PERIODIC SYSTEM OF ELEMENTS

ground state electron configurations and terms, atomic weights and ionization potentials

Periods \ Groups	I (a, b)	II (a, b)	III (b, a)	IV (b, a)	V (b, a)	VI (b, a)	VII (b, a)	VIII (a)	VIII	VIII
1	1.008 $1s$ $_{1}$H $^2S_{1/2}$ Hydrogen 13.598							$1s^2$ 4.003 $_{2}$He $^1S_0$ Helium 24.59; 54.42		
2	6.941(2) $2s$ $_{3}$Li $^2S_{1/2}$ Lithium 5.39; 75.6 122.4	9.012 $2s^2$ $_{4}$Be $^1S_0$ Beryllium 9.32; 18.2 153.9	$2s^22p$ 10.81 $_{5}$B $^2P_{1/2}$ Boron 8.30; 25.16 37.9	$2s^22p^2$ 12.01 $_{6}$C $^3P_0$ Carbon 11.26; 24.4 47.9	$2s^22p^3$ 14.01 $_{7}$N $^4S_{3/2}$ Nitrogen 14.53; 29.6 47.4	$2s^22p^4$ 16.00 $_{8}$O $^3P_2$ Oxygen 13.62; 35.1 54.9	$2s^22p^5$ 19.00 $_{9}$F $^2P_{3/2}$ Fluorine 17.42; 35.0 62.7	$2s^22p^6$ 20.18 $_{10}$Ne $^1S_0$ Neon 21.56; 41.0 62.6		
3	22.99 $3s$ $_{11}$Na $^2S_{1/2}$ Sodium 5.14; 47.3 71.6	24.30 $3s^2$ $_{12}$Mg $^1S_0$ Magnesium 7.65; 15.0 80.1	$3s^23p$ 26.98 $_{13}$Al $^2P_{1/2}$ Aluminium 5.99; 18.8 28.4	$3s^23p^2$ 28.09 $_{14}$Si $^3P_0$ Silicon 8.15; 16.3 33.5	$3s^23p^3$ 30.97 $_{15}$P $^4S_{3/2}$ Phosphorus 10.49; 19.8 30.2	$3s^23p^4$ 32.07 $_{16}$S $^3P_2$ Sulfur 10.36; 23.3 34.8	$3s^23p^5$ 35.45 $_{17}$Cl $^2P_{3/2}$ Chlorine 12.97; 23.8 39.6	$3s^23p^6$ 39.95 $_{18}$Ar $^1S_0$ Argon 15.76; 27.6 43.4		
4	39.10 $4s$ $_{19}$K $^2S_{1/2}$ Potassium 4.34; 31.6 45.8	40.08 $4s^2$ $_{20}$Ca $^1S_0$ Calcium 6.11; 11.9 50.9	44.96 $3d4s^2$ $_{21}$Sc $^2D_{3/2}$ Scandium 6.56; 12.8 24.8	47.88(3) $3d^24s^2$ $_{22}$Ti $^3F_2$ Titanium 6.83; 13.6 27.5	50.94 $3d^34s^2$ $_{23}$V $^4F_{3/2}$ Vanadium 6.75; 14.7 29.3	52.00 $3d^54s$ $_{24}$Cr $^7S_3$ Chromium 6.77; 16.5 31.0	54.94 $3d^54s^2$ $_{25}$Mn $^6S_{5/2}$ Manganese 7.43; 15.6 33.7	55.85 $3d^64s^2$ $_{26}$Fe $^5D_4$ Iron 7.90; 16.2 30.6	[illegible] 27 Cobalt 7.88; 17.1 33.5	58.69 $3d^84s^2$ $_{28}$Ni $^3F_4$ Nickel 7.64; 18.2 35.3
	$3d^{10}4s$ 63.55 $^2S_{1/2}$ $_{29}$Cu 7.73; 20.3 36.8 Copper	$3d^{10}4s^2$ 65.39(2) $^1S_0$ $_{30}$Zn 9.39; 18.0 39.7 Zinc	$4s^24p$ 69.72 $^2P_{1/2}$ $_{31}$Ga 6.00; 20.5 30.7 Gallium	$4s^24p^2$ 72.61(2) $^3P_0$ $_{32}$Ge 7.90; 15.9 34.2 Germanium	$4s^24p^3$ 74.92 $^4S_{3/2}$ $_{33}$As 9.79; 18.6 28.4 Arsenic	$4s^24p^4$ 78.96(3) $^3P_2$ $_{34}$Se 9.75; 21.2 30.8 Selenium	$4s^24p^5$ 79.90 $^2P_{3/2}$ $_{35}$Br 11.81; 21.8 35.9 Bromine	$4s^24p^6$ 83.80 $^1S_0$ $_{36}$Kr 14.00; 24.4 38.6 Krypton		
5	85.47 $5s$ $_{37}$Rb $^2S_{1/2}$ Rubidium 4.18; 27.3 39.2	87.62 $5s^2$ $_{38}$Sr $^1S_0$ Strontium 5.69; 11.03 42.9	88.91 $4d5s^2$ $_{39}$Y $^2D_{3/2}$ Yttrium 6.22; 12.2 20.5	91.22 $4d^25s^2$ $_{40}$Zr $^3F_2$ Zirconium 6.63; 13.1 23.1	92.91 $4d^45s$ $_{41}$Nb $^6D_{1/2}$ Niobium 6.76; 14.3 25.0	95.94 $4d^55s$ $_{42}$Mo $^7S_3$ Molybdenum 7.09; 16.2 27.2	[98] $4d^55s^2$ $_{43}$Tc $^6S_{5/2}$ Technetium 7.28; 15.3 29.5	101.1 $4d^75s$ $_{44}$Ru $^5F_5$ Ruthenium 7.36; 16.8 28.5	102.9 $4d^85s$ $_{45}$Rh $^4F_{9/2}$ Rhodium 7.46; 18.1 31.1	106.4 $4d^{10}$ $_{46}$Pd $^1S_0$ Palladium 8.34; 19.4 32.9
	$4d^{10}5s$ 107.9 $^2S_{1/2}$ $_{47}$Ag 7.58; 21.5 34.8 Silver	$4d^{10}5s^2$ 112.4 $^1S_0$ $_{48}$Cd 8.99; 16.9 37.5 Cadmium	$5s^25p$ 114.8 $^2P_{1/2}$ $_{49}$In 5.79; 18.9 28.0 Indium	$5s^25p^2$ 118.7 $^3P_0$ $_{50}$Sn 7.34; 14.6 30.5 Tin	$5s^25p^3$ 121.8 $^4S_{3/2}$ $_{51}$Sb 8.64; 16.5 25.3 Antimony	$5s^25p^4$ 127.6 $^3P_2$ $_{52}$Te 9.01; 18.6 28.0 Tellurium	$5s^25p^5$ 126.9 $^2P_{3/2}$ $_{53}$I 10.45; 19.1 33.0 Iodine	$5s^25p^6$ 131.3 $^1S_0$ $_{54}$Xe 12.13; 21.0 31.0 Xenon		
6	132.9 $6s$ $_{55}$Cs $^2S_{1/2}$ Caesium 3.89; 23.1 33.4	137.3 $6s^2$ $_{56}$Ba $^1S_0$ Barium 5.21; 10.0 35.8	138.9 $5d6s^2$ $_{57}$La $^2D_{3/2}$ * Lanthanum 5.58; 11.1 19.2	178.5 $5d^26s^2$ $_{72}$Hf $^3F_2$ Hafnium 6.83; 14.9 23.3	180.9 $5d^36s^2$ $_{73}$Ta $^4F_{3/2}$ Tantalum 7.89	183.8 $5d^46s^2$ $_{74}$W $^5D_0$ Tungsten 8.0	186.2 $5d^56s^2$ $_{75}$Re $^6S_{5/2}$ Rhenium 7.9	190.2 $5d^66s^2$ $_{76}$Os $^5D_4$ Osmium 8.7	192.2 $5d^76s^2$ $_{77}$Ir $^4F_{9/2}$ Iridium 9.1	195.1 $5d^96s$ $_{78}$Pt $^3D_3$ Platinum 9.0; 18.6
	$5d^{10}6s$ 197.0 $^2S_{1/2}$ $_{79}$Au 9.23; 20.5 34 Gold	$5d^{10}6s^2$ 200.6 $^1S_0$ $_{80}$Hg 10.44; 18.8 34.2 Mercury	$6s^26p$ 204.4 $^2P_{1/2}$ $_{81}$Tl 5.11; 20.4 29.8 Thallium	$6s^26p^2$ 207.2 $^3P_0$ $_{82}$Pb 7.42; 15.0 31.9 Lead	$6s^26p^3$ 209.0 $^4S_{3/2}$ $_{83}$Bi 7.29; 16.7 25.6 Bismuth	$6s^26p^4$ [209] $^3P_2$ $_{84}$Po 8.42 Polonium	$6s^26p^5$ [210] $^2P_{3/2}$ $_{85}$At 9.0 Astatine	$6s^26p^6$ [222] $^1S_0$ $_{86}$Rn 10.75 Radon		
7	[223] $7s$ $_{87}$Fr $^2S_{1/2}$ Francium 4.07	[226] $7s^2$ $_{88}$Ra $^1S_0$ Radium 5.29; 10.1	[227] $6d7s^2$ $_{89}$Ac $^2D_{3/2}$ ** Actinium 5.2; 11.7	[261] $6d^27s^2$ $_{104}$Db $^3F_2$ Dubnium						

Atomic Weights 1989
Abridged to Four Significant Figures
The values of $A_r(E)$ are considered reliable to ± 1, or less, in the last digit except when a larger single-digit uncertainty is inserted in parentheses

6.941(2)	2s
$_{3}$Li	$^2S_{1/2}$
Lithium	5.39; 75.6 122.4

Symbol of the element; Atomic number; Valence electron configuration; Ground state term; Ionization potential of Li I, II and III, respectively [eV]

* LANTHANIDES

140.1 $4f5d6s^2$ $_{58}$Ce $^1G_4$ Cerium 5.54; 10.8 20.2	140.9 $4f^36s^2$ $_{59}$Pr $^4I_{9/2}$ Praseodymium 5.46; 10.6 21.6	144.2 $4f^46s^2$ $_{60}$Nd $^5I_4$ Neodymium 5.52; 10.7 22.1	[145] $4f^56s^2$ $_{61}$Pm $^6H_{5/2}$ Promethium 5.55; 10.9 22.3	150.4 $4f^66s^2$ $_{62}$Sm $^7F_0$ Samarium 5.64; 11.1 23.4	152.0 $4f^76s^2$ $_{63}$Eu $^8S_{7/2}$ Europium 5.67; 11.2 24.9	157.2 $4f^75d6s^2$ $_{64}$Gd $^9D_2$ Gadolinium 6.15; 12.1 20.6	158.9 $4f^96s^2$ $_{65}$Tb $^6H_{15/2}$ Terbium 5.86; 11.5 21.9	162.5 $4f^{10}6s^2$ $_{66}$Dy $^5I_8$ Dysprosium 5.94; 11.7 22.8	164.9 $4f^{11}6s^2$ $_{67}$Ho $^4I_{15/2}$ Holmium 6.02; 11.8 22.8	167.3 $4f^{12}6s^2$ $_{68}$Er $^3H_6$ Erbium 6.11; 11.9 22.7	168.9 $4f^{13}6s^2$ $_{69}$Tm $^2F_{7/2}$ Thulium 6.18; 12.1 23.7	173.0 $4f^{14}6s^2$ $_{70}$Yb $^1S_0$ Ytterbium 6.25; 12.2 25.0	175.0 $4f^{14}5d6s^2$ $_{71}$Lu $^2D_{3/2}$ Lutetium 5.43; 13.9 21.0

** ACTINIDES

232.0 $6d^27s^2$ $_{90}$Th $^3F_2$ Thorium 6.1; 11.9 18.3	231.0 $5f^26d7s^2$ $_{91}$Pa $^4K_{11/2}$ Protactinium 5.9	238.0 $5f^36d7s^2$ $_{92}$U $^5L_6$ Uranium 6.19; 11.9 20	[237] $5f^46d7s^2$ $_{93}$Np $^6L_{11/2}$ Neptunium 6.27	[244] $5f^67s^2$ $_{94}$Pu $^7F_0$ Plutonium 6.06	[243] $5f^77s^2$ $_{95}$Am $^8S_{7/2}$ Americium 5.99	[247] $5f^76d7s^2$ $_{96}$Cm $^9D_5$ Curium 6.02	[247] $5f^97s^2$ $_{97}$Bk $^6H_{15/2}$ Berkelium 6.23	[251] $5f^{10}7s^2$ $_{98}$Cf $^5I_8$ Californium 6.30	[252] $5f^{11}7s^2$ $_{99}$Es $^4I_{15/2}$ Einsteinium 6.42	[257] $5f^{12}7s^2$ $_{100}$Fm $^3H_6$ Fermium 6.5	[258] $5f^{13}7s^2$ $_{101}$Md $^2F_{7/2}$ Mendelevium 6.6	[259] $5f^{14}7s^2$ $_{102}$No $^1S_0$ Nobelium	[260] $5f^{14}7s^27p$ $_{103}$Lr $^2P_{1/2}$ Lawrencium

Handbook of Physical Quantities

Handbook of

Physical Quantities

Edited by
Igor S. Grigoriev
and
Evgenii Z. Meilikhov
Russian Research Center
Kurchatov Institute
Moscow, Russia

Editor of the English Edition
A.A. Radzig
also of the
Russian Research Center
Kurchatov Institute
Moscow, Russia

CRC Press
Boca Raton New York London Tokyo

Library of Congress Cataloging-in-Publication Data

Handbook of physical quantities/ Igor S. Grigoriev and Evgenii Z. Meilikhov, editors.
p. cm.
Includes bibliographical references and index.
ISBN 0-8493-2861-6 (alk. paper)
1. Physics—Handbooks, manuals, etc. I. Grigoriev, Igor S. II. Meilikhov, Evgenii Zalmanovich.
QC61.H36 1995
530′.0212—dc 20 95-16928
CIP

Direct all inquiries to CRC Press LLC, 2000 N.W. Corporate Blvd., Boca Raton, Florida 33431.

International Standard Book Number 0-8493-2861-6
Library of Congress Card Number 95-16928
Printed in the United States of America 2 3 4 5 6 7 8 9 0
Printed on acid-free paper

About the Editors

Igor S. Grigoriev graduated from the Moscow Institute of Engineering and Physics in 1957 and went on to receive his Dr. Sci. degree, writing his thesis, "Neutron Propagation in Heterogeneous Media," in 1968. He was granted his Ph.D. in 1981, and wrote his thesis titled "Some Problems Concerning the Interaction of Laser Radiation with Substances."

Dr. Grigoriev is currently head of the laboratory for laser technology at the Russian Research Center *Kurchatov Institute* in Moscow and continues to follow his research interests in the physics of nuclear reactors, the interaction of laser radiation with gases and condensed substances, and laser technics.

Evgenii Z. Meilikhov is a 1962 graduate of the Moscow Institute of Physics and Technology. He wrote his theses, "Photoelectric Phenomena in Semiconductors at Microwave Frequencies" and "Plasma Phenomena in Semiconductors at Quantizing Magnetic Fields," in fulfillment of the requirements for a Dr. Sci. and a Ph.D. in 1967 and 1984, respectively.

Dr. Meilikhov now serves as chief researcher at the Russian Research Center *Kurchatov Institute*, as selected associate professor of the International Science Foundation, and as professor at the Moscow Institute of Physics and Technology. He is also associate editor of the Russian journal *Superconductivity* and the author of numerous scientific reviews and books.

His research interests include the physics of semiconductors at high magnetic fields, percolation phenomena in semi- and superconductors, and high-temperature conductivity.

Contributors

B.A. Aronzon
Russian Research Center
Kurchatov Institute
Moscow, Russia

A.P. Babichev
Russian Research Center
Kurchatov Institute
Moscow, Russia

N.A. Babushkina
Russian Research Center
Kurchatov Institute
Moscow, Russia

A.M. Bratkovskii
Russian Research Center
Kurchatov Institute
Moscow, Russia

M.E. Brodov
Institute of General Physics
Russian Academy of Sciences
Moscow, Russia

V.M. Bystrov
Scientific-Industrial Firm
Long-Range Communications
St. -Petersburg, Russia

V.M. Cherepanov
Russian Research Center
Kurchatov Institute
Moscow, Russia

A.G. Chertov
Likhachev Automobile
Construction Plant
Moscow, Russia

V.S. Egorov
Russian Research Center
Kurchatov Institute
Moscow, Russia

A.V. Eletskii
Russian Research Center
Kurchatov Institute
Moscow, Russia

E.B. Gel'man
Russian Research Center
Kurchatov Institute
Moscow, Russia

A.P. Geppe
Moscow Energy Institute
Moscow, Russia

I.S. Grigoriev
Russian Research Center
Kurchatov Institute
Moscow, Russia

K.G. Gurtovoy
Russian Research Center
Kurchatov Institute
Moscow, Russia

P.M. Imamov
Institute of Crystallography
Russian Academy of Sciences
Moscow, Russia

V.V. Ignat'ev
Russian Research Center
Kurchatov Institute
Moscow, Russia

A.V. Inyushkin
Russian Research Center
Kurchatov Institute
Moscow, Russia

V.Yu. Ivanov
Institute of General Physics
Russian Academy of Sciences
Moscow, Russia

V.L. Ivashintsova
A.F. Ioffe Physicotechnical
Institute
Russian Academy of Sciences
St. -Petersburg, Russia

N.V. Kadobnova
Institute for Aviation Materials
Moscow, Russia

I.I. Karasik
Institute of Industrial Standards
Moscow, Russia

M.N. Khlopkin
Russian Research Center
Kurchatov Institute
Moscow, Russia

I.N. Khlyustikov
Institute for Problems in
Microelectronics and
Superpure Materials
Chernogolovka, Moscow Region,
Russia

K.A. Kikoin
Russian Research Center
Kurchatov Institute
Moscow, Russia

V.A. Krivoruchko
Russian Research Center
Kurchatov Institute
Moscow, Russia

V.M. Kulakov
Russian Research Center
Kurchatov Institute
Moscow, Russia

S.D. Lazarev
Russian Research Center
Kurchatov Institute
Moscow, Russia

T.M. Lifshits
Institute of Radio Engineering
and Electronics
Russian Academy of Sciences
Moscow, Russia

Yu.E. Lyubarskii
Institute of Radioastronomy
Ukrainian Academy of Sciences
Khar'kov, Ukraine

A.A. Malyutin
Institute of General Physics
Russian Academy of Sciences
Moscow, Russia

S.V. Marin
Russian Research Center
Kurchatov Institute
Moscow, Russia

I.A. Maslov
Institute of General Physics
Russian Academy of Sciences
Moscow, Russia

E.Z. Meilikhov
Russian Research Center
Kurchatov Institute
Moscow, Russia

I.A. Migachev
Russian Research Center
Kurchatov Institute
Moscow, Russia

S.A. Mironov
A.F. Ioffe Physicotechnical
Institute
Russian Academy of Sciences
St. -Petersburg, Russia

A.L. Musatov
Institute of Radio Engineering
and Electronics
Russian Academy of Sciences
Moscow, Russia

V.A. Nikitin
Joint Institute for Nuclear
Research
Dubna, Moscow Region, Russia

L.A. Novitskii
Scientific Research Center
Technological Lasers
Shatura, Moscow Region, Russia

A.I. Obukhov
V.G. Khlopin Radium Institute
St. -Petersburg, Russia

V.I. Ozhogin
Russian Research Center
Kurchatov Institute
Moscow, Russia

R.V. Pisarev
A.F. Ioffe Physicotechnical
Institute
Russian Academy of Sciences
St. -Petersburg, Russia

V.S. Ptuskin
Institute for Terrestrial
Magnetism, Ionosphere, and
Wave Propagation
Troitsk, Moscow Region,
Russia

A.A. Radzig
Russian Research Center
Kurchatov Institute
Moscow, Russia

V.P. Rudakov
Russian Research Center
Kurchatov Institute
Moscow, Russia

V.G. Shapiro
Tel Aviv University
Tel Aviv, Israel

V.M. Shustryakov
Scientific Research Center for
Surface Properties and Vacuum
All-Russian Committee of
Standards
Moscow, Russia

B.D. Summ
Department of Physics
Moscow State University
Moscow, Russia

R.A. Syunyaev
Institute for Space Researches
Russian Academy of Sciences
Moscow, Russia

B.V. Vinogradov
Institute for High Pressures
Russian Academy of Sciences
Troitsk, Moscow Region, Russia

L.I. Vinokurova
Russian Academy of Sciences
Moscow, Russia

S.S. Yakimov
Russian Research Center
Kurchatov Institute
Moscow, Russia

V.P. Yanovskii
Institute of General Physics
Russian Academy of Sciences
Moscow, Russia

L.K. Zarembo
Department of Physics
Moscow State University
Moscow, Russia

Preface

For more than 20 years the universal reference book, *Tables of Physical Quantities* (published by Atomizdat, Moscow, 1976), edited by Acad. I.K. Kikoin, has been popular with specialists of different backgrounds. However, any reference book, no matter how carefully crafted, inevitably grows obsolete with time, despite its many benefits. Even the much-used *Tables* could not escape this fate. First it seemed that "cosmetic" changes and a few error corrections as well as minor additions would be enough to improve the book, but eventually it became obvious that a more profound and, in some cases, radical reworking of the material would be required, using fresh physical data and with a new team of authors. This gave rise to the idea of compiling a completely new and versatile reference work.

Throughout the course of the compilation, we have tried to keep in mind the potential users—specialists of different backgrounds (students, postgraduates, engineers, and scientists) who need specific numerical information in their everyday work and research. At the same time, it was our intention to present the material in such a way that it would be useful to nonspecialists as well. Toward that end, we have found it useful to present a summary of the principles that apply to the subject discussed in each chapter, as well as to list the units of the base physical quantities and, as far as possible, to provide references to in-depth studies.

In selecting numerical data, we limited ourselves, as a rule, to the most reliable of them (although in all fairness, it should be mentioned that the criteria of reliability are rather subjective and differ from one author to another). Wherever possible and convenient, the data are presented in charts and figures. Also, we have tried to avoid redundancy among the chapters. This consideration, coupled with the scope and variety of the material presented throughout the book, precluded the strict adherence to a uniform format. Nonetheless, we feel that the layout of the material is simple and allows the reader to find specific facts easily.

The number of inaccuracies in a printed work is usually in linear proportion to the sheer volume of the material presented. Bearing in mind that we have done our best to verify all the facts presented here, some things might have escaped our notice. We would be grateful to readers who point out inaccuracies so that we might improve future editions of this work.

I.S. Grigoriev
E.Z. Meilikhov
Russian Research Center
Kurchatov Institute
pl. Kurchatova 1
Moscow 123182
Russia

Contents

Acknowledgments

We are very grateful to Vladimir Borodin, Tanya Dmitrievskaya, Tanya Dumbrajs, Steven Frauzel, Janna Ivanenko, Irina Ivanova, Irina Kirienkova, Sergei Kryuchkov, Gil Pontecorvo, Natalia Shakura, Valerii Skobov, Georgii Smeshko, Artemii Ulantsev, and Evgenii Zhukovskii, who translated the original text into English.

The manuscript was prepared by a team of assiduous and highly professional typesetters, including Vitalii Arnold, Katya Goryacheva, Sergei Orlov, Irina Shompolova, German Solov'ev, Ilya Valuev, Alex Yustus, and Kirill Zhuravlev.

We are especially indebted to Igor Chaplygin, Vladimir Nikitin, and Veronica Vasendina for their highly skilled work, which resulted in a seemingly infinite number of camera-ready pages.

It gives us great pleasure to acknowledge the help and fruitful cooperation we have received from CRC Press during the preparation of this book. While many barriers along the way seemed insurmountable at first glance, the attentive and patient attitude of the CRC staff, including its Russian representative Svetlana Landau, helped bring our intentions to fruition.

We would especially like to express the warmest thanks to our project editor Gail Renard, who proofread the huge manuscript and contributed greatly to its improvement.

And finally, we offer our sincere thanks to Mark Licker, publisher at CRC, without whose active participation this book would hardly have reached the reader.

To all of them we are sincerely thankful.

I. Grigoriev
E. Meilikhov
Moscow, Russia
September 1996

1

Units of Physical Measure

A.G. Chertov

1.1 Units of the International System

1.1.1 SI Base Units and Measures

Length l is the unit of measure which characterizes the space and distance travelled by bodies or their parts along a given line. The dimension (dim) of length equals L that is symbolized by dim l = L; the unit of length is the meter (m).

The meter is the length of the path travelled by light in a vacuum during a time interval of $1/299,792,458$ of a second.

Mass m is the unit of measure which characterizes the inert and gravitational properties of material objects; dim m = M; the unit of mass is the kilogram (kg).

The kilogram is equal to the mass of the international prototype of the kilogram.

Time t is the unit of measure characterizing the successive change in phenomena and the states of matter which determines the duration of phenomenal being; dim t = T; the unit is a second (s).

The second is the duration of 9,192,631,770 periods of the radiation corresponding to the transition between the two hyperfine levels of the ground state of the cesium-133 atom.

Current strength I is a scalar measure equal to the time derivative from an electric charge sustained by a charge carrier through an observed surface; dim I = I, the unit of measure is the ampere (A).

The ampere is that constant current which, if maintained in two straight parallel conductors of infinite length, of negligible circular cross-section, and placed 1 meter apart in vacuum, would produce between these conductors a force equal to $2{\cdot}10^{-7}$ newton per meter of length.

Thermodynamic temperature T is the temperature calculated according to a thermodynamic temperature scale from absolute zero; dim $T = \Theta$; the unit is a kelvin (K).

The kelvin is the fraction 1/273.16 of the thermodynamic temperature of the triple point of water.

Notes:

(1) In addition to Kelvin temperature (designated as T) Celsius temperature (designated as t) can also be used, defined by the expression $t = T - T_0$, where T_0 = 273.15 K by definition. Kelvin temperature is expressed in kelvins, Celsius temperature in Celsius degrees (international designation °C). In terms of dimension, one degree Celsius is equal to a kelvin (1°C = 1 K).

(2) The interval or diversity of kelvin temperature is expressed in kelvins. The interval or diversity of Celsius temperature may be expressed both in kelvins and Celsius degrees.

(3) In addition to the thermodynamic scale, the International Practical Temperature Scale can also be used. This scale was recommended at the General Conference of Weights and Measures and is based on certain basic and supplementary reference points (see Chapter 8 "Thermometry") and is chosen so that all

0-8493-2861-6/97/$0.00+$.50

temperatures calculated on this scale will be close to thermodynamic temperature and so that the diversities will be within the limits of contemporary experimental error.

Amount of substance n is the measure equal to the number of elementary entities contained in a body (system of bodies); dim n = N; the unit is a mole (mol).

The mole is the amount of substance of a system which contains as many elementary entities as there are atoms in 0.012 kg of carbon-12.

When using moles the elementary entities must be specified as these may be atoms, molecules, ions, electrons, other particles, or specified groups of such particles.

Luminous intensity I is the measure equal to the radiant flux emitted by a source of radiation in a given direction inside a small solid angle in relation to this solid angle; dim I = J, the unit is a candela (cd).

The candela is the luminous intensity, in a given direction, of a source that emits monochromatic radiation of frequency $540 \cdot 10^{12}$ hertz and that has a radiant intensity in that direction of (1/683) watt per steradian.

1.1.2 Supplementary units

The plane angle α is a geometric figure formed by two rays (sides of the angle) extending from one point. The plane angle has no dimension, the unit is a radian (rad).

The radian is the plane angle between two radii of a circle which cut off on the circumference an arc equal in length to the radius.

The solid angle Ω is that part of space enclosed within one cavity of conical surface with a closed directrix. A solid angle has no dimension, the unit is a steradian (sr).

The steradian is the solid angle which, having its vertex in the center of the sphere, cuts off an area of the surface of the sphere equal to that of a square with sides of length equal to the radius of the sphere.

1.1.3 Derived SI units used in physics

1.1.3.1 Space and time

Area S is the measure which characterizes the plane and curved surfaces of geometric figures. Area is determined in the simplest cases by the single squares which fill a plane figure, i.e. the squares with one side equal to one unit of length; dim S =L^2, the unit is a square meter (m^2).

The square meter is equal to the area of a square with sides whose lengths are equal to 1 m.

Volume V is the measure which characterizes geometric bodies and is determined in the simplest cases by the number of single cubes which fit in a body, i.e. of edgewise cubes equal to a unit of length; dim V = L^3, the unit is a cubic meter (m^3).

The cubic meter is equal to the volume of an edgewise cube whose length is equal to 1 m.

Velocity v is the measure equal to the first time derivative from movement:

$$v = dr/dt;$$

dim v = LT^{-1}, the unit is a meter per second (m/s).

The meter per second is equal to the speed of a point moving in a straight and uniform manner when this point moves a distance of 1 m in a time interval of 1 s.

Acceleration a is the measure equal to the first time derivative from velocity:

$$a = dv/dt;$$

dim a = LT^{-2}, the unit is a meter per second squared ($\mathrm{m/s^2}$).

The meter per second squared is equal to the acceleration of a point moving in a straight and uniform manner when the speed of the point changes over 1 s by 1 m/s.

Angular velocity ω is the measure equal to the first time derivative from the deflection angle:

$$\omega = d\phi/dt;$$

dim $\Omega = \mathrm{T}^{-1}$, the unit is a radian per second (rad/s).

The radian per second is equal to the angular velocity of a uniformly rotating body all points of which in 1 s turn relative to the axis per 1 rad angle.

Angular acceleration ϵ is the measure determined by the first time derivative from the angular velocity:

$$\epsilon = d\omega/dt;$$

dim $\epsilon = \mathrm{T}^{-2}$, the unit is a radian per second squared (rad/s^2).

The radian per second squared is equal to the angular velocity of a uniformly rotating body when it changes its angular velocity over 1 s by 1 rad/s.

1.1.3.2 Periodic and related phenomena

Oscillation phase ϕ is the independent variable of a function describing the quantity which changes according to the law of harmonic vibrations. The oscillation phase is dimensionless, its unit is the radian (rad).

Period T is the time interval during which a cycle of periodic process is completed; dim $T = \mathrm{T}$, the unit of time is the second (s).

Periodic process frequency f, ν is the measure inverse to the period; dim $f, \nu = \mathrm{T}^{-1}$, the unit is a hertz (Hz).

The hertz is equal to the frequency of periodic process during which in 1 s one periodic process cycle is completed.

Rotational frequency n is the measure equal to the number of rotations completed in 1 s; dim $n = \mathrm{T}^{-1}$, the unit is a minus first degree second (s^{-1}).

The minus first degree second is equal to the frequency of uniform rotary motion during which over a time interval of 1 s a body performs a complete revolution.

The wave number ν is the measure inverse to the wavelength λ:

$$\nu = 1/\lambda;$$

dim $\nu = \mathrm{L}^{-1}$, the unit is a minus first degree meter (m^{-1}).

A minus first degree meter is equal to the wave number of the vibration with a wavelength of 1 m.

Damping coefficient δ is the unit of measure inverse to the interval τ during which the amplitude A decreases e times; dim $\delta = \mathrm{T}^{-1}$, the standard unit is the minus first degree second (s^{-1}).

The minus first degree second is equal to the damping coefficient during which over a 1 s time interval the amplitude decreases e times, where e is the base of the natural logarithm.

Attenuation coefficient μ is the unit of measure which characterizes the properties of a substance and is equal to the ratio of the relative decrease in radiant intensity dI/I in terms of the elemental path dx covered by the radiation in a given substance:

$$\mu = -\frac{\ln(dI/I)}{dx};$$

dim $\mu = \mathrm{L}^{-1}$, a unit is the minus first degree meter (m^{-1}).

The minus first degree meter is equal to the attenuation coefficient when over a distance of 1 m the amplitude decreases e times (e is the base of a natural logarithm).

1.1.3.3 Mechanics

Density ρ is the unit of measure equal to the ratio of mass dm of a body's element to the volume dV of this element:

$$\rho = dm/dV;$$

dim $\rho = \mathrm{ML}^{-3}$, the unit is a kilogram per cubic meter (kg/m^3).

The kilogram per cubic meter is equal to the density of a homogeneous substance whose mass at a volume of 1 m^3 is equal to 1 kg.

Specific volume v is the unit of measure equal to the ratio of the volume dV of a body's element to the mass dm of this element:

$$v = dV/dm;$$

dim $v = \mathrm{L}^3\mathrm{M}^{-1}$, a unit is the cubic meter per kilogram (m^3/kg).

The cubic meter per kilogram equals the specific volume of a homogeneous substance whose volume at a mass of 1 kg is equal to 1 m^3.

Momentum p of a material point is the unit of measure equal to mass of a material point at its velocity v:

$$p = mv;$$

dim $p = \mathrm{LMT}^{-1}$, a unit is the kilogram–meter per second (kg·m/s).

The kilogram meter per second is equal to the impulse of a material point whose mass is 1 kg, moving at a velocity of 1 m/s.

Angular momentum L of a point revolving around a fixed axis is the unit of measure equal to point momentum over its distance to the axis:

$$L = mvr;$$

dim $L = \mathrm{L}^2\mathrm{MT}^{-1}$, a unit is the kilogram–meter squared per second (kg·m^2/s).

The kilogram-meter squared per second is equal to an angular momentum of material point moving around a circumference with a radius of 1 m and having an impulse of 1 kg·m/s.

Moment of inertia (dynamic moment of inertia) J of a material point in relation to a certain axis is the unit of measure equal to mass m of a point per square distance r to the axis:

$$J = mr^2;$$

dim $J = \mathrm{L}^2\mathrm{M}$, a unit is the kilogram–meter squared (kg·m^2).

The kilogram–meter squared is equal to the moment of inertia of a point whose mass is 1 kg, located at a distance of 1 m from the pivot.

Force F is the vector measure which is the degree of mechanical influence exerted on a body by other bodies. It is determined by Newton's second law:

$$F = ma;$$

dim $F = \mathrm{LMT}^{-2}$, a unit is the newton (N).

The newton is equal to the force which gives a body whose mass is 1 kg an acceleration of 1 m/s^2, in the direction of the acting force.

Moment of force M in relation to a certain point is the unit of measure equal to the force F to the arm h, i.e. the distance between the direction of the force and this point:

$$M = Fh;$$

dim $M = \mathrm{L}^2\mathrm{MT}^{-2}$, a unit is the newton–meter (N·m).

The newton–meter is equal to the moment of force equal to 1 N in relation to a point located at a distance of 1 m from the action line of the force.

Impulse of force I is the unit of measure equal to force F over a time interval during which the force acts:

$$dI = F\,dt;$$

dim $I = \mathrm{LMT^{-1}}$, a unit is the newton–second (N·s).

The newton–second is equal to the moment of the force equal to 1 N acting within 1 s.

Pressure p is a unit of measure equal to the ratio between the force dF, acting on a surface element perpendicular to it, and the area dS of this element:

$$p = dF/dS;$$

dim $p = \mathrm{L^{-1}MT^{-2}}$, the unit is the pascal (Pa).

The pascal is equal to the pressure exerted by a force of 1 N equally distributed along its perpendicular surface with an area of 1 $\mathrm{m^2}$.

Note: Pascals also express normal and tangential stresses, as well as modules of longitudinal elasticity, shear and volume compression.

Dynamic viscosity η is the coefficient of proportionality in the internal friction force formula:

$$F = \eta \frac{dv}{dl} \Delta S,$$

where (dv/dl) is the velocity gradient; ΔS is the area of the surface layer on which the inner friction force is calculated; dim $\eta = \mathrm{L^{-1}MT^{-1}}$, the unit is the pascal–second (Pa·s).

The pascal–second is equal to the dynamic viscosity of the medium, where the tangential stress is equal to 1 Pa in the laminar flow and at various velocities of layers located 1 m along the direction perpendicular to the velocity of 1 m/s.

Kinematic viscosity ν is the coefficient equal to the ratio of dynamic viscosity of the medium to its density:

$$\nu = \eta/\rho;$$

dim $\nu = \mathrm{L^2T^{-1}}$, a unit is the meter squared per second ($\mathrm{m^2/s}$).

The meter squared per second is equal to the kinematic viscosity of the medium with a dynamic viscosity of 1 Pa·s and a density of 1 $\mathrm{kg/m^3}$.

Surface tension σ of the liquid is a measure equal to the ratio between the force dF, acting on the segment of the free surface contour perpendicular to the contour and tangentially to the surface, and length dl of this segment:

$$\sigma = dF/dl;$$

dim $\sigma = \mathrm{MT^{-2}}$, the unit is a newton per meter (N/m).

The newton per meter is equal to the surface tension of the liquid exerted by the force of 1 N which acts on the segment of the free surface contour at a length of 1 m perpendicular to the contour and tangentially to the surface.

Work. Elementary work, designated dA, is a measure equal to the scalar product of the force $\mathbf{F}$ into an elementary displacement ds:

$$dA = \mathbf{F}ds;$$

dim $A = \mathrm{L^2MT^{-2}}$, the unit is a joule (J).

The joule is equal to the work of 1 N force which moves a body the distance of 1 m in the direction of the force.

Note: Joules also express all forms of energy.

Power N, P is the measure equal to the work ratio dA in relation to an infinitely small time interval dt, during which the work is completed:

$$N = dA/dt;$$

dim $N = L^2MT^{-3}$, the unit is a watt (W).

The watt is equal to the power whereby the work of 1 J is completed for 1 second.

1.1.3.4 Heat

Temperature coefficient α is the measure equal to the ratio of relative change dX/X_0 in a physical quantity to the temperature change dT reckoned from the initial temperature:

$$\alpha = dX/(X_0 dT),$$

where X_0 is the quantity at initial temperature; dim $\alpha = \Theta^{-1}$, the unit is a minus first degree kelvin (K^{-1}).

The minus first degree kelvin is equal to the temperature rate of the relative change in the quantity during which the temperature change per 1 K from the initial temperature produces a relative change in the quantity equal to one unit.

Temperature gradient ∇T is the vector measure numerically equal to the temperature change for a unit of length and pointed in the direction of most rapid change in the heat field; dim $\nabla T = L^{-1}\Theta$, the unit is a kelvin per meter (K/m).

The kelvin per meter is equal to the temperature gradient of the field where the temperature changes by 1 K in a segment 1 m long in the direction of the temperature gradient.

Internal energy U of the system is the energy from the random thermal movement of all microparticles in the system (molecules, atoms, ions, etc.) and the energy of these particles interaction.

Internal energy, like any other form of energy, has a work dimension: dim $U = L^2MT^{-2}$ and is expressed in joules (J).

Heat, quantity of heat Q is the part of internal energy which is transferred spontaneously, with no external influence from more to less heated bodies by means of heat conductivity or radiation; dim $Q = L^2MT^{-2}$, the unit is a joule (J).

The joule is equal to the quantity of heat which is equivalent to the work of 1 J.

Note: Joules also express thermodynamic potentials (enthalpy, Helmholtz free energy, Gibbs function), phase transition heat, and chemical reaction heat.

Heat flow rate Φ through a given surface is the measure equal to the ratio of the quantity of heat dQ which passes through this surface to time dt during which this quantity of heat is conveyed:

$$\Phi = dQ/dt.$$

The heat flow rate dimension is equal to the power dimension: dim $\Phi = L^2MT^{-3}$, the unit is a watt (W).

The watt is equal to the heat flow rate which is equivalent to the mechanical power of 1 W.

Heat flux density q is the measure equal to the ratio of the heat flux $d\Phi$ in relation to the area dS of the surface through which the flux passes:

$$q = d\Phi/dS;$$

dim $\Phi = MT^{-3}$, the unit is a watt per square meter (W/m^2).

The watt per square meter is equal to the surface density of the heat flux in which the heat flux is equally distributed along the surface of an area 1 m^2.

Thermal conductivity λ is the measure equal to the density of the heat flux produced by heat conductivity at temperature gradient, which is equal to one unit. The formula defining the quantity of heat transferred over time t through the surface of area S in the direction perpendicular to this surface includes the thermal conductivity as a coefficient:

$$Q = \lambda \frac{dT}{dx} St,$$

where dT/dx is the temperature gradient; dim $\lambda = \mathrm{LMT}^{-3}\ \Theta^{-1}$, the unit is a watt per meter–kelvin (W/m·K).

The watt per meter–kelvin is equal to the thermal conductivity of the substance in which at a stationary regime with a surface density of the heat flux 1 W/m^2 the temperature gradient of 1 K/m is established.

Heat capacity C of a body (system) is the measure equal to the ratio of heat dQ required to warm a body (system of bodies), to the difference in temperature dT of the body:

$$C = dQ/dT;$$

dim $C = \mathrm{L}^2\mathrm{MT}^{-2}\ \Theta^{-1}$, the unit is a joule per kelvin (J/K).

Joule per kelvin is equal to the heat capacity of a system whose temperature rises by 1 K when 1 J heat is introduced to a system.

Specific heat capacity c of a substance is the measure equal to the ratio of the heat capacity C of a uniform body (system) to its mass:

$$c = C/m;$$

dim $c = \mathrm{L}^2\mathrm{T}^{-2}\ \Theta^{-1}$, the unit is joule per kilogram–kelvin (J/kg·K).

The joule per kilogram–kelvin is equal to the specific heat capacity of a substance which has a heat capacity of 1 J/K at a mass of 1 kg.

Thermal diffusivity a is the measure which defines the speed of temperature leveling under non-stationary heat conductivity and equals to the ratio of the thermal conductivity λ to the volume heat capacity $c_p\rho$ of a substance:

$$a = \lambda/(c_p\rho),$$

where c_p is the specific heat capacity of a substance at constant pressure; ρ is the density of the substance; dim $a = \mathrm{L}^2\mathrm{T}^{-1}$, the unit is a square meter per second (m^2/s).

The square meter per second is equal to the thermal diffusivity of a substance with its thermal conductivity 1 W/(m·K), with a specific heat capacity at constant pressure of 1 J/(kg·K), and a density of 1 kg/m^3.

Specific gas constant B is the measure equal to the ratio of work dA completed by the ideal gas at isobaric heating, to the mass of the gas m and the temperature interval dT at which the gas is heated:

$$B = dA/(mdT);$$

dim $B = \mathrm{L}^2\mathrm{T}^{-2}\Theta^{-1}$, the unit is a joule per kilogram–kelvin (J/kg·K).

The joule per kilogram–kelvin is equal to the specific gas constant of an ideal gas with a mass of 1 kg which completes the work of 1 J when raised by a temperature of 1 K at constant pressure.

Entropy S of a system is the single–valued function of the state of that system determined by the ratio

$$dS = dQ/T,$$

where dQ is an infinitely small quantity of heat transmitted to a system at temperature T; dim $S = \mathrm{L}^2\mathrm{MT}^{-2}\Theta^{-1}$, the unit is joule per kelvin (J/K).

The joule per kelvin is equal to the change in entropy of a system which, at a temperature of n K in the isothermic process receives a quantity of heat n J.

Specific entropy s is the measure equal to the ratio of entropy dS to the mass dm of the system:

$$s = dS/dm;$$

dim $s = \mathrm{L^2T^{-2}\Theta^{-1}}$, the unit is a joule per kilogram–kelvin (J/kg·K).

The joule per kilogram–kelvin is equal to the specific entropy of a substance in which at a mass of 1 kg the entropy change is 1 J/K.

The specific quantity of heat q is the measure equal to the ratio of heat dQ introduced to or removed from a system during a process, in relation to the system's mass dm:

$$q = dQ/dm;$$

dim $q = \mathrm{L^2T^{-2}}$, the unit is a joule per kilogram (J/kg).

The joule per kilogram is equal to the specific quantity of heat in a process during which a substance with a mass of 1 kg either gains or loses a quantity of heat equivalent to 1 J.

Note: Joules per kilogram also express specific thermodynamic potentials, specific heat of a phase transition, and specific heat of a chemical reaction.

1.1.3.5 Electricity and Magnetism

Quantity of electricity (electric charge) Q is the measure equal to the product of electric current I into time t during which the current flows:

$$Q = It;$$

dim Q = TI, the unit is the coulomb (C).

The coulomb equals the quantity of electricity transmitted through a cross section of a conductor at a current of 1 A over a time interval of 1 s.

Spatial density of an electric charge ρ is the measure equal to the ratio of the charge dQ located in an element of space, in relation to the volume dV of this element:

$$\rho = dQ/dV;$$

dim $\rho = \mathrm{L^{-3}TI}$, the unit is the coulomb per cubic meter ($\mathrm{C/m^3}$).

The coulomb per cubic meter is equal to the spatial density of an electric charge whereby at a volume of 1 $\mathrm{m^3}$ a charge of 1 C is uniformly distributed.

Surface density of an electric charge σ is the measure equal to the ratio of the charge dQ located in a surface element, in relation to the area dS of this element:

$$\sigma = dQ/dS;$$

dim $\sigma = \mathrm{L^{-2}TI}$, the unit is a coulomb per square meter ($\mathrm{C/m^2}$).

The coulomb per square meter is equal to the surface density of an electric charge when the charge of 1 C is uniformly distributed along a surface with an area of 1 $\mathrm{m^2}$.

Linear density of electric charge τ is the measure equal to the ratio of the charge dQ located in an element of a line, in relation to the length dl of this element:

$$\tau = dQ/dl;$$

dim $\tau = \mathrm{L^{-1}TI}$, the unit is a coulomb per meter (C/m).

The coulomb per meter is equal to the linear density of an electric charge when the charge evenly distributed along a line 1 m long is equal to 1 C.

Electric voltage U is a measure equal to the ratio of the power P of a direct current, in relation to the intensity of the current I:

$$U = P/I;$$

dim $U = \mathrm{L^2MT^{-3}I^{-1}}$, the unit is a volt (V).

The volt is equal to the electric voltage which generates in an electric circuit a direct current of 1 A at 1 W power.

Note: Volts also express electrical potential, the potential difference of an electric field, and the electromotive force.

Strength of an electric field E is a vector quantity equal to the ratio of force dF, acting on a positive charge dQ placed at a certain point in the electric field, in relation to this charge:

$$E = dF/dQ;$$

dim $E = \mathrm{LMT^{-3}I^{-1}}$, the unit is a volt per meter (V/m).

The volt per meter is equal to the strength of a uniform electric field created by the potential difference of 1 V between points located at a distance of 1 m on the line of field action.

Electric displacement flux Ψ through an enclosed surface is the measure equal to the algebraic sum of the electric charges contained in the inner space of this surface:

$$\Psi = \sum_{i=1}^{n} Q_i;$$

dim $\Psi = \mathrm{TI}$, the unit is the coulomb (C).

The coulomb is equal to the electric displacement flux connected with the total free charge of 1 C.

Electric displacement D is the measure equal to the ratio of the electric displacement flux $d\Psi$ to the area dS of a surface element through which this flux passes:

$$D = d\Psi/dS;$$

dim $D = \mathrm{L^{-2}TI}$, the unit is a coulomb per square meter (C/m^2).

The coulomb per square meter is equal to the electric displacement whereby a flux of electric displacement passed through a cross section having an area of 1 m^2, is equal to 1 C.

Absolute dielectric permittivity ϵ of a medium is the proportionality coefficient in a formula which joins both the displacement and the strength of an electric field:

$$D = \epsilon E;$$

dim $\epsilon = \mathrm{L^{-3}M^{-1}T^4I^2}$, the unit is a farad per meter (F/m).

The farad per meter equals the absolute dielectric permittivity of a medium in which the strength of the electric field of 1 V/m produces an electric displacement of 1 C/m^2.

Note: Farads per meter also express the electric constant ϵ_0.

Electric moment of a dipole p is a vector quantity which equals the charge Q of a dipole at its shoulder l:

$$p = Ql;$$

dim $p = \mathrm{LTI}$, the unit is a coulomb–meter (C·m).

The coulomb–meter equals the electric moment of a dipole whose charges are each equal to 1 C and arrange at a distance of 1 m one from another.

Density of an electric current j is the measure that equals the ratio of the current strength dI to the area dS of a cross section:

$$j = dI/dS;$$

dim j =$\mathrm{L}^{-2}\mathrm{I}$, the unit is an ampere per square meter (A/m^2).

The ampere per square meter equals the density of an evenly distributed current of 1 A along a cross section whose area is 1 m^2.

Linear density of an electric current A is the measure equal to the ratio of a current strength dI in a thin sheet conductor, in relation to width da of this conductor:

$$A = dI/da;$$

dim $A = \mathrm{L}^{-1}\mathrm{I}$, the unit is an ampere per meter (A/m).

The ampere per meter equals the linear density of an electric current evenly distributed along a cross section of a thin sheet conductor 1 m in width, whereby the current strength equals to 1 A.

Electric resistance R is a measure which defines the conductor and comprises the proportionality coefficient in the formula linking the voltage U and the current strength I:

$$U = RI;$$

dim $R = \mathrm{L}^2\mathrm{MT}^{-3}\mathrm{I}^{-2}$, the unit is an ohm (Ω or Ohm).

The ohm equals the resistance of the conductor between the ends of which arises a voltage of 1 V at a current of 1 A.

Electric conductance G is the measure inverse to the resistance:

$$G = 1/R;$$

dim $G = \mathrm{L}^{-2}\mathrm{M}^{-1}\mathrm{T}^3\mathrm{I}^2$, the unit is a siemens (S).

The siemens equals the electric conductance of a conductor with a resistance of 1 Ω.

Electric resistivity ρ of a substance is the quantity numerically equal to the resistance of the conductor with unit length and unit cross section area:

$$\rho = RS/l;$$

dim $\rho = \mathrm{L}^3\mathrm{MT}^{-3}\mathrm{I}^{-2}$, unit is ohm–meter (Ω·m).

The ohm–meter equals the electric resistivity of a conductor, which at a cross section area of 1 m^2 and a length of 1 m has an electric resistance equal to 1 Ω.

Electric conductivity g of a substance is the quantity inverse to the electric resistivity:

$$g = 1/\rho;$$

dim $g = \mathrm{L}^{-3}\mathrm{M}^{-1}\mathrm{T}^3\mathrm{I}^2$, the unit is a siemens per meter (S/m).

The siemens per meter equals the electric conductivity of a conductor which has an electric conductance of 1 S when the area of the cross section measures 1 m^2 and length, 1 m.

Strength of a magnetic field H is the measure which characterizes a magnetic field. Its dimension and unit may be determined by using a field strength formula in the center of a long solenoid:

$$H = nI;$$

dim $H = \mathrm{L}^{-1}\mathrm{I}$, the unit is ampere per meter (A/m).

The ampere per meter equals the strength of a magnetic field in the center of a long solenoid with an evenly distributed coil along which is transferred a current of $1/n$ A, where n is the number of loops in the segment of a solenoid whose length is 1 m.

Magnetomotive force F_m is the measure which characterizes the magnetizing effect of an electric current and is equal to the circulation of a magnetic field strength along an enclosed circuit:

$$F_m = \oint H_l dl;$$

dim F_m = I, the unit is an ampere (A).

The ampere equals the magnetomotive force along an enclosed circuit coupled with the circuit of a direct current of 1 A.

Note: amperes also express the difference of magnetic potentials.

Magnetic flux. The unit and dimension of magnetic flux Φ is determined by the formula

$$Q = \Phi/R,$$

where Q is the quantity of electricity transferred to an enclosed circuit of resistance R when the magnetic flux Φ, connected to this circuit, is changed to zero. From this formula it follows that dim Φ = $L^2MT^{-2}I^{-1}$, the unit is a weber (Wb).

The weber equals the magnetic flux which, when reduced to zero in its related electrical circuit with a resistance of 1 Ω, has 1 C of electricity transferred through the cross section.

Magnetic induction B is a measure equal to the ratio of magnetic flux $d\Phi$ to the area dS of the cross section through which this flux is passed:

$$B = d\Phi/dS;$$

dim B =$MT^{-2}I^{-1}$, the unit is a tesla (T).

The tesla equals the magnetic inductance at which a magnetic flux of 1 Wb is passed through a cross section with an area of 1 m^2.

Inductance L is a measure defining an enclosed contour and is the proportionality coefficient between a magnetic flux linked to this circuit and the current strength in it:

$$\Phi = LI;$$

dim L = $L^2MT^{-2}I^{-2}$, the unit is a henry (H).

The henry equals the inductance of an electric circuit with which a magnetic flux of 1 Wb is connected at a direct current of 1 A.

Note: The henry also expresses mutual inductance.

Absolute magnetic permeability μ is the proportionality coefficient between the magnetic induction and the strength of a magnetic field:

$$B = \mu H;$$

dim μ =$LMT^{-2}I^2$, the unit is henry per meter (H/m).

The henry per meter equals the absolute permeability of a medium in which the magnetic field of strength 1 A/m produces a magnetic induction of 1 T.

Note: The henry per meter also expresses the magnetic constant μ_0.

Magnetic moment (Ampere's) p_m of the circuit with a current is a quantity which equals the circuit current I times the area S bounded by the circuit:

$$p_m = IS;$$

dim p_m =L^2I, the unit is the ampere–square meter (A·m^2).

The ampere–square meter equals the magnetic moment of a 1 A electric current as it passes through a circuit with an area of 1 m^2.

Note: The dimension of the magnetic moment (Coulomb's) dim $p_m = \mathsf{L}^3\mathsf{M}\mathsf{T}^{-2}\mathsf{I}^{-1}$, the unit is weber–meter (Wb·m).

Magnetization M is a measure which equals the ratio of the sum of the magnetic moments of all the magnetic dipoles comprising the element of the magnetic substance, in relation to the volume dV of this element:

$$M = \sum_{i=1}^{N} \frac{p_{m,i}}{dV},$$

where $p_{m,i}$ is the magnetic moment of the i-th dipole; N is the number of dipoles comprising the magnetic substance; dim $M = \mathrm{L}^{-1}\mathrm{I}$, the unit is ampere per meter (A/m).

The ampere per meter equals the magnetization at which a substance with a volume of 1 m^3 has a magnetic moment of 1 A·m^2.

Magnetic resistance R_m is the measure that is the proportionality coefficient in a formula which expresses the dependence of a magnetic flux on the magnetomotive force F_m:

$$F_m = R_m \Phi;$$

dim $R_m = \mathrm{L}^{-2}\mathrm{M}^{-1}\mathrm{T}^2\mathrm{I}^2$, the unit is the minus first degree henry (H^{-1}).

The minus first degree henry equals the magnetic resistance of a magnetic circuit in which the magnetomotive force of 1 A produces a magnetic flux of 1 Wb.

Magnetic conduction Λ is a measure inverse to magnetic resistance:

$$\Lambda = 1/R_m;$$

dim $\Lambda = \mathrm{L}^2\mathrm{M}\mathrm{T}^{-2}\mathrm{I}^{-2}$, the unit is a henry (H).

The henry is equal to the magnetic conduction of a magnetic circuit with a magnetic resistance of 1 H^{-1}.

1.1.3.6 Optics

Energy exposure H_e is the measure equal to the ratio of energy dW of radiation falling on the surface to area dS of this surface:

$$H_e = dW/dS;$$

dim $H_e = \mathrm{M}\mathrm{T}^{-2}$, the unit is a joule per square meter ($\mathrm{J/m}^2$).

The joule per square meter equals the energy exposure at which a radiation of 1 J falls on a surface with an area of 1 m^2.

Radiant flux Φ_e is the measure equal to the ratio of radiation energy ΔW to the time interval Δt of the radiation transfer which significantly exceeds the period of oscillations:

$$\Phi_e = \Delta W/\Delta t;$$

dim $\Phi_e = \mathrm{L}^2\mathrm{M}\mathrm{T}^{-3}$, i.e. it corresponds to the dimension of the power, the unit is a watt (W).

The watt equals the radiant flux which is equivalent to the mechanical power of 1 W.

Energy luminosity (radiance) M_e is the measure equal to the radiant flux $d\Phi_e$ in relation to the area dS from which this radiation is emitted:

$$M_e = d\Phi_e/dS;$$

dim $M_e = \mathrm{M}\mathrm{T}^{-3}$, the unit is a watt per square meter ($\mathrm{W/m}^2$).

The watt per square meter is equal to the energy luminosity at which a surface with an area of 1 m^2 emits a radiant flux of 1 W.

Irradiance E_e is the measure equal to the radiant flux $d\Phi_e$ in relation to the area dS by which this radiation is absorbed:

$$E_e = d\Phi_e/dS;$$

dim $E_e = \mathrm{MT}^{-3}$, the unit is a watt per square meter ($\mathrm{W/m^2}$).

The watt per square meter equals the irradiance at which a surface with an area of 1 $\mathrm{m^2}$ absorbs a radiant flux of 1 W.

Radiant intensity I_e is the measure equal to the ratio of radiant flux $d\Phi_e$ of a source to a solid angle $d\Omega$, within which this radiation is propagated:

$$I_e = d\Phi_e/d\Omega;$$

dim $I_e = \mathrm{L^2MT^{-3}}$, the unit is a watt per steradian (W/sr).

The watt per steradian is equal to the radiant intensity of a point source which emits a radiant flux of 1 W in a solid angle of 1 sr.

Energy brightness B_e is the measure equal to the ratio of radiant intensity dI_e of the surface element to the area dS of the projection of this element on a plane perpendicular to the observed direction:

$$B_e = dI_e/dS;$$

dim $B_e = \mathrm{MT^{-3}}$, the unit is the watt per steradian–square meter ($\mathrm{W/sr{\cdot}m^2}$).

The watt per steradian–squared meter equals the energy brightness of uniformly radiating plane with an area of 1 $\mathrm{m^2}$ in a perpendicular direction to it at a radiant intensity of 1 W/sr.

Luminous flux Φ_v emitted by a source of light at a certain solid angle is the measure equal to the product of the light intensity I produced by a source and this solid angle Ω:

$$\Phi_v = I\Omega;$$

dim $\Phi_v = \mathrm{J}$, the unit is a lumen (lm).

The lumen equals the luminous flux emitted by a point source in a solid angle of 1 sr at a luminous intensity of 1 kd.

Luminous energy Q is the measure equal to the luminous flux Φ_v produced over time t during which this luminous flux is emitted (or received):

$$Q = \Phi_v t;$$

dim $Q = \mathrm{TJ}$, the unit is a lumen–second (lm·s).

The lumen–second equals the luminous energy of a luminous flux in 1 lm acting over 1 s.

Brightness B_ϕ of a luminous surface in a given direction ϕ is the measure equal to the luminous intensity ratio I in this direction in relation to area S of the projection of a luminous surface on a plane perpendicular to a given direction:

$$B_\phi = I_e/(S\cos\phi);$$

dim $B_\phi = \mathrm{L^{-2}J}$, the unit is a candela per square meter ($\mathrm{kd/m^2}$).

The candela per square meter equals the brightness of a luminous surface with an area of 1 $\mathrm{m^2}$ at a luminous intensity of 1 cd.

Luminance (radiance) R is the measure equal to the ratio of luminous flux $d\Phi_v$ emitted by a luminous surface, in relation to its area dS:

$$R = d\Phi_v/dS;$$

dim $R = \mathrm{L}^{-2}\mathrm{J}$, the unit is a lumen per square meter ($\mathrm{lm/m^2}$).

The lumen per square meter equals the luminance of a surface with an area of 1 m^2 which emits a luminous flux of 1 lm.

Illuminance (intensity of illumination) E is the measure equal to the ratio of luminous flux $d\Phi_v$ falling on a surface element, in relation to the area dS of this element:

$$E = d\Phi_v/dS;$$

dim $E = \mathrm{L}^{-2}\mathrm{J}$, the unit is a lux (lx).

The lux is equal to the illuminance of a surface with an area of 1 m^2 while a luminous flux of 1 lm falls on this surface.

Luminous exposure H is the measure equal to the luminance E produced over time t during which the radiation occurs:

$$H = Et;$$

dim $H = \mathrm{L}^{-2}\mathrm{TJ}$, the unit is the lux–second (lx·s).

The lux–second equals the luminous exposure generated over 1 s at an illuminance of 1 lx.

1.1.3.7 Acoustics

Acoustic pressure p is the pressure which arises supplementarily in a gaseous or liquid medium when sound waves are passed through it. Acoustic pressure, like all other types of pressure, has a dimension dim $p = \mathrm{L}^{-1}\mathrm{MT}^{-2}$ and is expressed in pascals (Pa).

Vibration velocity v is the measure equal to amplitude A of the particle vibrations in a medium through which passes a sound at the angular frequency ω:

$$v = A\omega;$$

dim $v = \mathrm{LT}^{-1}$, the unit is meter per second (m/s).

Volume speed of sound q is the measure equal to the product of vibration velocity v into an area S of the cross section of a channel where sound is propagated:

$$q = vS;$$

dim $q = \mathrm{L}^3\mathrm{T}^{-1}$, the unit is a cubic meter per second ($\mathrm{m^3/s}$).

The cubic meter per second is equal to the volume speed of sound at a vibration velocity of 1 m/s and an area of 1 m^2 channel cross section.

Acoustic energy W is the energy of the particles of a medium in which sound is propagated. Acoustic energy, like any other form of energy, has a dimension dim $W = \mathrm{L}^2\mathrm{MT}^{-2}$ and is expressed in joules (J).

Density of acoustic energy is the measure equal to the acoustic energy dW contained in the element of a channel, in relation to the volume dV of this element:

$$w = dW/dV;$$

dim $w = \mathrm{L}^{-1}\mathrm{MT}^{-2}$, the unit is a joule per cubic meter ($\mathrm{J/m^3}$).

The joule per cubic meter is equal to the density of the acoustic energy in a channel with a volume of 1 m^3 at acoustic energy of 1 J.

Acoustic energy flux (acoustic power) P is the measure equal to the acoustic energy dW transmitted through a surface in relation to the time interval dt needed for this energy passage:

$$P = dW/dt;$$

dim $P = L^2MT^{-3}$, the unit is a watt (W).

Acoustic (sound) intensity I is the measure equal to the ratio of acoustic energy flux dP through a surface perpendicular to the direction of the sound propagation, in relation to the area dS of this surface:

$$I = dP/dS;$$

dim $I = MT^{-3}$, the unit is a watt per square meter (W/m^2).

The watt per square meter equals the acoustic intensity in a channel at an acoustic energy flux of 1 W and an area of the 1 m^2 cross section.

Acoustic resistance Z_a of the channel is the proportionality coefficient in an equation connecting amplitude p_0 of acoustic pressure and the volume speed q of sound:

$$p_0 = Z_a q;$$

dim $Z_a = L^{-4}MT^{-1}$, the unit is a pascal–second per cubic meter (Pa·s/m^3).

A pascal–second per cubic meter equals the acoustic resistance of the channel in which a volume speed of 1 m^3/s is produced at an acoustic pressure of 1 Pa.

Specific acoustic resistance Z_s is the measure equal to the product of the acoustic resistance and the area S of the cross section of a channel:

$$Z_s = Z_a S;$$

dim $Z_s = L^{-2}MT^{-1}$, the unit is the pascal–second per meter (Pa·s/m).

The pascal–second per meter equals the specific acoustic resistance of a channel with the area of a cross section 1 m^2 and having an acoustic resistance of 1 Pa·s/m^3.

Mechanical resistance Z_m is the measure equal to the ratio of force F acting on the cross section of a channel where sound is propagated, to the mean vibration velocity $< v >$ in this section:

$$Z_m = F/ < v >;$$

dim $Z_m = MT^{-1}$, the unit is the newton–second per meter (N·s/m).

The newton–second per meter is equal to the mechanical resistance of a channel in which at a force of 1 N the vibration velocity of 1 m/s arises.

1.1.3.8 Physical Chemistry and Molecular Physics

Molar mass M is the measure equal to the ratio of mass m of a system (body) to the amount of substance n of the system:

$$M = m/n;$$

dim $M = MN^{-1}$, the unit is a kilogram per mole (kg/mol).

The kilogram per mole equals the molar mass of a substance having a mass of 1 kg when the amount of substance is 1 mole.

Molar volume V_m is the measure equal to the ratio of volume V of a system (body) to its amount of substance n:

$$V_m = V/n;$$

$\dim V_m = L^3N^{-1}$, the unit is a cubic meter per mole (m^3/mol).

The cubic meter per mole equals the molar volume of a substance which occupies a volume of 1 m^3 at an amount of substance of 1 mole.

Molar internal energy U_m is the measure equal to the ratio of internal energy dU of a system (body) to its substance amount dn:

$$U_m = dU/dn;$$

$\dim U_m = L^2MT^{-2}N^{-1}$, the unit is a joule per mole (J/mol).

The joule per mole is equal to the molar internal energy of a 1 mol substance whose internal energy is equal to 1 J.

Note: Joules per mole are also used to express molar enthalpy, chemical potential, chemical affinity and activation energy.

Molar heat capacity C_m is the measure equal to the ratio of heat capacity of a system (body) to its amount of substance:

$$C_m = C/n;$$

$\dim C_m = L^2MT^{-2}\Theta^{-1}N^{-1}$, the unit is a joule per mole–kelvin (J/mol·K).

The joule per mole–kelvin equals the molar heat capacity of a substance which has a heat capacity of 1 J/K when its amount of substance is 1 mol.

Note: Molar entropy is also expressed in joules per mole–kelvin.

Concentration (number density) of molecules n in a homogeneous system is the measure equal to the ratio of the molecular number dN in a system to its volume dV:

$$n = dN/dV;$$

$\dim n = L^{-3}$, the unit is a minus third degree meter (m^{-3}).

The minus third degree meter is equal to the concentration of molecules at which in an element of a system with a volume of 1 m^3 there is one molecule.

Molar concentration of a substance C_B in a mixture (solution, alloy) is the ratio of the amount of substance dn_B to the volume dV of the mixture (solution, alloy):

$$C_B = dn_B/dV;$$

$\dim C_B = L^{-3}N$, the unit is a mole per cubic meter (mol/m^3).

The mole per cubic meter is equal to the molar concentration of the substance in a solution when at a volume of 1 m^3 it contains an amount of soluble substance equal to 1 mol.

Note: For this measure the term "molarity" was previously used. At present this designation is not recommended in the terminology of physical chemistry.

Molarity b_B of the solution of a component is the ratio of the amount of substance dn of a soluble component to the mass dm of the solvent:

$$b_B = dn/dm;$$

$\dim b_B = M^{-1}N$, the unit is a mole per kilogram (mol/kg).

The mole per kilogram is equal to the molarity of the solution at which for a mass of the solvent of 1 kg there is an amount of substance equal to 1 mole.

Note: Specific adsorption is also expressed in moles per kilogram.

Mass concentration ρ_B of the component B in a mixture (solution, alloy) is the measure equal to the ratio of mass dm of component B to the volume dV of a mixture (solution, alloy):

$$\rho_B = dm_B/dV;$$

dim $\rho_B = \mathrm{L^{-3}M}$, the unit is a kilogram per cubic meter (kg/m^3).

The kilogram per cubic meter equals the mass concentration of the component when at volume of 1 m^3 the mixture (solution, alloy) contains a component with a mass of 1 kg.

Chemical reaction rate v is the measure equal to the ratio of change ΔC_B of the molar concentration of the initial substance in a solution over the time interval Δt of the reaction:

$$v = \Delta C_B/\Delta t;$$

dim $v = \mathrm{L^{-3}T^{-1}N}$, the unit is a mole per second–cubic meter (mol/s·m^3).

The mole per second–cubic meter is equal to the mean rate of a monomolecular chemical reaction when in 1 s the molecular concentration of the original substance in a solution changes by 1 mol/m^3.

1.1.3.9 Atomic and Nuclear Physics

Rest mass of a particle (atom, nucleus) m; dim m = M, unit is kilogram (kg).

Defect of a mass Δm; dim Δm = M, unit is kilogram (kg).

Elementary charge e; dim e = TI, unit is coulomb (C).

Magnetic moment of an atom, nucleus μ; dim $\mu = \mathrm{L^2I}$, unit is ampere–square meter (A·m^2).

Nuclear magneton μ_N; dim $\mu_N = \mathrm{L^2I}$, unit is ampere–square meter (A·m^2).

Gyromagnetic ratio γ; dim $\gamma = \mathrm{M^{-1}TI}$, unit is ampere–square meter per joule–second (A·m^2/J·s).

Nuclear quadrupole moment Q; dim $Q = \mathrm{L^2}$, unit is square meter (m^2).

Force constant for the vibrational spectrum of a molecule k_e; dim $k_e = \mathrm{MT^{-2}}$, unit is newton per meter (N/m).

Binding energy E_B; dim $E_B = \mathrm{L^2MT^{-2}}$, unit is joule (J).

Level width Γ; dim $\Gamma = \mathrm{L^2MT^{-2}}$, unit is joule (J).

Particle transfer (particle fluence); dim = $\mathrm{L^{-2}}$, unit is minus second degree meter (m^{-2}).

Particle flux density ϕ; dim $\phi = \mathrm{L^{-2}T^{-1}}$, unit is minus first degree second by minus second degree meter (s^{-1}·m^{-2}).

Energy transfer (energy fluence); dim = $\mathrm{MT^{-2}}$, unit is joule per square meter (J·m^{-2}).

Effective cross–section σ; dim $\sigma = \mathrm{L^2}$, unit is square meter (m^2).

Differential effective cross–section $d\sigma/d\Omega$; dim $(d\sigma/d\Omega) = \mathrm{L^2}$, unit is square meter per steradian (m^2/sr).

Spectral effective cross–section σ_s; dim $\sigma_s = \mathrm{M^{-1}T^2}$, unit is square meter per joule (m^2/J).

Linear attenuation factor μ_l; dim $\mu_l = \mathrm{L^{-1}}$, unit is minus first degree meter (m^{-1}).

Atomic attenuation factor μ_a; dim $\mu_a = \mathrm{L^2}$, unit is square meter (m^2).

Mass attenuation factor μ_m; dim $\mu_m = \mathrm{L^2M^{-1}}$, unit is square meter per kilogram (m^2/kg).

Mass energy transformation factor; dim = $\mathrm{L^2M^{-1}}$, unit is square meter per kilogram (m^2/kg).

Mass absorption coefficient; dim = $\mathrm{L^2M^{-1}}$, unit is square meter per kilogram (m^2/kg).

Mean free path $< l >$; dim $< l >=$ L, unit is meter (m).

Mean mass range; dim = $\mathrm{L^{-2}M}$, unit is kilogram per square meter (kg/m^2).

Linear ionization density; dim = $\mathrm{L^{-1}}$, unit is minus first degree meter (m^{-1}).

Half–attenuation layer thickness $d_{1/2}$; dim $d_{1/2}$ = L, unit is meter (m).

Deceleration equivalent; dim = L, unit is meter (m).

Stopping power (linear); dim = LMT^{-2}, unit is joule per meter (J/m).

Linear energy transformation factor; dim = LMT^{-2}, unit is joule per meter (J/m).

Mean energy of ion formation; dim = L^2MT^{-2}, unit is joule (J).

Atomic stopping power; dim = L^4MT^{-2}, unit is joule–square meter ($J{\cdot}m^2$).

Mass deceleration ability; dim = L^4T^{-2}, unit is joule–square meter per kilogram ($J{\cdot}m^2/kg$).

Mobility b; dim $b = M^{-1}T^2I$, unit is square meter per volt–second ($m^2/V{\cdot}s$).

Neutron flux Φ; dim $\Phi = T^{-1}$, unit is minus first degree second (s^{-1}).

Concentration (number density) of ions, neutrons n; dim $n = L^{-3}$, unit is minus third degree meter (m^{-3}).

Bulk velocity of neutrons; dim = $L^{-3}T^{-1}$, unit is minus first degree second–minus third degree meter ($s^{-1}{\cdot}m^{-3}$).

Slowing down density; dim = $L^{-3}T^{-1}$, unit is minus first degree second–minus third degree meter ($s^{-1}{\cdot}m^{-3}$).

Moderation ability of the medium; dim = L^{-1}, unit is minus first degree meter (m^{-1}).

Diffusion coefficient for neutron flux density; dim = L, unit is meter (m).

Neutron age; dim = L^2, unit is square meter (m^2).

1.2 Recommended Measures and Units of Ionizing Radiation

1.2.1 Units and Measures Which Characterize Ionizing Radiation and Its Field

Energy of ionizing particles E; dim $E = L^2MT^{-2}$, unit is joule (J).
Preferred units: eV, keV, MeV, GeV.

Energy of ionizing radiation w is the summary energy of ionizing photons (disregarding rest energy) emitted, transferred or absorbed; dim $w = L^2MT^{-2}$, unit is joule (J).
Preferred units: fJ, pJ, nJ, μJ, mJ, J, kJ, MJ.

Rest mass of a particle, atom or atomic nucleus m_a; dim m_a = M, unit is kilogram (kg).
Preferred unit: atomic mass unit (a.m.u.).

Ionizing particles flux F is the ratio of the number of ionizing particles dN which passes through a given surface over time interval to this interval dt:

$$F = dN/dt;$$

dim $F = T^{-1}$, unit is a minus first degree second (s^{-1}).

The minus first degree second equals the ionizing particles flux dN at which one particle passes through a given surface over 1 second.

Preferred units: s^{-1}, min^{-1}.

Transfer (fluence) of ionizing particles Φ is the ratio of the number of ionizing particles dN which penetrate an elementary sphere, in relation to the area dS of the central section of this sphere:

$$\Phi = dN/dS;$$

dim $\Phi = L^{-2}$, unit is a minus second degree meter (m^{-2}).

The minus second degree meter equals the transfer (fluence) of ionizing particles at which one particle penetrates a sphere with an area of central section equal 1 m^2.

Preferred unit: cm^{-2}.

Flux density of ionizing particles ϕ is the ratio of ionizing particles flux dF which penetrates into the elementary sphere to the area dS of the central section of this sphere:

$$\phi = dF/dS = d\Phi/dt = d^2N/(dS \cdot dt);$$

dim $\phi = \mathrm{L}^{-2}\mathrm{T}^{-1}$, unit is minus first degree second–minus second degree meter ($s^{-1}{\cdot}m^{-2}$).

The minus first degree second–minus second degree meter is equal to the density of the ionizing particles flux at which one particle penetrates a sphere with an area of the central section of 1 m^2 over 1 s.

Preferred units: $s^{-1}{\cdot}cm^{-2}$, $min^{-1}{\cdot}cm^{-2}$.

Energy flux density of ionizing particles $\phi(E)$ is the ratio of the flux density ϕ of ionizing particles with energy ranging from E to $E + dE$ to the energy interval dE:

$$\phi(E) = d\phi/dE = d^2F/(dS \cdot dE) = d^2\Phi/(dt \cdot dE) = d^3N/(dS \cdot dt \cdot dE);$$

dim $\phi(E) = \mathrm{L}^{-4}\mathrm{M}^{-1}\mathrm{T}$, unit is minus first degree second–minus second degree meter–minus first degree joule ($s^{-1}{\cdot}m^{-2}{\cdot}J^{-1}$).

The minus first degree second–minus second degree meter–minus first degree joule is equal to the energy flux density of ionizing particles at which one particle, whose energy is enclosed in the energy interval of 1 J, penetrates the sphere with an area of the central section of 1 m^2 over 1 s.

Preferred units: $s^{-1}{\cdot}cm^{-2}{\cdot}eV^{-1}$, $s^{-1}{\cdot}cm^{-2}{\cdot}MeV^{-1}$.

Angular flux density of ionizing particles $\phi(\Omega)$ is the ratio of the flux density of ionizing particles $d\phi$ emitted within an elementary solid angle $d\Omega$ oriented in the direction of Ω to this solid angle:

$$\phi(\Omega) = d\phi/d\Omega = d^2\Phi/(dt \cdot d\Omega) = d^2F/(dS \cdot d\Omega) = d^3N/(dS \cdot dt \cdot d\Omega);$$

dim $\phi(\Omega) = \mathrm{L}^{2}\mathrm{T}^{-1}$, unit is a minus first degree second–minus second degree meter–minus first degree steradian ($s^{-1}{\cdot}m^{-2}{\cdot}sr^{-1}$).

The minus first degree second–minus second degree meter–minus first degree steradian equals the angular density of an ionizing particles flux at which one ionizing particle moving in a solid angle of 1 sr intersects a surface with an area of 1 m^2 perpendicular to the direction of the particles movement over 1 second.

Preferred unit: $s^{-1}{\cdot}cm^{-2}{\cdot}sr^{-1}$.

Energy–angular flux density of ionizing particles $\phi(E,\Omega)$ is the ratio of density of the ionizing particles flux ϕ with energy ranging from E to $E + dE$ emitted within a solid angle $d\Omega$ oriented in the Ω direction, to energy interval dE and this solid angle:

$$\phi(E,\Omega) = \frac{d^2\phi}{dE \cdot d\Omega} = \frac{d^3F}{dS \cdot dE \cdot d\Omega} = \frac{d^3\Phi}{dt \cdot dE \cdot d\Omega} = \frac{d^4N}{dS \cdot dt \cdot dE \cdot d\Omega};$$

dim $\phi(E,\Omega) = \mathrm{L}^{-4}\mathrm{M}^{-1}\mathrm{T}$, unit is minus first degree second–minus second degree meter–minus first degree joule–minus first degree steradian ($s^{-1}{\cdot}m^{-2}{\cdot}J^{-1}{\cdot}sr^{-1}$).

The minus first degree second–minus second degree meter–minus first degree joule–minus first degree steradian equals the energy–angular density of an ionizing particles flux at which one ionizing particle with energy enclosed in the energy interval of 1 J, moving in the direction of a solid angle of 1 sr, penetrates the surface with an area of 1 m^2 perpendicular to the direction of the particle movement over 1 s.

Preferred units: $s^{-1}{\cdot}cm^{-2}{\cdot}eV^{-1}{\cdot}sr^{-1}$, $s^{-1}{\cdot}cm^{-2}{\cdot}keV^{-1}{\cdot}sr^{-1}$, $s^{-1}{\cdot}cm^{-2}{\cdot}MeV^{-1}{\cdot}sr^{-1}$.

Energy flux of ionizing radiation F_w equals the ratio of energy of ionizing radiation dw transferred through a given surface over time interval dt, to this interval:

$$F_w = dw/dt;$$

dim $F_w = L^2MT^{-3}$, unit is a watt (W).

The watt equals the energy flux of an ionizing radiation at which a radiation with an energy of 1 J passes through a given surface over 1 s.

Preferred units: nW, μW, mW, W, kW, MW.

Energy transfer (fluence) of an ionizing radiation Φ_w is the ratio of energy of ionizing radiation dw which penetrates the elementary sphere, to area dS of the central section of this sphere:

$$\Phi_w = dw/dS;$$

dim $\Phi_w = MT^{-2}$, unit is joule per square meter (J/m^2).

The joule per square meter is equal to the energy transfer (fluence) of ionizing radiation at which a radiation with an energy of 1 J penetrates the sphere with an area of central section equal to 1 m^2.

Preferred units: fJ/cm^2, pJ/cm^2, nJ/cm^2, $\mu J/cm^2$, mJ/cm^2, J/cm^2, kJ/cm^2, MJ/cm^2.

Flux density of an ionizing radiation energy ϕ_w is the ratio of energy flux of an ionizing radiation dF_w which penetrates the elementary surface, to the area dS of the central section of this sphere:

$$\phi_w = dF_w/dS = d\Phi_w/dt = d^2w/(dS \cdot dt);$$

dim $\phi_w = MT^{-3}$, unit is a watt per square meter (W/m^2).

The watt per square meter equals the energy flux density of the ionizing radiation at which a radiation with an energy of 1 J penetrates a sphere with an area of central section equal to 1 m^2 over 1 s.

Preferred units: nW/m^2, $\mu W/m^2$, mW/m^2, W/m^2, kW/m^2, MW/m^2.

1.2.2 Units and Measures Which Characterize the Interaction of Ionizing Radiation With a Substance

Cross–section of ionizing particles interaction (interaction cross–section) σ_i is the ratio of a number n_i of certain (i-th) type interactions between ionizing particles and target particles in an elementary volume during passing of the ionizing particles with fluence Φ, to the number N of the target particles in this volume and to this fluence:

$$\sigma_i = n_i/(\Phi N);$$

dim $\sigma_i = L^2$, unit is a square meter (m^2).

The square meter equals the interaction cross–section of the ionizing particles at which the fluence of falling particles 1 m^{-2} leads on the average to one act of an interaction of the certain type in 1 m^3 of a substance containing one target particle in 1 m^3.

Preferred unit: fm^2.

Total interaction cross–section of ionizing particles (total interaction cross–section) σ is the sum of all cross–sections σ_i for interacting ionizing particles of a given type which corresponds to various reactions and processes:

$$\sigma = \sum_i \sigma_i;$$

dim $\sigma = L^2$, unit is a square meter (m^2).

The square meter equals the total interaction cross–section of ionizing particles at which in a substance containing one target particle in 1 m^3, the 1 m^{-2} fluence of falling particles leads on the average to one interaction at 1 m^3.

Preferred unit: fm^2.

Macroscopic interaction cross–section of ionizing particles (macroscopic interaction cross–section) Σ_i is the product of the interaction cross–section σ_i by the concentration C of target particles in a substance:

$$\Sigma_i = \sigma_i C;$$

dim $\Sigma_i = L^{-1}$, unit is a minus first degree meter (m^{-1}).

Preferred unit: cm^{-1}.

Linear attenuation factor μ is the ratio of fraction dN/N of indirectly ionizing particles which undergo interaction while passing through the elementary path dl in a substance, to the length of this path:

$$\mu = \frac{1}{N}\frac{dN}{dl};$$

dim $\mu = L^{-1}$, unit is a minus first degree meter (m^{-1}).

The minus first degree meter equals the linear attenuation factor at which on a path of 1 m a density flux in a parallel beam of indirectly ionizing particles reduces e times (e is the base of a natural logarithm).

Preferred unit: cm^{-1}.

Mass attenuation factor μ_m is the ratio of a linear attenuation factor μ to the density ρ of a substance through which an indirectly ionizing particles pass:

$$\mu_m = \frac{\mu}{\rho} = \frac{1}{\rho N}\frac{dN}{dl};$$

dim $\mu_m = L^2M^{-1}$, unit is a square meter per kilogram (m^2/kg).

The square meter per kilogram equals the mass attenuation factor at which on a path of 1 m in a substance with a density of 1 kg/m^3 the flux density in a parallel beam of indirectly ionizing particles reduces e times (e is the base of a natural logarithm).

Preferred unit: cm^2/g.

Atomic attenuation factor μ_a is the ratio of a linear attenuation factor μ to the concentration C of the atoms of a substance through which an indirectly ionizing particles pass:

$$\mu_a = \frac{\mu}{C} = \frac{1}{CN}\frac{dN}{dl};$$

dim $\mu_a = L^2$, unit is a square meter (m^2).

Preferred unit: cm^2.

Linear coefficient of energy transfer μ_{tr} is the ratio of a fraction of energy dw/w of an indirectly ionizing particles (excluding the energy of rest particles) which transforms into kinetic energy of charged particles when they pass through the elementary path dl in a substance, to this path:

$$\mu_{tr} = \frac{1}{w}\frac{dw}{dl};$$

dim $\mu_{tr} = L^{-1}$, unit is a minus first degree meter (m^{-1}).

The minus first degree meter is equal to the linear coefficient of energy transfer at which in a substance on a path of 1 m the density of an energy flux of indirectly ionizing particles reduces e times (e is the base of a natural logarithm).

Preferred unit: cm^{-1}.

Mass coefficient of energy transfer $\mu_{tr,m}$ is the ratio of the linear coefficient of energy transfer μ_{tr} to the density ρ of a substance through which an ionizing radiation passes:

$$\mu_{\text{tr},m} = \frac{\mu_{\text{tr}}}{\rho} = \frac{1}{\rho w}\frac{dw}{dl};$$

dim $\mu_{\text{tr},m} = \text{L}^2\text{M}^{-1}$, unit is a square meter per kilogram (m^2/kg).

The square meter per kilogram equals the mass coefficient of energy transfer at which on a path of 1 m in a substance with a density of 1 kg/m^3 the flux density of indirectly ionizing radiation reduces e times (e is the base of a natural logarithm).

Preferred unit: cm^2/g.

Linear coefficient of energy absorption μ_{en} is the linear coefficient of energy transfer μ_{tr} produced by the difference between a unit and a fraction g of secondary charged–particles energy transferred into the bremsstrahlung in a given substance:

$$\mu_{\text{en}} = \mu_{\text{tr}}(1 - g);$$

dim $\mu_{\text{en}} = \text{L}^{-1}$, unit is a minus first degree meter (m^{-1}).

Preferred unit: cm^{-1}.

Mass coefficient of energy absorption $\mu_{\text{en},m}$ is the ratio of a linear coefficient of energy absorption μ_{en} to density ρ of a substance through which an indirectly ionizing radiation passes:

$$\mu_{\text{en},m} = \frac{\mu_{\text{en}}}{\rho} = \frac{\mu_{\text{tr}}(1-g)}{\rho} = \mu_{\text{tr},m}(1 - g);$$

dim $\mu_{\text{en},m} = \text{L}^2\text{M}^{-1}$, unit is a square meter per kilogram (m^2/kg).

Preferred unit: cm^2/g.

Mean linear path of a charged ionizing particle R is the average modulus of a radius vector between the beginning and end of the path of a charged ionizing particle in a given substance; dim $R =$ L, unit is a meter (m).

Preferred units: μm, mm, cm, m.

Mean mass path of a charged particle R_m is the mean linear path R of a charged ionizing particle in a given substance produced by the density ρ of this substance:

$$R_m = R\rho;$$

dim $R_m = \text{ML}^{-2}$, unit is kilogram per square meter (kg/m^2).

Preferred unit: g/cm^2.

Linear ionization density i is the ratio of the number dn of one-sign ions which were formed by a charged ionizing particle on an elementary path dl, to this path:

$$i = dn/dl;$$

dim $i = \text{L}^{-1}$, unit is a minus first degree meter (m^{-1}).

Preferred units: cm^{-1}, μm^{-1}.

Linear stopping power of substance S is the ratio of lost energy dE of a charged ionizing particle as it passes through an elementary path dl, to the length of this path:

$$S = dE/dl;$$

dim $S = \text{LMT}^{-2}$, unit is a joule per meter (J/m).

Preferred unit: keV/μm.

Mass stopping power of substance S_m is the ratio of the linear stopping power S of a substance to the density ρ of a substance:

$$S_m = \frac{S}{\rho} = \frac{1}{\rho}\frac{dE}{dl};$$

dim $S_m = \mathrm{L^4T^{-2}}$, unit is a joule–square meter per kilogram (J·m^2/kg).

Preferred unit: keV·cm^2/g, MeV·cm^2/g.

Atomic stopping power of substance S_a is the ratio of the linear stopping power S of a substance to the concentration C of the atoms of this substance:

$$S_a = \frac{S}{C} = \frac{1}{C}\frac{dE}{dl};$$

dim $S_a = \mathrm{L^4MT^{-2}}$, unit is a joule–square meter (J·m^2).

Preferred unit: eV·cm^2.

Linear transfer of energy (LTE) L_Δ is the ratio of the energy dE_Δ which a charged ionizing particle transfers to the substance due to collisions along the path dl, to the length of this path:

$$L_\Delta = dE_\Delta/dl;$$

dim L_Δ =LMT^{-2}, unit is a joule per meter (J/m).

Preferred unit: keV/μm.

Mean energy of ion generation W is the ratio of the initial kinetic energy E of a charged ionizing particle to the mean number of ion pairs N formed by this particle until the full loss of its kinetic energy in a given substance:

$$W = E/N;$$

dim $W = \mathrm{L^2MT^{-2}}$, unit is a joule (J).

Preferred unit: eV.

Mass surface density ρ_s is the ratio of mass dm of a layer element with a surface area dS, to this area:

$$\rho_s = dm/dS;$$

dim $\rho_s = \mathrm{ML^{-2}}$, unit is a kilogram per square meter (kg/m^2).

The kilogram per square meter equals the mass surface density at which over a surface layer of 1 m^2 area a mass of 1 kg is evenly distributed.

Preferred units: mg/cm^2, g/cm^2.

1.2.3 Dosimetric Units and Measures

Absorbed dose of ionizing radiation (radiation dose) D is the ratio of mean energy dw transferred by an ionizing radiation to a substance in its elementary volume, to the mass of a substance dm in this volume:

$$D = dm/dw;$$

dim $D = \mathrm{L^2T^{-2}}$, unit is a gray (Gy).

The gray equals the absorbed dose of an ionizing radiation at which an ionizing radiation of 1 J energy is transferred to a substance with a mass of 1 kg.

Preferred units: nGy, μGy, mGy, Gy, kGy, MGy.

Rate of the absorbed dose of an ionizing radiation (radiation dose rate) $\dot{D}$ is the ratio of increment of absorbed dose dD over a time interval dt, to this time interval:

$$\dot{D} = dD/dt;$$

dim $\dot{D} = \mathrm{L}^2\mathrm{T}^{-3}$, unit is a gray per second (Gy/s).

The gray per second equals the rate of an absorbed dose of radiation at which over 1 s a radiation dose of 1 Gy is formed.

Preferred units: mGy/min, Gy/min, mGy/s, kGy/s.

Kerma K is the ratio of the sum of initial kinetic energies dE_k of all the charged ionizing particles formed under the action of indirectly ionizing radiation in elementary volume, to the mass dm of a substance in this volume:

$$K = dE_k/dm;$$

dim $K = \mathrm{L}^2\mathrm{T}^{-2}$, unit is a gray (Gy).

The gray is equal to kerma for which the sum of initial kinetic energies of all the charged ionizing particles formed under the action of indirectly ionizing radiation in a substance with a mass of 1 kg, equals 1 J.

Preferred units: nGy, μGy, mGy, Gy, kGy, MGy.

Kerma rate $\dot{K}$ is the ratio of increment of a kerma dK over a time interval dt, to this time interval:

$$\dot{K} = dK/dt;$$

dim $\dot{K} = \mathrm{L}^2\mathrm{T}^{-3}$, unit is a gray per second (Gy/s).

The gray per second equals the kerma rate at which a 1 Gy kerma is formed over 1 s in a substance.

Preferred units: mGy/min, Gy/min, mGy/s, Gy/s, kGy/s.

Exposure dose of radiation (exposure dose) X is the ratio of total charge dQ of all the one–sign ions formed in air when all the electrons and positrons released by the photons in an elementary volume of air with a mass of dm are fully stopped in air, to the air mass at a given volume:

$$X = dQ/dm;$$

dim $X = \mathrm{M}^{-1}\mathrm{TI}$, unit is a coulomb per kilogram (C/kg).

The coulomb per kilogram equals the exposure dose at which all the electrons and positrons released by photons in air with a mass of 1 kg produce ions carrying an electrical charge of 1 C for each sign.

Exposure dose rate of radiation (exposure dose rate) $\dot{X}$ is the ratio of increment of the exposure dose dX over time interval dt, to this time interval:

$$\dot{X} = dX/dt;$$

dim $\dot{X} = \mathrm{M}^{-1}\mathrm{I}$, unit is an ampere per kilogram (A/kg).

The ampere per kilogram equals the exposure dose rate of radiation at which an exposure dose of 1 C/kg is formed over 1 s.

Note: In transferring physical quantities to SI units, the terms exposure dose and exposure dose rate are no longer in common usage.

Equivalent dose of ionizing radiation (dose equivalent) H is the absorbed dose D produced by the mean quality coefficient k for an ionizing radiation in a given volume element with biological tissue of standard composition:

$$H = Dk;$$

dim $H = \mathrm{L}^2\mathrm{T}^{-2}$, unit is a sievert (Sv).

The sievert equals the dose equivalent at which the absorbed dose in a biological tissue of standard composition produced by a mean quality coefficient is equal to 1 J/kg.

Preferred units: μSv, mSv.

Equivalent dose rate of ionizing radiation (dose equivalent rate) $\dot{H}$ is the ratio of increment dH of an dose equivalent over time interval dt, to this time interval:

$$\dot{H} = dH/dt;$$

dim $\dot{H} = \mathrm{L}^2\mathrm{T}^{-3}$, units is a sievert per second (Sv/s).

The sievert per second equals the dose equivalent rate at which over 1 s a dose equivalent of 1 Sv is formed.

Preferred unit: μSv/h.

1.2.4 Units and Measures Characterizing the Sources of Ionizing Radiation

Radionuclide activity in a source (sample) (activity) A is the numerical ratio dN of spontaneous transitions from a given nuclear–energy state of the radionuclide which occur in the source (sample) over time interval dt, to this time interval:

$$A = dN/dt;$$

dim $A = \mathrm{T}^{-1}$, unit is a becquerel (Bq).

The becquerel equals the activity of a nuclide in a radioactive source wherein one spontaneous transition from a given nuclear–energy state of this radionuclide occurs over 1 s .

Preferred units: Bq, kBq, MBq, GBq, TBq.

Specific source activity A_m is the ratio of activity A of a radionuclide in a source (sample) to the mass m of the source (sample), or the mass of the element or its compound:

$$A_m = A/m;$$

dim $A_m = \mathrm{M}^{-1}\mathrm{T}^{-1}$, the unit is a becquerel per kilogram (Bq/kg).

The becquerel per kilogram equals the specific source activity at which the activity of a radionuclide in a source (element, compound) with a mass of 1 kg is equal to 1 Bq.

Preferred units: Bq/g, kBq/g, MBq/g, GBq/g, TBq/g.

Volume activity of a source A_v is the ratio of activity A of a radionuclide in a source (sample) to its volume V:

$$A_v = A/V;$$

dim $A_v = \mathrm{L}^{-3}\mathrm{T}^{-1}$, unit is a becquerel per cubic meter (Bq/m^3).

The becquerel per cubic meter equals the volume activity of a source when the radionuclide activity in a source with a volume of 1 m^3 is 1 Bq.

Preferred units: Bq/ml, kBq/ml, MBq/ml, GBq/ml, Bq/l, kBq/l, MBq/l, Bq/m^3.

Molar activity of a source A_{mol} is the ratio of activity A of a radionuclide in a source (sample) to the number of moles N of a substance (compound) containing a given radionuclide:

$$A_{\mathrm{mol}} = A/N;$$

dim $A_{\mathrm{mol}} = \mathrm{T}^{-1}\mathrm{N}^{-1}$, unit is a becquerel per mole (Bq/mol).

The becquerel per mole equals the molar activity at which in a source (compound) containing 1 mole of radioactive substance (compound), the activity equals 1 Bq.

Preferred units: Mbq/mol, GBq/mol, MBq/mmol, GBq/mmol, TBq/mmol.

Surface activity of a source A_s is the ratio of activity A of a radionuclide in a source (sample) spread over the surface of a source, to the area S of this surface:

$$A_s = A/S;$$

dim $A_s = \mathrm{L}^{-2}\mathrm{T}^{-1}$, unit is a becquerel per square meter (Bq/m^2).

The becquerel per square meter equals the surface activity at which the activity of a radionuclide (or radionuclides) spread over a surface with an area of 1 m^2, is equal to 1 Bq.

Preferred units: Bq/cm^2, kBq/cm^2, MBq/km^2, GBq/km^2.

Rate constant of air kerma of a radionuclide (kerma constant) Γ_δ is the ratio of the air kerma rate $\dot{K}_\delta$ formed by photons with energy greater than the given threshold value δ from the point isotropic radiation source of a given radionuclide source located in a vacuum at distance l from a source, multiplied by the square of the distance, to the activity A of the source:

$$\Gamma_\delta = \dot{K}_\delta l^2/A;$$

dim $\Gamma_\delta = \mathrm{L}^4\mathrm{T}^{-2}$, the unit is a gray–square meter per second–becquerel (Gy·m^2/s·Bq).

The gray–square meter per second–becquerel equals the constant rate of the air kerma of a radionuclide at which the air kerma rate produced by photons with energy greater than δ and whose point isotropic radiation source is with an activity of 1 Bq in a vacuum at a distance of 1 m, is equal to 1 Gy/s.

Preferred unit: Gy·m^2/(s·Bq).

Kerma-equivalent of a source K_e is the air kerma rate of radiation with a photon energy greater than the given threshold value δ of a point isotropic radiation source located in a vacuum at distance l, multiplied by the square of this distance:

$$K_e = \dot{K}l^2;$$

dim $K_e = \mathrm{L}^4\mathrm{T}^{-3}$, unit is a gray–square meter per second (Gy·m^2/s).

The gray–square meter per second equals the kerma equivalent of a source at which the point isotropic source of photons with an energy greater than δ produces in a vacuum at a distance of 1 m an air kerma rate of 1 Gy/s.

Preferred units: nGy·m^2/s, μGy·m^2/s, mGy·m^2/s, Gy·m^2/s.

Radioactive decay constant of a radionuclide λ is the ratio of fraction of nuclei dN/N for a radionuclide as these nuclei decay over a time interval dt, to this time interval:

$$\lambda = \frac{1}{N}\frac{dN}{dt};$$

dim $\lambda = \mathrm{T}^{-1}$, unit is a minus first degree second (s^{-1}).

The minus first degree second equals the decay constant at which over 1 s the number of nuclei of a radionuclide as a result of radioactive decay decreases by e times (e is the base of a natural logarithm).

Preferred units: s^{-1}, min^{-1}, h^{-1}, d^{-1}, y^{-1}.

Half–life of a radionuclide $T_{1/2}$ is the time during which the number of nuclei in a radionuclide decreases two fold as a result of radioactive decay; dim $T_{1/2}$ = T, unit is a second (s).

Preferred units: s, min, h, d, y.

Average lifetime of a radionuclide τ is the time during which the number of nuclei in a radionuclide decreases e times (e is the base of a natural logarithm) as a result of radioactive decay; dim τ =T, unit is a second (s).

Preferred units: s, min, h, d, y.

1.3 Relationships Between Units of Physical Measure

1.3.1 Length

Angström (Å)	1 Å	$= 10^{-10}$ m
Astronomical unit (AU)	1 AU	$= 1.49597870 \cdot 10^{11}$ m
Inch (in.)	1 in.	= 0.0254 m (exact)
X–unit (X)	1 X	$= 1.00206 \cdot 10^{-13}$ m
Cable length (–)	1 cable length	= 185.2 m
Micron (μm)	1 μm	$= 1 \cdot 10^{-6}$ m
Nautical mile (n.mile)	1 n.mile	= 1852 m (exact)
Land mile (mi)	1 mi	= 1609.344 m (exact)
Parsec (pc)	1 pc	$= 3.08568 \cdot 10^{16}$ m
Light year (l.y.)	1 l.y.	$= 9.46053 \cdot 10^{15}$ m
Fermi (fm)	1 fm	$= 1 \cdot 10^{-15}$ m
Foot (ft)	1 ft	= 0.3048 m (exact)
Yard (yd)	1 yd	= 0.9144 m (exact)

1.3.2 Area

Are (a)	1 a	$= 100$ m^2
Barn (b)	1 b	$= 10^{-28}$ m^2
Hectare (ha)	1 ha	$= 1 \cdot 10^4$ m^2
Square inch (in.^{2})	1 in.^{2}	$= 6.4516 \cdot 10-4$ m^2
Square foot (ft^2)	1 ft^2	$= 9.29030 \cdot 10^{-2}$ m^2
Square yard (yd^2)	1 yd^2	$= 0.83613$ m^2

1.3.3 Volume

Barrel (UK), dry measure	1 bbl(UK)	$= 0.16365$ m^3
Barrel of oil (US)	1 bbl(US,oil)	$= 0.158987$ m^3
Dry barrel (US) [bbl(US)]	1 bbl(US)	$= 0.115628$ m^3
Bushel (UK)	1 bushel(UK)	$= 3.63687 \cdot 10^{-2}$ m^3
Bushel (US) (bu)	1 bu	$= 3.52393 \cdot 10^{-2}$ m^3
Gallon (UK) [gal(UK)]	1 gal(UK)	$= 4.54609 \cdot 10^{-3}$ m^3
Liquid gallon (US) [gal(US)]	1 gal(US)	$= 3.78543 \cdot 10^{-3}$ m^3
Gallon, dry measure (US)	1 gal(US,dry)	$= 4.405 \cdot 10^{-3}$ m^3
Cubic inch (in.^{3})	1 in.^{3}	$= 1.63871 \cdot 10^{-5}$ m^3
Liter (l)	1 l	$= 1 \cdot 10^{-3}$ m^3
Lambda (λ)	1 λ	$= 1 \cdot 10^{-9}$ m^3
Pint (liquid pint) (UK) [pt.(UK)]	1 pt.(UK)	$= 5.68261 \cdot 10^{-4}$ m^3
Liquid pint (US) [liq.pt.(US)]	1 liq.pt.(US)	$= 4.73179 \cdot 10^{-4}$ m^3
Dry pint (UK) [dry pt.(UK)]	1 dry pt.(UK)	$= 4.73179 \cdot 10^{-4}$ m^3
Dry pint (US) [dry pt.(US)]	1 dry pt.(US)	$= 5.50614 \cdot 10^{-4}$ m^3
Ounce (UK) [fl.oz.(UK)]	1 fl.oz.(UK)	$= 2.841 \cdot 10^{-5}$ m^3
Ounce (US) [fl.oz.(US)]	1 fl.oz.(US)	$= 2.957 \cdot 10^{-5}$ m^3
Cubic foot (ft^3)	1 ft^3	$= 2.83168 \cdot 10^{-2}$ m^3
Cubic yard (yd^3)	1 yd^3	$= 0.76455$ m^3

1.3.4 Plane Angle

Gon $(\ldots^g)$	1^g	= 0.01570796 rad
Degree $(\ldots^\circ)$	1°	= 0.01745329 rad
Minute $(\ldots')$	$1'$	$= 2.908882 \cdot 10^{-4}$ rad

Second (...″)	1″	$= 4.848137 \cdot 10^{-6}$ rad
Round angle, revolution (rev)	1 rev	= 6.283185 rad
Right angle (...∟)	1∟	= 1.570796327 rad
Rhumb in meteorology	1 Rhumb(met.)	= 0.392699 rad
Rhumb in sea navigation	1 Rhumb(nav.)	= 0.1963495 rad

1.3.5 Solid Angle

Square degree (□°)	1□°	$= 3.0462 \cdot 10^{-4}$ sr
Full solid angle	1 solid angle	$= 4\pi = 12.56637$ sr

1.3.6 Time

Minute (min)	1 min	= 60 s
Hour (h)	1 h	= 3600 s
Day (d)	1 d	= 86400 s
Year (a)	1 a	$= 3.15569259747 \cdot 10^7$ s (on 1900 year; in 100 years a year contracts of 0.5305 s)

1.3.7 Temperature

Degree Rankine (°R)	1°R	= 0.556 K
Degree Reaumur (°Re)	1°Re	= 1.25 K
Degree Fahrenheit (°F)	1°F	= 0.556 K
Degree Celsius (°C)	1°C	= 1 K

1.3.8 Velocity

Kilometer per hour (km/h)	1 km/h	= 0.2778 m/s
Mile per hour (mi/h)	1 mi/h	= 0.44704 m/s (exact)
Marine knot (kn)	1 kn	= 0.514444 m/s
Foot per second (ft/s)	1 ft/s	= 0.3048 m/s

1.3.9 Acceleration

Gal (Gal)	1 Gal	$= 0.01$ m/s^2

1.3.10 Angular Velocity

Degree per second (°/s)	1 °/s	= 0.0174533 rad/s
Round angle per minute	1 rnd angle p.m.	= 0.1047197 rad/s
Round angle per second	1 rnd angle p.s.	= 6.283185 rad/s
Right angle per second (∟/s)	1 ∟/s	= 1.57080 rad/s
Radian per minute (rad/min)	1 rad/min	= 0.0166667 rad/s

1.3.11 Rotational Frequency

Revolution per minute (rpm)	1 rpm	$= 0.016(6)$ s^{-1}
Revolution per second (rps)	1 rps	$= 1$ s^{-1}

1.3.12 Mass

Unified atomic mass unit (a.m.u. or u)	1 a.m.u.	$= 1.66054110^{-27}$ kg
Gamma (γ)	1 γ	$= 1 \cdot 10^{-9}$ kg
Gram (g)	1 g	$= 1 \cdot 10^{-3}$ kg
Grain (gr)	1 gr	$= 6.479891 \cdot 10^{-5}$ kg
Carat (ct)	1 ct	$= 2 \cdot 10^{-4}$ kg (exact)

Slug (slug)	1 slug	= 14.5939 kg
Technical mass unit (t.m.u.)	1 t.m.u.	= 9.8066 kg (exact)
Ton (t)	1 t	= 1000 kg
Ton (UK) [t(UK)]	1 t(UK)	= 1016.05 kg
Short ton (UK) (sh.t)	1 sh.t	= 907.185 kg
Apothecary ounce (oz apoth.)	1 oz apoth.	= $31.1035 \cdot 10^{-3}$ kg
Commercial ounce (avoirdupois)	1 oz avdp	= $28.3495 \cdot 10^{-3}$ kg
Troy ounce (oz t)	1 oz t	= $31.1035 \cdot 10^{-3}$ kg
Commercial pound (avoirdupois)	1 lb	= 0.45359237 kg
Pound (US) [lb(US)]	1 lb(US)	= 0.4535924277 kg
Centner, multiple SI unit (q)	1 q	= 100 kg
Centner (UK) (cwt)	1 cwt	= 50.8023 kg
Short centner (UK) (sh.cwt)	1 sh.cwt	= 45.3592 kg

1.3.13 Density

Gram per cubic inch (g/in.^{3})	1 g/in.^{3}	= 61.0 kg/m^3
Gram per cubic meter	1 g/m^3	= $1 \cdot 10^{-3}$ kg/m^3
Gram per cubic centimeter (g/cm^3)	1 g/cm^3	= $1 \cdot 10^{3}$ kg/m^3
Gram per cubic foot (g/ft^3)	1 g/ft^3	= $3.53 \cdot 10^{-2}$ kg/m^3
Gram per liter (g/l)	1 g/l	= 1 kg/m^3
Ounce per cubic inch (oz/in.^{3})	1 oz/in.^{3}	= $1.73 \cdot 10^{3}$ kg/m^3
Ounce per cubic centimeter (oz/cm^3)	1 oz/cm^3	= $2.835 \cdot 10^{4}$ kg/m^3
Ounce per cubic foot (oz/ft^3)	1 oz/ft^3	= 1.0014 kg/m^3
Pound per cubic foot (lb/ft^3)	1 lb/ft^3	= 16.0185 kg/m^3

1.3.14 Linear Density

Mass per unit length (tex)	1 tex	= $1 \cdot 10^{-6}$ kg/m (exact)

1.3.15 Force

Dyne (dyn)	1 dyn	= $1 \cdot 10^{-5}$ N
Kilogram–force (kgf)	1 kgf	= 9.80665 N (exact)
Kilopond (kpond)	1 kpond	= 9.80665 N (exact)
Poundal (pdl)	1 pdl	= 0.138255 N (exact)
Pond (pond)	1 pond	= $9.80665 \cdot 10^{-3}$ N (exact)
Sthene (sn)	1 sn	= $1 \cdot 10^{3}$ N
Ton–force (tf)	1 tf	= $9.80665 \cdot 10^{3}$ N (exact)
Pound–force (lbf)	1 lbf	= 4.44822 N

1.3.16 Pressure

Technical atmosphere (at)	1 at	= $9.80665 \cdot 10^{4}$ Pa (exact)
Standard atmosphere (atm)	1 atm	= $1.01325 \cdot 10^{5}$ Pa (exact)
Bar (bar)	1 bar	= $1 \cdot 10^{5}$ Pa
Dyne per square centimeter (dyn/cm^2)	1 din/cm^2	= 0.1 Pa
Inch of water column (in. H_2O)	1 in. H_2O	= 249.089 Pa
Inch of mercury column (in. Hg)	1 in. Hg	= 3386.39 Pa
Kilogram–force per square meter (kgf/m^2)	1 kgf/m^2	= 9.80665 Pa (exact)
Kilogram–force per square centimeter (kgf/cm^2)	1 kgf/cm^2	= $9.80665 \cdot 10^{4}$ Pa (exact)
Kilopond per square centimeter (kp/cm^2)	1 kp/cm^2	= $9.80665 \cdot 10^{4}$ Pa (exact)
Millimeter of water column (mm H_2O)	1 mm H_2O	= 9.80665 Pa (exact)
Conventional millimeter of mercury (mmHg)	1 mmHg	= 133.32239 Pa
Pieze (pz)	1 pz	= $1 \cdot 10^{3}$ Pa

Ton–force per square meter (tf/m^2)	1 tf/m^2	= $9.80665 \cdot 10^3$ Pa (exact)
Torr [Torr, mm Hg]	1 Torr	= 133.32237 Pa
Pound–force per square inch ($lbf/in.^2$)	1 $lbf/in.^2$	= $6.89476 \cdot 10^3$ Pa
Foot of water column (ft H_2O)	1 ft H_2O	= $2.98907 \cdot 10^3$ Pa

1.3.17 Momentum

Gram–centimeter per second (g·cm/s)	1 g·cm/s	= $1 \cdot 10^{-5}$ kg·m/s
Kilogram–force–second (kgf·s)	1 kgf·s	= 9.80665 kg·m/s
Ton–meter per second (t·m/s)	1 t·m/s	= $1 \cdot 10^3$ kg·m/s

1.3.18 Moment of Force

Dyne–centimeter (dyn·cm)	1 dyn·cm	= $1 \cdot 10^{-7}$ N·m
Kilogram–force–meter (kgf·m)	1 kgf·m	= 9.80665 N·m (exact)
Kilopond–meter (kpond·m)	1 kpond·m	= 9.80665 N·m
Pound–force–foot (lbf·ft)	1 lbf·ft	= 1.35582 N·m

1.3.19 Angular Momentum

Gram–square centimeter per second (g·cm^2/s)	1 g·cm^2/s	= $1 \cdot 10^{-7}$ kg·m^2/s
Kilogram–force–meter–second (kgf·m·s)	1 kgf·m·s	= 9.80665 kg·m^2·s
Ton–square meter per second (t·m^2/s)	1 t·m^2/s	= $1 \cdot 10^3$ kg· m^2/s

1.3.20 Stress

Kilogram–force per square millimeter (kgf/mm^2)	1 kgf/mm^2	= $9.80665 \cdot 10^6$ Pa
Kilopond per square millimeter ($kpond/mm^2$)	1 $kpond/mm^2$	= $9.80665 \cdot 10^6$ Pa (exact)

1.3.21 Work, Energy

Watt–hour (W·h)	1 W·h	= 3600 J
Kilowatt–hour (kW·h)	1 kW·h	= $3.6 \cdot 10^6$ J
Kilogram–force–meter (kgf·m)	1 kgf·m	= 9.80665 J
Kilopond–meter (kpond·m)	1 kpond·m	= 9.80665 J
Liter–atmosphere (l·atm)	1 l·atm	= 101.328 J
Horsepower–hour (hp·h)	1 hp·h	= $2.64780 \cdot 10^6$ J
Pound–force–foot (lbf·ft)	1 lbf·ft	= 1.35582 J
Erg (erg)	1 erg	= $1 \cdot 10^{-7}$ J

1.3.22 Power

Calorie per second (cal/s)	1 cal/s	= 4.1868 W
Kilogram–force–meter per second (kgf·m/s)	1 kgf·m/s	= 9.80665 W
Horsepower (hp)	1 hp	= 735.50 W
Horsepower (UK) [hp(UK)]	1 hp(UK)	= 745.700 W
Erg per second (erg/s)	1 erg/s	= $1 \cdot 10^{-7}$ W

1.3.23 Dynamic Viscosity

Kilogram–force–second per square meter (kgf·s/m^2)	1 kgf·s/m^2	= 9.80665 Pa·s
Poundal–second per square foot (pdl·s/ft^2)	1 pdl·s/ft^2	= 1.48816 Pa·s
Poise (P)	1 P	= 0.1 Pa·s
Pound–force–second per square meter (lbf·s/m^2)	1 lbf·s/m^2	= 47.8803 Pa·s

1.3.24 Kinematic Viscosity

Square meter per hour (m^2/h)	1 m^2/h	$= 2.77(7)\cdot 10^{-4}$ m^2/s
Square foot per second (ft^2/s)	1 ft^2/s	$= 0.0929030$ m^2/s
Square foot per hour (ft^2/h)	1 ft^2/h	$= 2.58064\cdot 10^{-5}$ m^2/s
Stokes (St)	1 St	$= 1\cdot 10^{-4}$ m^2/s

1.3.25 Bulk Flow Rate

Cubic inch per second ($in.^3/s$)	1 $in.^3/s$	$= 1.6387\cdot 10^{-5}$ m^3/s
Cubic centimeter per second (cm^3/s)	1 cm^3/s	$= 1\cdot 10^{-6}$ m^3/s
Cubic foot per second (ft^3/s)	1 ft^3/s	$= 0.0283168$ m^3/s
Liter per minute (l/min)	1 l/min	$= 1.66(6)\cdot 10^{-5}$ m^3/s
Liter per hour (l/h)	1 l/h	$= 2.77(7)\cdot 10^{-8}$ m^3/s

1.3.26 Heat

British thermal unit (Btu)	1 Btu	$= 1.05506\cdot 10^3$ J
International Table calorie (cal_{IT})	1 cal_IT	= 4.1868 J
Fifteen degree calorie (cal_{15})	1 cal_{15}	= 4.1855 J
Thermochemical calorie (cal_th)	1 cal_th	= 4.1840 J
Therm (th)	1 th	$= 4.1868\cdot 10^3$ J

1.3.27 Specific Heat Capacity

Calorie per gram–degree Celsius (cal/g·°C)	1 cal/g·°C	$= 4.1868\cdot 10^3$ J/(kg·K)
Kilocalorie per kilogram–degree Celsius (kcal/kg·°C)	1 kcal/kg·°C	$= 4.1868\cdot 10^3$ J/(kg·K)

1.3.28 Molar Heat Capacity

Calorie per mole–degree Celsius (cal/mol·°C)	1 cal/mol·°C	= 4.1868 J/(mol·K)

1.3.29 Thermal Conductivity

British thermal unit per second–foot–degree Fahrenheit [Btu/(s·ft·°F)]	1 Btu/(s·ft·°F)	$= 6.23064\cdot 10^3$ W/(m·K)
Calorie per second–centimeter–degree Celsius [cal/(s·cm·°C)]	1 cal/(s·cm·°C)	= 418.7 W/(m·K)
Kilocalorie per hour–meter–degree Celsius [kcal/(h·m·°C)]	1 kcal/(h·m·°C)	= 1.163 W/(m·K)
Erg per second–centimeter–kelvin [erg/(s·cm·K)]	1 erg/(s·cm·K)	$= 1\cdot 10^{-5}$ W/(m·K)

1.3.30 Electric Charge

Faraday (F)	1 F	= 96,484.56 C
Franklin (Fr)	1 Fr	$= 3.33564\cdot 10^{-10}$ C

1.3.31 Electric Field Strength

Volt per centimeter (V/cm)	1 V/cm	= 100 V/m

1.3.32 Electric Dipole Moment

Debye (D)	1 D	$= 3.33564\cdot 10^{-30}$ C·m

1.3.33 Current Density

Ampere per square millimeter (A/mm^2)	1 A/mm^2	$= 1\cdot 10^6$ A/m^2

1.3.34 Specific Electric Resistance

Ohm–square millimeter per meter [(Ω· mm²/m)]	1 $\Omega\cdot mm^2/m$	$= 1\cdot 10^{-6}\ \Omega\cdot m$
Ohm–centimeter [(Ω·cm)]	1 $\Omega\cdot cm$	$= 0.01\ \Omega\cdot m$

1.3.35 Specific Electric Conductance

Minus first degree ohm–minus first degree centimeter [($\Omega^{-1}\cdot cm^{-1}$)]	1 $\Omega^{-1}\cdot cm^{-1}$	= 100 S/m
Meter per ohm–square millimeter [m/(Ω·mm²)]	1 $m/(\Omega\cdot mm^2)$	$= 1\cdot 10^{6}$ S/m

1.3.36 Magnetic Induction

Gauss (G, Gs)	1 G	$= 1\cdot 10^{-4}$ T
Weber per square centimeter (Wb/cm^2)	1 Wb/cm^2	$= 1\cdot 10^{4}$ T

1.3.37 Magnetic Flux

Maxwell (Mx)	1 Mx	$= 1\cdot 10^{-8}$ Wb

1.3.38 Magnetic Field Strength

Örsted (Oe)	1 Oe	= 79.5775 A/m
Ampere per centimeter (A/cm)	1 A/cm	= 100 A/m
Ampere–turn per centimeter (At/cm)	1 At/cm	= 100 A/m

1.3.39 Magnetomotive Force

Gilbert (Gi, Gb)	1 Gi	= 0.795775 A
Ampere–turn (At)	1 At	= 1 A

1.3.40 Magnetic Moment

Bohr magneton (μ_B)	1 μ_B	$= 9.274010\cdot 10^{-24}\ A\cdot m^2$

1.3.41 Brightness

Apostilb (asb)	1 asb	$= 0.318310\ kd/m^2$
Lambert (Lb)	1 Lb	$= 0.318310\cdot 10^{4}\ kd/m^2$
Stilb (sb)	1 sb	$= 1\cdot 10^{4}\ kd/m^2$

1.3.42 Absorbed Dose

Rad (rad)	1 rad	= 0.01 Gy

1.3.43 Equivalent Dose

Röntgen equivalent, man (rem.)	1 rem.	= 0.01 J/kg

1.3.44 Nuclide Activity in a Radioactive Source

Curie (Ci)	1 Ci	$= 3.700\cdot 10^{10}$ Bq (exact)

1.3.45 Exposure Dose of X–Ray and γ–Radiation

Röntgen (R)	1 R	$= 2.58\cdot 10^{-4}$ C/kg (exact)

1.4 Fundamental Physical Constants

1.4.1 Universal Constants

Speed of light in vacuum	c	$= 299\ 792\ 458$ m/s (exact)
Magnetic constant	μ_0	$= 4\pi \cdot 10^{-7} = 12.566370614 \cdot 10^{-7}$ N·A^{-2} (exact)
Electric constant	ϵ_0	$= 8.854187817 \cdot 10^{-12}$ F/m (exact)
Gravitational constant	G	$= 6.67259(85) \cdot 10^{-11}$ m^3/kg·s^2
Planck constant	h	$= 6.6260755(40) \cdot 10^{-34}$ J·s
Planck mass	$(\hbar c/G)^{1/2}$	$= m_{\rm P} = 2.176\ 71(14) \cdot 10^{-8}$ kg
Planck length	$\hbar/(m_{\rm P}c)$	$= l_{\rm P} = 1.616\ 05(10) \cdot 10^{-35}$ m
Planck time	$l_{\rm P}/c$	$= t_{\rm P} = 5.390\ 56(34) \cdot 10^{-44}$ s

1.4.2 Electromagnetic Constants

Elementary charge	e	$= 1.60217733(49) \cdot 10^{-19}$ C
Magnetic flux quantum	Φ_0	$= h/2e = 2.06783461(61) \cdot 10^{-15}$ Wb
Josephson constant	$2e/h$	$= 4.8359767(14) \cdot 10^{14}$ Hz/V
Quantized Hall resistance	$R_{\rm H}$	$= h/e^2 = 25812.8056(12) = \Omega$
Bohr magneton	$\mu_{\rm B}$	$= e\hbar/2m_e = 9.2740154(31) \cdot 10^{-24}$ J/T
Nuclear magneton	$\mu_{\rm N}$	$= e\hbar/2m_p = 5.0507866(17) \cdot 10^{-27}$ J/T
	$\mu_{\rm N}/e$	$= 3.15245166(28) \cdot 10^{-8}$ eV/T

1.4.3 Atomic Constants

Fine–structure constant	α	$= \mu_0 c e^2/2h = 7.29735308(33) \cdot 10^{-3}$
	$1/\alpha$	$= 137.0359895(61)$
Rydberg constant	R_∞	$= m_e c\alpha^2/2h = 10973731.534(13)$ m^{-1}
	$R_\infty c$	$= 3.2898419499(39) \cdot 10^{15}$ Hz
	$R_\infty hc/e$	$= 13.6056981(40)$ eV
Bohr radius	a_0	$= \alpha/4\pi R_\infty = 0.529177249(24) \cdot 10^{-10}$ m
Hartree energy	$E_{\rm h}$	$= 2R_\infty hc = 4.3597482(26) \cdot 10^{-18}$ J
	$E_{\rm h}/e$	$= 27.2113961(81)$ eV

1.4.4 Electron

Electron mass	m_e	$= 9.1093897(54) \cdot 10^{-31}$ kg $= 5.485\ 799\ 03(13)$ u
	$m_e c^2/e$	$= 0.51099906(15)$ MeV
Electron specific charge	$-e/m_e$	$= -1.75881962(53) \cdot 10^{11}$ C/kg
Electron molar mass	M_e	$= 5.48579903(13) \cdot 10^{-7}$ kg/mol
Compton wavelength	$\lambda_{\rm C}$	$= h/m_e c = 2.42631058(22) \cdot 10^{-12}$ m
Classical electron radius	r_e	$= \alpha^2 a_0 = 2.81794092(38) \cdot 10^{-15}$ m
Thomson cross–section	σ_e	$= (8\pi/3)r_e^2 = 0.66524616(18) \cdot 10^{-28}$ m^2
Electron magnetic moment	μ_e	$= 928.47701(31) \cdot 10^{-26}$ J/T
	$\mu_e/\mu_{\rm B}$	$= 1.001159652193(10)$
Magnetic moment anomaly	a_e	$= \mu_e/\mu_{\rm B} - 1 = 1.159652193(10) \cdot 10^{-3}$
g–factor	g_e	$= 2(1 + a_e) = 2.002319304386(20)$

1.4.5 Muon

Muon mass	m_μ	$= 1.8835327(11) \cdot 10^{-28}$ kg
Muon magnetic moment	μ_μ	$= 4.4904514(15) \cdot 10^{-26}$ J/T
	$\mu_\mu/\mu_{\rm N}$	$= 8.8905981(13)$
Magnetic moment anomaly	a_μ	$= \frac{\mu_\mu}{(e\hbar/2m_\mu)} - 1 = 1.1659230(84) \cdot 10^{-3}$
g–factor	g_μ	$= 2(1 + a_\mu) = 2.002331846(17)$

1.4.6 Proton

Proton mass	m_p	$= 1.6726231(10) \cdot 10^{-27}$ kg$= 1.007276470(12)$ u
	m_pc^2/e	$= 938.27231(28)$ MeV
Proton–electron mass ratio	m_p/m_e	$= 1,836.152701(37)$
Proton specific charge	e/m_p	$= 9.5788309(29) \cdot 10^{7}$ C/kg
Compton wavelength	$\lambda_{C,p}$	$= 1.321\ 41002(12) \cdot 10^{-15}$ m
Magnetic moment	μ_p	$= 1.41060761(47) \cdot 10^{-26}$ J/T
	μ_p/μ_N	$= 2.792847386(63)$
Diamagnetic shielding correction, spherical sample of pure water at 25°C	σ_{H_2O}	$= 1 - \mu'_p/\mu_p = 25.689(15) \cdot 10^{-6}$
Magnetic moment uncorrected for diamagnetism	μ'_p	$= 1.410571\ 38(47) \cdot 10^{-26}$ J/T
Gyromagnetic ratio	γ_p	$= 26,752.2128(81) \cdot 10^{4}\ s^{-1}T^{-1}$

1.4.7 Neutron

Neutron mass	m_n	$= 1.6749286(10) \cdot 10^{-27}$ kg$= 1.008664904(14)$ u
Neutron–proton mass ratio	m_n/m_p	$= 1.001\ 378404(9)$
Compton wavelength	$\lambda_{C,n}$	$= h/m_nc = 1.319591\ 10(12) \cdot 10^{-15}$ m
Magnetic moment	μ_n	$= 0.96623707(40) \cdot 10^{-26}$ J/T

1.4.8 Physico–Chemical Constants

Avogadro constant	N_A	$= 6.0221367(36) \cdot 10^{23}\ mol^{-1}$
Atomic mass unit	1 u	$= m_u = m(^{12}C)/12 = 1.6605402(10) \cdot ^{-27}$ kg
Standard atmosphere	1 atm	$= 101,325$ Pa
Boltzmann constant	k	$= R/N_A = 1.380658(12) \cdot 10^{-23}$ J/K
First radiation constant	c_1	$= 2\pi hc^2 = 3.741\ 7749(22) \cdot 10^{-16}\ W{\cdot}m^2$
Second radiation constant	c_2	$= hc/k = 0.01438769(12)$ m·K
Molar volume (ideal gas at $T = 273.15$ K, $p = 101,325$ Pa)	V_m	$= RT/p = 22.41410(19)$ l/mol
Stefan–Boltzmann constant	σ	$= \pi^2k^4/(60\hbar^3c^2) = 5.67051(19) \cdot 10^{-8}\ W/(m^2{\cdot}K^4)$
Standard acceleration of free fall	g_n	$= 9.80665\ m{\cdot}s^{-2}$
Faraday constant	F	$= N_Ae = 96,485.309(29)$ C/mol
Electron volt	1 eV	$= 1.60217733(49) \cdot 10^{-19}$ J

1.5 Relationships between Units of Physical Measure

Table 1.1 Forming decimal and sub–multiples of SI units with SI prefixes

Factor	Prefix	Symbol	Factor	Prefix	Symbol	Factor	Prefix	Symbol	Factor	Prefix	Symbol
10^{18}	exa	E	10^{6}	mega	M	10^{-1}	deci	d	10^{-9}	nano	n
10^{15}	peta	P	10^{3}	kilo	k	10^{-2}	centi	c	10^{-12}	pico	p
10^{12}	tera	T	10^{2}	hecto	h	10^{-3}	milli	m	10^{-15}	femto	f
10^{9}	giga	G	10^{1}	deka	da	10^{-6}	micro	μ	10^{-18}	atto	a

Notes:

(1) In accordance with international standard ISO 31/0-74 decimal and fractional units are not SI units.

(2) The prefixes hecto, deka, deci and centi may only be used in designating widely accepted divisible and fractional units (e.g., centimeter, millimeter, etc.).

In complex designations of a compound SI unit the prefix is joined to first unit forming the series or numerator of the fraction, e.g., kPa·s/m, but not Pa·ks/m.

By way of exception to this rule there are certain well–established and widely accepted cases when a prefix may temporarily be joined to the unit of the fractional denominator, e.g., kV/cm, A/mm, Bq/ml, keV/μm.

The choice of a decimal or fractional unit from an SI unit is essentially a matter of practical convenience. While a multitude of decimal and fractional units can be formed with prefixes, only those units are selected which yield acceptable numerical values. By and large, decimal and fractional units are chosen in such a way that the numerical values range from 0.1 to 1,000. In Table 1.1 the recommended decimal and fractional units derived from SI units are illustrated.

In order to reduce the likelihood of errors in calculation, decimal and fractional decimal units should be prefixed only in the final instance, while in the computational process all quantities should be expressed in SI units, replacing all prefixes with degrees of the number 10.

Aside from standard decimal and fractional units, the use of the latter is allowed in cases involving nondecimal units of time, solid angles and relative quantities, e.g., in units of time (minute, hour, day), and a plane angle (degree, minute, second).

Table 1.2 List of relative and logarithmic quantities and their units

Name of measure	Unit			Note
	Name	Symbol	Definition	
1. A relative quantity (dimensionless ratio of physical quantity to an analogous physical quantity considered as the initial one): efficiency, relative elongation, relative density, relative dielectric permittivity and magnetic permeability, magnetic susceptibility, mass fraction, molar fraction, etc.	unity per cent pro mille parts per million	1 % ‰ ppm	1 10^{-2} 10^{-3} 10^{-6}	
2. A logarithmic quantity (logarithm of dimensionless ratio of a physical quantity to an analogous one taken as the reference quantity): acoustic pressure level, gain, attenuation, etc.[*1]	bel	B	$1\ \mathrm{B} = \lg(P_2/P_1)$ at $P_2 = 10P_1$ $1\ \mathrm{B} = 2\lg(F_2/F_1)$ at $F_2 = \sqrt{10}\,F_1$	P_1, P_2 are analogous energy quantities (power, energy, density of energy, etc.) F_1, F_2 are analogous "force" quantities (voltage, current, pressure, field intensity, etc.)
	decibel	dB	0.1 B	
3. The same: loudness level	phon	phon	1 phon equals the sound loudness level at which the acoustic pressure of the loudness-equivalent sound with the frequency of 1000 Hz is equal to 1 dB	
4. The same: frequency interval	octave		1 octave $= \log_2(f_2/f_1)$ at $f_2/f_1 = 2$	f_1, f_2 are frequencies
	decade		1 decade $= \lg(f_2/f_1)$ at $f_2/f_1 = 2$	

[*1] In accordance with Publication 27-3 of the International Electrotechnical Commission (IEC), when it is necessary to indicate the initial quantity, its value should be bracketed after stating the logarithmic quantity, e.g., for an acoustic pressure level L_p(re 20 μPa)=20 dB (re stands for reference, i.e. the initial value). For the short form of notation, the initial value is shown in parentheses after indicating the level values, viz. 20 dB (re 20 μPa).

Table 1.3 Other units which can be used with SI units

Quantity	Unit		Relation to SI unit
	Name	Symbol	
Mass	ton	t	10^3 kg
	(unified) atomic mass unit	u	$1.660\ 541 \cdot 10^{-27}$ kg (approx.)
Time*[1]	minute	min	60 s
	hour	h	3600 s
	day	d	86400 s
Plane angle	degree	$\ldots^\circ$	$(\pi/180)$ rad$=1.745329 \cdot 10^{-2}$ rad
	minute	$'$	$(\pi/10800)$ rad$=2.908882 \cdot 10^{-4}$ rad
	second	$''$	$(\pi/648000)$ rad$=4.848137 \cdot 10^{-6}$ rad
	gon	gon	$(\pi/200)$ rad
Volume*[2]	liter	l or L	10^{-3} m^3
Length	astronomical unit	A.U.	$1.45598 \cdot 10^{11}$ m (approx.)
	light year	l.y.	$9.4605 \cdot 10^{15}$ m (approx.)
	parsec	pc	$3.0857 \cdot 10^{16}$ m (approx.)
Vergence of optical systems	diopter		1 m^{-1}
Area	hectare	ha	10^4 m^2
Energy	electron–volt	eV	$1.60219 \cdot 10^{-19}$ J (approx.)
Total power	volt–ampere	V·A	
Reactive power	volt–ampere–reactive	var	

*[1] Other widely accepted units may be used including week, month, year, century, millenium, etc.

*[2] Not recommended for accurate calculations. When l (the letter el) may be confused with the digit 1 the designation L may be used.

Note: Units of time (minute, hour, day), plane angle (degree, minute, second), astronomical unit, light year, diopter and atomic mass unit may not be used with a prefix.

Table 1.4 Units temporarily permitted in usage

Quantity	Unit		Relation to SI	Note
	Name	Symbol		
Length	nautical mile	n. mile	1852 m (exact)	Marine navigation
Mass	metric carat	ct	$2 \cdot 10^{-4}$ kg (exact)	For precious stones and gems
Linear density	tex	tex	10^{-6} kg/m (exact)	Used in textile industry
Velocity	knot	kn	0.5144(4) m/s	Marine navigation
Pressure	bar	bar	10^5 Pa	
Rotational frequency	revolution per second	r.p.s.	1 s^{-1}	
	revolution per minute	r.p.m.	1/60 s^{-1}	
Natural logarithm of a dimensionless ratio of physical quantity to an analogous physical quantity taken as the reference one	neper	Np		1 Np = 0.8686 B = 8.686 dB

Note: Units illustrated in Table 1.4 are temporarily accepted until final international authorization is granted.

Table 1.5 The relation of other units to SI units

Quantity	Unit		Relation to SI unit
	Name	Symbol	
Length	micron	μm	10^{-6} m
	angström	Å	10^{-10} m
	X–unit	X	$1.00206 \cdot 10^{-13}$ m (approx.)

Table 1.5 The relation of other units to SI units *(continued)*

Quantity	Unit		Relation to SI unit
	Name	Symbol	
Area	are	a	100 m^2
Cross–section	barn	b	10^{-28} m^2
Deflection angle	revolution	r	2π rad=6.28... rad
Solid angle	square degree	□°	3.0462... 10^{-4} sr
Mass	centner	q	100 kg
Acceleration	gal	Gal	0.01 m/s^2
Force, weight	dyne	dyn	10^{-5} N
	kilogram–force	kgf	9.80665 N (exact)
	kilopond	kpond	9.80665 N (exact)
	gram–force	gf	$9.80665 \cdot 10^{-3}$ N (exact)
	pond	pond	$9.80665 \cdot 10^{-3}$ N (exact)
Pressure	kilogram–force per cm^2	kgf/cm^2	98,066.5 Pa (exact)
	kilopond per cm^2	$kpond/cm^2$	98,066.5 Pa (exact)
	millimeter of a water column	mm H_2O	9.80665 Pa (exact)
	standard atmosphere	atm	101325 Pa (exact)
	torr (millimeter of a mercury column)	Torr (mm Hg)	101325/760 = 133.32237 Pa
	conventional millimeter of mercury	mmHg	$13.5951 \cdot 9.80665 =$ 133.32239 Pa
Stress	kilogram–force per mm^2	kgf/mm^2	$9.80665 \cdot 10^6$ Pa (exact)
	kilopond per mm^2	$kpond/mm^2$	$9.80665 \cdot 10^6$ Pa (exact)
Energy, work	erg	erg	10^{-7} J
Heat, thermodynamic potentials (internal energy, enthalpy, etc.), phase transition heat, chemical reaction heat	IT calorie (int.)	cal_{IT}	4.1868 J (exact)
	thermochemical calorie	cal_{th}	4.1840 J (approx.)
	15–degree calorie	cal_{15}	4.1855 J (approx.)
Power	horse power	hp	735.499 W
Dynamic viscosity	poise	P	0.1 Pa·s
Kinematic viscosity	stokes	St	10^{-4} m^2/s
Specific electrical resistivity	ohm–square millimeter per meter	$\Omega \cdot mm^2/m$	10^{-6} $\Omega \cdot m$
Magnetic field strength	oersted	Oe	$10^3/(4\pi)$=79.5775... A/m
Magnetic flux	maxwell	Mx	10^{-8} Wb
Magnetic induction	gauss	G (or Gs)	10^{-4} T
Magnetomotive force,	gilbert	Gb	$10/(4\pi)$ A=0.795775... A
magnetic potential difference	ampere–turn	At	1 A
Brightness	nit	nt	1 kd/m^2
Absorbed dose of radiation	rad	rad, rd	0.01 Gy
Dose equivalent of radiation, radiation equivalent measure	rem	rem	0.01 Sv
Exposure dose of radiation (exposure dose of γ- and X-ray radiation)	röntgen	R	$2.58 \cdot 10^{-4}$ C/kg (exact)
Nuclide activity in radioactive source	curie	Ci	$3.700 \cdot 10^{10}$ Bq (exact)

Table 1.6 Relationships between electromagnetic measures in SGS and SI systems

Quantity	Relationship between units	
Electric current intensity	1 SGSE=$3.33564 \cdot 10^{-10}$ A	1 SGSM=10 A
Electric charge	1 SGSE=$3.33564 \cdot 10^{-10}$ C	1 SGSM=10 C
Surface density of an electric charge	1 SGSE=$3.33564 \cdot 10^{-6}$ C/m^2	1 SGSM=10^5 C/m^2
Spatial density of an electric charge	1 SGSE=$3.33564 \cdot 10^{-4}$ C/m^3	1 SGSM=10^7 C/m^3

Table 1.6 Relationships between electromagnetic measures in SGS and SI systems *(continued)*

Quantity	Relationship between units	
Electric field strength	1 SGSE=2.997925·10^4 V/m	1 SGSM=10^{-6} V/m
Electrical voltage, electrical potential, electromotive force	1 SGSE=2.997925·10^2 V	1 SGSM=10^{-8} V
Electric displacement	1 SGSE=2.65442·10^{-7} C/m^2	1 SGSM=0.795775·10^3 C/m^2
Electric displacement flux	1 SGSE=2.65442·10^{-11} C	1 SGSM=795,775 C
Electric capacity	1 SGSE=1.11265·10^{-12} F	1 SGSM=10^9 F
Absolute dielectric permittivity	1 SGSE=8.854187·10^{-12} F/m	1 SGSM=7.95775·10^9 F/m
Electric dipole moment	1 SGSE =3.33564·10^{-12} C·m	1 SGSM=0.1 C·m
Electric current density	1 SGSE =3.33564·10^{-6} A/m^2	1 SGSM=10^5 A/m^2
Linear density of electric current	1 SGSE=3.33564·10^{-8} A/m	1 SGSM=10^3 A/m
Magnetic field strength	1 SGSE=2.65442·10^{-9} A/m	1 SGSM=1 Oe=79.5775 A/m
Magnetomotive force, difference of magnetic potentials	1 SGSE=2.65442·10^{-11} A	1 SGSM=1 Gb=0.795775 A
Magnetic induction	1 SGSE=2.997925·10^6 T	1 SGSM=1 G= 10^{-4} T
Magnetic flux	1 SGSE=299.7925 Wb	1 SGSM=1 Mx=10^{-8} Wb
Inductance, mutual inductance	1 SGSE=8.98755·10^{11} H	1 SGSM=10^{-9} H
Absolute magnetic permeability	1 SGSE=1.12941·10^{15} H/m	1 SGSM=1.256637·10^{-8} H/m
Ampere's magnetic moment	1 SGSE=3.33564·10^{-14} A·m^2	1 SGSM=10^{-3} A·m^2
Coulomb's magnetic moment	1 SGSE=2.997925 Wb·m	1 SGSM=10^{-10} Wb·m
Magnetization	1 SGSE=3.33564·10^{-8} A/m	1 SGSM=10^3 A/m
Electric resistance	1 SGSE=8.98755·10^{11} Ω	1 SGSM=10^{-9} Ω
Electric conductance	1 SGSE=1.11265·10^{-12} S	1 SGSM=10^9 S
Specific electric resistance	1 SGSE=8.98755·10^9 Ω·m	1 SGSM=10^{-9} Ω·m
Specific electric conductance	1 SGSE=1.11265·10^{-10} S/m	1 SGSM=10^{11} S/m
Magnetic resistance	1 SGSE=8.854186·10^{-14} A/Wb	1 SGSM=79.5775·10^6 A/Wb
Magnetic conductance	1 SGSE=1.12941·10^{13} Wb/A	1 SGSM=1.256637·10^{-8} Wb/A

Table 1.7 Correspondence between units of radioactivity and ionizing radiation and SI units

Quantity	Out–of–system unit	Relation to SI unit
Flux density of ionizing particles	1/(cm^2·h)	2.778 m^{-2}s^{-1}
Radiant intensity	erg/(cm^2·s)	10^{-3} W/m^2
	erg/(cm^2·min)	1.667·10^{-5} W/m^2
	erg/cm^2·h)	2.778·10^{-7} W/m^2
Absorbed dose of radiation	erg/g	10^{-4} J/kg
	rad	0.01 Gy
Rate of absorbed dose of radiation	erg/(g·s)	10^{-4} W/kg
	rad/s	0.01 Gy/s
	rad/h	2.778·10^{-6} Gy/s
Exposure dose of X-ray and γ-radiation	R (röntgen)	2.58·10^{-4} C/kg
Rate of exposure dose of X-ray and γ-radiation	R/s	2.58·10^{-4} A/kg
	R/min	4.3·10^{-6} A/kg
	R/h	7.17·10^{-8} A/kg

Table 1.8 SI units relation with Hartree atomic units $e = m_e = \hbar = 1$ and relativistic units $c = m_e = \hbar = 1$

Quantity	Unit of system $\{e, m_e, \hbar\}$=1	Unit of system $\{c, m_e, \hbar\}$=1
Length	5.292·10^{-11} m	3.862·10^{-13} m
Time	2.419·10^{-17} s	1.288·10^{-21} s
Area	2.800·10^{-21} m^2	1.491·10^{-25} m^2
Velocity	2.188·10^6 m/s	2.998·10^8 m/s
Acceleration	9.044·10^{22} m/s^2	2.327·10^{29} m/s^2

Table 1.8 SI units relation with Hartree atomic units and relativistic units *(continued)*

Quantity	Unit of system $\{e,\ m_e,\ \hbar\}=1$	Unit of system $\{c,\ m_e,\ \hbar\}=1$
Mass	$9.109 \cdot 10^{-31}$ kg	$9.109 \cdot 10^{-31}$ kg
Force	$8.239 \cdot 10^{-8}$ N	$2.120 \cdot 10^{-1}$ N
Momentum	$1.993 \cdot 10^{-24}$ kg·m/s	$2.731 \cdot 10^{-22}$ kg·m/s
Moment of force	$4.360 \cdot 10^{-18}$ N·m	$8.187 \cdot 10^{-14}$ N·m
Angular momentum	$1.055 \cdot 10^{-34}$ J·s	$1.055 \cdot 10^{-34}$ J·s
Work, energy	$4.360 \cdot 10^{-18}$ J	$8.187 \cdot 10^{-14}$ J
Electric charge	$1.602 \cdot 10^{-19}$ C	$1.876 \cdot 10^{-18}$ C
Electric current	$6.624 \cdot 10^{-3}$ A	$1.456 \cdot 10^{3}$ A
Electric field strength	$5.142 \cdot 10^{11}$ V/m	$1.130 \cdot 10^{17}$ V/m
Potential	27.211 V	$4.365 \cdot 10^{4}$ V
Magnetic induction	$1.715 \cdot 10^{3}$ T	$3.771 \cdot 10^{8}$ T
Magnetic moment	$2.542 \cdot 10^{-21}$ A·m^2	$2.171 \cdot 10^{-22}$ A·m^2

References

[1] Burdun, G. D., Handbook of the international system of units, Izdatel'stvo standartov, Moscow, 1980 (in Russian).

[2] Units of physical quantities, GSSSD 8.417-81, Izdatel'stvo standartov, Moscow, 1981 (in Russian).

[3] Ionizing radiations and their measuring: terms and definitions. GSSSD 15484-81, Izdatel'stvo standartov, Moscow, 1981 (in Russian).

[4] Methodological directions: adoption and application of GSSSD 15484-81– Metrology. Units of physical quantities. RD 50-160-79, Izdatel'stvo standartov, Moscow, 1982 (in Russian).

[5] Methodological directions: adoption and application of GSSSD 8.417-81– Units of physical quantities for ionizing radiations. RD 50-454-84, Izdatel'stvo standartov, Moscow, 1982 (in Russian).

[6] Ivanov, V. I., Mashkovich, V. P., Tsenter, E. M., International system of units in the atomic science and engineering, Energoizdat, Moscow, 1981 (in Russian).

[7] Sena, L. A., Units of physical quantities and their dimensions, Nauka, Moscow, 1977 (in Russian).

[8] Chertov, A. G., Units of physical quantities, Vysshaya shkola, Moscow, 1977 (in Russian).

[9] Shirokov, K. P., Boguslavskii, M. G., International system of units, Izdatel'stvo standartov, Moscow, 1984 (in Russian).

[10] Fundamental physical constants. GSSSD 1-76 / Official edition, Izdatel'stvo standartov, 1976 (in Russian).

[11] Luther, G. G., Towler, W. R., Phys. Rev. Lett., 48, 121, 1982.

[12] Petley, B. W., Nature, 303 N 5916, 373, 1983.

[13] Symbols, units and nomenclature in physics: Document U.I.P. 20, IUPAP, S.U.N. Comission, 1978.

[14] Physical encyclopedic dictionary, Ed. Prokhorov, A. M., Sovetskaya Entsiklopedia, 1984 (in Russian).

[15] Cohen, E. R., Taylor, B. N., The 1986 CODATA recommended values of the fundamental physical constants, J. Res. Nat. Bureau Stand., 92, 85, 1987.

[16] Yost, G. P., Barnett, R. M., Hinchliffe I. et al.(Particle Data Group), Review of particle properties, Phys. Lett., B239, 1, 1990.

[17] Wapstra, A. H., Audi, G., The 1983 atomic mass evaluation, Nucl. Phys., A 432, 1, 1985.

2

Periodic Table

K.A. Kikoin

2.1 Periodic Table

Periodic Table of elements reflects the law of periodic variation of the physical and chemical properties of elements with changing the nuclear charge Z and the number of electrons in the outer electron shell (the Periodic Law). This table in its initial form proposed by D.I. Mendeleev (1869) contained the elements arranged in line with the increase of their masses and grouped according to the similarity of their chemical properties. Later on the explanation of Periodic Law was given within a framework of the quantum theory of atoms. It was found in this quantum-mechanical explanation that the sequence of the elements in the Periodic Table is determined by the nuclear charge Z and the periodicity of their physico-chemical properties is connected with existence of the electron shells which are filled up periodically with increasing Z.

The state of the electron in the atom depends on the principal quantum number n, orbital angular momentum l and its projection m, as well as spin quantum number s and its projection σ. Notice that all momenta in quantum mechanics are measured in units of $\hbar$ = Planck constant/2π, which is omitted for brevity. The electrons with various n and l form electronic configuration and they fill different atomic shells. Due to the Pauli principle, the number of electrons with given l cannot exceed $2(2l+1)$. The atomic shells are designated by the low-case letters $s, p, d, f, g, h, i, k, ...$ in accordance with the value of $l = 0, 1, 2, 3, 4, 5, 6, 7...$ and also carry information on the principal quantum number n of electrons involved.

The Periodic Table in its modern form is arranged in such a way that the nuclear charge and the number of electrons in the outer shell increase by unity when turning from the given element to the next one. The nuclear charge Z determines the ordinal number of the element in the table.

All the elements are divided into periods in accordance with the way of filling their atomic shells. The first period contains two elements (H, He) whose electrons fill up the first $1s$-shell. Both the second and the third periods contain eight elements. These periods begin with the elements filling up the ns-shells (n=2 and 3) and continue with the elements filling up the np shells (n=2 and 3). Three first periods are called usually the small periods.

The periods from the fourth to the seventh are called the large periods. Each of them occupy two rows in the Periodic Table. In the beginning of the fourth and fifth periods are the elements filling up the ns-shells (n=4 and 5). Then the nd- shells of the previous layers are filled (n=3 and 4, respectively), and finally the np-shells (n=4 and 5) are completed. Both these periods contain 18 elements.

The sixth and the seventh periods are organized in accordance with the same principle. Each of these periods begins with the elements filling up the outer ns-shells (n=6 and 7, respectively), then the first electron appears in a previous d-shell (La and Ac elements), but the following elements fill up mainly the deeper nf-shells (n=4 and 5). Only after that the nd-shells (n=5 and 6) are completed.

The elements with incomplete inner shells are called the transition elements. Thus the fourth period

0-8493-2861-6/97/$0.00+$.50

contains the transition elements from Sc to Ni which fill up the $3d$-shells, the $4d$-transition elements from Y to Pd enter the fifth period, the $5d$ and $4f$ transition elements from La to Pt are found in the sixth period, the $5f$ elements beginning with Ac are found at the end of the seventh period.

The latter period is still incomplete because the transuranium elements have no stable and even long-lived isotopes, and this instability increases with growing Z. Thus the table ends with the last of the artificially obtained elements with the number $Z = 105$. All the unsettled names, doubtful electron configurations and hypothetical mass numbers of the mostly long-lived isotopes of radionuclides are usually given in square brackets.

Due to electron interaction effects there are some deviations from the general trend described above. For example, the number of d-electrons in the $3d$-shells of Cr and Cu increases by two in comparison with preceding elements V and Ni due to the transition of one of $4s$-electron to the inner shell. The same anomaly is seen for Nb, Mo, Ru, Rh, Pd, and Ag. Moreover, both $5s$-electrons of Pd "turn" to d-state. The atoms of Os, Pt and Au have one less $6s$-electron as compared to other elements in the sixth period. The f and d states compete in the f-transition elements, and the next electron is found not in the $4f$ ($5f$)-shell but in the $5d$ ($6d$)-shell of some lanthanide and actinide elements (Gd, Tb, and Lu in the $4f$-group, Pa, U, Np, Cm, Bk, and Lw in the $5f$-group).

On a level with the periods and horizontal rows there exists a vertical division of the elements among the Periodic Table in the groups. The elements entering the same group have similar form of the external electron shells. In a short-period version of the Table, each of the large periods is divided into two rows placed one under another. In this version one finds the main and secondary groups. The main subgroups Ia and IIa of the first two groups are formed by the elements which have one and two s-electrons in the outer shell, respectively (the elements from $1, 2, 3, 4, 6, 8$ and 10th rows). The elements with the filled d-shells form the secondary subgroups Ib and IIb (the rows $5, 7, 9$). In the following groups from III to VII the main subgroups contain the elements with unfilled p-shells (the rows $2, 3, 5, 7$ and 9) and the secondary subgroups are formed by transition elements (the rows $4, 6, 8$ and 10). Hydrogen, the first element in the Table, can be placed both in the group Ia because it has one electron in the s-shell, and in the group VIIa because its $1s$-shell lacks one electron to be completed (see the broken line in Fig. 2.1 which illustrates this alternative). The inert gases forming the group VIIIa have completely filled shells. These elements terminate each period. For the sake of clarity the names of the elements in the main and secondary subgroups are shifted to the left and to the right, respectively. The transition elements with the nearly filled d-shells form special subgroups VIIIb (triads). These subgroups are called the iron, palladium and platinum subgroups. The elements with unfilled f-shells also form the special subgroups of lanthanides and actinides. Based on above considerations, we present short-period versions of the Periodic Chart (*by courtesy of* A. Radzig) in the front- and back-fly leaves.

Many properties of the elements in a free and condensed state are predetermined by their place in the Periodic Table. For example, the atomic masses of the elements increase with growing ordinal number of the element although there are several exclusions from this rule (the pairs Ar–K, Co–Ni, Te–I), all the elements with $Z > 83$ have no stable isotopes. Only the metals with unfilled $3d$ and $4f$ shells can be the magnets (the solid oxygen is exclusion from this rule). Among the superconducting metals only the paramagnetic elements from the periods 4 to 7 can be found. The elemental semiconductors are formed by the elements from the subgroup IVa and the semimetals are presented only with the elements from the subgroup Va. All the periods are terminated by the elements forming solid dielectrics. Practically all physical and chemical characteristics of the elements demonstrate the distinct periodical dependence on the atomic number Z.

To demonstrate the chemical trends in the properties of the elements, the long-period form of the Periodic Table is particularly convenient. In this version of the Table all the elements with the similar structure of the outer electron shell are placed within the same group [1]. Sometimes this table is presented in a compact step-wise form [2] with the periods placed one under another symmetrically relative to the axis of the table. The elements from the same group are connected with the lines (Fig. 2.1). Another form of the long-period Table is used here for presenting the periodical law for the physical properties of the elements (see Periodic Table in the double-page spread appeared below in the text). This table shows the atomic and thermodynamic properties of the elements in a condensed state at ambient pressure. The symmetry of the lattice is indicated for the solid phase existing at room temperature (if the room-temperature phase of the matter is liquid or gas, the symmetry of the solid phase corresponds to the state which can be obtained by cooling this matter at ambient pressure). The atomic and ionic radii (according to L. Pauling) are presented for the main valent states of the elements. The boiling and melting temperatures as well as the temperatures of magnetic and superconducting transitions are also shown. If the crystal is nonmagnetic, the type of magnetic susceptibility is also indicated (paramagnetic or diamagnetic). The Debye temperatures are taken for the most elements from the specific heat measurements at room temperature.

The references for the phase transition temperatures can be found in the corresponding Chapters of the present Handbook. The values of Debye temperatures are taken mainly from [3] and [4]. All other solid state characteristics are given in correspondence with [5, 6].

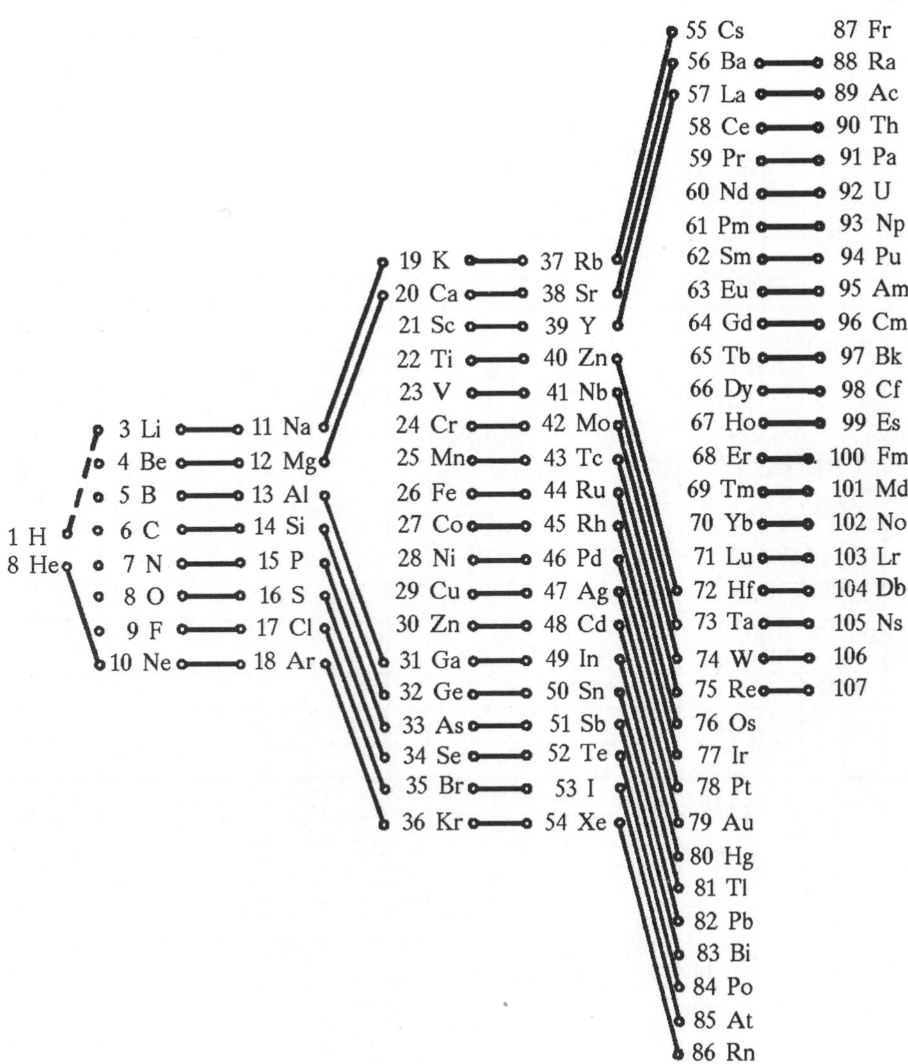

Figure 2.1 Long Periodic Table (step-wise form) of the elements.

PERIODIC

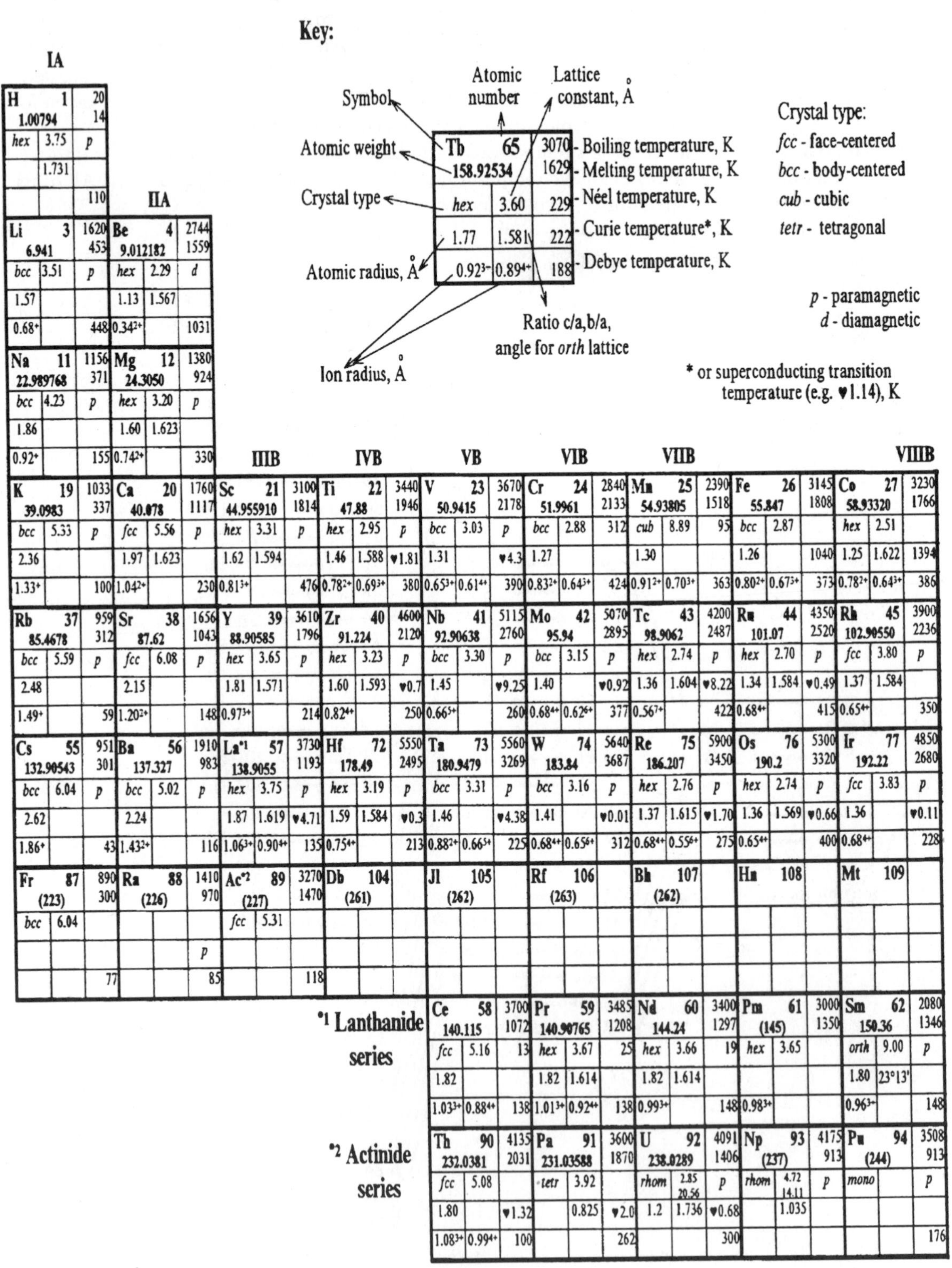
Key:
Symbol
Atomic number
Lattice constant, Å
Atomic weight
Crystal type
Atomic radius, Å
Ion radius, Å
Tb 65 158.92534 3070 1629 hex 3.60 229 1.77 1.581 222 0.92^{3+} 0.89^{4+} 188
- Boiling temperature, K
- Melting temperature, K
- Néel temperature, K
- Curie temperature*, K
- Debye temperature, K
Ratio c/a, b/a, angle for orth lattice
Crystal type:
fcc - face-centered
bcc - body-centered
cub - cubic
tetr - tetragonal
p - paramagnetic
d - diamagnetic
* or superconducting transition temperature (e.g. ♥1.14), K
IA
IIA
IIIB
IVB
VB
VIB
VIIB
VIIIB
H 1 1.00794 20 14 hex 3.75 p 1.731 110
Li 3 6.941 1620 453 bcc 3.51 p 1.57 0.68^{+} 448
Be 4 9.012182 2744 1559 hex 2.29 d 1.13 1.567 0.34^{2+} 1031
Na 11 22.989768 1156 371 bcc 4.23 p 1.86 0.92^{+} 155
Mg 12 24.3050 1380 924 hex 3.20 p 1.60 1.623 0.74^{2+} 330
K 19 39.0983 1033 337 bcc 5.33 p 2.36 1.33^{+} 100
Ca 20 40.078 1760 1117 fcc 5.56 p 1.97 1.623 1.04^{2+} 230
Sc 21 44.955910 3100 1814 hex 3.31 p 1.62 1.594 0.81^{3+} 476
Ti 22 47.88 3440 1946 hex 2.95 p 1.46 1.588 ♥1.81 0.78^{2+} 0.69^{3+} 380
V 23 50.9415 3670 2178 bcc 3.03 p 1.31 ♥4.3 0.65^{3+} 0.61^{4+} 390
Cr 24 51.9961 2840 2133 bcc 2.88 312 1.27 0.83^{2+} 0.64^{3+} 424
Mn 25 54.93805 2390 1518 cub 8.89 95 1.30 0.91^{2+} 0.70^{3+} 363
Fe 26 55.847 3145 1808 bcc 2.87 1.26 1040 0.80^{2+} 0.67^{3+} 373
Co 27 58.93320 3230 1766 hex 2.51 1.25 1.622 1394 0.78^{2+} 0.64^{3+} 386
Rb 37 85.4678 959 312 bcc 5.59 p 2.48 1.49^{+} 59
Sr 38 87.62 1656 1043 fcc 6.08 p 2.15 1.20^{2+} 148
Y 39 88.90585 3610 1796 hex 3.65 p 1.81 1.571 0.97^{3+} 214
Zr 40 91.224 4600 2120 hex 3.23 p 1.60 1.593 ♥0.7 0.82^{4+} 250
Nb 41 92.90638 5115 2760 bcc 3.30 p 1.45 ♥9.25 0.66^{5+} 260
Mo 42 95.94 5070 2895 bcc 3.15 p 1.40 ♥0.92 0.68^{4+} 0.62^{6+} 377
Tc 43 98.9062 4200 2487 hex 2.74 p 1.36 1.604 ♥8.22 0.56^{7+} 422
Ru 44 101.07 4350 2520 hex 2.70 p 1.34 1.584 ♥0.49 0.68^{4+} 415
Rh 45 102.90550 3900 2236 fcc 3.80 p 1.37 1.584 0.65^{4+} 350
Cs 55 132.90543 951 301 bcc 6.04 p 2.62 1.86^{+} 43
Ba 56 137.327 1910 983 bcc 5.02 p 2.24 1.43^{2+} 116
La*1 57 138.9055 3730 1193 hex 3.75 p 1.87 1.619 ♥4.71 1.06^{3+} 0.90^{4+} 135
Hf 72 178.49 5550 2495 hex 3.19 p 1.59 1.584 ♥0.3 0.75^{4+} 213
Ta 73 180.9479 5560 3269 bcc 3.31 p 1.46 ♥4.38 0.88^{2+} 0.66^{5+} 225
W 74 183.84 5640 3687 bcc 3.16 p 1.41 ♥0.01 0.68^{4+} 0.65^{6+} 312
Re 75 186.207 5900 3450 hex 2.76 p 1.37 1.615 ♥1.70 0.68^{4+} 0.55^{6+} 275
Os 76 190.2 5300 3320 hex 2.74 p 1.36 1.569 ♥0.66 0.65^{4+} 400
Ir 77 192.22 4850 2680 fcc 3.83 p 1.36 ♥0.11 0.68^{4+} 228
Fr 87 (223) 890 300 bcc 6.04 77
Ra 88 (226) 1410 970 p 85
Ac*2 89 (227) 3270 1470 fcc 5.31 118
Db 104 (261)
Jl 105 (262)
Rf 106 (263)
Bh 107 (262)
Hn 108
Mt 109
*1 Lanthanide series
Ce 58 140.115 3700 1072 fcc 5.16 13 1.82 1.03^{3+} 0.88^{4+} 138
Pr 59 140.90765 3485 1208 hex 3.67 25 1.82 1.614 1.01^{3+} 0.92^{4+} 138
Nd 60 144.24 3400 1297 hex 3.66 19 1.82 1.614 0.99^{3+} 148
Pm 61 (145) 3000 1350 hex 3.65 0.98^{3+}
Sm 62 150.36 2080 1346 orth 9.00 p 1.80 23°13' 0.96^{3+} 148
*2 Actinide series
Th 90 232.0381 4135 2031 fcc 5.08 1.80 ♥1.32 1.08^{3+} 0.99^{4+} 100
Pa 91 231.03588 3600 1870 tetr 3.92 0.825 ♥2.0 262
U 92 238.0289 4091 1406 rhom 2.85 20.56 p 1.2 1.736 ♥0.68 300
Np 93 (237) 4175 913 rhom 4.72 14.11 p 1.035
Pu 94 (244) 3508 913 mono p 176

TABLE

rhom - rhombic
hex - hexagonal
diam - diamond type
orth - orthorhomb
mono - monoclinic

			IB			IIB			IIIA			IVA			VA			VIA			VIIA			VIIIB		
																								He 2	4.002602	4.2
									B 5	10.811	4130 2348	C 6	12.011	3700	N 7	14.00674	77 63	O 8	15.9994	90 55	F 9	18.9984032	85 53.5	Ne 10	20.1797	27 24
									tetr	8.73	*d*	*diam*	3.57	*d*	*hex*	4.04		*cub*	6.83	24	*mono*		*d*	*fcc*	4.46	*d*
									0.97			0.77			0.71	1.651		0.74						1.6		
									0.2^{3+}		1250	0.20^{4+}	2.6^{-}	1860	0.11^{5+}	1.71^{3-}	79	1.36^{2-}		46	1.33^{-}					63
									Al 13	26.981539	2770 934	Si 14	28.0855	2870 1690	P 15	30.973762	554 317	S 16	32.066	718 386	Cl 17	35.4527	239 172	Ar 18	39.948	87 20
									fcc	4.05	*p*	*diam*	5.34	*d/p*	*cub*	7.17	*d*	*rhom*	10.47	*d*	*rhom*	6.24	*d*	*fcc*	5.26	*d*
									1.43		▼1.14	1.18			1.3			1.04	2.339 1.229		1.04	1.324 0.718		1.92		
									0.57^{3+}		394	0.39^{4+}		625	0.35^{5+}	1.86^{3-}					1.82^{-}	0.26^{7+}				85
Ni 28	58.69	3570 1726	Cu 29	63.546	2870 1356	Zn 30	65.39	1179 692	Ga 31	69.723	2500 303	Ge 32	72.61	2970 1210	As 33	74.92159	1090	Se 34	78.96	958 490	Br 35	79.904	332 267	Kr 36	83.80	120 89
fcc	3.52		*fcc*	3.61	*d*	*hex*	2.06	*d*	*hex*	3.19	*d*	*diam*	5.66	*d*	*orth*	4.13	*p*	*hex*	4.36	*d*	*rhom*	6.67	*d*	*fcc*	5.72	*d*
1.24		628	1.28			1.37	1.856	▼0.79	1.39	1.58	▼1.07	1.39			1.48	54°10'		1.6	1.136		1.19	1.307 0.672		1.98		
0.79^{+}	0.72^{2+}	375	0.98^{+}	0.80^{2+}	315	0.83^{2+}		234	0.62^{3+}		240			360	0.47^{5+}	1.91^{3-}	285	0.69^{4+}	1.98^{2-}	150	1.96^{-}					73
Pd 46	106.42	4250 1825	Ag 47	107.8682	2485 1234	Cd 48	112.411	1040 594	In 49	114.82	2348 438	Sn 50	118.710	2540 505	Sb 51	121.757	1910 904	Te 52	127.60	1263 723	I 53	126.90447	457 386	Xe 54	131.29	165 161
fcc	3.89	*p*	*fcc*	4.09	*d*	*hex*	2.98	*d*	*tetr*	4.58	*d*	*tetr*	5.82	*p*	*orth*	4.51	*d*	*hex*	4.45	*d*	*rhom*	7.27	*d*	*fcc*	6.20	*d*
1.37			1.44			1.56	1.886	▼0.54	1.66	0.108	▼3.37	1.58	0.546	▼3.69	1.61	57°6'		1.7	1.330		1.36	1.347 0.659		2.18		
0.65^{4+}		275	1.13^{+}		215	1.03^{2+}		120	1.30^{3+}	0.92^{3+}	129	1.02^{2+}	0.67^{4+}	170	0.9^{3+}	2.08^{3-}	200	0.89^{4+}	2.11^{2-}	135	2.20^{-}					
Pt 78	195.08	4800 2042	Au 79	196.96654	3220 1337	Hg 80	200.59	630 234	Tl 81	204.3833	1730 577	Pb 82	207.2	1998 600	Bi 83	208.98037	1830 544	Po 84	(209)	1435 527	At 85	(210)	575	Rn 86	(222)	477 223
fcc	3.92	*p*	*fcc*	4.08	*d*	*orth*	2.99	*d*	*hex*	3.46	*d*	*fcc*	4.05	*d*	*orth*	4.74	*d*	*cub*	3.35					*fcc*		*p*
1.39			1.44			1.57	70°45'	▼4.12	1.71	1.599		1.75		▼7.2	1.82	57°14'		1.19						1.34		
0.65^{4+}		230	1.37^{+}		170	1.1^{2+}		100	1.49^{3+}	1.05^{3+}	96	1.26^{2+}	0.76^{4+}	88	0.12^{3+}	0.74^{5+}	120							0.68^{4+}		

Eu 63	151.965	1710 1099	Gd 64	157.25	3270 1585	Tb 65	158.92534	3070 1629	Dy 66	162.50	2870 1680	Ho 67	164.93032	2870 1734	Er 68	167.26	3170 1770	Tm 69	168.93421	2000 1818	Yb 70	173.04	1700 1097	Lu 71	174.967	3600 1925
bcc	4.61	91	*bcc*	4.06		*hex*	3.60	229	*hex*	3.59	179	*hex*	3.58	131	*hex*	3.56	86	*hex*	3.54	56	*fcc*	5.49	*p*	*hex*	3.51	*p*
2.04			1.79		293	1.77	1.581	222	1.77	1.573	85	1.76	1.570	19	1.76	1.570	20	1.75	1.570	25	1.93				1.585	
0.95^{3+}		107	0.94^{3+}		176	0.92^{3+}	0.89^{4+}	188	0.91^{3+}		186	0.89^{3+}		191	0.88^{3+}		195	0.87^{3+}		200	0.86^{3+}		118	0.85^{3+}		207
Am 95	(243)	2880 1268	Cm 96	(247)	1610	Bk 97	(247)	1259	Cf 98	(251)	1270	Es 99	(252)		Fm 100	(257)		Md 101	(258)		No 102	(259)		Lw 103	(260)	

2.2 Names and Symbols of Transfermium Elements According to IUPAC Recommendation 1994

101 - Mendelevium (Md)	104 - Dubnium (Db)	107 - Bohrium (Bh)
102 - Nobelium (No)	105 - Joliotium (Jl)	108 - Hahnium (Ha)
103 - Lawrencium (Lr)	106 - Rutherfordium (Rf)	109 - Meitnerium (Mt)

References

[1] Lexikon der Physik, Bd. 3, Stuttgart, 1969.

[2] Handbuch der Physik, Bd. 36, Springer, Berlin, 1956.

[3] Gschneider K.A.J. Solid State Physics, Eds. Ehrenreich, H., Turnbull, D., and Seitz, F. Academic Press, New York, 1964, vol.14, p.275.

[4] Ashcroft, N.W., Mermin, N.D., Solid State Physics Holt, Rinehart and Winston, New York, 1976.

[5] Physical encyclopaedic glossary, Sovetskaya Entsiklopediya, Moscow, 1960–1969 (in Russian).

[6] Encyclopaedia of inorganic materials, Vyschaya shkola, Kiev, 1977 (in Russian).

3

The Mechanical Properties of Materials

N.V. Kadobnova and A.M. Bratkovskii

3.1 Introduction

Depending on the mode of loading (tension, compression, bending, torsion, or shear) and application conditions (temperature, rate, frequency and period) it is conventional to characterize the materials by various measures of their resistance to deformation and failure, i.e. by the characteristics of mechanical properties.

The mechanical properties can be arbitrarily subdivided into three main groups.

The first group includes a set of characteristics determined in the single short-time loading of the specimen. Among them are the elastic properties: modulus of elasticity E, shear modulus G, and Poisson's ratio μ. The resistance to low elastoplastic strains is defined by the elastic limit σ_e, proportional limit σ_p, and yield stress $\sigma_{0.2}$. The ultimate strength σ_b, the resistance to cut τ_c and resistance to shear τ_{sh}, the indentation resistance HB (Brinell hardness) and scratch resistance (on Mohs' scale), along with the breaking length L_b provide the characteristics of materials within the range of high strains up to a fracture. The plasticity is characterized by the specific elongation δ and contraction ratio ψ after breaking, as well as the deformability of some nonmetallic materials – by the extension on breaking δ_b. Also the notch impact strength KCU of the specimen is evaluated on impact bending.

The second group involves the parameters estimating the resistance of materials to varied and long-time static loads. Under repeated loading within the regime of high-cycle fatigue, the endurance limit is defined by the number of cycles to failure which reaches 10^7–$2{\cdot}10^7$ cycles. The low-cycle fatigue is separated from the high-cycle one by the arbitrary chosen number of cycling tests ($N \geq 5 \cdot 10^4$ cycles) and is differed by lower frequency of loading (f= 0.1–5 Hz). The resistance to low-cycle fatigue is estimated by longevity at the fixed level of repeated stresses or by the low-cycle fatigue limit at a fixed number of cycling tests. The resistance to long-term static loads is evaluated as a rule at a temperature above 20°C. The creep stress $\sigma_{0.2/\tau}$ and long-term ultimate strength σ_τ are the criteria of the material resistance to sustained constant stresses and temperatures. The long-term ultimate strength is determined at a given base of cycling tests, usually lasting 100 and 1000 hours, the creep limit – to tolerance for residual (usually 0.2%) or total strain at a given number of cycling tests.

The third group concerns the fracture properties. In engineering practice these characteristics are of comparatively recent usage. The fracture properties are obtained from the specimens containing initially induced cracks and the following parameters serve to estimate them: the fracture toughness, the critical stress intensity factor under plain deformation K_{IC}, the conventional critical stress intensity factor under the plane stress K_C, the specific propagation energy KCT of a cracked specimen and, finally, fatigue crack growth rate ($FCGR$) at a given range of stress intensity ΔK.

Of all mechanical properties it is only the elastic properties of the metallic materials that are structurally insensitive characteristics related to the parameters of crystal lattice and essentially independent

0-8493-2861-6/97/$0.00+$.50

of conditions of heat treatment and machining, if the latter do not lead to the allotropic transformations. The elastic constants for the practically isotropic polycrystalline metallic materials are related by the following equation: $E = 2G(1+\mu)$. The elastic properties are derived from the static tests (E_{st}, G_{st}) or by the dynamic method (E_{dyn}, G_{dyn}) through a resonance frequency of oscillations of a uniform cross-section thin rod under the action of low stresses. The values of elastic constants obtained by both methods at room temperature and near to it are practically the same. Under the static tests at elevated temperatures, the creep has an effect and as a result the static tests give progressive reduction in the values of elastic constants relative to those obtained by the dynamic method.

All other mechanical properties to a greater or lesser extent are structurally sensitive and anisotropic. Clearly defined anisotropy of elastic and other mechanical characteristics is native to many nonmetallic materials due to their oriented structure. Certain anisotropy is also inherent to the great majority of metallic materials. The strength of a material, the degree of plasticity, the fatigue level and the failure strength in the longitudinal (relative to strain axis) direction are usually higher than those in the transverse direction. However, for some alloys, e.g. titanium based, the reverse anisotropy is typical. The significant difference in tensile and compressive yield points is evident for the majority of wrought magnesium alloys ($\sigma_{0.2comp} \ll \sigma_{0.2}$).

Correlations between certain mechanical characteristics are experimentally obtained which allow (with a sufficient degree of accuracy) the estimation of the ultimate strength of a material in terms of hardness values to be made, and the shear resistance – in terms of the ultimate strength. The correlations between the fatigue limit and the ultimate strength, as well as between various fracture properties also exist.

3.2 Mechanical Properties at Room Temperature

In Tables 3.1–3.32 the following characteristics are presented:

E – modulus of elasticity specifies the ratio of the normal stress to the relative elongation (for tensile or compressive stresses less than the proportional limit of the material);

G – shear modulus is the ratio of the tangential stress to the relative shear (for tangential stresses below the shear proportional limit of the material);

μ – Poisson's ratio determines the absolute value of the ratio between the transverse deformation and the corresponding axial strain within the range of elasticity;

σ_p – proportional limit defines the limit of stress on a body below which the deformations are proportional to the stress applied; in practice, the conventional limit σ_p is adopted at which the deviation of deformation increment from linear dependence reaches a definite value, usually 50%;

$\sigma_{0.2}$ – yield stress (nominal) fits the stress at which the residual deformation after unloading is 0.2%;

σ_b – temporary resistance strength (ultimate strength), breaking strength (for nonmetallic materials) is peculiar to the stress complying with a maximum load the specimen can endure under the test;

δ – specific elongation governs the ratio of the absolute value of residual specimen elongation after breaking to the initial gage length;

ψ – contraction ratio decides on the ratio of the reduction in the specimen cross-section area after breaking to the initial area;

δ_b – specific elongation on breaking (for nonmetallic materials) settles the total change of the specimen gage length at the instant of breaking related to the initial gage length;

L_b – breaking length is characteristic of the strength of filaments, fibers, fabrics and other materials for which the cross-section area cannot be defined precisely; it is calculated by the expressions $L_b = P_b(l/Q)$, $L_b = P_b/gb$, where P_b is the breaking load; Q, l and b are, respectively, the mass, length and width of a specimen between the clamps; g is the mass of a material of 1 sq. meter area;

KCU – impact strength specifies the work done on a failure of circumferentially notched round specimen tested in impact bending divided by the notch cross-section area; for nonmetals, KCU is also determined for unnotched specimens;

HB – Brinell hardness determined by pressing the steel ball into a material under test, defines the average stress per unit area of spherical indentation surface;

H_μ – microhardness fits the resistance of a material to the indentation of the diamond indenter under very low loading which results in the impressions of small depth and size;

Mohs' hardness – the resistance of the minerals and other materials to the mechanical action determined by scratching. The measure of hardness is the number related to the hardest mineral not making a mark on scratching. The primary standards of hardness are talc (1), gypsum (2), calcium (3), fluorite (4), apatite (5), orthoclase (6), quartz (7), topaz (8), corundum (9), and diamond (10).

3.2.1 Metals

Table 3.1 Elastic properties of metals

Metal	E, GPa	G, GPa	μ	References	Metal	E, GPa	G, GPa	μ	References
Aluminum	69–72	25–26.5	0.31	[1,5]	Nickel	200–220	73–77	0.3–0.4	[1,2,4]
Antimony	57–78			[1,2]	Niobium	91–160		0.39	[1,6]
Beryllium	300	145	0.03	[1,3]	Osmium	575	225	0.28	[3,8]
Bismuth	32	12	0.33	[1,4]	Palladium	115–125	49–52	0.39	[1–4,8]
Boron	345			[1]	Platinum	150–175	61–68.5	0.36	[1–3,8]
Cadmium	50–53	20	0.3	[1,4,7]	Praseodymium	35–98	14	0.3	[2,3]
Caesium	1.75			[1]	Rhenium	475			[2]
Calcium	26			[1,2]	Rhodium	385	150	0.26	[3,8]
Cerium	44		0.25	[3]	Rubidium	2.5			[1]
Chromium	280–315	110	0.31	[1–3]	Ruthenium	420–500	175	0.31	[3,8]
Cobalt	206	78.5	0.32	[1,2]	Samarium	34–55	13	0.35	[2,3]
Copper	110–130	41.5–44	0.38	[1–3,7]	Selenium	55	19	0.45	[1,4]
Dysprosium	64–98	26	0.24	[2,3]	Silicon	110–160			[1,2]
Erbium	73–115	30	0.24	[2,3]	Silver	72–83.5	27–29.5	0.37	[1,3,4,8]
Gadolinium	56–98	23	0.26	[2,3]	Tantalum	190	70	0.35	[1,3,6]
Germanium	82			[1]	Tellurium	44			[4]
Gold	78–83	28.5	0.4	[3,8]	Terbium	57.5			[2]
Hafnium	79–150		0.29	[1,2,6]	Thallium	8			[1,2]
Holmium	67			[2]	Thorium	74–80			[1,2]
Indium	10.5		0.46	[1,2]	Tin	41-55	16–19	0.33	[1,2,4]
Iridium	520–590	220	0.28	[2,6,8]	Titanium	110	41.5	0.33	[3,10]
Iron	195–205	77–80	0.28	[1,3,4,7]	Tungsten (Wolfram)	350–400	125–155	0.3	[1,4,6]
Lanthanum	38	15	0.26	[2,3]					
Lead	14–18	5.5–8	0.45	[1,2,4,14]	Uranium	210			[1]
Lithium	5		0.42	[2,4]	Vanadium	139–170	47–60	0.36	[1,3,6]
Magnesium	42.5–45	16–18	0.35	[1,3,4,7]	Ytterbium	18			[2]
Manganese	200			[1]	Yttrium	66	26	0.27	[2,3]
Molybdenum	300–330	120	0.31	[3,6]	Zinc	100–130	37	0.3–0.35	[1,2,7]
Neodymium	38	15	0.28	[2,3]	Zirconium	84–97	33	0.35	[1,2,6]

Table 3.2 Strength and ductility of metals

Metal	Condition, degree of purity	$\sigma_{0.2}$, MPa	σ_b, MPa	δ, %	ψ, %	HB, MPa	References
Aluminum	Annealed, 99.95% Al	22	50	45	90	150	[1,3,4,5]
	Annealed, 99.5% Al	30	80	35	80	250	[1,3,4,5]
Antimony	Annealed		8	0	0		[1,4]
Barium	Annealed		12		40		[4,11]
Beryllium	Annealed	230	320	2.5			[1,3,4]

Table 3.2 Strength and ductility of metals *(continued)*

Metal	Condition, degree of purity	$\sigma_{0.2}$, MPa	σ_b, MPa	δ, %	ψ, %	HB, MPa	References
	Cast		140	0			[1,3,4]
	Hot-pressed	300	450	3.5		1500	[1,3,4]
Bismuth	Annealed		5–20	0	0	90	[1,4]
Boron	Annealed		250($\sigma_{b,\text{comp}}$)				[1]
Cadmium	Annealed	10	75	20	50	200	[1,4]
Calcium	Annealed	13.5	60	10	58	300	[1,4]
Cerium	Cast	90	105	24		250	[1,3]
	Forged	110	150	17		300	[1,3]
Chromium	Annealed	190	300	2	5	1000	[3,6]
Cobalt	Annealed, 99.4% Co	300	470	3.5	4.5	1550	[1,3,4]
	Cast, 99.4% Co		240			1250	[1,3,4]
	Quenched, 99.4% Co	210	280	4	8	1500	[1,3,4]
Copper	Annealed	70	215	60	75	400	[3]
	Cold-hardened	380	440	6	35	1000	[3]
Dysprosium	Cast	230	250	6		550	[3]
	Forged	330	435	3		1030	[3]
Erbium	Cast	295	300	4		600	[3]
	Forged	295	320	7		950	[3]
Gadolinium	Forged	270	395	7		700	[3]
Gallium	Annealed		40	40		60	[1]
Germanium	Annealed		68($\sigma_{b,\text{comp}}$)				[4,11]
Gold	Annealed	40	150	40	90	220	[1,8]
Hafnium	Annealed	500	700	6	25	1600	[1,6]
Holmium	Cast	220	260	5		500	[3]
Iridium	Annealed	90–100	400–500	6–10	10–15	1700–2200	[8]
Iron	Annealed, of peculiar purity	170	290	50	90	800	[3]
Lanthanum	Cast	125	130	8		400	[3]
	Forged	190	220	4			[3]
Lead	Annealed	5	14–18	50	100		[1,14]
Lithium	Annealed		115	50			[1,4]
Magnesium	Annealed	40	185	15	20		[1,3,4]
	Cast	24	115	8	9	300	[1,3,4]
	Wrought	90	195	10	12	400	[1,3,4]
Molybdenum	Annealed	570	670	25	50	1800	[1,3]
Neodymium	Cast	165	170	11			[3]
	Forged		215	2		400	[3]
Nickel	Annealed	80	400	40	70	800	[1,3]
Niobium	Annealed, of high purity	210	275	28	80	500	[1,3,4]
	Annealed, of industrial purity	280	330	20	80	700	[1,3,4]
	Cast	500	600	10	20		[1,3,4]
Osmium	Annealed					3000–4000	[1,6]
Palladium	Annealed	60	195	40	85	480	[3,8]
Platinum	Annealed	70	145	45	95	470	[3,8]
Praseodymium	Cast	100	110	10			[3]
	Forged	200	215	7		400	[3]
Rhenium	Annealed		500	20		2000	[1]
Rhodium	Annealed	70–100	400–570	8–15	20–25	1000–1300	[3,8]
Ruthenium	Annealed	360–400	500–600	3–10		2000–3000	[3,8]
Samarium	Cast	115	125	3			[3]

Table 3.2 Strength and ductility of metals *(continued)*

Metal	Condition, degree of purity	$\sigma_{0.2}$, MPa	σ_b, MPa	δ, %	ψ, %	HB, MPa	References
Samarium	Forged	180	190	8		550	[3]
Scandium	Annealed		400	10		550	[2,3]
Selenium	Annealed					710	[4]
Silicon	Annealed		700	0	0		[1,4]
Silver	Annealed	20–30	140–180	45	90	500	[1,3,8]
Strontium	Annealed		50	2	10		[4,8]
Tantalum	Annealed	400	500	30	75	1250–1400	[1,3,6]
	Cold-hardened		950	4	6	2250	[1,3,6]
Thallium	Annealed		9	35	100		[1]
Tellurium	Annealed		10	35	100	270	[1,4]
Thorium	Annealed	180	220	60	70		[1]
Tin	Annealed		30	40	75	60	[4]
Titanium	Annealed, of peculiar purity	100	250	70	85	600	[2,3,6,10]
	Annealed, 99.6% Ti	300	350	30	60		[2,3,6,10]
Tungsten	Recrystallized, 99.95% W		500	0	0	3200	[1,3]
	Wrought, 99.95% W	760	1000	0	0	4150	[1,3]
Uranium	Annealed	200	300	12	15		[1]
Vanadium	Annealed, 99.98% V	105	220	30	75	800	[3,4,6]
	Annealed, 99.6% V	540	590	10	25		[3,4,6]
Ytterbium	Annealed	67	75	6			[2]
Yttrium	Annealed	280	300	4	8	900–1650	[2,3]
Zinc	Annealed	100	125	12			[1]
Zirconium	Annealed	80–115	230–280	25	40	570	[1,4,11]

3.2.2 Steels

The steels used in machine building are classified as structural (carbon and alloy) and high-alloy stainless steels. The mark of a structural carbon steel contains two-digit number indicating the mean carbon content in 100th parts of percent by weight.

In designating the marks of structural alloy steels, the first two digits indicate the mean carbon content in 100th parts of percent by weight, the cyrillic letters after digits are: P–boron, Ю–aluminum, C–silicon, T–titanium, Ф–vanadium, X–chromium, Г–manganese, H–nickel, M–molybdenum, and B–tungsten. The digits after a letter indicate approximate percentage of alloying component in integer units; no digit implies that the steel contains less than 1.5% of this alloy element. For high-quality steels, the letter A is put at the end of designation. For extra high-quality steels, the letter Ш is put after dash at the end of designation.

In the stainless high-alloy steels, the alloy components are denoted as follows: A–nitrogen, B–tungsten, Д–copper, M–molybdenum, P–boron, T–titanium, Ю–aluminum, X–chromium, Б–niobium, Г–manganese, E–selenium, H–nickel, C–silicon, Ф–vanadium, K–cobalt, and Ц–zirconium. The mark digits after the letters in the same way as for designations of structural steels indicate the percentage of alloy element in integer units. The digit before letter indicates the mean or (when there is no lower limit) maximum carbon content for the steel in 100th parts of percent. The designation of casting steel is ended by letter Л.

The mechanical properties of structural steels are dependent on carbon content, for carbon steels the carbon percentage is a decisive factor (Fig. 3.1, Table 3.3).

The empirical relationships are established between some mechanical properties of steels. Brinell hardness is approximately proportional to the temporary resistance (ultimate strength): for low and medium-strength carbon steels and heat-treat steels $\sigma_t = 0.3 - 0.4\text{HB}$. For low- and medium-strength steels, the resistance to cut τ_c amounts to 65–80% of their ultimate strength, for high-strength steels, to 55–65%. The yield compression strength approximates the yield tensile limit $\sigma_{0.2}$.

Table 3.3 Mechanical properties of carbon and low-alloy steels of low and medium-strength after normalizing [3,24] [$E = 200 - 210$ GPa; $G = 77 - 81$ GPa; $\mu = 0.28 - 0.31$.]

Steels	$\sigma_{0.2}$, MPa	σ_b, MPa	δ, %	ψ, %	BH, MPa	KCU, MJ/m^2
Carbon steels of ordinary quality						
Ст0		320	22			
Ст1		320–400	33		1100	
Ст2	190–220	340–420	31		1160	
Ст3	210–240	380–500	25–27	55–65	1310	
Ст4	240–260	420–520	23–23		1430	
Ст5	260–280	500–640	19–21	45–55	1700	
Ст6	300–310	600–710	14–16	40	1970	
Quality and low alloy carbon steels						
08	200	330	33	60		
10	210	340	31	55	1160	
15	230	380	27	55	1300	
20, 15Г, 10г2	250	430	25	55	1430	
25, 20Г	280	460	24	50		0.9
30, 25Г	300	500	22	50	1970	0.9
35, 30Г	320	550	20	45		0.8
40, 35Г	340	580	19	45		0.7
45, 40Г, 30г2	360	610	16	45	2290	0.6
50, 45Г, 35г2	380	640	14	40	2410	0.5
55, 50Г, 40г2	390	670	12	40	2550	0.4
60, 45Г2	410	700	12	40	2550	
65, 60Г	420	710	11	35	2550	
70, 65Г, 50г2	440	750	10	30	2690	
70Г	460	800	8			
75*1	900	1100	7	30		
80*1	950	1100	6	30		
85*1	1000	1150	6	30		

*1 After hardening and tempering.

Table 3.4 Mechanical properties of structural wrought alloy steels after hardening and tempering [3,24] [$E = 200 - 210$ GPa; $G = 77 - 81$ GPa; $\mu = 0.28 - 0.31$.]

Steels	$\sigma_{0.2}$, MPa	σ_b, MPa	δ, %	ψ, %	BH, MPa	KCU, MJ/m^2
Low-strength steels*1						
09Г2, 09Г2Д, 12ГС, 14Г2, 19Г, 12ХГ	450–480	300–400	18			
15ГС, 18Г2С, 10Г2СД, 14ХГС	500	350	18			
10ХСНД, 15ХГН, 15ХСНД	520–540	360–400	16–18			
25Г2С, 35ГС	600	400	14			
Medium-strength steels						
20ХГСА	800	650	12	45	2070	0.7
39ХГН, 33ХС, 20ХН4ФА2*2	900	700	12–13	45–50	2290	0.8–1.0
35Х, 30ХРА, 38ХА, 30ХМ, 38ХС, 20ХН3А	950	750–800	11–12	45–55	1970–2290	0.7–1.0
40Х, 40ХР, 45Х, 35ХМ, 20ХГР, 40ХГ, 40ХГР, 40ХН, 18ХНВА	1000-1050	800–850	9–12	45–50	2070–2400	0.6–0.9
50Х, 50ХН, 35ХМФА, 30ХГНА, 25ХГСА, 30ХГС, 30ХГСА, 40ХНМА*2, 40ХНВА*2, 25ХНВА*2, 30ХН3А	1100	850–950	9–12	40–50	2170–2290	0.5–0.8

Table 3.4 Mechanical properties of structural steels *(continued)*

Steels	$\sigma_{0.2}$, MPa	σ_b, MPa	δ, %	ψ, %	BHN, MPa	KCU, MJ/m^2
High-strength steels						
35ХГСА*3, 30ХГСНА*3	1500–1650	1200–1300	10–13	45–55	4100–4500	0.7–0.9
30Х2ГН2СВМА*3						
30ХГНСМА	1750	1400	10	45	4600	0.6
40ХГСН3ВА*3	1900	1400	11	50	4900	0.6
40ХГСН3ВА	2000	1500	10	40	5150	0.55
Н18К9М5Т	2050	1950	7	50		0.5
Heat-resistant steels						
30ХМА, 20Х3МВФ, 33ХН3МА	900–950	750–850	12–19	50–60	2290	0.9–1.3
30ХА, 30ХГСА, 30Х3ВА, 40ХНВА	1050–1100	950	14–16	55–65	2290	0.6–1.4
30Х2Н2ВФА	1150	1100	15	60	2400	0.8
23Х2НВФА	1350	1150	13	55		0.6
30ХГСНА	1600	1350	9	45		0.6

*1 After normalizing.
*2 HB≈2690 MPa.
*3 Isothermal hardening.

Table 3.5 Strength and ductility of construction casting steels [3] [$E = 190 - 200$ GPa.]

Steels	σ_b, MPa	$\sigma_{0.2}$, MPa	δ, %	ψ, %	BH, MPa	KCU, MJ/m^2
Hardened-and-tempered alloy steels						
27ГЛ, 30ГСЛ, 40ХЛ	650	400–500	10–14	20–30	1800–2290	0.35–0.5
35ХНЛ, 35ХМЛ, 30ХГСТЛ	700	500–550	12	25	2070–2600	0.4
30ХНМЛ, 30ХНВЛ	800	600–650	10–12	20–30	2170–2690	0.4
35ХГСМЛ, 30ДХСНЛ, 27ХГСНЛ	900–1000	700–850	8–10	20–30		0.25–0.4
27ХГСНЛ, after isothermal hardening	1300–1500	1000–1200	5–6	20		0.25
Heat-resistant steels, normalized and tempered						
20МЛ, 25МЛ, 20ХМЛ, 20ХМФЛ	450–500	250–315	16–20	30–40		0.3–0.4
30ХМЛ, 23Х5МЛ	650–700	400–450	16–18	30–40		0.6
Х6Н2МВФ, after hardening and tempering	990	700	10	20		

Table 3.6 Mechanical properties of stainless steels [3,9]

Steels	σ_b, MPa	$\sigma_{0.2}$, MPa	δ, %	ψ, %	BH, MPa	KCU, MJ/m^2
Hardened austenite steels						
04Х18Н10, 08Х18Н10Т, 08Х18Н12Б, 10Х17Н13М3Т	420–520	180–200	40	55	1350–2000	
08Х18Н12Т, 12Х18Н9, 08Х17Н16М3Т, 20Х18Н9, 06Х23Н28М2Т	550–600	200–220	35–40	55	1350–2000	
20Х13Н4Г9, 12Х17Г9АН4, 12Х14Г14Н	650–700	250–300	35–40	45–50	1600–2000	
Х18Н9Л	450	200	25	35		
Hardened austenite-ferrite steels						
08Х20Н14С2, 20Х20Н14С2	550–600	250–300	35–40			1
08Х21Н5Т, 10Х21Н5Т, 08Х21Н6М2Т	650–700	300–450	20–30			0.6

Table 3.6 Mechanical properties of stainless steels [3,9] *(continued)*

Steels	σ_b, MPa	$\sigma_{0.2}$, MPa	δ, %	ψ, %	BH, MPa	KCU, MJ/m^2
Tempered ferrite and semi-ferrite steels						
12X17, 08X17T, 15X28, 15X25T	450–500	250–300	18–20	40–55	1400–1900	
10X13Л, 20X13Л	550–650	400–450	12–16	40–50		0.6–0.8
Hardened-and-tempered martensite steels						
10X13, 20X13	1200	1050	10–18	45–60	2550	0.5–0.7
30X13, 40X13	1650–1750	1400–1550	3–8			
23X13HBHΦ	1500	1100	10	40		0.4
10X17H2, 10X12H2BMΦ	1100–1300	900–1000	8–12	50–60		0.6–1.2
13X14HBPA						
08X17H3CЛ	850–950	650–750	6–8	10–20		0.25
10X13H3BΦЛ	1200	100	7	20		0.19
Transient class steels after hardening, cold treatment and age hardening						
08X15H9Ю, 08X16H6, 08X17H5M3	1200	900	10			0.5
10X15H4AM3	1450–1600	1200	15–20	50–60		1–1.5
Quenched-and-tempered martensite-aging steels						
08X15H5Д2TБ	1150–1300	1000–1300	18–20	60–70		1.4–2
03X12H10Д2TБ	1600	1500	4	60		0.8
Harden heat-resistant steels after age hardening						
45X14H14B2M, tempered	720	320	20	35		0.5
12XH35BMTP, 12XH35BT	800–950	450–550	25	40		0.6
40X15H7Г7Φ2MC, 40X12H8Г8MΦБ,						
10X12H20T3P, 10X12H22T3MP	900	600	15	25		0.3
10XH35BTЮ	1320	950–1000	20–25	40		0.75

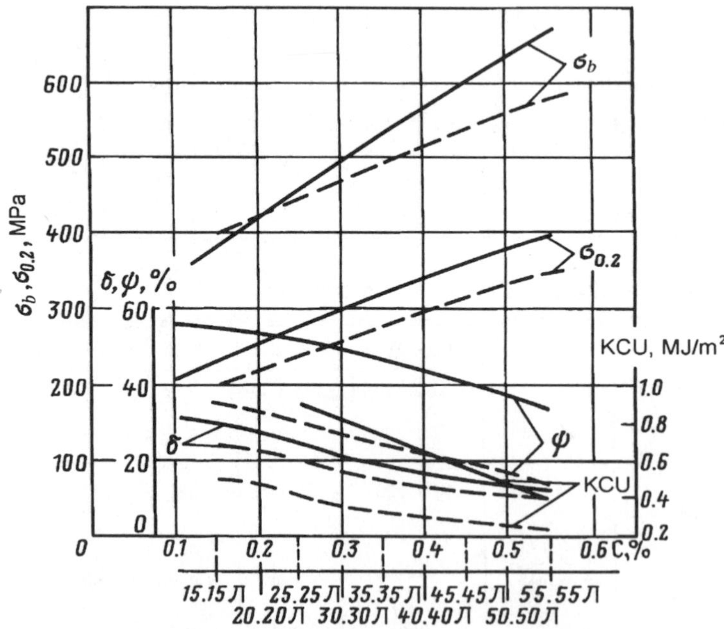

Figure 3.1 Mechanical properties of structural wrought and casting carbon steels with different carbon content [3,24]; solid lines are for wrought steels; dashed lines are for casting steels; the steel marks are displayed alongside x-axis in accordance with the mean carbon content.

3.2.3 Aluminum alloys

In designating the wrought aluminum alloys, the following codes are adopted: M – soft, annealed; П – cold-semihardened; H – cold-hardened; T – hardened and self-aged; T1 – hardened and artificially aged for high strength; T2 – hardened and artificially aged by the process providing higher fracture toughness and corrosion resistance under stress in comparison with those for T1 regime; T3 – hardened and artificially aged by process providing the highest corrosion resistance under stress and the high fracture toughness. The letter "ч" in the designation of alloy mark indicates higher than usual purity of an alloy (with respect to impurities).

The conditions of heat treatment of casting alloys are in addition to alloy mark designated as follows: T1 – artificial aging without prehardening; T2 – high-temperature aging. The casts not subjected to heat treatment are not supplied with additional code in alloy mark.

Table 3.7 Mechanical properties of wrought aluminum alloys of low and medium strength [3,5] [$E = 70 - 72.5$ GPa; $G = 27 - 28$ GPa; $\mu = 0.31 - 0.33$ (for thin clad sheets $E = 68.5 - 70.0$ GPa).]

Alloying system	Alloy, condition	Semifinished product	σ_b, MPa	$\sigma_{0.2}$, MPa	δ, %	τ_c, MPa	HB, MPa
Al, technical	АД1М	Rod, sheet	80	35	35	55	250
	АД1Н	same	150	100	6	70	320
Al–Mn	АМцМ	Sheet	110	60	23	80	300
	АМцП	same	170	130	10	100	400
	АМцН	same	220	180	5	110	550
Al–Mg	АМг1М	Sheet, rod	120	50	28	100	300
	АМг2М	same	190	100	23	125	450
	АМг2П	Sheet	250	210	8	140	680
	АМг2Н	same	280	230	5		770
	АМг3М	Sheet, rod	230	120	25	155	580
	АМг3Н	Sheet	270	230	8		750
	АМг4М	Sheet, rod	280	140	23	175	
	АМг4П	Sheet	320	240	12		
	АМг5М	same	300	150	20		650
	АМг5М	Rod, stamped	300	160	14	180	
	АМг6М	Sheet, shape	340	170	20	210	
	АМг6М	Forging	300	150	14		
	АМг6Н	Sheet	400	300	9	250	
Al–Mg–Si	АД31Т	Shape, rod	170	80	20		
	АД31Т1	same	250	210	13	150	800
	АД33Т	same	240	140	20	160	650
	АД33Т1	same	320	280	12	190	950
	АД35Т	same	270	200	15	155	600
	АД35Т1	same	330	300	10	180	950
	АВТ1	same	350	290	12	210	950
		Stamped	310	260	10	190	850
Al–Cu–Mg	Д1Т, Д1чТ	Stamped	410	280	25	270	
	Д16чТ, Д16Т	Sheet, plate	450	320	18	290	
	Д16чТ, Д16Т	Shape, rod	480	350	12	290	
	Д16Т1, Д16чТ1	Sheet, plate	470	400	8		
	Д19Т, Д19чТ	Sheet	440	310	16		
	Д19Т, Д19чТ	Shape, rod	480	350	10	290	
	ВАД1Т	Sheet	440	280	18		
		Shape, rod	500	360	13		
	ВД17Т1	Stamped	520	340	17		
	АК4-1Т1	Sheet, plate	420	360	7		
		Shape, rod	420	370	7		
		Forged	420	320	8		

Table 3.7 The mechanical properties of wrought aluminum alloys *(continued)*

Alloying system	Alloy, condition	Semifinished product	σ_b, MPa	$\sigma_{0.2}$, MPa	δ, %	τ_c, MPa	HB, MPa
Al–Zn–Mg	В92Т1	Sheet, plate	400	300	10		
		Shape, rod	470	350	8		
	1915Т1	Sheet, plate	360	280	11		
		Shape, rod	380	320	10		
	1911Т1	Sheet, plate	420	350	12		
		Shape, rod	520	420	15		
Al–Cu–Mg–Si	АК6Т1	Forged, stamped	420	340	10	260	
	АК8Т1	Stamped	420	310	10		
		Rod	480	380	10	290	

Table 3.8 Mechanical properties of high-strength wrought aluminum alloys [5]

Alloying system	Alloy, condition	Semifinished product	σ_b, MPa	$\sigma_{0.2}$, MPa	δ, %	ψ, %	τ_c	HB, MPa
Al–Zn–Cu–Mg	В95пчТ1	Sheet	570	500	12	27	320	
		Shape, slab	600	550	12	20	330	1600
	В95пчТ2	All forms	540	460	12	30	320	1500
	В95пчТ3	same	500	410	12	34	310	1400
	В93пчТ2	Forged, stamped	470	440	10	30	310	1250
	В93пчТ3	same	440	360	10	40	290	1150
	В96Ц3Т1	same	630	600	10	25		1750
	В96Ц3Т3	same	540	470	12			
	В96Ц1Т1	Shape	720	680	6			
	В96Ц1Т2	Shape	660	630	8	14		
Al–Cu–Li	ВАД23Т1	Sheet	560	500	5			
		Shape, rod	590	550	5		340	1500

Table 3.9 Mechanical properties of rivet aluminum alloys (wire) [5]

Alloying system	Alloy, condition	τ_c, MPa	σ_b, MPa	$\sigma_{0.2}$, MPa	δ, %	ψ, %	HB, MPa
Al–Mg	АМГ5пМ	190	270	150	23		700
Al–Cu–Mg	Д18Т	210	300	170	24	50	700
	В65Т	260	400	250	20	50	900
	Д19пТ	290	460	280	20	40	1200
	Д16пТ	290	450	290	18	35	1200
Al–Zn–Mg–Cu	В95пТ1	340	580	510	8		1500
	В95пТ3	310	500	360	10	45	

Table 3.10 Mechanical properties of powder and high-modulus aluminum alloys [5]

Alloying system	Alloy	Semifinished product	E, GPa	$\sigma_{0.2}$, GPa	σ_b, GPa	δ, %
Al–Al_2O_3	САП-1	Shape	72	210	300	20
	САП-1	Sheet	72	300	340	10
	САП-2	Shape	75	280	350	7
	САП-2	Sheet	75	290	370	7
	САП-3	Rod, strip	77	300	400	4
Al–Si–Ni	САС-1-50	Rod	100	210	340	2.5
		Tube	100	190	320	1.5
	САС-1-400	Rod	100	170	270	2.5
Al–Be–Mg	АБМ1	Sheet	135	300	450	16
	АБМ3	Sheet	200	480	580	7

Table 3.11 Mechanical properties of cast aluminum alloys [3,5] [Designation of casting methods: Z – sand casting; K – chill casting; O – casting in skin-dry molds; V – investment casting; D – pressure-die casting.]

Alloying system	Alloy, condition	Casting method	σ_b, MPa	$\sigma_{0.2}$, MPa	δ, %	τ_c, MPa	HB, MPa
		Heat-resistant alloys					
Al–Cu–Ni–Mg	АЛ1, АЛ1Т7	Z	200–220	170–180	1	170	800–900
	АЛ1Т5	Z	260	220	0.5	220	1000
	АЛ1Т5	K	300	260	1		1200
Al–Si–Cu–Mg	АЛ3Т5	Z,O	250	170	1.5		750
	АЛ3Т5	Z,K,V,O	210		1		700
	АЛ5Т5	Z	260	210	1		800
	АЛ5Т7	Z,K	240	180	1.5		900
	АЛ32Т5	Z,K	240–260	170–190	2		600–700
	АЛ32Т6	Z,K	250–270	180–220	1.5–2		600–700
	АЛ4МТ5	Z	340	250	3	240	1000
	АЛ4МТ5	K	370	270	4	250	1100
	В124Т6	K	420	365	2.5	330	1200
	В124Т6	Z	360	350	0.5		1100
Al–Cu–Mn	АЛ19Т4	Z,K,V,O	320	210	10		850
	АЛ19Т5	Z,K,V,O	375	280	5		1100
Al–Cu–Mn–Ni	АЛ33Т5	Z	280	180	2		900
	АЛ33Т6	Z	300		1		1000
		Leak-tight alloys					
Al–Si (Silumines)	АЛ2	Z	175	80	6		550
		D	215	115	2		
	АЛ4Т6	Z	260	200	4	145	700
	АЛ9Т4, Т5	Z,K	290–210	110–120	2–4		600
	АЛ9-1Т5	Z	270	150	5		
		K	270	220	4		
	АЛ9-1Т6	K	320	240	7		
	ВАЛ8Т5	K	400	340	4		
	АЛ34Т3	K	325	275	6		
		Corrosion-resistant alloys					
Al–Mg	АЛ8Т4	Z,V,K,O	300	170	12	230	600
	АЛ13Т4	Z,V,K,O	160	100	1–3		550
	АЛ22Т4	Z,O,V,K	240	180	1–3		900
	АЛ23Т4	K	250	140	10		600
	АЛ23-1	K	260	140	12	200	750
	АЛ23-1Т4	K	230	130	7	180	600
	АЛ27Т4	Z,V,K	360	180	16	250	800
	АЛ27-1Т4	Z,V,K	380	190	18	260	850
	АЛ28	Z,V,K	205		4–5		550
	АЛ31Т4	Z,K	360	160	25		

3.2.4 Titanium alloys

Table 3.12 Mechanical properties of titanium alloys [3,10] [$E = 110 - 120$ GPa; $G = 42 - 45$ GPa; $\mu = 0.31 - 0.34$.]

Alloy	Alloying system	σ_b, MPa	$\sigma_{0.2}$ MPa	δ, %	ψ, %	τ_c, MPa	KCU, MJ/m^2
			Annealed low-strength alloys				
ВТ1-0	99.28% Ti	350–500	300–420	30	60		
ВТ1-1	99.04% Ti	450–600	380–500	25	50	300	

Table 3.12 Mechanical properties of titanium alloys [3,10] *(continued)*

Alloy	Alloying system	σ_b, MPa	$\sigma_{0.2}$ MPa	δ, %	ψ, %	τ_c, MPa	KCU, MJ/m^2
			Annealed low-strength alloys				
ПТ7М	Ti–Al–Zr	480–680		20			0.8
ОТ4-0	Ti–0.8Al–Mn	500–650		30			
ОТ4-1	Ti–2Al–1.5Mn	600–750	470	20	35–70	400	0.5–1.0
АТ2	Ti–Zr–Mo	600–750		20			
ВТ1-1Л	99.04 % Ti	450–600		20	40		0.5
			Annealed medium-strength alloys				
ОТ4	Ti–Al–Mn	700–900	600	15	45		0.35–1.0
ВТ5	Ti–Al	750–950	650–700	15	25–40	600–650	0.4–0.5
ВТ4	Ti–Al–Mn	850–1000		12			
ВТ6С	Ti–Al–V	850–1000	800–900	10			
ВТ6	Ti–Al–V	950–1050	900	8	30	640	0.3–0.5
ВТ14	Ti–Al–Mo–V	930–1100	850–1000	10	20		0.6–1.2
ВТ20	Ti–Al–Mo–V	950–1100	850–1000	18			
ВТ5Л	Ti–Al	700–900	620	7–12	18–25		0.35
ВТЛ-1	Ti–Al–Si	1000–1100	900–1000	5–10	10–20		0.15
ВТ6Л	Ti–Al–V	$\geq$ 850		$\geq$5			$\geq$0.25
ВТ20Л	Ti–Al–Zr–Mo	$\geq$ 900	850	$\geq$5	20		$\geq$0.3
ВТ21Л	Ti–Al–Zr	$\geq$ 1000		$\geq$4			$\geq$0.2
			Hardened-and-aged high-strength steels				
ВТ6	Ti–Al–V	1150	1050	8	30		0.25
ВТ14	Ti–Al–Mo–V	1150–1400	1080–1300	7	20		0.25
ВТ22	Ti–Al–Mo–V–Fe–Cr	1250–1400	1180–1300	5–8	35–50		0.3
		1080*[1]	1000	18	40		
ВТ23	Ti–Al–Mo–V–Fe–Cr	1300–1500	1200–1400	5			0.6
ВТ15	Ti–Al–Mo–Cr	1300–1500	1160–1400	3			0.2
ТС6	Ti–Al–V–Cr						
			Annealed heat-resistant alloys				
ВТ8	Ti–Al–Mo–Si	1050	900	10		650	0.35
ВТ9	Ti–Al–Mo–Si–Zr	1150	1000	6		685	0.35
ВТ3-1	Ti–Al–Mo–Cr–Si–Fe	1000	950	12		635	0.4
ВТ9Л	Ti–Al–Mo–Fr	$\geq$950		$\geq$5			0.2

*[1] Annealed.

3.2.5 Magnesium alloys

In designating the wrought magnesium alloys, the next codes are adopted: M – annealed; H – cold-hardened; T1 – artificially aged; T4 – hardened; T6 – hardened and artificially aged; T8 – hardened, cold-hardened and artificially aged.

Cast alloys involve additional codes in alloy mark depending on heat treatment conditions of ingots: T1 – artificial aging; T2 – annealing; T4 – hardening; T6 – hardening and aging.

Magnesium alloys whose semifinished products after hot deformation or casting are not subjected to heat treatment, do not possess additional code in alloy designation.

Table 3.13 Mechanical properties of wrought magnesium alloys [3,24] [E = 43 GPa; G = 16 GPa; μ = 0.33.]

Alloying system	Alloy, conditions	Semifinished product	σ_b, MPa	$\sigma_{0.2}$, MPa	σ_p, MPa	δ, %	ψ, %	$\sigma_{0.2comp}$, MPa	τ_c, MPa	KCU, MJ/m^2
Alloys of low and medium strength										
Mg–Mn	MA1M	Sheet	210	120		8				0.05
	MA1M	Bar	240	150		4	6	85	130	0.06
Mg–Al–Zn–Mn	MA2	Bar, stamped	265	165	80	10	30	110	160	0.12
	MA2M	Sheet	250	150		18				
Mg–Mn	MA8	Bar	260	150		7			150	0.1
	MA8M	Sheet	250	160		18				
	MA8H	same	270	190		15				
	MA9	Bar	270	240		10	11			0.6
	MA9M	Sheet	250	180	100	15				
High-strength alloys										
Mg–Al–Zn–Mn	MA2-1	Plate, forged, stamped forming, shape	275	170	100	12		85	140	0.08
	MA2-1M	Sheet	280	180		16				
	MA3, MA3M	Bar, forged, stamped	275	220	100	13	25		140	0.1
	MA5T4	same	315	220	130	13	20	140	180	
Mg–Zn–Zr	MA14T1	Bar	340	280	150	10	25	160	160	0.08
	MA14T1	Stamped, forged	315	255		13				
Mg–Al–Cd–Ag–Mn	MA10T6	Bar, shape	400	295	130	5	8	240		0.03
Heat-resistant alloys										
Mg–Ce–Mn	MA11T6	Sheet, bar	270	140	80	10	12	120	170	0.04
Mg–Th–Zn	MA13T8	same	240	180	80	6		125		
	ВМД1	Bar	300	250	180	5	10	160	160	0.07
Superlight alloys (density ρ=1400–1600 kg/m^3; E=45–46 GPa)										
Mg–Li–Zn	ВМД5	Bar, stamped	160–220	120–180		15–40		130–180		0.25–0.9
Mg–Li–Al–Cd	ИМВ-2	same	210–280	160–250		8–20		180–240		0.05–0.1

Table 3.14 Mechanical properties of cast magnesium alloys [3,24] [E = 42 GPa; G = 15.5 GPa; μ = 0.34.]

Alloying system	Alloy, condition	σ_b, MPa	$\sigma_{0.2}$, MPa	σ_p, MPa	δ, %	ψ, %	$\sigma_{0.2comp}$, MPa	τ_c, MPa	KCU, MJ/m^2	HB, MPa
Medium-strength alloys										
Mg–Al–Zn	МЛ3	180	55		8	12		110	0.05	450
	МЛ7-1	180	70		6	8			0.03	550
High-strength alloys										
Mg–Al–Zn	МЛ4	170	95		4	6		130	0.02	500
	МЛ4Т4	235	85	18	7	15		135	0.04	580
	МЛ4Т6	240	115	45	4	6		145	0.02	675
	МЛ5	155	95		1.2	2.5				600
	МЛ5Т4	240	85	30	7	15	80	135	0.05	580
	МЛ5Т6	240	120	45	3	5		140	0.05	675
Mg–Zn–Zr	МЛ12	210	105	75	8	9		155	0.05	550

Table 3.14 Mechanical properties of cast magnesium alloys [3,24] *(continued)*

Alloying system	Alloy, condition	σ_b, MPa	$\sigma_{0.2}$, MPa	σ_p, MPa	δ, %	ψ, %	$\sigma_{0.2comp}$, MPa	τ_c, MPa	KCU, MJ/m^2	HB, MPa
				High-strength alloys						
	МЛ12Т1	230	130	85	7	7	130	155	0.04	550
Mg–Zn–La–Zr	МЛ12Т6	250	150	85	6.5	7	130	155	0.04	675
	МЛ15Т1	215	140	80	3.5	5	140	150	0.02	550
				Heat-resistant alloys						
Mg–Nd–Zr	МЛ9Т6	240	145	75	3	6	140	170	0.03	650
	МЛ10Т6	240	120	70	5	7.5	100	170		650
Mg–Ce–Zn–Zr	МЛ11	130	100	40	3	3.5	100			600
	МЛ11Т4	150	90	40	5	7	90	120	0.03	600
	МЛ11Т6	160	105	45	3	5	105	120	0.03	650
Mg–Th–Zn–Zr	МЛ14Т1	200	95	50	8					600
Mg–Th–Zr	ВМЛ1Т6	200	95	50	6					
	ВМЛ2Т6*[1]	260	120	60	6	8	120	170	0.07	650

*[1] Does not contain toxic and radioactive additives.

3.2.6 Copper alloys and tin-base and lead-base alloys

The copper alloys are subdivided into two main groups: the brasses and bronzes. The brasses refer to metals alloyed with zinc (Tables 3.15 and 3.16). Common and special marks of brasses are recognized. The common brasses (binary alloys) are designated with a letter Л followed by copper percentage. In designating the special brasses, a letter Л is followed by the capital letters of alloying elements and copper percentage and then – through the dash – by the percentage of each alloying element in its order.

Bronzes are defined as metals alloyed with various elements, less zinc (Tables 3.17 and 3.18). A bronze takes its name from a dominant alloying element: so, the alloy of copper and aluminum is called aluminum bronze, the copper-lead alloy is called lead bronze, etc. The bronzes are marked with the letters Бр, the rest being the same codes as in brasses. The alloys, in which nickel is a basic alloying element, are known as cupronickel alloys and go under special names (Table 3.19). Wrought copper alloys are supplied in soft (annealed–and–hardened), semi-hard (10–30% reduction), hard (30–50% reduction) and extra-hard (reduction over 60%) conditions.

Tin-base or lead-base alloys (babbitts) are marked with a letter Б followed by a digit denoting the tin content in an alloy (Table 3.20).

Table 3.15 Strength and ductility of wrought brasses [$E = 102 - 115$ GPa.]

Brass type	Brass mark	Soft condition			Hard condition		
		σ_b, MPa	δ, %	HB, MPa	σ_b, MPa	δ, %	HB, MPa
Common	Л96, Л90	240–260	50	550	450–470	2.5	1350
	Л85	280	50	570	550	3.5	1400
	Л80, Л70	320	50–55	600	655	4	1450
	Л68, Л62	300–320	45–55	600	700	4	1550
	Л63, Л60	400	45	640	705	3	1600
Aluminum	ЛА85-0.5	300	35		500	10	
	ЛА77-2	380	50	550	600	9	1550
	ЛАЖ60-1-1	450	50	550	700	8	1700
	ЛАН59-3-3	550	45	1150	700	9	1800
Silicon	ЛК80-3	300	55	1000	650	4	1800
Manganese	ЛЖМц59-1-1, ЛМц58-2, ЛМцА-57-3-1	380–450	40–60	800–950	680–750	5–10	1600–1850

Table 3.15 Strength and ductility of wrought brasses *(continued)*

Brass type	Brass mark	Soft condition			Hard condition		
		σ_b, MPa	δ, %	HB, MPa	σ_b, MPa	δ, %	HB, MPa
Nickel	ЛН65-5,	400	60	600	700	4.5	1650
Tin	ЛО90-1	280	45	570	520	4.5	1450
(marine)	ЛО70-1	350	60	600	700	4	1500
	ЛО62-1	400	40	800	700	7	1450
	ЛО60-1	380	40	770	580	4	1500
Lead	ЛС74-3, ЛС64-2, ЛС63-3	300–400	40–60	500-700	550–700	2–6	1000–1200
Muntzmetal	ЛС59-1, ЛС60-1	300–450	35–55	500–800	500–650	5	1200–1600

Table 3.16 Mechanical properties of cast brasses [3,24]

Brass	σ_b, MPa	$\sigma_{0.2}$, MPa	δ, %	KCU, MJ/m^2	HB, MPa	Brass	σ_b, MPa	$\sigma_{0.2}$, MPa	δ, %	KCU, MJ/m^2	HB, MPa
ЛА67-2,5	380	150	15		900	ЛКС80-3-3	340	140	20	0.4	950
ЛАЖМц66-6-3-2	640	350	7			ЛМцОС58-2-2-2	340	240	8		950
ЛАЖ60-1-1Л	400	250	20		900	ЛМцЖ52-4-1	550	300	20		
ЛК80-3Л	400	160	20	1.2	1050	ЛМцЖ55-3-1	550	250	10		1050
ЛМцС58-2-2	360	240	10	0.7	850	ЛС59-1Л	340	150	40	0.26	800

Table 3.17 Mechanical properties of wrought bronzes [3,24]

Bronze	E, GPa	Soft condition			Hard condition		
		σ_b, MPa	δ, %	BHN, MPa	σ_b, MPa	δ, %	BHN, MPa
БрОФ3-0.8, БрОФ7-0.2	116.5	400–500	60	850–1000	1100–1200	1	2000–2400
БрОФ6.5-0.4	112	450	60	900	800	7	2200
БрОФ4-0.25	100	340	50	650	600	8	1650
БрОЦ4-3	124	400	35	700	600	4.5	1600
БрОЦС4-4-2.5, БрОЦС4-4-4	73.5	350	30	700	600	2	1600
БрА5	120	400	65	650	800	4	2100
БрА7	120	470	70	750	1000	3	2200
БрАМц9-2	92	450	30	1100	800	4	1800
БрАЖ9-4	115	450	40	1100	700	4	2000
БрЖМц10-3-1.5	100	500	20	1400	700	9	2000
БрАЖН10-4-4	115	550	35	1500	830	9	2200
БрБ2, БрБНТ1.9, БрБНТ1.7	130	500	40	1400	900	2	
БрМц5	105	330	40	800	600		1700
БрКМц3-1	115	400	50	900	750	6	1700
БрКН1-3		450	25	1000	600	6	2000
БрХ0.7	135	250	25	700	500	4.5	1400

Table 3.18 Mechanical properties of cast bronzes [3,24]

Bronze	Chill casting			Sand casting		
	σ_b, MPa	δ, %	HB, MPa	σ_b, MPa	δ, %	HB, MPa
БрОЦСН3-7-5-1	210	5	600	180	8	600
БрОЦС3-12-5,	180	4	600	150	6	600

Table 3.18 Mechanical properties of cast bronzes [3,24] *(continued)*

Bronze	Chill casting			Sand casting		
	σ_b, MPa	δ, %	HB, MPa	σ_b, MPa	δ, %	HB, MPa
БрОЦС5-5-5						
БрОЦС3-5-7-5	250	3	900	220	3	800
БрОФ10-1						
БрОЦ10-2	230	5	750	220	8	700
БрОЦ5-25	140	6	500	120	4	400
БрАЖ9-4Л	400	10	1000	500	12	1000
БрАЖМЦ10-3-1.5	400	12	1200			
БрАЖН11-6-6	600	2	2500	600	2	2500
БрСуН6-2	260	6	810			
БрСуФ6-1	220	5	800			
БрСуСФ6-12-0,3	150	2	600			

Table 3.19 Mechanical properties of cupronickel alloys [3,24]

Name	Alloy	E, GPa	Soft condition		Hard condition	
			σ_b, MPa	δ, %	σ_b, MPa	δ, %
German silver	МНЖМц30-0,8-1	145	400	45	600	4.0
	МН19	140	350	40	550	4.0
Argentan	МНЦ15-20	140	425	45	650	2.5
	МНЦС16-29-1,8	125	400	42	650	3.0
Cumial	МНА13-3		420	12	900	3.0
	МНА6-1,5		400	35	700	5.0
Copel	МНМц43-0,5	120	420	38	650	3.5
Constantan	МНМц40-1,5	165	430	28	670	2.5

Table 3.20 Mechanical properties of babbitts [3]

Alloy	σ_b, MPa	$\sigma_{0.2}$, MPa	δ, %	σ_{bcomp}, MPa	$\sigma_{0.2comp}$, MPa	Compressive shortage, %	HB, MPa	KCU, MJ/m^2
Tin-base babbitts								
В89	80		9	100	50	40	250	0.08
Б83	90		6	115	72	38	300	0.06
Lead-tin babbitts								
Б16	78	76	0.2	123		15	300	0.014
БН*[1]	70	70	1.0	107		25	290	0.03
БТ*[2]	67	54	11.5	103			220	0.05
Б6	68		0.2	120		23	300	0.015
Calcium-base babbitts								
ДК	118	100	2.5	160		19	320	0.08
БК2	95	80	8				200	0.12

*[1] Nickel is included.
*[2] Tellurium is included.

3.2.7 Heat-resistant alloys and refractory metals-base alloys

Table 3.21 Mechanical properties of heat-resistant nickel-base and cobalt-base alloys [3,24]

Alloy, alloying system	σ_b, MPa	$\sigma_{0.2}$, MPa	δ, %	ψ, %
1.Nickel-base alloys:				
(a) wrought, hardened-and-aged				
ХН77ТЮ	1000	600	25	28
ХН77ТЮР	1000	650	20	21
ХН70ВМТЮ	1150	750	14	15
ХН70МВТЮБ	1050	700	16	16
ХН67МТЮБ	1050	650	25	25
(b) cast, normalized				
ВЛ7-45У	500		5	9
ЖС3	740		8	14
ЖС3ДК	1050		8	13
ЖС6	1050		1	2
ЖС6К	1000		2.5	6
ЭП23	1250		2.5	11
(c) wrought heat-resistant, normalized				
ХН78Т	780	275	40	
ХН75МБТЮ	850	400	45	
ХН60В	800	320	60	
ХН70Ю	720		75	
ХН60Ю	710		60	
2. Wrought cobalt-base alloys, aged				
ЛК4*1	700		8	10
ЛК4Я	900		5	5
4К66Я	900		6	7

*1 Cast condition.

Table 3.22 Strength and ductility of refractory metals-base alloys [3,6,24]

Base	Alloy, alloying system, condition	σ_b, MPa	δ, %
Chromium	ВХ2, ВХ2И	400	
	ВХ4	1100	
Molybdenum	ВМ1 (Mo–Ti–Zr–Nb), ЦМ2А (Mo–Ti–Zr), ЦМ6 (Mo–Zr) ВМ2 (Mo–Ti–Zr–Nb) annealed	800	
	МР47ВП (Mo–47Re)	1750	
Niobium	ВН-2 (Nb–Mo), wrought	750	23
	ВН-2А (Nb–Mo–Zr), wrought	850	4.5
	ВН-3 (Nb–Mo–Zr), wrought	780	18
	ВН-4 (Nb–Mo–Zr–La or Ce), recrystallized	800	16
	ВН-5А (Nb–Mo–Zr–La or Ce), recrystallized	600	27
	РН-6 (Nb–Mo–W–Zr), recrystallized	890	
	МН-1, recrystallized	670	18
	МН-1, hardened	760	17
	МН-2, recrystallized	1050	4
Tantalum	Ta-10W, recrystallized	950	25
	Ta-10W, wrought	1350	3
Tungsten	W-30Re, recrystallized	1400	5
	W-30Re, nonrecrystallized	2200	6

3.2.8 Composite materials

The composite materials offer oriented structure and can be divided into the fibrous materials, whose matrix contains unidimensional strengthening fillers (fibers, wires, threadlike crystals – whiskers), the lamellar composites representing an assembly of alternate layers of rigidly bound two-dimensional reinforcing elements (sheets, foils, etc.), and the dispersion-strengthened materials containing uniformly distributed ultrafine particles unsolved in a carrying matrix.

Table 3.23 Mechanical properties of unidirectional-structure composite materials based on polymer matrix [12,13]

Filler	Matrix	Filler content per volume, %	ρ, kg/m^3	σ_b, MPa	$\sigma_{b,comp}$, MPa	$\sigma_{b,bend}$ MPa	τ_{sh}^{*1}, MPa	E, GPa	KCU MJ/m^2	δ_b, %
Boron fiber	Epoxy resin	50	2000	1600	2500	1600	100	210	0.06	0.4
	Polyamide resin	50	2000	1000	1250	1550	60	250	0.11	
Carbon strip	Epoxy resin	50	1400	800	400	1000	30	120	0.05	0.5
	Polyamide resin	52.5	1300	400	300	800	26	80	0.04	
Carbon yarn	Epoxy resin	60	1500	1100	450	1300	45	180	0.045	
	Polyamide resin	57.5	1400	910	400	1100	30	140	0.04	
Organic fiber	Epoxy resin	70	1350	1800	290	675	49	80		2.3
Glass fiber	Epoxy resin	75	2000	1800	700	900	30	55		3

*1 Under interlayer shear.

Table 3.24 Mechanical properties of inidirectional-structure composite materials based on metal matrix [5,13,14,24]

Filler	Matrix	Filler content per volume, %	ρ, kg/m^3	σ_b, MPa	E, GPa
Boron fiber	Aluminum	50	2600	1350	240
	Alloy Al-Mg-Si	50	2600	1450	240
	Magnesium	50	2150	1200	220
	Magnesium cast	75	2400	1350	335
Carbon fiber	Aluminum	40–50	2200–2400	650–1250*1	125–260*1
	Nickel	50	5300	560	240
	Zinc	35	5260	780	120
	Lead	40	7350	730	205
Steel wire	Aluminum	40	4740	1500	120
		15	3470	850	90
Beryllium wire	Alloy Al–Mg–Si	50	2300	680	195
Molybdenum wire	Alloy Ti–Al–V	30	6250	1400	195
Tungsten wire	Nickel alloy	50	14000	700	
Threadlike SiO_2 crystals	Aluminum	50	2450	800	
HfO_2^{*2} particles	Nickel*1	2.5	8900	510	
SiC fibers	Titanium	25	4300	940	210

*1 Depending on type of carbon fiber.
*2 Dispersion-strengthened material.

3.2.9 Plastics, metal ceramics and other materials

Table 3.25 Mechanical properties of plastic [3,24]

Plastic	ρ, kg/m^3	σ_b, MPa	σb, bend, MPa	$\sigma_{b,\text{comp}}$, MPa	E, GPa	δ_b, %	KCU, 10^2 MJ/m^2
Asbestos textolite	1700	65	100	145	15		3
Caprone	1140	55	95	90	1.0	100	11
Caprone, polyamide, anid glass-filled	1350	130	200	115	6.5	2	2.5
Caprone, secondary	1150	35	45	60			2.5
Diflon	1200	70	105	85		50	13
Etrol, cellulose acetate	1400	55	50	55	2.25	7	2
Etrol, ethyl-cellulose	1200	12	34	17	2.2	50–100	1.5
Etrol, nitrocellulose	1900	30	35	20	2.3	25–150	0.6
АГ fiber glass	1700	405	410	260	22		15
FCBAM fiber glass-reinforced plastic	1900	460	460	420	35		30
ПС-1 foam plastic	100			1	0.05		0.11
ПХВ-1 foam plastic	200	4.5	4	2.6	0.18	6	0.15
ФК-20-СТ foam plastic, glass-filled	450	10		24			0.7
KAST-B glass textolite	1900	320	150	110	20	0.8	8
ВФТ-С glass textolite	1800	350	330	170	21		15
ЭФ 32-301 glass textolite	1800	400	400	255	21		15
Laminated paper plastic (getinaks)	1350	160	140		1.5		1.4
Molded laminate	1400	50	60	200	5		1.2
Organic glass:							
unoriented	1200	75	105		3	4	1.3
oriented СОЛ, СТ-1	1200	80	105		3	20	3
oriented 2-55, T2-55	1200	105	150		4	3	3
ПУ101Т Polyurethane foam	200			3.3			0.04
П-68 Polyamide	1110	50	80	90	1.5	100	15
Polyethylene BD	920	14	15	12	0.2	150	
Polyethylene HD	950	22–32	20–35	20–35	0.7	200	
Polypropylene plastic	910	32	75	100	1.15	600	
Polystyrene	1050	35	100	100	2.5	1.5	2
Polytetrafluoroethylene (PTFE)	2200	20	12.5		0.45	300	10
Polytrifluorochloroethylene (PTCE)	2120	37	70	50	1.15	30	2–16
Polyurethane	1200	55	75		0.01	55	1.1
Polyvinyl chloride plastic	1400	50	110	90	4.0	30	15
Talc- and graphite-filled polyamide	1140	60	80	80	1.15	100	10
ПТ textolite	1300	85	140	130	10	1	3.5

Table 3.26 Mechanical properties of metal ceramics*[1] [3,24]

Alloy type	Metal ceramics, composition	σ_b, MPa, in			ψ, %	HB, MPa
		tension	compression	bending		
Iron-base alloys	Sintered iron	195	300		4	540
	Ditto, 69–70% strained	275	370		6	760
	Sintered carbon steel	300	540		5	960

Table 3.26 Mechanical properties of metal ceramics*[1] [3,24] *(continued)*

Alloy type	Metal ceramics, composition	σ_b, MPa			ψ, %	HB, MPa
		tension	compression	bending		
Iron-base alloys	Sintered stainless steel 60–70% strained	565	720		16	1600
	Alloy 22Fe+20Ni+Mo	1000	1650		7	2700
Hard alloys	WC–Co: ВК2, ВК3, ВК8			1000		900
	ВК10, ВК11, ВК15			1500		900
	WC–TiC–Co: Т5К10, Т14К8, Т15К8			1100		900
	Т30К4, Т60К6			750–900		900
High-density alloys	W–Co–Ni (Cu, Cr)	1000	2800			2250
Filters	Bronze Cu–Sn	30–40	100–120		3	
	Low-carbon steel	1–7	30–250			
	Stainless steel	5–120			0.8	
Cermets	$70Al_2O_3$+30Cr	250	2300	390		
	80TiC+20Co		3150	1050		

*[1] Manufactured by powder metallurgy technique.

Table 3.27 Mechanical properties of ceramics [3,24]

Ceramics	ρ, kg/m^3	σ_b, MPa			E, GPa
		tension	bending	compression	
Fireclay of rough-grain structure	2100	6–10	10–20	30–90	
Fireclay of fine-grain structure	2170	25	25–70	80–150	
Porcelain	2350	30–60	50–120	400–500	
Stone casting	2900		40–65	330–450	90
Oxide ceramics					
Al_2O_3 (corundum)	3990	260	150	4000	382
Be_2O_3	3020	105		2100	272
MgO	3580	98	110	1400	214
ZrO_2	5600	148	233	2100	172
ThO_2	9690	100		1500	140

Table 3.28 Mechanical properties of inorganic glasses and sitalls [3,24]

Glass, sitall	ρ, kg/m^3	σ_b, MPa			E, GPa
		tension	bending	compression	
Glass:					
of quartz, nontransparent	2100	40	45	350	0.6
of quartz, transparent	2200	60	110	650	0.65
vacuum electrical	2100–2500	50–100		800–200	0.65
insulating	2500		40–65		0.7
Sitall:					
magnesian	2500–2850		160		0.12
pyroxenic	2900		400		0.14

Table 3.29 Mechanical properties of rubbers [3]

Rubber	σ_b, MPa	δ_b,	Rubber	σ_b, MPa	δ_b, %
Shock-absorbing	10–16	400–600	ХП-based	18–24	200–500
Sponge	0.02–0.1		Oil-resistant soft	4-12	250–600
Expanded	0.01–5	20–300	medium hard	4-16	200–350
Acid- and alkali-resistant			hard	4-20	120–300
СКБ-based	4.5–10	200–350	Cold-resistant on the base of СК МС-10	15	200
СКС-based	20–25	300–500	Resistant to hydraulic liquids on base of СКН	13	

3.2.10 Minerals and fibers

Table 3.30 Hardness of minerals

Mineral	Chemical composition	Hardness on the Mohs' scale	Other properties
Agate	SiO_2	6.5	$E = 98$ GPa
Almandine (garnet)	$Fe_3Al_2[SiO_4]_3$	8	$H_\mu = 1230$ GPa
Andalusite	Al_2SiO_5	7	
Andradite (garnet)	$Ca_3Al_2[SiO_4]_3$	7	
Azurite	$2CuCO_3{\cdot}Cu(OH)_2$	3.5	
Barite	$BaSO_4$	3.5	
Beryl	$Be_3Al_2[Si_6O_{18}]$	8	$H_\mu = 1120 - 1450$ GPa
Calcite	$CaCO_3$	3	
Corundum	Al_2O_3	9	$H_\mu = 2050$ GPa
Crocidolite	$3H_2O{\cdot}2Na_2O{\cdot}6(Fe,Mg)O{\cdot}2Fe_2O_3{\cdot}17SiO_2$	4	
Crysoberyl	Al_2BeO_4	8.5	
Crysotil-asbestos	$H_4Mg_3Si_2O_9$		$E = 158 - 210$ GPa, $\sigma_{b,comp} = 590 - 785$ MPa
Diamond	C	10	$E = 740 - 1000$ GPa, $\sigma_b = 240 - 480$ MPa, $H_\mu = 10000$ GPa
Fluorite	CaF_2	4	
Galenite	PbS	2.5	
Graphite	C	1.5	$\sigma_b = 4.5 - 6$ MPa, $\sigma_{b,comp} = 15 - 29$ MPa, $H_\mu = 6.8 - 11.8$ GPa
Grossularite (garnet)	$Ca_3Al_2[SiO_4]_3$	7	
Gypsum	$CaSO_4{\cdot}2H_2O$	2	
Halite	NaCl	2	
Hematite	$FeSiO_2$	6	
Kyanite	Al_2SiO_5	6	
Lazurite	$(Na,Ca)_{4-8}[AlSiO_4]_6[SO_4Cl,S]_{1-2}$	5.5	
Magnesite	$MgCO_3$	4	$\sigma_{b,comp} = 88.5$ MPa
Malachite	$Cu_2[CO_3](OH)_2$	4	
Millerite	NiS	4	
Mullite	$3Al_2O_3{\cdot}2SiO_2$	6	$E = 35$ GPa
Muscovite	$KAl_2(AlSi_3O_{10})(OH)_2$	2.5	$\sigma_b = 170 - 355$ MPa, $\sigma_{b,comp} = 420 - 540$ MPa
Opal	$SiO_2{\cdot}H_2O$	6	
Perovskite	$CaTiO_3$	6	
Phlogopite	$KMg_2[Si_3AlO_{10}][Fe,OH]_2$	2.5	$\sigma_{b,comp} = 205 - 265$ MPa
Pyrolusite	MnO_2	5.5	
Pyrope (garnet)	$Mg_3Al_2[SiO_4]_3$	7	
Pyrophyllite	$Al_2[Si_4O_{10}][OH]_2$	1.5	
Quartz	SiO_2	7	$H_\mu = 1100$ GPa
Realgar	AsS	2	
Rock crystal	SiO_2	7	
Rutile	TiO_2	6	
Scheelite	$CaWO_4$	5	
Schorlomite	$Ca_3(Al,Fe,Ti)[(Si,Ti)O_4]_3$	7	

Table 3.30 Hardness of minerals *(continued)*

Mineral	Chemical composition	Hardness on the Mohs' scale	Other properties
Sepiolite	$2MgO{\cdot}3SiO_2 \cdot nH_2O$	2.5	
Sillimanite	Al_2SiO_5	7	$\sigma_{b,\text{comp}} = 45 - 60$ MPa
Spessartine (garnet)	$Mn_3Al_2[SiO_4]_3$	7	
Talc	$Mg_3[Si_{14}O_{10}][OH]_2$	1	
Topaz	$Al_2[SiO_4][FeOH]_2$	8	
Tourmaline	$[Ni,Ca](NaAl)_6[Si_6Al_3B_3(O,OH)_{30}]$	7	
Uvarovite (garnet)	$Ca_3Cr_2[SiO_4]_3$	7	
Willemite	$Zn_2[SiO_4]$	5.5	
Wollastonite	$CaSiO_2$	4.5	$\sigma_{b,\text{comp}} = 1120 - 1450$ GPa
Zircon	$ZrSiO_4$	7.5	

Table 3.31 Mechanical properties of fibers [3]

Fiber	$L_b{\cdot}10^{-3}$, m under dry condition	Strength loss under wet conditions, %	Elongation, %		σ_b, MPa	E, GPa
			dry	wet		
Natural fibers						
Asbestos					300	155–205
Bamboo					345	33
Cotton:						
medium-stapled	24	0	6–8	7–9	245–390	
fine-stapled	35	0	6–8	7–9	390–540	
Flax:						
technical-glade	40	0	2–3	2–3	490–590	
individual	63	0	2–2.5	2–2.5	785–980	
Silk	35	20–30	15–20	20–25	440–490	
Wool:						
coarse	12	30–35	25–30	25–35	145–195	
fine	14	30–35	30–50		195–245	
Chemical fibers						
Acetate	11–14	40–45	22–30	28–35	175–215	
Albuminous (casein)	7–13	40–60	30–50	50–70	90–155	
Alginate	10–14	60–70	10–14	25–26		
Vinylon, vinyl	80–100	10–15	8–10	13–18	1030–1270	2.5–10.8
Viscose rayon cord:						
ordinary	27–30	35–40	10	16	340–390	3.9–5.9
high-strength	40–55	25–30	15	25	640–735	11.8–24.5
Viscose rayon:						
cuprammonium	15–23	35–40	10–17	15–30	225–315	
staple	15–20	55	20–25	20–30	225–300	
Hydrated cellulose	30–40	20	8–10	12–13		8–9.8
Nylon, capron	40–50	10	20–25		480–625	
Lavsan	45–55	0	9–12	15–18	450–540	
Orlon, acrylon	45–55	2–6	14–17	14–16	510–610	
Triacetate	10–15	20–25	20–23	35–40	135	4.5
Fluorlon (ftorlon)	45–55	0	6–9	6–9	880–1050	14.7

Table 3.32 Mechanical properties of fibers, wires and threadlike crystals used for reinforcement of high-strength and high-modulus composites [14,15,24]

Fiber, wire, crystal	ρ, kg/m^3	σ_b, GPa	E, GPa	δ_b, %
Fiber:				
boron	2630	2.5–3.5	380–450	0.7–0.8
carbon	1700	2–3.2	200–500	0.7–1.0
glass	2540	3.9–4.6	95–100	4–5
organic	1350	2.8–3.5	120–130	2–2.5
silicon carbide	3210	2–4	400–500	0.3–0.5
aluminum oxide	3960	2.1–2.6	500	
zirconium oxide	6270	2.4–2.7	480	
Wire:				
beryllium	1840	1–1.5	290–320	
tungsten	19300	2.4–4.2	400	2–3
steel	7800	3.5–4	200	2–2.5
titanium	4500	1.5–2	120	1.8–2
molybdenum	10000	1.75	350	1.3
Whiskers:				
of graphite	2260	21	100	
aluminum oxide	3960	28–42	500	
aluminum nitride	3300	15	380	
silicon carbide	3210	21–37	580	
silicon nitride	3180	15	495	

3.3 Mechanical Properties at Low and High Temperatures

For an each class of materials, the temperature dependences of mechanical properties are closely allied. Quantities characterizing the resistance to plastic deformation (hardness, ultimate strength and yield stress), as well as impact strength are most temperature sensitive. Elastic properties of metals and alloys vary only slightly with temperature. On the contrary, the modulus of elasticity for some nonmetallic materials can be reduced more than by a factor of two (with drop in temperature to –60°C).

In Tables 3.33–3.46 and in Figs.3.2–3.21, those characteristics as E, $\sigma_{0.2}$, σ_b , δ, ψ and KCU, whose definitions were given above in Section 3.1, are presented in the temperature range from –253 to +2400°C, along with $\sigma^T_{0.2/\tau}$ and σ^T_τ at elevated temperatures (up to +1600°C). Here, $\sigma^T_{0.2/\tau}$ is the creep stress, i.e. the highest sustained stress which over the definite period of time τ (usually 100 or 1000 h) at constant temperature t results in the residual strain not exceeding a given level (as a rule 0.2%); σ^T_τ is the long-term ultimate strength, i.e. the highest sustained stress causing the fracture over the definite period of time τ at a given constant temperature t.

Table 3.33 Mechanical properties of metals at low temperatures [3,6,11]

Metal	σ_b, MPa at t, °C			$\sigma_{0.2}$, MPa at t, °C			δ, % at t, °C			ψ, % at t, °C		
	20	–196	–253	20	–196	–253	20	–196	–253	20	–196	–253
Aluminum	120	210	350				30	42	45	85	75	65
Beryllium	300	270		200	230							
Niobium	350	1030	1120				30		7			
Copper	240	380	460				29	41	46	70	72	74
Iron	360	830	1060*[1]	260	780	1060*[1]	27	4	0.6*[1]	88	54	73*[1]

Table 3.33 Mechanical properties of metals at low temperatures *(continued)*

Metal	σ_b, MPa, at T, °C			$\sigma_{0.2}$, MPa at t, °C			δ, % at t, °C			ψ, % at t, °C		
	20	−196	−253	20	−196	−253	20	−196	−253	20	−196	−253
Lead	28	45	70				26	34	36			
Magnesium	120	160	210				5	5	5	10	7	8
Molybdenum	490	540	540				3	0	0	3	3	4
Nickel	390	600	750[*1]	125	145	145[*1]	54	61	62[*1]	85	80	75[*1]
Silver	180	290	360				39	82	88	90	88	80
Sodium	14	19	40				19	13	60			
Tantalum	620	1030	1150[*1]	560	1030	1150[*1]	13	6	0.6[*1]	69	74	77[*1]
Titanium	750	1190	1340[*1]	660	1080	1340[*1]	18	12	1[*1]	48	16	4[*1]
Titanium iodide	250	630		154	440		50	11.5		78	28	
Tin	36	71	73									
Vanadium	420	1080					14	3.5				
Zirconium iodide	260	470	680[*1]	140	230	160[*1]	35	49	47[*1]	52	58	52[*1]

[*1] At a temperature of −269°C.

Table 3.34 Strength of steels and alloys at low temperatures [3,5,19]

Steel, alloy, heat treatment	σ_b, MPa at t, °C				$\sigma_{0.2}$, MPa at t, °C			
	20	−70	−196	−253	20	−70	−196	−253
Steels								
Ст.45, normalizing	600		1050	1400	400		950	1400
Ст.45, hardening plus tempering	1000	1050	1320		890	960	1280	
30ХГСА, hardening plus tempering	1200	1300	1580		1100	1180	1500	
40ХНМА, hardening plus tempering	1100	1300	1550		980	1080	1400	
30ХГСНА, isothermal hardening	1600	1700	1900		1200	1230	1450	
40ХГСН3ВА, hardening plus tempering	2100	2200	2400		1450	1480	1700	
12Х18Н9Т, hardening	620	1150	1650	1900	250	300	370	420
08Х15Н5Д2Т, hardening plus tempering	1150	1230	1500	1750	1000	1080	1350	1650
10Х12Н20Т3Р, hardening plus aging	1080		1100	1450	620		1100	880
12Х14Г14Н3Т, hardening	750		1400	1500	500		500	600
08Х16Н6, hardening, cold treating, aging	1350	1500	1750	1850	1250	1350	1650	1800
45Л, tempering	770	850						
27ХГСНЛ, isothermal hardening	1450	1500			1250	1350		
35ХГСЛ, hardening plus tempering	950	1030	1350					
Titanium alloys								
ВТ5-1, annealing	750	950	1350	1600	650	850	125	1550
ОТ4, annealing	810	870	1220	1500	660	830	115	1300
ВТ6, annealing	100	120	1650	1850	950	1150	155	1750
ВТ5Л	850	1000	1350	1500	820	950	1300	1300
ВТ20Л	1000	1100						
ВТ21Л	1100	1200						
Aluminum alloys								
АД1М	80	95	170	260	30	30	40	54
АМГ6М	320	350	470	545	170	175	185	195
АД33Т1	310	350	410	530[*1]	270	310	350	390[*1]
Д16Т	440	470	550	700	290	330	420	520
Д16Т1	495	535	685	720[*1]	455	490	575	625[*1]
1201Т1	430	460	530	680	320	380	410	490

Table 3.34 Strength of steels and alloys at low temperatures [3,5,19] *(continued)*

Steel, alloy, heat treatment	σ_b, MPa at t, °C				$\sigma_{0.2}$, MPa at t, °C			
	20	–70	–196	–253	20	–70	–196	–253
Aluminum alloys								
В92Т1	440	470	510	610	320	330	360	400
В93Т1	490	520	580		450	470	530	
В95Т1	600	620	750	810	550	560	640	730
САП-1	320	380	480	560	220	260	350	450
АБм1	400	420	530		280	280	400	
АЛ4Т6	350	420	370	470	320	320	360	450
АЛ19Т4	320	320	370	390	210	210	250	330
АЛ9Т5	210	245	270		90			
ВАЛ8Т5	400	430	480		340			
АЛ27Т1	350	430	300	270	250	230	280	260
Magnesium alloys								
МА2М	250	300	370	430	140	180	210	230
МА2-1М	270	290	380		180		315	
МА14Т1	320	410	470	480	260	360	380	400
МЛ4	200	200	220					
МЛ5Т6	240	240	240					
МЛ10Т6	240	300	320					
МЛ15Т1	210	210	220					
Copper-base alloys								
ЛС59, soft	450		590	680				
БрОФ6.5-0.4, hard	630		840	950				
Nickel-base alloys								
ХН77ТЮР, quenching plus aging	1300		1360		690		800	
ХН77ТЮ, quenching plus aging	1200		1600	1050	600		650	500
ЖСКП	1400		1650	1630	1000		1200	1280

*[1] At a temperature of –269°C.

Table 3.35 Ductility and toughness of steels and alloys at low temperatures [3,5,19]

Steel, alloy, heat treatment	δ, % at t, °C				ψ, % at t, °C				KCU, MJ/m^2 at t, °C		
	20	–70	–196	–253	20	–70	–196	–253	20	–70	–196
Steels											
45, normalizing					55		38	0			
45, hardening plus tempering	13		13		52	52	48		1	0.6	0.1
30ХГСА, hardening plus tempering	14	14	7		50	47	13		0.7	0.4	0.15
40ХНМА, hardening plus tempering	17	13	12		55	50	27		1.1	6	0.4
30ХГСНА, isothermal hardening	13	14	2.5		52	53			0.9	0.45	0.1
40ХГСН3ВА, hardening plus tempering	11	13	10		43	45	14		0.55	0.55	0.15
12Х18Н9Г, hardening	40	37	30	25					1.3	1.6	2.0
08Х15НД2Т, hardening plus tempering	10	11	13	8	50	48	47	35			
10Х12Н20Т3Р, hardening plus aging	30		30	10							2.2
12Х14Г14Н3Т, hardening	45		40	35							
08Х16Н6, hardening, cold treating, aging					58	53	45	40	1.5	2.3	2.2

Table 3.35 Ductility and toughness of steels and alloys at low temperatures [3,5,19] *(continued)*

Steel, alloy, heat treatment	δ, % at *t*, °C				ψ, % at *t*, °C				KCU, MJ/m^2, at *T*, °C		
	20	−70	−196	−253	20	−70	−196	−253	20	−70	−196
Steels											
45Л, tempering	18	17			35	20			0.75	0.35	
27ХГСНЛ, isothermal hardening	8	10			45	40			0.6	0.35	
35ХГСЛ, hardening plus tempering	15	15	6		50	50	10				
Titanium alloys											
ВТ5-1, annealing	10	8	6	3							
ОТ4, annealing	24	15	12						0.5	0.3	0.2
ВТ6, annealing	16	15	12	3	33	33	30	4	0.4		0.3
ВТ5Л	9	11	8	3	25	25	14	14			
ВТ20Л	10	5									
ВТ21Л	10	2									
Aluminum alloys											
АД1М	35	45	50	45							
АМг6М	17	23	31	34	29	50	33	27			
АД33Т1	15		20	23[*1]	42		39	34[*1]			
Д16Т	14	13	10	16	16	16	15		0.24	0.24	0.24
Д16Т1	8	6	7.5	9.5[*1]	17	14	13	14[*1]			
1201Т1	10	11	13	10							
В92Т1	14	19	17	11							
В93Т1	7	8	6								
В95Т1	10	9	7	5	13	15	9		0.15	0.1	0.1
САП1	5	7	8	10							
АБМ1	15	16	35		20	16					
АЛ4Т6	3.5	2.5	2.5	1	5	6.5	3.5	3	0.03	0.03	0.03
АЛ19Т4	9.5	9.5	6	5							
АЛ9Т5	2	2.5	2.8								
ВАЛ8Т5	4	4	3.5								
АЛ27Т1	15	25	2	0.1	20	28	3	0.5			
Magnesium alloys											
МА2М	25	23	10	8							
МА2-1М	15	9	2.5								
МА14Т1	13	8	2.5								
МЛ4	8	6	5		12	9	4				
МЛ5Т6	5.5	4	2		7	5	4		0.035	0.025	0.02
МЛ10Т6	5	5	4								
МЛ15Т1	3.5	1	1								
Copper-base alloys											
ЛС59, soft	32		37	34	35		38	35			
БрОФ6.5-0.4, hard	12		29	29	61		54	51			
Nickel-base alloys											
ХН77ТЮР, hardening plus aging	25		18		20		15		0.5		0.35
ХН77ТЮ, hardening plus aging	28		30	18	35		28	8			
ЖС6КП, hardening	20		15	13							

[*1] At a temperature of −269° C.

Table 3.36 Mechanical properties of steels at elevated temperatures [3,24]

Steel	σ_b, MPa at t, °C						$\sigma_{0.2}$, MPa at t, °C						δ, % at t, °C					
	20	200	300	400	500	600	20	200	300	400	500	600	20	200	300	400	500	600
10	430	495	525	380	260	110	265	225	180	170	160	95						
20	480	440	450	360	220	130	270	230	180	150	130	90						
45	640	700	730	575	385	220	365	355	265	230	180	80						
38ХА, 40Х	950	900	890	700	500													
40ХФА	950	925	860		505	385	860	825	760	720	420		26	22	19	29	30	50
30ХМА	950	800		740	570		750	650		600	500		12	20		19	19	
30ХГСА, 40ХНМА	1100		1000	920	700	550[*1]	950		840	800	650	500[*1]	13		11	16	21	27[*1]
12Х2НВФА	1150		1020	1050	1020		950		900	920	920		17		14	15	12	
30Х2Н2ВФА	1180			1050	920		1090			930	860		15			12	12	
30ХГСН2А	1600	1600		1500	1150		1350	1350		1250	1050		9	9		11	9	
12Х18Н10Т	650		450	450	450	400	300		200	180	180	180	40		30	30	30	25
14Х17Н2	1200		1120		950	380	900		840				8		8		16	30
09Х15Н9Ю	1200	1130	1050	1000	750		950	900	850	800	500		10	8	8	8	10	
08Х17Н5М3	1200			1100	950	750[*1]	1000			850	600	500[*1]	10					10[*1]
20Л	440	430	450	370	230		260	240	200	170	130		30	22	21	32	45	
20ХМФЛ	600	580	550	520	480	350	400	400	390	370	340	280	18	14	13	14	15	20
30ХНВЛ	850	790	820	720	580	330	700	630	600	560	500	300	18	16	16	14	15	19
12Х13Н3ВФЛ	900			780	620	400	750			610	480	270	10			6	9	17
23Х6Н2МВФА	920		730	650	550	370	700		610	570	520	340	12		14	15	16	24
08Х17Н3СЛ	1200			1130	1040		1100			970	910		8			6.5	7	

[*1] Tested at a temperature of 550°C.

Table 3.37 Mechanical properties of titanium alloys at elevated temperatures [3,10]

Alloy	σ_b, MPa at t, °C						$\sigma_{0.2}$, MPa at t, °C						δ, % at t, °C					
	20	200	300	400	500	600	20	200	300	400	500	600	20	200	300	400	500	600
ВТ1-1	450		220	150			380		180	150			25		30	35		
ОТ4-1	650		380	310			470		290	260			20		23	17		
ОТ4	750		490	460			600		430	390								
ВТ5	800		480				700		380	350			15					
ВТ5-1	750	680	550	510	490	380	650		460				15	10	17	15	20	26
АТ4	900		680	600			850		600	550			15		17	20		
ВТ6	1000			580	530		900			500	420		8			9	16	
ВТ6С	920	670			510		870	580			400		10	11			15	
ВТ8	1070	860	830	770	730	600	920	710	690	630	570	400	10	10	9	8	8	9
ВТ9	1150	950	920	850	800	720	1030		740	720	660	550	6		10	7	7	9
ВТ3-1	1000	910	840	760	700	530	950	740	660	630	560	250	12	11	9	8	10	18
ВТ20	1000	780	730	700	650	470												
ВТ5Л	900		400	350	300								8		8	10	13	
ВТ6Л	900		550	500									9		9	9		
ВТ9Л	950		710	560	640	250[*1]							9		9	7	8	5[*1]
ВТ20Л	1000		670	610	560								10		10	10	12	
ВТ21Л	1100		700	670	630								10		10	11	12	

[*1] At a temperature of 800°C.

Table 3.38 Mechanical properties of aluminum alloys at elevated temperatures [3,5]

Alloy, condition	σ_b, MPa, at T, °C						$\sigma_{0.2}$, MPa at t, °C						δ, %, at T, °C					
	20	100	150	200	250	300	20	100	150	200	250	300	20	100	150	200	250	300
АД1М	80		60	42	28	18	30		29	25	17	10	35		55	65	75	80
АД1Н	150	135	110	42	28	18	100	70	45	28	14	10	6	7	11	55	65	80
АМг2М	190	170	160	130	110	70	80	80	70	60			23	26	35	50	60	75
АМг2П	260	260	220	160	80	50	210	210	190	100	50	35	14	16	25	40	80	100
АМг2Н	290		250	160	90	50	260		210	100	70	30	8		25	40	60	100
АМг6М	320	300	250	190	160	130	170	150	130	120	100	80	24	30	37	43	45	48
АД33Т1	310		240	200		90	270		200	180		80	2		11	11		8
АК8Т1	490		410	340	230		450		370	310	220		1		14	13	14	
Д1Т	410		280	150	90	50	250		210	110	65	35	15		16	28	45	95
Д16Т*1	440/520		380/440	330/420	220/290	150/190	290/380		265	255	195	115	19/16		19/14	11/9	13/10	13/12
ВАД1Т	430			340	270	170	280			230	210	140	18			20	16	18
АК4-1Т1	420		360	320	250		360		330	290	230		7		7	10	11	
Д20Т1	420		350	300	230		320		280	250	190		10		13	14	15	
1915Т1	380	325	290	250			310	280	250				10	14	18			
В92Т1	480	410	330	280			370	300	270	240			7	10	16	14		
В93Т1	490	430	380				450	400	340				7	6	4			
В95Т1*1	520/600	480/530	410/430	280/330	150/160		440/550	410/500	350/400	240/310	120/150		14/8	14/8	15/7	11	16/16	
АЛ1Т5				180	175	130										1.5	1.9	4
АЛ3Т5	220			180	150	100							0.8	1	1	1.4	1.5	4
АЛ32Т5, Т6	245		225	210	150		180		175	170	125		2		2.5	4	4	
АЛ4МТ5	340	300	280	260	180	145	250						3	2	2.5	3	3.5	6
АЛ4Т6	240	220	190	160	110								3	3	3.5	4	5.4	5
В124Т6	360			300	220	130							0.5			1	1.5	2.5
В124Т6К*2	415			310	210	140							2.5			3.5	4	6
АЛ19Т5	370			280	200	150	280			220	160	100	5			3	4	5
АЛ33Т5*3	280				210	160	180				150	100	2				2	2.5
АЛ9-1Т5	270		210	210	140		150		150	170	130		5		15	6	4	
АЛ9-1Т6К*2	320		240	200	130		240		200	180	120		7		12	6	6	
ВАЛ8Т5К*2	400		340	300	220	130							4		8	9	10	11

*1 Numerator for rolled, denominator for pressed semifinished products.

*2 Chill casting, the rest for sand casting.

*3 At a temperature 350°C: σ_b=100 MPa; $\sigma_{0.2}$=70 MPa; δ=5%.

Table 3.39 Strength of powder aluminum alloys and aluminum matrix and magnesium matrix composites at elevated temperatures [5,14,24]

Materials	σ_b, MPa at t, °C							
	20	100	200	250	300	350	400	500
Sintered aluminum powder (SAP-1)	300	260		180		120		60
Sintered aluminum powder (SAP-2)	350			210		130		85
Sintered aluminum powder (SAP-3)	400			240		180		90
Aluminum alloy + 50% boron fiber	1200	1100	1050	1000	950		900	800
Aluminum + 40% carbon fiber	1200	1180	1100	1150	1080	1050	1000	
Aluminum + 40% beryllium wire	560	510	420	390	340			
Aluminum + 50% quartz fiber (SiO_2)	840	820	780		730		470	390*1
Magnesium + 25% boron fiber	920		900		770		730	650
Magnesium + 50% boron fiber	1200							800*1

*1 At a temperature of 450°C.

Table 3.40 Mechanical properties of magnesium alloys at elevated temperatures [3,24]

Alloy, condition	σ_b, MPa at t, °C							$\sigma_{0.2}$, MPa at t, °C							δ, % at t, °C						
	20	100	150	200	250	300	350	20	100	150	200	250	300	350	20	100	150	200	250	300	350
МА2	270	210	165	115	75			170	115						10	30	45	60	75		
МА2-1	300	225	190	130	90	70		200	115	95	75	45	40		14	19	30	35	45	50	
МА5	310	220	170	125	85	70		220	130	100	70	55	35		11	22	30	38	45	85	
МА8	240	180	150	130	110	70		150	100	70	60	50			15	26	30	32	34	62	
МА11	280			210	180	140	100	145			110	90	80	60	10			13	15	19	45
МА14	330	260	210	150	105	70		255							10	20	28	50	58	62	
ВМД1[*1]	300			170	150	130	110	250				130	105	90	5			12	12	13	20
МЛ3	180	160	145	105		60		55	50	45	40		30		8	10	11	12		11	
МЛ4Т4	250	240	210	150	75			85	80	75	68				9	7	15	25	26		
МЛ5Т4	250	230	185	155	120	90		90	80	60	50	40			9	10	12	15	15		
МЛ6Т6	260	230	210	150	110	80		140		85		70			1.5	5	8	15			
МЛ9Т6	250			210	170	120		145			135	120	105	60	6			5	10	20	25
МЛ10Т6	240			190	165	135		150			140	130	110	2	5			8	13	17	
МЛ11Т6	160			140	130	105	75	100			80	75	60	45	3			6	8.5	30	30
МЛ12Т6	250		160	125	85	55		150		100	75	50	30		6		8	10	12	15	
МЛ15Т1	220		150	125	100	75		140		105	85	65	50		3.5		5	13	16	16	
ВМЛ1Т6[*2]	200					145	110	95					75	55	6					10	20

[*1] At a temperature of 400°C: σ_b=70 MPa; δ=24 %.

[*2] At a temperature of 400°C: σ_b=65 MPa; $\sigma_{0.2}$=35MPa, δ=25 %.

Table 3.41 Strength and ductility of heat-resistant steels and alloys at elevated temperatures [3,24]

Steel, alloy	σ_b, MPa at t, °C								δ, % at t, °C						
	20	500	600	700	800	900	1000	1100	20	500	600	700	800	900	1000
Steels															
45Х14Н1УВ2М	720	630	540	340	220				20	18	20	32	40		
40Х12Н8Г8МФБ	900	680	630	460	390	250			18	11	10	9	9	8	
10Х12Н22Т3МР	1000	900	850	700	470				9	9	10	10	17		
10ХН35ВТЮ	1300	1200	1110	1000	700	400[*1]			15	12	12	9	15	22[*1]	
40Х12Н8Г8МФЛ	800	550	500	370	280				19	32	32	29	20		
15Х12Н25Т3М3В4Л	900		800		500				12		5		5		
Nickel-base alloys															
ХН78Т	780			400	180	110	65	45	40			35	70	90	100
ХН60В[*2]	800			530	400	230	140	85							
ХН77ТЮР	1000		950	840	560	400[*1]			24		24	22	15	18[*1]	
ХН70ВМТЮ	1150		1000	900	740	500			20		17	10	9	15	
ХН55ВМТФКЮ	1200		950	850	650		300		12		10	10	10	17	20
ХН62МВТЮ	1250		980	980	930	620	300		18		18	10	8	12	17
ВЛ7-45У	500			400	300				7			5.5	10		
ЖС3	750				560	300			8.5				3	4	
ЖС6	1050				850	780	480		1				0.5	2.5	6
ЖС6К	1000				920	780	530		2.5				2	2	4.5
ЭП23	1250					940	720	450	2					6	13
Cobalt-base alloys															
ЛК4	700		520	470	400	270			8		12	10	6.5	9	
ЛК4Я	900		750	700	510	350	260[*3]		5.5		11	12.5	14	18	25[*3]
4К66Я	900			750	550	375	295[*3]		6			10.5	16	25	19[*3]

[*1] At a temperature of 850°C.

[*2] σ_b^{1200}=43 MPa.

[*3] At 950°C.

Table 3.42 Strength of refractory metal base alloys, ceramics and nickel matrix composites at elevated temperatures [3,14,24]

Material	σ_b, MPa at t, °C										
	20	1000	1100	1200	1400	1500	1800	1900	2000	2200	2400
Niobium alloys											
ВН2	650	400	300			80					
ВН3	780		450	270		125					
ВН4	810		700	550		170					
РН6	890			290			90				
Molybdenum alloys											
ВМ1	860			500	190	140	100				
ВМ2	750			450	300	160	90		80		
МР-47ВП	780					150	60				
ЦМВ-30						160	63				
Tungsten alloys (powder)											
Tungsten									80	60	30*4
Tungsten-50Mo							140*1	45		35	13
W-1ThO$_2$							260*1	190		90	
W-2ThO$_2$							200*1	180		120	70
W-30Re, cast	2180		1195				210*1	140			
Chromium-base alloys											
ВХ2	400			165*2		45					
ВХ4	1080	210		100*2							
Composites											
Nickel–2.5% HfO$_2$	525	150	105	85							
Nickel–50% carbon fiber	560	240									
ЖСК6–50% tungsten wire	580	540	380		290*3						
Ceramics ($\sigma_{t,\mathrm{bend}}$ in bending)											
30Cr–70Al$_2$O$_3$	390	230			175*3						
Of oxide Al$_2$O$_3$	150	95		76	60	36					
Hard alloy	1500			800							

*1 At a temperature of 1650°C.
*2 At 1150°C.
*3 At 1300°C.
*4 At 2500°C.

Table 3.43 Long-term ultimate strength and creep stress of steels and titanium alloys [3,10,24]

Steel, alloy	σ_{100}, MPa at t, °C							$\sigma_{0.2/100}$, MPa at t, °C					
	300	350	400	450	500	550	600	300	350	400	450	500	550
Steels													
30ХМ			740*1	580*1	380*1	260*1						142*2	59*2
30Х2Н2ВФА	980		800	750	540	320				640	440	370	240
30ХГСА			600*1	460*1	260*1	120*1				160*1	110*1	55*1	22*1
08Х15Н9Ю	1050		850										
08Х17Н5М3				950	650	400							
10Х12Н2ВМФ				730		440	270				580		200
23Х13НВМФА			1400	1000	700					870	620	210	
13Х14НВФРА	950			720		300		760			380		180
10Х13Н3ВФЛ					450	300	220			460	310	215	180

Table 3.43 Creep and creep rupture properties of steels and titanium alloys [3,10,24] *(continued)*

Steel, alloy	σ_{100}, MPa at t, °C							$\sigma_{0.2/100}$, MPa at t, °C						
	300	350	400	450	500	550	600	300	350	400	450	500	550
Titanium alloys														
BT5-1		450						400	330				
BT4		480	390						300	190	50		
BT6C		500			140				320				
BT6			580		380					360	140	60	
BT3-1			650	550	360					300	160	50	
BT8			750	650	450	380				370	230	120	
BT9					650	450	230					280	120
BT14			680	540					530	350			
BT5Л	400		350							280		170	
BT6Л			530		470								
BT9Л	650		620		500	350		500		460		280	200
BT20Л		600			430				450			160	
BT21Л			600		440					430		180	

*1 For 200 h.
*2 For 10000 h.

Table 3.44 Long-term ultimate strength and creep stress of wrought aluminum alloys [3,5,24]

Alloy, condition, semifinished product	σ_{100}, MPa at t, °C						$\sigma_{0.2/100}$, MPa at t, °C					
	125	150	175	200	250	300	125	150	175	200	250	300
D16T, D16T1, sheet	340	300	230	180			180	180	120	85		
D16чT, 1163T, plate	345	315	270				300	270	235			
D16чT, 1163T, press-formed shape	340	320	270	210				260	210			
BD17T1, press-formed strip				190	100	55				160	75	35
AK4-1T1, sheet	280	250	210	170			250	220	145	90		
AKA-1T1, plate	350	330	240				280	250	160			
1201T1, forged, press-formed shape, plate	290	250	220	180			240	200	140	130		
D20T2, press-formed strip		300	210	180	125	80		200	170	120	80	65
D21T1, sheet	370	330	240	170			290	250	170	105		
D21T1, plate	370	330	260	200	120		310	280	220	160	95	
B95T1, sheet		230		100								

Table 3.45 Long-termed ultimate strength of heat-resistant steels and alloys [3,24]

Steel, alloy	σ_{100}, MPa at t, °C						Steel, alloy	σ_{100}, MPa at t, °C					
	600	700	800	900	950	1000		600	700	800	900	950	1000
Steels							**Nickel-base alloys**						
45X14H14B2M	300	180	80				XH77TЮ	580	400	150			
38X12H8Г8MФБ	450	250	150				XH35BTЮ	670	390	230			
10X12H20T3P	580	390	280*1				XH77TЮP	680	420	200			
10X14H35M3TЮ	700	450	250				XH70BMTЮ		500	290	190*2		
15X10H22BMTЛ		400	200				XH55BMTФKЮ		700	440	200		80
30X12H10Г8BФЛ	400	280	140				XH62MBKЮ		750	450	200	120	
							XH78T		105	45	15		

Table 3.45 Long-term ultimate strength of heat-resistant steels and alloys [3,24] *(continued)*

Steel, alloy	σ_{100}, MPa at t, °C					
	600	700	800	900	950	1000
Nickel base alloys						
ХН75МБТЮ		165	80	29		
ЖС3			300	100		
ЖС6			500	280		140
ЖС6К			520	320		155

Steel, alloy	σ_{100}, MPa at t, °C					
	600	700	800	900	950	1000
Cobalt base alloys						
ЛК4	300	250	150	70	50	
ЛК4Я		300	180	140	110	
4К66Я			250	150	120	

*1 At a temperatures of 750°C.
*2 At 850°C.

Table 3.46 Long-term ultimate strength of refractory metal-base alloys and nickel-matrix composites [3,14,24]

Material	σ_{100}, MPa at t, °C				
	1100	1200	1300	1400	1500
Niobium-base alloys					
ВН2	90				
ВН3	160				
ВН4	280				
ВН5А	140				
РН6		110			
МН1	250	110			
МН2	360	260			
Molybdenum alloys					
ЦМ2А		90	65	40	
ВМ1		90			
ВМ2		155		50	
ЦМ6		130	40		
ЦМВ-30		250			5*1

Material	σ_{100}, MPa at t, °C				
	1100	1200	1300	1400	1500
Chromium-base alloys					
ВХ1, ВХ2		45*2			
ВХ3		150*2			
Tungsten alloys					
Tungsten, technical				70	60
Composites					
ЖС6–50% tungsten wire	150	80			
Nickel–2.5% ThO_2	105	80			
Nickel–2.5% HfO_2	70	50			

*1 At temperatures 1800 °C.
*2 At 1150 °C.

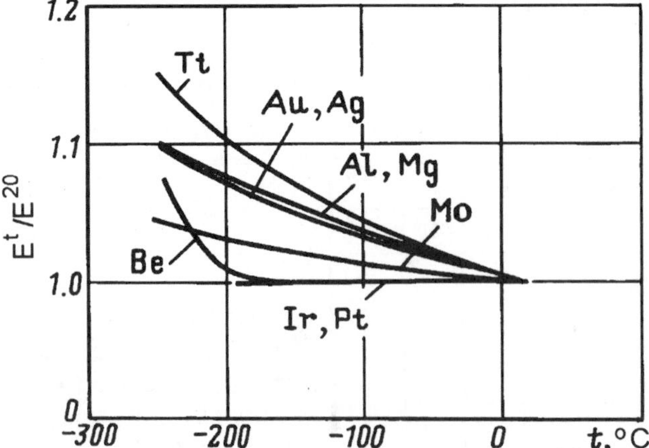

Figure 3.2 Effect of temperature on modulus of elasticity for selected materials [3,6] (for data at room temperature, see Table 3.1).

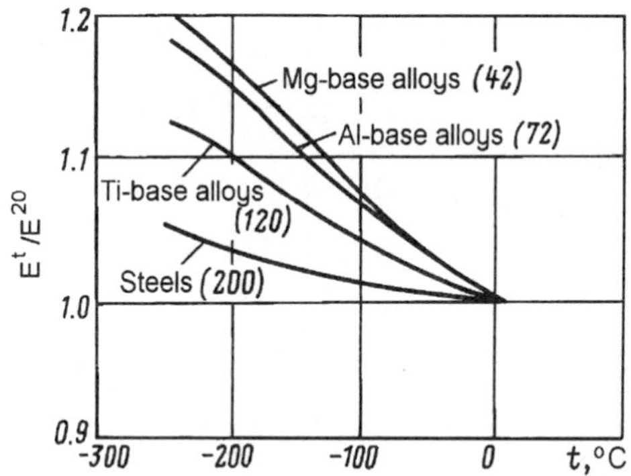

Figure 3.3 Effect of temperature on the modulus of elasticity for steels and alloys [3,5,10]. (In parentheses, the modulus of elasticity E_{st}^{20}, GPa at 20°C are given.)

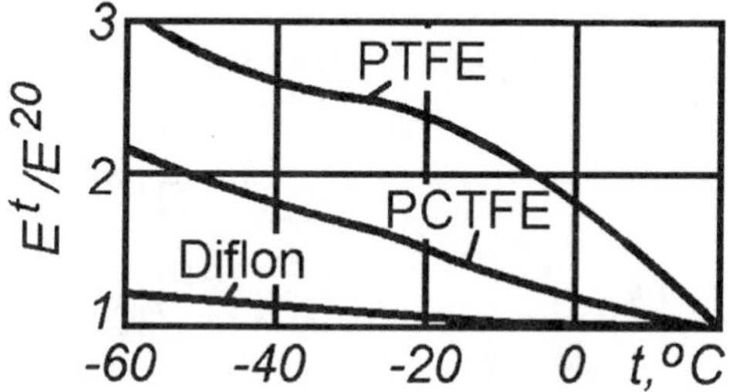

Figure 3.4 Effect of temperature on modulus of elasticity [24] for diflone ($E^{20} = 2.45$ GPa), polytrifluorochloroethylene (PTFCE, $E^{20} = 1.45$ GPa) and polytetrafluoroethylene (PTFE, $E^{20} = 850$ MPa.)

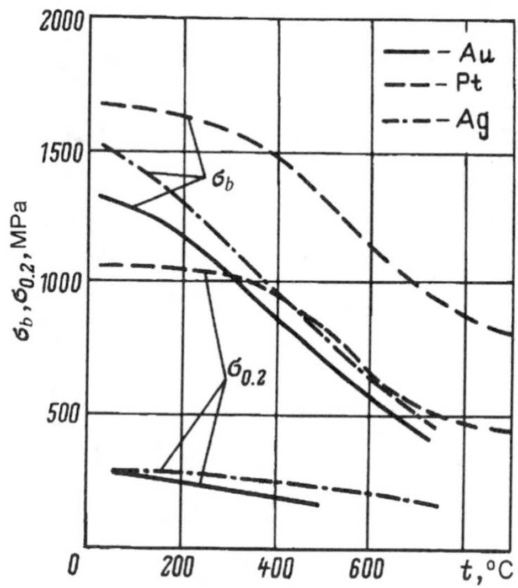

Figure 3.5 Ultimate and yield strengths of gold, platinum and silver at elevated temperatures [3,6,8].

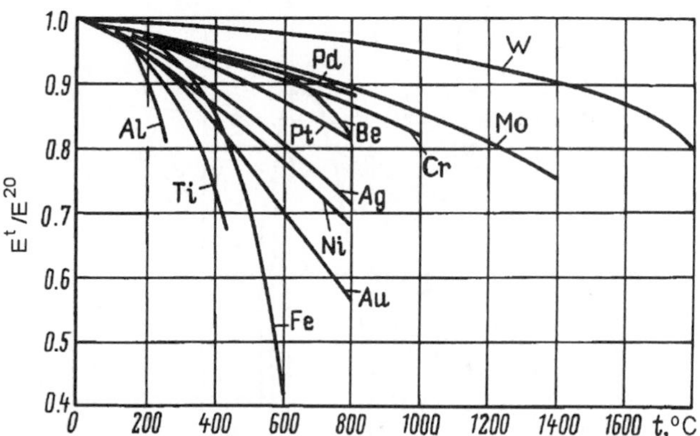

Figure 3.6 Effect of temperature on modulus of elasticity for metals [3,6] (for data at room temperatures, see Table 3.1).

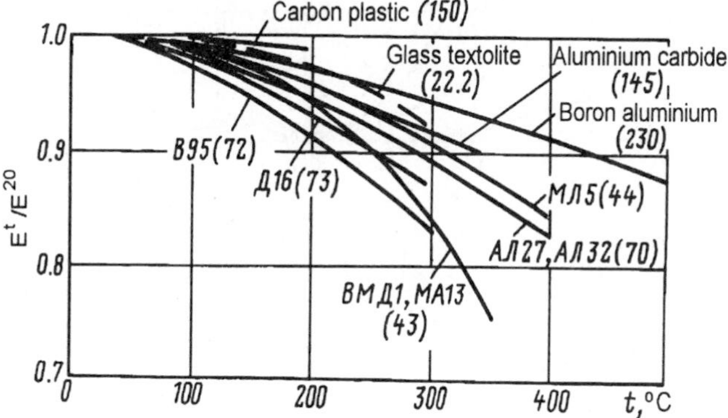

Figure 3.7 Effect of temperature increasing (up to 500°C) on modulus of elasticity for some materials [3,14], [24]. (In parentheses, E_{dyn}^{20}, GPa at 20°C are given.)

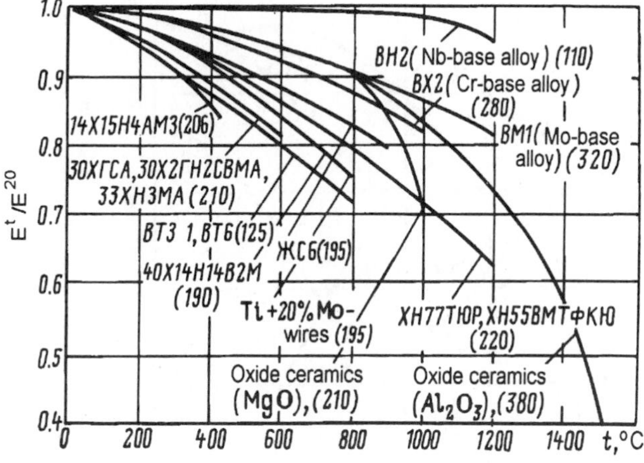

Figure 3.8 Effect of temperature increasing (up to 1200°C) on modulus of elasticity for some materials [3, 24]. (In parentheses, E_{dyn}^{20}, GPa at room temperature are given.)

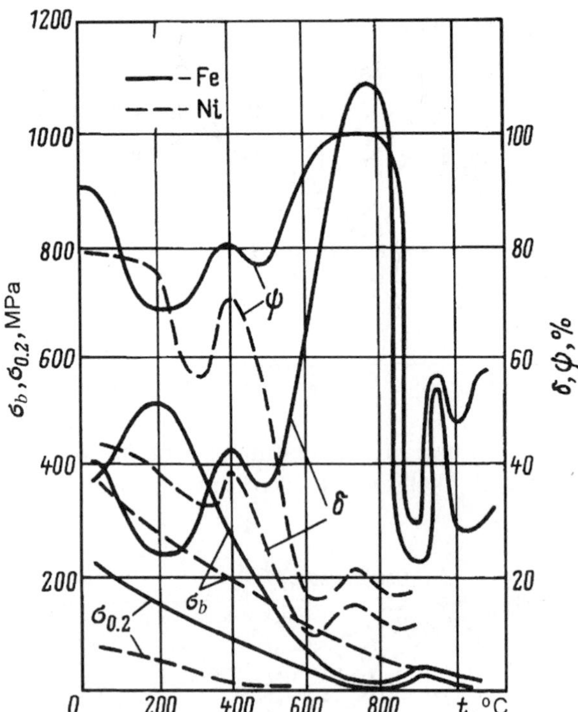

Figure 3.9 Mechanical properties of iron and nickel at elevated temperatures [6,10].

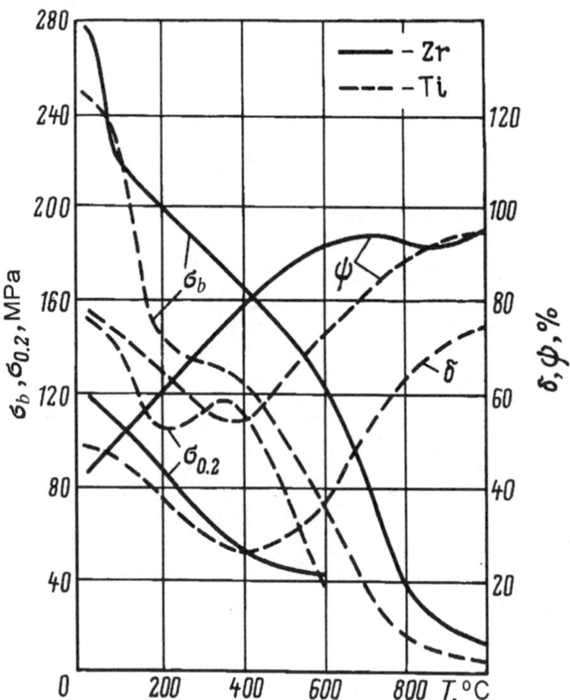

Figure 3.10 Mechanical properties of ultra-pure zirconium and titanium at elevated temperatures [6].

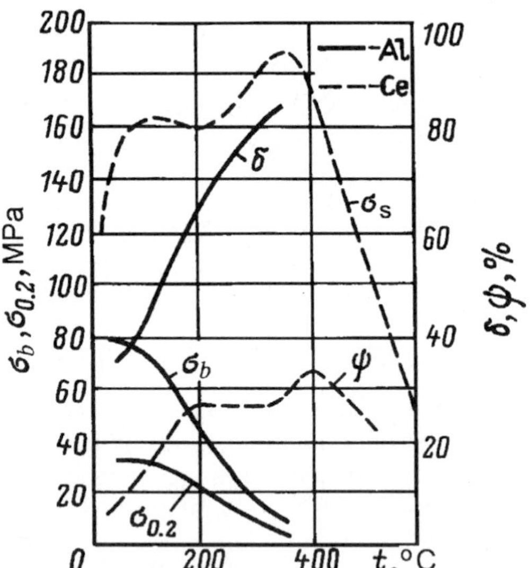

Figure 3.11 Mechanical properties of aluminum and cerium at elevated temperatures [4,6,18].

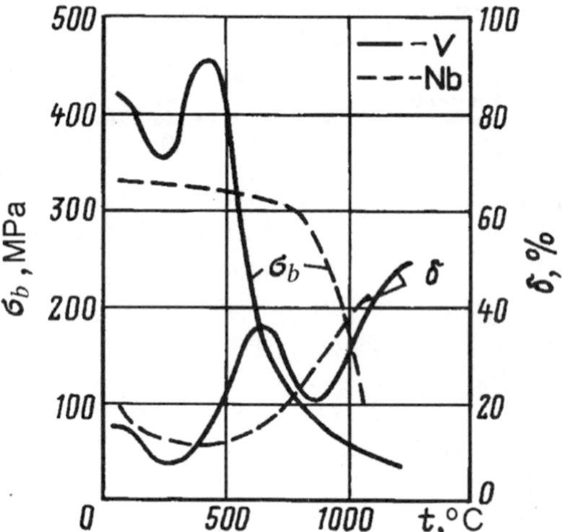

Figure 3.12 Mechanical properties of vanadium and niobium at elevated temperatures [3,6].

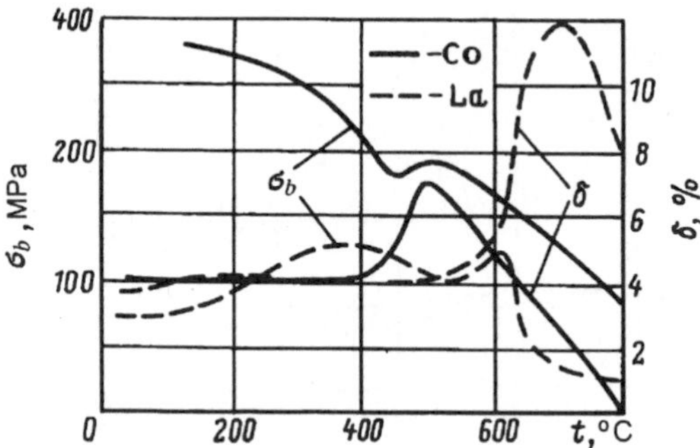

Figure 3.13 Mechanical properties of lanthanum and cobalt at elevated temperatures [3,6].

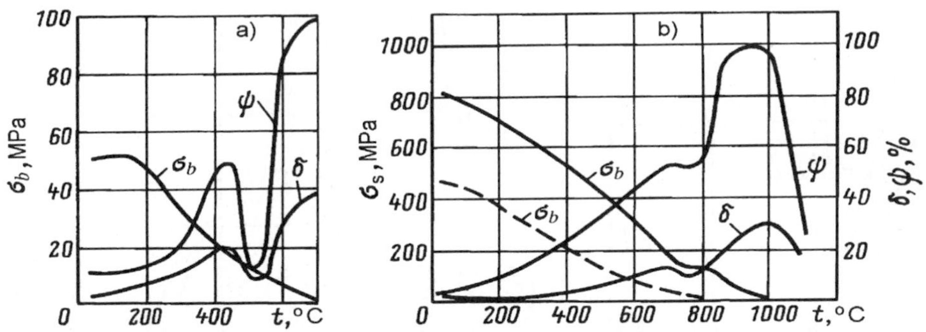

Figure 3.14 Mechanical properties of strontium, uranium and beryllium at elevated temperatures [6]: *a* – cast strontium; *b* — uranium; - - - - beryllium.

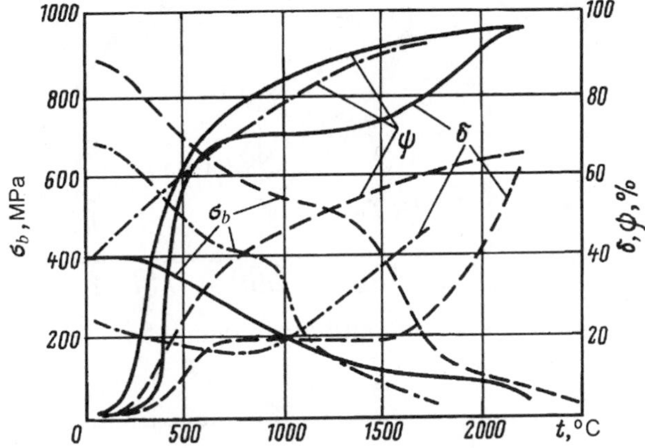

Figure 3.15 Mechanical properties of tungsten and molybdenum [3,6]: solid line – tungsten after annealing, dashed line – cold-hardened tungsten, Dotted-and-dashed line – molybdenum after annealing.

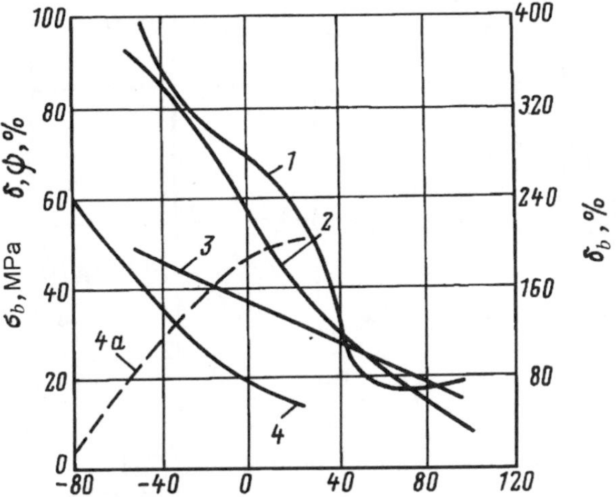

Figure 3.16 Dependence of ultimate strength on test temperature for selected nonmetallic materials [24]: *1*- capron (in bending); *2* – polychlorotrifluoroethylene (PCTFE, in bending); *3* – polyethylene; *4* – cold-resistant rubber; *4 a* – elongation at rupture of cold-resistant rubber.

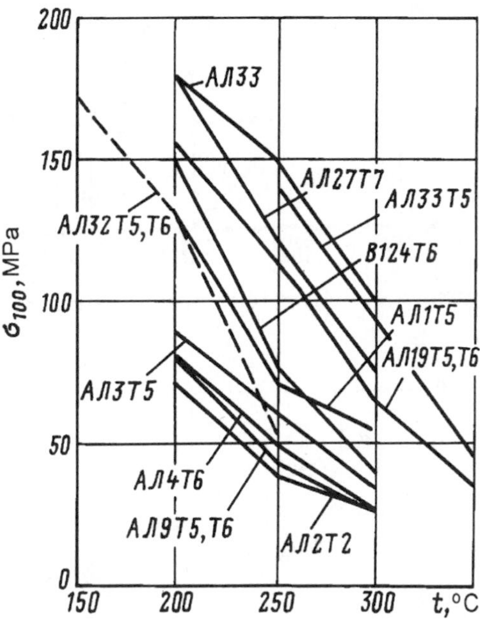

Figure 3.17 Long-term ultimate strength of cast aluminum alloys [3,5].

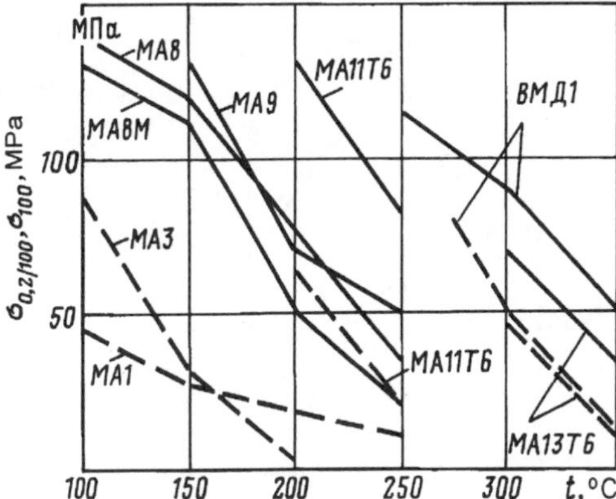

Figure 3.18 Long-term ultimate strength (solid lines) and creep stress (dashed lines) of wrought magnesium alloys [3].

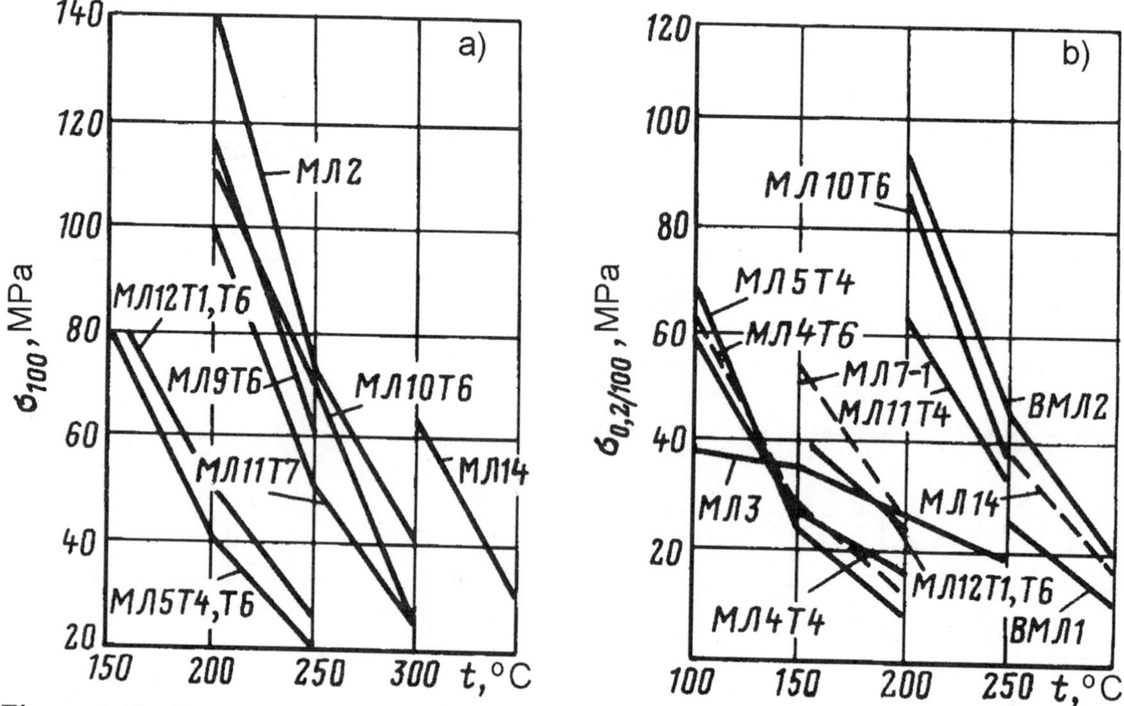

Figure 3.19 Long-term ultimate strength (*a*) and creep stress (*b*) of cast magnesium alloys [3].

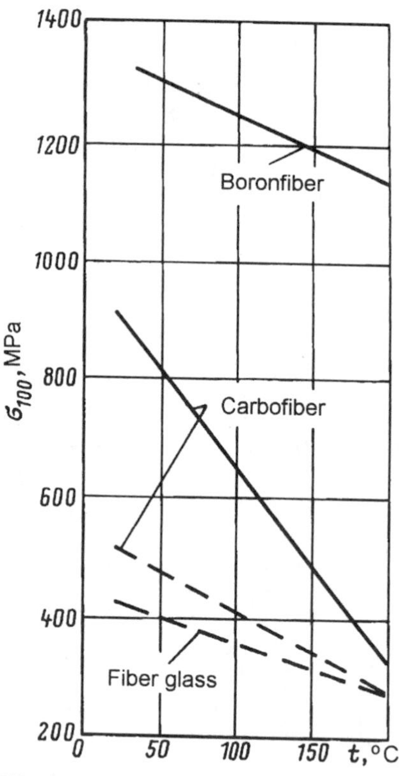

Figure 3.20 Long-term ultimate strength of polymer matrix composites in tension (solid lines) and bending (dashed lines) [13].

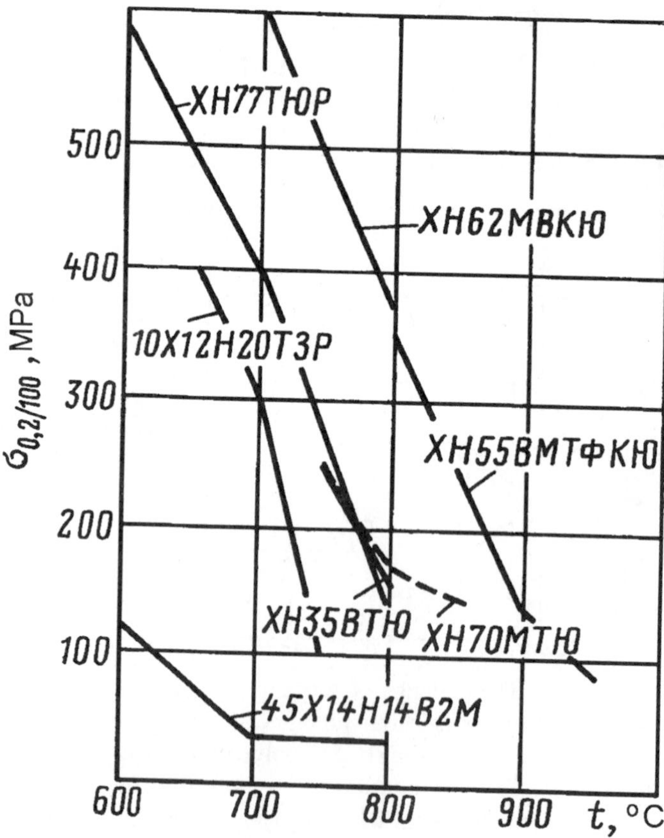

Figure 3.21 Creep stress of heat-resistant steels and nickel-based alloys [3].

3.4 Fatigue Properties

Conventional high-cycle and low-cycle fatigue properties of materials are assessed by various techniques. Low-cycle fatigue is characterized by a number of cycles to failure of $N \geq 5{\cdot}10^4$ cycles and the lowered frequencies of cyclic loading ranged from f =0.1 to 5 Hz, whereas the high-cycle fatigue, by $N \geq 10^7$ and f = 20–50 Hz. Damage or fracture within the high-cycle region occurs mainly at the elastic stresses, while that within the low-cycle region occurs at the elastoplastic stresses.

Fatigue properties are dependent on the shape and size of a specimen, as well as the manner and frequency of loading. The specimen resistance to fatigue is reduced with decreasing frequency of loading and increasing absolute size of the specimen. The fatigue limit of metals depends significantly on the surface conditions and is defined as a rule for polished specimens; the plastics are less sensitive to the surface quality.

The greater part of high-cycle fatigue data is derived from symmetrical bending tests to define σ_{-1}. To estimate the fatigue points under other modes of stress, the following correlations can be used: for structural steels, the fatigue limit in stress-strain tests is σ^s_{-1} = (0.8–0.9)σ_{-1}; in torsion tests τ_{-1} = (0.5–0.6)σ_{-1}; for aluminum alloys these ratios are 0.85–0.95 and 0.55–0.65, respectively.

For certain groups of materials, the relations between the fatigue and strength limits were established. For steels, the ratio σ_{-1}/σ_b is equal to 0.35–0.55 at the number of cycles to failure of $N = 2 \cdot 10^7$ cycles (for titanium alloys at the same conditions: σ_{-1}/σ_b is equal to 0.45–0.55); in this case, the lesser value of σ_{-1}/σ_b corresponds to the material of higher strength; for nonmetallic materials (textolite, organic glass, etc.) σ_{-1}/σ_b = 0.2–0.3 ($N = 10^7$ cycles).

The fibrous composites feature higher than usual ratio of σ_{-1}/σ_b = 0.6–0.8 ($N = 10^7$ cycles).

In Tables 3.47–3.53 and Figs. 3.22–3.24, the following fatigue properties are represented:

σ_{-1} – the fatigue limit at a symmetric bending, i.e. the greatest cycle stress at which the fatigue fracture of the specimen after action of arbitrary large numbers of cycles or fixed number of cycles to failure (10^7–$2{\cdot}10^7$ cycles) does not occur;

σ^H_{-1} – the fatigue limit of a specimen containing the notch or the hole at K_t = 2.2–2.6, with K_t being theoretical value of the stress concentration factor;

$\sigma_{\max}$ – the endurance limit (low-cycle fatigue) of a smooth (K_t = 1) or notched (K_t = 2.2–2.6) specimen at the nonsymmetrical tension, i.e. the greatest cycle stress at which the fatigue fracture of the specimen at the fixed number of cycles to failure does not occur;

N – the durability, i.e. the number of cycles to failure for a specimen, at fixed stress and test base.

Table 3.47 Fatigue limits for steels and titanium alloys on reversed bending

Steel, alloy	σ_b, MPa	σ_{-1}, MPa	σ^H_{-1}, MPa (K_t = 2.2)	Steel, alloy	σ_b, MPa	σ_{-1}, MPa	σ^H_{-1}, MPa (K_t = 2.2)
Steels, $N = 10^7$ cycles [3,10,24]							
20	400	210		23X2HBΦA	1350	645	380
35	600	330		25X2ΓHTA	1500	610	360
45	950	530	280	25X2ΓHTA	1600	700	440
30XMA	950	420	260	30XΓCHA	1650	700	490
30XMA	1600	620		30XΓCHA	1750	730	
30XΓCA	900	480	220	40XH2CMA	1800	700	
20X3MBΦ	920	530	200	40XH2CMA	2000	830	550
12X2HBΦA	960	520	310	12X18H9T	580	285	245
38XA	1050	500	330	20X13	950	480	240
30X3BA	1050	610	350	09X15H8IO	1350	620	
30X2H2BΦA	1180	560	360	13X12H2BMΦA	1200	600	400
37XH3A	1250	560	310	23X13HBMΦA	1600	600	450

Table 3.47 Fatigue limits for steels and titanium alloys on reversed bending *(continued)*

Steel, alloy	σ_b, MPa	σ_{-1}, MPa	σ_{-1}^H, MPa ($K_t = 2.2$)	Steel, alloy	σ_b, MPa	σ_{-1}, MPa	σ_{-1}^H, MPa ($K_t = 2.2$)
Steels, $N = 10^7$ cycles [3,10,24]							
40ХН2МА	1300	580		12Х13Н3ВФЛ	900	420	280
40ХН2МА	1600	660	370	10Х17Н3СЛ	1000	410	260
Titanium alloys, $N = 2 \cdot 10^7$ cycles [3,10,24]							
ВТ1-0, sheet	415	185		ВТ3-1, bar	1000	480	
ВТ1-1, same	450	230		ВТ3-1, same	1100	560*1	390*1
0Т4-1, same	650	340		ВТ3-1, same	1200	620*1	
0Т4, same	750	380		ВТ22, same	1100	530	330
ВТ5-1, ОТ4, sheet	750	400		ВТ8, same	1050	500	
ВТ5, bar	850	450	310	ВТ9, same	1150	540	
ВТ6С, sheet	950	500*1	420*1	ВТ5Л	800	280	250
ВТ6, bar	1100	550*1	300*1	ВТ14Л	950	270	230
ВТ20, sheet	950	420		ВТ6Л	850	200	
ВТ14, same	900	400		ВТ9Л	1000	180	180
ВТ14, same	1200	420		ВТ3-1Л	1000	220	220

*1 Number of cycles to failure $N = 10^7$ cycles.

Table 3.48 Fatigue limits on reversed bending of aluminum alloys, $N = 2 \cdot 10^7$ cycles [3,5,20]

Alloy, condition, semifinished product	σ_b, MPa	σ_{-1}, MPa	σ_{-1}^H, MPa ($K_t = 2.2$)
АД1М, sheet	80	35*1	
АД1Н, same	150	50	
АМг3М, same	230	80	
АМг5М, same	300	110	
АМг6М, sheet, pressed shape	340	100	
АД31Т1, pressed shape	250	90	50
АД33Т1, same	320	110	60
АД35Т1, same	330	110	
АВТ1, same	350	115	65
Д1Т, Д1чТ, stamped	410	140	
Д16чТ, plate, shape	440	140	80
Д16чТ1, sheet	470	120	
Д16чТ1, pressed panel	480	140	
Д19Т, Д19чТ, pressed shape	480	160	
ВАД1Т, sheet	440	120	
Д20Т1, sheet, pressed shape	420	130	70
Д21Т1, forged, stamped	430	110	
АК4-1Т1, plate, pressed shape	420	120	80
АК6Т1, stamped	420	130	80
АК8Т1, pressed shape	480	135	85
ВД17Т1, pressed strip	520	155	90
В92Т1, pressed bar	450	150	
В95пчТ1, sheet, plate	570	160	90
В95пчТ1, pressed panel	600	180	
В95пчТ2, sheet, plate, forged, pressed panel	540	160	
В95пчТ3, pressed panel	500	160	
В95пчТ3, forged	510	140	

Table 3.48 Fatigue limits on reversed bending of aluminum alloys, $N = 2 \cdot 10^7$ cycles *(continued)*

Alloy, condition, semifinished product	σ_b, MPa	σ_{-1}, MPa	σ_{-1}^H, MPa ($K_t = 2.2$)
В93пчТ1, same	520	140	80
В93пчТ2, same	470	140	
В93пчТ3, same	440	130	
В96Ц3Т1, pressure forming	630	170	
САП1, pressure formed strip	300	80	60
САП2, pressure formed shape	350	90	60
САП3, pressure formed bar	400	115	75
САС1-400	270	80	
САС1-50	340	90	
АЛ1Т5	260	56*1	
АЛ4Т6	260	70	
АЛ4МТ5	350	90*2	
АЛ5Т5	260	65*1	
АЛ8Т4	300	50*1	
АЛ9Т5	210	45*1	
АЛ9-1Т5*3	270	80	
АЛ19Т5	360	70	
АЛ19Т5*3	400	90	
АЛ20Т2, АЛ21Т2	210	70	
АЛ20Т7, АЛ21Т7	240	75	
АЛ23-1Т4	230	40	
АЛ27Т4	360	60	
АЛ27-1	390	80	50
АЛ27-1Т4	380	70	
АЛ33Т5	280	75	
ВАЛ8Т5*3	400	110	

*1 Test base of $N = 5 \cdot 10^8$ cycles.
*2 $N = 10^7$ cycles.
*3 Chill casting, the rest sand casting.

Table 3.49 Fatigue limits on reversed bending of magnesium alloys, $N = 2 \cdot 10^7$ cycles [3]

Alloy, condition, semifinished product	σ_b, MPa	σ_{-1}, MPa	σ_{-1}^H, MPa ($K_t = 2.2$)	Alloy, condition, semifinished product	σ_b, MPa	σ_{-1}, MPa	σ_{-1}^H, MPa ($K_t = 2.2$)
МА2-1, plate	270	105*1	70*1	МЛ5, МЛ6	150	85	70
МА3, strip	280	115		МЛ5Т4	240	100	80
МА5Т4, bar	320	130	100	МЛ5Т6, МЛ6Т6	240	85	70
МА14-Т1, strip	340	120	80	МЛ6Т4	235	95	75
МА10Т6, bar	430	125	80	МЛ12	210	50	50
МА11Т6, bar	280	85	50	МЛ12Т1	230	75	70
МА13Т8, sheet	240	70		МЛ12Т6	250	80	
ВМД1, bar	280	70	50	МЛ15Т1	215	90	70
МА1М, same	240	75		МЛ9Т6	240	80	
МА2, same	270	100*1		МЛ10Т6	240	70	
МА8, strip	260	80		МЛ11Т4, Т6	155	70	
МЛ4	170	80	65	МЛ14	200	50	
МЛ4Т4	235	90	80	ВМЛ2	260	60	
МЛ4Т6	240	80	70	МЛ3, МЛ7-1	180	55	

*1 Test base of $N = 5 \cdot 10^7$ cycles.

Table 3.50 Effect of temperature on fatigue limits in reversed bending of heat-resistant steels and nickel base alloys, $N = 2 \cdot 10^7$ cycles [3]

Alloy	σ_b, MPa	σ_{-1}, MPa at t, °C					σ_{-1}^H (K_t=2.2), MPa at t, °C			
		20	600	700	800	900	600	700	800	900
10Х12Н20Т3Р	900		330	280						
10Х13Н35МТ3Ю	1000			300	260					
08Х35ВТ10	1150			300*1	260					
ХН77ТЮ	1000	360	390	360	300					
ХН77ТЮР	1000	370	360	390	300		240	250	230	
ХН55ВМТФКЮ	1100			350	360	300		270	270	270
ХН62МВКЮ	1180			300	305	300		255	255	255
ХН70ВМТЮ	1150		350	370	365	250		290	270	190
ЖС3	750				300					
ЖС6*2	1040	270			270	260				
ЖС6К*2	1000	280			280	290				
ВЛ7-45У	500	260			160					

*1 At a temperature of 750 °C. *2 At 1000°C σ_{-1} = 220 MPa.

Table 3.51 Fatigue limits on reversed bending of selected composite and plastic materials, N=10^7 cycles [3,12,14,24]

Material	σ_b, MPa	σ_{-1}, MPa	Material	σ_b, MPa	σ_{-1}, MPa
Aluminum boride	1250	650	Boron plastic	1300	420
	1150	600	Carbon-filled plastic	1020	500
	1100	550		650	300
Magnesium boride	1200	550	Fiber glass plastic	350	90
Aluminum carbide	1200	400	Organic glass	62.5	17–30
	650	170			

Table 3.52 Low-cycle fatigue in asymmetric tension *1 [3,5,10,20,24,27,30,31]

Steel, alloy, semifinished product	σ_b, MPa	K_t	Test base N, 10^3 cycles	σ_{max} MPa	Steel, alloy, semifinished product	σ_b, MPa	K_t	Test base N 10^3 cycles	σ_{max} MPa
30ХГСА, bar	1000	1	50	650	Д16Т1, sheet	480	2.6	50	160
same	1000	1	20	800	Д19Т, same	470	2.6	50	180
same	1000	2.6	20	400	В95Т1, same	500	1	20	320
30ХГСНА, bar	1600	2.6	100	400	same	500	2.5	20	260
same	1750	1	50	900	В95пчТ2, sheet, plate	530	2.6	50	160
same	1750	1.6	50	650	Д16чТ, plate	470	2.6	50	200
same	1750	2.2	50	500	same	470	2.6	100	190
same	1750	4	50	250	АК4-1Т2, plate	430	2.6	50	185
03Х18К8М5Т, forged	1700	2.6	100	450	1201Т1, same	465	2.6	50	180
ВТ1-0, sheet	400	2.6	20	300	В93пчТ1, Т2, Т3, stamped	420–500	2.6	50	150
ВТ5-1, same	800	2.6	20	600	В95очТ2, forged	540	2.6	50	190
ВТ6С, same	980	2.6	20	750	Д16Т, bar	480	2.2	20	430
ВТ22, forged	1100	2.6	100	350	В95Т1, same	650	2.2	20	400
ВТ23, plate	1100	2.6	100	330	В95пчТ1, Т2, Т3, panel	480–565	2.6	50	180
ВТ5, bar	950	2.2	20	630	Д16чТ, pressed	440	2.6	20	250
Д16Т, sheet	480	1	50	280	В95Т1, same	600	2.6	20	200
same	480	1	20	400	ВАЛ8Т5, cast	400	1	50	200
same	480	2.5	20	280	same	400	2.2	50	180
same	480	2.6	20	250	Aluminum boride, sheet	1250	1	800	950
same	480	2.6	50	180	Magnesium boride, same	1200	1	800	800

*1 $R = \sigma_{min}/\sigma_{max}$ = 0–0.1.

Table 3.53 Effect of test temperature t on low-cycle fatigue resistance under asymmetric tension for strip containing a hole*1 (K_t=2.5) [3,20]

Alloy, condition, semifinished product	σ_b, MPa	Test base N, 10^3 cycles	t, °C	σ_{max}, MPa	Alloy, condition, semifinished product	σ_b, MPa	Test base N, 10^3 cycles	t, °C	σ_{max}, MPa
Д16T, sheet	470	20	20	260	AK4-1T1, plate	430	50	20	190
			200	220				150	170
B95T1, sheet	550	20	20	220	BT1-1, bar	700	20	20	390
			200	180				300	210

*1 $R = \sigma_{min}/\sigma_{max} = 0$.

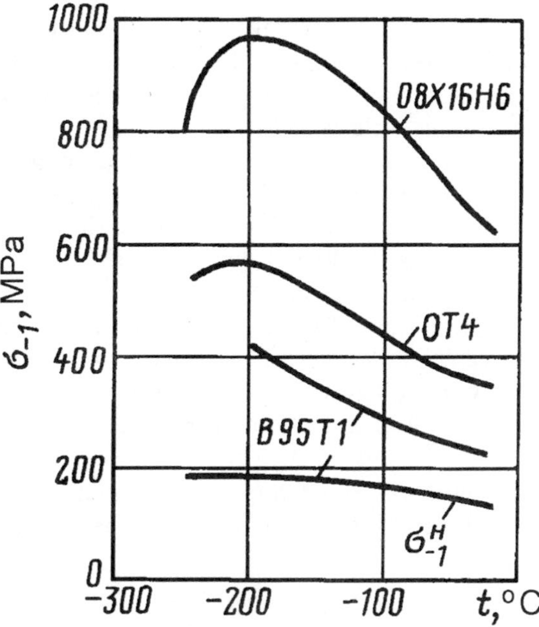

Figure 3.22 Effect of low temperatures on fatigue limit ($N = 10^6$ cycles) for selected alloys [3].

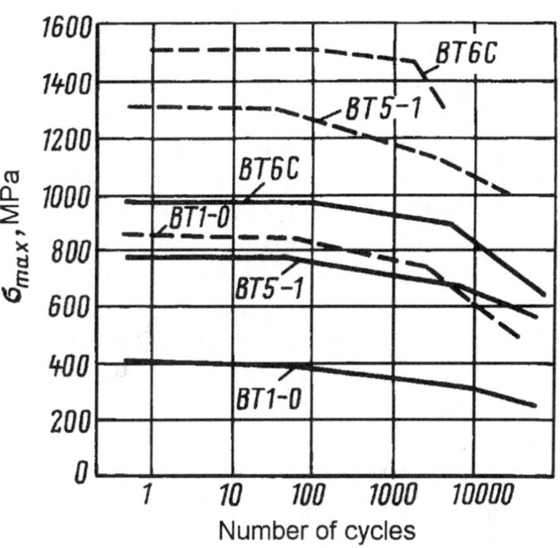

Figure 3.23 Low-cycle fatigue curves for titanium alloys at room temperature (solid lines) and t = 196°C (dashed lines) [10].

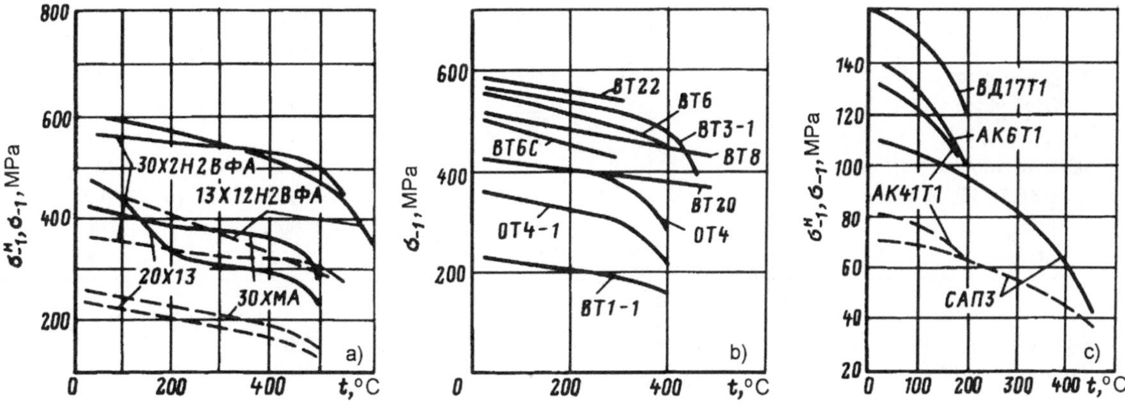

Figure 3.24 Effect of high temperatures on fatigue limit for (a) steels [3,31], (b) titanium alloys [3,10,24] and (c) aluminum alloys [31,5,3]: solid lines — σ_{-1}; dashed lines — σ_{-1}^{H}.

3.5 Fracture Properties

Among the basic properties of material fracture are the fracture toughness or critical stress intensity factor and fatigue crack growth rate. The fracture properties under single loading are assessed from testing specimens containing pregrown fatigue cracks. Stress intensity factor K characterizes the elastic stress singularity at the crack tip; in a general way, $K = \sigma(\pi l)^{1/2}y$, with σ being the stress at gross cross-section, l – the half length of the crack, and y – the function dependent on the geometries of the specimen and the crack. The critical stress intensity factor is defined at a point in time whereat the crack growth instability occurs.

The fracture toughness for plane strains K_{1C} is estimated by the data derived from the eccentric tension testing of compact specimens with thickness of $t = 2.5(K_{1C}/\sigma_{0.2})^2$, which imposes high constraints on plastic strain in front of the crack.

The fracture toughness K_C^y in a state of plain strain is defined as a rule from extension of wide, relatively thin plates containing a central slot (width of $B = 100 - 400$ mm, $t = 2 - 15$ mm). The length of slot together with induced fatigue cracks at its ends amounts to about $0.3B$. The initial crack size (together with the slot) must be taken into account in evaluating K_C^y.

The fatigue crack growth rates are derived from the plates containing the central slot as long as $2l \doteq 6 - 10$ mm and tested in cyclic tension. Tangent to the graph for the dependence of the crack increment $\Delta \cdot 2l$ on the cycle number N gives the rate of crack growth dl/dN in relation to the stress intensity factor range $\Delta K = \Delta\sigma(\pi l)^{1/2}y$, $\Delta\sigma = \sigma_{\max} - \sigma_{\min}$ being the stress range of a cycle.

The magnitudes of fracture parameters are dependent on the strength level and structure of a material, the specimen and crack geometry, as well as the conditions of loading. The data in Tables 3.54–3.58 were derived from the specimens with the through cracks. The plane strain fracture toughness K_C^y is substantially dependent on the specimen geometry, in particular on its width. Approximately, the magnitude of K_C^y is proportional to $\sqrt{B}$ (B is the plate width); in fact, however, it rises with B as B^n, where $n < 1/2$.

The correlations between the values of the plane strain fracture toughness K_{1C} and the specific work KCT delivered by the specimen containing a crack is established for some groups of materials tested in impact and static bending [5,21].

Table 3.54 Plane strain fracture toughness

Material, condition, semifinished product	Direction of specimen slot*1	$\sigma_{0.2}$, MPa	K_{1C}, MPa·m$^{1/2}$
	Steels [25–27]		
20, normalizing	DP	400	140
15X2HMΦA, normalizing	DP	640	148
40X, hardening and tempering	DP	1250	115
40XHM, ditto	DP	1500	80
30XΓCHA, isothermal hardening	DP	1550	85
03H18K8M5T, hardening and aging	DP	1550	130
03H18K9M5T, ditto	DP	1700	100
Ditto	DP	1800	75
Ditto	DP	2000	70
	Titanium alloys [21,26,28]		
OT4, annealing, stamping	DP	700	105
BT5-1, annealing, bar	DP	750	98
BT20, annealing, forging	DP	760	104
BT6ч, annealing, plate	DP	820	85
BT3-1, annealing, forging	DP	950	76

Table 3.54 Plane strain fracture toughness *(continued)*

Material, condition, semifinished product	Direction of specimen slot*[1]	$\sigma_{0.2}$, MPa	K_{1C}, MPa·m$^{1/2}$
Titanium alloys [21,26,28]			
BT3-1, hardening and aging, stamping	PD	1200	58
BT9, ditto	DP	1050	78
Ditto	DP	1150	58
BT22, annealing, stamping	DP	1080	73
BT22, annealing, plate	DP	1050	79
BT22, hardening and aging, bar	DP	1250	54
BT23, aging, plate	PD	1080	73
BT9Л, hardening and aging		1000	78
BT20Л, annealing		800	81
Aluminum alloys [5,20,22,28]			
Д16T, forging	PD	290	35
Д16T, plate	PD	340	27
Д16T, pressed strip	DP	340	37
Д16чT, forging	DP	290	38
Д16T1, pressed strip	DP	345	31
Д16чT1, forging	PD	375	24
Д16чT1, stamped	DP	465	31
Д16чT1, plate	DP	400	38
Ditto	DP	450	30
Ditto	PD	460	26
Д16очT1, plate	PD	430	30
1201T1, plate	DP	350	35
Ditto	PD	345	34
Ditto	VD	350	25
AK4-1T1, plate	DP	330	31
Ditto	DP	400	24
Ditto	PD	360	22
AK4-1T1, stamping	DP	400	25
AK6T1, forging	DP	320	37
AK8T1, forging, stamping	DP	420	29
B95T1, plate	DP	485	33
Ditto	PD	475	28
B95очT1, plate	PD	490	32
B95пчT2, plate	DP	460	36
Ditto	PD	455	29
B95пчT3, plate	DP	440	37
Ditto	PD	435	30
B93пчT1, stamping	DP	480	29
B93пчT2, ditto	DP	440	33
B93пчT3, ditto	DP	360	35
B96Ц3T3, ditto	DP	460	30
B96ЦT1, ditto	DP	560	22
Polymer matrix composites of unidirectional structure [12]			
Carbon fiber epoxyplastic	DP	1050	2.45
Boron epoxyplastic	DP	1200	2.7
Fiber-glass epoxyplastic	DP	1800	2.9

*[1] The first letter denotes the direction of specimen axis with reference to fibers; the second one, the direction of crack propagation plane.

Table 3.55 Plane strain fracture toughness for aluminum and titanium alloys[*1]

Alloy, condition, semifinished product	$\sigma_{0.2}$, MPa	B, m	K_c^y, MPa·m$^{1/2}$
Aluminum alloys [5,20,22,28]			
AMr6H, cold-rolled sheet	355	0.1	69
Д16T, ditto	330	0.1	60
Ditto	330	0.4	75
Д16T, hot-rolled sheet	350	0.1	70
Ditto	370	0.4	85
Д16T1, cold-rolled sheet	420	0.1	55
Д16чT, ditto	300	0.2	65
Д16чT, hot-rolled sheet	300	0.2	80
Ditto	360	0.2	75
Д16чT, plate	350	0.2	73
Д16чT, panel pressed	420	0.2	83
AK4–1T1, plate	380	0.2	52
Ditto	400	0.2	45
1201T1, cold-rolled sheet	380	0.2	50
1201T1, hot-rolled sheet	330	0.2	63
Ditto	330	0.3	75
Ditto	330	0.4	82
1201T1, plate	365	0.2	73
B95T1, hot-rolled sheet	480	0.1	52
Ditto	540	0.1	40
B95пчT1, hot-rolled sheet	530	0.1	60
B95пчT2, cold-rolled sheet	475	0.2	60
B95пчT2, hot-rolled sheet	460	0.2	73
Ditto	480	0.1	63
Ditto	480	0.2	80
B95пчT2, plate	465	0.2	80
B95пчT2, panel pressed	480	0.2	85
Titanium alloys [21,27,28]			
0T4, sheet, annealing	660	0.1	140
BT5-1, ditto	800	0.1	100
BT14, sheet, annealing	950	0.1	88
BT14, sheet, hardening and aging	1175	0.1	62
BT20, sheet, annealing	950	0.1	115
BT6, sheet, hardening and aging	1050	0.1	71–85

[*1] DP specimens.

Table 3.56 Effect of low temperatures on the plane strain fracture toughness for selected steels and alloys [22,25,26,27]

Steel, alloy	$\sigma_{0.2}$, MPa	K_{1C}, MPa·m$^{1/2}$ at t, °C: 20	−70	−130	−196	Steel, alloy	$\sigma_{0.2}$, MPa	K_{1C}, MPa·m$^{1/2}$ at t, °C: 20	−70	−130	−196
Steel 20	285			47	22	BT6ч, plate	820	85			60
25Г2HM	350	175	50	40		Д16T, pressed strip	340	37			55
15X2HMФA	640	148	83	68		Д16T1, ditto	350	31			35.5
30XГCHA	1550	85	45			1201T1, plate[*1]	350	35			35
03X18K8H5T	1550	130	130	120		AK4-1T1, ditto[*1]	330	31			36
03X18K9M5T	1700	100	79	63	47	AK4-1T1, ditto[*1]	405	27			30
03X18K9M5T	2000	70	70			B95T1, ditto[*1]	495	28			29

[*1] At a temperature −253°C, $K_{1C} = 40$MPa·$m^{1/2}$.

Table 3.57 Fatigue crack growth rate in aluminum alloys [5,20,27][*1]

Alloy, condition, semifinished product	$\sigma_{0.2}$, MPa	$dl/dN \cdot 10^6$, m/cycle at ΔK, MPa·m$^{1/2}$			
		12.4	15.5	18.6	21.7
Д16Т, sheet	325	0.37	0.6	1.0	
Д16чТ, ditto	350	0.2	0.35	0.5	0.9
Д16Т1, ditto	470	0.35	0.55	1.0	1.7
Д16чТ1, ditto	480	0.35	0.5	0.9	1.5
Д19Т, ditto	345	0.3	0.5	0.65	1.3
Д19Т1, ditto	450	0.35	0.75	1.7	3
1911Т1, ditto	355	0.4	0.8	1.5	2
В95Т1, ditto	500	0.5	0.8	1.3	1.6
В95пчТ1, ditto	530	0.3	0.65	1.0	1.2
В95пчТ2, ditto	480	0.4	0.7	0.9	
АМг6М, ditto	170	1.5	2.5		
АК4-1Т1, plate	450	0.4	0.75	1.5	2.2
1201Т1, ditto	345	0.3	0.5	1.1	1.4
В95пчТ1, ditto	500	0.65	1.25	1.95	
В95пчТ2, ditto	460	0.5	0.8	1.15	
Д16чТ, ditto	350	0.2	0.4	0.75	
Д16чТ1, ditto	470		0.6		
Д16чТ, stamped	325	0.3		0.7	
АК6Т1, ditto	350	0.4		0.85	
В93пчТ2, Т3, ditto	360–440	0.25		0.55	
Д16чТ, pressed panel	400	0.2	0.4	0.45	0.6
Д19Т, ditto	345	0.25	0.45	0.75	1.3
В95Т1, pressed strip	560	0.5	0.8	1.0	1.3
В95пчТ1, pressed panel	560	0.4	0.65	1.05	
В95пчТ2, ditto	470	0.3	0.5	0.95	
В95пчТ3, ditto	430	0.25	0.4	0.9	

[*1] Specimen makes the center-slotted 100-200 mm thick plate tested under asymmetric tension.

Table 3.58 Fatigue crack growth rate in steels and titanium alloys [21,26–28]

Alloy, steel, condition, semifinished product	$\sigma_{0.2}$, MPa	$dl/dN \cdot 10^6$, m/cycle at ΔK, MPa·m$^{1/2}$				
		15.5	25.4	31	46.6	54.4
ОТ4-1, annealing, sheet	570			0.45	1.05	1.8
ОТ4, ditto	660			0.7	2	
ВТ20, ditto	950			0.5	1.2	1.8
ВТ23, hardening and aging, sheet	1150			0.7	3	
ВТ6ч, annealing, plate[*1]	850	0.16	0.5	1	1.9	2.6
ВТ6ч, hardening and aging, plate[*1]	1000	0.15	0.6	0.9	1.1	2.3
ВТ22, annealing, plate[*1]	1100	0.15	0.4	0.9	1.8	
ВТ23, aging, plate[*1]	1100	0.15	0.5	0.8	1.5	
ВТ23, hardening and aging, forging[*1]	1050	0.22	0.8		2	
ВТ22, annealing, pressed forming	1000					2
ВТ23, annealing, plate	1000					2.5
30ХГСНА, isothermal hardening	1550					0.65
03Н18К8М5Т, hardening and aging	1550					0.55
Ст3	265		1.5			
18Г	240		1.4			
09Г2С	360		1.3			

[*1] Compact specimens in eccentric tension.

3.6 Mechanical Properties of Amorphous Metals and Alloys

Tables 3.59–3.62 present numerical data on the ultimate tensile strength σ_b , yield stress $\sigma_{0.2}$, Vickers hardness number of a material (HV), Young's modulus E, shear modulus G, bulk modulus B, Poisson's ratio μ, and crystallization temperature at annealing from amorphous state T_{cr}. The generally adopted names for certain alloys are listed in the notes.

The moduli of elasticity for amorphous materials are related as follows: $B = (\frac{E}{3})/(1 - 2\mu) = (\frac{EG}{3})/(3G - E)$.

The data in tables, unless otherwise stated, correspond to 20°C conditions; the mechanical properties of certain metals are given for polycrystalline or liquid state for comparison.

The amorphous alloys (AA) are produced from the melt by ultrafast quenching with a rate up to 10^5–10^7 K/s. The AA may be considered as an ideal elastoplastic material with vanishingly small strain hardening. Depending on temperature, two types of plastic flow are observed in AA. At temperatures below $T_p = (0.7\text{–}0.8)\mathrm{T}_{cr}$, a high local plasticity occurs with macroscopic-embrittle fracture mode. The sliding happens within localized strain bands (heterogeneous strain). At temperatures above T_p, the plastic deformation is uniform and proceeds by viscous flow (homogeneous deformation).

Table 3.59 Elastic properties of amorphous binary alloys [33–39,41]

Material (composites, %)	σ_b, GPa	σ_T, GPa	HV, GPa	E, GPa	G, GPa	B, GPa	μ	T_{cr}, K
$Mg_{70}Zn_{30}$				35				358.6
Mg (polycrystalline)				45	17.5	34	0.28	
$Fe_{88}B_{12}$				137				
$Fe_{80}B_{20}$ *1		3.63	10.8	168	64.9	141	0.3	
$Fe_{75}B_{25}$			12.89	175.5			0.32	
$Fe_{100-x}P_x$ $(14 \leq x \leq 21)$				120–130				
Fe (polycrystalline)				211	82.4		0.28	
Fe ("whiskers")		12			79.2			
$Co_{75}B_{25}$			11.3	176.5			0.34	
$Co_{85}P_{15}$				120	39			
$Co_{17}Sm_2$					42.6	133		
$Ni_{80}P_{20}$				103	36.7	161	0.394	
$Cu_{50}Ti_{50}$				96.7				
$Cu_{60}Zr_{40}$	1.96		5.3	74.5				753
$Cu_{57}Zr_{43}$	1.96		5.3	74.5				
$Cu_{50}Zr_{50}$			5.69	85.3			0.36	
$Cu_{50}Ti_{50}$			5.98	96.7			0.36	
Cu (polycrystalline)				123	45.5	136	0.35	
$Pd_{80}Si_{20}$	1.33	0.86	3.19	66.7	35.5	182	0.416	653
Pd (polycrystalline)				123	44.1	139	0.30	
$Pt_{75}P_{25}$			3.37	91.2			0.43	

*1 Metglas 2605.

Table 3.60 Elastic properties of amorphous multi-component alloys [33–42]

Material (composites, %)	σ_b, GPa	σ_T, GPa	HV, GPa	E, GPa	G, GPa	B, GPa	μ	T_{cr}, K	Notes
$Ti_{50}Be_{40}Zr_{10}$	1.86		7.16	105					Metglas 2204
$Fe_{78}Si_{10}B_{12}$	3.33		8.92	125.5				773	Amomet 26
$Fe_{80}P_{13}C_7$	3.04		7.45	121.6				693	
$Fe_{80}P_{14}C_4B_1$		2.44	8.19	135.3					Metglas 2615
$Fe_{60}Cr_6Mo_6B_{28}$		4.5							
$Fe_{40}Co_{40}B_{20}$				174.2	65.0	184	0.34		

Table 3.60 Elastic properties of amorphous multi-component alloys [33–42] *(continued)*

Material (composite, %)	σ_b, GPa	σ_T, GPa	HV, GPa	E, GPa	G, GPa	B, GPa	μ	T_{cr}, K	Notes
$Fe_{60}Ni_{20}B_{20}$				166	61.0	206	0.365		
$Fe_{40}Ni_{40}B_{20}$	2.4			159.7	59.6	167	0.341		Vitrovac 0040
$Fe_{40}Ni_{40}P_{14}B_6$	1.72	2.4	6.3–7.4	126–144					Metglas 2826
$Fe_{22}Ni_{36}Cr_{14}P_{12}B_6$	1.91		7.35	144					Metglas 2826A
$Fe_{29}Ni_{49}P_{14}B_6Si_2$		2.38	7.77	132	48.0	169	0.37		Metglas 2826B
$Fe_{82}Zr_{12}B_6$			8.29					868	
$Fe_{82}Nb_{12}B_6$			7.75					820	
$Fe_{78}Mo_2B_{20}$	2.60		9.95	144				885	
$Fe_{82}Hf_{12}B_6$			9.02						
$Co_{75}Si_{15}B_{10}$	2.94		8.92	104				763	Amomet 27
$Co_{82}Ti_{12}B_6$			8.73					833	
$Co_{74}Fe_6B_{20}$				175	66.7	132	0.32		
$Co_{82}Zr_{12}B_6$			8.34					648	
$Co_{82}Nb_{12}B_6$			6.77					705	
$Co_{82}Hf_{12}B_6$			9.17					820	
$Co_{82}Ta_{12}B_6$			7.50					723	
$Ni_{75}Si_8B_{17}$	2.65		8.43	103				733	Amomet 28
$Ni_{78}Si_8B_{14}$	1.85								Vitrovac 0080
$Ni_{82}Zr_{12}B_6$			6.62					703	
$Ni_{82}Hf_{12}B_6$			6.18					725	
$Ni_{82}Ta_{12}B_6$			5.74					653	
$Ni_{64}Pd_{16}P_{20}$	1.77		5.31	104			0.40		
$Pd_{77.5}Cu_6Si_{16.5}$		1.47	4.46	82.5	34.8	182			
$Pd_{64}Ni_{16}P_{20}$	1.57		4.43	92.2			0.41		
$Pd_{40}Ni_{40}P_{20}$				98.1					
$Pt_{60}Ni_{15}P_{25}$				96.1	33.8	202	0.421		

Table 3.61 Bulk moduli of amorphous (B_a) and liquid (B) metals, their ratios to B values in liquid phase, and volume jump at crystallization ($\Delta V/V$) [33]

Material (composition, %)	B_a , GPa	B/B_a	$\Delta V/V$, %
$Pd_{80}Si_{20}$	182	1.062	
$Pd_{77.5}Cu_6Si_{16.5}$	182	1.065	1.6
$Co_{17}Sm_2$	133	1.045	1.6
Na (liquid at T_K)	54	1.08	2.5
Cs (liquid at T_K)	16	1.13	2.6

Table 3.62 Young's E_a and shear G_a moduli for amorphous alloys, and those (E_c and G_a) for crystallized alloys [33]

Material (composition, %)	E_a, GPa	E_c/E_a	G_a, GPa	G_c/G_a
$Pd_{80}Si_{20}$			35.5	1.34
$Pd_{81}Si_{19}$	80	1.26		
$Co_{89.5}Sm_{10.5}$			42.6	1.36
$Co_{85}P_{15}$			39	1.34
$Co_{80}P_{20}$	105	1.23		
$Fe_{80}B_{20}$	168	1.24	64.9	1.35
$Ni_{76}P_{24}$ (quenched)	95	1.32	33.5	1.46
$Pd_{77.5}Cu_6Si_{16.5}$ (annealed)	94	1.24	34.8	1.35
SiO_2			31	1.38

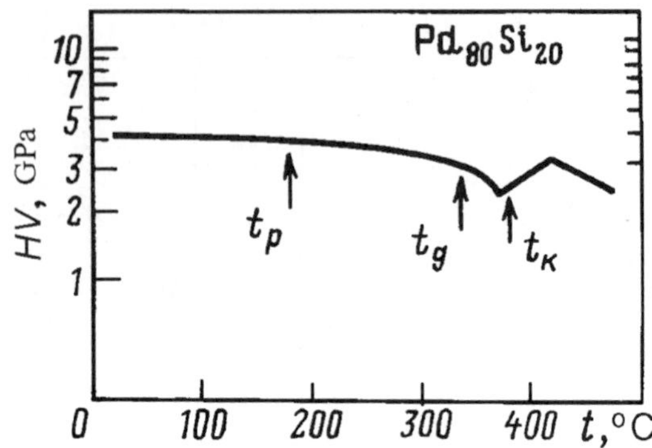

Figure 3.25 Effect of temperature on HV for $Pd_{80}Si_{20}$ amorphous alloy [34]: T_g – vitrification temperature, t_{cr} – crystallization temperature.

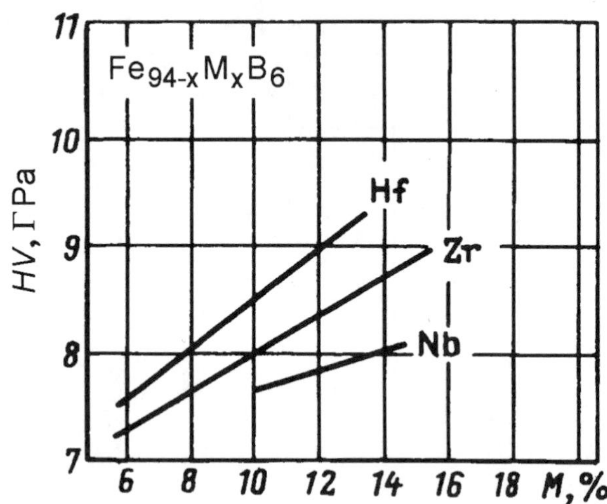

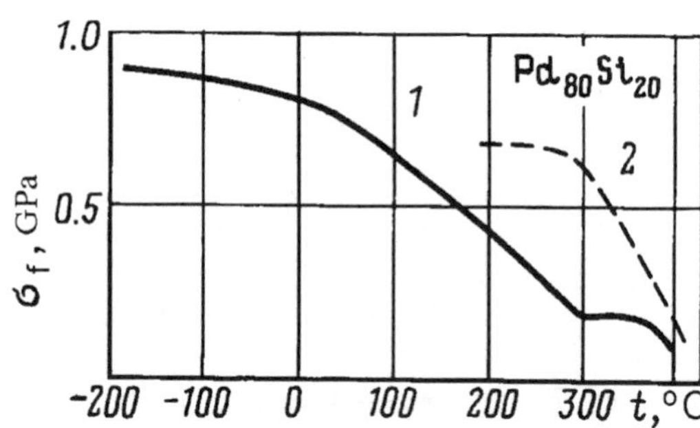

Figure 3.26 Effect of temperature on yield stress of $Pd_{80}Si_{20}$ amorphous phase (*1*) and Pd_3Si crystalline orthorhombic phase (*2*) [36].

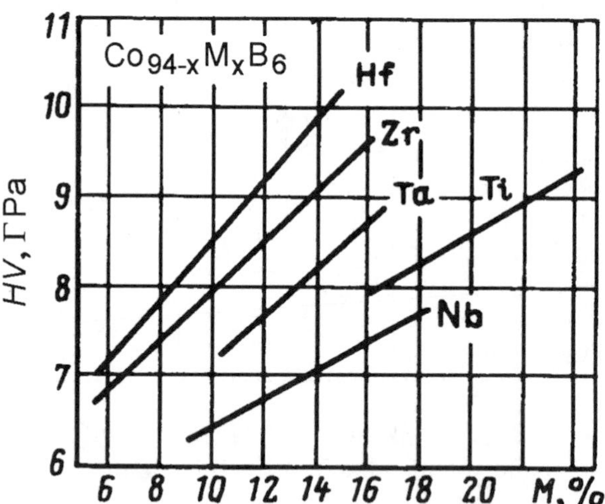

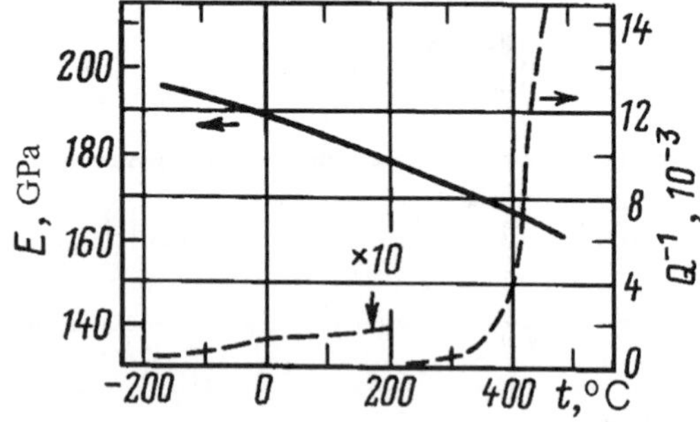

Figure 3.27 Effect of temperature on Young's modulus E and ultrasonic absorption coefficient Q^{-1} for amorphous $Co_{70.4}Fe_{4.6}Si_{15}B_{10}$ alloy. Measurements have been made at sound frequency of 140 Hz. Exponential growth of internal friction (Q^{-1}) at approaching the vitrification temperature (about 500°C in this case) is common to all amorphous materials [33].

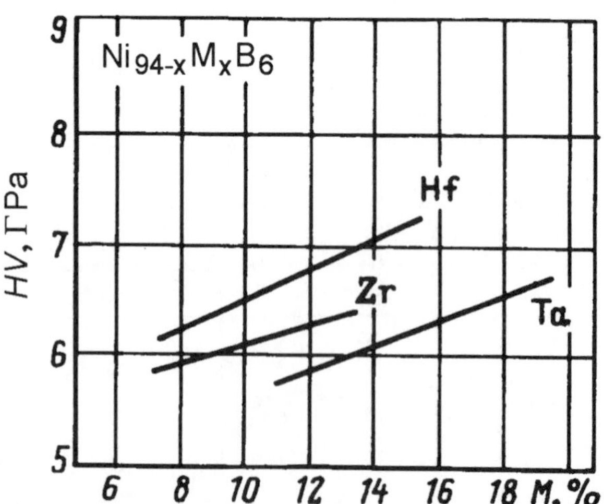

Figure 3.28 HV for amorphous alloys $Fe_{94-x}M_xB_6$, $Co_{94-x}M_xB_6$ and $Ni_{94-x}M_xB_6$ (where M=Ti, Zr, Hf, Nb, and Ta) versus the mass content of M–component.

References

[1] Metal science and thermal treatment of steels. Handbook, 2nd ed., Izdatel'stvo literatury po chernoi i tsvetnoi metallurgii, Moscow, 1961, vol. 1 (in Russian).

[2] Mechanical properties of rare metals, Ed. Sokolov, L. D., Metallurgiya, Moscow, 1972 (in Russian).

[3] Structural materials (in 3 volumes), Ed. Tumanov, A. T., Sovetskaya entsiklopediya, Moscow, 1965 (in Russian).

[4] Burkhardt, A., Mechanical and processing properties of pure metals, Metallurgizdat, Moscow–Leningrad, 1942 (in Russian).

[5] Industrial aluminum alloys. Handbook, Ed. Kvasov, F. I. and Fridlyander, I. N., 2nd ed., Metallurgiya, Moscow, 1972 (in Russian).

[6] Savitskii, E. M., Burkhanov, G. S., Metal science of alloys of refractory and rare metals, 2nd ed., Nauka, Moscow, 1971 (in Russian).

[7] Fridman, Ya. B., Mechanical properties of metals (in 2 parts), 3rd ed., Mashinostroenie, Moscow, 1974 (in Russian).

[8] Noble metals. Handbook, Ed. Savitskii, E. M., Metallurgiya, Moscow, 1984 (in Russian).

[9] Potak, Ya. M., High-strength alloys, in series "Uspekhi sovremennogo metallovedeniya", Metallurgiya, Moscow, 1972 (in Russian).

[10] Titanium alloys in machine building, Mashinostroenie, Moscow, 1974 (in Russian).

[11] Savitskii, E. M., Temperature effects on mechanical properties of metals and alloys, Izdatel'stvo Akad. Nauk SSSR, Moscow, 1957 (in Russian).

[12] Gunyaev, G. M., Structure and properties of polymer fibrous composite materials, Khimia, Moscow, 1981 (in Russian).

[13] Zabolotskii, A. A., Manufacturing and application of composite materials. Science and engineering results, in series "Kompozitsionnye materialy", VINITI Akad. Nauk SSSR, Moscow, 1976 (in Russian).

[14] Portnoi, K. I., Salibekov, S. E., Svetlov, I. L., Chubarov, V. M., Structure and properties of composite materials, Mashinostroenie, Moscow, 1979 (in Russian).

[15] Modern composite materials, Eds. Broutman L.J., Krock, H., Addison–Wesley Publ. Co., Reading, Massachusetts, 1967.

[16] Zdorik, T. B., Matias, V. V., Timofeev, I. N., Fel'dman, L. G., Minerals and rocks of the USSR, Mysl', Moscow, 1970 (in Russian).

[17] Lebedev, S. I., Determination of microhardness of minerals, Izdatel'stvo Akad. Nauk SSSR, Moscow, 1963 (in Russian).

[18] Aluminum–Taschenbuch, 13 Auflage, Eds. Nielsen, H., Hufnagel, W., Ganoulis, G., Aluminum–Verlag GmbH, Dusseldorf, 1974.

[19] Koshelev, P. F., Mechanical properties of alloys for cryogenic engineering, Mashinostroenie, Moscow, 1971 (in Russian).

[20] Kishkina, S. I., Failure strength of aluminum alloys, Metallurgiya, Moscow, 1981 (in Russian).

[21] Drozdovskii, B. A., Prokhodtseva, L. V., Novosil'tseva, N. I., Crack resistance of titanium alloys, Metallurgiya, Moscow, 1983 (in Russian).

[22] Kudryashov, V. G., Smolentsev, V. I., Fracture ductility of aluminum alloys, Metallurgiya, Moscow, 1976 (in Russian).

[23] Methods of testing, control and analysis of materials for machine building. Guidebook (in 3 volumes), Eds. Kishkina, S. I. and Sklyarov, N. M., Mashinostroenie, Moscow, 1974, vol. 2 (in Russian).

[24] Handbook of metal worker (in 5 volumes), Eds. Rakhshtadt, A. G., Brostrem, V. A., Mashinostroenie, Moscow, 1976, vol. 2 (in Russian).

[25] Romanov, O. N., Fracture ductility of structural steels, in series "Dostizheniya otechestvennogo metallovedeniya", Metallurgiya, Moscow, 1979 (in Russian).

[26] Resistance to fatigue crack development in metal alloys used in railway transport. Proceedings, Eds. Bushe, N. A. and Georgiev, M. N., Transport, Moscow, 1984 (in Russian).

[27] Cyclic crack resistance of metal materials and construction elements of transport facilities and structures, Proceedings, Eds. Danilov, V. N. and Mezhova, N. Ya., Transport, Moscow, 1984 (in Russian).

[28] Miklyaev, P. G., Neshpor, G. S., Kudryashov, V. G., Kinetics of failure, Metallurgiya, Moscow, 1979 (in Russian).

[29] Stepnov, M. N., Giatsintov, E. V., Fatigue of light structural alloys, Mashinostroenie, Moscow, 1979 (in Russian).

[30] Marin, N. I., Statistical strength of elements of aviation structures, Mashinostroenie, Moscow, 1973 (in Russian).

[31] Ratner, S. I., Failure under repeated loads, Oborongiz, Moscow, 1959 (in Russian).

[32] Serensen, S. V., Kogaev, V. P., Shneiderovich, R. M., Carrying capacity and calculations of machine elements' strength. Handbook and manual, Mashinostroenie, Moscow, 1975 (in Russian).

[33] Kunzi, H.-U., in: Glassy metals II, Eds. Guntherodt, H.-J., Beck, H.. Topics in Applied Physics, vol.52, Springer, New York, Berlin, Heidelberg, 1982.

[34] Masumoto, T., Sci. Rep., RITU A26, 246, 1977.

[35] Tonizawa, S., Masumoto, T., Sci. Rep. RITU, A26, 263, 1977.

[36] Masumoto, T., Maddin, R,. Acta Metallurgica, 19, 725, 1971.

[37] Davis, L. A., Ray, R., Ohou, C.-P., O'Handley, R. C., Scripta Metallurgica, 10, 541, 1976.

[38] Chen, H. S., Rept. Progr. Phys., 43, 353, 1980.

[39] Gilman, J. J., J. Phys. et radium, 41 c.8, 811, 1980.

[40] Inoue, A., Kobayashi, K., Nose, M., Masumoto, T, J. Phys. et radium, 41, c.8, 831, 1980.

[41] Ishio, S., Sato, Y., Ikeda, T., Takahashi, M., J. Noncrystalline Sol., 61, 62, 955, 1984.

[42] Bengus, V. Z., Tabachnikova, E. D., Startsev, V. I., Phys. Stat. Sol. (a), 81, K11, 1984.

4

Compressibility

B.V. Vinogradov

4.1 Introduction

The compressibility (or bulk elasticity) determines the reversible decrease of substance volume under a pressure action. The compressibility is described by the quantity

$$k = -\frac{1}{V}\left(\frac{\partial V}{\partial p}\right) = \frac{1}{\rho}\left(\frac{\partial \rho}{\partial p}\right), \tag{4.1}$$

where V is the volume; p is the pressure, and ρ is the density. The bulk modulus (or compressibility modulus) is called the reciprocal compressibility K: $K = 1/k$.

The compressibility depends on the conditions of pressure loading. The isothermal k_T and adiabatic k_S compressibilities are different. (Sometimes the isothermal compressibility is denoted by letter β, the adiabatic one — by κ, the bulk modulus — by B.) The compressibilities k_T and k_S are connected by the relationship [1]

$$k_T = \frac{c_p}{c_V} k_S = k_S + \frac{T\alpha^2}{c_p \rho}, \tag{4.2}$$

where c_p and c_V are the specific heats at constant pressure and constant volume, T is the absolute temperature, α is the volume thermal expansion. At room temperatures the differences between k_T and k_S for solids are usually less than several per cent (in Tables 4.1, 4.2 the adiabatic compressibility is denoted by the index S, whereas the isothermal one — by the index T).

The dimension changes under pressure for anisotropic solids depend on the direction and are characterized by linear compressibilities along the directions of principal axes for the crystals:

$$k_a = -\frac{1}{a}\left(\frac{\partial a}{\partial p}\right); k_b = -\frac{1}{b}\left(\frac{\partial b}{\partial p}\right); k_c = -\frac{1}{c}\left(\frac{\partial c}{\partial p}\right). \tag{4.3}$$

The compressibility is connected with the elastic constants s_{ij} according to formula

$$k = \sum_{i,j=1}^{3} s_{ij} = s_{11} + s_{22} + s_{33} + 2(s_{23} + s_{13} + s_{12}). \tag{4.4}$$

For the isotropic solids

$$\frac{1}{k} = K = \frac{E}{3\,(1 - 2\sigma)}, \tag{4.5}$$

where E is the Young's modulus, σ is Poisson's ratio.

0-8493-2861-6/97/$0.00+$.50

The compressibility depends essentially on the pressure. Experimental results for the dependence of solids volume on the pressure are presented in Tables 4.1 – 4.4 in a polynomial form, i.e.

$$(V - V_0)/V_0 = -ap + bp^2 - cp^3. \tag{4.6}$$

In this case the compressibility k (at pressure p) can be calculated from formula

$$k = (a - 2bp + 3cp^2)/(1 - ap + bp^2 - cp^3), \tag{4.7}$$

where the initial compressibility $k_0 = k(p = 0) = a$.

The parameters of the Murnaghan's equation [2]

$$p = \frac{K}{K'}\left[\left(\frac{V_0}{V}\right)^{K'} - 1\right] \tag{4.8}$$

or its modification [3]

$$p = \frac{3K}{2}\left[\left(\frac{V_0}{V}\right)^{7/3} - \left(\frac{V_0}{V}\right)^{5/3}\right]\left\{1 - \xi\left[\left(\frac{V_0}{V}\right)^{2/3} - 1\right]\right\}, \tag{4.9}$$

($K' = \frac{\partial K}{\partial p}$ is the pressure derivative of bulk modulus, ξ is a dimensionless parameter) are given in Tables 4.1 – 4.4 for some solids.

For the ideal gas $v = RT/p$ and $k = 1/p$ (v is the molar volume, m^3/mol, R is the universal gas constant, T is the temperature). The properties of real gases and liquids are described by more complicated $p - v$ relationships [4], for example, the Tait's equation

$$v = v_0\left(1 - C\ln\frac{B + p}{B + p_0}\right) \tag{4.10}$$

or the logarithmic equation

$$v = \frac{A}{\lg(p/p_0) - A/v_0}. \tag{4.11}$$

The parameters of the Tait's equation for several liquids and dense gases are presented in Tables 4.7, 4.12. Here, the compressibility k (at pressure p) can be calculated from

$$k(p) = \frac{C}{B + p}\left(1 - C\ln\frac{B + p}{B + p_0}\right). \tag{4.12}$$

The $p - v - T$ properties of liquids and gases are often characterized by a dimensionless quantity $Z = pv/RT$ which is known as the compressibility factor. Fig. 4.1 shows the $(p - v - T)$ relationships for liquids and gases as the generalized plot [5]: $Z = Z(T_r, p_r)$, where $T_r = T/T_c$; $p_r = p/p_c$; T_r and p_r are the reduced temperature and pressure, respectively; T_c and p_c are the critical temperature and pressure. The compressibility factor Z shown in Fig.4.1 in a wide range of pressures and temperatures differs from the experimental one less than 4 – 6 per cent for a great majority of substances.

The compressibility of pure liquids and gases, as well as its mixtures can be estimated using the data and procedures outlined in [6]. The information on the compressibility of solids and liquids is given also in [7]. The data on the compressibility and bulk modulus of solids are collected in [8–11].

4.2 Compressibility of Solids

Table 4.1 The compressibility and bulk modulus of elements [The data are given for $p = 10^5$ Pa and room temperature except where indicated; a, b, c – coefficients of the equation (4.6). The letters S and T mark the adiabatic and isothermal values. The numbers given in parentheses present the uncertainties in the units of the last significant figure.]

Element	$a(=k_0)$, 10^{-12} Pa^{-1}	b, 10^{-22} Pa^{-2}	c, 10^{-30} Pa^{-3}	K, 10^9 Pa	K'	Ref.
Aluminum *(T)*	12.676	3.347		78.877	3.165	[12]
Antimony *(T)*	23.535	9.78		40.43	4.282	[13]
Argon, 4 K *(T)*				2.86	7.2	[14]
Arsenic	31.6					[8]
Barium *(T)*	105.69	152.21	1.124	9.462	1.725	[12]
Beryllium *(T)*	9.97					[15]
Bismuth *(T)*	32.286	26.529		30.973	4.09	[12]
Boron	5.58	0.8				[8]
Cadmium *(T)*	21.140	11.358		47.304	4.083	[12]
Cesium *(T)*	500					[15]
Calcium *(T)*	54.757	54.206	0.395	18.263	2.616	[12]
Carbon (diamond)	1.8 [8]			560 *(T)*	4.0 *(T)*	[9]
Carbon (graphite)	30 [8]			33.7 *(T)*	12 *(T)*	[9]
Cerium	54.25	46.32	0.585			[16]
Chromium	5.25	0.9				[8]
Cobalt *(T)*	6.02	2.599		167.1	17.327	[13]
Copper *(T)*	6.621	1.099		151.03	4.01	[12]
Dysprosium *(T)*	23.749	11.259		40.327	5.113	[13]
Erbium *(T)*	22.009	8.3045		44.913	3.537	[13]
Europium	86.62	107.79	1.345			[16]
Gadolinium *(T)*	27.271	14.029		35.479	4.77	[13]
Gallium *(T)*	20					[8]
Germanium	12.93 [8]	3.6 [8]		75.02 *(S)*	4.55 *(S)*	[17]
Gold *(T)*	6.01	1.36		166.4	6.51	[18]
Hafnium *(T)*	8.8745			108.95	0.594	[13]
Holmium *(T)*	25.2					[15]
Hydrogen*[1], 4 K	5000					[8]
n-Hydrogen, 4.2 K *(T)*				0.170(6)	7.0(3)	[19]
p-Hydrogen, 6 – 10 K	5470(180)					[20]
Indium *(T)*	25.709	20.143	0.125	38.897	5.095	[12]
Iodine *(T)*	114.89	337.83	6.331	8.37	6.047	[13]
Iridium *(T)*	2.82					[15]
Iron *(T)*	5.8441	1.5008		171.11	7.789	[12]
Krypton*[2], 77 K	560					[8]
Krypton, 115 K *(T)*				1.34	7.94	[22]
Lanthanum *(T)*	40.623	31.874	0.183	24.617	2.863	[12]
Lead *(T)*	23.611	13.222		42.353	3.744	[12]
Lithium *(T)*	80.037	100.97	0.675	11.8(2)	3.33(9)	[21]
Lutetium *(T)*	24.3					[15]
Magnesium *(T)*	28.396	14.78		33.561	4.759	[13]
Manganese, 303 K	8.03	4.2				[8]
α-Mercury, 234 K	37.0					[8]

Table 4.1 The compressibility and bulk modulus of elements *(continued)*

Element	$a(=k_0)$, 10^{-12} Pa^{-1}	b, 10^{-22} Pa^{-2}	c, 10^{-30} Pa^{-3}	K, 10^9 Pa	K'	Ref.
β-Mercury, 50 K	21.7					[8]
Molybdenum *(T)*	3.951	1.115		253.1	13.288	[12]
Neodymium *(T)*	29.958	12.794		32.552	3.016	[13]
Neon*3, 4 K	1006					[8]
Nickel *(T)*	5.5475	2.7045		180.26	16.58	[12]
Niobium *(T)*	6.778	2.549		144.2	14.5	[13]
Nitrogen*4, 65 K	801					[8]
Palladium	5.34	0.9				[8]
Phosphorus (black)	29.6					[8]
Phosphorus (red)	55.6					[8]
Phosphorus (white)	205					[8]
Platinum	3.59					[8]
Plutonium	19.8					[8]
Potassium *(T)*	297.13	1440.4	48.26	3.40(1)	2.99(5)	[21]
Praseodymium *(T)*	33.824	19.278	0.132	30.223	1.612	[13]
Rhenium *(T)*	2.69					[15]
Rhodium	3.67					[8]
Rubidium *(T)*	369.76	2351.7	111.3	2.66(2)	3.23(2)	[21]
Ruthenium *(T)*	3.11					[15]
Samarium	26.38	10.46	0.0415			[16]
Scandium	15.21	3.083	0.006			[16]
Selenium (glass)	173					[8]
Selenium(cryst.) *(T)*	122.29	394.66	8.66	7.90	5.828	[13]
Silicon *(T)*	10.211	2.96		100.75	4.72	[13]
Silver *(T)*	9.031	2.317		110.73	4.681	[12]
Sodium *(T)*	162.4	547.7	15.66	6.2(1)	3.5(1.0)	[21]
Strontium *(T)*	84.756	137.98	1.811	12.08	2.498	[13]
Sulphur *(T)*	103.5	276.14	4.89	8.843	6.55	[13]
Tantalum *(T)*	4.953	0.594		201.9	3.842	[12]
Tellurium *(T)*	47.716	74.735	0.66	18.242	8.404	[12]
Terbium *(T)*	25.1					[15]
Thallium *(T)*	26.646	14.623		37.53	3.119	[12]
Thorium *(T)*	18.4					[15]
Thulium *(T)*	25.2					[15]
α-Tin *(T)*	9.01					[15]
β-Tin *(T)*	18.155	6.123		55.081	2.715	[12]
Titanium *(T)*	9.1598	1.745		109.35	3.355	[13]
Tungsten *(T)*	3.475	1.123		300.09	19.10	[13]
Uranium	10.2	5				[8]
Vanadium *(T)*	7.011	3.245		139.42	18.24	[13]
Xenon, 4 K				3.63	7.2	[14]
Ytterbium	78.1	93.01	1.108			[16]
Yttrium *(T)*	22.251	7.108		44.93	2.2	[13]
Zinc *(T)*	16.271	5.644		61.459	3.264	[12]
Zirconium *(T)*	9.676	1.77		102.84	3.142	[13]

*1 ξ in Eq. 4.9 is equal −1.9 for hydrogen (4 K).
*2 ξ in Eq. 4.9 is equal −5.1 for krypton (77 K).
*3 ξ in Eq. 4.9 is equal −4 for neon (4 K).
*4 ξ in Eq. 4.9 is equal −2.3 for nitrogen (65 K).

Table 4.2 The compressibility and bulk modulus for some inorganic compounds [The data are given for $p = 10^5$ Pa and room temperature except where indicated. a, b, c are coefficients of the equation (4.6). The letters S and T mark the adiabatic and isothermal values. The numbers given in parentheses present the uncertainties in the units of the last significant figure.]

Substance	$a(=k_0)$, 10^{-12} Pa^{-1}	b, 10^{-22} Pa^{-2}	c, 10^{-30} Pa^{-3}	K, 10^9 Pa	K'	Ref.
AgBr *(T)*	24.981	13.451		40.7 *(S)* [10]		[23]
AgCd	9.75 – 9.86					[11]
AgCl *(T)*	22.951	11.8		44.1 *(S)* [10]		[23]
AgF *(T)*	24.665	38.967				[23]
AgIn	10.18					[11]
AgMg	10.02 – 10.38			86 *(T)* [9]	3.6 *(T)* [9]	[11]
AgPd	9.4					[11]
AgSn	9.82					[11]
AgZn	9.69 – 9.87					[11]
AlB_{12} *(S)*	7.19			139		[24]
Al_2O_3 *(S)*				255.07	4.19	[11]
AlSb	16.9 *(T)*			59.2 *(S)* [10]		[25]
β-As_4S_3 *(T)*				17.0	5.5	[26]
As_2Se_3 (glass)	69.6			14.37 *(S)*		[27]
BN (cub.)	5					[28]
BN (hex.)*(T)*				25.5	15	[9]
BaF_2 *(T)*	17.59	9.6		56.82		[29]
$Ba(NO_3)_2$	30.74 – 30.9					[11]
BaO *(T)*				74.06	5.67	[30]
$BaSO_4$ *(S)*				22.3		[10]
$BaTiO_3$ (cub.)	8.91					[11]
$BaTiO_3$ (tetr.)	5.36					[11]
Be_2B *(S)*	8.06			124		[24]
Be_4B *(S)*	8.13			123		[24]
BeO *(T)*	4.13	4.3				[31]
Bi_2Te_3	27.3					[32]
CaF_2 *(T)*	12.26	4.5		81.7		[29]
CaO *(T)*				112	3.9	[9]
CaS *(T)*				56.7	4.9	[9]
CdO *(T)*				108	9	[9]
Cd_3As_2 *(T)*	20					[33]
CdS *(T)*	26.3	80	1.43			[31]
CdSe *(T)*	35.8	13.9	2.98			[31]
CdTe *(T)*	39.8	88	1.09			[31]
CeAs *(T)*	7.9					[34]
CeP *(T)*	6.6					[34]
CeSb *(T)*	10					[34]
CoF_2	9.2					[35]
CoO [9]	20(1)			190.5	3.9	[36]
CrAs	28.3					[37]
Cr_2FeO_4 *(S)*				203.7		[10]
CrN	4.36 – 4.38					[11]
α-Cr_2O_3	3.92					[11]
CsBr *(T)*	65.652	96.190	73.963	18.0 *(S)* [10]		[23]
CsCl *(T)*	56.401	73.842	54.194			[23]
CsF *(T)*	42.5					[38]
CsI *(T)*	78.488	130.42	115.48	12.4 *(S)* [10]		[23]
Cu_3Au	5.25 – 6.30					[11]
CuBr *(T)*	29.325	25.314	22.872			[23]

Table 4.2 The compressibility and bulk modulus for some inorganic compounds *(continued)*

Substance	$a(=k_0)$, 10^{-12} Pa^{-1}	b, 10^{-22} Pa^{-2}	c, 10^{-30} Pa^{-3}	K, 10^9 Pa	K'	Ref.
CuGa	7.34 – 7.44					[11]
CuGe	7.34 – 7.51					[11]
Cu_2O *(S)*				111.96(27)		[39]
CuSi	7.17 – 7.53					[11]
CuZn	7.47 – 7.87			116.2 *(S)* [10]		[11]
EuO				110(5)		[40]
EuS				61(5)		[40]
EuSe				52(5)		[40]
EuTe				40(5)		[40]
FeAl *(T)*				138	5.6	[9]
Fe_3Al *(T)*				136	4.6	[9]
Fe+5%Ni (bcc)				155(1)	4.2(8)	[41]
Fe+5%Ni (hcp)				209.0(1.5)	4.0	[41]
FeO *(T)*				154	3.4	[9]
Fe_2O_3 *(S)*				98		[10]
Fe_3O_4 *(S)*				161.6		[10]
Fe+8%Si *(T)*				174	4.6	[9]
Fe_3Si *(T)*				250	−2.0	[9]
GaAs *(T)*				74.66	4.67	[17]
GaN	5					[25]
GaP *(S)*	11.27(6)					[42]
GaSb *(T)*				56.14	4.78	[43]
GeO_2 *(S)*				25.9	6.15	[44]
GeO_2 *(T)*				39.1(4)	2.2(5)	[45]
H_2O, 4.2 K	91.743					[11]
H_2O-VII, 2.2 GPa				22.7	5.3	[46]
D_2O-VII, 2.2 GPa				24.2	4.6	[46]
$HfC_{0.967}$ *(S)*	4.12(2)			2.42		[47]
HgS	26.7					[25]
HgSe	25.4					[25]
HgTe	24.0					[25]
$HoZn_2$				6.43		[48]
InAs	17.2					[32]
InBi	29.1					[32]
InP	13.8					[25]
InSb	22.0			46.9 *(S)* [10]		[25]
In+10%Tl	25.0					[49]
K_3Al	63.6					[11]
KBr *(T)*	60.351	53.606		15.5 *(S)* [10]		[23]
KCl *(T)*	54.906	57.969				[23]
KF *(T)*	61.814	78.118				[23]
KI *(T)*	78.86	112.85		12.38 *(S)* [10]		[23]
LiBr *(T)*	38.951	21.239		25.66 *(S)* [10]		[23]
LiCl *(T)*	30.342	13.705		31.53 *(S)* [10]		[23]
LiF *(T)*	15.491	6.5364		69.8 *(S)* [10]		[23]
LiH *(S)*				32.35	3.80(15)	[50]
^{6}LiH	29.5				4	[51]
^{7}LiH	29.8				4	[51]
^{6}LiD	28.5				4	[51]
^{7}LiD	29.4				4	[51]
LiI *(T)*	57.466	66.674	43.154	18.83 *(S)* [10]		[23]
$MgCu_2$ *(S)*				87.99		[52]
MgF_2 *(T)*	10.0					[35]

Table 4.2 The compressibility and bulk modulus for some inorganic compounds *(continued)*

Substance	$a(=k_0)$, 10^{-12} Pa^{-1}	b, 10^{-22} Pa^{-2}	c, 10^{-30} Pa^{-3}	K, 10^9 Pa	K'	Ref.
MgO	6.45 – 6.52			178 *(T)* [9]	4 [9]	[11]
Mg_2Sn *(S)*				41.66		[53]
MnAs, 273 K	27.9					[54]
315 K	57.9					[54]
MnBi (ferromagn.)	21.0					[54]
MnBi (paramagn.)	26.4					[54]
MnF_2 *(T)*	11.5					[35]
MnO *(T)*				144.0	3.3	[9]
MnS *(T)*				81.0	3.3	[9]
$MoSi_2$	3.57					[11]
Mo_3Si_2	9.79					[11]
$(NH_4)_3Al$	65.1					[11]
NH_4Br *(T)*	62.162	102.73	98.655			[23]
NH_4Cl *(T)*	59.396	104.55	96.573	17.8*(S)* [10]		[23]
$NH_4H_2PO_4$	34.75 – 45.00					[11]
NaBr *(T)*	48.734	51.871	34.037	21.07 *(S)* [10]		[23]
$NaBrO_3$	32.4			30.8 *(S)* [10]		[11]
NaCl *(T)*	42.730	46.578	32.499	26.4 [9]	3.9 [9]	[23]
$NaClO_3$	36.63 – 39.11			25.6 *(S)* [10]		[11]
NaF *(T)*	20.648	9.0621		45.6 [9]	5.7 [9]	[23]
NaI *(T)*	66.622	96.809	81.299	15.0 [9]	4.1 [9]	[23]
$NaNO_2$				21.9(2)	4.3(8)	[55]
$NaNO_3$				25.8(6)	6.6(1.5)	[55]
NbC [9]	4.38			320 *(T)*	14 *(T)*	[11]
NbO_2 *(S)*				236(6)		[56]
NiF_2 *(T)*	8.4 [35]			119.75	5.07	[57]
NiO *(T)*				199	4.1	[9]
α-P_4S_3 *(T)*				5.9		[58]
$Pb(NO_3)_2$	27.93			37 *(S)* [10]		[11]
PbS	16.08 – 16.85			62 *(S)* [10]		[11]
PbSe *(S)*				54.1		[59]
PrSb *(T)*	19.8					[60]
RbBr *(T)*	76.9			13.81 *(S)* [10]		[38]
RbCl *(T)*	64.0			16.2 *(S)* [10]		[38]
RbF *(T)*	38.1			27.33 *(S)* [10]		[38]
RbI *(T)*	94.8			11.1 *(S)* [10]		[38]
SiC	4.26					[11]
β-SiC	4.6					[11]
α-SiO_2				36.4 – 37.7		[11]
β-SiO_2	17.71			56.5 *(S)* [10]		[11]
SiO_2-stishovite				481.5 *(S)* [10]		
SiP_2				128.4(3)	8.2(5)	[61]
SmS				15.1		[40]
SmSe				40(5)		[40]
SmTe				40(5)		[40]
SnI_4	201 [62]			7.9 *(T)*	5.8 *(T)*	[9]
SnO_2 *(S)*				212.3	5.13	[63]
SrO *(T)*				89.28	5.23	[30]
$SrTiO_3$	5.46 [11]			176 *(T)*	4.4 *(T)*	[64]
TaC *(T)*				330	3.3	[9]
TeO_2	22.33					[65]
TiB_2 *(S)*	3.98			251		[24]
$TiBe_{12}$ *(S)*	7.46			134		[24]

Table 4.2 The compressibility and bulk modulus for some inorganic compounds *(continued)*

Substance	$a(=k_0)$, 10^{-12} Pa^{-1}	b, 10^{-22} Pa^{-2}	c, 10^{-30} Pa^{-3}	K, 10^9 Pa	K'	Ref.
TiC	4.38 [11]			220 *(T)*	8 *(T)*	[9]
TiN	3.42 – 3.47					[11]
TiO *(T)*	3.68					[66]
$TiSi_2$	6.12					[11]
TlAl	54					[11]
TlBr	44.51			22.5 *(S)* [10]		[11]
TlBrCl	43.92					[11]
TlBrI	45.45 – 50.42					[11]
TlCl	38.75					[67]
TlI *(T)*				45	5	[9]
TmSb *(T)*	17.8					[60]
TmTe				46.5		[40]
UN	5.35					[11]
UO_2 *(T)*				210(10)	7(2)	[68]
VB_2	3.47					[11]
VC	2.65					[11]
V_3Ge *(S)*				169.2(3)	3.80(22)	[69]
V_3Si *(S)*				176.37(7)	4.30(4)	[70]
WC	1.58 – 1.73					[11]
$WC_{1.007}$ *(S)* t	3.03(5)			3.29		[47]
WSi_2	2.96					[11]
XeF_2				9.8(1)	5.8(3)	[71]
YbO				130	4	[72]
YbS				72(5)		[40]
YbSe				61(5)		[40]
YbTe				46(5)		[40]
ZnF_2 *(T)*	9.6					[35]
ZnO *(T)*	22.2	3.7				[31]
ZnS *(T)*	13.1	4.10				[31]
$ZnSO_4$	45.34					[11]
ZnSe *(T)*	24.69					[31]
ZnTe *(T)*	24.27					[31]
ZrC *(T)*	5.9(5) *(S)* [47]			195	8	[9]
ZrN	3.74 – 3.77					[11]
$ZrSi_2$	6.12					[11]

Table 4.3 The compressibility of minerals [8] [The coefficients of equation (4.6) are presented.]

Substance	$a(=k_0)$, 10^{-12} Pa^{-1}	b, 10^{-22} Pa^{-2}	Substance	$a(=k_0)$, 10^{-12} Pa^{-1}	b, 10^{-22} Pa^{-2}
Aegirine	9.4		Bromyrite	27.4	
Albite	20.2	21.6	Bismuthinite	33.2	
Almandine	5.45 – 5.7		Calcite	13.67	3.9
Analcinite	19.70 – 36.73		Celestite	15.7 – 16.3	
Anglesite	19.4		Cerargyrite	24	
Anhydrite	18.4		Cerussite	19.1	
Apatite	10.91	4.1	Chalcopyrite	12.9	
Aragonite	15.5		Chromite	4.9	
Argentite	25.1	33.8	Cobaltite	7.67	1.9
Arsenopyrite	9.9		Cuprite	19.4	19
Augite	10.2		Diopside	9.3	3.1
Barite	17.7 – 18.1		Dolomite	12.2	
Beryl	5.403	0.94	Enstatite	10.1	

Table 4.3 The compressibility of minerals [8] *(continued)*

Substance	$a(=k_0)$, 10^{-12} Pa^{-1}	b, 10^{-22} Pa^{-2}	Substance	$a(=k_0)$, 10^{-12} Pa^{-1}	b, 10^{-22} Pa^{-2}
Epsomite	22.9		Phlogopite	23.4	18.2
Fayalite	9.1		Pyrite	6.8 – 7.1	
Fluorite	12.2 – 12.6	6.5	Rhodochrosite	13	
Forsterite	7.9 – 8.2		Rutile	4.83	0.92
Galena	18.7 – 19.6		Siderite	10	
Gypsum	25		Sphalerite	13.03	1.28
Ilmenite	5.6		Spodumene	7.03	1.5
Iodargyrite	41.1		Staurolite	8	
Jadeite	7.5 – 11.1		Strontianite	17.5	
Labradorite	13.9 – 15.0		Thenardite	23.7	23.7
Marcasite	8.2		Topaz	6.11	1.1
Microcline	19.2	13	Tourmaline	8.16	1.95
Muscovite	12		Witherite	20.3	
Nepheline	20.5	5.2	Wurtzite	13.6	
Orthoclase	21.23	14.5	Zincite	7.8	
Periclase	5.98	1	Zircon	8.6	

Table 4.4 The compressibility and bulk modulus for some organic compounds and polymers

Substance	k, 10^{-12} Pa^{-1}	K, 10^9 Pa	K'	K'', 10^{-10} Pa^{-1}	Ref.
Adamantane	243				[73]
Anthracene		6.786	8.960	–9.261	[74]
Anthraquinone		8.092	11.016	–15.62	[74]
Anthrone		7.344	9.061	–10.027	[74]
Benzyl		6.129	7.168	–7.063	[74]
Benzophenone		6.421	10.328	–12.413	[74]
p-Diiodobenzene		7.8	7.0		[9]
Diphenyl		4.570	9.942	–13.088	[74]
9,10-Diphenylanthracene		7.316	9.467	–11.258	[74]
n-Heptacosane	193				[73]
Hexabromobenzene		8.089	11.067	–13.51	[74]
Hexachlorobenzene		8.388	8.223		[74]
Hexamethylenebenzene		5.600	10.272	–21.083	[74]
Hexamethylenetetramine	117				[73]
Naphthalene	65(5)				[75]
Nylon-6(γ)	179				[73]
Pentaerythritol	83				[73]
Perinaphtene		7.706	7.778	0.99	[74]
Perylene		10.141	7.701		[74]
Poly-1-butene	202				[73]
α-Poly(ethylene p-oxybenzoate)	138				[73]
Polyethylene (high dens.)	156				[73]
Polyethylene (low dens.)	192				[73]
Poly(4-methyl-1-pentane)	367				[73]
Polyoxymethylene	117				[73]
Polypropylene	207				[73]
Polytetrafluoroethylene, 297 K	250				[73]
Polytetrafluoroethylene-2, 283 K	189				[73]
Poly(tetramethylene oxide)-1	178				[73]
Poly(vinylidene fluoride)-2	94.7				[73]
p-Terphenyl		5.428	9.609	–8.844	[74]

Table 4.5 The compressibility of anisotropic minerals and organic compounds [The linear compressibilities k_i in the directions of principal axes are presented.]

Substance	k_a, 10^{-12} Pa^{-1}	k_b, 10^{-12} Pa^{-1}	k_c, 10^{-12} Pa^{-1}	Ref.
Apatite	4.23		2.45	[8]
Barite	5.026	6.816	5.760	[8]
Beryl	1.664		2.075	[8]
Calcite	2.73		8.22	[8]
Celestite	6.380	4.553	4.615	[8]
Graphite	-1.65		33.1	[76]
Naphthalene	55	31	26	[75]
Orthoclase	10.13	5.59	4.68	[8]
n-Paraffin ($C_{30}H_{62}$), $3 \cdot 10^8$ Pa	67	63		[77]
n-Paraffin ($C_{32}H_{66}$), $5 \cdot 10^8$ Pa	78	67		[77]
Polyethylene (high dens.)	87.7	67.2	1.3	[73]
Quartz	9.95		7.18	[8]
Rutile	1.90		1.05	[8]
Spodumene	1.83	2.50	2.03	[8]
Topaz	2.176	1.504	2.429	[8]
Tourmaline	1.65		4.86	[8]

Table 4.6 The compressibility of anisotropic elements and inorganic compounds [The linear compressibilities k_i along the directions of principal axes are presented. The numbers given in parentheses present the uncertainties in the units of the last significant digit.]

Substance	k_a, 10^{-12} Pa^{-1}	k_b, 10^{-12} Pa^{-1}	k_c, 10^{-12} Pa^{-1}	Ref.
As	-4.6		26.4	[78]
β-As_4S_3	22	14	15	[26]
BN (hex.)			34	[79]
$BaTiO_3$	1.9		4.0	[80]
Be	2.87		2.23	[8]
Bi	6.8(4)		18.7(7)	[81]
Bi_2Te_3	7.3		12.7	[32]
Cd	2.299		15.88	[8]
CdS	5.9		3.5	[82]
Co	1.72		1.81	[8]
CrAs, 190 K	4.7	8.5	5.8	[54]
293 K	1.04(80)	28.0(8)	0.73(80)	[54]
In	9.1(7)		7.0(7)	[83]
InBi	2.5		24.1	[32]
Mg	9.99		9.84	[8]
MnAs, 273 K	8.7(9)		9.6(3)	[54]
315 K	16.6(6)	23.4(8)	17.9(8)	[54]
MnBi (ferromagn.)	7.9(5)		10.6(2)	[54]
MnBi (paramagn.)	5.2(6)		10.5(2)	[54]
MnP	1.53	1.54	5.18	[84]
$NaNO_2$	26.3(1.0)	12.2(6)	7.3(7)	[55]
$NaNO_3$	7.18(6)		25(1)	[55]
$PbTiO_3$	1.4		14.3	[85]
$(SN)_x$, 10^5 Pa	55(5)	22(2)	50(5)	[86]
$2 \cdot 10^9$ Pa	7	3.1	7	[86]

Table 4.6 The compressibility of anisotropic elements and inorganic compounds *(continued)*

Substance	k_a, 10^{-12} Pa^{-1}	k_b, 10^{-12} Pa^{-1}	k_c, 10^{-12} Pa^{-1}	Ref.
Sb(As-struct.)	4.9		14.8	[87]
Sb(cub.)	3.2		20.6	[87]
Se	-3.3		40	[88]
β-Sn	6.13		6.84	[8]
Te	28.01		-4.23	[8]
TeO_2	8.11		6.11	[65]
Zn	1.59		13.76	[8]
α-Zr	5.09(46)		5.99(93)	[89]
ω-Zr	3.08(29)		2.97(27)	[89]
$ZrSiO_4$	0.92(7)		0.83(21)	[90]

4.3 Compressibility of Liquids and Gases

Table 4.7 The parameters of the Tait's equation (4.10) (B, C) and the compressibility of liquids

Liquid	T, K	Pressure, 10^5 Pa	C	B, 10^5 Pa	k, 10^{-12} Pa^{-1}	Ref.
Acetone	273	1–500			809	[91]
	273	500–1000			582	[91]
Almond oil	290				543	[91]
Benzene ($C_{7.1}H_{14.5}$)	296.5		0.111	874		[92]
	373		0.111	585		[92]
	423		0.111	200		[92]
Benzol	286	0.4–18			859	[91]
Bromine	293	100–200			570	[93]
Butyl alcohol	290.4	8			888	[91]
Carbon bisulfide	273	1–500			652	[91]
Carbon tetrachloride	293	100–200			895	[91]
Castor oil					470	[94]
Ethyl acetate	286.3	8–37			1027	[95]
Ethyl alcohol	273	1–50			948	[91]
	293	1–50			1106	[91]
Glycerin	288	1–10			217	[91]
Kerosene	289.5				687	[95]
Mercury	293				40.4	[8]
Methyl alcohol	273	1–500			784	[91]
	293	1000–6000	0.1018	937		[96]
	323	1000–6000	0.1018	774		[96]
Olive oil	293.5				625	[95]
Paraffin oil	287.8				619	[95]
Propyl alcohol	293	1000–8000	0.0870	913		[96]
	323	1000–7000	0.0870	731		[96]
Toluene	298				915	[94]

Table 4.8 The compressibility of water [97].

p, 10^7 Pa	Compressibility k, 10^{-14} Pa^{-1} at temperature t,°C									
	0	10	25	40	50	60	70	80	90	100
0	50885	47810	45246	44240	44174	44496	45161	46143	47430	49018
1	49479	46563	44106	43113	43021	43296	43891	44783	45959	47413
2	48122	45362	43012	42037	41925	42158	42693	43505	44581	45916
5	44313	42014	39989	39091	38942	39082	39476	40101	40943	41991
10	38745	37167	35684	34969	34820	34890	35155	35597	36199	36943

Table 4.9 The compressibility of D_2O [98]

p, 10^7 Pa	k, 10^{-14} Pa^{-1} at t,°C				p, 10^7 Pa	k, 10^{-14} Pa^{-1} at t,°C			
	5	25	50	75		5	25	50	75
0	51547	46480	44859	46121	5	45077	41095	39529	40179
1	50177	45326	43685	44789	10	39470	36546	35332	35702
3	47553	43137	41506	42351					

Table 4.10 The compressibility of liquid metals at the melting point

Metal	k_S, 10^{-12} Pa^{-1}	k_T, 10^{-12} Pa^{-1}	Ref.	Metal	k_S, 10^{-12} Pa^{-1}	k_T, 10^{-12} Pa^{-1}	Ref.
Ba	170	179	[97]	Mg	38.1	50.6	[97]
Bi	36.5		[99]	Na	175		[99]
Ca	82.7	110	[97]	Pb	28.8		[99]
Cd	25.5		[99]	Pr	40.8	41.7	[100]
Ce	52.2	53.0	[100]	Rb	421		[99]
Cs	573		[99]	Sn	27.5		[99]
Ga	21.6		[99]	Sr	116	131	[97]
Hg	9.2		[99]	Yb	99.1	116.9	[100]
In	28.6		[99]	Zn	19.2		[99]
La	41.0	42.4	[100]				

Table 4.11 The compressibility of melted substances

Substance	T, K	k, 10^{-12} Pa^{-1}	Ref.	Substance	T, K	k, 10^{-12} Pa^{-1}	Ref.
AgBr	722	96	[101]	KNO_3	623	185	[102]
	791	111			723	257	
AgCl	731	84	[101]	$K_2Cr_2O_7$	684	167	[101]
	783	98			741	187	
Al_2I_6	467	860	[101]	KSCN	457	246	[101]
	489	890			495	265	
CdI_2	661	430	[101]	$LiClO_4$	529	251	[101]
CsCl	944	410	[101]		561	230	
	1021	484		$LiNO_3$	573	197	[102]
CsI	923	650	[101]		673	241	
	973	690		$NaClO_3$	538	291	[101]
GaI_3	500	783	[102]	$NaNO_2$	573	180	[101]
I_2	393	452	[101]	$NaNO_3$	605	202	[101]
	413	527			673	230	
	443	646		Na_2S_4	595	171	[101]
ICl	309	540	[101]	Na_2S_5	595	180	[101]
	341	630		$RbNO_3$	623	203	[102]
InI_3	500	596	[102]		686	230	

Table 4.12 The parameters of the Tait's equation (4.10) (B, C, p_0) and of the logarithmic equation (4.11) (A, p_0) for dense gases [4]

Substance	t,°C	$p, 10^8$ Pa	C	$B, 10^5$ Pa	$A, 10^{-6}\text{m}^3$
Ammonia	50			673	87.7
($p_0 = 10^8$ Pa in Tait's equation	100			142	74.67
and in logarithmic equation)	150	1–10	0.3084	-184	66.76
	200			-280	
Argon	100			−1610	45.33
($p_0 = 3 \cdot 10^8$ Pa in Tait's equation,	200	3–10	0.3678	−1852	42.16
$p_0 = 2 \cdot 10^8$ Pa in logarithmic equation)	400			-2165	39.78
Helium	20		0.4776	−1995	
($p_0 = 3 \cdot 10^8$ Pa)	50	3–7	0.5101	−1925	
	100		0.5198	−1965	
	150		0.5276	-2005	
Hydrogen	25		0.4771	−1625	
($p_0 = 3 \cdot 10^8$ Pa)	50	3–7	0.4804	−1685	
	100		0.4852	−1766	
	150		0.4922	−1843	
Nitrogen	0			−1200	
($p_0 = 3 \cdot 10^8$ Pa in Tait's equation,	25			−1326	
$p_0 = 2 \cdot 10^8$ Pa in logarithmic equation)	50			−1420	51.13
	100	3–12	0.3678	−1576	49.45
	200			−1818	47.39
	300			−1970	46.71
	400			-2100	46.16

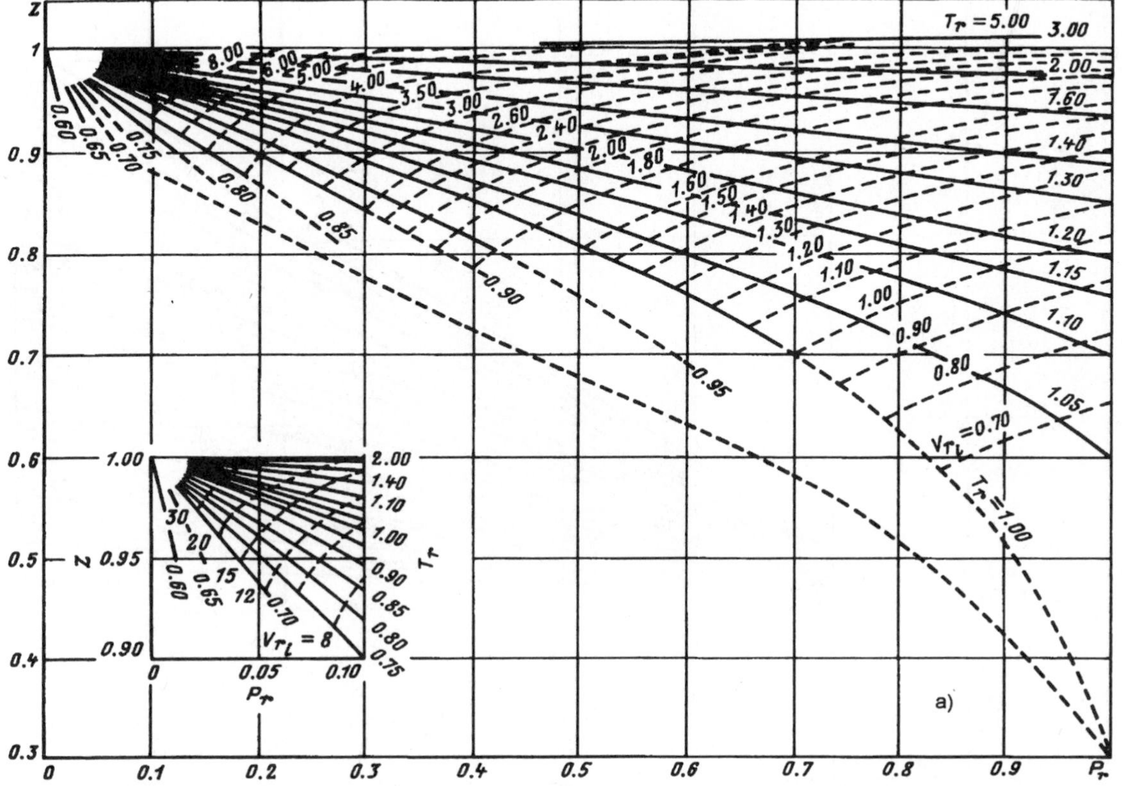

Figure 4.1 The compressibility factor Z for dense gases and liquids versus the reduced pressure p_r (a — $0 \leq p_r \leq 1$; b — $0 \leq p_r \leq 10$; c — $0 \leq p_r \leq 40$) at various reduced temperatures T_r [5].

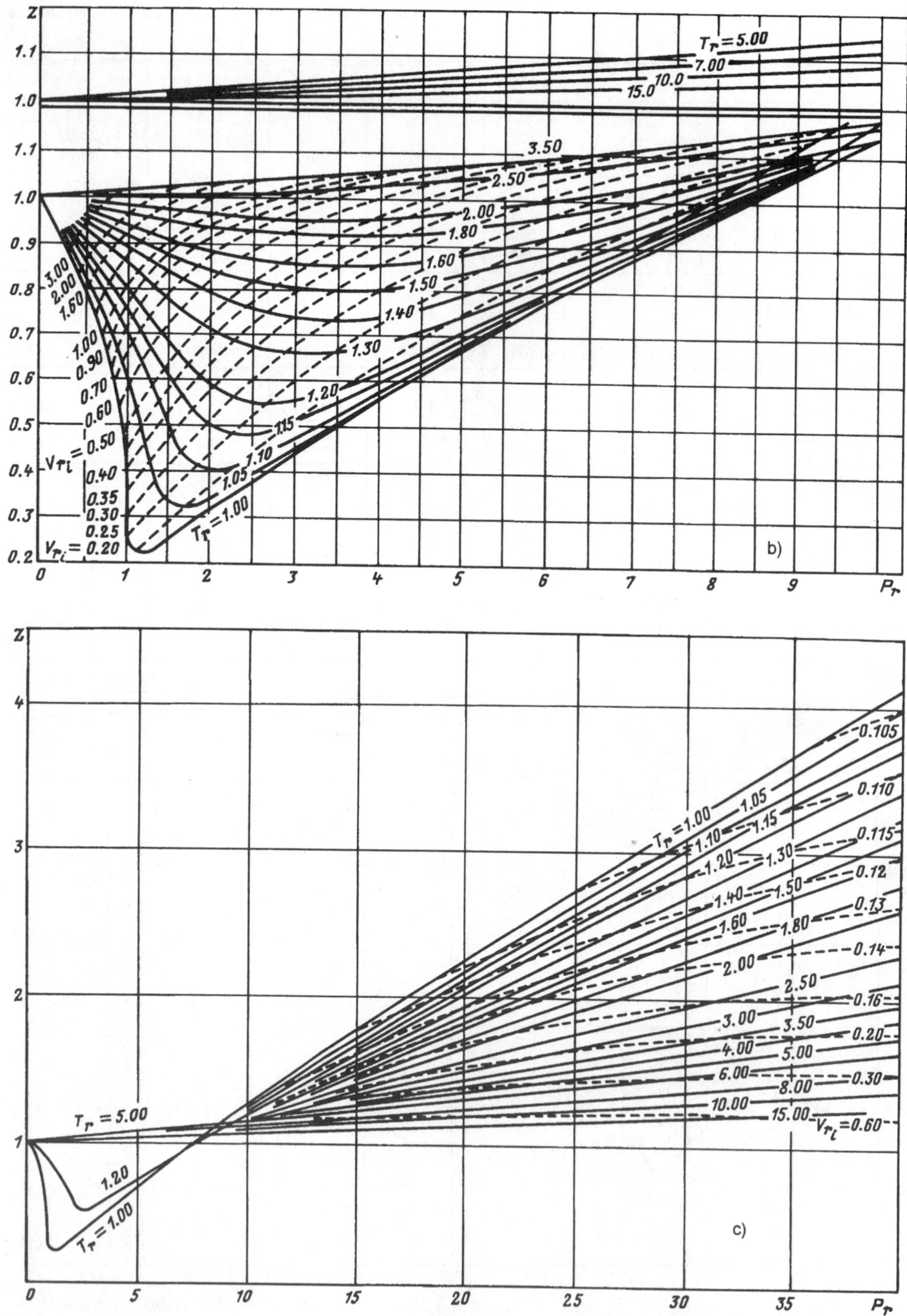

Figure 4.1 The compressibility factor Z for dense gases and liquids *(continued)*.

References

[1] Landau, L.D., Lifshitz, E.M., Statistical Physics. Vol. 1, 3rd ed., Pergamon Press, Oxford, 1980.

[2] Murnaghan, F.D., Proc. Nat. Acad. Sci., 30, 244, 1944.

[3] Birch, F., J. Geophys. Res., 57, 227, 1952.

[4] Tsyklis, D.S., Dense gases, Khimiya, Moscow, 1976 (in Russian).

[5] Nelson, L.C., Obert, E.F., Trans. ASME, 76, 1057, 1954.

[6] Reyd, R., Prausnitz, J., Sherwood, T., Properties of liquids and gases, 3rd ed. Rus. transl. Ed. Sokolov, B.I., Khimiya, Leningrad, 1982 (in Russian).

[7] Landolt – Börnstein, Zahlenwerte und Functionen aus Naturwissenschaften und Technik. Neue Serie. Gruppe IV. Band 4. Eigenschaften der Materie bei hohen Drucken, Berlin, Springer – Verlag, Heidelberg, New York, 1980.

[8] Birch, F. in: Handbook of Physical Constants, Geolog. Soc. of America, New York, 1966.

[9] Drickamer, H. G., Lynch, R. W., Clendenen, R.L., Perez – Albuerne, E. A., Solid State Physics, 19, 135, 1966.

[10] Anderson, O. in: Physical Acoustics, Ed. Mason, W.P., Vol. III, Part B, Academic Press, New York–London, 1965.

[11] Frantsevich, I. N., Voronov, F. F., Bakuta, S. A., Elastic constants and moduli of metals and non–metals. Handbook, Naukova dumka, Kiev, 1982 (in Russian).

[12] Vaidya, S. N., Kennedy, G. C., J. Phys. Chem. Solids, 31, 2329, 1970.

[13] Vaidya, S. N., Kennedy, G. C., J. Phys. Chem. Solids, 33, 1377, 1972.

[14] Anderson, M. S., Swenson, C. A., J. Phys. Chem. Solids, 36, 145, 1975.

[15] Gschneidner, K., Jr., Solid State Physics, 16, 275, 1964.

[16] Reynolds, C. L., Jr., Barker, R. E., Jr., J. Chem. Phys., 61, 2548, 1974.

[17] McSkimin, H. J., Jayaraman, A., Andreatch, P., Jr., J. Appl. Phys., 38, 2362, 1967.

[18] Barsch, G. R., Chang, Z. P., Phys. Status Solidi, 19, 139, 1967.

[19] Anderson, M. S., Swenson, C. A., Phys.Rev. B, 10, 5184, 1974.

[20] Udovichenko, B. G., Manzhelii, V. G., J. Low Temp. Phys., 3, 429, 1970.

[21] Vaidya, S. N., Gettings, I. C., Kennedy, G.C., J. Phys. Chem. Solids, 32, 2545, 1971.

[22] Stewart, J. W., J. Phys. Chem. Solids, 29, 641, 1968.

[23] Vaidya, S. N., Kennedy, G. C., J. Phys. Chem. Solids, 32, 951, 1971.

[24] Gust, W. H., Holt, A. C., Royce, E. B., J. Appl. Phys., 44, 550, 1973.

[25] Champhausen, D. L., Connel, G. A., Paul, W., Phys. Rev. Lett., 26, 184, 1971.

[26] Chattopadhyay, T., Werner, A., v. Schnering, H. G., J. Phys. Chem. Solids, 43, 919, 1982.

[27] Soga, N., Kunugi, M., Ota, R., J. Phys. Chem. Solids, 34, 2143, 1973.

[28] Frantsevich, I. N., Gnesin, G. G., Kurdyumov, A. V., Sverkhtverdye materialy, Ed. Frantsevich, I. N., Naukova dumka, Kiev, 1980 (in Russian).

[29] Wong, C., Shuele, D. E., J. Phys. Chem. Solids, 29, 1309, 1968.

[30] Chang, Z. P., Graham, E. K., J. Phys. Chem. Solids, 38, 1355, 1977.

[31] Cline, C. F., Stephens, D. R., J. Appl. Phys., 36, 2869, 1965.

[32] Akgöz, Y. C., Farley, J. M., Saunders, G.A., J. Phys. Chem. Solids, 34, 141, 1973.

[33] Banus, M. D., Lavine, M. C., High Temp.–High Pressures, 1, 276, 1969.

[34] Bartholin, H., Florence, D., Parisot, G., e.a., Phys. Lett. A, 60, 47, 1977.

[35] Gerlich, D., Hart, S., Whittal, D., Phys. Rev. B, 29, 2142, 1984.

[36] Sawaoka, A., Inoue, K., Saito, Sh., Jpn. J. Appl. Phys., 13, 579, 1974.

[37] Kamenev, V. I., Zavadskii, E. A., Fizika Tverdogo Tela, 20, 933, 1978.

[38] Tosi, M. P., Solid State Physics, 16, 1, 1964.

[39] Manghani, M. H., Brower, W. S., Parker, H.S., Phys. Status Solidi a, 25, 69, 1974.

[40] Jayaraman, A., Singh, A. K., Chatherjee, A., Usha Devi, S., Phys. Rev. B, 9, 2513, 1974.

[41] Takahashi, T., Bassett, W. A., Mao, H.-K., J. Geophys. Res., 73, 4717, 1968.

[42] Weil, R., Groves, W. O., J. Appl. Phys., 39, 4049, 1968.

[43] McSkimin, H. J., Jayaraman, A., Andreatch, P., Jr., Bateman, T. B., J. Appl. Phys., 39, 4127, 1968.

[44] Wang, H., Simmons, G., J. Geophys. Res., 8, 1262, 1973.

[45] Jorgensen, J. D., J. Appl. Phys., 49, 5473, 1978.

[46] Holzapfel, W., Drickamer, H. G., J. Chem. Phys., 48, 4798, 1968.

[47] Brown, H. L., Armstrong, P. E., Kempter, C.P., J. Chem. Phys., 45, 547, 1966.

[48] Michel, D. J., Earle, Ryba, Chang, Z. P., J. Appl. Phys., 39, 5547, 1968.

[49] Chandrasekhar, B. S., Rayne, J. A., Phys. Rev., 124, 1011, 1961.

[50] Gerlich, D., Smith, C. S., J. Phys. Chem. Solids, 35, 1587, 1974.

[51] Stephens, D. P., Lilley, E. M., J. Appl. Phys., 39, 177, 1968.

[52] Cheng, C. H., J. Phys. Chem. Solids, 28, 413, 1967.

[53] Davis, L. S., Whitten, W. B., Danielson, G.C., J. Phys. Chem. Solids, 28, 439, 1967.

[54] Zavadskii, E. A., Kamenev, V. I. in: Physics and technique of high pressures, No. 1, Naukova dumka, Kiev, 1980, p. 29 (in Russian).

[55] Hazen, R. M., Finger, L. W., J. Appl. Phys., 50, 6826, 1979.

[56] Boyle, W. F., Bennett, J. C., Shin, S. H., Sladek, P. J., Phys. Rev. B, 14, 526, 1976.

[57] Wu, A.Y., Phys. Lett. A, 60, 260, 1977.

[58] Chattopadhyay, T., Carlone, C., Jayaraman, A., v. Schnering, H. G., Phys. Rev. B, 23, 2471, 1981.

[59] Lippmann, G., Kästner, P., Wanniger, W., Phys. Status Solidi a, 6, K159, 1971.

[60] Mullen, M. E., Lüthy, B., Wang, P. S., e.a., Phys. Rev. B, 10, 186, 1974.

[61] Chattopadhyay, T., Werner, A., v. Schnering, H. G., J. Phys. Chem. Solids, 44, 699, 1983.

[62] Peercy, P. S., Samara, G. A., Morosin, B., J. Phys. Chem. Solids, 36, 1123, 1975.

[63] Chang, E., Graham, E. K., J. Geophys. Res., 80, 2595, 1975.

[64] Edwards, L. R., Lynch, R. W., J. Phys. Chem. Solids, 31, 573, 1970.

[65] Peercy, P. S., Fritz, I. J., Samara, G. A., J. Phys. Chem. Solids, 36, 1105, 1975.

[66] Banus, M. D., Lavine, M., High Temp.-High Pressures, 2, 671, 1970.

[67] Lewis, G. K., Perez-Albuerne, E. A., Drickamer, H. G., J. Chem. Phys., 45, 598, 1966.

[68] Benjamin, T. M., Zou, G., Mao, H.-K., Bell, P. M. in: Annu. Rept. Dir. Geoph. Lab., Washington D. C.: Carnegie Institution, s. a., 1980–1981, p.280.

[69] Carcia, P. F., Barsch, G. R., Phys. Rev. B, 8, 2505, 1973.

[70] Carcia, P. F., Barsch, G. R., Phys. Status Solidi b, 59, 595, 1973.

[71] Schwarz, W., Syassen, K., High Press. Res., 9/10, 47, 1992.

[72] Jayaraman, A., Rev. Mod. Phys., 55, 65, 1983.

[73] Ito, T. in: High Pressure Science and Technology. Vol. 1, Eds. Timmerhaus, K. D. and Barber, M. S., Plenum Press, New York–London, 1979, 482.

[74] Vaidya, S. N., Kennedy, G. C., J. Chem. Phys., 55, 987, 1971.

[75] Hamamsy, M. E., Elnahwy, S., Damask, A. C., e.a., J. Chem. Phys., 67, 5501, 1977.

[76] Losty, H. H. W. in: Modern Aspects of Graphite Technology, Ed. Blackman, L. C. F., Academic Press, London, 1970, p. 204.

[77] Kabalkina, S. S., Doklady AN SSSR, 125, 114, 1959 (in Russian).

[78] Pace, N. G., Saunders, G. A., Sümengen, Z., J. Phys. Chem. Solids, 31, 1467, 1970.

[79] Kabalkina, S. S., Vereshchagin, L. F., Doklady AN SSSR, 134, 330, 1960 (in Russian).

[80] Kabalkina, S. S., Vereshchagin, L. F., Shulenin, B. M., Doklady AN SSSR, 144, 1019, 1962 (in Russian).

[81] Morosin, B., Schriber, J. E., Solid State Commun., 10, 249, 1972.

[82] Kabalkina, S. S., Troitskaya, Z. V., Doklady AN SSSR, 151, 1068, 1963 (in Russian).

[83] Vereshchagin, L. F., Kabalkina, S. S., Troitskaya, Z. V., Doklady AN SSSR, 158, 1061, 1964 (in Russian).

[84] Iwata, N., Okamoto, T., Jpn. J. Appl. Phys., 14, 248, 1975.

[85] Kabalkina, S. S., Vereshchagin, L. F., Doklady AN SSSR, 143, 818, 1962 (in Russian).

[86] Clarce, T., Solid State Commun., 25, 333, 1978.

[87] Kolobyanina, T. N., Kabalkina, S. S., Vereshchagin, L. F., e.a., Zh. Exper. Teor. Fiz., 59, 1146, 1970 (in Russian).

[88] Vereshchagin, L. F., Kabalkina, S. S., Shulenin, B. M., Doklady AN SSSR, 165, 297, 1965 (in Russian).

[89] Olinger, B., Jamieson, J. C., High Temp.– High Pressures, 5, 123, 1973.

[90] Worlton, T. G., Cartz, L., Niravath, A., Ozkan, H., High Temp.–High Pressures, 4, 463, 1972.

[91] CRC Handbook of Chemistry and Physics, 37th ed. Ed. Hodgman, Ch.D., Chemical Rubber Co. Publ., Cleveland–Ohio, 1955 – 1956.

[92] Maslennikova, V. Ya., Nikiforova, M. B., Doklady AN SSSR, 273, 871, 1983 (in Russian).

[93] Technical encyclopaedia, Vol. 5, Ed. Martens, L.K., e.a. Sovetskaya entsiklopediya, Moscow, 1930 (in Russian).

[94] Gaivoronskii, A. T., Yakovlev, Yu. A., in: Physics and technique of high pressures, No. 7, Naukova dumka, Kiev, 1982 (in Russian).

[95] Kaye, G.W, Laby, T.H., Tables of Physical and Chemical Constants, Longmans, Green & Co., London, 1959.

[96] Atanov, Yu. A., Borzunov, V. A., Razumikhin, V. I., e.a. in: High pressure studies, Ed. Zolotykh, E.V., Izdatel'stvo Standartov, Moscow, 1969, Issue 104 (164), p. 99 (in Russian).

[97] Chen, C. T., Fine, R. A., Millero, F. J., J. Phys. Chem., 66, 2142, 1977.

[98] Fine, R. A., Millero, F. J., J. Chem. Phys., 63, 89, 1975.

[99] McAlister, S. P., Crozier, E. D., Cochran, J. F., Canad. J. Phys., 52, 1847, 1974.

[100] Wilson, D. R., Structure of liquid metals and alloys, Metallurgiya, Moscow, 1972.

[101] Cleaver, B., Spencer, P. N., High Temp.–High Pressures, 7, 539, 1975.

[102] Cleaver, B., Zani, P., High Temp.–High Pressures, 4, 463, 1972.

5

Density of Matter

A.P. Babichev

5.1 Introduction

1. The densities of matter summarized below in Tables 5.1 – 5.8 check with experimental conditions at a temperature of 20°C and atmospheric pressure except when another temperature is outlined in the parentheses.

2. The following notation is used throughout : (1) am. – amorphous, crys. – crystal, glass. – glassy, liq. – liquid, sol. – solid; (2) cub. – cubic, hex. – hexagonal, monocl. – monoclinic, rhomb. – rhombic, tetr. – tetragonal, tricl. – triclinic; (3) R – radioactive; (4) calc. – calculated result.

3. The densities of nitrogen, bromine, hydrogen, iodine, oxygen, phosphorus, fluorine and chlorine are given for their natural molecular states.

4. For some elements in Sect. 5.2, the abridged atomic masses (enclosed in parentheses) of most stable nuclides are exemplified for convenience sake. In Sect. 5.3, the Roman figures in parentheses indicate the state of oxidation for an element in the given compound.

5. It should be recommended to look at the following sources for more detailed information concerning the elements [1,2], inorganic materials [3], organic materials [7], liquids [12], mercury and water at various temperatures [12], and, in closing, plastic, alloys, minerals, wood and some other solids [8,9,14].

5.2 Elements

Table 5.1 Density of elements [1, 2]

Element	Atomic mass, amu	Density, 10^3 kg/m^3	Element	Atomic mass, amu	Density, 10^3 kg/m^3
$_{89}$Ac Actinium	(227)	10.07	$_{56}$Ba Barium	137.33	3.594
$_{13}$Al Aluminum	26.98154	2.6889	$_{97}$Bk Berkelium	(247)	14.78
$_{95}$Am Americium	(243)	13.67	$_{4}$Be Beryllium	9.01218	1.848
$_{51}$Sb Antimony	121.75 [3]	6.691	$_{83}$Bi Bismuth	208.9804	9.78
$_{18}$Ar Argon	39.948	$1.7837 \cdot 10^{-3}$	$_{5}$B Boron	10.811 [5]	
(−186°C, liq.)		1.40	(crys.)		2.34
$_{33}$As Arsenic	74.9216		(am.)		2.37
(blue, hex., α)		5.73	$_{35}$Br Bromine	79.904	3.119
(black, am.)		4.7 – 5.1	$_{48}$Cd Cadmium	112.41	8.65
(yellow, rhomb., γ)		1.97	$_{20}$Ca Calcium	40.078 [4]	1.55
$_{85}$At Astatine	(210)		$_{98}$Cf Californium	(251)	

0-8493-2861-6/97/$0.00+$.50

Table 5.1 Density of elements *(continued)*

Element		Atomic mass, amu	Density, 10^3 kg/m^3
$_{6}$C	Carbon (am.)		1.8 – 2.1
	(graphite)		1.9 – 2.3
	(diamond)		3.15 – 3.53
$_{58}$Ce	Cerium (hex., γ)	140.12	6.77
$_{55}$Cs	Cesium	132.9054	1.873
$_{17}$Cl	Chlorine	35.453	$3.214 \cdot 10^{-3}$
	(−35°C, liq.)		1.557
$_{24}$Cr	Chromium	51.9961 [6]	7.18 – 7.20
$_{27}$Co	Cobalt	58.9332	8.90
$_{29}$Cu	Copper	63.546 [3]	8.96
$_{96}$Cm	Curium	(247)	13.51 calc.
$_{66}$Dy	Dysprosium (25°C)	162.50 [3]	8.550
$_{99}$Es	Einsteinium	(254)	
$_{68}$Er	Erbium (25°C)	167.26 [3]	9.066
$_{63}$Eu	Europium (25°C)	151.96	5.243
$_{100}$Fm	Fermium	(257)	
$_{9}$F	Fluorine	18.998403	$1.696 \cdot 10^{-3}$
$_{87}$Fr	Francium	(223)	
$_{64}$Gd	Gadolinium	157.25 [3]	7.895
$_{31}$Ga	Gallium	69.723 [4]	
	(29.6°C, sol.)		5.904
	(29.8°C, liq.)		6.095
$_{32}$Ge	Germanium (25°C)	72.59 [3]	5.323
$_{79}$Au	Gold	196.9665	19.32
$_{72}$Hf	Hafnium	178.49 [3]	13.31
$_{2}$He	Helium	4.000602 [2]	$1.785 \cdot 10^{-4}$
	(−268.9°C, liq.)		0.1221
$_{67}$Ho	Holmium (25°C)	164.9304	8.795
$_{1}$H	Hydrogen	1.00794 [7]	$8.988 \cdot 10^{-5}$
	(−252.8°C, liq.)		0.0708
$_{49}$In	Indium	114.82	7.31
$_{53}$I	Iodine	126.9045	4.93
$_{77}$Ir	Iridium (17°C)	192.22 [3]	22.42
$_{26}$Fe	Iron	55.847 [3]	7.874
$_{36}$Kr	Krypton	83.80	$3.733 \cdot 10^{-3}$
	(−153.2°C, liq.)		2.155
$_{57}$La	Lanthanum (25°C)	138.9055 [3]	6.145
$_{82}$Pb	Lead	207.2	11.336
$_{3}$Li	Lithium	6.941 [2]	0.534
$_{71}$Lu	Lutetium	174.967	9.840
$_{12}$Mg	Magnesium	24.305	1.738
$_{25}$Mn	Manganese ($\alpha, \beta, \gamma, \delta$)	54.9380	7.21 – 7.44
$_{101}$Md	Mendelevium	(258)	
$_{80}$Hg	Mercury (liq.)	200.59 [3]	13.5461
	(−38.9°C, sol.)		14.193
$_{42}$Mo	Molybdenum	95.94	10.22
$_{60}$Nd	Neodymium	144.24 [3]	
	(hex., α)		7.007
	(cub., β)		6.80
$_{10}$Ne	Neon	20.179	$9.0035 \cdot 10^{-4}$
	(−246°C, liq.)		2.205
$_{93}$Np	Neptunium	237.0482	20.25
$_{28}$Ni	Nickel	58.69	8.91
$_{7}$N	Nitrogen	14.0067	$1.2506 \cdot 10^{-3}$
	(−195.8°C, liq.)		0.808
$_{41}$Nb	Niobium	92.9064	8.57
$_{76}$Os	Osmium	190.2	22.57
$_{8}$O	Oxygen	15.9994 [3]	$1.429 \cdot 10^{-3}$
	(−182.9°C, liq.)		1.14
$_{46}$Pd	Palladium	106.42	12.02
$_{15}$P	Phosphorus	30.97376	
	(white, cub.)		1.82
	(brown, tricl.)		2.0 – 2.4
	(black, rhomb.)		2.25 – 2.69
$_{78}$Pt	Platinum	195.08 [3]	21.45
$_{94}$Pu	Plutonium (25°C, α)	(244)	19.84
$_{84}$Po	Polonium	(209)	
	(cub., α)		9.32
	(hex., β)		9.4
$_{19}$K	Potassium	39.0983	0.862
$_{59}$Pr	Praseodymium	140.9077	
	(hex., α)		6.773
	(cub., β)		6.44
$_{61}$Pm	Promethium (25°C)	(145)	7.22
$_{91}$Pa	Protactinium	231.0359	15.37 calc.
$_{88}$Ra	Radium	226.0254	5–6
$_{86}$Rn	Radon	(222)	$9.73 \cdot 10^{-3}$
	(−62°C, liq.)		4.4
$_{75}$Re	Rhenium	186.207	21.02
$_{45}$Rh	Rhodium	102.9055	12.41
$_{37}$Rb	Rubidium	85.4678 [3]	1.532
$_{44}$Ru	Ruthenium	101.07 [2]	12.41
$_{62}$Sm	Samarium	150.36 [3]	
	(hex., α)		7.536
	(cub., β)		7.40
$_{21}$Sc	Scandium (25°C)	44.95591 [1]	2.989
$_{34}$Se	Selenium	78.96 [3]	
	(black, glass.)		4.28
	(brown, monocl., β)		4.46
	(blue, hex., α)		4.79
	(am.)		4.82
$_{14}$Si	Silicon (25°C, crys.)	28.0855 [3]	2.33
	(25°C, am.)		2.0
$_{47}$Ag	Silver	107.8682 [3]	10.50
$_{11}$Na	Sodium	22.98977	0.971
$_{38}$Sr	Strontium	87.62	
	(cub., α)		2.63
$_{16}$S	Sulphur	32.066 [6]	
	(yellow, am., γ)		1.92
	(yellow, monocl., β)		1.96
	(yellow, rhomb., α)		2.07
$_{73}$Ta	Tantalum	180.9479	16.654
$_{43}$Tc	Technetium	98.9062	11.49
$_{52}$Te	Tellurium (crys.)	127.60 [3]	6.24
	(am.)		6.00
$_{65}$Tb	Terbium	158.9254	8.272

Table 5.1 Density of elements *(continued)*

Element	Atomic mass, amu	Density, 10^3 kg/m^3	Element	Atomic mass, amu	Density, 10^3 kg/m^3
$_{81}$Tl Thallium (hex., α)	204.383	11.85	$_{92}$U Uranium (rhomb., α)	238.0289	19.040
$_{90}$Th Thorium	232.0381	11.72	$_{23}$V Vanadium	50.9415	5.96
$_{69}$Tm Thulium (25°C)	168.9342	9.321	$_{54}$Xe Xenon	131.29 [3]	$5.8971 \cdot 10^{-3}$
$_{50}$Sn Tin	118.710 [7]		(−109°C, liq.)		3.52
(blue, cub., α)		5.85	$_{70}$Yb Ytterbium (α)	173.04 [3]	6.965
(white, tetr., β)		7.29	$_{39}$Y Yttrium (25°C)	88.9058	4.469
$_{22}$Ti Titanium (hex., α)	47.88 [3]	4.505	$_{30}$Zn Zinc	65.39 [2]	7.133
$_{74}$W Tungsten	183.85 [3]	19.35	$_{40}$Zr Zirconium (hex., α)	91.224[2]	6.45

5.3 Inorganic Substances

Table 5.2 Density of inorganic substances [1, 3]

Name	Formula	Density, 10^3 kg/m^3	Name	Formula	Density, 10^3 kg/m^3
Aluminum			**Antimony**		
boride, di-	AlB_2	3.19	(III) bromide	$SbBr_3$	4.15
bromide	$AlBr_3$	3.01 (25°C)	(III) chloride	$SbCl_3$	3.14
carbide	Al_4C_3	2.36	(V) chloride	$SbCl_5$	2.34 liq.
chloride	$AlCl_3$	2.44 (25°C)	(III) fluoride	SbF_3	4.38 (25°C)
fluoride	AlF_3	3.07	(V) fluoride	SbF_5	2.99 (23°C) liq.
hydroxide	$Al(OH)_3$	2.42	hydride	SbH_3	$2.26 \cdot 10^{-3}$
hydroxide oxide	$AlO(OH)$	3.01	(III) iodide	SbI_3	4.92 (17°C)
iodide	AlI_3	3.95	(III) oxide (α)	Sb_2O_3	5.19 (25°C)
nitrate	$Al(NO_3)_3 \cdot 9H_2O$	1.72	(IV) oxide	Sb_2O_4	4.07
nitride	AlN	3.13	(V) oxide	Sb_2O_5	3.78
oxide (corundum)	Al_2O_3	3.97	stibine	SbH_3	$4.36 \cdot 10^{-3}$ (15°C)
phosphate, ortho-	$AlPO_4$	2.57	(antimonous hydride)		
sulfate	$Al_2(SO_4)_3$	2.71	sulfate	$Sb_2(SO_4)_3$	3.62 (4°C)
sulfide	Al_2S_3	2.02 (13°C)	(III) sulfide	Sb_2S_3	4.64
			(V) sulfide	Sb_2S_5	4.12
Americium			(III) telluride	Sb_2Te_3	6.50 (13°C)
(IV) oxide	AmO_2	11.7			
			Arsenic		
Ammonium			arsenic acid, ortho-	$H_3AsO_4 \cdot \frac{1}{2}H_2O$	2.0–2.5
amide	NH_4N_3	1.346	arsine	AsH_3	$43.502 \cdot 10^{-5}$
ammonia	NH_3	$0.771 \cdot 10^{-3}$	(hydrogen arsenide)		
bromide	NH_4Br	2.40	(III) bromide	$AsBr_3$	3.54 (25°C)
chlorate	NH_4ClO_3	1.80	(III) chloride	$AsCl_3$	2.16 liq.
chloride	NH_4Cl	1.53	(III) fluoride	AsF_3	2.67 liq.
fluoride	NH_4F	1.015	(V) fluoride	AsF_5	$7.71 \cdot 10^{-3}$
iodate	NH_4IO_3	3.31	(III) iodide	AsI_3	4.39 (13°C)
iodide	NH_4I	2.51	(V) iodide	AsI_5	3.93
nitrate	NH_4NO_3	1.725	(III) oxide	As_2O_3	3.74
nitrite	NH_4NO_2	1.69	(V) oxide	As_2O_5	4.09
phosphate, hydro-	$(NH_4)_2HPO_4$	1.62	(III) selenide	As_2Se_3	4.75
sulfate	$(NH_4)_2SO_4$	1.77	(II) sulfide	As_2S_2	3.20
sulfide, hydro-	$(NH_4)HS$	1.17	(III) sulfide	As_2S_3	3.43

Table 5.2 Density of inorganic substances *(continued)*

Name	Formula	Density, 10^3 kg/m^3
Barium		
azide	$Ba(N_3)_2$	2.936
boride, hexa-	BaB_6	4.36 (16°C)
bromide	$BaBr_2$	4.781 (24°C)
carbide, di-	BaC_2	3.75
carbonate (γ)	$BaCO_3$	4.43
chloride	$BaCl_2$	3.92
fluoride	BaF_2	4.83
hydride	BaH_2	4.21
iodate	$Ba(IO_3)_2$	5.00
iodide	BaI_2	4.92
nitrate	$Ba(NO_3)_2$	3.24
nitride	Ba_3N_2	4.78 (25°C)
peroxide	BaO_2	4.96
sulfate	$BaSO_4$	4.50 (15°C)
sulfide	BaS	4.25 (15°C)
oxide	BaO	5.72
Beryllium		
bromide	$BeBr_2$	3.47
carbide	Be_2C	1.90 (15°C)
chloride	$BeCl_2$	1.90 (25°C)
fluoride	BeF_2	1.99 (25°C)
hydroxide	$Be(OH)_2$	1.909
iodide	BeI_2	4.33 (25°C)
nitrate	$Be(NO_3)_2 \cdot 3H_2O$	1.56
nitride	Be_3N_2	2.71
oxide	BeO	3.01
sulfate	$BeSO_4$	2.44
sulfide	BeS	2.36
Bismuth		
(III) bromide	$BiBr_3$	5.60
(III) chloride	$BiCl_3$	4.75
(III) fluoride	BiF_3	8.75
hydroxide	$Bi(OH)_3$	4.36
iodide oxide	$BiIO$	7.92
(III) iodide	BiI_3	5.64
oxide chloride	$BiClO$	7.72 (15°C)
oxide fluoride	$BiFO$	7.5
(II) oxide	BiO	7.15–7.30
(III) oxide	Bi_2O_3	8.9
(III, V) oxide	$Bi_2O_4 \cdot 2H_2O$	5.6
(V) oxide	Bi_2O_5	5.10
phosphate, ortho-	$BiPO_4$	6.32 (15°C)
sulfate	$Bi_2(SO_4)_3$	5.08 (15°C)
(II) sulfide	BiS	~7.7
(III) sulfide	Bi_2S_3	7.6
Bismuthic acid	$Bi_2O_5 \cdot n\,H_2O$	5.75
Boron		
boric acid, meta-	HBO_2	2.49
boric acid, ortho-	H_3BO_3	1.435 (15°C)
bromide	BBr_3	2.65
tetra-, carbide	B_4C	2.52
chloride	BCl_3	1.434 (0°C)
chloride	BCl_3	1.434 (0°C)
decaborane	$B_{10}H_{14}$	0.94 (25°C)
diborane	B_2H_6	0.447 (−112°C) liq.
fluoride	BF_3	$2.99 \cdot 10^{-3}$
hexaborane	B_6H_{10}	0.69 (0°C)
iodide	BI_3	3.35 (50°C)
nitride	BN	2.34
oxide	B_2O_3	1.844 cryst.
pentaborane	B_5H_9	0.66 (0°C)
di-, sulfide, tri-	B_2S_3	1.55
tetraborane	B_4H_{10}	0.56 (−35°C) liq.
Bromine		
fluoride, tri-	BrF_3	2.49
fluoride, penta-	BrF_5	2.47 (25°C) liq.
hydrogen bromide	HBr	$3.645 \cdot 10^{-3}$
Cadmium		
bromide	$CdBr_2$	5.19
carbonate	$CdCO_3$	4.26 (4°C)
chloride	$CdCl_2$	4.047
fluoride	CdF_2	6.64
hydroxide	$Cd(OH)_2$	4.79 (15°C)
iodide	CdI_2	5.67 (30°C)
nitrate	$Cd(NO_3)_2 \cdot 6H_2O$	2.45 (17°C)
(II) oxide	CdO	8.15
selenide	$CdSe$	5.81 (15°C)
silicate, meta-	$CdSiO_3$	4.93
sulfate	$CdSO_4$	4.69
sulfide	CdS	4.82
telluride	$CdTe$	6.20 (15°C)
Calcium		
boride, hexa-	CaB_6	2.3
bromate	$Ca(BrO_3)_2 \cdot H_2O$	3.33
bromide	$CaBr_2$	3.35 (25°C)
carbide	CaC_2	2.22
carbonate (argonite)	$CaCO_3$	2.93
carbonate (calcite)	$CaCO_3$	2.71 (25°C)
chloride (hydrophilite)	$CaCl_2$	2.51 (25°C)
fluoride (fluorite, α)	Ca_3F_2	3.18
hydride (α)	CaH_2	1.7
hydroxide	$Ca(OH)_2$	2.24
iodide	CaI_2	3.96 (25°C)
nitrate	$Ca(NO_3)_2$	2.36
nitride	Ca_3N_2	2.63 (17°C)
oxalate	CaC_2O_4	2.2 (4°C)
oxide	CaO	3.37
phosphate, meta-	$Ca(PO_3)_2$	2.82
phosphate, ortho- (whitlockite, α)	$Ca(PO_4)_2$	3.14
phosphide	Ca_3P_2	2.51

Table 5.2 Density of inorganic substances *(continued)*

Name	Formula	Density, 10^3 kg/m^3
selenide	$CaSe$	3.57
silicate, meta-	$CaSiO_3$	
(wollastonite)		
(α)		2.5
(β)		2.90
sulfate (anhydrite, α)	$CaSO_4$	2.90–2.99
sulfate (gypsum)	$CaSO_4 \cdot 2H_2O$	2.32
sulfide (oldhamite)	CaS	2.18 (15°C)
telluride	$CaTe$	4.87
Carbon		
bromide	CBr_4	3.42
carbonyl bromide	$COBr_2$	2.44
carbonyl chloride	$COCl$	$1.392 \cdot 10^{-3}$
(phosgene)		
carbonyl fluoride	COF_2	1.139 (−114°C) liq.
carbonyl selenide	$COSe$	1.81 (4°C) liq.
carbonyl sulfide	COS	$1.073 \cdot 10^{-3}$
chloride, tetra-	CCl_4	1.587
cyanogen	$(CN)_2$	$2.335 \cdot 10^{-3}$
fluoride	CF_4	1.96 (−184°C) liq.
hydrogen cyanide	HCN	0.688 liq.
iodide	CI_4	4.34
(II)oxide	CO	$1.25 \cdot 10^{-3}$ (0°C)
(IV)oxide	CO_2	$1.977 \cdot 10^{-3}$ (0°C)
selenide	CSe_2	2.66 (25°C)
sulfide	CS	1.66
sulfide, di-	CS_2	1.263
tri-,sulfide, di-	C_3S_2	1.27
thiocarbonyl chloride	$CSCl_2$	1.51 (15°C)
thiocarbonyl selenide	$CSSe$	1.99
tri-,oxide, di-	C_3O_2	1.11 (0°C) liq.
Cerium		
(IV) boride	CeB_4	5.74
bromide	$CeBr_3$	5.18
carbide	CeI_2	5.23
chloride	$CeCl_3$	3.92 (0°C)
fluoride	CeF_3	6.16
iodide	CeI_3	2.27
(III) oxide	Ce_2O_3	6.86
(IV) oxide	CeO_2	7.3
phosphate, ortho-	$CePO_4$	5.22
selenate	$Ce_2(SeO_4)_3$	4.46
silicide	$CeSi_2$	5.67 (17°C)
(III) sulfate	$Ce_2(SO_4)_3$	3.91
(IV) sulfate	$Ce(SO_4)_2$	3.91 (18°C)
sulfide	Ce_2S_3	5.91
Cesium		
amide	$CsNH_2$	3.44 (25°C)
bromate	$CsBrO_3$	4.11 (16°C)
bromide	$CsBr$	4.44
chlorate	$CsClO_3$	5.57
chloride	$CsCl$	3.97
cyanide	$CsCN$	2.93
fluoride	CsF	3.59
hydride	CsH	3.41
hydroxide	$CsOH$	3.68
iodate	$CsIO_3$	4.85
iodide	CsI	4.51 (25°C)
iodide, tri-	CsI_3	4.47
nitrate	$CsNO_3$	3.69
oxide	Cs_2O	4.36
perchlorate	$CsClO_4$	3.33 (4°C)
peroxide	Cs_2O_2	4.25
selenate	$CsSeO_4$	4.45
sulfate	Cs_2SO_4	4.24
superoxide	CsO_2	3.77
Chlorine		
chlorosulfonic acid	HSO_3Cl	1.77
fluoride	ClF	1.62 (−108°C) liq.
fluoride, tri-	ClF_3	1.866 (10°C)
hydrate, octa-	$Cl_2 \cdot 8H_2O$	1.23
hydrogen chloride	HCl	$1.639 \cdot 10^{-3}$
(I) oxide	Cl_2O	$3.89 \cdot 10^{-3}$ (0°C)
(IV) oxide	ClO_2	1.64 (0°C) liq.
(VII) oxide	Cl_2O_7	1.86 (0°C)
perchloric acid	$HClO_4$	1.768
Chromium		
arsenide	$CrAs$	6.35 (16°C)
boride	CrB	6.17
(II) bromide	$CrBr_2$	4.37
(III) bromide	$CrBr_3$	4.25 (25°C)
carbide	Cr_3C_2	6.68
carbonyl, hexa-	$Cr(CO)_6$	1.77
(II) chloride	$CrCl_2$	2.88 (25°C)
(III) chloride	$CrCl_3$	3.03
(II) fluoride	CrF_2	4.11
(III) fluoride	CrF_3	3.78
(II) iodide	CrI_2	5.20
(III) iodide	CrI_3	4.92 (25°C)
nitride	CrN	5.8
(III) oxide	Cr_2O_3	5.21
(VI) oxide	CrO_3	2.70
oxide, di-, chloride, di-	$CrCl_2O_2$	1.91
phosphide	CrP	5.7 (15°C)
silicide	Cr_2Si_3	5.5 (0°C)
sulfate	$Cr_2(SO_4)_3$	3.01
sulfide	CrS	4.85
Cobalt		
(II) bromide	$CoBr_2$	4.91 (25°C)
carbonate	$CoCO_3$	4.13
(II) chloride	$CoCl_2$	3.36
(III) chloride	$CoCl_3$	2.94

Table 5.2 Density of inorganic substances *(continued)*

Name	Formula	Density, 10^3 kg/m^3
(II) fluoride	CoF_2	4.46 (25°C)
(III) fluoride	CoF_3	3.88
(II) hydroxide	$Co(OH)_2$	3.60 (15°C)
iodate	$Co(IO)_3$	5.01
iodide	CoI_2	
(α)		5.68
(β)		5.45
nitrate	$Co(NO_3)_2 \cdot 6H_2O$	1.87–2.13
(II) oxide	CoO	5.7
(II,III) oxide	Co_3O_4	6.07
(III) oxide	Co_2O_3	5.18
phosphide	Co_2P	6.4 (16°C)
(II) selenide	$CoSe$	7.65
silicate, ortho-	$CoSiO_4$	4.63
sulfate	$CoSO_4$	3.71 (25°C)
(II) sulfide	CoS	5.45
(III) sulfide	Co_2S_3	4.8
sulfide, di-	CoS_2	4.27
tri-, sulfide, tetra-	Co_3S_4	4.86
Copper		
(I) arsenide	Cu_3As	8.0
boride	Cu_3B_2	8.116
(I) bromide (α)	$CuBr$	4.72 (25°C)
(II) bromide	$CuBr_2$	4.77 (25°C)
carbide	Cu_2CO_3	4.40
(I) chloride (α)	$CuCl$	3.7
(II) chloride	$CuCl_2 \cdot 2H_2O$	2.38
(II) chloride	$CuCl_2$	3.05
(I) cyanide	$CuCN$	2.92
(II) fluoride	CuF_2	4.23
hydride	CuH	5.30
(II) hydroxide	$Cu(OH)_2$	3.37
(II) iodate	$Cu(IO_3)_2$	5.24 (15°C)
(I) iodide	CuI	5.65
(II) nitrate	$Cu(NO_3)_2 \cdot 3H_2O$	2.32
(I) nitride	Cu_3N	5.84 (25°C)
(I) oxide	Cu_2O	6.0
(II) oxide	CuO	6.45
(I) selenide	Cu_2Se	6.75 (30°C)
(II) selenide	$CuSe$	5.99
silicide	Cu_4Si	7.53
(II) sulfate (blue vitriol)	$CuSO_4 \cdot 5H_2O$	2.28
(II) sulfate	$CuSO_4$	3.60
(I) sulfide (copper glance) (α)	Cu_2S	5.5–5.8
(II) sulfide (covellite)	CuS	4.68
(I) thiocyanate	$CuSCN$	2.84
Curium		
(III) bromide	$CmBr_3$	6.87
(III) chloride	$CmCl_3$	5.81
(III) fluoride	CmF_3	9.70
(IV) fluoride	CmF_4	7.49
(III) iodide	CmI_3	6.37
Dysprosium		
bromide	$DyBr_3$	4.78
chloride	$DyCl_3$	3.67 (0°C)
fluoride	DyF_3	7.46
iodide	DyI_3	3.21
oxide	Dy_2O_3	7.81 (27°C)
Erbium		
boride	ErB_2	4.61
bromide	$ErBr$	4.93
chloride	$ErCl_3$	4.1
fluoride	ErF_3	7.81
iodide	ErI_3	3.28
oxide	Er_2O_3	8.64
sulfate	$Er_2(SO_4)_3$	3.68
sulfide	Er_2S_3	6.21
Europium		
(III) bromide	$EuBr_3$	5.40
(III) chloride	$EuCl_3$	4.89
(II) fluoride	EuF_2	6.50
(III) fluoride	EuF_3	6.79
(II) iodide	EuI_2	5.50
(III) oxide	Eu_2O_3	7.42
(II) sulfate	$EuSO_4$	4.99
Fluorine		
hydrogen fluoride	HF	0.991 (−19.9°C) liq.
oxide	F_2O	1.90 (−224°C)
oxide di	F_2O_2	1.45 (−57°C)
Gadolinium		
bromide	$GdBr_3$	4.57
chloride	$GdCl_3$	4.52 (0°C)
fluoride	GdF_3	7.05
iodide	GdI_3	3.14
nitrate	$Gd(NO_3)_3 \cdot 6H_2O$	2.33
oxide	Gd_2O_3	7.41 (15°C)
sulfate	$Gd(SO_4)_3$	4.14
sulfide	Gd_2S_3	6.15
Gallium		
(III) arsenide	$GaAs$	5.35
(III) bromide	$GaBr_3$	3.69 (25°C)
(III) chloride	$GaCl_3$	2.47 (25°C)
(III) fluoride	GaF_3	4.47
(III) iodide	GaI_3	4.15 (25°C)
(I) oxide	Ga_2O	4.77 (25°C)
(III) oxide	Ga_2O_3	
(α)		5.88
(β)		6.48

Table 5.2 Density of inorganic substances *(continued)*

Name	Formula	Density, 10^3 kg/m^3
(I) sulfide	Ga_2S	4.2
(III) sulfide	Ga_2S_3	3.65 (25°C)
Germanium		
(IV) bromide	$GeBr_4$	3.13 (29°C)
(IV) chloride	$GeCl_4$	1.87 (25°C)
digermane	Ge_2H_6	1.98 (−100°C)
(IV) fluoride	GeF_4	$6.65 \cdot 10^{-3}$
germane	GeH_4	$3.42 \cdot 10^{-3}$
(II) iodide	GeI_2	5.37
(IV) iodide	GeI_4	4.32 (25°C)
(IV) nitride	Ge_3N_4	5.25 (25°C)
(II) oxide	GeO	1.83
(IV) oxide	GeO_2	
(α)		6.239
(β)		4.703
(II) sulfide	GeS	4.01
(IV) sulfide	GeS_2	2.94 (14°C)
trigermane	Ge_3H_8	2.2
Gold		
(I) bromide	$AuBr$	7.90
(I) chloride	$AuCl$	7.4
(III) chloride	$AuCl_3$	3.9
(I) iodide	AuI	8.25
(III) sulfide	Au_2S_3	8.75
Hafnium		
carbide	HfC	12.7
nitride	HfN	13.9
oxide	HfO_2	9.68
sulfide	HfS_2	6.0
Holmium		
(III) bromide	$HoBr_3$	4.86
fluoride	HoF_3	7.83
(III) iodide	HoI_3	3.24
oxide	Ho_2O_3	8.35
Hydrogen		
bromide	HBr	2.16 (−68°C)
chloride	HCl	1.19 (−85°C)
fluoride	HF	0.991 (19.4°C)
iodide	HI	2.80 (−35°C)
peroxide	H_2O_2	1.442
selenide	H_2Se	2.12 (−42°C)
sulfide	H_2S	0.96 (−60°C)
telluride	H_2Te	2.57 (−20°C)
Indium		
(I) bromide	$InBr$	4.98
(III) bromide	$InBr_3$	4.75
(I) chloride	$InCl$	4.19
(III) chloride	$InCl_3$	3.45
(III) fluoride	InF_3	4.39 (25°C)

Name	Formula	Density, 10^3 kg/m^3
(I) iodide	InI	5.31
(III) iodide	InI_3	4.69
(I) oxide	InO_2	6.99 (25°C)
(III) oxide	In_2O_3	7.18
(III) sulfate	$In_2(SO_4)_3$	3.44
(I) sulfide	In_2S	5.87 (25°C)
(III) sulfide	In_2S_3	4.90
Iodine		
(I) bromide	IBr	4.42 (0°C)
chloride	ICl	
(α)		3.18 (0°C)
(β)		3.24 (34°C) liq.
chloride tri-	ICl_3	3.19
fluoride, hepta-	IF_7	2.8 (6°C) liq.
hydrogen iodide	HI	$5.789 \cdot 10^{-3}$
iodic acid	HIO_3	4.63 (0°C)
(IV) oxide	IO_2 or I_2O_4	4.2
(V) oxide	I_2O_5	4.80
fluoride, penta-	IF_5	3.5 liq.
Iridium		
(III) chloride	$IrCl_3$	5.30
(VI) fluoride	IrF_6	6.0
(IV) oxide	IrO_2	3.15
(IV) sulfide	IrS_2	8.43 (25°C)
Iron		
arsenide	$FeAs$	7.83
boride	FeB	7.15
(II) bromide	$FeBr_2$	4.64
tri-, carbide	Fe_3C	7.4–7.67
(II) carbonate	$FeCO_3$	3.8–3.9
carbonyl, penta-	$Fe(CO)_5$	1.457
carbonyl, tetra-	$Fe(CO)_4$	2.00 (18°C)
(II) chloride	$FeCl_2$	2.98
(III) chloride	$FeCl_3$	2.90 (25°C)
(II) fluoride	FeF_2	4.09 (25°C)
(III) fluoride	FeF_3	3.87
(II) hydroxide	$Fe(OH)_2$	3.4
(III) hydroxide	$Fe(OH)_3$	3.9
(II) iodide	FeI_2	5.31
(III) nitrate	$Fe(NO_3)_3 \cdot 9H_2O$	1.68
di-, nitride	Fe_2N	6.35
(II) oxide	FeO	5.7
(II, III) oxide (magnetite)	Fe_3O_4	5.18
(III) oxide (hematite)	Fe_2O_3	5.25
phosphide	Fe_2P	6.56
silicide	$FeSi$	6.1
(II) sulfate (iron vitriol)	$FeSO_4 \cdot 7H_2O$	1.90 (18°C)
(III) sulfate	$Fe_2(SO_4)_3$	3.10
(II) sulfide	FeS	4.76
(III) sulfide	Fe_2S_3	4.3
sulfide, di- (α)	FeS_2	5.03

Table 5.2 Density of inorganic substances *(continued)*

Name	Formula	Density, 10^3 kg/m^3	Name	Formula	Density, 10^3 kg/m^3
Lanthanum			iodide	LiI	4.06
boride	LaB_6	2.61	nitrate	$LiNO_3$	2.36
bromide	$LaBr_3$	5.07	oxide	Li_2O	2.01 (25°C)
carbide	LaC_2	5.02	perchlorate	$LiClO_4$	2.43
chloride	$LaCl_3$	3.84	peroxide	Li_2O_2	2.36
fluoride	LaF_3	5.94	phosphate, ortho-	Li_3PO_4	2.54
hydride	LaH_3	5.83	silicate, meta-	Li_2SiO_3	2.52 (25°C)
iodide	LaI_3	5.63	silicate, ortho-	Li_4SiO_4	2.39 (25°C)
oxide	La_2O_3	6.51 (15°C)	sulfate	Li_2SO_4	2.22
sulfate	$La_2(SO_4)_3$	3.60 (15°C)	sulfide	Li_2S	1.66
sulfide	La_2S_3	4.91 (11°C)			
			Lutetium		
Lead			bromide	$LuBr_3$	5.17
arsenate	$Pb(AsO_4)_2$	7.80	chloride	$LuCl_3$	3.98
(II) bromide	$PbBr_2$	6.67	fluoride	LuF_3	8.33
carbonate	$PbCO_3$	6.56	iodide	LuI_3	3.39
chlorate	$Pb(ClO_3)_2$	3.89	oxide	Lu_2O_3	9.42
(II) chloride	$PbCl_2$	5.85			
(IV) chloride	$PbCl_4$	3.18 (0°C)	**Magnesium**		
fluoride	PbF_2		arsenide	Mg_3As_2	3.15 (25°C)
(α)		8.37	borate, ortho	$Mg_3(BO_3)_2$	2.99
(β)		7.68	bromide	$MgBr_2$	3.72 (25°C)
iodate	$Pb(IO_3)_2$	6.16	carbonate (magnesite)	$MgCO_3$	2.96
iodide	PbI_2	6.16	chloride	$MgCl_2$	2.32
nitrate	$Pb(NO_3)_2$	4.53	fluoride	MgF_2	3.13
nitride	$Pb(N_3)_2$		hydroxide (brucite)	$Mg(OH)_2$	2.35–2.46
(α)		4.71	iodate	$Mg(IO_3)_2 \cdot 4H_2O$	3.3
(β)		4.93	iodide	MgI_2	4.43 (25°C)
(I) oxide	Pb_2O	8.342	nitrate	$Mg(NO_3)_2 \cdot 6H_2O$	1.464
(II) oxide	PbO		nitride	Mg_3N_2	2.71 (25°C)
(α)		9.51	oxide (periclase)	MgO	3.58 (25°C)
(β)		8.70	phosphate, di-	$Mg_2P_2O_7$	2.559
(II , IV) oxide (minium)	Pb_3O_4	8.79	phosphide	Mg_3P_2	2.05
(IV) oxide	PbO_2		silicate, meta-	$MgSiO_3$	3.18
(α)		9.67	silicate, ortho- (forsterite)	$MgSiO_4$	3.21
(β)		9.33	silicide	Mg_2Si	1.94
phosphate, ortho-	$Pb_3(PO_4)_2$	6.9–7.3	sulfate	$MgSO_4$	2.66
selenate	$PbSeO_4$	6.37	sulfide	MgS	2.84
selenide	$PbSe$	8.10 (15°C)	telluride	$MgTe$	3.86
silicate, meta-	$PbSiO_3$	6.49			
sulfate	$PbSO_4$	6.35	**Manganese**		
sulfide	PbS	7.59	arsenide	$MnAs$	6.18
telluride	$PbTe$	8.16	boride	MnB	6.2
			boride,di	MnB_2	6.9
Lithium			bromide	$MnBr_2$	4.38
arsenate, ortho-	Li_3AsO_4	3.07 (15°C)	tri-, carbide	Mn_3C	6.89 (17°C)
borate, meta-	$LiBO_2$	1.40 (42°C)	carbonate	$MnCO_3$	3.12
bromide	$LiBr$	3.46	(II) chloride	$MnCl_2$	2.98 (25°C)
carbonate	Li_2CO_3	2.11 (0°C)	(II) fluoride	MnF_2	3.98
chlorate	$LiClO_3$	1.12	hydroxide	$Mn(OH)_2$	3.26
chloride	$LiCl$	2.07 (25°C)	iodide	MnI_2	5.0
fluoride	LiF	2.63	(II) oxide	MnO	5.44
hydride	LiH	0.78 (25°C)	(II,III) oxide (α)	Mn_3O_4	4.72
hydroxide	$LiOH$	1.46 (25°C)	(III) oxide (braunite)	Mn_2O_3	4.50

Table 5.2 Density of inorganic substances *(continued)*

Name	Formula	Density, 10^3 kg/m^3	Name	Formula	Density, 10^3 kg/m^3
(IV) oxide (β)	MnO_2	5.03	oxide fluoride, tetra-	MoF_4O	3.00 (25°C)
(VII) oxide	Mn_2O_7	2.40	silicide, di-	$MoSi_2$	6.31
phosphate, di-	$Mn_2P_2O_7$	3.707 (25°C)	(IV) sulfide	MoS_2	4.80
phosphate, ortho-	$Mn_3(PO_4)_2 \cdot 3H_2O$	3.102	di-, sulfide, tri-	Mo_2S_3	5.91 (15°C)
phosphide	MnP	5.39	Molybdic acid	H_2MoO_4	3.11
selenide	MnSe	5.55 (15°C)			
silicate, meta-	$MnSiO_3$	3.72 (25°C)	**Neodymium**		
silicide	MnSi	5.90 (15°C)	bromide	$NdBr_3$	5.35
silicide, di-	$MnSi_2$	5.24 (13°C)	carbide, di-	NdC_2	5.15
di-, silicide	Mn_2Si	6.20 (15°C)	chloride	$NdCl_3$	4.17
(II) sulfate	$MnSO_4$	3.25	iodide	NdI_3	2.34
sulfide (α)	MnS	3.99	oxide	Nd_2O_3	7.24
			sulfate	$Nd_2(SO_4)_3 \cdot 8H_2O$	2.85
Mercury			sulfide	Nd_2S_3	5.18 (11°C)
(I) bromide	Hg_2Br_2	7.31			
(II) bromide	$HgBr_2$	6.11 (25°C)	**Neptunium**		
(I) chlorate	$Hg_2(ClO_3)_2$	6.41	azide	NpN	14.19
(I) chloride	Hg_2Cl_2	7.15	(III) bromide	$NpBr_3$	6.62
(II) chloride	$HgCl_2$	5.53	(III) chloride	$NpCl_3$	5.58
(II) cyanide	$Hg(CN)_2$	4.00	(IV) chloride	$NpCl_4$	4.95
(I) fluoride	Hg_2F_2	8.73 (15°C)	(III) fluoride	NpF_3	9.12
(II) fluoride	HgF_2	8.95 (15°C)	(IV) fluoride	NpF_4	6.8
(I) iodide	Hg_2I_2	7.70	(VI) fluoride	NpF_6	5.0
(II) iodide	HgI_2		(III) iodide	NpI_3	6.82
(α)		6.36	(II) oxide	NpO	13.35
(β)		6.09	(IV) oxide	NpO_2	11.1
(I) nitrate	$Hg_2(NO_3)_2 \cdot 2H_2O$	4.78	sulfide	Np_2S_3	8.9
(II) nitrate	$Hg(NO_3)_2 \cdot \frac{1}{2}H_2O$	4.39			
(I) nitrite	$Hg_2(NO_2)_2$	7.33	**Nickel**		
(I) oxide	Hg_2O	9.8	antimonide	NiSb	7.54
(II) oxide	HgO	11.1	arsenide	NiAs	7.57 (0°C)
(II) selenide	HgSe	8.27	boride	NiB	7.39
(I) sulfate	Hg_2SO_4	7.56	bromide	$NiBr_2$	4.64
(II) sulfate	$HgSO_4$	6.47	carbonyl, tetra-	$Ni(CO)_4$	1.32
(II) sulfide (cinnabar)	HgS		chloride	$NiCl_2$	3.55
(α)		8.10	fluoride	NiF_2	4.63
(β)		7.73	hydroxide	$Ni(OH)_2$	4.15
			iodate	$Ni(IO_3)_2$	5.07
Molybdenum			iodide	NiI_2	5.83
di-, boride	Mo_2B	9.26	nitrate	$Ni(NO_3)_2 \cdot 6H_2O$	2.05
boride	MoB	8.65	oxide	NiO	7.45
boride, di-	MoB_2	7.12	phosphide	Ni_2P	6.31 (15°C)
bromide	$MoBr_2$	4.88 (17.5°C)	selenide	NiSe	8.46
di-, carbide (α)	Mo_2C	8.9	silicide	Ni_2Si	7.2 (17°C)
(IV) carbide	MoC	8.40	sulfate	$NiSO_4$	3.68
carbonyl, hexa-	$Mo(CO)_6$	1.96	sulfide	NiS	5.3–5.65
(II) chloride	$MoCl_2$	3.71 (25°C)			
(III) chloride	$MoCl_3$	3.58 (25°C)	**Niobium**		
(V) chloride	$MoCl_5$	2.93 (25°C)	boride, di-	NbB_2	6.97
(VI) fluoride	MoF_6	2.55 (17.6°C) liq.	(V) bromide	$NbBr_5$	4.44
(II) iodide	MoI_2	5.28 (25°C)	carbide	NbC	7.74–8.2
(IV) oxide	MoO_2	6.47	(V) chloride	$NbCl_5$	2.75
(VI) oxide	MoO_3	4.69	(V) fluoride	NbF_5	3.29
oxide, di-, fluoride, di-	MoF_2O_2	3.49 (25°C)	(V) iodide	NbI_5	5.11

Table 5.2 Density of inorganic substances *(continued)*

Name	Formula	Density, 10^3 kg/m^3
nitride	NbN	8.40
(II) oxide	NbO	7.26
(IV) oxide	NbO_2	5.9
(V) oxide	Nb_2O_5	4.95
Nitrogen		
chloride	NCl_3	1.653
fluoride	NF_3	1.54 (−129°C) liq.
hydrazine	N_2H_4	1.01
hydrazoic acid	HN_3	1.09
(I) oxide	N_2O	$1.978 \cdot 10^{-3}$
(II) oxide	NO	$1.340 \cdot 10^{-3}$
(III) oxide	N_2O_3	1.45 (2°C)
(IV) oxide	NO_2	1.49 (0°C) liq.
(V) oxide	N_2O_5	1.64
nitric acid	HNO_3	1.513
nitrile chloride	NO_2Cl	$2.57 \cdot 10^{-3}$
nitrile fluoride	NO_2F	$2.90 \cdot 10^{-3}$
di-, sulfide, tetra-	N_2S_4	1.90
tetra-, sulfide, tetra-	N_4S_4	2.22 (15°C)
Osmium		
(VIII) fluoride	OsF_8	3.87
(IV) oxide	OsO_2	7.91
(VIII) oxide	OsO_4	4.91 (22°C)
(IV) sulfide	OsS_2	9.47
Oxygen		
oxygen fluoride	OF_2	1.90 (-223.8°C) liq.
ozone	O_3	$2.144 \cdot 10^{-3}$
Palladium		
bromide	$PdBr_2$	5.17 (16°C)
(II) chloride	$PdCl_2$	4.08
(II) fluoride	PdF_2	5.80
(III) fluoride	PdF_3	5.06
hydride	PdH_2	10.76
iodide	PdI_2	6.00
oxide	PdO	8.31
sulfide	PdS	6.60 (25°C)
Phosphorus		
(III) bromide	PBr_3	2.85 (15°C)
(III) chloride	PCl_3	1.57
(V) chloride	PCl_5	2.11
(III) fluoride	PF_3	$3.907 \cdot 10^{-3}$
(V) fluoride	PF_5	$5.805 \cdot 10^{-3}$
fluoride, tri-, chloride, di-	PCl_2F_3	$5.4 \cdot 10^{-3}$
hypophosphorous acid	$H(H_2PO_2)$	1.49
(III) iodide	PI_3	3.89
nitride chloride, di-	$(PCl_2N)_3$	1.98
(III) oxide	P_2O_3	2.135
(IV) oxide	PO_2	2.54
(V) oxide	P_2O_5	2.30
phosphine	PH_3	$1.529 \cdot 10^{-3}$
phosphine, di-	P_2H_4	1.012
phosphonium iodide	PH_4I	2.86
phosphoric acid, meta-	HPO_3	2.2–2.5
phosphoric acid, ortho-	H_3PO_4	1.83 (18°C)
phosphorous acid	$H_2(HPO_3)$	1.65
phosphoryl bromide	$POBr_3$	2.82
phosphoryl chloride	$POCl_3$	1.675 liq.
phosphoryl fluoride	POF_3	$4.8 \cdot 10^{-3}$
tetra-, selenide, tri-	P_4Se_3	1.31
tetra, sulfide, hepta-	P_4S_7	2.19 (17°C)
tetra-, sulfide, tri-	P_4S_3	2.03 (17°C)
tetra-, sulfide, deca-	P_4S_{10}	2.03
thiophosphoryl bromide	$PSBr_3$	2.85 (17°C)
thiophosphoryl chloride	$PSCl_3$	1.64
Platinum		
arsenide, di-	$PtAs_2$	11.8
(II) bromide	$PtBr_2$	6.65 (25°C)
(IV) bromide	$PtBr_4$	5.69 (25°C)
(II) chloride	$PtCl_2$	5.87
(III) chloride	$PtCl_3$	5.26 (25°C)
(IV) chloride	$PtCl_4$	2.43
(II) iodide	PtI_2	6.40 (25°C)
(III) iodide	PtI_3	7.41 (25°C)
(IV) iodide	PtI_4	6.06 (25°C)
(II) oxide	PtO	14.9
(IV) oxide	PtO_2	10.2
phosphide, di-	PtP_2	9.01 (25°C)
sulfide	PtS	10.04 (25°C)
(III) sulfide	Pt_2S_3	5.52
sulfide, di-	PtS_2	7.66 (25°C)
Plutonium		
arsenide	PuAs	10.39
(III) bromide	$PuBr_3$	6.69
(IV) carbide	PuC	13.5
(III) chloride	$PuCl_3$	5.70
(III) fluoride	PuF_3	9.32
(IV) fluoride	PuF_4	7.00
(VI) fluoride	PuF_6	4.86
(II) hydride	PuH_2	10.40
(III) hydride	PuH_3	9.61
(III) iodide	PuI_3	6.92
nitride	PuN	14.25
(II) oxide	PuO	13.89
(IV) oxide	PuO_2	11.44
sulfide	PuS	10.6
Potassium		
azide	KN_3	2.04
bromate	$KBrO_3$	3.24
bromide	KBr	2.76
carbonate (potash)	K_2CO_3	2.43

Table 5.2 Density of inorganic substances *(continued)*

Name	Formula	Density, 10^3 kg/m^3	Name	Formula	Density, 10^3 kg/m^3
chlorate	$KClO_3$	2.32	(VII) sulfide	Re_2S_7	4.87
chloride	KCl	1.99			
cyanide	KCN	1.56	**Rhodium**		
fluoride	KF	2.48	fluoride	RhF_3	5.38
fluoride, hydro-	KHF_2	2.37	(III) oxide	Rh_2O_3	8.20
hydride	KH	1.47	sulfide	Rh_2S_3	6.40 (25°C)
hydroxide	KOH	2.04			
iodate	KIO_3	3.89	**Rubidium**		
iodide	KI	3.12	bromate	$RbBrO_3$	3.68
nitrate (Indian saltpeter)	KNO_3	2.11 (16°C)	bromide	$RbBr$	3.36
oxide	K_2O	2.32 (0°C)	carbonate	Rb_2CO_3	3.47
perchlorate	$KClO_4$	2.52	chloride	$RbCl$	2.76
periodate	KIO_4	3.62	fluoride	RbF	3.56
permanganate	$KMnO_4$	2.70	hydride	RbH	2.60
peroxide	K_2O_2	2.18	hydroxide	$RbOH$	3.203 (11°C)
phosphate, ortho-	K_3PO_4	2.26	iodide	RbI	3.55
selenide	K_2Se	2.85 (15°C)	iodide, tri-	RbI_3	4.03
sulfate	K_2SO_4	2.66	nitrate	$RbNO_3$	3.11
sulfate, hydro-	$KHSO_4$	2.32	nitride	RbN_3	2.79
sulfide	K_2S	1.81 (14°C)	oxide	Rb_2O	3.72
superoxide	KO_2	2.14	perchlorate	$RbClO_4$	2.80
telluride	K_2Te	2.51	peroxide	Rb_2O_2	3.65 (0°C)
			selenate	Rb_2SeO_4	3.90
Praseodymium			sulfate	Rb_2SO_4	3.61
bromide	$PrBr_3$	5.26	sulfide	Rb_2S	2.91
carbide, di-	PrC_2	5.10	di, sulfide, penta-	Rb_2S_5	2.62 (15°C)
chloride	$PrCl_3$	4.02 (25°C)	superoxide	RbO_2	3.80
fluoride	PrF_3	6.14			
iodide	PrI_3	2.31	**Ruthenium**		
(II, IV) oxide	Pr_2O_3	7.07	(III) chloride	$RuCl_3$	3.11
(IV) oxide	PrO_2	6.82	(V) fluoride	RuF_5	2.96 (1.65°C)
sulfate	$Pr_2(SO_4)_3$	3.72 (16°C)	(IV) oxide	RuO_2	6.97
sulfide	Pr_2S_3	5.24	(VIII) oxide	RuO_4	3.29
			(IV) sulfide	RuS_2	6.99
Protactinium					
(IV) chloride	$PaCl_4$	4.72	**Samarium**		
(IV) fluoride	PaF_4	6.36	(II) bromide	$SmBr_2$	5.1
(II) oxide	PaO	13.43	(III) bromide	$SmBr_3$	5.40
(V) oxide	Pa_2O_5	9.0	carbide	SmC_2	5.86
			(II) chloride	$SmCl_2$	4.56 (25°C)
Radium			(III) chloride	$SmCl_3$	4.46
bromide	$RaBr_2$	5.78	(III) fluoride	SmF_3	6.64
chloride	$RaCl_2$	4.91	(III) iodide	SmI_3	3.14
			(II) nitrate	$Sm(NO_3)_2 \cdot 6H_2O$	2.38
Rhenium			(III) oxide	Sm_2O_3	7.43 (15°C)
(V) chloride	$ReCl_5$	4.9	(III) sulfate	$Sm_2(SO_4)_3 \cdot 8H_2O$	2.93
(IV) fluoride	ReF_4	5.38 (26°C)	sulfide	Sm_2S_3	5.73
(VI) fluoride	ReF_6	3.62 liq.			
oxide fluoride, tetra-	ReF_4O	3.717 liq.	**Scandium**		
(IV) oxide	ReO_2	11.4 (25°C)	boride	ScB_2	3.65
(VI) oxide	ReO_3	6.9–7.3	bromide	$ScBr_3$	3.91
(VII) oxide	Re_2O_7	8.2	chloride	$ScCl_3$	2.39 (25°C)
oxide, tri-, chloride	$ReClO_3$	3.87	oxide	Sc_2O_3	3.86
(IV) sulfide	ReS_2	7.51	sulfate	$Sc_2(SO_4)_3$	2.58

Table 5.2 Density of inorganic substances *(continued)*

Name	Formula	Density, 10^3 kg/m^3	Name	Formula	Density, 10^3 kg/m^3
Selenium			selenide	Ag_2Se	8.0
(I) bromide	Se_2Br_2	3.60 (15°C)	sulfate (α)	Ag_2SO_4	5.45 (29.2°C)
carbide	SeC_2	2.68	sulfide	Ag_2S	7.32
(I) chloride	Se_2Cl_2	2.91 (17.5°C)	telluride	Ag_2Te	8.5
(IV) chloride	$SeCl_4$	3.80			
(IV) fluoride	SeF_4	2.75 (25°C) liq.	**Sodium**		
(VI) fluoride	SeF_6	$2.26 \cdot 10^{-3}$ (−35°C)	acetate	$NaCH_3COO$	1.53
hydride	H_2Se	2.00 (−42°C) liq.	arsenate, meta-	$NaAsO_3$	2.30
hydrogen selenide	H_2Se	$3.670 \cdot 10^{-3}$	azide	NaN_3	1.85
(IV) oxide	SeO_2	3.95 (16°C)	borate, meta-	$NaBO_2$	2.46
(VI) oxide	SeO_3	3.6	borate, tetra-	$Na_2B_4O_7$	2.37
oxide chloride, di-	$SeCl_2O$	2.42	bromate	$NaBrO_3$	3.34 (17.5°C)
oxide fluoride, di-	SeF_2O	2.67	bromide	$NaBr$	3.21
selenic acid	H_2SeO_4	2.95 (15°C)	carbide	Na_2C_2	1.58 (15°C)
selenious acid	H_2SeO_3	3.00 (15°C)	carbonate	Na_2CO_3	2.53
sulfide	SeS	3.06 (0°C)	carbonate, hydro-	$NaHCO_3$	2.16
			chlorate	$NaClO_3$	2.49 (15°C)
Silicon			chloride	$NaCl$	2.165 (25°C)
(III) boride	SiB_3	2.52	cyanate	$NaOCN$	1.94
(IV) bromide	$SiBr_4$	2.77	cyanide	$NaCN$	1.60
carbide	SiC	3.22	fluoride	NaF	2.79
chloride	$SiCl_4$	1.48	hydride	NaH	1.38
fluoride	SiF_4	$4.684 \cdot 10^{-3}$	hydroborate, tetra-	$Na(BH_4)$	1.07
iodide	SiI_4	4.2	hydroxide (α)	$NaOH$	2.13
nitride	Si_3N_4	3.44	iodate	$NaIO_3$	4.40
(II) oxide	SiO	2.13	iodide	NaI	3.665 (4°C)
(IV) oxide (crystobalite)	SiO_2	2.32	nitrate	$NaNO_3$	2.26
(IV) oxide (quartz)	SiO_2	2.65	nitride	$NaNO_2$	2.17 (0°C)
(IV) oxide (tridymite)	SiO_2	2.26	oxide	Na_2O	2.27
silicic acid	H_2SiO_3	3.17	perchlorate	$NaClO_4$	2.50
sulfide	SiS_2	2.02	periodate	$NaIO_4$	4.17
			peroxide	Na_2O_2	2.60
Silver			phosphate, meta-	$NaPO_3$	2.48
arsenate, ortho-	Ag_3AsO_4	6.66 (25°C)	phosphate, ortho-	Na_3PO_4	2.54
bromate	$AgBrO_3$	5.21	selenate	Na_2SeO_4	3.21 (17°C)
bromide	$AgBr$	6.47 (25°C)	selenide	Na_2Se	2.63 (10°C)
carbonate	Ag_2CO_3	6.08	silicate, meta-	Na_2SiO_3	2.4
chlorate	$AgClO_3$	4.43	sulfate (Glauber's salt)	$Na_2SO_4 \cdot 10H_2O$	1.46
chloride	$AgCl$	5.56	sulfate	Na_2SO_4	2.70
cyanide	$AgCN$	3.95	sulfate, di- (pyrosulfate)	$Na_2S_2O_7$	2.66 (25°C)
(I) fluoride	AgF	5.85 (15°C)	sulfate, hydro-	$NaHSO_4$	2.74
(II) fluoride	AgF_2	4.57	sulfide	Na_2S	1.86 (14°C)
iodate	$AgIO_3$	5.53 (16.5°C)	sulfite	Na_2SO_3	2.63 (15°C)
iodide	AgI		telluride	Na_2Te	2.90
(α)		5.71			
(β)		5.61–5.67	**Strontium**		
nitrate	$AgNO_3$	4.35	boride	SrB_6	3.39 (15°C)
nitrite	$AgNO_2$	4.45	bromide	$SrBr_2$	4.22 (24°C)
(I) oxide	Ag_2O	7.22	carbide	SrC_2	3.2
(II) oxide	AgO	7.44	carbonate	$SrCO_3$	3.70
perchlorate	$AgClO_4$	2.81	chlorate	$Sr(ClO_3)_2$	3.15
periodate	$AgIO_4$	5.57	chloride	$SrCl_2$	3.05
phosphate, ortho-	Ag_3PO_4	6.37 (25°C)	fluoride chloride	$SrClF$	4.18
selenate	Ag_2SeO_4	5.72	fluoride	SrF_2	4.24

Table 5.2 Density of inorganic substances *(continued)*

Name	Formula	Density, 10^3 kg/m^3
hydride	SrH_2	3.72
hydroxide	$Sr(OH)_2$	3.63
iodate	$Sr(IO_3)_2$	5.05 (15°C)
iodide	SrI_2	4.55 (25°C)
nitrate	$Sr(NO_3)_2$	2.99
nitrite	$Sr(NO_2)_2 \cdot H_2O$	2.41 (0°C)
oxide	SrO	4.7
peroxide	SrO_2	4.56
selenate	$SrSeO_4$	4.23
selenide	$SrSe$	4.38
sulfate	$SrSO_4$	3.96
sulfide	SrS	3.70 (15°C)
telluride	$SrTe$	4.83
Sulphur		
(I) bromide	S_2Br_2	2.63
(I) chloride	S_2Cl_2	1.67 (25°C)
(II) chloride	SCl_2	1.62 (15°C)
disulfuric acid	$H_2S_2O_7$	1.9
(I) fluoride	S_2F_2	1.5 (−100°C) liq.
(IV) fluoride	SF_4	1.92 (−73°C) liq.
(VI) fluoride	SF_6	$6.50 \cdot 10^{-3}$
di-, fluoride, deca-	S_2F_{10}	2.08 (0°C)
hydride	H_2S	0.96 (−60°C)
hydrogen sulfide	H_2S	$1.538 \cdot 10^{-3}$ (25°C)
(IV) oxide	SO_2	$2.927 \cdot 10^{-3}$
(VI) oxide	SO_3	1.97 liq.
di-, oxide chloride, tetra-	S_2OCl_4	1.66 (0°C)
sulfuric acid	H_2SO_4	1.8305
sulfuryl chloride	SO_2Cl_2	1.67
sulfuryl fluoride chloride	SO_2ClF	$1.623 \cdot 10^{-3}$ (0°C)
sulfuryl fluoride	SO_2F_2	$3.72 \cdot 10^{-3}$
thionyl chloride	$SOCl_2$	1.655 (10.4°C)
thionyl fluoride	SOF_2	$2.93 \cdot 10^{-3}$
Tantalum		
boride	TaB_2	12.38
(V) bromide	$TaBr_5$	4.67
carbide	TaC	14.4
chloride	$TaCl_5$	3.76
(V) fluoride	TaF_5	4.74
(V) iodide	TaI_5	5.80
nitride	TaN	14.36
(V) oxide (α)	Ta_2O_5	8.53
Tellurium		
(II) bromide	$TeBr_2$	5.24
(IV) bromide	$TeBr_4$	4.31 (15°C)
(II) chloride	$TeCl_2$	7.05
(IV) chloride	$TeCl_4$	3.26
(VI) fluoride	TeF_6	2.56 (−26°C) liq.
hydride	TeH_2	2.68 (−12°C)
hydrogen telluride	H_2Te	$5.81 \cdot 10^{-3}$
(IV) iodide	TeI_4	5.40
(IV) oxide	TeO_2	
(α)		6.02
(β)		5.87
(VI) oxide	TeO_3	5.08
telluric acid, ortho-	H_6TeO_6	
(α)		3.07
(β)		3.17
tellurous acid	H_2TeO_3	3.05
Terbium		
bromide	$TbBr_3$	4.67
chloride	$TbCl_3$	4.35
fluoride	TbF_3	7.24
iodide	TbI_3	3.16
oxide	Tb_2O_3	7.81
Thallium		
(I) bromide	$TlBr$	7.56 (17°C)
carbonate	Tl_2CO_3	7.2
(I) chloride	$TlCl$	7.00
(I) fluoride (α)	TlF	8.36
(III) fluoride	TlF_3	8.36 (25°C)
(I) iodide	TlI	
(α)		7.07
(β)		7.30
(III) iodide	TlI_3	7.557
nitrate	$TlNO_3$	5.56 (21.4°C)
(I) oxide	Tl_2O	9.52 (16°C)
(III) oxide	Tl_2O_3	9.65
perchlorate	$TlClO_4$	4.89
phosphate, ortho-	Tl_3PO_4	6.89 (10°C)
selenide	Tl_2Se	9.05 (25°C)
sulfate (α)	Tl_2SO_4	6.67
(I) sulfide	Tl_2S	8.46
thiocyanate	$TlSCN$	4.96
Thorium		
(IV) boride	ThB_4	8.45
(VI) boride	ThB_6	6.4 (15°C)
bromide	$ThBr_4$	5.67
carbide	ThC_2	8.96
chloride	$ThCl_4$	4.59 (15°C)
fluoride	ThF_4	6.19
hydride	ThH_2	5.92
iodide	ThI_4	6.00
oxide	ThO_2	9.69
phosphate, meta-	$Th(PO_3)_4$	4.08 (16.4°C)
phosphide	Th_3P_4	8.56
silicate, ortho-	$ThSiO_4$	6.82 (16°C)
silicide	$ThSi_2$	7.96 (16°C)
sulfate	$Th(SO_4)_2$	4.22
sulfide	ThS_2	7.36

Table 5.2 Density of inorganic substances *(continued)*

Name	Formula	Density, 10^3 kg/m^3
Thulium		
bromide	$TmBr_3$	5.02
fluoride	TmF_3	7.97
iodide	TmI_3	3.32
oxide	Tm_2O_3	8.77
Tin		
(II) bromide	$SnBr_2$	5.18 (17°C)
(IV) bromide	$SnBr_4$	3.34 (35°C) liq.
(II) chloride	$SnCl_2$	3.95 (25°C)
(IV) chloride	$SnCl_4$	2.23 liq.
(IV) fluoride	SnF_4	4.78
(II) iodide	SnI_2	5.28 (25°C)
(IV) iodide	SnI_4	4.47 (0°C)
(II) oxide	SnO	6.45 (0°C)
(IV) oxide (cassiterite) (α)	SnO_2	6.95
(II) sulfide	SnS	5.08 (0°C)
(IV) sulfide	SnS_2	4.51
(II) telluride	$SnTe$	6.48
Titanium		
(I) boride	TiB	5.09
(II) boride	TiB_2	4.50
(II) bromide	$TiBr_2$	4.31
(IV) bromide (β)	$TiBr_4$	3.24
carbide	TiC	4.92
(II) chloride	$TiCl_2$	3.13
(III) chloride	$TiCl_3$	2.64
(IV) chloride	$TiCl_4$	1.73
(III) fluoride	TiF_3	3.40
(IV) fluoride	TiF_4	2.79
hydride	TiH_2	3.9 (12°C)
(II) iodide	TiI_2	4.99
(IV) iodide	TiI_4	4.40 (25°C)
nitride	TiN	5.43
(II) oxide	TiO	4.93
(III) oxide	Ti_2O_3	4.6
(IV) oxide	TiO_2	3.84
phosphide	TiP	3.95 (25°C)
sulfide	Ti_2S	4.12
sulfide, di-	TiS_2	3.22
di-, sulfide, tri-	Ti_2S_3	3.58
Tungsten		
boride	W_2B	17.7
boride, di-	WB_2	10.77
(VI) bromide	WBr_6	6.9
di-, carbide	W_2C	16.06–17.3
carbide	WC	15.7
carbonyl, hexa	$W(CO)_6$	2.65
(V) chloride	WCl_5	3.875
(VI) chloride	WCl_6	3.52
(VI) fluoride	WF_6	3.44 (15°C) liq.
(II) iodide	WI_2	6.9
(IV) iodide	WI_4	5.2 (18°C)
(IV) oxide	WO_2	12.11
(VI) oxide	WO_3	7.16
silicide, di	WSi_2	9.4
sulfide,di	WS_2	7.5 (10°C)
Tungstic acid	H_2WO_4	5.5
Uranium		
boride	UB_2	12.70
(III) bromide	UB_3	6.53
(IV) bromide	UBr_4	5.35
carbide	UC	13.63
carbide, di	UC_2	11.68
(III) chloride	UCl_3	5.35
(IV) chloride	UCl_4	4.87
(V) chloride	UCl_5	3.81
(VI) chloride	UCl_6	3.56
(III) fluoride	UF_3	8.96
(IV) fluoride	UF_4	6.7–6.9
(V) fluoride	UF_5	5.81
(VI) fluoride	UF_6	5.06
hydride	UH_3	10.95
(III) iodide	UI_3	6.38
(IV) iodide	UI_4	5.6 (15°C)
nitride	UN	14.31
(IV) oxide	UO_2	10.95
(VI) oxide	UO_3	
(α)		8.34
(β)		8.02
tri-, oxide, octa-	U_3O_8	8.30
sulfide	US	10.87
sulfide, di	US_2	7.96 (25°C)
Uranyl acetate	$UO_2(CH_3COO)_2 \cdot 2H_2O$	2.89 (15°C)
Uranyl chloride	UO_2Cl_2	5.28
Uranyl fluoride	UO_2F_2	5.8 calc.
Uranyl nitrate	$UO_2(NO_3)_2 \cdot 2H_2O$	3.35
Uranyl sulfate	$UO_2SO_4 \cdot 3H_20$	3.28 (16.5°C)
Vanadium		
boride	VB_2	5.10
bromide	VBr_3	4.00
carbide	VC	5.77
(II) chloride	VCl_2	3.23
(III) chloride	VCl_3	3.00
(IV) chloride	VCl_4	1.87
(III) fluoride	VF_3	3.36
(IV) fluoride	VF_4	2.975
(V) fluoride	VF_5	2.18
iodide	VI_2	5.44
nitride	VN	6.13
(II) oxide	VO	5.76
(III) oxide	V_2O_3	4.87
(IV) oxide	VO_2	4.34
(V) oxide	V_2O_5	3.36

Table 5.2 Density of inorganic substances *(continued)*

Name	Formula	Density, 10^3 kg/m^3	Name	Formula	Density, 10^3 kg/m^3
oxide chloride, tri-	VCl_3O	1.83	carbonate	$ZnCO_3$	4.42
oxide fluoride, tri-	VF_3O	2.46 (19°C)	chloride	$ZnCl_2$	2.91 (25°C)
(II) sulfide	VS	4.20	cyanide	$Zn(CN)_2$	1.85
(III) sulfide	V_2S_3	4.72	fluoride	ZnF_2	4.90
(V) sulfide	V_2S_5	3.0	hydroxide	$Zn(OH)_2$	3.05
			iodide	ZnI_2	4.73 (25°C)
Water	H_2O	1.0000 (4°C)	nitrate	$Zn(NO_3)_2 \cdot 6H_2O$	2.07 (14°C)
heavy water	D_2O	1.104	nitride	Zn_3N_2	6.22 (25°C)
			oxide	ZnO	5.7
Xenon			phosphate, ortho-	$Zn_3(PO_4)_2$	4.00 (15°C)
(II) fluoride	XeF_2	4.32	phosphide	Zn_3P_2	4.55
(IV) fluoride	XeF_4	4.04	selenide	ZnSe	5.42 (15°C)
(VI) fluoride	XeF_6	3.6	silicate, meta-	$ZnSiO_3$	3.42
(VI) oxide	XeO_3	4.6	silicate, ortho-	Zn_2SiO_4	4.103
			sulfate	$ZnSO_4$	3.74
Ytterbium			sulfide	ZnS	
(II) bromide	$YbBr_2$	5.91 (25°C)	(α)		4.09
(III) bromide	$YbBr_3$	5.10	(β)		3.98–4.08
(II) chloride	$YbCl_2$	5.08	telluride	ZnTe	6.34 (15°C)
(III) fluoride	YbF_3	8.17			
(II) iodide	YbI_2	5.40 (25°C)	**Zirconium**		
(III) iodide	YbI_3	3.33	boride	ZrB_2	6.09
(III) oxide	Yb_2O_3	9.17	carbide	ZrC	6.73
(III) sulfate	$Yb_2(SO_4)_3$	3.79	(II) chloride	$ZrCl_2$	3.16 (18°C)
			(III) chloride	$ZrCl_3$	3.00 (18°C)
Yttrium			(IV) chloride	$ZrCl_4$	2.80
bromide	YBr_3	3.95	(IV) fluoride	ZrF_4	4.43
carbide	YC_2	4.13	hydride	ZrH_2	5.74
chloride	YCl_3	2.8	hydroxide	$Zr(OH)_4$	3.25
fluoride	YF_3	4.01	nitride	ZrN	7.09
nitrate	$Y(NO_3)_3 \cdot 6H_2O$	2.68	oxide (α)	ZrO_2	5.68
oxide (α)	Y_2O_3	4.84	phosphide	ZrP_2	4.77 (25°C)
sulfate	$Y_2(SO_4)_3$	2.61	selenite	$Zr(SeO_3)_2$	4.3
			silicate, ortho-	$ZrSiO_4$	4.56
Zinc			silicide	$ZrSi_2$	4.88 (22°C)
antimonide	Zn_3Sb_2	6.33	sulfate	$Zr(SO_4)_2 \cdot 4H_2O$	3.22 (16°C)
arsenide	Zn_3As_2	5.53	sulfide	ZrS_2	3.87
bromide	$ZnBr_2$	4.20 (25°C)			

5.4 Organic Substances

Table 5.3 Density of organic substances [5,7]

Name	Formula	Density, 10^3 kg/m^3	Name	Formula	Density, 10^3 kg/m^3
Acetaldehyde	C_2H_4O	0.7834	Acetic acid	$C_2H_4O_2$	1.0492
– tribromo-	Br_3C_2HO	2.665 (25°C)	– ,allyl ether	$C_5H_8O_2$	0.9277
– trichloro-	Cl_3C_2HO	1.5121	– ,amide	C_2H_5ON	0.9986 (25°C)

Table 5.3 Density of organic substances *(continued)*

Name	Formula	Density, 10^3 kg/m^3	Name	Formula	Density, 10^3 kg/m^3
– ,amino-(glycine)	$C_2H_5O_2N$	1.1607	– 3-chloro-	C_7H_5ClO	1.2410
– ,anhydride	$C_4H_6O_3$	1.0820	– 4-chloro-	C_7H_5ClO	1.196 (61°C)
– ,bromanhydride	C_2H_3OBr	1.6625 (16°C)	– 2-fluoro-	C_7H_5FO	1.16 (15°C)
– bromo-	$C_2H_3O_2Br$	1.9335 (50°C)	– 3-fluoro-	C_7H_5FO	1.176 (24°C)
– ,butyl ether	$C_6H_{12}O_2$	0.8825	– 4-fluoro-	C_7H_5FO	1.181
– ,chloranhydride	C_2H_3OCl	1.1051	– 2-hydroxy-	$C_7H_6O_2$	1.1674
– chloro-	$C_2H_3O_2Cl$	1.4034 (40°C)	– 4-hydroxy-	$C_7H_6O_2$	1.129 (130°C)
– ,chloromethyl ether	$C_3H_5O_2Cl$	1.194	– 2-nitro-	$C_7H_5NO_3$	1.2844 (50°C)
– dibromo-	$C_2H_2O_2Br_2$	1.0921	– 3-nitro-	$C_7H_5NO_3$	1.2792
– dichloro-	$C_2H_2O_2Cl_2$	1.5634	– ,oxime-	C_7H_7NO	1.1111
– difluoro-	$C_2H_2O_2F_2$	1.5255	Benzene	C_6H_6	1.0880
– ethoxy-	$C_4H_8O_3$	1.1021	– azid-	$C_6H_5N_3$	1.0709 (22°C)
– ,ethyl ether	$C_4H_8O_2$	0.9003	– 1-azid-3-methyl-	$C_7H_7N_3$	1.0527 (22°C)
(ethylacetate)			– 1-azid-4-methyl-	$C_7H_7N_3$	1.0655 (25°C)
– ,fluoranhydride	C_2H_3OF	1.002 (15°C)	– bromo-	C_6H_5Br	1.4950
– fluoro-	$C_2H_3O_2F$	1.3693 (36°C)	– 1-bromo-2-chloro-	C_6H_4BrCl	1.6444
– ,iodanhydride	C_2H_3OI	2.0674	– 1-bromo-2-fluoro-	C_6H_4BrF	1.7038
– ,isobutyl ether	$C_6H_{12}O_2$	0.8710	– 1-bromo-2-iodo-	C_6H_4BrI	2.2571 (25°C)
– ,isopropenyl ether	$C_5H_8O_2$	0.9090	– 1-bromo-2-methoxy-	C_7H_7BrO	1.5018
– ,isopropyl ether	$C_5H_{10}O_2$	0.8718	– 1-bromo-2-methyl-	C_7H_7Br	1.4327
– mercapto-	$C_2H_4O_2S$	1.3253	– 1-bromo-2-nitro-	$C_6H_4BrNO_2$	1.6245 (80°C)
– methoxy-	$C_3H_6O_3$	1.1768	– 1-bromo-3-fluoro-	C_6H_4BrF	1.7081
– ,methyl ether	$C_3H_6O_2$	0.933	– chloro-	C_6H_5C	1.10630
– ,nitrile	C_2H_3N	0.7843	– 1-chloro-2-(chloromethyl)	$C_7H_6Cl_2$	1.2743
– ,pentyl ether	$C_7H_{14}O_2$	0.8745	– 1-chloro-2-(trichloromethyl)	$C_7H_4Cl_4$	1.5187
– ,propyl ether	$C_5H_{10})_2$	0.8867	– 1,2-dibromo-	$C_6H_4Br_2$	1.9843
– trichloro-	$C_2HO_2Cl_3$	1.62	– 1,3-dibromo-	$C_6H_4Br_2$	1.9523
– trifluoro-	$C_2HO_2F_3$	1.5331 (0°C)	– 1,2-dibromo-4-methyl-	$C_7H_6Br_2$	1.812
– ,vinyl ether	$C_4H_6O_2$	0.9342	– 1,2-dibromo-4-nitro-	$C_6H_3Br_2NO_2$	2.354 (8°C)
Aniline	C_6H_7N	1.02173	– 1,2-dichloro-	$C_6H_4Cl_2$	1.3048
– 2-bromo-	C_6H_6BrN	1.578	– 1,3-difluoro-	$C_6H_4F_2$	1.1572
– 3-bromo-	C_6H_6BrN	1.5793	– 1,4-difluoro-	$C_6H_4F_2$	1.1701
– 4-bromo-	C_6H_6BrN	1.4970 (100°C)	– 1,2-diiodo-	$C_6H_4I_2$	2.54
– 2-chloro-	C_6H_6ClN	1.21251	– 1,2-dinitro-	$C_6H_4N_2O_4$	1.3119
– 3-chloro-	C_6H_6ClN	1.217	– 1,4-dinitro-	$C_6H_4N_2O_4$	1.625
– 4-chloro-	C_6H_6ClN	1.175 (70°C)	– 2,4-dinitro-1-chloro-	$C_6H_3ClN_2O_4$	1.4717
– 2-fluoro-	C_6H_6FN	1.1478	– fluoro-	C_6H_5F	1.0225
– 3-fluoro-	C_6H_6FN	1.1580 (17.5°C)	– (fluoromethyl)-	C_7H_7F	1.0228 (25°C)
– 4-fluoro-	C_6H_6FN	1.1725	– 1-fluoro-2-(chloromethyl)-	C_7H_6ClF	1.2162 (24°C)
– 2-methoxy-	C_7H_9NO	1.0923	– 1-fluoro-2-(trichloromethyl)	$C_7H_4Cl_3F$	1.4523 (18.6°C)
– 3-methoxy-	C_7H_9NO	1.096	– 1-fluoro-2-(trifluoromethyl)	$C_7H_4F_4$	1.293 (26°C)
– 4-methoxy-	C_7H_9NO	1.071 (57°C)	– 1-fluoro-2-chloro-	C_6H_4ClF	1.2233 (30°C)
– 2-methyl-	C_7H_9N	0.99843	– 1-fluoro-3-chloro-	C_6H_4ClF	1.221 (25°C)
– 3-methyl-	C_7H_9N	0.98912	– hexadeutero-	C_6D_6	0.9465
– 4-methyl-	C_7H_8N	0.9619	– hexafluoro-	C_6F_6	1.612
– N-methyl-	C_7H_9N	0.98912	– iodo-	C_6H_5I	1.81548
– 2-nitro-	$C_6H_6N_2O_2$	1.442 (15°C)	– 1-iodo-2-(trifluoromethyl)-	$C_7H_4F_3I$	1.896 (26°C)
– 4-nitro-	$C_6H_6N_2O_2$	1.424	– 1-iodo-2-chloro-	C_6H_4ClI	1.9515 (25°C)
– 3-nitro-	$C_6H_6N_2O_2$	1.1747 (160°C)	– 1-iodo-2-methyl-	C_7H_7I	1.7090
Arabinose (α, β)	$C_5H_{10}O_5$	1.585	– 1-iodo-2-nitro-	$C_6H_4INO_2$	1.9786 (75°C)
Azetidine	C_3H_7N	0.8465	– 1-iodo-3-(trifluoromethyl)-	$C_7H_4F_3I$	1.887
Aziridine	C_2H_5N	0.8349 (25°C)	– 1-iodo-3-methyl-	C_7H_7I	1.6981
Benzaldehyde	C_7H_6O	1.0447	– 1-iodo-4-fluoro-	C_6H_4FI	1.9523 (15°C)
– 2-chloro-	C_7H_5ClO	1.2483	– methoxy-	C_7H_8O	0.99402

Table 5.3 Density of organic substances *(continued)*

Name	Formula	Density, 10^3 kg/m^3	Name	Formula	Density, 10^3 kg/m^3
– 1-methoxy-2,3-dinitro-	$C_7H_6N_2O_5$	1.524	– 4-fluoro-	$C_7H_5FO_2$	1.479 (25°C)
– 1-methoxy-2-chloro-	C_7H_7ClO	1.1911	– 2-hydroxy-	$C_7H_6O_3$	1.443
– 1-methoxy-2-fluoro-	C_7H_7FO	1.1293 (17.5°C)	– 4-hydroxy-	$C_7H_6O_3$	1.482 (25°C)
– 1-methoxy-2-nitro-	$C_7H_7NO_3$	1.2540	– ,nitrile	C_7H_5N	1.0052
– methyl-(toluene)	C_7H_8	0.86694	– 2-nitro-	$C_7H_5NO_4$	1.575
– 1-methyl-2,3,4-trinitro	$C_7H_5N_3O_6$	1.62	– 3-nitro-	$C_7H_5NO_4$	1.494
– 1-methyl-2,3-dinitro	$C_7H_6N_2O_4$	1.2625 (111°C)	Benzylamine	C_7H_9N	0.9813
– 1-methyl-2,4-dichloro-	$C_7H_6Cl_2$	1.2459	Bioxirane	$C_4H_6O_2$	1.1157
– 1-methyl-2,4-dinitro-	$C_7H_6N_2O_4$	1.321 (70°C)	1,2-Butadiene	C_4H_6	0.652
– 1-methyl-2-chloro-	C_7H_7Cl	1.0826	1,3-Butadiene	C_4H_6	0.6211
– 1-methyl-2-fluoro-	C_7H_7F	1.003	1,3-Butadiene, 1-bromo-	C_4H_5Br	1.4174
– 1-methyl-2-nitro-	$C_7H_7NO_2$	1.16296	– 2-bromo-	C_4H_5Br	1.397
– 1-methyl-2-nitro-4-chloro-	$C_7H_6ClNO_2$	1.256 (80°C)	– 1-bromo-2-methyl	C_5H_7Br	1.3271
– 1-methyl-3-fluoro-	C_7H_7F	0.9974	– 1,4-dichloro-	$C_4H_4Cl_2$	1.2692
– 2-methyl-1,3-dichloro-	$C_7H_6Cl_2$	1.2686	– hexachloro-	C_4Cl_6	1.682
– 1-nitro-2,3-dichloro-	$C_6H_3Cl_2NO_2$	1.721 (14°C)	– hexafluoro-	C_4F_6	1.553 (−20°C)
– 1-nitro-2-chloro-	$C_6H_4ClNO_2$	1.348 (45°C)	– 2-iodo-	C_4H_5I	1.7278
– 1-nitro-2-fluoro-	$C_6H_4FNO_2$	1.3289	1,3-Butadiene, 1-chloro-	C_4H_5Cl	0.9606
– nitropentachloro-	$C_6Cl_5NO_2$	1.718 (25°C)	– 2-chloro- (chloroprene)	C_4H_5Cl	0.9583
– pentachloro-	C_6HCl_5	1.8342 (16.5°C)	1,2-Butadiene, 3-methyl-	C_5H_8	0.6804
– pentafluoro- (trifluoromethyl)	C_7F_8	1.660 (25°C)	1,2-Butadiene, 4-bromo-	C_4H_5Br	1.4255
			1,2-Butadiene, 4-chloro-	C_4H_5Cl	0.9891
– 1,2,4-trichloro-	$C_6H_3Cl_3$	1.4542	2,3-Butadiene-1-ol	C_4H_6O	0.9164
– trifluoromethyl- (benzotrifluoride)	$C_7H_5F_3$	1.1886	1,3-Butadiene-2-methyl- (isoprene)	C_5H_8	0.6805
– 1-(trifluoromethyl)-2-chloro-	$C_7H_4ClF_3$	1.364 (25°C)	– 1-methoxy-	C_5H_8O	0.8296
– 1,3,5-trinitro-	$C_6H_3N_3O_6$	1.76	– 2-fluoro- (fluoroprene)	C_4H_5F	0.843 (4°C)
– 1,2,3-tribromo-4-methyl-	$C_7H_5Br_3$	2.456	Butadyine (diacetylene)	C_4H_2	0.7364 (0°C)
Benzenesulfonic acid			Butenal	C_4H_8O	0.8040
– ,chloranhydride	$C_6H_5ClO_2S$	1.3766 (25°C)	– 2-bromo-	C_4H_7OBr	1.469
– ,fluoranhydride	$C_6H_5FO_2S$	1.3286	– 4-chloro-	C_4H_7OCl	1.107 (8.5°C)
– ,methyl ether	$C_7H_8O_3S$	1.2734 (17°C)	– 2,3-dichloro-	$C_4H_6OCl_2$	1.266
1,2,3-Benzenetriole	$C_6H_6O_3$	1.453	– 3-ethoxy-	$C_6H_{12}O_2$	0.897
Benzenethiazole	C_7H_5NS	1.2460	– 4-methoxy	$C_5H_{10}O_2$	0.942
– 2-chloro-	C_7H_4ClNS	1.3715	– 2-methyl-	$C_5H_{10}O$	0.803
1,4-Benzoquinone	$C_6H_4O_2$	1.318	– ,oxime	C_4H_9ON	0.923
– 2-methyl-	$C_7H_6O_2$	1.08 (75°C)	– 2-oxo-	$C_4H_6O_2$	1.0285
Benzo (*c*) isoxazole	C_7H_5NO	1.1827	Butane	C_4H_{10}	0.5788
Benzoic acid	$C_7H_6O_2$	1.2659 (15°C)	– 1-bromo-3, 3-dimethyl-	$C_5H_{10}Br$	1.1556
– ,amide	C_7H_7ON	1.0792 (130°C)	– 1-chloro-	C_4H_9Cl	0.8862
– 2-amino-	$C_7H_7NO_2$	1.412	– 1,4-dibromo-	$C_4H_8Br_2$	1.7890
– 3-amino-	$C_7H_7NO_2$	1.5104 (4°C)	– 1,4-dichloro-	$C_4H_8Cl_2$	1.1408
– 4-amino-	$C_7H_7NO_2$	1.374 (25°C)	– 1,4-diiodo-	$C_4H_8I_2$	2.349 (26°C)
– ,azide	$C_7H_6ON_3$	1.1680 (35°C)	– 2,3-dimethyl-	C_6H_{14}	0.6616
– ,bromanhydride	C_7H_5OBr	1.5461	– 2,2-dimethyl-3, 3-dichloro-	$C_6H_{12}Cl_2$	0.8767
– 3-bromo-	$C_7H_5BrO_2$	1.845	– 1-fluoro-	C_4H_9F	0.7789
– 4-bromo-	$C_7H_5BrO_2$	1.894	– 1-fluoro-4-chloro-	C_4H_8FCl	1.0627 (25°C)
– ,chloranhydride	C_7H_5OCl	1.212	– 1-iodo-	C_4H_9I	1.6154
– 2-chloro-	$C_7H_5ClO_2$	1.544	– 2-iodo-	C_4H_9I	1.592
– 3-chloro-	$C_7H_5ClO_2$	1.496 (25°C)	– 1-iodo-2-methyl-	$C_5H_{11}I$	1.5253
– 2,3-dihydroxy-	$C_7H_6O_4$	1.542	– 1-methoxy-	$C_5H_{12}O$	0.7443
– ,fluoranhydride	C_7H_5OF	1.155	– 2-methyl (isopentane)	C_5H_{12}	0.6201
– 2-fluoro-	$C_7H_5FO_2$	1.460 (25°C)	– 2-methyl-2-chloro-	$C_5H_{11}Cl$	0.8653
– 3-fluoro-	$C_7H_5FO_2$	1.474 (25°C)	– 1-nitro-	$C_4H_9O_2N$	0.9710

Table 5.3 Density of organic substances *(continued)*

Name	Formula	Density, 10^3 kg/m^3
Butanedial	$C_4H_6O_2$	1.064
Butanedioic acid (succinic acid)	$C_4H_6O_4$	1.563
– ,dichloranhydride	$C_4H_4O_2Cl_2$	1.3748
– ,diethyl ether	$C_8H_{14}O_4$	1.041
– ,dimethyl ether	$C_6H_{10}O_4$	1.1197
– ,dinitrile	$C_4H_4N_2$	0.9867 (60°C)
– 2-hydroxy-	$C_4H_6O_5$	1.45
1,2-Butanediol	$C_4H_{10}O_2$	1.0024
1-Butanesulfonic acid	$C_4H_{10}O_3S$	1.1906 (25°C)
2-Butanesulfonic acid	$C_4H_{10}O_3S$	1.227 (25°C)
1-Butanethiol	$C_4H_{10}S$	0.8416
2-Butanethiol	$C_4H_{10}S$	0.830
1-Butanethiol, 2-methyl-	$C_5H_{12}S$	0.842
2-Butanethiol, 2-methyl-	$C_5H_{12}S$	0.812
1,2,4-Butanetriol	$C_4H_{10}O_3$	1.02
Butanoic acid (butyric acid)	$C_4H_8O_2$	0.9577
– ,amide	C_4H_9ON	0.885 (120°C)
– ,bromanhydride	C_4H_7OBr	1.4162 (17°C)
– ,butyl ether	$C_8H_{16}O_2$	0.8709
– 4-chloro-	$C_4H_7O_2Cl$	1.2236
– ,chloranhydride	C_4H_7OCl	1.0277
– 2,2-difluor-	$C_4H_6O_2F_2$	1.22 (25°C)
– 2,3-dioxo- ,-ethyl ether	$C_6H_8O_4$	1.151 (26°C)
– 2-ethyl	$C_6H_{12}O_2$	0.9239
– ,ethyl ether	$C_6H_{12}O_2$	0.8790
– ,fluoranhydride	C_4H_7OF	0.944 (11°C)
– 2-methoxy-	$C_5H_{10}O_3$	1.0486
– 4-methoxy-	$C_5H_{10}O_3$	1.0596
– 2-methyl	$C_5H_{10}O_2$	0.9340 (25°C)
– ,nitrile	C_4H_7N	0.7911
– 2-oxo-	$C_4H_6O_3$	1.20 (17°C)
– ,propyl ether	$C_7H_{14}O_2$	0.8722
1-Butanol	$C_4H_{10}O$	0.8096
2-Butanol	$C_4H_{10}O$	0.8063
2-Butanol, 1-amino-	$C_4H_{11}ON$	0.927 (17°C)
2-Butanol, 1-chloro-	C_4H_9OCl	1.0738
2-Butanol, 1-nitro-	$C_4H_9O_3N$	1.1303 (25°C)
1-Butanol, 2,2-dimethyl-	$C_6H_{14}O$	0.8273
2-Butanol, 2,3-dimethyl-	$C_6H_{14}O$	0.8277
1-Butanol, 2-amino-	$C_4H_{11}ON$	0.939 (26°C)
1-Butanol, 2-methyl	$C_5H_{12}O$	0.8193
2-Butanol, 3-iodo-	C_4H_9OI	1.7980 (15°C)
2-Butanone	C_4H_8O	0.8054
2-Butanone, 1-chloro-	C_4H_7OCl	1.085
2-Butanone, 1-hydroxy	$C_4H_8O_2$	1.026
2-Butenal (crotonaldehyde)	C_4H_6O	0.8495 (25°C)
2-Butenal, 2-chloro-	C_4H_5OCl	1.1404 (23°C)
3-Butenal, 2,2-dimethyl-	$C_6H_{10}O$	0.8184 (25°C)
2-Butenoic acid (*cis*) (isocrotonic acid)	$C_4H_6O_2$	1.0267
2-Butene (*trans*)	C_4H_8	0.6042
1-Butene	C_4H_8	0.5951
1-Butene, 1-bromo-3-methyl-	C_5H_9Br	1.2819
2-Butene, 1-chloro- (*trans*)	C_4H_7Cl	0.9205 (15°C)
2-Butene, 1-ethoxy-	$C_6H_{12}O$	0.7846
1-Butene, 2-bromo-3-methyl-	C_5H_9Br	1.2328
1-Butene, 2-ethyl-	C_6H_{12}	0.6894
2-Butene, 2-methoxy-	$C_5H_{10}O$	0.8054 (15°C)
2-Butene, 2-methyl-	C_5H_{10}	0.6623
2-Butene, 2-methyl-1-chloro-	C_5H_9Cl	0.9327
1-Butene, 2-nitro-	$C_4H_7O_2N$	1.0188
1-Butene, octafluoro-	C_4H_8	1.5443 (0°C)
2-Butene, octafluoro-	C_4H_8	1.5297 (0°C)
2-Butene-1,4-diol (*trans*)	$C_4H_8O_2$	1.080
3-Butene-1,2-diol	$C_4H_8O_2$	1.0466
2-Butene-1-ol (*cis*)	C_4H_8O	0.8662
2-Butene-1-ol (*trans*) (propyl ether)	C_4H_8O	0.8532
3-Butene-1-ol	C_4H_8O	0.8454
3-Butene-2-ol	C_4H_8O	0.8413
3-Butene-2-ol, 1-bromo-	C_4H_7OBr	1.5205
3-Butene-2-ol, 2,3-dimethyl-	$C_6H_{12}O$	0.8396
3-Butene-1-ol, 2-bromo-	C_4H_7OBr	1.5130
3-Butene-1-ol, 2-chloro-	C_4H_7OCl	1.1044
2-Butene-1-ol, 2-methyl-	$C_5H_{10}O$	0.863 (24°C)
2-Butene-1-thiol	C_4H_8S	0.883 (23°C)
3-Butene-1-thiol	C_4H_8S	0.9087
3-Butene-2-ol, 1-chloro-	C_4H_7OCl	1.1123
3-Butene-2-ol, 2-methyl-	$C_5H_{10}O$	0.825
3-Butene-2-on	C_4H_6O	0.8459
1-Butene-3-in	C_4H_4	0.7095 (0°C)
Butenedioic acid (*cis*) (maleic acid)	$C_4H_4O_4$	1.590
– ,dinitryl (*trans*)	$C_4H_2N_2$	0.9416 (111°C)
Butenedioic acid (*trans*) (fumaric acid)	$C_4H_4O_4$	1.625
2-Butenoic acid (*trans*) (crotonic acid)	$C_4H_6O_2$	0.973 (72°C)
3-Butenoic acid	$C_4H_6O_2$	1.0091
3-Butenoic acid, 2,2 dimethyl-	$C_6H_{10}O_2$	0.963
3-Butenoic acid, 2-chloro-	$C_4H_5O_2Cl$	1.237
2-Butenoic acid, 2-ethyl-	$C_6H_{10}O_2$	0.9578 (50°C)
sec- Butylamine	$C_4H_{11}N$	0.724
tert- Butylamine	$C_4H_{11}N$	0.6598
Butylamine	$C_4H_{11}N$	0.7414
Butylamine,N,N-dimethyl	$C_6H_{15}N$	0.7206
– N-isopropyl-	$C_7H_{17}N$	0.741
– N-methyl-	$C_5H_{13}N$	0.7377
– I-methyl-	$C_5H_{13}N$	0.7424
tert-Butylhydroperoxide	$C_4H_{10}O_2$	0.8930
tert-Butylhypochlorite	C_4H_9OCl	0.9583
tert-Butylisothiocyanate	C_5H_9NS	0.9187 (10°C)
Butylisothiocyanate	C_5H_9NS	0.9546
sec-Butylnitrate	$C_4H_9O_3N$	1.0264
Butylnitrate	$C_4H_9O_3N$	1.0153
sec-Butylnitrite	$C_4H_9O_2N$	0.8728

Table 5.3 Density of organic substances *(continued)*

Name	Formula	Density, 10^3 kg/m^3	Name	Formula	Density, 10^3 kg/m^3
tert-Butylnitrite	$C_4H_9O_2N$	0.867	– 1-methyl-	$C_7H_{14}O$	0.9194
Butylnitrite	$C_4H_9O_2N$	0.8823	Cyclohexanone	$C_6H_{10}O$	0.9455
1-Butyne	C_4H_6	0.6784 (0°C)	Cyclohexene	C_6H_{10}	0.8102
2-Butyne	C_4H_6	0.6510	1-Cyclohexene-1-carboxylic acid	$C_7H_{10}O_2$	1.109
1-Butyne, 3,3-dimethyl-	C_6H_{10}	0.6695	2-Cyclohexene-1-ol	$C_6H_{10}O$	0.9923 (15°C)
1-Butyne, 3-methyl-	C_5H_8	0.666	2-Cyclohexene-1-on	C_6H_8O	0.9931
2-Butyne,1.4-dimethoxy-	$C_6H_{10}O_2$	0.9575	1-Cyclohexenecarbaldehyde	$C_7H_{10}O$	0.9694
– 1,4-dichloro-	$C_4H_4Cl_2$	1.258	Cyclohexylamine	$C_6H_{13}N$	0.8671
Carbonic acid, benzyl ether, chloranhydride	$C_8H_7O_2Cl$	1.20	Cyclohexylisocyanate	$C_7H_{11}ON$	0.9852
– ,butyl ether, chloranhydride	$C_5H_9O_2Cl$	1.0513	Cyclopentadiene	C_5H_6	0.8021
– ,dichloranhydride (phosgene)	$COCl_2$	1.381	– hexachloro-	C_6Cl_6	1.7119
– ,diethyl ether	$C_3H_{10}O_3$	0.9764	Cyclopentane	C_5H_{10}	0.7454
– ,dimethyl ether	$C_3H_6O_3$	1.0706	– acetyl-	$C_7H_{12}O$	0.9172
– ,dipropyl ether	$C_7H_{14}O_3$	0.9435	– bromo-	C_5H_9Br	1.3873
– ,isobutyl ether, chloranhydride	$C_5H_9O_2Cl$	1.0426	– chloro-	C_5H_9Cl	1.0053
– ,isopentyl ether, chloranhydride	$C_6H_{11}OCl$	1.049	– ethyl-	C_7H_{14}	0.7665
			– methyl-	C_6H_{12}	0.7486
– ,isopropyl ether, chloranhydride	$C_4H_7O_3Cl$	1.0777	Cycopentanecarbaldehyde	$C_6H_{10}O$	0.9371
			Cyclopentane	C_7H_{14}	0.8098
– ,methyl ether, chloranhydride	$C_2H_3O_2Cl$	1.2298	Cyclopentanecarboxylic acid	$C_6H_{10}O_2$	1.0527
– ,propyl ether, chloranhydride	$C_4H_7O_2Cl$	1.0902	Cyclopentanol	$C_5H_{10}O$	0.9478
Cyclobutane	C_4H_8	0.720 (5°C)	Cyclopentanone	C_5H_8O	0.9450
– ethyl-	C_5H_{12}	0.728	Cyclopentene	C_5H_8	0.7720
– methyl-	C_5H_{10}	0.688	Cyclopropanecarboxylic acid	$C_4H_6O_2$	1.0889
Cyclobutanecarboxylic acid	$C_5H_8O_2$	1.057	– ,chloranhydride	C_4H_5OCl	1.1518
Cyclobutanol	C_4H_8O	0.9226 (15°C)	– ,ethyl ether	$C_6H_{10}O_2$	0.9638
Cyclobutanone	C_4H_6O	0.9548 (0°C)	– ,methyl ether	$C_5H_8O_2$	0.9972
Cyclobutene	C_4H_8	0.733 (0°C)	– ,nitryle	C_4H_5N	0.8946
1,2-Cycloheptanedione	$C_7H_{10}O_2$	1.0607 (22°C)	Cyclopropane	C_3H_6	0.6769 (−30°C)
Cycloheptanol	$C_7H_{14}O$	0.9478	– acetyl-	C_5H_8O	0.8984
Cycloheptanone	$C_7H_{12}O$	0.9491	– 1,1-dimethyl-	C_5H_{10}	0.6589
1,3,5-Cycloheptatriene	C_7H_8	0.8875	– methyl-	C_4H_8	0.6912 (−20°C)
2,4,6-Cycloheptatrienone	C_7H_6O	1.095 (22°C)	– vinyl-	C_5H_8	0.7160
Cycloheptene	C_7H_{12}	0.8255	Di-*tert*-butylperoxide	$C_8H_{18}O_2$	0.796
1,3-Cyclohexadiene	C_6H_8	0.8406	Diallylsulfide	$C_6H_{10}S$	0.8877 (27°C)
Cyclohexane	C_6H_{12}	0.7786	Diallylsulfone	$C_6H_{10}SO_2$	1.1215
– bromo-	$C_6H_{11}Br$	1.3360	Diallylsulfoxide	$C_6H_{10}SO$	1.0261
– chloro-	$C_6H_{11}Cl$	0.9891	Dibutyl ether	$C_8H_{18}O$	0.7689
– 1,2-dibromo- (*trans*)	$C_6H_{10}Br_2$	1.7759	Dibutylamine	$C_8H_{19}N$	0.7670
– 1,2-dichloro- (*trans*)	$C_6H_{10}Cl_2$	1.2021	Dibutylsulfide	$C_{18}H_{18}S$	0.8386
– ethyl-	C_8H_{16}	0.7880	Diethyl ether (ethyl ether)	$C_4H_{10}O$	0.71378
– fluoro-	$C_6H_{11}F$	0.9279	– 1-chloro-	C_4H_9OCl	0.9493
– iodo-	$C_6H_{11}I$	0.7694	– 2-chloro-	C_4H_9OCl	0.9894
– methoxy-	$C_7H_{14}O$	1.0610	– 1,2-dibromo-	$C_4H_8OBr_2$	1.7315
– methyl-	C_7H_{14}	0.8074	– 1,1-dichloro-	$C_4H_8OCl_2$	1.1285
– methylene-	C_7H_{12}	0.8752	Diethylamine	$C_4H_{11}N$	0.7059
– nitro-	$C_6H_{11}NO_2$	0.7794	Diethyldisulfide	$C_4H_{10}S_2$	0.9931
Cyclohexanecarbaldehyde	$C_7H_{12}O$	0.9235 (25°C)	*N*, *N*-Diethylhydrazine	$C_4H_{12}N_2$	0.8004
Cyclohexanecarboxylic acid	$C_7H_{12}O_2$	1.0274 (30°C)	Diethylperoxide (ethyl peroxide)	$C_4H_{10}O_2$	0.8240
1,-2-Cyclohexanedione	$C_6H_8O_2$	1.1187			
Cyclohexanethiol	$C_6H_{12}S$	0.9786	Diethylsulfate	$C_4H_{10}O_4S$	1.1842 (15°C)
Cyclohexanol	$C_6H_{12}O$	0.9624	Diethylsulfide	$C_4H_{10}S$	0.8362
– 2-chloro- (*trans*)	$C_6H_{11}OCl$	1.1461 (16°C)	Diethylsulfite	$C_4H_{10}O_3S$	1.0829

Table 5.3 Density of organic substances *(continued)*

Name	Formula	Density, 10^3 kg/m^3	Name	Formula	Density, 10^3 kg/m^3
Diisopropyl ether	C_6H_8O	0.7241	– hexafluoro-	C_2F_6	1.590 (−78°C)
Diisopropylamine	$C_6H_{15}N$	0.7163	– iodo-	C_2H_5I	1.9358
Diisopropyldisulphide	$C_6H_8S_2$	0.9435	– methoxy	C_3H_8O	0.7252 (0°C)
N, *N*-Diisopropylhydrazine	$C_6H_{10}N_2$	0.7929	– nitro	$C_2H_5NO_2$	1.0448 (25°C)
Diisopropylsulfide	C_6H_8S	0.8142	– pentachloro	C_2HCL_5	1.6796
Dimethylamine	C_2H_7N	0.6804 (0°C)	– pentabromo	C_2HBr_5	3.312
Dimethyldichlorogermanium	$C_2H_6GeCl_2$	1.492	– 1,1,1,2-tetrabromo-	$C_2H_2Br_4$	2.8748
Dimethyldisulfide	$C_2H_6S_2$	1.0647	– 1,1,1,2-tetrachloro-	$C_2H_2Cl_4$	1.54064
N,N-Dimethylhydrazine	$C_2H_8N_2$	0.7911	– 1,1,2-tribrom-	$C_2H_3Br_3$	2.6211
Dimethylsulfate	$C_2H_6O_4S$	1.3332 (15°C)	– 1,1,1-trichlor-	$C_2H_3Cl_3$	1.3390
Dimethylsulfite	$C_2H_6O_3S$	1.2129	(methylchloroform)		
Dimethylsulfide	C_2H_6S	0.8483	– 1,1,1-trifluoro-2,2,2-trichlor-	$C_2F_3Cl_3$	1.5790
Dimethylsulfoxide	C_2H_6OS	1.1014	– 1,1,1-trifluoro-	$C_2H_3F_3$	$3.78 \cdot 10^{-3}$
1,3-Dioxane	$C_4H_8O_2$	1.0319	Ethanedial (glyoxal)	$C_2H_2O_2$	1.14
1,3-Dioxolane	$C_3H_6O_2$	1.060	1,2-Ethanediol	$C_2H_6O_2$	1.1088
– 2-methyl-	$C_4H_8O_2$	0.9811	(etheleneglycol)		
1,3-Dioxolane-2-on	$C_3H_4O_3$	1.3218 (39°C)	Ethanesulfonic acid	$C_2H_6O_3S$	1.3341 (25°C)
(ethylene carbonate)			Ethanethiol	C_2H_6S	0.8391
– 4-methyl	$C_4H_6O_3$	1.2057	(ethyl mercaptan)		
Dipropyl ether	$C_6H_{14}O$	0.736	Ethanol (ethyl alcohol)	C_2H_6O	0.7983
Dipropyldisulfide	$C_6H_{14}S_2$	0.9599	– 2-amino-(ethanolamine)	C_2H_7ON	1.0159
Dipropylsulfate	$C_6H_{14}O_4S$	1.1064	– 2-azid-	$C_2H_5ON_3$	1.1454 (25°C)
Dipropylsulfide	$C_6H_{14}S$	0.8377	– 2-bromo-	C_2H_5OBr	1.7629
Dipropylsulfone	$C_6H_{14}O_2S$	1.0278 (50°C)	– 2-chloro	C_2H_5OCl	1.2003
Dipropylsulfoxide	$C_6H_{14}OS$	0.9654	– 2,2-dichloro-	$C_2H_4OCl_2$	1.4040 (25°C)
1,3-Dithiolane	$C_3H_6S_2$	1.259 (17°C)	– 2-ethoxy	$C_4H_{10}O_2$	0.9301
Divinyl ether	C_4H_6O	0.773	– 2-fluor-	C_2H_5OF	1.1040
Divinylsulfide	C_4H_6S	0.9174 (15°C)	– 2-iodo-	C_2H_5OI	2.1968
Ethane	C_2H_6	0.509 (−60°C)	– 2-mercapto-	C_2H_6OS	1.1143
– azid-	$C_2H_5N_3$	0.8765 (25°C)	– 2-methoxy-	$C_3H_8O_2$	0.9647
– bromo-	C_2H_5Br	1.4604	– 2-nitro-	$C_2H_5O_3N$	1.270 (15°C)
– 1-bromo-2-chloro-	C_2H_4BrCl	1.7392	– 2,2-oxydi-(diethyleneglycol)	$C_4H_{10}O_2$	1.1160
– 1-bromo-2-fluoro-	C_2H_4BrF	1.744 (25°C)	– 2,2,2-trifluoro-	$C_2H_3OF_3$	1.3816 (25°C)
– chloro-	C_2H_5Cl	0.8978	Ethylamine	C_2H_7N	1.0259
– 1,2-dibromo	$C_2H_4Br_2$	2.1792	Ethylene	C_2H_4	$1.26 \cdot 10^{-3}$ (0°C)
(ethylene bromide)			– bromo	C_2H_3Br	1.4933
– 1,1-dibromo-	$C_2H_4Br_2$	2.0555	– chloro	C_2H_3Cl	0.9106
(ethylidenebromide)			– 1,1-dichloro-	$C_2H_2Cl_2$	1.2129
– 1,2-dichloro	$C_2H_4Cl_2$	1.2351	– ethoxy	C_4H_8O	0.7589
(ethylene chloride)			– fluor	C_2H_3F	0.853 (−26°C)
– 1,1-dichloro-	$C_2H_4Cl_2$	1.1757	– iodo-	C_2H_3I	2.037
(ethylidenechloride)			– methoxy-	C_3H_6O	0.7725 (0°C)
– 1,2-difluoro-	$C_2H_4F_2$	1.024 (10°C)	– tetrachloro-	C_2Cl_4	1.6227
– 1,1-difluoro-	$C_2H_4F_2$	0.95	– tetrafluoro	C_2F_4	1.519 (−76.3°C)
(ethylidenefluoride)			– tetraiodo	C_2I_4	2.983
– 1,2-diiodo-	$C_2H_4I_2$	3.325	– tribrom-	C_2HBr_3	2.708
– 1,1-diiodo-	$C_2H_4I_2$	2.84 (0°C)	– trichlor	C_2HCl_3	1.4642
(ethylideneiodide)			Ethylenediamine	C_2H_8N	0.9006
– 1,2-dimethoxy-	$C_4H_{10}O_2$	0.86285	Ethylhydrazine	$C_2H_8N_2$	0.9079
– 1,1-dinitro-	$C_2H_4O_4N_2$	1.3503 (24°C)	Ethylhydroperoxide	$C_2H_6O_2$	1.013 (−6°C)
– fluor-	C_2H_5F	0.7182	*N*-Ethylhydroxylamine	C_2H_7ON	0.8827 (0°C)
– 1-fluoro-2-chloro-	C_2H_4FCl	1.1747	*O*-Ethylhydroxylamine	C_2H_7ON	0.9332
– hexabromo-	C_2Br_6	2.823	Ethylhypochlorite	C_2H_5OCl	0.384 (−10°C)
– hexachloro-	C_2Cl_6	2.091	Ethylisocyanate	C_3H_5ON	0.9031

Table 5.3 Density of organic substances *(continued)*

Name	Formula	Density, 10^3 kg/m^3	Name	Formula	Density, 10^3 kg/m^3
Ethylisocyanide	C_3H_5N	0.7402	1-Heptanol	$C_7H_{16}O$	0.8219
Ethylisothiocyanate	C_3H_5NS	0.9962	2-Heptanone	$C_7H_{14}O$	0.8111
Ethylnitrate	$C_3H_5O_3N$	1.1076	1-Heptene	C_7H_{14}	0.6970
Ethylnitrite	$C_2H_5O_2N$	0.9017 (14°C)	3-Heptene-2-on (*trans*)	$C_7H_{12}O$	0.8496
Ethyne (acetylene)	C_2H_2	0.6181 (−22°C)	1-Heptene-3-in	C_7H_{10}	0.7603
– bromo-	C_2HBr	$4.7 \cdot 10^{-3}$	2-Heptene-4-ol	$C_7H_{14}O$	0.8445
– chloro-	C_2HCl	$2.0 \cdot 10^{-3}$	1-Heptine	C_7H_{12}	0.7328
– ethoxy-	C_4H_6O	0.800	2-Heptine	C_7H_{12}	0.7480
– methoxy-	C_3H_4O	0.8001	3-Heptine	C_7H_{12}	0.7527
Formaldehyde (formic aldehyde)	CH_2O	0.815 (−20°C)	2-Heptine-1-ol	$C_7H_{12}O$	0.884 (27°C)
Formic acid	CH_2O_2	1.22	Heptylamine	$C_7H_{17}N$	0.7754
– ,allyl ether	$C_4H_6O_2$	0.946	2,4-Hexadienal	C_6H_8O	0.898
– ,amide (formamide)	CH_3ON	1.1334	1,2-Hexadiene	C_6H_{10}	0.7149
– ,butyl ether	$C_5H_{10}O_2$	0.8265	1,3-Hexadiene, 5-chloro-	C_6H_9Cl	0.9275
– ,ethyl ether	$C_3H_6O_2$	0.9168	1,4-Hexadiene-3-on	C_6H_8O	0.8959
– ,fluoranhydride	CHOF	1.099 (0°C)	1,3-Hexadiene-5-in	C_6H_6	0.7806
– ,heptyl ether	$C_8H_{16}O_2$	0.8784	2,4-Hexadienol	$C_6H_{10}O$	0.8967
– ,hexyl ether	$C_7H_{14}O_2$	0.8813	1,4-Hexadiine	C_6H_6	0.825 (0°C)
– ,isobutyl ether	$C_5H_{10}O_2$	0.8854	1,5-Hexadiine	C_6H_6	0.8049
– ,isopentyl ether	$C_6H_{12}O_2$	0.8857	Hexanal (caproic aldehyde)	$C_6H_{12}O$	0.8139
– ,isopropyl ether	$C_4H_8O_2$	0.8774	Hexane	C_6H_{14}	0.6603
– ,methyl ether	$C_2H_4O_2$	0.9705	– 1-bromo-	$C_6H_{13}Br$	1.1744
– ,nitrile (hydrogen cyanide)	CHN	0.6876	– 1-chloro-	$C_6H_{13}Cl$	0.8785
– ,pentyl ether	$C_6H_{12}O_2$	0.8853	– 1,2-dibromo-	$C_4H_{12}Br_2$	1.5774
– ,propyl ether	$C_4H_8O_2$	0.9039	– 1,2-dichloro-	$C_6H_{12}Cl_2$	1.085 (15°C)
Furan	C_4H_4O	0.9378	– 1,6-diiodo-	$C_6H_{12}I_2$	2.03
– 2-acetyl-	$C_6H_6O_2$	1.098 (20°C)	– 2,2-dimethyl-	C_6H_{18}	0.6953
– 2-bromo-	C_4H_3OBr	1.650	– 1-fluor-	$C_6H_{13}F$	0.7995
– 2-bromomethyl-	C_5H_5OBr	1.560	– 1-fluoro-6-chloro-	$C_6H_{12}FCl$	1.015
– 2-chloro-	C_4H_3OCl	1.1923	– 1-iodo-	$C_6H_{13}I$	1.4397
– 2,5-dibromo-	$C_4H_2OBr_2$	2.27	– 2-methyl-	C_7H_{16}	0.6787
– 2,5-dichloro-	$C_4H_2OCl_2$	1.371 (25°C)	1,6-Hexanedial (adipaldehyde)	$C_6H_{10}O_2$	1.003
– 2,5-dimethyl-	C_6H_8O	0.90	Hexanedioic acid (adipic acid)	$C_6H_{10}O_4$	1.360 (25°C)
– 2-ethoxy-	$C_6H_8O_2$	0.9849 (23°C)	– ,dimethyl ether	$C_8H_{17}O_4$	1.060
– 2-ethyl-	C_6H_8O	0.9013	– ,dinitrile	$C_6H_8N_2$	0.9676
– 2-iodo-	C_4H_3IO	2.024	2,5-Hexanediol	$C_6H_{14}O_2$	0.9610
– 2-methoxy-	$C_5H_6O_2$	1.0646 (25°C)	2,5-Hexanedione	$C_6H_{10}O_2$	0.9737
– 2-methyl-	C_5H_6O	0.9132	1-Hexanethiol	C_6H_{14}	0.8424
– tetrahydro-	C_4H_8O	0.8886	Hexanoic acid (caproic acid)	$C_6H_{12}O_2$	0.9274
2-Furancarbaldehyde	$C_5H_4O_2$	1.1594	– ,methyl ether	$C_7H_{14}O_2$	0.8846
Furfurylamine	C_5H_7ON	1.0533	– ,nitrile	$C_6H_{11}N$	0.8052
Glyoxal	$C_2H_2O_2$	1.14	– ,chloranhydride	$C_6H_{11}OCl$	0.9754
Heptanal	$C_7H_{14}O$	0.8495	1-Hexanol	$C_6H_{14}O$	0.8136
Heptane	C_7H_{16}	0.68376	2-Hexanone	$C_6H_{12}O$	0.8113
– 1-bromo-	$C_7H_{15}Br$	1.140	1,3,5-Hexatriene (*cis*)	C_6H_8	0.7175
– 1-chloro-	$C_7H_{15}Cl$	0.8758	1,3,5-Hexatriene (*trans*)	C_6H_8	0.7420
– 1,1-dichloro-	$C_7H_{14}Cl_2$	1.0008	2-Hexenal (*trans*)	$C_6H_{10}O$	0.8491
– 1-iodo-	$C_7H_{15}I$	1.3791	2-Hexene (*cis*)	C_6H_{12}	0.6869
– perfluoro-	C_7F_{16}	1.7333	3-Hexene (*cis*)	C_6H_{12}	0.6796
Heptanedioic acid	$C_7H_{12}O_4$	1.329 (15°C)	2-Hexene (*trans*)	C_6H_{12}	0.6784
1,7-Heptanediol	$C_7H_{16}O_2$	0.9569 (25°C)	3-Hexene (*trans*)	C_6H_{12}	0.6772
2,4-Heptanedione	$C_7H_{12}O_2$	0.9411 (15°C)	1-Hexene	C_6H_{12}	0.6731
1-Heptanethiol	$C_7H_{16}S$	0.8427	1-Hexene, 1-chloro-	$C_6H_{11}Cl$	0.8872
Heptanoic acid	$C_7H_{14}O_2$	0.920	2-Hexene, 2,5-dimethyl-	C_7H_{13}	0.720

Table 5.3 Density of organic substances *(continued)*

Name	Formula	Density, 10^3 kg/m^3	Name	Formula	Density, 10^3 kg/m^3
1-Hexene, 2-methyl-	C_7H_{14}	0.7030	– fluorodichloro-	$CHCl_2F$	1.405 (9°C)
1-Hexene-3-in	C_6H_8	0.7492	– fluorotrichloro-(Freon-11)	CCl_3F	1.4995 (15°C)
1-Hexene-3-ol	$C_6H_{12}O$	0.834	– iodo-(methyl iodide)	CH_3I	2.279
1-Hexine	C_6H_{10}	0.7155	– iodochloro-	CH_2ClI	2.422
2-Hexine	C_6H_{10}	0.7315	– iododichloro-	$CHCl_2I$	2.392
3-Hexine	C_6H_{10}	0.7236	– iododifluoro	CHF_2I	3.238 (−19°C)
Hexylamine	$C_6H_{15}N$	0.776	– iodofluoro-	CH_2FI	2.366
Hydroquinone	$C_6H_6O_2$	1.358	– iodotrichloro-	CCl_3I	2.355
Imidazole	$C_3H_4N_2$	1.0303 (101°C)	– nitro-	CH_3NO_2	1.1382
– 1-methyl-	$C_4H_6N_2$	1.0325	– nitrochloro-	CH_2ClNO_2	1.466 (15°C)
Indene	C_9H_8	0.9957	– nitrophenyl-	$C_4H_9NO_2$	1.1598
Indole	C_8H_7N	1.0718 (56°C)	(α-nitrotoluene)		
Indoline	C_8H_9N	1.069	– nitrotrichloro-	$CHCl_3NO_2$	1.6566
Isobutylamine	$C_4H_{11}N$	0.7346	(chloropicrin)		
Isobutylisothiocyanate	C_5H_9NS	0.9638 (14°C	– tetrabromo-	CBr_4	
Isopentylamine	$C_5H_{13}N$	0.7491	(α-phase)		2.9609 (100°C)
Isopentylnitrate	$C_5H_{11}O_3N$	0.9961 (22°C)	(β-phase)		3.273
Isopropenylisocyanate	C_4H_5ON	0.8776	– tetrachloro-	CCl_4	1.5940
Isoxazole	C_3H_3NO	1.078	– tetrafluoro-	CF_4	1.619 (−129.8°C)
Ketene	C_2H_2O	$1.45 \cdot 10^{-3}$	– tetraiodo-	CI_4	4.23
– ,diethylacetal	$C_4H_{12}O_2$	0.8776 (25°C)	– tetranitro-	$C(NO_2)_4$	1.6380
Malonic acid	$C_3H_4O_4$	0.619 (16°C	– tribromo-(bromoform)	$CHBr_3$	2.8899
– ,dichloranhydride	$C_3H_2O_2Cl_2$	1.4509	– tribromofluoro-	CBr_3F	2.7648
– ,dietyl ether	$C_7H_{12}O_4$	1.0551	– tribromonitro-	CBr_3NO_2	2.7930
– ,dimethyl ether	$C_5H_8O_4$	1.1528	– trichloro-(chloroform)	$CHCl_3$	1.4832
– ,dinitrile	$C_3H_2N_2$	1.1910	– trifluoro-(fluoroform)	CHF_3	1.52 (−100°C)
– methyl-(isosuccinic acid)	$C_4H_6O_4$	1.455	– triiodo-(iodoform)	CHI_3	4.008
Methane	CH_4	0.466 (−164°C)	– trinitro-(nitroform)	$CH(NO_2)_3$	1.648 (13°C)
– azid-	CH_3N_3	0.869 (8°C)	– trinitrochloro-	$CCl(NO_2)_3$	1.6769
– bromoiodo-	CH_2BrI	2.926 (17°C)	– tribromochloro-	CBr_3Cl	2.71 (15°C)
– bromo-	CH_3Br	1.6755	Methane-sulfonic acid	CH_4O_3S	1.4812
– bromodichloro-	$CHBrCl_2$	1.98 (16°C)	Methanethiol	CH_4S	0.8665
– bromodifluoro-	$CHBrF_2$	1.55	(methylmercaptan)		
– bromofluorochloro-	$CHBrClF$	1.9771 (0°C)	Methanol	CH_4O	0.7914
– bromotrichloro-	$CBrCl_3$	2.0122	Methanol-D	CH_3OD	0.8127
– bromotrinitro-	CO_6N_3Br	2.0313	Methyl nitrate	CH_3O_3N	1.2075
– chloro-(methyl chloride)	CH_3Cl	0.9159	Methyl nitrite	CH_3O_2N	0.991
– dibromo-	CH_2Br_2	2.4970	Methylamine	CH_5N	0.6628
(methylene bromide)			Methylhydrazine	CH_6N_2	0.8733
– dibromochloro-	$CHBr_2Cl$	2.451	Methylhydroperoxide	CH_4O_2	0.9967 (15°C)
– dibromodichloro-	CBr_2Cl_2	1.42 (25°C)	*N*-Methylhydroxysilamine	CH_5ON	1.0003
– dibromofluoro-	$CHBr_2F$	2.421	Methylisocyanate	C_2H_3ON	0.9744 (15°C)
– dichloro-	CH_2Cl_2	1.3266	Methylisothiocyanate	C_2H_3NS	1.0691 (37°C)
(methylene chloride)			Methylthiocyanate	C_2H_3NS	1.0678 (25°C)
– difluor-	CH_2F_2	0.909	Methyltrichlorogermanium	CH_3GeCl_3	1.73 (24.5°C)
(methylene fluoride)			Morpholine	C_4H_9ON	1.0005
– difluorochloro-(Freon-22)	$CHClF_2$	1.4909 (−69°C)	Naphthalene	$C_{10}H_8$	1.0253
– difluorodichloro-	CCl_2F_2	1.64 (−73°C)	Neopentylamine	$C_5H_{13}N$	0.7455
(Freon-12)			Norbornadiene	C_7H_8	0.9064
– diiododifluoro-	$CHFI_2$	3.1969 (26°C)	Octane	C_8H_{18}	0.8756
– diiodo-(methylene iodide)	CH_2I_2	3.3240	Orthoacetic acid,	$C_5H_{12}O_3$	0.9438 (25°C)
– dinitrochloro-	$CHCl\ (NO_2)_2$	1.6123	trimethyl ether		
– diphenyl-	$C_{13}H_{12}$	1.0060	Orthocarbonic acid,	$C_9H_{20}O_4$	0.9186
– fluoro-(methyl fluoride)	CH_3F	0.5786	tetraethyl ether		

Table 5.3 Density of organic substances *(continued)*

Name	Formula	Density, 10^3 kg/m^3	Name	Formula	Density, 10^3 kg/m^3
Orthoformic acid, trimethyl ether	$C_4H_{10}O_3$	0.9676	2-Pentanone	$C_5H_{10}O$	0.8089
			2-Pentenal	C_5H_8O	0.860
Oxalic acid	$C_2H_2O_4$	1.653	1-Pentene	C_5H_{10}	0.6405
Oxetane	C_3H_6O	0.9001	1-Pentene, 1-bromo-	C_5H_9Br	1.2391
Oxetane-2-on	$C_3H_4O_2$	1.146	2-Pentene (*cis*)	C_5H_{10}	0.6556
Oxirane	C_2H_4O	0.8971 (0°C)	2-Pentene (*trans*)	C_5H_{10}	0.6482
– chloromethyl-	C_3H_5OCl	1.180	1-Pentene-3-in (pyrilene)	C_5H_6	0.7401
– 2,2-dimethyl-	C_4H_8O	0.8112	1-Pentene-3-ol	$C_5H_{10}O$	0.8395
– ethyl-	C_4H_8O	0.837 (17°C)	1-Pentene, 2-ethyl-	C_7H_{14}	0.7079
– hydroxymethyl-	$C_3H_6O_2$	1.1143 (25°C)	1-Pentene, 2-methyl-	C_6H_{12}	0.6799
– methyl-	C_3H_6O	0.8311	1-Pentene, 3-chloro-	C_5H_9Cl	0.8988
– phenyl	C_8H_8O	1.0592	1-Pentene, decafluoro-	C_5F_{10}	1.2571 (25°C)
– 2,3-tetramethylene	$C_6H_{10}O$	0.9718	2-Pentene, 2,3-dimethyl-	C_7H_{14}	0.7227
– vinyl-	C_4H_6O	0.8745	2-Pentene, 2-methyl-	C_6H_{12}	0.6863
Oxiranecarbonic acid, ethyl ether	$C_5H_8O_3$	1.085	4-Pentene-1-ol	$C_5H_{10}O$	0.8457
			4-Pentene-2-ol	$C_5H_{10}O$	0.8367
1,2-Pentadiene	C_5H_8	0.6926	2-Pentenic acid	$C_5H_8O_2$	0.9809
1,4-Pentadiene-3-on	C_5H_6O	0.8811	Pentylamine	$C_5H_{13}N$	0.7561
1,3-Pentadiine	C_5H_4	0.7909	*tert*-Pentylamine	$C_5H_{13}N$	0.7320
1,2-Pentadiol	$C_5H_{12}O_2$	0.9802	Pentylnitrite	$C_5H_{11}O_2N$	0.8958
2,3-Pentadione	$C_5H_8O_2$	0.9565	1-Pentyne	C_5H_8	0.6901
Pentamethylenediamine	$C_5H_{14}N_2$	0.867 (25°C)	2-Pentyne	C_5H_8	0.7107
Pentanal (valeraldehyde)	$C_5H_{10}O$	0.8095	1-Pentyne, 5-chloro-	C_5H_7Cl	0.968
Pentane	C_5H_{12}	0.6262	1-Pentyne-3-ol	C_5H_8O	0.8859
– 1-bromo-	$C_5H_{11}Br$	1.2182	2-Penthynic acid	$C_5H_6O_2$	0.978
– 3-bromo-3-methyl-	$C_6H_{13}Br$	1.1835	Peracetic acid (acetyl hydroperoxide)	$C_2H_4O_3$	1.226 (15°C)
– 1-chloro-	$C_5H_{11}Cl$	0.8818			
– 1,2-dichloro-	$C_5H_{10}Cl_2$	1.0872	Phenol (carbolic acid)	C_6H_6O	1.0576
– 3,3-diethyl-	C_9H_{20}	0.7536	– 2-chloro-	C_6H_5ClO	1.2634
– 1,5-diiodo-	$C_5H_{10}I_2$	2.1903 (15°C)	– 4-mercapto-	C_6H_6OS	1.1285 (25°C)
– 2,2-dimethyl-	C_7H_{16}	0.6739	– 2-methyl-	C_7H_8O	1.0273
– 3-ethyl-	C_7H_{16}	0.6982	– 2-nitro-	$C_6H_5NO_3$	1.2942 (40°C)
– 1-fluoro-	$C_5H_{11}F$	0.7907	Phenylhydrazine	$C_6H_8N_2$	1.0986
– 1-iodo-	$C_5H_{11}I$	1.5161	Phenylisocyanate (carbanil)	C_7H_5NO	1.0946
– 1-methoxy-	$C_6H_{14}O$	0.7606	Phenylisothiocyanate	C_7H_5NS	1.1303
– 2-methyl-	C_6H_{14}	0.6532	Piperidine	$C_5H_{11}N$	0.8698
– 1-nitro-	$C_5H_{11}NO_2$	0.9525	– 1-methyl-	$C_6H_{13}N$	0.8159
Pentanedioic acid (glutaric acid)	$C_5H_8O_4$	1.192 (106°C)	– 1-nitroso-	$C_5H_{10}NO$	1.0631
			Propanal (propionic aldehyde)	C_3H_6O	0.8058
– ,dichloroanhydride	$C_5H_6O_2Cl_2$	1.3241	– 2-bromo-	C_3H_5OBr	1.592
– ,dimethyl ether	$C_7H_{12}O_4$	1.0876	– 2-chloro-	C_3H_5OCl	1.182 (15°C)
– ,dinitrile	$C_5H_6N_2$	0.9911 (15°C)	– 2,3-dibromo-	$C_3H_4OBr_2$	2.198 (16°C)
1-Pentanethiol	$C_5H_{12}S$	0.8421	– 2,3-dichloro-	$C_3H_4OCl_2$	1.40
Pentanoic acid (valerianic acid)	$C_5H_{10}O_2$	0.9391	– ,diethylacetal-	$C_5H_{16}O_2$	0.8232
			– 2,2-dimethyl-	$C_5H_{10}O$	0.7832
– 2-bromo-	$C_5H_9O_2Br$	1.381	– 2-methyl-	C_4H_8O	0.7904
– ,chloranhydride	C_5H_9OCl	1.0155 (15°C)	– ,oxime-	C_3H_7ON	0.9258
– 3-chloro-	$C_5H_9O_2Cl$	1.1484	– 2-oxo-	$C_3H_4O_2$	1.0455 (24°C)
– ,ethyl ether	$C_7H_{14}O_2$	0.8770	Propane	C_3H_8	0.5853 (−45°C)
– ,methyl ether	$C_6H_{12}O_2$	0.8947	– 1-bromo-	C_3H_7Br	1.3537
– 2-methyl-	$C_6H_{12}O_2$	0.9230	– 2-bromo-	C_3H_7Br	1.3140
– ,nitrile	C_5H_9N	0.7992	– 1-chloro-	C_3H_7Cl	0.8904
– 4-oxo- (levulinic acid)	$C_5H_8O_3$	1.1335	– 1,2-diacetoxy-	$C_7H_{12}O_4$	1.059
1-Pentanol (amyl alcohol)	$C_5H_{12}O$	0.8144	– 1,1-dibromo-	$C_3H_6Br_2$	1.982

Table 5.3 Density of organic substances *(continued)*

Name	Formula	Density, 10^3 kg/m^3	Name	Formula	Density, 10^3 kg/m^3
– 1,1-dichloro-	$C_3H_6Cl_2$	1.1321	2-Propanol (isopropyl alcohol)	C_3H_8O	0.7855
– 1,3-diiodo-	$C_3H_6I_2$	2.5755	2-Propanol, 1,3-dimethoxy-	$C_5H_{12}O_3$	1.0085
– 2,3-dimethyl (neopentane)	C_5H_{12}	0.6135	2-Propanol, 1-amino-	C_3H_9ON	0.9611
– 1-ethoxy-	$C_5H_{12}O$	0.7386	2-Propanol, 1-bromo-	C_2H_7OBr	1.5585 (30°C)
– 1-fluoro-	C_3H_7F	0.7956	2-Propanol, 1-chloro-	C_3H_7OCl	1.115
– 1-iodo-	C_3H_7I	1,7489	2-Propanol, 1-ethoxy-	$C_5H_{12}O$	0.9028
– 1-methoxy-	$C_4H_{10}O$	0.738	2-Propanol, 2,2-dimethyl-	$C_5H_{12}O$	0.812
– 2-methyl-(isobutane)	C_4H_{10}	0.551 (25°C)	1-Propanol, 2,3-dibromo-	$C_3H_6OBr_2$	2.0739
– 1-nitro-	$C_3H_7NO_2$	0.9955 (25°C)	1-Propanol, 2,3-dichloro-	$C_3H_6Cl_2$	1.3506 (25°C)
– 1,2,3-tribromo-	$C_3H_5Br_3$	2.4209	1-Propanol, 2-chloro-	C_3H_7OCl	1.103
– 1,2,3-trichloro-	$C_3H_5Cl_3$	1.3889	1-Propanol, 2-ethoxy-	$C_5H_{12}O$	0.9044
– 1-vinyloxy-	$C_5H_{10}O$	0.7680	2-Propanol, 2-methyl- (*tert*-butyl alcohol)	$C_4H_{10}O$	0.7887
1,2-Propanediol (propylene glycol)	$C_3H_8O_2$	1.0361	1-Propanol, 2-nitro-	$C_3H_7O_3N$	1.1841 (25°C)
1,3-Propanediol, 2-chloro-	$C_3H_7O_2Cl$	1.3219	2-Propanone (acetone)	C_3H_6O	0.7899
1,3-Propanediol, 2-methoxy-	$C_4H_{10}O_3$	1.124 (25°C)	– ,azine	$C_6H_{12}N_2$	0.8389
1,2-Propanediol, 3-mercapto- (tioglycerin)	$C_3H_8O_2S$	1.2455	– ,1-chloro-	C_3H_5OCl	1.15
			– ,hexachloro	C_3OCl_6	1.7444 (12°C)
1,3-Propanedithiol	$C_3H_8S_2$	1.0783	– ,1-hydroxy (acetol)	$C_3H_6O_2$	1.0824
2-Propanesulfonic acid	$C_3H_5O_3S$	1.187 (25°C)	Propenal (acrylaldehyde)	C_3H_4O	0.8410
1-Propanesulfonic acid, chloranhydride	$C_3H_7O_2SCl$	1.2826 (15°C)	– ,diacetate	$C_7H_{10}O_4$	1.0760
			– ,diethylacetal	$C_7H_{14}O_2$	0.8543 (15°C)
1,2,3-Propanethiol (glycerin)	$C_3H_8O_3$	1.2613	– 2-methyl-	C_4H_6O	0.837
1-Propanethiol (propylmercaptan)	C_3H_8S	0.8411	Propene (propylene)	C_3H_6	0.5193
			1-Propene,3-azid-	$C_3H_5N_3$	0.924 (25°C)
2-Propanethiol (isopropylmercaptan)	C_3H_8S	0.8143	– 1-bromo- (*trans*)	C_3H_5Br	1.428
			– 2-bromo-	C_3H_5Br	1.362
1-Propanethiol,2-methyl- (*tert*-butylmercaptan)	$C_4H_{10}S$	0.8002	– 3-bromo-	C_3H_5Br	1.398
			– 1-chloro- (*trans*)	C_3H_5Cl	0.9350
1-Propanethiol,2-methyl- (isopbutylmercaptan)	$C_4H_{10}S$	0.8339	– 1,2-dichloro-	$C_3H_4Cl_2$	1.1818
			– hexafluoro-	C_3F_6	1.583 (-40°C)
Propanoic acid (propionic acid)	$C_3H_6O_2$	0.993	– 3-iodo-	C_3H_5I	1.8494
			– 3-methoxy-	C_4H_8O	0.77 (11°C)
– 2-acetoxy-	$C_5H_8O_4$	1.1758	– 2-methyl-(isobutylene)	C_4H_8	0.5942
– ,anhydride	$C_6H_{10}O_3$	1.0110	– 1-nitro-	$C_3H_5O_2N$	1.065
– ,bromanhydride	C_3H_5OBr	1.521 (16°C)	– 3-ethoxy-	$C_5H_{10}O$	0.7651
– 2-bromo-	$C_3H_5O_2Br$	1.70	– 1,2,3-trichloro-	$C_3H_3Cl_3$	1.414
– ,butyl ether	$C_7H_{14}O_2$	0.8754	2-Propene-1-ol (allyl alcohol)	C_3H_6O	0.8540
– ,chloranhydride	C_3H_5OCl	1.0646	– 2-methyl (methallyl alcohol)	C_4H_8O	0.8515
– 2-chloro-	$C_3H_5O_2Cl$	1.2585	2-Propene-1-thiol (allylmercaptan)	C_3H_6S	0.925 (23°C)
– 2,2-dichloro-	$C_3H_4O_2Cl_2$	1.389 (23°C)			
– ,ethyl ether	$C_5H_{10}O_2$	0.9359	Propenoic acid (acrylic acid)	$C_3H_4O_2$	1.0511
– ,fluoranhydride	C_3H_5OF	0.972 (15°C)	– ,allyl ether	$C_6H_8O_2$	0.9441
– 3-fluoro-	$C_3H_5O_2F$	1.1815	– ,butyl ether	$C_7H_{12}O_2$	0.8898
– 2-iodo-	$C_3H_5O_2I$	2.073	– ,chloranhydride	C_3H_3OCl	1.1136
– ,isobutyl ether	$C_7H_{14}O_2$	0.8687	– ,ethyl ether	$C_5H_8O_2$	0.9234
– ,isopropyl ether	$C_6H_{12}O_2$	0.8660	– ,isobutyl ether	$C_7H_{12}O_2$	0.8896
– 3-mercapto-	$C_3H_6O_2S$	1.218	– ,methyl ether	$C_4H_6O_2$	0.9535
– ,methyl ether	$C_4H_8O_2$	0.9150	– 2-methyl- (methacrylic acid)	$C_4H_6O_2$	1.0153
– 2-methyl- (isobutyric acid)	$C_4H_8O_2$	0.9682	– ,nitrile	C_3H_3N	0.8060
– ,nitrile	C_3H_5N	0.7818	Propylnitrite	$C_3H_7O_2N$	0.935
– 2-oxo- (pyruvic acid)	$C_3H_4O_3$	1.2272	Propylamine	C_3H_9N	1.1064
– ,propyl ether	$C_6H_{12}O_2$	0.8809	– *N,N*-dimethyl-	$C_5H_{13}N$	0.7173
1-Propanol	C_3H_8O	0.8035	– 1-ethyl-	$C_3H_{13}N$	0.7487

Table 5.3 Density of organic substances *(continued)*

Name	Formula	Density, 10^3 kg/m^3	Name	Formula	Density, 10^3 kg/m^3
– 3-methoxy	C_3H_9ON	0.7204 (17°C)	2-Pyrrolidone	C_4H_7NO	1.120
– *N*-methyl-	$C_4H_{11}N$	0.7574	– 1,5-dimethyl-	$C_6H_{11}NO$	1.0242
– 1,2,2-trimethyl	$C_6H_{15}ON$	0.7668	– 1-methyl-	C_5H_9NO	1.0328
Propylenediamine	$C_3H_{10}N_2$	0.8584 (25°C)	3-Pyrroline	C_4H_7N	0.9097
Propylhydrazine	$C_3H_{10}N_2$	0.8406	Silanol, trimethyl-	$C_3H_{10}OSi$	0.8112
Propylisocyanide	C_4H_7N	0.7599 (17°C)	Tetraethyl germanium	$C_4H_{20}Ge$	0.9932
Propylisothiocyanate	C_4H_7NS	0.9781 (16°C)	Tetraethyl lead	$C_4H_{20}Pb$	1.6524
Propylnitrate	$C_3H_7O_3N$	1.0538	1,4-Thiazine, tetrahydro-	C_4H_9NS	1.0882
Propynal (propioaldehyde)	C_3H_2O	0.9152	Thiazole	C_3H_3NS	1.1998 (17°C)
Propyne	C_3H_4	0.7062 (−50°C)	1,2,3-Triazole	$C_2H_3N_3$	1.1861 (25°C)
– 3-bromo-	C_3H_3Br	1.579	Thiethane	C_3H_6S	1.0020
– 3-chloro-	C_3H_3Cl	1.0297	Thiirane	C_2H_4S	1.0130
– 1-ethoxy-	C_5H_8O	0.8276	– 2,2-dimethyl-	C_4H_8S	0.9013
– 3-iodo-	C_3H_3I	2.0177 (0°C)	– ethyl-	C_4H_8S	0.927
– 3-methoxy-	C_4H_6O	0.83 (12°C)	– methyl-	C_3H_6S	0.944
2-Propyne-1-ol (propargyl alcohol)	C_3H_4O	0.9485	– phenyl-	C_8H_8S	1.1044 (25°C)
			Thioacetic acid	C_2H_4OS	1.064
Propynoic acid (propargylic acid)	$C_3H_2O_2$	1.1380	Thiophene	C_4H_4S	1.0648
			– 2-acetyl-	C_6H_6OS	1.1679
Pyran-2-on, tetrahydro-	$C_5H_8O_2$	1.0794	– 2-bromo-	C_4H_3BrS	1.684
2*N*-Pyran, 3,4-dihydro-	C_5H_8O	0.9261	– 2-chloro-	C_4H_3ClS	1.2863
2*N*-Pyran-2-on (α-pyrone)	$C_5H_4O_2$	1.1972	– 2,5-dibromo-	$C_4H_2Br_2S$	2.147 (23°C)
4*N*-Pyran-4-on (γ-pyrone)	$C_5H_4O_2$	1.190	– 2,3-dimethyl-	C_6H_8S	1.0021
Pyrazine	$C_4H_4N_2$	1.0254 (60°C)	– 2-ethyl-	C_6H_8S	0.9928
Pyrazole	$C_3H_4N_2$	1.001 (99.8°C)	– 2-hydroxy-	C_4H_4OS	1.255
2-Pyrazoline	$C_3H_6N_2$	1.020 (17°C)	– 2-iodo-	C_4H_3IS	2.0595 (25°C)
Pyridazine	$C_4H_4N_2$	1.111	– 2-methyl-	C_5H_6S	1.0193
Pyridine	C_5H_5N	0.9835	– tetrachloro-	C_4Cl_4S	1.7036 (30°C)
– 2-acetyl-	C_7H_7NO	1.0776 (15°C)	– tetrahydro-	C_4H_8S	0.9987
– 2-bromo-	C_5H_4BrN	1.6337	Thiophene-1, 1-dioxide,tetrahydro-	$C_4H_8O_2S$	1.2615 (30°C)
– 2-chloro-	C_5H_4ClN	1.2029			
– 2,3-dimethyl-	C_7H_9N	0.9453	2-Thiophenecarbaldehyde	C_5H_4OS	1.224
– 2-ethyl-	C_7H_9N	0.9304	Thiophenol	C_6H_6S	1.0766
– 2-methoxy-	C_6H_7NO	1.0457	– 2-chloro-	C_6H_5ClS	1.2752
– 2-methyl-	C_6H_7N	0.9443	– 4-methoxy-	C_7H_8OS	1.1313 (25°C)
– 1,2,3,6-tetrahydro-	C_5H_9N	0.9153	– 2-methyl-	C_7H_8S	1.041
– 2-vinyl-	C_7H_7N	0.9757	Thiopyran-2-on, tetrahydro-	C_5H_8OS	1.155
Pyridine-4-carbaldehyde (isonicotine aldehyde)	C_6H_5NO	1.1339 (25°C)	Triethylaluminum	$C_6H_{15}Al$	0.825
			Triethylamine	$C_6H_{15}N$	0.7257
2-Pyridylamine, *N,N*–dimethyl-	$C_7H_{10}N_2$	1.0192	Triethylarsine	$C_6H_{15}As$	1.150
2-Pyridylamine, *N*-methyl-	$C_6H_8N_2$	1.0707	Triethylborate	C6H15O3B	0.8546 (28°C)
Pyrimidine	$C_4H_4N_2$	1.1293	Triethylphosphate	$C_6H_{15}O_4P$	1.0695
Pyrrole	C_4H_5N	0.9698	Triethylphosphine	$C_6H_{15}P$	0.8006
– 1-acetyl-	C_6H_7NO	1.044	Triethylphosphite	$C_6H_5O_3P$	0.9629
– 2,4-dimethyl-	C_6H_9N	0.9208	Trimethylaluminum	C_3H_9Al	0.752
– 1-ethyl-	C_6H_9N	0.9009	Trimethylamine	C_3H_9N	0.6356
– 2-isopropyl-	$C_7H_{11}N$	0.906 (25°C)	Trimethylarsine	C_3H_9As	1.144 (15°C)
– 2-methyl-	C_5H_7N	0.9295	Trimethylborate	$C_3H_9O_3B$	0.915
– 1-propyl-	$C_7H_{11}N$	0.8833	Trimethylenediamine	$C_3H_{10}N_2$	0.884 (25°C)
Pyrrolidine	C_4H_9N	0.8586	Trimethylphosphate	$C_3H_9O_4P$	1.2144
– 1-ethyl-	$C_6H_{13}N$	0.8263	Trimethylphosphite	$C_3H_9O_3P$	1.052
– 1-methyl-	$C_5H_{11}N$	0.8188	Trimethylstibine	C_3H_9Sb	1.523 (15°C)
– 1-propyl-	$C_7H_{15}N$	0.8172	Xylose	$C_5H_{10}O_5$	1.525)

5.5 Alloys, Minerals, Wood and Some Other Solids

Table 5.4 Density of alloys, minerals, wood and other solid substances

Name	Density, 10^3 kg/m^3
Alloys	
alumel	8.48
brass	8.2–8.85
bronze	7.5–9.1
cast iron:	
gray	7.0–7,2
white	7.6–7.8
chromel	8.7
constantan	8.88
copel	8.9
cunial	8.5–8.7
cupronickel	8.9
duralumin	2.6–2.9
Electron	1.8
ferronichrome	8.4
invar	7.9
magnalium	2.50
manganin	8.4
Monel metal	8.8
nichrome	8.4
nickeline	8.8
argentan	8.7–8.82
platiniridium	21.62
steel:	
stainless, heat-proof, heat-resistant	7.9–8.2
unalloyed, low- and medium-alloyed	7.77–7.85
Wood's alloy	9.7
Minerals	
agate	2.5–2.8
alabaster:	
carbonaceous	2.69–2.78
sulfate	2.26–2.32
albite	2.62–2.65
andesite	2.2–2.7
anorthite	2.74–2.76
asbestos	2.1–2.8
asbestos shale	1.8
basalt	2.6–3.25
beryl	2.69–2.70
beschtaunite	2.4–2.5
calcite	2.6–2.8
chalk	1.8–2.6
clay	1.6–2.9
coal:	
anthracite	1.4–1.8
bituminous	1.2–1.5
coke	1.0–1.7
copal	1.04–1.14
corundum	3.9–4.0
diamond	3.01–3.52
dolomite	2.84
emery	4.0
feldspar	2.55–2.75
flint	2.63
fluorite	3.18
galenite	7.3–7.6
garnet	3.15–4.3
gaseous coal	1.88
granite	2.34–2.76
quartz:	
melted	2.65
nontransparent	2.07
transparent	2.21
gypsum	2.31–2.33
hematite	4.9–5.3
hornblende	2.9–3.2
volcanic tuff	0.75–1.4
limestone	2.68–2.76
magnetite	4.9–5.2
malachite	3.7–4.1
marble	2.6–2.84
marl	2.3–2.5
mica:	
black	2.7–3.1
normal	2.6–3.2
white	2.76–3.00
opal	2.2
peat, dry	0.5
porphyry	2.6–2.9
pumice	0.4–0.9
pyrite	4.95–5.1
rock salt	2.18
sandstone	1.9–2.65
serpentine	2.5–2.65
shale	2.6–3.3
slaked lime	1.15–1.25
talk	2.7–2.8
thorianite (R)	9.32–9.33
thorite (R)	4.5–5.4
topaz	3.5–3.6
tourmaline	3.0–3.2
trogerite (R)	3.3
uranite (R)	6.5–9.7
uranite:	
calcium (R)	3.05–3.19
copper (R)	3.22–3.60
Other substances	
amber	1.1
asphalt	1.1–2.8
beeswax	0.96

Table 5.4 Density of alloys, minerals, wood and other solid substances *(continued)*

Name	Density, 10^3 kg/m^3	Name	Density, 10^3 kg/m^3
bitumen	~1	white	1.6–2.77
blast furnace slag	2.6–3.0	slate	2.7–2.8
bone	1.7–2.0	snow (soft)	0.12
brick:		soapstone	2.6–2.8
refractory	1.7–2.0	soil	1.3–2.0
normal	1.4–1.6	starch	1.5
camphor	0.99	stearin	1.0
carbon black	1.8–1.9	stone casting	2.9–2.95
cardboard	0.69	sugar	1.59
celluloid	1.4	tar	1.02
ceramics	2.1–2.3	vaseline	0.8–0.9
chamotte	1.85–2.2	**Wood (dry)**	
charcoal:		alder	0.42–0.68
oak	0.57	apple tree	0.66–0.84
pine	0.28–0.44	ash tree	0.65–0.85
cinnabar	8.12	aspen	0.50
colophony	1.07	balsa (cork-like)	0.11–0.14
common salt	2.2	bamboo	0.31–0.40
concrete	1.8–2.5	beech	0.70–0.90
cork	0.22–0.26	birch	0.51–0.77
dry skin	0.86	box tree	0.95–1.16
ebonite	1.15	Canadian spruce	0.48–0.70
faience	1.9–1.96	carob tree	0.67–0.71
gamboge	1.2	cedar	0.49–0.57
gelatin	1.27	cherry tree	0.70–0.90
glass ceramic	2.5	cornel	0.76
glass:		ebony (black)	1.11–1.33
borosilicate thermostable	2.2–2.4	elm	0.54–0.60
flint glass	3.9–5.9	fir	0.40
quartzitic	2.2	hickory	0.60–0.93
standard	2.4–2.8	holly	0.76
gum arabic	1.3–1.4	ironwood (guaiac)	1.17–1.33
hardened cement	2.6–3.2	juniper	0.56
ice (0°C)	0.917	larch	0.67
ivory	1.83–1.92	lime	0.32–0.59
linoleum	1.18	mahogany (Honduras)	0.66
natural resin	1.0–1.1	mahogany (Spain)	0.85
ocher	3.5	maple	0.62–0.75
paper	0.7–1.15	nut tree	0.64–0.70
paraffin	0.87–0.91	oak	0.60–0.90
plant rubber	0.91	pear tree	0.61–0.73
porcelain	2.3–2.5	pine:	
Pyrex	2.25	customary	0.37–0.60
red fiber	1.45	white	0.50–0.55
red lead	8.9–9.1	plane tree	0.40–0.60
rubber:		plum tree	0.66–0.78
pure	0.91–0.93	poplar	0.35–0.5
soft	1.1	sandalwood	0.91
solid	1.19	satinwood	0.95
sand:		sweet gum	1.00
damp	1.9–2.1	teakwood:	
dry	1.2–1.6	African	0.98
sealing wax	1.8	Indian	0.66–0.88
slag glass ceramic:		willow	0.40–0.60
gray	2.6–2.75		

5.6 Plastics

Table 5.5 Density of plastics [8]

Name	Density, 10^3 kg/m^3
Acrylnitrilebutadiene-styrene plastic	1.03–1.05
Acrylovinyl mass	1.13
Aminoplast	1.4–2.0
Angelite	1.7–1.95
Anid	1.14–1.16
Anid glass felled	1.4
Antifriction plastic	1.74–1.80
Asbestos textolite electrotechnical, sheet	1.5–1.7
Caprolite	1.2
Caprolone	1.15–1.16
Caprone	1.1–1.2
Copolymer:	
ethylene-propylene	0.90–0.96
ethylene-vinyl acetate	0.92–0.96
styrene	1.12–1.14
Dacryl	1.19
Difsane	1.32
Etrol:	
acetopropionate cellulose	1.16–1.25
acetyl cellulose	1.27–1.34
nitro cellulose	1.8–2.0
Fiber glass	1.65–1.78
Fiber glass material	1.7–1.9
Fiberglass anisotropic material	1.8–2.0
Flan	1.2–2.6
Fluoroplast	1.65–1.80; 2.02–2.23
Foam plastic K-40	0.2–0.4
Furanite	1.60–1.85
Getinaks electrotechicial, sheet	1.28–1.45
Glass organic:	
construction	1.18–1.19
technical	1.18
Glass textolite:	
construction	1.77–1.9
electrotechnical	1.6–1.9
Isodine	1.35–1.45
Laminated-wood plastic	1.25–1.30
Niplone	1.3–1.34
Pentoplast	1.32–1.40
Phenolic pressing mass	1.4–1.85

Name	Density, 10^3 kg/m^3
Phenolite	1.5–1.6
Phenoplast:	
cast	1.4–1.5
insulating	1.85
general-purpose	1.40–1.45
heat-resistant	1.75–1.90
moisture-chemical resistant	1.5–1.6
high-impact	1.45
Phenylon	1.35
Polyacrylate	1.2
Polyamide	1.02–1.13
Polyamide glass-filled	1.35–1.38
Polycarbonate (diflone)	1.2
Polyethelene:	
high-pressure	0.90–0.94
low-pressure	0.95–0.96
Polyethylenepyreftalate (Dacron)	1.32–1.53
Polymethyl methacrylate:	
casting	1.18–1.2
suspension	1.19
Polypropylene	0.9–0.92
Polysilfone	1.25
Polystyrene:	
general-purpose	1.05–1.10
high-impact	1.06
Polystyrolic plastic	1.2
Polyurethane	1.21
Polyvinyl chloride	
suspension	1.34–1.43
emulsion	0.5–0.63
Polyvinyl chloride foam plastic	0.1–0.3
Polyvinyl chloride plastic:	
sheet	1.38–1.43
high-impact	1.40–1.45
Retinax	2.4–2.7
Resin:	
epoxy	1.13–1.2
epoxydianil	1.2–1.3
polyamide	1.11–1.12
polyepoxy	1.1–1.2
polyester	1.08–1.3
Textolite:	
constructional	1.3–1.4
electrotechnical sheet	1.25–1.45
graphitized	1.3–1.4

5.7 Liquids

Table 5.6 Density of liquids [5, 17]

Name	Density, 10^3 kg/m^3	Name	Density, 10^3 kg/m^3
Acetone	0.80	Mazut	0.91–0.99
Benzene	0.88	Milk	1.03
Benzine	0.7–0.8	Nitric acid (100%)	1.50
Coal tar	1.05–1.25	Oil:	
Copper sulfate:		castor	0.96
10%	1.107	transformer	0.84–0.89
20%	1.230	vegetable	0.91–0.97
Diesel fuel	0.86	Petroleum	0.73–0.93
Drying oil	0.93–0.95	Sodium chloride solution in water:	
Ether	0.72	10%	1.071
Ethyl alcohol	0.79	20%	1.148
Glycerin	1.26	Sulfuric acid (concentrated)	1.83
Hydrochloric acid (20%)	1.10	Turpentine	0.87
Kerosene	0.81–0.84	Water glass (liq.)	1.35–1.53
Creosote	1.04–1.10		

5.8 Mercury at Atmospheric Pressure and Various Temperatures

Table 5.7 Density of mercury under various temperatures and atmospheric pressure [13]

t,°C	Density, 10^3 kg/m^3	t,°C	Density, 10^3 kg/m^3	t,°C	Density, 10^3 kg/m^3	t,°C	Density, 10^3 kg/m^3
0	13.59503						
10	13.57039	210	13.08814	410	12.6082	610	12.104
20	13.54583	220	13.06431	420	12.5838	620	12.078
30	13.52133	230	13.04048	430	12.5593	630	12.051
40	13.49689	240	13.01665	440	12.5348	640	12.025
50	13.47251	250	12.99282	450	12.5101	650	11.998
60	13.44819	260	12.96898	460	12.4854	660	11.972
70	13.42393	270	12.94514	470	12.4607	670	11.945
80	13.39971	280	12.92127	480	12.4358	680	11.918
90	13.37554	290	12.89739	490	12.4109	690	11.891
100	13.35142	300	12.87350	500	12.386	700	11.863
110	13.32734	310	12.8496	510	12.361	710	11.836
120	13.30330	320	12.8256	520	12.336	720	11.809
130	13.27929	330	12.8016	530	12.310	730	11.781
140	13.25531	340	12.7776	540	12.285	740	11.753
150	13.23137	350	12.7536	550	12.259	750	11.725
160	13.20754	360	12.7294	560	12.234	760	11.697
170	13.18356	370	12.7053	570	12.208	770	11.669
180	13.15968	380	12.6811	580	12.182	780	11.641
190	13.13582	390	12.6569	590	12.156	790	11.612
200	13.11197	400	12.6326	600	12.130	800	11.584

5.9 Distilled Water at Atmospheric Pressure and Various Temperatures

Table 5.8 Density of distilled water under various temperatures and atmospheric pressure [13]

t,°C	Density, 10^3 kg/m^3	t,°C	Density, 10^3 kg/m^3	t,°C	Density, 10^3 kg/m^3	t,°C	Density, 10^3 kg/m^3	t,°C	Density, 10^3 kg/m^3
0	0.999841								
1	0.999900	6	0.999941	11	0.999606	16	0.998943	21	0.997994
2	0.999941	7	0.999909	12	0.999498	17	0.998775	22	0.997772
3	0.999965	8	0.999849	13	0.999377	18	0.998596	23	0.997540
4	0.999973	9	0.999782	14	0.999244	19	0.998406	24	0.997299
5	0.999965	10	0.999701	15	0.999099	20	0.998205	25	0.997047

References

[1] Moses, J., The practicing scientist's handbook. A guide for physical and terrestrial scientists and engineers, Van Nostrand Reinhold Company, New York, 1978.

[2] Properties of elements. Handbook, Ed. Samsonov, G. V., Metallurgiya, Moscow, 1976, part 1 (in Russian).

[3] Properties of inorganic compounds. A handbook, Ed. Efimov, A.I., Khimiya, Leningrad, 1983 (in Russian).

[4] Nuclear wallet cards, Ed. Shirley, V. S. and Lederer, C. M., Produced by the Isotopes Project, Lawrence Berkeley Laboratory, on behalf of the U. S. Nuclear Data Network, 1979.

[5] Rabinovich, V. A., Khavin, Z. Ya., Concise chemical handbook, 2nd ed., Khimiya, Leningrad, 1978 (in Russian).

[6] Gordon, A., Ford, R., The chemist's companion. A handbook of practical data. Techniques and references, Wiley, New York, 1972

[7] Properties of organic compounds. A handbook, Ed. Potekhin, A. A., Khimiya, Leningrad, 1984 (in Russian).

[8] Kantsel'son, M. Yu., Balaev, G. A., Plastics. Properties and applications. Handbook, 3rd ed., Khimiya, Leningrad, 1978 (in Russian).

[9] Smiryagin, A. P., Smiryagina, N. A., Belova, A. V., Commercial nonferrous metals and alloys, 3rd ed., Metallurgiya, Moscow, 1974 (in Russian).

[10] Kaye, G., Laby, T., Tables of physical and chemical constants and some mathematical functions, Longmans, Green & Co, London, 1941.

[11] Kaye, G.W., Laby, T.H., Tables of physical and chemical constants, Longmans, Green & Co, London, 1959.

[12] Vargaftik, N. B., A guidebook for thermal properties of gases and liquids, Nauka, Moscow, 1972 (in Russian).

[13] Kuchling, H., Physik, Fachbuchverlag, Leipzig, 1980.

[14] Chirkin, V. S., Thermal properties in nuclear engineering. A handbook, Atomizdat, Moscow, 1968 (in Russian).

[15] Thermal and atomic power plants. Handbook, Eds. Grigor'ev, V. A., Zorin, V. M., Energoizdat, Moscow, 1982 (in Russian).

[16] A handbook for chemist. Basic properties of inorganic and organic compounds, Ed. Nikol'skii, B. P., 3rd ed., Khimiya, Leningrad, 1971 (in Russian).

[17] Perel'man, V. I., A concise handbook for chemist, 6th ed., Gosudarstvennoye nauchno–tekhnicheskoe izdatel'stvo khimicheskoi literatury, Moscow, 1963 (in Russian).

6

Friction

I.I. Karasik

6.1 Introduction

External friction (or, simply, friction) is an appearance of resistance to a relative motion arising between two bodies at the zones of contacting surfaces tangentially to them, which is attended with the energy dissipation. Static friction of rest is to be observed as predisplacement on a microscopic level before transition to relative motion on a macroscopic level, after which the dynamic friction occurs (when applied shearing force is increased). For friction of sliding the velocities of the bodies at point of contact are different in absolute value and/or direction as distinct from friction of rolling when these velocities are equal in modulus and direction.

The main characteristic of friction is a force of friction that is the force of resistance at relative motion of a body on the surface of another under the action of an external force. The former is tangential to common boundary between the bodies. In this connection one could recognize the greatest frictional force of rest within the preliminary microdisplacements (as a rule, named merely frictional static force or adhesive force) and a dynamic friction force. Similarly, the coefficient of cohesion and coefficient of friction f are distinguished, respectively, as a ratio of indicated forces of friction to a force (the load) normal to the surfaces of friction and pressing the bodies against each other.

In the general case, the dynamic friction force is given as

$$F = \frac{1}{v}\frac{dA}{dt}, \tag{6.1}$$

where v is the velocity of relative motion of mutually rubbing bodies; dA/dt is the power of friction losses related with mechanical (deformation, adhesive interaction), physical (adsorption, sound and electromagnetic effects, etc.), chemical (chemisorption, chemical modification of a surface layer) dissipative processes (A is the work done by a friction, i.e. dissipated energy).

For typical laboratory conditions in the inactive media and low sliding velocities, as well as for friction of rolling, the dissipation components related with mechanical interaction at zones of friction are of decisive importance.

Friction force is a result of a number of interaction events distributed in time and over small discrete sites of a real contact zone:

(1) direct contact of rubbing bodies materials in the initial state;

(2) contact through the layer of liquid oil lubricant (hydrostatic and hydrodynamic lubrication);

(3) contact through materials adsorbed and chemisorbed from surroundings and lubricant (in particular, through the boundary layers of lubricating material, the dust particles, etc.);

(4) contact through the secondary (produced in the course of friction) structures in the surface layers of rubbing bodies.

0-8493-2861-6/97/$0.00+$.50

The contributions of these interconnected components to total force of friction are different at various loadings P and sliding velocities v so that the coefficient of kinetic friction in the general case is a function of P and v, the ambient conditions of friction and lubrication (particularly, the heat removal conditions), shape and size of rubbing bodies.

When the indicated parameters and ambient conditions are fixed, after a long enough adaptation period, the rubbing bodies surface microconfiguration of a specific nature is established for the given system of parameters, composition structure of surface layers, the magnitudes of frictional force and coefficient of friction.

During the adaptation period, as a rule, the relative contribution of plastic deformation reduces and that for elastic deformation rises at the microcontacts, while the contact through protecting layers of lubricant and the secondary structures become more pronounced and direct contact is lowered to give reduction of frictional forces as a whole.

The experimental values of f in the strict sense are only true for the specific test conditions and for other ones they may be used as an approximation only. The available data of immediate interest in practice are concerned with a wide variety of different nonstandard test conditions.

Using the tables one should take into account that the relative f-data sets or the values for the ratio of f-magnitudes peculiar to various materials are dependent on test conditions to a lesser extent.

In vacuum, the higher friction is observed due to difficulties in formation of protective adsorption layers and the ensuing rise in adhesive interaction. During the running-in-period, the stabilization of f occurs at a higher value than the initial one and terminates after the wear of surface structures developed in the course of the prior surface machining.

A rise in v generally shows itself as a variation in f due to change of rheological characteristics of the materials at the friction zone and heat-up, and the latter exerts an effect on f because adsorption/desorption processes and hardness are varied.

The materials and their conjugations are conventionally classed as antifriction ones (f=0.15–0.12 without oil, f=0.1–0.05 with oil) used in the slider bearings, and friction ones (f = 0.3–0.35, rarely f = 0.5–0.6) used in the coupling friction devices (the brakes, the clutches, frictional transmission etc.).

For the estimation to be suitable with materials of the friction conjugations in the controlled initial state, they try to determine the experimental f-value under some specific conditions: (1) homogeneous contact over the friction surface (uniform pressure distribution on microscopic level), (2) constant area of friction surface and sliding velocity (not causing an essential heat-up), and (3) the certain pressure magnitudes (for example, at pressure equal to the material hardness or to its given portion).

Conversion to other conditions is performed on the basis of analytical expressions taking into account the real contact pressure.

The magnitude of coefficient of sliding friction f is determined as a sum of adhesive f_a and deformation f_d components, i.e. $f = f_a + f_d$. The latter components caused by redeformation losses of the thin surface layers are essential to extremely rough surfaces and polymers (Table 6.1). The adhesive/deformation components ratio is such that the value of f_d can be neglected at 1% accuracy.

The strength of adhesive bond is given by the expression [4]

$$\tau = \tau_0 + \beta p_r, \tag{6.2}$$

where p_r is the real pressure equal to the ratio of the load P to the total area of spots for a real discrete friction contact, resisting the load (with regard to microroughness, waviness and the macroanomalies of surface geometry); τ_0 , β are the parameters characterizing the properties of the surface layer materials.

On the basis of the above relationship one gets

$$f_a = \frac{\tau}{p_r} = \frac{\tau_0}{p_r} + \beta. \tag{6.3}$$

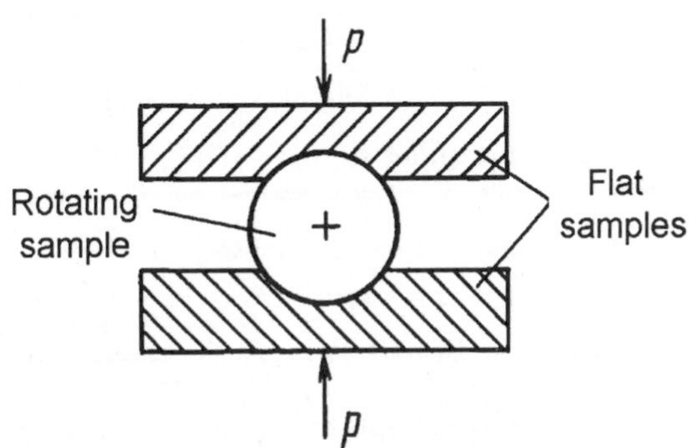

Figure 6.1 Arrangement of the samples for the standard method of τ_0 and β determination.

The standard experimental technique for determining the parameters τ_0 and β is based on the following procedure (see also Fig. 6.1):

(1) pressing in a ball-shaped specimen (made of more hard material) normal to the plane surface of the counter specimen (made of less hard material) with a force resulting in a plastic deformation of the latter;

(2) rotating the specimen about the axis normal to the counter-specimen surface;

(3) measuring the momenta developed by the frictional forces at the interface of specimen and counter-specimen, unloading of the specimens and measuring the size of counter-specimen indentations;

(4) repeating the experiments with the load by two orders of magnitude lesser than before; it is considered that a real contact area is equal to the spherical contour area of the pressing-in zone.

In Table 6.2, along with τ_0 and β, the values of frictional coefficients f_{ap} are also presented. They are taken at real pressure corresponding to hardness for specimen of less hard material of tested conjunction (p_r =HB).

The real pressure p_r depends on mechanical properties of a material at the surface and the surface microgeometry. The data of Table 6.2 can be used for a comparison of material friction conditions at the same values of real pressure. At very low pressures, the comparison of adhesive bond strength is performed using the values of τ_0. At high pressures close to HB, the coefficient of friction is estimated by f_{ap}, at high p and low τ_0 – by β.

The real pressure effect on the coefficient of friction is estimated using τ_0 and β.

Coefficient of rolling friction f_r, as well as that of sliding friction f_s, is determined as a sum of adhesive f_{ra} and deformation f_{rd} constituents: $f_r = f_{ra} + f_{rd}$. The adhesive constituent at rolling friction without sliding is related to the repeated (in the course of the friction process) breaking of adhesive bonds along their direction.

Deformation constituent f_{rd} at rolling friction is given by

$$f_{rd} = \frac{3}{16}\frac{a}{r}c \tag{6.4}$$

for sphere of radius r rolling over a plane, and

$$f_{rd} = \frac{2}{3\pi}\frac{a}{r}c \tag{6.5}$$

for cylinder of radius r rolling over a plane, where a is the radius of a contact area (for sphere) and the length of contact zone in the direction of motion (for cylinder) calculated according to Hertz formula; c is the coefficient of hysteresis (deformation) losses which is a measure of the internal friction for the near-surface material.

For rolling friction, the larger effect of deformation (hysteresis) losses is exhibited by more loaded bodies, materials having lower modulus of elasticity, smaller hardness (e.g., acrylic plastics). Therewith the higher values of f_r are evident in comparison with materials for which the adhesive constituent is prevailing (e.g., for glass). For other materials (e.g., steel, copper, etc.) the contributions of both constituents to f_r are substantial.

6.2 Adhesive Bond under Sliding Friction

Table 6.1 The ratio of adhesive f_a and deformation f_d component under sliding friction

Substance	Fluoroplastic	High-density polyethylene	Polyamides	Polycarbonates	Rubbers	Metals
f_a/f_d	0.2–1.0	1–2	2–9	4–9	2–3	$\leq$100

Table 6.2 Shearing strength of adhesive bond for various materials under friction over ШХ-15 steel [6]

Substance	HB, 10^7 Pa	f_a	$\tau_0, 10^7$ Pa	β
		Metals		
Aluminum	23	0.124	3.00	0.043
Antimony	27.0	0.127	0.73	0.100
Armco–iron	65.0	0.160		
Beryllium			0.45	0.250
Bismuth	7.70	0.175	0.454	0.116
Cadmium	23.0	0.096	0.943	0.055
Chromium	200	0.095		
	100	0.135	1.50	0.120
Cobalt	130.0	0.092		
Copper	85.0	0.10	1.70	
	52.0	0.115	1.82	0.080
	28.5	0.139	1.68	
Indium	0.80	0.20	0.107	0.066
Iron	130.0	0.097		
Lead	3.3	0.140	0.274	0.057
Magnesium	44.0	0.082	8.00	0.020
Molybdenum	186.0	0.095	2.79	0.080
	110.0	0.105	1.87	0.088
Nickel	180.0	0.095	3.48	
	105.0	0.130	1.47	0.116
	70.0	0.123	0.49	
Niobium	32.0	0.142	0.896	0.114
Platinum			9.50	0.10
Rhenium	105	0.095		
Silver	55	0.096	0.77	0.081
		0.096	1.00	0.080
Tantalum	78.0	0.115	2.42	0.084
Tin	4.40	0.170	0.449	0.068
Titanium	190	0.085	2.90	0.080
	128	0.10	2.82	0.078
Tungsten	285	0.082		
Vanadium	110.0	0.103		
Zinc	33.0	0.088		
Zirconium	74.0	0.121		
		Bearing alloys		
БН	25	0.102		
Б83	24	0.150		
ACC-6-5			1.0	0.065
A-20			1.6	0.050
Beryllium bronze	150	0.095		
A-9			0.75	0.12

Table 6.2 Shearing strength of adhesive bond *(continued)*

Substance	HB, 10^7 Pa	f_a	$\tau_0, 10^7$ Pa	β
Bearing alloys				
ЦАМ			0.50	0.05
Бр. Б2		0.095	2.45	0.085
Бр. АЖ Мц 10-3-1.5		0.120	2.00	0.100
Rigid PVC (clear)	12.0	0.091	0.372	0.06
Caprolan	13.0	0.065	0.15–0.195	0.05
Caprone industrial	7.0	0.063	0.161	0.04
Low-density polyethylene	2.0	0.080	0.044	0.058
	2.6	0.090	0.130	0.040
High-density polyethylene	3.8	0.080	0.114	0.050
Polypropylene	3.70	0.380	0.011	0.035
Polycaproamide	7.50	0.088		
68П Polyamide	16.0	0.085		
Polymethyl methacrylate		0.220	1.0	0.150
Acrylic plastic	16.0	0.220		
Fluoroplastic	3.10	0.028	0.341	0.017
Phenylone antipyzine	31.0	0.065		
Wood				
Compressed (unimpregnated)			0.050	0.061
Impregnated with industrial 45 oil			0.100	0.080
Impregnated with motor oil			0.025	0.076
Impregnated with F-4 fluoroplastic			0.70	0.074
Impregnated with ceresin			0.070	0.038
Impregnated with ceresin and stearic acid			0.135	0.028
Rubber				
Rubber			0.250	0.010
Tanning semifinished product				
Raw material			$0.22 \cdot 10^{-4}$	0.016
Depilated hide			$0.20 \cdot 10^{-4}$	0.014
Tanned			$1.7 \cdot 10^{-4}$	0.080
Synthetic diamond on steel				
30ХГСА	340	0.125	20.06	0.066
45	270	0.119	20.39	0.044
08Х18Н9Т	159	0.15	3.18	0.130
40Х	341	0.109	18.41	0.055
45	324	0.112	12.96	0.072
Friction with lubrication				
30ХГСА steel with grease				
ЦИАТИМ-201		0.045	0.1	0.04
АК-6		0.095	0.1	0.10
Petrolatum		0.100	1.0	0.12
Kerosene		0.123	1.65	0.11
Bronze Бр.Б2 with grease				
ЦИАТИМ-201		0.032	0.05	0.03
АК-6		0.063	0.10	0.06
Petrolatum		0.070	0.10	0.07
Kerosene		0.073	0.70	0.06
Bronze Бр.АЖМЦ 10-3-1.5 with grease				
ЦИАТИМ-201		0.340	0.05	0.032
АК-6		0.075	0.10	0.073
Petrolatum		0.08	0.10	0.078
Kerosene		0.10	0.10	0.10

6.3 Static Friction

Table 6.3 Coefficient of static friction (adhesion) [1]

Substance	f_a	Substance	f_a
Metals on metal:		Graphite on graphite:	
thoroughly cleaned of oxide scale	100	cleaned and degasified	0.5–0.8
unlubricated in the open air	1.0	in open air, cleaned	0.1
lubricated with petroleum oil	0.2–0.4	in open air, lubricated	0.1
lubricated with animal and vegetable oils	0.1	graphite (cleaned and lubricated) on steel	0.1
Alloys on steel:		cleaned rock salt on rock salt	0.8
copper-lead alloy, unlubricated	0.2	Soda nitrate on soda nitrate:	
copper-lead alloy, lubricated with petroleum oil	0.1	cleaned	0.5
White metal, Wood's alloy:		lubricated	0.12
unlubricated	0.7	Ice on ice:	
lubricated with petroleum oil	0.1	below −50°C	0.5
Phosphor bronze, brass:		within 0 and −20°C	0.05–0.1
unlubricated	0.35	Tungsten carbide on steel:	
lubricated with petroleum oil	0.15–0.2	both cleaned	0.4–0.6
Ordinary iron:		both lubricated	0.1–0.2
unlubricated	0.4	Plastics:	
lubricated with petroleum oil	0.1–0.2	Perspex, polyethylene on perspex, polyethylene, both cleaned	0.8
High-hard steel lubricated with:			
vegetable and animal oil	0.08–0.1	on steel, cleaned	0.3–0.5
petroleum oil	0.12	nylon on nylon	0.5
molybdenum disulfide	0.1	PTFE on PTFE	0.04–0.1
oleic acid	0.08	on steel	0.04–0.1
spirit, gasoline	0.4	Natural fibers:	
glycerin	0.2	wool fiber on horn (f_2 – against the nap, f_1 – with the nap)	$f_2 = 0.8$–0.1 $f_1 = 0.4$–0.6
unlubricated	0.6		
Thin metal layers on the solid base:		lubricated	$f_2 = 0.5$–0.3
indium film 10^{-3}–10^{-4} cm thick	0.08	same as above	$f_1 = 0.3$–0.4
lead films	0.15	Natural fibers, supplies:	
copper films	0.3	cotton on cotton (fiber)	0.3
Nonmetallic materials:		on cotton (cotton wool)	0.6
glass on glass, cleaned	1	silk on silk	0.2–0.3
lubricated with liquid hydrocarbons, fatty acids	0.3–0.6	Wood on wood:	
		cleaned, dry	0.25–0.5
lubricated with solid hydrocarbons, fatty acids	0.1	cleaned, wet	0.2
Precious stones:		Wood on metal:	
diamond on diamond (cleaned and degasified)	0.4	cleaned, dry	0.2–0.6
		cleaned, wet	0.2
cleaned in open air	0.1	Wood (cleaned, dry) on brick	0.3–0.4
lubricated	0.05–0.1	Leather on metal:	
Sapphire on sapphire:		cleaned, dry	0.6
cleaned and degasified	0.6	cleaned, wet	0.4
in open air, cleaned	0.2	cleaned, lubricated	0.2
in open air, lubricated	0.15–0.2	Braking material on cast iron:	
		cleaned	0.4
		wet	0.2
		lubricated	0.1

Table 6.4 Static friction of dry alloys on steel [11]

Alloy	f
Aluminum bronze	0.45
Brass	0.35
Cast steel	0.4
Constantan	0.4
Lead-base babbitt	0.55
Lead-copper (nondendritic)	0.22
Lead-copper (dendritic)	0.22
Phosphor bronze	0.35
Steel	0.8
Tin-base babbitt	0.8
Wood's alloy	0.7

Table 6.5 Static friction of lubricated metals on steel [11]

Carrying surface	Rapeseed oil	Castor oil	Petroleum oil	Fatty acids
Brass		0.11	0.19	0.13
Bronze	0.12	0.12	0.16	
Cast iron	0.11	0.15	0.21	
Gun steel	0.15	0.16	0.21	
Hard steel	0.14	0.12	0.16	0.09
Lead-base babbitt			0.1	0.08
Pure tin			0.6	0.21
Pure lead			0.5	0.22
Sintered bronze			0.13	
Tin-base babbitt			0.11	0.07

Table 6.6 Static friction of ice on ice [11]

t, °C	0	−12	−71	−82	−110
f	0.05 − 0.15	0.3	0.5	0.5	0.5

6.4 Sliding Friction without Lubrication

Table 6.7 Approximate coefficients of slipping friction for unlubricated materials [5]

Substance	f	Substance	f	Substance	f
Bronze on bronze	0.2	Ferrodo on metal	0.3–0.55	Steel on ice	0.014
Cast iron on cast iron	0.10–0.21	Fiber on metal	0.4–0.8	Steel on:	
Cellophane on:		Iron on:		steel	0.15–0.18
cast iron	0.30–0.37	bronze	0.15	cast iron	0.15–0.18
cellophane	0.38–0.40	oak	0.48	bronze	0.16–0.20
graph paper	0.35–0.38	Iron on:		Wood on wood:	
leather	0.43–0.44	iron	0.44	with grain	0.48
packing paper	0.3–0.4	cast iron	0.17–0.18	across grain	0.34
rubber	0.95	Leather on cast iron	0.2–0.4	Wood on bronze	0.3
steel	0.27–0.29	Paper on cast iron	0.15–0.4	Wood on cast iron	0.25–0.5
writing paper	0.45	Rubber on cast iron	0.5–0.8		

Table 6.8 Coefficient of sliding friction for unlubricated smooth surfaces [5]

Substance	f	Substance	f	Substance	f
Structural steel on:		structural steel	0.18	Chromium-nickel steel on:	
aluminum	0.18	textolite	0.22	chromium-nickel steel	0.18
duralumin	0.20	tin bronze	0.16	duralumin	0.21
ebonite	0.25	Quenched tool steel on:		L62 brass	0.16
electron	0.15	agate	0.22	structural steel	0.15
fiber	0.30	corundum	0.16	35XMЮA steel (nitrated) on:	
Getinaks (paper-based laminate)	0.20	electron	0.14	fiber	0.30
grey cast iron	0.19	fiber	0.22	getinaks	0.24
L62 brass	0.19	quenched steel	0.17	quenched steel	0.24
quenched steel	0.16	red copper	0.15	textolite	0.36
red copper	0.15	textolite	0.27	Brass (L59, L62) on:	
		tin bronze	0.15	aluminum	0.17

Table 6.8 Coefficient of sliding friction for unlubricated smooth surfaces [5] *(continued)*

Substance	f	Substance	f	Substance	f
brass	0.17	Tin bronze on:		duralumin	0.22
copper	0.30	ebonite	0.43	fiber	0.33
duralumin	0.22	fiber	0.27	red copper	0.30
fiber	0.32	grey cast iron	0.21	textolite	0.40
grey cast iron	0.16	textolite	0.23	Grey cast iron on:	
textolite	0.30	tin bronze	0.20	fiber	0.31
tin bronze	0.16	Duralumin on:		grey cast iron	0.22

Table 6.9 Coefficient of dry friction of reinforced phenol plastics on steel

Substance	Reinforcement substance	f
Textolite	Cotton fabric	0.3
Fiber reinforced plastic	Cotton fiber	0.33
Fiber-glass plastic	Glass fiber	0.34
Asbestos-cloth laminate	Asbestos fabric	0.34–0.38
Phenolite	Graphite, caprone, etc.	0.2

Table 6.10 Coefficient of friction of polymer and wood materials on steel [9]

Substance	f
Polymeric materials:	
thermoplastic	0.15–0.40
thermoreactive	0.1–0.4
Carbon graphite materials	0.2–0.35
Metal-ceramic materials	0.25–0.40
Laminated materials	0.1–0.15
Compressed wood impregnated with ceresine	0.08–0.1
Carbon-filled plastics, carbon graphites:	
without lubricant	0.1–0.3
with lubricant	0.01–0.1
Wood filled with:	
solution of polyethylene in MC–20, АРД–1 oil	0.08–0.12
synthesis binder	0.08–0.14
stearate in motor oil	0.06–0.12
stearate in organosilicon compound	0.09–0.12
copper salts and their oxides and АРД–2 glycerin	0.03–0.04

Table 6.11 Coefficient of friction for steel from experiments with inclined surface [5]

Substance	f
On agate	0.1
garnet	0.22
sapphire	0.25
glass	0.12–0.14
synthetic ruby	0.16
ruby	0.29

Table 6.12 Relationship between coefficient of friction and slipping speed [9]

Substance	v, m/s					
	50	100	200	300	400	500
Aluminum oxide	0.17	0.15	0.13	0.12	0.11	0.1
Antimony	0.25	0.18	0.14	0.1	0.07	
Bismuth	0.12	0.07	0.04			
Copper	0.23	0.17	0.12	0.09	0.08	0.07
Steel	0.12	0.10	0.08	0.06	0.05	0.04
Steel 10	0.05	0.035	0.03	0.025	0.022	0.02
Steel (magnet rail brake)	0.04					
Titanium carbide	0.22	0.16	0.11	0.08	0.06	0.05

6.5 Friction of Lubricated Surfaces

Table 6.13 Coefficient of friction of lubricated steel surfaces at room temperature [11]

Lubricant	Length of molecular chain	Melting point, °C	f	Phase transition temperature, °C
Paraffins				
Nonane	C_9	–54	0.26	
Decane	C_{10}	–30	0.23	
Hexadecane	C_{16}	17	0.16	17
Docosane	C_{22}	44	0.11	44
Triacontane	C_{30}	66	0.11	66
Alcohols				
Butyl	C_4	–89	0.3	
Octyl	C_8	–16	0.23	
Decanyl	C_{10}	7	0.16	
Cetyl	C_{16}	49	0.10	
Fatty acids				
Acetic	C_2	16	0.5	
Propionic	C_3	–22	0.4	
Valerianic	C_5	–35	0.17	
Caproic	C_6	–2	0.12	80
Pelargonic	C_9	12	0.11	90
Capric	C_{10}	31	0.11	95
Lauric	C_{12}	44	0.11	120
Myristic	C_{14}	58	0.11	125
Palmitic	C_{16}	64	0.11	130
Stearic	C_{18}	69	0.10	140

Table 6.14 Approximate coefficient of friction at main forms of lubricated contact [8]

Forms of frictional interaction	f
Shift of lubricant film	0.0003
Interfacial lubricant	0.02–0.01
Hydrodynamic radial bearings	0.005

Table 6.15 Coefficient of friction of selected materials on stainless steel in cryogenic liquid medium

Substance	Liquid medium	
	Nitrogen	Hydrogen
Graphite-fluorine metal	0.18	0.22
Graphite+phenol	0.04	0.06
Graphite (15%)+fluoroplastic–4 (85%)	0.09	0.16
Graphite (5%)+nylon (95%)	0.06	0.15

Table 6.16 Approximate coefficients of friction of materials for plain bearings on steel at faulty and mixed lubrication [7]

Substance	HB, 10^7Pa	f	Substance	HB, 10^7Pa	f
Bronzes			Бр. ОФ 10-1	80–100	0.008–0.180
Бр. ОЦСН-3-7-5-1	65	0.013–0.016	Бр. ОФ 7-0.2	75	0.01
Бр. ОЦС 3-12-5	60	0.01–0.015	Бр. ОФ 6-5-0.15	70–90	0.12
Бр. ОЦС 5-5-5	60–75	0.009	Бр. нМЦ 3-1	90	0.015
Бр. ОЦС 6-6-3	60	0.009	Бр. АМЦ 9-2	160	0.006–0.012
Бр. ОЦС 4-4-2.5	60	0.016	Бр. АМЦ 10-2	110	0.006
Бр. ОЦС 4-4-4	62	0.016	Бр. АЖ 9-4Л	120–140	0.012–0.040
Бр. ОЦ 10-2	75–90	0.008–0.060	Бр. АЖ 9-4	110–160	0.012–0.040

Table 6.16 Approximate coefficients of friction of materials *(continued)*

Substance	HB, 10^7Pa	f	Substance	HB, 10^7Pa	f
Бр. АЖМЦ 10-3-1.5	120–140	0.012	ЖГР-3-Д	70–100	0.04–0.07
Бр. АМЦ 9-2Л	90–120	0.012	АЖГр-6-3	20–24	0.005–0.008
Бр. АЖН 10-4-4Л	140–160	0.006–0.012	Бр. ОГ10-2	18–20	0.004–0.008
Бр. АЖН 11-6-6	250	0.011	**Self-lubricating polymers**		
Бр. С30	25	0.008	АМАН-2	29–31	0.12
ЛС 59-1Л	90–40	0.014	АМАН-4	27–29	0.1
ЛК 80-3Л	90–110	0.01–0.19	Esteran 33	22	0.08
Бр. Б2	100	0.016–0.05	АМАН-7	18–20	0.05
Babbitts			АМАН-13	14–16	0.16
Б6	32	0.005	АФ-3ам	28–30	0.12
Б16	30	0.006	($p = 0.2$ MPa; $v = 2$ m/s)		
Б83	27–30	0.005	**Plastics**		
БН	27–29	0.006	Polyamides П-610	10–15*1	0.26–0.32
БТ	26	0.009	Caprone	8–12	0.1–0.15
БКА	32	0.004	П-АК-93/7	10–12*1	0.24–0.25
Carbon materials			Polyimides		
АМС-1	40	0.1	ПМ-67	20–30	0.35*2
АМС-3	35	0.065	ПМ-69	20–27	0.40*2
АФ-3Т	46	0.087	ПАМ 15-69	33	0.18*2
АГ 1500СО5	65–70	0.063	Fluoroplastic-4	3–6	0.03–0.1
Powder and ceramic materials			Pentaplast		0.12
ЖГР-1-20	60–100	0.06–0.09	Polycarbonates	10–11	0.3*2
ЖГР-2-20	50–80	0.06–0.09	Polyacrylate	20	0.4*2

*1 Hardness is estimated by loading 365H indenter.
*2 The figures are determined without lubricant.

Table 6.17 Coefficients of friction in running water [5] [$p = 0.06$ MPa, $v = 6$ m/s.]

Substance	f
Beach and maple on bronze in motion	0.25–0.26
parallel to grains	0.12–0.23
perpendicular to grains	0.08–0.12
Ebonite on bronze	0.25–0.35
Brass on bronze	0.07–0.10
Steel on bronze	0.06
Chrome-plated steel on bronze	0.04
Bronze on bronze	0.04
Rubber	0.22–0.26
Textolite on bronze	0.12–0.14

Table 6.18 Coefficient of friction of graphite and boron nitride used as solid-film lubricants on steel [9]

Substance	f	
	Air	Vacuum
Natural graphite	0.19	0.44
Pyrolytic graphite	0.18	0.50
Boron nitride, hot pressed	0.25	0.70

Table 6.19 Coefficients of static f_{st} and kinematic f friction of materials with working solid-film lubricant on steel [9]

Substance	Conditions	10^5 Pa		$3 \cdot 10^{-4}$Pa	
	of friction (pressure, speed)	f_{st}	f	f_{st}	f
АМАН-2	0.2 MPa; 4 m/s	0.44	0.18	0.1	0.1
M-801 (Mo–MoS_2–Mo)	0.78 MPa; 0.265 m/s	0.4–0.5	0.1	0.15	0.5
Mo–$MoSe_2$	0.03–0.3 MPa; 0.02m/s	0.13	0.12	0.1	0.1

6.6 Friction on Ice and Snow

Table 6.20 Coefficient of friction on ice and snow [8]

Type of friction	f
On soggy ice	0.02
On smooth ice	0.06
Ski on snow	0.10–0.30
On rough ice	0.12
On dense snow	0.20

Table 6.21 Coefficient of friction on ice at speed of 4 m/s [11]

Substance	t,°C					
	0	−10	−20	−40	−60	−80
Ice	0.02	0.035	0.050	0.075	0.085	0.09
Ebonite	0.02	0.050	0.065	0.085	0.100	0.11
Brass	0.02	0.075	0.085	0.115	0.140	0.15

Note: At −3°C for unwaxed ski $f = 0.08$ and for waxed ski $f = 0.03$.

Table 6.22 Coefficient of friction of waxed ski on snow at speed of 0.1 m/s [11]

t,°C	0	−3	−10	−40
f	0.04	0.09	0.18	0.4

Table 6.23 Coefficient of friction of 12X18H10T steel in conjunction with other steels under vacuum [9]

Steel	Pressure P, Pa	Evacuation system	t,°C	τ_0, 10^7Pa	β
P9				6.43	0.232
У7	$6.5{\cdot}10^{-4}$	Oil		2.54	0.32
40X13			20	3.95	0.26
ШX15	$6.5{\cdot}10^{-1}$			5.75	0.159
P9	$6.5{\cdot}10^{-7}$	Oilless		9.65	0.575
P9	$4{\cdot}10^{-5}$		200	9.65	0.545

Table 6.24 Friction of pure metals at one-fold slipping in air and under vacuum of 10^{-3}Pa [9]

Friction pairs	Coefficient of friction	
	In air	Under vacuum, after outgasing
Cu – Ni	0.45	1.50
Ta – Ni	0.23	0.90
W – Ni	0.21	1.36
Cu – Fe	0.51	0.75
Ta – Cu	0.44	0.43
W – Cu	0.34	0.41

Table 6.25 Effect of residual pressure in test chamber on coefficient of friction of stainless steel on itself

Residual pressure, Pa	10^{-5}	$5{\cdot}10^{-5}$	$2{\cdot}10^{-6}$	10^{-6}
Coefficient of friction	0.47	1.22	2.74	2.94

Table 6.26 Coefficient of friction of metals degasified under vacuum [11]

Metal	Degasified metal	Metal after working	Metal after contact with O_2
Nickel on tungsten	6	6	1
Nickel on nickel	5–8	5	2.5
Copper on copper	4.8	4.8	0.7
Gold on gold	4.5	4.5	2.8

6.7 Rolling Friction

Table 6.27 Approximate coefficient of rolling friction [8]

Rubbing conjugations	f_r
Rolling bearings	0.0025
Resistance to free tire rolling	0.01
Rolling of ball on rubber	0.02–0.10
Friction between wheel and rail:	
under moist weather	0.09–0.15
under rainy weather	0.20
under dry weather	0.25–0.30
Rolling cylinder on rubber	0.10–0.30
Tires:	
with wet roads	0.40
with dry roads	0.70

Table 6.28 Coefficients of rolling friction determined with standard method [12] by damping of oscillations of pendulum having spherical bearing of ШХ15 steel [14]

Substance	f_r	Substance	f_r
Copper	$6 \cdot 10^{-4}$	ШХ15 steel	$2.2 \cdot 10^{-6}$
Glass	$5 \cdot 10^{-5}$	ШХ15 steel, uncleaned	$2 \cdot 10^{-5}$
Silicon	$3 \cdot 10^{-5}$		

Table 6.29 The values of hysteresis loss factor [13] determined with standard [12] method for selected materials

Substance	$c \cdot 10^3$	
	initial	annealed
Quartz	0.006	
Rubber	0.1	
Acrylic plastic	0.2	
Manganin	0.31	0.016
Copel	0.31	0.1
Rhenium	0.33	0.024
Ni – Ti alloy	0.41	0.43
Alumel	0.9	0.07
Chromel	0.95	0.87
Tungsten	1.0	0.18–0.35
Beryllium bronze	1.0	0.57
Nichrome	1.2	0.01
Steel	1.2	0.42
Elenvar	1.3	3.1
Silver	1.6	31
Aluminum	6.3	2.1
Copper	7.8	3.1
Constantan	31.4	10

References

[1] Bouden, F.P., Tabor, D., The friction and lubrication, Mashgiz, Moscow, 1960 (in Russian).

[2] Voronkov, B. D., Dry friction bearings, Mashinostroenie, Leningrad, 1971 (in Russian).

[3] USSR State Standard (code ГОСТ 23.002–78).

[4] USSR State Standard (code ГОСТ 23.202–78).

[5] Zaitsev, A. K., Fundamentals of the science of machine friction, wear and lubrication, Mashgiz, Moscow – Leningrad, 1947 (in Russian).

[6] Kragel'skii, I. V., Dobychin, M. N., Kombalov, V. S., Essentials of friction and wear analysis, Mashinostroenie, Moscow, 1977 (in Russian).

[7] Kragel'skii, I. V., Mikhin, N. M., Friction units of machines. Handbook, Mashinostroenie, Moscow, 1984 (in Russian).

[8] Moore, D, Principles and applications of tribology, Pergamon Press, Oxford (UK), 1975.

[9] Friction, wear and lubrication, A handbook, Mashinostroenie, Moscow, 1978, v.1 (in Russian).

[10] Physics encyclopedic dictionary, Ed. Prokhorov A. M., Sovetskaya entsiklopedia, Moscow, 1984 (in Russian).

[11] Bowden, F. P., Tabor, D., The friction and lubrication of solids, Clarendon Press, Oxford, 1954 (Part 1).

[12] USSR State Standard (code ГОСТ 23.214–83).

[13] Karagioz, O. V., Kocheryan, E. G., Izmailov, V. P., Improvement of the Q–factor of evacuated torsion system through plumb line annealing, Fizicheskaya i khimicheskaya obrabotka materialov, No.1, 87, 1972 (in Russian).

[14] Silin, A. A., Karagioz, O. V., Markachov, V. V., Izmailov, V. P., On the unity of the energy dissipation at rolling friction and other kinds of solid–state elastic deformation, Trenie i iznos, 1, 957, 1980 (in Russian).

7

Acoustics

L.K. Zarembo

7.1 Introduction

Acoustic waves refer to mechanical perturbations (displacement, density, pressure) of medium equilibrium state. They propagate in all media (plasma, gases, liquids, solids, etc.). For periodic perturbations, the following ranges are singled out according to the frequency f(in Hz): a sound range (range of hearing) (20 Hz – 20 kHz), an ultrasound range (20 kHz – 10^9 Hz), a hypersound range (10^9 – $\approx 10^{13}$ Hz). At higher frequencies, the discrete structure of solids affects the phenomenon.

Sound pressure (SP) $p(t)$ makes up the variable excessive pressure in a sound field (unit of measure is the pascal: 1 Pa = 1 kg/m·s^2; in CGS, the unit is 1 din/cm^2 = 0.1 Pa; 1 atm = 10^6 din/cm^2). The effective SP, $p_{\text{eff}} = \sqrt{\overline{p^2}}$, is often measured too. Here, the bar marks a time-average variable. For periodic wave $p_{\text{eff}} = p_{\max}/\sqrt{2}$, where $p_{\max}$ is the sound pressure amplitude.

With the displacement amplitude A(m) in the periodic wave of circular frequency $\omega = f \cdot 2\pi$, the vibration velocity amplitude is $v_v = \omega \cdot A$. In the plane wave $v_v = p/\rho_0 v$, where ρ_0 is the unperturbed density and v is the sound velocity in the medium.

Density of sound energy W is the time-average energy of the unit volume in the sound field: $[W] = \text{J/m}^3$. In CGS, the unit is 1 erg/cm^3 = 0.1 J/m^3. In the plane periodic wave $W = \rho_0 v_v^2/2 = p^2/2\rho_0 v^2$.

Sound intensity J is the sound energy flux for 1 s through the area of 1 m^2 that is perpendicular to the sound propagation direction: $\vec{J} = vW\vec{n}$ (W/m^2), here v is the sound velocity (group velocity in the dispersive medium), $\vec{n}$ is a unit vector of wave normal. In CGS, the unit is 1 erg/cm^2·s = 10^{-3} W/m^2.

Sound intensity level $B = 10\lg(J/J_0) = 20\lg(p/p_0) = 20\lg(v_v/v_{v0})$, where $J_0 = 5 \cdot 10^{-12}$ W/m^2 is the intensity, $p_0 = 2 \cdot 10^{-5}$ Pa is the sound pressure, and $v_{v0} = 5 \cdot 10^8$ m/s is the vibration velocity at the hearing threshold. The measure unit of sound intensity level is the decibel (dB). Due to variation of sensitivity of the human ear in the hearing range of frequencies, the sound intensity level does not indicate the subjective interpretation of loudness.

Loudness level of a sound or noise is a name for sound intensity level at the frequency of 1 kHz, which is equally loud with a given sound or noise. The unit for loudness level is the phone. Correlation of loudness level and sound intensity level is shown in Fig. 7.1. The approximate loudness levels of some noises (in phones) are listed below:

Explosions, noises near airplane jets	130 – 140	Factory noises	75
Generator room of electric power station, hearing threshold of pain	120	Street noises	50 – 70
Metro train noise	90 – 100	Conversation	50 – 70
Noise of passing automobile	85	Leaves rustling	20

Sound velocity. The propagation velocity for a given phase of periodic wave (of SP, for example) is called the sound phase velocity v. In a medium with dispersion of sound velocity (with dependence

0-8493-2861-6/97/$0.00+$.50

of the phase velocity on frequency), maximum perturbation caused by a packet of waves with close frequencies propagates with the group velocity v_g. In a medium without dispersion $v = v_g$.

In gases and liquids $v = \sqrt{\kappa\rho_0}$, where κ is the adiabatic compressibility. In the ideal gas $v \propto T^{1/2}$, where T is the gas temperature. In liquids, $dv/dT = -(2-6)$ m/s · K (water and some metal melts are the exceptions).

In unbounded isotropic solids, the longitudinal wave velocity (L-wave) v_l is

$$\rho_0 v_l^2 = \tilde{\lambda} + 2G = K + \frac{4}{3}G = \frac{G(4G-E)}{3G-E} = \frac{E(1-\sigma)}{(1+\sigma)(1-2\sigma)} = \frac{3K(1-\sigma)}{1+\sigma}; \tag{7.1}$$

and that of transverse waves (shear or S-waves) v_s is given as follows:

$$\rho_0 v_s^2 = G = \frac{E}{2(1+\sigma)} = \frac{3K(1-2\sigma)}{2(1+\sigma)}, \tag{7.2}$$

where $\tilde{\lambda}$ is the Lame constant; G is the shear modulus; K is the modulus of volume elasticity; E is the Young modulus; σ is the Poisson's ratio (all parameters are the adiabatic ones). The ratio of velocities takes the form: $v_s/v_l = \{(1-2\sigma)/[2(1-\sigma)]\}^{1/2} < 1$.

In crystals, the sound velocity depends on propagation direction, on wave polarization (the direction of vibrational displacement with respect to wave vector), on crystal symmetry, etc. For arbitrary direction in a crystal, three types of waves with mutually orthogonal polarizations can generally propagate. Among these are the quasilongitudinal waves (QL), for which polarization vector has minimum angle with respect to wave vector, and two quasitransverse waves: fast wave (QFS) and slow wave (QSS). Though there are directions (called special acoustic directions) along which pure transverse S and/or longitudinal L waves propagate. Table 7.1 presents effective modules $M = \rho_0 v^2$, where v is the wave velocity for corresponding type and special direction. The dash in the table means that for the particular crystal point group and for the particular wave type, the direction is not special.

In the tables below, for velocities and absorption of longitudinal (L) and transverse (S) waves, subscripts stand for Miller indexes of propagation direction and superscripts are the Miller indexes of wave polarization.

The above relations and velocity magnitudes in the tables below are presented mainly for the case of unbounded media, when the relation $\lambda \ll h$ is valid. Here $\lambda = v/f$ is the sound wavelength, h is the minimum dimension of a body. In the opposite case, the multiple reflection from the sample boundaries occurs, which leads to the waveguide character of sound propagation. Sound perturbations in this case can be represented as superposition of normal waves (modes) of waveguide, which are distinguished by spatial distribution of amplitudes. Modes, as a rule, possess dispersion. Fig. 7.2 illustrates two types of transverse waves in plates. Phase and group velocities of SH-modes are:

$$v_{SH} = v_S \left[1 - \frac{\pi^2 l^2}{(\omega h/2v_S)^2}\right]^{1/2}, \qquad v_{gSH} = v_S^2/v_{SH}, \tag{7.3}$$

where h is the plate thickness. Modes with inphase vibrations of opposite plate surfaces are called symmetrical (s), and those with antiphase vibrations are called antisymmetrical (a). For symmetrical ones $l = n$, and for antisymmetrical modes $l = (2n+1)/2$ ($n = 0, 1, 2, \ldots$). Phase and group velocities of SV waves (Lamb waves) are shown in Fig. 7.3. In thin plates with $\omega h/v_S \ll 1$, the zero-symmetrical mode of the Lamb wave displays the velocity (plate velocity):

$$v_{pl} = [E/\rho_0(1-\sigma^2)]^{1/2}. \tag{7.4}$$

Zero-antisymmetrical mode of the Lamb wave (bending wave in a thin plate) offers the velocity

$$v_b = [Eh^2\omega^2/12\rho_0(1-\sigma^2)]^{1/4}. \tag{7.5}$$

Phase velocities of the Lamb modes tend to the Rayleigh surface wave velocity v_R (see below) with the growth of wave thickness of a plate.

Velocity of a zero-longitudinal mode in the thin rods ($\omega a/v_r \ll 1$, a is the rod radius) equals $v_r = (E/\rho_0)^{1/2}$. In the nonthin rods, dispersion takes place (see Fig. 7.4). Zero mode of torsion waves in rods has velocity v_s; higher modes possess dispersion.

On a free surface of a solid, the Rayleigh surface waves (SAW) can propagate with a velocity of $v_R = bv_s$, where $b = (0.87 + 1.12\sigma)/(1+\sigma) < 1$. The medium particle trajectories in this wave are elliptic in the plane, which is orthogonal to the surface and contains wave normal. The penetration depth of *SAW* is on the order of wavelength.

On the boundary of media with different characteristic wave impedances $z_1 = \rho_1 v_1$, $z_2 = \rho_2 v_2$, **reflection** of incident and **refraction** of the passed wave occur.

The energy factor of the plane wave reflection from the flat media interface takes the form

$$R = \left(\frac{z_1\cos\theta_2 - z_2\cos\theta_1}{z_1\cos\theta_2 + z_2\cos\theta_1}\right)^2, \tag{7.6}$$

where θ_1 is the incidence angle, θ_2 is the refraction angle ($v_1\sin\theta_2 = v_2\sin\theta_1$). Transmission factor equals $T = 1 - R$. Amplitude factors are $r = R^{1/2}$ and $t = T^{1/2}$, accordingly. In media with spatial inhomogeneity of sound velocity wave, the normal direction changes during propagation (refraction). The sound ray declines to the regions with lower sound velocity.

Attenuation of sound intensity, as the wave is propagated through the media, occurs owing to diffraction (increase of the sound volume), to scattering on the medium inhomogeneities, and to dissipative losses (conversion of sound energy into heat). The two last types of losses are called the damping, and only dissipative losses are called the sound absorption. Damping (which comprises the absorption in homogeneous medium) is governed by the relation:

$$p(l) = p(0)\mathrm{e}^{-\alpha l}, \tag{7.7}$$

where l is the path of a plane wave, α (m^{-1}) is the damping (absorption) factor. Damping is often characterized by attenuation of the sound intensity level. Damping factor in this case is α' (dB/m) $= (20/l)\lg[p(l)/p(0)] = 8.686\alpha$.

Damping (absorption) may be characterized not only by spatial coefficient, but also by the time coefficient, which describes damping on the length passed by the wave for 1 s or for 1 μs: $\Gamma(1/\mathrm{s}) = \alpha\cdot v$; $\Gamma'(\mathrm{dB/s}) = \alpha'\cdot v = 8.686\alpha\cdot v$; $\Gamma''(\mathrm{dB}/\mu\mathrm{s}) = 10^{-6}\Gamma'$.

Damping in resonant systems is characterized by dimensionless quality factor $Q = \omega/2\alpha v$, as well as by logarithmic decrement of damping $\delta = \pi/Q$, which defines the diminishing of free vibrations amplitude during one period.

Absorption is usually caused by the medium viscosity. This part of absorption is $\alpha \sim \omega^2$ and rapidly grows with sound frequency. However, sound perturbations may often initiate more complex process leading to sound dissipation. In the frequency regions, where characteristic times of these processes are close to sound period, a pronounced declination from quadratic dependence of α on ω can be observed.

The accuracy of the best absorption measurements ranges within several per cent, but in general it is about 10%. Damping in structurally inhomogeneous media (for example, in polycrystalline metals) may differ even more.

The data of this chapter refer mainly to quantities measured in the linear acoustic region, i.e. at low perturbations when $v_v/v \sim p/\rho_0 v^2 \ll 1$. It is impossible to mark the precise boundary of transition to nonlinear acoustics. But at $v_v/v \geq 10^{-4}$, nonlinear acoustic phenomena appear, e.g. the compression wave front steepening during its propagation, which for the periodic wave can be represented as higher harmonic generation and the related growth of absorption. In gases and liquids, unidirectional flows are formed (acoustic flows); in liquids, in the regions with negative excessive pressure, formation of gaseous and vapor cavities (degassing and cavitation), heating of medium and self-focusing or self-defocusing, etc., can occur. The wave front steepening and harmonic generation depend (besides intensity) on nonlinearity of equation of state. Namely, for gases it depends on $(\gamma+1)/2$ (where $\gamma = c_p/c_v$ is the ratio of heat capacities); for liquids it depends on the nonlinearity parameter $B/2A$; and for solids it is due to parameters which determine declination from the linearity of the Hook's law.

For strong perturbation with $v_v \sim v$, shock wave is formed. There is a jump in the main parameter's values on its front: in pressure, density and velocity. The front velocity (shock wave velocity) is higher than sound velocity and depends on the jump level.

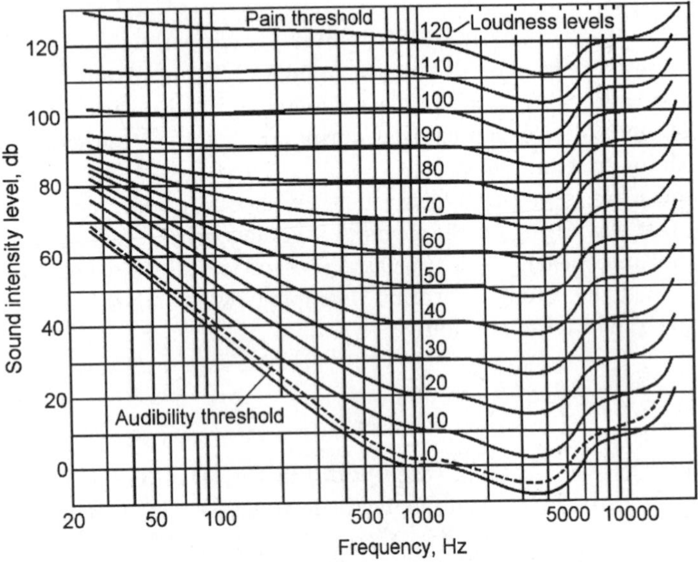

Figure 7.1 Relation between the loudness level and the sound intensity level (for example, at frequency of 100 Hz, sound with the intensity level of 60 dB has loudness 35 phones).

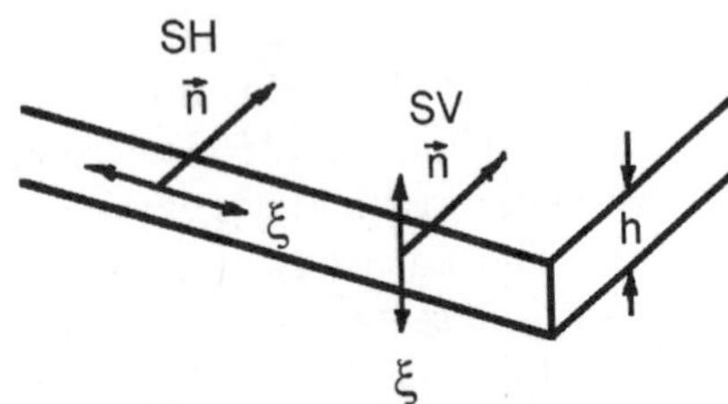

Figure 7.2 Transverse waves in a plate: SH-wave and the Lamb wave (SV-wave). $\vec{n}$ is the wave normal (direction of propagation), $\vec{\xi}$ is the direction of vibrational displacement.

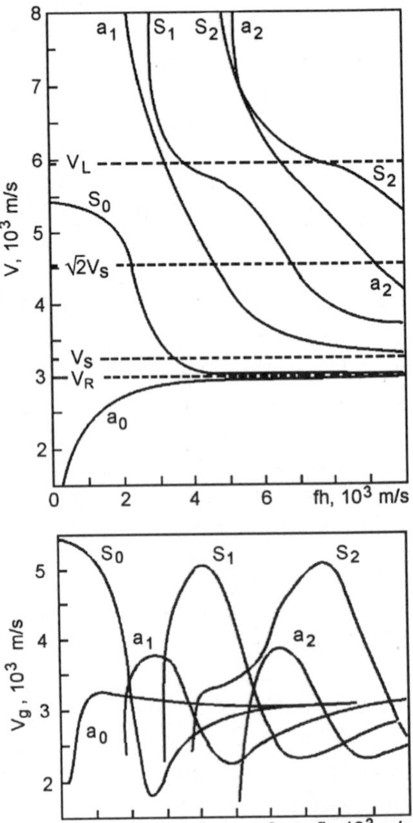

Figure 7.3 Theoretical dependence of phase (**a**) and group (**b**) velocities of symmetrical (s) and antisymmetrical (a) modes of the Lamb wave in a steel plate ($v_l = 5.96 \cdot 10^3$ m/s; $v_s = 3.22 \cdot 10^3$ m/s) placed in water on the product (frequency×plate thickness) $\sigma = 0.28$ [7].

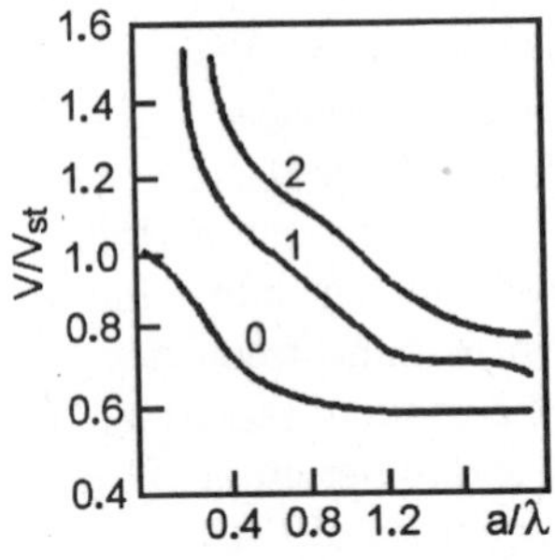

Figure 7.4 Theoretical phase velocity of the zero and two first modes of longitudinal waves in cylindrical rod ($\sigma = 0.28$).

Table 7.1 Effective elastic modulus for the acoustic waves propagating into peculiar acoustic direction in a crystal [2]

				Crystal system ⟶	Monoclinic		Orthorhombic		
				Point groups ⟶	2	$m, 2/m$	$mm2$	222	mmm
No.	Wave vector	Wave type	Polarization	Designation					
1	1, 0, 0	L	1, 0, 0	L_{100}	—	—	c_{11}		
2	0, 1, 0	L	0, 1, 0	L_{010}	$c_{22} + l_{22}^2/\varepsilon_0\varepsilon_{22}$	c_{22}			
3	0, 0, 1	L	0, 0, 1	L_{001}	—	—	$c_{33} + l_{33}^2/\varepsilon_0\varepsilon_{33}$	c_{33}	
4	$1/\sqrt{2}$, $1/\sqrt{2}$, 0	L	$1/\sqrt{2}$, $1/\sqrt{2}$, 0	L_{110}	—	—	—	—	—
5	$1/\sqrt{3}$, $1/\sqrt{3}$, $1/\sqrt{3}$	L	$1/\sqrt{3}$, $1/\sqrt{3}$, $1/\sqrt{3}$	L_{111}	—	—	—	—	—
6	$\cos\varphi$, $\sin\varphi$, 0	L	$\cos\varphi$, $\sin\varphi$, 0		—	—	—	—	—
7	1, 0, 0	S	1, 0, 0	S_{100}^{010}	$c_{66} + l_{16}^2/\varepsilon_0\varepsilon_{11}$	c_{66}			
8	1, 0, 0	S	0, 0, 1	S_{100}^{001}	—	—	$c_{55} + l_{14}^2/\varepsilon_0\varepsilon_{11}$	c_{55}	
9	0, 1, 0	S	1, 0, 0	S_{010}^{100}	—	—	c_{66}		
10	0, 1, 0	S	0, 0, 1	S_{010}^{001}	—	—	$c_{44} + l_{24}^2/\varepsilon_0\varepsilon_{22}$	c_{44}	
11	0, 0, 1	S	1, 0, 0	S_{001}^{100}	—	—	c_{55}		
12	0, 0, 1	S	0, 1, 0	S_{001}^{010}	$c_{44} + l_{34}^2/\varepsilon_0\varepsilon_{33}$	c_{44}			
13	$1/\sqrt{2}$, $1/\sqrt{2}$, 0	S	0, 0, 1	S_{110}^{001}	—	—	c)	d)	$(c_{55} + c_{44})/2$
14	$1/\sqrt{2}$, $1/\sqrt{2}$, 0	S	$-1/\sqrt{2}$, $1/\sqrt{2}$, 0	$S_{110}^{\bar{1}10}$	—	—	—	—	—
15	0, 0, 1	S	$\cos\varphi$, $\sin\varphi$, 0		—	—	—	—	—
16	$\cos\varphi$, $\sin\varphi$, 0	S	$-\sin\varphi$, $\cos\varphi$, 0		—	—	—	—	—
17	$\cos\varphi$, $\sin\varphi$, 0	S	0, 0, 1		—	—	f)	g)	g*)
18	$\sin\theta\cos\varphi$, $\sin\theta\cos\varphi$, $\cos\theta$	S	$-\sin\varphi$, $\cos\varphi$, 0		—	—	—	—	—
19	$\sin\theta$, 0, $\cos\theta$	S	0, 1, 0		n)	n*)	l)	o)	o*)
20	0, $\sin\theta$, $\cos\theta$	S	1, 0, 0		—	—	p*)	p)	p*)
21	$\sin\theta/\sqrt{2}$, $\sin\theta/\sqrt{2}$, $\cos\theta$	S	$-1/\sqrt{2}$, $1/\sqrt{2}$, 0		—	—	—	—	—
22	$1/\sqrt{3}$, $1/\sqrt{3}$, $1/\sqrt{3}$	S	u)		—	—	—	—	—

Table 7.1 Effective elastic modulus for the acoustic waves propagating into peculiar acoustic direction in a crystal *(continued)*

	Tetragonal						Trigonal			
	$\bar{4}$	4	$4mm$	$\bar{4}2m$	422	$4/m$, $4/mmm$	3	32	$3m$	$\bar{3}$, $\bar{3}m$
1	—	—	c_{11}				—	$c_{11} + l_{11}^2/\varepsilon_0\varepsilon_{11}$	c_{11}	
2	—	—	c_{11}				—	—	—	—
3	c_{33}	$c_{33} + l_{33}^2/\varepsilon_0\varepsilon_{33}$		c_{33}			$c_{33} + l_{33}^2/\varepsilon_0\varepsilon_{33}$	c_{33}	$c_{33} + l_{33}^2/\varepsilon_0\varepsilon_{33}$	c_{33}
4	—	—	$c_{11} + c_{22} + 2c_{66}/2$				—	—	—	—
5	—	—	—	—	—	—	—	—	—	—
6	—	—	—	—	—	—	—	—	—	—
7	—	—	c_{66}				—	—	—	—
8	$c_{44} + l_{15}^2/\varepsilon_0\varepsilon_{11}$			c_{44}			—	—	—	—
9	—	—	c_{66}				—	$c_{66} + l_{11}^2/\varepsilon_0\varepsilon_{11}$	c_{66}	
10	$c_{44} + l_{15}^2/\varepsilon_0\varepsilon_{11}$			c_{44}			—	—	—	—
11	c_{44}									
12	c_{44}									
13	$c_{44} + l_{14}^2/\varepsilon_0\varepsilon_{11}$	$c_{44} + l_{15}^2/\varepsilon_0\varepsilon_{11}$		$c_{44} + l_{14}^2/\varepsilon_0\varepsilon_{11}$	c_{44}		—	—	—	—
14	—	—	$(c_{11} - c_{12})/2$				—	—	—	—
15	c_{44}									
16	—	—	—	—	—	—	—	—	—	—
17	h)	$c_{44} + l_{15}^2/\varepsilon_0\varepsilon_{11}$		$c_{44} + l_{14}^2 \sin^2 2\varphi/\varepsilon_0\varepsilon_{11}$	c_{44}		—	—	—	—
18	—	—	—	—	—	—	—	—	—	—
19	—	—	l)	o)	p)	p*)	—	—	—	—
20	—	—	l)	o)	q)	q*)	—	r)	r)	r*)
21	—	—	s)	s)	t)	t*)	—	—	—	—
22	—	—	—	—	—	—	—	—	—	—

Table 7.1 Effective elastic modulus for the acoustic waves propagating into peculiar acoustic direction in a crystal *(continued)*

<table>
<tr><td></td><td colspan="6">Hexagonal</td><td colspan="2">Cubic</td></tr>
<tr><td></td><td>6</td><td>$\bar{6}$</td><td>622</td><td>$6mm$</td><td>$\bar{6}m2$</td><td>$6/m$, $6/mmm$</td><td>23, $\bar{4}3m$</td><td>$m3$, 432, $m3m$</td></tr>
<tr><td>1</td><td>c_{11}</td><td>$c_{11} + l_{11}^2/\varepsilon_0\varepsilon_{11}$</td><td colspan="6">$c_{11}$</td></tr>
<tr><td>2</td><td colspan="4">c_{11}</td><td>$c_{11} + l_{22}^2/\varepsilon_0\varepsilon_{11}$</td><td colspan="3">$c_{11}$</td></tr>
<tr><td>3</td><td colspan="2">$c_{33} + l_{33}^2/\varepsilon_0\varepsilon_{33}$</td><td>$c_{33}$</td><td>$c_{33} + l_{33}^2/\varepsilon_0\varepsilon_{33}$</td><td colspan="2">$c_{33}$</td><td colspan="2">$c_{11}$</td></tr>
<tr><td>4</td><td>—</td><td>—</td><td>—</td><td>—</td><td>—</td><td>—</td><td colspan="2">$(c_{11} + c_{12} + 2c_{44})/2$</td></tr>
<tr><td>5</td><td>—</td><td>—</td><td>—</td><td>—</td><td>—</td><td>—</td><td>a)</td><td>$c_{11} + 4c_{44} + 2c_{12}/3$</td></tr>
<tr><td>6</td><td>c_{11}</td><td>b)</td><td colspan="2">c_{11}</td><td>$c_{11} + l_{22}^2 \sin^2 3\varphi/\varepsilon_0\varepsilon_{11}$</td><td>$c_{11}$</td><td>—</td><td>—</td></tr>
<tr><td>7</td><td>c_{66}</td><td>$c_{66} + l_{22}^2/\varepsilon_0\varepsilon_{11}$</td><td colspan="2">$c_{66}$</td><td>$c_{66} + l_{22}^2/\varepsilon_0\varepsilon_{11}$</td><td>$c_{66}$</td><td colspan="2">$c_{44}$</td></tr>
<tr><td>8</td><td>$c_{44} + l_{15}^2/\varepsilon_0\varepsilon_{11}$</td><td colspan="2">$c_{44}$</td><td>$c_{44} + l_{15}^2/\varepsilon_0\varepsilon_{11}$</td><td colspan="4">$c_{44}$</td></tr>
<tr><td>9</td><td>c_{66}</td><td>$c_{66} + l_{11}^2/\varepsilon_0\varepsilon_{11}$</td><td colspan="4">$c_{66}$</td><td colspan="2">$c_{44}$</td></tr>
<tr><td>10</td><td>$c_{44} + l_{15}^2/\varepsilon_0\varepsilon_{11}$</td><td colspan="2">$c_{44}$</td><td>$c_{44} + l_{15}^2/\varepsilon_0\varepsilon_{11}$</td><td colspan="4">$c_{44}$</td></tr>
<tr><td>11</td><td colspan="8">c_{44}</td></tr>
<tr><td>12</td><td colspan="8">c_{44}</td></tr>
<tr><td>13</td><td>$c_{44} + l_{15}^2/\varepsilon_0\varepsilon_{11}$</td><td colspan="2">$c_{44}$</td><td>$c_{44} + l_{15}^2/\varepsilon_0\varepsilon_{11}$</td><td colspan="2">$c_{44}$</td><td>$c_{44} + l_{14}^2/\varepsilon_0\varepsilon_{11}$</td><td>$c_{44}$</td></tr>
<tr><td>14</td><td>c_{66}</td><td>$c_{66} + (l_{11} + l_{22})^2/2\varepsilon_0\varepsilon_{11}$</td><td>$c_{66}$</td><td>$c_{66}$</td><td>$c_{66} + l_{22}^2/\varepsilon_0\varepsilon_{11}$</td><td>$c_{66}$</td><td colspan="2">$(c_{11} - c_{12})/2$</td></tr>
<tr><td>15</td><td colspan="8">c_{44}</td></tr>
<tr><td>16</td><td>c_{66}</td><td>e)</td><td colspan="2">c_{66}</td><td>$c_{66} + l_{22}^2 \cos^2 3\varphi/\varepsilon_0\varepsilon_{11}$</td><td>$c_{66}$</td><td>—</td><td>—</td></tr>
<tr><td>17</td><td>$c_{44} + l_{15}^2/\varepsilon_0\varepsilon_{11}$</td><td colspan="2">$c_{44}$</td><td>$c_{44} + l_{15}^2/\varepsilon_0\varepsilon_{11}$</td><td colspan="2">$c_{44}$</td><td>$c_{44} + l_{14}^2 \sin^2 2\varphi/\varepsilon_0\varepsilon_{11}$</td><td>$c_{44}$</td></tr>
<tr><td>18</td><td>i)</td><td>j)</td><td>k)</td><td>l)</td><td>m)</td><td>l)</td><td>—</td><td>—</td></tr>
<tr><td>19</td><td>i)</td><td>j) $\varphi = 0$</td><td>k)</td><td>l)</td><td>m) $\varphi = 0$</td><td>l)</td><td colspan="2">c_{44}</td></tr>
<tr><td>20</td><td>i)</td><td>j) $\varphi = \pi/2$</td><td>k)</td><td>l)</td><td>m) $\varphi = \pi/2$</td><td>l)</td><td colspan="2">c_{44}</td></tr>
<tr><td>21</td><td>i)</td><td>j) $\varphi = \pi/4$</td><td>k)</td><td>l)</td><td>m) $\varphi = \pi/4$</td><td>l)</td><td>t)</td><td>t*)</td></tr>
<tr><td>22</td><td>—</td><td>—</td><td>—</td><td>—</td><td>—</td><td>—</td><td colspan="2">$(c_{11} + c_{44} - c_{12})/3$</td></tr>
</table>

Table 7.1 Effective elastic modulus for the acoustic waves propagating into peculiar acoustic direction in a crystal *(footnotes)*

a) $\frac{1}{3}(c_{11} + 4c_{44} + 2c_{12}) + 4l_{14}^2 \cdot (3\varepsilon_0\varepsilon_{11})^{-1}$
b) $c_{11} + (e_{11}\cos 3\varphi - e_{22}\sin 3\varphi)^2 \cdot (\varepsilon_0\varepsilon_{11})^{-1}$
c) $\frac{1}{2}(c_{44} + c_{55}) + (e_{15} + e_{24})^2[2\varepsilon_0(\varepsilon_{11} + \varepsilon_{22})]^{-1}$
d) $\frac{1}{2}(c_{44} + c_{55}) + (e_{14} + e_{25})^2[2\varepsilon_0(\varepsilon_{11} + \varepsilon_{22})]^{-1}$
e) $c_{66} + (e_{11}\sin 3\varphi + e_{22}\cos 3\varphi)^2 \cdot (\varepsilon_0\varepsilon_{11})^{-1}$
f) $c_{55}\cos^2\varphi + c_{44}\sin^2\varphi + (e_{15}\cos^2\varphi + e_{24}\sin^2\varphi)^2[\varepsilon_0(\varepsilon_{11}\cos^2\varphi + \varepsilon_{22}\sin^2\varphi)]^{-1}$
g) $c_{55}\cos^2\varphi + c_{44}\sin^2\varphi + (e_{14} + e_{25})^2\sin^2 2\varphi[4\varepsilon_0(\varepsilon_{11}\cos^2\varphi + \varepsilon_{22}\sin^2\varphi)]^{-1}$
h) $c_{44} + (e_{15}\cos 2\varphi + e_{14}\sin 2\varphi)^2(\varepsilon_0\varepsilon_{11})^{-1}$
i) $c_{44}\cos^2\theta + c_{66}\sin^2\theta + (e_{14}\sin 2\theta)^2[4\varepsilon_0(\varepsilon_{11}\sin^2\theta + \varepsilon_{33}\cos^2\theta)]^{-1}$
j) $c_{44}\cos^2\theta + c_{66}\sin^2\theta + (e_{11}\sin^2\theta\sin 3\varphi + e_{22}\sin^2\theta\cos 3\varphi)^2[\varepsilon_0(\varepsilon_{11}\sin^2\theta + \varepsilon_{33}\cos^2\theta)]^{-1}$
k) $c_{44}\cos^2\theta + c_{66}\sin^2\theta + (e_{14}\sin 2\theta)^2[\varepsilon_0(\varepsilon_{11}\sin^2\theta + \varepsilon_{33}\cos^2\theta)]^{-1}$
l) =i*) =j*) =k*) =m*) =o*) =q*) $c_{44}\cos^2\theta + c_{66}\sin^2\theta$
m) $c_{44}\cos^2\theta + c_{66}\sin^2\theta + (e_{22}\cos 3\varphi\sin^2\theta)^2[\varepsilon_0(\varepsilon_{11}\sin^2\theta + \varepsilon_{33}\cos^2\theta)]^{-1}$
n) $c_{44}\cos^2\theta + c_{66}\sin^2\theta + c_{46}\sin 2\theta + [e_{16}\sin^2\theta + e_{34}\cos^2\theta + \frac{1}{2}(e_{14} + e_{36})\sin 2\theta]^2[\varepsilon_0(\varepsilon_{11}\sin^2\theta + \varepsilon_{33}\cos^2\theta + \varepsilon_{31}\sin 2\theta)]^{-1}$
o) $c_{44}\cos^2\theta + c_{66}\sin^2\theta + [(e_{14} + e_{36})\sin 2\theta]^2[4\varepsilon_0(\varepsilon_{11}\sin^2\theta + \varepsilon_{33}\cos^2\theta)]^{-1}$
p) $c_{55}\cos^2\theta + c_{66}\sin^2\theta + [(e_{25} + e_{36})\sin 2\theta]^2[4\varepsilon_0(\varepsilon_{22}\sin^2\theta + \varepsilon_{33}\cos^2\theta)]^{-1}$
q) $c_{44}\cos^2\theta + c_{66}\sin^2\theta + (e_{14}\sin 2\theta)^2[4\varepsilon_0(\varepsilon_{11}\sin^2\theta + \varepsilon_{33}\cos^2\theta)]^{-1}$
r) $c_{44}\cos^2\theta + c_{66}\sin^2\theta + c_{14}\sin 2\theta + (e_{11}\sin^2\theta + \frac{1}{2}e_{14}\sin 2\theta)^2[\varepsilon_0(\varepsilon_{11}\sin^2\theta + \varepsilon_{33}\cos^2\theta)]^{-1}$
s) $(c_{11} - c_{12})\sin^2(\theta/2) + c_{44}\cos^2\theta$
t) $\frac{1}{2}(c_{11} - c_{12})\sin^2\theta + c_{44}\cos^2\theta + (e_{14}\sin 2\theta)^2[4\varepsilon_0(\varepsilon_{11}\sin^2\theta + \varepsilon_{33}\cos^2\theta)]^{-1}$
u) $\frac{1}{a}\cos\theta,\ \frac{1}{a}\sin\theta,\ -\frac{1}{a}(\sin\theta + \cos\theta);\ a = (2 + \sin 2\theta)^{1/2}$

Notes:
1) $\varepsilon_0 = 8.85 \cdot 10^{-2}$ F/m is the electrical constant;
c_{ik}, l_{ik}, ε_{ik} are the tensor components of elastic constants, piezoelectric and dielectric coefficients;
φ, θ are the angles between the direction of the wave vector and [100], [001] axes, accordingly.
2) For notes marked by asterisks, the moduli are taken at $e_{ik} = 0$.
3) In the table data on directions and/or polarizations of waves Nos. 6, 15–22 are displayed at arbitrary φ and θ.
Waves marked by dash do not exist in the given class of crystals. This does not exclude the existance of special acoustic directions at certain values of φ and θ. In the table they are marked by other numbers (for example, the wave with number 6 in the class 2 exists only at $\varphi = \pi/2$. Its effective modulus is given in the second row.)

7.2 Sound Propagation in Gases and Vapors

Table 7.2 Sound velocity in gases and vapors

Substance	t,°C	f, kHz	v, m/s	$\Delta v/\Delta t$, m/(K·s)	Ref.
Acetal	30	59.6	257		[4]
Acetone	18		327		[4]
	97.1		238.6	0.32	[6]
	134		251.2		[6]
Air	0		331.45	0.59	[9]
Amyl alcohol	136		218.8		[4]
Ammonia	0		415		[4]
	18		428.2		[4]
Argon	0		319.0	0.56	[4]
	20		321		[4]
	30.2	83.8	325.23		[5]
Benzene	97.1		202.0	0.3	[8]
	134	95	212.6		[4]
Bromine	0		135		[4]
Butyl alcohol					
secondary	134	95	215		[4]
ternary	134	95	180		[4]
Carbon tetrachloride	97.1	95	145.2		[8]
	134		153.6		[4]
Carbon tetrafluoride	22		178.2		[4]
Carbon dioxide	0	43.2	256.7		[4]
	20	58.5	274.6		[4]
	25	53	282		[13]
	100	53	311		[13]
	500	53	425		[13]
Carbon disulphide	55	59.5	205		[4]
	97.1	50	220.1		[8]
Carbon monoxide	0		338	0.6	[4]
	1000	27.4	717		[12]
	1800	27.4	909		[12]
Chlorine	0		206		[4]
Chloroform	97.1	95	171.4		[8]
	134	95	179.7		[4]
Cyclohexane	97.1		191.3	0.3	[8]
	134	95	201.9		[4]
Deuterium	0		890	1.6	[8]
Dichloroethane	97.1		181	0.24	[8]
	134		190		[8]
Diethyl ether	97.1		206	0.3	[8]
	134		217		[8]
Dimethyl ether	97.1		274		[8]
Dipropyl ether	97.1		194		[8]
Ethane	10		308		[8]
Ethyl alcohol	97.1	95	269.1	0.4	[4]
	134	95	284.4		[4]
Ethyl chloride	18		428.2		[4]
Ethyl ether	97.1		206.5	0.3	[4]
	134		217.4		[4]
Ethyl iodide	76	465	162		[4]
Ethylacetate	97.1		189.2	0.27	[8]
	134	95	198.8		[8]
Ethylene	0		317		[8]
	20	59.5	329		[4]
Ethylmethyl ketone	134		223		[8]
Gas natural	0		453		[8]
Helium	0		965	0.8	[4]
	30	83.8	1056		[5]
Hexane	134	94	199.6		[4]
Hydrogen	0		1284	2.2	[4]
	18		1301		[4]
	100		1463		[4]
Hydrogen bromide	0		200		[8]
Hydrogen chloride	0		296		[4]
Hydrogen iodide	0		157		[4]
Hydrogen sulphide	0		289.3		[8]
Iodine	0		108		[4]
Isopropyl alcohol	97.1	95	255	0.4	[4]
	134	95	270.2		[4]
Krypton	30.3	83.8	224.4		[5]
Methane	−87	588	353		[11]
	−20	588	411		[11]
	0		430		[8]
Methyl alcohol	97.1	95	335	0.46	[4]
	134		352.6		[4]
Methyl ether	97.1	95	273.9		[4]
Methyl iodide	43	465	154		[4]
Methylcyclohexane	134		185		[4]
Methylene chloride	97.1	95	204	0.24	[8]
	134	95	213		[8]
Neon	0		433.4		[1]
	30.5	83.8	461.3		[5]
Nitrogen	0		334.0	0.6	[4]
	19.1		349.0		[4]
Nitrogen monoxide	10		324		[4]
	16		332.4		[4]
Nitrogen oxide	0		263	0.5	[8]
	19	93.8	273		[10]
	109	93.8	310		[10]
Oxygen	0	43.2	314		[1]
Pentane	134		220		[4]
Propionic acid	146		232		[4]
Propyl alcohol	134	95	243.9		[4]
Propyl ether	97.1	50	194		[4]
Silicon fluorite	0		167		[8]
Sulphur dioxide	0		213	0.47	[4]
	20	111	221.5		[10]
	100	111	248.5		[10]
Vinylacetate	134		203		[4]
Water vapor	134		494		[8]

Table 7.3 Sound velocity in dry air versus temperature at atmospheric pressure [14]

t,°C	−30	−20	−10	0	10	20	30	100	500
v, m/s	313	319	325	331.45 [9]	338	344	350.70 [5]	386	553

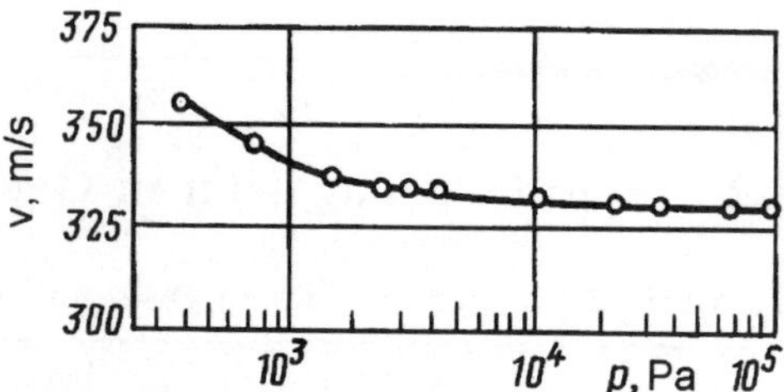

Figure 7.8 Sound velocity in air (free of CO_2) at atmospheric and lower pressure: $t = 0$°C; $f = 971$ kHz [15]

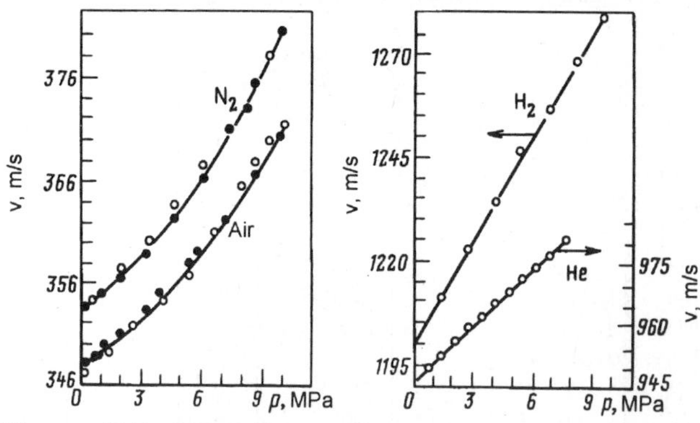

Figure 7.5 Dependence of sound velocity in air, nitrogen and helium on pressure [16]: ∘ – $f = 486$ kHz; • – $f = 286$ kHz.

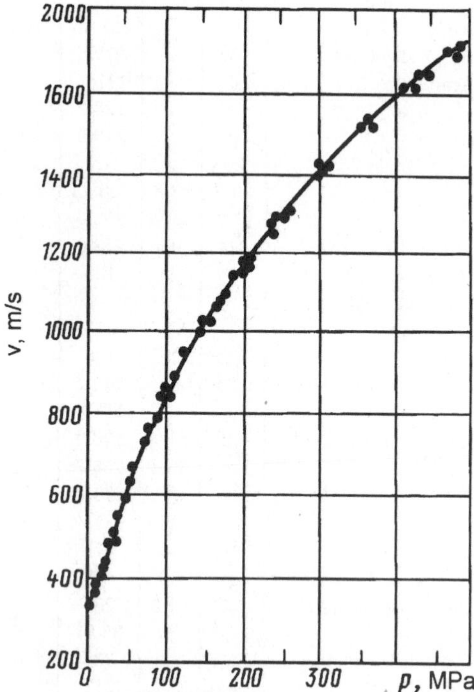

Figure 7.9 Dependence of sound velocity in nitrogen on pressure: $t = 20$°C, $f = 310$ kHz [17].

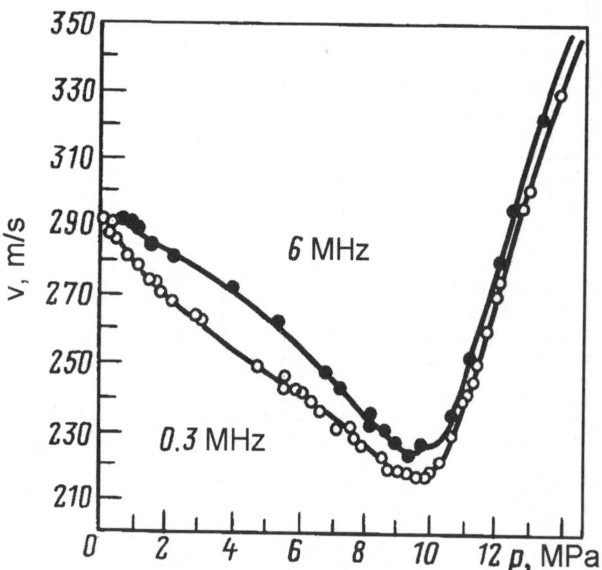

Figure 7.6 Dependence of sound velocity in carbon dioxide on pressure: $t = 50.8$°C [18].

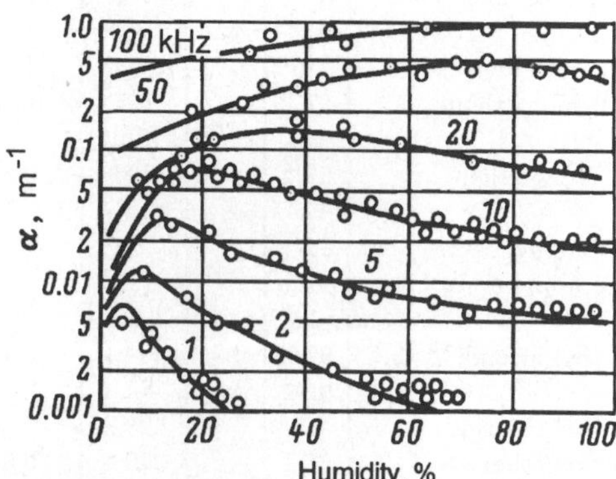

Figure 7.10 Sound damping in air at different frequencies as a function of relative moisture content [26].

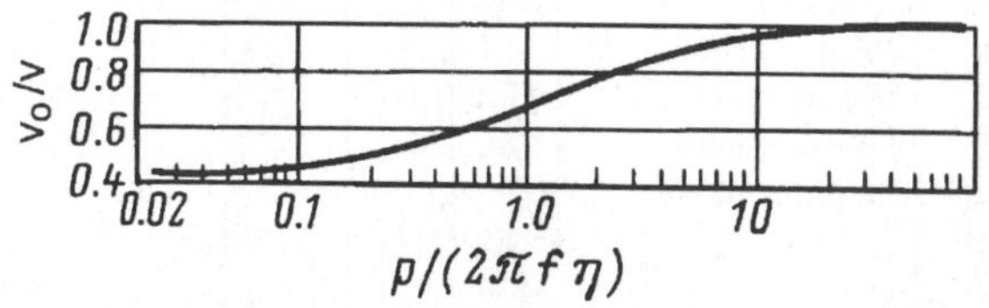

Figure 7.7 Sound velocity dispersion in air ($f = 100-200$ kHz) [19]: P_0 is the static pressure, η is the viscosity, v_0 is the sound velocity at atmospheric pressure.

Table 7.4 Sound damping coefficient in gases and vapors

Substance	t, °C	f, kHz	p_0, 10^5 Pa	α, m^{-1}	α/f^2, 10^{-11} $\mathrm{s^2/m}$	Reference
Acetone		97.8		50	525	[4]
Air		1940		69.7	1.85	[24]
Argon	20	4250	1.01	3.42	1.9	[21]
		500	0.58	8.34		[22]
		500	0.184	25.5		[22]
Benzene		97.8		100	1050	[4]
Carbon dioxide	16.6	304.4	0.99	24	27.1	[23]
Carbon disulphide		97.8		120	1270	[4]
Carbon oxide	18.7	304.4	0.86		5.78	[23]
Carbon tetrachloride		97.8		35	370	[4]
Chloroform		97.8		70	740	[4]
Ethyl alcohol		97.8		70	740	[4]
Ethyl ether		97.8		20	210	[4]
Ethyl iodide		97.8		45	474	[4]
Helium	17.5	598.9	1	10.7	2.96	[20, 23]
Hydrogen	19.9	598.9	1.01	12.9	3.58	[23]
Methyl alcohol		97.8		5	52.5	[4]
Methyl iodide		97.8		15	158	[4]
Methylene chloride		97.8		70	740	[4]
Neon	19	304.4	0.66		5.82	[23]
Nitrogen	19.9	598.9	0.98	4.9	1.35	[20]
Nitrogen oxide	16.3	598.9	0.96	6.6	1.83	[23]
Oxygen	19.6	598.9	1	6.0	1.68	[23]
Xenon		500	0.55		0.15	[22]
		500	0.091		0.93	[22]

Table 7.5 Sound damping in air-water fog, dB/s [25]

f, kHz	Water concentration, g/m³				
	0.26	0.46	0.52	0.86	1.03
0.5	0.2	0.4	0.3	0.2	1.0
1.0	0.5	0.8	1.5	3.0	3.0
2.0	2.0	4.6	6.0	11.0	22.6
4.0	7.0	15.5	16.0	30.5	37.0
6.0	11.0	25.0	26.0	51.0	58.0

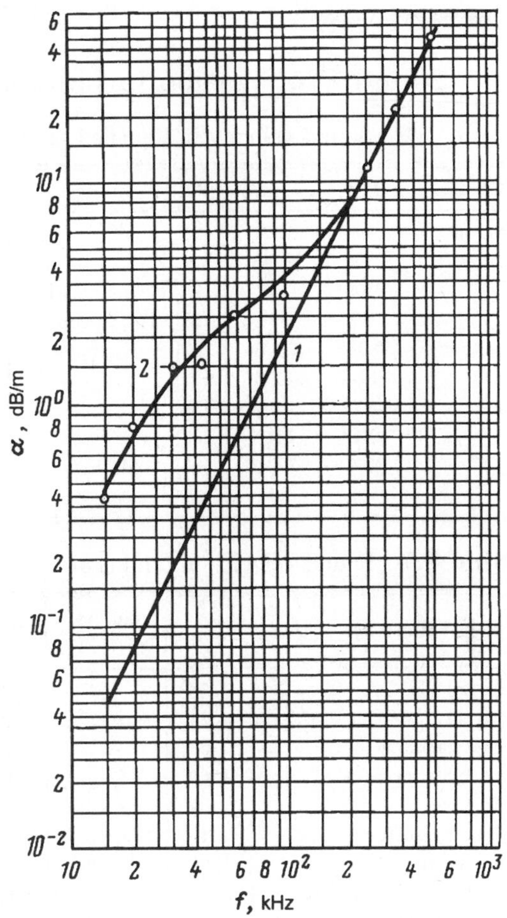

Figure 7.11 Sound damping at high and ultrasound frequencies in a dry (*1*) and humid (relative moisture content is 37%) (*2*) air: $t = 26.5$°C [27].

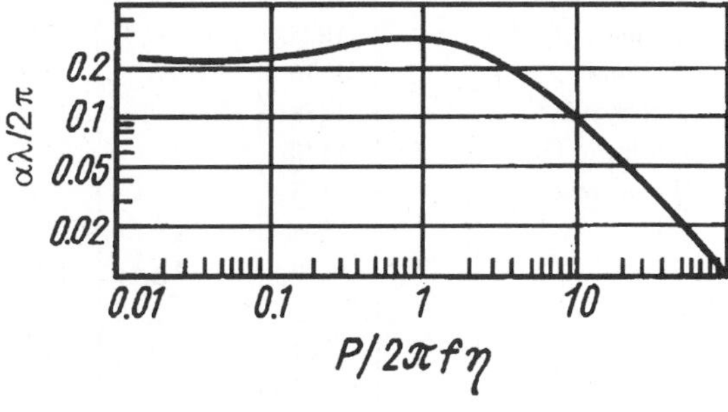

Figure 7.12 Dispersion of absorption factor in air [19]: p_0 is the static pressure, η is the viscosity, λ is the wavelength.

7.3 Sound Propagation in Liquids

Table 7.6 Sound velocity in liquefied gases near the boiling temperature

Substance	T, K	v, m/s	Ref.	Substance	T, K	v, m/s	Ref.	Substance	T, K	v, m/s	Ref.
Argon	64	1322	[29]		4.22	179.8	[33]	Nitrogen	65	1007	[1]
	78	1255	[29]		4.223	180.63	[32]		70	929	[28]
	84	863	[28]	Helium II	1.63	20	[35]		76	869	[28]
	87	837	[28]	(2nd sound)	2.18	3.4	[35]		77	867	[1]
Helium	0.985	237.66	[32]	Hydrogen	14.8	1340	[30]	Oxygen	53	1130	[36]
	1.08	237.67	[32]		16.5	1265	[30]		60	1119	[37]
	1.76	231.4	[33]		18.0	1260	[30]		70	1094	[37]
	2.18	221.7	[33]		20.3	1127	[31]		80	988	[1]
	2.5	223.3	[33]	Methane	94.9	1545	[38]		85	948	[1]
	3.0	220	[34]		111.4	1418	[38]		89.6	911	[28]

Table 7.7 Sound velocity in liquids at 20°C

Substance	v, m/s	$\Delta v/\Delta t$, m/(K·s)	Ref.	Substance	v, m/s	$\Delta v/\Delta t$, m/(K·s)	Ref.
Acetal (24°C)	1378		[40]	*n*-decane	1402	−3.7	[39]
Acetic anhydride (24°C)	1384		[40]	decyl	1413	−4.4	[52]
Acetylacetone	1383	−5.2	[41]	dodecyl (22.3°C)	1433		[52]
Acetyl chloride	1060		[41]	ethyl	1165	−3.4	[39]
Acetone	1189	−4.3	[42]	furyl (25°C)	1450	−3.5	[45]
Acetonylacetone	1416	−4.8	[41]	*n*-heptyl	1341	−3.7	[40]
Acetonitryle	1304	−5.0	[41]	*n*-hexyl	1322	−3.5	[39]
Acetate ether (25.5°C)	1417		[40]	methyl	1123	−3.3	[41]
Acetophenone	1496	−3.7	[40]	*n*-nonyl	1391	−4.3	[41]
Acid:				*n*-octyl	1358	−3.6	[39]
acetate	1150		[41]	*n*-pentyl	1294	−3.5	[39]
butyric	1203	−4.8	[41]	β-phenylethyl (30°C)	1512		[55]
caprylic	1331	−5.0	[41]	γ-phenylpropyl (30°C)	1523		[55]
caproic	1280	−5.0	[52]	*n*-propyl	1223	−3.7	[39]
elaidin (45°C)	1346	−3.5	[54]	*iso*-propyl	1170	−4.0	[41]
enanthic	1312	−5.0	[41]	tetradecyl (38.4°C)	1404		[52]
formic	1287		[41]	Allyl chloride (28°C)	1088		[40]
fumaric	1303	−3.5	[54]	Amyl acetate (26°C)	1168		[40]
maleic	1352	−3.8	[41]	Amyl bromide	981		[41]
oleic (45°C)	1333	−3.3	[51]	Amylformate	1223	−4.2	[39]
palmetic (62.1°C)	1328		[52]	Aniline	1659	−4.0	[42]
propionic	1176	−5.0	[41]	Benzaldehyde	1479	−4.0	[41]
pyroracemic	1471		[41]	Benzene	1324	−4.6	[42]
sulphuric (15°C)	1440		[53]	Benzene heavy	1238	−4.25	[43]
thioacetate	1168		[41]	Benzoyl chloride (28°C)	1318		[40]
valeric	1244	−4.8	[41]	Benzylacetone	1514		[41]
Acrolein	1190	−4.1	[39]	Benzyl chloride	1420	−4.6	[41]
Alcohol:				Bromal	966	−3.4	[41]
n-amyl	1294		[41]	Bromobenzene	1170	−3.12	[44]
iso-amyl	1260	−3.4	[39]	Bromoform	931	−2.2	[44]
tert-amyl	1204		[40]	α-Bromonaphthalene	1372	−3.1	[42]
benzyl	1540	−3.2	[41]	*n*-Butyl acetate	1226	−4.1	[39]
n-butyl	1263	−3.6	[39]	*iso*-Butyl acetate	1182	−4.0	[39]
iso-butyl	1212	−3.5	[39]	Butyl bromide	990	−4.0	[41]

Table 7.7 Sound velocity in liquids at 20°C *(continued)*

Substance	v, m/s	$\Delta v/\Delta t$, m/(K·s)	Ref.
Butyl chloride	1133	−4.6	[41]
Butyl iodide	977	−3.6	[41]
2,3-Butylene glycol (25°C)	1484		[45]
Butylformate	1215	−4.1	[39]
Carbon disulphide	1158	−3.2	[40]
Carbon tetrachloride	938	−3.0	[41]
Chlorobenzene	1289	−3.7	[42]
α-Chloronaphthalene	1481		[41]
Chloroform	1005	−3.6	[41]
meta-Chlorotoluene	1326		[41]
ortho-Chlorotoluene	1344		[41]
n-Chlorotoluene	1316		[41]
Cinnamic aldehyde (25°C)	1554	−3.1	[45]
Citral	1442	−3.8	[41]
ortho-Cresol (25°C)	1506	−3.5	[45]
Crotonaldehyde	1288	−4.1	[39]
Cyclohexane	1277	−4.6	[39]
Cyclohexanol	1493	−4.9	[41]
Cyclohexanone	1449	−5.4	[41]
Cyclohexene	1305	−5.4	[41]
Cyclohexylamine	1435		[41]
Cyclohexyl chloride	1319	−4.8	[41]
Cyclopentadiene	1421		[41]
Cyclopentane (30°C)	1182		[39]
Cyclopentanone (24°C)	1474		[40]
1-Decene	1250		[50]
Decyl chloride	1318	−4.2	[41]
Diacetyl (25°C)	1236	−4.5	[45]
Diamyl ether (26°C)	1153		[40]
Dibromoethylene	1009		[41]
cis-Dibromoethylene	957	−2.4	[51]
trans-Dibromoethylene	936	−2.2	[51]
meta-Dichlorobenzene	1232		[40]
ortho-Dichlorobenzene	1295		[40]
Dichloroethane (23°C)	1240		[40]
cis-Dichloroethylene	1090	−3.7	[51]
trans-Dichloroethylene	1031	−3.7	[51]
Diethylaniline	1482	−4.0	[41]
Diethyl carbonate (28°C)	1173		[40]
Diethyl ether	1008	−5.4	[41]
Diethyl ketone (25°C)	1218		[39]
Diethylene glycol (25°C)	1586	−2.4	[45]
Diethylphthalate (23°C)	1471		[40]
Dimethylaniline	1509	−3.6	[41]
Dioxane	1389	−6.2	[41]
Dipentene (23.8°C)	1328		[40]
Diphenylmethane	1501		[40]
Diphenyl ether (24°C)	1469		[40]
Dipropyl ether	1112		[41]
Ethyl acetate	1177	−4.5	[39]
Ethyl bromide	900	−3.4	[44]
Ethyl butyrate	1197	−4.1	[39]
Ethyl ether:			
chloroacetate (25.5°C)	1234		[40]
ortho-cresol (25°C)	1315		[40]
diethylene glycol (25°C)	1458		[45]
phenol (26°C)	1153		[40]
Ethyl iodide	876	−2.7	[44]
Ethylbenzylaniline	1586		[41]
Ethylbenzene	1338	−4.4	[41]
Ethylcaprylate (28°C)	1263		[40]
Ethylene bromide	1009	−2.6	[44]
Ethylene chloride	1216	−3.9	[44]
Ethylene glycol	1667	−2.5	[39]
Ethylformate (24°C)	1121		[40]
Ethylphenol ketone	1498		[41]
Ethylpropionate (23.5°C)	1185		[40]
Formamide	1550		[41]
Geranylacetate (28°C)	1328		[40]
Glycerol	1895	−1.9	[39]
Hemellitol	1372		[41]
Heptane	1162	−4.5	[41]
2-Heptanone	1207		[41]
1-Heptene	1128		[50]
Hexane	1083		[41]
Hexyl chloride	1221	−4.2	[41]
Hexyl iodide	1081	−3.2	[41]
Hydrindene	1403		[41]
Indene	1475		[41]
Linalool	1341		[41]
Mesityl oxide	1310		[45]
Mesitylene	1362	−4.8	[41]
Methylacetate	1182	−4.7	[39]
n-Methylaniline	1586		[41]
2-Methylbutanol (30°C)	1225		[55]
Methylcyclohexane	1247	−5.6	[41]
2-Methylcyclohexanol (25.5°C)	1421		[40]
4-Methylcyclohexanol (25.5°C)	1387		[40]
2-Methylcyclohexanone	1353		[40]
Methylene bromide	963	−2.6	[44]
Methylene chloride	1093	−3.9	[44]
Methylene iodide	973	−1.9	[44]
Methylethyl ketone	1207	−5.0	[41]
Methylhexalin (22.5°C)	1528		[40]
Methylhexyl ketone (24°C)	1324		[40]
Methyl iodide	834		[41]
Methylisopropylbenzene (28°C)	1308		[40]
Methyl propionate (24.5°C)	1215		[40]
Methyl salicylate (28°C)	1408		[40]
Monochlornaphtalene (27°C)	1462		[40]
Morpholine (25°C)	1442	−3.7	[45]
Nicotine	1491	−5.0	[41]
Nitrobenzene	1475	−3.4	[39]
Nitromethane	1346	−4.1	[41]
meta-Nitrotoluene	1481	−3.6	[39]
ortho-Nitrotoluene	1473	−3.7	[44]
Nonane	1248	−4.4	[41]
1-Nonene	1218		[50]

Table 7.7 Sound velocity in liquids at 20°C *(continued)*

Substance	v, m/s	$\Delta v/\Delta t$, m/(K·s)	Ref.	Substance	v, m/s	$\Delta v/\Delta t$, m/(K·s)	Ref.
n-Octane	1192	−4.2	[42]	Quinoline	1600	−4.8	[41]
iso-Octane	1111		[39]	Salicyl aldehyde (27°C)	1474		[42]
1-Octene	1184		[50]	Styrene	1354		[39]
Octyl bromide	1182	−4.8	[41]	Tetrabromoethane	1041		[42]
Octyl chloride	1280	−4.2	[41]	Tetrachloroethane	1171		[41]
Paraldehyde	1192	−4.2	[39]	Tetrachloroethylene	1053	−2.9	[39]
Pentachlorethane	1113		[41]	Tetraethylene glycol (25°C)	1586	−3.0	[45]
1-Pentadecene	1351		[50]	Tetralin	1492	−4.5	[41]
Pentane	1008	−4.2	[41]	Tetranitromethane	1039	−4.0	[41]
Perchloroethylene	1066		[41]	Thiophene	1300	−4.2	[41]
α-Picoline (28°C)	1453		[40]	Toluene	1328	−4.3	[42]
β-Picoline (28°C)	1419		[40]	*meta*-Toluidine	1594	−3.5	[39]
Phenol (100°C)	1274	−3.2	[56]	*ortho*-Toluidine	1618	−3.9	[39]
Phenylhydrazine	1738		[41]	1,2,4-Trichlorobenzene	1301		[41]
Pinene (24°C)	1247		[40]	Trichloroethylene	1049	−4.4	[41]
Piperidine	1400		[41]	1-Tridecene	1313		[50]
Propyl chloride	929	−4.0	[41]	Triethylene glycol (25°C)	1608	−3.8	[45]
Propyl iodide	1091	−4.4	[41]	Trimethylenebromide (23.5°C)	1144		[40]
iso-Propylbenzene	1342		[41]	Triolein	1482		[41]
n-Propylacetate	1198	−4.8	[39]	1-Undecene	1275		[50]
iso-Propylacetate	1133	−4.4	[39]	Water	1482.7	+3.1	[46]
Propyonitrile	1271		[41]	Water heavy (25°C)	1399	+2.8	[49]
Pseudobutyl-*meta*-xylene	1354		[41]	*meta*-Xylene	1344	−4.1	[39]
Pseudocumene	1368		[41]	*n*-Xylene	1330	−4.8	[41]
Pyridine	1441	−4.1	[44]	*ortho*-Xylene	1364	−3.8	[39]
Quinaldine	1575		[41]				

Table 7.8 Sound velocity in oils, petroleum and petroleum products [57]

Oils

Substance	t,°C	v, m/s	Substance	t,°C	v, m/s	Substance	t,°C	v, m/s
Oil:			geranium	27	1192	paraffin	33.5	1420
anise	28.5	1451	ionone	34	1331	peanut	31.5	1562
cassia	28.5	1460	lavender	28.5	1310	pine	31	1468
castor	21	1500 [36]	lemon	29	1076	rape	30.8	1450
cedar	29	1406	linseed	31.5	1772	spermaceti	33	1210
coconut	31.5	1490	linaloe	32	1397	turpentine	27	1280
eucalyptus	29.5	1276	mustard	31.5	1825	verbena	32.5	1432
gasoline	34	1250	olive	32.5	1381	xanthorrhiza	29	1394

Petroleum and petroleum products

Substance	t,°C	v, m/s	$\Delta v/\Delta t$, m/(°C·s)	Substance	t,°C	v, m/s	$\Delta v/\Delta t$, m/(°C·s)
Aviation oil:				Diesel fuel:			
MS–20	20	1506	−3.9 [58]	summer	20	1357	−4.15 [58]
Compressor				winter	20	1332	−3.9 [58]
KS–19	20	1503	−4.0 [58]	Gasoline A–66	20	1081	−4.0 [58]
Transformer	20	1445	−3.8 [58]	Gasoline A–72	19.6	1158	
	25	1415		Kerosene	34	1295	
Crude oil		1335–1379	−(3.88–4.09) [58]	Pitch oil	20	1512	−3.7 [58]

Table 7.9 Sound velocity in distilled water versus temperature at p_0=0.1 MPa [46]*

t,°C	v, m/s	t,°C	v, m/s	t,°C	v, m/s	t,°C	v, m/s
	1400 +		1400 +		1500 +		1500 +
0	2.7	26	99.6				
1	7.7		1500 +	51	43.9	76	55.4
2	12.6	27	2.2	52	44.9	77	55.3
3	17.3	28	4.7	53	45.9	78	55.2
4	22.0	29	7.1	54	46.8	79	55.0
5	26.5	30	9.4	55	47.7	80	54.8
6	30.9	31	11.7	56	48.5	81	54.6
7	35.2	32	13.9	57	49.3	82	54.3
8	39.5	33	16.1	58	50.0	83	54.0
9	43.6	34	18.1	59	50.7	84	53.6
10	47.6	35	20.1	60	51.3	85	53.2
11	51.5	36	22.1	61	51.9	86	52.8
12	55.3	37	23.9	62	52.4	87	52.4
13	59.1	38	25.7	63	52.9	88	51.9
14	62.7	39	27.5	64	53.4	89	51.3
15	66.3	40	29.2	65	53.8	90	50.8
16	69.7	41	30.8	66	54.1	91	50.2
17	73.1	42	32.4	67	54.4	92	49.6
18	76.4	43	33.9	68	54.7	93	48.9
19	79.6	44	35.3	69	54.9	94	48.2
20	82.7	45	36.7	70	55.1	95	47.5
21	85.7	46	38.1	71	55.3	96	46.8
22	88.6	47	39.3	72	55.4	97	46.0
23	91.5	48	40.6	73	55.4	98	45.1
24	94.3	49	41.7	74	55.5	99	44.3
25	97.0	50	42.9	75	55.5	100	43.4

* Measurements were made at frequencies f = 0.75 and 3.5 MHz. See also [59, 60].

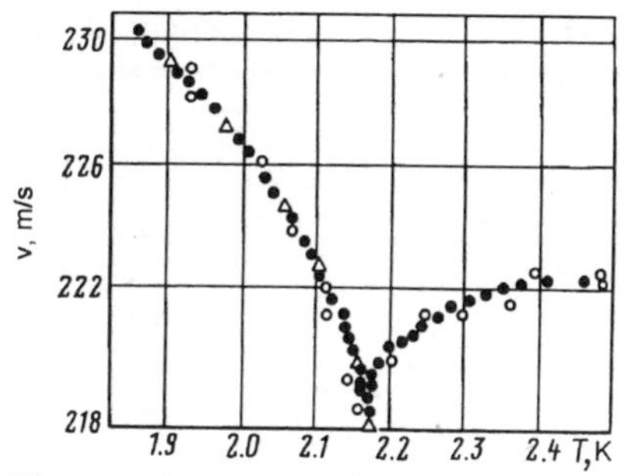

Figure 7.13 Sound velocity in liquid helium near the λ-point (the point of He I transition into a superfluid He II state) [71].

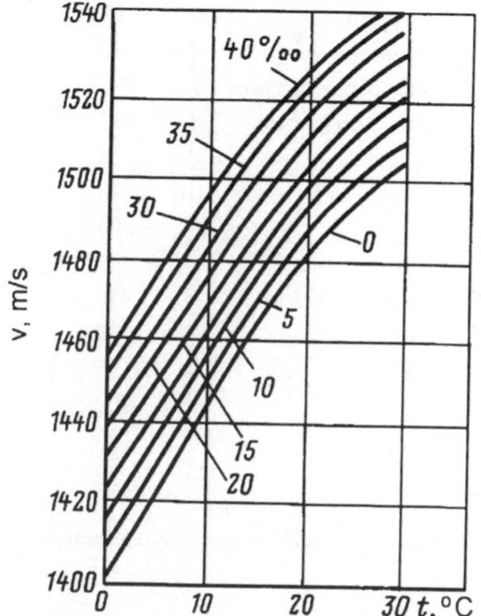

Figure 7.14 Dependence of sound velocity in a sea water on temperature at different salt concentrations [58].

Table 7.10 Sound velocity (m/s) in distilled water versus temperature and pressure [As an example, the sound velocity v = 1629.9 m/s at t = 50°C and P = 50 MPa.]*

P,	t,°C								
MPa	16	20	30	40	50	60	70	80	90
0.1	1469.3	1482.3	1509.1	1528.8	1542.5	1550.9	1554.7	1554.4	1550.4
10	485.7	499.0	526.1	546.1	560.3	569.3	573.7	574.0	570.6
20	502.2	515.6	543.0	563.4	578.0	587.5	592.5	593.4	590.7
30	518.8	532.2	559.9	580.6	595.5	605.4	610.9	612.4	610.4
40	535.6	548.9	576.7	597.6	612.7	622.9	628.8	631.0	629.5
50	552.7	565.8	593.5	614.6	629.9	640.3	646.6	649.1	648.3
60	570.0	582.9	610.3	631.4	646.8	657.5	664.0	666.9	666.6
70	587.1	599.9	627.1	648.0	663.6	674.5	681.2	684.5	684.6
80	603.9	616.8	643.7	664.5	680.1	691.2	698.3	701.8	702.5

* Calculation [47] with Barlow and Yazgan interpolation formula [48].

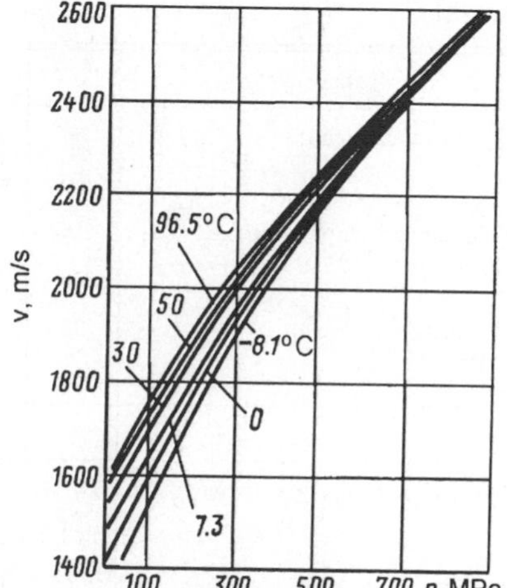

Figure 7.15 Dependence of sound velocity in water on pressure at different temperatures [86].

Table 7.11 Sound velocity in molten substances

Substance	t,°C	Velocity near the melting point	Temperature interval of measurement	$-\Delta v/\Delta t$, m/K·s	Reference
Aluminum	660	4673–4730		0.16–0.47	[61, 62]
Antimony	630	1980	630–880	0	[79]
Bismuth	280	1649–1663	289–356	0–0.08	[64, 65]
Cadmium	321	2200–2256	321–750	0.29–0.62	[66–68]
Cesium	29	967	29–130	0.3	[66]
Copper	1100	3460	1083–1500	0.46	[67]
Gallium	30	2740–2872	30–275	0.23	[66, 67]
Indium	156	2215–2315	156–950	0.27–0.5	[66, 67]
Lead	327	1790–1820	327–1000	0.3–0.53	[64, 66–68]
Mercury	20	1450–1452	0–204	0.31–0.7	[74, 76–78]
Potassium	64	1820–1880	64–160	0.5–0.53	[66, 69, 70]
Potassium bromide	742	1770	745–1010	0.67	[63]
Potassium chloride	770	2275	785–1020	0.88	[63]
Potassium iodide	682	1555	690–1020	0.64	[63]
Rubidium	39	1260	39–160	0.4	[66]
Silver	970	2710–2770	961–1540	0.41–0.47	[67, 79]
Sodium	100	2395–2653	98–700	0.3–0.66	[66, 69, 70, 72]
Sodium bromide	745	1798	750–1010	0.63	[63]
Sodium chloride	800	2483	810–1010	0.92	[63]
Sodium iodide	660	1502	670–1030	0.54	[63]
Sulphur	115	1315			[66]
Thallium	302	1625			[66]
Tellurium	460	920	460–950	change sign	[67]
Tin	232	2270–2480	232–1000	0.21–0.7	[66, 67, 73–75]
Zinc	420	2700–2850	419–850	0.27–0.31	[66, 67, 79, 80]

Table 7.12 Sound velocity and its anisotropy in liquid crystals

Substance	t, °C	v, 10^3 m/s		Ref.
Cholesterylinoleate	30	1.54		[88]
	50	1.44		[88]
Cholesterylpropionate	100	1.268		[89]
	124	1.204		[89]
Cholesterylaurate	73	1.384		[89]
	98.5	1.268		[89]
		$\theta^* = 0°$	$\theta^* = 90°$	
Diethyl-p-p-nitrooxydibenzoate	117.4	1.275	1.23	[90]
	113.2	1.30	1.255	[90]
Ethyl-p-metoxybenzylidene	88.2	1.65	1.56	[91]
aminocinnamate		1.475 ($\theta = 45°$)		[91]

* θ is the angle between propagation direction and helicoid axis.

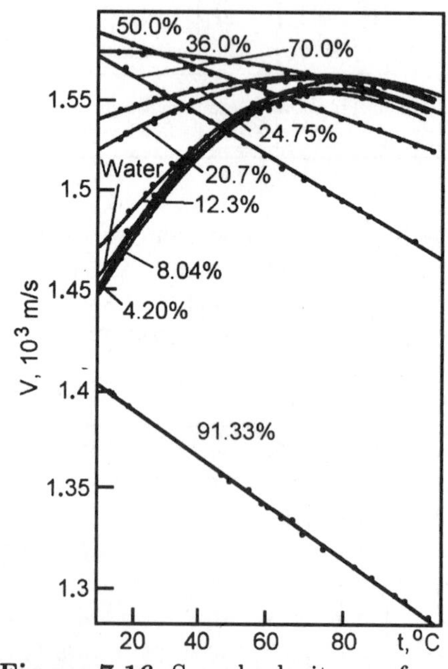

Figure 7.16 Sound velocity as a function of temperature in water solutions of H_2SO_4 with different concentrations (in weight %) [66].

Table 7.13 Shock wave velocity in liquids [92]

Substance	t,°C	Density ratio before and after the jump	Pressure jump, 10^2 MPa	Front velocity, m/s	Substance	t,°C	Density ratio before and after the jump	Pressure jump, 10^2 MPa	Front velocity, m/s
Acetone	30	0.623	45.7	3970	Water	20	0.716	31.4	3354
Benzene	16	0.647	51.6	4100			0.658	57.4	4126
Carbon							0.620	86.6	4813
tetrachloride	22	0.622	72.9	3510			0.622	84.3	4757
Glycerin	18	0.710	75.6	4580			0.577	131	5604
Methyl							0.488	329	8070
alcohol	15	0.625	46.0	3950			0.456	383	8450
Toluene	15	0.650	51.5	4120			0.444	395	8490
							0.450	414	8740

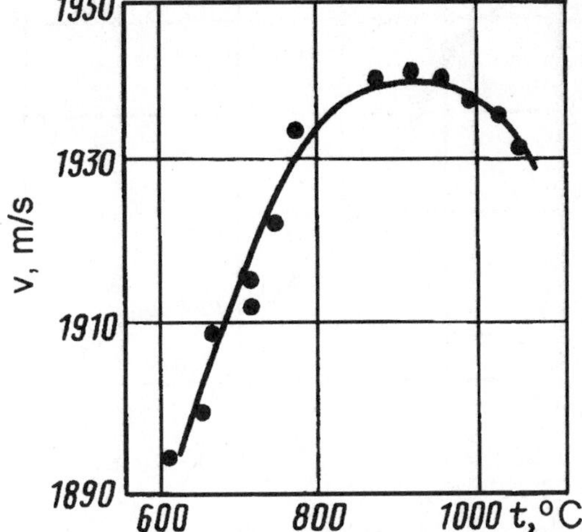

Figure 7.17 Dependence of sound velocity on temperature in liquid bismuth [67].

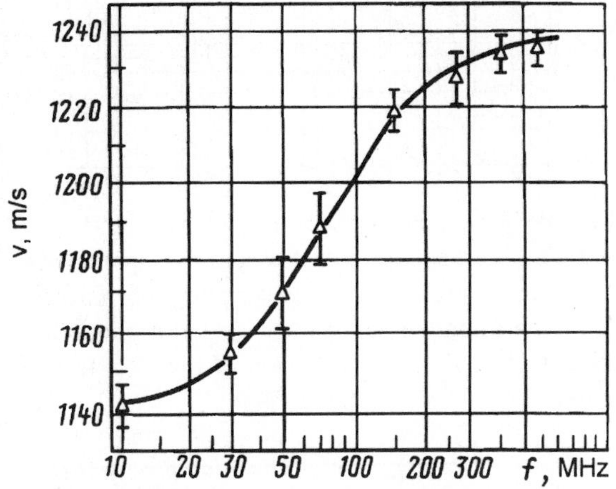

Figure 7.18 Sound velocity dispersion in carbon disulfide at 25°C (vibrational relaxation of C-S bonds) [87].

Table 7.14 Nonlinear adiabatic parameter B/A of liquids ($t \approx 20$°C)

Substance	B/A	References	Substance	B/A	References
Acetone	8.6–9	[97, 98]	Ethyl ether	3.1	[93]
Alcohol:			Gasoline A-70	10.2	[97, 98]
n-butyl	8.6	[97]	Glycerol	9.4	[102]
n-decyl	8.6	[97]	Heptane	10.16	[103]
ethyl	9.6–10.0	[97, 98]	Hydrogen (−259°C)	5.59	[96]
n-hexyl	9.7	[97]	Indium (160°C)	4.5	[95]
methyl	7.6–8.7	[93, 97, 98]	Mercury (30°C)	7.8	[95]
n-nonyl	9.0	[97]	Methane (−163°C)	17.95	[96]
n-octyl	8.7	[97]	Nitrogen (−195°C)	3.1–7.8	[93, 96]
n-propyl	8.9	[97]	Nonane	10.35	[103]
Amylacetate	5.1	[93]	Octane	10.34	[103]
Argon (−187°C)	5.01	[96]	Parahydrogen (−258°C)	7.12	[102]
Benzene	8.4–9.5	[98, 99]	Sodium (110°C)	2.7	[95]
Bismuth (318°C)	7.1	[95]	Tin (240°C)	4.4	[95]
Carbon tetrachloride	10.4–10.8	[97, 98]	Toluene	9.4–10.2	[97, 103]
Chloroform	10.5–10.6	[97, 98]	Transformer oil	6.5	[97]
Dichloroetane	7.6–7.7	[97, 98]	Water	5.06–6.6	[97, 98, 100, 101]
Dodecane	10.41	[103]	Water heavy	4.52	[102]
Ethylacetate	5.0	[93]	m-Xylene	8.7–9.6	[97, 100]

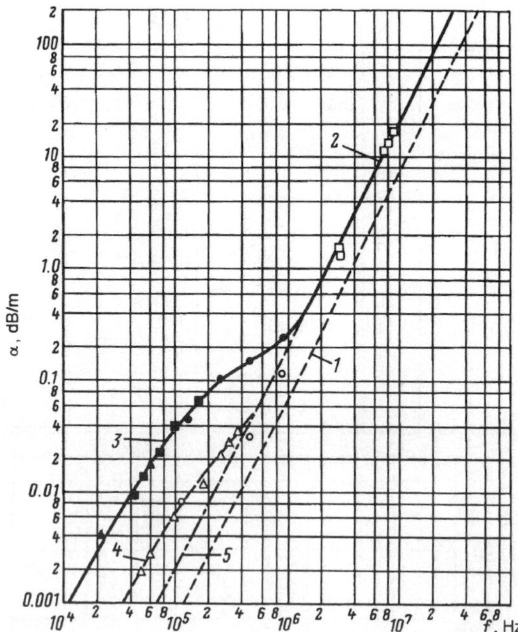

Figure 7.19 Sound damping in fresh and sea water at high sound and ultrasound frequencies: *1* – theory; *2* – experiment in sea and fresh water; *3* – experiment in sea water; *4* – experiment in fresh water; *5* – extrapolation of data on fresh water [105].

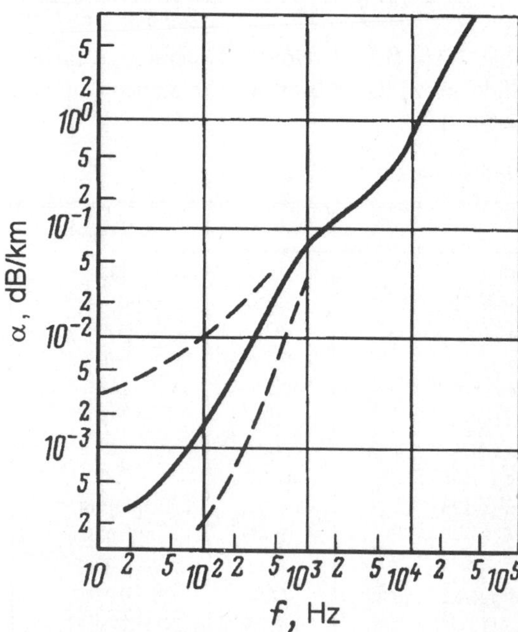

Figure 7.20 Sound damping in sea water at low frequencies:
solid line is for averaged results of measurements at long ocean paths [106];
dashed line shows the region of experimental data dispersion.

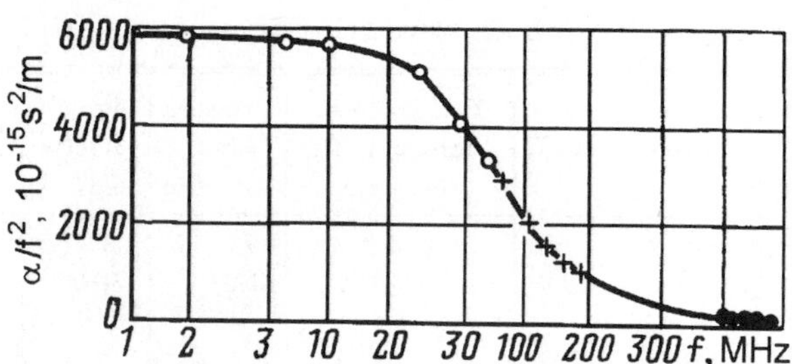

Figure 7.21 Dependence of α/f^2 on frequency in carbon disulfate at $t = 25°C$ (vibrational relaxation of C - S bonds) [115]: ∘ – data of [112].

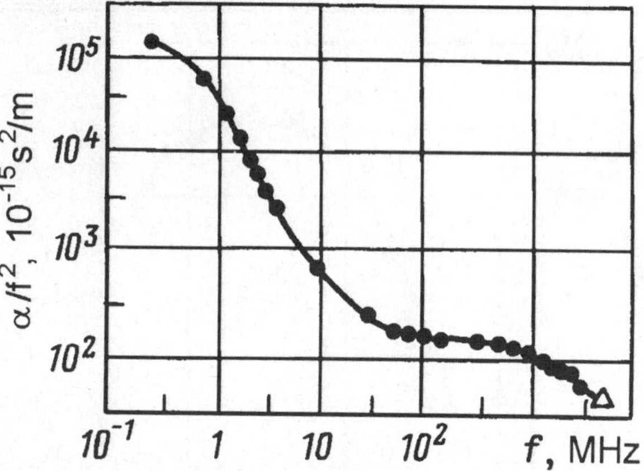

Figure 7.22 Dependence of α/f^2 on frequency in acetic acid at t=20–25°C [123]: △ – data of [124] obtained from the fine structure of scattered light.

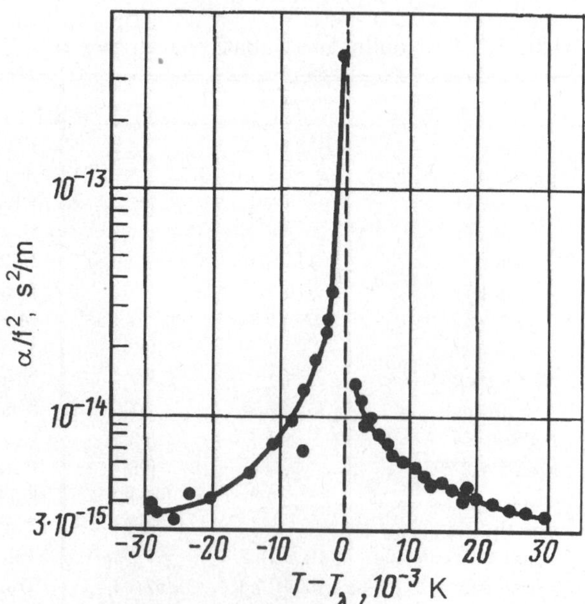

Figure 7.23 Dependence of α/f^2 on temperature near the λ-point in liquid helium (f = 1–12.1 MHz) [126].

Table 7.15 Sound absorbtion coefficient in liquids

Substance	t,°C	f, MHz	α/f^2, 10^{-15} s^2/m	Ref.
Acetal	25	100–200	57–43	[39]
Acetic anhydride	30	0.7–1.2	445	[109]
	24.6	104	58	[108]
Acetone	25	20–1900	31–28	[39, 112]
	25	1–4	70	[45]
Acetonitryle	25	7–10	80	[39]
Acetophenone	25	15–20	50	[39]
Acetyl chloride	24.6	104	82.8	[39]
Acid:				
acetate	18	0.5	30000	[116]
	18	67.5	158	[39]
	25	100–200	140	[39]
dichloracetate	25	100–200	214–201	[39]
formic	17.5	4	2270	[115]
	20.5	9.8	1170	[115]
furfuric	25	100–200	91	[39]
isovaleric	25	100–200	170	[39]
Alcohol:				
allyl	25	100–200	45	[39]
n-amyl	25	21	102	[39]
iso-amyl	25	100–200	131	[39]
benzyle	25	50–200	79	[39]
n-butyl	25	4–200	81	[39]
iso-butyl	25	100–200	140	[39]
ethyl	25	8–220	53	[39]
furfuryl	25	100–200	91	[39]
methyl	25	7–250	33	[39]
n-propyl	25	15–280	75	[39]
iso-propyl	23	15	92	[39]
n-Amyl acetate	20	20–200	65	[39]
ortho-Anisidine	24.8	104.3	58.8	[108]
Anisaldehyde	24.6	104.1	63.6	[108]
Aniline	25	0.3–0.6	50	[39]
Anisole	24.8	104	43.8	[108]
Argon	−188	44.4	10.1	[107]
Benzyl chloride	25	20–200	74	[39]
Benzylacetone	25	10–200	58	[39]
Benzylmethyl ketone	25	10–200	53	[39]
Benzene	25	0.5–30	880–870	[109]
	20	482	445	[111]
	20	1200–1900	184–95	[112]
Benzene bromide	25	20–100	145	[39]
	25	1000–1900	128–94	[112]
Benzene iodide	25	20–260	210	[111]
Benzotrichloride	25	100–200	107	[39]
n-Bromoanisole	25	100–200	62	[39]
ortho-Bromoanisole	25	100–200	70	[39]
Bromoform	25	0.2–200	250	[39]
n-Butyl acetate	20	40–200	46	[39]
iso-Butyl acetate	25	20–200	55	[39]
Butyl bromide	2	15	49	[109]
Butyl chloride	2	15	108	[109]
Butyl iodide	2	15	48	[109]
Carbon disulphide	20	1–10	6000	[115]
	25.5	189.2	776	[108]
	25	1000–1900	37–17	[112]
Carbon tetrachloride	30	0.5–1.2	560	[109]
	25	3–200	540–517	[39, 111]
	20	482	480	[111]
	20	1000–1900	405–226	[112]
ortho-Chloroaniline	25	100–200	54	[39]
Chlorobenzene	25	1–200	140	[39]
	25	1000–1900	125–96	[112]
Chloroform	25	0.2–200	418–380	[39]
	20	1000–1900	220–106	[112]
ortho-Chlorophenol	25	100–200	95	[39]
2-Chloroethanol	24.4	104.1	59	[108]
Cinnamic aldehyde	24.8	104.2	96	[108]
Crotonaldehyde	24.6	104.2	82.7	[108]
Cyclohexane	30	0.6–1.2	250	[109]
	25	100–200	101	[39]
Cyclohexanone	25	100–200	71	[39]
Cyclohexene	25	100–200	102	[39]
Cyclohexylamine	25	100–200	64	[39]
Decahydronaphthalene	25	100–200	121	[39]
1,2-Dibromomethane	25	30	311	[114]
1,1-Dimethyl-cyclohexane	24.1	104.1	127	[108]
1,4-Dioxane	25	100–200	114	[39]
ortho-Dichlorobenzene	24	100–200	131	[39]
1,1-Dichloroethane	25	30	100	[39]
1,2-Dichloroethane	30	0.7–1.2	105	[109]
trans-Dichloroethylene	25	30	360	[39]
Diethylamine	25	100–200	35	[39]
Diethylketone	25	10–200	25	[39]
Epychlorohydrin	24.4	104.1	68	[108]
Ethanolamine	24.8	100–200	152	[39]
Ether *iso*-propyl	24	104	53	[39]
Ether ethyl	25	10–100	45	[39]
Ethyl acetate	30	0.8–1.2	258	[109]
	25	1.0	500	[39]
	20	40–200	35	[39]
Ethyl bromide	25	1–200	70–62	[39]
Ethyl iodide	2	15	40	[39]
Ethylbenzene	25	100–200	55	[39]
Ehtylbutyrate	24.4	104	39	[108]
Ethylene bromide	25	0.2–200	300	[39]
Ethylene chloride	25	1–30	140	[39]
Ethylene glycol	25	5–15	120	[39]
Ethylformate	25	20–100	50	[39]
Fluorobenzene	25	30	278	[39]
	25	1000–1900	221–140	[112]
Formamide	25	10–100	40	[39]

Table 7.15 Sound absorption coefficient in liquids *(continued)*

Substance	t,°C	f, MHz	α/f^2, 10^{-15} s^2/m	Ref.
Glycerol	−18.8	30	12200	[113]
	20–27	0.15–4	2500	[45]
Helium	−270.2	15	110	[34]
n-Heptane	25	1–15	80	[39]
n-Hexane	25	20–200	60	[39]
Hydrogen	−256.2	44.4	5.6	[39]
Indene	25	100–200	133	[39]
Isoprene	15	10–190	58	[39]
Mesityloxyde	24.4	104.1	35.7	[108]
Methyl bromide	2	15	300	[39]
Methyl iodide	25	15–200	310	[39]
	24.4	104.2	36	[108]
Methylacetate	25	1	468	[118]
Methylal	24.4	104.1	39.2	[108]
Methylbenzene	20	27–482	80	[111]
Methylbutyrate	24.5	104.1	35.3	[108]
Methylcyclohexane	25	100–200	95	[39]
Methyldisulphide	24.9	104	58.7	[108]
Methylene bromide	2	15	304	[39]
	25	30	560	[39]
	25	1000–1900	104–40	[112]
Methylene chloride	20	21.6	920	[111]
	20	482	171	[111]
Methylene iodide	25	1–4	816–820	[39]
	25	30	250	[39]
Methylethylketone	25	10–200	30	[39]
Methylformamide	24.6	104.1	32	[108]
Methylformate	25	5–25	50	[39]
Methylsalicylate	24.5	104	57.5	[108]
ortho-Nitroanisole	25	100–200	70	[39]
Nitrobenzene	25	0.3–200	73	[39]
Nitrogen	−199.3	44.4	10.5	[107]
Nitromethane	25	7–10	90	[39]
Oil:				
castor	18.6	3.16	10900	[113]
	20	1.43	6000	[117]
linseed	20.5	3.1	1470	[117]
olive	21–25	1–4	1250	[118]
Oxygen	−213.2	44.4	8.6	[39]
Propyl bromide	2	15	39	[110]
Propyl chloride	2	15	42	[110]
Propyl iodide	2	15	54	[110]
n-Propylacetate	20	40–200	40	[39]
iso-Propylacetate	25	20–200	65	[39]
iso-Propylbenzene	25	2–200	64	[39]
β-Picoline	24.5	100–200	65	[39]
Pyridine	0	0.25–1.0	334	[121]
Thiophene	20	301	775	[111]
	20	482	469	[111]
	20	843–1900	160–46	[111, 112]
Toluene	27	0.15	205	[120]
	25	100–200	86	[39]
	20	1000–1900	81–66	[112]
Toluidine	25	27–482	50–60	[114]
Tribromoethane	25	10–110	200	[39]
Trichloroethane	30	1.0	245	[109]
	25	10–110	130	[39]
Trichloroethylene	25	0.2–0.6	200	[39]
Tetrachloroethylene	25	0.2–0.4	304	[39]
Water	20	7–1900	25	[39, 111, 112]
	20	843	24.2	[111]
ortho-Xylene	25	20–200	65	[39]
para-Xylene	25	1–20	60	[39]

Table 7.16 Sound absorption coefficient in molten metals

Substance	t,°C	α/f^2, 10^{-15} s^2/m	Ref.
Bismuth	280	8.05	[80]
	305	9.3	[78]
Cadmium	360	14.5	[80]
Gallium	30	1.58	[81]
Lead	340	9.4	[80]
Mercury	25	5.7–6.2	[78, 81]
Potassium	75	29.9	[78]
Sodium	100	14.5	[78]
Tin	240	5.63	[80]
Zinc	450	3.7	[80]

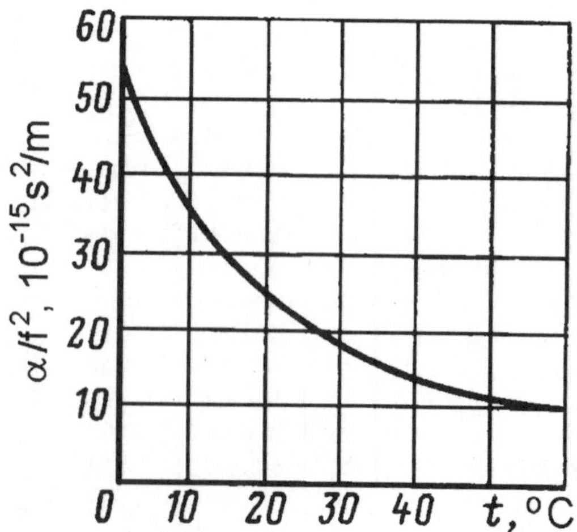

Figure 7.24 Dependence of α/f^2 on temperature in water (f = 0.1–100 MHz) [59,107].

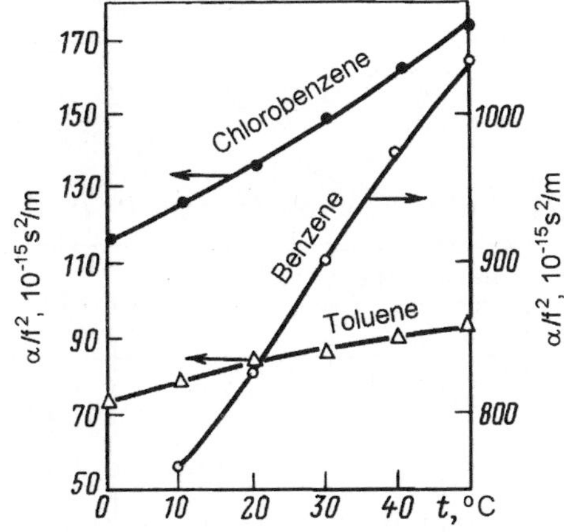

Figure 7.25 Dependence of α/f^2 on temperature in chlorobenzene ($f = 15-30$ MHz), in toluene ($f = 12-16$ MHz) and benzene ($f = 6-12$ MHz) [128].

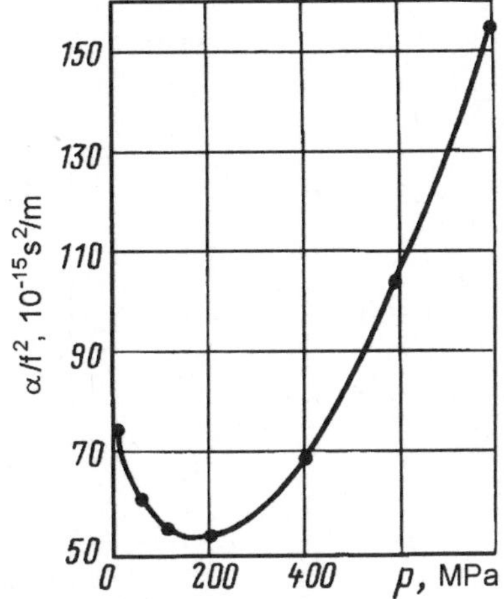

Figure 7.26 Dependence of α/f^2 on pressure in n-butyl alcohol [85].

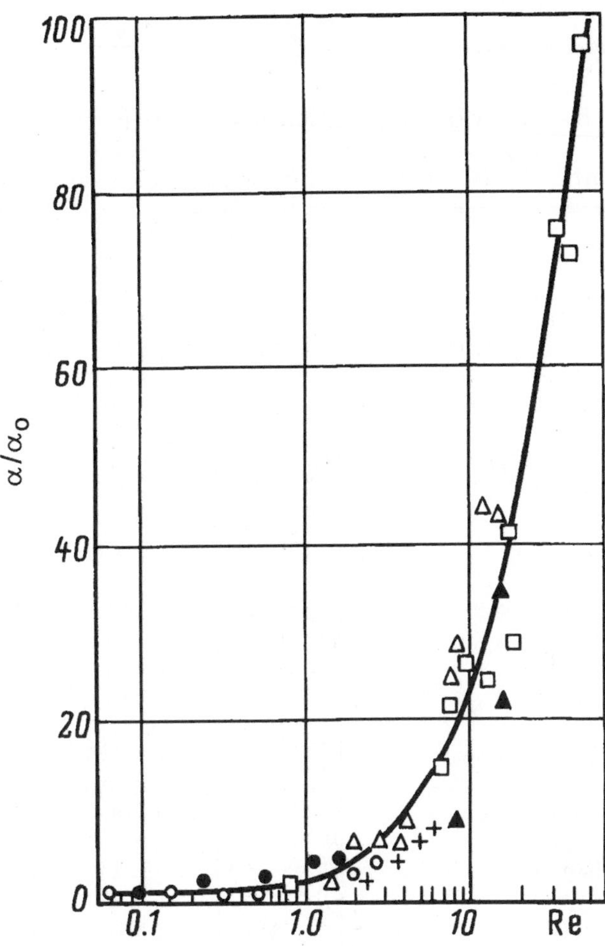

Figure 7.27 Nonlinear absorption of ultrasound in water as a function of the acoustic Reynolds number $Re = P/2\pi f b$: p is the amplitude of sound pressure; α_0 is the absorption factor for the low-amplitude wave; $b = \alpha_0 \rho_0 v^3/2\pi^2 f^2$ (at $t = 20°$C, $b = 4.2 \cdot 10^{-3}$ Pa·s).

7.4 Sound Propagation in Biological Tissues

Table 7.17 Velocity and dumping of ultrasound (0.5–7 MHz) in biological tissues ($t = 37°$C, $p_0 = 0.1$ MPa) [347]

Tissue	ρ, 10^3 kg/m^3	v, m/s	Z, 10^6 kg/m^2·s	α^*, m^{-1} (f, MHz)
Bone	1.700	3.360	6.0	
Brain	1.030	1.510	1.56	$(7-24)f^{(1.18-1.14)}$
Kidney	1.050	1.560	1.64	$(2.8-10)f^{(1.02-1.00)}$
Liver				$(2.6-8)f^{(1.17-1.13)}$
Fat	0.970	1.440	1.40	
Muscle	1.070	1.570	1.68	$(14-56)f^{(1.17-0.74)}$

* Absorption strongly depends on the state of tissue; given values must be treated as approximate ones.

7.5 Sound Propagation in Solids

Table 7.18 Sound velocity and specific impedance of polycrystal and amorphous solids ($t = 20°C$)

Substance	Sound velocity, 10^3 m/s			Specific impedance, 10^5 kg/m^2·s		Ref.
	In rod (v_r)	Longitudinal (v_l)	Shear (v_s)	Z_l	Z_s	
Aluminum	5.08	6.26	3.08	169	83	[6]
(372°C)	4.34					[136]
Antimony	3.40					[139]
Bakelite		1.59		36		[3]
Basalt		5.4–6.2	3.1–3.3	140–170	80–90	[146]
Beryllium (27°C)		12.55	8.83	232	162	[137]
Bismuth	1.79	2.18	1.10	214	108	[6]
Brass	3.49	4.43	2.12	361	172	[6]
Cadmium	2.40	2.48	1.50	240	129	[6]
Carbon						
pyrolitical		3.3		73		[3]
glassy		4.26	2.68	62.6	38.2	[3]
Cast iron	3.85	4.50	2.40	350	187	[143]
Chromium		6.65	4.03	466	282	[3]
Coal						
coke (Donbass)		$v_{\|} = 1.54$; $v_{\perp} = 1.50$		$Z_{\|} \approx Z_{\perp} = 23$		[147]
charcoal		$v_{\|} = 1.78 - 2.0$; $v_{\perp} = 1.23$		$Z_{\|} = 23 - 26$; $Z_{\perp} = 16$		[147]
Hard coal*		$v_{\|} = 2.52$; $v_{\perp} = 2.15$		$Z_{\|} = 40$; $Z_{\perp} = 34$		[147]
Constantan	4.30	5.24	2.64	460	232	[6]
Copper	3.71	4.70	2.26	418	201	[6]
Cork	0.50					[6]
Ebonite	1.57	2.40		28		[6]
Epoxy resin ED–5		2.58	1.22	29	13	[58]
Glasses:						
beryllium fluorite (17°C)		4.70	3.90	221	183	[53]
boron B_2O_3 (17°C)		3.47	1.25	62	23	[53]
crown glass	5.30	5.66	3.42	141	86	[6]
flint glass		4.26	2.56	154	92	[6]
fused quartz		5.935	3.740	132	83	[144]
germanate GeO_2 (17°C)		3.61	2.21	130	80	[53]
glassy selenium (17°C)		1.84	0.96	79	41	[53]
chalcogenide:						
As_2S_3 (17°C)		2.58	1.49	85	50	[53]
As_2Se_3 (17°C)		2.23	1.29	103	100	[53]
heavy crown glass	4.71	5.26	2.96			[6]
heavy flint glass	3.49	3.76	2.22	173	102	[6]
light flint	4.55	4.80	2.95			[6]
Gold	2.03	3.24	1.20	626	232	[6]
Granite	3.95	4.4–5.6	2.78	160–210	103	[58, 139]
Gypsum		4.79	2.37	111	55	[53]
Hydrogen solid (4.2 K)		2.19		1.9		[3]
Ice (0°C)	3.28	3.98	1.99	37	19	[6]
Indium		2.56	0.81	187	59	[138]
Iridium	4.79					[6]
Iron	5.17	5.85	3.23	456	252	[6]

Table 7.18 Sound velocity and specific impedance of polycrystal and amorphous solids *(continued)*

Substance	Sound velocity, 10^3 m/s			Specific impedance, 10^5 kg/m^2·s		Ref.
	In rod (v_r)	Longitudinal (v_l)	Shear (v_s)	Z_l	Z_s	
Lead	1.20	2.16	0.70	246	80	[6]
Limestone		4.6–6.2	2.3–3.2	120–165	60–86	[146]
Magnesium	4.90	5.78	3.05	101	53	[3, 139]
Manganese	3.83	4.66		346		[139]
Manganin	3.83	4.66	2.35	393	197	[6]
Marble	3.81	4.9–6.2	3.26	130–170	88	[139, 146]
Molybdenum		5.67	3.51	511	316	[140]
Nylon		2.65	1.1	29	12.3	[3, 145]
Neizilber (German silver)	3.58	4.76	2.16	400	181	[6]
Nickel	4.79	5.63	2.96	495	260	[6]
Niobium		4.92	2.10	422	180	[3]
Paraffin		1.5		2.3		[3]
Platinum	2.80	3.96	1.67	846	357	[6]
Plexiglass		2.67	1.12	30	13	[6]
Polyethylene		2.48				[145]
Polystyrene		2.35	1.12	25	12	[6]
Porcelain	4.88	5.34	3.12	128	75	[6]
Rubber		1.48		13		[6]
Rubidium		1.26		19.3		[6]
Schist	4.51					[139]
Silver	2.64	3.60	1.59	380	167	[6]
Steel (different types)	5.0–5.2	5.68–6.10	3.2	445–477	249	[3, 58, 141]
Tantalum	3.35	4.10	2.9	548	388	[3, 139]
Teflon		1.34		30		[145]
Thorium		2.40	1.56	332	216	[3]
Tin	2.73	3.32	1.67	242	122	[6]
Titanium	5.04	6.33	3.11	285	139	[140]
Tungsten	4.31	5.46	2.62	1042	500	[6]
Uranium		3.4	2.0	630	371	[3]
Vanadium		6.0	2.78	362	168	[3]
PVC plastic		2.30				[58]
Wood (oak)	4.05					[143]
Zinc	3.81	4.17	2.41	296	171	[6]
Zirconium		4.65	2.25	301	146	[3]

* $v_{\parallel}$ – velocity in the layer direction, $v_{\perp}$ – velocity in the direction orthogonal to the layer.

Table 7.19 Sound velocity in cubic crystals (t = 20–25°C)

Crystal, formula, point group	Density ρ	v, 10^3 m/s						References
	10^3 kg/m^3	L_{100}	S^{*1}_{100}	L_{110}	$S^{1\bar{1}0}_{110}$	L_{111}	S^{*1}_{111}	
Aluminum, Al, $m3m$	2.70	6.284	3.26	6.47	2.86	6.473	3.00	[53, 148, 153]
Alumomagnesium spinel, $MgAl_2O_4$, $\bar{4}3m$	3.58	9.02	6.54	10.0	6.55	10.5	5.0	[142]
Ammonium bromide, NH_4Br, $m3$	2.44	3.72	1.68	3.41	2.25	3.30	2.08	[53]
Barium fluoride, BaF_2, $m3m$	4.89	4.27	2.2	4.29	2.3	4.26	2.3	[142]
					2.26(S^{001}_{110})			[160]
Barium oxide, BaO, $m3m$	5.72	4.6921	2.4422	4.5326	2.60	4.61	2.57	[53]
Bismuth germanate, $Bi_{12}GeO_{20}$, 23	9.23	3.65	1.68	3.39	1.67	3.36		[163, 166]
Bismuth silicate, $Bi_{12}SiO_{20}$, 23	9.21	3.83						[176]

Table 7.19 Sound velocity in cubic crystals (t = 20–25°C) *(continued)*

Crystal, formula, point group	Density ρ	v, 10^3 m/s						References
	10^3 kg/m^3	L_{100}	S^{*1}_{100}	L_{110}	$S^{1\bar{1}0}_{110}$	L_{111}	S^{*1}_{111}	
Cadmium telluride, CdTe, $\bar{4}3m$	5.85	3.02	1.87	3.34	1.20	3.44	1.46	[53]
Cesium bromide, CsBr , $m3m$	4.46	2.636	1.30	2.47	1.299	2.41	1.50	[53, 148, 155]
Cesium chloride, CsCl, $m3m$	3.99	3.04	1.42	2.78	1.85	2.69	1.71	[53, 155]
Cesium iodide, CsI, $m3m$	4.51	2.326	1.17	2.21	1.39	2.16	1.32	[53]
Cobalt–zinc ferrite,								
$Co_{0.32}Zn_{0.22}Fe_{2.2}O_4$	5.43	7.00	3.80	7.28	3.23	7.37	3.42	[53]
Copper, Cu, $m3m$	8.94	4.34	2.90	4.96	1.62	5.16	2.14	[148]
Diamond, C, $m3m$,	3.51	17.5	12.80	18.3	11.6	18.6	12.0	[53, 148]
Fluorite, CaF_2, $m3m$	3.18	7.19	3.30	6.68	4.24	6.50	3.95	[148]
					3.3(S^{001}_{110})			[142]
Galenite, PbS,$m3m$	7.5	4.11	1.82	3.71	2.55	3.56	2.33	[53]
		3.69		3.56	2.07	3.52	1.99	[148]
Gallium antimonide, GaSb, $\bar{4}3m$	5.62	3.96	2.77	4.38	2.07	4.51	2.33	[53, 148]
Gallium arsenide, GaAs, $\bar{4}3m$	5.31	4.71	3.34	5.24	2.47	5.40	2.79	[53, 167]
Gallium phosphide, GaP, $\bar{4}3m$	4.13		4.13	6.32				[157]
Garnets:								
alumoyttrium, $Y_3Al_5O_{12}$, $m3m$	4.55	8.5630	5.0293	8.6016	4.9496	8.0384		[142, 168]
			5.0311	8.6167				[169, 170]
gallium-yttrium, $Y_3Ga_5O_{12}$, $m3m$	5.79	7.08	4.06					[142]
iron-yttrium, $Y_3Fe_5O_{12}$, $m3m$	5.18	7.0800	4.0601	7.172	3.902			[169, 171]
			3.843					[172]
Germanium, Ge, $m3m$	5.33	4.92	3.55	5.41	2.38	5.56	3.04	[148]
					2.75			
Gold, Au, $m3m$	19.3	3.10	1.47	3.33	0.866	3.39	1.10	[148]
Indium antimonide, InSb, $\bar{4}3m$	5.79	3.42	2.26	3.77	1.63	3.89	1.87	[149, 167]
Indium arsenide, InAs, $\bar{4}3m$	5.66	3.84	2.64	4.29	1.83	4.42	2.14	[148, 167]
α–Iron, Fe, $m3m$	7.86	5.55	3.77	6.24	2.46	6.46	2.97	[53]
Iron silicate, Fe_2Si, $m3m$	7.19	5.68	4.33	6.76	2.30	7.09	3.13	[53]
KRS–5, $m3m$	7.37	2.15		2.00	v_s=0.874^{*2}	1.96		[142, 175]
KRS–6, $m3m$	7.19	2.416		2.2170	1.0278			[175]
Lead, Pb, $m3m$	11.34	2.03	1.13	2.25	0.57	2.32	0.80	[148]
Lithium , Li, $m3m$	0.55	5.19	4.43	6.67	1.41	7.09	2.82	[148]
Lithium fluoride, LiF, $m3m$	2.64	6.63	4.90	7.40	3.54	7.60	4.16	[148, 151]
Lithium–indium , Li–In	5.16	3.29	2.27	3.82	1.17	3.76	1.62	[53]
Magnesium fluoride, MgF_2, $m3m$	3.98	6.64	7.72		4.16			[53, 142]
Magnesium oxide, MgO, $m3m$	3.58	8.94	6.43	9.66	5.27	9.89	5.68	[148]
Mercury telluride, HgTe, $\bar{4}3m$	8.08	2.58	1.62	2.87	1.03	2.96	1.26	[53]
Molybdenum, Mo, $m3m$	10.19	6.72	3.29	6.48	3.73	8.19	2.39	[148]
Nickel, Ni, $m3m$	8.90	5.26	3.74	6.01	2.36	6.24	2.90	[148]
Palladium, Pd, $m3m$	12.13	4.28	2.43	4.73	1.42	5.06	1.86	[53]
Potassium, K, $m3m$	0.91	2.24	1.70	2.73	0.68	2.88	1.13	[148]
Potassium bromide, KBr, $m3m$	2.75	3.55	1.36	3.02	2.29	2.84	2.19	[148]
Potassium chloride, KCl, $m3m$	1.99	4.52	1.77	3.89	2.9	3.61	2.59	[148, 151]
Potassium iodide, KI, $m3m$	3.13	2.92	1.16	2.51	1.89	2.33	1.68	[148]
Silicon, Si, $m3m$	2.33	8.43	5.84	9.175	4.67	9.35	5.09	[319]
					5.86 (S^{001}_{110})			[150]
Silver bromide, AgBr, $m3m$	6.48	2.95	1.05	2.83	1.34	2.79	1.25	[148]
Silver chloride, AgCl, $m3m$	5.57	3.29	1.06	3.13	1.47	3.07	1.35	[148]
Silver, Ag, $m3m$	10.49	3.41	2.07	3.79	1.20	3.92	1.55	[148]
Sodium, Na, $m3m$	1.01	2.44	2.41	3.31	0.84	3.58	1.55	[148]
Sodium bromate, $NaBrO_3$	3.34	4.08	2.13	3.93	2.41	3.87	2.32	[53]
Sodium bromide, NaBr, $m3m$	3.20	3.48	1.74	3.26	2.13	3.18	2.01	[148]

Table 7.19 Sound velocity in cubic crystals (t = 20–25°C) *(continued)*

Crystal, formula, point group	Density ρ	v, 10^3 m/s						References
	10^3 kg/m^3	L_{100}	S^{*1}_{100}	L_{110}	$S^{1\bar{1}0}_{110}$	L_{111}	S^{*1}_{111}	
Sodium chlorate, $NaClO_3$	2.49	4.47	2.17	4.18	2.68	4.09	2.52	[148]
Sodium chloride, NaCl, $m3m$	2.17	4.79	2.44	4.50	2.92	4.45	2.78	[151, 152]
Sodium fluoride, NaF, $m3m$	2.79		v_L=5.66*2		v_s=3.33*2			[173]
Sodium iodide, NaI, $m3m$	3.67	2.667	1.304	2.374	1.73	2.60	1.62	[53, 174]
Strontium titanate, $SrTiO_3$, $m3m$	5.12	7.876	4.910	8.098	4.918	8.141	4.703	[53]
Strontium fluoride, SrF_2, $m3m$	4.24				2.9(S^{001}_{110})	5.3		[142]
Strontium nitrate, $Sr(NO_3)_2$	2.99	3.98	2.21	4.05	2.07	3.79	2.12	[53]
Tallium bromide, TlBr, $m3m$	7.45	2.40	1.14	2.12	1.28	2.07	1.16	[156]
Tallium vanado–sulphate, Tl_3VS_4, $\bar{4}3m$	6.22	2.81	0.873	2.46	1.6			[177]
Thorium, Th, $m3m$	11.66	2.54	2.02	3.07	1.06	3.22	1.45	[148]
Tungsten, W, $m3m$	19.2	5.11	2.81	5.11	2.81	5.11	2.81	[148]
Vanadium, V, $m3m$	6.02	6.15	2.66	5.99	3.011	5.93	2.90	[53, 148]
Zinc selenide, ZnSe, $\bar{4}3m$	5.42	4.07	2.73	4.55	1.82	4.75	2.19	[53, 159]
Zinc sulphide, ZnS, $\bar{4}3m$	4.10	4.90	3.31	5.5	2.09	5.74	2.56	[53]

*1 Arbitrary polarization.
*2 The direction of the wave propagation is not shown.

Table 7.20 Sound velocity in hexagonal crystals ($t \approx 20$°C)

Crystal, formula, point group	Density ρ,	v, 10^3m/s						References
	10^3kg/m^3	L_{100}	L_{011}	L_{001}	S^*_{001}	S^{001}_{100}	S^{010}_{100}	
Beryllium, Be, $6/mm$	1.87	12.25		13.41	9.11	9.1	9.04	[148]
Cadmium selenide, CdSe, $6/mm$	5.68	3.630		3.856	1.521	1.592		[181]
Cadmium sulphide,	4.82	4.181		4.414	1.757	1.7565		[158]
CdS, $6/mm$								[182]
Cadmium,	8.64	3.56	2.33	3.74	1.46	1.46	2.01	[53]
Cd, $6/mm$			2.44				2.05	[148]
Cancrinite, $NaAlSiO_4$	2.44			5.81	3.12			
Cobalt, Co, $6/mm$	8.79	5.91	6.38		2.93	2.93	2.84	[148]
Ice, H_2O, (−16°C), $6/mm$	0.94	3.83		3.99	1.84	1.84	1.90	[148]
Lead germanate	7.29	3.01		3.47	1.67 (S^{100}_{010})			[198]
$Pb_5Ge_3O_{11}$, 6								
Lead vanado-germanate,	7.15	3.11		3.45	1.54			[178]
$Pb_5(GeO_4)(VO_4)_2$, $6/m$				4.427				[53]
Lead vanado-silicate,	7.02			3.62	1.73			[53]
$Pb_5(SiO_4)(VO_4)_2$								
Lithium iodate,	4.5	4.13		4.3	1.99	2.57		[142]
$LiIO_3$, 6								[179]
Magnesium, Mg, $6/mm$	1.79	5.84	5.94		3.06	3.06	3.09	[148]
β-Quartz,	2.53		6.61	3.78				[53]
SiO_2, (580°C), 32								
Rhenium, Re, $6/mm$	20.53			5.92	2.87			[53]
Ruthenium, Ru, $6/mm$	12.1			7.28	3.95			[53]
Yttrium, Y, $6/mm$	4.48			4.14	2.33			[53]
Zinc , Zn, $6/mm$	7.18	4.73	2.92		2.31	2.31	2.97	[148]
Zinc oxide, ZnO, $6/mm$	5.64	6.0776		6.0961	2.7353	2.7350		[180]
Zinc sulphide, ZnS, $6/mm$	4.1	5.512		5.582	2.647	2.652		[181]

* Arbitrary polarization.

Table 7.21 Sound velocity in tetragonal crystals ($t \approx 20°C$)

Crystal, formula, point group	Density ρ, $10^3 kg/m^3$	v, $10^3 m/s$										References
		L_{100}	L_{010}	L_{001}	L_{110}	S_{100}^{010}	S_{100}^{001}	S_{010}^{100}	S_{001}^{110}	$S_{110}^{1\bar{1}0}$	S_{110}^{001}	
Ammonium dihydrophosphate, NH_6PO_4, $\bar{4}2m$	1.80	6.15		4.35		1.83						[158]
Barium titanate, $BaTiO_3$, $4mm$	6.02			5.50					2.79*			[184]
Bastron, barium–strontium niobate, $Sr_{(0.75-0.5)}Ba_{(0.25-0.5)}Nb_2O_6$, $4mm$	5.4			5.5								[158]
Calomel, $HgCl_2$, $4/mmm$	6.97	1.6224		3.3434		1.3054				0.3471		[185]
Lead molybdate, $PbMoO_4$, $4/m$	6.9	3.98		3.632– −3.75		2.20	1.961– −1.98					[186–190]
Lead titanate, $PbTiO_3$, $4mm$	7.95			4.19								[197]
Magnesium fluoride, MgF_2, $4/mmm$				7.43					4.16			[142]
Paratellurite, TeO_2, 422	5.9	3.051		4.14– −4.3		3.317	2.100		2.08	0.616– −0.618		[189, 193] [194, 196]
Potassium dihydrophosphate, KH_2PO_4, $\bar{4}2m$	2.34	5.50		4.86		2.40*						[142, 158]
Rutile, TiO_2, $4/mmm$	4.28	7.929– −8.014	7.929– −7.958	10.94	9.935– −9.950	6.756	5.424	6.700	5.399	3.300	5.389	[189, 191] [192]
$TlGaS_2$, $2/m$	5.72	3.0		2.5		1.0*						[177]
$TlGaSe_2$, $4/mmm$	6.34	3.25		2.67		0.767*						[177]
$TlInSe_2$, $4/mmm$	7.18	2.48	2.48	2.48		1.35*						[177]

* Polarization is not shown.

Table 7.22 Sound velocity in trigonal crystals ($t \approx 20°C$)

Crystal, formula, point group	Density ρ, $10^3 kg/m^3$	v, $10^3 m/s$									References
		L_{100}	QL_{010}	L_{001}	QS_{100}^{010}	QS_{100}^{001}	S_{010}^{100}	QS_{010}^{001}	$S_{001}^{100,010}$	QS_{010}^{*}	
Berlinite, $AlPO_4$, 32		4.501		5.732	2.827	4.337					[199]
Boron, B, $\bar{3}m$	2.29			14							[142]
Cinnabar, α–HgS, 32	8.10			2.450							[206]
Lithium niobate, $LiNbO_3$, $3m$	4.62	6.54– −6.54873	6.829– −6.8822	7.271– −7.3328	4.034– −4.0593	4.76– −4.8012	3.94– −3.9615	4.457– −4.4943	3.57– −3.59	4.46– −4.46667	[207–212, 214, 219]
Lithium tantalate, $LiTaO_3$, $3m$	7.45	5.550– −5.5522	5.6917	6.160– −6.476	3.3556	4.210	3.529				[154, 157] [207–209]

Table 7.22 Sound velocity in trigonal crystals ($t \approx 20°C$) *(continued)*

Crystal, formula, point group	Density ρ, $10^3 kg/m^3$	v, $10^3 m/s$									References
		L_{100}	QL_{010}	L_{001}	QS_{100}^{010}	QS_{100}^{001}	S_{010}^{100}	QS_{010}^{001}	$S_{001}^{100,010}$	QS_{010}^{*}	
Proustite, Ag_3AsS_3, $3m$	5.63	3.2	2.98	2.60			1.50		1.28	1.50	[214]
Quartz, SiO_2, 32	2.65	5.747–-5.7509	6.0061–-6.0070	6.318–-6.325	3.2978	5.1145	3.9158–-3.9169		4.687–-4.6895	4.3207–-4.3249	[144, 200, 203, 205]
Sapphire, α–Al_2O_3, $\bar{3}m$	3.97	11.03–11.235		11.15	5.72	6.780					[215–217]
Tellurium, Te, 32	6.25	2.41		2.3			1.47*				[142, 218]
Thallium orthoselenoarsenide, Tl_3AsSe_3, 32 or 3	7.83			2.10					1.21		[177]
Tourmaline, $3m$	3.10			7.3							[142]

* Polarization is not shown.

Table 7.23 Sound velocity in orthorhombic crystals

Crystal, formula, point group	Density ρ, $10^3 kg/m^3$	v, $10^3 m/s$							References
		L_{100}	L_{010}	L_{001}	S_{100}^{010}	S_{100}^{001}	S_{010}^{001}	S_{001}^{*}	
Antimony sulphoiodide, SbSI, mmm	5.25	2.1–2.7			$L_{\perp c}$=1.7		$S_{\perp c}$=1.0		[231]
Banan, $Ba_2NaNb_5O_{15}$, $mm2$	5.41	6.15			3.65		3.65		[142]
Cesium biphtalate, $CsC_8H_7O_4$, $mm2$	2.18	3.27	2.51	2.54					[222]
Gadolinium molybdate, $Gd_2(MoO_4)_3$, $\bar{4}2m$	4.58	3.368–3.900	3.853–3.995	4.646	2.340–2.708		2.349–2.360	2.360–2.383	[227–229]
Iodic acid, α–HIO_3, 222	4.63–5.0	3.65	2.89	2.44	1.84				[226]
Lithium gallate, $LiGaO_2$, $mm2$	4.19			6.10				3.31	[142]
Lithium germanate, Li_2GeO_3, $mm2$	3.50	6.5–6.66				3.72–3.75	4.04–4.10		[223, 224]
Lithium metagermanate, Li_2GeO_3, $mm2$	3.49			6.66		3.72	4.04		[142]
Potassium biphtalate, $KC_8H_7O_4$, $mm2$	1.64	3.382	2.892	3.347	1.996	2.137	1.774		[221]
Rubidium biphtalate, $RbC_8H_7O_4$, $mm2$	1.93	2.68	2.85						[222]
Sodium–cobalt germanate, Na_2CoGeO_4, $mm2$		5.54	5.26	5.43	3.23			2.24	[225]
Thaliumorthoselenophosphate, Tl_3PSe_4	6.1	2.2	2.2	2.2	1.1				[177]
Thaliumorthosulphatoarsenide, Tl_3AsS_4		2.15						1.21	[224, 230]
Thallium–lead triiodide, $TlPbI_3$	6.39	2.145	1.807	2.289					[213]
Topaz, $Al_2(F,OH)_2 \times SiO_4$, mmm	3.50			9.3					[142]

* Polarization is not shown.

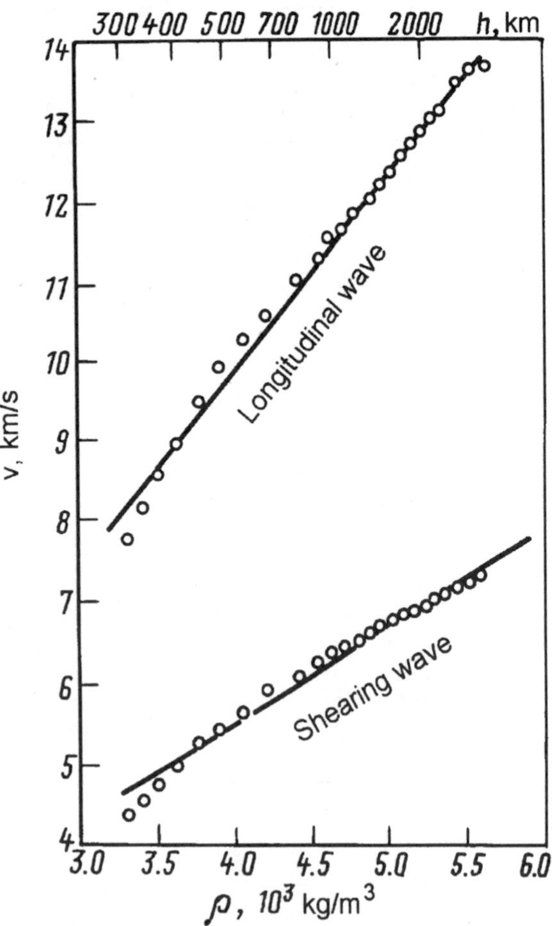

Figure 7.28 Dependence of sound velocity in the Earth's mantle on depth (density) [135].

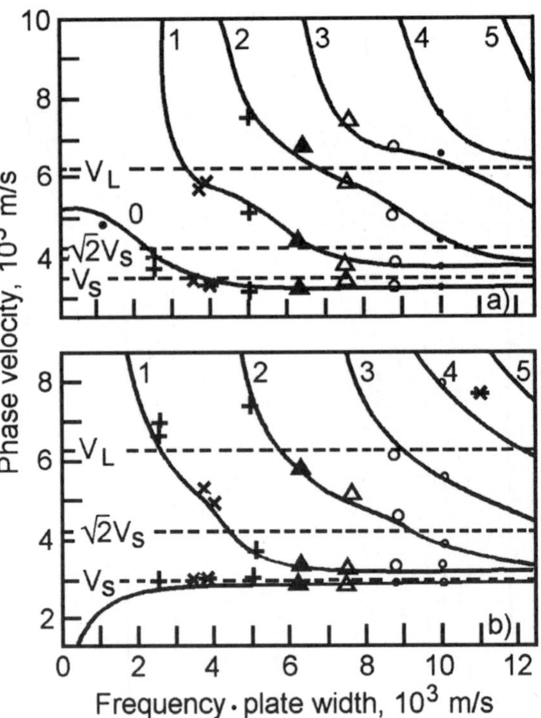

Figure 7.30 Theoretical and experimental dependence of the Lamb wave (SV-wave) phase velocity on the product of frequency and thickness of aluminum plates: a is for symmetrical modes; b is for antisymmetrical modes [348]; v_s and v_l are the velocities of bulk transverse and longitudinal waves. Numbers *0–5* mark numbers of modes.

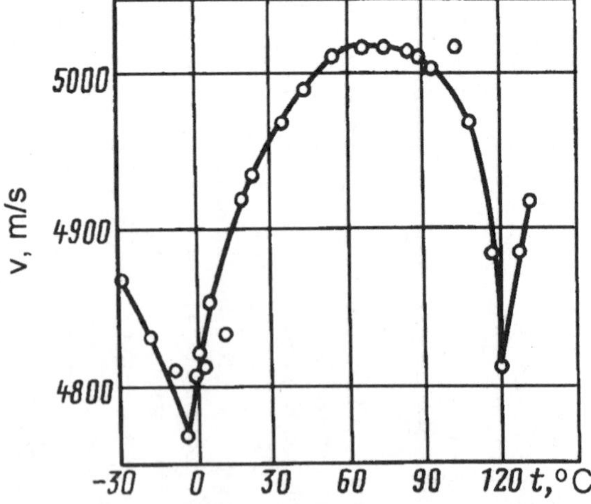

Figure 7.29 Dependence of sound velocity on temperature in polarized ceramics of barium titanate (longitudinal vibrations in the thickness of plate) [183].

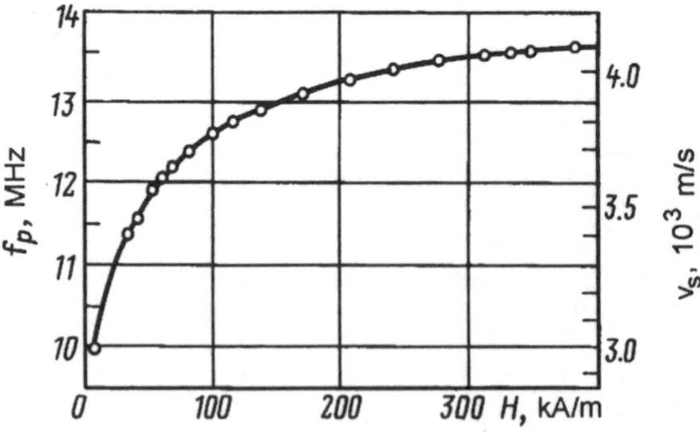

Figure 7.31 Magnetic field dependence of resonant frequency of shear vibrations in a plate made from natural crystal of hematite (α-Fe_2O_3): v_s is the effective velocity. Field direction is normal to the plate plane and coincides with a trigonal axis; "easy plane" is in the plane of the plate [260] (see also [261]).

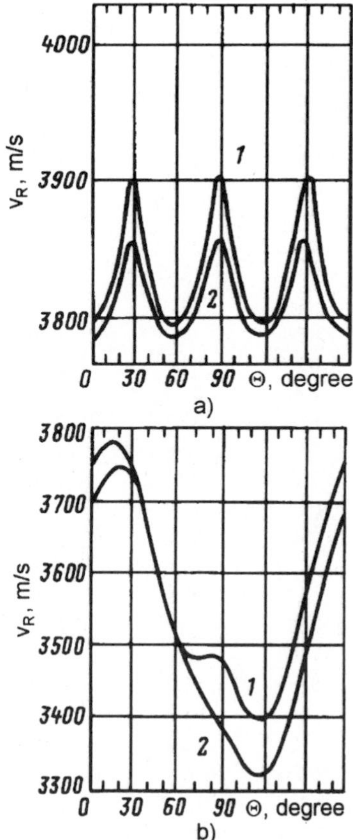

Figure 7.32 Anisotropy of velocity of SAW in $LiNbO_3$. (a) is for z–cut; θ is the angle between the X axis and wave normal; (b) is for x–cut, θ is the angle between the Y axis and wave normal. *1* corresponds to free surface and *2* corresponds to metallized surface [262].

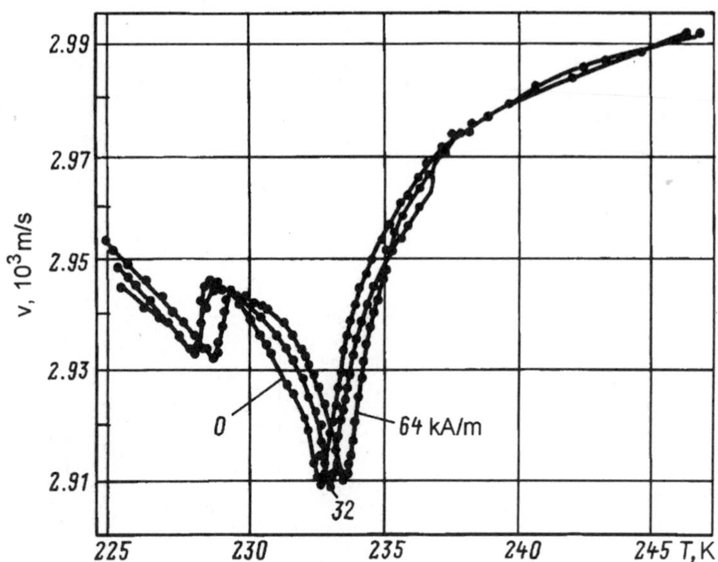

Figure 7.33 Dependence of longitudinal velocity (30 MHz) of the wave propagating along hexagonal axis of terbium on temperature at different field intensities. The field intensities in oersteds are, accordingly: 0 for *1*; 400 for *2*; 800 for *3*; in the vicinity of the Curie (~228 K) and Néel (~233 K) points [204].

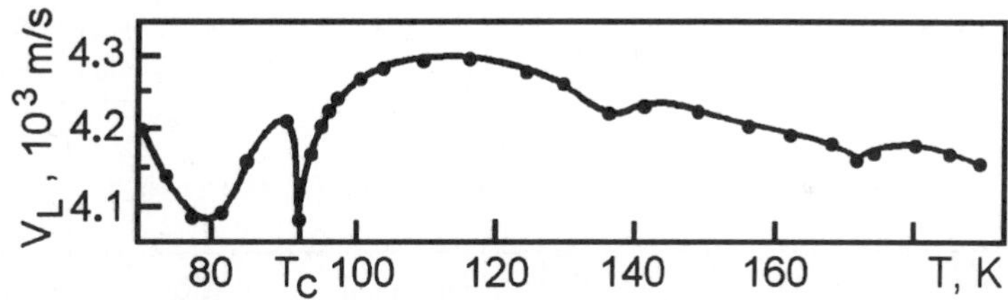

Figure 7.34 Dependence of the longitudinal ultrasound wave velocity (20–24 MHz) on temperature in $YBa_2Cu_3O_{6.9}$ ceramics: T_c is the point of transition into superconducting state [220].

Table 7.24 Materials for piezoelectrical transducers of bulk waves [142]

Material, formula, point group	ρ, 10^3 kg/m^3	Cut	Wave type	K^{*1}	$\varepsilon/\varepsilon_0$	v, 10^3 m/s	Z, 10^6kg/m^2·s	fh^{*2}, 10^{-3} Hz·m
Ammonium dihydrophosphate, $NH_4H_2PO_4$, $\bar{4}2m$	1.80	Z	L	0.28	15.4	4.35	7.8	2.17
Barium titanate*3,	5.7	0°	L	0.384	1260	5.47	31	2.73
$BaTiO_3$		90°	S	0.48	1115	3.16	18	1.58
Barium–sodium niobate,	5.41	Z	L	0.57	32	6.15	33.3	3.075
$Ba_2NaNb_5O_{15}$, $mm2$		X	S	0.21	222	3.65	19.3	1.823
		Y	S	0.25	227	3.65	19.3	1.825
Bismuth germanate,	9.23	111	L	0.15	38.6	3.36	30.4	1.68
$Bi_{12}GeO_{20}$, 23		110	S	0.24		1.76	16.2	0.88
Bismuth silicate,	9.14		L	0.20	42	3.83	35.0	1.92
$Bi_{12}SiO_{20}$, 23			S	0.31				

Table 7.24 Materials for piezoelectrical transducers of bulk waves *(continued)*

Material, formula, point group	ρ, 10^3 kg/m^3	Cut	Wave type	K^{*1}	$\varepsilon/\varepsilon_0$	v, 10^3 m/s	Z, 10^6kg/m^2·s	fh^{*2}, 10^{-3} Hz·m
Cadmium selenide,	5.68	Z	L	0.194	10.2	3.86	22	1.93
CdSe, 6*mm*		Y	S	0.13	9.53	1.54	8.8	0.77
Cadmium sulphide,	4.82	Z	L	0.262	9.53	4.41	21.2	2.20
CdS, 6*mm*		Y	S	0.188	9.02	1.8	8.7	0.9
		39.7°Y	S	0.212	9.33	2.1	10.1	1.05
Lithium gallate,	4.19	Z	L	0.30	8.5	6.10	25.6	3.05
$LiGaO_2$, *mm*2		X	S	0.17	7.0	3.31	13.9	1.65
Lithium germanate,	3.49	Z	L	0.34	10.8	6.66	23.2	3.33
Li_2GeO_3, *mm*2		X	S	0.089	7.1	3.72	13	1.86
Lithium niobate,	4.64	Z	L	0.17	275	7.3306	34.0	3.66
$LiNbO_3$, 3*m*		35°Y	L	0.49	38.6	7.4	34.3	3.7
		X	S	0.68	44.3	4.79	22.3	2.40
		163°Y	S	0.62	42.9	4.48	20.8	2.24
Lithium tantalate,	7.45	Z	L	0.19	45	6.1607	46	3.08
$LiTaO_3$, 3*m*		X	S	0.44	42.6	4.22	31.4	2.11
		47°Y	L	0.29	42.7	7.4	55.2	3.70
Lithium iodate,	4.5	Z	L	0.49	6.6	4.01	18.0	2.0
$LiIO_3$, 6		Y	S	0.61	8.4	2.57	11.5	1.28
		27°Y	L	0.65	6.5	4.23	19	2.11
Quartz,	2.65	X	L	0.098	4.58	5.7499	15.3	2.875
SiO_2, 32		Y	S	0.137	4.58	3.9169	10.2	1.958
Sodium–potassium niobate,	4.51	0°	L	0.46	306	6.94	31.3	3.47
$Na_{0.5}K_{0.5}NbO_3$		90°	S	0.59	435	3.10	13.6	1.55
Zinc oxide,	5.68	Z	L	0.282	8.84	6.40	36.4	3.2
ZnO, 6*mm*		Y	S	0.316	8.33	2.88	16.4	1.44
		39°Y	S	0.35	8.63	3.24	18.4	1.62

*1 Electromechanical coupling coefficient.
*2 Frequency factor: f first resonance frequency of plate thickness h.
*3 Ceramics.

Note: In column "Cut" directions of electrical polarisation are included. As example, for L–wave 1.5 MHz quartz resonance transducer it is need the X–cut plate of thickness $h = 2.875 \cdot 10^3$ Hz·m$/(1.5 \cdot 10^6$ Hz$) =$ 1.98 mm.

Table 7.25 Propagation velocity of surface acoustic waves

Crystal, formula, point groups	Cut (plane)	Propagation direction	Velocity v_R, 10^3 m/s	$\Delta v_R/v_R^*$, 10^{-4}	References
Aluminum, Al, *m*3*m*	Z	100	2.941		[252]
	110	001	2.962		[252]
	111	110	2.828		[252]
Banan, barium–potassium niobate,	X	Y+45°	3.32	12	[238]
$Ba_2NaNb_5O_{15}$, *mm*2	Y	001	3.177	5	[237]
		100	3.46	0	[237]
Berlinite, $AlPO_4$, 32	X	Y+19°	2.754	24	[239]
		Y+175.2°	2.865	26.4	[239]
Bismuth gallate, $Bi_{40}GaO_{63}$	X	110	1.61	$\bar{\kappa}$=1.76%	[240]
Bismuth germanate, $Bi_{12}GeO_{20}$, 423	Z	110	1.681	68	[242]
		100	1.620	0	[233]
		X+30°	1.654	55	[233]
	XY	001	1.5935	22	[237]
			1.620	37	[243]
		$1\bar{1}0$	1.7812	0	[237]

Table 7.25 Propagation velocity of surface acoustic waves *(continued)*

Crystal, formula, point groups	Cut (plane)	Propagation direction	Velocity v_R, 10^3 m/s	$\Delta v_R/v_R^*$, 10^{-4}	References
Bismuth silicate, $Bi_{12}SiO_{20}$, 23	Z	110	1.66		[176]
	111	110	1.61		[176]
	XY	110	1.748	$\bar{\kappa}$=1.5%	[142]
		001	1.60	4	[142]
Bismuth titanate, $Bi_{12}TiO_{20}$	XY	110+55°	1.72	$\bar{\kappa}$=1.58%	[256]
Cadmium suphide, CdS, $6mm$	X	010	1.7302	0	[254]
		001	1.7177	26.1	[254]
	Y	100	1.700		[232]
		001	1.716	2.6	[233]
		⊥001	1.720	31	[245]
	Z	Any	1.7308	23.6	[254]
Ceramics:					
CTS–23			2.253		[248]
PKR–2			1.8507		[248]
PZT–4			2.2	220	[142]
Copper, Cu, $m3m$	Z	100	2.012		[252]
	110	001	2.265		[252]
	111	110	1.613		[252]
Diamond, C, $m3m$	Z	100	10.971		[252]
	110	001	11.063		[252]
	111	110	10.756		[252]
Fresnoite, $Ba_2Si_2TiO_8$, $\bar{4}mm$	X	010	2.827	23	[142]
	Y	001	2.50	5	[249]
		100	2.678	80	[249]
	Z+45°	100	2.655	75	[142]
Fused quartz, SiO_2			3.411		[233]
Gallium arsenide, GaAs, $\bar{4}3m$	Z	100	2.719	0; 1.18	[232]
		110	2.863		[233]
		X+25°	2.750	3	[234]
		X+23°	2.760	2.85	[235]
		110+45°	2.480		[235]
	X	100	3.332		[236]
		010	2.720	0	[237]
		110	2.773	8	[236]
		011	2.8645	3.40	[237]
		210	2.725	70	[236]
	XY	001	2.819	1.16	[232]
		100	2.542	13	[236]
			2.820	0.95	[237]
		110	2.399		[233]
			2.512	19	[236]
		111	2.694	62	[236]
	111	110	2.428		[232]
		112	2.525		[236]
		$2\bar{1}\bar{1}$	3.0826	0	[237]
	211	111	2.621	1.2	[259]
Gallium phosphide, GaP, $\bar{4}3m$	Z	100	3.37	0	[234]
		X+23°	3.42	1.1	[234]
		X+45°	3.09	0	[234]
Germanium, Ge, $m3m$	Z	100	2.934	0.1	[162]
	XY	001	3.015	0.1	[162]
	111	110	2.683		[162]
Gold, Au, $m3m$	Z	100	1.124		[252]
	110	001	1.254		[252]
	111	110	0.902		[252]

Table 7.25 Propagation velocity of surface acoustic waves *(continued)*

Crystal, formula, point groups	Cut (plane)	Propagation direction	Velocity v_R, 10^3 m/s	$\Delta v_R/v_R^*$, 10^{-4}	References
Indium antimonide, InSb, $\bar{4}3m$	Z	100	1.836		[238, 252]
	110	001	1.920		[252]
	111	110	1.601		[252]
Iron, Fe, $m3m$	Z	100	3.014		[252]
	110	001	3.212		[252]
	111	110	2.559		[252]
Lead, Pb, $m3m$	Z	100	0.734		[252]
	110	001	0.872		[252]
	111	110	0.570		[252]
Lead molybdate, $PbMoO_4$, $4/m$	X	010	1.457		[142]
	Y	100	1.458793		[259]
Lead sulphide, PbS, $m3m$	Z	100	1.773		[251]
	110	001	1.762		[252]
	111	110	1.788		[252]
Lithium borate, LiB_4O_7	X	001	3.51	40	[142]
Lithium fluorite, LiF, $m3m$	Z		3.897		[162]
	XY		4.060		[162]
	111		3.458		[162]
Lithium gallate, $LiGaO_2$, $mm2$	X	010	3.173	55	[241]
Lithium iodate, $LiIO_3$, 6	$11\bar{1}$	$1\bar{1}0$	1.708	82	[242]
	X	001	1.909	26.2	[244]
	Z	100	2.258	443	[244]
Lithium niobate, $3m$	Z	100	3.798	260	[237, 249, 250]
	X	001	3.483	252	[237, 249, 250]
	Y	001	3.488	241	[237, 249, 250]
	Y+16.5°	100	3.503	268	[242]
	Y+41.5°	100	4.000	277	[242]
	Y+127.86°	100	3.950		[235]
Lithium tantalate, $LiTaO_3$, $3m$	Y	100	3.148	3.7	[242]
		001	3.230	33	[237, 242]
	Z	100	3.205	11.4	[242]
		010	3.329	59	[242]
	Z+22°	100	3.302	27	[237, 242]
Magnesium oxide, MgO, $m3m$	Z	100	5.513	120	[251]
	XY	001	5.640	118	[251]
	111	$\bar{1}10$	5.114		[251]
Molybdenum, Mo, $m3m$	Z	100	3.130		[252]
	110	001	3.116		[252]
	111	110	3.174		[252]
Nickel, Ni, $m3m$	Z	100	2.910		[252]
	110	001	3.113		[252]
	111	110	2.459		[252]
Paratellurit, TeO_2, 422	Z	Y+40°	1.4		[253]
	X	110	1.673	83	[142]
Potassium chloride, KCl, $m3m$	Z	100	1.753		[251]
	110	001	1.744		[252]
Quartz, SiO_2, 32	X	010	3.256		[245, 246]
	Y	100	3.154		[245, 247]
			3.159	9	[242]
	Z	010	3.258		[242]
			3.2617		[238]
	YX		3.1616		[248]
	ST	100	3.158	5.8	[242]
				6.7	[258]

Table 7.25 Propagation velocity of surface acoustic waves *(continued)*

Crystal, formula, point groups	Cut (plane)	Propagation direction	Velocity v_R, 10^3 m/s	$\Delta v_R/v_R^*$, 10^{-4}	References
Selen, Se	Y	100	0.813	$\bar{\kappa}$=1.7%	[142]
		001	1.54	$\bar{\kappa}$=0.8%	[142]
Silicon, Si, $m3m$	Z	100	4.92		[238]
	100	001	5.029		[252]
	111	110	4.539		[252]
Silver, Ag, $m3m$	Z	100	1.493		[252]
	110	001	1.671		[252]
	111	110	1.210		[252]
Sodium, Na, $m3m$	Z	100	1.047		[252]
	110	001	1.604		[252]
	111	110	0.786		[252]
Thallium tantaloselenate, $\bar{4}3m$	XY	$1\bar{1}0$	0.85		[255]
		001	0.77		[255]
Thallium vanadosulphate, Tl_3VS_4, $\bar{4}3m$	X	100	0.870		[177]
Tungsten, W, $m3m$	Z	100	2.646		[252]
	110	001	2.645		[252]
	111	110	2.647		[252]
Zinc oxide, ZnO, $6mm$	X	010	2.8378	17.44	[237]
		001	2.7524	0	[237]
	Z	Any	2.80		[237]
			2.64	45	[142]
Zinc sulphide, ZnS, $m3m$	Z	100	2.360		[252]
	110	001	2.475		[252]
	111	110	2.051		[252]

* velocity reduction after surface metallization.
κ – the electromechanical coupling factor.

Table 7.26 Sound dumping in nature material

Material	Frequency range, Hz	Excitation type	Q	References
Clay shale silvane	$(3.4–12.8)\cdot10^3$	Longitudinal resonance	73	[268]
Covering stratum	$(1.1–6.6)\cdot10^3$	Longitudinal resonance	45	[268]
	$100–2\cdot10^3$	Shear resonance	52	[268]
Granite (Kwinsley)	140–1600	Longitudinal resonance	100	[264]
	140–1600	Torsion, bending resonance	150–200	[264]
Granite	50–120	Bend	57	[265]
	$5\cdot10^4–4\cdot10^5$	Rayleigh waves	79	[266]
Diorite	50–120	Bend	125	[265]
Dolorite	50–120	Bend	90	[265]
Limestone:	$5\cdot10^6–10^7$	Longitudinal pulses	190–110	[267]
	$3\cdot10^6–15\cdot10^7$	Shear pulses	400	[267]
(Hunton)	$(2.8–10.6)\cdot10^3$	Longitudinal resonance	65	[268]
(Solenhofen)	$(3–9)\cdot10^6$	Shear pulses	190	[265]
	50–120	Bend	63	[265]
(oolite)	50–120	Bend	45	[268]
Sandstone	50–120	Bend	21	[265]
Sandstone amherste	$930–12.8\cdot10^3$	Longitudinal resonance	52	[265]

Table 7.27 Sound dumping in metals, glasses and plastics

Substance	Freuqency range, Hz	Excitation type	Q, 10^3	α/f, 10^{-7} s/m	References
Aluminum	$(3.1\text{–}7.5)\cdot10^6$	Longitudinal pulses	5.9	0.85	[269]
	$(5\text{–}15)\cdot10^6$		7.63	0.64	[269]
	$(3.5\text{–}4.5)\cdot10^6$	Shear pulses	19.4	0.53	[269]
	$(3\text{–}6.8)\cdot10^6$		17.2	0.59	[269]
	$(1\text{–}200)\cdot10^5$	Longitudinal resonance	150	0.03	[270]
single crystal ([110])	$(1.5\text{–}6)\cdot10^7$	Longitudinal pulses	1.09	4.6	[271]
Brass	10^7	Longitudinal pulses		70	[278]
	$9\cdot10^6$	Shear pulses		160	[278]
Copper	10–18	Longitudinal	2.14	3.10	[274]
	$(2.5\text{–}30)\cdot10^3$	Longitudinal resonance	2.18	3.10	[275]
	$(2.5\text{–}30)\cdot10^3$	Shear resonance	4.38	3.20	[275]
	1–6	Bend	0.98		[276]
	11–25	Bend	0.64		[277]
nontempered	$(1.5\text{–}6.5)\cdot10^7$	Longitudinal pulses	1.77	3.6	[271]
tempered	$(2.5\text{–}7.5)\cdot10^7$	Longitudinal pulses	5.83	1.1	[271]
Fused quartz	$(0.2\text{–}1.5)\cdot10^7$	Shear waves		0.19	[269]
	$5\cdot10^8$	Shear waves		0.1	[157]
	$(0.2\text{–}1.5)\cdot10^7$	Longitudinal waves		1.23	[269]
	$5\cdot10^8$	Longitudinal waves		0.7	[157]
Glasses:					
pyrex	$(2\text{–}15)\cdot10^6$	Longitudinal waves		4.89	[269]
flint glass	$(2\text{–}15)\cdot10^6$	Longitudinal waves		3.21	[269]
Lead	$(1.6\text{–}15)\cdot10^3$	Longitudinal	$3.6\cdot10^{-2}$	400	[275]
	$(1\text{–}9)\cdot10^3$	Shear	$3.4\cdot10^{-2}$	1400	[275]
Magnesium	$(2\text{–}100)\cdot10^3$	Longitudinal waves	4.9	1.08	[269]
	$(7\text{–}76)\cdot10^6$	Longitudinal pulses	0.965	5.5	[273]
Molybdenum	8–30	Bend	0.465		[277]
Monel	8–32	Bend	1.40		[277]
Nickel:					
monocrystal	10^7	Longitudinal waves		1.0	[279]
polycrystal	$8\cdot10^6$	Longitudinal pulses		102	[278]
	$2.5\cdot10^6$	Shear pulses		233	[278]
	12–33	Bend	0.96		[277]
Plexyglas	10^6	Longitudinal waves		250	[267]
Polystyrene	10^6	Longitudinal waves		170	[267]
Polyethylene	10^6	Longitudinal waves		520	[267]
Steel:	5–10	Longitudinal	5.0	1.0	[274]
	2–8	Bend	1.85		[276]
+3.5% Ni	8–25	Bend	1.36		[277]
tungsten carbon	$(2\text{–}100)\cdot10^3$	Longitudinal		0.38	[269]
molybdenum	$(2\text{–}100)\cdot10^3$	Longitudinal		1.42	[269]

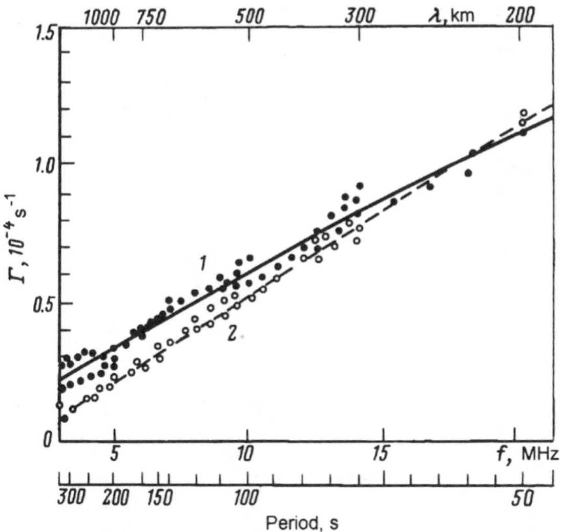

Figure 7.35 The damping factor of the Lave's X waves (*1*) and of the Rayleigh waves (*2*) as a function of period (frequency, wavelength). The data were obtained after developing records for a series of earthquakes [263].

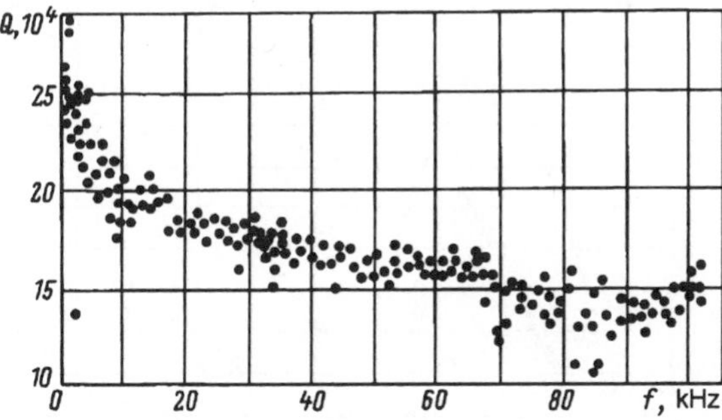

Figure 7.36 Frequency dependence of Q-factor of longitudinal vibrations of aluminum rod [270].

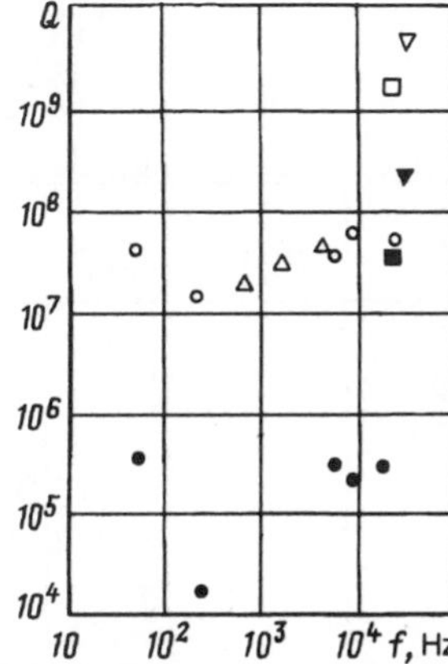

Figure 7.37 Q-factors of low frequency resonators made from different materials: ∘ for aluminum alloy (Al - 5056); △ for [280]; □ for Si [281]; ▽ for Al_2O_3 [282]. Dark marks are for $T = 300$ K and light marks are for $T = 4.2$ K.

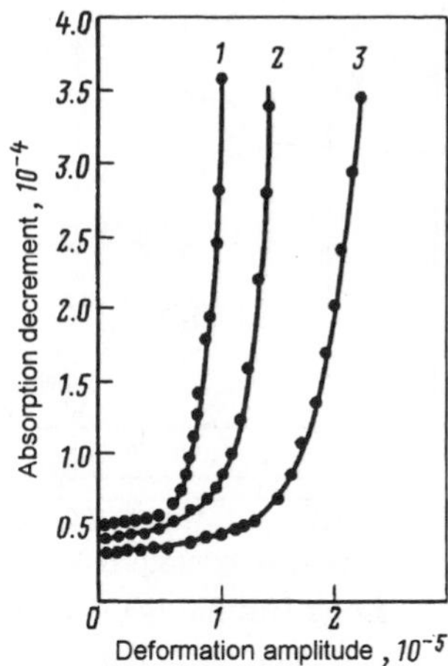

Figure 7.38 Amplitude-dependent damping in aluminum at $t = 84$°C: *1* is for not annealed sample; *2* is for sample after 3.5 hours; *3* is for the one after 13.5 hours (return time at annealing) [283].

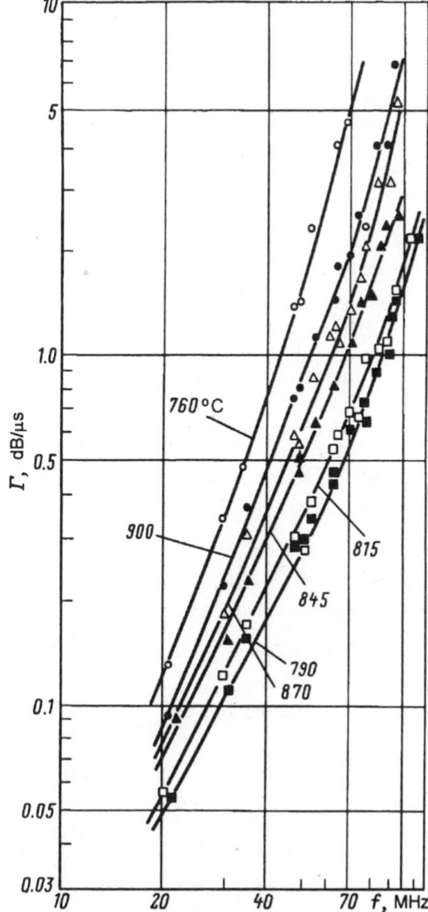

Figure 7.39 Damping of longitudinal waves in a hardened stamped steel. (Samples were austenized at different temperatures, accordingly: *1* – 760°C; *2* – 790°C; *3* – 815°C; *4* – 845°C; *5* – 870°C; *6* – 900°C [284].

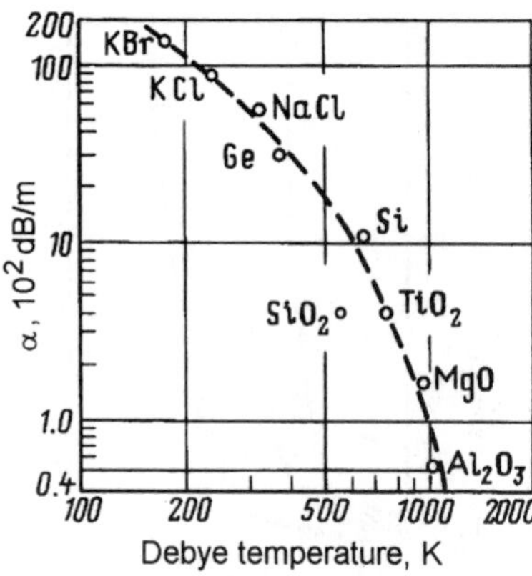

Figure 7.40 Longitudinal wave damping factor at 1 GHz at T = 300 K as a function of the Debye temperature. The propagation directions of L-wave was [100] in cubic crystals, X in SiO_2 and along c-axis in TiO_2 and Al_2O_3 [330].

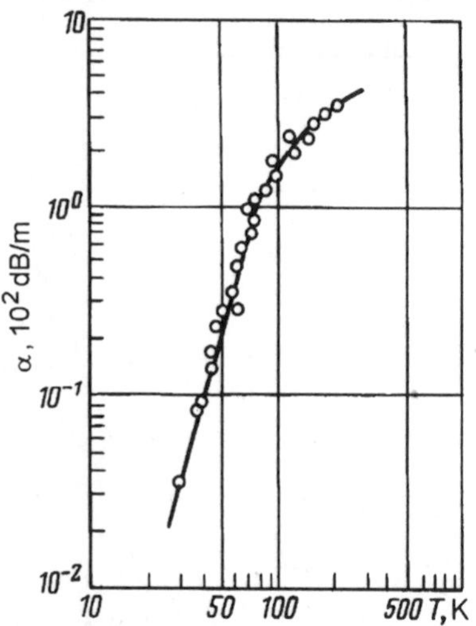

Figure 7.41 Dependence of damping in MgO crystal on temperature: S_{100}-wave at 3 GHz [330].

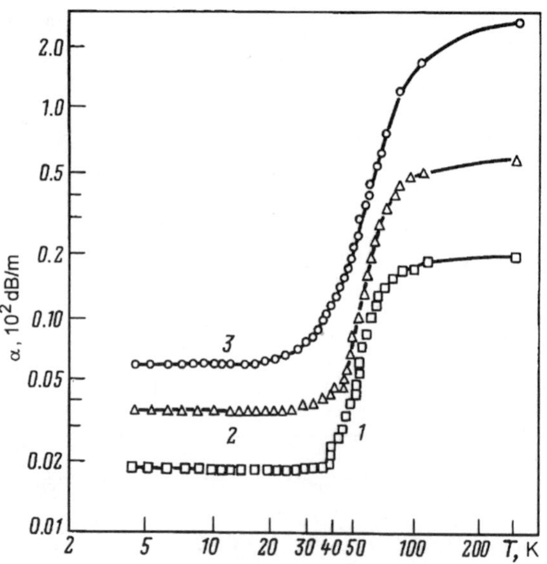

Figure 7.42 Temperature dependence of sound damping at frequency of 1 GHz in Al_2O_3 for waves propagating along the a axis: *1* is for longitudinal; *2*, *3* are for fast and slow transverse waves [331].

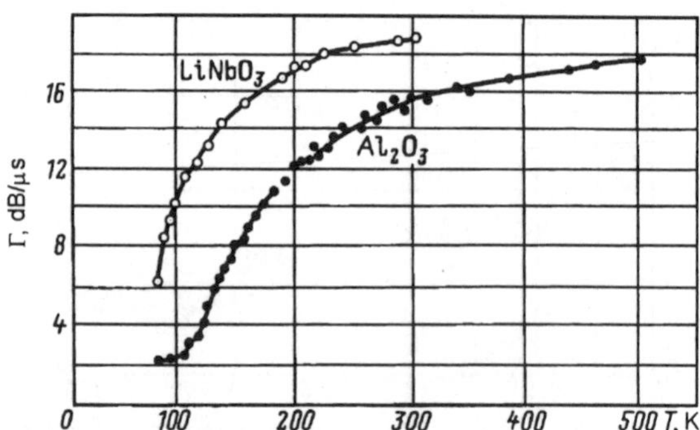

Figure 7.43 Damping of L-wave along the trigonal axis of the ruby and lithium niobate crystals at 9.4 GHz [332].

Table 7.28 Sound absoption in crystals ($t \approx 20°C$)

Crystal, formula, point group	Wave type*3	Γ, dB/μs (f, GHz)	α, 10^2 dB/m (f, GHz)	α/f^2, 10^2 dB/(m·GHz²)	References
Banan, $Ba_2NaNb_5O_{15}$, orthoromb., $mm2$	‖ **c**		0.1(0.077)		[287]
			0.4(0.5) (250 K)		
	‖ **a**		0.4(0.5) (250 K)		
Barium fluorite, BaF_2, cub., $m3m$	L_{100}	0.16(0.2)			[142, 160]
	L_{110}	0.16(0.2)			
	L_{111}	0.9(0.5)			
	S_{110}^{001}	0.5(0.5)			
Bastron, $Sr_{(0.75-0.5)}Ba_{(0.25-0.5)}Nb_2O_6$, tetragon., $4mm$	L_{001}		61(0.1)		[142, 288]
			≤1.0(0.5)		
Beryllium oxide, BeO, hex., $6mm$	L_{001}	1.3(1)			[142]
Bilead molybdate, Pb_2MoO_5, monocl., $2/m$	QL_{100}		4.2(0.3)		[142, 307, 308]
	L_{010}		~ 4(0.1)		
	QL_{001}		2.5(0.5)		

Table 7.28 Sound absorption in crystals ($t \approx 20°C$) *(continued)*

Crystal, formula, point group	Wave type*3	Γ, dB/μs (f, GHz)	α, 10² dB/m (f, GHz)	α/f², 10² dB/(m·GHz²)	References
Bismuth germanate, $Bi_{12}GeO_{20}$, cub., 23	L_{100}		2(0.025)		[142, 294]
	L_{001}		4(0.8)		
	L_{010}	0.11(0.5)	4(0.8)		
Bismuth orthosilicate, $Bi_4(SiO_4)_3$,	L_{111}			0.95	[294]
cub., $\bar{4}3m$	S_{110}^{001}			0.75	
Bismuth silicate, $Bi_{12}SiO_{20}$, cub., 23	L_{111}		5(0.8)		[294]
	S_{110}^{001}		2(0.8)		
Boron, B, trig., $\bar{3}m$	L_{001}	0.15(1)			[142]
Cadmium selenide, CdSe, hex., 6mm	L_{001}	3.0(0.2)			[142, 286]
	S_{1010}	0.5(0.2)			
		2.0(0.2)			
Cadmium sulphide, CdS, hex., 6mm	L_{100}			90 [158]	[142, 286]
	L_{001}	3.5(0.2)			
		80(1)			
	L_{1120}	17(1)			
	S_{1010}	0.6(0.2)			
		4.5(1)			
Cinnabar, α–HgS, trig., 32	L_{001}	7(1)	4.55(0.4)		[142, 206]
			7.1(0.5)		
Fluorite, CaF_2, cub., m3m	L_{110}	0.4(0.5)			[160]
	L_{111}	0.4(0.5)			
	SS_{110}	0.6(0.5)			
	FS_{110}	0.3(0.5)			
Gadolinium, Gd, hex.	L_{001}		13(0.05)		[324]
	S_{001}		0.4(0.05)		
Gallium arsenide, GaAs, cub., $\bar{4}3m$	L_{110}	0.6(0.2)		30 [237]	
	L_{111}	4.0(0.5)			
	S_{100}	0.7(0.2)			
Gallium phosphide, GaP, cub., $\bar{4}3m$	L_{110}	0.63(0.4)		6 [322]	[157, 293]
		1(0.5)			
Garnets:					
alumoittrium, $Y_3Al_5O_{12}$, cub., m3m	L_{100}	0.07(0.5)	0.2–0.32(1)		[142, 157, 216, 290, 293, 295, 296]
		0.6(1.5)			
		1.8(1.8)			
		5(3)			
	L_{110}	15(9.4)		0.25	
	L_{111}	14(9.4)			
	S_{110}	7(9.4)	0.4(0.9)	1.1	
	S_{100}	0.36(1.5)			
gallium-yttrium, $Y_3Ga_5O_{12}$, cub., m3m	L_{100}	0.7(0.1)			[142]
	S_{100}	0.1(0.1)			
	S_{110}	5.9; 18(9)			
iron-yttrium, $Y_3Fe_5O_{12}$, cub., m3m	$L_{110} = L_{111}$	1.0(1)			[142, 158, 290, 297, 323]
	L_{100}	1.0(1)	2(1)	1.1	
		0.6(1.5)	3.2(1.12)		
	S_{100}	0.13(1)	0.34(1)	0.35	
Germanium, Ge, cub., m3m	L_{100}	1.3(0.3)	23(1)	29.6	[142, 158, 290–293, 328]
		11(1)			
	L_{110}	10(1)			
	L_{111}	4.2(0.5)		30	
	S_{010}^{001}, $S_{110}^{1\bar{1}0}$	2(1)			
	S_{100}	0.29(0.3)	10(1)	9	

Table 7.28 Sound absoption in crystals ($t \approx 20°C$) *(continued)*

Crystal, formula, point group	Wave type*3	Γ, dB/μs (f, GHz)	α, 10^2 dB/m (f, GHz)	α/f^2, 10^2 dB/(m·GHz^2)	References
Indium antimonide, InSb, cub., $\bar{4}3m$	L_{100}	40(1)	4.8(0.11)		[142, 285, 328]
	L_{110}	8(0.5)			
		40(1)			
	S_{100}		5.1(0.58)		
	S_{110}^{001}	3.8(1)	4.5(0.58)		
		5(1)			
	$S_{1\bar{1}0}$	8(1)	8.8(0.58)		
	S_{111}		6.2(0.58)		
Iodic acid, α–HIO_3, orthoromb., 222	L_{010}	0.73(0.5)	2.5(0.5)		[142, 226]
	L_{001}	0.61(0.5)			
Lead germanate, Pb_5GeO_{11}, hex., 6	L_{100}			150	[158, 198]
	L_{010}		2(0.1)		
	L_{001}		~30(0.4)		
Lead germanovanadate, $Pb_5(GeO_4)(VO_4)_2$, hex., $6/m$	L_{010}		0.8(0.1)		[178]
			2.1(0.3)		
	L_{001}		0.5(0.1)		
			2.1(0.3)		
Lead molybdate, $PbMoO_4$, tetrag., $4/m$	S_{100}	1.4(1)	2.2(1)		[188, 189, 306]
	FS_{110}	1.6(1)	2.55(1.03)		
	L_{100}		2.5(0.5)		
	L_{001}	1.0(0.5)	0.3(0.1)		
	S_{100}^{001}		0.16(0.2)		
Lithium fluorite, LiF, cub., $m3m$	L_{100}	3.5(0.9)			[148, 328]
		10(1)			
	L_{110}	1.3(0.9)			
		2.5(1)			
	L_{111}	0.8(0.9)			
		1.5(1)			
	S_{100}	0.8(0.9)			
	S_{110}	0.8(0.9)			
	S_{111}	0.5(0.9)			
Lithium iodate, $LiIO_3$, orthoromb., 222	L_{100}	1.1(0.15)			[142, 179, 298]
	L_{001}	9.1(0.15)	20(0.002)		
			3(0.008)		
			7.5(0.14)		
	$S_{100,010}^{001}$	1.6(0.15)			
	$S_{100,010}^{\perp 001}$	0.29(0.15)			
Lithium niobate, $LiNbO_3$, trig., $3m$	L_{100}		0.29(0.4)		[142, 165, 212, 219, 305]
			1.39(1)		
			0.86(1)		
	QL_{010}	0.065(0.5)	0.16(0.4)		
			0.78(1)		
			0.71(1)		
	L_{001}	0.03(0.5)	0.03(0.4)		
		1.3(2)	0.17(1)		
			0.32(1)		
			0.34(1)		
	QS_{100}	0.07(0.5)	FS { 0.17(0.4)		
			FS { 1.04(1)		
			FS { 0.81(1)		
			FS { 1.71(1.1)		
			FS { 4.98(2)		
			SS 0.66(1)		

Table 7.28 Sound absorption in crystals ($t \approx 20°C$) *(continued)*

Crystal, formula, point group	Wave type[*3]	Γ, dB/μs (f, GHz)	α, 10^2 dB/m (f, GHz)	α/f^2, 10^2 dB/(m·GHz²)	References
	S_{010}		FS { 0.17(0.4), 0.96(1), 0.73(1), 1.05(1.1), 3.50(2)		
			SS { 0.14(0.4), 1.13(1), 1.24(1), 2.31(1.1), 5.81(1.74)		
	S_{001}	5.0(2.3)	0.39(0.4)		
			2.44(1)		
			4.3(1)		
Lithium tantalate, $LiTaO_3$, trig., $3m$	L_{100}	0.02(0.5)	0.1(0.5)	0.1 [157]	[142, 154, 157, 189, 305]
	L_{001}	0.01(0.5)	0.02(0.5)		
	S_{001}	4.4–6(2.3)			
Magnesium fluorite, MgF_2, tetrag., $4/mmm$	L_{001}	0.7(0.5)			[142]
	S^{100}_{001}	0.45(0.5)			
Magnesium oxide, cub., $m3m$	L_{100}	3(1)	3.30(1)		[142, 286, 293, 309, 310, 328]
		4.5(1)			
	L_{110}	1.5(1)			
	L_{111}	1.0(1)			
	L_{001}	1.0(0.2)	0.04(0.1)		
		4.4(1)	15(1)		
	S_{100}	0.21(1)	0.44(1)		
	$S^{001}_{1\underline{1}0}$	0.6(1)			
	$S^{1\bar{1}0}_{110}$	6.5(1)			
Mercury selenide, HgSe	L_{110}		~ 3 (0.075)		[329]
	$S^{1\bar{1}0}_{110}$		~ 3.2 (0.075)		
Nickel, Ni, cub., $m3m$	L_{110}		~ 0.7(0.0058)		[326]
			~ 5(0.09)		
	S^{001}_{110}		1.8(0.0095)		
			4.6(0.029)		
Paratellurite, TeO_2, tetrag., 422	L_{001}	3(1)		6.5[*2] [308]	[142]
	$S^{1\bar{1}0}_{110}$	18(1)		16[*2] [308]	
Pyrargirite, Ag_2SbS_3, trig., $3m$	L_{100}		1.3(0.003)		[311]
Potassium biphtalate, $KC_8H_7O_4$, orthoromb., $mm2$	L_{100}	1.1(0.4)			[222]
	L_{010}	1.5(0.4)			
	L_{001}	0.68(0.4)			
	FS_{100}	0.35(0.4)			
	FS_{010}	0.56(0.4)			
Potassium bromide, KBr, cub., $m3m$	L_{100}			36[*1]	[293]
	L_{110}			25[*1]	
	L_{111}			17[*1]	
	$S^{001}_{1\underline{0}0}$, S^{001}_{110}			6[*1]	
	$S^{1\bar{1}0}_{1\underline{1}0}$			40[*1]	
	$S^{1\bar{1}0}_{111}$			38[*1]	
Potassium chloride, KCl, cub., $m3m$	L_{100}			36[*1]	[289]
	L_{110}			25[*1]	
	L_{111}			18[*1]	
	$S^{001}_{1\underline{0}0}$, S^{001}_{110}			5[*1]	
	$S^{1\bar{1}0}_{110}$			40[*1]	
Potassium tantalate, $KTaO_3$, trig., $3m$	L^{*4}	0.45(0.18)			[325]

Table 7.28 Sound absorption in crystals ($t \approx 20°C$) *(continued)*

Crystal, formula, point group	Wave type*3	Γ, dB/μs (f, GHz)	α, 10^2 dB/m (f, GHz)	α/f^2, 10^2 dB/(m·GHz^2)	References
Prustite, Ag_3AsS_3, trig., $3m$	L_{100}	1.5(0.56)			[142, 312, 313]
	L_{001}	0.25(0.056)	~ 8(0.5)	4.5*2 [308]	
		2.5(0.56)			
Quartz, SiO_2, trig., 32	L_{001}	1.3(1)		3.25 [300]	[142, 216, 299, 301, 302]
	L_{100}	1.7(1)	1.3(0.73)	4.2 [300]	
		3.5(1.8)	9.0(1.96)	3.0 [158]	
			26.3(3.38)		
	QS_{100}^{001}		0.2(0.5)		
	S_{010}	3.5(1.8)			
	QS_{001}		2(1.01)		
	S_{AC}	1.65(1.5)			
Rubidium biphtalate, $RbC_8H_7O_4$, orthoromb., $mm2$	L_{100}	0.29(0.4)			[222]
	L_{010}	0.90(0.4)			
	L_{001}	9.36(0.4)			
Ruby, β–Al_2O_3, trig., $\bar{3}m$	L_{100}	19(9.4)			[317, 318]
	L_{001}	15(9.4)			
Rutile, TiO_2, tetrag., $4/mmm$	L_{100}	1.9(1)			[142, 148, 157, 158, 216, 314]
	L_{110}	2.5(1)			
	L_{001}	0.1(0.5)	0.5(0.5)		
		0.6(1)			
		7(1.8)			
	S_{001}	0.22(1)	3.4(2.8)	0.55 [158]	
Sapphire, α–Al_2O_3, trig., $\bar{3}m$	L_{100}	0.05(0.5)			[142, 157, 216]
		1(1.8)			
		16.5(9)			
	$L_{11\bar{2}0}$			0.2 [158, 320]	
	L_{001}	0.22(1)		0.2 [158, 320]	
	FQS‖a	1.4(1)	0.5(1)		
	SQS‖a		2.3(1)		
	QL‖a		0.2(1)		
Silicon, Si, cub., $m3m$	L_{100}	7(1)	7.6(1.03)		[142, 285, 328]
		7.1(1)			
	L_{110}	6(1)			
	L_{111}	2.3(0.5)	5.9(1.03)	6.5 [158]	
		6(1)			
Silicon carbide, SiC, hex., $6mm$	L_{001}	56(9.4)			[142]
	L_{1010}	50(9.4)			
	S_{001}	22(9.4)			
	S_{1010}	19(9.4)			
Sodium chloride, NaCl, cub., $m3m$	L_{100}	3.6(0.2)		18*1	[142, 289, 323]
		30(1)			
	L_{110}	0.56(0.2)		11*1	
		20(1)			
	L_{111}	0.28(0.2)		7*1	
		10(1)			
	S_{100}^{001}, S_{110}^{001}			1.9*1	
	S_{110}^{110}			26*1	
Sodium fluorite, NaF, cub., $m3m$	L_{100}	14(1)			[328]
	L_{110}	6.5(1)			
	L_{111}	10(1)			
	S_{110}^{001}	1.0(1)			
	S_{110}^{110}	12(1)			
Sphalerite, β–ZnS, cub., $\bar{4}3m$	L_{001}			27	[158]
	S_{001}			130	

Table 7.28 Sound absorption in crystals ($t \approx 20°C$) *(continued)*

Crystal, formula, point group	Wave type*3	Γ, dB/μs (f, GHz)	α, 10^2 dB/m (f, GHz)	α/f^2, 10^2 dB/(m·GHz²)	References
Strontium fluorite, SrF_2, cub., $m3m$	L_{111}	0.7(0.5)			[160, 323]
		2.5(1)			
	S_{110}^{001}	0.5(0.5)			
	$S_{110}^{1\bar{1}0}$	0.6(0.5)			
Strontium titanate, $SrTiO_3$, cub., $m3m$	L_{100}, L_{010}, L_{001}	1(1)	1.7(0.894)		[162, 321]
	L_{111}		1(0.972)		
	S_{100}		1.5(0.894)		
	S_{110}		1.1(0.77)		
Tallium bromide-chloride, KRS-6, TlBr–TlCl, cub., $m3m$	L*4		0.8(0.1)		[304]
Tallium bromide-iodide,	L_{111}		~8(0.5)		[142, 303]
KRS-5, TlBr–TlI, cub., $m3m$	L_{110}	0.3(0.1)			
Tellurium, Te, trig., 32	L_{100}		3 5(0.5)		[142]
Tourmaline, trig., $3m$	L_{001}	16(9)			[142]
Zinc oxide, ZnO, hex., $6mm$	S*4	0.5(0.2)			[286]

*1 Data on Γ/f^2 are in $dB/(\mu s \cdot GHz^2)$.
*2 Data on Γ/f^2 are in $dB/(\mu s \cdot MHz^2)$.
*3 Four-digit indices are the Brave indices.
*4 Wave propagation direction and polarization are not shown.

Table 7.29 Rayleigh surface waves damping in crystals ($t \approx 20°C$)

Crystal, formula, point group	Cut (plane)	Wave direction	Γ, dB/μs (f, GHz)	α, 10^2 dB/m at the frequency, GHz				Ref.
				0.1	0.2	0.4	1	
Banan, barium-sodium niobate, $Ba_2NaNb_5O_{15}$, orthoromb., $mm2$	Y*2	001	3.7(1)					[142]
Bismuth germanate, $Bi_{12}GeO_{20}$,	001*1	110	1.64(1)	0.199	0.571	1.832	9.756	[259]
cub., 23	111	$1\bar{1}0$	1.64(1)					[259]
	111	110		0.199	0.571	1.832	9.756	
Gallium arsenide, GaAs, cub., $\bar{4}3m$	211	111	3.62(1)					[259]
	110	100	4.22(1)					[142]
	001*1	110	12.8(1)					[142]
Lead molybdate, $PbMoO_4$, tetrag., $4/m$	X	010	7.5(1)					[142]
Lithium iodate, $LiIO_3$, hex., 6	Z	any	450(1)					[142]
Lithium niobate, $LiNbO_3$, trig., $3m$	X	001	1.1–1.78(0.9) 3.84(1.95)	0.079	0.210	0.621	3.06 3.04	[343] [343]
	Y	100	0.93(1)					[237]
	Y*3	001	1.07(1)					[237]
	Y+16.5°	100		0.086	0.227	0.669	3.28	
	Y+41.5°	100		0.093	0.225	0.600	2.62	
	Z−41.5°	100	1.05(1)					[237]
Lithium tantalate, $LiTaO_3$, trig., $3m$	Z	010	1.0(1)	0.092	0.230	0.646	3.004	[142]
	Y	001	1.14(1)					[142]
Piezoceramic PZT-4		any	6(0.036)					[142]
Quartz, SiO_2, trig., 32	Y*2	100	0.41(0.316) 4.1(1.047) 2.6(1)	0.211	0.558	1.661	8.244 8.220	[343] [343] [259]
	ST	100	3.09(1)	0.232	0.630	1.897	9.733 9.780	[259]
Zinc oxide, ZnO, hex., $6mm$	Z	any	37(1)					[142]

*1 $\Gamma = 1.45f^{1.9} + 0.19f$ [142]. *2 $\Gamma = 2.15f^2 + 0.45f$ [142]. *3 $\Gamma = 0.88f^{1.9} + 0.19f$ [142]. Here f is given in GHz.

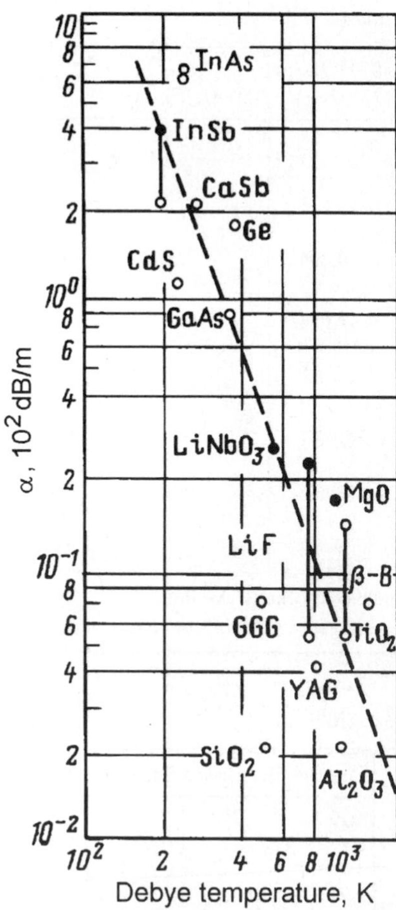

Figure 7.44 Residual absorption (absorption in the liquid helium temperature region) at 1 GHz as a function of the Debye temperature [333]: *L*-waves and *S*-waves.

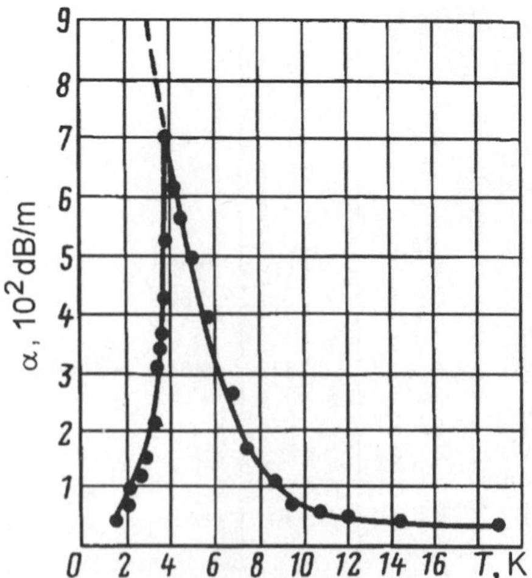

Figure 7.45 Absorption of *L*-wave (10.3 MHz) in pure tin in the region of transition into superconducting state; dashed line shows absorption in normal state [334].

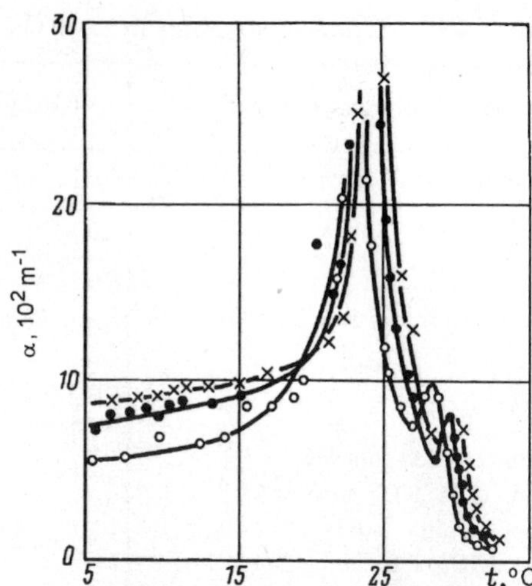

Figure 7.46 Absorption of *L*-wave (10 MHz) along the *c* axis of the antimony sulfoiodide crystal at different resistances of the sample [335].

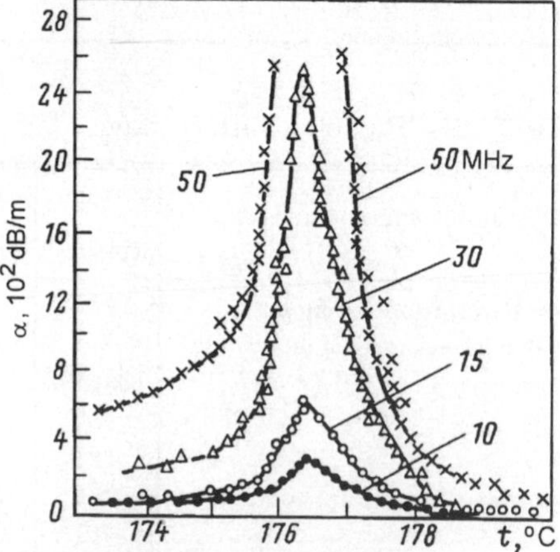

Figure 7.47 Temperature dependence of absorption factor of *L*-wave along the *Y* axis. Frequencies in MHz are marked by *1* – 10, *2* – 15, *3* – 30, *4* – 50 MHz, accordingly [336].

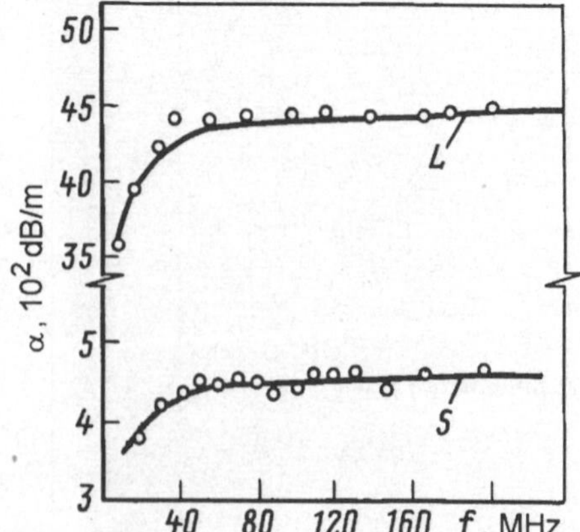

Figure 7.48 Damping of piezoactive share (*S*) and longitudinal (*L*) waves in lithium iodide [337].

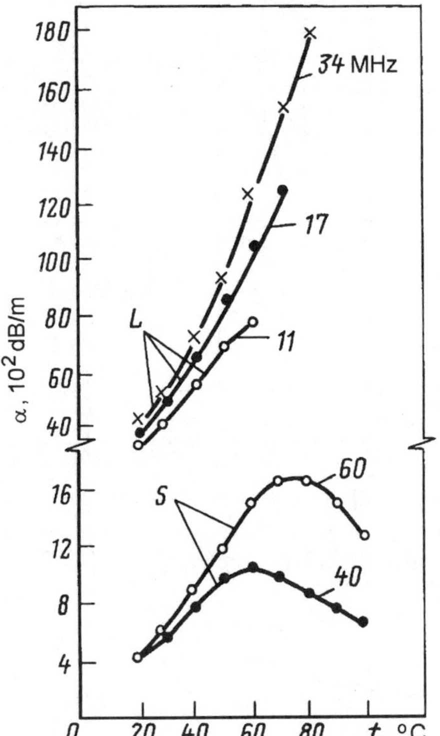

Figure 7.49 Damping of piezoactive share (S) and longitudinal (L) waves in lithium iodide. S-waves propagated perpendicular to the c axis, L-waves propagated parallel to c axis [337] (see also [338]).

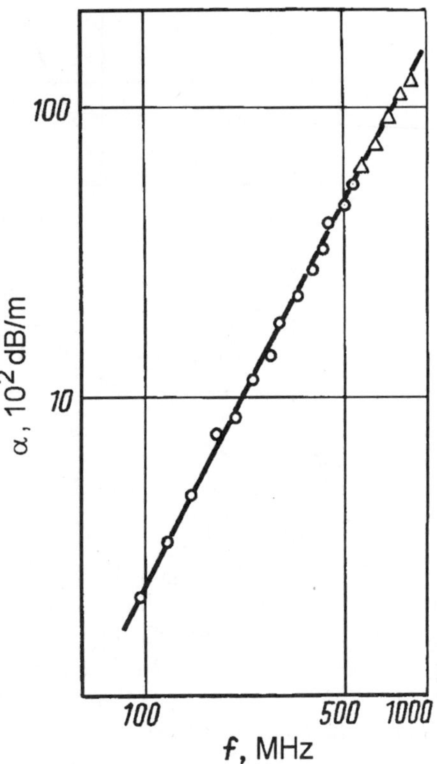

Figure 7.51 Frequency dependence of SS_{110}-wave damping in paratellurite ($\alpha = \alpha_0 f^{1.9}$, where $\alpha_0 = 3.6 \cdot 10^{-4}$ dB/cm; f is in MHz; $t = 10°C$) [340].

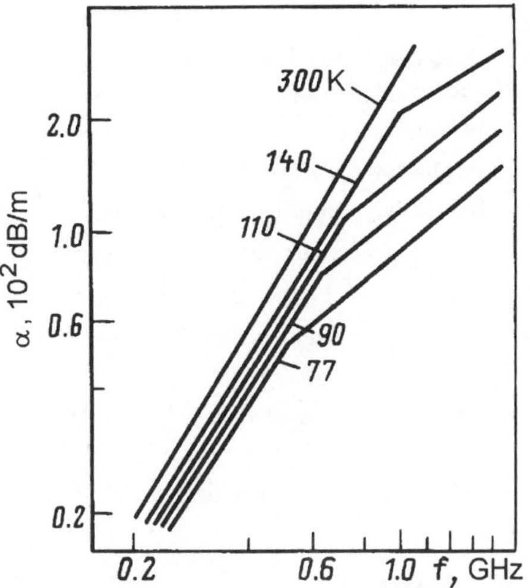

Figure 7.50 Frequency dependence of SS_{110}-wave absorption in n-silicon. The line-break corresponds to transition from the Akhiezer damping to the Landau–Rumer damping [339] (i.e. from sound modulation of heat phonons distribution to direct interaction of sound phonons with the thermal ones).

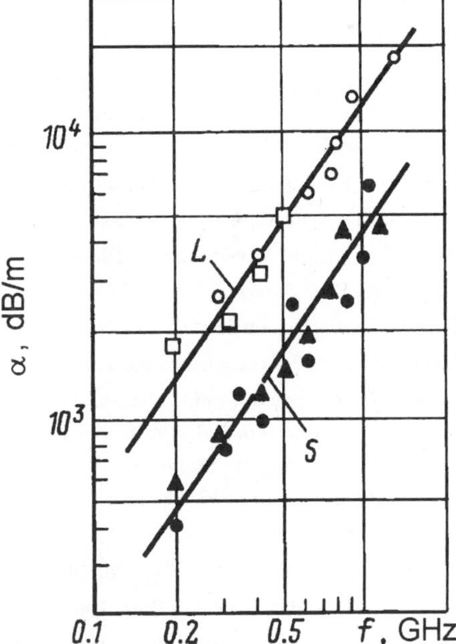

Figure 7.52 Nonelectron damping of L- and S-waves in cadmium sulfide at $T = 300$ K. Waves propagate at different angles θ with respect to the c axis: ○, □ 40°; • 0°; △ 90° [341].

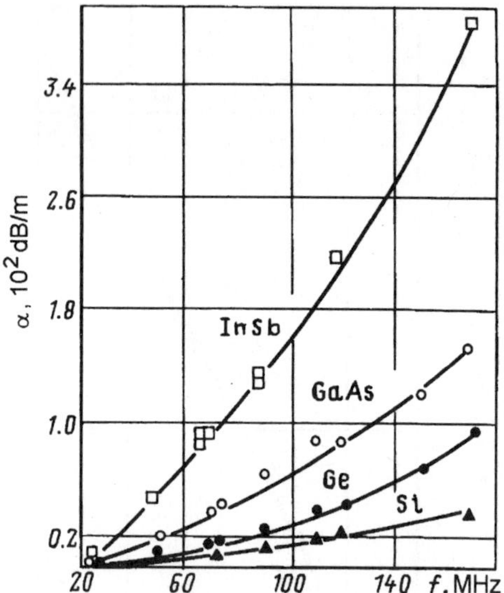

Figure 7.53 Frequency dependance of absorption in silicon, germanium, gallium arsenide, indium antimonide. T = 300 K; for L_{110}-waves. Impurity concentrations were: 10^{14} cm^{-3} of As in Si and Ge; $3 \cdot 10^{17}$ cm^{-3} of Te in GaAs and $2 \cdot 10^{16}$ cm^{-3} of Te in InSb [167].

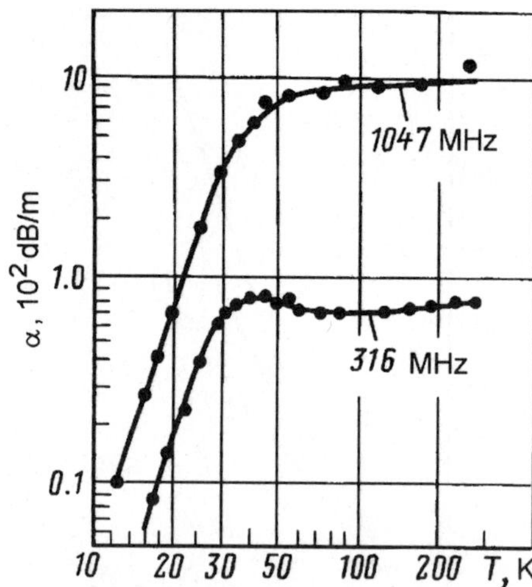

Figure 7.54 Absorption of the surface wave in the y-section of silicon. The wave propagates along the [100]-direction [344].

References

[1] Isakovich, M. A., General acoustics, Fizmatgiz, Moscow, 1973 (in Russian); Vinogradova, M. B., Rudenko, O. V., Sukhorukov, A. P., Theory of waves, Nauka, Moscow, 1990 (in Russian); Krasil'nikov, V. A., Krylov, V. V., Introduction into physical acoustics, Nauka, Moscow, 1985 (in Russian).

[2] Baranskii, K. N., Physical acoustics of crystals, Izdatel'stvo MGU, Moscow, 1991 (in Russian).

[3] Kino, G. S., Acoustic waves: devices, imaging and analog signal processing, Prentice Hall, New Jersey, 1987.

[4] Eder, F. X., Mod. Messmethoden Phys., Bd. 1, 1960; Petrallia, S., Nuovo Cimento Supp., 9, 1, 1952.

[5] Blythe, A. R., Acoustica, 16, 118, 1965-1966.

[6] Bergmann, L., Der ultraschall und seine Anwendung in Wissenschaft und Technik, S.Hirzel, Zurich, 1954.

[7] Merkulov, L. G., Akusticheskii zhurnal, 10, 163, 1964.

[8] Kneser, H. O., Ergeb. Exakt. Naturwiss., Bd. 22, 121, 1949; Nomoto, O., Ultrasonic waves, Tokyo, 1940; Petrallia, S., Nuovo cimento, 9, 1, 1951.

[9] Hardy, H. C., Telfair, D., Pielemeier, W. H., J. Acoustic Soc. Amer., 14, 226, 1942; Pielemeier, W. H., Hardy, H. C., J. Acoustic Soc. Amer., 13, 80, 1941.

[10] Penman, H. L., Proc. Phys. Soc., 47, 543, 1935.

[11] Quigley, Th. H., Phys. Rev., 67, 298, 1946.

[12] Sherratt, G. G., Griffiths, E., Proc. Royal Soc., Series A, 147, 292, 1934.

[13] Overbeck, C. J., Kendall, H. C., J. Acoustic Soc. Amer., 13, 26, 1941.

[14] Bachinskii, A. I., Putilov, V. V., Suvorov, N. P., A handbook in physics, Uchpedgiz, Moscow, 1951 (in Russian).

[15] Beyer, R. T., J. Acoustic Soc. Amer., 23, 176, 1951.

[16] Hodge, A. H., J. Chem. Phys., 5, 974, 1937.

[17] Volarovich, M. P., Balashova, D. B., Supersonics application for substance analysis, Izdanie MOPI, 13, 63, 1961.

[18] Henderson, M. C., Peselnick, L., J. Acoustic Soc. Amer., 29, 1074, 1957.

[19] Meyer, E., Sessler, G., Z. Phys., 149, 15, 1957.

[20] Itterbeek, A., Mariens, P., Physica, 4, 609, 1937.

[21] Keller, H. H., Z. Phys., 41, 386, 1940.

[22] Vaitonis, V. V., Yakovlev, V. F., Akusticheskii zhurnal, 12, 296, 1966 (in Russian).

[23] Itterbeek, A., Mariens, P., Physica, 5, 1533, 1938.; Itterbeek, A., Thys, L., Physica, 5, 889, 298, 640, 1938.

[24] Krasnushkin, P. E., Doklady Akad. Nauk SSSR, 27, 214, 1940 (in Russian); Phys. Rev., 65, 190, 1944.

[25] Mednikov, E. P., Acoustic coagulation and aerosol deposition, Izdatel'stvo Akad. Nauk SSSR, 1963 (in Russian).

[26] Evans, E. J., Bazley, E. N., Acoustica, 6, 238, 1956; Pohlmenn, W.,Proc. 3d Intern. Congr. Comm. Acoustics, Stuttgart, 1, 532, 1959.

[27] Sivian, L. J., J. Acoustic Soc. Amer., 19, 914, 1947.

[28] Liepmann, H. W., Helv. Phys. Acta, 11, 381, 1938; Helv. Phys. Acta, 12, 421, 1939.

[29] Gupill, E. W., Hoyt, C. K., Robinson, D. K., Canad. J. Phys., 33, 397, 1955.

[30] Beyer, R. T., J. Chem. Phys., 19, 788, 1951.

[31] Pitt, A., Jackson, W. J., Canad. J. Phys., 12, 686, 1935.

[32] Itterbeek, A., Forrez, G., Teirlinck, M., Physica, 23, 905, 1957.

[33] Findley, J. C., Pitt, A., Grayson, S. H., Phys. Rev., 54, 506, 1939; Chase, C. E., Proc. Royal Soc. (Lond.), A220, 116, 1953.

[34] Pellam, J. R., Squire, C. F., Phys. Rev., 72, 1245, 1947; 71, 477, 1947.

[35] Peshkov, V. P., Doklady Akad. Nauk SSSR, 45, 385, 1944; Livshits, E. M., Peshkov, V. P., Doklady Akad. Nauk SSSR, 47, 117, 1945.

[36] Mikhailov, I. G., Propagation of ultrasonic waves in liquids, Gostekhizdat, Moscow, 1949 (in Russian).

[37] Galt, J. K., J. Chem. Phys., 16, 505, 1948.

[38] Noury, J., Lacam, A., J. Phys. Rad., 15, 301, 1954.

[39] Landolt-Bornstein, Zahlenwerte und Funktionen aus Naturwissenschaften und Technik. Neue serie. Gruppe II, Springer-Verlag, Heidelberg, 1967 (Bd. 5).

[40] Parthasarathy, S., Proc. Indian Akad. Soc., A2, 497, 1935; A3, 285, 482, 519, 544, 1935; A4, 59, 213, 1936; Current Sci., 6, 213, 1937.

[41] Schaffs, W., Z. Physik, 114, 110, 1939; Z. Phys. Chem., 194, 28, 1944; Z. Naturforsch., 3, 396, 1948; Z. Phys. Chem., 196, 397, 1951.

[42] Freyer, E., Hubbard, J. C., Andrews, D., J. Amer. Chem. Soc., 51, 759, 1929.

[43] Schaffs, W., Nutsch-Kuhnkles, R., Acoustica, 10, 66, 1965-1966.

[44] Lagemann, R. T., McMillan, D. R., Woolf, W. E., J. Chem. Phys., 17, 369, 1949.

[45] Willard, G. W., J. Acoust. Soc. Amer., 19, 438, 1947.

[46] Greenspan, M., Tschiegg, C., Acoust. Soc. Amer., 31, 75, 1959.

[47] Sekoyan, S. S., Volkov, I. K., Tables of sound propagation velocities in distilled water within 16–94°C at pressures up to 800 bar, Izdatel'stvo VNIIFTRI, Moscow, 1969 (in Russian).

[48] Batlow, A. J., Yazgan, E., British J. Appl. Phys., 18, 645, 1967.

[49] Zeifen, N., Z. Physik, 108, 681, 1938.

[50] Lagemann, R. T., McMillan, D. R., Woosley, M., J. Chem. Phys., 16, 247, 1948.

[51] Baccaredda, M., Giacomini, A., Ricerca Sci., 16., 611, 1964; Att. Akad. Naz. Lincei, 1, 401, 1946.

[52] Tsvetkov, V., Marinin, V., Doklady Akad. Nauk SSSR, 68, 49, 1949.

[53] Shutilov, V. A., Fundamentals of physics of ultrasound, Izdatel'stvo LGU, Leningrad, 1980 (in Russian).

[54] Zeifen, N., Z. Physik, 108, 681, 1938.

[55] Weissler, A., J. Amer. Chem. Soc., 70, 1634, 1948.

[56] Kuhnkles, R., Schaffs, W., Acoustica, 13, 407, 1963.

[57] Pancholy, M., Paude, A., Parthasarathy, S., J. Sci. Ind. Res., 3, 5; 111; 159; 263; 354, 1944.

[58] Brazhnikov, N. I., Physical and physical-and-chemical control methods of substance composition and properties. Ultrasonic techniques, Ed. Shumilovskii, N. N., Energiya, Moscow-Leningrad, 1965 (in Russian).

[59] Wilson, W. D., J. Acoust. Soc. Am., 31, 1067, 1959; Sekoyan, S. S., Izmeritel'naya tekhnika, No 4, 51, 1963; Del Grosso, V. A., Mader, C. W., J. Acoust. Soc. Amer., 52, 1442, 1972; Ilgunas, V., Yaronis, E., Sukatskas, V., Ultrasonic interferometers, Mokslis, Vilnius, 1983 (in Russian).

[60] Mikhailov, I. G., Shutilov, V. A., Vestnik Leningradskogo universiteta, Ser. phys. chem., 3, No 16, 16, 1956.

[61] Seemann, H. J., Klein, F. K., Z. Angew. Phys., 19, 368, 1965.

[62] Vyugov, P. N., Gumenyuk, V. S., Ukrainskii fizicheskii zhurnal, 11, 440, 1966.

[63] Bockris, O. M., Richards, N. E., Proc. Royal Soc. (Lond.), A241, 44, 1957.

[64] Khodov, Z. L., Fizika matallov i metallovedenie, 10, 722, 1960.

[65] Webber, J., Stefens, R. in: Physical acoustics, Ed. Mason, W. P., v.IV, Part B, Academic Press, New York, London, 1968.

[66] Kleppa, O. J., J. Chem. Phys., 18, 1331, 1950.

[67] Gitis, M. B., Mikhailov, I. G., Akusticheskii zhurnal, 11, 434, 1965; 12, 17, 1966; 12, 145, 1966.

[68] Pronin, L. A., Filippov, S. I., Izvestiya vuzov, Ser. Ferrous metallurgy, 5, 10, 1963.

[69] Ilgunas, V., Yaronis, E., Akusticheskii zhurnal, 12, 145, 1966.

[70] Abowitz, G., Gordon, R. B., J. Chem. Phys., 37, 125, 1962.

[71] Barmatz, M., Rudnick, I., Phys. Rev., 170, 224, 1968.

[72] Trelin, Yu. S., Vasil'ev, I. V., Ultraacoustics application for substance investigation, Izdanie MOPI, 13, 3, 1961.

[73] Gordon, R. B., Acta Met., 7, 1, 1959.

[74] Polotskii, I. G., Khodov, Z. L., Voprosy fiziki metallov, 6, 70, 1955.

[75] Polotskii, I. G., Khodov, Z. L., Akusticheskii zhurnal, 4, 184, 1958.

[76] Hubbard, J. S., Loomis, A. L., Philos. Mag., 5, 1177, 1928.

[77] Kleppa, O. J., J. Chem. Phys., 17, 668, 1949.

[78] Jarzinski, J., Proc. Phys. Soc., 81, 745, 1963; Jarzinski, J., Litovitz, T. A., J. Chem. Phys., 41, 1290, 1964.

[79] Pronin, L. A., Filippov, S. I., Izvestiya vuzov, Ser. Ferrous metallurgy, 11, 11, 1963; Kazakov, N. B., Pronin, L. A., Filippov, S. I., J. Chem. Phys., 12, 11, 1964.

[80] Plass, K. G., Acoustica, 13, 240, 1963; Akust. Beih., 13, 240, 1963.

[81] Hunter, J. L., Hovan, K. S., J. Chem. Phys., 41, 4013, 1964; J. Acoust. Soc. Amer., 36, 1040, 1964; Hunter, J. L., Welch, T. J., Montrose, C. J., J. Acoust. Soc. Amer., 35, 1568, 1963.

[82] Otpushchennikov, N. F., Sysoev, I. V., Doklady I Vsesoyuznogo symp. po akustoopticheskoi spektroskopii, Tashkent, 1976 (p. 109).

[83] Wilson, W., Bradley, D., J. Acoust. Soc. Amer., 36, 333, 1964.

[84] Hawley, S., Allegra, J., Holton, Y., Acoust. Soc. Amer., 47, 258, 1970.

[85] Avgibaev, B., Ikramov, Sh. Kh., Shin, I. B., Doklady I Vsesoyuznogo symp. po akustoopticheskoi spektroskopii, Tashkent, 1976 (p. 148).

[86] Smith, A. H., Lawson, A. W., J. Chem. Phys., 22, 351, 1954.

[87] Hunter, J. L., Dardy, H. D., Bucaro, J. A., Preprint ND26 Congress Intern. Acoustic Liege, 1965 (p. 165).

[88] Pyro, J. F., Edmonds, P. D., Mol. Cryst. Liq. Cryst., 25, 175, 1974.

[89] Wetsel, G. C., Speer, R. S., Lowry, B. A., J. Appl. Phys., 43, 1495, 1972.

[90] Miyano, K., Ketterson, J. B., Phys. Rev. Lett., 34, 1047, 1973.

[91] Miyano, K., Ketterson, J. B., Phys. Rev., A12, 615, 1975.

[92] Walsh, J. M., Rice, M. H., J. Chem. Phys., 26, 815, 1957.

[93] Zarembo, L. K., Krasil'nikov, V. A., Introduction to nonlinear acoustics, Fizmatgiz, Moscow, 1966 (in Russian); Gun-Siu-Fang, Zarembo, L. K., Krasil'nikov, V. A., Akusticheskii zhurnal, 9, 382, 1963.

[94] Naugol'nykh, K. A., Ostrovskii, L. A., Nonlinear wave processes in acoustics, Nauka, Moscow, 1990 (in Russian).

[95] Beyer, R. T., Coppens, A. B., Proc. 5th International Acoustic Congr. Liege, 1965 (p. 515).

[96] Swamy, K. M., Naryan, K. L., Swamy, P. S., Acoustica, 32, 339, 1975.

[97] Shklovskaya-Kordi, V. V., Akusticheskii zhurnal, 9, 107, 1963.

[98] Mikhailov, I. G., Shutilov, V. A., Akusticheskii zhurnal, 6, 340, 1960.

[99] Emery, J., Kammoun, C., C. R. Akad. Sci. (Paris), B283, 131, 1976.

[100] Adler, L., Hiedemann, E. A., J. Acoust. Soc. Amer., 34, 410, 1962.

[101] Zverev, V. A., Kalachev, A. I., Akusticheskii zhurnal, 4, 321, 1958.

[102] Swamy, K. M., Naryan, K. L., Swamy, P. S., Acoustica, 34, 48, 1976.

[103] Thakur, K. P., Acoustica, 34, 270, 1976.

[104] Hagelberg, M. P., Holton, G., Kao, S., J. Acoust. Soc. Amer., 41, 564, 1967.

[105] Markham, J., Beyer, R., Lindsay, R. B., Rev. Mod. Phys., 23, 353, 1951.

[106] Thorp, W. H., J. Acoust. Soc. Amer., 42, 272, 1967.

[107] Lacam, A., Noury, J., J. Phys. Rad., 14, 272, 1953.

[108] Heaseall, L., Lamb, J., Proc. Phys. Soc., 69, 869, 1956.

[109] Mallikarjuna Rao, S. P., Suryanarayana, M., Ind. J. Pure and Appl. Phys., 11, 824, 1973.

[110] Pellam, J. R., Galt, J. K., J. Chem. Phys., 14, 608, 1946.

[111] Berdyev, A. A., Izvestiya Akad. Nauk Turkmenskoi SSR, Ser. Phys.-Techn., No 3, 16, 1965; Berdyev, A. A., Lezhnev, N. B., Akusticheskii zhurnal, 12, 247, 1966.

[112] Parpiev, K., Khabibulaev, P. K., Khaliulin, M. G., Akusticheskii zhurnal, 15, 466, 1969.

[113] Hunter, J. L., J. Acoust. Soc. Amer., 13, 36, 1941.

[114] Sette, D., J. Chem. Phys., 19, 1337, 1951.

[115] Bazhulin, P. A., Zh. eksper. teor. fiz., 8, 457, 1938.

[116] Lamb, J., Pinkerton, M. M., Proc. Royal Soc. (Lond.), 199, 114, 1949.

[117] Nomoto, O., Kishimoto, T., Ikeda, T., Bull. Kobayaci Inst. Phys. Res., 2, 72, 1952.

[118] Pinkerton, M. M., Proc. Phys. Soc., 62, 129, 1949.

[119] Lawley, L. E., Reed, R. D., Acoustica, 5, 316, 1955.

[120] Matthias, B. T., Nature, 161, 325, 1948.

[121] Pancholy, M., Parthasarathy, S., Chapgar, A. F., Nuovo Cimento, 10, 111, 1958.

[122] Andreas, J. H., Lamb, J., Proc. Phys. Soc., B64, 1021, 1951; Rapuano, R. A., Phys. Rev., (2)72, 78, 1947.

[123] Davidovich, L. A., Ivanov, A. A., Makhkamov, S. I., Akusticheskii zhurnal, 19, 26, 1973.

[124] Shoroshev, Yu. G., Lanshina, L. V., Shakhporonov, M. I., Doklady Akad. Nauk SSSR, 173, 70, 1967.

[125] Makhkamov, S. I., Khabibulaev, P. K., Khaliulin, M. G., Akusticheskii zhurnal, 20, 643, 1974.

[126] Chase, C. E., Proc. Royal Soc. (Lond.), A220, 116, 1953; Phys. Fluids, 1, 193, 1958.

[127] Fox, F. E., Rock, G. D., Phys. Rev., 2, 68, 1946; Smith, M. C., Beyer, R. T., Phys. Fluids, 27, 654, 1948; Pinkerton, J. M. M., Nature, 1960, 128, 1947.

[128] Koshkin, N. I., Ultraacoustics application for substance investigation, Izdanie MOPI, 1, 101, 1955.

[129] Fox, F. E., Wallace, W., J. Acoust. Soc. Amer., 26, 994, 1954.

[130] Towle, D. M., Lindsay, R. B., Acoust. Soc. Amer., 27, 530, 1955.

[131] Zarembo, L. K., Krasil'nikov, V. A., Shklovskaya-Kordi, V. V., Doklady Akad. Nauk SSSR, 109, 731, 1956.

[132] Zarembo, L. K., Phys. and math. Cand. thesis, MGU, Moscow, 1958.

[133] Naramsimhan, V., Beyer, R. T., J. Acoust. Soc. Amer., 28, 1233, 1956.

[134] Naugol'nykh, K. A., Romanenko, E. V., Akusticheskii zhurnal, 4, 200, 1958.

[135] Birch, F., in: Earth today, Eds. Cook, A. H. and Gaskell, T. F., Royal Astron. Soc., London, 1961.

[136] Salclanu, C., C. R. Akad. Sci., 252, 3021, 1961.

[137] Overton, W. C., J. Chem. Phys., 18, 113, 1950.

[138] Winder, D. R., Smith, C. S., Phys. Chem. Solids., 4, 128, 1958.

[139] Beranek, L. L., Acoustic measurements, Wiley, New York, 1949.

[140] Belyanin, V. A., Novikov, I. I., Proskurin, V. B., Trudy IX Vsesoyuznoi akusticheskoi conferentsii, V-III, No 2, 107, 1977.

[141] Otpushchennikov, N. F., Zhurnal eksperimental'noi i teoreticheskoi fiziki, 22, 782, 1952; Kuz'menko, V. A., Application of ultrasound oscillations for property studies, quality control and processing of metals and alloys, Izdatel'stvo Akad. Nauk USSR, Kiev, 1960 (in Russian).

[142] Morozov, A. I., Proklov, V. V., Stankovskii, B. A., Piezoelectric transducers for radioelectronic devices, Radio i svyaz', Moscow, 1981 (in Russian).

[143] Krasil'nikov, V. A., Sound and ultrasound waves in air, water and solids, Fizmatgiz, Moscow, 1960 (in Russian).

[144] Ivanov, V. B., Phys. and math. Cand. thesis, LETI, Leningrad, 1971.

[145] Auberger, M., Reinehart, J. S., J. Appl. Phys., 32, 219, 1961.

[146] Bacher, K., Verh. Deutsch. Phys. Ges., 3, 68, 1939; Z. Erdol und Kohle, 2, 125, 1949; Volarovich, M. P., Bayuk, E. I., Ultraacoustics application for substance investigation, Izdanie MOPI, 11, 1961.

[147] Martynov, E. G., Matveev, A. K., Ultraacoustics application for substance investigation, Izdanie MOPI, 10, 147, 1960.

[148] Truell, R., Elbaum, Ch., Chick, B. B, Ultrasonic methods in solid state physics, Academic Press, New York, London, 1969.

[149] Drichko, I. L., Ilisavskii, Yu. V., Gal'perin, Yu. M., Fizika tverdogo tela, 11, 2463, 1969.

[150] McSkimin, H. J., Andreatch, P. Jr., J. Appl. Phys., 35, 3312, 1964.

[151] Merkulov, L. G., Akusticheskii zhurnal, 5, 432, 1959.

[152] Papadakis, E. P., J. Appl. Phys., 34, 2168, 1963.

[153] Chatak, S. K., Sinha, S. K., Indian J. Phys., 48, 66, 1974.

[154] Meitzler, A. H., in: Ultrasonic transducer materials, Ed. Mattiat, O. E., Plenum Press, New York, 1971.

[155] Kor, K. S., Tandon, U. S., Mishra, P. K., J. Appl. Phys., 45, 2396, 1974.

[156] Vallin, J., Marklund, K., Sikstrom, J. O., Arkiv fur Physik, 33, 345, 1966.

[157] Dixon, R. W., J. Appl. Phys., 38, 3634, 1967.

[158] Uchida, N., Nizeki, N., Proc. IEEE, 61, 1073, 1973.

[159] Lee, B. H., J. Appl. Phys., 41, 2984, 1970.

[160] Lemanov, V. V., Avdonin, V. Ya., Pavlenko, A. V., Fizika tverdogo tela, 11, 3635, 1969.

[161] Beattic, A. G., Samara, G. A., J. Appl. Phys., 42, 2376, 1971.

[162] Farnell, G. W. in: Physical acoustics, Ed. Mason, W. P., Thurston, R. N., vol. VI, Academic Press, New York, London, 1970.

[163] Bairamov, B. Kh., Zakharchenya, B. P., Pisarev, R. V., Fizika tverdogo tela, 13, 3366, 1971.

[164] Melngailis, J., Vetelino, J. F., Jhunjhunwala, A., Appl. Phys. Lett., 32, 203, 1978.

[165] Sever, G. A., Baranskii, K. N., Izvestiya Akad. Nauk SSSR, Ser. Phys., 35, 935, 1971.

[166] Anisimkin, V. I., Zemlyanitsyn, M. A., Morozov, A. I., Fizika tverdogo tela, 17, 1513, 1975.

[167] Krasil'nikov, O. M., Vekilov, Yu. Kh., Kadyshevich, A. E, Fizika tverdogo tela, 11, 1200, 1969.

[168] Alton, W. J., Barlow, A. J., J. Appl. Phys., 38, 3023, 1967.

[169] Spencer, E. G., Denton, R. T., J. Appl. Phys., 34, 3059, 1963; Belt, R. F., Laser Focus, 6, 44, 1970.

[170] Roy, B., Ind. J. Pure and Appl. Phys., 25, 45, 1987.

[171] Bateman, T. B., J. Appl. Phys., 37, 2194, 1966.

[172] Clark, A. E., Strakna, R. E., J. Appl. Phys., 32, 1172, 1961.

[173] Landolt-Bornstein, Zahlenwerte und Funktionen aus Naturwissenschaften und Technik. Neue serie, Bd. 7, Kristallstrukturdaten anorganischer Verbindungen, Springer-Verlag, Berlin, 1971.

[174] Belyaev, L. M., Vitovskii, B. V., Dobrzhanskii, G. F., in: Crystal growth, v.1, Ed. Shubnikov, A. V., Izdatel'stvo Akad. Nauk SSSR, Moscow, 1957 (in Russian); Tyrbu, I. A., Ul'yanov, V. L., Botaki, A. A., Fizika tverdogo tela, 15, 3389, 1973.

[175] Voronkova, E. M., Grechushnikov, B. N., Distler, G. I., Optical materials for infrared equipment, Nauka, Moscow, 1965 (in Russian).

[176] Pratt, R. G., Simpson, G., Grossley, W. A., Electron. Lett., 8, 127, 1972.

[177] Gottlieb, M., Isaaks, T. J., Feichtner, J. D., J. Appl. Phys., 45, 5145, 1974; Isaaks, T. J., Cottlieb, M., Daniel, M., J. Electr. Mat., 4, 67, 1975.

[178] Yano, T., Nabeta, J., Watanabe, A., Appl. Phys. Lett., 18, 570, 1971.

[179] Warner, A. W., Pinnow, D. A., Acoust. Soc. Amer., 47, 791, 1970.

[180] Bateman, T. B., J. Appl. Phys., 33, 3309, 1962.

[181] Cline, C. F., Dunegan, H. L., Henderson, G. W., J Appl. Phys., 38, 1944, 1967.

[182] McFee, J. H., J. Appl. Phys., 34, 1548, 1963; Hutson, A. R., Phys. Rev. Lett., 4, 505, 1960.

[183] Fedotov, I. I., Kuznetsov, V. N., Ultraacoustics application for substance investigation, Izdanie MOPI, 14, 269, 1961.

[184] Aleksandrov, K. S., Anistratov, A. T., Rez, I. S., Fizika tverdogo tela, 19, 1863, 1977.

[185] Sil'vestrova, I. M., Barta, Ch., Dobrzhanskii, G. F., Kristallographiya, 20, 1062, 1975.

[186] Pinnow, D. A., Van Uitert, L. G., Warner, A. W., Appl. Phys. Lett., 15, 83, 1969.

[187] Badikov, V. V., Bogdanov, S. V., Godovikov, A. A., Akusticheskii zhurnal, 17, 300, 1971.

[188] Mozhaiskii, V. N., Pribory i tekhnika eksperimenta, 2, 200, 1974.

[189] Dieulesant, E., Royer, D., Ondes elastiques dans les solids. Application au traitement du signal, Masson, Paris, 1974.

[190] Coquin, G. A., Pinnow, D. A., Warner, A. W., J. Appl. Phys., 42, 2163, 1971.

[191] Lange, J. N., Phys. Rev., 176, 1030, 1968.

[192] Kleszczewski, Z., Archiwum Akustiyki, 13, 235, 1978.

[193] Ohmachi, Y., Uchida, N., J. Appl. Phys., 41, 2307, 1970.

[194] Dutoit, M., IEEE Trans. Sonics and Ultrasonics, SU-20, 279, 1973.

[195] Sosov, Yu. M., Yushin, N. K., Kudzin, A. Yu., Pis'ma Zh. Eksper. Teor. Fiz., 3, 475, 1977.

[196] Pisarevskii, Yu. V., Popolitov, V. I., Sil'vestrova, I. M., Materialy XII Vsesoyuznoi conf. po akustoelektronike i kvantovoi akustike, Saratov, 1983 (pt. 2, p. 363).

[197] Bondarenko, V. S., Gavrilyachenko, V. G., Chkalova, V. V., Nauchnye trudy vuzov LitSSR, Ultrasound, 9, 105, 1977.

[198] Ohmachi, Y., Uchida, N., J. Appl. Phys., 43, 3583, 1972.

[199] Sidek, A. A., Sounders, G. A., Wang Hong, Phys. Rev., 36, 7612, 1987.

[200] Ludanov, A. G., Fotchenkov, A. A., Yakovlev, L. A., Akusticheskii zhurnal, 22, 612, 1976.

[201] Zobnin, O. P., Yakovlev, L. A., Akusticheskii zhurnal, 22, 234, 1976.

[202] Weidner, D., Geophys. Res. Lett., 2, 189, 1975.

[203] Zagrai, N. P., Maksimov, V. N., Prikladnaya akustika, 1, 187, Izdatel'stvo TRTI, Taganrog, 1975.

[204] Anikeev, D. I., Zarembo, L. K., Karpachev, S. N., Fizika tverdogo tela, 24, 2938, 1982.

[205] Levites, A. F., Minaeva, K. A. Strukov, B. A., Pribory i tekhnika eksperimenta, 2, 187, 1974.

[206] Sapriel, J., Appl. Phys. Lett., 19, 533, 1971.

[207] Spenser, E. G., Lenzo, P. V., J. Appl. Phys., 38, 423, 1967.

[208] Smith, R. T., Welsh, F. S., J. Appl. Phys., 42, 6, 1971.

[209] Chkalova, V. V., Bondarenko, V. S., Klyuev, V. P., Elektronnaya tekhnika, Ser. 14, No 4, 158, 1971.

[210] Kludzin, V. V., Fizika tverdogo tela, 13, 651, 1971.

[211] Lezhnev, N. B., Karabash, V. I., Materialy XI Vsesoyuznoi conf. po akustoelektronike i kvantovoi akustike, Dushanbe, 1981 (pt. 2, p. 79).

[212] Baranskii, K. N., Berdyev, A. A., Pisarevskii, Yu. V., Izvestia Akad. Nauk Turkm. SSR, Ser. phys. techn., No 1, 37, 1975.

[213] Tokuro, I., Itsuro, I., Jpn. J. Appl. Phys., 15, 1451, 1976

[214] Carleton, H. R., Soref, R. A., Appl. Phys. Lett., 9, 109, 1966.

[215] Ultrasonic transducer materials, Ed. Mattias, O. E., Plenum Press, New York, 1971.

[216] Rodrigue, G. P., Proc. IEEE, 53, 1428, 1965

[217] Mustel', E. R., Parygin, V. N., Methods of light modulation and scanning, Nauka, Moscow, 1970 (in Russian).

[218] D'yakov, A. M., Ilisavskii, Yu. V., Farbshtein, I. I., Pis'ma Zh. Eksper. Teor. Fiz., 3, 564, 1977.

[219] Bajak, I. L., 2 pracovha konferencia cs. fyzikow, Bratislava, 1971 (p.184).

[220] Chernozatonskii, L. A., Golovashkin, A. I., Ivanenko, O. M., Fizika tverdogo tela, 30, 882, 1988.

[221] Moiseeva, N. A., Sil'vestrova, I. M., Pisarevskii, Yu. V, Acoustic properties of crystals. Potassium biphthalate, Preprint of Inst. of Crystallography Akad. Sci. USSR, Moscow, 1977 (in Russian).

[222] Belikova, G. S., Belyaev, L. M., Golovei, M. I., Kristallographiya, 19, 566, 1974; Belikova, G. S., Pisarevskii, Yu. V., Sil'vestrova, I. M., Fizika i khimiya tverdogo tela, 3, 18, 1973.

[223] Tokuru, I., Itsuro, I., Jpn. J. Appl. Phys., 15, 1451, 1976.

[224] Gottlieb, M., Laser Focus, 8, 24, 1972.

[225] Duderov, N. G., Demianets, L. N., Lobachev, A. M., J. Crystal Growth, 44, 483, 1978.

[226] Andrews, R. A., IEEE J. Quant. Electron., QE-6, 68, 1970.

[227] Epstein, D. J., Herrick, W. H., Turek, R. F., Solid State Commun., 8, 1491, 1970.

[228] Busch, M., Toledant, I. C., Torres, J., Opt. Commun., 10, 243, 1974.

[229] Hochli, U. T., Phys. Rev., B6, 1814, 1972.

[230] Roland, G. W., Appl. Phys. Lett., 21, 52, 1970.

[231] Kunigelis, V. F., Samulionis, V. I., Nauchnye trudy vuzov Lit. SSR, Ultrasound, 14, 11, 1972.

[232] White, R. M., Proc. IEEE, 58, 1238, 1970.

[233] Yakovkin, I. B., Petrov, D. V., Light diffraction on acoustic surface waves, Nauka, Novosibirsk, 1979 (in Russian).

[234] Pennury, O., Lakin, K. M., Proc. Ultrasonic Symp., Los Angeles, 1975 (p.478).

[235] Shibayama, K., Yamanouchi, K., Sato, H., Proc. IEEE, 64, 595, 1976.

[236] Levin, M. D., Lobanova, G. A., Pashchin, N. S., Akusticheskii zhurnal, 21, 68, 1975.

[237] Campbell, J. J., J. Appl. Phys., 41, 2797, 1970; Hutson, A. R., White, D. L., J. Appl. Phys., 33, 40, 1962; Szabo, E. L., Slobodnik, A. J., IEEE Trans. Sonics and Ultrasonics, SU-20, 240, 1973.

[238] Microwave acoustics handbook, Ed. Slobodnik, A., Conway, E. D., Office of Aerospace Research United State Air Force, New York, 1970 (v. 1).

[239] O'Connell, R. M., Carr, P. H., IEEE Trans. Sonics and Ultrasonics, SU-24, 376, 1977.

[240] Scott, B. A., Ingebrigsten, K. A., Tseng, C. C., Mat. Res. Bull., 5, 1045, 1970; Aniger, F. W., Electron. Lett., 7, 13, 1971.

[241] Jaffe, H, Berlincourt, D. A., Proc. IEEE, 53, 1372, 1965; Namatsu, S., Doi, K., Takahasti, M., Japan J. Appl. Phys., 11, 816, 1972; Gupta, S. N., Vatelino, J. F., Lipson, V. B., J. Appl. Phys., 47, 858, 1970.

[242] Slobodnik, A. J., Proc. IEEE, 64, 581, 1976.

[243] Karinskii, S. S., Ultrasonic acoustic wave devices for signal processing, Sovetskoe radio, Moscow, 1975 (in Russian).

[244] Yakovkin, I. B., Pashchin, N. S., Report abstracts for VIII Vsesoyuznaya conf. po akustoelektronike i kvantovoi akustike, Kazan, 1974 (p. 38).

[245] Ingebrigsten, K. A., Tonning, A., Appl. Phys. Lett., 9, 16, 1966.

[246] Coldren, L. A., Shaw, H. J., Proc IEEE, 64, 598, 1976.

[247] Coquin, G. A., Tiersten, H. F., J. Acoust. Soc. Amer., 41, 921, 1967.

[248] Zalesskii, V. V., Mel'tser, Ya. E., Stremovskii, E. V., Pribory i tekhnika eksperimenta, No 6, 107, 1976.

[249] Kraut, E. A., Tittman, B. R., Graham, L. J., Appl. Phys. Lett., 17, 271, 1970; Melngailis, J., Vetelino, J. F., Jhunjhunwala, A., Appl. Phys. Lett., 32, 203, 1978.

[250] Slobodnik, A. J., Conway, E. D., Electron. Lett., 6, 171, 1970; Slobodnik, A. J., Szabo, T. L., Electron. Lett., 7, 257, 1971.

[251] Farnell, G. W., in: Physical acoustics, Eds. Mason, W. P., Thurston, R. N., vol. VI, Academic Press, New York, london, 1970.

[252] Farnell, G. M., in: Acoustical surface waves, Ed. Oliner, A. A., Springer-Verlag, Berlin - Heidelberg - New York, 1978.

[253] Simpson, G., Electron. Lett., 9., 21., 1973.

[254] Bateman, T. B., McSkimin, H. J., Whelan, K., J. Appl. Phys., 30, 544, 1959; Hutson, A. R., Phys. Rev. Lett., 4, 505, 1960; Kino, G. S., Proc IEEE, 64, 724,1976.

[255] Isaaks, T. J., Weinert, R. W., J. Electron. Mater., 5, 13, 1976.

[256] Molokhia, N. H., Issa, M. A., Pramana, 11, 289, 1978.

[257] Dieulesaint, E., Royer, D., Piezoelectricite, 54, 180, 1976.

[258] Schulz, M. B., Holland, M. G., Proc. Conf. Component performance and systems applications of SAW devices, Scotland, 1973, p.180.

[259] Rouvaen, J. M., Bridox, E., Waxin, G., J. Appl. Phys., 49(4), 2306, 1978.

[260] Seavy, M. H., Solid State Commun., 10, 219, 1972.

[261] Berezhkov, V. V., Preobrazhenskii, V. L., Ekonomov, I. A., Radiotekhnika i elektronika, 28, 376, 1983.

[262] Campbell, J. J., Jones, W. R., IEEE Trans. Sonics and Ultrasonics, SU-15, 209, 1968.

[263] Knopff, L., in: Physical acoustics, Ed. Mason, W. P., vol. III, Part B, Academic Press, New York, London, 1965.

[264] Birch, F., Bancroft, D., Bull. Seismol. Soc. Amer., 28, 243, 1938.

[265] Bruckshaw, J., Mahanta, P., Petroleum (Lond.), 17, 14, 1954.

[266] Knopoff, L., Porter, L. D., J. Geophys. Res., 68, 6317, 1963.

[267] Peselnick, L., Zietz, I., Geophys., 24, 285, 1959.

[268] Born, W. T., Geophys., 6, 132, 1941.

[269] Mason, W. P., McSkimin, H. Y., J. Acoust. Soc. Amer., 19, 466, 1947.

[270] Zemanek, J. Jr., Rudnick, I., J. Acoust. Soc. Amer., 33, 1283, 1961.

[271] Lucke, K., J. Appl. Phys., 27, 1433, 1956.

[272] Auberger, M., Rinehart, J. S., J. Appl. Phys., 32, 219, 1961.

[273] Roth, W., J. Appl. Phys., 19, 901, 1948.

[274] Lindsay, G., Phys. Rev., 3, 397, 1914.

[275] Wegel, R. L., Walther, H., Physics, 6., 141, 1935.

[276] Gemant, A., Jackson, W., Philos. Mag., 23, 960, 1937.

[277] Kimball, A. L., Lovell, D. E., Phys. Rev., 30, 948, 1927.

[278] Papadakis, E. P., J. Acoust. Soc. Amer., 37, 711, 1965.

[279] West, F. G., J. Appl. Phys., 29, 480, 1958.

[280] Blair, D. G., Buckingham, M. I., Edwards, C., Proc. 2nd Marcel Grossmann Conf. General Relativity, Triest, 1980 (p. 185).

[281] McGuigan, D. G., Lam, C. C., Douglass, D. H., J. Low Temp. Phys., 30, 621, 1978.

[282] Braginskii, V. B., Mitrofanov, V. P., Panov, V. N., Low dissipation systems, Nauka, Moscow, 1981 (in Russian).

[283] Granato, A., Lucke, K., in: Physical acoustics, Ed. Mason, W. P., vol. IV, Part A., Academic Press, New York, London, 1966.

[284] Papadakis, E. P., in: Physical acoustics, Ed. Mason, W. P., vol. IV, Part B., Academic Press, New York - London, 1968.

[285] Helme, B. G., King., P. J., Phys. Status Solidi, 45, K33, 1978.

[286] Hickernell, F. S., IEEE Trans., Microwave theory and Techn., MTT-17, 957, 1969.

[287] Landolt-Bornstein, Zahlenwerte und Funktionen aus Naturwissenschaften und Technik. Neue serie, Bd. 3. Ferro- und Antiferroelectrische Substanzen, Springer-Verlag, Berlin, 1969.

[288] Venturini, E. L., Spencer, E. G., Ballmann, A. A., J. Appl. Phys., 40, 1622, 1969.

[289] Merkulov, L. G., Kovalenok, R. V., Konovodchenko, E. V., Fizika tverdogo tela, 11, 2769, 1969; 13, 1171, 1971; 14, 340, 1972.

[290] Auld, B. A., Acoustic fields and waves in solids, John Wiley and Sons, New York, 1973.

[291] Holland, M. G., IEEE Trans. Sonics and Ultrasonics, SU-15, 18, 1968.

[292] Gubanov, A. I., Davydov, S. Yu., Fizika tverdogo tela, 14, 2187, 1972.

[293] Oliver, D. W., Slack, G. A., J. Appl. Phys., 37, 1542, 1966.

[294] Rehwald, W., J. Appl. Phys., 44, 3017, 1973.

[295] Schloemann, E., Joseph, R. I., Kohane, T., Proc IEEE, 53, 1495, 1965.

[296] Belyaev, L. M., Bagdasarov, Kh. S., Sil'vestrova, I. M., Izvestiya Akad. Nauk SSSR, Ser. fiz., 35, 941, 1971.

[297] Grishmanovskii, A. P., Yushin, N. K., Bogdanov, V. L., Fizika tverdogo tela, 13, 1833, 1971.

[298] Abramovich, A. A., Khromova, N. N., Shutilov, V. A., Akusticheskii zhurnal, 22, 278, 1976.

[299] Wilkinson, C. D., Caddes, D. E., J. Opt. Soc. Amer., 40, 498, 1966.

[300] Broussand, G., Optoelectronique, Masson, Paris, 1974.

[301] Merkulov, L. G., Yakovlev, L. A., Akusticheskii zhurnal, 5, 374, 1959.

[302] King., P. J., J. Phys., 3, 500, 1970..

[303] Mastikhin, V. M., Bogdanov, S. V., Darvoid, T. I., Optiko-mekhanicheskaya promyshlennost', No 8, 36, 1977 (in Russian).

[304] Semenov, V. I., Sapozhnikov, V. K., Avdienko, K. I., Fizika tverdogo tela, 18, 2805, 1976.

[305] Smith, A. B., Kedzie, R. W., McManon, D. H., J. Appl. Phys., 40, 2687, 1976.

[306] Vasilevskaya, A. S., Sonin, A. S., Rez, I. S., Izvestiya Akad. Nauk SSSR, Ser. fiz., 31, 1159, 1967.

[307] Ohmachi, C., Uchida, N., J. Appl. Phys., 42, 521, 1971.

[308] Mastikhin, V. M., Sapozhnikov, V. K., Serbulenko, M. G., Avtometriya, 3, 31, 1975.

[309] Lewis, I. T., Lehoczky, A., Briscoe, C. V., Phys. Rev., 161, 877, 1967.

[310] Hemphill, R., Appl. Phys. Lett., 9, 35, 1966.

[311] Holovey, M. I., Gurzan, M. J., Olexeyk, J. D., Kristall und Techn., 8, 453, 1973.

[312] Esayan, S. Kh., Lemanov, V. V., Rez, I. S., Fizika tverdogo tela, 15, 907, 1973.

[313] Bogdanov, S. V., Zubrinov, I. I., Sheloput, D. V., Izvestiya Akad. Nauk SSSR, Ser. fiz., 35, 2013, 1971.

[314] Dutoit, H., IEEE Trans. Sonics and Ultrasonics, SU-20, 279, 1973.

[315] Midford, T. A., Wanuga, S., J. Appl. Phys., 36, 3362, 1965.

[316] Fitzgerald, T. M., Chick, B. B., Truell, R., J. Appl. Phys., 35, 2647, 1964.

[317] Grigoriev, M. A., Zyuryukin, Yu. A., Nayanov, V. I., Fizika tverdogo tela, 12, 3033, 1970.

[318] Grigor'ev, M. A., Zaitsev, B. D., Pylaeva, G. I., Fizika tverdogo tela, 15, 1398, 1973.

[319] Mason, W. P., in: Physical acoustics, Ed. Mason, W. P., vol. IV, Part A., Academic Press, New York - London, 1966.

[320] Fitzgerald, T. M., Silverman, B. D., Phys. Rev. Lett., 25A, 245, 1967.

[321] Reinjes, J., Schulz, M. B., J. Appl. Phys., 39, 5254, 1968.

[322] Smakula, A., Einkristalle, Wachtsum, Herstellung und Anwendung, Springer-Bergman, Berlin-Munchen, 1962.

[323] Lemanov, V. V., Phys. and math. Doctor Thesis, LFTI, Leningrad, 1973.

[324] Luthi, B., Pollina, R. J., Phys. Rev., 167, 482, 1968.

[325] Barret, H. H., Phys. Rev., 178, 743, 1969.

[326] Levy, S., Truell, R., Rev. Mod. Phys., 25, 140, 1953.

[327] Goncharov, K. V., Klyuev, V. P., Lyamov, V. E., Trudy vsesoyuznoi akusticheskoi conf., EII3, Moscow, 1968.

[328] Avdonin, V. Ya., Lemanov, V. V., Smirnov, I. A., Fizika tverdogo tela, 14, 877, 1972.

[329] Velikov, Yu. Kh., Kadyshevich, A. E., Krasil'nikov, O. E., Fizika tverdogo tela, 13, 1310, 1971.

[330] Dransfeld, K., J. Phys. Colloq. C1, Suppl. 2, 28, 157, 1967.

[331] De Klerck, J, in: Physical acoustics, Ed. Mason, W. P., vol. IV, Part A., Academic Press, New York, London, 1966.

[332] Grigor'ev, M. A., Zyuryukin, Yu. A., Nayanov, V. I., Elektronnaya tekhnika, Ser. 11, No 1(19), 121, 1970.

[333] Ivanov, S. N., Kotelyanskii, I. M., Medvedev, V. V., Zh. Eksper. Teor. Fiz., 70, 281, 1976.

[334] Bommel, H., Phys. Rev., 100, 758, 1955.

[335] Samulionis, V. I., Kunigelis, V. F., Garshka, E. P., Pis'ma Zh. Eksper. Teor. Fiz., 9, 459, 1969.

[336] Maishchik, E. P., Strukov, B. A., Sinyakov, E. V., Fizika tverdogo tela, 19, 335, 1977.

[337] Bogdanov, S. V., Balakirev, M. K., Ivanov, E. V., Materialy XII Vsesoyuznoi conf. po akustoelektronike i kvantovoi akustike, Saratov, 1983 (pt. 1, p. 15).

[338] Abramovich, A. A., Shutilov, V. A., Levitskaya, T. D., Fizika tverdogo tela, 14, 2585, 1972.

[339] Ilisavskii, Yu. V., Sternin, V. M., Materialy XII Vsesoyuznoi conf. po akustoelektronike i kvantovoi akustike, Saratov, 1983 (pt. 2, p. 281).

[340] Antonov, S. N., Proklov, V. V., Mirgorodskii, V. I., Materialy XI Vsesoyuznoi conf. po akustoelektronike i kvantovoi akustike, Dushanbe, 1981 (pt. 1, p. 174).

[341] Keller, O., Phys. Rev. Lett., 39A, 235, 1972; Siebert, F., Keller, O.,Wettling, W., Phys. Status Solidi, 4, 67, 1971.

[342] Farnell, G. M., Wave electronics, 2, 1, 1976.

[343] Carr, P. H., IEEE Trans., Microwave Theory and Techn., MTT, 458, 1969.

[344] Salzmann, E., Plieninger, T., Dransfeld, K., Appl. Phys. Lett., 13, 14, 1968.

[345] Acoustic crystals. Handbook, Ed. Shaskol'skaya, M. P., Nauka, Moscow, 1982 (in Russian).

[346] Biryukov, S. V., Gulyaev, Yu. V., Krylov, V. V., Plesskii, V. P., Surface acoustic waves in nonuniform media, Nauka, Moscow, 1991 (in Russian).

[347] Goss, S. A., Jonston, R. L., Dunn, F., J. Acoust. Soc. Amer., 64, 423, 1978; Goss, S. A., Frizzel, L. A., Dunn, F., Ultrasound Med. Biol., 5, 181, 1979; Bamber, J. S., Hill, C. R., Ultrasound Med. Biol., 5, 149, 1979; Chivers, R. C., Parry, R. J., J. Acoust. Soc. Amer., 63, 940, 1978.

[348] Worlton, D., J. Appl. Phys., 32, 967, 1961.

8

Thermometry

A.V. Inyushkin

8.1 Introduction

Temperature is the basic physical quantity describing the equilibrium thermodynamic state of a macroscopic system. The prerequisite for the temperature measurement is a temperature scale. The scale is based upon a realization of specified set of reproducible equilibrium states (defining fixed points) to each of which is assigned a numerical value of the temperature, and assumes using standard interpolation instruments between them.

The 11th General Conference of Weights and Measures in 1960 adopted the "International Thermodynamic (Kelvin) Scale" as a basic. The thermodynamic temperature scale is defined by the relation derived from consideration of the reversible Carnot cycle,

$$Q_1/Q_2 = T_1/T_2, \tag{8.1}$$

where Q_1 is the quantity of heat which the Carnot engine extracted from the hot reservoir in reversible isothermal process at the temperature T_1; Q_2 is the quantity of heat which the engine returned to the cold reservoir in the second isothermal process at the temperature T_2. The thermodynamic scale is independent of the working substance.

The unit of a thermodynamic temperature, symbol T, is the kelvin, symbol K, defined as the fraction 1/273.16 of the thermodynamic temperature of the triple point of water [1]. A thermodynamic temperature, T, expressed in terms of its difference from 273.15 K, the ice point, is known as a Celsius temperature, symbol t, defined by:

$$t/^\circ\mathrm{C} = T/\mathrm{K} - 273.15. \tag{8.2}$$

The unit of Celsius temperature is the degree Celsius, symbol °C, which is by definition equal in magnitude to the kelvin.

The direct harnessing of the Carnot cycle for temperature measurements results usually in low experimental accuracy. Due to this, practical methods for realization of thermodynamic temperatures have been developed in which a relation between the measured physical quantity and temperature is derived using the thermodynamic laws or statistical mechanics. In the list of such relations are the equation of state for a gas, the Curie law for a paramagnet, the temperature dependence of the sound velocity in a gas, the temperature dependence of electrical noise in a resistor (Nyquist relation), and the Planck radiation law. Temperature scales based on these relations depend upon the properties of a thermometric substance. This results in the appearance of such characteristics of the scale as reproducibility and accuracy. In addition, some scales are based upon approximate relations; the term – instrumental temperature (magnetic, color, etc.) different from the thermodynamic one – arises.

0-8493-2861-6/97/$0.00+$.50

The technical difficulties emerging on the direct realization of thermodynamic scale are overcome by introduction of practical temperature scales. These were formulated so as to allow measurements of temperature to be made precisely and reproducibly and with temperatures measured on these scales being as close an approximation as possible to the corresponding thermodynamic temperature. The first practical temperature scale, the International Temperature Scale of 1927 (ITS-27), was adopted by the 7th General Conference on Weights and Measures in 1927. A revised version, ITS-48, was adopted by 9th General Conference in 1948; an amended edition, International Practical Temperature Scale of 1948 (IPTS-48), was adopted by 11th General Conference in 1960. In 1968 the International Committee of Weights and Measures promulgated the IPTS-68, having been empowered to do so by the 13th General Conference of 1967-1968. The amended edition of the IPTS-68 was adopted by the 15th General Conference in 1975. In 1976, the International Committee of Weights and Measures approved the 1976 Provisional 0.5 K to 30 K Temperature Scale (EPT-76) for the low-temperature region. The present version of the scale is the International Temperature Scale of 1990.

Various aspects of thermometry are discussed in Refs. [2–12]. A full description of all international temperature scales can be found in Ref. [10]. A huge collection of papers detailing the research results, techniques, and findings is in the proceedings of international conferences on thermometry: Temperature, Its Measurement and Control in Science and Industry (TMCSI) [13].

8.2 The International Temperature Scale of 1990

The International Temperature Scale of 1990 (ITS-90) was adopted by the International Committee of Weights and Measures at its meeting in 1989, in accordance with the request embodied in Resolution 7 of the 18th General Conference of Weights and Measures in 1987. This scale supersedes the IPTS-68 (amended edition of 1975) [14] and the EPT-76 [15].

The ITS-90 defines both International Kelvin Temperatures, symbol T_{90}, and International Celsius Temperatures, symbol t_{90}. The relation between T_{90} and t_{90} is the same as that between T and t, i.e.:

$$t_{90}/°\mathrm{C} = T_{90}/\mathrm{K} - 273.15. \tag{8.3}$$

The unit of the physical quantity T_{90} is the kelvin, symbol K, and the unit of the physical quantity t_{90} is the degree Celsius, symbol °C, as in the case of the thermodynamic temperature T and the Celsius temperature t.

The ITS-90 extends upwards from 0.65 K to the highest temperature practically measurable in terms of the Planck radiation law with monochromatic radiation. The ITS-90 comprises a number of ranges and sub-ranges throughout each of which T_{90} temperatures are defined. Several of these ranges or sub-ranges overlap, and where such overlapping occurs, differing definitions of T_{90} exist: these differing definitions have equal status.

The defining fixed points of the ITS-90 are listed in Table 8.1. The effects of pressure, arising from significant depths of immersion of the sensor or from other causes, on the temperature of most of these points are given in Table 8.2.

Between 0.65 K and 5.0 K, T_{90} is defined in terms of the vapor-pressure p of ^{3}He and ^{4}He using equations of the form

$$T_{90}/\mathrm{K} = A_0 + \sum_{i=1}^{9} A_i\left[(\ln(p/\mathrm{Pa}) - B)/C\right]^i. \tag{8.4}$$

The constants A_0, A_i, B and C are given in Table 8.3 for ^{3}He in the range from 0.65 K to 3.2 K, and for ^{4}He in the ranges from 1.25 K to 2.1768 K (the λ point) and 2.1768 K to 5.0 K.

Between 3.0 K and the triple point of neon (24.5561 K), T_{90} is defined by means of an ^{3}He or ^{4}He gas thermometer of the constant-volume type that has been calibrated at three defining fixed points and using specified interpolation procedures. These points are the triple point of neon, the triple point of equilibrium hydrogen, and the temperature between 3.0 K and 5.0 K. This last temperature is determined using an ^{3}He or ^{4}He vapor pressure thermometer.

From 4.2 K to Triple Point of Neon (24.5561 K) with ^{4}He as the thermometric gas T_{90} is defined by the relation

$$T_{90} = a + bp + cp^2, \tag{8.5}$$

where p is the pressure in a gas thermometer, and a, b and c are the numerical coefficients obtained from measurements made at three defining fixed points given above, but with the further restriction that the lowest one of these points lies between 4.2 K and 5.0 K.

From 3.0 K to Triple Point of Neon (24.5561 K) with ^{3}He or ^{4}He as the thermometric gas. For a helium gas thermometer used below 4.2 K, the non-ideality of the gas must be accounted for explicitly, using the appropriate second virial coefficient $B_x(T_{90})$. In this range T_{90} is defined by the relation

$$T_{90} = \frac{a + bp + cp^2}{1 + B_x(T_{90})N/V}, \tag{8.6}$$

where p is the pressure in a gas thermometer, and a, b and c are the numerical coefficients obtained from measurements made at three defining temperatures as given above, N/V is the gas density with N being the quantity of gas and V the volume of the bulb, x is 3 or 4 according to the isotope used, and the values of the second virial coefficients are given by the relations:
for ^{3}He,

$$\begin{aligned} B_3(T_{90})(\mathrm{m}^3/\mathrm{mol}) = \big[&16.69 - 336.98(T_{90}/\mathrm{K})^{-1} \\ &+ 91.04(T_{90}/\mathrm{K})^{-2} - 13.82(T_{90}/\mathrm{K})^{-3}\big]10^{-6}; \end{aligned} \tag{8.7a}$$

for ^{4}He,

$$\begin{aligned} B_4(T_{90})(\mathrm{m}^3/\mathrm{mol}) = \big[&16.708 - 374.05(T_{90}/\mathrm{K})^{-1} \\ &- 383.53(T_{90}/\mathrm{K})^{-2} + 1799.2(T_{90}/\mathrm{K})^{-3} \\ &- 4033.2(T_{90}/\mathrm{K})^{-4} + 3252.8(T_{90}/\mathrm{K})^{-5}\big]10^{-6}. \end{aligned} \tag{8.7b}$$

Between the triple point of equilibrium hydrogen (13.8033 K) and the freezing point of silver (961.78°C), T_{90} is defined by means of platinum resistance thermometers calibrated at specified sets of defining fixed points, and using specified reference and deviation functions for interpolation at intervening temperatures. Temperatures are determined in terms of the ratio of the resistance $R(T_{90})$ at a temperature T_{90} and the resistance $R(273.16\ \mathrm{K})$ at the triple point of water. This ratio, $W(T_{90})$, is

$$W(T_{90}) = R(T_{90})/R(273.16\ \mathrm{K}). \tag{8.8}$$

An acceptable platinum resistance thermometer must be made from pure, strain-free platinum, and it must satisfy at least one of the following two relations:

$$W(29.7646°\mathrm{C}) \geq 1.11807, \tag{8.9a}$$

$$W(-38.8344°\mathrm{C}) \leq 0.844235. \tag{8.9b}$$

An acceptable platinum resistance thermometer that is to be used up to the freezing point of silver must also satisfy the relation

$$W(961.78°\mathrm{C}) \geq 4.2844. \tag{8.9c}$$

In each of the resistance thermometer ranges, T_{90} is obtained from $W_r(T_{90})$ as given by the appropriate reference function {Eqs.(8.10b) or (8.11b)}, and the deviation $W(T_{90}) - W_r(T_{90})$. At the defining fixed points this deviation is obtained directly from the calibration of the thermometer; at intermediate temperatures it is obtained by means of the appropriate deviation function.

(i) – In the range from 13.8033 K to 273.16 K the following reference function is defined:

$$\ln[W_r(T_{90})] = A_0 + \sum_{i=1}^{12} A_i \left[\frac{\ln(T_{90}/273.16\ \mathrm{K}) + 1.5}{1.5}\right]^i. \tag{8.10a}$$

An inverse function, equivalent to Eq.(8.10a) to within 0.1 mK, is

$$T_{90}/273.16\ \mathrm{K} = B_0 + \sum_{i=1}^{15} B_i \left[\frac{W_r(T_{90})^{1/6} - 0.65}{0.35}\right]^i. \tag{8.10b}$$

The constants A_0, A_i, B_0 and B_i are given in Table 8.4.

A thermometer may be calibrated for use throughout this range or, based on progressively fewer calibration points, for ranges with low-temperature limits of 24.5561 K, 54.3584 K and 83.8058 K, all having an upper limit of 273.16 K.

(ii) – In the range from 0°C to 961.78°C the following reference function is defined:

$$W_r(T_{90}) = C_0 + \sum_{i=1}^{9} C_i \left[\frac{T_{90}/\mathrm{K} - 754.15}{481}\right]^i. \tag{8.11a}$$

An inverse function, equivalent to Eq.(8.11a) to within 0.13 mK, is

$$T_{90}/\mathrm{K} - 273.15 = D_0 + \sum_{i=1}^{9} D_i \left[\frac{W_r(T_{90}) - 2.64}{1.64}\right]^i. \tag{8.11b}$$

The constants C_0, C_i, D_0 and D_i are given in Table 8.4.

A thermometer may be calibrated for use throughout this range or, based on fewer calibration points, for ranges with upper limits of 660.323°C, 419.527°C, 231.928°C, 156.5985°C or 29.7646°C, all having a lower limit of 0°C.

(iii) – A thermometer may be calibrated for use in the range from 234.3156 K (−38.8344°C) to 29.7646°C, the calibration being made at these temperatures and at the triple point of water. Both reference functions {Eqs.(8.10) and (8.11)} are required to cover this range.

The defining fixed points and deviation functions for the various ranges are given in Table 8.5.

Above the freezing point of silver, the temperature T_{90} is defined by the equation

$$\frac{L_\lambda(T_{90})}{L_\lambda[T_{90}(X)]} = \frac{\exp(c_2[\lambda T_{90}(X)]^{-1}) - 1}{\exp(c_2[\lambda T_{90}]^{-1}) - 1}, \tag{8.12}$$

where $T_{90}(X)$ refers to any of the silver $\{T_{90}(\mathrm{Ag}) = 1234.93\ \mathrm{K}\}$, the gold $\{T_{90}(\mathrm{Au}) = 1337.33\ \mathrm{K}\}$ or the copper $\{T_{90}(\mathrm{Cu}) = 1357.77\ \mathrm{K}\}$ freezing points and in which $L_\lambda(T_{90})$ and $L_\lambda[T_{90}(X)]$ are the spectral concentrations of a blackbody radiance at the wavelength (in vacuo) λ at T_{90} and at $T_{90}(X)$, respectively, and $c_2 = 0.014388$ m · K.

The preceding description of the ITS-90 has for the most part been taken from the English version of the text. For a complete description of the ITS-90 the reader should be refered to the official French text or the English version [16] which has been authorized by the Comité Consultatif de Thermométrie and approved by the Comité International des Poids et Mesures.

The apparatus, methods and procedures that will serve to realize the ITS-90 are given in "Supplementary Information for the ITS-90" [17]. This document also gives an account for the earlier International Temperature Scales and the numerical differences between successive scales that include, where practicable, mathematical functions for the differences $T_{90} - T_{68}$. A number of useful approximations to the ITS-90 are given in "Techniques for Approximating the ITS-90" [18]. These two documents have been prepared by the Comité Consultatif de Thermométrie and are published by the BIMP; they are revised and updated periodically.

The differences $T_{90} - T_{68}$ are listed in Table 8.6. Equations providing an analytic realization of the differences $T_{90} - T_{68}$ are given in Table 8.7. The helium vapour-pressure equations for the ITS-90 are those originally derived for the EPT-76 [19, 20], thus in the range below 4.2 K (omitted from Table 8.6) the differences $T_{90} - T_{76}$ can be disregarded.

The estimated uncertainties of the thermodynamic temperatures at most of the defining fixed points of the ITS-90 are given in Table 8.8 together with the very much smaller uncertainties estimated to be currently attainable under the best experimental conditions, in their realization on the ITS-90 itself.

Given in Table 8.9 are secondary fixed points. These points are fixed points that are *not* used in the definition of the ITS-90, but which have been well-characterized, and the temperatures of which have been carefully determined. The values of temperatures, which are on the IPTS-68, can be converted to the ITS-90 using the Table 8.6 or Table 8.7.

In some countries the Rankine, Fahrenheit and Reaumur scales are used besides the Kelvin and Celsius scales. The relations between the units of these scales and the kelvin are:

$$\text{the degree Rankine } (^\circ\text{Ra}) = \frac{5}{9}\text{ K},$$

$$\text{the degree Fahrenheit } (^\circ\text{F}) = \frac{5}{9}\text{ K},$$

$$\text{the degree Reaumur } (^\circ\text{R}) = \frac{5}{4}\text{ K}.$$

The following formulas are used to determine the temperatures on these scales:

$$n(^\circ\text{Ra}) = \frac{5}{9}n(\text{K}) = \left(\frac{5}{9}n - 273.15\right)(^\circ\text{C});$$

$$n(^\circ\text{F}) = \left[\frac{5}{9}(n-32) + 273.15\right](\text{K}) = \frac{5}{9}(n-32)(^\circ\text{C});$$

$$n(^\circ\text{R}) = \left(\frac{5}{4}n + 273.15\right)(\text{K}) = \frac{5}{4}n(^\circ\text{C});$$

$$n(\text{K}) = \frac{9}{5}n(^\circ\text{Ra});$$

$$n(^\circ\text{C}) = \frac{9}{5}(n + 273.15)(^\circ\text{Ra});$$

$$n(\text{K}) = \left[\frac{9}{5}(n - 273.15) + 32\right](^\circ\text{F});$$

$$n(^\circ\text{C}) = \left(\frac{9}{5}n + 32\right)(^\circ\text{F});$$

$$n(\text{K}) = \frac{4}{5}(n - 273.15)(^\circ\text{R});$$

$$n(^\circ\text{C}) = \frac{4}{5}n(^\circ\text{R}).$$

8.3 Liquid-in-Glass Thermometry

Liquid-in-glass thermometry is based upon the thermal expansion of liquid. Under variation of temperature the length of a liquid column in glass (quartz) capillary varies due to a difference between the thermal expansion of liquid and glass reservoir. The temperature is read from meniscus location relative to the scale engraved on the capillary directly or on a plate which is firmly affixed to the capillary. Liquid-in-glass thermometers have been used to measure temperatures from −200°C to 1200°C, but they are most frequently used within the range from −40°C to 250°C.

The temperature reading depends upon immersion depth of liquid-in-glass thermometer into measurable medium. For precise measurements, the thermometer is immersed to a prescribed depth (partial immersion) or to the top of the liquid column (total immersion). If this is impossible, the correction Δt due to the emergent column (that part of the liquid column between the point of immersion and the top of the column) is necessary. The corrections are

$$\Delta t = \beta n(t_1 - t_2) \tag{8.13}$$

and

$$\Delta t = \beta n(t_0 - t_2) \tag{8.14}$$

for thermometers calibrated at total and partial immersion, respectively, where t_2(°C) is the average temperature of the emergent column, t_1(°C) is the temperature reading recorded from thermometer, t_0(°C) is the column temperature under the thermometer calibration, n is the length of the emergent column expressed in the thermometer scale units, and β(°C^{-1}) is the differential coefficient of thermal expansion of the thermometer liquid relative to the thermometer glass. Properties of the most important thermometric liquids and glasses are given in Tables 8.11 and 8.12. Liquid-in-glass thermometry is discussed in Refs. [6, 10, 18, 23–30].

8.4 Resistance Thermometers

The principle of resistance thermometer (RT) action is based upon the temperature variation of electrical resistance R of metals, alloys and semiconductors. To obtain a temperature starting from the measured value of electrical resistance, empirical relations or tables are used. Resistance thermometers are used from about 0.01 K to 1100°C. The thermometers for precise measurements (at about 0.001 K level of uncertainty) are calibrated individually. Standard platinum RTs are used for interpolating between defining fixed points of the ITS-90. Standard platinum, germanium and rhodium-iron thermometers are used as transfer standards for maintaining the ITS-90 [18].

The industrial platinum resistance thermometers are widely used for temperature measurement and control in the range from −200°C to 750°C (rarely from −260°C to 1100°C) owing to very good thermometric properties of platinum. For industrial platinum RTs the standard calibrations have been developed (Tables 8.15, 8.16). In the range from −200°C to 200°C copper and nickel resistance thermometers are used rarely.

Semiconducting RTs (germanium, carbon, carbon-glass, gallium-arsenide) are used commonly below 0°C. At low temperatures their sensitivity is greater than that for metallic RTs. This is the main advantage of semiconducting RTs. Semiconducting thermometers have the complex $R(T)$ characteristics with poor specimen-to-specimen repeatability. Because of this, a standard description of the $R(T)$ is not possible.

Silicon, gallium-arsenide and germanium diodes have been developed as temperature sensors in cryogenic thermometry. Thermometric property is the junction voltage of a diode forward-biased at a constant current.

Different types of RTs and diode thermometers are discussed in Refs. [10, 11, 18, 28–36].

8.5 Thermoelectric Thermometry

Thermoelectric thermometry is based upon temperature dependence of the thermoelectric voltage E generated in a thermocouple formed by two dissimilar electrical conductors (metals or semiconductors) joined at one end to form a measuring junction. The measuring junction must be at the temperature to be determined, the reference is at fixed temperature, usually the ice point. The thermocouples are used in the 1 K to 3000 K range.

The uncertainties of thermocouple calibration are usually about a few kelvins but can be about 0.01 K. The accuracy of thermocouple (differential sensor) depends upon how accurately the reference junction temperature can be realized and determined.

Temperature is defined in terms of the thermoelectric voltage E using empirical equations or tables. The relations $E(T)$ presented here are the standard ones for calibration of individual thermocouples. The calibration tables of standard thermocouples correspond to the real sensors within known tolerances. The correction function of the polynomial type is found by measuring the deviation of E from tabulated value at a small number of fixed points. If the desired accuracy is higher than the tolerances of the standard reference tables, the thermocouple should be calibrated individually.

For most of thermocouples the $E(T)$ in working temperature range (for in the part of this range) can be approximated by using a function of the type $E = \sum_i a_i T^i$. The interpolation polynomials for standard thermocouples are given in footnotes to the reference tables.

Base-metal thermocouples are identified by letter designations accepted internationally.

Useful accounts of thermoelectric thermometry are given in Refs. [10, 11, 28, 29]. An extensive catalogue of thermocouples and discussion of their characteristics are given in Refs. [37–41]. Thermocouples for use at high temperatures are reviewed in Refs. [42–45].

8.6 Vapor-Pressure Thermometry

Vapor-pressure thermometry is based upon the temperature dependence of the vapor pressure of a liquid. Thermometric substances are usually liquefied gases: helium, hydrogen, neon, argon, oxygen, etc. To derive temperature from measured values of the vapor pressure, tables or empirical formulas are used. The low-temperature limit of vapor-pressure thermometers is determined by freezing point of thermometric liquid, and the upper limit – by the critical point. Temperature can be measured within uncertainty of 0.001 K using precise thermometers.

In measurements of vapor pressure with liquid-column pressure gauge a knowledge of the local value of the acceleration due to gravity g and the density of the liquid is required. Accurate local values of g may be obtained by using the Réseau Gravimétrique International Unifié 1971 (IGSN-71) de L'Union Géodésique et Géophysique Internationale.

The mean density of pure mercury at t_{90} in a barometric column supported by the pressure p being measured is given with sufficient accuracy, over the temperature range from 0°C to 40°C and for pressures relevant to these measurements, by the relation

$$\rho\left(t_{90}, \frac{p}{2}\right) = \frac{\rho(20°\mathrm{C}, p_0)}{[1 + A(t_{90} - 20°\mathrm{C}) + B(t_{90} - 20°\mathrm{C})^2] \times [1 - \chi(p/2 - p_0)]}, \tag{8.15}$$

where $A = 1.8120 \cdot 10^{-4}\ °\mathrm{C}^{-1}$, $B = 8 \cdot 10^{-9}\ °\mathrm{C}^{-2}$, $\chi = 4 \cdot 10^{-11}\mathrm{Pa}^{-1}$, $p_0 = 101325$ Pa and $\rho(20°\mathrm{C}, p_0) = 13545.854\ \mathrm{kg} \cdot \mathrm{m}^{-3}$ [50].

An account of vapor-pressure thermometry is given in Refs. [18, 34, 51]. The important for low-temperature thermometry technique of helium vapor pressure measurement is discussed in Refs. [35, 52, 53].

8.7 Optical Pyrometry (Radiation Thermometry)

Temperature measurements made with optical pyrometers are based upon the radiation laws of a blackbody. The radiation of real body differs from that of a blackbody. Because of this the measured temperature is not equal to the real temperature. The brightness (spectral), color (spectral ratio) and radiation temperatures are introduced in pyrometry.

Brightness (spectral) pyrometry is based upon the measurement of radiation intensity (brightness) of a body at fixed wavelength. If at the wavelength λ the radiation intensities of the body in question and of blackbody are the same, then the temperature of blackbody is equal to the brightness temperature T_b of the body. The relation of the T_b to the thermodynamic temperature T is given by equation

$$\frac{1}{T} = \frac{1}{T_b} + \frac{\lambda \ln[\epsilon(\lambda, T)]}{c_2}, \tag{8.16}$$

where $c_2 = 0.014388\ \mathrm{m \cdot K}$ is the second radiation constant in the Planck law, $\epsilon(\lambda, T)$ is the spectral coefficient of thermal radiation of the body at wavelength λ/m and temperature T/K. The equation (8.16) is valid under the Wien law approximation. In the case of precise measurements one must take into account the fact that Eq. (8.16) gives the underestimated values of T at temperatures above 3000 K.

Color (spectral ratio) pyrometry is based upon the comparison of the ratios of spectral concentrations of the radiance at two wavelengths λ_1 and λ_2 for a blackbody and for the body in question. If these ratios are the same then the color temperature T_c is equal to the temperature of blackbody. The relation between T_c and thermodynamic temperature is

$$\frac{1}{T} = \frac{1}{T_c} + \frac{\ln[\epsilon(\lambda_1, T)/\epsilon(\lambda_2, T)]}{c_2(1/\lambda_1 - 1/\lambda_2)}. \tag{8.17}$$

Color temperature of so called gray-body, for which $\epsilon(\lambda_1, T) = \epsilon(\lambda_2, T)$, is equal to the thermodynamic temperature. The spectral ratio technique is insensitive to a medium with transmissivity satisfying the equality $\tau(\lambda_1) = \tau(\lambda_2)$, which commonly present between a target and pyrometer (e.g., dust, vapors, windows, etc.).

Radiation (total radiation) pyrometry is based upon the measurement of the total radiation energy of a body (over a whole spectral interval). If the total radiant excitances of the body and a blackbody are the same then the radiation temperature T_r of body in question is equal to the temperature of a blackbody. The relation between T_r and thermodynamic temperature is

$$T = T_r[\epsilon_t(T)]^{-1/4}, \tag{8.18}$$

where $\epsilon_t(T)$ is the radiant total emissivity of a body. The total radiation pyrometers are most effective for measuring low temperatures, where the spectral emissivity is small.

Emissivity of metals and alloys strongly depends on the surface conditions (roughness, oxidation, etc.). In Tables 8.41–8.43 the data for clean, polished surfaces are listed. The total emissivity data for some oxides are presented in Table 8.44 (see also Ch.31 of this handbook).

An account of optical pyrometry is given in Refs. [10, 28, 29, 56–59].

Table 8.1 Defining fixed points of the ITS-90 [14]

Number	Temperature		Substance[*1]	State[*2]	$W_r(T_{90})$
	T_{90}, K	t_{90}, °C			
1	3 to 5	−270.15 to −268.15	He	V	
2	13.8033	−259.3467	e-H_2	T	0.001 190 07
3	≈ 17	≈ −256.15	e-H_2 (or He)	V (or G)	
4	≈ 20.3	≈ −252.85	e-H_2 (or He)	V (or G)	
5	24.5561	−248.5939	Ne	T	0.008 449 74
6	54.3584	−218.7916	O_2	T	0.091 718 04
7	83.8058	−189.3442	Ar	T	0.215 859 75
8	234.3156	−38.8344	Hg	T	0.844 142 11
9	273.16	0.01	H_2O	T	1.000 000 00
10	302.9146	29.7646	Ga	M	1.118 138 89
11	429.7485	156.5985	In	F	1.609 801 85
12	505.078	231.928	Sn	F	1.892 797 68
13	696.677	419.527	Zn	F	2.568 917 30
14	933.473	660.323	Al	F	3.376 008 60
15	1234.93	961.78	Ag	F	4.286 420 53
16	1337.33	1064.18	Au	F	
17	1357.77	1084.62	Cu	F	

[*1] All substances except ^{3}He are of natural isotopic composition, e-H_2 is hydrogen at the equilibrium concentration of the *ortho-* and *para-*molecular forms.

[*2] For complete definitions and advice on the realization of these various states, see "Supplementary Information for the ITS-90". The symbols have the following meanings: V: vapor pressure point; T: triple point (temperature at which the solid, liquid and vapor phases are in equilibrium); G: gas thermometer point; M, F: melting point, freezing point (temperature, at a pressure of 101 325 Pa, at which the solid and liquid phases are in equilibrium).

Table 8.2 Effect of pressure on the temperature of some defining fixed points [14] [The reference pressure for melting and freezing points is the standard atmosphere (p_0 = 101325 Pa). For triple points (T) the pressure effect is a consequence only of the hydrostatic head of liquid in the cell.]

Substance	Assigned value of equilibrium temperature T_{90}, K	Temperature variation with pressure p^{*1}, dT/dp 10^{-8}K·Pa^{-1}	Temperature variation with depth l^{*2}, dT/dl 10^{-3}K·m^{-1}	Substance	Assigned value of equilibrium temperature T_{90}, K	Temperature variation with pressure p^{*1}, dT/dp 10^{-8}K·Pa^{-1}	Temperature variation with depth l^{*2}, dT/dl 10^{-3}K·m^{-1}
e-Hydrogen (T)	13.8033	34	0.25	Indium	429.7485	4.9	3.3
Neon (T)	24.5561	16	1.9	Tin	505.078	3.3	2.2
Oxygen (T)	54.3584	12	1.5	Zinc	692.677	4.3	2.7
Argon (T)	83.8058	25	3.3	Aluminum	933.473	7.0	1.6
Mercury (T)	234.3156	5.4	7.1	Silver	1234.93	6.0	5.4
Water (T)	273.16	−7.5	−0.73	Gold	1337.33	6.1	10
Gallium	302.9146	−2.0	−1.2	Copper	1357.77	3.3	2.6

*1 Equivalent to millikelvins per standard atmosphere.
*2 Equivalent to millikelvins per metre of liquid.

Table 8.3 Constants for the helium vapor pressure Eqs.(4), and the temperature range for which each equation, identified by its set of constants, is valid [14]

	^{3}He 0.65 K to 3.2 K	^{4}He 1.25 K to 2.176 K	^{4}He 2.176 K to 5.0 K		^{3}He 0.65 K to 3.2 K	^{4}He 1.25 K to 2.176 K	^{4}He 2.176 K to 5.0 K
A_0	1.053 447	1.392 408	3.146 631	A_6	0.006 596	-0.017 976	-0.004 325
A_1	0.980 106	0.527 153	1.357 655	A_7	0.088 966	0.005 409	-0.004 973
A_2	0.676 380	0.166 756	0.413 923	A_8	-0.004 770	0.013 259	0
A_3	0.372 692	0.050 988	0.091 159	A_9	-0.054 943	0	0
A_4	0.151 656	0.026 514	0.016 349	B	7.3	5.6	10.3
A_5	-0.002 263	0.001 975	0.001 826	C	4.3	2.9	1.9

Table 8.4 The constants A_0, A_i; B_0, B_i; C_0, C_i; D_0 and D_i in the reference functions of Eqs. (10a), (10b), (11a), and (11b) respectively [14]

A_0	-2.135 347 29	B_0	-0.183 324 722	C_0	2.781 572 54	D_0	439.932 854
A_1	3.183 247 20	B_1	0.240 975 303	C_1	1.646 509 16	D_1	472.418 020
A_2	-1.801 435 97	B_2	0.209 108 771	C_2	-0.137 143 90	D_2	37.684 494
A_3	0.717 272 04	B_3	0.190 439 972	C_3	-0.006 497 67	D_3	7.472 018
A_4	0.503 440 27	B_4	0.142 648 498	C_4	-0.002 344 44	D_4	2.920 828
A_5	-0.618 993 95	B_5	0.077 993 465	C_5	0.005 118 68	D_5	0.005 184
A_6	-0.053 323 22	B_6	0.012 475 611	C_6	0.001 879 82	D_6	-0.963 864
A_7	0.280 213 62	B_7	-0.032 267 127	C_7	-0.002 044 72	D_7	-0.188 732
A_8	0.107 152 24	B_8	-0.075 291 522	C_8	-0.000 461 22	D_8	0.191 203
A_9	-0.293 028 65	B_9	-0.056 470 670	C_9	0.000 456 24	D_9	0.049 025
A_{10}	0.044 598 72	B_{10}	0.076 201 285				
A_{11}	0.118 686 32	B_{11}	0.123 893 204				
A_{12}	-0.052 481 34	B_{12}	-0.029 201 193				
		B_{13}	-0.091 173 542				
		B_{14}	0.001 317 696				
		B_{15}	0.026 025 526				

Table 8.5 Deviation functions and calibration points for platinum resistance thermometers in the various ranges in which they define T_{90} [14]

Ranges with an upper limit of 273.16 K			
Section	Lower temperature limit T, K	Deviation functions $W(T_{90}) - W_r(T_{90})$	Calibration points (see Table 1)
1	13.8033	$a[W(T_{90})-1]+b[W(T_{90})-1]^2+\sum_{i=1}^{5} c_i[\ln W(T_{90})]^{i+n}, n=2$	2–9
1.1	24.5561	As for 1 with $c_4 = c_5 = n = 0$	2, 5–9
1.2	54.3584	As for 1 with $c_2 = c_3 = c_4 = c_5 = 0, n = 1$	6–9
1.3	83.8058	$a[W(T_{90})-1]+ b[W(T_{90})-1]\ln W(T_{90})$	7–9
Ranges with a lower limit of 0°C			
Section	Upper temperature limit t,°C	Deviation functions $W(T_{90}) - W_r(T_{90})$	Calibration points (see Table 1)
2	961.78	$a[W(T_{90})-1]+b[W(T_{90})-1]^2+c[W(T_{90})-1]^3+d[W(T_{90}) -W(660.323°\text{C})]^2$	9, 12–15
2.1	660.323	As for 2 with $d = 0$	9, 12–14
2.2	419.527	As for 2 with $c = d = 0$	9, 12, 13
2.3	231.928	As for 2 with $c = d = 0$	9, 11, 12
2.4	156.5985	As for 2 with $b = c = d = 0$	9, 11
2.5	29.7646	As for 2 with $b = c = d = 0$	9, 10
Range from 234.3156 K (−38.8344°C) to 29.7646°C			
3		As for 2 with $c = d = 0$	8–10

*1 Calibration points 9, 12–14 are used with $d = 0$ for $t_{90} \leq 660.323$°C; the values of a, b and c thus obtained are retained for $t_{90} > 660.323$°C, with d being determined from calibration point 15.

Table 8.6 Differences between ITS-90 and EPT-76, and between ITS-90 and IPTS-68 for specified values of T_{90} and t_{90} [14]

$(T_{90} - T_{76})$, mK										
T_{90}, K	0	1	2	3	4	5	6	7	8	9
0						−0.1	−0.2	−0.3	−0.4	−0.5
10	−0.6	−0.7	−0.8	−1.0	−1.1	−1.3	−1.4	−1.6	−1.8	−2.0
20	−2.2	−2.5	−2.7	−3.0	−3.2	−3.5	−3.8	−4.1		
$(T_{90} - T_{68})$, K										
T_{90}, K	0	1	2	3	4	5	6	7	8	9
10					−0.006	−0.003	−0.004	−0.006	−0.008	−0.009
20	−0.009	−0.008	−0.007	−0.007	−0.006	−0.005	−0.004	−0.004	−0.005	−0.006
30	−0.006	−0.007	−0.008	−0.008	−0.008	−0.007	−0.007	−0.007	−0.006	−0.006
40	−0.006	−0.006	−0.006	−0.006	−0.006	−0.007	−0.007	−0.007	−0.006	−0.006
50	−0.006	−0.005	−0.005	−0.004	−0.003	−0.002	−0.001	0.000	0.001	0.002
60	0.003	0.003	0.004	0.004	0.005	0.005	0.006	0.006	0.007	0.007
70	0.007	0.007	0.007	0.007	0.007	0.008	0.008	0.008	0.008	0.008
80	0.008	0.008	0.008	0.008	0.008	0.008	0.008	0.008	0.008	0.008
90	0.008	0.008	0.008	0.008	0.008	0.008	0.008	0.009	0.009	0.009

Table 8.6 Differences between ITS-90 and EPT-76, and between ITS-90 and IPTS-68 *(continued)*

T_{90}, K	0	10	20	30	40	50	60	70	80	90
100	0.009	0.011	0.013	0.014	0.014	0.014	0.014	0.013	0.012	0.012
200	0.011	0.010	0.009	0.008	0.007	0.005	0.003	0.001		
$(t_{90} - t_{68})$/°C										
t_{90},°C	0	−10	−20	−30	−40	−50	−60	−70	−80	−90
−100	0.013	0.013	0.014	0.014	0.014	0.013	0.012	0.010	0.008	0.008
0	0.000	0.002	0.004	0.006	0.008	0.009	0.010	0.011	0.012	0.012
t_{90},°C	0	10	20	30	40	50	60	70	80	90
0	0.000	−0.002	−0.005	−0.007	−0.010	−0.013	−0.016	−0.018	−0.021	−0.024
100	−0.026	−0.028	−0.030	−0.032	−0.034	−0.036	−0.037	−0.038	−0.039	−0.039
200	−0.040	−0.040	−0.040	−0.040	−0.040	−0.040	−0.040	−0.039	−0.039	−0.039
300	−0.039	−0.039	−0.039	−0.040	−0.040	−0.041	−0.042	−0.043	−0.045	−0.046
400	−0.048	−0.051	−0.053	−0.056	−0.059	−0.062	−0.065	−0.068	−0.072	−0.075
500	−0.079	−0.083	−0.087	−0.090	−0.094	−0.098	−0.101	−0.105	−0.108	−0.112
600	−0.115	−0.118	−0.122	−0.125*1	−0.08	−0.03	0.02	0.06	0.11	0.16
700	0.20	0.24	0.28	0.31	0.33	0.35	0.36	0.36	0.36	0.35
800	0.34	0.32	0.29	0.25	0.22	0.18	0.14	0.10	0.06	0.03
900	−0.01	−0.03	−0.06	−0.08	−0.10	−0.12	−0.14	−0.16	−0.17	−0.18
1000	−0.19	−0.20	−0.21	−0.22	−0.23	−0.24	−0.25	−0.25	−0.26	−0.26
t_{90},°C	0	100	200	300	400	500	600	700	800	900
1000		−0.26	−0.30	−0.35	−0.39	−0.44	−0.49	−0.54	−0.60	−0.66
2000	−0.72	−0.79	−0.85	−0.93	−1.00	−1.07	−1.15	−1.24	−1.32	−1.41
3000	−1.50	−1.59	−1.69	−1.78	−1.89	−1.99	−2.10	−2.21	−2.32	−2.43

*1 A discontinuity in the first derivative of $(t_{90} - t_{68})$ occurs at a temperature of $t_{90} = 630.6$°C, at which $(t_{90} - t_{68}) = -0.125$°C.

Equations giving the differences $T_{90} - T_{68}$ and $t_{90} - t_{68}$ shown in Table 8.6 [15] are as follows: from 13.8 K to 83.8 K (accuracy is approximately ±1 mK)

$$(T_{90} - T_{68})(\mathrm{K}) = a_0 + \sum_{i=1}^{12} a_i \left[(T_{90} - 40\ \mathrm{K})/40\ \mathrm{K}\right]^i; \tag{8.19}$$

from 83.8 K to 630.6 °C (accuracy is approximately ±1.5 mK up to 0°C, and ±1 mK above 0°C)

$$(t_{90} - t_{68})(°\mathrm{C}) = \sum_{i=1}^{8} b_i (t_{90}/630°\mathrm{C})^i; \tag{8.20}$$

from 630.6 °C to 1064.18 °C (accuracy is approximately ±10 mK)

$$(t_{90} - t_{68})(°\mathrm{C}) = c_0 + \sum_{i=1}^{7} c_i \left[(t_{90} - 900°\mathrm{C})/300°\mathrm{C}\right]^i; \tag{8.21}$$

above 1064.18 °C

$$(T_{90} - T_{68})/\mathrm{K} = \Delta T(\mathrm{Au}) \cdot \left[\frac{T_{90}}{T(\mathrm{Au}}\right]^2 \cdot \frac{L_\lambda(T(\mathrm{Au}))}{L_\lambda(T_{90})} \cdot \frac{\exp[c_2/\lambda(T(\mathrm{Au}))]}{\exp[c_2/\lambda T_{90}]}, \quad (8.22)$$

where $L_\lambda(T_{90})$ and $L_\lambda(T(\mathrm{Au}))$ are given in Eq.(8.12), $T(\mathrm{Au}) = 1337.33$ K and $\Delta T(\mathrm{Au}) = -0.25$ K. The differences given by this equation are wavelength-dependent. The values given in Table 8.6 are those for the domain in which the Wien equation is a close approximation to the Planck equation.

The coefficients a_i, b_i and c_i are presented in Table 8.7.

Table 8.7

i	a_i	b_i	c_i	i	a_i	b_i	c_i
0	-0.005903	0	-0.00317	7	1.411912	7.438081	5.04151
1	0.008174	-0.148759	-0.97737	8	25.277595	-3.536296	0
2	-0.061924	-0.267408	1.25590	9	-19.183815	0	0
3	-0.193388	1.080760	2.03295	10	-18.437089	0	0
4	1.490793	1.269056	-5.91887	11	27.000895	0	0
5	1.252347	-4.089591	-3.23561	12	-8.716324	0	0
6	-9.835868	-1.871251	7.23364				

Table 8.8 Estimates of the $(1\,\sigma)$ uncertainty in the values of thermodynamic temperature ΔT_1, and in the current (1990) best practical realization ΔT_2 of the defining fixed points of the ITS-90 [15]

Fixed points	T_{90}, K	ΔT_1, mK	ΔT_2, mK	Fixed points	T_{90}, K	ΔT_1, mK	ΔT_2, mK
^{4}He vp*1	4.2221	0.3	0.1		t_{90},°C		
H_2 tp	13.8033	0.5	0.1	Ga mp	29.7646	1	0.05
H_2 vp	≈ 17	0.5	0.2	In fp	156.5985	3	0.1
H_2 vp	≈ 20.3	0.5	0.2	Sn fp	231.928	5	0.1
Ne tp	24.5561	0.5	0.2	Zn fp	419.5985	13	0.1
O_2 tp	54.3584	1.0	0.1	Al fp	660.323	25	0.3
Ar tp	83.8058	1.5	0.1	Ag fp	961.78	40	1*2, 10*3
Hg tp	234.3156	1.5	0.05	Au fp	1064.18	50	10*3
H_2O tp	273.16	0.0	0.02	Cu fp	1084.62	60	15*3

*1 ^{4}He vp: boiling point at a pressure of 101 325 Pa of ^{4}He (not a defining fixed point of the ITS-90).

*2 for platinum resistance thermometry.

*3 for radiation thermometry.

Table 8.9 Recommended values of temperature for a selected set of secondary reference points [21]

Equilibrium state*1	T_{68}, K	t_{68},°C	Uncertainty*2, K	Purity of material*3 wt. % or vol. %
Part A: First-quality points				
Equilibrium hydrogen*4 B	13.81	−259.34		
Equilibrium hydrogen*5 LV	13.81	−259.34	0.001	99.99
$p = p_0 \exp\left[\sum_{i=-1}^{1} a_i T_{68}^i\right] + \sum_{i=0}^{5} b_i T_{68}^i$,				
$a_{-1} = -101.80328$ K, $a_0 = 4.0306106$,				
$a_1 = 4.8778611 \cdot 10^{-2}$ K^{-1},	to	to		
$b_0 = 984.37489$, $b_1 = -293.52587$ K^{-1},				

Table 8.9 Recommended values of temperature *(continued)*

Equilibrium state*1	T_{68}, K	t_{68}, °C	Uncertainty*2, K	Purity of material*3 wt. % or vol. %
Part A: First-quality points *(continued)*				
$b_2 = 34.778619$ K^{-2}, $b_3 = -2.0464883$ K^{-3},				
$b_4 = 5.9799811 \cdot 10^{-2}$ K^{-4},				
$b_5 = 6.9416492 \cdot 10^{-4}$ K^{-5}	20.28	−252.87		
Neon isotope ^{20}Ne T	24.546	−248.604	0.001	99.95*4
Neon*4 B	27.102	−246.048		
Neon LV	24.562	−248.588	0.002	99.99
$\ln(p/p_0) = \sum_{i=-1}^{2} a_i T_{68}^i$,				
$a_{-1} = -252.5938$ K, $a_0 = 11.53431$,	to	to		
$a_1 = -1.216916 \cdot 10^{-1}$ K^{-1},				
$a_2 = 1.475653 \cdot 10^{-3}$ K^{-2}	27.102	−246.048		
Nitrogen T	63.1462	−210.0038	0.0007	99.999
Nitrogen B	77.344	−195.806	0.002	99.999
Nitrogen LV	63.146	−210.004	0.002	99.999
$\ln(p/p_0) = \sum_{i=-1}^{2} a_i T_{68}^i$,				
$a_{-1} = -874.467650$ K, $a_0 = 14.974170$,	to	to		
$a_1 = -6.655806 \cdot 10^{-2}$ K^{-1},				
$a_2 = 2.473924 \cdot 10^{-4}$ K^{-2}	77.344	−195.806		
Argon B	87.2946	−185.8554	0.0003	99.999
Argon LV	83.798	−189.352	0.001	99.999
$\ln(p/p_0) = \sum_{i=-1}^{1} a_i T_{68}^i$,				
$a_{-1} = -864.70028$ K, $a_0 = 10.593120$,	to	to		
$a_1 = -7.87611 \cdot 10^{-3}$ K^{-1}	87.294	−185.856		
Oxygen*4 B	90.199	−182.962		
Oxygen LV	65	−208.15	0.001	99.999
$\ln(p/p_0) = \sum_{i=-2}^{3} a_i T_{68}^i$,				
$a_{-2} = 9846.8897$ K^2, $a_{-1} = -1724.5240$ K,				
$a_0 = 33.935826$,	to	to		
$a_1 = -3.1312576 \cdot 10^{-1}$ K^{-1},				
$a_2 = 1.9275091 \cdot 10^{-3}$ K^{-2},				
$a_3 = -4.7205984 \cdot 10^{-6}$ K^{-3}	90.188	−182.962		
Methane (CH_4) T	90.6854	−182.4646	0.0003	99.995
Mercury F	234.3137	−38.8363	0.0005	99.9999
ICE POINT	273.15	0		
Gallium T	302.9240	29.7740	0.0002	99.99999
Water*4 B	373.15	100		
Indium T	429.7795	156.6295	0.0004	99.9999
Bismuth F	544.592	271.442	0.001	99.9999
Cadmium F	594.2582	321.1082	0.0003	99.9999
Lead F	600.652	327.502	0.001	99.9999
Antimony F	903.905	630.755	0.003	99.9999
Copper/71.9% silver autectic M	1052.70	779.55	0.07	99.999
Copper/71.9% silver autectic M	1053.03	779.88	0.05	99.999
Palladium F	1828.5	1555.3	0.2	99.999

Table 8.9 Recommended values of temperature *(continued)*

Equilibrium state*1	T_{68}, K	t_{68},°C	Uncertainty*2, K	Purity of material*3 wt. % or vol. %
Part A: First-quality points *(continued)*				
Platinum F	2041.9	1768.7	0.4	99.99
Rhodium F	2236	1963	5	99.99
Iridium F	2720	2447	6	99.99
Molybdenum F	2897	2624	5	99.9
Tungsten F	3693	3420	10	99.9
Part B: Second-quality points				
Normal hydrogen T	13.958	−259.192	0.002	99.99
Normal hydrogen B	20.397	−252.753	0.002	99.99
Oxygen $\alpha-\beta$ transition of solid phase	23.873	−249.277	0.005	99.999
Neon SV	20	−253.15	0.002	99.99
$\ln(p/p_0) = \sum_{i=-1}^{2} a_i T_{68}^i$,				
$a_{-1} = -266.9172$ K, $a_0 = 11.03809$,	to	to		
$a_1 = -7.875422 \cdot 10^{-2}$ K^{-1},				
$a_2 = 1.516646 \cdot 10^{-3}$ K^{-2}	24.562	−248.588		
Nitrogen $\alpha-\beta$ transition of solid phase	35.621	−237.529	0.006	99.999
Oxygen $\beta-\gamma$ transition of solid phase	43.8021	−229.3479	0.0006	99.999
Nitrogen SV	56	−217.15	0.002	99.999
$\ln(p/p_0) = \sum_{i=-1}^{1} a_i T_{68}^i$,				
$a_{-1} = -861.621597$ K, $a_0 = 12.189891$,	to	to		
$a_1 = -1.006522 \cdot 10^{-2}$ K^{-1}	63.146	−210.004		
Argon SV	81	−192.15	0.001	99.999
$\ln(p/p_0) = \sum_{i=-1}^{0} a_i T_{68}^i$,	to	to		
$a_{-1} = -955.992$ K, $a_0 = 11.02251$	83.788*6	−189.362*6		
Krypton T	115.764	−157.386	0.001	99.995
Xenon T	161.3897	−111.7603	0.0006	99.995
Carbon dioxide S	194.673	−78.477	0.003	99.99
Carbon dioxide T	216.580	−56.570	0.001	99.99
Bromobenzene T	242.417	−30.733	0.010	99.998
Phenoxybenzene (diphenyl ether) T	300.021	26.871	0.002	99.9999
Succinonitrile T	331.230	58.080	0.005	99.9995
Sodium F	370.969	97.819	0.005	99.99
Benzoic acid T	395.520	122.370	0.004	99.998
Benzoic acid F	395.532	122.382	0.004	99.998
Mercury B	629.811	356.661	0.004	99.9999
Mercury LV	622.15	349	0.010	99.9999
$t_{68} = \sum_{i=0}^{3} a_i (p/p_0 - 1)^i$,				
$a_0 = 356.657$, $a_1 = 55.552$,	to	to		
$a_2 = -23.03$, $a_3 = 14.0$	636.15	363		
Sulphur B	717.824	444.674	0.002	99.995
Sulphur LV	622.15	349	0.010	99.9999
$t_{68} = \sum_{i=0}^{3} a_i (p/p_0 - 1)^i$,				

Table 8.9 Recommended values of temperature *(continued)*

Equilibrium state[*1]	T_{68}, K	t_{68},°C	Uncer-tainty[*2], K	Purity of material[*3] wt. % or vol. %
Part B: Second-quality points *(continued)*				
$a_0 = 444.674, a_1 = 69.010,$	to	to		
$a_2 = -27.48, a_3 = 19.14$	726.15	453		
Copper/66.9% aluminum eutectic M	821.406	548.256	0.010	99.9999
Iron RT (653 nm)	1670	1397	6	99.95
Palladium RT (653 nm)	1688	1415	5	99.99
Titanium RT (997 nm)	1711	1438	6	99.9
Nickel F	1728	1455	3	99.9
Cobalt F	1768	1495	3	99.9
Titanium RT (653 nm)	1800	1527	3	99.9
Iron F	1811	1538	3	99.99
Vanadium RT (993 nm)	1875	1602	7	99.9
Zirconium RT (650 nm)	1940	1667	8	99.98
Titanium M	1943	1670	3	99.9
Vanadium RT (653 nm)	1990	1717	5	99.9
Zirconium M	2128	1855	8	99.98
Ruthenium RT (650 nm)	2294	2021	8	99.98
Aluminium oxide (Al_2O_3) M	2327	2054	6	99.9
Molybdenium RT (995 nm)	2331	2058	8	99.95
Niobium RT (650 nm)	2429	2156	5	99.9
Molybdenium RT (653 nm)	2529	2256	6	99.95
Ruthenium M	2607	2334	10	99.98
Tantalum RT (995 nm)	2620	2347	8	99.99
Yttria (Y_2O_3) M	2712	2439	12	99.999
Niobium M	2746	2473	7	99.9
Tantalum RT (653 nm)	2846	2573	8	99.99
Tungsten RT (653 nm)	3208	2935	10	99.95

[*1] The equilibrium states in this table are at a pressure p_0 = 101325 Pa (one standard atmosphere), except for the triple points, the radiance temperatures, and those cases where a range of pressure is explicitly allowed.

[*2] The uncertainties listed are (where possible) the standard deviation of the consensus values.

[*3] The minimum purity of the material to which the listed values of temperature and uncertainty apply is given in percent by volume when the material is liquid or gaseous at 0°C and one standard atmosphere and in percent by weight when the material is solid at 0°C and one standard atmosphere. (There are a few exceptions: for example, mercury).

[*4] This point is a defining fixed point of the IPTS-68 (Amended Edition of 1975) [14].

[*5] The term $\sum_{i=0}^{5} b_i T^i$ in the equation adds to the value of p a pressure amounting to the equivalent of up to 1 mK.

[*6] This equation does not apply within the melting range of argon, so the maximum temperature of application is (arbitrary) set 10 mK below the triple point.

Normal hydrogen is hydrogen at concentration of the *ortho-* and *para-*molecular forms at room temperature (about 75% *ortho-* and 25% *para-*molecular forms).

The symbols have the following meanings: LV: equilibrium between the liquid and vapor phases; SV: equilibrium between the solid and vapor phases; T: triple point; M: melting point; F: freezing point; B: boiling point; S: sublimation point; RT: radiance temperature. Radiance temperature values are given for the particular wavelength indicated. The radiance temperatures are determined with the specimens in vacuum or inert-gas environment at atmospheric pressure. Pressure corrections are negligible, if compared with measurement uncertainties.

Table 8.10 Superconductive fixed points [22]

Element	$T_{90}^{(c)}$, K	Uncertainty*1, mK	Element	$T_{90}^{(c)}$, K	Uncertainty*1, mK
Cd	0.5200	±3.0	In	3.4145	±2.5
Zn	0.8500	±3.0	Pb	7.1997	±2.5
Al	1.1810	±2.5	Nb	9.2880	±2.5

*1 The uncertainties given include the temperature measurement uncertainties, the non-uniqueness of the EPT-76, the differences of the $T_{76}^{(c)}$ values published for high-quality samples of different origin and different parameters, and the effect on the superconducting transition temperatures of the various factors that can be expected to influence the parameters of the investigated samples (see also [18]).

Table 8.11 Properties of liquids used in liquid-in-glass thermometers [28]

Substance	Formula	$\beta^{*1}, 10^{-6}$ K^{-1}	Freezing point, °C	Boiling point, °C
Acetone	CH_3COCH_3	1310	from −93.9 to −94.9	56.2
Carbon sulphide	CS_2	1220	−113.0	46.2
Complex amalgam		180	−90	from 360 to 2000
Ethanol	C_2H_5OH	1120	from −111.8 to −117.3	from 77.7 to 78.4
Gallium	Ga	110	29.77	2100
Kerosene		1150	from −50 to −15	from 200 to 290
Mercury	Hg	182.5	−38.84	356.66
Methanol	CH_3OH	1200	from −93.8 to −97.8	from 64.2 to 66.0
Pentane	C_6H_{12}	1600	below −200	from 30 to 40
Thallium amalgam	8.5 mass % Tl	182	−60	from 360 to 1460
Toluene	C_7H_8	1140	from −92.4 to −102.0	from 109.2 to 110.6

*1 The coefficient of thermal volume expansion.

Table 8.12 Coefficient of emergent column correction for some liquid-in-glass thermometers [28]

Glass	Liquid	$\beta^{*1}, 10^{-6}$ K^{-1}	Glass	Liquid	$\beta^{*1}, 10^{-6}$ K^{-1}
360	Mercury	160	quartz	Mercury	180
500	Mercury	165	glass at $t > 0$°C	Organic	1300
650	Mercury	170	glass at $t < 0$°C	Organic	800

*1 The coefficient of thermal volume expansion.

Table 8.13 Maximum error (°C) of liquid-in-glass thermometers [28]

Temperature range, °C		Graduation interval, °C							
from	to	0.01	0.02	0.05	0.1; 0.2	0.5	1	2	5; 10
−35	0	±0.05	±0.08	±0.10	±0.3	±1.0	±1	±2	±5
0	100	±0.04	±0.08	±0.10	±0.2	±1.0	±1	±2	±5
100	200		±0.10	±0.25	±0.4	±2.0	±1	±2	±5
200	300			±0.40	±1.0	±2.0	±3	±4	±5
300	400				±1.0	±3.0	±4	±4	±10
400	500					±3.0	±5	±5	±10
500	600						±6	±4	±10
600	650						±6	±6	±15

Table 8.14 Resistance ratios, $W(t_{90}) = R(t_{90})/R(0.01°\text{C})$, for standard platinum resistance thermometer

t_{90},° C	0	2	4	6	8	t_{90},° C	0	2	4	6	8
−260	0.00104					250	1.95970	1.96709	1.97446	1.98183	1.98920
−250	0.00684	0.00494	0.00345	0.00235	0.00157	260	1.99656	2.00392	2.01127	2.01862	2.02597
−240	0.02345	0.01917	0.01536	0.01203	0.00920	270	2.03331	2.04064	2.04797	2.05530	2.06262
−230	0.05126	0.04494	0.03897	0.03338	0.02820	280	2.06993	2.07724	2.08455	2.09185	2.09915
−220	0.08703	0.07942	0.07201	0.06483	0.05790	290	2.10644	2.11373	2.12102	2.12830	2.13557
−210	0.12731	0.11903	0.11084	0.10276	0.09482	300	2.14284	2.15011	2.15737	2.16462	2.17187
−200	0.16975	0.16117	0.15262	0.14412	0.13568	310	2.17912	2.18636	2.19360	2.20083	2.20806
−190	0.21301	0.20433	0.19566	0.18701	0.17837	320	2.21529	2.22250	2.22972	2.23693	2.24413
−180	0.25642	0.24775	0.23907	0.23038	0.22170	330	2.25133	2.25853	2.26572	2.27291	2.28009
−170	0.29967	0.29104	0.28240	0.27375	0.26509	340	2.28727	2.29444	2.30161	2.30877	2.31593
−160	0.34264	0.33407	0.32549	0.31689	0.30829	350	2.32309	2.33024	2.33738	2.34452	2.35166
−150	0.38529	0.37679	0.36827	0.35974	0.35119	360	2.35879	2.36592	2.37304	2.38015	2.38727
−140	0.42765	0.41920	0.41074	0.40227	0.39379	370	2.39438	2.40148	2.40858	2.41567	2.42276
−130	0.46972	0.46133	0.45292	0.44451	0.43608	380	2.42985	2.43693	2.44400	2.45107	2.45814
−120	0.51155	0.50320	0.49485	0.48648	0.47811	390	2.46520	2.47226	2.47931	2.48636	2.49340
−110	0.55315	0.54484	0.53653	0.52821	0.51988	400	2.50044	2.50747	2.51450	2.52153	2.52855
−100	0.59454	0.58628	0.57801	0.56973	0.56144	410	2.53556	2.54257	2.54958	2.55658	2.56358
−90	0.63575	0.62752	0.61929	0.61105	0.60280	420	2.57057	2.57756	2.58454	2.59152	2.59849
−80	0.67679	0.66860	0.66039	0.65219	0.64397	430	2.60546	2.61242	2.61938	2.62634	2.63329
−70	0.71767	0.70951	0.70134	0.69316	0.68498	440	2.64023	2.64717	2.65411	2.66104	2.66797
−60	0.75840	0.75027	0.74213	0.73398	0.72583	450	2.67489	2.68181	2.68872	2.69563	2.70253
−50	0.79899	0.79088	0.78277	0.77465	0.76653	460	2.70943	2.71632	2.72321	2.73009	2.73697
−40	0.83944	0.83136	0.82327	0.81518	0.80709	470	2.74385	2.75072	2.75758	2.76444	2.77130
−30	0.87975	0.87170	0.86364	0.85558	0.84751	480	2.77815	2.78500	2.79184	2.79867	2.80551
−20	0.91995	0.91192	0.90388	0.89585	0.88780	490	2.81233	2.81915	2.82597	2.83279	2.83959
−10	0.96001	0.95201	0.94400	0.93599	0.92797	500	2.84640	2.85320	2.85999	2.86678	2.87356
− 0	0.99996	0.99198	0.98400	0.97601	0.96801	510	2.88034	2.88712	2.89389	2.90065	2.90741
0	0.99996	1.00793	1.01590	1.02387	1.03183	520	2.91417	2.92092	2.92766	2.93441	2.94114
10	1.03978	1.04773	1.05568	1.06362	1.07156	530	2.94787	2.95460	2.96132	2.96804	2.97475
20	1.07949	1.08741	1.09533	1.10325	1.11116	540	2.98146	2.98816	2.99486	3.00155	3.00824
30	1.11907	1.12697	1.13487	1.14276	1.15065	550	3.01492	3.02160	3.02828	3.03494	3.04161
40	1.15853	1.16641	1.17428	1.18215	1.19001	560	3.04827	3.05492	3.06157	3.06822	3.07486
50	1.19787	1.20572	1.21357	1.22142	1.22926	570	3.08149	3.08812	3.09474	3.10136	3.10798
60	1.23709	1.24492	1.25275	1.26057	1.26838	580	3.11459	3.12120	3.12780	3.13439	3.14098
70	1.27619	1.28400	1.29180	1.29959	1.30738	590	3.14757	3.15415	3.16073	3.16730	3.17386
80	1.31517	1.32295	1.33073	1.33850	1.34627	600	3.18042	3.18698	3.19353	3.20008	3.20662
90	1.35403	1.36179	1.36954	1.37729	1.38503	610	3.21316	3.21969	3.22622	3.23274	3.23926
100	1.39277	1.40051	1.40824	1.41596	1.42368	620	3.24577	3.25228	3.25878	3.26527	3.27177
110	1.43139	1.43910	1.44681	1.45451	1.46221	630	3.27825	3.28474	3.29121	3.29769	3.30416
120	1.46990	1.47758	1.48527	1.49294	1.50061	640	3.31062	3.31708	3.32353	3.32998	3.33642
130	1.50828	1.51594	1.52360	1.53126	1.53890	650	3.34286	3.34929	3.35572	3.36214	3.36856
140	1.54655	1.55419	1.56182	1.56945	1.57708	660	3.37497	3.38138	3.38778	3.39418	3.40058
150	1.58470	1.59231	1.59992	1.60753	1.61513	670	3.40697	3.41335	3.41973	3.42610	3.43247
160	1.62272	1.63032	1.63790	1.64549	1.65306	680	3.43883	3.44519	3.45154	3.45789	3.46424
170	1.66064	1.66820	1.67577	1.68333	1.69088	690	3.47058	3.47691	3.48324	3.48956	3.49588
180	1.69843	1.70597	1.71351	1.72105	1.72858	700	3.50219	3.50850	3.51481	3.52111	3.52740
190	1.73610	1.74363	1.75114	1.75865	1.76616	710	3.53369	3.53997	3.54625	3.55253	3.55880
200	1.77366	1.78116	1.78865	1.79614	1.80363	720	3.56506	3.57132	3.57757	3.58382	3.59007
210	1.81110	1.81858	1.82605	1.83351	1.84097	730	3.59631	3.60254	3.60877	3.61499	3.62121
220	1.84843	1.85588	1.86333	1.87077	1.87821	740	3.62743	3.63364	3.63984	3.64604	3.65224
230	1.88564	1.89306	1.90049	1.90791	1.91532	750	3.65843	3.66461	3.67079	3.67697	3.68314
240	1.92273	1.93013	1.93753	1.94493	1.95232	760	3.68930	3.69546	3.70162	3.70777	3.71391

Table 8.14 Resistance ratios for standard platinum resistance thermometer *(continued)*

t_{90},° C	0	2	4	6	8	t_{90},° C	0	2	4	6	8
770	3.72005	3.72619	3.73232	3.73844	3.74456	870	4.02084	4.02673	4.03262	4.03850	4.04438
780	3.75068	3.75679	3.76290	3.76900	3.77509	880	4.05025	4.05612	4.06198	4.06784	4.07370
790	3.78118	3.78727	3.79335	3.79943	3.80550	890	4.07955	4.08539	4.09123	4.09707	4.10290
800	3.81157	3.81763	3.82368	3.82974	3.83578	900	4.10873	4.11455	4.12037	4.12618	4.13199
810	3.84183	3.84786	3.85389	3.85992	3.86594	910	4.13779	4.14359	4.14938	4.15517	4.16095
820	3.87196	3.87798	3.88398	3.88999	3.89599	920	4.16673	4.17251	4.17828	4.18405	4.18981
830	3.90198	3.90797	3.91395	3.91993	3.92590	930	4.19556	4.20132	4.20706	4.21281	4.21854
840	3.93187	3.93784	3.94380	3.94975	3.95570	940	4.22428	4.23001	4.23573	4.24145	4.24717
850	3.96165	3.96759	3.97352	3.97945	3.98538	950	4.25288	4.25858	4.26428	4.26998	4.27567
860	3.99130	3.99722	4.00313	4.00904	4.01494	960	4.28136	4.28705			

Note: Reference functions are given by Eqs.(8.10)–(8.11) and Table 8.4.

Table 8.15 Resistance ratios, $W(t_{68}) = R(t_{68})/R(0°\text{C})$, for industrial platinum resistance thermometers with $R(100°\text{C})/R(0°\text{C}) = 1.3850$ [18]

t_{68},°C	0	5	10	15	20	25	30	35	40	45
−200	0.1849									
−150	0.3971	0.3763	0.3553	0.3343	0.3132	0.2920	0.2708	0.2494	0.2280	0.2065
−100	0.6025	0.5822	0.5619	0.5415	0.5211	0.5006	0.4800	0.4594	0.4387	0.4179
−50	0.8031	0.7832	0.7633	0.7433	0.7233	0.7033	0.6833	0.6631	0.6430	0.6228
−0	1.0000	0.9804	0.9609	0.9412	0.9216	0.9019	0.8822	0.8625	0.8427	0.8229
0	1.0000	1.0195	1.0390	1.0585	1.0779	1.0973	1.1167	1.1361	1.1554	1.1747
50	1.1940	1.2132	1.2324	1.2516	1.2707	1.2898	1.3089	1.3280	1.3470	1.3660
100	1.3850	1.4039	1.4229	1.4417	1.4606	1.4794	1.4982	1.5170	1.5358	1.5545
150	1.5731	1.5918	1.6104	1.6290	1.6476	1.6661	1.6846	1.7031	1.7216	1.7400
200	1.7584	1.7768	1.7951	1.8134	1.8317	1.8499	1.8682	1.8863	1.9045	1.9226
250	1.9407	1.9588	1.9769	1.9949	2.0129	2.0308	2.0488	2.0667	2.0845	2.1024
300	2.1202	2.1380	2.1557	2.1735	2.1912	2.2088	2.2265	2.2441	2.2617	2.2792
350	2.2967	2.3142	2.3317	2.3491	2.3665	2.3839	2.4013	2.4186	2.4359	2.4531
400	2.4704	2.4876	2.5048	2.5219	2.5390	2.5561	2.5732	2.5902	2.6072	2.6242
450	2.6411	2.6580	2.6749	2.6918	2.7086	2.7254	2.7422	2.7589	2.7756	2.7923
500	2.8090	2.8256	2.8422	2.8587	2.8753	2.8918	2.9083	2.9247	2.9411	2.9575
550	2.9739	2.9902	3.0065	3.0228	3.0391	3.0553	3.0715	3.0876	3.1038	3.1199
600	3.1359	3.1520	3.1680	3.1840	3.1999	3.2159	3.2318	3.2476	3.2635	3.2793
650	3.2951	3.3108	3.3266	3.3423	3.3579	3.3736	3.3892	3.4047	3.4203	3.4358
700	3.4513	3.4668	3.4822	3.4976	3.5130	3.5283	3.5437	3.5590	3.5742	3.5894
750	3.6047	3.6198	3.6350	3.6501	3.6652	3.6802	3.6953	3.7103	3.7252	3.7402
800	3.7551	3.7700	3.7848	3.7997	3.8144	3.8292	3.8440	3.8587	3.8733	3.8880
850	3.9026									

Interpolation equation for the temperature range from −200°C to 0°C:

$$R(t_{68})/R(0°\text{C}) = 1 + At_{68} + Bt_{68}^2 + C(t_{68} - 100)t_{68}^3, \tag{8.23}$$

from 0°C to 850°C:

$$R(t_{68})/R(0°\text{C}) = 1 + At_{68} + Bt_{68}^2, \tag{8.24}$$

where $A = 3.90802 \cdot 10^{-3}$ °C^{-1}, $B = -5.8020 \cdot 10^{-7}$ °C^{-2}, $C = -4.274 \cdot 10^{-12}$ °C^{-4}.

Table 8.16 Resistance ratios, $W(t_{68}) = R(t_{68})/R(0°C)$, for industrial platinum resistance thermometers with $R(100°C)/R(0°C) = 1.3910$ [18]

t_{68},°C	0	5	10	15	20	25	30	35	40	45
-200	0.1729									
-150	0.3880	0.3668	0.3456	0.3243	0.3029	0.2814	0.2599	0.2382	0.2165	0.1947
-100	0.5964	0.5758	0.5552	0.5345	0.5137	0.4929	0.4721	0.4511	0.4302	0.4091
-50	0.8000	0.7798	0.7596	0.7394	0.7191	0.6987	0.6784	0.6580	0.6375	0.6170
-0	1.0000	0.9801	0.9603	0.9403	0.9204	0.9004	0.8804	0.8604	0.8403	0.8202
0	1.0000	1.0198	1.0396	1.0594	1.0791	1.0989	1.1185	1.1382	1.1578	1.1774
50	1.1970	1.2165	1.2360	1.2555	1.2749	1.2944	1.3137	1.3331	1.3524	1.3717
100	1.3910	1.4102	1.4295	1.4486	1.4678	1.4869	1.5060	1.5251	1.5441	1.5631
150	1.5821	1.6010	1.6200	1.6389	1.6577	1.6765	1.6954	1.7141	1.7329	1.7516
200	1.7703	1.7889	1.8075	1.8261	1.8447	1.8632	1.8818	1.9002	1.9187	1.9371
250	1.9555	1.9739	1.9922	2.0105	2.0288	2.0470	2.0652	2.0834	2.1016	2.1197
300	2.1378	2.1559	2.1739	2.1919	2.2099	2.2278	2.2458	2.2637	2.2815	2.2994
350	2.3172	2.3349	2.3527	2.3704	2.3881	2.4057	2.4234	2.4410	2.4585	2.4761
400	2.4936	2.5111	2.5285	2.5459	2.5633	2.5807	2.5980	2.6153	2.6326	2.6499
450	2.6671	2.6843	2.7014	2.7186	2.7357	2.7527	2.7698	2.7868	2.8038	2.8207
500	2.8376	2.8545	2.8714	2.8882	2.9051	2.9218	2.9386	2.9553	2.9720	2.9886
550	3.0053	3.0219	3.0384	3.0550	3.0715	3.0880	3.1044	3.1209	3.1373	3.1536
600	3.1700	3.1863	3.2026	3.2188	3.2350	3.2512	3.2674	3.2835	3.2996	3.3157
650	3.3317	3.3477	3.3637	3.3797	3.3956	3.4115	3.4274	3.4432	3.4590	3.4748
700	3.4906	3.5063	3.5220	3.5376	3.5533	3.5689	3.5844	3.6000	3.6155	3.6310
750	3.6465	3.6619	3.6773	3.6926	3.7080	3.7233	3.7386	3.7538	3.7691	3.7842
800	3.7994	3.8145	3.8297	3.8447	3.8598	3.8748	3.8898	3.9047	3.9197	3.9346
850	3.9494									

Interpolation equation for the temperature range from −200°C to 0°C:

$$R(t_{68})/R(0°C) = 1 + At_{68} + Bt_{68}^2 + C(t_{68} - 100)t_{68}^3, \tag{8.25}$$

from 0°C to 850°C:

$$R(t_{68})/R(0°C) = 1 + At_{68} + Bt_{68}^2, \tag{8.26}$$

where $A = 3.96868 \cdot 10^{-3}$ °C^{-1}, $B = -5.8677 \cdot 10^{-7}$ °C^{-2}, $C = -4.141 \cdot 10^{-12}$ °C^{-4}.

Table 8.17 Letter-designation for thermocouples [41]

Type	Thermoelement: positive (nominal composition, mass %)	Thermoelement: negative (nominal composition, mass %)
T	Copper	Copper-nickel alloy (55Cu–45Ni: constantan)
J	Iron (99.5Fe)	Copper-nickel alloy (55Cu–45Ni: constantan)
K	Nickel-chromium alloy (90Ni–10Cr: chromel)	Nickel-aluminium alloy (95Ni–2Al–2Mn–1Si: alumel)
E	Nickel-chromium alloy (90Ni–10Cr: chromel)	Copper-nickel alloy (55Cu–45Ni: constantan)
N	Nickel-chromium-silicon alloy (Ni–14.2Cr–1.4Si: nichrosil)	Nickel-silicon alloy (Ni–4.4Si–0.1Mg: nisil)
R	Platinum13%Rhodium	Platinum
S	Platinum10%Rhodium	Platinum
B	Platinum30%Rhodium	Platinum6%Rhodium
	Tungsten5%Rhenium	Tungsten20%Rhenium

Table 8.18 Working characteristics of thermocouples [28, 38, 42]

Thermocouple	Temperature range*1, K	Maximum emf, mV	Accuracy, K	Ambient
Cu–Au + 1.9%Co	2–270	10.2	0.1–1.5	Oxidizing, air
Chromel–Au + 1.9%Co	2–270	13.8	0.1–1.5	same as above
Cu–Au + 0.07%Fe	2–270	1.7	0.05–0.3	same as above
Chromel–Au + 0.07%Fe	2–270	5.2	0.05–0.3	same as above
Type T	10–670 (870)	21 (34.3)	0.1–5	Oxidizing, air up to 470 K
Type J	70–1000 (1500)	41 (70)	0.5–3	Oxidizing up to 1000 K, neutral
Type K	220–1270 (1570)	41 (52)	4–10	Oxidizing, neutral
Type E	220–870 (1070)	49 (66)	1–3	same as above
Type N	20–1570	47.5	< 10	Oxidizing, air
Type S	270–1570 (1870)	13 (16.6)	0.5–6	Oxidizing
Type B	570–1870 (2070)	11 (14)	0.5–7	same as above
W + 5%Re–W + 20%Re	270–2070 (2770)	27 (33.6)	8–30	Vacuum, neutral, weakly-reducing
Nickel–nichrome	250–1500	49	1–10	Oxidizing
Ir+60%Rh–Ir	400–2370	11.6	10–20	Vacuum, oxidizing, weakly-reducing
Ir+50%Rh–Ir+10%Ru	400–2370	17.3	10–20	Vacuum, oxidizing, neutral
W–Mo	1650–2100 (2400)	3.2	10	Vacuum, neutral, reducing

*1 In the brackets, maximum temperature of short-time usage of thermocouple is given.

Table 8.19 Reference table for low-temperature thermocouples [46, 47] [Temperature of reference junction is 0°C; constantan: 59.9% Cu, 40 ± 1% Ni, 0.5 ± 0.1% Mn; chromel: 90–91% Ni, 9–10% Cr.]

	Copper–Constantan		Copper–Au + 1.9 at.%Co		Chromel–Au + 1.9 at.%Co		Copper–Au + 0.07 at.%Fe		Chromel–Au + 0.07 at.%Fe	
T_{68}, K	E, μV	dE/dT, μV/K	E, μV	dE/dT, μV/K	E, μV	dE/dT, μV/K	E, μV	dE/dT, μV/K	E, μV	dE/dT, μV/K
2	6029.25	-0.10	10263.0	-2.81	13813.0	-2.87	1729.9	-10.88	5262.7	-11.05
3	6029.00	-0.41	10259.7	-3.72	13809.5	-3.98	1718.6	-11.68	5251.2	-12.08
4	6028.43	-0.72	10255.5	-4.61	13805.0	-5.08	1706.6	-12.38	5238.6	-12.97
5	6027.56	-1.02	10250.5	-5.49	13799.4	-6.14	1693.9	-12.97	5225.3	-13.73
6	6026.39	-1.32	10244.6	-6.35	13792.7	-7.19	1680.7	-13.48	5211.2	-14.38
7	6024.91	-1.62	10237.8	-7.20	13785.0	-8.21	1667.0	-13.90	5196.5	-14.93
8	6023.15	-1.91	10230.2	-8.03	13776.3	-9.21	1652.9	-14.23	5181.4	-15.39
9	6021.09	-2.20	10221.7	-8.85	13766.6	-10.19	1638.5	-14.48	5165.8	-15.77
10	6018.7	-2.49	10212.5	-9.65	13755.9	-11.14	1624.0	-14.67	5149.8	-16.08
12	6013.2	-3.05	10191.6	-11.22	13731.8	-13.00	1594.4	-14.86	5117.2	-16.52
14	6006.6	-3.60	10167.6	-12.72	13704.0	-14.77	1564.7	-14.85	5083.9	-16.77
16	5998.8	-4.13	10140.7	-14.17	13672.7	-16.47	1535.1	-14.70	5050.2	-16.88
18	5990.0	-4.66	10111.0	-15.56	13638.2	-18.10	1505.9	-14.45	5016.4	-16.90
20	5980.2	-5.17	10078.6	-16.89	13600.4	-19.67	1477.3	-14.14	4982.6	-16.85
22	5969.4	-5.66	10043.5	-18.16	13559.5	-21.17	1449.4	-13.81	4949.0	-16.77
24	5957.6	-6.15	10006.0	-19.38	13515.7	-22.62	1422.1	-13.47	4915.6	-16.67
26	5944.8	-6.62	9966.0	-20.55	13469.1	-24.00	1395.5	-13.16	4882.4	-16.56
28	5931.1	-7.08	9923.8	-21.66	13419.8	-25.34	1369.5	-12.87	4849.3	-16.46
30	5916.5	-7.53	9879.4	-22.72	13367.8	-26.62	1344.0	-12.59	4816.5	-16.37
32	5900.9	-7.97	9833.0	-23.73	13313.3	-27.85	1319.1	-12.31	4783.9	-16.30
34	5884.6	-8.40	9784.5	-24.69	13256.4	-29.04	1294.8	-12.05	4751.3	-16.25
36	5867.3	-8.82	9734.2	-25.61	13197.2	-30.18	1270.9	-11.80	4718.9	-16.22
38	5849.3	-9.23	9682.1	-26.48	13135.7	-31.28	1247.6	-11.55	4686.4	-16.20
40	5830.4	-9.63	9628.3	-27.31	13072.1	-32.34	1224.7	-11.32	4654.0	-16.21
45	5779.8	-10.60	9486.9	-29.21	12904.1	-34.82	1169.5	-10.77	4572.8	-16.28

Table 8.19 Reference table for low-temperature thermocouples [46, 47] *(continued)*

	Copper–Constantan		Copper–Au + 1.9 at.%Co		Chromel–Au + 1.9 at.%Co		Copper–Au + 0.07 at.%Fe		Chromel–Au + 0.07 at.%Fe	
T_{68}, K	E, μV	dE/dT, μV/K	E, μV	dE/dT, μV/K	E, μV	dE/dT, μV/K	E, μV	dE/dT, μV/K	E, μV	dE/dT, μV/K
50	5724.5	–11.50	9336.6	–30.87	12724.3	–37.09	1116.9	–10.28	4491.1	–16.43
55	5664.8	–12.37	9178.6	–32.33	12533.5	–39.17	1066.7	–9.82	4408.5	–16.62
60	5601.0	–13.18	9013.7	–33.61	12332.9	–41.06	1018.6	–9.40	4324.8	–16.86
65	5533.1	–13.96	8842.8	–34.73	12123.2	–42.81	972.62	–9.01	4239.9	–17.08
70	5461.4	–14.71	8666.6	–35.71	11905.1	–44.41	928.37	–8.63	4153.8	–17.34
75	5386.0	–15.43	8485.9	–36.58	11679.3	–45.87	886.12	–8.27	4066.4	–17.60
80	5307.1	–16.12	8301.0	–37.35	11446.5	–47.23	845.60	–7.94	3977.8	–17.84
85	5224.8	–16.79	8112.5	–38.04	11207.2	–48.47	806.70	–7.62	3888.0	–18.06
90	5139.2	–17.44	7920.7	–38.65	10962.0	–49.62	769.33	–7.33	3797.2	–18.27
95	5050.4	–18.08	7726.1	–39.21	10711.2	–50.67	733.39	–7.05	3705.4	–18.46
100	4958.4	–18.71	7528.8	–39.71	10455.4	–51.65	698.82	–6.78	3612.6	–18.65
110	4765.2	–19.93	7127.2	–40.58	9929.9	–53.40	633.48	–6.29	3424.3	–19.01
120	4559.9	–21.13	6717.7	–41.31	9388.2	–54.91	572.79	–5.85	3232.5	–19.35
130	4342.6	–22.33	6301.3	–41.93	8832.4	–56.22	516.31	–5.45	3037.4	–19.67
140	4113.4	–23.51	5879.3	–42.46	8264.3	–57.39	463.64	–5.09	2839.2	–19.97
150	3872.3	–24.70	5452.5	–42.90	7685.1	–58.43	414.41	–4.76	2638.0	–20.26
160	3619.4	–25.88	5021.6	–43.27	7096.1	–59.36	368.30	–4.46	2434.2	–20.52
170	3354.7	–27.05	4587.2	–43.59	6498.2	–60.21	325.64	–4.19	2227.8	–20.75
180	3078.4	–28.21	4150.0	–43.86	5892.2	–60.97	284.37	–3.94	2019.2	–20.97
190	2790.7	–29.34	3710.1	–44.10	5279.0	–61.66	246.10	–3.71	1808.5	–21.16
200	2491.8	–30.43	3268.1	–44.30	4659.3	–62.27	210.06	–3.50	1596.0	–21.34
210	2182.2	–31.48	2824.3	–44.46	4033.9	–62.80	176.12	–3.29	1381.7	–21.52
220	1862.3	–32.50	2379.1	–44.57	3403.5	–63.26	144.17	–3.10	1165.7	–21.68
230	1532.4	–33.47	1933.0	–44.63	2768.9	–63.65	114.10	–2.92	948.13	–21.83
240	1193.1	–34.41	1486.6	–44.64	2130.9	–63.96	85.76	–2.75	729.14	–21.96
250	844.3	–35.35	1040.2	–44.65	1490.0	–64.20	58.93	–2.62	509.07	–22.04
260	486.1	–36.31	593.3	–44.76	847.2	–64.36	33.24	–2.53	288.58	–22.04
270	117.8	–37.36	143.8	–45.24	203.3	–64.40	8.12	–2.51	68.87	–21.88
273	5.2	–37.70	7.7	–45.51	10.1	–64.38			3.37	–21.79

Table 8.20 Reference table for type T thermocouple [Thermal emf in mV; temperature of reference junction is 0°C.]

t_{68},°C	0	5	10	15	20	25	30	35	40	45
–250	–6.181	–6.209	–6.232	–6.248	–6.258					
–200	–5.603	–5.680	–5.753	–5.823	–5.889	–5.950	–6.007	–6.059	–6.105	–6.146
–150	–4.648	–4.758	–4.865	–4.969	–5.069	–5.167	–5.261	–5.351	–5.439	–5.522
–100	–3.378	–3.519	–3.656	–3.791	–3.923	–4.051	–4.177	–4.299	–4.419	–4.535
–50	–1.819	–1.987	–2.152	–2.315	–2.475	–2.633	–2.788	–2.939	–3.089	–3.235
–0	0.000	–0.193	–0.383	–0.571	–0.757	–0.940	–1.121	–1.299	–1.475	–1.648
0	0.000	0.195	0.391	0.589	0.789	0.992	1.196	1.403	1.611	1.822
50	2.035	2.250	2.467	2.687	2.908	3.131	3.357	3.584	3.813	4.044
100	4.277	4.512	4.749	4.987	5.227	5.469	5.712	5.957	6.204	6.452
150	6.702	6.954	7.207	7.462	7.718	7.975	8.235	8.495	8.757	9.021
200	9.286	9.553	9.820	10.090	10.360	10.632	10.905	11.180	11.456	11.733
250	12.011	12.291	12.572	12.854	13.137	13.421	13.707	13.993	14.281	14.570
300	14.860	15.151	15.443	15.736	16.030	16.325	16.621	16.919	17.217	17.516
350	17.816	18.118	18.420	18.723	19.027	19.332	19.638	19.945	20.252	20.560
400	20.869									

Interpolation equation for the temperature range from −270°C to 0°C [18]:

$$E = \sum_{i=0}^{14} a_i \cdot t_{68}^i, \tag{8.27}$$

where $a_0 = 0.0$, $a_1 = 3.8740773840 \cdot 10^{-2}$, $a_2 = 4.4123932482 \cdot 10^{-5}$,
$a_3 = 1.1405238498 \cdot 10^{-7}$, $a_4 = 1.9974406568 \cdot 10^{-8}$, $a_5 = 9.0445401187 \cdot 10^{-10}$,
$a_6 = 2.2766018504 \cdot 10^{-11}$, $a_7 = 3.6247409380 \cdot 10^{-13}$, $a_8 = 3.8648924201 \cdot 10^{-15}$,
$a_9 = 2.8298678519 \cdot 10^{-17}$, $a_{10} = 1.4281383349 \cdot 10^{-19}$, $a_{11} = 4.8833254364 \cdot 10^{-22}$,
$a_{12} = 1.0803474683 \cdot 10^{-24}$, $a_{13} = 1.3949291026 \cdot 10^{-27}$, $a_{14} = 7.9795893156 \cdot 10^{-31}$;
from 0°C to 400°C:

$$E = \sum_{i=0}^{8} b_i \cdot t_{68}^i, \tag{8.28}$$

where $b_0 = 0.0$, $b_1 = 3.8740773840 \cdot 10^{-2}$, $b_2 = 3.3190198092 \cdot 10^{-5}$,
$b_3 = 2.0714183645 \cdot 10^{-7}$, $b_4 = -2.1945834823 \cdot 10^{-9}$, $b_5 = 1.1031900550 \cdot 10^{-11}$,
$b_6 = -3.0927581898 \cdot 10^{-14}$, $b_7 = 4.5653337165 \cdot 10^{-17}$, $b_8 = -2.7616878040 \cdot 10^{-20}$.

Table 8.21 Reference table for type J thermocouple [Thermal emf in mV; temperature of reference junction is 0°C.]

t_{68},°C	0	5	10	15	20	25	30	35	40	45
-200	-7.890	-7.996	-8.096							
-150	-6.499	-6.663	-6.821	-6.974	-7.122	-7.265	-7.402	-7.533	-7.659	-7.778
-100	-4.632	-4.836	-5.036	-5.233	-5.426	-5.615	-5.801	-5.982	-6.159	-6.331
-50	-2.431	-2.663	-2.892	-3.120	-3.344	-3.566	-3.785	-4.001	-4.215	-4.425
-0	0.000	-0.251	-0.501	-0.748	-0.995	-1.239	-1.481	-1.722	-1.960	-2.197
0	0.000	0.253	0.507	0.762	1.019	1.277	1.536	1.797	2.058	2.321
50	2.585	2.849	3.115	3.381	3.649	3.917	4.186	4.455	4.725	4.996
100	5.268	5.540	5.812	6.085	6.359	6.633	6.907	7.182	7.457	7.732
150	8.008	8.284	8.560	8.837	9.113	9.390	9.667	9.944	10.222	10.499
200	10.777	11.054	11.332	11.609	11.887	12.165	12.442	12.720	12.998	13.275
250	13.553	13.830	14.108	14.385	14.663	14.940	15.217	15.494	15.771	16.048
300	16.325	16.602	16.879	17.155	17.432	17.708	17.984	18.260	18.537	18.813
350	19.089	19.364	19.640	19.916	20.192	20.467	20.743	21.019	21.295	21.570
400	21.846	22.122	22.397	22.673	22.949	23.225	23.501	23.777	24.054	24.330
450	24.607	24.884	25.161	25.438	25.716	25.994	26.272	26.551	26.829	27.109
500	27.388	27.668	27.949	28.230	28.511	28.793	29.075	29.358	29.642	29.926
550	30.210	30.496	30.782	31.068	31.356	31.644	31.933	32.222	32.513	32.804
600	33.096	33.389	33.683	33.977	34.273	34.569	34.867	35.165	35.464	35.764
650	36.066	36.368	36.671	36.975	37.280	37.586	37.893	38.201	38.510	38.819
700	39.130	39.442	39.754	40.068	40.382	40.697	41.013	41.329	41.647	41.965
750	42.283	42.602	42.922	43.242	43.563	43.885	44.207	44.529	44.852	45.175
800	45.498	45.821	46.144	46.467	46.790	47.112	47.434	47.755	48.076	48.397
850	48.716	49.036	49.354	49.672	49.989	50.305	50.621	50.936	51.249	51.562
900	51.875	52.186	52.496	52.806	53.115	53.422	53.729	54.035	54.341	54.645
950	54.948	55.251	55.553	55.854	56.155	56.454	56.753	57.051	57.349	57.646
1000	57.942	58.238	58.533	58.827	59.121	59.415	59.708	60.001	60.293	60.585
1050	60.876	61.168	61.459	61.749	62.039	62.330	62.619	62.909	63.199	63.488
1100	63.777	64.066	64.355	64.644	64.933	65.222	65.510	65.799	66.087	66.375
1150	66.664	66.952	67.240	67.527	67.815	68.103	68.390	68.677	68.964	69.250
1200	69.536									

Interpolation equation for the temperature range from $-200°C$ to $760°C$ [18]:

$$E = \sum_{i=0}^{7} a_i \cdot t_{68}^i, \tag{8.29}$$

where $a_0 = 0.0$, $a_1 = 5.0372753027 \cdot 10^{-2}$, $a_2 = 3.0425491284 \cdot 10^{-5}$,
$a_3 = -8.5669750464 \cdot 10^{-8}$, $a_4 = 1.3348825735 \cdot 10^{-10}$, $a_5 = -1.7022405966 \cdot 10^{-13}$,
$a_6 = 1.9416091001 \cdot 10^{-16}$, $a_7 = -9.6391844859 \cdot 10^{-20}$;
from 760°C to 900°C:

$$E = \sum_{i=0}^{5} b_i \cdot t_{68}^i, \tag{8.30}$$

where $b_0 = 2.9721751778 \cdot 10^{2}$, $b_1 = -1.5059632873 \cdot 10^{0}$, $b_2 = 3.2051064215 \cdot 10^{-3}$,
$b_3 = -3.2210174230 \cdot 10^{-6}$, $b_4 = 1.5949968788 \cdot 10^{-9}$, $b_5 = -3.1239801752 \cdot 10^{-13}$.

Table 8.22 Reference table for type K thermocouple [Thermal emf in mV; temperature of reference junction is 0°C.]

t_{68},°C	0	5	10	15	20	25	30	35	40	45
–250	-6.404	-6.425	-6.441	-6.452	-6.458					
–200	-5.891	-5.965	-6.035	-6.099	-6.158	-6.213	-6.262	-6.306	-6.344	-6.377
–150	-4.912	-5.029	-5.141	-5.249	-5.354	-5.454	-5.550	-5.642	-5.730	-5.813
–100	-3.553	-3.704	-3.852	-3.997	-4.138	-4.276	-4.410	-4.541	-4.669	-4.792
–50	-1.889	-2.067	-2.243	-2.416	-2.586	-2.754	-2.920	-3.082	-3.242	-3.399
–0	0.000	-0.197	-0.392	-0.585	-0.777	-0.968	-1.156	-1.342	-1.527	-1.709
0	0.000	0.198	0.397	0.597	0.798	1.000	1.203	1.407	1.611	1.817
50	2.022	2.229	2.436	2.643	2.850	3.058	3.266	3.473	3.681	3.888
100	4.095	4.302	4.508	4.714	4.919	5.124	5.327	5.531	5.733	5.936
150	6.137	6.338	6.539	6.739	6.939	7.139	7.338	7.538	7.737	7.937
200	8.137	8.336	8.537	8.737	8.938	9.139	9.341	9.543	9.745	9.948
250	10.151	10.355	10.560	10.764	10.969	11.175	11.381	11.587	11.793	12.000
300	12.207	12.415	12.623	12.831	13.039	13.247	13.456	13.665	13.874	14.083
350	14.292	14.502	14.712	14.922	15.132	15.342	15.552	15.763	15.974	16.184
400	16.395	16.607	16.818	17.029	17.241	17.453	17.664	17.876	18.088	18.301
450	18.513	18.725	18.938	19.150	19.363	19.576	19.788	20.001	20.214	20.427
500	20.640	20.853	21.066	21.280	21.493	21.706	21.919	22.132	22.346	22.559
550	22.772	22.985	23.198	23.411	23.624	23.837	24.050	24.263	24.476	24.689
600	24.902	25.114	25.327	25.539	25.751	25.964	26.176	26.387	26.599	26.811
650	27.022	27.234	27.445	27.656	27.867	28.078	28.288	28.498	28.709	28.919
700	29.128	29.338	29.547	29.756	29.965	30.174	30.383	30.591	30.799	31.007
750	31.214	31.422	31.629	31.836	32.042	32.249	32.455	32.661	32.866	33.072
800	33.277	33.482	33.686	33.891	34.095	34.299	34.502	34.705	34.909	35.111
850	35.314	35.516	35.718	35.920	36.121	36.323	36.524	36.724	36.925	37.125
900	37.325	37.524	37.724	37.923	38.122	38.320	38.519	38.717	38.915	39.112
950	39.310	39.507	39.703	39.900	40.096	40.292	40.488	40.684	40.879	41.074
1000	41.269	41.463	41.657	41.851	42.045	42.239	42.432	42.625	42.817	43.010
1050	43.202	43.394	43.585	43.777	43.968	44.159	44.349	44.539	44.729	44.919
1100	45.108	45.297	45.486	45.675	45.863	46.051	46.238	46.425	46.612	46.799
1150	46.985	47.171	47.356	47.542	47.726	47.911	48.095	48.279	48.462	48.645
1200	48.828	49.010	49.192	49.374	49.555	49.736	49.916	50.096	50.276	50.455
1250	50.633	50.812	50.990	51.167	51.344	51.521	51.697	51.873	52.049	52.224
1300	52.398	52.573	52.747	52.920	53.093	53.266	53.439	53.611	53.782	53.954
1350	54.125	54.296	54.466	54.637	54.807					

Interpolation equation for the temperature range from −270°C to 0°C [18]:

$$E = \sum_{i=0}^{10} a_i \cdot t_{68}^i, \tag{8.31}$$

where $a_0 = 0.0$, $a_1 = 3.9475433139 \cdot 10^{-2}$, $a_2 = 2.7465251138 \cdot 10^{-5}$, $a_3 = -1.6565406716 \cdot 10^{-7}$, $a_4 = -1.5190912392 \cdot 10^{-9}$, $a_5 = -2.4581670924 \cdot 10^{-11}$, $a_6 = -2.4757917816 \cdot 10^{-13}$, $a_7 = -1.5585276173 \cdot 10^{-15}$, $a_8 = -5.9729921255 \cdot 10^{-18}$, $a_9 = -1.2688801216 \cdot 10^{-20}$, $a_{10} = -1.1382797374 \cdot 10^{-23}$;
from 0°C to 1372°C:

$$E = \sum_{i=0}^{8} b_i \cdot t_{68}^i + 0.125 \exp\left[-\frac{1}{2}\left(\frac{t_{68} - 127}{65}\right)^2\right], \tag{8.32}$$

where $b_0 = -1.8533063273 \cdot 10^{-2}$, $b_1 = 3.8918344612 \cdot 10^{-2}$, $b_2 = 1.6645154356 \cdot 10^{-5}$, $b_3 = -7.8702374448 \cdot 10^{-8}$, $b_4 = 2.2835785557 \cdot 10^{-10}$, $b_5 = -3.5700231258 \cdot 10^{-13}$, $b_6 = 2.9932909136 \cdot 10^{-16}$, $b_7 = -1.2849848798 \cdot 10^{-19}$, $b_8 = 2.2239974336 \cdot 10^{-23}$.

Table 8.23 Reference table for type E thermocouple [Thermal emf in mV; temperature of reference junction is 0°C.]

t_{68},°C	0	5	10	15	20	25	30	35	40	45
-250	-9.719	-9.762	-9.797	-9.821	-9.835					
-200	-8.824	-8.947	-9.063	-9.172	-9.274	-9.368	-9.455	-9.534	-9.604	-9.666
-150	-7.279	-7.458	-7.631	-7.800	-7.963	-8.121	-8.273	-8.420	-8.561	-8.696
-100	-5.237	-5.460	-5.680	-5.896	-6.107	-6.314	-6.516	-6.714	-6.907	-7.095
-50	-2.787	-3.048	-3.306	-3.560	-3.811	-4.058	-4.301	-4.541	-4.777	-5.009
-0	0.000	-0.292	-0.581	-0.868	-1.151	-1.432	-1.709	-1.983	-2.254	-2.522
0	0.000	0.295	0.591	0.890	1.192	1.495	1.801	2.109	2.419	2.732
50	3.047	3.364	3.683	4.005	4.329	4.655	4.983	5.314	5.646	5.981
100	6.317	6.656	6.996	7.339	7.683	8.029	8.377	8.727	9.078	9.432
150	9.787	10.143	10.501	10.861	11.222	11.585	11.949	12.314	12.681	13.049
200	13.419	13.789	14.161	14.534	14.909	15.284	15.661	16.038	16.417	16.797
250	17.178	17.559	17.942	18.325	18.710	19.095	19.481	19.868	20.256	20.644
300	21.033	21.423	21.814	22.205	22.597	22.989	23.383	23.777	24.171	24.566
350	24.961	25.357	25.754	26.151	26.549	26.947	27.345	27.744	28.143	28.543
400	28.943	29.343	29.744	30.145	30.546	30.948	31.350	31.752	32.155	32.557
450	32.960	33.364	33.767	34.170	34.574	34.978	35.382	35.786	36.190	36.595
500	36.999	37.403	37.808	38.213	38.617	39.022	39.426	39.831	40.236	40.640
550	41.045	41.449	41.853	42.258	42.662	43.066	43.470	43.874	44.278	44.681
600	45.085	45.488	45.891	46.294	46.697	47.099	47.502	47.904	48.306	48.708
650	49.109	49.510	49.911	50.312	50.713	51.113	51.513	51.913	52.312	52.711
700	53.110	53.509	53.907	54.305	54.703	55.100	55.498	55.894	56.291	56.687
750	57.083	57.478	57.873	58.268	58.663	59.057	59.451	59.844	60.237	60.630
800	61.022	61.414	61.806	62.197	62.588	62.978	63.368	63.758	64.147	64.536
850	64.924	65.312	65.700	66.087	66.473	66.859	67.245	67.630	68.015	68.399
900	68.783	69.166	69.549	69.931	70.313	70.694	71.075	71.456	71.835	72.215
950	72.593	72.972	73.350	73.727	74.104	74.480	74.857	75.232	75.608	75.983
1000	76.358									

Interpolation equation for the temperature range from −270°C to 0°C [18]:

$$E = \sum_{i=0}^{13} a_i \cdot t_{68}^i, \tag{8.33}$$

where $a_0 = 0.0$, $a_1 = 5.8695857799 \cdot 10^{-2}$, $a_2 = 5.1667517705 \cdot 10^{-5}$, $a_3 = -4.4652683347 \cdot 10^{-7}$, $a_4 = -1.7346270905 \cdot 10^{-8}$, $a_5 = -4.8719368427 \cdot 10^{-10}$, $a_6 = -8.8896550447 \cdot 10^{-12}$, $a_7 = -1.0930767375 \cdot 10^{-13}$, $a_8 = -9.1784535039 \cdot 10^{-16}$, $a_9 = -5.2575158521 \cdot 10^{-18}$, $a_{10} = -2.0169601996 \cdot 10^{-20}$, $a_{11} = -4.9502138782 \cdot 10^{-23}$, $a_{12} = -7.0177980633 \cdot 10^{-26}$, $a_{13} = -4.3671808488 \cdot 10^{-29}$;
from 0°C to 1000°C:

$$E = \sum_{i=0}^{9} b_i \cdot t_{68}^i, \tag{8.34}$$

where $b_0 = 0.0$, $b_1 = 5.8695857799 \cdot 10^{-2}$, $b_2 = 4.3110945462 \cdot 10^{-5}$, $b_3 = 5.7220358202 \cdot 10^{-8}$, $b_4 = -5.4020668025 \cdot 10^{-10}$, $b_5 = 1.5425922111 \cdot 10^{-12}$, $b_6 = -2.4850089136 \cdot 10^{-15}$, $b_7 = 2.3389721459 \cdot 10^{-18}$, $b_8 = -1.1946296815 \cdot 10^{-21}$, $b_9 = 2.5561127497 \cdot 10^{-25}$.

Table 8.24 Reference table for type N thermocouple [Thermal emf in mV; temperature of reference junction is 0°C; wire diameter 1.6 mm.]

t_{68},°C	0	5	10	15	20	25	30	35	40	45
0	0.000	0.130	0.261	0.392	0.525	0.658	0.793	0.928	1.064	1.201
50	1.339	1.479	1.619	1.760	1.902	2.045	2.188	2.333	2.479	2.626
100	2.774	2.922	3.072	3.222	3.374	3.526	3.679	3.833	3.988	4.144
150	4.301	4.458	4.617	4.776	4.936	5.097	5.258	5.420	5.584	5.747
200	5.912	6.077	6.243	6.410	6.577	6.745	6.914	7.084	7.254	7.424
250	7.596	7.767	7.940	8.113	8.287	8.461	8.636	8.811	8.987	9.163
300	9.340	9.517	9.695	9.874	10.053	10.232	10.412	10.592	10.772	10.954
350	11.135	11.317	11.499	11.682	11.865	12.049	12.233	12.417	12.602	12.787
400	12.972	13.158	13.344	13.530	13.717	13.904	14.091	14.279	14.467	14.655
450	14.844	15.033	15.222	15.411	15.601	15.791	15.981	16.172	16.362	16.553
500	16.744	16.936	17.127	17.319	17.511	17.704	17.896	18.089	18.282	18.475
550	18.668	18.861	19.055	19.249	19.443	19.637	19.831	20.025	20.220	20.415
600	20.609	20.804	20.999	21.195	21.390	21.585	21.781	21.976	22.172	22.368
650	22.564	22.760	22.956	23.152	23.348	23.544	23.740	23.937	24.133	24.329
700	24.526	24.722	24.919	25.115	25.312	25.508	25.705	25.902	26.098	26.295
750	26.491	26.688	26.885	27.081	27.278	27.474	27.671	27.867	28.063	28.260
800	28.456	28.652	28.849	29.045	29.241	29.437	29.633	29.829	30.025	30.221
850	30.417	30.612	30.808	31.004	31.199	31.394	31.590	31.785	31.980	32.175
900	32.370	32.565	32.760	32.955	33.149	33.344	33.538	33.732	33.926	34.121
950	34.315	34.508	34.702	34.896	35.089	35.283	35.476	35.669	35.862	36.055
1000	36.248	36.441	36.633	36.826	37.018	37.210	37.402	37.594	37.786	37.978
1050	38.169	38.361	38.552	38.743	38.934	39.125	39.315	39.506	39.696	39.886
1100	40.076	40.266	40.456	40.645	40.835	41.024	41.213	41.401	41.590	41.778
1150	41.966	42.154	42.342	42.530	42.717	42.904	43.091	43.278	43.464	43.650
1200	43.836	44.022	44.207	44.393	44.577	44.762	44.947	45.131	45.315	45.498
1250	45.682	45.865	46.048	46.231	46.413	46.595	46.777	46.959	47.140	47.321
1300	47.502									

Interpolation equation for the temperature range from 0°C to 1300°C [18]:

$$E = \sum_{i=0}^{9} a_i \cdot t_{68}^i, \tag{8.35}$$

where $a_0 = 0.0$, $a_1 = 2.5897798582 \cdot 10^{-2}$, $a_2 = 1.6656127713 \cdot 10^{-5}$, $a_3 = 3.1234962101 \cdot 10^{-8}$, $a_4 = -1.7248130773 \cdot 10^{-10}$, $a_5 = 3.6526665920 \cdot 10^{-13}$, $a_6 = -4.4390833504 \cdot 10^{-16}$, $a_7 = 3.1553382729 \cdot 10^{-19}$, $a_8 = -1.2150879468 \cdot 10^{-22}$, $a_9 = 1.9557197559 \cdot 10^{-26}$.

Table 8.25 Reference table for type R thermocouple [Thermal emf in mV; temperature of reference junction is 0°C.]

t_{68},°C	0	5	10	15	20	25	30	35	40	45
-50	−0.226									
−0	0.000	−0.026	−0.051	−0.076	−0.100	−0.123	−0.145	−0.167	−0.188	−0.207
0	0.000	0.027	0.054	0.082	0.111	0.141	0.171	0.201	0.232	0.264
50	0.296	0.329	0.363	0.397	0.431	0.466	0.501	0.537	0.573	0.610
100	0.647	0.685	0.723	0.761	0.800	0.839	0.879	0.919	0.959	1.000
150	1.041	1.082	1.124	1.166	1.208	1.251	1.294	1.337	1.380	1.424
200	1.468	1.512	1.557	1.602	1.647	1.692	1.738	1.784	1.830	1.876
250	1.923	1.970	2.017	2.064	2.111	2.159	2.207	2.255	2.303	2.351
300	2.400	2.449	2.498	2.547	2.596	2.646	2.695	2.745	2.795	2.845
350	2.896	2.946	2.997	3.048	3.099	3.150	3.201	3.252	3.304	3.355
400	3.407	3.459	3.511	3.563	3.616	3.668	3.721	3.774	3.826	3.879
450	3.933	3.986	4.039	4.093	4.146	4.200	4.254	4.308	4.362	4.416
500	4.471	4.525	4.580	4.634	4.689	4.744	4.799	4.854	4.910	4.965
550	5.021	5.076	5.132	5.188	5.244	5.300	5.356	5.413	5.469	5.526
600	5.582	5.639	5.696	5.753	5.810	5.867	5.925	5.982	6.040	6.098
650	6.155	6.213	6.272	6.330	6.388	6.447	6.505	6.564	6.623	6.682
700	6.741	6.800	6.860	6.919	6.979	7.039	7.098	7.158	7.218	7.279
750	7.339	7.399	7.460	7.521	7.582	7.642	7.703	7.765	7.826	7.887
800	7.949	8.010	8.072	8.134	8.196	8.258	8.320	8.383	8.445	8.508
850	8.570	8.633	8.696	8.759	8.822	8.885	8.949	9.012	9.076	9.140
900	9.203	9.267	9.331	9.395	9.460	9.524	9.589	9.653	9.718	9.783
950	9.848	9.913	9.978	10.043	10.109	10.174	10.240	10.305	10.371	10.437
1000	10.503	10.569	10.636	10.702	10.768	10.835	10.902	10.968	11.035	11.102
1050	11.170	11.237	11.304	11.372	11.439	11.507	11.574	11.642	11.710	11.778
1100	11.846	11.914	11.983	12.051	12.119	12.188	12.257	12.325	12.394	12.463
1150	12.532	12.600	12.669	12.739	12.808	12.877	12.946	13.016	13.085	13.154
1200	13.224	13.293	13.363	13.433	13.502	13.572	13.642	13.712	13.782	13.852
1250	13.922	13.992	14.062	14.132	14.202	14.272	14.343	14.413	14.483	14.554
1300	14.624	14.694	14.765	14.835	14.906	14.976	15.047	15.117	15.188	15.258
1350	15.329	15.399	15.470	15.540	15.611	15.682	15.752	15.823	15.893	15.964
1400	16.035	16.105	16.176	16.247	16.317	16.388	16.458	16.529	16.599	16.670
1450	16.741	16.811	16.882	16.952	17.022	17.093	17.163	17.234	17.304	17.374
1500	17.445	17.515	17.585	17.655	17.726	17.796	17.866	17.936	18.006	18.076
1550	18.146	18.216	18.286	18.355	18.425	18.495	18.564	18.634	18.703	18.773
1600	18.842	18.912	18.981	19.050	19.119	19.188	19.257	19.326	19.395	19.464
1650	19.533	19.602	19.670	19.739	19.807	19.875	19.944	20.012	20.080	20.148
1700	20.215	20.283	20.350	20.417	20.483	20.550	20.616	20.682	20.748	20.813
1750	20.878	20.942	21.006	21.070						

Interpolation equation for the temperature range from −50°C to 630.74°C [18]:

$$E = \sum_{i=0}^{7} a_i \cdot t_{68}^i, \tag{8.36}$$

where $a_0 = 0.0$, $a_1 = 5.2891395059 \cdot 10^{-3}$, $a_2 = 1.3911109947 \cdot 10^{-5}$, $a_3 = -2.4005238430 \cdot 10^{-8}$, $a_4 = 3.6201410595 \cdot 10^{-11}$, $a_5 = -4.4645019036 \cdot 10^{-14}$, $a_6 = 3.8497691865 \cdot 10^{-17}$, $a_7 = -1.5372641559 \cdot 10^{-20}$;
from 630.74°C to 1064.43°C:

$$E = \sum_{i=0}^{3} b_i \cdot t_{68}^i, \tag{8.37}$$

where $b_0 = -2.6418007025 \cdot 10^{-1}$, $b_1 = 8.0468686747 \cdot 10^{-3}$, $b_2 = 2.9892293723 \cdot 10^{-6}$, $b_3 = -2.6876058617 \cdot 10^{-10}$;
from 1064.43°C to 1665°C:

$$E = \sum_{i=0}^{3} c_i \cdot t_{68}^i, \tag{8.38}$$

where $c_0 = 1.4901702702 \cdot 10^{0}$, $c_1 = 2.8639867552 \cdot 10^{-3}$, $c_2 = 8.0823631189 \cdot 10^{-6}$, $c_3 = -1.9338477638 \cdot 10^{-9}$;
from 1665°C to 1769°C:

$$E = \sum_{i=0}^{3} d_i \cdot t_{68}^i, \tag{8.39}$$

where $d_0 = 9.5445559010 \cdot 10^{1}$, $d_1 = -1.6642500359 \cdot 10^{-1}$, $d_2 = 1.0975743239 \cdot 10^{-4}$, $d_3 = -2.2289216980 \cdot 10^{-8}$.

Table 8.26 Reference table for type S thermocouple [Thermal emf in mV; temperature of reference junction is 0°C.]

t_{68},°C	0	5	10	15	20	25	30	35	40	45
−250	−6.181	−6.209	−6.232	−6.248	−6.258					
−200	−5.603	−5.680	−5.753	−5.823	−5.889	−5.950	−6.007	−6.059	−6.105	−6.146
−150	−4.648	−4.758	−4.865	−4.969	−5.069	−5.167	−5.261	−5.351	−5.439	−5.522
−100	−3.378	−3.519	−3.656	−3.791	−3.923	−4.051	−4.177	−4.299	−4.419	−4.535
−50	−0.236	−1.987	−2.152	−2.315	−2.475	−2.633	−2.788	−2.939	−3.089	−3.235
−0	0.000	−0.027	−0.053	−0.078	−0.103	−0.127	−0.150	−0.173	−0.194	−0.215
0	0.000	0.027	0.055	0.084	0.113	0.142	0.173	0.203	0.235	0.266
50	0.299	0.331	0.365	0.398	0.432	0.467	0.502	0.537	0.573	0.609
100	0.645	0.682	0.719	0.757	0.795	0.833	0.872	0.910	0.950	0.989
150	1.029	1.069	1.109	1.149	1.190	1.231	1.273	1.314	1.356	1.398
200	1.440	1.482	1.525	1.568	1.611	1.654	1.698	1.741	1.785	1.829
250	1.873	1.917	1.962	2.006	2.051	2.096	2.141	2.186	2.232	2.277
300	2.323	2.368	2.414	2.460	2.506	2.553	2.599	2.646	2.692	2.739
350	2.786	2.833	2.880	2.927	2.974	3.022	3.069	3.117	3.164	3.212
400	3.260	3.308	3.356	3.404	3.452	3.500	3.549	3.597	3.645	3.694
450	3.743	3.791	3.840	3.889	3.938	3.987	4.036	4.086	4.135	4.184
500	4.234	4.283	4.333	4.382	4.432	4.482	4.532	4.582	4.632	4.682
550	4.732	4.782	4.832	4.883	4.933	4.984	5.034	5.085	5.136	5.186
600	5.237	5.288	5.339	5.391	5.442	5.493	5.544	5.596	5.648	5.700
650	5.751	5.803	5.855	5.907	5.960	6.012	6.064	6.117	6.169	6.222

Table 8.26 Reference table for type S thermocouple *(continued)*

t_{68},°C	0	5	10	15	20	25	30	35	40	45
700	6.274	6.327	6.380	6.433	6.486	6.539	6.592	6.645	6.699	6.752
750	6.805	6.859	6.913	6.966	7.020	7.074	7.128	7.182	7.236	7.291
800	7.345	7.399	7.454	7.508	7.563	7.618	7.672	7.727	7.782	7.837
850	7.892	7.948	8.003	8.058	8.114	8.169	8.225	8.281	8.336	8.392
900	8.448	8.504	8.560	8.617	8.673	8.729	8.786	8.842	8.899	8.956
950	9.012	9.069	9.126	9.183	9.240	9.298	9.355	9.412	9.470	9.527
1000	9.585	9.642	9.700	9.758	9.816	9.874	9.932	9.990	10.048	10.107
1050	10.165	10.224	10.282	10.341	10.400	10.459	10.517	10.576	10.635	10.694
1100	10.754	10.813	10.872	10.931	10.991	11.050	11.110	11.169	11.229	11.288
1150	11.348	11.408	11.467	11.527	11.587	11.647	11.707	11.767	11.827	11.887
1200	11.947	12.007	12.067	12.128	12.188	12.248	12.308	12.369	12.429	12.489
1250	12.550	12.610	12.671	12.731	12.792	12.852	12.913	12.973	13.034	13.094
1300	13.155	13.216	13.276	13.337	13.397	13.458	13.519	13.579	13.640	13.701
1350	13.761	13.822	13.883	13.943	14.004	14.065	14.125	14.186	14.247	14.307
1400	14.368	14.429	14.489	14.550	14.610	14.671	14.731	14.792	14.852	14.913
1450	14.973	15.034	15.094	15.155	15.215	15.275	15.336	15.396	15.456	15.516
1500	15.576	15.637	15.697	15.757	15.817	15.877	15.937	15.997	16.057	16.116
1550	16.176	16.236	16.296	16.355	16.415	16.474	16.534	16.593	16.653	16.712
1600	16.771	16.830	16.890	16.949	17.008	17.067	17.125	17.184	17.243	17.302
1650	17.360	17.419	17.477	17.536	17.594	17.652	17.711	17.769	17.826	17.884
1700	17.942	17.999	18.056	18.113	18.170	18.226	18.282	18.338	18.394	18.449
1750	18.504	18.558	18.612	18.666						

Interpolation equation for the temperature range from −50°C to 630.74°C [18]:

$$E = \sum_{i=0}^{6} a_i \cdot t_{68}^i, \tag{8.40}$$

where $a_0 = 0.0$, $a_1 = 5.3995782346 \cdot 10^{-3}$, $a_2 = 1.2519770000 \cdot 10^{-5}$, $a_3 = -2.2448217997 \cdot 10^{-8}$, $a_4 = 2.8452164949 \cdot 10^{-11}$, $a_5 = -2.2440584544 \cdot 10^{-14}$, $a_6 = 8.5054166936 \cdot 10^{-18}$;
from 630.74°C to 1064.43°C:

$$E = \sum_{i=0}^{2} b_i \cdot t_{68}^i, \tag{8.41}$$

where $b_0 = -2.9824481615 \cdot 10^{-1}$, $b_1 = 8.2375528221 \cdot 10^{-3}$, $b_2 = 1.6453909942 \cdot 10^{-6}$;
from 1064.43°C to 1665°C:

$$E = \sum_{i=0}^{3} c_i \left(\frac{t_{68} - 1365}{300} \right)^i, \tag{8.42}$$

where $c_0 = 1.3943438677 \cdot 10^{1}$, $c_1 = 3.6398686553 \cdot 10^{0}$, $c_2 = -5.0281206140 \cdot 10^{-3}$, $c_3 = -4.2450546418 \cdot 10^{-2}$.

Table 8.27 Reference table for type B thermocouple [Thermal emf in mV; temperature of reference junction is 0°C.]

t_{68},°C	0	5	10	15	20	25	30	35	40	45
0	0.000	−0.001	−0.002	−0.002	−0.003	−0.002	−0.002	−0.001	−0.000	0.001
50	0.002	0.004	0.006	0.009	0.011	0.014	0.017	0.021	0.025	0.029
100	0.033	0.038	0.043	0.048	0.053	0.059	0.065	0.071	0.078	0.085
150	0.092	0.099	0.107	0.115	0.123	0.132	0.140	0.149	0.159	0.168
200	0.178	0.188	0.199	0.209	0.220	0.231	0.243	0.254	0.266	0.279
250	0.291	0.304	0.317	0.330	0.344	0.358	0.372	0.386	0.401	0.415
300	0.431	0.446	0.462	0.477	0.494	0.510	0.527	0.544	0.561	0.578
350	0.596	0.614	0.632	0.650	0.669	0.688	0.707	0.727	0.746	0.766
400	0.786	0.807	0.827	0.848	0.870	0.891	0.913	0.935	0.957	0.979
450	1.002	1.025	1.048	1.071	1.095	1.119	1.143	1.167	1.192	1.216
500	1.241	1.267	1.292	1.318	1.344	1.370	1.397	1.423	1.450	1.477
550	1.505	1.532	1.560	1.588	1.617	1.645	1.674	1.703	1.732	1.762
600	1.791	1.821	1.851	1.882	1.912	1.943	1.974	2.005	2.036	2.068
650	2.100	2.132	2.164	2.197	2.230	2.263	2.296	2.329	2.363	2.396
700	2.430	2.465	2.499	2.534	2.569	2.604	2.639	2.674	2.710	2.746
750	2.782	2.818	2.855	2.892	2.928	2.966	3.003	3.040	3.078	3.116
800	3.154	3.192	3.231	3.269	3.308	3.347	3.387	3.426	3.466	3.506
850	3.546	3.586	3.626	3.667	3.708	3.749	3.790	3.831	3.873	3.915
900	3.957	3.999	4.041	4.083	4.126	4.169	4.212	4.255	4.298	4.342
950	4.386	4.430	4.474	4.518	4.562	4.607	4.652	4.697	4.742	4.787
1000	4.833	4.878	4.924	4.970	5.016	5.063	5.109	5.156	5.202	5.249
1050	5.297	5.344	5.391	5.439	5.487	5.535	5.583	5.631	5.680	5.728
1100	5.777	5.826	5.875	5.924	5.973	6.023	6.073	6.122	6.172	6.223
1150	6.273	6.323	6.374	6.424	6.475	6.526	6.577	6.629	6.680	6.732
1200	6.783	6.835	6.887	6.939	6.991	7.044	7.096	7.149	7.202	7.255
1250	7.308	7.361	7.414	7.467	7.521	7.575	7.628	7.682	7.736	7.790
1300	7.845	7.899	7.953	8.008	8.063	8.118	8.172	8.227	8.283	8.338
1350	8.393	8.449	8.504	8.560	8.616	8.671	8.727	8.783	8.839	8.896
1400	8.952	9.008	9.065	9.121	9.178	9.235	9.291	9.348	9.405	9.462
1450	9.519	9.577	9.634	9.691	9.748	9.806	9.863	9.921	9.979	10.036
1500	10.094	10.152	10.210	10.268	10.325	10.383	10.441	10.500	10.558	10.616
1550	10.674	10.732	10.790	10.849	10.907	10.965	11.024	11.082	11.141	11.199
1600	11.257	11.316	11.374	11.433	11.491	11.550	11.608	11.667	11.725	11.784
1650	11.842	11.901	11.959	12.018	12.076	12.134	12.193	12.251	12.310	12.368
1700	12.426	12.485	12.543	12.601	12.659	12.718	12.776	12.834	12.892	12.950
1750	13.008	13.066	13.124	13.181	13.239	13.297	13.354	13.412	13.470	13.527
1800	13.585	13.642	13.699	13.756	13.814					

Interpolation equation for the temperature range from 0°C to 1820°C [18]:

$$E = \sum_{i=0}^{8} a_i \cdot t_{68}^i, \tag{8.43}$$

where $a_0 = 0.0$, $a_1 = -2.4674601620 \cdot 10^{-4}$, $a_2 = 5.9102111169 \cdot 10^{-6}$, $a_3 = -1.4307123430 \cdot 10^{-9}$, $a_4 = 2.1509149750 \cdot 10^{-12}$, $a_5 = -3.1757800720 \cdot 10^{-15}$, $a_6 = 2.4010367459 \cdot 10^{-18}$, $a_7 = -9.0928148159 \cdot 10^{-22}$, $a_8 = 1.3299505137 \cdot 10^{-25}$.

Table 8.28 Reference table for W+5%Re –W+20%Re thermocouple [48] [Thermal emf in mV; temperature of reference junction is 0°C.]

t_{68},°C	0	10	20	30	40	50	60	70	80	90
0	0.000	0.124	0.250	0.378	0.508	0.640	0.774	0.910	1.048	1.188
100	1.330	1.475	1.622	1.771	1.922	2.075	2.230	2.387	2.546	2.707
200	2.869	3.032	3.195	3.359	3.523	3.688	3.853	4.019	4.185	4.352
300	4.519	4.687	4.855	5.024	5.193	5.362	5.531	5.700	5.869	6.039
400	6.209	6.379	6.549	6.719	6.889	7.059	7.229	7.399	7.569	7.739
500	7.909	8.078	8.247	8.416	8.585	8.754	8.923	9.092	9.261	9.430
600	9.598	9.765	9.933	10.101	10.269	10.437	10.605	10.772	10.939	11.106
700	11.273	11.440	11.606	11.772	11.938	12.104	12.269	12.434	12.599	12.764
800	12.929	13.093	13.257	13.421	13.584	13.747	13.910	14.072	14.234	14.395
900	14.556	14.716	14.876	15.035	15.194	15.352	15.510	15.667	15.824	15.980
1000	16.136	16.291	16.446	16.600	16.754	16.907	17.060	17.212	17.364	17.515
1100	17.666	17.816	17.966	18.115	18.264	18.412	18.560	18.707	18.854	19.000
1200	19.146	19.291	19.436	19.580	19.724	19.867	20.010	20.152	20.294	20.435
1300	20.576	20.716	20.856	20.996	21.135	21.274	21.413	21.551	21.689	21.826
1400	21.963	22.099	22.235	22.370	22.505	22.639	22.773	22.906	23.039	23.171
1500	23.303	23.434	23.565	23.695	23.825	23.954	24.083	24.211	24.338	24.464
1600	24.590	24.715	24.840	24.964	25.088	25.211	25.334	25.456	25.578	25.699
1700	25.820	25.940	26.060	26.179	26.298	26.416	26.534	26.651	26.768	26.884
1800	26.999	27.113	27.226	27.339	27.451	27.563	27.675	27.786	27.897	28.007
1900	28.117	28.226	28.334	28.442	28.549	28.656	28.762	28.867	28.971	29.074
2000	29.177	29.279	29.381	29.483	29.584	29.685	29.786	29.886	29.986	30.086
2100	30.185	30.284	30.382	30.479	30.575	30.670	30.765	30.859	30.952	31.044
2200	31.136	31.227	31.318	31.408	31.498	31.587	31.676	31.764	31.852	31.939
2300	32.026	32.112	32.197	32.281	32.365	32.448	32.531	32.613	32.695	32.776
2400	32.857	32.937	33.017	33.096	33.175	33.253	33.331	33.408	33.485	33.561
2500	33.636									

Table 8.29 Reference table for type Au/Pt thermocouple [49] [Thermal emf in mV; temperature of reference junction is 0°C.]

t_{90},°C	0	5	10	15	20	25	30	35	40	45
0	0.000	0.031	0.062	0.095	0.128	0.163	0.198	0.234	0.271	0.309
50	0.347	0.387	0.427	0.468	0.510	0.553	0.596	0.641	0.685	0.731
100	0.778	0.825	0.872	0.921	0.970	1.020	1.071	1.122	1.174	1.226
150	1.279	1.333	1.388	1.443	1.498	1.555	1.611	1.669	1.727	1.785
200	1.845	1.904	1.965	2.026	2.087	2.149	2.211	2.275	2.338	2.402
250	2.467	2.532	2.598	2.664	2.731	2.798	2.865	2.934	3.002	3.072
300	3.141	3.211	3.282	3.353	3.425	3.497	3.569	3.642	3.716	3.790
350	3.864	3.939	4.014	4.090	4.166	4.243	4.320	4.398	4.476	4.554
400	4.633	4.712	4.792	4.872	4.953	5.034	5.115	5.197	5.279	5.362
450	5.445	5.529	5.613	5.697	5.782	5.868	5.953	6.039	6.126	6.213
500	6.300	6.388	6.476	6.565	6.654	6.743	6.833	6.923	7.014	7.105
550	7.197	7.289	7.381	7.474	7.567	7.660	7.754	7.849	7.943	8.039
600	8.134	8.230	8.327	8.423	8.521	8.618	8.716	8.815	8.914	9.013
650	9.112	9.212	9.313	9.414	9.515	9.617	9.719	9.821	9.924	10.027
700	10.131	10.235	10.340	10.445	10.550	10.656	10.762	10.868	10.975	11.082
750	11.190	11.298	11.407	11.516	11.625	11.735	11.845	11.955	12.066	12.178
800	12.289	12.402	12.514	12.627	12.740	12.854	12.968	13.083	13.198	13.313
850	13.429	13.545	13.661	13.778	13.896	14.013	14.131	14.250	14.369	14.488
900	14.608	14.728	14.848	14.969	15.090	15.212	15.334	15.456	15.579	15.702
950	15.826	15.950	16.074							

Interpolation equation for the temperature range from 0°C to 962°C [49]:

$$E = \sum_{i=1}^{8} a_i \cdot t_{90}^i, \tag{8.44}$$

where $a_1 = 6.0310705 \cdot 10^{-3}$, $a_2 = 1.9377291 \cdot 10^{-5}$, $a_3 = -2.2104563 \cdot 10^{-8}$, $a_4 = 3.1103354 \cdot 10^{-11}$, $a_5 = -3.5645621 \cdot 10^{-14}$, $a_6 = 3.1611501 \cdot 10^{-17}$, $a_7 = -1.7549464 \cdot 10^{-20}$, $a_8 = 4.2602445 \cdot 10^{-24}$.

Table 8.30 Temperature dependence of thermoelectric voltage for some high-temperature thermocouples [38, 42] [Thermal emf in mV; temperature of thermocouple reference junction is 0°C.]

t, °C	Nickel–nichrome	Ir+60%Rh–Ir	Ir+50%Rh–Ir+10%Ru	W–Mo	t, °C	Nickel–nichrome	Ir+60%Rh–Ir	Ir+50%Rh–Ir+10%Ru	W–Mo
100	3.7	0.36	0.7	–0.400	1100	40.6	5.85	10.1	
200	7.8	0.82	1.5		1200		6.39	11.0	
300	11.8	1.33	2.3		1300		6.92	11.8	–0.225
400	15.1	1.89	3.3		1400		7.46	12.7	+0.359
500	18.4	2.46	4.3		1500		8.01	13.5	0.997
600	21.9	3.05	5.3		1600		8.58	14.2	1.689
700	25.5	3.63	6.3		1700		9.16	15.0	2.435
800	29.2	4.20	7.3		1800		9.74	15.8	3.201
900	33.0	4.76	8.3		1900		10.36	16.5	
1000	36.7	5.31	9.2		2000		11.00	17.3	

Table 8.31 Temperature range of some vapor-pressure thermometers

	^3He	^4He	e-H_2	n-H_2	Ne	N_2	O_2	CF_4
T, K	0.2—3.2	0.5—5.2	14—23	14—30	19—40	56—84	54—94	90—173

Symbols: e-H_2 is equilibrium hydrogen, n-H_2 is normal hydrogen (see the definition in Table 8.9).

Table 8.32 ^3He vapor pressure as a function of temperature

T, K	0.00	0.01	0.02	0.03	0.04	0.05	0.06	0.07	0.08	0.09
					Pressure in Pa					
0.2	0.00147	0.00298	0.00569	0.01030	0.01783	0.02964	0.04754	0.07385	0.11150	0.16406
0.3	0.2358	0.3320	0.4584	0.6220	0.8305	1.0928	1.4186	1.8187	2.3050	2.8902
0.4	3.588	4.414	5.383	6.512	7.820	9.324	11.045	13.002	15.217	17.712
0.5	20.510	23.633	27.106	30.955	35.203	39.877	45.003	50.608	56.718	63.363
0.6	70.569	78.365	86.780	95.842	105.581	115.906	127.091	139.049	151.807	165.393
0.7	179.84	195.16	211.41	228.59	246.75	265.91	286.11	307.37	329.72	353.20
0.8	377.83	403.65	430.68	458.97	488.53	519.40	551.61	585.19	620.17	656.58
0.9	694.45	733.82	774.70	817.14	861.16	906.79	954.06	1002.99	1053.63	1105.99
					Pressure in kPa					
1.0	1.160	1.216	1.274	1.333	1.395	1.458	1.523	1.590	1.660	1.731
1.1	1.804	1.879	1.957	2.036	2.118	2.202	2.288	2.376	2.466	2.559
1.2	2.654	2.752	2.851	2.954	3.058	3.165	3.275	3.387	3.501	3.618
1.3	3.738	3.860	3.985	4.112	4.242	4.375	4.511	4.649	4.790	4.934
1.4	5.081	5.230	5.383	5.538	5.696	5.858	6.022	6.189	6.360	6.533
1.5	6.709	6.889	7.071	7.257	7.446	7.639	7.834	8.033	8.235	8.440

Table 8.32 ^{3}He vapor pressure as a function of temperature *(continued)*

T, K	0.00	0.01	0.02	0.03	0.04	0.05	0.06	0.07	0.08	0.09
					Pressure in kPa					
1.6	8.649	8.861	9.076	9.295	9.517	9.742	9.972	10.204	10.440	10.680
1.7	10.923	11.170	11.421	11.675	11.933	12.194	12.459	12.728	13.001	13.277
1.8	13.558	13.842	14.130	14.422	14.717	15.017	15.321	15.628	15.940	16.256
1.9	16.575	16.899	17.227	17.559	17.895	18.235	18.579	18.928	19.281	19.638
2.0	19.999	20.365	20.735	21.109	21.488	21.871	22.258	22.650	23.046	23.447
2.1	23.852	24.262	24.677	25.095	25.519	25.947	26.380	26.817	27.259	27.706
2.2	28.158	28.614	29.075	29.541	30.011	30.487	30.967	31.453	31.943	32.438
2.3	32.938	33.443	33.953	34.468	34.988	35.513	36.044	36.579	37.120	37.666
2.4	38.217	38.773	39.334	39.901	40.473	41.051	41.633	42.221	42.815	43.414
2.5	44.018	44.628	45.244	45.865	46.491	47.123	47.761	48.404	49.053	49.708
2.6	50.368	51.035	51.707	52.384	53.068	53.758	54.453	55.154	55.862	56.575
2.7	57.294	58.019	58.751	59.488	60.232	60.982	61.738	62.500	63.268	64.043
2.8	64.824	65.611	66.405	67.206	68.012	68.825	69.645	70.471	71.304	72.144
2.9	72.990	73.843	74.702	75.569	76.442	77.322	78.209	79.102	80.003	80.911
3.0	81.826	82.748	83.677	84.613	85.556	86.506	87.464	88.429	89.402	90.382
3.1	91.369	92.364	93.366	94.376	95.394	96.419	97.452	98.493	99.541	100.598
3.2	101.662									

In the range of 0.65 K to 3.2 K the temperature T_{90} is defined in terms of the ^{3}He vapor-pressure equation (see Eq.(8.4) and Table 8.3). From 0.5 K to 0.65 K and from 3.2 K to 3.3162 K (critical point) the vapor pressure at temperature T_{76} is given by the relation [19, 20]:

$$\ln(p/\mathrm{Pa}) = \sum_{i=-1}^{4} a_i T_{76}^i + b \ln T_{76}, \tag{8.45}$$

where $a_{-1} = -2.50943$ K, $a_0 = 9.70876$, $a_1 = -0.304433$ K^{-1}, $a_2 = 0.210429$ K^{-2}, $a_3 = -0.0545145$ K^{-3}, $a_4 = 0.0056067$ K^{-4}, $b = 2.25484$.

This equation is valid down to 0.2 K, the T_{76} being replaced by T.

Table 8.33 ^{4}He vapor pressure as a function of temperature

T, K	0.00	0.01	0.02	0.03	0.04	0.05	0.06	0.07	0.08	0.09
					Pressure in Pa					
0.5	0.00206	0.00287	0.00396	0.00539	0.00726	0.00969	0.01280	0.01676	0.02176	0.02803
0.6	0.0358	0.0454	0.0573	0.0716	0.0891	0.1101	0.1353	0.1653	0.2009	0.2429
0.7	0.2922	0.3500	0.4172	0.4952	0.5854	0.6892	0.8082	0.9444	1.0995	1.2757
0.8	1.475	1.700	1.954	2.239	2.558	2.914	3.310	3.751	4.240	4.782
0.9	5.379	6.038	6.762	7.556	8.427	9.379	10.418	11.550	12.781	14.119
1.0	15.570	17.140	18.839	20.672	22.649	24.777	27.066	29.523	32.159	34.983
1.1	38.005	41.234	44.682	48.359	52.276	56.445	60.877	65.585	70.580	75.875
1.2	81.483	87.418	93.693	100.322	107.318	114.734	122.505	130.687	139.294	148.344
1.3	157.85	167.83	178.31	189.29	200.80	212.85	225.47	238.66	252.45	266.86
1.4	281.91	297.61	313.98	331.04	348.82	367.33	386.59	406.62	427.44	449.07
1.5	471.54	494.85	519.05	544.13	570.13	597.06	624.95	653.82	683.69	714.58
1.6	746.51	779.50	813.58	848.77	885.09	922.56	961.21	1001.05	1042.12	1084.42
					Pressure in kPa					
1.7	1.128	1.173	1.219	1.267	1.315	1.366	1.417	1.470	1.525	1.581
1.8	1.638	1.697	1.758	1.820	1.883	1.949	2.015	2.084	2.154	2.226

Table 8.33 ^{4}He vapor pressure as a function of temperature *(continued)*

T, K	0.00	0.01	0.02	0.03	0.04	0.05	0.06	0.07	0.08	0.09
					Pressure in kPa					
1.9	2.299	2.374	2.451	2.530	2.610	2.692	2.776	2.862	2.949	3.039
2.0	3.130	3.223	3.317	3.414	3.512	3.613	3.715	3.819	3.924	4.032
2.1	4.141	4.253	4.366	4.480	4.597	4.716	4.836	4.958	5.082	5.207
2.2	5.335	5.465	5.597	5.731	5.867	6.005	6.146	6.288	6.433	6.580
2.3	6.730	6.881	7.036	7.192	7.351	7.512	7.675	7.841	8.009	8.180
2.4	8.354	8.529	8.708	8.889	9.072	9.258	9.447	9.638	9.832	10.028
2.5	10.228	10.430	10.634	10.842	11.052	11.265	11.481	11.700	11.921	12.145
2.6	12.373	12.603	12.836	13.072	13.311	13.552	13.797	14.045	14.296	14.550
2.7	14.807	15.068	15.331	15.597	15.867	16.140	16.416	16.695	16.977	17.263
2.8	17.552	17.844	18.140	18.438	18.741	19.046	19.355	19.667	19.983	20.302
2.9	20.625	20.951	21.281	21.614	21.951	22.291	22.635	22.982	23.333	23.688
3.0	24.046	24.408	24.774	25.143	25.517	25.894	26.274	26.659	27.047	27.439
3.1	27.835	28.235	28.638	29.046	29.458	29.873	30.292	30.716	31.143	31.575
3.2	32.010	32.450	32.893	33.341	33.793	34.249	34.709	35.173	35.641	36.114
3.3	36.591	37.072	37.557	38.047	38.541	39.039	39.542	40.049	40.560	41.076
3.4	41.596	42.121	42.650	43.184	43.722	44.264	44.811	45.363	45.919	46.480
3.5	47.045	47.615	48.190	48.769	49.353	49.942	50.536	51.134	51.737	52.345
3.6	52.957	53.574	54.197	54.824	55.456	56.092	56.734	57.381	58.032	58.689
3.7	59.351	60.017	60.689	61.366	62.047	62.734	63.426	64.123	64.825	65.533
3.8	66.245	66.963	67.686	68.414	69.148	69.886	70.631	71.380	72.135	72.895
3.9	73.660	74.431	75.208	75.989	76.777	77.569	78.368	79.171	79.981	80.796
4.0	81.616	82.442	83.274	84.111	84.954	85.803	86.658	87.518	88.384	89.256
4.1	90.133	91.017	91.906	92.801	93.702	94.609	95.522	96.441	97.366	98.296
4.2	99.233	100.176	101.125	102.080	103.041	104.009	104.982	105.962	106.948	107.940
4.3	108.94	109.94	110.95	111.97	113.00	114.03	115.06	116.10	117.15	118.21
4.4	119.27	120.34	121.42	122.50	123.59	124.68	125.78	126.89	128.01	129.13
4.5	130.26	131.40	132.54	133.69	134.84	136.01	137.18	138.35	139.54	140.73
4.6	141.93	143.13	144.35	145.57	146.79	148.03	149.27	150.52	151.77	153.04
4.7	154.31	155.58	156.87	158.16	159.46	160.77	162.09	163.41	164.74	166.08
4.8	167.42	168.78	170.14	171.51	172.89	174.27	175.66	177.06	178.47	179.89
4.9	181.32	182.75	184.19	185.64	187.10	188.56	190.04	191.52	193.01	194.51
5.0	196.02									

In the range from 1.25 K to 5.0 K, the temperature T_{90} is defined in terms of the ^{4}He vapor-pressure equation (see Eq.(8.4) and Table 8.3). From 0.5 K to 1.25 K and from 5.0 K to $T_{cr} = 5.1953$ K (critical point) the vapor pressure at temperature T_{76} is given by the relations [19, 20]:
from 0.5 K to 1.25 K

$$\ln(p/\text{Pa}) = \sum_{i=-1}^{6} a_i T_{76}^i, \tag{8.46}$$

where $a_{-1} = -7.41816$ K, $a_0 = 5.42128$, $a_1 = 9.903203\ \text{K}^{-1}$, $a_2 = -9.617095\ \text{K}^{-2}$, $a_3 = 6.804602\ \text{K}^{-3}$, $a_4 = -3.0154606\ \text{K}^{-4}$, $a_5 = 0.7461357\ \text{K}^{-5}$, $a_6 = -0.0791791\ \text{K}^{-6}$;
from 5.0 K to 5.1953 K

$$\ln(p/\text{Pa}) = \sum_{i=-1}^{8} a_i (T_{76}/T_{cr})^i + b(1 - T_{76}/T_{cr})^{1.9}, \tag{8.47}$$

where $a_{-1} = -30.93285$, $a_0 = 392.47361$, $a_1 = -2328.04587$, $a_2 = 8111.30347$, $a_3 = -17809.80901$, $a_4 = 25766.52747$, $a_5 = -24601.4$, $a_6 = 14944.65142$, $a_7 = -5240.36518$, $a_8 = 807.93168$, $b = 14.53333$.

Table 8.34 e-H_2 vapor pressure as a function of temperature [Pressure in kPa.]

T_{68}, K	0.0	0.2	0.4	0.6	0.8	T_{68}, K	0.0	0.2	0.4	0.6	0.8
14	7.842	8.773	9.787	10.888	12.082	19	67.879	72.480	77.303	82.354	87.641
15	13.372	14.765	16.266	17.879	19.609	20	93.169	98.944	104.973	111.263	117.818
16	21.463	23.445	25.562	27.818	30.220	21	124.65	131.75	139.15	146.83	154.82
17	32.772	35.481	38.352	41.392	44.606	22	163.11	171.71	180.62	189.87	199.44
18	47.999	51.579	55.351	59.321	63.495	23	209.35				

From 13.81 K to 23 K the vapor pressure as a function of temperature T_{68} is given by the equation [14]:

$$\ln(p/p_0) = \sum_{i=-1}^{2} a_i T_{68}^i, \tag{8.48}$$

where $a_{-1} = -101.33782$ K, $a_0 = 3.940796$, $a_1 = 5.43201 \cdot 10^{-2}$ K^{-1}, $a_2 = -1.10563 \cdot 10^{-4}$ K^{-2}.

Table 8.35 n-H_2 vapor pressure as a function of temperature [Pressure in kPa.]

T_{68}, K	0.0	0.2	0.4	0.6	0.8	T_{68}, K	0.0	0.2	0.4	0.6	0.8
14	7.384	8.271	9.238	10.289	11.430	23	203.35	213.39	223.77	234.50	245.59
15	12.665	13.999	15.437	16.984	18.646	24	257.05	268.88	281.09	293.68	306.67
16	20.428	22.335	24.373	26.546	28.862	25	320.05	333.84	348.05	362.67	377.72
17	31.325	33.940	36.715	39.654	42.763	26	393.20	409.12	425.49	442.31	459.59
18	46.049	49.517	53.173	57.023	61.073	27	477.33	495.55	514.25	533.44	553.13
19	65.330	69.798	74.486	79.397	84.540	28	573.31	594.01	615.22	636.96	659.22
20	89.920	95.543	101.415	107.543	113.934	29	682.02	705.37	729.27	753.73	778.75
21	120.59	127.53	134.74	142.24	150.04	30	804.35				
22	158.14	166.54	175.26	184.29	193.66						

From 13.956 K to 30 K the vapor pressure as a function of temperature T_{68} is given by the equation [54]:

$$\ln(p/p_0) = \sum_{i=-1}^{2} a_i T_{68}^i, \tag{8.49}$$

where $a_{-1} = -102.74982$ K, $a_0 = 3.994505$, $a_1 = 5.33898 \cdot 10^{-2}$ K^{-1}, $a_2 = -1.10563 \cdot 10^{-4}$ K^{-2}.

Table 8.36 Neon vapor pressure as a function of temperature [Pressure in kPa.]

T_{68}, K	0.0	0.2	0.4	0.6	0.8	T_{68}, K	0.0	0.2	0.4	0.6	0.8
19	1.932	2.227	2.560	2.934	3.354	30	223.09	234.27	245.85	257.84	270.26
20	3.823	4.348	4.932	5.581	6.302	31	283.11	296.40	310.13	324.32	338.98
21	7.099	7.980	8.952	10.021	11.196	32	354.11	369.72	385.82	402.43	419.54
22	12.484	13.894	15.435	17.117	18.949	33	437.18	455.34	474.05	493.30	513.10
23	20.942	23.107	25.456	27.999	30.751	34	533.47	554.42	575.95	598.08	620.82
24	33.723	36.929	40.385	44.022	47.356	35	644.16	668.14	692.75	718.00	743.91
25	50.882	54.605	58.534	62.676	67.037	36	770.48	797.74	825.67	854.31	883.66
26	71.626	76.449	81.515	86.831	92.404	37	913.73	944.52	976.07	1008.36	1041.42
27	98.243	104.355	110.747	117.429	124.408	38	1075.26	1109.89	1145.33	1181.57	1218.65
28	131.69	139.29	147.21	155.45	164.04	39	1256.56	1295.33	1334.96	1375.47	1416.88
29	172.97	182.25	191.89	201.91	212.31	40	1459.19				

The vapor pressure as a function of temperature T_{68} is given by the equations [54]:
from 19 K to 24.561 K

$$\ln(p/p_0) = \sum_{i=-1}^{2} a_i T_{68}^i, \tag{8.50}$$

where $a_{-1} = -261.18205$ K, $a_0 = 10.275895$, $a_1 = -4.54082 \cdot 10^{-2}$ K^{-1}, $a_2 = 1.035289 \cdot 10^{-3}$ K^{-2};
from 24.561 K to 40 K

$$\ln(p/p_0) = \sum_{i=-1}^{2} b_i T_{68}^i, \tag{8.51}$$

where $b_{-1} = -244.96075$ K, $b_0 = 10.618417$, $b_1 = -8.481135 \cdot 10^{-2}$ K^{-1}, $b_2 = 9.78350 \cdot 10^{-4}$ K^{-2}.

Table 8.37 Nitrogen vapor pressure as a function of temperature [Pressure in kPa.]

T_{68}, K	0.0	0.2	0.4	0.6	0.8	T_{68}, K	0.0	0.2	0.4	0.6	0.8
56	2.360	2.487	2.621	2.761	2.907	71	44.596	45.877	47.187	48.525	49.893
57	3.060	3.219	3.386	3.560	3.742	72	51.291	52.719	54.178	55.669	57.190
58	3.931	4.129	4.334	4.549	4.773	73	58.744	60.331	61.950	63.603	65.290
59	5.006	5.248	5.501	5.764	6.037	74	67.012	68.768	70.559	72.386	74.250
60	6.321	6.617	6.924	7.243	7.575	75	76.150	78.088	80.063	82.076	84.129
61	7.919	8.277	8.648	9.032	9.432	76	86.220	88.351	90.522	92.733	94.986
62	9.846	10.275	10.720	11.181	11.658	77	97.280	99.616	101.995	104.417	106.883
63	12.153	12.651	13.123	13.609	14.109	78	109.39	111.95	114.55	117.19	119.88
64	14.624	15.154	15.699	16.260	16.838	79	122.62	125.41	128.24	131.12	134.05
65	17.431	18.041	18.668	19.312	19.974	80	137.03	140.06	143.14	146.27	149.45
66	20.655	21.353	22.070	22.807	23.562	81	152.68	155.97	159.31	162.70	166.15
67	24.338	25.134	25.950	26.787	27.646	82	169.65	173.20	176.82	180.48	184.21
68	28.526	29.428	30.353	31.300	32.271	83	187.99	191.83	195.73	199.69	203.70
69	33.265	34.284	35.326	36.394	37.487	84	207.78				
70	38.605	39.750	40.921	42.118	43.344						

The vapor pressure as a function of temperature T_{68} is given by the equations [54]:
from 56 K to 63.146 K

$$\ln(p/p_0) = \sum_{i=-1}^{1} a_i T_{68}^i, \tag{8.52}$$

where $a_{-1} = -861.621597$ K, $a_0 = 12.189891$, $a_1 = -1.006552 \cdot 10^{-2}$ K^{-1};
from 63.146 K to 84 K

$$\ln(p/p_0) = \sum_{i=-1}^{2} b_i T_{68}^i + c \ln(T_{68}/T_0), \tag{8.53}$$

where $b_{-1} = -930.153333$ K, $b_0 = 13.569758$, $b_1 = -3.288437 \cdot 10^{-2}$ K^{-1}, $b_2 = 1.671382 \cdot 10^{-4}$ K^{-2}, $c = -2.36680$, $T_0 = 77.344$.

Table 8.38 Argon vapor pressure as a function of temperature [Pressure in kPa.]

T_{68}, K	0.0	0.2	0.4	0.6	0.8	T_{68}, K	0.0	0.2	0.4	0.6	0.8
81	46.454	47.825	49.228	50.666	52.138	85	78.968	80.749	82.561	84.404	86.279
82	53.646	55.189	56.769	58.387	60.042	86	88.186	90.126	92.098	94.103	96.142
83	61.736	63.469	65.242	67.056	68.911	87	98.214	100.320			
84	70.514	72.146	73.806	75.497	77.217						

The vapor pressure as a function of temperature T_{68} is given by the equations [21]:
from 81 K to 83.798 K

$$\ln(p/p_0) = \sum_{i=-1}^{0} a_i T_{68}^i, \tag{8.54}$$

where $a_{-1} = -955.992$ K, $a_0 = 11.02251$;
from 83.798 K to 87.294 K

$$\ln(p/p_0) = \sum_{i=-1}^{1} b_i T_{68}^i, \tag{8.55}$$

where $b_{-1} = -864.70028$ K, $b_0 = 10.593120$, $b_1 = -7.87611 \cdot 10^{-3}$ K^{-1}.

Table 8.39 Oxygen vapor pressure as a function of temperature

T_{68}, K	0.0	0.2	0.4	0.6	0.8	T_{68}, K	0.0	0.2	0.4	0.6	0.8
	Pressure in Pa						Pressure in kPa				
54			148.364	157.985	168.147	74	12.409	12.814	13.229	13.656	14.093
55	178.87	190.19	202.13	214.72	227.98	75	14.542	15.002	15.474	15.958	16.455
56	241.95	256.65	272.13	288.41	305.53	76	16.964	17.485	18.019	18.567	19.127
57	323.52	342.42	362.27	383.11	404.97	77	19.702	20.290	20.892	21.509	22.140
58	427.91	451.96	477.16	503.57	531.23	78	22.786	23.447	24.124	24.816	25.524
59	560.18	590.48	622.18	655.34	689.99	79	26.247	26.988	27.745	28.518	29.309
60	726.21	764.04	803.56	844.80	887.85	80	30.118	30.944	31.788	32.650	33.531
61	932.75	979.59	1028.41	1079.30	1132.32	81	34.431	35.350	36.288	37.246	38.223
	Pressure in kPa					82	39.222	40.240	41.280	42.340	43.422
62	1.188	1.245	1.305	1.367	1.432	83	44.526	45.652	46.800	47.971	49.165
63	1.499	1.569	1.642	1.718	1.796	84	50.382	51.623	52.888	54.176	55.490
64	1.878	1.963	2.051	2.142	2.236	85	56.828	58.192	59.581	60.996	62.437
65	2.334	2.436	2.542	2.651	2.764	86	63.904	65.399	66.921	68.470	70.047
66	2.881	3.002	3.128	3.258	3.392	87	71.652	73.286	74.948	76.640	78.361
67	3.531	3.675	3.824	3.977	4.136	88	80.113	81.894	83.706	85.549	87.424
68	4.300	4.470	4.644	4.825	5.011	89	89.330	91.268	93.239	95.242	97.278
69	5.204	5.402	5.607	5.818	6.035	90	99.348	101.452	103.590	105.762	107.970
70	6.259	6.491	6.729	6.974	7.226	91	110.21	112.49	114.81	117.16	119.54
71	7.487	7.754	8.030	8.314	8.606	92	121.97	124.43	126.93	129.47	132.05
72	8.906	9.214	9.532	9.858	10.194	93	134.67	137.32	140.02	142.75	145.53
73	10.539	10.893	11.257	11.631	12.015	94	148.35				

The vapor pressure as a function of temperature T_{68} is given by the equations [14]:
from 54.361 K to 94 K

$$\ln(p/p_0) = \sum_{i=-1}^{2} b_i T_{68}^i + c\ln(T_{68}/T_0), \tag{8.56}$$

where $b_{-1} = -1076.356664$ K, $b_0 = 13.726967$, $b_1 = -3.042408 \cdot 10^{-2}$ K^{-1}, $b_2 = 1.169807 \cdot 10^{-4}$ K^{-2}, $c = -1.6645120$, $T_0 = 90.188$.

Table 8.40 CF_4 vapor pressure as a function of temperature

T_{68}, K	0	1	2	3	4	T_{68}, K	0	1	2	3	4
	Pressure in Pa						Pressure in kPa				
90	119.70	147.49	180.76	220.41	267.46	130	30.379	33.215	36.261	39.527	43.024
95	323.00	388.31	464.77	553.90	657.42	135	46.763	50.757	55.017	59.556	64.386
100	777.18	915.22	1073.77	1255.24	1462.26	140	69.520	74.972	80.755	86.882	93.368
	Pressure in kPa					145	100.23	107.47	115.12	123.18	131.68
105	1.698	1.965	2.266	2.606	2.988	150	140.62	150.03	159.91	170.29	181.18
110	3.416	3.895	4.428	5.021	5.680	155	192.59	204.55	217.07	230.16	243.85
115	6.408	7.213	8.099	9.074	10.144	160	258.15	273.08	288.65	304.89	321.81
120	11.316	12.597	13.995	15.517	17.171	165	339.43	357.77	376.84	396.67	417.28
125	18.967	20.912	23.017	25.289	27.740	170	438.67	460.88	483.91	507.80	

From 89.6 K to 173 K the vapor pressure at temperature T_{68} is given by the equation [55]:

$$\ln(p/p_{cr}) = (A_1\tau + A_2\tau^{1.5} + A_3\tau^3 + A_4\tau^6)/T_R, \tag{8.57}$$

where $\tau = 1 - T_R$; $T_R = T_{68}/T_{cr}$; $p_{cr} = 3.742 \cdot 10^6$ Pa is the critical pressure; $T_{cr} = 227.5$ K is the critical temperature; $A_1 = -6.7651056$, $A_2 = 1.0777437$, $A_3 = -1.7490971$, $A_4 = -2.5845119$.

Table 8.41 Spectral emissivity of substances at $\lambda = 650$ nm [60–62]

Substance (density, ρ/ρ_{theor})	T, K	ϵ_λ^{*1}
Chromium	1733	0.39
Cobalt	1180–1530	0.39–0.37*2
Germanium	1000–2000	0.50–0.53
Graphite, synthetic	1200–3200	0.90–0.83
Hafnium	1500–1800	0.445–0.453
Hafnium carbide	1300–2900	0.73–0.60
Iridium	1300–2500	0.272–0.215
Iron	1130–1430	0.38–0.35*2
Molybdenum	1100–2800	0.409–0.352
Nickel	1080–1500	0.36–0.32
Nickel, oxidized	1100–1500	0.86–0.82
Niobium	1000–2600	0.368–0.348
Niobium carbide	1300–3400	0.66–0.49
Osmium	1200–1800–2500	0.55–0.38–0.39
Palladium	1000–1700	0.400–0.306
Platinum	1100–2000	0.292–0.300
Platinum, cold-rolled	1100–1500	0.32–0.42
Rhenium	1000–2900	0.432–0.370
Rhodium	1100–2100	0.269–0.196
Silicon	1000–1700	0.64–0.46
Tantalum	1200–2400	0.445–0.414
Tantalum, tempered	1100–1600–2800	0.49–0.44–0.41
Tantalum carbide, fused (0.95) [63]	300–3000	0.514*3
Thorium dioxide, fused (0.96) [63]	1500–3440	0.863–0.876*3
Titanium	1000–1900	0.503–0.444
Tungsten	1200–2600	0.453–0.418
Uranium carbide, fused (0.965) [63]	300–1400–2780	0.615–0.556–0.539*2
Uranium dioxide, fused (0.96) [63]	300–3120	0.824–0.836*3

Table 8.41 Spectral emissivity of substances at $\lambda = 650$ nm [60–62] *(continued)*

Substance (density, ρ/ρ_{theor})	T, K	ϵ_λ^{*1}
Vanadium	1000–2000	0.419–0.370
Zirconium	1000–2100	0.450–0.403
Zirconium dioxide	1155–1800	0.40–0.55*2

*1 Spectral emissivity corresponds to indicated temperatures. Linear interpolation between points is rather good.
*2 $\lambda = 630$ nm.
*3 $\lambda = 665$ nm.

Table 8.42 Spectral emissivity ϵ_λ of substances at different wavelength [60]

Substance	T, K	Type*1	Wavelength, μm											
			0.3	0.4	0.5	0.6	0.7	0.8	0.9	1.0	2.0	3.0	4.0	5.0
Aluminum	293	n	0.081	0.083	0.089	0.098	0.111	0.14	0.11	0.068	0.030	0.026	0.024	0.023
Gold	293	n	0.64	0.63	0.540	0.130	0.040	0.03	0.02	0.02	0.02	0.017	0.014	0.012
Copper	293	n	0.67	0.50	0.400	0.200	0.030	0.02	0.02	0.02	0.018	0.016	0.014	0.012
Graphite*2	293	h*3		0.97	0.960	0.950	0.925	0.90	0.89	0.87	0.72	0.600	0.500	
Iron	293	h*3			0.440	0.430	0.420	0.39	0.37	0.36	0.22	0.160	0.110	0.080
Molybdenum*4	293	n*3		0.480	0.475	0.463	0.461	0.459	0.420	0.405	0.125	0.082	0.073	0.060
	1000	n		0.458	0.438	0.417	0.394	0.367	0.333	0.302	0.106	0.063	0.046	0.035
	1400	n		0.440	0.422	0.403	0.383	0.361	0.333	0.310	0.151	0.096	0.073	0.059
	2000	n		0.419	0.403	0.387	0.370	0.352	0.333	0.317	0.191	0.133	0.102	0.084
Palladium	295	n		0.48	0.42	0.36	0.33	0.31	0.29	0.27	0.17	0.14	0.11	0.08
Silver	293	h*3			0.08	0.06	0.05	0.045	0.04	0.03	0.025	0.02	0.02	0.02
Steel, carbon	293	n	0.50	0.45	0.42	0.39	0.38	0.37	0.36	0.35	0.22	0.16	0.12	0.10
Tantalum*4	293	n*3			0.56*5	0.493*6				0.210	0.110	0.080	0.070	0.065
	1200	n		0.525	0.510	0.473	0.421	0.363	0.304	0.262	0.148	0.123	0.108	
	2400	n		0.498	0.464	0.432	0.399	0.366	0.338	0.317	0.220	0.190	0.168	
Tungsten*4	293	h*3									0.070	0.059	0.051	0.046
	1200	n	0.486	0.482	0.474	0.461	0.446	0.428	0.408	0.386	0.186	0.112	0.086	0.078
	2600	n	0.465	0.461	0.447	0.426	0.411	0.394	0.376	0.360	0.248	0.191	0.163	0.146

*1 n denotes the spectral emissivity in direction normal to surface; h denotes the spectral emissivity within hemisphere.
*2 Synthetic porous graphite.
*3 Data from Ref.[62].
*4 At high temperatures, at a fixed wavelength, the linear interpolation between given points is accurate.
*5 $\lambda = 467$ nm.
*6 $\lambda = 665$ nm.

Table 8.43 Radiant total emissivity ϵ_t of substances [60–62]

Substance	T, K	Type*1	ϵ_t^{*2}
Aluminum:	50–800	h	0.008–0.062
strong-oxidized	360–800	h	0.20–0.33
oxidized electrolytically 4–10 microns	310	h	0.72–0.83
Antimony	300–350	n	0.28–0.31
Beryllium	1100–1300–1480	n	0.41–0.57–0.87
Brass:	373	n	0.06
polished	500–610	h	0.02
oxidized	450–590	n	0.56–0.64

Table 8.43 Radiant total emissivity ϵ_t of substances [60–62] *(continued)*

Substance	T, K	Type*1	ϵ_t^{*2}
Bronze:			
aluminum	450–1270	n	0.03–0.06
oxidized	450–1270	n	0.08–0.16
Cadmium	80–300	h	0.03
Chromium	270–600–750–1220	n	0.06–0.06–0.10–0.42
Cobalt	1100–1500	h	0.21
Copper	50–300–800–1100	h	0.022–0.024–0.05–0.061
same as above	80–380–1160	n	0.02–0.01–0.02
Copper, oxidized	300–600–800–1100	h	0.38–0.47–0.59–0.87
Duralumin D16	500–900	n	0.016–0.03
Gafnium	1200–2000	h	0.284–0.324
Glass	293	n	0.94
Gold	50–1150	h	0.014–0.063
Graphite	1200–3200	h	0.77–0.83
Iridium	1300–2500	h	0.21–0.23
same as above	295	n	0.04
Iron	160–1100	h	0.081–0.254
same as above	600–1100	n	0.20–0.56
Iron oxide Fe_2O_3	310–1350	h	0.82–0.89
Lead:	310–530	h	0.04–0.08
gray, oxidized	270–470	h	0.28
oxidized at 473 K	473	h	0.63
Magnesium	410–490	h	0.12
Manganese, rolled	391	n	0.048
Mercury	273–373	n	0.09–0.12
Molybdenum	1100–2100–2400–2800	h	0.105–0.225–0.254–0.282
same as above	293	n	0.009
same as above	1100–2800	n	0.096–0.275
Nichrome:			
rolled	800–950–1100	n	0.20–0.24–0.36
oxidized	480–900–1200	h	0.62–0.67–0.78
Nickel	300–1500	h	0.068–0.196
same as above	500–1400	n	0.061–0.182
Nickel, oxidized	420–700–980	n	0.07–0.39–0.47
Niobium	1000–1700–2400	h	0.116–0.187–0.244
same as above	1000–1800–2600	n	0.085–0.170–0.232
Palladium	1000–1300–1600	h	0.10–0.15–0.18
same as above	400–1520	n	0.02–0.17
Platinum	300–1000–1400–1900	h	0.041–0.128–0.167–0.200
same as above	400–800–1600–2000	n	0.029–0.080–0.162–0.184
Rhenium	1000–1400–1900–	h	0.164–0.201–0.255–
	2300–2600–2900	h	0.290–0.309–0.322
Rhodium	900–1200–1600–	h	0.068–0.112–0.150–
	1900–2200	h	0.169–0.183
same as above	900–1600	n	0.053–0.100
Silver	100–1100	h	0.012–0.046
	80–300	n	0.020–0.075
Steel:			
carbon	100–1400	h	0.06–0.31
carbon, oxidized	573–1073	h	0.86–0.91
07KH16N6	250–900	n	0.13–0.28
07KH16N6, oxidized at 1173 K	300–1000	n	0.84
12KH18N10T	180–700–1200	n	0.13–0.24–0.38
08KH18N12B	100–1000–1390	n	0.17–0.25–0.65

Table 8.43 Radiant total emissivity ϵ_t of substances [60–62] *(continued)*

Substance	T, K	Type[*1]	ϵ_t^{*2}
Tantalum	1000–2300–2700–3300	h	0.132–0.251–0.282–0.316
same as above	80–300	n	0.020–0.075
same as above	1300–1600–2000	n	0.132–0.170–0.260
Tantalum nitride	800–1500–2600	n	0.74–0.80–0.60
Tellurium	295	n	0.22 7–0.323
Tin	310–360	h	0.05
Titanium:	900–1400–1900	n	0.217–0.274–0.323
electropolished	250–370	h	0.10–0.13
carbide	1200–3000	h	0.60–0.75
Tungsten	400–800–2000–3400	h	0.039–0.081–0.249–0.345
same as above	1200–1800–2200–3000	n	0.116–0.201–0.247–0.312
Uranium	1200	h	0.35
Uranium carbide	1600–2000	h	0.42
same as above	80	h	0.07
Vanadium	1000–1300–2000	h	0.145–0.190–0.257
Water:	272–373	n	0.92–0.96
ice, smooth	273	n	0.96
ice, frost-coated	273	n	0.985
Zinc	300–530	n	0.02–0.06
Zinc oxidized	300–470–800	n	0.28–0.14–0.11
Zirconium	1100–1600–2100	h	0.204–0.248–0.278
Zirconium carbide	1400–2900	h	0.40–0.55
Zirconium diboride	1200–2500	h	0.53–0.75

*1 n denotes the total emissivity in direction normal to surface; h denotes the total emissivity within hemisphere.
*2 Values of total emissivity correspond to indicated temperatures. Linear interpolation between points is accurate.

Table 8.44 Radiant total emissivity ϵ_t of some oxides [60]

Oxide	Temperature, K												
	200	400	600	800	1000	1200	1400	1600	1800	2000	2200	2400	2600
Al_2O_3		0.79	0.71	0.60	0.52	0.46	0.42	0.40	0.39				
MgO	0.73	0.69	0.57	0.52	0.42	0.35	0.30	0.28	0.29	0.35	0.49		
ThO_2			0.61	0.52	0.43	0.38	0.37	0.40	0.49	0.61	0.68	0.71	
ZrO_2	0.81	0.75	0.65	0.53	0.42	0.37	0.37	0.39	0.46	0.55	0.62	0.66	0.69

References

[1] Comptes Rendus des Séances de la Treizième Conférence Générale des Poids et Mesures (1967-1968), Resolutions 3 and 4, p. 104.

[2] Krichevskii, I. R., Notions and fundamentals of thermodynamics, 2-nd ed., Khimia, Moscow, 1970 (in Russian).

[3] Wolfe, H. G., TMCSI, 2, 3, 1955.

[4] de Boer, J., Metrologia, 1, 158, 1965.

[5] Quinn, T. J., Compton, J. P., Rep. Progr. Phys., 38, 151, 1975.

[6] Popov, M. M., Thermometry and calorimetry, Izdatel'stvo MGU, 2-nd ed., Moscow, 1954 (in Russian).

[7] Gordov, A. N., Temperature scales, Izdatel'stvo standartov, Moscow, 1966 (in Russian).

[8] Hudson, R. P., Rev. Sci. Instrum., 51, 871, 1980.

[9] Hall, J. A., J. Sci. Instrum., 43, 511, 1966.

[10] Quinn, T. J. Temperature, 2-nd ed., Academic Press, London, 1990.

[11] Orlova, M. P., Pogorelova, O. F., Ulybin, S. A., Low-temperature thermometry, Energoatomizdat, Moscow, 1987 (in Russian).

[12] McGee, T. D., Principles and methods of temperature measurements, Wiley, New York, 1988.

[13] Temperature, Its Measurement and Control in Science and Industry. Vol. 1, (Reinhold, New York, 1941); Vol. 2, ed. by H. C. Wolfe (Reinhold, New York, 1955); Vol. 3, ed. by C.M. Hertzfeld (Reinhold, New York, 1962); Vol. 4, ed. by H.H. Plumb (Instrument Society of America, Pittsburg, 1972); Vol. 5, ed. by J.F. Schooley (American Institute of Physics, New York, 1982); Vol. 6, ed. by J.F. Schooley (American Institute of Physics, New York, 1992).

[14] Preston-Thomas H., Metrologia, 12, 7, 1976.

[15] The 1976 Provisional 0.5 K to 30 K temperature scale, Metrologia, 15, 65, 1979.

[16] Preston-Thomas, H., The International Temperature Scale of 1990 (ITS-90), Metrologia, 27, 3, 1990; *ibid* 107.

[17] Supplementary Information for the ITS-90, Bureau International des Poids et Mesures, Sèvres, 1990.

[18] Techniques for approximating the International Temperature Scale of 1990, 1st ed., Bureau International des Poids et Mesures, Sèvres, 1990.

[19] Durieux, M., Rusby, R. L., Metrologia, 19, 67, 1983.

[20] Rusby, R. L., Durieux, M., Cryogenics, 4, 363, 1984.

[21] Bedford, R. E., Bonnier, G., Maas, H., Pavese, F., Metrologia, 20, 145, 1984.

[22] Fellmuth, B. and Maas, H., Recommended values of superconducting transition temperatures as reference temperatures for a selected set of materials, Comité Consultatif de Thermométrie, 16e Session, Document CCT/87-32, 1987.

[23] Wise, J. A., Liquid-in-glass thermometers, NBS Monograph, 150, 1976.

[24] Thompson, R. D., ISA Trans., 7, 87, 1968.

[25] Ween, S., ISA Trans., 7, 93, 1968.

[26] Annual Book of Standards: Section 14: Volume 14.01: Standard E 1-86: Standard Specification for ASTM Thermometers, American Society for Testing and Materials, Philadelphia, 1987. Standard E 77-84: Standard Method for Verification and Calibration of Liquid-in-glass Thermometers, ASTM, Philadelphia, 1987, 56-136.

[27] Guide to Selection and the Use of Liquid-in-glass Thermometers, BS 1041, Part 2, Section 2.1, British Standards Institution, London, 1985.

[28] Gerashchenko, O. A., Gordov, A. N., Lakh, V. I., Temperature measurements, Handbook, Naukova Dumka, Kiev, 1984 (in Russian).

[29] Sosnovskii, A. G., Stolyarova, N. I., Temperature measurement, Izdatel'stvo standartov, Moscow, 1970 (in Russian).

[30] Vepřek, I. J., Elekrická měřeni nízkých teplot, SNTL - Nakladatelstvi technické literatury, Praha, 1977.

[31] Riddle, J. L., Furukawa, G. T., Plumb, H. H., Platinum resistance thermometry, NBS Monograph, 126, 1973.

[32] Johnston, J. S., Resistance thermometry, Temperature-75, Conference Series No.26, Institute of Physics, London, 1975, 80.

[33] Sachse, H. B., Semi-conducting temperature sensors and their applications, Wiley, London, 1975.

[34] White, G. K., Experimental techniques in low-temperature physics, 3-d ed., Clarendon Press, Oxford, 1979.

[35] Astrov, D. N., Belyanskii, L. B., Low temperature measurement (Review), Fizika nizkih temperatur, 2, 821, 1967.

[36] Rubin, L. G., Brandt, B. L., Sample, H. H., Cryogenics, 22, 491, 1982.

[37] Kinzie, P. A., Thermocouple Temperature Measurement, John Wiley, New York, 1973.

[38] Rogel'berg, I. L., Beilin, V. M., Alloys for thermocouples. Handbook, Metallurgia, Moscow, 1983 (in Russian).

[39] Manual on the use of thermocouples in temperature measurement; ASTM special technical publication 470B, American Society for Testing and Materials, Philadelphia, 1981.

[40] Annual Book of ASTM Standards: Section 14: Volume 14.01: Standard E 230-87: Temperature-electromotive Force (emf) Tables for Standardized Thermocouples, American Society for Testing and Materials, Philadelphia, 242-372, 1987.

[41] Thermocouples, Part 1: Reference tables; International Electrotechnical Commission, IEC Standard, Publication 584-1, Bureau Central de la Commission Electrotechnique Internationale, Geneva, 1977.

[42] Danishevskii, S. K., Svede-Shvets, N. I., High temperature thermocouples, Metallurgia, Moscow, 1977 (in Russian).

[43] Samsonov, G. V., Kislii, P. S., High temperature non-metallic thermocouples and heads, Naukova Dumka, Kiev, 1965 (in Russian).

[44] Guildner, L. A., Burns, G. W., High Temp. - High Press., 11, 173, 1979.

[45] Bedford, R. E., High Temp. - High Press., 4, 241, 1972.

[46] Medvedeva, L. A., Orlova, M. P., Aleksakhin, I. A., Dukhovlinova, N. D., Low temperature measurement with thermocouples based on gold-cobalt alloys, Trudy VNIIFTRI, 4 (34), 154, 1973 (in Russian).

[47] Medvedeva, L. A., Orlova, M. P., Rabin'kin, A. G., Low temperature measurement with thermocouples based on gold-iron alloys, Trudy VNIIFTRI, 4 (34), 181, 1973 (in Russian).

[48] Industrial Thermocouples, Standard Reference Tables, Standard 3044-77, Gosudarstvennyj Komitet Standartov (USSR), 1977.

[49] Gotoh, M., Hill, K. D., Murdock, E. G., Rev. Sci. Instrum., 62(11), 2778, 1991.

[50] Ambrose, D., Metrologia, 27, 245, 1990.

[51] Hudson, R. P., Experimental Cryophysics, Butterworths, London, 1961 (chap. 9).

[52] Cataland, G., Edlow, M. H., Plumb, H. H., TMCSI, 3 (1), 413, 1962.

[53] Sydoriak, S. G., Rogerst, T. R., Sherman, R. H., J. Res. NBS, 68A, 547, 1964.

[54] Crovini, L., Bedford, R. E., Moser, A., Metrologia, 13, 197, 1977.

[55] Lobo, L. Q., Staveley, L. A. K., Cryogenics, 19, 335, 1979.

[56] Theory and practice of radiation thermometry, Ed. by DeWitt, D. P. and Nutter, G. N., Wiley, New York, 1989.

[57] Gordov, A. N., Pyrometry fundamentals, 2-nd ed., Metallurgia, Moscow, 1971 (in Russian).

[58] Kirenkov, I. I., Metrological principles of optical pyrometry, Izdatel'stvo standartov, Moscow, 1976 (in Russian).

[59] Optical methods of temperature measurement in metallurgy. Theory, systems, elements, Ed. Svet, D. Ya., Nauka, Moscow, 1979 (in Russian).

[60] Radiating properties of solid materials. Handbook, Ed. Sheindlin, A. E., Energia, Moscow, 1974 (in Russian).

[61] Novitskii, L. A., Stepanov, B. M., Optical properties of materials at low temperatures. Handbook, Mashinostroyeniye, Moscow, 1980 (in Russian).

[62] Kaspar, J., Radiometry, American Institute of Physics Handbook, 3-d ed., McGraw-Hill Company, New York, 1972.

[63] Boder, M., Karow, H. U., Muller, K., High Temp. - High Press., 12, 161, 1980.

9

Heat Capacity

M.N. Khlopkin

9.1 Introduction

9.1.1 Definitions and heat capacity units

Heat capacity of a body (C) is the ratio of the quantity of heat ΔQ supplied to a body at given conditions to the temperature increase ΔT. According to the third law of thermodynamics, heat capacity of any body tends to zero with nearing the temperature absolute zero.

The specific (c, J/(kg·K)) and molar (C, J/(mol·K)) heat capacities of a compound are defined as the heat capacities of unit mass and of a mole of the given compound. They are related by the equation $C = Mc$, where M is the mass of one mole (kg).

The heat capacity of a body depends on heating conditions. The most frequently used are the heat capacity at constant pressure (isobaric heat capacity) c_p, the heat capacity at constant volume (isochoric heat capacity) c_v, and the heat capacity under the saturated vapor pressure (along the line of the phase coexistence) c_s. The c_p and c_v quantities (J/(mol·K)) are related by the equation

$$c_p - c_v = \alpha^2 TV/k_T, \tag{9.1}$$

where α is the temperature expansion factor (K^{-1}), T is the temperature (K), V is the volume per mole (m^3/mol), and k_T is the isothermic compressibility factor (N/m^2 (Pa)).

9.1.2 Heat capacity of solids

The $c_p - c_v$ difference is usually not high in the case of solids. The heat capacity of solids (excluding solid helium) weakly depends on pressure. The most important parameter, characterizing the temperature-dependent heat capacity of solids, is the Debye characteristic temperature (Debye parameter) θ (K), which is determined by the relation $k\theta = h\nu$. Here k is the Boltzmann constant (J/K), h is the Planck constant (J·s), ν is the maximum vibrational frequency of an atom in the crystal (Hz).

According to the quantum Debye theory, the molar vibrational heat capacity of a solid c_v (J/mol·K) is determined by the relation

$$c_v = 3RnD(\theta/T), \tag{9.2}$$

where R is the gas constant (J/(mol·K)), n is the number of atoms in the molecule, D is the Debye function (it is tabulated in Table 9.7), θ is the Debye parameter (K), T is the temperature (K).

At temperatures higher than the Debye one ($T > \theta$), the Dulong-Petit law is obeyed

$$c_v = 3Rn \approx 25n, \text{ J/(mol} \cdot \text{K)}. \tag{9.3}$$

0-8493-2861-6/97/$0.00+$.50

At temperatures lower than the Debye parameter ($T < 0.1\theta$), the molar heat capacity of a solid c_v (J/(mol·K)) is usually represented in the following form:

$$c_v = 1944n(T/\theta)^3 + \gamma T, \tag{9.4}$$

where T and θ are the same as in the previous relations, γ is the electron heat capacity factor (J/(mol·K^2)). For insulators and semiconductors $\gamma = 0$. Significant deviations of heat capacity temperature dependence from the above formula are observed in lanthanides, actinides and in chemical compounds containing them, as well as in substances with magnetic ions, when thermal excitation of the atomic inner-shell electrons and of ion magnetic moments substantially contributes to heat capacity. The ordering processes (structural, magnetic, superconducting, etc.), which occur in solids, also lead to substantial anomalies in heat capacity.

9.1.3 Heat capacity of low-density gases

The isobaric and isochoric heat capacities difference $c_p - c_v$ (J/(mol·K)) in low-density gases is described by the relation $c_p - c_v = R$, where R is the universal gas constant.

According to the equipartition law, every translational and rotational molecular degree of freedom contributes $R/2$ to isochoric heat capacity, and every excited vibrational degree of freedom contributes R. Vibrational degrees of freedom become excited and contribute to heat capacity only at high temperatures ($T > h\nu/k$, where ν is the vibrational frequency of atoms in the molecules). When considering translational and rotational degrees of freedom and neglecting the vibrational ones, for molar isochoric (c_v) and isobaric (c_p) heat capacities (J/(mol·K)), the equipartition law gives values: $c_v = 1.5R,\ c_p = 2.5R$ in a monatomic gas; $c_v = 2.5R,\ c_p = 3.5R$ in a diatomic gas; $c_v = 3R,\ c_p = 4R$ in a polyatomic gas.

These relations are well obeyed in the monatomic gases, worse obeyed in the diatomic gases, and are poorly obeyed in the polyatomic gases.

The di– and polyatomic gas heat capacities at low temperature are lower than it follows from the equipartition law. This is owing to the decrease of the rotational degrees of freedom contribution. At high temperatures, heat capacity of these gases is higher than is predicted by the equipartition law, which is due to the thermal excitation of vibrational degrees of freedom and to partial dissociation of molecules.

9.1.4 Heat capacity of high-density gases and liquids

In the high-density gases, unlike the case of low-density gases, the molar heat capacities' difference $c_p - c_v$ may substantially differ from the universal gas constant R. The molar heat capacity of high-density gas is, as a rule, higher than that of a low-density one. At nearing critical point, heat capacity of the gas grows and it becomes infinite in the critical point. There are no simple relationships for treating the heat capacities of liquids.

9.1.5 Heat capacity of alloys, solutions, and mixtures

Heat capacity of alloys, solutions, and mixtures for a number of components with low deviation from ideality c_{mix} approximately meets the Kopp - Neumann additivity rule:

$$c_{\text{mix}} = \sum_j z_j C_j, \tag{9.5}$$

where z_j is the mole (mass) fraction, and C_j is the molar (specific) heat capacity of j-th mixture component, accordingly. The additivity rule is well obeyed for low-density gas, and worse for high-density gas. Substantial deviation from additivity rule is observed in electrolyte solutions, in a number of intermetal compounds and in the presence of phase transitions.

9.1.6 Heat capacity in the vicinity of phase transition point

Jump-like change in enthalpy is characteristic of the first-kind phase transitions (vaporization, melting, sublimation, crystal type modification, and so on), and it leads to corresponding transition hidden heat ΔH. Heat capacity usually changes in the phase transition of the first kind, the high-temperature phase heat capacity being either higher or lower than that of the low-temperature phase.

During the second-kind phase transition, enthalpy changes continuously, hidden heat release does not occur, and heat capacity undergoes the jump, accompanied by sharp maximum. As a rule, low-temperature phase heat capacity is higher than that of the high-temperature phase in the second-kind phase transition.

9.1.7 Reference books on heat capacity

Thermodynamic functions (heat capacities, entropy, and enthalpy) of individual substances (of elements and chemical compounds) at a temperature of 298.15 K are presented in [1]. Thermodynamic functions of individual substances (presumably in gaseous state) in a wide range of temperatures are presented in [2]. Heat capacities of elements and of binary alloys in a wide temperature range are contained in [3,15]. Heat capacities of elements, of inorganic and organic compounds are presented in [4]. Heat capacities of metals at temperatures lower than 900 K can be found in [5].

9.2 Heat Capacity of Elements

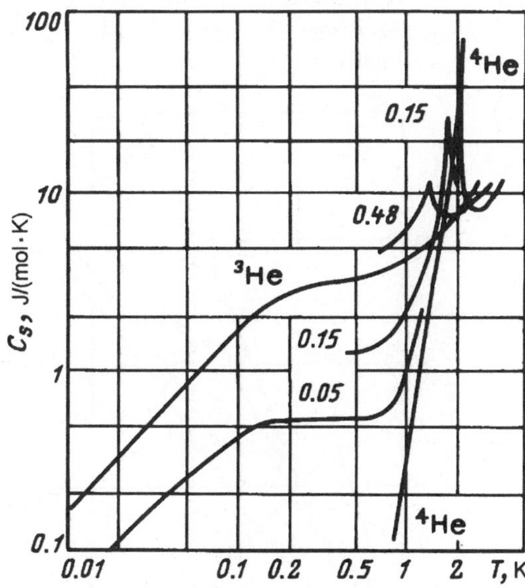

Figure 9.1 Molar heat capacity C_s of liquid helium isotopes ^{3}He and ^{4}He and of mixtures ^{3}He – ^{4}He with the molar fraction of ^{3}He equal 0.05, 0.15, 0.48 at the saturated vapor pressure [51].

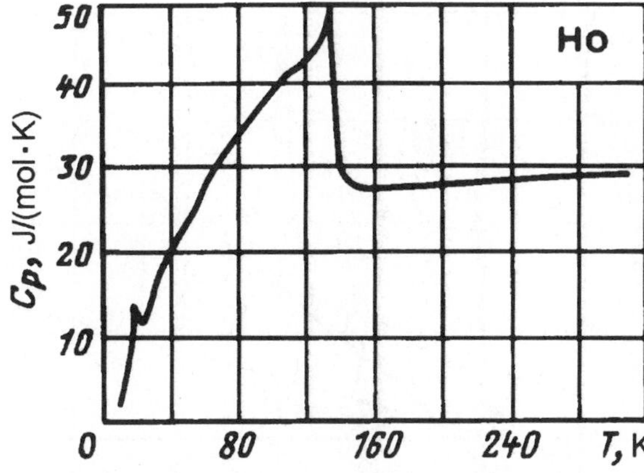

Figure 9.2 Molar heat capacity of holmium [4].

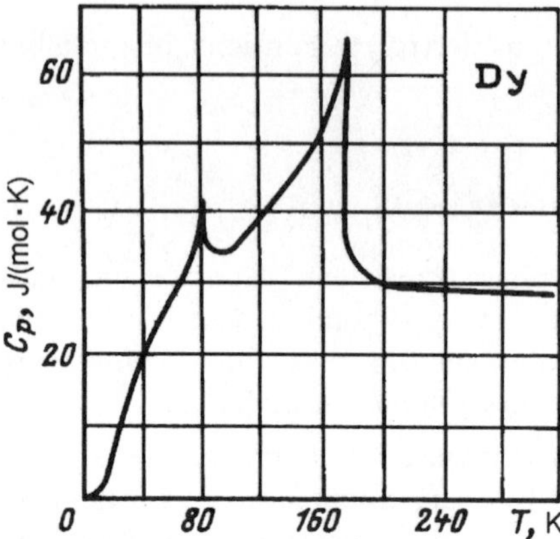

Figure 9.3 Molar heat capacity of dysprosium [4].

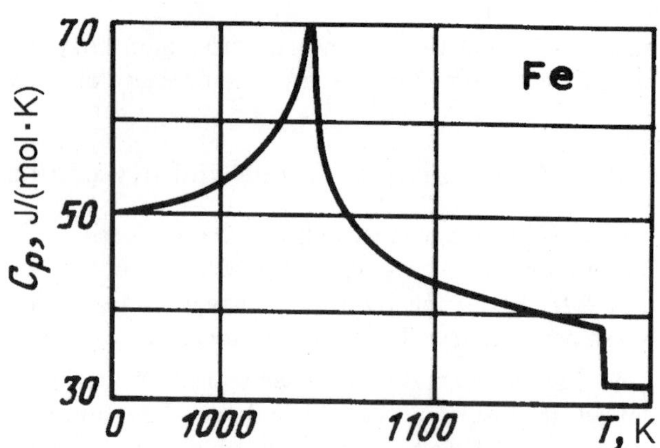

Figure 9.4 Molar heat capacity of iron near the ferromagnetic phase transition [52].

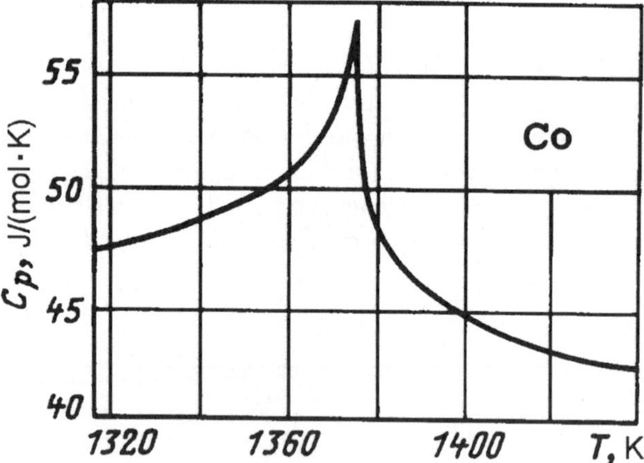

Figure 9.5 Molar heat capacity of cobalt near the ferromagnetic phase transition [53].

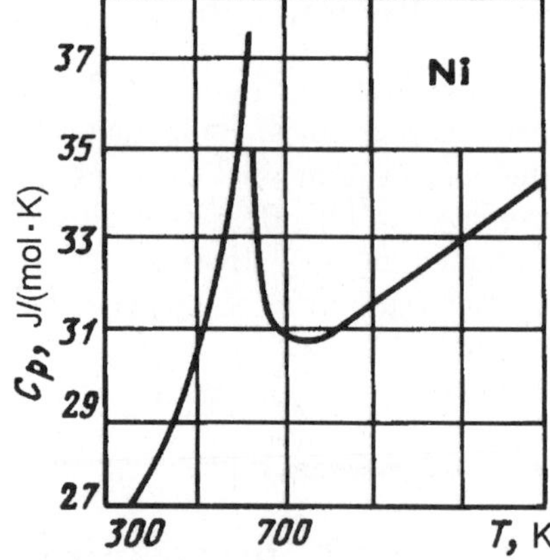

Figure 9.6 Molar heat capacity of nickel near the ferromagnetic phase transition [54].

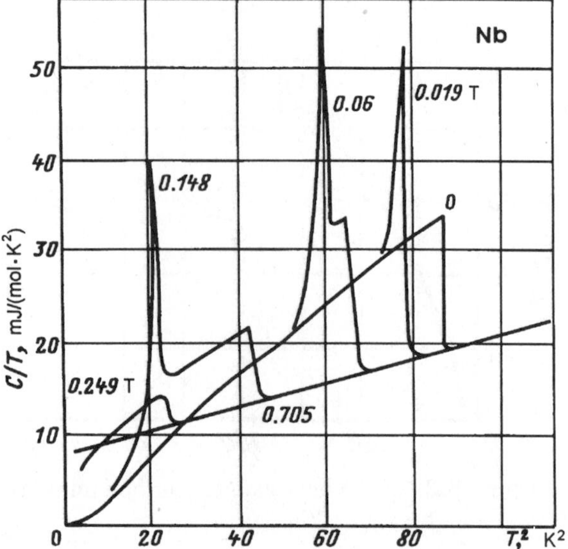

Figure 9.7 Molar heat capacity of niobium at different values of magnetic induction [55].

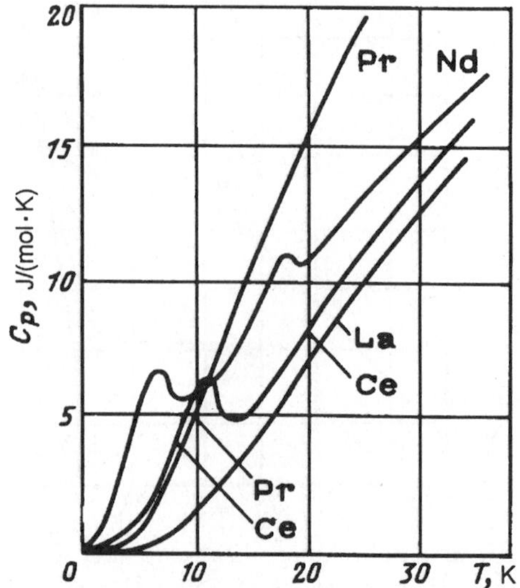

Figure 9.8 Molar heat capacity of lanthanum, cerium, praseodymium and neodymium [4].

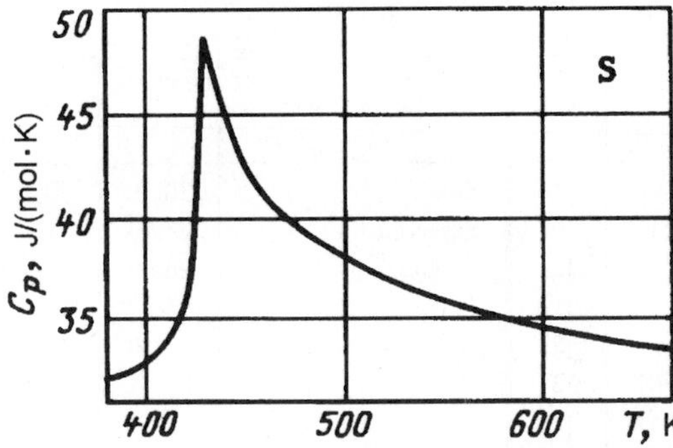

Figure 9.9 Molar heat capacity of liquid sulfur [4].

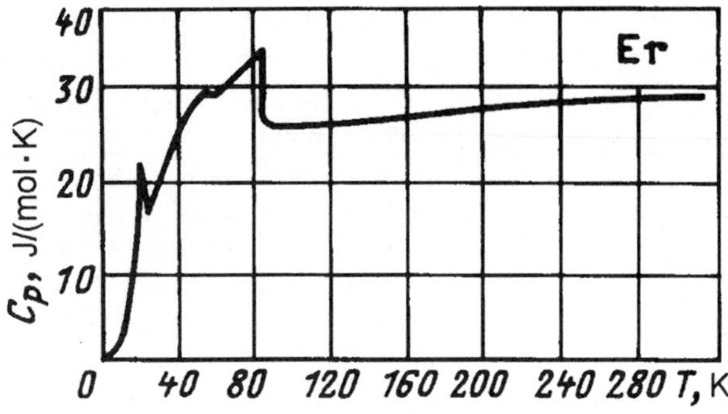

Figure 9.10 Molar heat capacity of erbium [4].

Table 9.1 Molar heat capacity C_p of elements at temperature 298.25 K and constant pressure [Substances in liquid and gaseous states are marked by letters "l" and "g". Different crystal modifications are marked by the greek letters $\alpha, \beta, \gamma, \delta$. Solids and liquids are at pressure of 0.1 MPa and gases are in the state of ideal gas (i.e. at utmost low pressure). Chemical formula is presented only for nonmonatomic molecules.]

Substance, formula	C_p, $\frac{J}{mol \cdot K}$	Ref.	Substance, formula	C_p, $\frac{J}{mol \cdot K}$	Ref.	Substance, formula	C_p, $\frac{J}{mol \cdot K}$	Ref.
Actinium	27.20	[1]	Gallium	26.07	[2]	Oxygen O_2 (g)	29.38	[2]
Aluminum	24.35	[2]	Germanium	23.22	[2]	Oxygen O_3 (g)	39.37	[2]
α-Americium	25.86	[1]	Gold	25.40	[1]	Palladium	25.86	[1]
Antimony	25.23	[1]	Hafnium	25.69	[2]	Phosphorus (black)	21.59	[1]
Antimony Sb_2 (g)	25.40	[1]	Helium γ	20.79	[3]	Phosphorus (red *IV*)	21.26	[3]
Argon (g)	20.79	[3]	α-Holmium	27.15	[1]	Phosphorus (red *V*)	21.19	[3]
α-Arsenic	24.68	[1]	*n*-Hydrogen H_2 (g)	28.83	[2]	Phosphorus (white)	23.82	[2]
Arsenic, As_2 (g)	35.10	[1]	*ortho*-Hydrogen H_2 (g)	28.46	[2]	Phosphorus P_2 (g)	32.05	[1]
Arsenic, As_4 (g)	77.40	[1]	*para*-Hydrogen H_2 (g)	29.95	[2]	Phosphorus P_4 (g)	67.15	[1]
Barium	28.10	[2]	Indium	26.90	[2]	Platinum	25.86	[1]
α-Beryllium	16.44	[2]	Iodine I_2	54.44	[2]	α-Plutonium	31.20	[1]
Bismuth	26.02	[1]	Iodine I_2 (g)	36.89	[2]	Polonium	26.40	[1]
Boron (amorph.)	11.95	[2]	Iridium	25.10	[1]	Potassium	29.60	[2]
Boron (crystal)	11.09	[2]	α-Iron	24.98	[1]	α-Praseodymium	27.45	[1]
Bromine Br_2 (g)	36.06	[2]	β-Iron	26.76	[4]	Radon (g)	20.79	[2]
Bromine Br_2 (l)	75.68	[2]	Krypton (g)	20.79	[2]	Rhenium	25.23	[1]
Cadmium	26.02	[1]	Lanthanum	27.11	[2]	Rhodium	24.98	[1]
Calcium	25.94	[2]	Lead	26.65	[2]	Rubidium	31.05	[1]
Carbon (diamond)	6.109	[2]	β-Lithium	24.85	[1]	Ruthenium	24.06	[1]
Carbon (graphite)	8.536	[2]	Lutecium	26.50	[2]	Samarium	29.54	[1]
β-Cerium	26.94	[1]	Magnesium	24.90	[2]	Scandium	25.50	[1]
Cesium	32.17	[1]	α-Manganese	26.28	[1]	Selenium (amorph.)	29.30	[1]
Chlorine Cl_2 (g)	33.94	[2]	β-Manganese	26.50	[1]	γ-Selenium (hex.)	25.36	[1]
Chromium	23.55	[2]	γ-Manganese	27.57	[1]	α-Selenium (monocl.)	27.20	[1]
α-Cobalt	24.8	[1]	Mercury (l)	27.98	[1]	Selenium Se_2 (g)	35.40	[1]
Copper	24.43	[1]	Molybdenum	23.93	[1]	Silicon	19.79	[2]
n-Deiterium D_2 (g)	29.20	[2]	Neodymium	27.40	[1]	Silver	25.36	[1]
α-Dysprosium	28.16	[1]	Neon (g)	20.79	[3]	β-Sodium	28.20	[1]
Erbium	25.44	[1]	α-Neptunium	29.46	[1]	Strontium	25.51	[2]
Europium	27.66	[1]	Nickel	26.07	[1]	Sulfur (monocl.)	23.64	[1]
Fluorine F_2 (g)	31.30	[2]	Niobium	24.44	[2]	Sulfur (rhombic)	22.69	[2]
Francium (l)	28.40	[1]	Nitrogen, N_2 (g)	29.12	[2]	Sulfur S_2 (g)	32.50	[2]
α-Gadolinium	37.07	[1]	Osmium	24.70	[1]	Sulfur S_8 (g)	160.70	[1]

Table 9.1 Molar heat capacity C_p of elements *(continued)*

Substance, formula	C_p, $\frac{J}{mol \cdot K}$	Ref.	Substance, formula	C_p, $\frac{J}{mol \cdot K}$	Ref.	Substance, formula	C_p, $\frac{J}{mol \cdot K}$	Ref.
Tantalum	25.29	[2]	α-Tin (grey)	25.78	[1]	Xenon (g)	20.79	[2]
Technetium	24.00	[1]	β-Tin (white)	27.11	[2]	Ytterbium	26.74	[1]
Tellurium	25.78	[1]	Titanum	25.06	[2]	Yttrium	26.52	[2]
Terbium	28.91	[1]	n-Tritium T_2 (g)	29.20	[1]	Zinc	25.44	[1]
α-Thallium	26.32	[2]	Tungsten	24.27	[2]	Zirconium	25.40	[2]
α-Thorium	26.23	[2]	α-Uranium	27.67	[2]			
Thulium	27.03	[1]	Vanadium	24.48	[2]			

Table 9.2 Isobaric specific heat capacity c_p (J/(kg·K)) of elements at temperatures from 20 to 1500 K [The substance state is marked before and after the phase transition: "s" for solid, "l" for liquid, "g" for gaseous. Different crystal modifications of solid state are marked by the greek letters α, β, γ and others. Solids and liquids are at pressure of 0.1 MPa and gases are in the state of ideal gas (i.e. at utmost low pressure).]

Substance	Temperature, K								Ref.
	20	50	100	200	400	600	1000	1500	
α-Arsenic	15.6	106	222	300	339	354	383		[3]
Aluminum	8.9	142	481	797	951	1037(s)	1177(l)	1177	[2,3,5]
Antimony	25.8	100	163	198	213	223	258(l)		[3]
Argon*[1]	306	627(s)	521(g)	521	521	521	521	521	[3,7]
Barium	67.4	140	177	192	259	300	328(s)	281(l)	[2,5]
Beryllium	1.61	19.2	199	1110	2179	2559	3060	3604	[2,5]
Bismuth	35	84.6	111	120	127(s)	141(l)	131		[3,5]
Boron (amorph.)	8.9	15.9	128	600	1463	1892	2337		[3]
Boron (cryst.)	2.2	7.08	99	561(γ)	1416(β)	1931	2306	2597	[3]
Bromine	79.6		273	336(γ)	230(g)	233	236	239	[2]
Cadmium	46	141	196	222	241(s)	264(l)			[3]
Calcium	36	271	488	613	670	758(α)	1066(β)	773(l)	[3,70]
Carbon (diamond)	0.21	1.88	20.5	194	854	1342	1799		[2,3]
Carbon (graphite)	6.3	42.2	138	411	994	1409	1799	2019	[2,3,58]
Cerium*[2]	62(α)				202(γ)	223	268(δ)	269(l)	[3]
Cesium	147	183	194	209(s)	240(l)	224	230	273	[9]
Chlorine	108		596	942(l)	634(g)	642	627	605	[3]
Chromium	2.1	35.8	193	385	482	516	614	806	[2,5]
Cobalt*[2]	5.4	73.1	236	379	450	503(α)	627(β)	674	[3]
Copper	7.29	96.9	252.2	355.5	397.5	416.7	451.1(s)	513(l)	[3,11]
α-Dysprosium*[2]	34.5	142	214	179	173	175	195	274	[3]
Erbium*[2]	126	170	147	161	170	174	192	229	[3]
Europium	65				184	199	250(s)	257(l)	[3,70]
Fluorine	347(α)	1299(β)	766(g)		869	928	979	1009	[3]
Gadolinium	25.4	123	450	230	179	185	207	243	[3]
Gallium	32.1	147	266	342(s)	394(l)	382	376	377	[3]
Germanium	12.5	85.3	190	290	337	347	375(s)	380(l)	[3]
Gold	15.9	72.6	108	124	131	135	146(s)	159(l)	[3,5]
Hafnium	9.5	62.8	115	136	153	156	169	197	[5,70]
Helium (g)*[1]	5242	5203	5195	5193	5193	5200	5200		[16,64]
Holmium*[2]	58	149	237	161	169	172	193	271	[3]
Hydrogen (eq.)*[1]	9520(l)	18890(g)	14070	13720	14590	14660	15100	16150	[8]
Indium	60.8	162	203	225	250(s)	245(l)	238	236	[2,5]
Iodine	64		180	203(s)	313(l)	148(g)	150	153	[2]
Iridium	2.0	38	90	122	129.5	135	150	172	[3,70]

Table 9.2 Isobaric specific heat capacity c_p (J/(kg·K)) of elements *(continued)*

Substance	Temperature, K								Ref.
	20	50	100	200	400	600	1000	1500	
α-Iron[*2]	4.6	54	216	384	490	574	975	706	[3]
γ-Iron						519	579	657	[3]
Krypton[*1]	188	295	389(s)	252(g)	249	248	248		[7]
Lanthanum[*2]	46.4	133	170		197(α)	197(β)	238(s)	247(l)	[3]
Lead	53	103	118	125	134	144(s)	142(l)	138	[2,5]
β-Lithium	57	585	1924	3107	3975(s)	4253(l)	4157	4205	[9]
Lutecium	20.9	90.6	129	147	153	156	173		[3]
Magnesium	15	235	646	932	1065	1110(s)	1382(l)	1115	[5,70]
α-Manganese	8.9	87	271	418	515	581(α)	686(β)	857(δ)	[3]
β-Manganese	21.7	106	279	431			686		[3,70]
γ-Manganese		92	269	436					[3]
Mercury	51.5	99.5	121	136(s)	137(l)	137	136		[12,13]
Molybdenum	2.87	41.0	139	222	264	276	294	330	[2,5,70]
Neodymium[*2]	71	150			200	222	291(s)	338(l)	[3,70]
Neon[*1]	945(s)	1042(g)	1033	1031	1030	1030	1030	1030	[7]
α-Neptunium					147				[3]
Nickel[*2]	5.8	68.2	232	383	482	592	561	616	[3,5]
Niobium	11.3	99	202	254	270	281	304	333	[3,5]
Nitrogen[*1]	681(α)	1476(β)	1069(g)	1043	1045	1075	1168	1244	[3,6]
Osmium	1.43				132	136	144	154	[3]
Oxygen (O_2)[*1]	429(α)	1440(γ)	968(g)	915	941	1003	1090	1137	[2,3,10]
Oxygen (ozone)					916(g)	1048	1166	1243	[2,10]
Palladium	9.5	76	168	226	251	261	281	307	[3,5]
Phosphorus (black)	13.5	110	293	560	850				[3]
Phosphorus (red *IV*)	30.9	122	284	554					[3]
Phosphorus (red *V*)	27.3	119	282	550	748	832			[3]
Phosphorus (white)	151	339	443(α)	681(β)	843(l)				[2]
Platinum	7.4	55.1	100	127	136	141	152	165	[3,5]
Plutonium	15	77	78	109(α)	138(β)	154(δ)	171(l)		[3]
Potassium	251	533	631	691(s)	805(l)	770	792	926	[9]
Praseodymium[*2]	94.4	183	185	192	202	224	286(s)	305(l)	[3]
Rhenium	2.8	42.8	97	127	139	145	158	171	[3,5,70]
Rhodium	2.7	49	147	220	253	273	310	349	[3,12]
Rubidium	189	275	299	321(s)	374(l)	362	360	424	[9]
Ruthenium	1.7	36.3	134	214	241	251	278	315	[3,12]
Samarium	49	150	253	180	221	271	301(s)	334(l)	[3,70]
Scandium	13.9	145	364	520	585	611	694	848	[2,5]
γ-Selenium (hex.)	43.5	136	231	297	354(s)	445(l)			[3]
Silicon (cryst.)	3.37	78.5	259	556	794	871	946	1013	[2,5]
Silver	15.5	108	187	225	239	250	277(s)	310(l)	[3,5]
α-Sodium	157								[3]
β-Sodium	170	670	977	1130(s)	1370(l)	1296	1257	1345	[3,9,70]
Strontium	54		268	293	313	343(α)	441(β)	411(l)	[2]
β-Sulfur (monocl.)[*2]			405	626(s)	1004(l)	1068	1004		[3]
α-Sulfur (rhomb.)	80.6	234	403	607					[3,14]
Tantalum	8.23	60.4	111	134	141	145	152	163	[2,5]
Technetium					211	225	290	324	[3]
Tellurium	33.5	113	167	184	219	253(s)	295(l)		[3,5]
Terbium	27.4	134	199	295	179	188	226	286	[3]
α-Thallium	51	103	120	126	134(s)	148(l)	141		[3,70]
α-Thorium	20	73	98	109	117	124	140	159	[3,70]
Thulium	63	226	150	157	161	163	186	209	[3]
α-Tin (grey)	32.3	94	163						[3]

Table 9.2 Isobaric specific heat capacity c_p (J/(kg·K)) of elements *(continued)*

Substance	Temperature, K								Ref.
	20	50	100	200	400	600	1000	1500	
β-Tin (white)	40	130	187	215	243(s)	242(l)	240		[3,5,13]
Titanum	7.0	99.2	302	465	5548	597	684	687	[2,5]
Tungsten	1.9	33.2	88.8	125	136	140	148	158	[3]
Uranium	13.5	65.9	93.4	108	125	146(α)	180(β)	201(l)	[3]
Vanadium	7.1	72.3	258	430	515	540	597	714	[3]
Xenon*1	133	186	214(s)	165(g)	159	159	158		[3,7]
Ytterbium	45.7	116	139	148	160	172	184		[3]
Yttrium	21.3	137	233	282	305	321	354	397	[3]
Zinc	26	171	293	367	402	436(s)	480(l)		[3,5]
Zirconium	11.7	101	205	264	300	321	362(α)	344(β)	[3]

*1 See also Table 9.5.

*2 See also Figures 9.2 to 9.10.

Table 9.3 Specific heat capacity c_p (J/(kg·K)) of elements at temperatures from 1 K to 15 K and constant pressure ["n" is for normal state and "s" is for superconducting state.]

Substance	Temperature, K								Ref.
	1	2	3	4	5	8	10	15	
Aluminum		0.11		0.30		0.9	1.41	4.6	[5]
Antimony	0.002	0.016	0.055	0.11	0.3		3.5	12.3	[3]
Argon	0.06	0.49		4.35		42.5	83.3		[7]
Arsenic	0.004	0.014	0.04	0.09	0.18		1.5	6.1	[3]
Barium		0.12	0.34	0.78	1.54		14	38.5	[3]
Beryllium	0.025	0.051	0.079	0.109	0.144	0.27	0.39	0.84	[5]
Bismuth	0.006	0.046	0.17	0.49		5.47	10.4	23.8	[12]
Bromine					2		16	45	[3]
Cadmium	0.008	0.029	0.087	0.234	0.577	3.98	8.62	25.9	[17]
α-Calcium	0.08	0.18	0.32	0.53	0.86		5.3	16.2	[3]
Carbon (graphite)	0.003	0.018	0.052	0.11	0.17		1.05	3.5	[3,5]
α-Cerium	0.15	0.33	0.55	0.83					[3]
β-Cerium		1.1	2.3	4.6	7.9		43	47	[3]
Cesium	0.28		8.3	16		61	83		[9]
Chlorine							20	54	[3]
Chromium	0.03	0.06	0.10	0.14	0.18		0.49	1.09	[3]
Cobalt	0.008	0.16	0.25	0.35	0.47		1.22	2.53	[3]
Copper	0.0116	0.0278	0.0530	0.0916	0.1482	0.4729	0.8709	2.907	[11,18]
α-Dysprosium	0.24	0.17	0.22	0.32	0.54		3.9	15.5	[3]
Erbium	0.26	0.24	0.38					40	[3]
Europium	0.24	1.3	2.9	4.9	7.4		26	57	[3]
Gadolinium	0.08	0.16	0.27	0.4	0.56		3.9	12.1	[3]
Gallium		0.024	0.051	0.10	0.19		3.46	14.8	[3]
Germanium	0.0006	0.0046	0.0155	0.036	0.069		0.86	4.43	[3]
Gold	0.006	0.025	0.07	0.16	0.29	1.2	2.2	7.4	[5]
Holmium	18.7	6.24	3.37	2.74	3.22		16.4	39.8	[3]
para-Hydrogen		4.98	17.4	43	89	450	1030	7100	[12,13]
Indium (n)	0.029	0.138	0.41	0.95	2.27	8.55	15.5	326.7	[5]
Indium (s)	0.019	0.142	0.464						[12]
Iodine					2.3		16	40	[3]
Iridium		0.034	0.053	0.075	0.10		0.31	0.83	[3]
α-Iron	0.09	0.18	0.28	0.38	0.5	0.9	1.24	2.49	[5]

Table 9.3 Specific heat capacity c_p (J/(kg·K)) of elements *(continued)*

Substance	Temperature, K								Ref.
	1	2	3	4	5	8	10	15	
Krypton	0.064	0.53		4.82		42.0	71.5		[7]
α-Lanthanum (n)	0.07	0.15	0.31	0.54	0.91		8.2	25	[3]
α-Lanthanum (s)	0.006	0.10	0.36	0.83	0.97				[3]
β-Lanthanum (n)	0.09	0.20	0.38	0.66	1.11				[3]
β-Lanthanum (s)	0.006	0.08	0.34	0.83	1.56				[3]
Lead (n)		0.094	0.281	0.684	1.48	7.37	13.7	33.5	[5]
Lead (s)		0.067	0.251	0.672	1.51				[20]
Lithium	0.24	0.54	0.90	1.39	2.02		9.3	25.7	[3]
Lutecium	0.07	0.14	0.22	0.33	0.49	1.43	2.75		[3]
Magnesium	0.055	0.117	0.19	0.29	0.44	1.08	1.9	5.8	[5]
α-Manganese	0.25	0.50	0.75	1.01	1.29	2.16	2.82	5.21	[5]
Mercury (n)	0.036	0.48	2.07	4.09		17.5	23.5	38	[12]
Mercury (s)	0.029	0.48	2.09	4.17					[12]
Molybdenum	0.02	0.04	0.065	0.09	0.13		0.48	1.00	[3]
Neodymium	1.07	5.0	11.0	18.9	27.2	41.5	36.5	51.9	[3]
Neon	0.24	1.96		17.7		159	278		[7]
Nickel	0.12	0.24	0.37	0.50	0.67	1.19	1.62	3.1	[5]
Niobium (n)	0.09	0.18	0.28	0.40		1.4	2.2	5.5	[12]
Niobium (s)*	0.001	0.014	0.087	0.27	0.56	2.28			[19]
α-Nitrogen			3.5	8.7	17.3		167	422	[3]
Osmium		0.025	0.04	0.06					[3]
α-Oxygen				3.9	7.8		69	224	[3]
Palladium	0.09	0.18	0.29	0.42	0.56		1.9	4.6	[3]
Phosphorus (white)					3.5		36.2	97	[2]
Platinum	0.03	0.07	0.12	0.18	0.27		1.17	3.5	[3]
α-Plutonium							3	9	[3]
Potassium	0.12		2.24	5.45		42.0	71.5		[9]
Praseodymium	0.35	0.63	1.3	2.4	4.3	18.1	29.2	60	[3]
Rhenium (n)	0.015	0.025	0.04	0.06	0.08		0.29	0.9	[3]
Rhodium	0.048	0.097	0.147	0.201		0.47	0.65	1.3	[12]
Rubidium	0.16		4.36	10.4		54	81		[9]
Ruthenium	0.03	0.06	0.10	0.13	0.17		0.41	0.87	[3]
Samarium	0.14	0.23	0.43	0.78	1.45		17.5	37	[3]
Scandium	0.267	0.505	0.747	1.00	1.3	2.5	3.31	7.0	[5]
γ-Selenium	0.007	0.056	0.19	0.49	1.0		7.9	23.8	[3]
Silicon (cr.)	0.0003	0.002	0.007	0.018	0.03		0.28	1.1	[3]
Silver	0.0072	0.0239	0.0595	0.124		0.91	1.8	6.4	[12]
α-Sodium	0.08	0.30	0.76	1.6	2.8		22.7	74	[3]
β-Sodium				1.78	3.42		28.4	85	[3]
Strontium	0.05	0.14	0.31	0.61	1.09		8.45	27	[3]
Sulfur (rhombous)					1.69		15.4	46	[14]
Tantalum (n)	0.032	0.068	0.112	0.171	0.26	0.65	1.17	3.6	[5]
Tantalum (s)	0.006	0.054	0.178	0.352					[12]
Tellurium	0.005	0.039	0.14	0.33	0.65		7.01	20.3	[3]
Terbium	1.45	0.50	0.43	0.55	0.68		3.7	14.3	[3]
Thallium (n)		0.06	0.43	1.02	2.0		15.3	33	[3]
α-Thorium		0.05	0.11	0.19	0.42		2.9	9.4	[3]
Thulium	0.25	0.38	0.79	1.44	2.35		11.6	32.7	[3]
α-Tin (grey)	0.0002	0.0014	0.005	0.12	0.46		6.0	12.0	[3]
β-Tin (white) (n)	0.017	0.047	0.109	0.245		4.2	8.1	22.6	[12]
β-Tin (white) (s)	0.004	0.048	0.151						[12]
Titanium	0.071	0.146	0.226	0.317	0.42	0.84	1.26	3.3	[5]
α-Uranium	0.044	0.097	0.17	0.26	0.38		1.76	5.9	[3]

Table 9.3 Specific heat capacity c_p (J/(kg·K)) of elements *(continued)*

Substance	Temperature, K								Ref.
	1	2	3	4	5	8	10	15	
Vanadium (n)	0.19	0.38	0.58	0.80	1.04		2.34	4.24	[3]
Vanadium (s)	0.005	0.20	0.70	1.45	2.43				[3]
Xenon	0.057	0.49		4.74		38.1	58.3	103	[7]
Ytterbium	0.02	0.087	0.23	0.50	0.98		8.5	25.3	[3]
Yttrium	0.10	0.20	0.32	0.46	0.64	1.5	2.6	8.6	[17]
Zinc	0.010	0.027	0.055	0.105	0.191	0.96	2.36	11.2	[17]
Zirconium	0.03	0.07	0.12	0.18	0.26		1.30	4.36	[3]

* See also Figure 9.7.

Table 9.4 Specific heat capacity c_s (J/(kg·K)) of liquid helium ^{4}He under the saturated vapor pressure at different temperatures (K) [21] [See also Figure 9.1.]

T, K	c_s	T, K	c_s	T, K	c_s	T, K	c_s	T, K	c_s
1.8	2.81	2.1	7.51	2.4	2.38	3.4	2.97	4.4	5.11
1.85	3.26	2.15	9.35	2.6	2.27	3.6	3.26	4.6	5.94
1.9	3.79	2.17	12.60	2.8	2.34	3.8	3.60	4.8	7.53
2.0	5.18	2.2	3.98	3.0	2.49	4.0	3.99	5.0	11.5
2.05	3.26	2.3	2.64	3.2	2.69	4.2	4.48	5.05	13.5

Table 9.5 Isobaric specific heat capacity c_p (J/(kg· K)) of elements in liquid and gaseous states at various pressures and temperatures [Data over the horizontal line correspond to liquid phase.]

T, K	P, MPa				T, K	P, MPa			
	0.1	1	10	100		0.1	1	10	100
				Argon [7]					
85	1.057	1.055			200	0.522	0.556	1.232	0.858
100	0.542	1.152	1.094		300	0.521	0.532	0.645	0.815
120	0.532	0.722	1.161	0.942	500	0.521	0.521	0.552	0.657
150	0.526	0.629	1.549	0.903	1300	0.521	0.521	0.526	0.556
				Helium ^{4}He [16,64]					
3	2.35	1.95			20	5.25	5.72	5.18	3.52
4	4.08	2.72	1.82		50	5.20	5.28	5.60	5.42
5	6.78	3.70	2.25		100	5.20	5.22	5.32	5.26
10	5.42	7.62	3.62		1400	5.20	5.20	5.20	–
				***n*-Hydrogen [8]**					
15	6.98	6.80			200	13.53	13.59	14.07	14.82
20	9.53	9.17	7.61		300	14.31	14.33	14.54	15.14
25	11.16	12.58	9.12		600	14.55	14.53	14.59	14.97
30	10.83	23.26	10.60		1000	14.98	14.98	14.66	15.01
50	10.48	12.12	15.53	9.64	2000	18.27	17.41	17.14	17.04
100	11.22	11.50	13.51	12.13					
				***para*-Hydrogen [8]**					
15	6.98	6.80			100	13.43	13.70	15.71	14.34
20	9.53	9.17	7.51		200	16.08	16.14	16.52	17.36
25	11.16	12.58	9.12		500	14.52	14.53	14.59	14.82
30	10.83	23.26	10.60		1000	14.98	14.98	14.99	15.01
50	10.54	12.18	15.59	9.7	2000	18.19	17.33	17.06	16.96

Table 9.5 Isobaric specific heat capacity c_p (J/(kg·K)) of elements *(continued)*

T, K	P, MPa				T, K	P, MPa			
	0.1	1	10	100		0.1	1	10	100
				Krypton [7]					
120	0.263	0.543	0.541		300	0.249	0.259	0.424	0.404
150	0.258	0.557	0.520	0.441	500	0.248	0.251	0.281	0.353
200	0.252	0.291	0.776	0.502	1300	0.248	0.248	0.251	0.268
				Neon [7]					
30	1.129	2.027	1.852		70	1.037	1.103	1.973	1.467
40	1.066	1.818	2.030	1.500	100	1.033	1.062	1.339	1.426
50	1.042	1.267	2.557	1.490	200	1.031	1.037	1.089	1.235
60	1.039	1.147	2.603	1.479	1000	1.030	1.030	1.031	1.035
				Nitrogen [6]					
70	2.023	2.012	1.903		200	1.043	1.083	1.619	1.484
80	1.022	2.089	1.997		300	1.041	1.056	1.197	1.355
90	1.068	2.099	1.968	1.701	400	1.045	1.053	1.121	1.271
100	1.069	2.269	1.998	1.774	600	1.075	1.078	1.104	1.201
120	1.057	1.309	2.220	1.685	1000	1.168	1.168	1.176	1.216
150	1.048	1.144	2.851	1.555	1500	1.244	1.244	1.247	1.266
				Oxygen [10]					
70	1.696	1.694	1.676		200	0.915	0.957	1.904	1.383
90	1.686	1.681	1.645		300	0.920	0.934	1.084	1.260
100	0.968	1.709	1.651		400	0.942	0.949	1.015	
110	0.947	1.780	1.681		600	1.003	1.006	1.031	
120	0.935	1.311	1.742		1000	1.090	1.091	1.099	
150	0.920	1.039	2.204	1.444					
				Xenon [7]					
170	0.168	0.348	0.339		500	0.159	0.163	0.208	0.248
200	0.165	0.356	0.338	0.294	1300	0.158	0.159	0.163	0.183
300	0.160	0.178	0.699	0.286					

Table 9.6 The dependence of Debye function $D(\theta/T)$ on θ/T [32]

θ/T	0	1	2	3	4	5	6	7	8	9	10	20
$D(\theta/T)$	1	0.952	0.825	0.663	0.503	0.369	0.266	0.191	0.138	0.101	0.076	0.0097

Table 9.7 The Debye parameter θ and the Sommerfeld constant γ for elements [22]

Substance	θ, K	γ, mJ/(mol·K^2)	Substance	θ, K	γ, mJ/(mol·K^2)
Aluminum	433	1.35	Boron (crystal) [4,22]	1219	0
Americium	121	27	Bromine [4]	111	0
Antimony	220	0.12	Carbon (diamond)	2250	0
Argon	92	0	Carbon (graphite)	413	0.014
Arsenic (amorph.) [27]	159		Cadmium	210	0.69
Arsenic (crystal)	282	0.19	Calcium	229	2.73
Barium	111	2.7	α-Cerium	179	12.8
Beryllium	1481	0.17	Cesium	40	3.97
Bismuth	120	0.0085	Chlorine [4]	115	0
Boron (amorph.) [4]	1102	0	Chromium	606	1.42

Table 9.7 The Debye parameter θ and the Sommerfeld constant γ for elements *(continued)*

Substance	θ, K	γ, mJ/(mol·K^2)	Substance	θ, K	γ, mJ/(mol·K^2)
Cobalt	460	4.4	Osmium	467	2.05
Copper	347	0.69	α-Oxygen [28]	126	0
Curium	123		γ-Oxygen [29]	46	0
ortho-Deiterium	114	0	Plutonium	206	25
α-Dysprosium	183		Potassium	91	2.08
Erbium [22,31]	188	10	Praseodymium	152	
Europium	118	6	Protactinium	185	5
Fluorine [4]	78	0	Rhenium	418	2.29
Gadolinium	182	6.38	Rhodium	512	4.65
Gallium	326	0.60	Rubidium	56	2.63
Germanium (amorph.) [27]	315	0	Ruthenium	555	3.1
Germanium (crystal)	373	0	Samarium	169	13.5
Gold	162	0.69	Scandium	346	10.3
Hafnium	252	2.15	Selenium (amorph.) [29]	123	0
Helium (hcp.) [26]	27	0	γ-Selenium	152	0
Holmium	190	6	Silicon (amorph.) [27]	528	0
para-Hydrogen [12]	122	0	Silicon (crystal)	645	0
Indium	112	1.66	Silver	227	0.64
Iodine	109	0	Sodium	156	1.38
Iridium	420	3.14	Strontium	147	3.64
α-Iron	477	4.9	Sulfur [4,12]	165 – 180	0
Krypton	72	0	Tantalum [22,30]	245	5
α-Lanthanum	150	9.45	Technetium	454	4.0
β-Lanthanum	140	11.5	Tellurium	152	0
Lead	105	2.99	Terbium	176	4.1
Lithium	344	1.65	Thallium	78	1.47
Lutecium	183	8.19	Thorium	160	4.08
Magnesium	403	1.26	Thulium	200	
α-Manganese	409	12.8	α-Tin (grey) [12]	212	0
Mercury	72	1.86	β-Tin (white)	200	1.78
Molybdenum	423	1.83	Titanum	420	3.36
Neodymium	163		Tungsten	383	1.01
Neon	75	0	Uranium	248	8.14
Neptunium	259	1.37	Vanadium [29]	382	9.82
Nickel	477	7.04	Xenon	64	0
Niobium	276	7.8	Ytterbium	118	2.9
Nitrogen [23]	81	0	Yttrium	248	8.2
Palladium	271	9.45	Zinc	329	0.64
Phosphorus (black)	137		Zirconium	290	2.77
Platinum	237	6.54			

9.3 Heat Capacity of Inorganic Compounds

Table 9.8 The Debye parameter θ and the Sommerfeld constant γ for inorganic compounds

Substance	Formula	θ, K	γ, mJ/(mol·K^2)	Ref.
Aluminum oxide	Al_2O_3	1042	0	[38]
Calcium fluoride	CaF_2	475	0	[12]

Table 9.8 The Debye parameter and the Sommerfeld constant for inorganic compounds *(continued)*

Substance	Formula	θ, K	γ, mJ/(mol·K^2)	Ref.
Chromium germanide	Cr_3Ge	473 – 670	14 – 16	[35]
Chromium silicide	Cr_3Si	620 – 670	10 – 12	[35]
Gallium antimonide	GaSb	269	0.002	[18]
Gallium arsenides	GaAs	345	0.0008	[18]
Indium antimonide	InSb	206	0	[18]
Indium arsenide	InAs	251	0	[18]
Iron sulfide	FeS_3	645	0	[12]
Lithium fluoride	LiF	736	0	[12]
Magnesium oxide	MgO	800	0	[12]
Niobium aluminide	Nb_3Al	280	32	[35]
Niobium stanide	Nb_3Sn	228	44	[36]
Potassium chloride	KCl	230	0	[12]
Silver bromide	AgBr	144	0	[12]
Silver chloride	AgCl	183	0	[12]
Sodium chloride	NaCl	275	0	[12]
Titanium oxide	TiO_2	778	0	[37]
Vanadium gallide	V_3Ga	302	97	[34]
Vanadium germanide	V_3Ge	392	30	[35]
Vanadium silicide	V_3Si	501	62	[35]
Water (Ice)	H_2O	192		[67]
Zinc sulfide	ZnS	270	0	[12]

Table 9.9 Molar isobaric heat capacity C_p (J/(mol·K)) of inorganic compounds at temperature 298.15 K [1] [Solids and liquids are at pressure of 0.1 MPa and gases are in the state of ideal gas (i.e. at utmost low pressure). Substances in liquid and gaseous states are marked by letters "l" and "g". Different crystal modifications of solids are marked by the greek letters $\alpha, \beta, \gamma, \delta$.]

Substance and state	C_p	Substance and state	C_p	Substance and state	C_p	Substance and state	C_p
AgBr	52.3	AlN [2]	30.1	BF_3 (g) [2]	50.5	$BiCl_3$	109
AgBr (g)	36.6	α-Al_2O_3 [2]	79.0	BH (g) [2]	29.2	Bi_2O_3	113.8
Ag_2CO_3	112.5	γ-Al_2O_3 [2]	82.7	B_2H_6 (g) [2]	57.6	Bi_2Tl	73.6
AgCl [4]	50.8	α-$Al_2O_3 \cdot H_2O$	105.5	$B_{10}H_{14}$	217.9	BrCl (g) [2]	35.0
AgCl (g) [33]	35.8	γ-$Al_2O_3 \cdot H_2O$	131.2	BN [2]	19.7	BrF (g) [2]	33.0
$AgClO_2$	87.3	$Al_2(SO_4)_3$ [2]	259.4	BN (g) [2]	29.6	BrF_3 (l)	124.5
AgH (g)	29.3	Al_2TiO_5	136.4	BO (g)	29.2	BrF_3 (g) [2]	67.4
α-AgI	57.0	$AsCl_3$ (l)	133.5	B_2O_2 (g) [2]	60.3	BrF_5 (g) [2]	101.3
$AgIO_3$	102.9	AsF_3 (l)	126.6	B_2O_3 [2]	62.8	BrI (g) [2]	36.5
AgMg	50.2	AsF_3 (g)	64.6	$Ba(OH)_2$	81.3	CCl_4 (l)	131.7
$AgNO_2$	79.1	AsH_3 (g)	38.6	$BaCO_3$	86.0	CCl_4 (g) [2]	82.9
$AgNO_3$	102.9	AsN (g)	30.5	$BaCl_2$ [2]	75.1	CN (g) [2]	29.2
Ag_2O	65.9	As_4O_6	203.8	$BaCl_2 \cdot 2H_2O$	203.3	C_2N_2 (g) [2]	57.1
Ag_2S	76.5	As_2O_5	116.5	BaF_2 [2]	71.1	CO (g) [2]	29.1
Ag_2SO_4	131.4	AuCl (g) [33]	35.44	BaH (g)	30.2	CO_2 (g) [2]	37.1
Ag_2Se	81.8	$AuSb_2$	79.5	$Ba(NO_3)_2$	151.6	C_3O_2 (g) [2]	62.2
Ag_2Te	88.7	AuSn	50.5	BaO [2]	47.0	COS (g) [2]	41.6
AlBr (g) [2]	35.6	BBr (g) [2]	32.8	$BaSO_4$ [2]	102.1	CP (g) [2]	29.9
$AlBr_3$	100.5	BBr_3 (g) [2]	67.8	$BaTiO_3$	102.5	CS (g) [2]	29.8
Al_4C_3	116.8	B_4C [2]	53.1	BeH (g) [2]	29.2	CS_2 (l)	75.7
AlCl (g) [2]	34.7	BCl (g) [2]	31.7	Be_3N_2 [2]	64.8	CS_2 (g) [2]	45.5
AlF (g) [2]	31.9	BCl_3 (g) [2]	62.6	BeO [2]	25.6	CaC_2	62.7
AlF_3 [2]	75.1	BF (g) [2]	29.6	$BeSO_4$ [2]	85.7	$CaCO_3$ [2]	83.5

Table 9.9 Molar isobaric heat capacity of inorganic compounds at temperature 298.15 K *(continued)*

Substance and state	C_p	Substance and state	C_p	Substance and state	C_p	Substance and state	C_p
$CaCl_2$ [2]	72.8	Cu_3N	90.8	HS (g) [2]	32.5	LiCl (g) [2]	33.2
CaF_2 [2]	67.0	CuO	42.3	H_2S (g) [2]	34.2	LiD	34.6
$CaFe_2O_4$	153.6	Cu_2O	63.6	H_2SO_4 (l)	138.9	LiF [2]	41.8
α-Ca_3N_2	114.6	CuS	47.8	H_2SO_4 (g) [2]	84.4	LiF (g) [2]	31.3
$Ca(NO_3)_2$	149.6	Cu_2S	76.3	$H_2SO_4{\cdot}H_2O$ (l)	214.8	$LiFeO_2$	82.8
CaO [2]	42.0	α-$CuSO_4$	98.9	H_2Se (g)	34.7	LiH [2]	29.0
$Ca(OH)_2$ [2]	82.5	Cu_2Sb	76.6	H_2Te (g)	35.6	LiH [2] (g)	29.7
$Ca(PO_3)_2$	145.1	D_2O [2]	84.3	$HfCl_4$	120.5	LiI	51.0
α-$Ca_3(PO_4)_2$	231.6	D_2O (g) [2]	34.4	HfO_2 [2]	60.3	$LiNO_3$	89.1
β-$Ca_3(PO_4)_2$	236	F_2O (g) [2]	43.5	$HgBr_2$ (g)	76.1	Li_2O [2]	54.1
CaS [2]	47.4	Fe_3C	105.8	HgCl (g)	36.4	LiOH [2]	49.6
$CaSO_3$	91.7	$FeCO_3$	83.3	$HgCl_2$ (g)	57.7	$LiOH{\cdot}H_2O$	79.5
$CaSO_4$	99.7	$Fe(CO)_5$ (l)	240.6	HgI_2 (red)	78.2	LiT	37.0
α-$CaSiO_3$	85.2	$FeCl_2$	76.4	HgI_2 (g)	61.1	Li_2TiO_3	111
β-$CaSiO_4$	128.6	$FeCl_3$	94.9	HgO	44.1	MgB_2	47.8
$CaTiO_3$	97.6	$FeCo_2O_4$	143.3	HgS	44.2	MgB_4	70.3
$CdCl_2$	76.6	$FeCr_2O_4$	133.7	Hg_2SO_4	132.0	$MgCO_3$ [2]	76.1
CdI_2	78.7	FeF_2	68.1	IBr (g)	34.5	$MgCl_2$ [2]	71.4
CdMg	51.6	ε-Fe_3N	122.6	ICl	56.2	$MgCl_2{\cdot}H_2O$	71.4
CdO	43.6	FeO	49.9	ICl (g)	35.6	$MgCr_2O_4$	126.8
CdS	47.3	α-Fe_2O_3	103.8	IF (g)	33.5	MgF_2 [2]	61.6
α-$CdSO_4$	99.6	Fe_3O_4	150.8	IrO_2	57.3	Mg_3N_2	92.1
CdSb	45.7	α-FeS	50.5	KBH_4	96.2	MgO [2]	37.2
Cd_2SiO_3	88.6	FeS_2	62.2	KBO_2	67.1	MgO (g) [2]	32.1
CeN	46.4	FeSi	48.5	KBr [2]	52.5	$Mg(OH)_2$ [2]	80.7
CeO_2	61.6	Fe_2SiO_4	132.9	KBr (g) [2]	36.9	$MgSO_4$ [2]	96.4
$Ce_2(SO_4)_3$	270.3	$FeTiO_3$	99.5	$KBrO_3$ [33]	104.9	$MgSiO_3$	81.9
Cl_2CO (g) [2]	57.8	Ga_2O_3	92.1	KCl [2]	51.3	$MgSiO_4$	118.7
ClF (g) [2]	32.1	$GeBr_4$ (g) [2]	101.7	KCl (g) [2]	36.5	$MgTiO_3$	91.9
ClF_3 (g) [2]	64.1	GeF_4 (g) [2]	82.0	$KClO_3$	100.3	$MgZn_2$	74.1
Cl_2O (g) [2]	47.8	GeH_4 (g) [2]	45.0	$KClO_4$	110.2	α-Mn_3C	93.3
ClO_2 (g) [2]	42	$GeHCl_3$ (g) [2]	54.8	K_2CrO_4	146.0	$MnCO_3$	94.8
$CoCl_2$	78.5	GeI_4 (g) [2]	104.2	KF [2]	49.0	$MnCl_2$	72.9
CoF_2	68.8	GeO_2 [2]	52.1	KH [2]	37.9	MnF_2 (g)	52.7
CoO	55.2	H_3BO_3 [2]	81.3	KH (g) [2]	31.0	MnO	44.1
Co_3O_4	122.8	HBr (g) [2]	29.1	KH_2AsO_4	126.7	β-MnO_2	54.0
β-Co_2O_4	103.2	HCN (l)	70.6	α-KHF_2	76.9	Mn_2O_3	107.5
Cr_3C_2	85.4	HCN (g) [2]	35.9	KH_2PO_4	116.6	Mn_3O_4	139.3
$Cr(CO)_6$	240.1	HCl (g) [2]	29.1	KI [2]	52.8	MnS	49.9
$CrCl_3$	91.8	HD (g) [2]	37.5	KI [2] (g)	37.1	$MnSO_4$	100.2
CrF_3	78.7	HF (g) [2]	30.2	γ-KIO_3	106.1	MnSe	51.3
Cr_2O_3 [2]	128.1	HI (g) [2]	28.8	$KMnO_4$	119.2	α-$MnSiO_3$	86.4
CsBr [2]	52.9	HN_3 (g) [2]	43.2	KNO_3 [2]	95.1	MnTe	72.3
CsCl [2]	52.5	HNO_2 (g) [33]	45.8	K_2O [2]	72	$Mo(CO)_6$	242.2
$CsClO_4$	108.4	HNO_3 (l)	109.9	KOH [2]	64.9	MoO_2 [2]	55.9
CsF [2]	51.1	HNO_3 (g)	54.2	$KReO_4$ [2]	122.6	MoO_3 [2]	75.1
CsH [2]	40.6	$HNO_3{\cdot}H_2O$ (l)	182.4	K_2SO_4 [2]	131.5	α-MoS_2	63.6
CsH (g) [2]	31.6	H_2O (l) [2]	75.3	La_2O_3 [2]	108.8	$MoSi_2$	64.8
CsI [2]	52.5	H_2O (g) [2]	33.6	$LiAlO_2$	67.8	Mo_3Si	93.0
γ-CuBr	54.7	H_2O_2 (l)	89.3	LiBr [2]	49.8	NF_3 (g) [2]	53.5
α-CuCl	48.5	H_2O_2 (g)	43.1	LiBr (g) [2]	33.9	NH_3 (g) [2]	35.6
α-$CuCl_2$	71.9	HOCl (g)	37.1	Li_2CO_3	98.3	$NH_3{\cdot}0.5\ H_2O$ (l)	117.9
α-CuI	54.0	H_3PO_4	106.1	LiCl [2]	48.0	$NH_3{\cdot}H_2O$ (l)	154.9

Table 9.9 Molar isobaric heat capacity of inorganic compounds at temperature 298.15 K *(continued)*

Substance and state	C_p	Substance and state	C_p	Substance and state	C_p	Substance and state	C_p
N_2H_4 [2]	48.4	NiO	44.3	$SbCl_3$ (g)	77.4	TiO_2 (anatase) [2]	55.3
NH_4Cl	84.1	γ-NiS	47.1	Sb_4O_6 (cubic)	209	TiO_2 (rutile) [2]	55.1
NH_4F [2]	65.3	$NiSO_4$	97.7	Sb_4O_6 (rhombic)	206	Ti_2O_3 [2]	95.8
$NH_4H_2PO_4$	142.2	$NiTe_2$	75.9	Sc_2O_3	94.1	Ti_3O_5 [2]	150.8
NH_4HSO_4	135.5	NpO_2	66.2	SeF_6 (g)	110.4	TiS_2	67.9
NH_4NO_3 [2]	139.1	PB_3 (g)	76.1	SeO (g)	31.1	ThBr	52.5
$(NH_4)_2SO_4$	187.4	PCl_3 (g)	71.8	$SiBr_4$ (g)	97	ThCl [2]	50.9
NH_4VO_3 [2]	129.3	PCl_5 (g) [2]	113.3	α-SiC [2]	26.8	ThCl (g) [2]	36.4
NO (g) [2]	29.9	PF_3 (g) [2]	58.7	β-SiC [2]	26.8	ThF [2]	53.4
N_2O (g) [2]	38.6	PH_3 (g)	37.1	$SiCl_4$ (l)	145.2	$ThNO_3$	99.6
NO_2 (g) [2]	37.2	PN (g) [2]	29.7	$SiCl_4$ (g)	90.4	UCl_3	102.9
N_2O_4 (g) [2]	79.2	P_3N_5 (g) [2]	150.6	SiF_4 (g)	73.6	UCl_4	122.0
N_2O_5	143.1	$POCl_3$ (g) [2]	84.4	SiH_4 (g)	42.9	α-UF_4 [2]	116.0
N_2O_5 (g) [2]	95.3	$POCl_3$ (l)	138.84	SiO_2 (quartz) [2]	44.6	UF_6 [2]	166.8
NOCl (g)	44.7	P_4O_{10}	215.6	SiO_2 (glass) [2]	44.0	UF_6 (g) [2]	129.5
NO_2Cl (g)	53.2	P_4O_{10} (g)	191	$SnBr_4$ (g)	103.0	β-UH_3	49.3
NOF (g) [2]	41.3	$PbBr_2$ (l)	56.9	$SnCl_2$ [2]	78.0	UO_2 [2]	63.6
NS (g) [2]	31.8	$PbCl_2$ [2]	77.0	$SnCl_2$ (g) [2]	54.6	UO_3 [2]	81.7
$NaAlO_2$	73.3	α-PbF_2	72.3	β-$SnNi_3$	99.6	U_3O_8 [2]	238
$NaBH_4$	86.7	PbO (red) [2]	46.4	SnO [2]	47.8	$UOCl_2$	95.1
$NaBO_2$ [2]	65.9	PbO (yellow) [2]	45.8	SnO (g) [2]	31.8	UO_2F_2 [2]	103.2
NaBr [2]	51.4	PbO (g) [2]	32.5	SnO_2 [2]	53.2	VC [4]	33.3
NaBr (g) [2]	36.3	PbO_2 [2]	61.0	SnPt	50.2	α-V_2C	55.9
Na_2CO_3 [2]	112.3	Pb_2O_3 [2]	107.7	SnS [2]	49.2	VCl_2	72.2
NaCl [2]	50.5	Pb_3O_4 [2]	146.7	SnS (g) [2]	34.5	VCl_3	93.2
NaCl (g) [2]	35.8	PbS [2]	49.5	SnS_2	70.1	V_2O_3 [2]	101.0
$NaClO_4$	110.4	$PbSO_4$	103.2	$SrBr_2$ [2]	76.5	V_2O_4 [2]	112.1
NaF [2]	46.8	PtS	43.4	$SrCl_2$ [2]	75.6	V_2O_5 [2]	127.8
NaH [2]	36.4	$PtSb_2$	68.6	$SrCO_3$ [2]	82.4	V_3Si	90.8
NaH (g) [2]	30.2	PuO_2 [2]	66.2	SrO [2]	45.0	α-WC	35.1
$NaHCO_3$	88.2	Pu_2O_3 [2]	117.0	SrO (g) [2]	33.1	WO_3 [2]	79.7
NaI [2]	52.1	RbBr [2]	52.8	$SrSO_4$ [2]	102.0	Y_2O_3	102.5
NaI (g) [2]	36.6	RbCl [2]	52.3	$SrTiO_3$	98.4	$ZnCO_3$	80.1
$NaNO_3$ [2]	93.0	$RbClO_3$	103.2	TaC	36.8	$ZnCl_2$	71.3
NaO_2 [2]	72.1	RbF [2]	50.6	TaN	42.7	α-ZnF_2	65.6
Na_2O [2]	69.1	RbI [2]	52.6	TeF_6 (g)	117	$ZnFe_2O_4$	138.0
Na_2O_2 [2]	89.3	Rb_2O[2]	74.0	TeO_2	64	ZnO	40.2
NaOH [2]	59.5	RhO	48.1	ThF_4	110.7	ZnS	45.5
Na_2SO_3	120.1	Rh_2O	73.2	ThN	45.2	α-$ZnSO_4$	99.1
Na_2SO_4 [2]	128.0	Rh_2O_3	103.8	ThO_2 [2]	61.8	Zn_2SiO_4	123.3
Na_2SiO_3	111.9	SF_6 (g) [2]	97.0	TiB_2	45.6	Zn_2TiO_4	137.3
Na_2TiO_3	125.6	SO (g) [2]	30.2	TiC	34.3	$ZrCl_2$	73.6
θ-NbN	37.5	SO_2 (g) [2]	39.8	TiCl (g)	37.2	ZrN	40.4
NBO_2 [2]	57.7	SO_3 (g) [2]	50.6	$TiCl_2$	69.8	Zr_3N_2	127.2
Nb_2O_5 [2]	132.0	$SOCl_2$ (l)	120.5	$TiCl_3$	97.1	ZrO_2 [2]	55.9
Nb_3Sn [39]	72.5	$SOCl_2$ (g)	66.7	$TiCl_4$ (l)	145.2	Zr_2SiO_4	98.3
$NiCl_2$	71.7	SOF_2 (g)	57.1	δ-TiN	37.1		
NiF_2	64.1	$SbCl_3$	110.5	TiO [2]	39.9		

Table 9.10 Molar isobaric heat capacity C_p of inorganic compounds at temperatures from 20 K to 1500 K [2, 4] [Solid and liquids are at the pressure of 0.1035 MPa and gases are in the state of ideal gas (i.e. at utmost low pressure). The substance state is shown before and after the phase transition: "s" is for solid, "l" is for liquid, "g" is for gaseous. Different crystal modifications are marked by the greek letters α, β, γ, δ and others.]

Substance	Temperature, K							
	20	40	100	200	400	600	1000	1500
AgBr		22.28	45.31	50.63	58.95	71.84(s)	37.34(g)	37.36
AgCl	8.71	22.30	41.67	49.71	58.86	61.63(s)	37.24(g)	37.28
AgI	15.31	28.03	45.77	52.26	64.68(s)		37.36(g)	37.38
$AgNO_3$	11.30	27.82	59.91	79.08	112.3			
Ag_2SO_4	12.72	38.49	82.76	112.5	144	167		
AlBr					36.40	37.11	37.62	37.96
AlF_3			24.64	56.74	86.28	97.31	100.83	105.86
Al_2O_3 (cor.) [68]	0.076	0.691	12.84	51.14	96.19	112.85	124.53	132.42
AsF_3	11.40	29.12	54.85	80.17(s)	71.9(g)	77.2	82.9	
AsH_3	18.28	44.97	51.07(s)	60.52(l)	45.4(g)	53.1	63.8	75.7
B_4C			5.10	27.32	77.87	96.76	112.33	128.61
BF_3	12.13	29.29	53.39(s)		57.55(g)	67.07	75.80	79.50
B_2H_6	5.02	21.30	54.02(s)		74.07(g)	101.36	136.44	157.57
BN	0.272	1.07	5.36	12.53	27.0	35.23	44.31	48.58
B_2O_3	0.38	3.22	20.87	43.93	78.52	97.83(s)	130.18(l)	128.44
$BaCO_3$			50.71	73.14	100.18	111.39	147.93	
BaF_2	2.51	12.47	44.06	65.86	75.07	81.67	93.92	
BaO			31.67	43.64	49.96	53.20	57.17	60.55
$BaSO_4$	5.48	21.76	57.53	84.18	118.76	137.01	161.69	
$BaTiO_3$			43.18	82.84	112.90	121.3	128.1	134.6
BeO			2.89	14.31	33.63	42.31	48.72	52.30
Bi_2O_3			62.80	96.69	116.90	123.60		
BrF_3(g)					72.72	78.03	81.20	
BrF(g)					34.52	36.15	37.32	37.95
CO	14.02	34.14(s)			29.34(g)	39.44	33.18	35.21
CO_2	5.13	19.62	39.87(s)		41.33(g)	47.33	54.32	58.40
COS	11.0	27.70	44.06(s)	70.92(l)	45.85(g)	51.16	56.77	59.86
CS_2	11.97	27.82	46.11(s)	75.14(l)	49.46(g)	54.35	58.90	61.04
$CaCO_3$ (calcite)			39.54	66.61	96.98	109.86	124.53	139.30
$CaCl_2$			48.83	67.28	75.30	78.81	84.34	
CaF_2			29.71	55.98	74.86	78.35	91.25	125.40
CaO	0.08	1.26	16.15	34.81	46.98	50.72	53.36	55.10
$Ca(OH)_2$	0.92	5.69	32.54	68.58	96.04	106.26	121.34	
$CaSO_4$			46.28	79.96	117.02	135.38	159.27	
$CaSiO_3$	0.79	7.20	30.50	67.15	100.4	112.9	123.8	
$CaTiO_3$			35.56	77.03	112.70	123.1	130.3	134.80
$CdCl_2$	7.49	19.75	51.13	69.20	77.32	85.35		
Cl_2O (g)					49.00	53.1	56.1	57.30
ClO_2 (g)					45.86	51.13	55.61	57.49
Co_3O_4			34.23	87.49	142.6	165.1	198.1	
Cr_3C_2	0.84	3.43	31.57	75.10	113.1	127.9	147.3	
CrF_3	3.75	15.29	36.11	63.93				
Cr_2O_3			24.39	75.65	112.3	120.5	127.0	132.5
CuBr		19.50	40.47	50.29	36.63(g)	37.04	37.24	37.31
CuO	0.59	3.51	16.52	34.79	46.82	50.84	58.87	
Cu_2O		18.95	39.69	54.04	71.88	76.65	86.19	
Cu_2S			50.42	68.95	97.28(α)	97.28		
D_2O	2.23	6.38	16.93	33.68(s)	35.62(g)	38.82	45.41	51.12
Fe_3C			43.60	898.03	115.6(α)	114.7(β)	119.7	126.0
$FeCl_2$			50.92	70.71	79.66	83.09		

Table 9.10 Molar isobaric heat capacity of inorganic compounds *(continued)*

Substance	Temperature, K							
	20	40	100	200	400	600	1000	1500
Fe_2O_3	0.38	3.99	31.50	76.6	129.1	140.8(α)	150.6(β)	143.7(γ)
Fe_3O_4			56.61	116.5	172.2	212.5(α)	200.8(β)	200.8
FeS	1.1	6.99	26.63	43.01	65.9(α)	57.02((γ)	61.0	
β-Ga_2O_3	0.46	5.56	32.69	69.45	105.53	116.78	126.98	135.78
HBr	11.05	20.50	43.26(s)	59.66(l)	29.21(g)	29.87	32.33	34.74
HCN	3.49	12.48	33.51	49.79	39.42	44.19	50.79	56.40
HCl	5.86	15.69	40.00(l)		29.16(g)	29.71	31.76	
HF	2.51	9.04	19.79(s)	43.10(l)	29.19(g)	29.23	30.17	32.33
HI	14.14	23.26	43.68	47.86(s)	29.29(g)	30.12	32.34	35.27
HNO_3	5.18	18.69	42.09	61.50(s)	63.40(g)			
H_2O	1.98	6.13	15.98	28.22(s)	34.16(g)	36.29	41.54	
H_2O_2	1.23	6.90	25.58(s)	43.02(l)	46.48(g)	53.52	62.62	
H_3PO_4	4.23	16.4	42.8	75.9				
H_2S	5.23	14.90	39.16(s)	68.03(l)	35.66(g)	39.05	45.97	51.78
H_2SO_4	7.24	21.84	45.06	75.72				
H_2Se			43.18	59.45	(s)36.28(g)	40.21	47.24	
HfO_2					66.97	73.13	79.73	86.13
HgCl	8.03	29.00	39.50	47.61	36.74(l)	37.11	37.28	37.36
HgO (red)	5.44	15.02	28.43	38.17				
IBr (g)					36.99	37.40	37.78	38.07
ICl (g)					36.40	37.11	37.66	37.99
KBr	6.94	23.18	43.34	50.20	53.32	56.64	64.62	
KCl	2.84	15.64	39.08	48.46	52.98	55.89	64.03	
KF	1.16	7.92	31.61	45.19	51.72	55.29	60.99	
KH_2PO_4	3.21	17.67	57.82	91.80				
$KMnO_4$	12.89	35.23	70.25	100.2				
K_2SO_4			79.12	110.0	146.84	174.46		
La_2OP_3	3.37	15.66	54.14	91.46	117.3	124.7	132.3	
Li_2CO_3	0.96	6.74	34.94	72.68				
LiF	0.08	0.75	13.14	32.72	46.38	51.57	59.18	
Li_2O	0.063	0.54	17.39	37.20	63.98	73.23	88.73	117.4
$MgCO_3$			24.27	58.03	89.87	107.70	135.88	
$MgCl_2$			40.25	63.51	75.62	79.85		
MgF_2			21.67	48.83	69.75	75.94	80.31	83.20
MgO	0.088	0.53	8.36	27.35	42.77	47.30	50.87	53.34
$Mg(OH)_2$	0.34	2.83	21.57	56.99	81.00	94.22		
Mn_3C			42.38	76.90	104.4	115.0	127.4(α)	159.0(β)
$MnCl_2$			47.11	67.03	77.15	81.84		
MnF_2	4.97	17.02	35.44	58.07				
MnO			32.71	38.07	47.45	50.33	54.14	58.49
Mn_2O_3			42.22	79.29	109.0	1120.8	137.2	
MnS			39.16	47.78	50.71	52.22	55.23	58.99
MoO_3	1.63	6.99	31.63	59.33	82.56	92.27	107.63	
Mo_3Si			46.67	71.11	97.02	102.3	110.7	120.5
NH_3	1.54	7.61	26.07(s)	73.47(l)	38.67(g)	45.23	56.24	65.94
N_2H_4	1.46	8.91	30.86	51.02(s)	63.2(g)	76.5	93.3	107
$NH_4H_2PO_4$	4.18	17.77	59.25	106.9				
NH_4NO_3	4.21	20.00	59.62	105.9				
NO	7.03	17.28	35.69(s)		29.95(g)	31.24	33.99	35.79
N_2O	6.32	21.46	41.42(s)		42.87(g)	48.86	56.02	60.45
N_2O_4	8.49	28.70	60.71	91.71(s)	90.46(g)	105.45	118.6	124.0
$NaBH_4$	1.59	10.54	40.21	71.42	94.56	108.6		
NaCl	1.42	10.04	35.06	46.86	51.92	55.53	65.52	
N_2O_5 (g)					106.8	121.9	136.4	143.0

Table 9.10 Molar isobaric heat capacity of inorganic compounds *(continued)*

Substance	Temperature, K							
	20	40	100	200	400	600	1000	1500
NaF			22.72	40.63	51.03	54.58	61.35	
$NaNO_3$	3.77	17.32	52.09	75.65	113			
Na_2O					73.93	80.64	91.85	
NaOH			28.77	49.20	66.64	86.04	84.57	84.56
Na_2SO_4	2.43	17.06	66.65	105.6	145.3	176.88	209.0	
Na_2SiO_3			48.20	89.83	130.6	148.0	168.9	
Na_2TiO_3			48.62	99.12	140(α)	151.3(β)	179.7	
Nb_3Al [39]	3.08	26.0	58.6	86.5	100.1			
Nb_3Sn [39]	5.04	32.7	65.8	90.0	99.8			
$NiCl_2$	3.29	13.08	43.76	64.94	75.4	79.79	85.94	
NiO			13.97	33.14	52.13(α)	51.84(β)	55.23	59.45
PH_3	10.29	40.88	46.90(s)		41.80(g)	50.92	64.31	72.84
$PbCl_2$	12.55	33.64	59.51	72.34	80.72	85.73		
PbO (red)			26.87	40.01	48.73	52.92		
PbO (yellow)			27.46	40.35			57.95	
RbI	15.06	434.73	47.70	51.38	53.5	57.35		
SF_6	22.64	42.93(s)	58.32(l)		117.9(g)	133.5	147.2	158.1
SO_2	6.95	24.18	48.07	87.74(s)	43.47(g)	48.99	54.52	57.11
Sb_2O_4			48.03	91.20(s)	121.7(g)	135.2	162.3	
SeF_6 (g)					127.9	141.3	150.7	156.6
SiC			4.18	16.32	34.19	41.24	47.72	51.26
SiH_4	9.41	26.19(s)	60.67(l)		51.42(g)	65.84	84.49	95.46
SiO_2 (cristobalite)			15.77	32.72	53.14	65.35	72.97	73.05
SiO_2 (glass)					53.1	61.3	70.0	
SiO_2 (quartz)	1.00	3.93	15.44	32.76	53.42	64.52	68.94	74.00
SiO_2 (tridymite)			16.28	33.72	61.50	63.68	68.12	73.64
SnO_2			20.48	42.26	65.09	74.28	81.18	86.10
SnS			34.31	45.61	51.00	56.08(s)	56.57(l)	74.90
$SrCO_3$			44.06	68.91	97.09	106.29	125.75	
SrO			24.02	40.04	48.45	52.04	56.31	60.14
$SrSO_4$	2.88	16.31	53.18	82.43	117.94	135.91	161.00	
$SrTiO_3$			43.35	80.71				
Ta_2O_5			57.91	107.4	150.3	167.0	179.6	195
ThO_2					67.69	72.74	77.61	82.29
TiC			10.79	27.53	41.46	47.32	51.34	53.85
TiO			12.68	30.63	46.58	50.73	57.76	
Ti_2O_3			26.69	72.30	119.24	136.38	143.47	158.02
UCl_3 [33]					97.15	105.1	115.2	118.4
UF_4	5.13	20.49	65.96	103.3	121.75	125.80	133.38	
UF_6	17.0	40.31	94.07	136.60(s)	139.8(g)	148.98	154.54	156.42
UO_2	4.19	11.03	29.11	52.17	73.43	79.13	82.70	90.97
VC			10.54	20.59	43.51	48.49	53.64	58.53
V_3Ga [40]			54.6	83.5	99.0			
V_3Ge [41]	1.62	16.42	53.3	84.8				
V_3Si [40,41]	1.72	12.92	44.84	778.68	93.8			
V_2O_3			28.70	80.92	117.05	127.46	133.09	139.29
WC [33]					37.03	38.83	42.47	47.03
WO_3 [4,33]			32.38	63.18	83.62			
ZnF_2	0.98	6.95	31.58	55.86				
ZnO				32.30	45.33	49.50	53.18	56.23
ZnS	1.67	8.20	24.52	39.66	49.41	52.38	55.48	
ZrN			18.54	31.67	44.77	48.66	52.76	
ZrO_2			19.00	43.51	63.56	70.04	76.07	78.1

Table 9.11 Isobaric specific heat c_p (kJ/(kg·K)) of ammonia, carbon dioxide, water and steam at different pressures and temperatures [Data over the horisontal line correspond to liquid phase.]

Ammonia NH_3 [42]

T, K	P, MPa				T, K	P, MPa			
	0.1	1	10	50		0.1	1	10	50
200	4.42	4.42	4.40	4.32	400	2.29	2.45	22.3	4.71
230	4.46	4.46	4.43	4.33	500	2.47	2.53	3.41	6.14
250	2.18	4.51	4.47	4.34	600	2.66	2.69	3.05	4.78
300	2.15	2.94	4.64	4.35	750	2.92	2.94	3.08	3.70

Carbon dioxide CO_2 [44,45]

T, K	P, MPa				T, K	P, MPa			
	0.1	1	10	100		0.1	1	10	100
220	0.786	1.820	1.745		350	0.900	0.936	1.930	1.561
240	0.798	1.018	1.864	1.466	400	0.941	0.966	1.327	1.530
260	0.814	0.945	2.039	1.609	600	1.077	1.084	1.161	1.422
280	0.832	0.920	2.273	1.648	1000	1.234	1.236	1.254	
300	0.852	0.916	2.977	1.630	1500	1.327	1.328	1.334	

Water and steam H_2O [43]

T, K	t, °C	P, MPa							
		0.1	1	2	5	10	20	50	100
273.15	0	4.217	4.212	4.207	4.196	4.165	4.117	3.993	3.890
323.15	50	4.181	4.179	4.176	4.170	4.158	4.137	4.080	4.038
237.15	100	2.038	4.213	4.211	4.205	4.194	4.173	4117	4.071
423.15	150	1.979	4.308	4.305	4.296	4.281	4.252	4.178	4.121
473.15	200	1.974	2.433	4.494	4.477	4.450	4.402	4.284	4.200
523.15	250	1.988	2.21	2.580	4.854	4.791	4.684	4.463	4.328
573.15	300	2.011	2.141	2.321	3.199	5.70	5.33	5.788	4.538
623.15	350	2.038	2.125	2.235	2.670	4.043	8.10	5.45	4.897
633.15	360	2.044	2.127	2.231	2.625	3.769	11.37	5.64	4.984
643.15	370	2.050	2.128	2.222	2.578	3.546	18.38	5.84	5.06
653.15	380	2.056	2.127	2.212	2.528	3.356	10.19	6.10	5.18
663.15	390	2.061	2.125	2.202	2.486	3.201	7.65	6.43	5.29
673.15	400	2.068	2.126	2.197	2.451	3.078	6.33	6.81	5.39
723.15	450	2.099	2.141	2.191	2.360	2.726	3.959	9.48	5.94
773.15	500	2.132	2.164	2.201	2.324	2.569	3.257	7.20	6.42
873.15	600	2.200	2.219	2.240	2.311	2.445	2.770	4.082	5.01
973.15	700	2.270	2.283	2.299	2.346	2.429	2.613	3.279	3.881
1073.15	800	2.341	2.352	2.364	2.401	2.465	2.598	3.024	3.415

9.4 Heat Capacity of Organic Compounds

Table 9.12 Isobaric molar heat capacity C_p (J/(mol·K)) of organic compounds at temperatures from 40 K to 1000 K [4] [Solids and liquids are at pressure of 0.1 MPa and gases are in the state of ideal gas (i.e. at utmost low pressure).]

Substance	Formula	Temperature, K							
		40	100	200	298.15	298.15	400	600	1000
		Solid or liquid				Gas			
Acetaldehyde	C_2H_4O	21.8	47.8	81.4		56.6	65.8	85.9	327.4
Acetic acid	$C_2H_4O_2$		50.3	67.2	123.4	66.5	81.7	105.2	133.8
Acetone	C_3H_6O	30.2	65.6	117	125	74.9	92.0	122.8	163.8
Aniline	C_6H_7N		71.6		191				
Benzene	C_6H_6	26.5	50.4	83.7	136.1	81.7	111.9	157.9	209.9
Benzoic acid	$C_7H_6O_2$ [61]	32.1	64.0	102.7					
Bromobenzen	C_6H_5Br		58.5	99.8	155.4				
Bromoform	$CHBr_3$				134.8	71.0	78.7	88.0	96.7
Bromomethane	CH_3Br	26.2	49.9	77.4		42.4	49.9	62.6	79.5
n-Butane	C_4H_{10}	27.4	66.7	119.6		98.7	124.7	169.1	226.8
1,3-Butadiene	C_4H_6	25.8	83.4	112.1		79.5	101.6	133.2	169.5
1-Butanol	$C_4H_{10}O$		64.9	138	179				
1-Butene	C_4H_8	24.5	107.7	110.1		89.3	112.7	149.9	197.7
Butylbenzene [21]	$C_{10}H_{14}$				175				
1-Butyne	C_4H_6	27.0	58.0	117.2		81.4	99.9	129.0	166.7
2-Butyne	C_4H_6	25.4	56.0	81.6	125.1	78.0	94.6	124.2	1645.4
Chlorobenzen	C_6H_5Cl		55.5	98.2	150.1	97.1	126.8	170.7	218.4
Chlorodifluoromethane*	$CHClF_2$	37.7	57.1	92.5		55.8	65.4	78.9	92.5
Chloroethane	C_2H_5Cl	29.6	59.7	95.8		62.3	77.7	102.4	132.5
Chloroethene	C_2H_3Cl					53.7	65.1	82.0	101.9
Chlorofluorormethane	CH_2ClF					47.1	55.6	69.3	85.3
Chloroform	$CHCl_3$				116.3	66.8	76.5	89.2	102.6
Chloromethane	CH_3Cl	22.5	49.0	74.1		40.7	48.2	61.3	78.9
2-Chloropropane	C_3H_7Cl					87.6	106.6	138.9	
Cumene	C_9H_{12}				189	134			
Cyclobutane	C_4H_8	23.4	53.4	91.1		72.2	99.9	145.4	200.6
Cyclobutene	C_4H_6					67.1	90.3	126.8	169.6
Cyclohexane	C_6H_{12}	28.1	58.6	109.5	156.5	106.3	149.9	225.2	317.1
Cyclohexene	C_6H_{10}	26.8	57.1	121.5	140.2	105.0	153.3	206.9	278.7
Cyclopentane	C_5H_{10}	27.8	59.9	102.3	126.8	82.9	118.2	177.2	250.0
Cyclopentene	C_5H_8	30.7	76.7	100.9	122.4	75.1	104.9	155.6	217.3
Cyclopropane	C_3H_6	22.6	51.3	76.4		55.9	76.7	109.4	147.9
Decane	$C_{10}H_{22}$	46.8	119.7	191.9	314.5	243.1	303.7	406.7	537.3
Dibromodichloromethane	CBr_2Cl_2					87.0	94.2	101.0	105.3
Dibromodifluoromethane	CBr_2F_2					77.0	85.6	95.4	102.8
Dibromomethane	CH_2Br_2				127.4	54.5	63.0	74.9	88.0
Dichlorodifluoromethane	CCl_2F_2				126	87.8	96.7	112.0	123.4
Dichloroethane	$C_2H_4Cl_2$	33.3	58.3	120.2	126.3	76.2	92.2	112.4	138.3
Dichloroethene	$C_2H_2Cl_2$	32.8	55.0	101.6	111.3	67.0	78.7	93.9	110.0
Dichlorofluoromethane	$CHCl_2F$					61.0	70.2	82.4	94.1
Dichloromethane	CH_2Cl_2				100	50.9	59.6	72.4	86.8
Difluoromethane	CH_2F_2					42.9	51.1	65.8	83.6
Dimethil ether	C_2H_6O	24.9	56.6	98.8		65.9			
p-Dioxane	$C_4H_6O_2$		52.8	86.4	152.9				

Table 9.12 Isobaric molar heat capacity C_p (J/(mol·K)) of organic compounds *(continued)*

Substance	Formula	Temperature, K							
		40	100	200	298.15	298.15	400	600	1000
		Solid or liquid				Gas			
Ethane*	C_2H_6	24.9	68.5			52.5	65.4	89.1	122.5
Ethanethiol	C_2H_6S	27.7	58.4	111.6	117.8	72.7	88.2	113.8	148.0
Ethanol cryst.	C_2H_6O	17.2	47.0	91.9	111.4	65.8	81.2	107.6	142.0
Ethanol amorph.	C_2H_6O	24.1	78.7						
Ethene (ethylene)*	C_2H_4	23.7	71.3			42.9	53.1	71.6	94.4
Ethine (acetylene)	C_2H_2					44.0	50.4	58.1	68.0
Ethyl ethanoate	$C_4H_8O_2$		77.4	160.2	170				
Ethylbenzene	C_8H_{10}	33.7	70.8	160	186	128.4	170.5	236.1	3123.8
Ethylene glycol	$C_2H_6O_2$		46.2	74.4	151				
Ethylene oxide	C_2H_4O	21.8	47.8	81.4	151	48.3	65.9	86.3	114.9
Fluorobenzen	C_6H_5F	30.1	54.1	95.2	146.4	94.4	125.5	171	220
Fluoroethane	C_2H_5F					59.0	73.6	98.1	
Fluoroform	CHF_3					51.1	61.1	76.4	92.6
Fluoromethane	CH_3F					37.5	44.0	57.7	77.1
Formaldehyde	CH_2O					35.4	39.2	48.1	61.9
Formic acid	CH_2O_2	14.2	34.8	50.8	99.1	45.8	60.0	69.7	89.7
Furan	C_4H_4O	29.3	48.6	100.5	114.6	65.4	88.7	122.6	158.5
Glycerin [21]	$C_3H_8O_3$				217				
Glycine	$C_2H_5O_2N$		42.8	72.7	100.3				
Heptane	C_7H_{16}	38.2	92.8	201.3	224.7	170.8	214.1	288.0	382.0
Hexachlourobenzen	C_6Cl_6		46.0	103	201				
Hexane*	C_6H_{14}	36.6	82.8	171.5	195	146.7	184.3	248.4	330.3
Hexanol	$C_6H_{14}O$	32	81.8		238				
1-Hexene	C_6H_{12}	33.8	79.1	158.5	183	138.4	173	229	301
1-Hexyne	C_6H_{10}					130.5	160	208	270
Iodomethane	CH_3I				127.2	44.1	51.6	63.9	80.1
Ketene	C_2H_2O				56.1	47.9	56.1	69.6	83.8
β-Lactose (anhydr.)	$C_{12}H_{22}O_{11}$		144	282	411				
Methane*	CH_4	28.7	54.8			35.7	4.6	52.7	73.7
Methanethiol	CH_4S	22.7	49.3	86.9		50.7	59.0	73.7	94.3
Methanol	CH_4O	18.3	43.6	70.7	81.6	44.1	51.8	67.5	89.7
Methyl amine	CH_5N					51.7	62.4	81.0	107.1
Methylcyclopentane	C_6H_{12}	32.	68.2	131.2	158.7	109.8	151	219.4	303.1
Naphthalene	$C_{10}H_8$	30.1	60.2	106.7	165.7	134.2	180.7	250.8	328.8
Nitrobenzen	$C_6H_5NO_2$		67.1	107	187.3				
Nonane	C_9H_{20}	43.3	111.4	184.8	284.4	219.0	273.8	367.1	485.6
Octane	C_8H_{18}	40.8	100.8	164	254.1	194.9	244.0	327.5	433.8
Pentane	C_5H_{12}	31.0	74.1	144.2	171.5	122.6	154.4	208.7	278.5
1-Pentanol	$C_5H_{12}O$		74.1	164	209.2				
1-Pentene	C_5H_{10}	32.9	73.0	133.5	155.3	114.6	143.1	189.8	249.4
1-Pentyne	C_5H_8					106.7	130.1	169.0	218.4
Phenol	C_6H_6O		53.0	85.1	134.7				
Phosgene	$COCl_2$	33.4	55.2	99.7		57.8	66.0	72.1	77.8
Propadiene	C_3H_4					59.0	72.0	92.0	117.2
Propane*	C_3H_8	25.1	85.0	93.5		73.5			
Propanol	C_3H_8O	24.7	53.5	110.3	148.6				
Propene	C_3H_6	29.0	90.8	89.2		63.9	79.9	107.5	144.2
Propyne	C_3H_4					60.7	72.5	91.2	115.9
Saccharose	$C_{12}H_{22}O_{11}$		146	282	425				
Styrene (styrol)	C_8H_8	11.7	64.8	135.4	182.9	122.1	160.3	218.2	284.2
Tetrabromomethane	CBr_4				128	91.0	96.9	102.4	105.9

Table 9.12 Isobaric molar heat capacity C_p (J/(mol·K)) of organic compounds *(continued)*

Substance	Formula	Temperature, K							
		40	100	200	298.15	298.15	400	600	1000
		Solid or liquid				Gas			
Tetrachloromethane	CCl_4	40.9	66.9	103.3	131.7	82.9	91.2	99.4	104.7
Tetrafluoromethane	CF_4	42.8	78.0			61.1	72.4	86.8	98.8
Toluene	C_7H_8		61.9	134	160	104	139.1	194.9	260.2
Tribromofluoromethane	CBr_3F					84.2	91.7	99.2	104.5
Trichlorofluoromethane	CCl_3F	40.0	62.3	112	121.5	78.1	87.2	96.8	103.5
1,2,2-Trichloro-1,1,2-trifluoroethane	$C_2Cl_3F_3$		41.3	67.0	176	117.4	135.1	155.1	
Urea	$(NH_2)_2CO$	18.1	41.3	67.0	93.1				
o-Xylene [21,4]	C_8H_{10}	31.8	71.7	119.9	183	133.3	171.7	234.2	311.1

* See also Table 9.13.

Table 9.13 Isobaric specific heat capacity c_p (kJ/(kg·K)) of organic liquids and gases at different pressures and temperatures [Data over the horizontal line correspond to liquid phase.]

T, K	P, MPa				T, K	P, MPa			
Chlorodifluoromethane (difluorochloromethane, freon-22) $CHClF_2$ [46]									
	0.1	1	10	20		0.1	1	10	20
273.15	0.621	1.161	1.125	1.105	403.15	0.768	0.807	2.169	1.334
303.15	0.657	0.842	1.166	1.114	453.15	0.813	0.830	1.422	1.465
353.15	0.720	0.804	1.378	1.225	473.15	0.830	0.841	1.228	1.454
Ethane (bimethyl, methylmethane, dimethyl) C_2H_6 [49]									
	0.1	1	10	70		0.1	1	10	70
120	2.296	2.293	2.274	2.250	220	1.511	2.607	2.500	2.301
140	2.356	2.354	2.335	2.277	240	1.561	2.834	2.633	2.360
160	2.362	2.359	2.332	2.222	300	1.765	1.910	3.457	2.527
180	2.392	2.387	2.347	2.205	400	2.182	2.227	3.211	2.838
200	2.392	2.387	2.347	2.205	500	2.597	2.620	2.914	3.136
Ethene (ethylene) C_2H_4 [50]									
	0.1	1	10	70		0.1	1	10	70
130	2.400	2.398	2.382	2.354	300	1.544	1.639	4.299	2.279
150	2.381	2.379	2.346	2.186	350	1.716	1.767	3.308	2.367
200	1.321	2.529	2.424	2.190	400	1.894	1.927	2.510	2.456
250	1.397	1.644	2.773	2.209	450	2.07	2.091	2.414	2.547
Fluoroform (trifluoromethane, freon-23) CHF_3 [46]									
	0.1	1	10	20		0.1	1	10	20
233.15	0.663	0.824	0.727	0.661	353.15	0.812	0.856	1.973	1.581
253.15	0.683	0.890	1.195	1.112	403.15	0.879	0.904	1.298	1.510
303.15	0.745	0.828	1.796	1.421	453.15	0.940	0.956	1.162	1.343
***n*-Hexane C_6H_{14} [66]**									
	0.1	1	10	20		0.1	1	10	20
180	1.976	1.969	1.911		350	1.944	2.503	2.464	2.395
200	2.000	1.995	1.973	1.740	400	2.144	2.769	2.681	2.568
250	2.087	2.086	2.074	2.022	500	2.526	2.679	3.282	2.995
300	2.290	2.288	2.268	2.223	600	2.870	2.922	3.675	3.096

Table 9.13 Isobaric specific heat capacity c_p (kJ/(kg·K)) of organic liquids and gases *(continued)*

T, K	P, MPa				T, K	P, MPa			
	Methane CH_4 [47]								
	0.1	1	10	100		0.1	1	10	100
100	3.258	3.255	3.226		200	2.106	2.294	5.321	2.962
110	3.390	3.379	3.289		300	2.236	2.290	3.018	3.047
120	2.192	3.549	3.431	3.046	400	2.534	2.559	2.808	3.149
130	2.163	3.683	3.523	3.202	600	3.273	3.282	3.367	3.612
140	2.141	3.822	3.580	3.230	800	3.959	3.964	4.006	4.175
150	2.125	2.941	3.644	3.187	1000	4.539	4.542	4.568	4.691
	Propane C_3H_8 [48]								
	0.1	1	10	70		0.1	1	10	70
90	1.911	1.911	1.908		300	1.696	2.751	2.556	2.317
100	1.923	1.922	1.919	1.903	350	1.916	2.050	3.012	2.544
150	1.998	1.997	1.988	1.953	400	2.136	2.209	3.915	2.795
200	2.125	2.122	2.099	2.028	500	2.552	2.586	3.238	3.052
250	1.496	2.336	2.277	2.145	700	3.239	3.252	3.394	3.573

9.5 Heat Capacity of Alloys, Mixtures, Solutions and Technical Materials

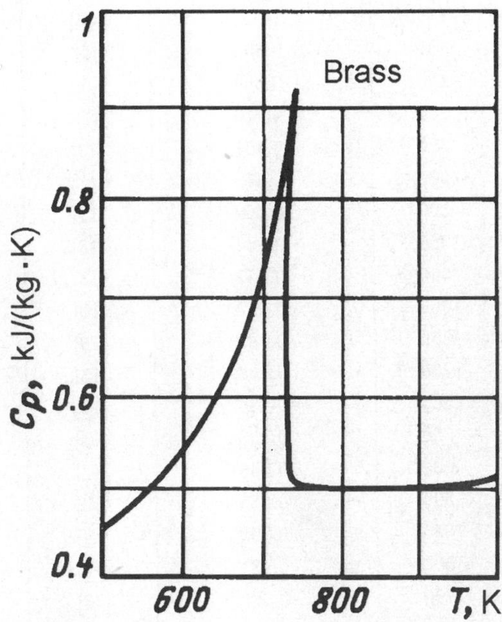

Figure 9.11 Specific heat of β–brass (mass ratios were 51.8% of Cu, 48.2% of Zn) near the ordering point [4].

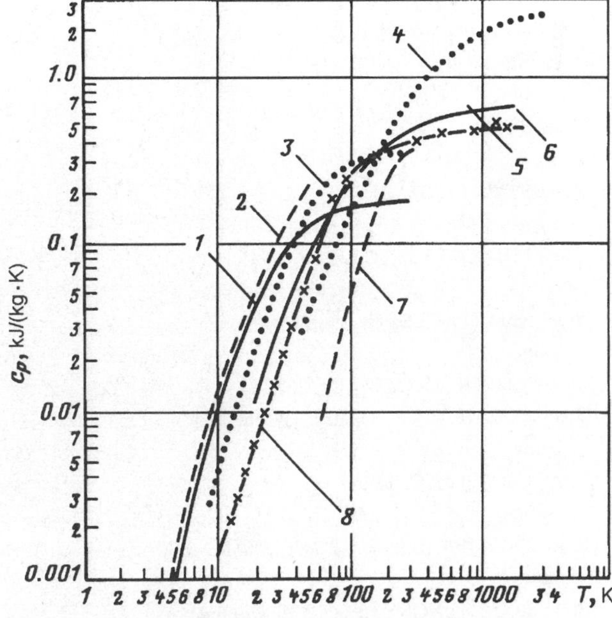

Figure 9.12 *1* – solder (Pb - 50%, Sn - 50%) [5]; *2* – Wood's alloy [5, 33]; *3* – brass L-62 [5]; *4* – graphite [5, 13, 58]; *5* – ST-45 [13]; *6* – ST-X18H9T, ST-X18H10T; *7* – constantan [5, 33]; *8* – Monel metal [5, 33].

Table 9.14 Isobaric specific heat capacity c_p (kJ/(kg·K)) of air at different pressures and temperatures [56, 57]

T, K	P, MPa				T, K	P, MPa			
	0.1	1	10	100		0.1	1	10	100
110	1.028	1.495	2.011	1.683	400	1.014	1.022	1.089	1.244
120	1.020	1.280	2.114	1.704	600	1.052	1.054	1.080	1.176
150	1.011	1.107	2.847	1.575	1000	1.141	1.142	1.150	1.191
200	1.007	1.048	1.641	1.433	1500	1.211	1.212	1.214	1.233
300	1.007	1.021	1.163	1.331	2000			1.252	1.261

Table 9.15 Specific heat capacity c_p of carbonized steels, alloyed steels and cast iron at different temperatures

Steel or iron index	T, K	c_p, kJ/(kg·K)	Ref.
Cast iron white	293	0.54	[5]
Cast iron (liquid)	1600	0.85	[71]
Grey iron	293	0.50	[5]
Stainless steel 12X18H9T (0.12%C, 18%Cr, 12%Ni, 1%Mn)	100	0.262	[5]
same	200	0.41	[5]
same	400	0.491	[5]
same	600	0.543	[5]
same	1000	0.616	[5]
Steel ball-bearing ШX18 (1%C, 1.5%Cr)	300	0.50	[5]
Steel high-speed R18	373	0.42	[63]
Steel high-speed R18	973	0.69	[63]
Steel high-speed 11P3AM3F2 (0.11%C, 3%B, 3%Mo, 2%V)	373	0.43	[63]
same	973	1.01	[63]
Steel sheet electrotechnical (3%Si)	80	0.358	[5]
same	150	0.389	[5]
same	250	0.426	[5]
Steel ST 20 (0.2% C)	300	0.461	[13]
same	1000	0.673	[13]
Steel ST 35 (0.35% C)	300	0.462	[13]
same	1000	0.564	[13]
Steel 13H2XA (0.13%C, 2%Ni, 1%Cr)	300	0.452	[13]
same	1000	0.612	[13]
Steel X5M (1%C, 5%Cr, 1%Mo)	300	0.482	[13]
same	1400	0.660	[13]
Steel X22H26 (1%C, 22%Cr, 26%Ni)	300	0.485	[13]
Steel 1X11MF (0%C, 11%Cr, 1%Mo, 1%V)	300	0.483	[13]
same	800	0.955	[13]
Steel 4X13 (0.4%C, 13%Cr)	300	0.485	[13]
same	1000	0.598	[13]
Steel 25X2MFA (0.25%C, 2%Cr, 1%Mo, 1%V)	300	0.481	[13]
same	800	0.506	[13]
Steel 30XM, 30XGS (0.3%C, 1%Cr, 0.6%Mn)	300	0.46	[13]
same	800	0.49	[13]
Steel Y8 (0.8% C, 0.2% Mn)	300	0.462	[13]
same	1200	0.662	[13]

Table 9.16 Specific heat capacity c_p of multicomponent alloys

Substance	Composition	T, K	c_p, J/(g·K)	Ref.
Alloys	92%Al, 8%Mg	293	1.15	[33]
	60%Al, 40%Zn	293 – 373	0.56	[33]
	80%Au, 20%Cu	137	0.164	[33]
	80%Au, 20%Cu	285	0.183	[33]
	78%K, 22%Na (liquid)	300	0.97	[13]
	78%K, 22%Na (liquid)	600	0.88	[13]
	78%K, 22%Na (liquid)	1000	0.89	[13]
	44%K, 56%Na (liquid)	300	1.16	[13]
	44%K, 56%Na (liquid)	600	1.06	[13]
	44%K, 56%Na (liquid)	1000	1.05	[13]
Bell metal		288 – 371	0.36	[33]
Bismuth tin	50%Bi, 50%Sn	273 – 373	0.182	[33]
Bismuth tin	25%Bi, 75%Sn	273 – 373	0.209	[33]
Brass	x%Cu, (100-x%)Zn ($10 < x < 90$)	288 – 373	0.162	[33]
Bronze	75%Cu, 25%Al	273	0.43	[33]
Bronze	67%Cu, 33%Al	373	0.46	[33]
Bronze	67%Cu, 33%Al	773	0.53	[33]
Bronze	50%Cu, 50%Al	123	0.35	[33]
Bronze	50%Cu, 50%Al	273	0.51	[33]
Bronze	67%Cu, 33%Mg	123	0.30	[33]
Bronze	67%Cu, 33%Mg	673	0.57	[33]
Bronze	90%Cu, 10%Ni	285	0.380	[33]
Bronze	80%Cu, 20%Sn	288 – 371	0.36	[67]
Constantan	40%Cu, 60%Ni	273	0.171	[33]
German silver		273	0.39	[67]
Invar	64%Fe, 36%Ni	288 – 373	0.50	[67]
Lead bismuth	44.5%Pb, 55.5%Bi	273	0.125	[33]
Lead bismuth	44.5%Pb, 55.5%Bi	423	0.146	[33]
Lead bismuth tin	31%Pb, 50%Bi, 19%Sn	273	0.138	[13]
Lead bismuth cadmium	26%Pb, 48%Bi, 13%Cd	273	0.13	[33]
Lead tin	64%Pb, 36%Sn	94 – 194	0.151	[33]
Lead tin	64%Pb, 36%Sn	194 – 291	0.163	[33]
Lead tin	64%Pb, 36%Sn	285 – 372	0.171	[33]
Lead tin	64%Pb, 36%Sn	285 – 372	0.171	[33]
Lead tin	50%Pb, 50%Sn	20	0.046	[5]
Lead tin	50%Pb, 50%Sn	100	0.152	[5]
Lead tin	50%Pb, 50%Sn	300	0.179	[5]
Lead thalium	36%Pb, 64%Th	285		[33]
Lipovitz alloy		298 – 323	0.144	[33]
Lipovitz alloy		373 – 423	0.178	[33]
Manganin		291	0.405	[33]
Platinum iridium	90%Pt, 10%Ir	293 – 373	0.135	[67]
Rose alloy		196 – 293	0.147	[33]
Rose alloy		293 – 362	0.23	[33]
Wood's alloy		278 – 423	0.147	[33]

Table 9.17 Specific heat capacity c_p (kJ/(kg·K)) of nonmetallic technical materials

Substance	T, K	c_p	Ref.	Substance	T, K	c_p	Ref.
Building materials							
Ashes	298	0.75	[13]	Brick silicated	298	0.84	[13]
Asphalt	298	1.68	[13]	Cartboard	298	1.51	[13]
Brick red	298	0.88	[13]	Cement, powder	293	0.84	[67]

Table 9.17 Specific heat capacity c_p (kJ/(kg·K)) of nonmetallic technical materials *(continued)*

Substance	T, K	c_p	Ref.	Substance	T, K	c_p	Ref.
Cement-sand solution	298	0.84	4[13]	Limestone	298	0.92	[33]
Chalk	298	0.88	[33]	Marble	298	0.92	[13]
Clay	273	0.75	[33]	Paper	298	1.51	[13]
Clay	673	1.13	[33]	Peat	298	1.7	[13]
Clay	1073	1.51	[33]	Plywood	273	2.5	[33]
Concrete	298	0.84	[13]	Sand	298	0.84	[13]
Cork	298	1.8	[33]	Slag	298	0.75	[13]
Felt	298	1.88	[13]	Soil	298	0.84	[13]
Glass	298	0.84	[13]	Wood	298	1.2	[33]
Granite	298	0.92	[13]				
Fuel and oil							
Coal, bituminous	293 – 1313	1.31	[33]	Oil MC-20	423	2.44	[21]
Glucerine	293	2.35	[21]	Oil, transformer	223	1.70	[21]
Glucerine	513	3.60	[21]	Oil, transformer	373	2.04	[21]
Kerosene T-1	293	2.00	[21]	Oil, vegetable	293	1.5 – 2.0	[33]
Kerosene T-1	423	2.63	[21]	Peat	298	1.7	[13]
Kerosene T-1, vapor	423	2.37	[21]	Petrol B-70	293	2.06	[21]
Kerosene T-1, vapor	473	2.47	[21]	Petrol B-70	423	2.74	[21]
Mazut (black oil)	293	2.18	[21]	Petrol B-70, vapor	423	2.28	[21]
Oil BM-4	243	1.44	[21]	Petrol B-70, vapor	523	2.58	[21]
Oil BM-4	373	1.62	[21]	Petroleum	293 – 333	2.10	[33]
Oil MC-20	273	1.98	[21]				
Glasses, natural silicates							
Albit	273 – 373	0.83	[33]	Glass, quartz	293	0.89	[33]
Albit	273 – 1173	1.1	[33]	Glass, quartz	873	1.00	[33]
Anorthite	273 – 373	0.83	[33]	Glass, quartz	1473	1.14	[33]
Anorthite	273 – 1173	1.10	[33]	Glass, sodium	293 – 373	0.80	[33]
Crown-glass	285	0.67	[33]	Glass, sodium	293 – 1273	1.12	[33]
Diopside	273 – 373	0.81	[33]	Glass, thermometric 16″	292 – 373	0.83	[33]
Diopside	273 – 973	1.02	[33]	Glass, window	273 – 373	0.67	[33]
Flint-glass	303	0.49	[33]	Microcline	273 – 373	0.80	[33]
Glass, Pyrex	173 – 273	0.60	[33]	Microcline	273 – 1373	1.09	[33]
Glass, Pyrex	273 – 573	0.86	[33]	Wollastonite	273 – 373	0.78	[33]
Glass, Pyrex	313 – 1273	1.2	[33]	Wollastonite	273 – 973	0.98	[33]
Minerals							
Apatite	288 – 372	0.79	[33]	Graphite natural	300	1	[5]
Asbestos	293	1.1	[5]	Soil	300	0.85	[13]
Beryl	323	0.84	[33]	Dolomite	293 – 372	0.93	[33]
Basalt	273	0.85	[33]	Kaoline	273	1.0	[33]
Basalt	1473	1.49	[33]	Kaoline	673	1.35	[33]
Gypsum	73	0.32	[33]	Lava	296 – 373	0.84	[33]
Gypsum	273	1.06	[33]	Lava	304 – 1049	1.09	[33]
Gneiss	273	0.74	[33]	Malachite	288 – 372	0.74	[33]
Gneiss	473	1.02	[33]	Mica	293	0.88	[33]
Granite	273	0.65	[33]	Spinel	282 – 371	0.81	[33]
Granite	1073	1.30	[33]	Talc	332	0.87	[33]
Plastics, polymers							
Bakelite	300	0.82	[5]	Paraffin	253 – 276	1.6	[33]
Ebonite	293	1.43	[33]	Paraffin (liquid)	333	3.0	[33]
Kapron	293	2.3	[33]	Plexiglas	100	0.55	[5]
Nylon-6	100	0.6	[5]	Plexiglas	200	1.05	[5]
Nylon-6	200	1.0	[5]	Polyethylene	100	0.62	[33]
Nylon-6	300	1.5	[5]	Polyethylene	293	2.5	[33]

Table 9.17 Specific heat capacity c_p (kJ/(kg·K)) of nonmetallic technical materials *(continued)*

Substance	T, K	c_p	Ref.	Substance	T, K	c_p	Ref.
Plexiglas	300	1.50	[5]	Teflon	5	0.006	[5]
Press-material AΓ-4C				Teflon	20	0.08	[5]
(phenol formaldehyde)	293	1.17	[33]	Teflon	50	0.21	[5]
Polyester porous	100	0.44	[5]	Teflon	100	0.39	[5]
Polyester porous	300	0.34	[5]	Teflon	200	0.70	[5]
Rubber	300	1.9	[5]	Teflon	300	1.16	[5]
				Textolite	298	1.95	[33]
Refractory materials							
Alumina	373	0.84	[33]	Coal, electrode	300 – 810	0.83	[33]
Alumina	1773	1.15	[33]	Coal, electrode	300 – 1723	1.62	[33]
Brick, chamotte	273	0.88	[13]	Porcelain, adjusting	300	0.92	[13]
Brick, chamotte	1273	1.14	[13]	Porcelain, high-volt.	300	0.75	[13]
Brick, magnesite	273	1.05	[13]	Porcelain, low-volt.	300	0.85	[13]
Brick, magnesite	1273	1.32	[13]	Sillimanite	273	0.90	[33]
Carborundum	273	0.93	[13]	Sillimanite	1273	1.16	[33]
Carborundum	1273	1.06	[13]	Zircon	273	0.55	[13]
Coal, electrode	300 – 350	0.70	[33]	Zircon	1273	0.68	[13]

References

[1] Thermal constants of substances. Handbook, Ed. Glushko, V. P., VINITI, Moscow, 1965–1981 (issues 1–10) (in Russian).

[2] Thermodynamic properties of individual substances. Reference 3-d ed., Ed. Glushko, V. P., Nauka, Moscow, 1977–1982 (vol. 1–4) (in Russian).

[3] Selected values of the thermodynamic properties of the elements, Hultgren, R., Desai, R. D., Hawkins, D. T., Gleiser, M., Kelley, K. K., Wagman, D. D., Amer. Soc. of Metals, Ohio, 1960.

[4] Landolt, H., Bornstein, R., Zahlenwerte und funktionen aus physik chemie, astronomie, geophysik und technik, Springer, Berlin, 1960 (Bd. 2).

[5] Kozhevnikov, I. G., Novitskii, L. A., Thermophysical properties of materials at low temperatures. Handbook, 2-nd ed., Mashinostroenie, Moscow, 1982 (in Russian).

[6] Thermodynamic properties of nitrogen, Sychev, V. V., Wasserman, A. A., Kozlov, A. D., Spiridonov, G. A., Kimarnii, V. A., Izdatel'stvo standartov, Moscow, 1977 (in Russian).

[7] Thermophysical properties of neon, argon, krypton and xenon, Eds. Rabinovich, V. A., Wasserman, A. A., Nedostup, V. I., Veksler, L. S., Izdatel'stvo standartov, Moscow, 1976 (in Russian).

[8] McCarty, R. D., Hord, J., Roder, H. H., Selected properties of hydrogen, Govern. Print. Off., Washington, DC, 1983.

[9] Thermophysical properties of alkaline metals, Ed. Kirillin, V. A., Izdatel'stvo standartov, Moscow, 1970 (in Russian).

[10] Thermodynamic properties of oxygen, Sychev, V. V., Wasserman, A. A., Kozlov, A. D., Spiridonov, G. A., Kimarnii, V. A., Izdatel'stvo standartov, Moscow, 1981 (in Russian).

[11] Ribkin, N. P., Orlova, M. P., Baranyuk, A. K., Nurullaev, N. G., Zhuchkov, E. S., Izmeritel'naya tekhnika, 3, 37, 1976 (in Russian).

[12] A handbook on physical and technical principles of cryogenics, 3rd ed., Ed. Malkov, M. P., Energoatomizdat, Moscow, 1985 (in Russian).

[13] Chirkin, V. S., Thermophysical properties of nuclear engineering materials. Handbook, Atomizdat, Moscow, 1978 (in Russian).

[14] Berezovskii, G. A., Paukov, I. E., Zhurnal fizicheskoi khimii, 52, 2677, 1978 (in Russian).

[15] Selected values of the thermodynamic properties of binary alloys, Hultgren, R., Desai, R. D., Hawkins, D. T., Gleiser, M., Kelley, K. K., Amer. Soc. of Metals, Ohio, 1973.

[16] Angus, S., de Reuck, K. M., McCarty, R. D., International thermodynamic tables of the fluid state – 4. Helium, Pergamon Press, Oxford, 1977.

[17] Cetas, T. C., Holste, J. C., Swenson, C. A., Phys. Rev., 182, 679, 1969.

[18] Holste, J. C., Cetas, T. C., Swenson, C. A., Rev. Sci. Instrum., 43, 670, 1972.

[19] Leopold, H. A., Boorse, H. A., Phys. Rev., 134, 1322, 1964.

[20] Heighbor, J. E., 155, 384, 1967.

[21] Vargaftik, N. B., Handbook on thermophysical properties of gases and liquids, 2nd ed., Nauka, Moscow, 1972 (in Russian).

[22] Stewart, G. R., Rev. Sci. Instrum., 54, 1, 1983.

[23] Bagatskii, M. I., Kucheryavy, V. A., Manzhelii, V. G., Popov, V. A., Phys. Status Solidi, 26, 453, 1968.

[24] Radebaugh, R., Keesom, H., Phys. Rev., 149, 209, 1966.

[25] Holste, J. C., Phys. Rev., B 6, 2495, 1972.

[26] Ahlers, G., Phys. Rev., A 2, 1505, 1970.

[27] Mertig, M., Pompe, G., Hegenbarth, E., Solid State Comm., 49, 369, 1984.

[28] Kostryukova, M. O., Strelkov, P. G., Doklady Akad. Nauk SSSR, 162, 543, 1965 (in Russian).

[29] Ashcroft H. W., Mermin N. D., Solid state physics, Holt, Rinehart and Winston, New York, 1976.

[30] Leopold, H. A., Iafrate, G. J., Rothwart, F., J. Low- Temp. Phys., 28, 241, 1977.

[31] Hell, R. W., Cesier, P., Hukin, D. A., J. Phys. F: Metal Phys., 14, 1265, 1984.

[32] Girifalco L. A., Statistical physics of materials, John Wiley and Sons, New York - London - Sydney - Toronto, 1973.

[33] Tables of physical quantities. Handbook, Ed. Kikoin, I. K., Atomizdat, Moscow, 1976 (in Russian).

[34] Junod, A., Staudeman, J.-L., Muller, J., Spitzli, P., J. Low-Temp. Phys., 5, 25, 1971.

[35] Gel'd, P. V., Kalishevich, G. I., Surikov, V. I., Doklady Akad. Nauk SSSR, 215, 833, 1974 (in Russian).

[36] Khlopkin, M. N., Sov. Phys. JETP, 63, 164, 1986.

[37] Sandine, T. R., Keesom, P. H., Phys. Rev., 177, 1370, 1969.

[38] Wiswanathan, R., J. Appl. Phys., 46, 4086, 1975.

[39] Knapp, G. S., Bader, S. D., Fisk, Z., Phys. Rev., B 13, 3783, 1976.

[40] Knapp, G. S., Bader, S. D., Culbere, H. V., 11, 4331, 1975.

[41] Surikov, V. I., Kalishevich, G. I., Gel'd, P. V., Zhurnal fizicheskoi khimii, 49, 555, 1975 (in Russian).

[42] Thermophysical properties of ammonia, Golubev, I. F., Kiyasheva, V. P., Perel'shtein, I. I., Parushin, E. B., Izdatel'stvo standartov, Moscow, 1978 (in Russian).

[43] Rivkin, S. L., Aleksandrov, A. A., Thermophysical properties of water and water vapor, Energiya, Moscow, 1980 (in Russian).

[44] US NBS Circular 564. Properties of gases, US Govern. Print. Off., Washington, DC, 1955.

[45] Angus, S., Armstrong, B., de Reuck, K. M., International thermodynamic tables of the fluid state – 3. Carbon dioxide, Pergamon Press, Oxford, 1976.

[46] Thermophysical properties of freons. Reference book. 1. Freons of the methane series, Eds. Altunin, V. V., Geller, V. Z., Petrov, E. K., Izdatel'stvo standartov, Moscow, 1980 (in Russian).

[47] Thermodynamic properties of methane, Eds. Sychev, V. V., Wasserman, A. A., Zagoruichenko, V. A. et al, Izdatel'stvo standartov, Moscow, 1979 (in Russian).

[48] Goodwin, R. D., Haynes, W. H., US NBS monograph. 170. Thermophysical properties of propane from 85 to 700 K at pressures to 79 MPa, US Govern. Print. Off., Washington, DC, 1982.

[49] Thermodynamic properties of ethane, Eds. Sychev, V. V., Wasserman A. A., Zagoruichenko, V. A., Kozlov, A. D., et al, Izdatel'stvo standartov, Moscow, 1982 (in Russian).

[50] Thermodynamic properties of ethylene, Eds. Sychev, V. V., Wasserman, A. A., Golovskii, E. A., et al, Izdatel'stvo standartov, Moscow, 1981 (in Russian).

[51] Peshkov, V. P., Usp. Fiz. Nauk, 94, 607, 1968 (in Russian).

[52] Physical and mechanical and thermophysical properties of metals, Ed. Rykalin, N. N., Nauka, Moscow, 1976 (in Russian).

[53] Kraftmakher, Ya. A., Romanishina, T. Yu., Fizika tverdogo tela, 8, 1966, 1966 (in Russian).

[54] Novikov, I. I., Roshchupkin, V. V., Mozgovoi, A. G., Semashko, N. A., Teplofizika vysokikh temperatur, 19, 858, 1981 (in Russian).

[55] Fereira da Silva, D., Burgemeister, E. A., Dokoupil, Z., Phys. Lett., 25A, 354, 1967.

[56] Thermodynamic properties of air, Eds. Sychev, V. V., Wasserman, A. A., Kozlov, A. D., et al., Izdatel'stvo standartov, Moscow, 1978 (in Russian).

[57] Air. Density, compressibility, enthalpy, entropy, isochoric and isobaric specific heat, velocity of sound and adiabatic exponent at temperatures 1300–2500 K and pressures from 5 to 100 MPa. Tables of recommended reference data. GSSSD R 32–81, Izdatel'stvo standartov, Moscow, 1982 (in Russian).

[58] Graphite UPV–1T. Enthalpy and isobaric specific heat at temperatures 1200–2900 K. Tables of standard reference data. GSSSD 25–81, Izdatel'stvo standartov, Moscow, 1982 (in Russian).

[59] Selected values of chemical thermodynamic properties, Eds. Rossini, F. D., Wagman, D. D., Evans, W. H., Levine, S., Jaffe, I., US Govern. Print. Off., Washington, DC, 1952.

[60] Selected values of chemical thermodynamic properties. Wagman, D. D., Evans, W. H., Parker, V. B., Hallow, J., Baily, S. M., Schumm, R. H., US Govern. Print. Off., Washington, 1965 - 1973.

[61] Ribkin, N. P., Orlova, M. P., Baranyuk, A. K., Nurullaev N G., Rozhnovskaya L. N., Izmeritelnaya tekhnika, 7, 29, 1974 (in Russian).

[62] Stainless steels 12X18H9T and 12X18H10T. Enthalpy and specific heat at temperatures 400-1380 K and atmospheric pressure. Tables of standard reference data. GSSSD 32–82, Moscow, 1983 (in Russian).

[63] High-speed tool steel. Physical properties. Tables of standard reference data. GSSSD 27–81, Moscow, 1982 (in Russian).

[64] Thermodynamic properties of helium-4, Eds. Sychev, V. V., Wasserman, A. A., Kozlov, A. D., Spiridonov, G. A., Kimarnii, V. A., Izdatel'stvo standartov, Moscow, 1984 (in Russian).

[65] Niobium. Physical properties. Tables of standard reference data. GSSSD 121–88, Izdatel'stvo standartov, Moscow, 1989 (in Russian).

[66] N-hexane. Thermodynamic properties in the ranges of 180 to 630 K and 0.1 to 100 Pa. Tables of standard reference data. GSSSD 90-85, Izdatel'stvo standartov, Moscow, 1986 (in Russian).

[67] CRC Handbook of chemistry and physics, 54th ed., Chemical Rubber Corporation Publ., Cleveland, Ohio, 1973-1974.

[68] Orlova, M. P., Korolev, Ya. A., Zhurnal fizicheskoi khimii, 52, 2576, 1978 (in Russian).

[69] Gorbunov, V. E., Gavrichev, K. S., Sharpataya, G. A., Zhurnal neorganicheskoi khimii, 26, 547, 1981 (in Russian).

[70] Zinov'ev, V. E., Thermophysical properties of metals at high temperatures. A handbook, Metallurgiya, Moscow, 1989 (in Russian).

10

Temperature Coefficients of Expansion

E.B. Gel'man

10.1 Introduction

A change in size of a body at constant pressure due to varying temperature is referred to as thermal expansion. This phenomenon is governed by the potential of interaction between atoms of a substance in a lattice being unsymmetrical, which leads to anharmonic oscillations of the atoms with respect to their mean positions.

The fractional change of the volume of a body, on being heated through one kelvin is called the temperature coefficient of volume expansion

$$\beta = \frac{1}{V}\left(\frac{\partial V}{\partial T}\right)_P, \tag{10.1}$$

where V is the volume of the body, and T is the temperature. The letter under the parentheses indicates that the derivative is taken at a constant pressure.

The thermal expansion of anisotropic substances is characterized by the temperature coefficient of linear expansion (TCLE) α, which is determined as

$$\alpha = \frac{1}{l}\left(\frac{\partial l}{\partial T}\right)_P, \tag{10.2}$$

where l is the size of a body in a given direction.

In the general case α is a symmetric second-rank tensor $\alpha = [\alpha_{ij}]$. If $[\alpha_{ij}]$ is reduced to the principal axes of symmetry, it will be entirely determined by the principal values of the temperature coefficient of linear expansion $\alpha_1, \alpha_2, \alpha_3$. Thus, a sphere separated in a substance, on being heated by ΔT, is transformed at $\Delta T \to 0$ into an ellipsoid with the axes proportional to $1 + \alpha_1\Delta T$, $1 + \alpha_2\Delta T$, $1 + \alpha_3\Delta T$, while $\beta = \alpha_1 + \alpha_2 + \alpha_3$. For cubic syngony $\alpha_{11} = \alpha_{22} = \alpha_{33}$ and the tensor is reduced to a scalar. For hexagonal (trigonal) syngony $\alpha_{11} = \alpha_{22} = \alpha_\perp$, $\alpha_{33} = \alpha_\parallel$, where the subscript $\parallel$ denotes the direction of sixth (third) order axis and $\perp$ denotes the direction perpendicular to this axis. For rhombic syngony, one should know the temperature coefficient of linear expansion in the directions of second-order axes. In the crystals of monoclinic and triclinic syngonies, the principal axes are not unambiguously determined by the crystallographic coordinate system and depend on temperature;

0-8493-2861-6/97/$0.00+$.50

therefore, the thermal expansion of such structures is described in terms of the temperature coefficients of linear expansion α_a, α_b, α_c for the directions of the crystallographic coordinate axes a, b, c.

Besides the coefficients α and β, which are called the true coefficients, in order to characterize the thermal expansion of a body one uses the average coefficients of expansion determined as the values of the temperature coefficients averaged over a finite range of temperature ΔT:

$$\overline{\alpha} = \frac{1}{l}\left(\frac{\Delta l}{\Delta T}\right)_P; \quad \overline{\beta} = \frac{1}{V}\left(\frac{\Delta V}{\Delta T}\right)_P, \tag{10.3}$$

as well as the temperature coefficient of linear expansion averaged over directions (for anisotropic substances)

$$\alpha_{\text{av}} = \frac{\beta}{3} = \frac{\alpha_{11} + \alpha_{22} + \alpha_{33}}{3}. \tag{10.4}$$

All the temperature coefficients are measured in K^{-1}.

There exists a phenomenological relation between the specific heat C_V of a substance at constant volume and its isothermal compressibility $æ_T$, which is described by the Gruneisen law

$$\beta = \gamma\left(\frac{C_V}{V}\right) æ_T, \quad æ_T = -\frac{1}{V}\left(\frac{\partial V}{\partial P}\right)_T, \tag{10.5}$$

where P is the pressure, γ is the so-called Gruneisen constant. There are also various empirical relations, e.g.,

$$\alpha T_{\text{m}}^{1.17} = A, \tag{10.6}$$

where T_{m} is the melting temperature of a substance, $A = 7.24 \cdot 10^{-2}$ for the substances with metallic bonds, $A = 11.5 \cdot 10^{-2}$ for alkali-halide compounds.

Generally α and β depend on pressure, temperature, chemical composition, the structure of a body, and its phase state. The monotonous character of the $\alpha(T)$ variation is violated close to the points of phase transitions, and also due to the superposition of electronic, magnetic, and lattice contributions to the thermal expansion, which can be of different signs and of comparable magnitudes in certain temperature ranges.

The temperature coefficients of linear expansion values are given in Tables 10.1–10.13 for some individual substances (elements and inorganic compounds) as well as for technical materials in a solid state. The values of temperature coefficients of volume expansion for certain liquids and gases are presented in Table 10.14.

10.2 Temperature Coefficients of Linear Expansion of Solids

Table 10.1 True temperature coefficients α (in 10^{-6} K^{-1}) of linear expansion of crystalline elements

Element	Temperature, K													Ref.
	5	10	30	50	100	200	300	400	500	600	800	1000	1200	
Ac							14.9 (estimate)							
Ag	0.0177	0.111	3.36	7.94	14.7	17.8	18.9	19.5	20.2	21.0	23.1	25.6	28.1	
Al		0.051	1.04	3.62	12.3	20.2	23.3	24.5	26.2	28.1	32.6	37.8 (900 K)		[2,3]
Am (α_{av})							7.1		9					
Ar (α_{av})	5.1	46.2	287	457	667 (80 K)									
As ($\alpha_{\parallel}$)	0.073	0.514	10.7	18.7	31.25 (85 K)		43.447 (283 K)							[2]
($\alpha_{\perp}$)							3.0 (279 K)							[2]
Au	0.026	0.228	4.22	7.65	11.5	13.3	14.0	14.5	15.0	15.5	16.5	17.7	19.1	
B							8 (273 K)			8.3 (658 K)				[2]
Ba	0.200	1.73	10.55	13.7	17.10 (85 K)		16.4	20.5	24.6					[2]
Be ($\alpha_{\parallel}$)					0.66	5.25	9.20	11.5	12.9	14.0	15.9	17.6	19.5	
($\alpha_{\perp}$)					1.48	7.89	12.4	14.9	16.9	18.3	20.2	21.4	23.4	
Bi ($\alpha_{\parallel}$)	0.26	2.07	13.1	16.0	16.85	17.1	17.2	17.4	17.5					
($\alpha_{\perp}$)	0.026	0.20	3.20	6.84	9.96	11.2	11.7	11.8	11.9					
Br (α_a)						133.1								
(α_b)						86.44								
(α_c)						29.55								
C (diamond)		$4{\cdot}10^{-5}$	$8{\cdot}10^{-4}$	0.004	0.05	0.45	1.0	1.80	2.53	3.09	3.83	4.32	4.83	[3]
(graphite, $\alpha_{\parallel}$)			3.8	8.7	17.6	24.9	26.7	27.5	28.1	28.5	28.9	29.2 (950 K)		[4]
(graphite, $\alpha_{\perp}$)			−0.09	−0.50	−1.07	−1.33	−1.22	−0.91	−0.53	−0.18	0.39	0.66 (950 K)		[4]
Ca	0.010	0.094	3.3	9.6	16.8	20.4	22.4	24.1	25.8	27.6	33.6			[2]
Cd ($\alpha_{\parallel}$)	0.28	7.2	50.0	58.9	59.7	56.5	54.0	50.5	43.4					
($\alpha_{\perp}$)	−0.01	−1.05	−2.95	2.6	10.6	16.7	19.6	25.8	37.0					
Ce (α_{av})	−4.0	−8.6	5.0 (15 K)					6.2	6.2	6.2	7.6	9.1 (950 K)		

Table 10.1 True temperature coefficients α (in 10^{-6} K^{-1}) of linear expansion of crystalline elements *(continued)*

Element	Temperature, K													Ref.
	5	10	30	50	100	200	300	400	500	600	800	1000	1200	
Cl (α_{av})			100 (39 K)											
Co (α_{av})	0.014	0.035	0.45	1.93	6.9	11.7	12.2	13.3	14.3	15.1	14.2	15.4 (1100 K)		[5]
Cr (α_{av})	−0.14	−0.29	−0.38	0.15	2.25	5.18	5.00	8.3	8.7	9.1	9.9	10.7	11.8	[3]
Cs							97.0	97.0 (370 K)						
Cu	0.009	0.021	1.00	3.87	10.5	15.2	16.7	17.3	17.9	18.6	20.1	21.8	23.8	[2,58,61]
Dy (α_{av})	($\alpha_a = 33$, $\alpha_b = 24$, $\alpha_c = 4.7$ at 50 K)				3.0	7.3	7.9	9.5	10.5	11.1	12.2			
($\alpha_{\parallel}$)								16.6 ($\alpha_{\perp} = 6.0$)						
Er (α_{av})		1.5	5.0		7.1	9.0	9.5	9.9	10.3	10.7	11.8	15.2	17.8 (1100 K)	[6]
Er ($\alpha_{\parallel}$)							($\alpha_{\parallel} = 19.3$, $\alpha_{\perp} = 6.6$, 400 K)							[6]
Eu (α_{av})							34.2	25.7	20.2	19.4				
Fe	0.017	0.040	0.22	1.00	5.09	9.96	12.0	13.2	14.4	15.5	16.5	14.7	22.5	
Fr							102 (estimate)							
Ga (α_{av})					15.7	18.8	20.7							[3]
(α_a)								16.2 ($\alpha_b = 11.1$, $\alpha_c = 30.2$)						[7]
Gd (α_{av})	−2	−3			5	6	6.3	6.6	7.8	8.3	9.2	11.2	17.6	
Ge	0.0011	0.0045	−0.071	0.12	2.29	4.93	5.82	6.28	6.71	7.12	7.83	8.73 (1100 K)		[2]
H_2			4800 (at triple point 13.96 K)											
Hf (α_{av})							6.0	6.7	6.9	7.11	7.54	7.98	8.4	[2]
Hg (α_{av})					36.9	47.6								
($\alpha_{\parallel}$)					42.9 ($\alpha_{\perp} = 34.0$, 100 K)									
Ho ($\alpha_{\parallel}$)			−100 (60 K)		−70	9	13	15.8	16.8	17.8	19.9	22.2 (950 K)		
($\alpha_{\perp}$)					32		5		5.0	5.0	5.4	7.0 (950 K)		
I (α_{av})				53	70	83	88							
(α_a)								133.4 ($\alpha_b = 95$, $\alpha_c = 35.1$)						
In ($\alpha_{\parallel}$)	−1.26	4.01	23.0	23.0	18.0	5.9	−16.2	−42.0						
($\alpha_{\perp}$)	0.99	0.94	7.2	17.0	28.0	37.9	53.9	79.5						

Table 10.1 True temperature coefficients α (in 10^{-6} K^{-1}) of linear expansion of crystalline elements *(continued)*

Element	Temperature, K													Ref.
	5	10	30	50	100	200	300	400	500	600	800	1000	1200	
Ir			0.93	2.0	4.11	6.09	6.40	6.51	6.73	7.02	7.66	8.72 (1100 K)		
K	0.95	6.01	35.5	49.1	59.5	68.3	79.6							[8]
Kr (α_{av})	8.1	57	216	280	430									
La (α_{av})	0.16				3.3	4.4	5.0	5.7	6.8		10.3	12.1		
Li					22.5	39.0	47.1							
Lu ($\alpha_{\parallel}$)	1.3	2.5	7.4	11.5	17.5	18.1	20.0	19.1	17			18		
($\alpha_{\perp}$)	0.3	0.6	2.0	3.5	7.3	6.2	5.8	5	7	9	12	12		
Mg ($\alpha_{\parallel}$)	0.0075	0.03	1.35	5.91	16.1	23.8								
($\alpha_{\perp}$)	0.011	0.05	1.35	5.34	15.0	22.3								
(α_{av})							25.8	27.3	29.0	31.0	35.0			
α-Mn	−0.14	−0.28	−0.9	−0.7	11.9	19.0	22.8	25.8	28.4	30.9	35.9			[3]
β-Mn					14.8	21.6	25.6					43.2	48.2	[3]
γ-Mn	0.049	0.130	1.40	4.01	8.10	12.3	15.1					45.2 (1400 K)		[3]
Mo	0.007	0.02	0.17	0.74	2.68	4.57	5.27	5.45	5.63	5.82	6.20	11.43 (2400 K)		[3,60]
N_2	12.5	30	550	692.0 (44 K)										
Na					45.7	64.7	71.5							
Nb	0.003*	0.04	0.64	2.14	4.84	6.39	7.10	7.30	7.50	7.70	8.09	10.39 (2400 K)		
Nd (α_{av})	−2.0	2.9			7.9	7.0	7.0	7.0	7.2	7.6	8.9	13.4 (1100 K)		
Ne (α_{av})	46	337	1723 (23K)											
Ni	0.02	0.05	0.26	1.64	6.61	11.4	13.0	13.7	14.9	16.9	16.7	18.2	20.3	[3]
Np (α_{av})							27.66			42.7 (700 K)				
O_2 (α_{av})	200 (21 K)		535	870 (48 K)										
Os ($\alpha_{\parallel}$)							5.84	6.17	6.55	7.01	8.18			
($\alpha_{\perp}$)							3.97	4.16	4.42	4.76	5.65			

Table 10.1 True temperature coefficients α (in 10^{-6} K^{-1}) of linear expansion of crystalline elements *(continued)*

Element	Temperature, K													Ref.
	5	10	30	50	100	200	300	400	500	600	800	1000	1200	
P (white)							124.5							
(red)							66.5 (estimate)							
Pa ($\alpha_{\parallel}$)								4.4 ($\alpha_{\perp} = 21.4$)				$-21.7(\alpha_{\perp} = 85)$		
Pb	0.11	3.02	17.0	21.8	25.4	27.3	28.5	29.63	31.73	33.3				
	(4 K)									(550 K)				
Pd	0.5		1.12	3.57	7.95	10.84	11.75	12.48	13.20	13.90	15.3	17.40		
	(20 K)											(1100 K)		
Pm							9.0 (estimate)							
Po ($\overline{\alpha}$)						23 (179–300 K)								
Pr					6.5	5.6	5.0	5.0	5.2	5.7	7.3	7.8		
												(850 K)		
Pt	0.0155	0.071	1.43	3.80	6.77	8.55	8.99	9.24	9.46	9.70	10.2	15.6		[2,61]
												(2100 K)		
Pu (α_{av})						39.5	48.7	41.3		8.61		−16.3		
								(440 K)		(650 K)		(750 K)		
Ra ($\overline{\alpha}$)								17.1 (300–600 K)						[5]
Rb							90							
Re (α_{av})	0.0053	0.0158	0.31	1.81	4.10	4.50	4.70	3.95		3.83	3.58	3.33	3.13	[9]
($\overline{\alpha}_{\parallel}$)								6.25 ($\overline{\alpha}_{\perp} = 7.0$, 293–1200 K)						[62]
Rh			0.40	1.70	4.99	7.57	8.50	8.93	9.38	9.84	10.80	12.40		[3]
												(1100 K)		
Ru ($\alpha_{\parallel}$)						7.0	8.70	9.09	9.49	9.92	10.83	11.81	12.87	
($\alpha_{\perp}$)						4.8	5.8	5.94	6.16	6.45	7.15	7.95	8.80	
S (rhomb.)					42	57	64.1							[2]
(monocl.)							80							[2]
Sb ($\alpha_{\parallel}$)		0.04	5.19	13.17	15.8	16.15	16.2							[4]
($\alpha_{\perp}$)		−0.01	0.47	2.31	5.99	7.72	8.31							[4]
(α_{av})								11.4	11.7	11.8	11.8			[4]
Sc ($\alpha_{\parallel}$)					15.0	15.0	15.1	15.1	15.1	15.1	15.4	16.1	17.3	[11]
($\alpha_{\perp}$)					7.2	7.4	7.61	8.07	8.53	8.99	9.93	10.9	11.9	[11]
Se (monocl., $\alpha_{\parallel}$)		−1.75	−16.1	−19.9	−13.8	−13.3	−13.4							[12]

Table 10.1 True temperature coefficients α (in 10^{-6} K^{-1}) of linear expansion of crystalline elements *(continued)*

Element	Temperature, K													Ref.
	5	10	30	50	100	200	300	400	500	600	800	1000	1200	
(monocl., $\alpha_\perp$)		2.4	29.7	42.0	53.5	62.4	69.8							[12]
(glass)		3.7	21.6	28.1	31.1	38.4	53.9							[12]
(hexagon., α_{av})				20.2 (140 K)		22.5	26.4	40.5	45.2 (478 K)					[13]
Si	0.0001	0.0009	−0.052	−0.24	−0.34	1.43	2.54	3.05	3.39	3.68	4.19	4.65		[14]
Sm (α_{av})							10.4							
α-Sn			−0.82	0.38	3.48	5.21 (220 K)								
β-Sn ($\alpha_\parallel$)	0.18	1.65	13.7	18.5	23.2	26.5	31.4	37.2	41.4					
($\alpha_\perp$)	−0.025	−0.10	1.2	4.9	11.8	14.8	16.0	18.6	20.3					
Sr	−0.037	0.27	8.2	13.7	19.4		22.45 (283 K)			22.2	20.0 (850 K)			[2]
Ta	0.009	0.048	0.92	2.40	5.02	6.25	6.6	6.72	6.84	6.95	7.12	7.32	7.53	
Tb (α_{av})							7.0	8.6	9.5	10.0	11.1	16 (1100 K)		[15]
Tc (α_{av})							8.1							
Te ($\alpha_\parallel$)	−0.274	−1.81	−9.5	−8.1	−4.2	−2.3	−2.0	−2.69 (450 K)						
($\alpha_\perp$)	0.34	2.48	17.2	23.3	25.9	28.0	29.3	31.9 (450 K)						
Th			7.3 (70 K)		9	10.3	11.2	12.3	13.1	13.7	14.8			
Ti		0.03	0.5	1.76	4.59	7.23	8.3	8.82	9.34	9.86	10.96	12.08	11.95	
Tl ($\alpha_\parallel$)	5.3 (7 K)		20.6	24.6	29.3	33.9								
($\alpha_\perp$)	1.3 (7 K)		14.2	21.5	23.2	25.0								
(α_{av})							29.6	30.3	41.5 (530 K)					
Tm ($\alpha_\parallel$)								20	20	20	20	22 (1050 K)		
($\alpha_\perp$)								12	12	12	12	14 (1050 K)		

Table 10.1 True temperature coefficients α (in 10^{-6} K^{-1}) of linear expansion of crystalline elements *(continued)*

Element	Temperature, K													Ref.
	5	10	30	50	100	200	300	400	500	600	800	1000	1200	
(α_{av})		−0.7	2.2	−1.3 (56 K)	3.0	5.0	5.8							
U (α_a)		−136 (36 K)		23.9 (90 K)		25.7	26	26.2	30.2	35.1	46.6	54.7 (900 K)		
(α_b)		−28.2 (36 K)		2.9 (90 K)		1.6	0.4	−0.8	−2.5	−5.6	−17.2	−25 (900 K)		
(α_c)		52.1 (36 K)		10.9 (90 K)		15.0	18.0	21.6	25.8	30.7	41.6	47.2 (900 K)		
V	0.025 (6 K)	0.045	0.35	1.13	3.95	7.49	7.84	8.5	9.3	9.8	10.7	11.7	12.7	
W	0.00044	0.007	0.2	0.8	2.3	4.1	4.6	4.6	4.6	4.7	4.8	4.9	5.1	[3]
									6.5 (2000 K)			8.8 (3200 K)		[3]
Xe	8.7	40	140.3	183.0	263.0	420.7 (155 K)								
Yb	0.17	1.23						25.7	27.0	29.1	33.1	33.7 (850 K)		
Yt ($\alpha_{\parallel}$)		0.1	2.1	6.2	13.7	18.4	19.2	19.4	19.6	19.7	20.1	22.3 (1050 K)		
($\alpha_{\perp}$)			0.3	0.8	2.9	4.3	4.6	5.0	5.4	5.6	6.7	10.1 (1050 K)		
Zn ($\alpha_{\parallel}$)	0.05	0.76	33.4	54.1	65.4	65.6	63.5	61.0	58.9	50.3 (650 K)				
($\alpha_{\perp}$)	−0.001	−0.06	−4.3	−3.0	4.1	10.4	13.2	15.7	17.7	27.9 (650 K)				
Zr ($\alpha_{\parallel}$)						6.55	7.36	8.20	9.07	9.97	11.84	14.87 (1100 K)		
($\alpha_{\perp}$)						4.87	4.99	5.09	5.16	5.20	5.19	4.94 (1100 K)		
(α_{av})		0.02	0.75	1.90	4.09									

* In the superconducting phase at 6 K.

Table 10.2 Temperature coefficients of linear expansion of individual substances [The true values of TCLE α (at a given temperature T) or mean TCLE $\bar{\alpha}$ (within the range ΔT) are presented.]

Substance	T; ΔT, K	α; $\bar{\alpha}$, 10^{-6} K^{-1}	Ref.	Substance	T; ΔT, K	α; $\bar{\alpha}$, 10^{-6} K^{-1}	Ref.
AgBr	300	34.3	[16]	$BaTiO_3$ (rhombohedr.)	113–174	8.8	[17]
AgCl	300	30.1	[13]	(cubic)	200	6.5	[18]
	473	34.54	[13]		300	7.5	[17]
	698	69.99	[13]		400	8.6	[18]
$AgGaSe_2$	300	7.1	[33]		773	12.3	[17]
α-, β-AgI	100	−0.3	[1]	(tetragon.)	300	11.4	[18]
	200	−0.9	[1]		293–393	3.5	[17]
	300	−1.5	[1]	(ortorhomb.)	174–277	11.4	[18]
	400	−3.2	[1]	Be_2C	300	5.6	[16]
	600	36.83	[1]		300–473	7.7	[16]
	750	44.6	[1]		300–673	9.5	[16]
γ-AgI	120–300	−1.6	[1]	BeO	300–373	5.42	[34]
$AgInSe_2$	300	1.9	[33]		300–673	7.08	[34]
$AgInTe_2$	300	4.3	[33]	Bi_2Te_3 ($\alpha_{\parallel}$)	50	13	[19]
AlAs	300	3.5	[1]		100	18	[19]
	400	4.1	[1]		200	20	[19]
	288–1113	5.2	[1]		300	21.3	[19]
AlN	300–873	4.8	[22]		400	22.6	[19]
Al_2O_3 ($\alpha_{\parallel}$)	300	6.7	[31]		600	24	[19]
	400	6.46	[31]	($\alpha_{\perp}$)	30	5	[19]
	600	7.38	[31]		50	8	[19]
	800	7.99	[31]		100	11	[19]
($\alpha_{\perp}$)	300	5.0	[31]		200	12	[19]
	400	5.82	[31]		300	12.9	[19]
	600	6.68	[31]		400	15	[19]
	800	7.23	[31]		600	17	[19]
AlSb	30	−0.90	[1]	CBr_4	200	1.03	[3]
	50	−0.94	[1]		300	1.50	[3]
	100	0.49	[1]	$(CH_2NH_2COOH_3)BeF_4$			
	200	3.36	[1]	(triglycinefluoroberyllate)			
	300–1323	4.88	[1]	(α_{100})	100	15	[3]
As_2O_3	300–1273	8.4	[16]		200	0	[3]
As_2S_3	300–438	24.62	[13]		300	20	[3]
B_4C	300–1073	4.5	[34]	(α_{010})	100	48	[3]
BN	300	7.5	[29]		200	64	[3]
BP	400	4.0	[35]		300	50	[3]
	600	5.0	[35]	(α_{001})	100	10	[3]
	800	6.2	[35]		200	16	[3]
BaB_6	300–1073	6.8	[16]		300	10	[3]
BaF_2	30	0.7	[1,3]	$(CH_2NH_2COOH_3)H_2SO_4$			
	50	2.9	[1,3]	(triglycinesulfate)			
	100	9.9	[1,3]	(α_{100})	100	5	[3]
	200	16.1	[1,3]		200	3	[3]
	300	18.4	[1,3]		300	−30	[3]
$Ba(NO_3)_2$	3–77	3.3	[16]	(α_{010})	100	40	[3]
	77–200	6.7	[16]		200	41	[3]
	300	15.2	[16]		300	−10.5	[3]
$BaNaNb_5O_{15}$	300	10.9	[36]	(α_{001})	100	10	[3]
$BaSi_2$	300–1373	8.4	[16]		200	16	[3]
$BaSrNb_2O_6$ (α_a)	300	10	[36]		300	8.5	[3]
(α_c)	200–900	9	[36]				

Table 10.2 Temperature coefficients of linear expansion of individual substances *(continued)*

Substance	T; ΔT, K	α; $\bar{\alpha}$, 10^{-6} K^{-1}	Ref.	Substance	T; ΔT, K	α; $\bar{\alpha}$, 10^{-6} K^{-1}	Ref.
$(CH_2NH_2COOH_3)H_2SeO_4$				CdS (α_{av})	50	−2.4	[1,13]
(triglycineselenate)				($\alpha_{\parallel}$)	77–298	2.1	[1]
(α_{100})	100	5	[3]		300	4.1	[13]
	200	0	[3]		600	4.2	[1]
	300	−30	[3]		800	4.8	[13]
(α_{010})	100	12	[3]	($\alpha_{\perp}$)	77–298	4.0	[1,13]
	200	35	[3]		300	6.5	[1]
	300	−30	[3]		600	6.6	[13]
(α_{001})	100	10	[3]		800	7.3	[1]
	200	16	[3]	CdSe ($\alpha_{\parallel}$)	77–298	2.45	[1,13]
	300	95	[3]	($\alpha_{\perp}$)	77–298	4.4	[1,13]
CaB_6	300	6.5	[38]	$CdSnAs_2$	300	4.7	[25]
$CaCO_3$ ($\alpha_{\parallel}$)	200	22.4	[3]	CdTe	30	−2.75	[1]
	300	24.4	[13]		50	−1.80	[13]
	323	26.6	[3]		100	1.38	[1]
($\alpha_{\parallel}$)	200	5.41	[13]		200	4.09	[13]
	300	5.68	[3]		300	4.96	[1]
	323	5.2	[13]		400	5.10	[13]
	638	−3.8	[3]		600	5.45	[1]
$CaCl_2$	300–463	22.3	[16]		800	5.8	[13]
CaF_2	30	0.3	[1,3]	$CdTl_2Te_4$	200	1.66	[3]
	50	1.6	[1,3]		300	2.35	[3]
	100	7.52	[1,3]	CeB_6	300–1073	7.3	[16]
	200	15.6	[1,3]	CeO_2	300	8.5	[25]
	300	18.80	[1,3]		373–773	8.6	[25]
	320	19.09	[1,3]	Ce_2S_3 ($\alpha_{\parallel}$)	300	13.2	[25]
$CaMoO_4$ (α_a)	300	19.4	[20]	($\alpha_{\perp}$)	300	10.5	[25]
(α_c)	300	25.5	[29]	Ce_2Se_3 (α_{av})	300	12.6	[25]
CaO	300–573	10.2	[31]	CmO_2	300	8.1	[31]
	300–673	13.2	[37]		400	8.1	[31]
$Ca_5(PO_4)_3$	300	9	[20]		600	8.1	[31]
$CaTiO_3$	300	10	[24]	Cm_2O_3	773–1073	6.6	[31]
$Ca_3(VO_4)_2$ (α_a)	300	5.5	[20]	CoF_2 ($\alpha_{\parallel}$)	300	3.6	[28]
(α_c)	300	3.5	[20]	($\alpha_{\perp}$)	300	10.7	[28]
$CaWO_4$ (α_a)	80–300	7.9	[17]	CoO	300	14.0	[31]
	300	11.2	[20]	CoSb	100	8.3	[3]
	400	11.2	[17]		200	11.4	[3]
	773–1073	13.8	[20]		300	12.5	[3]
(α_c)	80–300	12.7	[17]	CoSi	300–1073	10.6	[16]
	300	18.7	[20]	Cr_3C_2	300–373	8.8	[27]
	400	18.7	[17]		300–873	8.0	[34]
	773–1073	22.0	[20]	CrN	300–1073	2.3	[34]
Cd_3As_2	300	11.4	[3]	CrO_2 ($\alpha_{\parallel}$)	300	−10.3	[31]
CdB_6	300–1173	8.7	[16]	($\alpha_{\perp}$)	300	16.2	[31]
CdF_2	300	27.0	[13]	CrS	300–1300	12.3	[34]
$CdGeAs_2$	300	11.4	[3]	Cr_2S_3	300	12.3	[25]
$CdGeP_2$	300	3.2	[3]	CsBr	30	20.0	[3]
(glass)	300	5.2	[3]		50	28.7	[13]
CdI_2	323	29.1	[16]		100	38.6	[3]
Cd_2O_3	300–373	10.5	[37]		200	43.1	[13]
	300–1073	10.0	[37]		300	46.6	[3]
Cd_3P_2	300	9.63	[3]		400	52.24	[13]
					600	61	[3,13]

Table 10.2 Temperature coefficients of linear expansion of individual substances *(continued)*

Substance	T; ΔT, K	α; $\bar{\alpha}$, 10^{-6} K^{-1}	Ref.	Substance	T; ΔT, K	α; $\bar{\alpha}$, 10^{-6} K^{-1}	Ref.
CsCl	300	44.8	[16]	GaAs	10	0.0045	[1]
CsI	300	48.6	[17]		30	−0.177	[1]
Cu_3AsS_4	300	3.2	[25]		50	−0.15	[1]
Cu_3AsSe_4	300	9.5	[25]		100	2.05	[1]
CuB_6	300–1073	6.5	[16]		200	4.93	[1]
CuBr	293–423	19	[13]		300	5.82	[1]
CuCl	293–413	10	[13]		400	6.23	[1]
$CuGaSe_2$	300	5.4	[25]		600	6.98	[1]
$CuGaTe_2$	300	6.9	[33]		800	7.4	[1]
$CuGeS_3$	300	7.8	[25]	GaP	300	5.6	[1]
$CuGeSe_3$	300	8.4	[25]		400	5.7	[1]
CuI	300–875	25.2	[1]		800	6.1	[1]
$CuInSe_2$	300	6.6	[25]	GaSb	10	−0.026	[1]
$CuInTe_2$	300	7.1	[25]		30	−0.34	[1]
Cu_3SbSe_4	300	7.1	[25]		50	0.0	[1]
Cu_2SnSe_3	300	8.9	[25]		100	2.81	[1]
Cu_2SnS_3	300	7.8	[25]		200	5.81	[1]
DyBi	300	10.2	[3]		300	6.36	[1]
Dy_2O_3	300–1113	8.3	[34]		400	6.40	[1]
ErBi	300	10.8	[3]		600	6.40	[1]
Er_2O_3	373–573	5.7	[31]	Ga_2Se_3	30	1.70	[1]
Eu_2O_3	300–373	10.4	[31]		50	3.02	[1]
EuB_6	300	6.9	[38]		100	5.83	[1]
EuS	300	14.6	[25]		200	8.14	[1]
EuSe	300	18.6	[25]		300	8.99	[1]
FeF_2 ($\alpha_\parallel$)	300	−0.4	[28]		400	9.30	[1]
($\alpha_\perp$)	300	16.8	[28]	Ga_2Te_3	77	9.2	[1]
Fe_4N	300	7.9	[16]		300	8.3	[1]
FeS	293–1273	22.1	[34]	GdB_6	300–1060	11.86	[1]
FeS_2	320	8.9	[16]	GdBi	300	8.7	[38]
Fe_3Se_4	30	−0.30	[3]	Gd_2O_3	300	9.2	[3]
	50	−2.15	[3]	GeO_2 ($\alpha_\parallel$)	300–1073	10.0	[31]
	100	−3.80	[3]		300	2.0	[28]
	200	−0.90	[3]	($\alpha_\perp$)	300	6.0	[28]
	300	1.85	[3]	H_2O	23	−6.1	[16]
Fe_3Si	293–1273	14.4	[34]		83	3.3	[16]
Fe_5Si_3	70	0.5	[3]		193	39.2	[16]
	100	4.3	[3]		273	52.7	[16]
	200	8.3	[3]	HfB_2	300–1273	6.3	[31]
	300	8.2	[3]	HfC	300–473	6.0	[27]
$FeTe_2$ (α_a)	300	21	[1]		300–873	6.5	[27]
	400	24	[1]		300–873	6.5	[27]
	600	38	[1]	HfN	300–1373	6.9	[34]
	800	61	[1]	HfO_2	530–1120	6.84	[34]
(α_b)	300	46	[1]	HgBr (α_{100})	100	53.4	[3]
	400	48	[1]		200	42.8	[3]
	600	58	[1]		300	45.0	[3]
	800	80	[1]	(α_{010})	100	45.5	[3]
(α_c)	300	3	[1]		200	42.8	[3]
	400	7	[1]		300	45.0	[3]
	600	15	[1]	α_{001})	100	4.8	[3]
	800	22	[1]		200	3.2	[3]
					300	3.7	[3]

Table 10.2 Temperature coefficients of linear expansion of individual substances *(continued)*

Substance	T; ΔT, K	α; $\bar{\alpha}$, 10^{-6} K^{-1}	Ref.	Substance	T; ΔT, K	α; $\bar{\alpha}$, 10^{-6} K^{-1}	Ref.
HgCl ($\alpha_{100}, \alpha_{010}$)	100	44.1	[3]	In_2Te_3	30	2.97	[1]
	200	43.0	[3]		50	5.32	[1]
	300	47.0	[3]		100	8.03	[1]
(α_{001})	100	1.6	[3]		200	9.94	[1]
	180	10.0	[3]		300	10.10	[1]
	200	0.3	[3]	IrO_2 ($\alpha_{\parallel}$)	300	1.7	[28]
	300	1.4	[3]	($\alpha_{\perp}$)	300	3.8	[28]
HgSe	300	1.48	[1]	KBr	30	6.83	[3]
	400	1.64	[1]		50	16.7	[13]
	500	1.66	[1]		100	29.3	[3]
HgTe	300	4.80	[1]		200	34.6	[13]
	400	4.80	[1]		320–953	38	[13]
HoBi	300	10.5	[1]	KCl	30	3.11	[3]
Ho_2O_3	300–1073	8.44	[37]		50	11.4	[22]
InAs	10	−0.058	[1]		100	25.4	[3]
	30	−0.86	[1]		200	33.5	[22]
	50	−0.4	[1]		300	37.0	[3]
	100	2.05	[1]		400	38.8	[22]
	200	3.80	[1]		600	45.4	[3]
	300	4.41	[1]		800	52.45	[22]
	400	5.07	[1]	KF	200–300	45.0	[16]
	600	5.67	[1]		300–352	33.3	[16]
	800	5.92	[1]	KH	300–673	36	[39]
In_2O_3	300–1273	7.2	[31]	KH_2PO_4 ($\alpha_{\parallel}$)	120	34.3	[3]
InP	300	4.3	[1]		200	34.3	[3]
	400	4.5	[1]		300	34.3	[3]
	600	4.9	[1]	($\alpha_{\perp}$)	120	21.6	[3]
	800	5.9	[1]		200	21.6	[3]
InS ($\alpha_{\parallel}$)	150	7.05	[3]		300	21.6	[3]
	200	8.76	[3]	KI	30	10.3	[3]
	300	10.8	[3]		50	19.3	[23]
($\alpha_{\perp}$)	150	9.48	[3]		100	30.3	[3]
	200	11.5	[3]		200	35.8	[23]
	300	14.2	[3]		300	41.2	[3]
InSb	10	−0.24	[1]	$KNaC_4H_4O_6 \cdot 2H_2O$	100–200	38.0	[3]
	30	−1.5	[21]	(Seignett's salt)	200–300	40	[3]
	50	−0.4	[1]		293	40.2	[3]
	100	2.60	[21]	LaB_6	300	6.4	[38]
	200	4.67	[1]	LaF_3	300	20.0	[29]
	300	5.15	[21]	LaH_2	300–923	10.6	[39]
	400	5.3	[1]	La_2O_3	300–473	8.45	[31]
	473	5.4	[21]		300–1273	12.01	[37]
InSe ($\alpha_{\parallel}$)	150	8.94	[3]	LaS	300	11.6	[25]
	200	10.8	[3]	La_2S_3	300	9.9	[25]
	300	12.4	[3]	La_2Se_3	300	11.9	[25]
($\alpha_{\perp}$)	150	11.25	[3]	La_2Si_3	300–1373	8.7	[16]
	200	13.7	[3]	LiBr	100–200	39.3	[16]
	300	16.0	[3]		200–300	46.6	[16]
InTe ($\alpha_{\parallel}$)	150	8.82	[3]		273–350	46.6	[16]
	200	11.25	[3]	LiCl	273–350	40.6	[16]
	300	14.0	[3]		30	0.24	[3]
($\alpha_{\perp}$)	150	10.75	[3]		50	2.96	[3]
	200	13.55	[3]		100	15.2	[3]
	300	16.42	[3]		200	28.1	[3]
					300	33.7	[3]

Table 10.2 Temperature coefficients of linear expansion of individual substances *(continued)*

Substance	T; ΔT, K	α; $\bar{\alpha}$, 10^{-6} K^{-1}	Ref.	Substance	T; ΔT, K	α; $\bar{\alpha}$, 10^{-6} K^{-1}	Ref.
LiI	100–200	47	[16]	$NH_4H_2PO_4$ (α_a)	200	87	[3]
	200–300	55.7	[16]		290	80	[26]
	273–350	56	[16]	(α_b)	200	40	[3]
$LiNbO_3$ ($\alpha_{\parallel}$)	100	1.0	[24]		290	35	[26]
	200	3.8	[24]	(α_c)	200	41	[3]
	300	4.0	[24]		290	61	[26]
	400	2.0	[24]	NaBr	100–200	31.0	[16]
	600	2.0	[24]		200–300	36.7	[16]
($\alpha_{\perp}$)	100	1.9	[24]		288–333	45.5	[16]
	200	8.5	[24]	NaCl	100	33.9	[16]
	300	15.7	[24]		200	37	[16]
	400	17.5	[24]		300	39.6	[16]
	600	19.0	[24]	NaF	90–300	23.0	[3]
$LiTaO_3$ (α_a)	300	12.0	[24]		300	33.13	[3]
(α_c)	300	4.2	[24]		468	35.9	[16]
Lu_2O_3	300	7.8	[31]	NaH	300–673	64.0	[39]
MgF_2 ($\alpha_{\parallel}$)	300	13.42	[1]	$NaNO_3$	323	11	[16]
	400	14.49	[1]		400	12.4	[16]
	600	16.86	[1]	Na_2WO_4	300	18.2	[16]
	800	20.24	[1]	NbB_2	300–1273	8.0	[34]
($\alpha_{\perp}$)	300	9.44	[1]	NbC	300–773	6.25	[32]
	400	10.17	[1]	NbN	300–1273	10.1	[34]
	600	12.59	[1]	Nb_2O_5	300–473	0.0	[31]
	800	15.77	[1]		473–673	−1.2	[31]
MgO	293–373	11.7	[31]		300–1073	5.8	[31]
	293–873	12.9	[37]	Nb_2S_3	300–1273	10.0	[34]
Mg_2Si	30	0.08	[1]	NdB_6	300–1273	7.3	[16]
	50	0.74	[1]	Nd_2O_3	473	5.53	[37]
	100	5.49	[1]		1073	11.37	[25]
	200	12.31	[1]	Nd_2O_2Te	300	12.5	[25]
	300	13.96	[1]	NdS	300	15.4	[25]
Mg_2Sn	30	0.85	[1]	Nd_2S_3	300	12.9	[25]
	50	3.2	[1]	Nd_2Se_3	300	13.5	[25]
	100	9.2	[1]	NiO	300	10	[31]
	200	14.9	[1]		473	13.0	[37]
	300	16.0	[1]		675	13.5	[37]
MnF_2 ($\alpha_{\parallel}$)	300	12.6	[28]	Ni_2P	300–1073	15.05	[34]
($\alpha_{\perp}$)	300	3.5	[28]	Ni_3P	300–1073	12.65	[34]
MnS	100	17.3	[1]	NiS	100	0.03	[3]
	156.7	49.9	[1]		200	−13.5	[3]
	200	11.6	[1]		300	1.1	[3]
	300	17.3	[34]	NiSe	100	8.7	[3]
MnS_2	293–343	11.1	[16]		200	29.5	[3]
MnO_2 (α_a)	300–773	6.69	[37]		300	20.0	[3]
(α_c)	300–773	6.93	[37]	Ni_2Si	300–1143	16.5	[34]
MoC	285–463	7.8	[16]	NiTe	100	7.5	[3]
	300–1073	6.15	[16]		200	24.0	[3]
Mo_2N	293–1373	6.2	[16]		300	15.0	[3]
MoS_2	300–1273	10.7	[34]	$PbCl_2$	293–393	31	[13]
MoSi	293–1073	16.3	[16]	$PbNO_3$	348–453	31.6	[16]
$MoSi_2$	293–1723	5.1	[16]	PbO_2 ($\alpha_{\parallel}$)	300	9.3	[28]
NH_4Br	300	59.3	[16]	($\alpha_{\perp}$)	300	8.4	[28]
NH_4Cl	313	62.5	[16]				

Table 10.2 Temperature coefficients of linear expansion of individual substances *(continued)*

Substance	T; ΔT, K	α; $\bar{\alpha}$, 10^{-6} K^{-1}	Ref.	Substance	T; ΔT, K	α; $\bar{\alpha}$, 10^{-6} K^{-1}	Ref.
PbS	30	7.54	[1]	Sb_2O_3	313	19.63	[16]
	50	12.4	[1]	Sb_2S_3	573–773	17	[1]
	100	17.6	[1]	SbSI ($\alpha_{\parallel}$)	300	15	[3]
	200	19.8	[1]	($\alpha_{\perp}$)	300	42	[3]
	300	20.3	[1]	Sb_2Se_3	573–773	16	[1]
PbSe	30	7.65	[1]	Sb_2Te_3	573–773	12	[1]
	50	12.9	[1]	ScB_2 (α_c)	300–873	7.6	[34]
	100	17.4	[1]	(α_a)	300–873	6.8	[34]
	200	18.9	[1]	Sc_2O_3	300–673	8.5	[31]
	300	19.5	[1]	α-SiC ($\alpha_{\parallel}$)	100	0.3	[1]
PbTe	30	9.02	[1]		200	1.2	[1]
	50	14.3	[1]		300	2.8	[1]
	100	17.7	[1]		400	4.0	[1]
	200	19.6	[1]		600	4.6	[1]
	300	19.8	[1]		800	4.7	[1]
$PbTiO_3$	800	8.3	[13]	($\alpha_{\perp}$)	400	3.9	[1]
$PbTi_{0.48}Zr_{0.52}O_3$ (α_1)	300	1.0	[40]		600	4.2	[1]
	400	2.0	[40]		800	4.3	[1]
	600	4.0	[40]	β-SiC	100	0.3	[1]
(α_3)	300	4.0	[40]		200	1.2	[1]
	400	2.0	[40]		300	2.8	[1]
	600	−5	[40]		400	4.0	[1]
$PbZrO_3$	600	11.0	[18]		600	4.5	[1]
PrB_6	300	7.5	[38]		800	4.5	[1]
Pr_2O_3	293–373	8.0	[31]	Si_3N_4	300	2.7	[29]
	293–1273	8.3	[37]	SiO_2 (melt.)	4	−0.2	[1]
$Pr_2O_2Te_2$	300	10.4	[25]		10	−2.75	[1]
PrS	300	14.3	[25]		30	−8.54	[1]
Pr_2S_3	300	11.3	[25]		50	−8.46	[1]
Pr_2Se_3	300	13.0	[25]		100	−6.10	[1]
PuC	300–1053	10.7	[34]		200	0.45	[1]
Pu_2C_3	300–1053	14.7	[34]		300	4.50	[1]
PuO_2	300–773	9	[31]		400	6.10	[1]
	400–1184	15.2	[31]		600	6.25	[1]
RbBr	100–200	32.3	[16]		800	4.80	[1]
	200–300	34.7	[16]	($\alpha_{\parallel}$)	300–673	4.1	[31]
RbCl	100–200	30.5	[16]	($\alpha_{\perp}$)	300–673	6.6	[31]
	200–300	32.8	[16]	Sm_2O_3	300–1073	10.8	[37]
$RbHSO_4$ (α_a)	100	50	[3]	Sm_2O_2Te	300	13.5	[25]
	200	60	[3]	SmS	100	7.2	[3]
	300	60	[3]		200	11.3	[3]
(α_b)	100	30	[3]		300	11.8	[3]
	200	50	[3]	Sm_2Se_3	300	14.8	[25]
	300	33	[3]	SnO_2 (α_{av})	300	3.4	[31]
(α_c)	100	35	[3]	($\alpha_{\parallel}$)	300–680	5.6	[31]
	200	49	[3]	($\alpha_{\perp}$)	300–680	4.3	[31]
	300	70	[3]	α-SnS (α_{av})	300	14.1	[1]
RbI	100–200	37.3	[16]	β-SnS (α_{av})	300	15.0	[1]
	200–300	39.7	[16]	α-SnSe (α_{av})	300	16.9	[1]
Rh_2O_3 ($\alpha_{\parallel}$)	300	5.25	[31]	(α_a)	300	−26.6	[1]
($\alpha_{\perp}$)	300	5.35	[31]	(α_b)	300	26.7	[1]
RuO_2 ($\alpha_{\parallel}$)	300	−1.4	[28]	(α_c)	300	35.5	[1]
($\alpha_{\perp}$)	300	7.0	[28]	β-SnSe (α_{av})	300	22.1	[1]

Table 10.2 Temperature coefficients of linear expansion of individual substances *(continued)*

Substance	T; ΔT, K	α; $\bar{\alpha}$, 10^{-6} K^{-1}	Ref.
SnTe	30	6.4	[1]
	50	14.0	[1]
	100	17.6	[1]
	200	20.4	[1]
	300	20.8	[1]
SrB_6	300	6.7	[38]
$(SrBi)TiO_3$	300	9	[24]
SrF_2	30	0.4	[1]
	50	2.1	[1]
	100	8.5	[1]
	200	15.7	[1]
	273	17.5	[1]
$Sr(NO_3)_2$	348	32.2	[16]
SrO	300	13.9	[31]
SrO_2 ($\alpha_{\parallel}$)	300	3.6	[28]
($\alpha_{\perp}$)	300	3.3	[28]
$SrTiO_3$	300	9.4	[13]
TaB_2	300–1273	8.2	[34]
TaC	300–773	6.29	[27]
TaN	300–973	3.6	[34]
Ta_2N	300–1273	5.2	[16]
Ta_2O_5	300–673	0.8	[31]
TaS_2	300	13.9	[25]
$TaSi_2$	300–1273	8.85	[16]
TbB_6	300	7.8	[38]
TbBi	300	9.8	[3]
$TbO_{1.81}$	300–1273	3	[31]
ThB_6	300	5.8	[34]
	300–1073	7.8	[38]
ThC	300	6.53	[34]
ThO_2	300–673	7.1	[31]
	300–873	7.7	[31]
ThS	300–1053	10.2	[34]
TiB_2	290–673	5.5	[16]
TiC	100	6.2	[3]
	200	6.8	[27]
	300	7.3	[3]
TiI_4	243–398	74	[16]
TiN	300	4.7	[34]
	300–1373	9.35	[34]
TiO	300	6.73	[31]
TiO_2 ($\alpha_{\parallel}$)	300	9.80	[31]
	300–673	10.1	[31]
	670–1073	10.8	[31]
($\alpha_{\perp}$)	300	7.44	[31]
	300–673	8.0	[31]
	670–1073	8.3	[31]
Ti_6O	300–400	4.5	[31]
TiP	300	8.2	[34]
Ti_2S_3	300–1273	17.0	[34]
TlB_2	300–1573	8.1	[16]
TlBr	243–398	51.2	[16]
TlCl	288–333	54.6	[17]
TlI	300	45.0	[16]
Tm_2O_3	373–573	7.1	[31]
UB_2 (α_a)	300–573	9.0	[34]
(α_c)	300–573	8.0	[34]
UB_4	300–1273	7.0	[34]
UC	300–1273	10.4	[34]
UN	318–1273	8.61	[34]
UO_2	300–673	9.2	[31]
	673–1073	10.8	[41]
US	300–1253	11.9	[34]
VC	100	5.7	[28]
	200	6.2	[28]
	300	6.6	[3]
	300–573	7.2	[27]
V_3N	300–1373	8.1	[34]
WB_4	300	5.8	[34]
W_2B	300	6.7	[34]
WC	300–673	3.84	[27]
WSi_2	300–700	6.25	[34]
YbB_6	300	5.8	[34]
Yb_2O_3	373–573	4.9	[31]
	300–1273	9.30	[31]
$YAlO_3$	300	~ 7	[29]
$Y_3Al_5O_{12}$	100	4.25	[20]
	200	5.8	[29]
	300	7.5	[20]
$Y_3AlFe_4O_{12}$	200	6.3	[3]
	300	8.5	[30]
	600	11.1	[3]
YB_2 (α_a)	300	9.4	[34]
(α_c)	300	8.5	[34]
YB_6	300	6.2	[38]
$Y_3Fe_5O_{12}$	200	7.0	[29]
	300	8.3	[30]
	600	11.0	[30]
$Y_{1.5}Gd_{1.5}Fe_5O_{12}$	200	6.3	[30]
	300	8.1	[3]
	623	10.4	[30]
Y_2O_3	473	6.9	[13]
	1073	8.1	[13]
YVO_4	300	20	[29]
Zn_3As_2	300	10.4	[3]
ZnCl	300–433	29	[16]
ZnF_2 ($\alpha_{\parallel}$)	300	11.3	[28]
($\alpha_{\perp}$)	300	8.6	[28]
$ZnGeAs_2$	300	1.0	[25]
ZnO ($\alpha_{\parallel}$)	30	−0.48	[3]
	50	−0.88	[31]
	100	−0.62	[3]
	200	1.58	[31]
	300	2.92	[3]
	300–673	5.0	[34]

Table 10.2 Temperature coefficients of linear expansion of individual substances *(continued)*

Substance	T; ΔT, K	α; $\bar{\alpha}$, 10^{-6} K^{-1}	Ref.	Substance	T; ΔT, K	α; $\bar{\alpha}$, 10^{-6} K^{-1}	Ref.
ZnO ($\alpha_{\perp}$)	30	−0.27	[3]	ZnSe	400	8.17	[1]
	50	−0.50	[31]		600	9.97	[1]
	100	−0.04	[3]		800	11.69	[1]
	200	2.88	[31]	$ZnSnAs_2$	300	2.3	[25]
	300	4.75	[3]	ZnTe	30	−0.72	[1]
	300–673	5.5	[34]		50	0.36	[1]
Zn_3P_2	300	8.33	[3]		100	4.17	[1]
ZnS (cubic)	50	−0.34	[1]		200	7.58	[1]
	100	1.72	[1]		300	8.29	[1]
	200	5.17	[1]		400	8.7	[1]
	300	6.0	[1]		700	9.65	[1]
	600	6	[1]	ZrB_2	300–1073	5.9	[34]
	800	6	[1]	ZrC	100	5.1	[3]
ZnS (hexagon.)	100	2.0	[1]		200	5.8	[32]
	200	5.0	[1]		300	6.1	[27]
	300	6.1	[1]		300–873	6.80	[3]
	800	6	[1]	ZrN	300–1373	7.24	[34]
ZnSe	30	−2.87	[1]	ZrO_2	300–453	4.5	[31]
	50	−1.51	[1]		300–873	8.4	[37]
	100	2.57	[1]	ZrS_2	300–1273	11.7	[34]
	200	5.86	[1]	$ZrTiO_4$	300	7.5	[24]
	300	7.14	[1]				

Table 10.3 Temperature coefficients of linear expansion of inorganic compounds at temperatures above 1000°C [The true values of TCLE α (at a given temperature t) or mean TCLE $\bar{\alpha}$ (within the range Δt) are presented.]

Substance	t; Δt, °C	α; $\bar{\alpha}$, 10^{-6}, K^{-1}	Ref.	Substance	t; Δt, °C	α; $\bar{\alpha}$, 10^{-6}, K^{-1}	Ref.
Al_2O_3	20–1000	8.1	[31]	Cd_2O_3	20–1000	10.5	[37]
	20–1200	9.1	[37]		20–1400	10.5	[37]
	20–1400	9.8	[31]	CeO_2	20–1000	8.5	[37]
	20–1600	10.3	[37]		20–1200	12.5	[37]
	20–1800	10.6	[31]	CrB	20–1000	12.3	[34]
	20–2000	9.14	[37]		1000–2000	12.6	[34]
($\alpha_{\parallel}$)	1027	11.04	[37]	Cr_3C_2	20–1000	11.7	[27]
	1727	11.99	[37]		20–1200	11.7	[27]
($\alpha_{\perp}$)	1027	9.97	[37]	Cr_2O_3	20–1400	9.6	[31]
	1727	10.75	[37]	Eu_2O_3	20–1000	10.3	[37]
B_4C	80–1000	4.6	[34]		20–1200	10.3	[37]
	1200	3.1	[34]		20–1600	10.4	[37]
BeO	20–1000	9	[31]	HfB_2	20–1000	6.3	[34]
	20–1200	10.1	[37]	HfC	20–1000	6.6	[15]
	20–1400	10.3	[31]		20–1200	6.67	[32]
	20–1600	10.7	[37]		20–1400	6.8	[15]
	20–1800	10.85	[31]		20–1600	6.87	[32]
	20–2000	10.95	[37]		20–1800	6.66	[15]
	1200–2000	13.4	[37]		20–2000	6.8	[32]
CaO	20–1000	14.2	[31]		20–2200	6.9	[15]
	20–1200	14.75	[37]		20–2900	8.0	[32]
	20–1600	15.71	[31]	HfO_2	250–1300	5.8	[37]
	20–1800	16.02	[37]		20–1700	6.45	[37]
	20–2000	16.29	[31]				

Table 10.3 Temperature coefficients of linear expansion of inorganic compounds *(continued)*

Substance	t; Δt, °C	α; $\bar{\alpha}$, 10^{-6}, K^{-1}	Ref.	Substance	t; Δt, °C	α; $\bar{\alpha}$, 10^{-6}, K^{-1}	Ref.
MgO	20–1000	14.2	[31]		20–1200	10.3	[31]
	20–1200	15.38	[31]		20–1400	10.4	[31]
	20–1400	15.95	[31]		1400–1970	12.31	[37]
	20–1600	16.47	[31]		2000–2300	13.3	[37]
MgO	20–1800	17.08	[31]	TiB_2	20–1000	4.6	[34]
	20–2000	17.49	[31]		1000–2000	5.2	[34]
	20–2200	18.60	[31]	TiN	20–1000	7.5	[34]
MoC	20–1000	6.4	[27]		20–1200	7.5	[34]
	20–1200	6.7	[27]		20–1400	7.9	[34]
	20–1400	7.0	[27]		20–1600	8.1	[34]
	20–1600	7.35	[27]		20–1800	8.3	[34]
	20–1800	7.6	[27]		20–2000	8.5	[34]
	20–2100	9.75	[27]		20–2200	9.1	[34]
NbB_2	20–1000	8.0	[34]		20–2700	9.6	[34]
	20–2000	8.5	[34]	UO_2	800–1260	12.9	[41]
NbC	20–1000	6.88	[27]	VB_2	20–1000	7.6	[34]
	20–1200	6.90	[27]		20–2000	8.3	[34]
	20–1400	6.92	[27]	VC	20–1000	6.25	[27]
	20–1600	7.15	[27]		20–1200	6.60	[27]
	20–1800	7.41	[27]		20–1400	6.95	[27]
	20–2000	7.57	[27]		20–1600	7.45	[27]
	20–2900	8.02	[27]		20–1800	8.10	[27]
Nd_2O_3	1000	11.37	[27]		20–2000	8.95	[27]
	100–1000	11.8	[37]		20–2200	9.60	[27]
NiO	1000	14.1	[37]	WC	20–1000	4.87	[27]
	1400	14.9	[37]		0–2400	7.3	[27]
	1800	16.3	[37]	Yt_2O_3	1000	8.3	[31]
Sc_2O_3	20–900	7.6	[37]		1200	8.6	[31]
	20–1800	7.7	[37]		1400	8.8	[31]
Sm_2O_3	20–1000	9.9	[37]		1600	9.1	[31]
	20–1400	9.3	[37]		1800	9.3	[31]
TaB_2	20–1000	8.2	[34]	ZrB_2	20–1000	5.9	[34]
	1000–2000	8.8	[34]		1000–2000	6.5	[34]
TaC	20–1000	6.64	[27]	ZrC	20–1000	6.99	[27]
	20–1200	6.67	[27]		20–1300	7.20	[27]
	20–1400	6.89	[27]		20–1800	8.1	[27]
	20–1600	7.29	[27]		20–2400	8.77	[27]
	20–1800	7.59	[27]	ZrO_2	20–1000	9.1	[31]
	20–2000	7.94	[27]		20–1200	9.5	[31]
	20–2900	8.81	[27]		20–1400	9.47	[37]
ThO_2	20–1000	9.4	[31]		20–2100	13.0	[37]

Table 10.4 Average temperature coefficients of linear expansion of quartz and optical glasses $\bar{\alpha}$, 10^{-7} K^{-1} [4,42,43,61]

Glass specification	ΔT, K				Glass specification	ΔT, K			
	170–293	210–293	293–300	293–400		170–293	210–293	293–300	293–400
Quartz glasses									
КЛР−1	−1.5 (223–293)		−1.0 (273–293)		КУ-2, КВ	−6.4 (10); −8.6 (30); −8.6 (50); −7.1 (100)			
КЛР−2	−1.9 (223–293)		−1.0 (273–293)			−1.10 (170); 1.40 (210); 4.7 (293);			
КЧГ	+4.2 (223–293)		+4.5 (273–293)			6.1 (473); 5.3 (673); 3.4 (1073);			
					Pyrex	28.2	29.2	39.2	

Table 10.4 Average temperature coefficients of linear expansion of glasses $\bar{\alpha}$, 10^{-7} K^{-1} *(continued)*

Glass	ΔT, K				Glass	ΔT, K			
specification	170–293	210–293	293–300	293–400	specification	170–293	210–293	293–300	293–400
Optical glasses									
ЛК1	110	111	112	113	ТК12	57.0	58.0	61.2	65
ЛК3	85	86	88.4	92	ТК13	60.0	61.0	64.1	67
ЛК4	49	50	51.5	52	ТК14	61.0	63.0	65.1	69
ЛК5	32.0	33.0	34.4	35	ТК16	63.0	66.0	69.1	71
ЛК6	78	80.0	81.4	83	ТК17	68.0	68.0	72.2	75.5
ЛК7	38.0	40.0	41.3	44	ТК20	66.0	67.0	70.1	73
ЛК8		54.0	55.3	56	ТК21	72.0	73.0	76.1	81
ЛК103		84.0	86.4		ТК23	51.0	52.0	55.2	58
ЛК105		32.0	34.5		ТК104		58.0	61.3	
ЛК107		40.0	41.2		ТК109		68.0	71.2	
ФК1	83.0	84.0	87.2	91	ТК114		60.0	63.3	
ФК3				78	ТК115		63.0	66.3	
ФК4				107	ТК121		72.0	75.2	
ФК13	61.0	62.0	68.2	73	ТК123		50.0	54.4	
ФК14		88.0	92.0	98	СТК3	69.0	71.0	74.0 (273–293)	
ФК113		63.0	64.0 (273–293)		СТК7	82.0	84.0	88.0 (273–293)	
ФК114		87.0	92.0 (273–293)		СТК8	83.0	84.0	87.0 (273–293)	
К1	59.0	60.0	62.7	65	СТК9	50.0	51.0	54.0 (273–293)	
К2	56.0	57.0	60.3	64	СТК10	43.0	45.0	49.0 (273–293)	
К3	72.0	74.0	77.3	83	СТК12	49.0	57.0	61.0 (273–293)	
К5	66.0	67.0	70.3	74	СТК15		57.0	59.0 (273–293)	
К8	66.0	68.0	73.2	76	СТК16		58.0	61.0 (273–293)	
К14	61.0	62.0	66.0	70	СТК19		51.0	53.0 (273–293)	
К15	79.0	81.0	89.0	93	КФ1	62.0	63.0	64.2	66
К17	64.0	66.0	69.4	74	КФ3		83.0		91
К18	62.0	64.0	67.4	71	КФ4	62.0	63.0	66.2	71
К19	72.0	74.0	76.5	80	КФ5	59.0	59.0	60.1	62
К20	68.0	69.0	71.3	75	КФ6	60.0	62.0	65.3	68
К100		70.0	73.5 (273–293)		КФ7	54.0	55.0	57.3	60
К108		66.0	69.3 (273–293)		КФ8	70.0	72.0	74.1	76
К119		70.0	73.5 (273–293)		БФ1	65.0	66.0	69.2	72.5
ОК1				136	БФ4	69.0	70.0	72.1	74
БК4	73.0	74.0	76.2	80	БФ6	76.0	77.0	80.1	83
БК6	73.0	74.0	77.1	83	БФ7	67.0	68.0	71.2	75
БК8	55.0	56.0	59.1	63	БФ8	76.0	77.0	79.2	82
БК9	67.0	68.0	72.1	76	БФ11	61.0	63.0	67.2	70
БК10	63.0	65.0	67.0	72.5	БФ12	82.0	83.0	85.2	87.5
БК11	59.0	60.0	62.1	66.5	БФ13	59.0	61.0	65.2	68
БК12	70.0	71.0	73.1	77	БФ16	76.0	78.0	81.2	84
БК13	61.0	62.0	65.1	68.9	БФ18	71.0	72.0	75.2	79
БК104		71.0	74.2		БФ19	69.0	70.0	73.2	76
БК106		75.0	78.3		БФ21	70.0	71.0	74.1	77
БК110		65.0	67.2		БФ23	69.0	70.0	71.2	73
ББК1				99	БФ24	72.0	74.0	76.2	79
ББК2				94	БФ25	65.0	66.0	69.1	73
ТК1	58.0	59.0	60.2	61	БФ26	63.0	64.0	67.2	70
ТК2	63.0	64.0	67.2	70	БФ27	74.0	75.0	77.1	80
ТК3		54.0		60	БФ28	58.0	59.0	62.2	65
ТК4	58.0	59.0	62.2	66	БФ32		103	107	
ТК7		67.0		73	БФ101		66.0	69.8	
ТК8	60.0	62.0	66.2	70	БФ104		69.0	72.0	
ТК9	67.0	68.0	72.1	76	БФ106		76.0	79.0	

Table 10.4 Average temperature coefficients of linear expansion of glasses $\bar{\alpha}$, 10^{-7} K^{-1} *(continued)*

Glass specification	ΔT, K 170–293	210–293	293–300	293–400	Glass specification	ΔT, K 170–293	210–293	293–300	293–400
БФ112		78.0	82.0		Ф101		69.0	70.5	
БФ125		65.0	69.0		Ф104		70.0	71.3	
ТБФ3,4	63.0	65.0	69.0 (273–293)		Ф106		68.0	70.2	
ТБФ5		54.0	58.0	64	Ф108		91.0	95.0	
ТБФ7		58.0	61.0 (273–293)		Ф109		90.0	92.5	
ТБФ8		55.0	58.0 (273–293)		Ф113		68.0	70.4	
ТБФ25		55.0	59.0 (273–293)		ТФ1	81.0	82.0	84.3	86.5
ЛФ1	79.0	80.0	83.2	84.5	ТФ2	71.0	73.0	75.3	79
ЛФ5	66.0	67.0	69.4	72	ТФ3	76.0	77.0	81.0	82.5
ЛФ7	70.0	71.0	73.3	74.5	ТФ4	76.0	77.0	80.4	82
ЛФ8	83.0	84.0	87.5	90	ТФ5	77.0	78.0	81.3	83
ЛФ9	80.0	81.0	85.5	89	ТФ7	92.0	93.0	95.3	96
ЛФ10	71.0	72.0	74.6	78	ТФ8	76.0	77.0	79.3	82
ЛФ11	70.0	71.0	72.8	74	ТФ10	74.0	75.0	77.4	82
ЛФ12	85.0	86.0	88.3	91	ТФ11	95.0	97.0	104	108
ЛФ105		67.0	70.0		ТФ103		75.0	77.3	
Ф1	70.0	71.0	72.2	75	ТФ107		88.0	90.4	
Ф2	69.0	70.0	72.2	74	ТФ110		75.0	77.3	
Ф4	72.0	73.0	74.1	75.5	ОФ1	58.0	59.0	61.4	63
Ф6	69.0	70.0	72.2	74	ОФ2	69.0	70.0	72.2	74
Ф7	68.0	70.0	72.2	74.5	ОФ3	46.0	48.0	51.3	55.5
Ф8	94.0	95.0	98.1	101	ОФ4	46.0	48.0	51.3	52
Ф9	89.0	90.0	93.1	96.5	ОФ5	44.0	46.0	49.0	53
Ф13	69.0	71.0	72.2	74.5	ОФ101		58.0	60.4	
Ф18		106	109						

Note: Temperatures different from those specified are shown in parentheses.

Table 10.5 Temperature coefficients of linear expansion of technical glasses [The true values of TCLE α (at a given temperature) or mean TCLE $\bar{\alpha}$ (within the temperature range) are presented.]

Glass	α, $\bar{\alpha}$, 10^{-7} K^{-1}	Glass	α, $\bar{\alpha}$, 10^{-7} K^{-1}	Glass	α, $\bar{\alpha}$, 10^{-7} K^{-1}	Glass	α, $\bar{\alpha}$, 10^{-7} K^{-1}
Laser glasses (Russian) 293-400 K [20,29]							
ГСЛ-1	94	ГСЛ-3	121	ГСЛ-21 – ГСЛ-24	106*1	КГЗ-3	105*1
ГСЛ-2	119	ГСЛ-4	103	ЛГС-247-2	116	КГС-5	104*1
Laser glasses (world) 293 K [20]							
FG_1	80	LG-52	78	LG_1	99.5	LG_4 – LG_6	54.9
FG_2	70	LG-54	95	LG_2	102.0		
FG_3	60	LG-55	107	LG_3	104.0	Ba–crown	110.0
Optical fibers 293 K [43]							
cover, BO	53	core, BC	92	Optical fiber plate, ВОП	32–73		
Optical ceramics, 293–393 K [43,44]							
КО1	113	КО3	109.9	КО4	77	КЭО10	24
КО2	69						
Infrared glasses, 293–393 K [3,17,43]							
ИКС22	226	ИКС27	177	Irtran-2	66*2	KRS-5	580
ИКС23	246	ИКС28	220	Irtran-3	202*2	KRS-6	560
ИКС24	182	ИКС30	122	Irtran-4	77*2	KRS-13	391*1
ИКС25	220	К515	82	Irtran-5	120*2		
ИКС26	166	Irtran-1	107*2	Irtran-51	115*2		

Table 10.5 Temperature coefficients of linear expansion of technological glasses *(continued)*

Glass	α, $\bar{\alpha}$, 10^{-7} K^{-1}	Glass	α, $\bar{\alpha}$, 10^{-7} K^{-1}	Glass	α, $\bar{\alpha}$, 10^{-7} K^{-1}	Glass	α, $\bar{\alpha}$, 10^{-7} K^{-1}
Resistor glasses, 293–573 K [29]							
C27-1	27	C41-1	41	C74-1	74	C84-2	84
C36-1	36.5	C63-1	63	C77-1	77		
Insulator glasses, 293–373 K [29]							
13B	58	N1	94				
Electron-tube glasses [29], 293–373 K							
C37-1	37.5	C51-1*6	51.0	C40-1*3	40.0	C89-6	89.0
C38-1*3	38.0	C51-2*6	51.0	C47-1*4	47.0	C90-1	90.0
C39-1*3	39.5	C87-1*7	87.0	C48-1*4	48.0	Quartz	
C49-1*5	49.0	C88-1*7	88.0	C89-1*7	89.0	(300–350 K)	4.0
C49-2*5	49.0	C88-2	88.0	C89-2	89.0	(400–700 K)	6.0
Thermometrical glasses, 273–373 K [42]							
GSSD 1224–41	80	59^{III}	56.5	600	33	700	29.5
Chemical glasses, 300 K [42]							
Heat-resistant	33.2	T28	40	Щ23	65	G20	49
N13	50	Ц32	88.6	N51-A	48	Multal	50
N23	89	КС34	91	Uninost	98	Sial	50.6
N29	89.3	Щ14	87.1	Palex	64.9		
T16	50	Щ26	57.3	Murano 1922 N	48		
Doped glasses, 373–573 K [29,45]							
Aluminoboro-silicate	30–60	Yttrium-doped	171	Sodium-aluminosilicate	87	СЛК-5	
same alkali-free	8–38	Calcium-aluminosilicate	80–100	Fluoroberyllium	160	Fluorophos-phorous №436	165
Structural glasses, 300–573 K [42]							
Cast	80–95	Powder C25-1	25	Glass for pipe	50	Optical colour	< 70
Crystal (lead)	94–105	C48-2	48	same, pyrex	32	Glass enamel	70–120
Foam	5–6	C84-8	89	Glass cement СЦ90-1	97	Micales	80–90
Opal (opaque)	83	Unicolour-grade	90–99				

*1 At 230 K.
*2 In the range of 293–573 K.
*3 Fused with W, Н30К13Д.
*4 Fused wuth Mo, 29НК.
*5 Fused with 29НК.
*6 Fused with Mo.
*7 Fused with Fe–Cr–Ni-alloy.

Table 10.6 Average temperature coefficients of linear expansion of devitrified glasses $\bar{\alpha}$ (in 10^{-7} K^{-1}) [29,42,44].

Specification	ΔT, K					
	213–293	293–393	393–473	473–573	573–673	673–773
СТЛ-1	−12.5	−8	−5.4	−2.1	1.2	3.6
СТЛ-2	−10	−6	−2.5	2.7	5.3	5.6
СТЛ-3	−2.5	0.6	3	0.5	6.5	9
СТЛ-4	−1.3	2	5.3	7.3	13.6	15.9
СТЛ-5	−0.8	5	5.6	6	7	19.5
СТЛ-6	2.5	7	8.8	11	13	14
СТЛ-7	6.2	12	12.5	13	15	16
СТЛ-8	10	16	16.2	16.5	17	18
СТЛ-9		38.1	38.1	38.1	38.1	38.1
СТЛ-10	80	100	106.2	125	120	90
СТМ-1	11	27	31.2	32	34	37
СТМ-2		44	51.2	56	56	57

Table 10.6 Mean TCLE of devitrified glasses $\bar{\alpha}$ (in 10^{-7} K^{-1}) *(continued)*

Specification	ΔT, K					
	213–293	293–393	393–473	473–573	573–673	673–773
СТБ-1	27.5	29	32.5	37	38	60
СТБ-3	21	30	31.2	33	48	88
СО115М		0.0–1.5 (213–313 K)		2.5–6 (293–693 K)		
СО-21			−2.5 (293–693 K)		5.3	
СО-15					6.5	
С-15-12					1.2	

Note: Temperatures different from the specified ones are shown in parentheses.

Table 10.7 Temperature coefficients of linear expansion of cast iron [3,4,5]. [The true values of TCLE α (at given temperature t) or mean TCLE $\bar{\alpha}$ (within the range Δt) are presented.]

Specification or name	t; Δt, °C	α; $\bar{\alpha}$, 10^{-6} K^{-1}	Specification or name	t; Δt, °C	α; $\bar{\alpha}$, 10^{-6} K^{-1}
СЧ 00, СЧ 10	20	10.0	same	20–300	12.3
СЧ 12-28, СЧ 15-32,СЧ 18-36, СЧ 21-40	20–200	11.8	same	20–500	13.6
			same	20–700	14.7
СЧ 32-52	−77	11.2	austenitic	20–100	17
ЖЧН15Д7Х2	20	20.0	nickel doped	20–600	18.0
АВЧ-1, АЧК-1	20	11	chromium	20–600	11.0
ЧМ 1.3, ЧМ 1.8	20–100	12.0	Composed cast iron:		
ПЧ, ПЧИ, ХТВ, ХНВ	20–600	13.6	austenite	20–100	17–24
Pig iron:			ferrite	20–100	12–12.5
white cast iron	20	7–11	pearlite	20–100	10–11
same	20–100	10	cementite	20–100	6–8.5
grey cast iron	20	11	graphite	20–100	7.5–8
ferritic malleable	20–100	11			

Table 10.8 Temperature coefficients of linear expansion of steels [The mean values of TCLE $\bar{\alpha}$, 10^{-6} K^{-1}, in the range from 300 K up to the temperature indicated or the true TCLE α, 10^{-6} K^{-1}, are presented.]

Steel specification	Temperature, K										Ref.
	373	473	573	673	773	873	973	1073	1173	1273	
Carbon steels											
Ст.3, Ст.3кп(α)			5.50 (100 K)		10.1 (200 K)		11.8 (300 K)				[5]
08кп(α)	11.6	12.32	13.02	13.65	14.22	14.64	15.01				[16]
10	11.6	12.6	13.0		14.6						[63]
15	12.2	12.3	13.1	13.5	14.3	14.3	15.25	14.1	13.2	13.3	[16]
20, 20кп	11.1	12.1	12.8	13.4	13.9	14.4	14.8	12.9			[16]
25	12.2	12.7	13.1	13.5	13.9	14.4	14.9				[16]
25Л(α)	11.5	12.9	13.0	13.2	13.5	13.8					[16]
30(α)	12.9	14.5	15.8	15.8	16.7	16.2		16.4 (923 K)			[16]
35	11.09	11.89	13.42	14.02	14.43			17.2 (α, 923 K)			[16]
40	11.21	13.0	13.0	13.58	14.05	14.58	14.58	11.85	12.65	13.59	[16]
45	11.59	12.32	13.09	13.71	14.18	14.68	15.08	12.50	13.56	14.45	[16]
50(α)	4.90 (100 K)		9.30 (200 K)		10.9 (300 K)						[16]
	12.0	12.4	12.9	13.3	13.7	14.1	14.3	11.9	12.9	14.0	[16]
У8(α)	12.1	14.0	12.8	16.1	16.8	17.8		19.8 (923 K)			[16]
У9(α)	12.1	13.7	15.3	16.4	17.3	17.2		17.6 (923 K)			[16]
У12(α)	11.7	13.3	15.4	16.2	17.1	18.3		18.9 (923 K)			[16]

Table 10.8 Temperature coefficients of linear expansion of steels *(continued)*

Steel specification	Temperature, K										Ref.
	373	473	573	673	773	873	973	1073	1173	1273	
					Chromium steels						
15Х, 15ХА, 20Х	11.3	11.6	12.3	13.2	13.7	14.2					[5]
30Х	12.45	13.0	13.4	13.75	14.15	14.55	14.8	12.0	12.8	13.8	[5]
38ХА, 40Х (α)	−0.002 (5 K)		0.018 (10 K)	0.18 (20 K)		1.7 (50 K)	6.6 (100 K)		11.5 (200 K)		[46]
40Х	13.4	13.3		14.8		14.8					[5]
0Х13	10.5	11.1	11.4	11.8	12.1	12.3	12.5	12.8			[5]
1Х13	10.15	11.2	11.4	11.8	12.2	12.4	12.7	13.0	10.8	11.7	[5]
2Х13	10.2	11.2	11.5	11.9	12.2	12.8	12.8	13.0			[5]
3Х13	10.2	10.95	11.1	11.7	12.0	12.3	12.5	12.6	10.6	12.2	[5]
4Х13	10.65	11.5	11.85	12.2	12.5	12.75	12.95	13.2			[5]
ШХ15 (α)	14.0	15.1		15.5		15.7					[5]
08Х17	10.4	10.5	10.8	11.2	11.4	11.6	11.9	12.1			[37]
9Х18, 9Х18Ш	11.7	12.1	12.4	12.9	13.3	11.8	12.1	12.4			[5]
Х28	10.0				11.1	11.3	11.5	12.0	12.4		[5]
				Nickel-chromium low- and medium alloyed steels							
40ХН, 45ХН	11.8	12.3	13.4	14.0							[5,46]
18ХНВА	11.7	12.2	12.7	13.1	13.5	13.9					[5]
40ХНМА	11.7	12.2	12.7								[5]
35ХНМ, 34ХН1М	11.7	12.2	12.7								[5]
12ХН2	12.6	13.8	14.8	14.3							[5]
30ХН2МФА (α)	5.01 (70 K)			7.40 (100 K)		9.30 (200 K)		12.3 (300 K)			[3]
30ХН3	11.2	11.8	12.4	12.7	13.2	13.5	13.9	10.8	12.1	13.1	[5]
12ХН3А	11.8	13.0	14.0	14.7	15.3	15.6					[5,46]
20ХН3А	11.0	12.0	13.0	13.5	14.0	14.5					[5]
33ХН3МА, 34ХН3М	10.8	11.6	12.5	13.3	13.5	13.7					[5]
35ХН3МФ	11.8	12.1	12.6	13.0	13.4	13.7					[5]
0ХН3М (α)	5.0 (70 K)			7.60 (100 K)		9.45 (200 K)		11.6(300 K)			[3]
12Х2Н4А	11.0	12.0	13.0	14.7		15.6					[5]
18Х2Н4МА (α)	−0.06 (5 K)		−0.064 (10 K)		0.32 (30 K)		6.5 (100 K)		10.0 (200 K)		[3,46]
		11.2	12.5	13.1	13.7	13.8	13.8				[50]
				Nickel-chromium steels (Ni<20%)							
13Х12НВ2МФ	11.0	11.3	11.6	12.0	12.3	12.5					[5]
4Х12Н8Г8НФБ	16.0	16.9	17.7	18.5	19.5	19.9	20.2				[5]
0Х12Н16БС4	16.45	16.7	17.4	17.65	17.85	18.1	18.3	18.55	18.65		[5]
2Х13Н2	10.54	10.64	10.64	10.84	11.09	11.34					[5]
1Х13Н2С2	10.78	11.41	11.82	12.26	12.71	13.12	13.32				[5]
03Х13Н8Д2ТМ			10.2 (77–300 K)				8.5 (20–300 K)				[51,52]
Х13Н12М2В2Б1К10	15.6	15.8	16.5	16.9	17.1	17.3	17.7	18.0		18.6	[5]
Х13Н13В2Б	16.8	17.3	17.9	18.3	18.7	18.9	19.1	19.3	19.6		[5]
Х13Н13В2М2Б3К10	15.6	15.8	16.5	16.9	17.1	17.3	17.7	18.0	18.3	18.6	[5]
1Х13Н16Б	16.05	16.9	17.7	18.25	18.55	18.75	19.0	19.25	19.6	19.65	[5]
4Х14Н14В2М			17.0		18.0		18.0		19.0		[5,37]
1Х14Н14В2М	17.0	17.8	18.3	18.8	19.0	19.2	19.4	19.9	20.5		[5]
1Х14Н14В2МТ	17.2	17.2	17.5	18.0	18.5	18.6	18.9	19.3			[5]
2Х14Н14В2СТ	16.1	16.7	17.2	17.4	17.8	18.2					[5]
Х14Н14МВФБ	15.1	15.9	16.9	18.0	18.2	18.4	19.0				[5]
Х14Н14М2В2ФБТ				16.7	17.1	17.4	17.8	18.1			[5]
09Х14Н16Б	15.2	16.5	17.1	17.55	17.96	18.41	18.91	20.6			[37]
1Х14Н18В2Б	16.5	17.4	17.6	18.0	18.1	18.2	18.5	19.0	19.0		[5]
Х14Н18В2БР	15.9	16.5	17.2	17.6	18.0	18.3	18.6	18.7	19.0		[37]
09Х14Н19В2БР1	15.2	16.3	17.2	17.6	18.0	18.1	18.6	18.6			[5,37]
4Х15НГ7Ф2МС	17.0	17.7	18.4	19.1	20.5	20.8					[5,36]
(α)							22.8	22.78	23.32	24.64	[5,36]

Table 10.8 Temperature coefficients of linear expansion of steels *(continued)*

Steel specification	Temperature, K										Ref.
	373	473	573	673	773	873	973	1073	1173	1273	
0X15H7M2Ю	9.9	10.8	11.1	11.5	11.7	11.4	10.3	11.2	11.9		[5]
1X15H9C3Б1 (α)	17.4	18.7	19.7	20.2	20.5	21.0	21.55	21.75			[5]
X15H15M2K3BT	15.75	16.6	17.15	17.6	18.25	18.6					[5]
0X15H15M3	16.9	17.7	18.1	18.5	18.8	19.1	19.45	19.7	19.9		[5]
0X15H15M3Б	16.35	17.05	17.4	17.65	17.7	17.85	18.25	18.6	18.75		[5]
X16H9M2	17.0	17.5	18.0	18.4	18.9	19.3	19.5				[5]
08X16H13M2Б				17.1	17.4	17.8	18.2	18.6			[37]
X16H14Б	16.0		18.0		18.0			19.0			[5]
1X16H16B3MБP	17.1	17.1	17.1	17.9	18.2	18.5	18.8	19.1	19.2		[5]
07X16H16		11.7	12.1	12.5	12.9						[5]
X16H16B3MБ	15.8	16.8	17.3	17.6	17.8	17.9	18.1	18.2			[5]
2X17H1	8.1	8.76	10.26	10.04	10.54	11.54					[52]
2X17H2	10.54	10.67	10.94	10.84	11.24	11.34					[5]
1X17H2	10.3	10.4	10.7	11.1	11.8						[5]
0X17H4	9.7	10.2	10.6	10.9	11.2	11.3	9.6	9.6	10.2		[5]
0X17H4M2	10.6	11.0	11.4	11.6	11.9	11.7	11.1	11.7	12.3		[5]
X17H5M2	12.14	13.69	14.28	14.64	14.79	14.84					[5]
0X17H13M2T	15.7	16.1	16.7	17.2	17.6	17.9	18.2				[3]
0X18H7Ю1	15.6	16.5	17.3	17.9	18.1	18.4	18.5	18.7			[5]
X18H9	16.5	17.2	17.7	18.1	18.3	18.6	18.9	19.3	19.7	20.2	[5,37]
X18H9B		16.5	17.1	17.6	18.0	18.4	18.8	19.0	19.2	19.4	[5]
1X18H9BM			16.65	17.15	17.45	17.8	18.0	18.2			[5]
3X18H9B2ΦT	15.35	15.6	15.9	16.25	16.6	17.0	17.4	17.8	18.4	18.95	[5]
X18H9M	17.3	17.5	17.8	18.0	18.3	18.5	18.8				[5]
X18H9M2C2	16.6	17.0	17.35	17.7	18.05	18.4	18.75	19.1	19.4	19.75	[5]
X18H9C2	16.2	17.1	17.8	18.6	19.2	19.2	20.5				[5]
12X18H9T (α)	0.8 (20 K)		3.3 (50 K)		8.40 (100 K)		14.30 (200 K)		16.7 (300 K)		[3,52]
12X18H10T (α)		17.55 (400 K)		19.36 (700 K)		21.10 (1000 K)		22.92 (1300 K)			[60]
X18H9T	16.6	17.0	17.6	18.0	18.3	18.55	18.9	19.25	19.5	20.1	[5]
X18H9TЛ	14.8	16.0	16.9	17.1	17.6	18.0	18.4				[5]
04X18H10 (α)	3.0 (20 K)		8.0 (50 K)		10.8 (100 K)		15.4 (200 K)		15.5 (300 K)		[46,5]
X18H11Б	16.8	17.65	18.2	18.5	18.85	19.0	19.4	19.65	19.9		[5]
X18H12	16.8	17.2	17.6	17.8	18.2	18.5	18.8				[37]
0X18H12Б	16.0	18.0	18.0	19.0							[5]
X18H12M2T	15.7	16.1	16.7	17.2	17.6	17.9	18.2				[5]
X18H12MΦTP	15.9	16.9	17.6	17.8	17.9	18.4	18.8	19.0	19.2		[5]
X18H12M3	16.0		16.2		17.5	18.6				20.0	[5]
X18H12T	16.6	17.0	17.2	17.5	17.9	18.2	18.6	18.9	19.3		[5]
X18H14M2Б1				17.6	17.8	18.2	18.7			[5]	[5]
X18H15M3БЮP2	14.95	16.15	16.8	17.1	17.5	17.75	18.15	18.35	18.25		[5]
X18H15M3БЮP4	15.05	15.75	16.4	17.15	17.35	17.6	17.85	18.1	18.45		[5]
1X18H15M3B2БK13			16.7	16.7	16.75	17.0	17.25	17.4			[5]
0X18H15P4	16.5	17.4	17.8	18.15	18.45	18.85	19.2	19.45	19.8		[5]
0X18H15P7	16.8	17.4	17.7	18.05	18.2	18.55	18.95	19.4	19.8		[5]
10X18H18Ю4Д	15.5	16.5	17.0	17.4	17.65	18.2	18.4	18.8	18.6		[37]
3X19H9MBБT	16.65	16.9	17.2	17.5	17.8	18.15	18.5	18.9	19.3	19.7	[5,63]
X19H14Б2	17.0	17.2	17.4	17.6	17.9	18.6	18.8				[5]
2X20H11	17.3		17.8		18.4	18.7					[5]
X20H14C2	16.0					18.1	18.3	18.5	18.8	19.0	[5]
03X20H16AΓ6 (α)	0.01 (4 K)		0.06 (20 K)		0.57 (40 K)		8.27 (100 K)		16.10 (300 K)		[51,52]
07X21H5AΓ7			9.3 (20–300 K)		11.5 (77–300 K)			14.6 (173–300 K)			[51]
07X21H5AΓ7	15.7	16.0	16.8	17.3	18.0	18.4	18.5				[52]
0X21H5T	9.6	13.8	16.0	16.0	16.4	16.2	16.5	16.7	17.1		[5]
12X21H5T (α)		7.90 (100 K)		10.4 (200 K)			11.0 (300 K)		17.1	19.0	[3]

Table 10.8 Temperature coefficients of linear expansion of steels *(continued)*

Steel specification	Temperature, K										Ref.
	373	473	573	673	773	873	973	1073	1173	1273	
1X21H5T	10.2	14.4	16.8	16.8	17.4	17.5	17.7	18.0	18.5	[51]	[5]
0X21H6M2T	9.5	13.8	16.0	16.0	16.3	16.7	17.1	17.1	17.4		[5]
X22H9				17.5				18.5			[5]
20X23H18	14.9	15.7	16.6	17.3	17.5	17.85	17.85				[3,37]
X25H13AT,X25H13T					17.1				18.1		[5]
12X25H16Г17AP	16.6	16.2	16.8	17.4	18.0	18.5	18.7	18.9			[37]
Nickel–chromium steels (Ni > 20%)											
10X11H20T3P	15.9	16.4	17.2	17.9	18.4	18.9	19.2	8.15 (173–300 K)			[5,51]
00X18H20C3M3Д3Б (α)	−0.25 (10 K)		0.61 (30 K)		5.95 (70 K)		14.3 (200 K)		15.8 (300 K)		[3]
3X20H20M4B4K45			11.8	12.4	12.7	13.0	13.5	14.0	14.4	15.0	[5]
X20H20Б1M3K20				15.7	16.1	16.65	17.2	17.6	17.8 (1223 K)		[5]
X25H20C2	16.1					17.8	17.8	18.1	18.5	18.8	[5]
X26H20Л			15.5	15.6	15.9	16.2	16.6	17.2			[5]
X18H22B2T2 (α)	15.34	15.95	17.01	17.11	17.23	17.52					[5]
10X11H23T3MP (α)	0.03 (5 K)		0.14 (30 K)		1.46 (50 K)		9.10 (100 K)		16.0 (300 K)		[46]
08X15H24B4TP (α)	−0.12 (4 K)		−0.42 (30 K)		5.18 (70 K)		8.48 (100 K)		15.6 (300 K)		[46]
X15H24B4T	14.5	15.5	16.3	16.8	17.2	17.3	17.9	18.5			[5]
X15H24T2 (α)		3.60 (70 K)		5.85 (100 K)		14.3 (200 K)			16.4 (300 K)		[3]
1X16H25M6	15.0	15.7	15.9	16.3	16.6	16.9	17.1				[5]
36X18H25C2	13.0	13.6	14.7	16.1							[5]
X25H25TP	12.95	14.35	15.2	15.65	16.15	16.5	17.25	17.30	17.75		[5]
XH30MБЮ (α)	−0.14 (4 K)		0.21 (10 K)		1.9 (50 K)		8.2 (100 K)		14.5 (293 K)		[51]
1X15H30T2				16.1	16.7	16.9	17.5				[5]
XH32T	13.7	15.6	17.2	18.0	18.0	18.4	18.9	19.0			[37]
XH35BT	14.8	15.1	15.5	15.9	16.1	16.6	16.9	17.7			[37]
XH35BTЮ	12.7	14.1	15.0	15.4	15.8	16.0	16.6	16.8	18.4		[5,37]
X15H35B2M2TP	13.0			15.0	16.0	16.2	16.85	16.9			[5]
X15H35B3TK4		14.8	15.2	15.65	16.1	16.5	17.05				[5]
X15H35B5TP	14.1			15.8	16.2	16.2	16.7	16.9			[5]
1X16H36MБTЮP	17.73	14.02 (373–473 K)			16.95 (673–773 K)			24.96 (973–1073 K)			[5]
X26H36K30MБ			13.5	13.9	14.3	14.7	15.1	15.5	16.0		[5]
X16H38B3T	13.7	14.5	15.7	16.2	16.3	16.6	17.1	17.1	17.3		[5]
0X17H39Б	14.2	16.0	17.2	17.2	17.2	17.5	17.8				[5]
04XH40MДTЮ	8.15	11.0	12.0	12.8	13.6	14.1	19.9	16.2			[51]
0X21H40БP	14.3	15.4	15.8	15.9	16.0	16.3	16.8	17.1			[5]
X12H42MБT3	14.0				15.1			17.0 (300–1023 K)			[5]
XH45Ю	15.8	15.9	15.95	16.55		16.6		19.0	19.7		[37]
X15H45K28M3T2Ю3	12.3				14.3				17.0		[5]
X22H45M9	13.9									16.3	[5]
0X20H46Б	13.33	14.2	16.0	17.7	17.7	18.5	19.45	19.8			[5]
X15H51K20M5Ю5	11.9				14.3				17.7		[5]
X11H54K2M5ЮT2	12.4	13.1	13.6	14.0	14.3	14.5	15.0	15.7	16.8	18.1	[5]
X20H54K16T2Ю1	12.4				14.7				17.4		[5]
X11H55M5K20T1	12.4									18.8	[5]
X19H55K10M10T2Ю1	14.0				14.7				16.0		[5]
XH55BMKЮ	11.8	12.2	12.7	13.1	13.4	13.6	14.1	14.8	15.7	17.0	[37]
XH55BMTKЮ	10.8	11.7	12.4	12.8	13.4	13.8	14.1	14.5	15.5		[37]
XH55MБЮ		10.0 (20 K)			12.0 (80 K)			13.6 (210 K)			[51]
XH55MБЮ			14.1	14.3	14.5	15.1	15.5	16.3	16.3		[51]
XH55MBЮ	13.4	14.2	14.9	15.2	15.8	16.2		17.3	18.8		[37]
XH56MBTЮ	11.7	12.0	12.4	12.7	13.0	13.5	13.8	14.5	15.2		[37]
X10H56K20M5T4Ю5	11.7				14.0				16.2		[5]
X20H56K18T2Ю1.5	11.6	12.6	12.7	13.5	13.7	14.2	15.0	16.0	17.0		[5]
XH57MBTЮ	12.6	13.6	14.7	15.0	15.3	15.8	16.2	17.3	18.2		[37]

Table 10.8 Temperature coefficients of linear expansion of steels *(continued)*

Steel specification	Temperature, K										Ref.
	373	473	573	673	773	873	973	1073	1173	1273	
Х16Н57М17В4					13.3				11.8		[5]
Х15Н58М16	11.3									15.3	[5]
Х20Н58М10К10	12.5				14.3				16.5		[5]
Х20Н59К16Т2Ю1	12.8				15.1				17.9		[5]
ХН60В	12.7	13.2	13.6	14.1	14.5	15.5	15.7	16.0	16.2	16.8	[37]
ХН60КМВЮБ	11.3	12.0	12.6	13.2	13.5	13.8	14.4	15.3	16.1		[37]
ХН60Ю	12.28	13.08	13.42	14.2	14.22	14.59	15.11	15.18	15.88		[37]
Х18Н60Ю3	12.3	13.1	13.4	14.2	14.40	14.6	15.1	15.2	15.9		[5]
0Х20Н60Б	12.55	13.6	14.4	14.8	15.1	15.5	15.6	16.2	16.7		[5]
ХН62ВМКТЮБ	13.5	13.5	13.5	13.6	13.6	13.9	14.5	15.6			[37]
ХН62МБВЮ	12.0	12.8	13.3	13.5	13.8	14.5	14.8	15.5	16.4	17.3	[37]
ХН62МЮК10	11.49	12.57	13.03	13.37	13.68	13.98	14.33	14.75	15.56	16.9	[37]
ХН65ВМТЮ	11.5		12.3	12.75	13.0	13.2	14.0	14.3	15.0		[37]
ХН67ВМТЮ	12.0	12.3	13.5	13.6	14.4	14.8	15.2	16.2			[37]
Х18Н67В5М5Т2ЮР	10.4	11.1	11.7	12.0	12.5	12.8	13.6	13.8	14.9		[5]
ХН70ВМТЮ	12.0	12.5	13.5	13.9	14.3	14.8	15.8	16.3	16.7		[37]
ХН70ВМТЮФ	10.4	11.7	12.4	12.9	13.2	13.6	14.0	14.5	15.0		[37]
ХН70ВМЮ	10.2	11.3	12.0	12.2	12.4	12.6	13.1	13.2	14.0	15.1	[37]
ХН70ВМЮТ	12.2	12.65	13.2	13.6	14.1	14.55	15.1	15.8			[37]
Х15Н70В6М3Т2ЮР	12.4	12.8	13.05	13.3	13.6	14.0	14.6	15.25	16.1	17.35	[5]
Х15Н70В5М4Ю2ТР	12.2	12.6	13.2	13.6	14.1	14.5	15.1	15.8	16.5		[5]
Х27Н70Ю3										18	[5]
Х13Н71М5Б2Т1ЮБ	11.9				12.6				14.5		[5]
Х15Н73Т2Ю1Б			13.1	14.0	14.5	14.9	15.6	16.4			[5]
Х15Н73Т3Ю1Б1	13.7	13.9	14.2	14.4	14.6	15.0	15.4	16.2	16.8		[5]
Х20Н75ТЮ	12.2	13.0	13.4	13.8	14.1	14.7	15.4	15.5	16.0		[5]
ХН77ТЮР	12.67	12.9	13.3	13.8	14.2	14.6	15.1	15.5	16.2	16.8	[37]
Х20Н77Т2Ю	11.9	12.7	13.0	13.5	13.7	14.0	14.5	15.1	15.8		[5]
ХН78Т	12.2	13.0	13.4	13.8	14.1	14.7	15.4	15.5			[37]
Х16Н80ТБЮ	13.3	13.5	13.85	14.2	14.5	14.9	15.5	16.15	16.7	17.8	[5]
Х20Н80Т	11.8	12.9	12.9	14.4	14.8	15.8	16.1	16.50	16.8		[5]
Х20Н80Т3	11.9	12.7	13.0	13.5	13.7	14.0	14.5	15.1	15.8		[5]
Х20Н80										17.6	[5]
Х22Н47М9	13.8				14.9				16.8		[5]
Chromium-molybdenum steels											
12ХМ, 12МХ	11.2	12.5	12.7	12.9	13.2	13.5	13.8				[5]
15ХМ	12.2	12.95	13.3	13.7	14.0	14.3	14.5	13.35	11.15	12.45	[5]
20ХМ	11.8	12.5	13.0	13.6	14.0	14.3					[5]
20ХМЛ	10.9	12.4	12.8	13.1	13.6	13.9					[5]
20ХМФЛ	10.0	11.9	12.9	13.1	13.5	13.8					[5]
30ХМ, 30ХМА	11.6	12.5	13.2	13.8	14.3						[5]
34ХМ, 35ХМ	12.3	12.6	13.3	13.9	14.3	14.6					[5]
35ХМФА	11.8	12.5	12.7	13.0	13.4	13.7	14.0				[5]
38ХМЮА	12.3	13.1	13.3	13.5	13.5	13.8					
12Х1МФ	12.4	13.0	13.6	14.0	14.4	14.7	14.9	14.8	12.0		[5]
15Х1М1Ф	11.2	11.7	12.5	13.0	13.5	13.7					[5,37]
15Х1М1ФК1Р	12.1	12.5	13.1	13.8	14.6	14.9	14.6				[5]
20Х1М1Ф1ТР		12.0	12.3	12.85	13.0	13.55					[5]
20Х1М1Ф	12.05	12.4	12.85	13.15	13.45	13.8	13.9	14.25	12.65		[5]
25Х1МФ	11.3	11.7	12.8	13.2	14.2	14.4					[5]
25Х1М1Ф	10.9	12.0	12.7	13.65	13.7	13.8	14.0				[5]
1Х2М	12.1	12.3	12.8	13.2	13.8	14.0					[5]
12Х2МБ, 1Х2МФБ	12.3	12.4	12.6	12.6	13.3	13.5					[5]
12Х2МФСР	11.85	12.9	13.55	13.95	14.35	14.8	15.15	15.5	12.4		[5]

Table 10.8 Temperature coefficients of linear expansion of steels *(continued)*

Steel specification	Temperature, K										Ref.
	373	473	573	673	773	873	973	1073	1173	1273	
15X2M2ФБС	11.5	12.3	12.9	13.4	13.8	14.3	14.6	14.9	13.0		[5]
25X2M1Ф	12.5	12.9	13.3	13.7	14.0	14.7					[5]
X3MBФБ	10.6	11.45	12.3	12.65	13.1	13.35	13.4	13.1	11.45	10.35	[5]
X5M	11.3	11.6	11.9	12.2	12.3	12.5					[5]
10X7MBФБР	10.25	11.6	12.4	12.75	13.25	13.6	13.4				[37]
15X11MФ	10.3	10.6	10.8	11.3	11.7	12.0	12.2	12.4			[37]
18X11MФБ	10.3	10.6	10.8	11.3	11.7	11.8	12.0	12.4			[37]
12X12MBФБР	11.15	11.13	11.42	11.8	12.0	12.15	12.15	12.65	11.6		[37]
12X12MBФБ	10.25	11.6	11.6	11.6	13.85	12.7	14.95	10.5	13.85		[37]
12X13M2C2	10.5	11.2	12.0	12.3	12.5	12.8	13.0	13.5	14.1	13.9	[5]
1X13M2C1	10.2	10.8	11.3	11.7	12.0	12.4	12.6	13.0	13.0	13.6	[5]
1X13M2ФБР	9.7	10.6	11.5	11.6	11.8	12.2	12.2	12.6	12.4	12.6	[5]
15M, 16M, 20M	12.0	12.6	13.2	13.7	14.2	14.7					[5]
Manganese and chromium-manganese steels											
20Г		12.3	13.2			14.9					[63]
20Г (α)		0.3 (30 K)		5.80 (100 K)		10.0 (200 K)			11.9 (300 K)		[3]
22ГК	10.5	11.8	12.6	13.1	13.7	14.0					[5]
50Г	11.6	11.8	12.8	13.8	14.2	14.6					[5]
65Г	11.11	11.9	12.9	13.5		14.6					[5]
30Г2	11.9	12.7	13.4	13.9	14.3	14.7	15.1	13.6	13.7	14.5	[5]
14Г2АФ (α)		0.035 (4.3 K)		0.23 (20 K)	9.5 (100 K)		14.35 (200 K)		15.85 (300 K)		[46]
Г13	18.0	19.35	20.7	21.7	20.75	19.85	20.5	21.85	23.45	23.15	[5]
20ХГНР		11.7	12.7	13.5	14.0	14.3	15.2				[50]
25ХГСА	13.15	13.9	14.05	14.3	14.85	14.9	15.1				[5]
30ХГСА (α)				7.60 (50 K)			9.47 (100 K)				[5]
30ХГСА	12.0	12.5	12.9	13.2	13.6	13.9					[5]
Х12Г20Ф	15.4	16.9	17.5	18.2	18.9	19.7	20.5				[5]
03Х13АГ19 (α)		0.05 (10 K)		0.4 (40 K)	4.5 (100 K)		7.4 (200 K)		14.4 (300 K)		[5]
	13.85	16.80	18.70	19.40	20.40	21.80	21.8	22.5			[51]
07Х13АГ20Н4 (α)			−0.2 (20 K)			3.0 (77 K)		15.3 (293 K)			[5]
Х14Г14Н3Т	16.0	16.8	17.6	18.3	19.0	19.5	20.1				[5]
10Х14Г14Н4Т (α)			0.502 (20 K)			5.05 (100 K)		9.20 (200 K)			[37]
	15.3	15.8	18.9	19.05	19.07	20.02	20.07	21.1			[37]
Х15Г21Т	15.3	16.6	17.8	18.9	19.6	20.3	20.9	21.5			[5]
12Х17Г9АН4 (α)		3.60 (50 K)		9.05 (100 K)		14.1 (200 K)			14.6 (300 K)		[3]
07Х21Г7АН5 (α)		−0.03 (4 K)		−0.13 (20 K)		3.05 (70 K)	5.75 (100 K)		15.7 (300 K)		[3,46]
Chromium-tungsten steels											
28ХВФЦ	11.8	12.1	12.8	12.85	13.2						[5]
38ХВФЮА	12.25	12.9	13.1	13.35	13.75	14.1	14.45	14.1	11.6		[5]
3ХВ8Ф (α)	11.2										[5]
20Х3ВМФ				12.3	12.75	13.8					[5]
4Х4ВМФС (α)	11.44										[5]
4Х4ВМФСШ (α)	11.62										[5]
Х11В2НМФ	10.5	11.2	11.9	12.6	12.8	13.1					[5]
18Х12ВМБФР	11.5	11.13	11.42	11.8	12.0	12.5	12.65	11.6			[5]
15Х12ВНМФ	10.0	10.5	10.7	11.0	11.2	11.5	11.6	11.8	10.7	11.65	[5]
Х12В2МФ	10.8					13.5					[5]
Chromium-vanadium steels											
15ХФ	11.9	12.4	13.1	13.7	14.2	14.5	14.9				[5]
40ХФА	11.0		12.9			14.5					[5]
50ХФА	11.8	12.4	13.1	13.6	13.9	14.1					[5]
12Х2ФБ	11.5	12.2	12.4	12.9	13.2	13.4	13.5				[5]
70Х4Ф1В18	11.25	11.7	12.0	12.2	12.4	12.6	12.8	12.9	11.35	12.45	[5]

Table 10.8 Temperature coefficients of linear expansion of steels *(continued)*

Steel specification	Temperature, K										Ref.
	373	473	573	673	773	873	973	1073	1173	1273	
	Cobalt and chromium-cobalt steels and alloys										
15К, 20К		12.0	12.8	13.2	13.5	13.85					[5]
22К, 25К (α)		12.2	13.0	13.3	13.9						[5]
В3К (stellite, fused to the surface of a cutting tool)						14.0				16.0	[5]
03Х9К14Н6М3Д (α)	0.03 (4 K) −0.08 (9 K) 0.3 (20 K) 4.4 (70 K) 7.4 (110 K) 13.2 (300 K)										[51,52]
40К19Х14Н20М4В4Б4			14.25	14.5	14.75	15.1	15.4	15.75			[5]
15К20Х20Н20М3В2Б			15.55	15.95	16.3	16.65	17.05	17.55			[5]
40К20Х20Н20М4В4Б4			15.15	15.25	15.35	15.45	15.7	16.1	16.65		[5]
40К30Х20Н20М8В4			13.7	14.2	14.65	15.05	15.45	15.95	16.55		[5]
40К44Х19Н10В15			14.5	14.65	14.85	15.25	15.7	16.25	16.9		[5]
40К44Х20Н20М4В4Б4			17.7	17.55	17.55	17.55	17.6	17.65	17.75		[5]
12К51Х20Н10В15			13.65		14.65	15.3	15.85	16.3			[5]
40К51Х24Н16М6			13.9	14.05	14.2	14.4	14.7	15.1	15.5		[5]
40К55Х25Н10В8			14.05	14.4	14.9	15.5	16.35	16.7	16.75		[5]
40К65Х23Н2В6			13.7	14.15	14.55	15.05	15.65	16.3			[5]
25К65Х25Н2М5			14.05	14.25	14.55	14.85	15.4				[5]
125К65Х27В4С3			14.5	14.65	14.85	15.25	15.7	16.25			[5]
	Silicon steels										
30ХС, 37ХС, 40ХС	11.7	12.7	13.4	14.0	14.4	14.5					[5]
15ХСМБ	10.11	11.72	12.17	12.64	13.98	13.65					[5]
15ХСМФ	11.84	12.12	12.43	12.60	12.93	13.13					[5]
СХ6М	11.0				12.0	12.2	12.35				[5]
60С2А	12	12.8		13.5	13.4						[5]
50С2Г	11.2	12.2	12.9	13.4	13.7	14.1	14.4	13.6	13.7	14.5	[5]
	Titanium and chromium-titanium steels										
48ТС-1, 48ТС-2, 48ТС-3	11.9–12.6 (300–573 K)										[5]
15Х25Т		10.0	10.6	10.8	11.3	11.5	11.6	11.6	12.2	12.2	[37]
ХВГ	11.0	12.0	13.0	13.5	14.0	14.5					[5]
Э(Armco)		11.7	12.99	13.71	14.15						[5]
Э1	10.8	12.6	13.8	14.6	14.8	14.3	14.3				[5]
Э5	10.6	11.8	13.0	14.0	14.7	15.3	15.6				[5]
Э14		11.7	12.7	13.7	14.7	15.6	15.4				[5]
Э16	7.0	10.7	13.1	14.3	14.6	14.3	13.2				[5]

Note: Steels are marked by the combinations of figures and letters (Cyrillic).

Carbon steels of ordinary quality are marked by the letter "Ст" and a number that increases with carbon content. The letters "кп" and "сп" at the end of the marking denote rimed and killed steel, respectively. High-grade steels are marked only by the letter У, and figures denoting the carbon content in tenths of percent.

The dopants in heavy-doped steels and alloys are designated by the letters (Н – nickel, Х – chromium, С – silicon, В – tungsten, М – molybdenum, Г – manganese, П – phosphorus, А – nitrogen, Р – boron, Д – copper, Т – titanium, Ю – aluminum, Б – niobium, Ц – zirconium, and Ш – magnesium) followed by the figures indicating the approximate content of a given element in integer percent; if there are no figures, the element content is less that 1 percent. In these steels, the carbon content is given in the hundredths of a percent by the first letter in the marking.

Some heavy-doped steels form the groups marked by a first letter (Cyrillic) in the steel specification: Ж – stainless, Я – chromium-nickel stainless, Р – fast-cutting, Ш – ball-bearing, А – free-cutting, Л – casting, and У – spring steels. The letter А at the end of the steel specification denotes a high-grade steel purified with respect to sulphur and phosphorus (the content of each element is not more than 0.03%) [64].

The steel specifications are listed in the order of dopant content increase.

Table 10.9 Temperature coefficients of linear expansion of nonferrous metals and alloys [The true values of TCLE α (at a given temperature t) or mean TCLE $\bar{\alpha}$ (within the range Δt) are presented.]

Substance	t; Δt, °C	α; $\bar{\alpha}$ 10^{-6} K^{-1}	Substance	t; Δt, °C	α; $\bar{\alpha}$ 10^{-6} K^{-1}
Bronzes [3,5,29,46]					
Бр. А5	20	18.2		20–500	19.0
Бр. А7	20–300	17.8	Бр. ОЦСН 3-7-5-1	20	17.1
Бр. АМц 9-2	20	17.0	Бр. Х0.8	−170	10.5
Бр. АЖ 9-4	−200	10.5		20	16.7
	20	17.1	Бр. Х0.5	−200	11.7
	20–300	19.0		20	17.6
Бр. КН1-3	20	16.1	Бр. ОФ 10-1	−200	12.5
Бр. Мг0.3	25–300	17.6		20	17
Бр. О10	20	18.5		400	22
Бр. ОС 8-12	20	17.1	Бр. ОФ 7-02	20	17.5
Бр. ОС 5-25	20	17.6		400	19
Бр. АЖМц 10-3-1.5	−200	11.9	Бр. ОФ 6.5-0.4	20	17.1
	20	17.0		20–300	19.1
	20–400	20	Бр. ОФ 4-0.25	20–100	17.6
Бр. АЖН 11-6-6	20	14.9		20–400	19.4
Бр. АЖН 10-4-4	−200	11.4	Бр. ОЦ 10-2	0–100	18.3
	20	17.0	Бр. ОЦ 8-4	20	16.6
Бр. Б2	−250	11.5		20–180	18
	−100	15.6	Бр. ОЦ 4-3	20	18
	25	17.0	Бр. ОЦС 6-6-3	20	17.1
	200–300	19.0		300	18.2
Бр. Кд1	20	17.6	Бр. С30	20–260	18.4
Бр. КМц 3-1	−200	11.1	Бр. Alloy XOT	20–300	17
	20	18.5	Бр. Цр 0.4	20–100	16.32
	20–300	18		20–300	17.90
Бр. ОЦС 4-4-25	20	18.2		20–600	19.80
Brasses [3,5,29,46]					
Л56	20–300	18.1	ЛАЖ 60-1-1	20	21.6
Л59	20	21	ЛАН 59-3-2	20	19
Л62	−200	14.1	ЛАНКМц75-2-2.5-0.5-0.5	20	18.3
	20–100	19.0	ЛАЖМц 66-6-3-2		19.8
	20–300	20.6	ЛЖМц 59-1-1	−200	13.0
Л63	−200	14.1		20	22
	20	20.4	ЛК80	20	19.2
	20–300	20.6	ЛК80-3Л	−190	13.8
Л66	25–300	20.1		20	19.1
Л68	−250–20	13.5		20–300	17
	−100–20	17.2	ЛН 65-5	20	18.2
	20–100	18.5	ЛМц 58-2	20	21.2
	20–300	19.9	ЛО 90-1	20–100	18.4
Л70	20–300	19.9	ЛО 70-1	20	19.7
Л75	20–300	19.6		20–300	19.9
Л80	20	18.8	ЛО 62-1	20	19.3
	20–300	19.1	ЛС 74-3	20	17.5
Л85	20–100	17.7		20–300	19.8
	20–300	18.7	ЛС 64-2	20	20.3
Л90	20–100	14.4	ЛС 63-3	20	20.5
	20–300	18.2	ЛС 60-1	20–300	20.8
Л96	20	17	ЛС 59-1	−200	13.9
	20–300	18		−100	18.3
ЛА 85-0.5	20	18.6		20	20.6
ЛА 77-2	20–300	18.5			

Table 10.9 Temperature coefficients of linear expansion of nonferrous metals and alloys *(continued)*

Substance	t; Δt, °C	α; $\bar{\alpha}$ 10^{-6} K^{-1}	Substance	t; Δt, °C	α; $\bar{\alpha}$ 10^{-6} K^{-1}
Magnesium alloys [47]					
ВМ 65-1	20–100	20.9	МЛ2	20–100	26.6
	100–200	22.6		200–300	27.7
ВМ17	20–100	22.67	МЛ4	20–100	26.4
ВМД 3	20–100	25.9		200–300	28.3
	200–300	30.6	МЛ5	20–100	26.8
МА 1	20–100	23.1		200–300	28.7
	200–300	32.1	МЛ6	20–100	26.1
МА 2	20–100	26.0		200–300	27.7
	200–300	29.5	МЛ9	20–100	23.4
МА 3	20–100	26.0		200–300	32.7
	200–300	31.2	МЛ10	20–100	27.4
МА 5	20–100	26.1		200–300	28.3
	200–300	28.5	МЛ11	20–100	21.9
МА 8	20–100	23.7		200–300	24.8
	200–300	32.1	МЛ12	20–100	26.2
МА 9	20–100	25.5		200–300	31.1
	200–300	32.3	МЛ15	20–100	25.9
МА 11	20–100	25.7		200–300	27.9
	200–300	30.1			
Nickel alloys [3,5,29,46]					
МН16 (ТБ)	20	15.3	42Н	−100–20	4.9
МН19 (German silver)	20	16	36Н (Invar)	−100–20	1.10
МН95-5	20	16.4	49НД (Kovar)	−100	10.6
МНЖ 5-1	20	13.7		20	4.50
МНМц 43-0.5	20	14.0	47НД	−269	0.005
	20–600	16.8		−263	0.02
	20–1000	18.8		−253	0.12
МНМц 40-1.5 (Constantan)		14.4		−233	1.20
МНМц 3-12 (Manganin)	100	16		−173	6.52
МНЦ 15-20 (German silver)	20–100	16.6		−73	10.0
НМц5		13.7		20	10.4
НМц2.5		13.4	38НК	−100–20	4.30
НМцАК 2-2-1 (Alumel)	20	13.7		−60–20	3.90
НМЖМц30-1-1 (german silver)	25–300	16	34НК	−100–20	8.00
НМЖМц28-2.5-1.5	20–100	14.2	33НК	−100–20	8.60
(Monel-metal)	25–300	14.9		−70–20	10.0
	20–600	16.1	29НК	−100–20	6.30
НХ9.5 (Chromel T)	0–1000	12.8		−70–20	6.50
НХ9 (Chromel A)	20	12.8	38НКД	−100–20	7.20
	0–1000	12.8	32НКД (Invar)	−60–20	0.90
Н65М28 (Hastelloy)	20–300	11.1	30НКД	−100–20	5.40
	20–500	11.8	37НКДП	20–100	10.9
	20–1000	15.2		20–500	12.2
Н64М28	20–100	10.0	40НКМП	20–100	11.0
	20–1000	14.6		20–500	12.1
78Н	20–100	12.0	35НКТ (Invar)	−100–20	2.3
50Н	20–100	8.9		−60–20	2.2
	20–500	9.4	35НКХСП	20–100	10.8
46Н	−100–20	7.7		20–500	11.8
45Н	20–100	8.2	80НМ	20–100	12.2
	20–400	7.7		20–500	14.5
	20–500	9.0			

Table 10.9 Temperature coefficients of linear expansion of nonferrous metals and alloys *(continued)*

Substance	t; Δt, °C	α; $\bar{\alpha}$ 10^{-6} K^{-1}	Substance	t; Δt, °C	α; $\bar{\alpha}$ 10^{-6} K^{-1}
79НМ	20–100	10.5	36НХ (Invar)	−269	−0.5
	20–500	12.7		−263	−1.3
77НМД	20–100	12.7		−253	−1.6
	20–500	14.9		−223	−0.2
81НМТ	20–100	11.8		−173	1.9
	20–500	13.5		−73	1.9
65НП	20–100	11.8		20	1.9
	20–500	13.2	76НХД	20–100	11.8
				20–500	14.5
Fusible alloys and solders [3,48,49]					
ПМЦ36		22	50Pb50Sn	−173	20
ПМЦ48, ПМЦ54		21		−73	23.5
99Pb1Sb	20	28.8		20	25.5
85Pb15Sb	20	24.8			
Refractory alloys [3,5,20,57]					
АТ2	20	8.72	WC + 5.9%Co	20–100	4.5
ВР20	200	4		20–400	5.2
ВТ3	20	8.55	W + 0.015%Mo+	20–100	4.3
ВТ3-1	20	8.52	+0.005%Cu+	20–500	4.6
ВТ5	20–700	10.5	+0.002%As		
ВТ5-1	20	9.15	60W40Cu	20–100	8.0
ВТ6	20	9.10		20–700	10.2
ВТ8	20	8.6	98.8Zr1.2Hf	20	6.28
	20–300	9.8	Al1	20	13.8
	20–800	10.25	Al2	20	14.1
ВТ14	20	8.48	Cr5	20	14.5
ВТ15	20	8.30	Cr50	20	10.4
ОТ4-1	25	8.0	Cr70	20	8.52
ТВ1	20	8.74	Zr5	20	14.5
ТВ-10	20	6.34	Zr70	20	8.40
40Nb60V	20	8.27	98% Al	40–600	28.1
50Nb25V25Zr	20	7.85	90% Al	40–600	25.7

Note: Bronzes are marked by the letters Бр; the next letters denote only the doping elements, and the figures indicate their percentage. Brasses are marked by the first letter Л and other letters denote the dopant content. The first figure is the copper percentage, the others stand for the dopants percentage. Manganese alloys are designated by letters and the number of an alloy (Мл is for casting manganese alloys, the other notation corresponds to deformable alloys). Copper-nickel alloys are designated by the letters МН; the letters following denote the elements of an alloy; the first number denotes the percentage of nickel and cobalt, and the rest stand for the percentage of other elements in the same order with the letters. Babbitts are coded by the letter Б and the number for a tin percentage. Hard tungsten-cobalt alloys are marked by the letters ВК and the number for a cobalt percentage. For titanium-tungsten alloys, the letter Т and titanium carbide percentage are used. Titanium–tantalum–tungsten alloys are denoted by the letters ТТ and a number which stands for the total content of titanium and tantalum carbides.

The element codes in the designation of nonferrous alloys are as follows: А – aluminum, Ж – iron, Зл – gold, И – iridium, К – cobalt, Кд – cadmium, Мг – magnesium, Мц – manganese, М – copper, Н – nickel, Пд – palladium, Пл – platinum, О – tin, Рд – rhodium, Ре – rhenium, С – lead, Ср – silver, Су – antimony, ТТ – tantalum, Т – titanium, Ф – phosphorus, Х – chromium, Ц – zinc, and Цр – zirconium [64].

Table 10.10 Temperature coefficients of linear expansion of aluminum alloys [The true values of TCLE α (at given temperature t) or mean TCLE $\bar{\alpha}$ (within the range Δt) are presented.]

Specification	t; Δt, °C	α; $\bar{\alpha}$, 10^{-6} K^{-1}
А132	20–100	19.0
	20–200	20.0
	20–300	21.0
АВ	20–100	23.4
	20–200	24.5
	20–300	25.4
АД, АД1	−200	6.10
	−100	18.6
	20–100	23.5
	20–200	24.6
	20–300	25.6
АД33	−200–20	18.0
	−100–20	20.9
	20–100	23.2
	20–300	25.0
АК4	−173.2	16.4
	20–100	22
	20–200	23.1
	20–300	24
АК6	−200	7.60
	−100	16.0
	20–100	21.4
АК7	−200–20	16.6
	−100–20	19.3
АК8	−200	3.02
	−100	17.7
	20–100	22.5
	20–200	23.4
	20–300	24.8
АЛ1	20–100	22.3
	20–200	23.2
	20–300	24.4
АЛ2	−200	7.20
	−100	15.7
	20–100	21.1
	20–200	22.1
	20–300	23.3
АЛ3	20–100	22.0
	20–200	23.2
	20–300	24.0
АЛ4	20–100	21.7
	20–200	22.5
	20–300	23.5
АЛ5	20–100	23.1
	20–200	23.9
	20–300	25.2
АЛ6	20–100	21.5
	20–200	22.5
	20–300	23.5
АЛ7	20–100	23.0
	20–200	24.0
	20–300	25.0
АЛ8	−173.2	18.3
	20–100	24.5
	20–200	25.6
	20–300	23.3
АЛ9	−200	8.60
	−100	18.2
	20–100	23.0
	20–200	24.0
	20–300	24.5
АЛ10В	20–100	22.3
	20–200	23.3
	20–300	25.4
АЛ11	20–100	24.0
	20–200	24.4
	20–300	26.6
АЛ12	20–100	22.0
	20–200	23.0
	20–300	23.5
АЛ13	20–100	20.0
	20–200	24.0
	20–300	27.0
АЛ19	−173.2	16.4
	20–100	19.5
	20–200	21.9
	20–300	25.6
АЛ21	20–100	22.9
	20–200	24.6
	20–300	27.8
АЛ22	20–100	24.5
	20–200	25.6
	20–300	27.3
АЛ25	20–100	19.0
	20–200	20.0
	20–300	21.0
АМг2	20–100	23.8
	20–200	24.5
	20–300	25.4
АМг3	−193.2	9.2
	−100	18.3
	20–100	23.5
	20–300	25.5
АМг4	−200	7.5
	−100	18.5
АМг5	−200	4.82
	−100	18.9
	20–100	23.9
	20–200	24.8
	20–300	25.9
АМц	−193.2	6.05
	−100	16.8
	20–100	24
	20–200	24.8
	20–300	25.9

Table 10.10 Temperature coefficients of linear expansion of aluminum alloys *(continued)*

Specification	t; Δt, °C	α; $\bar{\alpha}$, 10^{-6} K^{-1}	Specification	t; Δt, °C	α; $\bar{\alpha}$, 10^{-6} K^{-1}
АМцС	−200	7.01		20–300	24.8
	−100	17.7	Д16	−200	14.6
	27	22.6		−100	17.0
В92	−200	16.5		20–100	22.7
	−100	21.4		20–200	23.4
В93	20–100	24.1		20–300	24.8
В94	20–100	21.9	Д18, Д18П	−200	10.0
В95	−200	4.23		−100	18.9
	−100	18.4		20–100	22.7
	20–100	23.2		20–200	23.4
	20–200	24.3		20–300	24.8
	20–300	25.9	Д20	−200	7.10
В65	−193.2	19.9		−100	18.5
	−100	22.3	АМ8	20–100	22.8
В14А	20–100	18.1	АЖ6	20–100	23.0
В300	20–100	22.9	АМК2	20–100	25.7
	20–300	27.8	АСС-5	20–100	23.9
Ви11-3	20–100	25.5	АСМ	20–100	24.0
	20–200	25.6	АН-2.5	20–100	25.0
	20–300	27.3	САС1	−223	6.05
ВАД-1	−200	16.4		−73	11.8
	−100	19.3		20	12.9
ВД17	−200	12.0	Alloy 1915	−200	7.52
	−100	18.3		−100	8.2
Д1	−200–20	21.8		27	22.7
	−100–20	21.9	Alloy 1201	−200	5.7
	20–100	22.9		−100	17.7
	20–200	23.4		7	22.1

Note: Aluminum alloys are marked by Cyrillic letters and the number of an alloy (АЛ is for casting aluminum alloys, the other notation corresponds to deformable alloys).

Table 10.11 Metallic alloys for brazing with glasses and ceramics [63] [The maximum values of mean TCLE α in a given temperature range ΔT are presented.]

Alloy	ΔT, K	α, 10^{-6} K^{-1}	Use of alloy
36Н	213–373	1.5	Parts with constant size
32НКД	213–373	1.0	High-precision and constant size parts
36НХ	15–273	1.0–2.0	Constructions and pipelines for cryogenic temperatures
	273–373	1.0–2.0	
39Н	15–293	4.0	Constructions and pipelines for cryogenic temperatures
35НКТ	213–293	3.5	For working in raised load conditions
	293–333	3.5	
32НК-ВИ	213–293	1.5	For parts with polished surfaces and low form rigidity
	293–373	1.5	
29НК	203–693	4.5–6.5	Vacuum-tight soldering with the glasses C49-1, C52-1, C48-1
30НКД	213–673	3.3–4.6	Soldering with the refractory glass C38-1 and glass C48-1
38НКД	213–673	7–7.8	Soldering with sapphire and the glass C72-4
47НХ	203–723	8.9	Soldering with the thermal glasses C72-4 and others
48НХ	203–723	8.5–9.5	
33НК	203–723	6–9	Joint with ceramics, mica, and the glass C72-4
42Н, 42НА	203–613	4.5–5.5	Electrovacuum technology
34НК	293–823	5.2–7.0	Soldering with thin films of the glass C84-2 and ceramics

Table 10.11 Metallic alloys for brazing with glasses and ceramics *(continued)*

Alloy	ΔT, K	α, 10^{-6} K^{-1}	Use of alloy
47Н3Х	203–713	9.5–10.5	Soldering with thin glass films
47НД	203–713	9–11	Joint with the glasses С93-2, С93-4, ceramics, mica, and the springs of hermetic contacts
47НХР	203–603	8.5–11	Soldering with the glasses С90-1, С93-2, С94-1, С95-2, and others
18ХТФ, 18ХМТФ	203–823	11–11.4	Soldering with the glasses С90-1, С93-4, С95-2, etc.
52Н	203–773	11–11.5	Soldering with the glasses С90-1, С93-2, С94-1, С95-2; the alloy with high permeability and the saturation induction of 1.5 T
58Н-ВИ	293–393	$11.5^{+0.3}$	For length standards and precision rulers; the alloy of high stability of size

Table 10.12 Temperature coefficients of linear expansion of plastics [True values of TCLE α (at a given temperature t) or mean TCLE $\bar{\alpha}$ (within the range Δt) are presented.]

Name	t; Δt, °C	α; $\bar{\alpha}$, 10^{-5} K^{-1}	Ref.
Aminoplastics		2.5–5.3	[53]
Aniline-formaldehyde resin polimers		5–6	[29]
Anyde		10	[29]
Bakelite (fenolic resin plastic)	−200–20	4.5	[3]
	−70–20	6.7	[3]
	0	7.6	[3]
Diphlon		6	[53]
Epoxy adhesive, К-9		3.5–4.0	[53]
	−70–20	8.01	[3]
	−10–20	8.75	[3]
ЭПК-1	−200–20	5.16	[3]
	−170–20	5.64	[3]
	−70–20	7.05	[3]
	20	7.74	[3]
К-63А	−170	3.52	[3]
	−70	4.92	[3]
	20	6.0	[3]
Epoxy material КЭП	20–150	2.6	[53]
	150–200	5.6	[53]
Epoxy organotitanosilicon polymer filler glass ceramic	−200–20	1.4–4.5	[3]
Epoxy resin		5.5–6.5	[29]
Foam plastic		2.7–7.9	[56]
Foam plastic ПВ	30	3.9	[56]
ПС	20	5.2–8.4	[56]
ПС-1	30	5.05	[56]
ПС-4	30	6.2	[56]
ПСБ	30	5.5	[56]
ПСВ		5.5–6.8	[56]
Foam plastic polychlorovinyl ПХВ-1	30	4.6	[5]
Foam plastic polyurethane ПУ	20	5.0	[56]
ПУ-101	40	6.6	[56]
Glass cloth			
along thread	25–100	5.0	[48]
across thread	25–100	8.5	[48]
Glass-cloth-base laminate			
ВФТ-С	20–100	0.8	[48]
КАСТ-В	20–100	0.9	[48]
	−173	1.25	[3]
	−73	1.8	[3]
	0	2.5	[3]
Glass fibre	20–80	0.9–1.2	[53]
	80–160	0.4–0.8	[53]
Glass-fibre-base laminate В, pressed	25–200	1.24	[48]
Glass-reinforced plastics		0.5–1.2	[53]
Kaprolon		6.6–9.8	[53]
Kapron А,Б,В		12–14	[29]
Kapron glas-filler		10–12	[29]
Lavsan		2.6–2.7	[29]
Melamine-formaldehyde polymers		4.0	[29]
Metallic-polymer for mould-casting	27	3.0	[3]
Moulding material Nyplon		5.0	[53]
Phenylon		3.1–3.5	[53]
Phenoplast base		1–2.5	[53]
Nylon	−263–20	4.3	[3]
	−233–20	4.6	[3]
	−173–20	5.2	[3]
	−73–20	5.4	[3]
	−7–20	4.0	[3]
Organosilicon polymers		0.5–2.0	[29]
Penton		7.8–8.0	[29]
Phenolformaldehyde polymers		2.5–6.0	[29]
Phenolformaldehyde polymers withstood impact		3.0–3.5	[53]
Phenolformaldehyde adhesive ВФТ-С	−210–20	3.98	[3]
	−173–20	4.35	[3]
	−73–20	5.21	[3]
	20	5.97	[3]

Table 10.12 Temperature coefficients of linear expansion of plastics *(continued)*

Name	t; Δt, °C	α; $\bar{\alpha}$, 10^{-5} K^{-1}	Ref.
Polyamide-6		8.2–9.7	[55]
-12		9.6–10	[53]
-66		9.9	[55]
-68	20	6.0	[3]
-68 with graphit		10–20	[29]
-68 BC		10–12	[29]
-68 T-40		4.5–4.8	[29]
-68 T-60	20	3–3.5	[29]
-548, -54		13	[29]
-H, -C		12	[29]
-AK7		10–11	[29]
Polyarylate		6	[53]
Polybatylenterephthalate		13	[54]
Polycarbon		6–7	[55]
Polychlorostyrene		7.0	[29]
Polydimethylstyrene		7.9	[29]
Polyethylene:			
ВД	0–100	22–55	[29]
НД	0–100	11–50	[29]
СД		10–55	[29]
average means for	−263–20	7.2	[29]
different density	−233–20	8.0	[29]
	−173–20	9.96	[29]
	−73–20	18.3	[29]
	20	34.0	[29]
cable		20	[29]
glass tiller, pressed:			
10% glass fibre	20	7.0	[3]
20% glass fibre	20	5.2	[3]
40%glass fibre	20	3.0	[3]
Polyethylenterephthalate	−100	5.0	[3]
	0	6.6	[3]
Polymethylacrylate		8	[29]
Polymethylmethacrylate:			
acrylic plastic ПА-200	−200	2.7	[3]
	−50	5.7	[3]
	0	7.7	[3]
	20	8.8	[3]
acrylic plastic СОЛ	20	7.1	[53]
	80	12.5	[53]
acrylic plastic СТ-1	20	7.7	[53]
	100	11.1	[53]
acrylic plastic 2-55	20	7.3	[53]
Polymethylpentene		1.2	[55]
Polymethylphenylsiloxane	−200	5.4	[3]
ПМФС-1	−100	6.7	[3]
	20	10.0	[3]
ПМФС-2	−200	7.2	[3]
	−100	8.5	[3]
	20	14.0	[3]
Polymethylsiloxane	−200	6.8	[3]
ПМС	−100	8.5	[3]
	20	10.0	[3]
Polyoxymethylene		8.1	[29]
Polypropylene	0–100	11	[29]
glass-filler		1.9	[29]
Polystyrene		6–8	[29]
МС		7–8	[29]
МСН		6–9	[54]
СА		7.4	[29]
САМП		7.5	[29]
СВ		8.5	[29]
lighting engineering		6–7	[53]
СН		8.6–9.5	[29]
withstood impact СНП-2		8.6	[29]
withstood impact ПС-СУ2		7.0	[29]
withstoodn impact УП-1Э		7.0	[29]
plastic АСБ		8–10	[29]
Polysulphone		5.6	[55]
Polytetrafiuoroethylene	−263–20	7.1	[3]
Fluoropolymer-4	−233–20	7.7	[3]
($\rho = 2.3$ kg·m^{-3})			
	−173	3.6	[3]
	0	11.5	[3]
Fluoropolymer-4	120–200	15	[48]
standard (Teflon)	210–280	21	[48]
Fluoropolymer-4M		9	[29]
Fluoropolymer-4O		6.2–9	[29]
Fluoropolymer-42		9.7–26	[29]
Fluoropolymer-4ОП	0	11.3	[3]
Polytrichlorofluoroethylene	−263–20	3.96	[3]
(fluoropolymer-3)	−233–20	4.26	[3]
	−173–20	4.74	[3]
	−73–20	5.29	[3]
	20	5.57	[3]
	120	10	[3]
Polyurethane ПУ-1		13.5	[29]
Polyuphenoloxide		1.6–3.4	[29]
Polyuphenylsiloxane ПФС	−200	4.5	[3]
	−100	5.7	[3]
	20	7.0	[3]
Polyuformaldehyde		7.9–8.1	[55]
Polyuchlorostyrene		7.4	[29]
Polyvinyl butyl ether	−180	9.0	[3]
	−80	13.0	[3]
	0–20	22.0	[3]
Polyvinyl *sec*-butyl ether	−180	6.6	[3]
	−80	10.7	[3]
	20	21.4	[3]
	−80	19.5	[3]
	−60	29.5	[3]
Polyvinyl alcohol		7–12	[29]
Polyvinyl chloride		6–9	[53]
(vinyl plastic, PVS)			
sheet		6.7	[29]
ВМЛ-25	(−50) - (−10)	2.8	[53]
	10–30	3.9	[53]

Table 10.12 Temperature coefficients of linear expansion of plastics *(continued)*

Name	t; Δt, °C	α; $\bar{\alpha}$, 10^{-5} K^{-1}	Ref.	Name	t; Δt, °C	α; $\bar{\alpha}$, 10^{-5} K^{-1}	Ref.
Polyvinyl decyl ether	−180	8.2	[3]	Polyvinyl phthalate		5.4	[29]
	−80	15.8	[3]	Polyvinylbutylene		9.2	[29]
		22.6	[3]	Polyvinylbutylphthalate		13	[29]
Polyvinyl ethyl ether	−180	8.1	[3]	Polyvinylcarbasole		4.0	[29]
	−80	9.5	[3]		−80	6.9	[3]
	0	24.0	[3]		20	22.4	[3]
Polyvinyl fluoride		5	[54]		−80	12.5	[3]
Polyvinyl hexyl ether	−180	6.5	[3]	Polyvinylidene fluoride		8–12	[54]
Polyvinyl isobutyl ether	−180	7.0	[3]	Polyvinyltoluene		7.5	[29]
	−80	11.4	[3]	Premix		3.2–4.0	[53]
	20	21.0	[3]	Shellac	< 46	9	[53]
Polyvinyl isopropyl ether	−160	8.8	[3]		> 46	4.4	[53]
	−80	9.0	[3]	Textile laminate sheet A-50	−233	0.55	[3]
	20	22.1	[3]	Urea-formaldehyde polymers		2.7	[29]
Polyvinyl methyl ether	−180	5.2	[3]	"Volocknit"		−3–3.5	[29]
Polyvinyl octyl ether	−180	7.1	[3]				

Table 10.13 Temperature coefficient of linear expansion α (in 10^{-6} K^{-1}) of construction materials at 20°C [3,4,16]

Substance	α	Substance	α
Aerated concrete	6.2–7.4	Ebonite (hard rubber) expanded	~60
Alumina cement stone	6.7	Expanded clay aggregate	
Ash concrete	9.1	К-1	6.1
Asphalt	200	К-2	4.2
Bakelite (phenolic resin plastic)	28	Expanded-clay lightweight concrete	5.5–6.9
Bitumen БНД 90/130 (−30° C)	310	Fasing file ceramics	5.9
Brick-work	4.0	Fibre	9.0
Clay	8.1	Foam-glass	10.8
Coke		Granite	
calcinated КНПС	2.5–3	gray middle-grained	8.3
uncalcinate КНПС	15–19	red small-grained	7.1
semi-coke	19	red coarse-grained	5.2
Concrete		Laminated compressed wood plastic (ДСП)	3–30
heavy-weight concrete	10–14	Limestone	
concrete granite	9.5	С-3	6.0
basalt	8.6	С-4	5.4
limestone	6.8	К-1	7.4
expanded clay aggregate	8.8–9.5	Я-7	6.5
slag	7.8–9.5	Marble	15
pearlite	9.5	Mortar	
vermiculite	14.1	sand-cement №1,2,3	10.4
diatomite	11.3	expanded-clay, vol. comp. 1:2.5	8.1
Corrugated asbestos board	20	pearlite, vol. comp. 1:2.5	9.0
Diabase		Mortar portland's cement sand-base,	
small-grained	7.05	comp. mass	
coarse-grained	6.6	1:1	11–13.3
Dolomite		1:2	10.1
С-1	9.3	1:3	11.2
С-2	4.8	1:6	9.2–10.4
С-21	3.4	quartz-sand base	9.5–11.2
Б-10	9.3	dolomite	11.4
Б-12	9.1	limestone, comp.mass 1:2	9.5

Table 10.13 TCLE α (in 10^{-6} K^{-1}) of construction materials at 20°C *(continued)*

Substance	α	Substance	α
Paper hard	10	Sandstone	
Pearlite concrete	6.8	П-1	10.4
Pearlite expanded	5.6	П-2	10.2
Pith		Slad concrete	9.15
coal-tar	36–45	Tuf	7.6
petrolium	45	Wood	
Pumice (stone)		beech	2.57
natural	5.6	oak	4.92
slag	8.1–9.7	pine	5.41
		ash	9.5

10.3 Temperature Coefficients of Volume Expansion of Liquids and Gases

Table 10.14 Temperature coefficients of volume expansion of gases and liquids at normal pressure [16] [True values of coefficients of volume expansion β (at a given temperature T) or mean coefficients of volume expansion $\bar{\beta}$ (within the range ΔT) are presented (l – liquid, and g – gas).]

Substance	State	$T; \Delta T$, K	$\beta, \bar{\beta}$, 10^{-3} K^{-1}	Substance	State	$T; \Delta T$, K	$\beta; \bar{\beta}$, 10^{-3} K^{-1}
Elements and inorganic compounds							
Aluminum	l	935–1373	0.113	Hydrogen chloride	g	273	3.769
Ammonia	l	223	1.93	Hydrogen cyanide	l	259	1.95
	g	373	3.808	Iodine	l	380–423	0.80
Argon	l	84–90	4.5	Krypton	g	373	3.311
	g	373	3.676	Lead	l	601–1098	0.120
Bismuth	l	544–903	0.122	Lithium	l	458–508	0.174
Bromine	l	293	1.132	Magnesium	l	923–1073	0.380
Cadmium	l	595–817	0.137	Mercury	l	323.15	0.18190
Carbon dioxide	g	223–273	4.95		l	373.15	0.18245
	g	273–373	3.723		l	573.15	0.18677
Carbon disulfide	l	293	1.218		l	1073	0.188
Carbon oxide	l	68–81	4.91	Neon	g	373	3.659
	g	273–373	3.669	Nitrogen	l	68–89	5.88
Cesium	l	303–325	0.341		g	373	3.670
Chlorine	l	172–239.4	1.41	Oxygen	l	68–89	3.85
Cyanogen	g	273	3.96	Ozone	l	90	2.0
Fluorine	l	55–85	3.0	Phosphorus	l	323–333	0.52
Gallium	l	373	0.121	Phosphorus tribromide	l	293	0.868
	l	473	0.1143	Phosphorus trichloride	l	293	1.154
	l	673	0.1066	Phosphorus oxychloride	l	293	1.116
	l	1173	0.0973	Potassium	l	373	0.280
Helium	l	4.2	137	Rubidium	l	313–413	0.339
	g	373	3.658	Silicon tetrachloride	l	293	1.430
Helium-3	l	3.2	580	Silver	l	1236–1373	0.111
Hydrogen	l	14–20.39	12.6	Sodium	l	373–453	0.275
	g	373	3.659		l	373–973	0.390

Table 10.14 Temperature coefficients of volume expansion of gases and liquids *(continued)*

Substance	State	$T; \Delta T$, K	$\beta; \bar{\beta}$, 10^{-3} K^{-1}	Substance	State	$T; \Delta T$, K	$\beta; \bar{\beta}$, 10^{-3} K^{-1}
Sulfur	l	388	0.430	Tin	l	505–673	0.106
	l	425	0.490		l	505–1873	0.100
	l	438	0.170	Water	l	273	-0.064
	l	483	0.344	(at the saturation	l	283	0.070
	l	630	0.388	line)	l	293	0.182
Sulfur dioxide	l	223	1.70		l	323	0.449
Sulfuric acid	l	293	0.576		l	373	0.752
Thallium	l	573–626	0.150	Zinc	l	693–816	0.147
Organic compounds							
Acethylene (ethyne)	g	273–373	3.739	Ethyl alcohol	l	293	1.120
Acetic acid	l	293	1.071	benzene	l	293	0.961
Acetonitrile	l	293	1.301	benzoate	l	293	0.900
Allyl alcohol	l	293	1.049	bromide	l	293	1.418
bromide	l	293	1.241	chloride	l	293	1.709
chloride	l	293	1.475	(diethyl) ether	l	293	1.656
iodide	l	293	1.091	formate	l	293	1.417
Amyl acetate	l	293	1.162	iodide	l	293	1.179
alcohol	l	293	0.902	Ethylene glycol	l	293	0.6375
benzoate	l	293	0.848	Formatic acid	l	293	1.025
bromide	l	293	1.102	Glycyl alcohol	l	293	0.505
chloride	l	293	1.208	Isobutyric acid	l	293	1.068
iodide	l	293	0.986	Isohexane	l	293	1.445
Aniline	l	293	0.858	Isopentane	l	293	1.680
Benzene (benzol)	l	293	1.237	Isoprene	l	293	1.570
Benzoyl chloride	l	293	0.880	Isopropyl alcohol	l	293	1.094
Butane	l	253–303	2.0	Methyl acetate	l	293	1.427
Butyl alcohol	l	293	0.950	bromide	l	293	1.684
Butyric acid	l	293	1.063	iodide	l	293	1.273
Capronic acid	l	293	0.975	alcohol	l	293	1.199
m-Xylene	l	293	0.99	formate	l	293	1.563
o-Xylene	l	293	0.97	ethyl ketone	l	293	1.315
p-Xylene	l	293	1.02	Neopentane	l	293	1.6
Chloral	l	293	0.934	Oleic acid	l	293	0.721
Chloroform	l	293	1.273	Pentane	l	293	1.605
Diallyl	l	293	1.357	Phenol	l	293	1.090
Diallyl ether	l	293	1.346	Propane	l	205–243	1.9
Diethyl ketone	l	293	1.233	Propyl bromide	l	293	1.447
oxalate	l	293	1.136	iodide	l	293	1.102
sulfide	l	293	1.278	chloride	l	293	1.591
Diisopropyl ether	l	293	1.452	alcohol	l	293	0.956
Dimethyl sulfide	l	293	1.082	Propionic acid	l	293	1.102
Dipropyl ether	l	293	1.354	Toluene	l	323	1.138
Ethane	l	143–200	2.4				
Mixtures							
Air	g	273–373	3.670	Oil machine MC-20	l	303	0.638
Gasoline Б-70	l	313	1.496	MK-20	l	323	0.886
	l	423	1.752	transformer	l	273	0.680
Kerosene T-1	l	293	0.955		l	393	0.740
				Petroleum ether	l	293	2.26

References

[1] Novikova, S. I., Thermal expansion of solids, Nauka, Moscow, 1974 (in Russian).

[2] Properties of elements. Handbook, Ed. Samsonov, G. V., 2nd ed., Metallurgia, Moscow, 1976 (in Russian).

[3] Kozhevnikov, I. G., Novitskii, L. A., Thermal properties of materials at low temperatures. Handbook, 2nd ed., Mashinostroeniye, Moscow, 1982 (in Russian).

[4] Novitskii, L. A., Kozhevnikov, I. G., Thermal properties of materials at low temperatures. Handbook, Mashinostroeniye, Moscow, 1975 (in Russian).

[5] Physical properties of steels and alloys used in power engineering. Handbook, Ed. Neimark, B. E., Energia, Moscow, 1967 (in Russian).

[6] Petrenko, N. S., Popov, V. P., Puskarev, E. A., Finkel, V. A., Phys. Stat. Solidi (b), 68, K145, 1975.

[7] Geshko, E. I., Mikhal'chenko, V. P., Sharlai, B. M., Fizika tverdogo tela, 14, 1803, 1972.

[8] Shouten, D. R., Swenson, S. A., Phys. Rev. B10, 2175, 1974.

[9] Properties and applications of metals and alloys for electrical vacuum devices. Guidebook, Ed. Nilender, R. A., Energia, Moscow, 1973.

[10] Petrenko, N. S., Popov, V. P., Fizika nizkikh temperatur, 5, 301, 1979.

[11] Sirota, N. N., Zhabko, T. E., Doklady Acad. Nauk SSSR, 236, 1120, 1977.

[12] Grosse, R., J. Phys., C11, 45, 1978.

[13] Handbook of Optics, Ed. Driscoll, W. G., New York, 1978.

[14] White, G. K., High Temp. High Press., 11, 471, 1979.

[15] Finkel', V. A., Smirnov, Yu. I., Vorobiev, V. V., Zhurnal eksperim. i teoreticheskoi fiziki, 51, 32, 1966.

[16] Tables of physical quantities. Handbook, Ed. Kikoin, I. K., Atomizdat, Moscow, 1976 (in Russian).

[17] Voronkova, E. M., Grechushnikov, B. I., Distler, G. I., Petrov, I. P., Optical materials for infra-red equipment, Nauka, Moscow, 1965 (in Russian).

[18] Bland, J. A., Canad. J. Phys., 37, 417, 1959.

[19] Barnes, J. O., Ragene, J. A., Phys. Lett., 46A, 317, 1974.

[20] Ryabtsev, N. G., Materials for quantum electronics, Sovetskoye radio, Moscow, 1972 (in Russian).

[21] Materials used in semiconductor devices, Ed. Hogarth, C.A., Interscience, John Wiley & Sons, New York, 1965.

[22] Materials for optoelectronics (collection of papers), Mir, Moscow, 1976 (in Russian).

[23] Rapp, J. E., Merchaut, H. D., J. Appl. Phys., 44, 3919, 1973.

[24] Barium titanate. Collection of articles, Ed. Belov, N. V., Nauka, Moscow, 1973.

[25] Sirota, N. N., Chemical bond in semiconductors and thermal dynamics, Nauka i tekhnika, Minsk, 1966.

[26] Boiko, A. A., Golovin, V. A., Kristallografiya, 15, 186,1970.

[27] Krzhizhanovskii, R. E., Shtern, Z. Yu., Thermal properties of nonmetallic materials. (Carbides), Energia, Moscow, 1977 (in Russian).

[28] Mazony, Y., Perkins, H. K., J. Appl. Phys., 41, 5130, 1970.

[29] A handbook on electrotechnical materials, Eds. Koritskii, Yu. V., Pasynkov, V. V., Tareev, B. M., 2nd ed., Energia, Leningrad, 1976.

[30] Shchelkotunov, V. A., Danilov, V. I., Kalacheva, V. S., Izvestiya Acad. Nauk SSSR, ser. "Inorganic materials," 12, 1076, 1976.

[31] Physical and chemical properties of oxides. Handbook, Ed. Samsonov, G. V., 2nd ed., Metallurgia, Moscow, 1978 (in Russian).

[32] High-temperature inorganic compounds, Naukova dumka, Kiev, 1965 (in Russian).

[33] Maslennikov, S. B., Heat-resistant steels and alloys. Handbook, Metallurgia, Moscow, 1983 (in Russian).

[34] Samsonov, G. V., Vinitskii, I. M., Refractory compounds. Handbook, 2nd ed., Metallurgia, Moscow, 1976 (in Russian).

[35] Mizutani, T., Jap. J. Appl. Phys., 15, 1305, 1976.

[36] Kuz'minov, Yu. S., Ferroelectric crystals for laser radiation control, Nauka, Moscow, 1982 (in Russian).

[37] Krzhizhanovskii, R. E., Shtern, Z. Yu., Thermal properties of nonmetallic materials. (Oxides). Reference book, Energia, Leningrad, 1973 (in Russian).

[38] Zhuravlev, N. N., Kristallografiya, 6, 791, 1961.

[39] Antonova, M. M., Properties of metal hydrides. Handbook, Naukova dumka, Kiev, 1965 (in Russian).

[40] Cook, W. R., Berlincourt, D. A., Scholz, F. J., J. Appl. Phys., 34, 1392, 1963.

[41] Materials for nuclear reactor, Ed. Sokurski, Yu.N., Gosatomizdat, Moscow, 1963 (in Russian).

[42] Glass. Handbook, Ed. Pavlushkin, N. M., Stroiizdat, Moscow, 1973 (in Russian).

[43] A handbook for an optics technologist, Eds. Kuznetsov, S. I., Okatov, M. A., Mashinostroeniye, Moscow, 1983 (in Russian).

[44] A handbook for a designer of optomechanical devices, Ed. Panova, V. A., Mashinostroeniye, Leningrad, 1980 (in Russian).

[45] Properties and development of new optical glasses, Ed. Tsarevskii, E. I., Mashinostroeniye, Leningrad, 1977 (in Russian).

[46] Solntsev, Yu. N., Stepanov, G. A., Materials in cryogenic technology. Handbook, Mashinostroeniye, Leningrad, 1982 (in Russian).

[47] Smiryagin, A. P., Smiryagina, M. A., Belova, A. V., Commercial nonferrous metals and alloys, 3rd ed., Metllurgiya, Moscow, 1974 (in Russian).

[48] Anur'iev, V. I., A handbook of machine-building designer, 6th ed., Mashinostroeniye, Moscow, 1982 (in Russian).

[49] Aref'iev, Yu. M., Thermal properties of substances and materials, 15, p. 115, Izdatel'stvo standartov, Moscow, 1980 (in Russian).

[50] Pridantsev, M. V., Davydova, L. N., Tamarina, I. A., Construction steels. Handbook, Metallurgia, Moscow, 1980 (in Russian).

[51] Ul'yanin, E. A., Sorokina, N. A., Steels and alloys in cryogenic technology. Handbook, Metallurgia, Moscow, 1984 (in Russian).

[52] Technical encyclopedia, A Handbook of physical, chemical and technological quantities, OGIZ, Moscow, 1930, vol. 1 (in Russian).

[53] Katsnelson, N. Yu., Balaev, G. A., Plastics. Properties and applications. Handbook, 3rd. ed., Khimia, Leningrad, 1978 (in Russian).

[54] Kalinchev, E. L., Sakovtseva, M. B., Properties and processing of thermal plastics. Handbook, Khimia, Leningrad, 1983 (in Russian).

[55] Thermal plastics for structural application, Ed. Trostyanskii, E. B., Khimia, Leningrad, 1975 (in Russian).

[56] Romanenkov, I. G., Physical and mechanical properties of foamy plastics, Izdatel'stvo standartov, Moscow, 1970 (in Russian).

[57] Proc.of the 1st All-Union Conference on thermal properties of substances at low temperature, February 1971, Izdatel'stvo BNIIFTRI, Moscow, 1972 (in Russian).

[58] Tables of standard reference data. GSSSD 56–83. Izdatel'stvo standartov, Moscow, 1984 (in Russian).

[59] Tables of standard reference data. GSSSD 3–77. Izdatel'stvo standartov, Moscow, 1978 (in Russian).

[60] Tables of standard reference data. GSSSD 59–83. Izdatel'stvo standartov, Moscow, 1984 (in Russian).

[61] Tables of standard reference data. GSSSD 45–83, Izdatel'stvo standartov, Moscow, 1984 (in Russian).

[62] Petukhov, V. A., Chekhovskoi, V. Ya., Andrianova, V. G., High Temp. High Press., 11, 625, 1979.

[63] Zhuravlev, V.N., Nikolaeva, O.A., Mashine steels, Handbook, 3rd ed., Mashinostroenie, Moscow, 1981 (in Russian).

[64] Technical handbook, 4th ed., Ed. Skorokhodov, E.A., Mashinostroenie, Moscow, 1990 (in Russian).

11

Pressure of Saturated Vapors

A.P. Babichev, V.V. Ignat'ev, V.A. Krivoruchko, and A.I. Migachev

11.1 Water Vapor

Table 11.1 Water vapor pressure [1]

t, °C	p, Pa	t, °C	p, Pa	t, °C	p, Pa	t, °C	p, Pa	t, °C	p, Pa
0	$6.108 \cdot 10^2$	64	$2.3910 \cdot 10^4$	130	$2.7012 \cdot 10^5$	196	$1.4291 \cdot 10^6$	262	$4.8484 \cdot 10^6$
0.01	$6.112 \cdot 10^2$	66	$2.6148 \cdot 10^4$	132	$2.8668 \cdot 10^5$	198	$1.4910 \cdot 10^6$	264	$5.0066 \cdot 10^6$
2	$7.054 \cdot 10^2$	68	$2.8561 \cdot 10^4$	134	$3.0406 \cdot 10^5$	200	$1.5551 \cdot 10^6$	266	$5.1688 \cdot 10^6$
4	$8.129 \cdot 10^2$	70	$3.1161 \cdot 10^4$	136	$3.2227 \cdot 10^5$	202	$1.6212 \cdot 10^6$	268	$5.3349 \cdot 10^6$
6	$9.346 \cdot 10^2$	72	$3.3957 \cdot 10^4$	138	$3.4137 \cdot 10^5$	204	$1.6895 \cdot 10^6$	270	$5.5051 \cdot 10^6$
8	$1.0721 \cdot 10^3$	74	$3.6963 \cdot 10^4$	140	$3.6136 \cdot 10^5$	206	$1.7601 \cdot 10^6$	272	$5.6794 \cdot 10^6$
10	$1.2271 \cdot 10^3$	76	$4.0190 \cdot 10^4$	142	$3.8228 \cdot 10^5$	208	$1.8329 \cdot 10^6$	274	$5.8579 \cdot 10^6$
12	$1.4015 \cdot 10^3$	78	$4.3650 \cdot 10^4$	144	$4.0418 \cdot 10^5$	210	$1.9079 \cdot 10^6$	276	$6.0406 \cdot 10^6$
14	$1.5974 \cdot 10^3$	80	$4.7359 \cdot 10^4$	146	$4.2707 \cdot 10^5$	212	$1.9855 \cdot 10^6$	278	$6.2277 \cdot 10^6$
16	$1.8170 \cdot 10^3$	82	$5.1328 \cdot 10^4$	148	$4.5099 \cdot 10^5$	214	$2.0654 \cdot 10^6$	280	$6.4191 \cdot 10^6$
18	$2.0626 \cdot 10^3$	84	$5.5572 \cdot 10^4$	150	$4.7597 \cdot 10^5$	216	$2.1478 \cdot 10^6$	282	$6.6150 \cdot 10^6$
20	$2.3368 \cdot 10^3$	86	$6.0107 \cdot 10^4$	152	$5.0205 \cdot 10^5$	218	$2.2327 \cdot 10^6$	284	$6.8155 \cdot 10^6$
22	$2.6424 \cdot 10^3$	88	$6.4947 \cdot 10^4$	154	$5.2926 \cdot 10^5$	220	$2.3201 \cdot 10^6$	286	$7.0206 \cdot 10^6$
24	$2.9824 \cdot 10^3$	90	$7.0108 \cdot 10^4$	156	$5.5764 \cdot 10^5$	222	$2.4102 \cdot 10^6$	288	$7.2303 \cdot 10^6$
26	$3.3600 \cdot 10^3$	92	$7.5607 \cdot 10^4$	158	$5.8722 \cdot 10^5$	224	$2.5030 \cdot 10^6$	290	$7.4448 \cdot 10^6$
28	$3.7785 \cdot 10^3$	94	$8.1460 \cdot 10^4$	160	$6.1804 \cdot 10^5$	226	$2.5985 \cdot 10^6$	292	$7.6642 \cdot 10^6$
30	$4.2417 \cdot 10^3$	96	$8.7685 \cdot 10^4$	162	$6.5014 \cdot 10^5$	228	$2.6968 \cdot 10^6$	294	$7.8885 \cdot 10^6$
32	$4.7536 \cdot 10^3$	98	$9.4301 \cdot 10^4$	164	$6.8355 \cdot 10^5$	230	$2.7979 \cdot 10^6$	296	$8.1178 \cdot 10^6$
34	$5.3182 \cdot 10^3$	100	$1.01325 \cdot 10^5$	166	$7.1830 \cdot 10^5$	232	$2.9019 \cdot 10^6$	298	$8.3521 \cdot 10^6$
36	$5.9401 \cdot 10^3$	102	$1.0878 \cdot 10^5$	168	$7.5445 \cdot 10^5$	234	$3.0089 \cdot 10^6$	300	$8.5917 \cdot 10^6$
38	$6.6240 \cdot 10^3$	104	$1.1668 \cdot 10^5$	170	$7.9202 \cdot 10^5$	236	$3.1189 \cdot 10^6$	302	$8.8364 \cdot 10^6$
40	$7.3749 \cdot 10^3$	106	$1.2504 \cdot 10^5$	172	$8.3106 \cdot 10^5$	238	$3.2319 \cdot 10^6$	304	$9.0865 \cdot 10^6$
42	$8.1983 \cdot 10^3$	108	$1.3390 \cdot 10^5$	174	$8.7161 \cdot 10^5$	240	$3.3480 \cdot 10^6$	306	$9.3420 \cdot 10^6$
44	$9.0998 \cdot 10^3$	110	$1.4326 \cdot 10^5$	176	$9.1370 \cdot 10^5$	242	$3.4674 \cdot 10^6$	308	$9.6031 \cdot 10^6$
46	$1.0085 \cdot 10^4$	112	$1.5316 \cdot 10^5$	178	$9.5739 \cdot 10^5$	244	$3.5899 \cdot 10^6$	310	$9.8697 \cdot 10^6$
48	$1.1161 \cdot 10^4$	114	$1.6361 \cdot 10^5$	180	$1.0027 \cdot 10^6$	246	$3.7158 \cdot 10^6$	312	$1.0142 \cdot 10^7$
50	$1.2335 \cdot 10^4$	116	$1.7464 \cdot 10^5$	182	$1.0497 \cdot 10^6$	248	$3.8450 \cdot 10^6$	314	$1.0420 \cdot 10^7$
52	$1.3612 \cdot 10^4$	118	$1.8628 \cdot 10^5$	184	$1.0984 \cdot 10^6$	250	$3.9776 \cdot 10^6$	316	$1.0704 \cdot 10^7$
54	$1.5001 \cdot 10^4$	120	$1.9854 \cdot 10^5$	186	$1.1488 \cdot 10^6$	252	$4.1137 \cdot 10^6$	318	$1.0994 \cdot 10^7$
56	$1.6510 \cdot 10^4$	122	$2.1145 \cdot 10^5$	188	$1.2011 \cdot 10^6$	254	$4.2533 \cdot 10^6$	320	$1.1290 \cdot 10^7$
58	$1.8146 \cdot 10^4$	124	$2.2503 \cdot 10^5$	190	$1.2552 \cdot 10^6$	256	$4.3965 \cdot 10^6$	322	$1.1592 \cdot 10^7$
60	$1.9919 \cdot 10^4$	126	$2.3932 \cdot 10^5$	192	$1.3112 \cdot 10^6$	258	$4.5434 \cdot 10^6$	324	$1.1900 \cdot 10^7$
62	$2.1837 \cdot 10^4$	128	$2.5434 \cdot 10^5$	194	$1.3692 \cdot 10^6$	260	$4.6940 \cdot 10^6$	326	$1.2215 \cdot 10^7$

0-8493-2861-6/97/$0.00+$.50

Table 11.1 Water vapor pressure [1] *(continued)*

t, °C	p, Pa	t, °C	p, Pa	t, °C	p, Pa	t, °C	p, Pa	t, °C	p, Pa
328	$1.2537 \cdot 10^7$	338	$1.4245 \cdot 10^7$	348	$1.6135 \cdot 10^7$	358	$1.8228 \cdot 10^7$	368	$2.0555 \cdot 10^7$
330	$1.2865 \cdot 10^7$	340	$1.4608 \cdot 10^7$	350	$1.6537 \cdot 10^7$	360	$1.8674 \cdot 10^7$	370	$2.1053 \cdot 10^7$
332	$1.3199 \cdot 10^7$	342	$1.4978 \cdot 10^7$	352	$1.6947 \cdot 10^7$	362	$1.9129 \cdot 10^7$	372	$2.1562 \cdot 10^7$
334	$1.3541 \cdot 10^7$	344	$1.5356 \cdot 10^7$	354	$1.7365 \cdot 10^7$	364	$1.9594 \cdot 10^7$	374	$2.2084 \cdot 10^7$
336	$1.3889 \cdot 10^7$	346	$1.5742 \cdot 10^7$	356	$1.7792 \cdot 10^7$	366	$2.0069 \cdot 10^7$	374.12	$2.2115 \cdot 10^7$

11.2 Vapors of Inorganic Substances

The vapor pressures (in Pa) for inorganic substances compiled in Tables 11.2–11.4 were recalculated from the values expressed in the units of mmHg. Recalculation procedure followed the least-squares method by using the familiar relationship between the pressure p of saturated vapor and its temperature T: $p = B - A/T$. The neptunium pressure was calculated by the expression $\lg p = 5.1 - 2.06 \cdot 10^4 (1/T)$ developed in [2].

Table 11.2 Saturation temperature (in K) of inorganic substances for the pressure range $p = 10^{-8}$ — 10^2 Pa [3]

Substance	Vapor pressure, Pa										
	10^{-8}	10^{-7}	10^{-6}	10^{-5}	10^{-4}	10^{-3}	10^{-2}	10^{-1}	1	10	10^2
Aluminum	816.7	861.8	912.2	968.8	1033	1106	1190	1289	1405	1544	1713
(III) fluoride								966.2	1024	1090	1165
(I) fluoride							812.2	898.0	1004	1138	
Americium [4, 5]							1066	1164	1283	1427	1609
(III) fluoride [6]				1083	1149	1225	1310	1409	1524		
Ammonium			89.9	95.2	101.1	107.8	115.4	124.2	134.4	146.5	160.9
Antimony	486.0	511.8	540.4	572.3	608.4	649.2	695.9	749.9	812.9	887.5	977.2
Argon			26.5	28.3	30.3	32.7	35.5	38.7	42.6	47.4	53.5
Arsenic	339.5	356.1	374.5	394.8	417.5	442.9	471.6	504.3	541.9	585.6	636.8
Barium	492.6	521.6	554.1	591.0	633.1	681.7	738.4	805.3	885.6	983.7	1106
fluoride [6]					1116	1187	1268	1361	1469		
oxide [7]			1153	1226	1309	1404	1514				
Beryllium	864.4	911.4	963.8	1022	1089	1164	1251	1353	1471	1613	1785
chloride [10]							439.5	468.0	500.6	537.9	581.4
fluoride [3, 8, 9]					681.1	721.4	766.8	818.3	877.2	945.2	1024
Bismuth	460.7	485.3	512.7	543.4	578.0	617.3	662.4	714.5	775.5	848.0	935.3
(III) chloride [12]					327.1	345.4	365.8	388.8	414.8	444.6	479.1
Boron	1456	1531	1614	1706	1810	1926	2059	2212	2389	2597	2845
oxide [3, 11]							1234	1331	1445	1581	1744
Bromine			127.4	134.5	142.5	151.4	161.6	173.2	186.6	202.2	220.7
Cadmium	307.5	324.6	343.9	365.5	390.0	418.1	450.5	488.3	533.1	586.9	652.8
arsenide						510.3	543.7	581.7	625.5	676.3	
bromide [25]							486.0	528.6	579.5	641.1	717.5
chloride					534.9	567.9	605.2	647.8	696.8	753.8	821.0
selenide							791.1	852.6	924.5	1009	
telluride							712.5	770.2	838.0	919.0	
Calcium	486.2	514.0	545.2	580.5	620.6	666.7	720.2	783.0	857.8	948.4	1060
chloride [29, 30]				874.8	926.4	984.5	1050	1125	1212	1314	1434
fluoride [27, 28]				1125	1193	1269	1356	1455	1571	1706	1866

Table 11.2 Saturation temperature (in K) of inorganic substances *(continued)*

Substance	Vapor pressure, Pa										
	10^{-8}	10^{-7}	10^{-6}	10^{-5}	10^{-4}	10^{-3}	10^{-2}	10^{-1}	1	10	10^2
Carbon (graphite [51]		1855	1953	2063	2185	2323	2480	2659	2867	3109	3396
(I) oxide			75.6	80.0	85.0	90.7	97.1	104.6	113.3	123.6	136.0
(II) oxide			26.4	28.1	30.0	32.2	34.7	37.6	41.0	45.2	50.2
Cerium					1306	1391	1487	1598	1727	1879	2059
Cesium	225.4	239.1	254.7	272.4	292.8	316.5	344.3	377.5	417.8	467.6	531.1
bromide [55]							696.7	747.4	806.2	874.9	956.5
chloride [55, 56]				581.0	617.3	658.3	705.2	759.2	822.3	896.3	986.1
fluoride [26, 55]					610.7	649.9	694.4	745.6	804.8	874.3	957.0
iodide [55]						644.0	687.6	737.5	795.1	862.6	942.6
Chlorine			83.7	88.7	94.3	100.7	108.1	116.6	126.5	138.3	152.5
Chromium	979.1	1030	1086	1149	1220	1301	1392	1498	1621	1765	1938
(II) bromide [53]					698.0	742.3	792.6	850.2	916.9	994.9	1087
(II) iodide [54]							826.2	871.0	920.9	976.9	1040
(VI) oxide							1716	1840	1985	2154	2354
Cobalt	877.4	922.4	972.3	1027	1090	1160	1240	1332	1439	1564	1713
Copper	889.1	937.8	992.1	1053	1122	1200	1291	1396	1520	1668	1849
Dysprosium [19]					1009	1082	1165	1263	1378	1517	1686
Erbium [19]					1128	1207	1298	1403	1527	1674	1854
Europium	500.0	528.1	559.7	595.5	635.6	681.9	735.4	798.1	872.4	961.9	1072
Fluorine										42.4	49.4
Francium	203.7	216.3	230.5	246.7	265.3	286.9	312.5	343.0	380.1	426.2	485.0
Gadolinium					1338	1437	1550	1683	1842	2033	2268
Gallium	744.2	786.2	833.3	886.4	946.6	1015	1095	1189	1300	1434	1599
(III) iodide [14]						302.8	322.6	345.2	371.2	401.5	437.1
(I) oxide [15]								1646	1752	1873	2012
Germanium	972.4	1025	1085	1152	1228	1314	1414	1530	1666	1830	2028
selenide [16]								677.8	730.4	792.0	864.8
sulfide [3, 17]							572.1	609.6	652.3	701.5	758.7
telluride [16, 18]						620.7	661.1	707.1	760.0	821.4	893.6
Gold	964.3	1017	1076	1143	1219	1305	1404	1520	1656	1819	2018
Hafnium	1675	1765	1865	1978	2104	2248	2414	2605	2829	3096	3418
Helium								0.6	0.8	0.9	1.2
Holmium	810.0	870.0	915.0	985.0	1040.0	1160.0	1225.0	1350.2	1480.0	1556.0	1720.0
Hydrogen			3.7	4.0	4.3	4.8	5.3	5.9	6.8	7.9	9.4
bromide			67.1	71.4	76.4	82.1	88.7	96.4	105.6	116.8	130.7
chloride			63.5	67.4	71.8	76.9	82.7	89.4	97.3	106.8	118.3
fluoride										172.5	178.2
sulfide			72.8	77.3	82.4	88.2	94.9	102.7	111.9	122.9	136.3
Indium	658.0	696.2	739.0	787.4	842.6	906.1	980.0	1067	1170	1297	1454
(III) bromide [14]						400.5	424.0	450.4	480.3	514.5	553.9
(III) chloride [14]						455.2	482.3	512.8	547.5	587.1	633.0
(III) iodide [14]						350.7	372.3	396.7	424.6	456.7	494.1
(I) oxide [22]							1504	1590	1686	1795	1919
Iodine	160.0	168.0	176.9	186.9	198.0	210.5	224.6	240.9	259.6	281.6	307.5
Iridium	1556	1636	1724	1822	1932	2057	2198	2360	2548	2768	3030
Iron	1010	1063	1123	1189	1263	1347	1443	1554	1684	1837	2021
(II) bromide [20, 21]						596.4	633.6	675.8	724.1	779.7	844.6
(II) chloride [21]							645.8	692.1	745.4	807.7	881.4
(II) iodide [21]							614.3	653.3	697.6	748.3	807.0
Krypton			36.5	38.9	41.7	45.0	48.8	53.3	58.7	65.3	73.6
Lanthanum	1152	1216	1288	1368	1459	1564	1684	1825	1991	2190	2433
(III) chloride				803.0	847.4	897.1	953.0	1016	1088	1172	

Table 11.2 Saturation temperature (in K) of inorganic substances *(continued)*

Substance	Vapor pressure, Pa										
	10^{-8}	10^{-7}	10^{-6}	10^{-5}	10^{-4}	10^{-3}	10^{-2}	10^{-1}	1	10	10^2
Lead	546.3	578.4	614.6	655.6	702.4	756.5	819.5	894.1	983.6	1092	1229
fluoride							729.6	793.1	868.7	960.2	1073
(II) oxide							885.0	945.2	1014	1094	1187
selenide [40]						725.6	776.5	835.0	903.0	983.2	1078
telluride [41]						729.7	778.5	834.3	898.69	973.7	1062
Lithium	447.7	473.8	503.2	536.4	574.4	618.1	669.0	729.1	801.0	888.7	997.9
chloride [3, 31]								778.0	849.0	934.3	1038
fluoride [26, 31]					829.6	880.1	937.2	1002	1077	1163	1265
Lutetium [19]					1412	1510	1623	1755	1909	2094	2318
Magnesium	404.1	426.8	452.1	480.6	513.0	550.0	592.8	642.8	702.0	773.3	860.6
chloride [31]							790.1	842.1	901.5	969.8	1049
fluoride [6, 31]						1170	1245	1329	1425	1537	1667
Manganese	731.5	770.8	814.6	863.6	918.8	981.7	1053	1137	1235	1351	1491
Mercury	181.9	192.1	203.5	216.3	230.9	247.5	266.7	289.2	315.8	347.8	387.0
chloride [39]						275.7	294.4	315.8	340.5	369.3	403.6
Methane			31.7	33.9	36.5	39.4	42.9	47.1	52.1	58.3	66.3
Molybdenum	1664	1751	1848	1955	2076	2213	2370	2550	2759	3007	
(VI) oxide [13]							798.8	837.3	879.6	926.4	978.5
Neodymium [19]	899.3	951.2	1009	1076	1150	1236	1336	1454	1595	1766	1978
Neon			7.3	7.8	8.4	9.1	9.9	10.9	12.1	13.6	15.5
Neptunium					1548	1685	1849	2048			
Nickel	848.5	892.0	940.1	993.7	1053	1121	1198	1287	1389	1510	1653
fluoride [33]							949.3	1023	1109	1212	1335
Niobium	1802	1895	1997	2110	2238	2382	2545	2733	2951	3206	3510
Nitrogen			23.4	25.0	26.7	28.7	31.0	33.8	37.1	41.1	46.0
Osmium	1792	1882	1982	2094	2218	2359	2518	2700	2911	3158	3450
Oxygen			27.8	29.6	31.6	33.8	36.4	39.5	43.1	47.4	52.7
Palladium [36, 37]	991.8	1048	1111	1182	1262	1355	1462	1588	1738	1918	2140
Phosphorus red	296.9	310.3	325.0	341.2	359.0	378.8	400.9	425.7	453.8	485.9	522.9
Phosphorus white	156.8	166.0	176.2	187.8	201.1	216.3	234.1	255.0	280.0	310.5	348.4
Platinum [3, 36]	1404	1478	1560	1650	1752	1867	1999	2150	2326	2534	2782
Plutonium						1351	1463	1596	1755	1949	
(III) fluoride [6]						1192	1269	1357	1458		
Polonium	351.6	371.3	393.3	418.0	446.1	478.2	515.3	558.6	609.9	671.5	747.0
Potassium	259.9	275.6	293.3	313.4	336.5	363.3	394.7	432.1	477.3	533.0	603.4
chloride						723.7	774.7	833.3	901.7	982.2	1078
fluoride [26]					738.7	785.9	839.6	901.2	972.5	1056	1155
Praseodymium [19]					1274	1371	1483	1616	1774	1967	2208
Radium	457.8	484.7	515.0	549.3	588.5	633.7	686.5	748.8	823.6	915.0	1029
Rhenium	1966	2070	2186	2317	2463	2630	2820	3041	3299	3604	3972
Rhodium	1383	1455	1535	1624	1724	1837	1966	2114	2287	2491	2734
Rubidium	237.0	251.4	267.6	286.2	307.5	332.2	361.2	395.7	437.6	489.4	555.1
chloride								799.2	867.0	947.5	1044
fluoride [26]					693.9	738.2	788.6	846.3	913.1	991.4	1084
Ruthenium	1646	1729	1820	1922	2036	2165	2311	2478	2670	2896	3163
Samarium	573.9	605.7	641.3	681.4	726.8	778.7	838.5	908.4	990.9	1089	1210
(III) chloride [24]								751.7	814.9	889.7	979.7
Scandium [19,23,43-45]	951.4	1004	1064	1130	1206	1293	1393	1511	1649	1816	2020
Selenium		335.6	352.8	371.8	393.0	416.6	443.6	474.1	509.1	549.7	597.3
Silicon	961.8	1011	1066	1128	1197	1275	1365	1467	1586	1727	1894
(II) oxide						1472	1615	1789			
Silver	752.5	793.8	840.0	891.8	950.4	1017	1094	1183	1289	1415	1569
bromide [25]							1076	1149	1233	1329	1442

Table 11.2 Saturation temperature (in K) of inorganic substances *(continued)*

Substance	Vapor pressure, Pa										
	10^{-8}	10^{-7}	10^{-6}	10^{-5}	10^{-4}	10^{-3}	10^{-2}	10^{-1}	1	10	10^{2}
telluride [42]							1013	1073	1140	1216	1303
Sodium	306.0	324.2	344.7	367.9	394.5	425.2	461.2	503.7	554.9	617.7	696.5
bromide [25]							740.8	802.1	874.5	961.2	1067
fluoride [26]					847.9	901.0	961.2	1030	1109	1202	1312
Strontium	453.3	479.2	508.2	540.9	578.2	620.9	670.5	728.7	797.9	881.6	985.0
fluoride [6]				1160	1225	1298	1381	1474	1581		
Sulphur	232.2	244.0	257.0	271.4	287.6	305.8	326.5	350.2	377.6	409.7	447.7
Tantalum	2003	2106	2221	2348	2492	2653	2838	3050	3296	3585	3929
Tellurium	399.5	419.5	441.7	466.3	493.9	524.9	560.0	600.3	646.7	700.9	765.0
(IV) oxide [48]					716.4	757.6	803.9	856.2	915.8	984.3	
Terbium [19]					1295	1389	1498	1625	1776	1958	2181
Thallium	490.4	518.8	550.7	586.7	627.7	674.9	729.9	794.5	871.7	965.5	1081
selenide [46]					459.3	489.4	523.8	563.3	609.3		
sulfide [47]							390.8	428.1	473.2	529.0	
Thorium			1693	1799	1919	2056					
(IV) fluoride [50]						985.9	1047	1116	1195	1286	1392
Thulium	655.0	690.3	729.6	773.6	823.3	879.8	944.6	1019	1107	1212	
Tin	850.0	898.9	953.7	1015	1086	1167	1261	1372	1504	1664	1862
selenide [34, 35]							725.5	783.2	850.9	931.3	1028
telluride [34, 35]							757.5	815.1	882.1	961.2	1055
Titanium	1202	1264	1333	1409	1495	1591	1701	1828	1975	2147	2352
sulfide [49]						1692	1811	1948	2108	2296	
Tungsten	2141	2249	2367	2500	2647	2814	3002	3218	3467	3759	4103
(VI) oxide [13]							1252	1323	1401	1490	1590
Uranium	1252	1323	1401	1490	1591	1706	1839	1995	2180	2402	2675
(IV) chloride [52]							597.1	633.9	676.5	723.0	
(IV) fluoride									1070	1148	1237
(IV) oxide			1581	1669	1768	1879	2005				
Vanadium	1282	1348	1421	1502	1594	1697	1814	1949	2105	2289	2508
Water			143.5	151.8	161.0	171.4	183.2	196.8	212.6	231.2	253.3
Xenon			50.3	53.7	57.6	62.1	67.3	73.5	81.0	90.1	101.6
Ytterbium [19]					533.7	573.5	619.8	674.1	738.9	817.4	914.6
Yttrium [14, 23]					1375	1474	1590	1725	1885	2078	2314
(III) chloride [24]								751.3	821.4	906.0	1010
Zinc	354.9	374.6	396.6	421.3	449.4	481.4	518.3	561.4	612.3	673.3	747.9
arsenide [59]					517.9	555.1	598.2	648.5			
bromide [25]					394.9	421.9	452.9	488.7	530.7	580.6	640.8
selenide							911.6	974.1	1045	1128	
sulfide									1127	1252	1408
telluride							778.2	839.6	911.4	996.6	
Zirconium	1565	1648	1741	1846	1963	2096	2249	2426	2633	2878	3174

Table 11.3 Saturation temperature (in K) of inorganic substances for the pressure range $p = 0.1 - 100$ kPa [3]

Substance	Vapor pressure, kPa									
	0.1	0.2	0.5	1	2	5	10	20	50	100
Aluminum	1719	1780	1868	1941	2019	2133	2228	2332	2486	2616
borohydride		205.4	216.7	226.1	236.4	251.5	264.2	278.3	299.5	317.8
bromide	343.3	355.5	372.9	387.3	402.8	425.3	444.1	464.6	494.9	520.5
(III) chloride	369.8	376.8	386.4	394.0	401.9	412.9	421.6	430.6	443.3	453.3
(I) fluoride [61]	1253	1266	1283	1297	1310	1329	1343	1357		

Table 11.3 Saturation temperature (in K) of inorganic substances *(continued)*

Substance	Vapor pressure, kPa									
	0.1	0.2	0.5	1	2	5	10	20	50	100
(III) fluoride	1496	1521	1557	1585	1614	1654	1685	1718	1763	1799
iodide	439.7	454.5	475.7	493.0	511.7	538.7	561.1	585.4	621.0	650.9
oxide	2394	2459	2551	2624	2702	2813	2903	2998	3135	3247
Americium [4]	1649	1718	1820							
Ammonium	159.7	165.1	172.7	178.9	185.7	195.4	203.4	212.2	224.9	235.7
azide	299.3	307.4	318.9	328.2	338.1	352.0	363.4	375.5	392.8	407.0
bromide	465.2	479.9	500.8	517.9	536.1	562.3	583.9	607.2	641.1	669.3
carbamate	243.9	250.5	259.8	267.4	275.4	286.7	295.9	305.7	319.7	331.2
chloride	428.4	441.8	460.7	476.1	492.6	516.2	535.7	556.6	587.0	612.3
cyanide	219.4	225.6	234.5	241.7	249.4	260.3	269.1	278.7	292.3	303.6
hydrosulfide	218.8	225.2	234.2	241.6	249.4	260.5	269.7	279.4	293.5	305.1
iodide	473.4	488.4	509.8	527.2	545.9	572.8	594.9	618.8	653.6	682.5
Antimony	999.1	1051	1128	1195	1270	1385	1486	1604	1792	1967
(III) bromide	362.2	374.2	393.2	408.3	424.6	448.2	467.9	489.5	521.1	547.9
(III) chloride	312.4	324.0	340.8	354.6	369.6	391.5	409.9	430.1	460.1	485.7
(V) chloride	291.3	301.8	317.0	329.6	343.1	362.9	379.5			
(V) fluoride	271.9	281.7	295.7	307.3	319.8	338.0	353.2	369.9	394.4	415.3
(III) iodide	429.4	445.5	468.9	488.2	509.2	539.9	565.7	594.1	636.3	672.5
(III) oxide	778.0	821.1	886.1	942.5	1006	1105	1195	1300	1470	1633
Argon	53.2	55.3	58.4	61.1	63.9	68.1	71.7	75.7	81.7	86.9
Arsenic	635.4	653.8	679.8	700.8	723.3	755.2	781.3	809.3	849.5	882.6
(III) bromide	309.0	320.9	338.2	352.5	368.1	391.0	410.3	431.6	463.4	490.7
(III) chloride	256.7	266.4	280.3	291.8	304.3	322.6	337.9	354.8	379.9	401.4
(III) fluoride					255.3	269.5	281.3	294.2	313.1	329.1
(V) fluoride	151.1	155.8	162.4	167.8	173.6	181.9	188.7	196.0	206.6	215.4
(III) iodide [88]	405.2	420.5	442.7	461.2	481.2	510.5	535.1	562.2	602.7	637.3
(III) oxide	466.4	483.0	506.9	526.5	547.7	578.6	604.3	632.4	673.9	709.1
(III) sulfide [89]	533.8	556.9	590.8	619.2	650.6	697.3	737.3	782.2	850.7	911.0
Arsine [62]	143.3	150.4	160.9	169.9	180.0	195.3	208.7	224.1	248.2	270.3
Astatine	427.7	441.0	459.7	475.0	491.4	514.8	534.1	554.9	584.9	609.9
Barium	1113	1161	1232	1291	1357	1455	1538	1632	1775	1902
chloride [63]	1533	1590	1672	1740						
Beryllium	1793	1857	1949	2025	2108	2228	2328	2438	2599	2737
borohydride	272.3	279.4	289.4	297.4	305.8	317.8	327.4	337.7	352.7	364.3
bromide	556.5	571.0	591.5	608.1	625.5	650.2	670.2	691.5	721.8	746.5
chloride	557.6	572.8	594.1	611.4	629.7	655.6	676.7	699.1	731.3	757.6
fluoride [8, 9]	1022	1052	1094	1128	1164	1216	1258			
iodide	550.3	566.0	588.2	606.1	625.2	652.3	674.5	698.2	732.2	760.3
Bismuth	1030	1072	1134	1185	1241	1324	1395	1473	1592	1695
(III) bromide		501.3	525.6	545.6	567.2	598.5	624.5	653.0	694.8	730.2
(III) chloride		484.4	508.3	528.0	549.4	580.3	606.2	634.5	676.2	711.5
(V) fluoride	343.4	354.7	370.8	383.9	398.1	418.5	435.4	453.7	480.4	502.7
(III) iodide	519.6	539.2	567.4	590.8	616.3	653.5	684.7	719.1	770.3	814.1
Boron	2859	2950	3079	3185	3299	3461	3596	3741	3952	4128
(III) chloride	177.8	184.7	194.7	203.0	212.0	225.3	236.5	248.9	267.4	283.3
(III) fluoride	116.3	120.2	125.7	130.3	135.2	142.2	148.0	154.3	163.6	171.4
(III) iodide [64]	301.4	313.1	329.9	343.9	359.1	381.4	400.2	421.0	452.0	478.6
Bromine	216.5	224.1	234.9	243.8	253.4	267.4	279.0	291.7	310.3	326.1
dichlorofluorosilane	182.2	189.9	201.1	210.5	220.8	236.1	249.1	263.7	285.8	305.2
(V) fluoride [65]	197.8	205.4	216.4	225.5	235.4	250.0	262.2	275.7	295.8	313.1
silane		173.8	183.7	192.1	201.2	214.7	226.1	238.8	258.1	274.8
Cadmium	655.9	681.0	717.3	747.4	780.2	828.2	868.6	913.1	979.6	1036
bromide [82]	727.9	756.9	799.0	834.0	872.2	928.5				

Table 11.3 Saturation temperature (in K) of inorganic substances *(continued)*

Substance	Vapor pressure, kPa									
	0.1	0.2	0.5	1	2	5	10	20	50	100
chloride [3, 83]	804.9	834.0	875.9	910.6	948.0	1002	1048	1098	1172	1234
dimethyl [84, 85]	242.9	251.8	264.6	275.2	286.6	303.3	317.2	332.5	355.2	374.5
fluoride	1636	1683	1750	1805	1863	1945	2012	2084	2188	2274
iodide	678.1	703.8	741.0	771.8	805.3	854.4	895.6	941.0	1008	1066
oxide	1255	1295	1354	1401	1452	1526	1587	1653	1749	1829
Calcium	1060	1104	1168	1221	1280	1366	1440	1522	1646	1754
chloride [63]	1467	1516	1586	1643						
Carbon (graphite)	3353	3480	3663	3814	3978	4219	4421	4644	4975	5258
dioxide	135.8	140.1	146.1	151.0	156.3	163.8	170.0	176.7	186.4	194.5
oxide	49.8	51.8	54.7	57.0	59.7	63.5	66.7	70.4	75.8	80.4
suboxide [111]	173.7	180.5	190.4	198.5	207.5	220.6	231.6	243.8	262.1	277.9
Carbonylborine	131.2	136.3	143.6	149.7	156.4	166.1	174.3	183.3	196.8	208.5
Carbonylchloride (phosgene)	176.3	183.1	192.8	200.9	209.8	222.7	233.5	245.5	263.4	278.8
Cerium	2070	2152	2270	2369	2477	2636	2770	2918	3141	3334
Cesium	532.1	556.5	592.3	622.7	656.3	706.8	750.4	799.8	876.0	944.1
bromide	1003	1041	1095	1140	1189	1260	1320	1386	1484	1567
chloride	1000	1038	1092	1137	1186	1257	1317	1383	1482	1566
fluoride	966.5	1003	1055	1099	1146	1216	1274	1338	1433	1515
iodide	993.5	1030	1083	1127	1175	1245	1304	1368	1464	1546
Chlorine	149.9	155.6	163.9	170.7	178.1	189.0	198.1	208.2	223.2	236.1
(II) oxide	171.8	178.5	188.3	196.3	205.2	218.1	229.0	241.1	259.2	274.7
(IV) oxide			201.7	209.5	217.9	239.1	240.3	251.4	267.9	281.8
(VI) oxide	276.0	285.6	299.2	310.5	322.6	340.2	354.9	370.8	394.2	414.0
Chlorine anhydride	224.5	232.9	245.2	255.3	266.3	282.5	296.0	310.9	333.1	352.2
fluoride		122.2	127.7	132.2	137.0	144.0	149.8	156.0	165.1	172.7
trifluoride		178.5	188.9	197.5	207.0	221.1	233.1	246.4	266.5	284.1
Chlorosilane	151.7	157.6	166.0	173.0	180.6	191.7	201.1	211.4	226.8	240.1
Chlorosulfonic acid	300.8	309.8	322.6	333.0	344.1	359.9	372.9	386.8	407.0	423.6
Chlorotrifluorosilane	126.0	131.0	138.1	144.1	150.6	160.1	168.2	177.1	190.4	201.9
Chromium	1939	2002	2093	2167	2247	2362	2457	2559	2709	2835
carbonyl	305.6	314.5	327.0	337.1	347.8	363.2	375.7	389.1	408.4	424.3
Chromyl chloride	249.4	258.5	271.7	282.7	294.5	311.7	326.2	342.1	365.5	385.6
Cobalt	1717	1774	1855	1921	1992	2095	2180	2272	2406	2519
chloride					974.4	1038	1091	1151	1241	1319
nitrosyltricarbonyl				255.7	266.7	282.8	296.4	311.3	333.4	352.4
Copper	1850	1929	2044	2141	2247	2404	2539	2689	2918	3118
bromide	814.8	856.3	918.0	971.0	1030	1121	1201	1293	1439	1574
chloride	783.7	827.7	894.1	951.8	1017	1119	1211	1320	1497	1666
iodide		786.5	849.2	903.6	965.5	1061	1148	1249	1415	1573
Deuteroammonium						192.9	201.9	211.9	226.7	239.3
Deuterodiborane [74]	109.0	113.5	119.9	125.3	131.1	139.8	147.1	155.3	167.6	178.3
Diborane [3, 74, 75, 76]	109.2	113.7	120.3	125.7	131.7	140.5	148.1	156.4	169.0	180.0
Dibromochlorofluorosilane	203.6	211.8	223.7	233.7	244.5	260.6	274.2	289.2	312.0	331.6
Dibromodifluorosilane			193.4	201.9	211.2	224.9	236.6	249.5	268.8	285.6
Dibromosilane	208.1	216.6	229.0	239.3	250.7	267.4	281.6	297.4	321.2	342.0
Dichlorodifluorosilane	145.3	151.3	160.2	167.5	175.6	187.6	197.8	209.2	226.4	241.5
Difluorosilane	123.6	128.3	135.0	140.6	146.7	155.6	163.0	171.3	183.5	193.9
Disilane	154.9	161.4	170.8	178.6	187.3	200.1	211.0	223.1	241.5	257.6
Disilasene	200.4	208.2	219.5	228.9	239.2	254.2	266.9	280.9	301.9	319.9
Disiloxane	158.1	164.5	173.7	181.4	189.9	202.3	212.8	224.4	242.0	257.2
Dysprosium [77]	1750	1810	2000	2060	2140	2300	2430	2590	2720	2880
Europium	1074	1116	1178	1230	1286	1369	1440	1517	1634	1735
chloride [78]	1549	1608	1694	1765	1843	1957	2052	2158	2316	2451

Table 11.3 Saturation temperature (in K) of inorganic substances *(continued)*

Substance	Vapor pressure, kPa									
	0.1	0.2	0.5	1	2	5	10	20	50	100
Fluorine	49.7	51.9	55.1	57.7	60.7	65.1	68.8	73.0	79.4	85.1
oxide	76.5	79.7	84.5	88.5	92.9	99.4	105.0	111.2	120.7	129.0
Fluorosilane	116.6	120.5	126.2	130.8	135.7	142.9	148.8	155.2	164.7	172.6
Fluorotrichlorosilane	176.5	183.4	193.5	201.8	210.9	224.3	235.6	248.1	266.8	282.9
Francium	486.8	509.4	542.8	571.1	602.6	649.8	690.8	737.4	809.4	874.0
Gadolinium	2280	2420	2634	2823	3041	3386	3704	4088	4738	5385
Gallium	1596	1656	1743	1815	1894	2009	2105	2211	2369	2504
(III) chloride	309.8	320.6	336.1	348.8	362.6	382.5	399.1	417.3	443.9	466.4
Germanium	2044	2117	2222	2309	2403	2540	2654	2778	2963	3119
bromide		293.2	309.7	323.4	338.5	360.6	379.4	400.2	431.6	458.8
chloride	222.9	231.5	243.9	254.2	265.4	281.8	295.6	310.8	333.6	353.2
hydride	108.2	112.8	119.6	125.4	131.7	141.0	149.1	158.0	171.7	183.8
hydride, di-	180.8	188.5	199.6	208.9	219.2	234.4	247.4	261.9	283.9	303.2
Gold	2025	2099	2204	2291	2385	2522	2636	2762	2947	3105
Hafnium	3370	3518	3734	3916	4117	4416	4673	4961	5403	5792
(IV) chloride [3, 71, 72]	441.0	452.5	468.6	481.6	495.4	514.7	530.5	547.1	570.9	590.3
Helium	1.2	1.3	1.5	1.6	1.8	2.1	2.4	2.8	3.6	4.6
Hexaborane [73]	234.9	243.3	255.2	265.1	275.8	291.3	304.2	318.4	339.2	356.9
Hexachloridusilane	272.9	282.4	296.1	307.4	319.6	337.3	352.0	368.0	391.7	411.7
Hexachloridusiloxane	267.0	276.3	289.6	300.6	312.4	329.6	343.9	359.4	382.3	401.7
Hexafluoridusilane	190.2	195.1	202.1	207.6	213.5	221.9	228.6	235.8	246.0	254.3
Hydrocyanic acid	197.8	204.5	214.1	222.0	230.6	242.9	253.1	264.2	280.5	294.3
Hydrodecaborane, di-	324.1	334.4	349.0	360.9	373.7	392.1	407.2	423.5		
Hydrogen	9.3	9.9	10.7	11.4	12.2	13.4	14.6	15.9	18.1	20.2
arsenide	127.2	132.4	139.9	146.2	153.1	163.3	171.9	181.5	196.0	208.6
bromide	131.4	136.3	143.4	149.2	155.6	164.8	172.6	181.1	193.8	204.6
chloride	119.3	123.8	130.2	135.5	141.3	149.7	156.8	164.5	176.1	185.9
fluoride	175.4	182.7	193.4	202.3	212.2	226.7	230.1	252.9	273.9	292.2
peroxide	282.9	292.8	307.1	318.9	331.6	350.1	365.5	382.3	407.1	428.0
iodide	146.7	152.6	161.0	168.1	175.8	187.2	196.8	207.4	223.4	237.2
selenide	154.0	159.2	166.6	172.7	179.2	188.6	196.5	205.0	217.4	227.9
sulfide	136.0	141.0	148.3	154.3	160.8	170.4	178.4	187.2	200.2	211.3
sulfide, di-	226.5	234.2	245.2	254.3	264.0	278.1	289.8	302.6	321.2	336.9
telluride	171.9	178.2	187.4	194.9	203.1	215.1	225.1	236.1	252.5	266.4
Hydrogen (deuterium)	11.9	12.5	13.4	14.2	15.1	16.5	17.7	19.1	21.3	23.4
Hydrogen (deuterohydrogen)	10.7	11.3	12.2	12.9	13.8	15.1	16.3	17.6	19.9	22.0
Hydrogen (tritium)	13.3	14.0	15.0	15.8	16.8	18.3	19.6	21.1	23.5	25.7
Hydropentaborane, di-		217.1	228.7	238.3	248.7	264.1	277.0	291.2	312.5	330.7
Hydroxylamine		298.8	308.8	316.8	325.2	337.1	346.6	356.8	371.1	382.7
Indium	1452	1508	1589	1656	1730	1837	1928	2029	2178	2307
chloride [80]	602.4	623.6	654.1	679.3	706.4					
sulfide [81]	1265	1303	1358	1402	1449	1517	1572	1631		
Iodine	304.7	314.8	329.3	341.1	353.9	372.3	387.5	404.0	428.1	448.4
fluoride, penta-	245.8	254.5	267.0	277.3	288.5	304.6	318.1	332.9	354.7	373.1
Iodosilane		204.4	215.8	225.2	235.5	250.7	263.5	277.7	299.0	317.4
Iodosilane, di-		256.4	272.4	285.8	300.7	323.0	342.2	363.7	396.8	426.1
Iridium	3033	3133	3274	3390	3514	3693	3841	4001	4235	4431
Iron	2031	2101	2201	2283	2371	2499	2605	2721	2891	3034
bromide [79]	825.2	853.0	892.8	925.5	960.6	1011	1053	1099	1166	1222
chloride			917.1	953.3	992.5	1049	1097	1149	1226	1292
chlorine	465.5	475.5	489.4	500.5	512.1	528.3	541.2	554.7	573.7	589.0
Iron carbonyl		248.8	262.0	273.0	284.9	302.4	317.0	333.2	357.4	378.1
Krypton	73.6	76.6	80.8	84.4	88.3	94.1	98.9	104.3	112.5	119.5

Table 11.3 Saturation temperature (in K) of inorganic substances *(continued)*

Substance	Vapor pressure, kPa									
	0.1	0.2	0.5	1	2	5	10	20	50	100
Lanthanum	2428	2516	2642	2746	2858	3022	3158	3308	3529	3717
Lead	1233	1284	1357	1418	1485	1584	1668	1761	1902	2024
bromide	770.3	798.0	837.9	870.9	906.5	958.3	1001	1049	1119	1178
chloride	805.2	833.5	874.2	907.7	943.9	996.4	1040	1087	1158	1217
fluoride		1061	1114	1157	1204	1272	1329	1391	1483	1561
iodide	735.4	762.1	800.6	832.4	866.8	917.0	958.9	1004	1072	1130
oxide	1199	1238	1293	1338	1387	1456	1514	1576	1666	1742
sulfide	1105	1138	1185	1223	1263	1321	1368	1419	1492	1553
Lithium	999.4	1038	1095	1142	1193	1269	1332	1403	1508	1599
bromide	1003	1041	1096	1141	1190	1263	1323	1390	1490	1575
chloride	1038	1077	1135	1183	1235	1312	1377	1448	1554	1645
fluoride	1299	1344	1408	1461	1518	1601	1669	1744	1854	1947
iodide	981.3	1013	1060	1098	1138	1197	1246	1298	1375	1440
Lutetium	2258	2371	2538	2681	2841	3085	3298	3544	3931	4285
Magnesium	858.4	891.8	940.2	980.3	1024	1088	1142	1202	1292	1369
chloride	1030	1071	1131	1182	1237	1318	1387	1463	1578	1678
Manganese	1494	1551	1635	1704	1780	1891	1984	2087	2241	2373
chloride		936.3	988.3	1031	1079	1148	1208	1273	1371	1456
Mercury	399.2	414.3	436.1	454.1	473.8	502.5	526.6	553.2	592.7	626.6
bromide	399.9	412.9	431.5	446.7	463.0	486.5	505.9	526.9	557.6	583.2
chloride	403.5	415.8	433.3	447.5	462.7	484.5	502.3	521.6	549.4	572.4
dimethyl [84]	227.7	236.6	249.4	260.0	271.6	288.6	303.0	318.8	342.5	362.9
iodide	420.6	434.5	454.3	470.5	487.9	513.0	533.7	556.3	589.1	616.7
Molybdenum	3326	3444	3614	3755	3906	4126	4310	4511	4807	5058
(V) chloride	355.4	368.4	387.1	402.5	419.2	443.6	464.0	486.4	519.5	547.7
dioxychloride [86]	332.1	340.1	351.3	360.3	369.8	383.0	393.8	405.1	421.1	434.0
(VI) fluoride	203.7	210.7	220.7	229.0	237.8	250.7	261.4	273.0	290.1	304.5
oxide	971.0	1001	1045	1080	1118	1173	1218	1267	1337	1396
(VI) oxytetrachloride [68]	328.3	341.4	360.3	376.0	393.2	418.5	439.8	463.5	499.1	529.8
Monobromodiboran	176.8	184.0	194.4	203.1	212.7	226.7	238.6	251.9	271.8	289.1
Monogermane [87]	109.4	114.0	120.8	126.5	132.8	142.0	150.0	158.8	172.3	184.1
Neodymium	2001	2091	2223	2335	2458	2643	2803	2982	3259	3505
Neon	15.6	16.3	17.3	18.1	19.0	20.4	21.6	23.0	25.0	26.8
Neptunium [2]	2538	2636	2778	2895	3023	3211	3369	3543	3804	4028
(VI) fluoride	236.8	243.6	253.1	260.9	269.2	280.9	290.5	300.8	315.5	327.7
Nickel	1656	1710	1786	1849	1916	2012	2092	2178	2303	2408
carbonyl					233.7	248.8	261.6	275.8	297.0	315.4
chloride	932.8	957.5	992.3	1020	1050	1091	1125	1162	1213	1255
Niobium	3529	3642	3803	3934	4075	4277	4444	4625	4887	5106
(V) bromide [90]	451.4	464.2	482.3	497.0	512.5	534.7	552.8	572.1	599.8	622.6
(V) chloride [90, 91]			381.4	395.1	409.8	431.1	448.8	467.9	495.9	519.3
(V) fluoride			338.1	352.9	369.1	392.8	412.9	435.2	468.6	497.6
Nitrogen	46.0	47.9	50.7	53.0	55.5	59.3	62.5	66.1	71.5	76.2
anhydride	234.0	239.6	247.5	253.8	260.5	269.8	277.3	285.3	296.5	305.6
(III) fluoride [3, 60]	87.3	90.8	96.0	100.4	105.2	112.2	118.2	124.8	134.8	143.6
fluoroxide	127.0	131.9	138.8	144.6	150.8	160.0	167.7	176.2	188.9	199.7
(I) oxide	127.9	132.0	137.7	142.4	147.4	154.6	160.5	166.9	176.1	183.9
(II) oxide	85.6	88.2	91.7	94.6	97.7	102.1	105.7	109.6	115.2	119.8
(IV) oxide	213.1	218.9	227.0	233.5	240.4	246.2	258.1	266.6	278.7	288.6
Nitrosyl fluoride	137.9	143.0	150.4	156.6	163.2	172.9	181.1	190.1	203.4	214.7
Nitrosyl chloride					199.4	211.9	222.4	234.0	251.4	266.4
Octachlorotrisilane	314.7	326.2	342.7	356.4	371.2	392.8	410.8	430.6	459.9	484.9

Table 11.3 Saturation temperature (in K) of inorganic substances *(continued)*

Substance	Vapor pressure, kPa									
	0.1	0.2	0.5	1	2	5	10	20	50	100
Osmium	3475	3579	3727	3846	3974	4157	4306	4467	4699	4891
(VIII) oxide:										
white	261.2	270.6	284.1	295.2	307.2	324.7	339.3	355.2	378.8	398.8
yellow	268.3	277.2	290.0	300.5	311.8	328.0	341.5	356.2	377.6	395.5
Oxygen	53.0	55.3	58.6	61.3	64.4	68.9	72.8	77.1	83.7	89.5
Ozone	91.7	95.9	102.0	107.2	112.9	121.5	128.9	137.3	150.3	161.8
Palladium	2214	2293	2408	2502	2604	2752	2876	3012	3212	3382
Pentaborane [93]	207.9	215.9	227.6	237.3	247.9	263.4	276.5	290.9	312.5	331.1
Pentadeuteroborane [93]	207.1	215.1	226.8	236.5	247.0	262.5	275.5	289.9	311.5	330.1
Perchlorylofluoride [94, 95]	140.6	146.1	154.1	160.7	167.9	178.4	187.4	197.2	212.0	224.7
Phosphine					133.5	142.9	150.9	159.9	173.5	185.5
Phosphonium bromide	226.4	232.8	241.8	249.0	256.8	267.7	276.6	286.2	299.9	311.1
Phosphonium chloride	179.4	184.3	191.3	197.0	203.0	211.5	218.5	225.9	236.5	245.2
Phosphonium iodide	244.7	251.5	261.1	268.9	277.1	288.8	298.4	308.5	323.1	335.1
Phosphorus red	503.3	517.4	537.1	553.1	570.1	594.2	613.9	634.8	664.9	689.6
(III) bromide	275.4	286.3	302.3	315.6	330.1	351.4	369.5	389.5	419.6	445.7
(III) chloride	217.0	225.5	237.7	247.8	258.9	275.1	288.8	303.9	326.5	345.9
(V) chloride	323.0	331.3	343.0	352.4	362.3	376.4	387.7	399.8	416.9	430.9
(III) oxide		291.8	307.5	320.6	334.8	355.6	373.2	392.6	421.6	446.5
(V) oxide (metastable form)	456.6	469.6	487.9	502.8	518.6	541.0	559.4	579.0	607.2	630.4
(V) oxide (stable form)	650.1	666.3	688.8	707.0	726.1	753.0	774.7	797.7	830.3	856.8
oxychloride			258.5	269.6	281.6	299.4	314.3	330.9	355.6	376.9
thiobromide	318.6	328.0	341.4	352.3	363.9	380.4	393.9	408.5	429.4	446.7
thiochloride	249.9	259.5	273.2	284.7	297.2	315.4	330.7	347.7	372.9	394.6
Phosphorus yellow	341.3	354.6	373.8	389.9	407.3	432.9	454.5	478.4	514.2	545.0
Platinum	2772	2859	2983	3085	3193	3349	3478	3616	3817	3985
(VI) fluoride [96]	230.6	238.2	249.0	257.8	267.3	281.0	292.3	304.5	322.4	337.4
Plutonium	2193	2278	2402	2505	2617	2782	2921	3075	3305	3503
(VI) fluoride				261.7	270.4	282.9	293.2	304.2	320.1	333.3
Polonium	757.9	788.3	832.5	869.4	909.7	969.1	1019	1075	1159	1231
Potassium	606.6	633.1	672.0	704.7	740.8	794.5	840.7	892.5	971.7	1041
bromide	1046	1086	1143	1191	1243	1319	1383	1454	1559	1649
chloride	1075	1115	1172	1220	1272	1348	1411	1481	1584	1673
fluoride	1138	1180	1241	1291	1345	1425	1491	1564	1673	1766
hydroxide	972.1	1011	1068	1116	1168	1246	1311	1384	1493	1589
iodide	999.5	1038	1094	1140	1190	1265	1327	1396	1499	1588
Protactinium	2937	3042	3193	3317	3452	3648	3811	3990	4254	4478
Radium	1038	1084	1151	1208	1271	1365	1446	1537	1676	1800
Radon	125.4	130.7	138.3	144.7	151.7	162.1	170.9	180.8	195.8	208.8
Rhenium	3972	4107	4299	4456	4626	4871	5074	5295	5619	5891
(VI) fluoride [97]	209.6	216.2	225.7	233.4	241.6	253.5	263.3	273.9	289.2	302.0
(VII) fluoride [97]	226.5	234.6	246.1	255.6	265.8	280.7	293.1	306.7	326.7	343.7
(VII) oxide	475.0	486.5	502.6	515.6	529.2	548.3	563.7	580.0	603.1	621.8
oxitetrachloride [98]	307.1	319.4	337.4	352.5	368.9	393.1	413.6	436.4	470.8	500.5
Rhodium	2764	2851	2975	3075	3183	3338	3465	3603	3802	3968
Rubidium	555.4	580.3	616.9	647.8	681.9	733.0	777.0	826.6	902.9	970.6
bromide	1035	1074	1130	1176	1227	1300	1362	1430	1530	1616
chloride	1045	1085	1142	1189	1241	1317	1380	1451	1555	1644
fluoride	1149	1185	1235	1277	1321	1384	1436	1491	1572	1640
iodide	1002	1040	1094	1139	1188	1260	1320	1386	1485	1569
Ruthenium	3170	3267	3404	3516	3635	3805	3945	4096	4313	4494
(IV) oxide [90, 100]	261.1	270.5	284.0	295.1	307.1	324.6	339.3	355.3	378.9	399.0

Table 11.3 Saturation temperature (in K) of inorganic substances *(continued)*

Substance	Vapor pressure, kPa									
	0.1	0.2	0.5	1	2	5	10	20	50	100
Samarium	1208	1252	1316	1369	1426	1510	1580	1657	1771	1868
chloride [78]	1460	1512	1587	1648	1714	1811	1891	1980	2109	2219
Selenium	612.9	634.5	665.5	691.1	718.7	758.8	792.3	828.8	882.5	928.1
(IV) chloride	343.2	352.4	365.3	375.7	386.8	402.4	415.1	428.7	448.0	463.8
(IV) oxide	425.5	437.8	455.1	469.2	484.3	505.6	523.1	541.8	568.7	590.9
(VI) fluoride	152.6	157.8	165.3	171.5	178.1	187.7	195.6	204.3	217.0	227.7
oxychloride	304.4	314.2	328.2	339.7	352.0	369.7	384.3	400.1	423.1	442.3
Silane [3, 101]	96.2	100.2	106.0	110.9	116.2	124.1	130.9	138.3	149.7	159.6
Silicon	1908	1975	2071	2151	2236	2361	2465	2578	2745	2886
(IV) chloride	205.6	213.6	225.2	234.8	245.2	260.6	273.5	287.8	309.2	327.6
(IV) iodide [64]	357.4	371.3	391.4	408.2	426.4	453.1	475.7	500.6	537.9	570.0
(IV) fluoride	127.5	131.2	136.6	140.9	145.5	152.1	157.4	163.2	171.5	178.4
oxide			1918	1979	2043	2134	2209	2289	2404	2500
Silver	1578	1635	1718	1786	1861	1969	2059	2158	2305	2430
chloride	1164	1208	1272	1325	1383	1467	1538	1616	1732	1832
iodide	1071	1115	1179	1233	1292	1379	1453	1535	1660	1768
Sodium	697.8	726.9	769.2	804.6	843.5	901.0	950.0	1004	1087	1159
bromide	1060	1100	1157	1205	1256	1331	1395	1464	1567	1656
chloride	1119	1160	1220	1269	1322	1400	1465	1536	1642	1733
cyanide	1071	1115	1180	1233	1292	1380	1454	1537	1661	1770
fluoride	1329	1374	1439	1491	1548	1630	1698	1772	1881	1972
hydroxide	995.2	1036	1096	1147	1202	1283	1353	1430	1548	1650
iodide	1022	1059	1113	1157	1205	1274	1332	1396	1490	1570
Stibine [102]			169.4	177.1	185.6	198.2	208.9	220.8	238.9	254.6
Strontium	979.9	1020	1080	1130	1185	1266	1335	1412	1529	1631
chloride [63]	1553	1610	1693	1761						
oxide	2314	2370	2449	2512	2578					
Sulphur	445.8	463.0	487.7	508.3	530.6	563.4	591.0	621.5	666.9	706.0
chloride	260.3	270.0	284.1	295.8	308.5	327.1	342.7	359.8	385.3	407.2
(VI) fluoride	138.3	143.2	150.2	156.0	162.2	171.3	178.8	187.1	199.2	209.5
Sulfuric acid	412.6	426.0	445.1	460.8	477.6	501.8	521.8	543.5	575.0	601.4
Sulfuric anhydride (α)	230.8	237.1	246.1	253.4	261.1	272.1	281.0	290.5	304.1	315.3
Sulfuric anhydride (β)	236.3	242.4	251.0	258.0	265.3	275.7	284.1	293.0	305.8	316.1
Sulfuric anhydride (γ)	255.3	260.9	268.7	274.9	281.4	290.5	297.7	305.4	316.1	324.7
Sulfurous anhydride	172.7	178.7	187.4	194.5	202.2	213.3	222.6	232.8	247.7	260.3
Sulfuryl chloride		221.7	233.8	244.0	255.0	271.1	284.8	299.9	322.6	342.1
Tantalum	3925	4044	4213	4350	4496	4706	4878	5063	5331	5552
(V) bromide [90]	430.2	442.8	460.7	475.2	490.7	512.8	530.8	550.2	578.1	601.2
(V) chloride [91, 103]	341.8	354.0	371.6	386.2	401.9	424.7	443.8	464.7	495.5	521.7
(V) fluoride						367.5	392.0	419.8	463.4	502.9
Technetium	3365			3750			4360		4699	4967
(VI) fluoride [107]	221.0	228.2	238.5	246.9	255.9	268.8	279.6	291.2	308.1	322.3
(VII) oxide [108]	389.6	403.1	422.4	438.3	455.5	480.3	501.0	523.5	556.5	584.5
Tellurium [3,104–106]	775.0	806.1	851.4	889.1	930.4	991.2	1042	1099	1185	1260
(IV) chloride			479.6	497.5	516.9	544.8	568.1	593.4	630.6	662.0
(VI) fluoride	159.1	164.4	172.0	178.2	184.8	194.4	202.4	211.0	223.6	234.2
Terbium [77]	2350	2430	2520	2615	2730	2850	3060	3200	3410	3660
Tetraborane	179.1	186.2	196.5	205.1	214.5	228.3	239.9	252.8	272.2	288.9
Tetrafluorohydrazine	115.1	120.2	127.6	133.9	140.8	151.1	160.0	169.9	185.2	198.7
Tetrahydropentaborane	219.6	227.7	239.4	249.0	259.5	274.8	287.6	301.6	322.4	340.2
Tetramethylgermanium	195.9	203.6	214.8	224.1	234.3	249.2	261.9	275.9	296.8	314.9
Tetrasilane	241.2	250.0	262.7	273.1	284.5	301.0	314.8	330.0	352.5	371.6

Table 11.3 Saturation temperature (in K) of inorganic substances *(continued)*

Substance	Vapor pressure, kPa									
	0.1	0.2	0.5	1	2	5	10	20	50	100
Thallium	1086	1128	1190	1241	1297	1379	1448	1525	1640	1739
bromide		711.4	749.6	781.4	816.0	866.7	909.4	956.6	1027	1087
chloride		707.4	744.9	776.1	809.9	859.5	901.2	947.2	1015	1074
(V) fluoride	327.7	339.5	356.5	370.5	385.7	407.7	426.2	446.4	476.3	501.6
iodide	699.1	725.2	762.8	794.0	827.7	877.1	918.5	964.1	1031	1089
sulfide [89]	926.6	963.4	1016	1061	1110	1181	1242	1309	1409	1496
Thionylbromide	262.5	272.4	286.8	298.7	311.7	330.6	346.6	364.1	390.3	412.7
Thionylchloride	212.8	221.6	234.3	244.9	256.5	273.6	288.2	304.4	328.9	350.2
Thionylfluoride [109]	146.7	152.1	159.9	166.3	173.3	183.4	192.0	201.3	215.2	227.0
Thorium	3010			3340			3855		4308	4470
Tin	1853	1926	2031	2119	2215	2355	2474	2605	2802	2972
bromide		308.5	325.4	339.4	354.7	377.2	396.2	417.2	448.6	475.7
chloride	578.0	599.1	629.4	654.4	681.6	721.1	754.2	790.5	844.1	889.8
chlorine	245.4	254.6	267.8	278.7	290.6	308.0	322.6	338.6	362.4	382.7
hydride	130.9	136.5	144.6	151.4	158.9	170.1	179.6	190.3	206.5	220.7
iodide		399.7	421.9	440.4	460.6	490.4	515.5	543.4	585.3	621.5
sulfide [92]	1029	1062	1110	1149	1190	1250	1300	1354	1432	1497
Titanium	2450	2524	2630	2716	2807	2939	3046	3162	3330	3469
(IV) chloride [110]	255.4	265.3	279.5	291.4	304.3	323.1	339.1	356.6	382.9	405.4
Tribromofluorosilane	222.4	231.0	243.6	254.0	265.3	282.0	296.1	311.6	334.8	354.9
Tribromosilane	237.7	247.1	260.7	272.1	284.5	302.7	318.1	335.2	360.8	382.9
Trichlorogermane	227.3	235.4	247.0	256.5	266.9	281.8	294.3	308.0	328.1	
Trichlorosilane	188.7	196.2	206.9	215.8	225.6	239.9	252.0	265.4	285.5	302.8
Trifluorosilane	118.7	122.7	128.6	133.3	138.5	146.0	152.2	158.9	168.8	177.2
Trigermane	232.1	241.6	255.5	267.1	279.8	298.5	314.5	332.2	359.1	382.4
Trisilane	200.0	208.1	219.7	229.5	240.1	255.8	269.1	283.8	306.0	325.2
Tungsten	4098	4214	4377	4510	4650	4850	5013	5188	5438	5643
(V) bromide [69]	436.2	451.4	473.2	491.2	510.5	538.6	562.0	587.5	624.9	656.6
(V) chloride [70]		406.0	423.0	436.9	451.6	472.8	490.1	508.8	535.7	558.1
(VI) chloride [68]	414.6	428.1	447.5	463.4	480.4	504.9	525.2	547.2	579.2	606.0
(VI) fluoride	198.2	204.5	213.5	220.9	228.8	240.1	249.4	259.5	274.2	286.5
oxichloride, tetra- [68]	374.2	383.6	396.7	407.2	418.3	434.0	446.6	460.0	478.9	494.4
Uranium	2691	2788	2929	3045	3171	3354	3508	3676	3924	4136
(IV) fluoride [112]	1202	1239	1292	1335	1381	1447	1502	1560	1645	1716
(VI) fluoride	239.3	246.1	255.6	263.2	271.4	283.0	292.5	302.6	317.1	329.0
Vanadium	2519	2600	2716	2810	2912	3057	3177	3308	3497	3655
oxide fluoride, tri- [66]							336.4	349.6	368.7	384.6
oxide chloride, tri- [3, 67]	244.7	254.6	268.9	280.8	293.9	313.2	329.5	347.6	374.9	398.6
Water	250.8	259.2	271.2	281.1	291.7	307.1	319.8	333.7	353.9	370.9
Water (deuterium)			271.7	281.8	292.7	308.4	321.5	335.7	356.5	374.1
Water (deuterohydrogenous)			270.8	280.8	291.7	307.4	320.4	334.6	355.4	373.0
Water (heavy-oxygen)			269.7	279.8	290.7	306.4	319.5	333.8	354.7	372.4
Xenon	103.3	107.3	113.1	118.0	123.3	131.0	137.6	144.8	155.6	165.0
Ytterbium	912.7	945.9	993.8	1033	1076	1138	1190	1248	1332	1405
chloride [78]	1510	1571	1659	1733	1814	1932	2033	2145	2313	2459
Yttrium	2362	2443	2560	2657	2760	2910	3035	3172	3371	3540
Zinc	750.4	778.9	820.2	854.4	891.6	946.1	991.9	1042	1117	1182
bromide [113]	636.5	657.2	686.7	710.8	736.7	773.9	804.6	838.0	886.5	927.1
chloride [3, 83, 113]	684.3	705.6	735.9	760.5	786.9	824.7	855.8	889.3	937.9	978.3
dimethyl [84]	201.7	209.2	220.0	229.0	238.7	252.8	264.7	277.8	297.2	313.8
fluoride	1475	1516	1573	1618	1667	1736	1792	1852	1938	2008
sulfide [114]	1313	1352	1406	1451	1498	1566	1621	1680		

Table 11.3 Saturation temperature (in K) of inorganic substances *(continued)*

Substance	Vapor pressure, kPa									
	0.1	0.2	0.5	1	2	5	10	20	50	100
Zirconium	3190	3292	3436	3554	3680	3861	4011	4172	4407	4603
(IV) bromide	475.7	487.6	504.3	517.7	531.8	551.8	567.8	584.9	609.1	628.7
(IV) chloride	454.2	466.0	482.6	495.9	510.0	529.9	546.0	563.1	587.5	607.4
(IV) iodide	531.5	544.8	563.5	578.5	594.3	616.6	634.6	653.6	680.7	702.7

Table 11.4 Vapor pressure (higher than 10^5 Pa) of elements and some compounds under various temperatures [115]

T, K	p, 10^5 Pa	T, K	p, 10^5 Pa	T, K	p, 10^5 Pa	T, K	p, 10^5 Pa
Ammonia NH_3		136	26.31	295	59.802	1500	27.6
240	1.0258	138	29.04	296	61.205	1550	33.0
250	1.6536	140	31.64	297	62.639	1600	38.9
260	2.559	142	34.41	298	64.098	1650	45.4
270	3.819	144	37.36	299	65.598	1700	52.4
280	5.518	146	40.50	300	67.155	1750	60.0
290	7.753	148	43.83	301	68.661	1800	68.2
300	10.624	150	47.39	302	70.246	1850	77.0
310	14.249	**Carbon dioxide CO_2**		303	71.858	1900	86.2
320	18.66	216.55	5.18	304.19	73.815	1950	96.0
330	24.22	220	6.00	**Carbon oxide CO**		2000	106.3
340	30.82	225	7.34	(t, °C)		2050	117.0
360	38.70	230	8.91	−191.52	1.013	**Chlorine Cl_2** (t, °C)	
370	58.91	235	10.75	−187.79	1.520	−34.04	1.013
380	71.54	240	12.82	−184.90	2.026	−23.33	1.580
390	86.06	245	15.18	−182.55	2.532	−12.22	2.411
400	102.8	250	17.87	−180.53	3.039	−1.11	3.535
405.6	113.0	255	20.85	−177.04	4.052	10.00	5.014
Argon Ar		260	24.21	−174.17	5.065	21.11	6.909
88	1.091	265	27.87	−171.69	6.078	32.22	9.279
90	1.337	270	32.03	−169.49	7.091	43.33	12.207
92	1.622	273.15	34.839	−167.46	8.104	54.44	15.752
94	1.952	274	35.633	−163.98	10.13	65.56	19.997
96	2.329	275	36.576	−161.02	12.16	76.67	25.011
98	2.758	276	37.543	−158.32	14.18	87.78	30.886
100	3.243	277	38.521	−155.94	16.21	98.89	37.694
102	3.787	278	39.520	−153.65	18.23	110.00	45.544
104	4.397	279	40.547	−151.70	20.26	121.11	54.520
106	5.074	280	41.588	−147.18	25.32	132.22	64.751
108	5.825	281	42.654	−143.30	30.39	137.78	70.363
110	6.652	282	43.732	−140.23	34.98	143.33	76.340
112	7.562	283	44.831	**Cesium Cs**		144.00	77.089
114	8.557	284	45.956	950	1.086	**Fluorine F_2**	
116	9.643	285	47.096	1000	1.693	95.0	2.775
118	10.82	286	48.261	1050	2.527	97.5	3.465
120	12.11	287	49.450	1100	3.629	100.0	4.282
122	13.49	288	50.666	1150	5.038	102.5	5.236
124	14.99	289	51.895	1200	6.790	105.0	6.340
126	16.60	290	53.148	1250	8.889	107.5	7.602
128	18.33	291	54.432	1300	11.41	110.0	9.029
130	20.20	292	55.732	1350	15.0	112.5	10.63
132	22.19	293	57.066	1400	18.7	115.0	12.41
134	24.32	294	58.421	1450	22.9	117.5	14.39

Table 11.4 Vapor pressure of elements and some compounds *(continued)*

T, K	p, 10^5 Pa	T, K	p, 10^5 Pa	T, K	p, 10^5 Pa	T, K	p, 10^5 Pa
120.0	16.59	4.40	1.1927	152	7.218	740	72.10
122.5	19.03	4.45	1.2468	154	7.928	760	82.60
125.0	21.73	4.50	1.3026	156	8.687	780	94.17
127.5	24.70	4.55	1.3601	158	9.497	800	106.85
130.0	27.98	4.60	1.4193	160	10.36	**Molybdenum hexa-**	
132.5	31.59	4.65	1.4803	**Lithium Li**		**fluoride MoF_6** [118]	
135.0	35.57	4.70	1.5431	1650	1.283	323.2	1.73
137.5	39.93	4.75	1.6078	1700	1.771	333.2	2.36
140.0	44.69	4.80	1.6743	1750	2.399	343.2	3.15
142.5	49.87	4.85	1.7427	1800	3.191	353.2	4.15
144.0	53.25	4.90	1.8131	1850	4.179	363.2	5.33
Heavy water D_2O		4.95	1.8855	1900	5.397	373.2	6.79
(t, °C)		5.00	1.9600	1950	6.871	383.2	8.52
110	1.3728	5.05	2.0367	2000	8.639	393.2	10.50
120	1.9134	5.10	2.1158	2100	12.4	403.2	12.90
130	2.6170	5.15	2.1976	2200	18.3	413.2	15.63
140	3.518	5.19	2.2654	2300	26.1	423.2	18.72
150	4.653	**Hydrogen H_2**		2400	33.1	433.2	22.33
160	6.066	21	1.209	2500	49.1	443.2	26.25
170	7.802	22	1.584	2600	66.1	453.2	30.70
180	9.911	23	2.036	2700	87.1	463.2	35.90
190	12.445	24	2.574	2800	113.7	473.2	41.75
200	15.462	25	3.206	2900	146.5	483.2	48.55
210	19.028	26	3.942	3000	186.6	**Neon Ne**	
220	23.194	27	4.789	3100	235.4	28	1.3210
230	28.031	28	5.755	3200	295.2	29	1.7351
240	33.606	29	6.848	3300	365.8	30	2.2381
250	39.993	30	8.077	3400	449.9	31	2.8402
260	47.280	30.5	8.747	3500	550.5	32	3.5526
270	55.527	31	9.455	3600	668.2	33	4.3860
280	64.834	31.5	10.20	3700	806.7	34	5.3518
290	75.288	32	11.00	3800	968.3	35	6.4618
300	86.968	32.5	11.84	**Mercury Hg** (t, °C)		36	7.7282
305	93.323	33	12.73	360	1.0772	37	9.1637
310	100.01	33.23	13.16	380	1.5207	38	10.7820
315	107.07	**Krypton Kr**		400	2.1024	39	12.597
320	114.54	120	1.031	420	2.852	40	14.625
325	122.38	122	1.202	440	3.801	41	16.882
330	130.65	124	1.395	460	4.986	42	19.387
335	139.37	126	1.610	480	6.446	43	22.157
340	148.54	128	1.849	500	8.222	44	25.217
345	158.20	130	2.114	520	10.358	44.4	26.54
350	168.35	132	2.406	540	12.901	**Nitrogen N_2**	
355	178.98	134	2.728	560	15.899	78	1.093
360	190.16	136	3.080	580	19.403	80	1.369
365	202.12	138	3.465	600	23.46	82	1.694
370	214.68	140	3.884	620	28.14	84	2.074
371	217.22	142	4.339	640	33.47	86	2.515
Helium ^4He [117]		144	4.832	660	39.53	88	3.022
4.25	1.0401	146	5.364	680	46.36	90	3.600
4.30	1.0894	148	5.938	700	54.03	92	4.256
4.35	1.1402	150	6.556	720	62.59	94	4.995

Table 11.4 Vapor pressure of elements and some compounds *(continued)*

T, K	p, 10^5 Pa
96	5.824
98	6.748
100	7.775
102	8.910
104	10.16
106	11.53
108	13.03
110	14.67
112	16.45
114	18.36
116	20.47
118	22.72
120	25.15
122	27.77
124	30.57
126	33.57
126.25	33.96
Oxygen O_2	
90.18	1.013
92	1.221
94	1.486
96	1.793
98	2.145
100	2.546
102	3.002
104	3.515
106	4.090
108	4.731
110	5.443
112	6.229
114	7.095
116	8.045
118	9.083
120	10.21
122	11.44
124	12.78
126	14.22
128	15.77
130	17.44
132	19.24
134	21.17
136	23.24
138	25.45
140	27.82
142	30.34
144	33.04
146	35.91
148	38.97
150	42.23
152	45.69
154.77	50.87
Potassium K	
1050	1.217
1100	1.864
1150	2.745
1200	3.913
1250	5.415
1300	7.304
1350	9.628
1400	12.44
1450	16.1
1500	20.0
1550	24.6
1600	29.8
1650	35.6
1700	42.0
1750	49.2
1800	57.0
1850	65.4
1900	74.5
1950	84.3
2000	94.7
2050	105.6
2100	117.2
2150	129.4
2170	134.4
Rubidium Rb	
1000	1.467
1050	2.241
1100	3.295
1150	4.684
1200	6.466
1250	8.698
1300	11.43
1350	14.7
1400	18.6
1450	23.2
1500	28.5
1550	34.5
Sodium Na	
1200	1.504
1250	2.244
1300	3.216
1350	4.563
1400	6.256
1450	8.383
1500	11.014
1550	14.6
1600	18.6
1650	23.4
1700	29.0
1750	35.6
1800	43.0
1850	51.6
1900	61.2
1950	72.1
2000	84.1
2050	97.5
2100	112.1
2125	120.0
Sulphur hexafluoride SF_6 [118]	
278.65	14.86
285.22	17.61
293.31	21.51
303.37	27.26
313.29	33.79
317.10	36.87
Sulfurous anhydride SO_2 (t, °C)	
-10	1.01
10	2.23
20	3.30
30	4.62
40	6.30
50	8.48
60	11.15
70	14.26
80	18.02
90	22.49
100	27.77
110	33.96
120	41.12
130	49.40
140	58.91
150	69.74
157.5	78.82
Tungsten hexafluoride WF_6 [116]	
304.2	1.73
314.2	2.39
324.2	3.18
334.1	4.21
344.1	5.48
354.1	7.02
364.0	8.80
374.0	11.01
383.3	13.53
393.8	16.57
403.7	19.96
413.7	23.88
423.6	28.40
433.5	33.50
443.5	39.35
Uranium hexafluoride UF_6 [119]	
364.0	3.33
374.0	4.28
383.8	5.44
393.8	6.84
403.7	8.47
413.7	10.38
423.6	12.59
433.5	15.15
443.5	18.01
453.4	21.28
463.3	24.98
473.2	29.13
483.2	33.93
493.0	39.08
502.9	45.08
Xenon Xe	
166	1.071
168	1.199
170	1.337
172	1.488
174	1.651
176	1.827
178	2.017
180	2.222
182	2.442
184	2.678
186	2.930
188	3.200
190	3.487
192	3.794
194	4.119
196	4.465
198	4.832
200	5.220
202	5.631
204	6.064
206	6.522
208	7.004
210	7.511
212	8.045
214	8.605
216	9.194
218	9.810
220	10.46
222	11.13
224	11.84

11.3 Vapors of Organic Substances

Table 11.5 Saturation temperature (in K) of organic substances for the pressure range $p = 0.1$ — 100 kPa [3]

Substance	Vapor pressure, kPa									
	0.1	0.2	0.5	1	2	5	10	20	50	100
Acetaldehyde	187.9	194.8	204.8	213.1	222.0	235.1	246.1	258.1	276.0	291.2
Acetaldoxime	263.1	271.8	284.3	294.5	305.4	321.2	334.3	348.4	369.1	386.5
Acetamide	332.6	343.8	359.9	373.1	387.3	407.8	424.8	443.3	470.4	493.2
Acetanilide	380.9	394.2	413.4	429.2	446.3	471.0	491.6	514.1	547.2	575.2
Acetic acid	252.0	261.2	274.6	285.7	297.7	315.2	329.8	345.9	369.8	390.1
Acetic anhydride	270.1	279.6	293.3	304.7	316.9	334.6	349.4	365.6	389.4	409.6
Acetone	209.7	217.5	228.9	238.3	248.5	263.4	275.9	289.7	310.2	327.7
Acetonitrile	221.5	230.0	242.5	252.8	264.1	280.6	294.5	309.9	332.9	352.6
Acetophenone	304.5	315.8	332.1	345.5	360.1	381.5	399.4	419.0	448.2	473.0
Acetylene	127.8	132.1	138.2	143.2	148.5	156.3	162.7	169.6	179.8	188.4
Acrolein	204.3	212.1	223.4	232.8	243.0	258.0	270.6	284.5	305.3	323.1
Acrylic acid	272.5	282.1	296.0	307.5	319.8	337.8	352.8	369.2	393.3	413.8
Acrylonitrile	217.5	226.1	238.4	248.7	259.9	276.4	290.3	305.7	328.8	348.7
Adipic acid	425.8	438.9	457.6	472.8	489.1	512.4	531.6	552.3	582.2	607.1
Allene	148.9	154.6	162.8	169.7	177.2	188.1	197.4	207.5	222.7	235.8
Allyl alcohol	249.4	257.7	269.7	279.5	290.1	305.3	318.0	331.7	351.8	368.7
Allyldichloroethylsilane	265.1	275.4	290.1	302.4	315.8	335.3	351.8	370.1	397.2	420.6
Allylisopropyl ether	225.5	233.9	246.1	256.1	267.0	283.0	296.4	311.1	332.9	351.5
Allylisothiocyanate	267.8	278.1	293.0	305.4	318.8	338.5	355.1	373.5	400.8	424.3
Allylpropyl ether	229.4	238.2	250.8	261.3	272.6	289.3	303.3	318.8	341.8	361.5
Allyltrichlorosilane	248.8	258.1	271.5	282.6	294.6	312.2	327.0	343.3	367.4	388.0
Amyl alcohol	281.5	290.5	303.3	313.8	325.1	341.2	354.6	369.0	390.0	407.5
tert-Amyl alcohol	256.3	264.6	276.4	286.1	296.5	311.5	323.9	337.3	356.8	373.1
Amylisopropionate	276.7	287.1	302.0	314.4	327.8	347.5	364.0	382.1	409.0	432.1
Amyltrimethylsilane	258.8	268.8	283.1	295.0	308.0	327.0	342.9	360.6	386.9	409.5
Aniline	300.4	310.9	325.9	338.3	351.7	371.1	387.2	404.8	430.7	452.6
2-Anilineethanol	371.3	383.8	401.8	416.5	432.4	455.3	474.3	495.0	522.2	550.7
Anisaldehyde	340.5	352.7	370.2	384.6	400.3	422.9	441.9	462.6	493.2	519.2
Anisole	273.3	283.5	298.3	310.5	323.8	343.2	359.4	377.3	403.9	426.7
Benzaldehyde	292.6	303.1	318.1	330.6	344.0	363.5	379.9	397.7	424.1	446.4
Benzene	229.3	237.3	248.9	258.4	268.7	283.6	296.0	309.5	329.5	346.4
Benzenesulfonylchloride	333.1	345.7	363.8	378.9	395.2	419.1	439.2	461.3	494.2	522.4
Benzildichlorosilane	312.9	323.5	338.7	351.2	364.7	384.1	400.2	417.8	443.5	465.1
Benzoic acid	360.9	372.2	388.4	401.6	415.7	435.9	452.6	470.6	496.7	518.5
Benzonitrile	312.0	323.3	339.6	353.1	367.7	388.9	406.6	426.1	454.8	479.3
Benzoyl bromide	314.6	326.3	343.2	357.2	372.3	394.5	413.1	433.5	463.9	489.8
Benzoyl chloride	299.7	310.9	327.1	340.5	355.0	376.3	394.2	413.8	443.0	468.0
Benzyl alcohol	323.6	334.1	349.2	361.5	374.8	393.9	409.6	426.7	451.6	472.5
Benzylamine	296.8	307.5	323.0	335.7	349.5	369.6	386.4	404.9	432.1	455.2
Benzylisothiocyanate	347.0	358.8	375.5	389.3	404.1	425.5	443.2	462.5	490.8	514.6
Bistrichlorosilane			331.0	344.7	359.5	381.2	399.4	419.5	449.3	474.8
Borinecarbonyl	131.3	136.3	143.7	149.8	156.4	166.1	174.3	183.3	196.8	208.5
4-Bromanisole	316.3	328.1	345.2	359.3	374.7	397.1	416.0	436.7	467.5	493.8
Bromobenzene	270.0	281.1	295.9	308.2	321.6	341.2	357.8	375.8	402.7	425.9
2-Bromo-1,4-xylene	304.8	316.3	332.8	346.5	361.3	383.0	401.3	421.4	451.2	476.7
1-Bromo-1-chloroethane	230.6	238.7	250.4	260.1	270.5	285.6	298.2	311.9	332.2	349.3
1-Bromo-2-chloroethane	239.6	248.7	261.9	272.8	284.7	302.1	316.7	332.8	356.8	377.4

Table 11.5 Saturation temperature (in K) of organic substances *(continued)*

Substance	Vapor pressure, kPa									
	0.1	0.2	0.5	1	2	5	10	20	50	100
3-Bromo-2,4,6-trichlorophenol	380.2	393.8	413.2	429.3	446.6	471.8	492.8	515.8	549.7	578.5
1-Bromo-2-butanol	292.5	301.5	314.3	324.8	335.9	351.9	365.0	379.2	399.7	416.7
1-Bromo-2-butanone	274.6	284.4	298.4	310.0	322.6	340.8	356.0	372.7	397.2	418.1
cis-2-Bromo-2-butene	229.4	238.3	251.1	261.8	273.4	290.4	304.8	320.7	344.4	364.7
trans-2-Bromo-2-butene	223.5	232.2	244.8	255.2	266.6	283.3	297.4	312.9	336.2	356.2
2-Bromo-2-nitroisopropane	235.3	243.5	255.3	265.0	275.4	290.6	303.2	317.0	337.2	354.3
1-Bromo-3-methylbutane	247.7	257.2	270.8	282.2	294.5	312.5	327.8	344.5	369.5	390.9
2-Bromo-4,6-dichlorophenol	351.4	364.1	382.5	397.6	414.0	437.9	457.9	479.7	512.1	539.6
1-Bromo-4-ethylbenzene	292.9	305.0	322.6	337.3	353.5	377.3	397.7	420.3	454.5	484.3
Bromoacetic acid	322.7	333.6	349.3	362.2	376.0	396.1	412.7	430.8	457.3	479.6
1-Bromobutane	235.5	244.5	257.6	268.4	280.2	297.5	312.0	328.1	352.0	372.6
1,4-Bromochlorobenzene	299.7	310.9	327.1	340.5	355.1	376.3	394.2	413.8	442.9	467.9
2-Bromoethyl-2-chloroethyl ether	304.2	315.2	331.1	344.1	358.3	378.8	396.1	414.9	442.8	466.4
1-Bromoethylene	174.3	181.5	192.0	200.7	210.2	224.4	236.4	249.8	270.0	287.6
1-Bromopropane	215.6	226.9	235.9	245.8	256.7	272.5	285.9	300.7	322.7	341.6
2-Bromopropane	207.0	215.1	226.8	236.5	247.1	262.7	275.9	290.4	312.1	330.9
3-Bromopyridine	284.3	295.0	310.3	323.0	336.8	357.0	373.9	392.5	420.2	443.8
2-Bromotoluene	291.4	302.1	317.4	330.1	343.9	364.0	380.8	399.2	426.4	449.7
3-Bromotoluene	287.4	298.6	314.8	328.4	343.1	364.7	383.0	403.1	433.3	459.4
4-Bromotoluene	283.2	294.6	311.1	324.9	340.0	362.3	381.2	402.2	433.7	461.0
α-Bromotoluene	299.8	311.1	327.4	341.0	355.7	377.2	395.3	415.2	444.8	470.1
Butacrilate	267.4	277.4	291.9	303.9	316.9	335.9	351.9	369.4	395.5	417.9
sec-Butylglycolate	296.3	306.8	321.8	334.1	347.5	366.8	383.0	400.6	426.5	448.5
1,2-Butadiene	179.9	186.9	197.2	205.8	215.1	228.8	240.4	253.2	272.4	289.0
1,3-Butadiene	166.6	173.2	182.6	190.5	199.0	211.6	222.2	234.0	251.6	266.7
Butane	168.0	174.6	184.2	192.3	201.0	213.8	224.7	236.7	254.7	270.3
1,3-Butanediol	299.0	311.3	329.1	343.9	360.2	384.3	404.7	427.5	461.8	491.6
2,3-Butanediol	312.4	322.5	337.0	348.8	361.4	379.7	394.7	411.0	434.8	454.6
1,2,3-Butanetriol	370.3	382.3	399.3	413.2	428.1	449.5	467.2	486.4	514.2	537.5
2-Butanone	220.3	228.8	241.1	251.3	262.5	278.9	292.6	307.9	330.6	350.2
1-Butene	164.8	171.3	180.6	188.5	197.0	209.5	220.1	231.9	249.4	264.6
cis-2-Butene	172.6	179.3	188.9	196.9	205.6	218.4	229.2	241.1	258.8	274.1
trans-2-Butene	169.8	176.4	186.0	194.0	202.7	215.5	226.3	238.2	256.1	271.4
3-Butenenitrile	248.7	258.1	271.6	282.8	294.9	312.7	327.6	344.1	368.5	389.4
Butenine	176.1	182.7	192.3	200.2	208.8	221.5	232.1	243.7	261.1	275.9
1-Butine	176.7	183.5	193.2	201.4	210.2	223.2	234.1	246.1	264.0	279.4
2-Butine	194.7	201.5	211.4	219.5	228.3	241.1	251.7	263.3	280.4	294.8
Butyl alcohol	268.1	276.7	289.0	299.1	309.9	325.4	338.2	352.0	372.2	389.0
sec-Butyl alcohol	256.8	265.0	276.6	286.1	296.2	310.8	322.9	335.9	354.8	370.6
tert-Butyl alcohol	247.9	255.5	266.4	275.2	284.6	298.1	309.2	321.1	338.4	352.8
Butyl formate	242.2	251.2	264.2	275.0	286.7	303.8	318.1	333.9	357.3	377.3
sec-Butyl formate	234.3	243.0	255.6	266.0	277.4	293.9	307.8	323.0	345.7	365.1
tert-Butyl formate	235.6	244.4	257.2	267.8	279.3	296.1	310.3	325.8	348.9	368.7
sec-Butyl chloroacetate	285.0	295.3	310.3	322.7	336.1	355.6	371.9	389.8	416.3	438.9
sec-Butyl chloride	208.8	217.3	229.5	239.8	251.0	267.4	281.4	297.0	320.4	340.7
tert-Butyl chloride					236.9	252.8	266.3	281.3	303.9	323.6
Butyltrimethylsilane	244.8	254.1	267.5	278.7	290.8	308.5	323.5	339.9	364.4	385.4
Butyric acid	293.6	303.5	317.7	329.3	341.8	359.8	374.8	391.1	414.9	435.0
Butyronitrile	248.1	257.4	270.8	282.0	294.0	311.7	326.5	342.8	367.0	387.7
Caprilonitrile	308.4	319.5	335.5	348.7	362.9	383.7	401.0	420.0	448.1	471.9
Caproic acid	152.8	159.1	168.2	175.8	184.1	196.4	206.9	218.5	236.0	251.2
Capronitrile	276.9	287.4	302.5	315.0	328.6	348.5	365.2	383.7	411.1	434.6

Table 11.5 Saturation temperature (in K) of organic substances *(continued)*

Substance	Vapor pressure, kPa									
	0.1	0.2	0.5	1	2	5	10	20	50	100
Caprylic acid	355.5	366.2	381.4	393.7	406.8	425.6	441.1	457.6	481.6	501.4
Caprylic aldehyde	342.9	350.7	361.6	370.3	379.4	392.1	402.4	413.2	428.3	440.6
Carbon selenosulfide	221.4	230.1	242.8	253.3	264.7	281.6	295.8	311.6	335.2	355.6
Carbon disulfide	196.3	204.1	215.4	224.9	235.2	250.3	263.2	277.4	298.8	317.2
Carbon dioxide	135.8	140.1	146.1	151.0	156.3	163.8	170.0	176.7	186.4	194.5
Carbon oxide	49.9	51.8	54.7	57.1	59.7	63.5	66.8	70.4	75.7	80.4
Carbon oxysulfide	138.0	143.4	151.2	157.7	164.8	175.2	184.0	193.7	208.3	220.8
Carbon subdioxide	173.9	180.7	190.6	198.8	207.7	220.9	232.0	244.3	262.7	278.6
Carbon subsulfide	282.4	293.2	308.9	322.0	336.2	357.0	374.5	393.9		
Carbon tetrabromide					345.3	366.8	384.8	404.7	434.5	460.0
Carbon tetrachloride	218.8	227.3	239.5	249.6	260.6	276.8	290.4	305.4	327.8	347.1
Carbon tetrafluoride	88.3	91.9	97.3	101.7	106.7	113.9	120.1	127.0	137.4	146.6
Carbonylselenide	305.8	316.4	331.7	344.3	357.9	377.6	394.0	411.9	438.1	460.4
Chloralhydrate	260.0	268.0	279.2	288.3	298.0	312.0	323.4	335.8	353.6	368.3
2-Chloraniline	312.2	323.4	339.5	352.8	367.2	388.2	405.6	424.8	453.0	477.0
4-Chloraniline	326.1	338.0	355.0	369.1	384.4	406.6	425.2	445.6	475.7	501.3
2-(2-Chlorethoxy)-ethanol	321.1	331.5	346.3	358.5	371.5	390.3	405.8	422.6	447.0	467.5
2-Chlorethyl-2-chlor-isopropyl ether	292.3	303.0	318.4	331.1	344.8	364.9	381.6	400.0	427.2	450.4
2-Chlorethyl-2-chloropropyl ether	297.3	308.4	324.5	337.8	352.3	373.4	391.1	410.6	439.6	464.4
bis-2-Chlorethylacetal	324.3	335.4	351.4	364.5	378.6	399.1	416.1	434.6	461.8	484.7
2-Chlorethylchloracetate	313.3	324.4	340.4	353.6	367.8	388.5	405.7	424.6	452.4	475.9
1-Chlorethylene	164.1	170.3	179.2	186.6	194.7	206.4	216.3	227.2	243.4	257.3
2-Chloro-α,α,α-trifluorotoluene	267.7	277.9	292.7	305.0	318.4	338.0	354.5	372.7	399.9	423.2
1-Chloro-2-ethoxybenzene	313.1	324.4	340.6	354.1	368.6	389.7	407.3	426.7	455.2	479.5
1-Chloro-2-ethylbenzene	284.6	295.4	311.0	323.9	337.9	358.5	375.7	394.8	423.1	447.4
1-Chloro-3-ethylbenzene	286.0	297.0	312.8	325.9	340.1	361.0	378.6	398.0	426.9	451.7
1-Chloro-4-ethylbenzene	287.0	298.1	314.1	327.5	342.0	363.3	381.2	401.0	430.6	456.0
1-Chloro-1,2,2-trifluorethylene	153.7	159.5	168.0	175.1	182.8	194.0	203.4	213.9	229.4	242.7
Chloroacetic acid	311.0	321.5	336.5	348.8	362.1	381.2	397.1	414.4	439.6	460.9
Chloroacetic anhydride	335.9	346.9	362.5	375.3	389.1	408.9	425.3	443.1	468.9	490.6
Chlorobenzene	254.9	264.7	278.7	290.4	303.1	321.7	337.3	354.5	380.2	402.3
1-Chlorobutane	219.6	228.1	240.4	250.6	261.7	278.0	291.8	306.9	329.6	349.1
α-Chlorocrotonic acid	338.3	348.8	363.7	375.9	388.9	407.5	422.8	439.3	463.2	483.1
Chlorodimethylphenylsilane	296.7	307.8	323.9	337.2	351.7	372.8	390.6	410.1	439.2	464.1
Chlorodifluoromethane	147.5	153.0	161.0	167.6	174.7	185.2	194.0	203.6	217.9	230.2
Chloromethylsilane	174.2	181.0	191.0	199.2	208.2	221.5	232.7	245.0	263.6	279.6
4-Chlorophenityl alcohol	351.4	363.7	381.3	395.9	411.6	434.3	453.3	473.9	504.3	530.1
2-Chlorophenol	279.6	290.4	306.1	319.1	333.3	354.2	371.7	391.1	420.1	445.0
3-Chlorophenol	311.5	323.1	339.8	353.6	368.5	390.4	408.8	428.9	458.8	484.4
4-Chlorophenol	317.3	329.0	345.9	359.8	375.0	397.0	415.5	435.8	465.9	491.6
Chloropicrine	243.1	252.3	265.6	276.7	288.7	306.3	321.1	337.4	361.6	382.4
1-Chloropropane	200.8	208.4	219.5	228.7	238.7	253.3	265.6	279.2	299.4	316.7
2-Chloropropane	190.5	198.0	208.9	218.0	227.8	242.4	254.7	268.3	288.7	306.4
1-Chloropropylene	187.6	195.3	206.3	215.6	225.7	240.6	253.3	267.4	288.6	307.0
3-Chloropropylene	198.8	206.5	217.5	226.7	236.7	251.4	263.7	277.4	297.7	315.2
γ-Chloropropyltrichloro-silane			315.4	328.4	342.6	363.4	380.9	400.1	428.7	453.3
2-Chloropyridine	280.4	291.1	306.4	319.2	333.0	353.3	370.4	389.1	417.1	441.1
3-Chlorostyrene	292.3	303.4	319.5	332.8	347.3	368.6	386.5	406.2	435.5	460.7
4-Chlorostyrene	295.1	306.1	322.0	335.3	349.6	370.5	388.1	407.5	436.2	460.8

Table 11.5 Saturation temperature (in K) of organic substances *(continued)*

Substance	Vapor pressure, kPa									
	0.1	0.2	0.5	1	2	5	10	20	50	100
2-Chlorotoluene	272.9	283.3	298.3	310.8	324.4	344.2	360.9	379.4	406.8	430.4
3-Chlorotoluene	272.8	283.3	298.6	311.2	325.0	345.3	362.3	381.2	409.3	433.5
4-Chlorotoluene	273.7	284.2	299.5	312.1	325.8	346.0	362.9	381.6	409.5	433.5
α-Chlorotoluene	289.8	300.6	316.0	328.8	342.6	362.8	379.8	398.4	426.0	449.6
Chlorotriethylsilane	263.3	273.4	288.0	300.1	313.3	332.6	348.9	366.8	393.6	416.6
Chlorotrifluoromethane	121.0	125.5	132.1	137.5	143.4	152.1	159.4	167.4	179.3	189.4
Chlorotrimethylsilane	206.3	214.3	225.8	235.4	245.9	261.3	274.2	288.6	310.0	328.4
Citraconic anhydride	314.2	325.7	342.2	355.8	370.6	392.1	410.2	429.9	459.2	484.1
2-Cresol	318.0	328.6	343.9	356.4	369.9	389.3	405.4	422.9	448.5	470.1
3-Cresol	319.3	329.9	345.1	357.6	370.9	390.2	406.2	423.6	448.9	470.2
4-Cresol			335.4	347.5	360.6	379.3	394.9	411.7	436.4	457.1
α-Crotonic acid	301.4	311.4	325.6	337.3	349.9	367.9	382.9	399.1	422.8	442.7
β-Crotonic acid	238.8	248.0	261.4	272.5	284.5	302.2	317.2	333.7	358.3	379.4
cis-Crotononitrile	248.7	258.5	272.0	283.4	295.8	314.0	329.3	346.2	371.3	393.0
trans-Crotononitrile	323.5	334.9	351.3	364.8	379.4	400.5	418.2	437.5	465.9	490.0
Cyan bromide	234.1	241.2	251.4	259.7	268.5	281.1	291.5	302.6	318.8	332.2
Cyan chloride	193.1	199.5	208.6	216.0	224.0	235.5	245.0	255.4	270.4	283.1
Cyan fluoride	137.1	141.5	147.9	153.2	158.8	166.9	173.6	180.8	191.4	200.2
Cyan iodide	294.6	303.4	315.8	325.9	336.7	352.1	364.7	378.3	397.8	413.9
2-Cyano-2-butylacetate	310.4	321.2	336.7	349.4	363.2	383.1	399.8	417.9	444.5	467.0
Cyclobutane	177.0	183.9	194.0	202.3	211.4	224.8	236.1	248.7	267.4	283.5
Cyclobutene	170.4	177.2	186.9	195.0	203.8	216.8	227.8	240.0	258.2	273.9
Cyclohexane	222.9	231.4	243.5	253.6	264.5	280.6	294.0	308.8	330.9	349.8
Cyclohexanethanol	318.0	329.0	344.7	357.6	371.5	391.6	408.4	426.6	453.3	475.9
Cyclohexanol	288.7	298.7	312.9	324.7	337.3	355.6	370.8	387.4	411.7	432.2
Cyclohexanone	268.7	279.1	294.0	306.4	319.9	339.7	356.4	374.8	402.3	425.9
Cyclopentane	201.1	208.9	220.1	229.5	239.6	254.6	267.2	281.1	301.8	319.7
Cyclopropane	153.1	158.8	166.9	173.6	180.8	191.4	200.3	210.1	224.5	236.8
Dehydracetic acid	359.0	371.4	389.4	404.1	420.0	443.0	462.2	483.1	513.9	539.9
Di-(2-bromethyl)-ether	315.1	326.5	342.8	356.4	371.0	392.3	410.1	429.6	458.4	482.9
Di-(2-chloroethoxy)-methane	320.5	331.8	348.2	361.6	376.1	397.2	414.8	434.1	462.4	486.4
Di-(chloroethyl)-ether	291.1	301.7	317.1	329.8	343.5	363.5	380.3	398.7	425.9	449.1
Di-(2-methoxyethyl) ether	281.0	291.2	305.8	317.8	330.9	349.8	365.7	383.0	408.7	430.5
Di-(nitrosoethyl)-amine	285.8	296.6	312.1	324.9	338.9	359.3	376.5	395.4	423.4	447.5
Di-(nitrosomethyl)-amine	270.8	281.0	295.7	307.9	321.1	340.5	356.7	374.6	401.2	424.0
Diacetamide	336.7	347.8	363.6	376.7	390.6	410.8	427.5	445.5	471.9	494.1
Diallyldichlorosilane	277.2	287.6	302.8	315.4	329.0	349.0	365.8	384.3	411.8	435.4
Diallylsulfide	258.6	268.5	282.9	294.8	307.8	326.8	342.8	360.5	386.8	409.4
1,2-Dibromethane	245.6	255.9	270.9	283.5	297.3	317.7	335.2	354.6	384.1	409.9
2,3-Dibromo-1-propanol	324.5	335.9	352.2	365.7	380.2	401.3	418.9	438.1	466.3	490.2
1,2-Dibromo-2-methyl-propane	238.5	249.3	265.0	278.4	293.1	315.2	334.2	355.7	388.7	418.0
1,3-Dibromo-2-methyl-propane	282.0	292.6	308.0	320.7	334.6	354.8	371.8	390.5	418.3	442.2
1,4-Dibromobenzene	320.7	331.9	347.9	361.1	375.4	396.0	413.2	432.0	459.6	482.9
meso-2,3-Dibromobutane	269.6	280.0	295.0	307.5	321.1	341.0	357.8	376.3	404.0	427.8
dl-2,3-Dibromobutane	272.5	283.0	298.1	310.7	324.4	344.5	361.4	380.1	407.9	431.8
1,2-Dibromobutane	275.4	286.1	301.4	314.2	328.1	348.4	365.6	384.6	412.8	437.2
1,4-Dibromobutane	299.5	310.7	326.9	340.4	355.0	376.3	394.2	413.9	443.2	468.2
(1,2-Dibromoethyl)-benzene	353.8	365.8	382.9	397.0	412.1	434.0	452.2	471.9	500.8	525.1
α,β-Dibromomaleic anhydride	317.8	329.3	345.7	359.3	374.1	395.5	413.4	433.0	461.9	486.6
Dibromomethane	234.0	242.9	255.8	266.5	278.2	295.2	309.6	325.4	349.0	369.2
1,2-Dibromopentane	287.9	298.5	313.7	326.3	340.0	359.9	376.6	394.9	422.0	445.2
1,2-Dibromopropane	261.1	271.0	285.4	297.3	310.2	329.1	345.1	362.6	388.8	411.2

Table 11.5 Saturation temperature (in K) of organic substances *(continued)*

Substance	Vapor pressure, kPa									
	0.1	0.2	0.5	1	2	5	10	20	50	100
2,3-Dibromopropylene	262.1	272.1	286.4	298.4	311.3	330.3	346.2	363.8	389.9	412.3
Dibutyldisulfide	299.5	315.2	338.8	359.1	382.0	417.1	448.3	484.6	542.6	596.6
Dibutylsulfide	292.5	303.4	319.2	332.2	346.3	367.0	384.3	403.4	431.7	455.9
3,4-Dichloro-α,α,α-trifluorotoluene	279.8	290.7	306.5	319.7	334.0	355.0	372.8	392.4	421.8	447.1
1,2-Dichloro-1,1,2,2-tetrafluorethane	174.2	180.8	190.4	198.3	207.0	219.6	230.2	241.9	259.3	274.3
1,2-Dichloro-1,2-difluorethylene	187.5	194.5	204.5	212.8	221.9	235.0	246.1	258.2	276.2	291.6
1,1-Dichloro-2-methylpropane	237.9	247.1	260.3	271.4	283.4	301.0	315.9	332.3	356.7	377.8
1,2-Dichloro-2-methylpropane	242.7	251.7	264.8	275.6	287.3	304.4	318.8	334.6	358.1	378.1
1,3-Dichloro-2-methylpropane	265.7	275.3	289.0	300.4	312.7	330.7	345.6	362.0	386.3	406.9
1,4-Dichloro-2-ethylbenzene	306.3	318.3	335.6	350.0	365.6	388.7	408.1	429.6	461.8	489.5
1,3-Dichloro-2-propanol	295.6	305.9	320.7	332.8	346.0	365.0	380.9	398.2	423.6	445.1
1,2-Dichloro-3-ethylbenzene	313.7	325.6	342.8	357.0	372.5	395.2	414.3	435.4	466.7	493.5
1,2-Dichloro-4-ethylbenzene	315.0	327.1	344.7	359.2	375.0	398.2	417.8	439.4	471.5	499.2
Dichloro-4-tolylsilane	313.4	324.1	339.4	352.0	365.5	385.1	401.3	419.0	444.9	466.8
Dichloroacetic acid	311.6	322.3	337.6	350.2	363.8	383.4	399.7	417.5	443.6	465.6
1,2-Dichlorobenzene	287.7	298.4	314.0	326.9	340.9	361.3	378.5	397.4	425.5	449.5
1,3-Dichlorobenzene	280.3	291.0	306.6	319.5	333.5	354.1	371.4	390.6	419.1	443.6
1,4-Dichlorobenzene			309.2	322.1	336.1	356.6	373.9	393.0	421.3	445.7
2,3-Dichlorobutane	243.1	252.5	266.1	277.4	289.6	307.7	322.9	339.7	364.8	386.3
1,2-Dichlorobutane	244.1	253.8	267.9	279.6	292.3	311.1	327.0	344.6	371.0	393.9
Dichlorodiethylsilane	258.8	268.3	282.1	293.4	305.7	323.7	338.7	355.2	379.7	400.5
Dichlorodifluoromethane	151.7	157.6	166.1	173.1	180.8	192.1	201.6	212.1	227.8	241.3
Dichlorodiisopropyl ether	297.6	308.2	323.5	336.1	349.7	369.5	386.0	404.0	430.7	453.3
Dichlorodimethylsilane	216.0	224.2	236.1	246.0	256.7	272.4	285.6	300.2	321.9	340.6
1,2-Dichloroethane	224.6	233.1	245.4	255.7	266.7	283.0	296.6	311.7	334.0	353.2
Dichloroethoxymethylsilane	235.3	244.2	257.2	267.9	279.6	296.7	311.1	326.9	350.5	370.8
Dichloroethoxyphenylsilane	319.3	330.9	347.7	361.6	376.6	398.6	416.9	437.0	466.8	492.2
1,1-Dichloroethylene	192.1	199.4	210.0	218.7	228.2	242.0	253.7	266.6	285.7	302.1
cis-1,2-Dichloroethylene	211.2	219.1	230.5	239.9	250.2	265.2	277.8	291.6	312.1	329.7
trans-1,2-Dichloroethylene	204.0	211.6	222.6	231.6	241.5	255.9	268.0	281.3	301.0	317.9
Dichloroethylphenylsilane	315.4	327.4	344.8	359.2	374.8	397.7	417.0	438.2	469.9	497.0
Dichlorofluoromethane	178.5	185.2	195.0	203.1	211.9	224.8	235.6	247.5	265.3	280.5
1,5-Dichloro-hexamethyltrisiloxane	293.6	304.4	320.0	332.9	346.8	367.2	384.3	403.0	430.8	454.5
Dichloromethane	199.4	206.9	217.7	226.6	236.3	250.5	262.4	275.5	295.0	311.6
Dichloromethylarsane	256.4	266.2	280.4	292.1	304.9	323.6	339.4	356.7	383.6	404.8
Dichloromethylphenylsilane	302.9	314.4	330.9	344.6	359.6	381.4	399.7	419.9	450.0	475.8
Dichloromethylsilane	193.7	201.4	212.5	221.7	231.8	246.7	259.2	273.1	293.9	311.9
1,7-Dichlorocta-methyltetrasiloxane	320.4	332.1	348.8	362.7	377.7	399.6	417.9	438.0	467.7	493.0
2,4-Dichlorophenol	320.6	331.7	347.6	360.6	374.7	395.1	412.1	430.5	457.7	480.6
2,6-Dichlorophenol	327.0	338.3	354.6	368.0	382.4	403.3	420.7	439.6	467.5	491.0
α,α-Dichloro-phenylacetonitrile	323.4	335.0	351.7	365.5	380.4	402.1	420.2	440.0	469.3	494.2
Dichlorophenylarsine	332.9	345.8	364.5	380.1	397.0	421.8	442.8	465.9	500.5	530.3
1,2-Dichloropropane	230.3	239.2	252.2	262.9	274.7	291.9	306.4	322.5	346.5	367.1
2,3-Dichlorostyrene	328.1	340.1	357.4	371.6	387.1	409.6	428.5	449.2	479.8	505.9
2,4-Dichlorostyrene	320.7	332.5	349.5	363.5	378.7	400.9	419.5	439.9	470.1	495.9
2,5-Dichlorostyrene	322.3	334.1	351.2	365.3	380.5	402.8	421.4	441.9	472.2	498.0
2,6-Dichlorostyrene	315.3	326.9	343.7	357.5	372.5	394.4	412.8	432.9	462.7	488.2
3,4-Dichlorostyrene	324.3	336.2	353.4	367.5	382.9	405.3	424.0	444.6	475.0	501.0

Table 11.5 Saturation temperature (in K) of organic substances *(continued)*

Substance	Vapor pressure, kPa									
	0.1	0.2	0.5	1	2	5	10	20	50	100
3,5-Dichlorostyrene	321.3	333.0	350.0	363.9	379.1	401.2	419.7	440.0	470.0	495.6
1,3-Dichloro-tetramethyldisiloxane	260.8	270.6	284.7	296.4	309.2	327.8	343.4	360.6	386.1	408.0
α, α-Dichlorotoluene	302.3	314.1	331.3	345.5	361.1	383.9	403.2	424.6	456.5	484.1
1,2-Dichlorotrichlorosilane			314.9	328.0	342.1	362.9	380.3	399.5	428.0	452.5
Dicyan	174.9	180.4	188.1	194.4	201.2	210.8	218.8	227.4	239.8	250.2
Diethoxydimethylsilane	249.0	258.1	271.2	281.9	293.6	310.6	324.9	340.4	363.5	383.2
1,2-Diethoxyethane	234.7	244.4	258.5	270.3	283.2	302.3	318.5	336.6	363.9	387.7
1,3-Diethoxytet-ramethylsiloxane	282.8	292.9	307.3	319.3	332.2	351.0	366.6	383.8	409.0	430.5
Diethyl ether	195.3	202.7	213.3	222.0	231.5	245.4	257.1	270.0	289.1	305.4
Diethyl-*d*-tartrate	369.2	381.9	400.0	414.9	430.9	454.1	473.4	494.4	525.2	551.1
Diethyl-*dl*-tartrate	367.4	380.2	398.4	413.4	429.6	453.0	472.6	493.8	525.0	551.4
Diethylacetal	245.5	254.2	266.7	277.1	288.2	304.4	317.9	332.7	354.4	372.9
Diethylamine			226.1	235.7	246.0	261.2	274.1	288.2	309.3	327.4
Diethylcarbonate	258.2	267.5	280.9	292.0	304.0	321.5	336.1	352.1	375.7	395.8
Diethyldichlorosilane			276.1	287.9	300.7	319.4	335.3	352.8	379.0	401.5
Diethyldifluorosilane	212.9	220.7	231.9	241.2	251.2	265.8	278.1	291.5	311.4	328.4
Diethyldioxysuccinate	337.5	349.0	365.6	379.3	394.0	415.2	432.9	452.1	480.4	504.2
Diethyleneglycol	360.3	371.7	387.9	401.1	415.3	435.6	452.4	470.5	496.8	518.7
Diethyleneglycol, butyl ether	334.6	345.9	362.1	375.4	389.7	410.4	427.5	446.2	473.5	496.5
Diethyleneglycol, ethyl ether	313.2	324.2	340.1	353.1	367.3	387.7	404.8	423.5	450.9	474.2
Diethyleneglycol-*bis*-chloroacetate	417.3	429.8	447.4	461.8	477.1	499.0	516.9	536.2	564.0	587.0
Diethylfumarate	320.7	332.2	348.7	362.4	377.1	398.6	416.5	436.2	465.1	489.7
Diethylisosuccinate	307.2	318.3	334.4	347.6	361.9	382.8	400.3	419.4	447.7	471.8
Diethylmalate	347.7	359.9	377.4	391.7	407.3	429.8	448.6	469.0	499.2	524.6
Diethylmaleinate	324.6	336.2	353.0	366.8	381.8	403.6	421.8	441.7	471.1	496.0
Diethylmalonate	308.4	319.5	335.4	348.6	362.8	383.5	400.8	419.7	447.7	471.4
Diethyloxalate	315.2	325.3	339.7	351.4	364.1	382.2	397.2	413.3	436.9	456.5
Diethylselenide	242.9	252.0	265.0	275.8	287.5	304.6	319.0	334.7	358.2	378.1
Diethylsuccinate	322.5	333.8	350.1	363.6	378.1	399.1	416.7	435.8	464.0	487.9
Diethylsulfate	314.3	325.6	341.7	355.1	369.5	390.4	407.9	427.0	455.3	479.2
Diethylsulfide	229.7	238.4	250.8	261.1	272.3	288.6	302.3	317.4	339.9	359.0
Diethylsulfite	277.4	287.6	302.3	314.4	327.6	346.8	362.9	380.5	406.7	429.0
Diethyl zinc	245.9	255.3	268.9	280.2	292.5	310.6	325.7	342.5	367.4	388.9
1,1-Difluorethane	157.9	163.7	172.1	179.1	186.6	197.6	206.8	217.0	232.0	244.8
Diisobutylamine	263.1	273.0	287.2	298.9	311.7	330.4	346.1	363.3	388.9	410.8
Diisopropyl ether	211.7	219.9	231.8	241.8	252.6	268.4	281.8	296.6	318.7	337.8
Diisopropyloxalate	311.2	321.9	337.2	349.8	363.3	382.9	399.2	417.0	443.0	465.0
1,2-Dimethoxyethane	220.3	229.4	242.6	253.6	265.6	283.4	298.6	315.5	340.9	363.1
N, *N*-Dimethylaniline	297.4	308.6	324.6	338.0	352.4	373.6	391.3	410.9	439.9	464.7
Dimethyl ether	154.8	160.9	169.7	177.1	185.1	196.9	206.9	217.9	234.5	248.8
Dimethyl-*d*-tartrate	369.2	381.9	400.0	414.8	430.8	454.0	473.3	494.3	525.0	550.9
Dimethyl-*dl*-tartrate	367.2	380.0	398.5	413.7	430.0	453.8	473.6	495.2	527.0	553.9
Dimethyl-*l*-malate	342.5	354.4	371.4	385.4	400.5	422.4	440.6	460.5	489.6	514.3
3,5-Dimethyl-1,2-pyrone	346.1	357.9	374.8	388.7	403.6	425.1	443.1	462.6	491.1	515.2
3,3-Dimethyl-2-thiobutane	233.8	242.7	255.5	266.1	277.6	294.5	308.7	324.3	347.6	367.5
2,5-Dimethyl-3-pentanone	273.9	282.6	295.0	305.2	316.0	331.6	344.5	358.4	378.6	395.4
Dimethylamine	181.6	188.2	197.6	205.4	213.9	226.1	236.4	247.6	264.2	278.3
Dimethylantimony	311.9	323.4	339.9	353.6	368.5	390.1	408.3	428.2	457.7	482.9
Dimethylarsanilate	283.5	293.6	308.2	320.2	333.2	352.0	367.8	385.1	410.5	432.1
Dimethylborane	163.4	170.1	179.8	188.0	196.9	210.1	221.2	233.7	252.5	268.8

Table 11.5 Saturation temperature (in K) of organic substances *(continued)*

Substance	Vapor pressure, kPa									
	0.1	0.2	0.5	1	2	5	10	20	50	100
2,2-Dimethylbutane	199.6	207.4	218.8	228.2	238.5	253.6	266.4	280.5	301.6	319.9
2,3-Dimethylbutane	205.3	213.3	224.9	234.6	245.1	260.6	273.6	288.1	309.7	328.3
Dimethylcitraconate	318.6	329.8	345.9	359.1	373.4	394.1	411.3	430.1	457.9	481.3
1,1-Dimethylcyclohexane	244.2	253.6	267.3	278.7	291.2	309.4	324.7	341.7	367.1	388.9
cis-1,2-Dimethylcyclohexane	251.6	261.3	275.3	286.9	299.5	318.5	333.6	350.8	376.5	398.5
trans-1,4-Dimethyl-cyclohexane	243.7	253.2	266.9	278.3	290.8	309.0	324.4	341.5	366.9	388.8
trans-1,2-Dimethyl-cyclohexane	246.9	256.5	270.4	281.9	294.4	312.8	328.3	345.5	371.1	393.1
cis-1,3-Dimethylcyclohexane	248.4	257.9	271.7	283.2	295.7	314.0	329.4	346.4	371.8	393.7
trans-1,3-Dimethylcyclohexane	245.5	254.9	268.5	279.8	292.2	310.2	325.4	342.2	367.2	388.7
cis-1,4-Dimethylcyclohexane	247.9	257.5	271.4	282.9	295.4	313.7	329.2	346.3	371.8	393.8
Dimethyldichlorosilane			233.0	243.2	254.3	270.6	284.4	299.8	322.7	342.6
2,2-Dimethylhexane	238.8	247.9	261.0	272.0	283.8	301.2	315.9	332.0	356.1	376.7
2,3-Dimethylhexane	244.9	254.2	267.6	278.8	290.9	308.6	323.5	340.0	364.4	385.4
2,4-Dimethylhexane	239.6	248.9	262.2	273.3	285.4	303.1	318.0	334.5	359.0	380.2
2,5-Dimethylhexane	241.1	250.2	263.4	274.3	286.2	303.5	318.1	334.2	358.1	378.6
3,3-Dimethylhexane	241.7	250.9	264.3	275.3	287.3	304.9	319.8	336.1	360.4	381.3
3,4-Dimethylhexane	245.9	255.2	268.8	280.0	292.2	310.0	325.1	341.6	366.3	387.4
trans-Dimethylisaconate	315.0	326.1	342.1	355.2	369.4	390.1	407.3	426.0	453.7	477.1
Dimethylitaconate	337.6	347.9	362.7	374.7	387.5	405.9	420.9	437.2	460.7	480.2
Dimethylmaleate	313.8	324.9	340.9	354.1	368.3	388.9	406.1	425.0	452.7	476.2
Dimethylmalonate	302.9	313.3	328.2	340.4	353.5	372.6	388.4	405.6	430.9	452.2
Dimethyloxalate	288.0	298.1	312.6	324.5	337.4	356.0	371.6	388.6	413.5	434.7
2,2-Dimethylpentane	219.7	228.2	240.4	250.6	261.7	277.9	291.6	306.7	329.3	348.7
2,3-Dimethylpentane	226.6	235.3	248.0	258.5	269.9	286.7	300.8	316.5	339.8	359.8
2,4-Dimethylpentane	220.8	229.3	241.6	251.8	262.9	279.2	292.9	308.0	330.5	349.9
3,3-Dimethylpentane	222.7	231.4	244.0	254.5	265.9	282.6	296.8	312.4	335.9	356.0
Dimethylphenylsilane	272.7	283.1	298.0	310.5	324.0	343.8	360.4	378.8	406.1	429.6
2,2-Dimethylpropane	167.2	174.3	184.7	193.4	203.0	217.2	229.3	242.9	263.5	281.5
4,6-Dimethylresorcin	316.7	328.1	344.7	358.3	373.1	394.6	412.6	432.3	461.4	486.2
Dimethylsilane	154.6	160.8	169.8	177.4	185.6	197.7	208.0	219.5	236.6	251.5
Dimethylsulfide	193.8	201.2	212.0	220.9	230.6	244.7	256.7	269.9	289.5	306.4
α,α-Dimethylsuccinic anhydride	328.7	340.0	356.1	369.3	383.6	404.2	421.3	440.0	467.3	490.4
1,4-Dioxane	234.7	243.8	257.0	267.9	279.8	297.2	311.9	328.1	352.3	373.2
1,4-Dioxane-2,6-dion		352.5	369.4	383.4	398.4	420.3	438.4	458.2	487.3	511.9
1,2-Dipropoxyethane	228.0	239.8	257.3	272.4	289.3	315.2	338.1	364.6	406.8	445.8
Dipropyl ether	225.5	234.3	247.0	257.5	269.0	285.9	300.2	315.9	339.4	359.7
Dipropyloxalate	320.4	331.7	347.8	361.1	375.5	396.3	413.7	432.6	460.5	484.1
Dipropyleneglycol	342.4	353.8	370.0	383.3	397.6	418.2	435.3	453.8	480.9	503.6
Enanthic acid	343.7	354.3	369.3	381.4	394.5	413.1	428.4	444.9	468.8	488.6
Enanthic aldehyde	276.7	286.3	300.0	311.3	323.5	341.1	355.8	371.8	395.2	415.1
Enanthiochloride	302.9	311.5	323.5	333.3	343.7	358.4	370.4	383.3	401.7	416.9
Enanthionitrile	288.6	299.6	315.6	328.8	343.1	364.2	381.9	401.4	430.5	455.4
Epichlorohydrin	252.1	261.3	274.6	285.5	297.4	314.7	329.2	345.0	368.5	388.6
Ethane	111.1	115.7	122.3	127.9	134.0	143.0	150.6	159.1	172.0	183.2
Ethanol	238.6	246.5	257.8	267.0	276.9	291.2	303.0	315.9	334.6	350.3
1,2-Ethoxy-2-methylpropane	199.6	207.7	219.4	229.2	239.9	255.7	269.1	284.0	306.4	325.8
Ethoxytrimethylsilane	217.8	226.2	238.4	248.6	259.6	275.8	289.5	304.6	327.2	346.6
Ethyl acetate	226.0	234.2	246.1	255.9	266.5	281.9	294.9	309.1	330.0	347.9
Ethyl acetoacetate	296.3	306.9	322.2	334.8	348.4	368.2	384.7	402.8	429.5	452.2
Ethyl bromide	195.2	202.6	213.5	222.5	232.2	246.6	258.7	272.0	291.8	308.9
Ethyl butyrate	249.8	259.2	272.8	284.1	296.3	314.1	329.2	345.7	370.2	391.3
Ethyl chloride	179.2	186.0	195.9	204.0	212.9	226.0	236.9	249.0	267.0	282.4
Ethyl chloroacetate	269.6	279.4	293.7	305.4	318.1	336.7	352.2	369.3	394.5	416.0

Table 11.5 Saturation temperature (in K) of organic substances *(continued)*

Substance	Vapor pressure, kPa									
	0.1	0.2	0.5	1	2	5	10	20	50	100
Ethyl cyanacetate	337.0	347.3	362.0	373.9	386.7	405.0	420.0	436.2	459.6	479.1
Ethyl dichloracetate	277.3	287.5	302.1	314.2	327.2	346.3	362.3	379.8	405.7	427.8
Ethyl fluoride	152.3	158.1	166.3	173.2	180.7	191.6	200.7	210.8	225.8	238.7
Ethyl formate	207.6	215.4	226.4	236.0	246.2	261.0	273.5	287.2	307.6	325.0
Ethyl glycolate	282.8	292.9	307.3	319.1	332.0	350.6	366.1	383.1	408.1	429.3
Ethyl isobutyrate	244.3	253.4	266.6	277.4	289.2	306.5	320.9	336.8	360.5	380.6
Ethyl isocapronate	279.1	289.3	304.1	316.3	329.5	348.8	364.9	382.6	408.9	431.3
Ethyl isovalerate	262.0	271.7	285.5	297.0	309.5	327.6	342.8	359.5	384.2	405.3
Ethyl isothiocyanate	255.5	265.2	279.3	291.0	303.7	322.4	338.0	355.3	381.1	403.2
Ethyl iodide	214.7	223.0	235.1	245.1	256.0	272.1	285.6	300.6	322.9	342.2
Ethyl levulinate	314.8	325.9	341.8	355.0	369.2	389.8	406.9	425.7	453.3	476.7
Ethyl mercaptan	192.7	200.1	210.9	219.8	229.5	243.7	255.7	268.8	288.6	305.6
Ethyl trichloracetate	288.1	298.3	313.1	325.2	338.3	357.4	373.3	390.7	416.3	438.1
2-Ethyl bromide-benzene	315.4	327.1	344.0	357.9	373.0	395.1	413.7	434.0	464.2	490.0
2-Ethyl bromide-cyclohexane	305.6	317.2	334.1	348.0	363.2	385.5	404.2	424.8	455.6	482.0
Ethyl-α-bromisobutyrate	278.2	288.6	303.7	316.1	329.7	349.5	366.1	384.4	411.5	434.8
Ethyl-α-chloropropionate	274.9	284.6	298.5	309.9	322.3	340.3	355.2	371.6	395.6	416.0
Ethyl-α-ethylacetoacetate	307.9	318.9	334.7	347.7	361.8	382.3	399.4	418.1	445.7	469.1
Ethyl-2-furoate	304.7	315.7	331.5	344.5	358.6	379.1	396.2	414.9	442.6	466.1
Ethyl-*l*-leucinate	298.0	309.2	325.4	338.8	353.4	374.7	392.6	412.3	441.6	466.6
Ethylacrylate	238.4	247.2	259.9	270.5	281.8	298.5	312.4	327.7	350.4	369.8
α-Ethylacrylic acid	315.5	325.3	339.2	350.6	362.7	380.1	394.4	409.9	432.2	450.8
α-Ethylacrylnitrile	239.1	248.5	262.2	273.6	285.9	304.2	319.6	336.6	362.1	384.1
Ethylamine	187.4	194.2	204.0	212.1	220.9	233.6	244.3	256.0	273.2	287.9
N-Ethylaniline	306.2	317.6	334.1	347.6	362.4	383.9	402.0	421.8	451.3	476.4
4-Ethylaniline	319.5	331.0	347.6	361.2	375.9	397.4	415.3	434.9	463.9	488.5
Ethylbenzene	258.2	268.0	282.2	294.0	306.7	325.4	341.2	358.5	384.3	406.4
Ethylcyclohexane	253.6	263.3	277.4	289.1	301.8	320.5	336.2	353.5	379.4	401.6
Ethylcyclopentane	236.0	245.0	258.0	268.9	280.6	297.9	312.4	328.4	352.3	372.7
Ethyldichlorosilane			233.9	244.4	255.9	272.9	287.3	303.3	327.5	348.5
Ethyl-*N*, *N*-diethyloxamate	343.8	356.1	373.7	388.2	403.9	426.7	445.7	466.5	497.2	523.2
2-Ethyldisilazane	206.7	215.0	227.1	237.2	248.2	264.4	278.2	293.5	316.5	336.4
Ethylene	102.8	107.0	113.2	118.3	123.8	132.1	139.1	146.9	158.6	168.8
Ethylene oxide	179.9	186.7	196.4	204.5	213.3	226.1	236.9	248.7	266.4	281.4
1,2-Ethylene glycol	322.3	332.8	347.7	360.0	373.1	392.0	407.7	424.6	449.3	470.0
Ethylenechlorohydrin	264.9	274.2	287.6	298.6	310.5	327.8	342.2	357.9	381.0	400.6
Ethylene-*bis*-(chloracetate)	379.7	392.1	409.8	424.3	439.8	462.3	480.8	500.9	530.2	554.7
Ethylenediamine	257.3	266.4	279.3	289.9	301.4	318.0	331.9	347.0	369.3	388.1
3-Ethylhexane	247.8	257.1	270.6	281.8	293.9	311.6	326.5	342.9	367.3	388.2
Ethylmethyl ether	178.0	184.7	194.4	202.4	211.1	223.8	234.4	246.2	263.6	278.5
3-Ethylpentane	230.7	239.4	252.0	262.5	273.8	290.4	304.4	319.7	342.6	362.2
2-Ethylphenol	313.8	325.0	341.1	354.4	368.8	389.7	407.1	426.2	454.3	478.2
3-Ethylphenol	327.7	338.8	354.6	367.6	381.6	401.8	418.6	436.8	463.5	486.0
4-Ethylphenol	326.7	338.0	354.3	367.7	382.1	403.1	420.5	439.5	467.4	491.0
Ethylpropionate	240.5	249.2	261.7	272.1	283.3	299.6	313.3	328.2	350.3	369.2
Ethylpropyl ether	204.4	212.6	224.4	234.3	245.2	261.1	274.6	289.6	312.1	331.6
Ethyltrichlorosilane			254.3	265.1	276.9	294.2	308.8	324.9	349.0	369.7
Ethyltriethoxysilane			304.9	317.1	330.3	349.5	365.6	383.3	409.4	431.7
Ethyltrifluorosilane	174.6	180.8	189.7	197.1	205.1	216.6	226.3	236.8	252.4	265.6
Ethyltrimethyltin	238.5	247.8	261.1	272.2	284.3	302.0	317.0	333.5	358.1	379.3
Ethylvinyldichlorosilane			271.5	283.1	295.7	314.2	329.8	347.1	372.9	395.1
1-Fluorethylene	120.8	125.8	133.0	139.1	145.7	155.5	163.9	173.2	187.2	199.4
Fluorobenzene	225.6	234.2	246.6	256.8	268.0	284.3	298.0	313.2	335.7	355.0

Table 11.5 Saturation temperature (in K) of organic substances *(continued)*

Substance	Vapor pressure, kPa									
	0.1	0.2	0.5	1	2	5	10	20	50	100
2-Fluorotoluene	243.9	253.2	266.6	277.7	289.8	307.4	322.3	338.7	363.1	384.0
3-Fluorotoluene	246.0	255.3	268.7	279.9	292.0	309.7	324.6	341.0	365.3	386.2
4-Fluorotoluene	246.7	256.0	269.4	280.6	292.7	310.5	325.4	341.8	366.2	387.2
Formaldehyde			174.7	182.1	190.1	201.9	211.8	222.8	239.1	253.2
Formamide	339.1	349.6	364.5	376.7	389.7	408.4	423.8	440.3	464.3	484.3
Formic acid	242.8	251.3	263.4	273.5	284.3	300.0	313.0	327.3	348.3	366.0
trans-Fumarylchloride	282.7	292.8	307.4	319.5	332.6	351.5	367.4	384.8	410.4	432.2
Furfurol	286.0	296.1	310.6	322.5	335.4	354.1	369.7	386.7	411.8	433.0
Furfuryl alcohol	300.5	310.5	324.8	336.5	349.1	367.2	382.2	398.5	422.3	442.3
Glutaric acid	423.8	435.2	451.4	464.4	478.2	497.7	513.6	530.5	554.7	574.5
Glutaric anhydride	368.5	381.5	400.2	415.7	432.4	456.6	476.8	498.8	531.4	558.9
Glutaronitrile	357.9	371.2	390.4	406.2	423.5	448.6	469.7	492.9	527.3	556.7
Gluterylchloride	324.0	335.4	351.7	365.1	379.5	400.5	418.0	437.0	465.1	488.8
Glycerin	391.7	403.7	420.7	434.6	449.4	470.7	488.1	506.9	534.1	556.6
Glycolacetate	306.0	316.8	332.2	344.8	358.5	378.3	394.8	412.9	439.4	461.8
Heptane	234.4	243.3	256.1	266.7	278.3	295.2	309.5	325.1	348.5	368.5
1-Heptanol	308.9	318.7	332.5	343.9	356.0	373.4	387.7	403.1	425.6	444.3
2-Heptanone	289.1	298.7	312.3	323.4	335.4	352.6	366.9	382.4	405.0	424.0
4-Heptanone	291.7	300.7	314.3	323.8	334.9	350.8	363.9	378.0	398.4	415.3
2-Heptene	232.7	241.6	254.6	265.4	277.1	294.3	308.7	324.7	348.6	369.1
Hexachlorobenzene	382.9	396.7	416.5	432.8	450.4	476.1	497.5	521.0	555.6	585.0
Hexachloroetane	298.7	309.6	325.3	338.2	352.2	372.6	389.7	408.4	436.1	459.7
1,5-Hexadiene-3-in	223.5	232.1	244.6	255.0	266.2	282.8	296.7	312.1	335.1	354.9
Hexamethylcyclotrisiloxsane						330.5	345.4	361.8	386.0	406.6
Hexamethyldisiloxane	239.7	248.5	261.1	271.6	282.9	299.4	313.2	328.3	350.8	369.9
Hexane	214.6	222.8	234.6	244.4	255.1	270.8	284.0	298.5	320.1	338.7
1-Hexanol	292.5	302.1	315.8	327.0	339.0	356.3	370.6	386.1	408.7	427.6
2-Hexanol	281.7	290.8	303.7	314.3	325.6	342.0	355.4	370.0	391.3	409.0
3-Hexanol	271.3	280.7	294.2	305.3	317.2	334.6	349.0	364.7	387.8	407.3
2-Hexanone	276.4	285.2	297.7	308.0	318.9	334.6	347.6	361.6	382.0	399.0
1-Hexene	211.2	219.4	231.3	241.2	251.9	267.7	281.0	295.7	317.6	336.6
Hydrocyanic acid	198.2	204.9	214.4	222.2	230.7	242.8	252.9	263.9	280.0	293.5
Hydroquinone	392.4	404.0	420.5	433.9	448.1	468.5	485.2	503.1	528.9	550.2
1-Iodo-2-methylpropane	251.7	261.1	274.6	285.7	297.9	315.5	330.4	346.7	370.9	391.5
1-Iodo-3-methylbutane	265.4	275.5	290.1	302.2	315.3	334.6	350.7	368.6	395.1	417.9
Iodobenzene	291.1	302.2	318.2	331.4	345.8	366.9	384.6	404.2	433.3	458.2
1-Iodooctane	312.3	324.4	341.9	356.4	372.2	395.4	414.9	436.5	468.7	496.4
1-Iodopropane	233.0	242.1	255.3	266.3	278.2	295.8	310.6	327.0	351.6	372.8
2-Iodopropane	225.5	234.3	246.9	257.5	269.0	285.8	300.1	315.8	339.3	359.5
2-Iodotoluene	304.6	316.2	333.1	347.1	362.3	384.6	403.4	424.1	454.9	481.4
Isoamyl alcohol	278.5	287.2	299.7	309.9	320.7	336.4	349.2	363.1	383.2	400.0
Isoamylacetate	267.9	277.7	291.7	303.4	316.0	334.4	349.7	366.6	391.6	412.8
Isoamylformate	251.1	260.6	274.4	285.8	298.1	316.2	331.5	348.2	373.2	394.6
Isoamylnitrate	272.9	282.7	296.9	308.5	321.2	339.6	354.9	371.7	396.6	417.7
Isobutyl alcohol	260.0	268.5	280.6	290.5	301.1	316.3	328.9	342.6	362.5	379.2
Isobutylacetate	247.3	256.7	270.2	281.5	293.7	311.5	326.5	343.1	367.7	388.8
Isobutylamine	218.8	226.9	238.6	248.2	258.7	273.9	286.7	300.8	321.6	339.3
Isobutylbutyrate	273.1	283.4	298.3	310.6	324.0	343.6	360.0	378.1	405.1	428.1
Isobutylcarbamate		335.8	351.2	363.8	377.4	397.1	413.3	431.0	456.7	478.4
Isobutylchloride	215.3	223.5	235.3	245.1	255.7	271.3	284.4	298.9	320.4	338.8
Isobutyldichloroacetate	295.8	306.6	322.1	334.8	348.7	368.8	385.7	404.2	431.5	454.7
Isobutylformate	236.3	245.1	257.9	268.4	279.9	296.7	310.7	326.2	349.2	368.8
Isobutylisobutyrate	272.3	282.2	296.4	308.1	320.8	339.3	354.7	371.7	396.7	418.0

Table 11.5 Saturation temperature (in K) of organic substances *(continued)*

Substance	Vapor pressure, kPa									
	0.1	0.2	0.5	1	2	5	10	20	50	100
Isobutylpropionate	266.0	275.6	289.5	300.8	313.2	331.1	346.1	362.5	386.8	407.4
Isobutyric acid	283.7	293.6	307.6	319.2	331.7	349.8	364.8	381.2	405.3	425.6
Isocaproic acid	327.8	338.0	352.6	364.5	377.2	395.4	410.4	426.6	450.1	469.6
Isoprene	189.9	197.3	207.9	216.8	226.5	240.7	252.7	266.0	285.8	302.9
Isoprolactone	305.7	317.2	333.7	347.4	362.3	384.0	402.3	422.4	452.2	477.7
Isopropylacetate	230.7	239.3	251.7	262.0	273.2	289.5	303.1	318.2	340.5	359.5
Isopropylchloroacetate	272.3	282.3	296.7	308.6	321.5	340.4	356.1	373.5	399.1	421.0
Isopropylformate	217.2	225.4	237.1	246.8	257.4	272.8	285.7	300.0	321.1	339.2
Isopropylisobutyrate	252.0	261.3	274.7	285.7	297.7	315.2	329.9	346.0	369.9	390.2
Isopropyllevulinate	315.1	326.2	342.2	355.4	369.7	390.4	407.6	426.5	454.3	477.8
Isovalerianic acid	302.9	313.0	327.4	339.2	352.0	370.3	385.5	402.0	426.1	446.3
Levulenic acid	370.0	380.9	396.3	408.8	422.1	441.2	456.7	473.4	497.5	517.4
Levulenic aldehyde	295.8	306.7	322.4	335.4	349.5	370.1	387.4	406.3	434.4	458.4
Mercaptoacetic acid	328.2	339.4	355.3	368.4	382.5	402.9	419.8			
Mesityl oxide	259.3	268.8	282.5	293.8	306.1	324.0	339.0	355.5	379.8	400.6
Methacrylic acid	293.1	302.9	316.9	328.3	340.7	358.5	373.2	389.2	412.6	432.3
Methacrylnitrile	224.4	233.2	246.0	256.7	268.3	285.5	299.9	315.9	339.9	360.7
Methane	66.2	69.0	73.1	76.5	80.3	85.9	90.7	96.0	104.1	111.2
Methanol	225.6	233.3	244.4	253.5	263.2	277.4	289.2	302.0	320.7	336.5
Methoacetic acid	320.4	331.3	346.9	359.7	373.5	393.5	410.1	428.1	454.5	476.7
2-Methoxyaniline	328.8	340.1	356.1	369.3	383.6	404.1	421.2	439.8	467.1	490.0
2-Methoxyethanol	255.8	265.2	278.8	290.2	302.1	319.8	334.6	350.9	375.0	395.6
2-Methoxyphenol	320.3	331.2	346.8	359.6	373.4	393.4	409.9	428.0	454.4	476.6
3-Methyl-2-butanone	249.0	257.0	268.3	277.6	287.5	301.8	313.5	326.2	344.7	360.2
2-Methyl-1-butene	180.0	187.2	197.6	206.3	215.8	229.8	241.6	254.8	274.5	291.6
2-Methyl-2-butene	194.5	202.0	213.0	222.0	231.9	246.4	258.6	272.1	292.2	309.6
2-Methyl-3-ethylpentane	244.1	253.4	266.9	278.1	290.3	308.1	323.2	339.7	364.4	385.7
3-Methyl-3-ethylpentane	244.1	253.5	267.1	278.5	290.8	308.9	324.2	341.0	366.2	387.9
2-Methyl-2-heptene	252.2	261.6	275.1	286.4	298.6	316.4	331.3	347.7	372.1	392.9
6-Methyl-3-heptene-2-ol	309.5	319.3	333.3	344.8	357.0	374.6	389.1	404.8	427.6	446.6
6-Methyl-5-heptene-2-ol	310.4	320.2	334.2	345.6	357.8	375.4	389.9	405.5	428.2	447.1
2-Methyl-1-pentanol	283.4	292.9	306.5	317.7	329.6	346.9	361.3	376.9	399.6	418.8
2-Methyl-2-pentanol	264.1	273.0	285.7	296.2	307.4	323.6	337.0	351.6	373.0	391.0
2-Methyl-4-pentanol	268.2	277.5	290.8	301.8	313.6	330.8	345.0	360.6	383.4	402.7
4-Methyl-2-pentanone	267.6	276.3	288.7	298.8	309.7	325.4	338.3	352.3	372.7	389.7
Methylacetate	212.0	219.9	231.1	240.4	250.5	265.2	277.6	291.1	311.2	328.3
Methylacetylene	158.8	164.7	173.3	180.4	188.1	199.4	208.8	219.2	234.7	247.9
Methylacrylate	225.5	233.9	245.9	255.9	266.7	282.4	295.6	310.1	331.6	350.0
Methylamine	174.6	180.8	189.7	197.1	205.0	216.6	226.2	236.8	252.2	265.4
N-Methylaniline	303.5	314.5	330.3	343.4	357.5	378.1	395.3	414.1	441.9	465.6
Methylanthranilate	344.1	356.9	375.5	390.8	407.5	431.8	452.2	474.7	508.0	536.5
Methylbenzoate	305.3	316.3	332.2	345.2	359.4	380.0	397.2	416.1	443.9	467.6
2-Methylbenzothiazole	338.0	349.2	365.2	378.3	392.4	412.7	429.5	447.8	474.5	496.9
α-Methylbenzyl alcohol	316.5	327.4	343.0	355.9	369.7	389.7	406.4	424.5	451.2	473.6
Methylbromide	173.4	180.1	189.7	197.7	206.4	219.2	230.0	241.8	259.6	274.8
2-Methylbutane	186.9	194.1	204.6	213.3	222.8	236.8	248.5	261.5	280.9	297.6
Methylbutyrate	242.2	251.0	263.7	274.1	285.5	301.9	315.7	330.8	353.2	372.2
Methylcapronate	273.8	293.8	298.2	310.2	323.1	341.9	357.6	374.9	400.5	422.3
Methylchloride	159.8	165.5	173.8	180.6	188.0	198.7	207.7	217.5	232.0	244.3
Methylchloroacetate	265.0	274.3	287.5	298.4	310.2	327.3	341.5	357.0	379.7	399.0
Methylcyclohexane	233.0	242.0	255.0	265.8	277.6	294.9	309.5	325.5	349.5	370.2
Methylcyclopentane	215.7	224.0	236.0	246.0	256.8	272.6	286.0	300.7	322.7	341.6
Methylcyclopropane	173.4	180.0	189.7	197.7	206.4	219.2	230.0	241.9	259.7	275.0

Table 11.5 Saturation temperature (in K) of organic substances *(continued)*

Substance	Vapor pressure, kPa									
	0.1	0.2	0.5	1	2	5	10	20	50	100
Methyldichloroacetate	271.0	280.7	294.7	306.2	318.7	336.8	351.9	368.5	392.9	413.7
Methyldichlorosilane			211.7	221.2	231.7	247.3	260.4	275.1	297.2	316.5
2-Methyldisilasane	193.0	200.3	211.0	219.9	229.5	243.6	255.5	268.6	288.1	304.8
Methylfluoride	122.9	127.5	134.1	139.7	145.7	154.4	161.8	169.9	182.0	192.3
Methylformate	195.2	202.5	212.9	221.5	230.8	244.5	255.9	268.4	287.0	302.9
α-Methylglutaric anhydride	360.4	373.5	392.2	407.7	424.5	449.0	469.4	491.8	524.9	553.0
Methylglycolate	277.8	287.7	301.9	313.6	326.2	344.6	359.9	376.6	401.3	422.2
2-Methylheptane	246.8	256.2	269.6	280.7	292.8	310.5	325.4	341.7	366.1	386.9
3-Methylheptane	248.3	257.5	270.9	282.0	294.1	311.7	326.5	342.7	366.8	387.5
4-Methylheptane	247.3	256.6	270.0	281.1	293.2	310.8	325.7	342.0	366.2	387.0
2-Methylhexane	228.5	237.2	249.8	260.2	271.6	288.2	302.2	317.6	340.6	360.3
3-Methylhexane	229.3	238.0	250.6	261.0	272.4	289.0	303.0	318.4	341.3	360.9
Methyliodide		203.1	214.3	223.6	233.7	248.6	261.2	275.2	296.1	314.2
Methylisobutyrate	234.9	243.5	255.9	266.1	277.2	293.4	307.0	321.8	343.9	362.6
Methylisothiocyanate	236.3	246.3	260.8	273.0	286.4	306.2	323.2	342.1	370.8	395.9
Methylisovalerate	249.2	258.4	271.7	282.7	294.7	312.1	326.7	342.7	366.4	386.7
Methyllevulinate	307.4	318.4	334.2	347.2	361.3	381.8	398.9	417.6	445.2	468.6
Methylmercaptan	179.2	185.8	195.3	203.2	211.8	224.2	234.7	246.1	263.2	277.7
Methylmethacrylate	237.4	246.3	259.3	270.0	281.6	298.6	312.9	328.6	352.0	372.0
2-Methylpentane	208.6	216.5	228.1	237.7	248.2	263.5	276.4	290.6	311.8	330.0
3-Methylpentane	210.0	218.0	229.7	239.5	250.0	265.6	278.6	293.1	314.6	333.1
2-Methylpropane	160.7	167.1	176.3	184.1	192.5	204.8	215.3	226.9	244.3	259.4
2-Methylpropene	162.9	169.4	179.0	187.0	195.7	208.6	219.6	231.7	250.0	265.9
Methylpropionate	227.1	235.4	247.4	257.3	268.0	283.6	296.7	311.1	332.3	350.4
2-Methylpropionyl-bromide	281.1	291.5	306.3	318.6	331.9	351.3	367.6	385.4	411.9	434.4
Methylpropyl ether	197.4	204.8	215.6	224.6	234.3	248.5	260.5	273.6	292.3	310.0
Methylsalicylate	320.5	332.1	348.7	362.5	377.4	399.0	417.1	437.0	466.3	491.2
Methylsilane	131.8	137.0	144.7	151.1	158.1	168.4	177.1	186.7	201.3	213.8
Methylthiocyanate	254.0	263.0	278.0	289.8	302.6	321.5	337.3	354.9	381.0	403.5
2-Methylthiophene	241.0	250.3	263.7	274.9	287.0	304.9	319.9	336.5	361.3	382.6
2, 3-Methylthiophene	244.0	253.3	266.7	277.8	290.0	307.7	322.6	339.1	363.6	384.7
Methyltrichlorosilane			231.0	241.0	251.8	267.7	281.2	296.1	318.3	337.5
Methylurethane	294.5	304.6	319.2	331.2	344.2	362.9	378.5	395.5	420.5	441.5
4-Nitro-1,3-xylol	332.2	344.4	361.9	376.5	392.2	415.1	434.3	455.4	486.6	513.2
2-Nitroaniline	371.0	383.8	402.3	417.4	433.8	457.6	477.2	498.7	530.3	557.0
3-Nitroaniline	386.0	399.3	418.4	434.1	451.0	475.4	495.8	517.9	550.4	577.9
4-Nitroaniline	410.0	423.9	443.9	460.2	477.9	503.4	524.6	547.6	581.3	609.7
2-Nitrobenzaldehyde	352.8	365.7	384.3	399.6	416.2	440.5	460.7	483.0	515.9	543.9
3-Nitrobenzaldehyde	362.9	375.7	394.1	409.2	425.5	449.2	469.0	490.5	522.3	549.2
Nitrobenzene	310.8	322.1	338.5	352.0	366.6	388.0	405.8	425.4	454.4	479.1
Nitroethane	247.4	256.6	269.8	280.8	292.7	310.1	324.7	340.8	364.6	384.9
Nitroglycerin	393.3	409.4	432.8	452.5	473.9	505.7				
Nitromethane	239.4	248.2	261.0	271.6	283.1	299.8	313.9	329.3	352.2	371.7
2-Nitrophenol	316.5	328.0	344.5	358.1	372.8	394.2	412.1	431.7	460.7	485.4
2-Nitrophenylacetate	368.2	379.6	395.8	409.0	423.1	443.4	460.0	478.0	503.9	525.5
1-Nitropropane	258.3	267.9	281.8	293.2	305.7	323.9	339.2	356.0	380.9	402.2
2-Nitropropane	249.6	259.0	272.6	283.9	296.2	314.1	329.1	345.7	370.3	391.4
2-Nitrothiophene	315.5	327.4	344.7	359.0	374.6	397.4	416.5	437.6	469.0	495.9
2-Nitrotoluene	318.0	329.8	346.7	360.8	376.0	398.2	416.8	437.2	467.5	493.4
3-Nitrotoluene	318.2	330.4	348.1	362.8	378.7	402.1	421.9	443.6	476.1	504.0
4-Nitrotoluene	321.3	333.8	351.8	366.7	383.0	406.9	427.0	449.3	482.5	511.1
Octamethylcyclotetrasiloxane	289.7	300.1	315.0	327.2	340.5	359.8	375.9	393.5	419.4	441.5
Octamethyltrisiloxane	275.3	285.2	299.5	311.2	323.9	342.4	357.9	374.8	399.8	421.0

Table 11.5 Saturation temperature (in K) of organic substances *(continued)*

Substance	Vapor pressure, kPa									
	0.1	0.2	0.5	1	2	5	10	20	50	100
Octane	253.5	262.9	276.5	287.8	300.0	317.9	332.9	349.3	373.7	394.6
1-Octanol	320.1	330.4	345.0	356.9	369.6	388.0	403.1	419.5	443.3	463.1
2-Octanol	300.8	311.1	325.9	338.0	351.2	370.1	385.9	403.0	428.2	449.5
2-Octanone	290.9	301.3	316.3	328.7	342.1	361.6	377.8	395.6	421.9	444.3
4-Oxy-3-methyl-2-butanone	312.4	322.6	337.2	349.1	361.9	380.4	395.6	412.1	436.2	456.4
4-Oxy-4-methyl-2-pentanone	290.1	300.4	315.0	327.1	340.1	359.0	374.8	392.0	417.3	438.8
4-Oxybenzaldehyde	387.5	400.9	420.0	435.8	452.7	477.3	497.7	519.9	552.5	580.1
α-Oxybutyronitrile	309.9	320.0	334.3	346.1	358.6	376.7	391.7	407.9	431.5	451.2
α-Oxyisobuteric acid	341.2	351.6	366.3	378.3	391.2	409.5	424.6	440.8	464.2	483.6
3-Oxypropionitrile	326.9	338.4	355.0	368.6	383.4	404.8	422.6	442.1	470.9	495.2
Paraformaldehyde	259.3	268.6	281.9	292.8	304.6	321.8	336.1	351.8	374.9	394.6
Pentachlorobenzene	365.8	378.4	396.4	411.3	427.2	450.4	469.6	490.6	521.4	547.3
Pentachloroethane	268.8	279.4	294.7	307.5	321.4	341.9	359.2	378.3	407.0	431.7
Pentachloroethylbenzene	363.4	377.1	397.0	413.4	431.3	457.5	479.5	503.7	539.7	570.6
Pentachlorophenol				441.8	458.4	482.5	502.5	524.1	555.9	582.5
1,3-Pentadiene	197.0	204.6	215.6	224.7	234.7	249.2	261.5	275.1	295.3	312.7
1,4-Pentadiene	186.3	193.5	204.0	212.7	222.1	236.0	247.7	260.7	280.0	296.6
Pentane	191.8	199.3	210.2	219.2	229.1	243.5	255.7	269.2	289.4	306.8
2-Pentanol	270.3	278.8	291.0	300.9	311.5	326.8	339.4	353.0	372.7	389.1
2-Pentanone	256.9	265.3	277.2	286.9	297.3	312.3	324.7	338.1	357.7	374.1
3-Pentanone	256.1	264.4	276.3	286.1	296.5	311.6	324.0	337.5	357.1	373.6
2,3,4-Pentatriol	424.0	436.9	455.3	470.3	486.3	509.2	528.0	548.2	577.5	601.8
1-Pentene	188.7	196.0	206.6	215.4	225.0	239.1	251.0	264.1	283.7	300.6
2-Phenetedine	334.2	345.7	362.1	375.7	390.3	411.4	429.0	448.2	476.3	500.0
Phenetole	285.7	296.3	311.5	324.1	337.8	357.8	374.5	392.9	420.1	443.3
Phenetyl alcohol	325.7	337.1	353.4	366.9	381.4	402.5	420.0	439.1	467.3	491.1
Phenol	306.0	316.1	330.6	342.4	355.1	373.5	388.7	405.2	429.2	449.4
2-Phenoxyethanol	345.2	357.1	374.0	387.9	402.9	424.6	442.6	462.2	491.0	515.2
Phenyl acetate	305.8	316.7	332.5	345.5	359.6	380.0	397.1	415.8	443.3	466.8
Phenyl acetochloride	315.5	326.7	342.9	356.2	370.6	391.6	409.0	428.1	456.3	480.2
Phenyl acetonitrile	327.3	339.2	356.4	370.5	385.9	408.3	427.0	447.6	478.0	503.9
Phenyl dichlorophosphate	333.4	345.4	362.6	376.8	392.2	414.6	433.3	453.8	484.0	509.7
Phenyl isothiocyanate	315.1	326.9	343.8	357.8	373.0	395.2	413.8	434.3	464.7	490.7
Phenyl isocyanate	278.0	288.5	303.6	316.2	329.8	349.7	366.5	384.9	412.3	435.7
Phenyl isocyanide	279.8	290.1	305.1	317.5	331.0	350.6	367.1	385.2	412.1	435.0
Phenylacetic acid	364.5	376.6	394.0	408.2	423.4	445.5	463.7	483.5	512.5	536.8
1,3-Phenyldiamine	366.6	379.6	398.2	413.6	430.1	454.2	474.3	496.2	528.5	555.9
Phenylhydrazine	339.3	351.2	368.4	382.6	397.9	420.1	438.6	458.8	488.5	513.7
Phenylmercaptan	286.1	296.4	311.3	323.6	337.0	356.3	372.6	390.3	416.5	438.9
Phenyltrichlorosilane			327.7	341.4	356.2	377.8	396.0	416.1	446.0	471.6
Phosgene	176.8	183.6	193.3	201.4	210.2	223.0	233.8	245.8	263.5	278.8
Phthalic anhydride	357.1	370.2	389.0	404.6	421.5	446.2	466.8	489.4	523.0	551.5
Phthalide	361.7	375.0	394.3	410.2	427.5	452.7	473.8	497.0	531.4	560.7
Phthalyl chloride	353.2	366.2	384.9	400.4	417.3	441.8	462.3	484.9	518.3	546.8
2-Picoline	257.9	267.4	281.2	292.6	304.9	322.9	338.0	354.6	379.2	400.3
Pimelic acid	430.7	444.0	462.9	478.3	494.7	518.3	537.6	558.5	588.7	613.8
Piperidine		248.0	261.2	272.1	284.0	301.5	316.2	332.4	356.5	377.2
1,2-Propandiol	313.8	324.1	338.9	351.0	363.9	382.6	398.1	414.9	439.4	459.9
1,3-Propandiol	327.7	338.9	354.8	367.8	381.9	402.2	419.0	437.4	464.2	486.9
Propane	141.8	147.4	155.6	162.4	169.9	180.8	190.1	200.4	215.8	229.2
1-Propanol	254.3	262.5	274.2	283.7	294.0	308.7	320.9	334.1	353.3	369.3
2-Propanol	243.3	251.2	262.5	271.7	281.6	295.8	307.5	320.3	338.8	354.3
Propionamide	333.5	344.3	359.8	372.5	386.1	405.7	421.9	439.4	465.0	486.4

Table 11.5 Saturation temperature (in K) of organic substances *(continued)*

Substance	Vapor pressure, kPa									
	0.1	0.2	0.5	1	2	5	10	20	50	100
Propionic acid	273.5	283.1	296.8	308.2	320.4	338.1	352.9	369.0	392.7	412.7
Propionic anhydride	288.5	298.7	313.4	325.6	338.7	357.7	373.6	391.0	416.5	438.2
Propionitrile	233.4	242.3	255.2	265.9	277.5	294.4	308.7	324.5	347.9	368.1
Propylacetate	242.3	251.1	263.7	274.1	285.4	301.9	315.6	330.7	353.0	371.9
Propylamine	204.9	212.5	223.5	232.6	242.5	256.9	269.0	282.3	302.1	318.9
Propylbutyrate	266.3	276.2	290.4	302.2	314.9	333.5	349.2	366.3	391.7	413.5
Propylcarbamate	320.3	330.7	345.5	357.6	370.6	389.3	404.8	421.5	445.9	466.2
Propylchloroglyoxylate	277.1	286.7	300.4	311.7	323.9	341.5	356.1	372.1	395.6	415.4
Propylene	137.8	143.3	151.4	158.1	165.4	176.3	185.5	195.6	211.0	224.3
Propylene oxide	194.5	201.8	212.5	221.3	230.9	244.9	256.6	269.6	288.9	305.4
Propylformate	226.0	234.4	246.5	256.6	267.5	283.5	296.9	311.6	333.5	352.2
Propylisovalerate	276.2	286.3	300.9	313.0	326.1	345.2	361.2	378.8	404.8	427.0
Propyllevulinate	326.1	337.5	353.8	367.3	381.8	402.8	420.4	439.5	467.6	491.4
Propylmercaptan	213.0	221.3	233.2	243.1	253.9	269.7	283.0	297.7	319.7	338.6
Propylpropionate	254.0	263.3	276.7	287.9	299.9	317.5	332.2	348.3	372.3	392.7
Pyridine	248.9	258.1	271.3	282.2	294.1	311.3	325.8	341.7	365.2	385.3
Pyrocatechin		353.3	370.7	384.9	400.3	422.7	441.4	461.8	491.8	517.3
Pyrogallol		398.1	417.4	433.3	450.5	475.4	496.1	518.8	552.1	580.3
Pyroracemic aldehyde	337.0	349.2	366.8	381.3	397.0	419.9	439.0	460.0	490.9	517.3
Resorcione	375.5	387.7	405.1	419.4	434.7	456.8	475.0	494.7	523.5	547.6
Salicyl aldehyde	300.5	311.7	327.8	341.2	355.7	376.8	394.6	414.1	443.0	467.8
Salicylic acid	377.2	387.9	403.0	415.2	428.2	446.7	461.7	477.9	501.0	520.0
Selenophene	228.9	238.6	252.6	264.4	277.3	296.5	312.9	331.2	358.9	383.2
Styrene	261.8	271.9	286.6	298.8	312.2	331.7	348.1	366.3	393.5	416.9
Suberic acid	439.3	452.4	470.9	486.0	502.1	525.1	543.9	564.1	593.2	617.3
Succinamide	380.6	393.1	411.0	425.7	441.4	464.1	482.9	503.2	532.9	557.8
Succinchloride	306.9	317.7	333.2	346.0	359.7	379.7	396.4	414.6	441.4	464.0
Succinic anhydride	354.6	366.8	384.2	398.5	414.0	436.3	454.9	475.2	504.9	529.9
1,1,1,2-Tetrabromoethane	326.0	336.4	351.3	363.4	376.4	395.1	410.5	427.2	451.5	471.7
1,1,2,2-Tetrabromoethane	332.5	344.7	362.4	376.9	392.7	415.7	435.0	456.2	487.6	514.3
1,1,2,2-Tetrachloro-1,2-difluoroethane	231.5	240.3	253.0	263.5	274.9	291.7	305.8	321.3	344.4	364.2
3,4,5,6-Tetrachloro-1,2-xylene	361.2	373.8	391.9	406.8	422.9	446.2	465.6	486.7	517.9	544.2
1,2,3,5-Tetrachloro-4-ethylbenzene	344.7	357.8	376.6	392.3	409.3	434.2	455.2	478.3	512.6	542.1
Tetrachloroamine	340.6	348.2	358.7	367.1	375.9	388.2	398.1	408.4	423.0	434.8
1,2,4,5-Tetrachlorobenzene					394.1	417.5	437.1	458.7	490.8	518.2
1,2,3,5-Tetrachlorobenzene	325.1	337.7	356.0	371.2	387.7	412.0	432.5	455.2	489.1	518.2
1,2,3,4-Tetrachlorobenzene	335.5	348.0	366.1	381.0	397.3	421.0	440.9	462.7	495.2	523.0
1,1,1,2-Tetrachloroethane	251.9	261.6	275.8	287.5	300.3	319.0	334.8	352.2	378.3	400.7
1,1,2,2-Tetrachloroethane	264.0	274.0	288.6	300.7	313.9	333.2	349.4	367.3	393.9	416.8
Tetrachloroethylene	248.0	257.5	271.1	282.5	294.8	312.9	328.1	344.9	369.9	391.3
2,3,4,6-Tetrachlorophenol	367.1	379.6	397.5	412.2	428.0	450.9	469.9	490.6	520.8	546.4
Tetrachlorosilane			223.7	233.5	244.2	259.9	273.2	287.9	310.0	329.1
2-α,α,α-Tetrachlorotoluene	336.8	349.8	368.8	384.5	401.6	426.7	447.9	471.3	506.3	536.5
Tetraethoxysilane	282.8	293.2	308.2	320.7	334.1	353.8	370.3	388.4	415.2	438.1
Tetraethyl lead	305.8	316.1	330.8	342.9	355.9	374.7	390.3	407.3	432.1	453.0
Tetraethyldistibine	363.4	377.9	399.1	416.8	436.1	464.5	488.6	515.3	555.5	590.3
Tetraethyleneglycol	421.8	433.7	450.5	464.1	478.5	499.0	515.7	533.6	559.3	580.4
Tetraethyleneglycol-chlorohydrine	378.1	390.5	408.3	422.8	438.4	460.9	479.6	499.8	529.2	553.9
Tetraethylsilane	266.8	277.1	292.0	304.4	317.8	337.6	354.2	372.6	400.0	423.6
Tetramethyl tin	217.5	226.0	238.4	248.7	259.9	276.3	290.2	305.6	328.7	348.5

Table 11.5 Saturation temperature (in K) of organic substances *(continued)*

Substance	Vapor pressure, kPa									
	0.1	0.2	0.5	1	2	5	10	20	50	100
Tetramethyl lead	239.1	248.3	261.7	272.8	285.0	302.7	317.7	334.3	359.1	380.3
Tetramethylboron	210.6	219.0	231.2	241.4	252.5	268.8	282.6	297.9	320.9	340.8
2,2,3,3-Tetramethylbutane	251.1	259.7	272.1	282.3	293.3	309.2	322.4	336.8	357.9	375.8
Tetramethylpiperazine	290.7	301.5	317.0	329.8	343.6	363.9	380.9	399.5	427.1	450.7
Tetramethylsilane	185.5	192.8	203.3	212.1	221.7	235.8	247.7	260.9	280.7	297.7
Tetranitromethane			278.5	289.9	302.3	320.3	335.5	352.2	377.0	398.2
2,2-Thiodiethanol	305.9	326.8	359.1	388.2	422.5	478.2	531.2			
Thiophene	227.4	235.9	248.0	258.1	269.1	285.0	298.4	313.2	335.0	353.7
Tigl aldehyde	243.9	253.3	267.0	278.4	290.7	308.9	324.2	341.1	366.4	388.1
Toluene	241.5	250.6	263.9	274.9	286.9	304.4	319.1	335.3	359.5	380.2
Toluene-2,4-diamine	373.4	385.9	403.7	418.4	434.2	457.0	475.8	496.3	526.3	551.5
2-Toluidine	308.7	320.0	336.2	349.6	364.1	385.2	402.9	422.2	450.9	475.3
3-Toluidine	309.5	320.5	336.3	349.4	363.4	383.9	401.0	419.6	447.1	470.5
4-Toluidine	304.2	315.6	332.1	345.7	360.6	382.2	400.4	420.5	450.2	475.7
4-Tolunitrile	309.4	321.2	338.2	352.4	367.8	390.3	409.2	430.1	461.3	488.0
4-Tolylhydrazine	292.6	303.4	319.0	331.9	345.9	366.3	383.4	402.2	430.1	453.9
2-Tolylisocyanide	310.5	321.3	336.9	349.7	363.5	383.6	400.3	418.5	445.3	468.0
Tribromacetaldehyde	286.7	297.4	312.7	325.4	339.2	359.3	376.2	394.7	422.2	445.7
1,1,2-Tribromethane	299.3	310.1	325.7	338.5	352.4	372.6	389.5	408.0	435.3	458.5
1,1,2-Tribromobutane	312.3	324.0	340.8	354.8	370.0	392.2	410.8	431.3	461.7	487.7
1,2,2-Tribromobutane	308.2	319.8	336.7	350.6	365.8	388.0	406.6	427.1	457.7	483.9
2,2,3-Tribromobutane	305.6	317.0	333.5	347.2	362.0	383.7	401.8	421.8	451.5	476.9
Tribromomethane		274.4	289.3	301.8	315.3	335.2	352.0	370.6	398.4	422.3
1,2,3-Tribromopropane	314.2	326.0	343.0	357.0	372.3	394.7	413.4	434.1	464.8	491.0
1,1,2-Trichloro-1,2,2-trifluorethane	201.3	209.0	220.2	229.4	239.5	254.3	266.7	280.4	300.9	318.4
Trichloroacetaldehyde	231.2	240.1	253.1	263.9	275.7	292.9	307.4	323.5	347.5	368.1
Trichloroacetic acid	318.7	329.2	344.1	356.3	369.5	388.4	404.0	421.0	445.7	466.4
Trichloroacetic anhydride	324.0	335.7	352.4	366.1	381.0	402.7	420.8	440.6	469.9	494.7
Trichloroacetylbromide	260.5	270.5	285.1	297.1	310.2	329.4	345.6	363.5	390.1	413.1
2,4,6-Trichloroaniline	401.9	412.2	426.5	438.1	450.3	467.5	481.4	496.1	517.1	534.1
1,2,3-Trichlorobenzene	308.2	320.2	337.5	351.9	367.6	390.6	410.0	431.5	463.6	491.2
1,2,4-Trichlorobenzene	305.6	317.4	334.3	348.4	363.8	386.2	405.1	426.0	457.2	483.9
1,3,5-Trichlorobenzene		313.4	330.3	344.3	359.6	382.0	400.9	421.8	453.0	479.8
1,2,3-Trichlorobutane	268.0	278.9	294.7	308.0	322.5	343.8	362.0	382.2	412.6	439.0
1,1,1-Trichloroethane	243.9	253.1	266.6	277.7	289.8	307.5	322.4	338.9	363.4	384.4
1,1,2-Trichloroethane	225.5	234.1	246.6	257.0	268.3	284.9	298.9	314.3	337.2	357.0
Trichloroethoxysilane			113.0	120.3	128.7	141.8	153.5	167.5	190.2	212.0
Trichloroethylene	240.8	249.6	262.1	272.5	283.8	300.2	313.9	328.9	351.1	369.9
Trichloroethylsilane	236.3	245.3	258.3	269.1	280.8	298.0	312.4	328.4	352.1	372.4
Trichlorofluoromethane	216.7	225.0	237.1	247.2	258.2	274.2	287.8	302.7	325.0	344.2
Trichloroisopropylsilane	244.1	253.6	267.2	278.5	290.9	309.0	324.2	341.0	366.2	387.8
Trichloromethane	210.9	218.9	230.4	240.0	250.4	265.5	278.3	292.3	313.2	331.1
Trichloromethylsilane	208.8	217.1	229.1	239.1	250.0	266.0	279.6	294.7	317.2	336.7
2,4,5-Trichlorophenol	340.0	352.2	369.7	384.2	399.9	422.6	441.6	462.5	493.2	519.3
2,4,6-Trichlorophenol	185.0	192.2	202.5	211.1	220.5	234.3	245.9	258.7	277.8	294.2
Trichlorophenylsilane	300.3	311.6	328.0	341.5	356.2	377.7	395.8	415.7	445.4	470.7
1,1,1-Trichloropropane	239.8	248.9	262.1	273.1	285.0	302.4	317.1	333.3	357.4	378.0
1,2,3-Trichloropropane	276.8	287.0	301.7	313.9	327.1	346.4	362.6	380.3	406.7	429.1
Trichlorosilane			206.9	215.9	225.7	240.1	252.3	265.8	286.0	303.5
α,α,α-Trichlorotoluene	312.6	324.1	340.7	354.5	369.4	391.2	409.4	429.4	459.1	484.4
Triethoxymethylsilane	338.0	349.6	366.1	379.7	394.3	415.5	433.1	452.3	480.4	504.0
Triethylformate	273.3	283.1	297.2	308.8	321.3	339.6	354.8	371.5	396.1	417.0
Triethylphosphate	306.7	318.3	335.2	349.3	364.5	386.9	405.6	426.4	457.2	483.7
Triethyleneboron	382.0	394.2	411.4	425.5	440.5	462.2	480.0	499.3	527.3	550.6

Table 11.5 Saturation temperature (in K) of organic substances *(continued)*

Substance	Vapor pressure, kPa									
	0.1	0.2	0.5	1	2	5	10	20	50	100
Triethyleneglycol	249.8	259.4	273.4	285.1	297.7	316.3	332.0	349.3	375.2	397.4
Triethylmethylsilane	275.9	287.5	304.4	318.6	334.2	357.3	377.0	398.9	432.3	461.5
Trifluorophenylsilane	237.6	246.4	259.1	269.6	280.9	297.5	311.4	326.6	349.3	368.6
α, α, α-Trifluorotoluene	236.5	245.5	258.4	269.2	280.9	298.1	312.5	328.4	352.1	372.5
Trimethyl borane	196.1	203.9	215.3	224.8	235.2	250.4	263.3	277.7	299.2	317.8
Trimethyl boron	153.7	160.0	169.2	176.9	185.3	197.7	208.2	220.0	237.7	253.1
Trimethyl gallium	208.1	216.0	227.6	237.1	247.5	262.7	275.6	289.7	310.8	328.9
Trimethyl phosphate	294.0	305.2	321.4	334.9	349.6	371.0	389.1	409.0	438.6	464.1
2,2,4-Trimethyl-2-pentanone	283.0	291.9	304.5	314.9	325.9	341.8	354.8	368.9	389.4	406.4
Trimethylamine	172.5	179.1	188.8	196.8	205.5	218.3	229.1	241.0	258.7	274.0
2,2,3-Trimethylbutane			238.4	248.9	260.4	277.3	291.7	307.6	331.4	352.1
Trimethylchlorosilane			223.7	233.5	244.2	260.0	273.3	288.1	310.3	329.5
2,2,3-Trimethylpentane	239.0	248.2	261.6	272.7	284.7	302.4	317.3	333.8	358.4	379.5
2,2,4-Trimethylpentane	232.0	241.0	253.9	264.7	276.5	293.7	308.2	324.3	348.2	368.8
2,3,4-Trimethylpentane	241.9	251.2	264.6	275.7	287.9	305.7	320.6	337.2	361.8	383.0
Trimethylpropyl tin	255.1	264.7	278.7	290.2	302.8	321.1	336.6	353.6	378.9	400.6
Trimethylpropylsilane	222.7	231.6	244.5	255.2	266.9	284.2	298.8	315.0	339.3	360.3
Trimethylsuccinic anhydride	320.2	332.3	349.7	364.1	379.8	402.7	422.0	443.2	474.7	501.7
Urethan		318.8	333.6	345.7	358.7	377.5	393.1	410.1	434.9	455.7
Valerianic acid	310.9	321.2	335.9	347.9	360.9	379.5	395.0	411.7	436.2	456.7
α-Valerolactone	304.3	315.8	332.4	346.1	361.1	382.9	401.2	421.4	451.4	477.2
Valeronitrile	262.3	272.2	286.5	298.3	311.2	330.0	345.9	363.3	389.2	411.4
Vanillin	374.7	387.5	405.7	420.7	436.8	460.1	479.4	500.5	531.3	557.3
Vinyl acetate	221.2	229.4	241.4	251.2	261.9	277.5	290.7	305.1	326.5	344.8
Vinyl trichlorosilane			246.6	257.4	269.2	286.5	301.1	317.4	341.7	362.8
Vinyl triethoxysiloxane			304.3	316.5	329.8	349.2	365.4	383.2	409.6	432.2
2-Xylene	263.9	273.9	288.2	300.1	313.1	332.0	347.9	365.4	391.4	413.7
3-Xylene	260.9	270.7	284.9	296.6	309.3	327.9	343.6	360.8	386.3	408.2
4-Xylene	259.7	269.5	283.7	295.4	308.2	326.8	342.5	359.7	385.4	407.4
2,3-Xylenol	319.1	330.1	345.9	358.9	372.9	393.2	410.1	428.5	455.5	478.2
2,4-Xylenol	333.7	345.0	361.3	374.7	389.0	409.8	427.1	445.9	473.5	496.7
3,5-Xylenol	329.7	340.9	357.0	370.2	384.4	404.9	422.0	440.6	467.8	490.7
2,4-Xylidine	319.9	331.0	347.1	360.2	374.5	395.1	412.3	431.1	458.6	482.0
2,6-Xylidine	311.4	323.1	340.0	354.0	369.1	391.3	410.0	430.5	460.9	487.0

Table 11.6 Vapor pressure (higher than 10^5 Pa) of organic substances [115]

t, °C	p, 10^5 Pa	t, °C	p, 10^5 Pa	t, °C	p, 10^5 Pa	t, °C	p, 10^5 Pa
Acetone C_3H_6O		253.2	15.20	400.0	40.52	490	18.70
78.6	2.02	263.0	20.26	422.4	50.65	500	21.66
113.0	5.06	271.6	25.33	426.0	53.08	510	24.96
144.5	10.13	278.9	30.39	Benzene C_6H_6 (T, K)		520	28.62
181.0	20.26	284.9	35.45	360	1.243	530	32.69
205.0	30.39	290.4	40.52	380	2.164	540	37.19
214.5	40.52	300.0	50.65	400	3.536	550	42.18
235.0	47.61	307.8	60.78	410	4.428	560	47.71
Acetylene C_2H_2 (T, K)		308.7	62.45	420	5.479	562.6	49.24
192.4	1.283	Aniline C_6H_7N		430	6.704	n-Butane C_4H_{10}	
200.9	2.026	212.8	2.02	440	8.121	0	1.032
209.4	3.039	254.8	5.06	450	9.746	10	1.483
221.5	5.065	292.7	10.13	460	11.60	20	2.074
230.4	7.091	342.0	20.26	470	11.69	30	2.83
240.7	10.13	375.5	30.39	480	16.01	40	3.78

Table 11.6 Vapor pressure of organic substances *(continued)*

t, °C	p, 10^5 Pa
50	4.96
60	6.39
70	8.10
80	10.13
90	12.50
100	15.29
110	18.52
120	22.23
130	26.48
140	31.29
150	36.73
152.01	37.96
n-Butanol $C_4H_{10}O$	
139.8	2.03
172.5	5.06
203.0	10.13
237.0	20.26
259.0	30.39
277.0	40.52
288.0	49.03
Butene-1 C_4H_8	
0	1.285
10	1.838
20	2.557
30	3.472
40	4.62
50	6.02
60	7.63
70	9.59
80	11.93
90	14.68
100	17.89
110	21.59
120	25.84
130	30.68
140	36.15
146.4	40.2
n-Butylbenzene $C_{10}H_{14}$	
190	1.1917
200	1.5018
210	1.8714
220	2.3078
230	2.8185
240	3.411
250	4.094
Butyne-1 C_4H_6	
10	1.06
20	1.47
30	1.70
40	1.96
Carbon tetrachloride CCl_4	
90	1.482
110	2.506
130	4.001
150	6.071
170	8.852
190	12.41
210	17.01
230	22.73
250	29.57
270	38.65
283.05	46.49
Chlorobenzene C_6H_5Cl	
140	1.252
160	2.044
180	3.158
200	4.693
220	6.739
240	9.434
260	12.85
329.2	31.90
359.2	45.22
Cyclohexane C_6H_{12}	
90	1.3272
110	2.260
130	3.622
150	5.52
170	8.059
190	11.34
210	15.52
230	20.82
250	27.45
270	35.61
Cyclohexen C_6H_{10}	
90	1.2450
100	1.6443
110	2.1358
120	2.7324
130	3.447
140	4.294
150	5.287
Cyclopentane C_5H_{10}	
50	1.038
70	1.905
90	3.252
110	5.211
130	7.950
150	11.67
170	16.58
190	22.87
210	30.74
230	40.39
n-Decane $C_{10}H_{22}$	
180	1.1737
190	1.4921
200	1.8740
210	2.3274
220	2.8603
230	3.482
240	4.199
250	5.022
346	21.07
Diethyl ester $C_4H_{10}O$	
40	1.228
60	2.311
80	3.964
100	6.471
120	10.01
140	14.73
160	21.03
180	29.02
190	34.01
193.8	36.06
Difluorodichloromethane (Freon-12) CF_2Cl_2	
−30	1.0143
−20	1.527
−10	2.219
0	3.125
10	4.285
20	5.739
30	7.529
40	9.687
50	12.28
60	15.34
70	18.94
80	23.13
90	27.99
100	33.60
111.8	41.32
Difluoromonochloromethane (Freon-22) CHF_2Cl	
−40	1.054
−30	1.641
−20	2.456
−10	3.552
0	4.983
10	6.811
20	9.097
30	11.908
40	15.315
50	19.395
60	24.236
70	29.94
80	36.62
90	44.43
96.13	49.86
1,1-Dimethylcyclopentane C_7H_{15}	
90	1.0784
100	1.4263
110	1.8550
120	2.3756
130	2.9998
140	3.7397
150	4.6070
160	5.6137
Diphenyl $C_{12}H_{10}$	
260	1.08
280	1.74
300	2.38
320	3.38
340	4.47
360	6.55
380	8.8
n-Dodecane $C_{12}H_{26}$	
220	1.1059
230	1.3890
240	1.7258
250	2.1227
386	18.1
Ethane C_2H_6 (T, K)	
184.52	1.013
190	1.347
200	2.174
210	3.340
220	4.921
230	7.002
240	9.675
250	13.02
260	17.12
270	22.08
280	28.01
290	35.10
300	43.65
305.5	49.13
Ethanol C_2H_6O	
97.5	2.03
126.0	5.06
151.8	10.13
183.0	20.26
203.0	30.39
218.0	40.52
230.0	50.65
242.0	60.78
243.01	63.92
Ethyl acetate $C_4H_8O_2$	
80	1.110

Table 11.6 Vapor pressure of organic substances *(continued)*

t, °C	p, 10^5 Pa	t, °C	p, 10^5 Pa	t, °C	p, 10^5 Pa	t, °C	p, 10^5 Pa
100	2.022	100	2.460	112.5	5.06	40	8.516
120	3.447	110	3.153	138.0	10.13	50	10.92
140	5.552	120	3.986	167.8	20.26	60	13.75
160	8.489	130	4.973	186.5	30.39	Methylcyclohexane C_7H_{14}	
180	12.42	140	6.128	203.5	40.52	110	1.2986
200	17.60	150	7.468	214.0	50.65	130	2.146
220	24.33	160	9.036	224.0	60.78	150	3.362
240	33.06	170	10.85	240.0	79.72	170	5.01
250.1	38.49	180	12.94	Methylacetylene C_3H_4		190	7.23
Ethylbenzene C_8H_{10}		190	15.32	–20	1.17	210	10.10
140	1.1211	200	18.02	–10	1.84	230	13.83
150	1.4477	210	21.08	0	2.79	250	18.48
160	1.8440	220	24.52	10	4.05	270	24.21
170	2.319	230	28.37	20	5.77	290	31.16
180	2.883	234.7	30.31	30	6.79	Methylcyclopentane C_6H_{12}	
190	3.546	Hexene-1 C_6H_{12}		40	8.00	80	1.2946
200	4.317	70	1.2398	50	10.84	100	2.238
210	5.208	80	1.6626	60	14.38	120	3.632
220	6.229	90	2.1885	70	18.64	140	5.59
Ethylene C_2H_4		100	2.8326	2-Methylbutadiene-1,3 C_5H_8		160	8.22
0	40.94	110	3.610	40	1.2374	180	11.73
2	42.84	120	4.537	50	1.7012	200	16.28
4	44.79	130	5.629	60	2.2889	220	22.05
6	46.82	Isopropylbenzene C_9H_{12}		70	3.0202	240	29.24
8	48.92	160	1.2290	80	3.915	2-Methylpropene C_4H_8	
9.5	50.6	170	1.5656	90	4.996	0	1.315
n-Heptadecane $C_{17}H_{36}$		180	1.9698	100	6.282	10	1.87
310	1.1794	190	2.4498	2-Methylbutane C_5H_{12}		20	2.58
462	13.169	200	3.0146	30	1.090	30	3.50
n-Heptane C_7H_{16}		210	3.673	40	1.511	40	4.64
100	1.0606	220	4.434	50	2.048	50	6.06
120	1.8330	230	5.306	60	2.719	60	7.77
140	2.979	Methane CH_4 (*T*, K)		70	3.541	70	9.82
160	4.599	115	1.324	80	4.536	80	12.26
180	6.78	120	1.920	90	5.721	90	15.09
200	9.71	125	2.691	100	7.11	100	18.37
220	13.54	130	3.671	110	8.77	110	22.16
240	18.47	135	4.895	120	10.69	120	26.34
260	24.65	140	6.375	130	12.93	130	31.28
267.01	27.36	145	8.136	140	15.49	140	36.70
Heptene-1 C_7H_{14}		150	10.33	150	18.43	Monofluorodichloromethane	
100	1.2180	155	12.88	160	21.75	(Freon-21) $CHFCl_2$	
110	1.6046	160	15.88	170	25.51	10	1.0578
120	2.0797	165	19.38	180	29.71	20	1.5305
130	2.6556	170	23.38	187.8	33.3	30	2.1534
140	3.345	175	27.88	Methylchlorine CH_3Cl		40	2.9550
150	4.161	180	32.88	–20	1.188	50	3.9655
160	5.118	185	38.54	–10	1.772	60	5.2160
n-Hexane C_6H_{14}		190	45.52	0	2.557	70	6.7389
70	1.0537	190.55	46.41	10	3.582	80	8.5675
80	1.4239	Methanol CH_4O		20	4.893	90	10.737
90	1.8876	84.0	2.03	30	6.525	100	13.283

Table 11.6 Vapor pressure of organic substances *(continued)*

t, °C	p, 10^5 Pa	t, °C	p, 10^5 Pa	t, °C	p, 10^5 Pa	t, °C	p, 10^5 Pa
110	16.245	322	22.8	0	3.81	Toluene C_7H_8	
120	19.666	*n*-Octane C_8H_{18}		10	5.28	120	1.312
130	23.593	130	1.1408	20	7.14	140	2.180
140	28.081	150	1.9022	25	8.24	160	3.433
150	33.195	170	3.006	30	9.47	180	5.165
160	39.008	190	4.536	40	12.4	200	7.474
170	45.612	210	6.582	50	15.8	220	10.93
178.25	51.812	230	9.270	60	20.0	240	15.26
Monofluorotrichloromethane		250	12.79	70	24.8	260	20.50
(Freon-11) $CHFCl_3$		270	17.29	Propane C_3H_8 (*T*, K)		280	26.69
24	1.026	290	22.94	231.10	1.013	300	33.82
30	1.263	296.2	24.96	259.83	3.039	310	37.74
40	1.748	Octene-1 C_8H_{16}		275.24	5.065	*n*-Tridecane $C_{13}H_{28}$	
50	2.366	130	1.2843	286.90	7.091	240	1.1230
60	3.138	140	1.6619	296.30	9.117	404	17.2
70	4.088	150	2.1202	317.42	15.20	Trifluoromonochloromethane	
80	5.240	160	2.6698	341.71	25.33	(Freon-13) CF_3Cl	
90	6.619	170	3.322	359.61	35.45	–80	1.094
100	8.253	180	4.086	370.0	42.65	–70	1.797
110	10.168	*n*-Pentane C_5H_{12}		Propanol-1 C_3H_8O		–60	2.803
120	12.393	40	1.156	117.0	2.03	–50	4.183
130	14.959	50	1.591	149.0	5.06	–40	6.017
140	17.896	60	2.145	177.0	10.13	–30	8.393
Naphthalene $C_{10}H_8$		70	2.832	210.8	20.26	–20	11.40
220	1.060	80	3.676	232.3	30.39	–10	15.13
230	1.321	90	4.693	250.0	40.52	0	19.70
240	1.629	100	5.906	263.7	50.55	10	25.23
250	1.993	110	7.333	Propylene C_3H_6		20	31.84
260	2.53	120	8.985	–40	1.425	29.13	39.00
270	2.99	130	10.94	–30	2.132	*ortho*-Xylene C_8H_{10}	
280	3.57	140	13.20	–20	3.08	150	1.172
290	4.23	150	15.79	–10	4.30	160	1.505
300	4.97	160	18.74	0	5.82	170	1.907
310	5.82	170	22.08	10	7.74	180	2.388
320	6.79	180	25.84	20	10.10	190	2.956
330	7.86	190	30.04	30	12.98	200	3.622
340	9.05	196.62	33.74	40	16.45	210	4.390
350	10.33	Pentene-1 C_5H_{10}		50	20.54	220	5.280
360	11.85	40	1.4162	60	25.38	230	6.305
370	13.37	60	2.5797	70	31.00	240	7.500
n-Nonadecane $C_{19}H_{40}$		80	4.354	80	37.47	250	8.820
340	1.195	100	6.900	90	44.88	260	10.32
487	12.156	120	10.54	91.9	46.0	270	12.00
n-Nonane C_9H_2O		140	15.26	Tetrafluorodichloroethane		280	13.90
160	1.2846	160	21.56	(Freon-114) $C_2F_4Cl_2$		290	16.02
170	1.6410	180	29.76	5	1.069	300	18.35
180	2.070	200	40.26	10	1.289	310	20.95
190	2.579	201	40.4	20	1.836	320	23.79
200	3.179	Propadiene C_3H_4		30	2.584	330	26.87
210	3.879	–30	1.22	40	3.453	340	30.32
220	4.689	–20	1.84	50	4.583	350	34.36
230	5.618	–10	2.68	60	5.963	358.44	38.08

11.4 Separation Factor of Isotope Molecules

A separation factor of a two-component liquid mixture is defined as the ratio of relative component concentrations c_1 and c_2 in the vapor and liquid, respectively:

$$\alpha = \left(\frac{c_1}{c_2}\right)_v \Big/ \left(\frac{c_1}{c_2}\right)_l . \tag{11.1}$$

In the case of ideal solutions, to which the mixtures of isotope molecules might be pertinent, the separation factor comes out to a ratio of vapor pressures for pure components: $\alpha(t) = p_1(t)/p_2(t)$. The separation factors of isotope molecules are collected in Table 11.7.

Table 11.7 Separation coefficient of binary mixture of isotopic molecules under various temperatures

t, °C	α
Acetic acid	
$C_2D_4O_2$–$C_2H_4O_2$ [139]	
20	1.0457
30	1.0450
40	1.0440
50	1.0438
60	1.0432
70	1.0427
80	1.0423
90	1.0418
100	1.0414
110	1.0410
120	1.0406
125	1.0405
Aniline	
C_6D_7N–C_6H_7N [122]	
50	1.036
55	1.026
60	1.019
65	1.014
70	1.011
75	1.008
80	1.006
Benzene	
C_6D_6–C_6H_6 [123]	
10	1.0241
20	1.0249
30	1.0258
40	1.0262
50	1.0261
60	1.0254
70	1.0251
80	1.0249
82	1.0249
$^{13}C_6H_6$–$^{12}C_6H_6$ [124]	
34.6	1.00025
Boron trichloride	
$^{11}BCl_3$–$^{10}BCl_3$ [126]	
-85	0.9988
-80	0.9991
-70	0.9997
-60	1.0002
-50	1.0007
-40	1.0012
-30	1.0016
-20	1.0020
-10	1.0023
0.0	1.0026
10	1.0029
12.7	1.0030
Boron trifluoride	
$^{11}BF_3$–$^{10}BF_3$ [125]	
-120	1.0035
-115	1.0048
-110	1.0060
-105	1.0072
n-Butyl alcohol	
$C_4D_{10}O$–$C_4H_{10}O$ [121]	
20	1.063
30	1.059
40	1.054
50	1.050
60	1.046
70	1.042
80	1.038
90	1.034
100	1.031
Butyric acid	
$C_4H_8O_2$–$C_4D_8O_2$ [122]	
50	1.010
60	1.029
70	1.047
80	1.060
90	1.055
100	1.049
110	1.043
115	1.040
Carbon monoxide	
^{12}CO–^{13}CO [138]	
-204	1.0130
-200	1.0119
-195	1.0106
-190	1.0094
-185	1.0084
-180	1.0075
-175	1.0067
-170	1.0059
Carbon tetrachloride	
$^{13}CCl_4$–$^{12}CCl_4$ [137]	
34.6	1.0013
Chloroform	
$^{13}CH_3Cl$–$^{12}CH_3Cl$ [124]	
34.6	1.0008
Cyclohexane	
C_6D_{12}–C_6H_{12} [123]	
10	1.0933
20	1.0903
30	1.0877
40	1.0837
50	1.0809
60	1.0774
70	1.0744
80	1.0683
Diborane	
B_2D_6–B_2H_6 [74]	
-155	1.0525
-150	1.0495
-140	1.0456
-130	1.0439
-120	1.0442
-110	1.0449
-100	1.0478
-90	1.0516
Ethylene	
C_2H_3D–C_2H_4 [140]	
-154.88	1.0024
-150	1.0033
-140	1.0046
-130	1.0055
-120	1.0062
-110	1.0067
-100	1.0070
-91.85	1.0071
cis-$C_2H_2D_2$–C_2H_4 [141]	
-160	1.0030
-150	1.0064
-140	1.0095
-130	1.0121
-120	1.0136
-110	1.0146
-100	1.0153
-90	1.0157
trans-$C_2H_2D_2$–C_2H_4 [141]	
-160	1.0047
-150	1.0085
-140	1.0112
-130	1.0131
-120	1.0143
-110	1.0153
-100	1.0159
-90	1.0162
heme-$C_2H_2D_2$–C_2H_4 [141]	
-160	1.0032
-150	1.0084
-140	1.0096
-130	1.0123
-120	1.0138
-110	1.0148
-100	1.0155
-90	1.0159
Hydrogen chloride	
$H^{35}Cl$–$H^{37}Cl$ [131]	
-105	1.0021
-100	1.0019
-95	1.0017

Table 11.7 Separation coefficient of binary mixture of isotopic molecules *(continued)*

t, °C	α
-90	1.0015
Hydrogen sulfide	
$H_2\,^{32}S$–$H_2\,^{35}S$ [137]	
-76	1.0055
-65	1.0058
Isobutyric acid	
$C_4H_8O_2$–$C_4D_8O_2$ [122]	
50	1.015
60	1.022
70	1.029
80	1.035
90	1.041
Isopropyl alcohol	
C_3D_8O–C_3H_8O [122]	
15	1.065
20	1.062
30	1.055
40	1.048
50	1.042
60	1.037
70	1.033
80	1.029
Isovaleric acid	
$C_5H_{10}O_2$–$C_5D_{10}O_2$ [122]	
55	1.034
60	1.036
70	1.037
80	1.038
90	1.039
100	1.039
110	1.039
Krypton	
^{82}Kr–^{86}Kr [133]	
-152	1.00048
Lithium	
^{6}Li–^{7}Li [134]	
400	1.060
450	1.047
500	1.036
550	1.026
Methane	
CD_4–CH_4 [135]	
-182.7	1.0130
$^{12}CH_4$–$^{13}CH_4$ [131]	
-182.7	1.0114
-180	1.0112
-175	1.0108
-170	1.0105
-165	1.0102
-161.3	1.0098
Methyl alcohol	
$^{13}CH_4O$–$^{12}CH_4O$ [124]	
34.6	1.0002
Neon ^{20}Ne–^{22}Ne [136]	
-256.74	1.131
-256	1.115
-255	1.103
-254	1.092
-253	1.085
-252	1.078
-251	1.071
-250	1.066
-249	1.061
-248	1.046
-247	1.043
-246	1.040
-245	1.038
-244	1.036
-243	1.034
Nitric tetroxide	
$^{14}N_2O_4$–$^{14}N^{15}NO_4$ [121]	
21	1.0038
Oxygen	
$^{16}O_2$–$^{18}O_2$ [132]	
-210	1.0131
-205	1.0121
-200	1.0112
-195	1.0105
-190	1.0098
-185	1.0092
-182.97	1.0089
Pentaborane	
B_5D_9–B_5H_9 [93]	
-45	1.0656
-40	1.0634
-30	1.0593
-20	1.0555
-10	1.0521
0.0	1.0489
10	1.0460
20	1.0434
30	1.0409
40	1.0387
50	1.0366
60	1.0347
Sulphur dioxide	
$^{32}SO_2$–$^{35}SO_2$ [137]	
-35	1.0018
Water	
H_2O–D_2O [127]	
0.0	1.255
10	1.182
20	1.154
30	1.137
40	1.122
50	1.107
60	1.094
70	1.081
80	1.071
90	1.061
100	1.052
110	1.045
120	1.038
130	1.032
140	1.027
150	1.022
160	1.018
170	1.014
180	1.011
190	1.008
200	1.005
210	1.003
220	1.001
230	0.9989
240	0.9958
H_2O–T_2O [127]	
20	1.113
30	1.096
40	1.076
50	1.064
60	1.052
$H_2\,^{16}O$–$H_2\,^{18}O$ [129]	
20	1.0093
30	1.0085
40	1.0077
50	1.0069
60	1.0062
70	1.0055
80	1.0049
90	1.0043
100	1.0037
110	1.0032
120	1.0027
130	1.0022
140	1.0018
150	1.0013
160	1.0009
170	1.0005
180	1.0001
190	0.9998
200	0.9994
210	0.9991
$H_2\,^{16}O$–$H_2\,^{17}O$ [130]	
52	1.0039
73	1.0030
Xenon	
^{130}Xe–^{136}Xe [133]	
-108	1.0001

References

[1] Rivkin, S. L., Aleksandrov, A. A., Thermal properties of water and water vapor. Handbook, Energoatomizdat, Moscow, 1984 (in Russian).

[2] Chaikhorskii, A. A., Neptunium chemistry, Atomizdat, Moscow, 1978 (in Russian).

[3] Tables of physical quantities. Handbook, Ed. Kikoin, I. K., Atomizdat, Moscow, 1976 (in Russian).

[4] Carniglia, S. C., Cunningham, B. B., J. Am. Chem. Soc., 77, 1502, 1955.

[5] Erway, N. D., Simpson, O. C., J. Chem. Phys., 18, 983, 1960.

[6] Green, J. W., J. Phys. Chem., 41, 2245, 1964.

[7] Shchukarev, S. A., Semenov, G. A., Zhurnal neorganicheskoi khimii, 2, 1217, 1957 (in Russian).

[8] Blauer, J. A., Greenbaum, M. A., Farber, M., J. Phys. Chem., 69, 1069, 1965.
[9] Greenbaum, M. A., Foster, J. N., Arin, M. L., J. Phys. Chem., 67, 36, 1963.
[10] Greenbaum, M. A., Yates, R. E., Farber, M., J. Phys. Chem., 67, 1802, 1963.
[11] Soulen, J. R., Sthapitanonda, P., Margrave, J. L., J. Phys. Chem., 59, 132, 1955.
[12] Darnell, A. J., Yosim, S. J., J. Phys. Chem., 63, 1813, 1959.
[13] Blackburn, P. E., Hoch, M., Jonston, H. L., J. Phys. Chem., 62, 769, 1958.
[14] Smith, F. J., Barrow, R. F., Trans. Faraday Soc., 54, 826, 1958.
[15] Shchukarev, S. A., Semenov, G. A., Rat'kovskii, I. A., Zhurnal neorganicheskoi khimii, 6, 1973, 1961 (in Russian).
[16] Lu Tsun-Khua, Pashinkin, A. S., Novoselova, A. V., Zhurnal neorganicheskoi khimii, 7, 963, 1962 (in Russian).
[17] Davydov, V. I., Diev, N. P., Zhurnal neorganicheskoi khimii, 2, 2003, 1957 (in Russian).
[18] Chikara Hirama, J. Phys. Chem., 66, 1563, 1962.
[19] Habermann, C. E., Daane, A. H., J. Chem. Phys., 41, 2818, 1964.
[20] Maclaren, R. O., Gregory, N. W., J. Phys. Chem., 59, 184, 1955.
[21] Sime, R. J., Gregory, N. W., J. Phys. Chem., 64, 86, 1960.
[22] Shchukarev, S. A., Semenov, G. A., Rat'kovskii, I. A., Zhurnal obshchei khimii, 31, 2090, 1961 (in Russian).
[23] Ackermann, R. J., Rauh, E. G., J. Chem. Phys., 36, 448, 1962.
[24] Pashinkin, A. S., Drobot, D. B., Shevtsova, Z. N., Zhurnal neorganicheskoi khimii, 7, 2811, 1962 (in Russian).
[25] Bloom, H., Bockris, O'M., Richards, N. E., J. Am. Chem. Soc., 80, 2044, 1958.
[26] Pugh, A. C. P., Barrow, R. F., Trans. Faraday Soc., 54, Part 5, 425, 1958.
[27] Schulz, D. A., Searcy, A. W., J. Phys. Chem., 67, 103, 1963.
[28] Blue, C. D., J. Phys. Chem., 67, 877, 1963.
[29] Hildenbrand, D. L., Potter, N. D., J. Phys. Chem., 67, 2231, 1963.
[30] Baytista, R. G., Margrave, J. L., J. Phys. Chem., 67, 2411, 1963.
[31] Hildenbrand, D. L., Potter, N. D., J. Chem. Phys., 40, 2882, 1964.
[32] Eick, H. A., Mulford, R. N. R., J. Chem. Phys., 41, 1475, 1964.
[33] Farber, M., Meyer, R. T., Margrave, J. L., J. Phys. Chem., 62, 883, 1958.
[34] Nesterova, Ya. M., Pashinkin, A. S., Novoselova, A. V., Zhurnal neorganicheskoi khimii, 6, 2014, 1961 (in Russian).
[35] Hirayama, C., Ichikawa Y., De Roo A. M., J. Phys. Chem., 67, 1039, 1963.
[36] Dreger, L. H., Margrave, J. L., J. Phys. Chem., 64, 1323, 1960.
[37] Zavitsano, P. D., J. Phys. Chem., 68, 2899, 1964.
[38] Phipps, T. E., Report No 735 at the International Conference on Peaceful Use of Atomic Energy, Geneva, 1955.
[39] Ruf, R., Treadwell, W. D., Helv. Chim. Acta., 37, 1941, 1954.
[40] Zlomanov, V. P., Popovkin, B. A., Novoselova, A. V., Zhurnal neorganicheskoi khimii, 4, 2661, 1959 (in Russian).
[41] Pashinkin, A. S., Novoselova, A. V., Zhurnal neorganicheskoi khimii, 4, 2657, 1959 (in Russian).
[42] Zhiteneva, G. M., Rumyantsev, Yu. V., Bolondz', F. M., Trudy Vostochno–Sibirskogo filiala Sib. otd. Akad. Nauk SSSR, 41, 121, 1962 (in Russian).
[43] Spedding, F. H., Trans. Am. Soc. Metals, 218, 608, 1960.
[44] Krikorian, O. H., J. Phys. Chem., 67, 1586, 1963.
[45] Karelin, V. V., Nesmeyanov, A. N., Priselkov, Yu. A., Doklady Akad. Nauk SSSR, 144, 352, 1962 (in Russian).
[46] Gekov L. Kh., Nesmeyanov, A. N., Priselkov, Yu. A., Vestnik Moskovskogo universiteta, Ser. 2: Khimia, 5, 34, 1962 (in Russian).
[47] Shakhtakhtinskii, M. G., Kuliev, A. A., Doklady Akad. Nauk SSSR, 123, 1071, 1958 (in Russian).
[48] Zlomanov, V. P., Popovkin, B. A., Zhurnal neorganicheskoi khimii, 3, 1473, 1958 (in Russian).
[49] Franzen, H. F., Gills, P. W., J. Chem. Phys., 42, 1033, 1965.
[50] Darnell, A. J., Keneshca, F. J., J. Phys. Chem., 62, 1143, 1958.
[51] A handbook of a chemist, Ed. Nikolskii, B. P., Khimia, Leningrad, 1971 (vol. 1) (in Russian).
[52] Shchukarev, S. A., Vasil'kova, I. V., Efimov, A. I., Zhurnal neorganicheskoi khimii, 1, 2272, 1956 (in Russian).
[53] Sime, R. J., Gregory, N. W., J. Am. Chem. Soc., 82, 800, 1960.
[54] Allen, T. L., J. Am. Chem. Soc., 78, 5476, 1956.

[55] Scheer, M. D., Fine, J., J. Chem. Phys., 36, 1647, 1962.

[56] Nesmeyanov, A. N., Sazonov, L. A., Zhurnal neorganicheskoi khimii, 5, 519, 1960 (in Russian).

[57] Ahman, D. H., U. S. Atomic Energy Comission AECD No. 3205, 1951.

[58] Kruglikh, A. A., Pavlov, V. S., Ukrainskii fizicheskii zhurnal, 10, 1029, 1965 (in Russian).

[59] Nesmeyanov, A. N., Iofa, B. Z., Strel'nikov, A. A., Zhurnal fizicheskoi khimii, 32, 955, 1958 (in Russian).

[60] Pierce L., Pace, Ė. L., J. Chem. Phys., 23, 551, 1955.

[61] Baimakova, A. Yu., Trudy Leningradskogo politekhnichskogo instituta, 188, 156, 1957 (in Russian).

[62] Sharman, R. H., Glaugue, W. F., J. Am. Chem. Soc., 77, 2154, 1955.

[63] Novikov, G. I., Gavryuchenkov, F. G., Zhurnal neorganicheskoi khimii, 9, 475, 1964 (in Russian).

[64] Andersen, H. C., Belz, L. H., J. Am. Chem. Soc., 75, 4828, 1953.

[65] Rogers, M. T., Speirs, J. L., J. Phys. Chem., 60, 1462, 1956.

[66] Verne, L., Trevorrow, E., J. Phys. Chem., 62, 362, 1958.

[67] Ruban, I. N., Ponomarev, V. D., Trudy instituta metallurgii i obogashcheniya Akad. Nauk KazSSR, 5, 34, 1962 (in Russian).

[68] Shchukarev, S. A., Suvorov, A. V., Vestnik Leningrad. universiteta, Ser. "Fizika i khimia," 4, 87, 1961 (in Russian).

[69] Shchukarev, S. A., Novikov, G. I., Kokovin, G. A., Zhurnal neorganicheskoi khimii, 4, 2185, 1959 (in Russian).

[70] Shchukarev, S. A., Novikov, G. I., Andreeva, N. V., Vestnik Leningradskogo universiteta, Ser. "Fizika i khimia," 4, 120, 1959 (in Russian).

[71] Sun In-Chzhu, Morozov, I. S., Zhurnal neorganicheskoi khimii, 4, 492, 1959 (in Russian).

[72] Palko, A. A., Ryon, A. D., Kuhn, D. W., J. Phys. Chem., 62, 319, 1958.

[73] Gibbs, S. G., Shapiro, J., J. Chem. Phys., 30, 1483, 1959.

[74] Ditter, J. F., Perrine, J., Shapiro, J., J. Chem. Eng. Data, 6, 271, 1961.

[75] Clarke, J. T., Rifkin, E. B., Jonston, H. L., J. Am. Chem. Soc., 75, 781, 1953.

[76] Wirth, H. E., Palmer, E. D., J. Phys. Chem., 60, 911, 1956.

[77] Zaidel', A. N., Ostrovskaya, G. V., Ostrovskii, Yu. I., Spectroscopy methods and practice, Nauka, Moscow, 1976 (in Russian).

[78] Polyachenok, O. G., Novikov, G. I., Zhurnal neorganicheskoi khimii, 8, 2631, 1963 (in Russian).

[79] Maclaren, R. O., Gregory, N. W., J. Phys. Chem., 59, 184, 1955.

[80] Fadeev, V. N., Fedorov, P. I., Zhurnal neorganicheskoi khimii, 8, 2007, 1963 (in Russian).

[81] Isakova, R. I., Nesterov, V. N., Shendyapin, A. S., Zhurnal neorganicheskoi khimii, 8, 18, 1963 (in Russian).

[82] Tolmacheva, T. A., Andrinovskaya, T. L., Vestnik Leningradskogo universiteta, Ser. "Fizika i khimia," 10, 131, 1960 (in Russian).

[83] Bloom, H., Welch B. J., J. Phys. Chem., 62, 1594, 1958.

[84] Long, L. H., Cattanach, J., J. Inorg. Nucl. Chem., 20, 340, 1961.

[85] Li, J. C. M., J. Am. Chem. Soc., 78, 1081, 1956.

[86] Shchukarev, S. A., Vasil'kova, N. V., Sharunin, B. N., Vestnik Leningradskogo universiteta, Ser. "Fizika i khimia," 22, 130, 1961 (in Russian).

[87] Devyatykh, G. G., Frolov, I. A., Zhurnal neorganicheskoi khimii, 8, 265, 1963 (in Russian).

[88] Cubissiotti, D., Eding, H., J. Phys. Chem., 69, 2743, 1965.

[89] Isakova, R. I., Nesterov, V. N., Trudy instituta metallurgii i obogashcheniya Akad. Nauk KazSSR, 5, 29, 1962 (in Russian).

[90] Berdonosov, S. S., Lapitskii, A. V., Bakov, E. K., Zhurnal neorganicheskoi khimii, 10, 322, 1965 (in Russian).

[91] Tarasenkov, D. N., Komandin, A. V., Zhurnal obshchei khimii, 10, 1319, 1940 (in Russian).

[92] Klushin, D. N., Chernykh, V. Ya., Zhurnal neorganicheskoi khimii, 5, 1409, 1960 (in Russian).

[93] Shapiro, J., Ditter J. F., J. Chem. Phys., 26, 798, 1957.

[94] Koehler, J. L., Giaugue, W. F., J. Am. Chem. Soc., 80, 2659, 1958.

[95] Jarry, R. L., J. Phys. Chem., 61, 498, 1957.

[96] Weinstock, B., Malm, J., Wealer, E. E., J. Am. Chem. Soc., 83, 4310, 1961.

[97] Malm, J. G., Selig, H., J. Inorg. Nucl. Chem., 20, 189, 1961.

[98] Baryshnikov, N. V., Zelikman, A. N., Teslitskaya, M. V., Zhurnal neorganicheskoi khimii, 7, 2634, 1962 (in Russian).

[99] Nikol'skii, A. B., Zhurnal neorganicheskoi khimii, 8, 1045, 1963 (in Russian).

[100] Nikol'skii, A. B., Zhurnal neorganicheskoi khimii, 10, 290, 1965 (in Russian).

[101] Zorin, A. D., Devyatykh, G. G., Krupnova, E. F., Zhurnal neorganicheskoi khimii, 9, 2280, 1964 (in Russian).

[102] Berka, L., J. Inorg. Nucl. Chem., 14, 190, 1960.

[103] Shchukarev, S. A., Kurbanov, A. R., Vestnik Leningradskogo universiteta, Ser. "Fizika i khimia," 10, 144, 1962 (in Russian).

[104] Kudryavtsev, A. A., Ustyugov, G. P., Zhurnal neorganicheskoi khimii, 6, 2421, 1961 (in Russian).

[105] Kudryavtsev, A. A., Ustyugov, G. P., Trudy Moskovskogo khimiko–tekhnologicheskogo instituta, 38, 42, 1962 (in Russian).

[106] Machol, R. E., Westrum, E. F., J. Am. Chem. Soc., 80, 2950, 1958.

[107] Selig, H., Malm, J. G., J. Inorg. Nucl. Chem., 24, 641, 1962.

[108] Smith, W. T., Cobbie, J. W., Boyd, G. E., J. Am. Chem. Soc., 75, 5773, 1953.

[109] Pace, E. L., Turnbull, B. F., J. Chem. Phys., 43, 1953, 1965.

[110] Pike, F. P., Foster, C. T., J. Chem. Eng. Data, 4, 305, 1959.

[111] McDougall, L. A., Kilpatrick, J. E., J. Chem. Phys., 42, 2311, 1965.

[112] Langer, S., Blaukenspir, F. F., J. Inorg. Nucl. Chem., 14, 26, 1960.

[113] Kenesha, F. J., Cubissiotti, D., J. Chem. Phys., 40, 191, 1964.

[114] Nesterov, V. N., Ponomarev, V. D., Izvestia Akad. Nauk KazSSR, Ser. "Metallurgiya, obogashcheniye i ogneupory," 1(4), 80, 1959 (in Russian).

[115] Vargaftik, N. B., Handbook on thermal properties of gases and liquids, 2nd ed., Nauka, Moscow, 1972 (in Russian).

[116] Malyshev, V. V., Teplofizika vysokikh temperatur, 12, 743, 1974 (in Russian).

[117] Committe consultatif de termometrie, 14 Session, 1982. Recommendation T1, p. T85-T86.

[118] Malyshev, V. V., Teplofizika vysokikh temperatur, 11, 1010, 1973 (in Russian).

[119] Clegg, H. P., Rowlinson, J. S., Sutton J. R., Trans. Faraday Soc., 51, 1327, 1955.

[120] Malyshev, V. V., Atomnaya energia, 34, 42, 1973 (in Russian).

[121] Begun, G. M., J. Chem. Phys., 25, 1279, 1956.

[122] Rabinovich, N. B., Sokolov, N. N., Artyukhin, P. I., Doklady Acad. nauk SSSR, 105, 762, 1955 (in Russian).

[123] Davis, R. T., Schisser, R. W., J. Phys. Chem., 57, 966, 1953.

[124] Baerstschi, F., Nature, 171, 1018, 1953.

[125] Sevryugova, N. N., Uvarov, O. V., Zhavoronkov, N. M., Zhurnal fizicheskoi khimii, 34, 1004, 1960 (in Russian).

[126] Sevryugova, N. N., Uvarov, O. V., Zhavoronkov, N. M., Doklady Acad. nauk SSSR, 126, 1044, 1959 (in Russian).

[127] Thermal properties of substances. Handbook, Ed. Vargaftik, N. B., Gosenergoizdat, Moscow, 1956 (in Russian).

[128] Smith, H. A., Fith, K. R., J. Phys. Chem., 67, 920, 1963.

[129] Uvarov, O. V., Sokolov, N. M., Zhavoronkov, N. M., Zhurnal fizicheskoi khimii, 36, 2699, 1962 (in Russian).

[130] Borowitz, J. L., J. Phys. Chem., 66, 1412, 1962.

[131] Matveev, K. I., Uvarov, O. V., Zhavoronkov, N. M., Doklady Acad. nauk SSSR, 125, 580, 1959 (in Russian).

[132] Devyatykh, G. G., Zorin, A. D., Zhurnal fizicheskoi khimii, 30, 1133, 1956 (in Russian).

[133] Grigoriev, V. N., Zhurnal fizicheskoi khimii, 36, 1779, 1962 (in Russian).

[134] Katal'nikov, S. G., Andreev, B. M., Atomnaya energia, 11, 240, 1961 (in Russian).

[135] Armstrong, G., J. Chem. Phys., 21, 1293, 1953.

[136] Bigeleisen, J., Roth, E., J. Chem. Phys., 35, 68, 1961.

[137] Devyatykh, G. G., Suloev, Yu. N., Zorin, A. D., Trudy Moscovskogo khimiko–tekhnologicheskogo instituta po khimii i tekhnologii, 1, 24, 1958 (in Russian).

[138] Sevryugova, N. N., Zhavoronkov, N. M., Doklady Akad. Nauk SSSR, 134, 875, 1960 (in Russian).

[139] Potter, A. E., Ritter, H. L., J. Phys. Chem., 58, 1040, 1954.

[140] Bigeleisen, J., Stern, M. J., Von Hook, W. A., J. Chem. Phys., 38, 497, 1963.

[141] Bigeleisen, J., Ribinkar, S. V., Von Hook, W. A., J. Chem. Phys., 38, 489, 1963.

12

Melting and Boiling

E.B. Gel'man

12.1 Introduction

The transition of a substance from the solid state to the liquid one is referred to as melting; the transition from the condensed state (solid or liquid) to the gaseous state (vapor) is called evaporation. In natural conditions melting is a steady-state process; evaporation becomes a steady-state process if the external pressure is equal to vapor pressure. Steady-state evaporation from the liquid state is known as boiling, and from the solid state, sublimation.

Melting, boiling, and sublimation processes are accompanied by energy absorption, which is designated melting heat ΔH_m, boiling heat ΔH_b, and sublimation heat ΔH_s for the substance given.

12.2 Melting and Boiling at Fixed Pressure

Melting, boiling, and sublimation of pure substances at a fixed pressure occur at the certain temperatures characteristic of the substance at hand, which are referred to as melting temperature T_m, boiling temperature T_b, and sublimation temperature T_s, respectively. In Tables 12.1–12.3, the melting and boiling (or sublimation) temperatures and heats are listed for elements, and organic and inorganic compounds.

Table 12.1 Melting, boiling, or sublimation point (t, °C) and heat (ΔH, kJ/mol) of elements and some isotopes at normal pressure (101,325 Pa)

Element	t_m	ΔH_m	t_b, t_s	$\Delta H_b, \Delta H_s$	Ref.
Actinium	1050	12	3297 [10]	378	[1]
Aluminum	660.24 [10]	10.8	2520	293	[1]
Americium	1180	14.4	≈ **2400** [11]	230	[1]
Antimony	630.5	20.1	1634	124.4	[1]
Argon	−189.30	1.190	−185.9	6.5	[1]
Arsenic (grey)	Sublimated		615.0	31.8	[1]
Astatine	302	8.8 [6]	334	90.4 [4]	[11]
Barium	727	7.53 [6]	1640 [9]	139	[1]
Beryllium	1287	13	2471	291	[1, 11]
Bismuth	217.4	11.0	1552	177	[1]
Boron	2075	23	3700	512	[1]

0-8493-2861-6/97/$0.00+$.50

Table 12.1 Melting, boiling, or sublimation point and heat of elements *(continued)*

Element	t_m	ΔH_m	t_b, t_s	$\Delta H_b, \Delta H_s$	Ref.
Bromine, Br_2	7.25	10.57	59.2	29.5	[1]
Cadmium	320.9 [9]	6.2	766.5	99.6	[1]
Calcium	842	8.66 [6]	1484 [10]	152	[1]
Carbon (graphite)	Sublimated		4200		[10]
Cerium	799	5.2 [1]	$\approx$ 3340 [9, 10]	409 [1]	[9]
Cesium	28.5	2.10	672	67	[1]
Chlorine, Cl_2	−101.03	6.61	−34.1	20.41	[1]
Chromium	1877 [10]	16.3 [6]	2672 [9]	338	[1]
Cobalt	1492	16.3	2957 [11]	376	[1]
Copper	1083	13.0	2543	302	[1]
Curium	1340				[10]
Deuterium tritide, DT ($^2H^3H$)	−253.5		−248.9		[4]
Deuterium					
para–D_2	−254.42	0.197	−249.9 [10]	1.226 [4]	[1]
ortho–D_2			−249.56	1.223	[4]
Dysprosium	1409	10.9	2562 [9]	228	[1]
Einsteinium	860				[10]
Erbium	1525	19.9	$\approx$ 2690 [9, 10]	261	[1]
Europium	822 [10]	9.2	1597 [10]	147	[1]
Fluorine, F_2	−219.6	0.510	−188.13	6.54	[1]
Francium	26.84 [10]	2.1 [6]	677	64	[9]
Gadolinium	1313 [9]	10.0	3280	360	[1]
Gallium	29.78 [4]	5.59 [2]	$\approx$ 2300 [9, 11]	256	[1]
Germanium	237.2	37.6 [6]	2847 [10]	334	[4]
Gold	1063.4	12.6	2877 [10]	331	[1]
Hafnium	2230 [10]	23.9 [6]	$\approx$ **4900** [9, 11]	569	[1]
Helium, ^4He	−271.4 (3.00 MPa)	0.007 (3.00 MPa) Solid phase is absent at normal pressure	−268.934 [9]	0.0837 [4]	[1]
Helium-3, ^3He			−269.95	0.0478 (−271.5°C)	[4]
Holmium	1474 [9]	12.4 [6]	2695 [10]	240 [1]	
Hydrogen deuteride, HD	−256.5	0.1548	−251.02	1.109 (−256.5°C)	[4]
Hydrogen tritide, HT	−254.7		−249.6		[4]
Hydrogen, *n*-H_2	−259.19	0.117	−252.77	0.916	[1]
para-, *p*-H_2	−259.27	0.117	−252.87	0.900	[4]
ortho-, *o*-H_2	−254.52	0.197			[4]
Indium	156.4	3.34 [6]	2024	228	[1]
Iodine, I_2	113.6	15.77 [4]	184.35 [9]	41.8	[1]
Iridium	2447	26.3 [4]	4380	613	[1]
Iron	1538 [10]	13.8	2872 [11]	350	[1]
Krypton	−157.37	1.64	−153.22	9.046	[1]
Lanthanum	920	6.3 [6]	3454 [10]	413	[1]
Lead	327.44 [10]	4.77	1745	178	[1]
Lithium	180.54 [9]	3.0	1347 [9]	138	[1]
Lutetium	1663 [10]	18.8	3395 [9]	356	[1]
Magnesium	650	8.56 [6]	1107 [10]	128	[1]
Manganese	1245	12.1 [6]	2080	227	[1]
Mercury	−38.89	2.30	356.66	59.23	[1]
Molybdenum	2620	36	$\approx$ **4700** [1, 10]	582	[1]
Neodymium	1024	10.68	3080	255	[12]
Neon, Ne	−248.52	0.33	−245.93	1.79	[1, 10]
Neptunium	637	52	$\approx$ 4000 [1, 9]	422	[1]

Table 12.1 Melting, boiling, or sublimation point and heat of elements *(continued)*

Element	t_m	ΔH_m	t_b, t_s	$\Delta H_b, \Delta H_s$	Ref.
neon-20, ^{20}Ne	−248.49	0.331			[12]
neon-22, ^{22}Ne	−248.31	0.334			[12]
Nickel	1455	17.6	≈ 2800 [9, 11]	370	[1]
Niobium	2469 [10]	28	≈ 4800 [9, 11]	662	[1]
Nitrogen, N_2	−210.012	0.7207 [4]	−195.812	5.59 [10]	[10]
nitrogen-15, $^{15}N_2$	−210.96 [12]	0.7205	−195.76	5.592	[4]
nitrogen-14, $^{14}N_2$	−210.01				[12]
Osmium	3045	31.8 [6]	5027	748 [1]	[9]
Oxygen, O_2	−218.79 [3]	0.4459 [3]	−182.97	6.833	[1]
Palladium	1554	17	≈ 3000 [9, 10]	353	[1]
Phosphorus (white)	44.1	0.66	257	13.1	[1]
(red)	Sublimated		429	29.8	[1]
(black)	Sublimated		453		[1]
Platinum	1772	20 [1]	3827	511 [1, 2]	[10]
Plutonium	639.7	2.8	≈ 3300	351	[1, 10]
Polonium	254	12.6 [6]	962	59	[1]
Potassium	63.5 [10]	2.33	761	76.6	[1]
Praseodymium	932	6.9	≈ 3360 [9, 10]	297	[1]
Promethium	≈ 1100	10.5 [6]	≈ 2600		[9, 11]
Protactinium	1572 [10]	16.7	≈ 3780 [10, 11]	481.5	[2]
Radium	700	9.2 [6]	≈ 1320	132 [1]	[9, 11]
Radon	−71 [10]	2.89	−61.9	16.8	[1]
Rhenium	3180 [10]	33.1 [6]	5627 [9]	715 [8]	
Rhodium	1963	21.8 [6]	≈ 3680 [9, 11]	496	[1]
Rubidium	39.49	2.19	686.04	70	[1, 10]
Ruthenium	2250	24.3 [6]	≈ 4050 [9]	602	[1]
Samarium	1072	8.6	1778	166	[1]
Scandium	1541	14.2 [6]	2831 [9]	315	[1]
Selenium (grey)	217	6.7 [6]	685	29	
Silicon	1415 [10]	49.8	3300	356	[1]
Silver	960.5	11.3 [6]	2167	251	[1]
Sodium	97.9	2.60 [6]	886	90.1	[1]
Strontium	768	8	1384 [9]	134	[1]
Sulphur					
(β, monocl.)	119.3	1.72	444.6	9.2	[1]
(α, rhomb.)	112.8		444.6	9.2	[1]
Tantalum	2996 [9]	35	≈ 5350 [9, 11]	745	[1]
Technetium	2200	24	≈ 4260 [9, 11]	593	[1]
Tellurium	449.8	17.5	990	51.0	[1]
Terbium	1356 [9]	10.8	3200	331	[1]
Thallium	303.5	4.2	1457	162.4	[1, 9]
Thorium	1750	16.7 [1, 6]	4790 [9]	540	[1]
Thulium	1545	16.9	≈ 1840 [9, 11]	191	[1]
Tin (grey)	231.97 [9]	7.2	≈ 2450 [9, 11]	296.1	[1]
Titanium	1608 [10]	15.1	3287 [9]	410	[1]
Tritium, $T_2(^3H_2)$	−252.53	0.23	−248.11	1.39	[1]
Tungsten	3420	35.1 [6]	5680	770	[1]
Uranium	1134	9.2	≈ 4030 [1, 10]	494	[1]
Vanadium	1917	17.6 [6]	3392	445 [1]	[11]
Xenon	−111.85	2.29	−107.96 [10]	12.63	[1]
Ytterbium	824	7.7	1211	130	[1]
Yttrium	1528	11	3337 [10]	362	[1]
Zinc	419.5	7.2	906.2	115.3	[1]
Zirconium	1852 [10]	14.6	4340	558	[1]

Note: Values that differ more than 10% according to different references are presented on bold type.

Table 12.2 Melting, boiling, or sublimation point (t, °C) and heat (ΔH, kJ/mol) of inorganic compounds at normal pressure (101,325 Pa) [1]

Substance	t_m	ΔH_m	t_b, t_s	$\Delta H_b, \Delta H_s$	Ref.	Substance	t_m	ΔH_m	t_b, t_s	$\Delta H_b, \Delta H_s$	Ref.
$AcBr_3$		subl.	800			BF_3	−128	4.6	−100	17.1	
$AcCl_3$	802	36.0	1760	188	[2]	B_2H_6	−165.6	4.5	−92.5	14.3	[4]
AcF_3	1327	33.4	2277	251	[2]	B_4H_{10}	−121		18	25.5	
Ac_2O_3	1977	83.7			[2, 3]	B_5H_9	48.8		60.0	28.9	
$AgBr$	424	8.8	1502	198	[4]	$B_{10}H_{14}$	99.5	32.7	213	44	[2]
$AgCN$	350	11.5				BI_3	49.9		209.5	41.8	
$AgCl$	455	13.2	1557	184	[5]	BN (α −)	3000		5067		[7]
$AgClO_3$	231		270 (decomp.)			$B_3N_3H_6$	−58		55	29.3	
AgF	435	17	1147	142	[2]	B_2O_3	450	24.6	2124	356.3	[3]
AgF_2	690	18.4	1227	150.5	[2]	$BaBr_2$	857	32.2	1980	232	
AgI	554	9.4	1506	143.9	[4]	$BaCO_3$			1400 (decomp.)		
$AgNO_3$	209	12.1	300 (decomp.)			$BaCl_2$	960	15.9	2050	241	
AgO	110 (expl.)					BaF_2	1370	17.6	2250	271	
Ag_2O	187	15.5 (decomp.)			[3]	BaH_2	1200	25			
Ag_3PO_4	849					BaI_2	711		1900		
Ag_2S	838	14.1 (decomp.)				$BaMoO_4$	1460		1730		
Ag_2SO_4	657	17.9	> 1085 (decomp.)		[4]	BaO	1923	57.7	2727	259.6	[3, 4]
Ag_2Se	897	17.58 (decomp.)			[6]	BaO_2	450	23.9	837 (decomp.)		[2, 3]
Ag_2Te	960	21.4			[6]	Ba_2O	607	21.8	767	83.7	[3]
AlB_{12}	2159					$Ba(OH)_2$	408	15		290	
$Al(BH_4)_3$	−64.5		44.5	30.0	[4, 5]	BaS	>2200		3000	435	[2, 7]
$AlBr_3$	97.5	11.3	256.2	23.5	[4]	$BaSO_4$	1350	40.6			[4]
Al_4C_3	2100				[2]	$BaSe$	1780	38			
$AlCl_3$		subl.	180.2	56.1	[2]	$BaWO_4$	1475		1730		
$AlCl_4$		subl.	180.1	112	[5]	$BeAl_2O_4$	1870	171.5			
AlF	827	21.8	1402	155	[2]	$BeBr_2$	508	18	540	98	
AlF_3		subl.	1272	318	[2]	Be_2C	2400	75.4	2537		[7]
AlH_3	> 105 (decomp.)					$BeCl_2$	415	15.3	550	110	
AlI_3	191	15.9	382	67		BeF_2	800	21	1175	200	
AlN	2227				[4]	BeI_2	490	18	530	98	
Al_2O_3	2046.5	113.0	2980	485.7	[3]	Be_3N_2	2200	109			
$AlPO_4$	2000					BeO	2550	85.4	4120	489.9	[3, 4]
Al_2S_3	1100		1500 (decomp.)			$BeSO_4$	1100	6 (decomp.)			
AmF_3	1427	54.3	2067	276	[2]	$BiBr_3$	219	21.7	461	75	
Am_2O_3	2205	71.2	3127	356	[2, 3]	$BiCl_2$	163	14.7	577	83.7	[2]
$AsBr_3$	31.2	11.8	221	41.8	[4]	$BiCl_3$	233	23.6	439	74	
$AsCl_3$	−16	10.1	130	38.1		BiF_3	727	25.96	1127	117.2	[2]
AsF_3	−8.5	10.4	57.8	33.5		BiF_5	151.4		230	62.3	[4]
AsF_5	−79.8	11.5	−52.8		[4]	BiI_3	407.7	32	542 (decomp.)	78	
AsH_3	−116.9	1.2	−62.5	16.7	[4]	BiO	902	15.5	1647	226	[3]
AsI_3	141	21.8	371	56.5		Bi_2O_3					
As_4O_6						cubic.	825	16.3	1890		[3]
cubic.	278	48.6	457	56.1	[3]	monoclinic.	817	62.8	1890		[3]
monoclin.	314	45.2	465	96.3	[3]	Bi_2S_3	685	79.4			
octahedr.	314	49.8	465	109.3	[3]	Bi_2Te_3	586	118.6			[7]
As_2S_3	307		723	86	[2]	BrF	−33		20 (decomp.)	25.1	[4]
$AuCl_3$	288	12.98			[2]	BrF_3	8.8	12.03	126	42.7	
AuF	577	12.1	1200	142.0	[2]	BrF_5	−61.3	7.4	40.8	30.6	
$AuZn$	760	24.6		7	[5]	CBr_4	93.6	4.0	190	38	
BBr_3	−46		89.8	29.7		CCl_4	−22.88	2.5	76.73	30.0	
B_4C	2450	104.7	> 3500		[7]	CF_4	−183.7*1	0.695*1	−128.0	12.3	
BCl_3	−107.2	6.8	12.5	23.9	[4]	CI_4	171		decomp.		
B_2Cl_4	−92.9	10.8	65.5	33.6	[4]	C_2N_2	−34.4	8.1	−21.2	23.3	

*1 At 0.01 MPa.

Table 12.2 Melting, boiling, or sublimation point and heat of inorganic compounds *(continued)*

Substance	t_m	ΔH_m	t_b, t_s	$\Delta H_b, \Delta H_s$	Ref.
C_2N_2	−34.4	8.1	−21.2	23.3	
CNI		subl.	139.8	59.4	[5]
CO	−205.02	0.8386	−191.50	6.044	[3]
CO_2		subl.	−78.515	25.23	[3]
C_3O_2	−111.3		7.0	25	[4]
COS	−138.2	4.73	−50.2	18.51	
COSe	−124.4		−21.7	22.6	[4]
CP	−110	10.9			[4]
CS_2	−111.9	4.39	46.24	26.78	
CSSe	−85		84.5	31.62	[4]
CSe_2	−45.5		125	33	
CaB_6	2235				[7]
$CaBr_2$	742	29.1	1830	212	
CaC_2	2300	32.2			[7]
$CaCO_3$	50	0.19			[2]
aragonite					
(calcite decomp.)					
$CaCl_2$	772	28.0	1960	225	
CaF_2	1418	29.7	2530	305	
CaH_2	814	21	1000 (decomp.)		
CaI_2	783	41.8	1760	179	
$CaMg(CO_3)_2$	750 (decomp.)				
$CaMg(SiO_3)_2$	1392	128			
$CaMoO_4$	1450				
Ca_3N_2	1195				
$Ca(N_3)_2$	140 (expl.)				
$Ca(NO_3)_2$	561	21.3	decomp.		[4]
CaO	2630	52	3500	625.3	[3]
$CaO_2 \cdot 8H_2O$	200		250 (expl.)		
$Ca(OH)_2$	580				
$Ca(PO_3)_2$	984	82.9			
$Ca_2P_2O_7$	1358	100.8			
CaS	2525	67			[7]
$CaSO_4$	1460	28			
$CaSIO_3$	1544				
$CaTiO_3$	1960				
$CaWO_4$	1580				
$CrZrO_3$	2350				
Cd_3As_2	721	74.1			[6]
CdBr	565	20.9	863		[2]
$CdBr_2$	568	33	865	102.5	
$CdCl_2$	568	31.8	964	121	
CdF_2	1072	22.6	1747	225.2	[2]
CdI_2	388	20.7	796	122.7	[2]
$Cd(NO_3)_2 \cdot 4H_2O$	59.4	32.6	132		
CdO		subl.	1559	225	[5]
CdS		subl.	1380	213.5	[2]
$CdSO_4$	1135				
CdSe	1252	45.2			[6]
CdTe	1092	57.4			[6]
CeB_6	2190				
$CeBr_3$	732		1705	188.5	[4]
CeC_2	2250				
$CeCl_2$	822	54	1650	199	

Substance	t_m	ΔH_m	t_b, t_s	$\Delta H_b, \Delta H_s$	Ref.
CeF_3	1432	58.5	2180	274.1	
CeO_2	2600	80		565.2	[3]
Ce_2O_3	1920	83.7	3227	339.9	[3]
ClF	−155.6		−100.1	22	
ClF_3	−76.31	7.612	11.76	27.53	
$ClFO_2$	−115		−6	29.5	[4]
$ClFO_3$	−147.7	3.83	−46.67	19.33	[4]
ClO_2	−59.8 (expl.)		9.7 (expl.)	26.33	[3]
Cl_2O	−116 (expl.)		2.2 (expl.)	25.96	[3]
Cl_2O_7	−90 (expl.)		78 (expl.)	32.4	[3]
$CoBr_2$	678		927		
$CoCl_2$	740	38	1053	145	[4]
CoF_2	1202	30.1	1400	200.9	[2]
CoI_2	515		570 (decomp.)		
CoO	1810	50.24	2627	255.4	[3]
Co_2P	1386				[7]
CoS	1100				
$CoSO_4$	1089				[2]
CoSi	1400	67			[2]
$CoSi_2$	1327	100.2			[7]
Co_2Si	1332	64.05			
Co_2SiO_4	1415				[7]
CrB	2070				
Cr_3C_2	1895		3800		[7]
$CrCl_2$	824	37	1330	198	
$CrCl_2O_2$	−96.5		116.7	35.9	[4]
CrF_2	894	19	1820	251.2	[2]
CrF_3	1100	41.8	1425	201	[2]
CrF_2O	31.6	23.4			[4]
CrO_3	187	25.5	727 (decomp.)		[3]
Cr_2O_3	2234	104.7	3000		[3]
CrS	1565				[7]
CsBr	638	23.6	1290	148	
CsCl	645	20.3	1300	138	
Cs_2CrO_4	975	36.0			
CsF	703	21,7	1250	123	
CsH	682	10.26	1252		[2]
CsI	632	26	1280	133	
CsI_3	215	25.1	decomp.		
$CsNO_3$	409	13.8			
CsO_2	560	20	> 597 (decomp.)		[3]
Cs_2O	490	19.2	495 (decomp.)		[3]
Cs_2O_2	594	22	650 (decomp.)		[3]
CsOH	343	4.6			
$CsSO_4$	1015	36			
CuBr	489	7.2	1355	68.2	
$CuBr_2$	498				
CuCl	430	10.2	1212	151	[2]
$CuCl_2$	596				
CuF_2	770	73.3	1387	150.7	[2]
CuI	602	10.88	1320	130.2	[2, 6]
CuO	1336	37.3 (decomp.)			[3]
Cu_2O	1242	64.23	1800 (decomp.)		[3]
CuS	1100	316.1 (decomp.)			[2]

Table 12.2 Melting, boiling, or sublimation point and heat of inorganic compounds *(continued)*

Substance	t_m	ΔH_m	t_b, t_s	$\Delta H_b, \Delta H_s$	Ref.
Cu_2S	1129	11.3			
$CuSO_4{\cdot}5H_2O$	250				
Cu_2Se	1113	18.0			[6]
DBr	−87.63	2.402	−66.85	17.82	[4]
DCl	−114.7	1.908	−84.75		[4]
DF	−83.6		18.36		[4]
DI	−51.82	2.863	−36.2	19.72	[4]
D_2O	3.813	6.314	101.43	45.43	[3]
D_2O_2	1.5		25.0	52.38	[3]
$DyBr_3$	881	38	1085	199	[2]
$DyCl_3$	653		1539	187	
DyF_3	1160	58.6	>2200		
DyI_3	983	41.9	1320	172	
Dy_2O_3	2385				[3]
$ErBr_3$	950	41.9	1460	179.6	[2]
ErF_3	1146	27.5	2200		
ErI_3	1015	41.9	1280	167	[2]
$EuBr_2$	683	25.12	1880	218	[2]
$EuCl_2$	854	23	2060	246	
EuF_2	1416		>2400		
EuF_3	1276		2280		
EuI_2	580	21	1775	167	[2]
F_2O	−223.85		−145.05	11.10	[3]
F_2O_2	−163.4				[3]
FeB	1650				[7]
$FeBr_2$	688	54	968	125	
$FeBr_3$	297		627		
$FeCO_3$	>490 (decomp.)				
$Fe(CO)_5$	−21	13.2	105	40.2	
Fe_2CaO_4	1220	108			
$FeCl_2$	677	43.0	1012	125	
$FeCl_3$	307.5	38	315	30	
FeF_2	1102	33.5	1827	208.3	[2, 4]
FeF_3	1027	50.2	1327	167.5	[2]
FeH_2CO_4	−70.2		−30	25.5	[4, 5]
FeI_2	594	75	935	111.9	[4]
$Fe_{0.95}O$	1374	31.4	2512	230.3	[3]
Fe_2O_3	1562 (decomp.)				[3]
magnetite					
F_2O_3	1350				[3]
hematite					
Fe_3O_4	1583	138.16	2623 (decomp.)		[3]
FeS	1193	32.4			[7]
FeS_2	1700				
$FeSO_4{\cdot}7H_2O$	64		300		
FeSi	1405	69			
Fe_2SiO_4	1217	92			
GaAs	1238	97.1			[6]
$GaBr_3$	121	11.72	279	43.55	[2]
$GaCl_2$	164	16.8	535	79.6	[2]
$GaCl_3$	78	11.5	201	62.8	[2]
GaF_3		subl.	952	188.4	[2]
GaH_3	−21.4		139 (decomp.)		
GaI_3	212	22	345	69	

Substance	t_m	ΔH_m	t_b, t_s	$\Delta H_b, \Delta H_s$	Ref.
Ga_2O	652	35.6	727	83.74	[3]
Ga_2O_3	1725	92	2627	314	[3]
Ga_2S_3	1110				
GaSe	960	38.9			[6]
GaTe	835	52.8			[6]
GdBr	785	33.5	1490		[2]
$GdCl_3$	605	40.5	1600	203.8	
CdI_3	926	41.9	1340	167	[2]
Gd_2O_3	2322				[3]
$GeBr_4$	26.1	12.14	186.8	41	[2]
$GeCl_4$	−49.5		83.1	33	
GeF_2	112	21.8	1552	180	[2]
GeF_4		subl.	−36.6	31	
GeH_4	−165.8	0.84	−88.5	14.06	
Ge_2H_6	−109		31	24.7	
Ge_3H_8	−105.6		110.5	29	
$GeHCl_3$	−71		75.3	33.5	[4]
GeI_4	144		440 (decomp.)		[3]
GeO		subl.	710	209.3	[3]
GeO_2					
tetragon.	1086	43.96	> 2352	255.4	[3]
hexagon.	1116	17.2	2352	255	[3]
GeS	665	21	827		[4]
GeSe	670	32.6	decomp.		[6]
HBO_2	236				
HBr	−86.9	2.406	−66.8	17.61	[4]
HCN	−13.3	8.41	25.65	25.2	
HCl	−114.2	2.0	−85.08	16.2	
$HClO_4$	−101	6.93	25 (3 kPa)	39.7 (3 kPa)	
HDO			100.85	44.67 (25°C)	[3]
HF	−88.36	3.929	19.52	7.49	
HI	−50.8	2.871	−35.4	19.77	
HIO_3	110				
H_2MoO_4	70				
HN_3	−80		36	29.7	
HNO_3	−41.6	10.5	83	30.14	[2]
H_2O	0.00	6.013	100.00	40.683	[3]
H_2O_2	−0.42	12.51	152.0	51.6 (25°C)	[3]
H_3PO_2	26.5	13.0	>50 (decomp.)		
H_3PO_3	70.1	12.85			[4]
H_3PO_4	42.35	13.0	213 (decomp.)		
$H_4P_2O_7$	61	9.2			
H_2S	−85.6	2.38	−60.35	18.7	
H_2S_2	−89.5	7.5	70.7	35.1	[4, 5]
H_2SO_4	10.31	10.7	279.6 (decomp.)		
$H_2SO_4{\cdot}H_2O$	8.48	19.5	290		
H_2Se	−65.7	2.515	−41.4	19.9	
H_2SeO_4	62.4	14.4	>65 (decomp.)		
HTO			100.8	46.89 (25°C)	[3]
H_2Te	−51	4.2	−2	23	

Table 12.2 Melting, boiling, or sublimation point and heat of inorganic compounds *(continued)*

Substance	t_m	ΔH_m	t_b, t_s	$\Delta H_b, \Delta H_s$	Ref.
H_2WO_4	>100				
$Hf(BH_4)_4$	29.0	14.2	118	40.2	[4]
$HfBr_4$		subl.	322	100.4	[4]
HfC	3890		5400		[7]
$HfCl_2$	727	24.7	1475	147	[2]
$HfCl_4$		subl.	315	103	
HfF_2	1377	20.9	2030	203	[2]
HfF_4		subl.	974	226	
HfN	3000				[7]
HfO_2	2780	71			[3]
$HgBr_2$	238	17.9	319	59	
Hg_2Br_2		subl.	392.5		
HgBrI	229		360		
$HgCl_2$	280	19.2	302	57.8	
Hg_2Cl_2		subl.	383.7		
HgF_2	645	23.03	647	92.1	[2]
Hg_2F_2	570		decomp.		
HgI_2	256	18.8	353	60	
HgO (red)	500 (decomp.)				[3]
HgS	825				
$HgSO_4$	>550 (decomp.)				
$HoBr_3$	919	50.2	1336		
$HoCl_3$	720	30.4	1517	190	
HoF_3	1140	56.3	2200		
HoI_3	989	41.9	1300	172	[2]
IBr	42		119 (decomp.)		
ICl	27.2	11.1	98 (decomp.)		
IF_5	9.4	15.9	100	41.3	[4]
IF_7		subl.	3.4	30.7	[5]
I_2O_4	130				[3]
InBr	220	24	680	92	[4]
$InBr_2$	235		633	82.4	[4]
$InBr_3$		subl.	409	123	
InCl	225	9	590	97	
$InCl_2$	235	14.65	523	92.1	[2]
$InCl_3$		subl.	500	170	
InF	452	13.4	900	109	[2]
InF_3	1170	46.1	1377	163.3	[2]
InI	351	22	713	85	
InI_3	210	20	447	95	
InO	1052	16.75	1727	251.2	[3]
In_2O	327	18.84	527	67	[3]
In_2O_3	1910	83.74	3327	356	[3]
InS	692		850		
In_2S_3		subl.	850		[6]
In_2Se_3	890	64			[6]
In_2Te_3	670	81.6			
$IrCl_3$	442	28.1	687	100	[2]
IrF_2	1107	9.63	1730	184	[2]
IrF_6	44.1	5.0	53.6	30.5	
IrO_2	>1100 (decomp.)				[3]
Ir_2O_3	1177	41.87	1977	209.3	[3]
KBF_4	570	18.0			
KBr	734	25.5	1407	149	
KBr_2I	58		180 (decomp.)		[5]
KCN	623	14.6			[4]
K_2CO_3	900	27.9	decomp.		
KCl	776	26.3	1430	152	
$KClO_3$	357	21			
K_2CrO_4	973	33.0			
KF	857	29.4	1505	172	
KHF_2	238.7	6.62	>400 (decomp.)		
$KHSO_4$	222	18	decomp.		
KH_2PO_4	252.6				
KI	681	24.0	1340	138	
$KMnO_4$	> 200 (decomp.)				
$KMoO_4$	926	39	1400 (decomp.)		
KNO_2	438	17	decomp.		
KNO_3	334.5	9.80	>400 (decomp.)		
$KNaC_4H_4O_6 \cdot 4H_2O$	75		215		
KO_2	380	20.52	543 (decomp.)		[3]
K_2O	707	28.5	decomp.		[3]
K_2O_2	490	29.31	1527	188.4	[3]
K_2O_3	430	25.54	700	104.7	[3]
KOH	405	9.4	1320	129	[4]
KPO_3	813	18.4	1320		
K_3PO_4	1640				
$KReO_4$	553	32	1367		
K_2S	948	16.2			
KSCN	177.0	10	500 (decomp.)		
K_2SO_4	1069	36.8	1700		
K_2SeO_4	1020				
K_2SiO_5	1045	35.2			
KVO_3	522				
K_2WO_4	923	31.0			
KrF_2	20 (decomp.)				
LaB_6	2715				
$LaBr_3$	783	33.5	1580	202	[2, 4]
LaC_2	2365				[7]
$LaCl_3$	862	54.4	1710	209	
LaF_3	1493	41.8	2330	234	[2]
LaI_3	779	33.5	1580	167	[2]
$La_2(MoO_4)_3$	1015				
La_2O_3	2217	75.4	4200		[3]
La_2S_3	2080				
$La_2(SO_4)_3$	1150 (decomp.)				
$LiBO_2$	849	33.9			
$Li_2B_4O_7$	920	120			
LiBr	550	17.7	1290	107	
Li_2CO_3	732	38			
LiCl	610	19.7	1382	121	[4]
$LiClO_3$	129		300 (decomp.)		
$LiClO_4$	246.7	17.0	400 (decomp.)		
LiF	849	27.1	1681	147	[2]
Li_2GeO_3	1239				
LiH	691	22	1680	213.5	[2]
LiI	469	14.6	1170	97	

Table 12.2 Melting, boiling, or sublimation point and heat of inorganic compounds *(continued)*

Substance	t_m	ΔH_m	t_b, t_s	$\Delta H_b, \Delta H_s$	Ref.
Li_2MoO_4	703	49.0			
Li_3N	813				
$LiNH_2$	375		430 (decomp.)		
$LiNO_3$	253.0	25.5	>600 (decomp.)		[4]
Li_2O	1727	58.6	2600	234.5	[3]
$LiOH$	473	20.1	925 (decomp.)		
Li_3PO_4	1220				
Li_2S	1370				
Li_2SO_4	860	9.3			
Li_2SiO_3	1201	28			
Li_2WO_4	740	28.5			
$LuCl_3$	925		1420	191	
LuF_3	1184	30.2	2200		
LuI_3	1045	46.0	1200	201	[2, 4]
Lu_2O_3	2467				[3]
$MgAl_2O_4$	2115				
$Mg_3(BO_3)_2$	1410				
$MgBr_2$	711	39	1230	151	[4]
$MgCO_3$	500 (decomp.)				
$MgCl_2$	714	43.1	1370	166	
$MgCr_2O_4$	2350				
MgF_2	1263	58	2270	274	
MgI_2	633	26	1014	151	
MgO	2825	77.5	3600	544.3	[3]
$Mg(OH)_2$	350				
$Mg_2P_2O_7$	1395	134			
$MgSO_4$	1137	15			
Mg_3Sb_2	1250				
Mg_2Si	1085	85.83			[7]
Mg_2SiO_4	1890	66.1			
$MgWO_4$	1360				
$MnAs$	936				
MnB_2	1988				[7]
$MnBr_2$	698				
$MnCl_2$	650	37.7	1238	148	
MnF_2	860	14	1640	256	
MnF_4	387	23.8	412	67	[2]
MnI_2	638				
MnO	1842	43.96	3127 (decomp.)		[3]
MnO_2	847 (decomp.)				[3]
Mn_2O_3	1347 (decomp.)				[3]
Mn_2O_7	5.9		>55 (decomp.)		
α-Mn_3O_4	1590	138	2630	314	[2]
β-Mn_3O_4	1560	127.7	2627	314	[3]
MnO_3F	−38		60	34	[4]
MnP	1147				[7]
MnS	1530	26.1			[7]
$MnSO_4$	700		850 (decomp.)		
$MnSi$	1275	59.4			[7]
$MnSiO_3$	1323	34.3			[4]
$MnTiO_3$	1404	33.5			
MoB	2550				[7]
MoB_2	2350 (decomp.)				[7]
α-MoC	2600				[7]

Substance	t_m	ΔH_m	t_b, t_s	$\Delta H_b, \Delta H_s$	Ref.
β-MoC_2	2522				[7]
$MoCl_5$	194	18	269	54	
MoF_4	97	31.4	617	83.7	[2]
MoF_6	17.6	4.3	33.9	27.2	
MoF_4O	97.2	3.5	186	50	
MoO_2	1927	67	1977 (decomp.)		[3]
MoO_3	801	49.0	1257	138	[3]
$MoSi_2$	2020				
NCl_3	−27		71 (expl.)		[5]
NF_3	−206.8	0.398	−129.0	11.55	[4]
NH_3	−77.75	5.66	−33.41	23.33	
N_2H_4	1.54	12.66	113.5	41.6	[4]
NH_4NO_3	169.6	6.40	210 (decomp.)		
NH_4HSO_4	146.9	14.2	decomp.		
NH_2OH	32	16	58	47.7	[4]
NH_4SCH	149.6		170 (decomp.)		
NO	−163.7	2.30	−151.65	13.783	[3]
N_2O	−91.0	6.544	−88.5	16.571	[3]
N_2O_3	−101		−40 (decomp.)		
N_2O_4	−11.2	14.662	21	38.56	[3]
N_2O_5		subl.	33	56.94	[3]
$NOBr$	−55.5		−2		
$NOCl$	−59.6	4	−5.8	25.1	
NO_2Cl	−141		−14.3	25.7	
NO_3Cl	−107		18	305	[4]
NOF	−132.5		−59.9	19.2	
N_2S_4	21		100 (expl.)		
NSF	−79		4.8	22.2	[4]
$NaAlSiO_4$	1526				
$NaBF_4$	406	13.6			
$NaBO_2$	966	36	1434		
$Na_2B_4O_7$	742	81	>1000 (decomp.)		
$NaBr$	747	26.2	1392	128	[5]
$NaCN$	564	18	1497	156	[4]
Na_2CO_3	858	28.0	decomp.		
$NaCl$	801	28.2	1490	138	
$NaClO_3$	263	21.3			
Na_2CrO_4	794	24.7			
NaF	996	33.5	1704	177	[4]
NaH	200 (decomp.)				[2]
$NaHSO_4$	186	10.4			
NaI	661	23.7	1304	1597	[4]
$NaIO_3$	422	23.7			
Na_2MoO_4	688	22			
$NaNH_2$	208		400		
$NaNO_2$	284	14.9	>320 (decomp.)		
$NaNO_3$	306.5	15.1	380 (decomp.)		
NaO_2	552	26.0	1027	117	[2, 3]
Na_2O	920	29.7	1350		[3]
Na_2O_2	596	16.7	decomp.		[3]
$NaOH$	322	6.4	1378	144.3	[4]
$NaPO_3$	628	17.3			
Na_3PO_4	1510				
$Na_4P_2O_7$	998	59			

Table 12.2 Melting, boiling, or sublimation point and heat of inorganic compounds *(continued)*

Substance	t_m	ΔH_m	t_b, t_s	$\Delta H_b, \Delta H_s$	Ref.	Substance	t_m	ΔH_m	t_b, t_s	$\Delta H_b, \Delta H_s$	Ref.
Na_2S	1168	30.1				PF_5	−93.7	11.9	−84.55	17.2	
NaSCN	307.5					PH_3	−133.8	1.13	−87.42	14.6	
Na_2SO_3	911	26				PH_4Cl		subl.	−28		
Na_2SO_4	884	24.3	890	24.07	[2]	PH_4I	18.5		80		
$Na_2SO_4{\cdot}10H_2O$	32.4	79.0	decomp.			PI_3	61.0		227	43.9	[4]
$Na_2S_2O_3{\cdot}5H_2O$	48.5	23.4	100			$(PNCl_2)_3$	114	20.9	255	55.2	[4]
$Na_2S_2O_7$	402	40.8	460 (decomp.)			$(PNF_2)_3$	27.8	21.8	50.9	31.8	[4]
Na_3Sb	1010	77			[4]	P_2O_3	23.9	14.07	175.4	18.84	[3]
Na_2Se	875					P_2O_4		subl.	100		[3]
Na_2SiF_6	846	99.6				P_2O_5					
Na_2SiO_3	1088	54				hexagon.		subl.	359	65	
Na_2Te	1035	14				rhomb.	420	≈ 65	584	79.13	[3]
$NaVO_3$	630	28.3				$(P_2O_5)_n$	562		588		[3]
Na_2WO_3	696	31				$POBr_3$	55.7		192	45.6	
NbB_2	3050					$POCl_3$	1.2	13.1	107.2	34.5	
$NbBr_5$	267.5	36	362	75.7		POF_3		subl.	−39.5	37.7	
NbC	3613		4500		[7]	P_4S_3	172		407		
$NbCl_5$	205	33.9	247.5	52.3		α-$PSCl_3$	−40.8		125	63.2	
NbF_5	79.5	12.2	234.5	51.0		$PaCl_5$	301	39.4	380	63	[2, 4]
NbN	2300				[7]	PaF_3	1280	36.8	2280	251	[2]
NbO	1940	67			[3]	PaF_5	297	28.5	587	83.7	[2]
NbO_2	2080	67	3527	356	[3]	Pa_2O_5	1777	108.9	3077	398	[3]
Nb_2O_5	1490	103.3	2927	334.9	[3]	$PbBr_2$	370	21	893	118	
$NdBr_3$	683	33.5	1614	195.8	[2, 4]	$PbCl_2$	495	24	953	129	
NdC_2	2207				[7]	$PbCl_4$	−7	10.0	127	34.3	[2]
$NdCl_3$	760	50.2	1620	209					(expl. near 100°C)		
NdF_3	1380	54.7	2300	259	[2]	PbF_2	822	11.9	1292	160.2	
NdI_3	787	33.5	1350	172	[2]	PbI_2	412	21	872	100	
Nd_2O_3	2212	92.1	3000		[3]	$PbMoO_4$	1070				
$NiBr_2$		subl.	919	224.6		PbO	886	25.52	1472	214	[3]
$Ni(CO)_4$	−19.3	13.8	42.3	29.8		PbO_2	290 (decomp.)				[3]
$NiCl_2$		subl.	970	225		Pb_3O_4	830				[3]
NIF_2	1157	37.3	1877	230	[2]	$Pb(OH)_2$	145				
NiI_2	797					PbS	1077	36.4	1281	229.4	[2]
NiO	1957	50.7			[3]	$PbSO_4$	1087	40.2			[4]
$Ni(OH)_2$	230					PbSe	1065	49.4			[6]
Ni_2P	1110					PbTe	917	57.4			[6]
NiS	996	44.1			[4]	$PbWO_4$	1125				
$NiSO_4$	840 (decomp.)					$PdCl_2$	678	40.6	1030	126	[2]
NiSb	1160		1400 (decomp.)			PdF_2	952	27.6	1730	180	[2]
Ni_2Si	1318	50.24			[4]	PdI_2	350 (decomp.)				
$NpCl_3$	802	41.9	1530	167	[2]	PdO	877 (decomp.)				[3]
$NpCl_4$	538	46.1	847	117	[2]	PdS	950 (decomp.)				
NpF_4	827	50.2	1477	192.6	[2]	$PdSe_2$	1000				
NpF_6	54.8	17.5	55.9	29.5		$PmBr_3$	687	33.5	1520	188	[2]
NpO_2	2560	62.8			[3]	PoCl	294	21.8	480	79.6	[2]
O_3	−192.5	2.09	−111.9	15.193	[3]	$PoCl_2$		subl.	195		
$OsCl_4$	97	13.8	330	50	[2]	PoO_2	552	23	decomp.		
OsF_2	1080	15.9	1630	213	[2]	$PrBr_3$	693	47.3	1695	167	
OsF_8	33.4	7	47.5	27.6	[8]	PrC_2	2120		2500 (decomp.)		[7]
OsO_4 (yellow)	39.50	14.28	130	37.3	[3]	$PrCl_3$	786	50.6	1630	213	
PBr	−40.5		173.3	38.7		PrF_3	1370	33.5	2330	259.6	[2]
PCl_3	−90.3	4.52	75.3	30.5	[4]	Pr_2O_3	2127	92.1	3727	376.8	[3]
PF_3	−151.3	0.94	−101.4	14.6		$PtCl_3$	435		decomp.		

Table 12.2 Melting, boiling, or sublimation point and heat of inorganic compounds *(continued)*

Substance	t_m	ΔH_m	t_b, t_s	$\Delta H_b, \Delta H_s$	Ref.
PtF_3	852	37.3	1200	159	[2]
PtF_6	61.3	5	69.2	30	
PtO_2	450	19.26	477 (decomp.)		[3]
PtP_2	>1500				[7]
$PuBr_3$	681	51	1512	184	[4]
$PuCl_3$	765	56	1770	195.6	[4]
$PuCl_4$	457	39.8	790	126	[2]
PuF_2	1410	54	2280	259	[2]
PuF_3	1426	33			[4]
PuF_4	1037	58.6	1427	197	[2]
PuF_6	51.6	18.7	62.2	30.1	
PuI_3	777	50.2			[2]
PuO	1017	30.14	2052	196.8	[3]
PuO_2	2390	62.8	3327	376.8	[3]
Pu_2O_3	2085	67	2977	314	[3]
$RaBr_2$	728		900		
$RaCl_2$	900	25.1	1610	176	[2]
RaF_2	1330	12.6	1930	230	[2]
RbBr	692	23.3	1352	141.8	[4]
RbCl	723	23.7	1390	149	
Rb_2CO_3	835	29	decomp.		
RbF	795	25.8	1430	159	
RbI	656	22.0	1327	135	
$RbNO_3$	312	4.6			
RbO_2	412	17.2	567 (decomp.)		[3]
Rb_2O_2	570	30.5	1010 (decomp.)		[3]
RbOH	385	8.9			
Rb_2SO_4	1070	38.4	1700		
$ReCl_3$	660	50.2	830	113	[2]
$ReCl_5$	278	37.7	330	58.6	[2]
$ReCl_3O$	4.5		131		
ReF_4	125	30.1	500 (decomp.)		[2]
ReF_6	18.8	4.6	33.7	28.3	
ReO_2	1202	50.2	2977	334.9	[3]
ReO_3	160	21.8	620		[3]
ReO_4	147	17.58	187	38.94	[3]
Re_2O_7	301.5	66.2	359	69.9	[3]
$RhCl_2$	780	26.4	960 (decomp.)		[2]
RhF_3	1187	46.0	1427	167.5	[2]
Rh_2O_3	1115 (decomp.)				[3]
$RuCl_3$	>500 (decomp.)				
RuF_5	101	188.4	227	56	[2]
RuO_4	25.4	10.89	100 (decomp.)		[3]
RuS_2	1000 (decomp.)				
S_2Br_3	−40		90		
SCl_2	−123		59 (decomp.)		
S_2Cl_2	−82		137 (decomp.)		
SF_4	−121.0		−38	21.8	
SF_6		subl.	−63.6	22.8	[4]
S_2F_2	−165		−11	22.8	
SO_2	−75.46	7.406	−10.01	24.95	[3]
α-SO_3		subl.	62.2	30	
β-SO_3	32.5	12.4	44.7	40.82	[3]
γ-SO_3	16.79	5.61	44.7	40.82	[3]
$SOBr_2$	−52		138	42.7	[4]
$SOCl_2$	−104.5		75.6	31.8	
SOClF	−139.5		12.2	24.8	[4]
SOF_2	−110.5		−43.8	21.8	
SOF_6	−86.0		−35.1	21.8	[4]
$SbBr_3$	96.6	15	289	51.2	
$SbCl_3$	73.2	13	233	46	
$SbCl_5$	3.0	10.0	172 (decomp.)		
SbF_3	292	16.3	319	58.6	[2]
SbF_5	8.3		142.7	43	
SbH_3	−94		−18	21	
SbI_3	170.5	18	401.6	61	
Sb_2O_3	656	110	1456	37.35	[3]
Sb_2S_3	560	65.3	1160		[3]
Sb_2Se_3	612	77.4			[6]
Sb_2Te_3	621	100.1			[6]
$ScCl_3$	960		967		[4]
Sc_2O_3	2405	96.3			[3]
$Sc_2(SO_4)_3$	973				
$SeBr_2O$	41.6		217 (decomp.)		
SeC_2	45.5		125		
$SeCl_3$	956	67.4	975	157	
$SeCl_4$		subl.	196	87.9	[2]
SeF_4	−9.5		107.7	47.02	[4]
SeF_6		subl.	−46.6	18.3	
SeO	1102	31.82	1802	188.4	[3]
SeO_2		subl.	337	84.6	[3]
SeO_3	118.5	7.1	>185 (decomp.)		
$SeOCl_2$	10.8	4.23	117.6	42.7	
$SiBr_4$	5.4		152.6	35	
SiC	2540		2830 (decomp.)		[7]
$SiCl_4$	−68.9	7.7	57.0	28.6	
$SiCl_6$	−1.0		145	42	[4]
SiF_4		subl.	−95.25	15.4	
SiH_4	−185	0.67	−111.9	12.4	
Si_2H_6	−131		−14.5	21.3	
Si_4H_{10}	−84.3		108	34.3	[5]
SiH_2F_2	−122	4	−77.8	16.3	[4]
SiI_4	120.5		301	50.2	[4]
Si_3N_4		subl.	1900		
SiO	2277	50.24			[3]
SiO_2					
quartz	1610	8.541	1997 (decomp.)		[3]
cristobalite	1720	7.704	2950	573.6 (25° C)	[3]
tridymite	1680	9.002			[3]
SiS_2	1090		1130		[2]
$SmBr_2$	669	25.12	1880		[2]
$SmBr_3$	664		1645	192.9	[4]
$SmCl_2$	858	14	1950	218	
$SmCl_3$	678	33.5	1560	197	[2]
SmF_3	1400	33.5	2330	259	[2]
SmI_2	520	21	1660	167	[2]
Sm_2O_3	2320	83.74	3527	340	[3]

Table 12.2 Melting, boiling, or sublimation point and heat of inorganic compounds *(continued)*

Substance	t_m	ΔH_m	t_b, t_s	$\Delta H_b, \Delta H_s$	Ref.
Sm_2S_3	1900				[7]
$SnBr_2$	232	7	641	97	
$SnBr_4$	30	11	208	37	
$SnBr_2I_2$	50		225		
$SnCl_2$	247	12.6	670	71	
$SnCl_4$	−33	9.16	112	37	[4]
SnF_2	212	13.8	853	113	[2]
SnF_4	447	27.6	705	92.1	[2]
SnH_4	−150		51.8	19.05	[4]
SnI_2	320	12.6	718	105	[4]
SnI_4	144.5	19.2	348.6	50	
SnO	1042	26.8	1527	251.2	[3]
SnO_2	1625	47.7	2500	314	[2, 3]
SnS	881	31.6	1276	156	
$SnSO_4$	360 (decomp.)				
SnTe	780	45.2	decomp.		[6]
SrB_6	2235		5100		[7]
$SrBr_2$	657	10.1	1970	230	
SrC_2	>1700				
$SrCl_2$	874	16.1	2040	243	
SrF_2	1477	28.5	2460	324	
SrI_2	538	19.7	1900	194	
$Sr(NO_3)_2$	645	49			
SrO	2415	69.9	3000	534.2	[3]
SrS	> 2000				[7]
$SrSO_4$	1605				
SrSe	1600				
$SrSiO_3$	1580				
SrTe	1490	57			
$SrWO_4$	1535 (decomp.)				
T_2O	4.49		101.6	46.05 (25°C)	[3]
$TaBr_5$	265	45.6	348.8	62.3	[4]
TaC	3985		5500		[7]
$TaCl_2$	937	28.5	1377	167	[2]
$TaCl_5$	216.5	34	236	56.1	
TaF_5	96	12.6	229.2	51.9	
TaI_5	365	6.7	543	75.8	[4, 5]
TaN	3087 (decomp.)				[7]
Ta_2O_5					
rhomb.	1785	67	2227		[3]
triclinic	1872	67	2227		[3]
$TbCl_3$	588				[4]
$TcCl_5$	152	32.7	232	50.2	[2]
TcF_6	36	9.6	57	27.6	[2]
TcO_2	2127	75.4	3727	439.6	[3]
Tc_2O_7	120	47.5	311	58.8	[3]
$TeBr_4$	380		421		
$TeCl_2$	208	14.7	328	64.1	[2]
$TeCl_4$	224	18.9	390	71.1	
TeF_4	129.6	26.57	374.5	52.8	[4]
TeF_6		subl.	−38.6	27	
TeO_2	733	29.5	1257	205.2	[3]
ThB_6	2190				[7]

Substance	t_m	ΔH_m	t_b, t_s	$\Delta H_b, \Delta H_s$	Ref.
$ThBr_4$	679	34	857	140	
ThC	2635		5000		[2]
$ThCl_4$	770	61	921	129	
ThF_4	1110	18.8	1650	270.3	
ThI_4	566	33.5	837	125	[2]
ThO	1877	54.4	2977	272.1	[3]
ThO_2	3050	90	4400		[7]
$ThSiO_4$	1975	decomp.			
TiB_2	2790				[7]
$TiBr_4$	38	12.9	231	44.4	
TiC	3257		4300		[7]
$TiCl_2$	1035	37.7	1330	147	[2]
$TiCl_3$	730	20.9	960	124	[4]
$TiCl_4$	−24.1	10.0	136.3	35.7	
TiF_3	1227	50.2	1400	217.7	[2]
TiF_4		subl.	285.5	90	
TiI_4	155	19.8	379.5	56	
TiN	2950				
β-TiO	1780	54.43	2700	563.2	[3]
TiO_2	1870 (21.3 kPa)	67	2927	decomp.	[3]
Ti_2O_3					
monoclinic.	1830	110.5	3027		[3]
hexagon.	2130				[3]
Ti_3O_5	2177		3327		[3]
TiP	1990	decomp.			
TiS	1780				[7]
TlBr	460	16.4	824	100	
Tl_2CO_3	269	18.4			
TlCl	431	15.6	820	101	
TlF	322	13.9	840	94	
TlF_3	550	35.6	927	138	[2]
TlO_2	579	30.3	600 (1 kPa)	71.2	[3]
Tl_2O_3	717	51.9	decomp.		[3]
TlI	441	14.7	833	decomp.	
$TlNO_3$	206.5	9.5	433	decomp.	[5]
Tl_2S	448	12.6	1180	159	[2]
Tl_2SO_4	632	24			
$TmBr_3$	945	41.9	1440		[2]
$TmCl_3$	845				[4]
TmI_3	1020	41.9	1260	167	[2]
UB_2	2385				[7]
UBr_3	730	46			
UBr_4	519	72	765	126	
UC_2	2520		4370		[7]
UCl_3	841	37.7	1780	197	[2]
UCl_4	590	50.0	792	138.2	[2]
UCl_5	287	36	417	67.9	[2]
UCl_6	179	20.9	277	50.2	[2]
UF_3	1495	37	2277	255	[2]
UF_4	1036	47.0	1417	200	[2]
UF_5	348	46.9	527	75.4	[2]
UF_6		subl.	56.5	47.7	

Table 12.2 Melting, boiling, or sublimation point and heat of inorganic compounds *(continued)*

Substance	t_m	ΔH_m	t_b, t_s	$\Delta H_b, \Delta H_s$	Ref.	Substance	t_m	ΔH_m	t_b, t_s	$\Delta H_b, \Delta H_s$	Ref.
UH_3	> 250	decomp.				XeF_2	140			51 (25°C)	
UI_3	680		1750			XeF_4	114			64 (25°C)	
UI_4	520	23.6	769			XeF_6	46		76		
UO	2477	58.62			[3]	YBr_3	905	37.7	1324		[2]
UO_2	2840	136.1	decomp.		[3]	YCl_3	721	31.5	1482	210	
US	2462				[7]	YF_3	1155	23.4	1502	184	[2]
US_2	1600					YI_3	997	50.2	1300	172	[2]
VC	2810		3900			Y_2O_3	2430	84	4300		[3]
VCl_2	1000	33.5	1375	147	[2]	$YbBr_2$	673	25.1	1800		[2]
VCl_4	−20.5	9.6	153	36		$YbCl_2$	702		2033	258	
VCl_3O	−78	9.6	126.7	34.7		YbF_3	1162	29.7	2200		
VF_3	1127	54.4	1627	230		YbI_2	772	27	1330	155	[2]
VF_5	19.5	4.2	48.0	43.9		$ZnBr_2$	394	15.7	670	109.6	
VN	2050				[7]	$ZnCl_2$	317	10.2	732	119	
VO	1830	62.8	3127	293.1	[3]	ZnF_2	875	41.8	1505	185	
VO_2	1545	57	2700		[3]	ZnI_2	446	17	624	decomp.	
V_2O_3	1970	100.5	3027	decomp.	[3]	ZnO	1975		decomp.		
V_2O_5	680	65.15	2052	264		Zn_3P_2	1193	163.2			
WB	2800				[7]	ZnS		subl.	1182	250.2	[4, 7]
WBr_5	295	17	392	58		$ZnSO_4$	>600	decomp.			
WBr_6	309		>400	decomp.		ZnSe	1520	53.6			[6]
WBr_4O	322	61	331	55.2		Zn_2SiO_4	1511				
WC	2785		6000		[7]	ZnTe	1300	65			
W_2C	2795		6000		[7]	ZrB_2	3050				
WCl_5	248	24	287	49		$Zr(BH_4)_4$	28.7	18.0	123	38.9	[4]
WCl_6	283	8	340	61.5		$ZrBr_4$		subl.	355	108	
WCl_4O	209.5	33	224	43		ZrC	3530		5100		[7]
WF_5	107	18.8	250	48	[2]	$ZrCl_2$	727	26.8	1387	147	[2]
WF_6	2.0	4.1	17.3	25.9		$ZrCl_4$		subl.	333	103.1	
WF_4O	106	8	185.9	56		ZrF_4		subl.	906	261	
WO_2	1570	48.15	1850	decomp.	[3]	ZrI_4		subl.	418	120	
WO_3	1473	73.5	1670	180	[3]	ZrN	2980				[7]
W_2Re_3	2987				[5]	ZrO_2	2900	87.1	4300	639	[3]
WSi_2	2160					$ZrSiO_4$	2430				

Note: m = melting; b = boiling; subl. = sublimation; expl. = exploded; decomp. = decomposed.

Table 12.3 Melting, boiling, or sublimation point (t, °C) and heat (ΔH, kJ/mol) of elements and some isotopes at normal pressure (101,325 Pa)

Substance	t_m	ΔH_m	t_b, t_s	ΔH_b, ΔH_s	Substance	t_m	ΔH_m	t_b, t_s	ΔH_b, ΔH_s
Acetaldehyde	−121	3.22	20.8	27.2	Allyl alcohol	−129		97	39.96
Acetamide	82.3	14.2	221		*n*-Amyl alcohol	−79	9.795	137.3	
Acetylene (ethyne)		subl.	−84	21.3	*t*-Amyl alcohol	−8.6	4.49	102.35	
Acethylthiophen	10.4	15.0	213.9	47.5	Anethol	21.4	16.0	234.5	
Acetic acid	16.64	11.73	117.8	23.7	Aniline	−6.15	10.54	184.13	45.15
Acetone	−95.4	5.72	56.2	29.09	Anisoll	−37.5		153.80	39.4
Acetonitrile	−44.9	8.91	81.5	32.8	Anthracene	216.041	28.8	340	
Acetophenone	19.75	16.65	202	38.8	*trans*-Azobenzene	68.5	24.7	293.0	
Acrylaldehyde (acrolein)	−87.0		52.5	28.85	Azulen	99	19.08	234	55.84
Acrylonitride	−83.5		77.3	32.6	Benzaldehyde	−26		179.2	40.9
Acrylic acid	13	11.1	141.6		Benzene	5.51	9.837	80.099	30.76
Allene (propadiene)	−136		−34.5	20.92	Benzil	95	19.4	347.0	

Table 12.3 Melting, boiling, or sublimation point and heat of elements *(continued)*

Substance	t_m	ΔH_m	t_b, t_s	ΔH_b, ΔH_s
Benzoic acid	122.5	18.00	250.0	
Benzonitrile	−14	9.08	190.7	45.80
α-Benzophenone	48.21	16.7	305.9	
Benzotrifluoride	−29.11		102.3	
Benzoyl bromide [14]	0.0		218.5	
Benzoyl chloride [14]	−0.5		197.2	
Benzyl alcohol	−15.3	8.97	205.4	50.46
Brombenzene	−30.82	10.63	156.06	36.40
1-Bromo--1-chlorobenzene [14]	16.6		82.7	33.1
Bromoacetic acid	49.5		208	54
Bromotrifluoromethane	−143.2		−59	22.7
1,2-Butadiene	−136.19	6.961	10.85	
1,3-Butadiene (divinyl)	−108.9	7.985	−4.4	22.59
Butadyine	−36.4		10.3	
Butane	−138.35	4.660	−0.50	22.39
s-Butanol	−114.7		98.5	43.56
t-Butanol	25.5	6.79	82.25	39.97
2-Butanole	−86.9	8.48	79.53	32.8
1-Butanthiol	−115.7	10.46	98.0	32.23
1-Butene	−185.35	3.848	−6.3	21.92
cis-2-Butene	−139.91	7.308	3.7	23.35
trans-2-Butene	−105.55	9.757	0.88	22.76
Butyl acrylate	−64.6		197.4	
Butyl alcohol (butanol)	−89.53	8.98	117.726	43.97
Butyl bromide	−112.4	9.241	101.6	32.53
Butyl formate	−91.9		106.8	37.1
Butyl propionate	−89.6		144.5	39.6
Butylamine	−50.5		70	27.2
Butylbenzene	−87.99		183.35	
Butylvinyl ether	−92		93.5	32.3
1-Butyne	−125.7	6.031	8.1	24.52
2-Butyne	−32.26	9.235	27	
Butyric acid	−5.26	11.07	163.25	42.01
α-Camphor [14]	178.4	6.82	209.2	
Capric acid	31.5		270	
Caproic acid	−1.5	15.1	205	
Caprylic acid	16.5	21.4	239.3	
Carbamide	135	15.1	154 (decomp.)	
Carbonyl bromide			60	30.1
Carbonyl fluoride	−114		−83.3	16.15
1-Chloro-1,1-difluoro-ethane	−130.8	2.69	−9.2	22.4
2-Chloro-1-ethanol	−67.5		128	41.4
Chloroacetic acid [5]	61.2	15.85	189.35	54
Chloroacetylene	−126		−30	22.5
m-Chloroaniline	−10.4	11.63	230.6	
Chlorobenzene	−45.58	9.556	131.69	35.66
1-Chlorobutane	−123.1		78.44	30.9
Chlorodifluoromethane	−146.5	4.12	−40.8	20.22
Chloroethane	−136.4	4.45	12.27	24.65
Chloroethylene	−153.8	4.74	−13.7	20.8
Chlorofluoromethane			−9.1	23.4
Chloroform	−63.5	9.5	61.1	29.67
Chloromethane	−97.73	6.43	−24.2	21.63
1-Chloropentane	−99		107.8	
4-Chlorophenol	43.5	15.1	219.8	
1-Chloropropane	−122.8	5.544	46.6	27.6
Chlorotrifluoromethane	−181		−81.1	14.63
Cinnamic acid	135		300.0	
Cinnamic alcohol	34		275.5	
m-Cresol	11.5		202.2	49.38
o-Cresol	30.9		191.0	46.94
p-Cresol	34.8		201.9	49.53
α-Crotonic acid	71.6	13.0	180.5	
Cyclobutane	−50	1.088	12	24.19
Cycloheptane	−12	1.882	118.5	
Cyclohexane	6.55	2.665	80.74	30.8
Cyclohexanol	25.2	1.70	161.4	42.4
Cyclohexene	−103.5	3.293	83	
Cyclononane	9.7	19.3	178.4	40.5
Cyclooctane	14.3	2.410	149	
Cyclopentanol	−19	1.54	140.8	
Cyclopentene	−135.1	3.363	44.2	
Cyclopropane	−127.5	5.443	−32.7	20.5
Cyclopentane	−93.9	0.609	49.3	27.2
o-Cymene	−71.54		178.15	
cis-Decaline	−43.0	9.489	195.8	42.72
Decane	−29.7	28.71	174.1	
1-Decene	−66.3	13.81	170.6	
Decyl acid	31.20	28.0	270	61.34
α-Decyl acid	28.6	25.1	280	
Decyl alcohol	7		229	
Deuteromethane	−183.37	0.902	−161.2	8.276
1,4-Dibromobenzene	87.5	20.0	220.4	
1,2-Dibromobutane	−65.4		166	
1,2-Dibromoethane	9.79	10.94	131.36	36.2
Dibromomethane	−52.55	4.2	97	36.0
1,2-Dibromopropane	−55.2		140	
Dibutyl ether	−95.3		142.0	36.92
trans-1,2-Dichloro-ethylene	−50	11.98	47.5	28.88
Dichloroacetic acid	10.8	7.65	194.4	42.7
1,2-Dichlorobenzene	−14	12.9	180.05	
1,3-Dichlorobenzene	−24.76	12.6	173.0	
1,2-Dichlorobutane [14]	−80.4		124	
Dichlorodifluoromethane	−158.2	4.14	−29.8	19.61
1,2-Dichloroethane	−35.36	8.837	83.47	31.45
1,1-Dichloroethane	−96.98	7.870	57.28	28.71
Dichloroethyl ether	−46.7	8.66	178.6	45.22
1,1-Dichloroethylene	−122.1	6.51	37	26.4
cis-1,2-Dichloroethylene	−80.5	7.205	60.3	30.23
Dichlorofluoromethane [5]	−135		9	24.0
Dichloromethane	−95.1	4.6	40	28.0
1,1-Dichlorotoluene [14]	−16.1		214.0	
Diethyl (ethyl) ether	−116.2	6.90	34.51	26.60

Table 12.3 Melting, boiling, or sublimation point and heat of elements *(continued)*

Substance	t_m	ΔH_m	t_b, t_s	ΔH_b, ΔH_s
Diethyl carbonate [14]	−43		126	31.1
Diethyl oxalate	−40.6		186.1	41.58
N, N-Diethylanilene	−38.8		217.0	46.32
Diethylamine	−48.0		55.5	28.83
1,3-Diethylbenzene	−83.92	11.0	181.1	
Difluoroacetic acid	−1.1		134.2	
1,1-Difluoroethane	−117		−24.7	23.8
Diisoamyl ether			173.2	35.1
Diisopropyl ether	−85.9	11.24	68	29.2
1,2-Dimethoxybenzene	22.7	16.0	206.1	
Dimethoxymethane	−104.8	7.97	42.5	28.6
N, N-Dimethylaniline	2.45	11.42	194.5	44.35
Dimethyl ether	−138.5	4.94	−23	21.51
Dimethyl oxalate	49.5	21.0	163.3	30.58
cis-1,4-Dimethyl-cyclohexane	−87.4	9.307	124.3	33.76
trans-1,3-Dimethyl-cyclohexane	−90.1	9.86	123.5	38.85
3,3-Dimethyl-1-butene	−115.2	10.95	41.2	25.65
2,2-Dimethyl-1-butene	−140.0		55.6	27.40
2,3-Dimethyl-2-butene	−74.28	6.452	73.2	29.64
Dimethylamine	−92.2	5.939	7.4	26.10
2,3-Dimethylbutane	−128.53	0.801	58	27.28
2,2-Dimethylbutane	−99.87	0.579	49.74	26.30
cis-1,2-Dimethylcyclo-hexane	−50.0	1.645	129.7	34.23
1,1-Dimethylcyclohexane	−33.5	2.023	119.5	32.97
1,1-Dimethylcyclopentane	−69.8	10.79	87.8	30.29
2,2-Dimethylheptane	−113.0	8.95	130.5	34.8
2,2-Dimethylhexane	−121.18	6.799	106.84	32.3
2,5-Dimethylhexane	−91.2	12.86	109.0	32.8
1,1-Dimethylhydrazine	−57.2	10.07	62.5	35.0
2,2-Dimethylpentane	−123.8	4.356	79.2	29.16
3,3-Dimethylpentane	−134.5	7.067	86.1	29.64
1,4-Dioxane	11.8	12.84	101.5	35.8
Diphenyl (biphenyl)	71.0	18.6	255.2	47.95
Diphenyl ether	26.9	17.22	257.9	47.74
Diphenylamine	54.5	17.53	302	55.23
1,2-Diphenylethane	53	30.5	285	
Diphenylmethane	25.2	18.4	264.3	
Divinyl ether	−101	7.95	28.35	26.2
Dodecane	−9.6	36.84	216.3	
1-Dodecane	−35.2	19.91	213.4	
Dodecyl acid	43.75	36.6	302.3	57.49
Dodecyl alcohol [14]	24.0		259.0	
1-Dodecyne	−19		215	
Eicosanic acid	76.2	70.92		
Enanthic acid	−7.5	15.0	223	
Ethane	−183.3	2.857	−88.63	14.70
Ethanedial	15		50.4	38
Ethyl acetate	−83.578	10.98	77.06	32.26
Ethyl acrylate	−71.2		99.8	

Substance	t_m	ΔH_m	t_b, t_s	ΔH_b, ΔH_s
Ethyl alcohol (ethanol)	−113.3	5.02	78.5	38.74
Ethyl bromide	−118.6	5.9	38.40	26.82
Ethyl butyrate	−100.8		121.55	36.7
Ethyl chloroacetate	−26.0		141.3	41.4
Ethyl cyanide	−91.8	6.07	97.10	31.0
Ethyl formate	−80.5	9.2	54.5	30.31
Ethyl isobutyrate	−88.2		110.1	34.43
Ethyl isovalerate [14]	−99.3		135.4	
Ethyl mercaptan	−147.9	4.975	35.0	26.78
Ethyl methyl ether			7.5	24.73
Ethyl nitrate	−94.6	8.527	87.7	33.9
Ethyl propionate	−73.9	12.6	99.1	34.42
Ethyl propyl ether			61.4	28.90
Ethyl silane	−179.7		−13.7	22.3
Ethyl vinyl ether	−115.8		37.5	
N-Ethylamine	−63.6		204.72	
Ethylamone	−81		16.6	27.30
Ethylbenzene	−94.975	9.16	136.19	35.98
Ethylcyclohexane	−111.3	8.333	131.8	34.69
Ethylcyclopentane	−138.4	6.896	103.5	32.17
Ethylene (ethene)	−169.15	3.351	−103.71	13.54
Ethylene bromide	−139.54	23.0	15.80	25.9
Ethylene glycol	−11.5	11.6	198	57.03
Ethylene oxide	−112.5	5.173	10.5	25.53
Ethylenediamine	8	19.3	117.4	38.9
3-Ethylhexane			118.53	33.60
1-Ethylnaphthalene	−13.9	16.3	258.7 (decomp.)	
Ethylnitrile	−44.9	8.91	81.5	32.75
3-Ethylpentane	−118.6	9.456	93.5	30.95
4-Ethylphenol	47		219.0	
2-Ethyltoluene	−80.83	10.63	165.15	38.87
Fluorobenzene	−41.2	11.30	85.1	31.20
Fluorobutane	−121.4		25.1	26.05
Fluoroethane	−143.2		−37.7	21.07
2-Fluoroethanol	−26.45		103.5	43.9
Fluoroethylene	−160.5		72.2	17.1
Fluoromethane	−141.8		−78.4	17.56
Fluorooxymethane	−138		−42	22.2
Formaldehyde	−118		−19.3	23.3
Formamide	2.6	6.7	210 (decomp.)	
Formic acid	8.4	12.68	100.7	22.26
Furan	−85.6	3.80	31.4	27.04
Furfurol	−38.7		161.5	
Furfuryl alcohol	−14.6		171	
Glycine	247.2 (decomp.)			
Glycolic acid	80	8.4	78	8.8
Glycyl alcohol (glycerin)	20	18.47	290 (decomp.)	
Heptadecane	22.0	40.48	301.8	
Heptane	−90.6	14.02	98.42	31.69
1-Heptane	−119	12.40	93.64	
s-Heptanol	−35.5		151.45	39.5

Table 12.3 Melting, boiling, or sublimation point and heat of elements *(continued)*

Substance	t_m	ΔH_m	t_b, t_s	ΔH_b, ΔH_s
trans-2-Heptene	−109.48		98	
Heptyl alcohol (heptanol)	−34.1	18.17	176	
Heptylcyclohexane	−40.4	23.3	244	
1-Heptyne	−81		99.74	
Hexachlorobenzene		subl.	322	
Hexachloroethane [5]		subl.	184.4	51.0
Hexadecane	18.6	53.36	286.8	
1,5-Hexadiene	−140.7		59.46	28.58
Hexafluoroethane	−94	2.69	−79	16.15
Hexamethylbenzene	166.5	20.59	265	53.85
Hexane	−95	13.03	68.95	28.85
2-Hexanol	−57		128	34.5
trans-1,3,5-Hexatriene	−12		78.5	20.59
1-Hexene	−139.82	9.347	63.35	28.03
cis-2-Hexene	−139.82		67.87	29.12
cis-3-Hexene	−137.82		66.44	28.70
Hexyl alcohol (hexanol)	−46.7	15.38	158	50.63
1-Hexyne	−131.9		71.3	
Hydrocinnamic acid	101.5	17.7	279.8	
Hydroquinone	173.5		285	
Indene	−1.8		182.44	
Iodobenzene	−31.33	9.75	188.4	40.80
Iodoethane	−108		72.3	29.8
Iodomethane	−66.4		42.4	26.14
1-Iodopropane	−101.3		102.4	
Isoamil acetate			143.6	37.7
Isoamil alcohol			132	44.60
Isoamil formate	−93.5		123.2	35.8
Isobutane	−159.6	5.541	−11.73	21.3
Isobutene	−140.4	5.93	−6.9	22.12
Isobutyl acetate	−98.6		116.2	36.7
Isobutyl alcohol [14]	−108.0		107.2	43.47
Isobutyl butyrate	−80.7		157.5	39.0
Isobutyl chloride [14]	−131.2		68.9	
Isobutyl formate	−95.8		98.4	33.85
Isobutyl propionate	−71.4		136.8	34.8
Isobutyric acid	−46.1	5.0	153.2	41.1
Isocrotonic acid	15.5	9.12	169	
Isopentane	−159.9	5.156	27.85	24.59
Isoprene	−145.95	4.83	34.07	25.82
Isopropyl acetate	−73.4		88	33.1
Isopropyl alcohol (isopropanol)	−89.5	5.37	82.4	40.48
Isopropylbenzene (cumene)	−96.04		152.39	
Isovaleric acid	−29.3	7.32	176.50	43.1
Ketene	−134.1	7.53	−41	20.25
Margaric acid	62.5	51.09	227	
Menthol	43	12.3	216	50.42
Methacrylic acid	16		162.5	
Methane	−182.48	0.938	−164	8.18
Methyl acetate	−98.0		57	31.20
Methyl acrylate			80.5	40.29
Methyl alcohol (methanol)	−93.9	3.17	64.96	35.27
Methyl bromide	−93.6	5.98	3.56	23.90
Methyl butyrate	−84.8		102.5	34.42
Methyl formate	−99	6.7	31.5	27.89
Methyl isobutyrate [14]	−84.7		92.6	33.76
Methyl mercaptan	−123	5.9	5.96	24.6
Methyl nitrate	−83.0	8.24	64.6	32.3
Methyl propionate	−87.5		79.8	32.64
Methyl-*t*-butyl ether	−108.6		55.1	28.3
2-Methyl-1-butene	−168.5	5.36	20	
3-Methyl-1-butene	−133.7	7.912	38.57	25.50
2-Methyl-1-pentene	−135.7		60.7	28.20
2-Methyl-2-pentanol	−103		121.1	39.3
2-Methyl-2-pentene	−135.1		67.3	29.0
2-Methyl-3-ethylpentane	−115.0	11.32	115.6	32.95
Methylamine [5]	−93.5	0.134	−6.3	25.81
3-Methylanilene	−30.4	3.89	203.35	45.61
Methylcyclohexane	−126.6	6.75	100.9	31.7
Methylcyclopentane	−142.4	6.23	71.8	29.3
2-Methylheptane	−109.0	10.25	117.6	33.6
2-Methylheptyl alcohol	−112.0		175.4	
2-Methylhexane	−118.27	8.87	90.0	30.66
Methylhydrazine	−52.4	10.42	25.0	40.37
2-Methylpentane	−153.7	6.266	60.3	27.79
Methylsilane	−156.8		−57.5	18.4
3-Methylthiophene	−69.0	40.54	115.4	34.25
Myristic acid (tetradecanoic acid)	53.7	44.95	328	61.5.
Naphthalene	80.28	18.98	217.96	
1-Naphthol	96	23.47	288.0	72.27
2-Naphthol	122	22.6	295	68.32
m-Nitroaniline	114.0	23.7	305.7	65.93
o-Nitroaniline	71.5	16.1	284.1	63.93
p-Nitroaniline	149	22.2	313.7	77.4
Nitrobenzene	5.85	12.12	211.03	
Nitroglycerine	13	21.9	256 (expl.)	
Nitromethane	−28.55	9.7	101.2	34
2-Nitrophenol	45.5	17.4	216	
3-Nitrotoluene	15	13.7	231.9	50.17
Nonane	−51	15.47	150.8	37.76
1-Nonene	−81.37		146.87	
1-Nonyl alcohol	−5.5		213.5	
1-Nonyne	−50		150.8	
Octane	−56.79	20.74	125.66	34.57
1-Octene	−101.73	15.31	121.3	
2-Octyl alcohol	−31.6		180.8	51.5
1-Octyl alcohol	16.7		194.45	53.1

Table 12.3 Melting, boiling, or sublimation point and heat of elements *(continued)*

Substance	t_m	ΔH_m	t_b, t_s	ΔH_b, ΔH_s	Substance	t_m	ΔH_m	t_b, t_s	ΔH_b, ΔH_s
1-Octyne	−79.3		125.5		Tetrachloroethylene	−19.0	10.5	121.0	34.7
Oleic acid	16.3		369	67.15	Tetrachloromethane	−22.99	2.5	76.54	30.0
Oxalic acid	subl.		157	90.8	Tetradecane	−5.9	45.07	253.7	
Palmitic acid	63	54.89	360	63.14	Tetraethyl lead	−136.2	8.79	195 (decomp.)	
Paraldehyde	−12.6	13.8	128		Tetraethyl silane	−83.8	13.0	153.0	39.7
Pentachloroethane	−29.0		162	40.6	Tetrafluoroethylene	−142.5	7.71	−76.3	16.82
(pentalin)					Tetrafluoromethane	−183.7	0.70	−129	12.33
Pentachlorofluoroethane	−106	1.878	−38	19.41	Tetramethyl silane	−99.0	6.90	26.6	24.2
Pentadecane	10.0	34.59	270.6		1,2,3,4-Tetramethylbenzene	−6.25	7.046	205.0	45.02
Pentadecyl acid	52.5	43.1	158		1,2,4,5-Tetramethylbenzene	79.24	21.0	196.8	45.52
1,4-Pentadiene	−148.28	6.14	25.97	26.0	2,2,3,3-Tetramethylbutane	−100.7	7.54	106.5	31.6
Pentane	−129.7	8.41	36.07	25.77	Tetramethylmethane	−16.55	3.256	9.50	22.75
1-Pentanethione	−75.7	17.49	126.6	37.02	(neopentane)				
1-Penten	−138	5.81	30.0	25.20	Tetranitromethane	14.2		126	38.5
cis-2-Penten	−151.4	7.11	36.9		Thiobutane	−105.9	9.76	66.7	29.52
trans-2-Penten	−136	8.39	36.3		2-Thiopentane	−113.0	9.91	95.5	32.08
1-Pentyn	−90.0		40.2		Thiophene	−38.3	5.09	84.2	31.47
Perlagonic acid	15	20.3	254		2-Thiopropane	−98.39	7.99		
Phenathrene	101	18.62	340		Toluene (methylbenzene)	−95.01	6.62	110.62	33.47
Phenol	40.8	11.6	181.8	47.30	1,2,2-Tribromobutane [14]			213.8	
Phenylacetic acid	76.7	17.7	266.5		1,1,2-Tribromoethane	−29.3		188.93	
Phenylhydrazine	19.8	16.4	248	54	Tribromomethane	8.3	11.08	149.5	
Phosgene	−127.8	5.69	7.56	24.40	1,2,3-Tribromopropane	16.9		222.2	
α-Picolyne	−66.7	9.82	129.6	37.77	Trichloroacetic acid	59.1	5.90	195.0	56.2
Picric acid	−122.5	19.5	320	87.9	1,2,4-Trichlorobenzene	17.05	15.48	213.0	
Piperidine	−9.0		106.5	32.3	1,1,1-Trichloroethane	−30.41	1.88	74.1	
Propane	−189.7	2.52	−42.1	18.77	1,1,2-Trichloroethane	−30.5	11.38	113.77	36.8
1,3-Propanediol	-85.1		213.5	44.33	Trichloroethylene	−73		87	32.13
Propene (propylene)	−185.2	3.00	−47.4	18.42	Trichlorofluoromethane	−111	6.893	23.8	24.9
Propionic acid	−20.8	7.53	141	30.62	1,1,1-Trichloropropane	−77.7		108.2	
Propyl acetate	−95		101.8	34.74	Tridecane	−5.4	25.50	235.4	
Propyl alcohol (propanol)	−126.5	5.20	97.4	41.81	Tridecyl acid	45	33.6	299.0	
1-Propyl bromide	−109.8	6.53	71.0	29.9	Trifluoroacetic acid	−15.3		72.4	34.7
2-Propyl bromide	−89	6.53	59.4	28.4	1,1,1-Trifluoroethane	−111.3	6.19	−47.31	19.8
Propyl butyrate	−97.2		142.7	37.2	Trifluoromethane	−160		−82.2	18.4
Propyl formate	−92.9		81.3	31.93	Trimethyl oxide	−97	6.49	46	
Propyl propionate	−75.9		122.3	35.6	Trimethyl silane	−135.9		6.7	24.4
Propylamine	−83.0		47.8	29.05	Trimethylamine	−117.3		2.8	24.3
Propyne	−101.5		−23.2	23.27	1,2,4-Trimethylbenzene	−43.91	12.34	16.35	39.25
Pyridine	−41.7	8.28	15.2	35.11	2,2,3-Trimethylbutane	−24.19	22.01	80.88	28.95
Pyrolidine	−59	8.58	86.6	33.01	2,2,3-Trimethylpentane	−112.3	8.63	109.8	32.2
Quinoline	−15.2	10.79	236.6	47.35	Undecane	−25.59	25.18	195.9	
Salicylic acid [14]	159.5		256.0		Urethan	48.5	15.2	187	
Spiropentane	−107.0	6.435	39.0	33.39	*n*-Valeric acid	−33.83	7.74	186.05	44.0
Stearic acid	71		374	66.11	Vinyl acetate	−93.2		73	34.4
Styrene	−30.63	10.95	145.2		*m*-Xylene	−47.87	11.57	139.1	36.40
Succinic anhydride [14]	119.6		261.0		*o*-Xylene	−25.18	13.60	144.41	36.82
Tetrabromomethane	90.5	3.95	189.5	43.1	*p*-Xylene	13.26	17.1	138.35	36.07
1,1,1,2-Tetrachloroethane	−70.21		130.5	36.8	Yperite (Dichlorodiethyl	13.5		215	
1,1,2,2-Tetrachloroethane	−36		146.2	38.7	sulfide)				

Many industrial materials in the solid state are alloys, solid solutions, or amorphous substances; their melting processes differ from that of pure substances.

Alloy and solid solution melting proceeds usually within the certain temperature range – from the lower temperature (solidus) to the upper temperature (liquidus); within this temperature range, the substance exists in the heterogeneous state. The exceptions are eutectic alloys with coincident solidus and liquidus. Tables 12.4–12.7 list melting temperatures (solidus, as a rule) of technical materials – semiconducting, optical, and high-temperature materials, steels, industrial and special alloys.

Table 12.4 Melting point (in °C) of semiconducting, optical and high-temperature compounds [30–35]

Substance	t_m	Substance	t_m	Substance	t_m	Substance	t_m
Semiconducting and		$CuFeTe_2$	467	NaSb	192	$ThO_2{\cdot}ZrO_2$	2680
optical compounds		$CuGaSe_2$	767	Nd_2S_3	1737	YN	2670
$AgFeSe_2$	574	$CuGaTe_2$	697	$PbMoO_4$	1065	ThC_2	2656
$AgFeTe_2$	387	Cu_2GeS_3	932	Pr_2S_3	1827	UN	2650
$AgGaSe_2$	847	$CuInSe_2$	987	SmS	902	ScN	2650
$AgGaTe_2$	717	$CuInTe_2$	697	$SnSe_2$	657	ThN	2630
$AgInS_2$	850	Cu_2S	1130	$SrMoO_4$	1065	CaO	2600
$AgInSe_2$	777	Cu_3SbSe	555	SnTe	790	BeO	2550
$AgInTe_2$	677	$CuSbTe_2$	527	Y_2Te_3	1252	SmB_6	2540
$AgSbSe_2$	637	Cu_2SnS_3	837	YVO_4	1940	NdB_6	2540
$AgSbTe_2$	557	Cu_2SnSe_3	687	$ZnAs_2$	768	LaB_6	2530
AlAs	1597	β-Cu_2Te	1111	$ZnGeAs_2$	602	Ta_4Si	2510
AlP	1500	$CuTlSe_2$	407	$ZnIn_2Se_4$	977	Ta_5Si_3	2500
AlSb	1054	Er_2Se_3	1520	$ZnIn_2Te_4$	802	$CeO_2{\cdot}ZrO_2$	2500
Al_2Se_3	950	GaN	1227	ZnSb	546	MgO	2500
Al_2Te_3	895	GaP	1350	**High-temperature**		UB_4	2495
AsSb	807	GaSb	712	**compounds**		SrO	2460
As_2Se_3	360	Ga_2Se_3	1020	HfC	3890	CeO	2450
As_2Te_3	362	Ga_2Te_3	790	TaC	3880	Cr_2O_3	2440
BAs	1900	Gd_3P_2	700	ZrC	3530	TaB	2430
$Ba_2NaNb_5O_{15}$	1430	Gd_2Se_3	450	NbC	3480	TaS	2425
$BaSrNbO_6$	1480	Gd_2Te_3	1147	Ta_2C	3400	ThS	2425
BiSe	607	$GeSe_2$	740	HfB_2	3250	B_4C	2425
Bi_2Se_3	707	GeTe	725	TiN	3205	Nb_2N	2420
$CaMoO_4$	1430	$HgIn_2Se_4$	827	TiC	3147	$ZrO_2{\cdot}SiO_2$	2420
$Ca(NbO_3)_2$	1560	$HgIn_2Te_4$	707	TaB_2	3100	Y_2O_3	2410
$Ca_5(PO_4)_3$	1705	HgSe	765	TaN	3087	VB_2	2400
Ca_2Pb	110	HgTe	655	NbB_2	3000	AlN	2400
Ca_2Si	910	InAs	943	BN	3000	U_2C	2400
$Ca_3(VO_4)_2$	1430	InN	927	HfN	2982	α-WB	2400
$CdAs_2$	621	InP	1054	ZrN	2982	UB_2	2385
$CdGeAs_2$	397	InSb	525	TiB_2	2980	VN	2360
$CdIn_2Se_4$	917	InSe	660	ThO_2	2956	MoB	2350
$CdIn_2Te_4$	787	InTe	696	Nb_2C	2927	UC	2350
Cd_3P_2	740	KH_2PO_4	255	NbB_4	2900	Be_2C	2325
CdSb	456	KSb	332	α-SiC	2800	La_2O_3	2310
Cd_4Sb_3	460	K_3Sb	812	TiB_2	2790	BeB_6	2300
$CdSiP_2$	<1200	$LaTe_2$	727	HfO_2	2777	CaC_2	2300
$CdSnAs_2$	322	$LiNbO_3$	1250	W_2B	2770	Th_2S	2300
$CeGaSe_3$	1125	$LiTaO_3$	1650	W_2C	2730	Th_4S_7	2300
Ce_2S_3	1617	Mg_2Ge	1115	UO_2	2730	YC_2	2300
Ce_2Se_3	1162	Mg_2Pb	650	WC	2720	YB_6	2300
Cu_3AsS_4	655	Mg_2Sn	778	ZrO_2	2700	W_2B_5	2300
$CuFeSe_2$	574	$NH_4H_2PO_4$	190	ZrB_{12}	2680	NbB	2280

Table 12.4 Melting point of semiconducting, optical and high-temperature compounds *(continued)*

Substance	t_m	Substance	t_m	Substance	t_m	Substance	t_m
High-temperature compounds		Ba_3N_2	2220	WSi_2	2165	V_3B_2	2070
		ThB_4	2210	ZrSi	2150	Ce_3S_4	2050
UC_2	2250	BaS	2205	ThB_6	2150	CrB	2050
VB	2250	Be_3N_5	2205	Mo_2B	2140	$MoSi_2$	2030
ScB_2	2250	Be_3N_2	2200	NdS	2140	Al_2O_3	2030
Mo_3B_2	2250	CaO·ZrO_2	2200	Ti_5Si_3	2120	TiO	2020
UB_{12}	2235	Nd_2S_3	2200	Th_3N_4	2100	Al_2O_3·BaO	2000
SrB_6	2235	CrB_2	2200	MoB_2	2100	BeO·ZrO_2	2000
CaB_6	2230	Ti_2B	2200	GdB_6	2100	Al_2O_3·BeO	2000
BaB_6	2230	$TiSi_2$	2200	La_2S_3	2100	Al_2O_3·MgO	1930
BP	2227	CeB_6	2190	VO	2077		

Note: See also Table 12.2.

Table 12.5 Melting point (solidus) (t_s, °C) of steels and industrial alloys (liquidus values are presented in bold type)

Index	t_s	Index	t_s	Index	t_s
Steels [24, 25, 26]		Бр. ОСН10-2-3	**1000**	ЛАЖ Мц 66-6-3-2	**899**
Carbon	**1535**	Бр. ОФ4-0,25	1060	ЛАН 59-3-2	892
Х13, Х25Т, Х28	**1500**	Бр. ОЦ 4-3	**1045**	ЛАНКМц75-2-2,5-0,5-0,5	940
Х28Л, Х34Л	1350	Бр. ОЦ 8-4	854	ЛЖМц 59-1-1	885
Х18Н9, Х23Н18	**1410**	Бр. ОЦ 10-2	1015	ЛК80-3	**900**
Х18Н9Т	**1500**	Бр. ОЦС3-12-5	**1000**	ЛКС65-1,5-3	870
Х18Н10Т	**1400**	Бр. ОЦС 3,5-6-5	**980**	ЛКС80-3-3	900
ОХ18Н10	**1410**	Бр. ОЦС4-4-2,5	887	ЛМц 58-2	865
Х20Н14С2	**1400**	Бр. ОЦС4-4-17	**920**	ЛМцА 57-3-1	**920**
Х23Н13	**1440**	Бр. ОЦС5-5-5	**955**	ЛМцЖ 52-4-1	**940**
4Х10С2М	**1480**	Бр. ОЦС6-6-3	967	ЛМц 55-3-1	**930**
Bronze [27, 28]		Бр. ОЦС8-4-3	**1015**	ЛМцС 58-2-2	**900**
Бр. А5	1056	Бр. ОЦСН3-7-5-1	**990**	ЛМцОС 58-2-2-2	**900**
Бр. А7	1040	Бр. С30	**975**	ЛН 56-3	**890**
Бр. А10	1040	Silver	1082	ЛН 65-5	**960**
Бр. АЖ9-4	1040	Бр. СН60-2,5	**885**	ЛО 59-1	885
Бр. АЖМц10-3-1,5	1045	Бр. СуН7-2	**950**	ЛО 60-1	885
Бр. АЖН10-4-4	1084	Бр. Х0,5	1073	ЛО 62-1	885
Бр. АЖН11-6-6	**1135**	alloy ХОТ	1075	ЛО65-1-2	**920**
Бр. АЖС7-1,5-1,5	**1020**	Бр. Цр0,4	965	ЛО70-1	890
Бр. АМц 9-2	1060	**Brass** [24, 27, 28]		ЛО74-3	885
Бр. Б2	864	Л59	885	ЛО90-1	995
Бр. Б2,5	930	Л62	898	ЛС59-1	**900**
Cadmium	1040	Л63	900	ЛС60-1	**900**
Бр. КМц3-1	970	Л66	905	ЛС63-3	885
Бр. КН1-3	1050	Л68	909	ЛС64-2	**910**
Бр. КС3-4	**1020**	Л70	915	ЛС74-3	**965**
Бр. КЦ4-4	**1000**	Л75	980	**Copper-nickel** [24, 27, 28]	
Бр. Мг 0,3	1076	Л80	965	Alumel	1430
Бр. Мц 5	1007	Л85	990	НМцАК2-2-1	
Бр. О10	1020	Л90	1025	Constantan	**1260**
Бр. ОС5-25	899	Л96	1055	МНМц40-1,5	
Бр. ОС8-12	940	ЛА 67-2,5	**995**	Copel	1220
Бр. ОС10-5	**980**	ЛА 77-2	930	МНМц43-0,5	
Бр. ОС10-10	**925**	ЛА 85-0,5	**1020**	Cunial A	**1183**
Бр. ОС12-7	**930**	ЛАЖ 60-1-1	**904**	МНА 13-13	

Table 12.5 Melting point (solidus) of steels and industrial alloys *(continued)*

Index	t_s
Copper-nickel [24, 27, 28]	
Cunial B	1140
МНА 6-1,5	
Manganin	960
МНМц 3-12	
German silver	1170
МНЖМц 30-1-1	
German silver	1130
МН19	
Monel metal	1370
НМЖМц 28-2,5-1,5	
German silver	1080
МНЦ 15-20	
German silver	965
МНЦС 16-29-1,8	
Nichrome Х20Н80	**1390**
Ferronichrome	**1230**
Х15Н60	
Chromel T НХ9,5	1435
Chromel K НХ9	1435
МН0,6 (ТП)	**1084**
МН5	**1120**
МН16 (ТБ)	1120
МН95-5	1086
МНА 13-3	**1120**
МНЦС 17-18-1,8	**1120**
НК 0,2	**1384**
НМ8	**1190**
НМ 56,6	**1290**
НММц 3-12	960
НМц 2,5	1440
НМц 5	1370
Aluminum alloys [24]	
АВ (Avial)	543
АК2	509
АК8	**638**
АЛ1	535
АЛ10В	488
АМг	627
АМг3	568
АМг5	**638**
АМг 7	550
АМц	643
АН-2,5	640
АСМ	657
В-95	477
Д1	513
Д16	502
Д18П	510
Magnesium alloys [27]	
ВМ65-1	516
МЛ1	505
МЛ2	645
МЛ3	561
МЛ4	400
МЛ5	430
МЛ6	440
МЛ7	505
МЛ9, МЛ10	550
МЛ11	593
МЛ15	539
МА1	645
МА2	565
МА3	510
МА5	482
МА8	645
Zinc alloys [28]	
ЦА15	**443**
ЦАМ2-4	**470**
ЦАМ2-5	**450**
ЦАМ 4-1	**390**
ЦАМ 4-3	**410**
ЦАМ 8-2	**405**
ЦАМ 10–2	**419**
ЦАМ 10–5	**395**
ЦМ1	**422**
ЦМ4	**445**
ЦОС 3–3	**411**
Type alloys [27, 28]	
Лн1, Ст2	240
МЛн1, Мп1, У	245
Мн, Ш3, Нт	255
МСМ1, П2	260
Ш2	270
Гс, МШ3	300
Ш1	330
К	340
Solders [28, 29]	
ПМЦ 36	800
ПМЦ 48	850
ПМЦ 54	876
ПОС40÷ПОС90	183
ПОС10	268
ПОСК 50-18	142
ПОССу 4-6	244
ПОССу 5-1	275
ПОССу 10-2	268
ПОССу 40-2	185
ПОССу 18-0,5	183
ПОССу 95-5	234
ПСр 1	225
ПСр 1,5	265
ПСр 2	225
ПСр 2,5	300
ПСр 2,5С	304
ПСр 3	314
ПСр3 Кд	295
ПСр 10	822
ПСр12М	793
ПСр44	650
ПСр65	695
ПСр72	779
Babbitts [28]	
Б6	416
Б83	370
Б91	223
Б93А	260
Б93	304
БК	440
БН	400
БС6	280

Note: For alloy designations see Chapter 3.

Table 12.6 Melting point (solidus) t_s (in °C) of two-component alloys [36]

Component		Content of B component, %								
A	B	10	20	30	40	50	60	70	80	90
Ag	Sn	870	750	630	550	495	450	420	375	300
Ag	Zn	850	755	705	690	660	630	610	570	505
Al	Ag	625	615	600	590	580	575	570	650	750
Al	Au	675	740	800	855	915	970	1025	1055	
Al	Cu	630	600	560	540	580	610	755	930	1055
Al	Fe	860	1015	1110	1145	1145	1220	1315	1425	1500
Al	Sb	750	840	925	945	950	970	1000	1040	1010

Table 12.6 Melting point (solidus) t_s of two-component alloys [36] *(continued)*

Component A	Component B	10	20	30	40	50	60	70	80	90
		Content of B component, %								
Al	Sn	645	635	625	620	605	590	570	560	540
Al	Zn	640	620	600	580	560	530	510	475	425
Au	Ag	1062	1061	1058	1054	1049	1039	1025	1006	982
Au	Cu	910	890	895	905	925	975	1000	1025	1060
Au	Pt	1125	1190	1250	1320	1380	1455	1530	1610	1685
Cd	Ag	420	520	610	700	760	805	850	895	910
Cd	Tl	300	285	270	262	258	245	230	210	235
Cd	Zn	280	270	295	313	327	340	355	370	390
Cu	Ag	1035	990	945	910	870	830	788	814	875
Cu	Ni	1180	1240	1290	1320	1335	1380	1410	1430	1440
Cu	Sn	1055	890	755	725	680	630	580	530	440
Cu	Zn	1040	955	930	900	880	820	780	700	580
K	Tl	133	165	188	205	215	220	240	280	305
Na	Bi	425	520	590	645	690	720	730	715	570
Na	Cd	125	185	245	285	325	330	340	360	390
Ni	Sn	1380	1290	1200	1235	1290	1305	1230	1060	800
Pb	Ag	460	545	590	620	650	705	775	840	905
Pb	Cu	870	920	925	945	950	955	985	1005	1020
Pb	Na	360	420	400	370	330	290	250	200	130
Pb	Sb	250	275	330	395	440	490	525	560	600
Pb	Sn	295	276	262	240	220	190	185	200	216
Pb	Tl	710	790	880	917	760	600	480	410	425
Sb	Ag	595	570	545	520	500	505	545	680	850
Sb	Bi	610	590	575	555	540	520	470	405	330
Sb	Sn	600	570	525	480	430	395	350	310	255
Sb	Zn	555	510	540	570	565	540	525	510	470

Table 12.7 Melting point (solidus) t_s (in °C) of lo v-melting alloys [24,37–39]

Alloy	t_s	Alloy	t_s	Alloy	t_s
Hg97.2 Na2.8*[1]	−48.2	Na99 Tl1*[1]	64	Bi52.5 Pb32 Sn12.5	96
Cs94.5 Na5.5*[1]	−30	Bi50.1 Pb24.9 Sn14.6 Cd10.8*[2]	65.5	Bi47 Pb35.3 Sn17.7	98
Cs93 Na7*[1]	−28	Bi50.4 Pb25.1 Sn14.3 Cd10.2*[2]	67.5	Bi40 Pb20 Sn40	100
K78 Na22*[1]	−11.4	Bi50 Pb25 Sn12.5 Cd12.5*[2]	68	Bi50 Pb40 Sn10	100
K80 Na20	−10	Bi50.1 Pb22.6 Sn13.3 Cd10*[3]	68	Bi50 Pb28 Sn22*[6]	100
Rb91.8 Na8.2*[1]	−4.5	Na70 Hg30	70	Bi54.4 Pb25.8 Sn19.8*[1]	101
K70 Na30	−3.5	Bi49.5 Pb27.27 Sn13.13 Cd10.1*[1]	70	Bi48 Pb28.5 Sn14.5 Hg9	105
K60 Na40	5	Bi38.4 Pb30.8 Sn15.4 Cd15.4	71	Bi42.1 Pb42.1 Sn15.8	108
K50 Na50	11	Bi33.7 In65.3*[1]	72	Bi40 Pb40 Sn20*[7]	113
K90 Na10	17.5	Bi27.5 Cd34.5 Pb27.5 Sn10.5	75	Bi36.5 Pb36.5 Sn27	117
Na56 K44*[1]	19	Bi50 Pb34.5 Sn9.3 Cd6.2	77	Bi33.4 Pb33.3 Sn33.3	123
Na85.2 Hg14.8*[1]	21.4	Na90 K10	77	Bi55.5 Pb44.5*[1]	124
Na60 K40	26	Bi58 In17 Sn25*[1]	79	Bi56.5 Sn43.5*[1]	125
Na70 K30	41	Bi35.3 Cd9.5 Pb35.1 Sn20.1	80	Bi27.2 Pb44.5 Sn33.3	127
Na50 Hg50	45	Na80 Hg20	80	Bi43 Pb43 Sn13	128
Bi47.7 In19.1 Sn8.3 Cd5.3 Pb22.6*[1]	47	Na96.7 Au3.3*[1]	80	Bi56 Sn40 Zn4*[1]	130
Bi36 Hg30 Pb28 Cd6	48	Na90 Hg10	90	Bi28.5 Pb43 Sn28.5	132
Bi42 Pb32 Hg20 Cd6	50	Na50 Hg50	90	K90 Tl10	133
Hg70 Na30	55	Bi55.2 Pb33.3 Tl11.5*[1]	91	Hg70 K30	135
Bi49.4 In21 Pb18 Sn11.6*[1]	57	Bi51.6 Cd8.1 Pb40.3	91	Bi57 Sn43*[1]	138
Na80 K20	58	Bi50 Pb30 Sn20*[4]	92	Bi57 Tl43*[1]	139
Na60 Hg40	60	Bi50 Pb25 Sn25	93	Cd18.2 Pb30.6 Sn51.2	142
Bi53.5 Sn19 Pb17 Hg10.5	60	Bi50 Pb31.2 Sn18.8*[5]	94	Pb42 Sn37	143
		Bi50 Sn25 Cd25	95	Bi60 Cd40*[1]	144

Table 12.7 Melting point (solidus) t_s of low-melting alloys *(continued)*

Alloy	t_s	Alloy	t_s	Alloy	t_s
Bi50 Pb50	145	Bi10.5 Pb42 Sn47.5	160	Na70 Hg30	181
Bi19 Pb38 Sn43	148	Bi13.3 Pb46 Sn40.7	165	Sn62 Pb38*1	183
Bi25 Pb50 Sn25	149	K80 Tl20	165	Bi44.2 Pb9.8 Tl48*1	186
Bi18.1 Pb36.2 Sn45.7	151	Bi12.8 Pb49 Sn38.2	172	Bi47.5 Tl52.5*1	188
Bi16 Pb36 Sn48	155	Pb32 Sn68	177	Bi76.5 Tl23.5*1	198
Bi13.7 Pb44.8 Sn41.5*1	160	Cd32 Sn68*1	177		

*1 Eutectic alloy. *3 Lipowitz's alloy. *5 Newton's alloy. *7 Bismuth solder.
*2 Wood's alloy. *4 Lichtenberg's alloy. *6 Rose's alloy.

Melting (and solidification) of substances with an amorphous structure in the solid state is not exemplified by the clear transition temperature; it occurs gradually and is characterized by softening and solidification temperatures. Tables 12.8 and 12.9 present softening temperatures of amorphous solid materials of this type – glasses and plastics, and Table 12.10 gives solidification temperatures of some commercial liquids.

Table 12.8 Softening point (t_{soft}, °C) of glasses [40]

Glass	t_{soft}	Glass	t_{soft}	Glass	t_{soft}
Quartz glass		T16	680	Foam glass silica	1100
КВ, КУ	1160	T28	645	Glass for pipe 13в	725
КИ	1220	Щ23	710	Glass-ceramic for pipe	1100
I	1300	Ц26	730	Medical glass АБ-1	590
II	1100	Ceramic glass slag	950	Medical glass НС-1	630
Chemical glass		Ceramic glass СТЛ	980	Multal	670
N23	580	Ceramic glass СТМ, СТБ	930	N51-A	574
N29	565	Crystal (lead) glass	530	Pyrex	565
Ц32	590	Dish glass	560	Sial	590
N846	582	Fibre glass alkaliless	830	Simax	570
Heat-resistant glass		Fibre glass soda	710	Uninost	530
N13	620	Foam glass	<600	Window cast glass	600

Table 12.9 Softening point (t_{soft}, °C) of plastics and polymers [29, 41–44]

Plastic	t_{soft}	Plastic	t_{soft}	Plastic	t_{soft}
Aminoplast	100	Foams plastic, epoxy resin base	170	Polycarbon	220
Asbestos glass laminate	130	Foams plastics	140	Polyester	115
Asbestos textile laminate	130	Foams polyurethanes	230	Polyethylene	90
Caprolon	190	Glass laminate	250	Polyethyleneterephthalate	130
Capron	215	Laminated compressed	150	Polyformaldehyde	177
Cellophane	60	wood plastic		Polyimide	250
Celluloid	40	Dacron	155	Polymethylmethacrylate	120
Delan	150	Nylon	60	Polypropylene	152
Dyphlon	150	Nyplon	330	Polystyrene	90
Enanthic fiber	225	Organic glass 2-55	133	Polyurethane	85
Epoxy resin	150	Organic glass СОЛ	90	Polyvinyl chloride	60
Epoxy-silicones material	220	Paper-based laminate	150	Polyvinyl fluoride	196
Etrol (Plastics)	70	Pentaplast	160	Premix	130
Fluoropolymer-3	125	Phenolic plastics	135	Shellac (Lac)	80
(Trifluoroethylene)		impact resistant	140	Terylen	264
Fluoropolymer-3M	150	Phenylone	150	Textile laminate	140
Fluoropolymer-4 (Teflon)	260	Polyacrylate	65	Urea resin	75
Fluoropolymer-4M	220	Polyamide	190	Vinyl plastics	180
Foams plastic Ysolan	210	Polyarylate	190		

Table 12.10 Crystallization point of fuels, oils and pressure liquids (in °C) [44, 22, 38] [Substances are ordered according to t_m value growth.]

Substance	t_m	Substance	t_m	Substance	t_m
Fuel[*1]		ТК, ТКп	−45	И-20А, И-40А	−15
Diesel fuel 3	−35	С-220	−30	И-70А, И-100А	−10
Diesel fuel 3С	−45	ГХБД	−21	Vacuum ВМ-1	−12
Diesel fuel A	−55	ТХБ	−16	**Nigrols**[*2]	
Diesel fuel Л	−10	Sovol	+5	Winter	−20
Furnace oil		(chlorinated biphenil)		ТС-14,5	−15
Gasoline (petrol) Б-70	−60	**Engine oil**[*2]		Summer	−5
Jet-fuel Т8	−55	МТ-14П	−43	**Pressure liquids**[*2]	
Kerosene (coal oil)	−38	МТ-16П	−25	АМГ-10, МГЕ-10	−70
Marine residual oil	−6	МС-20П	−18	РМ, МГЗ	−60
Mazut 100, МП	+25	Engine Т	0	ЭШ	−50
Mazut 40	+10	**Lubricating oil** [*2]		АУ, Р, АУП	−45
Transformer oil[*2]		МП-609	−80	МГ-30	−35
ФМ-5	−110	МП-601	−70	ИС-12, ОМТИ	−30
ФС-5	−100	МПВ	−60	ИС-50	−20
ФС-56	−90	МАС-19Н	−49	МС-20	−18
АТМ-65	−70	И-12А	−30	ИС-20	−15
Hexol, ПМС-50	−60	И-5А (velosite)	−25		
Т-750, Т-1500	−53	В-8А (sewing)	−20		

[*1] Crystallization point.
[*2] Solidification point.

In Table 12.11 are enumerated melting and boiling temperatures of liquids used as heat-transport media and coolants.

Table 12.11 Melting (t_m, °C) and boiling (t_b, °C) points of coolants, antifreezes and heat carriers [37, 38, 45] [Coolants are given according to nomenclature; other substances are given according to t_m growth.]

Index or composition, %		t_m	t_b	Index or composition, %	t_m	t_b
Coolant				**Antifreeze**		
R11	Trichlorofluoromethane	−111.0	23.65	ТОСОЛ-А65	−65	115
R12	Dichlorodifluoromethane	−155.9	−29.74	ТОСОЛ-А40	−40	108
R12B1	Bromochlorodifluoromethane	−80.0	−3.83	ТОСОЛ-А	−35	170
R13	Trichloromethane	−180.0	−81.59	**Heat-transfer agent**		
R13B1	Bromotrifluoromethane	−143.2	−55.77	CaCl 29.9; water 70.1	−55	
R14	Tetrafluoromethane	−184.0	−128.02	Ethylene glycol 67; water 33	−73	
R21	Dichlorofluoromethane	−135.0	8.37	Propylene glycol 60; water 40	−50	
R22	Chlorodifluoromethane	−160.0	−40.81	Silicon МАИ	−100	191
R23	Trifluoromethane	−155.0	−82.14	$AlCl_3$ 22.5; $AlBr_3$ 77.5	67	200
R30	Dichloromethane	−96.7	40.10	$AlCl_3$ 80.7; NaCl 9.7; KCl 9.6	70	254
R40	Chloromethane	−97.8	−23.86	Bi 55.5; Pb 43.5	125	1670
R113	1,1,2-Trichlorotrifluoroethane	−35.0	46.82	Diphenyl	69.5	255.6
R113B2	1,1,2-Trichloro-2-chlorodibromoethane	−72.9	94.57	Diphenyl 30.9; *o*-Triphenyl 47.5; *m*-Triphenyl 21.6	10.5	297.5
R114	1,2-Dichlorotetrafluoroethane	−93.9	3.63	Diphenyl 40.2; *o*-Triphenyl 59.2; *p*-Triphenyl 0.6	22.7	287
R114B2	1,2-Dibromotetrafluoroethane	−110.5	47.15			
R115	Chloropentafluoroethane	−106.0	−38.97	Diphenyl 50.5; *o*-Triphenyl 59.5	23	286.5
R116	Hexafluoroethane	−100.6	−78.21	Diphenyl ether	27.0	258.5
R142	1-Chloro-1,1-difluoroethane	−138.0	−9.20	Diphenylbutane	−25	295.3
R143a	1,1,1-Trifluoroethane	−111.3	−47.58	Diphenylethane	−17.9	272.6
R152a	1,1-Difluoroethane	−117	24.54	Diphenylmethane	25.4	264.3
RC318	Octafluorocyclobutane	−41.4	−5.97	Ethylene glycol	−15.6	197.3

Table 12.11 Melting and boiling points of coolants, antifreezes and heat carriers *(continued)*

Index or composition, %	t_m	t_b	Index or composition, %	t_m	t_b
Heat-transfer agent			*p*-Triphenyl	212.7	384
Glycerine	−17.9	290	*o*-Triphenyl 66.3; *m*-Triphenyl 33.7	29.3	345.5
Na22,K78	−11	784	*o*-Triphenyl 65.8; *m*-Triphenyl 33.5; *p*-Triphenyl 0.7	28.9	346
Na56,K44	19	825			
NaCl 23.1; water 76.9	−21.2		ДКМ	−22	335
Naphthalene	80.2	281.8	ДТМ	−30	296
$TiBr_4$	39	230	МИПД	−47	290
$TiCl_4$	−23	153.9	ТКОС	−33	457
m-Triphenyl	87.45	379	ТКС	−36	440
o-Triphenyl	56.25	337.5			

12.3 Melting and Boiling as a Function of Pressure

Melting, boiling, and sublimation processes depend on pressure according to the Clapeyron–Clausius equation

$$dT/dP = T(\Delta V/\Delta H), \tag{12.1}$$

where T, ΔH are the temperature and the transition heat, respectively; ΔV is the volume change at the transition; P is the pressure.

Boiling and sublimation temperatures always increase with the pressure. Dependence $T_b(P)$ [and $T_s(P)$] is said to be a vapor pressure curve; corresponding data are given in Chapter 11.

Melting temperature also, as a rule, rises with pressure. There are deviations for some substances within the limited ranges of pressure, which are explained by an imbalance between atomic package density in the liquid state and the solid state structure. Dependences $T_m(P)$ for elements and some organic and inorganic compounds are given in Tables 12.12 and 12.13, along with the derivative dT_m/dP, which allows us to calculate T_m values at small deviations from the normal pressure (up to 100–1000 MPa).

Table 12.12 Melting point of elements (t_m, °C) as a function of pressure and derivatives dt_m/dP (10^{-2} °C/MPa) at normal pressure [6, 15]

Element	dt_m/dP	Pressure, MPa							Additional
		10	50	100	500	1000	3000	5000	references
Actinium	19.40	1052	1060	1069	1147	1244	1632	2020	
Aluminum	6.41	661	663	667	692	724	852	981	[17]
Americium	2.50	1175	1176	1177	1185	1200	1245		
Antimony	−0.58	630	630	630	628	624	610	589	[17]
Argon	24.89	−187	−177	−166.8	−95.8				[16]
Arsenic	5.7	818	819	823	846	860	920	940	[17]
Astatine	55.80	307	330	358					
Barium	4.30	727	729	731	748	770	715	590	
		500 (6 GPa)		525 (7 GPa)		380 (7.5 GPa)			
Beryllium	5.00	1287	1289	1292	1312	1337	1437	1537	
Bismuth	−3.57	271	269	267.5	252	228	250	425	[16, 17]
		180 (2 GPa)		500 (6 GPa)					
Bromine	~ 25				80	170			
Cadmium	5.3	321	324	326	347	374	480	586	

Table 12.12 Melting point of elements as a function of pressure and derivatives *(continued)*

Element	dt_m/dP	Pressure, MPa							Additional
		10	50	100	500	1000	3000	5000	references
Calcium	14.90	843	849	857	916	985	1090	1200 (4 GPa)	
Carbon	19.60	3764	3772	3781	3860	3958	4350	4742	[20]
	4200 (7.5 GPa)		3710 (8 GPa)						
Cerium	−5.0	798	797	794	774	750	670	700	
	750 (7.5 GPa)								
Cesium	25.93	32	42	52	98.5	160	190	170	[17, 23]
	100 (4 GPa)								
Chromium	15.90	1878	1885	1893	1956				
Cobalt	3.50	1492	1494	1495	1509	1527	1597	1667	
Copper	4.69	1084	1085	1088	1107	1130	1215	1310	[17]
Deuterium	24.50	−252	−244	−236	−197.9	−157.7			[19]
Europium	12.80	823	828	835	886	950	1000	980	
Gadolinium	5.7	1312	1315	1318	1340	1369	1450	1480 (4 GPa)	
Gallium	−2.11	30	29	28	19	8	50	100	[17]
Germanium	−3.80	937	936	934	922	902	826	750	
Gold	6.20	1064	1067	1070	1095	1126	1224	1336	
Hafnium	17.3	2232	2238	2247	2316	2403	2749	3095	
Helium	25.7	−269.8	−264	−258.4	−246 (0.3 GPa)				[17, 21]
helium-3	18.2	−270.7	−263	−258.4	−246.2 (0.3 GPa)				[22]
Holmium	14.30	1475	1481	1488	1545	1617	1903	2189	
Hydrogen	29.60	−256	−247	−239	−198	−160			[18, 19]
Indium	5.22	157	159	162	182	205	287	380	[17]
								540 (9 GPa)	
Iodine	21.7	116	124	135	222	331	575	670	
								(4.5 GPa)	
Iridium	6.2	2447	2450	2453	2478	2509	2633	2757	
Iron	3.00	1538	1539	1541	1553	1568	1628	1645	[17]
Krypton	30.13	−154	−143	−129	−40	48			
Lanthanum	2.50	920	921	922	932	945	995	1045	
Lead	7.73	328	331	335	364	398	516	612	[17]
Lithium	3.41	181	182	184	194	204	228	242	[17]
Lutetium	15.90	1664	1671	1679	1742	1822	2140	2458	
Magnesium	7.50	651	654	657	687	725	875	1025	[17]
Manganese	3.30	1245	1246	1248	1261	1278	1344	1410	
Mercury	5.21	−38	−36	−34	−13	12.06	100	190[*1]	[16, 17]
Molybdenum	0.96	2611	2611	2612	2616	2620	2640	2659	[20]
Neodymium	2.60	1024	1025	1026	1037	1050	1080	1080	
		1075 (6 GPa)							
Neon	13.16	−247	−242	−235	−216 (0.3 GPa)				[17]
Neptunium	2.30	637	638	639	648	670	720	760 (4 GPa)	
Nickel	3.70	1455	1457	1459	1473	1492	1566	1640	
Niobium	16.2	2470	2477	2485	2550	2631	2955	3279	
Nitrogen	21.95	−208	−200	−190.9	−149.2				[16, 17]
Osmium	6.50	3045	3048	3051	3077	3110	3240	3370	
Oxygen	11.42	−218	−213	−208	−190 (0.3 GPa)				[17]
Palladium	6.40	1554	1557	1560	1586	1618	1746	1874	
Phosphorus	29.92	47	59	72.7	171.6	950	1030 (2 GPa)		[17]
Platinum	6.20	1749	1752	1755	1780	1811	1935	2015	[20]
Plutonium	−7.9	638	635	631	599	560	500	515	
Potassium	17.71	64	71	78.5	127	167	240	275	[16, 17]
Praseodymium		934				940	935	920[*2]	
Promethium	31.30	1093	1105	1121	1246	1403	2029	2655	

Table 12.12 Melting point of elements as a function of pressure and derivatives *(continued)*

Element	dt_m/dP	Pressure, MPa							Additional
		10	50	100	500	1000	3000	5000	references
Protactinium	18.40	1575	1581	1590	1664	1756	2124	2492	
Radium	21.30	842	850	861	946	1053	1479		
Radon	34.60	−68	−54	−37					
Rhenium	3.20	3177	3178	3180	3193	3209	3273	3337	[20]
Rhodium	5.90	1963	1966	1969	1992	2022	2140	2258	
Rubidium	21.11	41	49	57.9	115	164	235	280	[16, 17]
Ruthenium	6.10	2250	2253	2256	2280	2311	2433	2555	
Samarium	6.0	1072	1075	1078	1102	1132	1180	1200	
Scandium	15.10	1542	1548	1556	1616	1692	1994	2296	
Selenium	20.76	224	233	243	317	390	605	680	[17]
							630 (5.5 GPa)		
Silicon	−5.81	1419	1417	1414	1391	1330	1230	1110*3	[17]
Silver	4.90	961	963	966	985	1010	1108	1206	
Sodium	8.77	99	102	105.9	136	175	245	285	[16, 17]
Strontium	16.50	769	776	784	850	933	1130	1160 (4 GPa)	
Sulphur (rhomb.)	31.30		135	148	226	270			[17, 18]
(monocline)	33.14	117	130	143		280	415	560	[17]
Tantalum	5.40	2976	2979	2981	3003	3030	3138	3246	[20]
Technetium	5.50	2200	2203	2205	2227	2255	2365	2475	
Tellurium	4.50	450	4052	454		460	450	510	
Terbium	7.40	1357	1360	1363	1393	1450	1500 (2.5 GPa)		
Thallium	6.53	304	306	310	336	368	489		[17]
Thorium	18.60	1752	1759	1768					
Tin	2.61	232	233	234	245	256	299	420	[17]
Titanium	15.40	1609	1616	1623	1685	1762	2070	2378	
Tritium	22.12	−250	−243	−235	−213 (0.3 GPa)				[17]
Tungsten	7.80	3347	3350	3354	3385	3424	3580	3736	[20]
Uranium	18.10	1136	1143	1152		1150	1170	1160	
	1150 (7 GPa)								
Vanadium	15.7	1918	1925	1933	1995	2074	2388	2702	
Xenon	38.90		−93	−75	43 (0.3 GPa)				[17]
Ytterbium	17.0	816	823	831	902	990	1215	1230 (4 GPa)	
Yttrium	16.5	1529	1536	1544	1610	1693	2023	2353	
Zinc	4.81	420	422	424	443	466	547	619	[17]
Zirconium	16.30	1880	1887	1895	1960	2042	2368	2694	

*1 230 (7 GPa) [46].
*2 930 (7.5 GPa) [46].
*3 800 (100 GPa), 710 (12 GPa) [46].

Table 12.13 Melting point of individual compounds (t_m,°C) as a function of pressure and derivatives dt_m/dP (10^{-2} °C/MPa) at normal pressure [6, 15]

Element	dt_m/dP	Pressure, MPa					
		10	50	100	500	1000	3000
		Inorganic compounds					
Aluminum antimonide	−6.90	1059	1056	1053	1025	991	853
Ammonia	8.82	−77	−74	−70	−56 (0.3 GPa)		
Cadmium telluride	−20.0	1043	1035	1025	945	845	
Carbon dioxide [16]	20.82	−55	−47	−5	21.4	75.4	93.5
Carbon disulfide [16]	13.06	−110	−105	−74	−52	0	170
Carbon tetrachloride [16]	40.50	−19	−3	15	128	192.1	

Table 12.13 Melting point of individual compounds *(continued)*

Element	dt_m/dP	Pressure, MPa					
		10	50	100	500	1000	3000
Inorganic compounds							
Cesium chloride	48.29			707	870	1039	
Cupric chloride	−1.16	430	429	429	424	417	381
Gallium antimonide	−5.0	699	697	695	675	650	550
Gallium arsenide	−3.4	1138	1136	1134	1121	1104	1036
Hydrocyanic acid	23.28	−12	−3	7	40 (300 MPa)		
Indium antimonide	−10.0	1024	1020	1015	975	925	403
	449 (5 GPa)						
Indium arsenide	−4.29	939	938	936	918	897	811
Indium phosphide	−2.91	1060	1058	1057	1045	1031	973
Lithium chloride	24.22	607	617	629	715	810	
Orthophosphoric acid	8.13	43	46	50			
Potassium tetrasilicate	−6.02	764	762	759			
Rubidium chloride	25.01	720	729	741	815	1128	1426
Silicon tetrachloride [16]	31.25	−65	−53	−38	67.7	183.8	
Silver nitrate	9.09	213	216	239	257	303	
Sodium bromide	28.66	744	755	769	868	973	
Sodium chloride	23.81	803	812	824	910	1004	
Sodium fluoride	16.08	993	1000	1007	1063	1120	
Sodium iodide	32.81	658	671	687	798	914	1245
Tungsten carbide [20]	0.0	2850	2850	2850	2850	2850	2850
Water (hydrogen oxide) [6]	−7.68	−0.8	−3.84	−8.8	−7.5	26	135
		−22.0 (214.3 MPa)		−17 (357.3 MPa)		0.16 (646.4 MPa)	
Water, heavy (deuterium oxide), D_2O	−7.78	3	0	−5	−11 (0.3 GPa)		
Organic compounds							
Acetamide [16]	14.20	83	88	93.1	128	166	
Acetic acid [16]	24.70	19	28	37.7	119	148.3	
Acetonitrile	21.20	−42	−33	−24	44	112	
Acetophenone	25.05	22	32	42	78 (300 MPa)		
t-Amyl alcohol	22.02	−6	2	3			
Anethole	21.42	25	33	43	77 (300 MPa)		
Aniline [16]	21.31	−4	4	13.1	81	143.2	
Benzene [16]	29.79	9	20	32.5	117	190.5	
Benzoic anhydride	26.43	46	56	69			
Benzonitrile	20.40	−11	−3	7	86	180	
Benzophenone [16]	28.93	51	62	74.6	163	248	
Benzyl alcohol	15.38	−14	−8	0			
Betol	37.15	99	112	126	166 (300 MPa)		
Bromoacetic acid	18.45	57	64	73			
Bromobenzene [16]	20.50	−29	−21	−12.1	50	107.6	
Bromoethane	12.58	−118	−113	−107	−67	−28	80
1-Bromonaphthalene	27.52	9	20	34			
m-Bromonitrophenol	23.41	56	65	77			
Bromopropane [16]					−56	0	138
Bromotoluene	31.12	41	53	67	116 (300 MPa)		
t-Butanol	41.35	30	44	58	97 (300 MPa)		
n-Butyl alcohol (butanol) [16]	14.81	−88	−83	−76	−30	12	132
Butyric acid	19.12	−4	4	14			
1,4-Dibromobenzene	38.51	91	106	123			
1,2-Dibromoethane [16]			22.45	34.0			
1,4-Dioxane	11.81	13	17	23			

Table 12.13 Melting point of individual compounds *(continued)*

Element	dt_m/dP	Pressure, MPa					
		10	50	100	500	1000	3000
		Organic compounds					
1,2-Diphenylethane [16]				68.4	83.2 (200 MPa)		
Camphor	127.3	189	214 (30 MPa)				
Caprylic acid	18.87	18	26	35			
Cetyl alcohol	23.84	51	61	72			
Chloroacetic acid [16]	17.34	63	71	80	128	165	
Chlorobenzene [16]	18.89	−44	−36	−28	25	84.5	222
Chlorobenzene [16]							(2.5 GPa)
m-Chloronitrophenol	24.27	47	57	68			
Chloroform [16]	16.56	−62	−55	−47	17	79	243
Chloroform [16]							(2.5 GPa)
Chlorotoluene	27.81	11	21	34	78 (300 MPa)		
m-Cresol	13.73	13	19	26			
o-Cresol [16]	18.13	32	39	69	86	218	
p-Cresol	23.08	36	45	56	97 (300 MPa)		
Crotonic acid	37.29	75	82 (30 MPa)				
Cyclobutanone	24.42	−49	−39	−28	13 (300 MPa)		
Cyclohexane	54.12	12	33	59			
Cyclohexanol	40.82	30	46	66			
Dibenzyl	36.42	55	68	84			
Dibromomethane [16]			27.5	34.0			
Dichloroacetic acid	21.74	117	126	137			
Dichloromethane [16]					−46	0	157
Diethyl (ethyl) ether	13.66	−115	−109	−102	−75 (300 MPa)		
Diphenylamine	28.0	57	68	80	165	213 (800 MPa)	
Diphenylmethane	60.65	33	52	71	119 (300 MPa)		
Dodecane	24.35	−7	2	13	79		
Enanthic acid	18.27	−6	2	11			
Ethyl acetone	11.34	−82	−78	−73	−35	5	119
Ethyl alcohol (ethanol) [16]	9.14	−116	−113	−108	−75	−38	82
Formamide	9.54	4	7	12			
Formic acid	13.21	10	15	21	40 (300 MPa)		
Iodotoluene	31.56	38	50	65	115 (300 MPa)		
Menthol	25.10	45	55	65	99 (300 MPa)		
Methane	27.30	−180	−169	−155	−101 (300 MPa)		
Methyl alcohol	6.13	−97	−95	−92	−79 (300 MPa)		
Methyl benzoate	20.16	−10	−2	8			
Methyl oxalate	22.73	57	65	76	151		
Methylcyclohexane	7.11	−126	−123	−119	−95	−69	
Methylcyclopentanol	18.34	37	45	54			
Myristic acid	25.60	61	70	81	116 (300 MPa)		
1-Naphthol	24.53	98	108	120	164 (300 MPa)		
Naphthalene	39.4	84	99				
Naphthylamine	4.93	51	52 (30 MPa)				
Nitrobenzene [16]	23.68	8	17	27.9	107	184.5	
m-Nitrotoluene	24.83	19	28	41			
o-Nitrotoluene	20.59	−1	7	17	92		
p-Nitrotoluene	29.05	55	66	79			
Nitromethane	14.0	−24	−22	−15			
1-Nitronaphthalene	4.35	59	61	63	72 (300 MPa)		
m-Nitrophenol	19.83	98	106	115	151 (300 MPa)		
o-Nitrophenol	22.14	48	57	68	110 (300 MPa)		

Table 12.13 Melting point of individual compounds *(continued)*

Element	dt_m/dP	Pressure, MPa					
		10	50	100	500	1000	3000
		Organic compounds					
p-Nitrophenol [16]	27.26	115	125.5	137.7	198.8 (400 MPa)		
Nonane	13.33	−52	−47	−40	8	96	205
Octadiene	25.73	30	40	51	119		
Palmitic acid	20.75	66	74	84			
Pentachloroethane	22.56	−27	−18	−6			
Pentadecane	31.77	13	25	124			
Phenol [16]	14.7	42	48	53.4	115.48	184.6	
Phenyl salicylate	26.90	44	55	68			
Piperidine	17.30	−9	−2	6			
Propionic acid	22.50	−19	−10	0			
Propyl bromide	12.76	−109	−104	−98	−53	−6	142
Tetracosane	25.31	53	58 (30 MPa)				
Thiophene	54.69	−33	−14	6	105		
Thyme camphor	22.85	54	62	71	97 (300 MPa)		
o-Toluidine	18.80	−14	−7	3	59		
p-Toluidine [16]	25.55	46	56	68	153	204.9 (800 MPa)	
Tribromomethane [16]	25.66	10	20	31.5	115	194	
Tridecane	24.78	12	21	32	101		
Triphenylmethane	38.05	96	110	126	178 (300 MPa)		
Urethan	114.6	59	103	156	81	176	
Valeric acid	16.23	−33	−26	−18			
m-Xylene	20.47	−43	−35	−25			
o-Xylene	23.53	−23	−13	−1			
p-Xylene	35.08	17	31	47			

References

[1] Properties of inorganic compounds. Handbook, Ed. Efimov, A. I., Khimiya, Leningrad, 1983 (in Russian).

[2] Thermodynamic properties of inorganic substances, Ed. Zefirov, A. N., Atomizdat, Moscow, 1965 (in Russian).

[3] Physical and chemical properties of oxides. A handbook, Ed. Samsonov, G. V., Metallurgiya, Moscow, 1978 (in Russian).

[4] Landolt–Bornstein, Zahlenwerte und functionen aus physik, chemie, geophysik, astronomie, technik, Neue Serie, Gruppe IV, Bd. 4, Eigenschaften der Materie bei hohen Drucken, Springer–Verlag, Berlin, 1980 (4 teil).

[5] Selected values of chemical thermodynamic properties. NBS Circular, No. 500, Washington, DC, 1952.

[6] Regel', A. R., Glazov, B. M., Periodic law and physical properties of electron melts, Nauka, Moscow, 1978 (in Russian).

[7] Samsonov, G. V., Vinitskii, I. M., Refractory compounds. Handbook, 2nd ed., Metallurgiya, Moscow, 1976 (in Russian).

[8] Nekrasov, B. V., Fundamentals of general chemistry, 3rd ed., Khimiya, Moscow, 1973 (vols. 1, 2) (in Russian).

[9] Nuclear wallet cards, Eds. Shirley, V. S., Lederer, C. U, New York, 1979.

[10] Properties of elements. Handbook, Ed. Druts, M. E., Metallurgiya, Moscow, 1985 (in Russian).

[11] Thermal constants of substances (issues 1–10), Ed. Glushko, V. P., Izdatel'stvo Akad. Nauk SSSR, Moscow, 1965–1982 (in Russian).

[12] Ubbelohde, A., Melting and crystal structure, Oxford, Clarendon Press, 1965.

[13] Properties of organic compounds. Handbook, Ed. Potekhin, A. A., Khimiya, Moscow, 1984 (in Russian).

[14] Kaye, G., Laby, T., Tables of physical and chemical constants and some mathematical functions, Longmans, Green & Co., London, 1941.

[15] Cannon, J. F., J. Phys. Chem. Ref. Data, 3, 781, 1974.

[16] Handbook of a chemist, 3rd ed., Ed. Nikolskii, B. P., Khimiya, Leningrad, 1971 (vol. 1) (in Russian).

[17] Babb, S., Rev. Mod. Phys., 35, 400, 1963.

[18] Vezzoli, G. C., Wilsh, P. J., High Temp. - High Press., 9, 345, 1977.

[19] Liebenberg, D. H., Phys. Rev., B 15, 4526, 1978.

[20] Vereshchagin, L. F., Fateeva, N. S., High Temp. - High Press., 9, 619, 1977.

[21] Swenson C., Physics at high pressures, New York, 1960.

[22] Technical grade petroleum products. Properties and application. Handbook, Ed. Shkol'nikov, V. M., Khimiya, Moscow, 1978 (in Russian).

[23] Vaidya,. S. N., High Temp. High Press., 11, 335, 1979.

[24] Tables of physical quantities. Handbook, Ed. Kikoin, I. K., Atomizdat, Moscow, 1976 (in Russian).

[25] Fokin, M. N., Ruskov, Yu. S., Mosolov, A. V., Titanium and its alloys in the chemical industry, Khimiya, Leningrad, 1978 (in Russian).

[26] Handbook of a founder, Ed. Rubtsov, N. N., Gostekhteorizdat, Moscow, 1962 (in Russian).

[27] Smiryagin, A. P., Smiryagina, M. A., Belova, A. V., Commercial nonferrous metals and alloys, 3rd ed., Metallurgiya, Moscow, 1974 (in Russian).

[28] Orlov, I. D., Mironov, V. M., Handbook of a founder, Gostekhteorizdat, Moscow, 1960 (in Russian).

[29] Anur'ev, V. I., Handbook of a machine building designer, 6th ed., Mashinostroenie, Moscow, 1982 (vols. 1–3) (in Russian).

[30] Crystal chemical, physical and chemical properties of semiconducting materials, Ed. Bokii, G. B., Izdatel'stvo standartov, Moscow, 1973 (in Russian).

[31] Ryabtsev, N. G., Quantum electronics materials, Radio Publ., Moscow, 1972 (in Russian).

[32] Kuz'minov, Yu. S., Ferroelectric crystals for the laser radiation control, Nauka, Moscow, 1982 (in Russian).

[33] Kuz'minov, Yu. S., Lithium niobate and tantalate, Nauka, Moscow, 1975 (in Russian).

[34] Krzhizhanovskii, R. E., Shtern, Z. Yu., Thermal properties of nonmetallic materials. (Oxides). Reference book, Energiya, Leningrad, 1973 (in Russian).

[35] Krzhizhanovskii, R. E., Shtern, Z. Yu., Thermal properties of nonmetallic materials. (Carbides). Reference book, Energiya, Leningrad, 1977 (in Russian).

[36] Smithsonian physical tables, 9th ed., Washington, DC, 1954.

[37] Kanaev, A. A., Kopp, I. Z., Anhydrous vapors in power machine building, Mashinostroenie, Moscow, 1973 (in Russian).

[38] Fire-resistant cooling agents and hydraulic liquids. Guidebook, Ed. Sukhotin, A. M., Khimiya, Leningrad, 1979 (in Russian).

[39] Kogan, V. A., A handbook of metals and alloys for printing workers, Kniga Publ., Moscow, 1976 (in Russian).

[40] Glasses. A handbook, Ed. Pavlushkin, N. M., Stroiizdat, Moscow, 1973 (in Russian).

[41] Kalinchev, E. L., Sakovtseva, M. B., Properties and processing of thermal plastics. Guidebook, Khimiya, Leningrad, 1983 (in Russian).

[42] Katsnel'son, N. Yu., Balaev, G. A., Plastics. Properties and applications. Handbook, 3rd. ed., Khimiya, Leningrad, 1978 (in Russian).

[43] Katsnel'son, N. Yu., Balaev, G. A., Polymer materials. A handbook, Khimiya, Leningrad, 1982 (in Russian).

[44] A handbook of electrotechnical materials, 2nd ed., Eds. Koritskii, Yu. V., Pasynkov, V. V., Taraev, B. M., Energiya, Leningrad, 1976 (in Russian).

[45] Perel'shtein, I. I., Parushin, E. B., Thermodynamic and thermal properties of working agents of refrigerating machines and heat pumps, Legkaya i pishchevaya promyshlennost', Moscow, 1984 (in Russian).

13

Equation of State and Critical Parameters of Substances

E.B. Gel'man

13.1 Equation of State of Solids

An equation that shows a functional relation between temperature T, pressure P, and volume V in the state of thermodynamic equilibrium is referred to as an equation of state. At present, theoretically proved forms of equations of state are available only for substances in the solid crystal and gaseous states.

The equation of state of solids is most developed in the Mie–Gruneisen form [1]:

$$PV = P_0V + 2\gamma E, \tag{13.1}$$

where P_0 is the pressure at $T = 0$ K; γ is the Gruneisen constant; E is the heat energy of crystal lattice.

The magnitude of P_0 is determined by the lattice structure and by the interaction potential of particles that enter into its composition. For some types of lattices, the P_0 value can be written in the following way:

$$P_0 = Ar^{-2}\exp\{b(1-r)\} - Kr^{-m}, \tag{13.2}$$

where $r = (V/V_0)^{1/3}$, V_0 is the body volume at normal conditions, m^3; the parameter $m = 4$ for molecular crystals, and $m = 9$ for ion crystals and for nonalkali and other metals with strong overlapping of electron shells; the A, K, and b parameters are determined from experiment. Their values for a number of substances are presented in Table 13.1.

The $2\gamma E$ term in the Mie–Gruneisen equation allows for the temperature influence; its magnitude can be comparable with that of P_0V, and at high temperatures, it can exceed the latter.

The Gruneisen constant is equal to $\gamma = \beta V/k_T c_V$ [2]. Here, β is the volume coefficient of thermal expansion; c_V is the heat capacity of a body at constant volume; k_T is the isothermal compressibility, which weakly depends on temperature and volume. The heat energy of lattice in the first approximation is equal to $E = (3/2)RT$ (for one mole of substance). Its magnitude can be specified more accurately within the theory of solids (e.g., that of Debye, Einstein, and some others).

0-8493-2861-6/97/$0.00+$.50

Table 13.1 The Gruneisen constant and the parameters of the Mie–Gruneisen equation of state for some substances [γ_s is the Gruneisen constant at the compression ratio $s = V/V_0$.]

Substance	γ_1	$\gamma_{0.6}$	A, 10^{10} Pa	b	K, 10^{10} Pa	Substance	γ_1	$\gamma_{0.6}$	A, 10^{10} Pa	b	K, 10^{10} Pa
Metals and semiconductors											
Ag	2.47	1.9	2.6887	13.5293	2.8691	Pd	2.18	1.4	5.4307	12.3419	5.6172
Al	2.13	1.6	2.4433	10.9916	2.6180	Pt	2.63	2.2	13.6590	8.3348	13.8758
Au	3.05	2.6	8.8721	7.9029	9.0935	Rh	2.26	2.0	27.3303	5.3840	27.5471
Be	1.17	1.0	1.7918	19.9553	2.0571	Sb	0.86	0.2	1.0437	13.0215	1.0790
Cd	2.27	1.6	24.9938	2.5933	24.9224	Si	0.74	0.3	4.3669	8.8513	4.4217
Co	1.99	1.6	90.2906	2.6438	90.5267	Sn	2.03	1.5	1.6923	11.6741	1.7868
Cu	2.04	1.6	5.2703	9.9448	5.4910	Ta	1.69	1.3	9.5038	7.5979	9.6194
Fe	1.68	1.3	9.9743	7.0985	10.1639	Th	1.12	0.6	2.6096	8.1274	2.6512
Ge	0.72	0.4	4.5765	7.1077	4.6185	Ti	1.18	0.8	1.3882	23.8180	1.4761
In	2.24	2.0	1.2274	11.5108	1.3320	U	1.83	1.4	4.4319	10.2740	4.5397
Mg	1.46	1.1	2.8386	5.7902	2.9191	V	1.29	1.0	9.2284	7.1227	9.3461
Mo	1.58	1.2	15.8570	7.0914	15.9899	W	1.55	1.2	17.3243	7.3234	17.4481
Nb	1.68	1.4	19.8079	4.6213	19.9267	Zn	2.38	1.6	1.8403	12.1914	2.0362
Ni	1.91	1.4	4.7646	13.9474	4.9922	Zr	0.77	0.6	3.8994	8.6963	3.9410
Pb	2.78	2.4	1.7719	9.3515	1.8836						
Ion crystals and minerals											
Al_2O_3	1.60	1.4	41.4459	4.2629	41.7202	NaI	1.59	1.0	0.4397	12.3490	0.4972
CsBr	1.93	1.5	0.8952	7.5090	0.9547	SiO_2	0.71		1.1129	12.6690	1.2123
CsCl	1.97	1.4	0.5567	11.4866	0.6259	Andradite	1.1	0.5	4.6282	11.9457	4.8305
CsI	2.01	1.5	0.5456	9.0538	0.5968	Garnet	1.4	1.2	29.4744	3.8243	29.7665
Fe_3O_4	1.40	1.0	26.6830	3.9333	26.8730	Grossularite	1.0	0.2	2.6384	17.6332	2.8428
FeS_2	1.50	0.9	4.1727	12.7322	4.3377	Diopside	0.9	0.2	2.2236	17.0914	2.3652
MgO	1.40	1.0	10.7374	6.7287	10.9796	Labradorite	0.4	0.2	10.0589	4.1678	10.1128
NaBr	1.56	1.0	0.6910	10.8196	0.7644	Olivine	1.2	1.0	25.5397	3.4413	25.7285
NaCl	1.55	1.0	1.0849	8.9488	1.1733	Orthoclase	0.5	0.1	3.4593	6.7857	3.5255

13.2 Equation of State of Gases

The only theoretically verified form of the equation of gaseous state is the virial equation [3]

$$PV/RT = 1 + B/V + C/V^2 + \cdots, \tag{13.3}$$

where V is the volume of one mole of gas, cm^3/mol; R is the gas constant; B (cm^3/mol) and C (cm^6/mol^2) are the second, the third, and so forth virial coefficients, which depend on temperature and are independent of the pressure and gas density. At low densities ($V \to \infty$), equation (13.3) is degenerate to the equation of state of an ideal gas: $PV = RT$. The second, third (and so on) terms describe corrections for the gas being not ideal owing to the binary, triple (and so forth) interactions of the gas particles. The second and the third virial coefficients for a number of gases are tabulated in Tables 13.2 and 13.3.

Besides equation (13.3), the expansion of PV/RT in terms of the pressure is sometimes of use also:

$$PV/RT = 1 + B'P + C'P^2 + \cdots. \tag{13.4}$$

The latter coefficients are connected with the virial coefficients by the relations: $B = RTB'$, $C = (RT)^2(C' + B'^2)$, etc.

Table 13.2 The second virial coefficient of gas B (in cm^3/mol) [6]

T, K	Ar	CF_4	CH_4	CH_3Cl	CH_3F	$CHCl_3$	C_2H_2	C_2H_4	C_2N_2
293.15	−16.85	−91.18	−45.19	−439.61	−193.28	−1430.49	−162	−145.92	−381.96
400	−0.82	−33.71	−15.69	−208.31	−103.88	−580.56	−87	−72.30	−171.68
500	7.17	−4.50	−0.81	−128.24	−58.66	−348.42	−51	−35.92	−66.09
600	12.25	13.57	8.67	−86.32	−34.70	−241.47	−28	−13.08	−0.75
700	15.67	26.03	15.00	−60.79	−20.16	−181.03	−12	2.47	44.75
800	18.09	34.94	19.63	−43.80	−10.48	−142.42	−1	13.85	77.11
900	19.84	41.47	22.96	−31.64	−3.64	−115.76	7	22.19	101.40
1000	21.19	46.55	25.59	−22.60	1.41	−96.34	14	28.64	120.22
1200	23.01	53.62	29.22	−10.04	8.04	−69.77	23	37.84	146.60
1400	24.03	58.25	31.60	−1.83	12.51	−52.51	29	44.04	164.09
1600	24.86	61.20	33.06	3.89	15.51	−40.45	34	48.36	176.27
1800	25.49	63.21	34.08	7.23	17.70	−31.57	37	51.53	184.50
2000	25.91	64.91	35.00	11.04	19.90	−24.80	39	53.78	190.26
2400	26.24	67.35	36.26	15.60	20.54	−15.12	43	56.55	199.06

T, K	CO	CO_2	F_2	H_2	H_2O	He	Kr	N_2[33]	NH_3
293.15	−10.04	−97.94	−20.97	12.16		11.15	−53.79	−5.47	−302.71
400	7.64	−49.07	−1.58	14.36	−333.23	10.94	−22.88	9.183	−120.11
500	16.57	−25.0	8.09	14.98	−163.47	10.72	−7.25	16.45	−68.30
600	22.14	−9.89	14.22	15.66	−98.83	10.51	2.56	21.01	−45.97
700	25.85	0.39	18.37	16.38	−66.30	10.32	9.23	24.03	−29.77
800	28.49	7.92	21.26	16.68	−47.22	10.13	14.01	26.12	−20.83
900	30.39	13.43	23.41	16.70	−34.85	9.97	17.60	27.63	−14.67
1000	31.80	17.70	25.03	16.72	−26.19	9.81	20.32	28.73	−10.19
1200	33.48	23.81	27.29	16.69	−15.07	9.58	24.20	30.08	−4.18
1400	34.76	27.92	28.53	16.55	−8.08	9.37	26.76	30.95	−0.38
1600	35.56	30.78	29.47	16.39	−3.83	9.15	28.53	31.59	2.18
1800	36.19	32.87	30.25	16.22	−0.68	8.93	29.65	31.87	3.93
2000	36.36	34.44	30.78	16.06	1.47	8.72	30.49	32.04	5.37
2400	36.70	36.28	31.21	15.69	4.69	8.44	31.84	32.13	7.30

T, K	NO	N_2O	Ne	O_2	O_3	SF_6	SO_2	SiF_4	Xe
293.15	−18.25	−137.07	11.02	−17.00	−106.87	−292.12	−573	−144.01	−134.59
400	−3.98	−65.42	12.47	0.03	−54.20	−139.43	−339	−50.25	−70.79
500	3.36	−29.82	13.19	8.53	−27.73	−63.54	−123.46	−3.09	−39.83
600	7.90	−7.83	13.62	13.91	−11.19	−15.95	−84.62	26.56	−20.40
700	10.98	7.46	13.77	17.49	−0.15	16.41	−61.30	46.95	−6.92
800	13.19	18.43	13.87	20.06	7.96	39.66	−45.85	61.35	2.55
900	14.81	26.63	13.91	21.92	14.03	57.05	−32.92	71.97	9.67
1000	16.04	32.98	13.88	23.33	18.71	70.50	−26.81	80.18	15.29
1200	17.77	41.96	13.75	25.28	25.37	89.57	−15.57	91.72	23.25
1400	18.79	47.94	13.61	26.35	29.83	102.28	−8.21	99.23	28.53
1600	19.47	52.13	13.46	27.21	32.95	111.16	−3.05	103.99	32.32
1800	20.02	53.13	13.31	27.86	35.24	117.59	0.73	107.21	35.10
2000	20.47	57.19	13.16	28.30	36.96	121.89	3.62	109.90	37.19
2400	20.95	59.97	12.87	28.62	38.90	127.80	7.63	113.92	39.91

Table 13.3 The third virial coefficient of gas C (in cm^6/mol) [6]

T, K	Ar	CO_2	F_2	H_2	H_2O	He	N_2	O_2	SF_6
293.15	960	3159	1368	297		98.78	1437	1127	35868
400	848	2499	1205	283		89.48	1332	1000	28388
500	807	2173	1146	271		82.90	1288	954	25044
600	785	2005	1113	260	4650	77.63	1258	928	23361
700	769	1911	1090	251	2840	73.29	1233	910	22427
800	756	1852	1072	242	2170	69.65	1210	894	21846
900	744	1814	1056	234	1570	66.57	1188	880	21442
1000	733	1786	1040	227	1090	63.85	1167	867	21131
1200	713	1744	1011	215	715	60.50	1128	841	20646
1400	693	1709	983	204	537	57.40	1092	819	20233
1600	675	1679	959	195	412	54.30	1059	797	19853
1800	658	1650	934	188	366	51.20	1028	777	19492
2000	642	1622	912	181	319	48.09	1000	758	19148
2400	613	1571	872	169	246	44.73	951	724	18514

Various empirical equations of state are often used for experimental data representation [4]. In the simplest case, such equations contain two parameters, which can be calculated through the known values of the critical temperature T_c, pressure P_c, and volume V_c that are presented in Tables 13.4 – 13.6. The most commonly employed two-parameter equations can be introduced as follows:

(1) the van der Waals equation [4, 5]:

$$(P + \frac{a}{V^2})(V - b) = RT, \qquad a = \frac{27}{24} RT_c V_c, \; b = \frac{1}{3} V_c; \tag{13.5}$$

(2) the Dieterici equation [4]:

$$P(V - b) = RT\exp\{-a/RTV\}, \qquad a = 2RT_c V_c, \; b = \frac{1}{2} V_c; \tag{13.6}$$

(3) the Berthelot equation [4]:

$$(P + a/TV^2)(V - b) = RT, \qquad a = \frac{16}{3} RT_c V_c^2, \; b = \frac{1}{4} V_c; \tag{13.7}$$

(4) the Redlich–Kwong equation [29]:

$$P = \frac{RT}{V - b} - \frac{a}{T^{1/2} V(V + b)}, \qquad a = 0.4278 R^2 \frac{T_c^{5/2}}{P_c}, \; b = 0.26 V_c. \tag{13.8}$$

Sometimes empirical equations of state with greater number of parameters are utilized as well. They involve the modified Redlich–Kwong equation with three parameters [30], the Beattie–Bridgeman equation with six parameters [31], the Benedict–Webb–Rubin equation with eight parameters [32], and some others.

The approximation accuracy of equations of state is individual for the gas under consideration and depends on the range of variables' variation. In some cases it reaches about a few parts of per cent. Among the two-parameter equations, the Redlich–Kwong equation is often most accurate. Tables 13.4 – 13.6 give the van der Waals constants for some elements, organic substances, and inorganic materials.

The a_{AB}, b_{AB} constants of a chemical compound AB can be approximately calculated through the constants a_A, b_A and a_B, b_B of the A and B components in this compound according to the relations:

$$\sqrt{a_{AB}} = \sqrt{a_A} + \sqrt{a_B}, \qquad b_{AB} = b_A + b_B. \tag{13.9}$$

In the same way, the a_{A+B}, b_{A+B} constants for the mixture $A+B$ of the A and B components can be estimated in the following way:

$$\sqrt{a_{A+B}} = x\sqrt{a_A} + (1-x)\sqrt{a_B}, \qquad b_{A+B} = xb_A + (1-x)b_B, \tag{13.10}$$

where x is the A component concentration in the mixture. The more precise methods of the van der Waals constants calculation can be found in Bretschneider [5].

13.3 Critical Parameters of Substances

For every substance such values of temperature T_c (K), pressure P_c (Pa), volume V_c (cm^3/mol), and density ρ_c (g/cm^3) exist that the equation of state of this substance meets the conditions: $(\partial P/\partial V)_T = 0$, $(\partial^2 P/\partial^2 V)_T = 0$.

They are spoken of as the critical temperature, pressure, volume and density, accordingly (or, in total, critical parameters of a substance). The state corresponding to critical parameters is referred to as the critical state or critical point of a substance. According to the up-to-date knowledge, the critical point constitutes the isolated point of the second-type phase transition and simultaneously the end-point of the first-type phase transition between a liquid and a vapor [7, 8]. The critical point is characterized by a number of physical phenomena (the disappearance of meniscus, so-called critical opalescence, infinite compressibility, infinite specific heat at constant volume, and so on) that proceed identically in all the substances and are determined by the most general properties of intermolecular forces. In Tables 13.4 – 13.6, the values of critical parameters of the elements, some inorganic, and organic substances are presented (figures in parentheses indicate an error of a given value, i.e., 126.25(4) means 126.25±0.04 and 69.4(17.2) means 69.4±17.2).

Table 13.4 Critical parameters and van der Waals constants of elements [9]

Substance	Formula	T_c, K	P_c, MPa	ρ_c, g/cm^3	V_c cm^3/mol	a, N·m^4/mol^2	b, cm^3/mol
Aluminum*3 [35]	Al	8000	447	0.64			
Argon	Ar	150.65(10)	4.86(1)	0.531(1)	75.2	0.1361	32.191
Bromine	Br_2	584(5)	10.3(5)	1.18	135	0.9624	58.724
Cadmium*3	Cd	>2480 [14]					
Cesium*3	Cs	2043(15)	11.8(3)	0.43(4)	308(30)	10.356	180.64
Chlorine	Cl_2	417(5)	7.71	0.573	124	0.6576	56.202
Chromium*3 [35]	Cr	9620	968	2.30			
Copper*3 [10]	Cu	>5390 [14]					
Deuterium:							
normal	n-D_2	38.350(2)	1.6650(5)	0.0623	60.3	0.02576	23.940
equilibrium	eq-D_2	38.2	1.65	0.0669	60.3	0.02577	24.037
para-	p-D_2	38.26(4)	1.6498(61)	0.0623	60.3	0.02588	24.104
Deuterium tritide	DT	39.5	1.75	0.0867	58.5	0.02596	23.418
Fluorine	F_2	144	5.6	0.574 [16]	66.2 [12]	0.1085	26.854
Gallium*3 [35]	Ga	7210	431	1.8			
Germanium*3 [35]	Ge	9170	490	1.64			
Gold*3 [10]	Au	>4820 [14]					
Helium	He	5.20(10)	0.229(1)	0.0693	57.5	0.00344	23.599
Helium-3	^3He	3.34(2)	0.116(3)	0.0414 [20]	72.5 [20]	0.00279	29.789
Hydrogen:							
normal	n-H_2	33.24(5)	1.297(3)	0.0310	65.5	0.02484	26.635
equilibrium	eq-H_2	32.98	1.293	0.0314	64.2	0.02604	27.313
para-	p-H_2	32.98(2)	1.293(3)	0.03116	64.7	0.02453	26.509

Table 13.4 Critical parameters and van der Waals constants of elements *(continued)*

Substance	Formula	T_c, K	P_c, MPa	ρ_c, g/cm^3	V_c cm^3/mol	a, N·m^4/mol^2	b, cm^3/mol
Hydrogen deuteride	HD	35.91(2)	1.484(6)	0.0484	62.8	0.02535	25.158
Hydrogen tritide	HT	38.3	1.66	0.0667	60.3	0.02574	23.953
Indium*3 [35]	In	6420	243	1.9			
Iodine [12]	I_2	826(20)	11.6 [35]	1.64 [35]	155 [20]	1.1976	51.667
Krypton	Kr	209.38(10)	5.50(1)	0.908	92.3	0.2324	39.549
Lead*3 [35]	Pb	4980	184	3.25			
Lithium*3 [10]	Li	3223(600)	69.4(17.2)	0.12	58(15)	4.3645	48.258
Magnesium*3 [35]	Mg	3590	198	0.56			
Mercury [10]	Hg	1763(15)	153.5(1.5)	5.3(5)	36.5	0.5905	11.936
Molybdenum*3 [15]	Mo	11150(550)	553.3(117.5)	2.62	36.5(3.5)	6.5534	20.945
Neon	Ne	44.45(10)	2.72(1)	0.484(1)	41.7	0.0211	16.948
Niobium*3 [10]	Nb	>9880 [14]					
Nitrogen	N_2	126.25(4)	3.399(2)	0.304(2)	92.1(6)	0.1368	38.607
Oxygen	O_2	154.78(5)	5.081(10)	0.41(2)	78	0.1375	31.662
Ozone	O_3	261.05 (10)	5.53(1)	0.537	89.4	0.3592	49.038
Phosphorus	P	968(20)	8.1(5)	0.144 [12]	215 [12]	3.3711	124.10
Platinum*3 [10]	Pt	>6450 [14]					
Potassium	K	2280(50)	15.8(1.0)	0.194(25)	202(26)	9.5908	149.90
Radon	Rn	377.5	6.326(2)	1.613 [11]	137 [11]	0.6570	62.019
Rubidium	Rb	2106(15)	16.2 [10]	0.347(35)	246(25)	7.9782	135.00
Selenium*1 [19]	Se	1590(20)	38.5(3.0)	1.235 [17]	147 [17]	1.9148	42.916
Silver*3 [10]	Ag	>4300 [14]					
Sodium	Na	2503(50)	25.6(1.5)	0.207(30)	111(16)	7.1270	101.47
Sulphur*2 [18]	S	1313(5)	18.21	0.563	158	2.7611	74.041
Tin*3 [35]	Sn	8200	335	2.05			
Titanium*3 [35]	Ti	11790	763	1.30			
Tritium normal	n-T_2	43.7	2.11	0.112	53.7	0.02642	21.549
Tungsten*3 [14,35]	W	13400	337	4.3			
Vanadium*3 [35]	V	12500	1078	1.90			
Xenon	Xe	289.74(1)	5.841(1)	1.099(3)	119.47	0.4192	51.557
Zinc*3 [10]	Zn	>2590 [14]					
Zirconium*3 [35]	Zr	16250	752	1.80			

*1 The mean number of atoms in molecular Se in the critical point equals 2.3 [17].
*2 The mean number of atoms in molecular S in the critical point equals 2.78 [18].
*3 Estimated.

Table 13.5 Critical parameters and van der Waals constants of inorganic substances [9]

Substance	Formula	T_c, K	P_c, MPa	ρ_c, g/cm^3	V_c cm^3/mol	a, N·m^4/mol^2	b, cm^3/mol
Air		413.8	3.77	0.35		1.3247	114.09
Aluminum:							
tribromide	$AlBr_3$	763(2)	2.89(6)	0.861(5)	310 [20]	5.8791	274.59
trichloride	$AlCl_3$	625.65(50)	2.64(5)	0.51(3)	261 [20]	4.3165	245.86
triiodide [12]	AlI_3	955 [20]		1.002	407	3.6356	135.67
Ammonia	NH_3	405.45(5)	11.283(5)	0.233	73.1	0.4249	37.347
Antimony:							
tribromide	$SbBr_3$	904 [20]					
trichloride	$SbCl_3$	794.05(50)		0.842	271	2.0128	90.333
triiodide [21]	SbI_3	718.2	0.23				

Table 13.5 Critical parameters and van der Waals constants of inorganic substances *(continued)*

Substance	Formula	T_c, K	P_c, MPa	ρ_c, g/cm^3	V_c cm^3/mol	a, N·m^4/mol^2	b, cm^3/mol
Arsenic:							
trichloride	$AsCl_3$	654.45(50)		0.720	252	1.5426	84.000
trideuteride	AsD_3	372.0 [20]					
trihydride (arsine) [20]	AsH_3	373.0					
Bismuth:							
tribromide	$BiBr_3$	1220(5)	8.41	1.487(15)	302	5.1612	150.76
trichloride	$BiCl_3$	1178(5)	11.97(40)	1.210(6)	261	3.3818	102.31
Boron:							
diborane	B_2H_6	289.85(2)	4.00(2)	0.14(3)	170(10)	0.6121	75.263
tribromide	BBr_3	573(5)		0.9(1)	278 [22]	1.4900	92.667
trichloride	BCl_3	451.95(10)	3.87(1)	0.7(1)	150(20)	1.5390	121.35
trifluoride	BF_3	260.85(10)	4.98(2)	0.59 [12]	115 [12]	0.3980	54.379
triiodide	BI_3	773.15		1.10	356	2.5745	118.67
Carbon:							
dioxide (carbonic acid gas)	CO_2	304.15(5)	7.387(5)	0.468(1)	94	0.3652	42.792
diselenide	CSe_2	612(20)	7.0(1.0)	0.85(8)	200	1.5623	90.972
disulfide	CS_2	552(1)	7.90(2)	0.44(1)	173	1.1243	72.585
monoxide	CO	132.92(3)	3.499(5)	0.301(3)	93	0.1473	39.482
oxyselenide (carbonyl selenide)	$COSe$	394.25(50)					
oxysulfide (carbonyl sulfide)	COS	375.40(10)	6.2(1)	0.44 [20]	140 [20]	0.6649	63.120
Carbonic acid	H_2CO_3	304.19	7.380	0.459	135	0.3658	42.856
Chlorine:							
pentafluoride [12]	ClF_5	415.75	5.26	0.56 [20]	231	0.9585	82.161
trifluoride	ClF_3	447.5	5.8			1.0112	80.523
Cyanogen	C_2N_2	400(1)	6.0(1)			0.7805	69.536
Deuterium:							
bromide	DBr	362.0 [20]					
chloride	DCl	323.4 [20]					
iodide	DI	421.8 [20]					
oxide (heavy water)	D_2O	644.05(10)	21.86(3)	0.363	55.1	0.5535	30.625
peroxide [12]	D_2O_2	717	13.9			1.0722	53.290
selenide	D_2Se	412.35					
sulfide	D_2S	372.25					
Difluoramine	NHF_2	403	9.4			0.5026	44.445
cis - Difluorodiazine	N_2F_2	272	5.8 [12]	0.559 [12]	118 [12]	0.3736	48.944
trans - Difluorodiazine	N_2F_2	260	5.6	0.584 [12]	113 [12]	0.3537	48.486
Germanium tetrachloride	$GeCl_4$	552(2)	3.85(5)	0.65(1)	330	2.3078	148.99
Hydrazine	N_2H_4	653	14.7	0.230 [12]	139 [12]	0.8464	46.190
tetrafluoro $\bar{\nu}$ hydrazine	N_2F_4	309	3.7 [20]	0.574 [12]	181 [12]	0.7427	85.656
Hafnium:							
tetrabromide	$HfBr_4$	746(6)	4.1(5)	1.20(10)	415	3.9066	186.62
tetrachloride	$HfCl_4$	724(2)	5.7(3)	1.05(2)	304 [20]	2.6940	132.60
tetraiodide	HfI_4	913(3)	3.9(5)	1.30(10)	530	6.1515	240.11
Hydrogen:							
bromide	HBr	362.95(20)	8.51(5)	0.807 [22]	100 [22]	0.4514	44.317
chloride	HCl	324.55(50)	8.26(1)	0.42	86.8	0.3720	40.844
cyanide	HCN	465.65(50)	5.39(2)	0.195	139	1.1281	88.039
fluoride	HF	461(3)	6.49(35)	0.29	69	0.9542	76.760
iodide	HI	423(1)	8.22(10)	1.09	117	0.6350	53.496
oxide (water)	H_2O	647.30(5)	22.12(1)	0.32(1)	56.3	0.5524	30.413
peroxide [12]	H_2O_2	708.5	15.54	0.349	97.5	0.9418	47.372
selenide	H_2Se	411(1)	8.9	0.349 [13]	97.7 [13]	0.5525	47.903
sulfide	H_2S	373.55(10)	9.01(1)	0.349	97.7	0.4518	43.097
Iron pentacarbonyl	$Fe(CO)_5$	563(10)					

Table 13.5 Critical parameters and van der Waals constants of inorganic substances *(continued)*

Substance	Formula	T_c, K	P_c, MPa	ρ_c, g/cm^3	V_c cm^3/mol	a, N·m^4/mol^2	b, cm^3/mol
Mercury:							
mercuric bromide	$HgBr_2$	1011(10)					
mercuric chloride	$HgCl_2$	972(5)	11.52(40)	1.555(60)	174.6	2.3916	87.682
mercuric iodide	HgI_2	1070(10)					
Molibdenum:							
hexafluoride	MoF_6	473(5)	4.8(4)	0.93(3)	226	1.3700	103.22
pentachloride	$MoCl_5$	850(3)	5.3(5)	0.74(3)	369	3.9989	167.66
Nickel tetracarbonyl	$Ni(CO)_4$	473(10)	2.0(5)	0.5(1)	341	3.2196	242.57
Niobium:							
pentabromide	$NbBr_5$	1009(5)		1.05(4)	469	4.4264	156.33
pentachloride	$NbCl_5$	803.5(2.0)	4.88(11)	0.68(4)	397	3.8551	170.98
pentafluoride [20]	NbF_5	737	6.3	1.21	155	2.5215	121.92
Nitrogen:							
chloride difluoride	NF_2Cl	337.4	5.15			0.6450	68.122
dichloride fluoride	$NFCl_2$	337.45(1)	5.15(5)			0.6452	68.1316
difluoramine oxide	NOF_2	349.45(50)					
oxide	NO_2	431(1)	10,1(1)	0.56	82	0.5346	44.206
nitric oxide	NO	180(1)	6.54(2)	0.52	58	0.1444	28.579
nitrile fluoride	NO_2F	349.5 [12]	9.3 [12]	0.602	108	0.3821	38.964
nitrosyl chloride	NOCl	440 [20]	9.36 [21]			0.6030	48.841
nitrosyl fluoride [12]	NOF	349.4 [20]	7.98	0.595	82.3	0.4459	45.478
nitrous oxide	N_2O	309.58(1)	7.255(1)	0.453(1)	97.27	0.3852	44.347
trideuteride	ND_3	405.45(30)					
trifluoramine oxide [12]	NOF_3	302.65	6.42	0.593 [20]	169 [20]	0.4158	48.962
trifluoride	NF_3	233.10(10)	4.531(12)	0.574 [12]	123.8 [12]	0.3497	53.462
Oxygen:							
chloride [12]	Cl_2O	465	6.58	0.520	167	0.9589	73.487
fluoride	F_2O	215.5(1)	4.96	0.553	97.6	0.2733	45.200
Perchloryl fluoride	$FClO_3$	368.65(40)	5.4	0.637	161	0.7380	71.341
Phosphonium chloride	PH_4Cl	322.25(20)	7.37(4)			0.4111	45.463
Phosphonitrilic fluoride:							
trimer	$[PNF_2]_3$	460.85					
tetramer	$[PNF_2]_4$	496.35					
pentamer	$[PNF_2]_5$	523.95					
Phosphorus:							
chloride difluoride	PF_2Cl	362.32(5)	4.520			0.8470	83.303
dichloride fluoride	$PFCl_2$	462.99(5)	4.99			1.2514	96.322
pentachloride	PCl_5	645.15 [22]					
trichloride	PCl_3	563.15		0.520	264	1.3906	88.000
trideuteride	PD_3	323.6 [20]					
trifluoride	PF_3	271.10(2)	4.325			0.4955	65.134
trihydride (phosphine)	PH_3	324.45(20)	6.54(4)	0.30 [21]	113.5 [21]	0.4697	51.593
Rhenium:							
oxide	Re_2O_7	942(3)	6.9(3)	1.45(3)	334	3.7558	142.08
oxide tetrachloride	ReO_4Cl	782(3)	4.91(25)	0.95(2)	362	3.6289	165.37
Selenium tetrafluoride	SeF_4	563 [12]					
Stannic chloride	$SnCl_4$	591.85(10)	3.744(5)	0.742(5)	351	2.7285	164.28
Silane:	SiH_4	270(1)	4.28(15)	0.309	104(10)	0.4972	65.623
bromide	SiH_3Br	454(10)	5.64(15)	0.627	177(10)	1.0650	83.599
chloride	SiH_3Cl	409(10)	4.81(15)	0.444	150(10)	1.0136	88.315
chloride trifluoride	SiF_3Cl	307.63(2)	3.465(5)			0.7964	92.258
dibromide	SiH_2Br_2	550(10)	5.30(15)	0.772	246(10)	1.6647	107.86
dichloride	SiH_2Cl_2	470(10)	4.53(15)	0.515	196(10)	1.4223	107.84
dichloride difluoride	SiF_2Cl_2	368.92(2)	3.500(5)			1.1341	109.55

Table 13.5 Critical parameters and van der Waals constants of inorganic substances *(continued)*

Substance	Formula	T_c, K	P_c, MPa	ρ_c, g/cm^3	V_c cm^3/mol	a, N·m^4/mol^2	b, cm^3/mol
diiode	SiH_2I_2	660(10)	6.68(15)	1.224	232(10)	1.8996	102.57
fluoride	$SiFCl_3$	438.41(2)	3.600(5)			1.5569	126.56
iodide	SiH_3I	515(10)	6.94(15)	0.988	160(10)	1.1144	77.112
tetrabromide	$SiBr_4$	656(10)	4.18(15)	0.872	398(10)	2.9989	162.91
tetrachloride	$SiCl_4$	506(10)	3.75(15)	0.584	291(10)	1.9916	140.26
tetrafluoride	SiF_4	259.00(2)	3.714(5)			0.5266	72.462
tetraiodide	SiI_4	850(10)	4.26(15)	1.417	378(10)	4.9393	207.08
tribromide	$SiHBr_3$	610(10)	4.70(15)	0.768	350(10)	2.3081	134.84
trichloride	$SiHCl_3$	495(10)	4.17(15)	0.533	254(10)	1.7117	123.23
triiodide	$SiHI_3$	760(10)	5.93(15)	1.326	309(10)	2.8417	133.25
Sulphur:							
chloride pentafluoride	SF_5Cl	117.7(2)					
dioxide	SO_2	430.65(20)	7.88(1)	0.524(5)	122	0.6821	56.774
fluoronitride	$NF_2{\cdot}SF_5$	443.35(30)	3.3			1.7143	137.79
hexafluoride	SF_6	318.70(1)	3.759	0.732	199	0.7879	88.107
tetrafluoride	SF_4	364 [20]					
trifluoronitride (trifluoramine sulfide)	NSF_3	385	6.9	0.615	167.5	0.6274	58.070
trioxide	SO_3	491.15(50)	8.21(5)	0.633	126	0.8571	62.192
Tantalum:							
pentabromide	$TaBr_5$	973(5)		1.26(3)	461	4.1956	153.67
pentachloride	$TaCl_5$	767(3)	4.3(5)	0.89(3)	401	3.9376	182.95
Titanium tetrachloride	$TiCl_4$	638(3)	4.7 [20]	0.57(2)	336	2.5468	142.25
Tungsten:							
hexachloride	WCl_6	923(3)	4.9(5)	0.94(3)	420	5.0040	193.20
hexafluoride	WF_6	444(5)	4.4(4)	1.28(3)	233	1.3195	105.90
oxide tetrachloride	$WOCl_4$	782(4)	5.3(3)	1.01	338	3.3847	154.24
oxide trichloride	$WOCl_3$	637(3)		0.60(3)	289	1.7220	96.333
Uranium hexafluoride	UF_6	504(1)	4.59(6)	1.39(2)	250 [20]	1.6139	114.11
Vanadium oxytrichloride	$VOCl_3$	636 [20]		0.60 [20]	290	1.7252	96.667
Zirconium:							
bromide	$ZrBr_4$	805(6)	4.3(5)	0.97(10)	420	4.3374	192.01
chloride	$ZrCl_4$	778(2)	5.91(15)	0.73(2)	319	2.9881	136.87
iodide	ZrI_4	959(6)	4.1(5)	1.13(10)	530	6.4560	239.90

Table 13.6 Critical parameters and van der Waals constants of organic substances

Substance	Formula	T_c, K	P_c, MPa	ρ_c, g/cm^3	V_c, cm^3/mol	a, N·m^4/mol^2	b, cm^3/mol
			Hydrocarbons [23]				
Amylbenzene [21]	$C_{11}H_{16}$	678.9	2.65	0.284		5.0629	265.77
Benzene	C_6H_6	562.6 [24]	4.92 [24]	0.304 [24]	259	1.8744	118.73
1,3-Butadiene	C_4H_6	425	4.33	0.245	221	1.2175	102.08
n-Butane	C_4H_{10}	425.16	3.797	0.228	255	1.3884	116.38
1-Butene	C_4H_8	419.6	4.02	0.234	240	1.2764	108.40
cis-2-Butene	C_4H_8	435.55	4.20	0.240	234	1.3156	107.64
trans-2-Butene	C_4H_8	428.61	4.10	0.236	238	1.3055	108.54
n-Butylbenzene	$C_{10}H_{14}$	660.4	2.887	0.270	497	4.4058	237.75
sec-Butylbenzene [21]	$C_{10}H_{14}$	645	2.72	0.264		4.4510	245.93
tert-Butylbenzene [21]	$C_{10}H_{14}$	639.6	2.72	0.274		4.3768	243.87
1-Butyne (ethylacetylene)	C_4H_6	463.6					
2-Butyne (dimethylacetylene)	C_4H_6	488.6					

Table 13.6 Critical parameters and van der Waals constants of organic substances *(continued)*

Substance	Formula	T_c, K	P_c, MPa	ρ_c, g/cm^3	V_c, cm^3/mol	a, N·m^4/mol^2	b, cm^3/mol
Cycloheptane [26]	C_7H_{14}	604.3	3.81		353	2.7952	164.84
Cyclohexane	C_6H_{12}	553.4	4.07	0.273	308	2.1926	141.19
Cyclohexene	C_6H_{10}	560.41	4.24 [21]	0.288 [21]		2.1624	137.50
Cyclooctane [26]	C_8H_{16}	647.2	3.55		410	3.4442	189.66
Cyclopentane	C_5H_{10}	511.6	4.508	0.27	260	1.6932	117.92
Cyclopentene [21]	C_5H_8	504	4.55	0.277	251	1.6282	115.13
Cyclopropane	C_3H_6	397.80	5.495			0.8398	75.236
Cymene (isopropyltoluene)	$C_{10}H_{14}$	658 [13]					
cis-Decahydronaphthalene [21]	$C_{10}H_{18}$	677	2.49	0.247		5.3621	282.26
trans-Decahydronaphthalene [21]	$C_{10}H_{18}$	664	2.61	0.254		4.9182	263.97
n-Decane [24]	$C_{10}H_{22}$	619.5	2.11	0.236	602	5.3104	305.48
Deuteromethane [9]	CD_4	189.2(1)	4.66(1)	0.205(2)	97.7	0.22397	42.1858
1,2-Diethylbenzene	$C_{10}H_{14}$	662.8	2.96	0.279		4.3299	232.81
1,3-Diethylbenzene [21]	$C_{10}H_{14}$	657.1	2.92	0.287		4.5035	243.95
1,4-Diethylbenzene	$C_{10}H_{14}$	657.88	2.803	0.281 [21]		4.5033	243.95
2,2-Dimethylbutane	C_6H_{14}	489.35	3.08	0.240	359	2.2671	165.10
2,3-Dimethylbutane	C_6H_{14}	499.93	3.127	0.241	358	2.3309	166.16
1,1-Dimethylcyclopentane [24]	C_7H_{14}	550.15	3.5	0.28		2.4889	161.22
cis-1,2-Dimethylcyclopentane [24]	C_7H_{14}	565.15	3.4	0.27		2.7037	170.48
trans-1,2-Dimethylcyclopentane [24]	C_7H_{14}	555.15	3.4	0.27		2.6089	167.47
cis-1,3-Dimethylcyclopentane [24]	C_7H_{14}	555.15	3.4	0.27		2.6089	167.47
trans-1,3-Dimethylcyclopentane [24]	C_7H_{14}	555.15	3.5	0.28		2.5343	162.68
2,2-Dimethylhexane	C_8H_{18}	549.80	2.529	0.239	478	3.4856	225.92
2,3-Dimethylhexane	C_8H_{18}	563.42	2.628	0.244	468	3.5221	222.77
2,4-Dimethylhexane	C_8H_{18}	553.45	2.556	0.242	472	3.4942	224.99
2,5-Dimethylhexane	C_8H_{18}	549.99	2.628	0.237	482	3.4088	220.87
3,3-Dimethylhexane	C_8H_{18}	561.95	2.654	0.258	443	3.4703	220.07
3,4-Dimethylhexane	C_8H_{18}	568.78	2.692	0.245	460	3.5043	219.56
2,2-Dimethylpentane	C_7H_{16}	520.44	2.773	0.241	416	2.8486	195.03
2,3-Dimethylpentane	C_7H_{16}	537.29	2.908	0.255	393	2.89496	192.01
2,4-Dimethylpentane	C_7H_{16}	519.73	2.737	0.240	418	2.8783	197.36
3,3-Dimethylpentane	C_7H_{16}	536.34	2.946	0.242	414	2.8480	189.23
2,2-Dimethylpropane (neopentane)	C_5H_{12}	433.75	3.199	0.238	303	1.7152	140.92
Diphenyl (biphenyl)	$C_{12}H_{10}$	789	3.8	0.307	502	4.7150	212.96
n-Dodecane [24]	$C_{12}H_{26}$	659.15	1.81	0.237	718	6.9856	377.69
Eicosane	$C_{20}H_{42}$	775	1.1	0.24	1176	15.715	722.62
Ethane [9]	C_2H_6	305.45(10)	4.87(1)	0.203(5)	148	0.5571	64.997
Ethene (ethylene) [9]	C_2H_4	282.36	5.066(2)	0.227(5)	124	0.4589	57.921
Ethylbenzene	C_8H_{10}	617.09	3.609	0.284	374	3.0769	177.69
Ethylcyclopentane	C_7H_{14}	569.45	3.398	0.262	375	2.7835	174.19
3-Ethylhexane	C_8H_{18}	565.42	2.608	0.251	455	3.5747	225.30
3-Ethylpentane	C_7H_{16}	540.57	2.891	0.241	416	2.9479	194.34
o-Ethyltoluene [21]	C_9H_{12}	653	3.1	0.28	430 [13]	3.9587	216.05
m-Ethyltoluene [21]	C_9H_{12}	636	3.1	0.28	430 [13]	3.7553	210.43
p-Ethyltoluene [21]	C_9H_{12}	636	3.1	0.28	430 [13]	3.7553	210.43
Ethyne (acetylene)	C_2H_2	308.33(1)	6.24(1)	0.232(1)	112	0.4442	51.338
n-Heptadecane [24]	$C_{17}H_{36}$	735.15	1.32	0.24	1001	11.965	580.01
n-Heptane	C_7H_{16}	540.2	2.735	0.232	432	3.1107	205.21
1-Heptene	C_7H_{14}	537.23					
n-Hexadecane [24]	$C_{16}H_{34}$	725.15	1.42	0.24	924	10.810	531.25
1,5-Hexadiene	C_6H_{10}	507					
Hexamethylbenzene	$C_{12}H_{18}$	767					
n-Hexane	C_6H_{14}	507.4	2.97	0.233	370	2.5289	177.62

Table 13.6 Critical parameters and van der Waals constants of organic substances *(continued)*

Substance	Formula	T_c, K	P_c, MPa	ρ_c, g/cm^3	V_c, cm^3/mol	a, N·m^4/mol^2	b, cm^3/mol
1-Hexene	C_6H_{12}	503.98					
Isobutylbenzene	$C_{10}H_{14}$	640	3.1	0.274 [21]		3.8028	211.75
Isopropylbenzene	C_9H_{12}	631.15	3.1	0.28		3.6984	208.82
Methane [9]	CH_4	190.60(5)	4.63(1)	0.160(2)	100	0.2288	42.777
2-Methylbutane (isopentane)	C_5H_{12}	460.39	3.381	0.236	306	1.8281	141.50
2-Methyl-1-butene	C_5H_{10}	465	3.4			1.8304	140.27
2-Methyl-2-butene	C_5H_{10}	470	3.4			1.8699	141.78
Methylcyclohexane	C_7H_{14}	572.12	3.471	0.267	368	2.7498	171.28
Methylcyclopentane	C_6H_{12}	532.73	3.785	0.264	319	2.1869	146.29
2-Methyl-3-ethylpentane	C_8H_{18}	567.02	2.700	0.258	443	3.4722	218.22
3-Methyl-3-ethylpentane	C_8H_{18}	576.51	2.808	0.251	455	3.4521	213.39
2-Methylheptane	C_8H_{18}	595.57	2.485	0.234	488	4.1634	249.12
3-Methylheptane	C_8H_{18}	563.60	2.546	0.246	464	3.6379	230.03
4-Methylheptane	C_8H_{18}	561.67	2.542	0.240	467	3.6188	229.60
2-Methylhexane	C_7H_{16}	530.31	3.038	0.238	421	2.6998	181.43
3-Methylhexane	C_7H_{16}	535.19	2.814	0.248	404	2.9685	197.67
1-Methylnaphthalene	$C_{11}H_{10}$	772					
2-Methylnaphthalene	$C_{11}H_{10}$	761					
2-Methylpentane	C_6H_{14}	497.45	3.010	0.235	367	2.3972	171.73
3-Methylpentane	C_6H_{14}	504.4	3.124	0.235	367	2.3751	168.80
2-Methylpropane (isobutane)	C_4H_{10}	408.13	3.648	0.221	263	1.3317	116.28
2-Methylpropene (isobutylene)	C_4H_8	417.89	4.000	0.235	239	1.2731	108.56
Naphthalene	$C_{10}H_8$	748.4	4.051	0.31	410	4.0321	192.00
n-Nonadecane [21]	$C_{19}H_{40}$	760	1.2	0.24	1118	13.853	649.58
n-Nonane [21]	C_9H_{20}	595.4	2.316	0.236	534	4.4632	267.14
n-Octadecane [21]	$C_{18}H_{38}$	753.2	1.3	0.24	1059	12.560	594.25
n-Octane	C_8H_{18}	568.76	2.487	0.232	492	3.7940	237.72
1-Octene	C_8H_{16}	566.6					
n-Pentadecane [21]	$C_{15}H_{32}$	710.6	1.596	0.24	888	9.2274	462.75
n-Pentane	C_5H_{12}	469.9	3.369	0.237	304	1.9113	144.95
1-Pentene	C_5H_{10}	464.74	4.05			1.5541	119.17
cis-2-Pentene	C_5H_{10}	476	3.6			1.8114	135.61
trans-2-Pentene	C_5H_{10}	475	3.6			1.8038	135.33
1-Pentyne (proriylacetylene)	C_5H_8	493.4					
Propadiene [24]	C_3H_4	393.85	5.25			0.8619	77.984
Propane	C_3H_8	369.82	1.514	0.217	203	2.6347	253.89
Propene (prorylene)	C_3H_6	365.0	4.62	0.233	181	0.8409	82.098
n-Propylbenzene	C_9H_{12}	638.30	3.200	0.273	440	3.7132	207.31
Propyne (methylacetylene)	C_3H_4	402.38	5.526	0.245	164	0.8544	75.670
n-Tetradecane	$C_{14}H_{30}$	695	1.678	0.24	826	8.3949	430.45
1,2,3,4-Tetrahydronaphthalene [21]	$C_{10}H_{12}$	719	3.52	0.309		4.2876	212.52
1,2,3,4-Tetramethylbenzene [21]	$C_{10}H_{14}$	700.1	3.27	0.308		4.3673	222.31
1,2,3,5-Tetramethylbenzene [21]	$C_{10}H_{14}$	686.8	3.21	0.308		4.2824	222.21
1,2,4,5-Tetramethylbenzene	$C_{10}H_{14}$	675	2.9	0.306 [21]		4.5219	238.73
2,2,3,3-Tetramethylbutane	C_8H_{18}	567.8	2.87	0.248	461	3.2788	205.78
2,2,3,3-Tetramethylheptane [25]	$C_{11}H_{24}$	607.6	2.741			3.9280	230.38
2,2,3,4-Tetramethylheptane [25]	$C_{11}H_{24}$	592.6	2.602			3.9358	236.68
2,2,4,4-Tetramethylheptane [25]	$C_{11}H_{24}$	574.6	2.485			3.8754	240.35
2,3,3,4-Tetramethylheptane [25]	$C_{11}H_{24}$	607.5	2.716			3.9634	232.49
2,2,3,3-Tetramethylhexane [25]	$C_{10}H_{22}$	623	2.510			4.5098	257.97
2,2,5,5-Tetramethylhexane [25]	$C_{10}H_{22}$	581.5	2.186			4.5119	276.50
Toluene (methylbenzene)	C_7H_8	591.72	4.109	0.292	316	2.4851	149.67
n-Tridecane [21]	$C_{13}H_{28}$	677.2	1.778	0.24	767	7.5208	395.77

Table 13.6 Critical parameters and van der Waals constants of organic substances *(continued)*

Substance	Formula	T_c, K	P_c, MPa	ρ_c, g/cm^3	V_c, cm^3/mol	a, N·m^4/mol^2	b, cm^3/mol
1,2,3-Trimethylbenzene	C_9H_{12}	664.45	3.454	0.28 [21]	430 [13]	3.7274	199.91
1,2,4-Trimethylbenzene	C_9H_{12}	649.05	3.232	0.28 [21]	430 [13]	3.8008	208.68
1,3,5-Trimethylbenzene	C_9H_{12}	637.28	3.121	0.28 [21]	430 [13]	3.7951	212.22
2,2,3-Trimethylbenzene	C_7H_{16}	531.11	2.954	0.252	398	2.7851	186.87
3,3,5-Trimethylheptane [25]	$C_{10}H_{22}$	609.5	2.310			4.6894	274.18
2,2,5-Trimethylhexane	C_9H_{20}	568.0					
2,2,3-Trimethylpentane	C_8H_{18}	563.43	2.730	0.262	436	3.3915	214.51
2,2,4-Trimethylpentane (isooctane)	C_8H_{18}	543.89	2.568	0.244	408	3.3599	220.14
2,3,3-Trimethylpentane	C_8H_{18}	573.49	2.820	0.251	455	3.4013	211.36
2,3,4-Trimethylpentane	C_8H_{18}	566.34	2.730	0.248	461	3.4266	215.62
o-Terphenyl	$C_{18}H_{19}$	891.0	3.901	0.306	769	5.9348	237.37
m-Terphenyl	$C_{18}H_{19}$	924.8	3.506	0.300	784	7.1142	274.14
p-Terphenyl	$C_{18}H_{19}$	926.0	3.324	0.302	779	7.5241	289.56
n-Undecane [21]	$C_{11}H_{24}$	642.6	1.958	0.237	659	6.1515	341.14
o-Xylene [24]	C_8H_{10}	632.15	3.6	0.28	379	3.1948	180.10
m-Xylene [24]	C_8H_{10}	616.97	3.541	0.282	370	3.1346	181.10
p-Xylene [24]	C_8H_{10}	616.2	3.511	0.280	379	3.1539	182.40
Oxygen-containing compounds [23]							
Acetaldehyde [9]	C_2H_4O	461(5)					
Acetone (2-propanone)	C_3H_6O	508.2	4.70	0.278	209	1.6020	112.34
Acetic (ethanoic) acid [9]	$C_2H_4O_2$	594.75(10)	5.786(5)	0.351(50)	171	1.7829	106.83
Acetic anhydrides	$C_4H_6O_3$	569	4.68			2.0169	126.32
n-Amyl alhohol (1-pentanol)	$C_5H_{12}O$	586		0.270	326	1.7869	108.67
Benzaldehyde	C_7H_6O	625	2.18	0.330 [27]		5.2291	298.16
Biphenyl (diphenyl) ether [24]	$C_{12}H_{10}O$	805.15	3.57				
n-Butyl acetate	$C_6H_{12}O_2$	579					
n-Butyl alcohol(1-butanol)	$C_4H_{10}O$	562.93	4.413	0.270	274	2.0942	132.58
sec-Butyl alcohol (2-butanol)	$C_4H_{10}O$	535.95	4.194	0.276	268	1.9974	132.81
tert-Butyl alcohol (2-methyl-2-propanol)	$C_4H_{10}O$	506.2	3.972	0.270	275	1.8813	132.45
n-Butyric (butanoic) acid	$C_4H_8O_2$	628	5.27	0.304	290	2.1828	123.87
o-Cresol	C_7H_8O	697.6	5.01	0.375 [21]		2.8353	144.84
m-Cresol	C_7H_8O	705.8	4.56	0.357 [21]	310	3.1861	160.87
p-Cresol	C_7H_8O	704.6	5.15	0.347 [21]		2.8127	142.26
Cyclohexanol	$C_6H_{12}O_2$	625	3.75			3.0386	173.25
Cyclohexanone	$C_6H_{10}O$	629	3.85			2.9966	169.77
n-Decyl alcohol (1-decanol)	$C_{10}H_{22}O$	700		0.264	600	3.9285	200.00
Deuteroacetic acid [9]	$C_2H_3DO_2$	594.1(1)					
Deuteroethyl alcohol [9]	C_2H_5DO	514.9(1)					
1,2-Dimethoxyethane	$C_4H_{10}O_2$	536	3.87	0.333	271	2.1646	143.91
Dimethoxymethane (methylal)	$C_3H_8O_2$	497					
Dimethyl oxalate	$C_4H_6O_4$	628	3.98				
1,1-Diethoxyethane (acetal)	$C_6H_{14}O_2$	527					
1,4-Dioxane	$C_4H_8O_2$	587	5.48	0.370	238	1.8331	111.29
Ethoxybenzene (phenetole)	$C_8H_{10}O$	647	3.42			3.5645	196.33
Ethyl acetate	$C_4H_8O_2$	523.2	3.83	0.308	286	2.0843	141.96
Ethyl alcohol (ethanol) [9]	C_2H_6O	516(2)	6.4(1)	0.276(5)	167	1.2164	84.006
Ethyl butyrate	$C_6H_{12}O_2$	566	3.0	0.276	410	3.0733	193.51
Ethyl (diethyl) ether	$C_4H_{10}O$	466.70	3.638	0.265	280	1.7462	133.33
Ethyl formate	$C_3H_6O_2$	508.4	4.74	0.323	229	1.5895	111.42
Ethyl isobutyrate	$C_6H_{12}O_2$	553	3.0	0.28	410	2.9938	189.06
Ethyl isovalerate	$C_7H_{14}O_2$	588					

Table 13.6 Critical parameters and van der Waals constants of organic substances *(continued)*

Substance	Formula	T_c, K	P_c, MPa	ρ_c, g/cm^3	V_c, cm^3/mol	a, N·m^4/mol^2	b, cm^3/mol
Ethyl nonanoate	$C_{11}H_{22}O_2$	674					
Ethyl octanoate	$C_{10}H_{20}O_2$	659					
o-Ethylphenol	$C_8H_{10}O$	703					
m-Ethylphenol	$C_8H_{10}O$	716.4					
p-Ethylphenol	$C_8H_{10}O$	716.4					
Ethyl propionate	$C_5H_{10}O_2$	546	3.362	0.296	345	2.5859	168.78
Ethyl propyl ether	$C_5H_{12}O$	500.6 [21]	3.25	0.36	240	2.2469	159.95
Ethyl valerate	$C_7H_{14}O_2$	570					
Ethylene oxide (epoxyethane) [9]	C_2H_4O	468(1)	7.19(1)	0.32(1)	138	0.8878	67.607
Furan	C_4H_4O	490.2	5.50	0.312	218	1.2737	92.592
n-Heptyl alcohol (1-heptanol)	$C_7H_{16}O$	633		0.267	435	2.3980	135.00
n-Hexyl alcohol (1-hexanol)	$C_6H_{14}O$	610		0.268	381	2.1739	127.00
Isoamyl acetate	$C_7H_{14}O_2$	509					
Isobutyl acetate [21]	$C_6H_{12}O_2$	561	3.1	0.281		2.9218	185.61
Isobutyl alcohol (2-methyl-1-propanol)	$C_4H_{10}O$	547.73	4.295	0.272	273	2.0369	132.53
Isobutyl butyrate	$C_8H_{16}O_2$	611					
Isobutyl formate	$C_5H_{10}O_2$	551	3.88	0.29	350	2.2815	147.56
Isobutyl isobutyrate	$C_8H_{16}O_2$	602					
Isobutyl propionate	$C_7H_{14}O_2$	592					
Isobutyl valerate	$C_9H_{18}O_2$	621					
Isobutyric acid	$C_4H_8O_2$	609	4.05	0.302	292	2.6686	156.16
Isovaleric acid	$C_4H_{10}O_2$	634					
Isopentyl butyrate	$C_9H_{18}O_2$	619					
Isopentyl formate	$C_6H_{12}O_2$	578					
Isopentyl propionate	$C_8H_{16}O_2$	611					
Isopropyl ether	$C_5H_{12}O$	500	2.88	0.265	386	2.5336	180.57
Methoxybenzene (anisole)	C_7H_8O	641	4.18			2.8703	159.57
Methyl acetate	$C_3H_6O_2$	506.8	4.69	0.325	228	1.5966	112.27
Methyl alcohol (methanol) [9]	CH_4O	513.15(30)	7.95(5)	0.272(5)	118	0.9654	67.047
2-Methyl-2-butanol	$C_5H_{12}O_2$	545					
3-Methyl-1-butanol	$C_5H_{12}O_2$	579.40					
3-Methyl-2-butanone	$C_5H_{10}O$	553.4	3.85	0.278	310	2.3195	149.37
Methyl butyrate	$C_5H_{10}O_2$	554.4	3.48	0.300	340	2.5791	165.78
Methyl (dimethyl) ether	C_2H_6O	400.05(10)	5.37(5)	0.238(15)	193	0.8691	77.418
Methyl ethyl ether	C_3H_8O	437.8	4.40	0.272	221	1.27107	103.46
Methyl ethyl ketone (2-butanone)	C_4H_8O	535.6	4.15	0.270	267	2.0138	133.99
Methyl formate [9]	$C_2H_4O_2$	487.15(10)	5.998(5)	0.349(5)	172	1.1538	84.400
2-Methylfuran	C_5H_6O	527	4.72	0.333	247	1.7153	115.99
Methyl isobutyrate	$C_5H_{10}O_2$	540.8	3.44	0.301	339	2.4830	163.62
Methyl laurate (dodecanoate)	$C_{13}H_{26}O_2$	712					
4-Methyl-2-pentanone	$C_6H_{12}O$	571	3.27			2.9052	181.32
Methyl prorionate	$C_4H_8O_2$	530.6	4.004	0.312	282	2.0503	137.71
2-Methyltetrahydrofuran	$C_5H_{10}O$	537	3.76	0.322	267	2.2371	148.46
Methyl valerate [21]	$C_5H_{12}O_2$	567	3.2	0.279		2.8914	181.73
n-Nonyl alcohol (1-nonapol)	$C_9H_{20}O$	677		0.264	546	3.4575	182.00
n-Octyl alcohol (1-octanol)	$C_8H_{18}O$	658		0.266	490	3.0158	163.33
sec-Octyl alcohol (2-octanol)	$C_8H_{18}O$	637					
Paraldehyde (2,4,6-Trimethyl-*sec*-trioxane)	$C_6H_{12}O_3$	563					
2-Pentanone (methyl *n*-propyl ketone)	$C_5H_{10}O$	564	3.89	0.286	301	2.3842	150.64
3-Pentanone (diethyl ketone)	$C_5H_{10}O$	561.0	3.74	0.256	336	2.4548	155.93
n-Pentyl formate (*n*-amyl formate)	$C_6H_{12}O_2$	576					

Table 13.6 Critical parameters and van der Waals constants of organic substances *(continued)*

Substance	Formula	T_c, K	P_c, MPa	ρ_c, g/cm^3	V_c, cm^3/mol	a, N·m^4/mol^2	b, cm^3/mol
Phenol	C_6H_6O	694.2	6.13	0.401 [21]		2.2926	117.69
1-Propanol (*n*-propyl alcohol)	C_3H_8O	536.71	5.170	0.275	218	1.6250	107.90
2-Propanol (isopropyl alcohol)	C_3H_8O	508.31	4.764	0.273	220	1.5816	110.88
Propionic (propanoic) acid	$C_3H_6O_2$	612	5.37	0.32	230	2.0339	118.43
n-Propyl acetate	$C_5H_{10}O_2$	549.4	3.33	0.296	345	2.6405	171.28
n-Propyl butyrate	$C_7H_{14}O_2$	600					
Propyl formate	$C_4H_8O_2$	538.0	4.06	0.309	285	2.0774	137.61
n-Propyl isobutyrate	$C_7H_{14}O_2$	589					
n-Propyl isovalerate	$C_8H_{16}O_2$	609					
n-Propyl propionate	$C_6H_{12}O_2$	578					
Propylene oxide (1,2-epoxypropane)	C_3H_6O	482.2	4.92	0.312	186	1.3769	101.76
Tetrahydrofuran	C_4H_8O	540.2	5.19	0.322	224	1.6404	108.22
n-Valeric (pentanoic) acid	$C_4H_{10}O_2$	651					
Vinyl ethyl ether	C_4H_8O	475	4.07			1.6154	121.19
2,3-Xylenol	$C_8H_{10}O$	722.8					
2,4-Xylenol	$C_8H_{10}O$	707.6					
2,5-Xylenol	$C_8H_{10}O$	723.0					
2,6-Xylenol	$C_8H_{10}O$	701.0					
3,4-Xylenol	$C_8H_{10}O$	729.8					
3,5-Xulenol	$C_8H_{10}O$	715.6					
Halogen-containing compounds [23]							
Benzotrifluoride [21]	$C_7H_5F_3$	562.6	3.56	0.427	342	2.5954	164.40
Bromobenzene	C_6H_5Br	670	4.52	0.485	324	2.8968	154.08
Bromochlorodifluoromethane [28]	CF_2ClBr	428	4.310	0.741 [34]	245.7 [12]	1.2396	103.22
Bromochlorofluoromethane [34]	$CHFClBr$	507.5	5.35	0.707			
Bromochloromethane [34]	CH_2ClBr	555.5	6.32	0.669			
Bromodichlorofluoromethane [9]	$CFCl_2Br$	520					
Bromodichloromethane [34]	$CHCl_2Br$	585.43	5.60	0.712			
Bromodifluoromethane [9,34]	CHF_2Br	409	5.18	0.750			
Bromofluoromethane [34]	CH_2FBr	468	6.11	0.674			
Bromopentafluorobenzene	C_6F_5Br	670	4.52			2.8968	154.08
Bromotrichloromethane [34]	CCl_3Br	602.46	4.693	0.697			
Bromotrifluoromethane	CF_3Br	340	3.97	0.76	200	0.8488	88.960
1-Chloro-1, 1-difluoroethane [9]	$C_2H_3F_2Cl$	410.25(50)	4.12(7)	0.435(10)	231	1.1902	103.38
2-Chloro-1, 1-difluoroethylene [9]	C_2HF_2Cl	400.6(5)	4.46(7)	0.499(10)	197	1.0497	93.382
Chlorobenzene	C_6H_5Cl	632.4	4.52	0.365	308	2.5808	145.43
Chlorodifluoromethane [9]	CHF_2Cl	369.55(50)	4.91(5)	0.525(10)	165	0.8104	78.151
Chloroethane (ethyl chloride) [9]	C_2H_5Cl	460.35(50)	5.27(5)	0.331	195	1.1729	90.800
2-Chloro-1-fluoroethane [9]	C_2H_4FCl	520(15)	5.37(30)	0.40(2)	207	1.4684	100.63
Chlorofluoromethane [34]	CH_2FCl	424.83	6.00	0.443			
Chloroform (trichloromethane) [9]	$CHCl_3$	536.55(50)	5.47(20)	0.50(3)	239	1.5344	101.91
Chloromethane [9]	CH_3Cl	416.25(10)	6.60(1)	0.353(10)	143	0.76660	65.581
Chloropentafluoroacetone	C_3F_5ClO	410.6	2.88			1.7086	144.29
Chloropentafluorobenzene	C_6F_5Cl	571.0	3.22			2.9509	184.17
Chloropentafluoroethane [9]	C_2F_5Cl	353.1(1)	3.157(51)	0.613(4)	252	1.1516	116.22
Chloropropane (*n*-propyl chloride)	C_3H_7Cl	503	4.58			1.6110	114.14
3-Chloropropene (allyl chloride)	C_3H_5Cl	514					
Chlorotrifluoroethylene [9]	C_2F_3Cl	379(1)	4.05(10)	0.55(1)	212	1.0335	97.181
Chlorotrifluoromethane [9]	CF_3Cl	301.95(20)	3.95(5)	0.58(1)	180	0.6728	76.410
Deuterochloroform [9]	$CDCl_3$	535.9(1)					
o-Dibromobenzene [21]	$C_6H_4Br_2$	761.75	4.22	0.645		4.0144	187.81
Dibromochloromethane [34]	$CHClBr_2$	654.4	5.75	0.83			

Table 13.6 Critical parameters and van der Waals constants of organic substances *(continued)*

Substance	Formula	T_c, K	P_c, MPa	ρ_c, g/cm^3	V_c, cm^3/mol	a, N·m^4/mol^2	b, cm^3/mol
Dibromochlorofluoromethane [9]	$CFClBr_2$	570					
Dibromodichloromethane [34]	CCl_2Br_2	668.21	4.84	0.793			
1,1-Dibromoethylene [9]	$C_2H_4Br_2$	582.95(1.5)	7.15(8)			1.3854	84.689
Dibromodifluoromethane [9]	CF_2Br_2	464	4.22	0.866 [34]		1.4867	114.18
Dibromofluoromethane [34]	$CHFBr_2$	543.5	5.49	0.880			
Dibromomethane (methylene bromide) [28]	CH_2Br_2	583	7.3	0.525	165	1.3778	84.219
o-Dichlorobenzene [21]	$C_6H_4Cl_2$	697 [28]	4.10	0.408	360	3.4522	176.51
m-Dichlorobenzene [21]	$C_6H_4Cl_2$	684	3.88	0.415		3.5156	183.17
p-Dichlorobenzene [21]	$C_6H_4Cl_2$	684.75	3.90	0.395		3.5050	182.42
Dichlorodifluoromethane [9]	CF_2Cl_2	384.65(20)	4.01(1)	0.555(5)	218	1.0753	99.626
1,1-Dichloroethane [9]	$C_2H_4Cl_2$	523(5)	5.1(3)	0.308	321	1.5745	107.28
1,2-Dichloroethane [9]	$C_2H_4Cl_2$	561(2)	5.4(3)	0.44(3)	225	1.7091	108.56
1,1-Dichloroethylene [12]	$C_2H_2Cl_2$	494.15	5.23		220.3	1.3619	98.223
cis-1,2-Dichloroethylene [12]	$C_2H_2Cl_2$	544.15	5.86		220.3	1.4743	96.559
trans-1,2-Dichloroethylene [9]	$C_2H_2Cl_2$	516.5	5.52		220.3 [12]	1.4088	97.202
Dichlorofluoromethane [9]	$CHFCl_2$	451.65(50)	5.17(10)	0.522(10)	197	1.1512	90.831
Dichloromethane [9]	CH_2Cl_2	510(2)	6.1(1)	0.472(25)	180	1.2477	87.181
Dichlorotetrafluoroethane [24]	$C_2F_4Cl_2$	418.95	3.38	0.583		1.5125	128.65
1,1-Dichloro-1,2,2,2-tetrafluoroethane [9]	$C_2F_4Cl_2$	418.65(5)	3.30(7)	0.582(10)	294	1.5474	131.72
1,2-Dichloro-1,1,2,2-tetrafluoroethane	$C_2F_4Cl_2$	418.8	3.26	0.582	294	1.5677	133.40
Difluorodiiodomethane [9]	CF_2I_2	477					
1,1-Difluoroethane [9]	$C_2H_4F_2$	386.65(70)	4.49(7)	0.365(10)	181	0.9713	89.519
1,1-Difluoroethylene [9]	$C_2H_2F_2$	302.85(2)	4.46(1)	0.414(2)	155	0.5993	70.515
Difluoroiodomethane [9]	CHF_2I	459					
Difluoromethane [9]	CH_2F_2	351.55(20)	5.83(5)	0.430(50)	120	0.6182	62.664
Diiodomethane [34]	CH_2I_2	605.7	6.47	0.840			
Ethyl bromide [9]	C_2H_5Br	503.85(1.5)	6.23(5)	0.507	215	1.1881	84.029
Fluorobenzene	C_6H_5F	560.09	4.551	0.269	357	2.0104	127.91
Fluoroethane (ethyl fluoride) [9]	C_2H_5F	375.31(50)	4.72(5)			0.8700	82.605
Fluoroethene (vinyl fluoride)	C_2H_3F	327.8	5.24	0.320	144	0.5982	65.031
Fluoromethane [9]	CH_3F	317.75(50)	5.88(5)	0.300(5)	113	0.5010	56.190
Hexafluoroacetone	C_3F_6O	357.2	2.84			1.3115	130.84
Hexafluorodichloropropane [28]	$C_3F_6Cl_2$	449	2.99	0.62	356	2.0077	159.35
Hexafluoroethane	C_2F_6	292.85(50)	3.3	0.617	224	0.7479	91.0196
Iodobenzene	C_6H_5I	721	4.52	0.581	351	3.3546	165.81
Iodomethane (methyl iodide) [9]	CH_3I	528		0.83	171	0.8445	57.000
Iodotrifluoromethane [9]	CF_3I	359					
Methyl bromide [12]	CH_3Br	464	6.94		206	0.9045	69.475
Methyl bromide triiodide [9]	$CBrI_3$	750					
1H-Pentadecafluoroheptane	C_7HF_{15}	495.8					
Pentafluorobenzene	C_6HF_5	532.0	3.52			2.3475	157.25
1,1,2-Pentafluoropropane	$C_3H_3F_5$	380.11	3.137	0.491	273	1.3431	125.92
Perfluorobenzene	C_6F_6	516.72	3.304	0.493 [29]		2.3565	162.52
Perfluoro-*n*-butane	C_4F_{10}	386.4	2.323	0.629	378	1.8740	172.84
Perfluoro-2-butyltetrahydrofuran	$C_8F_{16}O$	500.2	1.607	0.707	588	4.5404	323.48
Perfluorocyclobutane	C_4F_8	388.37	2.777	0.616	325	1.5838	145.32
Perfluorocyclohexane	C_6F_{12}	457.2	2.4			2.5067	195.39
Perfluorocyclohexene	C_6F_{10}	461.8					
Perfluoro-*n*-decane	$C_{10}F_{22}$	542	1.45			5.9125	388.75
Perfluoro-*n*-heptane	C_7F_{16}	474.8	1.62	0.584	664	4.0552	304.36

Table 13.6 Critical parameters and van der Waals constants of organic substances *(continued)*

Substance	Formula	T_c, K	P_c, MPa	ρ_c, g/cm^3	V_c, cm^3/mol	a, N·m^4/mol^2	b, cm^3/mol
Perfluoro-1-heptene	C_7F_{14}	478.2					
Perfluoro-n-hexane	C_6F_{14}	447.6	1.905			3.0671	244.19
Perfluoro-1-hexene	C_6F_{12}	454.4					
Perfluoromethylcyclohexane	C_6F_{12}	488.6	2.3			2.9872	217.88
Perfluoronaphthalene	$C_{10}F_8$	673.0					
Perfluoro-n-nonane	C_9F_{20}	524	1.56			5.1316	348.99
Perfluoro-n-octane	C_8F_{18}	502	1.66			4.4225	313.95
Perfluoro-n-pentane	C_5F_{12}	422	2.04		475 [30]	2.5500	215.34
Perfluoro-n-propane	C_3F_8	345.0	2.680	0.628	299	1.2952	133.78
Perfluoropropylene	C_3F_6	358	3.36	0.6 [31]		1.1110	110.60
Phosgene (carbon oxychloride) [9]	CCl_2O	455(1)	5.67(10)	0.52(1)	190	1.0640	83.335
Tetrabromomethane (perbromomethane)	CBr_4	715 [9]					
Tetrachloroacetylene [9]	$C_2H_2Cl_4$	642(10)	3.95(30)	0.503(15)	334	3.0417	168.84
1,1,2,2-Tetrachloro-1, 2-difluoroethane [34]	$C_2F_2Cl_4$	551	3.33	0.569			
Tetrachloroethylene [12]	C_2Cl_4	613.2	4.48		289	2.4483	144.29
Tetrachloromethane [9]	CCl_4	556.25(50)	4.56(2)	0.558(10)	276	1.9789	126.78
Tetrafluoroethylene (perfluoroethane) [9]	C_2F_4	306(1)	3.95(10)	0.58(1)	172	0.6910	80.475
Tetrafluoromethane	CF_4	227.7	3.745	0.647	136	0.4044	63.291
Tetraiodomethane	CI_4	740 [9]					
Tribromochloromethane [34]	$CClBr_3$	674.67	5.02	0.968			
Tribromofluoromethane [9]	$CFBr_3$	591					
Tribromomethane [34]	$CHBr_3$	684.94	5.91	0.989			
Trichloroacetic acid	$C_2HF_3O_2$	491.3	3.258	0.559	204	2.1608	156.74
1,2,4-Trichlorobenzene [28]	$C_6H_3Cl_3$	735.0	3.98	0.472	384	3.9563	191.82
1,1,1-Trichloroethane [9]	$C_2H_3Cl_3$	550(10)	4.46(30)	0.464(15)	288	1.9787	128.21
Trichloroethylene [12]	C_2HCl_3	571.15	4.89		256.1	1.9438	121.28
Trichloroiodomethane [9]	CCl_3I	607					
Trichlorofluoromethane [9]	$CFCl_3$	471.15(50)	4.38(5)	0.554(10)	248	1.4789	111.86
1,2,2-Trichloro-1,1, 2-trifluoroethane [9]	$C_2F_3Cl_3$	487.25(50)	3.415(20)	0.576(10)	325	2.0276	148.29
1H-Tridecafluorohexane	C_6HF_{13}	471.8					
1,1,1-Trifluoroethane [9]	$C_2H_3F_3$	346.25(50)	3.76(7)	0.434(10)	193	0.9301	95.723
Trifluoromethane [9]	CHF_3	299.05(50)	4.86(10)	0.525(10)	133	0.5362	63.901
Trifluoropropylene [31]	$C_3H_3F_3$	376	3.92	0.455		1.0514	99.651
Triiodomethane [9]	CHI_3	690					
1H-Undecafluorocyclohexane	C_6HF_{11}	477.6					
1H-Undecafluoropentane	C_5HF_{11}	444.0					
Nitrogen-containing compounds [23]							
Acetonitrile (ethanenitrile)	C_2H_3N	548	4.80	0.237	173	1.8120	117.83
Aniline (phenylamine)	C_6H_7N	699	5.31	0.34	270	2.6837	136.82
Benzonitrile	C_7H_5N	690.4	4.22			3.2977	170.22
n-Butylamine	C_4H_9N	524	4.15			1.9275	131.08
Butanenitrile (butyronitrile)	C_4H_7N	582.2	3.79			2.6084	159.67
Decanenitrile (caprylonitrile)	$C_{10}H_{19}N$	622.0	3.25				
Deuteronitromethane [9]	CD_3NO_2	587.0					
Dimethylamine [9]	C_2H_7N	437.65(50)	5.31(5)			1.0520	85.664
N,N-Dimethylaniline	$C_8H_{11}N$	687	3.63			3.7944	196.82
N,N-Dimethyl-o-toluidine	$C_9H_{14}N$	668	3.12			4.1698	222.45
Di-n-propylamine	$C_6H_{15}N$	550	3.1	0.24	300	2.8085	181.97

Table 13.6 Critical parameters and van der Waals constants of organic substances *(continued)*

Substance	Formula	T_c, K	P_c, MPa	ρ_c, g/cm^3	V_c, cm^3/mol	a, N·m^4/mol^2	b, cm^3/mol
Diethylamine [9]	$C_4H_{11}N$	496.35(50)	3.71(5)			1.9373	139.09
Ethylamine [9]	C_2H_7N	456.35(50)	5.62(5)	0.243	185	1.0800	84.335
Ethylenediamine [9]	$C_2H_8N_2$	539(5)	6.3(3)	0.29(1)	206	1.6324	98.099
Isoxazole	C_3H_3NO	552.0					
Isoquinoline	C_9H_9N	803				3.4688	198.74
2,3-Lutidine (2,3-dimethylpyridine)	C_7H_9N	655.4					
2,4-Lutidine	C_7H_9N	647					
2,5-Lutidine	C_7H_9N	644.2					
2,6-Lutidine	C_7H_9N	623.8					
3,4-Lutidine	C_7H_9N	683.8					
3,5-Lutidine	C_7H_9N	667.2					
Methylamine [9]	CH_5N	430.05(50)	7.46(5)	0.223 [28]	254 [28]	0.7232	59.930
N-Methylaniline	C_7H_9N	701	5.20			2.7569	140.15
Methylhydrazine	CH_2N_2	567	8.035	0.170	271	1.1668	73.335
Nitromethane [9]	CH_3NO_2	588(1)	6.31(7)	0.352(10)	173	1.5972	96.804
α-Picoline	C_6H_7N	621					
β-Picoline	C_6H_7N	645					
γ-Picoline	C_6H_7N	646					
Piperidine	$C_5H_{11}N$	594.0					
Propanenitrile (propionitrile)	C_3H_5N	564.4	4.185	0.240	230	2.2199	140.16
n-Propylamine	C_3H_9N	497.0	4.74			1.5191	108.92
Pyridine	C_5H_5N	620.0	5.63	0.312	254	1.9898	114.37
Pyrrole	C_5H_5N	639.8					
Pyrrolidine	C_4H_9N	568.6	5.61	0.286	249	1.6796	105.27
Quinoline	C_9H_7N	782					
o-Toluidine (2-methylaniline)	C_7H_9N	693	3.75			3.7357	192.10
m-Toluidine	C_7H_9N	709	4.15			3.5287	177.36
p-Toluidine	C_7H_9N	667	2.38			5.4487	291.11
Trimethylamine [9]	C_3H_9N	433.30(10)	4.078(5)	0.233(3)	254	1.3428	110.44
Triethylamine [9]	$C_6H_{15}N$	532(10)	3.0(5)	0.26	390	2.7152	181.88
Sulfur-containing compounds [23]							
Diallyl sulfide	$C_6H_{10}S$	653					
Diethyl sulfide (3-thiapentane) [9]	$C_4H_{10}S$	556.95(50)	3.96(5)	0.279	323	2.2833	146.10
Dimethyl sulfide (2-thiapropane) [9]	C_2H_6S	503.05(50)	5.53(2)	0.309	201	1.3339	94.498
Ethyl diulfide	$C_4H_{10}S_2$	642					
Ethyl mercaptan (ethylthiol) [9]	C_2H_6S	498.65(50)	5.49(5)	0.300	207	1.3204	94.362
Isopentyl sulfide	$C_{10}H_{22}S$	664					
Methyl ethyl sulfide (2-thiabutane)	C_3H_8S	553	4.26			2.0956	135.04
Methyl mercaptan (methanthiol)	CH_4S	470.0	7.24	0.332	145	0.8904	67.515
Tetrahydrothiophene	C_4H_8S	632.0					
Thiophene	C_4H_4S	579.4	5.695	0.385	219	1.7192	105.74
Silicon-containing compounds [32]							
Dimethyl dichlorosilane	$C_2H_6Cl_2Si$	520.35	3.49	0.369		2.2654	155.14
Triethyl chlorosilane	$C_2H_5Cl_3Si$	559.95	3.33	0.406		2.7429	174.56
Trimethyl dichlorosilane	C_3H_9ClSi	497.75	3.20	0.297		2.2565	161.56
Mixtures [28]							
48% Dibenzyl, 52% Naphthalene		801					
73.5% Diphenyl ether, 26.5% Diphenyl		801	4.15			4.5037	200.38

References

[1] Zharkov, V. N., Kalinin, V. A., Equation of state for solids under high pressures and temperatures, Nauka, Moscow, 1968 (in Russian).

[2] Novikova, S. I., Thermal expansion of solids, Nauka, Moscow, 1974 (in Russian).

[3] Mason, E., Sperling, T., The virial equation of state, Pergamon Press, New York, 1969.

[4] Vukalovich, M. P., Novikov, I. I., Equation of state for real gases, Gosenergoizdat, Moscow - Leningrad, 1948 (in Russian).

[5] Bretshnider, S., Properties of gases and liquids, Khimia, Moscow-Leningrad, 1966.

[6] Thermodynamic properties of individual substances, 2nd ed., Ed. Glushko, V. P., Isdatel'stvo Acad. Nauk SSSR, Moscow, 1962 (vols. 1,2) (in Russian).

[7] Fisher, M., The nature of critical points, Colorado, 1965.

[8] Equation of state for gases and liquids, Nauka, Moscow, 1975 (in Russian).

[9] Thermal constants of substances (issues 1–10), Ed. Glushko, V. P., Izdatel'stvo Acad. Nauk SSSR, Moscow, 1965–1982 (in Russian).

[10] Ohse, R. W., Tippelskirch, H., High Temp. High Press., 9, 367, 1977.

[11] Herreman, W., Cryogenics, 2, 133, 1980.

[12] Thermodynamic properties of individual substances. Guidebook, 3rd ed., Ed. Glushko, V. P., Nauka, Moscow, 1978–1982 (vols. 1–2) (in Russian).

[13] Kobe, K. A., Lynn, R. E., Chem. Rev., 52, 117, 1953.

[14] Martynyuk, M. M., Karimkhodzhaev, I., Zhurnal fizicheskoi khimii, 48, 1243, 1974 (in Russian).

[15] Seydel, U., Fucke, W., J. Phys., F8, L157, 1978.

[16] Goodwin, R. D., J. Res. NBS, 74A2, 221, 1970.

[17] Rau, H., J. Chem. Thermod., 6, 525, 1974.

[18] Rau, H., Kutty, T. R. N., Guedes de Carvalho, J., J. Chem. Thermod., 5, 291, 1973.

[19] Hoshino, H., Schmutzler, R. W., Hensel, F., Ber. Bunsenges. Phys. Chem., 8, 27, 1976.

[20] Mathews, J. F., Chem. Rev., 72, 71, 1972.

[21] Handbook of a chemist, 3rd ed., Ed. Nikolskii, B. P., Khimia, Leningrad, 1963 (vol. 1) (in Russian).

[22] Properties of inorganic compounds. Handbook, Ed. Efimov, A. I., Khimia, Leningrad, 1983 (in Russian).

[23] Kudchadker, A. J., Alani, G. H., Zwolinski, B. J., Chem. Revs. 68, 659, 1968.

[24] Vargaftik, N. B., Handbook of thermal properties of gases and liquids, Fizmatgiz, Moscow, 1963 (in Russian).

[25] Amrose, D., Townsend, R., Trans. Faraday Soc., 64, 2622, 1968.

[26] Young, C. L., Austral. J. Chem., 25, 1625, 1972.

[27] Amrose, D., J. Chem. Thermod., 7, 1143, 1975.

[28] Tables of physical quantities. Handbook, Ed. Kikoin, I. K., Atomizdat, Moscow, 1976 (in Russian).

[29] Redlich, O., Kwong, J. N. S., Chem. Rev., 44, 233, 1949.

[30] Redlich, O., Thermodynamics: fundamentals, application, Elsevier, Amsterdam, 1976.

[31] Beattie, J. A., Bridgeman, O. C., J. Amer. Chem. Soc., 49, 1665, 1927.

[32] Benedict, N., Webb, G. B., Rubin, L. C., J. Chem. Phys, 8, 334, 1940.

[33] Tables of standard reference data. GSSD 49–83. Izdatel'stvo standartov, Moscow, 1984 (in Russian).

[34] Perel'shtein, I. I., Parushin, E. B., Thermodynamic and thermal properties of working agents of refrigerating machines and heat pumps, Legkaya i pishchevaya promyshlennost', Moscow, 1984 (in Russian).

[35] Anisimov, M. A., Rabinovich, V. A., Sychev V. V., Critical state thermodynamics of individual substances, Moscow, Energoatomizdat, 1990 (in Russian).

14

Surface Tension

B.D. Summ

14.1 Introduction

Surface tension refers to the basic thermodynamic parameter of liquid and solid interfaces (liquid/gas, liquid/liquid, solid/gas, etc.). It is governed by intermolecular (interatomic) interactions, and arises from the surface molecule interactions not only with the neighboring molecules of the same phase, but also with the nearby molecules of the adjacent phase. Surface tension σ may be specified in two ways [1,2].

(1) The energy approach considers the surface tension as a specific free surface energy, so that dimension of σ is J/m^2.

(2) The power approach treats of the surface tension as a force conditioned by intermolecular interactions and exerted on a unit length contour bounding the surface of the phase at hand. This force is directed at a tangent to the surface and along the inner perpendicular to the contour, so that the dimension of σ is N/m.

The above definitions are in close agreement for liquids; however, they are nonequivalent in the case of solids.

The total surface energy is used in thermodynamics along with surface tension. The fundamental thermodynamic equation reads as follows: $\epsilon = \sigma + \eta T$, where η is the specific surface entropy (the entropy surplus per unit area), T is the temperature. According to Einstein's theory, the total surface energy of pure liquids does not depend on temperature T.

The temperature dependence $\sigma(T)$ away from the critical point is approximately linear for many single-component nonassociated liquids (water, organic substances, melted metals and salts, etc.). Namely, $\sigma = \sigma_0 - \alpha(T - T_0)$, here σ_0 is the surface tension at temperature of T_0, the temperature coefficient $\alpha \approx 0.1$ mJ/(m$^2\cdot$)K. This empirical relation is valid over a fairly wide temperature range. The departures are observed at several kelvin degrees below the critical point of liquid T_0, so that if $T \to T_c$, then $\sigma \to 0$.

The surface tension is of principal importance for many physical, chemical, biological and geological phenomena. The following physical and chemical quantities depend essentially on the surface tension of liquids or on the free surface energy of solids [1–4]:

(1) The capillary pressure P_c originates beneath the curved liquid interfaces (liquid/gas, liquid/liquid, liquid/solid). The Laplace equation (the first law of capillarity) manifests that

$$P_c = \sigma(1/r_1 + 1/r_2), \tag{1}$$

where r_1 and r_2 are the principal curvature radii. They are positive for a convex surface ($r > 0$ and the capillary pressure $P_c > 0$), and they are negative for a concave one (when $r < 0$ and $P_c < 0$). For spherical surfaces, $P_c = \pm 2\sigma/r$, where r is the sphere radius.

0-8493-2861-6/97/$0.00+$.50

(2) The contact angle of wettability θ formed by a liquid meniscus in the vicinity of three phase lines (liquid/solid/gas or liquid/solid/other liquid), is determined by the Young's equation (the second law of capillarity):

$$\cos\theta = (\sigma_S - \sigma_{S1})/\sigma_1 \tag{2}$$

for liquid/solid/gas systems; σ_S and σ_{S1} are the free surface energies of the solid/gas and solid/liquid interfaces, σ_1 is the surface tension of the wetting liquid.

For liquid 1/solid/liquid 2 systems, the Young's equation takes the form

$$\cos\theta = (\sigma_{S1} - \sigma_{S2})/\sigma_{12}, \tag{3}$$

where σ_{S1} and σ_{S2} are the free surface energies of solid/liquid 1 and solid/liquid 2 interfaces, σ_{12} is the interfacial tension between 1 and 2 liquids.

(3) The saturated vapor pressure P_r over the curved liquid surface is determined by the Lord Kelvin equation (the third law of capillarity):

$$P_r = \pm P_0 \exp(2\sigma V_m / rRT), \tag{4}$$

here V_m is the molar volume of liquid; P_0 is the saturated vapor pressure over the planar liquid surface; r is the curvature radius; $R = 8.314$ J/(mol·K) is the universal gas constant; T is the temperature.

(4) The specific adsorption of surfactants Γ (i.e. the mass of substance adsorbed per unit area) is determined by the Gibbs equation

$$\Gamma = -d\sigma/d\mu, \tag{5}$$

with μ being the chemical potential of the adsorbed substance in the bulk phase, for example, in solution. For diluted solutions, the Gibbs equation (5) can be transformed to the following form:

$$\Gamma = -(d\sigma/dc)c/RT, \tag{6}$$

where c is the molar surfactant concentration.

(5) The adsorption layer of nonsoluble surfactants on a liquid surface is described by the Frumkin–Volmer equation of the two-dimensional state of the substance:

$$(\pi + a/A^2)(A - b) = kT, \tag{7}$$

where $\pi = \sigma_0 - \sigma$ is the two-dimensional surface pressure; σ_0 and σ are the surface tensions of the poor liquid (σ_0) and the "dirty" liquid (σ), whose surface is covered by the surfactant monolayer; a is the constant that is analogous to the van der Waals constant; A is the area applied to one surfactant molecule in the adsorption layer, b is the intrinsic area of the molecule; k is the Boltzmann constant.

(6) The electrocapillary effect is determined by the $\sigma(\varphi)$ dependence, with φ being the electrode potential. The Lippmann equation predicts that

$$-d\sigma/d\varphi = \rho_s, \tag{8}$$

where ρ_s is the surface charge density.

(7) The differential capacitiy of a double electrical layer C is given by

$$C = d^2\sigma/d\varphi^2. \tag{9}$$

(8) The equilibrium crystal facetting is determined by the Gibbs–Curie–Wulf equation:

$$\sigma_1/h_1 = \sigma_2/h_2 = \ldots = \sigma_i/h_i, \tag{10}$$

where σ_i is the free surface energy of the crystal i-th facet; h_i is the distance between this facet and the crystal center.

(9) The work W_c controlling the process of homogeneous nucleation of a new phase critical "embryo" is defined by thermodynamic Gibbs equations (11), (12), (13) pertinent to different cases of phase transformation. By this means one describes the following processes.

(a) The vapor–liquid condensation at the pressure P,

$$W_c = (16/3)\pi\sigma^3 V_m^2/(RT \ln P/P_0)^2, \tag{11}$$

where σ is the surface tension of the condensating liquid; V_m is the molar volume; P_0 is the pressure of saturated vapor over the planar liquid–vapor interface; R is the gas constant; T is the temperature.

(b) The crystallization from the supersaturated solution with the concentration $c > c_0$, where c_0 is the equilibrium concentration at a given temperature T. The work of crystallization is equal to

$$W_c = (16/3)\pi\sigma^3 V_m^2/(RT \ln c/c_0)^2. \tag{12}$$

In this equation, σ defines the specific free surface energy of the liquid–solid interface.

(c) The crystallization from the supercooled melt is described by the equation

$$W_c = (16/3)\pi\sigma^3 (V_m T_m/L\Delta T)^2, \tag{13}$$

where σ is the specific free energy of the melt–crystal interface; T_m is the melting point; L is the molar fusion heat; $\Delta T = (T_m - T)$ is the supercooling of a melt below the melting point.

(10) The theoretical strength of ideal solids P_i, as well as the real strength P_r, depends on the specific free energy σ_s of the solid–gas interface or on the free energy σ_{sl} of the solid–liquid interface for solids that are destroyed in the presence of the wetting liquid. The correlations between $P_i(\sigma_s)$ and $P_r(\sigma_s)$ are described by the Polanyi–Smecal (14) and the Griffith (15) equations:

$$P_i = (2\sigma_s E/\delta)^{1/2}, \tag{14}$$

$$P_r = (2\sigma_s E/\pi l)^{1/2}. \tag{15}$$

Here E is the elastic (Young) modulus of solids; δ is the distance between the nearest neighbor molecules (atoms, ions); l is the "embryonic" crack length. The Griffith theory manifests that the microcrack will spread spontaneously and destroy the stress at $P > P_r$.

(11) The capillary waves at the liquid surface are characterized by the Lord Kelvin equation

$$\sigma = \lambda^3 \rho/2\pi\tau^2 - g\lambda^2\rho/4\pi^2, \tag{16}$$

where λ is the capillary wavelength; ρ is the liquid density; τ is the oscillation period; g is the acceleration due to gravity.

(12) The elasticity of the liquid thin-film is determined by the Gibbs equation

$$E_s = 2(d\sigma/d \ln s), \tag{17}$$

where s is the film area; E_s is the modulus of elasticity of a film.

(13) A Bond number (Bo) determines the correlation between the gravitation and capillary forces:

$$\text{Bo} = (\rho_L - \rho_E)gl^2/\sigma, \tag{18}$$

where ρ_L and ρ_E are the densities of the liquid phase and the environmental phase (gas or another liquid); l is the scaling dimension (for example, the vessel diameter).

(14) The Weber number (We) correlates the inertial and capillary forces:

$$\mathrm{We} = \rho v^2 l / \sigma, \tag{19}$$

where v is the liquid velocity; ρ is the liquid density; l is the scale factor.

The surface tensions were measured for many pure liquids and solutions within wide temperature ranges. The modern techniques provide the measurements with a rather high accuracy [1,2]. The measurements of solid–gas and solid–liquid free surface energies σ_s and σ_{sl} are much more complicated, due to the creation of a new solid surface (e.g., by the cleavage of crystal) that inevitably causes the plastic deformations of the surface layers. Thus, the work needed for the solid surface creation surpasses the thermodynamic values of σ_s and σ_{sl}. It should be noted that the surface tension is very sensitive to the presence of certain impurities in the liquid. The impurity effect depends to a great extent on the chemical composition of the liquid and the adjacent phase, as well as on the impurity concentration. These factors cause noticeable deviations in different data on surface tensions due to differences in the composition of various systems. Thus, the data selection presents a complicated problem. In this chapter, the major data were compiled from Russian publications issued from 1980 to the early 1990s.

Table 14.1 Surface tension σ (in mN/m) of liquefied hydrogen vs. temperature T (in K) [5]

T	σ	T	σ	T	σ	T	σ
Hydrogen $n-H_2$		Deuterium $n-D_2$		Parahydrogen *para* $-H_2$		Orthodeuterium *ortho* $-D_2$	
20.55	1.898	20.57	3.437	20.70	1.818	20.91	3.355
20.86	1.844	20.96	3.346	21.14	1.749	21.28	3.278
21.51	1.729	21.00	3.338	22.47	1.524	21.29	3.278
22.07	1.638	21.64	3.201	22.56	1.509	21.88	3.141
22.10	1.629	22.08	3.127	23.07	1.421	21.92	3.139
22.21	1.616	22.28	3.062	23.68	1.318	22.66	2.980
22.59	1.547	22.91	2.930	24.09	1.252	23.05	2.895
23.23	1.437	22.95	2.924	24.45	1.189	23.41	2.811
24.14	1.282	23.35	2.839	24.51	1.175	24.22	2.632
24.63	1.198	23.61	2.777	25.19	1.065	25.00	2.462
25.04	1.129	24.30	2.634	25.54	1.006	25.04	2.448
25.33	1.084	24.95	2.493	25.94	0.944	25.49	2.347
25.83	0.998	25.50	2.375	26.41	0.867	26.00	2.235
26.27	0.929	25.92	2.275	26.73	0.817	26.56	2.117
26.69	0.857	26.52	2.159	27.17	0.744	27.12	1.999
26.99	0.811	27.28	1.983	27.50	0.692	27.51	1.914
27.32	0.758	27.33	1.973	27.99	0.616	28.09	1.787
27.70	0.693	28.43	1.743	28.00	0.615	28.50	1.697
28.46	0.578	29.00	1.621	28.46	0.543	29.44	1.493
29.00	0.498	29.95	1.428	28.75	0.499	30.04	1.379
29.40	0.438	30.87	1.240	29.02	0.461	30.52	1.275
29.59	0.410	31.47	1.119	29.51	0.388	30.98	1.181
30.01	0.351	32.95	0.824	29.70	0.360	31.51	1.075
30.45	0.289	33.93	0.637	30.00	0.320	32.41	0.899
30.53	0.281	34.51	0.538	30.44	0.258	32.49	0.887
30.96	0.226	34.53	0.531	30.48	0.255	33.48	0.699
31.06	0.210	35.14	0.425	30.99	0.191	34.03	0.597
31.58	0.149	35.50	0.364	31.11	0.175	34.41	0.537
31.95	0.107	35.93	0.294	31.62	0.114	34.98	0.432
		36.32	0.235			35.53	0.342
		36.70	0.181			36.03	0.264
						36.05	0.261

Table 14.2 Water surface tension σ (mN/m) vs. temperature t (°C) [7]

t	σ	t	σ	t	σ	t	σ	t	σ	t	σ	t	σ
0	75.50	80	62.69	160	46.51	240	28.52	320	9.84	363	1.37	371	0.31
10	74.40	90	60.79	170	44.38	250	26.13	330	7.69	364	1.22	372	0.20
20	72.88	100	58.91	180	42.19	260	23.73	340	5.61	365	1.07	373	0.10
30	71.20	110	56.97	190	40.00	270	21.33	350	3.64	366	0.93	374.15	0
40	69.48	120	54.96	200	37.77	280	18.94	355	2.71	367	0.79		
50	67.77	130	52.90	210	35.51	290	16.60	360	1.85	368	0.66		
60	66.07	140	52.79	220	33.21	300	14.29	361	1.68	369	0.54		
70	64.36	150	48.68	230	30.88	310	12.04	362	1.53	370	0.42		

Table 14.3 Saturated hydrocarbons (CH_4 – C_8H_{18}) surface tension σ (mN/m) vrs. temperature t (in °C) [7]

Substance	t	σ	Substance	t	σ	Substance	t	σ	Substance	t	σ
Methane	–180	18.0	Butane	–60	22.1	Isopentane	25	14.46	Octane	30	20.79
CH_4	–170	15.8	$n-C_4H_{10}$	–50	20.88	$iso-C_5H_{12}$	30	13.93	$n-C_8H_{18}$	40	19.78
	–160	13.7		–40	19.65	Hexane	0	20.56		60	17.82
Ethane	–160	28.08		–30	18.43	$n-C_6H_{14}$	10	19.51		80	15.94
C_2H_6	–150	26.34	Isobutane	–100	25.2		20	18.46		100	14.13
	–140	24.62	$iso-C_4H_{10}$	–90	23.9		30	17.40		120	12.39
	–130	22.91		–80	22.6	Heptane	20	20.86		140	10.70
	–120	21.23		–70	21.4	$n-C_7H_{16}$	30	19.54		160	9.07
	–110	19.57		–60	20.14		40	18.47		180	7.50
	–100	17.93		–50	18.90		60	16.39		200	5.99
	–90	16.31		–40	17.68		80	14.35		220	4.52
Propane	–130	27.8		–30	16.48		100	12.47		230	3.80
C_3H_8	–120	26.3	Pentane	–20	20.5		120	10.63		240	3.10
	–110	24.9	$n-C_5H_{12}$	–10	19.3		140	8.87	Isooctane	0	20.58
	–100	23.4		0	18.2		160	7.19	$iso-C_8H_{18}$	10	18.77
	–90	22.0		10	17.1		180	5.59		25	18.32
	–80	20.6		20	16.00		200	4.07		30	17.88
	–70	19.2		25	15.48		220	2.63		40	16.99
	–60	17.85		30	14.95		230	1.94		50	16.11
	–50	16.49		40	13.80		240	1.29		60	15.24
	–40	15.15	Isopentane	–20	19.40	Octane	–40	27.50		70	14.40
Butane	–100	27.2	$iso-C_5H_{12}$	–10	18.27	$n-C_8H_{18}$	–20	25.50		80	13.50
$n-C_4H_{10}$	–90	25.9		0	17.17		0	23.70		90	12.70
	–80	24.6		10	16.08		10	22.73			
	–70	23.4		20	15.00		20	21.76			

Table 14.4 Normal saturated hydrocarbons C_nH_{2n+2} (C_9H_{20} – $C_{20}H_{42}$) surface tension σ (in mN/m) vs. temperature t (in °C) [7]

Substance	t								
	0	10	20	25	30	40	50	76	100
Nonane	24.84	23.90	22.96	22.49	22.01	21.07	20.13	17.78	15.42
Decane	25.73	24.81	23.89	23.43	22.98	22.06	21.14	18.84	16.54
Undecane	26.57	25.68	24.78	24.34	23.89	23.00	22.10	19.87	17.63
Dodecane	27.24	26.34	25.48	25.04	24.60	23.72	22.85	20.65	18.45
Tridecane	27.87	27.00	26.13	25.69	25.26	24.39	23.52	21.34	19.16
Tetradecane		27.56	26.69	26.26	25.83	24.97	24.11	21.94	19.79
Pentadecane		28.02	21.17	26.74	26.32	25.46	24.61	22.48	20.35
Hexadecane			27.64	27.22	26.79	25.95	25.11	23.00	20.90

Table 14.4 C_nH_{2n+2} (C_9H_{20}–$C_{20}H_{42}$) surface tension σ vs. t (in °C) *(continued)*

Substance	t								
	0	10	20	25	30	40	50	76	100
Eicosane			29.0	28.6	28.1	27.2	26.4		23.0
Heptadecane				27.64	27.22	26.38	25.54	23.45	21.35
Nonadecane			28.7	28.2	27.8	26.9	26.0		22.5
Octadecane					27.59	26.75	25.92	23.84	21.75

Table 14.5 Surface tension σ (in mN/m) vs. temperature t (in °C) for unsaturated hydrocarbons of ethylene and acetylene series, and diolefins [7]

Substance	t	σ	Substance	t	σ	Substance	t	σ	Substance	t	σ
Ethylene	-120	19.50	Heptene	0	22.40	Octene-1	20	21.97	Acetylene	-50	13.13
C_2H_4	-110	17.65	C_7H_{14}	10	21.41	C_8H_{16}	25	21.44	C_2H_2		
	-100	15.71		20	20.42		30	20.95	Propadiene	-50	21.99
Hexene-1	0	20.58		25	19.93		40	19.99	C_3H_4	-40	20.44
C_6H_{12}	10	19.55		30	19.43		50	19.03		-30	18.90
	20	18.52		40	18.44		75	16.64		-20	17.35
	25	18.00		50	17.45		100	14.24			
	30	17.49		75	14.98	Acetylene	-80	18.92			
	40	16.46	Octene-1	0	23.83	C_2H_2	-70	16.99			
	50	15.43	C_8H_{16}	10	22.87		-60	15.06			

Table 14.6 Cyclic hydrocarbons surface tension σ (in mN/m) vs. temperature t (in °C) [7]

Substance	t	σ	Substance	t	σ	Substance	t	σ	Substance	t	σ
Cyclopentene	0	25.25	Cyclohexene	20	24.95	Ethylcyclo-	30	24.57	Chloroben-	60	28.7
C_5H_8	10	23.94	C_6H_{12}	25	24.35	hexane C_8H_{16}	40	23.47	zene C_6H_5Cl	80	26.4
	20	22.65		30	23.75		50	22.37		100	24.0
	25	21.99		40	22.45		75	19.62		150	18.5
	30	21.32		50	21.35		100	16.87		200	13.2
	40	20.03		75	18.35	Aniline	0	45.0		250	8.3
Cyclopentane	13.5	23.30	Methylcyclo-	0	25.80	$C_6H_5NH_4$	10	44.0		300	3.9
C_5H_{10}			hexane C_7H_{14}	10	24.74		20	42.9		310	3.1
Cyclohexene	0	29.00		20	23.68		30	41.8		320	2.4
C_6H_{10}	10	27.80		30	23.15		40	40.7		330	1.7
	20	26.61		40	22.62		50	39.5	Nitrobenzene	0	46.4
	25	26.01		50	21.56		100	33.7	$C_6H_5NO_2$	10	45.2
	30	25.41		60	20.50		150	27.9		20	43.9
	40	24.22		70	17.58		180	24.6		30	42.7
	50	23.02		100	15.20	Chloroben-	0	36.0		40	41.5
	75	20.03	Ethylcyclo-	0	27.87	zene C_6H_5Cl	10	34.8		60	39.0
Methylcyclo-	13.5	24.10	hexane C_8H_{16}	10	26.77		20	33.5		100	34.4
pentane C_6H_{12}				20	25.67		30	32.3		150	29.0
Cyclohexane	10	26.15		25	25.12		40	31.1		200	23.6

Table 14.7 Aromatic hydrocarbons surface tension σ (in mN/m) vs. temperature t (in °C) [7]

Substance	t									
	0	10	20	25	30	40	50	60	80	100
Benzene C_6H_6		30.24	28.88	28.18	27.49	26.14	24.88	23.66	21.2	
Toluene C_7H_8	30.92	29.70	28.53	27.29	27.32	26.15	25.04	23.94	21. 8	19.6

Table 14.7 Aromatic hydrocarbons surface tension σ (in mN/m) vs. t *(continued)*

Substance	*t*									
	0	10	20	25	30	40	50	60	80	100
Ethylbenzene C_8H_{10}	31.38	30.18	29.04	28.48	27.93	26.79	25.74	24.74	22.7	20.7
ortho-Xylene C_8H_{10}	32.28	31.16	30.03	29.48	28.93	27.84	26.76	25.70	23.6	21.5
meta-Xylene C_8H_{10}	30.92	29.78	28.63	28.08	27.54	26.44	25.36	24.26	22 .2	20.1
para-Xylene C_8H_{10}			28.31	27.76	27.22	26.13	25.06	24.02	22.0	20.1
iso-Propyl benzene (cumene) C_9H_{12}			28.20	27.68	27.17	26.09	25.08	24.07	22.2	

Table 14.8 Halogenated hydrocarbons surface tension σ (in mN/m) vs. temperature t (in °C) [7]

Substance	*t*	σ	Substance	*t*	σ	Substance	*t*	σ	Substance	*t*	σ
Diflourodichloro-methane (Freon-12) CF_2Cl_2	20	9.0	Triflourochloro-methane (Freon-13) CF_3Cl	0	18.5	Carbon tetra-chloride CCl_4	0	29.5	Carbon tetra-chloride CCl_4	100	17.3
	30	7.5		10	17.3		10	28.2		150	11.7
Triflourochloro-methane (Freon-13) CF_3Cl	−30	23.5		20	16.0		20	26.9		200	6.5
	−20	22.0		30	14.0		40	24.5		250	2.1
	−10	20.5	Chloromethyl CH_3Cl	0	12.0		60	22.1		260	1.4
				10	11.0		80	19.7		270	0.7

Table 14.9 Alcohols, ethers, ketones, and organic acids surface tension σ (in mN/m) vs. temperature t (in °C) [7]

Substance	*t*									
	0	10	20	30	50	100	150	200	220	240
Acetic acid CH_3COOH			27.8	26.8	24.8	19.8	15.0	10.4	8.5	5.7
Acetone C_3H_6O	26.2	25.0	23.7	22.5	19.9					
Butanol-1 C_4H_9OH	26.2	25.4	24.6	23.8	22.1	17.8				
Diethyl ester $C_4H_{10}O$	19.4	18.2	17.0	15.8	13.5	8.0	3.1			
Ethyl acetoacetate $C_4H_8O_2$	27.0	25.6	24.3	23.0	20.5	14.4	8.7	3.7	2.0	0.5
Ethanol C_2H_5OH	24.4	23.6	22.8	21.9	20.1	15.5	10.1	4.3	2.2	0.1
Glycerol $C_3H_5(OH)_3$			59.4	59.0	58.0	54.2	48.8			
Isopropanol C_3H_7OH	25.5	24.6	23.8	23.0	21.2	17.2				
Methanol CH_3OH	24.5	23.5	22.6	21.8	20.1	15.7	10.4	4.5	2.1	

Table 14.10 Organic liquid–water interfacial tension σ (in mN/m) at 20°C [4]

Substance	σ	Substance	σ	Substance	σ	Substance	σ
Amyl acetate	12.0	Cyclohexane	50.2	Hexane	49.4	Octanol	8.5
Amyl alcohol	4.4	Cyclohexanol	3.9	Hexanol	6.8	Oleic acid	15.7
Aniline	5.8	Cymene	34.6	Iodobenzene	41.8	Pentane	49.0
Benzaldehyde	15.5	Decane	51.2	Isoamyl alcohol	4.8	Styrene	34.0
Benzene	35.0	Dibromoethane	36.5	Isobutanol	2.0	Tetrachloro-ethylene	47.5
Bromobenzene	38.1	Dichloroethane	31.0	Isooctane	51.0	Tetradecane	52.2
Bromoform	40.0	Diethyl ester	10.7	Isopentane	49.0	Toluene	36.1
Butanol	1.8	Dodecane	51.8	Isovalerianic acid	2.7	*meta*-Xylene	37.9
Capronic acid	5.2	Ethylbenzene	38.4	Mesitylene	38.7	*ortho*-Xylene	36.1
Caprylic acid	8.2	Ethylbromide	31.2	Methylbenzoate	16.6	*para*-Xylene	37.8
Carbon tetrachloride	45.0	Furfurol	5.1	Nitrobenzene	25.7	*ortho*-Nitrotoluene	27.2
Chlorobenzene	37.4	Heptane	50.2	*meta*-Nitrotoluene	27.7	Octane	50.8
Chloroform	31.6	Heptanol	7.7				
		Hexadecane	52.6				

Table 14.11 Liquid alkali metals' surface tension σ (in mN/m) vs. temperature t (in °C) [7]

t	Li	Na	K	Rb	Cs	t	Li	Na	K	Rb	Cs
29					71.3	700	325	137.8	67.0	49.2	39.1
39				87.5		800	311	127.8	60.4	43.4	34.3
64			109.0			900	297	117.8	53.8	37.6	29.5
100		197.8	106.6	84.0	67.9	1000	283	107.8	47.2	31.8	24.7
200	395	187.8	100.0	78.2	63.1	1100	269	97.8	40.6	26.0	19.9
300	381	177.8	93.4	72.4	58.3	1200	255	87.8	34.0	20.0	15.1
400	367	167.8	86.8	66.4	53.5	1300	241	77.8	27.4	14.4	10.3
500	353	157.8	80.2	60.8	48.7	1400	227	67.8	20.8	8.6	
600	339	147.8	73.6	55.0	43.9	1500	223	57.78	14.2		

Table 14.12 Liquid metals' surface tension σ (in mN/m) vs. temperature t (in °C) [8]

Metal	t	σ	Metal	t	σ	Metal	t	σ	Metal	t	σ
Aluminum	660	915	Dyspro-			Manganese	1350	1070	Antimony	700	347.6
	800	850	sium	1407	490		1550	1030		750	346.2
	1000	830	Gadolinium	1350	670	Mercury	25	465		800	345.0
	1100	820	Gallium	50	706		50	462	Strontium	1373	286.1
	1600	725	Gallium	100	705		100	452		1473	280.3
	1830	680		200	704		150	439		1573	274.4
Barium	1273	267.3		300	699	Mercury	200	429		1673	268.6
	1373	259.7		400	694		250	416		1773	262.7
	1473	253.0	Germanium	1100	604.5		300	402		1873	256.8
	1573	246.3		1100	594	Molybdenum	2622	2225		1973	251.0
	1673	239.6		1550	558	Neodymium	840	688		2073	245.1
	1773	232.9	Gold	1100	1125	Nickel	1550	1735		2173	239.3
	1873	226.2	Hafnium	2200	1630		1850	1620		2273	233.4
	1973	219.4	Indium	200	556.0	Niobium	2468	2040	Tantalum	2996	2140–
	2073	212.7		300	547.4	Osmium	2700	2500			–2150
	2173	206.0		400	535.0	Palladium	1554	1500	Tellurium	452	179
	2273	199.3		500	531.2		1600	1460	Terbium	1356	700
Beryllium	1500	1100		600	522.5	Platinum	1760	1800	Thallium	302	464.5
		±35		700	512.8		1800	1699	Tin	300	530
Bismuth	280	372.1	Iridium	2454	2504.6	Praseo-				350	537
	300	371.0	Iron	1550	1850	dymium	940	706		450	535
	400	364.6		1600	1830	Rhenium	3180	2700		500	531
	500	357.9		1650	1790	Rhodium	1966	2000		600	525
	800	343		1700	1760	Ruthenium	2500	2250	Titanium	1672	1558
	900	336	Lanthanum	810	745	Selenium	220	105		1727	1460
	1000	328	Lead	350	440		240	104		1130	1550
Boron	2200	1060		450	428		280	101	Tungsten	3395	2316
Cadmium	330	560		750	420		340	87	Uranium	1130	1550
Calcium	851	370		800	412		380	80	Vanadium	1919	1950
Chromium	1830	1540		1000	388	Silicon	1550	750	Yttrium	1450	877
Cobalt	1550	1830	Magne-	670	552	Silver	1000	907	Zinc	420	767
	1600	1770	sium [11]	700	542		1100	894	Zirconium	1855	1455
Copper	1150	1370		720	534		1200	876		2460	1395
	1550	1265		740	528	Antimony	650	350.2			

Table 14.13 Specific free surface energy of solid metals [12]

Metal	t,°C	Atmosphere	σ, mJ/m^2	Metal	t,°C	Atmosphere	σ, mJ/m^2
Aluminum	180	Vacuum	1140± 200	Molybde-	1600	Argon	2100 ± 200
Beryllium	700	Helium	1000	num	2500	Argon	1920 ± 200
Bismuth	239	Vacuum, Argon	521 ± 6	Nickel	1343	Argon	1820 ± 180
Cobalt	1354	Hydrogen	1970 ± 175		1219	Vacuum	1860 ± 190
Copper	1006	Helium, Hydrogen	1720	Niobium	2250	Vacuum	2100 ± 100
Gold	970	Vacuum	1450 ± 80	Platinum	1310	Vacuum	2340 ± 800
	1025	Helium	1400 ± 65		1673	Air	1800 ± 200
	968	Air	1390 ± 80	Silver	750	Nitrogen	1140 ± 35
Indium	142	Vacuum, Argon	633 ± 7		909	Helium	1140 ± 90
Iron	1460	Argon	1910 ± 190	Tantalum	1500	Vacuum	2680 ± 500
	1410	Hydrogen	2320 ± 80	Thallium	272	Vacuum, Argon	562 ± 6
γ-Iron	1380	Argon	2170 ± 300	Titanium	1600	Vacuum	1700
Lead	309	Vacuum, Argon	560 ± 6	Zinc	480	Helium	830
Molybdenum	1427	Vacuum	2200 ± 200	Zinc (0001)	-195	Nitrogen (liquid)	410

Table 14.14 Influence of oxygen and carbon mass concentration ρ (in %) on the iron melts surface tension σ (in mN/m) [8]

ρ	σ	ρ	σ
Oxygen, $t = 1570$°C		Carbon, $t = 1570$°C	
0.0006	1717	0.42	1847
0.0077	1632	2.25	1822
0.02	1541	3.15	1793
0.041	1362	3.50	1805
0.07	1151	4.15	1788

Table 14.15 Melted salts' surface tension σ (in mN/m) vs. temperature t (in °C) [1]

Substance	t	σ
$BiCl_3$	271	66.0
$KClO_3$	368	81.0
KNCs	175	101.5
$NaNO_3$	308	116.6
$K_2Cr_2O_7$	397	129.0
$Ba(NO_3)_2$	595	134.8

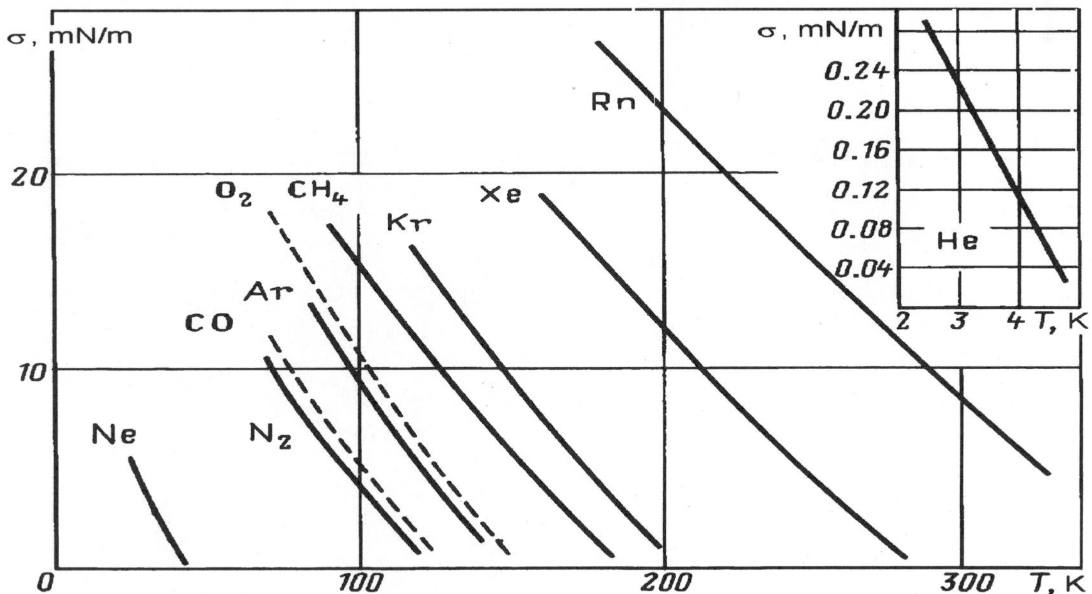

Figure 14.1 Surface tension of liquefied gases.

References

[1] Adamson, A. W., Physical chemistry of surfaces, 3rd ed., John Wiley and Sons, New York, 1976 (Russian transl., Mir, Moscow, 1979).

[2] Shchukin, E. D., Pertsov, A. V., Amelina, E. A., Colloid chemistry, Izdatel'stvo MGU, Moscow, 1982 (in Russian).

[3] Gibbs, J. W., The collected works in two volumes, Vol. I, Thermodynamics, Longmans, Green and Co., New York, 1948.

[4] Abramson, A. A., Surfactants, 2nd ed., Khimia, Leningrad, 1981 (in Russian).

[5] Baidakov, V. G., Khvostov, K. V., Skripov, V. P., Capillary constant and surface tension of neon, hydrogen and its isotopes, Izdatel'stvo Ural'skogo nauchnogo tsentra AN SSSR, Sverdlovsk, 1981 (in Russian).

[6] Baidakov, V. G., Khvostov, K. V., Muratov, G. N., Capillary constant and surface tension of argon, krypton, xenon, methane, oxygen and nitrogen, Izdatel'stvo Ural'skogo nauchnogo tsentra AN SSSR, Sverdlovsk, 1981 (in Russian).

[7] Vargaftik, N. B., Handbook of thermal properties of gases and liquids, 2nd ed., Nauka, Moscow, 1972 (in Russian).

[8] Nizhenko, V. I., Floka, L. I., Surface tension of liquid metals and alloys. A handbook, Metallurgiya, Moscow, 1981 (in Russian).

[9] Naidich, Yu. V., Contact phenomena in metallic melts, Naukova dumka, Kiev, 1972 (in Russian).

[10] Shpil'rain, E. E., Fomin, V. A., Kachalov, V. V., Sokol, G. F., Skovorod'ko, S. N., Thermal properties of alkaline-earth metals in the liquid phase (density, surface tension, viscosity). A review on thermal properties of substances, Izdatel'stvo nauchno-informatsionnogo tsentra po teplofizicheskim svoistvam chistykh veshchestv Akad. Nauk SSSR, Vol. 2 (40), Moscow, 1983 (in Russian).

[11] Andronov, V. N., Chekin, B. V., Nesterenko, S. V., Liquid metals and slags, Metallurgiya, Moscow, 1977 (in Russian).

[12] Missol, W., Energia powierzchni rozdzatu faz w metalach, Widawnitztwo "Slask," Katowice, 1975.

15

Thermal Conductivity

A.V. Inyushkin

15.1 Introduction

Thermal conductivity refers to a molecular heat transport (heat conduction) in a continuous medium caused by a temperature gradient. A thermal conductivity coefficient (it is adopted now as "a thermal conductivity") is defined by the Fourier equation

$$\mathbf{q} = -\lambda \nabla T, \tag{15.1}$$

where $\mathbf{q}$ is the density of heat flux, $\mathrm{W/m^2}$; λ is the thermal conductivity, $\mathrm{W/(m{\cdot}K)}$; ∇T is the temperature gradient, K/m. The Fourier equation is justified provided the temperature gradient is small enough (when the deviation of a system from the equilibrium state is small) and the mean free path of heat carriers (particles or quasiparticles) is small with respect to the system dimensions. For crystalline solids, the thermal conductivity is a symmetric tensor of second rank.

The heat conduction is described by the heat-transfer equation, the simplest form of which is

$$dT/dt = (\lambda/c\gamma)\nabla^2 T, \tag{15.2}$$

where $T(x, y, z, t)$ is the temperature, K, in the point with coordinates x, y, z; t is a time, s; c is a specific heat, $\mathrm{J/(kg{\cdot}K)}$; γ is the specific density, $\mathrm{kg/m^3}$; $\nabla^2 T = d^2T/dx^2 + d^2T/dy^2 + d^2T/dz^2$, $\mathrm{K/m^2}$.

Experimental methods of thermal conductivity measurements are based upon the solutions of the heat-transfer equation [1, 2]. Thermal conductivity depends on a state of aggregation of a substance, its composition, purity, temperature, pressure and other conditions. For most of the substances, the thermal conductivity of the liquid phase is about 10 times higher than that of the gaseous state, and for the solid state the thermal conductivity is much higher than for the liquid state near the melting point (excluding liquid bismuth, tin, and tellurium).

In practice, the thermal conductivities of a material near the surface and deep into it are different. This difference is due to the change of heat-transfer conditions and the variation of material structure (as a result of heat treatment, hardening, etc.). In the tables of this chapter, the thermal conductivity values are given for the body portion far away from its surface.

External factors such as irradiation, the change in pressure, and magnetic field can have an essential influence upon the thermal conductivity.

In a semitransparent medium, the thermal conductivity is accompanied by the radiative transport. The observed effective thermal conductivity of such a medium is a sum of the thermal conductivity proper and the radiative heat-transfer. The contribution of radiative heat-transfer increases with

0-8493-2861-6/97/$0.00+$.50

temperature and becomes essential at temperatures about several hundred degrees Celsius. The uncertainty of thermal conductivity values given in this chapter changes with materials, temperature ranges (as a rule, it is greater at low temperatures), and pressure (increases with pressure). In general, the uncertainty of thermal conductivity is about 10 to 20%. For some much–studied materials, the uncertainty is within 1 to 2%. The number of significant digits in the given values of λ generally agrees with their uncertainty. However, in those cases where it is advisable to give the characteristic variation of the thermal conductivity (for example, λ as a function of material composition or temperature), the number of significant digits is increased up to three.

The foundation of the theory of thermal conductivity as well as experimental methods of thermal conductivity investigation are discussed in [3].

The conversion factors for thermal conductivity are given below:

1 W/(cm·K)	= $1 \cdot 10^2$ W/(m·K)	1 W/(m·K)	= $1 \cdot 10^{-2}$ W/(cm·K)
1 erg/(cm·s·°C)	= $1 \cdot 10^5$ W/(m·K)	1 W/(m·K)	= $1 \cdot 10^5$ erg/(cm·c·°C)
1 cal_{IT}/(cm·s·°C)	= $4.1868 \cdot 10^2$ W/(m·K)	1 W/(m·K)	= $2.38846 \cdot 10^{-3}$ cal_{IT}/(cm·s·°C)
1 cal_{th}/(cm·s·°C)	= $4.184 \cdot 10^2$ W/(m·K)	1 W/(m·K)	= $2.39006 \cdot 10^{-3}$ cal_{th}/(cm·s·°C)
1 $kcal_{th}$/(m·h·°C)	= 1.16222 W/(m·K)	1 W/(m·K)	= 0.860421 $kcal_{th}$/(m·h·°C)
1 Btu_{IT}/(ft·h·°F)	= 1.73073 W/(m·K)	1 W/(m·K)	= $5.77789 \cdot 10^{-2}$ Btu_{IT}/(ft·h·°F)
1 Btu_{IT}·in/(ft^2·h·°F)	= 0.144228 W/(m·K)	1 W/(m·K)	= 0.693347 Btu_{IT}·in/(ft^2·h·°F)
1 Btu_{th}/(ft·h·°F)	= 1.72958 W/(m·K)	1 W/(m·K)	= $5.78176 \cdot 10^{-2}$ Btu_{th}/(ft·h·°F)
1 Btu_{th}·in/(ft^2·h·°F)	= 0.144131 W/(m·K)	1 W/(m·K)	= 0.693811 Btu_{th}· in/(ft^2·h·°F)
1 CHU/(ft·h·°F)	= 3.1152 W/(m·K)	1 W/(m·K)	= 0.32100 CHU/(ft·h·°F)

Remarks:
The international calorie: 1 cal_{IT} = 4.1868 J (exactly).
The thermochemical calorie: 1 cal_{th} = 4.1840 J.
The British thermal units: 1 Btu_{IT} = $1.05505 \cdot 10^3$ J; 1 Btu_{th} = 1.0543510^3 J.
1 CHU= 1899.1 J.

15.2 Thermal Conductivity of the Elements

In Table 15.1, the *generalized* values of thermal conductivity of the elements are mainly given over the temperature range from 4 K up to 1000 K. In a few cases, the data of specific experiments are given.

At moderate and high temperatures, the thermal conductivities of solid elements are close to that of a well-annealed high-purity (99.99 + %) sample. At low temperatures, the thermal conductivities greatly depend upon the crystal lattice imperfection, impurity concentration, and other defects. The values of λ at temperatures near and below the temperature at which the maximum in the thermal conductivity occurs, apply to a sample of a highest purity perfection. For the metals, the values of residual electrical resistivity ρ_0 (or ratio $\rho(300\ K)/\rho_0$), which characterize the quality of a metallic sample, are given.

The thermal conductivity of gaseous elements is given for a pressure of 10^5 Pa.

In total, among different elements, the uncertainty of thermal conductivity values varies from 2 to 10% at moderate temperatures and increases up to 20% at low and high temperatures.

Detailed information about the thermal conductivity of the elements can be found in Ho et al. [4, 5], Vargaftic et al. [6], and Okhotin et al. [7].

Table 15.1 Thermal conductivity of the elements, W/(m·K) [4]

Element	*T*, K											
	4	10	20	40	80	150	200	300	400	600	800	1000
Aluminum: $\rho_0 = 0.594 \cdot 10^{-9}\ \Omega \cdot$ cm	15700	23500	11700	2400	430	248	237	237	240	230	220	**93L**
Antimony: polycrystalline	190	480	240	110	55	36	30	24	21	18	17	**27L**
Argon: gas [6]					0.006*[1]	0.0096	0.0125	0.0178	0.0224	0.0304	0.0378	0.0434
Arsenic: polycrystalline							69	50	41			
Barium: $\rho_0 = 0.25 \cdot 10^{-6}\ \Omega \cdot$ cm	39					21	19	18				
Beryllium: polycrystalline,	720	1800	3500	4600	1600	450	300	200	160	126	106	91
$\rho_0 = 13.5 \cdot 10^{-9}\ \Omega \cdot$ cm												
Bismuth: polycrystalline			90	41	20.3	11.8	9.7	7.9	7.0	**13L**	**15L**	**17L**
‖ to trigonal axis			70	31	14.8	8.3	6.7	5.3	4.7			
⊥ to trigonal axis	1700	290	100	47	23.0	13.6	11.2	9.2	8.2			
Boron: polycrystalline	38	180	350	430	260	94	55	27	17	10.6	9.6	9.9
Bromine: liquid								0.12				
Cadmium: polycrystalline,	30000	1200	230	130	106	101	99	97	95	88*[2]	**42L**	
$\rho_0 = 0.112 \cdot 10^{-9}\ \Omega \cdot$ cm												
‖ to *c*-axis,	26000	1150	210	113	91	86	85	83	81	75*[2]		
$\rho_0 = 0.134 \cdot 10^{-9}\ \Omega \cdot$ cm												
⊥ to *c*-axis,	31000	1250	240	140	113	108	106	104	101	94*[2]		
$\rho_0 = 0.103 \cdot 10^{-9}\ \Omega \cdot$ cm												
Calcium: $\rho(237\ \mathrm{K})/\rho_0 = 70$ [26]				280	190	190	190	180				
Carbon: amorphous	0.016	0.071	0.16	0.31	0.56	0.94	1.2	1.6	1.9	2.2	2.4	2.5
diamond type I	11	140	790	2900	3500	2000	1400	900	650			
diamond type IIa	26	320	1700	6600	11700	6000	4000	2300	1500			
diamond type IIb	16	200	1100	4400	6600	3300	2300	1350	930			
graphite pyrolytic,												
‖ to *c*-axis		1.2	4.0	12	18	13	9.2	5.7	4.1	2.7	2.0	1.6
⊥ to *c*-axis	10	81	420	1600	4300	4500	3200	2000	1400	890	670	530
Cerium: polycrystalline,	0.48	1.1	1.9	3.2	5.2	7.7	9.0	11	13	17	19	22
$\rho(293\ \mathrm{K})/\rho(20\ \mathrm{K}) = 1.93$												
Cesium: $\rho_0 = 41.8 \cdot 10^{-9}\ \Omega \cdot$ cm	110	69	55	47	41	38	37	36	**20L**	**21L**	**19L**	**17L**
same as above [6]								18.2**L**	18.5**L**	17.3**L**	15.3**L**	
Chlorine: gas [6]							0.0054	0.0088	0.0124	0.0188		
Chromium: polycrystalline,	160	390	590	430	180	129	111	94	91	81	71	65
$\rho_0 = 60.8 \cdot 10^{-9}\ \Omega \cdot$ cm												
Cobalt: polycrystalline,	110	260	440	380	190	140	122	100	85	67	58	52
$\rho_0 = 90.75 \cdot 10^{-9}\ \Omega \cdot$ cm												

Table 15.1 Thermal conductivity of the elements, W/(m·K) [4] *(continued)*

Element	T, K											
	4	10	20	40	80	150	200	300	400	600	800	1000
Copper: $\rho_0 = 0.579 \cdot 10^{-9}\ \Omega \cdot$ cm	16200	24000	10800	2170	560	429	413	401	393	379	366	352
Deuterium: gas [6]			0.0120	0.0262	0.049	0.081	0.101	0.133	0.163	0.216	0.27	0.32
Dysprosium: polycrystalline, $\rho_0 = 4.93 \cdot 10^{-6}\ \Omega \cdot$ cm	4.4	10	14	14	12	9.0	9.6	11	11	12	14	15
ǁ to C-axis, $\rho_0 = 5.77 \cdot 10^{-6}\ \Omega \cdot$ cm		8.4	12	12.4	11.7	8.7	10	12				
⊥ to C-axis, $\rho_0 = 4.59 \cdot 10^{-6}\ \Omega \cdot$ cm		11	16	15	12	9.7	9.3	10				
Erbium: polycrystalline	3.6	7.1	7.8	9.3	11	13.7	14.6	14.3	14	14	15	16
ǁ to C-axis		7.2	6.4	8.6	12	17.4	18.5	18.4				
⊥ to C-axis		7.0	8.6	9.6	10	12	12.6	12.6				
Europium	2.5*3					17	15	14		10*4	11.5*4	
Fluorine: gas [6]						0.0135	0.0182	0.027	0.035	0.049		
Gadolinium: polycrystalline, $\rho_0 = 3.71 \cdot 10^{-6}\ \Omega \cdot$ cm	4.9	12	17	17	15	13	12	10.5				16*5
ǁ to C-axis, $\rho_0 = 2.62 \cdot 10^{-6}\ \Omega \cdot$ cm	5.9	14	17.5	16	14	12	11.2	10.8		14.9*6	16.4*6	17.8*6
⊥ to C-axis, $\rho_0 = 4.43 \cdot 10^{-6}\ \Omega \cdot$ cm	4.4	11	17	18	16	13.3	11.9	10.4		12.3*7	14.3*7	16.3*7
Gallium*8 : ǁ to a-axis, $\rho_0 = 0.100 \cdot 10^{-9}\ \Omega \cdot$ cm	9900	1200	270	80	50	44	42	41	34.5**L**	46.2**L**		
ǁ to b-axis, $\rho_0 = 0.034 \cdot 10^{-9}\ \Omega \cdot$ cm	27000	3300	1700	200	98	92	90	88				
ǁ to C-axis, $\rho_0 = 0.425 \cdot 10^{-9}\ \Omega \cdot$ cm	3100	350	84	33	20	16.7	16.3	15.9				
Germanium	880	1800	1500	800	330	132	97	60	43	27	20	17
Gold: $\rho_0 = 5.50 \cdot 10^{-9}\ \Omega \cdot$ cm	2100	3200	1580	520	332	325	323	317	311	298	284	270
Hafnium: polycrystalline, $\rho_0 = 4.23 \cdot 10^{-6}\ \Omega \cdot$ cm	3.5	9.5	18	24	26	25	24	23.0	22.3	21.3	20.8	20.7
Helium: gas [6]	0.0081	0.0162	0.0255	0.040	0.064	0.098	0.118	0.156	0.190	0.253	0.309	0.362
Holmium: polycrystalline, $\rho_0 = 2.67 \cdot 10^{-6}\ \Omega \cdot$ cm	5.8	12	14	16	15	13.5	15	16	13.5*9	14*9	15*9	16.5*9
ǁ to C-axis, $\rho_0 = 3.21 \cdot 10^{-6}\ \Omega \cdot$ cm		13	16	18	17	18	20	22				

Table 15.1 Thermal conductivity of the elements, W/(m·K) [4] *(continued)*

Element	T, K											
	4	10	20	40	80	150	200	300	400	600	800	1000
⊥ to c-axis, $\rho_0 = 2.82 \cdot 10^{-6}\ \Omega \cdot$ cm		12	14	15	14	12	13	14				
Hydrogen: gas, normal [6]			0.0163	0.0311	0.056	0.102	0.134	0.187	0.231	0.308	0.379	0.450
para-molecular form			0.0165	0.0310	0.060	0.128	0.157	0.193	0.232	0.309	0.380	0.450
Indium: polycrystalline, $\rho_0 = 0.587 \cdot 10^{-9}\ \Omega \cdot$ cm	5000	590	190	109	99	94	90	82	75	43**L**	48**L**	
Iodine								0.45**S**		0.007**G**		
Iridium: $\rho_0 = 19.1 \cdot 10^{-9}\ \Omega \cdot$ cm	520	1300	1900	750	210	160	153	147	144	138	132	126
Iron: $\rho_0 = 14.3 \cdot 10^{-9}\ \Omega \cdot$ cm	680	1480	1540	620	175	104	94	80	70	55	43	32
Krypton: gas [6]					0.0040*[10]	0.0050	0.0066	0.0096	0.0122	0.0168	0.0209	0.0245
Lanthanum: polycrystalline, $\rho_0 = 1.29 \cdot 10^{-6}\ \Omega \cdot$ cm	8.8	18	17	10	9.4	10.9	11.8	13.5	14.9	18	21	23
Lead: $\rho_0 = 0.862 \cdot 10^{-9}\ \Omega \cdot$ cm	2200	180	59	45	41	38	37	35	34	31	19**L**	22**L**
Lithium: $\rho_0 = 37.2 \cdot 10^{-9}\ \Omega \cdot$ cm	260	610	720	340	120	95	90	85	80	48**L**	54**L**	60**L**
same as above [6]										46.8**L**	52.4**L**	57.6**L**
Lutetium: polycrystalline, $\rho_0 = 1.45 \cdot 10^{-6}\ \Omega \cdot$ cm	7.9	20	25	22	20	18	18	16		13*[11]	14.5*[11]	16*[11]
‖ to c-axis, $\rho_0 = 0.76 \cdot 10^{-6}\ \Omega \cdot$ cm	16	36	41	32	29	26	25	23				
⊥ to c-axis, $\rho_0 = 2.65 \cdot 10^{-6}\ \Omega \cdot$ cm	5.3	15	19	18	16	15	15	14				
Magnesium: polycrystalline, $\rho_0 = 2.61 \cdot 10^{-9}\ \Omega \cdot$ cm	3800	5600	2700	720	200	161	159	156	153	149	146	84**L**
Manganese: $\rho_0 = 11.3 \cdot 10^{-6}\ \Omega \cdot$ cm	0.96	1.6	2.4	3.6	5.3	6.6	7.2	7.8			12*[12]	14*[12]
Mercury: polycrystalline	230	46	40	36	33	30	29	8.3**L**	9.8**L**	12.0**L**	12.8**L**	11.7**L**
‖ to trigonal axis	280	58	50	45	40	36	34					
⊥ to trigonal axis	200	40	35	32	29	27	26					
Molybdenum: $\rho_0 = 0.167 \cdot 10^{-6}\ \Omega \cdot$ cm	61	150	290	360	210	149	143	138	134	126	118	112
Neodymium: polycrystalline							17	17	17	18	20	22
same as above [30]	1.1	2.1	3.5									
Neptunium: polycrystalline							6.3					
Neon: gas [6]				0.010	0.018	0.030	0.037	0.049	0.059	0.078	0.094	0.109
Nickel: $\rho_0 = 11.2 \cdot 10^{-9}\ \Omega \cdot$ cm	860	1800	1650	580	210	122	107	91	80	66	68	72

Table 15.1 Thermal conductivity of the elements, W/(m·K) [4] *(continued)*

Element	T, K											
	4	10	20	40	80	150	200	300	400	600	800	1000
Niobium: $\rho_0 = 67.9 \cdot 10^{-9}\ \Omega \cdot cm$	140	290	250	95	58	53	53	54	55	58	61	64
Nitrogen: gas [6]					0.0077	0.0143	0.0186	0.0261	0.0327	0.0448	0.0564	0.0676
Osmium: polycrystalline,	420	1020	1600	640	140	96	91	88	87	87	87	87
$\rho_0 = 23.4 \cdot 10^{-9}\ \Omega \cdot cm$												
∥ to C-axis,	590	1400	2200									
$\rho_0 = 16.7 \cdot 10^{-9}\ \Omega \cdot cm$												
⊥ to C-axis,	350	870	140									
$\rho_0 = 27.8 \cdot 10^{-9}\ \Omega \cdot cm$												
Oxygen: gas [6]					0.0072	0.0143	0.0187	0.0269	0.0346	0.0495	0.0633	0.0760
Palladium: $\rho_0 = 12.3 \cdot 10^{-9}\ \Omega \cdot cm$	760	1150	600	170	81	73	72	72	74	80	87	94
Phosphorus:												
black, polycrystalline	0.51	6.5	27	44	35	23	18	12				
white, amorphous							0.31	0.24	0.18**L**	0.16**L**		
Platinum: $\rho_0 = 10.6 \cdot 10^{-9}\ \Omega \cdot cm$	880	1200	500	140	82	74	73	72	72	73	76	79
Plutonium:					2.9	3.6	4.1	5.2	7.3	11	12	
polycrystalline [31]												
Potassium: $\rho_0 = 2.20 \cdot 10^{-9}\ \Omega \cdot cm$	1900	460	170	115	108	105	104	102	52**L**	44**L**	37**L**	31**L**
same as above [6]									52.2**L**	44.5**L**	37.6**L**	31.2**L**
Praseodymium: polycrystalline					6.9	9.3	11	13	14	16	18	22
Promethium								18	18	19	20	21
Radium [23]								19			-	
Radon: gas							0.0026[*13]	0.0036	0.0048	0.0069	0.0087	0.0104
Rhenium: polycrystalline,	2500	3600	1200	160	63	54	51	48	46	44	44	45
$\rho_0 = 3.66 \cdot 10^{-9}\ \Omega \cdot cm$												
Rhodium: $\rho_0 = 8.40 \cdot 10^{-9}\ \Omega \cdot cm$	1200	2800	3600	1020	240	158	154	150	146	136	127	121
Rubidium: $\rho_0 = 38.4 \cdot 10^{-9}\ \Omega \cdot cm$	190	110	69	64	61	59	59	58	32L	29**L**	25**L**	22**L**
same as above [6]									31.8**L**	28.0**L**	24.4**L**	20.8**L**
Ruthenium: polycrystalline,	620	1500	2300	950	190	128	118	117	114	108	102	98
$\rho_0 = 15.8 \cdot 10^{-9}\ \Omega \cdot cm$												
Samarium: polycrystalline,	5.6[*14]	6.1	6.9	7.5	7.1	9.2	11	13	13	14	16[*15]	
$\rho_0 = 6.73 \cdot 10^{-6}\ \Omega \cdot cm$												
Scandium: polycrystalline,	2.8	6.8	12	14	14	15	15.5[*16]	16[*16]	16.2[*16]	16.7[*16]	17.2[*16]	17.7[*16]
$\rho_0 = 10.6 \cdot 10^{-6}\ \Omega \cdot cm$												
Selenium: ∥ to C-axis	140	140	59	26	13	7.6	6.1	4.5	5.4			
⊥ to C-axis	27	36	17	7.4	3.6	2.2	1.7	1.3	1.5			

Table 15.1 Thermal conductivity of the elements, W/(m·K) [4] *(continued)*

Element	T, K											
	4	10	20	40	80	150	200	300	400	600	800	1000
Selenium amorphous	0.032	0.042	0.056	0.079	0.13	0.20	0.26	0.53				
Silicon	300	2300	5000	3500	1340	410	260	150	99	62	42	31
Silver: $\rho_0 = 0.621 \cdot 10^{-9}\ \Omega \cdot$ cm	14700	16800	5100	1050	470	432	430	429	425	412	396	379
Sodium: $\rho_0 = 1.47 \cdot 10^{-9}\ \Omega \cdot$ cm	4900	2200	610	190	135	140	142	141	**87L**	**76L**	**67L**	**58L**
same as above [6]									**86.2L**	**74.6L**	**64.7L**	**55.5L**
Strontium						45	41	35	32	28	28	26
Sulfur: polycrystalline	10.6	8.2	2.4	1.1	0.65	0.43	0.36	0.27	0.13**L**	0.17**L**		
amorphous					0.16*1	0.175	0.185	0.206				
Tantalum: $\rho_0 = 0.214 \cdot 10^{-6}\ \Omega \cdot$ cm	46	107	140	87	60	58	58	58	58	59	59	60
Technetium: polycrystalline [28]								55	53	49	51	55
Tellurium: $\parallel$ to *c*-axis	670	310	95	32	12.2	5.9	4.6	3.4	2.8	2.3	4.2**L**	6.5**L**
$\perp$ to *c*-axis	250	130	41	15	6.2	3.3	2.6	2.0	1.7	1.4		
Terbium: polycrystalline, $\rho_0 = 2.19 \cdot 10^{-6}\ \Omega \cdot$ cm	8.4	19	23	19	15	12	10	11		14.7*17	15.3*17	16.7*17
$\parallel$ to *c*-axis, $\rho_0 = 1.87 \cdot 10^{-6}\ \Omega \cdot$ cm	18*14	26	29	23	19	15	13	15		15*18	16.4*18	18*18
$\perp$ to *c*-axis, $\rho_0 = 2.37 \cdot 10^{-6}\ \Omega \cdot$ cm	11*14	16	20	17	14	11	9.0	9.6		13*18	14.9*18	17*18
Thallium: polycrystalline, $\rho_0 = 0.24 \cdot 10^{-9}\ \Omega \cdot$ cm	1800	190	81	65	58	52	49	46	44			
Thorium: $\rho_0 = 26.8 \cdot 10^{-9}\ \Omega \cdot$ cm	360	470	170	84	63	56	55	54	55	56	57	58
Thulium: polycrystalline, $\rho_0 = 1.8 \cdot 10^{-6}\ \Omega \cdot$ cm	82	23	18	11	13	15	16	17		15*19	16*19	18*19
$\parallel$ to *c*-axis, $\rho_0 = 3.5 \cdot 10^{-6}\ \Omega \cdot$ cm	14*14	21	14	10.5	19	22	23.5	24				
$\perp$ to *c*-axis, $\rho_0 = 1.7 \cdot 10^{-6}\ \Omega \cdot$ cm	14*14	24	20	11	10.5	12.6	13.4	14.1				
Tin: polycrystalline, $\rho_0 = 0.132 \cdot 10^{-9}\ \Omega \cdot$ cm	1800	1900	320	130	92	78	73	67	62	**32L**	**36L**	**41L**
$\parallel$ to *c*-axis, $\rho_0 = 0.170 \cdot 10^{-9}\ \Omega \cdot$ cm	14000	1500	250	104	71	60	57	52	48			
$\perp$ to *c*-axis, $\rho_0 = 0.118 \cdot 10^{-9}\ \Omega \cdot$ cm	20000	2200	360	150	102	87	82	74	69			

Table 15.1 Thermal conductivity of the elements, W/(m·K) [4] *(continued)*

Element	T, K											
	4	10	20	40	80	150	200	300	400	600	800	1000
Titanium: polycrystalline, $\rho_0 = 1.90 \cdot 10^{-6}\ \Omega \cdot$ cm	5.8	14	28	39	33	27	25	22	20	19	20	21
Tungsten: $\rho_0 = 1.7 \cdot 10^{-9}\ \Omega \cdot$ cm	560	9700	4100	690	230	192	185	174	159	137	125	118
Uranium: polycrystalline, $\rho_0 = 2.14 \cdot 10^{-6}\ \Omega \cdot$ cm	4.4	9.8	16	18	21	24	25	28	30	34	39	44
Vanadium: $\rho_0 = 1.72 \cdot 10^{-6}\ \Omega \cdot$ cm	5.6	14	26	39	39	32	31	30.7	31.3	33.3	36	36
Xenon: gas [6]							0.0038	0.0056	0.0073	0.0103	0.0128	0.0154
Ytterbium: polycrystalline [33]						20	21	47				
same as above, $\rho(293\ \text{K})/\rho_0 = 4.9$ [34]		12	13	11	10							
Yttrium: polycrystalline, $\rho_0 = 5.54 \cdot 10^{-6}\ \Omega \cdot$ cm	2.7	7.0	13	15.2	15.7	16.4	16.6	17.2	18	21	23	25
‖ to *c*-axis, $\rho_0 = 2.30 \cdot 10^{-6}\ \Omega \cdot$ cm	5.3	13	23	24	24	25						
⊥ to *c*-axis, $\rho_0 = 8.70 \cdot 10^{-6}\ \Omega \cdot$ cm	1.9	5.0	9.4	12.1	13.3							
Zinc: polycrystalline, $\rho_0 = 1.28 \cdot 10^{-9}\ \Omega \cdot$ cm	7100	4700	1000	280	130	117	118	116	111	103	**56L**	**67L**
Zirconium: polycrystalline	44	100	110	59	37	28	25	23	22	21	22	24

*1 $T = 90$ K.
*2 $T = 294.258$ K – melting point.
*3 Data from Ref. [27].
*4 $\rho(300\ \text{K}) = 86 \cdot 10^{-6}\ \Omega$·cm [28].
*5 $\rho(300\ \text{K}) = 130 \cdot 10^{-6}\ \Omega$·cm [29].
*6 $\rho(300\ \text{K}) = 127 \cdot 10^{-6}\ \Omega$·cm [28].
*7 $\rho(300\ \text{K}) = 136 \cdot 10^{-6}\ \Omega$·cm [28].
*8 The values for thermal conductivity along *a*-axis are also good for polycrystalline gallium.
*9 $\rho(300\ \text{K}) = 90 \cdot 10^{-6}\ \Omega$·cm [28].
*10 $T = 120$ K.
*11 $\rho(300\ \text{K}) = 65 \cdot 10^{-6}\ \Omega$·cm [28].
*12 $\rho(300\ \text{K}) = 165 \cdot 10^{-6}\ \Omega$·cm [29].
*13 $T = 211$ K – melting point.
*14 $T = 6$ K.
*15 Data from Ref. [32].
*16 $\rho(300\ \text{K}) = 52 \cdot 10^{-6}\ \Omega$·cm [28].
*17 $\rho(300\ \text{K}) = 129 \cdot 10^{-6}\ \Omega$·cm [29].
*18 $\rho_{\parallel c}(300\ \text{K}) = 98 \cdot 10^{-6}\ \Omega$·cm, $\rho_{\perp c}(300\ \text{K}) = 127 \cdot 10^{-6}\ \Omega$·cm [28].
*19 $\rho(300\ \text{K}) = 78 \cdot 10^{-6}\ \Omega$·cm [28].

Note: The symbols have the following meanings: ‖ (⊥) denotes the thermal conductivity for the direction parallel (perpendicular) to the specified crystal axes; **G**: gas phase; **L**: liquid phase; **S**: solid phase.

15.3 Thermal Conductivity of Gases and Vapors

Thermal conductivity of rarefied gas weakly depends upon temperature (as $T^{0.6}$) and pressure (see Tables 15.2 and 15.3). A gas can be considered as rarefied for pressures on the order of 10^6 Pa at room temperature and for pressures up to about $4 \cdot 10^6$ Pa at $T = 1000$ K [8]. To estimate the thermal conductivity of rarefied gas, the Eucken formula can be used:

$$\lambda = 10^2 \eta c_V / M[1 + (9/4)R/c_V], \tag{15.3}$$

where η is the dynamic viscosity, Pa·s; c_V is the molar specific heat at constant volume, J/(mol·K); M is the molar mass, kg/mol; R is the universal gas-constant, 8.31 J/(mol·K).

At high pressures, the λ depends upon the pressure. In Table 15.4, the pressure dependence of thermal conductivity is given. At low pressures, when the mean free path of molecules is compared with the size of the gas container (for most systems it is the case for $p < 10$ Pa), the thermal conductivity is proportional to the gas pressure and vanishes as pressure decreases. Under these conditions, the thermal conductivity is not a specific property of the gas, but in addition depends upon the energy transfer through the container boundary, which is characterized by the accommodation coefficient.

The results of theoretical and experimental investigations on the thermal conductivity of gases are given in Hirshfelder et al. [9], Ferziger and Kaper [10], and Reid et al. [11]. The heat conduction in a gap filled with very rarefied gas is discussed in Kogan [12]. Experimental data on thermal conductivity for a number of substances in gaseous state are systematized in Vargaftic et al. [6], Vargaftic [13], and Vargaftic et al. [14].

Table 15.2 Thermal conductivity of inorganic gases at a pressure of 10^5 Pa [6, 14]

T, K	λ, 10^{-3} W/(m·K)	T, K	λ, 10^{-3} W/(m·K)	T, K	λ, 10^{-3} W/(m·K)	T, K	λ, 10^{-3} W/(m·K)
Air		1200	80.6	600	34.4	1370	64.4
80	7.5	1400	92.0	Nitric oxide NO		Sulfur hexafluoride SF_6	
100	9.3	1600	103	120	10.2	200	4.53
150	13.8	2000	122	200	17.6	300	13.2
200	18.0	Carbon monoxide CO		300	25.9	400	20.3
300	26.2	80	6.91	400	33.6	600	32.7
400	33.8	100	8.71	600	47.8	800	43.2
600	46.9	150	13.2	800	62.7	1000	54.0
800	57.5	200	17.4	900	71.2	Water vapour H_2O	
1000	67.6	300	25.1	Nitrous oxide N_2O		373	25.1
1200	77.7	400	32.2	200	9.75	473	33.3
Ammonia NH_3		600	44.2	250	13.4	573	43.4
200	13.3	800	55.3	300	17.3	673	54.8
300	24.4	1000	65.6	400	25.3	773	67.0
400	37.4	1200	75.6	600	41.1	873	79.9
600	66.8	1400	85.4	700	48.1	973	93.4
800	98.5	1600	95.4	Sulfur dioxide SO_2		1073	107
Carbon dioxide CO_2		2000	118	270	8.20	Heavy water vapour D_2O	
200	9.47	Hydrogen sulfide H_2S		300	9.75	423	29.0
300	16.6	180	3.36	400	15.1	473	33.8
400	24.3	200	6.08	600	26.1	573	44.5
600	39.8	250	11.2	800	37.1	673	56.7
800	54.5	300	14.8	1000	47.6	773	70.1
1000	68.0	400	20.3	1200	57.2	823	77.2

Table 15.3 Thermal conductivity of organic gases at a pressure of 10^5 Pa [6]

T, K	λ, 10^{-3} W/(m·K)	T, K	λ, 10^{-3} W/(m·K)	T, K	λ, 10^{-3} W/(m·K)	T, K	λ, 10^{-3} W/(m·K)
Acetone CH_3COCH_3		600	49.0	450	14.7	600	86.2
300	11.0	640	54.1	500	16.4	700	107
350	15.5	Ethane C_2H_6		Freon-12 CF_2Cl_2		800	127
400	20.3	200	11.4	250	8.0	900	148
450	25.4	250	15.6	300	10.2	1000	169
480	28.6	300	21.4	350	12.5	Methanol CH_3OH	
n-Amino alcohol $C_5H_{11}OH$		350	28.5	400	15.0	380	23.8
420	23.6	400	36.1	450	17.5	400	26.1
450	26.8	450	44.3	Freon-13 CF_3Cl		450	31.6
500	32.6	500	53.0	250	9.4	500	37.5
600	47.2	600	71.5	300	12.5	570	46.8
Benzene C_6H_6		700	90.7	350	15.6	n-Octane C_8H_{18}	
360	15.7	Ethanol C_2H_5OH		400	18.6	420	22.2
400	19.7	360	22.0	450	21.7	450	25.4
450	24.8	400	26.4	Freon-21 $CHFCl_2$		500	31.1
500	30.1	450	32.0	300	8.6	600	43.7
550	35.5	500	38.1	350	11.3	680	54.8
600	41.0	570	47.7	400	13.9	n-Pentane C_5H_{12}	
660	47.8	Ethyl chloride C_2H_5Cl		450	16.4	320	16.6
n-Butane C_4H_{10}		260	8.85	Freon-22 CHF_2Cl		350	20.0
280	13.8	300	11.6	250	8.9	400	26.0
300	16.1	350	15.3	300	11.2	450	32.5
350	22.1	400	19.4	350	14.2	460	33.8
400	28.4	450	23.8	400	17.5	Propane C_3H_8	
450	35.2	500	28.6	430	19.5	260	13.5
500	42.4	540	32.7	n-Heptane C_7H_{16}		300	18.2
600	58.0	Ethyl ether $(C_2H_5)_2O$		380	19.6	400	31.1
Carbon tetrachloride CCl_4		350	19.9	400	21.7	500	45.9
300	6.71	400	25.9	450	27.4	600	62.6
350	8.34	450	31.5	500	33.5	700	81.3
400	9.95	500	36.7	600	46.6	800	101
450	11.6	Ethyl formate $HCOOC_2H_5$		700	61.2	Propylene C_3H_6	
500	13.2	350	16.9	n-Hexane C_6H_{14}		230	10.9
600	16.4	400	22.7	360	19.1	250	12.6
620	17.1	450	29.0	400	23.6	300	17.6
Carbon tetrafluoride CF_4		500	36.8	450	29.5	350	23.5
200	8.98	600	58.0	500	36.0	400	30.0
250	12.6	Ethylene C_2H_4		600	50.0	450	36.7
300	16.5	180	9.7	680	62.2	500	43.5
350	20.4	200	11.0	Isopropyl alcohol C_2H_7OH		600	57.9
400	24.2	250	15.2	400	25.4	Toluene C_7H_8	
450	27.8	300	20.6	450	31.1	400	18.7
500	31.2	350	27.0	500	37.3	450	24.5
600	37.5	400	34.2	560	45.4	500	30.2
700	43.5	450	42.4	Methane CH_4		600	41.7
Cyclohexane C_6H_{12}		500	51.0	120	12.7	680	51.2
300	12.2	600	66.2	150	16.2	p-Xylene C_8H_{10}	
350	16.8	Freon-11 $CFCl_3$		200	21.8	460	25.4
400	22.4	300	8.7	300	34.1	500	29.7
450	28.7	350	10.7	400	49.2	600	41.2
500	35.4	400	12.7	500	66.8	680	50.4

Table 15.4 Thermal conductivity of gases at various temperatures and pressures [6, 14]

Gas	T, K	λ, 10^{-3} W/(m·K) at p, 10^5 Pa					Gas	T, K	λ, 10^{-3} W/(m·K) at p, 10^5 Pa				
		1	100	300	500	1000			1	100	300	500	1000
Air	100	9.3	124	143	158			100	68.4	94.0	147	192	292
	150	13.8	53.6	92.3	123			150	102	118	153	187	264
	200	18.0	29.4	61.9	82.8			200	134	145	170	197	258
	300	26.2	31.4	45.6	59.0	87.9		300	187	195	211	228	272
	400	33.8	37.3	46.4	55.4	76.9		400	231	238	249	261	295
	600	46.9	49.0	54.3	59.8	72.9		600	308	312	320	328	349
	800	57.5	59.5	63.5	67.6	77.0		800	379	382	388	394	410
	1000	67.6	69.7	73.2	76.4	83.5		1000	450	453	458	462	474
	1100	72.6	74.7	78.1	80.9	86.9		1500	630	632	635	638	646
Ammonia NH_3	200	13.2	658	662			*para*	20	16.5	121			
	300	24.4	479	501			Hydrogen p-H_2	50	37.5	107	177	223	285
	400	37.4	123	321				100	80.3	105	157	201	289
	600	66.8	76.9	108				150	128	139	178	212	286
Argon Ar	100	6.5	118	131				200	157	165	192	219	280
	150	9.6	61.5	84.3	98.6			300	193	198	214	233	279
	200	12.5	23.5	55.0	70.9			400	232	235	246	260	297
	300	17.8	22.5	34.8	47.0	69.6		600	309	310	316	325	350
	400	22.4	25.4	33.0	41.5	59.2		800	380	381	384	390	408
	600	30.4	32.9	36.7	41.4	52.8		1000	450	450	453	457	471
	800	37.3	39.0	42.0	45.3	53.7		1500		620	621	623	631
	1000	43.4	44.9	47.2	49.7	56.2	Methane CH_4	100		213	230	232	
Carbon dioxide	300	16.6	87.4	118	137	171		150	61.2	143	168	188	
CO_2	400	24.3	30.9	60.9	83.3	119.5		200	21.8	84.4	121	145	
	600	39.8	42.9	53.2	64.1	87.3		300	34.1	45.1	75.7	97.7	138
	800	54.5	56.6	63.2	69.8	85.6		400	49.2	56.3	72.7	88.4	120.5
	1000	68.0	69.6	74.1	78.8	90.6		600	86.2	90.9	100	109	129
Carbon	200	8.98	62.2	74.8	87.6			700	107	111	119	126	143
tetrafluoride	300	16.5	29.0	51.0	62.0		Nitrogen N_2	100	9.7	109	130	147	183
CF_4	400	24.2	30.9	48.5	59.9			150	14.3	48.5	80.0	100	139
Ethane C_2H_6	200	11.4	171	186	198			200	18.6	30.2	57.1	75.4	112
	300	21.4		111	130			300	26.1	31.1	44.6	57.4	85.1
	400	36.1	48.3	81.8	102	135		400	32.7	36.2	44.8	53.8	75.1
	600	71.5	75.3	87.7	99.7	124		600	44.8	47.1	52.2	57.5	71.2
	700	90.7	93.2	102	110	130		800	56.4	58.1	61.7	65.5	75.3
Ethylene C_2H_4	150	211^{*1}	222	233	244			1000	67.6	69.0	71.8	74.7	82.2
	200	11.0	166	188	205			1500	93.1	94.0	95.9	97.7	102.5
	300	20.6		107	129	168		2000	115	116	117	119	122
	400	34.2	45.0	77.1	97.3	133		2500	135.5	136	137	138	141
Helium He	10	16.2	52.8	78.2	94.7		Oxygen O_2	100	9.3	143	155	166	193
	20	25.5	55.7	91.0	119	171		150	14.3	79.8	106	125	164
	50	46.6	61.9	90.5	119	217		200	18.7	31.6	67.7	87.8	128
	100	74.1	81.9	102	119	178		300	26.9	31.8	46.2	60.2	88.5
	150	97.5	102	118	132	166		400	34.6	38.4	47.4	57.1	77.6
	200	118	122	134	146	173		600	49.5	52.1	57.8	63.8	77.9
	300	156	159	167	177	197		800	63.3	65.4	69.5	73.9	84.8
	400	190	192	199	206	224		1000	76.0	77.7	81.1	84.6	93.4
	600	253	254	258	264	277		1500	104	106	108	110	116
	800	309	311	314	318	330	Water vapour	273.15	561^{*1}	566^{*1}	577	588	612
	1000	362	364	367	370	379	H_2O	373.15	25.1	684^{*1}	695	706	732
	1500	480	481	483	484	490		473.15	33.3	671^{*1}	688	704	741
normal	20	16.3	121					573.15	43.4	551^{*1}	589	619	675
Hydrogen n-H_2	50	37.4	109	183	231	307							

Table 15.4 Thermal conductivity of gases at various temperatures and pressures [6, 14] *(continued)*

Gas	T, K	λ, 10^{-3} W/(m·K) at p, 10^5 Pa					Gas	T, K	λ, 10^{-3} W/(m·K) at p, 10^5 Pa				
		1	100	300	500	1000			1	100	300	500	1000
Water vapour	673.15	54.8	67.9	331	451	548	H_2O	973.15	93.4	100	121	151	236
H_2O	773.15	67.0	75.6	114	203	395		1073.15	107	113	131	156	225

*1 Liquid at pressure below critical.

15.4 Thermal Conductivity of Liquids

Thermal conductivity of liquids at 0°C for a pressure of one atmosphere can be estimated, within about ±10%, using the semi-empirical formula [15]

$$\lambda(0°\text{C}) = 9 \cdot 10^{-3} N^{-1/4} (T_b \gamma)^{1/2} c_p, \tag{15.4}$$

where N is the number of atoms in the molecule of the liquid; T_b is the boiling temperature, K; γ is the specific density of the liquid, g/cm^3; c_p is the specific heat at constant pressure, J/(cm^3·K).

Thermal conductivity of liquid increases with decreasing temperature (water and glycerin are an exception). In the temperature range from −50°C to 50°C, the dependence $\lambda(t)$ can be approximated by the relationship [15]

$$\lambda(t) = \lambda(0°\text{C})[1 + 0.01t(T_b^{1/2}/23.5 - 1)], \tag{15.5}$$

where t is the temperature, °C. Thermal conductivity of liquids is listed in Tables 15.5 and 15.6.

Thermal conductivity of liquids increases with pressure. This increasing is unessential for the pressures up to $5 \cdot 10^6$ Pa. At high pressures, up to $1.2 \cdot 10^9$ Pa, the thermal conductivity at temperature of 0°C can be calculated using the formula [15]:

$$\lambda(p) = \lambda(0\,\text{Pa})[1 + (T_b \gamma_0)^{-1/2} (p/1.44 \cdot 10^7)^{2/3}], \tag{15.6}$$

where γ_0 is the density at $p = 0$ Pa, g/cm^3; p is the pressure excess, Pa.

Thermal conductivity of liquids is discussed elsewhere in further detail [9, 11, 15, 16]. Several books provide compilations of experimental data [6, 13–15].

Table 15.5 Thermal conductivity of saturated liquids [6, 14]

T, K	λ, 10^{-3} W/(m·K)	T, K	λ, 10^{-3} W/(m·K)	T, K	λ, 10^{-3} W/(m·K)	T, K	λ, 10^{-3} W/(m·K)
Acetone CH_3COCH_3		400	230	350	128.5	350	91.5
200	197	Argon Ar		400	113.5	400	80.4
250	178	85	126	450	99	450	69.5
300	160	90	120	Carbon dioxide CO_2		470	65.1
350	141	100	108	220	181	Diphenyl $C_{12}H_{10}$	
400	122	110	96.3	250	140	350	137
450	104	120	84.2	270	113	400	130
500	84	130	71.8	290	90.0	450	123
Ammonia NH_3		140	59.2	295	86.0	500	116
200	657	150	40.4	300	85.0	550	109
250	562	Benzene C_6H_6		Carbon tetrachloride CCl_4^{*1}		580	105
300	467	290	148	250	114		
350	360	300	143	300	103		

Table 15.5 Thermal conductivity of saturated liquids [6, 14] *(continued)*

T, K	λ, 10^{-3} W/(m·K)	T, K	λ, 10^{-3} W/(m·K)	T, K	λ, 10^{-3} W/(m·K)	T, K	λ, 10^{-3} W/(m·K)
Ethanol C_3H_5OH		173	143	26	98.1	250	166
160	207	223	119	28	93.6	300	154
180	202	273	96.0	30	87.7	350	144
200	196	323	73.0	32	82.5	400	136
250	181	353	61.0	Methanol CH_3OH		450	128
300	166	Glycerol $C_3H_8O_3$		180	243	480	124
350	153	280	282	200	236	*m*-Terphenyl $C_{18}H_{14}$	
400	143	300	285	250	219	350	136
420	140	400	300	300	202	400	133
Ethylene glycol $(CH_2OH)_2$		500	309	350	189	450	130
280	250	Helium-I He-I		400	176	490	127
300	255	2.2	12.7	450	167	Toluene C_7H_8	
350	261	2.6	14.6	480	162	180	159
400	260	3.0	16.0	Nitrogen N_2		200	157
450	255	3.4	17.0	65	155	250	145
Freon-11 $CFCl_3$		3.8	17.8	70	149	300	132
173	120	4.2	18.5	80	133	350	120
223	108	4.6	19.8	90	114	400	108
273	95.8	4.8	21.4	100	95.3	450	97
323	82.5	5.0	25.3	110	77.3	500	85
373	69.5	normal Hydrogen n-H_2		120	59.9	550	76
413	57.5	14	81.1	124	54.9	Water H_2O	
Freon-12 CF_2Cl_2		16	91.6	Oxygen O_2		273.15	561
123	129	18	98.6	55	289	323.15	644
173	112	20	102.5	60	238	373.15	679
223	93.1	22	103.5	65	206	423.15	682
273	76.0	24	102.2	70	186	473.15	663
323	61.5	26	98.7	80	163	523.15	622
363	48.0	28	93.5	90	149	573.15	548
Freon-21 $CHFCl_2$		30	87.0	100	137	623.15	445
203	134	32	82.0	110	124	643.15	424
223	127	*para* Hydrogen p-H_2		120	109	Heavy water D_2O	
273	109	14	75.9	130	94.2	277.95	565
323	91	16	88.0	140	77.7	323.15	618
373	74.5	18	95.7	150	60.9	373.15	636
413	65.2	20	100	*n*-Propyl alcohol C_3H_7OH		423.15	625
Freon-22 CHF_2Cl		22	102	160	182	473.15	592
123	163	24	101	200	176	503.15	564

*[1] Recommended in Ref. [6] as a reference material.

Table 15.6 Thermal conductivity of oils and petroleum fractions [13, 15, 35]

Substance	t, °C	λ ,W/(m·K)*[1]	Substance	t, °C	λ ,W/(m·K)*[1]
Benzin (liquid)	−50 ··· 50 ··· 200	0.13 ··· 0.11 ··· 0.080	Oil:		
Diesel fuel	20 ··· 100	0.12 ··· 0.11	olive	20	0.17
Gasoline	30	0.14	red	30	0.14
Kerosene (liquid)	−50 ··· 50 ··· 300	0.13 ··· 0.11 ··· 0.074	transformer	70 ··· 100	0.18
Oil:			Paraffin	30	0.12
castor	20	0.18	Petroleum ether	30	0.13
light heat transfer	30 ··· 100	0.13	Vaseline	25	0.18

*[1] Thermal conductivity corresponds to the indicated temperatures.

15.5 Thermal Conductivity of Solids

In most cases, two mechanisms of heat transfer are responsible for the thermal conductivity of solids: the transport of thermal energy by the motion of conduction electrons (electronic thermal conductivity) and the heat transport by the lattice waves (phonon thermal conductivity). The first mechanism dominates in metals, the second determines the thermal conductivity of nonmetals. For some semiconductors, semi-metals, and strongly disordered alloys, both the conduction mechanisms give comparable contributions to the total conductivity.

Thermal conductivity of a solid is very sensitive to lattice imperfections (both the impurity atoms and lattice defects) at low temperatures. This is a result of the strong scattering of conduction electrons by imperfections of atomic scale in metals, and of the strong scattering of phonons by defects with dimensions of a few hundred in interatomic length in dielectrics. In nearly ideal dielectric crystals, at temperatures about 1 K, the mean free path of phonons becomes comparable to the sample width (usually about 5 mm). In this case, the thermal conductivity depends upon the character of phonon scattering by sample boundaries and upon sample sizes.

At high temperatures, the mean free path of heat carriers is limited by electron–phonon interaction in metals and by phonon-phonon interaction in electrically insulating crystals. Therefore, at high temperatures the thermal conductivity is weakly dependent on impurities and defects.

In amorphous dielectrics, the phonon mean free path is limited by the structural defect scattering over a wide temperature range. Thermal conductivity of amorphous solids is very much smaller than the thermal conductivity of crystals. Polycrystalline materials have thermal conductivity of intermediate magnitude between those of single crystals and amorphous solids.

The electronic thermal conduction of metals and alloys can be estimated by means of the Wiedemann–Franz law:

$$\lambda = L_0 \sigma T, \tag{15.7}$$

where L_0 is the Lorentz ratio, assumed to be a fundamental constant given by the Sommerfeld value: $L_0 = (\pi^2/3)(k_B/e)^2 = 2.445 \cdot 10^{-8}$ W·Ω/K^2; k_B is the Boltzmann constant and e is the electronic charge; σ is the electrical conductivity, $(\Omega\text{·m})^{-1}$; T is the temperature, K. For most of metals and alloys, the Wiedemann-Franz law is approximately true at high and very low temperatures. For pure metals, at intermediate temperatures, this law gives an overestimated value of thermal conductivity. For alloys it gives underestimated value (up to 10 times at temperatures near 20 K for highly disordered multicomponent alloys).

In this chapter, the values of thermal conductivity of some steels and alloys (Tables 15.7–15.17), nearly ideal dielectric single crystals (Table 15.18), glasses (Table 15.19), refractory materials and high-temperature composition of nuclear fuels (Tables 15.20–15.24), building and thermal insulating materials, woods, rocks and other materials (Tables 15.25–15.29) are given.

The theory of thermal conductivity of solids and experimental data are observed in other works [17–20]. An introduction to the theory of heat transport in solids and liquids under pressure, methods of measurements, and experimental data are given in Ross et al. [21]. A comprehensive set of thermal conductivity data for solids are found elsewhere [7, 22–25].

Table 15.7 Thermal conductivity of steels [22, 23]

Steel (composition in mass %)*1	T, K	λ, W/(m·K)*2
Armco iron, well-annealed,	10···20···40···80···150···	36.2···71.2···113···105···85.5···
$\rho_0 = 0.69 \cdot 10^{-6}$ Ω·cm [4]	300···600···1000···1183···1800	72.7···53.1···32.3···29.6···33.8
Armco iron*3, ρ(296 K) $= 14.5 \cdot 10^{-6}$ Ω·cm [36]	300···600···1000···1183···1673	71···52···32···32···38
St 0.8*4 (0.065 C, 0.4 Mn)	300···600···900	88···58···33
St 10*4 (0.1 C, 0.4 Mn, 0.17–0.37 Si)	300···600···800	83···57···44
St 15*4 (0.15 C, 0.35–0.65 Mn, 0.17–0.37 Si)	300···600···900	86···54···32
St 20*4 (0.15–0.25 C, 0.35–0.65 Mn, 0.17–0.37Si)	300···600···800···900	86···54···38···31

Table 15.7 Thermal conductivity of steels [22, 23] *(continued)*

Steel (composition in mass %)*1	T, K	λ, W/(m·K)*2
St 35*4 (0.32–0.4 C, 0.5–0.8 Mn, 0.17–0.37 Si)	300···600···800	85···50···36
St 45*4 (0.4–0.5 C, 0.5–0.8 Mn, 0.17–0.37 Si)	300···600···800	79···43···30
St 65G (0.62–0.7 C, 0.9–1.2 Mn, 0.17–0.37 Si)	300···600···1000	45···28···24
15KH (0.12–0.18 C, 0.4–0.7 Mn, 0.17–0.37 Si, 0.7–1 Cr)	300···600···800···1200	39···35···33···30
15KHA (0.12–0.17 C, 0.4–0.7 Mn, 0.17–0.37 Si, 0.7–1 Cr)	300···600···800···1200	39···35···33···30
20KH (0.17–0.23 C, 0.5–0.8 Mn, 0.17–0.37 Si, 0.7–1 Cr)	300···600···800···1200	39···35···33···30
30KH*4 (0.24–0.32 C, 0.5–0.8 Mn, 0.17–0.37 Si, 0.8–1.1 Cr)	300···600···900	48···38···28
35KH*4 (0.31–0.39 C, 0.5–0.8 Mn, 0.17–0.37 Si, 0.8–1.1 Cr)	300···600···900	48···38···28
40G*4 (0.37–0.45 C, 0.7–1 Mn, 0.17–0.37 Si)	300···500···700	65···51···46
35G2*4 (0.31–0.39 C, 1.4–1.8 Mn, 0.17–0.37 Si)	400···700	38···36
50G2*4 (0.46–0.55 C, 1.4–1.8 Mn, 0.17–0.37 Si)	300···600···800	43···36···35
40KHS*4 (0.37–0.45 C, 0.3–0.6 Mn, 1.2–1.6 Si, 1.3–1.6 Cr)	300···600···900	47···35···34
15KHM (0.11–0.18 C, 0.4–0.7 Mn, 0.17–0.37 Si, 0.8–1.1 Cr, 0.4–0.55 Mo)	300···600···800···1200	42···39···37···31
15KHMA (0.26–0.33 C, 0.4–0.7 Mn, 0.17–0.37 Si, 0.8–1.1 Cr, 0.15–0.25 Mo)	300···600···800···1200	42···39···37···31
30KHM (0.26–0.34 C, 0.4–0.7 Mn, 0.17–0.37 Si, 0.8–1.1 Cr, 0.15–0.25 Mo)	300···600···800···1200	39···38···37···350
30KHMA (0.26–0.33 C, 0.4–0.7 Mn, 0.17–0.37 Si, 0.8–1.1 Cr, 0.15–0.25 Mo)	300···600···800···1200	39···38···37···350
30KHGS (0.28–0.35 C, 0.8–1.1 Mn, 0.9–1.2 Si, 0.8–1.1 Cr)	300···600···800···1200	39···38···37···350
30KHGSA (0.28–0.34 C, 0.8–1.1 Mn, 0.9–1.2 Si, 0.8–1.1 Cr)	300···600···800···1200	39···38···37···350
15L–55L*5 (0.15–0.55 C depending upon the sort, 0.3–0.9 Mn, 0.2–0.4 Si, 0.45–0.6 S, 0.04–0.08 P)	300···600···800···1200	46···41···38···33
U8*6 (1.15–1.24 C, 0.15–0.35 Mn, 0.15–0.3 Si)	300···1200	50···26
U12 (1.15–1.24 C, 0.15–0.35 Mn, ≤ 0.2 Cr, 0.15–0.35 Si) [37, pp.197–198]	300···600···900···1200	45···37···32···25
R18 (0.7–0.8 C, 3.8–4.4 Cr, 17.5–18.5 W, 1–1.4 V, ≤ 1 Mo)	300···600···900···1200	22···26···26···24
R12 (0.8–0.9 C, 3.1–3.6 Cr, 12–13 W, 1.5–1.9 V, ≤ 1 Mo)[38]	300···500···700	16···19···26
12KH13*6 (0.09–0.15 C, 12–14 Cr, ≤ 0.8 Si, ≤ 0.8 Mn)	200···300···600···900···1400	31···31···33···34···33
07KH21G7AN5*6 (≤ 0.7 C, 6–7.5 Mn, 19.5–21 Cr, 5–6 Ni, 0.15–0.25 N) [24]	10···20···40···80··· 300	1.7···3.5···5.9···10.2··· 170
15KH12V2MF*6 (0.1–0.17 C, 0.5–0.8 Mn, 11–13 Cr, 1.7–2.2 W, 0.6–0.9 Mo, 0.15–0.3 V)	200···300···900···1400	30···31···33···32
18KH12VMBFR (0.15–0.22 C, 0.4–0.6 Mo, 0.5 Si, 11–13 Cr, 0.4–0.7 W, 0.15–0.3 V, 0.2–0.4 Nb)	200···300···600···900···1400	33···33···34···32···30
12KH18N9T*6 (≤ 0.12 C, 1–2 Mn, 17–19 Cr, ≤ 0.8 Si, 8–9.5 Ni, 5·C–0.8 Ti)	200···300···600···900···1400	13.5···14.5···19···23···28
12KH18N10T*6 (≤ 0.12 C, ≤ 2 Mn, ≤ 0.8 Si, 17–19 Cr, 9–11 Ni, 5·C–0.8 Ti) [24, 39]	10···20···40···80··· 150···300	1.5···3.7···5.5···8.2··· 11.0···15.1
12KH18N10T*7 (same as above) [40]	300···600···900···1100	15.0···19.8···26.6···27.8
20KH23N18 (≤ 0.1 C, ≤ 2 Mn, ≤ 1 Si, 22–25 Cr, 17–20 Ni)	200···300···600···1400	13.5···14···15···18
08KH16N13M2B (0.06–0.12 C, 15–17 Cr, 12.5–14.5 Ni, 2–2.5 Mo, 0.9–1.3 Nb)	200···300···600···1000··· 1400	14···15···16···17··· 18
08KH18N12B (≤ 0.08 C, 0.8 Si, 1–2 Mn, 17–19 Cr, 11–13 Ni, 8·C–1.2 Nb)	200···300···600···1000··· 1200	14···15···19···23··· 26
10KH18N9TL (≤ 0.14 C, ≤ 1 Si, 1–2 Mn, 17–20 Cr, 8–11 Ni, (C–0.03)*5–0.08 Ti)	200···300···600···1000··· 1400	13···14···18···25··· 28
KH35VT (≤ 0.12 C, 1–2 Mn, 14–16 Cr, 34–38 Ni, 2.8–3.5 W, 1.1–1.15 Ti)	200···300···600···1000··· 1400	13···14···15···17··· 19
KH35VTR (≤ 0.1 C, ≤ 1 Mn, 14.4–16 Cr, 35–38 Ni, 2.8–3.5 W, 1.1–1.5 Ti)	200···300···600···1000··· 1400	13···14···15···17··· 19
E11–E13, E1300 [41]	300	29
E41–E43A [41]	300	12

Table 15.7 Thermal conductivity of steels [22, 23] *(continued)*

Steel (composition in mass %)*1	T, K	λ, W/(m·K)*2
E310–E330 [41]	300	15
E45–E46 [41]	300	13
Cast iron:		
grey, medium strength	300	42–50
alloyed	400	29–58

*1 In the steel name, as a rule, number next to letter indicates the rounded mass percent of a component (when the component content is less than 1%, the number is omited).
*2 Thermal conductivity corresponds to the indicated temperatures.
*3 Chemical composition in mass percent: 99.5 Fe, 0.035 C, 0.12 S, 0.14 Mn, 0.025 S, 0.005 P, 0.20 Cu.
*4 Annealed cast steel.
*5 Annealed steel.
*6 Hardened steel.
*7 This steel is recommended as a standard reference material for thermal conductivity.

Table 15.8 Thermal conductivity of copper-based alloys [22, 24]

Alloy (composition in mass %)	T, K	λ, W/(m·K)*1
Brass:		
Cu62 (60.5–63.5 Cu, balance Zn)	300···600···900	110···150···200
Cu68 (67–70 Cu, balance Zn), deformed	80···150···300···900	71···84···110···120
Cu80 (79–81 Cu, balance Zn), rich low	300···600···900	110···120···140
CuAlNi59–3–2 (57–60 Cu, 2.5–3.5 Al, 2–3 Ni, balance Zn)	300···600···900	84···120···150
CuMn58–2 (57–60 Cu, 1–2 Mn, balance Zn)	300···600···900	70···100···120
CuSn70–1 (69–71 Cu, 1–1.5 Sn, balance Zn)	300···600	92···140
CuPb59–1 (57–60 Cu, 0.8–1.9 Pb, balance Zn), annealed	4···10···20···40···80	3.4···10···19···34···54
same as above	300	120
CuPb59–1V (57–61 Cu, 0.8–1.9 Pb, balance Zn)	300···600···900	110···140···180
Red brass:		
Cu90 (88–91 Cu, balance Zn), rolling	300···900	110···200
Cu96 (95–97 Cu, balance Zn), dragging	300···800	240···260
Bronze:		
Al5 (Cu, 4–6 Al)	300···600···900	105···105···150
AlFeMn10–3–1.5 (Cu, 9–11 Al, 3–4 Fe, 1–2 Mn)	300···600···800	59···77···84
Be2 (Cu, 1.8–2.1 Be), annealed 2 hrs at 573 K, soft	4···10···20···40···80	2.3···5.0···11···21···37
Cd1 (Cu, 0.9–1.2 Cd)	293	340
SiMn3–1 (Cu, 1–1.5 Mn, 2.75–3.45 Si)	300···500···700	42···55···54
Mn5 (Cu, 4.5–5.5 Mn)	300···500···700	94···110···130
SnP10–1 (Cu, 9–11 Sn, 0.4–1.1 P)	300···900	34···52
SnZn4–3 (Cu, 3.5–4 Sn, 2.7–3.3 Zn)	300···600···900	84···110···120
beryllium (Cu, 3.0 Be, $\leq$ 0.1 Fe), tempered	20···80···150···300	18···65···110···170
phosphorus (Cu, 18.07 Sn, 1.86 Zn, 0.16 Pb, 0.013 P), annealed	20···80···150···300	6.0···20···77···190
*Constantan**2 NiMn40–1.5 (Cu, 39–41 Ni, 1–2 Mn)	273···473···573···673	21···26···31···37
*Constantan**2 (55 Cu, 45 Ni)	4···10···20···40···80	0.8···3.5···8.8···13···18
same as above	300	23
Copel NiMn43–0.5*2 (Cu, 42.5–44 Ni, 0.1–0.5 Mn) [42]	473···1273	25···58
Cufenloy NiFeMn30–0.8–1 (Cu, 29–33 Ni, 0.5–1 Fe, 0.8–1.3 Mn)	300	29–37
Manganin NiMn3–12 (Cu, 2.5–3.5 Ni, 11.5–13.5 Mn)	273···573	22···36
same as above, [43, p. 320]	4···10···40···80···150···300	0.5···2···7···13···16···22

*1 Thermal conductivity corresponds to the indicated temperatures.
*2 Thermoelectric alloy.

Table 15.9 Thermal conductivity of nickel-based alloys [22, 24, 41]

Alloy (composition in mass %)	T, K	λ, W/(m·K)*1
Alumel MnAlCo–2–2–1*2 (Ni, 1.6–2.4 Al, 1.8–2.7 Mn, 0.85–1.5 Si, 0.6–1.2 Co) [42]	293···1073	27···44
Chromel NKH9.5*2 (Ni, 9.0–10 Cr, 0.6–1.2 Co) [42]	293···1073	18···34
Invar (35 Ni, 65 Fe)	273···573	11···13
Monel CuFeMn28–2.5–1.5(Ni, 27–29 Cu, 2–3 Fe, 1.2–1.8 Mn)	273···673	22···34
Nichrome KH20N80T (Ni, 20–23 Cr, 0.4–1.5 Si)	273···873	12···23
Ferronichrome KH15N60 (Ni, 15–18 Cr, 20–25 Fe, 0.4–1.5 Si)	273···673	12···18
Mn2.5 (Ni, 2.3–3.3 Mn)	293	53
Mn5 (Ni, 4.6–5.4 Mn)	293	48
44NKHTYU (43.5–45.5 Ni, 0.4–0.8 Al, 5.2–5.8 Cr, 2.2–2.7 Ti, balance Fe)	293	16
41NKHTA (41.5–43.5 Ni, 0.5–0.8 Al, 4.9–5.7 Cr, 2.2–3.0 Ti, balance Fe)	293	16
19NKH (18–20 Ni, 10–12 Cr, 0.3–0.6 Mn, 0.2–04 Si, balance Fe)	293	16
65NP (64.5–66 Ni, balance Fe)	293	29
79NM (76.5–80 Ni, 0.3–0.5 Si, 3.8–4.1 Mo, 0.6–1.1 Mn, balance Fe)	293	13

*1 Thermal conductivity corresponds to the indicated temperatures.
*2 Thermoelectric alloy.

Table 15.10 Thermal conductivity of aluminum-based alloys [24, 44]

Alloy (composition in mass %)	T, K	λ, W/(m·K)*1
AD1 (Al, admixtures $\leq$ 0.7), strain hardened	4···10···20···40	50···130···260···400
same as above	80···150···300	250···220···210
AV (Al, 0.45–0.9 Mg, 0.5–1.2 Si, 0.1–0.5 Cu, 0.15–0.35 Mn)	300···573	180..210
AD31 (Al, 0.1 Fe, 0.65 Mg, 0.1 Mn, 0.38 Si), temper-hardened	4···10···20···40	35···87···170···270
same as above	80···200	230···200
same as above	300–673	190
AD33 (Al, 0.8–1.2 Mg, 0.4–0.8 Si, 0.15–0.4 Cu, 0.15–0.35 Cr)	300···573	140···170
AK4 (Al, 1.9–2.5 Cu, 1.4–1.8 Mg, 1–1.5 Ni, 1.1–1.6 Fe, 0.5–1.2 Si) [22]	300···500···600···700	145···160···170···170
AK8 (Al, 3.9–4.8 Cu, 0.4–0.8 Mg, 0.6–1.2 Si), temper-hardened	20···40···80···150	50···72···10···125
same as above	300···573···673	160···180···180
AL1 (Al, 3.75–4.5 Cu, 1.75–2.25 Ni, 1.25–1.75 Mg) [22]	300···400···600	130···140···150
AL4 (Al, 8–10.5 Si, 0.6–1.2 Fe, 0.25–0.5 Mn, 0.17–0.3 Mg) [22]	300···473···673	150···160···155
AL5 (Al, 4.5–5.5 Si, 1–1.5 Cu, 0.35–0.6 Mg)	300···473···573	160···170···180
AL8 (Al, 9.5–11.5 Mg)	300···473···673	92···100···110
AMg2 (Al, 1.8–2.8 Mg, 0.2–0.6 Mn), annealed 1 hr at 623 K in vacuo	4···10···20···40	4.6···12···25···49
same as above	80···150···300	77··· 100···130
AMg5 (Al, 4.8–5.8 Mg, 0.5–0.8 Mn, 0.02–0.1Ti), annealed	10···20···40···80···	10···20···40···66···
same as above	150···300···473···673	92···130···130···150
AMn (Al, 1–1.6 Mn, 0.7 Fe, 0.6 Si, 0.2 Ti, 0.1 Zn, 0.05 Mg), strain hardened	4···10···20···40···80···	11···28···58···110···140···
	150···300···473···673	150···170···180···190
V93 (Al, 6.3–7.3 Zn, 1.6–2.2 Mg, 0.8–1.2 Cu, 0.2–0.45 Fe)	300···473···673	160···170···160
V95 (Al, 5–7 Zn, 1.8–2.8 Mg, 1.4–2 Cu, 0.2–0.6 Mn, 0.1–0.23 Cr)	300···473···673	155···160···160
VAL1 (Al, 5.5–6.2 Cu, 0.8–1.2 Ni, 0.6–1 Mn, 0.15–0.3 Cr, 0.05–0.2 Zn)	300···473···673	130···150···160
VAL5 (Al, 6.5–8.5 Si, 0.35–0.55 Mg, 0.1–0.3 Ti, 0.15–0.4 Be)	300···573···673	150···160···160
VAD1 (Al, 3.8–4.5 Cu, 2.3–2.7 Mg, 0.5–0.8 Mn, 0.08–0.15 Ti), annealed	20···80···300	30···61···160
VD17 (Al, 2.6–3.2 Cu, 2–2.4 Mg, 0.45–0.7 Mn)	300···673	130···170
D1 (Al, 3.8–4.8 Cu, 0.4–0.8 Mg, 0.4–0.88 Mn), annealed	20···40···80···150···293	30···55···97···125···170

Table 15.10 Thermal conductivity of aluminum-based alloys [24, 44] *(continued)*

Alloy (composition in mass %)	T, K	λ, W/(m·K)[*1]
D16 (Al, 3.8–4.8 Cu, 1.2–1.8 Mg, 0.3–0.9 Mn), temper-hardened	10···20···40···80···150··· 300···573	9···19···37···61···90··· 120···160
D20 (Al, 6–7 Cu, 0.25–0.45 Mg, 0.4–0.8 Mn, 0.1–0.2 Ti), temper-hardened	20···40···80···150 300···673	27···38···61···85 140···160

[*1] Thermal conductivity corresponds to the indicated temperatures.

Table 15.11 Thermal conductivity of titanium-based alloys [45]

Alloy (composition in mass %)	T, K	λ, W/(m·K)[*1]
VT1 (Ti, 0.014 C, 0.16 Fe, 0.045 Si, 0.028 N, 0.0092 H), annealed 40 min at 873 K in air [24]	10···20···40···80··· 150···300	5.5···10···14···18··· 20···16
VT5 (Ti, 4.3–6.2 Al)	300···673···973	8.8···12···17
VT6S (Ti, 5.3–6.8 Al, 3.5–5 V)	300···673···973···1173	8.4···13···17···20
VT8 (Ti, 6–7 Al, 2.8–3.8 Mo, 0.2–0.4 Si) [22]	300	7.1
VT14 (Ti, 3.5–6.3 Al, 2.5–3.8 Mo, 0.9–1.9 V)	300···673···973···1173	8.4···13···17···20
VT16 (Ti, 1.8–3.8 Al, 4.5–5.5 Mo, 4–5.5 V)	300···673···973···1173	10···15···18···21
VT3-1 (Ti, 5.5–7 Al, 2–3 Mo, 0.8–2.3 Cr, 0.15–0.4 Si) [22]	300	7.9
OT4 (Ti, 3.5–5 Al, 0.8–2 Mn)	300···673···873	9.6···13···16
OT41 (Ti, 1–2.5 Al, 0.7–2 Mn)		
OT40 (Ti, 0.2–1.4 Al, 0.2–1.3 Mn)	300···673···973···1173	13···15···18···20
OT42 (Ti, 5.7–6.7 Al, 1–2.3 Mn)	300···673···973···1173	7.1···13···18···21

[*1] Thermal conductivity corresponds to the indicated temperatures.

Table 15.12 Thermal conductivity of magnesium-based alloys [21, 46]

Alloy (composition in mass %)	T, K	λ[*1], W/(m·K)
MA1 (Mg, 1.3–2.5 Mn)	300···473···673	126···138···134
MA2 (Mg, 3–4 Al, 0.15–0.5 Mn, 0.2–0.8 Zn)	300···473···673	96···105···113
MA2-1 (Mg, 3.8–5 Al, 0.4–0.8 Mn, 0.8–1.5 Zn, 0.1 Ca), annealed 30 min at 533 K	20···80···300	13···26···96
MA5 (Mg, 7.8–9.2 Al, 0.15–0.5 Mn, 0.2–0.8 Zn)	300	59
MA8 (Mg, 1.5–2.5 Mn, 0.15–4.35 Ca), annealed 30 min at 623 K	300	130
MA11 (Mg, 1.5–2.5 Mn, 2.5–3.5 Nd)	300···473···673	110···117···117
MA15 (Mg, 2.5–3.5 Zn, 0.45–0.9 Zr) [47]	300	110
VM65-1 (Mg, 5–6 Zn, 0.3–0.9 Zr)	300···473···673	117···126···126
VMD1 (Mg, 1.2–2 Zn, 2.5–3.5 Th)	300	130
VMS1 (Mg, 3–4 Al, 1.2–2 Zn, 0.4–0.9 Zr)	293	96
ML2 (Mg, 1–2 Mn), as cast	293	130
ML3 (Mg, 2.5–3.5 Al, 0.15–0.5 Mn, 0.5–1.5 Zn), as cast	293	105
ML4 (Mg, 5–7 Al, 0.15–0.5 Mn, 2–3 Zn), hardened	293	180
ML5 (Mg, 7.5–9 Al, 0.15–0.5 Mn, 0.2–0.8 Zn), hardened	293	77
ML6 (Mg, 9–10.2 Al, 0.1–0.5 Mn, 0.6–1.2 Zn), as cast	293	77
ML8 (Mg, 5.5–6.6 Zn, 0.7–1.1 Zr, 0.2–0.8 Cd) [47]	293	120
ML10 (Mg, 0.1–0.7 Zn, 0.4–1 Zr, 2.2–2.8 Nd), temper-hardened	293	110
ML12 (Mg, 4–5 Zn, 0.6–1.1 Zr), tempered	293	130
ML14 (Mg, 1.7–2.3 Al–2.3 Al, 0.5–1 Zr, 2.6–3.8 Th), tempered	293	110
ML15 (Mg, 4–5 Zn, 0.7–1 Zr, 0.6–1.2 La), tempered	293	140
VML1 (Mg, 0.5–1 Zr, 2.5–4 Th), temper-hardened	293	110

[*1] Thermal conductivity corresponds to the indicated temperatures.

Table 15.13 Thermal conductivity of heat-resistant and refractory alloys on the base of refractory metals [7, 22]

Alloy (composition in mass %)	T, K	λ, W/(m·K)*1
KHN60V (Ni, $\leq$ 0.1 C, 23–26.5 Cr, 0.3–0.5 Ti, $\leq$ 0.5 Al, 13–16 W, 4 Fe)	200···300···600···1000	9.0···9.8···14···23
KHN60YU ($\leq$ 0.1 C, 15–18 Cr, 2.6–3.5 Al, $\leq$ 0.3 Mn, $\leq$ 0.8 Si, 55–58 Ni)	200···300···800···1400	8.0···9.6···19···29
KHN70VMYUT (Ni, $\leq$ 0.12 C, 13–16 Cr, 2–4 Mo, 0.1–0.5 V, 5–7 W, 1.8–2.3 Ti, $\leq$ 5 Fe, 1.7–2.3 Al)		
KHN70VMTYU (Ni, 0.1–0.16 C, 14–16 Cr, 1–1.4 Ti, 1.7–2.2 Al, 4–6 W, 3–5 Mo)	200···300···800···1400	7···8···18···24
KHN77TYU, KHN77TYUR (Ni, 19–22 Cr, 2.4–2.8 Ti, 0.6–1 Al)	200···300···600···1000	11···12···16···24
KHN78T (Ni, 0.12 C, 19–22 Cr, 0.15–0.35 Ti, 0.15 Al, 1 Fe)	300···900	13···23
KHN80TBYU (Ni, $\leq$ 0.08 C, $\leq$ 1 Mn, 15–18 Cr, $\leq$ 3Fe, 1–1.5 Nb, 1.8–23 Ti, 0.5–1 Al)	200···300···800···1400	11···12···21···27
WCu50 (W, 47.5–50 Cu)	573···773···973···1173	90···86···86···92
WRe20*2 (W, 20+0.5) Re)	523···773···973···1173	50···54···63···75
VM1 (Mo, 0.01 C, 0.08–0.25 Zr, 0.4 Ti)	373···773···1473···2173	130···120···110···90
VM2 (Mo, 0.02 C, 0.25–0.4 Zr, 0.2 Ti)		
VM3 (Mo, 0.25–0.5 C, 0.3–0.6 Zr, 0.8–1.3 Ti, 1–1.8 Nb)		
MoW30 (Mo, 0.01 C, 30 W)	673···973	120···100
Nb_3Sn (70.5 $\pm$ 0.05 Nb), superconductor	4···6···10···	0.038···0.13···0.65···
T_c = 18.3 K [48]	20···40···80	2.6···2.8···2.4

*1 Thermal conductivity corresponds to the indicated temperatures.
*2 Thermoelectric alloy.

Table 15.14 Thermal conductivity of alloys on the base of noble metals [24]

Alloy (composition in mass %)	T, K	λ, W/(m·K)*1
Ag–Cu (92.5 Ag, 7.5 Cu)	300	350
(90 Ag, 10 Cu)	300	345
(80 Ag, 20 Cu)	300	340
(50 Ag, 50 Cu)	300	310
Ag-Pb (95 Ag, 5 Pb)*2	300	220
(90 Ag, 10 Pb)*2	300	140
(80 Ag, 20 Pb)*2	300	92
(50 Ag, 50 Pb)*2	300	35
Ag-Au (99.63 Ag, 0.37 Au), [7]*3,*4	20···40···80···100	290···330···350···360
(90 Ag, 10 Au)	300	200
(30 Ag, 70 Au)	300	290
Au–Fe (Au, 0.03 Fe), annealed, ρ(300 K)/ρ_0 = 9.0, [7]*3,*4	4···10···20···40···80···300	50···130···210···240···250···310
Au–Co (Au, 2.1 Co), cold-drawing [7]*3,*4	4···10···20···40···80···100	1.0···4.2···8.5···14···20···24
Pt–Rh (90 Pt, 10 Rh)*3 [37]	300–400	30
Pt–Ir (90 Pt, 10 Ir)*5	300	30
(80 Pt, 20 Ir)*5	300	18
(70 Pt, 30 Ir)*5	300	16

*1 Thermal conductivity corresponds to the indicated temperatures.
*2 Machine processed in cold conditions.
*3 Thermoelectric alloy.
*4 Composition in atomic percent.
*5 Punched in cold conditions.

Table 15.15 Thermal conductivity of alloys on the base of radioactive metals [22]

Alloy (composition in mass %)	T, K	λ, W/(m·K)[*1]
U–Al (99 U, 1 Al)	200· · ·300· · ·900	14· · ·14· · ·13.6
(90 U, 10 Al)	200· · ·300· · ·900	10· · ·10· · ·9.8
(70 U, 30 Al)	200· · ·300· · ·800	31· · ·30· · ·28
(50 U, 50 Al)	200· · ·300· · ·500· · ·700	68· · ·66· · ·54· · ·34
(10 U, 90 Al)	300	190
UAl_3, $\gamma = 6.8$ g/cm^3	300	26
UBi	300	21
U_3Bi_4	300	19
UBi_2	300	17
U–Cr (95 U, 5 Cr)	300	21
(90 U, 10 Cr)	300	17
(70 U, 30 Cr)	300	10
(50 U, 50 Cr)	300	8.5
(30 U, 70 Cr)	300	16
U–Fe (95 U, 5 Fe)	300	24
(90 U, 10 Fe)	300	22
(70 U, 30 Fe)	300	17
(50 U, 50 Fe)	300	20
(30 U, 70 Fe)	300	33
U–Mo (95 U, 5 Mo)	300	25
(90 U, 10 Mo)	300	23
(70 U, 30 Mo)	300	27
(50 U, 50 Mo)	300	40
(10 U, 90 Mo)	300	110
U_3Si, $\gamma = 15.6$ g/cm^3	300	20
U_3Si_3, $\gamma = 12.2$ g/cm^3	300	14
U–Zr (95 U, 5 Zr)	300	19
(90 U, 10 Zr)	300	14
(70 U, 30 Zr)	300	6
(50 U, 50 Zr)	300	5
(30 U, 70 Zr)	300	6
(65 U, 20 Pu, 5 Mo, 10 fission products)	479· · ·808· · ·868	15· · ·23· · ·22
(70 U, 20 Pu, 10 fission products)	323· · ·573· · ·873· · ·1073	9.6· · ·16· · ·24· · ·30
Th–U (90 Th, 10 U)	323· · ·1073	36· · ·44
(80 Th, 20 U)	323· · ·1073	35· · ·43
(Zr, 8 U)	373· · ·723	14· · ·17
(Zr, 8 U, 1 H)	373· · ·473· · ·673	20· · ·18· · ·17

[*1] Thermal conductivity corresponds to the indicated temperatures.

Table 15.16 Thermal conductivity of antifriction alloys and solders [47]

Alloy (composition in mass %)	T, K	λ, W/(m·K)[*1]
Aluminum antifriction alloys		
AlNi2.5 (Al, 2.7–3.3 Ni)	300	130
AlSn6–1 (Al, 5.5–7 Sn, 0.7–1.3 Cu, 0.7–1.3 Ni)	300	180
Babbitts		
BCaAl (Pb, 0.95–1.15 Ca,0.7–0.9 Na, 0.05–0.2 Al)	300	21
B16 (Pb, 15–17 Sn, 15–17 Sb,1.5–2 Cu)	300	25
B83 (Sn, 10–12 Sb,5.5–6.5 Cu)	300	33
B88 (Sn, 7.3–7.8 Sb,2.5–3.5 Cu, 0.8–1.2 Cd, 0.15–0.25 Ni)	300	39

Table 15.16 Thermal conductivity of antifriction alloys and solders [47] *(continued)*

Alloy (composition in mass %)	T, K	λ, W/(m·K)*1
Solders		
Sn61Pb (Pb, 60–62 Sn)	300	50
(60 Sn, 40 Pb) [24]	4···10···40···80···300	18···40···51···49···50
Sn18Pb (Pb, 18 Sn, 2.5 Sb) [22]	300	38
Sn18Sb2Pb (Pb, 17–18 Sn, 1.5–2.0 Sb)	300	34
Sn40Sb2Pb (Pb, 39–41 Sn, 1.5–2.0 Sb)	300	42
Wood metal (50 Bi, 25 Pb, 12.5 Cd, 12.5 Sn) [24]	4···10···20···40···80···100	4.0···11···18···20···23···24
Rose metal (56.1 Bi, 28 Pb, 15.9 Sn) [24]	10···20···80···150···273	3.4···5···8.4···11···16
Ag25 (Zn, (25 ± 0.3) Ag, (40 ± 1) Cu)	300	105
Ag44 ((44 ± 1) Ag, $(27{+}1)$ Cu, (16 ± 2) Zn, (8 ± 1) Cd, (2 ± 0.5) Ni, (3 ± 0.5) Mn)	300	38
Ag70 (Zn, (70 ± 0.5) Ag, (26 ± 0.5) Cu)	300	170

*1 Thermal conductivity corresponds to the indicated temperatures.

Table 15.17 Thermal conductivity of semiconductors, W/(m·K) [7, 25]

Semiconductor	T, K					
	10	20	40	80	150	300
Cd_3As_2: undoped, $n = 2 \cdot 10^{18}$ cm^{-3}					2.7	2.8
CdSb: p-type, $n = (3-5) \cdot 10^{15}$ cm^{-3}, $\parallel$ to [001]				4.9 (100 K)	3.0	1.9
ZnSb: 99.9999% pure initial material	260	210	40	11.5	5.2	5.0
ZnO*1: $n_{imp} < 2.5 \cdot 10^{18}$ cm^{-3} [49]	300	520	450	260	134	54
ZnS: hexagonal lattice, $n_{imp} < 5 \cdot 10^{17}$ cm^{-3} [49]	300	380	310	155	70	27
CdS: $n_{imp} \geq 10^{16}$ cm^{-3}, $\parallel$ to c-axis	540	360	200	97	43	20
CdSe: n-type, > 99.99% pure initial material, undoped, $\parallel$ to c-axis	230	200	72	32		
CdTe: $n_{imp} < 2 \cdot 10^{18}$ cm^{-3} [49]	520	250	117	44	18.4	7.5
HgSe: n-type, n(4.2 K)$=2.1 \cdot 10^{17}$ cm^{-3} [50]	120	93	41	11	4.4	1.7
HgTe: p-type, acceptor concentration $10^{18} - 10^{19}$ cm^{-3}			25(60 K)	14	5.0	2.6
BN*2 [51]		2.0	10	43	112	180
AlN*3 [51]	65	100	175	290	330	200
AlSb: p-type, R(300 K) = 7.0 cm^3/C	72	280	330	210	115	69(250 K)
GaP: p-type, R(300 K) = 75 cm^3/C	190	590	700	450	210	140(250 K)
GaAs: n-type, n(77 K) $= 2 \cdot 10^{16}$ cm^{-3}	1400	2500	780	270	105	58
GaSb: p-type, $n = 1.5 \cdot 10^{17}$ cm^{-3}	140	340	320	180	85	36
InP: n-type, n(77 K) $= 2 \cdot 10^{16}$ cm^{-3}	180	2700	1200	470	190	70
InAs: n-type, n(77 K) $= 2 \cdot 10^{16}$ cm^{-3}	2900	1700	600	170		
InSb: n-type, $n = 7 \cdot 10^{13}$ cm^{-3}	2000	1100	370	90	42	
SiC: n-type, nitrogen concentration 10^{17} cm^{-3}, $\parallel$ to c-axis	350	1900	5100	4100	1500	490
PbS: p-type, n(300 K) $= 1.7 \cdot 10^{18}$ cm^{-3}	70	48	13	8.0		
PbS: natural, n(300 K) $= 1.48 \cdot 10^{17}$ cm^{-3}				5.5(100 K)	3.9	2.6
PbSe: p-type, n(300 K) $= 5.4 \cdot 10^{18}$ cm^{-3}	70	37	11.5	5.2		
PbSe: n-type, $n = 6.4 \cdot 10^{17}$ cm^{-3}, Cu doped				5.0	3.0	1.8
Bi_2Te_3: n-type, n(77 K) $= 3 \cdot 10^{17}$ cm^{-3}				6.4	3.5	2.9

*1 The mean values of thermal conductivity for the hexagonal lattice of ZnO are given: $\lambda_m = 1/3(2\lambda_a + \lambda_c)$, where λ_a and λ_c are the thermal conductivity values for the a and c crystal axes, respectively; in the range 30 K $< T <$ 300 K, $\lambda_a/\lambda_c = 1.2$.

*2 Hot-pressed polycrystalline ceramics, 97% of theoretical density, 20 microns grain size, oxygen and carbon impurity concentration $< 2 \cdot 10^{19}$cm^{-3}.

*3 Synthetic single crystal, oxygen concentration (1–5)$\cdot 10^{20}$ cm^{-3}.

Symbols: n is the carrier concentration; n_{imp} is the impurity concentration; R is the Hall constant.

Table 15.18 Thermal conductivity of dielectric single crystals, W/(m·K) [7, 25]

Crystal (composition in at.%)	T, K						
	4	10	20	40	80	150	300
BeO [51, 52]	47	750	5100	13500	6700	1360	370
MgO	270 (6 K)	1100	3000	2200	460	135	58
$Al_2O_3^{*1}$	240	3000	12500	13500	1100	125	40
Sapphire^{*2}, ∥ to *c*-axis	125	2300	10000	11500	1100	155	47
SiO_2^{*3}, ∥ to *c*-axis	470	2700	530	150	60	27.3	14.3
TiO_2	180	1750	600	65	18		
NiO_2	5	60	220	430	280	120	38
UO_2^{*4}	8.6	9.6	1.65	1.4	4.2	7.7	8.4
LiF:							
(0.02 ^{6}Li)	510	6000	10000	1380			
(7.4 ^{6}Li)^{*5}	440	2700	2800	930	127		
(50.1 ^{6}Li)	370	1900	1800	700	115	37	
NaF^{*6}, ∥ to [100] [53]	750	10000	14000	620	130		
CaF_2, ∥ to [100]	700	2600	2800	570	60	22	12.5
MnF_2^{*7}, ∥ to *a*-axis	67	430	450	135	40	16	7.7
CoF_2^{*8}	4.2	13	27	19.5	25	15.5	
NaCl	530	900	300	80	34	15.5	7.4
KCl	630	510	225	70	20		
RbCl	310	190	85	30	11		
NaBr	95	160	70	27	13		
KBr	480	250	90	34	15.5		4.8
KI	700	360	130	39	12		5.02
$SrTiO_3^{*9}$, ∥ to (100)	1.5	8.5	20	19	18	16	12
$Y_3Fe_5O_{12}^{*10}$, ∥ to [100]	13.5	125	220	112	37	15	7.4

^{*1} Sample dimensions: diameter of 5 mm, length of 50 mm.
^{*2} Impurity (Cr, Fe, Mg, Si) concentration $< 10^{-5}$cm^{-3}.
^{*3} Sample dimensions: 5 × 5 × 40 mm.
^{*4} Antiferromagnetic at $T < 30$ K.
^{*5} Natural isotopic composition.
^{*6} Impurity concentration $< 1 \cdot 10^{-6}$ cm^{-3}; square cross-section of 5.1 mm width.
^{*7} Antiferromagnetic at $T < 67$ K.
^{*8} Antiferromagnetic at $T < 38$ K; rod axis in direction of 26° to *c*-axis.
^{*9} Paraelectric, transition temperatures are 35, 65, and 110 K.
^{*10} Ferromagnetic at $T < 560$ K; mean diameter of 3.9 mm, length of 11.3 mm.

Table 15.19 Thermal conductivity of glasses [24, 54]

Material (composition in mass %)	T, K	λ, W/(m·K)^{*1}
Crown glass:		
(50–75 SiO_2, various other oxides)	173	0.50–0.86
	273	0.80–1.09
	373	0.88–1.21
light	90· · ·150· · ·300	0.52· · ·0.87· · ·1.2
barium (47 SiO_2, 4 B_2O_3, 1 Na_2O, 7 K_2O, 8 ZnO, 32 BaO)	173· · ·273· · ·373	0.63· · ·0.75· · ·0.88
borosilicate (60–65 SiO_2, 15–20 B_2O_3)	173	0.67–0.73
	273	0.88–0.94
	373	1.00–1.07
borosilicate (70–75 SiO_2, 5–10 B_2O_3)	173	0.80–0.86
	273	1.03–1.09
	373	1.15–1.21
Cut glass, $\gamma = 2.6 - 2.85$g/cm^3 [15]	293	0.88–0.91

Table 15.19 Thermal conductivity of glasses [24, 54] *(continued)*

Material (composition in mass %)	T, K	λ, W/(m·K)*1
Faience	300	0.93–1.3
Flint glass:		
(20–55 SiO_2, various other oxides)	173	0.38–0.63
	273	0.54–0.88
	373	0.63–0.96
light (65 SiO_2, 25 PbO, 10 others)	173	0.67–0.71
	273	0.88–0.92
	373	1.00–1.04
ordinary (45 SiO_2, 45 PbO, 10 others)	173	0.50–0.59
	273	0.67–0.75
	373	0.79–0.84
heavy (35 SiO_2, 60 PbO, 5 others)	173	0.46–0.50
	273	0.59–0.63
	373	0.71–0.75
barium (50 SiO_2, 24 BaO, 6 PbO, 8 K_2O, 3 Na_2O, 8 ZnO, 1 Sb2O3)	173···273···373	0.59···0.71···0.84
borosolicate (33 SiO_2, 31 B_2O_3, 25 PbO, 7 Al_2O_3, 3 K_2O, 1 Na_2O)	173···273···373	0.63···0.84···0.96
Pyrex glass (80 SiO_2, 13 B_2O_3, 2 Al_2O_3, 4 (Na_2O + K_2O))	20···80···150···300	0.14···0.54···0.84···1.34
Porcelain [55]	80···150···300···500···	0.96···1.35···1.68···1.72···
	700···1140	1.81···2.25
Quartz glass [fused quartz]	4···10···20···40···80	0.13···0.135···0.16···0.25···0.52
same as above [56]	60···80···160···300···	0.41···0.52···0.96···1.36···
	500···800···1100	1.63···1.81···1.98
same as above [39]	60···70···80···90	0.45···0.52···0.58···0.64
Soda-lime glass	20···80···150···300	0.14···0.46···0.70···1.05
same as above (60 SiO_2, 20 Na_2O, 20 CaO)	173···273···373	0.75···0.92···1.00
same as above (75 SiO_2, 12 Na_2O, 13 CaO)	173···273···373	0.88···1.09···1.17
Vycor glass (96 SiO2, 3 B2O3)	173···273···373	1.00···1.26···1.42

*1 Thermal conductivity corresponds to the indicated temperatures.

Table 15.20 Thermal conductivity of dense (porosity of 0%) sintering oxides [57]*1

Oxide	γ, g/cm^3	T, K	λ, W/(m·K)*2
Al_2O_3	3.7–3.8	400···700···1100···1500···2000	30···13···7···5···7
BeO [52]	3.01	400···700···1000···1500···2000	230···89···48···22···15
CaO	3.0–3.1	400···700···1500···2200	15···9···6···8
MgO	3.3–3.5	400···700···1100···1700···2000	36···16···8···6···9
NiO	5.0–5.1	400···700···1700	13···6···5
SiO_2	2.3–2.6	400···700···1300···1500	13···9···6···7
TiO_2	4.0–4.1	400···700···1500	8···5···3
ZrO_2	5.2–5.3	400···900···1500···2000	1.7···1.8···2.0···2.0

*1 Thermal conductivity data for oxides are given in Ref. [5].
*2 Thermal conductivity corresponds to the indicated temperatures.

Table 15.21 Thermal conductivity of pressed and sintered carbides [22, 58]

Carbide	γ, g/cm^3	T, K	λ, W/(m·K)*1
B_4C	2.32–2.5	300···110	28···13
Be_2C (in argon)	2.2	300···900	32···18
Cr_3C_2	6.68	293	13

Table 15.21 Thermal conductivity of pressed and sintered carbides [22, 58] *(continued)*

Carbide	γ, g/cm³	T, K	λ, W/(m·K)*1
$Cr_{23}C_6$	6.97	293	18
HfC	11.8–12.6	573···1473	9···17
Mo_2C	9.18	293	7
NbC	7.82	400···1200···2500	14..44···44
Nb_2C	7.85	400	19
SiC	3.2	200···300···600···	36···31···23···
		1000···1800···2500	16···12···14
TaC	14.5	573···1673···2673	34···40···38
Ta_2C	15.5	293	36
TiC	4.72–4.92	523···973	33···11
VC	5.48	293	25
WC	15.65	293	29
W_2C	17.3	293	36
ZrC	6.7	573···1673···2673	53···35···38

*1 Thermal conductivity corresponds to the indicated temperatures.

Table 15.22 Thermal conductivity of pressed and sintered borides, nitrides, and silicides at room temperature [22]

Compound	γ, g/cm³	λ, W/(m·K)	Compound	γ, g/cm³	λ, W/(m·K)
BeB_2		36	NbN	8.4	10
CaB_6		23	SiN [59]	2.4	16
CrB_2	5.6	22	TaN	13.8	23
Mo_2B_5	8.0	27	TiN	5.21	29
NbB_2	6.0	17	VN	6.04	31
SrB_6		26	ZrN	7.35	14
TaB_2	11.7	11	$CrSi_2$	4.4	6
TiB_2	4.5	24	$MoSi_2$	6.3	29
W_2B_5	11.0	32	$NbSi_2$	5.45	26
ZrB_2	6.1	23	$TaSi_2$	8.83	12
CrN	5.9	32	TSi_2	4.35	13
Cr_2N	6.5	60	WSi_2	9.33	19
MoN	8.6	16	$ZrSi_2$	4.88	18
Mo_3N	9.9	50			

Table 15.23 Thermal conductivity of dense fired refractory at a pressure of 1 atm, W/(m·K) [60]

Material (composition in mass %)	γ, g/cm³ at $t = 20°C$	Π*1,%	t, °C				
			200	400	800	1200	1600
Baddeleyite (> 90 ZrO_2)	5.80–5.51	0–3	1.8	1.7	1.7	1.9	2.3
Corundum (alumina) (> 90 Al_2O_3)	3.90–3.70	0–3	20	12.8	6.7	5.2	6.2
	3.12–2.89	20–24	4.1	3.5	2.7	2.4	2.5
Dinas (> 93SiO_2)	2.10–1.93	10–16	1.6	1.7	1.9	2.3	3.0
Dinas with additions (80 < SiO_2 < 93)	2.14–1.99	16–20	1.25	1.4	1.8	2.1	
Fireclay (28 < Al_2O_3 < 45)	2.16–1.92	16–20	0.88	0.94	1.1	1.2	1.1
Forsterite (50 < MgO < 65; 25 < SiO_2 < 40)	3.36–3.19	0–3	4.5	3.9	3.1	2.9	3.3*2
	2.69–2.49	20–24	3.3	2.8	2.0	1.8	
Magnesia (> 90 MgO)	3.55–3.37	0–3	26	17	8.3	5.6	6.2
	2.84–2.63	20–24	5.7	4.8	3.4	2.7	2.7
Malm (28 < Al_2O_3; < 85 SiO_2)	1.95–1.80	24–30	0.85	0.92	1.1	1.2	
Mullite (62 < Al_2O_3 < 72)	3.00–2.85	0–3	5.6	4.9	4.1	3.8	3.9
	2.52–2.34	20–24	1.2	1.3	1.5	1.7	1.8

Table 15.23 Thermal conductivity of dense fired refractory at a pressure of 1 atm *(continued)*

Material (composition in mass %)	γ, g/cm^3 at $t = 20$°C	Π[*1],%	t, °C				
			200	400	800	1200	1600
Mullite-corundum ($72 < Al_2O_3 < 90$)	2.89–2.68	16–20	2.7	2.6	2.4	2.3	2.3
Mullite-silica ($45 < Al_2O_3 < 62$)	2.45–2.26	16–20	1.4	1.4	1.4	1.4	1.4
Quartz glass (> 97 SiO_2)	2.20–2.09	0–3	1.4	1.9	3.0	4.5	
	1.67–1.50	24–30	0.42	0.34	0.46	0.94	1.8
Silicon-carbide with various	3.13–2.97	0–3	35	32	27	22	19
bonding (> 70 SiC)	2.50–2.32	20–24	13	12	9.7	8.1	7.2
Zirconia (> 50 ZrO_2, > 25 SiO_2):							
fused	3.40–3.14	20–24	1.8	1.6	1.4	1.3	1.25[*3]
polycrystalline	4.60–4.37	0–3	5.7	5.1	4.3	3.9	4.0
	3.86–3.59	16–20	4.6	4.2	3.5	3.2	3.1

[*1] Open porosity.
[*2] At $t = 1500$°C.

Table 15.24 Thermal conductivity of high-temperature composition of nuclear fuel [22]

Material (composition in mass %)	γ, g/cm^3	T, K	λ, W/(m·K)[*1]
PuO_2, sintered, Π = 0%	11.46	200···300···700···1100···1500	7.2···6.3···4.2···3.1···2.2
ThO_2, sintered, Π = 0%	9.96	400···500···700···1100···1800	8.2···6.9···4.6···3.0···2.0
UO_2, sintered [61]	10.97	300···600···1000···	8.2···5.6···3.7···
		1600···2600···3120	2.4···3.3···3.4
(50 UO_2 + 50 PuO_2)	11.1	300	4.3
(60 UO_2 + 40 Al), sintered mixture	4.9	300	7.0
(47 UO_2 + 53 BeO), sintered mixture	3.2–3.6	300	23–35
(19.9 UO_2 + 80.1 C), sintered mixture	1.93	300	18
(60 UO_2 + 40 MO)[*2]		373···1273	32···27
(50 UO_2 + 50 W)[*2]		373···773···1273	52···44···42
(20 UO_2 + 80 polyethelene)		273···363	0.31···0.24
(20 UO_2 + 60 polyethelene + 20 soot)		273···363	0.37···0.33
U_3O_8	7.90	300	6.8
UC		373···673···1100	25···22···27
UC_2	10.8	473···1873	13···20
(30.3 UC + 69.7 C), sintered mixture	2.22	300	57
UN (94.3 U, 5.32 N, 0.034 C, 0.047 O)	14.02	473···1073	16···21
US	10.87	300	11
$PuBe_{13}$	4.36	300	90

[*1] Thermal conductivity corresponds to the indicated temperatures.
[*2] Composition in volume %.

Symbols: Π is a porosity.

Table 15.25 Thermal conductivity of building materials at $t = (25 \pm 5)$°C and at a pressure of 1 atm [22, 23]

Material	γ, g/cm^3	λ, W/(m·K)
Alabaster slab	1.25	0.47
	0.84	0.27
Asbestos slate	1.9	0.35
Ash of wood fuel	0.7	0.15
Asphalt	1.8	0.72

Table 15.25 Thermal conductivity of building materials *(continued)*

Material	γ, g/cm^3	λ, W/(m·K)
Building gypsum [62]	1.25	0.35
Brick, dry:		
clinker	2.25	1.6
red [common], dense	1.8	0.67
red, deaerated	1.2	0.44
silica	0.5	0.12
silicate	1.9	0.81
tripoli	1.13	0.27
slag	1.4	0.58
Cardboard:		
common	0.7	0.17
dense	1.0	0.23
multilined corrugated, thickness of 5 mm	0.09	0.07
Concrete, dry:		
crushed-stone	2.0	1.3
foam	0.3	0.12
	0.8	0.33
reinforced [62]	2.4	1.6
sand and crushed-bricks	1.82	0.70
Dolomite, dense, dry	2.48	1.7
Grout:		
lime and sand	1.8	0.87
portland cement (without sand)	1.8	0.47
sand and cement	1.8	1.2
Limestone, dense, dry	2.32	1.9
Ruberoid	0.6	0.17
Sand, dry	1.6	0.87
Slag:		
blast-furnace, granulated	0.5	0.15
boiler	1.0	0.29
	0.7	0.19
Soil, dry (wet) [62]:		
clay and clay soil	1.6	0.87 (1.7)
	2.0	1.7 (2.6)
rocky	2.0	2.0 (2.7)
	2.4	2.3 (3.5)
sand and sandstone	1.6	1.1 (1.9)
	2.0	2.0 (3.2)
Tar	0.95	0.30
Tar paper	0.5	0.23

Table 15.26 Thermal conductivity of woods [15, 22]

Material	γ, g/cm^3	t, °C	λ, W/(m·K)*[1]
Balsa	0.11–0.13	30	0.043–0.052
Birch	0.72	20	0.15
Cedar, red	0.47	20	0.095
Cork	0.113	30	0.045
Fir	0.45	60	0.11
Larch	0.6	20	0.13
Maple	0.72	30	0.19
Oak	0.825	15	0.20

Table 15.26 Thermal conductivity of woods [15, 22] *(continued)*

Material	γ, g/cm^3	t, °C	λ, W/(m·K)*1
Pine:			
$\parallel$ to grains	0.545	15	0.15
$\perp$ to grains	0.545	15	0.40
Plywood	0.6	20	0.15
Poplar	0.58	50	0.17
Sawdust, insulating fill	0.25	20	0.093
Shaving, insulating packing	0.30	20	0.12

*1 Thermal conductivities of woods are given for direction perpendicular to the grains. Thermal conductivity along the grains is 2 to 3 times higher than that perpendicular to the grains. Wood wetness ranges from 7 to 10%. Thermal conductivity of dry wood can be estimated using the formula $\lambda = 0.0232 + 0.174\gamma$, where γ is the wood density, g/cm^3.

Table 15.27 Thermal conductivity of building thermal insulation at a pressure of 1 atm [62]

Material	γ, g/cm^3	t_{max},*1 °C	λ, 10^{-3} W/(m·K)
Aluminum foil:			
corrugated	0.02–0.04	350	$59 + 0.26t$
smooth	0.28	450	$87 + 0.12t$
Asbestos fabric, multilayer	0.5–0.6	450	$120 + 0.26t$
Asbestos mattress,			
filled in:			
fiber glass	0.20	450	$58 + 0.23t$
sovelite	0.28	450	$87 + 0.12t$
vermiculite	0.22	450	$81 + 0.23t$
Asbestos-vermiculite	0.25	600	$81 + 0.23t$
products (slabs, segments, shells)	0.30	600	$88 + 0.23t$
Diatomite foam-brick:			
mark PD-350	0.35	900	$81 + 0.19t$
mark PD-400	0.40	900	$93 + 0.19t$
Diatomite products	0.5	900	$100 + 0.23t$
	0.6	900	$140 + 0.23t$
Felt:			
constructional	0.20	100	$44 + 0.19t$
elastic, mineral wool	0.115	600	$43 + 0.22t$
warming	0.10	100	$60 + 0.23t$
Fiberglass (mats, strips)	0.20	450	$40 + 0.26t$
Foam concrete products	0.40	400	$110 + 0.30t$
	0.50	400	$130+0.3t$
Foam plastic FRP-1:	0.065–0.085	130	$41 + 0.23t$
series 75	0.065–0.085	130	$41 + 0.23t$
series 100	0.086–0.11	150	$43 + 0.19t$
Lime silica products (slabs, segments, shells)	0.2	600	$69 + 0.15t$
Linen, fiberglass VV-G	0.10	180	$38 + 0.15t$
Linens, ultramicrofine rock staple-fibers	0.03–0.07	700	$41 + 0.29t$
Mastic materials:			
asbothermite	0.57	500	$130 + 0.1t$
asbozurite-600	0.60	900	$160 + 0.17t$
newvel	0.37	350	$77 + 0.10t$
sovelite	0.50	500	$99 + 0.10t$
Mineral wool:			
mark 75	0.115	600	$44 + 0.29t$
mark 150	0.23	600	$53 + 0.19t$

Table 15.27 Thermal conductivity of building thermal insulation *(continued)*

Material	γ, g/cm^3	t_{max},*1 °C	λ, 10^{-3} W/(m·K)
Peat (slabs, segments, shells)	0.275	100	$64 + 0.15t$
	0.35	100	$76 + 0.15t$
Perlite, fine-pored:			
mark 75	0.09	875	$52 + 0.12t$
mark 150	0.18	875	$58 + 0.12t$
Perlite products (slabs, segments, shells):			
mark 250	0.25	900	$70 + 0.19t$
mark 400	0.40	900	$99 + 0.19t$
Perlite-cement products (slabs, segments, semicylinders):			
mark 250	0.25	600	$70 + 0.19t$
mark 350	0.35	600	$81 + 0.19t$
Slabs, mineral-wool with synthetic bond:			
mark 50	0.055–0.075	400	$40 + 0.29t$
mark 175	0.15–0.21	400	$52 + 0.20t$
Sovelite products (slabs, segments, semicylinders):			
mark 350	0.35	500	$75 + 0.15t$
mark 400	0.40	500	$78 + 0.15t$
Vermiculite, bulged, as fill	0.23	900	$70 + 0.27t$
Volcanic products (slabs, segments, shells)	0.35	600	$79 + 0.15t$
	0.40	600	$84 + 0.15t$

*1 Maximum temperature up to which a material can be used.

Table 15.28 Thermal conductivity of thermal insulating and auxiliary materials [15, 24, 63]

Material (composition in mass %)	γ, g/cm^3	t, °C	λ, 10^{-3} W/(m·K)*1
Asbestos, fibrous	0.47	$-200 \cdots 20$	$81 \cdots 160$
	0.70	$-200 \cdots 20$	$150 \cdots 240$
Cardboard, heat-insulating		$-260 \cdots 0$	$23 \cdots 39$
		750	84
Cellulose acetate	1.32	20	200–300
(60 aerogel-B + 40 aluminum powder)	0.18	$-180 < t < 20$	0.35
Cork, granulated ($d < 3$ mm)	0.037	$-200 \cdots 20$	$9 \cdots 34$
	0.161	$-70 \cdots 27$	$36 \cdots 50$
Ebonite [22]	1.2	20	160
Ebonite, bulged	0.064	$-100 \cdots 20$	$19 \cdots 32$
Fabric:			
quartz fiber ($d = 6 - 8\mu$, layer thickness 0.4 mm, $\gamma = 0.34$ kg/m^2)		$200 \cdots 600 \cdots 1000$	$120 \cdots 180 \cdots 290$
graphitizated, carbonic (argon, $d = 5\mu$, layer thickness 0.45 mm, $\gamma = 0.33$ kg/m^2)		$200 \cdots 1000 \cdots 1800$	$130 \cdots 270 \cdots 460$
Felt:			
graphitizated (vacuum, $d = 10 - 20\mu$)	0.04–0.12	$1000 \cdots 1800$	$400 \cdots 1200$
graphitizated, carbonic (argon)	0.05–0.10	$20 \cdots 800$	$80 \cdots 280$
zirconium carbide (argon, $d = 16\mu$, $\Pi = 80\%$)		$200 \cdots 1800$	$800 \cdots 1200$
glass	0.05	$-200 \cdots 20$	$14 \cdots 44$
Fiber:			
basalt	0.12	$20 \cdots 400$	$38 \cdots 78$
kaolin ($d = 4\mu$)	0.10	1100	$60 \cdots 380$
Fiber-glass laminate, reinforce layer		$-223 \cdots 20$	$190 \cdots 390$
Foam alumina (argon)	0.80	$200 \cdots 1400$	$400 \cdots 480$
Foam carbamide resin, white, soft	0.015	20	38
Foam glass	0.17	$-180 \cdots 20$	$49 \cdots 62$
	0.25	$-180 \cdots 20$	$61 \cdots 80$

Table 15.28 Thermal conductivity of thermal insulating and auxiliary materials *(continued)*

Material (composition in mass %)	γ, g/cm^3	t, °C	λ, 10^{-3} W/(m·K)[*1]
Foam plastic:			
PS-1 (d_{pore} = 0.5 mm)	0.10	$-170\cdots20$	$15\cdots40$
PS-4 (d_{pore} = 1.5 mm)	0.07	$-170\cdots20$	$18\cdots44$
Foam polyurethane:			
PPU-104B	0.39	$-170\cdots20$	$38\cdots67$
PPU-305A	0.14	$-100\cdots20$	$25\cdots43$
Ice	0.9	0	2200
Leather	1.0	20	170
Maylar [polyethylene terephthalate] [64]	1.39	$-200\cdots30$	$97\cdots125$
Mica:			
muscovite,			
$\parallel$ to cleavage plane	2.8	30	550
$\perp$ to cleavage plane	2.8	30	3400
phlogopite,			
$\parallel$ to cleavage plane	2.8	$20\cdots600$	$450\cdots570$
Paper, thickness of 75μ [22]	0.73	20	96
Perlite (air, $p = 0.13$ Pa)	0.20	$-190 < 180 < t < 20$	0.83
Perspex [plexiglas, polymethylmethacrylate], amorphous, transparent	1.174	$-190\cdots30$	$150\cdots200$
Polyethylene, high-pressure		30	330
Polystyrene [65]		30	140
Porolon	0.034	$-190\cdots20$	$13\cdots40$
Powder:			
Al_2O_3 ($d = 0.21$ mm, $\Pi = 51\%$)		$200\cdots800$	$360\cdots600$
MgO ($d = 0.18$ mm, $\Pi = 42\%$)		$200\cdots800$	$480\cdots780$
ZrO_2 ($d = 0.20$ mm, $\Pi = 42\%$)		$200\cdots800$	$250\cdots450$
Rubber:			
natural		$-180\cdots30$	$150\cdots150$
synthetic		$-180\cdots-100\cdots30$	$170\cdots210\cdots165$
Rubber [caoutchouc]:			
natural	0.86	50	42
fluorine-containing FK-20	0.18	$-200\cdots20$	$23\cdots59$
foaming, solidified	0.082	$-190\cdots20$	$15\cdots33$
Silk	0.10	0	40
Snow [66]	0.25	0	160
Talc (air, $p = 0.13$ Pa)	1.2	$-180 < 180 < t < 20$	1.6
Teflon [polytetrafluoroethylene]	2.12	$-190\cdots30$	$230\cdots260$
Thermal-insulating shield:			
one-side aluminized Maylar film of 5–12μ thickness, 3 × 3 mm corrugation, fiber-glass mat (15 shields/cm, $p = 10^{-3}$ Pa)	0.022	$-200 < 180 < t < 20$	< 0.1
niobium foil of 15μ thickness, sphere segment corrugation of 0.05–0.1 mm in height (20 shields/cm, in vacuo)		$800\cdots1000\cdots1200\cdots$	$42\cdots77\cdots130$
Wool:			
mineral	0.10	$-200\cdots20$	$16\cdots47$
	0.25	$-100\cdots20$	$36\cdots57$
glass	0.15	$-173\cdots27$	$20\cdots52$
cotton	0.05	$-200\cdots20$	$26\cdots59$
	0.081	$-200\cdots20$	$32\cdots59$

[*1] Thermal conductivities of non-monolithic materials are for the case where an interstitial gas is air at a pressure of 1 atm, unless otherwise noted.

Symbols: d is a diameter of fibers, grains, pores; p is a gas pressure; Π is a porosity.

Table 15.29 Thermal conductivity of rocks, W/(m·K) [67]

Rock	λ_{mean}	$\lambda_{min} - \lambda_{max}$	Rock	λ_{mean}	$\lambda_{min} - \lambda_{max}$
Magmatic rocks			Chalk [68]	1.6	0.8–2.2
Basalt	2.0	1.4–5.3	Clay	1.5	0.6–2.7
Diabase [dolerite]	2.6	1.6–4.4	Claystone	2.5	1.7–3.4
Granite	3.0	1.3–4.5	Coal	0.29	0.07–1.5
Lava	2.5	0.2–4.6	Dolomite	3.6	1.6–6.5
Obsidian [68]	1.5	1.4–1.6	Limestone	2.3	0.65–4.4
Tuff [68]	2.3	1.3–4.0	Marl and Marlstone	2.2	0.55–4.0
Metamorphic rocks			Peat [68]	0.07	
Gneiss	2.4	1.2–4.8	Salt	4.0	1.3–7.2
Marble [68]	2.6	1.6–4.0	Sand	1.44	0.1–2.8
Quartzite	4.6	3.1–8.0	Sandstone	2.5	0.9–6.6
Shale	2.1	0.6–4.2	Sediments, ocean and lake bottom	1.0	0.4–1.7
Sedimentary rocks					
Argillite	2.0	0.8–3.8	Siltstone	2.0	0.6–4.0

References

[1] Carslaw, H. S. and Jaeger, J. C., Conduction of heat in solids, 2nd ed., Oxford University, London, 1959.

[2] Lykov, A. V., Theory of heat conduction, Vysshaya shkola, Moscow, 1967 (in Russian).

[3] Thermal conductivity, Ed. Tye, R. P., Academic Press Inc., London, 1969 (vol. 1, 2).

[4] Ho, C. Y., Powell, R. W., Liley, P. E., J. Phys. Chem. Ref. Data, 2, 279, 1972.

[5] Ho, C. Y., Powell, R. W., Liley, P. E., J. Phys. Chem. Ref. Data, 3, Suppl. 1, 1974.

[6] Vargaftic, N. B., Filippov, L. P., Tarzimanov, A. A., Totskii, E. E., Handbook of thermal conductivity of liquids and gases, Energoatomizdat, Moscow, 1990 (in Russian).

[7] Okhotin, A. S., Borovikova, R. P., Nechaeva, T. V., Pushkarskii, A. S., Thermal conductivity of solids. Reference book. Ed. Okhotin, A. S., Energoatomizdat, Moscow, 1984 (in Russian).

[8] Childs, G. E., Hanley, H. J. M., Cryogenics, 8, 94, 1968.

[9] Hirschfelder, J. O., Curtiss, C. F. and Bird, R.B., Molecular theory of gases and liquids, Wiley, New York, 1966.

[10] Ferziger, J. H. and Kaper, H. G., Mathematical theory of transport processes in gases, North-Holland Publ., Amsterdam, London, 1972.

[11] Reid, R. C., Prausnitz, J. M., Sherwood, T. K., The properties of gases and liquids, 3d ed., McGraw-Hill, New York, 1977.

[12] Kogan, M. N. Dynamics of rarefied gas. The kinematic theory, Nauka, Moscow, 1967 (in Russian).

[13] Vargaftic, N. B. Handbook of thermophysical properties of gases and liquids, 2nd ed., Nauka, Moscow, 1972 (in Russian).

[14] Thermal conductivity of polyatomic liquids and gases: a digest, Eds. Vargaftik, N. B., Filippov, L. P., Tarzimanov, A. A., Totskii, E. E., Izdatel'stvo standartov, Moscow, 1981 (in Russian).

[15] Missenard A., Conductivité thermique des solides, liquides, gaz et de leurs mélanges, Editions Eyrolles, Paris, 1965.

[16] Filippov, L. P., Investigations of thermal conductivity of liquids, Izdatel'stvo MGU, Moscow, 1970 (in Russian).

[17] Ziman J. M., Electrons and phonons, Oxford University Press, London, 1960.

[18] Berman R., Thermal conduction in solids, Clarendon Press, Oxford, 1976.

[19] Mogilevskii, B. M., Chudnovskii, A. F., Thermal conductivity of semiconductors, Nauka, Moscow, 1972 (in Russian).

[20] Anderson, A. C., Amorphous solids. Low-temperature properties, Ed. Phillips, W. A., Springer-Verlag, Berlin, 1981, p.S65.

[21] Ross, R. G., Andersson, P., Sundqvist, B., Backstrom, G., Rep. Progr. Phys., 47, 1347, 1984.

[22] Chirkin, V. S., Thermophysical properties of nuclear engineering materials. Handbook, Atomizdat, Moscow, 1968 (in Russian).

[23] Chirkin, V. S., Thermal conductivity of industrial materials, 2nd ed., Mashgiz, Moscow, 1962 (in Russian).

[24] Kozhevnikov, I. G., Novitskii, L. A., Thermophysical properties of materials at low temperatures. Handbook, 2nd ed., Mashinostroeniye, Moscow, 1982 (in Russian).

[25] Childs, G. B., Ericks, L. Y., Powell, R. W., Thermal conductivity of solids at room temperature and below. A review and compilation of the literature, NBS (US) Monogr., 131, 1973.

[26] Cook, J. G., van der Meer, M. P., Laubitz, M. J., Canad. J. Phys., 50, 1386, 1972.

[27] Feger, A., Yanosh, Sh., Petrovich, P., et al., Fizika nizkikh temperatur, 4, 1305, 1978.

[28] Zinoviev, V. E., Kinetic properties of metals at high temperatures. Handbook, Metallurgia, Moscow, 1984 (in Russian).

[29] Zinoviev, V. E., Korshunov, I. G., Thermal conduction and temperature conduction of transition metals at high temperatures. Reviews on thermal properties of substances, IVT Acad. Nauk SSSR, Moscow, 1978 (part 1, No.1), 1979 (part 2, No.4) (in Russian).

[30] Campos Tome, M. A., J. Low Temp. Phys., 20, 677, 1975.

[31] Andrew, J.F., Klemens, P. G., Thermal conductivity and Lorenz number of plutonium and plutonium-gallium alloys, Proc. 17-th Internat. Thermal Conductivity Conf., Ed. Hust, J. G., Plenum Press, New York, 209, 1983.

[32] Vedernikov, M. V., Kizhaev, S. A., Petrov, A. V., et al., Fizika tverdogo tela, 17, 340, 1975.

[33] Chuah, D. G. C., Ratnalingam, R., J. Low Temp. Phys., 14, 257, 1974.

[34] Aliev, N. N., Vol'kenshtein, N. V., Zhurnal experim. i teoretich. fisiki, 49, 1450, 1965.

[35] Handbook of Chemistry and Physics. 70th ed., CRC Press, Boca Raton, FL, 1989–1990.

[36] Amasovich, E. S., Peletskii, V. Z., Teplofizika vysokikh temperatur, 20, 891, 1982.

[37] A handbook of a metal worker, 2nd ed., Ed. Acherkan, N. S., Mashinostroeniye, Moscow, 1965, vol. 1 (in Russian).

[38] Geller, Yu. A., Tool steels, Metallurgiya, Moscow, 1975 (in Russian).

[39] Zhdanovich, V. A., Chashkin, Yu. R., Izmeritel'naya tekhnika, No.3, 28, 1976.

[40] Sergeev, O. A., in Thermal properties of substances and materials, Izdatel'stvo standartov, Moscow, 1979, issue 13, p.133 (in Russian).

[41] Materials in instrument making and automatics. Handbook, Ed. Pyatin, Yu. M., Mashinostroeniye, Moscow, 1969 (in Russian).

[42] Rogel'berg, I. L., Beilin, V. M., Alloys for thermocouples. Handbook, Metallurgiya, Moscow, 1983 (in Russian).

[43] White, G. K., Experimental techniques in low-temperature physics, 3d ed., Clarendon Press, Oxford, 1979.

[44] Aluminum alloys. Guidebook, Metallurgiya, Moscow, 1972 (in Russian).

[45] Glazunov, S. G., Moiseev, V. N., Constructional titanium alloys, Metallurgiya, Moscow, 1974 (in Russian).

[46] Smiryagin, A. P., Smiryagina, M. A., Belova, A. V., Industrial non-ferrous metals and alloys, 3d ed., Metallurgiya, Moscow, 1974 (in Russian).

[47] A handbook of a metal worker, 3d ed., Eds. Rakhshtadt, A. G. and Bromstrem, V. A., Mashinostroeniye, Moscow, 1976, vol. 2 (in Russian).

[48] Cody, G. D., Cohen, R. W., Rev. Mod. Phys, 362, No.1 (part 1), 121, 1964.

[49] Slack, G. A., Phys. Rev., B6, 3791, 1972.

[50] Whitsett, C. R., Nelson, D. A., Broerman, J. G., Phys. Rev., B7, 4625, 1973.

[51] Slack, G. A., J. Phys. Chem. Solids, 34, 321, 1973.

[52] Slack, G. A., Austerman, S. B., J. Appl. Phys., 42, 4713, 1971.

[53] Jackson, H. E., Walker, C. T., Phys. Rev., B3, 1428, 1971.

[54] Ratcliffe, E. H., Thermal conductivities of glasses between −150 and 100°C, in Handbook of Chemistry and Physics, 70th ed., CRC Press, Boca Raton, FL, 1989–1990, pp. E7–E9.

[55] Shadrichev, E. V., Smirnov, I. A., Pribori i tekhnika eksperimenta, 5, 218, 1968.

[56] Sergeev, O. A., Shashkov, A. G., Umanskii, A. S., Inzhenerno-fizicheskii zhurnal, 43, 960, 1982.

[57] Krzhizhanovskii, R. E., Shtern, Z. Yu., Thermophysical properties of nonmetallic materials. (Oxides). Reference book, Energia, Leningrad, 1973 (in Russian).

[58] Krzhizhanovskii, R. E., Shtern, Z. Yu., Thermophysical properties of nonmetallic materials. (Carbides). Reference book, Energia, Leningrad, 1977 (in Russian).

[59] Touloukian, Y. S., Ho, C. Y., Thermophysical properties of selected aerospace materials. Part II: Thermophysical properties of seven materials, TERIAC CINDA, Purdue Univ., 1976.

[60] Litovskii, E. Ya., Puchkelevich, N. A., Thermal properties of regractory materials. Handbook, Metallurgia, Moscow, 1982 (in Russian).

[61] Fink, J. K., Chazanov, M. G., Leibovtz, L., J. Nucl. Mater., 102, 17, 1981.

[62] Thermal insulation. A handbook of a building worker, 4-th ed., Ed. Kuznetsov, G. F., Stroiizdat, Moscow, 1985 (in Russian).

[63] Kharlamov, A. G., Thermal conductivity of high-temperature thermal insulators, Atomizdat, Moscow, 1980 (in Russian).

[64] Hertz, J. and Haskins, J. F., Thermal conductivity of reinforced plastics at cryogenic temperatures, Adv. Cryogenic Engineering, Plenum Press, New York, 1965, vol.10, ed. K.D. Timmerhaus, p.163.

[65] van Krevelen, D. W. and Hoftyzer, P. J., Properties of polymers, their estimation and correction with chemical structure, 2nd ed., Elsevier Scientific, New York, 1976.

[66] Powell, R. L., Childs, G. E., American Institute of Physics Handbook, 3d ed., McGraw-Hill, New York, 1972 (ch. 4g).

[67] Čermak, V. and Rybach, L., Thermal conductivity and specific heat of minerals and rocks. In: Landolt-Börnstein: numerical data and functional relationships in science and technology. New series, Group V., Vol.1: Physical properties of rocks. Subvol. a. Ed. Angenheister, G., Springer-Verlag, Berlin-Heidelberg-New York, 1982, p.305.

[68] Physical properties of rocks and minerals (petrophysics). A handbook of a geophysicist, 2nd ed., Ed. Dortman, N. B., Nedra, Moscow, 1984 (in Russian).

16

Viscosity

A.V. Eletskii

16.1 Introduction

The viscosity of a fluid matter η_{xy} is determined by Newton's expression

$$F_x = -\eta_{xy} \nabla v_y \tag{16.1}$$

relating the internal friction F_x, acting on a unit area in the direction opposite to the velocity gradient of the viscid substance (gas, liquid) flow, and this velocity gradient ∇v_y. If the considered volume, filled with the fluid, displays no particular directions that could be due either to the presence of external fields or to the existence of some preferential orientation of the particles in the fluid, then all the components of the viscosity tensor are identical and the viscosity is a scalar. Precisely this case will be examined here.

Sometimes, the coefficient η is termed dynamic viscosity to distinguish it from the coefficient $\nu = \eta/\rho$ (ρ is the density of the fluid), known as the kinematic viscosity. In SI, the unit of dynamic viscosity is Pa·s = N·s/m^2 = kg/(m·s); the unit of kinematic viscosity is m^2/s.

16.2 The Viscosity of a Gas

Within a wide range of gas pressures, the viscosity of a gas is determined by pair collisions of its constituent particles. The lower limit of this range depends on the condition setting the characteristic path of particles in the gas much smaller than the dimensions of the volume of gas being examined. If the dimensions of the volume ~10 cm, the indicated limit corresponds to a pressure of ~1 Pa (10^{-2} mmHg). The upper limit is determined by the condition that the gas be ideal, in accordance with which the mean free path of particles exceeds significantly the mean distance between them, $n^{-1/3}$. The indicated condition, which if satisfied results in the role of triple and other multiple collision processes being insignificant as compared to pair collisions, may be expressed in the form $n \ll \sigma^{-3/2}$ (σ is the particle scattering cross-section). This condition restricts the pressure to values on the order of several tens of megapascals (several hundreds of atmospheres). Moreover, the condition requiring the gas to be ideal depends on the temperature and may be violated essentially as the triple point is approached.

In accordance with the elementary kinetic theory of gases, the expression for the viscosity of a gas simulated by small balls, with a cross-section independent of velocity, has the form

0-8493-2861-6/97/$0.00+$.50

$$\eta = \frac{1}{3}\frac{m\langle v\rangle}{\sigma}, \tag{16.2}$$

where m is the mass of the gas particle; $\langle v\rangle \approx \sqrt{2kT/m}$ is the mean thermal velocity; σ is the scattering cross-section. In real situations, the scattering cross-sections for atoms and molecules depend on the velocity. In this case, the expression for the viscosity determined by the solution of the Boltzmann kinetic equation, has a significantly more complicated form; nevertheless, from this expression, as from the one of an elementary theory (16.2), it follows that the viscosity is practically independent of pressure within a wide range of pressures[1], and that its enhancement with temperature is sharper than the $T^{1/2}$ law.

In Tables 16.1 and 16.2, viscosities of gaseous media are presented. The values given are obtained by averaging a large amount of experimental data and correspond to conditions where changes in the gas pressure do not result in a change of viscosity, within the experimental uncertainty (0.1 to 1%). Such a situation holds true at pressures lower than atmospheric pressure. The nature of the pressure dependence of viscosity is seen in Tables 16.3 and 16.4, in which the viscosities are presented for nitrogen and hydrogen at various temperatures and pressures [2]. The uncertainty in the data presented in Tables 16.1–16.4 does not exceed several percent.

[1] The dependence of viscosity on pressure arises when association processes changing the number of particles in the system take place in the gas.

Table 16.1 Viscosity of gases (in 10^{-6} Pa·s) at atmospheric pressure and various temperatures [1, 2, 4, 6, 7] [The uncertainty in the data ranges from to 1 to 10%.]

T, K	He	Ne	Ar	Kr	Xe	H_2	D_2	N_2	O_2	F_2	Cl_2	CO	CO_2
60	7.06	9.63	5.34			2.91	3.86						
80	8.41	12.1	6.83			3.60	4.88	5.59	6.27			5.40	
100	9.63	14.4	8.34	9.29		4.21	5.79	6.87	7.68	8.56		6.70	
150	12.3	19.4	12.3	13.4	12.2	5.57	7.77	10.0	11.3	12.9		9.84	
200	15.0	23.9	16.0	17.6	15.8	6.78	9.55	12.9	14.6	16.8		12.7	10.2
250	17.5	28.0	19.5	21.6	19.6	7.90	11.2	15.5	17.8	20.3		15.4	12.6
300	19.9	31.7	22.7	25.5	23.3	8.94	12.7	17.9	20.7	23.6	13.7	17.8	15.0
400	24.3	38.4	28.5	32.7	30.4	10.9	15.5	22.1	25.9	29.5	18.0	22.1	19.5
500	28.3	44.5	33.6	39.1	36.8	12.7	18.0	25.9	30.5	34.8	22.1	25.9	23.6
600	32.0	50.0	38.3	45.0	42.9	14.5		29.3	34.7		25.8	29.4	33.9
800	38.8	60.0	46.4	55.4	53.7	17.7		35.2	42.1		32.6	35.4	39.5
1000	45.0	68.9	53.5	64.5	63.2	20.7		40.4	48.5			40.6	42.9
1500	58.6	93.7	68.4	83.6	83.3	27.6		54.4	61.9			51.6	55.5
2000	70.7	114.0	80.7	108.0	106.0	33.6		66.8	73.1			66.5	67.8
2200	74.2	122.0	85.1	115.0	114.0			71.3	83.5			71.3	71.3
2500	90.1	133.0	103.0	125.0	124.0			78.2	91.0			77.8	78.0
3000	105.0	151.0	116.0	142.0	140.0			90.0	103.0			88.6	87.7
3500	118.0	168.0	129.0	157.0	156.0								
4000	131.0	184.0	141.0	172.0	171.0								
4500	144.0	200.0	153.0	186.0	185.0								
5000	156.0	215.0	165.0	200.0	199.0								
	H_2S	COS	CS_2	HCN	C_2N_2	SiH_4	Air	PH_3	CCl_4	CH_4	Br_2	T_2	NH_3
100							7.11						
150							10.3						
200							13.2			7.70			6.89
250							16.0			9.53			8.53
300	13.0	12.5	10.1	7.58	10.2	11.7	18.5	11.8	9.9	11.2	15.5		10.3

Table 16.1 Viscosity of gases *(continued)*

T, K	H_2S	COS	CS_2	HCN	C_2N_2	SiH_4	Air	PH_3	CCl_4	CH_4	Br_2	T_2	NH_3
400	17.3	16.6	13.6	10.8	13.7	15.3	23.0	15.6	13.0	14.3	20.3	18.6	13.9
500		20.4	16.9	13.9		18.9	27.0	19.1	16.0	17.0	25.1	23.0	17.6
600			20.2				30.6		19.0	19.5	29.9	27.2	21.4
800							37.0			23.9	39.2		28.8
1000							42.4			28.0			35.9
1500							53.0			44.6			
2000							63.0			44.6			
2200							67.0			47.3			
2500										51.6			
3000										58.2			
	BF_3	HCl	HI	H_2S	NO	NO_2	N_2O	SO_2	H_2O	CF_4	SF_6	C_2H_6	C_2H_4
150					10.5								
200	12.1				13.6		10.0	8.62		12.1			
250	14.6	12.1	15.9		16.6		12.6	10.8		14.9	12.8	7.77	8.60
300	17.1	14.6	19.0	12.6	19.3	13.0	15.0	13.0	9.13	17.6	15.3	9.36	10.4
400	21.7	19.6	25.1	16.9	24.1	21.3	19.5	17.3	13.2	22.2	19.9	12.2	13.6
500	26.1	24.3	31.0	20.8	28.4		23.6	21.3	17.3	26.5	23.9	14.9	16.6
600	30.2	28.8	36.7		32.3		27.3	25.1	21.3	30.3	27.7	17.3	19.3
800					39.0		34.1	32.1	29.5	37.3	34.3	21.7	24.1
1000					44.9		40.0	38.4	37.6	43.5	40.4	25.6	28.4
1500					57.3		52.5			57.4	54.1	34.2	38.1
2000					72.6					69.3	64.9	41.1	46.2
2200					77.7					73.3	68.9	44.0	48.9
2500					84.8					79.8	75.1		
3000					96.0					89.0	84.2		

Table 16.2 Viscosity (in 10^{-6} Pa·s) of gaseous hydrocarbons and their derivatives at atmospheric pressure [1, 2]

Gas	Temperature T, K							
	200	250	300	400	500	600	800	1000
Acetone C_3H_6O		6.78	7.77	10.1	12.8	15.5		
Acetylene C_2H_2			10.3	13.5	16.4	19.1		
Benzene C_6H_6			7.65	10.2	12.7	15.1		
Bromomethane CH_3Br		13.2	15.8	20.2	24.3			
iso-Butane *iso*-C_4H_{10}			7.60	10.0	12.3	14.0	17.5	
n-Butane n-C_4H_{10}			7.57	9.96	12.3	14.5	18.2	
Bromotrifluoromethane $CBrF_3$			15.0	19.8				
Chloromethane CH_3Cl		9.59	11.0	14.5	17.9	21.3		
Dichlorofluoromethane $CHCl_2F$			11.6	15.3	18.8			
Dichlorotetrafluoroethane $C_2Cl_2F_4$		10.0	11.6	15.0	18.4			
Diethyl ether $C_4H_{10}O$		6.3	7.6	10.1	12.5	14.7		
Difluorodicloromethane CF_2Cl_2		10.6	12.6	16.4	19.8			
Ethane C_2H_6	6.43	7.96	9.45	12.2	14.8	17.2	21.4	25.1
Ethanol C_2H_5OH			9.0	11.8	14.1	16.9		
Ethylene C_2H_4	7.1	8.8	10.4	13.5	16.3	18.8	23.4	27.5
n-Heptane n-C_7H_{16}			6.1	7.9	9.8	11.6		
n-Hexane n-C_6H_{14}			6.7	8.7	10.8	12.8	16.5	19.6
Methane CH_4	7.76	9.53	11.2	14.2	17.0	19.4	23.8	27.6
Methanol CH_3OH		8.3	9.9	13.1	16.5	19.8		
n-Octane n-C_8H_{18}				7.40	9.2	10.96		

Table 16.2 Viscosity of gaseous hydrocarbons and their derivatives *(continued)*

Gas	Temperature T, K							
	200	250	300	400	500	600	800	1000
Pentafluorochloroethane C_2F_5Cl			12.8	16.5	20.0			
iso-Pentane *iso*-C_5H_{12}				9.60	12.0	14.3	16.8	
n-Pentane *n*-C_5H_{12}				9.23	11.4	13.4	16.8	
n-Propane *n*-C_3H_8			8.25	10.8	13.3	15.6		
Propane C_3H_8		7.1	8.3	9.48	11.7	5.2	17.1	
iso-Propanol *iso*-C_3H_8O					13.1			
n-Propanol *n*-C_3H_8O				10.6	13.2			
Propylene C_3H_6		7.28	8.78	11.5	14.1	16.5		
Tetrachloromethane CCl_4			9.97	13.2	16.1	18.8	23.7	
Tetrafluoromethane CF_4		14.9	17.5	22.1	26.2			
Trichlorofluoromethane CCl_3F		9.42	11.0	14.3	17.8			
Trichlorotrifluoroethane $C_2Cl_3F_3$		9.13	10.4	12.5				
Trifluorochloromethane CF_3Cl		12.4	14.5	18.9	23.4			
Trifluoromethane CHF_3		12.4	14.9	19.6	23.8			

Table 16.3 Viscosity (in 10^{-6} Pa·s) of gaseous nitrogen at various temperatures and pressures [3]

T, K	Pressure, 10^5 Pa							T, K	Pressure, 10^5 Pa						
	1	10	50	100	150	200	400		1	10	50	100	150	200	400
80	5.52							500	25.8	25.9	26.3	26.8	27.4	28.1	31.1
100	6.88							600	29.1	29.2	29.5	29.9	30.4	30.9	33.2
120	8.21	8.68						700	32.1	32.2	32.4	32.8	33.2	33.6	35.5
140	9.45	9.83	15.5					800	34.9	35.0	35.2	35.5	35.8	36.2	37.8
160	10.6	11.0	13.4	22.1	31.3	38.4	59.0	900	37.5	37.6	37.8	38.1	38.4	38.6	40.0
180	11.8	12.1	13.8	18.3	24.5	30.1	49.0	1000	40.0	40.0	40.2	40.5	40.7	41.0	41.6
200	12.9	13.1	14.5	17.5	21.7	26.2	42.3	1100	42.3	42.4	42.5	42.8	43.0	43.2	44.3
250	15.5	15.7	16.6	18.2	20.4	22.7	33.5	1200	44.5	44.6	44.7	44.9	45.1	45.4	46.3
300	17.8	18.0	18.7	19.8	21.3	22.9	30.4	1300	46.6	46.7	46.8	47.0	47.2	47.4	48.2
400	22.0	22.2	22.7	23.4	24.2	25.2	29.6								

Table 16.4 Viscosity (in 10^{-6} Pa·s) of gaseous hydrogen at various temperatures and pressures [3]

T, K	Pressure, 10^5 Pa						T, K	Pressure, 10^5 Pa					
	1	10	20	50	100	500		1	10	20	50	100	500
30	1.6						300	8.96	8.96	8.98	9.02	9.10	10.3
50	2.49	2.54	2.80	4.20	6.25	14.4	500	12.6	12.6	12.6	12.7	12.7	13.3
100	4.21	4.23	4.24	4.42	5.0	9.95	750	16.6	16.6	16.6	16.6	16.6	17.0
200	6.81	6.82	6.85	6.91	7.06	9.13	1000	20.1	20.1	20.1	20.1	20.2	20.4

The viscosity of a mixture of two gases may depend in an irregular fashion on its partial composition. This follows both from straightforward experiments and from kinetic theory [3]. That nonmonotonous behavior is exhibited, for example, in the temperature and pressure dependences of the viscosity of partially dissociated molecular gases. A change in temperature and pressure of the gas results in a change in the degree of its dissociation, i.e. of its partial composition, and this, in turn, influences the viscosity. Tables 16.5–16.10 present the viscosities of some of the most common molecular gases at various pressures and temperatures in conditions when the respective gas is partially dissociated. In Tables 16.11–16.14, the viscosities are given of certain binary gas mixtures at different temperatures and partial compositions. The data presented are determined up to a precision on the order of 1%. Viscosity values for partially dissociated air are presented in Table 16.15.

Table 16.5 Viscosity (in 10^{-6} Pa·s) of partially dissociated hydrogen [3]

T, 10^3 K	Pressure, Pa 10^2	10^3	10^4	10^5	10^6	10^7	$2 \cdot 10^7$	T, 10^3 K	Pressure, Pa 10^2	10^3	10^4	10^5	10^6	10^7	$2 \cdot 10^7$
1.8	29.6	29.6	29.6	29.6	29.6	29.6	29.6	4.0	45.3	45.4	46.7	52.7	57.5	57.1	56.8
2.0	32.0	31.9	31.8	31.8	31.8	31.8	31.8	4.2	47.1	47.2	47.9	52.6	59.3	59.8	59.5
2.2	34.8	24.4	24.2	24.2	24.2	24.2	24.2	4.4	48.9	48.9	49.4	52.7	60.5	62.5	62.2
2.4	37.3	37.2	36.8	36.6	36.6	36.6	36.6	4.6	50.6	50.6	50.9	53.2	61.1	65.0	64.9
2.6	37.7	40.0	39.6	39.2	39.0	39.0	39.0	4.8	52.4	52.4	52.6	54.1	61.4	67.3	67.4
2.8	34.2	37.6	42.1	42.3	41.8	41.5	41.5	5.0	54.1	54.1	54.3	55.3	61.5	69.2	69.8
3.0	36.8	41.3	44.9	44.7	44.2	44.0	43.9	5.2	55.9	55.9	56.0	56.7	61.8	70.9	72.0
3.2	38.1	40.7	46.4	47.6	46.9	46.5	46.4	5.4	57.7	57.7	57.7	58.3	62.3	72.1	73.9
3.4	39.6	41.1	46.4	50.1	49.7	49.1	50.0	5.6	59.5	59.5	59.6	59.9	63.1	74.0	75.5
3.6	41.7	42.3	46.0	51.9	52.4	51.7	51.5	5.8	61.4	61.4	61.4	61.7	64.1	73.8	76.8
3.8	43.5	43.8	46.0	52.7	55.1	54.4	54.1	6.0	63.3	63.3	63.3	63.5	65.4	74.5	77.4

Table 16.6 Viscosity (in 10^{-6} Pa·s) of partially dissociated water vapour [3]

T, 10^3 K	Pressure, 10^5 Pa 0.1	0.4	1.0	5.0	20	40	60	100
1.8	6.12	6.12	6.12	6.12	6.12	6.12	6.12	6.12
2.0	6.57	6.58	6.59	6.59	6.60	6.60	6.60	6.60
2.2	6.98	7.02	7.03	7.04	7.05	7.06	7.06	7.06
2.4	7.31	7.39	7.42	7.46	7.49	7.49	7.50	7.50
2.6	7.52	7.67	7.75	7.84	7.89	7.90	7.91	7.92
2.8	7.54	7.84	7.98	8.15	8.24	8.28	8.29	8.31
3.0	7.56	7.90	8.10	8.38	8.54	8.60	8.63	8.65
3.2	7.63	7.78	8.14	8.52	8.77	8.86	8.91	8.96
3.4	7.95	7.96	8.16	8.59	8.93	9.06	9.13	9.21
3.6	8.44	8.22	8.25	8.63	9.03	9.20	9.30	9.41
3.8	9.03	8.67	8.51	8.68	9.08	9.29	9.41	9.55
4.0	9.53	9.22	8.95	8.82	9.14	9.36	9.49	9.65
4.2	9.97	9.76	9.49	9.09	9.22	9.42	9.55	9.73
4.4	10.3	10.2	10.0	9.50	9.39	9.52	9.63	9.80
4.6	10.8	10.7	10.6	9.99	9.65	9.68	9.75	9.89
4.8	11.2	11.1	11.0	10.6	10.0	9.93	9.94	10.0
5.0	11.6	11.6	11.4	11.0	10.4	10.2	10.2	10.2
5.2	11.9	11.9	11.8	11.6	10.9	10.47	10.6	10.5
5.4	12.3	12.2	12.2	12.0	11.5	11.2	11.0	10.8
5.6	12.6	12.6	12.6	12.4	12.0	11.6	11.4	11.2
5.8	13.0	13.0	13.0	12.8	12.5	12.2	12.0	11.7
6.0	13.4	13.4	13.4	13.2	13.0	12.6	12.4	12.2

Table 16.7 Viscosity (in 10^{-6} Pa·s) of partially dissociated carbon dioxide, CO_2 [3]

$T, 10^3$ K	Pressure, Pa 10^3	10^5	10^7	$T, 10^3$ K	Pressure, Pa 10^3	10^5	10^7
1.6	56.4	56.4	56.4	3.0	101.0	94.8	89.9
1.8	61.5	61.4	61.4	3.2	106.0	102.0	95.2
2.0	66.7	66.3	66.2	3.4	111.0	109.0	101.0
2.2	74.3	71.2	70.9	3.6	116.0	116.0	107.0
2.4	79.0	76.2	75.5	3.8	122.0	122.0	113.0
2.6	86.6	81.7	80.2	4.0	127.0	128.0	120.0
2.8	94.2	87.9	84.9				

Table 16.8 Viscosity (in 10^{-6} Pa·s) of partially dissociated nitrogen [3]

T, 10^3 K	Pressure, Pa						T, 10^3 K	Pressure, Pa					
	10^2	10^3	10^4	10^5	10^6	10^7		10^2	10^3	10^4	10^5	10^6	10^7
3.6	106	106	106	106	106	106	5.0	142	139	138	138	138	138
3.8	111	111	111	111	111	111	5.2	149	144	143	143	143	143
4.0	116	115	115	115	115	115	5.4	155	149	147	147	147	147
4.2	120	120	120	120	120	120	5.6	162	155	152	152	152	152
4.4	125	125	124	124	124	124	5.8	167	162	157	156	156	156
4.6	130	129	129	129	129	129	6.0	172	168	162	161	161	161
4.8	136	134	134	134	134	134							

Table 16.9 Viscosity (in 10^{-6} Pa·s) of partially dissociated oxygen [3]

T, 10^3 K	Pressure, Pa						T, 10^3 K	Pressure, Pa					
	10^2	10^3	10^4	10^5	10^6	10^7		10^2	10^3	10^4	10^5	10^6	10^7
2.0	73.7	73.6	73.5	73.5	73.5	73.5	4.2	132.0	132.0	133.0	139.0	138.0	129.0
2.2	79.1	78.6	78.4	78.3	78.3	78.3	4.4	139.0	137.0	138.0	142.0	146.0	136.0
2.4	85.9	84.0	83.3	83.0	82.9	82.9	4.6	142.0	142.0	142.0	145.0	152.0	145.0
2.6	93.9	90.6	88.6	87.8	87.6	87.5	4.8	146.0	146.0	147.0	149.0	156.0	153.0
2.8	100.0	98.6	94.8	92.0	92.2	92.0	5.0	151.0	151.0	151.0	153.0	160.0	161.0
3.0	104.0	106.0	102.0	98.6	97.0	96.5	5.2	156.0	156.0	156.0	157.0	163.0	168.0
3.2	108.0	111.0	111.0	105.0	102.0	101.0	5.4	160.0	160.0	160.0	161.0	166.0	174.0
3.4	112.0	114.0	118.0	113.0	108.0	106.0	5.6	165.0	165.0	165.0	166.0	170.0	180.0
3.6	117.0	118.0	122.0	121.0	115.0	111.0	5.8	170.0	170.0	170.0	170.0	173.0	184.0
3.8	122.0	123.0	126.0	129.0	122.0	116.0	6.0	174.0	174.0	174.0	174.0	177.0	188.0
4.0	127.0	127.0	129.0	135.0	130.0	133.0							

Table 16.10 Viscosity (in 10^{-6} Pa·s) of partially dissociated fluorine [3]

T, 10^3 K	Pressure, Pa			T, 10^3 K	Pressure, Pa		
	10^3	10^5	10^7		10^3	10^5	10^7
1.0	58.7	57.8	57.5	2.4	130.0	130.0	128.0
1.2	71.6	67.4	65.8	2.6	137.0	137.0	137.0
1.4	85.7	79.6	75.0	2.8	143.0	143.0	143.0
1.6	95.5	92.1	85.5	3.0	152.0	152.0	152.0
1.8	104.0	102.0	95.6	4.0	185.0	185.0	185.0
2.0	114.0	114.0	109.0	5.0	217.0	217.0	217.0
2.2	122.0	127.0	120.0	6.0	247.0	247.0	247.0

Table 16.11 Viscosity (in 10^{-6} Pa·s) of Ar–He mixture at atmospheric pressure, various temperatures and mole fractions of Ar [1]

Ar mole	Temperature, K								
fraction, %	72.0	90.2	192.5	229.5	293	373	456	473	523
0	7.98	9.12	14.5	16.4	19.7	23.2	16.9	27.2	29.0
0.05	8.23	9.40	15.4	17.6	20.6	23.8	28.3	27.8	29.8
0.10	8.34	9.60	16.0	17.9	21.3	24.4	29.2	28.4	30.6
0.15	8.37	9.70	16.4	18.4	21.8	25.0	29.9	29.0	31.4
0.20	8.37	9.74	16.6	18.8	22.2	25.6	30.5	29.5	32.1
0.25	8.32	9.70	16.8	19.0	22.5	26.0	31.0	30.0	32.7
0.30	8.24	9.60	16.2	19.2	22.7	26.4	31.3	31.4	33.2
0.35	8.13	9.47	16.9	19.2	22.8	26.8	31.6	31.8	33.6

Table 16.11 Viscosity of Ar–He mixture *(continued)*

Ar mole fraction, %	Temperature, K								
	72.0	90.2	192.5	229.5	293	373	456	473	523
0.40	8.00	9.32	16.8	19.1	22.9	27.0	31.9	32.2	34.0
0.45	7.85	9.17	16.7	19.0	22.9	27.2	32.1	31.4	34.3
0.50	7.70	9.03	16.6	18.9	23.0	27.4	32.2	31.7	34.5
0.55	7.54	8.90	16.4	18.8	23.0	27.5	32.4	31.9	34.7
0.60	7.40	8.76	16.3	18.8	22.9	27.5	32.4	32.0	34.8
0.65	7.24	8.63	16.2	18.7	22.9	27.5	32.5	32.1	34.9
0.70	7.09	8.49	16.1	18.6	22.8	27.5	32.5	32.2	35.0
0.75	6.95	8.35	16.0	18.5	22.7	27.4	32.5	32.2	35.0
0.80	6.82	8.21	15.9	18.3	22.6	27.3	32.5	32.2	34.9
0.85	6.70	8.07	15.8	18.2	22.5	27.2	32.4	32.2	34.8
0.90	6.57	7.93	15.7	18.0	22.4	27.1	32.4	32.2	34.8
0.95	6.50	7.80	15.6	17.9	22.3	27.0	32.4	32.1	34.6
1.00	6.35	7.68	15.5	17.7	22.1	26.8	32.3	32.1	34.5

Table 16.12 Viscosity (in 10^{-6} Pa·s) of Ar–Ne mixture at atmospheric pressure, various temperatures and mole fractions of Ar [1]

Ar mole fraction, %	Temperature, K							
	72.3	90.3	193.4	229	293	373	473	523
0	11.7	13.5	23.5	26.7	30.9	36.2	42.2	45.0
0.05	11.4	13.1	23.0	26.2	30.4	35.6	41.6	44.3
0.10	11.0	12.8	22.4	25.6	29.8	35.1	41.0	43.6
0.15	10.6	12.4	21.9	25.1	29.3	34.4	40.3	43.0
0.20	10.3	12.0	21.4	24.6	28.8	33.9	39.7	42.4
0.25	10.0	11.7	20.9	24.1	28.3	33.4	39.1	41.7
0.30	9.66	11.3	20.4	23.6	27.8	32.8	38.6	41.1
0.35	9.35	11.0	20.0	23.2	27.3	32.3	38.0	40.6
0.40	9.06	10.7	19.5	22.7	26.8	31.8	37.4	40.0
0.45	8.80	10.4	19.1	22.3	26.4	31.3	36.9	39.5
0.50	8.53	10.1	18.7	21.8	25.9	30.8	36.4	38.9
0.55	8.28	9.83	18.3	21.4	25.5	30.4	35.9	38.4
0.60	7.03	9.57	17.9	21.0	25.1	30.0	35.4	37.9
0.65	7.80	9.32	17.3	20.6	24.7	29.6	34.9	37.4
0.70	7.56	9.10	17.1	20.2	24.4	29.2	34.5	37.0
0.75	7.34	8.87	16.8	19.8	23.9	28.8	34.1	36.6
0.80	7.12	8.65	16.5	19.4	23.6	28.5	33.7	36.2
0.85	6.92	8.44	16.2	19.0	23.2	28.1	33.3	35.8
0.90	6.74	8.21	15.9	18.7	22.9	27.7	32.9	35.4
0.95	6.56	8.00	15.6	18.3	22.5	27.3	32.6	35.0
1.00	6.38	7.75	15.3	18.0	22.1	26.9	32.2	34.6

Table 16.13 Viscosity (in 10^{-6} Pa·s) of Ar–H_2 mixture at atmospheric pressure, various temperatures and mole fractions of Ar [1]

Ar mole fraction, %	Temperature, K				Ar mole fraction, %	Temperature, K			
	293	373	473	523		293	373	473	523
0.00	8.75	10.3	12.1	13.0	0.20	16.0	18.9	22.5	24.2
0.05	9.08	12.8	15.3	16.4	0.25	17.0	20.3	24.0	25.8
0.10	13.0	15.0	18.1	19.5	0.30	17.9	21.5	25.4	27.2
0.15	14.6	17.2	20.5	22.1	0.35	18.6	22.4	26.4	28.4

Table 16.13 Viscosity of Ar–H_2 mixture *(continued)*

Ar mole fraction, %	Temperature, K 293	373	473	523	Ar mole fraction, %	Temperature, K 293	373	473	523
0.40	19.2	23.2	27.4	29.4	0.75	21.6	26.2	30.9	33.4
0.45	19.7	23.9	28.2	30.2	0.80	21.7	26.3	31.2	33.7
0.50	20.2	24.5	28.9	30.9	0.85	21.8	26.5	31.5	33.9
0.55	20.4	24.9	29.4	31.6	0.90	22.0	26.6	31.7	34.2
0.60	20.9	25.3	29.9	32.2	0.95	22.1	26.8	31.9	34.3
0.65	21.8	25.6	30.3	32.7	1.00	22.1	26.8	32.1	34.5
0.70	21.4	25.9	30.7	33.1					

Table 16.14 Viscosity (in 10^{-6} Pa·s) of CO_2–H_2 mixture at atmospheric pressure, various temperatures and mole fractions of CO_2 [1]

CO_2 mole fraction, %	Temperature, K 300	400	500	550	CO_2 mole fraction, %	Temperature, K 300	400	500	550
0	8.9	10.8	12.6	13.4	0.55	15.0	19.3	23.1	25.1
0.05	10.9	13.4	15.4	16.9	0.60	15.1	19.4	23.3	25.2
0.10	12.1	15.0	17.4	18.7	0.65	15.1	19.4	23.4	25.3
0.15	12.9	16.1	19.0	20.1	0.70	15.1	19.5	23.5	25.4
0.20	13.6	16.9	20.0	21.4	0.75	15.0	19.5	23.6	25.4
0.25	14.0	17.6	20.8	21.6	0.80	15.0	19.5	23.6	25.4
0.30	14.3	18.1	21.5	23.4	0.85	15.0	19.5	23.6	25.5
0.35	14.6	18.5	22.0	24.0	0.90	15.0	19.5	23.6	25.5
0.40	14.8	18.8	22.4	24.4	0.95	15.0	19.5	23.6	25.5
0.45	14.9	19.0	22.7	24.7	1.00	14.9	19.4	23.5	25.6

Table 16.15 Viscosity (in 10^{-6} Pa·s) of partially dissociated air [3]

T, 10^3 K	Pressure, 10^5 Pa 0.001	1	10	100	T, 10^3 K	Pressure, 10^5 Pa 0.001	1	10	100
1.5	55.7	55.7	55.7	55.7	3.8	122.0	120.0	118.0	116.0
1.6	58.4	58.4	58.4	58.4	3.9	124.0	123.0	121.0	119.0
1.7	61.1	61.1	61.1	61.1	4.0	127.0	126.0	124.0	122.0
1.8	63.7	63.7	63.7	63.7	4.1	130.0	129.0	127.0	124.0
1.9	66.3	66.3	66.3	66.3	4.2	132.0	132.0	130.0	127.0
2.0	69.0	68.9	68.9	68.9	4.3	135.0	134.0	133.0	130.0
2.1	71.7	71.5	71.5	71.5	4.4	138.0	137.0	136.0	133.0
2.2	74.5	74.0	74.0	74.0	4.5	140.0	140.0	138.0	136.0
2.3	77.4	76.6	76.6	76.6	4.6	143.0	142.0	141.0	139.0
2.4	80.6	79.2	79.2	79.2	4.7	145.0	145.0	144.0	142.0
2.5	83.9	81.8	81.7	81.7	4.8	147.0	147.0	147.0	144.0
2.6	87.4	84.4	84.3	84.3	4.9	149.0	150.0	149.0	147.0
2.7	90.7	87.1	86.9	86.9	5.0	151.0	152.0	152.0	150.0
2.8	94.0	89.8	89.6	89.5	5.1	153.0	155.0	154.0	153.0
2.9	97.0	92.6	92.2	92.1	5.2	155.0	158.0	157.0	155.0
3.0	99.9	95.5	94.9	94.7	5.3	156.0	160.0	160.0	158.0
3.1	103.0	98.4	97.6	97.3	5.4	158.0	162.0	162.0	161.0
3.2	106.0	101.0	100.0	99.9	5.5	160.0	165.0	165.0	163.0
3.3	108.0	104.0	103.0	103.0	5.6	161.0	167.0	167.0	166.0
3.4	111.0	108.0	106.0	105.0	5.7	163.0	170.0	170.0	168.0
3.5	114.0	111.0	109.0	108.0	5.8	165.0	172.0	172.0	171.0
3.6	116.0	114.0	112.0	111.0	5.9	167.0	175.0	174.0	173.0
3.7	119.0	117.0	115.0	113.0	6.0	169.0	177.0	177.0	176.0

16.3 The Viscosity of Liquids

Experimental studies serve as the main source of information on the viscosities of liquids. Here, owing to the measurements being sensitive to the condition of the surface of the chamber, in which the experimental investigation of the viscosity is carried out, the uncertainty in the measurements of the viscosity of a liquid is somewhat larger than in the case of gases. In Tables 16.16–16.21, the viscosities are presented of liquefied gases and of some liquids, of liquid organic compounds, liquid metals, alloys, melts, salts, and bases at various temperatures.

Table 16.16 Viscosity η (in 10^{-3} Pa·s) of liquefied gases and some liquids at various temperatures [1, 2, 3] [The pressure corresponds to saturation conditions.]

T, K	η	T, K	η	T, K	η	T, K	η	T, K	η
He		18	0.016	70	0.22	290	0.15	CO	
1.28	0.00015	19	0.015	75	0.18	300	0.14	73.2	0.22
1.30	0.00016	20	0.014	80	0.15	310	0.12	75.2	0.20
1.34	0.00018	21	0.013	85	0.13	320	0.11	77.8	0.19
1.59	0.00023	22	0.012	90	0.11	330	0.10	82.8	0.16
1.76	0.00036	23	0.011	95	0.097	340	0.097	90.1	0.15
1.91	0.00068	24	0.010	100	0.087	350	0.088	99.6	0.12
2.00	0.00096	25	0.0096	105	0.078	360	0.080	112.0	0.10
2.09	0.0012	26	0.0088	110	0.071	370	0.070	130.0	0.066
2.11	0.0013	27	0.0082	115	0.060	380	0.061	H_2S	
2.14	0.0016	28	0.0076	120	0.048	390	0.051	191	0.52
2.16	0.0018	29	0.0070	125	0.032	400	0.040	193	0.51
2.18	0.0023	30	0.0064	126	0.019	405	0.025	198	0.49
2.32	0.0020	31	0.0058	O_2		H_2O		201	0.47
2.64	0.0024	32	0.0048	55	0.80	273	1.75	206	0.45
2.93	0.0024	33	0.0038	60	0.59	280	1.42	210	0.44
3.74	0.0028	Ne		65	0.46	300	0.82	HCl	
3.81	0.0029	25	0.15	70	0.37	320	0.56	161	0.57
4.02	0.0030	26	0.14	75	0.30	340	0.41	167	0.55
Ar		27	0.13	80	0.26	360	0.32	172	0.51
85	0.28	28	0.12	85	0.22	380	0.26	177	0.49
90	0.24	29	0.10	90	0.20	400	0.22	183	0.47
95	0.21	30	0.098	95	0.17	420	0.18	188	0.46
100	0.18	31	0.091	100	0.16	440	0.16	F_2	
105	0.16	32	0.084	105	0.14	460	0.14	69.2	0.41
110	0.15	33	0.078	110	0.13	480	0.13	73.2	0.35
115	0.13	34	0.072	115	0.12	500	0.12	75.3	0.33
120	0.12	35	0.067	120	0.11	520	0.11	78.2	0.30
125	0.11	36	0.062	125	0.10	540	0.10	80.9	0.28
130	0.10	37	0.056	130	0.096	560	0.094	83.2	0.26
135	0.089	38	0.052	135	0.088	580	0.086	CO_2	
140	0.075	39	0.047	140	0.078	600	0.079	260	0.12
145	0.060	40	0.043	145	0.066	620	0.071	265	0.11
150	0.045	41	0.039	154	0.026	647	0.042	270	0.10
151	0.028	42	0.034	NH_3		Air		275	0.096
H_2		43	0.031	240	0.28	90.1	0.13	280	0.091
14	0.026	44	0.027	250	0.25	107.2	0.094	285	0.086
15	0.022	N_2		260	0.22	111.0	0.090	290	0.079
16	0.020	60	0.36	270	0.19	125.1	0.082	295	0.070
17	0.018	65	0.27	280	0.17			300	0.060

Table 16.16 Viscosity η of liquefied gases and some liquids *(continued)*

T, K	η	T, K	η	T, K	η	T, K	η	T, K	η
CO_2		330	0.020	HI		232	1.34	191	0.85
304	0.032	350	0.020	223	1.42	236	1.30	194	0.84
310	0.022	400	0.021	227	1.38	HBr		199	0.82
		500	0.025			187	0.87		

Table 16.17 Viscosity (in 10^{-3} Pa·s) of liquid organic compounds at various temperatures [1–3]
[The pressure corresponds to saturation conditions.]

T, K	η	T, K	η	T, K	η	T, K	η	T, K	η	T, K	η
C_2H_6		340	0.11	130	0.33	340	0.062	380	0.49	290	0.26
285	0.74	360	0.088	140	0.26	360	0.041	390	0.44	300	0.24
300	0.58	380	0.069	150	0.22	369	0.021	400	0.41	310	0.22
320	0.45	400	0.055	160	0.19	$CHClF_2$		420	0.34	320	0.21
340	0.36	420	0.036	170	0.16	250	0.28	440	0.29	330	0.19
360	0.30	426	0.024	180	0.14	260	0.25	460	0.25	340	0.18
380	0.25	C_3H_6		190	0.13	270	0.23	C_8H_{10}		350	0.16
400	0.20	88.7	1.45	200	0.11	280	0.21	300	0.68	360	0.16
420	0.17	89.8	1.27	210	0.10	290	0.19	320	0.55	370	0.15
440	0.15	90.1	1.24	220	0.097	300	0.17	340	0.46	375	0.14
460	0.13	94.3	0.78	230	0.089	310	0.15	360	0.39	C_7H_{14}	
480	0.11	98.0	0.54	240	0.080	320	0.13	380	0.33	290	0.76
500	0.098	103	0.36	250	0.071	330	0.12	400	0.28	300	0.69
520	0.084	107	0.27	260	0.060	340	0.10	420	0.25	320	0.56
540	0.069	111	0.22	270	0.046	350	0.088	440	0.22	340	0.46
C_4H_{10}		119	0.16	280	0.030	360	0.072	460	0.19	360	0.38
190	0.75	C_2H_6		283	0.022	C_6H_{12}		480	0.17	380	0.38
200	0.61	100	0.88	CH_4		290	1.06	500	0.15	400	0.27
210	0.50	110	0.64	90	0.20	300	0.90	520	0.13	420	0.22
220	0.42	120	0.48	100	0.15	310	0.77	540	0.12	440	0.19
230	0.36	130	0.38	110	0.12	320	0.67	560	0.10	460	0.16
240	0.31	140	0.31	120	0.098	330	0.60	n-C_8H_{16}		$(C_8H_5)_2O$	
250	0.27	150	0.26	130	0.082	340	0.53	290	0.90	153	4.25
260	0.24	160	0.23	140	0.071	350	0.48	300	0.82	163	2.54
270	0.21	170	0.20	150	0.063	360	0.43	320	0.64	173	1.71
280	0.19	180	0.18	160	0.054	370	0.39	340	0.51	183	1.24
290	0.18	190	0.16	170	0.045	380	0.36	360	0.42	193	0.97
300	0.16	200	0.14	180	0.035	390	0.32	380	0.35	203	0.79
320	0.14	210	0.12	190	0.021	400	0.29	400	0.29	213	0.65
340	0.12	220	0.11	C_3H_8		420	0.25	420	0.25	223	0.55
360	0.095	230	0.099	80	19.2	440	0.22	440	0.21	233	0.47
380	0.074	240	0.089	100	3.8	460	0.19	460	0.18	243	0.41
400	0.051	250	0.079	120	1.5	480	0.17	n-C_7H_{14}		253	0.36
408	0.023	260	0.071	140	0.82	500	0.16	300	0.32	273	0.33
n-C_4H_{10}		270	0.052	160	0.54	n-$C_{12}H_{26}$		305	0.31	293	0.30
180	0.69	280	0.054	180	0.39	300	1.33	310	0.29	303	0.27
200	0.48	290	0.046	200	0.30	310	1.16	320	0.27	313	0.24
220	0.36	300	0.036	220	0.24	320	1.01	330	0.25	323	0.22
240	0.28	305	0.022	240	0.19	330	0.88	340	0.23	333	0.20
260	0.23	C_2H_4		260	0.16	340	0.77	350	0.21	343	0.166
280	0.19	100	0.80	280	0.13	350	0.68	360	0.20	353	0.140
300	0.16	110	0.56	300	0.11	360	0.61	n-C_6H_{12}		373	0.118
320	0.13	120	0.42	320	0.083	370	0.54	280	0.28		

Table 16.17 Viscosity of liquid organic compounds *(continued)*

T, K	η
C_6H_7	
273	10.2
283	6.5
293	4.4
303	3.12
313	2.30
323	1.80
333	1.50
343	1.28
353	1.10
363	0.94
373	0.80
383	0.69
393	0.59
$C_6H_5NO_2$	
273	3.1
283	2.5
293	2.0
303	1.7
313	1.4
323	1.2
333	1.1
353	0.87
373	0.70
$C_4H_8O_2$	
293	0.48
313	0.38
333	0.31
353	0.25
373	0.21
393	0.18
413	0.15
433	0.125
453	0.104
473	0.086
493	0.068
513	0.048
$C_2H_4O_2$	
303	0.79
313	0.69
323	0.62
333	0.55
353	0.45
373	0.38
393	0.32
403	0.30
n-C_5H_{12}	
153	2.31
173	1.25
193	0.77
203	0.64
213	0.55
223	0.47

T, K	η
233	0.42
243	0.38
253	0.34
263	0.31
273	0.28
283	0.26
293	0.24
303	0.22
n-$C_6H_{12}O_2$	
290	0.76
300	0.67
320	0.58
340	0.42
360	0.34
n-$C_{10}H_{22}$	
300	0.67
310	0.59
320	0.53
330	0.47
340	0.43
350	0.39
360	0.35
370	0.32
380	0.30
390	0.28
400	0.26
410	0.24
420	0.22
iso-C_8H_{18}	
100	0.232
110	0.214
120	0.198
130	0.183
140	0.169
150	0.156
160	0.146
170	0.135
180	0.125
190	0.115
200	0.106
210	0.097
220	0.089
230	0.080
240	0.072
250	0.064
260	0.054
270	0.037
n-C_8H_{18}	
213	2.47
223	1.83
233	1.43
243	1.16
253	0.97

T, K	η
263	0.83
273	0.71
283	0.62
293	0.55
303	0.49
313	0.44
323	0.39
343	0.32
363	0.28
383	0.23
403	0.19
423	0.16
443	0.14
463	0.12
483	0.10
503	0.084
523	0.070
543	0.056
563	0.041
n-C_9H_{20}	
263	1.15
273	0.97
283	0.83
293	0.71
303	0.63
313	0.56
323	0.50
333	0.44
343	0.40
353	0.37
363	0.34
373	0.31
383	0.28
393	0.26
403	0.24
413	0.23
423	0.21
n-$C_{10}H_{22}$	
243	2.51
253	1.93
263	1.55
273	1.27
283	1.07
293	0.91
Benzine	
223	1.7
233	1.4
253	0.99
273	0.73
293	0.52
313	0.41
333	0.33
353	0.27

T, K	η
373	0.22
393	0.19
413	0.16
433	0.15
453	0.13
473	0.11
493	0.10
513	0.089
533	0.081
553	0.073
573	0.067
Kerosene	
223	11.5
233	7.3
253	3.6
273	2.2
293	1.5
313	1.1
333	0.83
353	0.66
373	0.54
393	0.46
413	0.39
433	0.34
453	0.30
473	0.26
493	0.23
513	0.21
533	0.19
553	0.17
573	0.16
CCl_4	
263	1.7
273	1.4
283	1.1
293	0.97
303	0.84
313	0.74
323	0.65
333	0.59
343	0.52
353	0.47
363	0.43
373	0.39
383	0.35
393	0.32
413	0.28
433	0.23
453	0.20
CH_3OH	
173	16
193	5.7
213	3.0

T, K	η
233	1.8
253	1.2
273	0.82
293	0.58
313	0.45
333	0.35
353	0.27
373	0.21
393	0.17
413	0.14
433	0.11
453	0.088
473	0.072
493	0.058
513	0.046
C_2H_5OH (96%)	
273	1.8
293	1.2
313	0.82
333	0.59
353	0.43
373	0.32
393	0.24
413	0.19
433	0.15
453	0.12
473	0.095
493	0.072
513	0.049
$C_3H_8O_3$	
273	12100
283	3950
293	1480
303	600
313	330
323	180
333	102
343	59
353	35
363	21
373	13
393	5.2
413	1.8
433	1.0
453	0.45
473	0.22
$C_4H_{10}O$	
153	4.25
173	1.71
193	0.97
213	0.65
233	0.47
253	0.36

T, K	η
273	0.30
293	0.24
313	0.20
333	0.17
353	0.14
n-C_3H_7OH	
203	54.6
213	31.6
223	20.2
233	13.5
243	9.50
253	6.90
263	5.10
273	3.85
283	2.89
293	2.20
303	1.72
313	1.38
333	0.92
353	0.64
363	0.54
373	0.46
383	0.39
393	0.34
403	0.29
413	0.25
423	0.21
433	0.18
453	0.14
473	0.11
iso-C_3H_7OH	
213	66.1
223	37.6
233	23.2
243	14.9
253	10.1
263	6.8
273	4.6
283	3.26
293	2.39
303	1.77
313	1.33
323	1.03
333	0.80
343	0.65
353	0.52
n-C_3H_9OH	
223	34.7
233	22.4
243	14.6
253	10.3
263	7.4
273	5.19

Table 16.17 Viscosity of liquid organic compounds *(continued)*

T, K	η
283	3.87
293	2.95
303	2.28
313	1.78
323	1.41
333	1.14
343	0.93
353	0.76
363	0.63
373	0.54
383	0.46
n-$C_{11}H_{24}$	
263	2.16
273	1.74
283	1.42
293	1.18
303	1.01
313	0.87
323	0.76
333	0.67
343	0.60
353	0.54
363	0.48
373	0.44
393	0.36
413	0.31
433	0.27
453	0.23
473	0.20

T, K	η
n-$C_{12}H_{26}$	
263	2.90
273	2.26
283	1.83
293	1.49
303	1.25
313	1.06
323	0.92
333	0.80
343	0.71
353	0.63
363	0.57
373	0.51
n-$C_{13}H_{28}$	
273	2.96
283	2.34
293	1.88
303	1.56
313	1.31
323	1.11
333	0.97
343	0.85
353	0.75
363	0.67
373	0.60
393	0.49
413	0.41
433	0.35
453	0.30

T, K	η
473	0.26
n-$C_{14}H_{30}$	
283	2.96
293	2.32
303	1.89
313	1.56
323	1.32
333	1.14
343	0.99
353	0.87
363	0.77
373	0.68
n-$C_{15}H_{32}$	
293	2.84
303	2.29
313	1.87
323	1.57
333	1.34
343	1.16
353	1.01
363	0.89
373	0.79
n-$C_{16}H_{34}$	
293	3.45
303	2.75
313	2.32
323	1.85
333	1.56
343	1.34

T, K	η
353	1.16
363	1.01
373	0.89
393	0.72
413	0.58
433	0.49
453	0.41
473	0.35
493	0.30
513	0.26
n-$C_{17}H_{36}$	
303	3.29
313	2.65
323	2.17
333	1.83
343	1.56
353	1.34
363	1.16
373	1.01
393	0.79
413	0.66
433	0.55
453	0.46
473	0.39
493	0.34
513	0.30
533	0.26
553	0.23
573	0.20

T, K	η
n-$C_{18}H_{38}$	
303	3.81
313	3.06
323	2.49
333	2.06
343	1.75
353	1.48
363	1.30
373	1.13
$C_{20}H_{42}$	
313	4.07
323	3.26
333	2.66
343	2.22
353	1.19
363	1.16
373	1.14
393	1.09
413	0.88
433	0.72
453	0.60
373	0.50
493	0.43
C_5H_{10}	
273	0.56
283	0.49
293	0.44
303	0.39
313	0.35

T, K	η
n-$C_5H_{10}O_2$	
290	0.58
300	0.53
320	0.42
340	0.34
360	0.27
C_7H_8	
295	0.61
300	0.57
310	0.51
320	0.45
330	0.40
340	0.37
350	0.34
360	0.31
370	0.29
380	0.27
390	0.25
400	0.23
420	0.20
440	0.18
460	0.15
480	0.13
500	0.12
520	0.10
540	0.089
550	0.083

Table 16.18 Viscosity (in 10^{-3} Pa·s) of liquid metals [3]

T, K	Li	Na	K	Rb	Cs	Hg	Bi	Pb	Sn	Zn	Sb
300						1.6					
400		0.61	0.41	0.44	0.42	1.2					
500	0.53	0.42	0.30	0.32	0.32	1.0			1.9		
600	0.43	0.33	0.24	0.26	0.25	0.91	1.7	2.6	1.6		
700	0.36	0.27	0.20	0.22	0.22	0.84	1.2	1.3	1.3	3.3	
800	0.31	0.23	0.17	0.19	0.19	0.79		1.8	1.2	2.6	
900	0.28	0.20	0.15	0.17	0.17	0.76	1.0	1.5	1.0	2.1	1.6
1000	0.25	0.18	0.14	0.15	0.15	0.73		1.3	0.93	1.8	1.2
1100	0.22	0.16	0.12	0.14	0.14	0.71		1.2			1.1
1200	0.21	0.15	0.11	0.13	0.13						
1300	0.19	0.14	0.10	0.12	0.12						
1400	0.18	0.13	0.098	0.11	0.12						
1500	0.17	0.12	0.093	0.11							
1600	0.16	0.12	0.088								
1700	0.15	0.11	0.083								
1800	0.14	0.11	0.079								

Table 16.19 Viscosity η (in 10^{-3} Pa·s) of some molten salts and bases [5]

T, K	η
AgBr	
882	1.86
922	1.66
961	1.49
1043	1.22
1076	1.19
AgCl	
876	1.61
905	1.47
942	1.37
1007	1.19
AgI	
878	3.03
900	2.75
970	2.38
1000	2.12
1100	1.56
$CaCl_2$	
1070	4.94
$HgBr_2$	
520	3.0
530	2.0
KBr	
1020	1.5
1050	1.3
1080	1.2

T, K	η
KCI	
1060	1.4
1110	1.2
1200	1.0
1310	0.71
$K_2Cr_2O_7$	
670	13.2
690	11.7
710	10.4
730	9.2
750	8.1
770	7.0
KNO_3	
620	2.73
630	2.58
650	2.31
670	2.09
690	1.90
710	1.74
730	1.60
750	1.48
770	1.38
790	1.30
810	1.24
820	1.21

T, K	η
KOH	
670	2.3
720	1.7
770	1.3
820	1.0
870	0.8
$LiNO_3$	
530	6.48
560	5.48
580	4.70
600	4.32
630	3.63
660	2.49
700	2.05
$MgCl_2$	
1080	4.12
N_2H_4	
278	1.21
283	1.12
288	1.04
293	0.97
298	0.91
N_2O_4	
274	0.52
278	0.49
282	0.47
288	0.44

T, K	η
Na_3AlF_6	
1270	2.8
NaBr	
1035	1.42
1053	1.28
NaCl	
590	2.83
610	2.53
630	2.28
650	2.08
670	1.90
690	1.74
1100	1.43
1120	1.28
1150	1.14
1170	1.02
1200	0.91
1220	0.82
1250	0.75
1270	0.70
$NaNO_3$	
710	1.62
730	1.52
NaOH	
620	4.0
670	2.8
720	2.8
770	1.8
820	1.5

T, K	η
$NaPO_3$	
920	1250
970	700
1020	440
1070	300
1120	210
$PbBr_2$	
650	10.2
670	8.06
690	7.0
710	6.1
730	5.4
750	4.7
770	4.1
$PbCl_2$	
770	5.53
790	4.66
810	4.02
830	3.59
850	3.28
870	3.06
880	2.95
$SnCl_4$	
303	0.81
313	0.72
323	0.67

Table 16.20 Viscosity (in 10^{-3} Pa·s) of molten lead–tin alloy for various temperatures and molar fractions of lead [1]

T, K	Molar fraction of lead, %			
	0.000	0.025	0.300	0.382
460			4.2	
470			5.0	
475				2.6
480			3.0	2.1
500			2.5	2.1
505	2.6			
510	2.0	4.2		
520		1.9		
525			2.2	2.0
540	1.8	1.8		
550			2.2	2.2
560	1.7	1.6		
575			2.1	2.3
580	1.6	1.6		

T, K	Molar fraction of lead, %				
	0.000	0.025	0.300	0.382	1.000
600	1.6	1.5	2.0	2.3	3.0
612.5					2.8
620	1.5	1.5			
625			2.0	2.3	2.6
637.5					2.5
640	1.5	1.5			
650			2.0	2.3	2.4
660	1.5	1.4			
675			1.9	2.2	2.2
700			1.8	2.2	2.2
725			1.8		2.0
750					1.9
775					1.9

Table 16.21 Viscosity (in 10^{-3} Pa·s) of molten iron–carbon alloy for various temperatures and molar fractions of iron [1]

T, K	Molar fraction of iron, %								
	0.9514	0.9580	0.9715	0.9790	0.9870	0.9936	0.9960	0.9975	0.9992
1550		8.5							
1575		8.0							
1600		7.5							
1625		7.0	9.2						
1650	3.8	6.6	8.5						
1675	3.3	6.1	7.9						
1700	2.9	5.7	7.2	7.2					
1725	2.5	5.2	6.6	6.8	7.7				
1750	2.2	4.7	6.0	6.4	7.0				
1775			5.5	6.0	6.4				
1800		3.9	5.1	5.6	5.8				
1825		3.5	4.7	5.2	5.2		5.7	4.9	7.6
1850		3.2	4.4	4.9	4.8	4.7	4.9	4.4	6.4
1875			4.1	4.6	4.4	4.2	4.3	4.1	5.7
1900			3.9		4.1	3.9	3.9	3.9	5.2
1925					3.9	3.7	3.6	3.8	4.9
1950					3.7	3.6	3.4	3.6	4.6
1975					3.5	3.4	3.3	3.4	
2000						3.3			

References

[1] Touloukian, Y. S., Viscosity, NBS Edition, New York, 1974.

[2] Stephan, K., Lucas, K., Viscosity of dense fluids, Academic Press, New York, 1979.

[3] Vargaftik, N. B., Handbook on thermal properties of gases and liquids, Fizmatgiz, Moscow, 1972 (in Russian).

[4] Golubev, I. F., Gnezdilov, N. E., Viscosity of gas mixtures, Izdatel'stvo standartov, Moscow, 1971 (in Russian).

[5] Handbook of a chemist. Basic properties of inorganic and organic compounds, Ed. Nikol'skii, B.P., 3rd ed., Khimiya, Leningrad, 1971 (in Russian).

[6] Boushehri, A., Bzowski, J., Kestin, J., Meson, E. A., J. Phys. Chem. Ref. Data, 16, 445, 1987.

[7] Bich, E., Millat, J., Vogel, E., J. Phys. Chem. Ref. Data, 19, 1289, 1990.

17

Diffusion

A.V. Eletskii

17.1 Introduction

Diffusion is the process establishing equilibrium between the concentrations of particles (atoms, molecules, ions, electrons) in a medium. When there exists a gradient of concentration ∇N of particles in a medium, there arises a flux of these particles $\mathbf{j}$, which equalizes their concentrations. The flux and diffusion coefficient D are related by Fick's law:

$$\mathbf{j} = -D\nabla N. \tag{17.1}$$

This relation is valid when the size of the system along the gradient is much larger than the mean free path of the particles in the medium, and the change in concentration along the free path is much smaller than the distinctive particle concentration N. Moreover, the absence of external fields, temperature, and pressure gradients is implied.

17.2 The Diffusion of Atoms and Molecules in Gases

In perfect gases, where the density of neutral particles (atoms, molecules) N satisfies the condition

$$N \ll 1/a_0^3 \tag{17.2}$$

($a_0 \approx 10^{-8}$–10^{-7} cm is the typical range of intermolecular forces), the diffusion depends on pair collisions of the probe particle with atoms or molecules. Therefore, up to very high pressures[1] the diffusion coefficient is inversely proportional to the density of the gas particles and is expressed in terms of a characteristic of pair collisions of the probe particle with particles of the gas, namely, of the diffusion scattering cross section σ^*.

In accordance with the elementary kinetic theory of gases, the expression for the diffusion coefficient of small admixture in a gas consisting of one sort of particles, as well as for the coefficient of self-diffusion, has the form

$$D = \langle v \rangle \lambda/3, \tag{17.3}$$

[1]The range of pressures, where the indicated dependence holds with a sufficiently high accuracy (the uncertainty is 2 to 5%), depends on the temperature. Near the critical point, this dependence may be essentially violated.

0-8493-2861-6/97/$0.00+$.50

where $v = \sqrt{2kT/\mu}$ is the mean relative velocity between the colliding particles of the admixture and the gas; μ is the reduced mass of the colliding particles; T is the temperature of the gas; $\lambda = 1/N\sigma^*$ is the mean free path of the probe particles in the gas; N is the number density of gas particles. Relation (17.3) is exact if σ^* is independent of the energy of the colliding particles. Otherwise, the notion of mean free path loses sense, and the relation indicated is valid for effective quantities. The values of diffusion coefficients are generally referred to constant pressure, instead of constant temperature. In this case, taking into account the equation of state for a gas, $p = NkT$, we obtain the dependence

$$D = D_0 \left(\frac{T}{273}\right)^{3/2}, \tag{17.4}$$

where D_0 is the diffusion coefficient in normal conditions. This relation holds true for the same conditions as relation (17.3).

Self-diffusion coefficients of gases. The self-diffusion coefficients presented in Table 17.1 are appropriate for normal conditions ($T = 273$ K, $p = 0.1$ MPa). The diffusion coefficient can be approximated, within the temperature range given, by the power function

$$D(T) = D_0 \left(\frac{T}{273}\right)^{\alpha}; \tag{17.5}$$

the values of the parameter α are presented as well.

Table 17.1 The parameters entering into relationship (17.5) [1]

Gas	D_0, cm^2/s	α	T, K	Gas	D_0, cm^2/s	α	T, K	Gas	D_0, cm^2/s	α	T, K
Ar	0.156	1.92	77–353	HBr	0.179			N_2	0.17	1.9	77–353
CH_4	0.206			HCl	0.125			Ne	0.452		
CO	0.175			H_2O	0.277			O_2	0.18	1.92	77–353
CO_2	0.097			He	1.62	1.71	14–296	UF_6	0.0165	1.6	144-296
H_2	1.28			Kr	0.08			Xe	0.048		

The values of self-diffusion coefficients for molecular gases are presented in Table 17.2 within a broad range of temperatures.

Table 17.2 Self-diffusion coefficient (in cm^2/s) of molecular gases over a wide temperature range and at atmospheric pressure

T, K	N_2	O_2	NO	CO	N_2O	CO_2	CH_4	CF_4	SF_6	C_2H_4	C_2H_6
100	0.028			0.028							
150	0.062	0.061	0.059	0.062							
200	0.104	0.105	0.103	0.104			0.111	0.032			
250	0.156	0.158	0.155	0.156		0.083	0.170	0.048	0.025	0.089	0.075
273	0.182	0.185	0.182	0.182	0.099	0.099	0.200	0.057	0.030	0.107	0.090
293	0.206	0.210	0.207	0.206	0.114	0.114	0.228	0.066	0.034	0.122	0.103
300	0.214	0.185	0.215	0.214	0.119	0.119	0.238	0.068	0.036	0.128	0.108
313	0.231	0.236	0.232	0.231	0.129	0.129	0.257	0.073	0.039	0.139	0.117
333	0.257	0.263	0.259	0.257	0.146	0.146	0.288	0.082	0.044	0.157	0.132
353	0.284	0.291	0.287	0.284	0.163	0.163	0.319	0.091	0.049	0.175	0.179
373	0.313	0.321	0.316	0.313	0.181	0.181	0.352	0.100	0.054	0.194	0.203
423	0.388	0.399	0.393	0.388	0.230	0.228	0.440	0.125	0.168	0.246	0.256
473	0.469	0.483	0.477	0.469	0.282	0.280	0.536	0.152	0.083	0.301	0.298
523	0.556	0.574	0.566	0.556	0.339	0.336	0.638	0.181	0.100	0.361	0.333
573	0.648	0.670	0.661	0.648	0.400	0.396	0.747	0.212	0.117	0.426	0.363
623	0.746	0.772	0.762	0.746	0.465	0.459	0.862	0.245	0.136	0.494	0.388
673	0.849	0.879	0.868	0.849	0.533	0.526	0.984	0.279	0.155	0.566	0.409

Table 17.2 Self-diffusion coefficient of molecular gases *(continued)*

T, K	N_2	O_2	NO	CO	N_2O	CO_2	CH_4	CF_4	SF_6	C_2H_4	C_2H_6
723	0.958	0.992	0.980	0.958	0.605	0.596	1.11	0.315	0.175	0.641	0.427
773	1.07	1.11	1.10	1.07	0.680	0.670	1.25	0.353	0.197	0.720	0.443
873	1.31	1.36	1.34	1.31	0.841	0.827	1.53	0.434	0.242	0.889	0.468
973	1.58	1.63	1.61	1.58	1.01	0.996	1.88	0.521	0.291	1.07	0.488
1073	1.86	1.92	1.90	1.86	1.20	1.18	2.16	0.614	0.344	1.26	0.502
1173	2.15	2.23	2.21	2.15	1.40	1.37	2.51	0.713	0.400	1,47	0.514
1273	2.47	2.56	2.53	2.47	1.60	1.57	2.88	0.818	0.459	1.69	0.523
1773	4.33	4.49	4.44	4.33	2.80	2.75	5.02	1.42	0.800	2.95	0.545
2273	6.64	6.88	6.86	6.64	4.25	4.16	7.64	2.16	1.21	4.48	0.548
2773	9.38	9.70	9.73	9.38	5.93	5.81	10.7	3.02	1.69		
3273	12.5	12.9	13.0	12.5	7.83	7.66	14.2	3.98	2.23		

Note: The data were obtained by fitting the numerous calculation and experimental results with the estimated error within 1 to 4% range.

The values of self-diffusion coefficients for gases at high temperatures are presented in Table 17.3.

Table 17.3 Self-diffusion coefficient (in cm^2/s) of gases at high temperatures and atmospheric pressure

Gas	Temperature T, 10^3 K																				
	1	1.5	2	2.5	3	3.5	4	4.5	5	5.5	6	6.7	7	8	9	10	11	12	13	14	15
Ar	1.51	3.0	5.03	7.50	10.4	13.9	17.8	22.1	26.8	31.9	37.4	43.3	49.5	62.8	77.4	93.2	111	129	148	169	190
He	13.3	26.9	46.7	70.3	98.6	130	167	207	250	299	350	407	467	595	738	898	1070	1260	1460	1670	1840
Kr	0.85	1.72	2.83	4.14	5.7	7.4	9.3	11.4	13.8	16.4	19.2	22.2	25.5	32.6	40.7	49.6	59.6	70.2	81.5	93.7	107
N_2	1.56	3.08	5.05	7.5	10.4	13.8	17.6	21.9	26.6	31.8	37.3	43.2	49.6	62	79	96	113	133	153	174	197
Ne	3.9	7.76	12.7	18.5	25.2	32.6	41.1	50.4	60.7	72.0	83.8	96.5	110	139	172	207	244	284	325	318	414
Xe	0.50	1.01	1.65	2.43	3.32	4.32	5.42	6.6	8.0	9.4	10.9	12.6	14.4	18.2	22.5	27.2	32.4	37.7	43.4	49.4	55.7

Note: The calculated results were found using the interatomic interaction potential obtained on the basis of the atomic beam scattering measurements [1].

The coefficient of mutual diffusion in gases. The respective fluxes $\mathbf{j}_1$ and $\mathbf{j}_2$ of particles of a first and of a second sort in a two-component gas mixture in the absence of external fields, chemical reactions, as well as temperature and pressure gradients, are expressed by the relations

$$\mathbf{j}_1 = -nD_{12}\,\mathbf{grad}\,x_1, \qquad \mathbf{j}_2 = -nD_{21}\,\mathbf{grad}\,x_2, \tag{17.6}$$

where n is the total particle density; x_1 and x_2 are the relative concentrations of particles of the first and second sorts, respectively. These relations are valid in a reference frame in which no resulting particle flux is present: $\mathbf{j}_1 + \mathbf{j}_2 = 0$, since, also, $x_1 + x_2 = 1$, $D_{12} = D_{21}$.

The diffusion coefficient D_{AB} of particles of arbitrary sorts in a mixture of the two gases A and B is determined by the following relation (Blank's law), valid with a precision up to several percent:

$$D_{AB}^{-1} = \frac{x_A}{D_{1A}} + \frac{x_B}{D_{1B}}, \tag{17.7}$$

where D_{1A} and D_{1B} are the mutual diffusion coefficients of the probe particles in the gases A and B, respectively, at a pressure equal to the total pressure in the mixture under consideration.

Experimental studies serve as the main source of information on the coefficient of mutual diffusion in gases. The precision with which this coefficient is known depends essentially on the temperature range and on the sort of gases studied. The results of measurements of mutual diffusion coefficients, presented below for various pairs of gases, are divided into four groups depending on the class of accuracy (Tables 17.4, 17.5). The uncertainties characterizing the mutual diffusion coefficients of the first three groups (I, II, III) are seen in Figure 17.1, and the respective information for the fourth group is presented in Table 17.5.

Table 17.4 Estimated accuracy (in %) of the diffusion coefficients for interacting systems classified under three groups

Accuracy group	System of interacting particles	Temperature, K					
		1.75	65	300	500	10^3	10^4
I	He – Ne, He – Ar, He – Kr, He – Xe, Ne – Ar, Ne – Kr, Ne – Xe, Ar – Kr, Ar – Xe, Kr – Xe, H_2 – N_2		2	1	2	5	10
II	^3He – ^4He, He – H_2, He – N_2, He – CO, He – O_2, He – air He – CO_2, H_2 – Ne, H_2 – Ar, H_2 – Kr, H_2 – D_2, CO – air, CO – CO_2, N_2 – Ar, N_2 – CO, N_2 – CO_2	6	4	2	3	7	15
III	Ar – CH_4, Ar – CO, Ar – CO_2, Ar – air, Ar – SF_6, H_2 – Xe, H_2 – CH_4, H_2 – O_2, H_2 – SF_6, CH_4 – He, CH_4 – N_2, CH_4 – O_2, CH_4 – air, CH_4 – SF_6, N_2 – Ne, N_2 – Kr, N_2 – Xe, N_2 – CH_4, N_2 – O_2, N_2 – SF_6, O_2 – Ar, O_2 – CO_2, O_2 – SF_6, CO_2 – air, CO_2 – N_2O, CO_2 – SF_6, SF_6 – He, SF_6 – air, CO – Kr, CO – O_2, CO – air, CO – CO_2, CO – SF_6			3	4	10	20

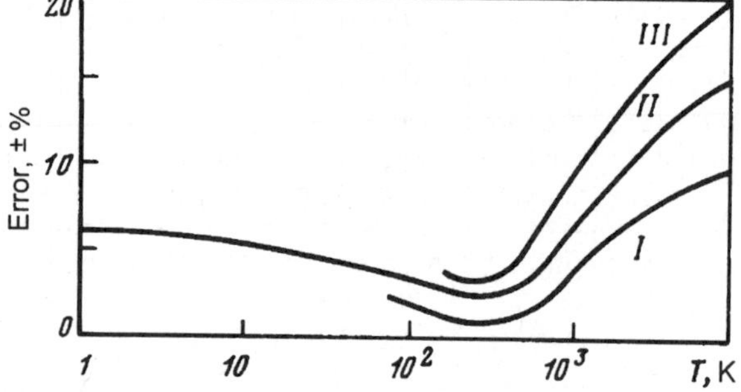

Figure 17.1 Temperature dependence of uncertainty in data presented in Table 17.4.

Table 17.5 Estimated accuracy of the mutual diffusion coefficients for interacting systems classified under the fourth group

System of interacting particles	Temperature range, K	Accuracy, %
H_2 – N_2	282–373	4
H_2 – O_2	282–1070	7
H_2O – air	282–1070	5–10
H_2O – CO_2	296–1640	7–10
CO_2 – Ne	195–625	3–5
CO_2 – C_3H_8	298–550	3–5
H – H_2	300	5
	1000	30
H – N_2; O – N_2	300	10
O – O_2	1000	25
H – He; H – Ar	300	15
O – He; O – Ar	1000	30

In Tables 17.6 and 17.7, the parameters are presented that enter into the empirical dependences of mutual diffusion coefficients at atmospheric pressure. For the systems presented in Table 17.6, this dependence takes the form

$$D = D_0 \left(\frac{T}{273}\right)^{\alpha} \exp\left(-\frac{S}{T} - \frac{S'}{T^2}\right) \tag{17.8}$$

where S, S' are the empirical parameters, while the exponential factor may be dropped without loss of accuracy at temperatures over 200 K.

The empirical dependence for the systems presented in Table 17.7 is of the form

$$D = D_0 \left(\frac{T}{273}\right)^{\alpha} \exp\left(-\frac{S}{T}\right), \tag{17.9}$$

and, here, unity can be substituted for the exponential factor at $T > 20S$.

Table 17.6 The parameters entering into relationship (17.8) and applying to the temperature interval ($T_1 - 10^4$ K) [2]

System of interacting particles	D_0, cm^2/s	α	S, K	S', K^2	T_1, K	Accuracy group
$^3He - ^4He$	1.55	1.501	−0.963	1.894	1.74	II
Ne − Ne	0.235	1.509	1.87		65	I
He − Ar	0.635	1.552	1.71		77	I
He − Kr	0.503	1.609	−32.65	2036	77	I
He − Xe	0.391	1.644	−68.87	5416	169	I
He − H_2	1.32	1.510			90	II
He − N_2	0.613	1.524			77	II
He − CO	0.613	1.524			77	II
Ne − Ar	0.278	1.546	1.82	1170	90	I
Ne − Kr	0.242	1.555	20.4		112	I
Ne − Xe	0.197	1.584	10.1		169	I
Ar − Kr	0.14	1.556	47.3		169	I
Ar − Xe	0.122	1.563	59.9		169	I
Ar − N_2	0.79	1.519	39.8		242	II
Kr − Xe	0.0812	1.608	52.7		169	I
Kr − H_2	0.664	1.564	26.4		77	II
$H_2 - D_2$	0.99	1.500	6.072	38.10	14	II
$H_2 - N_2$	0.66	1.548	−2.80	1067	65	I
H_2 − CO	0.66	1.548	−2.80	1077	65	II
N_2 − CO	0.175	1.576	36.2	3825	78	II

Table 17.7 The parameters entering into empirical relationship (17.9) [2]

System of interacting particles	D_0, cm^2/s	α	S, K	Temperature range, K	Accuracy group
He − CH_4	0.57	1.750		$298-10^4$	III
He − O_2	0.45	1.710		$244-10^4$	II
He − air	0.62	1.729		$244-10^4$	II
He − CO_2	0.52	1.720		200−530	II
He − SF_6	0.35	1.627		$290-10^4$	III
Ne − H_2	0.99	1.731		$90-10^4$	II
Ne − N_2	0.28	1.743		$293-10^4$	III
Ne − CO	0.22	1.776		195−625	III
Ar − CH_4	0.172	1.785		$307-10^4$	III
Ar − N_2	0.17	1.752		$244-10^4$	II
Ar − CO	0.17	1.752		$244-10^4$	III
Ar − O_2	0.167	1.736		$243-10^4$	III
Ar − air	0.165	1.749		$244-10^4$	III
Ar − CO_2	0.177	1.646	89.1	276−1800	III
Ar − SF_6	0.114	1.596	145.4	$328-10^4$	III
Kr − N_2	0.13	1.766		$248-10^4$	III
Kr − CO	0.13	1.766	2	$248-10^4$	III
Xe − H_2	0.54	1.712	16.9	$242-10^4$	III
Xe − N_2	0.106	1.789		$242-10^4$	III
$H_2 - CH_4$	0.62	1.765		$293-10^4$	III
$H_2 - O_2$	0.69	1.732		$252-10^4$	III
H_2 − air	0.66	1.750		$252-10^4$	II
$H_2 - CO_2$	0.56	1.750	11.7	200−550	II
$H_2 - SF_6$	0.52	1.570	102.5	$298-10^4$	III

Table 17.7 The parameters entering into empirical relationship (17.9) *(continued)*

System of interacting particles	D_0, cm^2/s	α	S, K	Temperature range, K	Accuracy group
$CH_4 - N_2$	0.2	1.750		298–10^4	III
$CH_4 - O_2$	0.22	1.695	44.2	294–10^4	III
CH_4 – air	0.186	1.747		298–10^4	III
$CH_4 - SF_6$	0.119	1.657	69.2	298–10^4	III
$N_2 - O_2$	0.182	1.724		285–10^4	III
$N_2 - H_2O$	0.204	2.072		282–373	IV
$N_2 - CO_2$	0.208	1.570	113.6	288–1800	II
$N_2 - SF_6$	0.122	1.590	119.4	328–10^4	III
$CO - O_2$	0.175	1.724		285–10^4	III
CO – air	0.182	1.730		285–10^4	III
$CO - CO_2$	0.142	1.803		282–473	III
$CO - SF_6$	0.129	1.584	139.4	297–10^4	III
$O_2 - H_2O$	0.207	2.072		282–450	IV
$O_2 - H_2O$	0.264	1.632		450–1070	IV
$O_2 - CO_2$	0.174	1.661	61.3	287–1083	III
$O_2 - SF_6$	0.138	1.522	129.0	297–10^4	III
H_2O – air	0.205	2.072		282–450	IV
H_2O – air	0.26	1.632		450–1070	IV
CO_2 – air	0.207	1.590	102.1	280–1800	IV
SF_6 – air	0.126	1.576	121.1	328–10^4	IV
$H_2O - CO_2$	0.41	1.500	307.9	296–1640	IV
$CO_2 - N_2O$	0.095	1.866		195–550	III
$CO_2 - SF_6$	0.069	1.886		328–472	III
H – He	2.35	1.732		275–10^4	IV
H – Ar	0.112	1.597		275–10^4	IV
$H - H_2$	0.184	1.728		190–10^4	IV
$N - N_2$	0.29	1.774		280–10^4	IV
O – He	0.84	1.749		280–10^4	IV
O – Ar	0.23	1.841		280–10^4	IV
$O - N_2$	0.28	1.774		280–10^4	IV
$O - O_2$	0.28	1.774		280–10^4	IV

The diffusion coefficients for metastable atoms in inert gases are presented in Tables 17.8 – 17.10.

Table 17.8 Diffusion coefficient (in cm^2/s) of metastable inert atoms in the parent gases at a pressure of 133 Pa (or 1 mm Hg) [Estimated uncertainty does not exceed 10%.]

Atom	Temperature, K					
	20	40	77	150	300	500
He ($2\,^3S$)	60	100	160	260	460	710
He ($2\,^1S$)	45	75	120	210	370	570
Ne ($^3P_2$)	19	33	56			
Ar ($^3P_2$)	8	13	20	34	64	
Kr ($^3P_2$)		12	12	20	38	51
Xe ($^3P_2$)			8	13	22	33

Table 17.9 Diffusion coefficient (in cm^2/s) of metastable inert atoms in the foreign gases measured at a temperature of $T = 300$ K and a pressure of 1 atm [13]

Atom	Gas				
	He	Ne	Ar	Kr	Xe
Ne ($3\,S$)	1.09		0.315	0.25	0.22
Ar ($4\,S$)	0.70	0.32			
Kr ($5\,S$)	0.63	0.26	0.14		
Xe ($6\,S$)	0.52	0.22	0.104		

Table 17.10 Diffusion coefficient (in cm^2/s) of metastable mercury atoms in the rare gases at various temperatures and pressures

Inert atom	Electronic state of mercury atom	Temperature, K							
		296	380	475	515	580	680	780	890
He	$6\,^3P_0$	0.59	1.1		2.1				5.8
	$6\,^3P_0$	0.51	1.0		1.9				5.3
Ne	$6\,^3P_0$	0.31	0.57	0.93		1.4	1.9	2.4	2.9
	$6\,^3P_0$	0.27	0.48	0.86		1.3	1.8	2.3	2.9
Ar	$6\,^3P_0$	0.16	0.28	0.46		0.74	0.97	1.2	1.6
	$6\,^3P_0$	0.12	0.21	0.39		0.56	0.80	1.0	1.4
Kr	$6\,^3P_0$	0.1	0.19	0.32		0.50	0.65		1.1
	$6\,^3P_0$	0.07	0.13	0.23		0.35	0.49		
Xe	$6\,^3P_0$	0.06	0.13	0.23		0.33	0.45	0.58	0.7
	$6\,^3P_0$	0.48	0.10	0.18		0.27	0.37	0.47	

17.3 Diffusion in Liquids

The diffusion of large molecules in a solvent. Diffusion in liquids is due to multiparticle processes involving interaction of the probe particle with particles of the liquid. Therefore, theoretical determination of diffusion coefficients in liquids is difficult, and, in practice, experimental studies serve as the only sources of reliable information. An exception is presented by the diffusion of large molecules in a solvent of low molecular mass, for the description of which one can apply the Einstein–Stokes formula

$$D_{12} = \frac{kT}{6\pi r\eta}, \tag{17.10}$$

where r is the radius of the large molecule, the shape of which is approximated by a sphere; η is the dynamic viscosity of the liquid. The quantity r in relation (17.10), which is valid for many systems, remains the undetermined parameter, usually known up to a factor of 2.

Diffusion in solutions. The diffusion coefficients for many weak solutions are well approximated by the following empirical expression [5]:

$$D_{12} = K_1 K_2, \tag{17.11}$$

where K_1 is a coefficient independent of the properties of the solvent, and K_2 is a coefficient independent of the properties of the dissolved substance. The coefficients K_1 and K_2 depend on the temperature; however, if D_{12} is known at temperature T_1, then $D_{12}(T_2)$ can be determined on the basis of the following relation:

$$D_{12}(T_2) = D_{12}(T_1)\frac{T_2}{T_1}\frac{\eta(T_1)}{\eta(T_2)}, \tag{17.12}$$

where $\eta(T_2)$, $\eta(T_1)$ are the dynamic viscosities of the solvent at temperatures T_2 and T_1, respectively. The uncertainty pertinent to relations (17.11) and (17.12) amounts to several tens percent. Such uncertainties generally characterize the available information on diffusion coefficients in liquids. Values of K_1 and K_2 are presented in Tables 17.11 and 17.12.

Table 17.11 Values of K_1 (in 10^{-3} cm·s$^{-1/2}$) for some solutes in diluted solutions [4]

Solute	T, K	K_1	Solute	T, K	K_1
Acetic acid	298	2.3	Formic acid	279	4.2
Acetone	288	3.5	Furfural	288	2.2
Allyl alcohol	288	2.3	Iodomethane	280	3.9
Aniline	288	2.2	Isoallyl alcohol	288	2.3
Benzaldehyde	288	2.4	Methyl alcohol	288	2.9
Benzene	298	3.5	Nitrobenzene	288	2.3
Bromoethane	288	4.0	Phenol	278	1.9
Bromobenzene	288	2.4	Propionic acid	288	3.0
Butyl alcohol	288	2.0	Propyl alcohol	288	2.2
Chloral hydrate	288	2.0	Pyridine	293	2.7
Chlorobenzene	288	2.6	Tetrachloromethane	298	3.6
1,2-Dibromoethane	288	2.9	Toluene	298	3.0
Diethyl ether	288	3.8	Trichloromethane	288	3.3
Ethyl alcohol	288	2.5	Water	298	2.6

Table 17.12 Values of K_2 (in 10^{-3} cm·s$^{-1/2}$) for various solvents [5]

Solvent	T, K	K_2	Solvent	T, K	K_2
Benzene	288	5.9	Methyl alcohol	288	6.7
Bromobenzene	298	5.1	Tetrachloromethane	298	4.2
Chlorobenzene	298	5.7	Toluene	298	6.1
Ethyl alcohol	298	3.8	Water	298	4.1

The diffusion coefficient in a strongly diluted electrolyte solution is given by the relation

$$D_{12} = \frac{[1/n^+ + 1/n^-]RT}{[1/\lambda_0^+ + 1/\lambda_0^-]F}, \tag{17.13}$$

where $R = 8.314$ J/(K·mol) is the gas constant; T is the temperature; λ_0^+ and λ_0^- are the respective conductivities of the solution due to the positive and negative ions within the limit of zero concentrations, which are measured in A·g-equiv/(cm^4·V); n^+ and n^- are the respective valences of the cation and anion; $F = 96485$ C/mol is the Faraday constant. The ionic conductivities $\lambda_0^\pm$ for some aqueous solutions are presented in Table 17.13. The diffusion coefficients for some substances dissolved in water are given in Tables 17.14 and 17.15. It must be noted that in the case of liquids, when multiparticle interactions play an essential part in the diffusion process, reciprocity relations cannot be applied, i.e., $D_{12} \neq D_{21}$.

Table 17.13 Limit ion conductivities (in A·g-equiv/(cm^4·V)) in water at $T = 300$ K [6]

Cation	λ_0^+	Anion	λ_0^-
H^+	350	OH^-	198
Li^+	38.7	Cl^-	76.3
Na^+	50.1	Br^-	78.3
K^+	73.5	I^-	76.8
NH_4^+	73.4	NO_3^-	71.4
Ag^+	61.9	ClO_4^-	68.0
Tl^+	74.7	HCl_3^-	44.5
$(1/2)Mg^{2+}$	53.1	HCO_2^-	54.6
$(1/2)Ca^{2+}$	59.5	$CH_3CO_2^-$	40.9
$(1/2)Sr^{2+}$	50.5	$ClCH_2CO_2^-$	39.8
$(1/2)Ba^{2+}$	63.6	$CNCH_2CO_2^-$	41.8

Table 17.13 Limit ion conductivities (in A·g-equiv/(cm^4·V)) in water at $T = 300$ K *(continued)*

Cation	λ_0^+	Anion	λ_0^-
$(1/2)Cu^{2+}$	54.0	$CH_3CH_2CO_2^-$	35.8
$(1/2)Zn^{2+}$	53.0	$CH_3(CH_2)_2CO_2^-$	32.6
$(1/3)La^{3+}$	69.5	$C_6H_5CO_2^-$	32.3
$(1/3)CO(NH_3)_6^{3+}$	102.0	$HC_2O_4^-$	40.2
		$(1/2)C_2O_4^{2-}$	74.2
		$(1/2)SO_4^{2-}$	80.0
		$(1/3)Fe(CN)_6^{3-}$	101.0
		$(1/4)Fe(CN)_6^{4-}$	111.0

Note: The data are given for a concentration of 1 g-equiv/cm^3.

Table 17.14 Diffusion coefficient D for inorganic substances dissolved in water [1]

Solute	Concentration, mol/l	T, K	D 10^{-5} cm^2/s	Solute	Concentration, mol/l	T, K	D 10^{-5} cm^2/s
Br_2	0.00105	300	1.25		0.05	298	2.9
	0.00173	300	1.24		0.2	283	2.1
	0.00183	300	1.24		0.2	298	3.0
	0.00193	300	1.23		0.35	283	2.1
	0.00230	300	1.22		0.35	298	3.1
	0.00309	300	1.20		0.43	273	1.6
	0.00429	300	1.19		1.0	298	3.6
	0.00501	300	1.18		2.0	273	1.8
CO_2	0	291	1.46		2.0	284	2.5
CO	0	300	2.67		5.0	273	2.2
$CaCl_2$	0.29	282	0.79		6.0	284	3.1
	0.37	282	1.09		8.0	273	2.7
	1.5	282	0.84		9.0	285	3.4
$CdSO_4$	0	290	0.35	HNO_3	0.84	278	1.7
	0.5	290	0.34		3.0	279	1.8
	1.0	290	0.33		3.0	280	2.4
	1.5	290	0.454		20	282	2.3
	2.0	290	0.43	H_2O	0	293	0.9
	3.0	290	0.44	I_2	0.05	298	1.25
	3.5	290	0.34	KBr	1.0	283	1.2
	4.5	290	0.36	K_2CO_3	3.0	283	0.7
	5.5	290	0.4	KCl	0.00125	298	1.96
	6.5	290	0.45		0.00194	298	1.95
	7.0	290	0.48		0.00325	298	1.95
Cl_2	0.1	285	1.4		0.00585	298	1.94
	0.1	290	1.3		0.00704	298	1.93
$CoCl_2$	0.0062	291	0.7		0.00980	298	1.92
	0.0127	284	0.73		0.02	298	1.91
$CuCl_2$	1.5	283	0.5		0.1	298	1.89
$CuSO_4$	0.1	290	0.45		0.5	298	1.82
	0.5	290	0.34		0	282	1.7
	0.95	290	0.27		0.05	282	1.56
D_2O	0	300	2.5		0.10	282	1.54
H_2	0	291	3.6		0.20	282	1.53
HCl	0	283	2.2		0.50	282	1.55
	0	298	3.1		1.00	282	1.61
	0.02	283	2.1		1.5	282	1.67
	0.02	298	3.0		2.00	282	1.73
	0.05	283	2.0				

Table 17.14 Diffusion coefficient D for inorganic substances dissolved in water *(continued)*

Solute	Concentration, mol/l	T, K	D 10^{-5} cm^2/s	Solute	Concentration, mol/l	T, K	D 10^{-5} cm^2/s
KI	0.01	273	1.3	NH_3	0.683	277	1.23
	0.01	291	1.7		3.55	277	1.23
	0.10	273	1.24	NaBr	2.9	283	1.0
	0.10	291	1.61	Na_2CO_3	2.4	283	0.45
	1.0	273	1.21	NaCl	0.05	291	1.26
	1.0	291	1.59		0.40	291	1.2
	5.5	273	1.37		1.00	291	1.24
	5.5	291	1.8		2.0	291	1.29
KNO_3	0.05	291	1.45		3.0	291	1.36
	0.20	291	1.39		4.0	291	1.43
	0.40	291	1.35		5.0	291	1.49
	0.80	192	1.28	NaI	1.0	283	0.93
	1.0	291	1.24		2.0	283	1.04
	2.0	291	1.15	$NaNO_3$	0.6	286	1.04
	2.5	291	1.17		3	284	0.88
KOH	0.1	287	2.2		5	284	0.96
	0.9	287	2.16		6	286	0.90
	3.9	287	2.82	NaOH	0.02	285	1.3
K_2SO_4	0.02	293	1.27		0.1	285	1.29
	0.05	293	1.12		0.9	285	1.21
	0.28	293	1.0		3.9	285	1.14
	0.95	293	0.92	Na_2SO_4	1.4	283	0.76
LiBr	2.3	283	0.93	$NaCH_3COO$	0.2	285	0.71
	4.4	283	1.04	$Ni(NO_3)_2$	0.0088	291	0.667
LiCl	0.01	282	0.88		0.0226	291	0.822
	0.01	291	1.16		0.068	291	0.93
	1.0	282	0.81	NO	0	289	1.54
	1.0	291	1.06	O_2	0	298	2.60
	4.2	282	0.84	$Pb(NO_3)_2$	0.22	285	0.82
	4.2	291	0.11		0.82	285	0.76
LiI	1.3	283	0.93	$ZnSO_4$	0.025	293	0.58
$MgSO_4$	0.5	288	0.54		0.050	293	0.54
	1.0	288	0.53		0.55	293	0.12
	3.0	288	0.59		2.95	293	0.38
	4.5	288	0.73	$Zn(CH_3COO)_2$	2	273	0.139
N_2	0	291	1.63		2	291	0.243

Table 17.15 Diffusion coefficient D of organic compounds in aqueous solutions for zero concentration [7]

Solute	T, K	$D, 10^{-5}$cm^2/s	Solute	T, K	$D, 10^{-5}$cm^2/s
Acetamide	293	1.05	ethyl	283	0.84
Acetic acid	293	1.19		288	1.00
	286	0.91*1		298	1.24
Acetonitrile	288	1.26	Alloxan	293	0.67
Alcohols:			Aniline	293	0.92
allyl	288	0.90	Arabinose	293	0.70
benzyl	293	0.82	Benzene	275	0.58
butyl	288	0.77		283	0.75
isoamyl	288	0.69		293	1.02
methyl	288	1.28		333	2.55
propyl	288	0.87	Benzoic acid	298	1.21

Table 17.15 Diffusion coefficient D of organic compounds in aqueous solutions *(continued)*

Solute	T, K	$D, 10^{-5}\text{cm}^2/\text{s}$	Solute	T, K	$D, 10^{-5}\text{cm}^2/\text{s}$
n-Butane	277	0.50		333	3.55
	293	0.89	Methylcyclopentane	275	0.48
	333	2.51		283	0.59
Caffeine	283	0.42		293	0.85
Diethylamine	288	0.97		333	1.92
Ethyl acetate	0	1.04	Nicotine	293	0.53
Ethyl glycol	298	1.16	Oxalic acid	293	1.53
Ethylbenzene	275	0.58	Pentaerythritol	293	0.69
	283	0.75	Pyridine	288	0.58
	293	1.02	Pyrogallol	288	0.56
	333	2.55	Propylene	298	1.44
Furfural	293	0.87	Raffinose	293	0.36[*3]
Glucose	288	0.52	Resorcin	293	0.77
Glycerin	288	0.72	Saccharose	285	0.36
	283	0.63[*2]	Tartaric acid	288	0.61
Hydrogen sulfide	288	1.43	Urea	285	0.99
Hydroquinone	293	0.78		293	1.20
Lactose	283	0.46		298	1.38
Maltose	283	0.328	Urethan	288	0.87
	293	0.426	Vinyl chloride	298	1.34
Mannitol	293	0.605		323	2.42
Methane	283	0.328		348	3.67
	293	0.85			

[*1] At concentration 0.001 mol/l.
[*2] At concentration 0.125 mol/l.
[*3] At concentration 0.005 mol/l.

The diffusion coefficients in various liquids and melts are given below in Tables 17.16 – 17.22.

Table 17.16 Diffusion coefficient D of water in organic solvents at $T = 300$ K [6]

Solvent	D, 10^{-5} cm^2/s	Solvent	D, 10^{-5} cm^2/s	Solvent	D, 10^{-5} cm^2/s
Acetone	4.56	n-propyl	0.61	Glycerin	0.0083
Alcohols:		Aniline	0.70	n-Hexadecane	3.76
benzyl	0.37	n-Butyl acetate	2.87	Nitrobenzene	2.80
iso-butyl	0.30	n-Butyric acid	0.79	Pyridine	2.73
n-butyl	0.56	Dichloromethane	6.53	Toluene	6.19
ethyl	1.24	Ethyl acetate	3.20	1, 1, 1-Trichloroethane	4.64
methyl	1.75	Ethylene glycol	0.18	Trichloroethylene	8.82
iso-propyl	0.38	Furfural	0.90	Triethylene glycol	0.19

Note: The water molar content $< 1\%$.

Table 17.17 Diffusion coefficient D of various solutes in methyl alcohol at $T = 288$ K [7]

Solute	D, 10^{-5} cm^2/s	Solute	D, 10^{-5} cm^2/s	Solute	D, 10^{-5} cm^2/s
Acetanilide	1.50	dichloroacetic	1.36	Allyl alcohol	1.80
Acetonitrile	2.64	iodopropionic	1.36	Aniline	1.49
Acetyl diphenylamine	0.98	lactic	1.36	Benzaldehyde	1.66
Acids:		phthalic	1.30	Bromoaniline	1.41
bromopropionic	1.35	propionic	1.62	Bromobenzene	1.75
chloroacetic	1.52	tribromoacetic	1.23	Bromonitrophenol	1.43
chlorobenzoic	1.29	trichloroacetic	1.45	3-Bromopropylene	2.22

Table 17.17 Diffusion coefficient D of various solutes in methyl alcohol at $T = 288$ K *(continued)*

Solute	D, 10^{-5} cm^2/s	Solute	D, 10^{-5} cm^2/s	Solute	D, 10^{-5} cm^2/s
Bromophenol	1.34	Dinitrobenzene	1.56	Isoamyl alcohol	1.34
Carbon tetrachloride	1.70	1, 1-Dinitrohydride	1.36	m-Nitrobenzaldehyde	1.24
Chloral hydrate	1.16	Dinitronaphthalene	1.32	Nitrophenol	1.38
Chloroaniline	1.34	2, 4-Dinitrophenol	1.40	Phenol	1.40
Chlorobromomethane	2.50	Ethyl nitrate	2.20	Phenolphthalein	0.78
Chloroform	2.07	Ethyl:		Phenotol	1.40
1-Chlorohydrin	1.30	bromide	2.40	Phenylacetate	1.62
Chloronitrobenzene	1.68	iodide	2.16	3-Propylene	1.72
Chlorophenol	1.32	Ethylene:		Pyridine	1.58
Diazodinitrophenol-	1.68	bromide	1.95	Salol	1.29
methane		iodide	1.56	2, 4, 6-Tribromophenol	1.12
Dibenzylamine	0.86	Furfural	1.70	2, 4, 6-Trichlorophenol	1.21
n-Dibromobenzene	1.55	Hydroquinone	1.25	2, 4, 6-Trinitrophenol	1.32
Dibromonaphthalene	1.33	Iodobenzene	1.65	Urethan	1.41
Dichloronaphthalene	1.52	Iodoform	1.83	Vanillin	1.00

Note: Molar concentrations of the solutions $< 1\%$.

Table 17.18 Diffusion coefficient D (in 10^{-5} cm^2/s) in various liquids in the case of infinite dilution [5, 6]

Solute	Solvent	T, K	D	Solute	Solvent	T, K	D
Bromobenzene	Acetic acid	293	1.68	Adipic acid	n-Butyl alcohol	303	0.40
Bromobenzene (0.1)		280	3.25	Benzene		298	1.0
Iodine		280	1.51	Butyric acid		303	0.51
Iodine (0.1)		289	3.60	n-Dichlorobenzene		298	0.82
				Diphenyl		298	0.63
Acetic acid	Acetone	288	2.92	Methyl alcohol		303	0.59
		298	3.31	Oleic acid		303	0.25
		313	4.04	Propane		273	1.02
Benzoic acid		298	2.62			298	1.57
Bromoform		293	2.73	n-Propyl alcohol		303	0.40
Formic acid		298	3.77	Tartaric acid		303	0.40
Water		298	4.56	Water		298	0.56
Iodine (0.1)	Anisole	293	1.13	Bromine	Carbon disulfide	292	3.40
				Phenol		292	0.20
Acetic acid	Benzene	298	2.09				
Aniline		298	1.96	Iodine (0.1)	Carbon tetrachloride	293	1.36
Benzoic acid		298	1.38				
Bromobenzene		280	1.45	Acetone	Chloroform	298	2.35
Formic acid		298	2.28			313	2.90
Heptane		298	2.10	Benzene		288	2.51
		318	2.75			298	2.89
		338	3.65			313	3.552
		353	4.25			325	4.25
Methyl ethyl ketone		303	2.09	n-Butyl acetate		298	1.71
Naphthalene		280	1.19	Diethyl ether		288	2.07
Oxygen		303	2.89			298	2.14
Toluene		298	1.85	Ethyl acetate		298	2.02
1, 2, 4-Trichlorobenzene		313	2.39	Ethyl alcohol		288	2.20
		280	1.34	Iodine (0.05)		283	1.93
Vinyl chloride		280	1.77	Methyl ethyl ketone		298	2.13
				Phenol		283	1.60

Table 17.18 Diffusion coefficient D (in 10^{-5} cm^2/s) in various liquids *(continued)*

Solute	Solvent	T, K	D	Solute	Solvent	T, K	D
Bromobenzene	Decalin	280	0.47	Resorcin		293	0.45
α-Bromonaphthalene		280	0.34	Saligenin		293	0.61
				Stearic acid (0.2)		293	0.59
α-Bromonaphthalene	Dibenzyl ether	280	0.149	Tetrachloromethane		298	1.50
				Urea		285	0.54
Acetic acid	Ethyl acetate	293	2.18	Water		298	1.24
Acetone		293	3.18				
Ethylbenzene		293	1.85	Bromobenzene	Ethylbenzene	280	1.44
Iodine (0.1)		293	2.15				
Methyl ethyl ketone		303	2.93	Bromoform	Ethyl ether	290	3.30
Nitrobenzene		293	2.25	Nitrobenzene		281	3.24
Water		298	3.20	Phenol		292	3.60
Acetal (1)	Ethyl alcohol	293	1.13	Benzene	Heptane	298	3.40
Acetamide		293	0.67			318	4.40
Acetoin		293	0.56	Iodine (0.1)		372	8.40
Aconicotine (0.04)		293	0.27				
Allyl alcohol (1)		293	0.95	Bromobenzene	n-Hexane	280	2.60
Amyl alcohol		293	0.98	n-Hexane		298	4.21
Azobenzene (0.2)		293	0.74	Dodecane		298	2.73
Bromonaphthalene (0.05)		293	0.74	Methyl ethyl ketone		303	3.74
Bromoform		293	0.97	Propane		298	4.87
Brucine (0.06)		293	0.28	Tetrachloromethane		298	3.70
Camphor		293	0.70	Toluene		298	4.21
Carbon dioxide		290	3.2				
Chloral (1)		293	0.61	Phenol	Isoallyl alcohol	292	0.20
Chloroform (1)		293	1.25				
Hydroquinone		293	0.49	Tetrabromoethane (0.06)	Tetrachloroethane	323	0.94
Iodine		298	1.32				
Iodobenzene		293	1.00	Bromobenzene	Tetralin	280	0.48
Isoamyl alcohol		293	0.81	α-Bromonaphthalene		280	0.36
Oxygen		303	2.64				
Phenol (0.1)		293	0.80	Bromobenzene	Toluene	280	1.59
Pyridine		293	1.10	Iodine (0.1)		293	1.59

Note: A solution concentration other than zero is indicated (in mol/l) in parentheses.

Table 17.19 Diffusion coefficient D for various substances in molten salts [7]

Diffusing substance	Medium	T, K	D, 10^{-5} cm^2/s	Diffusing substance	Medium	T, K	D, 10^{-5} cm^2/s
$AgNO_3$	KNO_3	630	4.56	KCl	KNO_3	630	2.96
	KNO_3	660	4.86	$NaNO_3$	KNO_3	630	5.22
	$NaNO_3$	600	4.57	$PbCl_2$	KCl, LiCl	800	2.06
	$NaNO_3$	630	5.06		KCl, LiCl	1000	4.40
AgBr	KBr	1050	4.92	$Sr(NO_3)_2$	KNO_3	630	2.81
AgCl	KCl	1000	4.63		$NaNO_3$	620	4.17
	KCl	1050	5.10	TlBr	KBr	1040	4.28
	KCl, LiCl	750	5.61	TlCl	KCl, LiCl	800	3.10
	KCL, LiCl	1000	6.61		KCl	800	3.14
$Ba(NO_3)_2$	KNO_3	640	2.06	$TlNO_3$	KNO_3	620	3.17
	$NaNO_3$	630	3.71	$TlNO_3$	$NaNO_3$	630	4.31
KBr	KNO_3	630	3.01				

Table 17.20 Diffusion coefficient D (in 10^{-5} cm^2/s) of various solutes in benzene ($T = 288$ K) [5]

Solute	D	Solute	D
Acetyldiphenylamine	0.90	m-Dinitrobenzene	1.54
Acids:		Dinitronaphthalene	1.23
benzoic	1.36	n-Dichlorobenzene	1.90
phthalic	1.37	Dichloronaphthalene	1.40
chloroacetic	1.48	Ethyl ether	2.21
Alcohols:		Ethylene:	
isoamyl	1.48	bromide	1.97
propyl	1.60	iodine	1.40
Benzaldehyde	1.73	Iodine (0.05 M, 293 K)	1.95
Bromine (0.1 M, 285 K)	2.00	Iodobenzene (1%, 281 K)	1.35
Bromoaniline	1.41	Iodoform	1.38
Bromobenzene	1.86	Methyl salicylate	1.56
α-Bromonaphthalene	1.30	Naphthalene (1%)	1.19
Bromonitrobenzene	1.33	Nitrobenzene	1.84
Bromoform (291 K)	1.62	α-Nitronaphthalene	1.39
Bromophenol	1.34	Phenanthrene (1%, 281 K)	0.95
Chloroaniline	1.56	Phenol (278 K)	1.27
Chlorobenzene	1.42	Quinone	1.68
Chloroform	2.11	Salicylaldehyde	1.78
α-Chloronaphthalene	1.20	1, 2, 4, 5-Tetrachlorobenzene (< 1%, 281 K)	1.24
Chloronitrobenzene	1.70	1, 2, 4-Trichlorobenzene (< 1%, 281 K)	1.34
Chlorophenol	1.42	1, 2, 3-Trichloropropane	1.72
Dexachlorobenzene (< 1%, 281 K)	1.02	2, 4, 6-Trinitrophenol	1.39
n-Dibromobenzene	1.37	Trinitrotoluene	1.39
Dibromonaphthalene	1.25		

Note: The molar concentration of the solution is 1%.

Table 17.21 Diffusion coefficient D (in 10^{-5} cm^2/s) of metal atoms in amalgams [7]

Atom	T, K	D	Atom	T, K	D	Atom	T, K	D	Atom	T, K	D
Ag	289	1.11	Ca	283	0.62	Li	298	0.93	Sn	298	2.1
Au	298	0.73	Cd	298	2.0	Na	298	0.86	Sr	283	0.54
Ba	281	0.60	Cs	298	0.64	Pb	298	2.1	Tl	298	1.18
Bi	298	1.5	K	298	0.71	Rb	280	0.53	Zn	298	2.4

Table 17.22 Diffusion coefficient D (in 10^{-5} cm^2/s) of metal atoms in the metallic melts [8]

Atom	Diffusion medium	T, K	D	Atom	Diffusion medium	T, K	D	Atom	Diffusion medium	T, K	D
Ag	Sn	800	4.8	Mg	Al	1000	7.54	Rh	Pb	800	3.52
Au	Bi	800	5.22	Pb	Sn	800	3.68	Si	Fe	1750	2.4
Au	Sn	800	5.37	Pt	Pb	760	1.95			1830	10.8

17.4 Diffusion in Solids

Experimental studies serve as the main source of information on diffusion coefficients both in solids and in liquids. Here, because the results of measurements are extremely sensitive to the purity of the sample being investigated, to the method with which it was produced, and to variations in temperature, it is common for the results of different measurements to differ within the limits on the order of magnitude. For this reason, the data presented in the tables are the results of an averaging over a large number of experimental data, and, because the averaging procedure is arbitrary, they can only claim to be valid within an order of magnitude.

The temperature dependence of the diffusion coefficient in a solid is well described by the semiempirical formula

$$D = D_0 \exp\left(-\frac{Q}{RT}\right), \qquad (17.14)$$

where D_0 is the diffusion factor; Q is the activation energy. Relation (17.14) is satisfied in a wide temperature range within the limits of the diffusion coefficient measurement precision. For substances having a crystalline structure, the diffusion coefficient is a tensor. The data presented below are mainly the result of an averaging of this tensor over various orientations. In certain cases, diffusion coefficients are presented for orientations parallel ($D_{\|}$) and perpendicular ($D_{\perp}$) to the principal axis of the crystal. Values of the parameters occurring in (17.14), as well as the temperature range for which measurements were performed, are presented in Tables 17.23 – 17.42.

Table 17.23 The parameters entering into relationship (17.14) for the self-diffusion coefficient of metals [8]

Metal	Temperature range, K	D_0, cm^2/s	Q, kJ/mol	Metal	Temperature range, K	D_0, cm^2/s	Q, kJ/mol
Ag	700 – 1200	0.81	191	Ni	1300 – 1700	1.9	285
Al	600 – 700	0.10	128	Pb	450 – 600	0.28	102
	700 – 900	1.7	143	Pd	1300 – 1800	0.20	267
Au	900 – 1200	0.031	165	Pr	1100 – 1200	0.087 [10]	123 [10]
	1100 – 1300	.11	177	Pt	1500 – 2000	0.22	279
Be, $\perp$	800 – 1300	0.52	158	δ-Pu	310 – 700	0.0045	100
Be, $\|$	800 – 1300	0.62	165	ϵ-Pu	800 – 900	0.027	77.4
Cd, $\perp$	380 – 490	0.10	80	Rb	208 – 520	0.23	39.4
Cd, $\|$	380 – 490	0.05	76	Sb, $\perp$	770 – 900	0.10	150
Co	1050 – 1300	0.5	274	Sb,	770 – 900	56	201
	1300 – 1600	0.2	260	β-Sn, $\perp$	430 – 500	10.7	110
Cr	1200 – 1500	10^{-4}	221	β-Sn, $\|$	430 – 500	7.7	107
	1500 – 1900	0.28	309	Ta	1500 – 2400	0.12	413
Cs	1100 – 1300	47	25.7		2100 – 2800	2	462
	1000 – 1200	11	23.9	α-Th	1400 – 1700	700	350
Cu	800 – 1300	70	235	β-Th	1700 – 1800	10^5	417
	1000 – 1200	120	278	α-Ti	960 – 1100	$6.4 \cdot 10^{-8}$	123
α-Fe	1200 – 1500	3.6	298	β-Ti	1170 – 2850	0.0016	146
γ-Fe	1350 – 1600	1.1	284	α-Tl, $\perp$	420 – 560	0.4	94.5
δ-Fe	1000 – 1400	0.48	276		470 – 690	0.39	104
Hf	2070 – 2300	0.0012	162	α-Tl, $\|$	470 – 690	0.08	92
	2100 – 2600	0.3	398	α-U	850 – 920	0.002	168
In, $\perp$	320 – 420	3.7	78.2	β-U	970 – 1020	0.014	176
In, $\|$	320 – 420	2.7	78.2	γ-U	1050 – 1300	0.002	113
K	220 – 330	0.16	3.92	V	1150 – 1600	0.36	308
La	930 – 1120	1.5 [10]	189 [10]		1600 – 2100	210	394
Li	208 – 520	0.24	55.2	W	1560 – 1730	$6.3 \cdot 10^{-7}$	570
Mg, $\perp$	740 – 900	1.5	136		2300 – 3000	0.54	502
Mg, $\|$	740 – 900	1.0	135		2900 – 3600	43	640
Mo	2000 – 2200	2.8	465	α-Zr	600 – 1000	$5 \cdot 10^{-8}$	92
Na	208 – 520	0.2	41.9		900 – 1100	$5 \cdot 10^{-2}$	218
Nb	1150 – 2700	1.1	402	β-Zr	1200 – 1800	0.0024	159

Table 17.24 The parameters entering into relationship (17.14) for the self-diffusion coefficient of some crystals

Crystal	T, K	D_0, cm^2/s	Q, kJ/mol
Ar	78	350	17.4
C (graphite)	2270–2620	10	680
Ge	1000–1200	10.8	291
P (white)	273–303	$1.1 \cdot 10^{-3}$	39.3
Xe	121–158	7.3	34.9

Table 17.25 The parameters entering into relationship (17.14) for the diffusion coefficient of minor impurities in nickel [8]

Impurity	T, K	D_0, cm^2/s	Q, kJ/mol
Al	1070–1250	1.1	249
	1400–1600	1.87	268
Au	1200–1400	2	272
Co	1020–1460	0.75	271
	1430–1650	1.11	272
Cr	940–1170	0.03	171
	1400–1600	1.1	273
Cu	1320–1620	0.57	258
Fe	1200–1800	0.8	255
Mg	1070–1250	$2.3 \cdot 10^{-5}$	131
	1400–1600	0.44	235
Mn	1400–1600	7.5	281
Mo	1400–1600	3	288
Si	1070–1250	10.6	271
	1400–1600	1.5	258
Ti	1400–1600	0.86	257
W	1400–1600	11.1	322

Table 17.26 The parameters entering into relationship (17.14) for the diffusion coefficient of minor impurities in copper [8]

Impurity	T, K	D_0, cm^2/s	Q, kJ/mol
Ag	<1356	0.63	195
As	<1356	0.12	176
Au	700–1000	0.15	192
Cd	<1356	0.93	189
Co	701–1077	1.93	226
Cu	500–1000	70	234
Fe	719–1074	1.4	217
Ga	<1356	0.55	192
Hg	<1356	0.35	184
Ni	743–1076	2.7	237
Pd	807–1056	1.74	228
Sb	600–1000	0.34	176
Tl	785–996	0.71	181
Zn	<1356	0.34	191

Table 17.27 The parameters entering into relationship (17.14) for the diffusion coefficient of minor impurities in silver [8]

Impurity	T, K	D_0, cm^2/s	Q, kJ/mol
Ag	450–900	0.81	191
Au	650–950	0.26	191
Cd	592–937	0.44	175
Co		104	251
Cu	700–950	1.23	193
Fe	718–927	2.42	205
Ge		0.084	153
Hg	650–950	0.08	160
In	592–937	0.41	170
Ni	750–950	21.9	230
Pb	700–825	0.22	160
Pd	735–939	9.57	238
Ru	794	180	276
Sb	468–942	0.17	160
Sn	592–937	0.25	164
Tl	650–800	0.15	159
Zn	640–925	0.54	174

Table 17.28 The parameters entering into relationship (17.14) for the diffusion coefficient of minor impurities in gold [8]

Impurity	T, K	D_0, cm^2/s	Q, kJ/mol
Ag	700–1000	0.072	168
Au	700–1050	0.091	175
Co	700–950	0.068	174
Fe	700–950	0.082	174
Hg	500–1030	0.116	157
Ni	700–950	0.034	176
Pt	800–1050	7.6	255

Table 17.29 The parameters entering into relationship (17.14) for the diffusion coefficient of atomic impurities in the alkalies [10]

Metal	Impurity	T, K	D_0, cm^2/s	Q, kJ/mol
K	Au	280–326	1300	12.4
Li	Cu	320–450	0.47	38.6
	Ag	340–430	0.12	52.8
	Au	320–420	0.21	46.1
	Zn	330–450	0.57	54.5
	Na	330–450	0.41	52.8
Na	Au	270–350	3.3	9.2

Table 17.30 The parameters entering into relationship (17.14) for the diffusion coefficient of different impurities in the metals of III and IV groups [8]

Metal	Impurity	T, K	D_0, cm^2/s	Q, kJ/mol
β-Al	Na^+	470–670	$2{\cdot}10^{-3}$	15.9
	Ag^+	470–670	$1.65{\cdot}10^{-3}$	16.8
	K^+	470–670	$0.78{\cdot}10^{-3}$	22.4
	Rb^+	470–670	$0.34{\cdot}10^{-3}$	30.1
	Li^+	470–670	$14.5{\cdot}10^{-3}$	36.4
	Tl^+	470–670	$0.65{\cdot}10^{-3}$	34.4
In	Au	300–420	$9{\cdot}10^{-3}$	281.0
	Ag	300–420	0.11	48.2
Pb	Cu	>500	$7.9{\cdot}10^{-3}$	33.5
	Ag	>400	$4.6{\cdot}10^{-3}$	60.3
		390–570	$7.5{\cdot}10^{-3}$	63.6
	Au	460–570	$4.1{\cdot}10^{-3}$	39.1
		460–600	$8.7{\cdot}10^{-3}$	41.9
	Zn	450–570	$1.6{\cdot}10^{-3}$	47.3
	Cd	>420	0.41	88.9
	Hg	470–570	1.05	95.0
	Tl	480–600	0.5	104.0
	Sn	510–560	0.16	96.5
	Na	520–490	6.3	119.0
	Ni	480–870	$9.4{\cdot}10^{-3}$	44.4
	Pd	480–860	$3.8{\cdot}10^{-3}$	36.0
Sn	Cu	410–500	$2.4{\cdot}10^{-3}$	33.1
	Au [001]	400–510	$5.8{\cdot}10^{-3}$	46.0
	[100]	400–510	0.16	74.1
	Ag [001]	400–510	$7.1{\cdot}10^{-3}$	51.5
	[100]	400–510	0.18	77.0
	Zn	360–510	$5.9{\cdot}10^{-3}$	49.4
	Cd [001]	460–500	220	118.0
	[100]	460–500	120	113.0
	Hg [001]	450–500	7.5	106.0
	[110]	450–500	30	112.0
	In [001]	450–500	12	107.0
	In [100]	450–500	34	108.0
	Sb [001]	470–500	79	122.0
	[100]	470–500	77	123.0
Tl	Ag	360–500	0.33	48.2
	Au	360–500	$5.3{\cdot}10^{-3}$	21.8

Note: The Miller's indices in brackets define a direction of diffusive flow [10].

Table 17.31 The parameters entering into relationship (17.14) for the diffusion coefficient of atomic impurities in the lanthanides and actinides [10]

Metal	Impurity	T, K	D_0, cm^2/s	Q, kJ/mol
La	Au	980–1070	0.022	75.8
Pr	Cu	930–1060	0.084	75.8
		1090–1190	0.057	74.5
	Co	880–1040	0.047	68.7
	Au	880–1020	0.043	82.5
		1070–1180	0.033	84.1
	Ag	880–1040	0.14	106.0
		1070–1190	0.032	90.0
	Zn	880–1040	0.18	104.0
		1100–1200	0.63	113.0
γ-U	Co	1050–1350	$3.5{\cdot}10^{-4}$	52.8
	Fe	1050–1350	$2.7{\cdot}10^{-4}$	50.2
	Ni	1050–1350	$5.4{\cdot}10^{-4}$	65.8
	Mn	1050–1350	$1.8{\cdot}10^{-4}$	58.2
	Cr	1050–1350	$5.4{\cdot}10^{-3}$	103.0
	Cu	1050–1350	$2.0{\cdot}10^{-3}$	101.0
	Nb	1050–1350	0.049	166.0

Table 17.32 The parameters entering into relationship* $D = D_{01}\exp(-Q_1/RT) + D_{02}\exp(-Q_2/RT)$ for the diffusion coefficient of minor impurities in β-Ti [8]

Impurity	T, K	D_{01}, 10^{-4} cm^2/s	Q_1, kJ/mol	D_{02}, 10^{-4} cm^2/s	Q_2, kJ/mol
Co	1200–1900	130	130	16	257
Cr	1220–1900	74	154	14	274
Fe	1200–1900	80	126	15	254
Mn	1200–1900	76	144	12	270
Mo	1170–1900	7	155	3.6	272
Nb	1270–1900	13	146	9.5	291
Ni	1200–1900	170	133	20	252
P	1220–1900	36.2	101	5	237
Sc	1200–1560	21	137	34	
Sn	1230–1870	3.8	132	9.5	291
Ta	1300–1900	7.2	151	50	335
Ti	1170–1800	2	125	1.0	250
V	1200–1820	10	146	3.4	258

* The first term dominates at low temperatures, while the second, at high temperatures.

Table 17.33 The parameters entering into relationship (17.14) for the self-diffusion coefficient of some metals and alloys [1] [The dashes imply that impurity concentration vanishes.]

Diffusing element	Diffusion medium	Atomic concentration of diffusing element, %	T, K	D_0, cm^2/s	Q, kJ/mol
Ag	Al	1.26	740–850	1.1	137
	Au	9	1120–1270	$2.9{\cdot}10^{-2}$	159
	Cu	3	100–1070	$2.9{\cdot}10^{-2}$	156
	Pb	0.12	500–560	$7.4{\cdot}10^{-2}$	64

Table 17.33 The parameters entering into relationship (17.14) for some metals and alloys *(continued)*

Diffusing element	Diffusion medium	Atomic concentration of diffusing element, %	T, K	D_0, cm^2/s	Q, kJ/mol
Al	Cu	15–21	770–1120	$7.1 \cdot 10^{-2}$	164
Au	Ag	100	500–870	$5.3 \cdot 10^{-4}$	125
		18.4	1040–1200	$1.1 \cdot 10^{-4}$	112
	Cu	2.4–3.5	670–1250	$6.8 \cdot 10^{-6}$	94
	Pb	0.03–0.09	370–570	0.35	59.5
B	α-Fe	—	<1800	10^{5}	260
Bi	Pb	2.0	500–560	$1.8 \cdot 10^{-3}$	77
C	Fe	Carburization	1100–1400	$1.67 \cdot 10^{-2}$	120
		1.1 mass.%	1200–1500	0.49	153
		0.1–1 mass.%	100–1500	0.12	137
	α-Fe	—	<1800	$7.9 \cdot 10^{-3}$	76
		—	<1800	$2 \cdot 10^{-2}$	84
Cd	Ag	2.0	920–1170	$4.9 \cdot 10^{-5}$	94
		—	<1200	0.44	175
	Cu	3	1000–1130	$3.04 \cdot 10^{-4}$	99
	Pb	1.6	440–520	$1.83 \cdot 10^{-3}$	65
Co	α-Fe	—	<1800	0.4	226
	γ-Fe	—	<1800	$1.2 \cdot 10^{5}$	435
	Steel	—	<1800	90	335
Cr	α-Fe	—	<1800	$3 \cdot 10^{4}$	343
	γ-Fe	—	<1800	$1.8 \cdot 10^{4}$	406
	Steel	—	<1800	10	314
Cs	W	First adsorption layer	300–700	0.2	58.6
		Second adsorption layer	300–700	$1.64 \cdot 10^{-2}$	96
Cu	Ag	2.0	920–1170	$5.9 \cdot 10^{-5}$	104
	Al	Eutectics	700–800	2.3	146
		0.085–0.17	730–840	$8.4 \cdot 10^{-2}$	137
	Au	100	570–890	$1.06 \cdot 10^{-3}$	115
		25.6	700–1000	$5.8 \cdot 10^{-4}$	115
	Ge	—	<1200	$1.9 \cdot 10^{-4}$	172
	Pt	13.9	1310–1730	$4.8 \cdot 10^{-2}$	233
	β-brass (45% Zn)	—	<1100	$3.8 \cdot 10^{-2}$	104
Fe	Au	18.3	1030–1280	$1.16 \cdot 10^{-4}$	102
	Fe-20%	—	<1800	$18 \cdot 10^{-0.92[C]}$*	314 [C]
	Ni-C	—	<1800	$71 \cdot 10^{-0.65[C]}$	314 [C]
	Ni	—	<1800	$8.4 \cdot 10^{-3}$	214
	W	0.04	2200–2800	11.5	586
Ge	Ge	Self-diffusion	<1200	87	308
Hg	Cd	4	430–475	2.6	82
In	Ag	20	920–1170	$7.3 \cdot 10^{-5}$	102
Li	Ge	—	<1200	$2.5 \cdot 10^{-3}$	49.4
	Si	—	<1700	$2.3 \cdot 10^{-3}$	63.6
Mg	Al	Eutectics	640–710	$1.5 \cdot 10^{-2}$	161
		5.5–11.0	670–850	0.110^{-1}	119
Mn	Cu	8–11.4	670–1120	$7.2 \cdot 10^{-6}$	96
Mo	W	Single crystal	1800–2530	$6.3 \cdot 10^{-4}$	33.7
		Polycrystal	1800–2530	$5 \cdot 10^{-3}$	33.7
N	Fe	—	<1800	$3.1 \cdot 10^{-3}$	75
		—	<1800	$3 \cdot 10^{-3}$	76
		—	<1800	$6.6 \cdot 10^{-3}$	78
	Th	—	<1970	$2.1 \cdot 10^{-3}$	94
	α-Ti	—	<1940	$1.2 \cdot 10^{-2}$	190
	β-Ti	—	<1940	0.35	142

Table 17.33 The parameters entering into relationship (17.14) for some metals and alloys *(continued)*

Diffusing element	Diffusion media	Atomic concentration of diffusing element, %	T, K	D_0, cm^2/s	Q, kJ/mol
Ni	Au	15.0	1070–1280	$1.74 \cdot 10^{-3}$	131
	Cu	7.5–11.8	820–1220	$6.5 \cdot 10^{-5}$	125
	Pt	14.9	1320–1670	$7.8 \cdot 10^{-4}$	181
O	β-Ti	—	<1940	1.6	202
Pd	Ag	20.2	720–1200	$6.4 \cdot 10^{-6}$	84.5
	Au	17.1	1000–1250	$1.13 \cdot 10^{-3}$	91.5
	Cu	4.3–6.2	760–1240	$1.6 \cdot 10^{-6}$	163
Pt	Au	20.1	1000–1250	$1.24 \cdot 10^{-3}$	163
	Cu	2.4–3.5	760–1230	$1.0 \cdot 10^{-6}$	91.5
Sb	Ag	2.0	920–1170	$5.3 \cdot 10^{-5}$	91
Si	Al	0.5	740–870	0.9	128
Sn	Ag	20	920–1170	$7.8 \cdot 10^{-6}$	89.5
	Cu	3.9–5.6	670–1120	$4.1 \cdot 10^{-3}$	131
β-Sn	Pb	2.0	510–560	4.0	110
Th	W	Diffusion along grain boundaries	2050–2500	1.13	394
		Bulk diffusion	2400	1.0	503
		Surface diffusion	1650	0.47	278
Tl	Pb	2.0	500–550	$2.5 \cdot 10^{-2}$	81
U	W	—	2000	1.14	420
W	α-Fe	—	<1800	380	293
	γ-Fe	—	<1800	10^3	377
	Steel	—	<1800	13	314
Y	W	—	2000	0.11	260
Zn	Zn	Self-diffusion in Zn of 99.999%-purity	<690	0.22	60
		Self-diffusion in Zn of 99.99%-purity	<690	0.38	61
	Al	0.84	690–730	12	116
	Cu	6.8–9.7	630–1150	$3 \cdot 10^{-6}$	82.5
		3	1000–1130	$3.7 \cdot 10^{-6}$	92
		0–28.6	910–1150	$3.2 \cdot 10^{-3}$	176
	β-brass (45% Zn)	—	<1100	0.024	95.5
Zr	W	—	2000	1.1	326

* [C] is the atomic carbon concentration (in %).

Table 17.34 The parameters entering into relationship (17.14) for the diffusion coefficient of ions in the salt crystals [10]

Ion	Salt crystal	T, K	D_0, cm^2/s	Q, kJ/mol	Ion	Salt crystal	T, K	D_0, cm^2/s	Q, kJ/mol
Cs^+	CsCl	630–730	1.1	130	Br^-	KBr	670–970	10^{-3}	196
		760–880	0.1	134	Br^-	NaBr	920–1020	1.0	164
Cs^+	CsI	570–770	80	158	Cl^-	CsCl	550–730	1.51	122
K^+	KBr	730–1000	0.01	122			760–880	0.7	152
K^+	KCl	800–950	137	207	Cl^-	KCl	770–1010	178	216
		860–1020	$1.8 \cdot 10^{-5}$	76	Cl^-	NaCl	770–1020	60.7	206
K^+	KF	870–1100	2	172	Cl^-	RbCl	870–970	33.3	193
K^+	KI	730–950	10^{-5}	62	I^-	CsI	680–830	0.39	122
Li^+	LiF	850–1000	0.8	192	I^-	KI	730–950	$1.2 \cdot 10^{-3}$	108
		1000–1100	5.6	289					
Na^+	NaBr	770–950	0.67	147					
Na^+	NaCl	550–820	$3.5 \cdot 10^{-6}$	81.2					
		860–1070	76.9	196					
Rb^+	RbCl	840–970	33.3	193					

Table 17.35 The parameters entering into relationship (17.14) for the diffusion coefficient of impurities in the salt crystals [10]

Salt crystal	Impurity	T, K	D_0, cm^2/s	Q, kJ/mol
KBr	Pb	550–670	$1.5 \cdot 10^{-3}$	87.5
	Tl	410–480	0.091	96.5
		620–770	50	192
KCl	Li	770–1000	20	142
	Na	860–1020	2.2	169
KCl	Ag	470–920		40.6
	Bi	670–950	$5.6 \cdot 10^{-3}$	94
	Ce	770–970	$1.1 \cdot 10^{-3}$	100
	Cd	620–770	$4.7 \cdot 10^{-5}$	52.5
	Cl	770–920	$1.5 \cdot 10^{-3}$	109
	Co	470–920		19.3
	Cs	840–1020	0.7	169
	Cu	620–920	10.6	118
	Eu	700–820	0.064	124
	I	770–920	50	193
	OH	670–890	$1.2 \cdot 10^{-3}$	193
	Pb	500–750	10^{-3}	96.5
	Rb	880–1040	26.8	195
	Tl	500–730	$5.8 \cdot 10^{-11}$	23
		730–1000	$1.3 \cdot 10^{-3}$	126
NaCl	Ag	850–1000	380	192
	Au		0.2	103
	Br	770–920	20	187
	Ca	700–1020	$6 \cdot 10^{-4}$	87
NaCl	Cd	800–930	$3.9 \cdot 10^{-3}$	251
	Co	880–1030	$8 \cdot 10^{-3}$	106
	Cs	870–970	1.62	192
	Cu	620–920	33.8	138
	Hg	720–820	$8.2 \cdot 10^{-5}$	55
	I	800–970	500	216
	K	670–930		48.3
	Mn	670–970	10^{-4}	65
	Ni	900–1020	0.02	125
	Pb	600–800	0.015	94.5
	Rb	870–1020	28.5	192
	Sr	390–1070	$1.7 \cdot 10^{-3}$	126
	Zn	810–1000	0.04	94.5
NaF	Li	850–1000	0.8	192
		1000–1100	5.6	290
NaI	Tl	670–910	$5 \cdot 10^{-3}$	113

Table 17.36 The parameters entering into relationship (17.14) for the diffusion coefficient of Ag^+ ions in the crystals [1]

Diffussion medium	T, K	D_0, cm^2/s	Q, kJ/mol
α-AgI	454–744		9.45
α-AgI	451–701	$1.6 \cdot 10^{-4}$	9.45
α-Ag_2I	473–653	$5.8 \cdot 10^{-3}$	19.1
α-Ag_2S	443–693	$4.6 \cdot 10^{-4}$	13.3
α-Ag_2Se	421–673	$1.6 \cdot 10^{-4}$	12.3
α-Ag_2Se		$6.7 \cdot 10^{-4}$	85.5
α-Ag_2Te	428–678	$3.8 \cdot 10^{-5}$	11.1
α-CuI	685–753	$4.5 \cdot 10^{-3}$	28.3
α-Cu_2S	586–1192	$3.3 \cdot 10^{-4}$	19.1
α-Cu_2Te	603–794	2.4	87
NaBr		50	199
NaCl	<550	1.6	75
	>550	3.1	174
$PbCl_2$	439–743	7.7	154
PbI_2	387–438	7.8	126
	528–588	10.6	127
PbS	733–1043	1.3	176

Table 17.37 Diffusion coefficient of ions in the crystals at various temperatures [1]

Diffusing ion	Diffusion medium	T, K	D, 10^{-6} cm^2/s
Ag^+	CuBr	518	51
Cu^+	α-Cu_2S	503	1.85
		603	4.6
		693	9.4
	α-Cu_2Te	603	0.359
		749	1.98
		794	4.66
	Cu_3Sb	723	0.139
Li^+	AgCI	511	0.24
	α-AgI	753	37–41
	α-Ag_2S	443	0.4
		503	4.1
		603	6.3
		693	16.5
Na^+	AgCl	723	0.116
	AgBr	573	0.023
Pb^{++}	$PbCl_2$	473	4.5
		523	11
		578	15
		653	23
Pb^{++}	PbI_2	387	$7.3 \cdot 10^{-11}$
		420	$1.35 \cdot 10^{-9}$
		438	$7.35 \cdot 10^{-9}$
Cl^-	AgI	453	$1.1 \cdot 10^{-4}$
Se^{2-}	Cu_2S	844	$1.1 \cdot 10^{-3}$
		967	$4.9 \cdot 10^{-3}$

Table 17.38 Mutual diffusion D coefficient of solid salts [1]

Diffusing substance	Diffusion medium	T, K	D, 10^{-11} cm^2/s
Ag	Na-glass	627	25.2
(from $AgNO_3$)	Na-permutite	627	14.8
Ag_4Sn	Cu_4Sn	723	231
$BaMoO_4$	$BaWO_4$	1223	9.61
$BaWO_4$	$BaMoO_4$	1223	3.47
$MgMoO_4$	$MgWO_4$	1073	2.31
$MgWO_4$	$MgMoO_4$	1073	1.39
$SrMoO_4$	$SrWO_4$	1223	9.14
$SrWO_4$	$SrMoO_4$	1223	3.01
$ZnMoO_4$	$ZnWO_4$	1073	3.94
		1123	6.36
$ZnWO_4$	$ZnMoO_4$	1073	1.00
		1123	2.89

Table 17.39 The parameters entering into relationship (17.14) for the diffusion coefficient of hydrogen isotopes in metals [9]

Atom	Metal	T, K	D_0, cm^2/s	Q, kJ/mol
H	Cu	600–1100	$3.5 \cdot 10^{-4}$	39.5
D	Cu	800–1100	$2.5 \cdot 10^{-4}$	39.2
T	Cu	800–1100	$2 \cdot 10^{-4}$	39.2
H	α-Fe	<1800	$7.5 \cdot 10^{-4}$	10
H	Nb	>273	$5.0 \cdot 10^{-4}$	10.2
		<223	$0.9 \cdot 10^{-4}$	6.5
D	Nb	<1000	$5.2 \cdot 10^{-4}$	12.3
T	Nb	<1000	$4.5 \cdot 10^{-4}$	13
H	Ni	>631	$6.9 \cdot 10^{-3}$	40.5
		<631	$4.8 \cdot 10^{-3}$	39.4
H	Pb	<900	$2.9 \cdot 10^{-3}$	22.2
H	Ta	>273	$4.4 \cdot 10^{-4}$	13.5
		<200	$2 \cdot 10^{-6}$	3.85
D	Ta	<600	$4.6 \cdot 10^{-4}$	15.5
H	V	<600	$3.1 \cdot 10^{-4}$	4.36
D	V	<600	$3.8 \cdot 10^{-4}$	7.12

Table 17.40 The parameters entering into relationship (17.14) for the diffusion coefficient of gases in solids [1]

Diffusing gas	Diffusion environment	T, K	D_0, 10^{-6}cm^2/s	Q, kJ/mol
H_2	SiO_2	473	8.5	42.7
		773	11	42.7
He	SiO_2	293	5.7	23
		773	2.9	23
	Pyrex	293	1.3	36.4
NH_3	Analcite*	773	5.5	

* Natural aluminosilicate.

Table 17.41 The parameters entering into relationship (17.14) for the diffusion coefficient of rare gases in salt crystals [10]

Crystal	Gas	Temperature range, K	D_0, cm^2/s	Q, kJ/mol
CsBr	Xe	530–850	100	140
CsCl	Xe	620–740	0.1	86.6
		740–920	0.1	83.5
CsF	Xe	720–870	10^2	19.3
CsI	Xe	420–770	0.57	96
KCl	Ar	>670	$7.9 \cdot 10^{-4}$	36.7
	Kr	450–520	$8 \cdot 10^{-5}$	107
		520–770	10^6	203
KBr	Ar	290–570	10^5	145
KBr	Xe	290–570	10	135
KI	Xe	420–770	1.5	98
RbBr	Kr	460–600	$4 \cdot 10^7$	163
		>600	$5 \cdot 10^{-4}$	28.9
RbCl	Kr	630–970	$7.9 \cdot 10^4$	43
RbF	Kr	570–1000	2.5	132
RbI	Kr	450–610	$1.3 10^6$	136
		>610	$1.6 \cdot 10^{-3}$	29.4
	Xe	420–770	0.082	89.5

Table 17.42 The parameters entering into relationship (17.14) for the diffusion coefficient of atomic impurities in semiconductors [11]

Semi-conductor	Diffusion media	D_0, cm^2/s	Q, kJ/mol	Temperature range, K	Semi-conductor	Diffusion media	D_0, cm^2/s	Q, kJ/mol	Temperature range, K
Ge	Ag	0.044	96.5	900–1200	Ge	Fe	0.13	106	1000–1200
	Al	0.05	261	1000–1200		Ga	20	320	1000–1200
	As	4.5	232	900–1200		Ge	49	209	1000–1200
	Au	0.055	242	1000–1200		H	0.0027	36.7	1000–1200
	B	0.084	444	1000–1200		He	0.0061	68	900–1200
	Be	0.5	242	900–1200		In	0.048	231	1000–1200
	Bi		234	900–1200		Li	0.0012	49.2	900–1200
	Co	0.16	106	900–1200		N		249	900–1200
	Cu	0.0033	17.4	900–1200		Ni	0.42	87	900–1200

Table 17.42 The parameters of (17.14) for diffusion of impurities in semiconductors *(continued)*

Semi-conductor	Diffusion media	D_0, cm^2/s	Q, kJ/mol	Temperature range, K	Semi-conductor	Diffusion media	D_0, cm^2/s	Q, kJ/mol	Temperature range, K
Ge	O	0.17	195	900–1200	CdS	Cd	$1.1{\cdot}10^{-5}$	60	670–1000
	P	1.3	240	1000–1200		Cu	2000	92.7	420–670
	Pb		348	1000–1200		Cu	0.0015	73.4	670–1020
	Sb	3.6	232	1000–1200		Li	$3{\cdot}10^{-6}$	65.6	880–1230
	Sn	0.017	183	1000–1200	CdSe	P		203	1070–1270
	Ta	$2.5{\cdot}10^{-6}$	112	900–1200		Se	$1.3{\cdot}10^{5}$	42.8	1220–1700
	Zn	3.3	247	900–1200	CdTe	Au	67	193	600–1000
Si	Ag	0.002	154	1400–1600		Cu	$3.7{\cdot}10^{-4}$	64.7	370–570
	Al	4.8	323	1400–1700		In	0.041	154	720–1300
	As	34	376	1300–1600		Se	$1.7{\cdot}10^{-4}$	130	950–1200
	Au	0.0011	106	1100–1500	GaAs	As	$4{\cdot}10^{4}$	980	1400–1500
	B	$6{\cdot}10^{-7}$	163	1000–1400		Au	10^{-3}	106	860–1330
	Bi	10^{3}	444	1300–1600		Cd	0.0013	212	1070–1370
	C	0.33	282	1350–1700		Ga	10^{7}	540	1400–1500
	Cr	0.01	96.5	1200–1500		Ge	7.5	348	1320–1410
	Cu	0.0047	41.5	670–970		Li	0.53	96.5	520–770
		0.4	96.5	1100–1400		Mg	$4{\cdot}10^{-5}$	118	1070–1270
	Fe	0.0062	83	1300–1600		Mn	0.65	240	1000–1300
	Ga	90	376	1300–1600		S	180	251	1100–1500
	Ge	$6.3{\cdot}10^{5}$	510	1400–1700		Sn	$6{\cdot}10^{4}$	241	1100–1500
	H	0.0094	46.3	1240–1480		Te	$2.6{\cdot}10^{-5}$	193	1270–1370
	He	0.11	121	1300–1600		Zn	$3{\cdot}10^{-7}$	96.5	1070–1370
	Ir	18	374	1300–1600	GaP	S	3200	453	1400–1700
	K	0.0011	72.5	800–1060	GaSb	Cd	$1.5{\cdot}10^{-6}$	69.5	770–900
	Li	0.0025	63.6	300–400		Ga	3200	304	930–970
		0.0023	62.7	700–1100		Li	$2.3{\cdot}10^{-4}$	184	1070
	Na	0.0016	69	800–1100		Sb	0.0087	109	600–920
	O	0.21	292	1300–1600		Sb	34000	333	920–970
	P	20	364	1300–1600		Sn	$2.4{\cdot}10^{-5}$	77.4	600–920
	S	0.92	212	1300–1600		Te	$3.8{\cdot}10^{-4}$	116	600–920
	Sb	9.2	380	1300–1600	HgTe	Cd	$3.1{\cdot}10^{-4}$	66.7	520–620
	Si	$5.4{\cdot}10^{3}$	483	1400–1700	InAs	Ag	$7.3{\cdot}10^{-4}$	25	730–1200
	Te	16.5	375	1300–1600		As	$3{\cdot}10^{7}$	432	1000–1200
	Zn	0.1	135	1250–1550		Au	0.0058	62.8	900–1200
Se	Fe	10^{-5}	37.5	490		Cd	$6{\cdot}10^{-4}$	112	900–1200
	Ge	$9.4{\cdot}10^{-6}$	38.6	490		Cu	0.036	51	1200
	In	$5.2{\cdot}10^{-6}$	30.9	490		Ge	$3.7{\cdot}10^{-6}$	113	900–1200
	S	4.9	26	490		In	$6{\cdot}10^{5}$	386	1000–1200
	Se	$2.8{\cdot}10^{-8}$	28	490		Mg	$2{\cdot}10^{-6}$	113	900–1200
	Sn	$4.8{\cdot}10^{-8}$	37.6	490		P	130	261	920–1170
	Te	$5.4{\cdot}10^{-6}$	39.6	490		S	6.8	212	900–1200
	Tl	$1.4{\cdot}10^{-6}$	33.8	490		Se	13	212	900–1200
	Zn	$3.8{\cdot}10^{-7}$	28	490		Sn	$1.5{\cdot}10^{-6}$	113	900–1200
Te	Hg	$3.4{\cdot}10^{-5}$	78	640–700		Te	$3.4{\cdot}10^{-5}$	124	900–1200
	Se	260	120	590–700		Zn	0.0037	104	900–1200
	Te	$3.5{\cdot}10^{-4}$	96.5	600–700	InP	Ag	$3.6{\cdot}10^{-4}$	57	800–1200
	Tl	320	172	630–700		Au	$1.3{\cdot}10^{-5}$	463	870–1100
AlSb	Al		174			Cd	$1.1{\cdot}10^{-7}$	70	1000–1200
	Cu	0.0035	34.5	420–770		Cu	30	66.5	900–1200
	Sb		145			In	10^{5}	373	1120–1270
	Zn	0.33	186	930–1130		P	$7{\cdot}10^{10}$	545	1120–1270
CdS	Ag	0.24	77.2	570–770	InSb	Ag	10^{-7}	24.2	700–800
	Au	200	174	770–1070		Au	$7{\cdot}10^{-4}$	30.9	400–800
	Cd	3	193	1000–1420		Cd	10^{-5}	106	520–770

Table 17.42 The parameters of (17.14) for diffusion of impurities in semiconductors *(continued)*

Semi-conductor	Diffusion media	D_0, cm^2/s	Q, kJ/mol	Temperature range, K	Semi-conductor	Diffusion media	D_0, cm^2/s	Q, kJ/mol	Temperature range, K
InSb	Co	$2.7 \cdot 10^{11}$	37.7	770–770	InSb	Se	0.016	125	470–720
	Cu	$3.5 \cdot 10^{-5}$	35.7	500–760		Te	$6.6 \cdot 10^{-5}$	115	630–770
	Fe	10^{-7}	24.2	700–780		Zn	$1.6 \cdot 10^{6}$	222	620–770
	Ge	$5 \cdot 10^{-6}$	91.5	600–770	ZnS	S		328	970–1160
	Hg	$4 \cdot 10^{-6}$	113	700–770		Zn	$3 \cdot 10^{-4}$	147	1170–1200
	Li	$7 \cdot 10^{-4}$	27	770		Zn	$1.5 \cdot 10^{-4}$	314	1210–1300
	In	$2 \cdot 10^{-9}$	27	600–800	ZnSe	Cu	$1.7 \cdot 10^{-5}$	64	470–840
	S	$4 \cdot 10^{5}$	101	470–720		Zn	1000	332	1000–1100
	Sb	$1.4 \cdot 10^{-4}$	363	600–800	ZnTe	Te	$2 \cdot 10^{-4}$	367	1000–1270
	Sn	$1.3 \cdot 10^{-6}$	62.7	570–770		Zn	0.01	183	1060–1220

References

[1] Handbook of a chemist, Vol. 3, Khimia, Moscow, 1966 (in Russian).

[2] Marrero, T. A., Mason, E. A., J. Phys. Chem. Ref. Data, 1, 3, 1972.

[3] Eletskii, A. V., Palkina, L. A., Smirnov, B. M., Transport phenomena in weakly ionized plasma, Atomizdat, Moscow, 1975 (in Russian).

[4] Kryukov, N. A., Interaction between mercury atoms and inert gas atoms, Author's abstract of candidate thesis, Leningrad State University, Leningrad, 1984.

[5] Reid, R. C., Sherwood, T. K., The properties of gases and liquids. Their estimation and correlation, 2nd ed., McGraw-Hill, New York, 1966.

[6] Reid, R. C., Prausnitz, J. M., Sherwood, T. K., The properties of gases and liquids, 3rd ed., McGraw-Hill, New York, 1977.

[7] Bretsznaider, S., Prediction of transport and other physical properties of fluids, Pergamon Press, Oxford, 1971.

[8] Adda, Y., Philibert, J., La diffusion dans les solids, Vol. 2, Press universitaires de France, Paris, 1966.

[9] Hydrogen in metals, Eds. Alefeld, G. and Volki, J., Springer, Berlin, 1974.

[10] Warburton, W. K., Turnbull, D., Diffusion in solids, North Holland, Amsterdam, 1975.

[11] Boltaks, B. I., Diffusion and point defects in semiconductors, Nauka, Leningrad, 1972 (in Russian).

[12] Bousheri, A., Bzowski, J., Kestin, J., Meson, E. A., J. Phys. Chem. Ref. Data, 16, 449, 1987.

[13] Devdariani, A. Z., Zagrebin, A. L., in: Plasma Chemistry, v.1, Ed. Smirnov, B. M., Energoatomizdat, Moscow, 1989 (in Russian).

18

Elementary Processes in Gases and Plasmas

A.V. Eletskii

18.1 Vibrational Relaxation of Molecules in Gases

The process resulting in the energy of molecular vibrations being transformed into the energy of translational motion of particles is described by the formula

$$AB(v) + M \rightarrow AB(v') + M + \Delta E, \tag{18.1}$$

where AB denotes the molecule; v and v' are vibrational quantum numbers; M denotes a gas particle (atom, molecule); ΔE represents the energy released in the form of the energy of translational motion of the colliding particles. The processes proceeding most effectively, in the case of vibrational relaxation, are those related to a unit change in the vibrational quantum number ($v - v' = 1$) [1, 34]. Precisely such processes are the ones best studied in detail experimentally. Table 18.1 presents the rate constants k for the vibrational relaxation of diatomic molecules measured at various temperatures for molecules in the first excited vibrational state ($v = 1$, $v' = 0$). The uncertainty in the presented data amounts to several hundreds percent. Rate constants of molecular vibrational relaxation measured at various temperatures and values of the vibrational quantum number v are reported in Tables 18.2 and 18.3. Parameters characterizing the process of vibrational relaxation and the exchange of vibrational quanta are illustrated in Figures 18.1 and 18.2 for the CO molecule.

Table 18.1 Vibrational relaxation rate constant k (in 10^{-16} cm^3/s) for diatomic molecules in the first excited vibrational state [1, 2]

Molecule	Admixture	Temperature, K			Molecule	Admixture	Temperature, K		
AB	M	300	800	2000	AB	M	300	800	2000
H_2	H_2	0.8			HCl	HCl	180	1000	10^4
D_2	D_2^{*1}	0.3 [5]			HBr	HBr	170	800	10^4
HD	HD	1.3			HI	HI		1000	10^4
N_2	N_2		0.56	14	DF	DF	$1.1 \cdot 10^4$	5100	8600
CO	CO	0.18		97	DCl	DCl	63		
NO	NO	640		$4.7 \cdot 10^5$	CS	CS	3400		
O_2	O_2	0.06	11	260	D_2	H_2	2.7		
F_2	F_2	18				HD	1.4		
Cl_2	Cl_2	89	3500			DF	8300		
Br_2	Br_2	1150			HF	He			1300
HF	HF	$1.7 \cdot 10^4$	10^4	$8.2 \cdot 10^3$		Ar			250

0-8493-2861-6/97/$0.00+$.50

Table 18.1 Vibrational relaxation rate constant k *(continued)*

Molecule AB	Admixture M	Temperature, K 300	800	2000
HF	F	10^6		$4.2\cdot10^5$
	H_2	6800		500
	D_2	1050	2600	
	N_2	36	1200	
	DF	$1.8\cdot10^4$	$1.2\cdot10^4$	10^4
	H	$3.6\cdot10^4$		
HCl	H_2	55	900	8000
	Ar	0.03		
	D_2	4.4		
	HD	23		
	Br	2600		
	Cl	$7\cdot10^4$		
	He	$5.5\cdot10^4$		
	Ne	$4.4\cdot10^4$		
	Kr	$4.2\cdot10^4$		
	H	$7.7\cdot10^4$		
	D	$1.1\cdot10^5$		
	Cl	55		
HBr	He	$4.5\cdot10^4$		
	HD	80		
	Ne	$4.8\cdot10^4$		
	DF	2800		
	Kr	$4.7\cdot10^4$		
	Br	$2.6\cdot10^4$		
	H_2	100		
	D_2	110		
DF	Ar			120
	N_2	230		1200
	H_2	180	1200	8200
	HF	10^4		$1.7\cdot10^5$
	D_2	6400		
DCl	He	0.62		
	Cl	$5.5\cdot10^4$		
	H_2	190		
	HD	780		
	Br	2300		
	D_2	17		

Molecule AB	Admixture M	Temperature, K 300	800	2000
	H	$1.8\cdot10^4$		
	D	$2\cdot10^4$		
H_2	HF	8300		
	He*2	0.2	3	2000
	Ar	0.1		
	D_2	0.24		
CO	He	0.27 [5]	2.1 [5]	300 [5]
	H			$2.7\cdot10^6$
	Ne	0.009		5.3
	H_2*3	4.0 [4]		
	HD	1.9 [5]		
	D_2	0.12 [5]		
	O_2	0.4		
	O	3600	5800	$5\cdot10^4$
NO	Ar	85		
	CO	340		
	N_2	10.3		
	He	2600		
Cl_2	He	10^4		
	H_2	$1.9\cdot10^4$		
	N_2	70		
N_2	O	45	360	2500
	He	14		
	H_2	150		
	D_2	12		
D_2	He	0.07		
	H_2	2.8		
O_2^+	He	<20 [6]		
	Ne	<130 [6]		
	Ar	10^4 [6]		
	Kr	$1.1\cdot10^5$ [6]		
	H_2	$2.5\cdot10^4$ [6]		
	D_2	6500 [6]		
	N_2	$1.9\cdot10^4$ [6]		
	CO	$4.4\cdot10^5$ [6]		
	O_2	$3\cdot10^6$ [6]		

*1 $k=0.011\cdot10^{-16}$cm^3/s at 77 K [5]. *2 $k=0.017\cdot10^{-16}$cm^3/s at 77 K [1, 2]. *3 $k=0.18\cdot10^{-16}$cm^3/s at 77 K [4].

Table 18.2 Vibrational relaxation rate constants k (in 10^{-12} cm^3/s) for HF (DF) molecules colliding with unexcited HF (DF) molecules [9, 10]

ν	T, K 300	500	1000	1800	2400	ν	T, K 300	500	1000	1800	2400
	HF (ν) + HF (0) → HF (ν-1) + HF(0)						DF (ν) + DF (0) → DF (ν-1) + DF(0)				
1	1.5	1.2	0.88	0.57	0.88	1	1.0	0.56	0.42	0.74	1.34
2	7.4	6.2	4.6	2.9	4.6	2	4.4	2.3	1.9	3.2	5.4
3	16	13	8.8	5.6	8.8	3	9.3	4.8	4.1	6.7	11
4	23	21	15	9.4	14	4	13	7.7	6.2	10	17
5	34	28	20	12	19	5	18	10.3	7.5	13	21
6	41	32	24	14	24	6	21	12	8.8	15	27

Table 18.2 Vibrational relaxation rate constant k *(continued)*

ν	T, K					ν	T, K				
	300	500	1000	1800	2400		300	500	1000	1800	2400
	DF (ν) + HF (0) $\to$ DF (ν-1) + HF(0)						HF (ν) + DF (0) $\to$ HF (ν-1) + DF(0)				
1	1.1	0.73	0.72	0.52	0.63	1	1.8	1.4	0.95	0.97	1.0
2	4.5	3.1	2.5	2.2	3.0	2	8.3	6.9	4.6	4.4	4.5
3	11	7.3	4.9	4.4	6.1	3	18	14	9.4	9.6	10
4	16	11	7.3	8.0	9.3	4	30	24	15	14	16
5	22	16	11	10	12	5	41	31	21	20	21
6	25	18	14	12	13	6	53	37	25	22	22

Table 18.3 Vibrational relaxation rate constant k (in 10^{-12} cm^3/s) for diatomic molecules in various excited vibrational states (T = 300 K) [1, 2]

Molecule	Admixture	ν						
AB	M	1	2	3	4	5	6	7
HCl	H [7]	50	85	110			130	
	Cl [7]	8.3	14.2	18.3			70	
	Br [8]	0.28	1.4					
	H_2	0.0044	0.0088	0.014	0.019			
HF	O_2	0.0011	0.0074	0.02	0.05	0.14	0.34	0.5
	N_2	0.0046	0.026	0.08	0.14	0.57	1.7	3.7
	D_2	0.09	0.57	1.4	3	5.7	8	12
	CO_2	1.1	5.7	10	20	30	43	43
	H_2	0.68	0.21	0.15	0.21	0.49	0.99	0.16
DF	H_2	0.18	0.06	0.14	0.29			
	N_2	0.023	0.074	0.17	0.3			

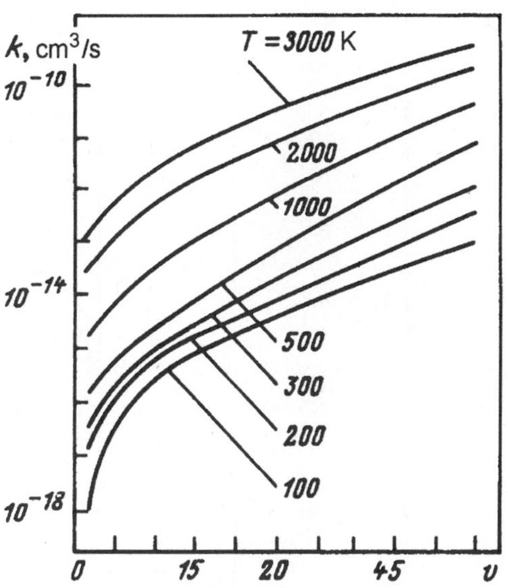

Figure 18.1 Dependence of the rate constant of vibrational relaxation of CO molecules colliding with He atoms upon the vibrational quantum number of the CO molecule [3]: CO(v) + He $\to$ CO($v-1$) + He + ΔE.

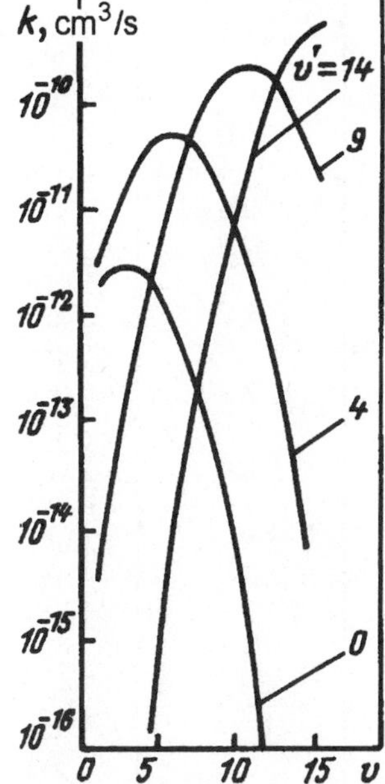

Figure 18.2 Dependence of the rate constant for the exchange of vibrational quanta in collisions between CO molecules upon the molecular vibrational quantum number [3]: CO(v) + CO(v') $\to$ CO($v-1$) + CO($v'+1$).

18.2 Rotational Relaxation of Molecules in Gases

Molecules which offer rotational degrees of freedom influence the propagation of perturbations in molecular gases, the formation and decay kinetics of inverse population densities via vibrational-rotational transitions of the molecules in molecular gas lasers, and, also, the thermal properties of molecular gases. The exchange of energy between the translational and rotational degrees of freedom of colliding molecules (rotational relaxation) is rather complicated and is related to the strong dependence of the energy of a molecular rotational quantum on its number. Since such exchanges usually involve the simultaneous participation of a significant number of molecular rotational states, a detailed level-by-level description of the exchange process encounters significant technical difficulties and, most likely, cannot be achieved with sufficient reliability at present. Therefore, in describing the intensity of energy exchange in rotational relaxation, one usually makes use of an average parameter Z_r, equal to the ratio of the mean time of rotational relaxation τ_r to the mean time τ_o required for the molecule to cover its mean free path in a gas-kinetic collision. Here, τ_r is defined as the time required for thermodynamic equilibrium to be established between the rotational and translational degrees of freedom of a molecule. Thus, parameter Z_r represents the average number of gas-kinetic collisions required for exchanging energy between the translational and rotational degrees of freedom of a molecule. Values of this parameter measured by various authors within a wide range of temperatures for certain colliding pairs of molecules are compiled in Table 18.4 [1, 33]. The data presented are characterized by uncertainties of 10 to 30%.

Table 18.4 Average number Z_r of gas-kinetic collisions required for the rotational relaxation of molecules [1, 33]

Relaxed molecule	Collision partner	T, K	Z_r	Relaxed molecule	Collision partner	T, K	Z_r
p-H_2	p-H_2	77	716			90.5	91
		170	325		Ar	77.3	79
		273	199			90.5	95
		293	134	HD	HD	27	10
	o-H_2	77	394			39.6	11
		170	194			77.3	7
	^{3}He	170	194			293	15
	^{4}He	90.5	369	n-H_2	n-H_2	79.5	560
		170	207			90	540
	Ne	90.5	235			207	215
		170	140			273	303
	Ar	90.5	225			298	215
		170	198	n-D_2	n-D_2	77.3	212
	Kr	111.5	277			273	205
		170	253			293	111
	Xe	170	356	N_2	N_2	77.1	4.6
o-H_2	o-H_2	170	790			150	4.0
		233	510			273	6.2
		293	530			293	4.8
o-D_2	o-D_2	31.2	370			400	6.6
		77	300			500	3.8
		293	111			700	8.1
	^{4}He	44.5	238			1000	9.8
		77.3	175			1300	11.4
		90.5	152	CO	CO	298	2.1
o-D_2	Ne	52.5	126			398	3.3
		77.3	96			500	1.9

Table 18.4 Average number Z_r of gas-kinetic collisions *(continued)*

Relaxed molecule	Collision partner	T, K	Z_r	Relaxed molecule	Collision partner	T, K	Z_r
		676	3.5	SO_2	SO_2	366	3.87
NO	NO	366	3.6	D_2O	D_2O	381	1.6
N_2	He	310	6.1			525	1.2
	Ne	310	2.5	H_2S	H_2S	300	3.0
	Ar	310	3.0			398	5.45
	Xe	310	4.6			500	1.6
O_2	O_2	162	1.7			398	4.32
		293	3.8			444	6.12
		400	5.5	NH_3	NH_3	300	2.1
		700	6.7			393	2.0
		1000	7.6			474	1.6
		1300	8.5	CO_2	CO_2	284	2.4
O_2	He	293	8.4			500	3.9
HCl	HCl	300	5.2			676	4.1
		400	1.7		He	298	0.8
		500	1.6	C_2H_2	C_2H_2	287	1.8
DCl	DCl	300	0.9	CH_4	CH_4	77	3.0
		400	0.2			293	8.6
		471	1.0			500	14.8
HBr	HBr	200	10.0	CD_4	CD_4	180	4.7
		300	6.0			293	7.6
		500	1.8	CF_4	CF_4	288	3.0
HF	HF	350	11.0	SF_6	SF_6	291	2.8
		400	8.0	CH_4	CH_4	308	3.0
		500	5.0		Ar	308	12.0
DF	DF	350	3.4		Xe	308	27.0
		400	2.5	CD_4	He	308	2.6
		450	1.8		Ne	308	4.1
H_2O	H_2O	350	2.2		Ar	308	7.3
		400	1.4		Xe	308	14.0
		450	0.9	C_2H_6	C_2H_6	283	4.0
		500	0.5	C_2H_4	C_2H_4	283	2.4

18.3 Electron Scattering on Atoms and Molecules

Table 18.5 Momentum transfer cross section (in 10^{-16}) cm^2 of electron scattering on atoms and molecules [27]

Electron energy, eV	He	Ne	Ar	Kr	Xe	H_2	N_2	D_2	O_2	CO_2	CO	H_2O
0.01	5.2		5.7	25	120	7.3	2.2	7.3	3.0	170	7.8	4000
0.02	5.4		4.0	20	80	8.0	2.9	8.0	3.0	120	5.9	1700
0.04	5.5	0.50	2.4	1	50	9.0	3.9	9.0	3.0	85	5.2	790
0.06	5.7	0.57	1.5	10	37	9.6	4.6	9.6	3.0		6.0	500
0.08	5.8	0.64	0.8	8	28	10.1	5.2	10.1	3.4			350
0.1	5.9	0.70	0.4	6.9	21	10.5	6.0	10.6	4.4	52	7.3	
0.30	6.4	1.09	0.16	1.0	4.0	13.0	9.6	14	4.6	18	12	
0.40	6.5	1.22	0.18	0.6	1.9	13.9	10	14.5	4.7		13.5	
0.50	6.6	1.32	0.22	0.56	1.4	14.7	10	15	5.1	7.3	14	
0.60	6.7	1.40	0.4	0.52	1.3	15.6	10	16	5.5	5.7	15	

Table 18.5 Momentum transfer cross section (in 10^{-16} cm^2) of electron scattering *(continued)*

Electron energy, eV	He	Ne	Ar	Kr	Xe	H_2	N_2	D_2	O_2	CO_2	CO	H_2O
0.70	6.7	1.47	0.54	0.54	1.4	16.3	10	17	6.1	5.0	15.5	
0.80	6.8	1.53	0.77	0.56	1.5	17	10	17.5	6.8		16	
1.0	6.8	1.62	1.0	0.60	2.1	18	10.5	18	7.6	4.1	17.0	
2.0	7.0	1.82	2.6	2.0	8.0	18.0	20		7.1	4.8		
2.5	7.0	1.86				17	32		6.3	6.4		
3.0	6.9	1.91	4.0	4.0	15	16	30		5.6	10.4		
4.0	6.6	1.98	5.9	7.0	23	14	13		5.2	17.1		
5.0	6.3	2.07	7.9	10.0	32	13	11		5.4	14		
6.0	6.0	2.14	9.0	13	34		10		5.8			
7.0		2.21	10.6	16	35		9.7					
8.0			11	18	35		9.6		7.2	11.7		
9.0			12	19	33		9.5					
10			14	21	32		9.8		8.4	12.9		
15			12	22	30		11		9.4			
20			10	19	22		13		9.1			

18.4 Ionization Processes Involving Excited Atoms

When an excited atom collides with another atom or molecule (excited or unexcited), there exist two kinds of processes that may result in ionization [13, 14]. In processes of the first kind, known as Penning ionization, the ionization can proceed if the excitation energy of the atom A* exceeds the ionization potential of the other atom or molecule B:

$$\mathrm{A}^* + \mathrm{B} \to \mathrm{A} + \mathrm{B}^+ + e. \tag{18.2}$$

In processes of the second kind, termed associative ionization, part of the energy needed for detaching the electron is released as a result of the production of a molecular ion:

$$\mathrm{A}^* + \mathrm{B} \to \mathrm{AB}^* + e. \tag{18.3}$$

If the binding energy of the molecular ion AB^+ exceeds the ionization potential of the molecule AB, then associative ionization may take place even if only unexcited particles are involved. Such a situation is realized in the case of associative ionization resulting from the collision of V and Th atoms with oxygen molecules. The cross sections and rate constants of the Penning and associative ionization processes presented in Tables 18.6 and 18.7 are characterized by uncertainties of several tens percent. Cross sections and rate constants for ionization due to the collisions of excited Na atoms with electronegative molecules are given in Table 18.8. These processes result in the production of negative ions [35].

Table 18.6 Cross-section (in 10^{-16} cm^2) for the Penning process $\mathrm{A}^* + \mathrm{B} \to \mathrm{A} + \mathrm{B}^+ + e$ at thermal energies [11, 13, 35] [The data are averaged over numerous measurements, the uncertainty in the data is 50 to 100%.]

B	He				Ne				Ar				Kr			
	2^3S	2^1S	2^1P	3^1P	$^3P_2$	$^1P_1$	$^3P_0$	$^3P_1$	$^3P_2$	$^1P_1$	$^3P_0$	$^3P_1$	$^3P_2$	$^1P_1$	$^3P_0$	$^3P_1$
H	22	33														
Ar	7	25	86	56	3.1											
Kr	8.8	34	89	50	1				1							

Table 18.6 Cross section in 10^{-16} cm^2 for the Penning process $A^* + B \rightarrow A + B^+ + e$ *(continued)*

B	He				Ne				Ar				Kr			
	2^3S	2^1S	2^1P	3^1P	$^3P_2$	$^1P_1$	$^3P_0$	$^3P_1$	$^3P_2$	$^1P_1$	$^3P_0$	$^3P_1$	$^3P_2$	$^1P_1$	$^3P_0$	$^3P_1$
Xe	12	40	84	73	12											
Hg	270								30							
He(2^3S)	100															
Na	14	17														
Zn	36				42				53				93			
Cd	72				46				65				110			
H_2	2.3	2.7	26		2											
N_2	5.3	11	69		6											
O_2	14	25	69		1.7	4	2.8	2.8	1.2	1.8	1.8	1.6	1.9	1.6	1.8	2.2
CO	8	16			7											
NO	17	36			21											
CO_2	51	70			11											
SF_6	28	53			38											
Ne				40												
Cs	6.5	22					400	400								
N_2O	36	38			25											
K	70															
Li	32															
Rb	35															

Table 18.7 Average cross section Q_{ai} (averaged over fine-structure levels) and rate constant k_{ai} of associative ionization in collisions involving excited atoms: $A^* + B \rightarrow AB^+ + e$ [11, 13, 35]

A^*	B	T, K	k_{ai}, 10^{-13} cm^3/s	Q_{ai}, 10^{-16} cm^2
Li($2\,^2P$)	Li($2\,^2P$)	900		0.005
Li($4\,^2P$)	Li($2\,^2S$)	1100	13000	
Li($5\,^2P$)	Li($2\,^2S$)	1100	33000	
Li($6\,^2P$)	Li($2\,^2S$)	1100	55000	
Li($7\,^2P$)	Li($2\,^2S$)	1100	64000	
Li($8\,^2P$)	Li($2\,^2S$)	1100	76000	
Li($9\,^2P$)	Li($2\,^2S$)	1100	65000	
Li($10\,^2P$)	Li($2\,^2S$)	1100	61000	
Li($11\,^2P$)	Li($2\,^2S$)	1100	55000	
Li($12\,^2P$)	Li($2\,^2S$)	1100	50000	
Li($13\,^3P$)	Li($2\,^2S$)	1100	42000	
Li($14\,^2P$)	Li($2\,^2S$)	1100	38000	
Li($15\,^2P$)	Li($2\,^2S$)	1100	33000	
Li($16\,^2P$)	Li($2\,^2S$)	1100	29000	
Li($17\,^2P$)	Li($2\,^2S$)	1100	28000	
Li($18\,^2P$)	Li($2\,^2S$)	1100	22000	
Li($19\,^2P$)	Li($2\,^2S$)	1100	20000	
Na($3\,^2P$)	Na($3\,^2P$)	500	380	3.8
Na($4\,^2P$)	Na($3\,^2S$)	700	830	
Na($5\,^2P$)	Na($3\,^2S$)	700	3600	
Na($6\,^2P$)	Na($3\,^2S$)	700	6000	
Na($7\,^2P$)	Na($3\,^2S$)	700	8400	
Na($8\,^2P$)	Na($3\,^2S$)	700	9000	
Na($9\,^2P$)	Na($3\,^2S$)	700	9600	
Na($10\,^2P$)	Na($3\,^2S$)	700	9400	
Na($11\,^2P$)	Na($3\,^2S$)	700	8300	
Na($12\,^2P$)	Na($3\,^2S$)	700	8100	
Na($13\,^2P$)	Na($3\,^2S$)	700	7500	

Table 18.7 Average cross-section Q_{ai} and rate constant k_{ai} *(continued)*

A^*	B	T, K	k_{ai}, 10^{-13} cm^3/s	Q_{ai}, 10^{-16} cm^2
$Na(14\ ^2P)$	$Na(3\ ^2S)$	700	6200	
$Na(15\ ^2P)$	$Na(3\ ^2S)$	700	5300	
$Na(16\ ^2P)$	$Na(3\ ^2S)$	700	4600	
$Na(17\ ^2P)$	$Na(3\ ^2S)$	700	4400	
$Na(18\ ^2P)$	$Na(3\ ^2S)$	700	4200	
$Na(19\ ^2P)$	$Na(3\ ^2S)$	700	3800	
$Na(20\ ^2P)$	$Na(3\ ^2S)$	700	3800	
$K(4\ ^2P)$	$K(4\ ^2P)$	500	9	0.13
$K(5\ ^2P)$	$K(4\ ^2S)$	570	56000	
$K(6\ ^2P)$	$K(4\ ^2S)$	570	98000	
$K(7\ ^2P)$	$K(4\ ^2S)$	570	130000	
$K(8\ ^2P)$	$K(4\ ^2S)$	570	150000	
$K(9\ ^2P)$	$K(4\ ^2S)$	570	160000	
$K(10\ ^2P)$	$K(4\ ^2S)$	570	120000	
$K(11\ ^2P)$	$K(4\ ^2S)$	570	77000	
$K(12\ ^2P)$	$K(4\ ^2S)$	570	29000	
$Rb(5\ ^2P)$	$Rb(5\ ^2P)$	470	3.2	0.7
$Rb(5\ ^2P)$	$Rb(5\ ^2S)$	520	62	
$Rb(6\ ^2P)$	$Rb(5\ ^2S)$	520	3000	
$Rb(7\ ^2P)$	$Rb(5\ ^2S)$	520	12000	
$Rb(8\ ^2P)$	$Rb(5\ ^2S)$	520	20000	
$Rb(9\ ^2P)$	$Rb(5\ ^2S)$	520	20000	
$Rb(10\ ^2P)$	$Rb(5\ ^2S)$	520	19000	
$Rb(11\ ^2P)$	$Rb(5\ ^2S)$	520	13000	
$Cs(6\ ^2P)$	$Cs(6\ ^2P)$	425	2	0.54
$Cs(6\ ^2P)$	$Cs(6\ ^2S)$	500	5400	
$Cs(7\ ^2P)$	$Cs(6\ ^2S)$	500	8200	
$Cs(8\ ^2P)$	$Cs(6\ ^2S)$	500	8700	
$Cs(9\ ^2P)$	$Cs(6\ ^2S)$	500	8700	
$Cs(10\ ^2P)$	$Cs(6\ ^2S)$	500	7800	
$Cs(11\ ^2P)$	$Cs(6\ ^2S)$	500	7300	
$Cs(12\ ^2P)$	$Cs(6\ ^2S)$	500	5300	
$He(3\ ^3S)$	He	320		0.07
$He(3\ ^3P)$	He	320		1.7
$He(3\ ^3D)$	He	320		3.2
$He(3\ ^1P)$	He	320		2.1
$He(3\ ^1D)$	He	320		17
U	O	2000		16.2
U	O_2	2000		0.17
U	O_3	2000		4
Th	O	2000		10.3
Th	O_2	2000		0.15
Th	O_3	2000		2000
$Hg(6^3P_0)$	Hg	400		460

Table 18.8 Ionization parameters of Na* excited atoms undergoing thermal collisions with electronegative molecules (T = 600 K) [35, 39]

Atom	Molecule	Rate constant, 10^{-10} cm^3/s	Cross section, 10^{-13} cm^2	Resulting negative ions
$Na(4^2D)$	SF_6	2.8	1.6	SF_6^-
$Na(4^2D)$	CH_3Br	3.5	8.4	Br^-
$Na(4^2D)$	CCl_2F_2	0.54	2.8	Cl_2^-, Cl^-, F_2^-

Table 18.8 Ionization parameters of Na^* excited atoms undergoing thermal collisions *(continued)*

Atom	Molecule	Rate constant, 10^{-10} cm^3/s	Cross section, 10^{-13} cm^2	Resulting negative ions
$Na(4^2D)$	C_6F_6	0.08	$1.2 \cdot 10^{-4}$	$C_6F_6^-$
$Na(4^2D)$	C_6F_5H	0.18	$2.7 \cdot 10^{-4}$	
$Na(4^2D)$	CH_3I	2.1		I^-
$Na(4^2D)$	O_2	2		O_2^-
$Na(5^2S)$	O_2	0.13		O_2^-
$Na(5^3S)$	SF_6	0.16		SF_6^-
$Li(4^2S)$	F_2		$2 \cdot 10^{-3}$	
$Li(4^2S)$	Cl_2		$2 \cdot 10^{-4}$	
$Li(4^2S)$	NO_2		$2 \cdot 10^{-5}$	
$K(4^2P)$	Br_2	0.94	0.24	
$Pb(5^2P)$	I_2	1.4	0.27	

Note: The uncertainty in the data ranges from 10 to 30%.

18.5 Resonance Charge Exchange Processes

The following processes belong to resonance charge exchange:

$$AB^+ + AB \to AB + AB^+; \tag{18.4}$$

$$AB^- + AB \to AB + AB^-; \tag{18.5}$$

$$A^+ + A \to A + A^+; \tag{18.6}$$

$$A^- + A \to A + A^-. \tag{18.7}$$

Most of the information on the parameters of these processes is obtained from calculations by the asymptotic theory [17] that exhibit uncertainties of 10 to 13%, which is essentially lower than the uncertainties pertaining to modern experiments. The cross sections of processes (18.4)–(18.7) are presented in Tables 18.9–18.11 for various collision energies [20].

Table 18.9 Cross-section (in 10^{-16} cm^2) of resonance charge exchange of positive and negative molecular ions on the parent molecules [20]

Ion	Collision energy, eV						Ion	Collision energy, eV					
	0.1	1.0	10	100	1000	10000		0.1	1.0	10	100	1000	10000
H_2^+		25	17	7	7	6	CO^+	82	67	52	39	29	
N_2^+	66	52	38	32	23		NO^+	49	37	25	17	5.5	
O_2^+	50	36	25	18	14		O_2^-	13	7	5.5	4.7	2.2	

Table 18.10 Cross-section (in 10^{-16} cm^2) of resonance charge exchange of ions and parent atoms [20]

Atom	Collision energy, eV						Atom	Collision energy, eV					
	0.1	1.0	10	100	1000	10000		0.1	1.0	10	100	1000	10000
H	67.1	49.2	36.7	26.2	17.2	9.5	Be	150	108	83.6	63.9	46.6	31.4
He	37.1	27.4	20.8	15.3	10.6	6.6	B	120	83.2	62.0	45.0	30.5	18.4
Li	313	228	180	140	103	70.8	C	95.6	58.7	40.7	28.2	18.3	10.5

Table 18.10 Cross-section (in 10^{-16} cm^2) of resonance charge exchange of atomic ions *(continued)*

Atom	Collision energy, eV						Atom	Collision energy, eV					
	0.1	1.0	10	100	1000	10000		0.1	1.0	10	100	1000	10000
N	60.5	39.5	28.5	20.4	13.7	8.1	Sb	137	114	92	72	55	40
O	60.2	41.3	30.3	21.6	14.4	8.4	Te	128	106	85	67	51	36
F	54.7	33.5	23.2	16.0	10.5	6.1	I	146	97	73	56	41	29
Ne	38.3	25.7	18.9	13.7	9.4	5.9	Xe	116	82	64	51	39	28
Na	350	264	212	168	128	92.0	Cs	548	427	359	292	231	174
Mg	211	155	123	96.0	72.4	51.2	Ba	400	296	240	194	152	114
Al	204	147	113	84.7	60.2	38.9	La	328	278	233	191	152	116
Si	169	106	76.1	55.0	37.9	23.7	Ce	218	182	149	118	91	66
P	110	75.3	57.1	43.0	31.0	20.8	Pr	352	299	249	202	160	120
S	99.3	66.0	48.7	55.6	24.5	15.3	Nd	244	204	167	133	102	74
Cl	100	54.1	42.9	31.0	21.5	13.8	Pm	244	204	168	133	102	74
Ar	71.9	44.2	36.6	27.7	20.2	13.8	Sm	362	307	256	209	165	125
K	457	347	282	225	174	127	Eu	297	250	207	167	130	97
Ca	301	218	174	138	105	76	Gd	264	222	184	148	115	85
Sc	281	206	165	131	101	74	Tb	218	182	149	118	90	64
Ti	273	196	157	124	96	70	Dy	214	179	146	115	87	62
V	279	208	168	135	105	78	Ho	210	174	142	112	85	61
Cr	239	181	144	114	86	61	Er	210	174	141	112	85	60
Mn	241	177	142	114	88	64	Tu	208	174	141	111	84	60
Fe	223	163	130	104	80	59	Yb	250	210	172	137	105	76
Co	227	171	139	112	87	65	Hf	247	210	175	143	113	86
Ni	235	182	148	120	94	70	Ta	208	176	146	118	93	70
Cu	207	161	130	104	79	58	W	241	174	140	115	89	66
Zn	174	128	102	81	62	45	Re	226	192	160	130	103	78
Ga	189	154	123	94	67	46	Os	179	151	125	101	79	59
Ge	145	118	94	72	52	36	Ir	172	145	120	97	78	57
As	105	86	69	53	39	27	Pt	158	132	108	86	66	48
Se	105	86	68	53	39	27	Au	158	132	109	87	67	49
Br	128	78	56	41	30	20	Hg	158	120	97	78	60	45
Kr	90	62	48	37	28	18	Tl	204	169	137	107	80	57
Rb	500	386	316	256	201	150	Pb	218	148	112	86	64	44
Sr	326	253	206	166	129	96	Bi	186	155	128	102	79	59
V	255	214	177	142	111	82	Po	117	95	75	57	41	28
Zr	235	198	164	132	103	76	Ac	348	297	249	205	164	127
Nb	218	182	149	119	91	66	Th	282	239	199	162	128	97
Mo	210	176	144	114	88	64	Pa	314	266	222	181	143	108
Tc	221	186	154	124	97	72	U	295	221	209	171	136	104
Ru	198	166	135	107	82	59	Np	455	391	330	273	221	172
Rh	195	163	133	105	80	58	Pu	376	320	269	220	176	135
Pd	50	37	25	17.1	10.2	5.8	Am	488	422	357	297	241	190
Ag	224	170	137	109	84	61	Cm	456	392	333	277	225	177
Cd	191	141	113	90	70	51	Bk	477	410	347	288	234	183
In	208	171	137	107	79	55	No	438	375	316	261	210	162
Sn	158	129	102	78	57	39							

Table 18.11 Cross-section (in 10^{-16} cm^2) of resonance charge exchange between negative atomic ions and parent atoms [20]

Atom	Collision energy, eV						Atom	Collision energy, eV					
	0.1	1.0	10	100	1000	10000		0.1	1.0	10	100	1000	10000
H	415	266	172	104	23		C	166	118	80	50	29	14.1
Li	640	480	375	233	145	79	O	123	85	55	33	17.6	8.1
W	323	239	140	72	30	9.2	F	85	59	41	27	16.6	9.2

Table 18.11 Cross-section (in 10^{-16} cm^2) of resonance charge exchange of negative ions *(continued)*

Atom	Collision energy, eV						Atom	Collision energy, eV					
	0.1	1.0	10	100	1000	10000		0.1	1.0	10	100	1000	10000
Na	822	628	461	322	209	123	Se	181	138	101	71	47	29
Al	380	292	202	126	62	36	Br	153	104	77	56	39	26
Si	222	166	120	82	53	31	Rb		906	691	507	253	229
P	293	212	145	93	54	28	Ag	487	388	302	227	163	110
S	148	110	78	53	34	19.4	Sn	313	242	181	130	89	57
Cl	122	84	61	43	29	18.5	Sb	326	254	187	132	88	54
K	939	723	537	380	252	152	Te	220	170	127	91	62	40
Fe		807	588	406	261	150	I	162	129	99	74	53	36
Co	605	475	361	263	182	117	Cs					773	505
Ni	463	368	284	211	150	100	Au	320	265	216	171	132	99
Cu	472	373	286	212	149	98							

18.6 Ion–Molecular Reactions

Table 18.12 Rate constant k (in 10^{-30} cm^6/s) of processes involving the molecular ions ($T = 300$ K) [21]

Process	k	Process	k
$He^+ + 2He \to He_2^+ + He$	1.0	$O^+ + N_2 + He \to N_2O^+ + He$	54 (80 K)
$Ne^+ + 2Ne \to Ne_2^+ + Ne$	0.60	$NO^+ + 2NO \to N_2O_2^+ + NO$	5
$Ar^+ + 2Ar \to Ar_2^+ + Ar$	2.2	$CO^+ + 2CO \to C_2O_2^+ + CO$	140
$Kr^+ + 2Kr \to Kr_2^+ + Kr$	2.3	$CO_2^+ + 2CO_2 \to C_2O_4^+ + CO_2$	320
$Xe^+ + 2Xe \to Xe_2^+ + Xe$	2.8	$C_2H_2^+ + 2C_2H_4 \to C_4H_8^+ + C_2H_4$	2000
$Cs^+ + 2Cs \to Cs_2^+ + Cs$	150	$O_2^+ + CO_2 + He \to CO_4^+ + He$	23 (200 K)
$Ar^+ + Ar + He \to Ar_2^+ + He$	1.0 (82 K)	$O_2^+ + N_2O + He \to N_2O_3^+ + He$	52 (200 K)
$Kr^+ + Kr + He \to Kr_2^+ + He$	0.6	$O_2^+ + SO_2 + He \to SO_4^+ + He$	600 (200 K)
$Xe^+ + Xe + He \to Xe_2^+ + He$	1.3	$Mg^+ + O_2 + Ar \to MgO_2^+ + Ar$	2.5
$Hg^+ + Hg + He \to Hg_2^+ + He$	1.7 (700 K)	$NO^+ + H_2O + NO \to NOH_2O^+ + NO$	150
$Ne^+ + 2He \to NeHe^+ + He$	0.21	$NO^+ + 2H_2O \to NOH_2O^+ + H_2O$	150
$Ar^+ + Ar + Ne \to Ar_2^+ + Ne$	3	$NOH_2O^+ + H_2O + H_2 \to NO(H_2O)_2^+ + He$	300
$HeH^+ + 2He \to He_2H^+ + He$	0.4 (200 K)	$NOH_2O^+ + H_2O + Ar \to NO(H_2O)_2^+ + Ar$	900
$H_3^+ + 2Ar \to H_3^+Ar + Ar$	1	$NOH_2O^+ + H_2O + N_2 \to NO(H_2O)_2^+ + N_2$	1100
$Li^+ + 2Ar \to LiAr + Ar$	1.8	$NOH_2O^+ + H_2O + O_2 \to NO(H_2O)_2^+ + O_2$	800
$H^+ + 2H_2 \to H_3^+ + H_2$	30	$NOH_2O^+ + H_2O + NO \to NO(H_2O)_2^+ + NO$	1100
$D^+ + 2D_2 \to D_3^+ + D_2$	30	$H_3O^+ + H_2O + He \to H_5O_2^+ + He$	120
$H_3^+ + 2H_2 \to H_5^+ + H_2$	0.65	$Hg^+ + 2Hg \to Hg_2^+ + Hg$	1 (700 K)
$D_3^+ + 2D_2 \to D_5^+ + D_2$	0.45	$He_2^+ + 2He \to He_3^+ + He$	1.7 (77 K)
$N^+ + 2N_2 \to N_3^+ + N_2$	30	$Ar_2^+ + 2Ar \to Ar_3^+ + Ar$	32 (77 K)
$N^+ + N_2 + He \to N_3^+ + He$	30 (80 K)	$Ar_2^+ + Ar + He \to Ar_3^+ + He$	5.5 (80 K)
$N_2^+ + 2N_2 \to N_4^+ + N_2$	85	$Cs_2^+ + Cs \to Cs_3^+ + Cs$	300
$O_2^+ + 2O_2 \to O_4^+ + O_2$	2.5	$Ne^+ + Ne + He + Ne_2^+ + He$	3.0
$O_2^+ + O_2 + He \to O_4^+ + He$	1.5	$Ca^+ + O_2 + Ar \to CaO_2^+ + Ar$	6.6
$O_2^+ + O_2 + Kr \to O_4^+ + Kr$	8 (80 K)	$Fe^+ + O_2 + Ar \to FeO_2^+ + Ar$	1.0
$O_2^+ + O_2 + H_2O \to O_2^+ + H_2O$	1.5	$Na^+ + O_2 + Ar \to NaO_2^+ + Ar$	0.2
$O_2^+ + 2N_2 \to O_2^+N_2 + N_2$	0.8	$K^+ + O_2 + Ar \to KO_2^+ + Ar$	<0.2
$O_2^+ + N_2 + He \to O_2^+ + He$	19 (80 K)	$Li^+ + 2N_2 \to LiN_2^+ + N_2$	2.0
$O_2^+ + H_2 + He \to O_2^+H_2 + He$	0.74 (80 K)	$Li^+ + 2O_2 \to LiO_2^+ + O_2$	1.1
$O_2^+ + O_3 + He \to O_5^+ + He$	100 (200 K)	$Na^+ + 2O_2 \to NaO_2^+ + O_2$	0.1 (193 K)

Table 18.12 Rate constant k (in 10^{-30} cm^6/s) of processes involving the molecular ions *(continued)*

Process	k	Process	k
$LiN_2^+ + 2N_2^+ \to Li(N_2)_2^+ + N_2$	2.2	$Na^+ + He + H_2O \to NaH_2O^+ + He$	4.7
$O_2N_2^+ + N_2 + He \to N_4O_2^+ + He$	10 (80 K)	$Na^+ + 2H_2O \to NaH_2O^+ + H_2O$	100
$O_4^+ + O_2 + He \to O_6^+ + He$	5 (80 K)	$K^+ + He + H_2O \to KH_2O^+ + He$	2.6
$O_4^+ + N_2 + He \to O_4N_2^+ + He$	10 (80 K)	$K^+ + 2H_2O \to KH_2O^+ + H_2O$	45
$O_4^+ + 2O_2 \to O_6^+ + O_2$	0.07	$Cs^+ + H_2O + N_2 \to CsH_2O^+ + N_2$	9
$O_6^+ + 2O_2 \to O_8^+ + O_2$	25 (90 K)	$Xe^+ + H_2O + He \to XeH_2O^+ + He$	15
$NO^+ + 2N_2 \to NO^+ + N_2 + N_2$	0.2	$NO^+ + H_2O + He \to NOH_2O^+ + He$	34
$NO^+ + 2O_2 \to NO_3^+ + O_2$	0.09	$NO^+ + H_2O + N_2 \to NOH_2O^+ + N_2$	150
$NO^+ + 2CO_2 \to NOCO_2^+ + CO_2$	24	$NO^+ + H_2O + O_2 \to NOH_2O^+ + O_2$	86
$NO^+ + CO_2 + He \to NOCO_2^+ + He$	4	$H_3O^+ + H_2O + Ar \to H_5O_2^+ + Ar$	40
$NO^+ + CO_2 + Ar \to NOCO_2^+ + Ar$	25 (200 K)	$H_3O^+ + H_2O + N_2 \to H_5O_2^+ + N_2$	340
$NO^+ + CO_2 + N_2 \to NOCO_2^+ + N_2$	30 (200 K)	$H_3O^+ + H_2O + O_2 \to H_5O_2^+ + O_2$	270
$NO^+ + 2NH_3 \to NONH_3^+ + NH_3$	54	$NO_2^+ + H_2O + N_2 \to NO_2H_2O^+ + N_2$	500
$NH_4^+ + NH_3 + O_2 \to N_2H_7^+ + O_2$	1800	$O_2^+ + H_2O + He \to H_2O_3^+ + He$	87
$Na^+ + 2CO_2 \to NaCO_2^+ + CO_2$	50	$O_2^+ + H_2O + N_2 \to H_2O_3^+ + N_2$	270
$Na^+CO_2 + 2CO_2 \to Na^+(CO_2)_2 + CO_2$	0.05	$O_2^+ + H_2O + O_2 \to H_2O_3^+ + O_2$	200
$Cs^+ + SO_2 + N_2 \to CsSO_2^+ + N_2$	30	$O_2^+ + H_2O + Ar \to H_2O_3^+ + Ar$	170

Table 18.13 Rate constant k (in 10^{-30} cm^6/s) of processes involving the molecular negative ions ($T = 300$ K) [21]

Process	k	Process	k
$O^- + 2O_2 \to O_3^- + O_2$	0.9	$O^- + H_2O + O_2 \to H_2O_2^- + O_2$	100
$O_2^- + 2O_2 \to O_4^- + O_2$	0.4	$O_2^- + H_2O + O_2 \to H_2O_3^- + O_2$	160
$O_2^- + O_2 + He \to O_4^- + He$	0.34 (200 K)	$O_3^- + H_2O + O_2 \to H_2O_4^- + O_2$	210
$O^- + CO_2 + He \to CO_3^- + He$	150	$H_2O_3^- + H_2O + O_2 \to H_4O_4^- + O_2$	540
$O^- + 2CO_2 \to CO_3^- + CO_2$	90	$NO^- + CO_2 + Ar \to NO^-CO_2 + Ar$	56
$O_2^- + CO_2 + He \to CO_4^- + He$	47 (200 K)	$NO^- + N_2O + Ar \to N_3O_2^- + Ar$	7.8
$O_2^- + 2CO_2 \to CO_4^- + CO_2$	9	$NO_2^- + H_2O + NO \to NO_2H_2O^- + NO$	150
$O_2^- + CO_2 + O_2 \to CO_4^- + O_2$	20	$Br^- + 2Br_2 \to Br_3^- + Br_2$	29
$O^- + N_2 + He \to N_2O^- + He$	0.04 (200 K)	$Cl^- + H_2O + NO \to ClH_2O^- + NO$	120
$O_2^- + N_2 + He \to N_2O_2^- + He$	0.04 (200 K)	$WO_3^- + O_2 + Ar \to WO_5^- + Ar$	10^3

18.7 Neutralization Processes of Charged Particles in a Plasma

Neutralization of charged particles in a volume of plasma proceeds via various recombination processes. These processes are characterized by recombination coefficients defined through the relationship

$$dN_-/dt = -\alpha_{\rm rec} N_- N_+, \tag{18.8}$$

where N_- and N_+ are the concentrations of negative and positive particles, respectively; $\alpha_{\rm rec}$ is the recombination coefficient, which depends on the sort of particles undergoing recombination, the state of the plasma, and the concrete recombination mechanism.

Three-body recombination. The process responsible for neutralization of a plasma with sufficiently high charged particle densities is described by the formula

$$A^+ + 2e \to A + e \tag{18.9}$$

(A stands for the atomic or molecular ion).

The expression for the coefficient of three-body recombination takes the form [12]

$$\alpha_{tr} \approx \frac{4e^{10}N_e}{m^{1/2}T_e^{9/2}} \approx \frac{0.6 \cdot 10^{-27}N_e}{T_e^{9/2}}(\mathrm{cm}^3/\mathrm{s}), \tag{18.10}$$

where e is the electron charge; N_e is the electron number density, cm^{-3}; T_e is the electron temperature, eV. Expression (18.10) is valid in the limit of high electron densities, when transitions between highly excited states of atoms (molecules) are due to inelastic collisions with electrons, while radiative processes are insignificant. Moreover, the electron temperature T_e is considered much smaller than the ionization potential of the atomic particle, so that the most probable transitions to occur in collisions are transitions between energy states that are close to each other. When the indicated conditions are not satisfied, the coefficient α_{tr} of three-body recombination depends on the sort of particle undergoing recombination. The dependence of the ratio α_{tr}/N_e on the temperature of the plasma is illustrated in Fig. 18.3 for K, Cs, N, Ar, He, and H atoms [12].

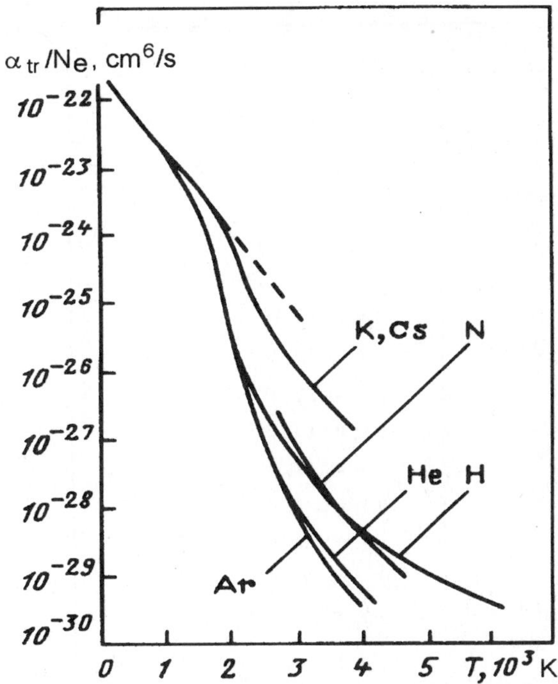

Figure 18.3 Coefficient of radiative-collisional electron recombination for some gases at various temperatures [12].

Dissociative recombination. This process represents an important channel for the neutralization of charged particles in a low-temperature plasma, where molecular ions make the main sort of positive ions [15]:

$$e + \mathrm{AB}^+ \rightarrow \mathrm{A} + \mathrm{B}. \tag{18.11}$$

Usually, one of the products of dissociative recombination turns out to be in an electron-excited state. Rate constants of dissociative recombination, obtained by averaging over a large body of experimental data, are presented in Table 18.14; the uncertainty ranges from 10 to 20%. Data on the energy dependence of cross sections and on the temperature dependence of dissociative recombination coefficients are introduced in Table 18.15, where also are indicated the temperature and energy ranges, within which the uncertainties in the approximations involved do not exceed 30%.

Table 18.14 Coefficient of dissociative recombination α_d (10^{-7} cm^3/s) at $T = 300$ K [15]

Ion	H_2^+	Ne_2^+	Ar_2^+	Kr_2^+	Xe_2^+	$HeNe^+$	C_2^+	O_2^+	N_2^+	NO^+
α_d	0.3	1.8	6.9	10.3	20	0.2	10 (100 K)	2.0	3.3	3.7
Ion	CH^+	CO^+	CO_2^+	CH_2^+		He_3^+	Cs_2^+		Cs_3^+	
α_d	5 (100 K)	6.8	3.6	8.7 (100 K)		34 (80 K)	1.7 (600 K)		0.2 (600 K)	
Ion	O_4^+	N_4^+	$N_2O_2^+$	H_3O^+	$H_5O_2^+$	$H_7O_3^+$	$H_9O_4^+$		NH_4^+	NaO_2^+
α_d	23 (200 K)	15	17	10 (540 K)	20 (540 K)	34	36		18	50
Ion	$NaCO_2^+$	$H_{11}O_5^+$	$H_{13}O_6^+$	$NH_4{\cdot}NH_3^+$	$NH_4(NH_3)_2^+$	$NH_4(NH_3)_3^+$	H_3^+	$CO{\cdot}CO^+$	H_5^+	$CO^+{\cdot}(CO)_2$
α_d	50	80	28	27	25	25	2.3	13	30	19

Table 18.15 Energy dependences of the cross-section σ_d and the rate constant α_d for dissociative recombination of electrons and molecular ions [16]

Ion	$\sigma_d, 10^{-15}$cm^2	$\alpha_d, 10^{-7}$cm^3/s	Comments
H_2^+	$8(0.1/\epsilon_e)^{0.92}$	$2.3(300/T_e)^{0.4}$	
HD^+	$8(0.1/\epsilon_e)^{0.92}$	$2.3(300/T_e)^{0.4}$	
D_2^+	$5.5(0.1/\epsilon_e)^{0.8}$	$1.6(300/T_e)^{0.3}$	
CH^+	$8(0.1/\epsilon_e)^{0.92}$	$3(300/T_e)^{0.4}$	
NH^+	$2.5(0.1/\epsilon_e)^{1.0}$	$0.86(300/T_e)^{0.5}$	
OH^+	$2(0.1/\epsilon_e)^{1.0}$	$0.75(300/T_e)^{0.5}$	
C_2^+	$14(0.1/\epsilon_e)^{1.0}$	$6(300/T_e)^{0.5}$	
N_2^+	$10(0.1/\epsilon_e)^{1.0}$	$3.5(300/T_e)^{0.5}$	
NO^+	$5(0.1/\epsilon_e)^{1.0}$	$2.3(300/T_e)^{0.5}$	
O_2^+	$4.5(0.1/\epsilon_e)^{1.0}$	$1.9(300/T_e)^{0.5}$	
Ne_2^+		$1.8(300/T_e)^{0.43}(300/T)^{1.1}$	
Ar_2^+		$9.1(300/T_e)^{0.61}(300/T)^{0.7}$	
Kr_2^+		$16(300/T_e)^{0.55}$	
Xe_2^+		$27(300/T_e)^{0.6}$	
H_3^+	$10(0.1/\epsilon_e)^{1.0}$	$2.3(300/T_e)^{0.5}$	
HD_2^+	$8(0.1/\epsilon_e)^{1.0}$		
D_3^+	$6(0.1/\epsilon_e)^{0.84}$		
CH_2^+	$13(0.1/\epsilon_e)^{1.0}$	$5.0(300/T_e)^{0.5}$	
CH_3^+	$16(0.1/\epsilon_e)^{1.0}$	$7.0(300/T_e)^{0.5}$	$T_e < 2000$ K, $\epsilon_e < 0.15$ eV
	$6(0.1/\epsilon_e)^{1.6}$		$\epsilon_e > 0.15$ eV
CH_4^+	$14(0.1/\epsilon_e)^{1.0}$	$7.0(300/T_e)^{0.5}$	$\epsilon_e < 0.07$ eV, $T_e < 1000$ K
	$5.6(0.2/\epsilon_e)^{1.5}$		$\epsilon_e > 0.07$ eV
CH_5^+	$14(0.1/\epsilon_e)^{1.0}$	$7.0(300/T_e)^{0.5}$	$\epsilon_e < 0.08$ eV, $T_e < 1000$ K
	$5.6(0.2/\epsilon_e)^{1.5}$		$\epsilon_e > 0.08$ eV
NH_2^+	$30(0.04/\epsilon_e)^{1.0}$		$\epsilon_e < 0.05$ eV
	$9(0.1/\epsilon_e)^{1.35}$		$\epsilon_e > 0.05$ eV
NH_3^+	$31(0.07/\epsilon_e)^{1.0}$		$\epsilon_e < 0.09$ eV
	$6.5(0.2/\epsilon_e)^{1.8}$		$\epsilon_e > 0.09$ eV
NH_4^+	$11(0.1/\epsilon_e)^{1.4}$	$13(410/T_e)^{0.5}$	$\epsilon_e < 0.5$ eV
	$0.5(0.5/\epsilon_e)^{3.43}$		$\epsilon_e > 0.5$ eV

Table 18.15 Energy dependences of the cross-section σ_{tr} and recombination coefficient *(continued)*

Ion	σ_{tr}, 10^{-15}cm^2	α_{tr}, 10^{-7}cm^3/s	Comments
H_2O^+	$18(0.1/\epsilon_e)^{1.0}$		$\epsilon_e < 0.15$ eV
	$6(0.2/\epsilon_e)^{2.0}$		$\epsilon_e > 0.15$ eV
$H_3O^+(D_3O^+)$	$18(0.08/\epsilon_e)^{1.0}$	$6.3(300/T_e)^{0.5}$	$\epsilon_e < 0.1$ eV, $T_e < 1000$ K
	$5(0.2/\epsilon_e)^2$		$\epsilon_e > 0.1$ eV
C_2H^+	$14(0.1/\epsilon_e)^{1.0}$	$5.4(300/T_e)^{0.5}$	
$C_2H_2^+$	$14(0.1/\epsilon_e)^{1.0}$	$5.4(300/T_e)^{0.5}$	$\epsilon_e < 0.1$ eV
	$60(0.2/\epsilon_e)^{1.2}$		$\epsilon_e > 0.1$ eV
$C_2H_3^+$	$20(0.1/\epsilon_e)^{1.0}$	$9(300/T_e)^{0.5}$	$\epsilon_e < 0.1$ eV
	$8.4(0.2/\epsilon_e)^{1.3}$		$\epsilon_e > 0.1$ eV
HCO^+		2.0	$T_e = 300$ K
$N_2H^+ \cdot (N_2O^+)$	$12(0.1/\epsilon_e)^{1.0}$	$7.5(300/T_e)^{2.5}$	$\epsilon_e < 0.1$ eV
	$4.4(0.2/\epsilon_e)^{1.3}$		$\epsilon_e > 0.1$ eV
CO_2^+		3.8	$T_e = 300$ K
H_5^+		36	$T_e = 205$ K
$H_3O^+ \cdot (H_2O)$		25	$T_e = 300$ K
$H_3O^+ \cdot (H_2O)_2$		22	$T_e = 415$ K
		38	$T_e = 300$ K
$H_3O^+ \cdot (H_2O)_3$		49	$T_e = 300$ K
$H_3O^+ \cdot (H_2O)_4$		60	$T_e = 205$ K
$H_3O^+ \cdot (H_2O)_5$		75	$T_e = 205$ K
$NH_4^+ \cdot (NH_3)$		$28(300/T_e)^{0.147}$	
$NH_4^+ \cdot (NH_3)_2$		$27(300/T_e)^{0.05}$	
$NH_4^+ \cdot (NH_3)_3$		30	$T_e = 200$ K
N_4^+		$14(300/T_e)^{0.41}$	
$CO^+ \cdot (CO)$		$13(300/T_e)^{0.34}$	
$CO^+ \cdot (CO)_2$		$19(300/T_e)^{0.33}$	
$O_2^+ \cdot (O_2)$		23	$T_e = 205$ K

Ion-ion recombination. An important mechanism of charged particle neutralization in plasmas of electronegative gases is due to ion–ion recombination:

$$A^+ + B^- \to A + B \tag{18.12}$$

or

$$A^+ + B^- + R \to A + B + R. \tag{18.13}$$

Rate constants for pair ion–ion recombination (18.12) at room temperature are presented in Tables 18.16. These data were obtained by averaging the results of a great number of experiments [17]. The uncertainty in the data given amounts to 50%. The coefficient of pair ion–ion recombination exhibits the following dependence on the temperature of the gas:

$$\alpha_i = \alpha_{i0}(T_0/T)^{1/2}, \tag{18.14}$$

where α_{i0} values represent the recombination coefficient at $T = T_0$.

Table 18.16 Rate constant α_{i0} (in 10^{-7} cm^3/s) of pair ion-ion recombination at T=300 K [17, 18]

Ion pair	α_{i0}	Ion pair	α_{i0}
$H^+ + H^-$	8.6	$CCl_3^+ + Cl^-$	0.45
$O^+ + O^-$	1.4	$CCl_2F^+ + Cl^-$	0.41
$N^+ + O^-$	1.4	$CCl_2F_2^+ + Cl^-$	0.41
$O_2^+ + O_2^-$	4.2	$NH_4^+ + Cl^-$	0.67
$N_2^+ + O_2^-$	1.6	$Cl_2^+ + Cl^-$	0.5
$NO^+ + NO_2^-$	3.4	$O_2^+ + CO_3^-$	0.95
$O_2^+ + NO_2^-$	4.1	$CF_3^+ + F^-$	0.58
$NO^+ + NO_3^-$	0.45	$NF_2^+ + F^-$	0.75
$O^+ + O_2^-$	2	$N_2F^+ + F^-$	0.85
$Na^+ + O^-$	1.3	$H_3O^+(H_2O)_3 + NO_3^-$	0.55
$O_2^+ + O^-$	1.2	$H_3O^+(H_2O)_3 + NO_3^- HNO_3$	0.57
$NO^+ + O^-$	1.2	$NO^+(NO_2)_2 + NO_3^-(HNO_3)_3$	0.35
$SF_3^+ + SF_5^-$	0.4	$NH_4^+(NH_3)_2 + Cl^-$	0.79
$SF_5^+ + SF_6^-$	0.4	$NH_4^+(NH_3)_2 + NO_2^-$	0.49
$CCl_2^+ + Cl^-$	0.45	$H_3O^+(H_2O)_3 + Cl^-$	0.68

At pressures exceeding several hundred pascals, pair recombination (18.12) is no longer the main ion–ion recombination mechanism, and this role is assumed by three-body recombination (18.13). The rate constant of this process exhibits a nonmonotonous dependence on pressure with a maximum at $p \approx 10^5$ Pa [17].

In Table 18.17, the dependences are presented of rate constants for triple ion–ion recombination of negative halogen ions with positive rare gas ions upon the density of the inert gas, the atoms of which play the role of the third particle [19].

Table 18.17 Coefficients (in 10^{-6} cm^3/s) of three-body ion-ion recombination at various concentrations of inert gas atoms and T=300 K [19]

Ion pair	Inert gas	Concentrations of inert gas atoms, 10^{19} cm^{-3}							
		0.27	0.81	2.7	5.4	10.8	27	54	135
$Kr^+ + F^-$	Ne	0.19	0.52	1.5	2.4	3.0	2.1	1.2	0.45
	Ar	0.33	0.84	2.2	2.7	2.2	1.0	0.49	0.20
	Kr	0.36	0.93	2.2	2.1	0.91	0.56	0.2	0.13
	Xe	0.47	1.2	2.4	2.0	0.8	0.47	0.26	0.09
$Kr_2^+ + F^-$	Ne	0.18	0.48	1.4	2.1	2.6	2.0	1.2	0.45
	Ar	0.29	0.73	1.8	2.5	2.0	0.95	0.48	0.19
	Kr	0.33	0.85	2.0	2.2	1.5	0.60	0.3	0.13
	Xe	0.37	0.96	2.0	1.8	1.1	0.44	0.23	0.09
$Xe^+ + Cl^-$	Ne	0.12	0.3	0.86	1.6	2.3	2.1	1.3	0.50
	Ar	0.26	0.68	1.7	2.2	1.8	0.8	0.41	0.17
	Kr	0.31	0.80	1.8	1.9	1.3	0.54	0.27	0.12
	Xe	0.36	0.88	1.7	1.4	0.8	0.34	0.17	
$Xe_2^+ + Cl^-$	Ne	0.11	0.30	0.80	1.4	2.0	20	1.2	0.48
	Ar	0.23	0.60	1.5	1.9	1.8	0.81	0.42	0.17
	Kr	0.27	0.70	1.6	1.8	1.3	0.52	0.27	0.11
	Xe	0.30	0.80	1.6	1.5	0.9	0.40	0.19	

18.8 Negative Ion Formation and Decay Processes

Negative ions may be formed as a result of pair processes of dissociative electron attachment [17]:

$$AB + e \rightarrow A + B^-, \tag{18.15}$$

due to electron captures yielding autoionization states of negative ions that are subsequently quenched in collisions with third particles:

$$e + AB \rightarrow (AB^*)^-; \quad (AB^*)^- + M \rightarrow AB^- + M. \tag{18.16}$$

Negative ions may be also produced as a result of the processes with molecules capturing electrons in triple collisions:

$$e + AB + M \rightarrow AB^- + M. \tag{18.17}$$

The rate constant of dissociative attachment (18.15) is a function of the mean electron energy in the gas. The rate constant of the process (18.16) is measured in units of cm^3/s, although it depends on the pressure and composition of the gas mixture. The rate constant of three-body electron attachment, cm^6/s, depends on the temperature and composition of the gas. Cross sections and rate constants of processes (18.15)–(18.17) are presented in Tables 18.18–18.22 and in Fig. 18.4.

Table 18.18 Cross-section (in 10^{-18} cm^2) of dissociative electron attachment to the SO_2 molecule, corresponding to the production of negative ions of various types [23]

Electron energy, eV	Negative ion			Electron energy, eV	Negative ion		
	O^-	S^-	SO^-		O^-	S^-	SO^-
2.5	0.01	0.0	0.0	5.4	1.12	0.044	2.80
3.0	0.10	0.013	0.02	5.6	0.69	0.028	1.21
3.1	0.16	0.019	0.0	5.8	0.45	0.013	0.62
3.2	0.22	0.034	0.03	6.0	0.43	0.015	0.39
3.3	0.36	0.056	0.03	6.2	0.81	0.008	0.33
3.4	0.58	0.079	0.16	6.3	0.86	0.018	0.46
3.5	0.89	0.122	0.16	6.4	1.17	0.017	0.40
3.6	1.34	0.167	0.36	6.5	1.32	0.020	0.58
3.7	2.20	0.225	0.54	6.6	1.59	0.022	0.44
3.8	3.16	0.276	0.84	6.7	1.80	0.019	0.52
3.9	4.38	0.298	1.41	6.8	2.10	0.017	0.58
4.0	5.58	0.313	2.05	6.9	2.33	0.025	0.41
4.1	6.92	0.310	3.27	7.0	2.56	0.027	0.42
4.2	7.52	0.295	4.58	7.1	2.68	0.026	0.40
4.3	8.08	0.271	6.94	7.2	2.41	0.030	0.42
4.4	8.02	0.246	8.51	7.3	2.47	0.031	0.51
4.5	7.22	0.231	10.10	7.4	2.41	0.031	0.44
4.6	6.48	0.191	10.80	7.5	2.14	0.036	0.51
4.7	5.48	0.162	10.98	7.6	1.89	0.030	0.43
4.8	4.69	0.164	10.53	7.7	1.56	0.029	0.33
4.9	3.79	0.131	9.07	7.8	1.33	0.027	0.28
5.0	2.14	0.114	7.76	7.9	1.06	0.016	0.21
5.1	2.56	0.091	5.86	8.0	0.93	0.015	0.21
5.2	1.90	0.065	4.84	8.1	0.76	0.015	0.13

Table 18.18 Cross-section (in 10^{-18} cm^2) of dissociative electron attachment *(continued)*

Electron energy, eV	Negative ion			Electron energy, eV	Negative ion		
	O^-	S^-	SO^-		O^-	S^-	SO^-
8.2	0.61	0.015	0.11	8.9	0.07	0.030	
8.3	0.50	0.013	0.05	9.0	0.04	0.027	
8.4	0.35	0.013	0.0	9.1	0.04	0.026	
8.5	0.27	0.014	0.01	9.2	0.02	0.017	
8.6	0.25	0.016	0.0	9.3	0.01	0.015	
8.7	0.14	0.026		9.5	0.01	0.002	
8.8	0.09	0.021					

Table 18.19 Rate constant (in 10^{-10} cm^3/s) of dissociative electron attachment to the molecules in an electric field [23–25]

Molecule	E/N, 10^{16} V·cm^2											
	0.2	0.5	1	5	10	15	20	25	30	40	50	60
SF_6				6.3	7	7.8	7.3	6.8	6.8	6.7	6.5	5.8
CF_4				0.39	0.68	0.68	0.64	0.5	0.46			
C_2F_6								2.7	2.6	1.5	0.65	
C_3F_8								3.0	2.4	1.8	1.0	
C_4F_{10}								3.3	2.3	1.0	0.6	
CCL_2F_2									4.6	4.6	3.9	2.5
H_2O				0.017	0.8	1.0	0.78					
N_2O	$6.5 \cdot 10^{-6}$	$1.2 \cdot 10^{-4}$	0.001									
HCl	1.6	2.6										
HBr	34											
HI	53	26										

Table 18.20 Rate constant (in 10^{-10} cm^3/s) of dissociative electron attachment to the molecules as a function of mean electron energy [23–25]

Molecule	Mean electron energy, eV							
	0.25	0.5	0.75	1	1.5	2	2.5	3
CCl_4	700	450	350					
$CHCl_3$	150	195	130	80				
CH_2Cl_2	35	90	120	110				
C_2HCl_3	40	80	95	70				
1-1-1-$C_2H_3Cl_3$	44	50	52	36				
1-1-2-$C_2H_3Cl_3$	16	35	40	30				
CH_3Br	0.48	0.2	0.1	0.022	0.0075	0.0016		
C_2H_5Br	0.19	0.65	0.84	0.63	0.34	0.16	0.075	
n-C_3H_7Br	0.23	0.67	1.0	0.65	0.32	0.17	0.085	
iso-C_4H_9Br	0.33	0.85	1.2	0.75	0.40	0.18	0.12	
iso-$C_5H_{16}Br$	0.45	1.0	1.4	0.8	0.42			
iso-$C_6H_{13}Br$	0.48	1.2	1.5	0.9	0.5	0.3	0.2	
cis-C_4F_6	320	130	68	40				
2-C_4F_6	370	420	400	300				
1,3-C_4F_6	570	300	190	135				
cis-C_4F_8	390	350	260	170				
2-C_4F_8	300	230	170	120				
cis-C_5F_8	1000	480	310	210				

Table 18.20 Rate constant of dissociative electron attachment to the molecules *(continued)*

Molecule	Mean electron energy, eV							
	0.25	0.5	0.75	1	1.5	2	2.5	3
cis-C_6F_{10}	980	480	300	200				
cis-C_6F_{12}	940	540	330	220				
C_7F_8	1000	540	330	220				
C_8F_{16}	930	540	330	220				
cis-C_7F_{14}	200	150	100	70				
cis-C_7F_{14}	1200	800	500	320				
CCl_3F	500	220	100					
CCl_2F_2	15	18	20	14				
$CClF_3$		0.027	0.16	0.3				
CO_2					0.0003	0.0011	0.009	
F	130	90	60	45	25	20	16	14
Br_2	0.65	1.3	1.5					
HBr	4	8.6	7.6	6.2	4.6			

Table 18.21 Rate constant of triple electron attachment to the molecules, $e+AB+M \rightarrow AB^- + M$, measured by various authors at $T = 300$ K [17, 37]

AB	M	Rate constant, 10^{-30} cm^6/s	Pressure range, torr	AB	M	Rate constant, 10^{-30} cm^6/s	Pressure range, torr
O_2	O_2	2.1	1–600	NO	NO	0.3	0.01–160
	He	0.046	1–425		N_2O	6	5–20
	Ne	0.023	90–300		CO_2	1.7	5–20
	Ar	0.05	100–300		NH_3	2.5	5–20
	Kr	0.05	60–200		C_2H_4	0.7	5–20
	Xe	0.085	35–170		C_6H_6	4.6	5–20
	N_2	0.21	1–300		CH_3OH	6.5	5–20
	H_2	0.48	100–300	N_2O	N_2O	0.005	5–300
	D_2	0.14	140–400		N_2	0.003	400–900
	H_2O	14	5–760		C_2H_6	0.0006	30–400
	CO_2	3.2	5–1900		C_3H_8	0.001	30–250
	H_2S	9	5–20		*n*-C_4H_{10}	0.0029	5–160
	NH_3	6.8	5–20		*i*-C_4H_{10}	0.0048	15–120
	CH_4	0.34	65–280		*n*-C_4H_8	0.019	20–140
	C_2H_4	2.6	5–850		*c*-2-C_4H_8	0.023	30–160
	C_2H_6	1.7	50–200		*t*-2-C_4H_8	0.014	40–170
	C_3H_8	3.3	50–240		*i*-C_4H_8	0.14	15–120
	C_6H_6	8.5	5–20		*n*-C_5H_{12}	0.013	10–100
	n-C_4H_{10}	5	100		He	1500	1–60
	n-C_5H_{12}	7.5	25–600		Ne	1000	1–60
	n-C_6H_{14}	8.1	30–240		Ar	1700	1–60
	CH_3OH	9.9	5–70		Kr	1100	1–60
	C_2H_5OH	18	10–50		Xe	1100	1–60
	CH_3COCH_3	27	5–20		H_2	1200	1–60
SO_2	CO_2	6	3–160		D_2	1200	1–60
	C_2H_4	55	8–30		N_2	1700	1–60
	C_2H_6	12	8–16		CO_2	1900	1–60
	CH_3OH	70	8–16	C_6H_6	N_2	0.001	2000–6000

Note: The uncertainties in data range from 10 to 30%.

Table 18.22 Rate constant (in 10^{-12} cm^3/s) for negative ion destruction due to collisions with atomic particles ($T = 300$ K) [17, 36]

Reaction	Energy defect, eV	Rate constant	Reaction	Energy defect, eV	Rate constant
$H^- + H \rightarrow H_2 + e$	3.8	1800	$CN^- + H \rightarrow HCN + e$	1.5	1000
$H^- + O_2 \rightarrow H_2O + e$	1.25	1200	$S^- + O_2 \rightarrow SO_2 + e$	3.8	30
$H^- + NO \rightarrow HNO + e$	1.4	500	$S^- + CO \rightarrow COS + e$	1.6	300
$H^- + CO \rightarrow HCO + e$	0.54	50	$C^- + CO \rightarrow C_2O + e$	2.0	400
$O^- + O \rightarrow O_2 + e$	3.6	200	$C^- + CO_2 \rightarrow 2CO + e$	4.3	500
$O^- + N \rightarrow NO + e$	5.1	200	$C^- + N_2O \rightarrow CO + N_2 + e$	8.2	900
$O^- + H_2 \rightarrow H_2O + e$	3.1	600	$C^- + N_2O \rightarrow CN + NO + e$	1.6	
$O^- + NO \rightarrow NO_2 + e$	1.4	200	$F^- + H \rightarrow HF + e$	2.5	1600
$O^- + O_3 \rightarrow 2O_2 + e$	2.6	300	$Cl^- + H \rightarrow HCl + e$	0.7	1000
$O^- + CO \rightarrow CO_2 + e$	4.0	600	$I^- + H \rightarrow HI + e$	~ 0	60
$O^- + SO_2 \rightarrow SO_3 + e$	2.1	2000	$Cl^- + N \rightarrow ClN + e$	0.7	10
$O^- + C_2H_4 \rightarrow C_2H_4O + e$	1.2	800	$Cl^- + O \rightarrow ClO + e$	0.9	10
$O^- + O_2(^1\Delta_g) \rightarrow O_3 + e$	0.5	300	$NO^- + He \rightarrow NO + He + e$	–0.02	0.27
$O_2^- + O_2 \rightarrow 2O_2 + e$	–0.43	$2.2 \cdot 10^{-6}$	$NO^- + He \rightarrow NO + He + e$	–0.02	0.035
		0.03 (600 K)	$NO^- + H_2 \rightarrow NO + H_2 + e$	–0.02	0.26
$O_2^- + N_2 \rightarrow O_2 + N_2 + e$	–0.43	0.00018	$NO^- + NO \rightarrow 2NO + e$	–0.02	6
		(600 K)	$NO^- + CO \rightarrow NO + CO + e$	–0.02	0.55
$O_2^- + O \rightarrow O_3 + e$	0.6	300	$NO^- + N_2O \rightarrow NO + N_2O + e$	–0.02	6.1
$O_2^- + N \rightarrow NO_2 + e$	4.0	400	$NO^- + CO_2 \rightarrow NO + CO_2 + e$	–0.02	9.5
$O_2^- + O_2(^1\Delta_g) \rightarrow 2O_2 + e$	0.6	200	$NO^- + NH_3 \rightarrow NO + NH_3 + e$	–0.02	22
$Cl^- + H \rightarrow HCl + e$	0.7	900	$CN^- + H \rightarrow HCN + e$	1.0	800
$OH^- + O \rightarrow HO_2 + e$	0.9	200	$HS^- + H \rightarrow H_2S + e$	1.7	1300
$OH^- + H \rightarrow H_2O + e$	3.2	1000			

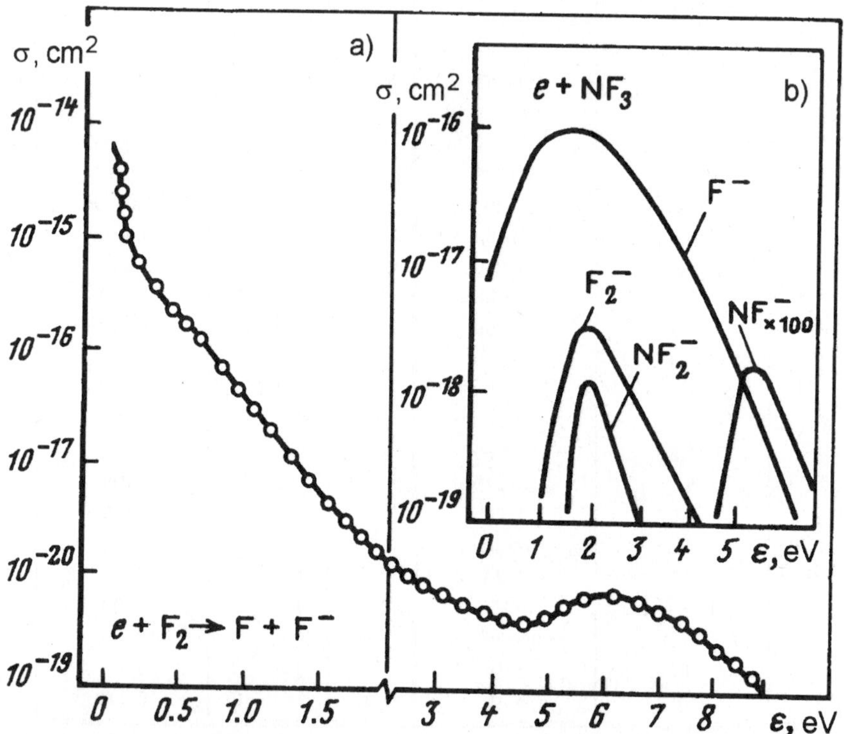

Figure 18.4 Cross section of dissociative electron attachment to molecules F_2 (*a*) and NF_3 (*b*) [23].

18.9 Production and Quenching Processes of Excited Atoms and Molecules in Collisions with Heavy Particles

Cross-sections and rate constants of such processes are presented in Tables 18.23–18.31. The uncertainties in the data involved amount to 50%.

Table 18.23 Cross-section (in 10^{-16} cm^2) of quenching the resonantly excited alkalis by molecules (T = 400–500 K) [14]

Quenching molecule	Na	K	K	K	Rb		Cs	
	^2P	$^2P_{1/2}$	$^2P_{3/2}$	^2P	$^2P_{1/2}$	$^2P_{3/2}$	$^2P_{1/2}$	$^2P_{3/2}$
N_2	39	35	39	26	47	40	66	67
H_2	17	7	4	7	6	3	7	5
HD	11	11	14	12	6	5	8	7
D_2	10	2	1	8	3	5	4	3
CO	88							

Table 18.24 Cross-section (in 10^{-16} cm^2) of quenching the resonantly excited atoms by molecules in flames (T = 1400–1800 K) [4]

Quenching molecule	Li(2^2P)	Na(3^2P)	K(4^2P)	Rb(5^2P)	Cs(6^2P)	Tl(7^2S)	Rb(7^2P_1)
H_2	16.3	8.6	3.3	3.0	5.2	0.2	1.3
N_2	12.2	22	19	22	48	20	18
O_2		36	52	84		41	47
CO	40	40	42	37		43	41
CO_2	29	51	67	75		102	91
H_2O	6	2	2.9	4.0	13	5.5	25

Table 18.25 Rate constant (in 10^{-11} cm^3/s) of quenching the excited inert gas atoms and molecules in collisions with atoms and molecules (T = 300 K) [14]

Quenching molecule	Ne($^3P_2$)	Ar($^3P_2$)	Ar($^3P_0$)	Kr($^3P_2$)	Xe($^3P_2$)	He$_2$($^3\Sigma_u^+$)	Ar($^3P_1$)	Ar($^1P_1$)
He*[1]								
Ne	0.00072					0.44		
Ar		0.00011	0.00057			3.1		
Kr							0.9	0.1
Xe					0.00023		22	33
H_2	5	8.7	7.8	3	1.6		20	25
D_2		6.5	7.8	2.5				
N_2	8.4	3.6	1.6	0.4	1.9	3.0		4.7
O_2		17	24	16	22		31	
CO	5.4	1.6	13	5,8	3.6	5.6		9.0
NO		20	25	19	27			56
Cl_2	41	71	72	73	72			
F_2	41	75	90	72	75			
HCl	17							
HI		75						
HBr		52						
CO_2		54	59	40	45	9.5	50	74

Table 18.25 Rate constant (in 10^{-11}cm^3/s) of quenching the excited inert gas atoms *(continued)*

Quenching molecule	$Ne(^3P_2)$	$Ar(^3P_2)$	$Ar(^3P_0)$	$Kr(^3P_2)$	$Xe(^3P_2)$	$He_2(^3\Sigma_u^+)$	$Ar(^3P_1)$	$Ar(^1P_1)$
N_2O	22	44	48	31	44		45	58
NF_3	7.8	14	7	12	9			
N_2F_4	3		33					
CH_4		50	55	37	33	6.1		
SF_6		27	17	18	23		63	65
C_2H_4							56	103

*1 The rate constant of process $He(2^1S) + He(2^1S_0) \rightarrow 2He(1^1S_0)$ equals $3.6{\cdot}10^{-15}$ cm^3/s.

Table 18.26 Rate constant (in 10^{-14} cm^3/s) of collisional quenching the metastable oxygen and nitrogen atoms and molecules (T = 300 K) [14, 38]

Quenching particle	$O_2(^1\Delta_g)$	$O_2(^1\Sigma_g^+)$	$O(^1S)$	$O(^1D)$	$N(^2D)$	$N(^2P)$	$N_2(A^3\Sigma_u^+)$
Ar	$8.3{\cdot}10^{-7}$	0.00058	0.039			0.07	
N_2	$3{\cdot}10^{-7}$	0.23		6100	1.5	6	
O_2	0.00019	0.02	30	5400	690	6	
H_2	0.00027	52	0.058	$2{\cdot}10^4$	290	0.19	0.27
D_2	0.004			$1.5{\cdot}10^4$			
NO	2500	5	$4{\cdot}10^4$	$1.7{\cdot}10^4$	9200	3300	6600
CO		0.3		7000	405	90	440
CO_2	$3.8{\cdot}10^{-5}$	19	30	1900	43	0.12	
N_2O			940	$1.3{\cdot}10^4$			
H_2O	0.00056	330	$4.2{\cdot}10^4$	$1.2{\cdot}10^4$			
O_3	0.47	1900	$6.9{\cdot}10^4$	$1.9{\cdot}10^4$			

Table 18.27 Cross-section σ_q of quenching the excited $Hg(6^3P_1)$ atoms colliding with atoms and molecules $T \approx 300$ K) [14]

Quenching particle	H_2	D_2	O_2	N_2	CO	CO_2	Xe
σ_q, 10^{-16} cm^2	22	22	57	0.9	19	9	0.002

Table 18.28 Quenching rate constant (in 10^{-10} cm^3/s) for some excited atoms, radicals, and molecules ($T \approx 300$ K) [14, 22]

Excited particle	Quenching particle										
	Ar	N_2	O_2	Cl_2	CO	CO_2	CF_4	SF_6	CCl_4	SiF_4	D_2
$Ar(4d\,[3/2])$	1.9	1.6	1.3	1.3	2.3	1.6	3.3	5.1	2.2	1.6	
$F(3p\ ^2P_{3/2})$	0.4	0.3	0.3	1.5	0.5	0.6	0.6	0.9	0.3	0.6	
$F(3p\ ^4D_{7/2})$	1.2	1.0	0.8	1.9	1.5	1.8	1.8	2.6	0.9	1.8	
$O(3p\ ^5P)$		2.0	1.7	3.7	3.0	3.5	4.2	4.8	3.1	3.1	
$CO(B^1\Sigma)$	1.3	1.1	0.9		1.7	1.8	2.5	3.3	3.1	2.0	
$N_2(C^3\Pi_u)$	0.4	0.1	0.3		0.6	0.6	0.9	1.1	1.0	0.7	
$CF_2(^1B_2)$		0.2					0.05				
$OH(A^2\Sigma+)$	0.22				3	1.3					
$CO(a^3\Pi)$	1.8	0.15	1.5		12.6	0.16					1.6
$NH(b^1\Sigma+)$	$2.5{\cdot}10^{-6}$	$4.5{\cdot}10^{-6}$	$2.4{\cdot}10^{-5}$								
$ND(b^1\Sigma+)$	$5{\cdot}10^{-8}$	$3.8{\cdot}10^{-7}$									
$CH(c^2\Sigma+)$		0.0007			0.048						

Table 18.28 Quenching rate constant for some excited atoms *(continued)*

Excited particle	Quenching particle										
	NO	O (^3P)	CH_4	C_2H_4	He	N (^4S)	H_2	NH_3	ND_3	H_2O	NO_2
Ar($4d[3/2]$)											
F($3p\ ^2P_{3/2}$)											
F($3p\ ^4D_{7/2}$)											
O($3p\ ^5P$)											
CO($B^1\Sigma$)											
$N_2(C^3\Pi_u)$											
$CF_2(^1B_2)$											
OH($A^2\Sigma^+$)							0.92			5	3.5
CO($a^3\Pi$)	2.4	1.9	3.1	5.9							
NH($b^1\Sigma^+$)		0.18			$7\cdot10^{-7}$	0.34	0.01	0.0039			
ND($b^1\Sigma^+$)					$5\cdot10^{-8}$		0.009		$5.2\cdot10^{-4}$		
CH($c^2\Sigma^+$)										0.45	

Table 18.29 Rate constant (in 10^{-10} cm^3/s) of excimer molecules production in the bimolecular reaction of substitution ($T \approx 300$ K) [26]

Reaction	Rate constant	Reaction	Rate constant
$Ar(^3P_2) + F_2 \to ArF^* + F$	9.0	$Kr(^3P_2) + Cl_2 \to KrCl^* + Cl$	7.3
$Ar(^3P_2) + NF_3 \to ArF^* + NF_2$	1.0	$Xe(^3P_2) + F_2 \to XeF^* + F$	7.5
$Ar(^3P_2) + Cl_2 \to ArCl^* + Cl$	7.1	$Xe(^3P_2) + NF_3 \to XeF^* + NF_2$	0.9
$Kr(^3P_2) + F_2 \to KrF^* + F$	6.2	$Xe(^3P_2) + OF_2 \to XeF^* + OF$	5.7
$Kr(^3P_2) + NF_3 \to KrF^* + NF_2$	1.0	$Xe(^3P_2) + Cl_2 \to XeCl^* + Cl$	7.2
$Kr(^3P_2) + OF_2 \to KrF^* + OF$	5.3	$Xe(^3P_2) + Br_2 \to XeBr^* + Br$	10

Table 18.30 Rate constant k (10^{-33} cm^6/s) of excimer molecule production in triple collisions: $A^* + B + C \to AB^* + C$ ($T \approx 300$ K) [26]

A*	B	C	Excimer molecule	k	A*	B	C	Excimer molecule	k
He(2^3S)	He	He	He_2^*	0.23	Hg($^3P_0$)	Hg	Hg	Hg_2^*	250 (470 K)
Ne($^3P_2$)	Ne	Ne	Ne_2^*	0.5	Hg($^3P_1$)	Hg	Hg	Hg_2^*	160
Ne($^3P_0$)	Ne	Ne	Ne_2^*	0.07 (77 K)	Cs(6^2P)	Cs	Xe	Cs_2^*	4200 (620 K)
Ne($^1P_1$)	Ne	Ne	Ne_2^*	5.8	Xe($^3P_2$)	Xe	He	Xe_2^*	160 (670 K)
Ar($^3P_2$)	Ar	Ar	Ar_2^*	10	Xe($^3P_2$)	Xe	Ne	Xe_2^*	14
Ar($^1P_1$)	Ar	Ar	Ar_2^*	14	Xe($^3P_2$)	Xe	Ar	Xe_2^*	26
Ar($^3P_1$)	Ar	Ar	Ar_2^*	12	Xe($^3P_2$)	Ar	Ar	XeAr*	0.7
Kr($^3P_2$)	Kr	Kr	Kr_2^*	36	Hg($^3P_0$)	Hg	N_2	Hg_2^*	1000 (430 K)
Kr($^3P_2$)	Ar	Ar	KrAr*	1.0	Hg(O_g^+)	Hg	N_2	Hg_3^*	200
Kr($^3P_0$)	K	Kr	Kr_2^*	54	ArF($B_{1/2}$)	Ar	Ar	Ar_2F^*	490
Kr($^3P_1$)	Kr	Kr	Kr_2^*	30	KrF($B_{1/2}$)	Ar	Ar	ArKrF*	90
Kr($^1P_1$)	Kr	Kr	Kr_2^*	1.6	KrF($B_{1/2}$)	Kr	Ar	Kr_2F^*	600
Xe($^3P_2$)	Xe	Xe	Xe_2^*	55	KrF($B_{1/2}$)	Kr	Kr	Kr_2F^*	600
Xe($^3P_0$)	Xe	Xe	Xe_2^*	40	XeF($B_{1/2}$)	Xe	Ne	Xe_2F^*	780
Xe($^3P_1$)	Xe	Xe	Xe_2^*	70	XeF($B_{1/2}$)	Xe	Xe	Xe_2F^*	260

Table 18.31 Rate constant k (in 10^{-12} cm^3/s) of quenching the excimer molecules ($T \approx 300$ K) [26]

Excimer molecule	Quenching particle	k	Excimer molecule	Quenching particle	k	Excimer molecule	Quenching particle	k
$ArF(B_{1/2})$	Ar	9	$XeCl(B_{1/2})$	Ne	1.0	$HgBr(B_{1/2})$	He	0.044
	Kr	1600		Xe	32		Ar	0.072
	Xe	4500		HCl	1400		Xe	0.31
	F_2	1900	$XeBr(B_{1/2})$	Br_2	800		N_2	0.13
$XeF(B_{1/2})$	He	1.2	$XeI(B_{1/2})$	Xe	9		Br_2	290
	Ne	0.77		I_2	500		HBr	130
	Ar	2.7		CH_3I	360		CF_3Br	87
	Xe	45	$HgCl(B_{1/2})$	He	0.041		CCl_3Br	180
	N_2	7.0		Ne	0.033	$HgI(B_{1/2})$	Ar	0.11
	F_2	470		Ar	0.05		Xe	0.22
	CO_2	250		Kr	0.073		CF_3I	290
	XeF	305		Xe	0.31	$HgI(C_{3/2})$	He	3.8
	NF_3	18		N_2	0.061		Ne	5.1
$XeF(C_{3/2})$	He	1.12		Cl_2	170		Ar	16
	Ne	0.3		CCl_4	160		Xe	41
	Ar	0.09	$KrF(B_{1/2})$	Ar	5		N_2	24
	Xe	1.0		Kr	3.6	$Ar_2F(^2B_2)$	He	0.005
	N_2	0.4		F_2	650		Ar	0.022
	F_2	80		KrF	370		F_2	210
	XeF_2	240		NF_3	52		NF_3	300
	NF_3	16				$Kr_2F(^2B_2)$	F_2	880

18.10 Photoionization and Photorecombination

In the case of the photoionization process

$$A + \hbar\omega \rightarrow A^+ + e, \tag{18.18}$$

the energy required for removing the electron from the atom or molecule is supplied by the radiation. Photoionization cross sections of atoms in the vicinity of the threshold are energy-independent; therefore, the threshold photoionization cross section serves as the main parameter of this process applied in describing phenomena in a low-temperature plasma. The dependence of the photoionization cross section of the helium atom upon the wavelength of the incident photon [28] is shown in Fig. 18.5. Threshold values of photoionization cross sections for certain atoms [29] are presented in Table 18.32.

Table 18.32 Threshold values of the photoionization cross-section for some atoms [28]

Atom	Threshold wavelength, nm	Cross-section, 10^{-18} cm^2	Atom	Threshold wavelength, nm	Cross-section, 10^{-18} cm^2
H	91.2	6.3	Mg	162	1.2
He	50.4	7.4	Ar	78.7	35
Li	230	2.5	K	286.0	0.012
Be	133	8.2	Ca	202.8	0.45
B	149	19	Ga	207	0.2
C	110	11	Kr	84.5	35
N	85.2	9	Rb	297	0.11
O	91	2.6	In	214	0.3
F	71.3	6	Cs	318.5	0.22
Ne	57.5	4	Tl	203	4.5
Na	241.2	0.12			

The cross-section of the hydrogen atom photoionization by photons of energies much higher than the binding energy of the electron in the atom (13.6 eV) is given (in cm^2) by the following expression [31]:

$$\sigma_{\mathrm{phi}} = 23.8\lambda^{7/2}, \tag{18.19}$$

where λ is the wavelength of the incident photon in cm. The photoionization cross section of a highly excited atom (in cm^2) with an effective value of the principal quantum number n^*, is determined by the Kramers formula [30]:

$$\sigma_{\mathrm{phi}} = 0.022\lambda^3/n^{*5}, \tag{18.20}$$

The inverse process of photoionization (18.18) is referred to as photorecombination. The cross-section (in cm^2) of an electron and an ion photorecombination into a highly excited atomic state with an effective value of the principal quantum number n^* is described by a relation following from the Kramers formula (18.20), i.e.

$$\sigma_{\mathrm{phr}} = \frac{2 \cdot 10^{-16}\lambda}{\varepsilon n^{*3}}, \tag{18.21}$$

where ϵ is the electron energy (in eV). The temperature dependence of the coefficient of an electron and a proton photorecombination in hydrogen plasma [32] is presented in Fig. 18.6.

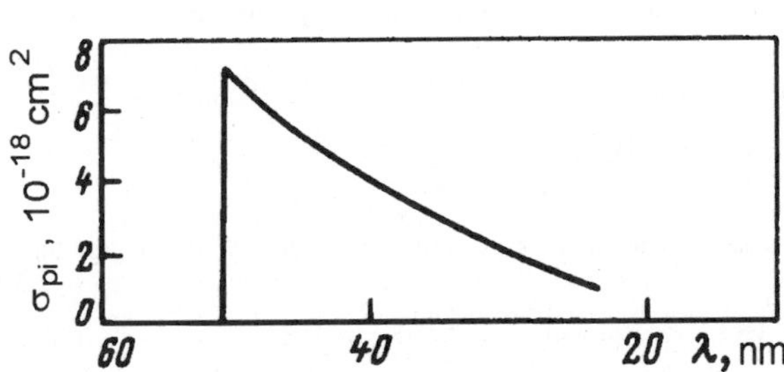

Figure 18.5 Dependence of photoionization cross section of the helium atom on the wavelength of the incident radiation [28].

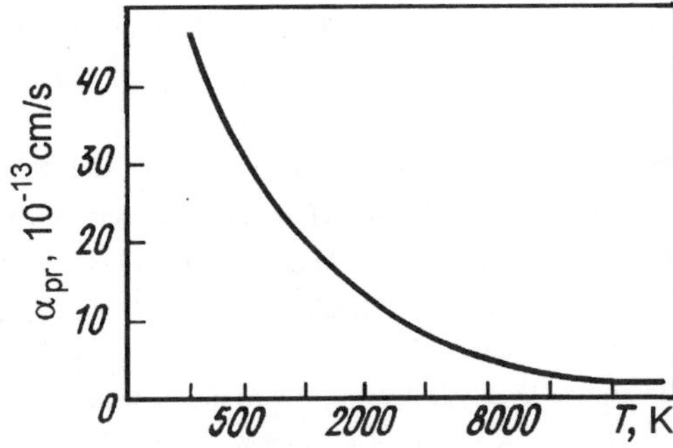

Figure 18.6 Temperature dependence of the photorecombination coefficient of an electron and a proton in hydrogen plasma [32].

References

[1] Eletskii, A. V., Palkina, L. A., Smirnov, B. M., Transport phenomena in weakly ionized plasma, Atomizdat, Moscow, 1975 (in Russian).

[2] Eletskii, A. V., Uspekhi fizicheskikh nauk, 134, 237, 1981 (in Russian).

[3] Rich, J. W., Appl. atomic collision physics, Vol. 3: Gas lasers, Academic Press, New York, 1982, p. 99.

[4] Andrews, A. J., Simpson, C. J. S. M., Chem. Phys. Lett., 36, 271, 1975.

[5] Drosdoski, W. S. e.a., J. Chem. Phys., 65, 1542, 1976.

[6] Bohringer, H. e.a., J. Chem. Phys., 79, 420, 1983.

[7] Wilkins, R. L., J. Chem. Phys., 63, 534, 1975.

[8] MacDonald, R. G., Moore, C. B., J. Chem. Phys., 73, 1681, 1980.

[9] Wilkins, R. L., Kwok, M. A., J. Chem. Phys., 73, 3198, 1980.

[10] Dzelkalns, L. S., Kaufman, F., J. Chem. Phys., 79, 3836, 1983.

[11] Smirnov, B. M., Uspekhi fizicheskikh nauk, 133, 569, 1981.

[12] Eletskii, A. V., Smirnov, B. M., Fundamentals of plasma physics, Eds. Sagdeev, R. Z., Rozenblut, M., Energoatomizdat, Moscow, 1983 (Vol. 1) (in Russian).

[13] Klyucharev, A. N., Bezyglov, N. N., Excitation processes and atomic ionization on light absorption, Izdatel'stvo LGU, Leningrad, 1983 (in Russian).

[14] Smirnov, B. M., Excited atoms, Energoatomizdat, Moscow, 1982 (in Russian).

[15] Eletskii, A. V., Smirnov, B. M., Uspekhi fizicheskikh nauk, 136, 25, 1982 (in Russian).

[16] Mitchel, J. B. A., McGowan, P. W., Physics of ion–ion and electron–ion collisions, Plenum Press, New York, 1983 (p. 279).

[17] Smirnov, B. M., Negative ions, Atomizdat, Moscow, 1978 (in Russian); Smirnov, B. M., Asymptotic methods in the theory of atomic collisions, Atomizdat, Moscow, 1978 (in Russian).

[18] Smith, D., Adams, N. G., Physics of ion–ion and electron–ion collisions, Plenum Press, New York – London, 1983 (p. 501).

[19] Flannery, M. R., Appl. atomic collision physics, Vol. 3: Gas lasers, Academic Press, New York, 1982, p. 141.

[20] Duman, E. L., Preprint of the IAE No 3532/12, Moscow, 1982 (in Russian).

[21] Smirnov, B. M., Uspekhi fizicheskikh nauk, 121, 231, 1977 (in Russian).

[22] Slovetskii, D. I., in: Plasma chemistry, Vol. 10, Ed. Smirnov, B. M., Energoatomizdat, Moscow, 1983 (in Russian).

[23] Chantry, P. J., Appl. atomic collision physics, Vol. 3: Gas lasers, Academic Press, New York, 1982, p. 35.

[24] Orient, O. J., Srivastava, S. K., J. Chem. Phys., 78, 2949, 1983.

[25] Gallagher, J. W., J. Phys. Chem. Ref. Data, 12, 109, 1983.

[26] Smirnov, B. M., Uspekhi fizicheskikh nauk, 139, 53, 1983 (in Russian).

[27] Huxley, L. G. H., Crompton, R. W., The diffusion and drift of electrons in gases, John Wiley and Sons, New York, 1974.

[28] Weissler, G. L., Handbuch der physik, XXI, Springer, Berlin, 1956.

[29] Hasted, J. B., Physics of atomic collisions, Butterworth, London, 1964.

[30] Kramers, H. A., Phil. Mag., 46, 836, 1923.

[31] Berestetskii, V. B., Lifshitz, E. M., Pitaevskii, L. P., Quantum theory of relativity, Part I, Nauka, Moscow, 1968 (in Russian).

[32] Bates, D. R., Dalgarno, A., Atomic and molecular processes, Ed. Bates, D. R., Academic Press, New York, 1962.

[33] Osipov, A. I., in: Plasma chemistry, Vol. 16, Ed. Smirnov, B. M., Energoatomizdat, Moscow, 1990, p. 3 (in Russian).

[34] Nikitin, E. E., Osipov, A. I., and Umanskii, S. Ya., in: Plasma chemistry, Vol. 15, Ed. Smirnov, B. M., Energoatomizdat, Moscow, 1990, p. 3 (in Russian).

[35] Klyucharev, A. N., Yanson, M. L., Elementary processes in alkali metals plasma, Energoatomizdat, Moscow, 1988 (in Russian).

[36] Lifschitz, C., e.a., J. Chem. Sol., 67, 2381, 1977.

[37] Aleksandrov, N. L., Uspekhi fizicheskikh nauk, 154, 177, 1988 (in Russian).

[38] Dvoryankin, A. N., e.a., in: Plasma chemistry, Vol. 14, Ed. Smirnov, B. M., Energoatomizdat, Moscow, 1987, p. 102 (in Russian).

[39] Beterov, I. M., Fateev, N. V., in: Plasma chemistry, Vol. 13, Ed. Smirnov, B. M., Energoatomizdat, Moscow, 1987, p. 40 (in Russian).

19

Atomic and Molecular Ionization

A.A. Radzig and V.M. Shustryakov

19.1 Introduction

The set of processes spanning the ionization phenomena in substances is rather widespread and varied. The ionization of atoms and molecules under the action of incident charged and neutral particles, photons, as well as on exposure to external fields may be of chief interest in one or another case being considered. Not attempting to cover the real diversity of the elementary processes, we will discuss below adequate information concerning the most important threshold parameter of the ionization process, i.e. the ionization potential of atoms, atomic ions, and molecules. Furthermore, the electron impact ionization cross-section for atomic and molecular species will allow the observation of the most common process of electron and ion formation in plasma media.

In addition, the binding energies of electrons in negative ions, along with the energies released on the addition of a proton to atomic and molecular species, are presented. The binding energies of electrons in the atomic inner-shells are also covered comprehensively.

It should be noted that atomic and molecular photoionization processes, as well as the Penning and associative ionization of atomic species at collisions with excited atoms, are touched upon in the preceding Chapter 18.

19.2 Ionization Potentials of Atoms and Molecules

The minimum energy expended in the transition of particle valence electron into the continuous spectrum is referred to as the ionization potential of the particle in study. Table 19.1 shows the ionization potentials of neutral atoms, which were found by extrapolating the optical transition series to the edge of a continuous spectrum. The associated electron transitions were initiated through the use of different excitation sources. The optical ionization potentials can be determined experimentally either by assuming an appropriate function (e.g., the Ritz formula) for the low l (l is the atomic electron quantum number of orbital angular momentum) energy levels of a long Rydberg series and extrapolating it to the infinite principal quantum number n of the electron, or by considering the high n and l energy levels whose binding energies are so close to the hydrogen value that the difference can be expressed entirely in terms of atomic core polarization [1]. Thus, apart from the atomic ionization potentials (in eV), the first ionization limit (in cm^{-1}) for a series of optical transitions with respect to the atomic ground level is also given. The data derived from the optical spectra are the most accurate and reliable.

0-8493-2861-6/97/$0.00+$.50

In addition, part of the numerical data was obtained by the techniques of photoionization-threshold measurement, photoelectron spectroscopy, electron-impact appearance potential measurement and surface ionization. For hydrogen and its isotopes, the theoretical values accounting for many recoil, relativistic and quantum-electrodynamic effects should be used in preference to experimental results. In a separate column in Table 19.1, the configurations of the valence electron shell and the ground state electronic terms are specified. Basic information on the atomic ionization potentials can be found in [2–6].

The ionization potentials of light and intermediate atomic ions are compiled in Table 19.2. These data feature all the stages of ionization for ions with nuclear charge $Z \leq 36$ and are of particular interest for high-temperature plasma physics. Most of the numerical data presented for the low degrees of atomic ionization were based on the observable spectra of optical transitions between highly excited ion levels. Whereas, in the case of multiply-ionized ions, the various procedures of potential extrapolation along the isoelectronic series were employed [2,5,6]. Table 19.3 lists the ionization potentials of single-, double-, and triple-charged atomic ions with $37 \leq Z \leq 92$, which were found using the limits of line convergence in the optical spectra [2,3,5,6]. The values listed in Tables 19.1—19.3 are truncated to the point where the estimated uncertainty in their determination is at most $\pm$ 1 in the last significant figure quoted.

Table 19.4 presents information on the ionization potentials (IP) of molecules. In this case, the minimum energy is correlated with the transition energy for the zero vibrational levels of ground electronic states of the molecule and the molecular ion formed. This energy can be termed as an adiabatic ionization potential of the molecule. Among the main experimental techniques of ionization potential determination in molecules, we shall highlight the spectroscopic method, and photoionization and electron impact ionization. The molecular spectroscopy procedure is based on finding the Rydberg series limits in molecular band spectra.

To give an indication of the experimental errors for the molecular IPs, we assigned numerical values to four accuracy classes: A– for uncertainties within 1%; B– for uncertainties within 3%; C– for uncertainties within 10%, and, finally, D– for uncertainties within 30%. The classification procedure relied on accuracy estimation in determining the ionization potentials of molecules. Data compiled in Table 19.4 were taken from numerous journal publications of the last decade as well as the monographs [7,8].

Table 19.1 Ionization potentials of neutral atoms

Atomic number Z	Atom (valence electron configuration – ground state term)	Optical limit, cm^{-1}; ionization potential, eV	Atomic number Z	Atom (valence electron configuration – ground state term)	Optical limit, cm^{-1}; ionization potential, eV
1	H	109678.7717	5	B	66928.1
	$(1s - {}^2S_{1/2})$	13.59844		$(2p - {}^2P^{\circ}_{1/2})$	8.29803
	D	109708.6146	6	C	90820.4
	$(1s - {}^2S_{1/2})$	13.60214		$(2p^2 - {}^3P_0)$	11.26030
	T	109718.5439	7	N	117225.7
	$(1s - {}^2S_{1/2})$	13.60337		$(2p^3 - {}^4S^{\circ}_{3/2})$	14.53413
2	^{4}He	198310.772	8	O	109837.0
	$(1s^2 - {}^1S_0)$	24.58741		$(2p^4 - {}^3P_2)$	13.61806
	^{3}He	198300.3	9	F	140524
	$(1s^2 - {}^1S_0)$	24.5861		$(2p^5 - {}^2P^{\circ}_{3/2})$	17.4228
3	Li	43487.15	10	Ne	173929.7
	$(2s - {}^2S_{1/2})$	5.39172		$(2p^6 - {}^1S_0)$	21.56455
4	Be	75192.6	11	Na	41449.451
	$(2s^2 - {}^1S_0)$	9.32270		$(3s - {}^2S_{1/2})$	5.139079

Table 19.1 Ionization potentials of neutral atoms *(continued)*

Atomic number Z	Atom (valence electron configuration – ground state term)	Optical limit, cm^{-1}; ionization potential, eV	Atomic number Z	Atom (valence electron configuration – ground state term)	Optical limit, cm^{-1}; ionization potential, eV
12	Mg	61671.05	33	As	78950
	$(3s^2 - {}^1S_0)$	7.64623		$(4p^3 - {}^4S^o_{3/2})$	9.7886
13	Al	48278.48	34	Se	78658.2
	$(3p - {}^2P^o_{1/2})$	5.985771		$(4p^4 - {}^3P_2)$	9.75238
14	Si	65747.8	35	Br	95284.8
	$(3p^2 - {}^3P_0)$	8.1517		$(4p^5 - {}^2P^o_{3/2})$	11.8138
15	P	84580.8	36	Kr	112914.4
	$(3p^3 - {}^4S^o_{3/2})$	10.48669		$(4p^6 - {}^1S_0)$	13.99961
16	S	83559	37	Rb	33690.881
	$(3p^4 - {}^3P_2)$	10.3600		$(5s - {}^2S_{1/2})$	4.177138
17	Cl	104591	38	Sr	45932.20
	$(3p^5 - {}^2P^o_{3/2})$	12.9676		$(5s^2 - {}^1S_0)$	5.69487
18	Ar	127109.88	39	Y	50144
	$(3p^6 - {}^1S_0)$	15.75962		$(4d5s^2 - {}^2D_{3/2})$	6.2171
19	K	35009.814	40	Zr	53506
	$(4s - {}^2S_{1/2})$	4.340665		$(4d^25s^2 - {}^3F_2)$	6.6339
20	Ca	49306.0	41	Nb	54514
	$(4s^2 - {}^1S_0)$	6.11316		$(4d^45s - {}^6D_{1/2})$	6.7589
21	Sc	52922	42	Mo	57204
	$(3d4s^2 - {}^2D_{3/2})$	6.5615		$(4d^55s - {}^7S_3)$	7.0924
22	Ti	55073	43	Tc	58700
	$(3d^24s^2 - {}^3F_2)$	6.8281		$(4d^55s^2 - {}^6S_{5/2})$	7.28
23	V	54413	44	Ru	59366
	$(3d^34s^2 - {}^4F_{3/2})$	6.7463		$(4d^7 - {}^5F_5)$	7.3605
24	Cr	54570	45	Rh	60160
	$(3d^54s - {}^7S_3)$	6.766		$(4d^85s - {}^4F_{9/2})$	7.4589
25	Mn	59959.4	46	Pd	67241.6
	$(3d^54s^2 - {}^6S_{5/2})$	7.43402		$(4d^{10} - {}^1S_0)$	8.3369
26	Fe	63737	47	Ag	61106.6
	$(3d^64s^2 - {}^5D_4)$	7.9024		$(5s - {}^2S_{1/2})$	7.57625
27	Co	63565	48	Cd	72540.1
	$(3d^74s^2 - {}^4F_{9/2})$	7.8810		$(5s^2 - {}^1S_0)$	8.99383
28	Ni	61619	49	In	46670.11
	$(3d^84s^2 - {}^3F_4)$	7.6398		$(5s^25p - {}^2P^o_{1/2})$	5.786358
29	Cu	62317.4	50	Sn	59232.7
	$(4s - {}^2S_{1/2})$	7.72638		$(5p^2 - {}^3P_0)$	7.34392
30	Zn	75769.3	51	Sb	69431.4
	$(4s^2 - {}^1S_0)$	9.39420		$(5p^3 - {}^4S^o_{3/2})$	8.60840
31	Ga	48387.63	52	Te	72670
	$(4p - {}^2P^o_{1/2})$	5.999304		$(5p^4 - {}^3P_2)$	9.010
32	Ge	63713.2	53	I	84295.1
	$(4p^2 - {}^3P_0)$	7.89944		$(5p^5 - {}^2P^o_{3/2})$	10.45125

Table 19.1 Ionization potentials of neutral atoms *(continued)*

Atomic number Z	Atom (valence electron configuration – ground state term)	Optical limit, cm^{-1}; ionization potential, eV	Atomic number Z	Atom (valence electron configuration – ground state term)	Optical limit, cm^{-1}; ionization potential, eV
54	Xe	97833.77	75	Re	64000
	$(5p^6 - {}^1S^0)$	12.12985		$(5d^56s^2 - {}^6S_{5/2})$	7.9
55	Cs	31406.4667	76	Os	70450
	$(6s - {}^2S_{1/2})$	3.893907		$(5d^66s^2 - {}^5D_4)$	8.7
56	Ba	42034.90	77	Ir	73000
	$(6s^2 - {}^1S_0)$	5.211665		$(5d^76s^2 - {}^4F_9)$	9.1
57	La	44980	78	Pt	72300
	$(5d6s^2 - {}^2D_{3/2})$	5.577		$(5d^96s - {}^3D_3)$	9.0
58	Ce	44670	79	Au	74409.7
	$(4f5d6s^2 - {}^1G^o_4)$	5.539		$(5d^{10}6s - {}^2S_{1/2})$	9.22563
59	Pr	44100	80	Hg	84185.0
	$(4f^36s^2 - {}^4I^o_{9/2})$	5.46		$(6s^2 - {}^1S_0)$	10.4376
60	Nd	44560	81	Tl	49266.7
	$(4f^46s^2 - {}^5I_4)$	5.525		$(6s^26p - {}^2P^o_{1/2})$	6.10829
61	Pm	45000	82	Pb	59819.6
	$(4f^56s^2 - {}^6H^o_{5/2})$	5.55		$(6p^2 - {}^3P_0)$	7.41668
62	Sm	45520	83	Bi	58761.7
	$(4f^66s^2 - {}^7F_0)$	5.644		$(6p^3 - {}^4S^o_{3/2})$	7.28552
63	Eu	45734.9	84	Po	67885.3
	$(4f^76s^2 - {}^8S^o_{7/2})$	5.6704		$(6p^4 - {}^3P_2)$	8.41671
64	Gd	49603	85	At	
	$(4f^75d6s^2 - {}^9D^o_2)$	6.150		$(6p^5 - {}^2P^o_{3/2})$	9.0
65	Tb	47300	86	Rn	86692
	$(4f^96s^2 - {}^6H^o_{15/2})$	5.864		$(6p^6 - {}^1S_0)$	10.749
66	Dy	47900	87	^{212}Fr	32848.88
	$(4f^{10}6s^2 - {}^5I_8)$	5.939		$(7s - {}^2S_{1/2})$	4.072743
67	Ho	48570	88	Ra	42573.4
	$(4f^{11}6s^2 - {}^4I^o_{15/2})$	6.022		$(7s^2 - {}^1S_0)$	5.27843
68	Er	49260	89	Ac	42000
	$(4f^{12}6s^2 - {}^3H_6)$	6.108		$(6d7s^2 - {}^2D_{3/2})$	5.2
69	Tm	49880	90	Th	49000
	$(4f^{13}6s^2 - {}^2F^o_{7/2})$	6.1843		$(6d^27s^2 - {}^3F_2)$	6.1
70	Yb	50441	91	Pa	47000
	$(4f^{14}6s^2 - {}^1S_0)$	6.2539		$(5f^26d7s^2 - {}^4K_{11/2})$	5.9
71	Lu	43762.6	92	U	49960
	$(4f^{14}5d6s^2 - {}^2D_{3/2})$	5.42587		$(5f^36d7s^2 - {}^5L^o_6)$	6.194
72	Hf	55048	93	Np	50540
	$(5d^26s^2 - {}^3F_2)$	6.8251		$(5f^46d7s^2 - {}^6L_{11/2})$	6.266
73	Ta	63600	94	Pu	49000
	$(5d^36s^2 - {}^4F_{3/2})$	7.89		$(5f^67s^2 - {}^7F_0)$	6.06
74	W	64000	95	Am	48300
	$(5d^46s^2 - {}^5D_0)$	8.0		$(5f^77s^2 - {}^8S^o_{7/2})$	5.99

Table 19.1 Ionization potentials of neutral atoms *(continued)*

Atomic number Z	Atom (valence electron configuration – ground state term)	Optical limit, cm^{-1}; ionization potential, eV	Atomic number Z	Atom (valence electron configuration – ground state term)	Optical limit, cm^{-1}; ionization potential, eV
96	Cm	48600	100	Fm	52000
	$(5f^7 6d7s^2 - {}^9D_2^o)$	6.02		$(5f^{12}7s^2 - {}^3H_6)$	6.5
97	Bk	50200	101	Md	53000
	$(5f^9 7s^2 - {}^6H_{15/2}^o)$	6.23		$(5f^{13}7s^2 - {}^2F_{7/2}^o)$	6.6
98	Cf	50800	102	No	54000
	$(5f^{10}7s^2 - {}^5I_8)$	6.30		$(5f^{14}7s^2 - {}^1S_0)$	6.6
99	Es	51800			
	$(5f^{11}7s^2 - {}^4I_{15/2}^o)$	6.42			

Table 19.2 Ionization potentials (in eV) of atomic ions ($2 \le Z \le 36$) [Xξ is a spectroscopic symbol of the ion, where $\xi = Z - N + 1$, and N is the number of ion electrons.]

Atomic number Z	Element X	ξ: II	III	IV	V	VI	VII	VIII
2	He	54.418						
3	Li	75.641	122.45					
4	Be	18.211	153.90	217.72				
5	B	25.155	37.931	259.38	340.23			
6	C	24.384	47.89	64.49	392.09	490.00		
7	N	29.602	47.45	77.47	97.891	552.07	667.05	
8	O	35.1211	54.936	77.414	113.90	138.12	739.34	871.42
9	F	34.971	62.71	87.14	114.24	157.164	185.19	953.91
10	Ne	40.963	63.46	97.12	126.2	157.9	207.28	239.10
11	Na	47.287	71.620	98.92	138.40	172.2	208.5	264.2
12	Mg	15.03528	80.144	109.265	141.27	186.5	224.9	266.0
13	Al	18.829	28.448	119.99	153.83	198.48	241.44	284.60
14	Si	16.346	33.493	45.142	166.77	205.3	246.5	303.5
15	P	19.770	30.2026	51.444	65.026	220.42	263.6	309.6
16	S	23.338	34.83	47.305	72.7	88.05	280.9	328.2
17	Cl	23.814	39.61	53.47	67.8	97.0	114.20	348.9
18	Ar	27.630	40.911	59.81	75.0	91.01	124.32	143.46
19	K	31.63	45.81	60.91	82.7	99.4	117.6	154.7
20	Ca	11.872	50.913	67.3	84.5	108.8	127.2	147.2
21	Sc	12.800	24.757	73.49	91.7	110.7	138.0	158.1
22	Ti	13.58	27.49	43.27	99.30	119.53	140.8	170.4
23	V	14.66	29.31	46.71	65.28	128.13	150.6	173.4
24	Cr	16.50	31.0	49.2	69.5	90.64	160.2	184.7
25	Mn	15.640	33.67	51.2	72.4	95	119.3	194.5
26	Fe	16.188	30.65	54.8	75	99	124.88	151.06
27	Co	17.084	33.5	51.3	79.5	102	129	158
28	Ni	18.169	35.3	54.9	76	108	133	162
29	Cu	20.293	36.84	57.4	80	103	139	166
30	Zn	17.964	39.72	59.57	83	108	134	174
31	Ga	20.515	30.726	64.2	90	117	147	179
32	Ge	15.935	34.2	45.715	93.5	120	151	183
33	As	18.59	28.4	50.1	62.6	128	154	187
34	Se	21.16	30.82	42.95	68	81.7	155.33	191
35	Br	21.81	35.90	47.3	59.7	88.6	103	192.8
36	Kr	24.3598	36.95	52.5	64.7	78.5	111	125.80

Table 19.2 Ionization potentials (in eV) of atomic ions ($2 \leq Z \leq 36$) *(continued)*

Atomic number Z	Element X	ξ IX	X	XI	XII	XIII	XIV	XV
9	F	1103.13						
10	Ne	1195.8	1362.21					
11	Na	299.87	1465.1	1648.71				
12	Mg	328.2	367.5	1761.8	1962.68			
13	Al	330.1	399.4	442.0	2086.0	2304.2		
14	Si	351.1	401.4	476.4	523.4	2437.7	2673.2	
15	P	372.1	424.4	479.5	560.8	611.7	2816.9	3069.87
16	S	379.1	447.1	504.8	564.7	651.6	707.2	3223.9
17	Cl	400.1	455.6	529.3	592.0	656.7	749.8	809.4
18	Ar	422.4	478.7	539.0	618.3	686.1	755.7	854.8
19	K	175.82	504	564.7	629.4	714.6	786.6	861.1
20	Ca	188.3	211.28	591.9	657.2	726.6	817.6	894.5
21	Sc	180.0	225.1	249.84	687.4	756.7	830.8	927.5
22	Ti	192.1	215.92	265	291.49	787.8	863.1	941.9
23	V	205.8	230.5	256	308	336.28	896.0	976
24	Cr	209.3	244.4	271	298	354.8	384.2	1011
25	Mn	221.8	248.3	286.0	314.4	343.6	403.0	435.6
26	Fe	234	262.1	290	331	361	392	457
27	Co	186.1	275	305	336	379	411	444
28	Ni	193	225	321	352	384	430	464
29	Cu	199	232	266	369	401	435	484
30	Zn	203	238	274	311	420	454	490
31	Ga	214	248	284	321	358	475	510
32	Ge	217	255	291	330	369	409	533
33	As	222	259	300	338	379	421	462
34	Se	227	264	304	347	388	431	475
35	Br	232	270	310	352	398	441	486
36	Kr	230.9	275	316	358	403	451	497

Atomic number Z	Element X	ξ XVI	XVII	XVIII	XIX	XX	XXI	XXII
16	S	3494.2						
17	Cl	3658.4	3946.3					
18	Ar	918.0	4120.7	4426.2				
19	K	968	1033.4	4610.9	4934.1			
20	Ca	974	1087	1157	5128.9	5469.9		
21	Sc	1009	1094	1213	1288.0	5674.9	6033.8	
22	Ti	1044	1131	1221	1346	1426	6249.1	6625.8
23	V	1060	1168	1260	1355	1486	1569.6	6851.29
24	Cr	1097	1185	1299	1396	1496	1634	1721
25	Mn	1133	1244	1317	1437	1539	1644	1788
26	Fe	489.3	1262	1360	1470	1582	1690	1800
27	Co	512.0	546.6	1397	1505	1603	1735	1846
28	Ni	499	571.3	607.0	1541	1648	1756	1894
29	Cu	520	557	633	671	1690	1793	1905
30	Zn	542	579	619	698	737	1846	1953
31	Ga	546	596	637	677	765	807	2010
32	Ge	568	607	658	701	744	836	880
33	As	594	630	670	724	769	813	910
34	Se	519	657	695	736	793	839	886
35	Br	533	579	724	762	806	864	913
36	Kr	545	593	642	794	833	878	939

Table 19.2 Ionization potentials (in eV) of atomic ions ($2 \leq Z \leq 36$) *(continued)*

Atomic	Element	ξ						
number Z	X	XXIII	XXIV	XXV	XXVI	XXVII	XXVIII	XXIX
23	V	7246.18						
24	Cr	7432	7894.8					
25	Mn	1880	8141	8571.9				
26	Fe	1960	2046	8828	9277.7			
27	Co	1962	2119	2219	9544	10,012.1		
28	Ni	2011	2131	2295	2399	10,290	10,775	
29	Cu	2045	2173	2298	2460	2585	11,062	11,568
30	Zn	2070	2216	2350	2479	2647	2780	11,865
31	Ga	2120	2242	2393	2533	2668	2840	2982
32	Ge	2180	2294	2421	2577	2723	2863	3041
33	As	957	2357	2474	2606	2768	2920	3065
34	Se	987	1036.3	2542	2661	2798	2965	3123
35	Br	961	1068	1119	2733	2855	2997	3170
36	Kr	989	1039	1151	1205	2931	3056	3203

Atomic	Element	ξ						
number Z	X	XXX	XXXI	XXXII	XXXIII	XXXIV	XXXV	XXXVI
30	Zn	12,389						
31	Ga	12,696	13,239					
32	Ge	3192	13,550	14,119				
33	As	3248	3409	14,440	15,029			
34	Se	3274	3463	3633	15,370	15,968		
35	Br	3334	3490	3684	3865	16,315	16,937	
36	Kr	3381	3551	3712	3912	4105	17,290	17,936

Table 19.3 Ionization potentials (in eV) of single-, double-, and triple-charged heavy atomic ions ($37 \leq Z \leq 92$) [Xξ is a spectroscopic symbol of the ion.]

Atomic	Element	ξ			Atomic	Element	ξ		
number Z	X	II	III	IV	number Z	X	II	III	IV
37	Rb	27.290	39.2	52.2	60	Nd	10.7	22.1	40.4
38	Sr	11.03028	42.88	56.28	61	Pm	10.9	22.3	41.0
39	Y	12.224	20.525	60.61	62	Sm	11.1	23.4	41.1
40	Zr	13.13	23.1	34.419	63	Eu	11.24	24.9	42.7
41	Nb	14.32	25.0	37.7	64	Gd	12.1	20.6	44.0
42	Mo	16.16	27.2	46.4	65	Tb	11.5	21.9	39.4
43	Tc	15.26	29.5		66	Dy	11.7	22.8	41.4
44	Ru	16.76	28.5		67	Ho	11.8	22.8	42.5
45	Rh	18.08	31.1		68	Er	11.9	22.7	42.7
46	Pd	19.43	32.9		69	Tm	12.1	23.7	42.7
47	Ag	21.49	34.8		70	Yb	12.18	25.05	43.6
48	Cd	16.908	37.47		71	Lu	13.9	20.96	45.25
49	In	18.87	28.0	57.0	72	Hf	14.9	23.3	33.4
50	Sn	14.632	30.50	40.74	78	Pt	18.56		
51	Sb	16.53	25.32	44.16	79	Au	20.5	34	43
52	Te	18.6	27.96	37.42	80	Hg	18.76	34.2	46
53	I	19.131	33.0		81	Tl	20.43	29.85	
54	Xe	20.98	31.0	45	82	Pb	15.033	31.94	42.33
55	Cs	23.1575	33.4	46	83	Bi	16.70	25.56	45.3
56	Ba	10.004	35.8	47	88	Ra	10.15		
57	La	11.1	19.18	49.9	89	Ac	11.75	20	
58	Ce	10.8	20.20	36.76	90	Th	11.9	18.3	28.7
59	Pr	10.6	21.62	38.98	92	U	11.9	20	37

Table 19.4 Ionization potentials (IP) of ground state molecules

Molecule	IP, eV	Molecule	IP, eV	Molecule	IP, eV	Molecule	IP, eV
Diatomic molecules		CsF	8.8 (*B*)	KLi	4.7 (*B*)	PrO	4.9 (*C*)
Ag_2	7.656 (*A*)	CsI	7.25 (*A*)	KNa	4.5 (*B*)	PtB	10 (*D*)
AgF	11.4 (*B*)	Cu_2	7.4 (*B*)	Kr_2	12.87 (*A*)	PtSi	7.9 (*C*)
Al_2	5.99 (*A*)	D_2	15.4665 (*A*)	LaO	4.9 (*B*)	Rb_2	3.45 (*B*)
AlCl	9.4 (*C*)	DBr	11.67 (*A*)	Li_2	5.145 (*A*)	RbBr	7.7 (*B*)
AlF	9.8 (*C*)	DCl	12.76 (*A*)	LiBr	10.0 (*C*)	RbCl	8.3 (*B*)
AlO	9.5 (*C*)	DF	16.06 (*A*)	LiCl	10.1 (*C*)	RbI	7.1 (*B*)
AlS	9.5 (*C*)	DI	10.39 (*A*)	LiD	7.7 (*C*)	RhO	9.3 (*C*)
AlTe	9.0 (*C*)	DT	15.475 (*A*)	LiF	11.3 (*C*)	RuO	8.7 (*C*)
Ar_2	14.5 (*A*)	DyF	6.0 (*C*)	LiH	7.85 (*B*)	S_2	9.356 (*A*)
ArKr	13.4 (*A*)	ErF	6.3 (*C*)	LiI	8.6 (*C*)	SH	10.4 (*A*)
ArXe	12.0 (*A*)	F_2	15.686 (*A*)	LiNa	4.9 (*B*)	SO	10.29 (*A*)
As_2	9.9 (*B*)	FeO	8.7 (*B*)	LiO	8.4 (*B*)	Sb_2	9.0 (*C*)
At_2	8.3 (*C*)	GaF	10.6 (*C*)	MgCl	7.5 (*B*)	Se_2	8.88 (*A*)
AuBr	9.2 (*C*)	GaO	9.4 (*C*)	MgF	7.8 (*C*)	SeH	9.85 (*A*)
AuSi	8.3 (*C*)	GaS	8.9 (*C*)	Mn_2	7.4 (*C*)	Si_2	7.4 (*C*)
BC	10.5 (*C*)	GaSe	8.8 (*C*)	MnF	8.7 (*C*)	SiC	10.2 (*B*)
BF	11.1 (*A*)	GaTe	8.4 (*C*)	MoO	8.0 (*C*)	SiF	7.26 (*A*)
BH	9.8 (*B*)	Ge_2	7.9 (*C*)	N_2	15.581 (*A*)	SiH	7.9 (*B*)
BO	13.5 (*C*)	GeBr	7.3 (*C*)	NF	12.3 (*B*)	SiO	11.4 (*A*)
BSi	7.8 (*C*)	GeC	10.3 (*C*)	NH	13.1 (*A*)	SnBr	7.4 (*C*)
BaCl	5.0 (*B*)	GeCl	7.2 (*C*)	NO	9.2643 (*A*)	SnCl	6.6 (*C*)
BaF	4.9 (*C*)	GeF	7.5 (*C*)	NS	8.9 (*B*)	SnF	7.4 (*C*)
BaO	6.9 (*C*)	GeO	11.1 (*B*)	Na_2	4.889 (*A*)	SnO	10.5 (*C*)
BeF	9.1 (*C*)	GeSi	8.2 (*C*)	NaBr	8.3 (*B*)	SnS	9.7 (*C*)
BeH	8.2 (*B*)	GeTe	10.1 (*C*)	NaCl	8.93 (*A*)	SnSe	9.7 (*C*)
BeO	10.1 (*C*)	H_2	15.4258(*A*)	NaI	7.64 (*A*)	SnTe	9.1 (*C*)
Br_2	10.52 (*A*)	HBr	11.67 (*A*)	NaLi	4.9 (*B*)	SrCl	5.6 (*B*)
BrCl	11.1 (*B*)	HCl	12.75 (*A*)	NaK	4.416 (*A*)	SrF	4.9 (*C*)
BrF	11.8 (*B*)	HD	15.4445 (*A*)	Nb_2	6.2 (*B*)	SrO	6.1 (*C*)
BrO	10.3 (*A*)	HF	16.04 (*A*)	NdO	5.0 (*D*)	T_2	15.487 (*A*)
C_2	12.15 (*A*)	HI	10.38 (*A*)	NiCl	11.4 (*B*)	TaO	6.0 (*C*)
CCl	8.98 (*A*)	HT	15.451 (*A*)	NiO	9.5 (*C*)	Te_2	8.29 (*A*)
CF	9.11 (*A*)	He_2	22.22 (*A*)	O_2	12.071(*A*)	TeO	8.7 (*B*)
CH	10.64 (*A*)	HfO	7.5 (*B*)	OD	12.9 (*A*)	Ti_2	6.3 (*C*)
CN	14.2 (*A*)	HoF	6.2 (*C*)	OH	12.9 (*A*)	TiO	6.4 (*C*)
CO	14.014 (*A*)	I_2	9.3 (*B*)	OT	12.9 (*A*)	TiS	7.1 (*C*)
CS	11.33 (*A*)	IBr	9.85 (*A*)	P_2	10.5 (*A*)	TlBr	9.14 (*A*)
CaF	5.83 (*A*)	ICl	10.08 (*A*)	PC	10.5 (*C*)	TlCl	9.70 (*A*)
CaH	5.9 (*B*)	IF	10.5 (*B*)	PH	10.6 (*B*)	TlF	10.5 (*B*)
CaO	6.5 (*C*)	InBr	9.1 (*B*)	PN	11.8 (*A*)	TlI	8.47 (*A*)
CeO	4.9 (*C*)	InCl	9.5 (*B*)	PO	8.2 (*B*)	V_2	6.4 (*C*)
Cl_2	11.50 (*A*)	InF	9.6 (*C*)	PbBr	7.8 (*C*)	VO	5.0 (*D*)
ClF	12.7 (*B*)	InI	8.5 (*B*)	PbCl	7.5 (*C*)	WO	9.1 (*D*)
ClO	11.0 (*B*)	InS	7.0 (*C*)	PbF	7.5 (*C*)	UC	6.2 (*C*)
CoO	9.0 (*C*)	InSe	7.1 (*C*)	PbO	9.0 (*C*)	UN	7.0 (*C*)
CrF	8.4 (*C*)	InTe	7.6 (*C*)	PbS	8.6 (*C*)	UO	5.7 (*B*)
CrO	8.4 (*C*)	IrC	9.5 (*D*)	PbTe	8.2 (*C*)	US	6.3 (*C*)
Cs_2	3.64 (*B*)	K_2	4.061 (*A*)	Pd_2	7.7 (*C*)	Xe_2	11.18 (*A*)
CsBr	7.72 (*A*)	KCl	8.4 (*B*)	PdO	9.1 (*C*)	ZrN	7.9 (*C*)
CsCl	8.3 (*B*)	KI	8.2 (*C*)	PbSi	8.4 (*C*)	ZrO	6.1 (*C*)

Table 19.4 Ionization potentials (IP) of ground state molecules *(continued)*

Molecule	IP , eV	Molecule	IP , eV	Molecule	IP , eV	Molecule	IP , eV
Triatomic molecules		HO_2	11.53 (*A*)	BH_3	11.4 (*C*)	$SnCl_4$	12.10 (*A*)
BH_2	9.8 (*C*)	H_2O	12.614 (*A*)	BI_3	9.40 (*A*)	XeF_4	12.65 (*A*)
$BaCl_2$	9.2 (*B*)	H_2S	10.46 (*A*)	CCl_3	8.3 (*C*)	**Hexatomic molecules**	
BaI_2	8.1 (*C*)	H_2Se	9.89 (*A*)	CH_3	9.840 (*A*)	C_2Cl_4	9.34 (*A*)
BeF_2	14.7 (*C*)	ICN	10.87 (*A*)	C_2H_2	11.406 (*A*)	C_2F_4	10.12 (*A*)
BrCN	11.84 (*A*)	Li_2O	6.8 (*B*)	H_2CN	11 (*C*)	C_2H_4	10.51 (*A*)
CCl_2	13.2 (*B*)	NF_2	12.11 (*A*)	H_2O_2	10.9 (*C*)	CH_3OH	10.85 (*A*)
CF_2	11.8 (*B*)	NH_2	10.15 (*A*)	HBO_2	12.6 (*B*)	N_2F_4	12.0 (*A*)
CH_2	10.396 (*A*)	NO_2	9.59 (*A*)	NH_3	10.15 (*A*)	N_2H_4	8.74 (*A*)
CO_2	13.78 (*A*)	N_2O	12.890 (*A*)	P_4	9.2 (*B*)	C_2H_5	8.12 (*A*)
COS	11.18 (*A*)	O_3	12.52 (*A*)	PF_3	9.71 (*A*)	SF_6	15.7 (*A*)
CS_2	10.07 (*A*)	PF_2	8.85 (*A*)	SO_3	12.83 (*A*)	**Some more complex**	
$CaCl_2$	10.3 (*A*)	SO_2	10.58 (*A*)	SiH_3	8.13 (*A*)	**polyatomic molecules**	
ClCN	12.34 (*A*)	SiC_2	10.2 (*B*)	**Pentatomic molecules**		UF_6	14.14 (*A*)
ClO_2	11.1 (*C*)	SiF_2	11.0 (*C*)	CBr_4	11.0 (*B*)	XeF_6	12.19 (*A*)
Cs_2O	4.45 (*A*)	SiO_2	11.7 (*C*)	CCl_4	11.47 (*A*)	B_2H_6	11.41 (*A*)
CsOH	7.21 (*B*)	$SnCl_2$	10.2 (*C*)	CF_2Cl_2	12.31 (*A*)	C_2H_6	11.50 (*A*)
FCN	13.32 (*A*)	$SrCl_2$	9.7 (*B*)	$CFCl_3$	11.77 (*A*)	C_2H_5Br	10.29 (*A*)
$GeBr_2$	9.5 (*C*)	UO_2	5.5 (*B*)	CH_4	12.66 (*A*)	C_2H_5OH	10.47 (*A*)
$GeCl_2$	10.4 (*C*)	XeF_2	12.42 (*A*)	CH_2Cl_2	11.35 (*A*)	C_6H_6	9.244 (*A*)
GeF_2	11.8 (*B*)	**Tetraatomic molecules**		CH_3Br	10.53 (*A*)	C_7H_7	7.25 (*A*)
H_3	3.665 (*A*)	BBr_3	10.68 (*A*)	CH_3I	9.54 (*A*)	C_{60}	7.60 (*A*)
HCN	13.73 (*A*)	BCl_3	11.73 (*A*)	SiH_4	11.4 (*C*)		
HCS	7.45 (*A*)	BF_3	15.95 (*A*)	$SnBr_4$	11.0 (*A*)		

19.3 Binding Energies of Electrons in the Atomic Inner-Shells

Table 19.5 represents the core-electron binding energies in the atoms, i.e. the minimum energies required to remove the inner-shell electron from a bound atomic level to the continuous spectra. The electron ejection can leave the ion in the different states. The interaction between the outer (valence) electron shell and the inner-shell with the vacancy results in a relatively small energy splitting for these states. The binding energies presented here correspond to the formation of the atomic ion in its lowest energy state.

In Table 19.5, the core-electron binding energies are given for the isolated neutral atoms, i.e. they refer to the vacuum zero-energy level. For elements in the solid state, the measured atomic binding energies refer, naturally, to the Fermi level of the respective conductor (for metals) or to the Fermi level of the spectrometer-tank material (for insulators and semiconductors). To compare the free-atom values, they must be augmented by the work function of each element, yielding the vacuum-referenced binding energies. The work function comprises, on the average, about 4 eV, varying for the polycrystalline specimens in the range from 1.81 (Cs) to 5.32 eV (Pt).

The most precise technique of measuring the core-electron binding energies in the atoms is based on the gas-phase X-ray photoelectron spectroscopy (uncertainties can reach $\pm$ 0.1 eV). Several experimental methods are also available for such studies, e.g., photoabsorption, X-ray emission spectroscopy, and Auger electron spectroscopy, etc. [9].

The two-way symbols of electronic subshells in an atom are placed in the top rows of Table 19.5. They refer to various quantum numbers of electrons in the inner-shell, their orbital and total angular momenta. The main numerical information on this subject can be found in [6, 9–16]. Most of the energy values were rounded off to give uncertainties of a few units in the last place.

Table 19.5 Binding energies (in eV) of electrons from the inner-shells of free atoms ($Z = 3 - 18$)

Z	Atom	Electron subshells, principal quantum numbers and angular momenta of ejected electrons					Z	Atom	Electron subshells, principal quantum numbers and angular momenta of ejected electrons				
		K_I	L_I	L_{II}	L_{III}	M_I			K_I	L_I	L_{II}	L_{III}	M_I
		$1s_{1/2}$	$2s_{1/2}$	$2p_{1/2}$	$2p_{3/2}$	$3s_{1/2}$			$1s_{1/2}$	$2s_{1/2}$	$2p_{1/2}$	$2p_{3/2}$	$3s_{1/2}$
3	Li	64.40					11	Na	1079	70.9	38.46	37.99	
4	Be	123.6					12	Mg	1311.4	96.5	57.8	57.6	
5	B	200.8	12.93				13	Al	1567	126	81	80	10.62
6	C	296.1	16.59				14	Si	1844	154	107	107	13.46
7	N	410	20.33				15	P	2150	195	137	136	16.15
8	O	544	28.48				16	S	2480	234	171	170	20.20
9	F	699	37.86				17	Cl	2829	279	209	208	24.59
10	Ne	870.27	48.47				18	Ar	3206	326.4	250.77	248.62	29.24

Table 19.5 ($Z = 19 - 36$) *(continued)*

Z	Atom	Electron subshells, principal quantum numbers and angular momenta of ejected electrons									
		K_I	L_I	L_{II}	L_{III}	M_I	M_{II}	M_{III}	M_{IV}	M_V	N_I
		$1s_{1/2}$	$2s_{1/2}$	$2p_{1/2}$	$2p_{3/2}$	$3s_{1/2}$	$3p_{1/2}$	$3p_{3/2}$	$3d_{3/2}$	$3d_{5/2}$	$4s_{1/2}$
19	K	3614	384	303.3	300.7	41	24.82	24.49			
20	Ca	4048	447	360	356	48	34.7	34.3			
21	Sc	4494	503	408	404	56	33	33			
22	Ti	4972	569	468	462	65	39	38			
23	V	5475	638	532	524	77	44	43			
24	Cr	5996	703	591	582	79	49	48			
25	Mn	6550	782	663	652	95	55	53			
26	Fe	7124	857	733	720	104	61	59			
27	Co	7725	940	810	795	115	68	66			
28	Ni	8348	1024	887	870	125	75	73			
29	Cu	8988	1106	960	942	130	82	80	11.0	10.4	
30	Zn	9667	1203	1052	1029	145	98.7	96.1	17.5	17.2	
31	Ga	10371	1302	1146	1119	162	111	107	27	26	11.87
32	Ge	11107	1413	1251	1220	184	133	131	41	40	14.28
33	As	11871	1531	1362	1330	212	151	145	56	55	18.96
34	Se	12662	1656	1479	1441	234	173	166	62	61	22.19
35	Br	13481	1787	1602	1556	263	197	190	79	78	23.80
36	Kr	14327	1923	1731	1678	293	221.8	214.5	95.0	93.8	27.51

Table 19.5 ($Z = 37 - 54$) *(continued)*

Z	Atom	Electron subshells, principal quantum numbers and angular momenta of ejected electrons														
		K_I	L_I	L_{II}	L_{III}	M_I	M_{II}	M_{III}	M_{IV}	M_V	N_I	N_{II}	N_{III}	N_{IV}	N_V	O_I
		$1s_{1/2}$	$2s_{1/2}$	$2p_{1/2}$	$2p_{3/2}$	$3s_{1/2}$	$3p_{1/2}$	$3p_{3/2}$	$3d_{3/2}$	$3d_{5/2}$	$4s_{1/2}$	$4p_{1/2}$	$4p_{3/2}$	$4d_{3/2}$	$4d_{5/2}$	$5s_{1/2}$
37	Rb	15203	2068	1867	1807	325	254.3	245.4	117.4	116.0	32	21.77	20.71			
38	Sr	16108	2219	2010	1943	361	288	278	144	142	40	29.2	28.2			
39	Y	17041	2375	2158	2083	397	315	304	163	161	48	30	29			6.48
40	Zr	18002	2536	2311	2227	434	348	335	187	185	56	35	33			
41	Nb	18990	2702	2469	2375	472	382	367	212	209	62	40	38			
42	Mo	20006	2872	2632	2527	511	416	399	237	234	68	45	42			
43	Tc	21050	3048	2800	2683	551	451	432	263	259	74	49	45			

Table 19.5 $(Z = 37 - 54)$ *(continued)*

Z	Atom	Electron subshells, principal quantum numbers and angular momenta of ejected electrons														
		K_I	L_I	L_{II}	L_{III}	M_I	M_{II}	M_{III}	M_{IV}	M_V	N_I	N_{II}	N_{III}	N_{IV}	N_V	O_I
		$1s_{1/2}$	$2s_{1/2}$	$2p_{1/2}$	$2p_{3/2}$	$3s_{1/2}$	$3p_{1/2}$	$3p_{3/2}$	$3d_{3/2}$	$3d_{5/2}$	$4s_{1/2}$	$4p_{1/2}$	$4p_{3/2}$	$4d_{3/2}$	$4d_{5/2}$	$5s_{1/2}$
44	Ru	22123	3230	2973	2844	592	488	466	290	286	81	53	49			
45	Rh	23225	3418	3152	3010	634	526	501	318	314	87	58	53			
46	Pd	24357	3611	3337	3180	677	565	537	347	342	93	63	57			
47	Ag	25520	3812	3530	3357	724	608	577	379	375.6	101	69	63	13.28	12.43	
48	Cd	26715	4022	3732	3542	775	659	625	419	412	112	78	71	18.28	17.58	
49	In	27944	4242	3943	3735	830	707	669	455	447	126	90	82	25	24	11.03
50	Sn	29204	4469	4160	3933	888	761	719	500	493	141	102	93	33	32	13.10
51	Sb	30496	4703	4385	4137	949	817	771	542	536.3	157	114	104	38	37	16.86
52	Te	31820	4945	4618	4347	1012	876	825	589	581.3	174	127	117	48	46	17.84
53	I	33176	5195	4858	4563	1078	937	881	638	628.3	195	141	131	60	58	20.61
54	Xe	34561	5453	5107	4787	1149	1002	941	689.3	676.7	213.3	157	145.5	69.5	67.5	23.40

Table 19.5 $(Z = 55 - 86)$ *(continued)*

Z	Atom	Electron subshells, principal quantum numbers and angular momenta of ejected electrons									
		K_I	L_I	L_{II}	L_{III}	M_I	M_{II}	M_{III}	M_{IV}	M_V	N_I
		$1s_{1/2}$	$2s_{1/2}$	$2p_{1/2}$	$2p_{3/2}$	$3s_{1/2}$	$3p_{1/2}$	$3p_{3/2}$	$3d_{3/2}$	$3d_{5/2}$	$4s_{1/2}$
55	Cs	35987	5717	5362	5014	1220	1068	1005	746	732	233
56	Ba	37442	5991	5626	5249	1293	1138	1063	797	788	254
57	La	38928	6269	5894	5486	1365	1207	1124	851	834	273
58	Ce	40446	6552	6167	5726	1437	1275	1184	903	885	291
59	Pr	41995	6839	6444	5968	1509	1342	1244	954	934	307
60	Nd	43575	7132	6727	6213	1580	1408	1303	1005	983	321
61	Pm	45188	7432	7017	6464	1653	1476	1362	1057	1032	335
62	Sm	46837	7740	7315	6720	1728	1546	1422	1110	1083	349
63	Eu	48522	8056	7621	6981	1805	1618	1484	1164	1135	364
64	Gd	50243	8380	7935	7247	1884	1692	1547	1220	1189	380
65	Tb	51999	8711	8256	7518	1965	1768	1612	1277	1243	398
66	Dy	53792	9050	8585	7794	2048	1846	1678	1335	1298	416
67	Ho	55622	9398	8922	8075	2133	1926	1746	1395	1354	434
68	Er	57489	9754	9267	8361	2220	2008	1815	1456	1412	452
69	Tm	59393	10118	9620	8651	2309	2092	1885	1518	1471	471
70	Yb	61335	10490	9981	8946	2401	2178	1956	1580	1531	490
71	Lu	63320	10876	10355	9250	2499	2270	2032	1647	1596	514
72	Hf	65350	11275	10742	9564	2604	2369	2113	1720	1665	542
73	Ta	67419	11684	11139	9884	2712	2472	2197	1796	1737	570
74	W	65529	12103	11546	10209	2823	2577	2283	1874	1811	599
75	Re	71681	12532	11963	10540	2937	2686	2371	1953	1887	629
76	Os	73876	12972	12390	10876	3054	2797	2461	2035	1964	660
77	Ir	76115	13422	12828	11219	3175	2912	2554	2119	2044	693
78	Pt	78399	13883	13277	11567	3300	3030	2649	2206	2126	727
79	Au	80729	14356	13738	11923	3430	3153	2748	2295	2210	764
80	Hg	83108	14845	14214	12288	3567	3283	2852	2390	2300	809
81	Tl	85536	15350	14704	12662	3710	3420	2961	2490	2394	852
82	Pb	88011	15867	15206	13041	3857	3560	3072	2592	2490	899
83	Bi	90534	16396	15719	13426	4007	3704	3185	2696	2588	946
84	Po	93105	16933	16242	13815	4160	3850	3300	2800	2685	992
85	At	95730	17485	16780	14213	4315	4000	3415	2905	2785	1043
86	Rn	98400	18053	17335	14615	4480	4155	3505	3015	2890	1095

Table 19.5 ($Z = 55 - 86$) *(continued)*

Z	Atom	Electron subshells, principal quantum numbers and angular momenta of ejected electrons											
		N_{II}	N_{III}	N_{IV}	N_V	N_{VI}	N_{VII}	O_I	O_{II}	O_{III}	O_{IV}	O_V	P_I
		$4p_{1/2}$	$4p_{3/2}$	$4d_{3/2}$	$4d_{5/2}$	$4f_{5/2}$	$4f_{7/2}$	$5s_{1/2}$	$5p_{1/2}$	$5p_{3/2}$	$5d_{3/2}$	$5d_{5/2}$	$6s_{1/2}$
55	Cs	174	164	81	79			25	19.07	17.21			
56	Ba	193	181	101	98.4			31	24.7	22.8			
57	La	210	196	105	103			36	22	19			
58	Ce	225	209	114	111			39	25	22			
59	Pr	238	220	121	117			41	27	24			
60	Nd	250	230	126	122			42	28	25			
61	Pm	261	240	131	127			43	28	25			
62	Sm	273	251	137	132			44	29	25			
63	Eu	286	262	143	137			45	30	26			
64	Gd	300	273	150	143			46	31	27			
65	Tb	315	285	157	150			48	32	28			
66	Dy	331	297	164	157			50	33	28			
67	Ho	348	310	172	164			52	34	29			
68	Er	365	323	181	172			54	35	30			
69	Tm	382	336	190	181			56	36	30			
70	Yb	399	349	200	190	10.17	8.91	58	37	31			
71	Lu	420	366	213	202	13	12	62	39	32			7.0
72	Hf	444	386	229	217	21	20	68	43	35			7.46
73	Ta	469	407	245	232	30	28	74	47	38			
74	W	495	428	261	248	40	38	80	51	41			
75	Re	522	450	278	264	52	50	86	56	45			
76	Os	551	473	295	280	56	54	92	61	49			
77	Ir	581	497	314	298	64	62	99	66	53			
78	Pt	612	522	335	318	78	75	106	71	57			
79	Au	645	548	357	339	91	87	114	76	61	12.90	11.09	
80	Hg	686	584	385	366	111	107	134	90	72	16.70	14.84	
81	Tl	726	615	411	391	130.4	125.7	139	98	79	21	19	6.13
82	Pb	769	651	441	419	148.5	143.6	153	111	90	28.2	25.3	14.60
83	Bi	813	687	472	448	170	165	167	125	101	34	32	16.73
84	Po	855	720	500	475	190	184	183	137	108	39	35	15
85	At	902	760	533	504	212	205	198	150	118	45	42	19
86	Rn	950	795	565	535	235	230	215	165	129	53	48	24

Table 19.5 ($Z = 87 - 96$) *(continued)*

Z	Atom	Electron subshells, principal quantum numbers and angular momenta of ejected electrons											
		K_I	L_I	L_{II}	L_{III}	M_I	M_{II}	M_{III}	M_{IV}	M_V	N_I	N_{II}	N_{III}
		$1s_{1/2}$	$2s_{1/2}$	$2p_{1/2}$	$2p_{3/2}$	$3s_{1/2}$	$3p_{1/2}$	$3p_{3/2}$	$3d_{3/2}$	$3d_{5/2}$	$4s_{1/2}$	$4p_{1/2}$	$4p_{3/2}$
87	Fr	101135	18640	17905	15030	4650	4320	3660	3135	3000	1155	1005	840
88	Ra	103920	19240	18490	15450	4825	4490	3790	3253	3110	1213	1062	885
89	Ac	106760	19850	19090	15875	5005	4660	3920	3375	3225	1275	1118	930
90	Th	109655	20475	19700	16305	5187	4835	4050	3495	3337	1335	1173	972
91	Pa	112600	21115	20320	16735	5372	5007	4180	3610	3445	1390	1228	1010
92	U	115605	21763	20953	17170	5553	5185	4308	3730	3555	1445	1277	1050
93	Np	118675	22432	21605	17615	5745	5370	4440	3855	3670	1505	1332	1090
94	Pu	121795	23110	22270	18062	5938	5555	4570	3975	3780	1565	1383	1125
95	Am	124984	23803	22950	18512	6135	5748	4710	4102	3900	1623	1440	1165
96	Cm	128261	24523	23655	18975	6313	5946	4830	4235	4015	1665	1495	1195

Table 19.5 ($Z = 87 - 96$) *(continued)*

Z	Atom	Electron subshells, principal quantum numbers and angular momenta of ejected electrons												
		N_{IV}	N_V	N_{VI}	N_{VII}	O_I	O_{II}	O_{III}	O_{IV}	O_V	P_I	P_{II}	P_{III}	Q_I
		$4d_{3/2}$	$4d_{5/2}$	$4f_{5/2}$	$4f_{7/2}$	$5s_{1/2}$	$5p_{1/2}$	$5p_{3/2}$	$5d_{3/2}$	$5d_{5/2}$	$6s_{1/2}$	$6p_{1/2}$	$6p_{3/2}$	$7s_{1/2}$
87	Fr	604	572	265	255	235	184	144	65	60	33	19	14	
88	Ra	640	608	293	285	260	205	158	75	70	40	25	19	
89	Ac	680	645	325	318	280	222	173	90	82	45	29	22	7.49
90	Th	718	682	358	349	295	238	185	100	92	50	33	25	
91	Pa	748	713	379	369	313	250	190	105	97	50	32	24	
92	U	785	743	403	392	328	265	200	110	100	52	34	24	
93	Np	820	775	425	414	345	280	210	115	105	54	35	25	
94	Pu	852	805	449	436	355	290	220	120	110	53	34	23	
95	Am	890	840	474	460	373	303	230	125	115	54	44	36	
96	Cm	920	865	498	483	388	315	240	130	120	60	39	27	

19.4 Electron Affinities of Atoms and Molecules

The binding energy of electrons in the negative ion is termed as the electron affinity (EA) of the appropriate particle. It relates to the minimum energy spent in the valence electron separation from the negative atomic or molecular ions. For molecules, EA is considered to mean energy fitted transition between the ground vibrational states of the molecular ion and molecule.

Table 19.6 presents electron affinities of atomic particles. Most reliable experimental studies of atomic EA are based on the threshold behavior of photodetachment cross-sections in laser induced processes, and also on laser photoelectron spectroscopy [17–20]. The designation of the outer electron shell and the electronic term of a negative ion are shown in the separate columns of Table 19.6.

The electron affinities of diatomic and more complex molecules are given in Table 19.7. The photoelectron spectroscopy, along with photodetachment, charge exchange and collisional ionization techniques are among the most reliable methods of molecular EA determination [17,18,21,22].

To give more insight into experimental errors in atomic and molecular EAs, data in the tables were grouped according to four accuracy classes: A– for uncertainties within 1%; B– for uncertainties within 3%; C– for uncertainties within 10%, and D– for uncertainties within 30%. This classification stems from error estimations in the applied measurement techniques.

Table 19.6 Electron affinities (EA) of atoms

Atomic number Z	Atomic negative ion (electron term)	Outer-shell electron configuration	EA, eV	Atomic number Z	Atomic negative ion (electron term)	Outer-shell electron configuration	EA, eV
1	H^- (^1S)	$1s^2$	0.75421 (A)	9	F^- (^1S)	$2s^22p^6$	3.40119 (A)
	D^- (^1S)	$1s^2$	0.7546 (A)	11	Na^- (^1S)	$3s^2$	0.5479 (A)
2	He^- (^4P)	$1s2s2p$	0.077 (B)	12	Mg^- (^4P)	$3s3p^2$	0.32 (B)
3	Li^- (^1S)	$1s^22s^2$	0.618 (A)	13	Al^- ($^3P_0$)	$3p^2$	0.44 (B)
4	Be^- (^4P)	$1s^22s2p^2$	0.28 (B)		Al^- ($^1D_2$)	$3p^2$	0.11 (B)
5	B^- (^3P)	$2s^22p^2$	0.267 (B)	14	Si^- (^4S)	$3p^3$	1.39 (A)
6	C^- (^4S)	$2s^22p^3$	1.263 (A)		Si^- (^2D)	$3p^3$	0.52 (B)
	C^- (^2D)	$2s^22p^3$	0.034 (B)		Si^- (^2P)	$3p^3$	0.03 (B)
8	O^- (^2P)	$2s^22p^5$	1.46112 (A)	15	P^- (^3P)	$3p^4$	0.746 (A)

Table 19.6 Electron affinities (EA) of atoms *(continued)*

Atomic number Z	Atomic negative ion (electron term)	Outer-shell electron configuration	EA, eV	Atomic number Z	Atomic negative ion (electron term)	Outer-shell electron configuration	EA, eV
16	S^- (^2P)	$3p^5$	2.07711 (A)	44	Ru^- (^4F)	$4d^75s^2$	1.1 (D)
17	Cl^- (^1S)	$3p^6$	3.6127 (A)	45	Rh^- (^3F)	$4d^85s^2$	1.14 (A)
19	K^- (^1S)	$4s^2$	0.5015 (A)	46	Pd^- (^2D)	$4d^95s^2$	0.56 (A)
20	Ca^- (^2P)	$4s^24p$	0.04 (D)	47	Ag^- (^1S)	$4d^{10}5s^2$	1.30 (A)
21	Sc^- (^1D)	$3d4s^24p$	0.19 (C)	49	In^- (^3P)	$5p^2$	0.3 (D)
	Sc^- (^3D)	$3d4s^24p$	0.04 (D)	50	Sn^- (^4S)	$5p^3$	1.11 (A)
22	Ti^- (^4F)	$3d^34s^2$	0.08 (D)	51	Sb^- (^3P)	$5p^4$	1.05 (A)
23	V^- (^5D)	$3d^44s^2$	0.53 (C)	52	Te^- (^2P)	$5p^5$	1.971 (A)
24	Cr^- (^6S)	$3d^54s^2$	0.67 (B)	53	I^- (^1S)	$5p^6$	3.0591 (A)
26	Fe^- (^4F)	$3d^74s^2$	0.15 (B)	55	Cs^- (^1S)	$6s^2$	0.4716 (A)
27	Co^- (^3F)	$3d^84s^2$	0.66 (B)	57	La^- (^3F)	$5d^26s^2$	0.5 (D)
28	Ni^- (^2D)	$3d^94s^2$	1.16 (A)	70	Yb^- (^2P)	$4f^{14}6s^26p$	0.01 (D)
29	Cu^- (^1S)	$3d^{10}4s^2$	1.23 (A)	73	Ta^- (^5D)	$5d^46s^2$	0.32 (C)
31	Ga^- (^3P)	$4p^2$	0.3 (D)	74	W^- (^6S)	$5d^56s^2$	0.82 (A)
32	Ge^- (^4S)	$4p^3$	1.23 (A)	75	Re^- (^5D)	$5d^66s^2$	0.15 (D)
33	As^- (^3P)	$4p^4$	0.80 (C)	76	Os^- (^4F)	$5d^76s^2$	1.4 (D)
34	Se^- (^2P)	$4p^5$	2.0207 (A)	77	Ir^- (^3F)	$5d^86s^2$	1.57 (A)
35	Br^- (^1S)	$4p^6$	3.36359 (A)	78	Pt^- (^2D)	$5d^96s^2$	2.123 (A)
37	Rb^- (^1S)	$5s^2$	0.4859 (A)	79	Au^- (^1S)	$5d^{10}6s^2$	2.3086 (A)
39	Y^- (^1D)	$4d5s^25p$	0.31 (C)	81	Tl^- (^3P)	$6p^2$	0.3 (D)
	Y^- (^3D)	$4d5s^25p$	0.16 (D)	82	Pb^- (^4S)	$6p^3$	0.37 (B)
40	Zr^- (^4F)	$4d^35s^2$	0.43 (B)	83	Bi^- (^3P)	$6p^4$	0.95 (B)
41	Nb^- (^5D)	$4d^45s^2$	0.89 (B)	84	Po^- (^2P)	$6p^5$	1.9 (D)
42	Mo^- (^6S)	$4d^55s^2$	0.75 (B)	85	At^- (^1S)	$6p^6$	2.9 (D)
43	Tc^- (^5D)	$4d^65s^2$	0.6 (D)				

Table 19.7 Electron affinities (EA) of molecules

Molecule	EA, eV	Molecule	EA, eV	Molecule	EA, eV	Molecule	EA, eV
Diatomic molecules		CaH	0.93 (C)	LiH	0.3 (D)	RbCl	0.54 (B)
Al_2	2.42 (C)	Cl_2	2.44 (B)	LiN	0.4 (D)	S_2	1.66 (B)
AlO	3.6 (C)	ClO	2.28 (A)	MgCl	1.59 (A)	SF	2.5 (C)
As_2	0.7 (D)	CoH	0.67 (B)	MgH	1.05 (C)	SH	2.31 (A)
AsBr	1.3 (C)	Cr_2	0.50 (B)	NH	0.3744 (A)	SO	1.12 (A)
AsCl	1.3 (C)	CrH	0.56 (B)	NO	0.025 (D)	Sb_2	1.28 (A)
AsF	1.3 (C)	CsCl	0.45 (B)	NS	1.19 (A)	Se_2	1.9 (C)
AsH	1.0 (D)	CuO	1.78 (A)	NaBr	0.79 (B)	SeH	2.2125 (A)
BO	3.0 (C)	F_2	2.96 (B)	NaCl	0.73 (B)	SeO	1.46 (B)
Be_2	0.3 (D)	FO	2.27 (A)	NaF	0.52 (B)	SiH	1.28 (A)
BeH	0.7 (D)	FeO	1.49 (B)	NaI	0.86 (B)	Sn_2	1.96 (A)
BeO	1.8 (C)	I_2	2.51 (B)	NaH	0.32 (D)	SnPb	1.57 (A)
Bi_2	1.27 (A)	IBr	2.6 (C)	NiH	0.48 (B)	Te_2	1.9 (C)
Br_2	2.6 (C)	ICl	2.8 (C)	O_2	0.45 (B)	TeH	2.10 (A)
BrO	2.35 (A)	IO	2.38 (A)	OH	1.8277 (A)	ZnH	0.95 (D)
C_2	3.39 (B)	KBr	0.64 (B)	OD	1.8255 (A)	**Triatomic molecules**	
CBr	1.7 (C)	KCl	0.58 (B)	P_2	0.59 (C)	AlF_2	2.3 (C)
CF	3.3 (C)	KI	0.73 (B)	PH	1.03 (A)	AlO_2	4.1 (C)
CH	1.24 (A)	Li_2	0.7 (D)	PO	1.09 (B)	$AsBr_2$	3.5 (C)
CN	3.86 (A)	LiCl	0.59 (B)	Pb_2	1.37 (A)	$AsCl_2$	2.2 (C)
CS	0.21 (C)	LiF	0.44 (C)	Pd_2	1.68 (A)	AsF_2	0.8 (D)

Table 19.7 Electron affinities (EA) of molecules *(continued)*

Molecule	EA, eV	Molecule	EA, eV	Molecule	EA, eV	Molecule	EA, eV
AsH_2	1.27 (*B*)	P_3	0.9 (*D*)	**Pentatomic molecules**		SF_5	3.7 (*C*)
BF_2	2.2 (*C*)	PH_2	1.27 (*A*)	CCl_3F	1.1 (*D*)	UF_5	4.0 (*C*)
BO_2	4.3 (*C*)	PF_2	1.4 (*C*)	CCl_2F_2	0.4 (*D*)	VF_5	4.5 (*C*)
Bi_3	1.6 (*B*)	PtO_2	2.7 (*C*)	CD_3O	1.55 (*B*)	**Some more complex**	
C_3	2.1 (*C*)	S_3	2.09 (*B*)	C_2F_3	2.0 (*D*)	**polyatomic molecules**	
C_2H	3.73 (*B*)	S_2O	1.88 (*A*)	CF_3Br	0.9 (*D*)	C_2Cl_5	1.6 (*D*)
C_2O	1.85 (*B*)	SCN	2.2 (*D*)	CF_2CO	2.4 (*D*)	C_2F_5	2.2 (*C*)
CCl_2	1.60 (*A*)	SH_2	1.1 (*C*)	CF_3I	1.6 (*C*)	C_2H_5	0.89 (*C*)
CF_2	2.1 (*D*)	SO_2	1.11 (*A*)	CF_3O	1.4 (*D*)	C_4H_2N	1.7 (*D*)
CH_2	0.65 (*A*)	Sb_3	1.8 (*B*)	CF_3S	1.8 (*D*)	Fe_2F_5	4 (*D*)
CNS	2.0 (*C*)	SeCN	2.6 (*C*)	CH_2CN	1.55 (*A*)	IrF_6	6.5 (*C*)
COS	0.5 (*D*)	SeO_2	1.82 (*B*)	CH_3Br	0.4 (*D*)	MoF_6	3.8 (*C*)
CS_2	0.85 (*D*)	SiCN	2.6 (*C*)	CH_3O	1.57 (*B*)	OsF_6	5.9 (*C*)
ClO_2	2.14 (*A*)	SiH_2	1.12 (*B*)	CH_3S	1.6 (*D*)	PtF_6	7.0 (*C*)
CoH_2	1.45 (*A*)	**Tetraatomic molecules**		CO_4	1.22 (*C*)	RuF_6	6.5 (*C*)
FCN	4.0 (*D*)	CCl_3	1.2 (*D*)	IrF_4	4.7 (*C*)	SF_6	1.0 (*C*)
FeH_2	1.05 (*B*)	CF_3	1.9 (*C*)	$LiCH_3$	0.24 (*D*)	SeF_6	3.0 (*C*)
GeF_2	1.3 (*D*)	CH_3	1.07 (*B*)	CeF_4	3.7 (*C*)	TeF_6	3.3 (*B*)
HCBr	1.45 (*A*)	CO_3	2.82 (*B*)	FeF_4	5.4 (*C*)	UF_6	5.1 (*B*)
HCCl	1.21 (*A*)	GeF_3	3.0 (*D*)	HNO_3	0.6 (*D*)	WF_6	4.3 (*C*)
HCF	0.54 (*A*)	FeF_3	3.6 (*C*)	MnF_4	5.5 (*C*)	C_3F_5	2.2 (*D*)
HCl	1.68 (*A*)	MnF_3	4.4 (*C*)	$OH{\cdot}H_2O$	1.95 (*C*)	C_2H_5O	0.6 (*D*)
HNO	0.34 (*C*)	NO_3	3.7 (*B*)	OsF_4	3.9 (*C*)	C_2H_5S	1.4 (*D*)
IO_2	2.58 (*A*)	PBr_3	1.6 (*D*)	$POCl_3$	1.4 (*D*)	Fe_2F_6	4.4 (*C*)
LiCN	0.74 (*D*)	PBr_2Cl	1.6 (*D*)	PtF_4	5.5 (*C*)	C_4F_5	1.0 (*D*)
LiNC	0.62 (*D*)	$PBrCl_2$	1.5 (*C*)	RhF_4	5.4 (*C*)	C_2H_6N	1.0 (*D*)
LiOH	0.22 (*D*)	PCl_3	0.8 (*D*)	RuF_4	4.8 (*C*)	C_3F_7	2.2 (*C*)
MnH_2	0.44 (*C*)	$POCl_2$	3.8 (*C*)	SF_4	2.35 (*B*)	C_3H_7	0.6 (*D*)
NCO	3.61 (*A*)	SF_3	2.9 (*C*)	UF_4	1.7 (*D*)	C_5H_5	1.79 (*B*)
NCS	3.54 (*A*)	SO_3	1.7 (*C*)	VF_4	3.6 (*C*)	C_6Cl_5	2.8 (*C*)
NH_2	0.77(*A*)	SO_2F	2.8 (*C*)	**Hexatomic molecules**		C_4F_7	3 (*D*)
NF_2	1.7 (*D*)	SiF_3	2.7 (*C*)	C_3F_3	4.0 (*D*)	C_6F_5	2.7 (*C*)
NO_2	2.27 (*A*)	SiH_3	1.4 (*D*)	C_3H_3	2.3 (*D*)	C_6H_5	2.3 (*C*)
N_2O	0.24 ()	UF_3	1.5 (*D*)	MoF_5	3.3 (*C*)	C_6F_6	0.5 (*D*)
NiH_2	1.93 (*A*)	VF_3	2 (*D*)	PtF_5	6.5 (*C*)	U_2F_{10}	4 (*C*)
O_3	2.10 (*A*)	WO_3	3.3 (*C*)	RuF_5	5 (*C*)	C_{60}	2.6 (*B*)

19.5 Proton Affinities of Atoms and Molecules

The energy released in proton (H^+) addition to neutral atomic and molecular species is called the proton affinity (PA) of the appropriate particle. It is the main characteristic of gas-phase protonation reactions under particle collisions in a low-temperature plasma. The general definition of PA is based on examination of the hypothetical reaction

$$X + H^+ \rightarrow XH^+. \qquad (19.1)$$

The proton affinity of X molecules therewith equals the enthalpy variation with opposite sign for the above reaction, or if we take advantage of the heat formation ΔH_f° for X, H and XH particles, the consideration gives

$$PA(\mathrm{X}) = \Delta H^{\circ}_{f0}(\mathrm{X}) + \Delta H^{\circ}_{f0}(\mathrm{H}^+) - \Delta H^{\circ}_{f0}(\mathrm{XH}^+), \tag{19.2}$$

where ΔH°_{f0}'s are the heats of formation of components in their standard states.

The basic experimental methods of PA determination start from measuring the equilibrium constant of protonation reactions with the use of high-pressure mass-spectrometers, ion-cyclotron resonance and continuous plasma afterglow, etc. [23–25].

Table 19.8 shows atomic PAs which correspond to the dissociation energy of a hydride molecular ion in the zero vibrational state. The values of PA for some gas-phase molecules are collected in Table 19.9. The key information about atomic and molecular PAs is given in [22–25]. The numerical values listed were grouped into the accuracy classes as follows: A– for uncertainties within 1%; B– for uncertainties within 3%; C– for uncertainties within 10%, and D– for uncertainties within 30%.

Table 19.8 Proton affinities (PA) of atoms

Atom	PA, eV	Atom	PA, eV	Atom	PA, eV	Atom	PA, eV
Ar	3.87 (A)	F	3.42 (A)	Kr	4.41 (B)	S	6.87 (B)
Br	5.73 (B)	H	2.651 (A)	N	3.4 (C)	Xe	5.15 (B)
Cl	5.34 (B)	He	1.845 (A)	Ne	2.08 (B)	Zn	6.77 (B)
Cs	7.59 (B)	I	6.31 (B)	O	5.1 (B)		

Table 19.9 Proton affinities (PA) of molecules

Molecule	PA, eV	Molecule	PA, eV	Molecule	PA, eV	Molecule	PA, eV
Diatomic molecules				N_2O	5.8 (B)	SO_2	7.0 (C)
Br_2	6.1 (B)	HF	4.09 (B)	PH_2	7.2 (C)	SiH_2	8.7 (B)
CO	6.15 (A)	HI	6.3 (C)	**Tetraatomic and more complex polyatomic molecules**			
CN	5 (D)	HS	7.2 (C)				
CS	8.2 (A)	N_2	4.8 (C)	C_2H_2	5.5 (C)	CF_4	5.2 (C)
H_2	4.40 (B)	NH	6.1 (C)	C_2N_2	7.0 (B)	CH_3Cl	6.9 (C)
D_2	4.56 (B)	NO	5.50 (B)	CH_3	5.4 (C)	CH_4	5.4 (B)
HBr	6.10 (A)	O_2	4.10 (B)	H_2CS	8.0 (A)	GeH_4	7.1 (B)
HCl	5.86 (B)	OH	6.2 (C)	HNO_2	8.1 (C)	HNO_3	7.5 (B)
Triatomic molecules				$(HF)_2$	5.2 (C)	SiH_4	6.6 (B)
CH_2	8.6 (B)	H_2S	7.4 (B)	NF_3	6.6 (C)	$TiCl_4$	7.6 (B)
CO_2	5.5 (B)	H_2Se	7.4 (B)	NH_3	8.9 (A)	C_2H_4	6.6 (B)
HCN	7.46 (A)	H_2Te	7.6 (B)	N_2H_2	7.9 (C)	$(HF)_3$	5.8 (C)
H_2O	7.23 (B)	NH_2	8.1 (B)	PH_3	8.07 (A)	SF_6	3.7 (C)
HO_2	6.4 (C)	NO_2	6.6 (B)	SO_3	6.2 (B)	C_2H_6	6.9 (B)

19.6 Effective Cross-Sections of Electron Impact Ionization of Atoms and Molecules

The cross-section σ^{i+} of i-fold ionization is defined as the ratio of appearance probability for an i-fold ion to the flux density of incident electrons. The total ionization cross-section σ_n is related to partial σ^{i+} through the summation: $\sigma_n = \sigma^+ + 2\sigma^{2+} + 3\sigma^{3+} + ...$ At electron energies less than the threshold of two-fold ionization, the cross-section σ_n agrees with σ^+. The total ionization cross-section defines the probability of total (net) positive charge production per unit time, and for the molecular ionization it covers also the dissociative ionization by electron impact.

At electron energies E far exceeding the ionization potential U_α of an atomic α-shell, the ionization cross-section is described by the Bethe formula [26,27]:

$$\sigma_\alpha^+ = a_\alpha q_\alpha \ln(E/U_\alpha)/EU_\alpha, \tag{19.3}$$

where a_α is the constant varied from 2.6 to $4.5 \cdot 10^{-16}$ $\mathrm{cm}^2 \cdot \mathrm{eV}^2$ upon the comparison of calculated and experimental results at $E \gg U_\alpha$ [27]; q_α is the number of equivalent electrons at the α-shell. The Bethe formula provides the foundation for many empirical relations which are used for approximating the cross-sections of atomic ionization. The Lotz approximation is most commonly invoked for cross-sections of one-fold ionization of ground state atomic particles [27,28]:

$$\sigma^+ = \sum_\alpha a_\alpha q_\alpha [\ln(E/U_\alpha)/EU_\alpha]\{1 - b_\alpha \exp[c_\alpha - (E/U_\alpha - 1)]\}, \tag{19.4}$$

here a_α, b_α, and c_α are the constants for a given electron shell, which are determined by fitting the experimental findings. Table 19.10 presents the above constants for some atomic shells. The approximation error comes to 30–40% [28] and can be over this range in the vicinity of ionization thresholds and close to maximum cross-sections. This approximation does not account for the contribution of the autoionizing atomic states to the total ionization cross-section. Figures 19.1–19.34 illustrate the cross-sections of atomic and molecular ionization by collisions with electrons.

The analytical fitting of computed (in Born approximation) cross-sections and rate constants of some ionization processes with participation of neutral and ionized atomic particles was produced in monographs [29,30]. These data are well suited for practice.

Table 19.10 Parameters a, b, and c entering into Eq.(19.4) for electron impact ionization of atomic inner-shells [n takes the numbers 4, 5, 6, and 7 for s-shells; 4, 5, and 6 for p-shells; 5 and 6 for d-shells; and, finally, 4 and 5 for f-shells; a is measured in units of 10^{-14} $\mathrm{cm}^2 \cdot \mathrm{eV}^2$.]

Parameter	$1s$	$1s^2$	$2p$	$2p^2$	$2p^3$	$2p^4$	$2p^5$	$2p^6$	$3d$	$3d^2$	$3d^3$	$3d^4$	$3d^5$	$3d^6$	$3d^7$	$3d^8$	$3d^9$	$3d^{10}$
a	4	4	3.8	3.5	3.2	3.0	2.8	2.6	3.7	3.4	3.1	2.8	2.5	2.2	2.0	1.8	1.6	1.4
b	0.60	0.75	0.6	0.7	0.8	0.85	0.90	0.92	0.6	0.7	0.8	0.85	0.90	0.92	0.93	0.94	0.95	0.96
c	0.56	0.50	0.4	0.3	0.25	0.22	0.20	0.19	0.4	0.3	0.25	0.20	0.18	0.17	0.16	0.15	0.14	0.13
Parameter	$2s$	$2s^2$	$3p$	$3p^2$	$3p^3$	$3p^4$	$3p^5$	$3p^6$	$4d$	$4d^2$	$4d^3$	$4d^4$	$4d^5$	$4d^6$	$4d^7$	$4d^8$	$4d^9$	$4d^{10}$
a	4	4	4	4	4	4	4	4	4	3.8	3.5	3.2	3.0	2.8	2.6	2.4	2.2	2.0
b	0.3	0.5	0.35	0.40	0.45	0.50	0.55	0.6	0.3	0.45	0.6	0.7	0.8	0.85	0.90	0.92	0.93	0.94
c	0.6	0.6	0.6	0.6	0.6	0.5	0.45	0.4	0.6	0.5	0.4	0.3	0.25	0.20	0.18	0.17	0.16	0.15
Parameter	$3s$	$3s^2$	np	np^2	np^3	np^4	np^5	np^6	nd	nd^2	nd^3	nd^4	nd^5	nd^6	nd^7	nd^8	nd^9	nd^{10}
a	4	4	4	4	4	4	4	4	4	4	3.8	3.6	3.4	3.2	3.0	2.8	2.6	2.4
b	0	0.3	0	0	0.2	0.3	0.4	0.5	0	0.2	0.3	0.45	0.6	0.7	0.8	0.85	0.90	0.92
c	0	0.6	0	0	0.6	0.6	0.6	0.5	0	0.6	0.6	0.5	0.4	0.3	0.25	0.20	0.18	0.17
Parameter	ns	ns^2	nf	nf^2	nf^3	nf^4	nf^5	nf^6	nf^7	nf^8	nf^9	nf^{10}	nf^{11}	nf^{12}	nf^{13}	nf^{14}		
a	4	4	3.7	3.4	3.1	2.8	2.5	2.2	2.0	1.8	1.6	1.4	1.3	1.2	1.1	1.0		
b	0	0	0.6	0.7	0.8	0.85	0.90	0.92	0.93	0.94	0.95	0.96	0.96	0.97	0.97	0.97		
c	0	0	0.4	0.3	0.25	0.20	0.18	0.17	0.16	0.15	0.14	0.13	0.12	0.12	0.11	0.11		

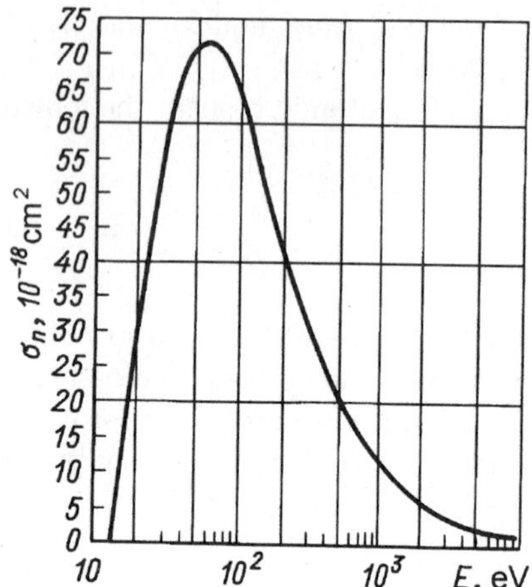

Figure 19.1 Cross-section of e–H ionization (relative error $\pm 7\%$) [31].

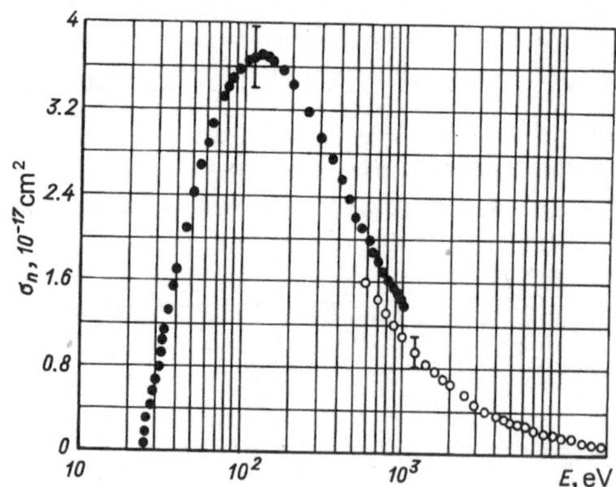

Figure 19.2 Cross-section of e–He ionization: • – [32]; ∘ – [33].

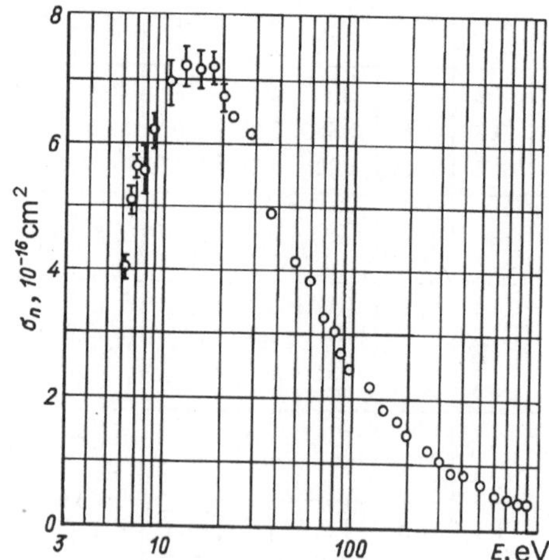

Figure 19.3 Total cross-section of e–He ionization from the metastable state $2\,^3S$ [34].

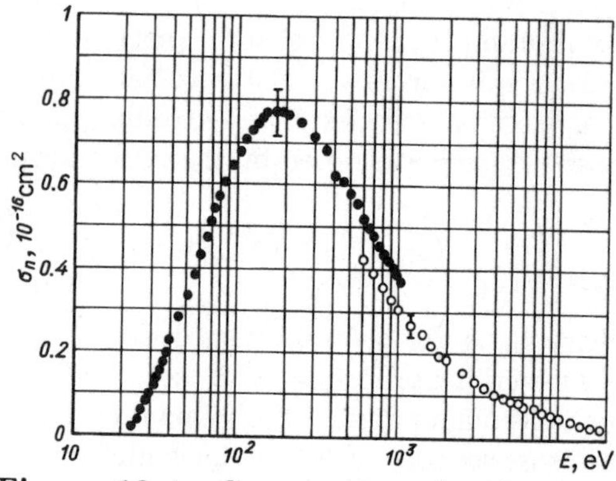

Figure 19.4 Cross-section of e–Ne ionization: • – [32]; ∘ – [33].

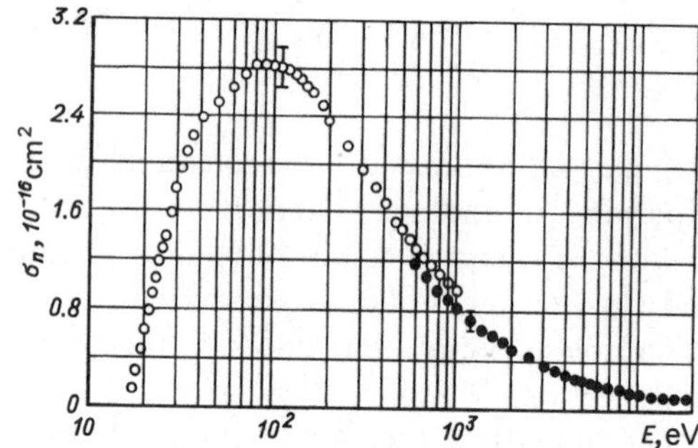

Figure 19.5 Cross-section of e–Ar ionization: ∘ – [32]; • – [33].

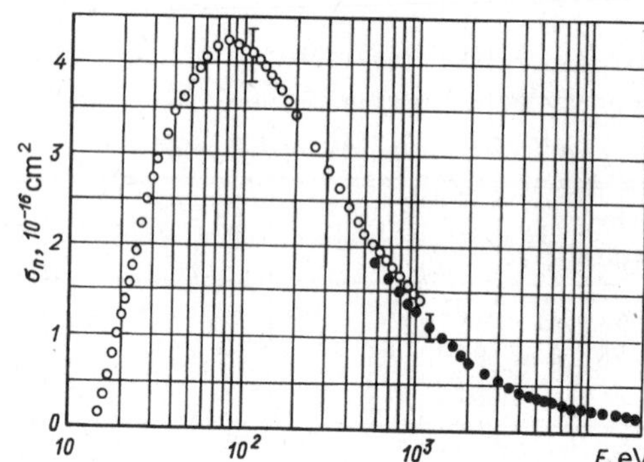

Figure 19.6 Cross-section of e–Kr ionization: ∘ – [32]; • – [33].

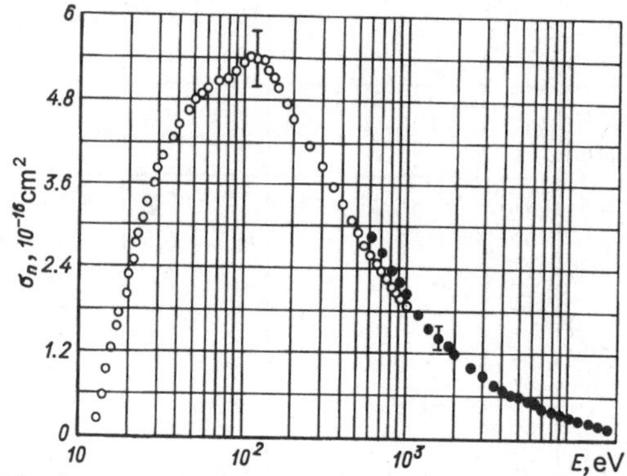

Figure 19.7 Cross-section of e–Xe ionization: ∘ – [32]; • – [33].

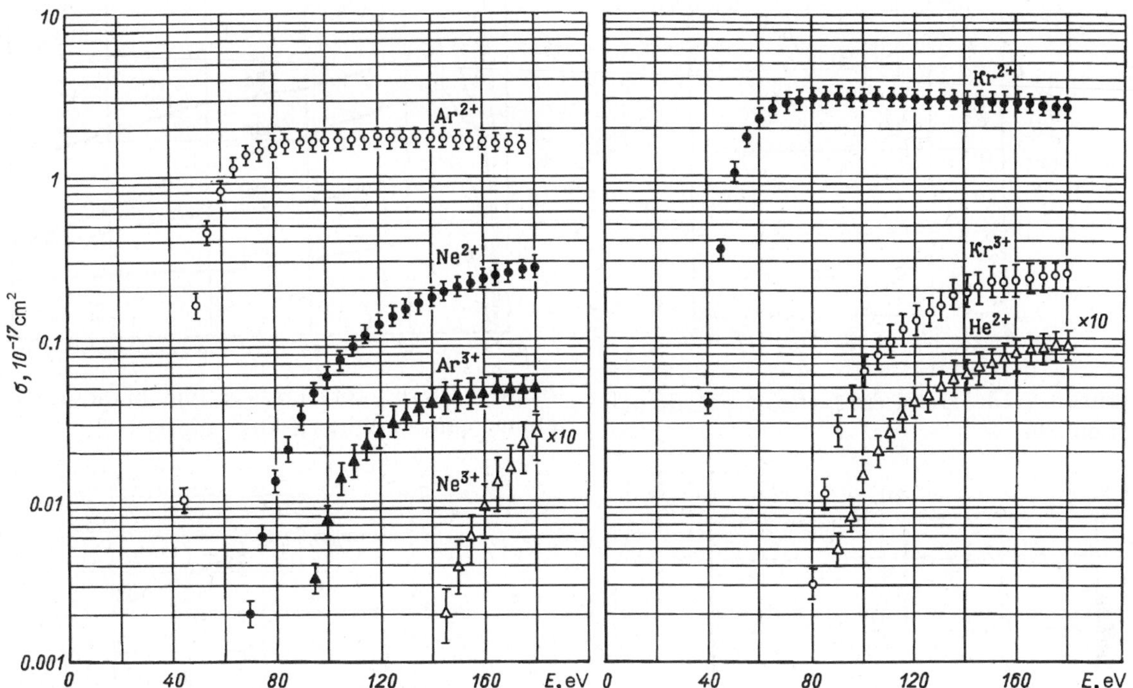

Figure 19.8 Cross-sections of rare gas multiple ionization by electron impact [35]. *Note*: data for Ne^{3+} and He^{2+} were multiplied by 10.

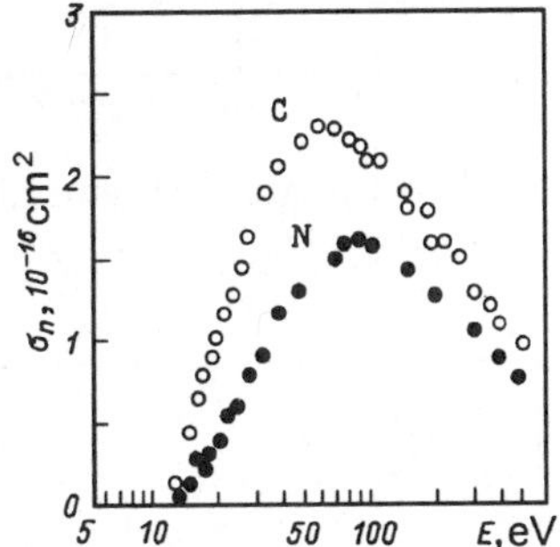

Figure 19.9 Total cross-sections of e–C and e–N ionization [36].

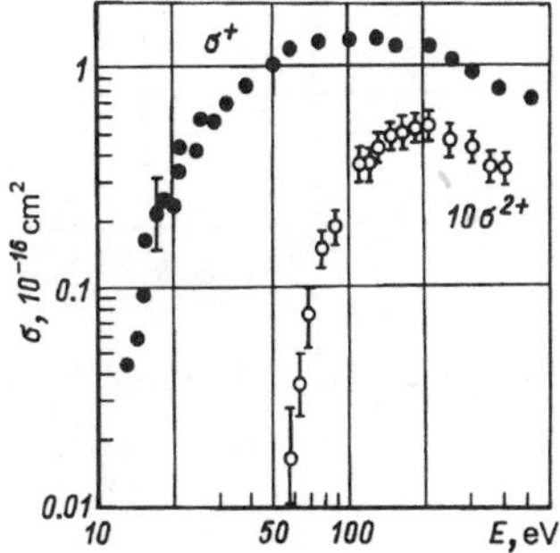

Figure 19.10 Cross-sections of e–O ionization: σ^+ – [36]; σ^{2+} – [37].

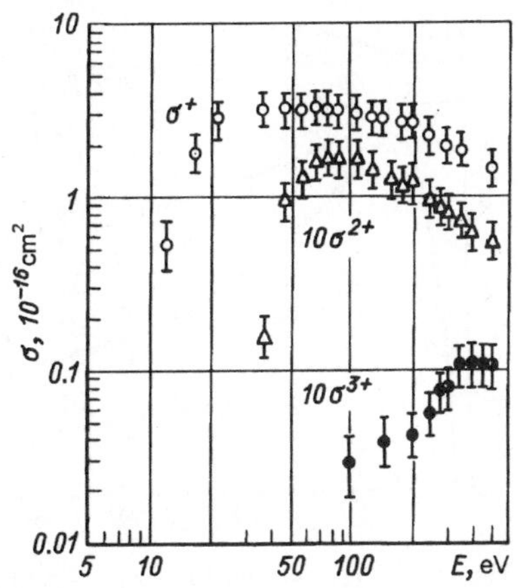

Figure 19.11 Cross-section of e–S ionization [38].

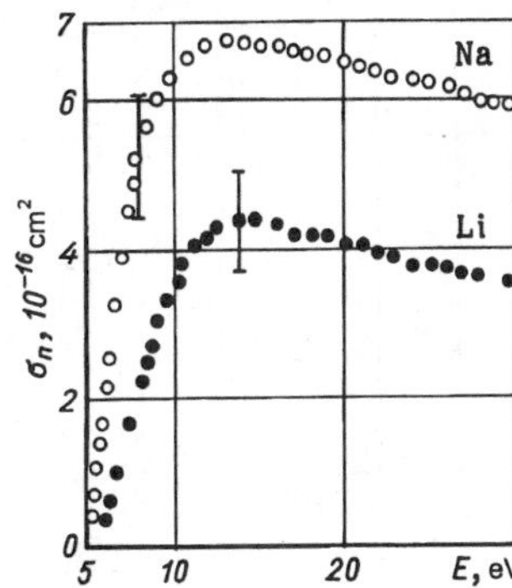

Figure 19.12 Total cross-sections of e–Li and e–Na ionization [39].

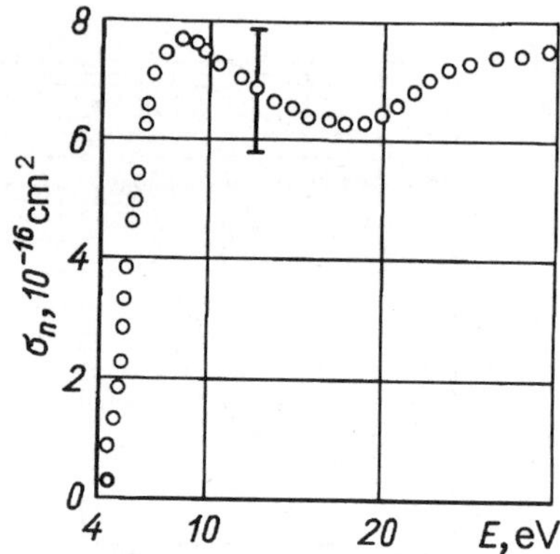

Figure 19.13 Total cross-section of e–K ionization [39].

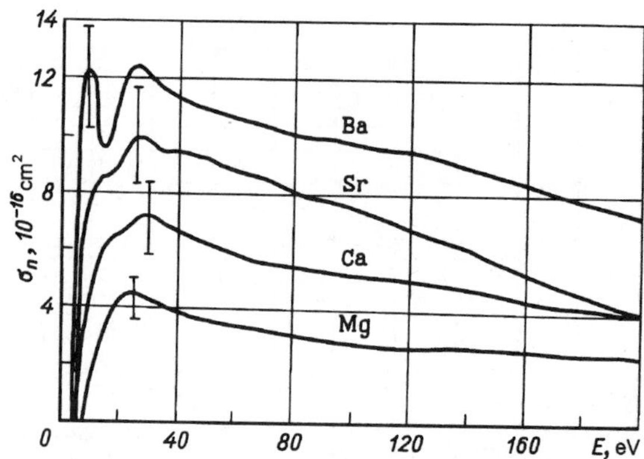

Figure 19.16 Total cross-sections of rare-earth atomic ionization by electron impact [42].

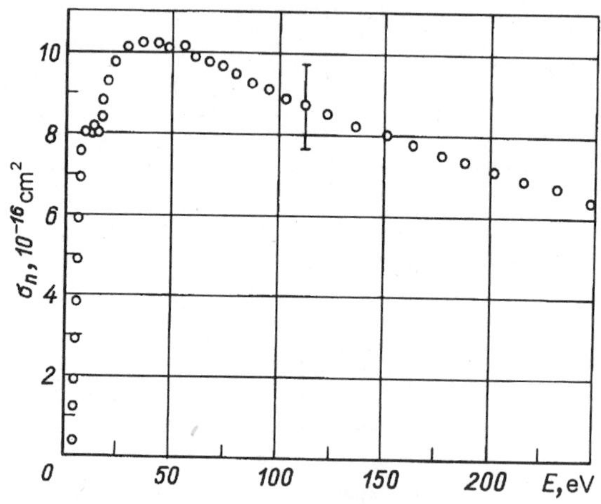

Figure 19.14 Total cross-section of e–Rb ionization [40].

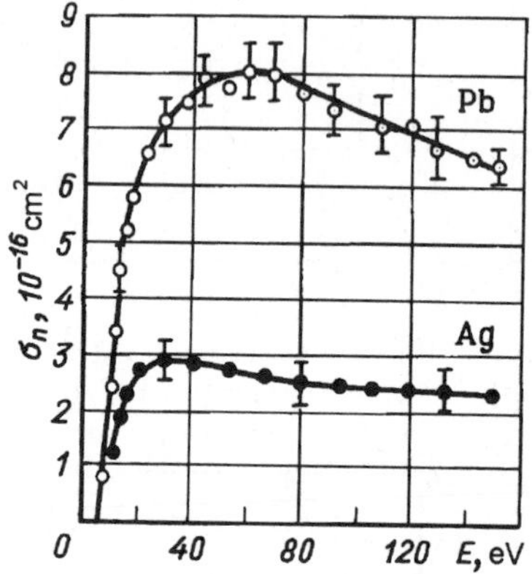

Figure 19.17 Total cross-sections of e–Pb and e–Ag ionization [45].

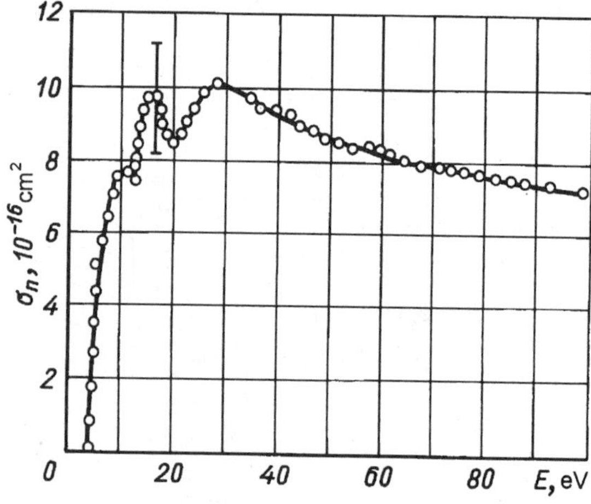

Figure 19.15 Total cross-section of e–Cs ionization [41].

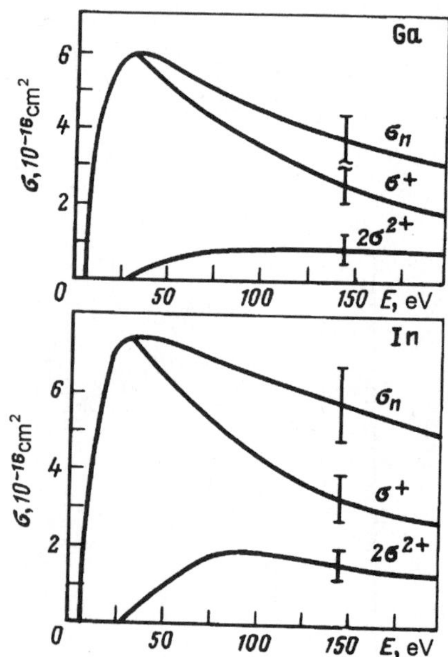

Figure 19.18 Cross-sections of e–Ga and e–In ionization [46].

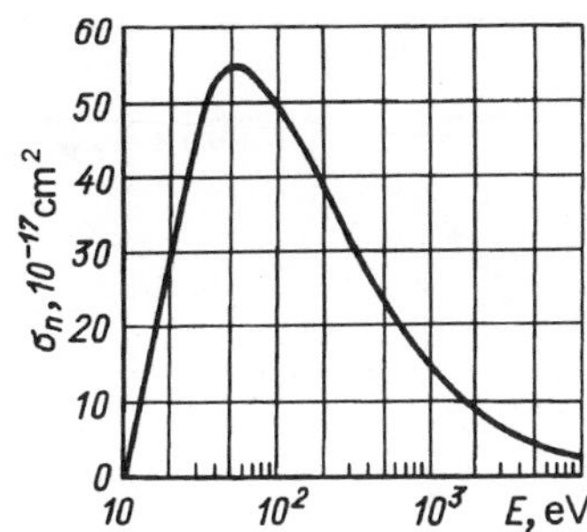

Figure 19.19 Total cross-section of e–Hg ionization (relative error ±20% [43,44].

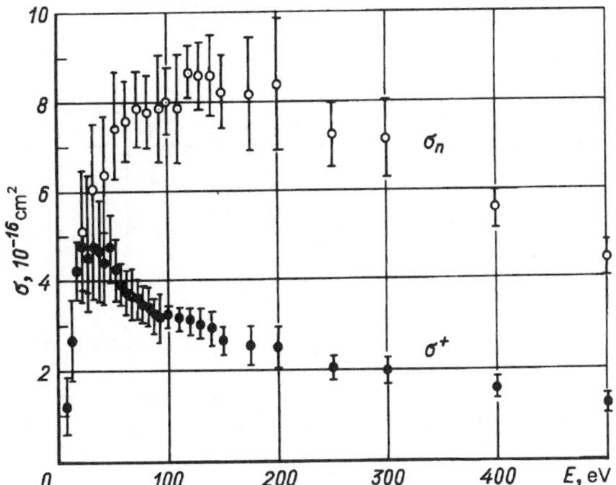

Figure 19.20 Total and one-fold cross-sections of e–U ionization [47].

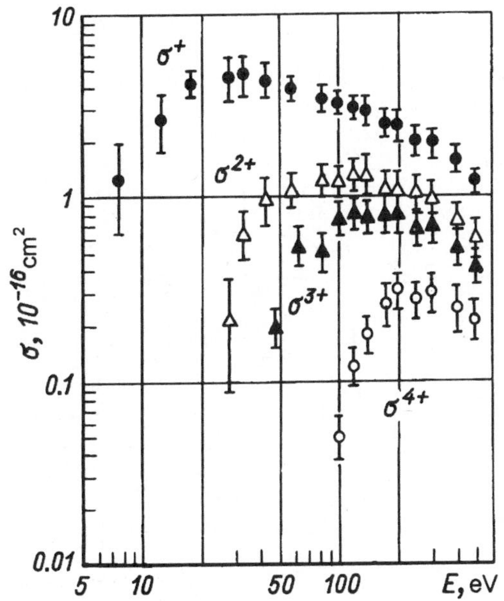

Figure 19.21 Cross-sections of multiple e–U ionization [47].

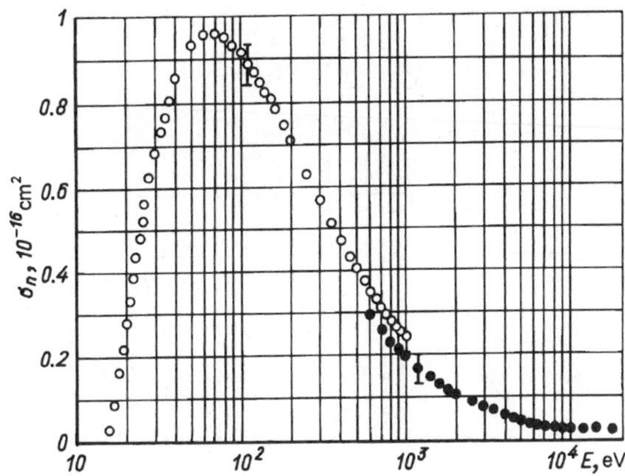

Figure 19.22 Total cross-section of e–H_2 ionization: ○ – [32]; ● – [33].

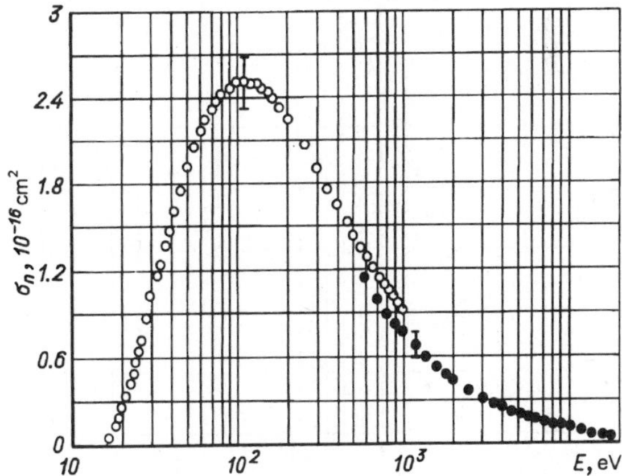

Figure 19.23 Total cross-section of e–N_2 ionization: ○ – [32]; ● – [33].

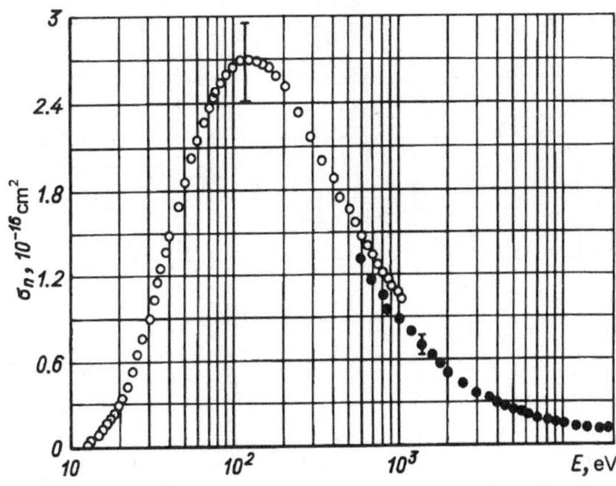

Figure 19.24 Total cross-section of e–O_2 ionization: ○ – [32]; ● – [33].

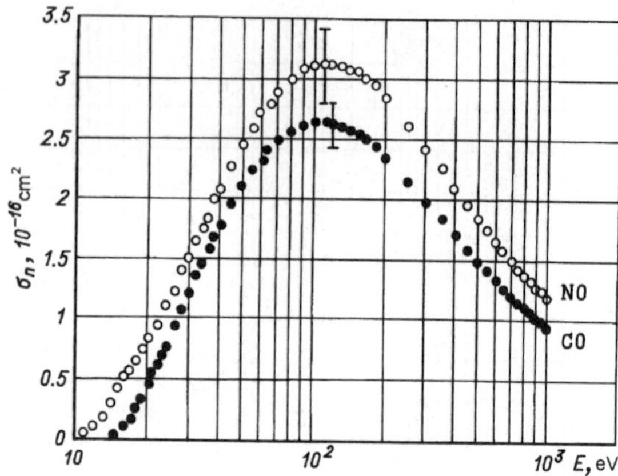

Figure 19.25 Total cross-sections of e–CO and e–NO ionization [32].

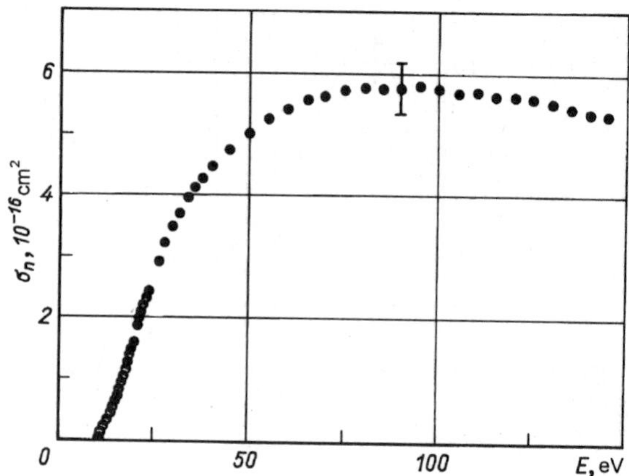

Figure 19.28 Total cross-section of e–C_2H_4 ionization [32].

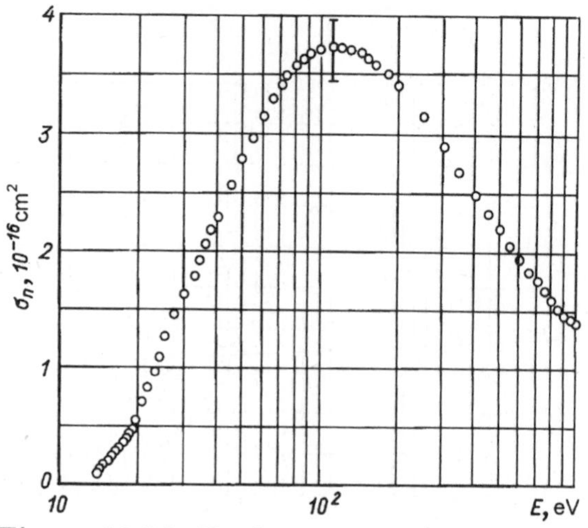

Figure 19.26 Total cross-section of e–N_2O ionization [32].

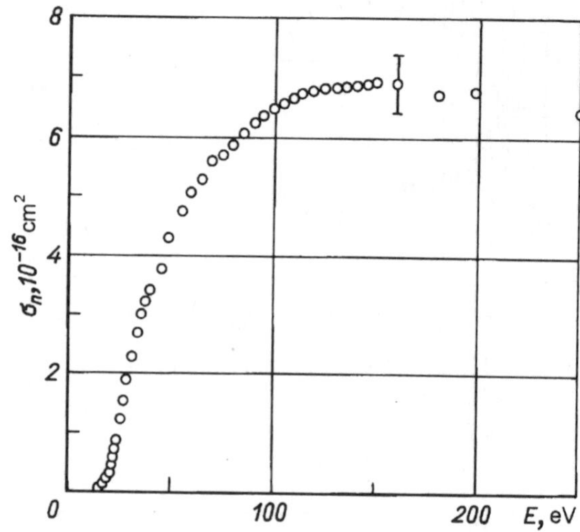

Figure 19.29 Total cross-section of e–SF_6 ionization [32].

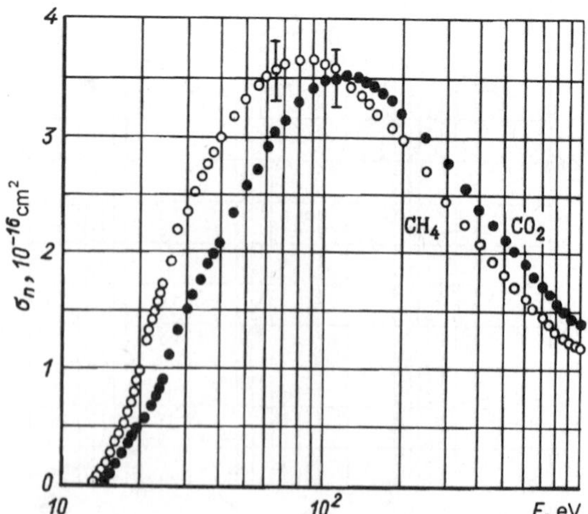

Figure 19.27 Total cross-sections of e–CO_2 and e–CH_4 ionization [32].

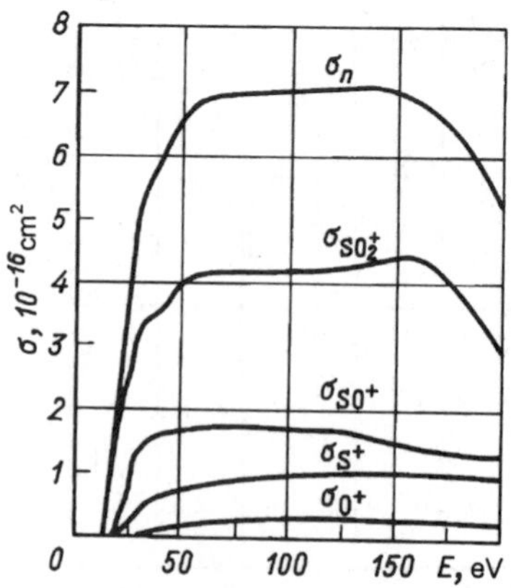

Figure 19.30 Cross-sections of total and dissociative SO_2 ionization by electron impact [47,48].

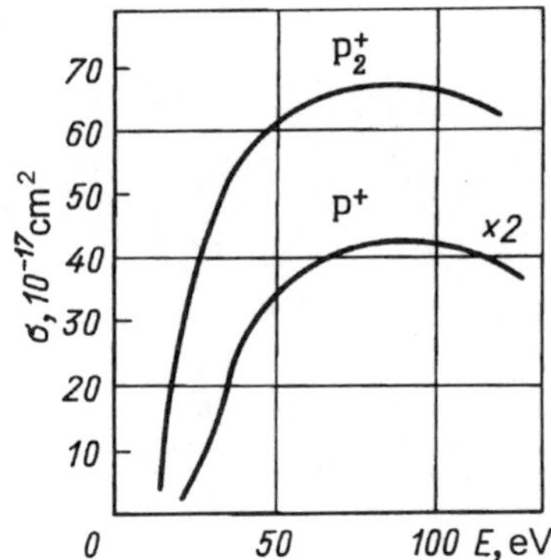

Figure 19.31 Cross-sections of e–P_2 ionization [49].

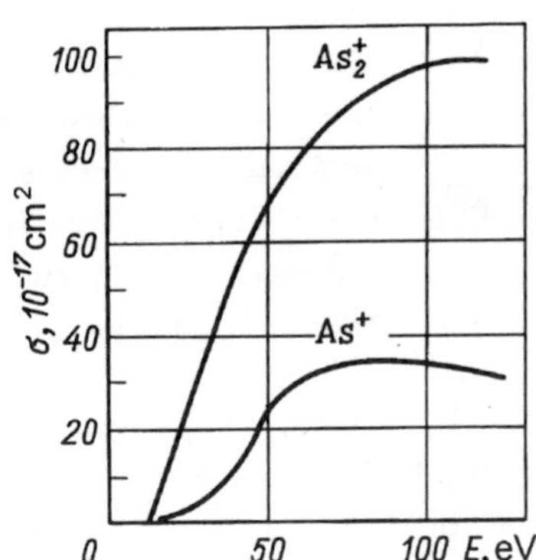

Figure 19.33 Cross-sections of e–As_2 ionization [49].

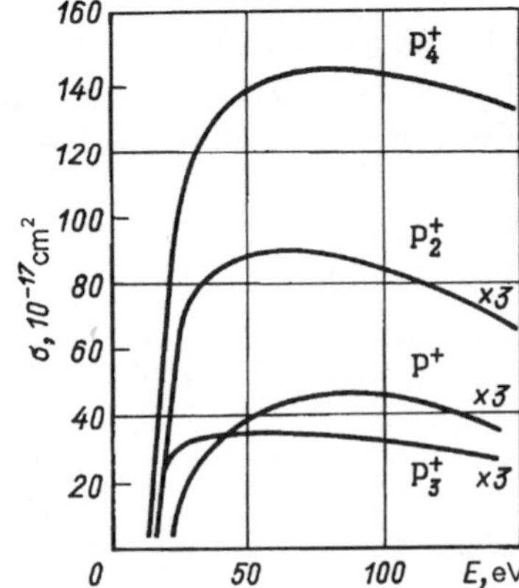

Figure 19.32 Cross-sections of e–P_4 ionization [49].

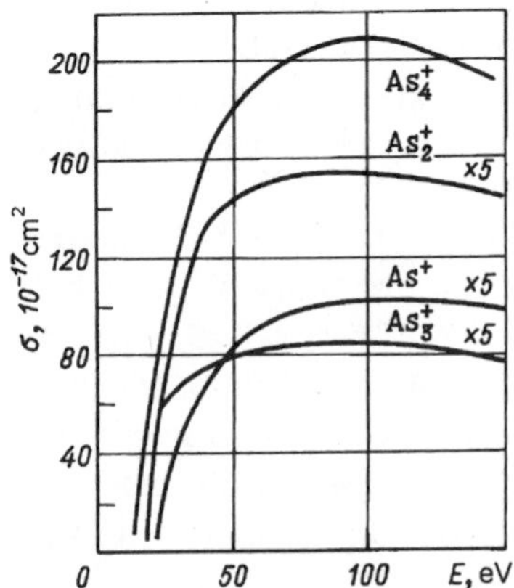

Figure 19.34 Cross-sections of e–As_4 ionization [49].

References

[1] Edlen, B., Atomic spectra. Spectroscopy I, Ed. Flugge, S., Encyclopedia of physics, Springer, Berlin, 1964 (vol. 27).

[2] Moore, C. E., Ionization potentials and ionization limits derived from optical spectra, Nat. Stand. Ref. Data Ser., Nat. Bur. Stand., NBS, Washington, 1970 (vol. 34).

[3] Martin, W. C., Zalubas, R., Hagan, L., Atomic energy levels. The rare-earth elements, Nat. Stand. Ref. Data Ser., Nat. Bur. Stand., NBS, Washington, 1978 (vol. 60).

[4] Worden, E. F., Conway, J. G., Multistep laser photoionization of the lanthanides and actinides. Lanthanide and actinide chemistry and spectroscopy, Amer. Chem. Soc. Symp. Series 131, Ed. Edelstein, N. M., Amer. Chem. Soc., Washington, 1980 (ch. 19).

[5] Kelly, R. L., Atomic and ionic spectrum lines below 2000 Å(H through Ar), Preprint ORNL-5922, Oak Ridge, Tennessee, Oct. 1982.

[6] Radzig, A. A., Smirnov, B. M., Parameters of atoms and atomic ions, Energoatomizdat, Moscow, 1986 (in Russian).

[7] Huber, K.-P., Herzberg, G., Molecular spectra and molecular structure. IV. Constants of diatomic molecules, Van Nostrand Reinhold, New York, 1979.

[8] Energies of chemical bond breakage. Ionization potentials and electron affinities, Ed. Kondratiev, V. N., Nauka, Moscow, 1974 (in Russian).

[9] Carlson, T. A., Photoelectron and Auger spectroscopy, Plenum Press, New York, 1975.

[10] Shirley, D. A., Martin, R. L., Kowalczyk, S. P., Phys. Rev., B15, 544, 1977.

[11] Mehlhorn, W., Breuckmann, B., Hausamann, D., Phys. Scripta, 16, 177, 1977.

[12] Porter, F. T., Freedman, M. S., J. Phys. Chem. Ref. Data, 7, 1267, 1978.

[13] Sevier, K. D., Atom. Data Nucl. Data Tabl., 24, 323, 1979.

[14] Fuggle, J. C., Martensson, N., J. Electron Spectr. Related Phenomena, 21, 275, 1980.

[15] Siegbahn, H., Karlsson, L., Photoelectron spectroscopy. Encyclopedia of physics, Ed. Mehlhorn, W., Springer, Berlin, 1982 (vol. 31).

[16] Jolly, W. L., Bomben, K. D., Eyermann, C. J., Atom. Data Nucl. Data Tabl., 31, 433, 1984.

[17] Massey, H., Negative ions, 3rd ed., Cambridge Univ. Press, Cambridge (UK), 1976.

[18] Smirnov, B. M., Negative ions, McGraw–Hill, New York, 1985.

[19] Hotop, H., Lineberger, W. C., J. Phys. Chem. Ref. Data, 4, 539, 1975; 14, 731, 1985.

[20] Mead, R. D., Stevens, A. E., Lineberger, W. C., Photodetachment in negative ion beams, in: Gas Phase Ion Chemistry, Ed. Bowers, M. T., Academic Press, New York, 1984 (vol. 3, ch. 22).

[21] Drzaic, P. S., Marks, J., Brauman, J. I., Electron photodetachment from gas-phase molecular anions, in: Gas Phase Ion Chemistry, Ed. Bowers, M. T., Academic Press, New York, 1984 (vol. 3, ch. 21).

[22] Christodoulides, A. A., McCorkle, D. L., Christophorou, L. G., Electron affinities of atoms, molecules, and radicals, in: Electron-Molecule Interactions and Their Applications, Ed. Christophorou, L. G., Academic Press, New York – London, 1984 (vol.2).

[23] Walder, R., Franklin, J. L., Intern. J. Mass Spectrom. Ion Phys., 36, 85, 1980.

[24] Lias, S. G., Liebman, J. F., Levin, R. D., J. Phys. Chem. Ref. Data, 13, 695, 1984.

[25] Raksit, A. B., Bohme, D. K., Intern. J. Mass Spectrom. Ion Phys., 57, 211, 1984.

[26] Bethe, H. A., Ann. Phys., 5, 325, 1930.

[27] Lotz, W., Z. Phys., 206, 205, 1967.

[28] Lotz, W., Z. Phys., 232, 101, 1970.

[29] Sobel'man, I. I., Vainshtein, L. A., Yukov, E. A., Excitation of atoms and broadening of spectral lines, Springer, Heidelberg, 1981.

[30] Vainshtein, L. A., Sobel'man, I. I., Yukov, E. A., Atomic excitation and broadening of spectral lines, Nauka, Moscow, 1979 (in Russian).

[31] Bell, K. L., Gilbody, H. B., Hughes, J. G., J. Phys. Chem. Ref. Data, 12, 891, 1983.

[32] Rapp, D., Englander–Golden, P., J. Chem. Phys., 43, 1464, 1965.

[33] Schram, B. L., de Heer, F. J., van der Wiel, M. J., Kistemaker, J., Physica, 31, 94, 1965.

[34] Dixon, A. J., Harrison, M. F. A., Smith, A. C. H., J. Phys. B: Atom and Molec. Phys., 15, 2617, 1976.

[35] Stephan, K., Helm, H., Mark, T. D., J. Chem. Phys., 73, 3763, 1980.

[36] Brook, E., Harrison, M. F. A., Smith, A. C. H., J. Phys. B: Atom. and Molec. Phys., 11, 3115, 1978.

[37] Ziegler, D. I., Newman, J. H., Smith, K. A., Stebbings, R. F., Planet. and Space Sci., 30, 451, 1982.

[38] Ziegler, D. I., Newman, J. H., Goeller, L. N., Planet. and Space Sci., 30, 1269, 1982.

[39] Zapesochnii, I. I., Aliksakhin, I. S., Zhurnal eksperimental'noi i teoreticheskoi fiziki, 55, 76, 1968 (in Russian).

[40] Nygaard, K. J., Hahn, Yu. B., J. Chem. Phys., 58, 3494, 1973.

[41] Nygaard, K. J., J. Chem. Phys., 49, 1995, 1968.

[42] Vainshtein, L. A., Ochkur, V. I., Rakhovskii, V. I., Stepanov, A. M., Zhurnal eksperimental'noi i teoreticheskoi fiziki, 61, 511, 1971 (in Russian).

[43] Kieffer, L. J., Dunn, G. H., Rev. Mod. Phys., 38, 1, 1966.

[44] Lotz, W., Astrophys. J. Suppl., 14, 207, 1967.

[45] Pavlov, S. I., Rakhovskii, V. I., Fedorova, G. M., Zhurnal eksperimental'noi i teoreticheskoi fiziki, 52, 21, 1967 (in Russian).

[46] Golovach, D. G., Rakhovskii, V. I., Shustryakov, V. M., Izmeritel'naya tekhnika, 2, 52, 1986 (in Russian).

[47] Halle, J. C., Lo, H. H., Fite, W. L., Phys. Rev., 23A, 1708, 1981.

[48] Orient, O. J., Srivastava, S. K., J. Chem. Phys., 80, 140, 1984.

[49] Monnom, G., Gancherel, Ph., Paparoditis, C., J. Phys., 45, 77, 1984.

20

Transport Phenomena in Weakly Ionized Plasma

A.V. Eletskii

20.1 Diffusion and Mobility of Charged Particles in Weakly Ionized Plasma

In an ideal weakly ionized plasma, where the ionization degree satisfies the condition $N_e/N_a \ll (kT_e/\mathrm{Ry})^2$ (N_e and N_a are the number densities of electrons and neutral particles, respectively; T_e is the electron temperature; Ry = 13.6 eV is a measure of the atomic energy scale), the diffusion of charged particles (electrons and ions) depends mainly on pair collisions between these particles and neutral particles (atoms and molecules). The number densities of neutral and charged particles must, here, satisfy the ideality criteria

$$N_a \ll 1/a_0^3, \tag{20.1}$$

$$N_e \ll (kT_e)^3/e^6, \tag{20.2}$$

where e is the electron charge; k is the Boltzmann constant, and a_0 is the characteristic size of the atom. When the electric field strength is low and the mean energies of charged and neutral particles differ insignificantly, the diffusion coefficient D of the charged particles and their mobility K are related by the Einstein formula

$$D = \frac{kT}{e}K. \tag{20.3}$$

The mobility K of the charged particles is determined by the relationship $K = w/E$, where w is the drift velocity of the charged particles in an electric field of strength E. In an electric field of high strength E, when the charged particle velocity distribution is not Maxwellian and the charged particle "temperature" has no straightforward physical meaning, relation (20.3) holds true approximately, with a precision up to 10–15%, if in this case the "temperature" of charged particles is understood to be a quantity related to their mean energy $\bar{\varepsilon}$ by the relation $\bar{\varepsilon} = 3kT/2$. In plasma, the conductivity of which depends mainly on the motion of electrons in the electric field, the electron mobility K_e is related to the conductivity σ of the plasma by the expression

$$K_e = \frac{\sigma}{eN_e}. \tag{20.4}$$

Relations (20.3) and (20.4) allow determination of the electron diffusion coefficient from data on the electric conductivity of weakly ionized plasma.

0-8493-2861-6/97/$0.00+$.50

If the typical distance, at which the density of charged particles in a plasma varies significantly, is much larger than the Debye radius

$$r_D = \sqrt{\frac{kT_e}{8\pi N_e e^2}}, \tag{20.5}$$

then the plasma is quasineutral, i.e. the densities of positive and negative charges coincide. In this case, joint diffusion of charged particles of differing signs (ambipolar diffusion) occurs with a sole diffusion coefficient D_a for electrons and ions, which is expressed through the ion diffusion coefficient D in the same gas:

$$D_a = D_i \left(1 + \frac{T_e}{T_i}\right), \tag{20.6}$$

where T_i and T_e are the ion and electron temperatures, respectively. Relation (20.5) holds true with the same precision as the Einstein relation (20.3).

The main quantities characterizing the motion of electrons in a gas under the influence of an electric field are the transverse diffusion coefficient $D_\perp$ and the electron drift velocity w_e. However, since the transverse diffusion coefficient of electrons in a gas depends on the density of the gas particles, w_e and the characteristic electron energy $\bar{\varepsilon}_e$, defined as the ratio between $D_\perp$ and the electron mobility K_e, are normally used as reference quantities. Both of these parameters are single-valued functions of the ratio of the electric field strength E and the number density N_a of the gas particles. The quantities $\bar{\varepsilon}_e$ and w_e measured for certain gases at various values of the ratio E/N_a are presented in Table 20.1. Diffusion coefficients D_i of ions in a zero electric field and at atmospheric pressure are given in Table 20.2.

Table 20.1 Drift velocity w_e (in 10^5 cm/s) and the ratio $eD_\perp/K_e$ (in eV) for electrons in some gases [1]

E/N, 10^{-17} V·cm^2	He		Ne		Ar		Kr		Xe [1, 13, 14, 17]	
	w_e	$eD_\perp/K_e$	w_e	$eD_\perp/K_e$	w_e	$eD_\perp/K_e$	w_e	$eD_\perp/K_e$	w_e	$eD_\perp/K_e$
0.03	0.7	0.034	1.3		1.2	0.94	1		0.16	0.21
0.05	1	0.042	1.6		1.4	1.2	1.2		0.5	1.1
0.07	1.3	0.051	1.8		1.5	1.3	1.2		0.74	1.9
0.1	1.6	0.065	2.1	0.4	1.6	1.5	1.3		0.88	2.0
0.3	2.8	0.15	3.4	0.91	2.2	2.3	1.6		1.1	4.0
0.5	3.7	0.24	4.3	1.4	2.5	2.8	1.7		1.2	5.0
0.7	4.3	0.32	5	1.8	2.7	3.2	1.9		1.3	
1	5.1	0.43	5.8	2.5	3	3.8	2		1.4	
3	8.8	1.2	15	5.2	4	7.2	3.2		1.8	
5	12	2.3	19	6	4.9	8.2			2.6	
7	16	3	26	6.4	7.2	8.2			4.0	
10	25	3.6	38	6.9	11	7.7			5.9	
30	68	5.3	100		26	7.7			21	
50	130	6.6	150		42	7.1			29	
70	190	7.3	200		55	6.4			49	
100	280	8.2	260		74	6.4			65	
200	580	11	460		120	6.2			130	
300	880	14	640			5.8			180	
500	1400	16				6.6			250	
800	1700	18				7.8				
1000						8.5				

E/N, 10^{-17} V·cm^2	H_2		D_2		N_2 [1, 17]		O_2		CO_2	
	w_e	$eD_\perp/K_e$	w_e	$eD_\perp/K_e$	w_e	$eD_\perp/K_e$	w_e	$eD_\perp/K_e$	w_e	$eD_\perp/K_e$
0.03	0.46	0.027	0.46	0.026	1.1	0.028	2.6		0.054	
0.05	0.74	0.028	0.74	0.027	1.6	0.031	3.7		0.095	

Table 20.1 Drift velocity w_e and the ratio $eD_\perp/K_e$ for electrons in some gases *(continued)*

E/N,	H_2		D_2		N_2 [1, 17]		O_2		CO_2	
10^{-17} V·cm^2	w_e	$eD_\perp/K_e$	w_e	$eD_\perp/K_e$	w_e	$eD_\perp/K_e$	w_e	$eD_\perp/K_e$	w_e	$eD_\perp/K_e$
0.07	1	0.029	1	0.029	2	0.035	4.1		0.14	0.026
0.1	1.4	0.032	1.4	0.031	2.4	0.042	4.3		0.19	0.026
0.3	3.1	0.046	3	0.048	3.1	0.1	5.4		0.54	0.026
0.5	4.3	0.059	4	0.067	3.5	0.16	7.3		0.89	0.026
0.7	5.2	0.074	4.7	0.088	3.9	0.20	9.2		1.2	0.026
1	6.2	0.095	5.4	0.12	4.4	0.28	10	0.14	1.8	0.026
3	9.8	0.23	8.9	0.28	7.5	0.61	22	0.3	5.4	0.03
5	13	0.33	11	0.39	11	0.74	26	0.49	9.1	0.035
7	15	0.41	14		14	0.84	28	0.73	13	0.043
10	19	0.51	17		18	0.93	31	1.1	21	0.065
30	37	1			42	1.2	72	2.4	120	0.99
50	57	1.7			61	1.4	110	2.9	130	1.7
70	83	2.2			79	1.5	120	3.0	130	2.4
100	130	2.7			100	1.9	160	3.4	130	3.2
200	340	3.8			200	3.4	260		180	4.4
300	470	4.5			290	4.2	330		240	5.0
500	730	5.4			420	5.2	450		320	6.5
800	960	6.5			530	6.2	640		420	7.1
1000	1200	7.1			590		710		490	7.9

E/N,	CO		NO		H_2O (vapor)		CH_4		CF_4 [2]		SF_6 [2]	
10^{-17} V·cm^2	w_e	$eD_\perp/K_e$	w_e	$eD_\perp/K_e$	w_e	$eD_\perp/K_e$	w_e	$eD_\perp/K_e$	w_e	$eD_\perp/K_e$	w_e	$eD_\perp/K_e$
0.3	7.4	0.045	7.8				14	0.055	30			
0.5	7.7	0.063	8.6				29	0.052	45			
0.7	9	0.08	9.1				39	0.061	57			
1	10	0.09	12				52	0.17	68			
3	18	0.18	26	0.2	0.7		100	0.15	100			
5	22	0.25	32	0.23	1.2		110	0.25	115			
7	25	0.32	38	0.25	1.6		95	0.44	130			
10	26	0.42	47	0.28	2.3		85	0.7	150	0.1	43	
30	39	0.82	66	0.63	7.5		75	1.5		0.6	53	
50	52	1.1	84	1.1	27		52	4.0		1.7	77	
70	66	1.3			58	0.022	62	4.0		3.0	86	
90	85	1.4			110	0.029	72	4.5	150	3.7	97	
100	96	1.6					74	4.9	170	4	100	
200	230	2.8					150	5.2	220	5.0	160	
300	290	3.7					190	5.1	300	5.5	190	5.1
500	430	5.1					280	5.3			260	5.5
800							400	6.6			340	6.8
1000							470	7.8			360	

E/N,	SiF_4 [12]		BF_3 [12]		Hg [15]		CH_4 [16]		NF_3 [18]		CCl_2F_2 [19]	
10^{-17} V·cm^2	w_e	$eD_\perp/K_e$	w_e	$eD_\perp/K_e$	w_e	$eD_\perp/K_e$	w_e	$eD_\perp/K_e$	w_e	$eD_\perp/K_e$	w_e	$eD_\perp/K_e$
0.3								0.055	2.4			
0.5	1.95		0.62					0.052	3.3			
0.7	2.7		0.89					0.061	3.7			
1	3.9		1.2					0.077	4.4			
3	11.6		3.8					0.15	7.9			
5	19		6.3					0.25	11			
7	26		8.9					0.44	14			
10	38		12,8		4.7	1.04		0.7	18			

Table 20.1 Drift velocity w_e and the ratio $eD_\perp/K_e$ for electrons in some gases *(continued)*

E/N,	SiF_4 [12]		BF_3 [12]		Hg [15]		CH_4 [16]		NF_3 [18]		CCl_2F_2 [19]	
10^{-17} V·cm^2	w_e	$eD_\perp/K_e$	w_e	$eD_\perp/K_e$	w_e	$eD_\perp/K_e$	w_e	$eD_\perp/K_e$	w_e	$eD_\perp/K_e$	w_e	$eD_\perp/K_e$
30					17	1.3		1.5	39			
50					24	1.5		4.0	56			
70					32	1.7		4.5				
90								4.5				
100					50	1.9		4.5				
200					110	2.5						
300					180	3.4					3.5	150
500					320	3.9					3.8	240
800					490	5.8						
1000					770	7.2						

E/N,	I_2		N_2O		C_3F_8		C_4F_{10}		CCl_2F_2		Cl_2		Br_2		NH_3	HCl
10^{-17} V·cm^2	w_e	$\frac{eD_\perp}{K_e}$	w_e	$\frac{eD_\perp}{K_e}$	w_e	$\frac{eD_\perp}{K_e}$	w_e	$\frac{eD_\perp}{K_e}$	w_e	$\frac{eD_\perp}{K_e}$	w_e	$\frac{eD_\perp}{K_e}$	w_e	$\frac{eD_\perp}{K_e}$	w_e	$\frac{eD_\perp}{K_e}$
0.1			0.7													
0.2			1.3													
0.4			2.8												0.17	
0.6			4												0.27	
0.8			6												0.36	
1			7.8												0.45	
2			18												0.85	
4			37												1.8	
6			53												2.7	
8			65												3.4	3
10			80	0.35											4.4	4
20				0.51										1.1	10	9
40		0.75		1.2							95	1.8	63	2.6	50	27
60		1.9		1.9							94	2.3	74	2.8	110	61
80	80	3.1		2.45							92	2.6	80	3.0		88
100	82	3.3		3.1							94	2.8	82	3.1		100
200											115	3.2	103	3.1		130
300					160	3.1	140	2.85		3.7						
400					190	3.8	150	3.4	190	3.7						
500					220	4.3	170	3.9	220	3.8						
600					240	4.6	200		250	3.9						
800										4.1						
1000																

Table 20.2 Diffusion coefficient D_i (in cm^2/s) of ions in various gases at atmospheric pressure and in the zero electric field [3–7] [For gas temperatures different from 300 K, its real magnitude is marked in parentheses.]

Ion	He	Ne	Ar	Kr	Xe	Rb	Cs	Hg	H_2	D_2
He^+	0.27	0.44		0.116						
He_2^+	0.43	0.24	0.098	0.076	0.059				0.28	
Ne^+	0.79	0.107	0.069	0.052	0.039				0.27	
$HeNe^+$	0.66	0.17								
Ne_2^+	0.45	0.16	0.056	0.040	0.030				0.26	
Ar^+	0.50	0.14	0.039	0.040	0.030				0.26	0.21
Kr^+	0.52	0.12	0.056	0.024	0.022				0.26	0.20
Xe^+	0.46	0.12	0.045	0.029	0.015				0.28	0.20
H^+	0.82	0.63		0.23					0.40	0.45

Table 20.2 Diffusion coefficient D_i (in cm^2/s) of ions in various gases *(continued)*

Ion	He	Ne	Ar	Kr	Xe	Rb	Cs	Hg	H_2	D_2
D^+	0.64	0.42		0.16						0.29
Hg^+	0.51	0.154	0.056	0.030				0.066 (350)		0.20
N^+	0.52	0.197		0.066						0.23
N_2^+	0.49	0.22	0.059	0.047	0.031				0.26	0.21
O^+	0.54	0.19	0.089	0.062						0.22
H_2^+	0.64	0.42		0.16						
Kr_2^+		0.14		0.028	0.021				0.29	0.20
Xe_2^+	0.36	0.12	0.043	0.026	0.019				0.025	0.20
Rb_2^+		0.13		0.030	0.021	0.016 (621)	0.0055	0.017	0.29	0.21
Cs_2^+		0.13		0.028	0.019	0.0066	0.011 (625)	0.015	0.029	0.20
Hg_2^+		0.13		0.027	0.018	0.0063	0.0047	0.019 (500)	0.29	0.20
H_3^+	0.78	0.35		0.13	0.104				0.39	0.30
D_3^+		0.26		0.096	0.074				0.33	0.21
N_3^+		0.15		0.043	0.032				0.30	0.21
N_4^+		0.15		0.039	0.028				0.029	0.21
Ba^+		0.14		0.031	0.022			0.018	0.018	0.20
O_4^+		0.14		0.038	0.027				0.29	0.21
H^-	1.02	0.58		0.23	0.18	0.053	0.047	0.16	0.50	0.45
O^-	0.75	0.19		0.062	0.047	0.014	0.0125	0.041	0.30	0.22
O_2^-	0.56	0.16		0.047	0.035	0.011	0.0093	0.030	0.33	0.22
O_3^-	0.48	0.15	0.63	0.041	0.030	0.0095	0.0080	0.026	0.29	0.21
O_4^-		0.14		0.038	0.027	0.0087	0.0072	0.023	0.29	0.20
CO_3^-		0.15	0.62	0.038	0.028	0.0089	0.0074	0.023	0.29	0.20
CO_4^-	0.43	0.14		0.036	0.026	0.0083	0.0068	0.021	0.29	0.20
O_2^+	0.55	0.21	0.067	0.045	0.032				0.26	
Li^+	0.63	0.25	0.12	0.096	0.072	0.021	0.018		0.32	0.25
Na^+	0.56	0.205	0.078	0.057	0.042	0.012	0.011		0.32	0.23
K^+	0.53	0.187	0.069	0.048	0.034	0.010	0.0086		0.33	0.24
Rb^+	0.52	0.169	0.058	0.038	0.026	0.0097 (621)	0.0066		0.33	
Cs^+	0.48	0.153	0.054	0.034	0.023	0.0073	0.0065 (628)		0.33	
U^+	0.41					0.0067				
SF_6^-	0.24						0.0057			
SF_5^-	0.28						0.0059			
Ar_2^+	0.37	0.12	0.0475	0.031	0.023				0.26	
H_2O^+	0.53	0.20	0.069							
H_3O^+	0.55	0.18	0.087 (337)						0.33	
$H_5O_2^+$	0.45	0.16	0.073 (337)							
$H_7O_3^+$	0.36	0.15	0.064 (337)							
$H_9O_4^+$		0.14	0.058 (337)				0.0076			
ReO_3^-		0.13	0.05				0.0051			
ReO_4^-		0.13	0.043				0.0051			
WO_3^-		0.13	0.051				0.0054			

Table 20.2 Diffusion coefficient D_i (in cm^2/s) of ions in various gases *(continued)*

Ion	He	Ne	Ar	Kr	Xe	Rb	Cs	Hg	H_2	D_2
NO^+	0.54	0.15	0.061	0.046	0.033				0.26	
CO^+	0.51		0.069	0.050	0.037					
$N_2O_2^+$			0.057	0.038	0.028					
$C_2O_2^+$			0.062	0.039	0.028					
C^+			0.092	0.070	0.054					
$(SF_6)_2^-$			0.047	0.028	0.019		0.0056			
$(SF_6)_3^-$			0.046	0.027	0.018		0.0047			
He_3^+	0.12 (77)		0.092	0.039						
Cl_2^-			0.055	0.037	0.026		0.0070			
Br_2^-			0.049	0.031	0.021		0.0055			
Br^-	0.49		0.058	0.035	0.025		0.0067			
N_2O^+	0.47	0.18	0.056	0.044	0.032					
NO_2^+	0.47		0.058	0.043	0.030					
SO_2^+	0.44		0.049	0.038	0.027					
ArH^+	0.50		0.062	0.043	0.032					
COH^+	0.52		0.064	0.049	0.036					
O_2H^+	0.50		0.066	0.047	0.035					
CO_2^+	0.51	0.17	0.055	0.042	0.031					
NH_3^+	0.60		0.81	0.060	0.046					
NH_4^+	0.57		0.80	0.059						
N_2OH^+	0.44	0.15	0.052		0.031					
$O_2H_2^+$	0.47	0.16	0.065		0.034					
$NO_2{\cdot}H_2O^+$	0.43	0.14	0.056		0.027					
CH_5^+	0.55	0.19	0.081		0.046					
$CH_3O_2^+$	0.44	0.15	0.055		0.030					
Cl^-	0.52	0.21	0.073	0.048	0.035	0.011	0.0090	0.029	0.30	0.21
OH^-	0.66	0.19			0.046	0.014	0.012	0.040	0.30	0.22
NO_2^-	0.49	0.15			0.035	0.0095	0.081	0.026	0.29	0.21
SO_2F^-	0.39	0.14			0.025	0.0081	0.066	0.021	0.29	0.20
$C_2H_2^-$	0.45	0.17			0.038	0.012	0.010	0.031	0.30	0.21
SO_3^-	0.40	0.14			0.026	0.0082	0.0067	0.021	0.29	0.20
HeH^+	0.61	0.28			0.081	0.024	0.022		0.34	0.27
F^-	0.76	0.18	0.086	0.060	0.043	0.013	0.012	0.038	0.30	0.22
I^-	0.42	0.14	0.059		0.022	0.0074	0.0059	0.018	0.29	0.20
N_2H^+		0.16			0.037	0.011			0.30	0.21

Ion	N_2	O_2	NO	CO	SF_6	Cl_2	Br_2	CH_4	CO_2	
He^+										
He_2^+	0.098	0.103		0.094					0.078	
Ne^+	0.073	0.074		0.070					0.054	
$HeNe^+$										
Ne_2^+	0.061	0.062		0.058					0.044	
Ar^+	0.061	0.062		0.058		0.033			0.044	
Kr^+	0.053	0.054	0.060	0.05		0.027			0.037	
Xe^+	0.046	0.103	0.057	0.094		0.024	0.016	0.059	0.035	
H^+	0.098	0.2	0.28	0.29		0.17			0.22	
D^+	0.20		0.19	0.20		0.12			0.16	
Hg^+	0.055	0.54	0.055	0.063		0.023	0.015	0.058	0.037	
N^+	0.078		0.16	0.17		0.049				
N_2^+		0.067	0.15	0.063		0.037				

Table 20.2 Diffusion coefficients D_i of ions in various gases *(continued)*

Ion	N_2	O_2	NO	CO	SF_6	Cl_2	Br_2	CH_4	CO_2
O^+	0.085		0.15	0.16		0.46			0.049
H_2^+	0.20		0.20	0.21		0.12			
Kr_2^+			0.056	0.052		0.024	0.016		
Xe_2^+	0.049	0.048	0.054	0.046		0.022	0.014		0.033
Rb_2^+			0.056	0.056		0.023	0.015		
Cs_2^+			0.054	0.051		0.022	0.014		
Hg_2^+			0.054	0.050		0.021	0.013		
H_3^+			0.17	0.16		0.098	0.082		
D_3^+			0.12	0.14		0.071	0.058		
N_3^+			0.068	0.062		0.032	0.024		
N_4^+	0.058		0.063	0.059		0.030	0.022		
Ba^+	0.06		0.057	0.053		0.024	0.016		
O_4^+	0.053	0.049	0.062	0.058		0.029	0.021		
H^-		0.056	0.28	0.26		0.17			
O^-		0.83	0.086	0.08		0.046			0.05
O_2^-		0.056	0.071	0.066		0.035			
O_3^-		0.066	0.065	0.061		0.031			
O_4^-		0.056	0.062	0.058		0.029			
CO_3^-		0.065	0.063	0.058		0.029			0.033
CO_4^-		0.063	0.060	0.056		0.027			0.032
O_2^+	0.063	0.058		0.060		0.035	0.027		0.048
Li^+	0.102	0.105		0.098		0.066	0.054		0.082
Na^+	0.073	0.071		0.066		0.04	0.031	0.067	0.057
K^+	0.065	0.069	0.058	0.059		0.033	0.025	0.066	0.037
Rb^+	0.058	0.062		0.05		0.027	0.019		0.032
Cs^+	0.057	0.057		0.052		0.025	0.016		0.029
U^+						0.023	0.014		
SF_6^-		0.052			0.014			0.056	
SF_5^-		0.057			0.015			0.056	
Ar_2^+	0.054	0.054		0.051		0.027	0.019	0.061	
H_2O^+						0.0444	0.035	0.077	
H_3O^+	0.071					0.043	0.034	0.076	
$H_5O_2^+$	0.079					0.038	0.026	0.067	
$H_7O_3^+$	0.055					0.030	0.022	0.063	
$H_9O_4^+$						0.028	0.020	0.062	
ReO_3^-						0.028	0.014	0.058	
ReO_4^-						0.024	0.014	0.058	
WO_3^-						0.027	0.014	0.058	
NO^+	0.065	0.067	0.049	0.062					0.048
CO^+				0.045					
$N_2O_2^+$			0.046						
$C_2O_2^+$				0.046					
C^+				0.070					
$(SF_6)_2^-$					0.012				
$(SF_6)_3^-$					0.011				
He_3^+									
N_2O^+	0.069						0.050		
COH^+	0.069								
CO_2^+	0.056								
Cl_2^-									
Br_2^-									
Br^-						0.083			

20.2 The Thermal Conductivity of an Ionized Gas

The heat transfer in an ionized gas is realized by both neutral and charged particles. In the case of a weakly ionized gas, the contribution of positive ions to the thermal conductivity is relatively small, since the amounts of energy carried by an ion and by a neutral atom are comparable, while the fraction of ions is relatively small. Electrons, the velocities of which exceed significantly the velocities of heavy particles, start to contribute noticeably to the thermal conductivity of a plasma already at low ionization densities, i.e. $N_e/N_a \geq \sqrt{m_e/M}$ (m_e is the electron mass; M is the mass of the atom, or of the ion). Here, since the energy exchange between the electrons and the atoms is small ($\sim m_e/M$), the contributions to the thermal conductivity of the neutral particles and of the electrons add up. Besides the above, a certain contribution to the thermal conductivity of an ionized gas in equilibrium is due to the process that establishes ionization equilibrium: $A \leftrightarrow A^+ + e$. The atoms enter the region of higher temperature and are ionized there, thus transferring an energy equal to the ionization potential. On the contrary, the charged particles passing from a hotter region to the region of lower temperature undergo recombination, thus transferring the amount of energy in the same direction. Thus, the thermal conductivity coefficient κ of a weakly ionized plasma in equilibrium is represented in the form

$$\kappa = \kappa_a + \kappa_e + \kappa_{\text{int}}, \tag{20.7}$$

where κ_a, κ_e, κ_{int} are the contributions to the thermal conductivity due to neutral particles, to electrons, and to the transfer of ionization energy, respectively. The contribution to the thermal conductivity of the last term reaches its maximum when $N_e/N_a \sim \sqrt{m_e/M}$, and in these conditions amounts to 30%. Since the equilibrium ionization degree depends not only on the temperature, but on the pressure as well, the thermal conductivity of an ionized gas also depends essentially on these two parameters. The uncertainties in the data presented in Table 20.3 amount to $\sim$ 10%. Data on the thermal conductivity of air plasma are compiled in Table 20.4.

Table 20.3 Thermal conductivity (in 10^{-2} W/(cm·K)) of rare gas plasma at atmospheric pressure and various temperatures [8]

Gas	Temperature, 10^4 K								
	2	3	4	5	6	7	8	9	10
He	0.51	0.66	1.0	1.2	1.5	1.7	1.9	2.1	2.3
Ne	0.17	0.22	0.29	0.32	0.35	0.4	0.45	0.54	0.63
Ar	0.055	0.08	0.1	0.12	0.14	0.17	0.23	0.37	0.84
Kr	0.032	0.045	0.052	0.07	0.09	0.12	0.19	0.40	0.89
Xe	0.024	0.038	0.038	0.05	0.06	0.08	0.14	0.40	1.1

Table 20.4 Thermal conductivity κ (in 10^{-2} W/(cm·K)) of air plasma at atmospheric pressure and various temperatures [10]

T, 10^3K	κ	T, 10^3 K	κ	T, 10^3 K	κ	T, 10^3 K	κ
2	0.12	7	3.42	12	2.57	17	2.95
3	0.51	8	1.75	13	3.35	18	2.44
4	0.74	9	1.38	14	3.94	19	2.21
5	0.77	10	1.44	15	4.0		
6	2.07	11	2.0	16	3.46		

20.3 The Viscosity of Weakly Ionized Plasma

The viscosity of a weakly ionized plasma composed of neutral particles, electrons, and positive ions may differ from the viscosity of the neutral gas. Here, the electrons, having a small mass, give practically no noticeable contribution to the momentum transfer, independent of the conditions, and their role in determining the viscosity of the plasma may be neglected. The contribution of ions to the viscosity starts to be significant already at low ionization degrees, since the cross section of momentum transfer occurring when an ion collides with an atom and due to resonance charge exchange of the ion on the parent atom exceeds essentially the cross section of momentum transfer due to collisions between atoms. In accordance with elementary kinetic theory, the dependence of the viscosity η of a plasma on its parameters is given by the expression

$$\eta \sim \frac{\sqrt{TM}}{\sigma_{\text{el}} + \sigma_{\text{res}} N_i/N_a}, \tag{20.8}$$

where T is the temperature of the plasma; M is the mass of an atom; σ_{el} is the cross section of elastic scattering between atoms resulting in a momentum transfer; σ_{res} is the cross section of resonance charge exchange between an ion and an atom; and N_i and N_a are the number densities of ions and atoms, respectively. In an equilibrium plasma, the ratio N_i/N_a is a function of the pressure and temperature of the plasma, which is precisely what determines the dependence of the plasma's viscosity on its parameters.

In the case of a plasma with a low ionization degree, when the contribution of ions to the viscosity may be neglected, the viscosity of the plasma increases with the temperature. Since usually $\sigma_{\text{res}} \gg \sigma_{\text{el}}$, when $N_i/N_a \sim \sigma_{\text{el}}/\sigma_{\text{res}} \ll 1$, the temperature dependence of the viscosity exhibits a maximum. The maximum values of viscosity for equilibrium plasmas of inert gases are presented in Table 20.5. The viscosities of equilibrium plasmas of inert gases, air, and Cs at various pressures and temperatures are given in Tables 20.6–20.9 and Fig. 20.1 [8–11]. The uncertainties in the data involved amount to 10–30%.

Table 20.5 Maximum values of viscosity $\eta_{\max}$ for rare gas plasma and of requisite temperature $T_{\max}$ [8, 9]

Gas	p, Pa	$T_{\max}$, K	N_i/N_a	$\eta_{\max}$, Pa·s
Ar	10^4	7800	0.069	$3.0 \cdot 10^{-4}$
	10^5	8600	0.058	$3.4 \cdot 10^{-4}$
	10^6	10200	0.048	$3.9 \cdot 10^{-4}$
Kr	10^5	10000	0.060	$3.3 \cdot 10^{-4}$
Xe	10^5	9000	0.067	$3.4 \cdot 10^{-4}$

Table 20.6 Viscosity (in 10^{-4} Pa·s) of the helium and argon plasmas at atmospheric pressure and various temperatures [8, 9]

T, 10^3 K	2	3	4	5	6	7	8	9	10	11	12
He	0.72		1.3	1.6	1.8	2.1	2.4	2.9	2.9	3.2	3.3
Ar	0.82	1.1	1.3	1.6	1.9	2.1	2.4	2.7	2.9	3.0	

Table 20.7 Viscosity (in 10^{-4} Pa·s) of the xenon plasma at various pressures and temperatures [8, 9]

p, Pa	Temperature, K				
	6000	7000	8000	9000	10,000
10^4	2.6	2.9	3.0	2.7	1.7
10^5	2.6	3.0	3.3	3.4	3.2
10^6	2.6	3.0	3.3	3.6	3.9

Table 20.8 Viscosity of air plasma at atmospheric pressure and various temperatures [10]

T, 10^3 K	2	4	6	8	10	12	14	16	18	20
η, 10^{-4} Pa·s	0.7	1.2	1.8	2.4	2.8	2.1	1.2	0.6	0.3	0.1

Table 20.9 Viscosity of partially ionized Cs vapor at various pressures and temperatures [11]

T, 10^3 K	Pressure, Pa						
	1	10	10^2	10^3	10^4	10^5	10^6
2	317	386	406	413	412	412	418
3	6.85	3.88	398	480	561	583	592
4	3.34	6.64	19.6	49.5	224	488	663
5	5.58	7.93	11.7	18.8	66.8	226	482
7	11.6	15.0	19.4	25.4	37.7	80.5	221
10	26.0	31.4	38.2	48.1	63.2	91.0	188

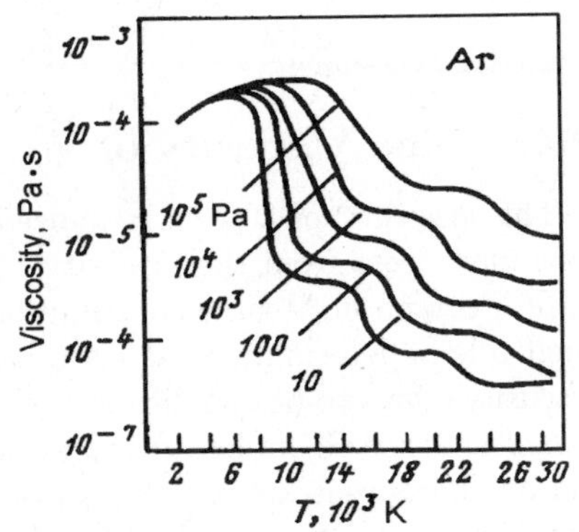

Figure 20.1 Viscosity of partially ionized argon at various pressures and temperatures

References

[1] Bychkov, V. L., Eletskii, A. V., Smirnov, B. M., in: Plasma chemistry, v.10, Ed. Smirnov, B. M., Energoatomizdat, Moscow, 1983 (issue 10) (in Russian).

[2] Gallagher, J., J. Phys. Chem. Ref. Data, 12, 109, 1983.

[3] Smirnov, B. M., Ions and excited atoms in plasma, Atomizdat, Moscow, 1974 (in Russian).

[4] McDaniel, E. W., Mason, E. A., The mobility and diffusion of ions in gases, John Wiley and Sons, New York, 1973.

[5] Linuma, K., J. Chem. Phys., 79, 3893, 1983.

[6] Lamm, D. R., J. Chem. Phys., 79, 1965, 1983.

[7] Bohringer, H., Durup-Ferguson, M., Fahey, D. W., J. Chem. Phys., 79, 1974, 1983.

[8] Eletskii, A. V., Palkina, L. A., Smirnov, B. M., Transport phenomena in a weakly ionized plasma, Atomizdat, Moscow, 1975 (in Russian).

[9] Palkina, L. A., Smirnov, B. M., Teplofizika vysokikh temperatur, 12, 3, 1974 (in Russian).

[10] Rolin, N. M., Inzhenerno–fizicheskii zhurnal, 34, 444, 1978 (in Russian).

[11] Vargaftik, N. B., Handbook of thermal properties of gases and liquids, Fizmatgiz, Moscow, 1972 (in Russian).

[12] Hunter, S. R., Carter, J. G., Christophorou, L. G., J. Appl. Phys., 65, 1858, 1989.

[13] Puech, V., Mizzi, S., J. Phys. D, 24, 1974, 1991.

[14] Patrick, E. L., Andrews, M. L., Garscadden, A., Appl. Phys. Lett., 59, 3239, 1991.

[15] Lin, J., Govinda Raju, G. R., J. Phys. D, 25, 167, 1992.

[16] Shimura, N., Makabe, T., J. Phys. D, 25, 751, 1992.

[17] Roznerski, W., Mechlinska–Dzewko, J., Leja, K., J. Phys. D, 23, 1461, 1990.

[18] Ushizoda, S., Kajita, S., Kondo, Y., J. Phys. D, 23, 47, 1990.

[19] Novak, J. P., Frechette, M. F., J. Appl. Phys., 57, 4368, 1985.

21

Electric Properties of Metals and Alloys

V.S. Egorov and I.N. Khlyustikov

21.1 Introduction

Metals are those substances that have, at absolute zero temperature, one or more energy bands which are not fully filled with electrons. A surface in the configurational (momentum) space separating the filled and the vacant states is referred to as the Fermi surface (FS). The presence of FS differentiates metals from all other substances, whereas the shape of FS gives rise to variations from one metal to another.

The resistance of metals to electric current is due to the scattering of conduction electrons. An expression for resistivity ρ is written as follows:

$$\rho = \frac{m}{Ne^2\tau} = \frac{mv_F}{Ne^2 l}, \tag{21.1}$$

where m and e are the electron mass and charge, N is the effective electron number density, v_F is the electron velocity on the Fermi surface, τ is the mean electron lifetime between two sequential collisions, and, finally, l is the corresponding mean free path.

Two possible scattering mechanisms manifest themselves: (1) electron collisions with local immobile centers – impurities, defects, etc.; (2) scattering on the lattice thermal oscillations – phonons.

As long as these processes are independent, the effective collision frequency is equal to the sum of collision frequencies for the first and second processes, that is $1/\tau_{\text{eff}} = 1/\tau_1 + 1/\tau_{\text{ph}}$. This leads to the Mathissen rule

$$\rho = \rho_{\text{res}} + \rho_{\text{id}}(T), \tag{21.2}$$

where ρ_{res} is the temperature independent residual resistivity, which is determined by pre-treatment of a particular metal sample, being lower for more perfect samples; $\rho_{\text{id}}(T)$ is the resistivity of the ideal lattice of the material, as determined by scattering on phonons. The latter is strongly temperature dependent and is described by the Bloch-Grueneisen formula

$$\rho_{id} = \frac{KT^5}{M\theta^6}\int_0^{\theta/T} \frac{\xi d\xi}{(\exp\xi - 1)(1-\exp(-\xi))}, \tag{21.3}$$

where K is a constant related to the specific metal volume, M is the atomic mass, and θ is the Debye characteristic temperature, i.e. the metal parameter defining the maximum oscillation frequency of the lattice ($\hbar\omega_{\max} = k\theta$). At high temperatures ($T/\theta \geq 1$)

$$\rho = \frac{KT}{4M\theta^2}, \tag{21.4}$$

0-8493-2861-6/97/$0.00+$.50

that is, the resistivity is a linear function of temperature, whereas at low temperatures ($T/\theta \leq 1$)

$$\rho = \frac{124.4KT^5}{M\theta^6}. \tag{21.5}$$

For metals with nearly spherical FS, the power law $\rho \propto T^5$ is coming at $T \leq 0.1\theta$. For metals with complex FS, this law is observed at much lower temperatures, when the momenta of thermal phonons become smaller than any size of the Fermi surface (for example, $T \leq 0.1$ K in bismuth).

The presence of some impurities can significantly influence $\rho_{\mathrm{id}}(T)$, thus leading to deviations from the Mathissen rule.

A scattering on magnetic impurities results in a resistivity addition that increases logarithmically with decreasing temperature (the Kondo effect). This phenomenon is used in thermometry. Some metals at low temperatures turn into the superconducting state. The shortest possible mean free path (on the order of interatomic spacing) determines the maximum metallic resistivity $\rho_{\max} \sim a\hbar^2/e^2 \cong 0.2 \cdot 10^{-3}$ Ω·cm (a is the interatomic distance).

21.2 Pure Metals

Table 21.1 lists the values of resistivity ρ_0 and the temperature coefficient of resistance $\alpha_0 = (1/\rho_0)(d\rho/dT)$ at 0°C [1]. The fourth column comprises the values of characteristic temperature θ obtained from heat capacity measurements [2]. According to Eq. (21.3), the reduced resistance $r \equiv \rho(T)/\rho_0$ should be a universal function of the reduced temperature T/θ for any metal:

$$r_T = 1.056\left(\frac{T}{\theta}\right) F\left(\frac{\theta}{T}\right). \tag{21.6}$$

Relying on data of Table 21.1, one can calculate values of the resistivity in a broad temperature range through the use of

$$\rho(T) = \rho_0 \frac{r_T}{r_0}, \tag{21.7}$$

where ρ_0 and r_0 are the resistivity and reduced resistance at 0°C, respectively, whereas the values of $F(\theta/T)$ can be found in Table 21.2 [3]. Furthermore, Table 21.3 [4, 5] gives resistivities of anisotropic monocrystals along and transverse to their principal axes. The data for graphite [6–8] correspond to both natural monocrystals of graphite–1 and artificially grown monocrystals of pyrolytic graphite–2. In addition, Table 21.4 [9] correlates the resistance of a wire and its length in the range of diameters 0.05–4.0 mm and of resistivities (0.016–1.4)$\cdot 10^{-4}$ $\Omega\cdot$ cm.

Liquid metals. Melting leads to an increase in resistance for the majority of metals. However, the resistance is diminished in the metals whose volume decreases after melting (see Tables 21.5, 21.6, and Figs. 21.1, 21.2).

Table 21.1 Resistivity ρ_0, temperature coefficient of resistance α_0 at 0°C, and the Debye temperature θ for pure metals [1, 2]

Metal	ρ_0, $10^{-6}\,\Omega\cdot$cm	α_0, 10^{-5} K^{-1}	θ, K	Metal	ρ_0, $10^{-6}\,\Omega\cdot$cm	α_0, 10^{-5} K^{-1}	θ, K
Aluminum	2.50	460	433	Bismuth	110.0	454	120
Antimony	39.0	511	220	Cadmium	7.07	462	210
Arsenic	26.0	475	282	Calcium	4.06	417	229
Barium	36.0	649	111	Cerium	72.7	97	179
Beryllium	3.2	900	1481	Cesium	18.1	503	40

Table 21.1 Resistivity ρ_0, temperature coefficient of resistance α_0 at 0°C, and Debye temperature θ *(continued)*

Metal	ρ_0, $10^{-6}\,\Omega\cdot$cm	α_0, $10^{-5}\,K^{-1}$	θ, K	Metal	ρ_0, $10^{-6}\,\Omega\cdot$cm	α_0, $10^{-5}\,K^{-1}$	θ, K
Chromium	14.1	301	606	Osmium	9.5	420	467
Cobalt	5.57	604	460	Palladium	9.77	377	271
Niobium	16.1	343	276	Platinum	9.81	396	237
Copper	1.55	433	347	Polonium	42.0	460	
Dysprosium	56.0	119	183	Potassium	6.1	673	91
Erbium	107.0	252	188	Praseodymium	65.8	171	152
Gadolinium	140.0	176	182	Rhenium	18.9	455	416
Gallium	40.0	369	325	Rhodium	4.35	462	512
Gold	2.06	402	162	Rubidium	11.29	637	56
Hafnium	30.0	440	252	Ruthenium	7.16	458	555
Holmium	87.0	171	190	Samarium	88.0	148	169
Indium	8.19	490	112	Silver	1.49	430	227
Iridium	4.93	411	420	Sodium	4.28	546	156
Iron	8.6	651	477	Strontium	30.3	383	147
Lanthanum	57.6	213	150	Tantalum	12.4	382	245
Lead	19.2	428	105	Thallium	16.2	517	78
Lithium	8.55	489	344	Thorium	13.0	275	160
Lutetium	79.0	240	183	Thulium	79.0	195	200
Magnesium	4.31	412	403	Tin	11.15	465	199
α−Manganese	278.0	50	409	Titanium	42.0	546	420
β−Manganese	91.0	136		Tungsten	4.89	510	383
γ−Manganese	39.2	628		Uranium	21.0	282	248
Mercury	94.07	99	72	Vanadium	18.2	390	398
Molybdenum	5.03	473	423	Ytterbium	30.0	130	118
Neodymium	71.0	200	163	Zinc	5.65	417	328
Nickel	6.14	692	477	Zirconium	41.0	440	290

Table 21.2 Values of the function $F(\theta/T)$ at different θ/T [3]

θ/T	F(θ/T)	θ/T	F(θ/T)	θ/T	F(θ/T)	θ/T	F(θ/T)	θ/T	F(θ/T)
0.0	1.0000	2.2	0.7733	4.4	0.4008	6.6	0.1725	8.8	0.07272
0.1	0.9994	2.3	0.7559	4.5	0.3867	6.7	0.1658	8.9	0.07000
0.2	0.9978	2.4	0.7383	4.6	0.3729	6.8	0.1593	9.0	0.06740
0.3	0.9950	2.5	0.7205	4.7	0.3595	6.9	0.1531	9.1	0.06490
0.4	0.9912	2.6	0.7026	4.8	0.3466	7.0	0.1471	9.2	0.06250
0.5	0.9862	2.7	0.6846	4.9	0.3340	7.1	0.1414	9.3	0.06021
0.6	0.9803	2.8	0.6666	5.0	0.3217	7.2	0.1359	9.4	0.05800
0.7	0.9733	2.9	0.6486	5.1	0.3098	7.3	0.1306	9.5	0.05589
0.8	0.9653	3.0	0.6307	5.2	0.2983	7.4	0.1255	9.6	0.05386
0.9	0.9563	3.1	0.6128	5.3	0.2871	7.5	0.1206	9.7	0.05192
1.0	0.9465	3.2	0.5950	5.4	0.2763	7.6	0.11599	9.8	0.05005
1.1	0.9357	3.3	0.5775	5.5	0.2658	7.7	0.11150	9.9	0.04826
1.2	0.9241	3.4	0.5600	5.6	0.2557	7.8	0.10719	10.0	0.04655
1.3	0.9118	3.5	0.5428	5.7	0.2460	7.9	0.10306	10.1	0.04490
1.4	0.8986	3.6	0.5259	5.8	0.2366	8.0	0.09909	10.2	0.04332
1.5	0.8848	3.7	0.5091	5.9	0.2275	8.1	0.09529	10.3	0.04181
1.6	0.8704	3.8	0.4927	6.0	0.2187	8.2	0.09165	10.4	0.04035
1.7	0.8554	3.9	0.4766	6.1	0.2103	8.3	0.08816	10.5	0.03896
1.8	0.8398	4.0	0.4608	6.2	0.2021	8.4	0.08480	10.6	0.03762
1.9	0.8238	4.1	0.4453	6.3	0.1942	8.5	0.08159	10.7	0.03633
2.0	0.8073	4.2	0.4301	6.4	0.1867	8.6	0.07851	10.8	0.03509
2.1	0.7905	4.3	0.4153	6.5	0.1795	8.7	0.07555	10.9	0.03390

Table 21.2 Values of the function $F(\theta/T)$ at different θ/T [3] *(continued)*

θ/T	$F(\theta/T)$	θ/T	$F(\theta/T)$	θ/T	$F(\theta/T)$	θ/T	$F(\theta/T)$	θ/T	$F(\theta/T)$
11.0	0.03276	12.4	0.02073	14.6	0.010915	17.4	0.005427	38.0	0.0002387
11.1	0.03167	12.5	0.02009	14.8	0.010344	17.6	0.005185	40.0	0.0001944
11.2	0.03061	12.6	0.01948	15.0	0.029805	17.8	0.004956	44.0	0.0001328
11.3	0.02960	12.7	0.01889	15.2	0.009302	18.0	0.004740	48.0	0.00009375
11.4	0.02863	12.8	0.01832	15.4	0.008831	19.0	0.003819	50.0	0.00007964
11.5	0.02769	12.9	0.01777	15.6	0.008389	20.0	0.003111	52.0	0.00006806
11.6	0.02680	13.0	0.01725	15.8	0.007974	22.0	0.002125	56.0	0.00005061
11.7	0.02593	13.2	0.01624	16.0	0.007584	24.0	0.001500	60.0	0.00003841
11.8	0.02510	13.4	0.01531	16.2	0.007218	26.0	0.001089	64.0	0.00002967
11.9	0.02430	13.6	0.01445	16.4	0.006873	28.0	0.0008097	68.0	0.00002328
12.0	0.02353	13.8	0.01364	16.6	0.006549	30.0	0.0006145	70.0	0.00002073
12.1	0.02279	14.0	0.01289	16.8	0.006243	32.0	0.0004747	72.0	0.00001852
12.2	0.02208	14.2	0.012185	17.0	0.005955	34.0	0.0003724	76.0	0.00001492
12.3	0.02139	14.4	0.011528	17.2	0.005683	36.0	0.0002963	80.0	0.00001215

Table 21.3 Resistivity of anisotropic metal crystals [4–8]

Metal	t,°C	Crystalline structure	$\rho_{\parallel}$, 10^{-6} Ω·cm	$\rho_{\perp}$, 10^{-6} Ω·cm	$\rho_{\perp}/\rho_{\parallel}$
Antimony	0	rhomb.	26.3	36.0	1.37
Arsenic	0	hex.	26	23.8	0.92
Beryllium	0	hex.	3.58	3.12	0.88
Bismuth	0	rhomb.	127	99	78
Cadmium	0	hex.	7.73	6.35	0.87
Gallium	20	rhomb.	55.5	17.3	0.31
Graphite–1	20	hex.	5400	41	$7.6 \cdot 10^{-3}$
	−195	hex.	2300	20	$8.7 \cdot 10^{-3}$
Graphite–2	20	hex.	$1.7 \cdot 10^{5}$	44.3	$2.6 \cdot 10^{-4}$
	−195	hex.	$3 \cdot 10^{5}$	25.9	$8.6 \cdot 10^{-5}$
Hafnium	0	hex.	32.7	32.0	0.98
Magnesium	0	hex.	3.58	4.22	1.21
Mercury	−45.5	rhomb.	17.8	23.5	1.32
Tin	0	tetr.	9.09	13.08	1.46
Tungsten	−183		0.892 [100]	0.843 [111]	0.945
Zinc	0	hex.	5.59	5.39	0.96

Note: Graphite–1 and Graphite–2 are the natural and artificial pyrolytic ones, respectively.

Table 21.4 Resistance and length of wires with diameter 0.05–4.0 mm and resistivity $(0.016 - 1.40) \cdot 10^{-4} \Omega \cdot$ cm

Diameter	0.016	0.017	0.018	0.019	0.020	0.025	0.027	0.030	0.040
↓	**Resistance of a 1-meter long wire, Ω**								
0.05	8.15	8.66	9.17	9.68	10.19	12.73	13.75	15.3	20.4
0.10	2.04	2.16	2.29	2.42	2.55	3.18	3.44	3.8	5.1
0.15	0.91	0.96	1.02	1.07	1.13	1.41	1.53	1.70	2.26
0.20	0.51	0.54	0.57	0.60	0.64	0.80	0.859	0.95	1.27
0.25	0.33	0.35	0.37	0.39	0.41	0.51	0.550	0.61	0.81
0.30	0.226	0.240	0.255	0.269	0.283	0.354	0.382	0.424	0.566
0.40	0.127	0.135	0.143	0.151	0.159	0.199	0.215	0.239	0.318

Table 21.4 Resistance and length of wires with diameter 0.05–4.0 mm *(continued)*

Diameter	0.016	0.017	0.018	0.019	0.020	0.025	0.027	0.030	0.040
↓	**Resistance of a 1-meter long wire, Ω**								
0.50	0.081	0.087	0.092	0.097	0.102	0.127	0.138	0.153	0.204
0.60	0.057	0.060	0.064	0.067	0.071	0.088	0.0955	0.106	0.141
0.70	0.042	0.044	0.047	0.049	0.052	0.065	0.0702	0.078	0.104
0.80	0.0318	0.0338	0.0358	0.0378	0.0398	0.0497	0.0537	0.060	0.080
1.00	0.0204	0.0216	0.0229	0.0242	0.0255	0.0318	0.0344	0.038	0.051
1.2	0.0142	0.0150	0.0159	0.0168	0.0177	0.0221	0.0239	0.0265	0.0354
1.4	0.0104	0.0110	0.0117	0.0123	0.0130	0.0162	0.0175	0.0195	0.0260
1.6	0.0080	0.0085	0.0090	0.0095	0.0099	0.0124	0.0134	0.0149	0.0199
2.0	0.0051	0.0054	0.0057	0.0060	0.0064	0.0080	0.00859	0.0095	0.0127
2.5	0.00326	0.00346	0.300367	0.00387	0.00407	0.0051	0.00550	0.0061	0.0081
3.0	0.00226	0.00240	0.00255	0.00269	0.00283	0.0035	0.00382	0.0042	0.0057
3.5	0.00166	0.00177	0.00187	0.00197	0.00208	0.00260	0.00281	0.00312	0.00416
4.0	0.00127	0.00135	0.00143	0.001581	0.00159	0.00199	0.00215	0.00239	0.00318
	Length of a wire with the resistance of 1 Ω, m								
0.05	0.123	0.115	0.109	0.103	0.098	0.079	0.073	0.065	0.049
0.10	0.49	0.46	0.44	0.41	0.39	0.314	0.291	0.262	0.196
0.15	1.11	1.04	0.98	0.93	0.88	0.71	0.655	0.59	0.44
0.20	1.96	1.85	1.75	1.65	1.57	1.26	1.16	1.05	0.79
0.25	3.07	2.89	2.73	2.58	2.45	1.96	1.82	1.64	1.23
0.30	4.4	4.2	3.9	3.7	3.5	2.83	2.62	2.36	1.77
0.40	7.9	7.4	7.0	6.6	6.3	5.03	4.65	4.19	3.14
0.50	12.3	11.5	10.9	10.3	9.8	7.85	7.27	6.54	4.91
0.60	27.6	16.6	15.7	14.9	14.1	11.3	10.5	9.4	7.1
0.70	24.0	22.6	21.4	20.3	19.2	15.4	14.3	12.8	9.6
0.80	31.4	29.6	27.9	26.4	25.1	20.1	18.6	16.8	12.6
1.00	49.1	46.2	43.6	41.3	39.3	31.4	29.1	26.2	19.6
1.2	71	66	63	59	57	45	41.9	37.7	28.3
1.4	96	90	86	81	77	62	57.0	51.3	38.5
1.6	126	118	112	106	101	80	74.5	67	50
2.0	196	185	175	165	157	126	116	105	79
2.5	307	289	273	258	245	196	182	164	123
3.0	442	416	393	372	353	283	262	236	177
3.5	601	566	535	506	481	385	356	321	241
4.0	785	739	698	661	628	503	465	419	314

Diameter	0.05	0.06	0.07	0.08	0.10	0.15	0.20	0.25	0.30
↓	**Resistance of a 1-meter long wire, Ω**								
0.05	25.5	30.6	36	41	51	76	102	127	153
0.10	6.4	7.6	8.9	10.2	12.7	19.1	25.5	31.8	38
0.15	2.83	3.4	4.0	4.5	5.7	8.5	11.3	14.1	17.0
0.20	1.59	1.91	2.23	2.55	3.18	4.8	6.4	8.0	9.5
0.25	1.02	1.22	1.43	1.63	2.04	3.06	4.1	5.1	6.1
0.30	0.707	0.849	0.99	1.13	1.41	2.12	2.83	3.54	4.24
0.40	0.398	0.477	0.56	0.64	0.80	1.19	1.59	1.99	2.39
0.50	0.255	0.306	0.36	0.41	0.51	0.76	1.02	1.27	1.53
0.60	0.177	0.212	0.248	0.283	0.354	0.53	0.71	0.88	1.06
0.70	0.130	0.156	0.182	0.208	0.260	0.39	0.52	0.65	0.78
0.80	0.099	0.119	0.139	0.159	0.199	0.298	0.398	0.497	0.60
1.00	0.064	0.070	0.080	0.102	0.127	0.191	0.255	0.318	0.38

Table 21.4 Resistance and length of wires with diameter 0.05–4.0 mm *(continued)*

Diameter	0.05	0.06	0.07	0.08	0.10	0.15	0.20	0.25	0.30
↓	**Resistance of a 1-meter long wire, Ω**								
1.2	0.0442	0.053	0.062	0.071	0.088	0.133	0.177	0.221	0.265
1.4	0.0325	0.039	0.045	0.052	0.065	0.097	0.130	0.162	0.195
1.6	0.0249	0.0298	0.0348	0.0398	0.0497	0.075	0.099	0.124	0.149
2.0	0.0159	0.0191	0.0223	0.0255	0.0318	0.048	0.064	0.080	0.095
2.5	0.0102	0.0122	0.0143	0.0163	0.0204	0.0305	0.0407	0.051	0.061
3.0	0.0071	0.0085	0.0099	0.0113	0.0141	0.0212	0.0283	0.035	0.042
3.5	0.0052	0.0062	0.0073	0.0083	0.0104	0.0156	0.0208	0.0260	0.0312
4.0	0.0040	0.0048	0.0056	0.0064	0.0080	0.0119	0.0159	0.0199	0.0239
	Length of a wire with the resistance of 1 Ω, m								
0.05	0.039	0.0327	0.0280	0.0245	0.0196	0.0131	0.0098	0.0079	0.0065
0.10	0.157	0.131	0.112	0.098	0.079	0.052	0.039	0.0314	0.0262
0.15	0.35	0.295	0.252	0.221	0.177	0.118	0.088	0.071	0.059
0.20	0.63	0.52	0.45	0.39	0.314	0.209	0.157	0.126	0.105
0.25	0.98	0.82	0.70	0.61	0.49	0.327	0.245	0.196	0.164
0.30	1.41	1.18	1.01	0.88	0.71	0.47	0.35	0.283	0.236
0.40	2.51	2.09	1.80	1.57	1.26	0.84	0.63	0.503	0.419
0.50	3.93	3.27	2.80	2.45	1.96	1.31	0.98	0.785	0.654
0.60	5.7	4.7	4.0	3.53	2.83	1.88	1.41	1.13	0.94
0.70	7.7	6.4	5.5	4.81	3.85	2.57	1.92	1.54	1.28
0.80	10.1	8.4	7.2	6.3	5.0	3.25	2.51	2.01	1.68
1.00	15.7	13.1	11.2	9.8	7.9	5.24	3.93	3.14	2.62
1.20	22.6	18.8	16.2	14.1	11.3	7.5	5.7	4.5	3.77
1.40	30.8	25.7	22.0	19.2	15.4	10.3	7.7	6.2	5.13
1.60	40	33.5	28.7	25.1	20.1	13.4	10.1	8.0	6.7
2.0	63	52.4	44.9	39.2	31.4	20.9	15.7	12.6	10.5
2.5	98	82	70	61	49	32.7	24.5	19.6	16.4
3.0	141	118	101	88	71	47.1	35.3	28.3	21.6
3.5	192	160	137	120	96	64	48	38.5	32.1
4.0	251	209	180	175	126	84	63	50.3	41.9

Diameter	0.40	0.50	0.60	0.75	0.85	1.00	1.10	1.20	1.40
↓	**Resistance of a 1-meter long wire, Ω**								
0.05	204	255	306	382	433	509	560	611	713
0.10	51	64	76.4	95.5	108	127	140	153	178
0.15	22.6	28.3	33.9	42.4	48	56.6	62.2	67.9	79.2
0.20	12.7	15.9	19.1	23.9	27.1	31.8	35.0	38.2	44.6
0.25	8.1	10.2	12.2	15.3	17.3	20.4	22.4	24.4	28.5
0.30	5.66	7.07	8.49	10.6	12.0	14.1	15.5	17.0	19.8
0.40	3.18	3.98	4.78	5.97	6.76	7.96	8.76	9.55	11.1
0.50	2.04	2.55	3.06	3.82	4.33	5.09	5.60	6.11	7.13
0.60	1.41	1.77	2.12	2.65	3.00	3.54	3.89	4.24	4.95
0.70	1.04	1.30	1.56	1.95	2.21	2.60	2.86	3.12	3.64
0.80	0.80	0.99	1.19	1.49	1.69	1.99	2.19	2.39	2.79
1.00	0.51	0.64	0.764	0.955	1.08	1.27	1.40	1.53	1.78
1.2	0.354	0.442	0.531	0.663	0.75	0.884	0.973	1.06	1.24
1.4	0.260	0.325	0.390	0.487	0.55	0.650	0.715	0.779	0.909
1.6	0.199	0.249	0.198	0.373	0.423	0.497	0.547	0.597	0.696
2.0	0.127	0.159	0.191	0.239	0.271	0.318	0.350	0.382	0.446
2.5	0.081	0.102	0.122	0.153	0.173	0.204	0.224	0.244	0.285

Table 21.4 Resistance and length of wires with diameter 0.05–4.0 mm *(continued)*

Diameter	0.40	0.50	0.60	0.75	0.85	1.00	1.10	1.20	1.40
↓	**Resistance of a 1-meter long wire, Ω**								
3.0	0.057	0.071	0.085	0.106	0.120	0.141	0.156	0.170	0.198
3.5	0.0416	0.052	0.062	0.078	0.088	0.104	0.114	0.125	0.146
4.0	0.0318	0.040	0.048	0.060	0.068	0.080	0.087	0.095	0.111
	Length of a wire with the resistance of 1 Ω, m								
0.05	0.0049	0.0039	0.0033	0.0026	0.0023	0.0020	0.0018	0.0016	0.0014
0.10	0.0196	0.0157	0.0131	0.0105	0.0092	0.0078	0.0071	0.0065	0.0056
0.15	0.044	0.035	0.0295	0.00236	0.0208	0.0177	0.0161	0.0147	0.0126
0.20	0.079	0.063	0.0524	0.0419	0.0370	0.0314	0.0286	0.0262	0.0224
0.25	0.123	0.098	0.0818	0.0655	0.0578	0.0491	0.0446	0.0409	0.0351
0.30	0.177	0.141	0.118	0.0943	0.083	0.0707	0.0643	0.0589	0.0505
0.40	0.314	0.251	0.209	0.168	0.148	0.126	0.114	0.105	0.0898
0.50	0.491	0.393	0.327	0.262	0.231	0.196	0.179	0.164	0.140
0.60	0.71	0.57	0.471	0.377	0.332	0.283	0.257	0.236	0.202
0.70	0.96	0.77	0.641	0.513	0.453	0.385	0.350	0.321	0.275
0.80	1.26	1.01	0.838	0.670	0.592	0.503	0.457	0.419	0.359
1.00	1.96	1.57	1.31	1.05	0.924	0.785	0.0714	0.655	0.561
1.20	2.83	2.26	1.89	1.51	1.33	1.13	1.03	0.943	0.808
1.40	3.85	3.08	2.57	2.05	1.81	1.54	1.40	1.28	1.10
1.60	5.0	4.0	3.35	2.68	2.37	2.01	1.83	1.68	1.44
2.0	7.9	6.3	5.24	4.19	3.7	3.14	2.86	2.62	2.24
2.5	12.3	9.8	8.18	6.55	5.8	4.91	4.46	4.09	3.51
3.0	17.7	14.1	11.8	9.43	8.3	7.07	6.43	5.89	5.05
3.5	24.1	19.2	16.0	12.8	11.3	9.62	8.75	8.02	6.87
4.0	31.4	25.1	20.9	16.8	14.8	12.6	11.4	10.5	8.98

Table 21.5 Resistivity and volume changes of some metals under melting [1]

Metal	t_{melt}, °C	ρ_{liq}, $10^{-6}\ \Omega \cdot cm$	ρ_{liq}/ρ_{sol}	$\Delta V/V_{liq}$
Aluminum	660.0	20.1	1.64	+0.048
Antimony	630.0	108.0	0.71	-0.09
Bismuth	271.0	123.0	0.43	-0.033
Cadmium	321.0		1.89	+0.05
Cesium	29.7		1.66	+0.026
Copper	1083.0	21.5	2.07	+0.042
Gallium	29.9	25.9	0.58	-0.03
Gold	1063.0	30.8	2.28	+0.051
Indium	156.0		2.12	
Iron	1535.0	139.0	1.09	+0.03
Lithium	180.0		1.68	+0.017
Magnesium	651.0	27.9	1.63	+0.041
Mercury	-38.9	90.0	3.36	+0.037
Potassium	62.5		1.56	+0.026
Rubidium	38.7		1.61	+0.028
Silver	961.0	16.4	1.9	+0.038
Sodium	97.6		1.45	+0.027
Thallium	302.0		2.0	+0.03
Tin	327.4	99.3	2.07	+0.035
Zinc	420.0	32.6	2.11	+0.042

Table 21.6 Resistivities of some liquid metals ρ (in 10^{-6} Ω·cm) [9]

t, °C	K	$K_{56}Na_{44}$	$K_{78}Na_{22}$	Cu	Ni	Sn	Sb	Zn	Cd	Al	Fe	Pb
100	15.49	41.61	45.63									
200	21.8	47.23	51.33									
300	28.2	54.33	58.58			49.4						
400		62.21	65.65			51.6			33.7			98
500		69.37	73.48			53.9		34.5(5)	34.1			103
600		78.29	82.61			56.0		35.5(5)	34.8			107
700		88.23	91.76			58.3	129	35.6(5)	35.8	27.8		112
800		99.68	104.51			60.5	131	35.7(0)		29.3		116
900						62.7	133	35.7(5)		30.8		121
1000						65.0	135			32.2		126
1100				21.5		67.2	138			33.7		
1200				22.4		69.5	140			35.2		
1300				23.3		71.7						
1400				24.2		74.0						
1500				25.0	109	76.2						
1550					110						133	
1600					110(5)						136	
1650					111(5)						138	

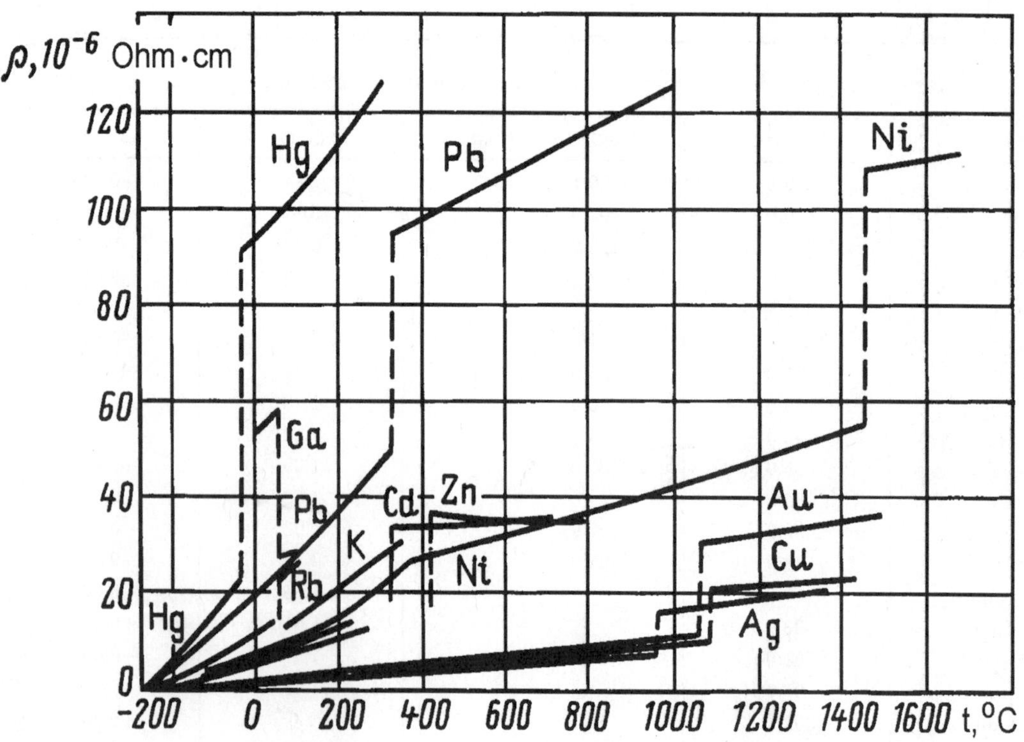

Figure 21.1 Plots of resistivity vs. temperature for some metals in solid and liquid states [9].

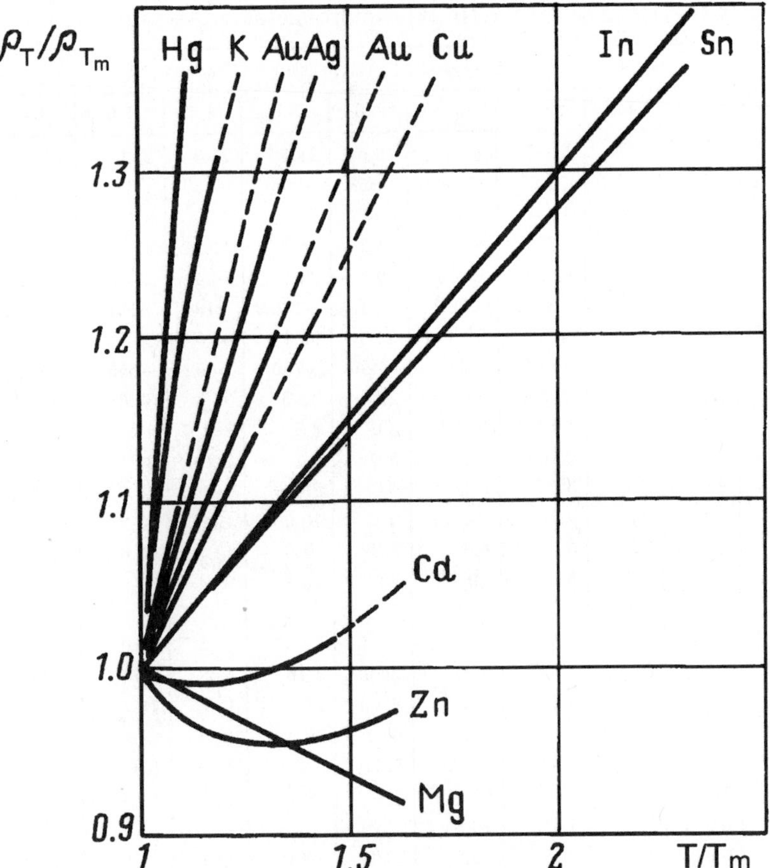

Figure 21.2 Temperature dependence of relative resistivity for various liquid metals (dashed curves are interpolated; T_m is the melting point) [4].

21.3 The Effect of Hydrostatic Compression on the Resistance of Metals

As a rule, the resistance of metals diminishes under the effect of external hydrostatic pressure P. Dependence $R(P)$ in some metals can be nonmonotonic, with breaks and steps as a result of phase transformations. The latter are used in high pressure physics as reference points. The relative variations of resistance (not the resistivity, which should take into account the sample size variations) with opposite sign are given in Table 21.7 in the pressure range $0 < P \leq 10$ GPa. All data correspond to measurements at 25–30°C [1].

Table 21.7 Relative resistance change (in %) of pure metals $-\Delta R/R_0 = -(R_p - R_0)/R_0$ under hydrostatic pressure [1]

Metal	Pressure, GPa										
	0.5	1.0	2.0	3.0	4.0	5.0	6.0	7.0	8.0	9.0	10.0
Aluminum	2.15	4.06	7.66	10.8	13.5		17.6		20.6		23.0
Antimony		−6.0	−10.4	−10.3	−7.8	−2.1	6.2	16.3	24.4	32.6	39.5
Antimony (41° to c-axis)	−5.5	−12.6	−25.6	−31.5							
Antimony (87° to c-axis)	−2.6	−4.5	−4.2	−4.7							

Table 21.7 Relative resistance change (in %) of pure metals *(continued)*

Metal	Pressure, GPa										
	0.5	1.0	2.0	3.0	4.0	5.0	6.0	7.0	8.0	9.0	10.0
Arsenic, $\perp c$		6.7	11.8	15.6	18.6	21.1	23.3	25.0	26.6	28.2	30.2
Barium*1	−24.5	−55.6	−137.7	−246.6	−358	−296	−137	−89	−81.5	−80.5	−81
Beryllium	0.85	1.6	3.1	4.6	5.9	7.1	8.3	9.4	10.4	11.6	12.4
Bismuth*2	−6.5	−15.2	−38.3	39.6	43.3	45.4	47.2	48.3	50.2	51.5	52.6
Cadmium	3.95	7.32	12.8	17.16	20.8	23.8	26.4	28.7	30.7	32.5	34.2
Calcium	−7.2	−15.2	−33.4	−56.7	−81.1	−106.9	−134	−161.8	−191.6	−224.4	−265.0
Cerium*11	−2.0	45.2	49.4	51.3	52.3	53.1	53.8	53.7	52.9	51.4	51.8
Cesium*10	18.8	−0.5	−91.7	−320	−496	−790	−832	−655	−555	−488	−433
Chromium*9	12.45	17.33	22.56	26.88	29.9	32.7	35.3	37.6	39.9	42.0	44.2
Cobalt	0.465	0.904	1.704	2.397	3.0	3.5	3.9	4.2	4.5	4.7	4.9
Niobium	0.69	1.366	2.662	3.888	5.0	6.2	7.2	8.2	9.1	9.9	10.6
Copper	0.9	1.8	3.5	5.1	6.6	8.0	9.3	10.5	11.6	12.5	13.4
Dysprosium		2.3	4.7	7.0	9.1	10.6	12.0	13.2	15.2	17.1	16.3
Erbium		2.7	5.0	6.8	8.2	11.0	13.7	15.6	16.8	17.4	17.6
Gadolinium		4.6	8.8	19.9	23.6	26.8	29.4	31.6	33.3	34.8	35.9
Gallium*3	2.65	5.31									
Gold	1.5	2.9	5.6	8.1	10.4	12.6	14.5	16.0	17.1	17.9	18.4
Hafnium		0.87	1.63	2.27	2.79	3.16	3.48	3.74	3.93	4.08	4.19
Holmium		2.2	4.4	6.6	9.0	11.6	13.8	15.7	17.3	18.4	19.2
Indium	1.7	12.2	21.4	28.3	34.1	39.0	43.5	47.1	50.4	53.5	56.2
Iridium	0.69	1.368	2.694	3.978	5.2	6.4	7.6	8.6	9.6	10.5	11.4
Iron	1.2	2.3	4.4	6.4	8.1	19.7	11.2	12.5	13.8	14.9	15.9
Lanthanum*4	0.9	1.7	3.5	10.6							
Lead	6.8	12.5	21.8	29.2	34.9	39.4	42.7	45.4	47.6	49.5	51.0
Lithium	−3.5	−7.2	−5.2	−23.9	−30.5	−36.6	−42.8	−48.7	−55.1	−12.3	−70.4
Lutetium		1.31	2.86	4.75	6.40	7.81	9.03	10.09	10.98	11.75	12.39
Magnesium	2.29	4.7	8.4	11.4	14.1	16.3	18.3	20.0	21.4	22.4	23.3
Manganese	1.92	3.54	6.08	7.89	9.1	10.0	10.8	11.7	12.8	14.4	15.9
Mercury*6	11.4	20.8	85.0	95.6	112.6	114.6	116.7	118.5	120.1	121.6	122.8
Molybdenum	0.65	1.289	2.532	3.729	4.9	6.0	7.1	8.1	9.1	10.0	10.8
Neodymium	0.9	1.5	3.1	4.0	4.5	4.7	4.7	4.7	4.6	4.3	3.8
Nickel*5	0.9	1.824	3.552	5.184	6.7	8.2	9.6	10.8	12.0	13.1	14.2
Osmium											
Palladium	1.05	2.10	4.2	6.3	7.5	9.1	10.6	11.9	13.1	14.2	15.3
Platinum	0.95	1.879	3.664	5.340	6.9	8.4	9.7	10.9	12.0	13.0	13.9
Polonium											
Potassium	50.9	69.7	81.8	82.2	79.7	74.8	69.1	63.0	56.1	48.6	40.4
Praseodymium	0.3	0.4	0.4	0.2	−0.1	0.6	6.9	11.1	12.0	11.8	11.0
Rhenium ($\Delta R/R_{20}$)		14.0		14.4		21.3		25.8		29.4	
Rhodium	0.819	1.622	3.182	4.680	6.1	7.5	8.8	10.0	11.1	12.0	12.8
Rubidium	52.9	62.9	62.4	48.4	30.5	7.6	−21.8	−59.4	−97.2	−138	−195
Ruthenium	1.24	2.48									
Samarium		3.6	6.1	8.2	9.5	10.5	11.2	11.8	12.3	12.6	12.8
Silver	1.7	3.4	6.4	9.0	11.5	13.6	15.4	16.9	18.1	19.1	19.8
Sodium	24.6	38.2	51.8	58.2	60.8	61.0	60.1	58.8	57.0	54.8	52.1
Strontium*7	−24.5	−55.6	−137.7	−246.6	−358	−229	−137	−89	−81.5	−80.5	−81
Tantalum	0.811	1.621	3.196	4.650	5.9	7.1	8.2	9.2	10.2	11.0	11.8
Thallium*8	6.38	11.80	20.5	27.3	33.5	56.4	60.9	64.8	68.1	71.0	73.5
Thorium		3.4	6.3	8.8	10.8	12.4	13.8	15.0	16.0	16.9	17.9
Thulium		2.6	4.7	6.4	7.7	8.7	9.x6	10.3	11.2	12.1	13.1

Table 21.7 Relative resistance change (in %) of pure metals *(continued)*

Metal	Pressure, GPa									
	0.5	1.0	2.0	3.0	4.0	5.0	6.0	7.0	8.0	9.0	10.0
Tin	4.9	9.2	16.5	22.5	27.6	31.7	35.3	38.2	40.8	43.1	45.2
Titanium	0.575	1.118	2.126	3.051	3.9	4.7	5.5	6.2	6.9	7.6	8.4
Tungsten	0.66	1.305	2.554	3.744	4.9	6.0	7.0	8.0	8.9	9.7	10.5
Uranium	2.355	4.56	8.634	12.31	16.3	18.8	21.0	23.0	24.8	26.3	27.6
Vanadium		1.6	3.1	4.6	5.9	7.1	8.4	9.4	10.5	11.5	12.2
Ytterbium		−97	−288	−624	−101.3	−11.85		21.14	23.04	24.12	24.67
Zinc		6.3	11.3	15.3	18.8	21.7	24.4	26.7	28.7	30.5	32.1
Zinc (87° to *c*-axis)	25	4.8	8.5	11.4							
Zirconium*12	0.195	0.39	0.78		0.8	1.0	1.2	1.4	1.6		

*1 At 8 GPa, a jump from 103.2 to 148.5%.
*2 At 2.47 GPa, a jump from +51.4 to −76.1%; at 2.6 GPa, from -76.4% to -38.3%.
*3 For solid gallium (0°C) at 1 GPa, $\Delta R/R_0 = -2.47\%$.
*4 Increasing pressure. Hysteresis is observed in the backward run; at 2, 1 and 0.5 GPa, $\Delta R/R_0$ =-8.4, -4.4, -1.7%.
*5 A break at 1 GPa.
*6 Liquid to solid transition at 1.36 GPa, a jump from −26.2 to 75.3%; at 4 GPa, a jump from −101.7 to −112%.
*7 At 4.6 GPa, a maximum of +405%.
*8 At 4.5 GPa, a jump from −36.1 to 54%.
*9 A break at 0.34 GPa, $\Delta R/R_0$ =9.76%.
*10 At 2.207 GPa, a jump from 120.3 to 143.9%; at 5.495 GPa, a maximum of +102%.
*11 At 0.7 GPa, a jump from +2.7 to −41.8%.
*12 Above 8 GPa, the resistance falls by 16−17%.

21.4 Electrical Properties of Some Metals and Alloys

Tables 21.8–21.9 and Figs. 21.3–21.10 show data on resistance of metals, metals with impurities and alloys that are widely used in electrical technology, device manufacturing, and other branches of industry.

Table 21.8 The resistivity ρ_{20} and the temperature coefficient of resistance α_{20} at 20°C for alloys exploited up to the temperature t_{max} [1]

Alloy	Component content, mass.%	t_{max},°C	$\rho_{20}\cdot 10^6, \Omega\cdot$cm	$\alpha_{20}, 10^{-5}$K^{-1}
Constantan	54Cu, 45Ni, 1Mn	400	50	-3
Manganin	86Cu, 12Mn, 2Ni	300	43	1-2
	85Cu, 15Mn	300	51	0.8
	84Cu, 13Mn, 3Al	400	50	-(0.2-2)
	85Cu, 9.5Mn, 5.5Al	400	45	1-3
Copper-aluminum	95Cu, 5Al	350	11	80
Copper-manganese	91Cu, 9Mn	350	33.4	-0.8
	90.5Cu, 9.5Mn	350	35.2	9.1
Copper-manganese-	93Cu, 5Mn, 2Al	400	22	5
aluminum	88Cu, 10Mn, 2Al	400	38	-3
Nickeline	67Cu, 30Ni, 2-3Mn	300	40	11
Nickeline-	58Cu, 22Ni, 20Zn	300	36	31
neusilber	54Cu, 26Ni, 20Zn	300	43	23
Neusilber	60Cu, 17Ni, 23Zn	300	30	35

Table 21.9 Resistivity of alloys (exploited up to the temperature t_{max}) at 20°C and 1000°C [1]

Alloy	Component content, mass.%	t_{max}, °C	ρ_{20}, $10^{-6}\Omega\cdot$cm	ρ_{1000}, $10^{-6}\Omega\cdot$cm
Nichrome	70–80Ni, 20Cr, 0–2Mn	1150	106	112
	70Ni, 8Fe, 20Cr, 2Mn	1150	110	120
	62Ni, 23Fe, 15Cr	1100	110	119
	63Ni, 20Fe, 15Cr, 2Mn	1150	112	125
	20Ni, 55Fe, 25Cr	1000	97	130
Chromium-aluminum	65Fe, 30Cr, 5Al	1350	140	142
	72Fe, 20Cr, 5Al, 3Co	1300	145	151
	86Fe, 12Cr, 2Al	1000	110	122

Table 21.10 Resistivity ρ_{20} and the temperature coefficient α_{20} for some alloys at 20°C [1]

Alloy	Component content, mass.%	ρ_{20}, $10^{-6}\Omega\cdot$cm	α_{20}, $10^{-5}K^{-1}$
Alumel	2Mn, 2Al, 1Si, Ni, balance Co	305	
Rose alloy	48Bi, 28Pb, 24Sn	67	190
Wood's alloy	56Bi, 14Pb, 14Sn, 16Cd	54	230
	64Fe, 36Ni	75	
	90Fe, 10Al	100	327
	88Fe, 12Mn	55	≈ 200
	96Fe, 4Si	50	≈ 90
Copel	44Ni+Co, 0.1–1Mn, balance Cu	465	
Monel	67Ni, 28Cu, balance Fe+Mn	48	≈120
Permalloy C	70–75Ni, balance Fe+ +Cu+Cr	55	
Platinum–iridium	90Pt, 10Ir	23.6	123
	85Pt, 15Ir	27.4	100
	80Pt, 20Ir	30	80
Platinum–rhodium	90Pt, 10Rh	21.7	139
Platinum–silver	33Pt, 67Ag	27	24

Table 21.11 Conductivity of conducting bronzes at 20°C [11, 12]

Alloy	State	Conductivity with respect % to copper, %	Alloy	State	Conductivity with respect % to copper, %
Bronze*1:			beryllium	annealed	17
cadmium	annealed	95	(2.25% Be)	aged at 350°C	30
(0.9% Cd)	hard-drawn	83–90	phosphorous	annealed	10–15
tin	annealed	55–60	(7% Sn, 0.1% P)	hard-drawn	10–15
(0.8% Cd, 0.6% Sn)	hard-drawn	50-55	Chromium alloyed copper (0.5% Cr)	aged	80–85
aluminum	annealed	15–18			
(2.5% Al, 2% Sn	hard-drawn	15–18			

*1 Balance in all bronzes – copper.

Table 21.12 Electrical properties of various brasses at 20°C [11]

Alloy	Mark (composition, %)	ρ_{20}, 10^{-6} Ω·cm	α_{20}, 10^{-3} K^{-1}
Tombac	Л–96 (Cu96, Zn4)	4.3	2.7
Brass	Л–62 (Cu62, Zn38)	7.1	1.7
Iron-manganese-alloy brass	ЛЖМц 59–1, 1 (Cu59, Zn39, Fe 1, Mn 0.8)	8.9	1.8
Manganese-alloy brass	ЛМц 58–2 (Cu58, Zn40, Mn 2)	21.2	1.3
Lead-alloy brass	ЛС 59–1 (Cu59, Zn40, Pbl)	6.5	1.7

Table 21.13 Resistivity ρ of cast iron and its constituents at 20°C [13]

Cast iron	Cast iron			Cast iron component			
sort (component)	Gray	Malleable	White	Ferrite	Perlite	Cementite	Graphite
ρ, 10^{-6}Ω · cm	80±40	50±20	70±20	10.4	20	140	150–300

Table 21.14 Resistivity ρ of stainless steels at 20°C [14]

Steel mark*1	3X13	X17	1X18H9, 2X18H9, 1X18H9T	X25C3H	X18H25C2	X20H14C2	X10C2M
ρ, 10^{-6}Ω· cm	57	65	75	80	102	95	75

*1 Cf. Chapter 3.

Table 21.15 Electrical properties of a piece (1-cm long and 1-cm in diameter) of an infinite molybdenum thread*1 [11] [Notation: R – resistance, W – heating power; U – voltage, I – current.]

T, K	R, 10^{-6}Ω	W, W	U, 10^{-3} V	I, A	T, K	R, 10^{-6}Ω	W, W	U, 10^{-3} V	I, A
273	6.85				1800	59.8	35.5	46.1	772.0
300	7.36				2000	69.6	60.3	64.8	816.0
400	10.36				2200	75.4	96.5	85.2	1130.0
1000	30.4	1.73	7.24	248.5	2400	83.3	147.8	11.0	1330.0
1200	36.9	4.5	12.9	349.0	2600	91.8	218.5	141.4	1547.0
1400	44.6	10.0	21.1	473.0	2800	99.8	304.0	174.0	1745.0
1600	52.4	19.8	32.2	614.0	2895	103.4	364.4	194.0	1875.0

*1 The thread is infinite in the sense of vanishing longitudinal heat removal. When recalculating to a diameter d(cm), one should divide R by d^2, multiply W by d, divide U by $\sqrt{d}$, and multiply I by $d^{3/2}$.

Table 21.16 Electrical properties of a piece (1-cm long and 1-cm in diameter) of an infinite tantalum thread*1 [11] [Notation: T – true temperature, R – resistance, W – heating power, U – voltage, I – current, I_e – emission current.]

T, K	R, 10^{-6}Ω	W, W	U, V	I, A	I_e, A/cm
300	17.65				
100	56.2	2.495	0.0118	211	
1100	60.2	3.87	0.0152	254	
1200	64.8	5.78	0.0193	299	
1300	69.7	8.58	0.0244	352	
1400	74.0	12.42	0.0304	408	
1500	79.3	17.2	0.0368	469	
1600	83.7	23.3	0.0438	528	$4.4 \cdot 10^{-5}$
1700	88.1	31.7	0.0527	602	$3.14 \cdot 10^{-4}$

Table 21.16 Electrical properties of a piece of an infinite tantalum thread*1 *(continued)*

T, K	R, $10^{-6}\Omega$	W, W	U, V	I, A	I_e, A/cm
1800	92.2	41.7	0.0617	676	$1.57 \cdot 10^{-3}$
1900	96.4	53.8	0.0716	751	$6.28 \cdot 10^{-3}$
2000	100.3	68.0	0.0821	828	$2.8 \cdot 10^{-2}$
2100	104.3	85.2	0.0936	910	$8.8 \cdot 10^{-2}$
2200	108.3	107.5	0.1072	1002	$2.7 \cdot 10^{-1}$
2300	112.4	132.9	0.1212	1095	$7.85 \cdot 10^{-1}$
2400	116.2	161.3	0.1357	1189	2.04
2500	120.1	196.2	0.1522	1288	4.4
2600	124.0	236.8	0.1699	1394	
2700	127.6	282.8	0.1880	1502	
2800	131.0	332.0	0.2064	1606	
2900	134.5	387.0	0.2257	1715	
3000	138.5	454.0	0.2479	1830	
3100	142.0	526.0	0.2700	1948	
3200	145.0	611.0	0.2940	2075	
3269	147.0	674.0	0.3110	2164	

*1 See footnote to Table 21.15.

Table 21.17 Properties of a piece (1-cm long and 1-cm in diameter) of an infinite tungsten thread*1 [5, 11] [Notation: T_l – luminance temperature; see other designations in Table 21.16*2.]

T, K	T_l, K	R, $10^{-6}\Omega$	W, W	U, 10^{-3} V	I, A	I_e, A/cm^2
273		6.37				
293		6.99				
300		7.20				
400		10.26	0.0062	0.253	24.67	
500		13.45	0.0305	0.64	47.62	
600		16.85	0.0954	1.268	75.25	
700		20.49	0.240	2.218	108.2	
800		24.19	0.530	3.581	148.0	
900		27.94	1.041	5.393	193.1	
1000	966	31.74	1.891	7.749	244.1	$3.36 \cdot 10^{-15}$
1100	1058	35.58	3.223	10.71	301.0	$4.77 \cdot 10^{-13}$
1200	1149	39.46	5.210	14.34	363.4	$3.06 \cdot 10^{-11}$
1300	1240	43.40	8.060	18.70	430.9	$1.01 \cdot 10^{-9}$
1400	1330	47.37	12.01	23.85	503.5	$2.08 \cdot 10^{-8}$
1500	1420	51.40	17.33	29.85	580.6	$2.87 \cdot 10^{-7}$
1600	1509	55.46	24.32	36.73	662.2	$2.91 \cdot 10^{-8}$
1700	1597	59.58	33.28	44.52	747.3	$2.22 \cdot 10^{-5}$
1800	1684	63.74	44.54	53.28	836.0	$1.4 \cdot 10^{-4}$
1900	1771	67.94	58.45	63.02	927.4	$7.15 \cdot 10^{-4}$
2000	1857	72.19	75.37	73.75	1022.0	$3.15 \cdot 10^{-3}$
2100	1943	76.49	95.69	85.57	1119.0	$7.23 \cdot 10^{-2}$
2200	2026	80.83	119.8	98.40	1217.0	$4.17 \cdot 10^{-2}$
2300	2109	85.22	148.2	112.4	1310.0	$1.28 \cdot 10^{-1}$
2400	2192	89.65	181.2	127.5	1422.0	0.364
2500	2274	94.13	219.3	143.6	1526.0	0.935
2600	2356	98.66	263.0	161.1	1632.0	2.25
2700	2437	103.22	312.7	179.7	1741.0	5.12
2800	2516	107.85	368.9	199.5	1849.0	11.11
2900	2595	112.51	432.4	220.6	1961.0	22.95
3000	2673	117.24	503.5	243.0	2072.0	44.4
3100	2750	121.95	583.0	266.7	2187.0	83.0
3200	2827	126.76	671.5	291.7	2301.0	150.2
3300	2903	131.60	769.7	318.3	2418.0	265.2

Table 21.17 Properties of a piece of an infinite tungsten thread*1 [5, 11] *(continued)*

T, K	T_1, K	R, $10^{-6}\Omega$	W, W	U, 10^{-3} V	I, A	I_e, A/cm^2
3400	2978	136.49	878.3	346.2	2537.0	446.0
3500	3053	141.42	998.0	375.7	2657.0	732.0
3600		146.40	1130.0	406.7	2777.0	1173
3655	3165	149.15	1202.0	423.4	2838.0	1505

*1 See footnote to Table 21.15.

*2 I_e is the current normalized per unit length of a thread of 1-cm in diameter, and not per unit area.

Table 21.18 Resistivity and the temperature coefficient of resistance ($\alpha_0 = \rho^{-1} \cdot d\rho/dT$) for rhenium [15]

t, °C	ρ, $10^{-6}\Omega\cdot$ cm	α_0, 10^{-3}K^{-1}	t, °C	ρ, $10^{-6}\Omega\cdot$ cm	α_0, 10^{-3}K^{-1}
−253	0.015		900	72.5	3.13
−200	2.3		1100	80.5	2.94
−100	10.0		1300	87.0	2.78
0	17.5		1500	93.0	2.58
20	19.3		1700	98.5	2.44
100	25.4	3.95	1900	103.0	2.31
300	40.0	3.83	2100	106.5	2.17
500	52.0	3.58	2300	109.0	2.04
700	63.0	3.33			

Table 21.19 Resistivity of tungsten-rhenium alloys vs. temperature [15]

4.75% Re		21% Re		27% Re	
T, K	ρ, $10^{-6}\Omega\cdot$cm	T, K	ρ, $10^{-6}\Omega\cdot$cm	T, K	ρ, $10^{-6}\Omega\cdot$cm
300	10.4	300	24.2	1400	67.3
630	20.4	550	34.0	1600	73.9
850	27.3	810	42.0	1800	79.9
1020	32.4	980	47.0	2000	85.7
1160	36.7	1120	51.6	2400	91.1
1270	40.0	1250	54.6	2600	96.8
1370	43.2	1360	57.4	2800	102.8
1465	45.9	1460	59.8	3000	109.2
1540	48.5	1540	62.6		
		1610	64.4		

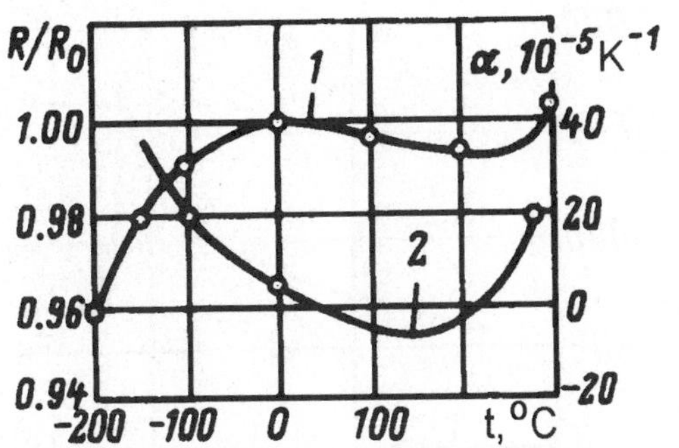

Figure 21.3 The relative resistance (*1*) and the temperature coefficient (*2*) of manganin vs. temperature [11].

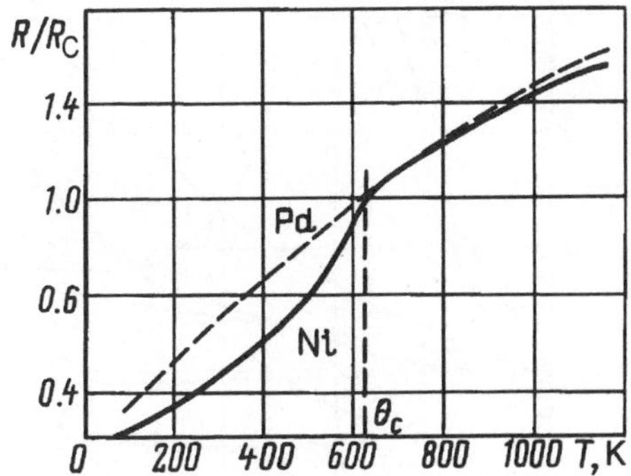

Figure 21.4 The relative resistance of nickel and palladium as a function of temperature. The resistance is normalized to Curie point resistance for nickel (631 K) [4].

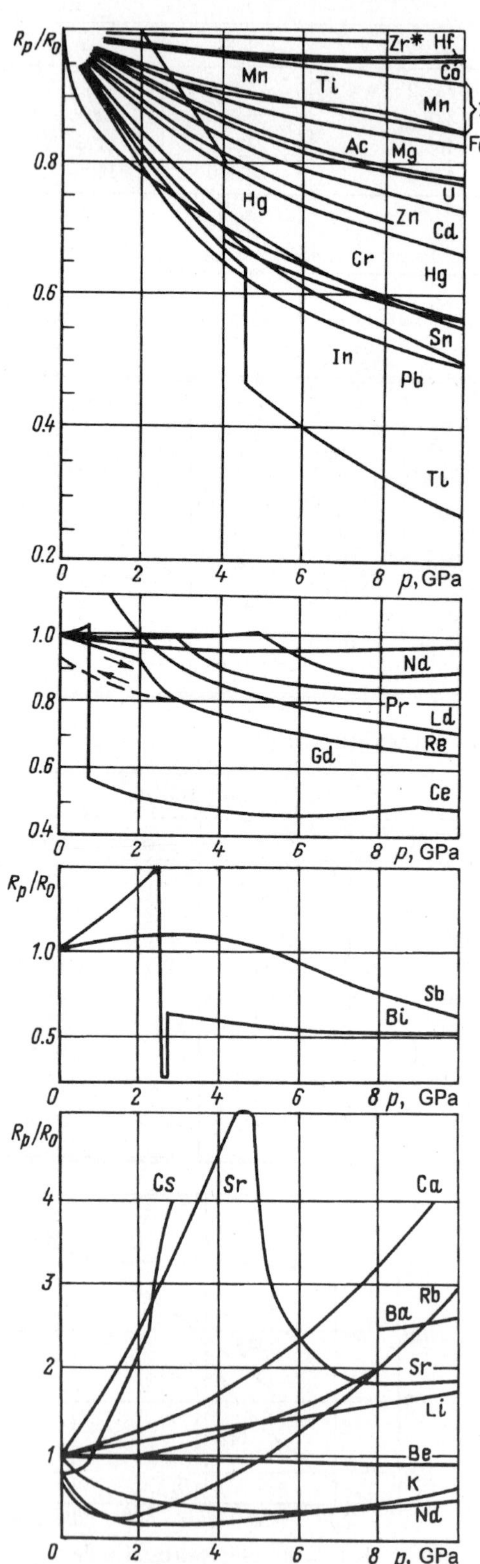

Figure 21.5 The effect of pressure on relative resistance of metals [4].

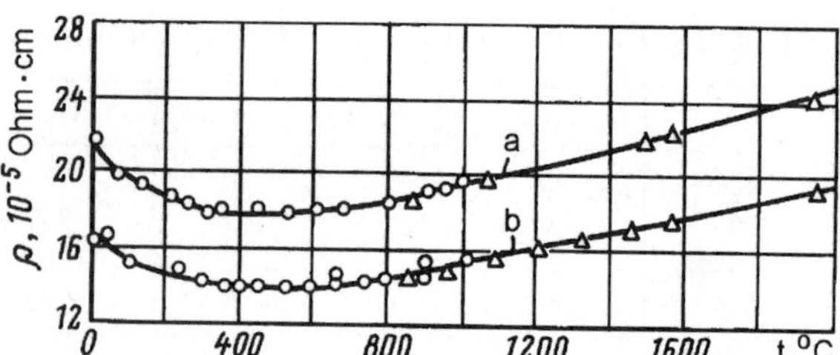

Figure 21.6 The resistivity of electrode graphite transversely to (*a*) and along (*b*) pressing direction vs. temperature [10].

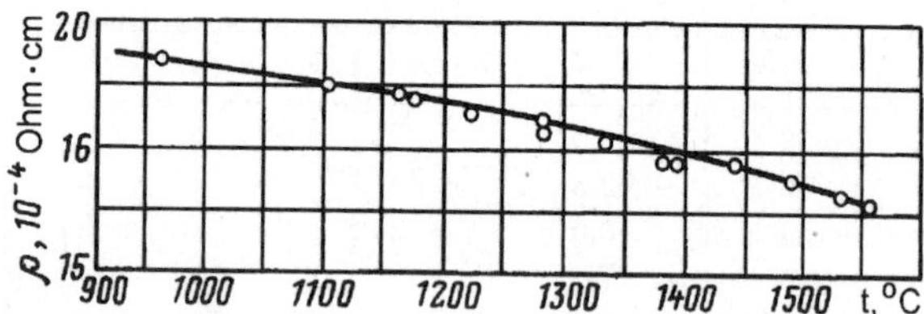

Figure 21.7 The resistivity of a carbon material produced from petroleum coke at temperatures above annealing point [10].

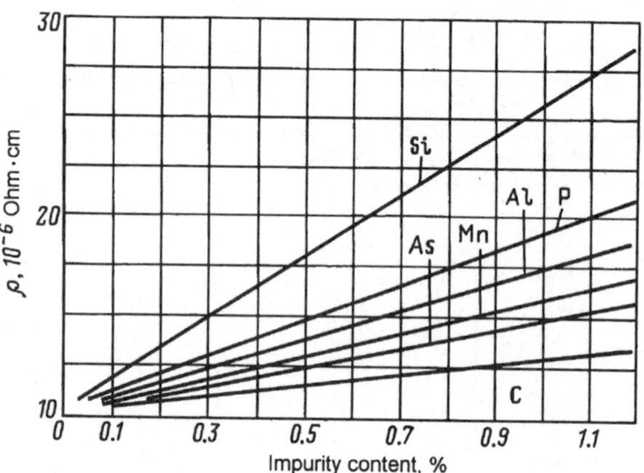

Figure 21.8 The resistivity of steel as a function of impurity content [11].

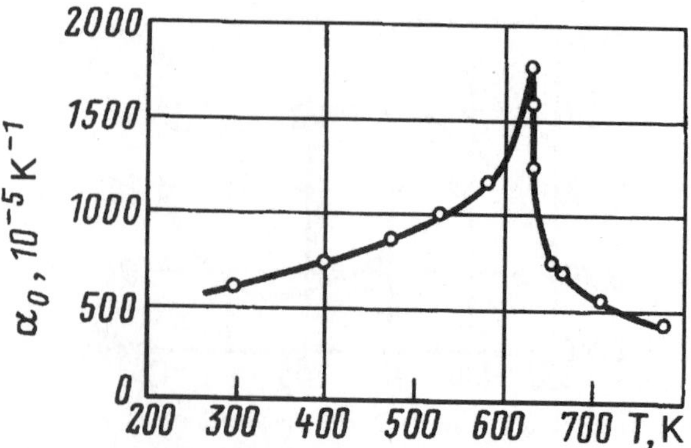

Figure 21.9 The temperature coefficient of resistance for nickel vs. temperature [4].

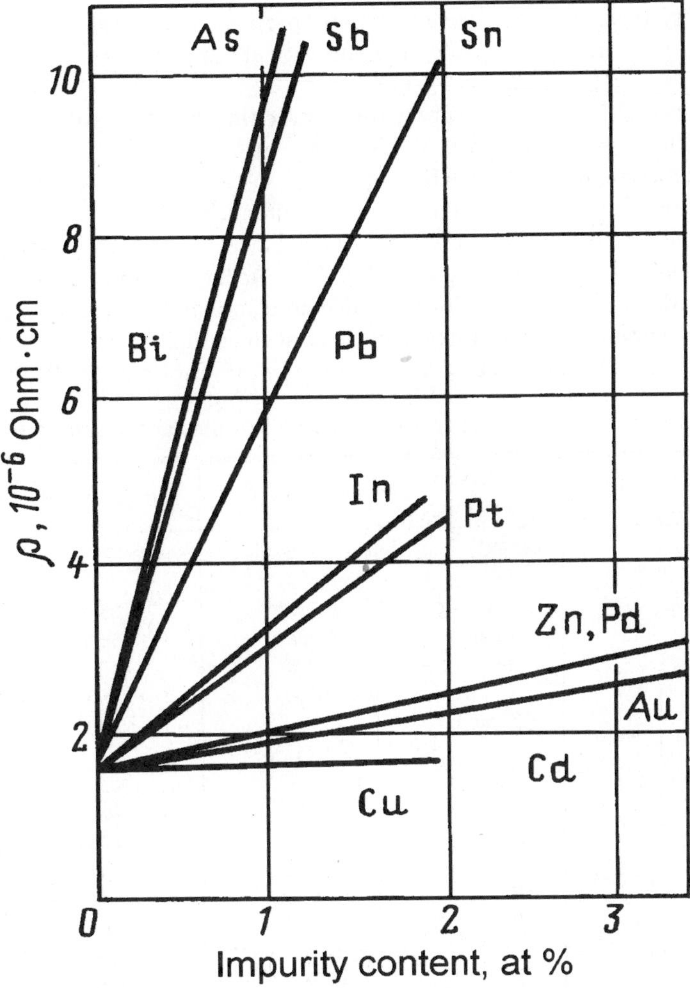

Figure 21.10 The effect of impurities on the resistivity of silver [11].

21.5 Superconductors

Superconductivity refers to a quantum phenomenon originating from a Bose-condensation of conduction electron pairs. Two most important macroscopic indications of the superconducting state are as follows: (1) the absence of resistance to the direct electric current at temperatures below some critical value T_c, and (2) pushing out of the magnetic field from the bulk of a superconductor (Meissner effect). There exist a critical magnetic field H_c and a critical current density j_c, above which a superconductivity disappears. The dependence of critical magnetic intensity on temperature is well fitted by the expression

$$H_c(T) = H_{c0}\left[1 - \left(\frac{T}{T_c}\right)^2\right], \tag{21.8}$$

where T is the absolute temperature, and H_{c0} is the critical magnetic field strength at zero temperature.

The conduction electrons in a metal are paired due to the electron-phonon interactions, so that the superconductivity turns out to be sensitive to properties of the crystal lattice. Various crystalline modifications of the same material possess different critical temperatures; T_c also depends on external pressure P.

Microscopic characteristics of a superconductor include the characteristic size of an electron pair xi_0 and the magnetic field penetration depth λ.

The majority of superconducting alloys belongs to the so-called type II superconductors, wherein the superconductivity can coexist with a magnetic field (Shubnikov phase). In the bulk of such superconductors, magnetic field creates thin threads of the normal metal (Abrikosov vortices) with characteristic size $\sim \lambda$, each bearing a quantum of magnetic flux $\Phi_0 = \hbar c/2e$, where $\hbar$ is the Planck's constant, c is the light velocity, and e is the electron charge. Since no complete Meissner effect occurs in the type II superconductors, the superconductivity exists in them at much higher magnetic fields $H_{c1} < H < H_{c2}$.

Various characteristics of superconductors are presented in Tables 21.20–21.26 and in Fig. 21.11.

Table 21.20 Superconductivity of elements (massive samples at atmospheric pressure)

	Element	Crystal structure*1	T_c, K [16]	H_{c0}, Oe [16]	dT_c/dP, 10^{-11}K/Pa [17]
Al	Aluminum	FCC	1.1796*2	104.9	−25.5
Be	Beryllium	Hex.	0.0237*2		
Cd	Cadmium	Hex.	0.519*2	29.6	−18.2
Ga	Gallium	Rhomb.	1.091	58.9	−18
Hf	Hafnium	Hex.	0.015?		
Hg	Mercury α	Rhomb.	4.153	390	−36.6
	Mercury β		3.95	340	−47.5
	Mercury γ		3.74		
In	Indium	Tetr.	3.4145*2	289	−39.1
Ir	Iridium	FCC	0.100*2	20.1	−0.5
La	Lanthanum α	Hex.	4.88	808	190
	Lanthanum β		6.0	1096	113
Lu	Lutetium	Hex.	0.1		
Mg	Magnesium	Hex.	0.0005		
Mo	Molybdenum	BCC	0.92	98	−1
Nb	Niobium	BCC	9.3*2	1980	0
Os	Osmium	Hex.	0.66	65	−1.8
Pa	Protactinium	BCC	1.4		
Pb	Lead	FCC	7.1999*2	803.4	−38.6
Re	Rhenium	Hex.	1.698	198	−2
Rh	Rhodium	FCC	0.002?		
Ru	Ruthenium	Hex.	0.493	47	0
Sn	Tin	Tetr.	3.722	308	−46.3
Ta	Tantalum	BCC	4.46	831	−2.6
Th	Thorium	FCC	1.374	162	−17
Ti	Titanium α	Hex.	0.42	56	5.5
	Titanium β		5?		
Tl	Thallium	Hex.	2.38	179.5	−23
Tc	Technetium	Hex.	7.8*2	1410	−12.5
U	Uranium γ	Rhomb.	1.8		
V	Vanadium	BCC	5.46	1167	11
W	Tungsten	BCC	0.015*2	1.07	
Zn	Zinc	Hex.	0.851*2	52	−25
Zr	Zirconium α	Hex.	0.53	47.7	14
	Zirconium β		1.5		
	Zirconium ω		0.72		

*1 FCC – face-centered cubic; BCC – body-centered cubic; tetr. – tetragonal; hex. – hexagonal; rhomb. – rhombic.
*2 Critical temperatures recommended as reference points [19–21].

Table 21.21 Superconductivity of elements under pressure [16]

Element		T_c^{max}, K	P, 10^8Pa	dT_c/dP, 10^{-11}K/Pa
As II	Arsenic	0.25	100	
As III		0.5	140	
Ba II	Barium	3.2	83	1000
Ba III		3.0	88	−150
Ba IV		5.3	200*1	13
Bi	Bismuth	3.916	25	
Bi II		8.2	25	−32
Bi III		7.25	27	−40
Bi IV		7.0	43	−46
Bi V		6.7	68	−30
Bi VI		8.55	80	−23
Ce	Cerium	1.7	50	
Cs V	Cesium	1.6	125	
Ga II	Gallium	6.24	20	−30
Ge	Germanium	5.35	115	
Hf	Hafnium	0.24	160*1	
α−La	Lanthanum	8.2	23	140
β−La		12	250*1	
Lu	Lutetium	0.7	150*1	
P	Phosphorus	5.8	170	
Pb II	Lead	3.6	160	−2.2
Sb	Antimony	3.55	85	
Se	Selenium	6.9	130*1	
Si	Silicon	7.1	120	
Sn II	Tin	5.3	113	−49.5
Te II	Tellurium	4.2	70	80
Te III		4.3	70	−10
Te IV		4.3	80	−42
Tl II	Thallium	2.395	2	−10
Tl III		1.45	35	−49.5
Tl IV		2.32	40	−20
α−U	Uranium	2.1	10	
β−U		1.15	9.5	
Y	Yttrium	2.7	170	
α−Zr	Zirconium	0.69	45	3.5

*1 Maximum experimentally reached pressure.

Table 21.22 Superconductivity of elements in thin films [16]

Element		T_c^{max}, K	Element		T_c^{max}, K
Al	Aluminum	5.8	Nb	Niobium	10.0
Be	Beryllium	9.6	Re	Rhenium	7
Bi	Bismuth	8	Sn	Tin	6.0
Cd	Cadmium	0.9	Ta	Tantalum	4.51
Cr	Chromium	1.52	α-Ti	Titanium	1.3
Ga	Gallium	8.5	Tl	Thallium	3.15
α-Hf	Hafnium	1.3	V	Vanadium	6.02
In	Indium	7	W	Tungsten	4.55
β-La	Lanthanum	6.74	Zn	Zinc	1.9
Mo	Molybdenum	6.7			

Table 21.23 Microscopic parameters of some superconductors [18]

Element		ξ_0, nm	λ, nm
Al	Aluminum	1600−1360	16−50
Cd	Cadmium	760	110
Hg	Mercury		38−45
In	Indium	354−240	64−39
Nb	Niobium	38	39
Pb	Lead	96−51	39−63
Sn	Tin	300−100	34−75
Tl	Thallium	420	

Table 21.24 Superconductivity of binary compounds [16]

Composition	T_c, K
Cubic structure A15	
Ti_3Ir	4.63–5.40
Ti_3Pt	0.48–0.58
Ti_3Sb	6.5–5.7
Zr_3Au	0.92
Zr_3Pb	0.76
Zr_3Sn	0.94–0.76
V_3Al	11.65
V_3Au	3.15–2.97
V_3Ga	14.6–16.8
V_3Ge	6.3
V_3In	13.9
V_3Ir	1.71
VOs	5.15–4.49
V_3Pd	0.082
V_3Pt	3.27–3.20
V_3Sb	0.80
V_3Si	17.2–17.0
V_3Sn	7–3.8
Nb_3Al	18.5–18.8
Nb_3Au	10.6–11.5
Nb_3Bi	3.05
Nb_3Ga	14.5
Nb_3Ge	6.9 (23.2 in film)
Nb_3In	9.2
Nb_3Ir	1.76–1.63
Nb_3Os	1.05–0.94
Nb_3Pb	9.6
Nb_3Pt	10.9–8.18
Nb_3Rh	2.64–2.4
Nb_3Sb	1.9–0.4
Nb_3Sn	18.1–18.5
Ta_3Au	0.51–10
Ta_3Pb	17.0
Ta_3Pt	0.40
Ta_3Sb	0.72–0.59
Ta_3Sn	6.4
Cr_3Ir	0.83
Cr_3Os	4.25–4.03
Cr_4Rh	0.072
Cr_3Ru	3.43
Mo_3Al	0.58
Mo_3Ga	0.76
Mo_3Ge	1.43
Mo_3Ir	9.6
Mo_3O	4–6.7
Mo_3Os	11.76
Mo_4Pt	4.53–5.6
Mo_3Re	15.
Mo_3Si	1.34
Mo_3Tc	13.5–14
W_3O	0.4–3.35
Cubic Laves–phase C15	
$BaRh_3$	6.0
$BiAu_2$	1.84
$CaIr_2$	6.15
$CaRh_2$	6.40
$CeCo_2$	1.5
$CeRu_2$	6.20
$CsBi_2$	4.75
$HfMo_2$	0.07
HfV_2	9.4
KBi_2	3.58
$LaAl_2$	3.24
$LaIr_2$	0.48
$LaOs_2$	6.5
$LaRu_2$	4.1
$LaPt_2$	0.54
$LuIr_2$	2.89
$LuRh_2$	1.27
$NbBe_2$	2.15
$PbAu_2$	1.18
$RbBi_2$	4.25
$RhHf_2$	1.98
$ScIr_2$	2.46
$ScRu_2$	2.24
$SrIr_2$	5.7
$SrPt_2$	0.7
$SrRh_2$	6.2
$ThIr_2$	6.50
$ThRu_2$	3.56
YIr_2	2.18
YPt_2	1.70
YRh_3	1.07
$ZrIr_2$	4.10
ZrV_2	8.8
ZrW_2	2.7
$CaAu_5$(C 15*b*)	0.34–0.38
Hexagonal Laves-phase C14	
$HfOs_2$	2.69
$HfRe_2$	5.61
$HfTe_2$	5.6
$LuOs_2$	3.49
$LuRu_2$	0.86
$ScOs_2$	4.60
$ScRe_2$	4.2
$ScRu_2$	2.24
TaV_2	10.0
$ThRe_2$	5.05
$ThTc_2$	5.3
YOs_2	4.7
YRe_2	1.83
YRu_2	2.42
$ZrOs_2$	3.0
$ZrRe_2$	6.8
$ZrRu_2$	1.84
$ZrTe_2$	7.6
Zr_3Ir	2.13
VRe_3	6.26
Nb_3Ir_2	9.8
Nb_3Os_2	1.86
Nb_xPd_y	2.0
Nb_2Pt	4.2
NbRe	3.8
Nb_3Rh_2	4.1
Ta_7Pt_3	1.5
Ta_2Re_3	1.4
Ta_3Rh_2	2.35
Cr_2Re_3	2.50
Cr_5Ru_3	2.10
Mo_7Ir_3	6.8
Mo_5Os_3	5.65
Mo_2Re_3	6.5
Mo_5Ru_3	9.2
Mo_3Tc_7	15.8
$W_{0.72}Ir_{0.28}$	4.46
W_xOs_y	4.4
W_xRe_y	5.2
W_3Ru_2	4.67
W_2Tc_3	7.88
$ReFe_3$	6.55
Cubic χ–phase A12	
Al_5Re_{24}	3.35
Hf_5Re_{24}	5.86
$NbOs_2$	2.86
Nb_3Pd_2	2.47
NbRe	9.7
NbTc	10.5
$MoRe_3$	9.89
Sc_5Re_{24}	2.2
TaOs	1.95
Ta_5Re_{24}	6.78
Ti_5Re_{24}	6.6
WRe_3	9.0
Zr_5Re_{24}	7.40
$ZrTc_6$	9.7

Table 21.24 Superconductivity of binary compounds [16] *(continued)*

Structure type	Composition	T_c, K
	Cubic structures	
B1	GeTe	0.04–0.31
	AsSn	3.41–3.65
	As_3Sn_4	1.23–1.16
	SbSn	1.3–2.37
B2	InPd	0.70
	LaAg	0.92–0.96
	LiPb	7.2
	$Mg_{0.47}Tl_{0.53}$	2.75
	OsTi	0.46
	RuTi	1.07
	TiCo	0.71
	TiO	2.3–0.58
	UCo	1.70
B20	AuBe	2.64
	GaPt	1.74
Cl	$AuAl_2$	0.095–0.074
	$AuGa_2$	1.12–1.05
	$AuIn_2$	0.096–0.093
	$CoSi_2$	1.22–1.4
	$PtGa_2$	1.7–1.9
	PRh_2	1.3
	$PtAl_2$	0.48–0.55
C2	$AuSb_2$	0.58
	$\alpha-PdBi_2$	1.45
	$IrTe_2$	0.3–1.18
	$PdSb_2$	0.35–1.25
	$RhSe_2$	6.0
	$\alpha-RhT_2$	0.51
$E9_3$	$CoHf_2$	0.56
	$CoTi_2$	3.44
	$RhHf_2$	1.98
	$RhZr_3$	11
LI_2	$AlLa_3$	6.16
	$AlZr_3$	0.73
	$CaPb_3$	0.65
	$CaTl_3$	2.0
	$BiTl_3$	4.15–4.4
	$InLa_3$	10.40
	$InLu_3$	0.14–0.24
	$LaIn_3$	0.70–0.71
	$LaPb_3$	4.07–4.10
	$LaSn_3$	6.02–6.55
	$LaTl_3$	1.51–1.63
	$NaPb_3$	5.62
	$LuCa_3$	2.30
	$SiNb_3$	1.5
	$SrBi_3$	2.62–5.70
	$SrPb_3$	1.85
	$ThPb_3$	5.5
	$ThSn_3$	3.33
	$ThTl_3$	0.87
	YIn_3	0.78
	YPb_3	4.72
LI_2	YTl_3	1.52
	$YbAl_3$	0.94
	$YbPb_3$	0.23
	$ZrHg_3$	3.28
$D8_2$	Ag_5Hg_8	0.64
DO_3	Cu-Sb	0.127–1.84
$D2_3$	$RuBe_{13}$	1.3
$D7_3$	La_3S_4	6.5–8.25
	La_3Se_4	8.6–1.25
	La_3Te_4	2.45–3.75
Complex	$Be_{22}Re$	9.65
	$Be_{22}Tc$	5.25
	$Be_{22}Mo$	2.529–5.545
	$Be_{22}W$	4.038–4.12
	$AgAu_4$	0.4–0.7
	$AgTl_3$	2.6
	$AuZn_3$	1.21
	Hg_5Tl_2	3.14–3.8
	$In_{19}Cd$	3.0–3.55
	$Pd_{22}S$	1.63
	$Ph_{17}S_{15}$	5.8
	Sb_2Tl_7	5.20
	γ-SnTl	4.2–6.4
Nitrides	TiN	4.86–5.8
	ZrN	8.9–10.7
	HfN	6.2–8.7
	VN	7.5–8.2
	NbN	16.1
	TaN	4.84–6.5
	Mo_2N	5.0
	$ReN_{0.34}$	4–5
	ThN	3.2
	UN	5.6
	$\alpha-MC_x$	9.26–14.3
Carbides	NbC	1.05–11.7
	ReC	3.4
	$Sc_{13}C_{10}$	8.5
	TaC	1.05–11.2
	TcC	3.85
	TiC	3.32–3.42
	VC	0.03
	$\beta-WC_x$	5.2–10.0
	PdH	5
Hydrides	PdH_2	16
	HfB	3.1
Borides	ZrB	2.8–3.4
	Hexagonal structures	
A3	Ag_2Al	1.28–0.088
	Ag_2Ga	6.5–8
	Ag_4Ge	0.85
	Ag_5Sb	0.019–0.065
	Ag_5Sn	0.025–0107
	Au_5In	0.035–0.331
	Au_5Sn	0.4–1.1

Table 21.24 Superconductivity of binary compounds [16] *(continued)*

Structure type	Composition	T_c, K
	Hexagonal structures	
A_3	Cu_3Ge	0.025–0.26
	Cu_5Si	0.050–0.058
	Pb_2Bi	8.2–8.5
	RhW	2.64–3.37
	RhMo	1.97
$B8_1$	AuSn	1.25
	BiNi	4.25
	$BiPd_2$	4.0
	BiPt	1.21–2.4
	BiRh	2.06
	PdSb	1.44–1.67
	PdTl	3.85
	PtSb	2.10
	PtSn	0.37
C32	$BiIn_2$	5.6
	MoB_2	1.0–6.4
	$\beta-ThSi_2$	2.41
	YGa_2	1.68
DO_{19}	AlTh	0.75
	Cd_3Mg	0.185
	Hg_3Li	1.7
D2d	$LaIr_5$	2.13
	$BaAu_5$	0.7–0.35
$D8_8$	Zr_5Ga_3	2.5–4.0
	Zr_5Pb_3	4.60
	Zr_5Sb_3	1.74
$D10_2$	B_3Ru_7	2.58
	Co_3Th_7	1.83
	Fe_3Th_7	1.86
	Ir_3La_7	2.24
	Ir_3Lu_7	0.72
	Ir_3Th_7	1.52
	Ni_3Th_7	1.98
	Os_3Th_7	1.51
	Pt_3Th_7	0.98
	Rh_3La_7	2.58
	Rh_3Th_7	2.15
$B8_2$	Pd_3Sn_2	0.47–0.64
	Zn_2Rh	8.2–6.4
C7	NbS_2	5.4–6.15
	$NbSe_2$	5.4–7.5
	$NbTe_2$	0.5–0.74
	TaS_2	0.71–2.1
	$TaSe_2$	0.13–0.22
B18	CuS	1.62
	Ag_2F	0.066
Complex	AsRh4 − 1.6	0.03–0.56
	$\beta-Bi_2Pt$	0.155
	$\gamma-B_4Rh$	2.70
	Hg_4Na	3.05
	MoPd	3.52
	$\epsilon-MoRh_2$	1.97
	Nb_3Te_4	1.49
Complex	$\alpha-Pd_2As$	0.66
	Pd_5As_2	0.46
	Re_2B	2.8
	TaSi	4.25–4.38
	Ti_3Bi_5	6.4–6.6
	$HgSn_6$	5.1
Nitrides	Ta_2N	10.6
	MoN	12.0
Carbides	$\beta-Mo_2C$	2.4–7.2
	$\eta-MoC_x$	7.4–9.26
	$\gamma-MoC$	7.6–8.3
	Nb_2C	1.98–9.11
	RuC	1.9–2.0
	Ta_2C	3.2–3.3
	$\alpha-W_2C$	2.74–3.6
	$\gamma-W_2C$	2.85–3.05
Borides	Mo_2B_5	8.1
	NbB_2	1.0–6.4
	Nb_2B_5	6.4
	Re_2B	2.80
	Ru_7B_3	2.58
	Tetragonal structures	
C16	$AgZn_2$	2.11–2.46
	$AgTh_2$	2.19–2.26
	$AlTh_2$	0.09
	$AuPb_2$	3.10–3.15
	$AuTh_2$	3.08–3.65
	$AuTl_2$	4.2
	$CoTa_2$	0.82
	$CoZr_2$	5.0–6.30
	$CuAl_2$	0.65
	$CuTh_2$	3.44–3.49
	$GaHf_2$	0.21
	$GaZr_2$	0.38
	$FeZr_2$	0.17
	$NiHf_2$	0.87
	$NiTa_2$	0.90
	$NiZr_2$	1.52–1.6
	$PdPb_2$	2.95–3.01
	$PdTh_2$	0.75–0.85
	$PdTl_2$	1.32
	$PtTl_2$	1.58
	$RhPb_2$	1.32
	$RhSn_2$	0.60
	$RhZr_2$	10.8–11.1
	$ZrTh_2$	0.67
DO_c	Mo_3P	5.31
	W_3P	2.26
$D2_c$	CoU_6	2.29–2.4
	FeU_6	3.86
	MnU_6	2.32
	NiU_6	0.41
$L1_o$	$\alpha-LiBi$	2.47
	NaBi	2.25

Table 21.24 Superconductivity of binary compounds [16] *(continued)*

Structure type	Composition	T_c, K
	Tetragonal structures	
C_c	$CaSi_2$	1.58
	$LaGe_2$	1.57–3.49
	$LaSi_2$	2.3–2.5
	$ScGe_2$	1.3–1.31
	Sr_2Si_3	0.55
	$ThGa_2$	2.56
	$\alpha - ThSi_2$	3.16–3.20
	YGe_2	2.4–3.8
Complex	Nb_3Be_2	2.30
	Ta_3Be_2	1.0
	Th_3Al_2	2.6
	$BaBi_3$	5.8
	Tl_5Te_3	2.28–2.078
	$\beta - In_3Sn$	7.3
	$AuPb_3$	4.40
	GeP	1.8–4.2
	Pd_4Se	0.42–0.66
Borides	Ta_2B	0.06–3.12
	W_2B	3.10–3.22
	Mo_2	4.74–5.86
	Rhombic structures	
B31	AsRh	0.58
	GeIr	4.70
	GePt	0.40
	GeRh	0.96
	SiPd	0.93
	SiPt	0.88
	SnPd	0.41
B31	SbRu	0.35–1.27
Orthorhombic	PtTh	0.44
	RhTh	0.36
	$AuSn_4$	2.38
	$PtPb_4$	2.8
	$CoLu_3$	0.35
	$CoLa_3$	4.01
	Bi_3Ni	4.06
	Bi_3Rh	3.2
Orthorhombic	$AlRe_6$	1.85
	Pd_2Sn	0.41
	BiPd	3.74–3.42
	KHg_2	1.2
	PtTe	0.59
	Rh_5P_4	1.22
Rhombohedral	In_3Te_4	1.15–1.25
	Ge_3Te_4	1.80–1.55
	AsS_4	1.16–1.21
	P_3Sn_4	1.24–1.10
	Nb_2Be_17	1.47
	$\alpha - Pd_7P_3$	0.70
	$\beta - Pd_7P_3$	1.00
Complex	Zr_3Au	0.98
	AuSn	2.48
	$AsSn_4$	1.16–1.19
	Ge_3Rh_5	2.12
Carbides	$\alpha - Mo_2C$	2.4–7.2
	$\beta - W_2C$	3.1–3.90
Borides	TaB	4.0

Table 21.25 Other structure types and compounds with yet undefined structure

Composition	T_c, K	Composition	T_c, K	Composition	T_c, K
$PdTe_2$	1.53–1.69	In_3Ru	2.68	Rh_5Th	1.07
$\beta - Pd_2As$	1.71	Ir_2Y	1.61	$RuBe_2$	1.35
Bi_2Pd	1.7	$LaIr_3$	2.46	Ta_2Ge	1.60
Os_4Al_{13}	5.9	$LaRh_5$	1.65	Y_3Co	0.34
AuIn	0.4–0.6	$LaRh_3$	2.60	YIr_{3-4}	3.50
$NbRu_3$	1.2	$LuIr_2$	0.84	CsC_8	0.020–0.135
AsGe	3–3.5	$LuRh_5$	0.49	KC_8	0.39–0.55
Be_5Ca	6.7	Mo_3Sb_4	2.10	La_2C_3	5.9–11.0
Be_5Ir	1.5	Nb_3Se_4	1.61	La_2C_2	1.61–3.33
Be_5Os	9.2	$OsBe_2$	3.07	RbC_8	0.023–0.151
Be_5Pt	2.3	PPb	7.8	Th_2C_3	4.1
CaPb	7.0	$P_{2.65}Sn_4$	1.21–1.10	V_2C_3	8.2
$CeIr_3$	3.34	$Pd_{2.5-2.8}Se$	2.20–2.30	WC_x	2.4–4.05
$CeIr_5$	1.82	Pd_3Te	0.76	Y_2C_3	6.0–11.5
CoSe	0.35	Pd_5Te	0.40	YC_2	3.35–3.88
Ge_2Se	1.3–1.31	Pt_5Th	3.13	NbB	8.25
Hg_3K	3.18	$RhBe_2$	1.37	NbB_6	3?
Hg_4K	3.27	$RhSc_3$	0.32–0.92	ZrB_{12}	5.9–6.0
Hg_8K	3.42	Rh_2Te	0.49		

Table 21.26 Superconductivity of engineering-relevant compounds [16] (see also Fig. 21.11):

Compound	T_c, K	j_c, A/cm^2 at 4.2 K	B_{c2}, T
Nb_3Sn	18.1–18.5	(1–8)·10^5 (B_{ext}=0)	24.5–28 (T=0)
NbTi	9.5–10.5	(3–8)·10^4 (B_{ext}=5 T)	12.5–16.5 (T=1.2 K) 12 (T=4.2 K)
NbN	14.5–17.8	(2–5)·10^7 (B_{ext}=18 T)	25 (T=1.2 K) 8–13 (T=4.2 K)

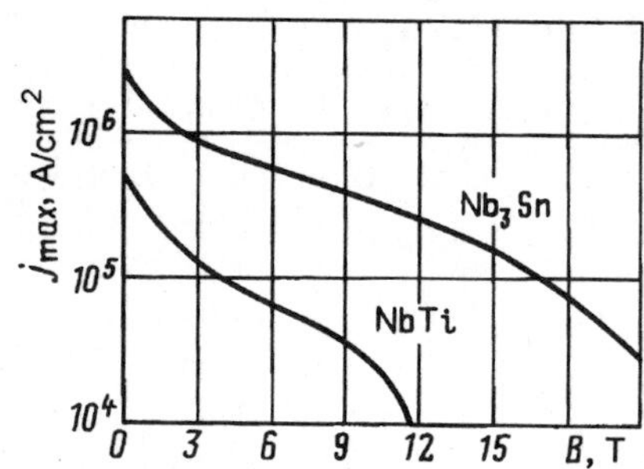

Figure 21.11 The critical current density vs. magnetic induction at 4.2 K [22].

References

[1] Landolt–Börnstein, Zahlenwerte und funktionen, II Bd., 6 Teil, Springer–Verlag, Berlin–Gottingen–Heidelberg, 1959.

[2] Stewart, G. R., Rev. Sci. Instrum., 54, 1, 1983.

[3] Encyclopedia of physics, Ed. Flugge, S., v.XIV–XV, Springer–Verlag, Berlin–Gottingen–Heidelberg, 1956.

[4] Handbuch der Physik, Bd. XIX, Springer–Verlag, Berlin–Gottingen-Heidelberg, 1956.

[5] Smithells, C.J., Tungsten, Chapman & Hall, Ltd., London, 1952

[6] Soule, D. E., Phys. Rev., 112, 698, 1958.

[7] Primak, W., Fuchs, L. H., Phys. Rev., 95, 22, 1954.

[8] Spain, J. L., Ubbelohde, A. R., Young, D. A., Phil. Trans. Royal Soc., 262, N 1128, 345, 1967.

[9] Landolt–Börnstein, Zahlenwerte und funktionen, IV Bd., 3 Teil, Springer–Verlag, Berlin–Gottingen–Heidelberg, 1957.

[10] High-temperature technology, Ed. Campbell, I., Wiley, NY–Chapmen & Hall, London, 1956

[11] Handbook of electrotechnical materials, Ed. Andrianov K. A., Gosenergoizdat, Moscow–Leningrad, 1960 (in Russian).

[12] Gotman, P. E., Berezin, V. B., Khaikin, A. M., Electrotechnical materials, Energiya, Moscow, 1969 (in Russian).

[13] Metal science and thermal treatment, Ed. Gudtsov, N. G., Metallurgiya, Moscow, 1966 (in Russian).

[14] Concise handbook of a metal worker, Ed. Malov, A. M., Mashinostroenie, Moscow, 1965.

[15] Savitskii, E. M., Tylkina, M. L., Povarova, K. B., Rhenium alloys, Nauka, Moscow, 1965 (in Russian).

[16] Superconducting materials, Ed. Savitskii E. M., Metallurgiya, Moscow, 1976 (in Russian).

[17] Superconductivity, Ed. Parks, R., Marcel Dekker, New York, 1969.

[18] Kittel, Ch., Introduction to solid state physics, Chapman & Hall, Ltd, London, 1953.

[19] Schooley, J. F., J. Phys., 39, 1169, 1978.

[20] Utton, D. B., Soulen, R. J., Marshak, H., Low-Temp. Phys., 4, 76, 1975.

[21] Durieux, M., Astrov, D. N., Kemp, W. R. G., Swenson, C. A., Metrologia, 15, 57, 1979.

[22] Tables of physical quantities. Handbook, Ed. Kikoin, I. K., Atomizdat, Moscow, 1976 (in Russian).

22

Electrophysical Properties of Semiconductors

S.D. Lazarev, E.Z. Meilikhov, and B.A. Aronzon

22.1 Introduction

The substances in which, at $T = 0$ K, the uppermost of the energy bands occupied by electrons (the valence band) and the lowermost of the unoccupied energy bands (the conduction band) do not overlap are referred to as semiconductors or dielectrics. The dividing line between the two classes of substances is rather arbitrary. Unlike dielectrics, in semiconductors the energy gap between the conduction band and the valence band is not too wide. This brings into existence the noticeable number of free charge carriers in the bands at $T \neq 0$ K.

The electrical resistivity of semiconductors (at room temperature) usually lies in the range 10^{-3}–10^{10} Ω·cm (in metals, $\rho = 10^{-6}$–10^{-4} Ω·cm).

The characteristic features of this class of substances are an increase in electrical conduction with temperature, low (as compared with metals) densities of current carriers, a high sensitivity of the electrical properties to radiations and impurities, as well as the nonohmic behavior of contacts.

The forbidden band width E_g represents an energy gap between the absolute maximum of the valence band and the absolute minimum of the conduction band. It is determined from the temperature dependence of resistance or by optical methods (absorption edge, long-wave threshold of photoconductivity). The value of E_g depends on temperature and pressure; its functional form is governed by the coefficients $\alpha_T = \partial E_g/\partial T$ and $\alpha_P = \partial E_g/\partial P$.

Carrier mobility and conductivity. The drift mobility of carriers is defined as

$$\mu_d = v_d/E,$$

where v_d is the drift velocity of carriers in the electric field of strength E, and is determined by direct experiments from the time of propagation of an injected pulse of minority carriers in a sample. The electrical conductivity σ is related to the drift mobilities of electrons and holes μ_n, μ_p and to their densities n and p by the formula

$$\sigma = e\,(n\mu_n + p\mu_p)\,.$$

The measurement of the Hall effect allows the determination of the Hall mobility

$$\mu_H = \mid R\sigma \mid,$$

where R is the Hall coefficient.

Hall effect. In a semiconductor placed in a magnetic field perpendicular to a current flowing in the semiconductor, there occurs an electric field perpendicular to both the current and the magnetic field.

0-8493-2861-6/97/$0.00+$.50

This phenomenon is called "Hall effect" described by the relationship

$$E_H = RjH,$$

where E_H is the strength of the Hall electric field; j is the current density; H is the strength of the magnetic field; R is the Hall coefficient.

In the case of one sort of carriers (with a density n)

$$R = \pm \frac{r}{ne}.$$

Here R is measured in cm^3/C; n in cm^{-3}; $e = 1.6 \cdot 10^{-19}$ C; r is the numerical factor (so-called Hall factor) whose value is determined by the mechanism of carrier momentum relaxation. The sign of R correlates with that of the carrier charge ($R < 0$ in n-type samples, $R > 0$ in p-type samples).

When two sorts of carriers (for example, electrons and holes) are present, R depends on the magnetic field strength:

$$R = \frac{R_n \sigma_n^2 \left(1 + \sigma_p^2 R_p^2 H^2\right) + R_p \sigma_p^2 \left(1 + \sigma_n^2 R_n^2 H^2\right)}{\left(\sigma_n + \sigma_p\right)^2 + \left[\sigma_n \sigma_p \left(R_n + R_p\right) H\right]^2}.$$

Here $\sigma_n = ne\mu_n$; $\sigma_p = pe\mu_p$; $R_n = -r/ne$; $R_p = r/pe$.

For $H \to \infty$ and $p \neq n$

$$R \to \frac{1}{e\,(p - n)}.$$

For $n = p$ and arbitrary magnetic field

$$R = \frac{1}{ne} \frac{\mu_p - \mu_n}{\mu_p + \mu_n}.$$

The Hall mobility of carriers μ_H is defined by the relationship $\mu_H = \mid R\sigma \mid$.

The lifetime of carriers τ represents a time for which the nonequilibrium density of carriers decreases down to the equilibrium value through their recombination. The basic recombination mechanisms involve radiative (the energy of recombining electron-hole pair is emitted, as a photon), phonon (the energy is transferred to the lattice), and collisional ones (the energy of the pair is transferred to a third particle).

More frequently the recombination occurs not directly, but through recombination centers (impurities, defects).

The theoretical estimation of the radiative recombination time τ_R yields the upper limit of the carrier lifetime. For $T = 300$ K and a near-intrinsic density, the values of τ_R are given in Table 22.1 along with the real τ values.

Table 22.1 The recombination lifetime τ and the radiative recombination lifetime τ_R for some semiconductors [162] (near-intrinsic carrier density, $T = 300$ K)

Semiconductor	Si	Ge	InSb	InAs	PbS	PbSe	PbTe
τ_R, s	3	0.3	$3 \cdot 10^{-7}$	10^{-5}	10^{-5}	$3 \cdot 10^{-6}$	$2 \cdot 10^{-6}$
τ, s	$2 \cdot 10^{-3}$	$3 \cdot 10^{-3}$	$3 \cdot 10^{-7}$	10^{-7}	10^{-5}		

Surface recombination. Besides the recombination in the bulk of a semiconductor, the carriers can recombine at its surface. The surface recombination rate s is defined as the velocity of particle flow from the bulk to the surface such that is required to maintain an excess of nonequilibrium carriers. The rate s strongly depends on the method of surface treatment. For example, for germanium $s \approx 10$ cm/s upon etching its surface by boiling H_2O_2, and $s \approx 10^5$ cm/s and more upon lapping. Usually one finds $s \approx 10^2$–10^3 cm/s.

The diffusion length L_D is a distance characterizing the spatial decrease in the nonequilibrium carrier density down to the equilibrium value. The L_D quantity is defined through the diffusion coefficient D and the lifetime τ by the relationship

$$L_D = \sqrt{D\tau}.$$

The diffusion coefficient and the mobility are related by the well-known Einstein relation

$$D = kT\mu/e$$

(in a nondegenerate semiconductor). The maximum diffusion length characterizes the degree of crystal perfection and purity. At $T = 300$ K, $L_D \approx 0.5$ cm in Ge, ≈ 0.3 cm in Si, and $\approx 10^{-2}$ – 10^{-3} cm in InSb [162].

Band structure and effective masses. The effective mass of a carrier characterizes its motion in the crystal lattice. The reverse effective mass $(m^*)^{-1}$ furnishes the tensor quantity defined by the dependence $E(\mathbf{p})$ of the carrier energy E on its quasimomentum $\mathbf{p}$.

$$(m^*)^{-1}_{ij} = \frac{\partial^2 E(\mathbf{p})}{\partial p_i \partial p_j}.$$

Usually it will suffice to know the behavior of $E(\mathbf{p})$ only in the vicinity of the extreme points: the energy minima or maxima. The isoenergetic surfaces near the extrema are often represented as spheres (with the effective masses, for example, for several subbands of the valence band m_{p1}, m_{p2}, etc.) or ellipsoids (with the effective masses for the conduction band $m_{n\|}$, $m_{n\perp 1}$, $m_{n\perp 2}$.[1]

To analyze experimental data, the concept of a scalar density-of-states effective mass (m_{dn} and m_{dp}, for electrons and holes, respectively) is in frequent use. In the case of the ellipsoidal isoenergetic surfaces, this quantity is determined from the relationship

$$m_d = N^{2/3}\left(m_{\|} \cdot m_{\perp 1} \cdot m_{\perp 2}\right)^{1/3},$$

where N is the number of extrema in the band.

For the energy bands having the degenerate spherical surfaces of constant energy with the effective masses m_{p1}, m_{p2}, etc., the density-of-states effective mass is defined as follows:

$$m_{dp} = \left(m_{p1}^{3/2} + m_{p2}^{3/2} + \ldots\right)^{2/3}.$$

The concepts of an ohmic effective mass m_σ and a cyclotron effective mass m_c are also introduced. They are defined by the relationships

$$\frac{1}{m_\sigma} = \frac{1}{3}\left(\frac{1}{m_{\|}} + \frac{1}{m_{\perp 1}} + \frac{1}{m_{\perp 2}}\right)$$

[1]Often $m_{\perp 1} = m_{\perp 2}$, then the symbol $m_\perp = m_{\perp 1} = m_{\perp 2}$ is used.

(for ellipsoidal isoenergetic surfaces), and

$$m_c = \frac{\hbar}{2\pi}\frac{\partial S}{\partial E}$$

(S is the area of the section of the isoenergetic surface by a plane normal to a magnetic field).

The general expressions for m_d, m_σ, and m_c can be found, for example, in [84].

The cyclotron resonance constitutes a direct method for determining $E(\mathbf{p})$ and the effective masses. The valuable information about the band structure and the carrier effective masses can be obtained in measuring the anisotropy of magnetoresistance, the Shubnikov-de Haas type effects, and the magnetooptical effects.

The intrinsic carrier density n_i corresponds to an ideally pure material and can be calculated, if the band structure and the carrier effective masses are known, by the formula

$$n_i = 4.82 \cdot 10^{15} T^{3/2} \left(\frac{m_{dn} m_{dp}}{m_0^2}\right)^{3/4} \exp\left(\frac{E_g - \alpha_T T}{2kT}\right),$$

where m_{dn} and m_{dp} are the density-of-states effective masses of electron and holes, respectively; m_0 is the mass of a free electron; k is the Boltzmann constant; α_T is the temperature coefficient of the forbidden band width.

As a rule, n_i is determined experimentally from the measurements of the Hall effect and the conduction in the temperature involved.

The Debye temperature T_D is defined through the limiting lattice vibration frequency ω_m by the relationship

$$kT_D = \hbar\omega_m.$$

The different Debye temperatures correspond, generally speaking, to the different branches of lattice vibrations. The values of T_D determined from thermal measurements are averaged over the vibration branches that are essential at the temperature of measurements. More comprehensive information can be obtained, for example, in measuring the elastic constants.

The limiting frequency of optical phonons ω_l, or ω_t, is the frequency of respective (longitudinal or transverse) optical lattice vibrations with wavelengths considerably longer than the interatomic distance. It is determined from the infrared absorption and reflection spectra as well as by neutron spectroscopy. For elementary semiconductors (Ge, Si, etc.) one has $\omega_l = \omega_t = \omega_0$.

The spectroscopic splitting factor g in the conduction (or valence) band characterizes the splitting of the carrier energy levels in a magnetic field due to their magnetic moments:

$$E = (n + 1/2)\,\hbar\omega_H \pm (1/2)g\hbar\omega_H, \quad n = 1, 2\ldots,$$

where ω_H is the cyclotron frequency of a charge carrier in the crystal which is determined from n easuring the oscillation effects in the magnetic field by the para- and nuclear magnetic resonance methods.

Dielectric constant. The values ε_0 and ε_∞ of the dielectric constant are determined from static and high-frequency (or optical) measurements, respectively. In the tables throughout this chapter, except as otherwise indicated, the values presented belong to $T = 290$ K.

The tables thereafter contain mainly the data on semiconductors with $E_g < 3$ eV. Ternary and more complex semiconducting compounds are not described here.[2] Data on the parameters of semiconductor devices are also lacking.

[2]For the properties of some classes of ternary compounds, see monographs [16, 121, 123, 145, 289, 290].

22.2 Elementary Semiconductors

22.2.1 Silicon and germanium

Silicon and germanium comprise the widely used and best-investigated semiconductors. They crystallize into a diamond lattice and have a complex band structure.

In Si, six equivalent absolute minima of the conduction band are located on the [100] axes inside the Brillouin zone. In the vicinity of each of these minima, the isoenergetic surfaces represent ellipsoids of revolution (six ellipsoids).

In Ge, eight equivalent absolute minima of the conduction band are located on the [111] axes at the boundary of the Brillouin zone.

In the vicinity of each of these minima, the isoenergetic surfaces represent ellipsoids of revolution (the equivalent number of ellipsoids equals four).

The valence bands in Si and Ge are split into three subbands: two of them are degenerate at k = 0 and the third subband is split off due to spin-orbit interaction.

The electronic band structures of Si and Ge (with the symbols used in the tables) are shown in Figures 22.1 and 22.2, respectively.

The electrophysical properties of Si and Ge are presented in Tables 22.2–22.4 and Figures 22.3–22.22.

Table 22.2 Electrophysical properties of Si and Ge

Element	Crystal structure: System, group	Crystal structure: a, nm [123]	ρ, g/cm^3 [123]	T_{melt}, K [123]	T_D, K [117]	$\hbar\omega_0$, eV [123]	ϵ_0 [28]	$\epsilon_\infty(\lambda)$
Si	Cubic, O_h^7	0.543	2.33	1690	689 539 (80 K)	0.063	11.7	12.7 (1μm)[28] 13.7 (8 mm)[117]
Ge	Cubic, O_h^7	0.566	5.32	1210	406 353 (80 K)	0.037	16	16.5 (2.5μm)[28]

Element	n_i, cm^{-3} [292]	ρ_i, Ω·cm [11]	E_g, eV [292]	$E_{\Gamma\Lambda}$, eV	$E_{\Gamma\Delta}$, eV	Δ_{so}, eV	$\partial E_g/\partial T$, 10^{-4}eV/K	$\partial E_g/\partial P$, 10^{-6}eV/bar
Si	$1.02 \cdot 10^{10}$ (300 K)	$2.3 \cdot 10^5$	1.11 1.17(0 K)	2.3 [84]	1.5 [84]	0.044 [123]	−2.8 [292]	−1.4 [292] −3.8 (80K)
Ge	$2.33 \cdot 10^{13}$ (300 K)	47	0.664 0.744 (1.5 K)	0.05 [123] 0.15 [84]	0.02 [123]	0.28 [117] 0.30 [123]	−3.7 [292]	5 [123] 3 [292]

Element	$\partial E_{\Gamma\Lambda}/\partial T$, 10^{-4}eV/K	$\partial E_{\Gamma\Lambda}/\partial P$, 10^{-6}eV/bar	$\partial E_{\Gamma\Delta}/\partial P$, 10^{-6}eV/bar	m_n/m_0 [84]	m_{dn}/m_0	$m_{n\sigma}/m_0$ [117]
Si				0.9163 (‖) 0.1905 (⊥)	0.33	0.26
Ge	−4 [123]	7 [123]	−(12–14) [123]	1.588 (‖) 0.0815 (⊥)	0.22	0.12

Element	m_{p1}/m_0	m_{p2}/m_0	m_{p3}/m_0	m_{dp}/m_0	μ_n, cm^2/(V·s)	μ_p, cm^2/(V·s)
Si	0.537	0.153	0.25	0.81 [292]	3000 ($\sim T^{-2.6}$)	500 [123] ($\sim T^{-2.3}$) [117]
Ge	0.34	0.043	0.08	0.39 [123]	3800 ($\sim T^{-2.3}$)	1820 [123] ($\sim T^{-1.67}$) [28]

Table 22.3 Properties of impurities in Si [292]

Impurity	Ag		Al	As	Au		B	Bi	Cu
Type*	*A*	*D*	*A*	*D*	*A*	*D*	*A*	*D*	*A*
E_i, eV	0.93 0.86 0.40	0.59 0.83 0.89	0.068	0.054	0.6–0.7	0.8	0.045	0.071	0.24 0.37 0.52

Impurity	Fe	Ga	In	Li	O	P	S	Sb	Tl	Zn
Type*	*A*	*A*	*A*	*D*	*D*	*D*	*D*	*D*	*A*	*A*
E_i, eV	0.4	0.071	0.155	0.034	0.06	0.045	0.31	0.043	0.246	0.316 0.617

* *D*–donor, *A*–acceptor.

Note: The ionization energy E_i is reckoned from the bottom of the conduction band for donors and from the valence-band edge, for acceptors.

Table 22.4 Properties of impurities in Ge [117, 123, 292]

Impurity	Ag	Al	As	Au	B	Be	Bi	Cd	Co	Cr	Cu	Fe
Type*	*A*	*A*	*D*	*A*	*A*	*A*	*D*	*A*	*A*	*A*	*A*	*A*
E_i, eV	0.13 0.5 0.7	0.0111	0.0142	0.16 0.59 0.75	0.0108	0.07	0.0128	0.05 0.15	0.09 0.25 0.48	0.07 0.12	0.33 0.4 0.53	0.35 0.52

Impurity	Ga	In	Li	Mn	Ni	O	P	Pt	S	Sb	Se	Te
Type*	*A*	*A*	*D*	*A*	*D*	*D*	*D*	*A*	*D*	*D*	*D*	*D*
E_i, eV	0.0113	0.0120	0.0100	0.16	0.3	0.017 0.04 0.20	0.0129	0.2 0.4	0.18	0.0103	0.14	0.11 0.30

* *D*–donor, *A*–acceptor.

Note: The ionization energy E_i is measured from the bottom of the conduction band for donors and from the valence-band edge, for acceptors.

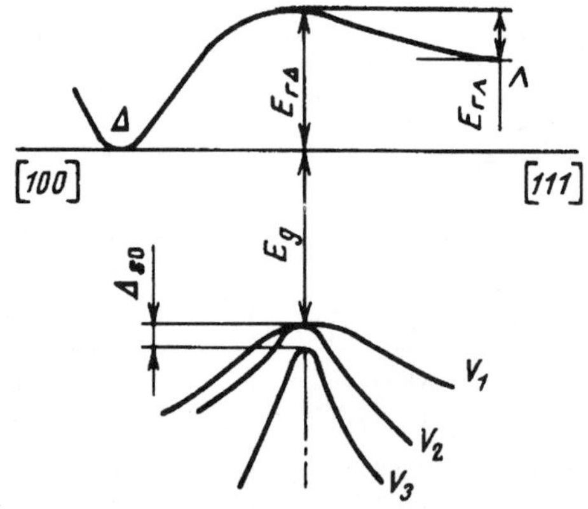

Figure 22.1 The band structure of Si [117].

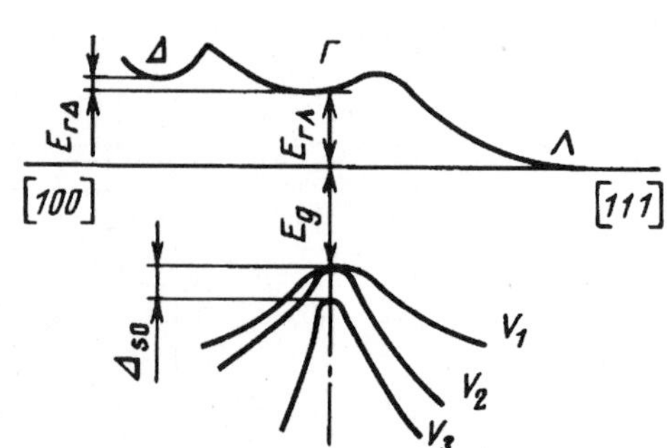

Figure 22.2 The band structure of Ge [117].

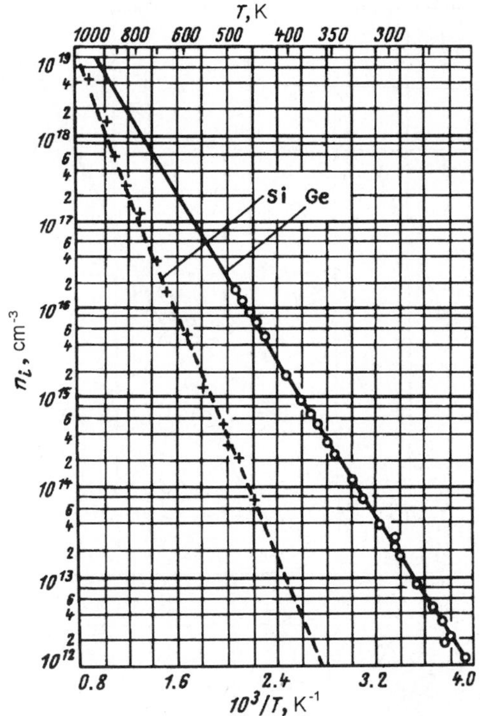

Figure 22.3 The temperature dependence of the intrinsic carrier density in Ge and Si [221].

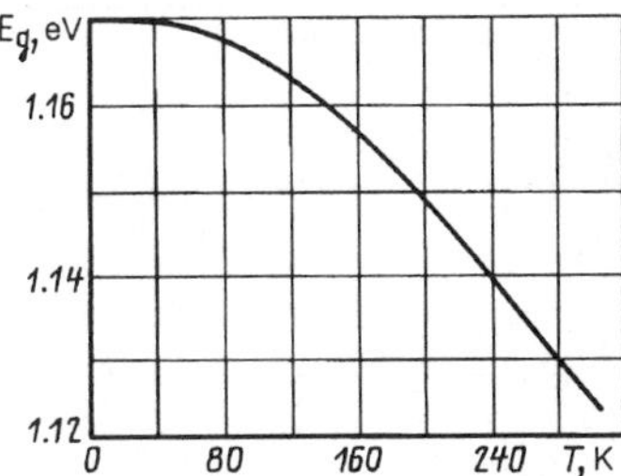

Figure 22.4 The temperature dependence of the forbidden band width for Si [58].

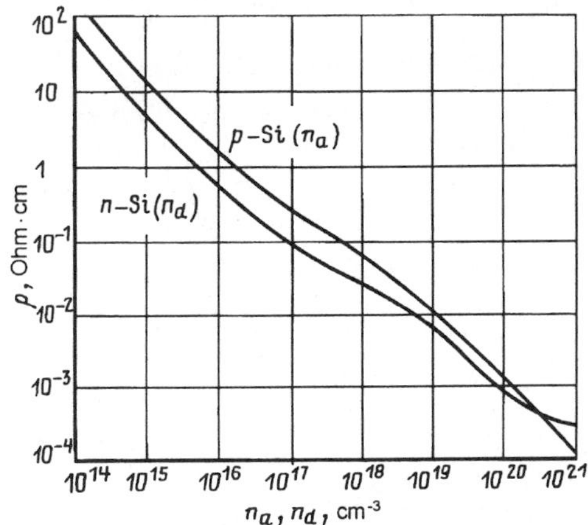

Figure 22.5 The dependence of the Si resistivity at 300 K on the concentration of donor or acceptor impurities [63].

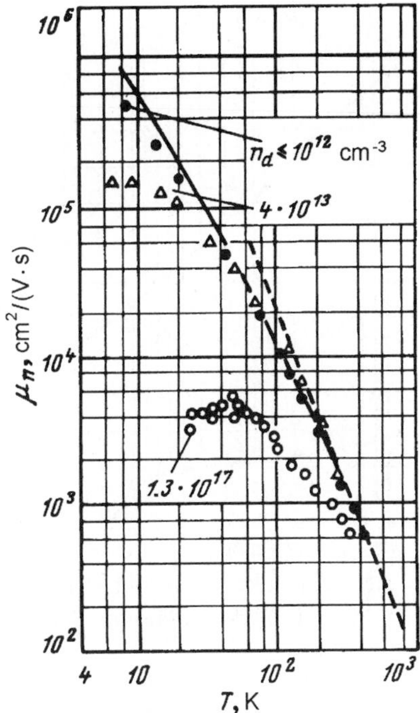

Figure 22.6 The temperature dependence of the electron mobility in Si at different donor densities [64]: solid line is a calculation with allowance for the phonon scattering of electrons; dashed line fits the asymptotic $T^{-2.42}$ approximation.

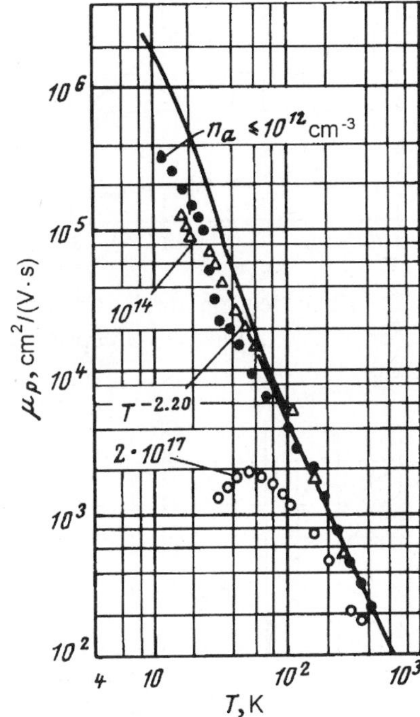

Figure 22.7 The temperature dependence of the hole mobility in Si at different acceptor densities [64]: solid line is a calculation with allowance for the phonon scattering of holes.

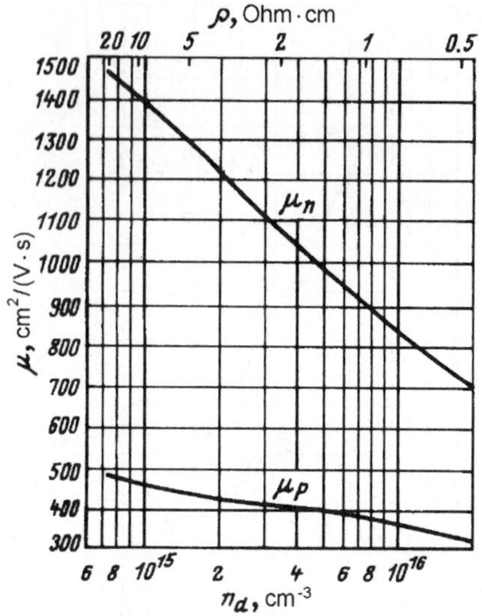

Figure 22.8 The electron and hole mobilities in Si at 300 K vs. the donor density [226].

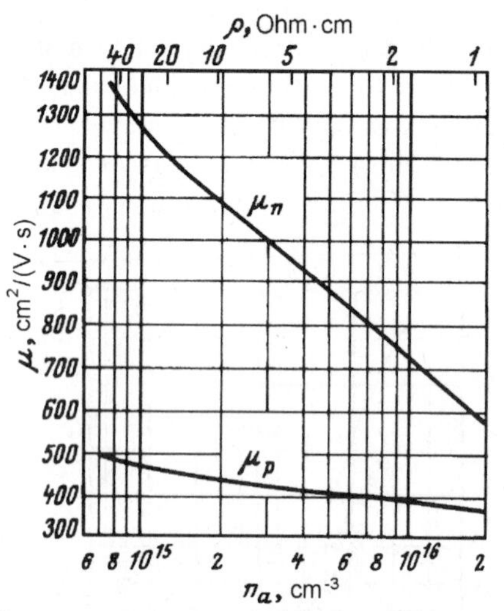

Figure 22.9 The electron and hole mobilities in Si at 300 K vs. the acceptor density [226].

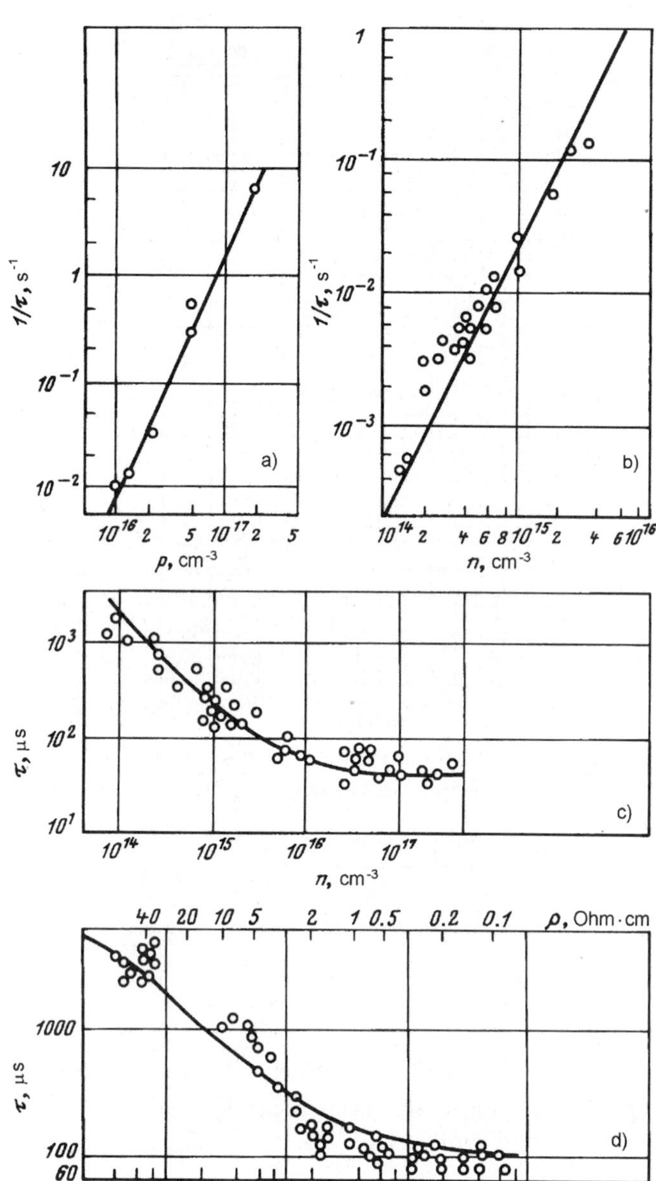

Figure 22.10 The lifetime of minority carriers in n-Si (a), p-Si (b), n-Ge (c) and p-Ge (d) vs. the majority carrier density at 300 K [10,237].

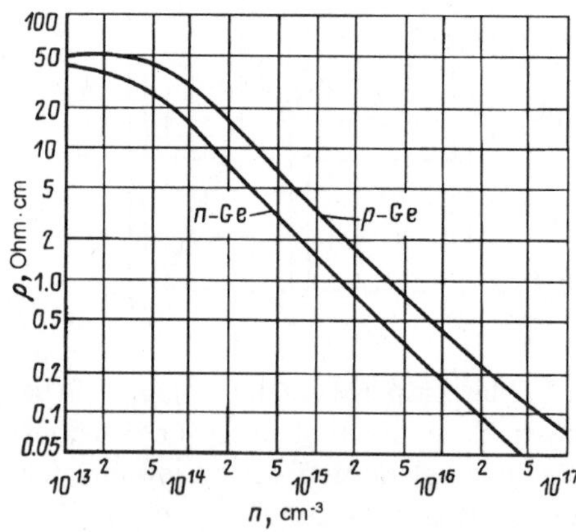

Figure 22.11 The resistivity of n- and p-Ge vs. the carrier density at 300 K [172].

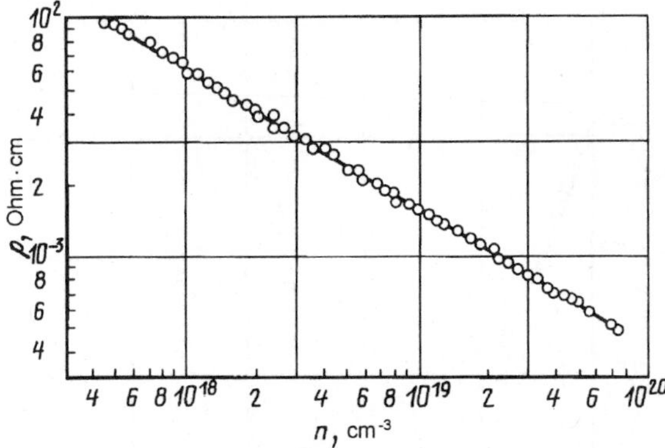

Figure 22.12 The resistivity of heavy-doped n-Ge crystals vs. the impurity concentration at 300 K [225].

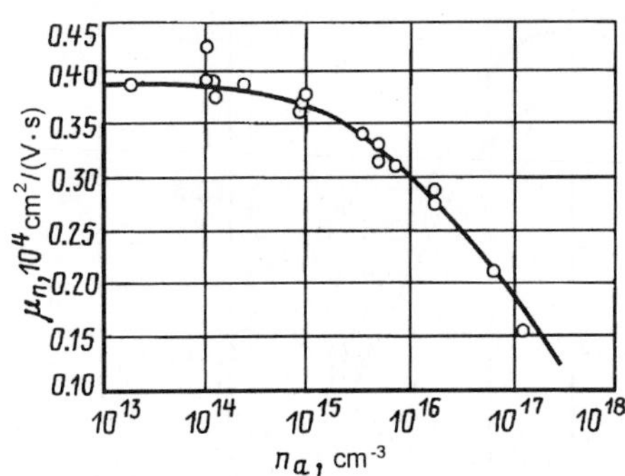

Figure 22.14 The drift mobility of electrons in p-Ge at 300 K vs. the acceptor density: solid line – calculated results [72].

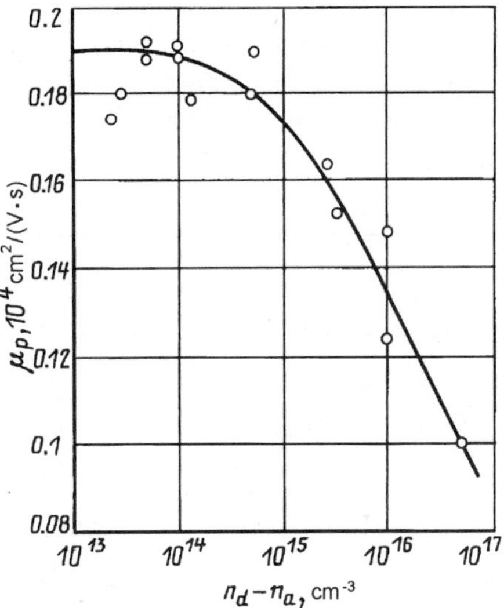

Figure 22.15 The drift mobility of holes in n-Ge at 300 K vs. the carrier density: solid line – calculated results [72].

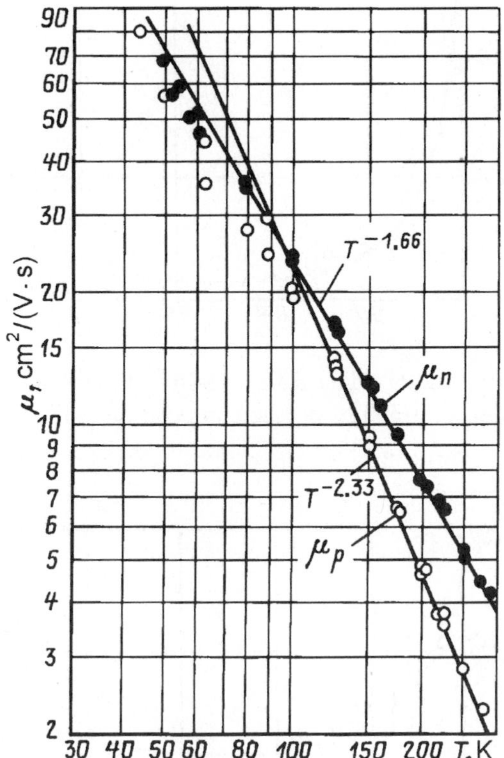

Figure 22.13 Temperature dependences of electron and hole mobilities in pure Ge at the constant carrier density [71].

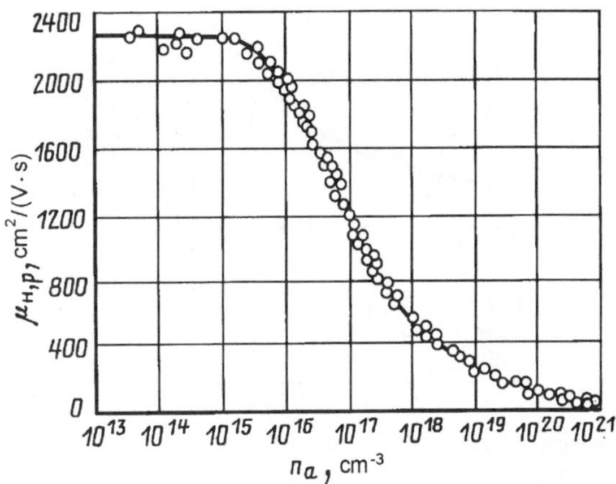

Figure 22.16 The Hall mobility of holes in p-Ge at room temperature vs. the acceptor density [73].

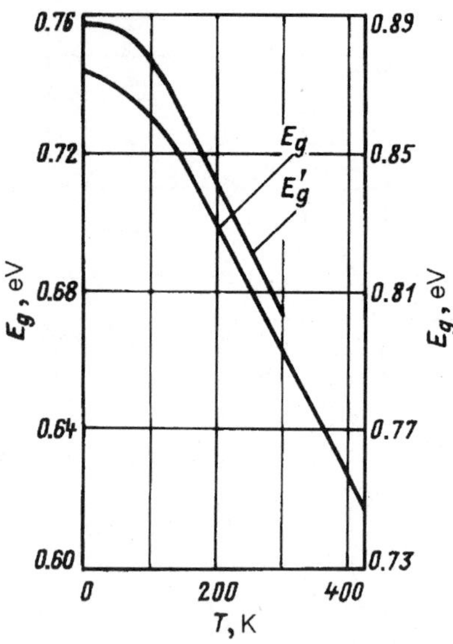

Figure 22.17 The temperature dependence of the direct-gap (E_g') and indirect-gap (E_g) transitions in Ge [65].

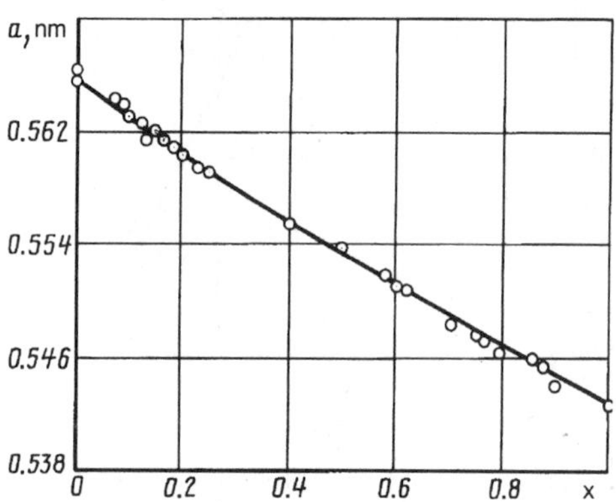

Figure 22.18 The lattice parameter a of Si_xGe_{1-x} vs. the composition [71].

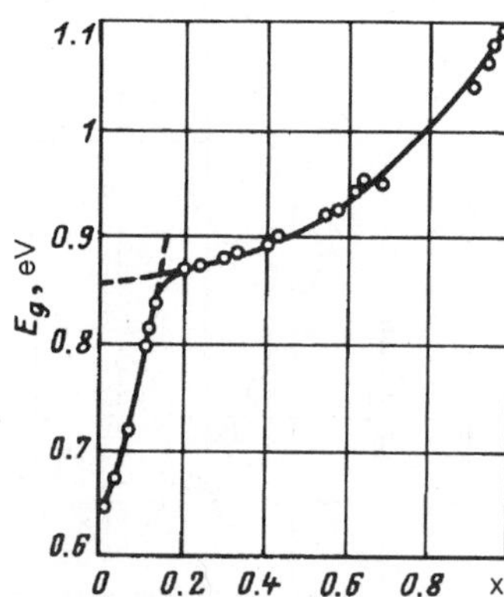

Figure 22.19 The forbidden band width of Si_xGe_{1-x} at 300 K vs. the composition. At $x \cong 0.15$ the transition from the band structure of Si to that of Ge occurs [76].

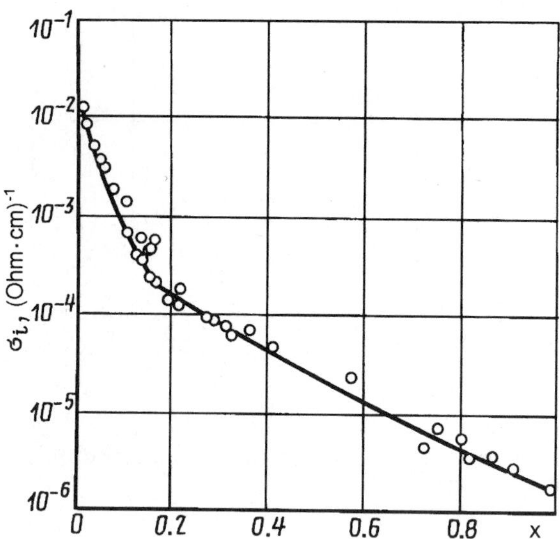

Figure 22.20 Intrinsic conductivity of Si_xGe_{1-x} at 300 K vs. the composition [80].

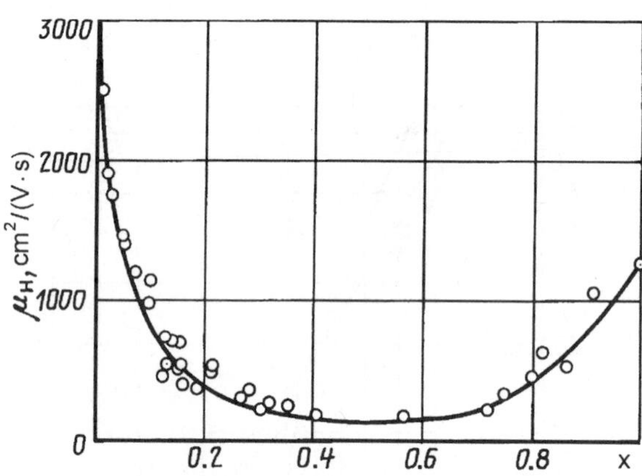

Figure 22.21 The Hall mobility of electrons in Si_xGe_{1-x} at 300 K vs. the composition [80].

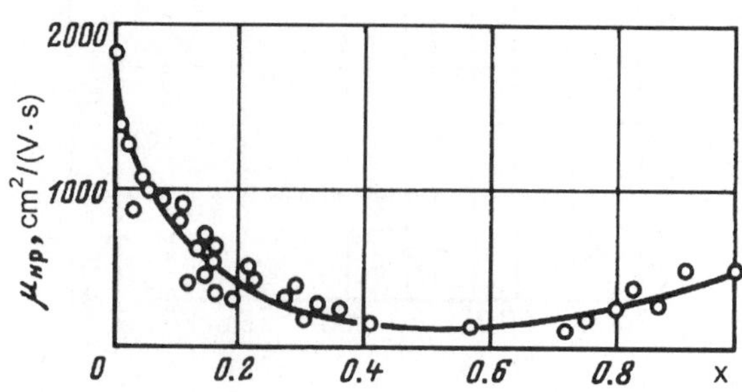

Figure 22.22 The Hall mobility of holes in Si_xGe_{1-x} at 300 K vs. the composition [80].

22.2.2 Other elementary semiconductors

Table 22.5 Electrophysical properties of elementary semiconductors [293]

Element	Crystal structure: System, group	Crystal structure: a, b, c, nm	ρ, g/cm^3	T_{melt}, K	E_g, eV	$\partial E_g/\partial T$, 10^{-4} eV/K	n_i, cm^{-3}
α-B	Rhombohedral, D_{3d}^5	0.5057 (a), α=58°06′	2.45	2350–2400	1.97 (**E**$\parallel c$) 1.90 (**E**$\perp c$)		
β-B	Rhombohedral, D_{3d}^5	1.014 (a), α=65°17′	2.33	2350–2400	1.63 (**E**$\parallel c$) 1.61 (**E**$\perp c$) 1.56 (0 K)	−2	
C (diamond)	Cubic, O_h^5	0.3566 [1]	3.51 [123]	4300 [123]	5.4 [28] 5.48 [292]	−0.5 [292]	
P (black)	Orthorhombic, D_{2h}^{18}	0.331 (a); 0.437 (b); 1.047 (c)	2.70	1300	0.33 (0 K)	−2.8	
α-S	Orthorhombic, D_{2h}^{24}	1.046 (a); 1.288 (b); 2.448 (c)	2.069		3.8		
As (grey)	Rhombohedral, D_{3d}^5	0.376 (a); 1.076 (c)	5.72	1090	0.175		$2.2 \cdot 10^{20}$
Se	Trigonal, D_3^4, D_3^6	0.435 (a); 0.494 (c)	4.819	490	1.9 (5 K)	−2.7	10^{14}
α-Se	Monoclinic	0.905 (a); 0.907 (b); 1.161 (c)	4.9		2.48 (5 K)		
α-Sn	Cubic, O_h^7	0.649 [292]	7.28 [292]	505*1 [121]	0.09 [122]	−5 [122]	$> 10^{17}$ (77 K)
Sb	Rhomboheral, D_{3d}^5	0.4308 (a); 1.1274 (c)	6.69	903	0.1 (1.15 K)		$4.2 \cdot 10^{19}$
Te	Trigonal, D_3^4, D_3^6	0.446 (a); 0.595 (c)	6.24	723	0.335 (4.2 K)	+1.8 ($T < 4.2$ K) −0.4 (100–300 K)	$5.6 \cdot 10^{15}$

Table 22.5 Electrophysical properties of elementary semiconductors *(continued)*

Element	T_D, K	m_n/m_0	m_p/m_0	μ_n, cm^2/(V·s)	μ_p, cm^2/(V·s)	ϵ_0	ϵ_∞
α-B	1430 (0 K)				120		
β-B	1220 (100 K) 1300 (0 K)	4.5	1.8	1 [11]	300	11.55 (**E**$\perp$); 10.24 (**E**$\parallel$)	9.12 (**E**$\perp$); 8.41 (**E**$\parallel$)
C (diamond)	2240*2 [123]	1.4 (m_l) [313] 0.36 (m_t)	1.1*3 [213] 2.1 [123]	1800 [11] ($\sim T^{-3/2}$)	2100 [213] ($\sim T^{-3/2}$)	5.7 [292]	5.7 [121]
P (black)	400	0.16 (m_{n1}) 0.81 (m_{n2}) 0.24 (m_{n3})	0.17 (m_{p1}) 0.71 (m_{p2}) 0.59 (m_{p3})	220	350		
α-S	250			7.5	10	3.6–4.6 (anisotropic)	
As (gray)	250 282 (4.2 K)	0.134 (m_{n1}) 1.252 (m_{n2}) 0.141 (m_{n3})	0.146 (m_{p1}) 0.104 (m_{p2}) 0.166 (m_{p3})	40–550 (anisotropic)	50–1210 (anisotropic)	50	
Se	152.5 171 (0 K)		1.4 (m_d)	1 [234]	40	8.5 [234]	
α-Se	128 (<20 K)			4 ($\sim T^{-3/2}$)	0.2	7.39	6.1
α-Sn	230 [121]	0.023 [122] (000) 0.21 [213] (111)	0.26 [122] 0.20*4 [292]	2500 [11] ($\sim T^{-1.65}$) 10^5 (78K) [238] 10^5 (4.2K) [238]	2400 [11] ($\sim T^{-2}$) 10^4 [213] (100K)		24 [292]
Sb	168 (10 K) 211 (0 K)	0.068 (m_{n1}) 0.63 (m_{n2}) 0.34 (m_{n3})	0.093 (m_{p1}) 1.14 (m_{p2}) 0.093 (m_{p3})	0.11–2.74 (anisotropic)	0.18–3.63 (anisotropic)	80 (4.2 K)	
Te	140 ($\perp c$) [220] 290 ($\parallel c$) [120]	0.06 ($m_\perp$) 0.05 ($m_\parallel$)	0.114 ($m_\perp$) 0.109 ($m_\parallel$)	2380 ($\parallel c$) 1150 ($\perp c$)	1260 ($\parallel c$) 650 ($\perp c$)		23 ($\perp c$) [28] 39 ($\parallel c$) [28]

*1 Transforms into β-Sn at 285 K.
*2 $\hbar\omega_t = \hbar\omega_l = 0.165$ eV [123].
*3 Heavy holes; for light holes $m_p/m_0 = 0.36$ [23], 0.7 [195].
*4 Heavy holes; for light holes $m_p/m_0 = 0.06$ [292].

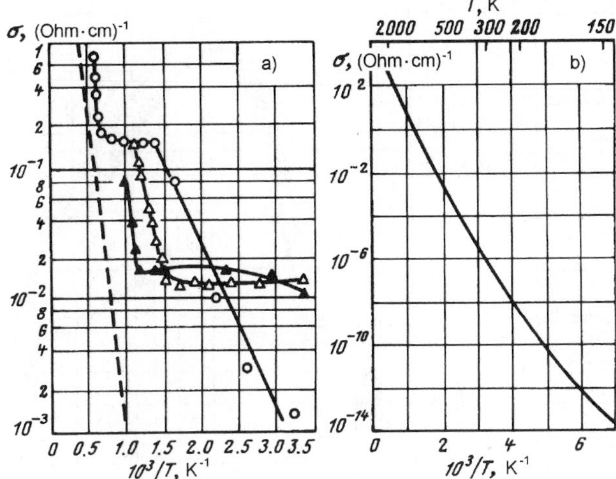

Figure 22.23 The temperature dependence of the conductivity for α-B (a) and β-B (b). The dashed line fits the asymptotic approximation at $E_a = 1$ eV: △ - granules; ▲ - crystals; ○ - polycrystal.

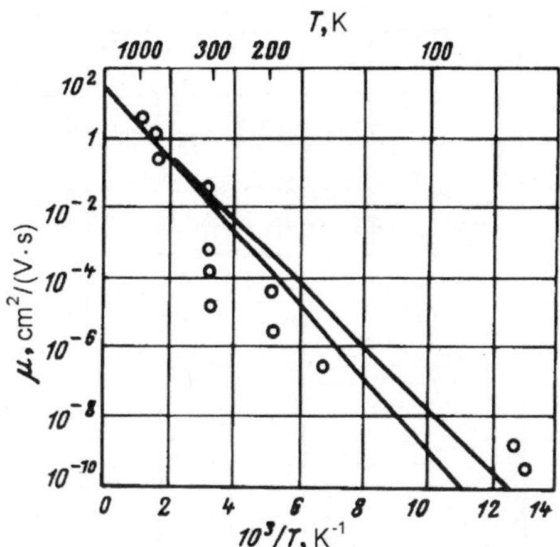

Figure 22.25 The temperature dependence of the drift mobility of electrons in β-B [293]. The solid lines correspond to calculations based on different models.

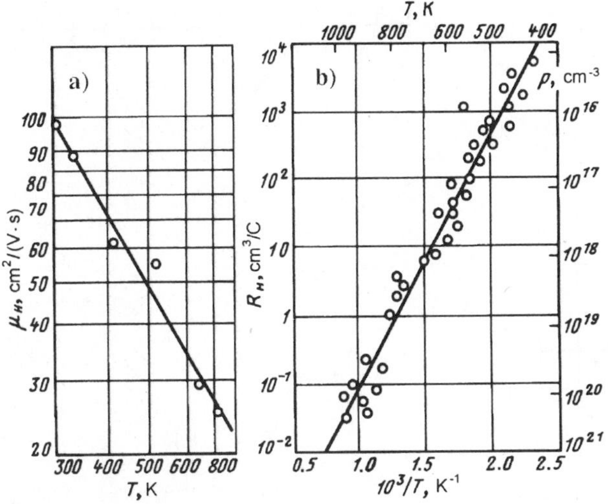

Figure 22.24 (a) High-temperature Hall mobility of holes in α-B [6] and (b) the Hall coefficient and the hole density in β-B [293].

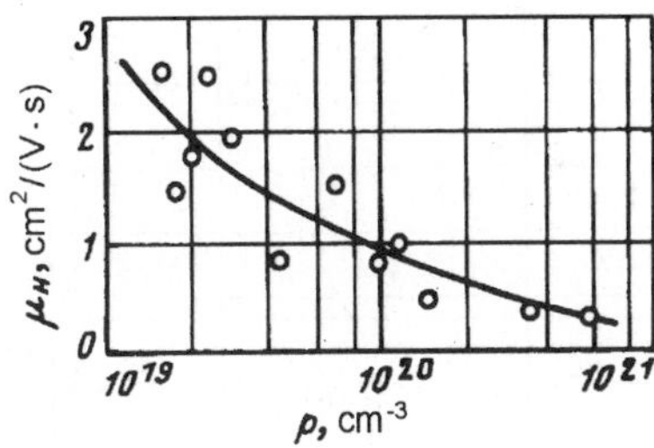

Figure 22.26 The electron Hall mobility in polycrystalline β-B vs. the carrier density [7,14].

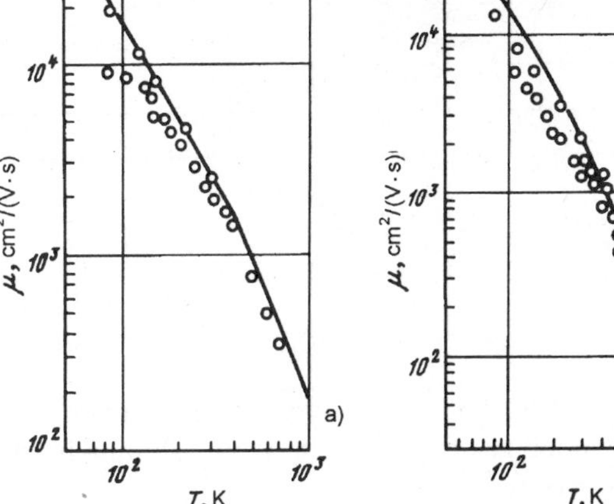

Figure 22.27 The temperature dependences of the electron (a) and hole (b) mobilities in diamond [313].

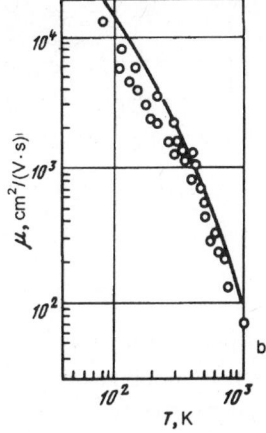

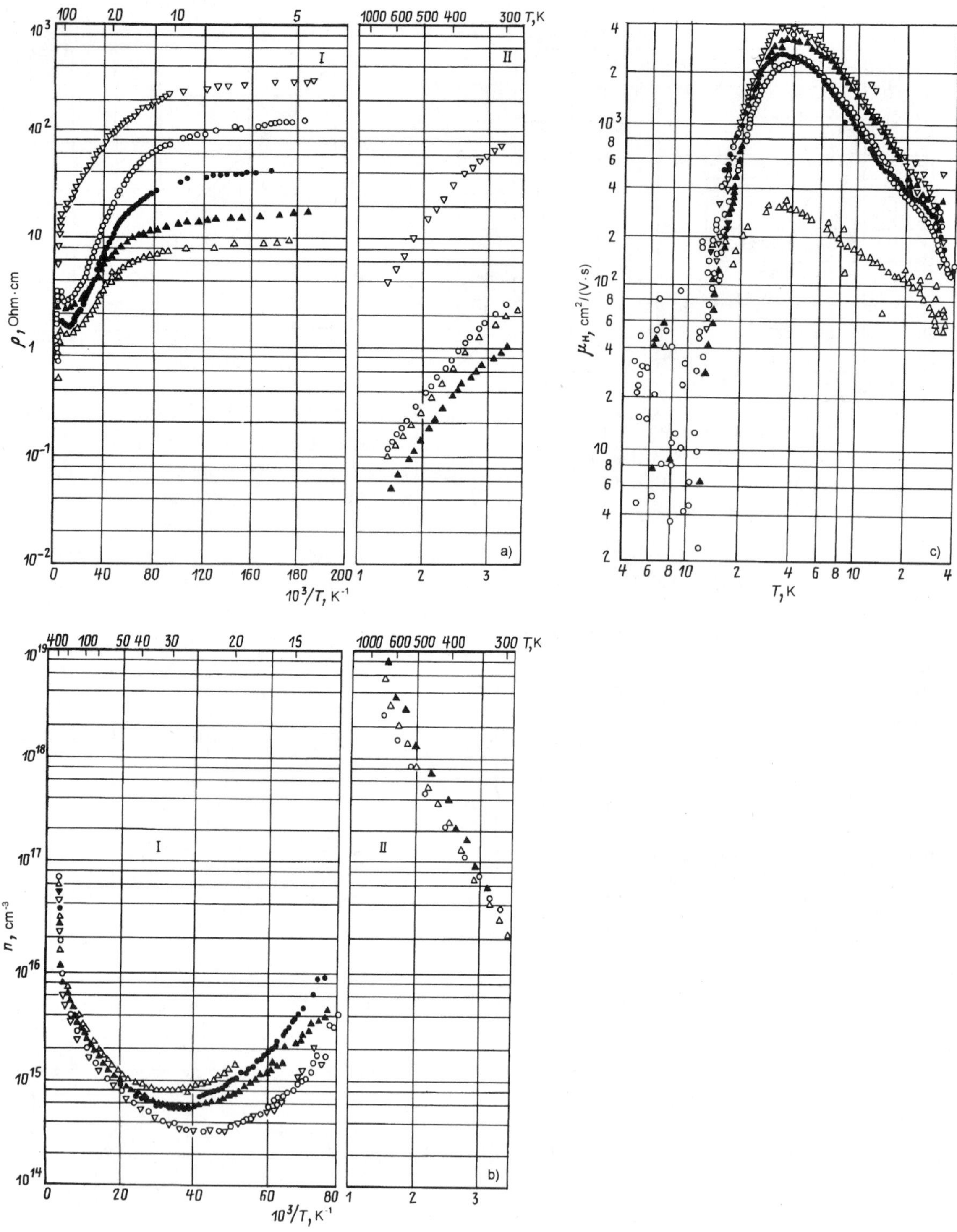

Figure 22.28 The temperature dependences of the resistivity (a), the total carrier density (b) and the Hall mobility of holes (c) in polycrystalline black phosphorus [17]: I – low temperatures; II – high temperatures.

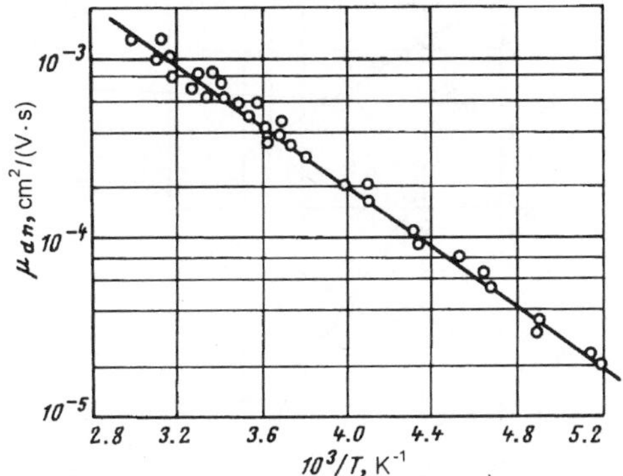

Figure 22.29 The temperature dependence of the electron drift mobility in α-S [37].

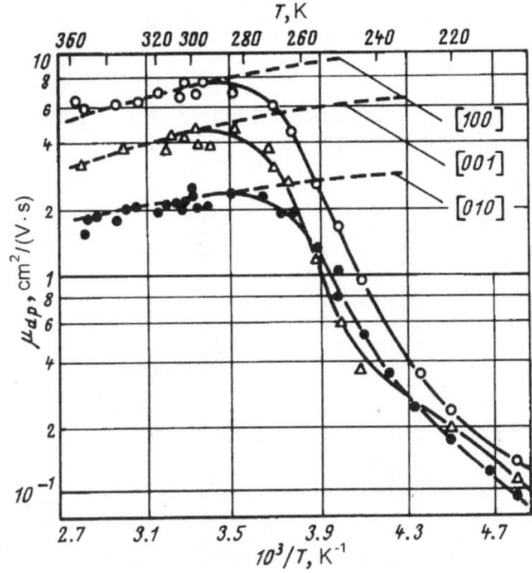

Figure 22.30 The temperature dependence of the hole drift mobility along different crystal axes in α-S [39]. The dashed line shows the dependence $\mu \propto T^{-n}$, where n = 1.6 [100], n=1.1 [010], and n=1.7 [001].

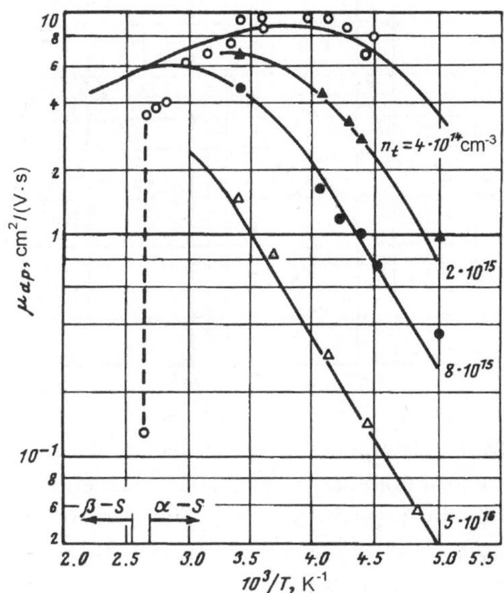

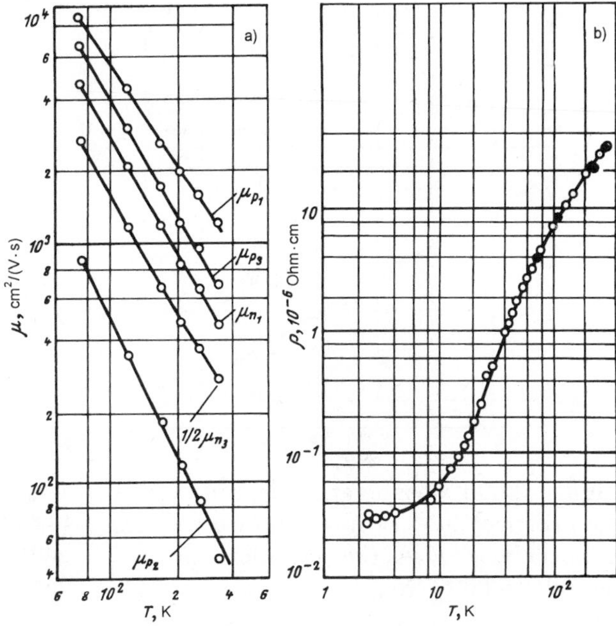

Figure 22.32 Temperature dependences of the electron and hole mobilities (a) and the resistivity (b) in As: μ_{n1}, μ_{n2} and $\mu_{n3} \propto T^{-1.7}$; $\mu_{p1} \propto T^{-1.5}$; $\mu_{p2} \propto T^{-2.0}$ [57].

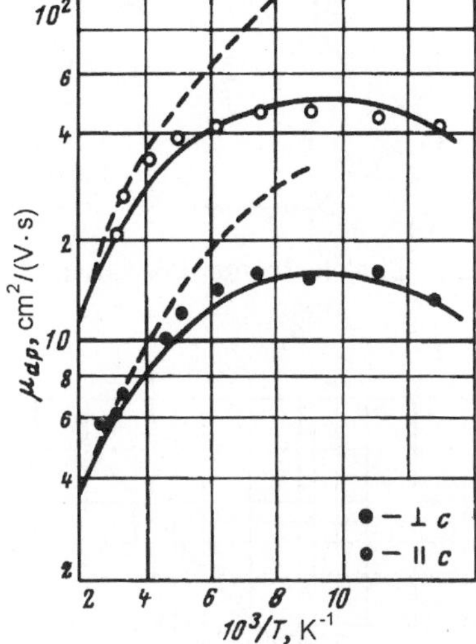

Figure 22.33 The temperature dependence of the hole drift mobility in trigonal Se [19]. Dashed lines correspond to the $T^{-3/2}$ dependence.

Figure 22.31 The temperature dependence of the hole drift mobility in α-S samples with different hole trap densities n_t. The mobility abruptly decreases on the phase transition α-S $\rightarrow$ β-S [53].

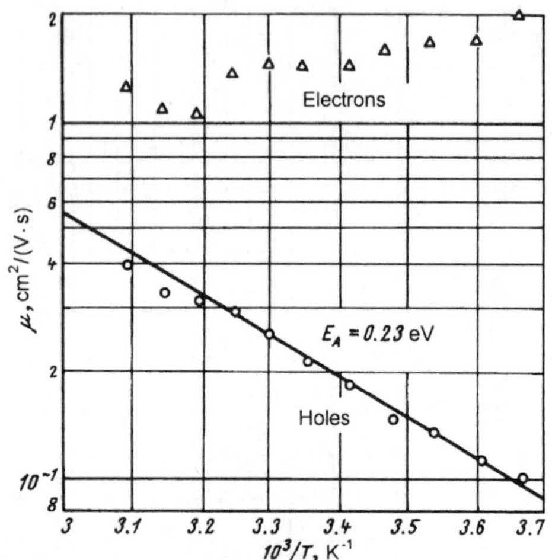

Figure 22.34 Temperature dependences of the drift mobilities of holes and electrons in monoclinic Se: E_A is the activation energy of the hole mobility [20].

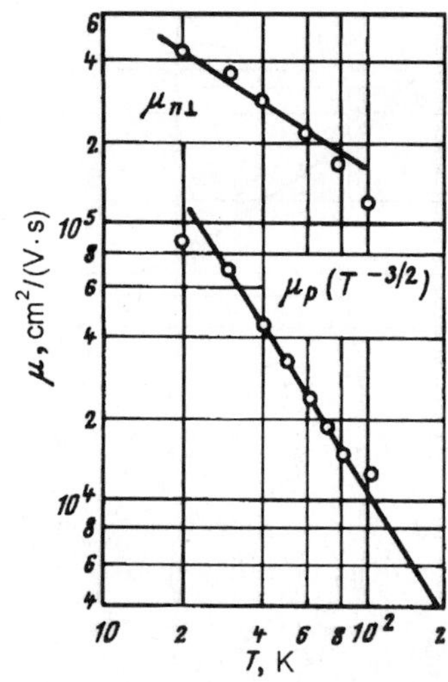

Figure 22.37 Temperature dependences of the mobilities of light electrons and holes in α-Sn. The solid lines check with the calculations [24].

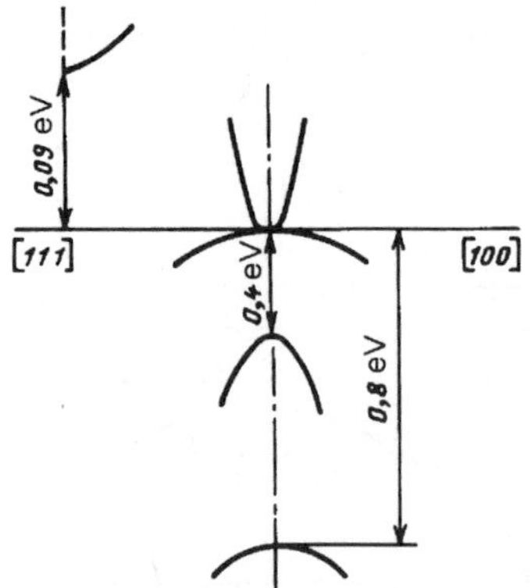

Figure 22.35 The electronic band structure of α-Sn [17].

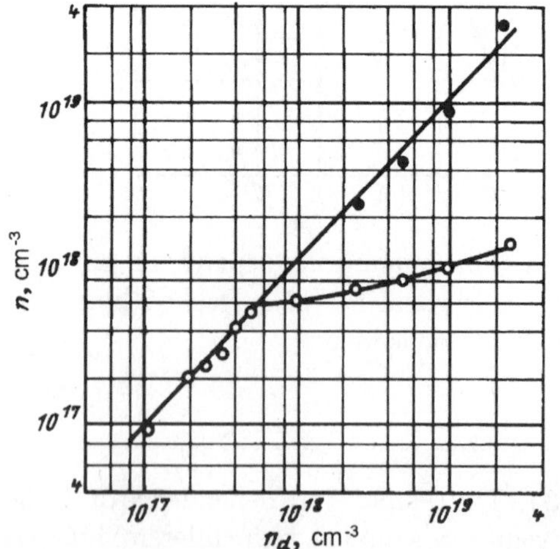

Figure 22.36 The electron density in α-Sn at T = 4.2 K vs. the donor density [24]: ○ – light electrons; ● – all electrons.

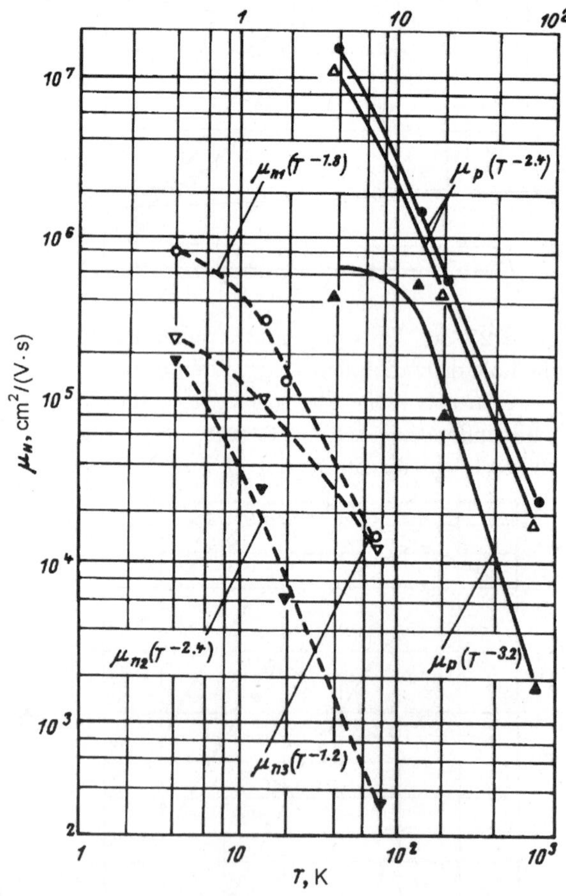

Figure 22.38 Temperature dependences of the mobility tensor components for electrons (dashed lines) and holes (solid lines) in Sb [29]. The electron and hole densities are $4{\cdot}10^{19}\text{cm}^{-3}$: the lower temperature scale – for n, the upper one – for p.

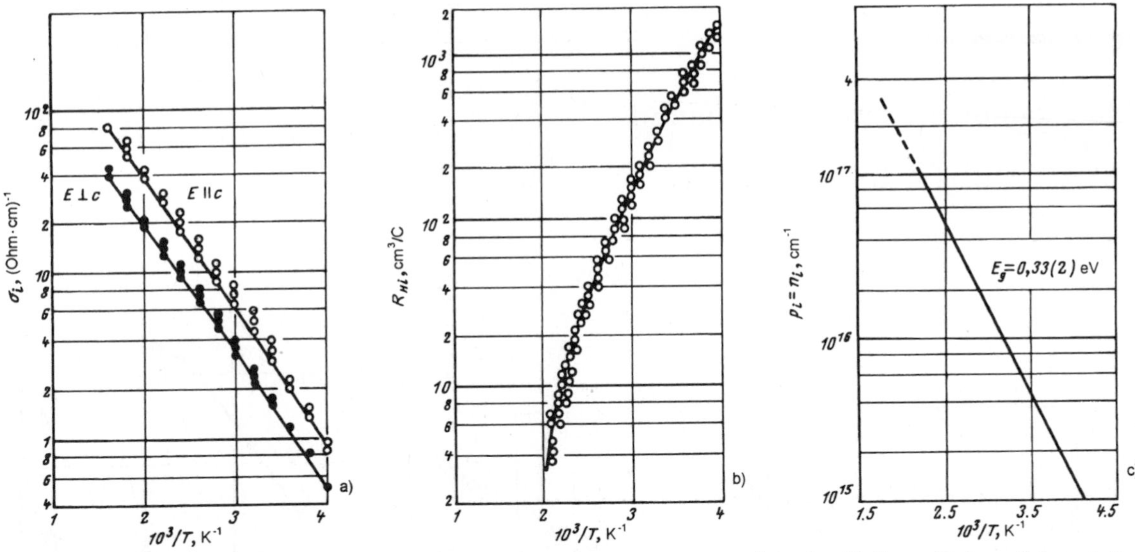

Figure 22.39 The temperature dependences of the conductivity (a), the Hall coefficient (b) and the intrinsic carrier density (c) in Te [30].

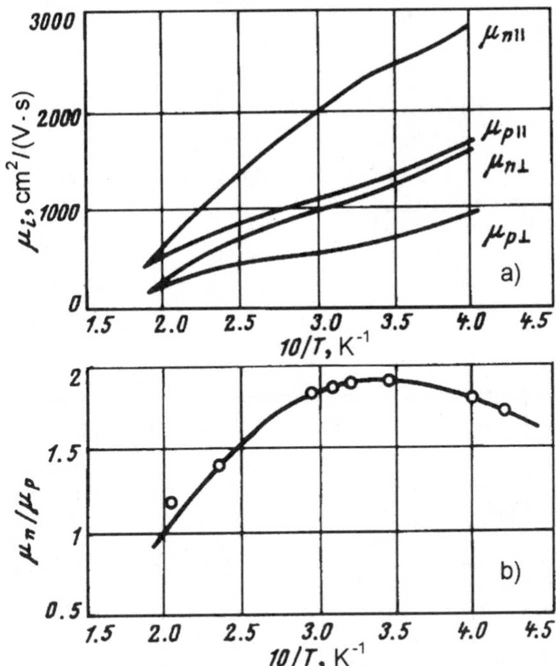

Figure 22.40 Temperature dependences of the electron and hole mobilities (a) and the mobility ratio (b) in the intrinsic region of Te [30].

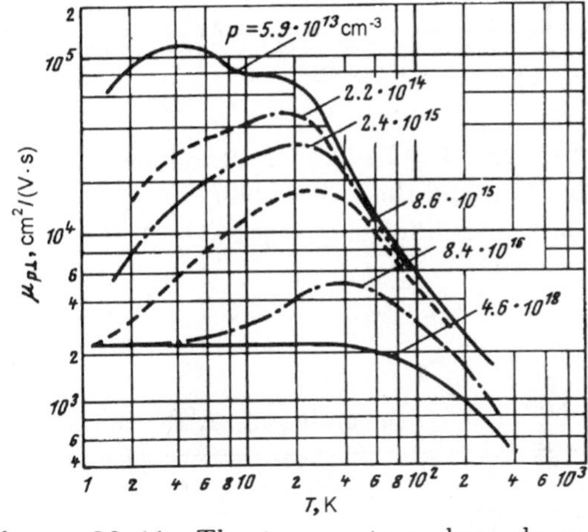

Figure 22.41 The temperature dependence of the hole mobility in Te along the c-axis [35].

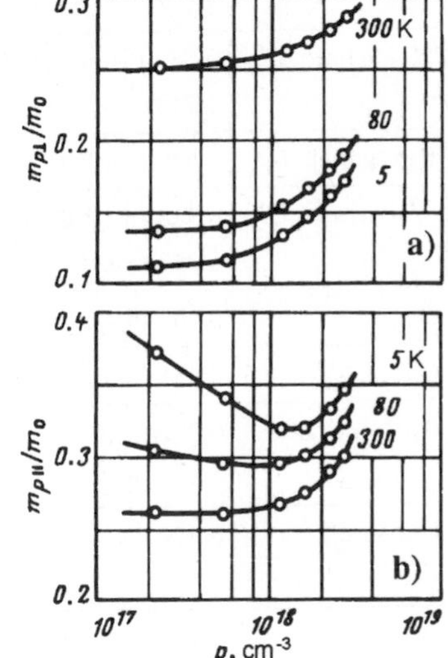

Figure 22.42 Dependences of the transverse (a) and longitudinal (b) effective masses of holes on their densities in Te [29].

22.3 Semiconducting Compounds

22.3.1 $A^IB^I - A^IB^{VII}$ compounds

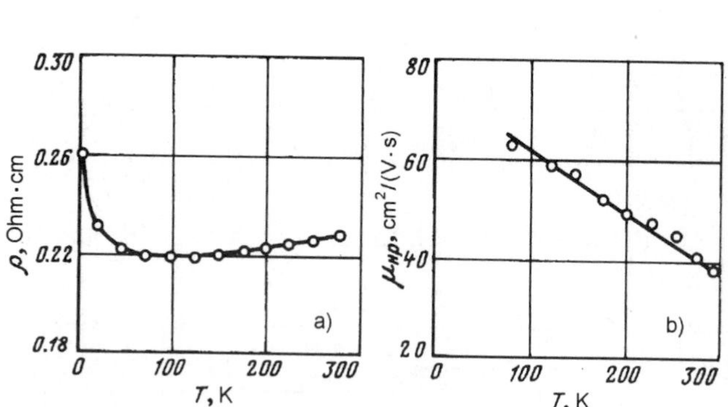

Figure 22.43 The temperature dependences of the resistivity (a) and the Hall mobility of holes (b) in CsAu (a 130-mm-thick film sample) [81].

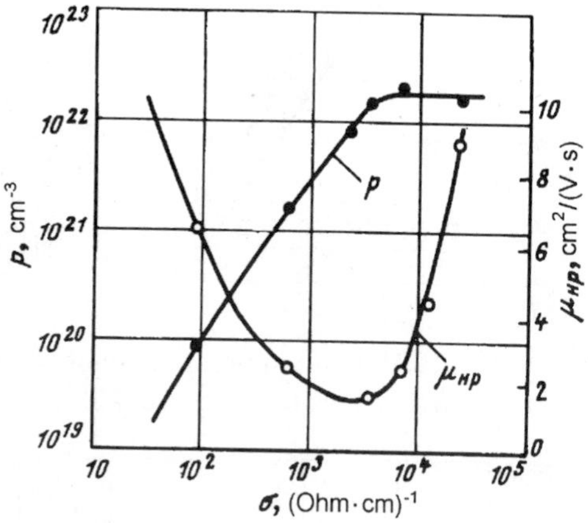

Figure 22.44 The Hall mobility of holes and the hole density vs. the conductivity of 270-nm-thick CsAu films with different degrees of oxidation [83].

Table 22.6 Semiconducting A^IB^V compounds [293]

Compound	Crystal structure		ρ, g/cm^3	T_{melt}, K	E_g, eV	μ_n, cm^2/(V·s)	μ_p, cm^2/(V·s)
	System, group	a, b, c, nm					
β-Li_3Sb	Cubic, O_h^5	0.656	3.29 [288]	923 [288]	1.4–1.6 [288] 1.25 [16]		
NaSb	Monoclinic, C_{2h}^5	0.680 (a); 0.634 (b); 1.248 (c); β=117.6°	4.03 [288]	738 [16]	0.82 [16]		
α-Na_3Sb	Hexagonal, D_{6h}^4	0.536 (a); 0.95 (c) [1]	2.67 [123]	1130 [123]	1.1 [123]		
KSb	Monoclinic, C_{2h}^5	0.718 (a); 0.697 (b); 1.340 (c); β=115.1°	3.52 [288]	878 [16]	0.9 [16]		
α-K_3Sb	D_{6h}^4	0.603 (a); 1.069 (c) [1]	2.35 [123]	1085 [123]	1.1 [123]		
β-Rb_3Sb	Cubic, O_h^5	0.884		1006	1.0 [10]		
CsSb	Rhombic, D_2^4	0.757 (a); 0.734 (b); 1.327 (c)		856	0.8 [10] 0.6 [288]	500 [16] ($T^{-3/2}$)	100 [16]
Cs_3Sb	Cubic, O_h^5	0.914 [1]	4.40 [288]	998	1.6 [123]	500 [10] ($T^{-3/2}$)	10 [10] 200–600 [288]
β-Cs_3Bi	Cubic, O_h^5	0.931	4.92 [288]	908	0.55 [16]		350 [288]

Table 22.7 Semiconducting $A^{I}B^{VI}$ compounds [293]

Compound	Crystal structure		ρ, g/cm^3	T_{melt}, K	E_g, eV	$\partial E_g/\partial T$, 10^{-4} eV/K
	System, group	a, b, c, nm				
Cu_2O	Cubic, O_h^4	0.426 [1]	6.1 [123]	1508 [16]	2.17 (4.2 K)	−2 [10]
α-Cu_2S	Rhombic, C_{2v}^{15}	1.18 (a); 2.728 (b); 1.349 (c)	5.6 [216] 5.8 [288]	1393 [288] 1373 [216]	1.3	
α-Cu_2Se	Cubic, O_h^5	0.575 [288]	6.8 [216] 7.1 [288]	1390 [216] 1398 [288]	1.25 [94]	
Cu_2Te	Hexagonal, D_{4h}^7	0.396 (a); 0.612 (c)	7.34 [216]	1173 [216]	0.5	−3
Ag_2O	Cubic, O_h^4	0.472	6.9	1361	1.2	−20
α-Ag_2S	Monoclinic	0.423 (a); 0.691 (b); 0.787 (c); β=99°35′	7.23	1098	1.0 [16] 1.3 (0 K)	−12
β-Ag_2S	Cubic, O_h^9	0.488 [288]	7.2 [288] 6.7	859 [288]		
β-Ag_2Se	Cubic, O_h^9	0.499	8.25 7.58 [288]	1170 [16]	0.15	
AgTe	Rhombic	0.890 (a); 2.01 (b); 0.462 (c) [288]	7.61 [288]	483 [288]	0.73 [288]	
α-Ag_2Te	Monoclinic, C_{2h}^1	0.818 (a); 0.448 (b); 0.809 (c)	8.41	[216]*	0.67	−0.86 [102]
β-Ag_2Te	Cubic, O_h^5	0.658	8.5 [216] 7.6 [288]	1228 [216] 963 [288]	0.2	

Compound	T_D, K	m_n/m_0	m_p/m_0	μ_n, cm^2/(V·s)	μ_p, cm^2/(V·s)	ϵ_0	ϵ_∞	n_i, cm^{-3}
Cu_2O	188	0.98	0.58	0.1 [288]	100 [123]	7.5 [23]		
α-Cu_2S	300 [288]		1.8	3.0 [242]	5.0			
α-Cu_2Se	235 [288]		0.5		1400 [288] 750 [16]	9 [288]	11.6	
Cu_2Te	220 [288]	0.9 [288]	0.8	200 [40] 400 [288]	900 [88] ($T^{-3.2}$)		17 [288]	
Ag_2O		0.7	1.9					
α-Ag_2S	70	4.55 (m_d) [15]	7.8 (m_d) [15]	63	18			$3.7 \cdot 10^{15}$
β-Ag_2S	160 [288]	0.24 [288]		60−120 [288]				
β-Ag_2Se	190 (80 K)	0.2	0.54	2000 [16]	520			$3.2 \cdot 10^{18}$
AgTe					200 [288]			
α-Ag_2Te	150 (80 K) [288]	0.026−0.034		10000	1000		16 [288]	
β-Ag_2Te	120 [288]	0.11	1.5	1100 4000 [16]	18			

* Transforms into β-Ag_2Te at $T \leq 423$ K [16].

Table 22.8 Semiconducting A^IB^{VII} compounds

Compound	Crystal structure		ρ, g/cm^3	T_{melt}, K	E_g, eV	T_D, K [290]
	System, group	a, nm				
γ-CuCl	Cubic,	0.541	4.136	695	3.39 [246]	164
CuBr	Cubic, T_d^2	0.568 [1]	4.72 [121]	750 [290]	2.9 [16] 3.08 [290]	163
γ-CuI	Cubic, O_h^5	0.604 [290]	5.67 [290]	642 ($\gamma \to \beta$) 681 ($\beta \to \alpha$) 875 [290]	3.1 [238]	165
AgCl	Cubic, O_h^5	0.555 [290]	5.56 [216]	728 [216]	3.0 [242] 4.1 [11]	161
AgBr	Cubic, O_h^5	0.577 [1]	6.47 [216]	703 [216]	2.0 [78]	120
AgI	Cubic, T_d^2	0.647 [1]	5.67 [121]	825 [121]	2.8 [10]	114

Compound	μ_n, cm^2/(V·s)	μ_p, cm^2/(V·s)	m_n/m_0	m_p/m_0	ε_0 [290]	ε_∞ [290]
γ-CuCl			2.5 [246]		7.9	3.6
CuBr	30 [16]		0.28 [290]	1.4 [290]	8.6	4.06
γ-CuI			0.33 [290]	1.4 [290] 2.4 [290]	6.5	4.6
AgCl	70 (200 K) 45000 (12 K) [109]	40 (200 K) [11]	0.30 [290] 0.36 [109]		11.1	3.9
AgBr	240 [10] 4000 [123] 600000 (1.7 K) [82]	2 [290] 30000 (4.2 K)	0.29 [291]	1.1 [246]	12.4	4.6
AgI	50 [16]				7.0	4.1

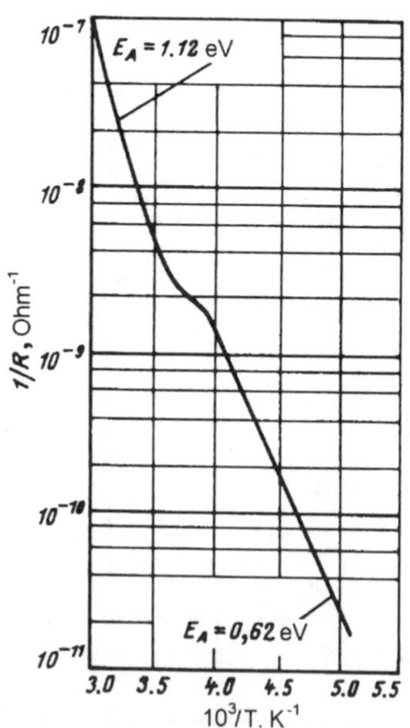

Figure 22.45 The temperature dependence of the reverse resistance in Cs_3Sb [89].

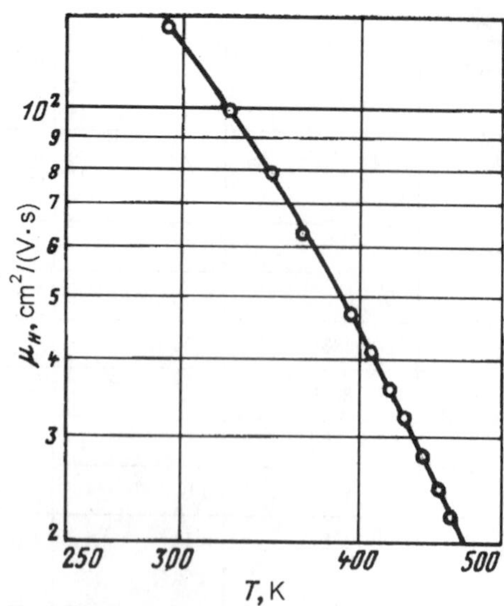

Figure 22.46 The temperature dependence of the hole mobility in Cu_2O [90].

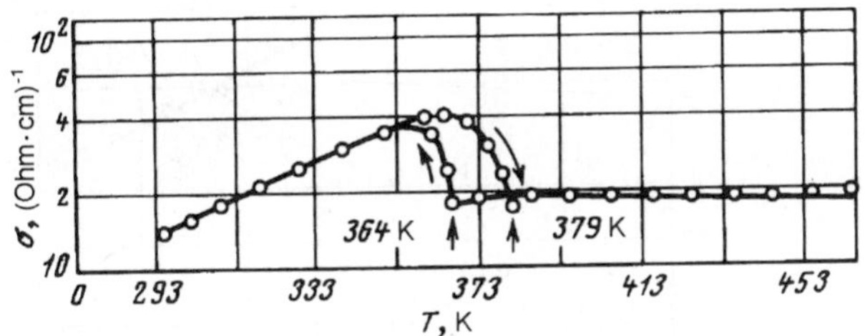

Figure 22.47 The temperature dependence of the conductivity in Cu_2S [91]: the arrows directed down and up – measurements for increasing and decreasing temperatures, respectively.

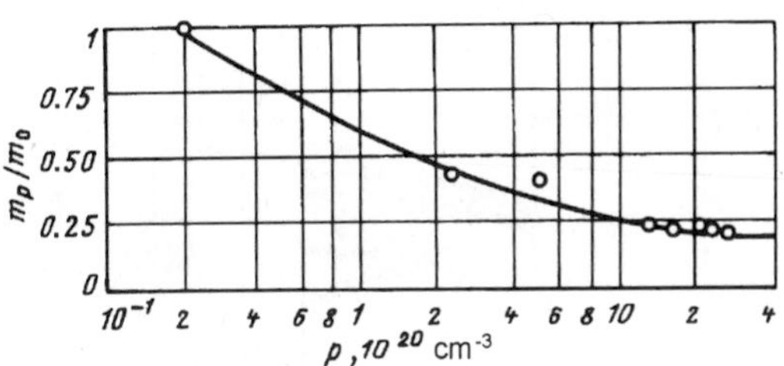

Figure 22.49 The effective mass of holes in Cu_2Se vs. the hole density [93].

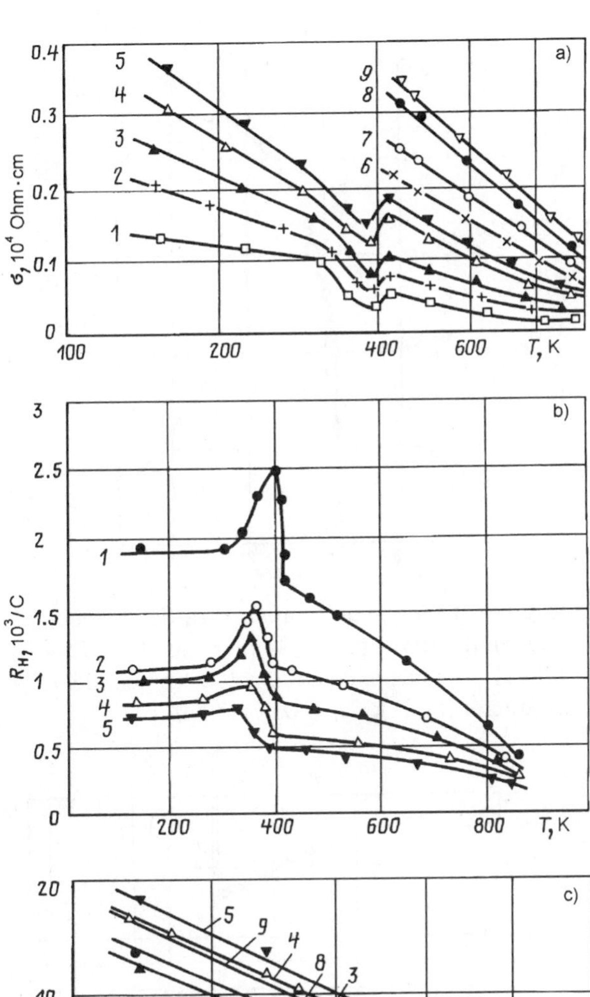

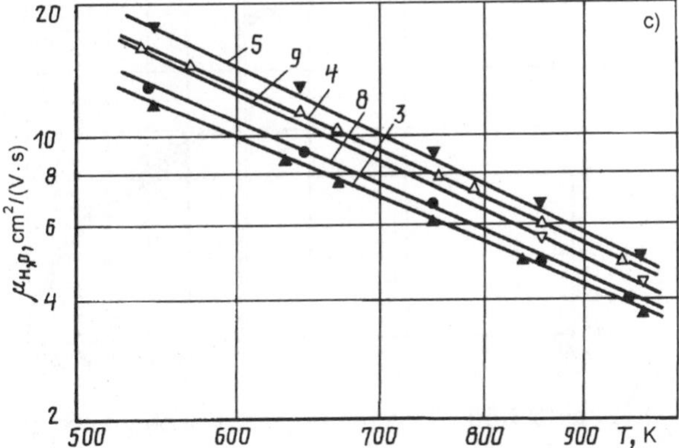

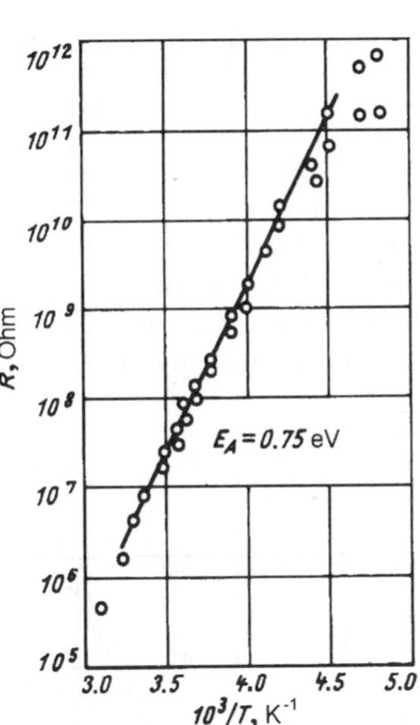

Figure 22.50 The temperature dependence of the resistance in Ag_2O (a film sample) [96].

Figure 22.48 The temperature dependences of the electron conductivity (a), the Hall coefficient (b) and the Hall mobility of holes (c) for Cu_2Se samples [92]. The Se content increases from sample *1* to sample *9* as does the carrier density (from $1.9 \cdot 10^{20}$ to $24.5 \cdot 10^{20} cm^{-3}$).

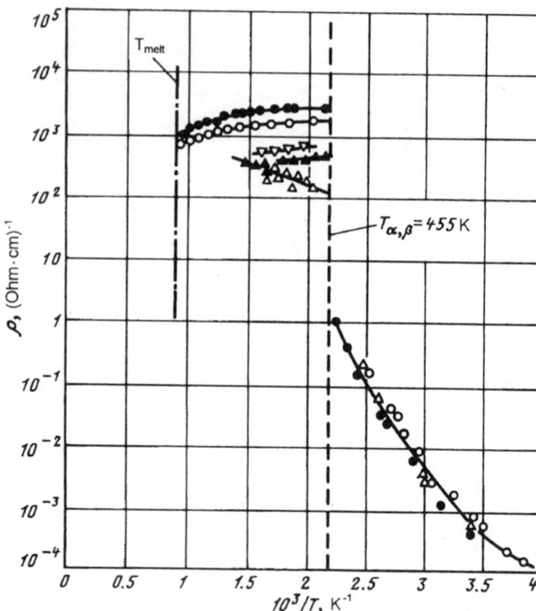

Figure 22.51 The temperature dependence of the conductivity of Ag_2S [97]. The points correspond to the samples of nominal purity and stoichiometry. (The electrochemical control of the Ag/S ratio was not made). For the α-phase ($T < T_{\alpha\beta}$), the thermal prehistory of the sample is important.

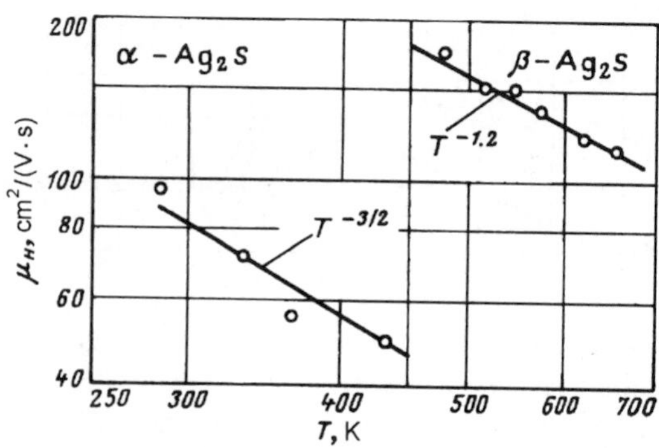

Figure 22.53 The temperature dependence of the electron Hall mobility in α- and β-Ag_2S [101].

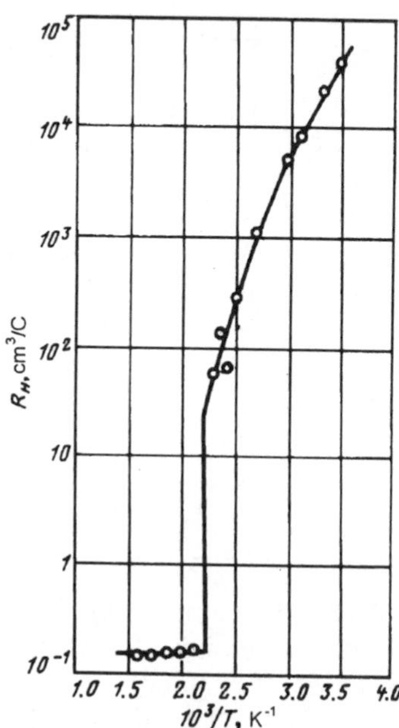

Figure 22.52 The temperature dependence of the Hall coefficient in Ag_2S [101].

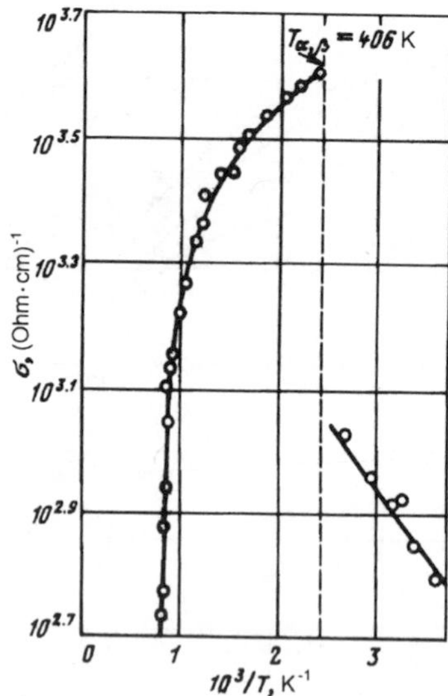

Figure 22.54 The temperature dependence of the conductivity in Ag_2Se [97]. At $T < T_{\alpha,\beta} =$ 406 K, the dependence is of a semiconducting character; at $T < T_{\alpha,\beta}$, $d\sigma/dT < 0$.

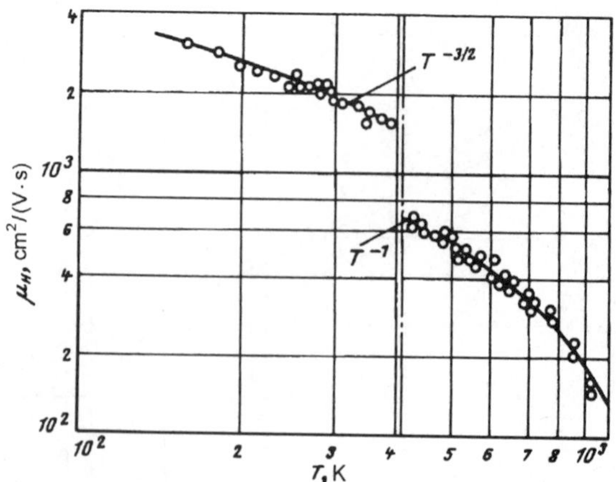

Figure 22.55 The temperature dependence of the electron Hall mobility in Ag_2Se [101].

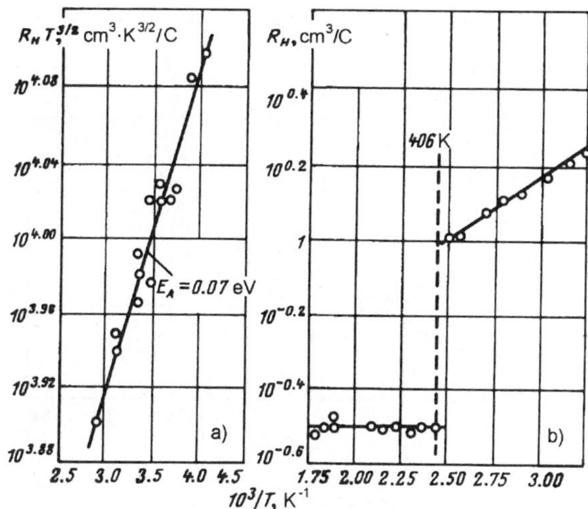

Figure 22.56 Temperature dependences of the Hall coefficient in Ag_2Se samples at low (a) and high (b) temperatures [101].

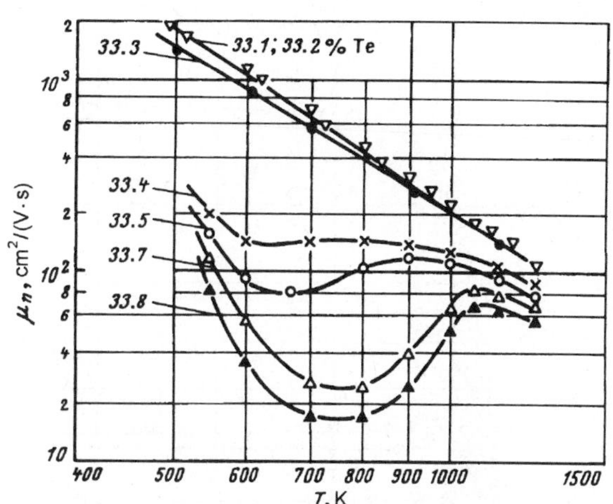

Figure 22.58 The temperature dependence of the electron mobility in Ag_2Te [93] at various Te concentrations.

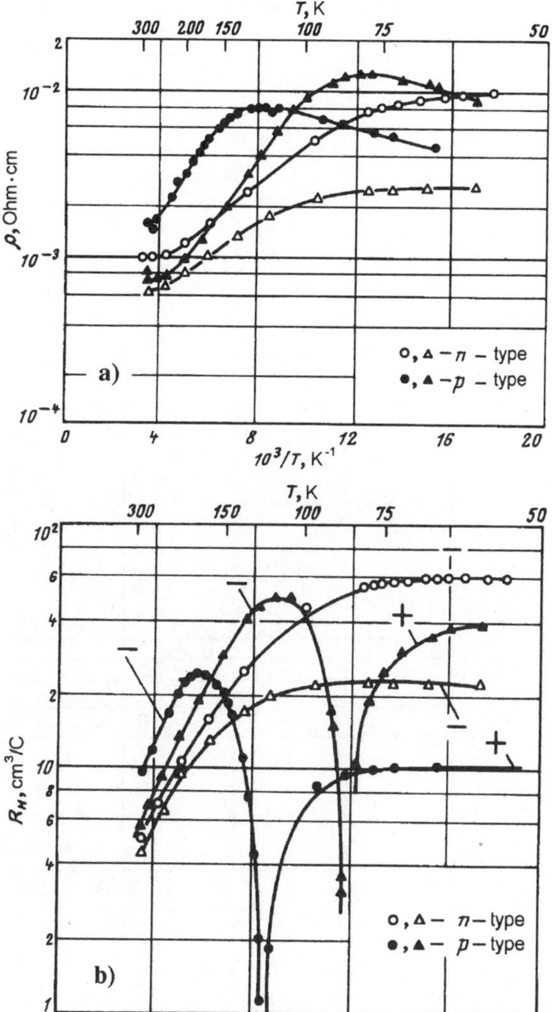

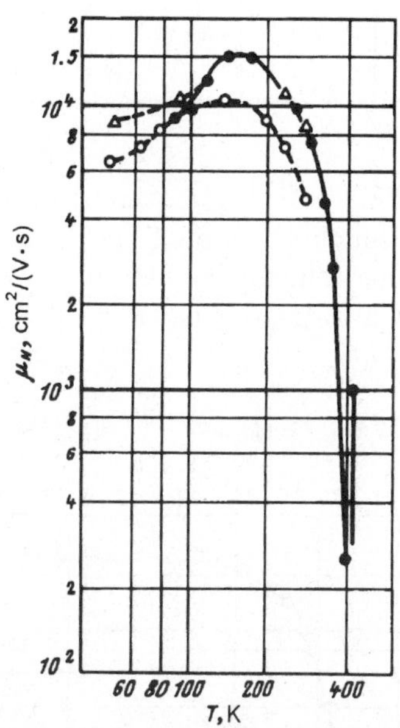

Figure 22.59 The temperature dependence of the electron Hall mobility in Ag_2Te [123]:
○, △ – bulk crystals; ● – 135 nm-thick films with a spherolitic structure deposited on the substrate at 120°C.

Figure 22.57 Temperature dependences of (a) the resistivity [108] and (b) the Hall coefficient [93] for n- and p-Ag_2Te:
+, – in (b) are signs of the Hall emf.

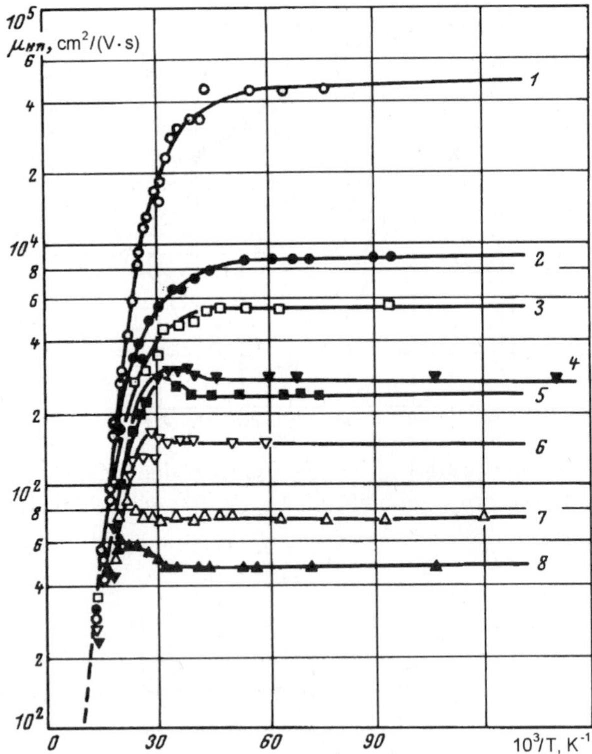

Figure 22.60 The temperature dependence of the electron Hall mobility in pure and doped AgCl crystals: *1* – zone refining; *2* – Cu impurity; *3–8* – Fe impurity, various treatments [109].

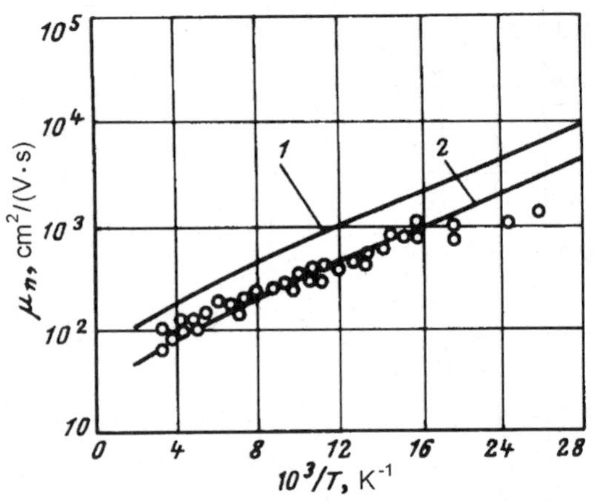

Figure 22.62 The temperature dependence of the electron drift mobility in AgBr: *1* – calculations with the use of the measured polaron mass; *2* – root-mean-square "fitting" curve [120].

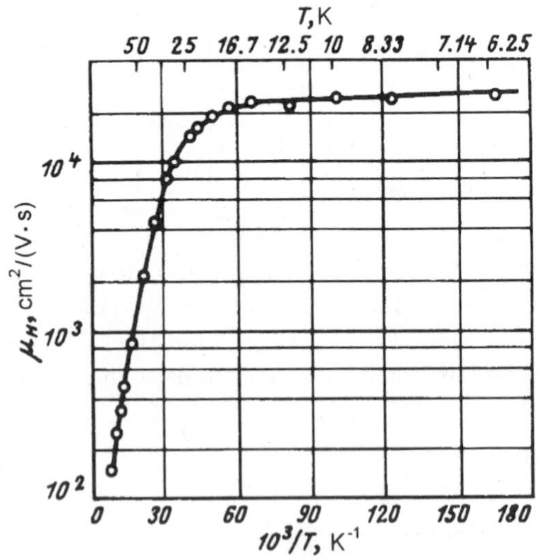

Figure 22.61 The temperature dependence of the electron mobility in AgBr: solid line is a calculation with allowance for scattering of electrons by acoustic and optic phonons and impurities [119].

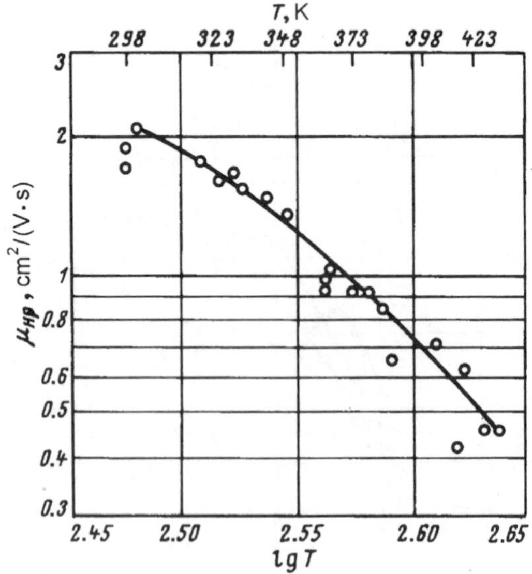

Figure 22.63 The temperature dependence of the Hall mobility of holes in AgBr [113].

22.3.2 $A^{II}B^{IV}$ – $A^{II}B^{VII}$ compounds

Amongst the $A^{II}B^{V}$ compounds, ZnSb and CdSb as well as their solid solutions, which are of interest for manufacturing the thermoelectric devices, were most extensively studied. ZnSb and CdSb crystallize into a rhombic structure (D_{2h}^{15}) giving rise to a strong anisotropy in their mechanical and electrical properties. The band structure is characterized by the indirect gap [18, 117]. The conduction-band minimum falls on the center of the Brillouin zone (at $\mathbf{k} = 0$). In the vicinity of this minimum, the constant-energy surface is similar to an ellipsoid of revolution with the major axis directed along the a axis. The extremum of the valence band is arranged at $\mathbf{k} \neq 0$. The review of the properties of $A^{II}B^{V}$ compounds can be found in [241].

All of the $A^{II}B^{VI}$ compounds represent the phases of variable composition. Polymorphism and polytypism inherent in many of them lead to a strong dependence of their structure-sensitive properties on crystal growth and thermal treatment conditions.

The best-investigated semiconductors crystallize into a sphalerite T_d^2 or a wurtzite C_v^4 lattice and have a direct-gap band structure (the extrema of the conduction and valence bands are at $\mathbf{k} = 0$). The cubic crystals (sphalerite) are isotropic; one of the subbands of their valence band is split out due to spin-orbit interaction Δ_{so} (see Figure 22.95a). The hexagonal (wurtzite) crystals are slightly anisotropic (this anisotropy is often ignored); the presence of an additional interaction Δ_{cr} (the crystal field of the noncubic crystal) causes the splitting of the valence band into three subbands (see Figure 22.95b). The splittings E_1 and E_2 determined experimentally by optical techniques are related to Δ_{so} and Δ_{cr} through the formulas [138]

$$E_{1,2} = \frac{1}{2}(\Delta_{so} + \Delta_{cr}) \pm \left(\frac{1}{4}(\Delta_{so} + \Delta_{cr})^2 - \frac{2}{3}\Delta_{so}\Delta_{cr}\right)^{1/2}.$$

The HgS, HgSe and HgTe compounds possess a complex band structure with the overlapping bands (see Figure 22.110). The overlapping energy E_t is of order 0.001 eV [199, 200] (according to other findings $E_t = 0.02$ eV for HgTe and 0.07 eV for HgSe [138]).

Table 22.9 Semiconducting $A^{II}B^{IV}$ compounds

Compound	Cubic, O_h^5, a, nm [123]	ρ, g/cm^3 [123]	T_{melt}, K [123]	E_g, eV [293]	$\partial E_g/\partial T$, 10^{-4} eV/K	n_i, cm^{-3} [293]	T_D, K [293]
Mg_2Si	0.634	1.95	1375	0.78 (0 K)	−6.4 [131]	10^{14}	450
Mg_2Ge	0.639 [293]	3.086	1388	0.57 [67]	−1.8 [67]	$2\cdot10^{14}$	363
Mg_2Sn	0.676	3.592	1051	0.23 0.36 (0 K)	−3.5 [293]	$2.7\cdot10^{17}$	240
Mg_2Pb	0.686 [293]	5.54 [203]	828 [293]	0.15			244

Compound	m_n/m_0	m_p/m_0	μ_n, cm^2/(V· s)	μ_p, cm^2/(V· s)	ϵ_0	ϵ_∞
Mg_2Si	0.46 (m_d) [75]	0.87 [10] 2.0 [288]	550 [293]	70 [123]	18.8 [293]	13.3 [293]
Mg_2Ge	0.18 (m_d) [55]	0.31	530 4335 (77 K) [293]	110 1180 (77 K) [293]	21.7 [293]	13.9 [293]
Mg_2Sn	0.8 ($m_\parallel$) [133] 0.3 ($m_\perp$) [133] 1.17 (m_d) [293]	1.28 [10]	320 ($T^{-2.2}$)	260 [123]	23.75 [219]	15.5 [219]
Mg_2Pb	0.04 [293]	0.35 [293]	12000 (4.2 K) [293]	14000 [293]		

Table 22.10 Semiconducting $A^{II}B^{V}$ compounds [293]

Compound	Crystal structure		ρ, g/cm^3	T_{melt}, K	E_g, eV	$\partial E_g/\partial T$,
	System, group	a, b, c, nm	[123]	[123]	[293]	10^{-4}eV/K
Mg_3As_2	Cubic	1.233		1073	2.6	−9
Mg_3Sb_2	Trigonal, D^3_{3d}	0.454(a); 0.723(c) [1]	4.09 [123]	1500 [123]	0.82 [151]	
α-Zn_3P_2	Tetragonal, D^4_4	0.508(a); 1.859(c)	3.54	1258 1313 [288]	2.18 2.0 [288]	−5.5
β-ZnP_2	Monoclinic, C^5_{2h}	0.885(a); 0.729(b); 0.756(c)	3.55	1265	1.37	
Zn_3P_2	Tetragonal, D^{15}_{4h}	0.81(a); 1.145(c) [2]	4.54 [16]	1466	1.15 [10]	
$ZnAs_2$	Monoclinic, C^5_{2h}	0.921(a); 0.764(b); 0.798(c); β=102°28′	4.94	1041 [133]	0.92 [16]	
Zn_3As_2	Tetragonal, D^{15}_{4h}	1.178(a); 2.364(c)	5.58 [123]	1288	1.0 (0 K)	−4.55
ZnSb	Rhombic, D^{15}_{4h}	0.622(a); 0.774(b); 0.812(c); [117]	6.38 [117]	819 [117]	0.5; 0.61 (4.2 K) [103]	
β-Zn_4Sb_3	Monoclinic, C	1.074(a); 1.22(b); 0.820(c)	6.81	836	1.2 [10]	
CdP_4	Monoclinic, C^5_{2h}	0.527(a); 0.519(b); 0.766(c); β=80°32′	2.04 3.90 [288]		1.0 [111]	−(2.7–3.7)
β-CdP_2	Tetragonal, D^{15}_{4h}	0.529(a); 1.974(c)	4.19	1057	2.02	−8.6
Cd_3P_2	Tetragonal, D^{15}_{4h}	1.256(a); 2.544(c)	5.60 [123]	970 [123] 1015 [288]	0.55 [54]	−1.8
$CdAs_2$	Tetragonal, D^{10}_4	0.795(a); 0.468(c)	5.88 5.64 [288]	894 [133]	1.13 [16] 1.0 [15]	
α-Cd_3As_2	Tetragonal, C^{12}_{4v}	1.265(a); 2.544(c)	6.25 [123]	994 [133]	0.12 (0 K)	−3.3
CdSb	Rhombic, D^{15}_{2h}	0.647(a); 0.824(b); 0.853(c) [117]	6.92 [117]	729 [133]	0.56 [117]	−5.4 [117]
β-Cd_4Sb_3	Monoclinic, D^{13}_{2h}	0.815(a); 0.816(b); 1.196(c)		703	1.25 [16]	−5.4 [288] −3.6

Compound	n_i, cm^{-3}	T_D, K	m_n/m_0	m_p/m_0	μ_n, cm^2/(V·s)	μ_p, cm^2/(V·s)	ϵ_0	ϵ_∞
Mg_3As_2					10 20			
Mg_3Sb_2			3.1(m_d)			100 [10] ($\sim T^{-3.2}$)		
α-ZnP_2		465 [288]			1800			
β-ZnP_2*1		288 [288]			3.5	20 [288] 1000 (77 K) [288]		
Zn_3P_2	$3.56 \cdot 10^{14}$	320 369 (160 K)				100 [288] 20		
$ZnAs_2$		234 284 (55 K)	0.35 [288]	2.45 [288]	500	100	12–15	
Zn_3As_2		290 275 (120 K)	1.7 [16]	0.65 [16]	$10 \cdot (T/300)^{-1.1}$ [135]	17 [16] ($\sim T^{-1.64}$)	11.8	
ZnSb	10^{15}	225 (80 K) [117]	0.175 ($m_\parallel$) [117] 0.146 ($m_\perp$)	0.53 [18]	800 (100 K)	700 [50] ($\sim T^{-3/2}$)		
β-Zn_4Sb_3			0.16 [288]	0.12	330	980 [16]		

Table 22.10 Semiconducting $A^{II} B^{V}$ compounds *(continued)*

Compound	n_i, cm^{-3}	T_D, K	m_n/m_0	m_p/m_0	μ_n, cm^2/(V·s)	μ_p, cm^2/(V·s)	ϵ_0	ϵ_∞
CdP_4					150 [111]	600 [128]		
β-CdP_2								
Cd_3P_2		251 235 (55 K)	0.09 [77] 0.05 [188] $g_n = 2.15$	0.5	3000 [54] ($\sim T^{-1.1}$) 5·10^4 (4.2 K)		37	14–17
Cd_3As_2		233 (55 K)	0.15 ($m_\parallel$) [133] 0.58 ($m_\perp$) 0.37 (m_d)	0.094 ($m_\parallel$) [133] 0.346 ($m_\perp$) 0.22 (m_d)	100–400 [133] 3000 [288]	400 [133]	15.4–17.4	11.5–13.8
α-Cd_3As_2		258 [288]	0.046 (m_d); g_n=17–30.2 [52]	0.12	2.6·10^4 2.8·10^4 (4.2 K)	430 [16]	42	16
CdSb		180 (80 K) [117]	0.72	0.34–0.4 1.07	660	2000 [117] ($\sim T^{-3/2}$)	16.4	
β-Cd_4Sb_3			0.1–0.2 [288]	0.35 [288]	1000 [288]	110 900 [288]		

*[1] Anisotropy of resistivity approaches $\rho_a : \rho_b : \rho_c = 5:35:1$.

Table 22.11 Semiconducting $A^{II}B^{VI}$ compounds

Compound	Crystal structure: System, group	Crystal structure: a, c, nm	ρ, g/cm^3	T_{melt}, K	E_g, eV	$\partial E_g/\partial T$, 10^{-4} eV/K
ZnO	Hexagonal, C_{6v}^4	w*[1], 0.325(a); 0.521(c) [290]	5.60 [121] 5.66 [123]	2248 [123]	3.35 [144] 3.43 [66]	−8 [144]
ZnS	Cubic, T_d^2	s*[2], 0.54(a) [138]	4.10 [140]	2103 [236]	3.54 [123] 3.6 [114] 3.67(w) [138] 3.91(w) [123]	−5.3 [114] −3.8(w)
ZnSe	Cubic, T_d^2	s, 0.5668(a) [138]	5.28 [316]	1788 [123]	2.7 [114] 2.8 [138]	−4.5 [280]
ZnTe	Cubic, T_d^2	s, 0.6104(a) [138]	5.68 [141]	1512 [123]	2.1 [47] 2.34 [142]	−4.1 [289] −5 [290]
CdO	Cubic, O_h^5	0.470 [1]	8.15 [216]	1099 [288]	2.3 [123] 2.68 [59]	−6 [59]
CdS	Hexagonal, C_{6v}^3	w*[1], 0.4136(a); 0.6713(c) [138]	4.82 [139]	2023 [123]	2.52 [142] 2.4(s) [214]	−4.4 [138]
CdSe	Hexagonal, C_{6v}^3	w*[1], 0.430(a); 0.701(c) [138]	5.81 [139]	1531 [123]	1.67 [138] 1.8 [114]	−4.6 [144]
CdTe	Cubic, T_d^2	s, 0.648(a) [138]	5.86 [139] 6.20 [121]	1371 [123]	1.5 [237] 1.6 (77 K)	−3.0 [68] −5.6 [288]
α-HgS	Trigonal, D_3^6	0.414 (a); 0.949 (c) [288]	8.09 [288]	618 [288]	2.1 [290]	–
β-HgS*[3]	Cubic, T_d^2	0.585(a) [290]	7.73 [290]	1098 [288]	0.15 [289]	–
HgSe*[3]	Cubic, T_d^2	0.608(a) [138]	8.26 [139]	1071 [123]	0.2 [148]; 0.22 [156];	5 [289]
HgTe*[3]	Cubic, T_d^2	0.646(a) [138]	8.09 [289]	943 [123]	0.115 [33] 0.250 (77 K) [133]; 0.302 (4.2 K) [38]	7 [289]

Table 22.11 Semiconducting $A^{II}B^{VI}$ compounds *(continued)*

Compound	g_n	g_p^{*4}	T_D, K	$\hbar\omega_l$, eV	$\hbar\omega_t$, eV	$< \epsilon_0 >^{*5}$	$< \epsilon_\infty >^{*5}$
ZnO	−1.94 [155]	−(1.24–1.74) [155]	416 [141]	0.0732 [254] 0.605 [220]	0.0512 [254] 0.0545 [220]	8.5 [159]	4.0 [159]
ZnS	1.88 [289] 2.0 (‖c) (w) 2.3(⊥c) [138]	1.04 [148] ~1.5 [289]	310 [140]	0.044 [138]	0.0367 [138] 0.0394(w)	8.54 [289] 8.32 [138]	5.2 [289] 5.13 [138]
ZnSe	1.46 [148] 1.14–1.37 [289]	5.25 [148]	400 (80 K) [144]	0.031 [138]	0.0263 [138]	9.1 [138] 9.73 [289]	5.9 [138] 7.18 [289]
ZnTe	−0.38 [289] −0.57	0.9–1.1 [289]	250 (80 K) [114]	0.025 [138]	0.024 [138] 0.0219 [254]	10.1 [138] 9.67 [142]	7.28 [138] 8.26 [114]
CdO	1.81 [201]	–	415 [288]	0.0654 [254]	0.0325 [254]	21.9 [254]	5.4 [254] 5.3 [290]
CdS	1.78(‖c) [138] 1.72(⊥c)	1.15(‖c) [138] 1.00 [148]	250–300 [114]	0.038 [138]	0.0324 [138] 0.0301 [254]	9.3 [138] 8.5 [289]	5.2 [138]
CdSe	0.6(‖c) [138] 0.51(⊥c)	$g_p - g_n$=1.8 (‖c) [280]	230 (80 K) [114]	0.027 [138]	0.0263 [138] 0.0172 [254]	9.4 [138] 9.63 [138]	5.98 [138] 6.10 [142]
CdTe	−1.1 [202]	−2(‖c) [138]	200 (80 K) [114] 160 [236]	0.0212 [138]	0.0173 [138]	10.6 [138] 10.29 [289] 11.01 [161]	7.6 [114] 7.21 [138] 7.19 [289]
α-HgS	–	–	–	–	–	–	–
β-HgS	–	–	–	–	0.022 [289]	18.2 [290]	11.3 [289]
HgSe	−14 [289] −36 [157]	–	242 [114]	–	0.015 [289]	25.6 [114]	7.2 [289] 11
HgTe	−22 [289] −25 [157] −41*6	6(‖[100]) 4(‖[110]) 3(‖[111]) [289]	143 [288]	0.0171 [252]	0.0146 [252]	21 [252]	14 [138]

Compound	$\partial E_g/\partial P$, 10^{-6} eV/bar	m_n/m_0	m_p/m_0	μ_n, $cm^2/(V\cdot s)$	μ_p, $cm^2/(V\cdot s)$
ZnO	0.6–1.9 [10]	0.27 [123]	0.59(v_1, v_2) [280] 0.31–0.55 (v_3)	180 [123] 1000 [288]	–
ZnS	5.6 [114] 9(w)	0.34–0.39 [289] 0.27 (w) [138]	≈0.2 (l), 1.5 (h)[316] 1.4(‖ c, w) [289] 0.49(⊥ c, w)	140 [138] 200 [16]	5 (700 K) [138]
ZnSe	6 [114]	0.15–0.17 [114] 0.18 [142]	0.6 [138] 0.75 [289]	260 (300–550 K) 530 [114, 138] 6000 (55 K) [153]	15 (200–400 K) [114] 23 [154]
ZnTe	6 (114)	0.12 [289]	0.65 [181] 0.15*7 [229]	350 ($\propto T^{-3/2}$) [114] 340 [138]	100 ($\propto T^{-3/2}$) [16] 110 [138]
CdO	–	0.1 [143] 0.3	–	120 [86] 500 [288]	–
CdS	3.3 [123]	0.18–0.20 0.18(s) [289]	5.0(‖ c) [138] 0.7(⊥ c) m_d/m_0=1.34 [21]	350 [138] $64000 \cdot T^{-3/2}$ [138] (50–200 K) 20 ($\propto T^{-3/2}$) (s) [114]	15 [138] 50 [46]

Table 22.11 Semiconducting $A^{II}B^{VI}$ compounds *(continued)*

Compound	$\partial E_g/\partial P$, 10^{-6} eV/bar	m_n/m_0	m_p/m_0	μ_n, $cm^2/(V\cdot s)$	μ_p, $cm^2/(V\cdot s)$
CdSe	4 [290]	0.11–0.13 [289] 0.10–0.12 (s)	$> 7(\parallel c, v_1)$ [138] $0.45(\perp c, v_1)$ [138] $0.9(\perp c, v_2)$ [22] $m_d/m_0{=}0.63(v_1)$ [138]	580 [114] 650 [138]	50 [114]
CdTe	3.0 [218] 7.9 [290]	0.10–0.12 [289]	0.1*7 [289] 0.4*8	1200 [121] 3000 (77 K) [121] 57000 (30 K) [149]	$80(\propto T^{-3/2})$ [114] 50 [121] –
α-HgS	–	–	–	45 ($\parallel c$) [289] 13 ($\perp c$)	
β-HgS	–	0.028 [289]	–	10000 (77 K) [289] 100000 (4.2 K)	–
HgSe	5 [289]	0.050 [156]	0.17 [84] 0.55 [112]	20000 [289]	
HgTe	10 [84]	0.017 0.03 (4.2 K) [84]	0.16 0.35 (4.2 K) [84]	33000 100000 (77 K) [289] 900000 (4.2 K)	–

Compound	ϵ_{011} [138]	ϵ_{033} [138]	$\epsilon_{\infty 11}$ [138]	$\epsilon_{\infty 33}$ [138]
ZnO	7.8($\perp c$) [290]	8.75($\parallel c$) [290]	3.7($\perp c$) [290]	3.75($\parallel c$) [290]
ZnS	–	–	5.13	–
ZnSe	–	–	5.9	–
ZnTe	–	–	7.28	–
CdO	–	–	–	–
CdS	9.02	9.53	5.17	5.23
CdSe	10.2	9.33	5.96	6.05
CdTe	–	–	7.21	–
α-HgS	18.2 [290]	23.5 [290]	6.25 [290]	7.9 [290]
β-HgS	–	–	–	–
HgSe	–	–	–	–
HgTe	–	–	–	–

*1 Can crystallize into a sphalerite structure.
*2 Can crystallize into a wurtzite structure.
*3 Cf. Figure 22.110.
*4 Heavy holes.
*5 The value averaged over the crystal directions.
*6 Light holes.
*7 Light hole mass.
*8 Heavy hole mass.

Note: The letters "s" and "w" denote the data for the sphalerite- and wurtzite-type modifications, respectively. Other data relate to the basic modification.

Table 22.12 Valence–band parameters for $A^{II}B^{VI}$ compounds [138, 289]

Compound	Δ_{so}, eV	Compound	Δ_{so}, eV	Δ_{cr}, eV	E_1, eV	E_2, eV
Sphalerite		Wurtzite				
ZnS	0.06–0.07	ZnO	—	—	0.007 [159]	0.052 [159]
ZnSe	0.43	ZnS	0.092	0.055	0.029	0.117
ZnTe	0.90–0.96	CdS	0.065	0.027	0.015	0.025
CdTe	0.8–0.9	CdSe	0.42	0.041	0.078	0.433
HgSe	0.40–0.45					
HgTe	0.94–1.10					

Table 22.13 Semiconducting $A^{II}B^{VII}$ compounds

Compound	Crystal structure: System, group	Crystal structure: a, b, c, nm	ρ, g/cm^3	T_{melt}, K	E_g, eV	$\partial E_g/\partial T$, 10^{-4} eV/K
CdF_2	Cubic, O_h^5	0.539 [1]	–	–	6.05 [26]	–
CdI_2	Hexagonal	0.424(a); 0.683(c)	5.67	660	3.47	−12
β-HgI_2	Rhombic, C_{2v}^{12}	0.467(a); 1.376(b); 0.732(c) [1]	6.09	532	2.39	−(7–14)

Compound	m_n/m_0	m_p/m_0	μ_n, cm^2/(V· s)	μ_p, cm^2/(V· s)	ϵ_0	ϵ_∞
CdF_2	0.4 [26]	–	19 [26]	–	–	–
CdI_2		–	–	–	5.9–12.9*	4.3–4.6*
β-HgI_2	0.25 ($m_\parallel$) 0.29 ($m_\perp$)	1.72 ($m_\parallel$) 0.56 ($m_\perp$)	100	–	8.5–25.9*	5.1–6.8*

* Anisotropic.

Figure 22.64 The band structure of Mg_2Si, Mg_2Ge, and Mg_2Sn [123, 129, 230, 293].

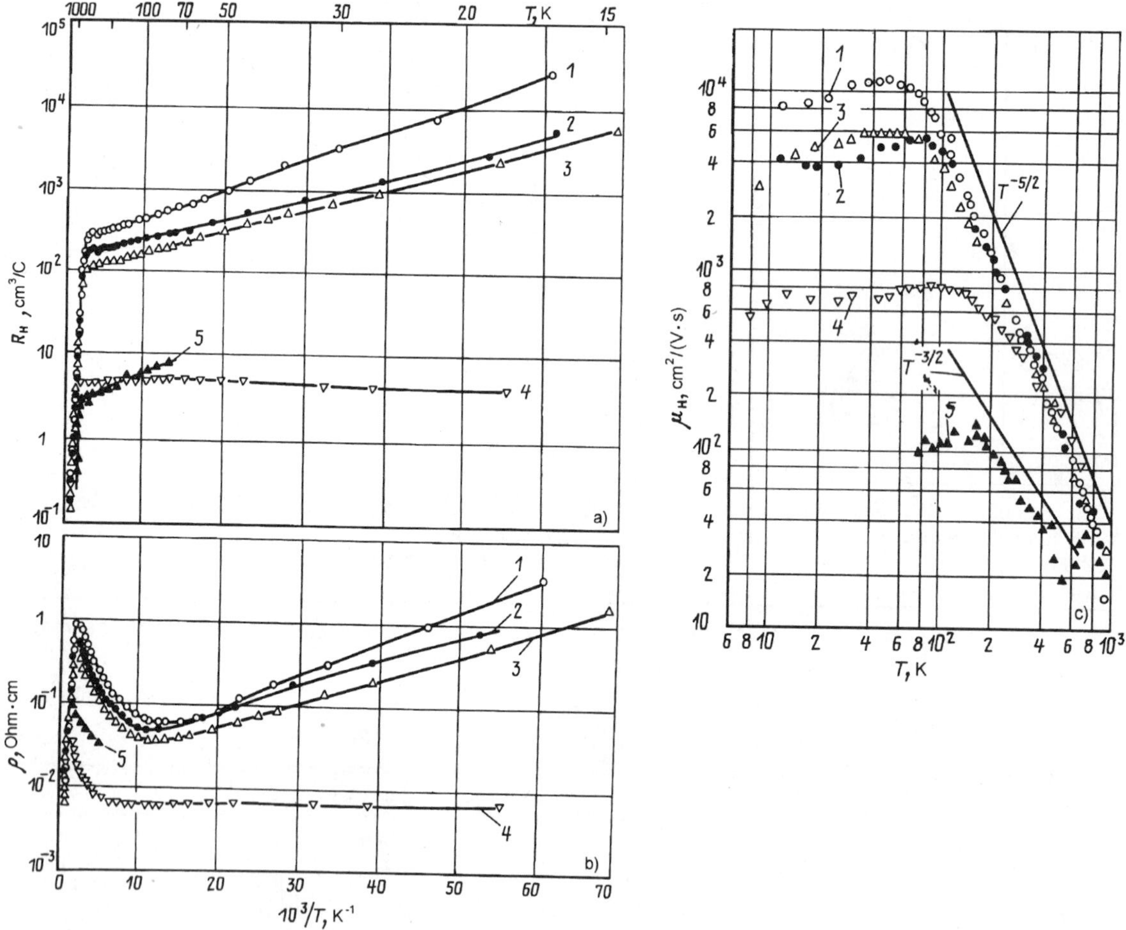

Figure 22.65 Temperature dependences of the Hall coefficient (a), the resistivity (b), and the Hall mobility of electrons and holes (c) in n-type (*1–4*) and p-type (*5*) Mg_2Si samples [134]. The carrier densities are in the range $n = 2.8 \cdot 10^{16}$–$1.7 \cdot 10^{18}$ cm^{-3}.

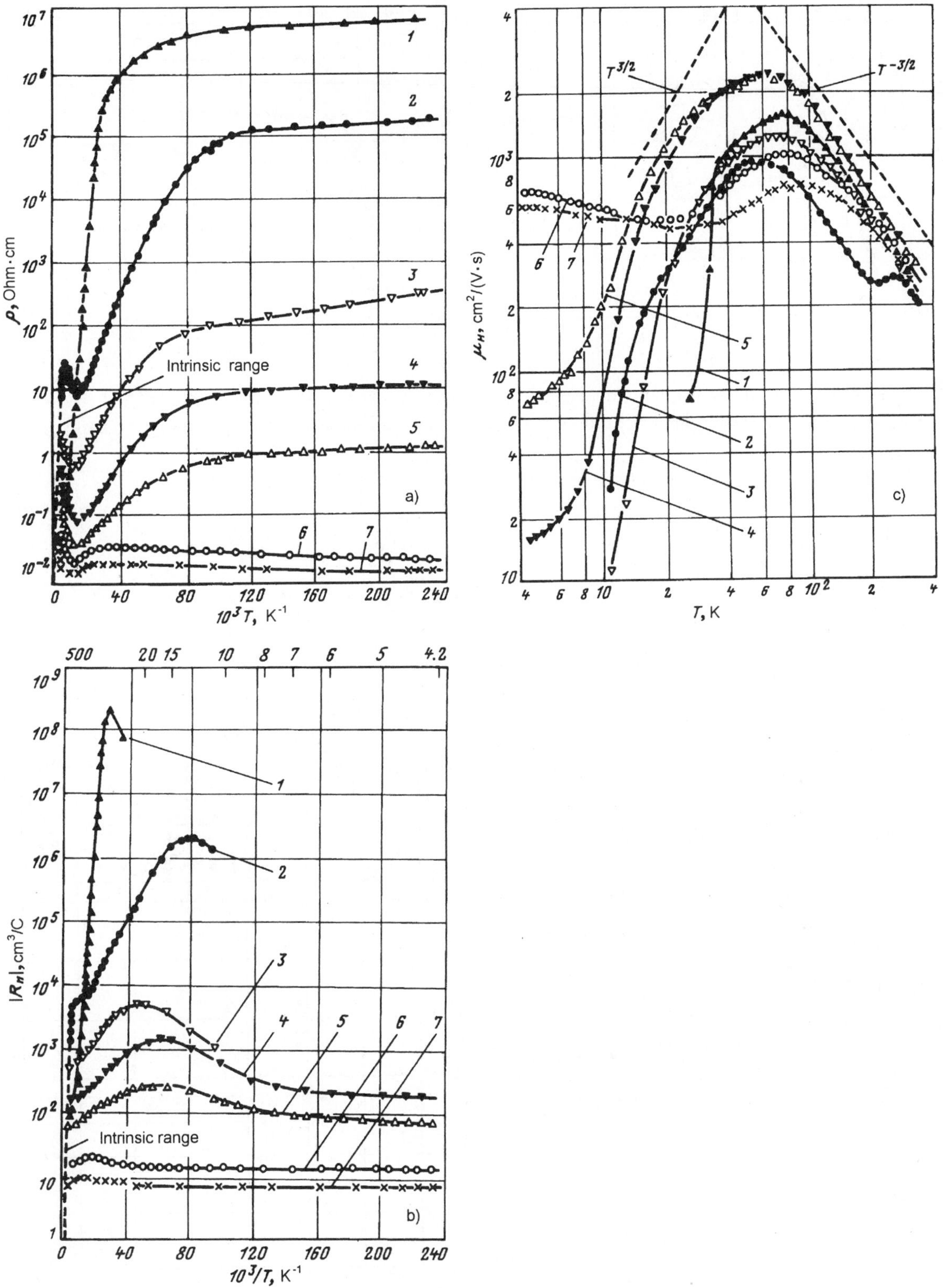

Figure 22.66 Temperature dependences of the resistivity (a), the Hall coefficient (b) and the Hall mobility of electrons (c) in different n-Mg_2Ge samples with the carrier density varying from $1.3 \cdot 10^{16}$ to $8.2 \cdot 10^{17} cm^{-3}$ (*1–7*).

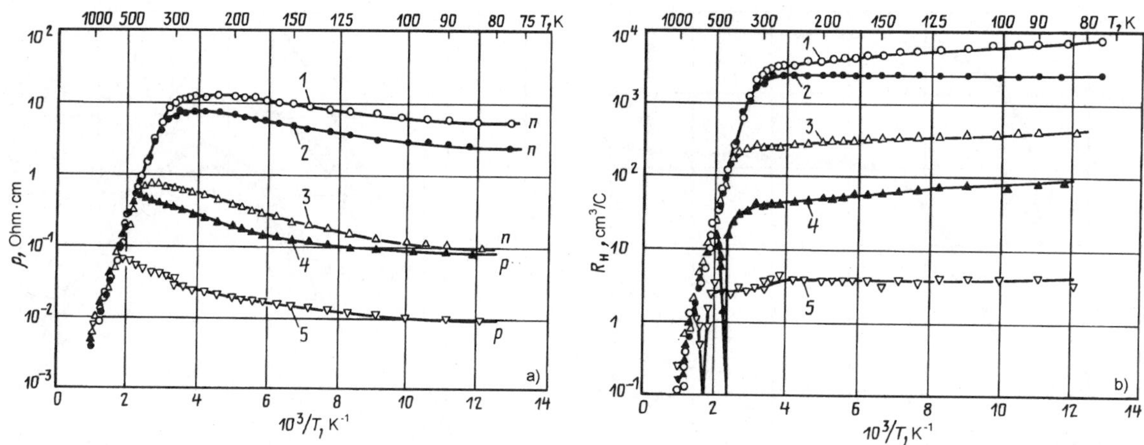

Figure 22.67 Temperature dependences of the resistivity (a) and the Hall coefficient (b) in n- and p-Mg_2Ge [135]: *1-5* – samples with different carrier densities increasing from sample *1* to sample *5*.

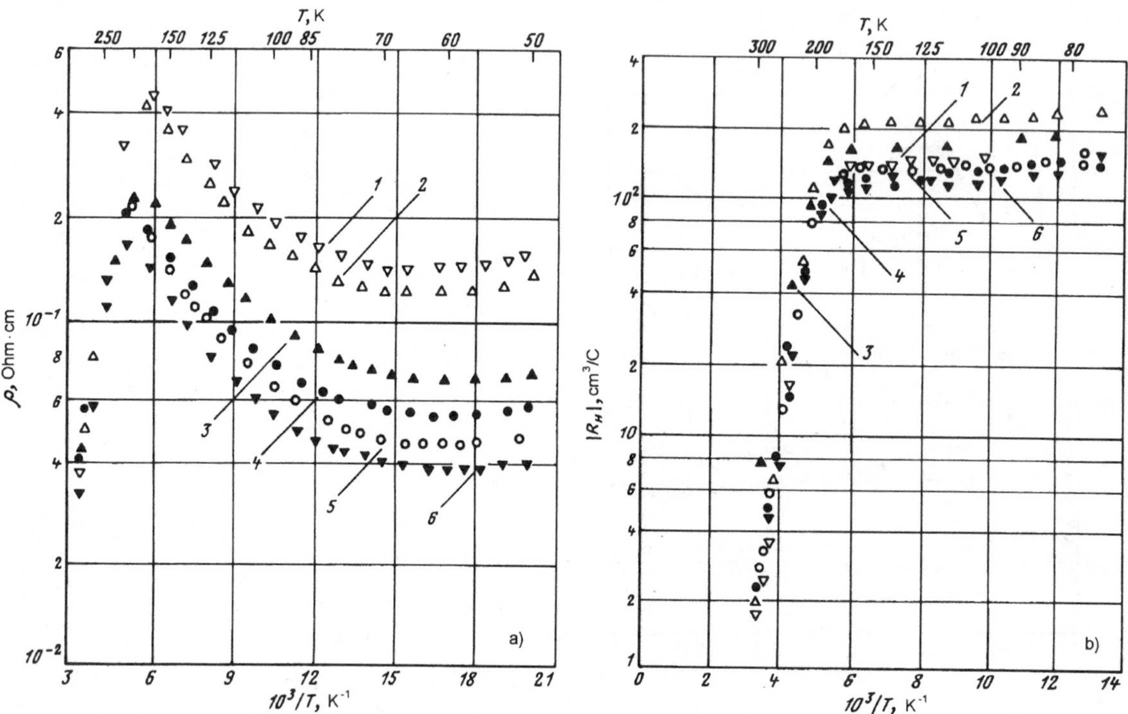

Figure 22.68 Temperature dependences of the resistivity (a) and the Hall coefficient (b) for n-Mg_2Sn samples with the carrier density varying from $3{\cdot}10^{16}$ to $6{\cdot}10^{16}\text{cm}^{-3}$ (*1-6*) in the mixed conduction region [137].

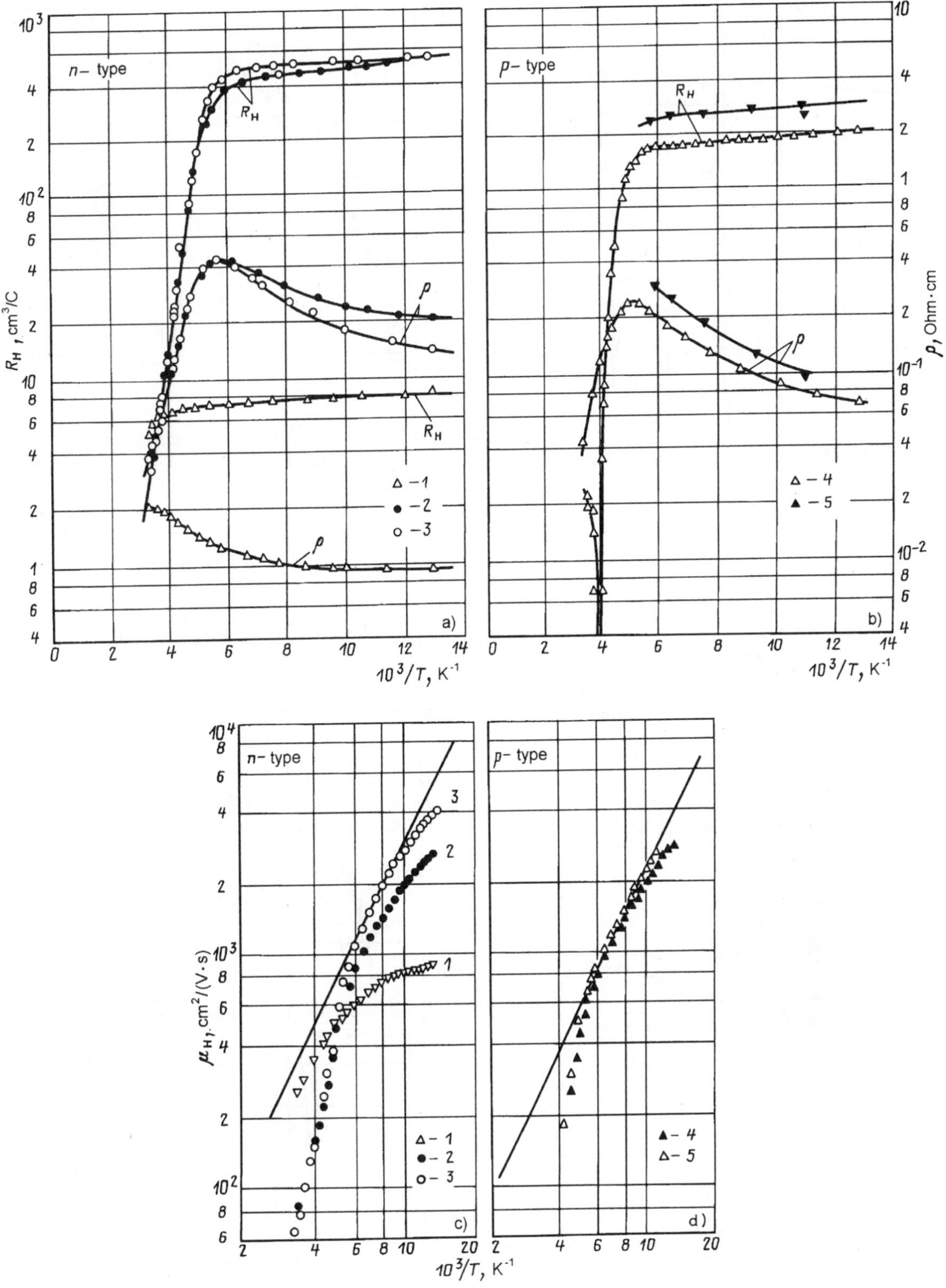

Figure 22.69 Temperature dependences of the resistivity and the Hall coefficient in n-Mg_2Sn (a) and p-Mg_2Sn (b) as well as the Hall mobility of electrons (c) and holes (d) in different n- and p-type samples [146]: carrier density, cm^{-3}: *1* – $9 \cdot 10^{17}$ (n); *2* – $1.5 \cdot 10^{16}$ (n); *3* – $1.3 \cdot 10^{16}$ (n); *4* – $4.3 \cdot 10^{16}$ (p); *5* – $3 \cdot 10^{16}$ (p).

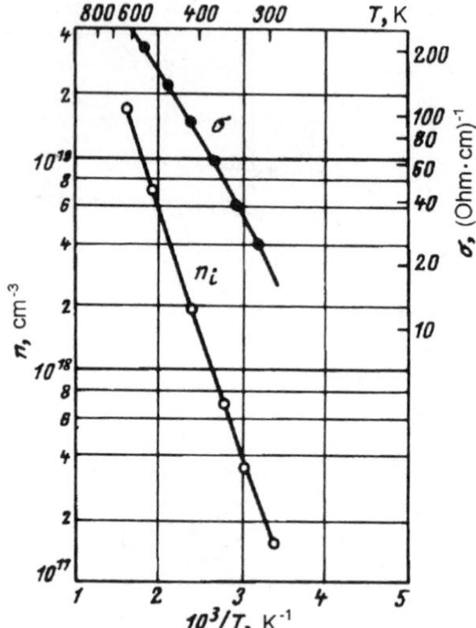

Figure 22.70 The temperature dependence of the intrinsic carrier density in Mg_2Sn determined from the data on the free carrier reflection (o) and on the conductivity (•) [147].

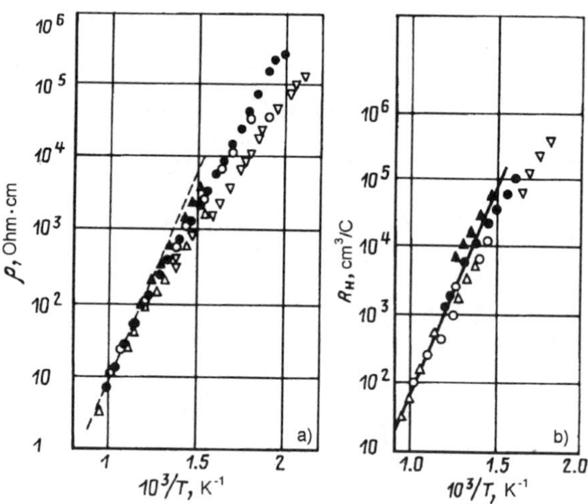

Figure 22.72 The temperature dependence of the resistivity (a) and the Hall coefficient (b) for polycrystalline Mg_3As_2 samples [160].

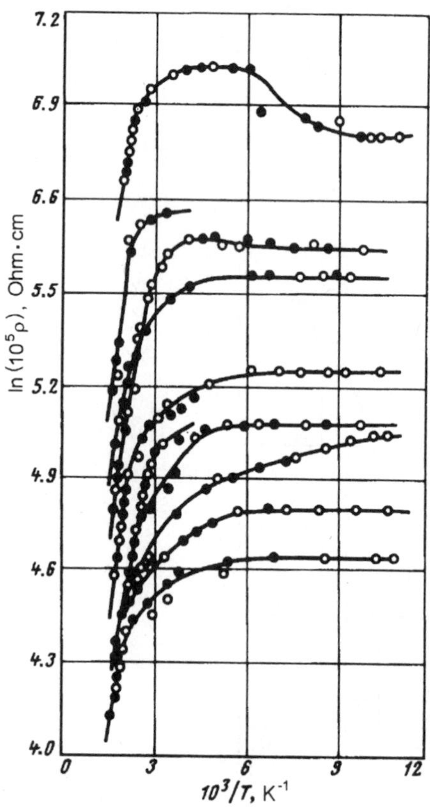

Figure 22.71 The temperature dependence of the resistivity for p-Mg_2Pb samples [158]: • – measurements for increasing sample temperature; o – measurements for decreasing sample temperature.

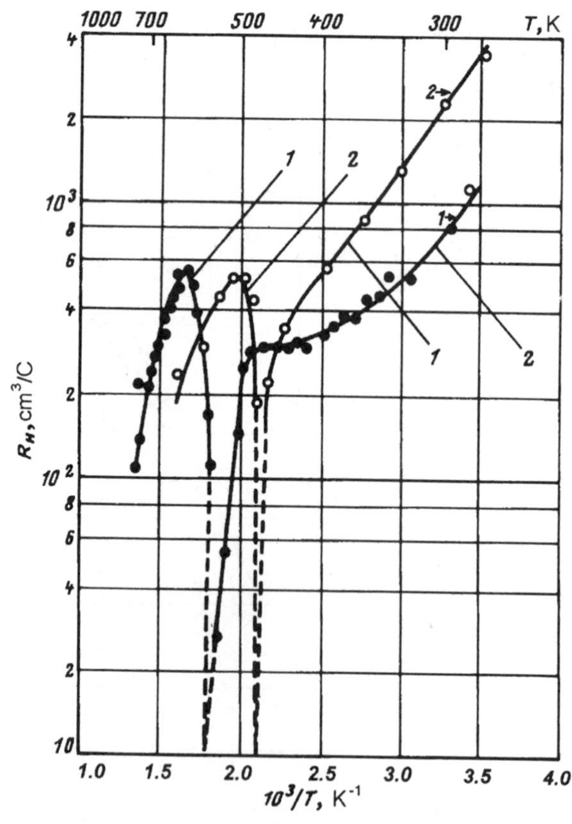

Figure 22.73 The temperature dependence of the Hall coefficient in β-ZnP_2 [166]: *1* – crystals grown from the vapor phase; *2* – crystals grown from the liquid phase.

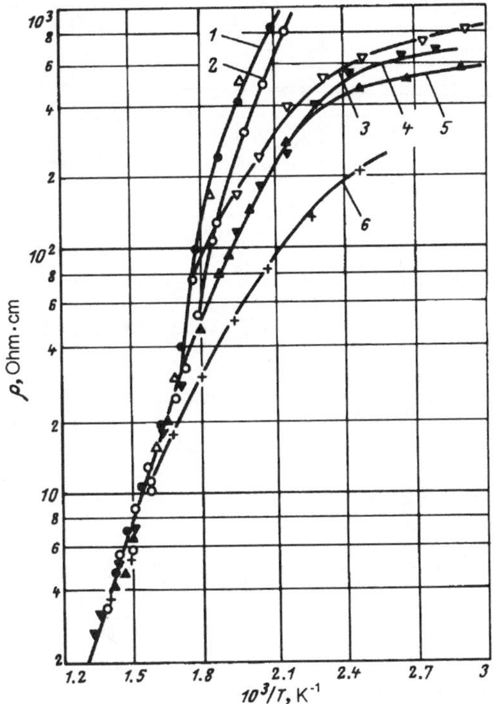

Figure 22.74 The temperature dependence of the resistivity in the intrinsic region for Zn_3P_2 [163]. The carrier density increases from sample *1* to sample *5*.

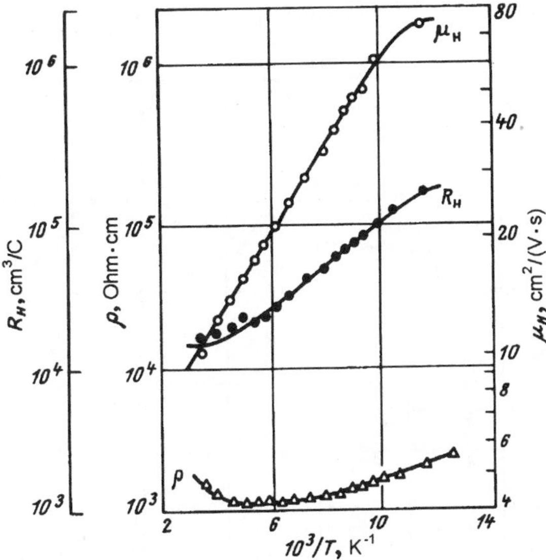

Figure 22.75 The temperature dependences of the resistivity, the Hall coefficient and the hole mobility in Zn_3P_2 [165].

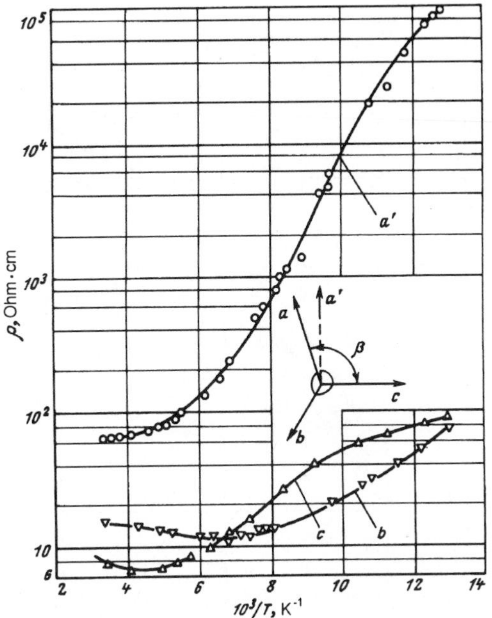

Figure 22.76 The temperature dependence of the resistivity for three samples cut from the same $ZnAs_2$ single crystal [167]. The orientation of the samples is shown in the insert; the b and c directions correspond to the crystallographic axes; the a' direction is perpendicular to b and c and makes an angle of 12° with the a monoclinic axis.

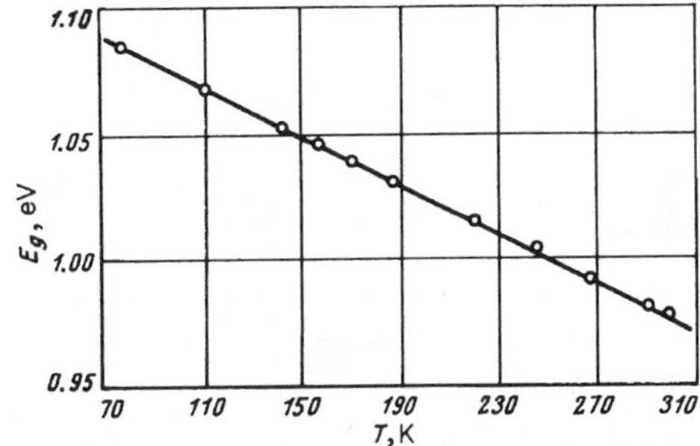

Figure 22.77 The temperature dependence of the forbidden band width in Zn_3As_2 [176].

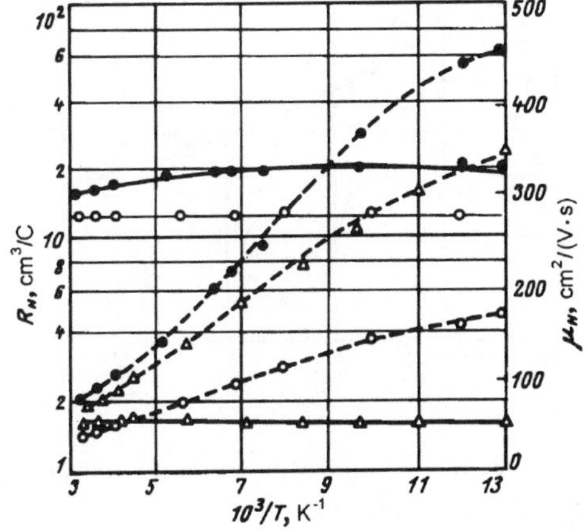

Figure 22.78 Temperature dependences of the Hall coefficient (solid lines) and the hole mobility (dashed lines) in Zn_3As_2 [177]: ○ – undoped sample; △ – Cu-doped; ● – Te-doped.

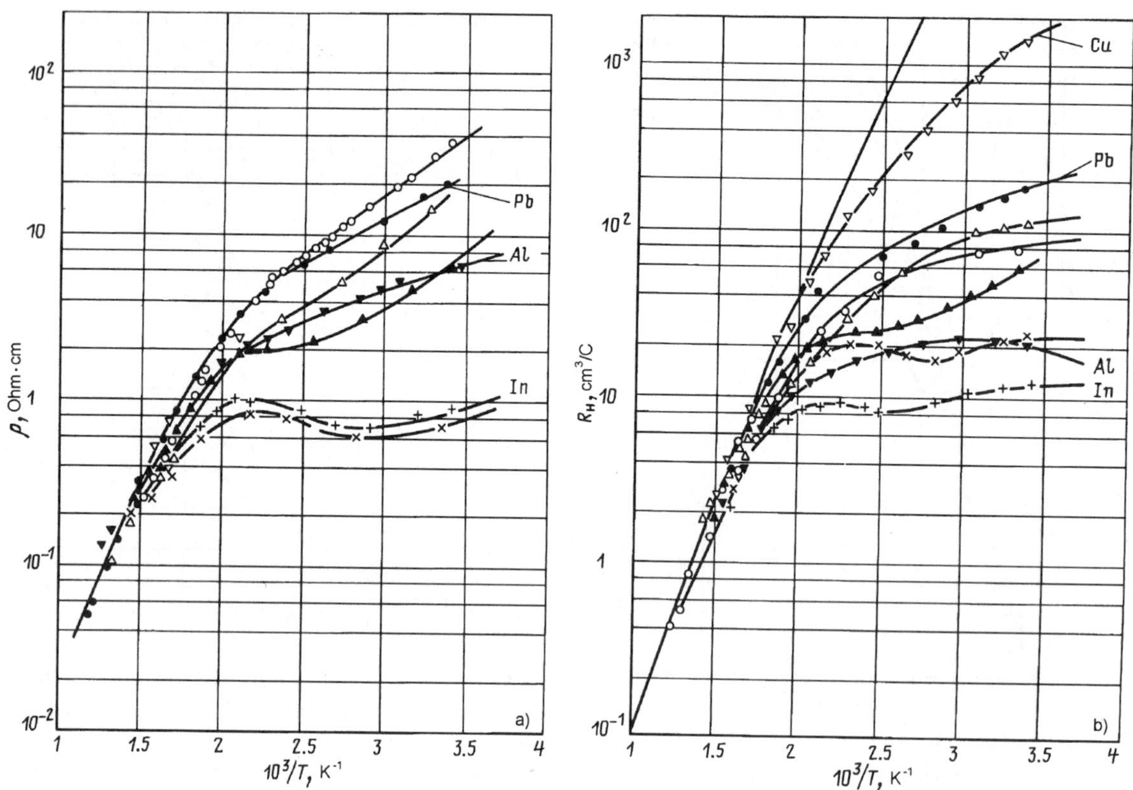

Figure 22.79 Temperature dependences of the resistivity (a) and the Hall coefficient (b) in Zn_3As_2 [179]. The doping elements are indicated for some curves; other curves relate to the pure material; the curve without experimental points displays the Hall coefficient in the intrinsic material [179].

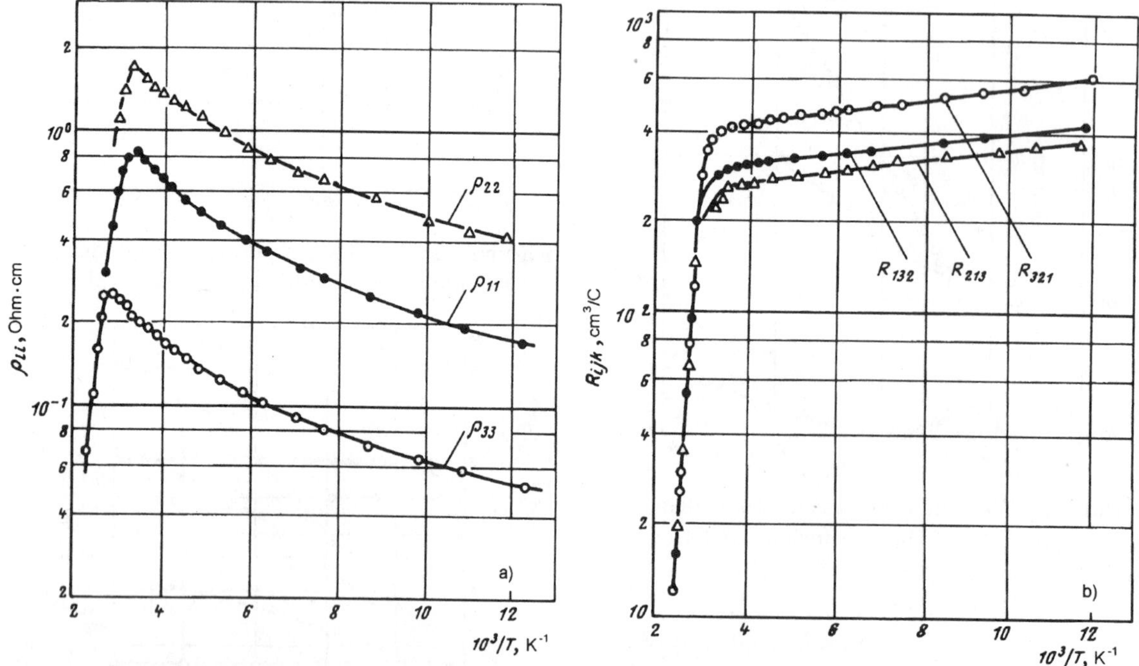

Figure 22.80 Temperature dependences of the tensor components for the resistivity (a) and the Hall coefficient (b) in ZnSb [180].

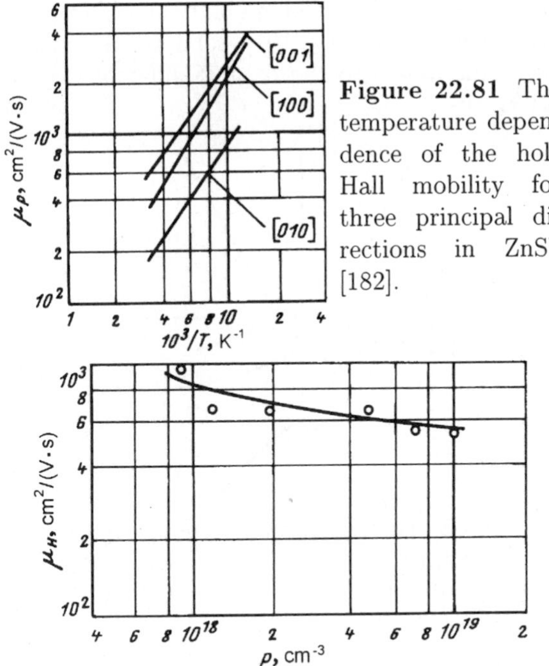

Figure 22.81 The temperature dependence of the hole Hall mobility for three principal directions in ZnSb [182].

Figure 22.82 The Hall mobility of holes vs. their density in Zn_4Sb_3 at 300 K [187].

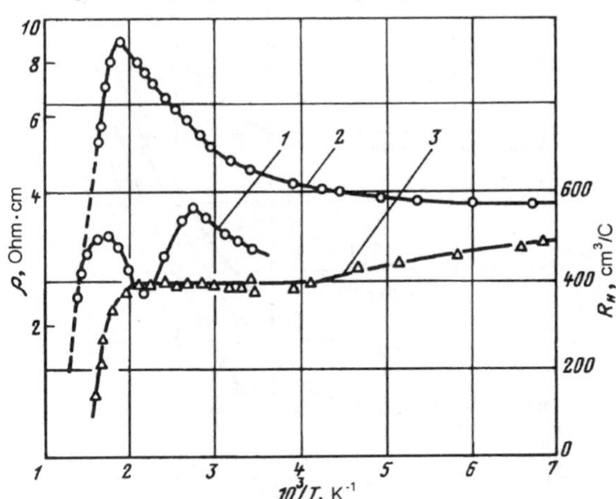

Figure 22.83 Temperature dependences of the resistivity (*1, 2*) and the Hall coefficient (*3*) in CdP_4 [189]: *1* – single crystal; *2, 3* – polycrystals.

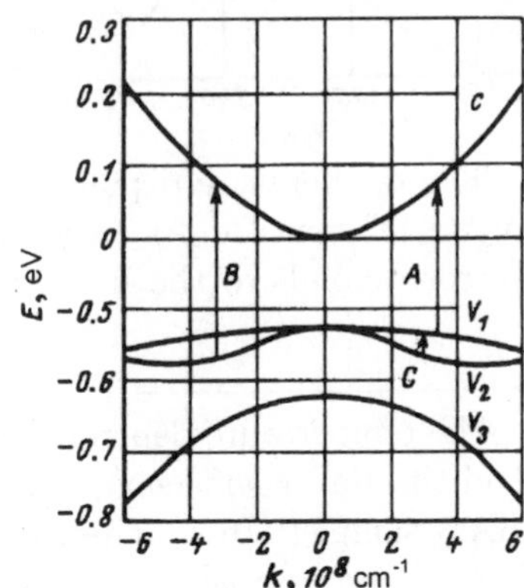

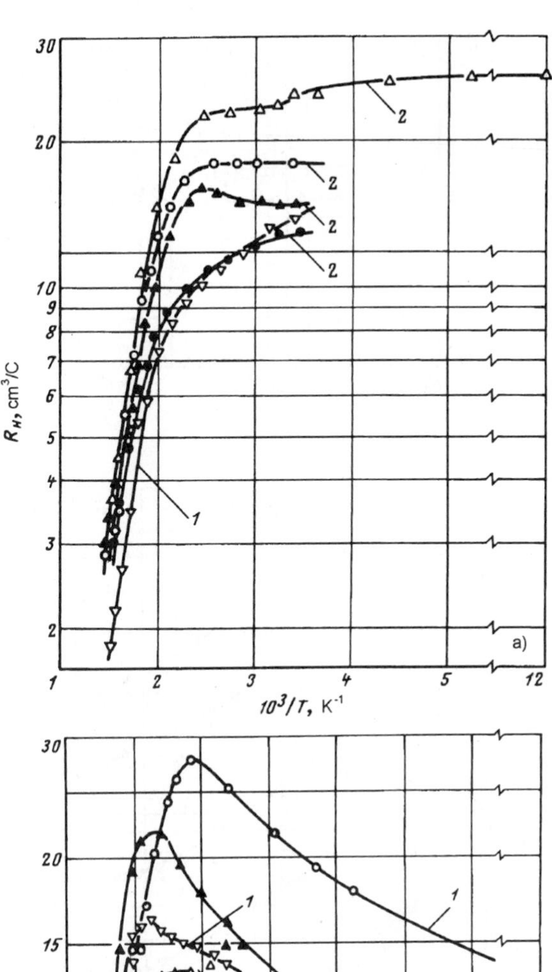

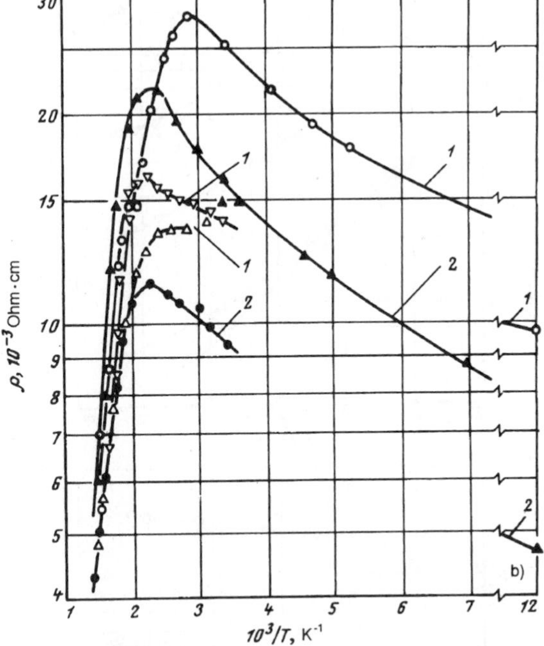

Figure 22.85 Temperature dependences of the Hall coefficient (a) and the resistivity (b) in monocrystalline (*1*) and polycrystalline (*2*) Cd_3P_2 samples [190].

Figure 22.84 The band structure for Cd_3P_2 ($E_g = 0.53$ eV, $\Delta_{so} = 0.1$ eV, $P = 6.7 \cdot 10^{-8}$ eV·cm, $m_{v1} = 0.5m_0$) [192]: A, B, C – allowed direct-gap interband transitions; V_1, V_2, V_3 – valence bands; c – conduction band.

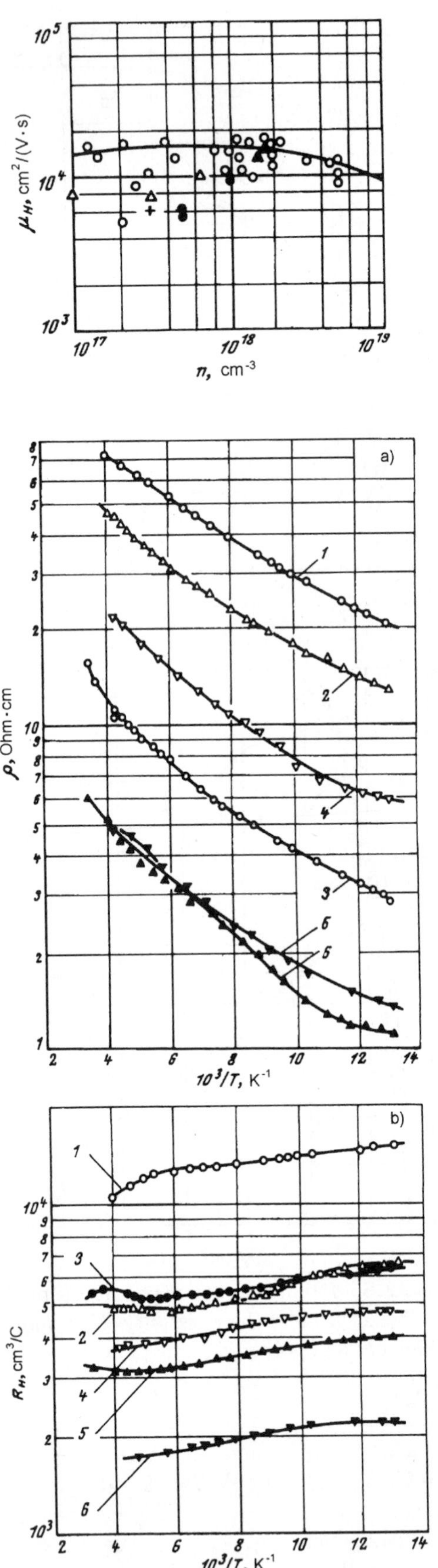

Figure 22.86 The electron mobility at 77 K vs. the electron density in Cd_3P_2 [191]: solid line is a calculation; points relate to the data of different experimental works.

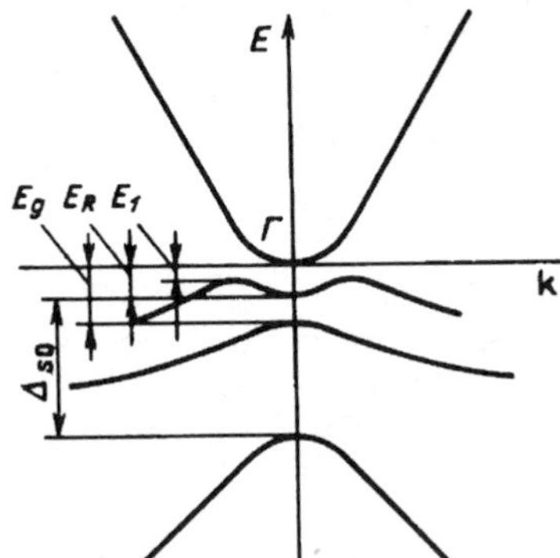

Figure 22.88 The energy band structure in Cd_3As_2 [293]: $E_g = 0.12 + 2.2 \cdot 10^{-4}T$ eV; $E_R = 0.03$ eV (10 K); $E_R = 0.05$ eV (4.2 K); $E_1 = 0.026$ eV (10 K); $\Delta_{so} = 0.21 - 0.31$ eV.

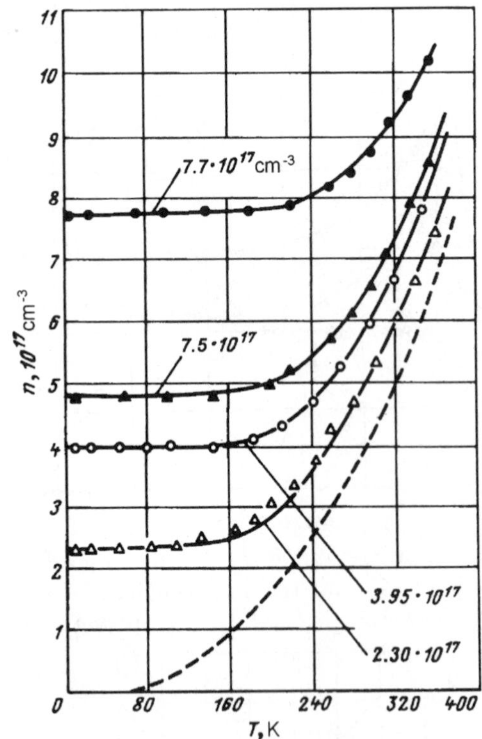

Figure 22.89 The temperature dependence of the electron density in different Cd_3As_2 crystals [196]. The dashed line corresponds to the intrinsic carrier density.

Figure 22.87 The temperature dependences of the resistivity (a) and the Hall coefficient (b) for monocrystalline n-$CdAs_2$ samples [167]. The carrier density increases from *1* to *6* samples.

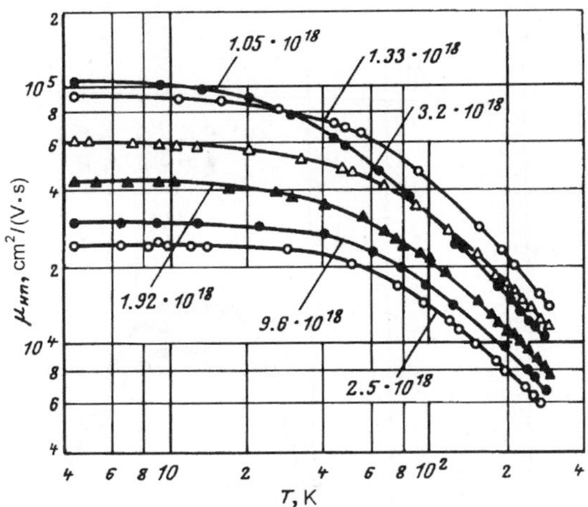

Figure 22.90 The temperature dependence of the electron Hall mobility in Cd_3As_2 [187].

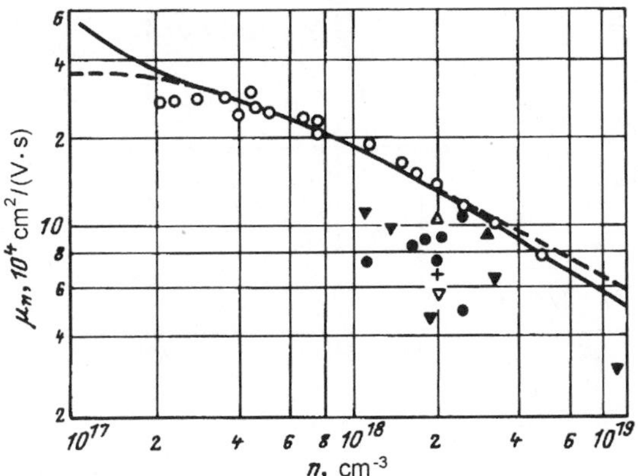

Figure 22.91 The mobility of electrons vs. their density in Cd_3As_2 at 4.2 K from the data of different authors [204]. The solid and dashed lines show the calculations.

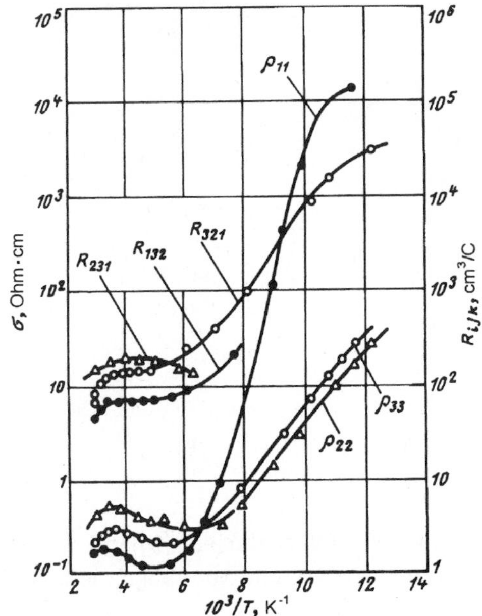

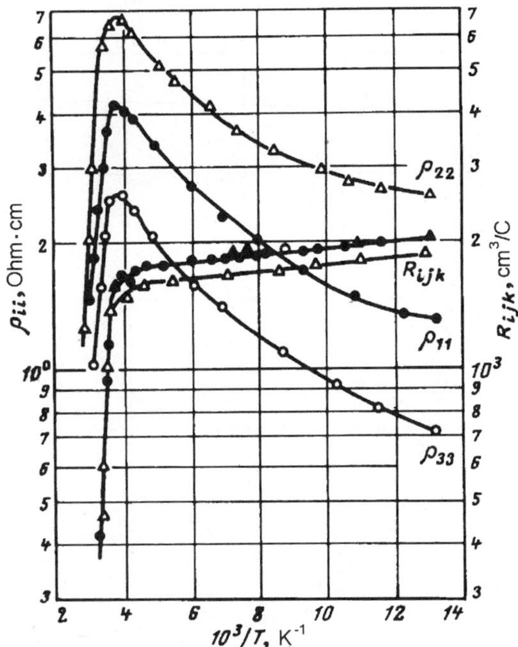

Figure 22.93 The temperature dependence of the tensor components for the resistivity and the Hall coefficient in p-CdSb [205].

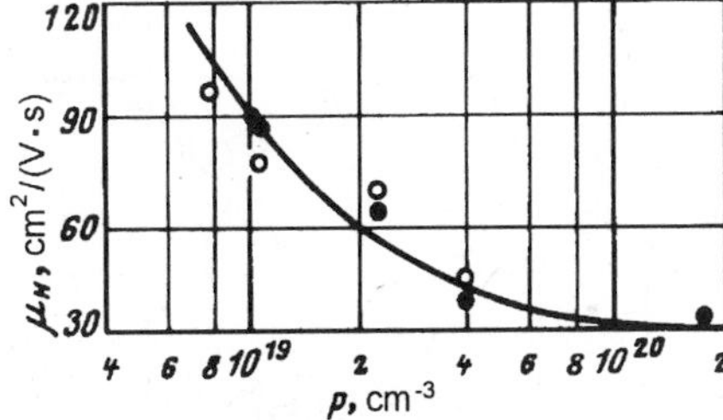

Figure 22.94 The Hall mobility of holes vs. their density in Cd_4Sb_3 [187].

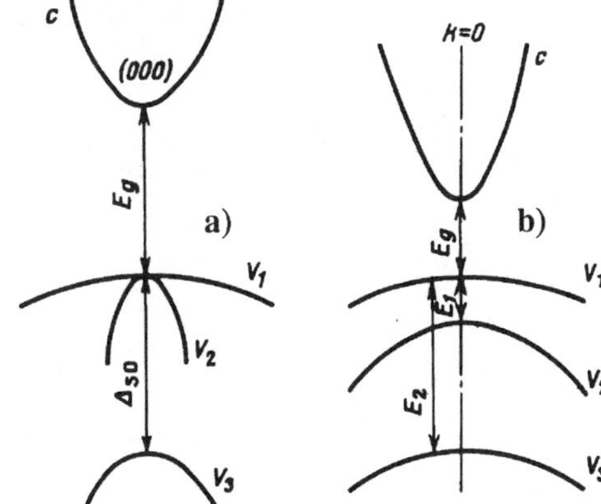

Figure 22.95 The band structure of $A^{II}B^{VI}$ semiconductors [138]: (a) – sphalerite, (b) – wurtzite.

Figure 22.92 Temperature dependences of the tensor components for the resistivity and the Hall coefficient in Ga-doped n-CdSb [205].

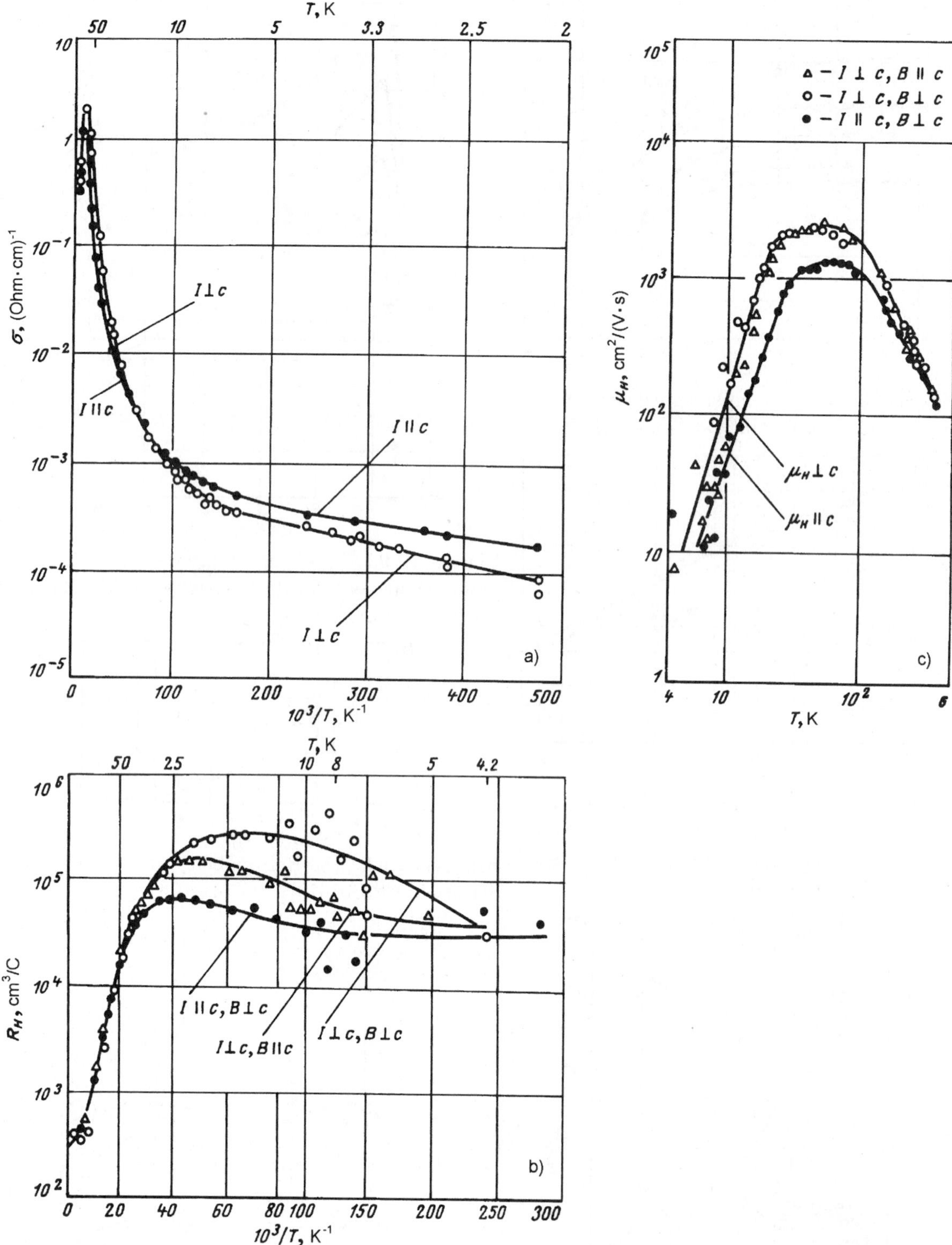

Figure 22.96 Temperature dependences of the conductivity (a), the Hall coefficient (b) and the Hall mobility of electrons (c) in ZnO for different orientations of the current I and the magnetic field B with respect to the c axis [207].

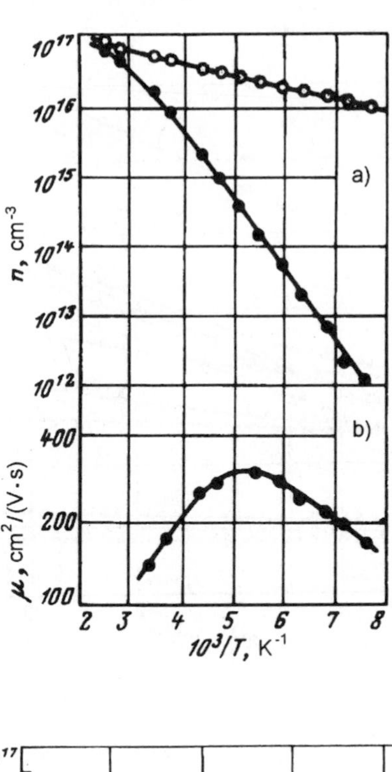

Figure 22.97 Temperature dependences of the electron density (a) and the Hall mobility of electrons (b) in ZnS [138]: • — hexagonal Al-doped ZnS annealed at 1050 K; ∘ — cubic I-doped ZnS annealed at 950 K.

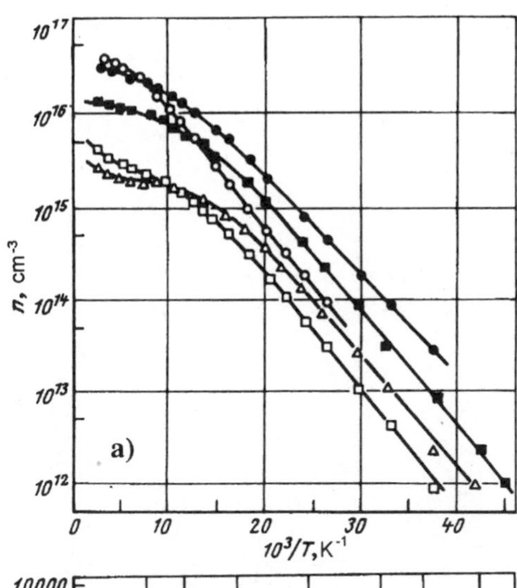

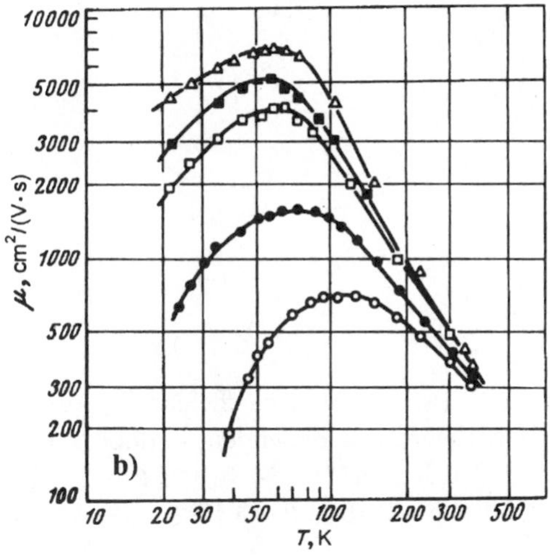

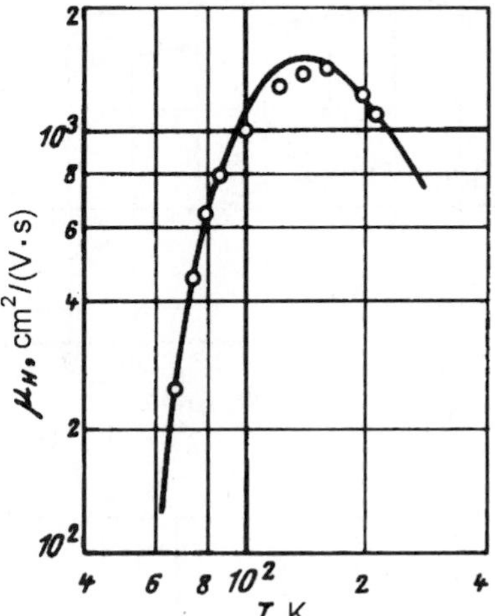

Figure 22.99 The temperature dependence of the electron drift mobility in ZnSe [208].

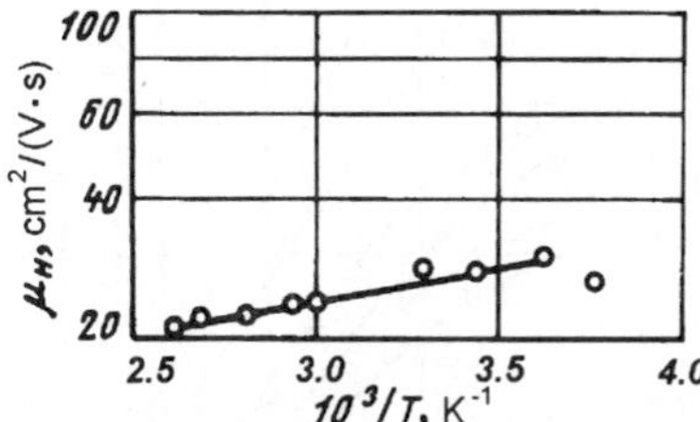

Figure 22.100 The temperature dependence of the hole Hall mobility in ZnSe [210].

Figure 22.98 Temperature dependences of the electron density (a) and the Hall mobility of electrons (b) in n-ZnSe crystals with different donor and acceptor densities [138]. N_D, N_A are measured in 10^{16}cm^{-3}:
△ – 0.34 (N_D), 0.13 (N_A); ■ – 1.8 (N_D), 0.5 (N_A);
□ – 1.05 (N_D), 0.75 (N_A); • – 3.7 (N_D), 0.5 (N_A);
∘ – 7.4 (N_D), 3.4 (N_A).

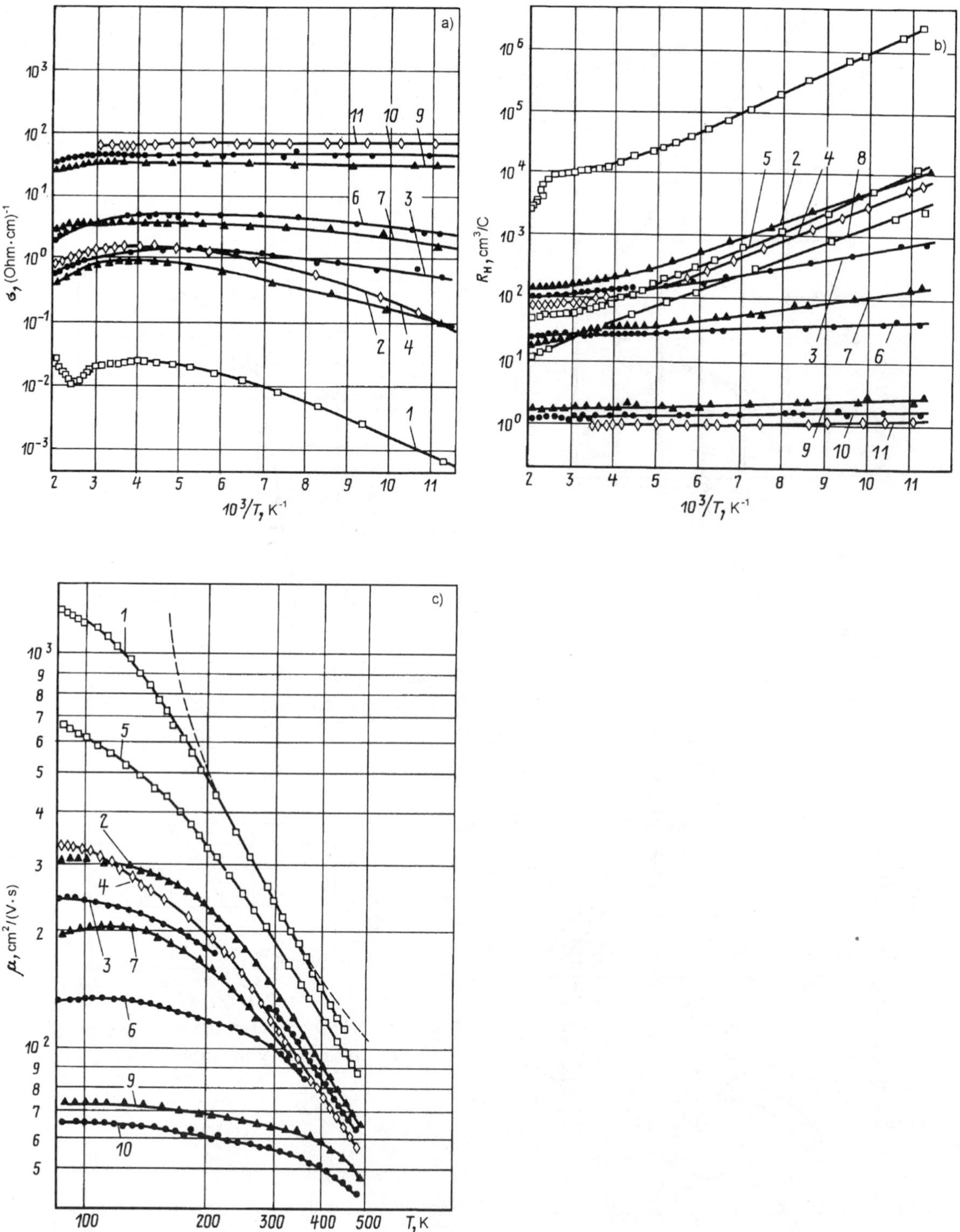

Figure 22.101 Temperature dependences of the conductivity (a), the Hall coefficient (b), and the hole mobility (c) in *p*-ZnTe crystals [235]: dashed line is a theoretical dependence; doping impurities: ◇ – Cs; □ – Te; △ – P; • – Li; carrier density at 300 K, cm^{-3}: *1* – $5{\cdot}10^4$; *2* – $3.2{\cdot}10^{16}$; *3* – $5.1{\cdot}10^{16}$; *4* – $6.9{\cdot}10^{16}$; *5* – $7.9{\cdot}10^{16}$; *6* – $2.8{\cdot}10^{17}$; *7* – $2.1{\cdot}10^{17}$; *8* – $2{\cdot}10^{17}$; *9* – $3{\cdot}10^{18}$; *10* – $4.5{\cdot}10^{18}$; *11* – $6.3{\cdot}10^{18}$.

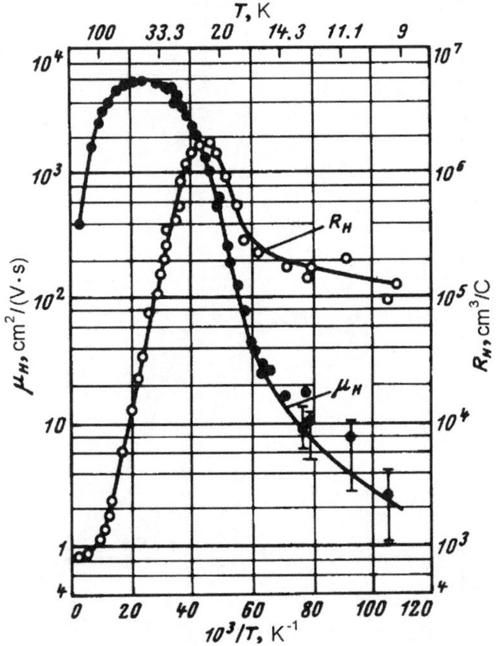

Figure 22.102 Temperature dependences of the Hall coefficient and the Hall mobility of electrons in an undoped CdS crystal [211].

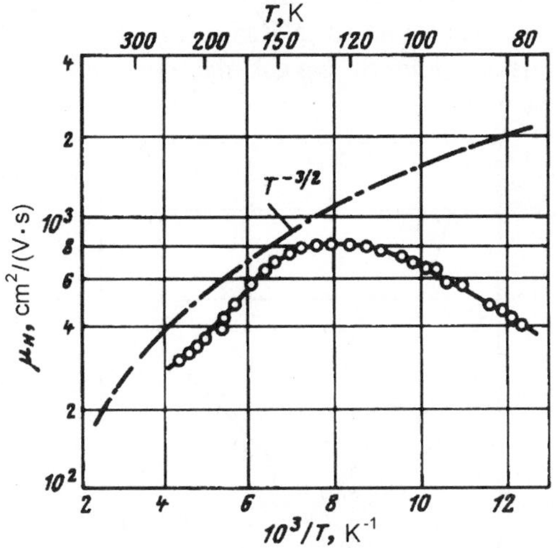

Figure 22.103 The temperature dependence of the electron drift mobility in different CdS crystals [215].

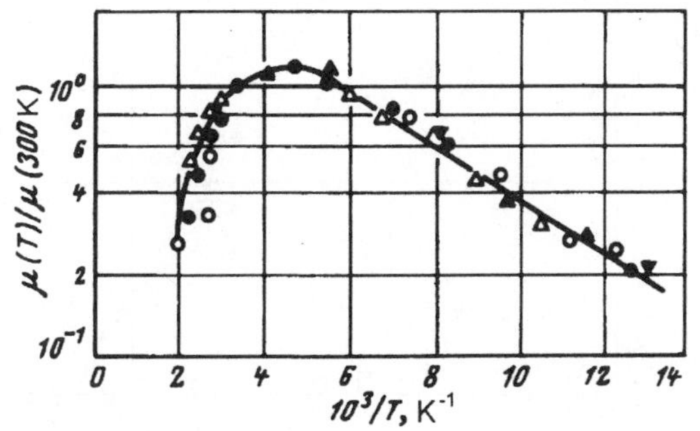

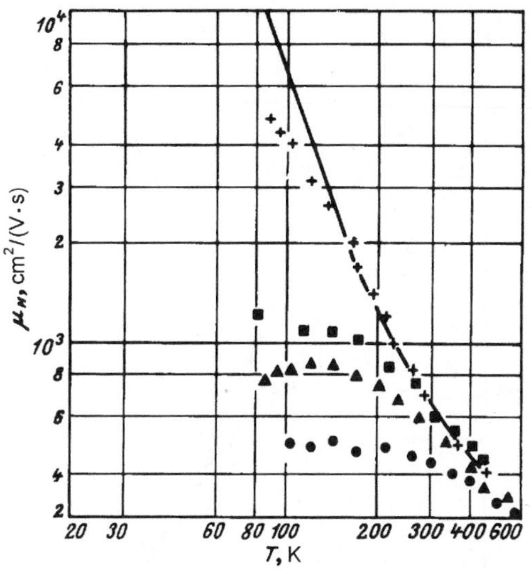

Figure 22.105 The temperature dependence of the electron Hall mobility in CdSe [138]: concentration of Ga donors, cm^{-3}: + – $1{\cdot}10^{16}$ (undoped sample); ■ – $1.4{\cdot}10^{17}$; ● – $2.9{\cdot}10^{18}$; curve checks with a calculation.

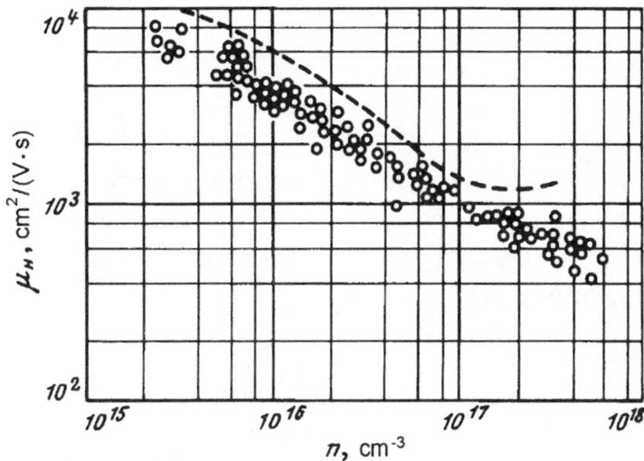

Figure 22.106 The Hall mobility of electrons in CdSe at 77 K vs. their density: dashed line is a calculation [218].

Figure 22.104 The temperature dependence of the hole drift mobility in different CdS samples [217].

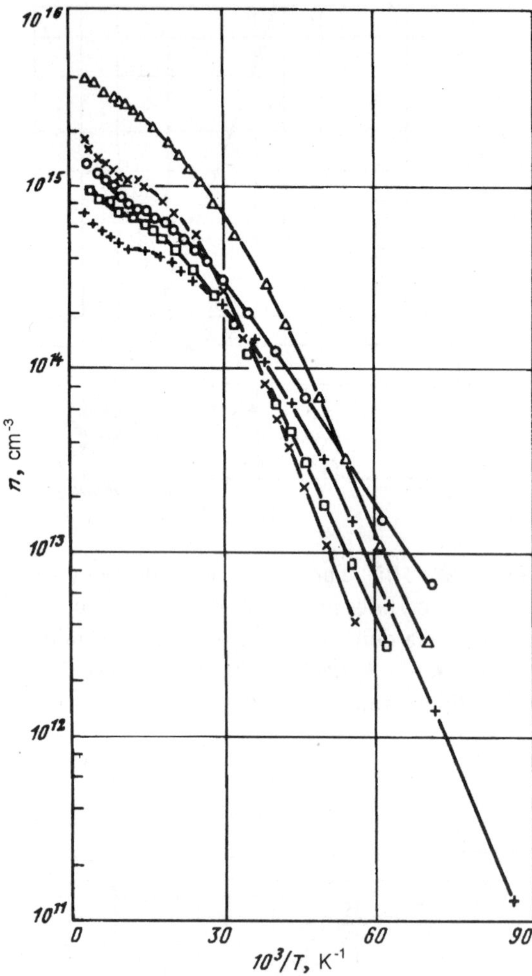

Figure 22.107 The temperature dependence of the electron density in n-CdTe [138]: +; ○, ×, □, △ – undoped samples after zone refining; ○– annealing with excess in Cd.

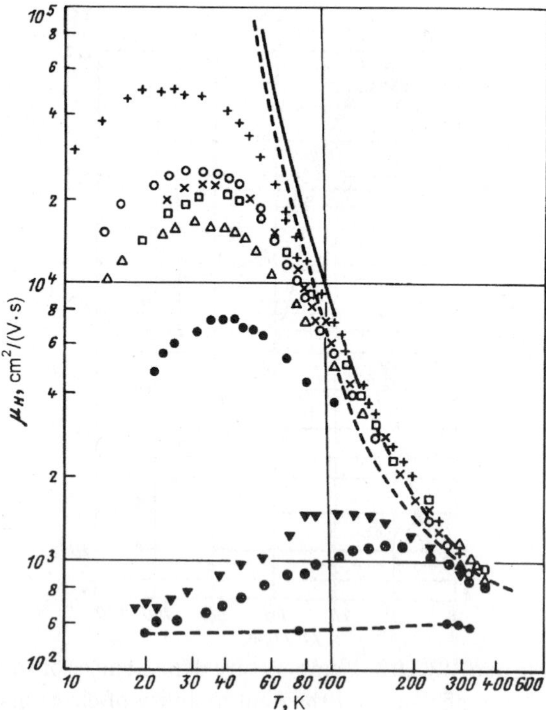

Figure 22.108 Temperature dependences of the electron Hall mobility in n-CdTe [138]. The symbols are the same as in Figure 22.107; the solid and dashed lines fit the calculated results for different models.

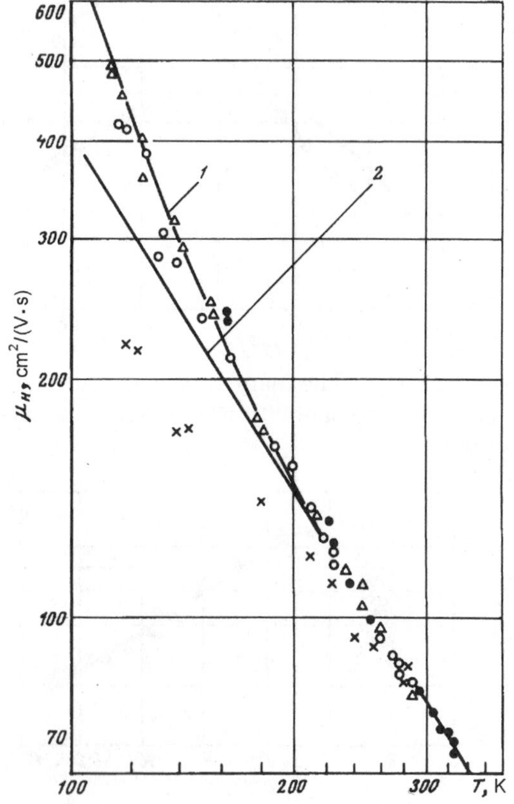

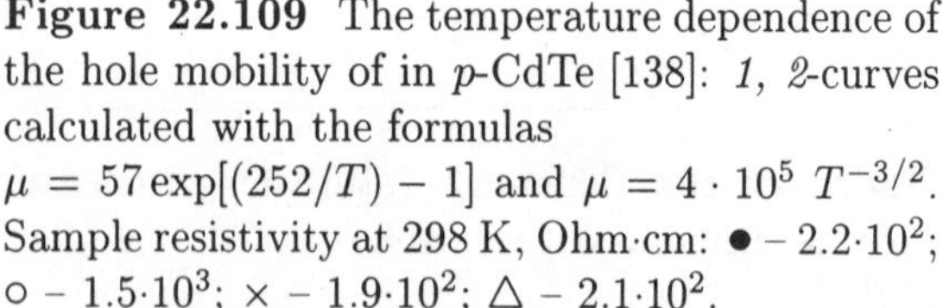

Figure 22.109 The temperature dependence of the hole mobility of in p-CdTe [138]: *1*, *2*-curves calculated with the formulas
$\mu = 57\exp[(252/T) - 1]$ and $\mu = 4 \cdot 10^5\ T^{-3/2}$.
Sample resistivity at 298 K, Ohm·cm: ● – $2.2{\cdot}10^2$; ○ – $1.5{\cdot}10^3$; × – $1.9{\cdot}10^2$; △ – $2.1{\cdot}10^2$.

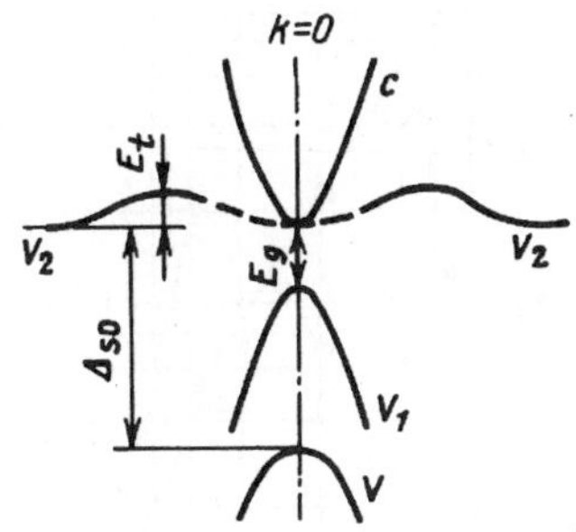

Figure 22.110 The band structure of β-HgS, HgSe, and HgTe [138, 199, 200].

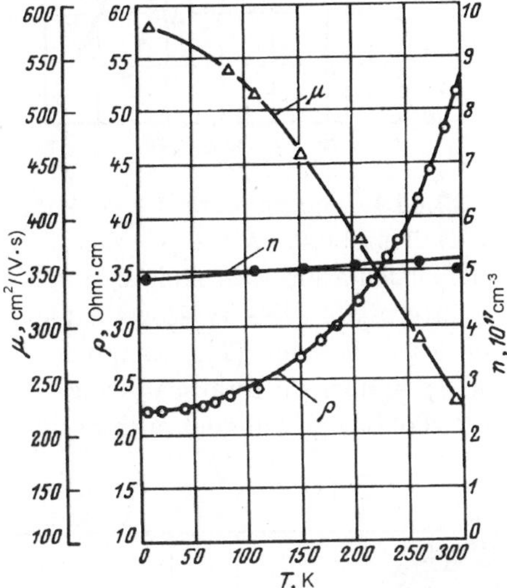

Figure 22.111 Temperature dependences of the resistivity, electron mobility and electron density in a 11.5 μm-thick β-HgS film [222].

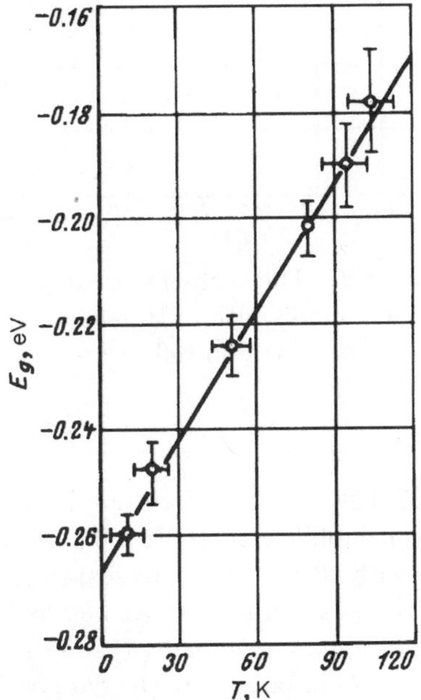

Figure 22.112 The temperature dependence of the forbidden band width in HgSe [223].

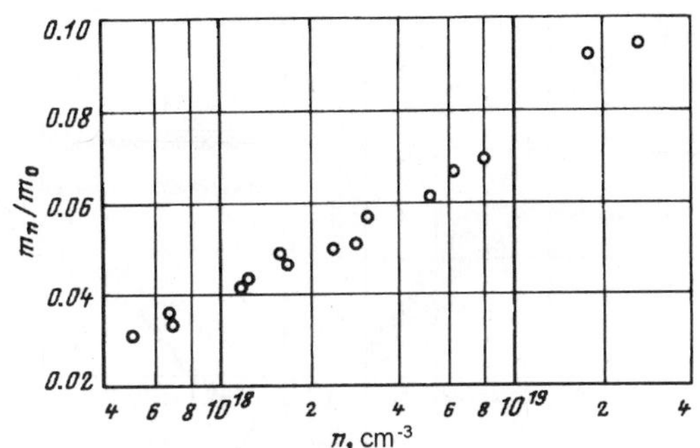

Figure 22.113 The effective mass of electrons vs. their density in HgSe at 300 K [224].

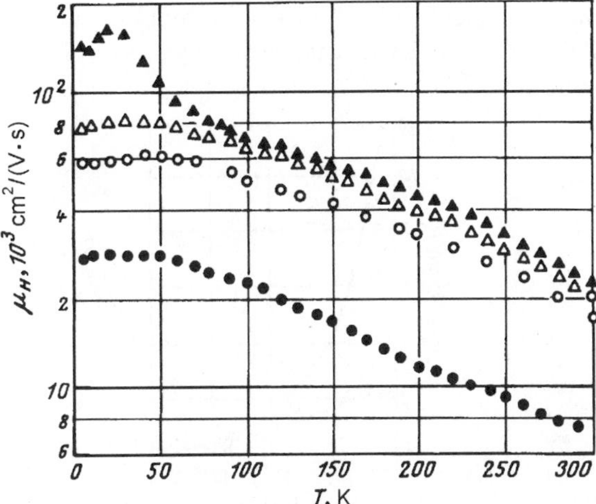

Figure 22.114 The temperature dependence of the electron mobility in HgSe [227]. The crystal was repeatedly annealed to obtain different electron densities n at T = 4.2 K: ▲ – $3.6 \cdot 10^{16}$; △ – $1.89 \cdot 10^{17}$; ○ – $3.78 \cdot 10^{17}$; ● – $3.92 \cdot 10^{18}$ cm^{-3}.

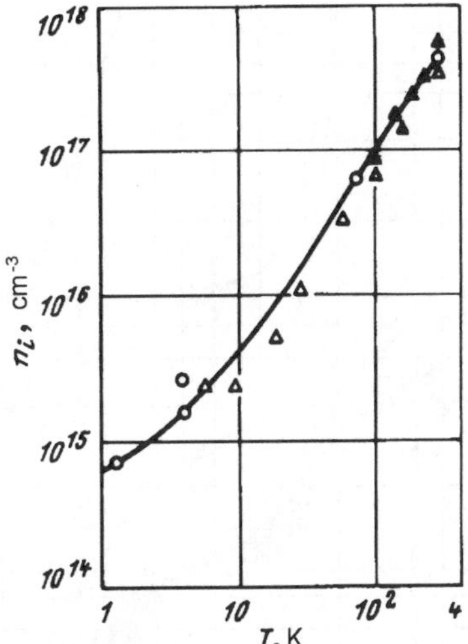

Figure 22.115 The temperature dependence of the intrinsic carrier density in HgTe: solid line fits a calculation [243]; points are the data of different authors.

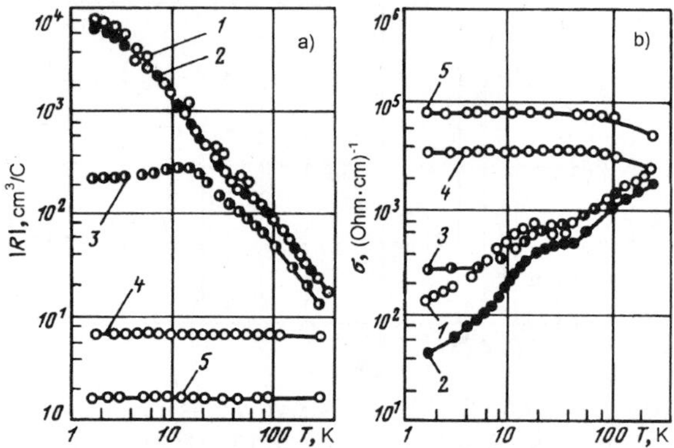

Figure 22.116 Temperature dependences of the Hall coefficient (a) and the conductivity (b) in HgTe [239]. The carrier density grows from sample *1* to sample *5*.

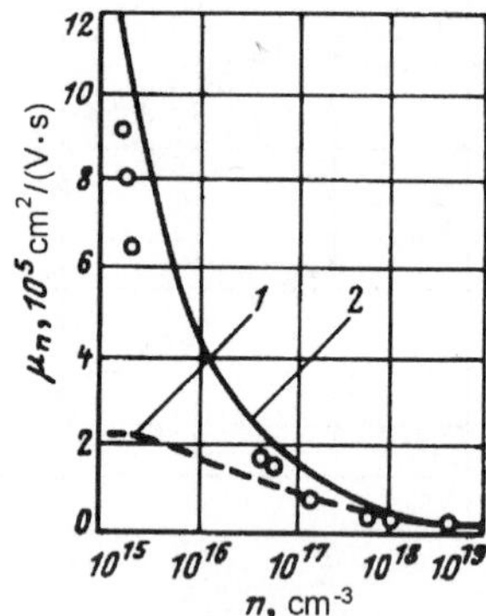

Figure 22.117 The mobility of electrons in HgTe at 4.2 K vs. their density. The calculation was made without (dashed curve *1*) and with (solid curve *2*) allowance for the screening of the charged centers due to interband transitions [240].

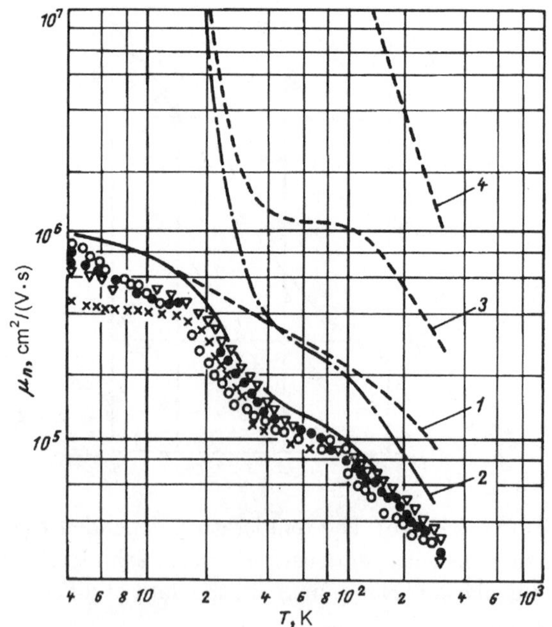

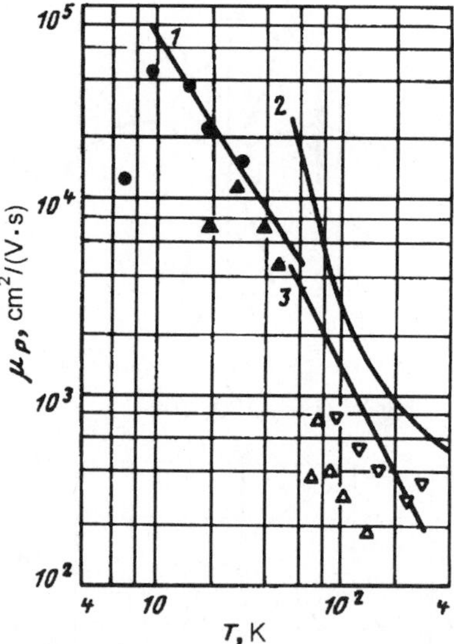

Figure 22.119 The temperature dependence of the hole Hall mobility in the intrinsic region of HgTe. The calculation was made with allowance for the scattering of holes by acoustical and nonpolar optical phonons (*1*), nonpolar optical phonons (*2*), and for the total scattering (*3*) [245]; points are the data of different authors.

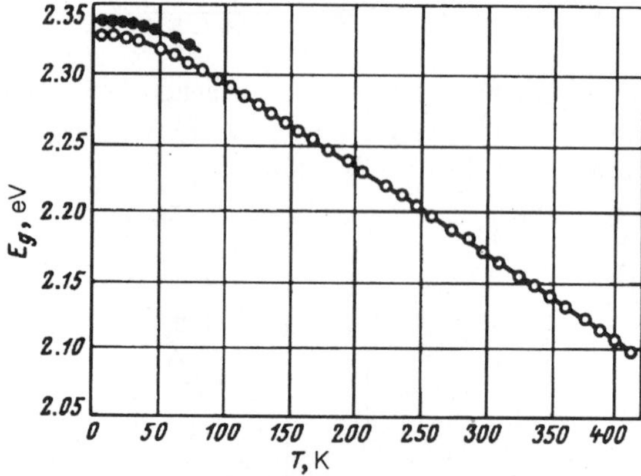

Figure 22.120 The temperature dependence of the forbidden band width in HgI_2 [9]: • – data on transmission; ○ – data on reflection.

Figure 22.118 The temperature dependence of the electron mobility in HgTe. The calculation was made with allowance for the scattering of electrons by charged centers (*1*) as well as by polar optical (*2*), nonpolar optical (*3*), and acoustical (*4*) phonons [244]; points are the data of different authors.

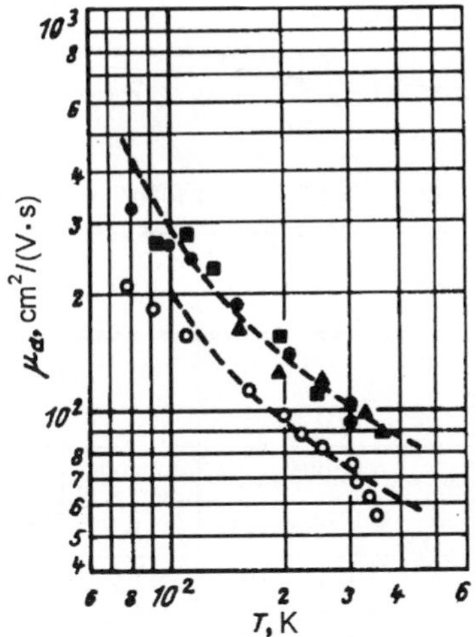

Figure 22.121 The temperature dependence of the electron drift mobility in HgI_2 [104]: ○– $E \perp c$; other symbols – $E \parallel c$ for samples with different thicknesses.

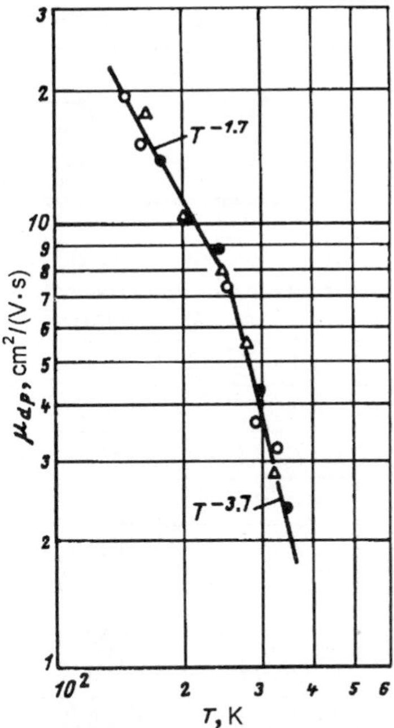

Figure 22.122 The temperature dependence of the hole drift mobility in $HgI_2$2 [110]: ○– $E \perp c$; other symbols – $E \parallel c$.

22.3.3 $A^{III}B^{IV}$ – $A^{III}B^{VII}$ compounds

$A^{III}B^{V}$ type semiconductors exhibit a high mobility and a small effective mass of electrons. Their properties have been thoroughly studied thus providing a wide use of these semiconductors. The latter crystallize into a sphalerite T_d^2 lattice. The conduction bands have, as a rule, several minima at $\mathbf{k} = 0$ and on the [111] and [100] axes.

In the GaN, GaAs, GaSb, InN, InP, InAs and InSb compounds the absolute minimum Γ is placed at the center of the Brillouin zone ($\mathbf{k} = 0$) (see Figure 22.123); in BP, AlP, AlAs, and AlSb, the absolute minimum Δ falls on the [100] axis (see Figure 22.124). The numerical values of the band structure parameters can be seen in Table 22.15.

The valence band is degenerate at $\mathbf{k} = 0$. The maximum of the light-hole band V_2 is arranged at $\mathbf{k} = 0$; the maxima of the heavy-hole band V_1 are somewhat (by 0.01-0.001 eV) higher on the [111] axes. There exists also a band V_3 which is split out due to spin-orbit interaction Δ_{so}.

The semiconducting compounds of the III and VI groups have the stoichiometric composition of two types: $A^{III}B^{VI}$ and $A_2^{III}B_3^{VI}$, and show many crystal modifications.

$A^{III}B^{VI}$ compounds fall mainly into two classes: (1) with the hexagonal structure D_{6h}^3 (GaS,.GaSe, InSe), and (2) with the tetragonal structure D_{4h}^{18} (InTe, TlS, TlSe, TlTe). The majority of the $A^{III}B^{VI}$ crystals possesses a complex structure.

$A_2^{III}B_3^{IV}$ compounds have a low mobility of carriers (as a result of the imperfect structure [150]) and, as a rule, show an n-type conductivity (due to the excess number of chalcogen atoms). These compounds are not very sensitive to the doping by other impurity atoms. Many of these compounds crystallize into lattices of several types.

The semiconducting properties of the rare-earth chalcogenides can be recognized in the collection book [62]. The UO_2 compound crystallizing as cubic O_h^5 with $a = 0.546$ nm [1] possesses the semiconducting properties too: $E_g = 1.3$ eV [87] and $\mu_p = 10$ cm^2/(V·s) [249].

Table 22.14 The electrophysical properties of boron carbide B_4C

Crystal structure		ρ, g/cm^3	T_{melt}, K	E_g, eV	ρ, Ohm·cm	T_D, K
System, group	a, c, nm					
Rhombohedral, D_{3d}^5	0.560 (a); 1.212 (c) [242]	2.52 [216]	2723 [16]	1.64 [16]	0.1–10 [16]	1300 [288]

Table 22.15 Semiconducting $A^{III}B^{V}$ compounds [292]

Compound	Crystal structure		ρ, g/cm^3	T_{melt}, K	T_D, K	$\hbar\omega_l$, eV	$\hbar\omega_t$, eV
	System, group	a, nm					
BP	Cubic, T_d^2	0.4538	2.89 [242]	1400 (T_{decomp})	985	0.103	0.102
AlP	Cubic, T_d^2	0.547	2.40	2823	588 [178]	0.062	0.0545 [254]
AlAs	Cubic, T_d^2	0.566	3.76 [213]	2013	417 [121]	0.050	0.0447 [254]
AlSb	Cubic, T_d^2	0.614	4.26 [213]	1327 [123]	292 [121]	0.042	0.0395 [254]
GaN	Hexagonal, C_{6v}^4	0.318(a); 0.5166 (c) [242]	6.10 [242]	1500 [242] 2000	600	—	—
GaP	Cubic, T_d^2	0.545	4.14	1740	446 [121]	0.0499	0.0455 [254]
GaAs	Cubic, T_d^2	0.569 [1]	5.32 [213]	1510 [123]	344 [121]	0.0362	0.0354 [251]
GaSb	Cubic, T_d^2	0.609 [1]	5.61	985 [123]	265 [121]	0.0298	0.0286 [254]
InN	Hexagonal, C_{6v}^4	0.353 (a); 0.569 (c) [242]	6.88 [242]	1373	—	0.086	0.059
InP	Cubic, T_d^2	0.586 [1]	4.79 [123]	1327 [123]	321 [121]	0.0435	0.038 [254]
InSb	Cubic, T_d^2	0.647	5.78 [121]	798 [123]	262 [121]	0.0244	0.0229 [254]
InAs	Cubic, T_d^2	0.606 [1]	5.68 [121]	1216 [123]	249 [121]	0.030	0.0272 [254]

Compound	ε_0	ε_∞	μ_n, cm^2/(V·s)	μ_p, cm^2/(V·s)	Absolute valence band minimum
BP	11.6 [242]		30–120	500 [213]	(100)
AlP	9.83 [254]	7.56 [254]	50 [28]	450 [246]	(100)
AlAs	10.06 [213]	8.16 [213]	75–300 1200 [115]	100 [234]	(100) [170]
AlSb	12.0 [213]	10.2 [254]	200 [121] 700(77 K)	420 [27] 3700(77 K)	(100) [170]
GaN	12.2 [214]	5.8 [214]	380 [214]		(000)
GaP	11.1 [213]	9.11 [213]	300 [123] 500(77 K) [27]	75 [27] 420(77 K)	(100) [170]
GaAs	12.53 [170] 12.9 [254]	11.6 [170] 10.9 [254]	8500 [27] 210000(77 K)	420 [27] 9000(77 K)	(000) [170]
GaSb	15 [27] 16.1 [254]	15.2 [170] 14.4 [254]	4000 [27] 6000(77 K)	1400 [27] 3600(77 K)	(000 [170]
InN		9.3	250(300 K) [213]		(000)
InP	12.4 [254]	9.6 [254]	4600 [27] 130000(77 K) [213]	150 [27] 1200(77 K)	(000) [170]
InSb	16.8 17 [27]	15.6 [254] 15.7	78000 [27] 1200000(77 K)	750 [27] 10000(77 K)	(000) [170]
InAs	14.5 [27] 14.9 [254]	11.6 [170] 12.3 [254]	33000 [27] 82000(77 K)	460 690(77 K) [27]	(000) [170]

Table 22.15 Semiconducting $A^{III}B^{V}$ compounds *(continued)*

Compound	E_g, eV	$E_{\Gamma\Delta}$, eV	$E_{\Gamma\Lambda}$, eV	Δ_{so}, eV	$m_n^{(000)}/m_0$	$m_n^{(111)}/m_0$
BP	2.0					
AlP	2.45 2.5(2 K)	1.1				
AlAs	2.14 2.23(2 K)	1.0 1.3 [172]		0.29 [173]	0.124	0.096(⊥) 1.9(‖) [213]
AlSb	1.63 1.69(4.2 K)	0.7 [171]		0.67 [213]	0.09 [123]	
GaN	3.44 3.50(1.6 K)				0.2 [213]	
GaP	2.27 2.35(0 K)	0.35 [170]	0.25 [185]	0.08 0.1 [171]	0.12 [123]	
GaAs	1.43 1.52(2 K)	0.38	0.5 [185]	0.33 [170]	0.067 [84]	
GaSb	0.70 0.811(2 K)	0.4 [170]	0.08	0.75 [84] 0.8 [171]	0.047 [170]	0.9 [170] 0.12(⊥) 1.3(‖)
InN	2.05(300 K) [213] 2.11(78 K)				0.11	
InP	1.34 1.42(2 K)	0.7 [170]	0.4 [84]	0.1 [171]	0.073 [84]	
InSb	0.180 0.236(4 K)		0.45 0.5 [84]	0.85 [213]	0.014 [170]	0.09 [213]
InAs	0.36 0.41(0 K)			0.38 [84] 0.43 [170]	0.023 [170]	

Compound	$m_n^{(100)}/m_0$	m_{p1}/m_0	m_{p2}/m_0	m_{dp}/m_0	$\partial E_g/\partial T$, 10^{-4} eV/K	$\partial E_g/\partial P$, 10^{-11} eV/Pa	g_n
BP							
AlP	0.13 [234]	0.51(h)(100) [246] 0.21(l)(100) 1.37(h)(111) 0.15(l)(111)	0.20 0.39 [173]	–	−3.5 [234]		
AlAs	0.19(⊥) [213] 1.1(‖) [213]	0.41(100) 1.02(111) [213]	0.15(100) [213] 0.11 [111]		−4	2.8 [246]	
AlSb	0.33 0.25(⊥) [84] 1;1.64(‖)	0.34(100) [213] 0.87(111)	0.12(100) 0.09(111) [213]	0.9 [173]	-3.5 -4 [123]	-1.6 [170] (100)	
GaN	0.2 [234]			0.6–1.0 [214]	-3.9 [242] -6.7	-4.2 (000)	
GaP	0.22(⊥) [84] 1.15(‖)	0.5 [170] 0.56 [105]	0.13 [170] 0.16		-5.5 [170]	-1.7 [170] (100)	
GaAs	1.2 [170] 0.3(⊥) 2.0(‖)	0.475 [231] 0.5 [170]	0.89 [232]	0.5 [173]	-5(000) [170] -2.4(100)	12.5(000) [186] -8.7(100) [170]	0.52–0.37 [84] 0.523 [170]

Table 22.15 Semiconducting $A^{III}B^V$ compounds *(continued)*

Compound	$m_n^{(100)}/m_0$	m_{p1}/m_0	m_{p2}/m_0	m_{dp}/m_0	$\partial E_g/\partial T$, 10^{-4} eV/K	$\partial E_g/\partial P$, 10^{-11} eV/Pa	g_n
GaSb	0.22(⊥) [213] 1.51(‖) [213]	0.23 [123] 0.36(⊥) [233] 0.26(‖) [233]	0.04 0.052 [233]	0.39 [173]	−3.5(000) [170] +2(111) [123]	14.5(000) −10(111)	−9 −7.8 [213]
InN		0.5 [213]	0.17 [213]				
InP		0.60 [105]	0.12 [105]		−2.9 [123] −4.6 [170]	4.6(000) [170]	1.3 [292] −0.6 [230]
InSb		0.4 [123]	0.015 [123] 0.020 [231]	0.2 [173]	−2.8 [170] −2.9 [123]	15.5(000) [170] 14.2(000)	−50 [170] −48 [84]
InAs		0.4 [170]	0.025 [170]	0.33 [173]	−2.2 [170] −3.7 [123]	4.8(000) 8.5(000) [170] 3.2(111)	−14.7 [84]

22.16 Properties of impurities in $A^{III}B^V$ compounds [292]

Compound	E_D, eV	E_A, eV
AlP		0.15 (?) 0.37 (?)
AlAs	0.05 (Mg) 0.06 (Zn) 0.07 (Si)	
AlSb	0.68 (Te) 0.147 (Se) 0.060 (?)	0.033(?) 0.041(?) 0.102(?)
GaN	0.017–0.042 (V_N)	0.225(V_{Ga}) 0.41(Hg) 0.750(Li) 0.37; 0.48; 0.65; 1.02; 1.42 (Zn)
GaP	0.061 (Li) 0.091 (Li) 0.072 (Sn) 0.085 (Si) 0.105 (Se) 0.107 (S) 0.204 (Ge) 0.897 (O) 0.930 (Te)	0.052 (?) 0.054 (C) 0.057 (Be) 0.060 (Mg) 0.070 (Zn) 0.102 (Cd) 0.210 (Si) 0.265 (Ge) 0.530 (Cu)
GaAs	0.058 (Si) 0.058 (Se) 0.058 (Pb) 0.058 (V_{Ga}) 0.059 (Ge) 0.059 (Si) 0.059 (C)	0.027 (C) 0.028 (Be) 0.029 (Mg) 0.031 (Zn) 0.035 (Si, Cd) 0.040 (Ge) 0.113 (Mn) 0.167 (Sn)
GaSb	0.145 (S) 0.085 (Se) 0.020 (Te) } (111) ∼ 0.3 (S) 0.2 (Se) $<$ 0.08 (Te) } (100)	0.0094 (Si) 0.013 – 0.015 (Ge) 0.008 – 0.130 (?)
InP	0.106 (Ti) 0.175 (Si) } (100)	0.031; 0.108 (Mg) 0.031; 0.143 (Be) 0.041 (C) 0.046 (Zn) 0.057 (Cd) 0.098 (Hg) 0.210 (Ge) 0.270 (Mn)
InAs	020 – 0.025 (?)	0.01 (Sn); 0.02; 0.035 (?) 0.014 (Ge) 0.02 (Si)
InSb	∼ 0.007 (S) ∼ 0.007 (Se) ∼ 0.007 (Te) } (000) 0.05 (Te) 0.15 (Se) 0.25 (S) 0.55 (?) } (111)	0.008 (Co) 0.0091; 0.0099 (Zn) 0.00925 (Ge) 0.0095 (Mn) 0.0099(Mg) 0.013 (Fe) 0.028; 0.056 (Cu) 0.030; 0.056 (Ag) 0.07 (Cr) 0.12 (?)

Table 22.17 Semiconducting $A^{III}B^{VI}$ compounds

Compound	Crystal structure		ρ, g/cm^3	T_{melt}, K	T_D, K	E_g, eV
	System, group	a, b, c, nm				
GaS	Hexagonal, D^4_{6h}	03585 (a); 1.55 (c) [150]	3.75 [288]	1235 [288]	215 [288]	3.06 (77 K)
GaSe	Hexagonal, D^4_{6h}	0.3755 (a); 1.594 (c) [150] β- and ϵ-GaSe [294]	5.03 [150]	1211 [294]	190 (0 K) [294]	2.09 [294]
GaTe	Monoclinic, C^1_2	1.744 (a); 1.046 (b); 0.4077 (c) [294]	5.44 [150]	1097 [294] 1108 [150]	158 [294]	1.7 [152] 1.8 (0 K) [294]
InS	Rhombic, D^{13}_{2h}	0.394 (a); 0.444 (b); 1.064 (c) [150]	5.18 [150]	965 [150]		1.9 [294] 2.07 (4.2 K)
InSe	Hexagonal, D^4_{6h}	0.400 (a); ϵ:1.670 (c); γ: 2.495 (c) [294]	5.55–5.72 [150]	888 [288] 933 [150]	190 [294]	1.18 1.32 (30 K) [294]
InTe	Tetrahedral, D^{18}_{4h}	0.8437 (a); 0.7139 (c) [150]	6.29 [150]	966 [150]		0.26–0.36 [288]
TlS	Tetrahedral, D^{18}_{4h}	0.779 (a); 0.679 (c) [150]	7.61 [150]	623 [150]		1.36 [294]
TlSe	Tetrahedral, D^{18}_{4h}	0.803 (a); 0.701 (c) [150]	8.2 [288] 8.15 [294]	607 [288] 623 [294]	180 [294]	0.75 [294] 0.7 [32]
TlTe	Tetrahedral, D^{18}_{4h}	1.294 (a); 0.6158(c) [150]	8.42 [150]	573 [288] 603 [150]		

Compound	$\partial E_g/\partial T$, 10^{-9} eV/Pa	m_n/m_0	m_p/m_0	μ_p, cm^2/(V·s)	μ_p, cm^2/(V·s)	ϵ_0 [294]	ϵ_∞ [294]
GaS	-7.2 [288]	5 [288]		12 [294]	80 [294]	5.9 (**E** $\parallel c$) 10.0 (**E** $\perp c$)	5.3 ($E \parallel c$) 6.7 (**E** $\perp c$)
GaSe	-4 [10]	0.51 [288] 0.5 ($\perp$)[294] 1.6 ($\parallel$) [294]	1.34 [288] 0.8 ($\perp$)[294] 0.2 ($\parallel$) [294]	80 ($\parallel c$) [294] 300 ($\perp c$)	210 (**E** $\parallel c$) [294] 60 ($\perp c$)	6.18 (**E** $\parallel c$) 10.6 (**E** $\perp c$)	5.76 (**E** $\parallel c$) 7.44 (**E** $\perp c$)
GaTe	-(4–5) [294]		1.34 [288]	50 [31] $\mu_n/\mu_p \simeq 10$ [288]	15 [152] 40 [150]	10.58(**E** $\parallel c$) 9.66 (**E** $\perp c$)	7.29 (**E** $\parallel c$) 6.97 (**E** $\perp c$)
InS	-6 [10] -7.9 [294]	0.4 [294]		50 [294]			15.2 [72]
InSe	-4.2 [288]	0.12–0.16 [294]	0.5 ($\perp$) [294] 1.5 ($\parallel c$)	900 [294]	$1.4{\cdot}10^7 T^{-2/3}$ ($\perp c$) [294]	5.4 (**E** $\parallel c$) 8.6 (**E** $\perp c$)	4.9 (**E** $\parallel c$) 6.2 (**E** $\perp c$)
InTe		0.09 [294]	0.26 [294]	$\mu_n/\mu_p = 0.75$ [294]	150 [294]		14.7 (**E** $\parallel c$) 14 (**E** $\perp c$)
TlS		0.07 [294]	0.11 (m_{dp}) [294]	$\mu_n/\mu_p = 0.4$ [294]	20 [294]		
TlSe	-4.5 [288] 3.9 [32]	0.3 (m_{dn}) [32]	0.6 (m_{dp}) [32]	$\mu_n/\mu_p \simeq 0.4$ [288]	15–150 [288]	15–20(**E** $\parallel c$) 34–44(**E** $\perp c$)	12
TlTe		0.03 (m_{dn}) [294]	0.5 (m_{dp}) [294]		60 [294] 1120 (4.2 K)		

Table 22.18 Semiconducting $A_2^{III}B_3^{VI}$ compounds

Compound	Crystal structure		ρ, g/cm^3	T_{melt}, K	E_g, eV	$\partial E_g/\partial T$,	μ_n,
	System, group	a, b, c, nm				10^{-4} eV/K	cm^2/(V·s)
α-Al_2O_3	Hexagonal, C_6^2	2.3 2.5[288]	2.32 [242]	1373 [242]	4.1 [242]	-11.5 [242]	
α-Al_2S_3	Hexagonal, D_{3h}^6	0.4758 (a); 1.295 (c) [242]		2323 [242]	2.5 [242]		
α-Al_2Se_3	Monoclinic, C_3^4	1.168(a); 0.673(b); 0.733 (c); $\beta = 121.1°$[150]	3.91 [150]	1253	3.1 [150]	-11.2 [10]	
Al_2Te_3	Hexagonal, C_{6v}^4	0.408 (a); 0.694 (c) [150]	4.5 [150]	1168 [150]	2.2 [150] 2.3 [123]		
β-Ga_2S_3	Hexagonal, C_{6v}^4	0.3685(a); 0.6028 (c) [288]	3.65–3.74 [150]	1398 [150]	2.84 [123]	- 7 [58]	28 [150]
α-Ga_2Se_3	Monoclinic, C_3^4	1.114 (a); 0.641(b); 0.704 (c); $\beta = 121.2°$ [150]	4.92 [150]	1293 [150]	1.75 [150]	- 5 [16]	10 [16]
α-Ga_2Te_3*1	Cubic, T_d^2	0.589 (a) [150]	5.57 [150]	1065 [150]	1.56 [31]	- 6 [242]	340 [123]
In_2O_3*2	Cubic	1.012 (a) [132]	7.04 [132]	2270 [132]	2.8 [132]	- 8[132]	270 [60]
β-In_2S_3	Cubic, O_h^5	1.072 (a) [150]	5.92 [150]	1363 [150]	1.1 [150]	- 7 [10]	< 100 [150]
α-In_2Se_3	Hexagonal	1.60 (a); 1.924 (c) [150]	5.67 [150]	1173 [150]	1.2 [10]	- 4 [10]	125 [150]
α-In_2Te_3*3	Cubic, T_d^2	1.840 (a) [121]	5.8 [121]	940 [150]	1.02 [150] 1.12 [16]	- 4[242]	10 [150] 15–70 [16]

*1 μ_p=26 cm^2/(V·s); $m_n/m_0 = 0.39$; $m_p/m_0 = 0.23$ [123].
*2 $m_{dn}/m_0 = 0.55$ [60].
*3 $m_n/m_0 = 0.7$; $m_p/m_0 = 1.1$ [150].

Table 22.19 Electrophysical properties of thallium halides [294]

Compound	Crystal structure		ρ, g/cm^3	T_{melt}, K	T_D, K	E_g, eV	m_n/m_0^*	m_p/m_0^*	μ_n, cm^2/(V·s)	μ_p, cm^2/(V·s)	ϵ_0 (ϵ_∞)
	System, group	a, nm									
TlCl	Cubic, O_h^5	0.384	7.02	704	393	3.2	0.5–0.7	0.6–1	20 5000 (4.2 K)	67000 (4.8 K)	32.7 (4.76)
TlBr	Cubic, O_h^5	0.399	7.45	733	290	2.64 (4.2 K)	0.5	0.7	30 40000 (1.8 K)	4 35000 (1.8 K)	30.6 (5.34)

* Polaron mass.

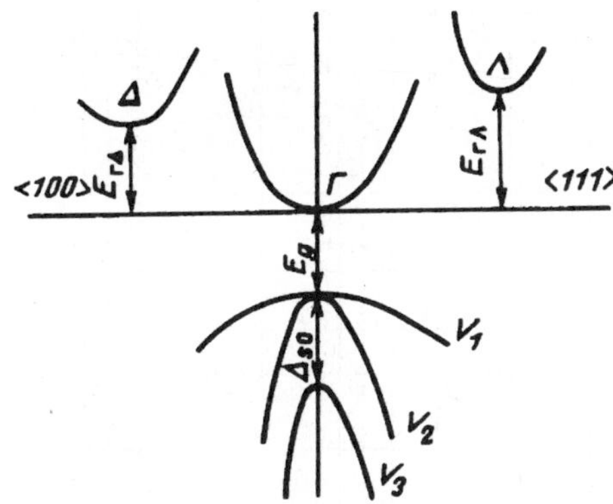

Figure 22.123 The band structure of GaN, GaAs, GaSb, InN, InP, InAs, and InSb [121, 170, 171].

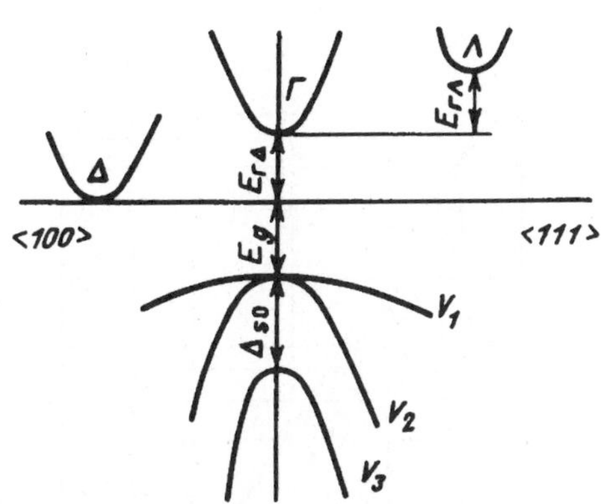

Figure 22.124 The band structure of BP, AlP, AlAs, and AlSb [121, 170-172].

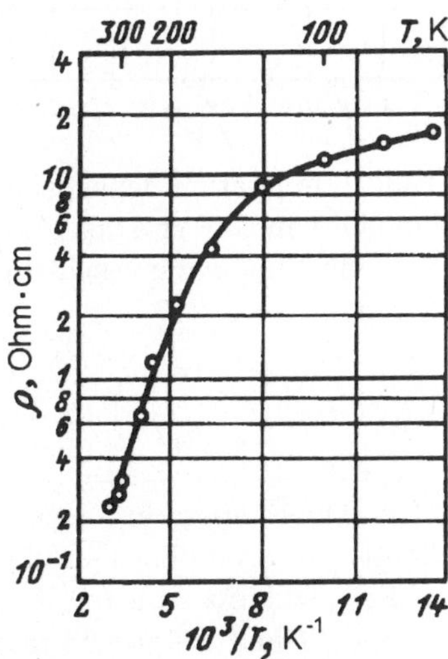

Figure 22.125 The temperature dependence of the resistivity in an n-BP single crystal [247].

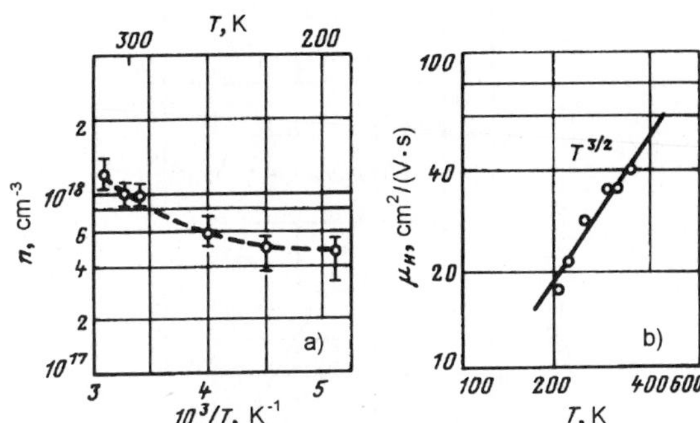

Figure 22.126 Temperature dependences of the density of electrons (a) and their Hall mobility (b) in a BP single crystal [247].

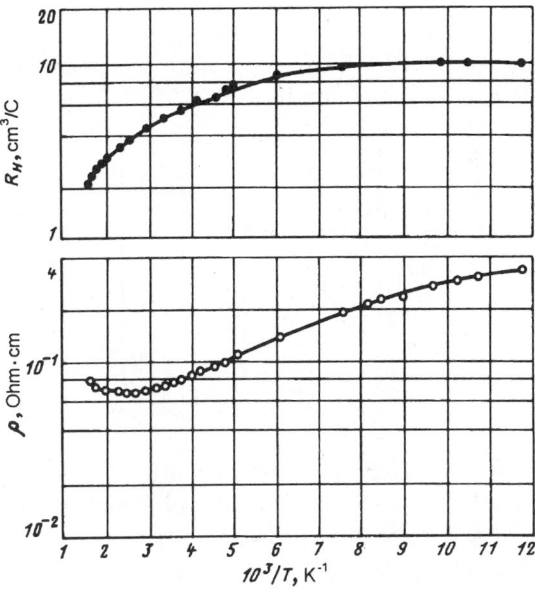

Figure 22.127 Temperature dependences of the Hall coefficient and the resistivity in n-AlP [248].

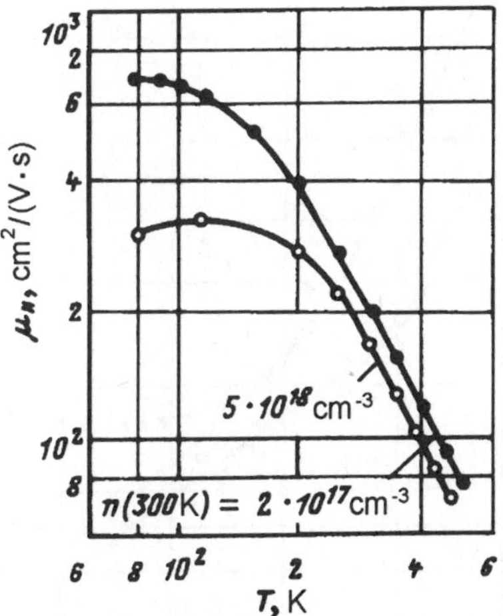

Figure 22.128 The temperature dependence of the electron Hall mobility in AlSb [250].

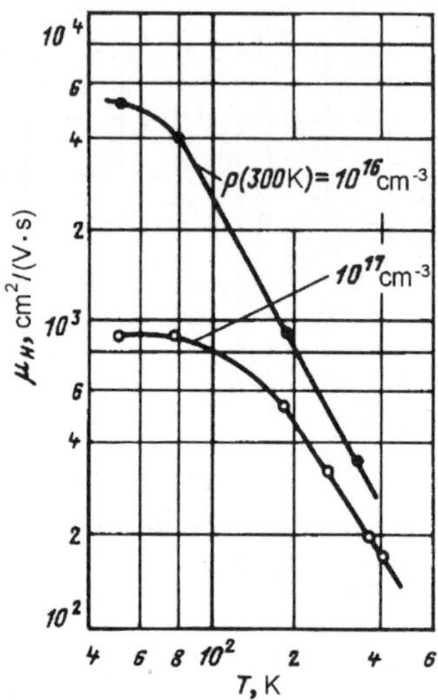

Figure 22.129 The temperature dependence of the hole Hall mobility in AlSb [251].

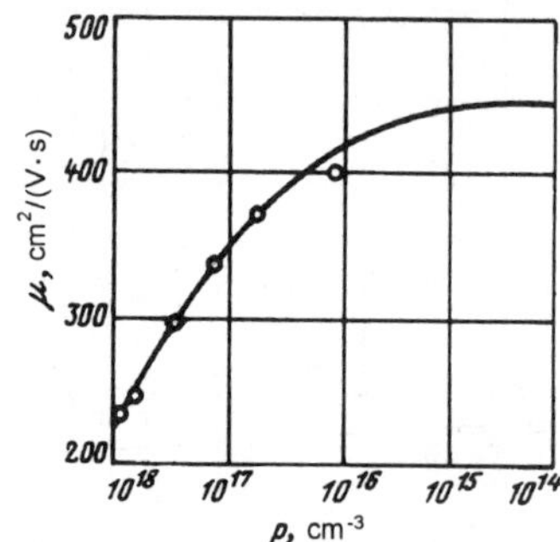

Figure 22.130 The hole mobility vs. the hole density in AlSb at 293 K [170]: solid line is a calculation for the combined scattering by acoustical phonons and ionized impurities.

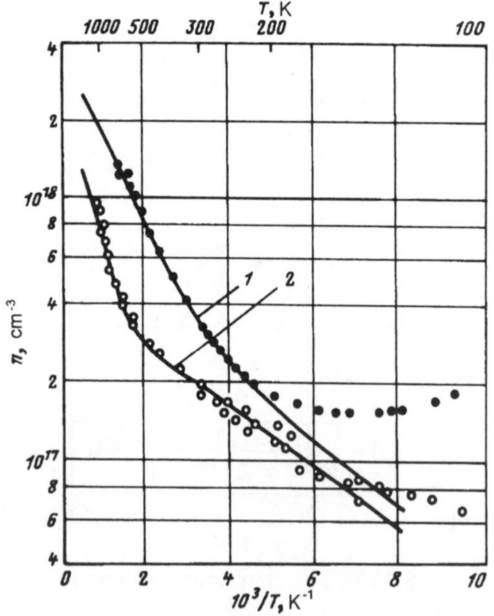

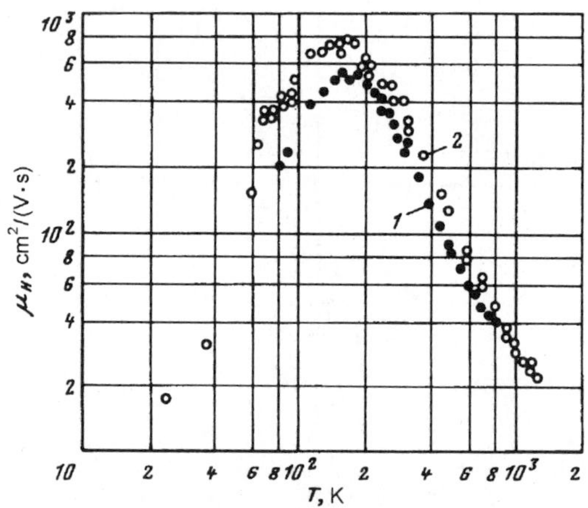

Figure 22.132 The temperature dependence of the electron Hall mobility for two GaN crystals [253]; the same crystals as in Fig. 22.131.

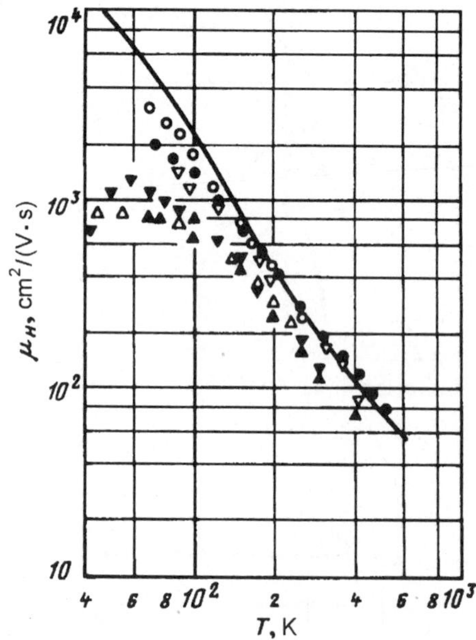

Figure 22.133 Temperature dependences of the electron Hall mobility in GaP crystals [256]: points are the data of different works; solid line fits a calculation.

Figure 22.131 The temperature dependence of the electron density for crystalline GaN films [253]: solid lines fit the calculations by the model with donors of two types; *1* and *2* – growth rate of 0.75 and 0.45 μm/min, respectively.

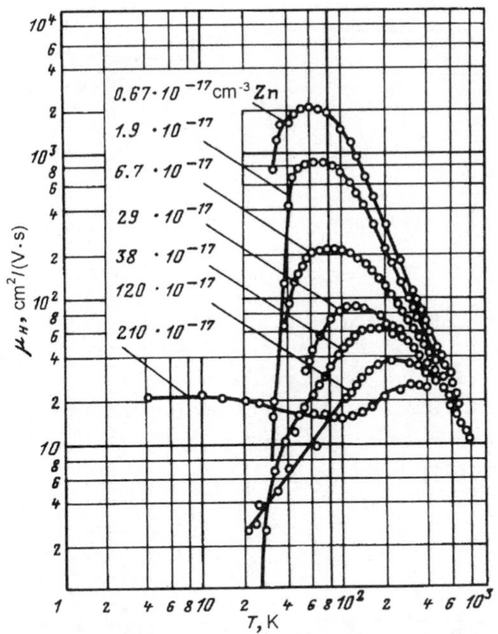

Figure 22.134 The temperature dependence of the hole Hall mobility in the Zn-doped GaP crystals with various Zn concentrations [256].

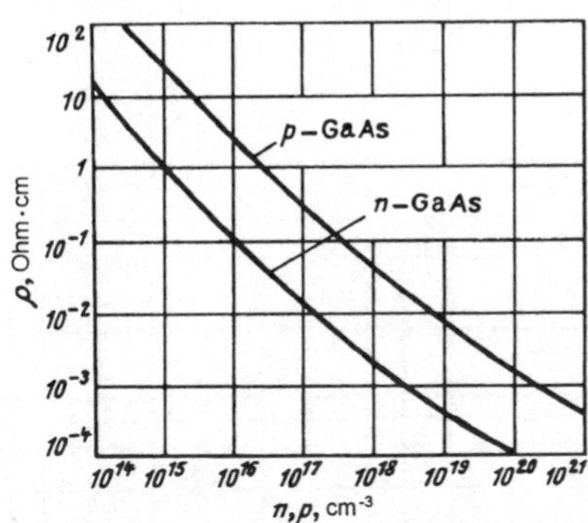

Figure 22.135 The resistivity of n- and p-GaAs at 300 K vs. the electron and hole densities [257].

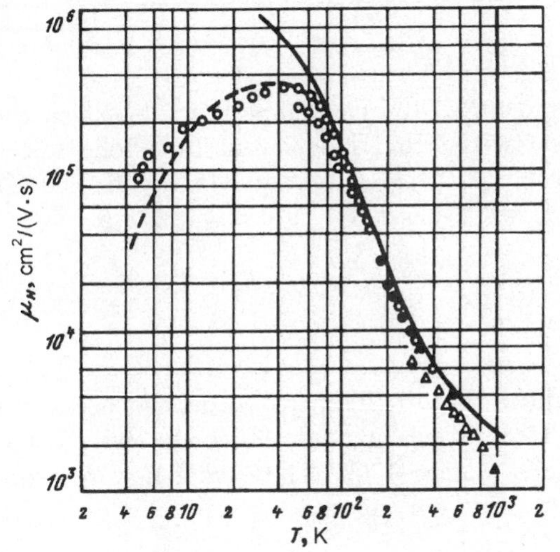

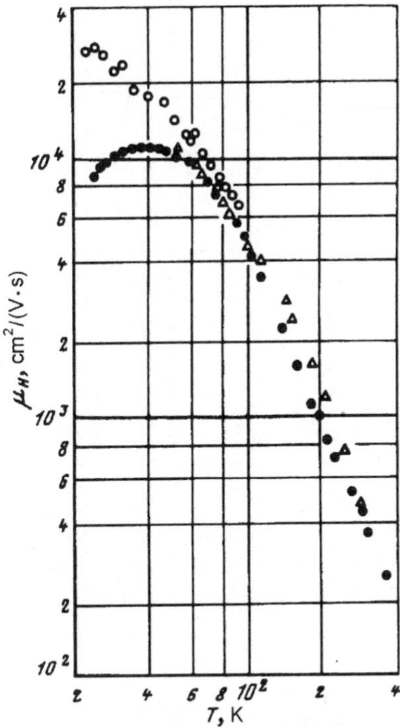

Figure 22.137 The temperature dependence of the hole Hall mobility in GaAs [259]. Acceptor and donor densities N_A and N_D, 10^{14}cm^{-3}: ○ – 3.3 and 1.8; △ – 11 and 3; ● – 7 and 5.5.

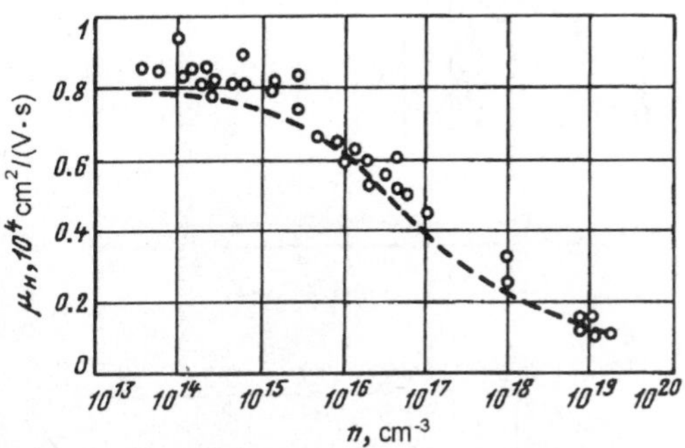

Figure 22.138 The Hall mobility of electrons at 300 K vs. the electron density in Sn-doped GaAs crystals. Dashed line fits a calculation [260].

Figure 22.136 The temperature dependence of the electron Hall mobility in GaAs. Lines are calculations without (solid) and with (dashed) allowance for the scattering of electrons by ionized impurities; points are the data of different works [258].

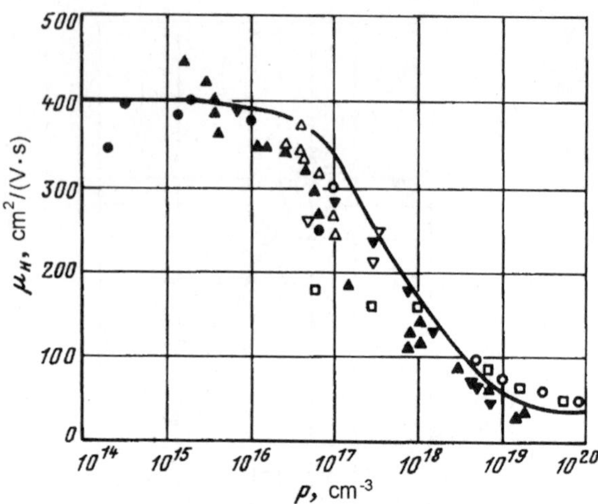

Figure 22.139 The Hall mobility of holes at 300 K vs. the hole density in GaAs. Solid line is a calculation [259]; points are the data of different works.

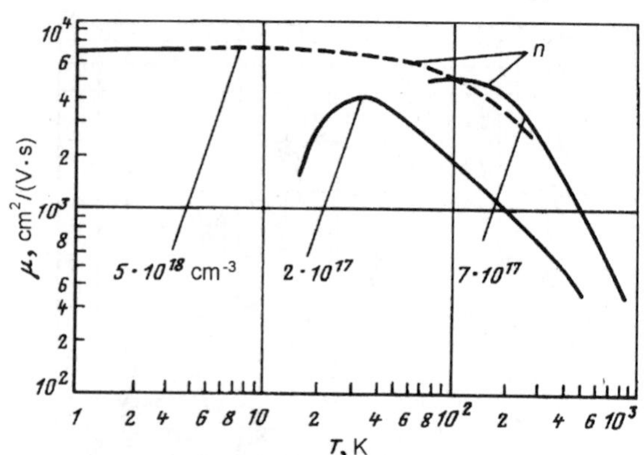

Figure 22.141 Temperature dependences of the carrier mobility in GaSb of the n- and p- (lower curve) types [127].

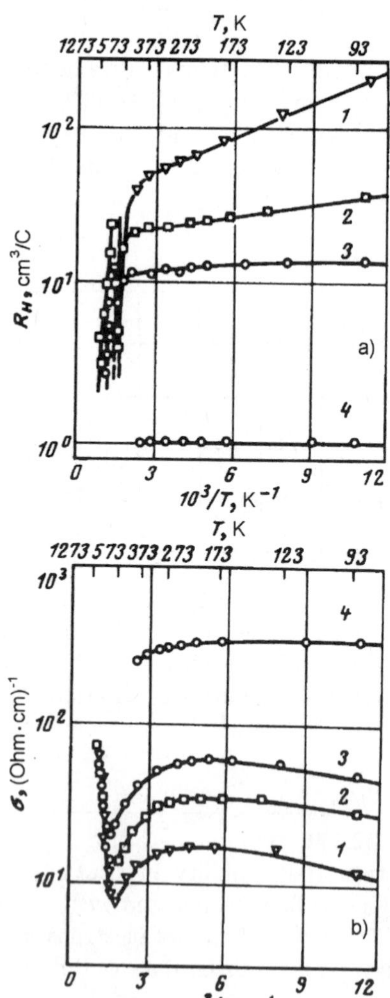

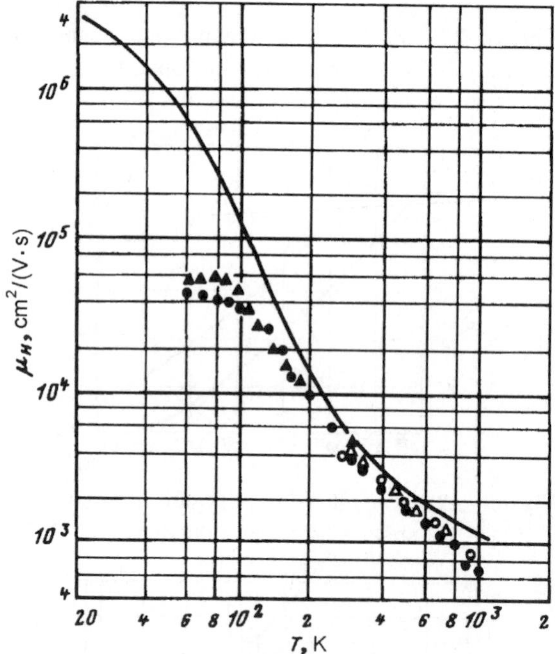

Figure 22.142 The temperature dependence of the electron Hall mobility in InP. Solid line is a calculation [261]; electron density n (300 K), 10^{15} cm^{-3}: ● – 2; ▲ – 1.7; ○ – 4; △ – 6.

Figure 22.140 Temperature dependences of the Hall coefficient (a) and the conductivity (b) in GaSb crystals [170]. The carrier density grows from curve *1* to curve *4*.

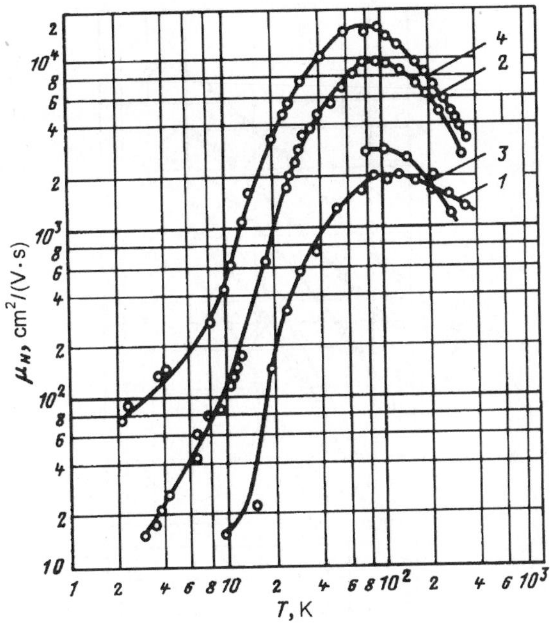

Figure 22.143 The temperature dependence of the electron Hall mobility in InP [262]. Electron density n (300 K), 10^{15}cm^{-3}: *1* – 2 (Cu); *2* – 5 (Fe); *3* – 5 (Te, Cu); *4* – 10 (Co).

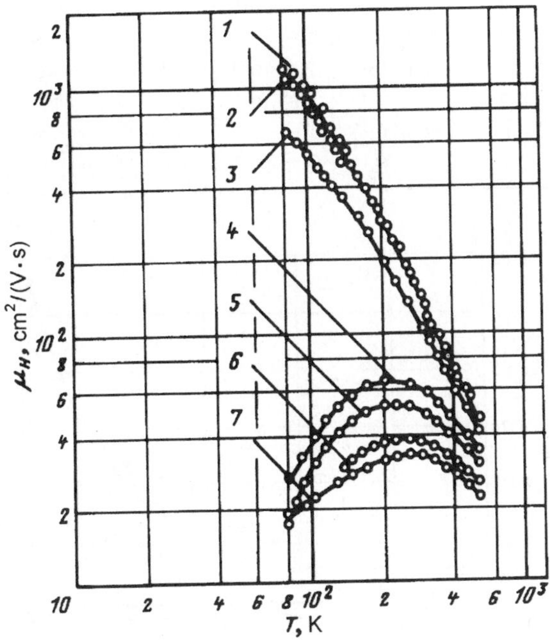

Figure 22.144 The temperature dependence of the hole Hall mobility in INP [263]. p(300 K), 10^{17} cm^{-3}: *1* – 0.7; *2* – 1; *3* – 3; *4* – 13; *5* – 30; *6* – 56; *7* – 85.

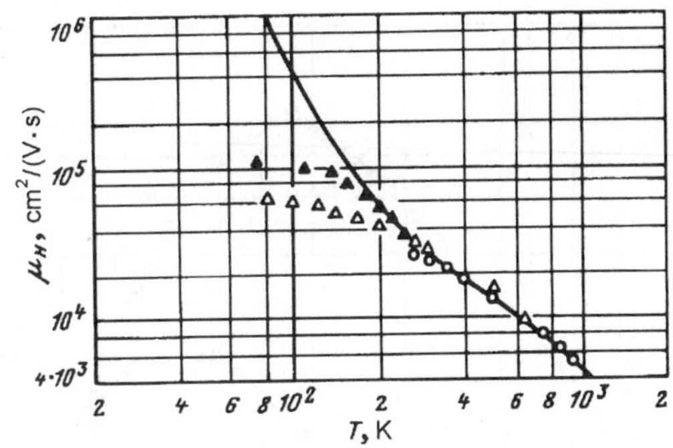

Figure 22.145 Temperature dependences of the Hall mobility of electrons in InAs: solid line is a calculation [255]; electron density n (300 K), 10^{16} cm^{-3}: △ – 1.7; ○ – 4; ▲ – 0.4.

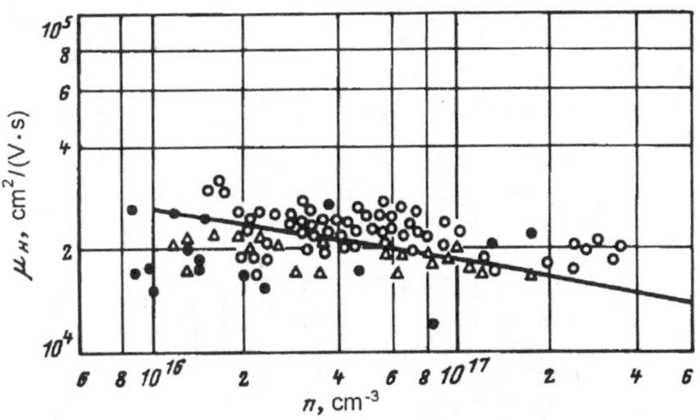

Figure 22.146 The Hall mobility of electrons at 300 K vs. the electron density in InAs crystals with various degrees of compensation $K = N_A/K_D$ [247]. Solid line is a calculation; ○ – $K < 0.15$; △ – $0.15 < K < 0.3$; • – $K > 0.3$.

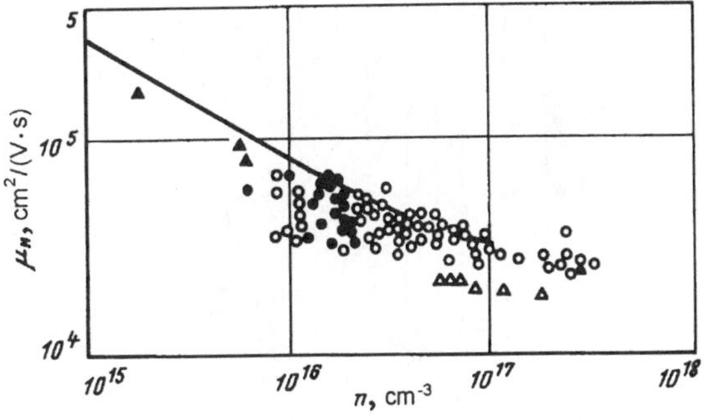

Figure 22.147 The Hall mobility of electrons at 77 K vs. the electron density in InAs crystals [247]. Solid line is a calculation; • – undoped sample; ○ – Cu - doped sample; △ – Sn-Zn-doped sample; ▲ – epitaxial films.

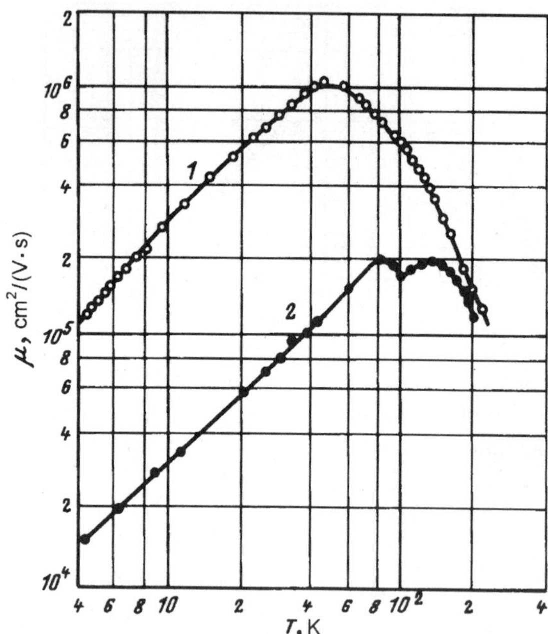

Figure 22.148 Temperature dependences of the electron mobility in pure (*1*) and doped (*2*) (impurity concentration $> 10^{14}\text{cm}^{-3}$) InSb crystals [265].

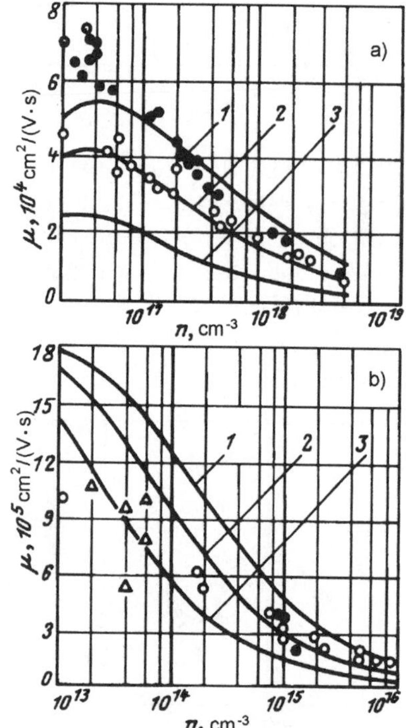

Figure 22.149 The electron mobility in InSb at 300 K (a) and 77 K (b) vs. the electron density [267]. Solid lines are the calculated dependences of the drift mobility at $(N_D^+ + N_A^-)/(n+p) = 1(\mathit{1})$; $2(\mathit{2})$ and $5(\mathit{3})$; points are the data from different works.

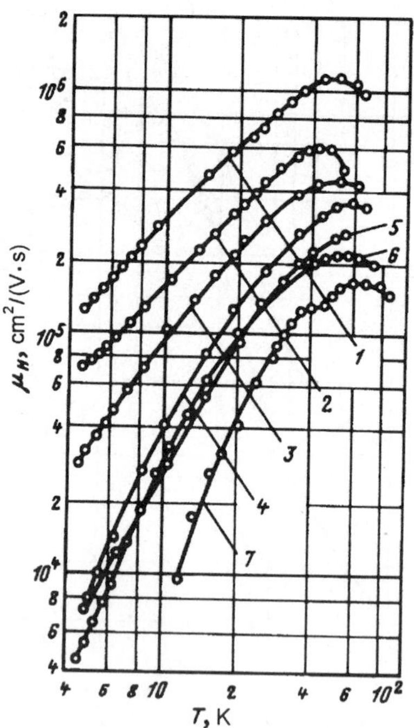

Figure 22.150 Temperature dependences of the electron mobility in InSb crystals with various impurity concentrations [266]:

No. of sample	$N_D - N_A$, cm^{-3}	$N_D + N_A$, cm^{-3}	N_A/N_D	$\mu_n(30\text{ K})$, $\text{cm}^2/(\text{V·s})$
1	$1.2 \cdot 10^{14}$	$6.5 \cdot 10^{14}$	0.50	830000
2	$1.9 \cdot 10^{14}$	$1.3 \cdot 10^{15}$	0.75	480000
3	$9.8 \cdot 10^{13}$	$1.6 \cdot 10^{15}$	0.88	330000
4	$1.4 \cdot 10^{13}$	$2.3 \cdot 10^{15}$	0.99	160000
5	$1.6 \cdot 10^{13}$	$2.6 \cdot 10^{15}$	0.99	150000
6	$3.6 \cdot 10^{12}$	$1.9 \cdot 10^{15}$	0.98	150000
7	$6.2 \cdot 10^{11}$	$2.7 \cdot 10^{15}$	0.99	92000

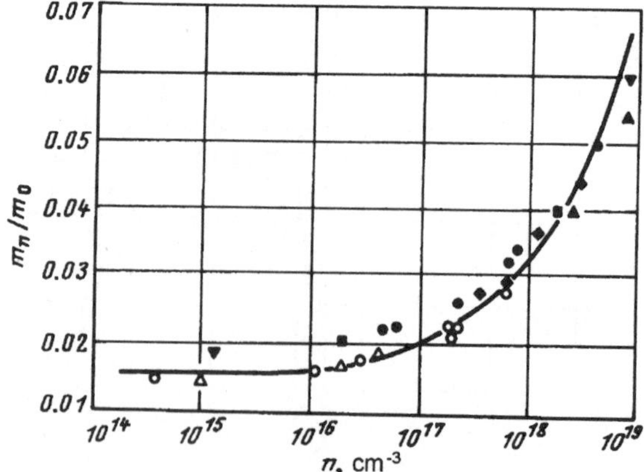

Figure 22.151 The effective mass of electrons in *n*-InSb vs. the electron density [264]. Solid line fits a calculation by the Kane theory; points are the data from different works.

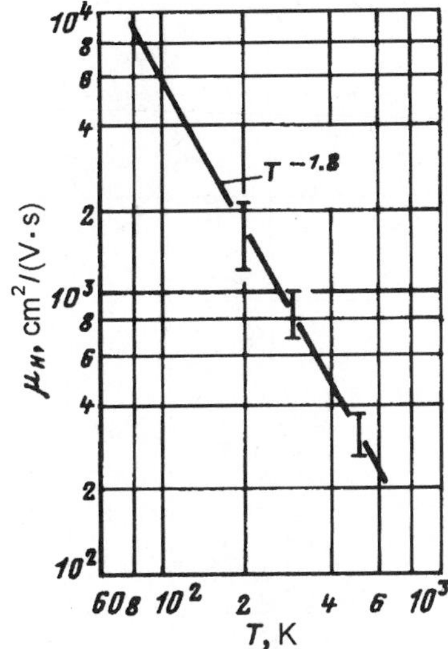

Figure 22.152 The temperature dependence of the hole Hall mobility in InSb [268].

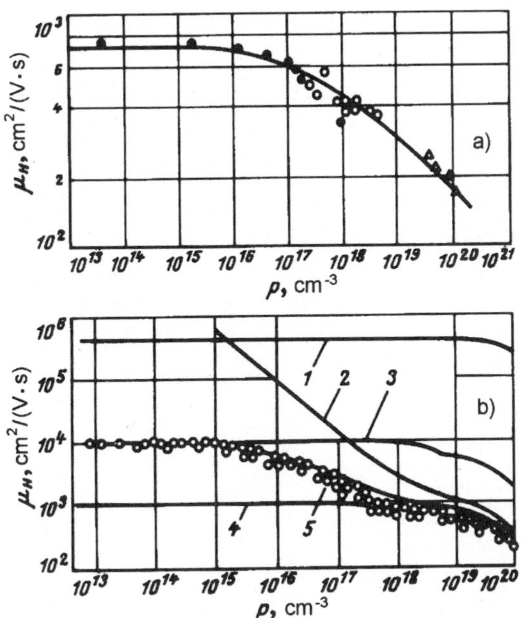

Figure 22.153 The hole mobility in InSb at 290 K (a) and 77 K (b) vs. the hole density [268–270]. Solid lines (b) fit the calculations with allowance for the hole scattering by optical phonons (*1*), impurity ions (*2*), acoustical phonons (*3* – E_1 = 7 eV; *4* – E_1 = 21 eV), impurities and acoustical phonons (*5*); points are the data of different works.

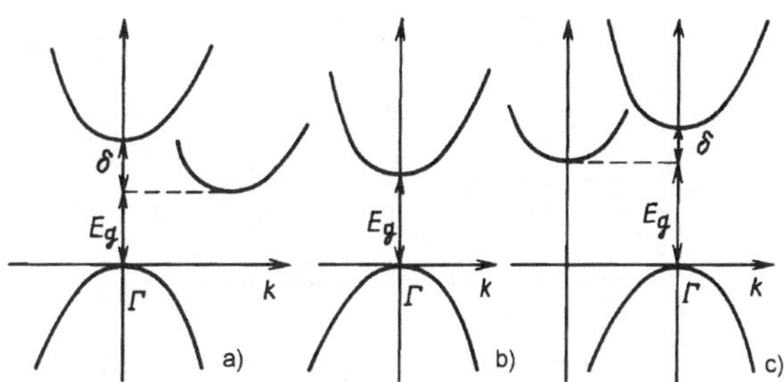

Figure 22.154 The band structure of GaS, GaSe, InS, and TlS (a), GaTe (b), and InSe (c) [294]:

Compound	GaS	GaSe	GaTe	InS	InSe	TlS
δ, eV		0.025		0.55	0.076	0.2

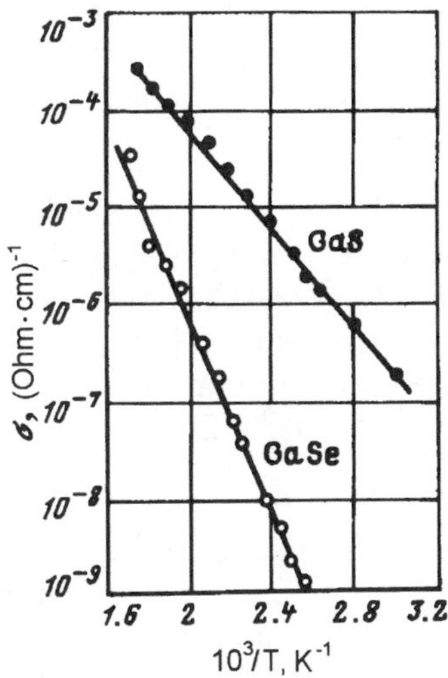

Figure 22.155 Temperature dependences of the conductivity of n-GaS and n-GaSe crystals in the layer plane [271].

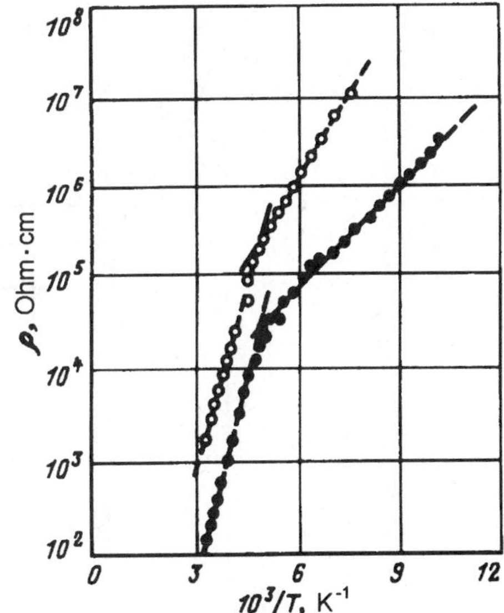

Figure 22.156 Temperature dependences of the resistivity in a GaS crystal along (•) and across (○) the layers [274].

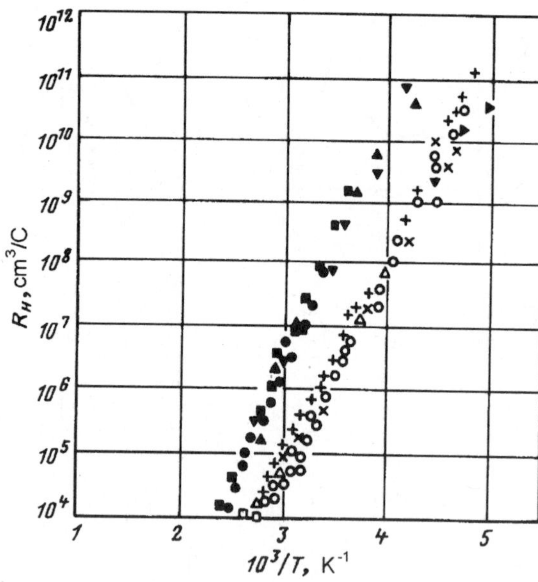

Figure 22.157 Temperature dependences of the Hall coefficient for various *p*-GaSe crystals. The current is directed along the layers, the magnetic field – across the layers [275]: ×, + , ○, △, □ – undoped GaSe crystals; dopant: ■, ● – 10% Ge; ▲, ▼– 5% Sn; ▹ – 10^{-3}% Zn.

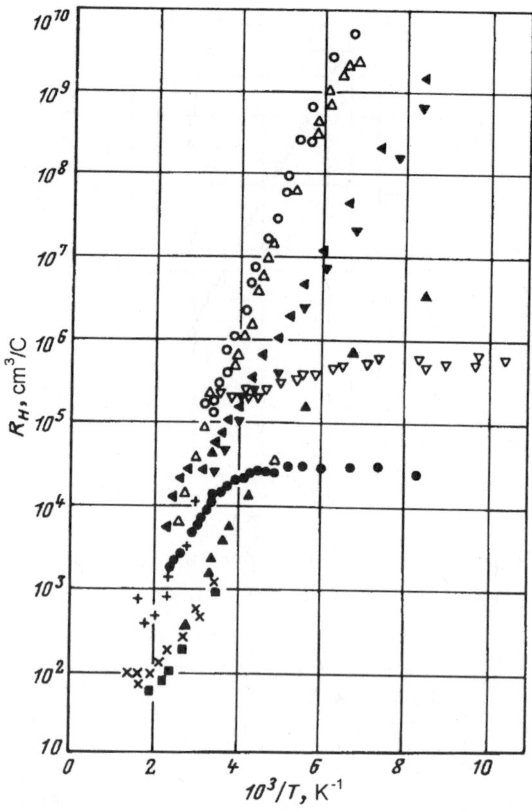

Figure 22.159 Temperature dependences of the Hall coefficient for various *p*-GaSe crystals. The current is directed along the layers, the magnetic field – across the layers [275]: ○, △, ▽ – undoped GaSe crystals; dopant: ●, ◃ – 10% Zn; ▼– 0.5% Zn; ▲ – 0.1% Zn; ■, × – 10^{-2}% Zn; + – $3{\cdot}10^{-3}$% Zn.

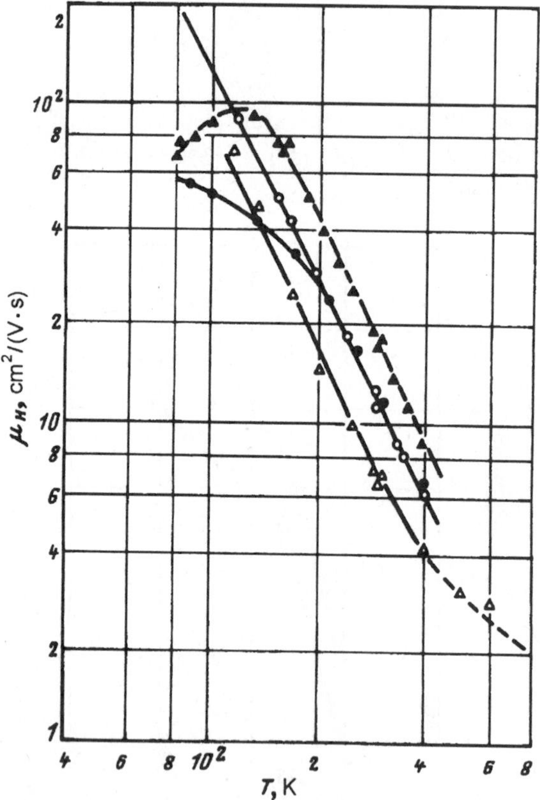

Figure 22.158 The temperature dependence of the carrier Hall mobility in GaS crystals. The current is directed along the layers, the magnetic field - across the layers. The method of production and the type of the crystals: ▲ – iodine transport, *n*-type; ● – gas-phase deposition, *n*-type; △, ○ – gas-phase deposition, *p*-type [272, 273].

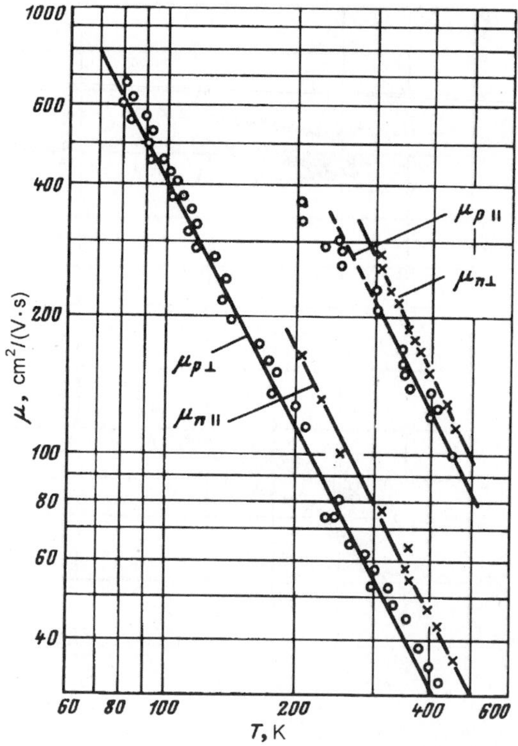

Figure 22.160 The temperature dependences of the Hall mobility of carriers in GaSe [276]: $\parallel$ and $\perp$ indicate mobilities along and across the layers, respectively.

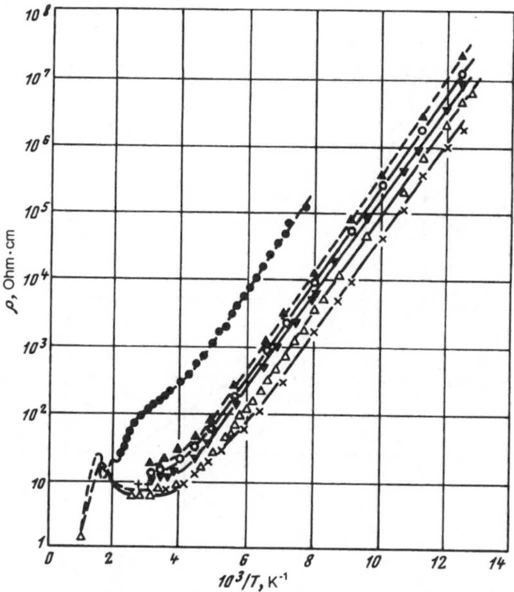

Figure 22.161 Temperature dependences of the resistivities for various GaTe crystals [277]: full symbols – across the layers; other symbols – along the layers; crystal treatment: ●, △ – etching; ▼, + – mild etching; ▲, ○ – cold processing; × – Cu doped.

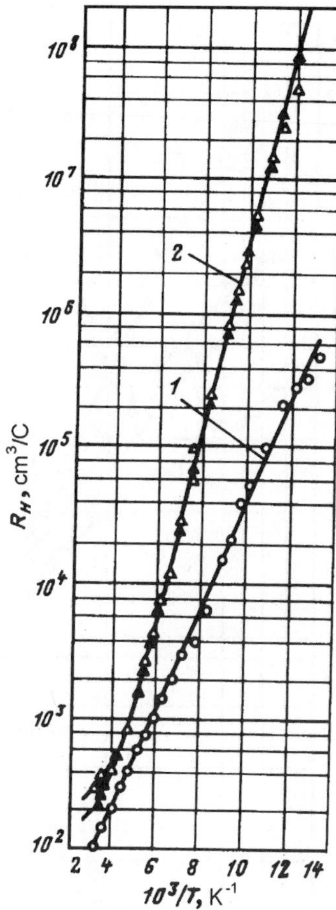

Figure 22.162 Temperature dependences of the Hall coefficient for various GaTe crystals. The current is directed along the layers, the magnetic field – across the layers [278]: samples *1* and *2* are from different batches.

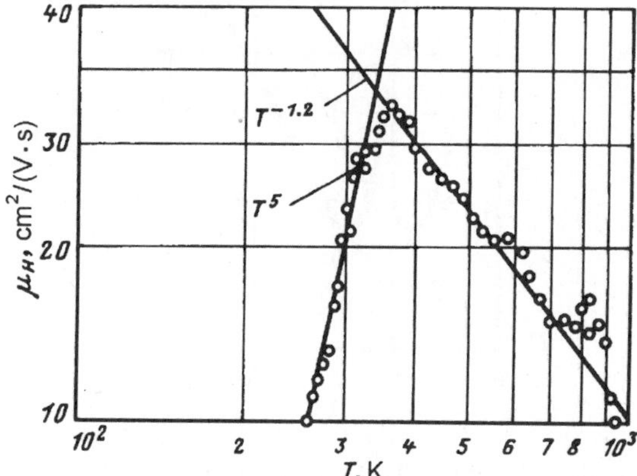

Figure 22.163 The temperature dependence of the electron mobility in polycrystalline GaTe samples consisting of large single-crystal regions. The current is directed along the layers [279].

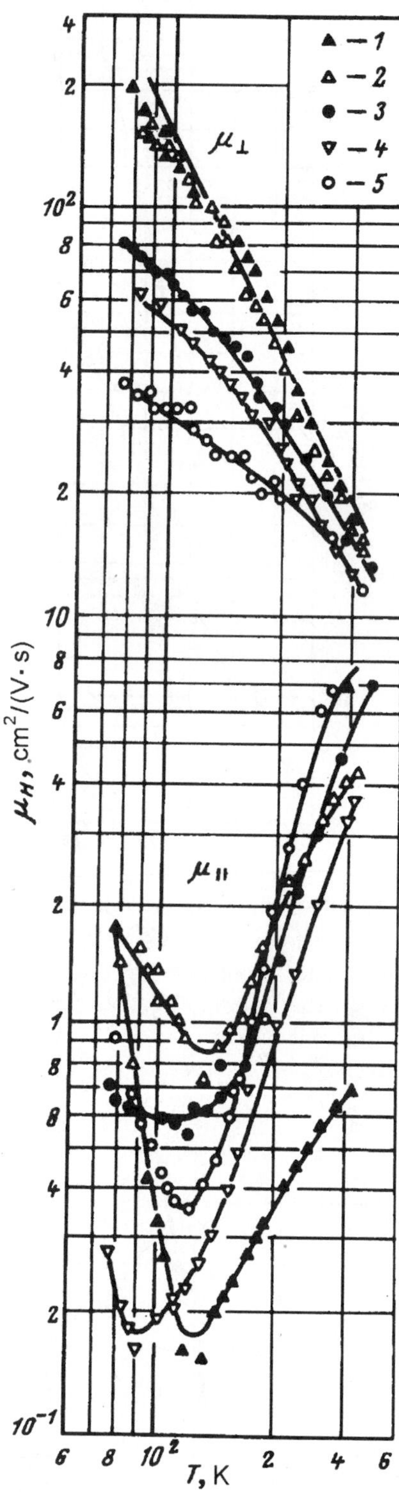

Figure 22.164 The temperature dependence of the hole Hall mobility in p-GaTe crystals [278]: $\perp$ and $\|$ indicate mobilities along and across the layers, respectively; the carrier density grows from sample 1 to sample 5.

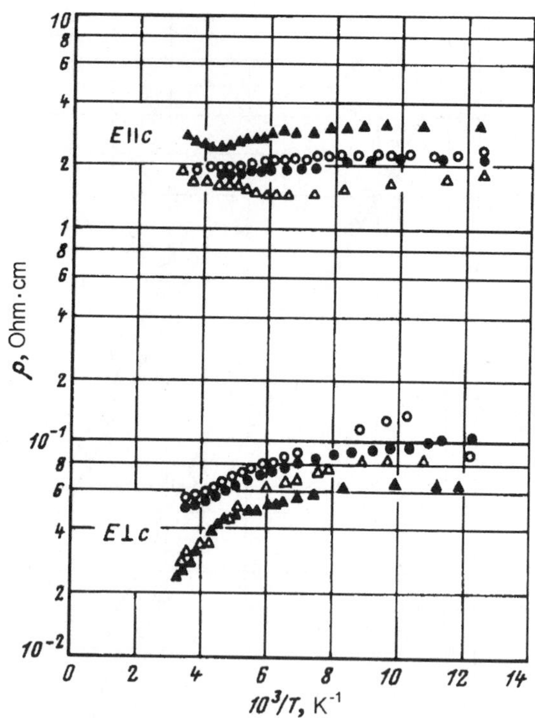

Figure 22.165 The temperature dependence of the resistivity for melt-grown InS crystals [280]: electron density n(300 K), 10^{18}cm^{-3}: ▲ – 9; △ – 6.5; ○ – 5.5; • – 4.5.

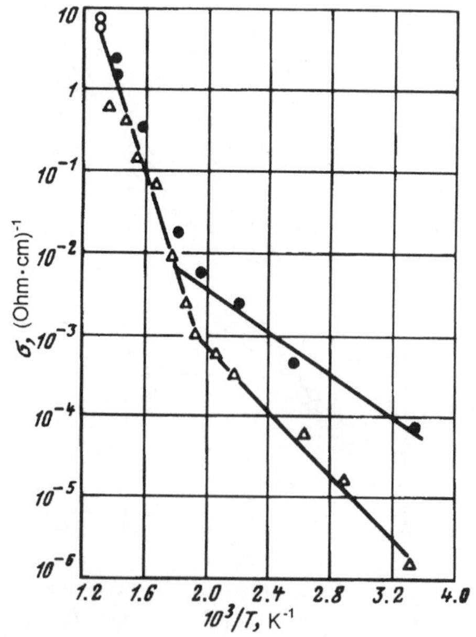

Figure 22.166 The temperature dependence of the conductivity for various InSe crystals along (△) and across (•) the c axis [281].

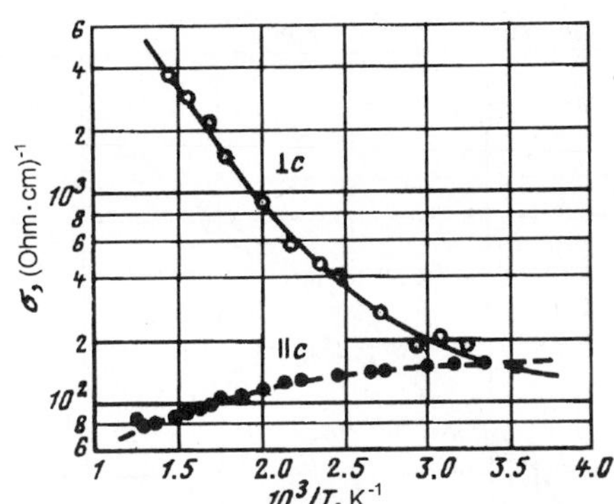

Figure 22.167 The temperature dependence of the conductivity for InTe along (||) and across (⊥) the c axis [282].

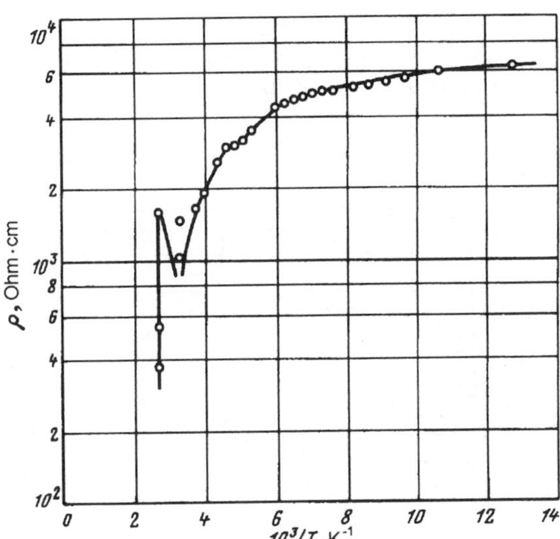

Figure 22.168 The temperature dependence of the resistivity for various TlS crystals (the current flows along the c axis) [283].

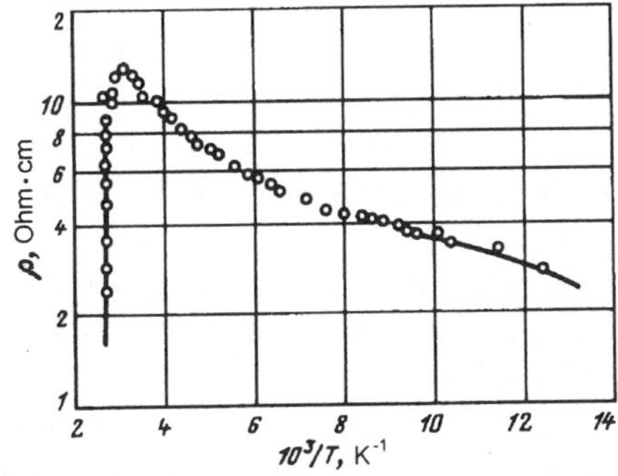

Figure 22.169 The temperature dependence of the resistivity for TlSe (the current flows along the c axis) [283].

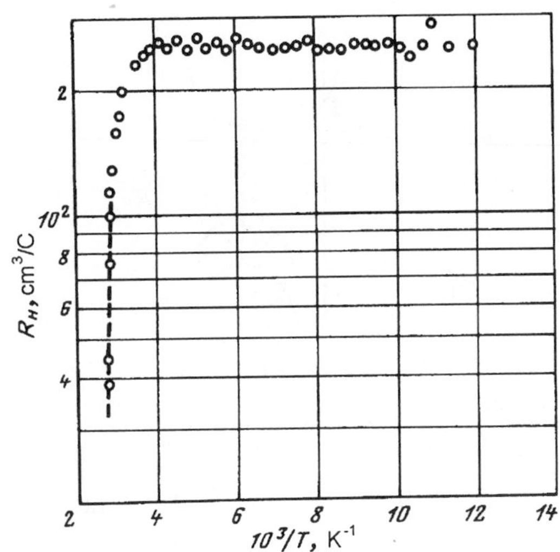

Figure 22.170 Temperature dependences of the Hall coefficient for various TlSe crystals in a field $B = 13$ kG (the current flows along the c axis; the magnetic field is in the [110] direction and transverse to the c axis) [283].

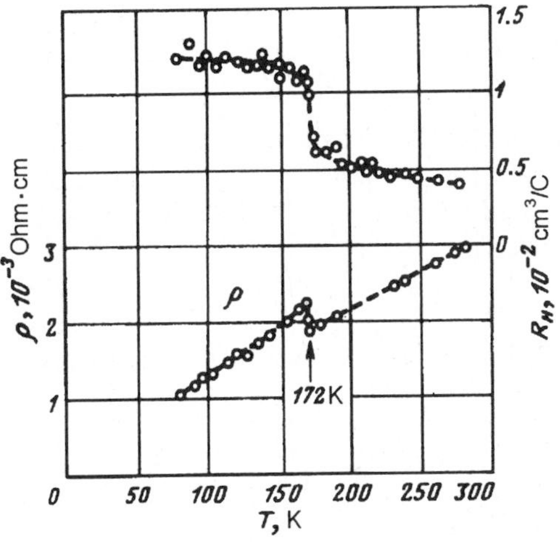

Figure 22.171 Temperature dependences of the resistivity and the Hall coefficient for a polycrystalline TlTe sample [284]. The tetragonal (above 172 K) – rhombic structure phase transition occurs at 172 K.

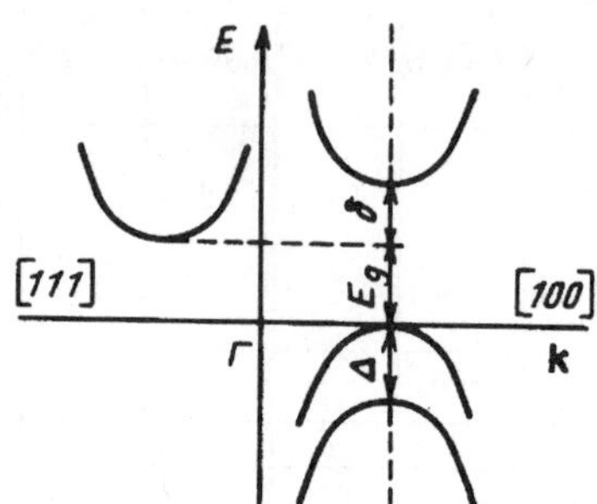

Figure 22.172 The band structure of TlCl and TlBr [294]:

Compound	TlCl	TlBr
Δ, eV	1.7	1.04
δ, eV	0.2	0.27

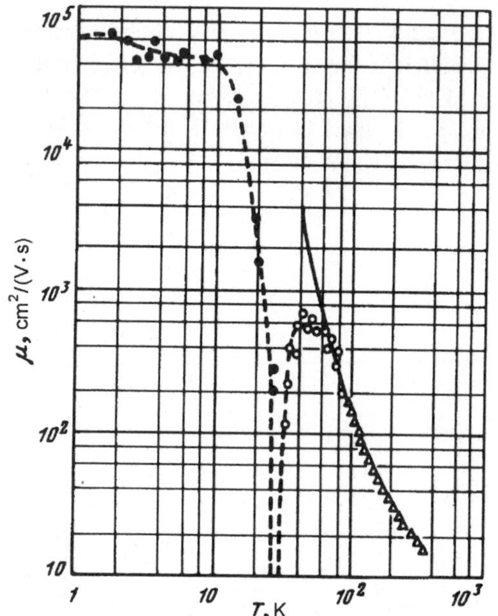

Figure 22.173 The temperature dependence of the Hall (•, ○) and drift (△) mobilities of photoexcited carriers in TlCl [285]: • – holes; ○, △ – electrons.

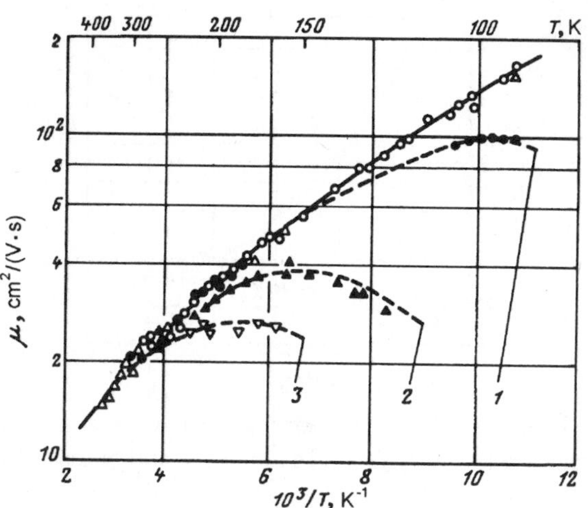

Figure 22.174 The temperature dependence of the electron drift mobility in TlCl crystals with various trap concentrations (cm^{-3}): *1* – $8.4{\cdot}10^{14}$; *2* – $1.2{\cdot}10^{16}$; *3* – $3.3{\cdot}10^{16}$; solid line corresponds to an undoped crystal [286].

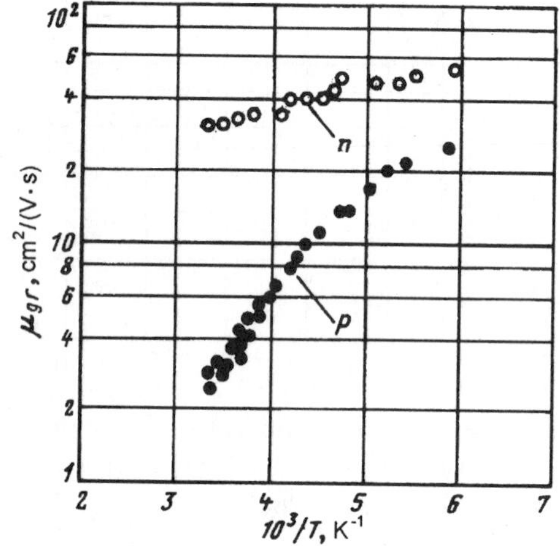

Figure 22.175 Temperature dependences of the drift electron and hole mobilities in TlBr [287].

22.3.4 $A^{IV}B^{IV}$ – $A^{IV}B^{VI}$ compounds

Amongst the semiconducting $A^{IV}B^{V}$ compounds, Sn_3As_2 is known to crystallize into a rhombic lattice (T_{melt} = 870 K, E_g = 0.47 eV, μ_p = 270 cm^2/(V·s) [100]).

The best-studied $A^{IV}B^{VI}$ compounds represent lead chalcogenides (PbS, PbSe, PbTe) which crystallize as face-centered O_h^5 cubes. Their band structure has a direct gap, the absolute extrema of the bands being positioned at the boundary of the Brillouin zone in the [111] direction (see Figure 22.178). In the vicinity of the extrema, the isoenergetic surfaces constitute ellipsoids of revolution (their equivalent number is four for each band). The valence band is split into two subbands; the upper one (the heavy-hole subband) has a maximum inside the Brillouin zone on the [111] axes and manifests itself in the *p*-type materials at elevated temperatures (in PbTe, at $T \geq 400$ K). Lead chalcogenides exhibit an anomalously high dielectric constant.

Some compounds of the type considered have an indirect-gap band structure, as, for example, in SnS_2 and $SnSe_2$ (see Figure 22.213), or a complex band structure (see Figure 22.208 for SnTe). The conduction type of the compounds under consideration is often determined by a deviation from stoichiometry.

Table 22.20 Electrophysical properties of SiC

Compound	Crystal structure: System, group	Crystal structure: a, c, nm	T_{melt}, K	ρ, g/cm^3	T_D, K	$\hbar\omega_\ell$, eV	$\hbar\omega_t$, eV	ϵ_0
α-SiC (2H–SiC)	Hexagonal, C_{6v}^4	0.308 (a); 1.511 (c) [292]	3073 [132]	3.21 [193]	1200 [193]	0.119 [194]	0.098 [194]	10.2 [124] 9.8 [194]
β-SiC (3C–SiC)	Cubic, T_d^2	0.436 (a) [1]	2600 [124] ($\beta \to \alpha$)	3.21 [132]	1080 [246]	0.103 [292]	0.094 [292]	9.7[288]

Compound	ϵ_∞	E_g, eV	$\partial E_g/\partial T$, 10^{-4}eV/K	m_n/m_0	m_p/m_0	μ_n, cm^2/(V·s)	μ_p, cm^2/(V·s)
α-SiC (2H–SiC)	6.9 [124] 6.73 [194]	3.33 [246]	−3.8 [85]	0.25 (⊥c); 1.5 (∥c) [292]	1.0[123] 1.2 [124]	400 [234]	70 [16]
β-SiC (3C–SiC)	6.5 [288]	2.4 (2K) [292]	−5.8 [15, 79]	0.24 (⊥c); 0.65 (∥c) [292]	0.59 [15]	1000 [288]	20 [213]

Table 22.21 Electrophysical properties of TiO_2, Ti_2O_3, and TiO

Compound	Crystal structure: System, group	Crystal structure: a, c, nm	T_{melt}, K	ρ, g/cm^3	ϵ_0	ϵ_∞	T_D, K	E_g, eV	μ_n, cm^2/(V·s)
TiO_2 (rutile)	Tetragonal, D_{4h}^{14}	0.454(a); 0.296(c)	2113	4.28 [209]	140 (⊥c) 260 (∥c)	7 [211]	670 [242]	3.0 [16]	0.16(⊥c) 0.57(∥c) [25]
					89(⊥c) 173 (∥c) [25]	9(⊥c) 6.8 (∥c) [25]			
Ti_2O_3	Tetragonal, D_{3d}^6	0.516(a); 1.357(c)	2400	4.93–5.13 [288]	–	–	–	0.02 [288]	~1000 (4.2 K)[288]
TiO	Cubic, O_h^5	0.4177(a)	2010	5.82 [288]	–	–	–	0.1 [288]	~1 [288]

Table 22.22 Electrophysical properties of lead oxide and chalcogenides [125, 294]

Compound	Crystal structure: System, group	Crystal structure: a, c, nm	ρ, $\frac{g}{cm^3}$	T_{melt}, K	E_g, eV	δ_n, eV	δ_p, eV	$\partial E_g/\partial T$, 10^{-4} eV/K	$\partial E_g/\partial P$, 10^{-6} eV/bar	$\frac{m_{n\parallel}}{m_0}$	$\frac{m_{n\perp}}{m_0}$	$\frac{m_{p\parallel}}{m_0}$	$\frac{m_{p\perp}}{m_0}$
PbO	Tetragonal, D_{4h}^7	0.397(a) 0.502(c)	9.53	1163	2.07 2.03^{*2}			1					
PbS	Cubic, O_h^5	0.594	7.6	1387	0.41 0.31^{*1} 0.29^{*2}	1.46	1.67	4; 5.2	−(8–9)	0.105	0.080	0.105	0.075
PbSe	Cubic, O_h^5	0.612	8.3	1355	0.278 0.176^{*1} 0.145^{*2}	1.36	1.72	4; 5.1	−(8–9)	0.070	0.040	0.068	0.034
PbTe	Cubic, O_h^5	0.650	8.2	1190	0.32 0.22^{*1} 0.19^{*2}	1.11	0.77	4.5	−(7–8)	0.22	0.024	0.24	0.025

Table 22.22 Electrophysical properties of lead oxide and chalcogenides *(continued)*

Compound	μ_n, $cm^2/(V{\cdot}s)$	μ_p, $cm^2/(V{\cdot}s)$	n_i (300 K), 10^{16} cm^{-3}	ϵ_0	ϵ_∞	$\hbar\omega_\ell$, meV	$\hbar\omega_t$, meV	T_D(200 K), K	g_n*3	g_p*3
PbS	610 11000*1 68500*2	620 15000*1 80000*2	2.0	175	17	26.3	8.2	227	12	13
PbSe	1000 15500*1 139000*2	1000 13700*1 57900*2	3.0	250	24	16.5	5.4	138	27(‖) 19.6(⊥)	32(‖) 17.1(⊥)
PbTe	1730 31600*1 800000*2	840 21600*1 250000*2	1.5	400 1000*1 3000*2	33	13.6	3.9	125	60(‖) 16(⊥)	58(‖) 19(⊥)

*1 $T = 77$ K.
*2 $T = 4.2$ K.
*3 ‖ (⊥) – magnetic field is parallel (perpendicular) to the principal axis of the energy ellipsoid.

Table 22.23 Semiconducting $A^{IV}B^{VI}$ compounds

Compound	Crystal structure		ρ, g/cm^3	T_{melt}, K	T_D, K	$\partial E_g/\partial T$, 10^{-4}eV/K
	System, group	a, c, nm [1]				
Si_2Te_3	Hexagonal, C_{3v}^4	0.743 (a); 1.347 (c) [288]	4.5 [288]	1165 [288]		
GeS	Rhombic, D_{2h}^{16}	1.044(a); 0.365(b); 0.430(c)	4.01 [114]	895 [35] 938 [294]		–(6–8) [294]
GeSe	Rhombic, D_{2h}^{16}	1.079(a); 0.382(b); 0.438(c)	5.52 [114]	943 [114]		–5 [294]
GeTe	Cubic, O_h^5	0.602(a) [114]	6.19 [114]	993 [114]	166 (<1 K)[164]	–4 [294]
SnS	Rhombic, D_{2h}^{16}	1.118(a); 0.398(b); 0.433(c)	5.08 [114]	1155 [216]	270 (80 K) [114]	
SnSe	Rhombic, D_{2h}^{16}	1.157(a); 0.419(b); 0.446(c)	6.18 [114]	1133 [216]	210 (80 K) [114]	
SnTe	Cubic, O_h^5	0.628(a); 0.632(a) [294]	6.45 [114]	1079 [114]	140 [114]	
SnO_2	Tetrahedral, D_{4h}^{14}	0.472(a); 0.317(c)	6.95 [216]	1400 [216]	570 [294]	12 [294]
SnS_2	Trigonal, D_{3d}^3	0.364(a); 0.587(c)	4.5 [216]	1038 [294] 1143 [288]		–8.6[294]
$SnSe_2$	Trigonal, D_{3d}^3	0.381(a); 0.614(c) [288]	6.01 [294]	923 [216]		

Compound	E_g, eV	m_n/m_0	m_p/m_0,	μ_n [294], $cm^2/(V{\cdot}s)$	μ_p, $cm^2/(V{\cdot}s)$	ϵ_0* [294]	ϵ_∞* [294]
Si_2Te_3	1.89 [294]	5.4 (‖c)	5.4 (‖c) [294]		$2 \cdot 10^{-3}$ (‖c) [294]		
GeS	1.65 1.74 (4.2 K) [294]		0.3–0.7 [294]		90 [294]	25–30	10–15
GeSe	1.0 [8] 1.16 [114]				70 ($\propto T^{-2}$) [114] 60 [8]	22–30	14–22
GeTe	0.1–0.2 [294]		1.15 (m_{p1}) [114] 5.0 (m_{p2})		150 [126]	40 [288]	36
SnS	1.07 [3]	0.45 (m_{dn})	0.95 (m_{dp}) [127] 0.2 ($m_\perp$) 1.0 ($m_\parallel$)		90 (⊥c) [127] $\mu_{\parallel a} \simeq \mu_{\parallel b}$ $\mu_{\parallel a}/\mu_{\parallel c} \simeq 5.5$	32–48	14–16
SnSe	0.9–0.95 [288]		0.15 (m_{dp})[15] 0.07 [288]		1000 [288]	42–62	13–17

Table 22.23 Semiconducting $A^{IV}B^{VI}$ compounds *(continued)*

Compound	E_g, eV	m_n/m_0	m_p/m_0,	μ_n [294], cm^2/(V·s)	μ_p, cm^2/(V·s)	ϵ_0* [294]	ϵ_∞* [294]
SnTe	0.19 [294] 0.26 [114]		0.13 (m_{p1}) 0.09 (m_{p2}) [294] 3 (m_{dp1})		1700 (μ_1) [181] 50 (μ_2) [114] 3500 (100 K) [114]	1770 1200	
SnO_2	3.54 [288] 3.97 [98]	0.30 ($m_\perp$) 0.23 ($m_\parallel$) 0.22 (m_{dn}) [69]		260 8800 (77 K)	300 [69]	9.6–13.5	3.8–4.2
SnS_2	2.07 [183]			50		6–20	5.6–8.8
$SnSe_2$	0.97 [183]	0.4 2.9 (m_{dn})		27 [183] 66 (77 K)		10–21	9.4–10.7

* ϵ_0 and ϵ_∞ depend on the mutual orientation of **E** and the c axis.

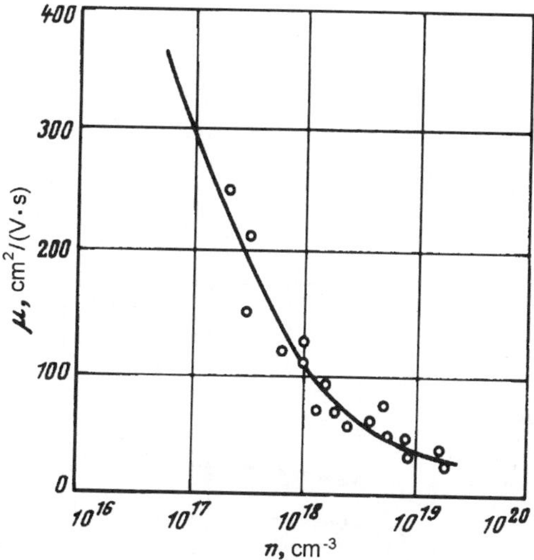

Figure 22.176 The electron drift mobility in α-SiC at 300 K vs. the electron density [295].

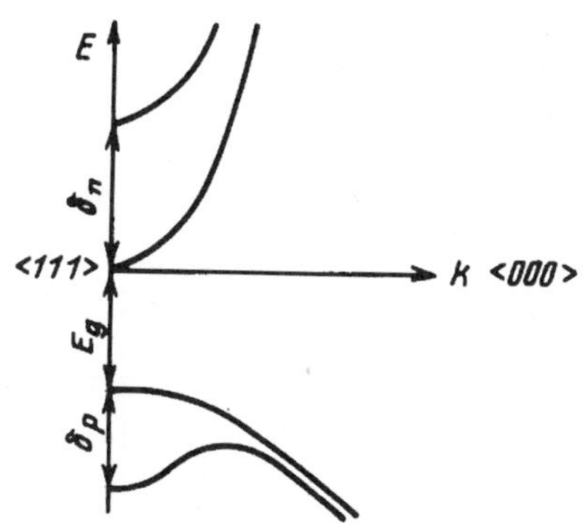

Figure 22.178 The band structure of PbS, PbSe, and PbTe [125]. The values of the parameters are listed in Table 22.22

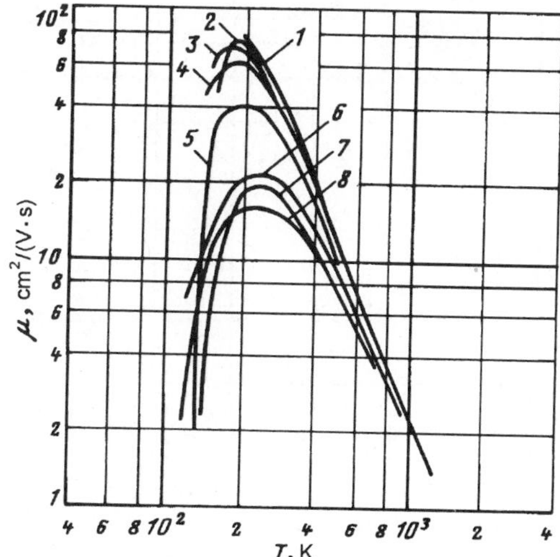

Figure 22.177 The temperature dependence of the hole drift mobility in α-SiC crystals [296]:

No.	N_A, 10^{17}cm^{-3}	N_D, 10^{17}cm^{-3}	Note
1	3.2	1.1	Ar atmosphere
2	3.7	1.9	The same
3	36	7.6	Ar atmosphere with an admixture of Cl
4	23	7.6	The same
5	850	140	The same
6	340	230	Ar atmosphere with an admixture of CO
7	3600	120	Ar atmosphere with an admixture of Cl
8	320	190	Ar atmosphere

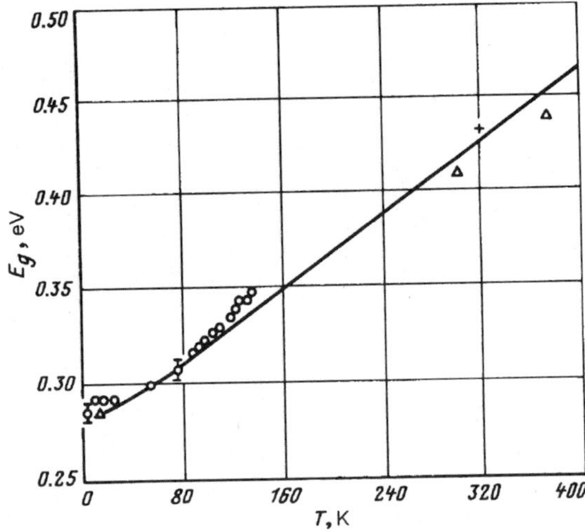

Figure 22.179 The temperature dependence of the forbidden band width for PbS [297].

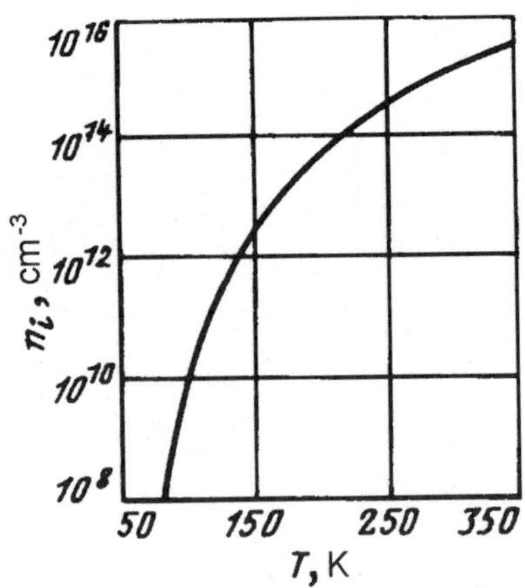

Figure 22.180 The temperature dependence of the intrinsic carrier density in PbS [297].

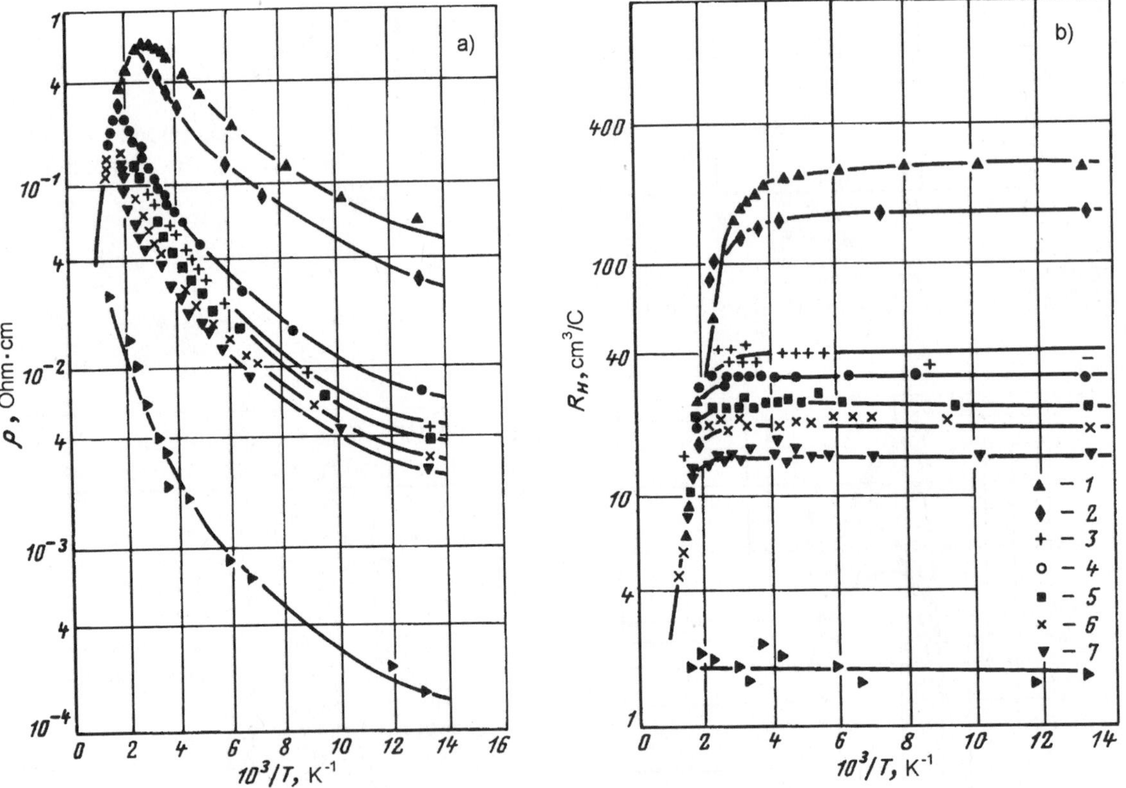

Figure 22.181 Temperature dependences of the resistivity (a) and the Hall coefficient (b) for n-PbS crystals [117]. The carrier density grows from sample *1* to sample *7*.

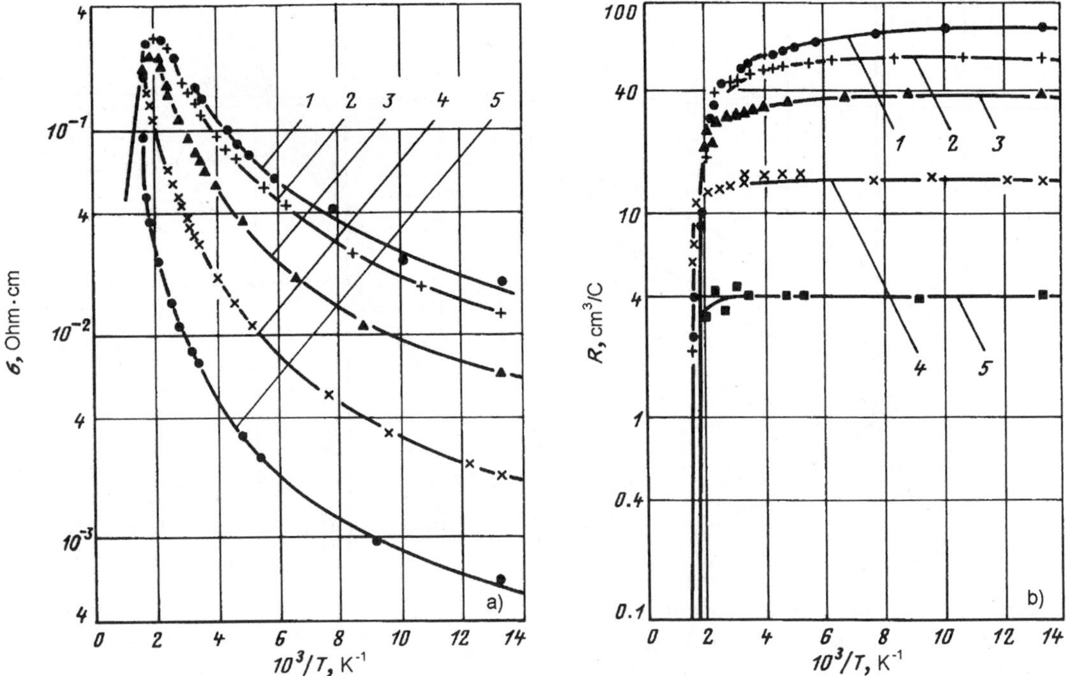

Figure 22.182 Temperature dependences of the resistivity (a) and the Hall coefficient (b) for *p*-PbS crystals [228]. The carrier density grows from sample *1* to sample *6*.

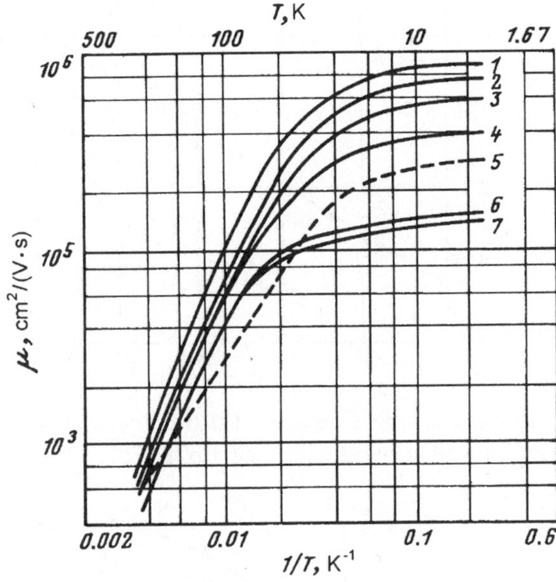

Figure 22.183 Temperature dependences of the Hall mobilities of electrons and holes in PbS crystals [117]: carrier density, 10^{18} cm^{-3}: *1* – 2.66 (*p*); *2* – 4.25 (*n*); *3* – 7.45 (*n*); *4* – 4.63 (*n*); *5* – 2.72 (*n*); *6* – 0.184 (*n*); *7* – 0.164 (*n*).

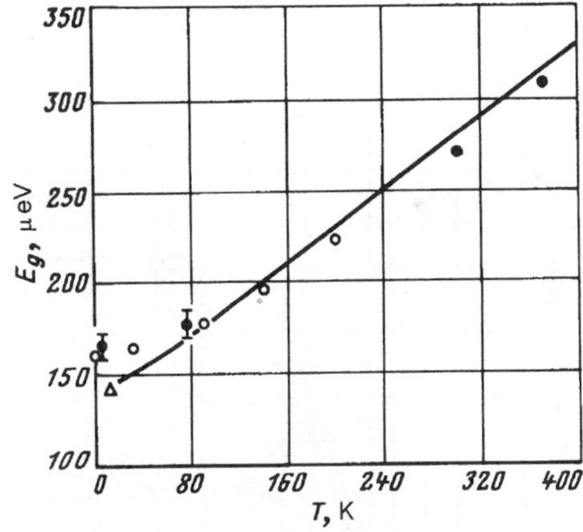

Figure 22.184 The temperature dependence of the forbidden band width for PbSe [297].

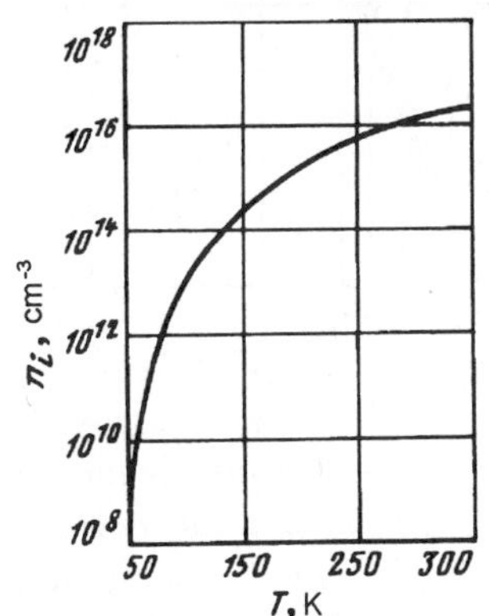

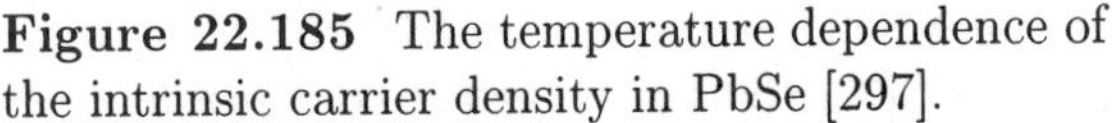

Figure 22.185 The temperature dependence of the intrinsic carrier density in PbSe [297].

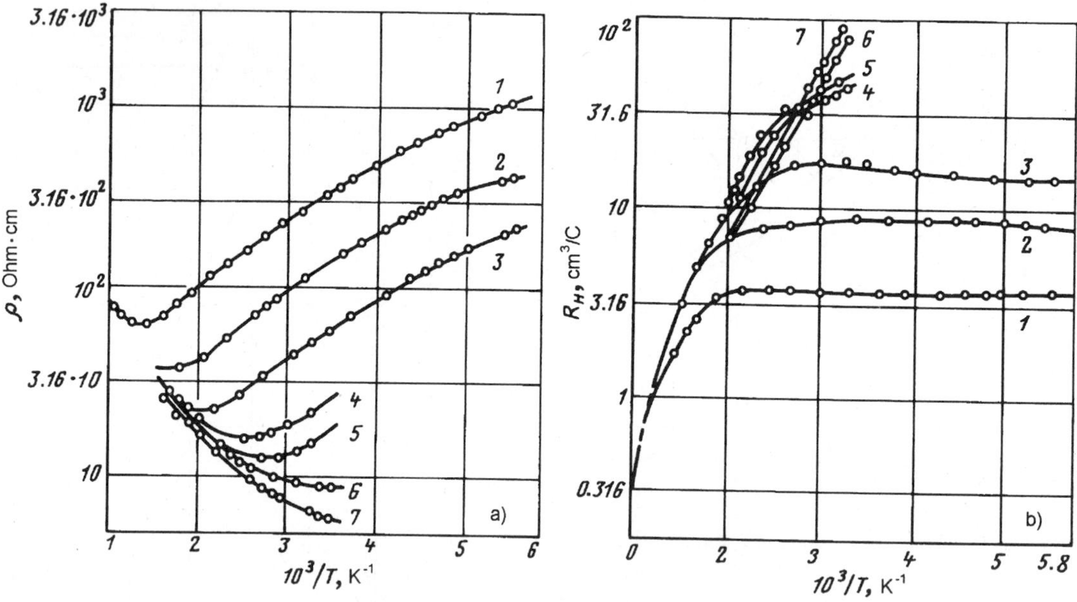

Figure 22.186 Temperature dependences of the conductivity (a) and the Hall coefficient (b) in n-PbSe crystals with various donor densities [117]. Donor density, cm^{-3}: *1* – $1.6 \cdot 10^{18}$; *2* – $7.8 \cdot 10^{17}$; *3* – $3.5 \cdot 10^{17}$; *4* – $1.58 \cdot 10^{17}$; *5* – $1.26 \cdot 10^{17}$; *6* – $4.9 \cdot 10^{16}$; *7* – $3 \cdot 10^{16}$.

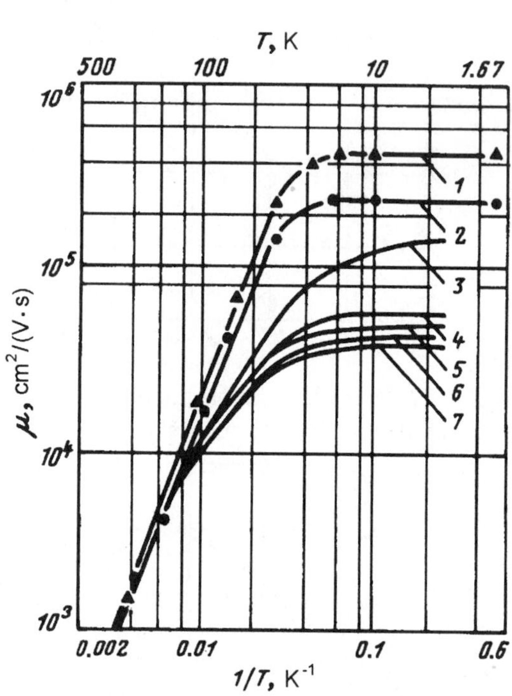

Figure 22.187 The temperature dependence of the Hall mobilities of electrons and holes in PbSe crystals [117]. Carrier density, 10^{18} cm^{-3}: *1* – 0.16 (n); *2* – 0.11 (p); *3* – 3.6 (n); *4* – 4.3 (p); *5* – 1.4 (p); *6* – 3.5 (p); *7* – 2.4 (n).

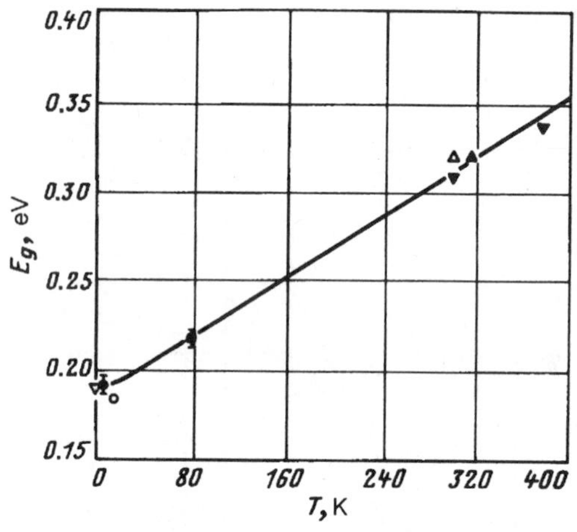

Figure 22.188 The temperature dependence of the forbidden band width for PbTe [297].

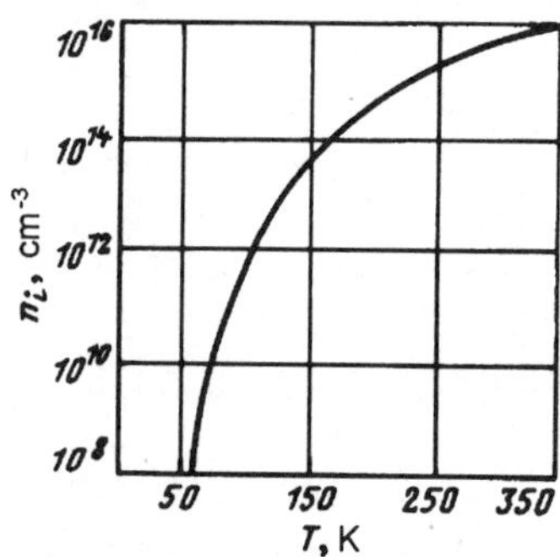

Figure 22.189 The temperature dependence of the intrinsic density in PbTe [297].

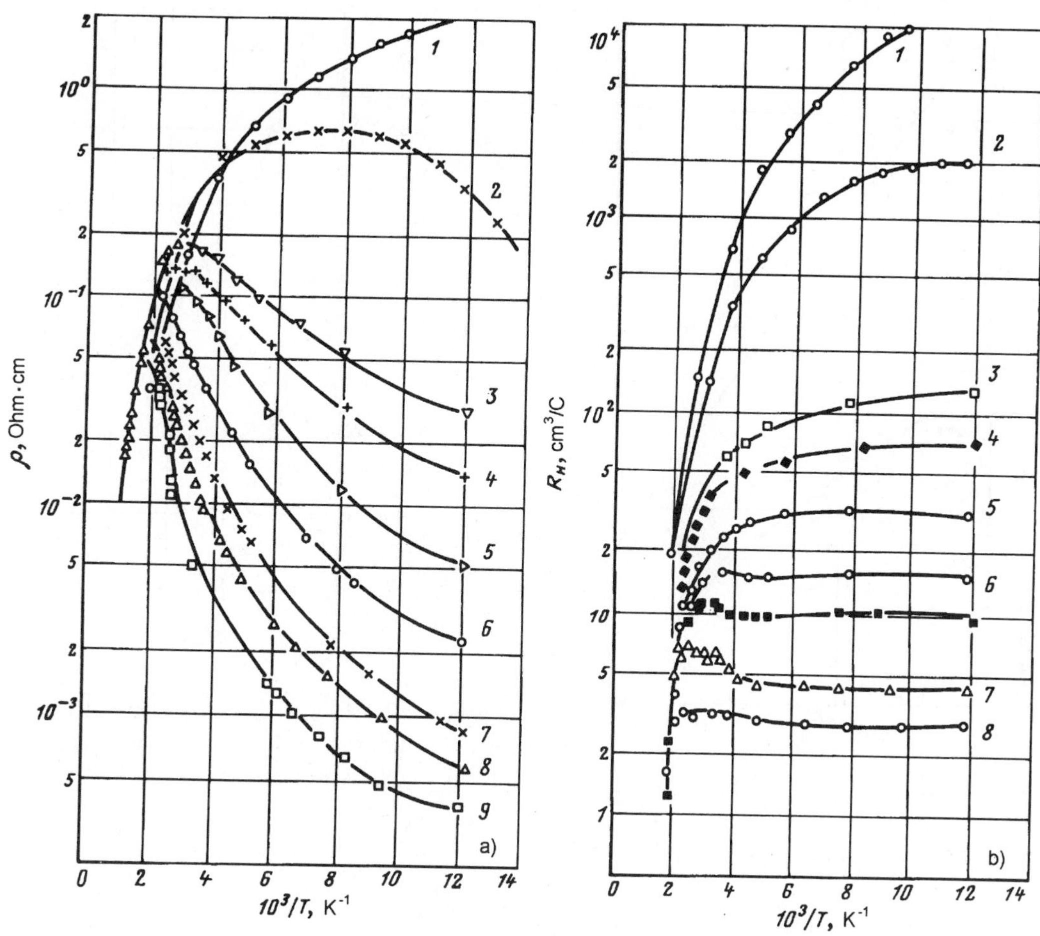

Figure 22.190 Temperature dependences of the resistivity (a) and the Hall coefficient (b) in PbTe crystals [117]: *1*, *2* – near-intrinsic samples; *3*-*9* – *p*-PbTe; the carrier density grows from sample *1* to sample *9*.

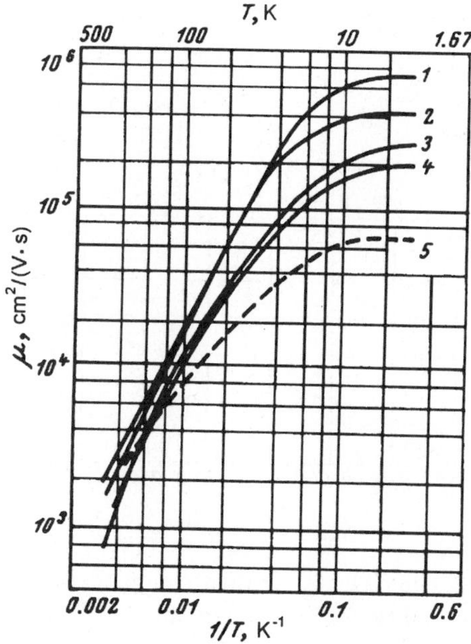

Figure 22.191 Temperature dependences of the Hall mobilities of electrons and holes in PbTe crystals with various carrier densities (10^{18} cm^{-3}): *1* – 1.0 (*n*); *2* – 0.58 (*n*); *3* – 2.1 (*p*); *4* – 3.0 (*p*); *5* – 9.5 (*n*) [117].

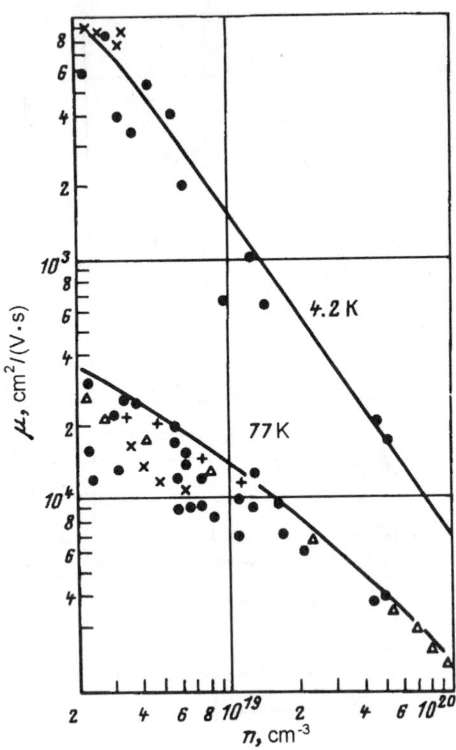

Figure 22.192 The electron mobility vs. the electron density for PbTe at 77 and 4.2 K [125]: points are the data of different authors.

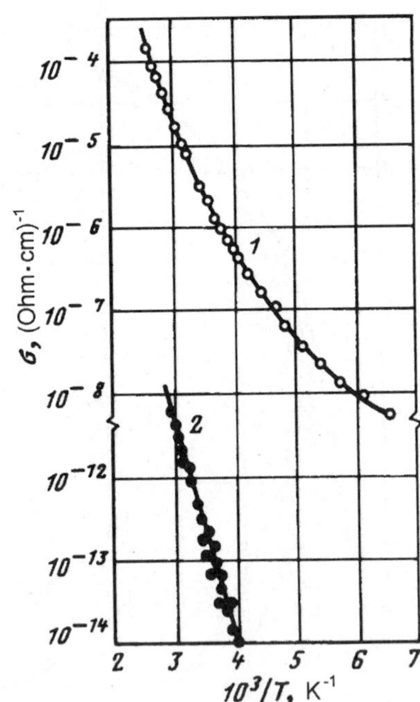

Figure 22.194 The temperature dependences of the conductivity for crystalline (*1*, transverse to the c axis) and amorphous (*2*) GeS [299].

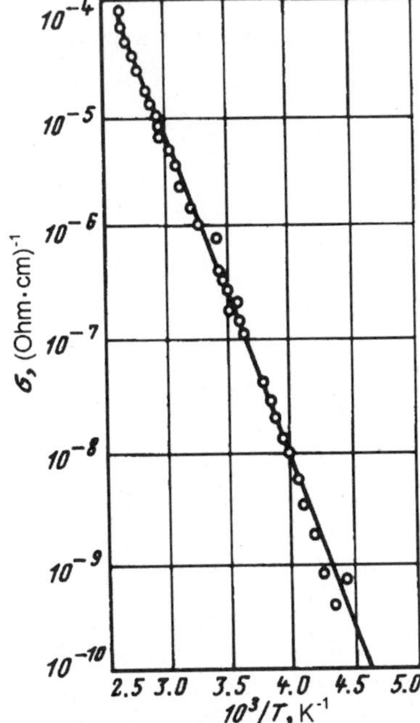

Figure 22.193 The temperature dependence of the conductivity for needle-like p-GeS crystals [298].

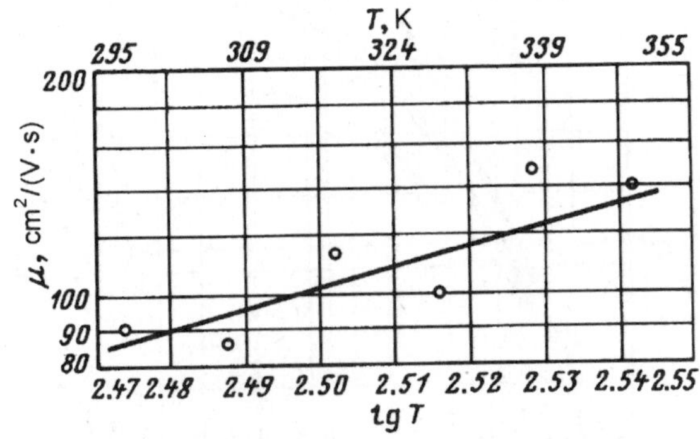

Figure 22.195 The temperature dependence of the hole mobility for a 730-nm-thick GeS film [300].

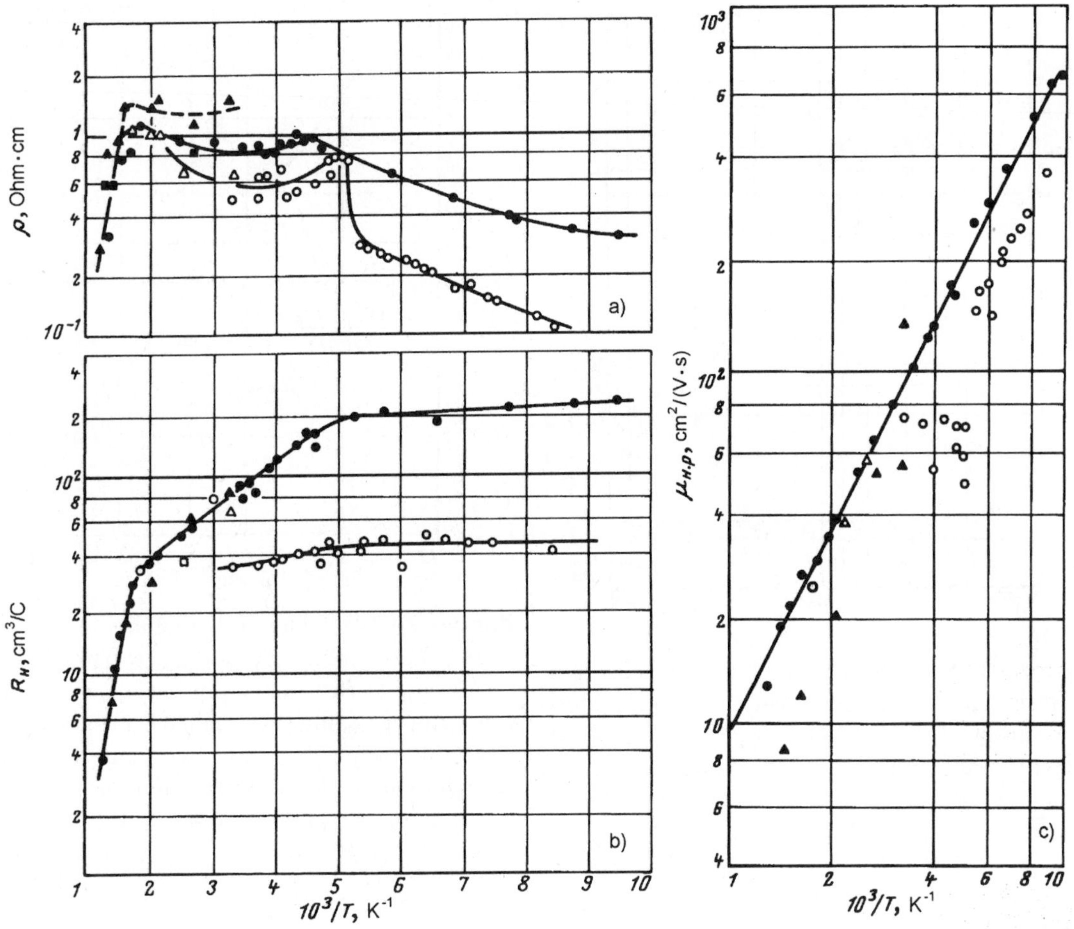

Figure 22.196 Temperature dependences of the resistivity (a), the Hall coefficient (b), and the hole mobility (c) transverse to the c axis for p-GeSe crystals of various stoichiometric compositions [301].

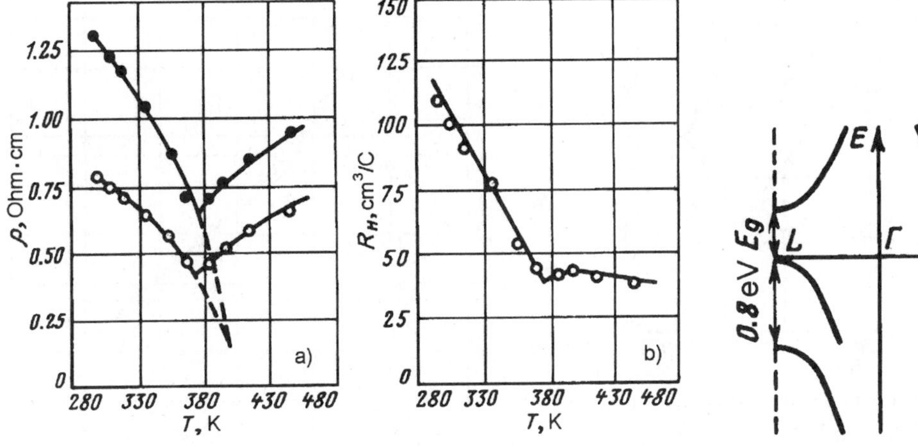

Figure 22.197 Temperature dependences of the resistivity (a) along the a (•) and b (○) axes and the Hall coefficient (b) in the (001) plane for GeSe [302].

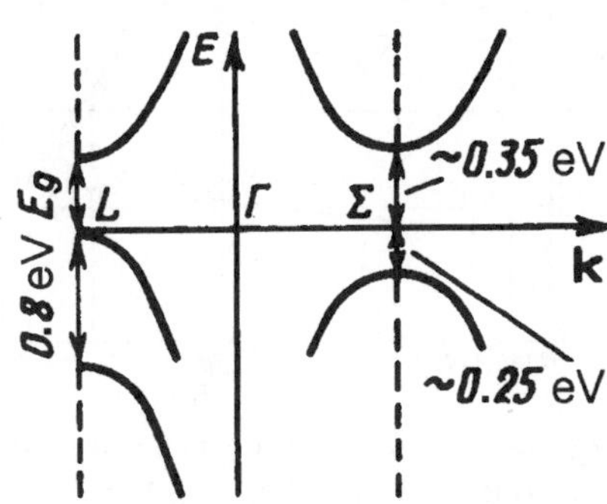

Figure 22.198 The band structure of GeTe [294].

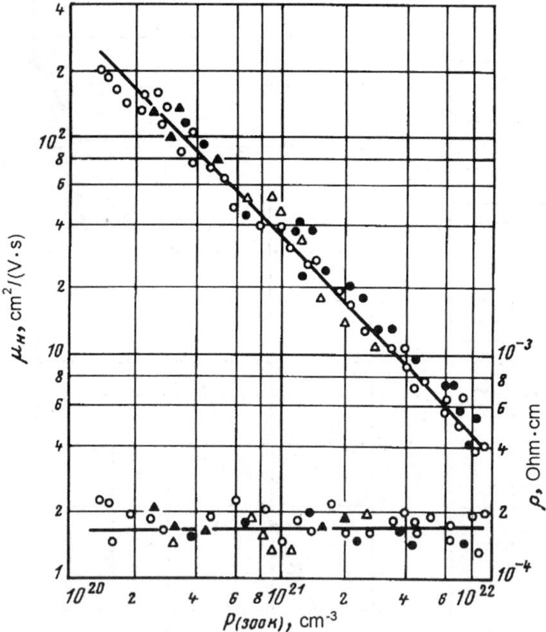

Figure 22.199 The Hall mobility of holes and the resistivity vs. the hole density in GeTe films with various Te contents: ○ – 54% (unannealed sample); other symbols – 51.5% (after several annealing cycles) [303].

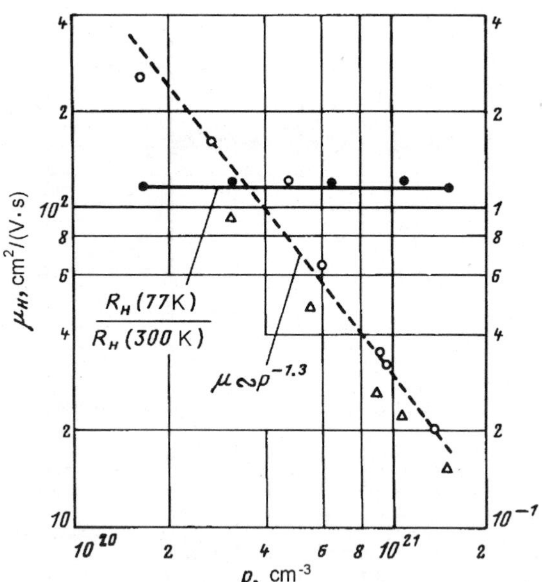

Figure 22.201 The Hall mobility of holes at 77 K (○) and 300 K (△) and the ratio of the Hall coefficients at the same temperatures vs. the hole density in crystalline GeTe films deposited on substrates at various temperatures [304]. The values of R_H (300 K) for different films are different; for bulk polycrystalline samples $R_H(300\text{ K}) = 6.5 \cdot 10^{-3}\text{ cm}^3/\text{C}$.

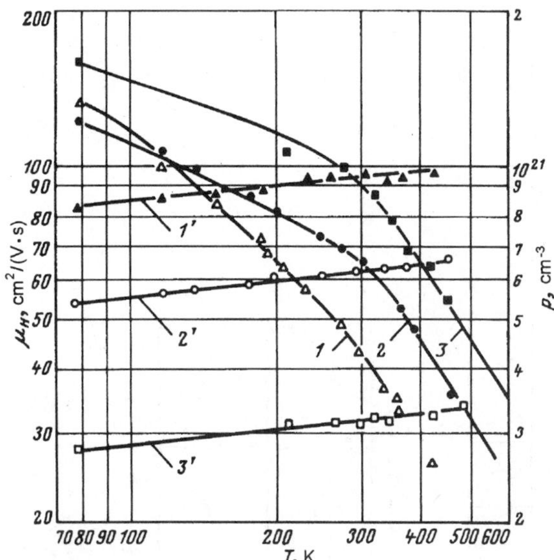

Figure 22.200 Temperature dependences of the hole Hall mobility (*2*, *3*) and the hole density (*2′*, *3′*) in polycrystalline samples and single-crystal films (*1*, *1′*) of GeTe [304]. The temperature of film deposition onto a glass substrate, °C: *2*, *2′* – 250; *3*, *3′* – 300.

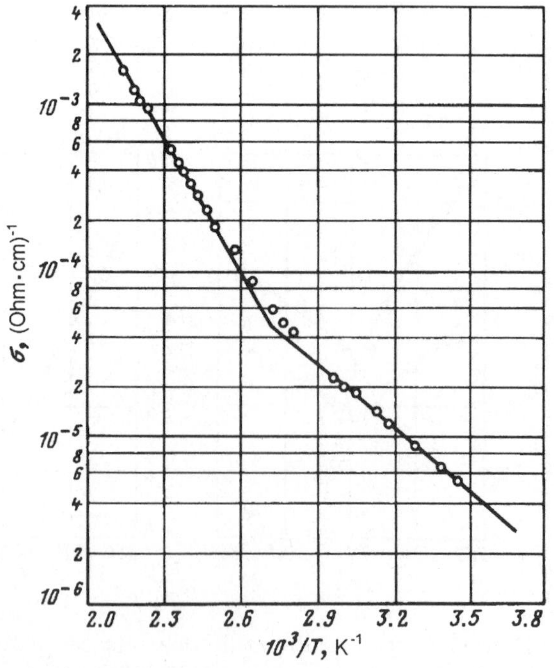

Figure 22.202 The temperature dependence of the conductivity in a SnS single crystal [305].

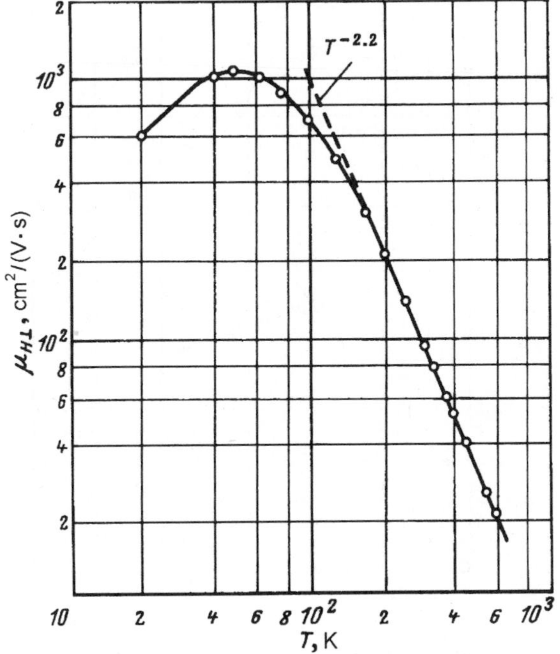

Figure 22.203 The temperature dependence of the hole Hall mobility in SnS across the c axis [306].

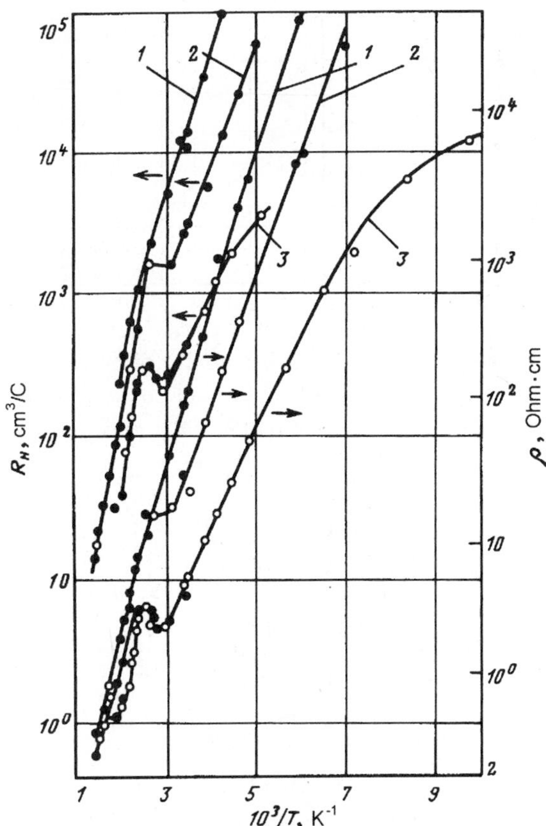

Figure 22.205 Temperature dependences of the resistivity and the Hall coefficient in as-grown (*1*), as-annealed (*2*) and as-quenched (*3*) n-SnSe crystals [307].

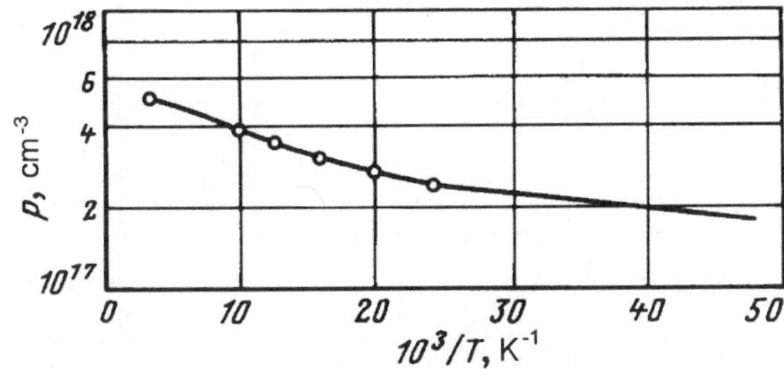

Figure 22.204 The temperature dependence of the hole density in SnS [306].

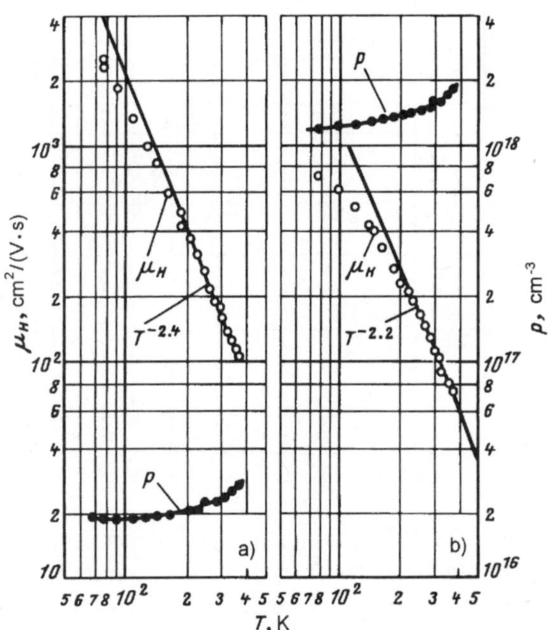

Figure 22.206 Temperature dependences of the hole mobility along the [001] axis and the hole density in SnSe crystals with a low (a) and a high (b) hole density [308].

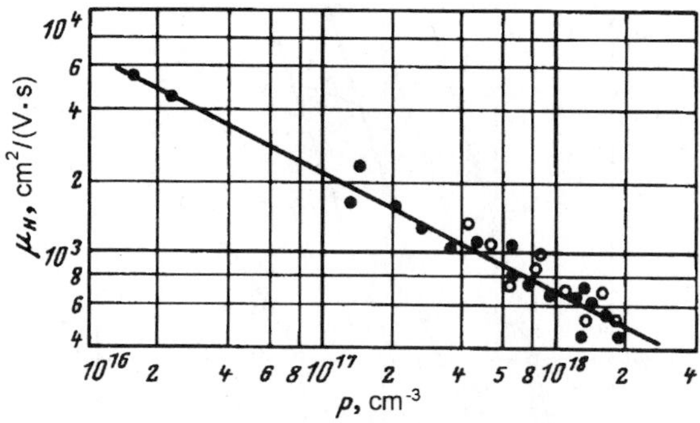

Figure 22.207 The Hall mobility of holes along the [001] axis at 77 K vs. the hole density in as-grown (○) and as-annealed (●) SnSe [308].

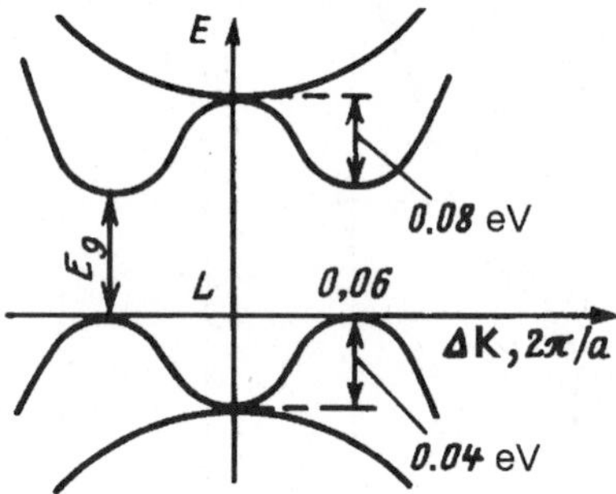

Figure 22.208 The band structure of SnTe in the vicinity of the L-point of the Brillouin zone [294].

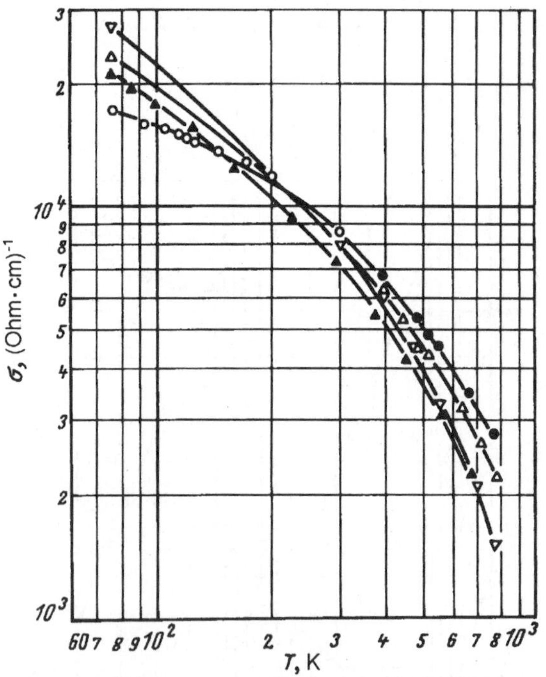

Figure 22.209 The temperature dependence of the conductivity for various SnTe crystals [309]: hole density at 77 K, 10^{20}cm^{-3}: ▽ – 2.5; ▲ – 3.48; △ – 8.05; ○ – 18.8; ● – 17.6.

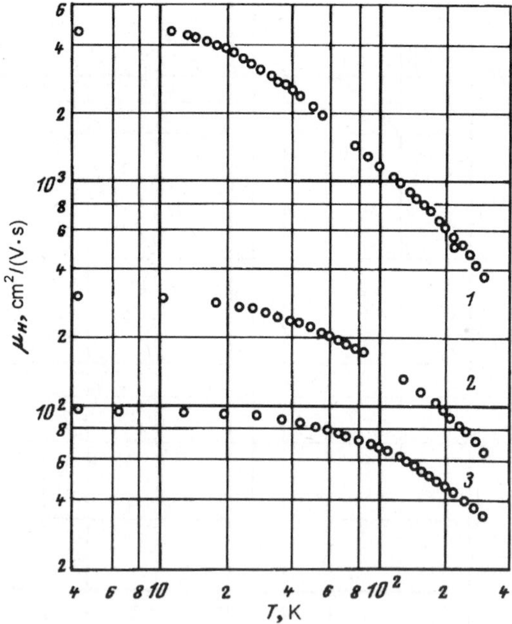

Figure 22.210 The temperature dependence of the hole Hall mobility in various SnTe crystals [310]: hole density at 4.2 K, 10^{20} cm^{-3}: *1* – 1.4; *2* – 7.73; *3* – 1.74.

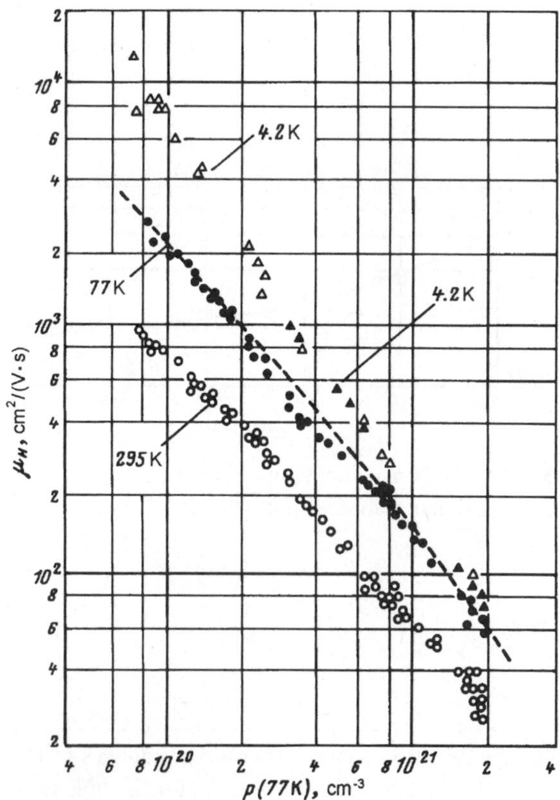

Figure 22.211 The Hall mobility of holes at different temperatures vs. the hole density at 77 K for SnTe crystals [309, 310].

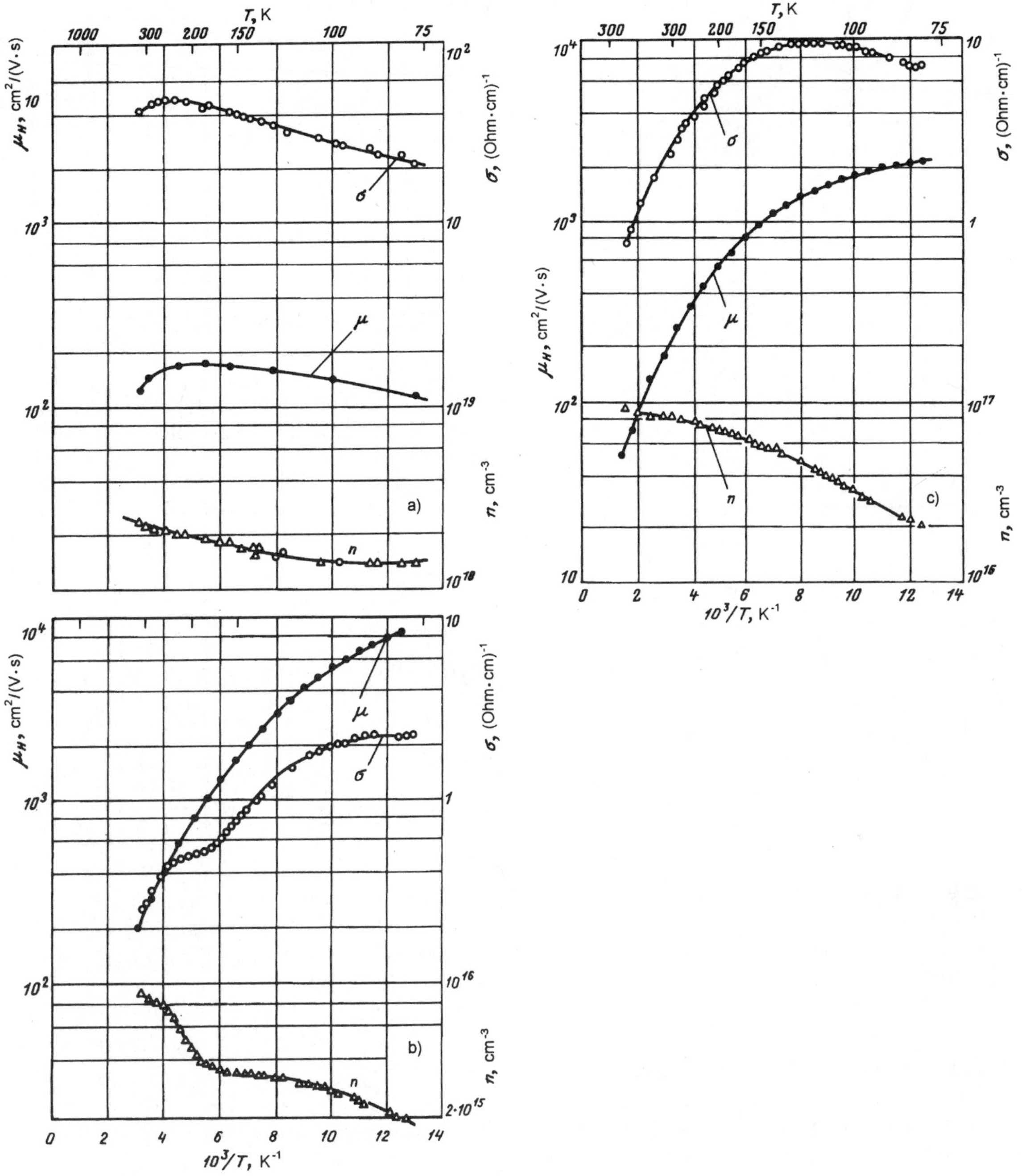

Figure 22.212 Temperature dependences of the conductivity, the electron density and the Hall mobility of electrons in SnO_2 crystals with high (a), medium (b) and low (c) electron densities [311].

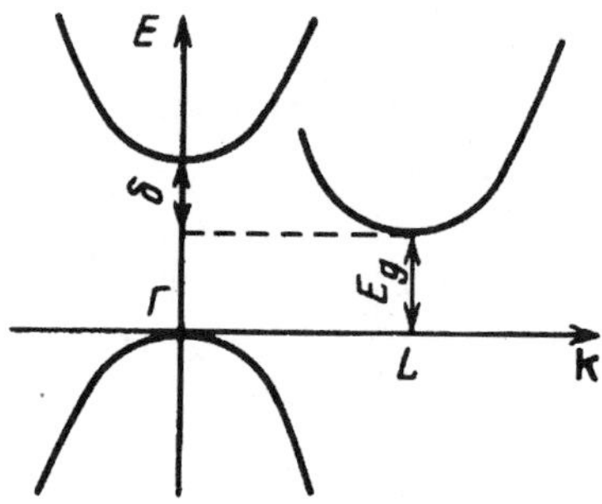

Figure 22.213 The band structure of SnS_2 and $SnSe_2$ [294]: δ, eV: SnS_2 – 0.8; $SnSe_2$ – 0.65.

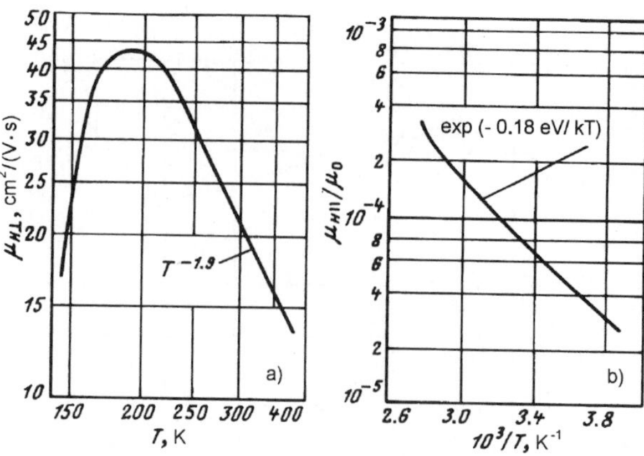

Figure 22.214 The temperature dependence of the electron Hall mobility across (a) and along (b) the c axis in SnS_2 [312].

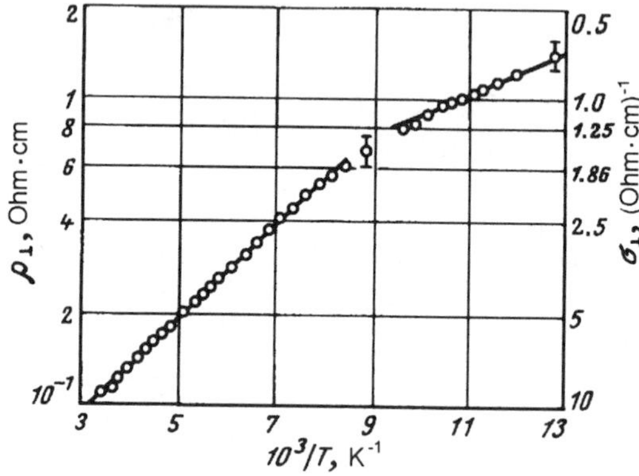

Figure 22.215 The temperature dependence of the resistivity (conductivity) across the c axis for a $SnSe_2$ crystal with an electron density $n = 1.57 \cdot 10^{18}$ cm^{-3} at 290 K [36].

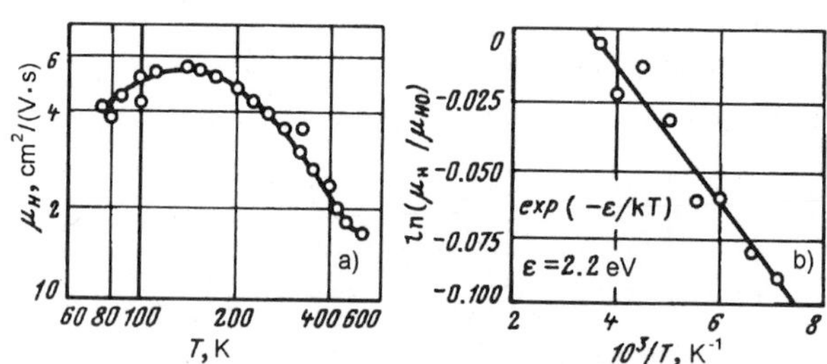

Figure 22.216 Temperature dependences of the electron mobility in the basal plane (a) and in the direction perpendicular to the basal plane (b) in a $SnSe_2$ crystal [270].

22.3.5 A^VB^V - A^VB^{VII} compounds

The only known semiconducting A^VB^{VII} compound is SbI which crystallizes into a hexagonal lattice ($a = 0.748$ nm, $c = 2.09$ nm, $E_g = 2.5$ eV [242]).

Table 22.24 Electrophysical properties of $Bi_{1-x}Sb_x$ alloys [293]

Crystal structure		T_D, K	E_g, eV	m_n/m_0	μ_n,	μ_p,	ϵ_0
System	a, b, c, nm				cm²/(V·s)	cm²/(V·s)	(4.2 K)
Rhombohedral	0.474–0.450 (a) (x=0–1); 1.186–1.154 (c) (x=0–0.5)	128.5 (x=0.1)	$6 \cdot 10^{-3}$ ($x = 0.06$) $8 \cdot 10^{-3}$ ($x = 0.063$)	0.025 (m_1) ($x = 0.05$) 0.75 (m_2) ($x = 0.05$) 0.0065 (m_3) ($x = 0.05$)	$3.97 \cdot 10^4$ ($x = 0.03$)	$4.73 \cdot 10^4$ ($x = 0.072$)	360 175 (x=0.092)

Table 22.25 Semiconducting $A^{V}B^{VI}$ compounds

Compound	Crystal structure		ρ, g/cm^3	T_{melt}, K	E_g, eV	$\partial E_g/\partial T$, 10^{-4}eV/K	m_n/m_0
	System, group	a, b, c, nm					
As_2S_3	Monoclinic, C_{2h}^5	1.147 (a); 0.957 (b); 0.424 (c) [1]	3.43 [123]	598 [123]	2.5 [123] 2.6 [294]	−5.6 [10] −7 [294]	
As_2Se_3	Monoclinic, C_{2h}^5	1.205 (a); 0.989 (b); 0.428 (c) [288]	4.75 [216]	633 [216]	1.8 [114] 1.85 [294] 1.9 [107]	−8 [288]	
As_2Te_3	Monoclinic, C_{2h}^5	1.44 (a); 0.992 (b); 0.405 (c) [1]	6.1 [123]	635 [116]	0.5 [288] 1.0 [123]	−3 [288]	0.36 (m_{dn}) [123]
Sb_2S_3	Rhombic, D_{2h}^{16}	1.12 (a); 1.128 (b); 0.386 (c) [1]	4.64 [123]	819 [116]	1.62 [12] 1.72 [114] 1.88 [294]	−5.7 [114] −9 [294]	
Sb_2Se_3	Rhombic, D_{2h}^{16}	1.158 (a); 1.168 (b); 0.398 (c) [1]	5.81 [123]	885 [116]	1.1 [294] 1.2 [123]	−5.5 [294] −7 [114]	2.2 [288]
Sb_2Te_3	Trigonal, D_{3d}^5	1.043 (a) [294]; $\alpha = 23°23'$	6.5 [198] 6.57 [114]	895 [118]	0.19 [48] 0.22 [294]	−1 [288]	0.37 (m_{dn}) [123]
Bi_2S_3	Rhombic, D_{2h}^{16}	1.113 (a); 1.127 (b); 0.397 (c) [1]	6.73 [123] 6.81 [294]	1036 [294] 1123 [116]	1.25 [16] 1.3 [116]	−8 [288]	
BiSe	Hexagonal, D_{3d}^1	0.418 (a); 2.28 (c) [288]	8.4 [288]	880 [123]	0.4 [123]		
Bi_2Se	Trigonal, D_{3d}^5	0.984 (a) [132]; $\alpha = 24°24'$	7.40 [114] 7.68 [294]	979 [116]	0.16 [294] 0.2 [288]	−2 [168]	0.02 ($m_\perp$) 0.13 ($m_\parallel$) 0.15 (m_{dn}) [294]
Bi_2Te_3	Trigonal, D_{3d}^5	1.048 (a); [117] $\alpha = 24°9'$	7.86 [117]	853 [116]	0.13 [117] 0.16 [45]	−0.95 [117] −1.5 [294]	0.27 [294] 0.32 (m_{dn}) [123]

Compound	m_p/m_0	μ_n, cm^2/(V·s)	μ_p, cm^2/(V·s)	ϵ_∞ [294]	ϵ_0 [294]	T_D, K
As_2S_3		∼ 1 [294]		5.7–8.8 7.5 [184]	5.9–12 9.7 [184]	590 [288]
As_2Se_3		20–80	∼ 10 [288]	8.8–10.5 6.0 [184]	12.4–13.9 8.9 [184]	418 [288]
As_2Te_3	0.5 (m_{dp}) [15]	170 [15]	80 [15]			260 [288]
Sb_2S_3		15 [115]	45 [115]	7.2 ($\perp c$) 9.5 ($\parallel c$)	180($\parallel c$) 15 ($\perp c$)	310 [132]
Sb_2Se_3	1.4 [288]	75 [16]	45 [16]	13.7 ($\perp c$) 15.1 ($\parallel c$)	120 ($\parallel c$)	240 [132]
Sb_2Te_3	0.3 [294] 0.34 (m_{dp}) [15]		400 [44] 10000 (4.2 K)[294]	32.5 ($\perp c$) 51 ($\parallel c$)	168 ($\perp c$) 36.5 ($\parallel c$)	160 [132]
Bi_2S_3		200 [123]		13 ($\parallel c$) 9 ($\perp c$)	38 ($\perp c$) 120 ($\parallel c$)	435 [288]
BiSe		20 [288]				
Bi_2Se_3	0.12 [294]	600–2000 ($\propto T^{-7/2}$)[114]	40 [294]	29 ($\perp c$)	113 ($\perp c$) 100 [288]	180 [132]
Bi_2Te_3	0.24 [123] 0.35 (m_{dp}) [294]	1200 [13] ($\propto T^{-1/7}$)	600 ($\propto T^{-2}$) [13]	50 ($\parallel c$) 85($\perp c$)	360 [288]	165 [294]

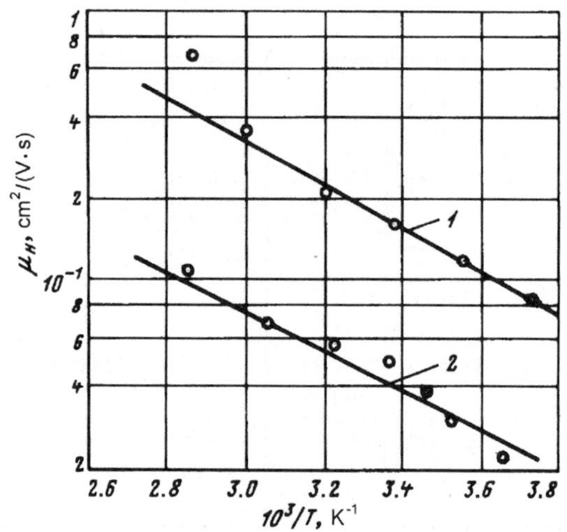

Figure 22.217 The temperature dependence of the hole drift mobility (across the layers) in two As_2S_3 crystals [268]. The trap concentration is higher in crystal *2* than in crystal *1*.

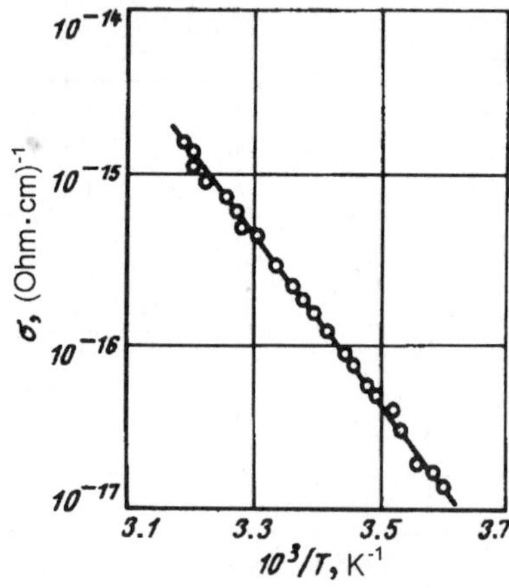

Figure 22.218 The temperature dependence of the conductivity (across the layers) for As_2Se_3 [261].

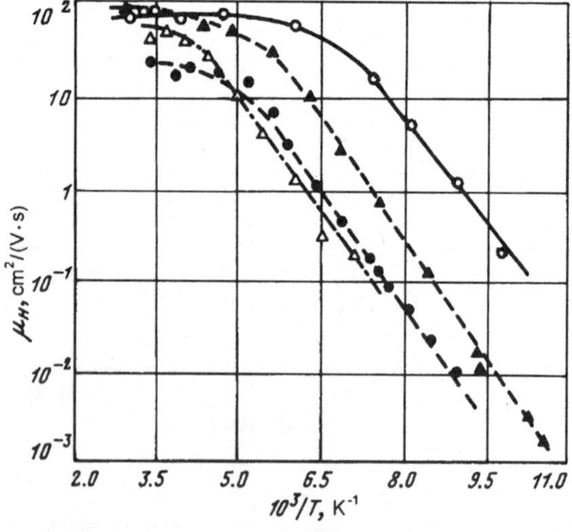

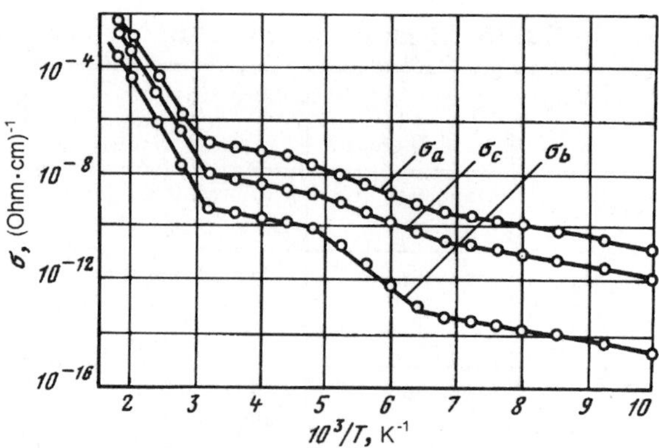

Figure 22.220 Temperature dependences of the conductivity along the *a*, *b*, and *c* axes in Sb_2S_3 [258].

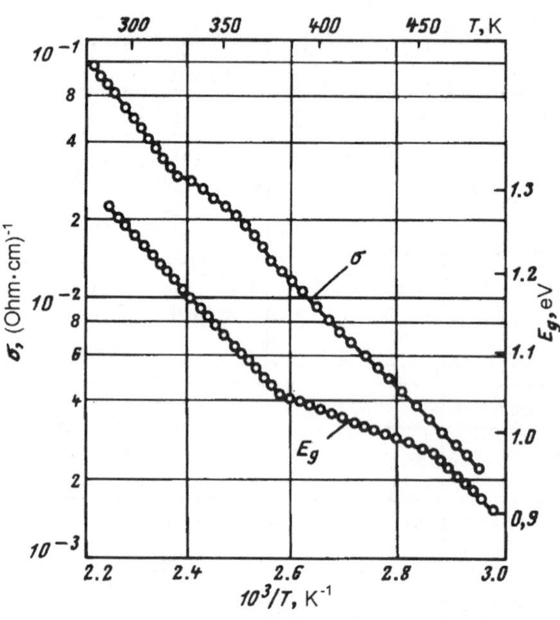

Figure 22.221 Temperature dependences of the conductivity (along the *c* axis) and the forbidden band width in Sb_2Se_3 [248].

Figure 22.219 Temperature dependences of the electron drift mobility (across the layers) in various As_2Se_3 crystals [261]. The ratio of the trap concentration to the effective density of states in the conduction band equals: ○ – 10^{-5}; ▲ – $2{\cdot}10^{-5}$; ● – $9{\cdot}10^{-5}$; △ – $3{\cdot}10^{-4}$.

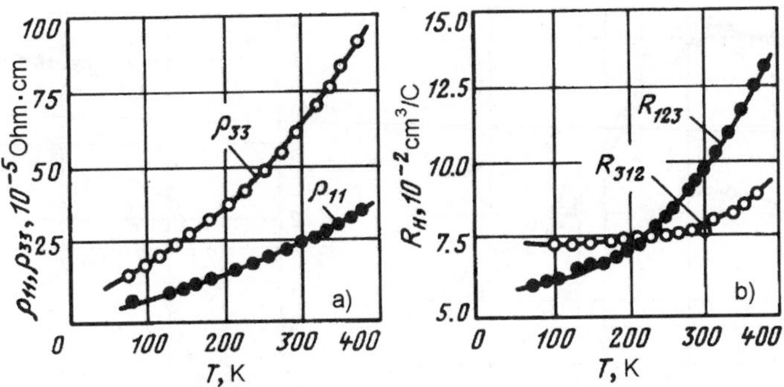

Figure 22.222 The temperature dependence of the resistivity (a) along (ρ_{33}) and transverse (ρ_{11}) to the c axis as well as the Hall coefficients (b) in the geometries $I \parallel c \perp B$ (*R312*) and $I \perp c \parallel B$ (*R123*) for Sb_2Te_3 [206].

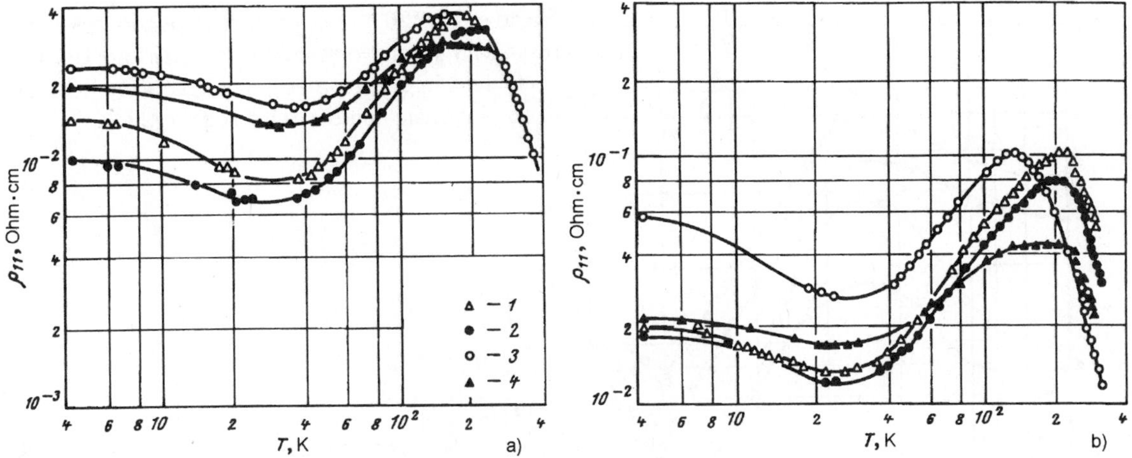

Figure 22.223 Temperature dependences of the resistivity (transverse to the c axis) for n-Bi_2Se_3 (a) and p-Bi_2Se_3 (b) crystals [203]. The carrier density grows from sample *1* to sample *4*.

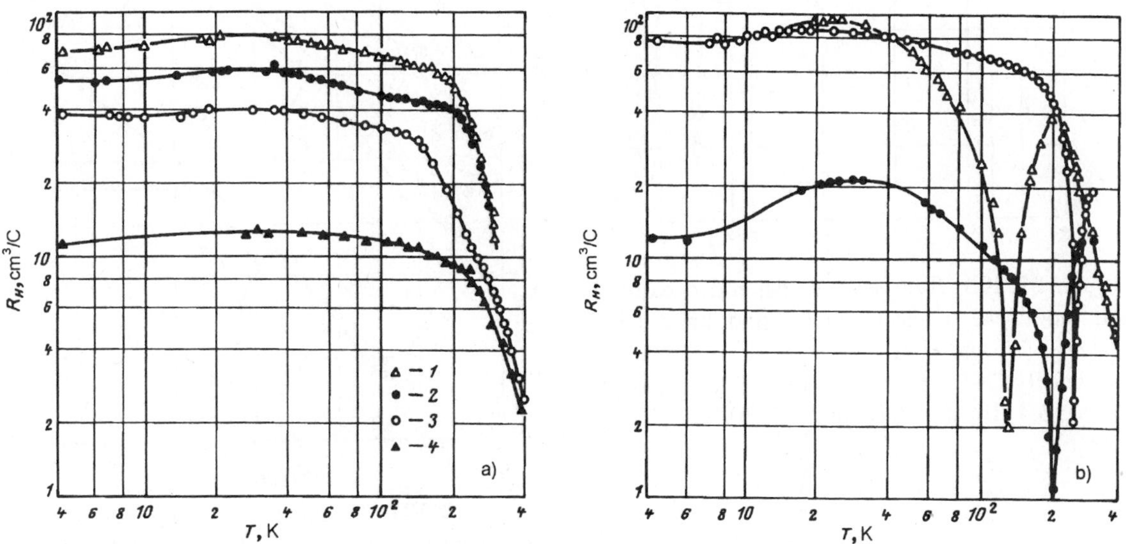

Figure 22.224 Temperature dependences of the Hall coefficient in the geometry $I \parallel c \perp B$ for n-Bi_2Se_3 (a) and p-Bi_2Se_3 (b) crystals [203]. The same crystals as in Figure 22.223.

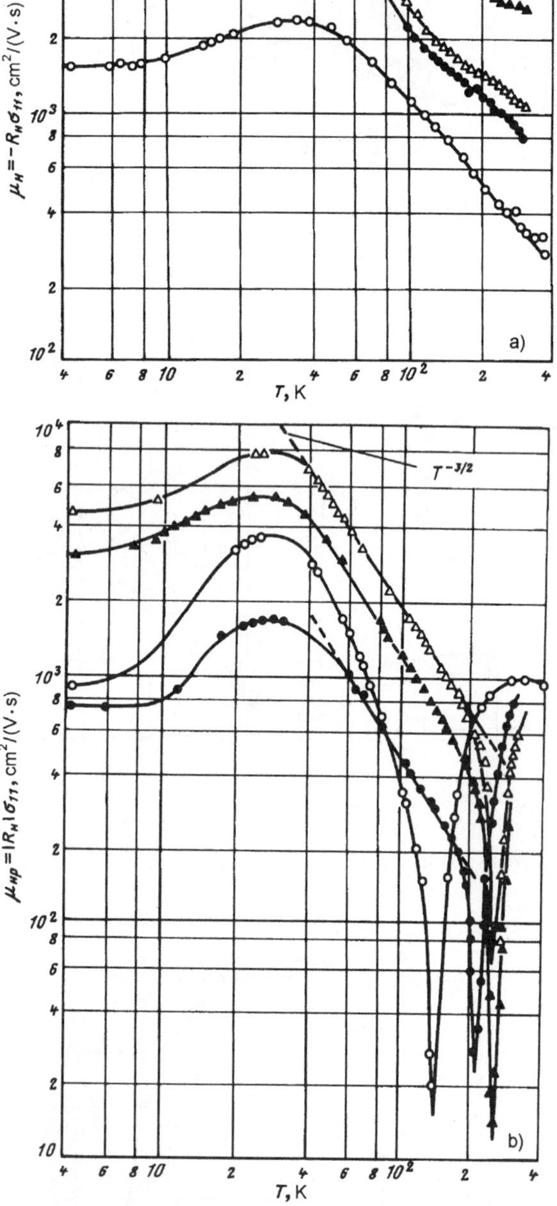

Figure 22.225 Temperature dependences of the Hall mobilities of electrons (a) and holes (b) in n-Bi_2Se_3 (a) and p-Bi_2Se_3 (b) crystals [203]. Carrier density, 10^{17} cm^{-3}: ○ – 1.6 (a); 0.69 (b); ● – 1.2 (a); 5.4 (b); △ – 0.87 (a); 6.7 (b); ▲ – 5.4 (a); 0.72 (b).

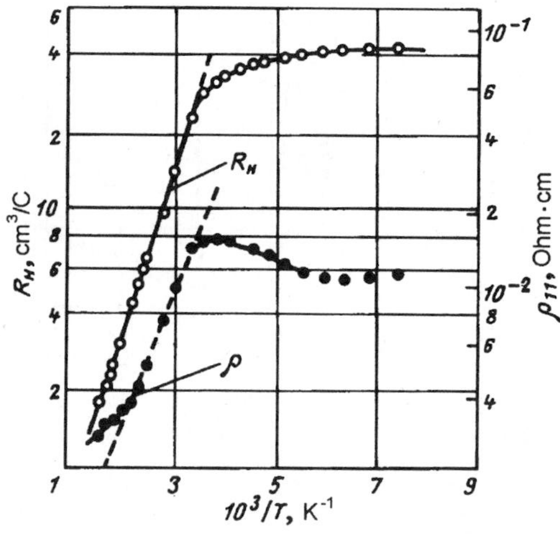

Figure 22.226 Temperature dependences of the resistivity (transverse to the c axis) and the Hall coefficient in the geometry $I \parallel c \perp B$ for n-Bi_2Se_3 in the intrinsic conductivity region [203].

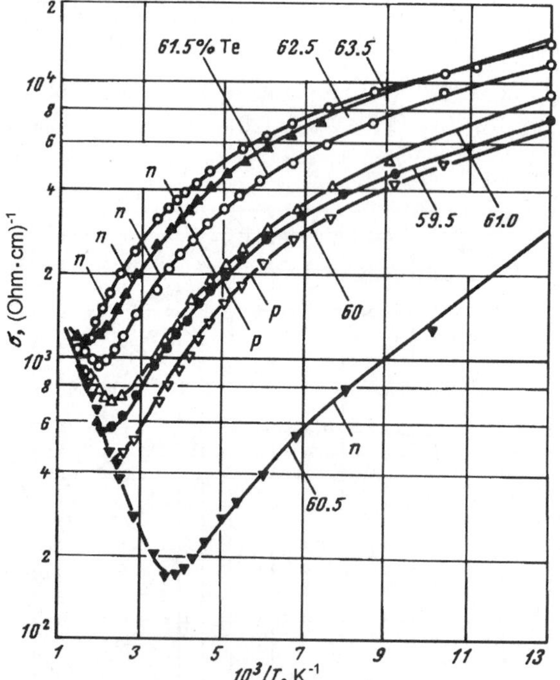

Figure 22.227 Temperature dependences of the conductivity (transverse to the c axis) for n- and p-Bi_2Te_3 crystals [204]. The numbers in the curves show the atomic Te content in at. %.

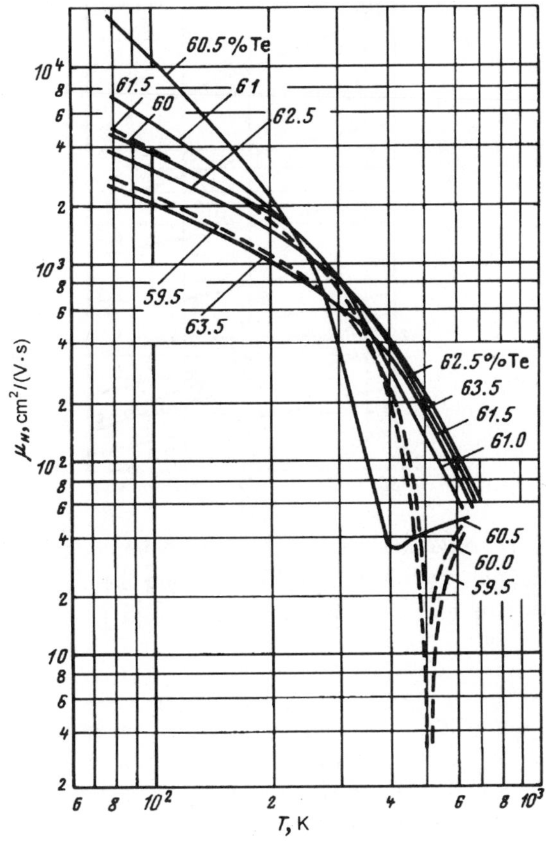

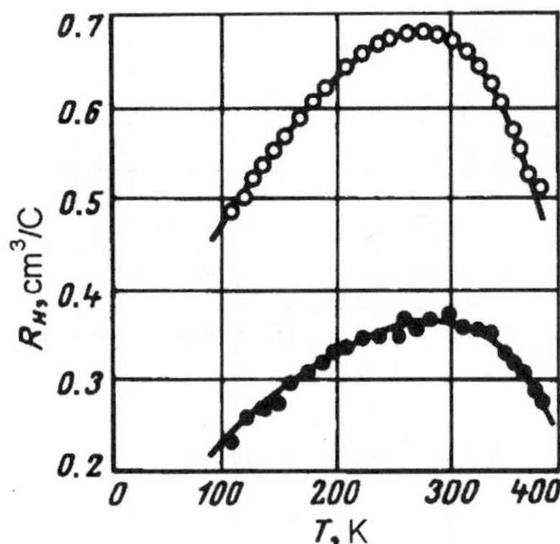

Figure 22.229 Temperature dependences of the Hall coefficient for Bi_2Te_3 in the geometries $I \perp c \parallel B$ (▲) and $I \perp c \perp B$ (○) [174].

Figure 22.228 The temperature dependence of the Hall mobility transverse to the c axis in n-Bi_2Te_3 (dashed lines) and p-Bi_2Te_3 (solid lines) crystals [128]. The numbers in the curves show the Te content in at.%.

22.3.6 $A^{VI}B^{IV}$ - $A^{VI}B^{VI}$ compounds

The only known semiconducting $A^{VI}B^{IV}$ compound is p-$CrSi_2$ which possesses a very low hole mobility (E_g = 1.3 eV [16]).

Table 22.26 Semiconducting $A^{VI}B^{VI}$ compounds

Compound	Crystal structure		E_g, eV	μ_n, μ_p, cm^2/(V·s)
	System, group	a, b, c, nm		
CrO_3	Cubic, C_{2v}^{16}	0.5743 (a); 0.8557 (b); 0.4789 (c) [242]	1.4 [242]	
Cr_2O_3	Hexagonal, D_{3d}^{6}	0.4954 (a); 0.13584 (c) [242]	1.59–1.63 [242]	
$CrTe_2$			0.17 [16]	12 (μ_n) [70]
MoS_2	Hexagonal, D_{6h}^{4}	0.315 (a); 0.1230 (c) [1]	1.2 [10]	200–300 (μ_p) [249]
TeO_2	Tetrahedral, D_{4h}^{14}	0.479 (a); 0.377 (c) [1]	1.5 [10]	

Table 22.27 Electrophysical properties of Se_xTe_{1-x} alloys [293]

Crystal structure* a, b, c, nm	ρ, g/cm^3	E_g, eV	$(m_n m_p)^{1/2}$ m_0	n_i, cm^3	μ_n (∥c), cm^2/(V·s)	μ_p (∥c), cm^2/(V·s)
x =0–1;	6.2	0.36 (x=0.046)	0.17	$10^{14}-10^{16}$	10000	1000 (x=0.1)
0.446–0.438 (a)	(x=0.05)	0.2 (x=0.55)	(x=0.05)	(x=0–1)	(x=0.1)	350 (x=0.2)
0.593–0.492 (c)	5.01	0.4 (x=0.85)	0.22			17 (x=0.5)

* Trigonal system.

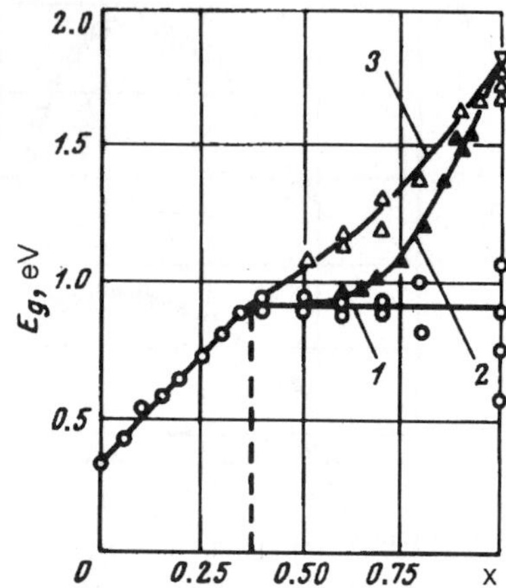

Figure 22.230 The forbidden band width E_g vs. x in Se_xTe_{1-x} [111]: *1* – intrinsic conductivity; *2* – photoconductivity: *3* – temperature dependence of conductivity.

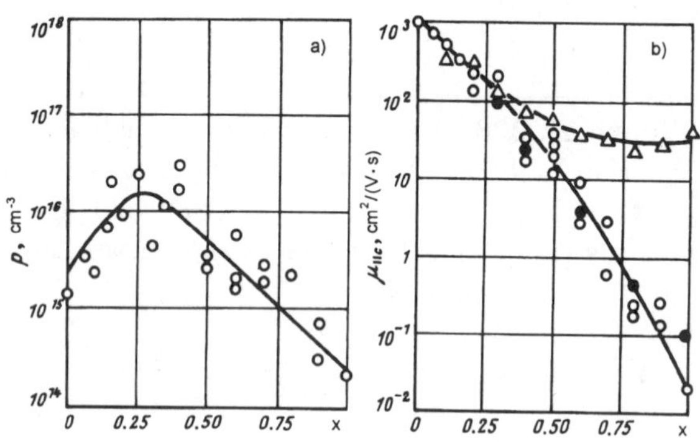

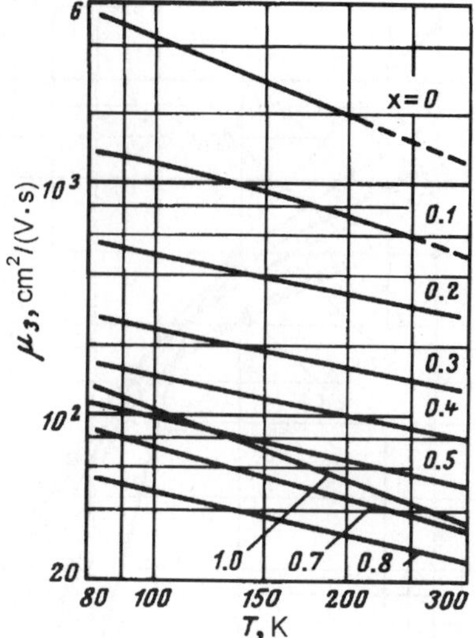

Figure 22.232 The temperature dependence of the hole mobility in Se_xTe_{1-x} along the c axis at various x (from magnetoresistance) [87].

Figure 22.231 The hole density in Se_xTe_{1-x} vs. x from the data of thermopower measurements [87] (a) and the hole mobility in Se_xTe_{1-x} vs. x at 300 K (b) determined by various methods [111]: △ – from magnetoresistance; ○ – from thermopower; ● – from the Hall effect.

22.3.7 $A^{VII}B^{III}$ - $A^{VII}B^{VI}$ compounds

Known are $MnAl_3$ as an $A^{VII}B^{III}$ semiconducting compound (E_g=0.45–0.58 eV, $\mu_n > 200$ cm^2/(V·s), $\mu_p \sim 200$ cm^2/(V·s) [242]), $ReSi_2$ as an $A^{VII}B^{IV}$ compound ($E_g = 0.13$ eV [16]), and MnSi crystallizing as cubes ($a = 0.456$ nm; $E_g = 0.5$-0.6 eV, μ_p=10–40 cm^2/(V·s) [242]).

Table 22.28 Semiconducting $A^{VII}B^{VI}$ compounds [288]

Compound	Crystal structure		ρ, g/cm^3	T_{melt}, K	T_D, K	T_N^{*1}, K	E_g, eV
	System	a, b, c, nm					
MnO_2^{*2}	Orthorhombic	0.927 (a); 0.287(b);				92	0.13–0.19
MnS	Cubic	0.522	3.83	1890		150	
MnS_2	Cubic, T^6	0.610		677		48	
MnSe	Cubic	0.545	5.35	1780	360	150	0.7
$MnSe_2$	Cubic, T^6	0.643		863		75	0.15
MnTe^{*3}	Hexagonal	0.415 (a); 0.671(c)		1470	160	310	1.3
$MnTe_2$	Cubic, T^6	0.695		1010		84	0.01–0.04
		0.453 (c)					

*1 All the compounds listed in Table 28 are antiferromagnetics; T_N is the Néel temperature.
*2 $\mu_n \simeq 1.2$ cm^2/(V·s) [242].
*3 $m_p/m_0 = 0.25$ ($\perp c$) and 0.47; $\mu_n \simeq 1$cm^2/(V·s) [106].

22.3.8 $A^{VIII}B^{V}$ – $A^{VIII}B^{VI}$ compounds

$A^{VIII}B^{V}$ compounds have been scantily studied. The conduction band in $PtSb_2$ represents eight ellipsoids of revolution whose major axes are directed along [111]. The valence zone has six ellipsoids of revolution positioned along the [100] axes [41]. The direct transition energy $E_d = 0.4$ eV; E_g corresponds to the indirect transition [175].

Table 22.29 Semiconducting $A^{VIII}B^{V}$ compounds

Compound	Crystal structure		T_{melt}, K	E_g, eV	m_n/m_0	m_p/m_0	μ_n,
	System, group	a, b, c, nm					$cm^2/(V\cdot s)$
$CoSb_2$	Rhombic	0.321 (a); 0.578 (b); 0.642 (c) [1]		0.2 [16]			300[16]
$CoSb_3$	Cubic, T_h^5	0.9034 [242] 0.9936 [242]	1131 [249]	0.5 [16, 249]			
$PtSb_2^*$	Cubic, T_h^6	0.643 [1]	1503 [41]	0.08 [41] 0.1 [288]	0.42–0.54 [41] 1.4 [41]	0.57–0.72 (m_d) [41] 0.168($m_\parallel$) [175] 0.098($m_{\perp 1}$) [175] 0.06($m_{\perp 2}$) [175]	$3.26\times 10^6 T^{-1.57}$ [288]

* $\epsilon_0 = 32$; $n_i = 1.32\cdot 10^{16} T^{3/2}\exp(0.11\text{eV}/2kT)$ [288].

Table 22.30 Semiconducting $A^{VIII}B^{VI}$ compounds

Compound	Crystal structure		T_{melt}, K	E_g, eV	μ_n,
	System, group	a, b, c, nm			$cm^2/(V\cdot s)$
Fe_2O_3	Rhombohedral	0.5427 [242]	1500 [242] 1838 [16]	1.6 [16] 2.2 [242]	
FeS_2	Cubic, T_h^5	0.5405 [242]	962 [249]	1.25 [16]	200 [16]
FeS_2	Rhombic, D_{2h}^{12}	0.358 (a); 0.479 (b); 0.572 (c) [1]	1183*1 [16]	0.95 [16]	
$FeTe_2$	Rhombic, D_{2h}^{12}	0.385 (a); 0.534 (b); 0.626 (c) [1]	1015*1 [16]	0.46 [16]	1 [51]
NiO*2	Cubic, O_h^5	0.418 [1]	3170 [42]	3.7*3 [95]	0.2 [25] 0.1 [95]

*1 The crystallization temperature is given, because these dichalcogenides melt with decomposition.
*2 $\rho = 6.6$ g/cm^3 [132]; $\epsilon_0 = 11.9$; $\epsilon_\infty = 4.75$ [29]. The polaron semiconductor; can have a magnetic order.
*3 Corresponds to the $3d \to 4s$ electron transition, but not to the energy gap width. The energy gap is much wider.

References

[1] Landolt–Börnstein, Zahlenwerte und funktionen aus physik, chemie, astronomie, geophysik, technik, Bd. 1, Springer–Verlag, Berlin, 1955.

[2] Wickoff, R W. G., Crystal structures, Intersci. Publ., New York, 1953.

[3] Khartsiev, V. E., Fizika tverdogo tela, 4, 433, 1962 (in Russian).

[4] Hoen, F. H., in: Boron, vol.1, Eds. Kohn, J. A., Nye, W. F. and Ganle, C. K., Plenum Press, New York, 1960.

[5] Golikova, O. A., Soloviev, N. E., Ugai, Ya. A., Feigel'man, V. A., Fizika i tekhnika poluprovodnikov, 13, 825, 1979 (in Russian).

[6] Golikova, O. A., J. Less-Common Metals, 82, 362, 1982.
[7] Golikova, O. A., Zhubanov, M. Zh., Klimashin, G. N., Fizika i tekhnika poluprovodnikov, 2, 548, 1968.
[8] Asanable, S., J. Phys. Soc. Jpn., 15, 989, 1960.
[9] Harbeke, G., Tosatti, E., Proc. of the 12th International Conf. on Phys. Semicond., Tuebner, Stuttgart, 1974, p. 626.
[10] Bube, R., Photoconductivity of solids, Wiley, New York, 1960.
[11] Smith, R., Semiconductors, Cambridge, Univ. Press, Cambridge, 1959.
[12] Ybuki, I., J. Phys. Soc. Jpn., 10, 549, 1955.
[13] Drabbe, G., J. Phys. Soc., B69, 1101, 1956; B72, 380, 1958.
[14] Golikova, O. A., Subanov, M. Z., Boron, Ed. Niemyski, T., Pol. Sci. Publ., Warsaw, 1970 (vol. 3).
[15] Rodot, M., Phys. Status Solidi, 3, 10, 1963.
[16] Ugai, Ya. A., Introduction to semiconductor chemistry, Vysshaya shkola, Moscow, 1965 (in Russian).
[17] Warschauer, D., J. Appl. Phys., 34, 1853, 1963.
[18] Kot, M. V., Kretsu, L. V., Izvestiya Akad. Nauk SSSR, Ser. "Physika", 281, 259, 1964 (in Russian).
[19] Mort, J., Phys. Rev. Lett., 18, 540, 1967.
[20] Caywood, J. M., Mead, C. A., J. Phys. Chem. Solids, 31, 983, 1970.
[21] Hopfield, J. J., Thomas, D. G., Phys. Rev., 122, 35, 1961.
[22] Dimmock, J. O., Wheeler, R. J., J. Appl. Phys., 32, 2271, 1961.
[23] Noguet, P. C., J. Phys., 26, 317, 1965.
[24] Lavine, C. F., Ewald, A. W., J. Phys. Chem. Solids, 32, 1121, 1971.
[25] Bogomolov, V. N., Kudinov, E. K., Firsov, E. A., Fizika tverdogo tela, 9, 3175, 1967 (in Russian).
[26] Prener, I., Woodbury, H. H., Intern. Conf. Semicond. Phys., Paris, 1964, p. 1231.
[27] Hilsum, C., Rose-Innes, A., Semiconducting III-V compounds, Oxford, Univ. Press, Oxford, 1961.
[28] Moss, T., Optical properties of semiconductors, Gordon and Breach, London, 1959.
[29] Bresler, M. S., Red'ko, N. A., Zhurnal eksperimental'noi i teoreticheskoi fiziki, 34, 149, 1972 (in Russian).
[30] Grosse, P., Springer tracts in modern phys., vol. 48, Ed. Hohler, G., J. Springer, Berlin - Heidelberg - New York, 1969.
[31] Gramatskii, V. I., Mushinskii, V. P., Izvestiya Akad. Nauk SSSR, Ser. "Physika", 28, 1077, 1964.
[32] Akhundov, A., Abdullaev, G. B., Gusseinov, G. D., Intern. Conf. Semicond. Phys., Paris, 1964, p. 1277.
[33] Veric, C., New Develop. Semiconductors, Leyden, 1973.
[34] Mirzabaev, M., Tuchkevich, V. M., Shmartsev, Yu. V., Izvestiya Akad. Nauk SSSR, Ser. "Physika", 28, 1300, 1964 (in Russian).
[35] Parfeniev, R. V., Pogarskii, A. M., Farbshtein, I. I., Fizika tverdogo tela, 4, 2630, 1963 (in Russian).
[36] Evans, R. L., Hazelwood, R. A., J. Phys. D2, 1507, 1969.
[37] Gibbons, Dy., Spear, W. E., J. Phys. Chem. Solids, 27, 1917, 1966.
[38] Fuldner, J., Proc. XI Int. Conf. Phys. Semicon., Warsaw, 1972, 351.
[39] Nitzki, V., Stossel, W., Phys. Status Solidi, 39(b), 309, 1970.
[40] Rodot, M., Compt. Rend. Acad. Sci., 260, 1908, 1965.
[41] Emtage, P. R., Phys. Rev., 138, 24, 1965.
[42] Botling, J. F., J. Chem. Phys., 33, 305, 1960.
[43] A handbook on semiconductor diodes, transistors and integrated circuits, Ed. Goryunov, N. N., Energiya, Moscow, 1972 (in Russian).
[44] Von Liebe, L., Ann. Phys., 7, 179, 1965.
[45] Salzer, O., Nieke, H., Ann. Phys., 7, 192, 1965.
[46] Onuki, J. Phys. Soc. Jpn., 20, 171, 1965.
[47] Kot, M. V., Panasyuk, L. M., Simashkevich, A. V., Tsurkan, A. E., Fizika tverdogo tela, 7, 1242, 1965 (in Russian).
[48] Yaschke, R., Ann. Phys., 7, 106, 1965.
[49] Lomakina, G.A., Fizika tverdogo tela, 7, 600, 1965 (in Russian).
[50] Komiga, H., Phys. Rev., 133, A1679, 1964.

[51] Suchet, J., Compt. Rend. Acad. Sci., 259, 3219, 1964.
[52] Resenman, M. J., Compt. Rend. Acad. Sci., 259, 2621, 1964.
[53] Adams, A. R., Spear, W. E., J. Phys. Chem. Solids, 25, 1113, 1964.
[54] Hacke, J., J. Appl. Phys., 35, 2484, 1964.
[55] Redin, A. D., Phys. Rev., 109, 1916, 1958.
[56] Rodot, M., Compt. Rend. Acad. Sci., 258, 5414, 1964.
[57] Jeavons, A. P., Sannders Y. A., J. Royal Phys. Soc., A310, 415, 1969.
[58] Bludau, W., Onton, A., Heinke, W., J. Appl. Phys., 45, 1846, 1974.
[59] Dunstadter, H., Z. Phys, 137, 383, 1954.
[60] Vol'fshtein, V. M., Fistul', V. I., Electronics and its applications, vol.4, Ser. "Itogi nauki i tekhniki," VINITI AN SSSR, Moscow, 1974 (in Russian).
[61] Sommer, A.H., Photoemissive materials. Preparation, properties and uses, Wiley, New York, 1968.
[62] Physical properties of chalcogenides – rare-earth elements, Ed. Zhuze, V. P., Nauka, Leningrad, 1974 (in Russian).
[63] Irvin, J. C., Bell Syst. Techn. J., 41, 387, 1962.
[64] Jacoboni, C., Canali, C., Ottaviani, G., Alberigi Quaranta, A., Solid State Electron., 20, 77, 1977.
[65] McLean, T. P., Progress in Semiconductors, vol. 5, Ed. Gibson, A. F., Heywood, London, 1960.
[66] Park, V. S., Phys. Rev., 143, 512, 1966.
[67] Lott, S. A., Synch, D. W., Phys. Rev., 141, 681, 1966.
[68] Camassel, J., Auvergne, D., Matthieu, H., Solid State Commun., 13, 63, 1973.
[69] Marby, J. A., Phys. Rev., 140, 304, 1965.
[70] Arumi, Z., Mancu, P., Rend. Semin. Fac. Sci. Univ. Gagliari, 33, No 3-4, 26, 1963.
[71] Morin, F. J., Phys. Rev., 93, 62, 1954.
[72] Prince, M. B., Phys. Rev., 92, 681, 1953.
[73] Golikova, O. A., Moizhes, B. Ya., Stil'bans, L. S., Fizika tverdogo tela, 3, 3105, 1961.
[74] Logan, R. A., Rowell, J. M., Trumbore, F. A., Phys. Rev., 136A, 1751, 1964.
[75] Whitaker, J., Solid State Electron., 8, 649, 1965.
[76] Braunstein, R., Moore, A. R., Herman, F., Phys. Rev., 109, 695, 1958.
[77] Zdanowich, W., Wojkowski, A., Phys. Status Solidi, 8, 569, 1965.
[78] Douglas, R., J. Phys. Chem. Solids, 21, 329, 1965.
[79] Richard, D., J. Phys. Chem. Solids, 26, 439, 1965.
[80] Busch, G., Vogt, O., Helv. Phys. Acta, 33, 437, 1960.
[81] Wooten, J., Condas, Y. A., Phys. Rev., 131, 657, 1963.
[82] Tamura, H., Masumi, T., Solid State Commun., 12, 1183, 1973.
[83] Hall, R. F., Wright, H. C., Brit. J. Appl. Phys., 18, 33, 1967.
[84] Tsidil'kovskii, I. M., Electrons and holes in semiconductors, Nauka, Moscow, 1972 (in Russian).
[85] Orlova, N. I., Shishkin, P. T., Izvestiya Akad. Nauk SSSR, Ser. "Physics–Mathematics," No 4, 53, 1964 (in Russian).
[86] Winker, H., Helv. Phys. Acta, 28, 633, 1955.
[87] Mell, H., Stuke, J., Phys. Status Solidi, 45, 163, 1971.
[88] Sorokin, G. P., Izvestiya vuzov, Ser. "Physika," No 4, 140, 1965 (in Russian).
[89] Wallis, Y., Ann. Phys., 17, 401, 1956.
[90] Pollack, Y. P., Trivich, D., J. Appl. Phys., 46, 163, 1975.
[91] Okamoto, K., Kawai, S., Jap. J. Appl. Phys., 12, 1130, 1973.
[92] Voskanyan, A. A., Inglizyan, P. N., Lalykin, S. P., Fizika i tekhnika poluprovodnikov, 12, 2096, 1978 (in Russian).
[93] Gorbachev, V. V., Putilin, I. M., Phys. Status Solidi (a), 16, 553, 1973.
[94] Sorokin, G. P., Penshev, Yu. M., Oush, P. T., Fizika tverdogo tela, 7, 2244, 1965 (in Russian).
[95] Ksendzov, Ya. M., Drabkin, I. A., Fizika tverdogo tela, 7, 1884, 1965 (in Russian).
[96] Fortin, E., Weichman, F. L., Phys. Status Solidi, 5, 515, 1964.
[97] Junod, P., Phil. Mag., 36, 941, 1977.

[98] Aliev, S. A., Suyunov, U. Kh., Arasly, D. G., Aliev, M. I., Fizika i tekhnika poluprovodnikov, 7, 1086, 1973.

[99] Aulich, E., Brebner, J. L., Mooser, E., Phys. Status Solidi, 31, 129, 1969.

[100] Ugai, Ya. A., Zaval'skii, Yu. P., Ugai, V. A., Doklady Akad. Nauk SSSR, 163, 663, 1965 (in Russian).

[101] Junod, P., Helv. Phys. Acta, 32, 567, 1959.

[102] Davlen, V. R. M., Phys. Rev., 143, 666, 1966.

[103] Komiga, U., Phys. Rev., A133, 679, 1964.

[104] Minder, R., Ottaviani, G., Canali, C., J. Phys. Chem. Solids, 37, 417, 1976.

[105] Leotin, J., Barbaste, R., Askenazy, S., Solid State Commun., 15, 693, 1974.

[106] Wassher, J. D., Seuter, R. M. J., Haas, C., Intern. Conf. Semicond. Phys., Paris, 1964, p. 1269.

[107] Kolomietz, B. T., Lebedev, E. A., Mazek, F. T., Intern. Conf. Semicond. Phys., Paris, 1964, p. 1283.

[108] Wood, C., Harrap, C., Kane, W. M., Phys. Rev., 121, 978, 1961.

[109] Masumi, T., Ahrenkiel, R. K., Brown, F. C., Phys. Status Solidi, 11, 163, 1965.

[110] Ponpon, L. P., IEEE Trans. Nucl. Sci., S-22, 182, 1975.

[111] Beyer, W., Mell, H., Stuke, J., Phys. Status Solidi (b), 45, 153, 1971.

[112] Krevs, V. E., Lutsiv, R. V., Pashovskii, M. B., Retrov, P.P., Phys. Status Solidi (b), 65, K43, 1974.

[113] Hanson, R. C., J. Phys. Chem., 66, 2376, 1962.

[114] Abrikosov, N. Kh., Semiconductor compounds, their manufacturing and properties, Nauka, Moscow, 1967 (in Russian).

[115] Reimherr, A., General Survey of the Semiconductor Field, NBS Technical Note, NBS, DC, 1962 (p. 153).

[116] Semiconductors, Ed.Hannay, N.B., American Chem. Soc. Monograph Ser., Reynolds Publ. Corp., New York, 1959.

[117] Materials used in semiconductor devices, Ed. Hogarth, C.A., Interscience Publishers, John Wiley and Sons, New York–London–Sydney, 1965.

[118] Thermophysical Properties of High-Temperature Solid Materials, Purdue Univ., 1967.

[119] Burnham, D. C., Brown, F. C., Knox, R. S., Phys. Rev., 119, 1560, 1960.

[120] Ahrenkiel, R. K., Phys. Rev., 18, 180, 1969.

[121] Goryunova, N. A., Complex diamond-like semiconductors, Sovetskoye radio, Moscow, 1968 (in Russian).

[122] Wargur, R. J., Ewald, K. A., J. Phys. Chem. Solids, 32, 697, 1971.

[123] Rodot, M., Les materiaux semi-conducteurs, Dunod, Paris, 19651

[124] Dobrolezh, S. A., Silicon carbide, Gosudarstvennoye izdatel'stvo tekhnicheskoi literatury UkrSSR, 1963 (in Russian).

[125] Ravich, Yu. I., Efimova, B. A., Smirnov, I. A., Methods of semiconductor analysis as applied to lead chalcogenides, Nauka, Moscow, 1968 (in Russian).

[126] Kolomietz, N. V., Lev, E. Ya., Sysoeva, L. M., Fizika tverdogo tela, 5, 2871, 1963 (in Russian).

[127] Albers, W., J. Appl. Phys., Suppl. 32, 2220, 1961.

[128] Champness, C. H., Kipling, A. L., Canad. J. Phys., 44, 769, 1966.

[129] Sobolev, V. V., Fizika tverdogo tela, 12, 2687, 1970 (in Russian).

[130] Cuhen, M. L., Au–Yang, M. T., Phys. Rev., 178, 1358, 1969.

[131] Winkler, H., Helv. Phys. Acta, 28, 663, 1955.

[132] Agrain, P., Balkansky, M., Table de constante des semiconducteurs, Pergamon Press, Paris, 1961.

[133] Turner, W. J., Fischer, A. S., Reese, W. E., J. Appl. Phys., Suppl. 32, 2241, 1961.

[134] Heller, M. W., Danielson, G. C., J. Phys. Chem. Solids, 23, 601, 1962.

[135] Redin, R. D., Morris, R. G., Danielson, G. C., Phys. Rev., 109, 1916, 1958.

[136] Li, P. W., Lee, S. N., Danielson, G. C., Phys. Rev., B6, 442, 1972.

[137] Crossman, L. D., Danielson, G. C., Phys. Rev., 171, 867, 1968.

[138] Physics and chemistry of II–VI compounds, Eds. Aven, M., Prener, J.S., North–Holland Publ. Company, Amsterdam, 1967.

[139] Hamilton, P. H., Semicond. Proc. Solid-State Technology, 7, No 6, 15, 1964.

[140] Betts, D. D., Canad. J. Phys., 39, 223, 1961.

[141] Robic, R. A., Edwards, J. L., J. Appl. Phys., 37, 2659, 1966.

[142] Rode, D. L., Phys. Rev., B2, 4036, 1970.

[143] Miloslavskii, V. K., Shklyarevskii, O. N., Fizika i tekhnika poluprovodnikov, 5, 926, 1971 (in Russian).

[144] Watanabe, H., Wada, H., Takahashi, T., Jpn. J. Appl. Phys., 3, 617, 1964.

[145] Rodot, H., J. Phys. Chem. Solids, 25, 85, 1964

[146] Umeda, J., J. Phys. Soc. Jpn., 19, 2052, 1964.

[147] Geik, R., Hakel, W. J., Perry, C. H., Phys. Rev., 148, 824, 1966.

[148] Balkansky, M., Zellag, B. A. M., Longen, D., J. Phys. Chem., 27, 299, 1966.

[149] Segall, B., Lorenz, M., Woodbury, H., Phys. Rev., 129, 2471, 1963.

[150] Medvedeva, Z. S., Chalcogenides of the IIIB subgroup elements of the Periodic table, Nauka, Moscow, 1968 (in Russian).

[151] Tagiev, B. F., Phys. Stat. Sol. (a), 3, K119, 1970.

[152] Catsuyama, C., Watanabe, Y., Hamaguchi, C., J. Phys. Soc. Jpn., 29, 150, 1970.

[153] Aven, M., J. Appl. Phys., 42, 1204, 1971.

[154] Park, Y. S., Hemenger, P. M., Chung, C. H., Appl. Phys. Lett., 18, 45, 1971.

[155] Reignold, D. G., Litton, C. W., Collins, T., Phys. Rev., 146, 1726, 1965.

[156] Seiler, D. G., Galazka, R. R., Becher, W. M., Phys. Rev., B3, 4274, 1971.

[157] Bashirov, R. I., Tadzhiev, R. M., Fizika tverdogo tela, 4, 1936, 1970 (in Russian).

[158] Bush, G., Moldanova, M., Helv. Phys. Acta, 35, 500, 1962.

[159] Dietz, R. F., Hopfield, J. J., Thomas, D. G., J. Appl. Phys., 32, 2282, 1961.

[160] Pigon, K., Helv. Phys. Acta, 41, 1104, 1968.

[161] Fisher, P., Fan, H., Bull. Amer. Phys. Soc., 4, 409, 1959.

[162] Ryvkin, S. M., Photoelectric phenomena in semiconductors, Fizmatgiz, Moscow, 1963 (in Russian).

[163] Zdanowicz, W., Kenkie, Z., Bull. Acad. Pol. Sci., Ser. Sci. Chem., 12, 729, 1964.

[164] Fingold, L., Phys. Rev. Lett., 13, 233, 1964.

[165] Shevchenko, V. Ya., Zhurnal neorganicheskikh materialov, 11, 1719, 1975 (in Russian).

[166] Zdanowicz, W., Wielzak, B., Zdanowicz, P., Acta Phys. Polon., A48, 27, 1975.

[167] Turner, W. C., Fischler, A. S., Reese, W. G., Phys. Rev., 121, 759, 1961.

[168] Black, J., J. Phys. Chem. Sol., 2, 240, 1957.

[169] Smirnov, I. A., Shadrichev, E. V., Kutasov, V. A., Fizika tverdogo tela, 11, 3311, 1969 (in Russian).

[170] Madelung, O., Physics of III–V compounds, John Wiley and Sons, New York, 1964.

[171] Long, D., Energy bands in semiconductors, Sydney Interscience Publ., a Division of John Wiley and Sons, New York - London, 1968.

[172] Minden, H. T., Appl. Phys. Lett., 17, 358, 1970.

[173] Kessler, F. R., Phys. Status Solidi, 6, 3, 1964.

[174] Stordeur, M., Kuhnberger, W., Phys. Status Solidi (b), 69, 377, 1975.

[175] Damon, D. H., Miller, R. C., Emtage, P. R., Phys. Rev. 5B, 2175, 1972.

[176] Becla, P., Gummieuny, Z., Misiewicz, J., Opt. Appl., 9, 143, 1979.

[177] Shevchenko, V. Ya., Marenkin, S. F., Ponomarev, V. F., Zhurnal neorganicheskikh materialov, 13, 1898, 1977 (in Russian).

[178] Steigmeier, E. F., Kudman,I., Phys. Rev., 141, 767, 1966.

[179] Pigon, K., Bull. Acad. Pol. Sci., Ser. Sci. Chem., 9, 751, 1961.

[180] Masumoto, K., Komiya, H., J. Jpn. Instrum. Met., 28, 273, 1964.

[181] Tsurkan, A. E., Maksimova, O. G., Verlan, V. I., Complex semiconductors and their physical properties, Shtinitsa, Kishinev, 1971 (in Russian).

[182] Naake, M. J., Belcher, S. C., J. Appl. Phys., 35, 3064, 1964.

[183] Domingo, G., Itoga, R. S., Kannewur, C. R., Phys. Rev., 143, 536, 1966.

[184] Markov Yu. F., Investigation of the optical properties of semiconducting compounds in the far IR spectrum range, Phys.-mat. candidate thesis, A. Ioffe Phys. Tech. Institute, Leningrad, 1972 (in Russian).

[185] Ehrenreich, H., Phys. Rev., 120, 1951, 1960.

[186] Panfilov, V. V., Subbotin, L. S., Vereshchagin, L. F., Doklady Akad. Nauk SSSR, 96, 559, 1971 (in Russian).

[187] Semiconductor compounds of groups II–V, Nauka, Moscow, 1978 (in Russian).

[188] Heller, M. W., Babiskin, J., Radoff, R. L., Phys. Rev., A36, 363, 1971.

[189] Zdanowicz, W., Wojakowski, A., Phys. Status Solidi, 16, K129, 1966.

[190] Zdanowicz, W., Wojakowski, A., Phys. Status Solidi, 8, 569, 1965.

[191] Arushanov, E. K., Lashul, A. V., Mashovets, D. V., Phys. Status Solidi (b), 102, K121, 1980.

[192] Gelten, M. J., Von Liehout, A., van Es, C., Blout, F. A. P., J. Phys., C11, 227, 1978.

[193] Slack, G. A., J. Appl. Phys., 143, 666, 1966.

[194] Moroz, A. I., Odarich, V. A., Ukrainskii fizicheskii zhurnal, 16, 1501, 1971 (in Russian).

[195] Kozlov, S. F., Phys.–math. doctorate thesis, Kurchatov Inst. of Atomic Energy, Moscow, 1985 (in Russian).

[196] Bloom, F. A. P., Gelten, M. J., Phys. Rev., B19, 2411, 1979.

[197] Kolomietz, B. T., Mazets, T. F., Sarsembinov, Sh. Sh., Fizika i tekhnika poluprovodnikov, 5, 2301, 1971 (in Russian).

[198] Shutov, S. O., Sobolev, V. V., Smeshlivyi, L. I., Semiconductor compounds and their solid solutions, Izdatel'stvo Akad. Nauk MoldSSR, Kishinev, 1970, p. 155 (in Russian).

[199] Whitsett, C. R., Nelson, D. A., Phys. Rev. 5B, 3125, 1972.

[200] Overhof, H., J. Phys. Chem. Solids, 43, 221, 1971.

[201] Miller, K. A., Schneider, J., Phys. Lett., 4, 288, 1963.

[202] Look, D. C., Moore, D. L., Phys. Rev. 5B, 3406, 1972.

[203] Kohler, H., Fabricins, A., Phys. Status Solidi (b), 71, 487, 1975.

[204] Bloom, F. A. P., Intern. Summ. School on Narrow-Gap Semicond. Physics and Applications, Nimes (France), 1979, p. 191.

[205] Kawasaki, T, Tanaka, T., J. Phys. Soc. Jpn., 21, 2475, 1966.

[206] Eichler, W., Simon, G., Phys. Status Solidi (b), 86, K85, 1978.

[207] Helbig, R., Wagner, P., J. Phys. Chem. Solids, 35, 327, 1974.

[208] Hagenberg, F., Ph. D. Thesis, D83, Techn. Univ., Berlin, 1980.

[209] Grant, F. A., Rev. Mod. Phys., 31, 646, 1959.

[210] Yu, P. Y., Cardona, M., J. Phys. Chem. Solids, 34, 29, 1973.

[211] Frederikse, H. P. R., J. Appl. Phys., Suppl. 32, 2211, 1961.

[212] Austin, I.G., Mott, N. E., Adv. Phys., 18, 41, 1969.

[213] Madelung, O., in: Semiconductors: group VI elements and III–IV compounds, Ed. Madelung, O., Springer–Verlag, Berlin, 1991.

[214] Kesamanly, F. P., Fizika i tekhnika poluprovodnikov, 8, 225, 1974 (in Russian).

[215] Kikuchi, Y., Chubachi, N., Iinuma, K., Sendai Symp. Acoustoelectron., Japan, 1968.

[216] Handbook of a chemist, Ed. Nikol'skii, B. P., Khimiya, Moscow, 1971 (in Russian).

[217] Spear, W. E., Mort, P., Proc. Phys. Soc., 81, 130, 1963.

[218] Krupishev, R. S., Abagyan, S. A., Davydov, A. A., Fizika i tekhnika poluprovodnikov, 6, 75, 1972 (in Russian).

[219] Kahan, A., Lipson, H. G., Loewinstein, E. V., Intern. Conf. Semicond. Phys., Paris, 1964, p. 1067.

[220] Mitra, S. S., Marshall, R., Intern. Conf. Semicond. Phys., Paris, 1964, p. 1085.

[221] Kireev, P. S., Semiconductor physics, Vysshaya Shkola, Moscow, 1969 (in Russian).

[222] Zallen, R., Phys. Rev., B1, 4058, 1970.

[223] Szusukiewicz, W., Phys. Status Solidi (b), 91, 361, 1979.

[224] Wright, G. B., Strauss, A. J., Harman, T. C., Phys. Rev., 125, 1534, 1962.

[225] Fistul', V. I., Strongly alloyed semiconductors, Nauka, Moscow, 1967 (in Russian).

[226] Prince, M. B., Phys. Rev., 93, 1204, 1954.

[227] Lehoczky, S. L., Broerman, J. G., Nelson, D. A., Whitsett, C. R., Phys. Rev., B9, 1598, 1974.

[228] Brebrick, R. P., Scanlon, W. W., Phys. Rev., 99, 598, 1954.

[229] Stradling, R., Solid State Commun., 6, 665, 1968.

[230] Cardona, M., Phys. III–V compounds, vol. 3, Academic Press, New York, 1966.

[231] Stradling, R., Electronic components, Pergamon Press, London, 1958, p. 1135.

[232] Walton, A., Mishra, U., Proc. Phys. Soc., 90, 1111, 1967.

[233] Stradling, R., Phys. Lett., 20, 217, 1966.

[234] Gavrilenko, V.I., Grekhov, A.M., Korbutyak, D.V., Litovchenko, V.G., Optical properties of semiconductors, Naukova Dumka, Kiev, 1987 (in Russian).

[235] Kot, M. V., Prilepov, V. D., Tsurkan, A. E., Semicondictor compounds and their solid solutions, Izdatel'stvo Akad. Nauk MoldSSR, Kishinev, 1970, p. 3 (in Russian).

[236] Mashkovskii, M. V., Foreign electronic technology, No 12, Izdatel'stvo CNII "Electronika," Moscow, 1974, p. 3 (in Russian).

[237] Kalashnikov, S. G., J. Phys. Chem. Solids, 8, 52, 1959.

[238] Ewald, A. W., Tutle, O. N., J. Phys. Chem. Solids, 8, 523, 1959.

[239] Ivanov-Omskii, V. I., Kolomietz, B. T., Ogorodnikov, V. K., Fizika i tekhnika poluprovodnikov, 4, 264, 1970 (in Russian).

[240] Galazka, R. R., Phys. Lett., 32A, 101, 1970.

[241] Nasledow, D. N., Shevchenko, V. Ya., Phys. Status Solidi (a), 15, 3, 1973.

[242] Crystal–chemical, physical and chemical properties of semiconductor materials, Izdatel'stvo Standartov, Moscow, 1973 (in Russian).

[243] Gel'mont, B. L., Ivanov-Omskii, V. I., Kolomietz, B. T., Fizika i tekhnika poluprovodnikov, 5, 266, 1971 (in Russian).

[244] Szymanska W., Physics of narrow-gap semiconductors, Proc. IIIth Internat. Conf., Warsaw, Sept. 1977, Eds. Rauluszkewicz, J., Gorska, M. and Kaczmarek, F., PWN–Polish–Sci. Publishers, Warsaw, 1978.

[245] Dziuba, Z., Wrobel, J., Phys. Status Solidi (b), 100, 379, 1980.

[246] Landolt–Börnstein, Numerical data and functional relationship in science and technology, vol. 22, subvol. a, Ed. Madelung, O., Springer–Verlag, Berlin, 1987.

[247] Kato, N., Kummura, M., Iwami, M., Kawabe, K., Jpn. J. Appl. Phys., 16, 1623, 1977.

[248] Grigas, J., Krist. Techn., 13, 683, 1978.

[249] Sominskii, M. S., Semiconductors, Nauka, Leningrad, 1967 (in Russian).

[250] Stirn, R. J., Becker, W. M., Phys. Rev., 148, 907, 1966.

[251] Reid, F. J., Willardson, R. K., J. Electron. Control, 5, 54, 1958.

[252] Baars, J., Jorger, F., Solid State Commun., 10, 875, 1972.

[253] Ilegems, M., Montgomery, H. C., J. Phys. Chem. Solids, 34, 885, 1973.

[254] Burstein, E., Pinczuk, A., Wallis, R. F., J. Phys. Chem. Solids, 32, Suppl. No 1, 251, 1971.

[255] Rode, D. L., in: Semiconductors and semimetals, vol. 10, Eds. Willardson, R. K. and Beers, A. C., Academ. Press, New York, 1975.

[256] Casey, H. C., Ermanis, F., Wolfstern, K. B., J. Appl. Phys., 40, 2945, 1969.

[257] Solomon, R., 2nd Intern. Symp. GaAs and Related Compounds, Inst. Phys., London, 1968, p. 11.

[258] Roy, B., Chakrabarty, B. R., Bhattacharya, R., Dutta, A. K., Solid State Commun., 25, 937, 1978.

[259] Wiley, J. D., in: Semiconductors and semimetals, vol. 10, Eds. Willardson, R. K. and Beers, A. C., Academic Press, New York, 1975 (vol. 10).

[260] Vilms, J., Garrett, J. P., Solid State Electron., 15, 443, 1972.

[261] Marshall, J. M., J. Phys. 10C, 1283, 1977.

[262] Dakhno, A. N., Emel'yanenko. O. V., Lagunova, T. S., Metreveli, S. G., Fizika i tekhnika poluprovodnikov, 10, 677, 1976 (in Russian).

[263] Golovanov, V. V., Metreveli, S. G., Snukaev, N. V., Starosel'tseva, S. P., Fizika i tekhnika poluprovodnikov, 3, 120, 1969 (in Russian).

[264] Stillman, G. E., Wolfe, C. M., Dimmock, J. O., in: Semiconductors and semimetals, vol. 12, Eds. Willardson, R. K. and Beers, A. C., Academic Press, New York, 1977.

[265] Trifonov, V. I., Yaremenko, N. G., Fizika i tekhnika poluprovodnikov, 5, 953, 1971 (in Russian).

[266] Yaremenko, N. G., Potapov, V. T., Ivleva, V. S., Fizika i tekhnika poluprovodnikov, 6, 1238, 1972 (in Russian).

[267] Rode, D. L., Phys. Rev., B3, 3287, 1971.

[268] Schein, L. B., Phys. Rev., B15, 1024, 1977.

[269] Filipchenko, A. S., Bol'shakov, L. P., Phys. Status Solidi, 77, 53, 1976.

[270] Likhter, A. I., Pel, E. G., Prisyazhnyuk, S. I., Phys. Status Solidi (a), 14, 265, 1972.

[271] Kipperman, A. H. M., Solid State Commun., 9, 1825, 1971.

[272] Kipperman, A. H. M., Van der Leeden, G. A., Solid State Commun., 6, 657, 1968.

[273] Kipperman, A. H. M., Vermij, C. J., Nuovo Cimento, B63, 29, 1969.

[274] Tatsuyama, C., Hamaguchi, C., Tomita, H., Nakai, J., Jpn. J. Appl. Phys., 10, 1698, 1971.

[275] Fivaz, R., Mooser, E., Phys. Rev., 163, 743, 1969.

[276] Fivaz, R., Shmid, Ph. E., in: Physics and Chemistry of Materials with Layered Structures, vol. 4: Optical and electrical properties, Ed. Lee, P. A., D. Reidel Publ. Comp., London, 1976, p. 343.

[277] Fisher, G., Brebner, J. L., J. Phys. Chem. Solids, 23, 1363, 1962.

[278] Gouskov, L., Gouskov, A., Solid State Commun, 28, 99, 1978.

[279] Fielding, P., Fisher, G., Mooser, E., J. Phys. Chem. Solids, 8, 434, 1959.

[280] Nishino, T., Hamakawa, Y., Jap. J. Appl. Phys., 16, 1291, 1977.

[281] Damon, R. W., Redington, R. W., Phys. Rev., 96, 1498, 1954.

[282] Sugaike, S., Minerals J., 2, 63, 1957.

[283] Itoga, R. S., Kannewurf, C. R., J. Phys. Chem. Solids, 32, 1099, 1971.

[284] Ikari, T., Hashimoto, K., Phys. Status Solidi (a)., 31, K115, 1975.

[285] Makita, Y., Kobayashi, K., Kanada, M., Kawai, T., J. Phys. Soc. Jpn., 25, 816, 1968.

[286] Kobayashi, K., Kawai, T., Kanada, M., J. Phys. Soc. Jpn., 23, 305, 1967.

[287] Kawai, T., Kobayashi, K., Kurita, M., Makita, Y., J. Phys. Soc. Jpn., 30, 1101, 1971.

[288] Physical and chemical properties of semiconductors. Handbook, Nauka, Moscow, 1979 (in Russian).

[289] Berchenko, N. N., Krevs, V. E., Sredin, V. G., Semiconductor solid solutions and their application, Voenizdat, Moscow (in Russian).

[290] Landolt–Börnstein, Numerical data and functional relationship in science and technology, vol. 17, subvol. b, Ed. Madelung, O., Springer–Verlag, Berlin, 1982.

[291] Conwell, E.M., High-field transport in semiconductors, Academic Press, New York – London, 1967

[292] Landolt–Börnstein, Numerical data and functional relationship in science and technology, vol. 17, subvol. a, Ed. Madelung, O., Springer–Verlag, Berlin, 1982.

[293] Landolt–Börnstein, Numerical data and functional relationship in science and technology, vol. 17, subvol. e, Ed. Madelung, O., Springer–Verlag, Berlin, 1983.

[294] Landolt–Börnstein, Numerical data and functional relationship in science and technology, vol. 17, subvol. f, Ed. Madelung, O., Springer–Verlag, Berlin, 1983.

[295] Kamath, G. S., Mater. Res. Bull., 4, S57, 1969.

[296] Van Daal, H. J., Knippenberg, W. F., Wasscher, J. D., J. Phys. Chem. Solids, 24, 109, 1963.

[297] Nimitz, G., Schlicht, B., Springer tracts in modern physics, vol. 98, Springer–Verlag, Berlin, 1983.

[298] Van den Dries, J., Lieth, R., Phys. Status Solidi (a), 5, K171, 1971.

[299] Stourac, L., Zavetova, M., Abraham, A., Proc. of the 12th Int. Conf. on Phys. Semicond., Tuebner, Stuttgart, 1974, p. 621.

[300] Stanchev, A., Vodenicharov, C., Thin Solid Films, 38, 67, 1976.

[301] Asanabe, S., Okazaki, A., J. Phys. Soc. Jpn., 15, 989, 1960.

[302] Kyriakos, D. S., Valassiades O., Economou, N. A., Instrum. Phys. Conf., 1979, ser.43, chapt.8.

[303] Anisimov, B. B., Gabedova, A. A., Dzhamagidze, Sh. Z., Izvestiya Akad. Nauk SSSR, Ser. "Inorganic materials," 14, 1417, 1978 (in Russian).

[304] Bahl, S. K., Chopra, K. L., J. Appl. Phys., 41, 2196, 1970.

[305] Yabumoto, T., J. Phys. Soc. Jpn., 13, 972, 1958.

[306] Albers, W., Haas, C., Vink, H., J. Appl. Phys., 32, 2220, 1961.

[307] Umeda, J., J. Phys. Soc. Jpn., 16, 124, 1961.

[308] Maier, H., Daniel, D. R., J. Electron. Mater., 6, 693, 1977.

[309] Sagar, A., Miller, R. C., Proc. of Int. Conf. on Phys. Semicond., July 1962, Exeter, Ed. Stickland, A. C., p.653.

[310] Allgaier, R. S., Houston, B., Phys. Rev., B5, 2186, 1972.

[311] Fonstad, C. J., Rediker, R. H., J. Appl. Phys., 42, 2911, 1971.

[312] Gowers, J. P., Lee, P. A., Solid State Commun., 8, 1447, 1970.

23

Dielectrics

A.P. Geppe

23.1 Introduction

The substances whose basic electrical property consists in their capacity to polarization and in which an electrostatic field can exist are referred to as dielectrics. The electrostatic field can be conserved for a long time only in the media that poorly conduct an electric current. The parameter of a substance determining quantitatively its electric conductance is the specific electric conductivity σ (in S/m), as well as the specific electric resistivity $\rho = 1/\sigma = nqu$ (in Ω·m), where n is the concentration of charge carriers (in m^{-3}), q is the electric charge of a carrier (in C), u is the carrier mobility (in $m^2/(V \cdot s)$): $u = v/E$, where v is the velocity of directed motion of a carrier (in m/s) in an electric field with a strength E (in V/m). The charge carriers in dielectrics can be ions, electrons, and charged colloid particles. The conduction current density is defined as $j = \sigma E$ (in A/m^2). In real dielectrics, $\rho = 10^7$–10^{18} Ω·m; because of a low concentration of carriers, the conductivity of dielectrics is so small that they are considered as materials which practically do not conduct a current. There does not exist any exact boundary between dielectrics and semiconductors, and, therefore, the lower limit of ρ-values for dielectrics is rather arbitrary.

When an electric voltage applied to a dielectric sample increases, the latter remains essentially non-conducting (holds a high ρ), until a high-conductance channel arises in the dielectric under the action of electric field. This gives rise to practically a short circuit between the electrodes, that is, to the breakdown of the dielectric. The minimum voltage that, when applied to a dielectric specimen, causes its breakdown is called the breakdown voltage U_b. The parameter that determines the resistance of a dielectric material to the breakdown is an electric strength E_b. The latter refers to the strength of an electric field in the dielectric at which its breakdown occurs. This parameter is defined as follows: $E_b = U_b/d$, where d is the thickness of a dielectric specimen at the breakdown site. The electric strength E_b is expressed usually in megavolts per meter (MV/m). It is more convenient in practice to express it in kilovolts per millimeter (1 kV/mm = 1 MV/m).

The important parameter of dielectrics comprises a dielectric constant characterizing the ability of a material to create a capacitance. In the SI, a distinction is made between a relative dielectric constant ϵ_r, which is a dimensionless quantity, and an absolute dielectric constant $\epsilon_r\epsilon_0$ (in F/m), where $\epsilon_0 = 8.854 \cdot 10^{-12}$ F/m is the electric constant.

Dielectrics consist essentially of bound charged particles: positive nuclei and electrons rotating around the nuclei in atoms, molecules, and ions as well as elastically-bound oppositely charged ions positioned in the lattice sites of ionic crystals. The polarization of dielectrics makes an ordered displacement of bound charges under the action of an applied electric field. The displacement l is small and ceases as soon as the electric field force responsible for the movement of charges relative to each other is balanced by a force of their interaction. As a result of polarization, each molecule (or any other

0-8493-2861-6/97/$0.00+$.50

particle) in a dielectric becomes an electric dipole – a system of two equal, opposite in sign charges q spaced at l. An electric moment $p = ql$ (in C·m) is induced in a particle as a result of its polarization. For linear dielectrics (to which most dielectrics belong), there exists a direct proportionality between the induced moment and the electric field strength E acting on the particle: $p = \alpha E$. The proportionality factor α (in F·m^2) is called the polarizability of the given particle. The relationship between ϵ_r and α is written as

$$\epsilon_r = 1 + N\alpha/\epsilon_0.$$

This formula is suitable for gaseous dielectrics, but in some cases it can also be applied with some degree of accuracy to liquid and solid dielectrics. The polarizability α of a particle depends on the mechanism of polarization which is in turn determined by the nature of a dielectric.

The basic mechanisms of polarization are the electronic, ionic, and dipole ones.

The electronic polarization constitutes an elastic displacement of electron orbits with respect to nuclei in atoms, molecules, and ions under the action of an external electric field. This polarization is fast, proceeding for about 10^{-14}–10^{-15} s, which is comparable with the light oscillation period.

The ionic polarization presents an elastic displacement of oppositely charged ions in the lattice sites of ionic crystals under the action of an external electric field. The polarization of this type is also fast, taking place in 10^{-12}–10^{-13} s. The electronic component of polarization occurs in the ionic crystals as well. For this reason, such dielectrics possess a higher value of ϵ_r than nonpolar ones.

The fast polarizations proceed without loss of energy.

The dipole polarization is typical of polar dielectrics. Polar molecules are asymmetrical in structure. The centers of gravity of their equivalent opposite in sign charges do not coincide, and, therefore, in the absence of an external field these molecules have an intrinsic electric moment, that is, they make rigid (constant) dipoles. The essence of the dipole polarization is that these rigid dipoles turn (orient) along the direction of an applied electric field. In a polar molecule, electronic polarization also occurs. Because of this, in polar dielectrics the values of ϵ_r are higher than in nonpolar ones. The dipole polarization belongs to the delayed (relaxation) types of polarization, taking about 10^{-2}–10^{-10} s for its completion. The relaxation-type polarizations fell behind the variation of an external electric field inducing these polarizations and are accompanied by an energy loss.

Along with the electronic and ionic polarizations, a spontaneous polarization is characteristic of nonlinear dielectrics – ferroelectrics. The spontaneous polarization, being a polarization of the relaxation type, develops spontaneously under the action of internal processes within a certain temperature range limited by the Curie ferroelectric points. As this takes place, the structure of a crystal unit cell becomes asymmetrical and gains an electric moment. The electric moments of the adjacent unit cells are parallel to each other within small areas (domains), so that the domain is spontaneously, very intensively polarized. But the adjacent domains are polarized in different directions, and initially the total polarization P of a ferroelectric specimen is equal to zero. In an external electric field, the polarization vectors of the domains orientate parallel, and the whole specimen turns out to be intensively polarized in a single direction. In ferroelectrics, the value of ϵ_r near the Curie point can amount to 10^6.

In a dielectric placed in an electric field there occurs an energy dissipation. The energy (power) dissipated per second is known as "dielectric losses." The lost energy is transformed to heat, resulting in a heating of the dielectric. Consequently, the electric and other important characteristics of the dielectric are impaired. The losses in dielectrics are observed at both an alternating and a direct voltage. However, by dielectric losses are meant a power dissipation in an alternating electric field. The phase of the current vector in the dielectric specimen to which an alternating voltage is applied leads the phase of the voltage vector by an angle $\phi < 90°$. The angle δ complementary to 90° is termed a dielectric loss angle. In the ideal (loss-free) dielectrics $\phi = 90°$ and $\delta = 0$. As for the dielectric parameter, the tangent of dielectric loss angle $\tan\delta$ is commonly used.

23.2 Gaseous Dielectrics

Gases under normal conditions exhibit a high resistivity and very low dielectric losses. Among the advantages of gases are the restoration of their insulating properties after breakdown and the absence of aging effects (deterioration of the properties with time). Their disadvantage furnishes a low (as compared with liquid and solid dielectrics) electric strength at normal pressure. To elevate the electric strength, use is made of both an increase in pressure of gases and their higher rarefaction. The electric strength of a gaseous insulation can also be enhanced by the employment of electronegative gases. The molecules of these gases, containing usually the atoms of fluorine, chlorine, and other halogens, can capture free electrons to turn into hardly mobile negative ions. The removal of the mobile electrons hinders the development of an electric discharge. As a result, the electric strength of the gas becomes higher.

Air, particularly under near-normal atmospheric conditions, is the most important gaseous dielectric in practical electrical engineering. The properties of some gases, compared to those of air, are given in Table 23.1. (The corresponding parameters of air are taken to be unity.)

The dielectric constant of gases is little different from unity. The values of ϵ_r for some gases are presented in Table 23.2. With increasing pressure (at constant temperature) in gases ϵ_r grows because of an enhancement in the concentration of polarizable particles. The ϵ_r values for some gases at different pressures are listed in Table 23.3.

In weak electric fields at moderate temperatures, gases provide a very low electric conductivity. Under these conditions the free charge carriers, electrons and ions, being few in number are produced only under the action of low-intensive external ionizers: cosmic rays and natural ionizing radiation. As a consequence, under the above conditions, gases make excellent dielectrics with a resistivity of about 10^{18} Ω·m and practically without dielectric losses ($\tan\delta$ is about 10^{-8}). The electric conductivity of gases increases at high temperatures beginning with 10^3–10^4 K when the thermal-motion energy of gas particles is high, so that they can ionize each other and a thermal ionization occurs. The thermal ionization of air is built up beginning with a temperature of 8,000 K. At 20,000 K, air is almost fully ionized. The process of thermal ionization plays a decisive role in a highly conducting electric-arc channel whose temperature reaches 4,000–15,000 K.

The mechanism of breakdown in gases is based on the process of impact ionization caused by free electrons which, when accelerated in an electric field, collide with neutral molecules of the gas and ionize them.

The breakdown in a uniform electric field begins almost immediately after attaining a certain voltage U_b. A spark jumps across the electrodes. At a sufficient power of the voltage source, the spark can transform into an electric arc. The Paschen's law is obeyed in gases: at constant temperature the breakdown voltage in a gas depends on the product of its pressure p and the distance d between the electrodes: $U_b = f(pd)$. This dependence is presented in Fig. 23.1 for air and hydrogen. For each gas there exists a minimum value of breakdown voltage at a certain value of pd (327 V for air at $pd = 665$ Pa·mm). To site an example, the minimum breakdown voltage is equal to 195 V for argon, 280 V for hydrogen, and 420 V for carbon dioxide. In regard to the breakdown at an alternating voltage, the above figures relate to the amplitude values.

The electric strength of a gas does not remain unchanged even in a uniform field at constant pressure and temperature. For small electrode spacings, E_b grows considerably because the formation of electron avalanches becomes complicated. The values of E_b for air under normal conditions with d varying from 0.006 to 1 cm are presented in Table 23.4.

In a nonuniform electric field, the breakdown voltage of a gas at the same electrode spacing is the lower, the greater a degree of field nonuniformity. The least value of U_b is observed for a gas gap between a rod and a plane electrode, because an electric field with the highest degree of nonuniformity is generated between such electrodes. Figures 23.2 and 23.3 show U_b in an air gap as a function of the electrode spacing for a uniform and a highly nonuniform field, respectively.

In asymmetric electric fields which are set up either between the electrodes of different shapes or between similar electrodes, if one of them is grounded (in so doing, the symmetry of the field is disturbed under the influence of the ground or other grounded parts), in the case of the breakdown of a gas gap at constant voltage, the polarity of the electrodes has a considerable effect on U_b (Fig. 23.4).

The breakdown of a gas at high frequencies proceeds with distinguishing features. Figure 23.5 displays the breakdown voltage of air in a uniform electric field at a pressure of 0.1 MPa as a function of frequency (considered is the ratio of $U_{b,f}$ at the frequency f to $U_{b,0}$ in a uniform electric field). At frequencies of up to 10^4 Hz, the breakdown voltage is frequency-independent and with further increase in frequency gradually decreases, passing through a minimum in the frequency range of 10^6–10^7 Hz.

A discharge in air along the surface of a solid dielectric is named "surface discharge," or "surface flashover." The insertion of a solid dielectric into an air gap essentially reduces its discharge voltage. The discharge develops in air along the surface of the solid dielectric, always at a lower voltage compared with that of the air gap without the dielectric insert. Figure 23.6 shows the voltage U_d of a surface discharge in air along insulating cylinders made of different solid dielectrics at a frequency of 50 Hz as a function of the cylinder height (the length of the discharge gap). A decrease in the discharge voltage results from the distortion of the electric field uniformity due to a varying thickness of a moisture film on the surface of the dielectric cylinder. By thoroughly drying hydrophilic materials, their flashover voltage can be increased practically up to the voltage of a pure air gap without a dielectric insert.

The flashover voltage in a nonuniform electric field is much lower than in a uniform field, with the hygroscopic properties of the solid dielectric having a weaker effect (Fig. 23.7).

Liquefied gases possess high insulating properties: their ϵ_r is slightly more than unity, whereas the electric strength amounts to 30–35 MV/m [7]. Some properties of liquefied gases are listed in Table 23.5.

Table 23.1 Properties of gases relative to the properties of air [1]

Characteristic	Nitrogen	Carbon dioxide	Hydrogen	SF_6 gas
Density	0.97	1.52	0.07	5.19
Thermal conductivity	1.08	0.64	6.69	0.70
Specific heat	1.05	0.85	14.35	0.59
Electric strength	1.00	0.90	0.60	2.3

Table 23.2 Dielectric constant ε_r of gases at 20°C and a pressure of 0.1 MPa [1]

Gas	ε_r (low frequencies)	ε_r (wavelength of 30 mm)
Helium	1.000072	
Hydrogen	1.00027	1.00036
Oxygen	1.00055	1.00053
Argon	1.00056	
Nitrogen	1.00058	1.00059
Carbon dioxide	1.00096	1.00099
SF_6 gas	1.00191	
Air	1.00058	1.00058

Table 23.3 Dielectric constant ε_r of gases at various pressures [1]

Gas	0.1 MPa	2 MPa	4 MPa	10 MPa
Air (at 19°C)	1.00058	1.0109	1.0218	1.0549
Nitrogen (at 20°C)	1.00058	1.0109		1.055
Carbon dioxide (at 15°C)	1.00098	1.020	1.050	

Table 23.4 The electric strength of air in a uniform electric field at 20°C and a pressure of 0.1 MPa for various electrode spacings [2]

d, cm	0.006	0.008	0.01	0.02	0.04	0.06	0.1	0.4	1.0
E_b, MV/m	12.5	10.7	9.7	7.4	5.9	5.3	4.5	3.6	3.1

Table 23.5 Properties of liquefied gases [8]

Characteristic	Nitrogen	Oxygen	Hydrogen	Neon	Helium
Boiling point, K	77.4	90.2	20.4	27.2	4.2
Dielectric constant ε_r	1.431	1.48	1.231		1.047

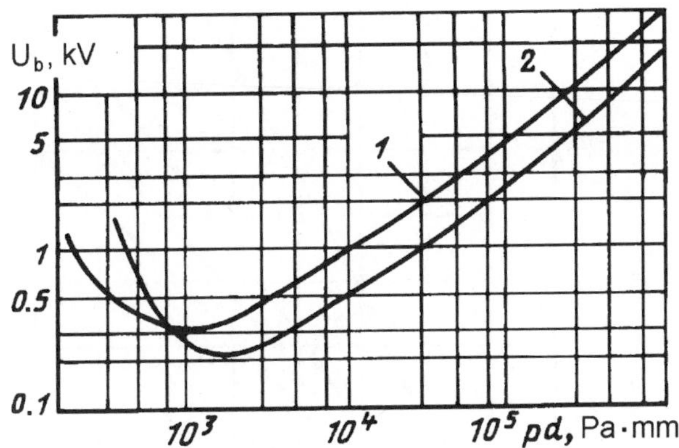

Figure 23.1 The dependence of U_b on the product of the pressure p by the electrode spacing d for air (*1*) and hydrogen (*2*) [1].

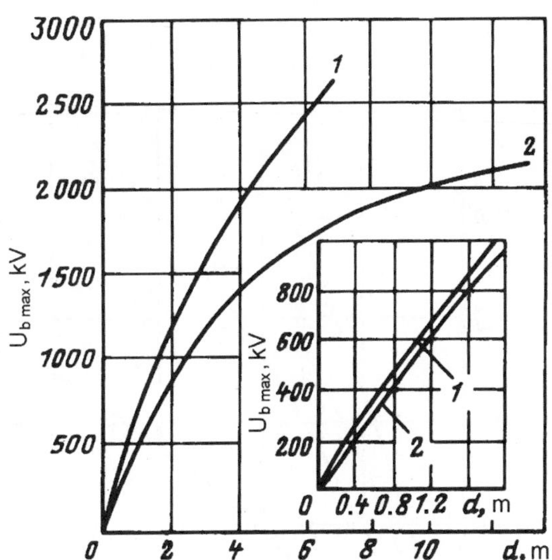

Figure 23.3 The dependence of the U_b amplitude values in an air gap on the distance between two rod electrodes (*1*) and between a rod and a plane electrode (*2*) at a frequency of 50 Hz, a temperature of 20°C and a pressure of 0.1 MPa [4].

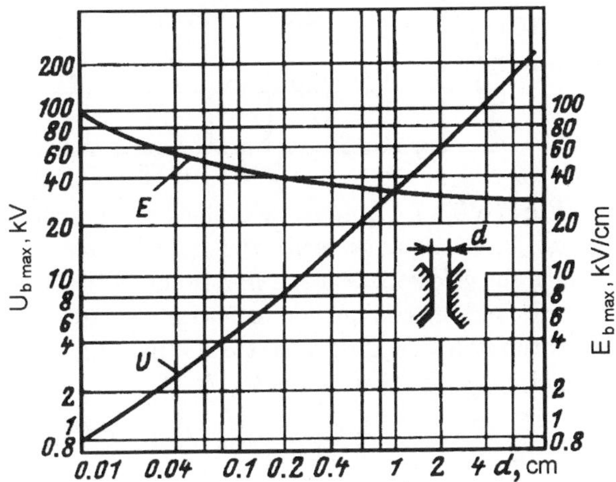

Figure 23.2 Dependences of U_b and E_b (in amplitude values) in a uniform electric field on the electrode spacing in air at a frequency of 50 Hz, a temperature of 20°C and a pressure of 0.1 MPa [3].

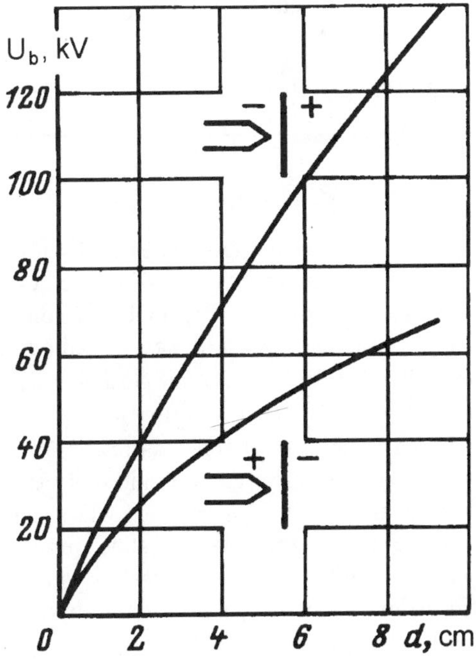

Figure 23.4 The dependence of U_b in an air gap on the distance between a rod and a plane electrode at different electrode polarities [5].

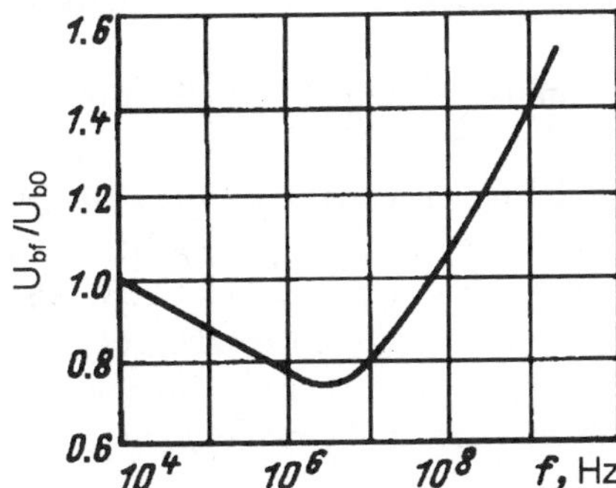

Figure 23.5 The a.c. to d.c. breakdown voltage ratio as a function of the frequency f for air under normal conditions in a uniform field [1].

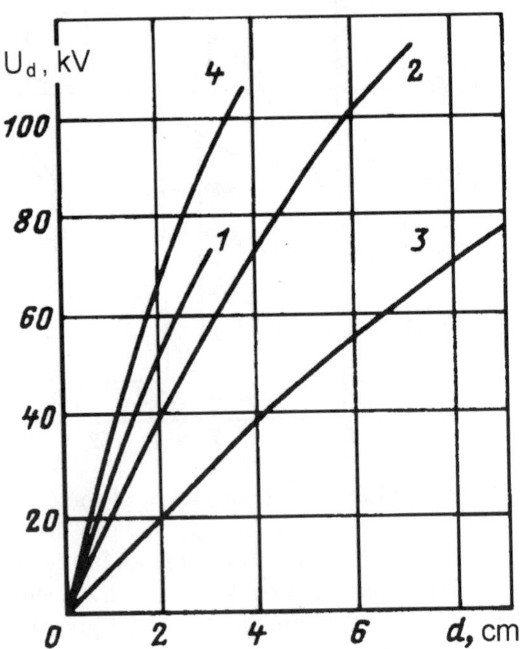

Figure 23.6 Amplitude of the discharge voltage in air along different insulating 50-mm-diameter cylinders in a uniform field at a frequency of 50 Hz versus the gap length (height of the cylinder d): *1* – paraffin; *2* – glass; *3* – glass in a loose contact with the electrodes; *4* – air gap without a dielectric insert [6].

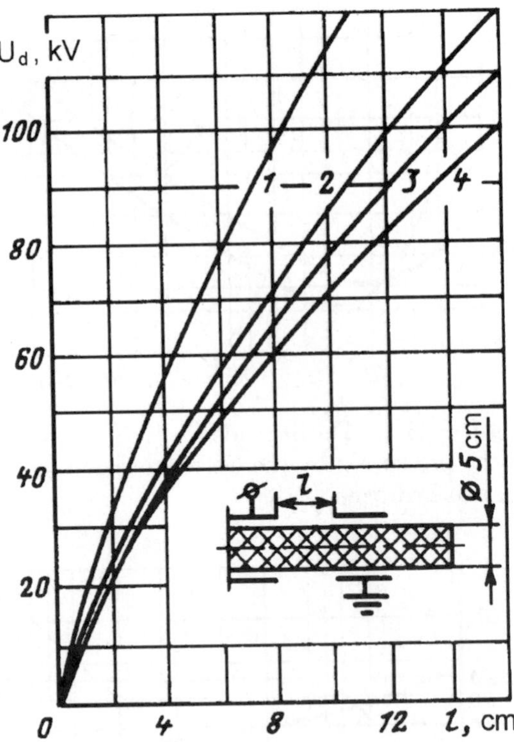

Figure 23.7 Amplitude of the discharge voltage in air along the solid dielectric surface versus the electrode spacing in a nonuniform electric field at a frequency of 50 Hz: *1* – air gap without a dielectric insert; *2* – paraffin; *3* – bakelite; *4* – porcelain and glass [6].

23.3 Liquid Dielectrics

Liquid dielectrics are characterized by ionic and molionic electric conductances. The ionic conductance results from dissociation of molecules of the liquid itself (intrinsic conduction) and impurities (impurity conductance). The molionic conductance is caused by charge carriers like charged colloid particles (molions) in colloid water, resinous substances, soaps, etc.

Impurities impair some other electrical characteristics of liquids ($\tan\delta$, E_b) as well. The value of ϵ_r is relatively less sensitive to impurities.

The characteristics of some natural and synthetic organic liquid dielectrics are presented in Table 23.6. Among natural organic-liquid dielectrics, mention may be made of petroleum oils: transformer, condenser and cable (low-viscosity MH-2, medium-viscosity C-20 and high-viscosity P-28) oils as well as castor oil and condenser Vaseline; among synthetic organic-liquid dielectrics are, for example, a polyolefin liquid – octol and diesters, such as dibutyl sebacinate. The characteristics of synthetic liquid dielectrics based on chlorinated hydrocarbons, silicone, and organofluoric compounds are listed in Tables 23.7, 23.8, and 23.9.

The resistivities of nonpolar liquids can be increased up to 10^{16}–10^{18} Ω·m, whereas E_b up to 140–260 MV/m using their special physical and chemical purification [11]. The dependence of the electric strength E_b on ϵ_r for high-purity liquid dielectrics used as solvents and for other purposes is presented in Table 23.10.

Table 23.6 Characteristics of organic liquid dielectrics [1, 9, 10]

Characteristic	Petroleum oils				
	Transformer oil	Condenser oil	Cable oils		
			MH-2	C-220	P-28
Solidification temperature, °C (no more than)	−45	−45	−45	−30	−10
Flash point, °C (no less than)	135	135	135	180	240
Kinematic viscosity at 20°C, 10^{-6} m^2/s	17–26	30–45	37	800	2000
ρ at 20°C, Ω·m	10^{11}–10^{12}	10^{12}–10^{13}	10^{12}–10^{13}	10^{13}–10^{14}	10^{11}–10^{12}
ε_r at 20°C	2.1–2.4	2.1–2.3	2.2–2.3	2.1–2.2	2.2–2.4
$\tan\delta$ at 50 Hz and 20°C	0.0005–0.002				
$\tan\delta$ at 50 Hz and 100°C		≤0.005	0.003	0.003	0.025
E_b at 50 Hz and 20°C, MV/m	12–26	20–25	18–20	20–24	14–18

Table 23.6 *(continued)*

Characteristic	Castor oil	Octol	Dibutyl sebacinate*1	Vaseline*2 (condenser petroleum)
Solidification temperature, °C (no more than)	(−15)–(−17)	−12	−8	37 (drop point)
Flash point*3, °C (no less than)		165	180	
Kinematic viscosity at 20°C, 10^{-6} m^2/s	130 (at 50°C)	100 (at 100°C)		
ρ at 20°C, Ω·m	10^{8}–10^{11}	10^{12}	10^{8}–10^{9} (50°C)	10^{12}–10^{13}
ε_r at 20°C	4.0–4.5	2.2–2.4	4.2–4.3	3.8–4.0
$\tan\delta$ at 50 Hz and 20°C	0.02	0.0012	0.80 (1 kHz, 50°C)	0.0002 (at 1 kHz)
$\tan\delta$ at 50 Hz and 100°C				
E_{br} at 50 Hz and 20°C, MV/m	14–16	16–17	16	20–22

*1 Dibutyl sebacinate is used in production of capacitors.
*2 Vaseline is a semiliquid material.
*3 The flash point indicates the flash temperature of the vapor of a liquid mixed with air and determined by the standard method.

Table 23.7 Characteristics of chlorinated liquid dielectrics [1]

Characteristic	Trichloro-diphenyl	Chlorinated biphenyl (sovol)	HCBD (hexachloro-butadiene)	TCB (trichloro-benzene)	Sovtol-10 (90% sovol+ 10% TCB)	Hexol (20% sovol + 80% HCBD)
Solidification temperature, °C	−18	5	−21	−16	−7	−60
Initial boiling point (°C) at 0.1 MPa	320	350	212	215	221	215
Kinematic viscosity at 20°C, 10^{-6} m^2/s	126	1800	2	1.5	650	4
ρ at 90°C, Ω·m	$>3{\cdot}10^{9}$	$>5{\cdot}10^{9}$	$>5{\cdot}10^{10}$	$>7{\cdot}10^{8}$	$>8{\cdot}10^{8}$	$>1.3{\cdot}10^{10}$
ε_r at 90°C	4.5	4.1	2.3	3.3	4.3	2.7
$\tan\delta$ at 50 Hz and 90°C	<0.025	<0.015	<0.003	<0.150	<0.100	<0.010
E_b, MV/m (no less than)	20 at 65°C	20 at 65°C	20 at 20°C	16 at 20°C	20 at 65°C	24 at 20°C

Table 23.8 Characteristics of organosilicon liquid dielectrics [1, 10]

Characteristic	Polymethylsiloxanes			Polyethyl-siloxanes		Polymethylphenyl-siloxanes		Polychloro(fluoro)organo-siloxanes	
	PMS–10	PMS–20	PMS–50	PES–D	No 3	PM–5	PM–1322	PC–5	PC–56
Solidification temperature, °C	−60	−60	−60	−60	−70	−110	−70	−100	−90

Table 23.8 Characteristics of organosilicon liquid dielectrics [1, 10] *(continued)*

Characteristic	Polymethylsiloxanes			Polyethyl-siloxanes		Polymethylphenyl-siloxanes		Polychloro(fluoro)organo-siloxanes	
	PMS–10	PMS–20	PMS–50	PES–D	No 3	PM–5	PM–1322	PC–5	PC–56
Flash point at 20°C	170	180	200	150	125	200	200	200	340
Kinematic viscosity at 20°C, 10^{-6}m^2/s	10	20	50	80–180	15	15.7	24.5	18	720
$\tan\delta$ at 20°C	0.0008–0.010			0.0003	–	0.0020	0.0016	0.0200	
ρ at 20°C, Ω·m	10^{12}			$2{\cdot}10^{12}$	10^{11}	10^{11}	10^{11}	$5{\cdot}10^{10}$	
ε_r at 20°C	2.6–2.7			2.4–2.8	–	2.8	2.7	5.4	6.3

Table 23.9 Characteristics of organofluoric liquid dielectrics [1, 10]

Name	Temperature, °C		E_b at 60 Hz and 25°C, MV/m		ε_r	$\tan\delta$, 10^{-4}	ρ at 20°C,
	boiling	freezing	liquid	vapor at 0.1 MPa	(25°C, 100 Hz)	(25°C, 100 Hz)	Ω·m
Freon-114	3.6	−94	19.3	9.8	2.26	1	10^{14}–10^{15} (0°C)
Freon-113	47.6	−35	12.2	11.0	2.41	1	10^{14}
Freon-215	74.0	−80	12.6	12.6	2.76	6	10^{13}
Freon-212	92.8	23.8	12.6		2.56	2	10^{12}
Freon-214	114.0	−92.8	11.8	13.0	2.78	2	10^{11}

Table 23.10 Electrical characteristics of high-purity liquid dielectrics at 20°C [10, 12]

Liquid	ε_r	E_b, MV/m	Liquid	ε_r	E_b, MV/m
n–Hexane	1.88	86.8	Diethyl ester	4.38	58.0
n–Heptane	1.93	84.0	Chloroform	4.89	54.5
Cyclohexane	2.04	83.0	*m*–Dichlorobenzene	4.90	53.8
Carbon tetrachloride	2.24	81.0	Bromobenzene	5.31	49.9
Benzene	2.28	78.4	Chlorobenzene	5.54	49.9
Toluene	2.39	78.6	1,1,2,2-Tetrachloroethane	8.08	35.5
Tetrachloroethylene	2.46	77.6	Methylene chloride	8.56	32.5
Hexachlorobutadiene	2.55	89.0	*ortho*-Dichlorobenzene	9.43	31.2
Trichloroethylene	3.44	67.0	1,2-Dichloroethane	10.03	27.5
1,2,4–Trichlorobenzene	3.98	62.0	Paraldehyde	15.06	18.2

23.4 Solid Dielectrics

Natural resins and synthetic polymers (high-molecular compounds) are used to produce insulating varnishes, enamels, compounds, and plastics as well as films, fibrous materials and so on. Natural resins and synthetic polymers can be thermoplastic (being heated they do not lose the capacity to melt and dissolve in appropriate solvents) and thermoreactive (become infusible and insoluble after heating). Synthetic polymers are produced by two types of reactions: polymerization and polycondensation. Some polymers of the latter type are also named resins (for example, phenol-formaldehyde, polyester, and epoxy resins) by analogy with natural materials. The three-volume *Encyclopedia of polymers* [19] contains comprehensive data on all divisions of chemistry, physics and technology of polymers and polymer materials (plastics, rubbers, chemical fibers, film materials, varnishes, paints, etc.). The properties of natural resins are listed in Table 23.11.

Plastics make composite materials with a polymer as a base component. This polymer determines the main properties of a composite and serves as a binder cementing together all components of the composite to form a monolith. The other components, such as fillers, plasticizers, stabilizers, etc., when introduced in nonpolar polymers, impair their insulating properties. For this reason, plastics based on such polymers, which are excellent dielectrics, involve practically a binder alone. Table 23.12 presents the properties of thermoplastic polymer-organic dielectrics and materials on their bases. The properties of thermoreactive plastics and laminated plastics with a sheet (roll) filler are listed in Tables 23.13 and 23.14, respectively.

Insulating compounds (composites) constitute hardening materials. When used for technological purposes (impregnation, potting), they are in a liquid state, being solid under the working conditions. The properties of composites are presented in Table 23.15.

The properties of **fibrous materials**, such as papers (including semiconducting paper), board and fiber, are listed in Table 23.16; the properties of varnished cloths (including semiconducting ones) are displayed in Table 23.17.

Mica refers to an inorganic dielectric. Table 23.18 exhibits the properties of the most important sorts of mica. Micanites are mica-based glued sheet materials which may also have fibrous substrates. The properties of some micanites and Micalex (a mica-based plastic) are presented in Table 23.19. Materials made of mica-loaded papers and mica plastics are the substitutes of micanites. The properties of some of these materials are listed in Table 23.20. The mica-loaded paper is fabricated from waste muscovite; mica plastics are made of waste phlogopite.

Glasses consist in inorganic amorphous substances representing usually systems of oxides. The properties of quartz glasses and some other insulating glasses are presented in Tables 23.21 and 23.22, respectively.

Devitrified glasses (such as Pyroceram) represent glass-ceramic materials made by crystallizing the glasses of specific composition. The properties of some technical glass ceramics are given in Table 23.23.

Electric-grade ceramic constitutes a stone-like material which is produced by sintering the mass of a required composition and consists of a crystalline and an amorphous phase. The properties of some commonly used electric-grade ceramics are presented in Table 23.24.

Table 23.11 The properties of natural resins at 20°C and 50 Hz [1]

Resin	ρ, $\Omega\cdot$m	ε_r	$\tan\delta$	E_b, MV/m	Solvents
Shellac	10^{13}–10^{14}	3.5	0.01 (0.001*1)	20–30	Ethanol
Colophony	10^{13}–10^{14}	2.8	0.003	10–15	Liquid hydrocarbons, vegetable oils, alcohol, turpentine
Amber	10^{17}	2.8	0.001		Insoluble in a cold state

*1 After prolonged heating.

Table 23.12 Characteristics of thermoplastic polymer dielectrics and materials on their bases at 20°C [9, 13, 14]

Name	ρ, $\Omega\cdot$m	ε_r	$\tan\delta$ at 50 Hz	E_b, MV/m
Polyethylene	10^{15}	2.2–2.3	$(2\text{–}6)\cdot10^{-4}$ at 10^6 Hz	25–60
Polypropylene	10^{14}–10^{15}	2.2	$(2\text{–}5)\cdot10^{-4}$ at 10^6 Hz	25–40
Polystyrene	10^{14}–10^{15}	2.4–2.6	$(2\text{–}8)\cdot10^{-4}$ at 10^6 Hz	20–25
Polystyrene shock-resistant	10^{13}–10^{14}	2.6–2.7	$(2\text{–}3)\cdot10^{-3}$ at 10^6 Hz	23–25
Polytetrafluoroethylene (PTFE)	10^{15}–10^{18} (up to 150°C)	1.9–2.2	$\leq 1\cdot10^{-4}$ (up to 10^{10} Hz)	25–27 at d = 4mm
Polychlorotrifluoroethylene (Kel-F)	10^{16}	3.0	$1.5\cdot10^{-2}$	13–15 at d = 4mm
Flexible PVC, insulating	10^{10}–10^{12}	4–8	$(5\text{–}8)\cdot10^{-2}$	20–50
Hard PVC	10^{12}–10^{14}	4.0	0.02	25–60
Polymethylmethacrylate (Plexiglas)	10^{10}–10^{11}	3.6	0.06	15–25
Polyformaldehyde	10^{12}–10^{13}	3.7	$(3\text{–}4)\cdot10^{-3}$	25

Table 23.12 Characteristics of thermoplastic polymer dielectrics *(continued)*

Name	ρ, Ω·m	ε_r	$\tan\delta$ at 50 Hz	E_b, MV/m
Polyethylene terephthalate (lavsan)	10^{13}–10^{14}	3.1–3.2	$(2–3)\cdot10^{-3}$	140–180 (film)
Polycarbonate (diflon)	10^{14}–10^{15}	3.0–3.1	$(2–3)\cdot10^{-3}$	20–25 (180 for film)
Polyamides:				
capron	10^{11}–10^{12}	4.5	0.06–0.1	20–22
nylon	10^{12}–10^{13}	4.6	0.04	20–25
phenylone	10^{12}–10^{13}	4.5	0.02	18–22 (>100 for film)
Polyimide (polypyromellitimide)	10^{14}–10^{16}	3.0–4.5	$(1–5)\cdot10^{-3}$	100–300 (film)
	10^{10}–10^{12} (at 250°C)		$(2–6)\cdot10^{-3}$ (at 250°C)	

Table 23.13 Properties of thermoreactive insulating plastics at 20°C [9, 13, 14]

Brand of plastic	ρ, Ω·m	ε_r	$\tan\delta$ at 50 Hz	E_b, MV/m	Composition
E1-340-02; E2-330-02; E8-361-63; E9-342-73; E10-342-63; E11-342-63; E15-121-02	10^{10}–10^{11}	7.5–9.5	0.08	13–20	Binder – novolac and resol phenol–formaldehyde resins (or their modifications) Filler – organic, mineral, their mixture
E3-340-65; E4-100-30; E5-101-30; E6-014-30 (high-frequency)	10^{11}–10^{12}	6–8	0.010–0.012 (10^6 Hz)	15–20	
B1, B2, B3, B4, B5 (with higher insulating properties)	10^{10}–10^{12}	4–6	0.03–0.05	12–14	Binder – urea- and melanino-formaldehyde resins. Filler – organic, mineral, their mixture
D1, E1 (with higher arc resistance, heat stability and mechanical strength)	10^{11}–10^{11}	4–5	0.05–0.06	5–10	
AG-4 (for work at temperatures of −196 to 200°C and under tropical condition)	10^{10}–10^{11}	5–8 (10^6 Hz)	0.02–0.05 (10^6 Hz)	13–18	Binder – modified phenol-formaldehyde resin Filler – glass fiber
KMK-218; ВПМ-1; ПК-9; КF-9; КF-10, etc. (for working temperatures of 200–350°C, tropicalized)	10^{10}–10^{13}	3–6	0.004–0.02	4–7 (up to 20) for КФ)	Organosilicone resins with mineral fillers
Ebonite	10^{12}–10^{13}	2.8–3.5	0.004–0.013 (10^6 Hz)	20–35	Hard rubber, vulcanized with 30–35% sulphur
Escapon	10^{14}–10^{15}	2.7–2.9	0.0005 (10^6 Hz)	30–35	Synthetic rubber SCB, polymerized, unvulcanized

Table 23.14 Properties of main-type laminated plastics at 20°C [1, 16]

Plastic	ρ, Ω·m	ρ_s, Ω	ε_r	$\tan\delta$ at 50 Hz	E_b, MV/m	Composition
Getinacs (paper-based laminate), low-voltage (brands I, II, III, IV, VI)	10^{10}–10^{11}	10^{12}	7–8	0.10–0.40	20–35	Binder – resol phenol-formaldehyde resins and epoxy resins Filler – cellulose paper
Getinacs, high-voltage (V)	10^{11}	10^{13}	7–8	0.015	27–30	
Getinacs, high-frequency (VII, VIII)	10^{10}–10^{11}	10^{12}–10^{13}	7–8	0.020–0.025 (10^6 Hz)	28–40	
Textolite (fabric-based laminate), low-voltage (brands A, B, D)	10^8–10^9	10^{11}–10^{12}	8	<0.6	4.5–12	Filler – cotton and lavsan fabric (LT) Binder – resol phenol-formaldehyde resins and epoxy resins (LT)
Textolite, high-frequency						
brand HF	10^{10}	10^{12}	7–8	0.005 (10^6 Hz)	8–16	
brand LT	10^{12}	10^{13}	4	0.003 (10^6 Hz)	25–32	

Table 23.14 Properties of main-type laminated plastics at 20°C [1, 16] *(continued)*

Plastic	ρ, Ω·m	ρ_s, Ω	ε_r	tan δ at 50 Hz	E_b, MV/m	Composition
Glass textolite low-voltage:						Filler – glass cloth Binder – phenol-formaldehyde, epoxy and silicone resins
working temperature 130°C (brands GT, GT-B, GT-1)	10^9–10^{11}	10^{11}–10^{13}		<0.4	10–25	
working temperature 155°C (brands GT-11)	10^{11}	10^{12}–10^{13}			12–20	
Glass textolite for working temperature of 155°C:						
high-voltage (GTEF-1)	10^{11}	10^{13}	5–6	0.03–0.05	20–30	
high-voltage and high-frequency (GTEF)	10^{11}–10^{12}	10^{13}–10^{14}	5–6	0.003–0.005	20–30	
Glass textolite for working temperature of 180°C, high-voltage (GTK)	10^{12}–10^{13}			0.001–0.005	18–25	

Table 23.15 Characteristics of insulating compounds at 20°C [1, 9, 15, 17]

Brand and type of compound	Softening temperature, °C	Cold endurance, °C (no higher than)	Volume shrinkage, % (no more than)	ρ, Ω·m	ε_r	tan δ at 50 Hz	E_b, MV/m
Impregnating compounds							
225 (bituminous)	97–102	−25	8.0–8.5	10^{11}–10^{12}	2.8–3.2	0.02	20–22
KGMS-1 (polyester, hot-setting	250	−60	6–7	10^{11}–10^{12}	6	0.02–0.04	18–22
KGMS-2 (the same, but more flexible)	250	−60	8–10	10^{11}–10^{12}	6	0.02–0.04	18–20
KP-18 (polyester, hot-setting)	unsoftenable	−50		10^{10}–10^{11}	4.4–5.0 at 10^3 Hz	0.02–0.03 at 10^3 Hz	28–33
D-1 (epoxy, hot-setting)	the same	−50	0.5–1	10^{11}–10^{12}	4.0 at 10^6 Hz	0.01–0.02 at 10^6 Hz	28–30
K-67, K-67F (organo-silicone, hot-setting)	the same	−60	5–8	10^{13}–10^{14}		0.005–0.015	20–22
Impregnating-potting compounds							
MBK-1 (methacrylic, rigid)	250	−60	0.5	10^{11}–10^{12}	3.1–3.5	0.05–0.07	20–25
MBK-2 (the same, less rigid)	250	−60	0.5	10^{10}–10^{11}	4.0–5.6	0.03–0.04	17–20
MBK-3 (the same, flexible)	250	−60	0.5	10^{10}–10^{11}	4.5–5.2	0.03–0.04	16–18
K-168 (epoxy, cold-setting)		−60	0.5–1.5	10^{10}–10^{12}		0.02–0.04	25–30
K-293 (the same, hot-setting)		−60	1–2	10^{10}–10^{12}		0.05–0.08	20–25
Potting compounds							
K-30 (polyurethane, hot-setting)	unsoftenable	−80	3–4	10^9–10^{10}	6	0.02–0.03	27–30
K-31 (the same)	the same	−80	3–4	10^{12}–10^{13}	5	0.05–0.07	27–30
Vicsynt K-18 (organo-silicone, cold-setting)		−60	2–3	10^{11}–10^{12}	3.0	0.01–0.02	15–18
MK-45 (oil-colophony)	45–48	−8	6–7	10^9–10^{10}			12–16
MB-70 (bituminous)	70–73	−10	8–9	10^{10}–10^{11}			14–16
MB-90 (bituminous)	90–92	−10	8–9	10^{10}–10^{11}			14–16
MBM (bitumen-oil)	55–60	−45	7–8	10^{10}–10^{11}			15–17

Table 23.16 Characteristics of fibrous materials [15]

Type of material	Brand	Thickness, mm	Mechanical characteristics	tan δ at 50 Hz	Layer breakdown voltage, V (no less than)	E_b, MV/m
			Breaking length*[1], m			
Capacitor paper, ordinary	CON-1	0.010–0.030	8000	$\leq$0.26 at 100°C	360–620	
	CON-2	0.004–0.030	8500	$\leq$0.32 at 100°C	240–680	
Capacitor paper, special, high–quality	SCON-1	0.010–0.030	8000	$\leq$0.20 at 100°C	380–620	
	SCON-2	0.004–0.022	8500	$\leq$0.24 at 100°C	270–590	
	SCON-3	0.005–0.022	8500	$\leq$0.32 at 100°C	300–610	
Capacitor paper, low–dielectric loss	MCON-08	0.010–0.020	8000	$\leq$0.12–0.16 at 120°C	320–470	
	MCON-1	0.008–0.030	8000	$\leq$0.16–0.20 at 120°C	340–620	
	MCON-2	0.006–0.030	8000	$\leq$0.22–0.28 at 120°C	300–680	
	MCON-3	0.006–0.050	8000	$\leq$0.26–0.32 at 120°C	310–520	
			Breaking load, N			
Cable paper	C-080	0.080	83.4	unstandardized	unstandardized	
	C-120	0.120	127.5			
	C-170	0.170	171.7			
Cable paper, laminated	CM-120	0.120	142.2	same	same	
	CM-170	0.170	186.4			
Cable paper, laminated, strengthened	CMP-120	0.120	152.0	same	same	
	CMP-170	0.170	196.2			
Cable paper, high–voltage, laminated	CHM-080	0.080	74	$\leq$0.0022 at 100°C	same	
	CHM-120	0.120	142	$\leq$0.0022 at 100°C		
	CHM-170	0.170	186	$\leq$0.0022 at 100°C		
Cable paper, high–voltage, laminated stabilized	CHMS-080	0.080	69	$\leq$0.0018 at 100°C	same	
	CHMS-120	0.120	108	$\leq$0.0019 at 100°C		
	CHMS-170	0.170	147	$\leq$0.0019 at 100°C		
Cable paper, high–voltage, laminated, stabilized, thickened	CHMSU-080	0.080	98	$\leq$0.0026 at 100°C	same	
	CHMSU-120	0.120	137	$\leq$0.0026 at 100°C		
Telephone cable paper	KT-050	0.050	60.8	unstandardized		
Dielectric cable paper	ECTM	0.44	44.2	same		$\geq$40 in oil at (90±5)°C
Cable paper, semiconducting, thickened, one-color, with carbon black content	CPU-80	0.080	78.4	$\rho=10^3$–$9\cdot10^4$ Ω·m		
	CPU-120	0.120	117.6			
Cable paper, semiconducting, heavyweight, two–color, with carbon black content of one ply	CPDU-80	0.080	94	$\rho_s=5\cdot10^4$–10^6 Ω		
	CPDU-120	0.120	127.4			
			Longitudinal tensile strength, MPa			
Insulating board for operation in air	EF	0.10–0.3	83–98	unstandardized		8–12
	EFG	0.20–0.40	127			12
	EFP	0.10; 0.20	127			12
	EFT	0.10–0.50	118			12–13
	EFA	0.15	118			11

Table 23.16 Characteristics of fibrous materials [15] *(continued)*

Type of material	Brand	Thickness, nm	Mechanical characteristics	tan δ at 50 Hz	Test voltage, kV	E_b, MV/m
Insulating board for operation in transformer oil	AM	2.0; 2.5; 3.0	39	same	40–50	
	A	2.0; 2.5; 3.0	39		40–50	
	B	1.0–6.0	49		31–90	
	C	2.0; 2.5; 3.0	59		55–70	
	D	0.5–3.0	34–39		19–57	
Fiber, sheet, electrical and technical	FE	0.4–3.0	69–74	same		≥3.5–7.0
	FT	0.6–25.0	29–49			≥2–4

*[1] The breaking length is a length at which paper is torn by gravity. The breaking load is applied to a paper strip 15 mm in width.

Table 23.17 Characteristics of varnished cloths [15]

Brand of varnished cloth	Warp (cloth)	Impregnating agent	Rated thickness, mm	Specific breaking load (on the warp), N/cm	ρ, Ω·m (no less than)	E_b, kV: at 15–35°C and an air relative humidity of 45–75%	E_b, kV: after 24-hour exposure at 20°C and an air relative humidity of (95±2)%	Heat resistance, °C
VCoM-105	Cotton	Oil varnish	0.15; 0.17; 0.20; 0.24; 0.30	38–45	10^{11}	4–9.5	1.5–4.0	105
VCoMC-105	Same	Same	0.17; 0.20; 0.24	44–51	10^{11}	4.5–7.4	1.9–3.5	105
VCoMM-105	Same	Same	0.17; 0.20; 0.24	50–64	10^{11}	4.8–9.2	2.0–4.0	105
VCoB-105	Same	Oil-bituminous varnish	0.17; 0.20; 0.24	58–80	10^{11}	4.5–9.2	2.0–3.6	105
VSM-105	Silk	Oil varnish	0.08; 0.10; 0.12; 0.15	7–24	10^{11}	2.3–8.5	1.4–3.8	105
VSMC-105	Same	Same	0.04; 0.05; 0.06; 0.10; 0.12	11–25	10^{11}	0.4–9.3	2.0–4.2	105
VCaM-105	Capron	Same	0.10; 0.12; 0.15	20–30	10^{11}	3.3–7.8	1.7–3.4	105
VCaMC-105	Same	Same	0.10; 0.12; 0.15	20–30	10^{11}	3.6–9.3	2.0–4.2	105
VGM-105/120	Glass	Same	0.15; 0.17; 0.20; 0.24	85–170	10^{11}	3.2–8.0	1.5–4.2	120
VGMM-105/120	Same	Same	0.17; 0.20; 0.24	95–170	10^{11}	4.0–9.2	2.3–4.6	120
VGL-105/120	Same	Butadiene-styrene latex	0.15; 0.17; 0.20	95–150	10^{12}	3.2–6.8	1.3–3.0	120
VGE-105/130	Same	Escapon varnish	0.12; 0.15; 0.17; 0.20; 0.24	70–170	10^{12}	2.6–9.6	0.7–3.6	130
VGB-120/130	Same	Bitumen-oil-alkyd varnish	0.12; 0.15; 0.17; 0.20; 0.24	70–170	10^{12}	2.6–10.8	0.9–4.1	130
VGP-130/155	Same	Epoxy-polyester varnish	0.08; 0.10; 0.12; 0.15; 0.17	35–130	10^{10}	0.8–9.0	0.6–3.5	155
VSK-155/180	Same	Organosilicone lacquer	0.05; 0.06; 0.08; 0.10; 0.12; 0.15; 0.17; 0.20	20–150	10^{12}	1.0–9.0	0.6–5.0	180
VSKR-180	Same	Organosilicone rubber	0.12; 0.15; 0.17; 0.20	70–150	10^{12}	0.7–4.9	0.6–2.7	180
VSKL-155	Same	Adhesive organo silicone varnish	0.12; 0.15	70–105	10^{12}	0.8–0.9		155
VSU	Same	Polyurethane varnish	0.13; 0.15; 0.17	90–130	10^{11}	5.5–9.5	3.0–5.5	155
VSK-2	Same	Organosilicone enamel	0.12; 0.15; 0.20	50–120	10^{11}	2.5–7.5	1.3–4.2	180
VSK-5	Same	Semiconducting silicone enamel	0.12; 0.15; 0.20	70–150	$\rho_s=10^3$–10^5 Ω			180

Table 23.18 Characteristics of insulating mica at 20°C [1]

Type of mica	Heat resistance, °C	ρ, Ω·m	ρ_s, Ω·m	ε_r	tan δ at 50 Hz	tan δ at 10^6 Hz	E_b, MV/m
Muscovite	500–600	10^{12}–10^{14}	10^{11}–10^{12}	6.1–8.4	0.0004–0.008	0.0001–0.0004	100–250
Phlogopite	800–1000	10^{11}–10^{12}	10^{10}–10^{11}	5.5–6.7	0.006–0.015	0.0002–0.006	70–150
Fluorophlogopite (synthetic mica)	1000–1050	10^{12}–10^{14}	10^{14}–10^{15}	6.1–7.5		0.0001–0.0003	100–250

Table 23.19 Characteristics of some micanites and Mycalex [1, 9, 15]

Name	Brand	Type of mica and adhesive	Thickness, nm	ρ at 20°C, Ω·m	E_b at 20°C, MV/m	Heat resistance, °C
Commutator micanite	CPS	Phlogopite ordinary; shellac	0.4–1.5	10^{11}–10^{12}	19–22	130
	CPG	Same; glyptal resin	0.4–1.5	10^{12}–10^{13}	19–22	130
	CPP	Same; polyester resin	0.7–1.5	10^{11}–10^{12}	19–22	155
	CPA	Phlogopite, heat-resistant; ammophos	0.7–1.2	10^{11}–10^{12}	19–22	>180
Plate micanite	PMG PPG PGG	Muscovite, phlogopite ordinary; their mixture; glyptal resin	0.5–1.0	10^{11}–10^{12}	16–23	130
	PPS	Phlogopite, heat-resistant; Organosilicone resin	0.15–1.0	10^{10}–10^{11}	16–38	180
Flexible micanite	GML GPL	Muscovite, phlogopite; oil-glyptal varnish (light)	0.15	10^{11}–10^{12}	20–30	130
	GMB GPB	Muscovite, phlogopite; oil-bitumen varnish (black)	0.2–0.5	10^{11}–10^{12}	20–30	130
	GML-PP GPL-PP GMB-PP GPB-PP	Muscovite, phlogopite; light and black varnishes. Micanite lined with paper on both sides	0.2–0.5	10^{11}–10^{12}	16–23	130
	GPS	Phlogopite; organosilicone varnish	0.2–0.5	10^{11}–10^{12}	19–25	180
Micalex		Filler – Mica powder, muscovite, Binder – low-fusible glass	4–15	10^{10}–10^{12}	13–18	300–350

Table 23.20 Characteristics of some mica-based composites and plastics [1, 9, 15]

Name	Brand	Type of adhesive, substrate	Thickness, mm	ρ at 20°C, Ω·m	E_b MV/m	Heat resistance, °C
Commutator mica composite	CMS	Shellac	0.45–1.2	10^{10}–10^{12}	28–35	130
Flexible mica composite	GGP	Polyester varnish	0.15–0.30	10^{11}–10^{12}	17–23	130
	G_1GP	Same; glass substrate on one side	0.10; 0.15	10^{11}–10^{12}	20–22	130
	G_2GP	Same; glass substrate on both sides	0.2–0.3	10^{11}–10^{12}	28–30	130
	GGP-PL-10	Polyester varnish; polyethylene terephthalate 10-μm-thick film on one side	0.15; 0.20	10^{11}–10^{12}	30–32	130
	GGP-PL-20	Same, 20 μm	0.20; 0.25	10^{11}–10^{12}	28–30	130
	GGP-PL-50	Same, 50 μm	0.25; 0.30	10^{11}–10^{12}	26–28	130
	G_1SS	Organosilicone varnish; glass substrate on one side	0.10; 0.15	10^{11}–10^{12}	20–22	155
	G_2SS	Same; glass substrate on both sides	0.2–0.3	10^{11}–10^{12}	28–30	155

Table 23.20 Characteristics of some mica-based composites and plastics [1, 9, 15] *(continued)*

Name	Brand	Type of adhesive, substrate	Thickness, mm	ρ at 20°C, $\Omega\cdot$m	E_b MV/m	Heat resistance, °C
	G_1SSH	Organosilicone varnish; heat-resistant; glass substrate on one side	0.10; 0.15	10^{11}–10^{12}	20–32	300
	G_2SSH	Same; glass substrate on both sides	0.2–0.3	10^{11}–10^{12}	26–28	300
Commutator mica plastic	CIPS	Shellac	0.4–1.5	10^{10}–10^{11}	22–26	130
	CIPP	Polyester resin	0.4–1.5	10^{10}–10^{11}	22–24	155
	CIPH-B	Heat-resistant binder	0.4–1.5	10^{11}–10^{12}	24–26	>180
	CIPS	Organosilicone resin	0.4–1.5	10^{11}–10^{12}	24–26	>180
Plate mica plastic	PIPS	Shellac	0.5–1.5	10^{11}–10^{12}	20–28	130
	PIPSA	Same	0.5–1.5	10^{11}–10^{12}	20–28	130
	SIPT	Aluminophosphate with organosilicone varnish	0.5–1.5	10^{11}–10^{12}	20–28	>180
Mica plastic compound	GIT-CG	Triethylene glyptal or glyptal varnish; substrates from fiber glass cloth and fiber glass gauze	0.25; 0.30	10^{11}–10^{12}	10–19	130
	GIP-CG GIP-GG	Polyester varnish; substrates from fiber glass cloth and fiber glass gauze or both substrates from fiber glass gauze	0.35; 0.45	10^{11}–10^{12}	10–19	155
	GIS-CG GIS-GG	Organosilicone varnish; substrates from fiber glass cloth and fiber glass gauze or both substrates from fiber glass gauze	0.35; 0.45	10^{11}–10^{12}	9–18	185

Table 23.21 Properties of electrotechnical quartz glasses [1]

Characteristic	Type of glass		Characteristic	Type of glass	
	Transparent	Opaque		Transparent	Opaque
Mean TC of linear expansion, °C^{-1}, within temperature ranges, °C:			1000	10^6	10^2
			ε_r	3.8	3.7
20–50	$4.0\cdot10^{-7}$		$\tan\delta$, 10^{-4} at 10^{10} Hz and 20°C	1.0	3.0
120–420	$6.0\cdot10^{-7}$		Same at 1000°C	5.0	
420–1200	$5.0\cdot10^{-7}$		E_b, MV/m at T, °C:		
ρ, $\Omega\cdot$m at T, °C:			20	44	32
20	10^{14}	10^{14}	200	32	21
200	10^{12}	10^{10}	600	6	3
600		10^4			

Table 23.22 Properties of some electrical insulating glasses [1]

Brand of glass	Mean TC of linear expansion within 20–300°C range, 10^{-7} °C^{-1}	ε_r at 20°C		$\tan\delta$, 10^{-4} at 20°C		ρ, $\Omega\cdot$m	
		at 10^6 Hz	at 10^{10} Hz	at 10^6 Hz	at 10^{10} Hz	at 150°C	at 300°C
C5-1	5	3.81	3.80	1	1		
C40-1	40.0±1.5	5.10	4.70	22	53		
C48-1	48.0±1	5.40	4.90	22	65		
C87-1	87.0±1	7.20	6.80	13	62		
C63-1	63.0±2	14.0	12.0	18	131	$3\cdot10^{11}$	$6.2\cdot10^7$
C77-1	77.0±2	13.0	11.0	12	78	$9\cdot10^{11}$	$6.3\cdot10^7$

Table 23.23 Properties of some technical brands of glass-ceramics [1]

Brand of glass-ceramics	Mean TC of linear expansion within 300–400°C range, 10^{-7} °C^{-1}	ε_r at 25°C		tan δ, 10^{-4} at 25°C		ρ, Ω·m	ρ_s, Ω	E_b, MV/m
		at 10^6 Hz	at 10^{10} Hz	at 10^6 Hz	at 10^{10} Hz			
C-15-12	1.2	9.7		461		$2.2\cdot10^{10}$	$1.8\cdot10^{11}$	70.7
C-12-14	7.0	7.4	7.05	32	153	$4.1\cdot10^{111}$	$1.9\cdot10^{12}$	27.0
TC-81	17	7.0	6.42	41	64	$3.3\cdot10^{10}$	$2.35\cdot10^{11}$	29.6
AC-05-C-023	120	5.7	5.40	21	120	$4.6\cdot10^{10}$	$1.15\cdot10^{11}$	85.3
CT-50-1	50	8.3	7.90	15	45	$7.1\cdot10^{10}$	$4.56\cdot10^{11}$	47.2
CT-50-2	50	5.6	5.50	200	5	$3.2\cdot10^{10}$	$1.67\cdot10^{11}$	27.9
AT-50-336	48	5.1	4.96	10	17	$4.9\cdot10^{10}$	$1.04\cdot10^{12}$	65.9

Table 23.24 Basic characteristics of electrical ceramic materials [1]

Material	Density, kg/m^3	Mechanical characteristics*1			TC of linear expansion, 10^{-6} °C^{-1}	Electrical characteristics at 20°C			
		Tensile strength, MPa	Static bending strength, MPa	Specific impact strength, kJ/m^2		ρ, Ω·m	ε_r	tan δ at 50 Hz	E_b, MV/m
Electric-grade porcelain	2200	30–55	60–110	1.8–2.2	3.5–5.0	10^{11}–10^{12}	5–8	0.022–0.025	30–32
Ultraporcelain UP-46, УФ-43	3200	50–60	200–250	2.5–2.8	5.0–5.5	10^{12}–10^{13}	8.0–8.8	0.0005–0.001	30–36
Steatite SK-4, TK-21	3000	60–70	170–190	3.0–3.5	6.0–6.4	10^{13}–10^{14}	6.5–7.0	0.001–0.003	40–42
Cordierite	2800		70–88	2.0–3.0	2.0–2.3	10^{9}–10^{10}	5–6		4.5–6.0

*1 Mechanical characteristics are given for unglazed samples. For glazed porcelain and steatite, the characteristics are 15–20% higher. Values for the TC of expansion are given within a 20–100°C range.

23.5 Active Dielectrics

Ferroelectrics refer to the dielectrics that exhibit a spontaneous polarization in a certain temperature range. Among the basic features of ferroelectrics we can mention a high and ultra-high value of ϵ_r, a drastic dependence of ϵ_r on temperature with a sharp peak at the Curie point, and an abrupt dependence of ϵ_r on the electric field strength. These properties are used in devices based on ferroelectric ceramic materials. Figure 23.8 shows the dependence of ϵ_r on temperature for barium titanate. It is seen from the figure that in this material the Curie point occurs at 125°C. The dependence of ϵ_r on the electric field strength E_{eff} for the same material is in Figure 23.9. Created on the base presented of ferroelectric ceramics are varicaps – nonlinear dielectric capacitors whose capacity varies abruptly under changes of alternating and constant voltages applied to them. The properties of the ferroelectric ceramics for varicaps are presented in Table 23.25. Among the basic characteristics of varicaps are: the Curie point T_C; the initial dielectric constant ϵ_i in a weak alternating electric field (2–5 kV/m); the nonlinearity factor in an alternating field $K_\sim = \epsilon_{\max}/\epsilon_i$, where $\epsilon_{\max}$ is the maximum value of the dielectric constant in an alternating field (at $E = E_{\max}$); the reversible nonlinearity factor in a constant field $K_- = \epsilon_\sim/\epsilon_{\min}$, where $\epsilon_\sim$ is the dielectric constant in an alternating field without any biasing constant field; $\epsilon_{\min}$ is the dielectric constant at the same alternating field and a significant biasing constant field, such that ϵ is already independent of the biasing field for practical purposes.

Table 23.25 Properties of ferroelectric ceramics for varicaps [8, 20]

Material	T_C, °C	ε_i at 20°C	$K_\sim$	$E_{\max}$, kV/m	K_-	tgδ at 20°C
VC-1	75	2300–2500	6–8	150–200	≥ 2	0.02
VC-2	75	2000–2500	15-20	120–150	≥ 2	0.01
VC-3	25	10,000–20,000	1–2	50–100	≥ 8	0.03
VC-4	105	1800–2000	10-16	250–300	≥ 2	0.015
VC-5	25	2000–3000	35-50	80–100	≥ 2	0.01
VC-6	200	400–500	20-50	500–600		0.03
VC-7	≤ 20	2000–4000			≥ 2	0.001

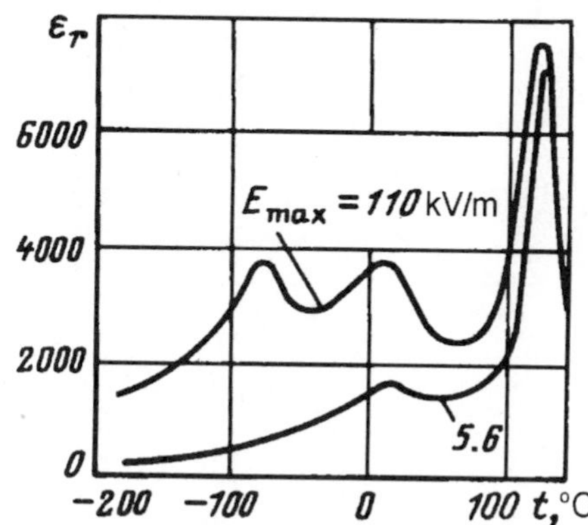

Figure 23.8 The temperature dependence of ε_r at different electric field strengths for barium titanate [18].

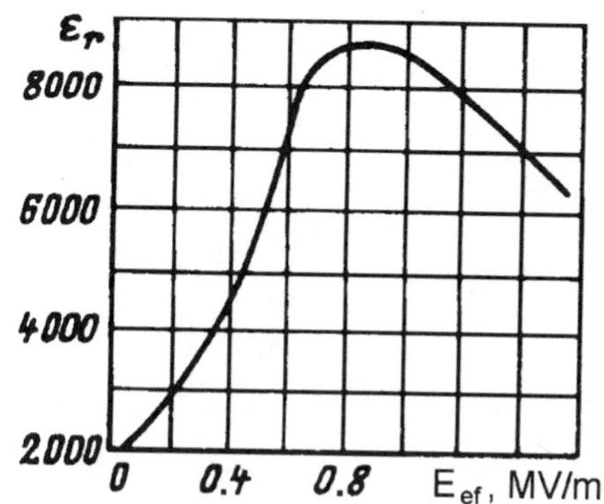

Figure 23.9 Dependence of ε_r on the electric field strength for barium titanate at 22°C [18].

References

[1] Handbook on electrotechnical materials, Eds. Koritskii, Yu. V., Pasynkov, V. V. and Tareev, B. M., 2nd ed., Energiya, Moscow, 1974 (Vols. 1, 2), 1976 (Vol. 3) (in Russian).

[2] Skanavi, G. I., Insulator physics (high field region), Fizmatgiz, Moscow, 1958 (in Russian).

[3] Babikov, M. A., Komarov, N. S., Sergeev, A. S., High voltage equipment, Gosenergoizdat, Moscow, 1955 (in Russian).

[4] Dolginov, A. I., High voltage equipment in electric power engineering, Energiya, Moscow, 1968 (in Russian).

[5] High voltage equipment, Ed. Razevig, D. V., Gosenergoizdat, Moscow, 1963 (in Russian).

[6] High voltage equipment, Ed. Razevig, D. V., 2nd ed., Energiya, Moscow, 1976 (in Russian).

[7] Kazarnovskii, D. M., Yamanov, S. A., Radiotechnical materials, Vysshaya Shkola, Moscow, 1972 (in Russian).

[8] Electrical radio materials, Ed. Tareev, B. M., Vysshaya Shkola, Moscow, 1978 (in Russian).

[9] Electrotechnical reference data, Eds. Gerasimov, V. G., Grudinskii, P. G., Zhukov, L. A., 6th ed., Energiya, Moscow, 1980 (Vol. 1) (in Russian).

[10] Shakhnovich, M. I., Synthetic liquids for electric devices, Energiya, Moscow, 1972 (in Russian).

[11] Adamchevskii, I., Electroconductivity of liquid insulators, Energiya, Leningrad, 1972 (in Russian).

[12] Balygin, I. E., Electrical strength of liquid insulators, Energiya, Moscow, 1964 (in Russian).

[13] A handbook on plastic materials, Eds. Kataev, V. M., Popov, V. A. and Sazhin, B. I., 2nd ed., Khimiya, Moscow, 1975 (Vols. 1, 2) (in Russian).

[14] Polyvinyl chloride plastic compounds and their application to cables, Ed. Troitskii, I. D., Energiya, Moscow, 1978 (in Russian).

[15] Electrotechnical materials. Handbook, Eds. Berezin, V.B., Prokhorov, N. S., Rykov, G. A., Khaikin, A. M., 3rd ed., Energoatomizdat, Moscow, 1983 (in Russian).

[16] Baranovskii, V. V., Dulitskaya, G. M., Laminated plastic materials for electrotechnical purposes, Energiya, Moscow, 1976 (in Russian).

[17] Gladkov, A. Z., Electrical isolating varnishes and compounds, Energiya, Moscow, 1973 (in Russian).

[18] Pasynkov, V. V., Electronic engineering materials, Vysshaya Shkola, Moscow, 1980 (in Russian).

[19] Encyclopedia of polymers, Sovetskaya Entsiklopedia, Moscow, 1972 (vol. 1), 1974 (vol. 2), 1977 (vol. 3) (in Russian).

[20] Varicaps in electronic pulsed circuits, Ed. Bulybenko, V.Yu., Sovetskoe Radio, Moscow, 1971 (in Russian).

24

Thermoelectric Phenomena

N.A. Babushkina

Thermoelectricity embraces a set of physical phenomena (among them are the Seebeck, Peltier, and Thompson effects) caused by a correlation between thermal and electrical processes in the solid conductors.

The *Seebeck effect* lies in the fact that in a closed electric circuit of heterogeneous metals, there arises a thermoelectromotive force (thermopower) E_{12}, if their junctions are maintained at different temperatures T_1 and T_2.

The thermopower E_{12} depends only on temperatures T_1 and T_2 of the connected conductors and on the nature of the materials composing the thermocouple. From the value of E_{12}, it is possible to estimate the temperature at the junction site. In a relatively narrow temperature range, the relationship $E_{12} = S_{12}(T_1 - T_2)$ holds, where S_{12} is the thermopower coefficient determined by the nature of the materials in the thermocouple and by the temperature range to which it is applied. The S_{12} coefficient is subject to drastic variations with temperature and is free even to change its sign.

Measured thermopowers always relate to a pair of metals and hence they do not provide the characteristics of individual metals. Usually tabulated values of thermopower are given with respect to lead, platinum, or copper (as, for example, in Table 24.3).

The *Peltier effect* consists of the following. As an electric current flows through the junction of different conductors, an amount of heat Q_Π apart from the Joule heat is released or absorbed (depending on the current direction), which is proportional to the electric charge passed through the junction (i.e., to current strength I and time t):

$$Q_\Pi = \Pi I t \tag{24.1}$$

where Π is the Peltier coefficient depending on the nature of materials in contact interior (see Table 24.7).

Both the Peltier and Seebeck effects can be observed only when two heterogeneous conductors are present in an electric circuit.

The *Thompson effect* refers to a reversible release (or absorption) of heat in a homogeneous conductor carrying an electric current, which occurs if a temperature gradient parallel to the current is also present:

$$Q_\mu = \mu(T_2 - T_1)It, \tag{24.2}$$

where μ is the Thompson coefficient depending on the nature of the material (see Table 24.8), I is the current strength, and t is the time.

Contrary to the Seebeck and Peltier phenomena, the Thompson effect relates to one homogeneous conductor, therefore coefficient μ for any conductor can be determined independently.

According to the thermodynamic approach, the Thompson, Seebeck, and Peltier coefficients (μ, S, and Π) are connected by the Kelvin relations in the following way:

$$\mu = T\frac{dS}{dT}, \quad \Pi = ST. \tag{24.3}$$

0-8493-2861-6/97/$0.00+$.50

From the known temperature dependence of the Thompson coefficient, one is able to calculate the thermopower and Peltier coefficients of individual metals (the so-called absolute thermopower and Peltier coefficients of metals):

$$S = \int_0^T \frac{\mu}{T} dT. \tag{24.4}$$

The differential thermopower of a pair of metals is presented as a difference in the absolute values of thermopower for the individual metals, namely:

$$S_{12} = S_1 - S_2. \tag{24.5}$$

The absolute thermopower of a metal at low temperatures can be measured if a thermocouple is composed of the metal and a superconductor, since the differential thermopower is produced in this by its normal branch only.

The absolute thermopower of lead (see Table 24.1) obtained in such a way by Christian et al. [1] makes up a universally recognized standard used in all thermoelectrical measurements. The values of absolute thermopower compiled in a later paper by Roberts [2] differ from Christian's data at temperatures above 20 K (Table 24.1).

At high temperatures, the precious metals – copper, silver, and gold – may also be used as "standards" (Table 24.2). They are more preferable compared to the transition metals (platinum and tungsten), whose absolute values of thermopower are considerably higher. In Tables 24.4–24.5, the thermopowers of transition metals are presented [38].

In addition, in Table 24.6 lists differential thermopower of metals with respect to platinum. In Figs. 24.1–24.18, temperature-dependent curves of absolute thermopower for a number of metals are illustrated.

The thermopower of pure metals is generally presumed to consist of two contributions: S_d defines a diffusion term arising due to conduction electrons, while S_g is the phonon-drag term i.e. a conduction electron contribution induced by nonequilibrium effects in the phonon system. In the simplest free electron model, the thermopower is directly proportional to temperature, the contributions are additive, and limiting functional forms take the form

$$\begin{aligned} S(T) &= AT + BT^3, \qquad T \ll \Theta_D; \\ S(T) &= AT + C/T, \qquad T \gg \Theta_D, \end{aligned} \tag{24.6}$$

where Θ_D is the Debye temperature. Phonon-phonon scattering alters the nonequilibrium condition in the phonon spectrum and causes changes in the limiting expressions for the phonon-drag term. These simple equations are modified to incorporate a variety of forms for more realistic models, and hence they are valid for a restricted number of elements. Nonetheless, experimental results are often reported in relation to the above cited equations.

For elements having magnetic properties, an additional thermopower contribution combined with modifications of the above components, is due to spin scattering. The effect of spin scattering is still under very active study and no simple formulae, either based upon theory or empirical data, have been generally adopted. Moreover, no metal is ever absolutely pure. Some impurities are always present in the specimen and thus impurity scattering effects on thermopower cannot be ignored.

Numbers in the tables that follow should be regarded as most plausible rather than absolutely accurate, as the thermopower of a substance is sensitive to a very small number of impurities, to crystalline-grain orientation, and to thermal or even cold treatment of the material. For this reason, a thermopower may arise (in the presence of a temperature gradient) in a circuit involving only one material, if various sections of the circuit were subjected to treatment of different types. In comparing the S values of materials investigated by a variety of researchers, a discrepancy might be revealed of the 10% range.

It should be remarked that graduating tables for thermocouples are given in Chapter 8.

The comprehensive theoretical and experimental data on the thermoelectric phenomena have been summarized in [3]. Integrated data on composition as well as physical, chemical, and metrological properties of alloys currently employed in thermocouples can be found in the handbooks [4,17].

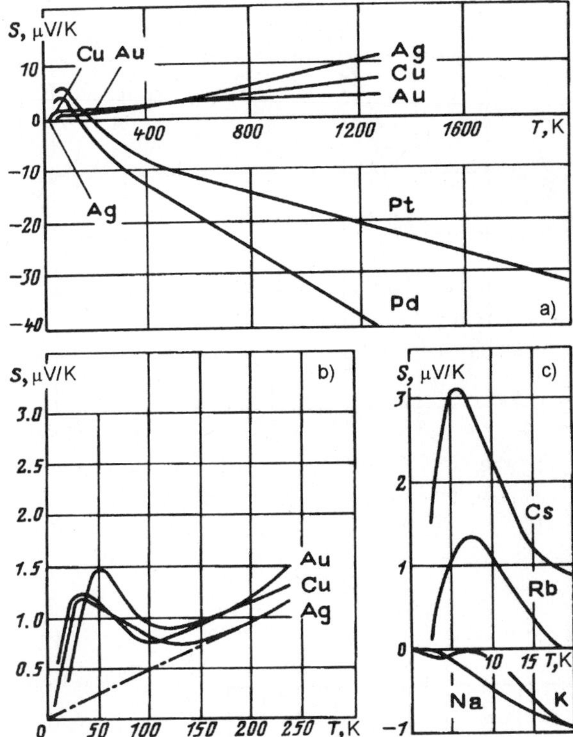

Figure 24.1 Absolute thermopower of metals: (**a**) refractory and precious metals in the temperature range above 300 K [5]; (**b**) copper, silver, and gold at temperatures from 0 to 250 K [7]; (**c**) alkali metals at low temperatures [7].

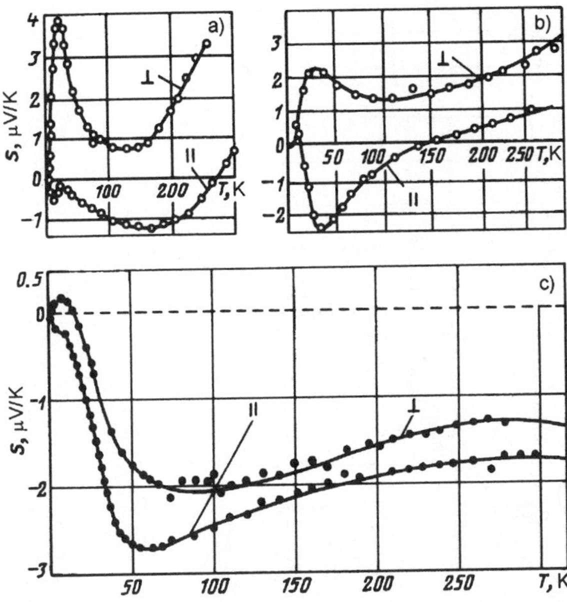

Figure 24.2 Temperature-dependent curves of absolute thermopower for cadmium (**a**), zinc (**b**), and magnesium (**c**) in the directions parallel (||) and perpendicular (⊥) to the hexagonal axes [9].

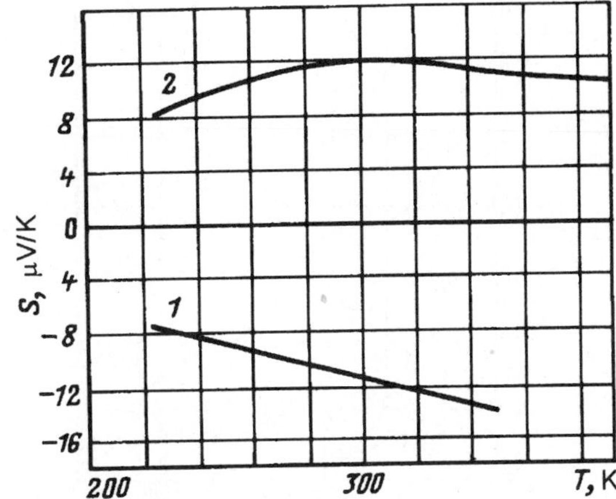

Figure 24.3 Temperature-dependent curves of absolute thermopower for alkali metals: (*1*) generalized curve for sodium, potassium, and rubidium; (*2*) for lithium [8].

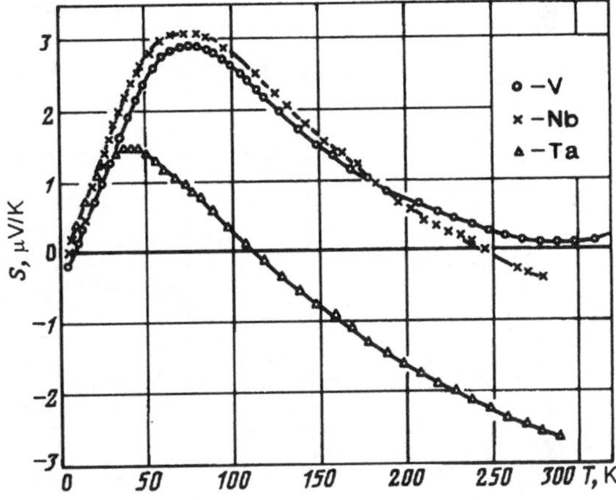

Figure 24.4 Temperature-dependent curves of absolute thermopower for vanadium, niobium, and tantalum [10].

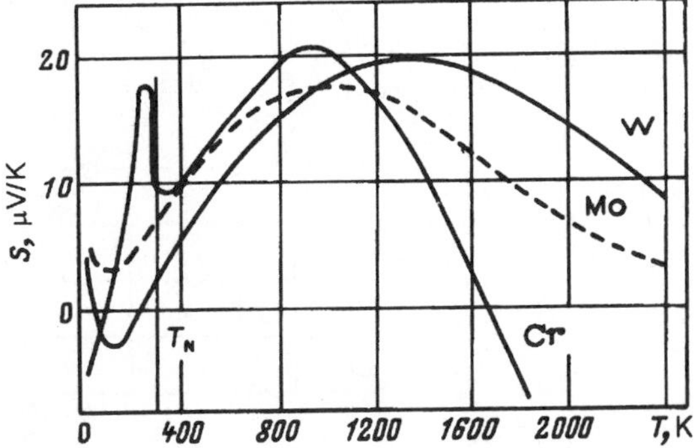

Figure 24.5 Temperature-dependent curves of absolute thermopower for tungsten, molybdenum, and chromium [8]; T_N is the Néel temperature for chromium.

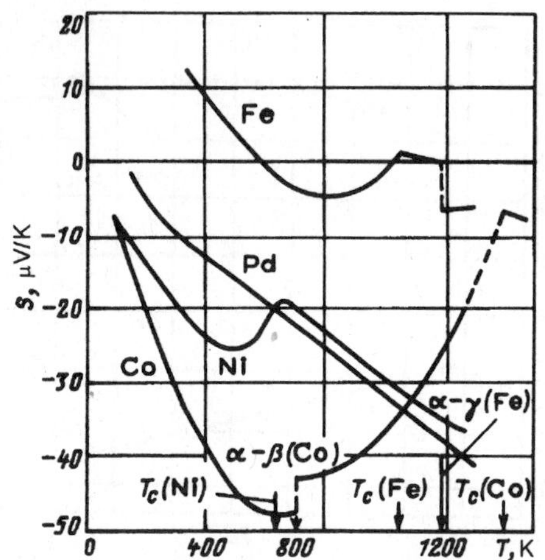

Figure 24.6 Temperature-dependent curves of absolute thermopower for iron, cobalt, nickel, and palladium [11]: T_C is the Curie point; $\alpha-\beta, \alpha-\gamma$ are the points of structural transitions for cobalt and iron.

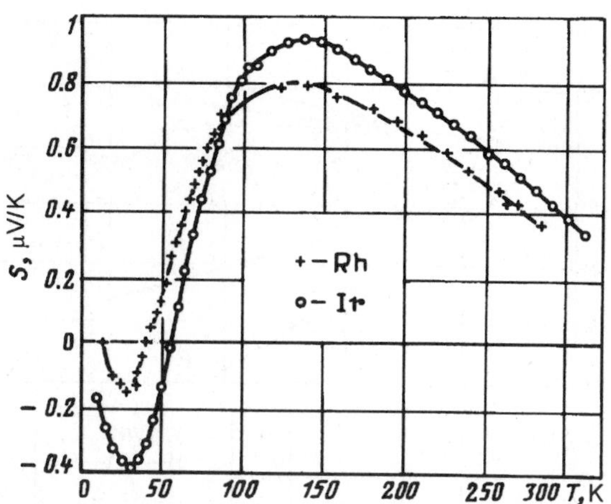

Figure 24.7 Temperature-dependent curves of absolute thermopower for rhodium and iridium [10].

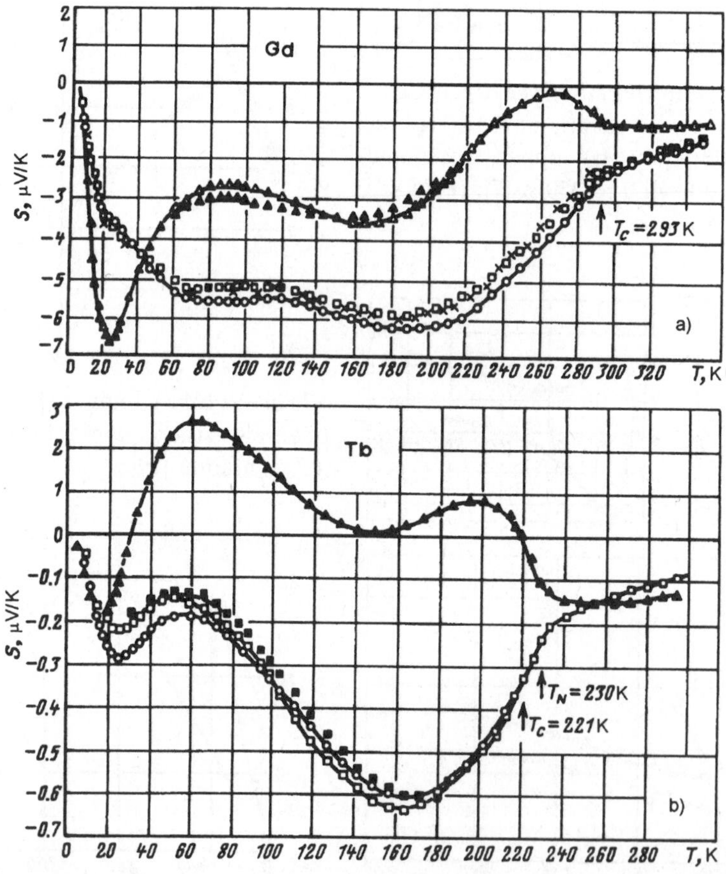

Figure 24.8 Temperature-dependent curves of absolute thermopower of single crystals: **(a)** gadolinium [12] (○ – in the basal plane, △ – along the c-axis, ▲– along the c-axis at $B = 0.1$ T, □ – in the a–c plane, 17° off the a-axis); **(b)** terbium [12] (○ – along the a-axis, □ – along the b-axis, ▲ – along the c-axis, ■ – along the b-axis at $B = 0.13$ T).

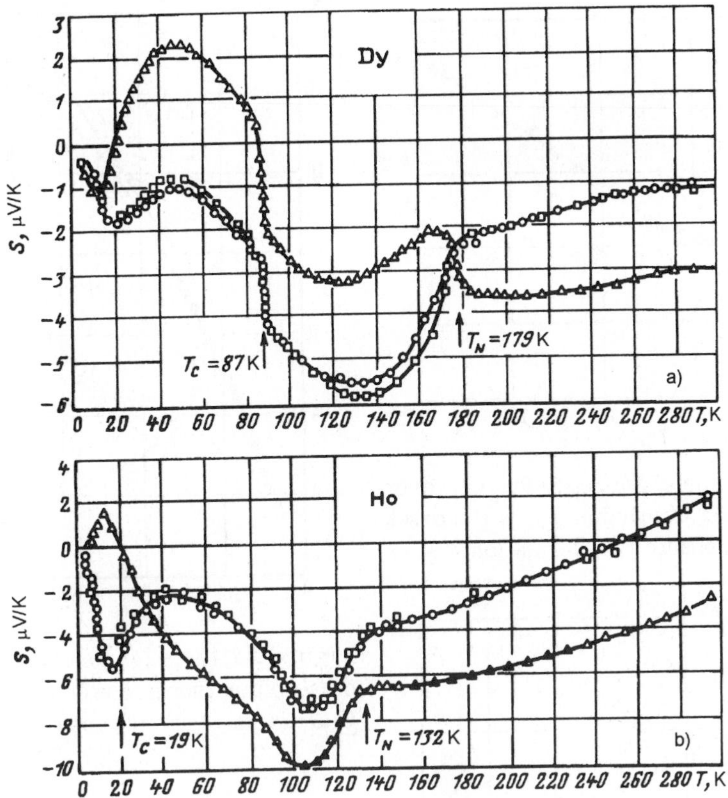

Figure 24.9 Temperature-dependent curves of absolute thermopower of dysprosium (**a**) and holmium (**b**) single crystals [12]: ○, □, and △ – along the *a*-, *b*-, and *c*-axes, respectively.

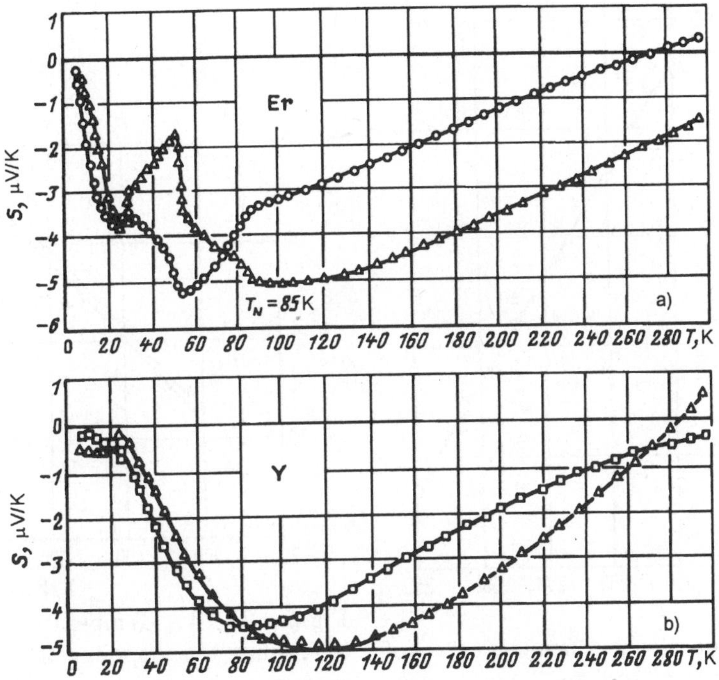

Figure 24.10 Temperature-dependent curves of absolute thermopower of erbium (**a**) and yttrium (**b**) single crystals [12]: ○ – in the basal plane, △ – along the *c*-axis.

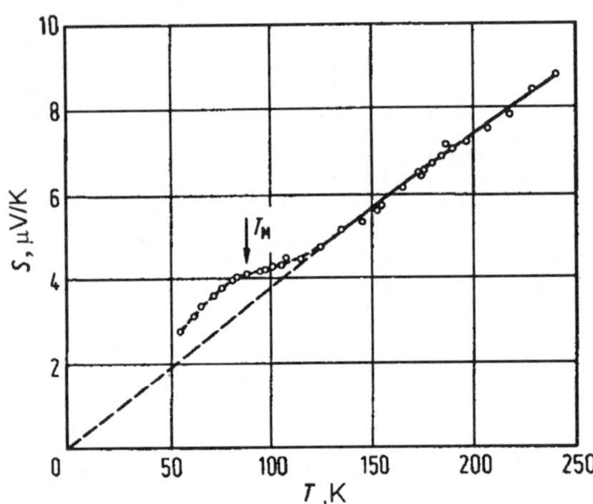

Figure 24.11 Temperature-dependent curves of absolute thermopower of lithium: T_M is the onset temperature of martensitic transformation [17].

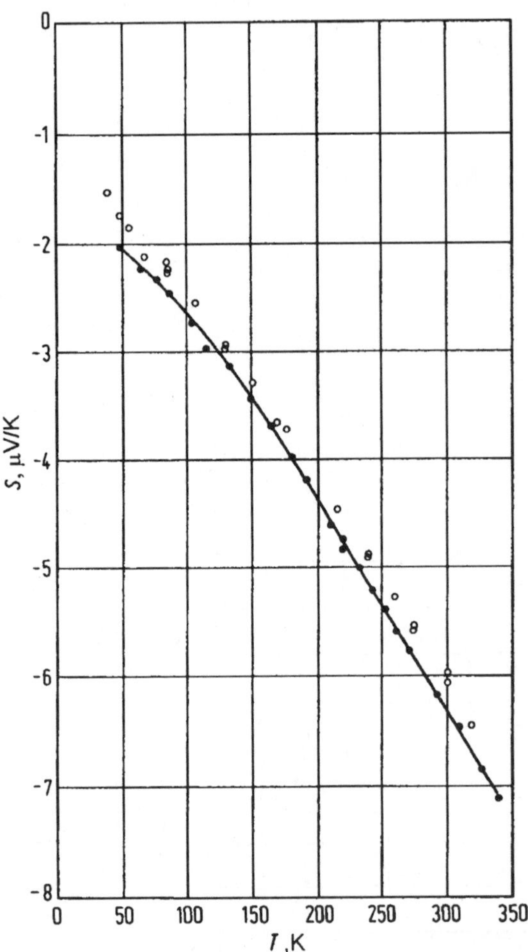

Figure 24.12 Temperature-dependent curves of absolute thermopower of sodium. Actual data points are for two samples of comparable purity [17].

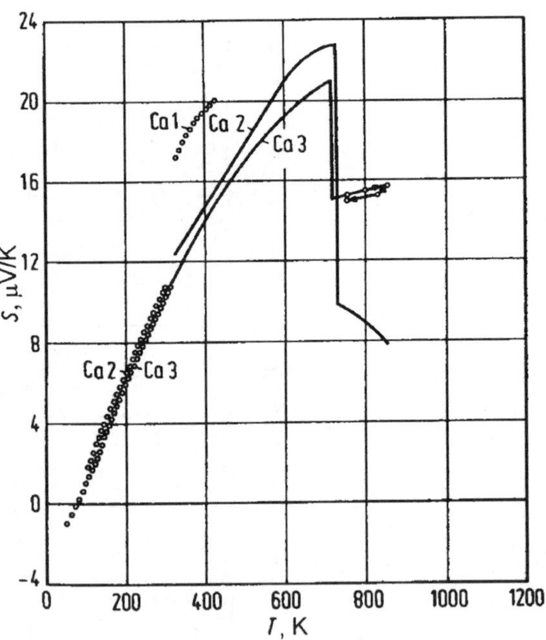

Figure 24.13 Temperature-dependent curves of absolute thermopower of calcium: Ca*2* and Ca*3* are purer samples as compared to Ca*1* [17].

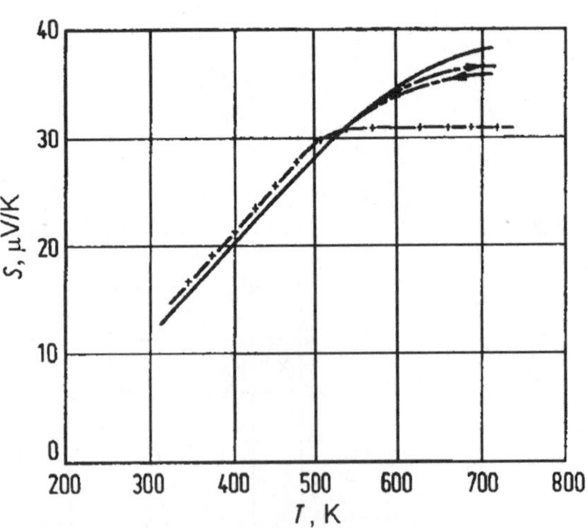

Figure 24.14 Temperature-dependent curves of absolute thermopower of barium. These data show the effects of sample purity: — very pure sample, –+– pure sample, - - - charged with H_2 at 535 K [17].

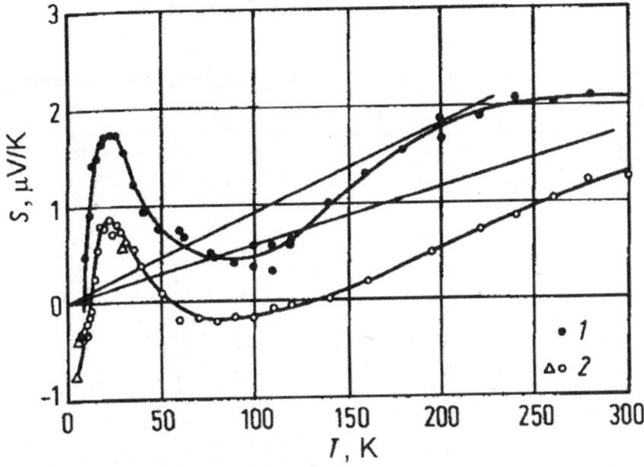

Figure 24.15 Temperature-dependent curves of absolute thermopower of indium single crystals. *1* – along the [100] direction, *2* – along the [001] direction, the straight lines illustrate possible diffusion contributions, in these preferred directions, deduced from the higher temperature data [17].

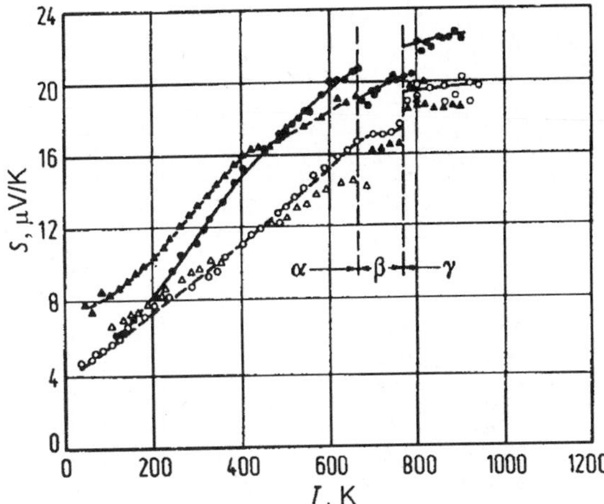

Figure 24.16 Temperature-dependent curves of absolute thermopower of uranium. Data illustrate that sample history effects (thermal hysteresis and annealing after mechanical deformation) are pronounced [17].

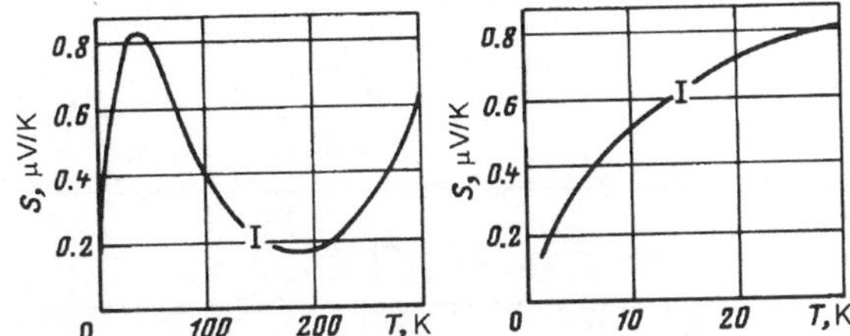

Figure 24.17 Temperature-dependent curves of absolute thermopower of manganin alloy (82.2% Cu + 13.3% Mn + 4.5% Ni) [13].

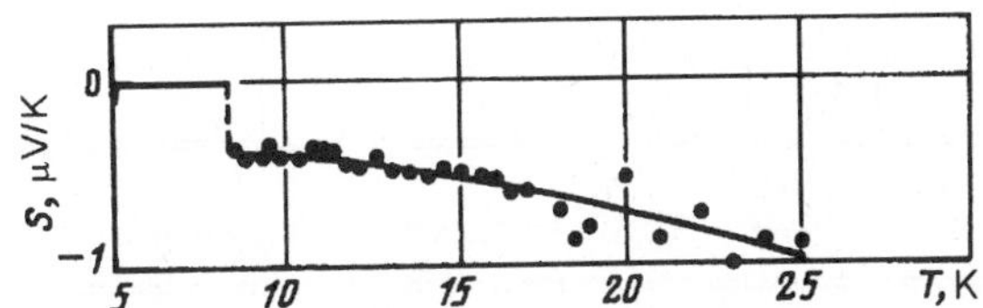

Figure 24.18 Temperature-dependent curve of absolute thermopower of superconducting alloy 48% Nb + 52% Ti (T_C = 9 K) [14].

Table 24.1 Absolute thermopower of pure lead after Christian et al. [1] and Roberts [2]

T, K	S_{Chr}, µV/K	S_{Rob}, µV/K	T, K	S_{Chr}, µV/K	S_{Rob}, µV/K	T, K	S_{Chr}, µV/K	S_{Rob}, µV/K	T, K	S_{Chr}, µV/K	S_{Rob}, µV/K
7.25	−0.204	−0.204	16.5	−0.777	−0.777	40.0	−0.764	−0.575	170.0	−1.054	−0.760
7.5	−0.221	−0.220	17.0	−0.781	−0.782	42.0	−0.766	−0.564	180.0	−1.075	−0.785
8.0	−0.257	−0.226	18.0	−0.785	−0.786	44.0	−0.768	−0.555	190.0	−1.098	−0.810
8.5	−0.297	−0.298	19.0	−0.785	−0.784	46.0	−0.770	−0.548	200.0	−1.120	−0.834
9.0	−0.343	−0.343	20.0	−0.784	−0.779	48.0	−0.772	−0.542	210.0	−1.143	−0.858
9.5	−0.390	−0.388	21.0	−0.783	−0.771	50.0	−0.774	−0.537	220.0	−1.161	−0.882
10.0	−0.434	−0.433	22.0	−0.782	−0.760	55.0	−0.777	−0.530	230.0	−1.175	−0.904
10.5	−0.475	−0.476	23.0	−0.781	−0.748	60.0	−0.779	−0.527	240.0	−1.191	−0.927
11.0	−0.516	−0.517	24.0	−0.780	−0.735	65.0	−0.782	−0.528	250.0	−1.205	−0.948
11.5	−0.556	−0.556	25.0	−0.779	−0.721	70.0	−0.784	−0.531	260.0	−1.224	−0.969
12.0	−0.593	−0.593	26.0	−0.778	−0.707	80.0	−0.794	−0.544	270.0	−1.243	−0.989
12.5	−0.628	−0.627	27.0	−0.777	−0.694	90.0	−0.824	−0.562	280.0	−1.259	−1.009
13.0	−0.658	−0.657	28.0	−0.776	−0.681	100.0	−0.865	−0.583	290.0	−1.271	−1.028
13.5	−0.683	−0.684	29.0	−0.775	−0.669	110.0	−0.899	−0.606	300.0		−1.047
14.0	−0.706	−0.707	30.0	−0.774	−0.657	120.0	−0.927	−0.631	310.0		−1.065
14.5	−0.728	−0.728	32.0	−0.772	−0.636	130.0	−0.952	−0.656	320.0		−1.084
15.0	−0.746	−0.745	34.0	−0.770	−0.617	140.0	−0.980	−0.682	330.0		−1.101
15.5	−0.760	−0.759	36.0	−0.768	−0.601	150.0	−1.010	−0.708	340.0		−1.119
16.0	−0.771	−0.770	38.0	−0.766	−0.587	160.0	−1.034	−0.734	350.0		−1.136

Table 24.2 Absolute thermopower for some metals [5] in the temperature range from 100 to 2400 K (based on data for Pb [1])

T, K	S, μV/K						
	Cu	Ag	Au	Pt	Pd	W	Mo
100	1.19	0.73	0.82	4.29	2.00		
150	1.12	0.85	1.02	1.32	−1.63		
200	1.29	1.05	1.34	−1.27	−4.85		
273	1.70	1.33	1.79	−4.45	−9.00		
300	1.83	1.51	1.94	−5.28	−9.99	1.07	5.57
400	2.34	2.08	2.46	−7.83	−13.00	4.44	8.52
500	2.83	2.82	2.86	−9.89	−16.03	7.53	11.12
600	3.33	3.72	3.18	−11.66	−19.06	10.29	13.27
700	3.83	4.72	3.43	−13.31	−22.09	12.66	14.94
800	4.34	5.77	3.63	−14.88	−25.12	14.65	16.13
900	4.85	6.85	3.77	−16.39	−28.15	16.28	16.86
1000	5.36	7.95	3.85	−17.86	−31.18	17.57	17.16
1100	5.88	9.06	3.88	−19.29	−34.21	18.53	17.08
1200	6.40	10.15	3.86	−20.69	−37.24	19.18	16.65
1300	6.91		3.78	−22.06	−40.27	19.53	15.92
1400				−23.41	−43.30	19.60	14.94
1600				−26.06	−49.36	18.97	12.42
1800				−28.06	−55.42	17.41	9.52
2000				−31.23	−61.48	15.05	6.67
2200						12.01	4.30
2400						8.39	2.87

Table 24.3 Absolute thermopower of pure metals near room temperature [17]

Metal	S, μV/K	Metal	S, μV/K	Metal	S, μV/K
Ag	1.51	In	1.68	Pu	≥10
Al	−1.66	Ir	≈1.35...≈0.35	Rb	−10.23
Au	1.89	K	−13.7	Re	−5.42
Ba	12.1	La	≈1.6...≈−1.0	Rh	≈1.3
Ca	10.3	Li	10.6*	Ru	≈1.0, ≈−1.5
Cd	2.56	Lu, sc	≈ −3.3...≈ −4.3	Sc	−5.3...−16.2*
Co	−30.8		b-axis: −8.0	Sm	≈1.2, ≈1.5
Cr	21.7		c-axis: 4.0	Sn	−0.92*
Cs	−0.9*	Mg	−1.46	Sr	1.1
Cu	1.85	Mn	−10.0*	Ta	−2.2*
Dy	≈ −0.3...≈ −1.8	Mo	5.57	Tb	≈−0.9, ≈−1.2
Er	≈0.5...≈-0.5	Na	−5.8*	Th	≈−3.0...≈−4.0
Eu	≈25...≈15	Nb	≈−0.6	Ti	4.5*
Fe	16.2*	Nd	≈ −2.3...≈ −4.0	Tm	≈1.8...≈1.3
Ga		Ni	−19.24, −19.52	U	>5
< 100 >	1.78	Np	≈−4	V	0.13*
< 010 >	0.56	Os	≈ −4.2*	W	1.1, ≈1.5
< 001 >	2.13	Pb	−1.29	Y	≈1.2
Gd	≈ −0.5...≈ −2.0	Pd	−10.7	Yb	≈28...≈14
Hf	≈8.3...≈6.0	Pr	≈−0.6...≈−3.1	Zn	positive
Ho	≈−0.5...≈−2.0	Pt	−5.14	Zr	≈10...≈5

* Values at 273 K.

Note: The data provide a concise summary of experimental findings for pure metals at 300 K. Entries preceded by ≈ were taken from graphical sources. Dual entries separated by a comma are supposed to be equally probable. All values are given on the basis of the absolute scale [1]. To convert these values to a new scale [16], add 0.55 μV/K to figures at 273 K, and add 0.24 μV/K to those at 300 K.

Table 24.4 Absolute thermopower (in μV/K) for pure transition metals at sub-room temperatures [3] (based on data for Pb [1])

Element	Thermopower at given temperature (in K)								
	10	20	50	80	100	150	200	250	273
Group III B									
Sc	−1.6	−3.0	−8.0	−12.7	−14.0	−15.6	−16.3	−16.5	−16.2
Y*1	−0.3	−0.4	−3.3	−4.5	−4.4	−3.1	−1.9	−0.9	−0.7
	−0.5	−0.4	−2.8	−4.5	−4.9	−4.5	−3.2	−1.4	−0.5
La					+0.3	+0.4	+0.7	+1.0	+1.3
Group IV B									
Ti			−3.0	−3.0	−2.6	0	+2.0	+4.0	+4.5
Zr			0	+3.0	+4.5	+7.5	+8.5	+9.5	+9.5
Hf				0	0	+2.5	+3.7	+4.7	+5.3
Group V B									
V*2	+0.19	+0.76	+2.45	+2.91	+2.65	+1.52	+0.72	+0.26	+0.13
Nb*2	+0.31	+0.98	+2.73	+3.09	+3.13	+1.42	+0.65	−0.04	−0.20
Ta*2	+0.36	+1.03	+1.41	+0.78	0	−0.8	−1.5	−2.0	−2.2
Group VI B									
Cr*3	+3.1	+6.7	+8.2	+5.0	+5.0	+7.0	+11.8	+17.5	+18.8
Mo*3	−0.02	−0.11	−0.48	−0.2	+0.1	+0.94	+2.50	+4.08	+4.57
W*3	+0.05	−0.28	−2.78	−3.70	−4.04	−2.45	−1.41	−0.10	+.056
Group VII B									
Mn		+12.5	+15.5	+6	−2.5	−7.0	−8.5	−9.7	−10.0
Re	+0.61	+1.32	+1.18	+0.08	−0.66	−2.21	−3.51	4.63	−5.03
Group VIII B									
Fe*4	+1.0	+2.5	+8.0	+12.0	+13.0	+16.0	+17.0	+15.5	+15.0
Ru				+0.2	0	−1.1	−1.5	−1.5	−1.5
Os				−2.2	−3.2	−3.8	−4.0	−4.0	−4.0
Co*4		−0.5	−1.0	−3.0	−4.0	−9.0	−12.0	−18.0	−19.0
Rh*5	−0.19	−0.33	−0.11	+0.54	+0.78	+0.92	+0.75	+0.58	+0.48
Ir*5		−0.11	+0.13	+0.57	+0.73	+0.77	+0.64	+0.46	+0.35
Ni*4	−2.0	−4.7	−7.2	−8.1	−11.0	−11.0	−13.5	−17.0	−18.0
Pd*4	+0.4	+1.6	+4.3	+3.7	+2.00	−1.63	−4.85	−7.42	−9.00
Pt	+0.6	+2.3	+5.8	+5.5	+4.29	+1.82	−1.27	−3.28	−4.45

*1 In the numerator, data relate to the basal plane; in the denominator, data for the c-axis.
*2 See also Fig. 24.4 [9].
*3 See also Fig. 24.5 [7].
*4 See also Fig. 24.7 [10].
*5 See also Fig. 24.6 [9].

Table 24.5 Ranges of linear dependence for transition–metals thermopower [3,8]

Element	Temperature range of linearity, K	Range of thermopower, mV
Group III B		
Sc	150 – 500	(−7.2) – (−5)
	750 – 1400	(−5) – (+9)
La	300 – 583	(−2) – (+2.3)
Group IV B		
Ti	400 – 800	(+3) – (−5)
Zr	400 – 800	(+7.5) – (−5)
	900 – 1400	(−6) – (+2)
Hf	500 – 1000	8 – 0
Group V B		
V	500 – 1100	1 – 6
Nb	600 – 1400	(−1.5) – (+6)
Ta	1100 – 1800	(−2) – (+9)
Group VI B		
Cr	500 – 800	10 – 12
Mo	300 – 600	6 – 13
	1200 – 2100	1.6 – 7
W	200 – 300	(−2) – (+2.5)
	1800 – 2400	17 – 8
Group VII B		
Mn	300 – 1000	(−1) – (−4)
Re	450 – 1500	(−7) – (−1)
Group VIII B		
Fe	300 – 700	(+12) – (−5)
Ru	500 – 1800	(−2) – (−9.5)
Os	600 – 1800	(−3.8) – (−10)
Co	100 – 500	(−5) – (−46)
	1400 – 1600	(−8) – (−8.5)
Rh	600 – 1700	(+0.5) – (−0.8)
Ir	300 – 1300	(+1) – (−0.4)
	1400 – 1800	(−4.5) – (−9)
Ni	200 – 500	(−15) – (−2.5)
	600 – 1500	(−20) – (+40)
Pd	300 – 1800	(−10) – (−30)
Pt	400 – 2000	(−8) – (+55)

Table 24.6 Values of thermopower (in mV) in reference to platinum for some materials at various temperatures in degrees of IPTS-68 [6]

Material	Temperature, K									
	73	173	373	473	673	873	1073	1273	1473	1673
Elements										
Aluminum	0.45	0.06	0.42	1.06	2.84	5.15				
Antimony			4.89	10.14	20.53	28.87				
Bismuth	−0.04	−0.31	0.90	2.35						
Cadmium	−0.04	−0.31	0.90	2.35						
Calcium			−0.51	−1.13						
Carbon			0.70	1.54	3.72	6.79	10.98	16.46		
Cerium	0.22	−0.13								
Cobalt			−1.33	−3.08	−7.24	−11.28	−13.99	−14.21		
Copper	−0.19	−0.37	0.76	1.83	4.68	8.34	12.81	18,16		
Germanium	−44.00	−26.62	33.9	72.4	82.3	43.9				
Gold	−021	−0.39	0.78	1.84	4.63	8.12	12.26	17.05		
Indium			0.69							
Iridium	−0.25	−0.35	0.65	1.49	3.55	6.10	9.10	12.57	16.45	20.47
Lead	0.24	−0.13	0.44	1.09						
Lithium	−1.12	−1.00	1.82							
Magnesium	0.37	−0.09	0.44	1.10						
Mercury			−0.60	−1.33						
Molybdenum			1.45	3.19	7.57	13.13	19.83	27.74	36.96	
Nickel	2.28	1.12	−1.48	−3.10	−5.45	−7.04	−9.83	−12.11		
Palladium	0.81	0.48	−0.57	−1.23	−2.82	−5.03	−7.96	−11.61	−15.86	−20.40
Potassium	1.61	0.78								
Rhodium	−0.20	−0.34	0.70	1.61	3.91	6.77	10.14	14.02	18.39	22.99
Rubidium	1.09	0.46								
Silicon	63.13	37.17	−41.56	−80.57						
Silver	−0.21	−0.39	0.74	1.77	4.57	8.41	13.33			
Sodium	1.00	−0.29								
Tantalum	0.21	−0.10	0.33	0.93	2.91	5.95	10.02	15.15	21.37	
Thallium			0.58	1.30						
Thorium			−0.13	−0.26	−0.50	−0.45	0.22	1.72	4.03	
Tin	0.26	−0.12	0.42	1.07						
Tungsten	0.43	−0.15	1.12	2.62	6.70	12.26	19.25	27.73	37.72	
Zinc	−0.07	−0.33	0.76	1.89	5.29					
Materials for thermocouples										
Alumel[*1]	2.39	1.29	−1.29	−2.17	−3.64	−5.28	−7.07	−8.78	−10.33	−11.72
Chromel[*2]	−3.36	−2.20	2.81	5.96	12.75	19.61	26.20	32.47	38.48	44.04
Constantan[*3]	5.35	2.98	−3.51	−7.45	−16.19	−25.46	−34.81	−43.85		
Alloys										
Beryllium bronze[*4]			0.67	1.26	4.19					
Ferronichrome[*5]			0.85	2.01	5.00	8.68	13.03	18.06		
Manganin[*6]			0.61	1.55	4.25	7.84				
Nichrome[*7]			1.14	2.62	6.25	10.53	15.41	20.87		
Phosphor bronze[*8]			0.55	1.34		3.50	6.30			

[*1] 94% Ni + 2% Al+2% Si + 2% Mn.
[*2] 90% Ni + 10% Cr.
[*3] 60% Cu + 40% Cr.
[*4] 98% Cu + 2% Be.
[*5] 60% Ni + 24% Fe + 16% Cr.
[*6] 84% Cu + 4% Ni + 12% Mn.
[*7] 80% Ni + 20% Cr.
[*8] 85% Cu + 13% Sn + 1.86% Zn + 0.16% Pb + 0.013% P.

Note: The temperature of the thermocouple-free ends was 273 K. The positive sign of the metal thermopower corresponds to a current flow through the thermocouple junction from platinum to the metal.

Table 24.7 Peltier coefficients for various pairs of metals [15]

Iron–constantan		Copper–nickel		Lead–constantan	
T, K	Π, mV	T, K	Π, mV	T, K	Π, mV
273	13	292	8.0	293	8.7
293	15	328	9.0	383	11.8
403	19	478	10.3	508	16.0
513	26	563	8.6	578	18.7
593	34	613	8.0	633	20.6
833	52	718	10.0	713	23.4

Note: Copper–constantan displays Π=11.0 mV at T=293 K.

Table 24.8 Thompson coefficients for various metals [15]

Metal	T, K	μ, μV/K	Metal	T, K	μ, μV/K
Ag	105	−0.10	Ni	100	−4.5
	150	+0.55		200	−12.1
	300	+1.31		260	−15.7
Al (99 %)	260	+0.11	Pb (99.99%)	120	−0.19
	300	−0.08		200	−0.45
	400	−0.56		400	−0.85
Au	105	+0.29	Pd	20	+1.9
	150	+0.96		100	−7.8
	300	+1.61		200	−12.1
Cd (single crystal) ‖[*1]	373	+6.9		300	−18.2
	473	+7.3	Pt	203	−9.6
⊥[*2]	373	+8.2		273	−9.1
Co	473	+8.7		393	−9.2
	100	−8.4	Sn (99.99%)	100	+0.86
	200	−19.6		200	+0.42
	300	−25.4		300	−0.07
Cu	70	−0.26		400	−0.45
	170	+0.47	Zn (single crystal) ‖[*1]	322.5	+1.6
	300	+1.52		398	+4.6
Cu+0.37% Au	20	+0.44	⊥[*2]	322.5	+4.1
	40	+2.17		398	+8.8
	60	+2.96	W	328	+8.5
	90	+2.20	Constantan	323	−24.6
	130	+1.71		384	−25.5
	300	+2.33		446	−26.0
Fe (Armco)	323	−15.3	German silver	328	−12.0
	423	−22.8			
	523	−26.3			

[*1] ‖ – the specimen is cut out along the hexagonal crystal axis (Thompson effect is observed along a specimen).
[*2] ⊥ – the specimen is cut out at the angle of 90° to the hexagonal crystal axis.

References

[1] Christian, J. W., Jan, J.-P., Pearson, W. B., Templeton, I.M., Proc. Royal Soc., London, A245, 213, 1958.

[2] Roberts, R. B., Phil. Mag., 36, 91, 1977.

[3] Blatt, F. J., Schroeder, P. A., Foiles, C. L., Greig, D., Thermoelectric power of metals, Plenum Press, New York, 1976.

[4] Rogel'berg, I. L., Beilin, V. M., Alloys for thermocouples. Handbook, Metallurgiya, Moscow, 1983 (in Russian).

[5] Cusack, N., Kendall, P., Proc. Phys. Soc., 72, 898, 1958.

[6] Physics Vade Mecum, AIP 50th Anniv., Ed. Anderson, H. L., AIP, New York, 1981.

[7] MacDonald, D. K. C., Thermoelectricity: an introduction to the principles, John Wiley and Sons, New York, 1962.

[8] Vedernikov, M. N., Burkov, A. T., Present state of experimental knowledge on thermopower of metals at high temperatures above 77 K, in: Thermoelectricity in metallic conductors, Eds. Blatt, F. J., Schroeder P. A., Plenum Press, New–York–London, 1978.

[9] Rowe, V. A., Schroeder, P. A., J. Phys. Chem. Solids, 31, 1, 1970.

[10] Carter, R., Davidson, A., Schroeder, P. A., J. Phys. Chem. Solids, 31, 2374, 1970.

[11] Greig, D., Thermoelectricity in transition metals, in: Thermoelectricity in metallic conductors, Eds. Blatt, F. J., Schroeder P.A., Plenum Press, New York–London, 1978.

[12] Sill, L. R., Legvold, S., Phys. Rev., 137, 1139, 1965.

[13] Rathanayaka, K. D. D., J. Phys. E: Sci. Instrum., 18, 380, 1985.

[14] Harmans, C., Cryogenics, 22, 39, 1982.

[15] Landolt–Bornstein, Zahlenwerte und funktionen aus physik, chemie, astronomie, 6 Teil, Elektrische Eigenschaften, Springer–Verlag, Berlin–Gottingen–Heidelberg, 1959.

[16] Rowland, T., Cusack, N. E., Ros, R. G., J. Phys. E: Sci. Instrum., 4, 2189, 1974.

[17] Landolt–Bornstein, Numerical data and functional relationships in science and technology, New series, Eds.–in–Chief: Hellwege, K.-H., Madelung, O., Group III: Crystal and Solid State Physics, vol. 15: Metals: Electronic Transport Phenomena, (subvol. *b*), Springer–Verlag, Berlin, 1985

25

Electron and Ion Emission

T.M. Lifshits and A.L. Musatov

25.1 Introduction

Electron and ion emission constitutes a process of electron or ion ejection by different bodies being subjected to external factors: heating, incident photon, electron and ion streams, or strong electric field. Depending on the type of the external action one can differentiate thermoelectron, thermionic, photoelectron, secondary electron, secondary ion, electron–ion, ion–electron and field emissions.

For all the types of emission except the field one, the role of external action consists of raising the energy of some electrons or ions in the body to a level that allows them to overcome attractive forces and pass to the vacuum or other medium. Positive as well as negative ions may be ejected in ion emission.

The body emitting electrons or ions is called an *emitter*. To observe and utilize electron or ion emission, one has to set up electric fields near the emitter surface that draw off the ejected particles. Usually it is sufficient to apply rather moderate field strengths to a body (tens or hundreds of volts per centimeter) in order to guarantee the saturation of the emission current. In the case of field emission, the external electric field transforms the surface energy threshold existing on the body boundary and preventing the electrons from leaving the body, into the finite-width barrier and reduces its height, which makes the quantum mechanical tunneling of electrons through the barrier possible. As this takes place, the electric field energy is spent only to accelerate the emitted electrons. To initiate field emission, one has to apply the strong electric field ($\mathscr{E} \sim 10^9$ V/m), and the current density may reach 10^{11} A/m^2. In still higher impulse fields, some sections of the emitter, like protrusions and cusps, blow up after being strongly warmed, mainly by field emission current. A part of the emitter material transforms from the condensed phase into dense plasma. This process is accompanied by the ejection of the intensive electron stream and burst electron emission results. The monographs and reviews on emissive electronics and characteristics of various emitters can be of use [1–4, 7, 12, 14–17, 27–30, 34].

25.2 Work Function

The *work function* $\Phi = e\phi$ (here e is the electron charge, ϕ provides the potential) is the most important characteristic of solids and equals the minimum energy necessary for transferring a Fermi surface electron from the body to the point in vacuum where field intensity dies away to practically zero [1]. If one reckons the potential from the level corresponding to an electron at rest in vacuum, then ϕ comprises the potential inside the crystal corresponding to the Fermi level. According to modern concepts, the main contributions to the surface potential barrier are made by exchange and correlation

0-8493-2861-6/97/$0.00+$.50

effects and also, though in lower degree, by the electric double layer near the body surface. The most generally employed experimental methods of determining the work function are the emission ones: by using temperature, spectral, or field dependence of thermo-, photo-, or field emission, and also measuring the contact potential difference between the body under study and another body (anode) with the known work function [1, 2]. Tables 25.1, 25.3, and 25.4 present the work functions for elements and some compounds. It is worth noting that the external electric field lowers a work function (Schottky effect). If the emitter surface is homogeneous, a work function decrease upon imposing an electric field $\mathscr{E}$ (in V/m) is equal to

$$\Delta\Phi = e(e\mathscr{E}/4\pi\epsilon_0)^{1/2} = 3.79\cdot 10^{-5}\mathscr{E}^{1/2}\,\text{eV}. \tag{25.1}$$

Thin layers of adsorbed alkaline or alkaline-earth metals on the sample surface essentially reduce a work function. Work functions of metals and semiconductors are particularly strongly decreased by adsorption of cesium, barium, and related oxide layers on prepurified surfaces of these samples. Carbon and oxygen being adsorbed on the sample surfaces increase, as a rule, their work functions (Table 25.2, Figs. 25.1 and 25.2).

Table 25.1 Work function of the elements [2] [The crystallographic indices of the single-crystal faces are given in parentheses.]

Element	Φ, eV	
	Poly-crystal [2]	Single crystal [3]
Ag	4.3	4.64(100)
Al	4.25	4.41(100)
As	3.75[3]	4.55(332)
Au	5.1[3]	5.47(100)
B	4.5	
Ba	2.49	
Be	3.92	
Bi	4.4	
C	4.7	
Ca	2.80	
Cd	4.1	
Ce	2.7	
Co	4.41	
Cr	4.58	
Cs	1.81	
Cu	4.40	4.59(100)
		4.48(110)
		4.98(111)
		4.53(112)
Dy	3.25	
Er	3.25	
Eu	2.5	
Fe	4.31	4.67(100)
	4.70α	4.81(111)
	4.62β	
	4.68γ	
Ga	3.96	
Gd	3.1	
Ge	4.76	
Hf	3.53	
Hg	4.52	
Ho	3.22	

Element	Φ, eV	
	Poly-crystal [2]	Single crystal [3]
In	3.8	
Ir	5.27[3]	5.42(110)
		5.76(111)
		5.67(100)
		5.00(210)
K	2.22	2.39(110)
La	3.3	
Li	2.38	
Lu	3.33	
Mg	3.64	
Mn	3.83	
Mo	4.3	4.53(100)
		4.95(110)
		4.55(111)
		4.36(112)
		4.50(114)
Na	2.35	
Nb	3.99	4.02(001)
		4.87(110)
		4.36(111)
		4.63(112)
		4.29(113)
		3.95(116)
		4.18(310)
Nd	3.2	
Ni	4.50	5.22(100)
		5.04(110)
		5.35(111)
Os	4.7	
Pb	4.0	
Pd	4.8	5.6(111)
Pr	2.7	

Element	Φ, eV	
	Poly-crystal [2]	Single crystal [3]
Pt	5.32	5.7(111)
Rb	2.16	
Re	5.0	5.75(1011)
Rh	4.75	
Ru	4.60	
Sb	4.08	4.7(100)
Sc	3.3	
Se	4.72	
Si	4.8	4.85(100)(n-type)
		4.91(100)(p-type)
		4.60(111)(p-type)
Sm	2.7	
Sn	4.38	
Sr	2.35	
Ta	4.12	4.15(100)
		4.80(110)
		4.00(111)
Tb	3.15	
Te	4.73	
Th	3.30	
Tl	3.7	
Tm	3.10	
U	3.3	3.73(100)
		3.90(110)
		3.67(113)
V	4.12	
W	4.54	4.63(100)
		5.25(110)
		4.47(111)
		4.18(113)
		4.30(116)

Table 25.2 Work function of the polycrystalline elements covered by adsorbate with optimal thickness [2]

Substance-adsorbate	Φ, eV	Substance-adsorbate	Φ, eV	Substance-adsorbate	Φ, eV
Ag-0,Cs	1.0–1.2	Mo-Cs	1.54–1.66	Re-Y	2.38
Ag-Cs	1.65	Mo-Na	2.64	Rh-Ba	2.1–2.2
Au-Ba	2.3–2.8	Nb-Ba	2.2	Ru-Ba	2.22
Au-C	4.05	Nb-Cs	1.37	Ta-Ba	2.2
Au-Cs	1.8	Ni-Ba	1.52	Ta-Cs	1.6–1.69
Au-O	5.66	Ni-Cs	1.37	Ta-Th	2.52
Be-Cs	1.94	Os-Ba	2.22	Ta-Y	3.02
C-Cs	1.37	Os-Cs	1.44–1.5	W-αU	3.37
Ce-Ba	2.2	Pd-Cs	1.51	W-βU	3.31
Co-Cs	1.79	Pt-Ba	2.05	W-γU	3.19
Cr-Cs	1.71–1.79	Pt-Cs	1.59	W-Ba	2.1
Cu-Ba	3.35	Pt-Na	2.10	W-Ba,O	1.96
Fe-Cs	1.82	Pt-O	5.7–6.55	W-Cs	1.62–1.78
Hf-Ba	2.3–2.4	Pt-Rb	1.57	W-K	2.0
Ir-Cs	1.79	Re-Ba	2.42	W-Th	3.0–3.3
Mn-Cs	1.80	Re-Cs	1.4–1.51	W-Y	3.0
Mo-Ba	2.2	Re-Th	2.58–3.15		

Table 25.3 Work function of some monocrystalline semiconductors [2] [Errors are given in parentheses.]

Semiconductor	Crystal face	Φ, eV	Notes	Semiconductor	Crystal face	Φ, eV	Notes
$AgAsS_2$	(100)	5.7	cleaved facet		(110)	4.01(.02)	n-type
AlSb	(110)	4.86	p-type, cleaved	InAs	(110)	4.9(.05)	cleaved facet
			facet	InSb	(110)	4.57–4.77	
Bi_2Te_3	(0001)	5.40			(111)	4.39–4.43	
GaAs	(100)	4.38(.05)	p-type	PbS	(100)	3.5 (.2)	cleaved facet
	(110)	4.05–4.45	n-type	PbSe	(001)	4.14–4.30	n- and p-types
	(110)	4.65–5.35	p-type	PbTe	(100)	5.14(.2)	n- and p-types
	(111)	5.13	n-type	V_2O_5	(010)	4.5 (.2)	
GaSb	(100)	4.0	p-type		$(0\bar{1}0)$	6.71 (.08)	
	(110)	4.55	p-type				

Table 25.4 Work functions of high-melting compounds of transition metals with nonmetals [4]

Metal	Φ, eV	Φ, eV							
		Borides		Carbides		Nitrides		Silicides	
Cr	4.58	CrB_2	3.36					$CrSi_2$	3.78
Fe	4.31	FeB_2	3.5–3.75						
Hf	3.53	HfB_2	3.85	HfC	2.04–4.15	HfN	3.85–3.90		
La	3.3	LaB_6	2.41–3.20						
Mn	3.83	MnB_2	4.14						
Mo	4.3	MoB_2	3.83–4.14	Mo_2C	3.80–4.74			$MoSi_2$	4.02–4.73
Nb	3.99	NbB_2	3.65	NbC	2.21–4.1	NbN	3.92	$NbSi_2$	4.34
Re	5.0							$ReSi_2$	4.02
Sc	3.3	ScB_2	2.29–3.76						
Ta	4.2	TaB_2	2.8–4.4	TaC	3.05–4.4	TaN	3.8–4.42	$TaSi_2$	4.42–4.71
Ti	3.95	TiB_2	3.80–3.95	TiC	2.35–4.12	TiN	2.92–3.75	$TiSi_2$	3.95
V	4.12	VB_2	3.88	VC	3.85	VN	3.56		
W	4.54	WB_2	2.62	W_2C	2.6–4.58			WSi_2	4.04–4.62
Y	3.3	YB_6	2.22–3.58					YSi_2	3.26
Zr	3.9	ZrB_2	3.60–4.48	ZrC	2.1–4.39	ZrN	2.92–3.90	$ZrSi_2$	3.95

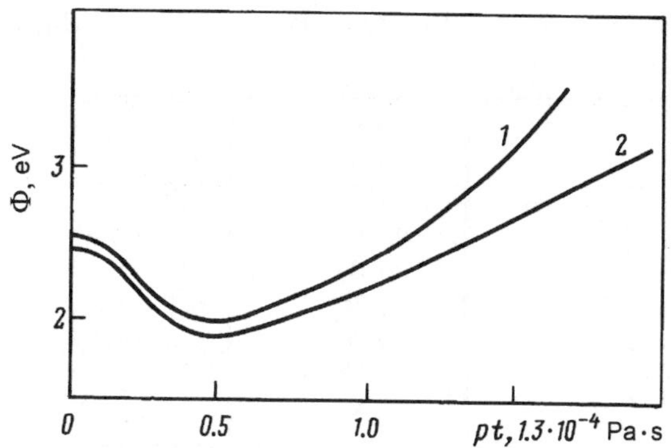

Figure 25.1 Change of the work function during the oxidation of the Ba-monolayer on the (100)–W (*1*) and (100)–Ir (*2*) surfaces. The exposition of surfaces to oxygen (*pt*) is given along the x-axis [5].

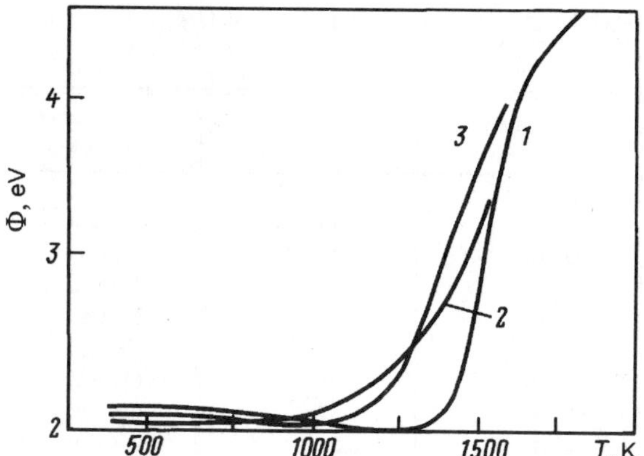

Figure 25.2 Change with heating in the work function of the (100)–W (*1*), (100)–Ir (*2*) and Os–Ir alloy (*3*) surfaces covered with the barium oxide film 0.8 monolayer thick [5].

25.3 Thermoelectron Emission

The saturation current density of *thermoelectron emission* (TE) for the emitter with the homogeneous surface in a weak external electric field having no significant effect on the work function is defined by the Richardson–Dushman equation [2]

$$j_e = (1 - \bar{r})A_0 T^2 \exp(-e\phi_T/kT), \tag{25.2}$$

where $\bar{r}$ is the coefficient of electron reflection from the potential barrier, averaged over electron energies, $A_0 = 4\pi mek^2/h^3 = 1.204 \cdot 10^6 \text{A·m}^{-2}\text{·K}^{-2}$ is the universal constant common to all solids, $\phi_T = \phi_0 + (d\phi/dT)T$, ϕ_0 is the value of ϕ (in eV) for T=0, k=1.38·10^{-23} J/K is the Boltzmann constant (Fig. 25.3).

The thermoemissive cathode (thermocathode) is a component of the electronic vacuum or gas–filled device serving as a source of electrons. The main types of thermocathodes include the metal, oxide, porous metal (dispensed), metallic alloy, or boride cathodes. Metal thermocathodes made of thoriated tungsten find only limited use nowadays.

Oxide thermocathodes consist of a mixture of metal oxides coated onto a metal kern. One uses the mixture of alkaline-earth metal oxides – barium, calcium, and strontium, in order that the low–temperature thermocathodes be operated in the temperature range from 900 to 1300 K. These oxides are produced at the decomposition of alkaline-earth carbonates coated onto the metallic cathode kern when it is heated directly inside a vacuum device, where the cathode has to operate. Oxide low-temperature cathodes are most widely used in electronic vacuum devices. For high-temperature oxide cathodes with the operation temperature 1400–1900 K, the yttrium and thorium oxides are commercially available. Such cathodes are employed in magnetrons.

Dispensed barium–tungsten thermocathodes constitute a barium film-coated porous tungsten sponge. The barium film reduces the work function and guarantees high TE current. In operation, the barium film disintegrates due to ionic bombardment and action of gases evolving from the device components. Regeneration of the film occurs as a result of the barium supply from the tungsten sponge on thermal decomposing of its active substance. There exist several types of porous metal cathodes: (a) a chambered or L–cathode consists of a chamber filled with the active substance (barium–strontium carbonate), and closed with the sponge wall, whose external side comprises an emitting surface; (b) an impregnated cathode consists of a porous tungsten, rhenium, or a molybdenum sponge whose voids are filled with the active substance (barium–calcium aluminate or tungstate); and, finally, (c) hot-pressed sintered cathodes are produced in the form of pellets or ceramic tubes by pressing the mixture of yttrium or thorium oxide powders and high-melting metal (tungsten, molybdenum, tantalum) powders. Cathodes of this type as well as thorium-oxide cathodes operate at temperatures 1700–1800°C and are intended for use in RF devices, mainly magnetrons.

Boride thermocathode refers to the cathode based on metal-like compositions of the MeB_6 type (where Me is alkaline-earth and rare-earth metals or thorium). One uses most extensively lanthanum hexaboride, less often yttrium and gadolinium hexaboride and chromium diboride as thermocathodes. Coating of the oxide layer by the osmium thin film reduces the work function of the cathode and increases its emission capacity. Lanthanum–hexaboride thermocathodes operate at a temperature of 1650 K and guarantee the TE current density up to $5{\cdot}10^5$ A/m^2. High mechanical strength and resistance of such cathodes to ion bombardment make it possible to use them in the operating conditions of thermic field emission (at the external electric field intensity 10^8 V/m, the substantial part of the emission current is caused by the electron tunneling through the barrier). In this regime, the lanthanum–hexaboride cathode may emit a current with density up to 10^7 A/m^2 at temperatures 1400–1500 K. The lanthanum–hexaboride cathodes are not poisoned in air and operate steadily in relation to poor vacuum. Their life does not depend on the residual gas pressure inside the device up to pressures of the order 10^{-2} Pa. These cathodes are employed in accelerators and different vacuum devices.

The main characteristics of thermocathodes involve the work function $e\phi$, operating temperature T, the TE saturation current density j_e and its temperature dependence, the rate of active substance vaporization v_{evap} at operating temperature, the cathode efficiency η, that is the ratio of the TE current density to the power dissipated at the cathode under its heating, the cathode performance criterion q, that is the ratio of work function to the active substance vaporization heat at given temperature, and, finally, the depth of an active layer d (or the diameter, for homogeneous cathodes). The characteristics of different cathodes are presented in Tables 25.5 – 25.14 and in Figs. 25.4 – 25.11.

Table 25.5 Thermoemission properties of the tungsten thermocathode [6] [Pure tungsten, ρ=1.93·10^4 kg/m^3, Φ=4.54 eV.]

T, K	j_e, A/m^2	v_{evap}, kg/(m$^2\cdot s$)	j_e/v_{evap}, C/kg	$d^{*1}_{1\text{h}}$, mm/h	Lifetime, h		
					d=0.1 mm	d=1 mm	Bulk cathode
2100	39.0	$1.6\cdot10^{-11}$	$2.5\cdot10^{12}$	$3.0\cdot10^{-9}$	$1\cdot10^6$	$1\cdot10^7$	$1\cdot10^8$
2200	$1.3\cdot10^2$	$1.2\cdot10^{-10}$	$1.1\cdot10^{12}$	$2.3\cdot10^{-8}$	$1.4\cdot10^5$	$1.4\cdot10^6$	$1.4\cdot10^7$
2300	$4.1\cdot10^2$	$7.8\cdot10^{-10}$	$5.0\cdot10^{11}$	$1.5\cdot10^{-7}$	$2\cdot10^4$	$2\cdot10^5$	$2.2\cdot10^6$
2400	$1.2\cdot10^3$	$4.4\cdot10^{-9}$	$2.7\cdot10^{11}$	$8.2\cdot10^{-7}$	$3\cdot10^3$	$3\cdot10^4$	$3\cdot10^5$
2500	$3.0\cdot10^3$	$2.0\cdot10^{-8}$	$1.5\cdot10^{11}$	$3.7\cdot10^{-6}$	830.0	8300.0	$8\cdot10^4$
2600	$7.0\cdot10^3$	$8.8\cdot10^{-8}$	$8.0\cdot10^{10}$	$1.6\cdot10^{-5}$	180.0	1800.0	$1.8\cdot10^4$
2700	$1.6\cdot10^4$	$3.2\cdot10^{-7}$	$5.0\cdot10^{10}$	$6.0\cdot10^{-5}$	50.0	500.0	$5\cdot10^3$
2800	$3.5\cdot10^4$	$1.1\cdot10^{-6}$	$3.1\cdot10^{10}$	$2.1\cdot10^{-4}$	14.0	140.0	1440.0
2900	$7.3\cdot10^4$	$3.5\cdot10^{-6}$	$2.1\cdot10^{10}$	$6.5\cdot10^{-4}$	4.6	46.0	460.0
3000	$1.4\cdot10^5$	$9.7\cdot10^{-6}$	$1.7\cdot10^{10}$	$1.8\cdot10^{-3}$	1.7	17.0	170.0

*1 The thickness of layer evaporating for 1 h at temperature T.

Table 25.6 Thermoemission properties of the dispensed barium-tungsten thermocathode [6] [Φ=1.8–2.0 eV, amount of BaO reaches 0.3 kg/m^2.]

T, K	j_e, A/m^2	v_{evap} kg/(m^{2}s)	j_e/v_{evap}, C/kg	$d_{1\text{h}}$, mm/h	Lifetime, h	T, K	j_e, A/m^2	v_{evap}, kg/(m^{2}s)	j_e/v_{evap}, C/kg	$d_{1\text{h}}$, mm/h	Lifetime, h
800	$1\cdot10^{-1}$					1300	$2\cdot10^4$	$1.7\cdot10^{-9}$	$1.2\cdot10^{13}$	$6\cdot10^{-4}$	$5\cdot10^4$
900	3.0					1400	$1\cdot10^5$	$2.0\cdot10^{-8}$	$5.0\cdot10^{12}$	$7\cdot10^{-3}$	4300
1000	50.0					1500	$3\cdot10^5$	$1.7\cdot10^{-7}$	$1.8\cdot10^{12}$	$6\cdot10^{-2}$	500
1100	$6\cdot10^2$					1600	$9\cdot10^5$	$1.0\cdot10^{-6}$	$9.0\cdot10^{11}$	0.4	75
1200	$4\cdot10^3$					1700	$3\cdot10^6$	$5.5\cdot10^{-6}$	$5.5\cdot10^{11}$	2.0	15

Table 25.7 Thermoemission properties of the BaO–SrO oxide thermocathode [6] [$\bar{\rho} = 5.5 \cdot 10^3$ kg/m^3, $\Phi = 1.6$ eV, the thickness of active layer is 20 μm (0.1 kg/m^2), the kern is made of nickel, $d^{*1}_{\text{crit}} = 6 \mu$m.]

T, K	j_e, A/m^2	v_{evap}, kg/(m^{2}s)	j_e/v_{evap}, C/kg	$d^{*2}_{1\text{h}}$, mm/h	Lifetime, h	T, K	j_e, A/m^2	v^{*2}_{evap}, kg/(m^{2}s)	j_e/v^2_{evap}, C/kg	$d^{*2}_{1\text{h}}$, mm/h	Lifetime, h
500	0.1					1000	$3 \cdot 10^4$	$5 \cdot 10^{-10}$	$6 \cdot 10^{13}$	$3.2 \cdot 10^{-7}$	5000
600	4					1100	$1.2 \cdot 10^5$	$2 \cdot 10^{-8}$	$6 \cdot 10^{12}$	$1.3 \cdot 10^{-5}$	115
700	70					1200	$5 \cdot 10^5$	$8 \cdot 10^{-7}$	$6 \cdot 10^{11}$	$5.2 \cdot 10^{-4}$	3
800	$4 \cdot 10^2$					1300	$2 \cdot 10^6$	$1 \cdot 10^{-5}$	$2 \cdot 10^{11}$	$6.5 \cdot 10^{-3}$	.23
900	$4 \cdot 10^3$										

*1 The minimum thickness of the active layer for which the cathode retains its efficiency.

*2 Values for BaO; for BaO–SrO thermocathodes, the evaporation rate is slightly less.

Table 25.8 Recommended work conditions for oxide thermocathodes [2]

j_e, A/m^2	T, K	Maximum lifetime, h	j_e, A/m^2	T, K	Maximum lifetime, h
Continuous mode			**Pulsed mode**		
500	1000–1040	20000	25 000*1	1070–1100	3 000
1500	1000–1070	5000	30 000	1000–1040	10000
2000	1070–1100	3000	40 000*1	1070–1100	2000
3000	1070–1100	2000	50 000	1000–1040	5000
			60 000*2	1070–1100	3000
			100 000*2	1070–1100	2000

*1 Pulse duration measures several hundreds of microseconds.

*2 Pulse duration measures microseconds.

Table 25.9 Thermoemission characteristics and the lifetime of pressed, dispenser Ni-oxide cathodes [7]

T, K	j_e, A/m^2	Lifetime, h	Working mode	T, K	j_e, A/m^2	Lifetime, h	Working mode
1200	5000	5000	stationary	1270	10000	5000	stationary
1220	5000	5000	stationary	1340	30000	3000	stationary

Table 25.10 Thermoemission constants of borides allowing for calculation of Φ_T by the formula $\Phi_T = \Phi_O + T(d\Phi/dT)$, and the secondary emission coefficient σ [7]

Boride	Φ_0, eV	$d\Phi/dT$, 10^{-4} eV/K	σ	Boride	Φ_0, eV	$d\Phi/dT$, 10^{-4} eV/K	σ
BaB_6	3.45	7.5		PrB_6	3.46	2.5	0.8
CeB_6	2.59	2.3	0.68	ScB_6	2.96	2.3	0.58
CrB_2	3.36		0.77	SrB_6	2.67	2.5	
DyB_6	3.53	1.5	0.8	TbB_6	3.26		
ErB_6	3.37	2.3		ThB_6	2.92	2.0	
EuB_6	4.9			TiB_2	3.88		0.82
GdB_6	2.05	4.0	0.8	$TmB_6 + TmB_4$	3.38		
HoB_6	3.42	1.6	0.7	VB_2	3.95		0.8
LaB_6	2.68	1.4	0.95	YB_6	2.22	2.18	1.0
LuB_6	3.0	1.6	0.8	YbB_6	3.13	3.0	
MnB_2	4.14			ZrB_2	3.67	2.0	0.85
NdB_6	3.97	1.6	0.8				

Table 25.11 Thermoemission properties of LaB_6 [6, 8] [$\rho = 2.61 \cdot 10^3$ kg/m^3, $\Phi = 2.68$ eV.]

T, K	j_e, A/m^2	v_{evap}, kg/(m^2·s)	j_e/v_{evap}, C/kg	d, mm	Lifetime for indirectly heated cathode, h	
					point cathode $d_{crit} = 3\ \mu$m	bulk cathode $d_{crit} = 100\ \mu$m
1600	$4 \cdot 10^3$	$1 \cdot 10^{-10}$	$4.0 \cdot 10^{13}$	$1.4 \cdot 10^{-7}$	$2 \cdot 10^4$	$7 \cdot 10^5$
1700	10^4	$1 \cdot 10^{-9}$	$1.0 \cdot 10^{13}$	$1.4 \cdot 10^{-6}$	2150	$7 \cdot 10^4$
1800	$3 \cdot 10^4$	$1 \cdot 10^{-8}$	$3.0 \cdot 10^{12}$	$1.4 \cdot 10^{-5}$	215	$7 \cdot 10^3$
1900	$8.5 \cdot 10^4$	$9 \cdot 10^{-8}$	$9.5 \cdot 10^{11}$	$1.2 \cdot 10^{-4}$	25	800
2000	$2.5 \cdot 10^5$	$7 \cdot 10^{-7}$	$3.6 \cdot 10^{11}$	$1.0 \cdot 10^{-3}$	3	100
2100	10^6	$6 \cdot 10^{-6}$	$1.7 \cdot 10^{11}$	$8.3 \cdot 10^{-3}$	0.35	12
2200	$4 \cdot 10^6$	$4 \cdot 10^{-5}$	$1.0 \cdot 10^{11}$	$5.55 \cdot 10^{-2}$	0.05	1.8

Table 25.12 Thermoemission properties of pressed cathodes with various compositions [7]

Composition (mass. %)	T, °C	j_e, A/m^2	
		Pulsed mode	Stationary mode
$ThH_2 + W_2C + W$ (0.5+10+89.5)	1700		
$W + ThO_2$ (96+4)	1500	$8 \cdot 10^3$	
$W + ThO_2$ (96+4)	1600	$2 \cdot 10^4$	
$W + ThO_2 + B$ (95+4+1)	1370		$4 \cdot 10^3$
$W + ThO_2 + B$ (97.5+2+0.5)	1370		$4 \cdot 10^3$
$W + La_2O_3$ (70+30)	1300		$1.4 \cdot 10^4$
$W + Nd_2O_3$ (70+30)	1300		$5 \cdot 10^3$
$W + Cd_2O_3$ (70+30)	1400		$4 \cdot 10^3$
$W + [75\%Cd_2O_3 + 25\%La_2O_3]$ (70+30)	1400		$8.6 \cdot 10^3$
W+Th (impregnated)	1600		$(2\text{–}3) \cdot 10^4$

Table 25.13 High-temperature thermocathodes [14]

Cathode	T, °C	j_e, A/m^2	Lifetime, h
Mo–La_2O_3	1580	70,000	7000
Mo–C–La_2O_3–Pt	1570	50,000	
Ir–La	1430	40,000	4500
Re–LaB_6	1400	30,000	1500
W–Re–Y	1440	10,000	800
W–Re–Nd	1730	20,000	500
ZrC–W	1780	10,000	10,000
ZrC–W, ZrC–Mo	2030	10,000	9000

Table 25.14 Work function Φ, heat of vaporization Q, and performance criterion $q = \Phi/Q$ for thermocathodes [7]

Type of cathode	Φ, eV	Q, eV	Φ/Q	T, K for endurance limit:	
				10000 h	1000 h
Oxide	1.6	4.0	0.4	1030	1100
W–Barium	2.1	4.7	0.45	1300	1420
Boride	2.8	6.8	0.41	1720	1870
Th–oxide	3.2	7.6	0.42	1770	1870
Tantalum	4.2	7.9	0.53	2000*[1]	2200*[1]
Tungsten	4.54	8.0	0.57	2300*[1]	2500*[1]

*[1] A rough measure, as the service life for the directly-heated cathode depends also on its diameter.

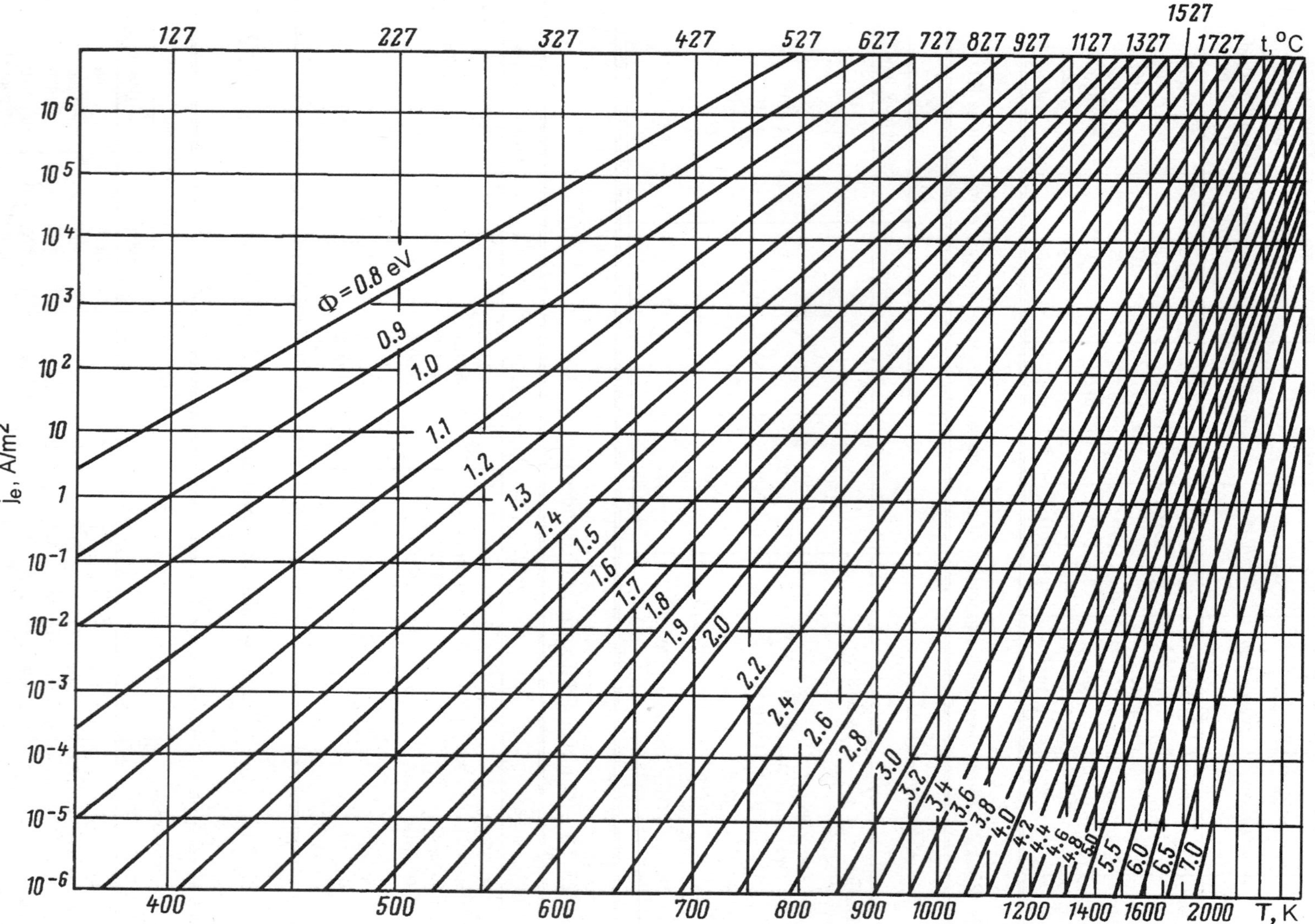

Figure 25.3 Temperature dependence of the TE current densities for cathodes with various work functions [6].

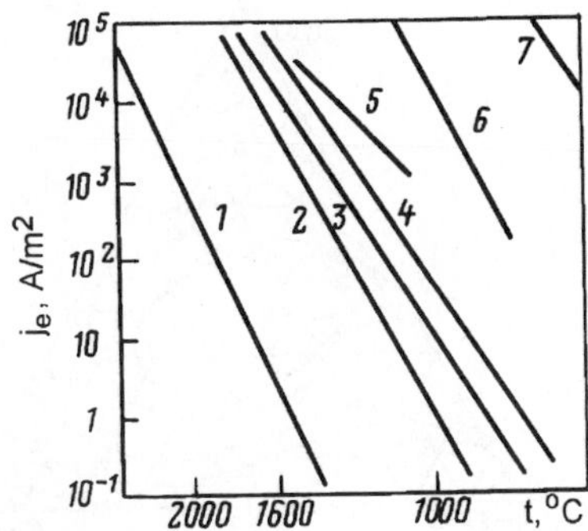

Figure 25.4 Temperature-dependent TE current densities for various cathodes [7]: *1* – tungsten; *2* – thoriated tungsten; *3* – thorium oxide; *4* – lanthanum hexaboride (see also [8]); *5* – yttrium-oxide cathode with tungsten, molybdenum or tantalum added; *6* – *L*-cathode; *7* – oxide cathode.

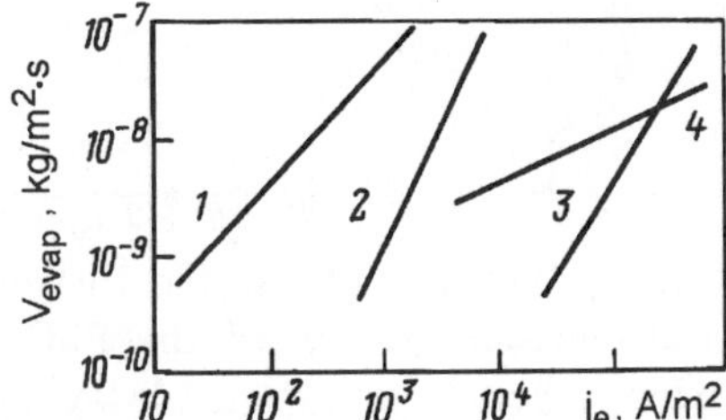

Figure 25.5 Evaporation rate of some cathode materials vs. the thermoelectron emission current density [7]: *1* – tungsten; *2* – thoriated tungsten; *3* – impregnated cathode; *4* – lanthanum hexaboride [7, 8].

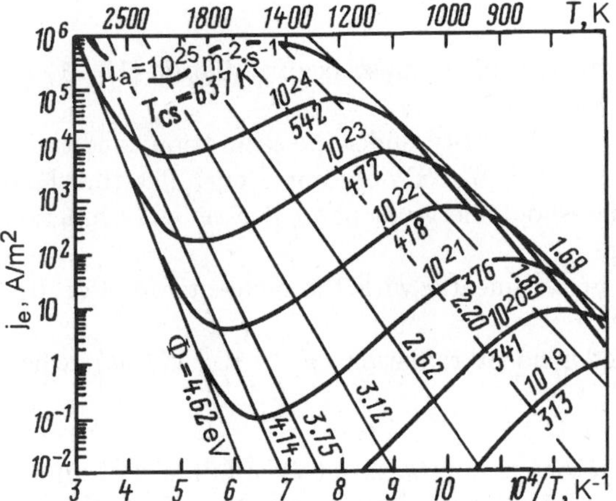

Figure 25.6 Thermoelectron emission of the polycrystalline tungsten in the Cs-vapor environment [9]. The temperature of the Cs liquid phase (T_{Cs}) is shown at the bottom of the curves, and the Cs flux density for a cathode – at the top of each curve. The inclined straight lines depict the lines of constant work function.

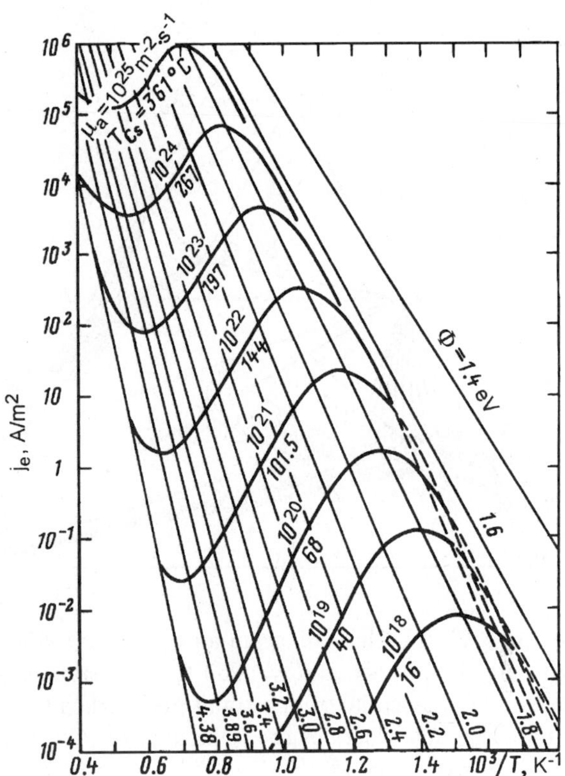

Figure 25.7 Thermoelectron emission of the polycrystalline molybdenum in Cs-vapor environment [9]. Designations are the same as in Fig. 25.6.

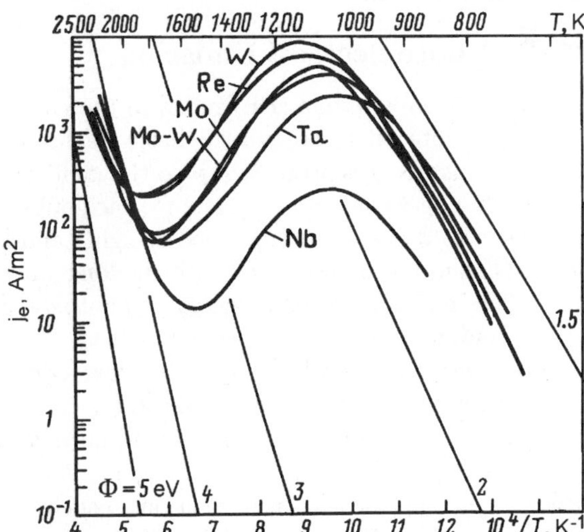

Figure 25.8 Thermoelectron emission of the five high-melting metals and tungsten-molybdenum alloy (in equal mass quantities) in the Cs–vapor environment [9]. The temperature of the Cs–liquid phase is 200°C. The inclined straight lines depict the lines of constant work function.

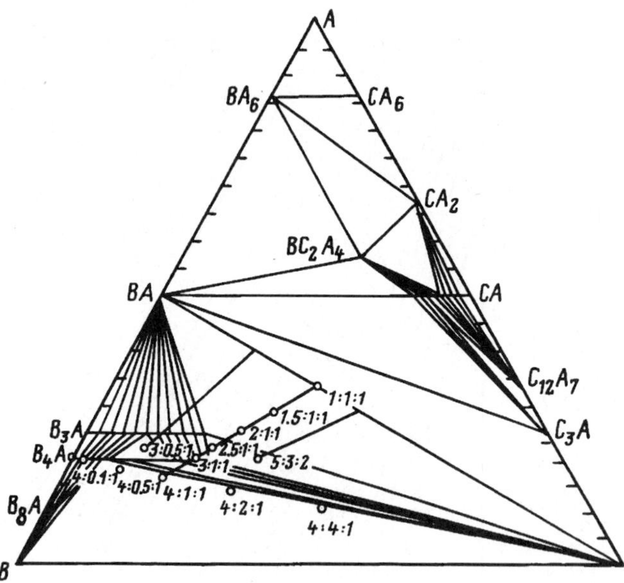

Figure 25.9 Phase diagram for BaO–CaO–Al_2O_3 oxide cathode determined at $T = 1250°C$. $B \equiv$ BaO; $C \equiv$ CaO; $A \equiv Al_2O_3$. Essential compositions are marked with circles, and their mole fractions are related as $B : C : A$. [10]

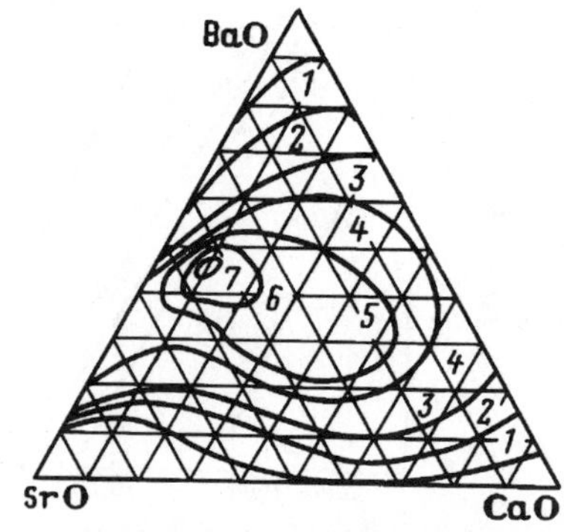

Figure 25.10 Emissivity of the various composition oxide cathodes [12]. The numbers denote the values of pulsed currents in 10^4 A/m^2.

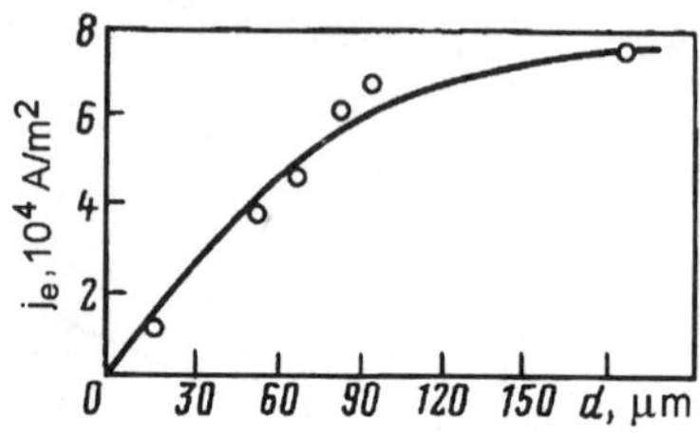

Figure 25.11 The stable thermoelectron emission current of the BaO–SrO–oxide cathode versus the oxide layer thickness d [13].

25.4 Photoelectron Emission

Photoelectron emission (PE) refers to the electron emission under substance irradiation [15–21, 40, 41]. It is of interest to present the basic PE laws.

(1) Photocurrent is proportional to the incident radiation intensity under the saturation conditions.

(2) The long-wave edge λ_0 exists for each substance. At $\lambda > \lambda_0$, PE does not occur. The threshold frequency $\nu_0 = c/\lambda_0$ (c is the speed of light) and the threshold energy or photoelectric work function $h\nu_0$ are connected immediately with the long-wave edge.

(3) The maximum kinetic energy of photoelectrons grows linearly with the incident radiation frequency and does not depend on its intensity.

These laws are broken at very high flux densities of incident radiation ($F \geq 10^4$ W/m^2), when multiphoton processes become significant.

Let us list the main characteristics of photocathodes.

Quantum yield $Y_\lambda = (I_\lambda/e)/(F_\lambda/h\nu)$ (in electron/photon) is appropriate to the number of photoelectrons passing to vacuum in relation to each photon incident on the photocathode surface. Here I_λ is the saturation photocurrent for a given wavelength, F_λ is the radiation flux for a given wavelength.

Spectral sensitivity $S_\lambda = I_\lambda/F_\lambda = 0.807\, Y_\lambda \lambda$ (in mA/W) fits the saturation photocurrent per power unit of the monochromatic radiation incident on the photocathode surface (usually, λ is in nm).

Spectral characteristics correspond to the S_λ or Y_λ dependence on the frequency or wavelength of the incident radiation.

Integral sensitivity S matches the ratio of the saturation photocurrent to the radiation flux of the reference light source usually taken in the form of the incandescent lamp with a filament temperature of 2850 K:

$$S = I/F = \int_0^{\lambda_0} S_\lambda F_\lambda d\lambda \Big/ 683 \int_{\lambda_1}^{\lambda_2} F_\lambda K_\lambda d\lambda\,, \tag{25.3}$$

where S is in μA/lm, S_λ is in μA/W, K_λ is the relative sensitivity of the human eye ($K_{\max} = K_{\lambda=555\,\text{nm}}$), and λ_1 and λ_2 are the edge wavelengths of the visible-light spectrum.

Dark current density j_d conforms to the TE current density for the unexposed photocathode at operation temperature. The dark current makes a main noise source in the photoelectron devices. The noise current root-mean-square in the absence of radiation is given by the formula

$$(\overline{i_n^2})^{1/2} = (2ej_d s\Delta f)^{1/2}, \tag{25.4}$$

where s is the photocathode surface area, and Δf is the frequency bandwidth of the recording device.

Photoemission from metals. The PE threshold energy for metals coincides with their work function. PE from metallic photocathodes at radiation frequencies not very distant from the threshold frequency ($\nu \leq 1.5\nu_0$) is well described by the phenomenological Fowler theory according to which

$$I = \begin{cases} \alpha A_0 T^2 \exp\left(-\dfrac{h\nu_0 - h\nu}{kT}\right), & \nu < \nu_0; \\ \dfrac{\alpha A_0}{2}\left[\dfrac{(h\nu - h\nu_0)^2}{k^2} + \dfrac{\pi^2}{3}T^2\right], & \nu > \nu_0, \end{cases} \tag{25.5}$$

where $\alpha \approx F/h\nu$ is the ratio of the density of electrons excited by light to the density of normal electron gas in the emitter, $A_0 = 4\pi emk^2/h^3$ is the Richardson constant. The PE threshold frequency ν_0 is strictly defined only at $T = 0$. At $T > 0$, PE is observed near the threshold also for frequencies $\nu < \nu_0$. The PE quantum yield from pure metal surfaces in the visible–light spectral range has the order of 10^{-4} electron/photon, and for $h\nu \leq 10$ eV it does not exceed 10^{-2} electron/photon (Fig. 25.12).

Photoemission from semiconductors. PE in semiconductors may result from the excitation of valence band electrons, electrons from impurity and defect levels, surface state electrons and conduction band electrons (in degenerate n-type semiconductors). The threshold frequency value differs for each of these cases. Usually, unless otherwise specified, the photoelectron work function is used in reference to the minimum photon energy such that PE from the semiconductor valence band sets in (Table 25.15). This value as a rule exceeds the work function. The spectral dependence of the PE quantum yield in semiconductors in the threshold vicinity takes the form

$$Y \sim (h\nu - h\nu_0)^m, \tag{25.6}$$

with m=1–3 depending on optical transition processes and the photoelectron scattering mechanism. The PE quantum yield in semiconductors depends on the electron affinity χ; that is, the energy necessary to transfer an electron from the valence band bottom to the vacuum level. All the effective photocathodes possess small χ values, so that $\chi/E_g < 1$, where E_g denotes the band gap. The photoemission quantum yield of semiconductors with $\chi/E_g > 1$ is small (Fig. 25.14). Reduction in the semiconductor work function due to the adsorption of electropositive atoms (cesium, barium, etc.) leads to a decrease in χ and to sharp increase of the quantum yield. In the case of GaAs, GaP, Si and a number of other semiconductors, the coadsorption of cesium and oxygen results in such a drastic fall of the work function that the condition of the negative electron affinity (NEA) is achieved. Semiconductors with NEA display the greatest quantum yields in the visible and infrared spectral ranges.

Effective photocathodes. All the effective photocathodes are semiconductors. Metals exhibit high quantum yield (about 0.1 electron/photon) only in the range $h\nu > 12$ eV and usually if the oxide films exist on their surfaces.

Alkali halides (CsI, CsBr and others), copper iodides and silver halides represent effective photocathodes for the spectral range $\lambda < 200$ nm ($h\nu > 6$ eV). High quantum yield in the range $\lambda = 200$–350 nm was obtained with cesium telluride (Cs_2Te) and rubidium telluride (Rb_2Te). It should be noted that these photocathodes fall in the category of so called "sun–blind", i.e. photocathodes sensitive to the ultraviolet radiation, but not responsive to sun radiation. The photocathodes for the visible spectrum discussed below show high quantum yield also in the ultraviolet part of the spectrum.

Photocathodes sensitive in the visible-light spectrum make up alkali metal antimonides. Some of them (e.g. $CsNa_2KSb$) are sensitive also to the near infrared spectral range up to λ = 850–900 nm. Until recently, the Ag–O–Cs photocathode was the only photocathode for the interval λ = 900–1100 nm. At present, a new class of photocathodes, namely photocathodes with NEA, has been developed. The photocathodes with NEA comprise highly doped p-type semiconductors (GaAs, Si, solid solutions like GaInAs, InGaAsP, and others), whose work function is strongly reduced through adsorption of cesium and oxygen, so that the vacuum level is situated below the valence band bottom in the semiconductor. They possess the highest sensitivity in the red and infrared spectral ranges. According to recent data [41], CsNaKSb is also a photocathode with NEA.

In the range $\lambda > 1.1\,\mu$m, only photocathodes based on semiconducting heterostructures with the external bias (InGaAs – InP – Ag – CsO) exhibit a high quantum yield. The long-wave edge in such photocathodes is determined by the forbidden band width of a narrow-band semiconductor (InGaAs), wherein the light absorption occurs. A strong electric field initiates photoelectron transfer from InGaAs to the upper layer of InP and augments its energy to the value exceeding the vacuum level. The photocathodes of this type are sensitive to the λ range up to 1.7 μm. The characteristics of most widely used photocathodes are presented in Tables 25.16, 25.17 and in Figs. 25.13, 25.15, and 25.16.

X-ray photoelectron emission (XRPE) occurs under the influence of X-ray radiation and is connected with the electron transitions from the deep atomic levels to a vacuum. The presence of sharp lines corresponding to photoelectrons which escape a body without energy scattering (Tables 25.18 and Figs. 25.28–25.30) is a feature of the XRPE spectrum. When using the long-wave X-ray radiation ($h\nu \approx 1$ keV), the energy of emitted electrons comprises several hundreds electronvolts. The mean free path of such electrons is equal to 0.5–2 nm (Fig. 25.27), so that the line XRPE spectra reflects the features of the near-surface layer with a width up to five monolayers. This specific feature of XRPE spectra allows their use in analyzing the surface composition in X-ray photoelectron spectroscopy (XRPS). The energies for elements in various compounds differ by several electronvolts. Thus, the energy of the carbon 1s line changes from 281 (HfC, TiC) to 292 eV (CO_2). This effect, usually referred to as the chemical shift, provides a possibility of obtaining (with XRPS) information not only on the surface composition, but also on chemical bonds in the subsurface area.

Table 25.15 The band gap E_g, electron affinity χ, and the photoemission threshold $h\nu_0$ of different semiconducting materials [16]

Material	E_g, eV	χ, eV	$h\nu_0$, eV	Material	E_g, eV	χ, eV	$h\nu_0$, eV
Barium oxide	3.7	1.3	5.0	Lead telluride	0.3	4.7	5.0
Barium titanate	2.7	2.6	5.3	Lithium fluoride	12.0	1.0	13.0
Bismuth telluride	0.1	5.2	5.3	Lithium iodide	5.9	1.4	7.3
Cadmium selenide	3.0	4.0	7.0	Magnesium antimonide	0.8	3.0	3.8
Cadmium sulfide	2.4	3.8–4.8	6.2–7.2	Magnesium oxide	8.7	1.0	<10
Cadmium telluride	1.5	4.4	5.9	Potassium iodide	6.2	1.1	7.3
Cesium auride	2.6	1.4	4.0	Rubidium iodide	6.1	1.2	7.3
Cesium iodide	>6	<0.5	6.4	Rubidium telluride	3.3	0.5	3.7
Cesium telluride	3.3	< 0.5	3.5	Selenium	2.3	3.7	6.0
Copper iodide	3.0	3.0	6.0	Silicon carbide			
Gallium antimonide	0.7	4.1	4.8	cubic	2.2	4.8	7.0
Gallium arsenide	1.4	4.1	5.5	hexagonal	2.9	4.1	7.0
Germanium	0.7	4.2	4.9	Silicon	1.1	4.0	5.1
Indium antimonide	0.2	4.6	4.8	Silver bromide	2.5	3.5	6.0
Indium arsenide	0.4	4.9	5.3	Silver chloride	3.0	3.0	6.0
Indium phosphide	1.3	4.4	5.7	Sodium iodide	5.8	1.5	7.3
Lead sulfide	0.4	<4.2	<4.6	Tellurium	0.3	4.6	4.9

Table 25.16 Properties of alkali antimonide photocathodes [15, 16, 40, 41]

Photocathode	Crystal structure	Y	λ_0, nm	S, μA/lm	E_g, eV	χ, eV	Conductivity type	j_d, A/m^2
Li_3Sb	Hexagonal		320		1.0	2.9	n	
Na_3Sb	Hexagonal	0.02	330		1.1	2.5	n	
K_3Sb	Cubic	>0.07	550	12	1.4	0.9	p	
K_3Sb	Hexagonal	0.07	460	2	1.1	1.6	n	
Rb_3Sb	Cubic	0.10	580	25	1.0	1.2	p	
Cs_3Sb	Cubic	0.15–0.25	620–700	40–80	1.6	0.45	p	10^{-12}
Na_2KSb	Cubic	0.30	600–670	30–60	1.0	1.0	p	10^{-13}
K_2CsSb	Cubic	0.30	650–700	50–100	1.0	1.1	p	10^{-13}
K_2CsSb (O)	Cubic	0.35	780	130	1.0	<1.1	p	10^{-12}
$CsNa_2KSb$	Cubic	0.30–0.40	870–940	300	1.4	≈ 0	p	10^{-12}
Rb_2CsSb	Cubic	0.30	680–750	130	1.45	0.2		10^{-13}

Table 25.17 Parameters of the main photocathodes for visible and near infrared regions of the spectrum [16, 18]

Photocathode	λ_0, nm	Y	S_{av}, μA/lm	S, μA/lm	j_d, A/m^2
Cs_3Sb	620–700	0.2–0.25	40–80	120	10^{-12}–10^{-}
K_2CsSb	650–700	0.3–0.4	55–65	200	10^{-13}
$CsNa_2KSb$	900–940	0.3–0.4	200–400	700	10^{-12}
Ag–O–Cs	1200	0.005	20–40	70	10^{-9}–10^{-7}
GaAsP–Cs–O	680	0.5	200–300	375	$< 10^{-10}$
GaAs–Cs–O	900	0.3	800–1400	2150	10^{-10}–10^{-12}
InGaAsP–Cs–O	1100–1150	0.2	200–500	1640	10^{-7}–10^{-9}
$In_{0.53}Ga_{0.47}As$–InP–Ag–Cs–O	1700	0.08			

Table 25.18 Energy of the principal X-ray photoemission lines (binding energy) for chemical elements ($h\nu$=1486.6 eV) and relative photoionization cross section for these lines [The photoionization cross section for the sodium 1s line is taken equal to unity [37, 39].]

Element	Mean peak energy, eV	Line energy range, eV	Relative photoionization cross section	Element	Mean peak energy, eV	Line energy range, eV	Relative photoionization cross section
	1s – transition			$_{17}$Cl	199	11	0.198
$_{3}$Li	56		0.009	$_{18}$Ar	241	0	
$_{4}$Be	113	4	0.033	$_{19}$K	293	1	0.424
$_{5}$B	191	8		$_{20}$Ca	347	2	0.377
$_{6}$C	287	12	0.127	$_{21}$Sc	402	6	
$_{7}$N	402	9	0.188	$_{22}$Ti	458	8	
$_{8}$O	531	4	0.338	$_{23}$V	515	6	0.421
$_{9}$F	686	6	0.480	$_{24}$Cr	577	6	0.522
$_{10}$Ne	863	0		$_{25}$Mn	641	4	0.478
$_{11}$Na	1072	2	1.000	$_{26}$Fe	710	8	0.865
$_{12}$Mg	1305	2		$_{27}$Co	781	6	0.777
	$2p_{1/2}$ – transition			$_{28}$Ni	855	6	0.836
$_{13}$Al	74	4	0.086	$_{29}$Cu	934	4	1.500
	$2p_{3/2}$ – transition			$_{30}$Zn	1022	2	0.135
$_{14}$Si	102	6		$_{31}$Ga	1117	2	0.200
$_{15}$P	133	8	0.141	$_{32}$Ge	1219	4	0.210
$_{16}$S	165	8	0.163	$_{33}$As	1329	7	0.180

Table 25.18 Energy of the principal X-ray photoemission lines for chemical elements *(continued)*

Element	Mean peak energy, eV	Line energy range, eV	Relative photoionization cross section	Element	Mean peak energy, eV	Line energy range, eV	Relative photoionization cross section
	$3d_{5/2}$ – transition			$_{61}$Pm	1034		
$_{34}$Se	57	8	0.239	$_{62}$Sm	1083		1.21
$_{35}$Br	69	7	0.262	$_{63}$Eu	1136		1.24
$_{36}$Kr	88	0		$_{64}$Gd	1186		1.91
$_{37}$Rb	110	1	0.284	$_{65}$Tb	1244		
$_{38}$Sr	133		0.380	$_{66}$Dy	1295		
$_{39}$Y	158		0.430		$4d_{5/2}$ – transition		
$_{40}$Zr	181	6	0.570	$_{67}$Ho	161		0.119
$_{41}$Nb	206	8	0.564	$_{68}$Er	169		0.237
$_{42}$Mo	230	6		$_{69}$Tm	180		0.189
$_{43}$Tc	253			$_{70}$Yb	185		0.282
$_{44}$Ru	282	4	0.846	$_{71}$Lu	197		0.236
$_{45}$Rh	309	4	0.990		$4f_{7/2}$ – transition		
$_{46}$Pd	337	5	0.894	$_{72}$Hf	17	6	0.427
$_{47}$Ag	368	2	1.170	$_{73}$Ta	25	8	0.660
$_{48}$Cd	405	2	1.410	$_{74}$W	34	6	
$_{49}$In	445	3	1.880	$_{75}$Re	43	6	0.885
$_{50}$Sn	486	3	2.35	$_{76}$Os	52	3	0.625
$_{51}$Sb	530	4	2.81	$_{77}$Ir	62	4	0.860
$_{52}$Tb	575	5	1.89	$_{78}$Pt	73	5	0.86
$_{53}$I	619	6	1.88	$_{79}$Au	85	3	1.04
$_{54}$Xe	672	4		$_{80}$Hg	100	2	1.14
$_{55}$Cs	724	2	3.12	$_{81}$Tl	118	2	1.17
$_{56}$Ba	780	2	2.70	$_{82}$Pb	138	3	1.507
$_{57}$La	834		1.00	$_{83}$Bi	159	4	1.67
$_{58}$Ce	882		1.00	$_{90}$Th	335	3	3.30
$_{59}$Pr	930		1.00	$_{92}$V	380	5	3.34
$_{60}$Nd	930		1.51				

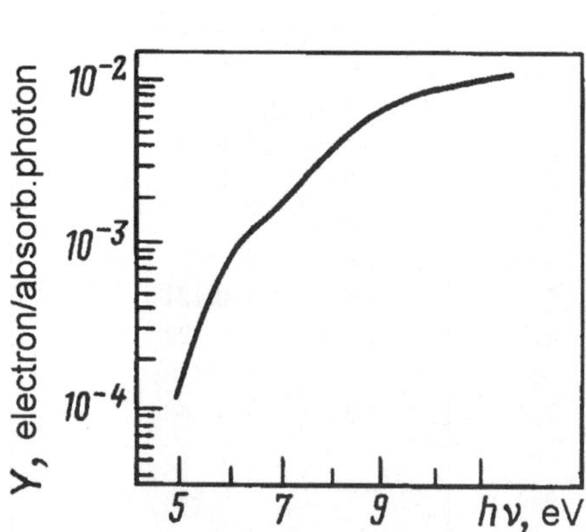

Figure 25.12 PE quantum yield of copper with clean surface [18].

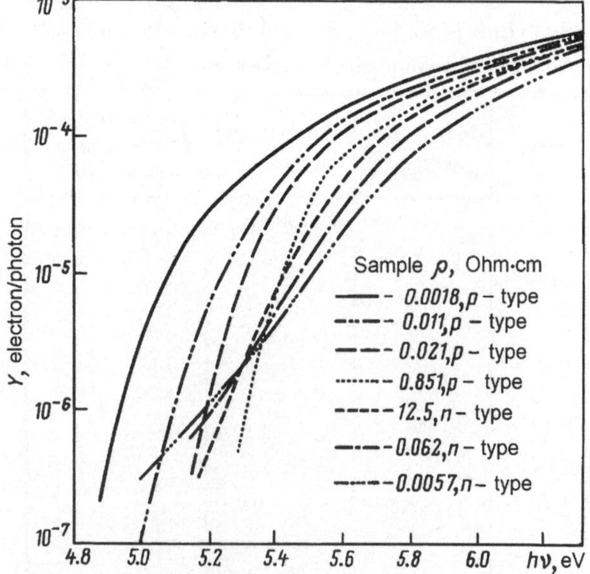

Figure 25.13 PE quantum yield of n- and p-type silicon with different doping ($\chi/E_g \approx 4$) [16].

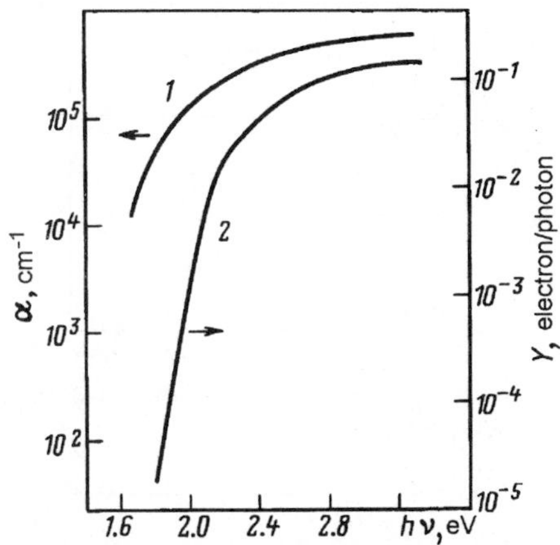

Figure 25.14 Spectral curves of absorption coefficient α (*1*) and quantum yield (*2*) for Cs_3Sb-photocathode [16].

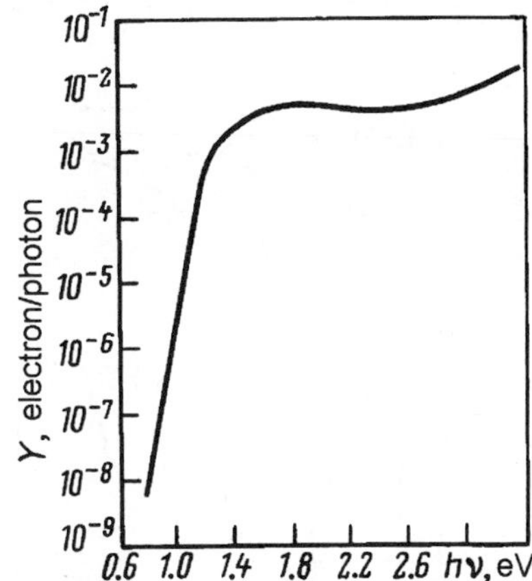

Figure 25.17 PE quantum yield of Ag–O–Cs photocathode [16].

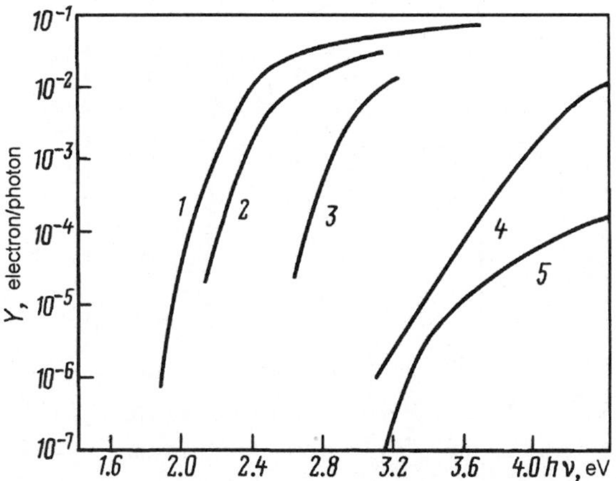

Figure 25.15 PE quantum yield of alkali antimonide photocathodes: *1* – Rb_3Sb; *2* – cubic K_3Sb; *3* – hexagonal K_3Sb; *4* – Na_3Sb; *5* – Li_3Sb [16].

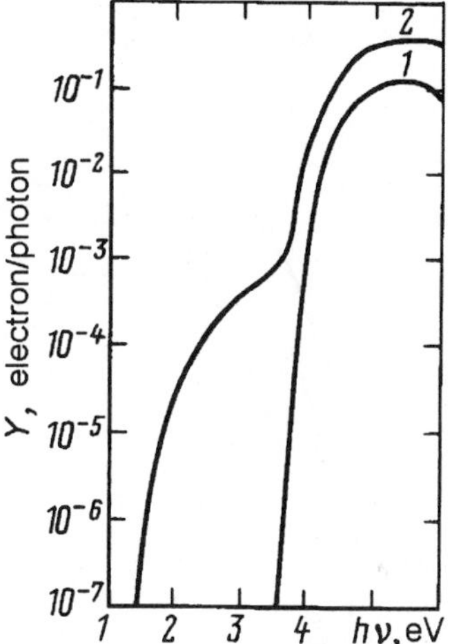

Figure 25.18 PE quantum yield of Cs_2Te photocathodes without (*1*) and with (*2*) Cs excess [16].

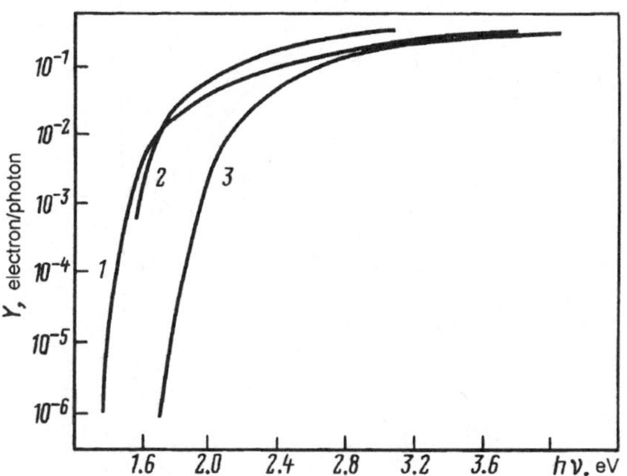

Figure 25.16 PE quantum yield of multialkali and bialkali photocathodes: *1* – $CsNa_2KSb(Cs)$; *2* – $K_2CsSb(O)$; *3* – Na_2KSb [16].

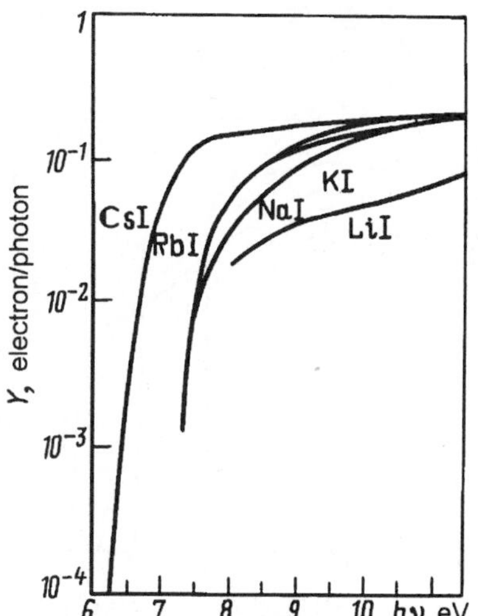

Figure 25.19 PE quantum yield of alkali iodides [16].

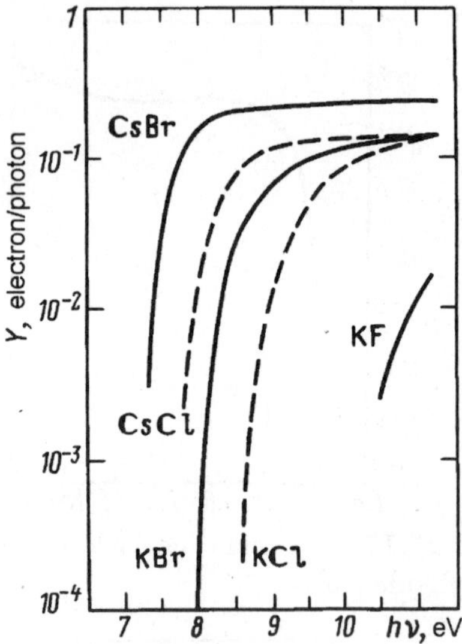

Figure 25.20 PE quantum yield of alkali halides [16].

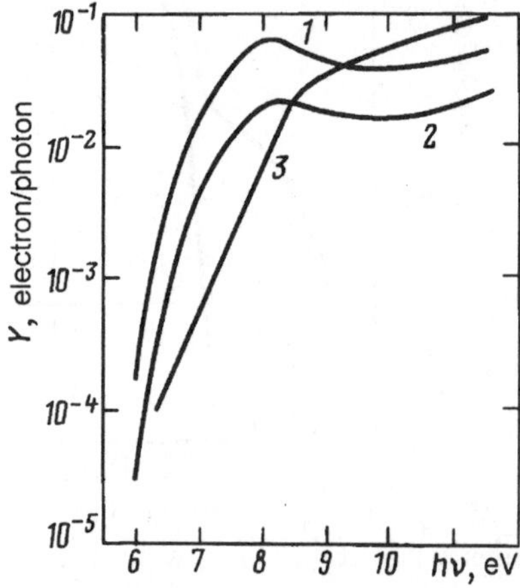

Figure 25.21 PE quantum yield of some Ag halides: *1* – single-crystal AgBr; *2* – melt AgBr; *3* – AgCl [16].

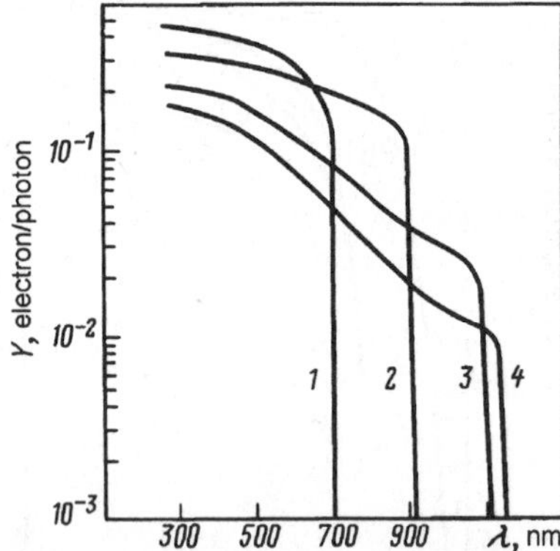

Figure 25.22 PE quantum yield of NEA photocathodes: *1* – GaAsP; *2* – GaAs; *3, 4* – InGaAsP with different band gaps [18].

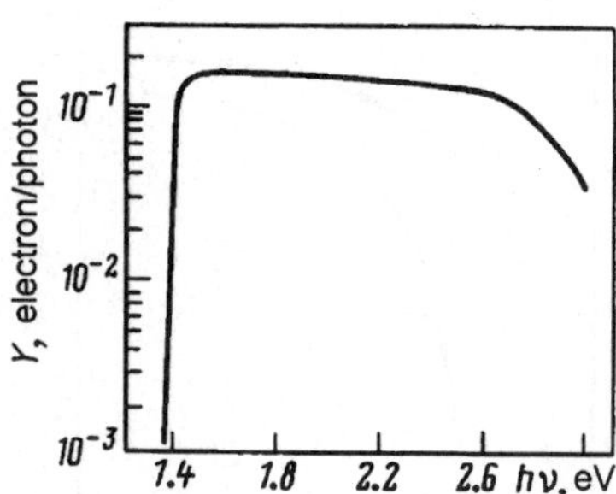

Figure 25.23 PE quantum yield of semitransparent GaAs photocathode [18].

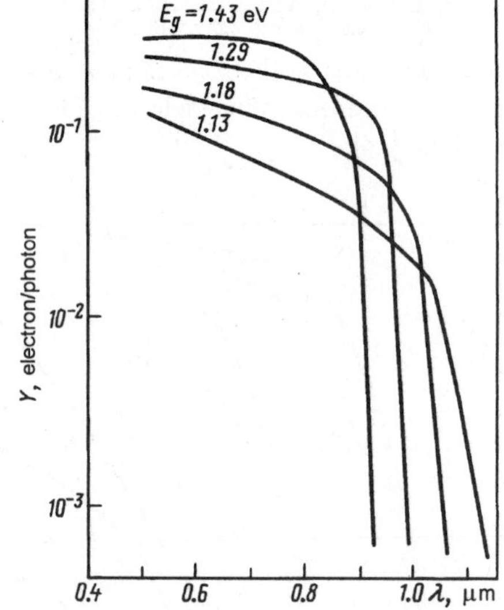

Figure 25.24 PE quantum yield of $Ga_{1-x}In_xAs$ photocathodes with various band gaps: $x = 0$, $E_g = 1.43$ eV; $x = 0.1$, $E_g = 1.29$ eV; $x = 0.17$, $E_g = 1.18$ eV; $x = 0.2$, $E_g = 1.13$ eV [20].

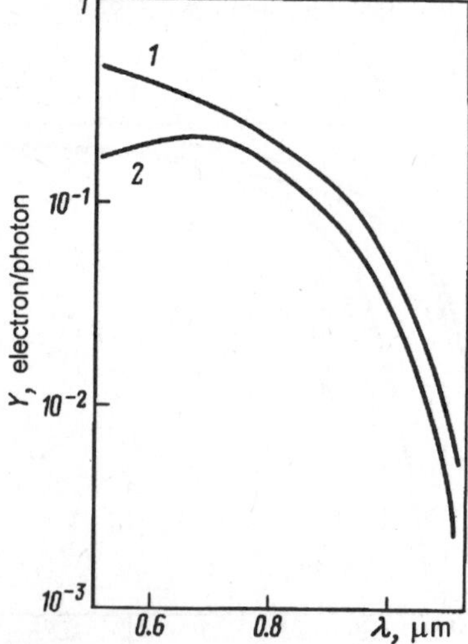

Figure 25.25 PE quantum yield of silicon photocathode in reflection (*1*) and transmission (*2*) modes [21].

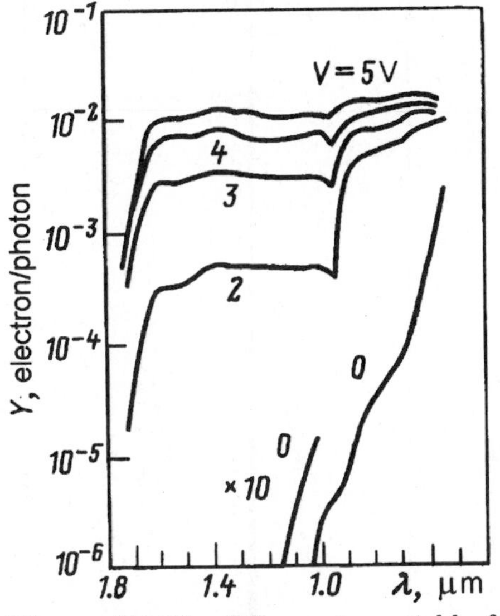

Figure 25.26 PE quantum yield of InGaAs–InP–Ag–Cs–O photocathodes at various bias voltages [19].

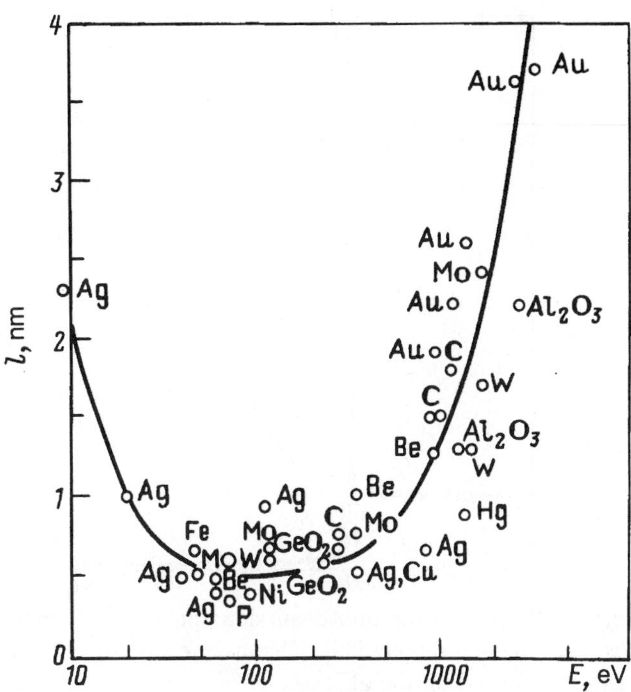

Figure 25.27 The mean depth l of electron exposure of solids as a function of electron energy [23].

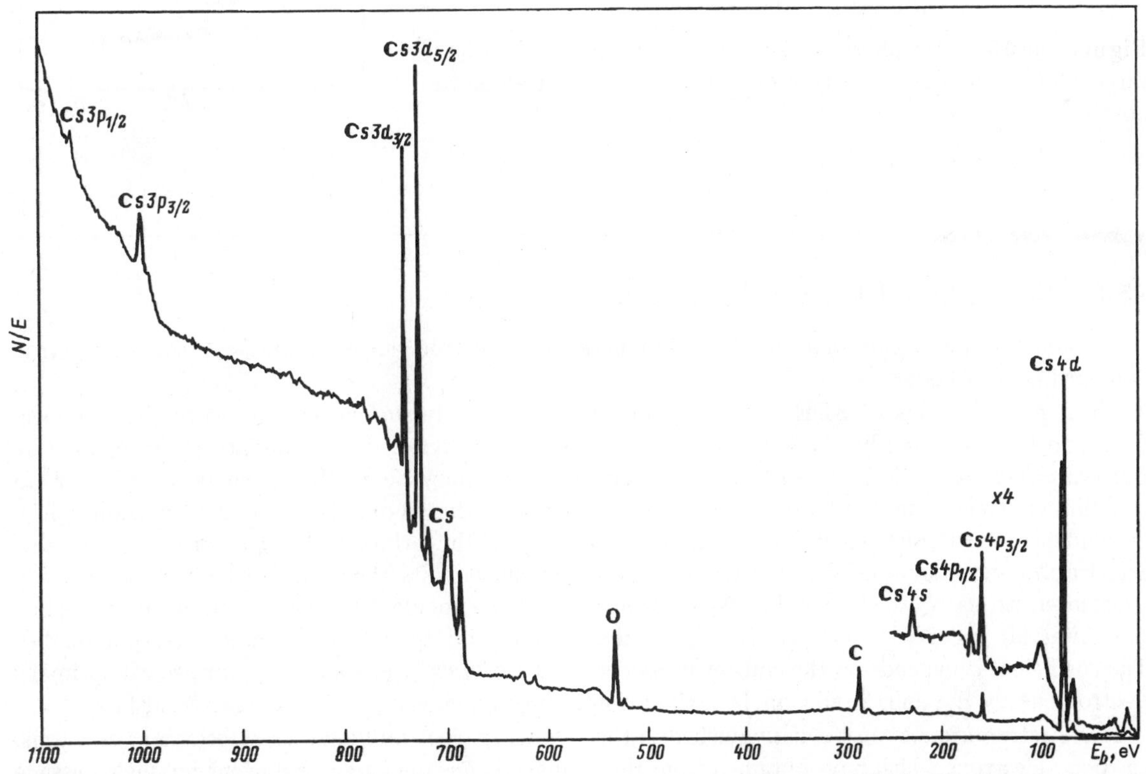

Figure 25.28 X-ray photoemission spectrum of cesium (for CsOH target) [39]. The atomic electron binding energy is plotted on the abscissa.

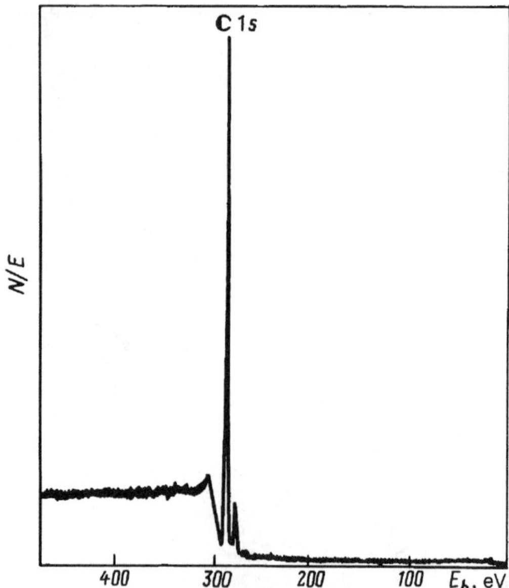

Figure 25.29 *X*-ray photoemission spectrum of carbon (for polyethylene target) [39]. The atomic electron binding energy is plotted on the abscissa.

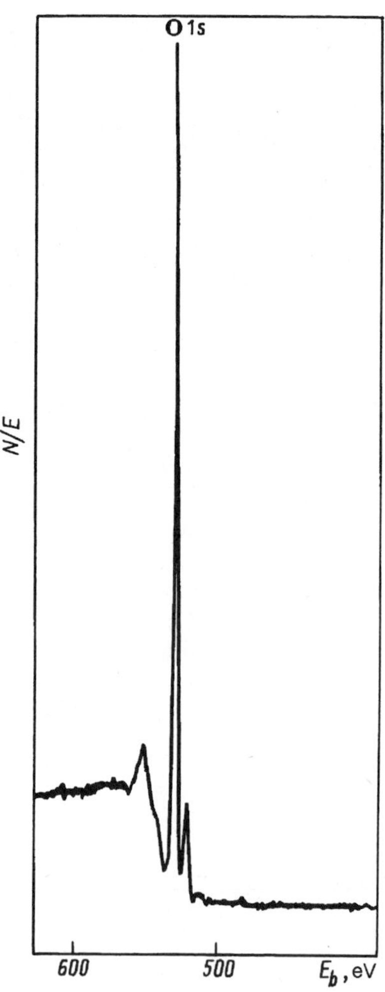

Figure 25.30 *X*-ray photoemission spectra of oxygen (for Al_2O_3 target) [39]. The atomic electron binding energy is plotted on the abscissa.

25.5 Secondary Electron Emission

Secondary electron emission (SEE) is designated the electron emission initiated by the electron bombardment of bodies.

Main peculiarities of SEE. Electrons bombarding the body surface are said to be primary electrons, electrons emitted by the body are called the secondary ones. The secondary electrons may be emitted either from the body surface irradiated by the primary beam (SEE "on reflection") or (in thin-film emitters) from the body surface opposite to the irradiated one (SEE "at shooting through"). Dependence of the SEE coefficient σ on the energy E_p of the primary electrons is the main characteristic of secondary-electron emitters. The SEE coefficient fits the ratio of the number of body emitted electrons N_2 to the number N_1 of primary electrons incident on its surface during the same time interval: $\sigma = N_2/N_1 = I_2/I_1$ (I_1, I_2 are the primary and secondary currents, correspondingly). The coefficient σ depends on the emitter properties and structure, the state of its surface, the primary electron energy E_p, and, finally, angle of the primary beam incidence upon the emitter surface.

Two electron groups can be distinguished in the secondary electron beam: true secondary electrons, i.e. body electrons which have obtained from the primary beam the energy sufficient for their passage to vacuum, and (elastically and inelastically) reflected electrons, i.e. a portion of the primary beam reflected from the body. At low E_p ($E_p < 10$ eV), the elastically reflected electrons constitute a main part of the secondary electrons. With a rise in E_p, the fraction of elastically reflected electrons diminishes and at $E_p > 0.1$ keV it forms only several percent of the total SEE. The true secondary electrons would acquire energy from 0 to about 50 eV, their most probable energy falling within the range 1.5–3.5 eV. By convention it is agreed to consider the secondary electrons with the energy exceeding 50 eV as the inelastically reflected ones. The ratio of a number of inelastically reflected electrons to a number of primary electrons is designated the inelastic reflection coefficient $\eta = N_2(E > 50\,\text{eV})/N_1$ (N_2 includes also the elastically reflected electrons, but their number is small and it does not influence η).

SEE from metals and semiconductors. In metals and semiconductors, the maximum value of σ (σ_m) usually lies in the range 0.5–1.8 (Tables 25.19–25.21 and Figs. 25.31–25.38). In some dielectrics (alkali–halide crystals, MgO) σ_m is much greater (10–35). This results from the great depth of the secondary electron exposure of these substances (20–100 nm). The presence of strong electric field in dielectrics directed from the emitting surface towards the layer depths (i.e. accelerating the secondary electrons) leads to a substantial increase in σ. The strong field is usually set up by the electron bombardment of the thin dielectric layer on the conducting substrate at energy E_p such that $\sigma > 1$. As a result, the surface of the dielectric charges positively with respect to the metal substrate. The SEE current arising in the presence of a strong electric field in the emitter consists of two components: the low-inertial one (this part is spoken of as the secondary electron emission amplified by the field; its time lag is less than 10^{-6} s), and the self–maintaining component existing also in the absence of the primary beam (once the initial charging of the layer has been taken).

Effective emitters of secondary electrons. Effective emitters of photoelectrons (antimony–cesium emitter, multi-alkaline emitter, photoemitters with NEA, and some others) are at the same time effective emitters of secondary electrons. Effective emitters of secondary electrons based on the alloys of magnesium, beryllium and some other elements have received wide acceptance. These emitters make the layer of corresponding metal oxide on the surface of the parent alloy (Ag–Mg, Al–Mg, Cu–Be, Ni–Be, and so on). Emitters of secondary electrons from conducting glasses are used in channel secondary electron multipliers.

Characteristics of some effective emitters of secondary electrons are presented in Tables 25.22–25.24 and in Figs. 25.39–25.43. Values of E_{pm} in Tables 25.22–25.24 relate to the primary electron energy such that the maximum value of the secondary electron emission coefficient σ_m is achieved.

Auger electron emission (AEE). There appear sharp lines in the SEE spectra corresponding to secondary electrons which originated from the Auger process. The latter takes place in the bombardment of a body by primary electrons, leaving the body without scattering (Table 25.25 and Figs. 25.44–25.46). This process is going on as follows. The transition of an electron from some internal level (for example, K) to a free state above the vacuum level takes place on the excitation of an atom by primary electrons. The level having become vacant is occupied by an electron from an above lying level (for example, L). The energy liberated in the process and approximately equal to $E_K - E_L$, where E_K and E_L are the electron binding energies, correspondingly, at K and L levels, is emitted in the form of an X-ray quantum or is transferred to another electron occupying an adjacent level. Such a process leads to the emission of an electron with the energy E_{KLL} approximately equal to $E_K - 2E_L$. Besides the KLL-series, the Auger series LMM, MNN, and others are also intensive.

The Auger peak energy characterizes the given atom, therefore the analysis of Auger electron spectra makes it possible to gain information on the composition of the near-surface zone of the solid, from whence the AEE originates. The energy of Auger electrons lies in the 30–2000 eV range. The mean free path of electrons with the same energies measures 0.5–2 nm, so that the Auger electron spectra reflect properties of the near-surface layer about five monolayers thick. The Auger peak amplitude is proportional to the concentration of atoms of a given sort on the solid surface and to Auger transition effectiveness characterized by the value called Auger sensitivity. The latter is determined by the ratio between the number of secondary Auger electrons with a given energy emitted by the specific element and the number of primary electrons and depends on the primary electron energy. The analysis of the Auger electron spectrum lies at the heart of Auger electron spectroscopy, i.e. the main method of studying the solid surfaces.

Table 25.19 SEE characteristics for different polycrystals [22]

Element	σ_m	E_{pm}, eV	Element	σ_m	E_{pm}, eV
Aluminum	0.88	300	Molybdenum	1.23	600
Antimony	1.19	600	Nickel	1.34	600
Barium	0.83	400	Niobium	1.2	370
Beryllium	0.55	200	Osmium	1.7	750
Bismuth	1.32	900	Palladium	1.75	550
Boron	1.2	150	Platinum	1.8	850
Cadmium	1.59	800	Potassium	0.53	170
Calcium	0.60	200	Rhenium	1.6	750
Carbon (graphite)	1.0	300	Scandium	0.83	200–250
Cesium	0.72	400	Selenium	1.23	300
Copper	1.40	700	Silicon	1.03	300
Gallium	1.08	500	Silver	1.70	800
Germanium	1.08	400	Strontium	0.72	400
Gold	1.81	900	Tellurium	1.22	450
Hafnium	1.16	700	Thallium	1.30	700
Indium	1.41	500	Thorium	1.14	600–1000
Iridium	1.8	800	Tin	1.43	600
Iron	1.3	350	Titanium	0.83	300
Lanthanum	1.03	500	Tungsten	1.4	700
Lead	1.4	700	Yttrium	0.93	350
Lithium	0.48	75	Zinc	1.41	700
Magnesium	0.88	300	Zirconium	1.1	350
Mercury	1.63	700			

Table 25.20 SEE characteristics for some oxides and chalcogenides [22]

Compound	σ_m	E_{pm}, eV	Compound	σ_m	E_{pm}, eV
Boron oxide	2.5–2.8	250	Antimony selenide	1.2–1.4	500–600
Antimony oxide	1.6–1.8		Antimony sulfide	1.1–1.35	450–500
Tellurium oxide	1.7–1.85	600	Arsenic sulfide	1.5–1.8	300
Lead oxide	1.8–2.0	600	Lead sulfide	1.25–1.3	500
Bismuth selenide	1.25–1.35	600–700	Germanium sulfide	1–1.05	400
Indium selenide	1.3–1.5	300–350	Arsenic telluride	1.1–1.3	400
Calcium selenide	1.4–1.6	300–350	Antimony telluride	1.2–1.35	700–800
Arsenic selenide	1.1–1.4	400			

Table 25.21 Anisotropy of secondary-electron emission properties of single crystals [4]

Metal	SEE parameters	Single-crystal planes						Polycrystals
		(100)	(111)	(110)	(116)	(112)	(102)	
Tungsten	σ_m	1.66	1.58	1.48	1.39			1.42
	E_{pm}, eV	780	670	550–720	820			700
Iron	σ_m			1.34				1.30
	E_{pm}, eV			450				350
Iridium	σ_m	2.11	1.85	1.95				1.80
	E_{pm}, eV	650	650	650				800
Molybdenum	σ_m	1.52–1.58	1.4–1.45	1.35		1.14	125	1.28
	E_{pm}, eV	350–400	350–510	400–460		720–760	550	400
Nickel	σ_m		1.52					1.34
	E_{pm}, eV		770					600
Niobium	σ_m	1.38	1.29	1.25–1.34				1.20
	E_{pm}, eV	350–820	360	380–880				370
Tantalum	σ_m	1.43				1.25		1.3
	E_{pm}, eV	750				740		700

Table 25.22 Maximum values of SEE parameters for some effective emitters [13]

Emitter	σ_m	E_{pm}, keV
$CsNa_2KSb$	39	1.8
Cs_3Sb	10	0.5
Cu–Be	6–8	0.6
Ag–Mg	12	0.6
GaP (100)(single crystal)–CsO	500	12
GaP (polycrystal)–CsO	200	4
GaAs (100)–CsO	540	20
Si (100)–CsO	950	20
$GaAs_{0.5}P_{0.5}$ (100)–CsO	110	1.7
CsCl	20	2
CsI	20–35	1.7–5.5
CaF_2	5	1.4
MgO	20	0.9

Table 25.23 SEE coefficient of effective emitters in the "shooting through" mode [24]

Emitter	σ (at E_p, keV)				
	2	5	10	15	20
KCl	0.5	4.2	2	2	2
MgO	2	5.0	4		
Si–CsO	0	50.0	200	520	725
GaAs–CsO	0	2.0	15	60	112

Table 25.24 SEE coefficient of effective emitters for low-energy primary electrons [24]

Emitter	σ (at E_p, keV)		
	300	600	1000
CuBe	2–4	4–6	7–12
CuAlMg	2–4	3–7	7–16
Cs_3Sb	2–6	4–15	3–12
$CsNa_2KSb$	3–5	7–11	10–19
GaP (polycrystal)–CsO	28	50	68
GaAs (polycrystal)–CsO	26	42	65

Table 25.25 Energy E_{Ae} of main peaks of Auger electrons for different elements and relative Auger sensitivity ξ [38]

Element	E_{Ae}, eV	ξ^{*1} (at E_p= 3 keV)	Element	E_{Ae}, eV	ξ^{*1} (at E_p= 3 keV)	Element	E_{Ae}, eV	ξ^{*1} (at E_p= 3 keV)
	KLL-transition		$_{27}$Co	775	0.27	$_{54}$Xe	532	
$_3$Li	43		$_{28}$Ni	848	0.26	$_{55}$Cs	563	0.16
$_4$Be	104	0.14	$_{29}$Cu	920	0.22	$_{56}$Ba	584	0.12
$_5$B	179	0.15	$_{30}$Zn	994	0.17	$_{57}$La	625	0.09
$_6$C	272	0.2	$_{31}$Ga	1070	0.13	$_{58}$Ce	661	0.065
$_7$N	379	0.3	$_{32}$Ge	1147	0.1	$_{59}$Pr	669	0.05
$_8$O	503	0.5	$_{33}$As	1228	0.08	$_{60}$Nd	730	0.04
$_9$F	647	0.48	$_{34}$Se	1315	0.065	$_{62}$Sm	814	0.028
$_{11}$Na	990	0.22	$_{35}$Br	1396	0.05	$_{63}$Eu	858	0.026
$_{12}$Mg	1180	0.11	$_{37}$Rb	1565	0.03	$_{64}$Gd	895	0.024
$_{13}$Al	1396	0.05	$_{38}$Sr	1649	0.025	$_{65}$Tb	1073	0.024
	LMM-transition		$_{39}$Y	1746	0.02	$_{66}$Dy	1126	0.024
$_{13}$Al	68	0.24		*MNN*-transition		$_{67}$Ho	1175	0.024
$_{14}$Si	92	0.35	$_{40}$Zr	147	0.22	$_{68}$Er	1393	0.025
$_{15}$P	120	0.55	$_{41}$Nb	167	0.26	$_{69}$Tm	1449	0.027
$_{16}$S	152	0.8	$_{42}$Mo	186	0.33	$_{70}$Yb	1514	0.03
$_{17}$Cl	181	1.0	$_{44}$Ru	273	0.5	$_{71}$Lu	1573	0.036
$_{18}$Ar	215		$_{45}$Rh	302	0.65	$_{72}$Hf	1624	0.045
$_{19}$K	252	0.8	$_{46}$Pd	330	0.8	$_{73}$Ta	1680	0.055
$_{20}$Ca	291	0.5	$_{47}$Ag	351	1.0	$_{74}$W	1736	0.05
$_{21}$Sc	340	0.35	$_{48}$Cd	376	1.0	$_{75}$Re	1799	0.04
$_{22}$Ti	418	0.45	$_{49}$In	404	0.95	$_{77}$Ir	1908	0.027
$_{23}$V	473	0.45	$_{50}$Sn	430	0.8	$_{78}$Pt	1967	0.022
$_{24}$Cr	529	0.32	$_{51}$Sb	454	0.6	$_{79}$Au	2024	0.019
$_{25}$Mn	589	0.23	$_{52}$Te	483	0.45	$_{80}$Hg	2078	
$_{26}$Fe	703	0.21	$_{53}$I	511	0.32			

*1 Relative Auger sensitivity of silver (E_p = 351 eV) is taken equal to unity.

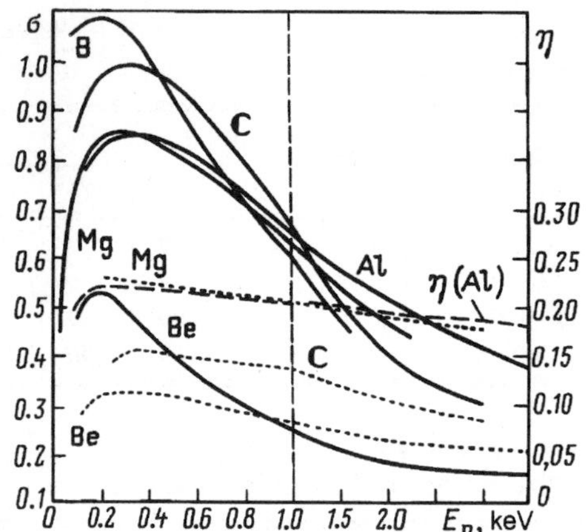

Figure 25.30 SEE coefficient σ (solid lines) and inelastic reflection coefficient η versus primary electron energy for boron, carbon, beryllium, magnesium, and aluminum [22].

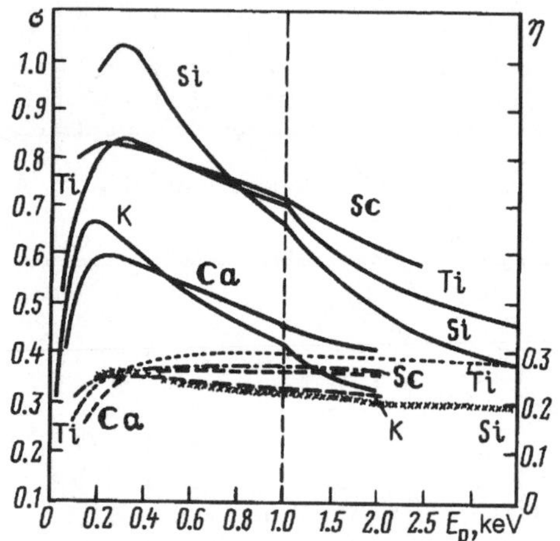

Figure 25.31 SEE coefficient σ (solid lines) and inelastic reflection coefficient η versus primary electron energy for potassium, calcium, scandium, titanium, and silicon [22].

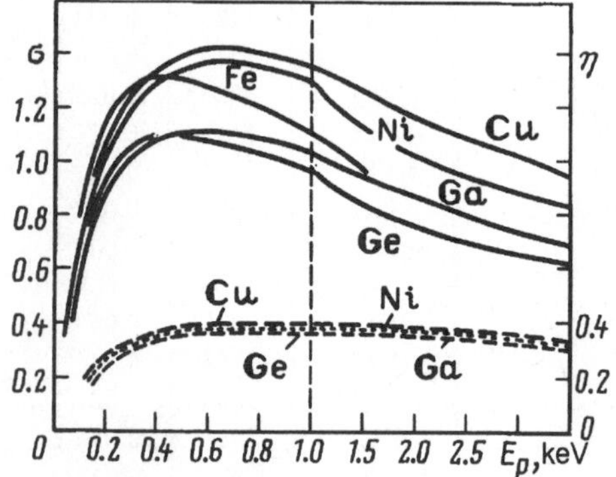

Figure 25.32 SEE coefficient σ (solid lines) and inelastic reflection coefficient η versus primary electron energy for iron, nickel, copper, gallium, and germanium [22].

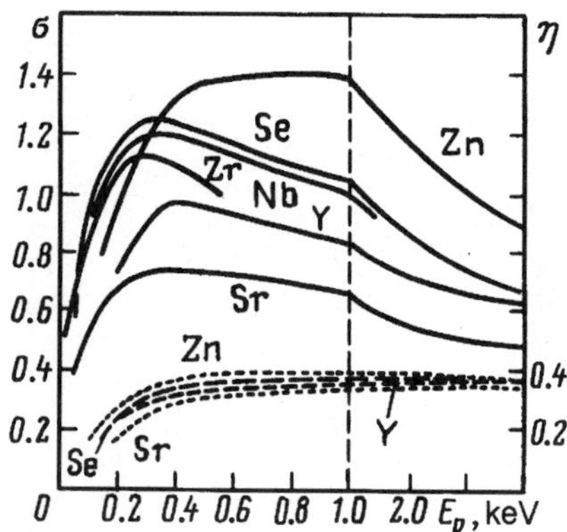

Figure 25.33 SEE coefficient σ (solid lines) and inelastic reflection coefficient η versus primary electron energy for zinc, selenium, strontium, yttrium, zirconium, and niobium [22].

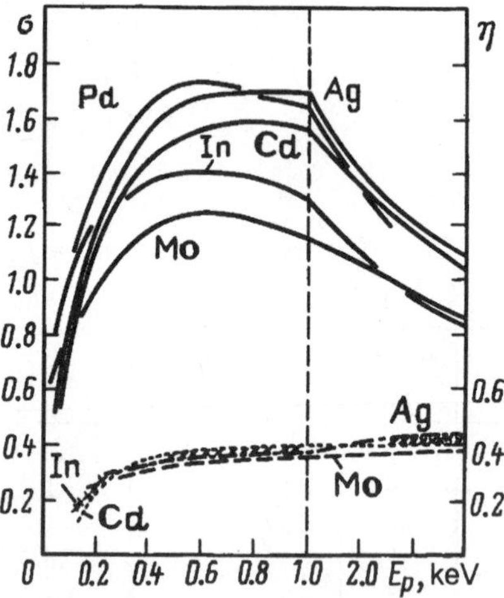

Figure 25.34 SEE coefficient σ (solid lines) and inelastic reflection coefficient η versus primary electron energy for molybdenum, palladium, silver, cadmium, and indium [22].

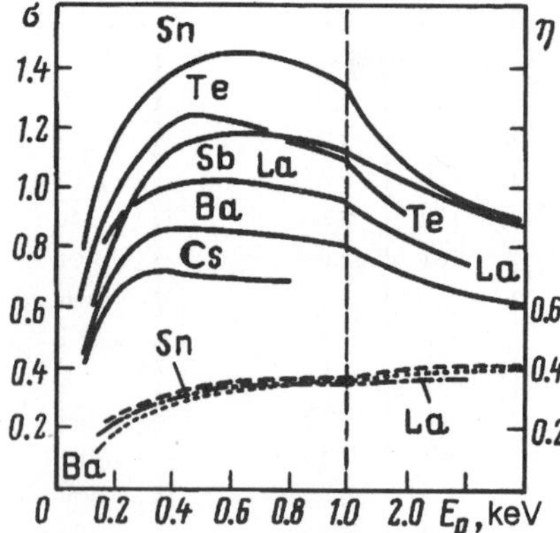

Figure 25.35 SEE coefficient σ (solid lines) and inelastic reflection coefficient η versus primary electron energy for tin, antimony, tellurium, cesium, barium, and lanthanum [22].

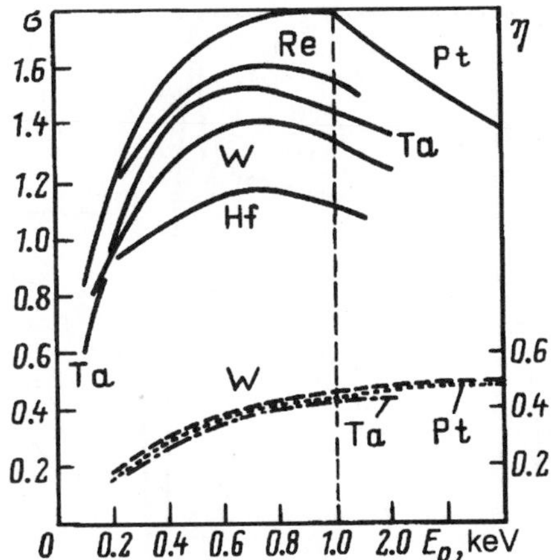

Figure 25.36 SEE coefficient σ (solid lines) and inelastic reflection coefficient η versus primary electron energy for hafnium, tantalum, wolfram, rhenium, and platinum [22].

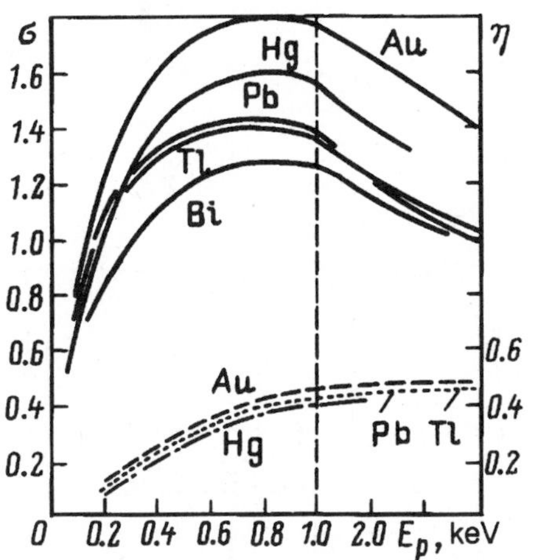

Figure 25.37 SEE coefficient σ (solid lines) and inelastic reflection coefficient η versus primary electron energy for gold, mercury, thallium, lead, and bismuth [22].

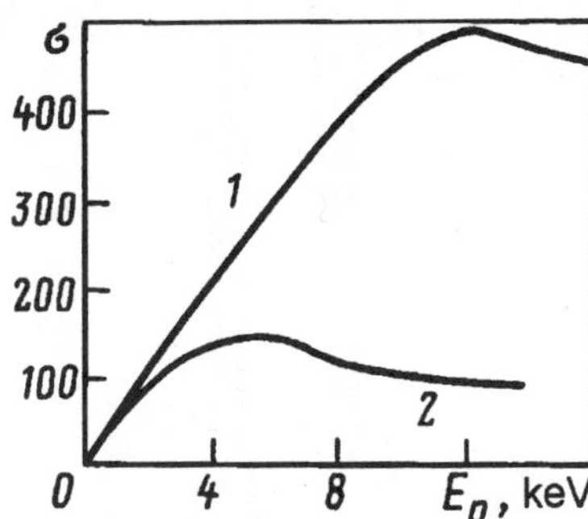

Figure 25.38 SEE coefficient versus primary electron energy for single-crystal (*1*) and polycrystal (*2*) GaP–CsO emitters [24].

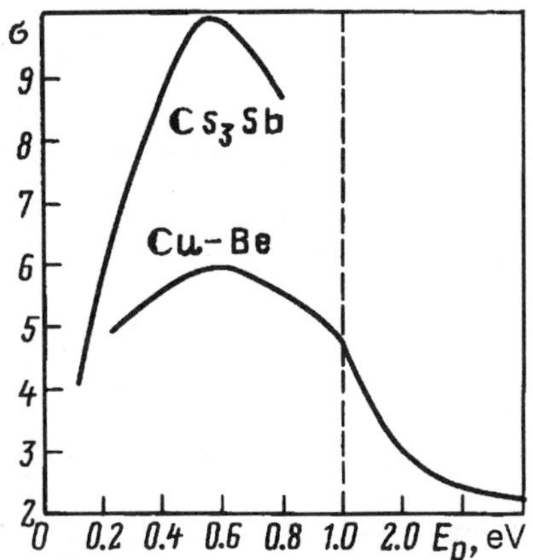

Figure 25.39 SEE coefficient as a function of primary electrons energy for Cs_3Sb and copper–beryllium alloy [15].

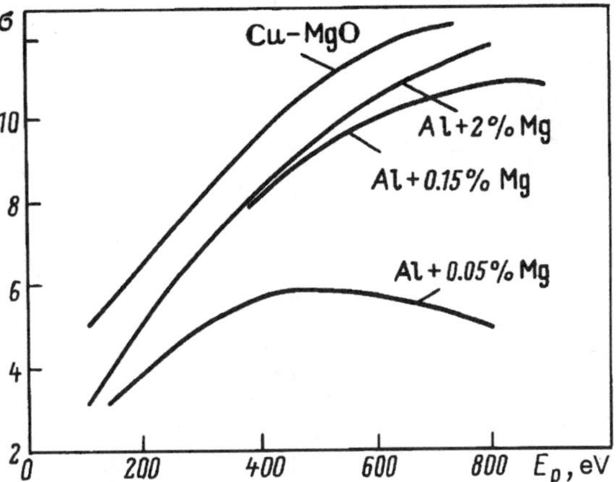

Figure 25.40 SEE coefficient as a function of primary electron energy for aluminum–magnesium and copper–magnesium alloys [15].

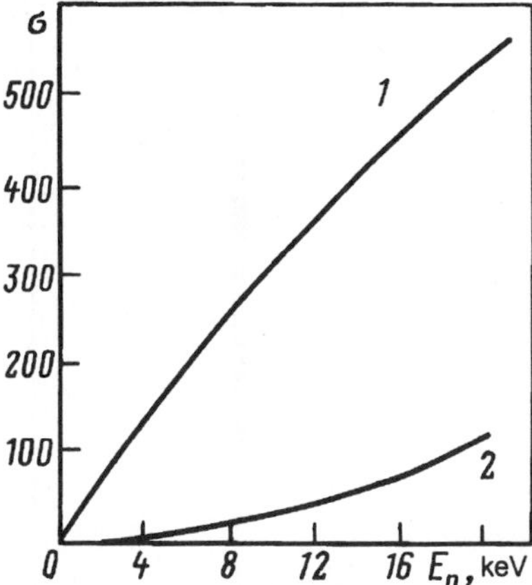

Figure 25.41 SEE coefficient versus primary electron energy for GaAs–CsO emitter in reflection (*1*) and "shooting through" (d=3–4 μm) (*2*) modes [26].

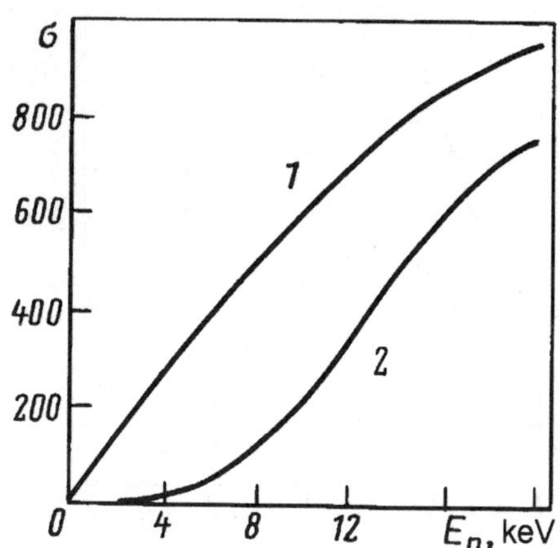

Figure 25.42 SEE coefficient versus primary electron energy for Si–CsO emitter in reflection (*1*) and "shooting through" (d=4–5 μm) (*2*) modes [25].

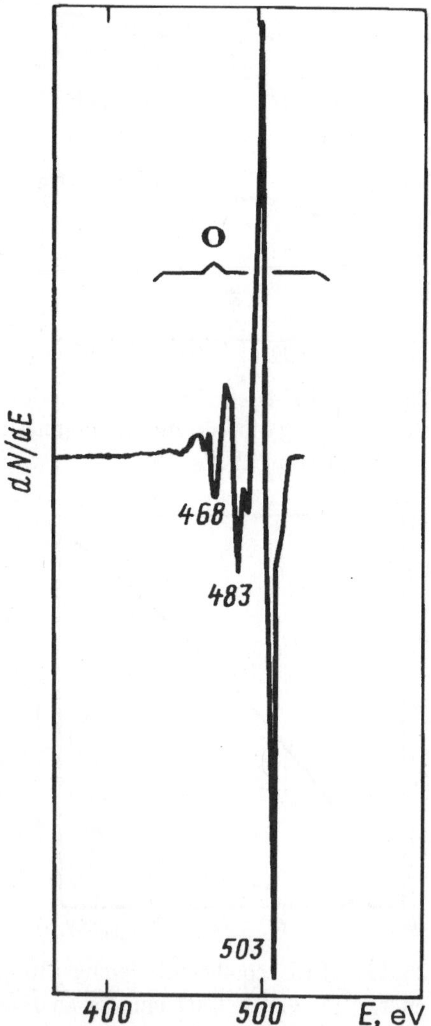

Figure 25.43 Auger spectrum of oxygen (for MgO target) [42].

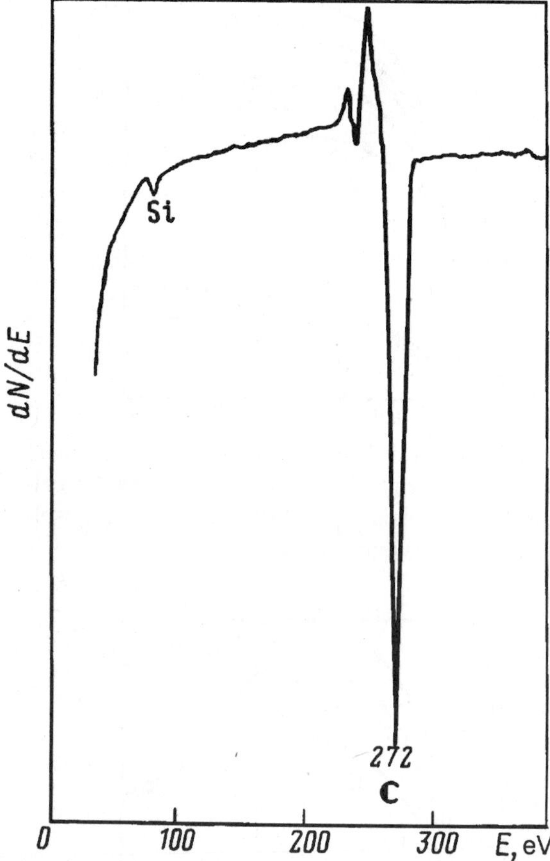

Figure 25.44 Auger spectrum of carbon [42].

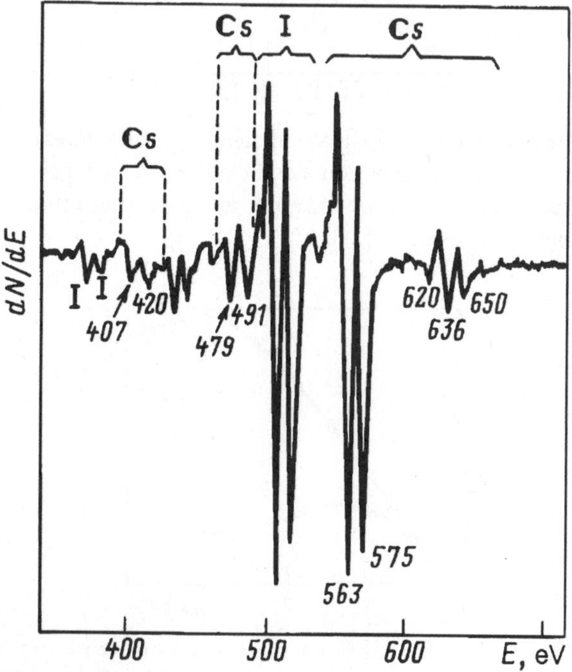

Figure 25.45 Auger spectrum of cesium (for CsI target) [42].

25.6 Field Electron Emission

Field (tunnel, autoelectron) *emission* (FE) makes up the electron ejection by bodies under the effect of a strong electric field near their surfaces. If the external electric field is strong enough for the potential threshold on the body boundary to become a barrier of the finite and even small width ($\mathcal{E} \geq 10^9$ V/m), the electrons escape the barrier (the quantum mechanical tunneling) and their passage to vacuum becomes possible. In this case, the electrons having penetrated the barrier possess the same energy as inside the body, and the electric field executes work only by accelerating the electrons in vacuum in the interelectrode space between emitter and anode and by heating the emitter with the current flowing in it. The FE current density for metals at low temperatures (practically at $T \leq 20°$C) is described in a good approximation by the formula ($\mathcal{E}$ is in V/cm) [27]:

$$j = \frac{e^3 \mathcal{E}^2}{8\pi h \Phi} \left[-\frac{8\pi}{3} \frac{(2m)^{3/2}}{h} \frac{(e)^{3/2}}{e\mathcal{E}} \theta(y) \right], \tag{25.7}$$

where $y = (e^3 \mathcal{E})^{3/2}/\Phi$ is the relative reduction of the work function by an external electric field of intensity $\mathcal{E}$ in accordance with the Schottky effect, $\theta(y)$ is the Nordheim function (Table 25.26). If the work function is measured in eV, and all the other values are taken in SI units, then

$$j = 1.54 \cdot 10^{-6} \frac{\mathcal{E}^2}{\Phi} \exp \left[-\frac{6.85 \cdot 10^9 \, \Phi^{3/2}}{\mathcal{E}} \theta\,(y) \right]. \tag{25.8}$$

It is seen that FE current density depends on the electric field strength in just the same way as TE varies with the temperature: $\ln(j/\mathcal{E}^2) = f(1/\mathcal{E})$ (Fig. 25.47). At high temperatures, the FE current density increases with T, especially strongly in the region of weak (but already giving rise to FE) fields. If FE occurs at low emitter temperatures, the energy distribution of electrons emitted from the metal starts with the electron energy corresponding to a Fermi level in the metal (taken to be zero) and extends to the negative energy range. The distribution half–width approximates 0.5 eV (Fig. 25.48). When the temperature increases, the energy spectrum of electrons emitted from the metal widens to the direction of positive energies. FE in semiconductors shows some features related to the specific electron energy distribution in them, to external electric field penetration through semiconductors, and to strong thermo- and photosensitivity of semiconductors influencing the FE current (Fig. 25.49) [28, 29]. It is possible to obtain high density FE currents from emitters having a form of the cusp. The limiting current density $j_{\rm cr}$ as yet not eroding the cusp increases with a rise in the emitting cone angle because this growth improves the heat removal from the cusp (Table 25.27, Fig. 25.50). In very strong electric fields, when the FE current density reaches 10^{12}–10^{13} A/m^2, local areas of the emitter, where the emission occurs from cusps, blow up due to intensive heating, thus producing dense plasma which expand with the velocity $v = 10^8$ m/s. This process is accompanied by the initiation of intensive electron emission (explosive electron emission, see Fig. 25.51) [30]. The current I (in A) of the explosive electron emission under the explosion of a unit cusp takes the form

$$I = 3.7 \cdot 10^{-5} U^{3/2} vt/(d - vt), \tag{25.9}$$

where U (in V) is the voltage between the anode and cathode during the explosive emission, d is the distance between them in m, and t is the time from the moment of the pulse application ($t < d/v$).

FE is widely used in some vacuum electronic devices, in field-electron and ion microscopy, whereas the explosive electron emission is employed in high-current electron accelerators and pulsed X-ray sources of high intensity [30].

Table 25.26 Nordheim function $\theta(y)$ [27]

y	$\theta(y)$
0	1.0000
0.05	0.9948
0.1	0.9817
0.15	0.9622
0.2	0.9370
0.25	0.9068
0.3	0.8718
0.35	0.8322
0.4	0.7888
0.45	0.7413
0.5	0.6900
0.55	0.6351
0.6	0.5768
0.65	0.5152
0.7	0.4504
0.75	0.3825
0.8	0.3117
0.85	0.2379
0.9	0.1613
0.95	0.0820
1.0	0

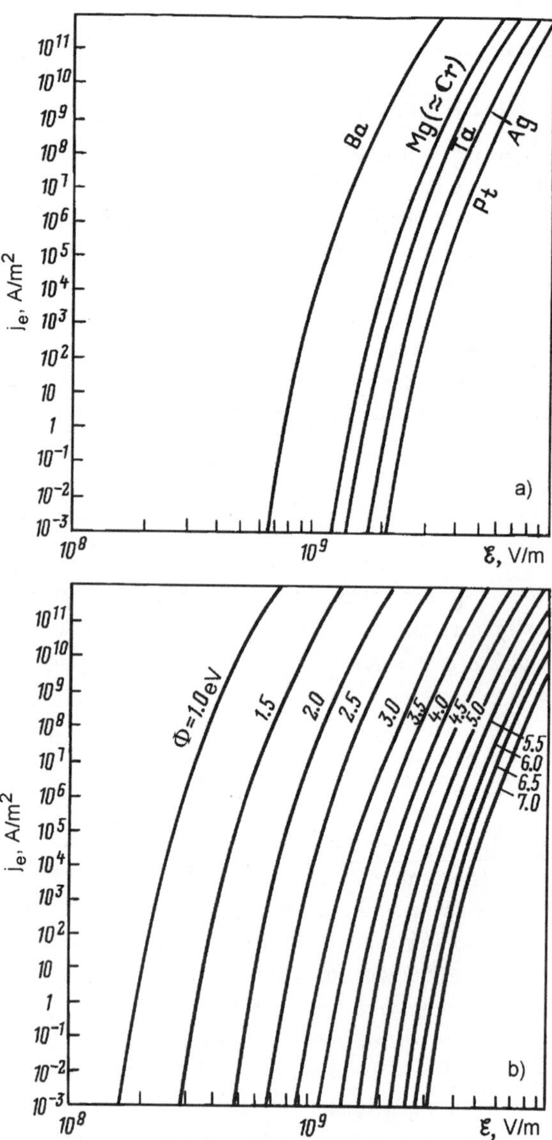

Figure 25.46 Dependences of FE current density vs. the electric field strength for some metals (**a**) and for emitters with various work functions (**b**) [6].

Table 25.27 Limiting FE current density j_{cr} for a single-cusp cathode (measured values [28])

Cathode material	Field-pulse duration, s	j_{cr}, A/m^2
W	stationary voltage	10^{11}
W	10–10^{-3}	$2{\cdot}10^{11}$
W	10^{-5}–10^{-6}	$5{\cdot}10^{11}$–10^{12}
W	10^{-7}	$3{\cdot}10^{12}$–$5{\cdot}10^{12}$
W	10^{-8}–10^{-9}	10^{13}
W-Zr	$4{\cdot}10^{-6}$	$1{\cdot}10^{13}$–$5{\cdot}10^{13}$
Ta, Re	$4{\cdot}10^{-6}$	$5{\cdot}10^{11}$
LaB_6	$3{\cdot}10^{-6}$	10^{11}–10^{12}
ZrC	$3{\cdot}10^{-6}$	10^{11}–10^{12}

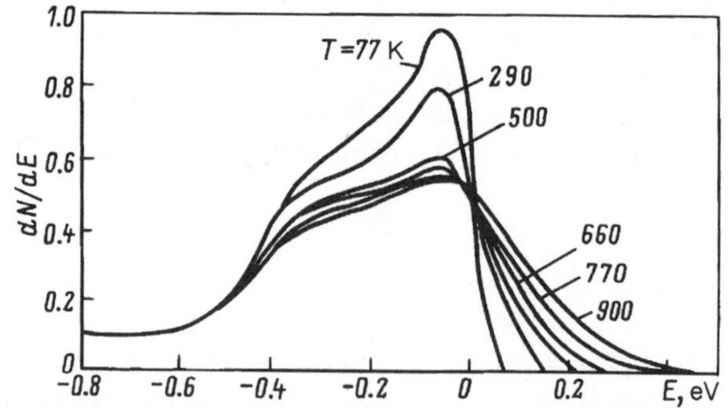

Figure 25.47 Energy distribution of FE electrons from ⟨100⟩-oriented tungsten point cathode at various temperatures [31].

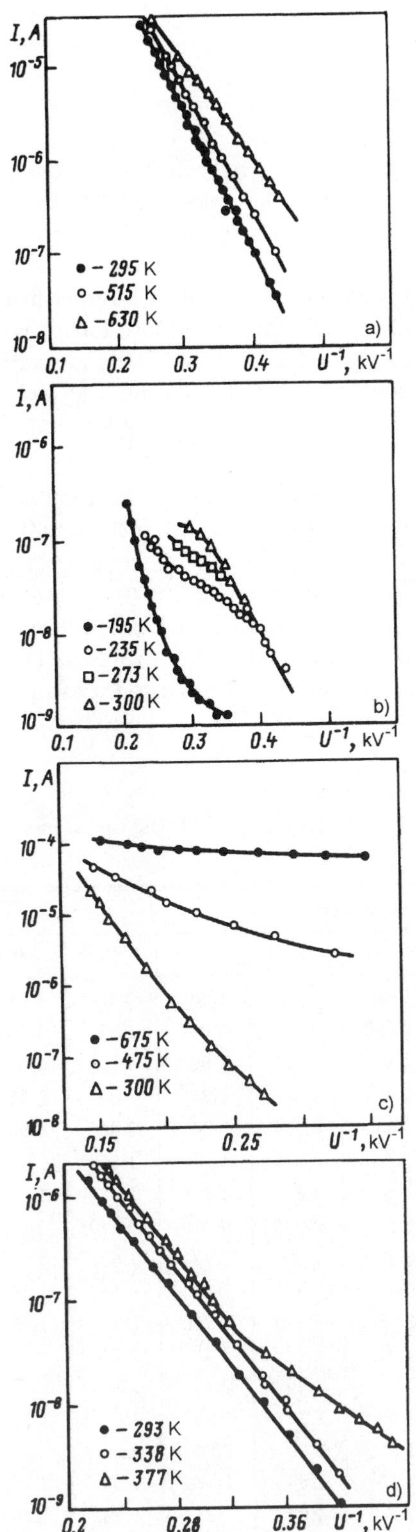

Figure 25.48 Current-voltage characteristics of FE from semiconductors at various temperatures. **a** – low-resistivity n-type Ge; **b** – high-resistivity p-type Ge; **c** – zinc sulfide; **d** – cadmium sulfide [29].

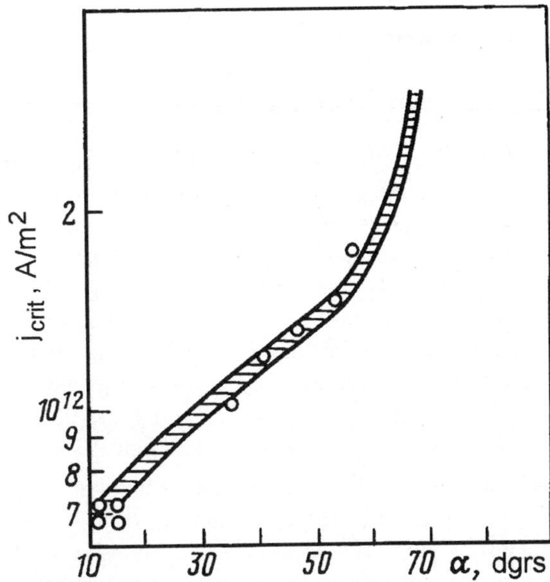

Figure 25.49 The critical (maximum nondestructive) FE current density versus the apex angle of the W-cathode cone [28]. Current impulses are 35 μs long with repetition frequency equal to 50 Hz. The circles indicate the experimental values, the shaded strip – calculated results.

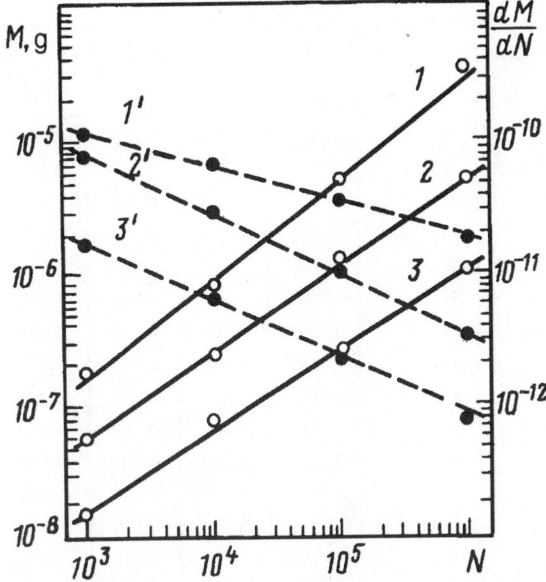

Figure 25.50 The mass consumption M of copper point cathodes during numerous switchings on of currents corresponding to explosive FE (N – number of current pulses; $U = 30$ kV; the distance between the cathode and anode is 3 mm; the apex angle of the cathode cone is 6°): *1, 3* – $M(N)$; *1′, 3′* – $\partial M/\partial N$ [30]. The current pulse duration is: *1, 1′* – $5 \cdot 10^{-8}$ s; *2, 2′* – $2 \cdot 10^{-8}$ s; *3, 3′* – $5 \cdot 10^{-9}$ s.

25.7 Ion and Ion–Electron Emission

Ion emission consists in the body ejection of ionized elements entering its composition on heating (thermionic emission) or on its surface bombardment by other particles: electrons (electron–ion emission), ions (secondary ion emission), or else on the photon irradiation (photodesorption). The electron emission may also arise under the ion bombardment (ion–electron emission, see Table 25.28 and Fig. 25.52). Bodies with a skeleton-cavity structure of crystal lattice (zeolites, alumosilicates, and others) exhibit intensive thermionic emission when alkali atoms are introduced into lattice voids. These atoms are loosely bound with the main lattice and may easily travel inside the body at heating, diffusing toward the surface and evaporating from it. Effective solid–ion sources are produced with the use of such materials (Table 25.29 and Figs. 25.52, 25.54). In the course of secondary ion emission, some portion of particles is emitted as neutral atoms.

Basic characteristics of ion emission involve: j_i, the ion current density; S, the atomization coefficient, i.e. the ratio of the total number of atomized particles (neutral and ionized) to the number of primary ions; K^+, the secondary ion emission coefficient, i.e. the ratio of the number of secondary ions with a given charge/mass ratio to the number of primary ions (for nonelementary targets, the secondary ion emission coefficient is taken to mean the value $\gamma_i^+ = K^+/C_i$, where C_i is the relative concentration of i-th kind atoms in the target); and, finally, $\beta^+ = K^+/S$, the ionization coefficient (Tables 25.30–25.32). Secondary ion emission is widely used in secondary ion mass-spectroscopy for analyzing the composition of solid near-surface layers [34].

Table 25.28 Ion–electron emission coefficients of metals γ (in electron/ion) [35] [Data presented were obtained by graphical interpolation of measured values.]

Target	Primary ion	Primary ion energy, keV									
		2	3	4	7	10	15	20	30	40	50
Li	H^+	0.475	0.560	0.650	0.830	0.965	1.12	1.23	1.36	1.46	1.55
	H_2^+	0.260	0.320	0.365	0.465	0.550	0.680	0.775	0.945	1.07	1.16
	He^+	0.465	0.525	0.590	0.740	0.880	1.10	1.27	1.54	1.77	1.95
Al	H^+	0.225	0.310	0.380	0.560	0.700	0.875	1.01	1.23	1.34	1.38
	H_2^+	0.087	0.132	0.156	0.230	0.310	0.450	0.555	0.745	0.885	0.955
	He^+	0.280	0.330	0.380	0.510	0.620	0.765	0.885	1.10	1.31	
Cr	H^+	0.280	0.360	0.430	0.580	0.710	0.880	1.04	1.26	1.36	1.42
	H_2^+	0.240	0.270	0.305	0.375	0.450	0.540	0.635	0.810	0.940	1.01
	He^+	0.310	0.420	0.485	0.650	0.780	0.960	1.10	1.42	1.69	1.94
Cu	H^+		0.420	0.480	0.650	0.780	0.960	1.08	1.29	1.45	1.57
	H_2^+	0.162	0.210	0.265	0.380	0.470	0.580	0.670	0.830	0.960	1.08
	He^+		0.390	0.455	0.620	0.800	1.04	1.18	1.44	1.70	
Ag	H^+		0.650	0.810	1.15	1.42	1.70	1.89	2.17	2.33	
	H_2^+	0.225	0.295	0.375	0.610	0.800	1.03	1.20	1.50	1.67	1.78
	He^+	0.360	0.450	0.570	0.910	1.21	1.65	2.00	2.50	2.89	
Au	H^+	0.360	0.540	0.680	0.980	1.22	1.46	1.64	1.77	2.03	2.15
	H_2^+	0.138	0.240	0.325	0.550	0.720	0.930	1.09	1.38	1.59	1.70
	He^+	0.315	0.410	0.510	0.810	1.09	1.52	1.83	2.34	2.68	
Target	Primary ion	Primary ion energy, keV									
		50	75	100	125	150	175	200	225		
W [36]	H^+	1.54	1.62	1.64	1.65	1.64	1.61	1.56	1.53		

Table 25.29 Thermoionic properties of zeolites and alumosilicates of alkali metals [33]

Substance	T, °C	Emitted ions	j_i, A/m^2
Zeolite	1100	Cs^+	120
Li-alumosilicate ($Li_2O \cdot Al_2O_3 \cdot 2SiO_2$)	1100	Li^+	6
Rb-alumosilicate ($RbO_2 \cdot Al_2O_3 \cdot 2SiO_2$)	1160	Rb^+	4
Rb-alumosilicate + Mo (10%)	1160	Rb^+	20–50
Na-alumosilicate ($Na_2O \cdot Al_2O_3 \cdot 2SiO_2$)	1200	Na^+	20
Na-alumosilicate + W (10%)	1200	Na^+	200

Table 25.30 The basic parameters of secondary ion emission for pure metal films [34]

Primary ion	Parameter	Metal							
		Al	Ti	Mn	Ni	Cu	Sn	Ta	Au
Ar^+	$\beta^+, 10^{-4}$ ion/atom	110	92	22	6.4	1.9	0.66	6.9	0.056
	S, atom/ion	3.7	2.3	3.5	6.1	8.5	8.3	2.9	12
	$K^+, 10^{-4}$ ion/ion	410	212	77	39	16	565	20	0.71
He^+	$\beta^+, 10^{-4}$ ion/atom		337	26.8	9.8	4.73	4.15	50	0.185
	S, atom/ion		0.8	1.1	1.5	1.9	2.8	0.68	3.2
	$K^+, 10^{-4}$ ion/ion		270	29.8	14.7	9	11.6	34	0.59
O_2^+	$\beta^+, 10^{-4}$ ion/atom	9600	5900	265	63.5	14	9.3	250	0.32
	S, atom/ion	1.6	1.1	2.3	3.3	4.3	5.1	1.1	6.5
	$K^+, 10^{-4}$ ion/ion	15300	5900	610	210	60	47.5	275	2.08

Table 25.31 Relative coefficients of the secondary ion emission for the most widespread metals under Ar^+, He^+, and O^+ ion bombardment [The energy of the primary ions is 8 keV; Fe bombarded by Ar^+ ions is taken as a standard [34].]

Metal	Primary ion			Metal	Primary ion		
	Ar^+	He^+	O^+		Ar^+	He^+	O^+
Ag	0.108	0.07	0.135	Nd	4.13	1.11	17.8
Al	9.62	62.2	447.0	Ni	0.98	0.415	4.42
Au	0.012	0.025	0.051	Pb	0.086	1.08	1.22
Be	2.19	15.5	163.0	Pd	0.042	0.033	1.55
Bi	0.358	0.13	0.464	Pt	0.112	0.013	0.076
Cd	0.019	0.016	0.106	Pr	1.03	2.61	22.2
Ce	0.40	0.773	14.7	Re	0.65	2.14	20.5
Co	1.22	0.37	5.97	Rh	0.89	0.556	24.0
Cr	1.01	2.95	41.6	Ru	0.76	0.724	20.6
Cu	0.262	0.235	1.26	Sc	7.75	31.0	205.0
Dy	1.88	2.84	23.1	Sm	0.275	1.84	19.5
Er	4.50	2.03	15.8	Sn	0.094	0.202	0.558
Fe	1.00	0.872	15.8	Ta	0.186	0.853	5.43
Gd	2.28	1.47	13.2	Tb	7.00	3.63	23.8
Hf	2.25	1.33	8.58	Ti	1.56	7.14	127.0
Ho	2.88	3.03	21.4	Tu	1.68	6.7	22.1
In	1.50	30.0	53.8	V	2.75	13.1	200.0
La	0.49	0.725	12.8	W	0.187	0.447	5.83
Lu	15.6	2.64	16.6	Y	2.87	8.4	51.2
Mg	11.4	13.3	86.0	Yb	1.05	2.8	25.2
Mn	2.15	0.95	13.5	Zn	0.034	0.107	0.918
Mo	0.385	1.86	25.2	Zr	0.296	2.08	28.2
Nb	1.16	4.05	35.6				

Table 25.32 Basic parameters of the secondary ion emission in carbides [The primary ions are Ar^+ with energies 8 keV; the density of the primary ion flux is 10 A/m^2 [34].]

Parameter	TiC	VC	CrC_2	ZrC	NbC	MoC	HfC	TaC	WC
S	3.0	2.9	3.3	2.3	2.9	2.9	4.0	2.8	2.5
γ^+_{Me}	21.0	24.0	14.0	2.7	6.7	1.9	1.3	0.82	0.52
γ^+_C	0.056	0.034	0.033	0.047	0.054	0.030	0.056	0.030	0.080
β^+_{Me}	0.20	0.24	0.12	0.033	0.067	0.018	0.009	0.008	0.006
β^+_C	0.007	0.005	0.004	0.008	0.007	0.004	0.005	0.004	0.012

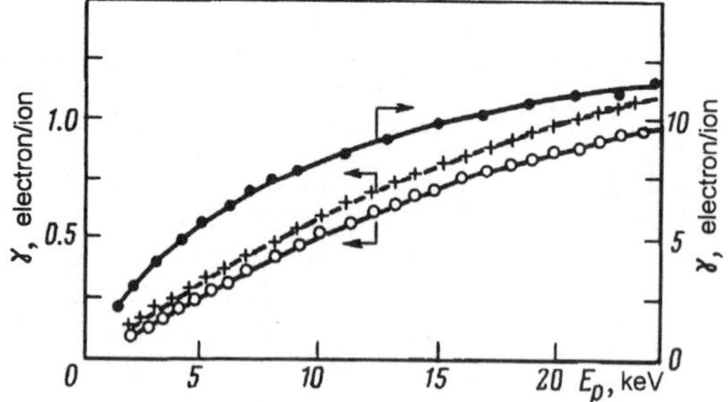

Figure 25.51 The ion–electron emission coefficient γ versus $^{40}Ar^+$ primary-ion energy E_p [23]. The targets are: polycrystalline Mo (+); (111) surfaces of single-crystal Ge (o); monocrystalline sodium chloride (•). Measurements were made at pressures $p < 10^{-8}$ Pa. The surfaces of the targets were ion etched in vacuum at $p \sim 10^{-7}$ Pa and $T = 350°C$ (Ge and NaCl); Mo target has been previously annealed at $T = 1700$ K and at the same pressure.

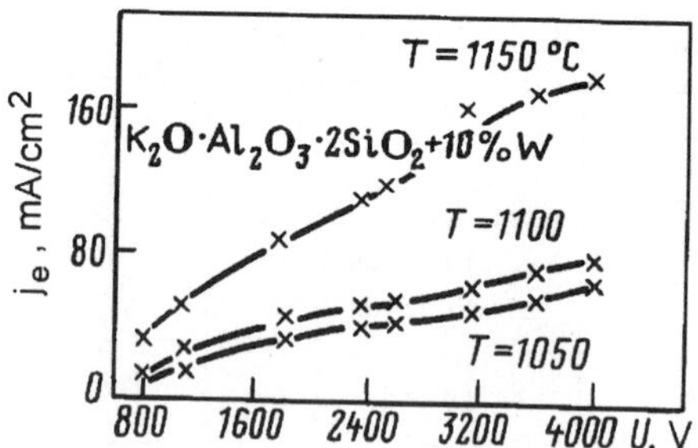

Figure 25.52 Current-voltage characteristics of the K^+ emission source from potassium alumosilicate with tungsten added ($K_2O \cdot Al_2O_3 \cdot 2SiO_2$ + 10%W) in the pulsed mode of operation [33]. The current values are averaged over the pulse duration. Current pulses are 700 μs long with repetition frequency equal to 10 Hz. The values of source temperatures are shown alongside the curves.

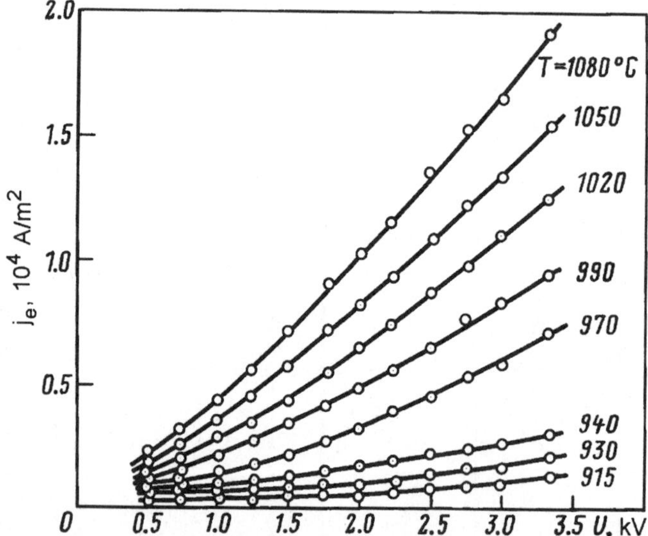

Figure 25.53 Current-voltage characteristics of the Cs-ion emission source from Cs-alumosilicate ($Cs_2 \cdot A_2O_3 \cdot 2SiO_2$) in the pulsed mode of operation at various cathode temperatures [32]. Distance between cathode and anode is 2 mm, current pulses are 10 μs long, with repetition frequency equal to 25 Hz.

References

[1] Dobretsov, L. N., Gomoyunova, M. V., Emission electronics, Nauka, Moscow, 1966 (in Russian).

[2] Fomenko, V. S., Emission properties of materials, 4th ed., Naukova Dumka, Kiev, 1981 (in Russian).

[3] Michaelson, H. B., J. Appl. Phys., 48, 4729, 1977.

[4] Savitskii, E. M., Burov, I. V., Pirogova, S. V., Litvak, L. N., Electrical and emission properties of alloys, Nauka, Moscow, 1978 (in Russian).

[5] Haas, G. A., Shin, A., Marrian, C. R. K., Appl. Surface Sci., 16, 139, 1983.

[6] Ardenne, M., Tabellen zur Angewandten physik auflage, Bd. 1, VEB Verlag der Wissenschaften, Berlin, 1962.

[7] Kudintseva, G. A., Mel'nikov, A. I.,Morozov, A. V., Nikonov, B. P., Thermal electronic cathodes, Energiya, Moscow, 1966 (in Russian).

[8] Futamoto, M., Nakazawa, M., Usami, K., J. Appl. Phys., 51, 3869, 1980; Cronin, J. L., IEE Proc., 128, 1, 1981 (pt. I).

[9] Houston, J. M., Webster, H. F., Advances in Electronics and Electr. Phys., 17, 125, 1962.

[10] Lipeles, R. A., Kan, H. K. A., Appl. Surface Sci., 16, 189, 1983.

[11] Kul'varskaya, B. S., Radiotekhnika i electronika, 15, 1717, 1970 (in Russian).

[12] Nikonov, B. P., Oxide cathode, Energiya, Moscow, 1979 (in Russian).

[13] Soukup, R. J., J. Appl. Phys., 48, 1098, 1977.

[14] Ashkinazi, L. A., Soboleva, N. A., Itogi nauki i tekhniki, Ser. "Elektronika," 15, 154, 1983 (in Russian).

[15] Soboleva, N. A., Melamid, A. E., Photoelectron devices, Vysshaya Shkola, Moscow, 1974 (in Russian).

[16] Sommer, A.H., Photoemissive materials, Wiley, New York–London, 1968.

[17] Bell, R.L., Negative electron affinity devices, Oxford Univ. Press (Clarendon), London–New York, 1973.

[18] Escher, J. S., Semiconductors and semimetals, 15, 195, 1981.

[19] Escher, J. S., Bell, R. L., Gregory, P. E., IEEE Trans. Electron Devices, ED–27, 1244, 1980.

[20] Fischer, D. G., Enstrom, R. E., Escher, J. S., J. Appl. Phys., 43, 3815, 1972.

[21] Howorth, J. R., Folkes, J. R., Palmer, J. C., J. Phys. D, 9, 785, 1976.

[22] Bronshtein, I. M., Fraiman, B. S., Secondary electron emission, Nauka, Moscow, 1969 (in Russian).

[23] Electron and ion spectroscopy of solids, Eds. Fiermans, L., Vennik, J., and Dekeyser, W., Plenum, New York, 1978.

[24] Gavanin, V. A., Kutenin, Yu. D., Itogi nauki i tekhniki, Ser. "Elektronika i prilozheniya," 12, 43, 1980.

[25] Martinelli, R. U., Appl. Phys. Lett., 17, 313, 1970.

[26] Martinelli, R. U., Schultz, M. L., Gossenberger, H. F., J Appl. Phys., 43, 4803, 1972.

[27] Elinson, M. I., Vasil'ev, G. F., Autoelectronic emission, Fizmatgiz, Moscow, 1958 (in Russian).

[28] Nonheated cathodes, Ed. Elinson, M. I., Sovetskoe Radio, Moscow, 1974 (in Russian).

[29] Fursei, G. N., and L'vov, O. I., Additional review in: Fisher, R., and Neuman, H., Autoelectron emission of semiconductors, Nauka, Moscow, 1971 (Russian transl.).

[30] Bugaev, S. P., Litvinov, E. A., Mesyats, G. A., Preobrazhenskii, D. I., Uspekhi fizicheskikh nauk, 115, 101, 1975 (in Russian).

[31] Swanson, L. W., Crouser, L. C., Phys. Rev., 163, 622, 1975.

[32] Kul'varskaya, B. S., Kan, Kh. S., Dotsenko, V. G., Theses of reports at the 5th All–Union Symp. on High Current Electronics, Part 1, Institut sil'notochnoi electroniki, Sibirskoe otdelenie Akad. Nauk SSSR, Tomsk, 1984 (in Russian).

[33] Kul'varskaya, B. S., Itogi nauki i tekhniki, Ser. "Elektronika," 13, 111, 1981 (in Russian).

[34] Cherepin, V. T., Ion probe, Naukova Dumka, Kiev, 1981 (in Russian).

[35] Baragiola, R. A., Alonso, E. A., Oliva Florio, A., Phys. Rev. B, 19, 121, 1979.

[36] Ewing, R. I., Phys. Rev., 139, 1840, 1965.

[37] Photoemission in solids, Eds. Cardona, M. and Ley, L. in: Topics in Applied Physics, Springer-Verlag, Berlin–Heidelberg, 1978.

[38] Palmberg, P. W., Riach, G. E., Weber R. E., Handbook of Auger electron spectroscopy, Phys. Electron Industry Inc., Edina, Minnesota, 1975.

[39] Wagner, C.D., Riggs, W. M., Davis, L. E., Handbook of X–ray photoelectron spectroscopy, Perkin–Elmer Corp., Minnesota, 1979.

[40] Hoene, E. L., Advances in Electronics and Electr. Phys., 33A, 369, 1972.

[41] Beguchev, V. P., Shefova, I. A., Musatov, A. L., J. Phys. D: Appl. Phys., 26, 1499, 1993.

26

Magnetic Properties of Dia- and Paramagnetics

V.Yu. Ivanov and L.I. Vinokurova

This chapter discusses magnetic characteristics of substances which do not possess atomic magnetic structures (in the absence of magnetic field): simple elements, the most known inorganic compounds, and some practically important compounds (silicides, selenides, tellurides, etc.)

In weak magnetic fields, $\mu_B H \ll kT$, the magnetization J of such substances increases proportionally to magnetic field intensity: $J = \chi_v H$, where χ_v is the magnetic susceptibility per unit volume, μ_B is the Bohr magneton, H is the magnetic field intensity, k is the Boltzmann constant, and T is the temperature).

The substances with negative magnetic susceptibility are referred to as diamagnetics ($\chi_v < 0$). The diamagnetism results from electromagnetic induction of in-molecule currents, this induction being generated by an external magnetic field in atomic electron shells. All the substances feature diamagnetic properties without exception.

Paramagnetism ($\chi_v > 0$) is typical for the substances whose particles (atoms, molecules, ions, atomic nuclei) exhibit their own magnetic moments, but in the zero external magnetic field, these moments are not oriented, so on the whole $J = 0$. In the external magnetic fields, the atomic magnetic moments of paramagnetic substances are oriented along the field. If the field is very strong, $\mu_B H \gg kT$, then all the magnetic moments of paramagnetic particles are oriented closely to the field (magnetic saturation).

Due to the disordering effect of particle thermal motion, in the simplest cases the magnetic susceptibility decreases with temperature rise according to Curie's law: $\chi_v = C/T$, where C is the Curie constant, $C = Np_{\text{eff}}^2\mu_B^2/3k$. Here p_{eff} is the effective magnetic moment per molecule, N is the number density of molecules. For substances with magnetic moment carriers interacting with each other and with the intracrystalline field, the paramagnetic susceptibility follows the Curie-Weiss law: $\chi_v = C'/(T - \theta)$. In most cases the constant C' practically coincides with C taken from Curie's law for free magnetic ions, and θ defines the interaction of magnetic ions between each other and with the intracrystalline field. The Curie-Weiss law is usually obeyed in a certain temperature range. Deviations from this law are observed at low temperatures (below 70 K), due to the influence of nonuniform electric fields of neighboring ions or oriented solvent molecular dipoles on the electron orbital momentum. The Curie-Weiss law is also true for ferro- and antiferromagnetics in a given temperature range higher than the magnetic ordering temperature.

Chemical compounds can also be paramagnetic even when they contain ions without magnetic moments in the ground state. Paramagnetism of such compounds is explained by a quantum correction due to an admixture of excited states with the magnetic moments. Such paramagnetism (polarizational or Van Fleck's paramagnetism) is not influenced by temperature.

The contribution resulting from the spin paramagnetism of conduction electrons (Pauli's paramagnetism) in metals is nearly independent of temperature.

Semiconductor paramagnetic susceptibility due to impurity carriers depends exponentially on temperature in the simplest cases: $\chi_v = AT^{1/2}\exp(-\Delta\varepsilon/(2kT))$, where A is the material constant, $\Delta\varepsilon$ is the width of energy gap for the semiconductor at hand.

The specific magnetic susceptibility, that is the magnetic susceptibility per unit mass, is usually used for characterizing the substance magnetic properties: $\chi = \chi_v/\rho$, where ρ is the substance density.

0-8493-2861-6/97/$0.00+$.50

The magnetic susceptibility is often referenced to one mole (χ_m). The relationship between χ and χ_m reads as $\chi_m = \chi M$, where M is the relative molecular mass.

Table 26.1 lists the values of dia– and paramagnetic specific susceptibility for those materials that do not follow the Curie-Weiss law.

Table 26.2 shows the room temperature values of susceptibility for some paramagnetics with temperature dependences conforming to the Curie-Weiss law. This table presents only those substances for which no magnetic ordering was found at low temperatures (higher than 4.2 K).

The values of χ are given below in SI units (m^3/kg). To get χ in CGSM units (cm^3/g), the SI value of χ should be multiplied by $10^3/4\pi$.

Tables which follow comprise data for a solid polycrystalline state if there are no special marks. In other cases the following abbreviations were used: s/c – single-crystalline state (when data for both the single-crystalline and polycrystalline states are included, the latter is specially marked with p/c); $\chi_\parallel$ and $\chi_\perp$ are the susceptibilities measured in fields which are parallel or perpendicular to the axis of high symmetry, respectively; χ_a, χ_b, χ_c are susceptibilities along elementary cell translation vectors in a crystal lattice; (g) – gas; (liq) – liquid; (ss) – solid state; (s) – solution; (p) – the hole concentration in a semiconductor; (n) – the electron concentration in a semiconductor; T_{melt} – the melting point; T_{ev} – the evaporation temperature; ΔT – the temperature interval where the Curie-Weiss law is valid, in so doing a dash means that the measurement temperature was not cited in the original work.

Table 26.1 Magnetic susceptibility of dia- and paramagnetics

Substance	T, K	χ, 10^{-9} m^3/kg	Ref.	Substance	T, K	χ, 10^{-9} m^3/kg	Ref.
H_2		−1.974	[1]	DH		−1.325 (g)	[8]
H_2	293	−1.9867	[2]	DHO	302	−0.682 (liq)	[1]
H_2	90	−1.97	[2]	D_2O	276.8	−0.626 (ss)	[1]
H_2	<20	−2.7 (liq)	[3]	D_2O	276.8	−0.632 (liq)	[1]
H_3AsO_3	293	−0.406 (ss)	[4]	D_2O	293	−0.637 (liq)	[1]
H_3BO_3	298	−0.55	[1]	He	298	−0.47	[1]
HBr	273	−0.407 (liq)	[5]	Li^{*2}	298	2.04	[10]
HCl	273	−0.62 (liq)	[1]	Li^{*2}	298	3.35	[9]
HF	287	−0.43 (liq)	[1]	LiBr	298	−0.400	[1]
HI	281	−0.373 (liq)	[1]	$LiBrO_3$	298	−0.289	[1]
HI	195	−0.370	[1]	$Li(C_2H_3O_2)$	298	−0.513	[1]
HIO_3	298	−0.27	[1]	Li_2CO_3	298	−0.365	[1]
HIO_4	298	−0.294	[1]	LiCl	298	−0.573	[1]
HNO_3	298	−0.316	[1]	$LiClO_3$	298	−0.319 (s)	[1]
H_2O^{*1}	273	−0.7019 (ss)	[1]	LiF	298	−0.389	[1]
H_2O^{*1}	273	−0.7177 (liq)	[1]	LiH	298	−0.79	[1]
H_2O^{*1}	293	−0.7199 (liq)	[1]	LiI	298	−0.37	[1]
H_2O^{*1}	273	−0.72454 (liq)	[1]	$LiIO_3$	298	−0.26	[1]
H_2O^{*1}	>273	−0.727 (g)	[1]	$LiNO_3$	292	−0.48	[7]
H_2O_2	298	−0.015	[1]	$LiNO_3{\cdot}3H_2O$	298	−0.504	[1]
H_3PO_3	298	−0.518 (s)	[1]	Li_2O	293	−0.57	[7]
H_3PO_4	298	−0.447 (s)	[1]	LiOH	298	−0.516	[1]
H_2S	298	−0.748	[1]	Li_2SO_4	298	−0.364	[1]
H_2SO_4	298	−0.406	[1]	$Li_2SO_4{\cdot}H_2O$	290	−0.43	[7]
H_2SeO_3	298	−0.352 (ss)	[1]	Be^{*3} (s/c) $\chi_\parallel$	293	−2.38	[11]
H_2SeO_4	298	−0.353	[1]	Be^{*3} (s/c) $\chi_\perp$	293	−0.80	[11]
H_2TeO_3	298	−0.24	[7]	Be (p/c)	103	−0.75	[12]
H_2TeO_4	291	−0.43	[7]	Be	291	−1.0	[1, 12]
D_2		−0.994 (g)	[8]	Be	573	−1.2	[12]

*1 Temperature coefficient $(T/\chi){\cdot}\mathrm{d}\chi/\mathrm{d}T$ decreases from $2.9{\cdot}10^{-4}$ at 278 K to $0.62{\cdot}10^{-4}$ at 343 K [6].
*2 When temperature decreases from 298 to 78 K, χ increases by 1.3% [9].
*3 See Figure 26.1.

Table 26.1 Magnetic susceptibility of dia- and paramagnetics *(continued)*

Substance	T, K	χ, 10^{-9} m^3/kg	Ref.	Substance	T, K	χ, 10^{-9} m^3/kg	Ref.
Be	1200	−1.3	[12]	NH_4I	298	−0.455 (ss)	[1]
$BeCl_2$	298	−0.332	[1]	NH_4IO_3	298	−0.323	[1]
$Be(NO_3)_2$	298	−0.308 (ss)	[1]	NH_4NO_3	298	−0.421 (ss)	[1]
$Be(NO_3)_2{\cdot}3H_2O$	298	−0.311	[4]	NH_4OH	298	−0.90	[1]
BeO*4	298	−0.476	[1]	$(NH_4)_2SO_4$	298	−0.51	[1]
$Be(OH)_2$	290	−0.537	[1]	$(NH_4)_2SO_8$	294	−0.440	[4]
$BeSO_4$	298	−0.354	[1]	NH_4SCN	298	−0.632	[1]
$BeSO_4{\cdot}4H_2O$	290	−0.51	[7]	NO	60-90	0.66 (ss)	[1]
B*5	90	−0.63	[13]	NO	117.64	3.81 (liq)	[1]
B*5	290	−0.62	[1, 13]	NO	146.9	77.45 (g)	[1]
B*5	373	−0.73	[14]	NO	298	48.7 (g)	[1]
B*5	503	−0.76	[14]	NO_2	408	3526 (g)	[1]
B*5	1373	−0.71	[14]	N_2O	285	−0.429 (liq)	[1]
α-B (rhombohedral)	300	−0.795	[15]	N_2O	289	−0.429 (g)	[1]
β-B (rhombohedral)	300	−0.78(0.10)	[16]	N_2O_3	291	−0.206 (liq)	[1]
B (amorphous)	300	−0.3	[15]	N_2O_4	257	−0.276 (ss)	[1]
B (amorphous)	300	−0.645	[16]	N_2O_4	295.1	−0.250	[1]
BCl_3	298	−0.511 (liq)	[1]	N_2O_5	289	−0.332 (ss)	[1]
B_2O_3	298	−0.56	[1]	β-O_2	33	118 (ss)	[24]
C (diamond)	103	−0.49	[12]	γ-O_2	54.3	319 (ss)	[1]
C (diamond)	293	−0.49	[1]	O_2	54.3	310 (liq)	[1]
C (diamond)	300	−0.98	[17]	O_2	70.8	271.4 (liq)	[1]
C (diamond)	473	−0.50	[14]	O_2	90.1	240.6 (liq)	[1]
C (diamond)	673	−0.51	[14]	O_2	293	107.8 (g)	[1]
C (diamond)	1173	−0.54	[14]	O_3	90	0.14*7 (liq)	[1]
C (diamond)	1473	−0.56	[14]	O_3	90	0.14 (g)	[25]
C (graphite) (p/c)	14	−4.7	[18]	Ne	293	−0.33 (g)	[1]
C (graphite) (p/c)	289	−3.0	[18]	Na*8	78	0.582	[9]
C (s/c), $\chi_{\parallel}$*6	293	−21	[19]	Na*8	298	0.610	[9]
C (s/c), $\chi_{\perp}$	293	−0.4	[19]	Na*8	303	0.600	[26]
CCl_4	291	−0.442	[4]	Na*8	523	0.620	[26]
CH_2	293	−0.4936	[20]	$Na_2B_4O_7$	298	−0.422	[1]
CN_2	298	−0.432 (liq)	[21]	$Na_2B_4O_7{\cdot}10H_2O$	293	−0.593	[27]
CNCl	298	−0.527	[1]	NaBr	298	−0.40	[1]
CO_2	298	−0.454	[1]	$NaBr{\cdot}2H_2O$	291	−0.420 (s)	[4]
CO	298	−0.35	[1]	$NaBrO_3$	298	−0.293	[1]
COS	298	−0.539	[1]	$NaC_2H_3O_2$	298	−0.458	[1]
$COCl_2$	298	−0.485	[1]	$NaC_2H_3O_2{\cdot}3H_2O$	282	−0.483	[4]
CS_2	293	−0.554 (liq)	[22]	Na_2CO_3	298	−0.39	[1]
N_2	88	−0.507	[23]	$Na_2CO_3{\cdot}10H_2O$	290	−0.58	[7]
N_2	293	−0.427	[1]	NaCl	123	−0.051 (ss)	[28]
NH_3	298	−1.06 (g)	[1]	NaCl	298	−0.518 (ss)	[1]
NH_4Br	298	−0.484 (ss)	[1]	NaCl	635	−0.499 (ss)	[28]
$NH_4C_2H_3O_2$	298	−0.533	[1]	NaCl	1206	−0.508	[28]
$(NH_4)_2CO_3$	298	−0.545	[1]	$NaClO_3$	298	−0.326	[1]
NH_4Cl	298	−0.686 (ss)	[1]	$Na_2Cr_2O_7$	298	−0.21	[1]
NH_4ClO_3	298	−0.415	[1]	NaF	298	−0.383 (ss)	[1]
NH_4F	298	−0.62 (ss)	[1]	$NaHCO_3$	293	−0.21	[29]

*4 χ is temperature-independent.
*5 See Figure 26.2.
*6 See Figure 26.3.
*7 χ is temperature-independent at $T<90$ K.
*8 See Figures 26.4, 26.5.

Table 26.1 Magnetic susceptibility of dia- and paramagnetics *(continued)*

Substance	T, K	χ, 10^{-9} m^3/kg	Ref.	Substance	T, K	χ, 10^{-9} m^3/kg	Ref.
$NaHPO_3$	293	−0.457	[30]	$AlBr_3$	292	−0.32	[7]
$NaHPO_4$	293	−0.399	[30]	$AlCl_3$	292	−0.60	[7]
$NaHPO_4 \cdot H_2O$	296	−0.438 (ss)	[4]	AlF_3	302	−0.16	[1]
NaH_2PO_2	293	−0.381	[30]	Al_2O_3*11	298	−0.36	[1]
$NaI \cdot 2H_2O$	300	−0.405 (s)	[4]	$Al_2O_3 \cdot 2H_2O$	293	−0.337	[84]
$NaIO_3$	298	−0.268	[1]	$Al_2(SO_4)_3$	298	−0.27	[1]
$NaNO_2$	298	−0.210	[1]	Si*12	300	−0.228	[17]
$NaNO_3$	298	−0.301	[1]	$SiBr_4$	298	−0.370	[1]
Na_2O	298	−0.319	[1]	SiC	300	−0.265	[35]
Na_2O_2	298	−0.36	[1]	$SiCl_4$	298	−0.520	[1]
NaOH	290	−0.59	[7]	SiO_2	293	−0.493	[1]
$NaPO_3$	298	−0.417	[1]	$Si(OH)_4$	293	−0.443	[1]
Na_2S	298	−0.50	[1]	$SiO(OH)_2$	293	−0.427	[36]
Na_2S_2	298	−0.48	[1]	$Si_2O(OH)_6$	293	−0.429	[36]
Na_2S_3	298	−0.48	[1]	P (black)	293	−0.86	[1]
Na_2S_4	298	−0.48	[1]	P (white)	103	−0.90	[2]
Na_2S_5	298	−0.48	[1]	P (white)	317	−0.90	[2]
Na_2SO_4	298	−0.37	[1]	P (red)	293	−0.67	[14]
$Na_2SO_4 \cdot 10H_2O$	298	−0.571	[1]	P	318	−0.90 (liq)	[14]
$Na_2SO_3 \cdot 7H_2O$	293	−0.462	[30]	P	393	−0.90 (liq)	[14]
$Na_2S_2O_3$	298	−0.246	[1]	PCl_3	292	−0.463	[1]
$Na_2S_2O_3 \cdot 3H_2O$	290	−0.368 (ss)	[4]	PCl_5	298	−0.490	[1]
$Na_2S_2O_4$	298	−0.200	[10]	P_2O_5	291	−0.46	[7]
Na_2Se	298	−0.48	[31]	$POCl_3$	286	−0.449	[4]
Na_2Se_2	298	−0.392	[31]	α-S	293	−0.487	[1]
Na_2Te	298	−0.432	[31]	α-S (rhombohedral)	293	−0.487	[1]
Na_2Te_2	298	−0.349	[31]	β-S (monoclinic)	293	−0.464	[1]
Mg*9 (p/c)	1.2	0.25	[32]	S*13	293	0.259–0.374	[37]
Mg*9 (p/c)	293	0.46	[33]	S	$T>T_{melt}$	−0.480 (liq)	[1]
Mg (s/c), $\chi_\perp$	293	0.46	[33]	S	828	22 (g)	[1]
$MgBr_2$	298	−0.39	[1]	S	1023	14.5 (g)	[1]
$Mg(C_2H_3O_2) \cdot 4H_2O$	298	−0.541	[1]	S (s/c), χ_a	298	−0.500	[38]
$MgCO_3$	298	−0.384	[1]	S (s/c), χ_b	298	−0.478	[38]
$MgCO_3 \cdot 3H_2O$	293	−0.525	[1]	S (s/c), χ_c	298	−0.474	[38]
$MgCl_2$	298	−0.498	[1]	S_2C	298	−0.554 (liq)	[10]
$MgCl_2 \cdot 6H_2O$	291	−0.57	[7]	SCl	298	−0.480	[1]
MgF_2	298	−0.364	[1]	SCl_3	298	−0.357	[1]
MgI_2	298	−0.399	[1]	S_2Cl_3	298	−0.365	[1]
$Mg(NO_3)_2 \cdot 6H_2O$	291	−0.287 (s)	[4]	SF_6	293	−0.301	[1]
MgO_2	290	−0.25	[1]	SH_2	298	−0.748	[10]
$Mg(OH)_2$	288	−0.378	[1]	SI	298	−0.332	[1]
$MgSO_4$	298	−0.415	[1]	SI_2	298	−0.184	[10]
$MgSO_4 \cdot H_2O$	298	−0.441	[1]	S_4N_4	298	−0.553	[10]
$MgSO_4 \cdot 5H_2O$	298	−0.518	[1]	SO_2	298	−0.284	[1]
$MgSO_4 \cdot 7H_2O$	298	−0.551	[1]	$SOCl_2$	298	−0.878	[10]
$Mg_3(PO_4)_3 \cdot 4H_2O$	298	−0.499	[1]	SO_2Cl_2	283	−0.402	[10]
Al*10	298	0.61	[1]	SOC	298	−0.539	[10]
Al*10	$T>T_{melt}$	0.44 (liq)	[1]	Cl	213	−0.57 (liq)	[20]

*9 See Figure 26.6.

*10 See Figure 26.7.

*11 χ depends on the substance preparation procedure [10].

*12 The dependence of χ on a concentration of electrons and holes is discussed in [98].

*13 χ depends on the preparation technique and thermal treatment.

Table 26.1 Magnetic susceptibility of dia- and paramagnetics *(continued)*

Substance	T, K	χ, 10^{-9} m^3/kg	Ref.	Substance	T, K	χ, 10^{-9} m^3/kg	Ref.
Cl	298	−0.57	[1]	Ca*15 (p/c)	291	1.1	[12]
ClF_3	298	−0.287	[1]	Ca*16 (s/c)	293	1.05	[33]
ClH	195	−0.647	[10]	$CaBr_2$	298	−0.369	[1]
ClH	273	−0.620 (liq)	[10]	$CaBr_2{\cdot}3H_2O$	298	−0.453	[1]
Ar	90	−0.49	[2]	$Ca(BrO_3)_2$	298	−0.287	[1]
Ar	293	−0.49	[1]	$Ca(C_2H_3O_2)_2$	283	−0.446	[1]
K*14	293	0.532	[1]	$CaCO_3$	298	−0.382	[1]
K*14	303	0.460	[26]	$CaCl_2$	290	−0.49	[1]
K*14	523	0.467 (liq)	[26]	$CaCl_2{\cdot}6H_2O$	288	−0.462	[4]
KBr	298	−0.413	[1]	CaF_2	293	−0.360	[1]
$KBrO_3$	298	−0.315	[1]	CaI_2	298	−0.371	[1]
$KC_2H_3O_2$	298	−0.458	[1]	$Ca(IO_3)_2$	298	−0.260	[1]
KCN	298	−0.568	[1]	$Ca(NO_3)_2$	298	−0.239 (s)	[10]
KCNO	298	−0.465	[39]	$Ca(NO_3)_2{\cdot}4H_2O$	298	−0.194	[1]
K_2CO_3	298	−0.427	[1]	CaO	289	−0.27	[1]
KCl	298	−0.523	[1]	CaO_2	298	−0.330	[1]
$KClO_3$	298	−0.349	[1]	$Ca(OH)_2$	298	−0.297	[1]
$KClO_4$	298	−0.342	[10]	$CaSO_4$	293	−0.364	[1]
K_2CrO_4	298	−0.020	[1]	$CaSO_4{\cdot}\frac{1}{2}H_2O$	298	−0.384	[10]
$K_2Cr_2O_7$	293	−0.100	[1]	$CaSO_4{\cdot}H_2O$	293	−0.384	[30]
KF	298	−0.406	[1]	$CaSO_4{\cdot}2H_2O$	298	−0.434	[1]
$K_4Fe(CN)_6$	298	−0.353	[1]	Sc*17	90	8.2	[1]
KHF_2	293	−0.428	[40]	Sc*17	292	7.0	[41]
KH_2PO_4	290	−0.435	[4]	Sc*17	293	6.45(0.05)	[42]
$KHSO_4$	298	−0.36	[1]	$Sc_2(C_2O_4)_3$	293	−0.25	[29]
KI	298	−0.384	[1]	$ScH_0.16$	293	6.38	[42]
KIO_3	298	−0.295	[1]	$ScH_0.26$	293	6.20	[42]
KIO_4	295	−0.292	[4]	$ScH_0.36$	293	3.61	[42]
$KMnO_4$	298	−0.127	[1]	Sc_2O_3	293	−0.018	[29]
KNO_3	293	−0.333	[1]	$Sc_2(SO_4)_3$	293	−0.33	[43]
KNO_2	298	−0.274	[1]	Ti*18 (p/c)	90	3.13	[1]
KOH	298.5	−0.390	[1]	Ti*18 (p/c)	293	3.19	[1]
K_2S	298	−0.54	[1]	Ti*18 (p/c)	293	3.36	[44]
K_2S_2	298	−0.50	[1]	Ti (s/c), $\chi_{\parallel}$	298	3.35	[45]
K_2S_3	298	−0.46	[1]	Ti (s/c), $\chi_{\parallel}$	573	3.65	[45]
K_2S_4	298	−0.43	[1]	Ti (s/c), $\chi_{\perp}$	298	3.07	[45]
K_2S_5	298	−0.41	[1]	Ti (s/c), $\chi_{\perp}$	573	3.38	[45]
KSCN	298	−0.49	[1]	TiC*19	293	0.6	[46]
K_2SO_3	298	−0.404	[1]	$TiCl_4$	293	−0.213	[47]
K_2SO_4	298	−0.385	[1]	TiI_2	288	5.93	[1]
$K_2S_2O_3$	299	−0.393	[4]	TiN*19	293	1.00	[48]
$K_2S_2O_5$	295	−0.389	[4]	TiO*20	79	1.38	[49]
$K_2S_4O_6$	293	−0.412	[30]	TiO_2	293	0.08	[50]
$K_2S_2O_7$	295	−0.398	[4]	Ti_3O_5	293	8.1	[51]
$K_2S_2O_8$	293	−0.371	[30]	TiO_3	248	1.38	[1]
K_2SiO_3	299	−0.383 (s)	[4]	TiP	297	0.25	[52]

*14 See Figures 26.4, 26.5; when temperature decreases from 298 to 78 K, χ increases by 0.2% [26].

*15 See Figure 26.6.

*16 χ is almost independent of T [48].

*17 See Figure 26.8.

*18 See Figure 26.9.

*19 χ is almost independent of T [48].

*20 χ does not depend on T in the range from 79 to 373 K [49].

Table 26.1 Magnetic susceptibility of dia- and paramagnetics *(continued)*

Substance	T, K	χ, 10^{-9} m^3/kg	Ref.	Substance	T, K	χ, 10^{-9} m^3/kg	Ref.
TiS	298	5.40	[1]	CoSi	293	−0.3	[76]
Ti_2S_3	293	0.91	[51]	CoSi	293	−0.44	[80]
TiS_2	296	0.408	[53]	$CoSi_2$	293	0.4	[76]
TiS_3	90–573	−0.09	[54]	NiAs	298	0.322	[1]
TiSi	293	0.20	[55]	β-$NiAs_2$	90–373	0.2309	[73]
$TiSi_2$	293	1.2	[55]	Ni_3P	4.2	8.8	[68]
Ti_5Si_3	293	2.54	[56]	Ni_3S_2	293	4.3	[81]
V*21	293	5.30	[57]	Ni_2Si	293	0.3	[79]
VCl_4NO	298	5.0	[58]	NiSi	293	−0.3	[76]
V_2Cl_7NO	298.5	10.504	[59]	NiSi	293	−0.08	[71]
$V_3Cl_8(NO)_5$	298.5	10.998	[59]	$N_{1.03}Si_{1.93}$	293	0.19	[82]
V_2O_5*22	293	1.1	[60]	Cu*24	296	−0.086	[1]
$VOCl_3$	293	−0.069	[61]	Cu*24	$T>T_{melt}$	−0.097 (liq)	[1]
VP	297	0.032	[52]	CuBr	298	−0.345	[1]
V_3Si	293	4.4	[62]	CuCN	298	−0.267	[1]
V_5Si_3	293	3.0	[63]	CuCl	298	−0.400	[1]
VSi_2	293	1.50	[56]	CuI	298	−0.330	[1]
VSi_2	293	1.32	[64]	Cu_2O	298	−0.216	[83]
$Cr(CO)_6$	298	0.050	[1]	Cu_3P	298	−0.149	[1]
$Cr(C_6H_6)_2$	298	−0.724	[1]	Cu_3P_2	291	−0.23	[7]
CrO_3	293	0.40	[65]	CuP_2	298	−0.28	[1]
CrO_2Cl	220.4	−0.094	[66]	CuS	293	0.05	[84]
CrO_2Cl	249.3	−0.040	[66]	Cu_2S	77–1073	0.30	[84]
CrO_2Cl	296.2	0.017	[66]	CuS_2	300	0.15	[85]
CrP	298	3.80	[67]	CuSCN	298	−0.398	[1]
Cr_3P	4.2	4.0	[68]	CuSe	293	−0.14	[86]
Cr_3Si	293	2.1	[70]	Cu_2Se	77–1073	0.25	[84]
Cr_3Si	293	2.3	[69]	$CuSe_2$	298	0.42	[87]
Cr_5Si_3	293	3.0	[69]	$CuTe_2$	298	−0.40	[87]
Cr_5Si_3	293	3.06	[63]	Zn (p/c)	298	−0.175	[1]
CrSi	293	5.1	[69]	Zn (p/c)	$T>T_{melt}$	−0.12 (liq)	[1]
$CrSi_2$	293	(−0.38)-(−0.94)	[71]	Zn*25 (s/c), $\chi_{\parallel}$	293	−0.190	[88]
Mn_nSi_{2n-m}*23	298	−0.4	[72]	Zn*25 (s/c), $\chi_{\perp}$	293	−0.145	[88]
$FeAs_2$	90–670	2.1834	[73]	Zn_3As_2	300	−0.434	[89]
FeB_2 (powder)	100–300	0.74	[74]	$ZnBr_2$	297	−0.40	[7]
FeB_2 (s/c)	115–300	−0.138	[74]	$Zn(C_2H_3O_2)_2{\cdot}2H_2O$	298	−0.46	[1]
$Fe(CN)_6H_4$	298	−0.329	[1]	$Zn(CN)_2$	298	−0.392	[1]
$Fe(CO)_4$	298	−0.019	[1]	$ZnCO_3$	298	−0.271	[1]
$Fe(CO)_5$	298	−0.480	[1]	$ZnCl_2$	296	−0.477	[1]
FeS_2	295	0.084	[75]	ZnF_2	299.6	−0.370	[1]
FeSi	293	7.9	[76]	ZnI_2	298	−0.307	[1]
CoP	298	0.6	[67]	$Zn(NO_3)_2$	298	−0.333	[1]
Co_2P	298	3.54	[67]	ZnO	298	−0.565	[1]
CoPS	298	0.17	[77]	$Zn(OH)_2$	298	−0.67	[1]
CoPSe	298	0.089	[77]	ZnP_2	300	−0.465	[89]
Co_3S_4	293	3.4	[78]	ZnS	77	−0.36	[10]
Co_2Si	293	6.4	[79]	ZnS	298	−0.26	[1]

*21 See Figure 26.12; when $T<$ 770 K, λ does not depend on T [10].
*22 χ does not depend on T [60].
*23 χ does not depend on T [72].
*24 See Figure 26.10.
*25 See Figure 26.13.

Table 26.1 Magnetic susceptibility of dia- and paramagnetics *(continued)*

Substance	T, K	χ, 10^{-9} m^3/kg	Ref.	Substance	T, K	χ, 10^{-9} m^3/kg	Ref.
Zn_3Sb_2	300	−0.466	[89]	As_2O_3	298	−0.208	[10]
$ZnSO_4$	298	−0.34	[10]	AsOCl	298	−0.240	[10]
$ZnSO_4 \cdot H_2O$	298	−0.351	[1]	As_2S_2	298	−0.327	[10]
$ZnSO_4 \cdot 7H_2O$	298	−0.497	[1]	As_2S_3	291	−0.03	[7]
$Zn_3(PO_4)_2$	298	−0.365	[1]	Se (hexagonal)*29	298	−0.32	[1]
Ga*26 (p/c)	80	−0.35 (ss)	[1]	Se (hexagonal)*29	1.6–77	0.272	[100]
Ga*26 (p/c)	290	−0.31 (ss)	[1]	Se (hexagonal)*29	293	0.188–0.271	[37]
Ga*26 (p/c)	303	0.036 (liq)	[1]	Se*29	900	−0.304 (liq)	[1]
Ga (s/c), χ_a	298	−0.12	[90]	Se (amorphous)	300	−0.279	[101]
Ga (s/c), χ_b	298	−0.42	[90]	Se_2Br_2	298	−0.356	[1]
Ga (s/c), χ_c	298	−0.23	[90]	Se_2Cl_2	298	−0.414	[1]
GaAs	300	−0.230	[17]	SeF_6	298	−0.264	[1]
$GaCl_2$	293	−0.45	[1]	SeO_2	298	−0.245	[1]
GaI_2	298	−0.46	[1]	SeO_3H_2	298	−0.352	[10]
GaN	293	−0.332	[6]	SeO_4H_2	298	−0.353	[10]
Ga_2O	298	−0.219	[1]	$SeOCl_2$	298	−0.293	[10]
GaP	300	−0.298	[17]	Br_2	103	−0.40 (ss)	[12]
Ga_2S	298	−0.210	[1]	Br_2	265	−0.40 (ss)	[12]
GaS	295	−1.715	[91]	Br_2	266	−0.40 (liq)	[12]
Ga_2S_3	298	−0.34	[1]	Br_2	291	−0.40 (liq)	[12]
GaSb	293	−0.201	[92]	Br_2	$T>T_{melt}$	−0.353 (liq)	[1]
GaSe	293	−0.20	[93]	Br_2	$T>T_{ev}$	−0.46 (g)	[1]
Ge*27	4.2	−0.1102	[94]	BrF_3	298	−0.246	[1]
Ge*27	77.3	−0.1095	[94]	BrF_5	298	−0.258	[1]
Ge*27	293	−0.1059	[1]	Kr	291	−0.38 (liq)	[102]
$GeCl_4$	298	−0.34	[1]	Kr	$T>T_{ev}$	−0.344 (g)	[1]
GeF_4	298	−0.34	[10]	Rb*30	298	0.21	[9]
GeI_4	298	−0.30	[1]	Rb*30	303	0.198	[1]
GeO	298	−0.325	[1]	RbBr	298	−0.341	[1]
GeO_2	298	−0.328	[1]	Rb_2CO_3	298	−0.326	[1]
GeS	298	−0.91	[1]	RbCl	298	−0.380	[1]
GeS_2	298	−0.390	[1]	RbF	298	−0.305	[1]
As (p/c)	84	0.46	[99]	RbI	298	−0.340	[1]
As (p/c)	181	0.092	[99]	$RbNO_3$	293	−0.281	[1]
As (p/c)	298	0.016	[99]	Rb_2S	298	−0.394	[1]
As (p/c)	461	−0.087	[99]	Rb_2S_2	298	−0.383	[1]
β-As (amorphous)	90	−0.30	[1]	Rb_2S_3	298	−0.374	[10]
β-As (amorphous)	293	−0.316	[1]	Rb_2S_4	298	−0.371	[10]
γ-As (amorphous)	293	−0.307	[1]	Rb_2S_5	298	−0.368	[10]
As (s/c) $\chi_{\parallel}$*28	82	1.005	[99]	Rb_2S_6	298	−0.363	[10]
As (s/c) $\chi_{\parallel}$*28	292	0.578	[99]	Rb_2SO_4	293	−0.331	[1]
As (s/c) $\chi_{\parallel}$*28	531	0.273	[99]	Sr	295	1.05	[1]
As (s/c) $\chi_{\perp}$	82	−0.238	[99]	Sr	533	0.73	[103]
As (s/c) $\chi_{\perp}$	292	−0.279	[99]	$SrBr_2$	298	−0.35	[1]
As (s/c) $\chi_{\perp}$	510	−0.302	[99]	$SrBr_2 \cdot 6H_2O$	298	−0.45	[1]
$AsBr_3$	298	−0.337	[1]	$Sr(BrO_3)_2$	298	−0.272	[1]
$AsCl_3$	298	−0.441	[10]	$Sr(C_2H_3O_2)_2$	284	−0.385	[1]
AsI_3	298	−0.312	[1]	$SrCO_3$	298	−0.32	[1]

*26 See Figure 26.11.

*27 See Figures 26.14–26.16; to find χ-dependences from the concentrations of electrons and holes, see [95–98].

*28 See Figure 26.17.

*29 See Figure 26.18.

*30 See Figure 26.4; when temperature decreases from 298 to 78 K, χ rises by 2.5% [9].

Table 26.1 Magnetic susceptibility of dia- and paramagnetics *(continued)*

Substance	T, K	χ, 10^{-9} m^3/kg	Ref.
$SrCl_2$	293	−0.40	[1]
$SrCl_2{\cdot}6H_2O$	298	−0.544	[1]
$Sr(ClO_3)_2$	298	−0.287	[1]
$SrCrO_4$	298	−0.025	[1]
SrF_2	293	−0.296	[1]
SrI_2	298	−0.328	[1]
$Sr(IO_3)_2$	298	−0.247	[1]
$Sr(NO_3)_2$	298	−0.270	[1]
$Sr(NO_3)_2{\cdot}4H_2O$	298	−0.374	[1]
SrO	298	−0.34	[1]
SrO_2	298	−0.27	[1]
$Sr(OH)_2$	298	−0.33	[1]
$Sr(OH)_2{\cdot}8H_2O$	298	−0.512	[1]
$SrSO_4$	298	−0.315	[1]
Y*31	90	2.43	[1]
Y*31	292	2.15	[1]
Y_2O_3	293	0.197	[1]
Y_2S_3	298	0.365	[1]
Zr*32	90	1.305	[1]
Zr*32	293	1.33	[1]
Zr*32	293	1.38	[57]
$ZrBr_4$	301	−0.261	[104]
ZrC	293	−0.22	[48]
$ZrCl_4$	303	−0.302	[104]
ZrF_4	301	−0.194	[104]
ZrI_4	302	−0.191	[104]
ZrN	293	0.57	[48]
Zr_3N_4	293	−0.03	[105]
$Zr(NO_3)_4{\cdot}5H_2O$	298	−0.179	[1]
ZrO_2	288	−0.112	[1]
ZrP	297	−0.21	[52]
ZrS_3	90–573	−0.19	[54]
$ZrSiO_4$	292	−0.215	[4]
Nb*33	14	2.34	[11]
Nb*33	289	2.28	[11]
Nb*33	293	2.34	[57]
$NbH_{0.86}$	81	0.62	[106]
$NbH_{0.86}$	291	0.70	[106]
NbF_4	297	0.580	[107]
NbO_2	295	0.05	[50]
Nb_2O_5	295	0.09	[108]
NbP	297	−0.52	[52]
Mo*34	20.4	1.56	[1]
Mo*34	63.8	1.13	[1]
Mo*34	298	0.93	[1]

Substance	T, K	χ, 10^{-9} m^3/kg	Ref.
Mo*34	293	0.82	[44]
Mo_3Br_6	290.5	−0.06	[1]
$Mo(CO)_6$	293	−0.28	[109]
MoF_6	293	−0.12	[1]
MoO_2	289	0.32	[1]
MoO_3	293	0.02	[1]
Mo_2O_3	298	−0.175	[1]
Mo_2O_5	293	0.23	[110]
Mo_3O_6	298	0.101	[1]
MoS_2	289	−0.48	[10]
MoS_3	289	−0.33	[1]
$MoTe_2$	293	−0.342	[111]
Tc	78	2.9	[1]
Tc	298	2.7	[1]
Tc	402	2.5	[1]
$TcO_2{\cdot}2H_2O$	78	3.4	[112]
$TcO_2{\cdot}2H_2O$	300	1.2	[1]
$TcO_2{\cdot}2H_2O$	398	1.34	[112]
Tc_2O_7	298	−0.129	[1]
$TcO_4(NH_4)$	78	0.077	[10]
$TcO_4(NH_4)$	298	0.050	[10]
Ru*35	18	0.43	[12]
Ru*35	293	0.34	[44]
Ru*35	723	0.496	[1]
$RuAs_2$	90–770	−0.1681	[73]
RuO_2	297	1.26	[113]
$RuOCl_6$	299	0.74	[114]
RuP_2	90–765	−0.2849	[73]
Rh*36	20	0.9269	[115]
Rh*36	293	1.03	[44]
Rh*36	723	1.195	[1]
$RhCl_3$	298	−0.036	[1]
RhF_4	293	2.79	[1]
Rh_2O_3	298	0.408	[1]
$Rh_2(SO_4)_3{\cdot}6H_2O$	298	−0.173	[1]
$Rh_2(SO_4)_3{\cdot}14H_2O$	298	−0.200	[1]
Rd*37	20	7.32	[115]
Rd*37	288	5.333	[1]
Rd*37	1503	1.7	[14]
$RdCl_2$	291.5	−0.214	[1]
$RdCl_2$	300	0.255	[116]
Ag*38	14	−0.19	[18]
Ag*38	293	−0.181	[1]
Ag*38	975	−0.1786	[117]
Ag*38	$T > T_{melt}$	−0.22 (liq)	[10]

*31 See Figure 26.8.
*32 See Figures 26.12, 26.19.
*33 See Figure 26.12.
*34 See Figure 26.20.
*35 See Figure 26.21.
*36 See Figure 26.22.
*37 See Figure 26.23.
*38 See Figure 26.24.

Table 26.1 Magnetic susceptibility of dia- and paramagnetics *(continued)*

Substance	T, K	χ, 10^{-9} m^3/kg	Ref.	Substance	T, K	χ, 10^{-9} m^3/kg	Ref.
AgBr	283	−0.318	[1]	$Cd(NO_3)_2$	298	−0.233	[1]
$AgC_2H_3O_2$	298	−0.362	[1]	$Cd(NO_3)_2 \cdot 4H_2O$	298	−0.454	[1]
AgCN	299	−0.322	[1]	CdO	300	−0.380	[119]
Ag_2CO_3	298	−0.292	[1]	CdO	300–900	−0.278	[120]
AgCl	290	−0.34	[1]	$Cd(OH)_2$	298	−0.28	[1]
$AgCrO_4$	298	−0.179	[1]	Cd_3P_2	300	−0.338	[89]
Ag_2F	297–350	−0.274	[10]	$Cd_3(PO_4)_2$	298	−0.302	[1]
AgF	298	−0.288	[1]	CdS	298	−0.436	[121]
AgI	298	−0.34	[1]	$CdSO_4$	298	−0.284	[1]
$AgNO_2$	298	−0.273	[1]	$3CdSO_4 \cdot 8H_2O$	286	−0.281	[4]
$AgNO_3$	298	−0.269	[1]	CdSb	293	−0.30	[122]
Ag_2O	298	−0.578	[1]	Cd_3Sb_2	300	−0.434	[89]
AgO	287	−0.158	[1]	CdSe (s/c), $\chi_{\parallel}$	298	−0.354	[123]
AgP_2	298	−0.318	[10]	CdSe (s/c), $\chi_{\perp}$	298	−0.304	[123]
AgP_3	298	−0.329	[10]	CdP_2*40		−0.415	[124]
Ag_3PO_4	298	−0.287	[1]	Cd_3P_2	300	−0.16	[125]
Ag_2S (rhombic)	77–1073	0.32	[84]	In*41 (p/c)	293	−0.077	[126]
AgSCN	298	−0.3726	[1]	In*41 (p/c)	298	−0.56	[1]
$AgSO_4$	299	−0.292	[4]	In (s/c), $\chi_{\parallel}$	293	−0.158	[127]
Ag_2SO_4	298	−0.298	[1]	In (s/c), $\chi_{\perp}$	293	−0.078	[127]
Ag_2Se	77–1073	0.35	[84]	InAs	293	−0.2714	[128]
Cd*39 (p/c)	14	−0.310	[10]	$InBr_3$	298	−0.30	[1]
Cd*39 (p/c)	293	−0.176 (ss)	[1]	InCl	298	−0.20	[1]
Cd*39 (p/c)	$T>T_{melt}$	−0.160 (liq)	[1]	$InCl_2$	298	−0.30	[1]
Cd (s/c), $\chi_{\parallel}$	14	0.679	[118]	$InCl_3$	298	−0.57	[1]
Cd (s/c), $\chi_{\parallel}$	283	−0.243	[118]	InF_2	298	−0.40	[1]
Cd (s/c), $\chi_{\perp}$	14	−0.130	[118]	InN*42	293	−0.320	[129]
Cd (s/c), $\chi_{\perp}$	293	−0.142	[118]	In_2O	298	−0.191	[1]
Cd_3As_2	300	−0.431	[89]	In_2O_3	298	−0.076	[1]
$CdBr_2$	298	−0.321	[1]	InP*43	293	−0.313	[130]
$CdBr_2 \cdot 4H_2O$	288	−0.407	[1]	$InP_{0.2}As_{0.8}$	293	−0.282	[130]
$Cd(C_2H_3O_2)_2$	298	−0.379	[1]	In_2S	298	−0.19	[1]
$Cd(C_2H_3O_2)_2 \cdot 3H_2O$	291	−0.365	[4]	InS	298	−0.19	[1]
$Cd(CN)_2$	298	−0.328	[1]	In_2S_3	298	−0.30	[1]
$CdCO_2$	288	−0.305	[4]	InS_3	293	−0.19	[7]
$CdCO_3$	298	−0.271	[1]	InSb*44	296	−0.281	[130]
$CdCl_2$	298	−0.375	[1]	InSb*45	293	−0.284	[131]
$CdCl_2 \cdot H_2O$	285	−0.368	[4]	InTe (cubic)	203	−0.14	[132]
$CdCl_2 \cdot 2H_2O$	298	−0.451	[1]	InTe (tetragonal)	293	−0.22	[132]
$CdCrO_4$	298	−0.074	[1]	α-Sn (grey)	273*46	−0.310	[130]
CdF_2	298	−0.270	[1]	α-Sn (grey)	273*46	−0.310	[130]
CdI_2	298	−0.320	[1]	α-Sn (grey)	273*47	−0.265	[130]
$Cd(IO_3)_2$	298	−0.235	[1]	α-Sn (grey)	100	−0.267	[1]

*39 See Figure 26.25.
*40 Lattice susceptibility.
*41 See Figure 26.26.
*42 Lattice part of χ.
*43 $n = 1.9 \cdot 10^{16}$ cm^{-3}.
*44 $n = 6.2 \cdot 10^{15}$ cm^{-3}.
*45 $n = 4 \cdot 10^{14}$ cm^{-3}.
*46 $n = 10^{17}$cm^{-3}.
*47 Lattice part of χ.

Table 26.1 Magnetic susceptibility of dia- and paramagnetics *(continued)*

Substance	T, K	χ, 10^{-9} m^3/kg	Ref.	Substance	T, K	χ, 10^{-9} m^3/kg	Ref.
β-Sn (white)*48 (p/c)	4.2	0.023	[94]	$TeBr_2$	298	−0.369	[1]
β-Sn (white)*48 (p/c)	293	0.026	[1, 94]	$TeCl_2$	298	−0.474	[1]
β-Sn (s/c), $\chi_\parallel$	293	0.026	[127]	TeF_6	298	−0.273	[1]
β-Sn (s/c), $\chi_\perp$	293	0.029	[127]	TeO_2	291	−0.14	[7]
Sn (s/c)	T>T_{melt}	−0.038 (liq)	[1]	TeO_2H_2	298	−0.216	[10]
$SnBr_4$	298	−0.340	[1]	$Te(CH_3)_2Cl_2$	298	−0.402	[10]
$Sn(CH_3)_4$	293	−0.738	[10]	$Te(CH_3)_2I_2$	298	−0.353	[10]
$Sn(C_2H_5)_4$	293	−0.144	[10]	I_2	14	−0.313 (ss)	[136]
$Sn(C_4H_9)_4$	293	−0.669	[10]	I_2	298	−0.35 (ss)	[1]
$SnCl_2$	298	−0.36	[1]	I_2	368	−0.38 (ss)	[14]
$SnCl_4$	298	−0.441 (liq)	[1]	I_2	386	−0.39 (liq)	[14]
$SnCl_2 \cdot 2H_2O$	298	−0.405	[1]	I_2	433	−0.33 (liq)	[14]
SnI_4	298	0.327	[10]	I_1 (atomic)	1303	6.85 (g)	[1]
SnO	298	−0.14	[1]	I_1 (atomic)	1440	8.82 (g)	[1]
SnO_2	298	−0.27	[1]	I_2 (s/c), χ_a	298	−0.354	[137]
Sn_2O_3	289	−0.33	[7]	I_2 (s/c), χ_b	298	−0.331	[137]
$Sn(OH)_4$	293	−0.321	[1]	I_2 (s/c), χ_c	298	−0.336	[137]
$SnO(OH)_2$	293	−0.278	[30]	ICl	285	−0.336 (liq)	[1]
$SnSO_4$	291	−0.29	[30]	ICl_2	288	−0.387 (s)	[4]
SnSe	293	−0.10	[93]	ICl_3	298	−0.387	[1]
Sb*49 (p/c)	293	−0.81 (ss)	[1]	IF_5	298	−0.262	[1]
Sb*49 (p/c)	T>T_{melt}	−0.02 (liq)	[1]	IH	195	−0.369 (ss)	[10]
Sb (s/c), $\chi_\parallel$	90	−1.73	[133]	IH	281	−0.373 (ss)	[10]
Sb (s/c), $\chi_\parallel$	293	−1.42	[133]	Xe	298	−0.33	[1]
Sb (s/c), $\chi_\perp$	90	−0.50	[133]	Cs*53	298	0.23	[9]
Sb (s/c), $\chi_\perp$	293	−0.50	[133]	Cs*53	T>T_{melt}	0.20 (liq)	[1]
$SbBr_3$	298	−0.318	[1]	CsBr	298	−0.316	[1]
$Sb(C_6H_5)_3$	298	−0.515	[10]	$CsBrO_3$	298	−0.288	[1]
SbCd	293	−0.213	[134]	Cs_2CO_3	298	−0.318	[1]
$SbCl_3$	298	−0.380	[1]	CsCl	298	−0.337 (ss)	[1]
$SbCl_5$	298	−0.401 (liq)	[1]	$CsClO_3$	298	−0.30	[1]
SbF_3	298	−0.256	[1]	CsF	298	−0.293	[1]
SbI_3	298	−0.293	[1]	CsI	298	−0.318	[1]
Sb_2O_3	298	−0.238	[1]	$CsIO_3$	298	−0.270	[1]
SbOCl	298	−0.214	[10]	$CsNO_3$	298	−0.279	[10]
Sb_2S_3	298	−0.25	[1]	Cs_2S	298	−0.349	[1]
Sb_2Se_3*50	130–500	−0.383	[135]	Cs_2S_2	298	−0.355	[10]
Sb_2Te_3	130–150	−0.398	[135]	Cs_2S_3	298	−0.354	[10]
SbZn	293	−0.285	[134]	Cs_2S_4	298	−0.353	[10]
$SbZn_3$	293	−0.261	[134]	Cs_2S_5	298	−0.352	[10]
$SbZn_3$	T>T_{melt}	−0.107 (liq)	[134]	Cs_2S_6	298	−0.349	[10]
Te*51	14.2	−0.293	[136]	Cs_2SO_4	298	−0.321	[1]
Te*51	293	−0.283(0.005)	[37]	Cs_2Se	298	−0.331	[10]
Te*51	T>T_{melt}	−0.05 (liq)	[1]	Ba	293	0.150	[1]
Te*52 (s/c), $\chi_\parallel$	293	−0.329	[10]	Ba	573	0.316	[138]
Te*52, $\chi_\perp$	293	−0.296	[10]	Ba	673	0.415	[138]

*48 See Figure 26.27.

*49 See Figure 26.28.

*50 p-conductivity.

*51 See Figure 26.30.

*52 When temperature rises, $\chi_\parallel$ decreases, but $\chi_\perp$ is almost invariable; when T=493 K, $\chi_\parallel=\chi_\perp$.

*53 When temperature falls from 298 to 78 K, χ increases by 12%.

Table 26.1 Magnetic susceptibility of dia- and paramagnetics *(continued)*

Substance	T, K	χ, 10^{-9} m^3/kg	Ref.	Substance	T, K	χ, 10^{-9} m^3/kg	Ref.
$BaBr_2$	298	−0.31	[1]	Ta_2O_5*57	293	−0.095	[145]
$BaBr_2{\cdot}2H_2O$	298	−0.357	[1]	TaON	293	−0.064	[145]
$Ba(BrO_3)_2$	298	−0.269	[1]	TaP	297	−0.62	[52]
$Ba(C_2H_3O_2)_2H_2O$	298	−0.366	[1]	$TaBr_4$	293	−0.15	[144]
$BaCO_3$	298	−0.298	[1]	W*58	293	0.29	[44]
$BaCl_2$	298	−0.35	[1]	WBr_4	297	−0.149	[147]
$BaCl_2{\cdot}2H_2O$	298	−0.409	[1]	WBr_6	293	0.03	[148]
$Ba(ClO_3)_2$	298	−0.288	[1]	WC	293	0.05	[1]
BaF_2	298	−0.291	[1]	WCl_2	293	−0.098	[1]
BaI_2	298	−0.317	[1]	WCl_4	303	−0.178	[147]
$BaI_2{\cdot}2H_2O$	298	−0.38	[1]	WCl_6	298	−0.179	[1]
$Ba(IO_3)_2$	298	−0.251	[1]	WF_6	298	−0.134 (liq)	[1]
$Ba(NO_3)_2$	298	−0.254	[1]	WO_2	298	0.264	[1]
BaO	298	−0.19	[1]	W_2O_5	298	0.20	[148]
BaO_2	298	−0.240	[1]	WO_3	298	−0.068	[1]
$Ba(OH)_2$	298	−0.31	[1]	Re*59	293	0.363	[1]
$Ba(OH)_2{\cdot}8H_2O$	298	−0.497	[1]	ReO_2 (rhombic)	77–300	0.25	[149]
BaS	291	−0.32	[7]	ReO_2 (monoclinic)	300	0.44	[149]
$BaSO_4$	298	−0.306	[1]	Re_2O_2	293	0.2	[150]
$BaS_2O_6{\cdot}2H_2O$	293	−0.359	[30]	ReO_3	298	0.088	[151]
La*54	293	0.73	[139]	Re_2O_7	298	−0.033	[1]
La*54	298	0.85	[1]	$ReO_2{\cdot}2H_2O$	295	−0.291	[1]
LaB_6	298	−0.294	[10]	ReS	293	0.2	[152]
$LaCl_3$	298	5.6	[7]	ReS_2	290.5	−0.50	[153]
$La(NO_3)_3$	298	−0.072	[10]	Re_2S_7	290–373	0.19	[153]
La_2O_3	298	−0.24	[1]	Os	298	0.052	[1]
La_2S_3	292	−0.099	[1]	Os	698	0.070	[10]
La_2S_4	293	−0.246	[1]	Os	$T>T_{melt}$	0.14 (liq)	[10]
LaS_3	293	−0.03	[140]	$OsAs_2$	90–800	−0.3279	[73]
LaS_4	293	−0.26	[140]	$OsCl_2$	298	0.158	[10]
$La_2(SO_4)_3$	293	−0.30	[43]	$OsCl_3$	295.5	3.04	[154]
$La_2(SO_4)_3{\cdot}9H_2O$	293	−23	[1]	$OsCl_4$ (rhombic)	300	3.25	[155]
Hf*55	4.2	0.46	[141]	$OsCl_4$ (cubic)	300	2.65	[155]
Hf*55	77	0.40	[141]	OsI_3	295	0.557	[156]
Hf*55	293	0.39	[142]	OsO_2	77	0.536	[157]
HfO_2	293	−0.109	[65]	OsO_2	300	0.528	[157]
HfP	297	−0.10	[52]	OsO_4	77	−0.045	[157]
HfS_2	293	0.005	[53]	OsO_4	300	−0.048	[157]
HfS_3	90–573	−0.21	[143]	OsP_2	90–760	−0.3845	[88]
Ta*56 (p/c)	293	0.85	[1]	Ir*60	293	0.14	[26]
Ta*56 (p/c)	2143	0.685	[1]	Ir*60	698	0.167	[1]
$TaBr_4$	293	−0.15	[144]	Ir*60	1423	0.31	[14]
$TaCl_5$	304	0.391	[1]	$IrBr_3$	293	−0.2	[158]
Ta_3N_5	293	−0.008	[145]	$IrCl_3$ (rhombic)	298	−0.183	[159]
TaO_2	293	0.32	[146]	$IrCl_3$ (hexagonal)	298	0	[159]

*54 See Figure 26.8.

*55 See Figure 26.9, 26.31.

*56 See Figure 26.29.

*57 At $H \to \infty$.

*58 See Figure 26.32.

*59 See Figure 26.32.

*60 See Figure 26.21.

Table 26.1 Magnetic susceptibility of dia- and paramagnetics *(continued)*

Substance	T, K	χ, 10^{-9} m^3/kg	Ref.	Substance	T, K	χ, 10^{-9} m^3/kg	Ref.
IrO_2 (s/c), χ_a	300	0.92 (0.02)	[160]	$Hg(CN)_2$	298	−0.265	[1]
IrO_2 (s/c), χ_c	300	0.83 (0.02)	[160]	HgCl	298	−0.22 (ss)	[1]
Pt*61	20	1.094	[115]	HgCl	$T>T_{melt}$	−0.248 (liq)	[10]
Pt*61	293	0.97	[44]	$HgCl_2$	298	−0.302	[1]
Pt*61	623	0.795	[161]	Hg_2Cl_2	298	−0.253	[4]
Pt*61	1493	0.30	[14]	$HgCrO_4$	298	−0.039	[1]
PtCl	298	−0.51	[161]	$HgCrO_4$	298	−0.122	[1]
$PtCl_2$	295	0.188	[162]	HgF	298	−0.241	[1]
$PtCl_3$	298	−0.221	[1]	HgF_2	302	−0.260	[1]
$PtCl_4$	298	−0.276	[1]	HgI	298	−0.253	[1]
$PtCl_2 \cdot CO$	293	−0.289	[109]	HgI_2	298	−0.283	[1]
$PtCl_2 \cdot 2CO$	293	−0.37	[109]	$HgIO_3$	298	−0.245	[1]
Pt_2O_3	298	−0.086	[1]	$HgNO_3$	298	−0.213	[1]
PtI_2	300	−0.225	[163]	$Hg(NO_3)_2$	298	−0.228	[1]
PtI_3	300	−0.234	[163]	$Hg(NO_3)_2 \cdot 2H_2O$	298	−0.225	[4]
PtI_4	300	−0.285	[163]	HgO	298	−0.20	[1]
PtS	195–723	−0.14	[164]	Hg_2O	298	−0.183	[1]
PtS_2	90–723	−0.28	[164]	$Hg_2(OH)_2$	298	−0.23	[1]
PtSe	195–723	−0.11	[164]	HgS	298	−0.238	[1]
$PtSe_2$	90–723	−0.13	[164]	$Hg(SCN)_2$	298	−0.305	[1]
PtTe	195–723	−0.12	[164]	$Hg_2(OH)_2$	298	−0.23	[1]
$PtTe_2$	90–723	−0.05	[164]	HgS	298	−0.238	[1]
Au*62	14	−0.132	[165]	$Hg(SCN)_2$	298	−0.305	[1]
Au*62	296	−0.142	[1]	$HgSO_4$	298	−0.233	[1]
Au*62	975	−0.1417	[117]	Hg_2SO_4	298	−0.247	[1]
Au*62	$T<T_{melt}$	−0.17	[1]	α-Tl (p/c) (hexagonal)	14.2	−0.258	[136]
AuBr	298	−0.22	[1]	α-Tl (p/c) (hexagonal)	298	−0.249	[1]
$AuCl_3$	298	−0.37	[1]	β-Tl (p/c) (cubic)	>508	−0.158	[1]
AuCl	298	−0.288	[1]	Tl	573	−0.131 (liq)	[1]
AuF_3	298	0.291	[1]	α-Tl*64 (s/c), $\chi_\perp$	293	−0.420	[10]
AuI	298	−0.28	[1]	α-Tl*64 (s/c), $\chi_\parallel$	293	−0.164	[168]
AuP_3	298	−0.369	[1]	TlBr	298	−0.225	[1]
Hg*63 (p/c)	80	−0.118 (ss)	[166]	$TlBrO_3$	298	−0.228	[1]
Hg*63 (p/c)	293	−0.167 (liq)	[1]	TlCN	298	−0.213	[1]
Hg*63 (p/c)	560.5	−0.1637 (liq)	[167]	TlCNS	298	−0.254	[1]
Hg*63 (p/c)	$T>T_{ev}$	−0.39 (g)	[1]	$TlCO_3$	298	−0.217	[1]
Hg (s/c), $\chi_\parallel$	80	−0.112	[166]	TlCl	298	−0.241	[1]
Hg (s/c), $\chi_\perp$	80	−0.121	[166]	$TlClO_3$	298	−0.228	[1]
$HgBr_2$	298	−0.261	[1]	$TlCrO_4$	298	−0.075	[1]
HgBr	298	−0.204	[1]	TlF	298	−0.199	[1]
$HgBrO_3$	298	−0.239	[10]	TlI	298	−0.248	[1]
$Hg(CH_3)_2$	293	−0.199	[36]	$TlIO_3$	298	−0.229	[1]
$Hg(C_2H_5)_2$	293	−0.158	[36]	$TlNO_2$	298	−0.203	[1]
$Hg(C_4H_9)_2$	293	−0.115	[36]	$TlNO_3$	298	−0.212	[1]
$Hg(C_5H_{11})_2$	293	−0.103	[36]	Tl_2O_3	298	−0.166	[1]
$Hg(C_2H_3O_2)_2$	301	−0.317	[1]	Tl_2PO_4	298	−0.304	[1]
$Hg_2(C_2H_3O_2)_2$	300	−0.276	[4]	Tl_2S	298	−0.201	[1]

*61 See Figure 26.33.
*62 See Figure 26.36.
*63 See Figure 26.34.
*64 See Figure 26.6.

Table 26.1 Magnetic susceptibility of dia- and paramagnetics *(continued)*

Substance	T, K	χ, 10^{-9} m^3/kg	Ref.	Substance	T, K	χ, 10^{-9} m^3/kg	Ref.
Tl_2O_4	298	−0.223	[1]	$Bi(NO_3)_3{\cdot}5H_2O$	298	−0.328	[1]
Pb (p/c)	14.2	−0.132	[109]	BiO	298	−0.49	[1]
Pb (p/c)	289	−0.111	[1]	Bi_2O_3	298	−0.18	[1]
Pb (p/c)	330	−0.075 (liq)	[1]	$Bi(OH)_3$	298	−0.253	[1]
Pb*65 (s/c)	293	−0.114	[127]	$BiPO_4$	298	−0.203	[1]
$PbBr_2$	298	−0.247	[1]	Bi_2S_3	298	−0.239	[1]
$Pb(C_2H_3O_2)_2$	298	−0.274	[1]	$Bi_2(SO_4)_3$	298	−0.282	[1]
$Pb(CNS)_2$	298	−0.254	[1]	Bi_2Se_3	130–600	−0.410	[135]
$PbCO_3$	298	−0.229	[1]	BiTe	293	0.57	[13]
$PbCl_2$	298	−0.265	[1]	Bi_2Te_3	130–600	0.402	[135]
$PbCrO_4$	298	−0.056	[1]	Th	90	0.662	[1]
PbF_2	298	−0.237	[1]	Th	298	0.57	[1]
PbI_2	298	−0.274	[1]	Th	130–300	0.410	[174]
$Pb(IO_3)_2$	298	−0.235	[1]	$ThCl_4{\cdot}8H_2O$	305.2	−0.348	[1]
$Pb(NO_3)_2$	298	−0.223	[1]	$Th(NO_3)_4$	298	−0.225	[1]
PbO	298	−0.19	[1]	ThO_2	83	−0.056	[175]
PbO_2	293	−0.09	[169]	ThO_2	298	−0.061	[1]
Pb_3O_4	291	−0.24	[7]	α-U*68	80	1.63	[176]
$Pb_3(PO_4)_2$	298	−0.224	[1]	α-U*68	293	1.72	[176]
PbS (powder)	293	−0.342	[170]	α-U*68	623	1.85	[1]
PbS	293	−0.390	[170]	β-U*69	288–1193	2.02	[176]
$PbSO_4$	298	−0.230	[1]	γ-U*70	288–1193	2.06	[176]
PbSe*66	293	−0.447	[135]	UF_6	90	0.01	[177]
PbTe	293	−0.446 (ss)	[135]	UF_6	293	0.12	[177]
PbTe	293	−0.051 (liq)	[134]	UO_3	293	4.5	[1]
Bi*67 (p/c)	14	−1.55	[171]	UOS	290.5	10.527	[178]
Bi*67 (p/c)	298	−1.34	[1]	U_2S_2	293	4.545	[179]
Bi*67 (p/c)	$T>T_{melt}$	−0.0502 (liq)	[1]	U_3Se_5	293	3.567	[180]
Bi (s/c), $\chi_{\parallel}$	14	−1.20	[171]	α-Np	300	2.34	[181]
Bi (s/c), $\chi_{\parallel}$	85	−1.295	[172]	α-Pu*71	293	2.5	[1]
Bi (s/c), $\chi_{\parallel}$	298	−1.053	[173]	α-Pu*72	78–793	2.35*71	[181]
Bi (s/c), $\chi_{\perp}$	14	−1.77	[171]	β-Pu*73	78–793	2.44*71	[182]
Bi (s/c), $\chi_{\perp}$	85	−2.04	[172]	γ-Pu*74	78–793	2.36*71	[182]
Bi (s/c), $\chi_{\perp}$	85	−1.482	[173]	δ-Pu*75	78–793	2.31	[182]
$BiBr_3$	298	−0.328	[1]	ε-Pu	78–793	2.36	[182]
$BiCl_3$	298	−0.084	[1]	PuF_6	295	0.483	[1]
$Bi_2(CrO_4)_3$	298	0.201	[1]	PuO_2	300	2.645	[1]
BiF_3	303	−0.23	[1]	Am	300	4.0	[1]
BiI_3	298	−0.34	[1]	Am	300	2.82	[181]
$Bi(NO_3)_3$	298	−0.23	[1]				

*65 See Figure 26.27; the anisotropy of χ in (100) plane is absent [130].
*66 Lattice part of χ.
*67 See Figure 26.35.
*68 See Figure 26.37.
*69 See Figure 26.37; α-U transforms to β-U at T=993 K.
*70 See Figure 26.37; β-U transforms to γ-U at T=1040 K.
*71 See Figure 26.38; α-Pu transforms to β-Pu at T=400 K.
*72 The average value of κ is given for the T–interval indicated.
*73 β-Pu transforms to γ-Pu at T=400 K.
*74 γ-Pu transforms to δ-Pu at T=590 K.
*75 δ-Pu transforms to ε-Pu at T=740 K.

Table 26.2 The magnetic susceptibilities of paramagnetics which possess the temperature dependences in accordance with the Curie-Weiss law

Substance	T, K	χ, 10^{-9} m^3/kg	ΔT, K	$p_{\rm eff}$, $\mu_{\rm B}$	θ, K	References
Sc (ss)	292	7.0		1.8	(−950)–(−1000)	[10]
$TiBr_2$	288	3.10	90–300	2.8	−1110	[10]
TiS_2	296	0.408		0.52		[53]
$Ti(SO_4)_3$			14–290	1.41		[10]
VBr_2	293	15.3		4.30	−400	[191]
VBr_3	293	9.95		2.72	−20	[191]
VCl_2	293	19.8	14–300	4.15	−565	[10]
VCl_4	293.8	5.86		1.62	±1	[192]
VF_3	293	2.53		2.55		[193]
VI_2 (red)	295	14.0		3.32	−140	[194]
VI_2 (red)	90	26.6		2.43		[194]
VI_2 (black)	295	13.8		3.27	−140	[194]
VI_2 (black)	90	26.3		2.42		[194]
VI_3	295	8.02		2.92		[194]
VI_3	90	39.0		3.50		[194]
VOCl	293	20.3		2.22		[195]
VTe	293	>0		1.7	0	[10]
CrB_2			100–600	2.07	−1550	[196]
CrH	295	13.63		1.77	−247	[197]
$CrSO_4$				4.80		[198]
$Cr_2(SO_4)_3$	293	30.1	90–900	3.84	−16	[10]
MnF_4			90–295	3.84	−10	[199]
$Fe(OH)_2$	301.4	124.6		5.22		[200]
$Fe_2(SO_4)_3$	283	30.6	65–300	5.8	−65	[10]
$Fe_2(SO_4)_3$			300–900	5.8	−75	[10]
Co_2O_3	293	27.5		2.3		[201]
$CoTe_2$	293	3.11	90–723	2.81	−540	[202]
NiI_2	293	12.4	83–603	3.25	−42	[10]
NiS_2			4–440	3.15	−2100	[203]
NiS_2			440–900	2.70	−1000	[203]
NiTe	293	0.844		0.61		[204]
NiTe			90–600	0.99	−1100	[205]
$NiTe_2$	293	0.445	100–400	0.57		[204]
CuS_2	298	0.29	4.2–100	0.58	−25	[206]
$ZrBr_3$	300	0.18		≈0.4		[207]
$ZrCl_2$	300	0.925		0.6		[207]
$ZrCl_3$	300	0.35		≈0.4		[207]
ZrS_2	296	0.535		0.61		[53]
NbF_3	293	0.95		0.7	−180	[208]
NbO	295	0.29	1.5–150	0.052		[108]
NbO_2	295	0.05		0.32		[209]
Nb_2O_5	295	0.09	1.5–150	0.237	0	[108]
NbS_2	293	0.39		0.98	−1100	[146]
$MoBr_2$	293	1.34		0.35		[210]
$MoBr_3$	293	1.56		1.24		[210]
$MoBr_4$	293	1.62		1.02–1.28		[210]
$MoCl_3$	290	0.21		0.67		[210]
$MoCl_4$	296	9.6		2.54	−34	[211]
α-$MoCl_4$	299.6	0.856		≈0.85		[212]

Table 26.2 The magnetic susceptibilities of paramagnetics *(continued)*

Substance	T, K	χ, 10^{-9} m^3/kg	ΔT, K	p_{eff}, μ_B	θ, K	References
$MoCl_5$	305	3.80		1.52	23	[213]
$MoCl_5$			195–293	≈1.6		[10]
$TcCl_4$	306	20.5		3.14	57	[213]
RuF_4			291	3.04	−74	[214]
RuF_5			292	3.60		[214]
RuO_2	298	1.22		0.62		[161]
RhF_4	293	2.99		1.1		[193]
Pd	288	5.333	450–1000	1.62	−228	[10]
Pd			>1000	1.82	−578	[10]
PdF_5	293	14.3		2.05		[193]
CeB_6	293	11.0	620–1030	2.91	−344	[10]
CeF_3	293	11.1	100–293	2.51	−62	[10]
Ce_2S_3	292	13.5		2.66	−57	[10]
Ce_2Sn_3				2.8		[215]
Pr_2O_3				3.59	−73	[215]
NdB_6		>0	620–1030	3.82	−455	[10]
NdF_3	293	24.7	153–373	3.75	−56	[10]
$Nd(NO_3)_3$	293	15.2	83–373	3.72	−49	[10]
$Nd_2(SO_4)_3$	293	17.3	83–373	3.69	−42	[10]
			300–673	3.47	−18	[10]
Nd_2Se_3		>0		3.58	−33	[10]
Nd_2O_3				3.66	−27	[215]
Sm_2O_3	293	5.15		1.50	−150	[10, 215]
$EuBr_2$	292	86.2	90–292	7.95	0	[10]
$EuCl_2$	292	119	90–292	7.91	0	[10]
Eu_2O_3	298	28.7	180–673	3.62	−135	[10]
$Eu_2(SO_4)_3$	293	17.6	293–628	3.62	−100	[10]
CdB_6			623–1033	7.63	−49	[10]
CdC_2			80–300	7.95	−44	[10]
CdH_2			80–300	7.83	11	[10]
$Cd_2(SO_4)_3$	285.5	89.9	83–700	7.92	−0.4	[10]
$CdSn_3$				8.0	−73	[215]
$Tb_2(SO_4)_3$	293	129	293–630	9.63	−2.3	[10]
$Er_2(SO_4)_3$	293–700	>0	293–700	9.53	−10.8	[10]
Er_2Se_3		>0		9.63	−5	[10]
Tm_2O_3	296.5	133	290–700	7.28	−41.6	[10]
Tm_2O_3				9.43	−25	[215]
Yb_2B_6			623–1033	4.58	−2	[10]
$Yb_2(SO_4)_3$	293	11.7	290–700	4.83	−82	[10]
Yb_2Se_3		>0	90–293	4.75	−63	[10]
HfS_2	296	0.005		0.44		[53]
TaF_3	293	3.34		1.4		[193]
TaS	293	0.11		0.57	−690	[205]
WBr_3	293	0.88		0.94		[216]
WBr_5	293	0.43	195–293	≈1	0	[10]
WBr_5	302.3	1.0		1.19	−7	[217]
WCl_5	293	1.07	90–293	≈1.1	13	[10]
WCl_5	300.6	1.20		1.02	20	[217]
$ReBr_3$			293	0.17		[218]
$ReCl_3$	305	1.38		2.04	−985	[213]

Table 26.2 The magnetic susceptibilities of paramagnetics *(continued)*

Substance	T, K	χ, 10^{-9} m^3/kg	ΔT, K	p_{eff}, μ_B	θ, K	References
$ReCl_4$	300	1.99	220–300	1.55	−108	[219]
$ReCl_5$	293	3.37	90–290	2.32	−265	[10]
$ReCl_5$	302	4.01	77–300	2.57	−266	[219]
$ReCl_5$			<150	2.21	−164	[219]
$ReCl_6$	297	4.25	98–297	2.07	−30	[219]
ReO_2	293	0.20		0.49		[220]
$ReCl_4$			90–300	1.51	−25	[10]
ReS_2	293	0.15		0.38		[220]
$OsCl_4$			294	2.02		[221]
OsF_5			295.4	2.31		[222]
OsF_6			297	1.50	−66	[223]
OsF_7	195	2.65	195	1.19		[224]
$IrCl_4$			293	1.98		[225]
IrF_6			293.6	≈2.90	−30	[226]
IrO_2	298	0.997	300–700	2.8	−4025	[10]
Pt	290.3	1.035	290–720	1.61	−1096	[10]
Pt			>90	1.81	−1617	[10]
PtF_4	292	1.68		1.1		[227]
UBr_3	294	9.92	290–480	3.29	25	[10]
UBr_4	293	6.33	77–570	3.12	−35	[10]
UCl_3	300	10.0	300–509	3.03	−29	[10]
UCl_4	294	9.69	90–550	3.29	−62	[10]
UF_4	300	11.24	77–500	3.28	−116	[10]
UI_3	293	7.2	200–394	3.31	5	[10]
$U(SO_4)_2$	298	0.072	150–300	3.46	−110	[10]
$PuCl_3$		>0	90–590	1.1–1.4		[10]
PuF_4	301	5.5	200–450	2.4	−100	[10]

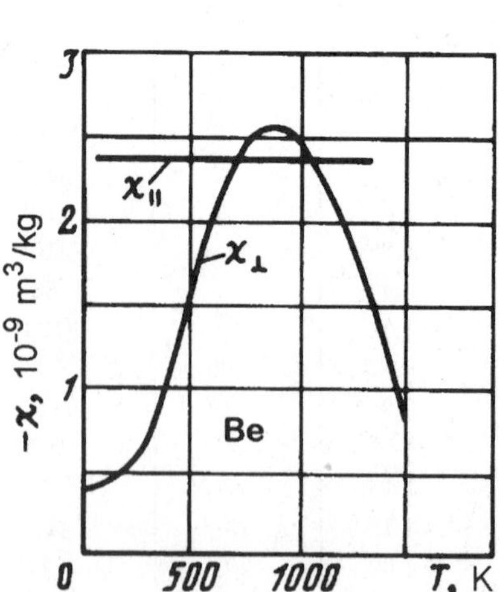

Figure 26.1 The temperature-dependent $\chi_{\|}$ and $\chi_{\perp}$ for Be [183].

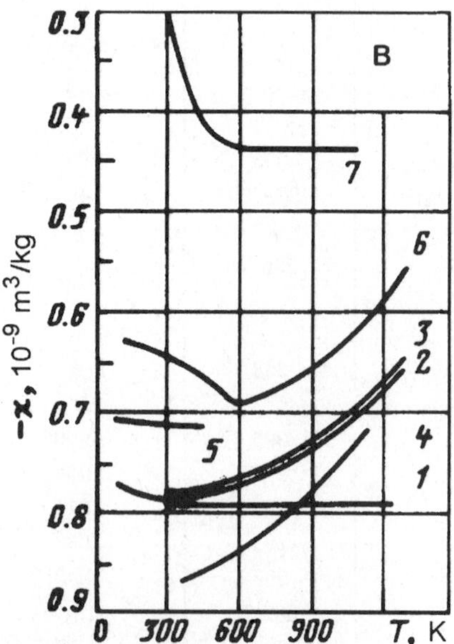

Figure 26.2 The temperature dependence of specific magnetic susceptibility for B (according to different works): *1* – α–B; *2-5* – β–B; *6,7* – amorphous B [184].

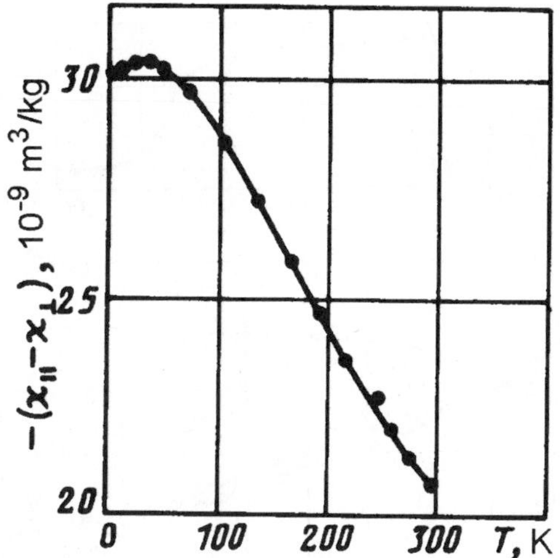

Figure 26.3 The temperature dependence of $\chi_{\perp} - \chi_{\parallel}$ for graphite [19].

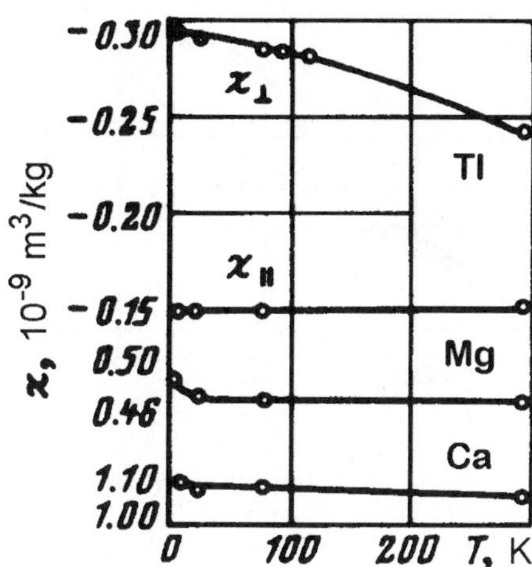

Figure 26.6 The temperature dependence of χ for Ca, Mg, Tl [33].

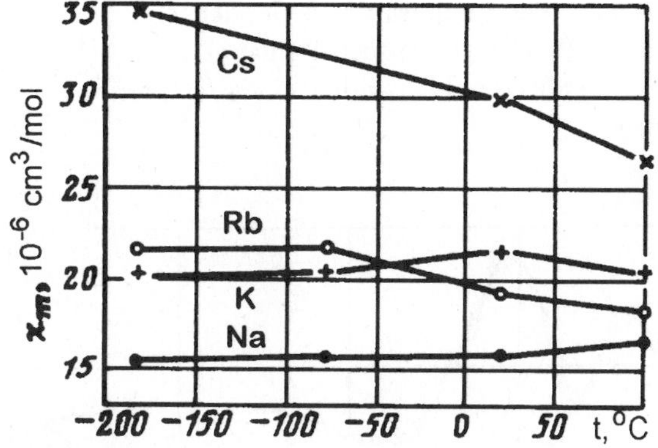

Figure 26.4 The temperature dependence of χ_m for Cs, Rb, Na, K [185, 186].

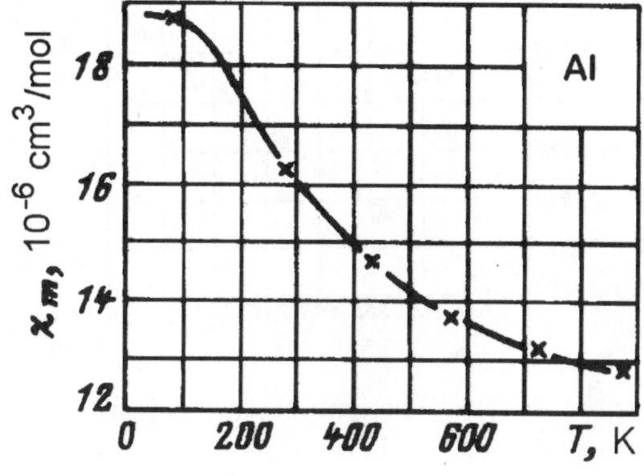

Figure 26.7 The temperature dependence of χ_m for Al [33].

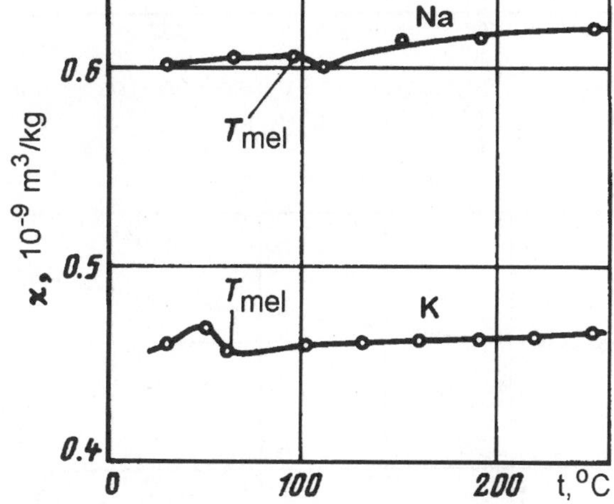

Figure 26.5 The temperature dependence of χ for Na and K [26, 185].

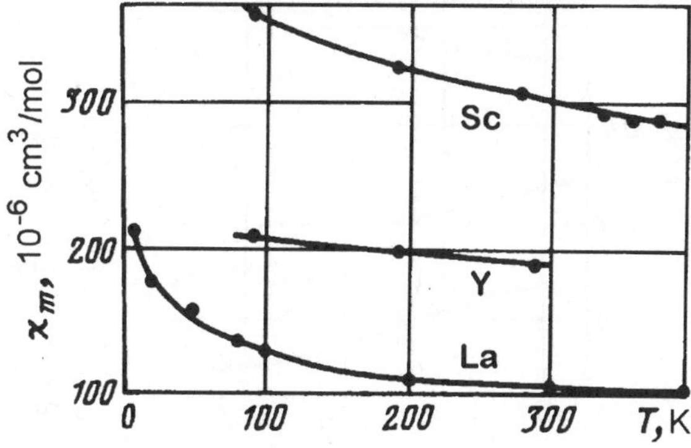

Figure 26.8 The temperature dependence of χ_m for La [41, 139], Sc, and Y [41, 185].

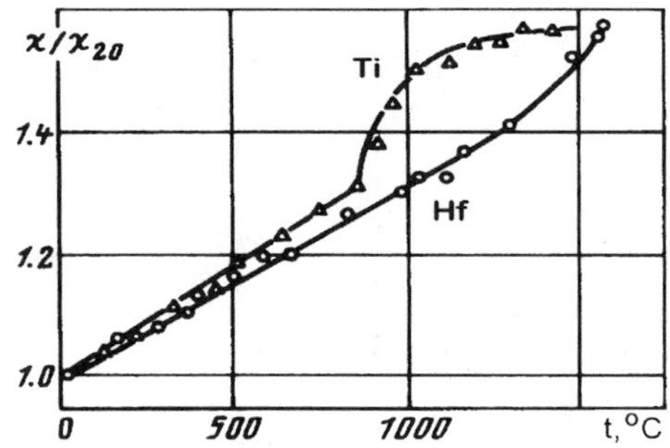

Figure 26.9 The temperature dependence of χ/χ_{20} for Ti and Hf (χ_{20} is the susceptibility at 20°C) [44].

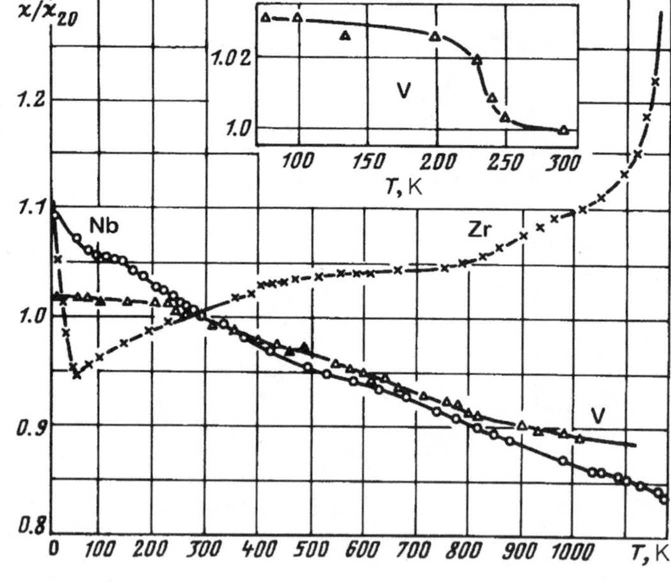

Figure 26.12 The temperature dependence of χ/χ_{20} for V, Zr, and Nb (χ_{20} is the susceptibility at 20°C) [57]. In the inset — the abnormal dependence of χ/χ_{20} for V at low temperatures.

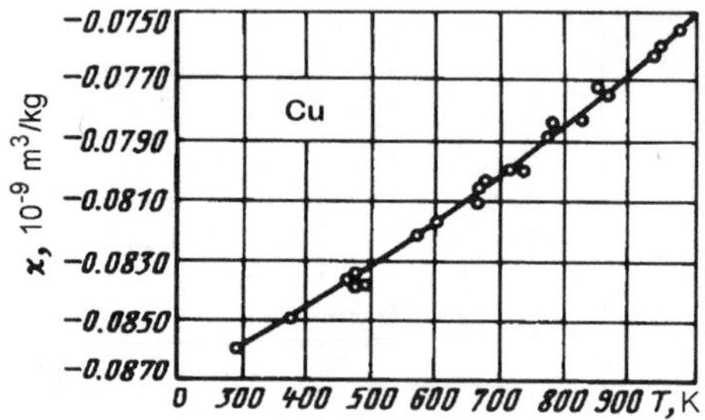

Figure 26.10 The temperature dependences of χ for Cu [117].

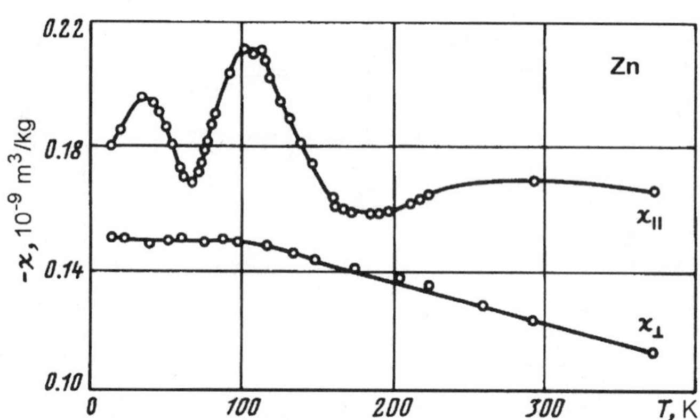

Figure 26.13 The temperature dependences of $\chi_{\parallel}$ and $\chi_{\perp}$ for Zn [18, 185] ($H = 65.7 \cdot 10^4$ A/m).

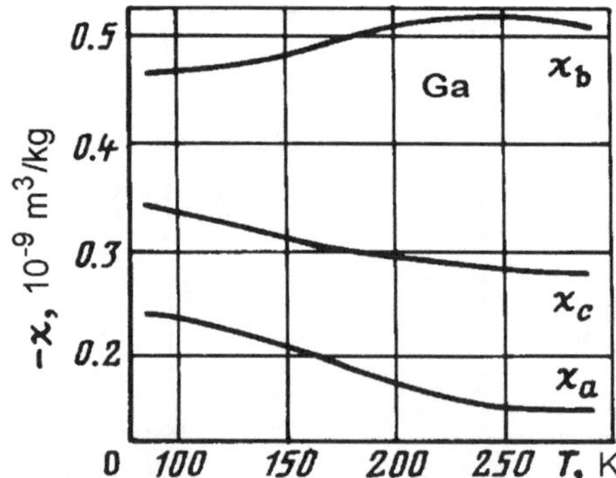

Figure 26.11 The temperature dependences of χ_a, χ_b, χ_c for Ga [187].

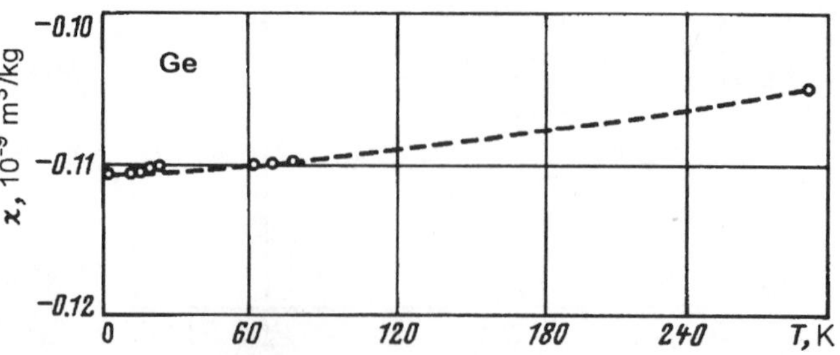

Figure 26.14 The temperature dependence of χ for Ge. Resistivity is equal to 44–46 Ohm·cm, the concentration of Sb is below 10^{14} cm^{-3} [94].

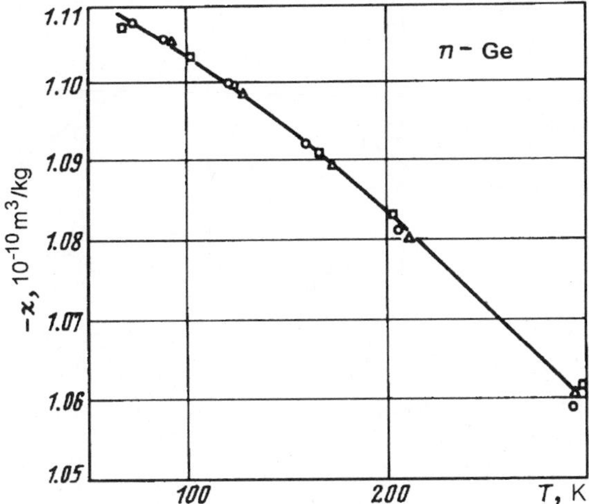

Figure 26.15 The temperature dependence of χ for n-Ge. Carrier concentrations in the samples are as follows: ○, △ – 10^{13} cm^{-3}, □ – 10^{14} cm^{-3} [94, 185].

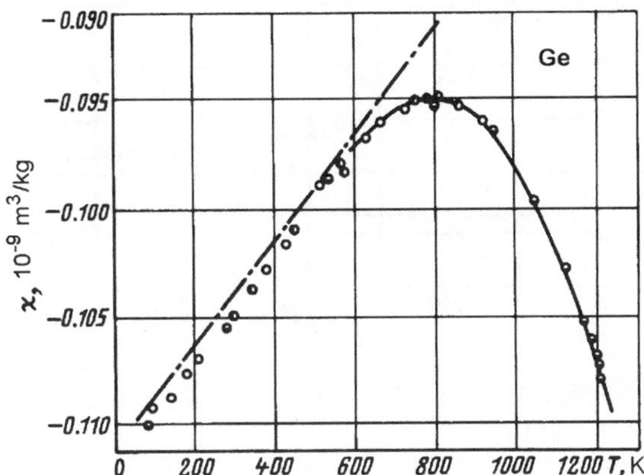

Figure 26.16 The temperature dependence of χ for Ge [130, 185].

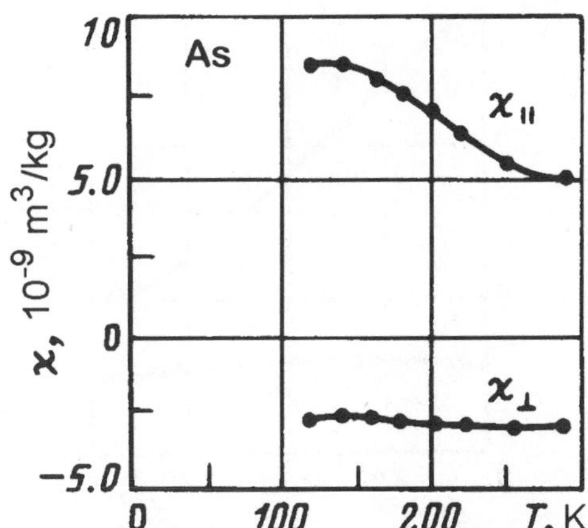

Figure 26.17 The temperature dependences of $\chi_{\perp}$ and $\chi_{\parallel}$ for As [188].

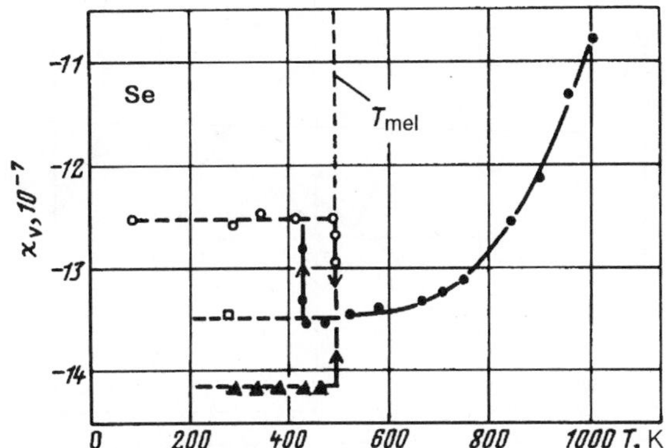

Figure 26.18 The temperature dependence of χ_v for solid and liquid Se [101]: ○– hexagonal Se (slow cooling from T_{melt}); □ – amorphous (quenching from T_m to 280 K); ▲– monoclinic (annealing at 450 K); ● – liquid.

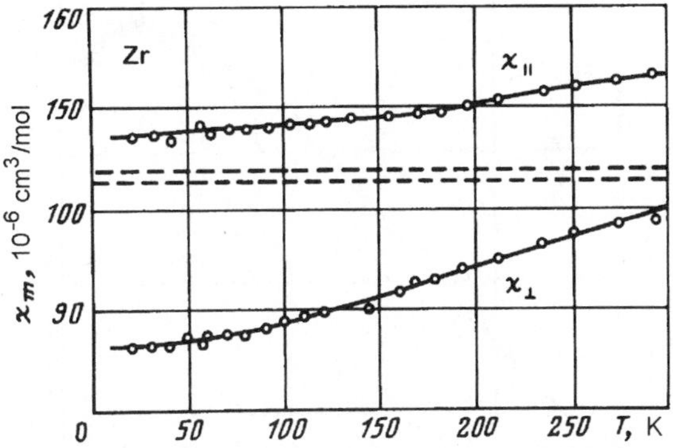

Figure 26.19 The temperature dependences of $\chi_{\perp}$ and $\chi_{\parallel}$ for Zr [189] ($H = 8.75 \cdot 10^5 A/m$).

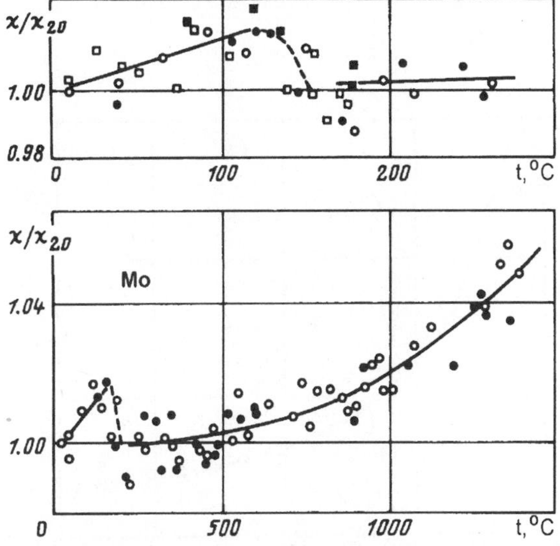

Figure 26.20 The temperature dependence of χ/χ_{20} for Mo (χ_{20} is the susceptibility at 20 °C) [44]. Different points correspond to various measurement cycles.

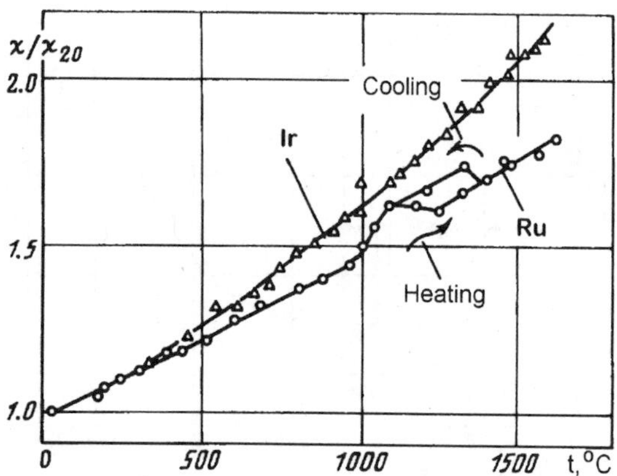

Figure 26.21 The temperature dependence of χ/χ_{20} for Ru and Ir (χ_{20} is the susceptibility at 20°C) [4].

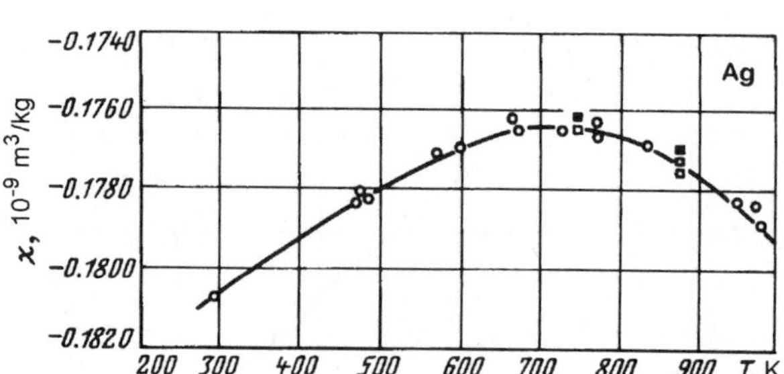

Figure 26.24 The temperature dependence of χ for Ag [117]: ○– normal sample, ■ – after 12 h of exposure at measuring temperature; □ – after sample degassing.

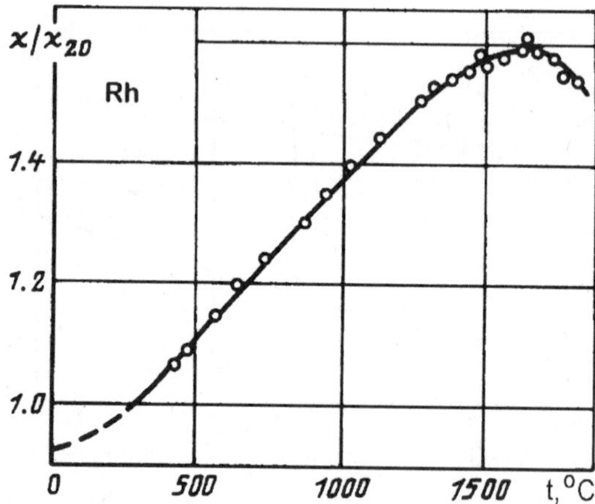

Figure 26.22 The temperature dependence of χ/χ_{20} for Rh (χ_{20} is the susceptibility at 20°C) [44].

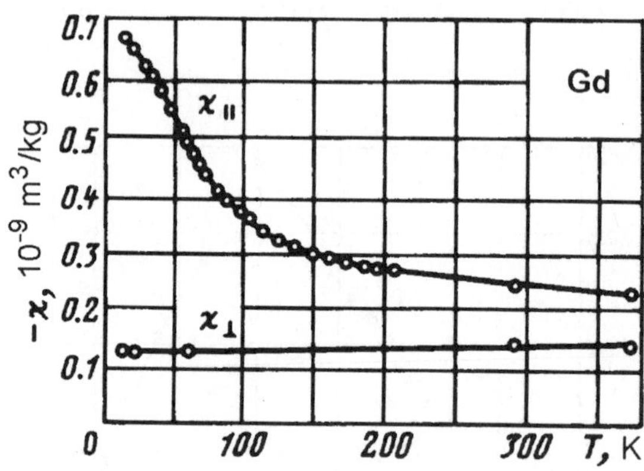

Figure 26.25 The temperature dependences of $\chi_{\perp}$ and $\chi_{\parallel}$ for Cd [118, 185].

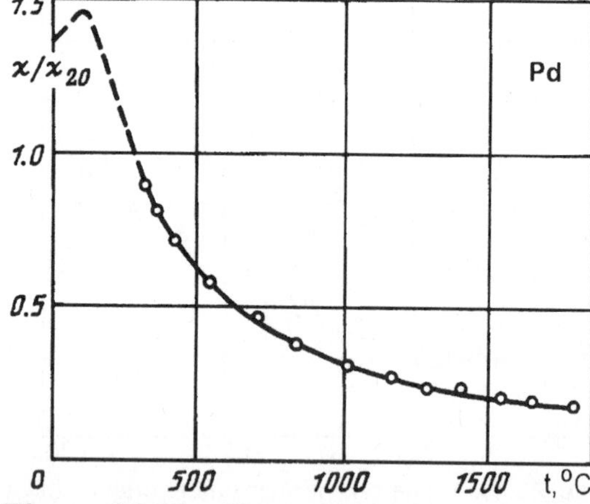

Figure 26.23 The temperature dependence of χ/χ_{20} for Pd (χ_{20} is the susceptibility at 20°C) [44].

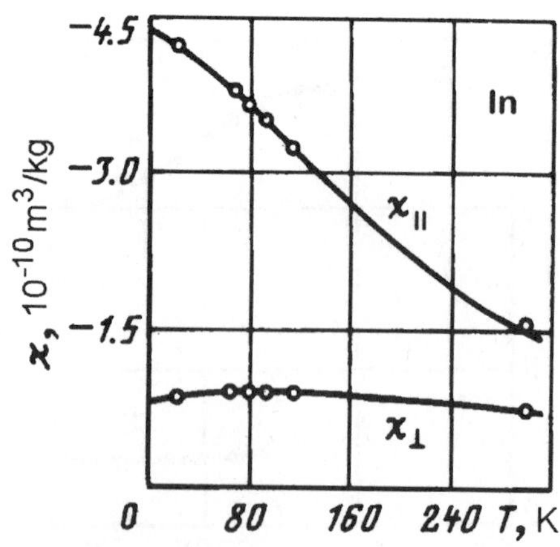

Figure 26.26 The temperature dependences of $\chi_{\perp}$ and $\chi_{\parallel}$ for In [127].

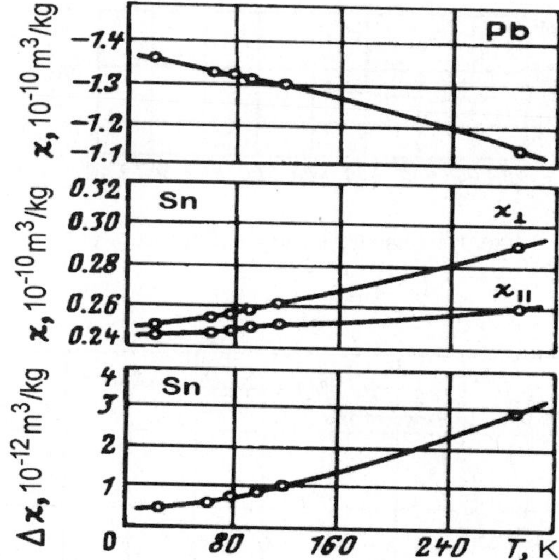

Figure 26.27 The temperature dependences of χ for Sn and Pb, and $\Delta\chi = \chi_{\perp} - \chi_{\parallel}$ for Sn [127].

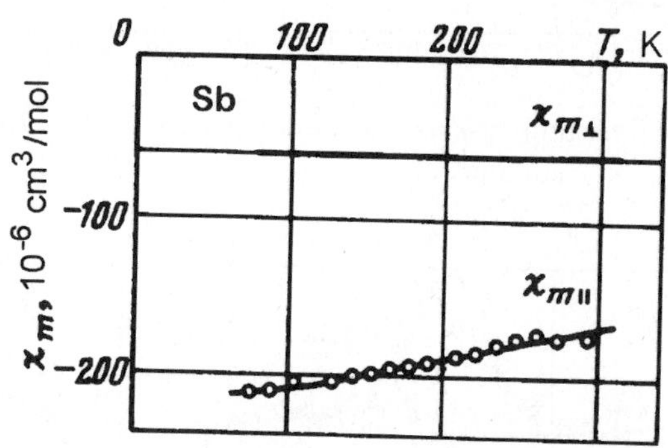

Figure 26.28 The temperature dependences of $\chi_{m\perp}$ and $\chi_{m\parallel}$ for Sb [190].

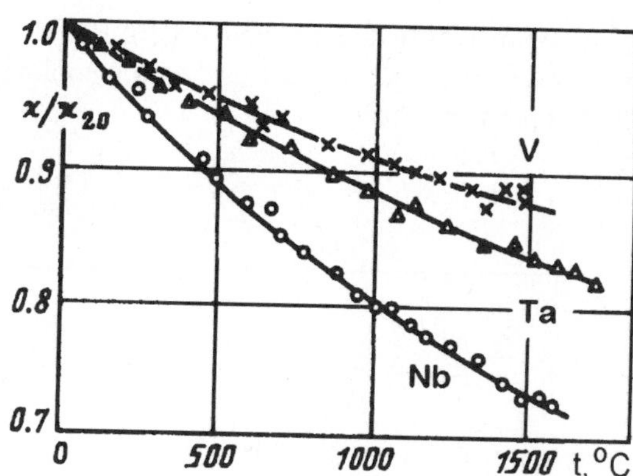

Figure 26.29 The temperature dependence of χ/χ_{20} for V, Ta, and Nb (χ_{20} is the susceptibility at 20°C) [44].

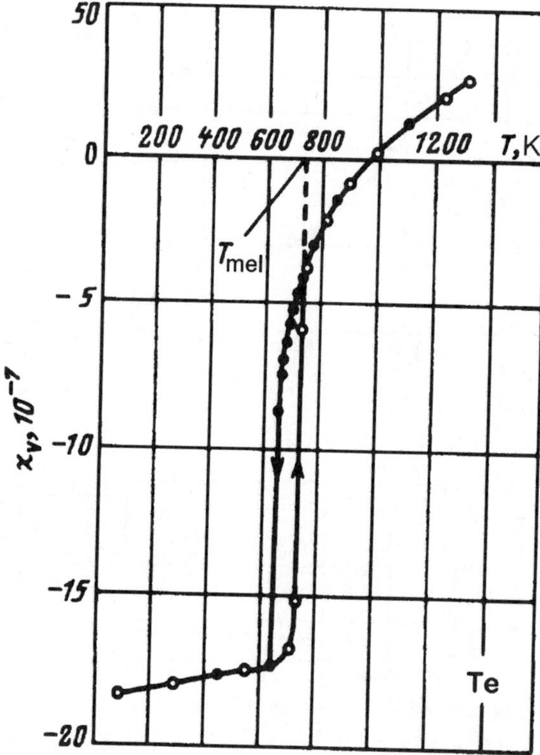

Figure 26.30 The temperature dependence of χ_v for solid and liquid Te [101]: ○– heating, ● – cooling.

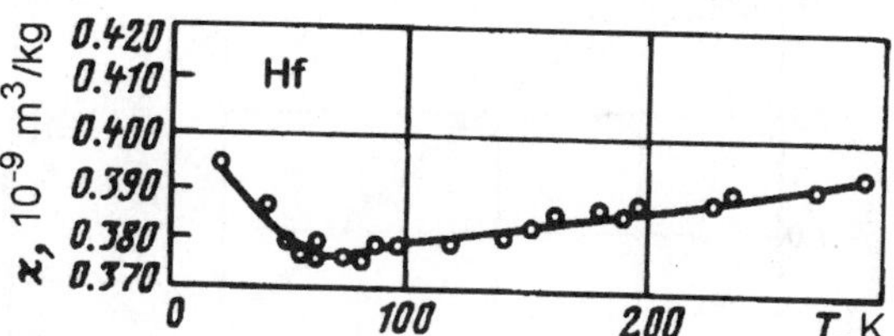

Figure 26.31 The temperature dependence of χ for Hf [142].

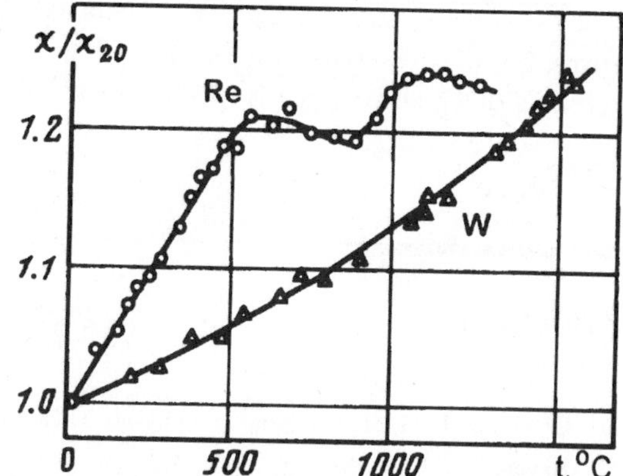

Figure 26.32 The temperature dependence of χ/χ_{20} for Re and W (χ_{20} is the susceptibility at 20°C) [44].

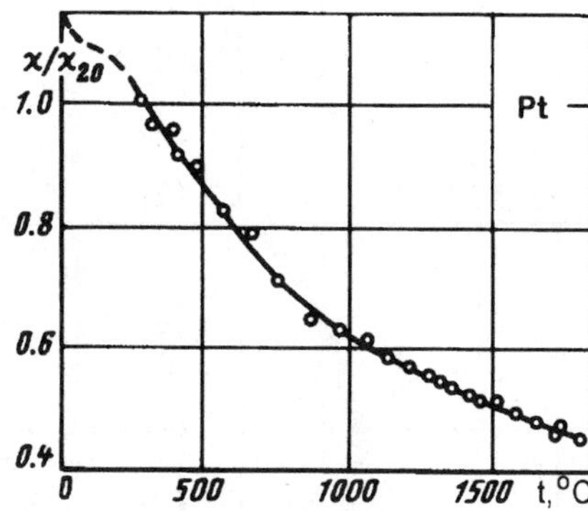

Figure 26.33 The temperature dependence of χ/χ_{20} for Pt (χ_{20} is the susceptibility at 20°C) [44].

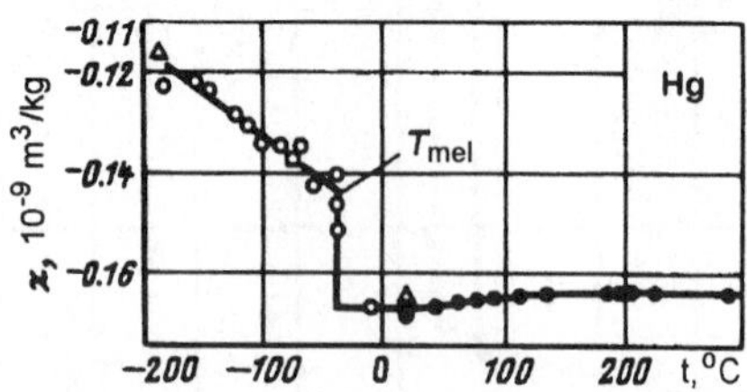

Figure 26.34 The temperature dependence of χ for Hg [185]: ○, △, □, ● — data from the different sources.

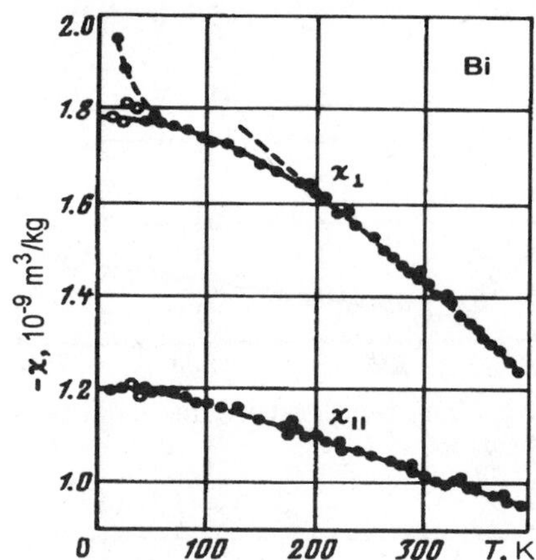

Figure 26.35 The temperature dependences of $\chi_{\|}$ and $\chi_{\perp}$ for Bi [185].

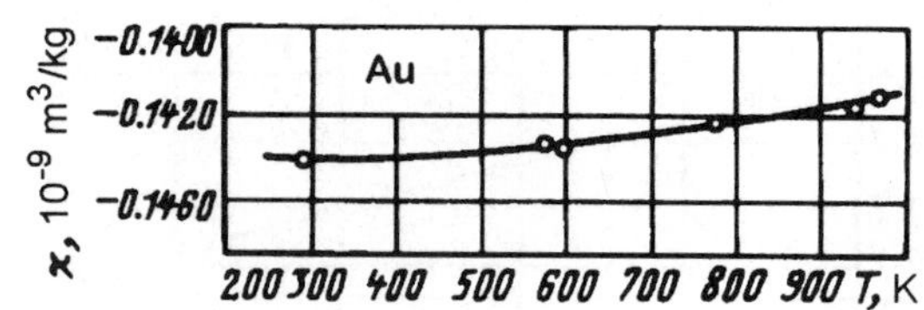

Figure 26.36 The temperature dependence of χ for Au [177].

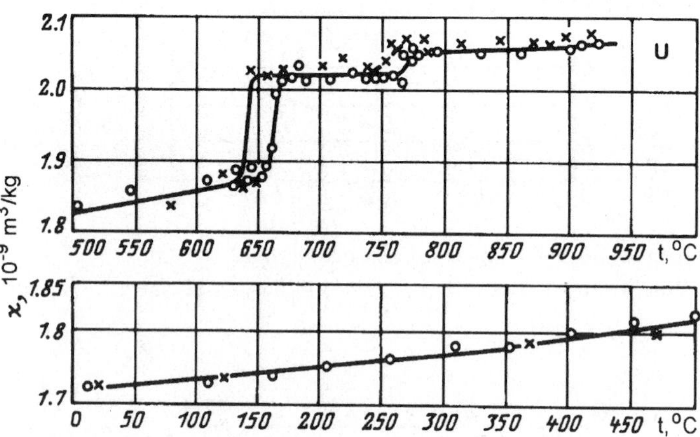

Figure 26.37 The temperature dependence of χ for U [176, 185]: ○– heating, × – cooling.

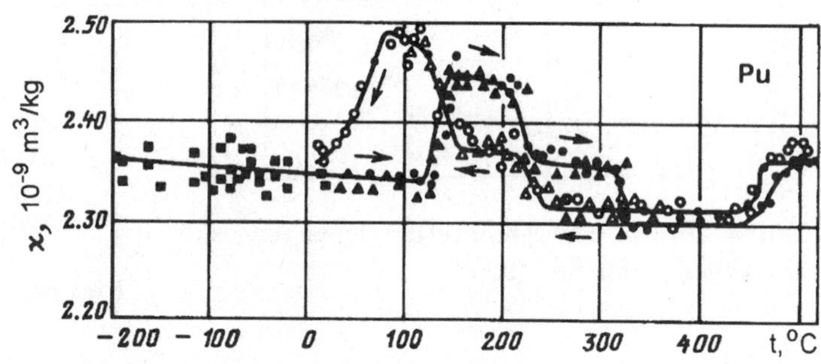

Figure 26.38 The temperature dependence of χ for Pu [181, 185]. Measurements were made during heating (▲, ●), cooling (△, ○) and at low temperatures (■).

References

[1] CRC handbook of chemistry and physics, 63rd ed., CRC Press Inc., Boca Raton, 1982–1983.

[2] Havens, G., Phys. Rev., 43, 992, 1933.

[3] Onnes, H. K., Perrior, A., Proc. Acad. Sci. Amsterdam, 14, 115, 1911.

[4] Kido, K., Sci. Reports Tohoku Univ., 21, 149; 288; 869, 1932.

[5] Ehrlich, P., Z. Anorgan. und Allgem. Chem., 249, 219, 1942.

[6] Cabrera, B., Fahlenbrach, H., Ann. Soc. Espan. Fos. Quim., 32, 525, 1934.

[7] Meyer, St., Phys. Z., 26, 51, 1925.

[8] Witmer, E. E., Phys. Rev., 61, 387, 1942.

[9] Collings, E. W., J. Phys. Chem. Solids, 26, 949, 1965.

[10] Foex, G., Tables de constantes et donnees numeriques 7. Constantes selectionnees diamagnetisme et paramagnetisme, Paris, 1957.

[11] Verkin, B. I., Dmitrenko, I. M., Svechkarev, I. V., Zhurnal eksperimental'noi i teoreticheskoi fiziki, 40, 670, 1961 (in Russian).

[12] Owen, M., Ann. Phys., 37, 657, 1910.

[13] Klemm, L., Z. Electrochem., 45, 354, 1939.

[14] Honda, K., Ann. Phys., 32, 1003, 1910.

[15] Soloviev, N. E., Makarov, V. S., Ugai, Ya. A., et al., J. Less-Common Metals, 67, 161, 1979.

[16] Kubler, L., Gewinner, G., Koulmann, H., et al., Phys. Status Solidi (b), 69, 323, 1975.

[17] Hudgens, S., Kastner, M., Fritzsche, H., Phys. Rev. Lett., 33, 1552, 1972.

[18] De Haas, W., van Alphen, P. M., Proc. Koninke Nederlands Acad. Wet. Amsterdam, 36, 263, 1933.

[19] Mauroubi, A., Flandorois, S., Coulon, C., J. Phys. Chem. Solids, 43, 1103, 1982.

[20] Pascal, P., Compt. Rend., 147, 56; 242; 742, 1908; Compt. Rend., 150, 1167, 1910; Ann. Chimie et de Phys., 19, 5, 1910.

[21] Pascal, P., Compt. Rend. Paris, 148, 413, 1909.

[22] Joussot-Dubien, J., Lemanceau, B., Pacault, A., J. Chim. Phys., 53, 198, 1956.

[23] Bitter, F., Phys. Rev., 35, 1572, 1930.

[24] De Fotis, G. C., Phys. Rev. B23, 4714, 1981.

[25] Laine, P., Compt. Rend., 196, 910, 1933.

[26] Venskateswarlu, K., Sriraman, S., Z. Naturforsch., 13a, 445, 1958.

[27] Prusad, M., Kanekar, C. R., Kamat, L. S., J. Chem. Phys., 19, 686, 1951.

[28] Ishiwara, T., Sci. Reports Tohoku Univ., Ser. I, 3, 303, 1914; 5, 53, 1916; 9, 233, 1920.

[29] Meslin, J., Ann. Chimie et de Phys., 7, 145, 1906.

[30] Pascal, P., Compt. Rend., 173, 144; 712, 1921; 174, 457; 1698, 1922; 175, 1063, 1923.

[31] Klemm, W., Sodomann, H., Langmesser, P., Z. Anorgan. und Allgem. Chem., 241, 281, 1939.

[32] Thomas, J. G., Mendoza, E., Philos. Mag., 43, 900, 1951.

[33] Verkin, B. I., Svechkarev, I. V., Ukrainskii fizicheskii zhurnal, 7, 322, 1962 (in Russian).

[34] Wilson, J., Proc. Royal Soc., A98, 274, 1921; 103, 185, 1923.

[35] Das, D., Indian J. Phys., 41, 525, 1967.

[36] Pascal, P., Compt. Rend., 156, 323, 1914; 158, 37, 1914.

[37] Suvorova, L. N. Baidakov, L. A., Izvestiya Akad. Nauk SSSR, Ser. Inorg. Mat., 13, 806, 1977 (in Russian).

[38] Nilakantan, P., Proc. Indian Acad. Sci. (A), 4, 419, 1936..

[39] Pascal, P., Compt. Rend., 176, 1887, 1923; 177, 765, 1923.

[40] Endo, H., Sci. Reports Tohoku Univ., Ser. I., 14, 479, 1925.

[41] Bommer, H., Z. Electrochem., 45, 357, 1939.

[42] Volkenshtein, N. V., Goloshina, E. V., Kost, M. E., et al., Phys. Status Solidi (b), 117, K47, 1983.

[43] Wedekind, E., J. Angew. Chemie, 37, 87, 1924.

[44] Kojima, H., Tebble, R. S., Williams, D. E. G., Proc. Royal Soc., Ser. A, 260, 237, 1961.

[45] Reekie, J., Yao, Y. L., Proc. Phys. Soc. (London), B69, 417, 1956.

[46] Munster, A., Lagel, K., Z. Physik, 144, 139, 1956.

[47] Gervais, D., Basso-Bert, M., Choukroum, R., et al., C. R. Acad. Sci. Paris, Ser. C, 269, 257, 1969.

[48] Klemm, W., Schuth, W., Z. Anorgan. und Allgem. Chem., 201, 24, 1931.

[49] Vasiliev, Yu. V., Khrishcheva, D. D., Aria, S. M., Zhurnal neorganicheskoi khimii, 8, 785, 1963 (in Russian).

[50] Rüdorff, W., Luginsland, H. H., Z. Anorgan. und Allgem. Chem., 334, 125, 1964.

[51] Wedekind, E., Horst, C., Ber. Chem. Ges., 45, 262, 1912; 48, 105, 1915.

[52] Scott, B. A., Eulenberger, G. R., Bernheim, R. A., J. Chem. Phys., 48, 263, 1968.

[53] Conroy, L. E., Park, K. C., Inorg. Chem., 7, 459, 1968.

[54] Haakon, H., Kjekshug, A., Rost, E., et al., Acta Chem. Scan., 17, 1283, 1963.

[55] Radovskii, I. Z., Rabinovich, B. S., Sidorenko, F. A., et al., Trudy Ural'skogo politekhnicheskogo instituta, 186, 155, 1970 (in Russian).

[56] Robins, D. A., Philos. Mag., 3, 313, 1958.

[57] Suzuki, H., Miyahara, S., J. Phys. Soc. Jpn., 20, 2102, 1965.

[58] Kreissman, C. J., Rev. Mod. Phys., 25, 122, 1953; Phys. Rev., 9, 837, 1954.

[59] Whitteker, A. G., Yost Don, M., J. Amer. Chem. Soc., 71, 3135, 1949.

[60] Roch, J., Compt. Rend., 249, 56, 1959.

[61] Gervais, D., Choukroum, R., J. Inorg. Chem., 36, 3679, 1974.

[62] Clogston, A. M., Gossard, A. C., Jaccarino, V., et al., Phys. Rev. Lett., 9, 262, 1962.

[63] Rykova, M. A., Sabirzyanov, A. V., Zagryazhskii, V. A., et al., Trudy Ural'skogo politekhnicheskogo instituta, 186, 37, 1970 (in Russian).

[64] Sidorenko, F. A., Radovskii, I. Z., Zelenin, L. P., et al., Poroshkovaya metallurgiya, No 9, 68, 1966 (in Russian).

[65] Tilk, W., Klemm, W., Z. Anorgan. und Allgem. Chem., 240, 355, 1939.

[66] Bartecki, A., Wajda, S., Proc. Symp. Wroclaw, Poland, 1962 (p. 305).

[67] Stein, B. F., Walmsley, R. H., Phys. Rev., 148, 933, 1966.

[68] Gambino, R. J., McGuire, T. R., Nakamura, Y., J. Appl. Phys., 38, 1253, 1967.

[69] Radovskii, I. Z.,Shubina, T. S., Gel'd, P. V., et al., Poroshkovaya metallurgiya, No 2, 33, 1965 (in Russian).

[70] Radovskii, I. Z., Bortnik, A. N., Trudy Ural'skogo politekhnicheskogo instituta, 167, 120, 1968 (in Russian).

[71] Gel'd, P. V., Sidorenko, F. A., Silicides of period IV transition metals, Metallurgiya, Moscow, 1971 (in Russian).

[72] Levinson, L. M., J. Solid State Chem., 6, 126, 1973.

[73] Holseth, H., Kjekshug, A., J. Less-Common Metals, 16, 472, 1968.

[74] Boda, G., Stenstrom, B., Lugredo, V., et al., Phys. Scr., 4, 132, 1971.

[75] Miyahara, S., Teranishi, T., J. Appl. Phys., 39, 896, 1968.

[76] Shinoda, D., Asanabe, S. T., J. Phys. Soc. Jpn., 21, 555, 1966.

[77] Nahigian, H., Steger, J., McKinzie, H. L., et al., Inorg. Chem., 13, 1498, 1974.

[78] Heidelberg, R. F., Luxem, A. H., Talhouk, S., Inorg. Chem., 5, 194, 1966.

[79] Frolov, A. A., Sidorenko, F. A., Krentsis, R. P., Trudy Ural'skogo politekhnicheskogo instituta, 186, 18, 1970 (in Russian).

[80] Yaccarino, V., Wertheim, G. K., Wernick, J. R., et al., Phys. Rev., 160, 476, 1967.

[81] Badger, E. H. M., Griffith, R. H., Newling, W. B. S., Proc. Royal Soc., A197, 184, 1949.

[82] Sidorenko, F. A., Miroshnikov, L. A., Gel'd, P. V., Izvestiya vuzov, Ser. Fizika, No 5, 114, 1969 (in Russian).

[83] Czanderna, A. W., J. Chem. Phys., 45, 3159, 1966.

[84] Adou, J. J., Baudet, J., J. Chim. Phys., 64, 1540, 1967.

[85] Gautier, F., Krill, G., Panissod, P., et al., J. Phys., 7, L170, 1974.

[86] Klemm, W., Schüth, W., Z. Anorgan. und Allgem. Chem., 203, 104, 1931.

[87] Bither, T. A., Bouchard, R. J., Cloud, W. H., Inorg. Chem., 7, 2208, 1968.

[88] MacClure, J. W., Marcus, J. A., Phys. Rev., 84, 787, 1951.

[89] Zyubina, T. A., Toroptsev, V. P., Toroptsev, Yu. P., et al., Izvestiya Akad. Nauk SSSR, Ser. Inorg. Mat., 13, 355, 1977 (in Russian).

[90] Pankey, T., J. Appl. Phys., 31, 1802, 1960.

[91] Rustamov, P., Aliev, O. M., Kurbanov, T. Kh., et al., Izvestiya Akad. Nauk SSSR, Ser. Inorg. Mat., 13, 1748, 1977 (in Russian).

[92] Busch, G. A., Kern, R., Helv. Phys. Acta, 32, 24, 1959.

[93] Dovletov, K., Markhuda, Yu. A., Anikin, A. V., et al., Izvestiya Akad. Nauk SSSR, Ser. Inorg. Mat., 14, 33, 1978 (in Russian).

[94] Van Itterbeek, A., Duchateau, W., Physica, 22, 649, 1956; 23, 169, 1957.

[95] Hedgcock, F. T., J. Electronics, 2, 513, 1957.
[96] Bowes, R., Phys. Rev., 108, 683, 1957.
[97] Stevens, D. K., Cleland, J., Crawford, L. H., et al., Phys. Rev., 100, 1084, 1955.
[98] Geist, D., Z. Physik, 157, 335; 490, 1959.
[99] Bennett, S., Heyding, R. D., J. Phys. Chem. Solids, 27, 471, 1966.
[100] Badley, B. G., Disalvo, F. J., Warszczak, J. V., Solid State Commun., 11, 89, 1972.
[101] Busch, G., Risi, M., Uuan, S., Helv. Phys. Acta, 33, 1002, 1960.
[102] Honda, K., Ishiwara, T., Sci. Reports Tohoku Univ., 4, 215, 1915.
[103] Rao, S. R., Sovithri, K., Proc. Indian Soc., 14A, 584, 1941.
[104] De Monsabert, W. R., Boudreaux, E. A., Phys. Chem., 62, 1422, 1958.
[105] Juza, R., Rabenau, A., Nitgshke, J., Z. Anorgan. und Allgem. Chem., 332, 1, 1964.
[106] Trzebiatowski, W., Stalinski, B., Bull. Acad. Polon. Sci., 1, 317, 1953.
[107] Gortseva, F. P., Didchenko, R., Neorganicheskaya khimia, 4, 182, 1965 (in Russian).
[108] Khan, H. R., Raub, C. J., Gardner, W. E., et al., Mater. Res. Bull., 9, 1129, 1974.
[109] Klemm, W., Yacobi, H., Tilk, W., Z. Anorgan. und Allgem. Chem., 201, 1, 1931.
[110] Fendins, H., Dissert., Hannover Univ., 1931.
[111] Morette, A., Compt. Rend., 215, 86, 1942.
[112] Nelson, C. M., Boyd, R. E., Smith, W. T., J. Amer. Chem. Soc., 76, 348, 1954.
[113] Fletcher, J. M., Gardner, W. R., Greenfield, B. F., et al., J. Chem. Soc. (A), 1968, 653, 1968.
[114] Fletcher, J. M., Gardner, W. R., Hooper, E. W., et al., Nature, 199, 1089, 1963.
[115] Hoare, F. E., Matthews, J. C., Proc. Royal Soc., A212, 137, 1952.
[116] Papatheodorou, G. N., Inorg. Nuclear Chem. Lett., 10, 115, 1974.
[117] Garber, M., Henry, W. G., Hoeve, H. G., Canad. J. Phys., 38, 1595, 1960.
[118] Marcus, J. A., Phys. Rev., 76, 413; 621, 1949.
[119] Mookherji, T., J. Electrochem. Soc., 117, 1201, 1970.
[120] Strakhov, L. P., Rimadkhanov, K., Korolev, V. V., Izvestiya Akad. Nauk SSSR, Ser. Neorganicheskie Mat., 9, 645, 1973 (in Russian).
[121] Singh, S., Singh, P., J. Phys. Chem. Solids, 41, 135, 1980.
[122] Pilat, I. M., Fizika metallov i metallovedenie, 4, 232, 1957 (in Russian).
[123] Strakhov, L. P., Krepkaya, V. P., Kazennov, B. A., Fizika tverdogo tela, 11, 3595, 1969 (in Russian).
[124] Potykevich, I. V., Bondar', G. I., Koval', V. S., et al., Ukrainskii fizicheskii zhurnal, 28, 1072, 1983 (in Russian).
[125] Sirota, N. N., Vitkina, Ts. Z., Antyukov, A. M., et al., Izvestiya Akad. Nauk SSSR, Ser. Fiz.–Mat., No 2, 126, 1976 (in Russian).
[126] Verharghe, H., Van der Meerssche, G., Le Conepte, C., Phys. Rev., 80, 758, 1959.
[127] Aleksandrov, B. N., Verkin, B. I., Svechkarev, I. V., Zhurnal eksperimental'noi i teoreticheskoi fiziki, 39, 37, 1960 (in Russian).
[128] Busch, G., Menth, A., Natterer, B., Z. Naturforsch., 19a, 542, 1964.
[129] Juza, R., Hahn, H., Z. Anorgan. und Allgem. Chem., 244, 111, 1940.
[130] Busch, G., Kern, R., Helv. Phys. Acta, 32, 24, 1959.
[131] Stevens, D. K., Crawford, J. H., Phys. Rev., 99, 487, 1955.
[132] Darnell, A. J., Libby, W. F., Phys. Rev., A135, 1453, 1964.
[133] Broniewski, W., Franczek, S., Witkowski, R., Ann. Phys., 10, 5, 1938.
[134] Endo, H., Sci. Reports Tohoku Univ., 16, 201, 1927.
[135] Matyás, M., Czechoslov. J. Phys., 8, 301, 1958.
[136] De Haas, W. J., van Alphen, P. M., Acad. Sci. (Amsterdam), 36, 158, 1933.
[137] Rao, S. R., Venkataramiah, H. S., J. Mysore Univ., 8, No 2, 39, 1948.
[138] Lane, C. T., Phys. Rev., 44, 43, 1933.
[139] Lock, J. M., Proc. Phys. Soc. (London), B70, 476; 566, 1957.
[140] Klemm, W., Meisel, K., von Vogel, H., Z. Anorgan. und Allgem. Chem., 190, 123, 1930.

[141] Kreissman, C. J., McGuire, T. R., Phys. Rev., 98, 936, 1955.

[142] Volkenshtein, N. V., Goloshina, E. V., Fizika metallov i metallovedenie, 18, 784, 1964 (in Russian).

[143] Hall, J. R., Marchunt, N. K., Plowman, R. A., Australian J. Chem., 16, 34, 1963.

[144] Schäfer, H., Gerken, R., Scholz, H., Z. Anorgan. und Allgem. Chem., 335, 96, 1965.

[145] Brauer, G., Weidlein, J., Strähle, J., Z. Anorgan. und Allgem. Chem., 348, 298, 1966.

[146] Krylov, E. I., Zhurnal neorganicheskoi khimii, 3, 1487, 1958 (in Russian).

[147] McCarley, R. E., Brown, T. M., Inorg. Chem., 3, 1232, 1964.

[148] Fendins, H., Dissert., Hannover Univ., 1931.

[149] Gibart, P., Compt. Rend., 259, 4237, 1964.

[150] Biltz, W., Z. Anorgan. und Allgem. Chem., 214, 227, 1930.

[151] Greiner, J. D., Shanks, H. R., J. Solid State Chem., 5, 262, 1972.

[152] Cheretien, A., Odent, G., Compt. Rend., 257, 2290, 1963.

[153] Thaore, K., Bull. Soc. Chim. (France), 1284, 1965.

[154] Belova, V. I., Semenov, I. N., Zhurnal neorganicheskoi khimii, 16, 2871, 1971 (in Russian).

[155] Machmer, P., Z. Naturforsch., 24B, 200, 1969.

[156] Schüber, H., Huneke, K. H., Brendel, C., Z. Anorgan. und Allgem. Chem., 383, 49, 1971.

[157] Belova, V. I., Syrkin, Yu. K., Zhurnal neorganicheskoi khimii, 3, 2016, 1958 (in Russian).

[158] Kolbin, N. I., Samoilov, V. M., Zhurnal neorganicheskoi khimii, 13, 906, 1968 (in Russian).

[159] Brodersen, K., Machmer, P., Z. Naturforsch., 176, 127, 1962.

[160] Ryden, W. D., Lawson, A. W., J. Chem. Phys., 52, 6058, 1970.

[161] Gouthrie, A. N., Bourland, L. T., Phys. Rev., 37, 303, 1931.

[162] Brodersen, K., Thiele, G., Schnering, H. G., Z. Anorgan. und Allgem. Chem., 337, 120, 1965.

[163] Argue, G. R., Banewicz, J. J., Inorg. Nuclear Chem., 25, 923, 1963.

[164] Gronvold, F., Haraldsen, H., Kjekshug, A., Acta Chem. Scand., 14, 1879, 1960.

[165] Rao, S. R., Sriraman, S., Proc. Indian Acad. Sci., 5A, 343, 1937.

[166] Vogt, E., Ann. Phys., 14, 1, 1932.

[167] Bates, L. F., Baker, C. I. W., Proc. Phys. Soc. (London), 50, 409, 1938.

[168] Rao, S. R., Narayanaswamy, A. S., Philos. Mag., 26, 1018, 1938.

[169] Pascal, P., Minne, P., Compt. Rend., 193, 1303, 1931.

[170] Mikhail, H., Mekkawy, J., Czechoslov. J. Phys., B28, 216, 1978.

[171] Shoenberg, D., Uddin, M. Z., Proc. Royal Soc., A156, 687, 1936.

[172] Kapitza, P., Proc. Royal Soc., A131, 224, 1931.

[173] Rao, S. R. J., J. Mysore Univ., 5, No 2, 69, 1945.

[174] Smith, J. F., Greiner, J. D., Phys. Rev., 115, 884, 1959.

[175] Trzebiatowski, W., Selwood, P. W., J. Amer. Chem. Soc., 72, 4501, 1950.

[176] Bates, L. F., Hughes, D., Proc. Phys. Soc. (London), B67, 28, 1954.

[177] Henkel, P., Klemm, W., Z. Anorgan. und Allgem. Chem., 222, 70, 1935.

[178] Picon, M., Flahaut, J., Compt. Rend., 237, 1160, 1953.

[179] Picon, M., Flahaut, J., Compt. Rend., 240, 784, 1955.

[180] Khodadad, P., Compt. Rend., 247, 1205, 1958.

[181] Brodsky, M. V., Proc. Inst. Phys. London, 1971, Conference digest on rare earths and actinides, No 3, p. 75.

[182] Seguin, M., Compt. Rend., 246, 3243, 1958.

[183] Grechnev, G. E., Svechkarev, I. V., Sereda, Yu. P., Zhurnal eksperimental'noi i teoreticheskoi fiziki, 75, 993, 1978 (in Russian).

[184] Landolt–Börnstein, Numerical data and functional relationships in science and technology, New series, Eds.-in-Chief: Hellwege, K.-H., Madelung, O., Group III: Crystal and Solid-State Physics, vol. 17, Springer-Verlag, Berlin, 1985 (subvol. *e*).

[185] Landolt–Börnstein, Zahlenwerte und funktionen, 6th ed., Bd. II, 9 teil, Springer–Verlag, Berlin–Göttingen–Heidelberg, 1962.

[186] Böhm, B., Klemm, W., Z. Anorgan. und Allgem. Chem., 243, 69, 1939.

[187] Marchand, A., Compt. Rend., 241, 468, 1955.

[188] Yamaguchi, Y., Solid State Commun., 8, 833, 1970.

[189] Volkenshtein, N. V., Goloshina, E. V., Shchegolikhina, N.I., Fizika metallov i metallovedenie, 25, 1840, 1968 (in Russian).

[190] Browne, S. H., Lane, C. T., Phys. Rev., 60, 895, 1941.

[191] Klemm, W., Hoschek, E., Z. Anorgan. und Allgem. Chem., 226, 359, 1936.

[192] Clark, J. K., Machin, D. J., J. Chem. Soc., 4430, 1963.

[193] Nyholm, R. S., Sharpe, A. G., J. Chem. Soc., 3579, 1952.

[194] Juza, D., Giegling, D., Schäber, H., Z. Anorgan. und Allgem. Chem., 366, 121, 1969.

[195] Vorob'ev, N. I., Pechkovskii, V. V., Kobets, L. V., Zhurnal neorganicheskoi khimii, 19, 3, 1974 (in Russian).

[196] Cadeville, M. C., J. Phys. Chem. Solids, 27, 667, 1966.

[197] Proskurnikov, A. A., Krylov, E. I., Zhurnal neorganicheskoi khimii, 10, 1017, 1965 (in Russian).

[198] Hume, D. N., Stone, H. W., J. Amer. Chem. Soc., 63, 1200, 1941.

[199] Hoppe, R., Daehne, W., Klemm, W., Ann. Chem., 658, 1, 1962.

[200] Zernicke, J., Rec. Trav. Chim., 72, 390, 1953.

[201] Williams, E. H., Phys. Rev., 28, 167, 1926.

[202] Haraldsen, H., Grönvold, F., Hurlen, T., Z. Anorgan. und Allgem. Chem., 283, 143, 1956.

[203] Furuseth, S., Kjekshug, A., Andersen, A. F., Acta Chem. Scand., 23, 2325, 1969.

[204] Vandenbempt, E., Pauwels, L., de Clippeleir, K., Bull. Soc. Chim. Belg., 80, 283, 1981.

[205] Uchida, E., Kondoh, H., J. Phys. Soc. Jpn., 11, 21, 1955.

[206] Munson, R. A., de Sorbo, W., Koubel, J. S., J. Chem. Phys., 47, 1769, 1967.

[207] Lewig, J., Machin, D. J., Newnham, I. E., et al., J. Chem. Soc., No 11, 2036, 1962.

[208] Ehrlich, P., Plöger, F., Pietzka, G., Z. Anorgan. und Allgem. Chem., 282, 19, 1955.

[209] Rüdolf, W., Luginsland, H. H., Z. Anorgan. und Allgem. Chem., 334, 125, 1964.

[210] Klemm, W., Steinberg, H., Z. Anorgan. und Allgem. Chem., 227, 193, 1936.

[211] Schäfer, H., v. Schnering, H. G., Tillack, J., Z. Anorgan. und Allgem. Chem., 353, 281, 1967.

[212] Kepert, D. L., Mandyszewsky, R., Inorg. Chem., 7, 2091, 1968.

[213] Knox, K., Coffey, G. E., J. Amer. Chem. Soc., 81, 5, 1959.

[214] Holloway, J. H., Peacock, R. D., J. Chem. Soc., p. 3892, 1963.

[215] Taylor, K., Darby, M., Physics of rare-earth compounds, Wiley, London, 1972

[216] Brown, D. A., Glass, W. K., O'Daly, C., J. Chem. Soc., Dalton Trans., No 12, 1311, 1973.

[217] Brisdon, B. L., Edwards, D. A., Machin, D. J., J. Chem. Soc. (A), p. 1825, 1967.

[218] Perakis, N., J. Phys. Radium, 15, 191, 1954.

[219] Brown, D., Colton, R., J. Chem. Soc., (A), p. 714, 1964.

[220] Schüth, W., Klemm, W., Z. Anorgan. und allgem. Chem., 220, 193, 1934.

[221] Colton, R., Farthing, R. H., Australian J. Chem., 21, 589, 1968.

[222] Hargreaves, G. B., Peacock, R. D., J. Chem. Soc., No 11, 2618, 1960.

[223] Hargreaves, G. B., Peacock, R. D., Proc. Chem. Soc., No 1, 85, 1959.

[224] Glemser, O., Roesky, H. W., Hellberg, K. H., et al., Chem. Ber., 99, 2652, 1966.

[225] Bose, D. M., Bhar, H. G., Z. Physik, 48, 716, 1928.

[226] Figgis, B. N., Lewis, J., Mabbs, F. E., J. Chem. Soc., (A), p. 3138, 1961.

[227] Nast, R., Hoerl, W., Chem. Ber., 95, 1470, 1962.

27

Magnetic Properties of Ferromagnetic Metals and Alloys

K.G. Gurtovoy

27.1 Introduction

Ferromagnets constitute, generally, those substances which have a spontaneous magnetic moment, that is a finite magnetization at low enough temperatures and vanishingly small applied magnetic field. Magnetic structure of these substances may be relatively complex (see "Cone" and "Ferri" structures in Fig. 27.16). Dependence of the magnetic permeability on the applied magnetic field and on the sample treatment prehistory is typical for ferromagnets, as well as the existence of a certain temperature, above which the substance passes to the paramagnetic state with zero spontaneous magnetic moment.

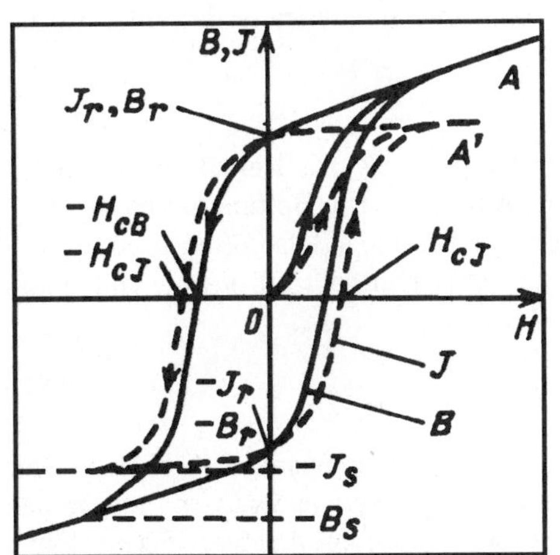

Figure 27.1 Magnetic induction B and magnetization J of a ferromagnetic material vs. field intensity H.

Magnetization and magnetic induction. The magnetic moment per unit volume of a magnetic substance is called magnetization. Magnetization $\mathbf{J}$, A/m, magnetic induction $\mathbf{B}$, T, and field strength $\mathbf{H}$, A/m, are related through the well-known equation

$$\mathbf{B} = \mu_0(\mathbf{H} + \mathbf{J}), \tag{27.1}$$

where $\mu_0 = 4\pi \cdot 10^{-7}$ H/m is the permeability of vacuum. If an ellipsoidal magnetic sample is placed into an external magnetic field $\mathbf{H}$, the field inside the sample $\mathbf{H}_i$ is defined by the formula

$$\mathbf{H}_i = \mathbf{H} - \hat{N}\mathbf{J}, \tag{27.2}$$

where $\hat{N}$ is the tensor of demagnetizing factors.

Magnetization of ferromagnets from the zero induction state is represented by the magnetization curve OA (or OA′) (Fig. 27.1) depicting the dependence of magnetic induction (or magnetization) on H. As the field increases, the magnetization reaches a limit J_s, which is called the saturation magnetization. On magnetic reversal the dependence $B(H)$ (or $J(H)$) forms a characteristic S-shaped curve — the hysteresis loop (Fig. 27.1). The points of a hysteresis loop intersection with the ordinate and the abscissa axes give the values of the remanent induction B_r (or the remanent magnetization J_r) and of the coercive force H_{cB} (or H_{cJ}), respectively. The specific magnet moment σ (or specific magnetization), i.e. the

0-8493-2861-6/97/$0.00+$.50

magnetic moment per unit mass, is often used to describe magnetic substances. The SI unit of the specific magnetic moment is A·m^2/kg, the CGSM unit — emu/g. Sometimes, an atomic magnetic moment n, that is, the average magnetic moment per an atom or per a substance formula unit, is referred to as the magnetic characteristic; it is measured in Bohr magnetons μ_B. The specific magnetic moment and the atomic magnetic moment are related by

$$n = (M/N_A\mu_B)\sigma, \tag{27.3}$$

where M is the molar mass, corresponding to a substance formula unit, kg/mol, and N_A is the Avogadro constant, mol^{-1}.

Curie temperature and compensation point. The temperature T_C of magnetic phase transition from the ferromagnetic to the paramagnetic state is called the Curie temperature or point. In some intermetallic compounds with complex magnetic structure, the spontaneous magnetization disappears at the so-called compensation temperature T_{comp} due to making up for the constituent magnetizations of magnetic sublattices.

Magnetic susceptibility and permeability. Dependence of J versus H takes the form $J = \chi H$, where χ is the magnetic (volume) susceptibility, whereas the dependence of B on H is accepted to be written as $B = \mu_a H$, where μ_a is the absolute magnetic permeability, H/m. Thus, the magnetic susceptibility and permeability are related via the expression

$$\chi = (\mu_a/\mu_0) - 1 = \mu_r - 1, \tag{27.4}$$

where $\mu_r = \mu_a/\mu_0$ is the relative magnetic permeability of a substance. Induction differentiation with respect to the field intensity results in the relative incremental magnetic permeability

$$\mu_{rd} = (1/\mu_0)(dB/dH). \tag{27.5}$$

In weak fields this quantity is called the relative initial permeability (μ_i), whereas its maximum value along the whole magnetization curve is spoken of as the relative maximum permeability ($\mu_{\max}$).

At temperatures $T > T_C$, the paramagnetic susceptibility of many magnetics is described rather well by the Curie-Weiss law

$$\chi = C/(T - \theta_p), \tag{27.6}$$

where $C = \mu_0(N_A/V)\mu_{\text{eff}}^2/3k$; θ_p is the paramagnetic Curie temperature, K. Here V is the molar volume, m^3/mol; μ_{eff} is the effective atomic magnetic moment, A·m^2; k is the Boltzmann constant.

In addition to the above defined magnetic susceptibility per unit volume χ, also often used are the specific mass susceptibility χ_ρ and the molar susceptibility χ_m, i.e. the susceptibilities per unit mass or one mole of the substance. These quantities are interrelated as

$$\chi_\rho = \chi/\rho \quad \text{and} \quad \chi_m = \chi_\rho/M, \tag{27.7}$$

where ρ is the substance density, kg/m^3. The volume susceptibility is dimensionless in both SI and CGSM systems, the specific mass susceptibility is expressed in m^3/kg and emu/g, and the molar susceptibility — in m^3/mol and emu/mol, respectively. The unit for the absolute magnetic permeability in SI is H/m. In the CGSM system, the magnetic permeability, as defined by the formula $B = \mu H$, is dimensionless (just like the relative permeability in SI).

Magnetic anisotropy. The difference in magnetic properties of ferromagnets along nonequivalent directions in the body, which is called magnetic anisotropy, is most pronounced in single crystals. A measure of magnetic anisotropy is a magnetization work necessary to turn the vector **J** from the position along the easy magnetization axis, where this vector is directed in the absence of a field, to a new position along the applied field. This work determines the density of magnetic-anisotropy free energy E_a, J/m^3, which is expressed in terms of the angles between the magnetization vector **J** and the crystallographic axes:

(1) for cubic crystals

$$E_a = K_1 s + K_2 \alpha_1^2 \alpha_2^2 \alpha_3^2 + K_3 s^2 + \dots, \tag{27.8}$$

where α_1, α_2 and α_3 are the direction cosines of the **J** vector; $s = \alpha_1^2\alpha_2^2 + \alpha_1^2\alpha_3^2 + \alpha_2^2\alpha_3^2$;

(2) for hexagonal crystals

$$E_a = K_1 \sin^2\theta + K_2 \sin^4\theta + \dots, \tag{27.9}$$

where θ is the angle between the **J** vector and the hexagonal c-axis.

Magnetostriction. A change of the size and the form of a body in the course of its magnetization is spoken of as magnetostriction. This is further distinguished by the volume magnetostriction, which is characterized by the relative change of the body volume $\omega = \Delta V/V$, and the anisotropic magnetostriction marked by the relative variation of the body size $\lambda = \delta l/l$ nearly without change in its volume. Sometimes the latter type of magnetostriction is said, though not quite correctly, to be linear magnetostriction. In the general case, the volume magnetostriction shows up as a sum of three terms:

$$\omega = \omega_I + \omega_K + \omega_F. \tag{27.10}$$

The first term ω_I depends only on the true magnetization $J > J_s$. In strong magnetic fields, when orientation of the magnetization vector no longer changes, ω_I is linear, to a first approximation, in magnetic field intensity. The terms ω_K and ω_F describe the dependence of the body volume on the direction of the magnetization vector and on the sample form, respectively.

The saturation magnetostriction, that is the magnetostriction corresponding to the material magnetization up to saturation, is defined for cubic crystals as

$$\begin{aligned}\lambda = {} & h_1(\alpha_1^2\beta_1^2 + \alpha_2^2\beta_2^2 + \alpha_3^2\beta_3^2 - 1/3) + 2h_2(\alpha_1\alpha_2\beta_1\beta_2 + \alpha_1\alpha_3\beta_1\beta_3 + \alpha_2\alpha_3\beta_2\beta_3) + B \\ & + h_4(\alpha_1^4\beta_1^2 + \alpha_2^4\beta_2^2 + \alpha_3^4\beta_3^2 + (2/3)s - 1/3) + 2h_5(\alpha_1^2\alpha_2\alpha_3\beta_2\beta_3 + \alpha_1\alpha_2^2\alpha_3\beta_1\beta_3 + \alpha_1\alpha_2\alpha_3^2\beta_1\beta_2) + \dots,\end{aligned} \tag{27.11}$$

where α_i and β_i are the direction cosines of the **J** vector and the vector along the magnetostriction measurement direction; $B = h_3 s$ if the [100] axis is aligned with the easy magnetization direction and $B = h_3(s - 1/3)$ when the easy magnetization direction is aligned with the [111] axis.

The longitudinal ($\alpha_i = \beta_i$) magnetostriction along the principal crystallographic axes of a cubic crystal is defined by

$$\lambda_{100} = (2/3)(h_1 + h_4), \quad \lambda_{110} = (1/12)(2h_1 + 6h_2 + h_4), \quad \lambda_{111} = (2/9)(3h_2 + h_5). \tag{27.12}$$

One can often set $h_3 = h_4 = h_5 = 0$ and use a simplified formula

$$\lambda = (3/2)\lambda_{100}(s - 1/3) + 3\lambda_{111}(\alpha_1\alpha_2\beta_1\beta_2 + \alpha_1\alpha_3\beta_1\beta_3 + \alpha_2\alpha_3\beta_2\beta_3). \tag{27.13}$$

For a polycrystalline material, the anisotropic magnetostriction can be deduced from the relation

$$\lambda = (3/2)\lambda_s(\cos^2\theta - 1/3), \tag{27.14}$$

where $\lambda_s = (2\lambda_{100} + 3\lambda_{111})/5$, and θ is the angle between the vector **J** and the measurement direction.

The saturation magnetostriction of hexagonal crystals takes the form

$$\begin{aligned}\lambda = {} & \lambda_A\left[(\alpha_1\beta_1 + \alpha_2\beta_2)^2 - (\alpha_1\beta_1 + \alpha_2\beta_2)\alpha_3\beta_3\right] + \lambda_B\left[(1 - \alpha_3^2)(1 - \beta_3^2) - (\alpha_1\beta_1 + \alpha_2\beta_2)^2\right] \\ & + \lambda_C\left[(1 - \alpha_3^2)\beta_3^2 - (\alpha_1\beta_1 + \alpha_2\beta_2)\alpha_3\beta_3\right] + 4\lambda_D(\alpha_1\beta_1 + \alpha_2\beta_2)\alpha_3\beta_3.\end{aligned} \tag{27.15}$$

The direction cosines α_i, β_i are defined in the Cartesian coordinate system with the 1 (x) and 2 (y) axes coinciding with the a and b directions, and the 3 (z) axis – with the hexagonal c-axis (see Fig. 27.35).

For a polycrystalline material, the longitudinal saturation magnetostriction can be obtained from the formula

$$\lambda_{\parallel} = (2/5)\lambda_A + (8/15)\lambda_D. \tag{27.16}$$

The transverse saturation magnetostriction (i.e. the magnetostriction measured at right angles Oto **J**) is expressed by

$$\lambda_{\perp} = (2/15)\lambda_A + (1/3)(\lambda_B + \lambda_C) - (4/15)\lambda_D. \tag{27.17}$$

For hexagonal crystals of rare-earth metals (see Figs. 27.32–27.34) the expressions for the saturation magnetostriction differ from those given above, namely

$$\begin{aligned}\lambda = {} & \left[\lambda_1^{\alpha,0} + \lambda_1^{\alpha,2}(\alpha_3^2 - 1/3)\right](\beta_1^2 + \beta_2^2) + \left[\lambda_2^{\alpha,0} + \lambda_2^{\alpha,2}(\alpha_3^2 - 1/3)\right]\beta_3^2 \\ & + (1/2)\lambda^{\gamma,2}\left[(\alpha_1\beta_1 + \alpha_2\beta_2)^2 - (\alpha_1\beta_2 - \alpha_2\beta_1)^2\right] + 2\lambda^{\epsilon,2}(\alpha_1\beta_1 + \alpha_2\beta_2)\alpha_3\beta_3\end{aligned} \tag{27.18}$$

or

$$\begin{aligned}\lambda = {} & A\left[2\alpha_1\alpha_2\beta_1 + (\alpha_1^2 - \alpha_2^2)\beta_2\right]^2 + B\alpha_3^2\left[(\alpha_1\beta_1 + \alpha_2\beta_2)^2 - (\alpha_1\beta_2 - \alpha_2\beta_1)^2\right] \\ & + C\left[(\alpha_1\beta_1 + \alpha_2\beta_2)^2 - (\alpha_1\beta_2 - \alpha_2\beta_1)^2\right] + D(1 - \alpha_3^2)(1 - \beta_3^2) + E\alpha_3^2\beta_3^2(1 - \alpha_3^2) + F\alpha_3^2(1 - \alpha_3^2) \\ & + G\beta_3^2(1 - \alpha_3^2) + H\alpha_3\beta_3(\alpha_1\beta_1 + \alpha_2\beta_2) + I\alpha_3^2\beta_3^2(\alpha_1\beta_1 + \alpha_2\beta_2) + J\alpha_3^2(1 - \beta_3^2) + K\alpha_3^2\beta_3^2.\end{aligned} \tag{27.19}$$

(Sometimes the notations $C = \lambda^{\gamma,2}/2$, $A = -\lambda^{\gamma,4}$ are also used.)

Soft magnetic materials. Soft magnetic materials denote those that are magnetized to saturation and reverse the magnetization in relatively weak magnetic fields $H \approx 10 - 10^3$ A/m. These materials are characterized by high values of the relative magnetic permeability – both the initial one $\mu_i = 10^2$–10^5 and the maximum one $\mu_{\max} = 10^3$–10^6. The coercive force H_c of magnetically soft materials falls usually within 1 to 10^2 A/m, whereas the core loss is very small — 1–10^3 J/m^3 per one cycle of magnetic reversal. For many materials, the specific loss is cited as a reference measure, that is the power loss P at the alternating magnetization field frequencies of 50 or 60 Hz for different values of the induction amplitude (e.g., $P_{1.0/50}$ is the power loss at the induction equal to 1.0 T and the field frequency of 50 Hz).

Magnetostrictive materials. The basic characteristics of the magnetostrictive materials (see Table 27.29) used for producing magnetostrictive transducers, include: magnetomechanical coupling factor K, the square of which equals the ratio of the energy (mechanical or magnetic) transduced to that (magnetic or mechanical, respectively) supplied; the dynamic magnetostrictive constant $a = (\partial\sigma/\partial B)_\epsilon$ and the magnetostrictive sensitivity constant $\Lambda = (\partial B/\partial\sigma)_H$, where σ is the mechanical stress, N/m^2, B – the magnetic induction, T, and the subscripts ϵ and H mean constant deformation and magnetic field. The value of a is essential for the performance of emitters, whereas the value of Λ – for that of receivers. The density ρ and the Young's modulus E determine the resonant frequency of transducers; the limiting intensity of magnetostrictive transducers depends on the mechanical strength, the saturation magnetostriction λ_s, and the saturation induction B_s; the mechanical quality Q, the electrical resistivity ρ_{el}, and coercive force H_c determines the eddy-current and hysteresis losses during the transducer performance. The values of K, a, and Λ depend substantially on the bias field intensity. The value H_{opt} of the latter, corresponding to the maximum K, is usually referred to as the optimal one.

Hard magnetic materials. Hard magnetic materials concern those that are magnetized to saturation and reverse the magnetism in comparatively strong magnetic fields $H \approx 10^3 - 10^5$ A/m. Magnetically hard materials are characterized by high values of the coercive force H_c, remanent induction B_r and maximum magnetic energy density $(BH)_{\max}$ on the demagnetization portion $B_r - H_{cB}$ of the hysteresis loop (Fig. 27.1). In Sections 27.2 and 27.3, each item comprising one or another group of metallic ferromagnets begins with the temperature-dependent paramagnetic susceptibility along with the dependences of magnetization, magnetic moment per unit mass or atomic magnetic moment on the temperature, magnetic field intensity, alloy composition, and the dependences of Curie temperatures of alloys on their composition. Then, data on magnetic anisotropy and, finally, on magnetostriction are presented. In selecting the material, we had no intention to reach overwhelming completeness, but an attempt was made to reflect the modern state and the trends in the physics of magnetic phenomena. The chapter also includes some information about the most widespread engineering magnetic materials.

The properties of metallic ferromagnetic materials are concerned in the manual [78] and in the encyclopedia [32]. The most complete information about magnetic properties of ferromagnetic metals and alloys can be found in monographs [12,39] as well as in a multivolume handbook [66].

27.2 Elements

27.2.1 Iron, cobalt, nickel

Among the transition d-metals, only Fe, Co and Ni ($3d$-metals) are ferromagnetic, these metals being the base of nearly all magnetic materials. Tables 27.1–27.5 and Figs. 27.2–27.15 comprise information about their magnetic properties. A review of magnetic properties for these metals can be found in [102].

Table 27.1 Principal magnetic properties of Fe, Co, and Ni

Element	σ_0, A·m^2/kg	n, μ_B	T_C, K	σ_s (4.2 K), A·m^2/kg	σ_s (287 K), A·m^2/kg	References
Fe	221.71	2.216	1044			[29]
		2.226		222.671	218.210	[84]
Co*	162.55	1.715	1388			[80]
		1.729 (**H** ∥ c) 1.721 (**H** ∥ [100] in basal plane)		163.862 (**H** ∥ c) 163.078 (**H** ∥ [100] in basal plane)	162.624	[84]
Ni	58.57	0.616	627.4			[29]
		0.619		58.872	55.370	[84]

* At temperatures near 700 K, a transition in Co occurs from the hexagonal close-packed (hcp) α–phase (stable at low temperatures) to the face-centered cubic (fcc) β–phase (stable at high temperatures) [102].

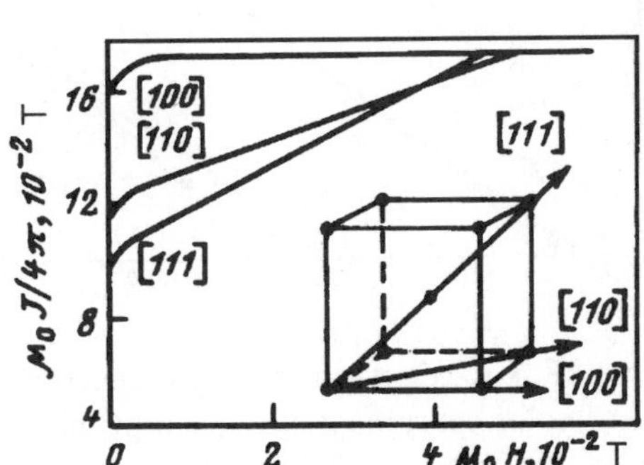

Figure 27.2 Magnetization curves for the main crystallographic directions in a Fe single crystal at T=20°C [53].

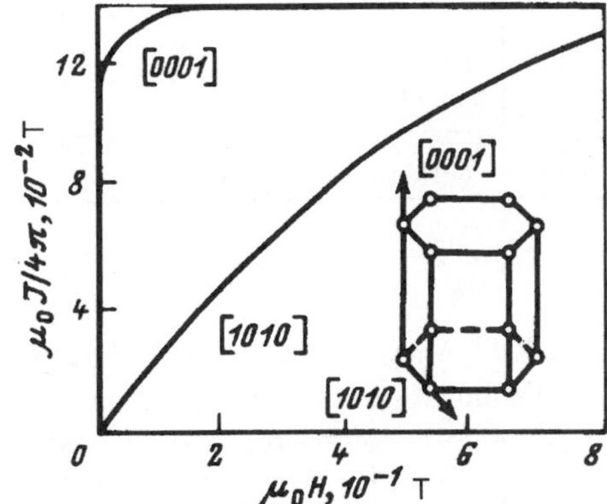

Figure 27.3 Magnetization curves for the main crystallographic directions in a Co single crystal at T=20°C [59].

Table 27.2 Dependence of the relative magnetization J_s/J_o for Fe, Co, and Ni on the relative temperature T/T_C [2]

T/T_C	Fe	Co, Ni
0	1	1
0.1	0.996	0.996*
0.2	0.99	0.99
0.3	0.975	0.98
0.4	0.95	0.96
0.5	0.93	0.94
0.6	0.90	0.90
0.7	0.85	0.83
0.8	0.77	0.73
0.85	0.70	0.66
0.9	0.61	0.56
0.95	0.46	0.40
1	0	0

* Only for Ni.

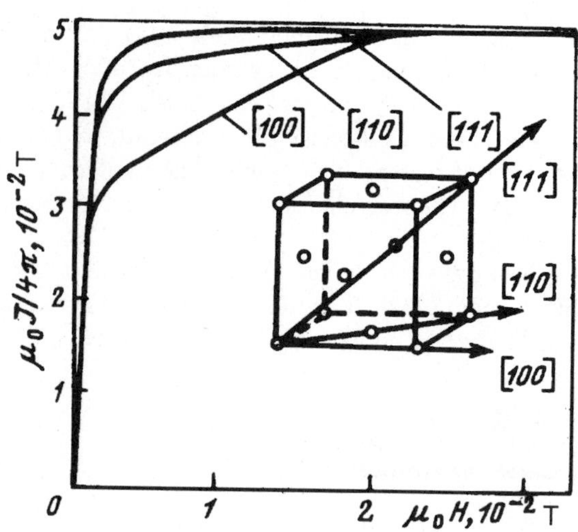

Figure 27.4 Magnetization curves for the main crystallographic directions in a Ni single crystal at T=20°C [58].

Table 27.3 The derivative $(1/\mu_0)(\partial\omega_I/\partial H)_T$ for elements of the iron group

Element	T, K	$(1/\mu_0)(\partial\omega_I/\partial H)_T$, 10^{-6} T^{-1}	References
Fe	1.5	45	[92]
	77	4.5	[92]
	293	4.5–6	[65,92]
β-Co, fcc*	4.2	4.8	[91]
α-Co, hcp*	4.2	3.2	[91]
		7.2	[102]
	293	6	[14]
Ni	4.2	1.2	[96]
	77	1.5	
	294	1.0	
	377	0.5	
	428	0.0	
	479	−0.7	
	568	−3.8	
	606	−11.0	

* See footnote to Table 27.1.

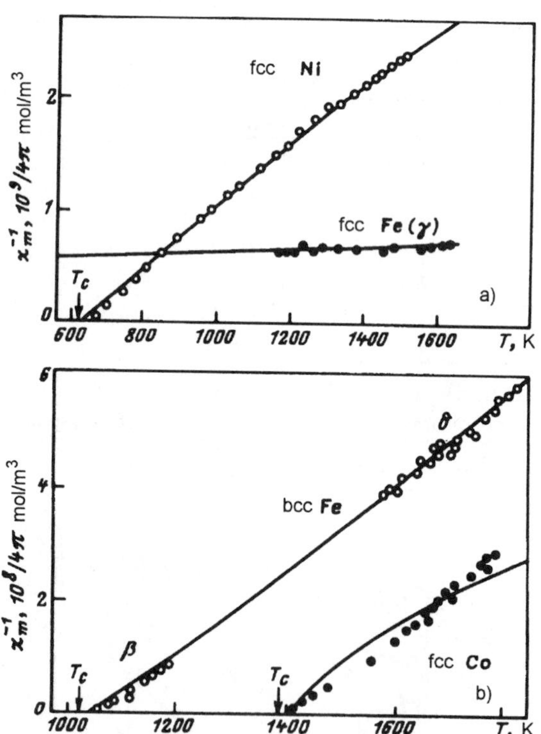

Figure 27.5 Temperature dependence of the inverse molar paramagnetic susceptibility for face-centered cubic Ni [31] and face-centered cubic γ-phase of Fe (**a**), as well as for body-centered cubic β- and δ-phases of Fe and face-centered cubic β-phase of Co (**b**) [81].

Table 27.4 Magnetostriction constants h_i (in 10^{-6}) of Fe and Ni at room temperature

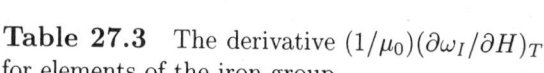

Element	h_1	h_2	h_3	h_4	h_5	References
Fe	36.2	−34.0	2			[42]
Ni	−94– −99	−42– −44	−0.7– +0.1	−1.4– +3.4	0.2– 1.5	[11,67]

Table 27.5 Magnetostriction constants of α-Co (hcp)* at various temperatures [55]

T,°C		λ_A, 10^{-6}	λ_B, 10^{-6}	λ_C, 10^{-6}	λ_D, 10^{-6}
−200		−66	−123	+126	−128
0		−52	−109	+126	−108
20		−50	−107	+126	−105
200		−32	−88	+120	−82
400		−16	−70	+105	−52
20	[14]	−45	−95	+110	−100

* See footnote to Table 27.1.

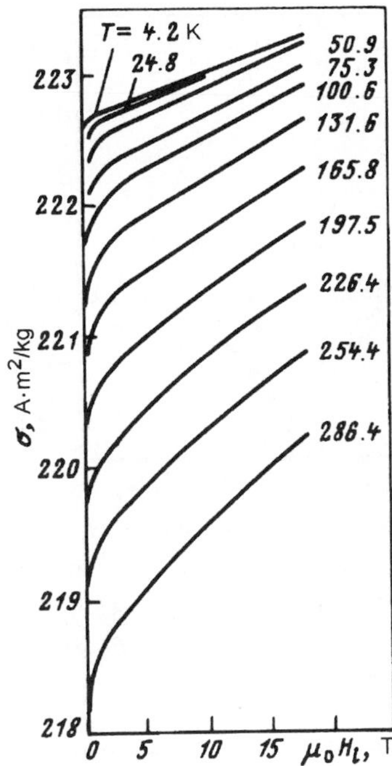

Figure 27.6 Magnetization curves of a Fe single crystal in strong magnetic fields at different temperatures. The magnetizations along the [100], [110] and [111] axes differ here by less than 0.05%. The curves are drawn through the average values of these magnetizations; $\mu_0 H_i$ — the induction of an internal magnetic field [84].

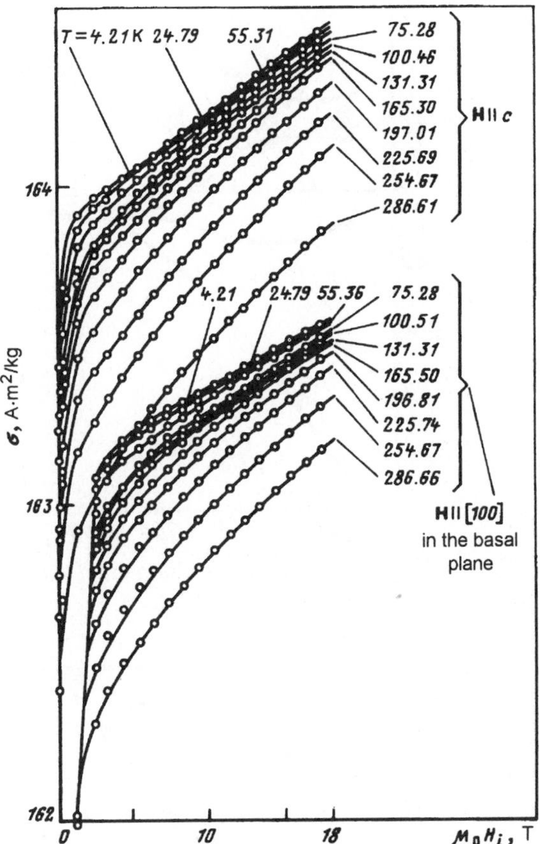

Figure 27.7 Magnetization curves of a hexagonal Co crystal in strong magnetic fields at different temperatures; H_i, H — the internal and applied magnetic fields, c — the hexagonal axis [84].

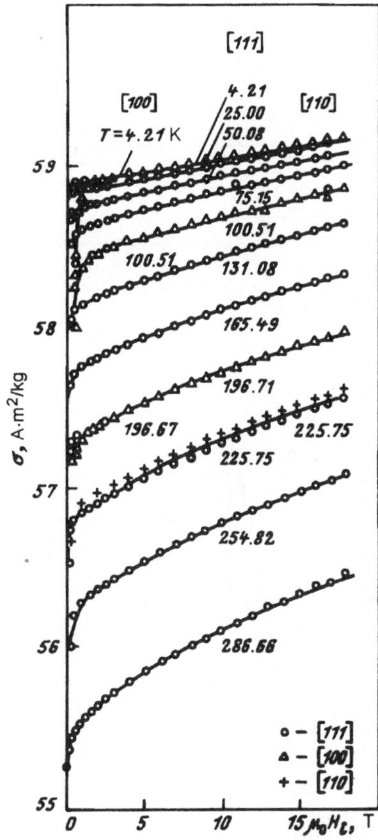

Figure 27.8 Magnetization curves of a Ni crystal in strong magnetic fields at different temperatures; $\mu_0 H_i$ — the induction of an internal magnetic field [84].

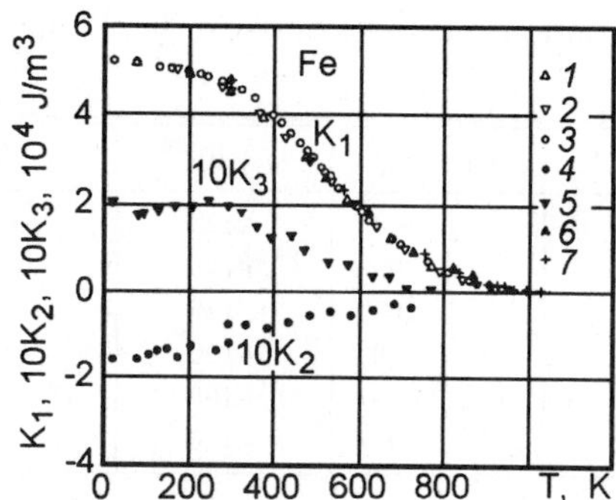

Figure 27.9 Temperature dependence of magnetocrystalline anisotropy constants of Fe [66] as obtained from torque measurements [60,41], ferromagnetic resonance [8], and ac susceptibility [50,60].

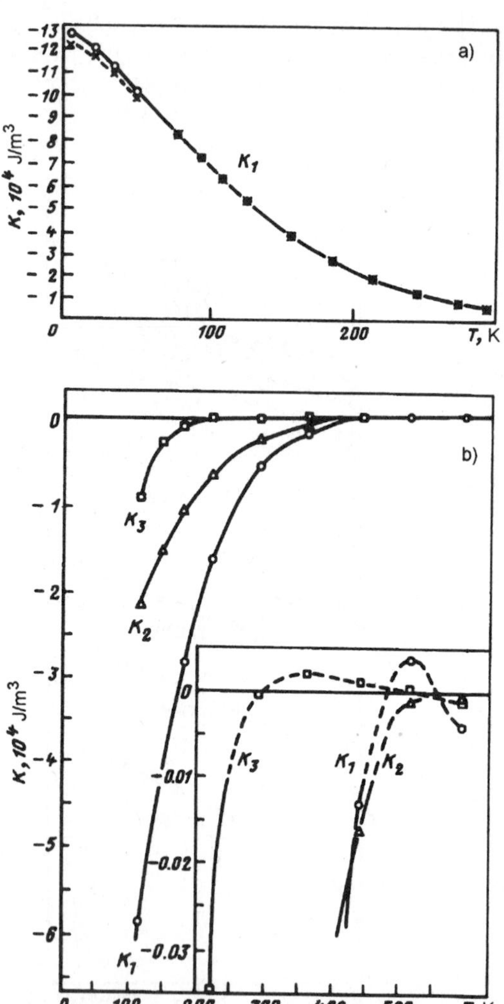

Figure 27.10 Temperature dependences of the magnetic anisotropy constants in Ni after [37] (a) (circles – the first paper in [37], crosses – the second one) and [4] (b).

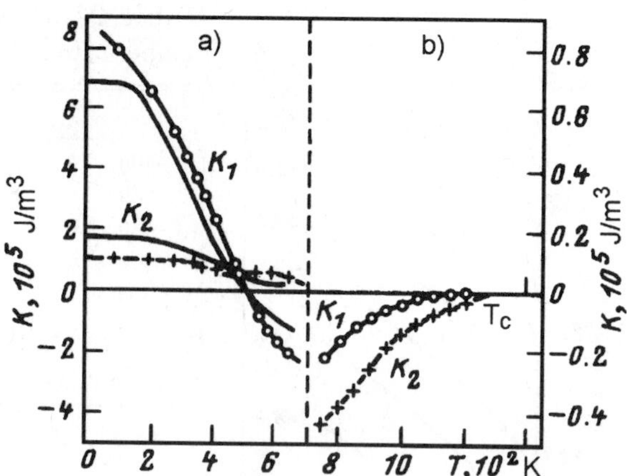

Figure 27.11 Temperature dependence of the magnetic anisotropy constants of Co for hexagonal (a) and face-centered cubic (b) phases: curves with data points – after [54,95], curves without data points – after [5].

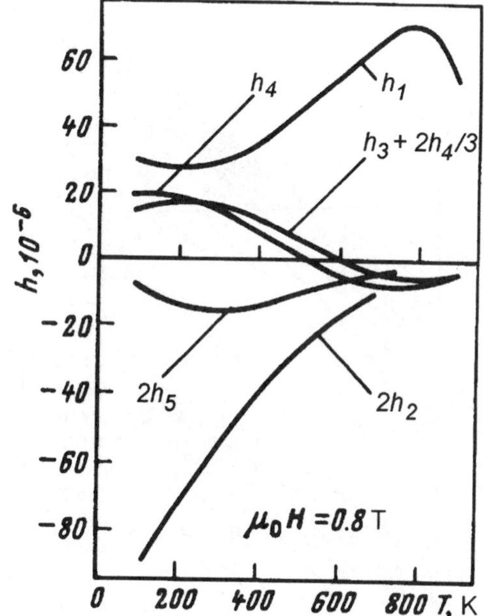

Figure 27.12 Temperature-dependent magnetostrictive coefficients of Fe [101].

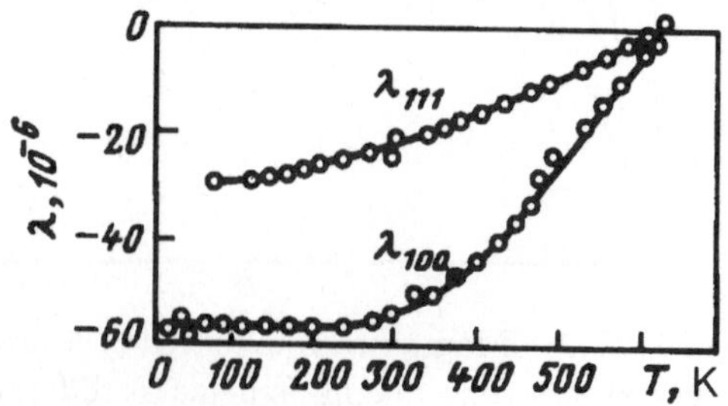

Figure 27.13 Temperature-dependent λ_{100} and λ_{111} magnetostrictive coefficients of Ni [65].

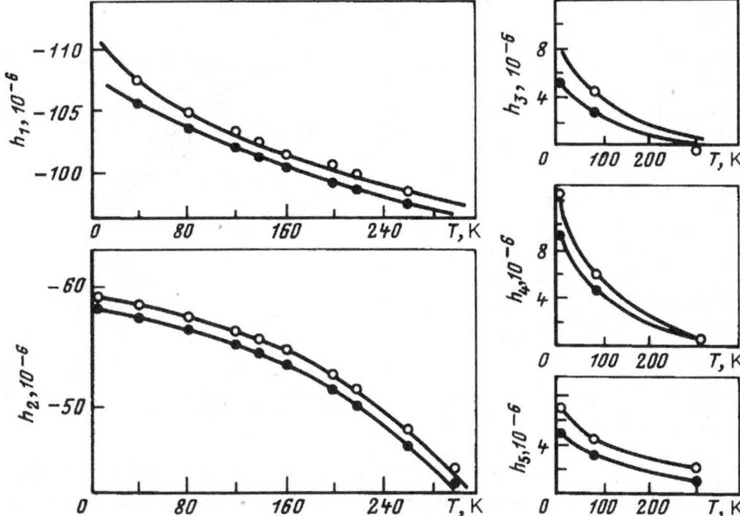

Figure 27.14 Temperature dependences of the magnetostrictive coefficients for two Ni samples (closed and open circles, respectively) [67].

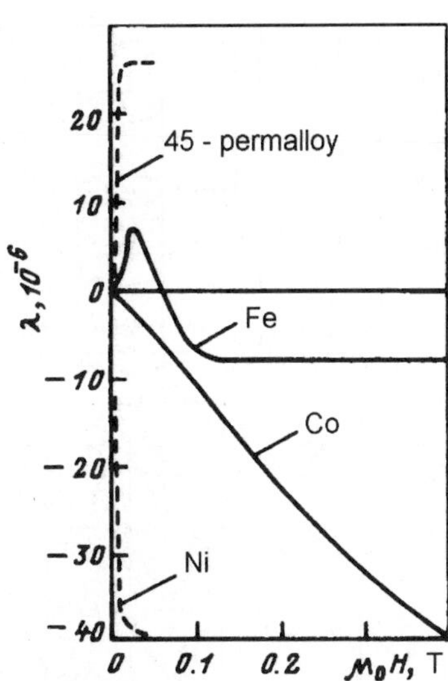

Figure 27.15 Longitudinal magnetostriction of polycrystalline samples as a function of applied magnetic field induction at $T = 20°C$ [12].

27.2.2 Rare-earth and actinoid elements

Ferromagnetic ordering in f-metals manifests itself more often in comparatively complex magnetic structures ("Cone", "Ferri" in Fig. 27.16). A transition to them from the paramagnetic state ("Para" in Fig. 27.16) occurs during the temperature decrease, as a rule through the antiferromagnetic (AF) phases ("Helix", "AF-Cone", and "CAM" in Fig. 27.16). There exists information about the ferromagnetic cubic phases of Pr, Nd, and Pm [69], which are metastable at and beyond the ambient temperature. Ferromagnetism was found [56] in the face-centered cubic phase of Cm (see Figs. 27.29, 27.30), which is also metastable at temperatures below the Curie temperature. Hexagonal (dhcp) Cf seems to have a complex ferromagnetic structure below $T_C = 51$ K [56]. Data on the magnetic properties of rare-earth metals are presented in Table 27.6. The temperature and field dependences of the spontaneous magnetic moment, as well as the temperature dependences of the inverse paramagnetic susceptibility of the elements from the groups discussed, are depicted in Figs. 27.17–27.30. In what follows the typical temperature dependences of anisotropy and magnetostriction coefficients are given for heavy rare-earth elements (Figs. 27.31–27.34) and, finally, the magnetostriction dependence on the magnetic induction is exemplified for Dy (Figs. 27.35, 27.36).

Table 27.6 Magnetic properties of ferro- and ferrimagnetic rare-earth metals (phase with the space group $P6_3/mmc$) [69]

Element	μ_{eff}, μ_B	n, μ_B	Direction of the easy-magnetization axis at $T = 0$ K	T_N, K	T_C, K	θ_p, K	
						$H \parallel c$	$H \perp c$
Gd	7.98	7.63			293.4	317	317
Tb	9.77	9.34	b	230.0	219.5	195	239
Dy	10.83	10.33	a	179.0	89.0	121	169
Ho	11.2	10.34	b	132.0	20.0	73.0	88.0
Er	9.9	9.1		85.0	20.0	61.7	32.5
Tm	7.61	7.14	c	58.0	32.0	41.0	−17.0

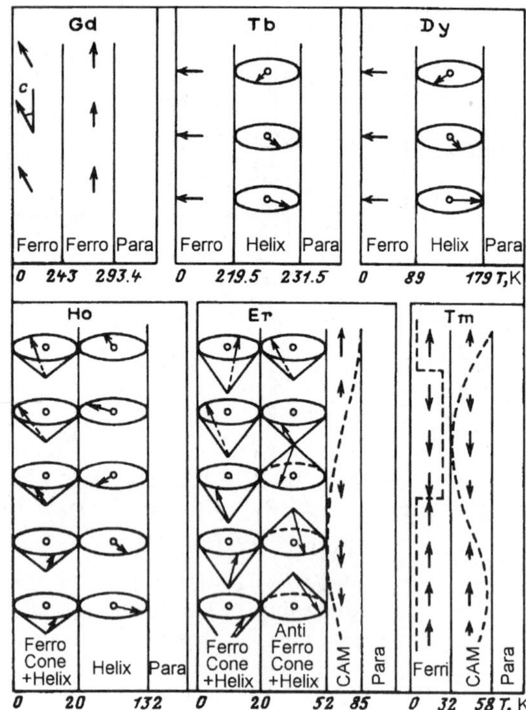

Figure 27.16 Magnetic ordering in the heavy rare-earths as found by neutron diffraction [69]. CAM is the c-axis modulated structure.

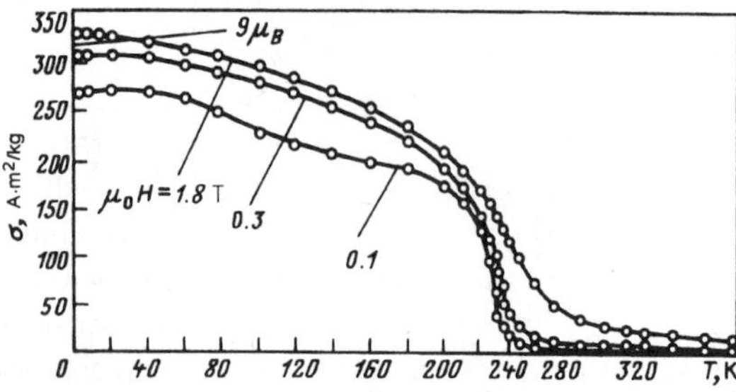

Figure 27.17 Temperature dependence of the specific magnetic moment (along b-axis) in Tb crystal at different values of the applied magnetic field induction [52].

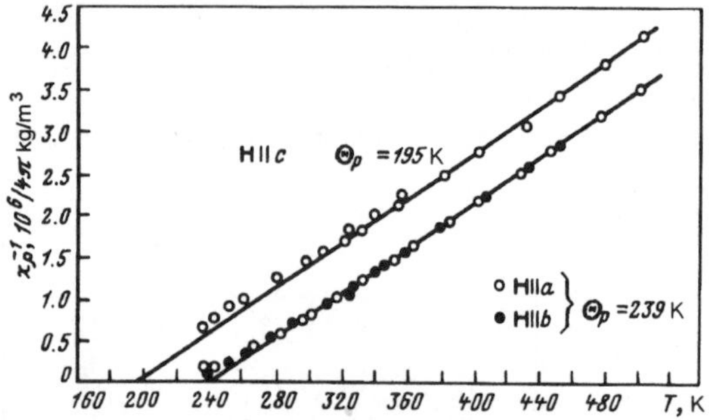

Figure 27.18 Reciprocal of the paramagnetic susceptibility (along a-, b- and c-axes) in Tb crystal vs. temperature [45].

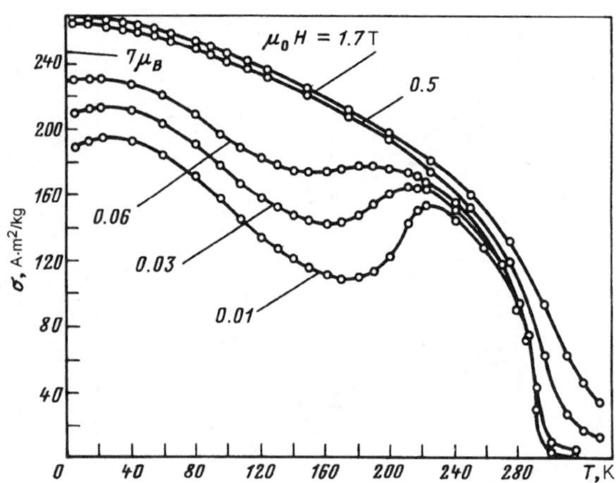

Figure 27.19 Temperature-dependent specific magnetic moment (along c-axis) of Gd crystal at different values of the applied magnetic field induction [82].

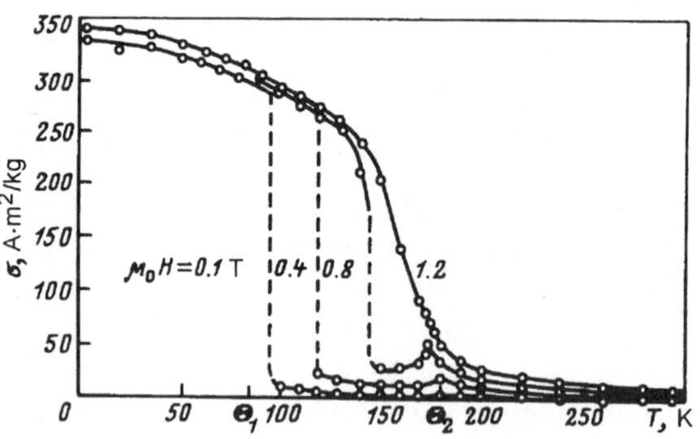

Figure 27.20 Temperature-dependent specific magnetic moment of Dy single crystal at different values of the induction of magnetic field applied along the a-axis in the basal plane [31].

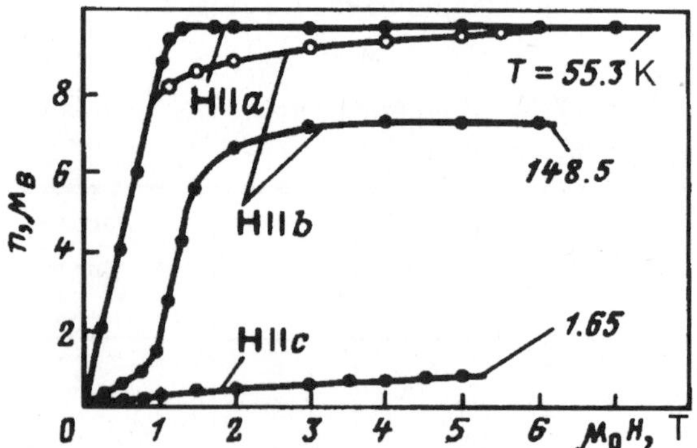

Figure 27.21 Magnetization isotherms for a Dy single crystal [38].

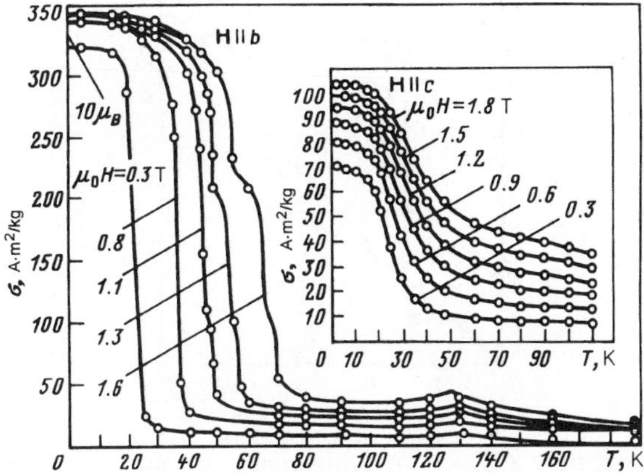

Figure 27.22 Temperature-dependent specific magnetic moment (along *b*- and *c*–axes) of Ho crystal at different values of the applied magnetic field induction [93].

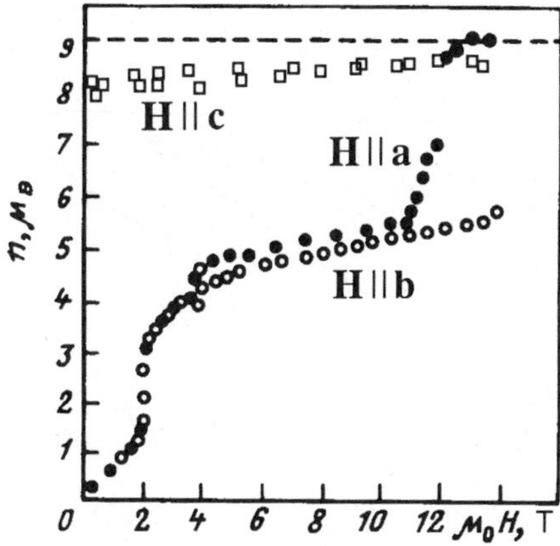

Figure 27.25 Atomic magnetic moment of a Er single crystal vs. induction of the magnetic field applied along *a*-, *b*- and *c*-axes at a temperature of 4.2 K [15]. Some of the *n* values for **H** ∥ **c** correspond to that from [87] multiplied by 0.963 [75]. Black/white circles – values of *n* from [87].

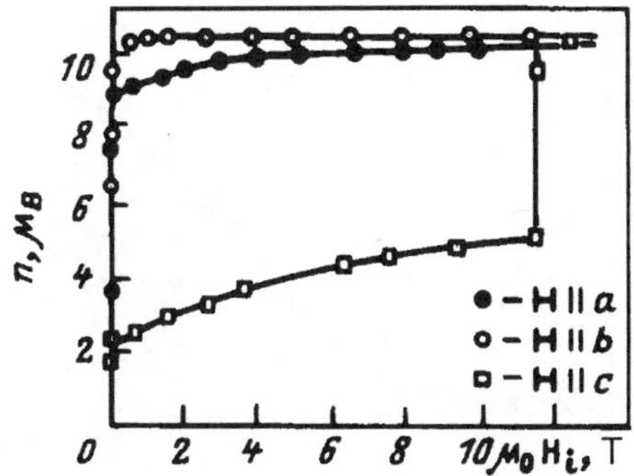

Figure 27.23 Atomic magnetic moment of a Ho single crystal vs. induction of the internal magnetic field. An external field was applied along *a*-, *b*- and *c*-axes at a temperature of 4.2 K [15].

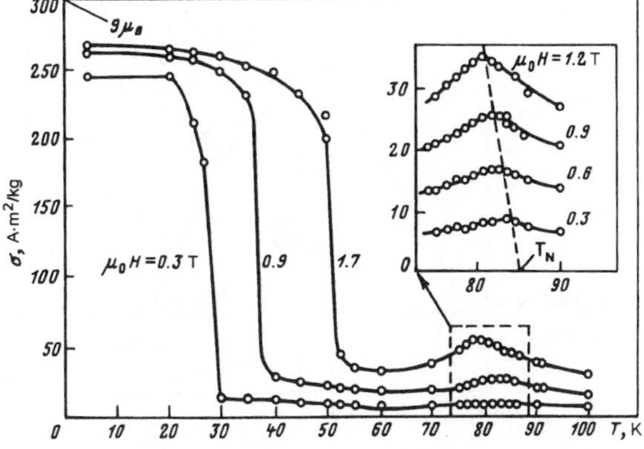

Figure 27.24 Temperature-dependent specific magnetic moment (along *c*–axis) of Er crystal at different values of the applied magnetic field induction [45].

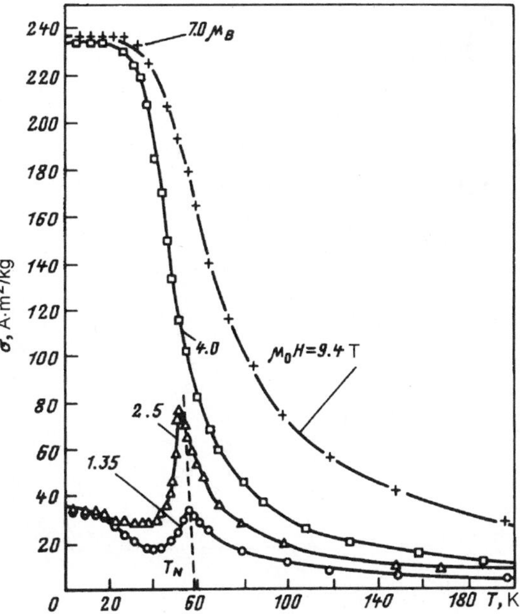

Figure 27.26 Temperature dependence of the magnetic moment per unit mass (along *c*–axis) of Tm crystal [88].

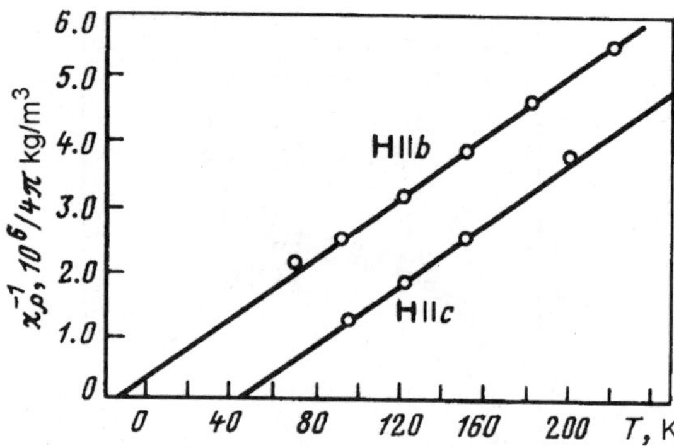

Figure 27.27 Temperature dependence of the inverse specific susceptibility in the paramagnetic region (along b- and c-axes) of Tm crystal [88].

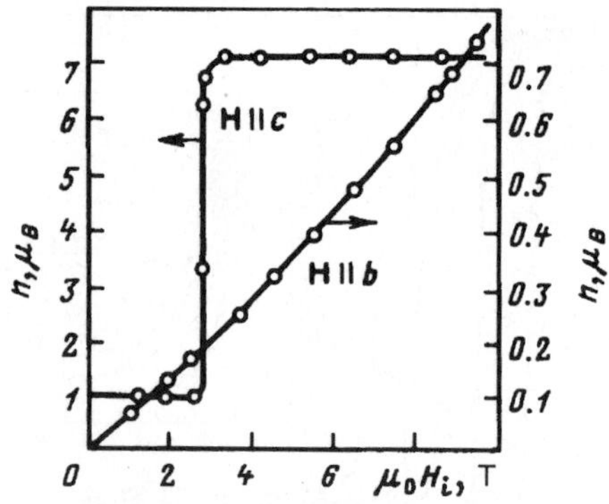

Figure 27.28 Atomic magnetic moment of a Tm single crystal vs. induction of internal magnetic field at a temperature 4.2 K [88].

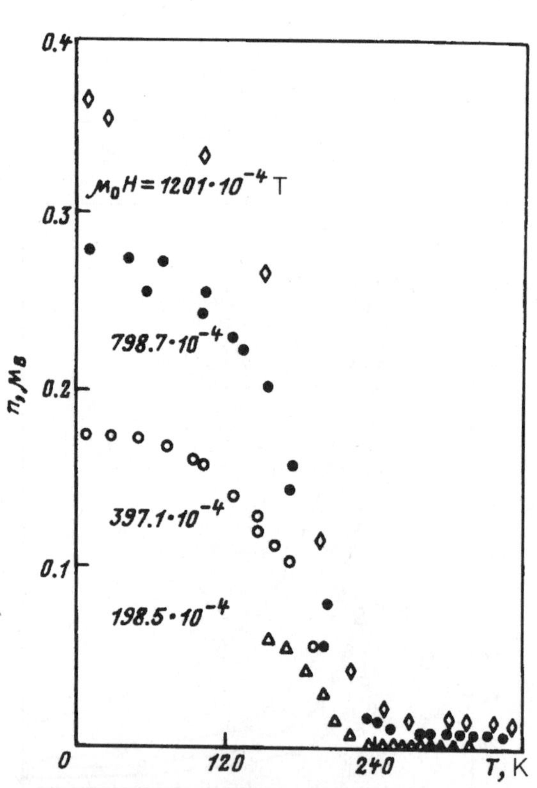

Figure 27.29 Temperature-dependent atomic magnetic moment for Cm in face-centered cubic β-phase at different values of applied magnetic field induction [56].

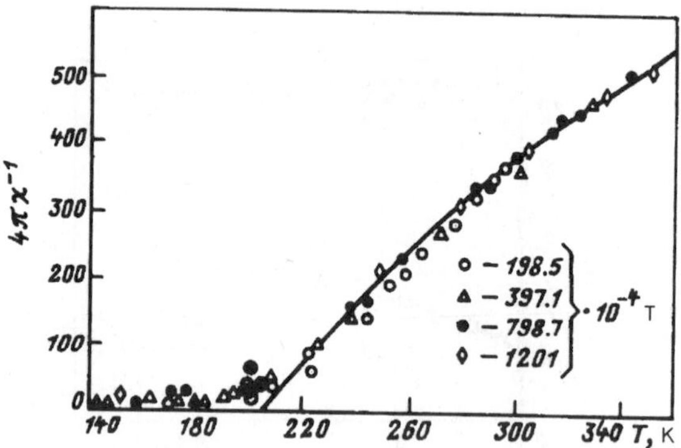

Figure 27.30 Temperature-dependent inverse bulk paramagnetic susceptibility of Cm in face-centered cubic β-phase at different values of applied magnetic field induction [56].

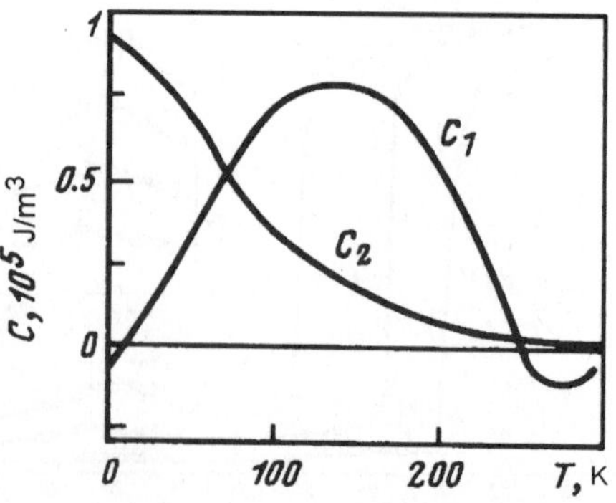

Figure 27.31 Temperature dependence of the magnetic anisotropy constants $C_1 = -K_1 + K_2$ and $C_2 = K_2/2$ in a hexagonal Gd crystal [76].

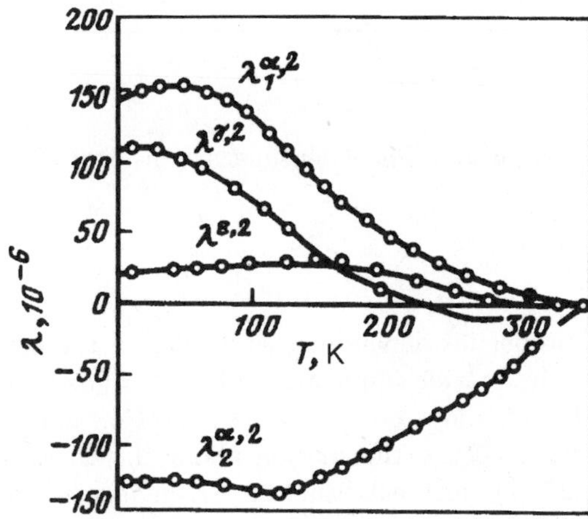

Figure 27.32 Temperature dependences of the magnetostriction constants in Gd [77].

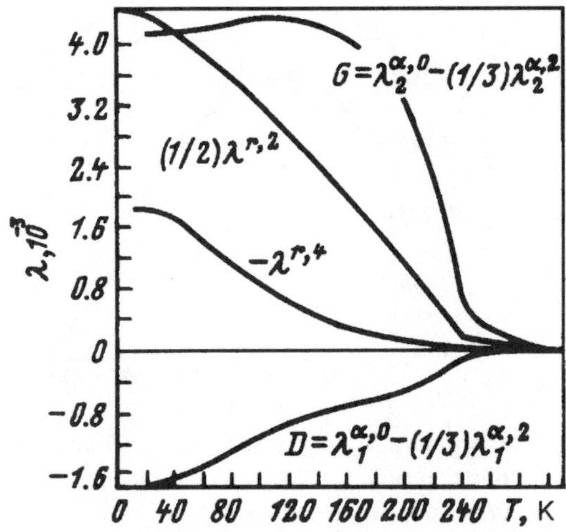

Figure 27.33 Temperature-dependent magnetostriction constants of Tb extrapolated to a zero applied field [86].

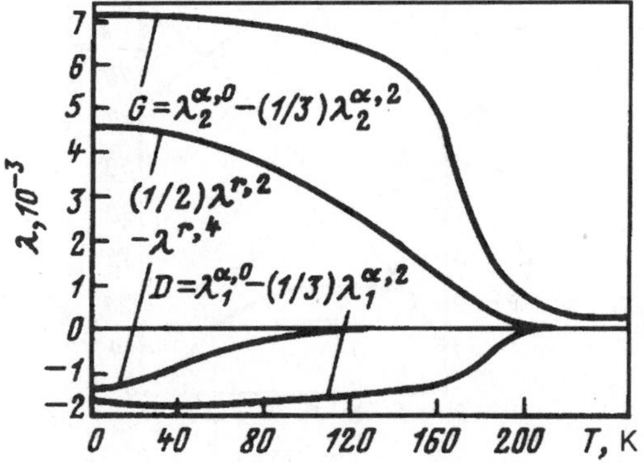

Figure 27.34 Temperature-dependent magnetostriction constants of Dy in an applied field with the induction of 3 T [85].

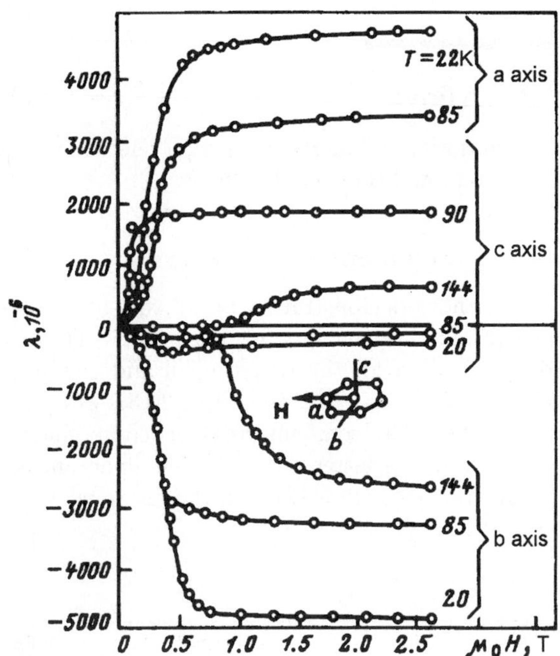

Figure 27.35 Magnetostriction along the axes of Dy single crystal vs. induction of a magnetic field applied along the a-axis [68].

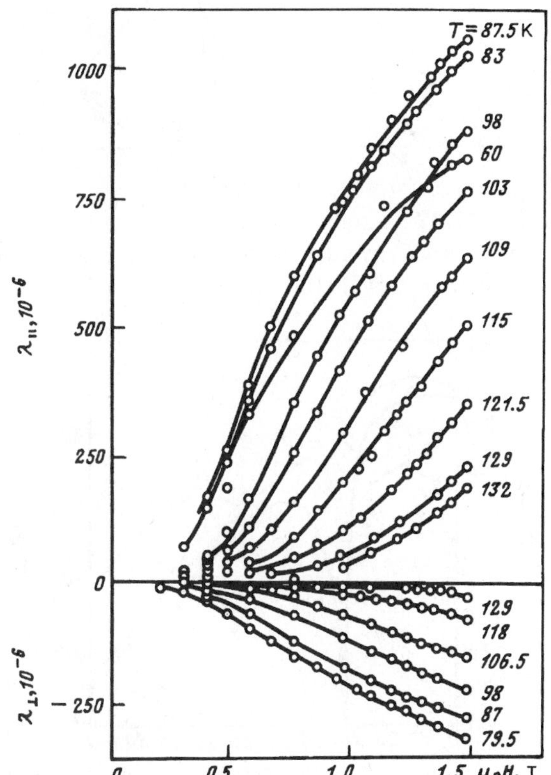

Figure 27.36 Isotherms of the longitudinal and transverse magnetostrictions in polycrystalline hexagonal Dy [6].

27.3 Alloys

The majority of data in this section refers to two-component alloys. The exclusions are the Fe–Co–Ni alloy system and the Geissler alloys.

27.3.1 *d*-element based alloys

These alloys demonstrate a broad variety of magnetic properties depending, as a rule, on the mechanical and thermomagnetic treatments. This fact provides a wide application of these alloys. In addition to data for the well-studied and engineering relevant alloys on the base of Fe, Co, and Ni (Tables 27.7, 27.8, 27.12 and Figs. 27.37–27.54), this item concerns information about the Geissler alloys (Table 27.9), some intermetallic compounds (Table 27.11) and weak itinerant ferromagnets (Table 27.10). In the latter alloys, a weak spontaneous magnetization ($n < 1$) arises due to the ordering of conduction electrons. Data on the dilute alloys demonstrating peculiar ferromagnetic properties can be found in [79].

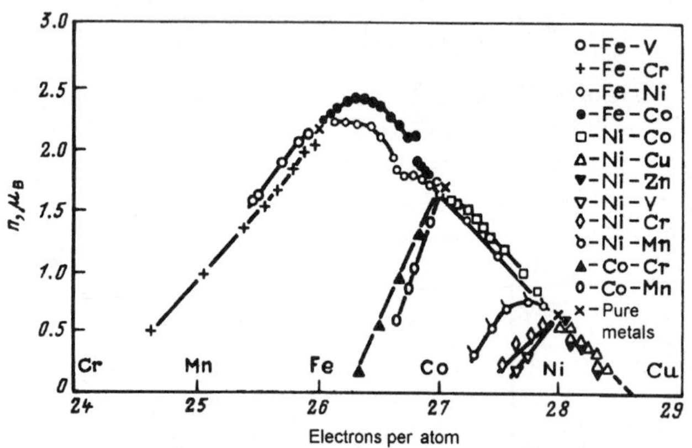

Figure 27.37 Atomic magnetic moment of alloys and pure metals vs. average number of electrons per atom (Slater–Poling curve) [12].

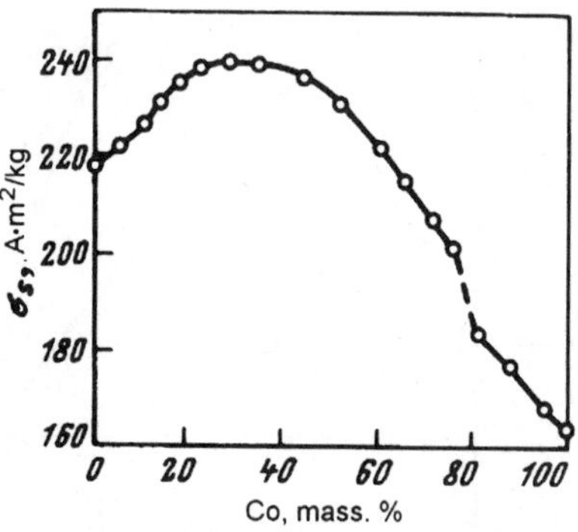

Figure 27.39 Specific saturation magnetic moment of Fe–Co alloys at T = 20°C vs. composition [100].

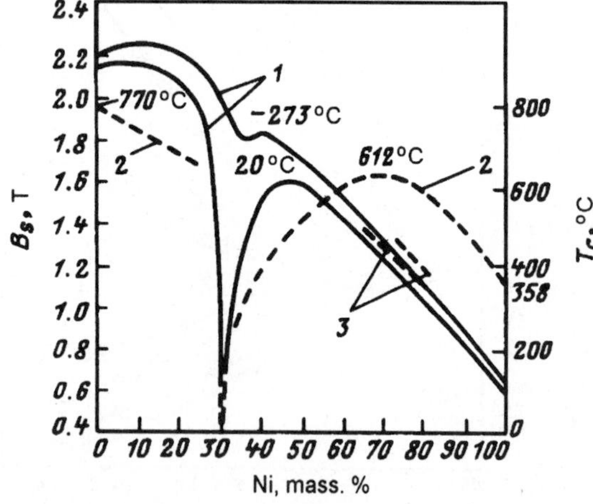

Figure 27.38 Saturation induction (*1*) and Curie temperature (*2*) of Fe–Ni alloys and saturation induction of ordered alloys with composition close to $FeNi_3$ (*3*) [12].

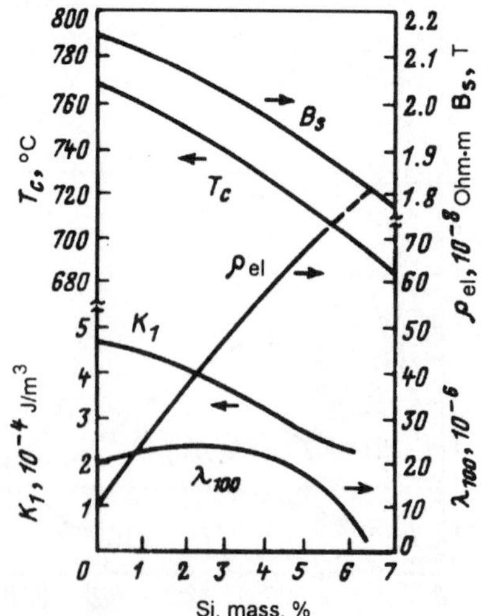

Figure 27.40 Magnetic properties and electrical resistivity of Fe–Si alloys at T = 20°C [70].

Table 27.7 Specific saturation magnetic moment and the Curie temperature of binary iron-based alloys [2]

Second component of alloy	Content, at.%	σ_s, A·m^2/kg (T=20°C)	n, μ_B	T_C, K
Al	7.1	207	2.05	1029
	19.7	184	1.74	937
	24.9	134	1.29	714
	26.0	149	1.40	767
Au	6.2	174	2.08	1040
	10.2	154	2.02	1041
Co	20	236	2.42	1223
	33	238	2.52	1243
	50	233	2.42	1253
	75	203	2.14	1143
	80	184	1.94	1183
Cr	17.7	196	1.70	951
	47.5	90	0.98	756
	67.8	35	0.53	541
Ir	4.0	200	2.25	1023
	15.0	120	1.67	
Os	8.1	158	1.97	
	12.5	50	0.69	
Pd	5.5	203	2.19	1027
	40.0	129	1.89	
	74.8	45	0.97	523
Pt	8.1	191	2.36	
	12.4	177	2.43	
	24.8	104	2.23	437
	50.0	32	0.75	
Rh	10.0	209	2.32	
	25.0	192	2.39	987
	40	161	2.26	897
Ru	7.0	200	2.18	933
	12.5	105	1.17	
Sn	2.3	208	2.18	1041
	6.0	197	2.16	1041
Si	8.3	204	2.00	993
	15.9	174	1.67	926
	23.5	141	1.32	860
V	5.9	204	2.09	1088
	10.6	184	1.91	1078
	18.6	149	1.58	1056

Table 27.8 Specific saturation magnetic moment and the Curie temperature of binary cobalt-based alloys [2]

Second component of alloy	Content, at.%	σ_s, A·m^2/kg (T=20°C)	n, μ_B	T_C, K
Cr	5.6	134	1.42	
	10.6	100	1.07	
	16.7	59.5	0.64	
	22.1	19	0.24	
Mn	4.2	144	1.53	
	11.9	109	1.16	
	17.3	84	0.89	
	22.5	48	0.57	
Ni	40	124	1.38	1173
	70	90	0.97	953

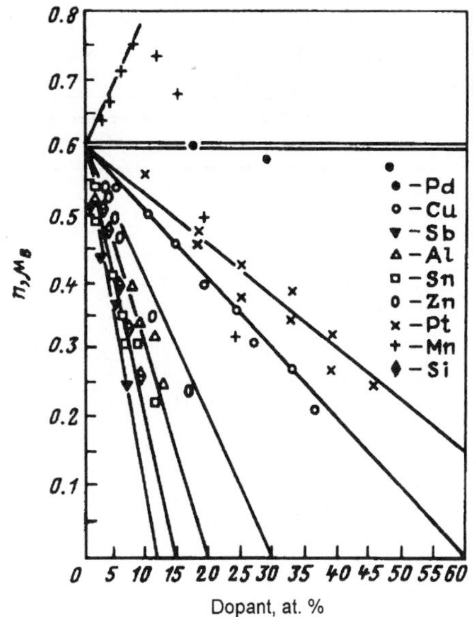

Figure 27.41 Atomic saturation magnetic moments vs. composition of Ni alloys with nonferromagnetic metals [12].

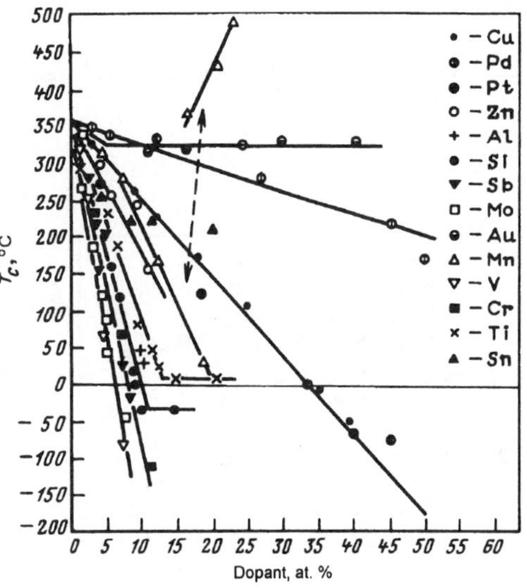

Figure 27.42 Curie temperatures of Ni-based alloys [12].

Table 27.9 The Curie temperature and atomic magnetic moments of Geissler alloys (simple cubic structure of CsCl) [97]

Compound	T_C, K	n, μ_B
Cu_2MnAl	600	3.6
Cu_2MnIn	520	4.0
Cu_2MnSn	530	4.1
Co_2MbSi	982	5.1
Co_2MnGa	694	4.1
Co_2MnGe	905	5.1
Co_2MnSn	829	5.1
Ni_2MnGa	379	4.2
Ni_2MnIn	323	4.4
Ni_2MnSn	344	4.1
Ni_2MnSb	360	3.3
Pd_2MnGe	170	3.2
Pd_2MnSn	189	4.2
Pd_2MnSb	247	4.4
Au_2MnAl	258	3.1

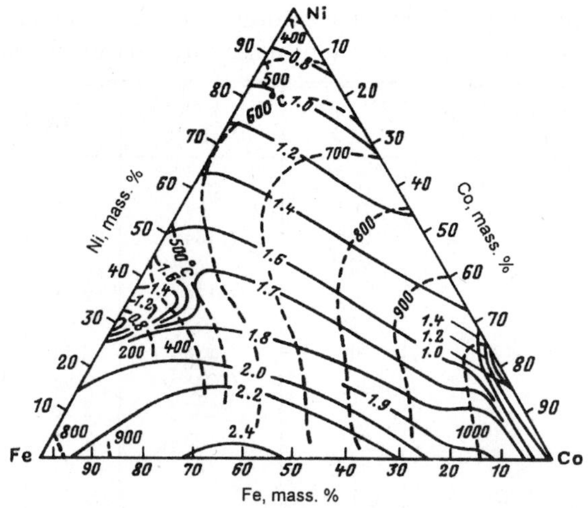

Figure 27.43 Induction (in T, solid curves) corresponding to magnetization J in a field $H = 1.19 \cdot 10^5$ A/m and Curie temperature (in °C), dashed curves) for Fe–Co–Ni alloys [57].

Table 27.10 Magnetic properties and crystalline structure of itinerant ferromagnets on the base of transition d-elements [7,91]

Compound	Crystal structure: Syngony, space group	Crystal structure: Lattice parameters, nm	ρ, 10^3kg/m^3	T_C, K	θ_p, K	μ_{eff}, μ_B	n, μ_B	References
$ZrZn_2$	Cubic, $Fd3m$	a=0.739	7.16	16–27	33	1.3–1.4 0.75	0.12–0.18	[83,85] [93]
Sc_3In	Hexagonal, $P6_3/mmc$	a=0.642–0.656 c=0.512–0.518	4.38	6–7.5	16	0.67–1.3	0.06–0.15 (Sc)	[60,84]
Au_4V	Body-centered tetragonal structure of Ni_4Mo type, $D1_a$	a=0.640 c=0.398		43–55		1.4–1.7	0.41–0.92 (V)	[51,53]
Ni_3Al	Orthorhombic		7.45	41.5–75	130	1.0	0.075–0.1	[35]

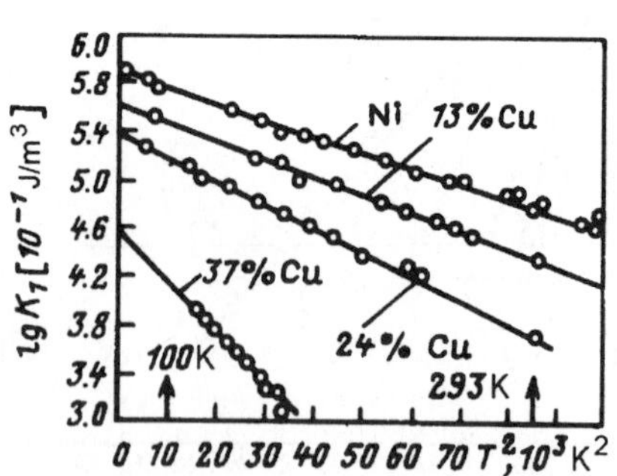

Figure 27.44 Temperature-dependent magnetic anisotropy constant of Ni-Cu alloys with different composition [12].

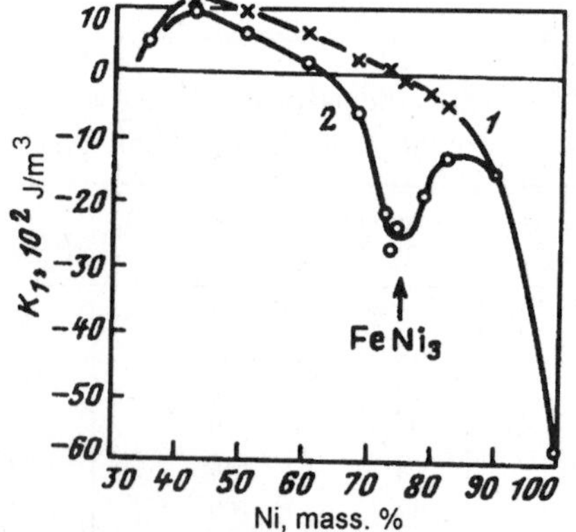

Figure 27.45 Magnetic anisotropy constants K_1 for Fe–Ni alloys at T = 20°C: *1* – quenched disordered alloys; *2* – slowly cooled down ordered alloys [13].

Table 27.11 Atomic magnetic moments per one atom of a 3d-element and Curie temperatures of some binary intermetallics [16]

Compound	Syngony (crystal lattice type)	T_C, K	n, μ_B
Au_4Mn	Body-centered tetragonal (Ni_4Mo)	363	4.15
CoB	Orthorhombic (FeB)	477	0.28
Co_2B	Tetragonal ($CuAl_2$)	429	0.76
Co_3B	Orthorhombic (Fe_3C)	747	1.11
CoPt	Tetragonal (AuCu)	813	0.17
CoS_2	FCC (FeS_2)	122–130	0.84–0.96
$CrBe_{12}$	Tetragonal ($MoBe_{12}$)	50	~0.2
$CrGe_2$		98	~0.1
$Cr_{1.2}Pt_{2.8}$	FCC (Cu_3Au)	>77	2.56(Cr)
$CrS_{1.19}$	Hexagonal (NiAs)	160 (T_N) 305 (T_C)	0.11
CrTe	Hexagonal (NiAs)	239–334	2.45
Cr_3Te_4	Monoclinic	80 (T_N) 329 (T_C)	2.3
FeAl	Cubic (CsCl)	623	~1.0
Fe_3Al	BCC (CsCl superlattice)	773	1.46 (Fe I) 2.14 (Fe II)
FeB	Orthorhombic (FeB)	598	1.12
Fe_2B	Tetragonal ($CuAl_2$)	1043	1.91
$FeBe_5$	FCC ($MnCu_2$)	75	~0.1
Fe_3C	Orthorhombic (Fe_3C)	483	2.01
Fe_3Cr	FCC (Cu_3Au)	993	~1.3
Fe_3Ge	Hexagonal (Ni_3Sn)	365	1.90
FeP	Orthorhombic (MnP)	215	0.36
Fe_2P	Hexagonal (Fe_2P)	266–278	0.77–1.32
Fe_3P	Tetragonal (Ni_3P)	716	1.84
$FePd_3$	FCC (Cu_3Au)	540	2.7 (Fe)
FePt	Tetragonal (AuCu)	743	~0.2
FeRh	Cubic (CsCl)	330 (T_N) 675 (T_C)	3.0 (Fe) 0.9 (Rh)
Fe_3Si	Cubic (Cu_2MnAl)	808	1.51 (Fe I) 2.15 (Fe II)
Fe_3Sn	Hexagonal (Ni_3Sn)	743	1.9
MnAs	Hexagonal (NiAs)	318 (heating) 306 (cooling)	3.4
MnB	Orthorhombic (FeB)	578	1.92
MnB_2	Hexagonal (AlB_2)	143–157	0.19–0.25
MnBi	Hexagonal (NiAs)	633	3.52
Mn_3Ga	Hexagonal	470	~0.02
Mn_3Ge	Hexagonal (Ni_3Sn)	28	0.38
Mn_5Ge_3	Hexagonal (Mn_5Si_3)	320	2.5
Mn_3In	Cubic (Cu_5Zn_8, γ-brass)	583	~0.1
$MnPt_3$	FCC (Cu_3Au)	<300	3.60 (Mn) 0.17 (Pt)
MnSb	Hexagonal (NiAs)	583	3.53
MnSi	Cubic (FeSi)	34	0.4
Mn_5Sn_3	Hexagonal (NiIn)	263	1.23
Mn_5Y	Orthorhombic ($GdMn_5$)	490	2.2
$MnZn_3$	Hexagonal (Ni_3Sn)	>400	~1.0
NiPt	Tetragonal (CuAu)	136	0.06
Ni_3Y	Orthorhombic ($CeNi_3$)	33	0.16

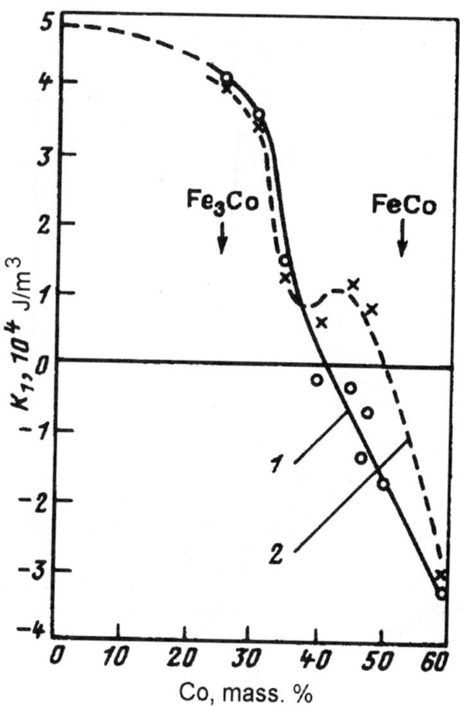

Figure 27.46 Magnetic anisotropy constants K_1 for Fe–Co alloys at $T = 20°C$: *1* – quenched disordered alloys; *2* – slowly cooled down ordered alloys [48].

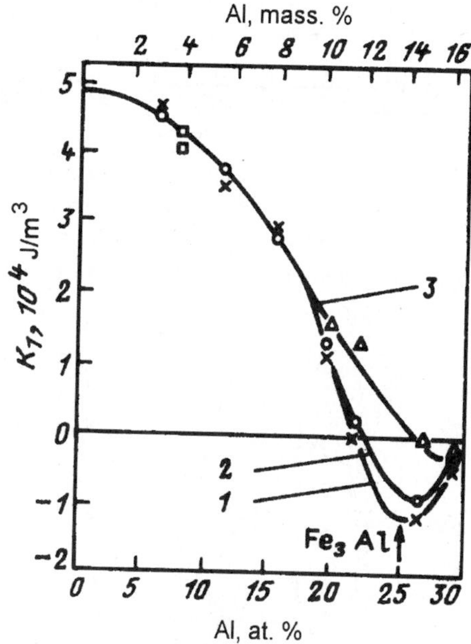

Figure 27.47 Magnetic anisotropy constants K_1 for Fe–Al alloys at $T = 20°C$: *1* – slowly cooled down ordered alloys; *2* – partially ordered alloys; *3* – quenched disordered alloys [47].

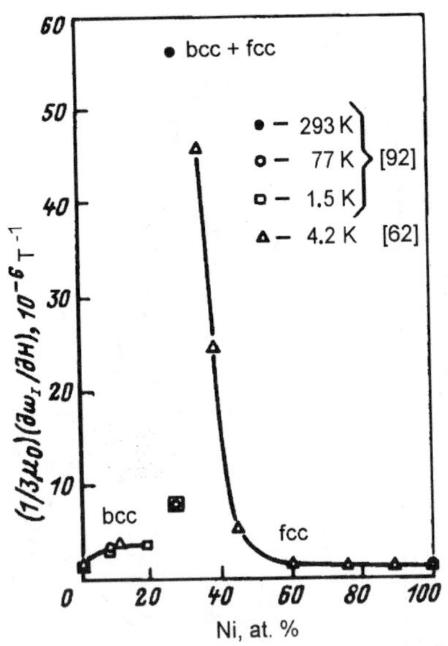

Figure 27.48 Derivative $\frac{1}{3\mu_0}\left(\frac{\partial\omega}{\partial H}\right)_T$ for Fe–Ni alloys.

Table 27.12 Dependence of magnetic anisotropy constants K_1 and K_2 (in 10^2 J/m^3) on temperature for Fe–Co–Ni and Co–Ni alloys [12, 65]

Composition, mass. %			T=20°C		T=200°C		T=398°C	
Fe	Co	Ni	K_1	K_2	K_1	K_2	K_1	K_2
50	10	40	61	−160	19	4	7	−60
25	25	50	4	16	4	2	−3	22
20	15	65	9	−110	−1	−18	−3	2
15	25	60	−26	34	−10	−45	−3	15
10	40	50	−72	−4	−54	41	−9	−102
10	30	60	−38	−80	−17	−50	−12	−37
10	20	70	−29	−17	−25	70	−14	29
10	10	80	−2	−39	−2	−20	−2	6
	65	35	−258	150				
	50	50	−108	−40				
	40	60	−74	40				
	20	80	−4	8				
	10	90	16	−40				
	3	97	−10	9				

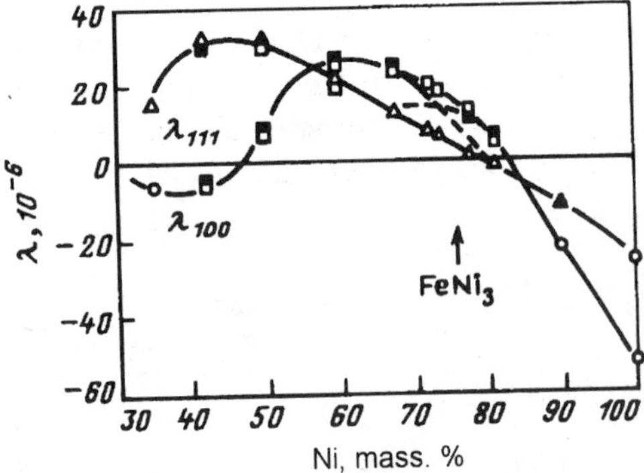

Figure 27.49 Magnetostriction constants of face-centered cubic Fe–Ni alloys at T = 20°C: solid curves – quenched disordered alloys; dashed curves – slowly cooled down ordered alloys [13].

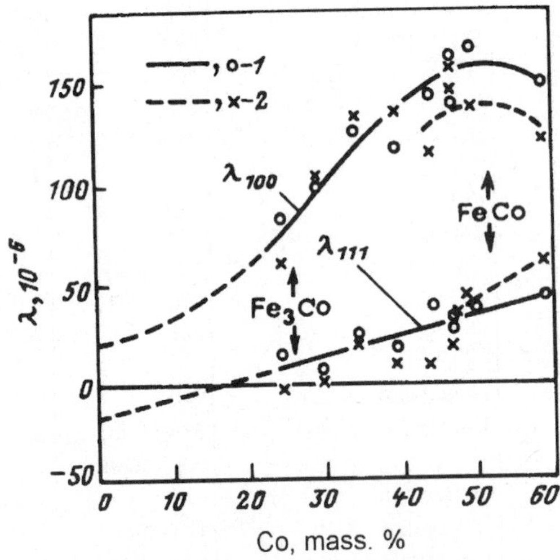

Figure 27.50 Saturation magnetostriction along crystal axes [100] and [111] for Fe–Co alloys at T = 20°C: *1* – quenched disordered alloys; *2* – slowly cooled down ordered alloys [48].

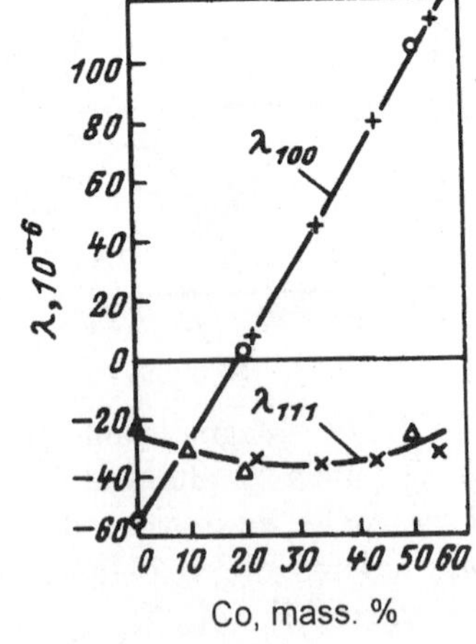

Figure 27.51 Magnetostriction constants λ_{100} and λ_{111} of face-centered cubic Co-Ni alloys at T = 20°C [20] in accordance with [47, 107].

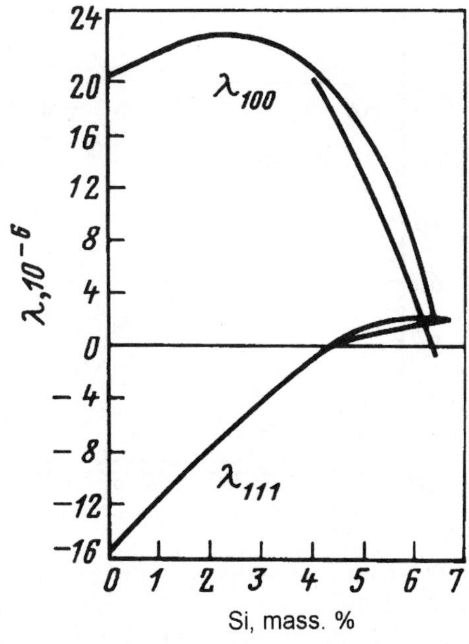

Figure 27.52 Magnetostriction constants of Fe–Si alloys at T=20°C: the lower curves in the region of 4–7 mass. % of Si are for more ordered slowly cooled down samples [47].

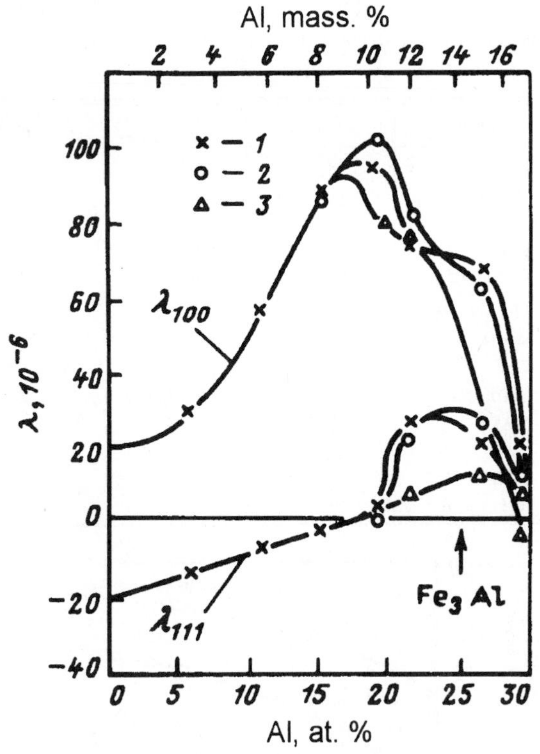

Figure 27.53 Magnetostriction constants λ_{100} and λ_{111} along crystal axes [100] and [111] for Fe–Al alloys at T = 20° C: *1* – slowly cooled down ordered alloys; *2* – cooled down at intermediate rate, partially ordered alloys; *3* – quenched disordered alloys [47].

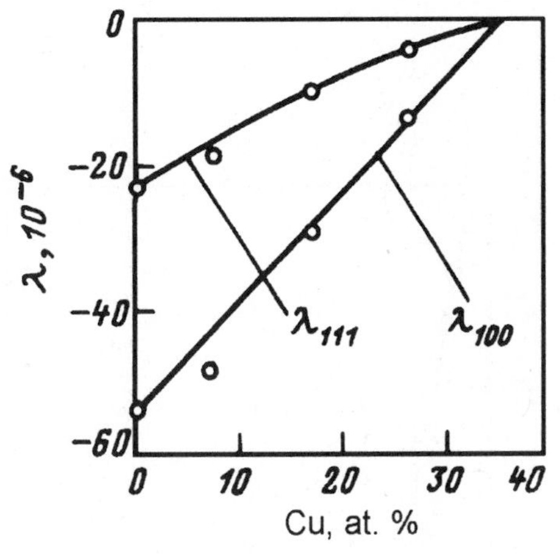

Figure 27.54 Magnetostriction constants λ_{100} and λ_{111} of Ni–Cu alloys at T = 20°C [65].

27.3.2 *f*-element based alloys

Despite a broad variety of alloys based on rare-earth elements, data for only three typical intermetallic systems are presented below (R being the rare-earth element): RFe_2, RCo_5, and R_2Co_{17} (Tables 27.13, 27.14 and Figs. 27.55–27.68). Materials based on them have found wide use in engineering practice. Data about some ferromagnetic actinoid compounds and alloys are summarized in Tables 27.15, 27.16.

Table 27.13 The saturation magnetic moments, Curie temperatures and compensation temperatures of intermetallics (space group $Fd3m$) [18]

Compound	n, μ_B	T_C, K	T_{comp}, K
YFe_2	2.78–3.1	535–554	
$CeFe_2$	2.38–2.59	221–240	
SmF_2	2.50–2.8	676–700	
$GdFe_2$	2.8–3.75	785–810	
$TbFe_2$*	3.68–5.8	696–711	
$DyFe_2$	4.9–7.31	633–638	
$HoFe_2$*	5.1–6.67	597–614	
$ErFe_2$*	4.75–5.85	590–595	480–490
$TmFe_2$	2.52–2.7	566–610	255–248
$YbFe_2$	1.8–2.3		31
$LuFe_2$	2.70–2.97	558–610	

* Ferrimagnetic structure.

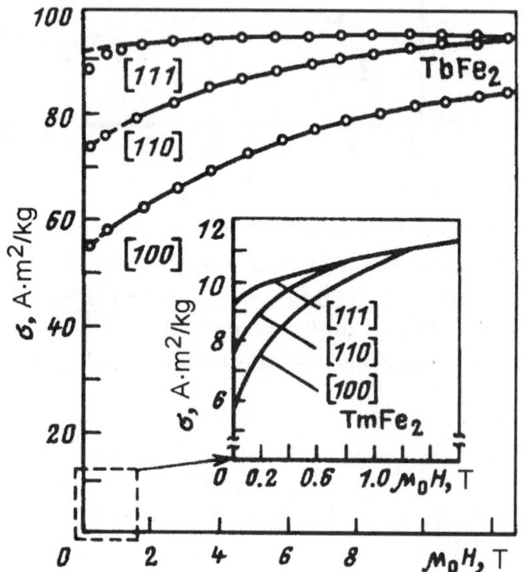

Figure 27.55 Magnetization curves of $TbFe_2$ and $TmFe_2$ at $T = 20°C$ [1, 22].

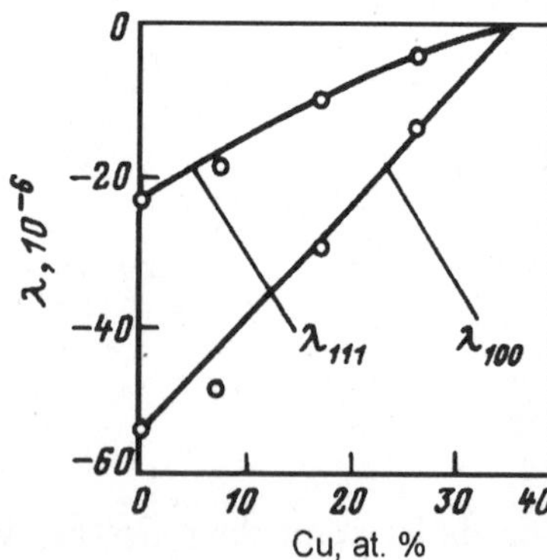

Figure 27.56 Magnetostriction constants λ_{100} and λ_{111} of Ni–Cu alloys at $T = 20°C$ [65].

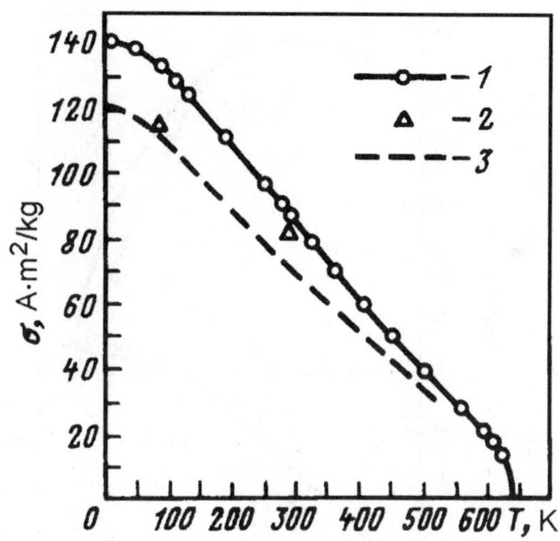

Figure 27.57 Temperature-dependent specific magnetic moment of $DyFe_2$ compound [24]: *1* – single crystal, [100] axis; *2* – polycrystal in a field with the induction of 12 T [24]; *3* – polycrystal [17].

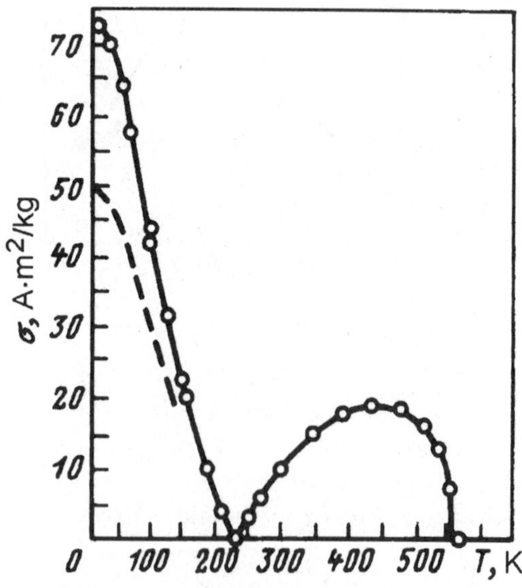

Figure 27.58 Temperature-dependent specific magnetic moment of $TmFe_2$ alloy [24]: solid curve – single crystal, [111] axis; dashed curve – polycrystal [17].

Table 27.14 Magnetic properties and lattice constants of intermetallic RCo_5 compounds (space group Pb/mmm) [35,18]

Compound	a, nm	c, nm	ρ, 10^3 kg/m^3	n, μ_B	T_C, K	T_{comp}, K
YCo_5*1	0.4941	0.3971	7.60	6.8–8.2	977–997	
$LaCo_5$	0.5052	0.3970	7.03	7.3	840	
$CeCo_5$*1	0.4923	0.4019	8.57	5.7–7.4	737	
$PrCo_5$	0.5013	0.3988	8.32	9.9–10.0	912	
$NdCo_5$*1	0.5015	0.3982	8.39	9.5–11.7	910	
$SnCo_5$	0.5001	0.3970	8.53	6.0–6.8	1020	
$GdCo_5$	0.4974	0.3971	8.86	1.2–1.5	1008–1020	
$TbCo_{5.1}$*2	0.4951	0.3978	8.91	0.57–0.8	980	99–120
$DyCo_{5.2}$*2	0.4910	0.3996	9.15	0.7–1.5	966	93–170
$HoCo_{5.5}$*2	0.4888	0.4003	9.13	1.1–2.0	1000–1066	71
$ErCo_{5.9}$	0.4873	0.4003	9.03	0.46–2.1	986–1123	
$TmCo_6$				1.5–1.9	1020	

*1 Ferromagnetic structure.
*2 Ferrimagnetic structure.

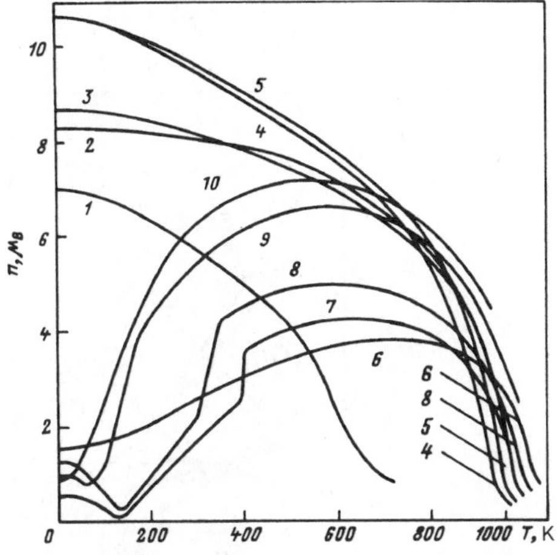

Figure 27.59 Temperature dependence of the atomic moment per formula unit of RCo_5 compounds [34]:
1 – $CeCo_5$; *2* – YCo_5; *3* – $SmCo_5$; *4* – $NdCo_5$; *5* – $PrCo_5$; *6* – $GdCo_5$; *7* – $TbCo_{5.1}$; *8* – $DyCo_{5.2}$; *9* – $HoCo_{5.5}$; *10* – $ErCo_6$.

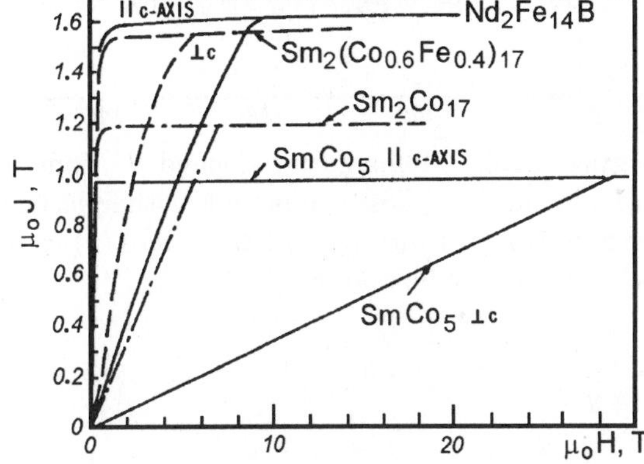

Figure 27.60 Easy- and hard-axis magnetization curves of several high-anisotropy compounds whereon practical permanent magnets are based: $SmCo_5$ and $Nd_2Fe_{14}B$ – single crystals, Sm_2Co_{17} and $Sm_2(Co_{0.6}Fe_{0.4})_{17}$ – powders [94].
1 – single crystals; *2* – powders

Table 27.15 Magnetic properties and lattice constants of ferromagnetic actinoid-based monochalcogenides and monopnictides (space group $Fm3m$) [98]

Compound	Lattice constant, nm	T_C, K	θ_p, K	μ_{eff}, μ_B	n, μ_B
US	0.5487–0.5490	172–180	173–180	2.22–2.31	1.20–1.76
USe	0.571–0.575	160.5–187	188	2.51	1.31–2.0
UTe	0.6146–0.6161	102–110	104	2.84	1.10–2.20
NpN	0.4898	82–100	82–100	2.13–2.44	1.4–2.2
PuP	0.5651	125	130	1.06	0.42–0.77
PuAs	0.586	129	129	0.97	0.35
PuSb	0.6240	85	90	1.0	0.57
CmN	0.5027–0.5041	109		7.02	
CmAs	0.5905	88–140		6.58	

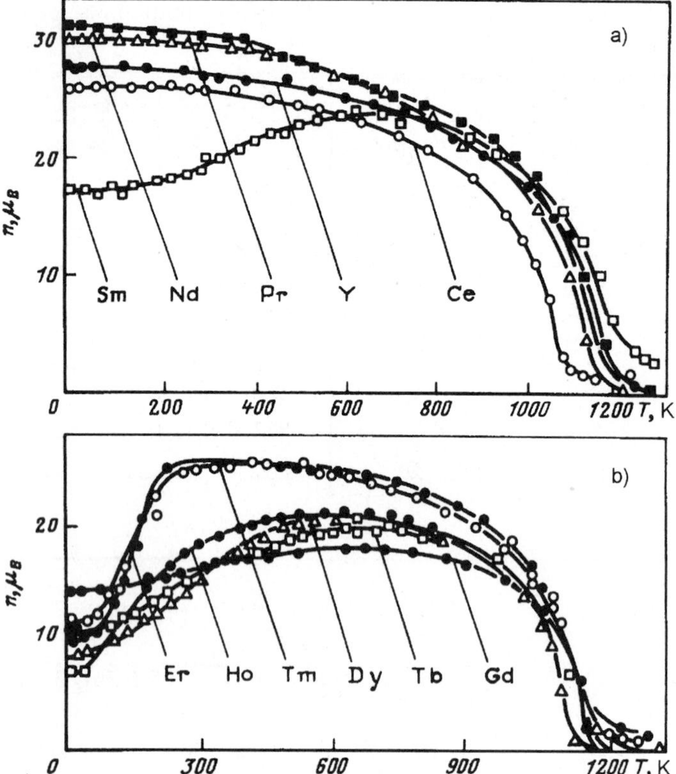

Figure 27.61 Temperature-dependent atomic moments per formula unit of R_2Co_{17} compounds with light lanthanides and yttrium (*a*), and with heavy lanthanides (*b*) as measured in a field with the induction of 2 T [64].

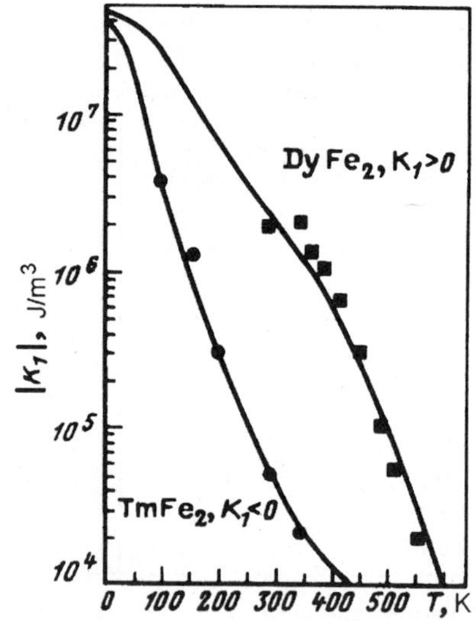

Figure 27.63 Temperature-dependent absolute magnetic anisotropy constant K_1 for $DyFe_2$ and $TmFe_2$ [24].

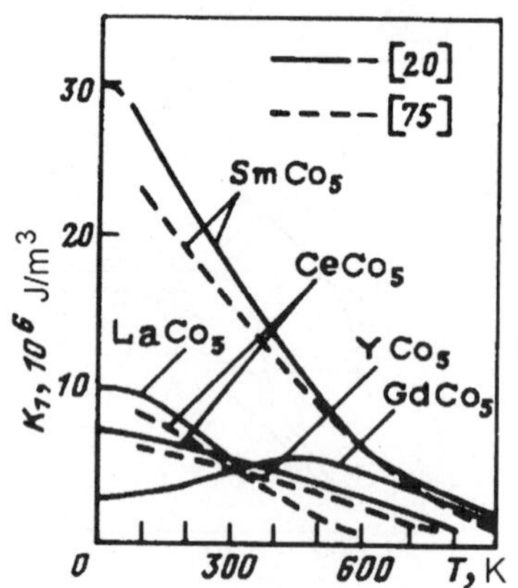

Figure 27.64 Temperature dependence of the magnetic anisotropy constant K_1 in RCo_5 compounds [18].

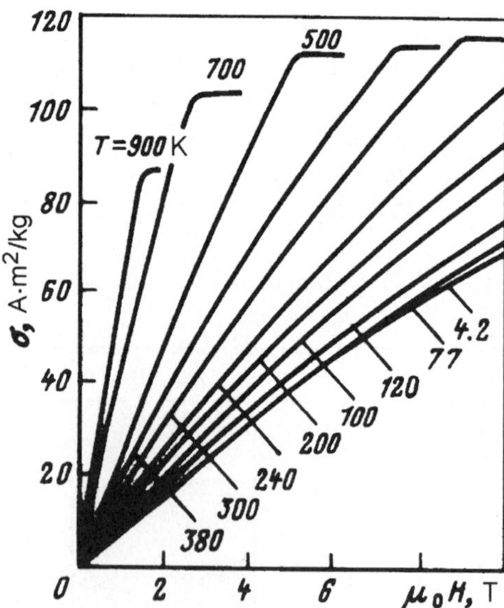

Figure 27.62 Magnetization curves of Sm_2Co_{17} single crystal along [100] crystal axis at different temperatures [63].

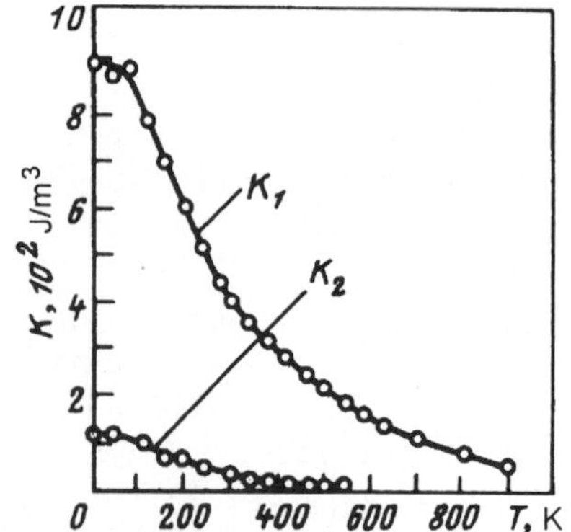

Figure 27.65 Temperature dependence of the magnetic anisotropy constants in the Sm_2Co_{17} compound [30].

Table 27.16 Magnetic properties and crystal structure of some ferromagnetic actinoid based intermetallic compounds [98]

Compound	Crystalline structure		T_C, K	θ_p, K	μ_{eff}, μ_B	n, μ_B
	Syngony, space group	Lattice parameters, nm				
$ThCo_5$	Hexagonal, $P6/mmm$		410–740			5–8
$UCo_{5.3}$	Rhombohedral, $R3m$	a=0.476 c=3.649	360			2.4
UFe_2	Cubic, $Fd3m$	0.758	160–162	170	2.0–3.0	0.06(U); 0.59(Fe); 1.09(Σ)
UNi_2	Hexagonal, $P6_3/mmc$	a=0.406 c=0.825	30	45	2.5	0.12
UPt	Orthorhombic, $Cmcm$	a=0.372 b=1.077 c=0.441	27–30		2.62	0.07–0.52
$NpFe_2$	Cubic, $Fd3m$	0.7144	492–500	523	4.2	1.09(Np); 1.35(Fe); 2.6(Σ)
$NpAl_2$	Cubic, $Fd3m$	0.7785	56	56	2.3	1.5
$NpMn_2$	Cubic, $Fd3m$	0.7230	18			0.3–0.4
$NpNi_2$	Cubic, $Fd3m$	0.7098	28–32	20		1.2
$NpOs_2$	Cubic, $Fd3m$	0.7528	7.5–8	−110	2.3	0.4
$PuFe_2$	Cubic, $Fd3m$	0.7190	564–600	599	3.7	0.45(Pu); 1.47(Fe); 2.3(Σ)
$AmFe_2$	Cubic, $Fd3m$	0.730	350–400		4	2.8

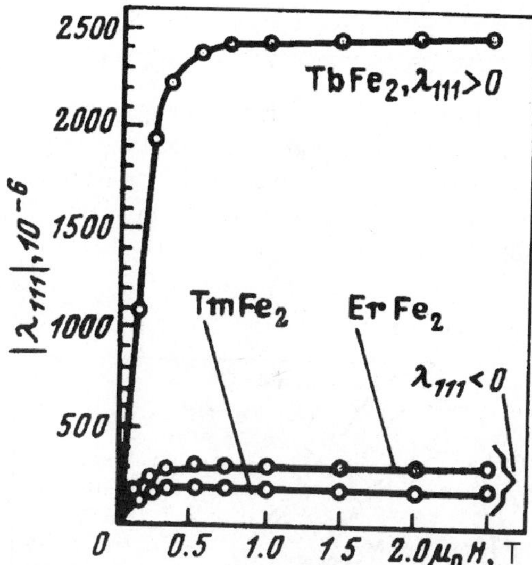

Figure 27.66 Absolute value of the magnetostriction constant $|\lambda_{111}|$ in $TbFe_2$, $ErFe_2$, and $TmFe_2$ single crystals at room temperature [1,22].

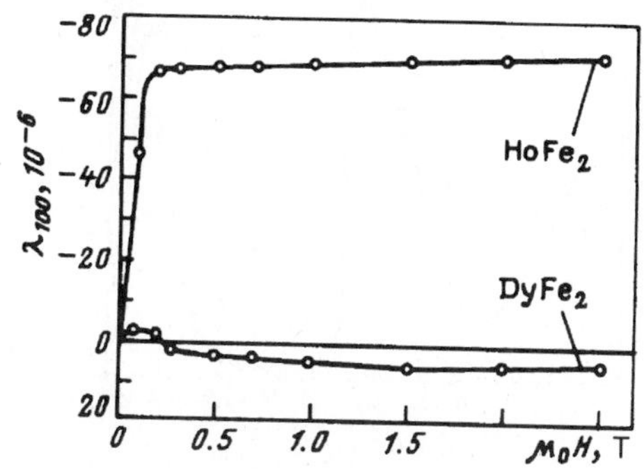

Figure 27.67 Magnetostriction constant λ_{100} in $DyFe_2$ and $HoFe_2$ single crystals at T=20°C [25].

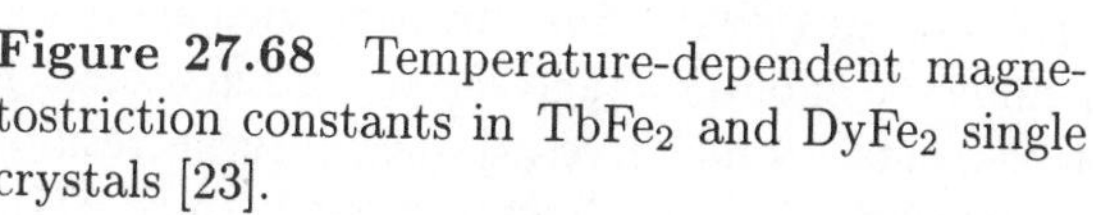

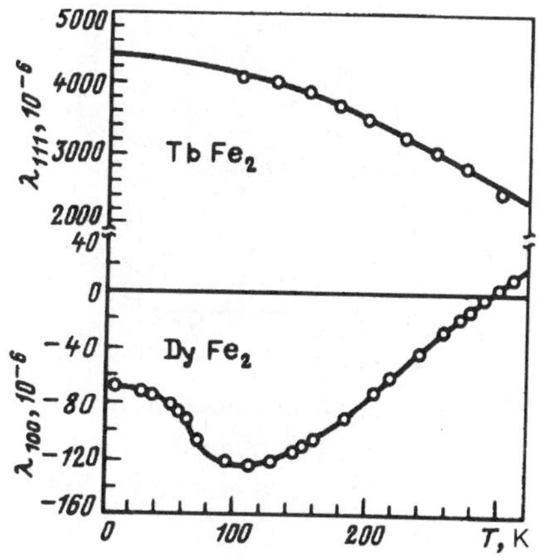

Figure 27.68 Temperature-dependent magnetostriction constants in $TbFe_2$ and $DyFe_2$ single crystals [23].

27.4 Metal Magnetic Materials

Data compiled in this section correspond to room temperature, i.e. 20°C, except as noted.

27.4.1 Soft magnetic materials

Materials with the largest saturation magnetization are referred to as soft magnetic materials. This group of materials includes the irons with minimum impurity content (Tables 27.17, 27.18), unalloyed electrical engineering steels (Tables 27.18, 27.19), Fe–Co based alloys (Figs. 27.69, 27.70, Table 27.20, see also Figs. 27.39, 27.46, 27.50, Table 27.7) including Permendur (composition: 49 wt.%Co, 2 wt.% V, Fe balance).

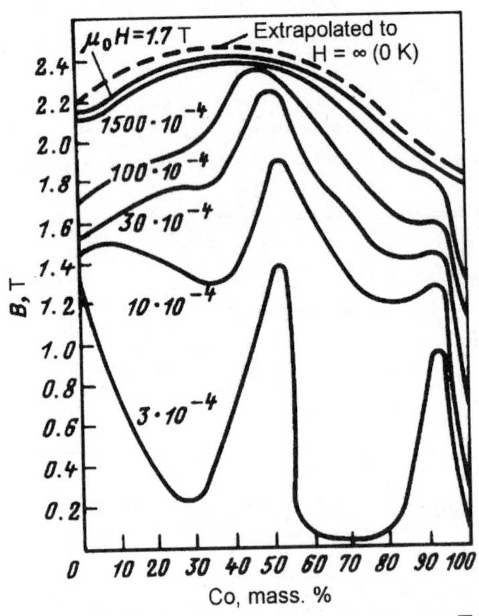

Figure 27.69 Magnetic induction in Fe–Co alloys vs. cobalt content and the applied magnetic field induction at $T = 20°C$ [12].

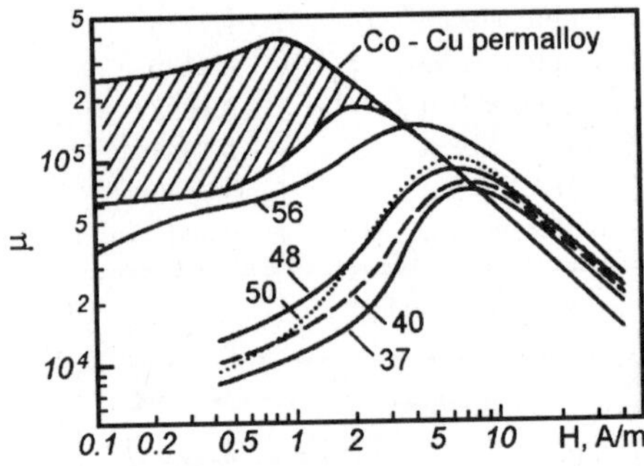

Figure 27.71 Permeability vs. field strength in Ni–Fe alloys with a round hysteresis loop (toroidal tape–wound cores, tape thickness 0.05–0.1 mm, 50 Hz). Numbers alongside the curves indicate Ni content (in mass. %) in the alloys [32].

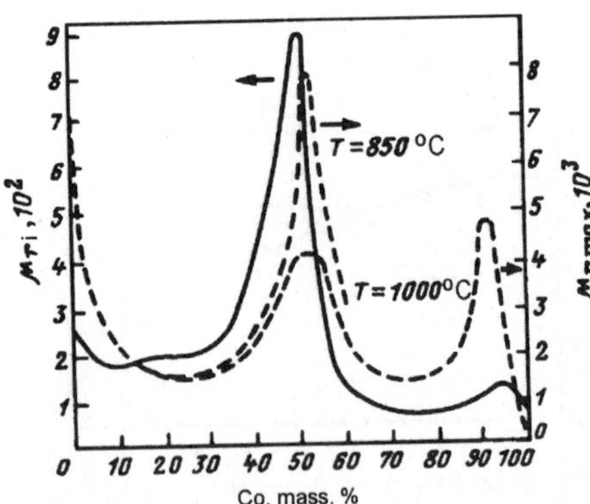

Figure 27.70 Initial and maximum relative magnetic permeabilities of Fe–Co alloys vs. cobalt content and the annealing temperature [12].

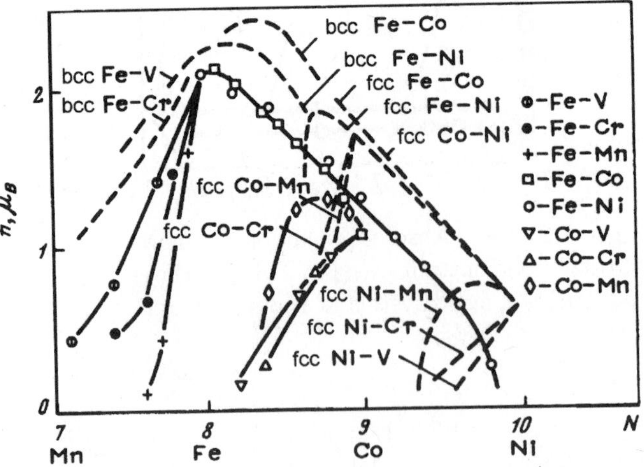

Figure 27.72 Atomic magnetic moments (per atom of transition element) of quasibinary amorphous alloys $(L_xM_{1-x})_{80}P_{10}B_{10}$ (L, M – 3*d* transition elements) vs. average number N of outer *s*- and *d*-electrons in atoms of transition elements – solid curves; dashed curves – results for corresponding crystalline alloys without P and B [71].

Table 27.17 Typical chemical composition (in weight percent) of some irons and steels [21,70]

Material	Form	Si	C	Mn	P	S	O	N	Al
Magnetic ingot iron (Armco)	Mill sheet	0.003	0.015	0.030	0.005	0.025	0.15	0.007	0.003
Electrolytical iron		0.004	0.01	0.01	0.003	0.003			0.009
Low-carbon steel (SAE 1008)	Mill sheet	0.005	0.08	0.40	0.020	0.025	0.05	0.007	0.005
Low-carbon steel	Decarburized	0.005	<0.01	0.40	0.020	0.025	0.05	0.007	0.005
M36 cold-rolled	Fully processed	2.2	<0.01	0.25	0.015	0.01	0.01	0.01	0.3
M22 cold-rolled	Fully processed	2.8	<0.01	0.25	0.015	0.01	0.01	0.01	0.3
M15 hot-rolled	Fully processed	3.3	<0.01	0.25	0.015	0.01	0.01	0.01	0.3
(110)[001]oriented	Fully processed	3.2	<0.002	0.07	0.010	<0.002	<0.003	0.001	0.002

Table 27.18 Typical magnetic properties of some irons and steels (fully annealed) [21,70]

Material	B_s, T	Density, 10^3 kg/m^3	Resistivity, 10^{-8} Ω·m	H_s, A/m (B_m^*=1.0 T)	Permeability		Core loss (1.5 T/60 Hz) at thickness shown, W/kg		
					H=79.6 A/m	H=796 A/m	0.36 mm	0.47 mm	0.64 mm
Mild steel, 0.2%C normalized	2.14	7.85		318					
Cast magnetic ingot iron	2.15	7.85	10.7	67.6	3500	1500			
Magnetic ingot Fe, 0.08 in sheet	2.15	7.85	10.7	88.3	1800	1575			13.2
Electromagnet Fe, 0.08 in sheet	2.15	7.85	12	81.2	2750	1575			
Low-carbon steel, decarburized	2.14	7.85	12.5	71.6	2000	1530	8.15	9.25	11.4
Oriented Fe	2.15	7.85	10.7	31.8		1925	4.40	5.29	8.59
M36, cold-rolled	2.04	7.75	41	35.8	7400	1485		3.85	4.73
M22, cold-rolled	1.98	7.65	49	31.0	8100	1450		3.63	4.29
M15, hot-rolled	1.97	7.65	53	27.8	7800	1390	2.75		
(110)[001] 3.2%Si-Fe	2.03	7.65	48	5.6	16000	1820	1.39		
(100)[001] 3.2%Si-Fe	2.03	7.65	48	5.6	14000	1600	2.20		

* B_m is the maximum induction reached at the minor hysteresis loop for which the cited data were obtained.

Table 27.19 Magnetic properties of some high-purity irons [21]

Iron	B_s, T	H_c, A/m	Permeability	
			μ_i	μ_{max}
Cast (vacuum melting)		22–27		17000–21000
Electrolytic (annealing)		12–25	44–60	8100–41500
Electrolytic (vacuum melting and annealing)		7.2	1150	61000
Puron (annealing in hydrogen)	2.16	4.0		100000

Table 27.20 Properties of Permendur (regular) and Supermendur (high-purity Co–Fe–2% V) [32]

Property	Permendur	Supermendur
Electrical resistivity (Ω·cm)	$25 \cdot 10^{-6}$	$25 \cdot 10^{-6}$
Saturation induction (T)	2.4	2.4
Remanent induction (T)	1.5	2.215
Magnetic permeability	4000–8000	92500
Coercive force (A/m)	$3.98 \cdot 10^2$	15.92
Magnetostriction coefficient*	$+60 \cdot 10^{-6}$	$+60 \cdot 10^{-6}$

* At $7.96 \cdot 10^3$ A/m.

Table 27.21 Properties of unoriented fully processed test specimens of flat-rolled electrical steel (ASTM A677) [21, 99]

AISI type	Si+Al content, wt.%	Thickness, nm	Max. core loss at 1.5 T, W/kg		B_s, T
			60 Hz	50 Hz	
M-15	3.50	0.36	3.20	2.53	1.98
M-15		0.47	3.70	2.93	1.98
M-19	3.30	0.36	3.48	2.75	1.99
M-19		0.47	3.84	3.03	1.99
M-19		0.64	4.59	3.62	1.99
M-22	3.20	0.36	3.70	2.93	2.00
M-22		0.47	4.08	3.22	2.00
M-22		0.64	4.81	3.80	2.00
M-27	2.80	0.36	3.97	3.13	2.02
M-27		0.47	4.19	3.31	2.02
M-27		0.64	4.96	3.92	2.02
M-36	2.65	0.36	4.19	3.31	2.02
M-36		0.47	4.52	3.57	2.02
M-36		0.64	5.29	4.18	2.02
M-43	2.35	0.47	5.07	4.01	2.04
M-43		0.64	5.95	4.70	2.04
M-45	1.85	0.47	6.72	5.31	2.07
M-45		0.64	7.94	6.27	2.07
M-47	1.05	0.47	10.14	8.01	2.11
M-47		0.64	12.65	10.01	2.11

Table 27.22 Maximum core losses of grain-oriented fully processed test specimens of flat-rolled electrical steel (ASTM A665-15 kG, ASTM A725-17 kG) [Nominal Si+Al content is 3.15 wt.%; B_{800}^{*1} = 1.8 T, B_s = 2.00 T] [21,99]

Designation		Thickness, nm	Maximum core loss, W/kg			
			at 1.5 T		at 1.7 T	
AISI*2	ASTM*3		60 Hz	50 Hz	60 Hz	50 Hz
M-3		0.28			1.57	
		0.30			1.64	
		0.35			1.68	
M-4	27G053	0.27	1.17	0.89		
	27H076	0.27			1.67	1.27
M-5	30G058	0.30	1.28	0.97		
	30H083	0.30			1.83	1.39
M-6	35G066	0.35	1.45	1.11		
	35H094	0.35			2.07	1.57

*1 B_{800} is the induction in a magnetic field of 800 A/m.
*2 American Society for Testing and Materials.
*3 American Iron and Steel Institute.

Table 27.23 Core loss of high-induction ($B_{800} \geq 1.88$ T, B_s = 2.00 T) grain-oriented silicon steel at 1.7 T (W/kg) [Nominal Si+Al content is 2.90–3.15 wt.%] [21,99]

AISI type	0.27 nm		0.30 nm		0.35 nm	
	50 Hz	60 Hz*1	50 Hz	60 Hz	50 Hz	60 Hz
M-0H	(0.99)*2	1.33	1.05	(1.41)		
M-1H	(1.04)	1.40	1.11	(1.49)	1.16	(1.55)
M-2H	(1.11)	1.49	1.17	(1.56)	1.22	(1.63)
M-3H	(1.17)	1.57	1.23	(1.65)	1.28	(1.72)
M-4H					1.37	(1.84)

*1 Data for 0.28 mm thick samples.
*2 Data in parentheses were calculated on the basis of Loss(60 Hz)/Loss(50 Hz) = 1.34, typical of high-induction materials.

Table 27.24 Properties of high-permeability alloys [21]

Alloy	Chemical composition (wt.%)	T_C, °C	B_s, T	K_1, 10^{-1} J/m^3	$\lambda_{100} \times 10^6$	$\lambda_{111} \times 10^6$	$\mu_i \times 10^{-3}$	ρ_{el}, 10^{-8} Ω·m
Supermalloy	79 Ni, 15.5 Fe, 5.0 Mo, 0.5 Mn	400	7.90				100	60
Supermalloy	79.7 Ni, 14.5 Fe, 5.1 Mo, 0.7 Mn	394	7.27	0	1.79	0.17	163	68
Mumetal	76 Ni, 16.9 Fe, 4.8 Cu,1.9 Cr, 0.4 Mn	400	8.00				100	55
Mumetal	76.0 Ni, 16.6 Fe, 4.7 Cu, 1.7 Cr	461	8.31	1000	5.5	0	90	60
1040	72 Ni, 11 Fe, 14 Cu, 3 Mo	290	6.00				40	56
Mo–Cr–Permalloy	80.5 Ni, 15.1 Fe, 2.6 Mo, 1.8 Cr	411	7.50				108	64
Mo–Cu–Cr–V–Permalloy	78.7 Ni, 15.3 Fe, 1.4 Mo, 2.4 Cu, 1.0 Cr, 1.2 V	405	7.64	0	3.01	0.44	150	72
V-Permalloy	82.3 Ni, 13.3 Fe, 3.9 V	371	6.72	0	0.77	−0.52	128	65
W-Permalloy	75.8 Ni, 15.1 Fe, 9.1 W	414	7.20	300	3.13	0.63	82	66
Cr-Permalloy	81.4 Ni, 15.0 Fe, 3.6 Cr	421	7.80				34	61
Cu-Permalloy	69.5 Ni, 13.9 Fe, 15.9 Cu	413	7.50				74	30
Ta-Permalloy	73.0 Ni, 12.0 Fe, 15.0 Ta		7.00				57	64
Nb-Permalloy	79.2 Ni, 11.8 Fe, 9.0 Nb		6.00			$0.35\lambda_s$	125	75

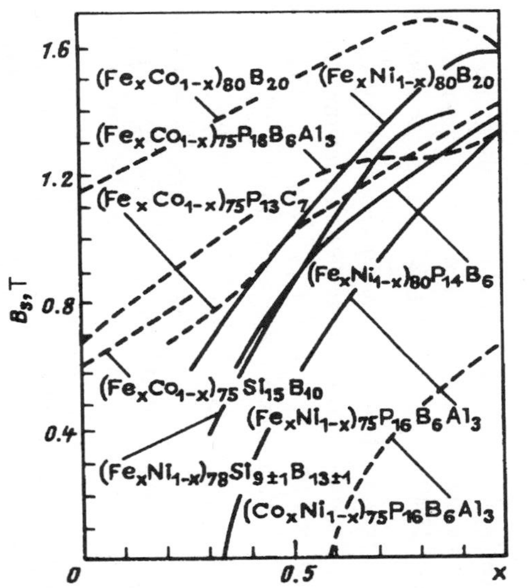

Figure 27.73 Saturation induction of Fe, Co, and Ni amorphous alloys at $T = 20$°C [71].

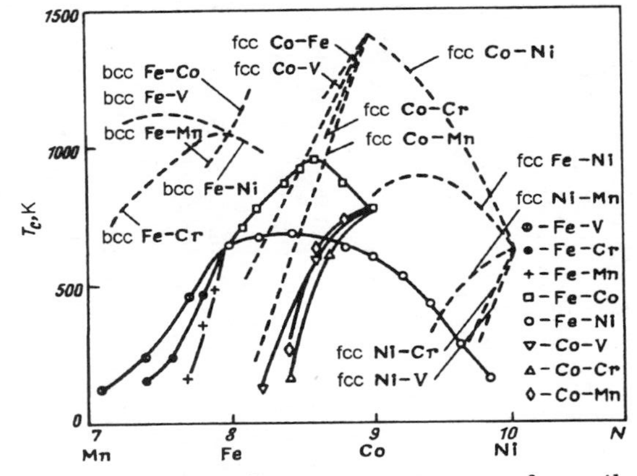

Figure 27.74 Curie temperatures of quasibinary amorphous alloys $(L_xM_{1-x})_{80}P_{10}B_{10}$ (L, M – 3d transition elements) vs. average number N of outer s- and d-electrons in atoms of transition elements – solid curves; dashed curves – results for corresponding crystalline alloys without P and B [71].

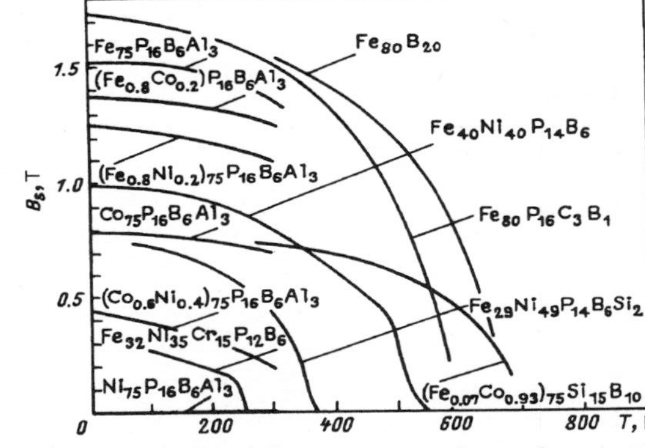

Figure 27.75 Temperature-dependent saturation induction for various amorphous alloys (see also Tables 27.26 and 27.27) [71].

Table 27.25 Materials for laminations, cut cores and tape wound cores [21]

Alloy	μ_i $\times 10^{-3}$	μ_{max} $\times 10^{-3}$	H_c, A/m	B_s, T	B_r, T	B_r/B_m*	T_C, °C	Resistivity, $10^{-8}\Omega\cdot$m	Density, kg/m^3	Notes
					High initial μ					
36Ni	3	20	16	1.3			250	75	8.15	1
48Ni	11	80	2.4	1.55			480	48	8.25	2
56Ni	30	125	1.6	1.5			500	45	8.25	3
4Mo–56Ni	40	200	1.2	0.80			460	58	8.74	4
4Mo–5Cu–77Ni	40	200	1.2	0.80			400	58	8.74	5
5Mo–80Ni	70	300	0.4	0.78			400	65	8.77	6
4Mo–5Cu–77Ni	70	300	0.4	0.80			400	60	8.74	7
					Square loop					
4Mo–80Ni			2.4	0.80	0.66	0.80	460	58	8.74	8
4Mo–5Cu–80Ni			2.4	0.80	0.66	0.80	400	58	8.74	9
3Mo–65Ni			2	1.25	1.05	0.94	520	60	8.50	10
50Ni			8	1.60	1.50	0.95	500	45	8.25	11
3Si			32	2.03	1.63	0.85	730	50	7.65	12
2V–49Co			16	2.30	2.01	0.90	940	26	8.15	13
					Skewed (flat) loop					
4Mo–5Cu–77Ni			1.2	0.8	0.12		400	58	8.74	14
3Mo–65Ni			10	1.25	0.15		520	60	8.50	15

* See footnote to Table 27.18.

1 Permenorm 3601 K2, Hyperm 36M, Radiometal 36 (0.3 mm/50Hz).

2 4750 Alloy, Hight Permeability 49, Hyperm 52, Alloy 48, Superperm 49, PB-2, Super Radiometal, Permenorm 5000 H2 (0.15 mm/60 Hz).

3 Permax M (0.15 mm/60 Hz).

4 4-79Mo-Permalloy, HyMu 80, Round Permalloy 80, Superperm 80 (0.1mm/60 Hz).

5 Mumetal Plus, Hyperm 900, Vacoperm 100 (0.1 mm/60 Hz).

6 Supermalloy, HyMu 800, Hyperm Maximum (0.1 mm/60 Hz).

7 Supermumetal, Ultraperm 10 (0.1 mm/60 Hz).

8 Square Permalloy, HyRa 80, Square Permalloy 80, Square 80 (0.05 mm/0.5 Oe dc).

9 Ultraperm Z, Othomumetal (0.015 mm/0.5 Oe dc).

10 Permax Z (0.05 mm/0.5 Oe dc).

11 Deltamax, HyRa 49, Hyperm 50T, Othonol, Square 50, PE-2, HCR Alloy, Permenorm Z (0.05 mm/0.5 Oe dc).

12 Silectron, Microsil, Oriented T-S, Hyperm 5T/7T, Thin-Gage Orientcore (0.1 mm/0.5 Oe dc).

13 Supermendur, Vacoflux Z, Hyperm Co50 (0.1 mm/3 Oe dc).

14 Ultraperm F.

15 Permax F.

Materials with low core loss. This group of materials includes a large number of electrical engineering silicon steels with the silicon content from 0.4 to 5 wt.% (Tables 27.17–19, 27.21–23, see also Figs. 27.40, 27.52 and Table 27.7); it includes also some amorphous magnetic materials (see below).

Materials with the largest permeability in weak fields. The largest permeability in weak fields is demonstrated by some electrical engineering steels (Table 27.18) and iron–nickel alloys, the so-called Permalloys (Tables 27.24, 27.25, Fig. 27.71, see also Figs. 27.38, 27.45, 27.48, 27.49), as well as some amorphous materials (see below).

Amorphous magnetic materials. A separate group of soft magnetic materials involves amorphous metallic materials obtained by special technologies. Two types of such materials are widely known: iron-based amorphous metals with 10–20 at.% of such metalloids as B, C, N, Si, P, and amorphous alloys of transition metals with the rare-earths. Only data for the first class of materials are cited below (Tables 27.26, 27.27), since they are employed as the materials with low core losses and high magnetic susceptibility in weak fields (see above). Data on the second-type materials can be found in [36]. The results listed in this section were borrowed from [71]. The spontaneous magnetic moment and the saturation induction as a function of alloy composition and temperature, as well as the dependence of the Curie temperature on alloy composition are depicted in Figs. 27.72–27.75. In Figs. 27.72 and 27.74, the abscissa axis N corresponds to the alloy composition. Fig. 27.76 gives an idea about magnetostriction in materials of various compositions. Figs. 27.77, and 27.78 include the practically relevant characteristics of initial permeability and losses on reversing the magnetism.

Table 27.26 Static magnetic properties of some amorphous alloys [71]

Alloy	Sample shape*	Annealed samples			Unannealed samples		
		$\mu_0 H_c$, 10^{-4} T	J_r/J_s	μ_{max}, 10^3	$\mu_0 H_c$, 10^{-4} T	J_r/J_s	μ_{max}, 10^3
$Fe_{80}B_{20}$	t				0.075	0.46	
	t	0.08	0.51	100	0.04	0.77	300
$Fe_{40}Ni_{40}P_{14}B_6$	t	0.06	0.45	58	0.02	0.71	275
	r	0.010	0.90		0.009	0.89	
	t	0.05	0.45	70	0.01	0.71	550
$Fe_{29}Ni_{49}P_{14}B_6Si_2$	r	0.01	0.8		0.03(A)	0.995(A)	
	t	0.057	0.54	46	0.11	0.70	310
$Fe_3Co_{72}P_{16}B_6Al_3$	t	0.023	0.45	120	0.013	0.71	340
$Fe_{4.7}Co_{70.3}Si_{15}B_{10}$	r	0.013	0.36	190	0.006	0.63	700
	r	0.01	0.35	230	0.015	0.82	370
$Fe_{80}P_{13}C_7$	r	0.08	0.42	74	0.018	0.36	280
	t	0.12	0.42	42	0.06	0.4	80
$(Fe_{0.8}Ni_{0.2})_{78}S_{10}B_{12}$	r	0.04	0.44	145	0.015	0.93	800
$(Fe_{0.8}Ni_{0.2})_{78}S_8B_{14}$	r	0.018	0.41	300	0.006	0.95	2000
$(Fe_{0.6}Ni_{0.4})_{78}S_8B_{14}$		0.022	0.44	210	0.006	0.97	1700
$(Fe_{0.4}Ni_{0.6})_{78}S_8B_{14}$	r	0.024	0.46	120	0.005	0.76	950
$Fe_{80}P_{16}C_3B_1$	r	0.080			0.068		
	t	0.062	0.40	96	0.050	0.42	130

* t – toroid, r – ribbon, A – sample is annealed under tension.

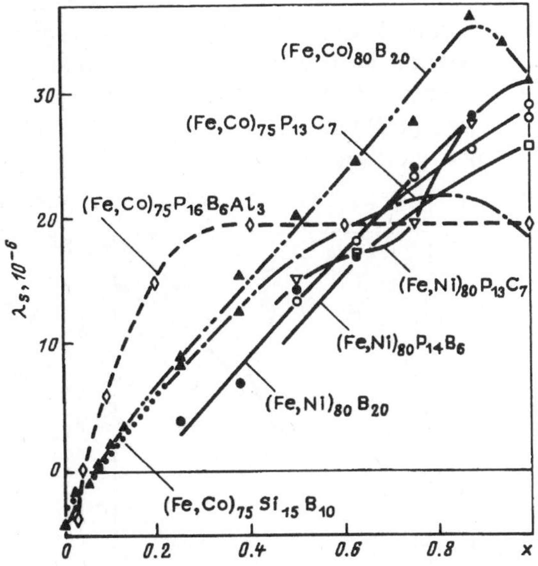

Figure 27.76 Saturation magnetostriction vs. composition of some amorphous alloys $(Fe_xM_{1-x})_{(75-80)}(P, B, C)_{(25-20)}$ (here, M is for Co or Ni) [71].

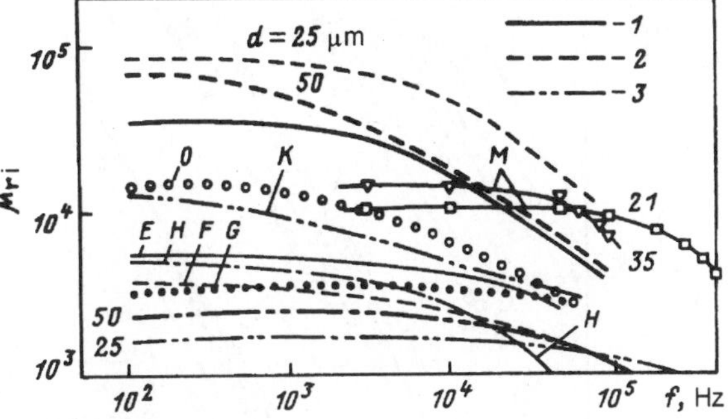

Figure 27.77 Typical dependences of the relative initial magnetic permeability on the frequency of remagnetization field for some amorphous and polycrystalline alloys at different sample thicknesses [71]: data for amorphous alloys are marked with letters corresponding to Table 27.27; *1* and *2* – Permalloy (composition: 4 mass. % Mo, 79 mass. % Ni, Fe balance) and Supermalloy (composition: 4 mass. % Mo, 80 mass. % Ni, Fe balance, round hysteresis loop); *3* – Silectron (textured electrical engineering steel with 3.2 mass. % Si).

Materials with special magnetic properties and magnetostrictive materials. Here we present data (Table 27.28) on the alloys with highly stabilized magnetic permeability in weak fields (μ_i is constant within 1–5% experimental error for H in the range from 0 to 300 A/m), the typical characteristics of one of the thermomagnetic materials (Fig. 27.79), as well as information about magnetostrictive materials (Table 27.29). The first group includes mainly the annealed (Perminvar) or thermomechanically treated (Isoperm) iron–nickel–cobalt alloys (see also Fig. 27.43 and Table 27.12) and the second one — the iron–nickel–chromium (Compensator) (Fig. 27.79), iron–nickel (Thermalloy) and nickel- copper (Calmalloy) alloys. Information about rare-earth–iron compounds (perspective magnetostrictive materials) is contained in Section 27.3.2 (see Table 27.13, Figs. 27.55, 27.56, 27.63, 27.66–27.68). A detailed description of magnetic materials with special properties can be found in the monograph [21].

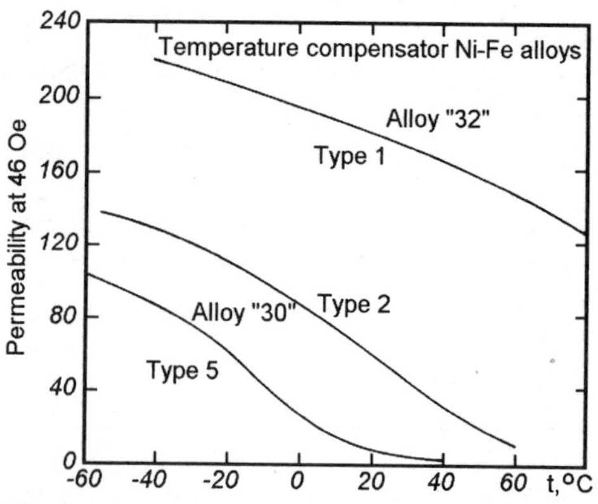

Figure 27.79 Permeability at 46 Oe as a function of temperature for three types of commercial temperature compensator Ni–Fe alloys. Type 1 contains ≈ 32% Ni, and types 2 and 5 contain ≈30% Ni [21].

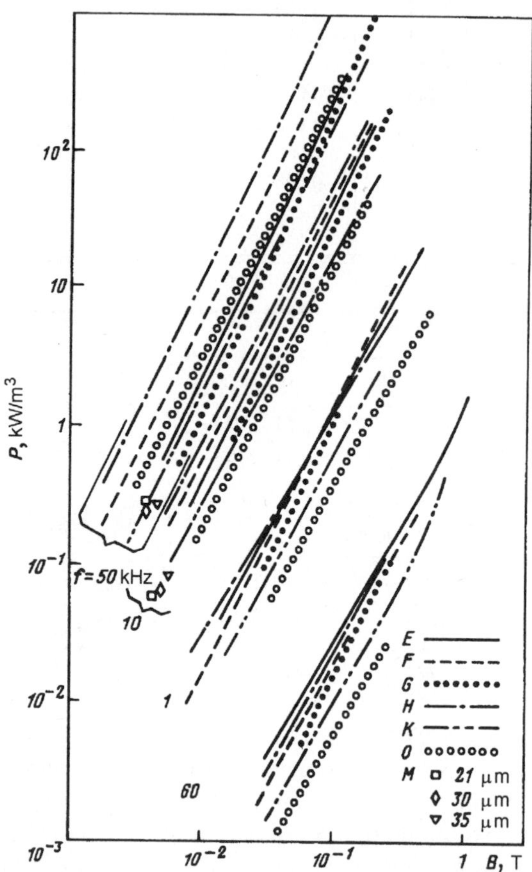

Figure 27.78 Core losses vs. magnetic induction amplitude for some amorphous alloys at different frequencies of the remagnetization field [71]. Letters denote alloys according to Table 27.27. Sample thickness equals 25–50 μm. The curves marked with 60 were obtained at a frequency of 60 Hz.

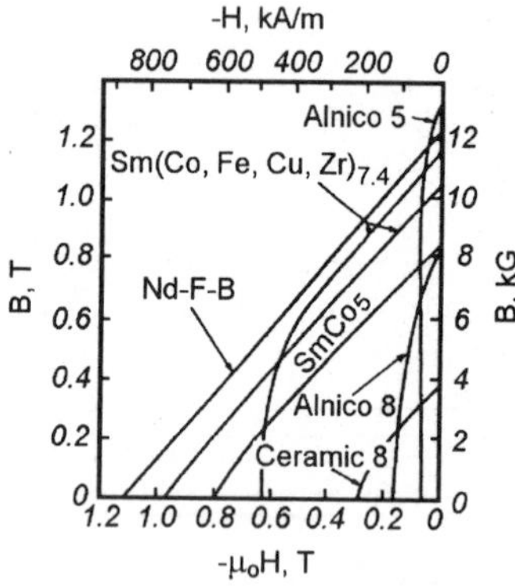

Figure 27.80 Comparison of rare-earth permanent magnets with older permanent magnets. B vs. H demagnetization curves are typical for high-grade commercial magnets indicated [94].

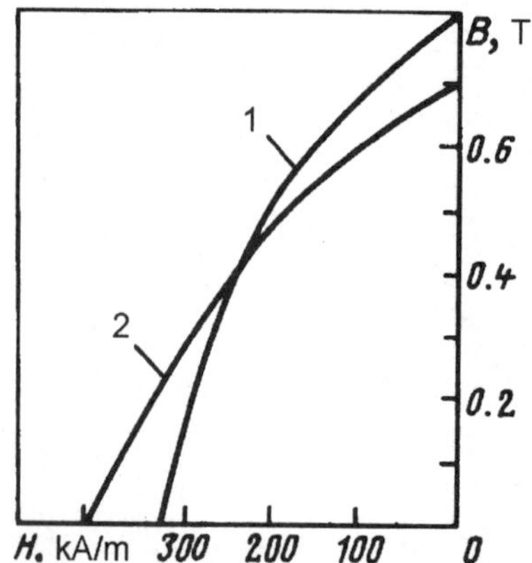

Figure 27.81 Typical demagnetization curves of Co–Pt alloys: *1* – 76 wt. % Pt, *2* – 78 wt. % Pt, Co balance [90].

Table 27.27 Specific core losses in unannealed and annealed samples of amorphous alloys [71]

Alloy	Notation in Figs. 27.77 and 27.78	ρ, 10^3 kg/m^3	Losses at B_{max} = 0.1 T, kW/m^3 — Unannealed samples at varying frequencies, kHz				Annealed samples at varying frequencies, kHz				Ribbon thickness, μm	Annealed samples	
			0.060	1	10	50	0.060	1	10	50		$\mu_0 H_c$, 10^{-4} T	J_r/J_s
$Fe_{80}B_{20}$		7.39	0.071	2.8	65		0.030	1.1	18		30–35	0.040	0.78
	E		0.10	5	350	1000	0.025	1.2	35	200	30	0.075	0.46
$Fe_{80}P_{16}C_3B_1$		7.7		8.5	310			6.3	160		40	0.050	0.42
$Fe_{40}Ni_{40}P_{14}B_6$		7.5–7.7		13	250			1.8	42		50	0.019	0.58
	K		0.07	4	400	3000	0.010	0.60	18	180	50	0.020	0.70
$Fe_{29}Ni_{49}P_{14}B_6Si_2$		7.51		10	160			0.75	20		70	0.011	0.70
$Fe_{40}Ni_{40}B_{20}$	*F*	7.14	0.07	4	450	2300	0.016	1.2	45	600	30	0.090	0.68
$Fe_{80}P_{14}B_6$	*G*	6.86	0.04	2.5	80	950	0.014	1.1	28	200	25	0.1	0.37
$Fe_{50}Ni_{30}P_{14}B_6$	*H*	7.65	0.09	5	380	1700	0.022	1.2	50	1000	30	0.050	0.84
$Fe_{40}Ni_{40}P_{14}B_6$		7.5–7.7	0.06	5.5	550	1300	0.014	0.92	49	820	36	0.035	0.85
$Fe_5Co_{70}S_{15}B_{10}$	*M*	8.0										0.010	0.85
$Fe_3Co_{72}P_{15}B_6Al_3$	*O*	7.3	0.025	1	28	190	0.006	0.35	16	230	20	0.015	0.82

Table 27.28 Magnetic properties of a cold-rolled ribbon from alloys with the low remanent induction and improved constancy of the magnetic permeability [44]

Alloy composition, wt.%	Thickness, mm	μ_i	μ_{max}/μ_i	α_S*1 at H=800 A/m	$d\mu/dT$*2, 10^{-2} °C^{-1} (T=(-60)–(+120)° C)
47% Ni, 23% Co,balance Fe	0.01–0.1	900–1100	1.15	0.05	0.06
47% Ni, 22% Co, 2% Cr, balance Fe	0.02–0.1	1500	1.2	0.05	0.03
64% Ni, balance Fe	0.01–0.1	2000–2200	1.2	0.07	0.06
40% Ni,25% Co, 4% Mo, balance Fe	0.01–0.1	1800	1.2	0.07	

*1 $\alpha_S = B_r/B_{max}$ – the hysteresis loop squareness factor.
*2 $d\mu/dT$ – the permeability temperature factor.

Table 27.29 Basic characteristics of metal magnetostriction materials [43] [for notation see Section 27.1*]

Material and its chemical composition	ρ, 10^3 kg/m^3	E, 10^{11} N/m^2	Sound velocity c_s, 10^3 m/s	μ_r	Magnetic loss tangent tgβ, 10^3	H_{opt}, 10^3 A/m	a, 10^7 N/(m^2·T)	K	Λ, 10^{-9} T·m^2/N
Ni more than 98%	8.9	2.15	4.9	35	60	1–2	2.3	0.26–0.3	4.2
4% Co, Ni balance	8.9	2.15	4.9		55	2	2.7	0.43–0.5	
2.3%Cr, 1.4% Co, Ni balance	8.8	2.25	5.0			2.5	2.9	0.37	
4% Co, 2% Si, Ni balance	8.8	1.9	4.8	210		0.4–0.6	1.8	0.49	28
49% Co, 2% V, Fe balance	8.2	2.05	5.2	200	400	0.4–0.6	2.2	0.48–0.54	27
65% Co, Fe balance	8.2	2.05	5.2	70	500	1–1.7	1.5	0.27–0.30	7
14% Al, Fe balance	6.6	1.63	5.0	110	60	0.3–0.6	1.15	0.26	15
12.4% Al, Fe balance	6.7	1.58	4.8	30	30	0.3–0.6	0.85	0.30	20

Material and its chemical composition	Q	Resistivity, 10^{-7} Ω·m	H_c, 10^2 A/m	B_s, 10^{-1} T	λ_s, 10^{-6}	T_C, °C	$\lvert\lambda_s E\rvert$, 10^5 N/m^2	Dynamic fatigue strength σ_{dyn}, 10^5 N/m^2	Tensile strength σ_t, 10^5 N/m^2	Yield strength σ_y, 10^5 N/m^2
Ni more than 98%	750	0.7	1.7	6.1	−37	360	40	1000	3600	1000
4% Co, Ni balance		1.0		7.0	−36	410	41	1000		
2.3% Cr, 1.4% Co, Ni balance		3.0		4.8	−33	260	37			
4% Co, 2% Si, Ni balance		1.8	0.17–0.25	6.2	−(25–27)	350	26–28		4200	850
49% Co, 2% V, Fe balance	600	3.4	1.4	24.0	+70	980	80	1100	5000	3500
65% Co, Fe balance	600	0.8	1.0	22.0	+90	980	90		6700	4500
14% Al, Fe balance	300	12	0.25	12.0	+40	500	34		7500	5500
12.4% Al, Fe balance	400	16	0.1	16.0	+40	600	34	300	8000	6000

* For references to the magnetostriction rare-earth based compounds see Section 27.4.1.

27.4.2 Hard magnetic materials

The following main groups of materials are applied to permanent magnets manufacturing: the alloyed carbon steels characterized by moderate values of the coercive force H_c and the maximum density of magnetic energy $(BH)_{max}$; the iron–nickel–aluminum precipitation hardening alloys (of the Alnico, Ticonal, Alni type, Table 27.30, Fig. 27.80); the deformed iron–cobalt–chromium (Table 27.31) and iron–cobalt–vanadium (Vicalloy, Table 27.32) alloys, having the improved values of the above-mentioned parameters; the platinum–cobalt alloys (Fig. 27.81) and the intermetallic $SmCo_5$-based alloys (Table 27.33, Figs. 27.60, 27.80) with large values of these parameters. The record characteristics are demonstrated by the materials developed on the base of a compound $Nd_2Fe_{14}B$ (Figs. 27.60, 27.80). This material possesses $T_C = 585$ K, forms a tetragonal crystal lattice with the periods $a = 0.880$ nm and $c = 1.221$ nm, and its density is 7.6 g/cm^3 [19].

Table 27.30 Compositions and magnetic properties of isotropic alnicos 1–4 and anisotropic alnicos 5–9*[1] [32]

Characteristics	Alnico	Composition (wt.%)*[2]					Magnetic properties		
		Ni	Al	Co	Cu	Others	B_r, T	H_{cB}, kA·m^{-1}	$(BH)_{max}$, kJ·m^{-3}
Isotropic, Co-free	3	24–30	12–14	0	0–3	0–1 Ti	0.5–0.6	54–40	10
Isotropic, low Co	1&4	21–28	11–13	3–5	2–4	0–1 Ti	0.55–0.75	56–36	11–12
Isotropic, medium Co	2	16–20	9–11	12–14	3–6	0–1 Ti	0.65–0.8	50–40	13–14
Isotropic, extra-high H_{cB}	2	18–21	8–10	17–20	2–4	4–8 Ti 0–1 Nb	0.6–0.7	72–60	14–16
Field-treated random grain	5	12–15	7.8–8.5	23–25	2–4	0–0.05 Ti 0–1 Nb	1.2–1.3	52–46	40–44
Field-treated random grain, higher H_{cB}	6	13–16	7.8–8.5	23–25	2–4	1–1.5 Ti 1–2 Nb	1.05–1.15	62–56	30–36
Field-treated random grain, high H_{cB}	7	18	8	24	3	5	0.74	85	24
Field-treated random grain, high H_{cB}	8	14–16	7–8	32–36	4	4–6 Ti 0–1 Nb	0.8–0.9	140–110	40–45
Field-treated random grain, extra-high H_{cB}	8	16–15	7–8	37–40	3	7–8 Ti	0.74–0.78	170–150	44–48
Field-treated directed-grain, form of alnico 5	5DG	13–15	7.8–8.5	24–25	2–4	0–1 Nb	1.3–1.4	62–56	56–64
Field-treated columnar grain form of alnico 8	9	14–16	7–8	32–36	4	4–6 Ti 0–1 Nb 0.3 S	1.0–1.1	140–110	60–75

*[1] Greater details can be found in [51].
*[2] Fe balance.

Table 27.31 Typical magnetic properties of Cr–Co–Fe alloys [32]

Alloy (wt.%, Fe balance)	B_r, T	H_c, kA·m^{-1}	$(BH)_{max}$, kJ·m^{-3}	Alloy (wt.%, Fe balance)	B_r, T	H_c, kA·m^{-1}	$(BH)_{max}$, kJ·m^{-3}
Isotropic magnets				5Co, 30Cr	1.34	42	42
23Co, 31Cr	0.84	50	18	2Co, 33Cr	1.24	38	34
10.5Co, 28Cr	0.98	32	14	**Anisotropic magnets (deformation aging)**			
7Co, 28Cr	0.97	26	11	23Co, 32Cr, 2Cu	1.30	86	78
Anisotropic magnets (field aging)				16Co, 32Cr, 2Cu	1.29	70	65
25Co, 30Cr, 3Mo	1.15	62	40	11.5Co, 32Cr, 2Cu	1.15	65	50
15Co, 24Cr, 3Mo	1.54	67	76	7Co, 32Cr, 2Cu	1.18	42	34
Anisotropic magnets [(100) columnar]				**Sintered magnets**			
15Co, 22Cr	1.56	51	66	25Co, 29.5Cr, 3Mo	0.95	69	27
12Co, 25Cr	1.45	50	61	12Co, 25Cr	1.40	44	42
9Co, 27Cr	1.30	46	50	5Co, 31Cr	1.23	40	35
7Co, 28Cr	1.20	41	42				

Table 27.32 Magnetic properties of Co-Fe-V alloys [32]

Alloy (wt.%, Fe balance)	B_r, T	H_c, kA·m^{-1}	$(BH)_{max}$, kJ·m^{-3}
10V, 52Co (Vicalloy I)	0.75	20	64
14V, 52Co (Vicalloy II wire)	1.00	42	28
8V, 53Co, 4Cr (Koerflex 300 wire)	1.15	30	25
8V, 53Co, 4Cr (Koerflex 300 wire)	1.05	24	15
5V, 53Co, 4Cr (Koerflex 200)	1.30	10	8.8

Table 27.33 Magnetic properties of various permanent magnets based on cobalt–rare-earth alloys [32]

Composition	B_r, T	H_{cB}, kA·m^{-1}	H_{cJ}, kA·m^{-1}	$(BH)_{max}$, kJ·m^{-3}
Co_5Sm*1	1.06	792	1360	224
Co_5Sm*2	0.91	716	1430	159
Co_5Sm*3	0.86	660	1430	143
$Co_5MM_{0.5}Sm_{0.5}$	0.90	605		159
$Co_5Sm_{0.4}Pr_{0.6}$	1.03	804	1320	207
$Co_5Sm_{0.5}Pr_{0.3}Nd_{0.2}$	1.05	770	1150	210
$Co_5Gd_{0.42}Sm_{0.58}$	0.64	454	3020	74
$(Co_{0.72}Fe_{0.14}Cu_{0.14})_5Ce$	0.72	398	421	99
$(Co_{0.73}Fe_{0.20}Cu_{0.05}Zr_{0.02})_{7.5}Ce_{0.5}Sm_{0.5}$	1.06	648	744	210
$(Co_{0.70}Fe_{0.16}Cu_{0.12}Mn_{0.02})_{6.6}Ce_{0.3}Sm_{0.7}$	0.97	557	597	179
$(Co_{0.73}Fe_{0.20}Cu_{0.05}Zr_{0.02})_{7.5}Sm$	1.07	760	1000	216
$(Co_{0.69}Fe_{0.20}Cu_{0.10}Zr_{0.01})_{7.4}Sm$	1.12		533	240
$(Co_{0.65}Fe_{0.28}Cu_{0.05}Zr_{0.02})_{7.8}Sm$	1.20		1110	263

*1 Laboratory best value.
*2 Commercial high grade.
*3 Commercial average.

References

[1] Abbundi, R., Clark, A. E., J. Appl. Phys., 49, 1969, 1978.

[2] American institute of physics handbook, Ed. Gray, D. E., McGraw-Hill, New York, 1963.

[3] Arajs, S., Colvin, R. V., J. Phys. Chem. Solids, 24, 1233, 1963.

[4] Aubert, G., J. Appl. Phys., 39, 504, 1968.

[5] Barnier, Y., Pauthenet, R., Rimet, G., Compt. Rend. Acad. Sci., 252, No 19, 2839, 1961.

[6] Belov, K. P., Levitin, R. Z., Nikitin, S. A., Ped'ko, A. V, Zhurnal eksperimental'noi i teoreticheskoi fiziki, 40, 1562, 1961 (in Russian).

[7] Berdyshev, A. A., Introduction into quantum theory of ferromagnetism, A course of lectures, Part 4, Izdatel'stvo Ural'skogo gosudarstvennogo universiteta, 1971 (in Russian).

[8] Bhagat, S. M., Rothstein, M. S., Solid State Commun., 11, 1535, 1972.

[9] De Boer, F. R., Biesterbos, J., Schinkel, C. J., Phys. Lett., 24A, No 7. 355, 1967.

[10] Booth, J. G., in: Ferromagnetic materials, Vol.4, Eds. Wohlfarth, E. P. and Buschow, K. H. J., Elsevier Science Publishers B. V., Amsterdam, 1988.

[11] Bower, D. I., Proc. Roy. Soc., A326, No 1564, 87, 1971.

[12] Bozorth, R. M., Ferromagnetism, Van Nostrand, Princeton, 1951.

[13] Bozorth, R. M., Rev. Mod. Phys., 25, 42, 1953.

[14] Bozorth, R. M., Phys. Rev., 96, 311, 1954.

[15] Bozorth, R. M., Clark, A. E., Van Vleck, J. H., Intern. J Magnetism, 2, 19, 1972.

[16] Bozorth, R. M., McGuire, T. R., Hudson, R. P., in: American institute of physics handbook, Ed. Gray, D. E., McGraw-Hill, New York, 1972 (pp. 5–139).

[17] Burso, E., Z. Angew. Phys., 32, 127, 1971.

[18] Buschow, K. H. J., in: Ferromagnetic materials, Vol.1, Ed. Wohlfarth, E. P., North Holland, Amsterdam, 1980.

[19] Buschow, K. H. J., in: Ferromagnetic materials, Vol.4, Eds. Wohlfarth, E. P. and Buschow, K. H. J, Elsevier Science Publishers B. V., Amsterdam, 1988.

[20] Carr, W. J., in: Handbook der physik, Vol.18/2, Ed. Flügge, S., Springer–Verlag, 1966.

[21] Chin, G. Y., Wernick, J. H., in: Ferromagnetic materials, Vol.2, Ed. Wohlfarth, E. P., North Holland, Amsterdam, 1980.

[22] Clark, A. E., Cullen, J., Sato, K., AIP Conf. Proc., 1975, N 24. Amer. Inst. Phys., New York (p. 192).

[23] Clark, A. E., Abbundi, R., Savage, H. T., McMasters, O. D, Physica, 86-88B, 73, 1977.

[24] Clark, A. E., Abbundi, R., Gillmor, W. G., IEEE Trans. Mag. MAG–14, 542, 1978.

[25] Clark, A. E., in: Ferromagnetic materials, Vol.1, Ed. Wohlfarth, E. P., North Holland, Amsterdam, 1980 (p. 531).

[26] Cohen, R. L., Sherwood, R. C., Wernick, J. H., Phys. Lett., 26A, 462, 1968.

[27] Corner, W. D., Tanner, B. K., J. Phys., 26A, 462, 1976.

[28] Creveling, L., Luo, H. L., Knapp, G. S., Phys. Rev. Lett., 18, 851, 1967.

[29] Danan, H., Herr, A., Meyer, A. J. P., J. Appl. Phys., 39, 669, 1968.

[30] Deryagin, A. V., Kudrevatykh, N. V., Bashkov, Yu. F., Proc. Intern. Conf. on Magnetism ICM–73, Vol.1, Nauka, Moscow, 1974 (p. 222) (in Russian).

[31] Elliot, J. F., Legvold, S., Spedding, F., Phys. Rev., 94, 1143, 1958.

[32] Encyclopedia of material science & engineering, Vol.4, Ed. Bever, M. B., Pergamon Press, New York, 1986.

[33] Ermolenko, A. S., Proc. Intern. Conf. on Magnetism ICM–73, Vol.1, Nauka, Moscow, 1974, (p. 231) (in Russian).

[34] Ermolenko, A. S., Physical properties of magnetic materials, Uralskii nauchnyi tsentr Akademii nauk SSSR, Sverdlovsk, 1982 (p. 32) (in Russian).

[35] Ermolenko, A. S., Ph. D. Thesis. Institut Fiziki Metallov AN SSSR, Sverdlovsk, 1983 (in Russian).

[36] Eschenfelder, A. H., in: Ferromagnetic materials, Vol.2, Ed. Wohlfarth, E. P., North Holland, Amsterdam, 1980 (p. 345).

[37] Escudier, P., Ph. D. Thesis, Grenoble Univ., 1973; Franse, J. J. M., de Vries, G., Physica, 39, 477, 1968.

[38] Feron, J. L., Hug, G., Pauthenet, R., Z. Angew. Phys., 31, 61, 1970; Les elements des terres rares, Vol.2, CNRS, Paris, 1970 (p. 19).

[39] Ferromagnetic materials, Vol.1–4, Eds. Wohlfarth, E. P. and Buschow, K. H. J., Elsevier Science Publishers B. V., Amsterdam, 1980–1988.

[40] Gardner, W. E., Smith, T. F., Howlett, B. W., Phys. Rev., 166, 577, 1968.

[41] Gengnagel, H., Hoffmann, U., Phys. Status Solidi, 29, 91, 1968.

[42] Gersdorf, R., Ph. D. Thesis, Amsterdam Univ., 1961.

[43] Golyamina, I. P., in: Ultrasound. Little encyclopedia, Ed. Golyamina, I. P., Sovetskaya Entsiklopedia, Moscow, 1979 (in Russian).

[44] GOST (National Standard of the Russian Federation) 10160–75. Soft Magnetic Alloys. Moscow, Izdatel'stvo Standartov, 1975 (in Russian).

[45] Graham, C. D., Jr., Phys. Rev., 112, 1117, 1958.

[46] Green, R. W., Legvold, S., Spedding, F. H., Phys. Rev., 122, 827, 1961.

[47] Hall, R. C., J. Appl., Phys., 30, 816, 1959.

[48] Hall, R. C., Trans. Metallurg. Soc. AIME, 218, 268, 1960.

[49] Handbook on precision alloys, Ed. Molotilov, B. V., Metallurgiya, Moscow, 1983 (in Russian).

[50] Hauham, S. D., Arrot, A. S., Heinrich, B., J. Appl. Phys., 52, 1941, 1981.

[51] Heck, C., Magnetic materials and their applications, Crane, Russak & Co., New York, 1974.

[52] Hegland, D. E., Legvold, S., Spedding, F. H., Phys. Rev., 131, 158, 1963.

[53] Honda, K., Kaya, S., Sci. Repts. Tohoku Univ., 15, 721 1926.

[54] Honda, K., Masumoto, H., Sci. Repts. Tohoku Univ., 20, 322, 1931.

[55] Hubert, A., Unger, W., Kranz, J., Z. Phys., 224, 148, 1969.

[56] Huray, P. G., Nave, S. E., Handbook on the physics and chemistry of the actinides, Vol.5, Eds. Freeman, A. J. and Lander, G. H., Elsevier Science Publishers B. V., Amsterdam, 1987.

[57] Kase, T., Sci. Repts. Tohoku Univ., 16, 491, 1927.

[58] Kaya, S., Sci. Repts. Tohoku Univ., 17, 639, 1927.

[59] Kaya,S., Sci. Repts. Tohoku Univ., 17, 1157, 1928.

[60] Klein, H.-P., Kneller, E., Phys. Rev., 144, 372, 1966.

[61] Klein, H. P., Menth, A., Perkins, R. S., Physica, 80B, 153, 1975.

[62] Kondorskii, E. I., Sedov, V. L., Zhurnal eksperimental'noi i teoreticheskoi fiziki, 35, 586, 1959 (in Russian).

[63] Kudrevatykh, N. V., Ph. D. Thesis, Uralskii Gosugarstvennyi Univ., Sverdlovsk, 1977 (in Russian).

[64] Laforest, J., Lemaire, R., Pauthene, R., Schweizer, J., Compt. Rend. Acad. Sci., B262, 1260, 1966.

[65] Landolt–Börnstein, Zahlenwerte und funktionen aus physik, chemie, geophysik, technik, Bd. 2, Teil 9, Springer–Verlag, Berlin, 1962.

[66] Landolt–Börnstein, New Serie, Group III (Crystal and solid state physics), Springer-Verlag, Berlin, Vol.19 (Magnetic Properties of Metals), subvol. *a*, 1986; subvol. *b*, *c*, 1987; subvol. *d*1, 1991; subvol. *d*2, *e*, 1990; subvol. *f*1, *f*2, 1991, subvol. *g*, 1988; subvol. *h*, 1991; subvol. *i*1, *i*2, 1992.

[67] Lee, E. W., Asgar, M. A., Proc. Roy. Soc., A326, No 1564, 73, 1971.

[68] Legvold, S., Alstad, J., Rhyne, J., Phys. Rev. Lett., 10, 509, 1963.

[69] Legvold, S., in: Ferromagnetic materials, Vol.1, Ed. Wohlfarth, E. P., North Holland, Amsterdam, 1980.

[70] Littman, M. F., IEEE Trans. Mag., MAG–7, 48, 1971.

[71] Luborsky, P. E., in: Ferromagnetic materials, Vol.1, Ed. Wohlfarth, E. P., North Holland, Amsterdam, 1980; Handrich, K., Kobe, S., Amorphe ferro- und ferrimagnetika, Akademie–Verlag, Berlin, 1980.

[72] Mattias, B. T., Bozorth, R. M., Phys. Rev., 169, 604, 1958.

[73] Mattias, B. T., Clogston, A. M., Williams, H. J., Phys. Rev. Lett., 7, 7, 1961.

[74] Mattocks, P. G., Melville, D., J. Phys. F: Metal Phys. 8, 1291, 1978.

[75] McEwen, K. A., Handbook on the physics and chemistry of rear-earths, Ch.6, Eds. Gschneider, K. A. and Eyring, L., North Holland, Amsterdam, 1978.

[76] Mihai, V., Franse, J. J. M., Rev. Roum. Phys., 21, 1041, 1976.

[77] Mishima, A., Fuju, H., Okamoto, T., J. Phys. Soc. Jpn., 40, 962, 1976.

[78] Morrish, A. H., The physical principles of magnetism, John Wiley & Sons, New York, 1965.

[79] Mydosh, J. A., Nieuwenhuys, G. J., in: Ferromagnetic materials, Vol.1, Ed. Wohlfarth, E. P., North Holland, Amsterdam, 1980.

[80] Myers, H. P., Sucksmith, W., Proc. Roy. Soc., A207, No 1091, 427, 1951.

[81] Nakagawa, Y., J. Phys. Soc. Jpn., 11, 855, 1956.

[82] Nigh, H. E., Legvold, S., Spedding, F. H., Phys. Rev., 132, 1092, 1963.

[83] Ogawa, S., Sakamoto, N., J. Phys. Soc. Jpn., 22, 1214, 1967.

[84] Pauthenet, R., in: High-field magnetism, Ed. Date, M., North Holland Publ. Comp., Amsterdam, 1983 (p. 77).

[85] Rhyne, J. J., Ph. D. Thesis, Iowa State Univ., Ames, Iowa, 1965.

[86] Rhyne, J. J., Legvold, S., Phys. Rev., 138A, 507, 1965.

[87] Rhyne, J. J., Foner, S., McNiff, E. J., Doclo, R., J. Appl. Phys., 39, 892, 1968.

[88] Richards, D. B., Legvold, S., Phys. Rev., 186, 508, 1969.

[89] Sechovsky, V., Havela, L., in: Ferromagnetic materials, Vol.4, Eds. Wohlfarth, E. P. and Buschow, K. H. J., Elsevier Science Publishers B. V., Amsterdam, 1988.

[90] Sergeev, V. G., in: Handbook of electrical engineering, Vol.1, Ed. Gerasimov, V. G., Energoatomizdat, Moscow, 1985 (in Russian).

[91] Shimizu, M., Rep. Prog. Phys., 44, 329, 1981.

[92] Stoelinga, J. H. M., Gersdorf, R., de Vries, G., Physica, 31, 349, 1965.

[93] Strandburg, D. L., Legvold, S., Spedding, F. H., Phys. Rev. 127, 2046, 1962.

[94] Strnat, K. J., in: Ferromagnetic materials, Vol.4, Eds. Wohlfarth, E. P. and Buschow, K. H. J., Elsevier Science Publishers B. V., Amsterdam, 1988.

[95] Sucksmith, W., Thompson, J. E., Proc. Roy. Soc., A225, No 1162, 362, 1954.

[96] Tange, H., Tokunaga, T., J. Phys. Soc. Jpn., 27, 554, 1969.

[97] Tebble, R. S., Craig, D. J., Magnetic materials. Wiley Intersci., London, 1969.

[98] Trzebiatowski, W., in: Ferromagnetic materials, Vol.1, Ed. Wohlfarth,E. P., North Holland, Amsterdam, 1980.

[99] Washko, S. D., Shen, T. H., in: Encyclopedia of material science & engineering, Vol.4, Ed. Bever, M. B., Pergamon Press, New York, 1986.

[100] Weiss, P., Forrer, R., Ann. Phys., 12, 279, 1929.

[101] Williams, G. M., Pavlovic, A. S., J. Appl. Phys., 39, 571, 1968; Benninger G. N., Pavlovic A. S., J. Appl. Phys., 38, 1325, 1967.

[102] Wohlfarth, E. P., in: Ferromagnetic materials, Vol.1, Ed. Wohlfarth, E. P., North Holland, Amsterdam, 1980, pp.1–70.

[103] Yamamoto, M, Nakamichi, T., J. Phys. Soc. Jpn., 13, 228, 1958.

28

Antiferromagnets

V.I. Ozhogin and V.G. Shapiro

28.1 Introduction

Electron exchange interaction determining an ordered magnetic structure of substances can result in both the parallel and antiparallel (and even more complex) mutual arrangement of magnetic moments of nearest neighbor atoms in crystals. The substances in which magnetic moments of separate ions are completely (or almost completely) compensated are referred to as *antiferromagnets.* Information about magnetic structures and properties of antiferromagnets can be found in monographs [1–16] and the handbook [17]. Experimental data on antiferromagnets' magnetic properties, as a rule, are well explained if their magnetic structure may be vizualized as a superposition of two (or more) sublattices inserted one into another, magnetic moments of atoms in each sublattice being parallel to each other. However, at present a great number of various noncollinear magnetic structures is coming to light, in particular, helicoidal (e.g., MnO_2, $MnAu_2$), sinusoidal-modulated (of the type $CeAl_2$) and others, in which magnetic moment compensation can take place over rather large, nearly macroscopic, volume of the crystal. Such substances are usually classed with the antiferromagnets too. More details on various noncollinear magnetic structures can be drawn in the monograph [12] and the review [18].

28.2 Selected Properties of Antiferromagnets

The numerical data given below basically deal with magnetic properties of insulators with antiferromagnetic ordering. Besides, the properties of certain semiconductors, metals, and metallic alloys[1] are also presented.

Table 28.1 does not cover all the antiferromagnets known at the time of this handbook edition. The authors aimed to present a general view on the properties of various types of antiferromagnetic crystals from the best-known and well-studied antiferromagnetic insulators to recently discovered superconducting compounds. An important, though not necessary, criterion of data choosing was the fact of the compound antiferromagnetic structure being established by neutron diffractometry. There was no possibility to give the most complete data on certain groups of compounds. Thus, a wide group (with the exception of two compounds) of antiferromagnets with garnet structure is not practically represented.[2] There appears only scarce information on many rare-earth alloys and compounds.[3]

[1] The main classes of antiferromagnetic insulators are in [56].
[2] The data on them can be found in [57].
[3] The antiferromagnetism of these compounds is considered in [58].

0-8493-2861-6/97/$0.00+$.50

Information on magnetic structures of many antiferromagnets studied by neutron diffraction is presented in [3]. Types of the magnetic ordering in cubic lattices are shown in Fig. 28.1.

The consecutive order in which the compounds are listed in Table 28.1 corresponds to the atomic number of a magnetic atom in Mendeleev's periodic system of elements. The compounds with the same magnetic atom are given in order of growing complexity (e.g., the compounds involving three elements of the $R_iA_kB_l$ kind with the magnetic ion R are arranged in order of growing succession $n = 100i + 10k + l$). The numbers of n being equal, the compounds are lined up in the order of A and B atoms position in the periodic system of elements.

Notation and abbreviations used in Table 28.1 are as follows:

I, II, III, IV – the possible types of magnetic order in a face-centered cubic lattice (Fig. 28.1b);

A, C, E, F, G – the possible types of magnetic order in a simple cubic lattice (Fig. 28.1a);

$[a]$, $[b]$, $[c]$ – the crystallographic axes in rhombohedral, monoclinic, orthorhombic, and hexagonal syngonies;

(ab), (bc), (ac) – the crystallographic planes in triclinic, monoclinic, and orthorhombic syngonies;

$\mathbf{a}^*$, $\mathbf{b}^*$, $\mathbf{c}^*$ – the reciprocal lattice vectors;

c – the highest symmetry axis in tetragonal, trigonal, and hexagonal syngonies;

D – the spin dimension defined by a number of spin orthogonal components;

d – the magnetic lattice dimension defined by spatial anisotropy of exchange interaction energy[4];

E_g – the width of energy gap in an electron excitation spectrum of substances which exhibit semi-conducting properties;

F, **G**, **C**, **A** – the vectors resulting from a linear combination of magnetic moments in a four-sublattice model of an antiferromagnet[5]:

$$\mathbf{F} = (1/4)(\mathbf{M_1} + \mathbf{M_2} + \mathbf{M_3} + \mathbf{M_4}), \tag{28.1}$$

$$\mathbf{G} = (1/4)(\mathbf{M_1} - \mathbf{M_2} + \mathbf{M_3} - \mathbf{M_4}), \tag{28.2}$$

$$\mathbf{C} = (1/4)(\mathbf{M_1} + \mathbf{M_2} - \mathbf{M_3} - \mathbf{M_4}), \tag{28.3}$$

$$\mathbf{A} = (1/4)(\mathbf{M_1} - \mathbf{M_2} - \mathbf{M_3} + \mathbf{M_4}); \tag{28.4}$$

H_A – the strength of effective field of magnetic anisotropy[6];

H_C – the external magnetic field in which one or another feature in the dependence of substance magnetization upon magnetic field is observed;

$H_{c2}(0)$ – the upper (second) critical field at $T = 0$ K for a type II superconductor;

H_D – the strength of effective field due to Dzyaloshinskii interaction;

H_E – the strength of effective exchange field;

H_{res} – the strength of effective magnetic field at which antiferromagnetic resonance is observed;

ΔH_{res} – the linewidth of antiferromagnetic resonance in magnetic field;

J – the constant characterizing the exchange interaction between neighbor magnetic ions in isotropic systems or between neighbor magnetic ions in linear chains in quasi-one-dimensional ($d = 1$) systems or inside the planes in quasi-two-dimensional ($d = 2$) systems;

J' – the constant characterizing the exchange interaction between magnetic ions which belong to neighbor linear chains in quasi-one-dimensional systems ($d = 1$) or neighbor planes in quasi-two-dimensional systems ($d = 2$);

[4]In a model description of antiferromagnet behavior, an exchange Hamiltonian is usually considered in the form $H = \sum_{ij} J^{\|}_{ij} S^z_i S^z_j + J^{\perp}_{ij}(S^x_i S^x_j + S^y_i S^y_j)$, the case of $J^{\|}_{ij} = J^{\perp}_{ij}$ corresponding to the Heisenberg model ($D = 3$), the case of $J^{\perp}_{ij} = 0$ corresponding to the Ising model ($D = 1$), and $J^{\|}_{ij} = 0$ – to XY - model. If $J^{\|} \neq J^{\perp}$, one can speak about spin anisotropy of exchange interaction. If $J(R_{ij})$ variously depends on R along different main directions of the crystal, the spatial anisotropy of exchange interaction is said to exist.

[5]Stable spin configuration (magnetic order) in antiferromagnetic crystals is often described by means of the invariants of the second order, which are generated from the vector **F**, **G**, **C**, **A** components and transformed according to the same irreducible representation of the crystal space group [11].

[6]Numerical values of magnetic induction corresponding to the magnetic field strength in SI (for a vacuum there is the relation 1 T= $\mu_0 \cdot$ 1 A/m, where $\mu_0 = 4\pi \cdot 10^{-7}$ H/m). If the temperature is not specified, the value is given at $T \ll T_N$, where T_N is the temperature of magnetic ordering (or Néel point).

la, mb, nc – the magnetic cell parameters (dimensions of the magnetic cell are enlarged compared to that of a crystallographic unit cell along the a, b and c axes by the factors of l, m, and n, respectively);

n_s – the number of Bohr magnetons per atom in the state of saturation;

p – the effective number of Bohr magnetons per atom in the expression determining C_M ($C_M = Np^2\mu_B/3k$);

T_N, $T_N(\mathrm{A})$ – the temperature of antiferromagnetic ordering, the temperature of A ions ordering[7];

T_r – the temperature of magnetic moment reorientation with respect to crystallographic axes;

$\alpha(\mu, [ikl])$ – the angle between the magnetic moment μ vector and the crystallographic axis $[ikl]$;

θ, C_M – Curie-Weiss constants, $\chi_M = C_M/(T - \theta)$ (as a rule, $\theta < 0$);

$\lambda(0)$ – the wavelength of electromagnetic radiation corresponding to antiferromagnetic resonance frequency in zero magnetic field (at $T \ll T_N$);

μ – the magnetic moment of an ion (atom);

$\mu_A(T)$[8] – the magnetic moment of the A ion (atom) at the temperature T (in Bohr magnetons μ_B);

${}^A\nu_z$ – the nuclear magnetic resonance frequency corresponding to the internal magnetic field at the nucleus ${}^AZ$, where A is the mass number of the element Z;

ρ – the resistivity;

$\sigma_m(T)$[9] – the molar spontaneous magnetic moment at the temperature T, A·m^2/mol (1 CGSM unit $= 10^{-3}$ SI unit);

$\sigma_s(T)$ – the specific spontaneous magnetic moment at the temperature T, A·m^2/kg (1 CGSM unit = 1 SI unit);

χ_M – the molar magnetic susceptibility, cm^3/mol (1 CGSM unit = 4π SI unit);

χ_s – the specific magnetic susceptibility, cm^3/g (1 CGSM unit = 4π SI unit);

AF – antiferromagnetic state;

AFI between (ikl) – antiferromagnetic interaction between magnetic atoms belonging to neighbor planes (ikl), which results in antiparallel arrangement of magnetic moments of atoms from neighbor planes (ikl);

AFMR – antiferromagnetic resonance;

SDW – spin density wave;

HS – helicoidal magnetic structure ($\mathbf{k}_0$ is the helicoid propagation vector);

CS – collinear magnetic structure;

EA – easy axis;

EP – easy plane;

MO – magnetic order;

NCS – noncollinear magnetic structure;

P – paramagnetic state;

SW – sinusoidal-modulated magnetic structure (spin wave);

CAF – canted antiferromagnetic state;

F – ferromagnetic state;

FI (AFI) in (ikl) – ferromagnetic (antiferromagnetic) exchange interaction between magnetic atoms belonging to the same plane (ikl), which results in parallel (antiparallel) arrangement of magnetic moments of atoms from the plane (ikl);

NMR – nuclear magnetic resonance.

Indices used in writing some physical quantities[10]:

$\parallel$ ($\perp$) – corresponds to the direction parallel (normal) to the chosen crystal axis;

$[ikl]$, $[a]$, $[b]$, $[c]$ – the symbol is associated with the corresponding crystallographic direction.

[7]The values of Néel temperature T_N presented in Table 28.1 are obtained mainly from magnetic susceptibility or neutron scattering measurements. The values obtained from specific heat measurements differ to some extent from the former.

[8]$\mu(\mathrm{A})$ is taken at $T \to 0$ K.

[9]$\sigma(0)$ is taken at $T \to 0$ K.

[10]The same indices written alongside the numerical values denote the fact that appropriate measurement was carried out in the corresponding direction.

Table 28.1 Certain properties of antiferromagnets

Substance	Structure: crystallographic	Structure: magnetic	T_N, K	Additional information	References
α-O_2	Monoclinic C_{2h}^3	CS, $\mu\|[010]$	23.9	$\mu_{O_2}(4.2K) = 2\mu_B$	[3]
α–KO_2^{*1}	Tetragonal D_{4h}^{17}	$\mu\|(001)$ FI in (001) AFI between (001)	7	$\theta = -3\,K,\ p = 1.03$ at $T < 150$ K; $\theta = -53\,K,\ p = 1.2$ at $T > 150$ K	[3,17]
$LaTiO_3$	Orthorhombic D_{2h}^{16}	G-type MO (Ti)	125	$\mu_{Ti^{3+}}(10\,K) = 0.45\mu_B$ Metallic conductivity at $T > 125$ K	[59]
α-VS	Hexagonal D_{6h}^4		1040 900*2	$\theta = -3000\,K$ $\chi_M(T_N)/4\pi =$ $= 6.6 \cdot 10^{-5}\ cm^3/mol$	[17]
α-VSe	Hexagonal D_{6h}^4		163	$\theta = -2570\,K$	[17]
VF_2	Tetragonal D_{4h}^{14}	HS, $\mathbf{k}\|\mathbf{c}$, $\mu\perp c$	7	$\theta = -80\,K$	[3,17]
VCl_3	Trigonal D_{3d}^6		30		[17]
V_2O_3	Trigonal D_{3d}^6	$\mu\|[111]$	150	$\rho(T_N+0)/\rho(T_N-0) \approx 10^{9*3}$ $\mu_{V^{3+}} < 0.05\mu_B$	[3,17]
VOCl	Orthorhombic D_{2h}^{13}	CS, $\mu\|[a]$	80.5	$\mu_V(4\,K) = 1.48\mu_B$ Magnetic properties correspond to Heisenberg model, $d = 2$	[60]
$LaVO_3$	Cubic O_h^5	$\mu\perp c$ FI in (010) AFI between (001)	133–144	$\theta = -665\,K$ $\mu_V(4.2\,K) = 1.31\mu_B$	[3]
V_2WO_6	Tetragonal D_{4h}^{14}	NCS, $\mu\perp c$	370		[3]
MgV_2O_4	Cubic O_h^7 (tetragonal distortions at $T < T_N$)	$\mu\| c$	45	$\theta = -750\,K,\ p = 3.43$ $\mu_{V^{3+}}(4.2\,K) = 1.0\mu_B$	[3]
CaV_2O_4	Orthorhombic D_{2h}^{16}	CS, $\mu\|[b]$, $(a,\ 2b,\ 2c)$		$\mu_{V^{3+}}(4.2\,K) = 1.0\mu_B$	[3]
ZnV_2O_4	Cubic O_h^7 (tetragonal distortions at $T < T_N$)	CS, $\mu\| c$	45	$\theta = -256$ K, $p = 2.18$ $\mu_{V^{3+}}(4.2\,K) = 0.8\mu_B$	[3]
α–Cr	Cubic O_h^9	$\mu\|[100]$ $T_{SF} \approx 122\,K^{*4}$	312	$\mu(77\,K) = 0.59\mu_B$	[1,3,61]
CrN	Cubic O_h^5 (orthorhombic distortions at $T < T_N$)	IV type MO, $\mu\|[110]$ $(2a,\ 2a,\ 2a)$	273	$\mu_{Cr}(77\,K) = 2.4\mu_B$	[3]

*1 Magnetic properties of α-KO_2 are due to unpaired electron in π-type molecular (2p-atomic orbital) of O_2^- ion.

*2 By specific heat measurements.

*3 Simultaneously with the P–AF transition there occurs metal–semiconductor transition, at $T = T_N$ conductivity changes by nine–ten order of magnitude. For the description of the electronic structure and the review of of V_2O_3 magnetic properties see the book by Bugaev A.A., Zakharchenya B.M., Chudnovskii E.A. *Metal-semiconductor phase transition and its applications*, Leningrad, Nauka, 1979 (in Russian).

*4 The magnetic structure constitutes a standing SDW with the wave vector $\mathbf{Q}$ and polarization vector η. At $T < T_{SF}$ – longitudinal SDW ($\mathbf{Q}\|\eta$), at $T > T_{SF}$ – transverse SDW ($\mathbf{Q}\perp\eta$).

Table 28.1 Certain properties of antiferromagnets *(continued)*

Substance	Structure		T_N, K	Additional information	References
	crystallographic	magnetic			
CrS	Monoclinic C_{2h}^6	FI in (011) AFI between (011)	460	$\theta = -695\,\text{K}$, $p = 4.22$ $\mu_{\text{Cr}}(77\,\text{K}) = 3.4\mu_B$	[3,17]
CrAs	Orthorhombic D_{2h}^{16}	NCS, $\mathbf{k}_0 \parallel [c]$	260–280	$\mu_{\text{Cr}}(77\,\text{K}) = 2.0\mu_B$ $\mu_{\text{Cr}}(4.2\,\text{K}) = 1.7\mu_B$	[3]
CrSe	Hexagonal D_{6h}^4	NCS $\alpha(\mu,\ c) = 45°$ $(3a,\ 3a,\ c)$	200–300	$p = 4.6$, $\mu_{\text{Cr}}(4.2\,\text{K}) = 2.9\mu_B$	[3,17]
CrSb	Hexagonal D_{6h}^4	CS, FI in (001) AFI between (001)	663–723	$p = 3.89$, $\chi_M(T_{\text{N}})/4\pi =$ $1.92 \cdot 10^{-3}\text{cm}^3/\text{mol}$ $\mu_{\text{Cr}^{3+}}(290\,\text{K}) = 2.84\mu_B$	[3,17]
CrRe (alloy with Re concentration of more than 17%)	FCC		160	Itinerant antiferromagnet; superconductor at $T \leq 3\,\text{K}$	[62]
CrF_2	Monoclinic C_{2h}^5	$\mu \parallel (010)$ $\alpha(\mu,\ [c]) = 32°$	53	$p = 4.3$, $\mu_{\text{Cr}^{2+}}(4.2\,\text{K}) = 3.96\mu_B$	[3,17]
$CrCl_2$	Orthorhombic D_{2h}^{12}	CS*5 FI in (011) AFI between (011) $(a,\ 2b,\ 2c)$	20	$\theta = -149\,\text{K}$ $\mu_{\text{Cr}}(4.2\,\text{K}) = 3.68\mu_B$	[3,17]
$CrSb_2$	Orthorhombic D_{2h}^{12}	$\mu \parallel [101]$, FI in (011) AFI between (011) $(a,\ 2b,\ 2c)$	273	$\mu_{\text{Cr}}(4.2\,\text{K}) = 1.49\mu_B$	[3]
CrF_3	Trigonal D_{3d}^6	CS, $\mu \parallel (111)$	80	$\theta = -124\,\text{K}$, $p = 4.1$, $\mu_{\text{Cr}}(4.2\,\text{K}) = 3\mu_B$ $\sigma(0\ \text{K}) = 169 \cdot 10^{-3}\ \text{A} \cdot \text{m}^2/\text{mol}$	[3,17]
$CrCl_3$	Trigonal D_3^3	CS, $\mu \perp c$ FI in (001) AFI between (001) $(a,\ a,\ 2a)$	16.8	$\theta = 29\,\text{K}$ $\chi_M(T_N)/4\pi = 6.0\,\text{cm}^3/\text{mol}$ $\mu_{\text{Cr}^{3+}}(4.2\,\text{K}) = 2.82\mu_B$ $\mu_0 H_{c\parallel}(1.7\,\text{K}) = 5\ \text{T}$ $\mu_0 H_{c\perp}(1.7\,\text{K}) = 0.26\ \text{T}$	[3,17]
$CrAu_4$	Tetragonal D_{4h}^{18}	NCS, $\mu \perp c$	380		[3]
$\beta AlCr_2$	Tetragonal D_{4h}^{17}	$\alpha(\mu,\ c) = 65°$	598	$\mu_{\text{Cr}} = 0.92\mu_B$	[3]
Cr_2As	Tetragonal D_{4h}^7	$\mu \parallel (001)$ $(a,\ a,\ 2c)$	393–438	$\theta = -2067\,\text{K}$, $p = 1.8$ $\mu_{\text{Cr I}}(77\,\text{K}) = 1.1\mu_B$ $\mu_{\text{Cr II}}(77\,\text{K}) = 1.2\mu_B$	[3,17]
Cr_2O_3	Trigonal D_{3d}^6	CS, $\mu \parallel c$	308	$p = 2.6$, $\mu_{\text{Cr}}(4.2\,\text{K}) = 2.8\mu_B$ $\mu_0 H_{c\parallel}(77\,\text{K}) = 5.9\ \text{T}$ $\lambda(0) = 189\ \mu\text{m}$ $^{59}\nu_{\text{Cr}}(1.6\,\text{K}) = 70.43\,\text{MHz}$*6	[3,17]

*5 Magnetic moments of ions are parallel to the Cr–Cl long bond.

*6 See Appendix II in [8].

Table 28.1 Certain properties of antiferromagnets *(continued)*

Substance	Structure: crystallographic	Structure: magnetic	T_N, K	Additional information	References
β-Cr_2S_3	Trigonal C_{3i}^2	CS, $\mu \perp c$	120–122	$\theta = -636$ K $\mu_{Cr\,I}(4.2\,K) = 1.72\mu_B$ $\mu_{Cr\,II}(4.2\,K) = 1.36\mu_B$ $\mu_{Cr\,III}(4.2\,K) = 1.82\mu_B$	[3]
Cr_2O_5			100	$\theta = -140$ K	[17]
Cr_3S_4	Monoclinic C_{2h}^3	$\alpha(\mu, [101]) = 29°$ (2a, b, 2c)	280	$\theta = -547$ K	[3,17]
Cr_3Se_4	Monoclinic C_{2h}^3	$\mu \parallel (10\bar{1})$ $\alpha(\mu, [b]) = 30°$ (2a, b, 2c)	80	$\theta = -60$ K $\mu_{Cr\,I}(4.2\,K) = 4\mu_B$ $\mu_{Cr\,II}(4.2\,K) = 3\mu_B$	[3,17]
Cr_5S_6	Hexagonal D_{3d}^2	HS, $\mathbf{k}_0 \parallel c$ at $T < T_N$	150–160	Ferrimagnet at $T_N < T < T_C = 305$ K $\Delta\sigma(T_N) = 1.85\ A \cdot m^2/kg$ $\rho = (77\,K) = 5 \cdot 10^4\ \Omega \cdot cm$ $\mu_{Cr\,I}(4.2\,K) = 2.98\mu_B$ $\mu_{Cr\,II}(4.2\,K) = 2.77\mu_B$ $\mu_{Cr\,III}(4.2\,K) = 2.78\mu_B$ $\mu_{Cr\,IV}(4.2\,K) = 2.75\mu_B$	[3,17]
$LiCrS_2$	Trigonal D_3^2 or D_{3d}^3	NCS	55	$p = 3.83$, $\mu_{Cr}(4.2\,K) = 2.26\mu_B$	[3]
$KCrF_3$	Tetragonal C_{4h}^{10}	$\mu \perp c$ A-type MO	40	$\mu_{Cr^{2+}}(4.2\,K) = 4.27\mu_B$	[3,17]
$YCrO_3$	Orthorhombic D_{2h}^{16}	CS, $\sigma \parallel c$ $\mu \parallel [a]$	141	$\mu_0 H_{c,[a]} = 3.3$ T*7 $\mu_{Cr^{3+}}(4.2\,K) = 2.96\mu_B$	[3,17]
$LaCrO_3$	Orthorhombic D_{2h}^{16}	G-type MO	282–320	$\theta = -430$ K, $p = 2.76$ $\chi_M(T_N)/4\pi = 1.9\times$ $\times 10^{-3} cm^3/mol$ $\mu_{Cr^{3+}}(4.2\,K) = 2.45\mu_B$	[3,17,63]
$CeCrO_3$	Orthorhombic D_{2h}^{16}		257(Cr)		[3]
$PrCrO_3$	Orthorhombic D_{2h}^{16}	CS, $\mu_{Cr} \parallel [a]$ $\mu_{Pr} \parallel [c]$*8	239(Cr)	$\theta = -133$ K, $p = 2.87$ $\mu_{Cr^{3+}}(4.2\,K) = 2.46\mu_B$ $\mu_{Pr^{3+}}(4.2\,K) = 0.5\mu_B$	[3,17,63]
$NdCrO_3$	Orthorhombic D_{2h}^{16}	CS(Cr), $\mu \parallel [a]$ at $T > T_r = 20$ K $\mu \parallel (001)$, $\alpha(\mu, [a]) = 60°$ at $T < T_r$; CS(Nd) $\mu \parallel [c]$	224(Cr) 10(Nd)	$\theta = -292$ K, $p = 2.81$ $\mu_{Cr}(4.2\,K) = 2.55\mu_B$ $\mu_{Nd}(4.2\,K) = 1.3\mu_B$	[3,17,63]
$SmCrO_3$	Orthorhombic D_{2h}^{16}	NCS, $\mu \parallel [a]$ at $T > T_r = 38$ K $\mu \parallel [c]$ at $T < T_r$	192(Cr)		[3,17]
$EuCrO_3$	Orthorhombic D_{2h}^{16}		181(Cr)		[17]
$GdCrO_3$	Orthorhombic D_{2h}^{16}		170(Cr)	$\theta = -20$ K, $p = 2.8$	[17,63]

*7 Spin-flop magnetic field.

*8 At $T = 4.2$, K the Pr magnetic moments are ferromagnetically ordered.

Table 28.1 Certain properties of antiferromagnets *(continued)*

Substance	Structure		T_N, K	Additional information	References
	crystallographic	magnetic			
$TbCrO_3$	Orthorhombic D_{2h}^{16}	CS(Cr) NCS(Tb) at $T > T_r$ CS(Tb) at $T < T_r$, $T_r = 3$ K	158(Cr) 4(Tb)	$\theta_{Cr} = -110$ K $\mu_{Cr^{3+}}(4.2\,K) = 2.85\mu_B$ $\mu_{Tb^{3+}}(1.5\,K) = 8.6\mu_B$	[3,17]
$DyCrO_3$	Orthorhombic D_{2h}^{16}	CS(Cr) $\mu_{Cr} \parallel [c]$ NCS(Dy)	146(Cr) 2.1(Dy)	$\theta_{Cr} = -35$ K $\mu_{Cr^{3+}}(1.5\,K) = 2.76\mu_B$ $\theta_{Dy} = -1$ K, $\mu_{Dy}(1.5\,K) = 9.6\mu_B$	[3,17]
$HoCrO_3$	Orthorhombic D_{2h}^{16}	Change in Cr structure at $T_r = 80$ K NCS(Ho)	141(Cr) 12(Ho)	$\theta_{Cr} = -28$ K $\mu_{Cr}(4.2\,K) = 2.94\mu_B$ $\mu_{Ho}(4.2\,K) = 7\mu_B$ (AF-component) $\mu_{Ho}(4.2\,K) = 3.4\mu_B$ (F-component)	[3,17]
$ErCrO_3$	Orthorhombic D_{2h}^{16}	CS(Cr) $\mu \parallel [a]$ at $T > T_r = 20$ K $\alpha(\mu, [a]) = 55°$ at $T < T_r$, CS(Er)	133(Cr) 16.8(Er)	$\mu_{Cr^{3+}}(4.2\,K) = 2.90\mu_B$ $\mu_{Er^{3+}}(4.2\,K) = 5.2\mu_B$	[3,17]
$TmCrO_3$	Orthorhombic D_{2h}^{16}	CS(Cr)	124(Cr) $T_C = 4$ K (Tm)*9	$\mu_{Cr^{3+}}(4.2\,K) = 2.58\mu_B$ $\mu_{Tm^{3+}}(4.2\,K) = 0.8\mu_B$	[3]
$YbCrO_3$	Orthorhombic D_{2h}^{16}	CS(Cr)	158(Cr), 3.0(Yb)	$\mu_{Cr}(4.2\,K) = 2.80\mu_B$	[3,17]
$LuCrO_3$	Orthorhombic D_{2h}^{16}	CS, $\mu \parallel (010)$ $\alpha(\mu, [a]) = 63°$	112(Cr)	$\mu_{Cr}(4.2\,K) = 2.51\mu_B$	[3,17]
$PbCrO_3$	Cubic O_h^1	G-type MO $(2a, 2a, 2a)$	210–250	$\theta = -240$ K, $p = 2.83$ $\mu_{Cr}(4.2\,K) = 1.9\mu_B$	[3,17]
$CrVO_4$	Orthorhombic D_{2h}^{17}	FI in (001) AFI between (001)	50	$\mu_{Cr^{3+}}(4.2\,K) = 2.1\mu_B$ $p = 4$	[3,17]
$CrUO_4$	Orthorhombic D_{2h}^{14}	CS, $\mu \perp [c]$	45	$\mu_{Cr} = 2.4\mu_B$	[3,17]
$CrNb_4S_8$	Hexagonal D_{6h}^4	CS FI in (001) AFI between (001)		$\theta = -4$ K at low temperatures $\theta = -243$ K at high temperatures $\mu_{Cr}(4.2\,K) = 1.5\mu_B$	[3]
Cr_2BeO_4	Orthorhombic D_{2h}^{16}	HS, $\mathbf{k}_0 \parallel [c]$ $\mu \parallel (010)$	28	$\mu_{Cr\,I}(4.2\,K) = 1.55\mu_B$ $\mu_{Cr\,II}(4.2\,K) = 2.75\mu_B$	[3]
$MgCr_2O_4$	Cubic O_h^7 (tetragonal distortions at $T < T_N$)	NCS, $\mu \perp (001)$ at 13.5 K*10 $< T_N$	16	$\theta = -350$ K, $p = 3.84$ $\mu_{Cr^{3+}}(4.2\,K) = 2.63\mu_B$	[3]
$CoCr_2O_4$	Cubic O_h^7	NCS at $T < T_N$ $(3a, 3a, a)$	31 F→AF	Ferrimagnet at $T_N < T < T_C = 97$ K	[3]
$ZnCr_2O_4$	Cubic O_h^7 (tetragonal distortions at $T < T_N$)	NCS $\mu \parallel (100)$ $\mu \parallel [010]$ $(2a, 2a, 2a)$	16	$\theta = -392$ K	[3,17]

*9 At $T < T_C$, the Tm magnetic moments are ferromagnetically ordered.
*10 At $T = 13.5$ K, there is AF-structure rearrangement.

Table 28.1 Certain properties of antiferromagnets *(continued)*

Substance	Structure		T_N, K	Additional information	References
	crystallographic	magnetic			
Cr_2TiS_4	Monoclinic C^3_{2h}	$(2a,\ b,\ 2c)$			[3]
Cr_2NiS_4	Monoclinic C^3_{2h}	FI in (101) AFI between (101) $(2a,\ b,\ 2c)$			[3]
$ZnCr_2S_4$	Cubic O^7_h		20	$\theta = 18\,K$	[65]
$HgCr_2S_4$	Cubic O^7_h	NCS at $T < 25\,K$	25 F→AF	Ferromagnet at $25\,K < T < T_C = 60\,K$ $\mu_{Cr^{3+}}(4.2\,K) = 2.71\mu_B$	[3]
$ZnCr_2Se_4$	Cubic O^7_h	NCS at $T<T_N$ $\mu \parallel (001)$	22 F→AF	$\theta = -115\,K$ $\mu_{Cr}(4.2\,K) = 3\mu_B$ Ferrimagnet at $T_N < T < T_C = 129\,K$	[3,17]
Cr_2TeO_6	Tetragonal D^{14}_{4h}	NCS, $\mu \perp c$	105	$\mu_{Cr^{3+}}(4.2\,K) = 2.45\mu_B$	[3,17]
Cr_2WO_6	Tetragonal D^{14}_{4h}	NCS, $\mu \perp c$	69	$\theta = -196\,K$ $\mu_{Cr^{3+}}(4.2\,K) = 2.14\mu_B$	[3]
$CrTiNdO_5$	Orthorhombic D^9_{2h}	NCS $\mu_{Cr\ I} \parallel [b]$ $\mu_{Cr\ II} \parallel [c]$ $\mu_{Nd} \parallel (001)$	30(Cr) 13(Nd)	$\mu_{Nd}(1.5\,K) = 2.89\mu_B$ $\mu_{Cr}(1.5\,K) = 2.95\mu_B$	[3]
$CrK(SO_4)_2\cdot$ $\cdot 12H_2O$	Cubic T^6_h		0.004	$\chi_M(T_N)/4\pi = 2 \cdot 10^3 cm^3/mol$	[17]
$Cr(CH_3NH_3)\cdot$ $\cdot(SO_4)_2\cdot 12H_2O$			0.02		[17]
α-Mn	Cubic T^3_d	Complex multisublattice	100	$\mu(0\,K) \approx 0.5\mu_B$ (averaged value)	[2,3]
γ-Mn	Cubic O^5_h	I type MO	1370	$\mu = 2.4\mu_B$, stable at $1352\,K < T < 1416\,K$	[3]
MnO	Cubic O^5_h (rhombohedric distortions at $T<T_N$)	$\mu \parallel [110]$ II type MO $(2a,\ 2a,\ 2a)$	113–120	$\theta = -698\,K$, $p = 5.97$ $\lambda(0) = 362\ \mu m$ $\mu_{Mn^{2+}} = 5\mu_B$ $^{55}\nu_{Mn}(1.5\,K) = 618\,MHz$*11	[3,17]
MnP	Orthorhombic D^{16}_{2h}	HS, $\mathbf{k}_0 \parallel [a]$ at $T < T_N$	47 F→AF	Ferromagnet at $T_N < T < T_C = 291\,K$ $\mu_0 H_{c,[c]}(4.2\,K) = 0.24$ T $\mu_0 H_{c,[b]}(4.2\,K) = 0.62$ T	[3,17]
αMnS	Cubic O^5_h	II type MO $\mu \parallel (111)$ $(2a,\ 2a,\ 2a)$	146–155	$\theta = -465\,K$, $\mu_{Mn^{2+}}(4.2\,K) = 5\mu_B$	[3]
βMnS	Cubic T^3_d	II type MO $(a,\ 2a,\ a)$	150	$\theta = -982\,K$ $\mu_{Mn^{2+}}(4.7\,K) = 5\mu_B$ $\chi_M(T_N)/4\pi = 6 \cdot 10^{-3} cm^3/mol$	[3,17]
MnSe	Cubic O^5_h	II type MO $\mu \perp [111]$ $(2a,\ 2a,\ 2a)$	147–150	$\theta = -740\,K$ $\chi_M(T_N)/4\pi = 19 \cdot 10^{-3} cm^3/mol$ Semiconductor, $E_g = 0.68$ eV	[3,17]

*11 See Appendix II in [8].

Table 28.1 Certain properties of antiferromagnets *(continued)*

Substance	Structure		T_N, K	Additional information	References
	crystallographic	magnetic			
MnTe	Hexagonal D_{6h}^4	FI in (001) AFI between (001) $\mu \parallel (001)$	307–310	$\theta = -692\,\mathrm{K}$, $p = 5.97$ $\chi_M(T_N)/4\pi = 4.6 \cdot 10^{-3}\,\mathrm{cm^3/mol}$ $\mu_{\mathrm{Mn}^{2+}}(290\,\mathrm{K}) = 4.6\mu_B$ Semiconductor, $E_g = 1$ eV	[3,17]
MnNi	Tetragonal D_{4h}^1	$\mu \perp c$	1070–1140	$\mu_{\mathrm{Mn}}(0\,\mathrm{K}) = 4\mu_B$ $\mu_{\mathrm{Ni}}(0\,\mathrm{K}) < 0.6\mu_B$	[3,17]
γ-MnNi	Cubic O_h^5		453	$\mu_{\mathrm{Mn}} = 1.5\mu_B$	[3]
MnCu	Tetragonal	$\mu \parallel c$	300	Alloy(69–85% Mn)	[17]
MnPd	TetragonalD_{4h}^1	*12	813–825	$\mu_{\mathrm{Mn}}(4.2\,\mathrm{K}) = 4.0\mu_B$ $\mu_{\mathrm{Pd}}(4.2\,\mathrm{K}) < 0.2\mu_B$	[3]
MnRh	CubicO_h^1 (tetragonal dis- tortions at $T<T_N$)		170	$\theta = -260\,\mathrm{K}$	[17]
MnPt	Tetragonal D_{4h}^1	Structure change at $T = 710\,\mathrm{K}$ $(2a,\ 2a,\ c)$	970	$\mu_{\mathrm{Mn}}(290\,\mathrm{K}) = 4.3\mu_B$	[3]
β-MnAu	Tetragonal O_h^1	A-type MO $(a,\ a,\ 2a)$	513	$\theta = -600\,\mathrm{K}$, $p = 3.9$ $\chi_M(T_N)/4\pi = 4.3 \cdot 10^{-3}\mathrm{cm^3/mol}$	[3,17]
MnHg	Cubic O_h^1	G-type MO $\mu \parallel [001]$ at $T < 198\,\mathrm{K}$ $\mu \parallel [111]$ at $T > 198\,\mathrm{K}$ $(2a,\ 2a,\ 2a)$	460	$\chi_M(T_N)/4\pi = 9.4 \cdot 10^{-3}\mathrm{cm^3/mol}$ $\rho(T_N) = 5 \cdot 10^{-7}\Omega{\cdot}\mathrm{m}$ $\mu_{\mathrm{Mn}}(4.2\,\mathrm{K}) = 3.7\mu_B$	[3,17]
IrMn	Tetragonal D_{4h}^1	AFI between (001)	> 1200	$\mu_{\mathrm{Mn}}(290\,\mathrm{K}) = 3.4\mu_B$ $\mu_{\mathrm{Ir}}(290\,\mathrm{K}) = 0.2\mu_B$ (49% Ir)	[3]
MnB_2	Hexagonal D_{6h}^1	$\mu \perp c$, FI in (001) AFI between (001) $(a,\ a,\ 2c)$	> 885	$\mu_{\mathrm{Mn}}(77\,\mathrm{K}) = 2.6\mu_B$	[3]
β-MnO_2	TetragonalD_{4h}^{14}	HS	84–94	$\theta = 1050\,\mathrm{K}$, $p = 5.52$	[3,17]
MnS_2	Cubic T_h^6	III-type MO $\mu \parallel [010]$, $(a,\ 2a,\ a)$	48	$\theta = -592\,\mathrm{K}$, $p = 6.3$ $\mu_{\mathrm{Mn}^{2+}}(4.2\,\mathrm{K}) = 5\mu_B$	[3,17]
$MnSe_2$	Cubic T_h^6	$\mu \parallel [010]$ $(a,\ 3a,\ a)$	75	$\theta = -483\,\mathrm{K}$, $p = 5.93$ $\mu_{\mathrm{Mn}^{2+}}(4.2\,\mathrm{K}) = 5\mu_B$	[3,17]
$MnTe_2$	Cubic T_h^6	I-type MO $\mu \parallel (001)$	84	$\theta = -528\,\mathrm{K}$, $p = 6.22$ $\mu_{\mathrm{Mn}^{2+}}(4.2\,\mathrm{K}) = 5\mu_B$	[3,17]
MnF_2	Tetragonal D_{4h}^{14}	$\mu \parallel c$	67.3	$\theta = -113\,\mathrm{K}$ $\chi_{M\perp}(0\,\mathrm{K})/4\pi =$ $= 25 \cdot 10^{-3}\mathrm{cm^3/mol}$, $\lambda(0) = 1.11$ mm $\mu_0 H_{c\parallel}(4.2\,\mathrm{K}) = 9.33$ T $\mu_0 \Delta H_{\mathrm{res}} = 3 \cdot 10^{-4}$ T at $\nu_{\mathrm{AFMR}} = 23$ GHz $^{55}\nu_{\mathrm{Mn}}(0\,\mathrm{K}) = 680$ MHz*13	[3,17]

*12 The possible magnetic structures are considered in [3].

*13 See Appendix II in [8].

Table 28.1 Certain properties of antiferromagnets *(continued)*

Substance	Structure		T_N, K	Additional information	References
	crystallographic	magnetic			
$MnCl_2$	Trigonal D^5_{3d}	$\alpha(\mu, c) = 57°$ at $T<T_N$ $\mu \parallel [a]$ at $T<T_r = 1.82$ K	1.96	$\theta = -3.3$ K, $p = 5.73$	[3,17]
$MnBr_2$	Trigonal D^3_{3d}	CS*14	2.16	$\theta = -4.66$ K, $\mu_{Mn}(1.35\ K) = 5\mu_B$	[3,17]
MnI_2	Trigonal D^3_{3d}	HS*15	3.4	$\mu_{Mn}(1.3\ K) = 4.6\mu_B$	[3,17]
$MnSn_2$	Tetragonal D^{18}_{4h}	Structure change at $T = 73$ K	323–325	$\mu_{Mn}(4.2\ K) = 2.33\mu_B$	[3,17]
$MnAu_2$	Tetragonal D^{17}_{4h}	HS	363–370	$\mu_{Mn^{2+}}(290\ K) = 3.04\mu_B$	[3,17]
MnF_3	Monoclinic C^6_{2h}	$\mu \parallel (101)$	47	$\theta = 8$ K, $p = 5$ $\mu_{Mn^{3+}}(4.2\ K) = 4\mu_B$	[3,17]
α-$MnZn_3$	Cubic O^1_h (tetragonal distortions at $T<T_N$)	A-type MO $(a,\ a,\ 2a)$	155		[3]
$MnPd_3$	Tetragonal D^{17}_{4h}	$\mu \perp c$	220	$\mu_{Mn}(77\ K) = 4.1\mu_B$	[3]
$MnAu_3$			145	$\theta = 200$ K	[17]
Mn_2N	Orthorhombic D^{14}_{2h}	NCS	301	$\mu_{Mn}(120\ K) = 1.6\mu_B$	[3]
Mn_2P	Hexagonal D^3_{3h}		103–110	$\mu_{Mn\ I}(4.2\ K) = 0.02\mu_B$ $\mu_{Mn\ II}(4.2\ K) = 1.68\mu_B$	[3]
Mn_2As	Tetragonal D^7_{4h}	$\mu \parallel (001)$ $(a,\ a,\ 2c)$	573	$\theta = -1947$ K, $p = 2.58$ $\mu_{Mn\ I}(190\ K) = 3.7\mu_B$ $\mu_{Mn\ II}(290\ K) = 3.5\mu_B$	[3,17]
YMn_2	Cubic O^7_h		100	$\mu_{Mn} = 2.7\mu_B$	[65]
Mn_2O_3	Cubic T^7_h		80		[17]
Au_5Mn_2	Monoclinic C^3_{2h}		353	$\theta = 120$ K, $p = 5.56$	[3]
Mn_2Hg_5	Tetragonal D^5_{4h}	FI in (001) AFI between (001) $\mu \perp c$	342	$\mu_{Mn}(4.2\ K) = 4.5\mu_B$	[3]
Mn_3Rh	Cubic O^1_h	NCS, $\mu \parallel (111)$	853	$\mu_{Mn}(4.2\ K) = 3.5\mu_B$	[3]
Mn_3Sn	Hexagonal D^4_{6h}	HS, $\mathbf{k}_0 \parallel c$ at $T<T_r = 270$ K	420	$\mu_{Mn}(4.2\ K) = 2.5\mu_B$	[3]
Mn_3Pt	Cubic O^1_h	CS at $T>T_r$ NCS at $T<T_r$ $T_r = 365$ K	475	$\mu_{Mn}(77\ K) = 3.0\mu_B$	[3]
Pd_3Mn_2			593	$\theta = -430$ K $\chi_M(T_N)/4\pi =$ $= 2 \cdot 10^{-5}$ cm^3/mol	[17]
Mn_3Ge_2			164 F→AF	Ferromagnet at $T_N < T < T_C = 283$ K $\Delta\sigma(164\ K) = 3.5$ A·m^2/kg	[17]

*14 Ferromagnetically ordered (011) planes are alternated consecutively as $++--$.

*15 Ferromagnetic ordering in (307) plane. Magnetic moments in neighbor (307) planes are turned by an angle of $\pi/16$.

Table 28.1 Certain properties of antiferromagnets *(continued)*

Substance	Structure		T_N, K	Additional information	References
	crystallographic	magnetic			
Mn_5Si_3	Hexagonal D_{6h}^3	$\mu \perp c$ $(2a,\ a,\ c)$	62–68	$\theta = -9.1\,K,\ p = 4.05$ $\mu_{Mn\ I}(4.2\,K) = 0.4\mu_B$ $\mu_{Mn\ II}(4.2\,K) = 1.2\mu_B$	[3]
MnOOH	Monoclinic C_{2h}^2	Modulated with a period of 0.488 nm	40		[3]
FeMnAs	Tetragonal D_{4h}^7	$\mu \perp c$ $(a,\ a,\ 2c)$	463	$\mu_{Fe}(290\,K) = 0.2\mu_B$ $\mu_{Mn}(290\,K) = 3.6\mu_B$	[3]
CuMnSb	Cubic T_d^2	III type MO $\mu \parallel [111]$ $(2a,\ 2a,\ 2a)$	38	$\mu_{Mn}(4.2\,K) = 3.9\mu_B$	[3]
$MnCO_3$	Trigonal D_{3d}^6	$\sigma \perp c$ (EP anisotropy)	32.5	$\theta = -64.5\,K$ $\sigma(0\,K) = 18.8 \cdot 10^2\ A{\cdot}m^2/mol$ $\lambda(0) = 2.44$ mm $\mu_0 H_D = 0.44$ T $^{55}\nu_{Mn}(4.2\,K) = 640$ MHz*16	[3,17,31]
$MnSiO_3$	Triclinic			$\theta = -45\,K,\ p = 5.91$	[17]
$CaMnO_3$	Cubic O_h^1	C-type MO $(2a,\ 2a,\ 2a)$	123–130	$\theta = -425\,K,\ \sigma(0\,K) \neq 0$ $\chi_M(T_N)/4\pi = 4.7 \cdot 10^{-3}\ cm^3/mol$ $\mu_{Mn^{4+}} = 2.43\mu_B$	[3,17]
Ca_2MnO_4	Tetragonal D_{4h}^{17}	$(\sqrt{2}a,\ \sqrt{2}a,\ 2c)$	114	$\mu_{Mn^{4+}}(4.2\,K) = 2.0\mu_B$	[3]
$ScMnO_3$	Hexagonal C_{6v}^3	NCS	120	$\mu_{Mn}(4.2\,K) \approx 4.0\mu_B$	[3,17]
$MnTiO_3$	Trigonal C_{3i}^2	$\mu \parallel [111]$	60–65	$\theta = -219\,K,\ p = 5.93$ $\mu_{Mn^{2+}}(4.2\,K) = 4.55\mu_B$ $\mu_0 H_{c\parallel} = 5.2$ T	[3,17]
$MnGeO_3$	Orthorhombic D_{2h}^{15}	$\mu \parallel [b]$	16	$\theta = -46\,K,\ p = 5.48$ $\mu_{Mn\ I}(1.1\,K) = 4.37\mu_B$ $\mu_{Mn\ II}(1.1\,K) = 4.27\mu_B$	[3,17]
$MnYO_3$	Hexagonal C_{6v}^3	$\mu \perp c$	46	$\theta = -475\,K,\ \sigma(0\,K) \neq 0$	[3,17]
$BaMnO_3$	Hexagonal D_{6h}^4	CS, $\mu \parallel c$ $(\sqrt{3}a,\ \sqrt{3}a,\ c)$	2.3	$\mu_{Mn}(1.8\,K) = 3.0\mu_B$	[3]
$SrMnO_3$	Cubic O_h^1	G-type MO $(2a,\ 2a,\ 2a)$	260	$\mu_{Mn^{4+}}(77\,K) = 2.6\mu_B$	[3]
$LaMnO_3$	Orthorhombic D_{2h}^{16}	NCS A_xF_y-type MO	100–150	$\theta = 40\,K$ $\sigma(0\,K) = 1.21\ A{\cdot}m^2/mol$ $\chi_M(T_N)/4\pi = 48.4 \cdot 10^{-3}\ cm^3/mol$ $\mu_{Mn^{2+}}(4.2\,K) = 3.7\mu_B$	[3,17]
$PrMnO_3$	Orthorhombic D_{2h}^{16}	CS, $\mu \parallel [b]$	91	$\mu_{Mn^{3+}}(1.5\,K) = 1.77\mu_B$	[3]
$NdMnO_3$	Orthorhombic D_{2h}^{16}	CS, $\mu \parallel (001)$ $\alpha(\mu, [b]) = 36°$	85	$\mu_{Mn^{3+}}(1.5\,K) = 1.71\mu_B$	[3]
$HoMnO_3$	Hexagonal C_{6v}^3	$\mu \parallel c$ at $T<T_r = 50$ K $\mu \perp c$ at $T>T_r$	76	$\mu_{Mn}(4.2\,K) = 3.5\mu_B$	[3,17]
$ErMnO_3$	Hexagonal C_{6v}^3	NCS, $\alpha(\mu, c) = 70°$	79	$\mu_{Mn}(4.2\,K) = 3.5\mu_B$	[3,17]

*16 See Appendix II in [8].

Table 28.1 Certain properties of antiferromagnets *(continued)*

Substance	Structure		T_N, K	Additional information	References
	crystallographic	magnetic			
$TmMnO_3$	Hexagonal C_{6v}^3	NCS $\alpha(\mu, c) = 45°$	86	$\mu_{Mn}(4.2\,K) = 3.8\mu_B$	[3,17]
$LuMnO_3$	Hexagonal C_{6v}^3	NCS $\alpha(\mu, c) = 55°$	91	$\mu_{Mn}(4.2\,K) = 3.7\mu_B$	[3,17]
$NaMnF_3$	Orthorhombic D_{2h}^{16}	G-type MO $\sigma\|\|[c]$	67	$\mu_0 H_E = 39.0\,T$ $\mu_0 H_D = 0.43\,T$ $\mu_0 H_A = 2 \cdot 10^{-2}\,T$	[3,17]
$KMnF_3$	Tetragonal D_{4h}^5 at $T < 91$ K	G-type MO (2a, 2a, 2a)	88	$\theta = -238\,K$ $\sigma_M(4.2\,K) = 19.3 \cdot 10^{-3}$ A·m^2/mol*17 $\chi_M(T_N)/4\pi = 17.7 \cdot 10^{-3}\,cm^3/mol$ $\mu_{Mn^{2+}}(4.2\,K) = 5.06\mu_B$ $\mu_0 H_D = 4.5 \cdot 10^{-2}\,T$ $\mu_0 H_E = 86.8\,T$ $^{55}\nu_{Mn}(4.2\,K) = 676$ MHz*18	[3,6,17]
$RbMnF_3$	Cubic O_h^1	G-type MO $\mu\|\|$ [111] (2a, 2a, 2a)	83	$\theta = -118\,K$ $\mu_0 H_{c\|\|} = 0.28\,T$ $\mu_0 H_A = 4 \cdot 10^{-4}\,T$ $\mu_0 H_E = 89\,T$ $\chi_M(T_N)/4\pi = 17.7 \cdot 10^{-3}\,cm^3/mol$	[3,6,17,66]
$CsMnF_3$	Hexagonal D_{6h}^4	EP anisotropy $\mu \perp c$	51–54	$\mu_0 H_{A\|\|} = 0.75$ T*19 $\mu_0 H_{A\perp} = 1.1 \cdot 10^{-4}\,T$ $\mu_0 H_E = 35\,T$ $^{55}\nu_{Mn}(4.2\,K) = 666$ MHz*20 Magnetic properties correspond to Heisenberg model, $d = 2$	[3,6,17,66]
$TlMnF_3$	Cubic O_h^1	G-type MO	77	$\theta = -125\,K$ $\mu_0 H_E = 68.0\,T$ $\mu_0 H_A = 0.67\,T$ $\mu(4.2\,K) = 4.9\,\mu_B$	[17]
NH_4MnF_3	Cubic O_h^1	G-type MO (2a, 2a, 2a)	84		[3]
$NaMnCl_3$	Trigonal C_{3i}^2	$\mu \perp c$ (EP anisotropy)	6.5	$\theta = 1\,K$	[6]
$KMnCl_3$	Tetragonal*21 D_{4h}^{17} ([c]/[a]) = 1.005		100	$\theta = -122\,K$ $\mu_0 H_E = 122\,T$	[6,17]
$KMnCl_3$	Orthorhombic D_{2h}^{16}	HS, $\mathbf{k}_0\|\|$ [b] with a period 3.17[b]	2.1	$\mu_{Mn}(1.4\,K) = 3.78\mu_B$	[67]

*17 Spontaneous moment in $KMnF_3$ appears at the temperature $T = 81.5$ K. The change of the structure at $T_C = 81.5$ K makes the phase transition of the first order.

*18 See Appendix II in [8].

*19 $\mu_0 H_A = 2.48 \cdot 10^{-1}$ T, where H_A is the field of anisotropy obtained by AFMR high-frequency branch measurements [117].

*20 See Appendix II in [8].

*21 According to [67], perovskite structure – space group D_{2h}^{16}.

Table 28.1 Certain properties of antiferromagnets *(continued)*

Substance	Structure		T_N, K	Additional information	References
	crystallographic	magnetic			
$RbMnCl_3$	Hexagonal*22 D_{6h}^4	$\mu \perp c$ (EP anisotropy)	95	$\theta = -204$ K Magnetic properties correspond to Heisenberg model, $d = 2$ $\Delta\sigma_{\|}(4.2\,K) = = 4 \cdot 10^{-3}$ A·m²/mol at $\mu_0 H_{\perp} = 0.63$ T	[6,17]
$CsMnCl_3$	Trigonal D_{3d}^5	$\mu \perp c$ (EP anisotropy)	67–70	$\theta = -145$ K $\mu_0 H_E = 70.0$ T $\mu_0 H_{c\perp} = 5.26$ T $\mu_0 H_{A\|} = 0.74$ T	[6,17]
$TlMnCl_3$	Orthorhombic C_{4v}^{10}	G-type MO	118*23	$\theta = -259$ K $\mu_{Mn}(77\,K) = 4.7\mu_B$ $\mu_0 H_D = 0.16$ T at $T_C = 108$ K*23 AF→CAF	[3,6]
NH_4MnCl_3	Cubic O_h^1 (orthorhombic distortions at $T<T_N$)	G-type MO (2a, 2a, 2a)	105	$\mu_{Mn}(4.2\,K) = 5\mu_B$	[3,17]
$CsMnBr_3$	Hexagonal D_{6h}^4	HS, $\mathbf{k}_0 \| c$ ($\sqrt{3}a$, $\sqrt{3}a$, c)	8.3	$\theta = -167$ K, $J/k = 9.9$ K Magnetic properties correspond to Heisenberg model, $d = 1$	[6]
$TlMnI_3$	Orthorhombic D_{2h}^{16}	HS, $\mathbf{k}_0 \| [b]$ $\mu \| (001)$	6.0	$\theta = 10.6$ K, $p = 5.9$	[68]
$MnSO_4$	Orthorhombic D_{2h}^{17}	HS, $\mathbf{k}_0 \| [a]$ with a period 3 nm	11	$\mu_{Mn^{2+}}(4.2\,K) = 4.8\mu_B$	[3,17]
$MnSeO_4$	Orthorhombic D_{2h}^{16}	CS, $\mu \| (001)$	20	$\mu_{Mn}(4.2\,K) = 4.95\mu_B$	[3]
$MnWO_4$	Monoclinic C_{2h}^4	(4a, 2b, 2c)	14–16	$\theta = -71$ K, $p = 5.83$	[3]
$MnUO_4$	Orthorhombic D_{2h}^{28}	$\mu \| [b]$	12	$\theta = -8$ K $\mu_{Mn}(4.2\,K) = 4.88\mu_B$ $\chi_M(T_N)/4\pi = 0.2$ cm³/mol	[3,17]
$BaMnF_4$	Orthorhombic C_{2v}^{12}*24	NCS	26	$\mu(4.7\,K) = 4.8\mu_B$ Magnetic properties correspond to Heisenberg model, $d = 2$	[6,120]
$Mn(OH)_2$	Trigonal D_{3d}^3	NCS ($\sqrt{3}a$, $\sqrt{3}a$, 2c)	12	$\theta = -28$ K $\mu_{Mn}(4.2\,K) = 4.9\mu_B$	[3,17]
$MnAl_2O_4$	Cubic O_h^7	$\mu \| [100]$	6	$\theta = -156$ K, $p = 5.13$ $\mu_{Mn^{2+}}(4.2\,K) = 3.91\mu_B$	[3]
$MnGa_2O_4$	Cubic O_h^7	$\mu \| [111]$	33	$\theta = -154$ K, $p = 5.77$ $\mu_{Mn^{3+}}(4.2\,K) = 3.6\mu_B$	[3,17]

*22 Hexagonal structure at room temperature. At $T = 272$ K, the structural transition $D_{6h}^4 \to C_{2h}^2$ is found by optical measurements [118].

*23 T_N and T_C are given according to [3], $T_N = T_C \approx 110$ K – according to [6].

*24 C_{2v}^{12} at $T > 247$ K; at $T = 247$ K – structural transition to incommensurate phase with unit cell doubling in *bc* plane and incommensurate modulated structure along the [a] axis appears.

Table 28.1 Certain properties of antiferromagnets *(continued)*

Substance	Structure		T_N, K	Additional information	References
	crystallographic	magnetic			
α-$MnGa_2S_4$	Monoclinic C_{2h}^6		23.5	$\theta = -50$ K	[69]
β-$MnGa_2S_4$	Orthorhombic C_{2v}^9		11	$\theta = -28$ K	[69]
K_2MnF_4	Tetragonal D_{4h}^{17}	$\mu \parallel c$ $(2a,\ 2a,\ c)$	42–45	$\mu_{Mn}(4.2\,K) = 4.54\mu_B$ $\lvert J'/J \rvert = 10^{-6}$ $kT_N/J = 10.1$ Magnetic properties correspond to Heisenberg model, $d = 2$	[3,6]
Rb_2MnF_4	Tetragonal D_{4h}^{17}	$\mu \parallel c$ $(2a,\ 2a,\ c)$	38	$T_N/\theta = 0.44$ $\lvert J'/J \rvert = 10^{-6}$ $kT_N/J = 10.3$ Magnetic properties correspond to Heisenberg model, $d = 2$	[3,6]
Rb_2MnCl_4	Tetragonal D_{4h}^{17}	$\mu \parallel c$ $(2a,\ 2a,\ c)$	56	$T_N/\theta = 0.39$ $\lvert J'/J \rvert = 10^{-6}$ $kT_N/J = 9.2$ $\mu_{Mn}(4.2\,K) = 4.4\mu_B$ Magnetic properties correspond to Heisenberg model, $d = 2$	[3,6]
Cs_2MnCl_4	Tetragonal D_{4h}^{17}	$\mu \parallel c$ $(2a,\ 2a,\ c)$	52	$p = 5.9$ Magnetic properties correspond to Heisenberg model, $d = 2$	[3]
α-Cs_2MnCl_4	Orthorhombic		0.93*25	$\theta = -4$ K	[17]
$MnNb_2O_6$	Orthorhombic D_{2h}^{14}	G_x-type MO $(a,\ 2b,\ c)$	4.4	$\theta = -18$ K	[3,17]
$MnTa_2O_6$	Orthorhombic D_{2h}^{14}	G_xG_z-type MO $\alpha(\mu, [a]) = 10°$			[3]
Pd_2MnAl	Cubic O_h^1	II type MO	240	$\mu_{Mn}(4.2\,K) = 4.4\mu_B$	[3]
Pd_2MnIn	Cubic O_h^5	II type MO $(2a,\ 2a,,\ 2a)$	142	$\mu_{Mn}(4.2\,K) = 4.3\mu_B$ $\theta = -50$ K	[3]
$Mn(C_2H_5)_2$			134	$\theta = -492$ K, $p = 7.18$	[17]
Mn_2SiO_4	Orthorhombic D_{2h}^{16}	CS*26 $\mu \parallel [a]$ at $T>T_r$ NCS at $T<T_r$ $T_r = 13$ K	50	$\theta = -163$ K, $p = 5.87$ $\mu_{Mn^{2+}}(4.2\,K) = 5\mu_B$	[3,17]
$CaMn_2O_4$	Orthorhombic D_{2h}^{11}	CS, $\mu \parallel [a]$ $(2a,\ b,\ c)$	225	$\theta = 294$ K, $p = 4.72$ $\mu_{Mn}(4.2\,K) = 3.6\mu_B$	[3]
Mn_2GeO_4	Orthorhombic D_{2h}^{16}	NCS	24	$\theta = -162$ K, $p = 5.85$	[3]
$ZnMn_2O_4$	Tetragonal D_{4h}^{19}		200		[64, p.325]

*25 By specific heat measurements.
*26 $\mu_{Mn\ I} \parallel [a]$, $\mu_{Mn\ II}$ form a canted AF-structure at $T < T_r = 20$ K; CS at $T_r < T < T_N = 47$ K [116].

Table 28.1 Certain properties of antiferromagnets *(continued)*

Substance	Structure		T_N, K	Additional information	References
	crystallographic	magnetic			
$BiMn_2O_5$	Orthorhombic D_{2v}^7	NCS $(2a,\ b,\ c)$	52	$\theta = -236$ K	[3]
Mn_2GeS_4	Orthorhombic D_{2h}^{16}	CS, G_y-type MO		$\mu_{Mn}(4.2\,K) = 4.7\mu_B$	[3]
$ThMn_2Si_2$	Tetragonal D_{4h}^{17}	CS, $\mu \parallel c$	483	$p = 2.4,\ \mu_{Mn} = 1.75\mu_B$	[3]
$ThMn_2Ge_2$	Tetragonal D_{4h}^{17}	NCS	400	$p = 2.4$	[3]
Cu_3Mn_2Al	Cubic O_h^7	$\mu \parallel (001)$	873	$\mu_{Mn}(4.2\,K) = 4.5\mu_B$	[3]
Mn_3B_4	Orthorhombic D_{2h}^{25}	CS at $226\,K < T < T_N$ HS at $110\,K < T < 226\,K$ $\mathbf{k}_0 \parallel [b]$, CS at $T < 110$ K $\mu \parallel c$	392	$\theta = -543$ K $\mu_{Mn\ I}(4.2\,K) = 2.92\mu_B$ $\mu_{Mn\ II}(4.2\,K) = 0.44\mu_B$	[3]
Mn_3GaC	Cubic O_h^1	CS, $\mu \parallel [111]$ $(2a,\ 2a,\ 2a)$	150–164 F→AF	$\mu_{Mn}(4.2\,K) = 1.8\mu_B$ Ferromagnet at $T_N < T < T_C = 248$ K	[3,17]
Mn_3NiN	Cubic O_h^1	NCS (structure change at $T = 180$ K)		$\mu_{Mn}(77\,K) = 0.98\mu_B$	[3]
Mn_3ZnN	Cubic O_h^1	NCS (structure change at $T = 140$ K)	$T_N = 183$ K	$\mu_{Mn\ I} = 1.21\mu_B$ at $140\,K < T < T_N$ $\mu_{Mn\ I}(4.2\,K) = 0.617\mu_B$ $\mu_{Mn\ II}(4.2\,K) = 1.0\mu_B$	[3,17]
Mn_3GaN	Cubic O_h^1	NCS	298	$\mu_{Mn}(4.2\,K) = 2.34\mu_B$	[3]
$CsMn_3Cl_5$	Tetragonal		0.59*27	$\theta = -0.9$ K	[17]
$Mn_3B_2O_6$	Orthorhombic D_{2h}^{12}		35	$\theta = -185\,K,\ p = 6.18$	[3]
$Nb_2Mn_4O_9$	Trigonal D_{3d}^4	CS $\mu \parallel c$	110*28 125	$\theta = -250$ K $\mu_{Mn^{2+}}(4.2\,K) = 5\mu_B$; magnetoelectric at $140\,K < T < T_N$	[3,17]
$Ta_2Mn_4O_9$	Trigonal D_{3d}^4		103*28	Magnetoelectric	[17]
$Dy_2Mn_4O_9$	Orthorhombic D_{2h}^9		8		[17]
$LiMnPO_4$	Orthorhombic D_{2h}^{16}	CS $\mu \parallel [100]$	35	$\theta = -80\,K,\ p = 5.2$ $\chi_M(T_N)/4\pi =$ $= 33 \cdot 10^{-3}\ cm^3/mol$	[3,17]
$CuMnSnS_4$	Tetragonal D_{2d}^{11}	CS, $\mu \parallel (101)$, $(2a,\ a,\ c)$		$\mu_{Mn^{2+}}(290\,K) = 4.3\mu_B$	[3]
$BaMnUO_6$	Cubic O_h^5	FI in (001) AFI between (001)	12.8	$\mu_{Mn} = 2.25\mu_B$	[3]
$MnCl_2\cdot$ $\cdot 4H_2O$	Monoclinic C_{2h}^5	CS $\alpha(\mu, [c]) = 7°$	1.62	$\theta = -1.79$ K $\mu_0 H_{c\parallel}(1.2\,K) = 0.25–0.5$ T $^{54}\nu_{Mn} = 500.4$ MHz*29	[17]

*27 By specific heat measurements.

*28 By magnetoelectric measurements.

*29 The strongest line ($M = -3 \to M = -2$) for quadrupole-split NMR spectrum in oriented nuclei of Mn [70].

Table 28.1 Certain properties of antiferromagnets *(continued)*

Substance	Structure		T_N, K	Additional information	References
	crystallographic	magnetic			
$MnBr_2 \cdot 4H_2O$	Monoclinic C_{2h}^5	CS, $\mu \parallel [c]$	2.136	$\mu_0 H_{c\parallel}(1.2\,K)$=0.75–1.0 T	[17]
$Mn_3Al_2Ge_3O_{12}$	Cubic O_h^{10}	NCS*30	6.65	$\theta = -28\,K$, $p = 5.89$ $c/\lambda(0) = 39$ GHz*31	[121,122]
α-$RbMnCl_3 \cdot$ $\cdot 2H_2O$ (α-RMC)		CS	4.56	Magnetic properties correspond to Heisenberg model, $d = 1$; $\lvert J/k \rvert = 3.0\,K$ $\lvert J'/J \rvert = 7 \cdot 10^{-3}$ $\mu_0 H_{c\parallel} = 1.31\,T$ $\mu_0 H_E = 20.0\,T$ $\mu_0 H_A = 4.27 \cdot 10^{-2}\,T$	[71]
$CsMnCl_3 \cdot 2H_2O$ (CMC)	Orthorhombic D_{2h}^8	CS	4.89	Magnetic properties correspond to Heisenberg model, $d = 1$; $\lvert J/k \rvert = 3.2\,K$ $\lvert J'/J \rvert = 8 \cdot 10^{-3}$ $\mu_0 H_{c\parallel} = 1.64\,T$ $\mu_0 H_E = 23.8\,T$ $\mu_0 H_A = 5.65 \cdot 10^{-2}\,T$	[71]
$CsMnBr_3 \cdot 2H_2O$ (CMB)	Orthorhombic	CS	5.75	Magnetic properties correspond to Heisenberg model, $d = 1$; $\lvert J/k \rvert = 3.0\,K$ $\lvert J'/J \rvert = 11 \cdot 10^{-3}$ $\mu_0 H_{c\parallel} = 2.05\,T$ $\mu_0 H_E = 22.3\,T$ $\mu_0 H_A = 9.41 \cdot 10^{-2}\,T$	[71]
$MnSiF_6 \cdot 6H_2O$	Trigonal C_{3i}^2		0.1		[17]
$Cs_2MnCl_4 \cdot 2H_2O$	Triclinic		1.80*32		[17]
$(CH_3)_4N \cdot$ $\cdot MnCl_3$ ($TMMC$)	Hexagonal D_{6h}^9 (monoclinic distortions at $T < 171\,K$), structure transition at $T = 45\,K$	CS	0.84	Magnetic properties correspond to Heisenberg model, $d = 1$; $\lvert J/k \rvert = 6.7\,K$ $\lvert J'/J \rvert = 10^{-4}$ $\mu_0 H_{c\parallel} = 1.14\,T$ $\mu_0 H_E = 49.9\,T$ $\mu_0 H_A = 1.30 \cdot 10^{-2}\,T$	[6,71]
$(CH_3NH_3) \cdot$ $\cdot MnCl_3 \cdot$ $\cdot 2H_2O$ (MMC)	Monoclinic C_{2h}^5		4.12	Magnetic properties correspond to Heisenberg model, $d = 1$; $\lvert J/k \rvert = 3.01\,K$ $\lvert J'/J \rvert = 6 \cdot 10^{-3}$	[72]

*30 Manganese magnetic moments lie in the (111) plane and are aligned with or opposed to the [211], [121], and [112] axes.

*31 The behavior of $Mn_3Al_2Ge_3O_{12}$ magnetic-resonance three branches is studied in [122].

*32 By specific heat measurements.

Table 28.1 Certain properties of antiferromagnets *(continued)*

Substance	Structure		T_N, K	Additional information	References
	crystallographic	magnetic			
$(CH_3)_2NH_2\cdot$ $\cdot MnCl_3$ (*DMMC*)		CS	3.60	Magnetic properties correspond to Heisenberg model, $d = 1$; $\|J/k\| = 6.5\,K$ $\|J'/J\| = 10^{-3}$ $\mu_0 H_{c\parallel} = 1.83\,T$ $\mu_0 H_E = 48.4\,T$ $\mu_0 H_A = 3.46 \cdot 10^{-2}\,T$	[71]
$(C_5H_5NH)\cdot$ $\cdot MnCl_3$ (*PMCA*)		CS	2.32	Magnetic properties correspond to Heisenberg model, $d = 1$; $\|J/k\| = 6.5\,K$ $\|J'/J\| = 4 \cdot 10^{-4}$ $\mu_0 H_{c\parallel} = 1.0\,T$ $\mu_0 H_E = 48.4\,T$ $\mu_0 H_A = 1.03 \cdot 10^{-2}\,T$	[71]
$Mn(C_2H_6O_2)\cdot$ $\cdot 4H_2O$	Monoclinic C_{2h}^5		3.2	$\theta = -5.2\,K$ σ_m(0 K)=2.79 A·m^2/mol	[3,17]
$Mn(HCOO)_2\cdot$ $\cdot 2H_2O$	Monoclinic C_{2h}^5	$\mu \parallel (100)$	3.7*33	Magnetic properties correspond to Ising model, $d = 2$; $\|J/k\| = 0.35$	[17,73]
$Mn(DCOO)_2\cdot$ $\cdot 2D_2O$	Monoclinic C_{2h}^5	$\sigma \parallel (101)$ at $T>T_r$ $\sigma \parallel [b]$ at $T<T_r$ $T_r = 1.7\,K$	3.7	$\mu_{Mn^{2+}}(2.5\,K) = 5.06\mu_B$ $\mu_{Mn^{2+}}(1.5\,K) = 5.06\mu_B$	[3]
$Mn(NH_4)_2\cdot$ $\cdot(SO_4)_2\cdot 6H_2O$	Monoclinic		0.14	$\sigma_m(0\,K) = 7.82$A·m^2/mol	[17]
FeO	Cubic O_h^5 (rhombohedric distortions at $T<T_N$)	II type MO $\mu \parallel [11]$ $(2a,\ 2a,\ 2a)$	198	$\theta = -570\,K,\ p = 4,6$ $\chi_M(T_N)/4\pi = 8 \cdot 10^{-3}\,cm^3/mol$ $\mu_{Fe^{2+}}(4.2\,K) = 3.32\mu_B$	[3,17]
FeSi	Cubic T_4		523	$\theta = -149\,K,\ p = 2.55$	[3,17]
FeP	Orthorhombic D_{2h}^{16}	HS, $\mathbf{k}_0 \parallel [c]$ $\mu \perp [c]$ $(a,\ b,\ 5c)$	125	$\theta = -50\,K$ $\mu_{Fe\ I}(4.2\,K) = 0.46\mu_B$ $\mu_{Fe\ II}(4.2\,K) = 0.37\mu_B$	[3,17]
FeS	Hexagonal D_{3h}^4	$\mu \parallel c$ at $T<T_r$ $\mu \perp c$ at $T>T_r$ $T_r = 390\,K$	593–600	$\theta = -917\,K$ $\chi_M(T_N)/4\pi =$ $= 2.2 \cdot 10^{-3}\,cm^3/mol$	[3,17]
FeCo	Cubic O^1h		1253	$\mu_{Fe} = 2.9\mu_B,\ \mu_{Co} = 1.9\mu_B$	[3]
FeGe	Hexagonal D_{6h}^1	CS, $\mu \parallel c$, $(a,\ a,\ 2c)$	400–412	$p = 3.19,\ \mu_{Fe}(4.2\,K) = 1.67\mu_B$	[3]
FeAs	Orthorhombic D_{2h}^{16}	HS, $\mathbf{k}_0 \parallel [c]$	77	$p = 3.1,\ \mu_{Fe}(4.2\,K) = 0.5\mu_B$	[3]
FeSe	Hexagonal D_{6h}^4		847	Ferromagnet at $T < 423\,K$	[17]
FeSn	Hexagonal D_{6h}^1	FI in (001) AFI between (001) $\mu \parallel (001)$, $(a,\ a,\ 2c)$	365–373	$\theta = -158\,K,\ p = 4.5$ $\mu_{Fe}(4.2\,K) = 1.55\mu_B$	[3]

*33 At this temperature only a part of magnetic moments is being ordered. The rest of the moments provide the paramagnetic properties. The Schottky anomalies in a specific heat were observed at $T = 0.23\,K$.

Table 28.1 Certain properties of antiferromagnets *(continued)*

Substance	Structure		T_N, K	Additional information	References
	crystallographic	magnetic			
FeRh	Cubic O_h^1	G-type MO $(2a,\ 2a,\ 2a)$	328 F→AF	Ferromagnet at $T_< T < T_C = 668$ K $\Delta\sigma_s = (328\text{ K}) = 117$ A·m²/kg $\rho(0\text{ K}) = 5 \cdot 10^{-7}\Omega$·m	[3,17]
FeF_2	Tetragonal D_{4h}^{14}	$\mu \parallel c$	78	$\theta = -117$ K $\mu_{Fe^{2+}}(23\text{ K}) = 4\mu_B$, $\lambda(0) = 189$ μm	[3,17]
$FeCl_2$	Trigonal D_{3d}^5	FI in (001) AFI between (001) $\mu \parallel c$, $(2a,\ 2a,\ 2a)$	23	$\theta = 48$ K, $p = 5.26$ $\mu_{Fe^{2+}}(4.2\text{ K}) = 4.2\mu_B$ Methamagnet: $H_A > H_E$ $\mu_0 H_{c\parallel} = 1.05$ T	[3,6,17]
$FeBr_2$	Trigonal D_{3d}^3	FI in (001) AFI between (001) $\mu \parallel c$, $(a,\ a,\ 2c)$	11	$\theta = 6$ K, $\mu_{Fe}(4.2\text{ K}) = 4.2\mu_B$ Methamagnet: $H_A > H_E$ $\mu_0 H_{c\parallel} = 3.15$ T	[3,6,17]
FeI_2	Trigonal D_{3d}^3	$\mu \parallel c$	10	$\theta = -23$ K Methamagnet: $H_A > H_E$ The behavior in a magnetic field is described by a model with 8–12 sublattices (4 phase transitions in magnetic field)	[3,6,17]
FeP_2	Orthorhombic D_{2h}^{12}		250*34	$\theta = -17$ K $\chi_M(T_N)/4\pi = 1.18 \cdot 10^{-3}$ cm³/mol	[17]
$FeGe_2$	Tetragonal D_{4h}^{18}	$\mu \parallel (001)$	270–315	$\theta = -510$ K, $\mu_{Fe}(4.2\text{ K}) \approx 1\mu_B$	[3,17]
$FeSn_2$	Tetragonal D_{4h}^{18}	$(2a,\ 2a,\ c)$	377–384	$\theta = -230$ K, $p = 3.36$ $\chi_M(T_N)/4\pi = 1.95 \cdot 10^{-3}$ cm³/mol	[3,17]
$FeTe_2$			85	$\theta = -320$ K, $p = 0.194$	[17]
FeF_3	Trigonal D_{3d}^6	CS, $\mu \parallel (111)$	362–365	$\mu_{Fe}(4.2\text{ K}) = 5\mu_B$	[3,17]
$FeCl_3$	Trigonal C_{3i}^2	HS	10	$\theta = -115$ K, $p = 5.73$ $\mu_{Fe^{2+}}(4.2\text{ K}) = 4.3\mu_B$	[3,17]
$FePt_3$	Cubic O_h^1	C-type MO $(2a,\ 2a,\ a)$	120	$\mu_{Fe} = 3.3\mu_B$	[3,17]
Fe_2As	Tetragonal D_{4h}^7	$\mu \perp c$ $(a,\ a,\ 2c)$	368	$\theta = -90$ K, $p = 4.66$ $\mu_{Fe\ I}(290\text{ K}) = 1.28\mu_B$ $\mu_{Fe\ II}(290\text{ K}) = 2.05\mu_B$	[3]
$\alpha-Fe_2O_3$	Trigonal D_{3d}^6	$\mu \parallel c$ at $T < T_M^{*35}$ $\mu \perp c$ at $T_M < T < T_N$	950–959	$\mu_{Fe}(4.2\text{ K}) = 4.9\mu_B$ $^{57}\nu_{Fe}(290\text{ K}) = 71.2$ MHz*36 $\sigma_m(300\text{ K}) = 3 \cdot 10^{-2}$ A · m²/mol $\mu_0 H_E = 900$ T $\mu_0 H_D = 2.2$ T $\mu_0 H_{c\parallel}(77\text{ K}) = 6.8$ T $\mu_0 H_{c\perp}(77\text{ K}) = 16$ T	[3,17]

*34 There is no certainty of the transition being of P→AF type.

*35 At $T = T_M = 262$ K, there occurs the transition of CAF→AF.

*36 See Appendix II in [8].

Table 28.1 Certain properties of antiferromagnets *(continued)*

Substance	Structure		T_N, K	Additional information	References
	crystallographic	magnetic			
FeOF	Tetragonal D_{4h}^{14}	$\mu \parallel c$	315	$\mu_{Fe^{3+}}(4.2\,K) = 4.8\mu_B$	[3]
FeOCl	Orthorhombic D_{2h}^{13}		92.3		[17]
$LiFeO_2$	Tetragonal D_{4h}^{19}	CS $\mu \parallel c$ at $T>T_r = 90\,K$ $\mu \parallel (111)$ at $T<T_r$	315	$\mu_{Fe^{3+}}(77\,K) = 4.5\mu_B$ $\mu_{Fe^{3+}}(4.2\,K) = 2.5\mu_B$ $\sigma(0\,K) \neq 0$	[3,17]
β-$FeNaO_2$	Orthorhombic C_{2v}^{9}	$\sigma \parallel [b]$, $\mu \parallel c$	723	$\mu_{Fe}(290\,K) = 4.2\mu_B$	[3,17]
$CuFeS_2$	Tetragonal D_{2d}^{12}	$\sigma \perp c$	815–823	$\mu_{Fe} = 3.85\mu_B$, $E_g = 2.5$ eV	[3,17]
$FeBO_3$	Trigonal D_{3d}^{6}	NCS $\mu \parallel (111)$ $\sigma \perp c$	348	$\mu_{Fe}(77\,K) = 4.7\mu_B$ $\mu_0 H_D(293\,K) = 6.2\,T$ $^{57}\nu_{Fe}(77K) = 76.5MHz$*37	[3,17]
$FeCO_3$	Trigonal D_{3d}^{6}	CS $\mu \parallel c$	38	$\theta = -14\,K$ $\mu_0 H_{c\parallel} = 15,3\,T$ Magnetic properties correspond to Ising model	[2,3,40,74]
$FeTiO_3$	Trigonal C_{3i}^{2}	$\mu \perp [111]$, $(2a,\ 2a,\ 2a)$	68	$p = 5.23$	[3,17]
$FeVO_3$	Trigonal D_{3d}^{6}				[17]
$YFeO_3$	Orthorhombic D_{2h}^{16}	$\sigma \parallel [001]$	640–650	$\sigma_m(4.2\,K) = 0.30\,A \cdot m^2/mol$ $\mu_0 H_c(4.2\,K) = 7.25\,T$*38	[3,17]
$LaFeO_3$*39	Orthorhombic D_{2h}^{16}	$\sigma \parallel [001]$ $(2a,\ 2b,\ 2c)$	738–750	$\theta = -480\,K$ $\sigma_M^{*40} = 0.24\ A{\cdot}m^2/mol$	[3,17]
$CeFeO_3$	Orthorhombic D_{2h}^{16}	$\sigma \parallel [001]$ at $T>T_r$, $T_r = 230\,K$	719		[17]
$PrFeO_3$	Orthorhombic D_{2h}^{16}	$\sigma \parallel [001]$	707–711	$p = 3.6$	[3,17]
$NdFeO_3$	Orthorhombic D_{2h}^{16}	$\mu \parallel [a]$ at $T > T_{r_1} = 160\,K$ $\mu \parallel c$ at $T < T_{r_2} = 70\,K$ $(2a,\ 2b,\ 2c)$	687–689	$p = 3.6$ $\mu_{Fe^{3+}}(43\,K) = 4.57\mu_B$	[3,17]
$SmFeO_3$	Orthorhombic D_{2h}^{16}	$\sigma \parallel [100]$ at $T<T_r$ $\sigma \parallel [001]$ at $T>T_r$ T_r =470–490 K	672–674	$\sigma_m = 0.21\,A \cdot m^2/mol$	[3,17]
$EuFeO_3$	Orthorhombic D_{2h}^{16}	$\sigma \parallel [001]$	666	$\sigma_m = 0.20\,A \cdot m^2/mol$	[17]
$GdFeO_3$	Orthorhombic D_{2h}^{16}	A_x-type MO at $T>T_r = 80\,K$	657–670 (Fe) 2.6(Gd)	$\theta = -4.3\,K$, $p = 8$ $\sigma_m = 0.26\,A \cdot m^2/mol$	[3,17]
$TbFeO_3$	Orthorhombic D_{2h}^{16}	$\sigma \parallel [001]$ at $T>T_r$ $T_r = 8.4\,K$ (change of Tb sublattices magnetic structure at $T = 3.1\,K$)	650–681 (Fe) 8.3(Tb)	$p = 9.7$ $\mu_{Fe}(97\,K) = 4.8\mu_B$ $\mu_{Tb^{3+}}(1.5\,K) = 8.6\mu_B$ $\sigma_m = 0.35\,A \cdot m^2/mol$	[3,17]

*37 See Appendix II in [8].

*38 At $H = H_c$, the second order phase transition takes place resulting in a magnetic symmetry change [75].

*39 The review of magnetic properties of orthoferrites ($RFeO_3$, where R is a rare-earth ion) is presented in [76]. Magnetic orientation transitions in orthoferrites are considered in [11].

*40 The σ magnitudes for $RFeO_3$ are given at $T < T_N(Fe)$, but $T > T_N(R)$.

Table 28.1 Certain properties of antiferromagnets *(continued)*

Substance	Structure		T_N, K	Additional information	References
	crystallographic	magnetic			
$DyFeO_3$	Orthorhombic D_{2h}^{16}	$\sigma \parallel [001]$ at $T>T_r$ $\sigma = 0$ at $T<T_r$ $T_r = 30$–40 K*41	645–648 (Fe) 3.7(Dy)	$\sigma_m = 0.33\ \text{A}\cdot\text{m}^2/\text{mol}$ Reorientation of magnetic structure at $T = T_r$ occurs step-wise	[3,17]
$HoFeO_3$	Orthorhombic D_{2h}^{16}	$\sigma \parallel [100]$ at $T<T_r$ $\sigma \parallel [001]$ at $T>T_r$ $T_r = 51$–63 K	639–647 (Fe) 6.5(Ho)	$p = 10.5$ $\mu_{Fe^{3+}}(43\,\text{K}) = 4.6\mu_B$ $\mu_{Ho^{3+}}(1.25\,\text{K}) = 7.25\mu_B$ $\sigma_m = 0.24\ \text{A}\cdot\text{m}^2/\text{mol}$	[3,17]
$ErFeO_3$	Orthorhombic D_{2h}^{16}	$\sigma \parallel [100]$ at $T<T_r$ $\sigma \parallel [001]$ at $T>T_r$ $T_r = 80$–100 K	636–643 (Fe) 3.9–4.5(Er)	$p = 9.5$ $\mu_{Fe^{3+}}(43\,\text{K}) = 4.6\mu_B$ $\mu_{Er^{3+}}(1.25\,\text{K}) = 5.8\mu_B$ $\sigma_m = 0.24\ \text{A}\cdot\text{m}^2/\text{mol}$	[3,17]
$TmFeO_3$	Orthorhombic D_{2h}^{16}	$\sigma \parallel [100]$ at $T<T_r$ $\sigma \parallel [001]$ at $T>T_r$ $T_r = 80$–92 K	630 – 632	$p = 7.3$ $\sigma_m = 0.32\ \text{A}\cdot\text{m}^2/\text{mol}$	[3,17]
$YbFeO_3$	Orthorhombic D_{2h}^{16}	$\sigma \parallel [001]$ at $T>T_r$ $T_r = 6$–8 K	627 – 634	$p = 4.5$ $\sigma_m = 0.35\ \text{A}\cdot\text{m}^2/\text{mol}$	[3,17]
$LuFeO_3$	Orthorhombic D_{2h}^{16}	$\mu \parallel [a]$	622–625	$\sigma_m = 0.28\ \text{A}\cdot\text{m}^2/\text{mol}$	[3,17]
$BiFeO_3$	Trigonal D_{3d}^6, C_{3v}^6 *42		643		[17]
$KFeO_3$	Cubic O_h^1 (trigonal distortions at $T<T_N$)	G-type MO $\mu \parallel [111]$ $(2a,\ 2a,\ 2a)$	113	$\mu_{Fe^{2+}}(4.2\,\text{K}) = 4.42\mu_B$	[3,17]
$RbFeF_3$	Cubic*43 O_h^1	G-type MO $\mu \parallel c$, $(2a,\ 2a,\ 2a)$	102–105	$\theta = -88\,\text{K}$, $p = 5.82$ $\mu_{Fe^{2+}} = 4.6\mu_B$	[3,17]
$RbFeCl_3$	Hexagonal D_{6h}^4	NCS	2.5	Magnetic phase transitions with structure change at $T = 2.35$ K and $T = 1.95$ K	[77]
$CsFeF_3$	Hexagonal D_{6h}^4	$\alpha(\mu, c) = 75°$	60–62	$\mu_{Fe^{2+}}(77\,\text{K}) = 4.4\mu_B$	[3]
$KFeCl_3$	Orthorhombic D_{2h}^{16}	CS, $\mu \parallel c$ FI in (001) AFI between (001)	15	$\mu_{Fe^{2+}}(77\,\text{K}) = 4.1\mu_B$	[3]
$TlFeI_3$	Orthorhombic D_{2h}^{16}	$\mu \parallel [b]$, FI in (100) AFI between (100)	21.5	$\theta = -44\,\text{K}$, $p = 6.7$	[68]
$FePO_4$	Trigonal D_{3d}^4	$\mu \parallel c$	25	$\theta = -60\,\text{K}$	[17]
$FeSO_4$	Orthorhombic D_{2h}^{17}	CS, $\mu \parallel [b]$ FI in (010) AFI between (010)	21–23	$\theta = -30.5\,\text{K}$, $p = 5,2$ $\chi_M(T_N)/4\pi = 78.5\cdot 10^{-3}\ \text{cm}^3/\text{mol}$ $\mu_{Fe^{2+}}(4.2\,\text{K}) = 4.1\mu_B$	[3,17]

*41 At low temperatures ($T < T_r$), a number of phase transitions induced by magnetic field is observed in $DyFeO_3$. Magnetic properties of $DyFeO_3$ correspond to the Ising model.

*42 Suggested structures.

*43 At $T > T_N$, there appears a cubic structure; at $87\,\text{K} < T < T_N$, tetragonal distortions; at T=40–87 K, orthorhombic structure, and at $T < 40\,\text{K}$, a monoclinic one.

Table 28.1 Certain properties of antiferromagnets *(continued)*

Substance	Structure		T_N, K	Additional information	References
	crystallographic	magnetic			
$FeNbO_4$	Monoclinic C_{2h}^4	$\mu \parallel (010)$ FI in (100) AFI between (100) $(2a,\ b,\ c)$			[3]
$FeTaO_4$	Tetragonal D_{4h}^{14}	$\mu \parallel c$	180	$\theta = -456\,\mathrm{K},\ p = 5.3$ Ferrimagnet at $T < 30\,\mathrm{K}$ $\mu_{Fe^{3+}}(77\,\mathrm{K}) = 2.96\mu_B$	[3]
$FeWO_4$	Monoclinic C_{2h}^4	$\mu \parallel (010)$ $\alpha(\mu, [a]) = 29°$ FI in (100) AFI between (100) $(2a,\ b,\ c)$	76	$\theta = 27\,\mathrm{K},\ p = 5.41$ $\mu_{Fe}(4.2\,\mathrm{K}) = 2.19\mu_B$	[3]
$FeUO_4$	Orthorhombic D_{2h}^4	CS*44, $\mu \parallel [b]$	55	$p = 5.42$ At $T < 42\,\mathrm{K}$, ferromagnet with a weak antiferro-magnetism, $\rho = 3.2 \cdot 10^7\,\Omega \cdot \mathrm{cm}$	[3,17]
$KFeF_4$	Orthorhombic D_{2h}^{17}	$\mu \parallel [c]$	137		[3]
$BaFeF_4$	Orthorhombic C_{2v}^{12}	$\mu \parallel c$ $(a,\ 2b,\ 2c)$	54	Magnetic properties are described by 2d-model $\mu_{Fe^{2+}}(4.2\,\mathrm{K}) = 4.1\mu_B$	[3,17]
$RbFeF_4$	Orthorhombic C_{2h}^5	$\mu \parallel [b]$, FI in (111) AFI between (111) $(a,\ 2b,\ c)$			[3]
$FeAl_2O_4$	Cubic O_h^7		8	$\theta = -76\,\mathrm{K},\ p = 4.14$	[3]
$FeSb_2O_4$	Tetragonal D_{4h}^{13}	NCS, $\mu \perp [a]$	46	$\mu_{Fe^{2+}}(4.2\,\mathrm{K}) = 3.8\mu_B$	[3,17]
Rb_2FeF_4	Tetragonal D_{4h}^{17}	$\mu \parallel [b]$ $(2a,\ 2a,\ c)$	50–56	$\lvert T_N/\theta \rvert = 0.54$ Magnetic properties correspond to XY-model, $H_A/H_E = 0.1$	[3,6]
$FeCr_2Se_4$	Monoclinic C_{2h}^3	$\alpha(\mu, [10\bar{1}]) = 55°$ FI in $(10\bar{1})$ AFI between $(10\bar{1})$ $(2a,\ b,\ 2c)$	4.2		[3]
$FeNb_2O_6$	Orthorhombic D_{2h}^{14}	CS, $\mu \parallel [a]$ $(a,\ 2b,\ c)$	< 25	$\mu_{Fe}(4.2\,\mathrm{K}) = 3.87\mu_B$	[3]
$FeTa_2O_6$	Tetragonal D_{4h}^{14}	$\mu \parallel (001)$ $\alpha(\mu, [a]) = 45°$ $(2a,\ 2b,\ 2c)$	14	$\mu_{Fe} = 4.08\mu_B$	[3]
Ba_2FeF_6	Tetragonal D_{4h}^{17}	$\mu \parallel c$ $(2a,\ 2a,\ c)$	48	Magnetic properties correspond to 2d-model $\mu_{Fe^{2+}}(4.2\,\mathrm{K}) = 3.46\mu_B$	[78]

*44 For AF-structure.

Table 28.1 Certain properties of antiferromagnets *(continued)*

Substance	Structure		T_N, K	Additional information	References
	crystallographic	magnetic			
$FeNb_3S_6$	Hexagonal D_6^6	$\mu\parallel c$		$\theta=-100$ K, $p=4.93$ $\mu_{Fe}(4.2\,K)=3.8\mu_B$	[3]
Fe_2SiO_4	Orthorhombic D_{2h}^{16}	$\mu\parallel[b]$*45 CS at $T>T_r=20$ K NCS at $T<T_r$	65	$\theta=-150$ K, $p=6.03$ $\mu_{Fe\ I}(4.2\,K)=4\mu_B$ $\mu_{Fe\ II}(4.2\,K)=4\mu_B$	[3,17]
Fe_2TiO_4	Cubic O_h^7 (tetragonal distortions at $T<T_N$		140–142	$\sigma\neq 0$ $\mu_{Fe}(4.2\,K)=4.2\mu_B$	[3]
$CaFe_2O_4$	Orthorhombic D_{2h}^{16}	CS, $\mu\parallel[c]$	180–200	$\theta=-580$ K, $p=5.4$ $\mu_{Fe^{3+}}(4.2\,K)=4.0\mu_B$	[3]
$ZnFe_2O_4$	Cubic O_h^7	NCS $(a,\ a,\ 2a)$	9–17	$\theta=-21$ K $\mu_{Fe^{3+}}(4.2\,K)=4.0\mu_B$	[3,17]
$BaFe_2O_4$	Orthorhombic C_{3v}^9	G_z-type MO	880	$\mu_{Fe}=4.58\mu_B$	[3]
Fe_2Mn_4P	Orthorhombic D_{2h}^{16}	NCS, $\mu\parallel(101)$ $(a,\ b,\ 2c)$	340	$\theta=-81$ K	[3]
Fe_2GeS_4	Orthorhombic D_{2h}^{16}	NCS (structure change at $T=69$ K)	108	$\mu_{Fe\ I}(4.2\,K)=3.98\mu_B$ $\mu_{Fe\ II}(4.2\,K)=3.60\mu_B$	[3]
Fe_2TeO_6	Tetragonal D_{4h}^{14}	$\mu\parallel c$	201–218	Magnetoelectric $\mu_{Fe^{3+}}(4.2\,K)=4.19\mu_B$	[3,17]
$Ca_2Fe_2O_5$	Orthorhombic D_{2h}^{16}	$\mu\parallel[c]$ $c<a<b$	720–730	$\theta=-615$ K $\mu_{Fe^{3+}}(4.2\,K)=4.5\mu_B$	[3,17]
$Sr_2Fe_2O_5$	Orthorhombic D_{2h}^{16}	CS, $\mu\parallel[c]$	600–700	$\mu_{Fe^{3+}}(77\,K)=4.5\mu_B$	[3,17]
$Fe_2Mo_3O_8$	Hexagonal C_{6v}^4	$\mu\parallel c$	59.5	$\theta_\perp<-200$ K $\theta_\parallel=-75$ K $p_\perp=4.4$, $p_\parallel=5.8$	[79]
$YBa_2Fe_3O_{7+x}$	Orthorhombic $Pmmm$	$\mu\perp c$	700		[123]
$Na_3Fe_5O_9$	Monoclinic C_{2h}^6	$\mu\perp[001]$	375–381	$\mu_0H_{eff}=51.0$ T*46	[3,17]
α-FeOOH	Orthorhombic D_{2h}^{16}	CS, $\mu\parallel[b]$	330–403	$p=6.4$ $\mu_{Fe^{3+}}(77\,K)=5\mu_B$	[3,17]
β-FeOOH	Tetragonal C_{4h}^5	$\mu\parallel c$	273–285	$\theta=-640$ K $\mu_{Fe^{3+}}(77\,K)=5\mu_B$	[3]
γ-FeOOH	Orthorhombic D_{2h}^{17}	CS, FI in (001) AFI between (001)	50–75	$p=4.9$	[3]
δ-FeOOH	Trigonal D_{3d}^4	$\mu\perp c$	450–460	$\mu_{Fe\ I}(77\,K)=5\mu_B$ $\mu_{Fe\ II}(77\,K)=5\mu_B$	[3]
$LiFePO_4$	Orthorhombic D_{2h}^{16}		53	$\mu_0H_E=43.0$ T	[17]
$FeTiNdO_5$	Orthorhombic D_{2h}^9	NCS, $(a,\ b,\ 2c)$			[3]
$FeCrWO_6$	Tetragonal D_{4h}^{14}	CS, $\mu\parallel c$			[3]
Ca_2FeAlO_5	Orthorhombic D_{2h}^{28}	$\mu\parallel[a]$	60		[17]

*45 $\mu_{Fe\ I}\parallel[b]$, $\mu_{Fe\ II}$ form a canted AF-structure at $T<T_N$ [116].

*46 Effective field of hyperfine interaction averaged over three nonequivalent Fe^{3+}-ion positions.

Table 28.1 Certain properties of antiferromagnets *(continued)*

Substance	Structure		T_N, K	Additional information	References
	crystallographic	magnetic			
$FeCl_2 \cdot 2H_2O$	Monoclinic C_{2h}^3	$\mu \parallel (101)$ $\alpha(\mu, [a]) = 58°$	23	Methamagnet $\mu_0 H_{c_1} = 3.92\,T$ $\mu_0 H_{c_2} = 4.56\,T$	[17]
$FeCl_2 \cdot 4H_2O$	Monoclinic	$\mu \parallel [b]$	1	$\chi_M(T_N)/4\pi =$ $= 0.105\,cm^3/mol$	[17]
$K_3Fe(CN)_6$	Monoclinic C_{2h}^5		0.129		[17]
$FeSO_4 \cdot H_2O$	Monoclinic C_{2h}^6	CS, $\mu \parallel [c]$			[3]
$FeC_2O_4 \cdot 2H_2O$			20		[17]
$Cs_2FeF_5 \cdot 4H_2$	Orthorhombic D_{2h}^7	$\mu \parallel [a]$	2.4	EA anisotropy	[80]
$Fe_3(PO_4)_2 \cdot 4H_2O$	Monoclinic C_{2h}^5	NCS, $\sigma \parallel [b]$	15–20	$\sigma_m = 4.46\,A \cdot m^2/mol$ $\mu_{Fe^{2+}}(4.2\,K) = 0.82\mu_B$	[3,17]
$Fe_3(PO_4)_2 \cdot 8H_2O$	Monoclinic C_{2h}^3	$\mu \parallel (101)$	8–12	$\mu_{Fe} = 4\mu_B$	[3,17]
$Fe(HCOO)_2 \cdot$ $\cdot 2H_2O$	Monoclinic C_{2h}^5	$\sigma \parallel [b]$	3.7	$\sigma \neq 0$, $\mu_{Fe\ I}(1.5\,K) = 3.7\mu_B$ Magnetic moments of Fe II are ordered at $T_{N_2} = 0.4\,K$	[3,17]
$Fe(NH_4) \cdot (SO_4)_2 \cdot$ $\cdot 12H_2O$			0.043	$\chi_M(T_N)/4\pi =$ $= 1\,cm^3/mol$	[17]
$[\{Fe(C_{13}H_{17}N_3)_2\} \cdot$ $\cdot SO_4 \cdot 6H_2O]$ (Organometallic polymer)			> 300	In a weak magnetic field it transforms into ferromagnetic phase ($\mu_0 H < 6 \cdot 10^{-2}$ T)	[81]
CoO	Cubic O_h^5 (tetragonal distortions at $T<T_N$)	$\alpha(\mu, [001]) = 27.4°$ $(2a,\ 2a,\ 2a)$	291–292	$\theta = -280\,K$, $p = 4.96$ $\chi_M(T_N)/4\pi =$ $= 5.3 \cdot 10^{-3}\,cm^3/mol$ $\mu_{Co^{2+}}(77\,K) = 3.52\mu_B$	[3,17]
CoS	Hexagonal D_{6h}^4		358	$\theta = -632\,K$, $p = 1.7$	[17]
CoF_2	Tetragonal D_{4h}^{14}	$\mu \parallel [001]$	37.7	$\theta = -52.7\,K$ ${}^{59}\nu_{Co}(0\,K) = 180\,MHz$*47 $\lambda_1(0) = 351\ \mu m$*48 $\lambda_2(0) = 278\ \mu m$ $\mu_0 H_{c\perp} = 12.0\,T$*49 $\mu_{Co^{2+}}(23\,K) = 3\mu_B$	[3,17]
$CoCl_2$	Trigonal D_{3d}^5	FI in (111) AFI between (111) $(2a,\ 2a,\ 2a)$	25	$\theta = 20\,K$ $\mu_{Co^{2+}}(4.2\,K) = 3.15\mu_B$ Magnetic properties at $T>T_N$ corresponds to 2d antiferromagnet model	[3,17]

*47 Central frequency of quadrupole-split spectrum consisting of seven lines ($I = 7/2$) [82].

*48 At $\mathbf{H} \parallel [001]$, g-factors for $\lambda_1(H)$ and $\lambda_2(H)$ amount to $g_1 = 1.18$ and $g_2 = 2.80$, respectively.

*49 In the range of $H_{c\parallel,1} < H < H_{c\parallel,2}$ ($\mu_0 H_{c\parallel,1} = 21\,T$, $\mu_0 H_{c\parallel,2} = 25.5\,T$), an angular phase is realized, in which sublattice moments are off by different angles with respect to c_4-axis [83].

Table 28.1 Certain properties of antiferromagnets *(continued)*

Substance	Structure: crystallographic	Structure: magnetic	T_N, K	Additional information	References
$CoBr_2$	Trigonal D_{3d}^3	$\mu\|$ (001) (a, a, $2a$)	19	$\mu_{Co}(4.2\,K) = 2.86\mu_B$ Magnetic properties at $T > T_N$ corresponds to 2d antiferromagnet model	[3,17]
α-CoI_2	Trigonal D_{3d}^3	$\mu\| c$	3		[17]
$CoSe_2$	Cubic T_h^6	(a, a, $2a$)	93	$\theta = -160\,K$, $p = 2.2$ $\mu_{Co}(4.2\,K) = 1\mu_B$ Semiconductor	[3,17]
CoF_3	Trigonal D_{3d}^6	CS, $\mu\|$ [111]	460	$\mu_{Co}(4.2\,K) = 4.4\mu_B$	[3,17]
Co_3O_4	Cubic O_h^7	$\mu\|$ [001]	40	$\theta = -53\,K$, $p = 4.14$ $\mu_{Co^{2+}}(4.2\,K) = 3.25\mu_B$	[3,17]
$CoCO_3$	Trigonal D_{3d}^6	$\mu \perp c$*50 $\sigma \perp c$	18.1	$\theta = -52\,K$, $p = 4.34$ EP anisotropy $\sigma_\perp(0\,K) = 1.4$ A·m²/mol $\mu_0 H_D = 2.7 - 5.1$ T*51	[2,3,86]
$CoTiO_3$	Trigonal C_{3i}^2	$\mu\|$ (111), ($2a$ $2a$ $2a$)	37–42	$\theta = -9.3\,K$, $p = 5.46$	[3,17]
$CoGeO_3$	Monoclinic C_{2h}^6		8.5	$\theta = -14\,K$, $p = 5.33$	[17]
$NaCoF_3$	Orthorhombic D_{2h}^{16}	CS, $\mu\|$ [b]	74–78	$\mu_{Co^{2+}}(4.2\,K) = 3.4\mu_B$	[3]
$KCoF_3$	Cubic O_h^1 (tetragonal distortions at $T < 78\,K$)	G-type MO ($2a$, $2a$, $2a$)	135–144	$\theta = -125\,K$ $^{59}\nu_{Co} = 372.7$ MHz*52 $\mu_{Co^{2+}}(4.2\,K) = 3.3\mu_B$	[3,17]
$RbCoF_3$	Cubic O_h^1 (tetragonal distortions at $T < 101\,K$ $c/a = 0.997$ at $T = 4.2\,K$)	G-type MO ($2a$, $2a$, $2a$)	98	$\theta = -180\,K$, $p = 5.53$ $\mu_{Co}(4.2\,K) = 3\mu_B$	[3,6,17]
$CsCoF_3$	Trigonal D_{3d}^5		8	$\theta = -62\,K$	[6]
$RbCoCl_3$	Hexagonal D_{6h}^4		18	Magnetic properties correspond to Ising model, $d = 1$	[6]
$CsCoCl_3$	Hexagonal D_{6h}^4	CS at $T > 9.2\,K$	21	$\theta = -100\,K$ Magnetic properties correspond to Ising model, $d = 1$ $\|J'/J\| = 1.6 \cdot 10^{-2}$ Ferrimagnet at $T < 9.2\,K$	[6]
$RbCoBr_3$	Hexagonal D_{6h}^4	CS, $\mu\| c$ ($\sqrt{3}a$, $\sqrt{3}a$, c)	36	Magnetic properties correspond to Ising model, $d = 1$ $\mu_{Co}(4.4\,K) = 3.4\mu_B$	[3,6,17]

*50 $\sigma_\perp \gg \sigma_\| = 7 \cdot 10^{-3}$ A·m²/mol [84].

*51 For a compendium of H_D values measured by different authors see [85].

*52 The central frequency of quadrupole-split spectrum which consists of seven lines with average interval between the lines of $\Delta\nu = 2.1$ MHz [87].

Table 28.1 Certain properties of antiferromagnets *(continued)*

Substance	Structure		T_N, K	Additional information	References
	crystallographic	magnetic			
$CsCoBr_3$	Hexagonal D_{6h}^4	CS at $T > 15\,K$ $\mu \| c$	28	Magnetic properties correspond to Ising model, $d = 1$ Ferrimagnet at $T < 15\,K$	[6]
α-$CoSO_4$	Orthorhombic D_{2h}^{17}	NCS	12–15	$\theta = -47\,K,\ p = 5.65$ $\mu_{Co^{2+}}(4.2\,K) = 3.3\mu_B$	[3]
β-$CoSO_4$	Orthorhombic D_{2h}^{16}	NCS	12–15	$\theta = -24\,K,\ p = 5.65$ $\chi_M(T_N)/4\pi = 62 \cdot 10^{-3}\,cm^3/mol$ $\mu_0 H_{c,[a]} = 1.2\,T$ $\mu_{Co}(4.2\,K) = 3.8\mu_B$	[3,17]
$CoSeO_4$	Orthorhombic D_{2h}^{16}	NCS	30	$p = 3.64$	[3]
$CoWO_4$	Monoclinic C_{2h}^4	$\mu \| (010)$ FI in (100) AFI between (100) $(2a,\ b,\ c)$	55	$\theta/T_N = -1.55$	[3]
$CoUO_4$	Orthorhombic D_{2h}^{28}	NCS, $\mu \| [a]$ and $\mu \| [c]$ $(2a,\ b,\ 2c)$	12	$\theta = -23\,K$ $\chi_M(T_N)/4\pi = 83 \cdot 10^{-3}\,cm^3/mol$ $\mu_{Co}(4.2\,K) = 4.06\mu_B$	[3,17]
$BaCoF_4$	Orthorhombic C_{2v}^{12}	$\mu \| [a]$	70	Magnetic properties correspond to Ising model, $d = 2$; $\mu_{Co^{2+}}(4.7\,K) = 3.4\mu_B$	[3]
$CoRh_2O_4$	Tetragonal		27	$\theta = -30\,K$	[17]
K_2CoF_4	Tetragonal D_{4h}^{17}	$\mu \| c$ $(2a,\ 2a,\ c)$	107–125	Magnetic properties correspond to Ising model, $d = 2$; $\|J'/J\| \approx 10^{-6}$ $H_A/H_E = 0.7$	[3,17,88]
Rb_2CoF_4	Tetragonal D_{4h}^{17}		101	Magnetic properties correspond to Ising model, $d = 2$; $\|J'/J\| \approx 10^{-6}$ $H_A/H_E = 0.8$	[88]
$CoNb_2O_6$	Orthorhombic D_{2h}^{14}	$\mu \| [a]$, $(a,\ b,\ 2c)$			[3]
$CoCs_3Cl_5$	Tetragonal		0.52		[17]
$CoCs_3Br_5$	Tetragonal				[17]
Co_2SiO_4	Orthorhombic D_{2h}^{16}	CS G_y-type MO	49	$\theta = -65\,K,\ p = 5.09$ $\mu_{Co^{2+}}(4.2\,K) = 3.3\mu_B$	[3,17]
$GeCo_2O_4$	Cubic O_h^7	$\mu \| (111)$ $(2a,\ 2a,\ 2a)$	20	$\theta = -90\,K,\ p = 4.7$ $\mu_{Co^{3+}}(4.2\,K) = 3.2\mu_B$	[3]
$Co_2Mo_3O_8$	Hexagonal C_{6v}^4	$\mu \| c$	40.8	$\theta_\perp = -185\,K,\ p_\perp = 5.9$ $\theta_\| = -95\,K,\ p_\| = 5.8$	[79]
$Co_3B_2O_6$	Orthorhombic D_{2h}^{12}	CS, $\mu \| [c]$ $(a,\ 2b,\ 2c)$	30–37	$\theta = -63\,K,\ p = 5.29$	[3]

Table 28.1 Certain properties of antiferromagnets *(continued)*

Substance	Structure		T_N, K	Additional information	References
	crystallographic	magnetic			
$Nb_2Co_4O_9$	Trigonal D_{3d}^4	CS, $\mu \parallel c$	30	$\theta = -10\,\mathrm{K}$, $p = 4$ $\mu_{Co^{2+}}(4.2\,\mathrm{K}) = 3\mu_B$	[3,17]
$Ta_2Co_4O_9$	Trigonal D_{3d}^4		206*53		[17]
$LiCoPO_4$	Orthorhombic D_{2h}^{16}	CS A_y-type MO	23	$\theta = -90\,\mathrm{K}$, $p = 5.7$ $\mu_0 H_E = 32\,\mathrm{T}$	[3,17]
NH_4CoF_3	Cubic O_h^1 (tetragonal distortions at $T<T_N$)	G-type MO	124	$\sigma \neq 0$ at $T<T_N$	[89]
Na_2CoSiO_4	Orthorhombic C_{2h}^5	$\mu \perp [b]$, $(2a,\ b,\ 2c)$	4	$\mu_{Co}(1.2\,\mathrm{K}) = 2.7\mu_B$	[3]
Na_2CoGeO_4	Orthorhombic C_{2h}^5	$\alpha(\mu, [b]) = 80°$ $(2a,\ b,\ c)$	4	$\mu_{Co}(1.2\,\mathrm{K}) = 3.0\mu_B$	[3]
$BaCoWO_6$	Cubic O_h^5	$\alpha(\mu, [111]) = 20°$			[17]
Ba_2CoWO_4	Cubic O_h^5	II type MO $\alpha(\mu, [111]) = 23°$ $(2a,\ 2a,\ 2a)$	17	$\mu_{Co}(4.2\,\mathrm{K}) = 2.0\mu_B$	[3]
$BaCoUO_4$	Cubic O_h^5	FI in (001) AFI between (001)	9	$\mu_{Co} = 1.35\mu_B$	[3]
$CoCl_2 \cdot 2H_2O$	Monoclinic C_{2h}^3	$\mu \parallel [b]$	18	Methamagnet $\mu_0 H_{c_1,[b]} = 3.2\,\mathrm{T}$ $(\Delta m = 1\mu_B)$ $\mu_0 H_{c_2,[b]} = 4.6\,\mathrm{T}$ $(\Delta m = 3\mu_B)$ $\chi_m(T_N) = 21\ \mathrm{cm^3/mol}$	[3,17]
$CoBr_2 \cdot 2H_2O$	Monoclinic C_{2h}^3	$\mu \parallel [b]$	9.5	$\theta = 0\,\mathrm{K}$, Methamagnet $\mu_0 H_{c_1,[b]} = 1.37\,\mathrm{T}$ $\mu_0 H_{c_2,[b]} = 2.98\,\mathrm{T}$	[17]
$CoCl_2 \cdot 6H_2O$	Monoclinic C_{2h}^3	$\mu \parallel [c]$ $(a,\ b,\ 2c)$	2.3	$\theta = -20\,\mathrm{K}$ $\chi_M(T_N)/4\pi =$ $= 17.8 \cdot 10^{-2}\,\mathrm{cm^3/mol}$ $\mu_{Co^{2+}}(1.5\,\mathrm{K}) = 3.8\mu_B$	[3,17]
$CoBr_2 \cdot 6H_2O$	Monoclinic C_{2h}^3		3.2		[17]
$Co(NC_5H_5)Cl_2$	Monoclinic		3.7	$\theta = -4\,\mathrm{K}$ $\chi_M(T_N)/4\pi = 2.8\,\mathrm{cm^3/mol}$	[17]
$Co(H_2O)_6 \cdot SiF_6$	Trigonal	$\sigma \perp c$	0.15	$\sigma = 3.5 \cdot 10^{-2} \cdot 2M_0^{*54}$	[17]
$CoCl_2[(NH_2)_2 \cdot CS]_4$	Tetragonal C_{4h}^4		0.92*55		[17]
$CoK_2(SO_4)_2 \cdot 6H_2O$			0.193		[17]
$Co(HCOO)_2 \cdot 2H_2O$	Monoclinic C_{2h}^5	$\mu \parallel (100)$	5.1	Magnetic properties correspond to Ising model, $d = 2$	[17]

*53 By magnetoelectric effect measurements.

*54 $2M_0 = Ng\mu_B s$, where N is the Avogadro number, n is the number of magnetic ions in a molecule, g is the Landé factor, s is the magnetic ion spin.

*55 By specific heat measurements.

Table 28.1 Certain properties of antiferromagnets *(continued)*

Substance	Structure		T_N, K	Additional information	References
	crystallographic	magnetic			
$Co(NH_4)_2\cdot$ $\cdot(SO_4)_2\cdot 6H_2O$	Monoclinic		0.084	$\sigma \approx 7.1$ A·m²/mol	[17]
$[Co(\gamma{-}CH_3\cdot$ $\cdot C_5H_4NO)_6]\cdot$ $\cdot(ClO_4)_2$	Monoclinic		0.49	Magnetic properties correspond to Ising model, $d = 2$	[90]
NiO	Cubic O_h^5 (rhombohedric distortions below T_N)	$\mu \parallel (111)$ II type MO	523	$\theta = -2470$ K, $p = 4.6$ $\lambda(0) = 274$ μm	[3,17]
NiS	Hexagonal D_{6h}^4	$\mu \parallel c$, FI in (001) AFI between (001)	263–265*56	$\mu_{Ni}(4.2\,K) = 1.8\mu_B$	[3,17]
NiAs	Hexagonal D_{6h}^4	$\mu \perp c$	150	$\theta = -300$ K	[17]
NiF_2	Tetragonal D_{4h}^{14}	NCS $\sigma \parallel [010]$	73	$\theta = -100$ K $\sigma = 0.35$ A·m²/mol $\lambda_1(0) = 3000$ μm, $\lambda_2(0) = 320$ μm $\mu_{Ni^{2+}}(23\,K) = 2\mu_B$	[2,3]
$NiCl_2$	Trigonal D_{3d}^5	$\mu \perp c$	52	$\theta = 67$ K	[17]
$NiBr_2$	Trigonal D_{3d}^5		60	$\theta = -20$ K, $p = 3.0$	[17]
$NiCO_3$	Trigonal D_{3d}^6	NCS, $\sigma \perp c$ $\alpha(\mu, [111]) = 63°$	25	$\sigma_m(4.2\,K) = 2.08$ A · m²/mol $\mu_0 H_D = 9.0$ T, $\mu_0 H_E = 24.0$ T	[3,17,84]
$NiTiO_3$	Trigonal C_{3i}^2	$\mu \perp [111]$ FI in (111) AFI between (111) $(a, a, 2a)$	23	$\theta = -11$ K, $p = 3.2$ $\lambda(0) = 1.6$ mm $\mu_{Ni}(4.2\,K) = 2.2\mu_B$	[3,17]
$NaNiF_3$	Orthorhombic D_{2h}^{16}	NCS $\mu \parallel (010)$ $\sigma \parallel [c]$	138–156	$\theta = -280$ K $\sigma_m = 0.335$ A · m²/mol $\mu_0 H_D = 16.2$ T, $\mu_0 H_A = 1.2$ T $\mu_0 H_E = 200$ T*57 $\mu_{Ni}(4.2\,K) = 2\mu_B$	[3,17]
$KNiF_3$	Cubic O_h^1	G-type MO $\mu \parallel [001]$ $(2a, 2a, 2a)$	275	$\theta = -234$ K $\chi_M(T_N)/4\pi = 2.05 \cdot 10^{-3}$ cm³/mol $\mu_{Ni^{2+}}(4.2\,K) = 2.2\mu_B$ Magnetic properties correspond to Heisenberg model, $d = 3$ $\mu_0 H_E = 3.6 \cdot 10^2$ T $\mu_0 H_A = 2.7 \cdot 10^{-2}$ T	[3,6,17,66]
$CsNiF_3$	Hexagonal D_{6h}^4	$\mu \perp c$	2.61	Magnetic properties correspond to XY-model, $d = 1$	[6]
$RbNiCl_3$	Hexagonal D_{6h}^4	HS	11	$\theta = -101$ K, $\mu_{Ni^{2+}}(4.2\,K) = 1.5\mu_B$ $\lvert J'/J \rvert = 2 \cdot 10^{-2}$ Magnetic properties correspond to Heisenberg model, $d = 1$	[3,6,17]

*56 The P→AF transition possesses the properties of the first order transition. The crystal unit cell parameters change at the point of the transition: $\Delta a/a = 3 \cdot 10^{-3}$; $\Delta c/c = 10^{-2}$.

*57 The effective field intensities are obtained by AFMR experiments with the assumption of sublattice g-factor anisotropy: $g_{xx} \approx g_{zz} = 2.14$; $g_{xz} = g_{zx} = -2.5 \cdot 10^{-2}$ (x, y, z are directed along the axes $[a]$, $[b]$, $[c]$, respectively).

Table 28.1 Certain properties of antiferromagnets *(continued)*

Substance	Structure: crystallographic	Structure: magnetic	T_N, K	Additional information	References
$CsNiCl_3$	Hexagonal D_{6h}^4	HS at $T>T_r = 4.4$ K $\mu \parallel (010)$ at $T<T_r$	4.85	$\theta = -69$ K, $p = 3.41$ $\chi_{\parallel}(0\,K)/4\pi = 3.5 \cdot 10^{-3}$ cm^3/mol $\chi_{\perp}(0\,K)/4\pi = 7.3 \cdot 10^{-3}$ cm^3/mol $\mu_{Ni^{2+}}(1.6\,K) = 1\mu_B$ Magnetic properties correspond to Heisenberg model, $d = 1$ $\lvert J'/J \rvert = 6 \cdot 10^{-2}$	[3,6,17,119]
$TlNiCl_3$	Hexagonal D_{6h}^4	NCS	13	Magnetic properties correspond to Heisenberg model, $d = 1$; $\lvert J'/J \rvert = 2 \cdot 10^{-2}$	[6]
$NiSO_4$	Orthorhombic D_{2h}^{17}	CS, $\mu \parallel [b]$ FI in (010) AFI between (010)	37	$\theta = -82$ K, $p = 3.82$ $\chi_M(T_N)/4\pi = 15 \cdot 10^{-3}$ cm^3/mol $\mu(4.2\,K) = 2.1\mu_B$	[3,17]
$NiCrO_4$	Orthorhombic D_{2h}^{17}	CS A_x-type MO	23	$\theta = -105$ K $\mu_{Ni^{2+}}(4.2\,K) = 1.28\mu_B$	[3]
$NiSeO_4$	Orthorhombic D_{2h}^{17}	$\mu \perp c$ A_{xy}-type MO	27	$p = 2.14$	[3]
$BaNiF_4$	Orthorhombic C_{2v}^{12}	CS, $\mu \parallel [b]$ $(a,\ 2b,\ 2c)$	150	$\mu_{Ni}(4.2\,K) = 1.96\mu_B$	[3,17]
$Ni(OH)_2$	Trigonal D_{3d}^3	CS, $\mu \parallel c$	28–35	$p = 3.2$, $\mu_{Ni}(78\,K) = 2.0\mu_B$	[3]
K_2NiF_4	Tetragonal D_{4h}^{17}	$\mu \parallel c$	97	$\theta = -600$ K $\mu_0 H_{c\parallel}(4.2\,K) = 18.0$ T $H_A/H_E = 2 \cdot 10^{-3}$ Magnetic properties correspond to Heisenberg model, $d = 2$; $\lvert J'/J \rvert \approx 10^{-6}$	[6,17]
Rb_2NiF_4	Tetragonal D_{4h}^{17}	$\mu \parallel c$	91	$\lvert T_N/\theta \rvert = 0.41$ $\mu_0 H_{c\parallel}(4.2\,K) = 35.0$ T $H_A/H_E = 10^{-2}$ Magnetic properties correspond to Heisenberg model, $d = 2$; $\lvert J'/J \rvert = 10^{-6}$	[6,17]
$NiRh_2O_4$	Tetragonal		18	$\theta = -20$ K	[17]
Tl_2NiF_4	Tetragonal D_{4h}^{17}		101	$T_N/\theta = -0.42$ Magnetic properties correspond to Heisenberg model, $d = 2$; $\lvert J'/J \rvert = 10^{-6}$	[6,17]
$NiSiO_4$	Orthorhombic		8.2	$\theta = 14$ K	[17]
Ba_2NiF_6	Tetragonal D_{4h}^{17}	$\mu \parallel c$ $(2a,\ 2a,\ c)$	93	$\mu_{Ni^{2+}}(4.2\,K) = 1.9\mu_B$ Magnetic properties correspond to Heisenberg model, $d = 2$	[78]
$GeNi_2O_4$	Cubic O_h^7	$\mu \parallel (111)$ $(2a,\ 2a,\ 2a)$	15–16	$\theta = -6$ K, $p = 3.24$ $\mu_{Ni^{3+}}(4.2\,K) = 2.2\mu_B$	[3,17]
$Ni_3B_2O_6$	Orthorhombic D_{2h}^{12}	NCS, $(a,\ 2b,\ 2c)$	49	$\theta = -5$ K, $p = 3.07$	[3]
$LiNiPO_4$	Orthorhombic D_{2h}^{16}	CS, A_z-type MO	23	$\theta = -79$ K, $p = 3.35$	[3]

Table 28.1 Certain properties of antiferromagnets *(continued)*

Substance	Structure		T_N, K	Additional information	References
	crystallographic	magnetic			
$SrNiMoO_6$			71.5		[17]
NH_4NiCl_3	Hexagonal D_{6h}^4	NCS	9	Magnetic properties correspond to Heisenberg model, $d = 1$	[6]
Na_2NiAlF_3	Orthorhombic C_{2v}^{20}	NCS $\sigma \parallel [b]$ at $T < 11$ K	90	Canted antiferromagnet at $T < 11$ K	[3]
$BaNiWO_6$	Cubic O_h^5	$\mu \parallel [111]$ II type MO $(2a,\ 2a,\ 2a)$	17	$\mu_{Ni}(4.2\,K) = 1.9\mu_B$	[3,17]
$NiCl_2{\cdot}6H_2O$	Monoclinic C_{2h}^3	$\alpha(\mu, [a]) = 22°$ $(a,\ b,\ 2c)$	5.34–5.8	$\theta = -7$ K	[3,17]
$NiBr_2{\cdot}6H_2O$			6.5*58		[17]
$Ni_3B_7O_{13}I$			120		[17]
$Ni(NO_3)_2{\cdot}2H_2O$	Monoclinic	$\mu \parallel [a]$	4.2	$\theta = -2.5$ K	[17]
$Ni(IO_3)_2{\cdot}2H_2O$			3.08	$\theta = -5$ K $\sigma_s(0\,K) = 1.45$ A·m²/kg	[91]
$Ni(IO_3)_2{\cdot}2D_2O$	Orthorhombic D_{2h}^{15}	NCS	3.1	$\sigma \neq 0$, $\mu_{Ni} = 2\mu_B$	[3]
$Ni(NO_3)_2{\cdot}6NH_3$	Cubic		1.35*58	$\theta = -3.3$ K	[17]
$Ni(HCOO)_2{\cdot}$ $\cdot 2H_2O$	Monoclinic C_{2h}^5	$\mu \parallel (100)$	15.7*59	Magnetic properties correspond to Heisenberg model, $d = 2$	[17]
$Ni(NH_3)_2{\cdot}$ $\cdot Ni(CN)_4{\cdot}2C_6H_6$	Tetragonal C_{4h}^1	$\mu \parallel c$	2.37	$\mu_0 H_{c\parallel} = 3.5$ T $\mu_0 H_A(1.46\,K) = 2.2$ T	[17]
CuO	Monoclinic C_{2h}^6		230	$\mu_{Cu} \leq 0.5\mu_B$	[17]
CuF_2	Monoclinic O_{2h}^5	NCS $\mu \parallel (a, b)$, $\sigma \parallel (010)$ $(2a,\ b,\ c)$	69–70	$\theta = -200$ K $\sigma_s \leq 1$ A·m²/kg	[3,17]
$CuCl_2$	Monoclinic C_{2h}^1		23.9	Magnetic properties correspond to Heisenberg model, $d = 1$	[92]
$CuBr_2$	Monoclinic C_{2h}^1		74	Magnetic properties correspond to Heisenberg model, $d = 1$ $\lvert J'/J \rvert = 7 \cdot 10^{-2}$	[92]
Cu_2S_6	Tetragonal		373	$\theta = -1694$ K, $p = 43$	[17]
$KCuF_3$	Tetragonal D_{4h}^{18} and D_{4h}^5 *60	$\mu \parallel (a, b)$ A-type MO	$38(D_{4h}^{18})$ $22(D_{4h}^5)$	$\mu_{Cu^{2+}}(4.2\,K) = 0.45\mu_B$ for D_{4h}^5 $\mu_{Cu^{2+}}(4.2\,K) = 0.54\mu_B$ for D_{4h}^{18}	[3,17]

*58 By specific heat measurements.

*59 Only the (100) plane magnetic moments are ordered at this temperature. The system of the (200) plane moments determines paramagnetic properties at the temperature below T_N. At $T = 3.1$ K there is Schottky anomaly in a specific heat.

*60 Both structures were observed simultaneously.

Table 28.1 Certain properties of antiferromagnets *(continued)*

Substance	Structure		T_N, K	Additional information	References
	crystallographic	magnetic			
$CsCuCl_3$	Hexagonal D_6^2 or D_6^3	HS, $\mathbf{k}_0 \parallel c$ $\mu \perp c$	10.7	$\mu_{Cu^{2+}}(4.2\,K) = 0.58\mu_B$	[93]
$CuSO_4$	Orthorhombic D_{2h}^{16}	$\mu \parallel [a]$	35	$\theta = -88\,K$ $\mu_{Cu^{2+}}(4.2\,K) = 0.83\mu_B$	[3,17]
$CuSeO_4$	Orthorhombic D_{2h}^{16}	CS, $\mu \parallel [c]$ FI in (001) AFI between (001)	34	$\mu_{Cu^{2+}}(7.2\,K) = 0.9\mu_B$	[3,17]
Nd_2CuO_4	Tetragonal D_{4h}^{17}	$\mu_{Cu} \perp c$	≈ 280		[124]
Eu_2CuO_4	Tetragonal D_{4h}^{17}	$\mu_{Cu} \perp c$	≈ 155		[125]
Gd_2CuO_4	Tetragonal D_{4h}^{17}	$\mu_{Cu} \perp c$	280(Cu); 6.5 (Gd)		[94]
$CuWO_4$	Monoclinic C_{2h}^4		90	$\theta/T_N = -1.89$	[3]
La_2CuO_4	Orthorhombic	$\mu \perp [b]$	240	$\mu = 1.1 \pm 0.3\mu_B$	[123]
$CuCl_2{\cdot}2H_2O$	Orthorhombic D_{2h}^7	$\mu \parallel [a]$, $(a,\ b,\ 2c)$	4.3	$\theta = 5\,K$, $\mu_0 H_{c\parallel} = 0.65\,T$	[2,3]
$CuCl_2{\cdot}2D_2O$	Orthorhombic D_{4h}^7	$\mu \parallel [a]$	4.3		[17]
$Cu(C_2H_3O_2)_2$			270		[17]
YBa_2CuO_6	Tetragonal	$\mu \parallel c$	500	$\mu = 0.48 \pm 0.05\mu_B$	[124]
$CuSiO_3{\cdot}2H_2O$			21^{*61}		[17]
$LiCuCl_3{\cdot}$ ${\cdot}2H_2O$	Monoclinic C_{2h}^5	$\mu \parallel (101)$	5–6	$\theta = -10\,K$, $p = 1.8$ $\chi_M(T_N)/4\pi = 30 \cdot 10^{-3}\,cm^3/mol$ $\mu_0 H_{c\parallel} = 1.0\,T^{*62}$ $\mu_{Cu^{2+}}(4.2\,K) = 1\mu_B$	[3,17]
$CuSO_4{\cdot}5H_2O$	Triclinic C_i^1		0.029^{*61}		[17]
$CuSeO_4{\cdot}5H_2O$			0.046^{*61}		[17]
$Cu(NO_3)_2{\cdot}$ ${\cdot}2.5H_2O$			0.45	$\chi_M(T_N)/4\pi = 65 \cdot 10^{-3}\,cm^3/mol$	[17]
$Cu_3(CO_3)_2{\cdot}$ ${\cdot}(OH)_2$	Monoclinic C_{2h}^5		1.86	$\theta = -10\,K$	[17]
$Cu(HCO_2)_2{\cdot}$ ${\cdot}4H_2O$	Monoclinic		17	$\theta = -175\,K$ $\sigma_m(0\,K) = 0.15\,A \cdot m^2/mol$ Magnetic properties correspond to Heisenberg model, $d = 2$	[17]
$CuK_2(SO_4)_2{\cdot}$ ${\cdot}6H_2O$			0.05		[17]
$Cu(C_2H_3O_2)_2{\cdot}$ ${\cdot}H_2O$			250–280		[17]
$(C_6H_{11}NH_3){\cdot}$ ${\cdot}CuCl_3$	Orthorhombic D_2^4	$\mu \parallel (a,b)$ $\alpha(\mu,[b]) = 17°$	2.21	Magnetic properties correspond to ferromagnetic chains model, $\|J'/J\| = 10^{-3}$	[95]

*61 By specific heat measurements.

*62 Spin-flop field.

Table 28.1 Certain properties of antiferromagnets *(continued)*

Substance	Structure		T_N, K	Additional information	References
	crystallographic	magnetic			
$(C_6H_{11}NH_3)\cdot$ $\cdot CuBr_3$	Orthorhombic D_2^4	$\mu\|\| (a, b)$ $\alpha(\mu, [b]) = 25°$	1.5	Magnetic properties correspond to ferromagnetic chains model $\|J'/J\| = 10^{-3}$	[95]
$(C_2H_5NH_3)_2\cdot$ $\cdot CuCl_4$	Orthorhombic D_{2h}^{15}	FI in (a, b) AFI between (a, b)		Magnetic properties correspond to Heisenberg model, $d = 2$ $\mu_0 H_{E_1} = 8 \cdot 10^{-2}$ T (between layers) $\mu_0 H_{E_2} = 50.0$ T (inside a layer) $\mu_0 H_{c\|\|} = 0.16$ T $\mu_0 H_D = 1.2 \cdot 10^{-2}$ T	[17,96]
$Cu(NH_3)_4SO_4\cdot$ $\cdot H_2O$			0.37*63	$\theta = -1.2$ K	[17]
$Cu\cdot$ $\cdot(C_6H_5COO)_2\cdot$ $\cdot 3H_2O$	Monoclinic		1.4	Magnetic properties correspond to Heisenberg model, $d = 1$	[17]
MoF_3	Cubic O_h^1		185		[17]
AgF_2	Orthorhombic D_{2h}^{15}	$\mu\|\| (100)$ $\alpha(\mu, [c]) = 0.5°$, $\sigma\|\| [b]$	163	$p = 2$ $\mu_{Ag^{2+}}(4.2\,K) = 0.66\mu_B$	[3]
Ce	Hexagonal D_{6h}^4	Ferrimagnetic ordering in the plane normal to c and AFI between the planes normal to c	13	$\theta = -46$ K $\mu \approx 0.6\mu_B$	[12,14]
CeS	Cubic O_h^5	$\mu\|\| [111]$ II type MO	7	$\theta = -45$ K, $p = 2.57$ $\mu_{Ce}(4.2\,K) = 0.57\mu_B$	[3]
CeZn	Cubic O_h^1		29–36	θ =(-18)–(-2) K, $p = 2.3$	[13]
CeSb	Cubic O_h^5 (tetragonal distortions at $T<T_N$)	Structure changes at $T = 15.5$; 14.9; 13.5; 8.4 K	16–18	$p = 2.58$ $\mu_{Ce}(4.2\,K) = 2.1\mu_B$	[3,17]
CeBi	Cubic O_h^5	$\mu\|\| [001]$ I type MO	25–26	$\theta = -12$ K, $p = 2.38$ $\mu_{Ce}(4.2\,K) = 2.0\mu_B$	[3,17]
CeC_2	Tetragonal D_{4h}^{17}	$\mu\|\| c$, I type MO	33	$\theta = 2.54$ K, $\mu_{Ce}(4.2\,K) = 1.74\mu_B$	[3]
$CeAl_2$	Cubic O_h^7	Incommensurate SW	3.9	$n_s = 0.71$*64	[97]
$CeZn_2$	Orthorhombic D_{2h}^{28}	$\mu\|\| [b]$	7.5	$\mu_{Ce}(4.2\,K) = 1.6\mu_B$	[3]
$CeCl_3$	Hexagonal		0.345		[17]
CeB_6	Cubic structure of CaB_6 type	NCS	2.3	$\mu_{Ce}(1.3\,K) = 0.7\mu_B$ Magnetic properties correspond to Kondo-lattice model	[99]

*63 By specific heat measurements.

*64 For magnetic characteristics of $CeAl_2$ see [58, pp. 74–75]. The results on magnetization measurements in strong magnetic field are presented in [98].

Table 28.1 Certain properties of antiferromagnets *(continued)*

Substance	Structure: crystallographic	Structure: magnetic	T_N, K	Additional information	References
$CeTiO_3$	Orthorhombic D_{2h}^{16}	G-type MO (Ti) F-type MO (Ce)	116	$\mu_{Ce^{3+}}(81\,K) = 0.4\mu_B$ $\mu_{Ti^{3+}}(81\,K) = 0.36\mu_B$ Metallic conductivity at $T > 100$ K	[59]
$Ce_2Zn_3\cdot$ $\cdot(NO_3)_{12}\cdot 24H_2O$	Trigonal		0.0063		[17]
Pr	Hexagonal D_{6h}^{4}		23–35	Pure monocrystal samples do not exhibit clear magnetic structure	[3]
PrAg	Cubic O_h^1	C-type MO, $(2a,\ 2b,\ c)$	11–14	$\mu_{Pr^{3+}}(4.2\,K) = 2.1\mu_B$	[3]
PrMg	Cubic O_h^1		45–47	$\theta = -10\,K,\ p = 3.4$	[13]
PrC_2	Tetragonal D_{4h}^{17}	I type MO, $\mu\Vert c$	15	$\mu_{Pr}(4.2\,K) = 1.14\mu_B$	[3]
$PrCl_3$	Hexagonal		0.7		[17]
$PrSn_3$	Cubic O_h^1	A-type MO, $(a,\ a,\ 2a)$	8.6	$\theta = -8\,K$	[3]
PrB_6	Cubic		6.9*65	Metal $\rho(300\,K) = 15\cdot 10^7\ \Omega\cdot m$	[17]
$PrAlO_3$	Orthorhombic D_{2h}^{16}			$\theta = -100\,K$	[17]
$PrCo_2Si_2$	Tetragonal D_{4h}^{17}	$\mu\Vert c$	31	$\mu_{Pr}(4.2\,K) = 3.19\mu_B$	[100]
Nd	Hexagonal	FI in the plane normal to c and AFI between the planes normal to c (structure change at $T = 7.5$ K)	19	$\theta = -16\,K$ $\mu_{Nd}(4.2\ K) = 2\ \mu_B$	[12,14]
NdMg	Cubic O_h^1		48–64	$\theta = -11\,K,\ p = 3.7$	[13]
NdAl	Orthorhombic D_{2h}^{11}	NCS, $\alpha(\mu, [a]) = 58°$ $(2a,\ b,\ c)$	29	$\theta = -4\,K$ $\mu_{Nd}(4.2\,K) = 2.7\mu_B$	[3]
NdP	Cubic O_h^5	I type MO, $\mu\Vert [100]$		$\mu_{Nd^{3+}}(4.2\,K) = 1.8\mu_B$	[3]
NdS, $\alpha(\mu, [111]) = 24°$	CubicO_h^5	NCS		$\theta = -24\,K,\ p = 3.62$	[3,17]
NdAs	Cubic O_h^5	I type MO, $\mu\Vert [100]$		$\mu_{Nd^{3+}}(4.2\,K) = 2.13\mu_B$	[3]
NdSe	Cubic O_h^5	II type MO $\mu\Vert [111]$, $(2a,\ 2a,\ 2a)$	14	$\theta = -9\,K,\ p = 3.5$ $\mu_{Nd}(4.2\,K) = 1.57\mu_B$	[3,17]
NdAg	Cubic O_h^1		22	$\theta = -3\,K,\ p = 3.6$	[13]
NdIn	Cubic O_h^1		66–148	$\theta = 35\,K,\ p = 3.3$	[13]
NdTe	Cubic O_h^5	II type MO $\mu\Vert [111]$, $(2a,\ 2a,\ 2a)$	13	$\theta = -14\,K,\ p = 9.8$ $\mu_{Nd}(4.2\,K) = 1.1\mu_B$	[3,17]
NdSb	Cubic O_h^5 (tetragonal distortions at low temperatures)		10–16	$\theta = -3\,K,\ p = 3.75$ $\mu_0 H_c = 10.5\,T$ $\mu_{Nd^{3+}}(4.2\,K) = 2.99\mu_B$	[3,17]

*65 By specific heat measurements.

Table 28.1 Certain properties of antiferromagnets *(continued)*

Substance	Structure		T_N, K	Additional information	References
	crystallographic	magnetic			
NdBi	Cubic O_h^5	I type MO, $\mu\parallel[001]$	25	$\theta = -1\,\text{K},\ p = 3.58$	[3,17]
NdC_2	Tetragonal D_{4h}^{17}	I type MO, $\mu\parallel c$	29	$\mu_{\text{Nd}}(4.2\,\text{K}) = 2.95\mu_B$	[3]
$NdCl_3$	Hexagonal		1.035		[17]
$NdIn_3$	Cubic O_h^1	C-type MO (2a, 2a, a)	7	$\theta = -17\,\text{K},\ p = 3.7$ $\mu_{\text{Nd}} = 2.1\mu_B$	[3,13]
$NdSn_3$	Cubic O_h^1	A-type MO (a, a, 2a)	4.7	$\theta = -22\,\text{K},\ p = 3.6$ $\mu_{\text{Nd}} = 1.53\mu_B$	[3,13]
$NdPb_3$	Cubic O_h^1		2.7	$\theta = -23\,\text{K},\ p = 3.6$	[13]
$Nd(OH)_3$	Hexagonal C_{6h}^2	$\mu\parallel c$	1.7*66		[17]
$NdVO_3$	Orthorhombic D_{2h}^{16}	C_x-type MO	132	$p = 3.32$	[3]
$NdFe_2Si_2$	Tetragonal D_{4h}^{17}	CS	16	Canted antiferromagnet at $T_N < T < T_C \approx 690$ K $\mu_{\text{Nd}} = 3.1\mu_B$	[3]
$NdRh_4B_4$	Tetragonal (crystal structure of $CeCo_4B_4$-type)		1.31	$\theta = -6.2\,\text{K},\ p = 3.58$ Superconductor at $T < 5.4$ K	[102]
Sm	Rhombohedral	$\mu\parallel c$ (structure change at $T = 14$ K	106	$\mu_{\text{Sm}} = 0.1\mu_B$	[12,14]
$SmCl_3$	Hexagonal		0.4		[17]
$SmRh_4B_4$	Tetragonal (crystal structure of $CeCo_4B_4$-type)		0.87	$\theta = -1.93\,\text{K},\ p = 0.63$ Superconductor at $T < 2.72$ K $\mu_0 H_{c2} = 0.185\,\text{T}$	
Eu	Cubic O_h^9	HS, $\mathbf{k}_0\parallel[100]$	87–91	$p = 8,\ \mu_{\text{Eu}}(4.2\,\text{K}) = 5.9\mu_B$	[3,12,14]
EuSe	Cubic O_h^5	NCS	4.6	Semiconductor, ferromagnet at $T < 2.8$ K	[3,17]
EuTe*67	Cubic O_h^5	II type MO $\mu\parallel[110]$, (2a, 2a, 2a)	8 – 11	$\theta = -7.5\,\text{K}$ $\mu_0 H_c = 7.5\,\text{T},\ n_s = 7$	[17]
EuF_2	Cubic O_h^5		2; 19.5*68		[17]
Eu_3O_4	Orthorhombic		5.3	$\mu_0 H_c = 0.2\,\text{T}$	[17]
$EuTiO_3$	Cubic		5.2	$\mu_0 H_c = 1.4\,\text{T}$ $n_s = 6.93$	[17]
$EuGd_2O_4$	Orthorhombic		4.5		[17]
GdP	Cubic		15	$\theta = -2\,\text{K}$ $\mu_0 H_c = 7.5\,\text{T}$	[17]
GdS	Cubic O_h^5	II type MO $\mu\parallel(111)$, (2a, 2a, 2a)	50		[3,17]

*66 $T_N = 0.265\,\text{K}$ [101].

*67 Magnetic properties of EuTe samples correlate with electric resistivity which can be changed in a wide range ($10^6 - 10^{-4}$ Ω·m at $T = 300$ K).

*68 By AFMR measurements of samples containing impurities [104].

Table 28.1 Certain properties of antiferromagnets *(continued)*

Substance	Structure		T_N, K	Additional information	References
	crystallographic	magnetic			
CdCu	Cubic O_h^1		41–140	θ =(−75)–(−26) K, $p = 8.4$	[13,17]
GdAs	Cubic (trigonal distortions at $T<T_N$)		19–25	$\theta = -12\,K$, $p = 8.2$ $\mu_0 H_c = 18.0\,T$, $n_s = 7.2$	[17]
GdSe	Cubic O_h^5	II type MO, $\mu \parallel (111)$ (2a, 2a, 2a)	60		[3,17]
GdAg	Cubic O_h^1		138–150	θ =(−84)–(−70) K $p = 8.2 - 8.8$ $\chi_M(T_N)/4\pi = 40 \cdot 10^{-3}\,cm^3/mol$	[3,17]
GdIn	Tetragonal		28	θ =(−66)–(−18) K, $p = 8.1$ $\chi_M(T_N)/4\pi = 73.5 \cdot 10^{-3}\,cm^3/mol$	[13,17]
GdSb	Cubic O_h^5	II type MO, $\mu \parallel (111)$ (2a, 2a, 2a)	28	$\theta = -42\,K$, $p = 8.1$	[3,17]
GdTe	Cubic O_h^5		80		[17]
GdBi	Cubic O_h^5	II type MO, $\mu \parallel (111)$ (2a, 2a, 2a)	28–32	$p = 8.3$	[3,17]
$GdCu_2$	Orthorhombic D_{2h}^{28}		41	$\theta = 11\,K$, $p = 8.4$	[17]
$GdCoO_3$	Orthorhombic D_{2h}^{16}	A-type MO (Gd)	2.9(Gd)	$\mu_{Gd} = 7\mu_B$	[3]
$GdAlO_3$	Orthorhombic D_{2h}^{16}	$\mu \parallel [b]$	3.87	$\theta = -4.81\,K$ $\mu_0 H_{c\parallel}(0.5\,K) = 1.1\,T$*69	[17]
$GdVO_3$	Orthorhombic		7.5	$p = 8.0$	[17]
$GdVO_4$	Tetragonal D_{4h}^{19}		2.49	At room temperature possesses ferromagnetic properties	[105]
$Gd(OH)_3$	Hexagonal	$\mu \parallel c$	2.0		[17]
$Gd_{1.2}Mo_6S_8$	Rhombohedral		0.8–0.9	Superconductor at $T < 1.4\,K$	[108,109]
$SrGd_2O_4$	Orthorhombic		2.8	$\theta = -6\,K$	[17]
$GdCl_3 \cdot 6H_2O$			0.182*70		[17]
Tb	Hexagonal D_{6h}^4	HS at $216\,K < T < T_N$ $\mu \perp c$	229–230	Ferromagnet at $T < 216\,K$	[3,12]
TbAl	Orthorhombic D_{2h}^{11}	NCS (2a, b, c)	72	$\theta = 10 - 24\,K$, $p = 10$ $\mu_{Tb}(4.2\,K) = 8.8\mu_B$	[3,13]
TbP	Cubic O_h^5	II type MO $\mu \parallel [111]$, (2a, 2a, 2a)		$\theta = 1\,K$, $p = 9.2$, $n_s = 8$ $\mu_0 H_c = 0.43\,T$, $\mu_{Tb^{3+}} = 6.2\mu_B$	[3,17]
TbCu	Cubic O_h^1	C-type MO $\mu \parallel (001)$, (2a, 2a, a)	177	$\theta = -20\,K$, $p = 9.6$ $\mu_{Tb} = 8.6\mu_B$	[3,13]
TbAs	Cubic O_h^5	II type MO $\mu \parallel [111]$ (2a, 2a, 2a)	10-12	$\theta = -4\,K$, $p = 9.7$ $n_s = 7.9$ $\mu_0 H_c = 2.8\,T$	[3,17]

*69 Spin-flop field.

*70 By specific heat measurements.

Table 28.1 Certain properties of antiferromagnets *(continued)*

Substance	Structure		T_N, K	Additional information	References
	crystallographic	magnetic			
TbSe	Cubic O_h^5	II type MO $\mu\|\|$ [111]	52	$\theta = -53.5$ K, $p = 9.8$ $\mu_{Tb}(4.2\,K) = 7.5\mu_B$	[3]
TbAg	Cubic O_h^1	C-type MO $\mu\|\|$ [001], (2a, 2a, 2a)	100–106	$\theta = (-36) - (-11)$ K $p = 9.4 - 10.1$, $\mu_{Tb} = 8.3\mu_B$	[3,13]
TbSb	Cubic O_h^5	II type MO $\mu\|\|$ [111] (2a, 2a, 2a)	14–17	$\theta = -14$ K, $p = 9.7$ $n_s = 7.5$, $\mu_0 H_c = 6.0$ T $\mu_{Tb^{3+}}(0\,K) = 8.2\mu_B$	[3,17]
TbBi	Cubic O_h^5	II type MO $\mu\|\|$ [111], (2a, 2a, 2a)	18	$\theta = -33$ K, $p = 9.52$ $\mu_{Tb}(4.2\,K) = 7.9\mu_B$	[3,17]
TbD_2	Cubic O_h^5	$\mu\|\|$ [001]*71	40	$\mu_{Tb}(4.2\,K) = 7.9\mu_B$	[3]
TbC_2	Tetragonal D_{4h}^{17}	HS, $\mathbf{k}_0\|\|$ [a] (structure change at $T \approx 30$ K)	66	$\mu_{Tb}(4.2\,K) = 5.2\mu_B$	[3]
TbO_2	Cubic O_h^5	II type MO $\mu\|\|$ (111), (2a, 2a, 2a)	3	$p = 7.8$ $\mu_{Tb^{4+}} = 6.25\mu_B$	[3]
$TbCu_2$	Orthorhombic D_{2h}^{28}		54	$\theta = -6$ K, $p = 9.8$	[13,17]
$TbAg_2$	Tetragonal D_{4h}^{17}	FI in (100) AFI between (100) (2a, 2a, c)	35	$\theta = -32$ K, $n_s = 8.95$	[3,13]
$TbZn_2$	Orthorhombic D_{2h}^{28}	NCS at $T_r < T < T_N$ CS at $T<T_r= 60$ K*72 $\mu\|\|$ [b], (a, b, 2c)	75	$\mu_{Tb}(4.2\,K) = 8.8\mu_B$	[3]
$TbAu_2$	Tetragonal D_{4h}^{17}	$\mu\|\|$ c at $T<T_r= 42$ K*73	55	$\theta = -21$ K, $p = 9.8$ $\mu_{Tb^{3+}}(4.2\,K) = 9\mu_B$ $\mu_{Tb}(42.6\,K) = 5.1\mu_B$	[3,13]
$TbPt_3$	Cubic O_h^1	FI in (111) AFI between (111) (2a, a, a)	20–22	$\theta = 17$ K, $p = 9.9$ $n_s = 8.4$	[3,13]
α-Tb_2C_3	Cubic T_d^6	$\mu\|\|$ [001]	33	$\mu_{Tb}(4.2\,K) = 6.9\mu_B$, $n_s = 9$	[3]
Tb_2O_3			2.4	$\theta = -13$ K, $p = 9.67$	[17,108]
Tb_3Ni	Orthorhombic D_{2h}^{16}	NCS, (2a, b, c)	62	$\theta = -5$ K, $p = 10$	[3,13]
Tb_3Nd	Trigonal D_{3d}^5	$\mu \perp c$	129	Ferromagnet at low temperatures	[3]
Tb_4La	Trigonal D_{3d}^5	$\mu \perp c$	124	$\mu_{Tb} = 6.8\mu_B$	[3]
Tb_4Pr	Trigonal D_{3d}^5	$\mu \perp c$	130	Ferromagnet at $T < 30$ K	[3]
TbOCl	Tetragonal D_{4h}^7	$(2\sqrt{2}a,\ 2\sqrt{2}a,\ 4c)$		$p = 9.84$	[3]
$TbAlO_3$	Orthorhombic D_{2h}^{16}	NCS, $\alpha(\mu, [a]) = 34°$	3–5	$\mu_{Tb}(1.5\,K) = 8.25\mu_B$	[3]
$TbVO_3$	Orthorhombic D_{2h}^{16}	CS (V) NCS (Tb)		$\mu_{Tb}(1.5\,K) = 7.6\mu_B$ $\mu_V(4.2\,K) = 1.3\mu_B$	[3]

*71 Commensurate SW propagating along [001] with the period $\tau = 2.4$ nm.
*72 $T_r = 55$ K [13].
*73 In the range $T_r < T < T_N$, incommensurate SW polarized along the [c] axis and propagating along the [a] axis.

Table 28.1 Certain properties of antiferromagnets *(continued)*

Substance	Structure: crystallographic	Structure: magnetic	T_N, K	Additional information	References
$TbCoO_3$	Orthorhombic D_{2h}^{16}	NCS, $\alpha(\mu, [a]) = 33°$	3.3	$\mu_{Tb}(1.5\,K) = 8.0\mu_B$	[3]
$BaTbO_3$	Trigonal D_{3d}^5	CS	36	$\mu_{Tb}(11\,K) = 6.7\mu_B$	[3]
$TbCo_2Si_2$	Tetragonal D_{4h}^{17}	CS, $\mu \parallel c$	46	$\mu_{Tb}(4.2\,K) = 9.12\mu_B$	[100]
$Tb_{1.2}Mo_6S_8$	Rhombohedral		1.0	Superconductor at $T < 2.05\,K$ Ferromagnet at $H > H_{c2}(0)$ $\mu_0 H_{c2}(0) = 0.19$ T	[106,107,109]
Tb_2O_2S	Trigonal D_{3d}^3	$\mu \parallel [110]$ $(a,\ 2a,\ 2c)$		$\theta = -17\,K$ $\mu_{Tb}(1.5\,K) = 8.36\mu_B$	[3]
Tb_2O_2Se	Trigonal D_{3d}^3	$\alpha(\mu,\ c) \approx 30°$ $(a,\ 2a,\ 2c)$	7	$\theta = -18\,K$ $\mu_{Tb^{3+}}(1.5\,K) = 6.5\mu_B$	[3]
$Tb_3Al_5O_{12}$	Cubic O_h^{10}	NCS	1.5		[11,17]
Dy	Hexagonal D_{6h}^4	HS at $T_C < T < T_N$ T_C =85–90 K, $\mu \perp c$	175–184	$\theta = 153\,K$ Ferromagnet at $T < T_C$	[3,12,14]
DyP	Cubic O_h^5	NCS at $T = 1.8\,K$		$\theta = 6\,K,\ p = 9.9$ $n_s = 7.8,\ \mu_0 H_c = 1.7\,T$	[17]
DyCu	Cubic O_h^9	a) CS, C-type MO [*74] $\mu \parallel [001]$; b) NCS[*74]	61–64	θ =(−26)–(−18) K, $p = 10.7,\ \mu_{Dy^{3+}} = 10.6\mu_B$	[3,13]
DyAs			8.5	$\theta = 2\,K,\ p = 10.4$ $n_s = 8.1,\ \mu_0 H_c = 3.2\,T$	[17]
DySb	Cubic O_h^5	FI in (111) AFI between (111) $\mu \parallel [001]$, $(2a,\ 2a,\ 2a)$	9.5	$\theta = -4\,K,\ p = 10.7$ $n_s = 7.7,\ \mu_0 H_c = 1.95\,T$ $\mu_{Dy}(5.8\,K) = 9.4\mu_B$	[3,17]
DyBi	Cubic O_h^5	II type MO $(2a,\ 2a,\ 2a)$	12–13	$\theta = -30\,K$ $\mu_{Dy}(4.2\,K) = 8.7\mu_B$	[3,17]
DyC_2	Tetragonal D_{4h}^{17}	$\mu \parallel C$[*75]	59	$\mu_{Dy}(4.2\,K) = 8.37\mu_B$ $n_s = 11.8$	[3]
$DyCu_2$	Orthorhombic D_{2h}^{28}		24	$\theta = 5\,K,\ p = 10.75$	[13,17]
$DyGa_2$	Hexagonal D_{6h}^1	CS, $\mu \parallel [b]$ $(a,\ a,\ 2\sqrt{3}c)$	15	$\theta = -6\,K,\ p = 10.7$ $\mu_{Dy}(4.2\,K) = 7.5\mu_B$	[3,13]
$DyAg_2$	Tetragonal D_{4h}^{17}	Incommensurate structure at $T_N > T > T_r = 9\,K$[*75] Commensurate structure at $T<T_r$	15	$\theta = -25\,K,\ p = 10.5$ $\mu_0 H_{c\parallel} = 3.5\,T$[*76] $\mu_{Dy}(1.7\,K) = 7.4\mu_B$	[3,13,17]
$DyAu_2$	Tetragonal D_{4h}^{17}	Incommensurate structure at $T_N > T > T_r = 25\,K$[*77] Commensurate structure at $T<T_r$		θ =(−13)–(−24) K $p = 10.5$ $\mu_{Dy}(1.7\,K) = 9.2\mu_B$	[3,13,17]

*74 Possible structures.
*75 Incommensurate transverse SW propagating along the [a] axis.
*76 Spin-flop field.
*77 Incommensurate transverse SW propagating along the [a] axis.

Table 28.1 Certain properties of antiferromagnets *(continued)*

Substance	Structure		T_N, K	Additional information	References
	crystallographic	magnetic			
$DyIn_3$	Cubic O_h^1	FI in (110) AFI between (110) $\alpha(\mu, [001]) = 27.5°$ $(a,\ a,\ 2a)$	23–24	$\theta = -35\,K,\ p = 10.8$ $\mu_{Dy}(4.2\,K) = 8.8\mu_B$	[3,13]
$DyPt_3$	Cubic O_h^1	II type MO, $(2a,\ 2a,\ 2a)$	13	$\mu_{Dy} = 9.0\mu_B$	[3,13]
DyB_6	Cubic (*bcc*)		20.5	$\theta = -21\,K,\ p = 9.43$	[110]
DyOOH	Monoclinic C_{2h}^2	FI in (010) AFI between (010) $\mu \parallel (010),\ \alpha(\mu, [a]) = 80°$		$\mu_{Dy^{3+}}(4.2\,K) = 9.4\mu_B$	[3]
DyOCl	Tetragonal D_{4h}^7	$\mu \parallel (001),\ (a,\ a,\ 2c)$	9	$\mu_0 H_c(4.2\,K) = 1.4$ T	[3,17]
$DyAlO_3$	Orthorhombic D_{2h}^{16}	NCS, $\alpha(\mu, [a]) = 57°$	3.4–3.5	$\mu_{Dy}(1.5\,K) = 8.2\mu_B$	[3,17]
$DyCoO_3$	Orthorhombic D_{2h}^{16}	$\alpha(\mu, [a]) = 60°$	3.6	The Co ions are diamagnetic due to strong crystal field; $\mu_{Dy}(1.5\,K) = 8.8\mu_B$	[3]
$DyPO_4$	Tetragonal D_{4h}^{19}	G-type MO $\mu \parallel c$	3.4	Magnetic properties correspond to Ising model, $d = 3$ $\mu_{Dy^{3+}}(1.8\,K) = 9.0\mu_B$	[3,111]
$DyVO_4$	Tetragonal D_{4h}^{19}	$\mu \parallel [b]$	3.0	$\mu_{Dy}(1.85\,K) = 9.0\mu_B$	[3,17]
$DyAsO_4$	Tetragonal D_{4h}^{19}	NCS $\mu \parallel (007)$	2.5–2.8	Ferromagnet at $T < T_C = 11\,K$	[3]
$Dy_{1.2}Mo_6S_8$	Rhombohedral	$\mu \parallel [111]$	0.4	Superconductor at $T < 2.05\,K$ $\mu_0 H_{c2}(0) = 0.12$ T Ferromagnetic component at $\mu_0 H > 2 \cdot 10^{-2}$ T $\mu_{Dy}(0.07\,K) = 8.77\mu_B$	[107,109,112]
Dy_2O_2S	Trigonal D_{3d}^3	$\mu \parallel [110]$ $(a,\ 2a,\ 2c)$	5.85	$\theta = -14.8\,K$ $\mu_{Dy}(1.5\,K) = 4.41\mu_B$	[3]
Dy_2O_2Se	Trigonal D_{3d}^3	$\mu \parallel c,\ (a,\ 2a,\ 2c)$	8.5	$\mu_{Dy^{3+}}(1.5\,K) = 9.0\mu_B$	[3]
$Dy_3Al_5O_{12}$	Cubic O_h^{10}	NCS*78	2.5	$\mu_{Dy^{3+}}(1.3\,K) = 9.0\mu_B$	[3,11,17]
Ho	Hexagonal D_{6h}^4	HS, $\mathbf{k}_0 \parallel c$ at $T_N > T > T_C = 20\,K$	130–133	Ferromagnetic spiral at $T < T_C$	[3,12,14]
HoSi	Orthorhombic D_{2h}^{17}	CS, $\mu \parallel [103]$ $(2a,\ b,\ 2c)$	25	$\theta = 2\,K,\ p = 10.6$ $\mu_{Ho^{3+}}(4.2\,K) = 7.0\mu_B$	[3,13]
HoAs			4.8	$\theta = 1\,K,\ p = 10.5$ $n_s = 9.0,\ \mu_0 H_{c1} = 0.18\,T$ $\mu_0 H_{c2} = 0.6\,T$	[17]
HoRh	Cubic O_h^1	A-type MO $\mu \parallel (001)$	3.2	$\theta = -3$ K, $p = 10.4$ $\mu_{Ho}(1.2\,K) = 5.2\mu_B$	[3]
HoSb	Cubic O_h^5	II type MO $\mu \parallel [100]$ $(2a,\ 2a,\ 2a)$	5–9	$\theta = -2.4\,K,\ p = 10.8$ $n_s = 8.7,\ \mu_0 H_c = 1.5\,T$ $\mu_{Ho^{3+}} = 9.3\mu_B$	[3,17]

*78 Magnetic structure consists of three groups of mutually orthogonal magnetic sublattices inserted one into the other and aligned with or opposed to the axes [100], [010], and [001].

Table 28.1 Certain properties of antiferromagnets *(continued)*

Substance	Structure		T_N, K	Additional information	References
	crystallographic	magnetic			
HoD_2	Cubic O_h^5	NCS	8	$p = 9.9$	[3]
HoC_2	Tetragonal D_{4h}^{17}	HS, $\mathbf{k}_0 \| [a]$	26	$\mu_{Ho}(4.2\,K) = 6.89\mu_B$	[3]
$HoCu_2$	Orthorhombic D_{2h}^{28}		9	$\theta = -6\,K,\ p = 10.5$	[13,17]
$HoGa_2$	Hexagonal D_{6h}^1	CS, $\mu \| [a]$ $(2a,\ \sqrt{3}a,\ c)$	10	$\theta = -2\,K,\ p = 10.7$ $\mu_{Ho}(4.2\,K) = 9.5\mu_B$	[3,13]
$HoAg_2$	Tetragonal D_{4h}^{17}	Incommensurate SW *79 propagating along the a-axis	5–8	$\theta = -13.5\,K$	[3,13]
$HoAu_2$	Tetragonal D_{4h}^{17}	NCS	9	$\theta = -8.0\,K,\ p = 10.97$	[3,13]
$HoIn_3$	Cubic O_h^1	FI in (110) AFI between (110) $(2a,\ a,\ 2a)$	11	$\theta = -18\,K,\ p = 10.65$ $\mu_{Ho^{3+}}(4.2\,K) = 9\mu_B$	[3,13]
α-Ho_2C_3	Cubic T_d^6	$\mu \| [111]$	19	$p = 8.7,\ \mu_{Ho} = 7.3\mu_B$	[3]
$HoCoO_3$	Orthorhombic D_{2h}^{16}	NCS $\alpha(\mu, [a]) = \pm 63°$	2.4	The Co ions are diamagnetic due to strong crystal field $\mu_{Ho}(1.5\,K) = 7.0\mu_B$	[3]
Ho_2O_2S	Trigonal D_{3d}^3	CS, $\mu \| c$ $(a,\ 2a,\ c)$	2.5	$\theta = -8\,K$ $\mu_{Ho}(1.5\,K) = 7.9\mu_B$	[3]
Ho_2O_2Se	Trigonal D_{3d}^3	CS, $\mu \| c$, $(\sqrt{3}a,\ a,\ c)$	4	$\mu_{Ho}(1.5\,K) = 9.3\mu_B$	[3]
Er	Hexagonal D_{6h}^4	At $T > 52$ K*80 $\mu \| c$, at 20 K $< T <$ 52 K $\mu \| c$ and $\mu \perp c$	79–86	Ferrimagnetic coil at $T < T_C = 20\,K$	[3,12,14]
ErAl	Orthorhombic D_{2h}^{11}	NCS $(a,\ 2b,\ c)$	10–13	$\theta = 25\,K,\ p = 9.7$ $\mu_{Er}(4.2\,K) = 7.0\mu_B$	[3,13]
ErSi	Orthorhombic D_{2h}^{17}	CS $(2a,\ b,\ 2c)$	10	$\theta = -5\,K,\ p = 9.38$ $\mu_{Er^{3+}}(4.2\,K) = 6.2\mu_B$	[3]
ErP	Cubic O_h^5	II type MO $\mu \perp [111]$	3–4	$\theta \approx 0\,K,\ p = 9.3$ $n_s = 8.5$ $\mu_0 H_c = 0.52\,T,\ \mu_{Er} = 5.7\mu_B$	[3,17]
ErAs			3.5	$\theta = -1.5\,K,\ p = 9.6$ $n_s = 8.4,\ \mu_0 H_c = 1.05\,T$	[17]
ErSb	Cubic O_h^5	$\mu \perp [111]$ $(2a,\ 2a,\ 2a)$	3.5–3.7	$\theta = -3\,K,\ p = 9.6$ $n_s = 7.3$ $\mu_0 H_c = 1.2\,T,\ \mu_{Er^{3+}} = 7\mu_B$	[3,17]
ErRh	Cubic O_h^1	A-type MO $\mu \| (001)$, $(a,\ a,\ 2a)$	3.3	$\theta = -4\,K,\ p = 9.4$ $\mu_{Er}(1.2\,K) = 6.1\mu_B$	[3,13]
ErC_2	Tetragonal D_{4h}^{17}	A set of incommensurate SW at $T > 10$ K*81	19	$\mu_{Er}(2\,K) = 7.9\mu_B$	[3]

*79 At $T > T_r = 47\,K$, longitudinal polarization occurs at $T < T_r$, transverse polarization parallel to the c axis.

*80 At $52\,K < T < T_N$, longitudinal SW propagates along c with a half-period of seven atomic layers. At $T < 52\,K$, a magnetization component in the plane normal to the c axis is added.

*81 At $T < 10\,K$, a commensurate AFSW contributes to the magnetic structure.

Table 28.1 Certain properties of antiferromagnets *(continued)*

Substance	Structure		T_N, K	Additional information	References
	crystallographic	magnetic			
$ErCu_2$	Orthorhombic D_{2h}^{28}		11	$\theta = 4\,K,\ p = 9.35$	[13,17]
$ErAu_2$	Tetragonal D_{4h}^{17}	$\mu \parallel [b]$*82	6.7	$\theta = -4\,K,\ p = 9.45$ $\mu_{Er} = 9.2\mu_B$	[3,13]
α-$ErAl_3$	Trigonal D_{3d}^5	CS, $\mu \parallel c$, $(2a,\ 2a,\ c)$	5–6	$\mu_{Er^{3+}}(2.15\,K) = 6.15\mu_B$	[3,13]
β-$ErAl_3$	Cubic O_h^1	C-type MO $\mu \parallel (001)$ $(2a,\ 2a,\ a)$	5	$\theta = -16\,K,\ p = 9.87$ $n_s = 6.25$ $\mu_{Er^{3+}}(4.2\,K) = 5.1\mu_B$	[3]
$ErIn_3$	Cubic O_h^1	C-type MO $(2a,\ 2a,\ a)$	6	$\theta = -10\,K,\ p = 9.75$ $n_s = 7.3$	[3,13]
Er_2O_3	Cubic T_h^7	NCS	3.4	$\theta = -11\,K,\ p = 10.5$ $\mu_{Er^{3+}\ I}(1.25\,K) = 6.06\mu_B$ $\mu_{Er^{3+}\ II}(1.25\,K) = 5.36\mu_B$	[3,17,108]
Er_3Co	Orthorhombic D_{2h}^{16}	NCS (structure change at $T = 7\,K$)	13	$\theta = 20\,K$ $\mu_{Er\ I}(4.2\,K) = 6.4\mu_B$ $\mu_{Er\ II}(4.2\,K) = 6.4\mu_B$	[3,13]
Er_3Ni	Orthorhombic D_{2h}^{16}	NCS $\alpha(\mu, [a]) = 29°$ $(2a,\ 2b,\ c)$	5–9	$\theta = -6\,K,\ p = 9.8$ $\mu_{Er\ I}(4.2\,K) = 6.9\mu_B$ $\mu_{Er\ II}(4.2\,K) = 6.9\mu_B$ $\mu_{Ni}(4.2\ K) = 0$	[3,13]
ErOOH	Monoclinic C_{2h}^2	CS, $\mu \parallel [b]$		$\mu_{Er}(1.6\,K) = 7.14\mu_B$	[3]
$ErVO_3$	Orthorhombic D_{2h}^{16}	$\mu_V \parallel [b]$ $\mu_{Er} \parallel [c]$	20(V) 16(Er)	$\mu_V(4.2\,K) = 1.2\mu_B$ $\mu_{Er}(4.2\,K) = 4.25\mu_B$	[3]
$LaErO_3$	Orthorhombic D_{2h}^{16}	NCS	2.4	$p = 9.45$ $\mu_{Er}(1.5\,K) = 6.34\mu_B$	[3]
Tm	Hexagonal D_{6h}^4	$\mu \parallel c$ at $40\,K < T < T_N^{*83}$	56	Ferrimagnetic ordering at $T \leq T_C = 40\,K$	[12,14]
TmAl	Orthorhombic D_{2h}^{11}	NCS $(2a,\ b,\ c)$	11	$\theta = -2\,K$ $\mu_{Tm\ I}(4.2\,K) = 5.5\mu_B$ $\mu_{Tm\ II}(4.2\,K) = 5.5\mu_B$	[3]
TmSi	Orthorhombic D_{2h}^{17}	CS $(a,\ b,\ 2c)$	10	$\theta = 10\,K,\ p = 7.45$ $\mu_{Tm^{3+}}(4.2\,K) = 5.0\mu_B$	[3,13]
$TmAu_2$	Tetragonal D_{4h}^{17}		3.2–3.5	$p = 7.62$	[3,13]
Yb_2O_3	Cubic T_h^7	NCS $\mu_{Yb\ I} \parallel [111]$, $\mu_{Yb\ II} \parallel [101]$	2.3	$\mu_{Yb\ I}(1.25\,K) = 1.05\mu_B$ $\mu_{Yb\ II}(1.25\,K) = 1.86\mu_B$	[3]
Yb_2O_2S	Trigonal D_{3d}^3	CS $\mu \parallel [110]$	3	$\theta = -6.5\,K$ $\mu(1.5\,K) = 1.66\mu_B$	[3,17]
Yb_2O_2Se	Trigonal D_{3d}^3	CS, $\mu \perp c$		$\mu_{Yb}(1.5\,K) = 1.6\mu_B$	[3]
K_2ReCl_6	Cubic O_h^5	I type MO	12	$\mu_{Re^{4+}}(4.2\,K) = 2.6\mu_B$ $n_s = 2.6$	[3,17]

*82 Incommensurate transverse SW spreads along the a-axis ; At $T < 4\,K$, a commensurate antiferromagnetic SW is added.

*83 At $T_C < T < T_N$, a transverse SW propagates along the c-axis with half-period of seven atomic layers.

Table 28.1 Certain properties of antiferromagnets *(continued)*

Substance	Structure		T_N, K	Additional information	References
	crystallographic	magnetic			
K_2ReBr_6	Cubic O_h^5	I type MO, $\mu \parallel (001)$	15.3		[3]
K_2IrCl_6	Cubic O_h^5	III type MO $(a,\ a,\ 2a)$	3.0	$\theta = -32\,\text{K}$	[3,17,113]
Rb_2IrCl_6	Cubic O_h^5		1.8	$\theta = -17\,\text{K}$	[113]
Cs_2IrCl_6	Cubic O_h^5		0.5	$\theta = -4\,\text{K}$	[113]
$(NH_4)_2IrCl_6$	Cubic O_h^5		2.15	$\theta = -20\,\text{K}$	[113]
UN	Cubic O_h^5	I type MO	50–55	$\theta =$(-200)–(-300) K, $p = 3$ $\mu_{U^{3+}}(4.2\,\text{K}) = 0.75\mu_B$	[1,3,114]
UP	Cubic O_h^5		122–125	$\theta = 49\,\text{K},\ p = 3.3$	[17,114]
UGa	Rhombic		27		[114]
UAs	Cubic O_h^5	I type MO, $\mu \parallel [001]$	128	$\theta = 32\,\text{K},\ \mu_U = 1.89\mu_B$	[1,3,17,114]
USb	Cubic O_h^5	I type MO	213–246	$\theta = 95\,\text{K},\ p = 3.85$ $\mu_U(4.2\,\text{K}) = 2.6\mu_B$	[1,3,114]
UBi	Cubic O_h^5		285–290	$\theta = 115\,\text{K}$	[114]
UO_2	Cubic O_h^5	$\mu \parallel [111]$*84	28–31	$\mu_{U^{4+}} = 1.7\mu_B$	[3,17]
UP_2	Tetragonal D_{4h}^7	$\mu \parallel c$ $(a,\ a,\ 2c)$	203–206	$\theta = 30\,\text{K},\ p = 2.4$ $\mu_U(80\,\text{K}) = 2.0\mu_B$	[3,17]
UMn_2	Cubic O_h^7		260		[114]
UAs_2	Tetragonal D_{4h}^7	$\mu \parallel c$ $(a,\ a,\ 2c)$	283	$\theta = 34\,\text{K},\ p = 2.9$ $\mu_U(4.2\,\text{K}) = 1.6\mu_B$	[3,114]
USe_2			11–13		[114]
USb_2	Tetragonal D_{4h}^7	$\mu \parallel c$ $(a,\ a,\ 2c)$	206	$\theta = 18\,\text{K},\ p = 3.0$ $\mu_U(80\,\text{K}) = 0.94\mu_B$	[3,114]
UHg_2	Tetragonal C_{4h}^6		70		[114]
UBi_2	Tetragonal D_{4h}^7	$\mu \parallel c$	183	$\theta = -53\,\text{K},\ p = 3.4$ $\mu_U(80\,\text{K}) = 2.1\mu_B$	[3,17]
UGa_3	Cubic O_h^1	G-type MO $(2a,\ 2a,\ 2a)$	70	$\mu_U(4.2\,\text{K}) = 0.72\mu_B$	[3,114]
UIn_3	Cubic O_h^1	G-type MO $(2a,\ 2a,\ 2a)$	95–100	$\theta = -215\,\text{K}$	[3,114]
UTe_3			56		[17]
UTl_3	Cubic O_h^1	G-type MO $(2a,\ 2a,\ 2a)$	80–90	$\theta = -150\,\text{K}$ $\mu_U(4.2\,\text{K}) = 1.6\mu_B$	[3,114]
UPb_3	Cubic O_h^1	FI in (001) AFI between (001) $\mu \parallel [001]$, $(a,\ a,\ 2a)$	32		[3,114]
UPd_4	Cubic O_h^1	CS, $\mu \parallel (001)$ $(2a,\ 2a,\ a)$	10–30	$\theta = -130\,\text{K},\ p = 3.47$ $\mu_U \approx 0.8\mu_B$	[3]

*84 Noncollinear three-rays magnetic structure with atomic uranium magnetic moments aligned with spatial diagonals of a cube [115] is the most probable.

Table 28.1 Certain properties of antiferromagnets *(continued)*

Substance	Structure		T_N, K	Additional information	References
	crystallographic	magnetic			
UCu_5	Cubic T_d^2	CS, $\mu\perp(111)$ FI in (111) AFI between (111)	15–16	$\mu_U(4.2\,K) = 0.9\mu_B$	[3,114]
U_2N_3	Cubic T_h^7		94–96		[114]
UOS	Tetragonal D_{4h}^7	$\mu\parallel c$, $(a,\ a,\ 2c)$	55	$\theta = -51\,K$, $\mu_U(4.2\,K) = 1.9\mu_B$	[3]
UOSe	Tetragonal D_{4h}^7	$\mu\parallel c$	72	$\theta =(-20)$–(-130) K $p = 2.87$, $\mu_U(4.2\,K) = 2.2\mu_B$	[3]
UOTe	Tetragonal D_{4h}^7	$\mu\parallel c$	157–160	$\theta = (-56) - (-60)$ K $\mu_U(78\,K) = 2.7\mu_B$	[3]
U_2N_2P	Trigonal D_{3d}^3	CS, $\mu\parallel c$	366	$\mu_U(4.2\,K) = 1.7\mu_B$	[3]
U_2N_2S	Trigonal D_{3d}^3	CS, $\mu\parallel c$	233	$\mu_U(4.2\,K) = 1.3\mu_B$	[3]
U_2N_2As	Trigonal D_{3d}^3	CS, $\mu\parallel c$, $(a,\ a,\ 2c)$	406	$\mu_U(4.2\,K) = 1.3\mu_B$	[3]
U_2N_2Se	Trigonal D_{3d}^3	CS, $\mu\parallel c$, $(a,\ a,\ 2c)$	245	$\mu_U(4.2\,K) = 2.2\mu_B$	[3]
NpC	Cubic O_h^5		310	$p = 3.37$, ferromagnet at $T < T_C = 200\,K$, $\mu_{Np} = 1.4\mu_B$[*85]	[17]
NpP	Cubic O_h^5		130		[114]
NpS	Cubic O_h^5	II type MO	23[*86]	$p = 2.1$, $\mu_{Np}(4.2\,K) = 0.9\mu_B$	[3]
NpAs	Cubic O_h^5	Longitudinal SW at $T > T_r = 150\,K$ I type MO $\mu\parallel[001]$ at $T < T_r$	177	$(\mu_{Np})_{max} = 1.4\mu_B$ at $T_r < T < T_N$ $\mu_{Np}(4.2\,K) = 2.5\mu_B$	[3]
NpSb	Cubic O_h^5	I type MO	207	$\mu_{Np}(4.2\,K) = 2.5\mu_B$	[3]
NpPt	Orthorhombic D_{2h}^{17}		27		[114]
NpO_2	Cubic O_h^5		25.3		[17]
$NpCo_2$	Cubic O_h^7		15	$\mu_{Np} = 0.5\mu_B$ $\mu_{Co} = 9.15\mu_B$	[3]
$NpAs_2$			180		[114]
$NpIr_2$	Cubic O_h^7		7.5		[114]
$NpPd_3$	Cubic O_h^1	CS, G-type MO $(2a,\ 2a,\ 2a)$	55	$\mu_{Np}(4.2\,K) = 2.0\mu_B$	[3,114]
$NpSn_3$	Cubic O_h^1		9.5		[114]
NpB_4			52.5		[114]
PuN	Cubic O_h^5		13		[114]
$PuRh_3$	Cubic O_h^1		6.2		[114]
$PuPd_3$	Cubic O_h^1		24		[114]
$PuBe_{13}$	Cubic O_h^6		11.5		[114]
Pu_3S_4	Cubic		10		[114]

[*85] In the ferromagnetic state.

[*86] At $T = T_N$, the phase transition of the first order occurs.

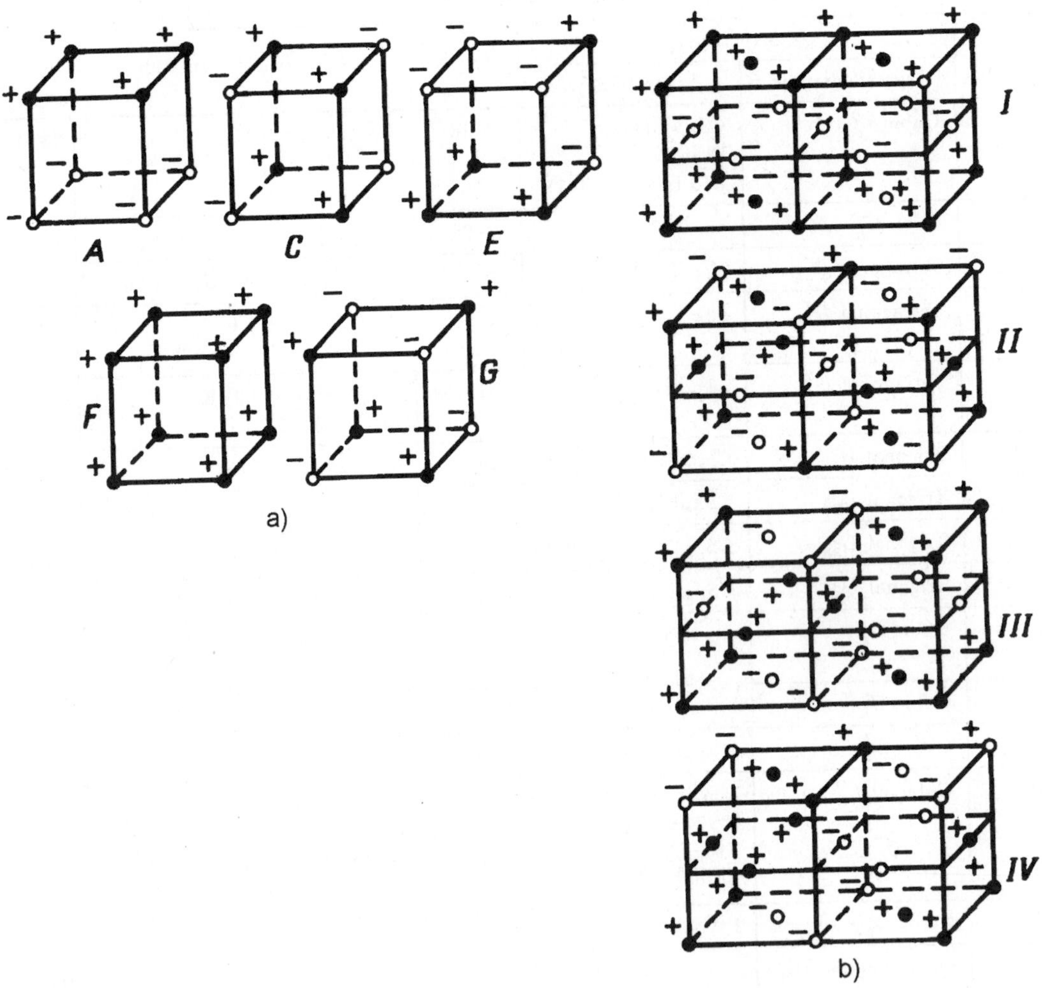

Figure 28.1 Possible types of magnetic order in a simple cubic (*a*) and face-centered cubic (*b*) lattices. Signs +, − correspond to parallel and anti-parallel orientation of ion magnetic moments.

References

[1] Vonsovskii, S. V., Magnetism, Nauka, Moscow, 1971 (in Russian).

[2] Borovik–Romanov, A. S., Itogi nauki, Ser. "Fiz.–mat. nauki", No 4, Izdatel'stvo Acad. nauk SSSR, Moscow, 1962 (in Russian).

[3] Oles, A., Kajzar, M., Kucab, W., Magnetic structures determined by neutron diffraction, Krakow panstwowe wydawnictwo naukowe, Warszawa, 1976.

[4] Krinchik, G. S., Physics of magnetic phenomena, Izdatel'stvo MGU, Moscow, 1985 (in Russian).

[5] Borovik–Romanov, A. S., Antiferromagnets with the easy-plane anisotropy, in: Problems of Magnetism, Nauka, Moscow, 1972 (p. 47) (in Russian).

[6] Aleksandrov, K. S., Fedoseeva, N. V., Spevakova, I. P., Magnetic phase transitions in halloid crystals, Nauka, Novosibirsk, 1983 (in Russian).

[7] Eremenko, V. V., Introduction to optical spectroscopy of magnetic substances, Naukova Dumka, Kiev, 1975 (in Russian).

[8] Turov, E. A., Petrov, M. P., Nuclear magnetic resonance in ferro- and antiferromagnets, Nauka, Moscow, 1969 (in Russian).

[9] Izyumov, Yu. A., Naish, V. E., Ozerov, R. P., Neutrons and solid state, vol.2, Atomizdat, Moscow, 1981 (in Russian).

[10] Turov, E. A., Physical properties of magnetically ordered crystals, Izdatel'stvo Akad. Nauk SSSR, Moscow, 1963 (in Russian).

[11] Zvezdin, A. K., Matveev, V. A., Mukhin, A. A., Popov, A. I., Rare-earth ions in magnetically ordered crystals, Nauka, Moscow, 1985 (in Russian); Belov, K. P.,Zvezdin, A. K., Kadomtseva, A. M., Levitin, R. Z., Orientational transitions in rare earth magnetic materials, Nauka, Moscow, 1979 (in Russian).

[12] Chikasumi, S., Physics of ferromagnetism, Syokabo, Tokyo,1980.

[13] Buschow, K. H. J., in: Ferromagnetic materials, vol.1, Ed. Wohlfarth, E. P., North-Holland Publ. Comp., Amsterdam–New York–Oxford, 1980, p. 297.

[14] Legvold, S., in: Ferromagnetic materials, vol. 1, Ed. Wohlfarth, E. P., North-Holland Publ. Comp., Amsterdam–New York–Oxford, 1980, p. 183.

[15] Keffer, F., Handbuch der Physik, Bd 18/2, Springer, Berlin, 1966.

[16] Morrish, A. H., The physical principles of magnetism, John Wiley and Sons Inc., New York, 1965.

[17] Ozhogin, V. I., Shapiro, V. G., in: Tables of physical quantities, Handbook, ch.30, Ed. Kikoin, I. K., Atomizdat, Moscow, 1976 (in Russian).

[18] Hurd, C. M., Contemp. Phys., 23, 469, 1982.

[19] Bizette, H., Tsai, B., Compt. Rend., 238, 1575, 1954.

[20] Stout, J. W., Adams, H. E., J. Amer. Chem. Soc., 64, 1535, 1942.

[21] Jacobs, J. S., J. Appl. Phys., 32S, 61, 1961.

[22] Shapira, Y., Foner, S., Phys. Rev. B1, 3083, 1970.

[23] Low, G. G., Okazaki, A., Stevenson, R. W. H., J. Appl. Phys., 35, 998, 1964.

[24] Johnson, F. M., Nethercot, A. H. Jr., Phys. Rev., 114, 705, 1959.

[25] Eremenko, V. V., Belyaeva, A. I., Uspekhi fizicheskikh nauk, 98, 27, 1969 (in Russian).

[26] Allen, S. J. Jr., Loudon, R., Richards, P. L., Phys. Rev. Lett., 16, 463, 1966.

[27] Green, R. L., Sell, D. D., Yen, W. M., Phys. Rev. Lett., 15, 656, 1965.

[28] Kharchenko, N. F., Eremenko, V. V., Physics of a condensed state, No XIII, Izdatel'stvo instituta nizkihk temperatur, Khar'kov, 1971 (in Russian).

[29] Borovik–Romanov, A. S., Zhurnal eksperimental'noi i teoreticheskoi fiziki, 36, 766, 1959 (in Russian).

[30] Ozhogin, V. I., Zhurnal eksperimental'noi i teoreticheskoi fiziki, 58, 2079, 1970 (in Russian).

[31] Borovik–Romanov, A. S., Prozorova, L. A., Zhurnal eksperimental'noi i teoreticheskoi fiziki, 55, 1727, 1968 (in Russian).

[32] Borovik–Romanov, A. S., Rudashevskii, E. G., Zhurnal eksperimental'noi i teoreticheskoi fiziki, 47, 2095, 1964 (in Russian).

[33] Borovik–Romanov, A. S., Kreines, N. M., Prozorova, L. A., Zhurnal eksperimental'noi i teoreticheskoi fiziki, 45, 64, 1963 (in Russian).

[34] Ozhogin, V. I., Maximenkov, P. P., Digests of Intermag. Conf., Kyoto, Japan, 1972, p. 494; Seavey, M. H., Solid State Commun., 10, 219, 1972.

[35] Borovik–Romanov, A. S., Kreines, N. M., Zhotikov, V. G., Problems of magnetic resonance, Nauka, Moscow, 1978 (in Russian).

[36] Borovik–Romanov, A. S., Prozorova, L. A., J. Physique, 32, C1, 1971.

[37] Ozhogin, V. I., Yakubovskii, A. Yu., Zhurnal eksperimental'noi i teoreticheskoi fiziki, 63, 2155, 1972 (in Russian).

[38] Ozhogin, V. I., Inyushkin, A. V., Babushkina, N. A., J. Magnetism and Magnetic Materials, 31–34, 147, 1983 (in Russian).

[39] Jacobs, J. S., Lawrence, P., J. Appl. Phys., 35, 996, 1964.

[40] Gillot, M., Eremenko, V. V., Marchand, A., Proc. Intern. Symp. on High-Field Magnetism, Ed. Date, M., Osaka, 1982, North–Holland Publ. Comp., Amsterdam–New York–Oxford, 1983, p. 63.

[41] Dzyaloshinskii, I. E., Zhurnal eksperimental'noi i teoreticheskoi fiziki, 32, 1547, 1957 (in Russian).

[42] Borovik–Romanov, A. S., Ozhogin, V. I., Zhurnal eksperimental'noi i teoreticheskoi fiziki, 39, 27, 1960 (in Russian).

[43] Ozhogin, V. I., Zhurnal eksperimental'noi i teoreticheskoi fiziki, 45, 1687, 1963 (in Russian).

[44] Ozhogin, V. I., Shapiro, V. G., Zhurnal eksperimental'noi i teoreticheskoi fiziki, 54, 96, 1968 (in Russian).

[45] Kreines, N. M., Zhurnal eksperimental'noi i teoreticheskoi fiziki, 40, 762, 1961 (in Russian).

[46] Borovik–Romanov, A. S., Zhurnal eksperimental'noi i teoreticheskoi fiziki, 38, 1088, 1960 (in Russian).

[47] Borovik–Romanov, A. S., Physical encyclopedic dictionary, Sovetskaya Entsiklopedia, 1983 (in Russian).

[48] Astrov, D. N., Zhurnal eksperimental'noi i teoreticheskoi fiziki, 40, 1035, 1961 (in Russian).

[49] Hornreich, R. M., Digests of Intermag. Conf., Kyoto, Japan, 1972, p. 404.

[50] Schmid, H., Intern. J. Magnetism, 4, 337, 1974.

[51] Ozhogin, V. I., Izvestiya Akad. Nauk SSSR, Ser. "Fizika," 42, 1625, 1973 (in Russian).

[52] Buzdin, A. I., Bulaevskii, L. N., Uspekhi fizicheskikh nauk, 144, 415, 1984 (in Russian).

[53] Mortensen, K., Tomkiewicz, Y., Bechgaard, K., Phys. Rev., B25, 3319, 1982.

[54] Scott. I. S., J. Appl. Phys., 53, 1845, 1982.

[55] Walsh, W. M. Jr., Wudl, F., Aharon-Shalom, E., Phys. Rev. Lett., 49, 885, 1982.

[56] Smolenskii, G. A., Nedlin, G. M., Physics of magnetic insulators, Nauka, Leningrad, 1974 (in Russian).

[57] Belov, K. P., Sokolov, V. I., Uspekhi fizicheskikh nauk, 121, 285, 1977 (in Russian).

[58] Handbook on the physics and chemistry of rare-earths, Eds.Gschneider, K., Eyring, Le Roy, North-Holland Publ. Comp., Amsterdam-New York-Oxford, 1979.

[59] Goral, J. P., Greedan, J. E., J. Magnetism and Magnetic Materials, 37, 315, 1983.

[60] Wiedenmann, A., Venien, J. P., Palvadeau, P., J. Phys., C16, 5339, 1983.

[61] Kondorskii, E. I., Kostina, T. I., Galkina, V. Yu., Zhurnal eksperimental'noi i teoreticheskoi fiziki, 69, 1753, 1975 (in Russian).

[62] Nishihara, Y., Yamaguchi Y., Waki, S., J. Phys. Soc. Jpn., 52, 2301, 1983.

[63] Tripathi, A. K., Lal, H. B., Indian J. Pure and Appl. Phys., 20, 271, 1982.

[64] Krupicka, S., Physik der ferrite und der verwandten magnetischen oxide, Academia, Praha, 1973.

[65] Yoshimura, K., Nakamura, Y., J. Magnetism and Magnetic Materials, 40, 55, 1983.

[66] Oleveira, N. F., Shapira, Y., J. Appl. Phys., 50, 1790, 1979.

[67] Gurewitz, E., Horowitz, A., Shaked, H., Phys. Rev., 20, 4544, 1979.

[68] Zandbergen, H. W., J. Solid State Chem., 37, 189, 1981.

[69] Rimet, R., Buder, R., Schlenker, C., Solid State Commun., 37, 693, 1981.

[70] Kotlicki, A., McLeod, B. A., Shott, M., Phys. Rev., 29, 26, 1984.

[71] Phaff, A. C., Swuste, C. H. W., Kopinga, K., J. Phys., C16, 6635, 1983.

[72] Simizu, S., Chen, I. Y., Fridberg, S. A., J. Appl. Phys., 55, 2398, 1984.

[73] Takeda, K., Koyama, K., J. Phys. Soc. Jpn., 52, 648, 1983.

[74] Lhao, M. G., Du, M. L., Phys. Rev., B28, 6481, 1983.

[75] Ozhogin, V. I., Shapiro, V. G., Gurtovoy, K. G., Zhurnal eksperimental'noi i teoreticheskoi fiziki, 62, 2221, 1972 (in Russian).

[76] White, R., Uspekhi fizicheskikh nauk, 103, 593, 1971 (in Russian).

[77] Wada, N., Ubukoshi, K., Hirakawa, K., J. Phys. Soc. Jpn., 51, 2833, 1982.

[78] Renaudin, J., Pannetier, J., Peland, S., Solid State Commun., 47, 445, 1983.

[79] McAlister, S. R., Strobel, P., J. Magnetism and Magnetic Materials, 30, 340, 1983.

[80] Belov, N. V., Golovastikov, N. I., Ivashchenko, A. N., Kristallographia, 27, 511, 1982 (in Russian).

[81] Sugano, T., Kinoshita, M., Shirotani, I., Solid State Commun., 45, 99, 1983.

[82] Jaccarino, V., Phys. Rev., B2, 163, 1959.

[83] Gurtovoy, K. G., Lagutin, A. S., Ozhogin, V. I., Zhurnal eksperimental'noi i teoreticheskoi fiziki, 83, 1941, 1982 (in Russian).

[84] Bazhan, A. N., Zhurnal eksperimental'noi i teoreticheskoi fiziki, 67, 1520, 1974 (in Russian).

[85] Kotyuzhanskii, B. Ya., Prozorova, L. A., Svistov, L. E., Zhurnal eksperimental'noi i teoreticheskoi fiziki, 88, 221, 1985 (in Russian).

[86] Brown, P. J., Welford, P. J., Forsyth, J. B., J. Phys., C6, 1405, 1973.

[87] Tsuda, T., Yasuoka, H., J. Phys. Soc. Jpn., 45, 115, 1978.

[88] Breed, D. I., Gilianse, K., Miedema, A. R., Physica, 45, 205, 1961.

[89] Bartolome, P., Navarro, R., Gonzales, D., Physica, 92B, 45, 1977.

[90] Carlin, R. L., van der Bilt, A., Jong, K. O., Physica, 111BC, 147, 1981.

[91] Burgiel, J. S., Jaccarino, V., Schawlow, A. L., Phys. Rev., 122, 429, 1961.

[92] Bastow, T. J., Whitfield, H. J., Bristow, G. K., Phys. Lett., 84A, 266, 1981.

[93] Adachi, A., Achiva, N., Mekata, M., J. Phys. Soc. Jpn., 49, 545, 1980.

[94] Smirnov, A.I., Khlustikov, I.N., Pis'ma v ZhETF, 59, 783, 1994.

[95] Phaff, A. C., Swuste, C. H. W., de Jonge, W. J. H., J. Phys., C17, 2583, 1984.

[96] Vasyukov, V. N., Zhuravlev, A. V., Lukin, S. N., Fizika tverdogo tela, 26, 1297, 1984 (in Russian).

[97] Barbara, B., Boucherle, J. X., Buevoz, J. L., Solid State Commun., 24, 481, 1977.

[98] Aarts, J., de Boer, F. R., Horn, S., Physika, 107B, 381, 1981.

[99] Hanzawa, K., Kasuya, T., J. Phys. Soc. Jpn., 53, 1809, 1984.

[100] Yakinthous, J. K., Routsi, Ch., Penelope Schobinger–Paramantellous, J. Phys. Chem. Solids, 45, 689, 1984.

[101] Ellingsen, O. S., Bratsberg, H., Mroczkowsky, S., J. Appl. Phys., 53, 7948, 1982.

[102] Hamaker, H. C., Woolf, L. D., MacKay, H. B., Solid State Commun., 31, 139, 1979.

[103] Hamaker, H. C., Woolf, L. D., Mac Kay, H. B., Solid State Commun., 32, 289, 1979.

[104] Lee, K., Muir, H., J. Appl. Phys., 36, 1043, 1965.

[105] Gaur K., Tripathi, A. K., Lal, H. B., J. Materials Sci. Lett., 2, 161, 1983.

[106] Ishikawa, M., Muller, J., Solid State Commun., 27, 761, 1978.

[107] Buzdin, A. I., Bulaevskii, L. N., Kulich, M. L., Uspekhi fizicheskikh nauk, 144, 597, 1984 (in Russian).

[108] Taylor, K., Darby, M., Physics of rare-earth solids, Wiley, London, 1972.

[109] Thomlinson, W., Shirane, G., Moncton, D. E., J. Appl. Phys., 50, 1981, 1970.

[110] Ali, N., Woods, S. B., J. Low-Temp. Phys., 56, 575, 1984.

[111] Schienle, M., Kasten, A., Muller, P. H., Phys. Status Solidi (b), 119, 611, 1983.

[112] Moncton, D. E., Shirane, G., Thomlinson, W., Phys. Rev. Lett., 41, 1133, 1978.

[113] Raaen, A. M., Svare, J., Pedersen, V., Physica, 121B, 89, 1983.

[114] Trezebiatowski, W., in: Ferromagnetic materials, Ed. Wohlfarth, E. P., North-Holland Publ. Comp., Amsterdam–New York–Oxford, 1980, p. 415.

[115] Izyumov, Yu. A., Uspekhi fizicheskikh nauk, 131, 387, 1980 (in Russian).

[116] Lottermoser, W., Muller, R., Fuess, M., J. Magnetism and Magnetic Materials, 54-57, 1005, 1986.

[117] Borovik–Romanov, A. S., Kotyuzhanskii, B. Ya., Prozorova, L. A., Zhurnal eksperimental'noi i teoreticheskoi fiziki, 58, 1911, 1970 (in Russian).

[118] Aleksandrov, K. S., Anistratov, A. T., Mel'nikova, S. V., Fizika tverdogo tela, 21, 1119, 1979 (in Russian).

[119] Buyers, W. J. L., Morrs, R. M., Armstrong, R. L., Phys. Rev. Lett., 56, 371, 1986.

[120] Cox, D. E., Shapiro, S. M., Cowley, R. A., Phys. Rev., B19, 5754, 1979.

[121] Valyanskaya, T. V., Plakhtii, V. P., Sokolov, V. I., Zhurnal eksperimental'noi i teoreticheskoi fiziki, 70, 2279, 1976 (in Russian).

[122] Prozorova, L. A., Marchenko, V. I., Krasnyak, Yu. V., Pis'ma ZhETF, 41, 522, 1985 (in Russian).

[123] Felner, I., et al., Phys. Rev., B48, 16040, 1993.

[124] Smirnov, A.I., Khlustikov, I.N., Zhurnal experimental'noi i teoreyicheskoi fiziki, 105, 1040, 1994.

[125] Golovenchitz, E.I., et al., Zhurnal expewrimental'noi i teoreticheskoi fiziki, 107, April, 1995.

29

Ferrites and other Magnetic Dielectrics

M.V. Bystrov, V.L. Ivashintsova, S.A. Mironov, and R.V. Pisarev

29.1 Introduction

29.1.1 Magnetic properties

Ferrimagnetic substances, or, alternatively, ferrites, form a separate group among the magnetically ordered materials. In contrast to the ordinary ferromagnetics or antiferromagnetics, which are characterized by the magnetic atoms being placed in the translationally-equivalent sites, ferrites include materials in which nonequivalent sublattices (in crystallographic or magnetic respect) exist. With this definition of ferrimagnetism, the ferromagnetic is a particular case of ferrimagnetic with one magnetic sublattice and the simple antiferromagnetic is the one with two equivalent sublattices. The presence of nonequivalent sublattices gives rise to a wide range of magnetic properties of ferrimagnetics, which are different from those of ferro- and antiferromagnetic ones, although in some circumstances one can find common features in these different groups of magnetic substances.

Magnetic properties of ferrimagnetics were first explained by Néel [1] on the basis of the two-sublattice model that he suggested for ferrites with the spinel structure (see below). In this model magnetic ions occupy the tetrahedral positions (sites A) and in the octahedral positions (sites B). The main interaction constitutes the antiferromagnetic (negative) interaction between the ions of different sublattices, which causes an antiparallel state of their magnetic moments. In this case the total magnetization (per 1 m^3 of the substance) can be represented by the difference of sublattices' magnetizations:

$$M_s(T,H) = M_B(T,H) - M_A(T,H). \tag{29.1}$$

The M_s quantity performance as a function of temperature and field intensity can be more complex than that of ferromagnetics and the character of the M_A and M_B changes with temperature and field intensity can be different. For example, when temperature rises, M_s can monotonously decrease and become zero at the Curie point T_C, above which the substance is paramagnetic, though the paramagnetic susceptibility varies with temperature according to the law different from the Curie law for the simple paramagnetics. When temperature rises in the range below T_C, the spontaneous magnetization increase is also possible in the definite temperature interval. For some ferrites (particularly for many rare-earth ferrites, which are spoken of as garnets; see Table 29.15 and Fig. 29.22) the compensation temperature $T_{\rm comp}$ exists at which magnetizations of sublattices become equal and total magnetization reduces to zero. The compensation point appearance is also possible when the composition of ferrimagnetics changes. Particularly, it is observed in the yttrium–ferro–gallium garnets.

Specific saturation magnetization, $\sigma_s = M_s/d$ (where d is the density), is also used for describing the magnetic properties of ferrites. The limiting value of σ_s when temperature tends to zero is designated by σ_s^0. The relation between σ_s^0 and the number of Bohr magnetons n_B per one formula unit of a

0-8493-2861-6/97/$0.00+$.50

ferrite is given by the expression

$$n_B = \frac{M}{N_A \mu_B} \sigma_s^0, \tag{29.2}$$

where μ_B is the Bohr magneton; M is the molecular mass corresponding to one formula unit; N_A is the Avogadro number. Magnetic moment p_m^0 per molecule at $T = 0$ is determined as follows:

$$p_m^0 = (n_B - n_A)\mu_M, \tag{29.3}$$

where n_A, n_B are the numbers of Bohr magnetons per atom in the A and B positions, accordingly.

The magnetic interaction energy that depends on the magnetization orientation with respect to crystallographic axes is referred to as the crystallographic magnetic anisotropy energy. For cubic crystals (to which ferrites with the structure of spinel and garnet belong) this energy is usually written in the form [2]:

$$E_a = K_1(\alpha_1^2\alpha_2^2 + \alpha_2^2\alpha_3^2 + \alpha_3^2\alpha_1^2) + K_2\alpha_1^2\alpha_2^2\alpha_3^2 + \dots, \tag{29.4}$$

where K_1, $K_2 \dots$ are the anisotropy constants; α_1, α_2, α_3 are the direction cosines of the magnetization vector with respect to the axes coinciding with the cube edges. Magnetic anisotropy can be characterized by an effective internal field with the intensity of $H_a = 2K_1/M_s$.

Under the influence of magnetostriction the crystal lattice deforms with the change of magnetization. In the particular case of ferrite with cubic structure, the relative magnetostrictive deformation can be represented in the following form:

$$\frac{\Delta l}{l} = \lambda_s = \frac{3}{2}\lambda_{100}(\alpha_1^2\beta_1^2 + \alpha_2^2\beta_2^2 + \alpha_3^2\beta_3^2 - \frac{1}{3}) + 3\lambda_{111}(\alpha_1\alpha_2\beta_1\beta_2 + \alpha_2\alpha_3\beta_2\beta_3 + \alpha_1\alpha_3\beta_1\beta_3), \tag{29.5}$$

where λ_{100} and λ_{111} are the constants of saturation magnetostriction in the [100] and [111] directions, respectively; β_1, β_2, β_3 are the direction cosines of the axis along which deformation is measured.

When ferrite is placed into a constant magnetic field with intensity H_0 and the RF magnetic field (which is perpendicular to the constant one) it absorbs the microwave energy. This absorption is of a resonant character (ferromagnetic resonance) and shows maximum at the frequency ω_0, which is related to the field H_0 in a particular way. The dependence of frequency ω_0 on H_0 is rather complicated and is determined by magnetic crystallographic anisotropy, by shape anisotropy, by the elastically stressed state of a sample, and so on [3]. In the simplest case of an isotropic sphere it takes the form:

$$\omega_0 = \gamma H_0. \tag{29.6}$$

Here γ is the gyromagnetic ratio

$$\gamma = g\frac{e}{2mc}, \tag{29.7}$$

where e is the electron charge; m is the electron rest mass; c is the speed of light; g is the spectroscopic splitting factor. For the spinel-structure ferrites with two sublattices, the g-factor value is given by the following expression [4]:

$$g = \frac{M_B - M_A}{M_B/g_B - M_A/g_A}, \tag{29.8}$$

where g_A and g_B are the spectroscopic splitting factors for ions in the A and B cites accordingly.

The amplitude and form of the resonant absorption curve is determined by the relaxation processes. Their presence leads to complex-valued tensor components of the magnetic permeability. When the external field is absent, the magnetic permeability is a scalar. The width of the ferromagnetic resonance curve ΔH is usually defined as the difference between the field strengths at which the imaginary part of the diagonal component of the permeability tensor μ'' becomes one half of its value μ''_{res} at the resonance point. The dependences of its real μ' and imaginary μ'' parts on frequency are known as magnetic spectra. Two regions of dispersion are characteristic of the magnetic spectra of ferrites. A low-frequency region is due to the shift of the domain boundaries, and a high-frequency region is caused by the "natural" ferromagnetic resonance in effective anisotropy fields and demagnetization fields.

29.1.2 Elastic properties and magnetoelastic interaction

Magnetoacoustic phenomena in ferrites appear as a result of interaction between magnetic ions spins and the elastic vibrations of crystal lattice, i.e. as a result of the same interactions, which determine magnetostrictive phenomena. The expression for the elastic and magnetoelastic energy can be written in the following way:

$$E_{\text{el.}} + E_{\text{m-el.}} = \frac{1}{2}c_{11}(e_{xx}^2 + e_{yy}^2 + e_{zz}^2)$$

$$+\frac{1}{4}c_{44}(e_{xy}^2 + e_{yz}^2 + e_{zx}^2) + c_{12}(e_{xx}e_{yy} + e_{yy}e_{zz} + e_{zz}e_{xx}) \tag{29.9}$$

$$+B_1(\alpha_x^2 e_{xx} + \alpha_y^2 e_{yy} + \alpha_z^2 e_{zz}) + B_2(\alpha_x\alpha_y e_{xy} + \alpha_y\alpha_z e_{yz} + \alpha_z\alpha_x e_{zx}),$$

where $c_{11} = c_{1111}$, $c_{12} = c_{1122}$, $c_{44} = c_{2323}$ are the coefficients of elastic rigidity; $B_1 = b_{1111}$, $B_2 = b_{2323}$ are the magnetoelastic coefficients.

The conditions for elastic and magnetoelastic isotropy are formulated as follows:

$$c_{11} - c_{12} = 2c_{44};\ B_1 = B_2. \tag{29.10}$$

The first condition is true, for example, in the yttrium garnet ferrite with an accuracy of 5%. The values of coefficients determining elastic and magnetoelastic properties are listed in Tables 29.18, 29.21, and 29.22.

29.1.3 Optical and magnetooptical properties

Ferrites possess relatively high transmittance in a number of near- and far-infrared spectral regions. Garnet ferrites are characterized by a higher transmittance than spinel ferrites. For example, in yttrium garnet ferrites the transparency windows appear at wavelengths of $1 < \lambda < 0.1$ mm and $1 < \lambda < 10$ μm. A strong lattice absorption is observed between these two regions. In the first region of transparency of the rare-earth garnet ferrites one can observe absorption due to ferromagnetic resonance (if the anisotropy field is strong) resulting from the exchange resonance of rare-earth sublattice in the field of iron sublattices as well as due to electronic transitions between the levels of the main multiplet in rare-earth ions. In the second region, electronic transitions are observed in the rare-earth ions and (at shorter wavelengths) in iron ions in octahedral and tetrahedral positions. Garnet ferrites display substantial Faraday effect in visible and near infrared spectral regions, when light propagates along the magnetization vector, and approximately the same in absolute value Cotton-Mouton effect (magnetic linear double refraction), when light propagates perpendicular to the magnetization vector [109, 110].

Below, the distinguishing features in the structure and magnetic properties of ferrites of different groups are considered. These groups are most interesting from the technical and scientific points of view and they involve ferrites with the structure of spinel, garnet, and hexaferrites. Also some data are presented on the chalcogenide spinels (which possess ferromagnetic and antiferromagnetic properties) as well as data on ferromagnetic and antiferromagnetic chalcogenides of europium and some other ferromagnets with various structure. Properties of a large and important class of orthoferrites were reviewed in the previous chapter on antiferromagnets.

29.2 Spinel Ferrites

Spinel ferrites exhibit a crystal structure of the spinel Al_2MgO_4 mineral and show a the chemical formula $Me^{3+}Fe_2^{2+}O_4$, where Me^{2+} is the bivalent metal ion and the iron ions Fe^{3+} are trivalent. In the case of simple ferrites, Me refers to one of the bivalent ions of the transition elements, for example Mn, Ni, Co, or Mg. Combination of these ions (solid solutions of ferrites or mixed ferrites) are also possible. Trivalent ions of iron in $MeFe_2O_4$ can be fully or partially substituted by other trivalent ions, for example, by Al^{3+} or Cr^{3+} (mixed ferrite-aluminates or ferrite-chromates).

In the the spinel-type structure, oxygen ions form the face-centered cubic lattice with the edge length a. Between oxygen ions, the ions of metals are placed, the latter being surrounded by four or six oxygen ions. Such surroundings are called the tetrahedral (or A) and octahedral (or B) positions, accordingly. Spinels in which Me^{2+} ions lie in tetrahedral positions and Fe^{3+} ions lie in octahedral positions are termed normal. If ions Me^{2+} and half of the Fe^{3+} ions are in the octahedral positions, then the spinel is described as inverted. At mixed distribution, bivalent metal ions are to be found both in A and B positions.

It is agreed that ions in tetrahedral positions are written in the ferrite formula before the square brackets, and ions in octahedral positions in brackets. Then, for example, the chemical formula of zinc ferrite with the structure of a normal spinel will be written in the form $Zn^{2+}[Fe_2^{3+}]O_4$, the formula of nickel ferrite with the structure of an inverted spinel has to be written in the form $Fe^{3+}[Ni^{2+}Fe^{3+}]O_4$, and, finally, ion distribution in mixed manganic spinel is described by the formula $Fe_{0.2}^{3+}Mn_{0.8}^{2+}[Mn_{0.2}^{2+}Fe_{1.8}^{3+}]O_4$.

A collection of properties of spinel ferrites is presented in Tables 29.1–29.9 and in Figs. 29.1–29.19.

29.2.1 Simple Ferrites

Table 29.1 Main parameters of simple ferrites with a spinel structure [pc – polycrystalline; sc – single crystal; tetr – tetragonal; cub – cubic.]

Parameter*1	Fe_3O_4	$MgFe_2O_4$	$MnFe_2O_4$	$CuFe_2O_4$
Lattice constant a, nm	0.839 [5]	0.836 [50]	0.8517 [23]	0.837 (cub)*2 [34] c=0.870 (tetr) [35] a=0.822 (tetr) [35]
Cation distribution	Fe[Fe^{2+}Fe] [6]	$Mg_{0.1}Fe_{0.9}\cdot 2.3$ [$Mg_{0.9}Fe_{1.1}$] [50,51]	$Mn_{0.8}Fe_{0.2}$ [$Mn_{0.2}Fe_{1.8}$] [23]	Fe[CuFe] [36]
Oxygen parameter u	0.379 [6]	0.381 [50]	0.385 [23]	0.380 [112]
X-ray density d_X, 10^3 kg/m^3	5.24 [6]	4.52 [50]	5.0 [23]	5.35 [12]
Curie temperature T_C, K	858 [5]	713 [5]	573 [5]	728 [5]
Magnetic moment per molecule p_m^0, μ_B	4.1 (0 K) [6]	1.1 (25 K) [52]	4.6 (0 K) [23]	2.3(0 K) (cub) [37] 1.3(0 K) (tetr) [37]
Specific saturation magnetization σ_s, A·m^2/kg	98 (0 K) [5] 92 (300 K) [5]	31 (0 K) [58] 27 (300 K) [5]	112 (0 K) [58] 80 (300 K) [5]	30 (0 K) [5] 25 (300 K) [5]
Saturation magnetization M_s, kA/m	510 (0 K) [5] 477 (300 K) [5]	143 (0 K) [5] 119 (300 K) [5]	557 (0 K) [5] 398 (300 K) [5]	158 (0 K) [5] 135 (300 K) [5]
The first anisotropy constant K_1, 10^3 J/m^3	−10.7 [7]	−3.5 [53] −2.5 [54]	−18.7 (77 K) [24] −2.8 (300 K) [24]	−20.6 (77 K) [38] −6.3 (300 K) [38]
The second anisotropy constant K_2, 10^3 J/m^3	−2.8 [7]		−0.3 (77 K) [24] −0.2 (300 K) [24]	
Magnetostriction constant λ, 10^{-6}	77.6[111]; 19.5[100]; 57.1[110] [8]	2.0[111]; −6(λ_s) [5] −10.6[100] [59]	−1 [111]; 5(λ_s) [5] −35[100] [14]	−15(λ_s) [39]
Spectroscopic splitting factor g	2.03 (123 K) [9] 2.12 (294 K) [9]	2.03–2.06 (pc) [55]	2.019 (77 K) [24] 2.004 (300 K) [24]	2.20 (77 K) [39] 2.04 (300 K) [39]
Ferromagnetic resonance line-width ΔH, kA/m		1.59 [53]	0.95 (423 K) [25] 1.03 (294 K) [26]	27.8 (pc) [41]
Relative initial magnetic permeability μ_H	70 (pc) [8]	36 (pc) [54]	250 [5]	70 [40]
Dielectric permeability $\varepsilon = \varepsilon' - i\varepsilon''$		ε'=9.66 (4.55 GHz) ε''=0.17 [42]		ε'=9.24 (4.55 GHz) (pc) [42]; ε''=0.52
Resistivity ρ_0, 10^4 Ω·m	5·10^{-9} [10,43]	~1 (T=373 K) [44]	2·10^{-3} [27] ($Mn_{1.16}Fe_{1.84}O_4$)	~1 (373 K) [44]

Table 29.1 Main parameters of simple ferrites with a spinel structure *(continued)*

Parameter*1	$CoFe_2O_4$	$NiFe_2O_4$	$Li_{0.5}Fe_{2.5}O_4$
Lattice constant a, nm	0.838 [17]	0.8337 [11, 13]	0.833 [28]
Cation distribution	Fe[CoFe] [45]	Fe[NiFe] [12]	Fe[$Li_{0.5}Fe_{1.5}$] [28]
Oxygen parameter u	0.381 [45]	0.381 [60]	0.382 [28]
X-ray density d_X, 10^3 kg/m^3	5.29 [45]	5.37 [13]	4.75 [28]
Curie temperature T_C, K	793 [5]	858 [13]	943 [5]
Magnetic moment per molecule p^0_m, μ_B	3.94 (0 K) [46]	2.3 (0 K) [5]	2.6 (0 K) [5]
Specific saturation magnetization σ_s, A·m^2/kg	94 (0 K) [58] 80 (300 K) [5]	56 (0 K) [58] 50 (300 K) [5]	69 (0 K) [58] 65 (300 K) [5]
Saturation magnetization M_s, kA/m	477 (0 K) [5] 422 (300 K) [5]	302 (0 K) [5] 255 (300 K) [61]	334 (0 K) [5] 285 (300 K) [61]
The first anisotropy constant K_1, 10^3 J/m^3	4.4·10^2 (77 K) [15] 2.9·10^2 (300 K) [15]	−8.7 (77 K) [14] −6.2 (300 K) [14]	−12.7 (77 K) [29] −8.4 (300 K) [29]
The second anisotropy constant K_2, 10^3 J/m^3		−3 [16]	−9.7 (77 K) [30] −0.2 (300 K) [30]
Magnetostriction constant λ, 10^{-6}	120[111] ($Co_{0.8}Fe_{2.2}O_4$) [15] -590[100]; −110(λ_s) [17]	-4[111]; -26(λ_s) [17] -36[100] [14]	2.7 [111] −8 (λ_s) [31] -28.7[100] [2]
Spectroscopic splitting factor g	2.7 (363 K) (sc) [48] 2.22 (373 K) (pc) [49]	2.198 (85 K) [18] 2.196 (298 K) [18]	2.012 (77 K) [30] 2.003 (300 K) [30]
Ferromagnetic resonance line-width ΔH, kA/m		0.75 [19] 1.19 [20]	0.06 (134 K) [32] 0.14 [32]
Relative initial magnetic permeability μ_H	25 (pc) [5]	80 (sc) [14] 39 (pc) [21]	33 (pc) [5]
Dielectric permeability $\varepsilon = \varepsilon' - i\varepsilon''$	ε'=10 (4.55 GHz, pc) [42]	19 (sc) [14] 21 (pc) [14]	
Resistivity ρ_0, 10^4 Ω·m	0.5 (373 K) [44]	2 [22]	~1·10^{-4} [33]

*1 Values are related (if not specially outlined) to single crystals at a temperature of 293 K.

*2 Structure of copper ferrite $CuFe_2O_4$ depends on thermal treatment. On slow cooling down below the temperature of transformation T_{tr}, transition of a cubic structure into a tetragonal one occurs: T_{tr}=1033 K [36], 688 K [56], and 663 K [57]. As a result of hardening at $T > T_{tr}$, the cubic structure remains at a temperature of 293 K.

29.2.2 Mixed ferrites

Table 29.2 Experimental data on neutron diffraction in a number of $Mg_xMn_{1-x}Fe_2O_4$ ferrites [11] [T_{hard} is the temperature of hardening; a is the lattice constant; u is the oxygen parameter.]

x	T_{hard}, K	a, nm	u	Fraction of ions at the tetrahedral sites	
				Mg^{2+}	Mn^{2+}
0.25	1673	0.8485	0.3850	0.12	0.89
0.25	1273	0.8479	0.3848	0.09	0.78
0.5	1673	0.8455	0.3839	0.18	0.92
0.5	1273	0.8450	0.3836	0.14	0.82
0.75	1673	0.8424	0.3834	0.24	0.90
0.75	1273	0.8421	0.3831	0.21	0.98
0.9	1673	0.8406	0.3831	0.25	1.00
0.9	1273	0.8403	0.3821	0.21	0.90

Table 29.3 Dependence of some parameters of cobalt–zinc $Co_xZn_{1-x}Fe_2O_4$ ferrites on cobalt content x [62, 63] [T_C – the Curie temperature; σ_s – specific saturation magnetization.]

x	T_C, K	p^0_m, μ_B	σ_s, A·m^2/kg at temperature T, K			
			285	77	20	0
0.2	298	4.06	18.5	62.0	72.6	95.0
0.3	355	4.90	39.3	91.0	95.5	114.0
0.4	410	5.78	61.0	119.2	126.0	132.0
0.6	548	6.03	87.5	128.5	133.2	143.0
0.8	662	4.87	93.7	110.0	110.0	115.5
0.9	733	4.29	90.0	93.9	94.1	102.0
1.0	788	3.67	83.6	83.0	82.6	87.5

Table 29.4 Dependence of saturation magnetization M_s and spectroscopic splitting factor g for polycrystalline $Co_{0.7}Zn_{0.3}Fe_2O_4$ ferrite on temperature [The values of g-factor and the first anisotropy constant K_1 for a single crystal of the same composition are also given. The Curie temperature is $T_C = 613$ K. Measurements in the temperature range from 223 to 283 K were performed at a frequency of 23.6 GHz, and in the range from 293 to 553 K, at a frequency of 9.25 GHz [64, 65].]

T, K	Polycrystal		Single crystal		T, K	Polycrystal		Single crystal	
	M_s, kA/m	g	g	K_1, 10^4 J/m^3		M_s, kA/m	g	g	K_1, 10^4 J/m^3
223	560		1.90	12.03	353	338	2.12	2.06	−0.21
248	545		1.90	8.41	393	269	2.15	2.12	−0.41
283	500		1.91	4.51	433	203	2.16	2.12	−0.34
293	506	2.07	1.91	2.36	473	134	2.17	2.17	−0.21
313	450	2.10	1.92	1.34	513	61	2.18	2.16	−0.07
333	381	2.11	1.96	0.16	553	8	2.19	2.15	−0.01

Table 29.5 Dependence of some parameters of nickel $NiFe_{2-x}Al_xO_4$ aluminate ferrites on nickel content x [66, 67] [a – lattice constant; T_C – the Curie temperature; p_m^0 – magnetic moment per molecule; g – spectroscopic splitting factor.]

x	a, nm	T_C, K	p_m^0, μ_B		g
			slow cooling	hardening at 1623 K	
0.00	0.8337	580	2.29	2.29	2.3
0.25	0.83062	506	1.30	1.59	2.7
0.45	0.82769	465	0.61	1.19	
0.50	0.82705	430	0.44	1.99	6.9
0.625	0.82521	360	0–0.045		3.8
0.75	0.82329	294	0.38	0.58	1.5
1.00	0.81951	198	0.64	0.42	
1.00	0.81951	198	0.64	0.42	

Table 29.6 Spectroscopic splitting factor g and magnetic moment per molecule p_m^0 for $LiFe_5O_8$ ferrite and its solid solutions with $CdFe_2O_4$ and $LiAl_5O_8$ [68]

Composition	g (T=300 K)	p_m^0, μ_B
$LiFe_5O_8$	1.96	2.38
$(LiFe_5O_8)_{0.75}(CdFe_2O_4)_{0.25}$	1.97	3.44
$(LiFe_5O_8)_{0.65}(CdFe_2O_4)_{0.35}$	1.96	4.10
$(LiFe_5O_8)_{0.50}(CdFe_2O_4)_{0.50}$	1.97	4.70
$(LiFe_5O_8)_{0.25}(CdFe_2O_4)_{0.75}$	1.99	4.53
$Li(Fe_{0.9}Al_{0.1})_5O_8$	1.97	1.99
$Li(Fe_{0.8}Al_{0.2})_5O_8$	1.99	0.62
$Li(Fe_{0.7}Al_{0.3})_5O_8$	2.01	0.10
$Li(Fe_{0.6}Al_{0.4})_5O_8$		0.13

Table 29.7 Dependence of lithium-chromic ferrites' parameters on chromium content [69]

Chromium content x	distribution of metal ions	a, nm	T_C, K	T_{comp}, K	p_m^0, μ_B
0.0	$Fe_{1.00}[Li_{0.50}Fe_{1.50}]O_4$	0.8331	953		2.47–2.60
0.50	$Fe_{1.00}[Li_{0.50}Fe_{1.00}Cr_{0.50}]O_4$	0.8306	773		1.50–1.62
0.75	$Fe_{0.98}Li_{0.02}[Li_{0.48}Fe_{0.77}Cr_{0.75}]O_4$	0.8296	683		1.35
1.00	$Fe_{0.96}Li_{0.04}[Li_{0.44}Fe_{0.56}Cr_{1.00}]O_4$	0.8292	588	478	0.84
1.25	$Fe_{0.91}Li_{0.09}[Li_{0.41}Fe_{0.34}Cr_{1.25}]O_4$	0.8290	487	311	0.61
1.50	$Fe_{0.80}Li_{0.20}[Li_{0.30}Fe_{0.20}Cr_{1.50}]O_4$	0.8287	392	257	0.55
1.60	$Fe_{0.64}Li_{0.36}[Li_{0.14}Fe_{0.26}Cr_{1.60}]O_4$	0.8288	440	284	0.42
1.70	$Fe_{0.54}Li_{0.46}[Li_{0.14}Fe_{0.26}Cr_{1.70}]O_4$	0.8290	428	293	0.22
2.00	$Fe_{0.50}Li_{0.50}[Cr_{2.00}]O_4$	0.8288	353±16	310±15	

Table 29.8 Main parameters of chrome chalcogenide spinels

Compound	Lattice constant, nm	Type of magnetic ordering	Curie or Néel temperature, K	Magnetic moment, μ_B/molec.	Absorption edge, μm		Type of conductivity and resistivity ρ, $\Omega{\cdot}m$ at 300 K	References
					300 K	4.2 K		
$CdCr_2Se_4$	1.0755	ferro	130	5.98	0.94	1.08	p–type semiconductor, 10^2–10^3	[76,–81]
$CdCr_2S_4$	1.0244	ferro	85	5.9	0.79	0.69	n-type semiconductor, 10^4–10^6	[78,79,82–84]
$HgCr_2Se_4$	1.0753	ferro	106	5.6	1.48	3.88	p-type semiconductor, $0.7{\cdot}10^{-2}$	[78,85–88]
$HgCr_2S_4$	1.0237	antiferro	36	5.5		1.27	p-type semiconductor, $0.7{\cdot}10^{-2}$	[89,90]
$ZnCr_2Se_4$	1.0443	antiferro	20		0.97	1.10	p-type semiconductor	[91,92]

Table 29.9 Magnetooptic parameters of chrome chalcogenide spinels

Compound	Faraday rotation, deg/cm	Magnetooptic Q-factor, deg/dB	Ref.	Compound	Faraday rotation, deg/cm	Magnetooptic Q-factor, deg/dB	Ref.
$CdCr_2S_4$	$8.0{\cdot}10^3$ ($\lambda = 0.8$ μm)	30 ($\lambda = 1$ μm)	[93]	$HgCr_2Se_4$	10^3	4.5	[95]
	T=80 K, H=480 kA/m				T=85 K, H=360 kA/m, λ=10.6 μm		
$CdCr_2Se_4$	$9.2{\cdot}10^3$ ($\lambda = 1.17$ μm)	63 ($\lambda = 1.3$ μm)	[94]	$CoCr_2Se_4$	$1.3{\cdot}10^6$ ($\lambda = 1$ μm, 80 K) (calculated by Kerr effect)		[96]
	T=82 K, H=1200 kA/m						

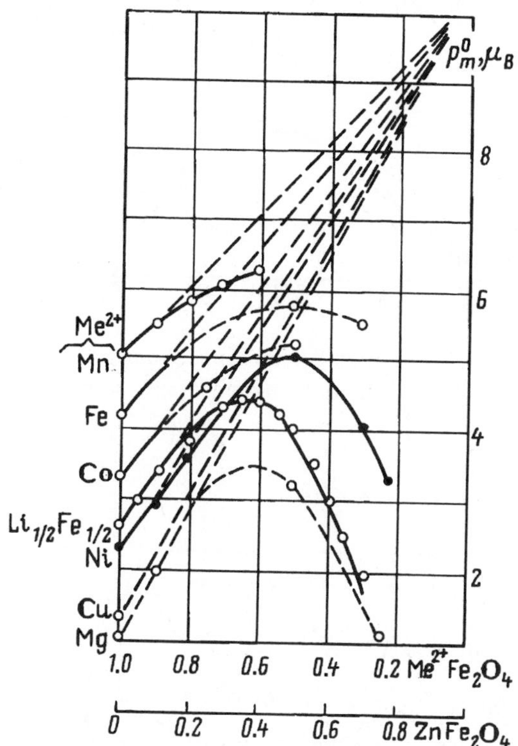

Figure 29.1 Saturation magnetic moments p_m^0 (at 0 K) of some mixed ferrites obtained by substitution of bivalent metal magnetic ions Me^{2+} with nonmagnetic Zn ions (Me^{2+} is one of the Mn, Fe, Co, Ni, Cu, Mg, or $Li_{0.5}Fe_{0.5}$ ions) [37].

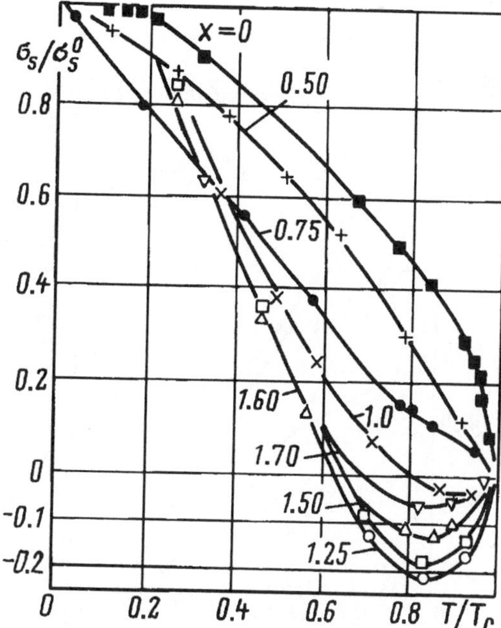

Figure 29.2 Dependence of relative saturation magnetization σ_s/σ_s^0 on normalized temperature T/T_c for lithium $Li_{0.5}Fe_{2.5-x}$ Cr_xO_4 chromite ferrites [70].

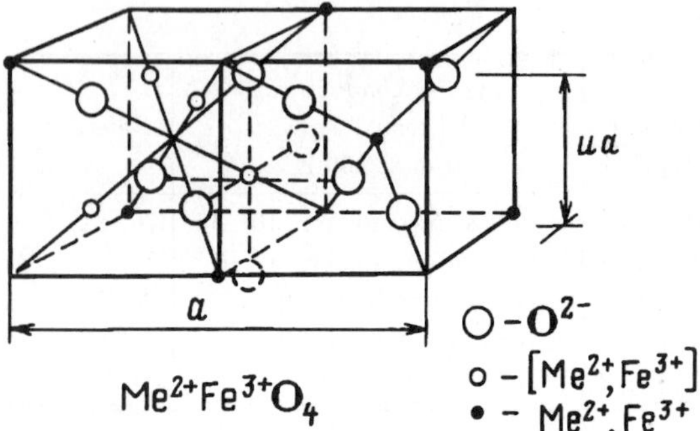

Figure 29.3 Two octants of spinel structure. Large open circles denote oxygen ions, little open and closed circles denote metal ions in octahedral and tetrahedral positions, respectively [5].

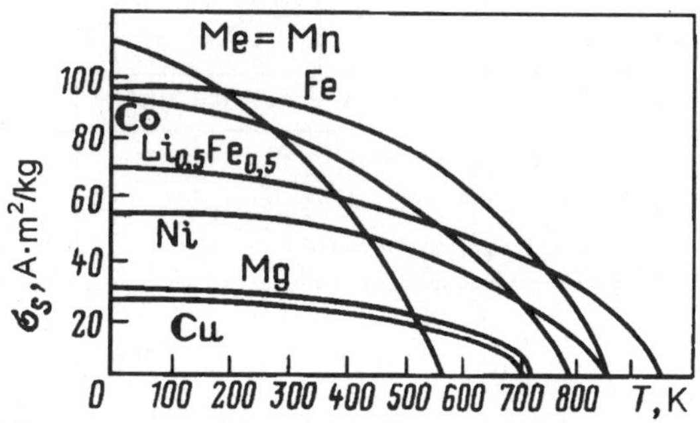

Figure 29.4 Dependence of specific saturation magnetization σ_s on temperature for some simple spinel ferrites.

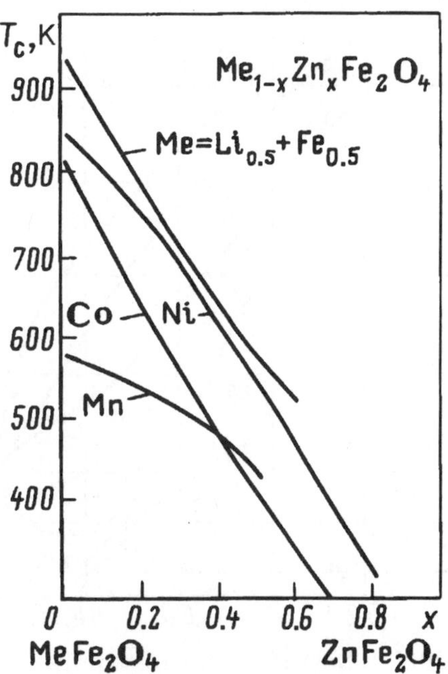

Figure 29.5 Dependence of Curie point T_C of some mixed ferrites of the $Me_{1-x}Zn_xFe_2O_4$ type on zinc concentration [5].

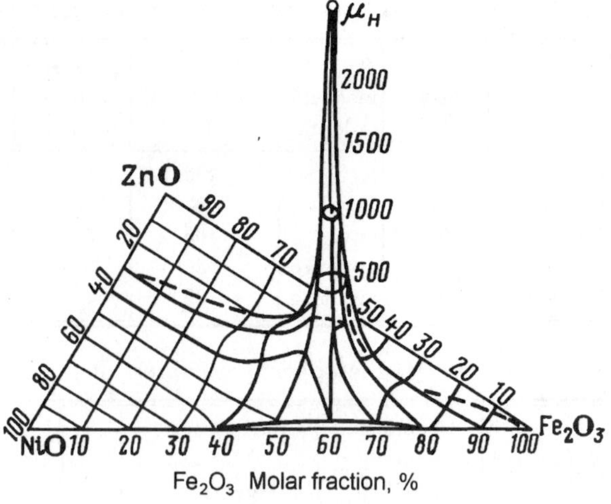

Figure 29.6 Dependence of initial magnetic permeability μ_i of ferrites in the system Ni–ZnO–Fe_2O_3 on their composition [71].

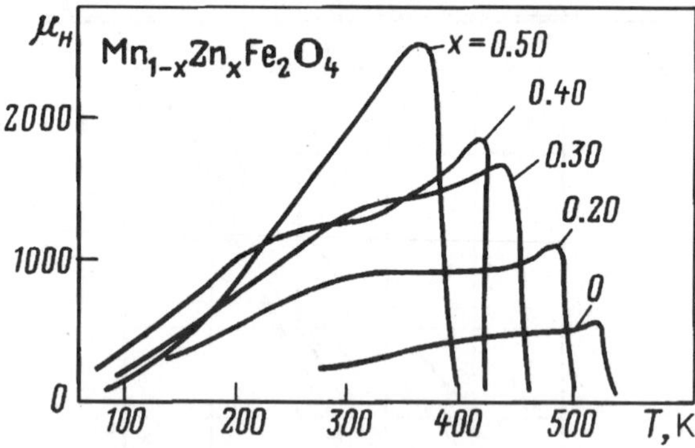

Figure 29.7 Dependence of initial magnetic permeability μ_i of mixed manganese–zinc ferrites on temperature [5].

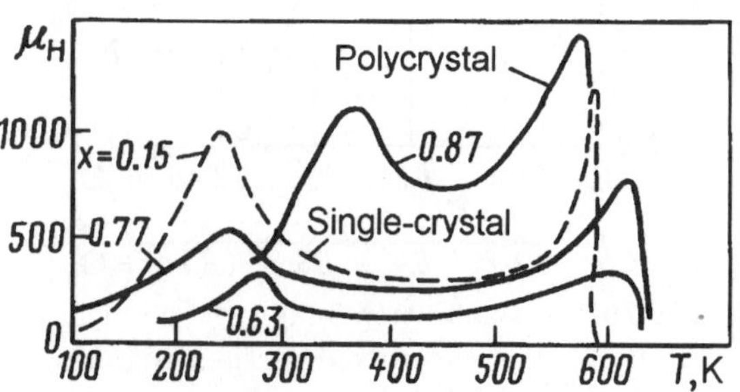

Figure 29.8 Dependence of initial magnetic permeability μ_i of $Mn_{1-x}Fe^{2+}{}_xFe_2O_4$ ferrites on temperature [72].

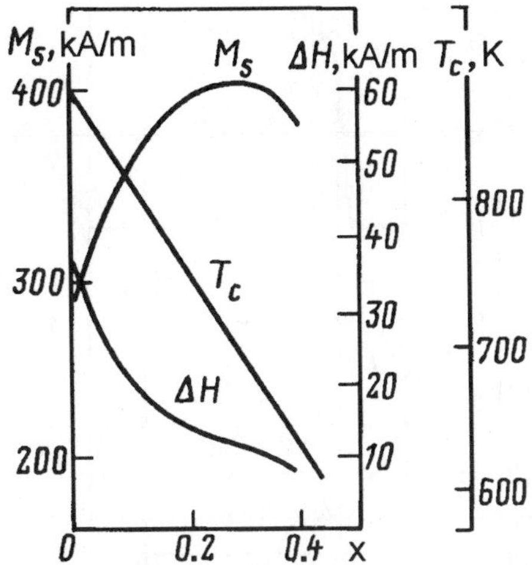

Figure 29.9 Dependence of saturation magnetization M_s, ferromagnetic resonance line-width ΔH (10 GHz) at temperature 293 K and Curie temperature T_C of lithium $Li_{0.5-x/2}Zn_xFe_{2.5-x/2}O_4$ ferrites on x [61].

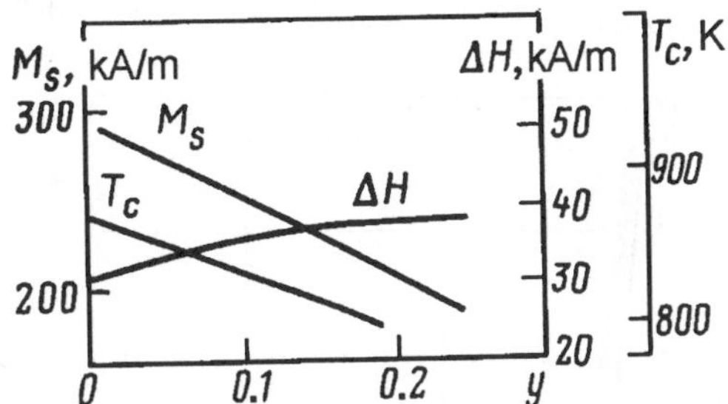

Figure 29.10 Dependence of saturation magnetization M_s, ferromagnetic resonance line-width ΔH (10 GHz) at temperature 293 K and Curie temperature T_C of lithium $Li_{0.5+y/2}Fe_{2.5-3y/2}Ti_yO_4$ ferrite on y [61].

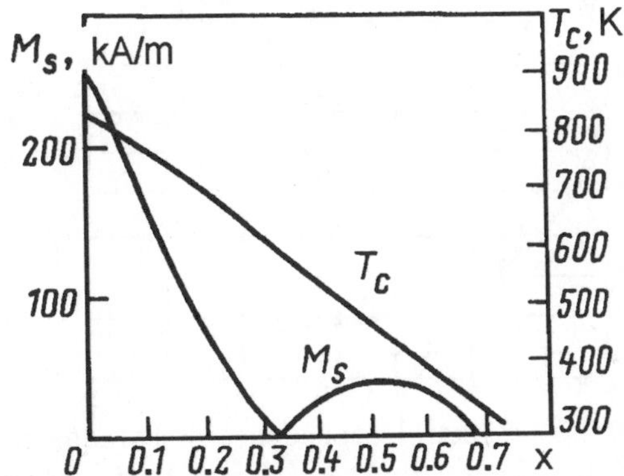

Figure 29.11 Dependence of saturation magnetization M_s at room temperature and Curie temperature T_C of nickel $NiFe_{2-2x}Al_{2x}O_4$ ferrite on x [61]. Sample was annealed.

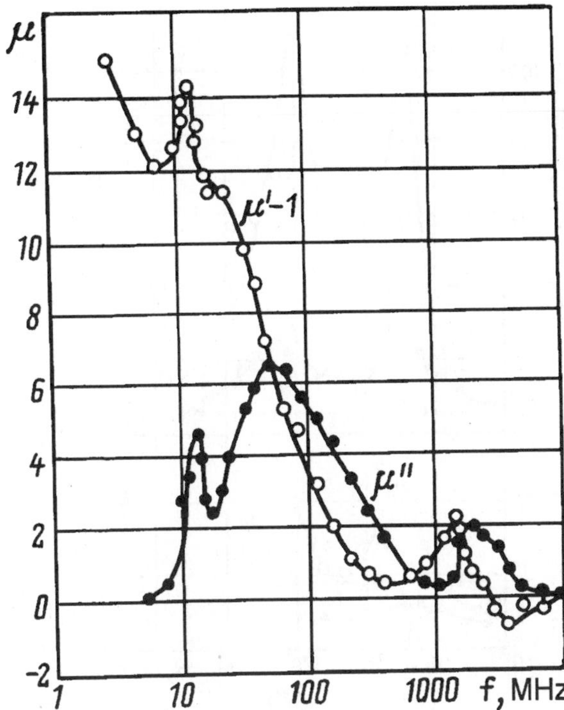

Figure 29.12 Dependence of real μ' and imaginary μ'' parts of initial magnetic permeability on frequency for polycrystalline $(MgO)_{0.81}(FeO)_{0.19}Fe_2O_3$ ferrite at temperature $T = 293$ K [54]; $d = 3.68 \cdot 10^3$ kg/m^3; $\sigma_s(299\text{ K}) = 26.4$ A·m^2/kg.

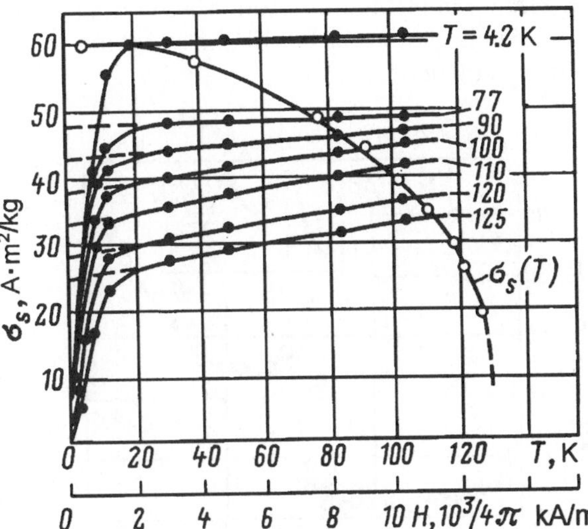

Figure 29.13 Dependence of spontaneous specific magnetization on the external field intensity at different temperatures and dependence of $\sigma_s(T)$ on temperature in $CdCr_2Se_4$ [76].

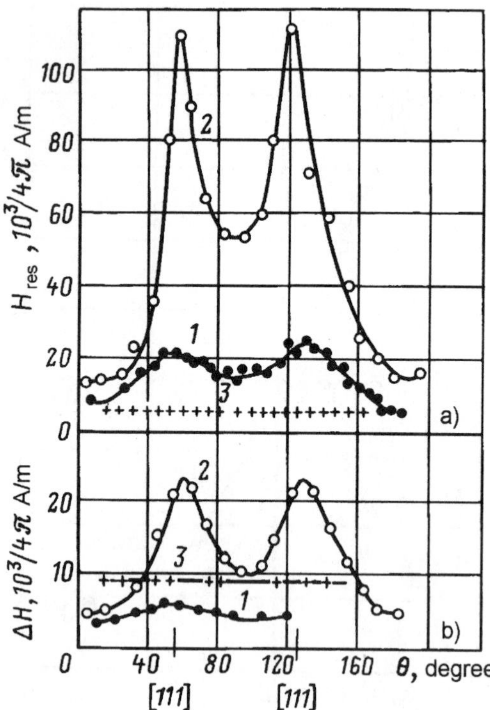

Figure 29.14 Angle dependences of resonance field (*a*) and resonance curve width (*b*) for $CdCr_2Se_4$ sphere [76]: $T = 4.2$ K; θ is an angle between constant magnetic field lines and [100]-axis; *1* – crystal grown by the crystallization method from a quasihomogeneous melt; *2* – crystal grown by the transport method in the liquid phase; *3* – crystal grown by the method of crystallization from a quasihomogeneous melt with molar admixture of Ag 0.08%.

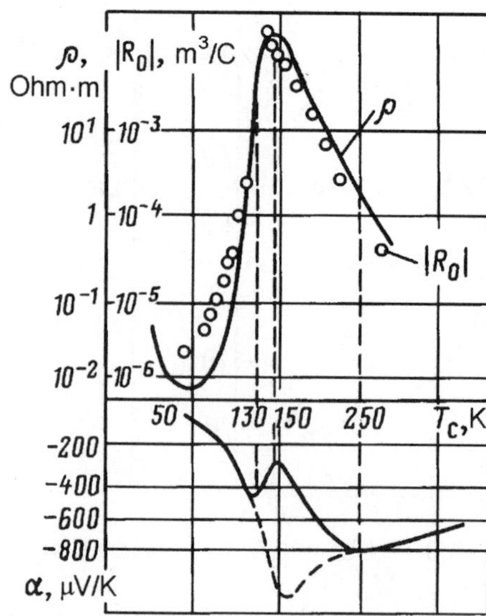

Figure 29.16 Temperature-dependent electric resistivity ρ, normal Hall coefficient R_0, and thermopower α for a $CdCr_2Se_4$ sample (1% admixture of In) [77]: —— experimental curve; - - - - calculation according to a one-zone model.

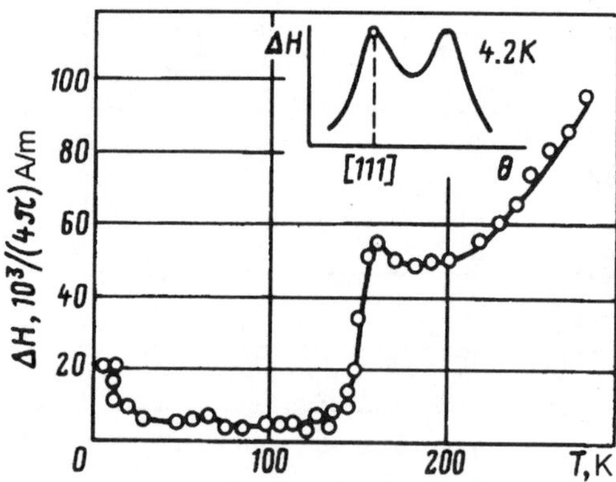

Figure 29.15 Temperature-dependent resonance curve width for $CdCr_2Se_4$ sphere [76]: T=4.2 K; f=8.9 GHz; constant magnetic field is directed along the [111] axis.

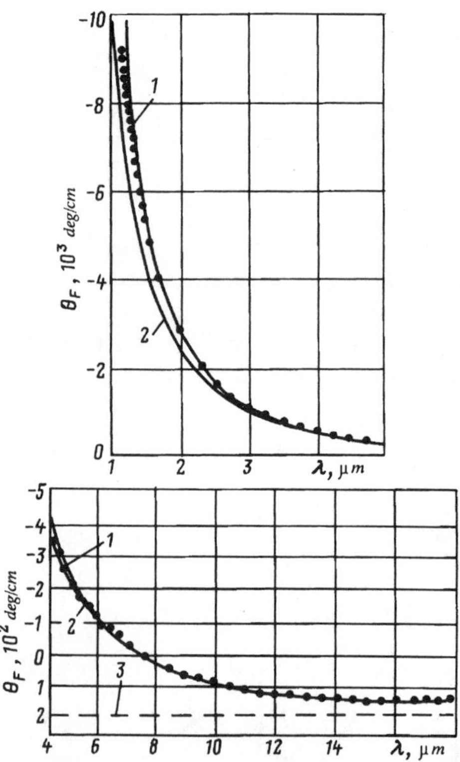

Figure 29.17 Spectral dependence of Faraday rotation in $CdCr_2Se_4$ at T=82 K, $H = 1.2 \cdot 10^3$ kA/m [81] (points refer to experimental data, and lines refer to calculated results): *1* – resonance wave-length; *2* – quadratic law; *3* – constant contribution from ferromagnetic resonance.

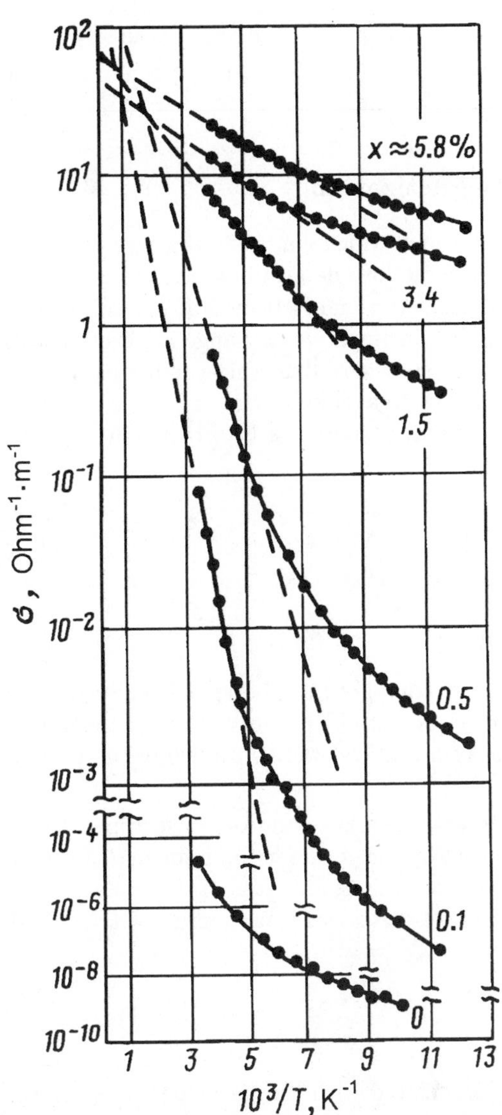

Figure 29.18 Temperature-dependent electrical conductivity for $Cd_{1-x}Ag_xCr_2Se_4$ single crystals of p type at different values of x [78]. Data on conductivity of a pure crystal ($x = 0$) are from [79].

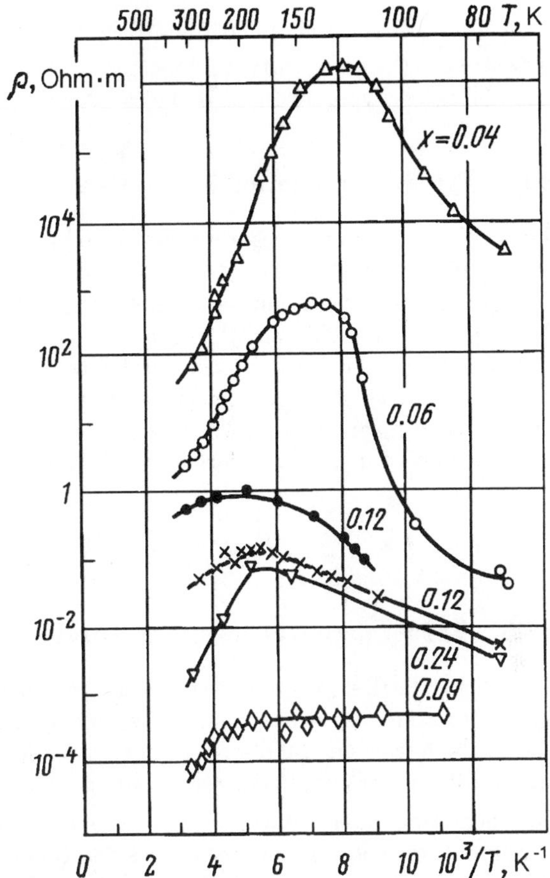

Figure 29.19 Temperature-dependent electrical resistivity for n-type $Cd_{1-x}In_xCr_2Se_4$ single crystals at different values of x [80].

29.3 Ferrites with a Garnet Structure

29.3.1 Crystallographic structure

Ferrimagnetic oxides of the garnet type crystallize in a structure which is isomorphic to the classical mineral of garnet $\{Ca_3\}[Al_2](Si_3)O_{12}$. The garnet structure is described by the cubic space group $Ia3d$ - O_h^{10}. An element of this structure is shown in Fig. 29.20. Cubic elementary cell of garnet contains eight formula units. Sixteen ions of Al^{3+} occupy the octahedral positions denoted as $16a$, twenty four Si^{4+} ions take up the central positions of tetrahedrons and are labelled by $24d$, and, finally, each of the twenty four Ca^{2+} ions is surrounded by eight oxygen ions, their positions being marked by 24*c*.

Attention to the garnet structure increased substantially after synthesizing the ferrimagnetic garnets of the $M_3Fe_5O_{12}$ type, where M is the rare-earth metal ion or yttrium ion.

29.3.2 Magnetic properties and saturation magnetization

Three magnetic sublattices were introduced in treating garnets, which contrasts the case of ferrites with the spinel structure. The strongest antiferromagnetic interaction (determining the Curie temperature T_C) occurs between ions of trivalent iron in the octahedral 16*a*- and tetrahedral 24*d*-sublattices. The rare-earth ions' sublattice 24*c* is most strongly coupled by the negative exchange interaction with the tetrahedral sublattice (with the octahedral one in the garnets where light rare-earth ions from Pr to Sm are present), this coupling being approximately 10 times weaker than the $(a - d)$ interaction. The saturation magnetization M_s in the case of heavy rare-earth garnets can be written through the magnetizations of separate sublattices in the following way:

$$M_s = M_{24d} - M_{16a} - M_{24c}. \qquad (29.11)$$

29.3.3 Ferromagnetic resonance and anisotropy

The garnet ferrites possess lower specific magnetization as compared to the spinel ferrites. High attention to them was caused mainly by their unique properties in the RF spectral range. Minimum values of the ferromagnetic resonance line-width $\Delta H \simeq 16$ A/m (0.2 Oe) were obtained in the yttrium garnet ferrite free from admixtures of rare-earth ions.

For theoretical interpretation of the results on ferromagnetic resonance and on anisotropy of rare-earth garnet ferrites a simultaneous consideration of ion levels splitting under the influence of crystal field, spin-orbital and exchange interaction is needed. These interactions are often of the same order of magnitude. Up-to-date information on electron levels of the rare-earth ions is not sufficient for reliable theoretical interpretation of experimental results.

29.3.4 Magnetostriction

Magnetostriction of rare-earth garnet ferrites is linearly associated with the rare-earth ions concentration and strongly increases with decrease in temperature. The record values of $\lambda_{111} = 2420 \cdot 10^{-6}$ and $\lambda_{100} = 1200 \cdot 10^{-6}$ in the field of intensity $H = 2000$ kA/m were obtained at a temperature of 4.2 K in the terbium garnet ferrite. They are comparable in an order of magnitude with the magnetostriction of rare-earth metals.

A collection of properties of ferrites with a garnet structure is presented in Tables 29.10–29.31 and in Figs. 29.20–29.30.

Table 29.10 Garnet structure [111]

Space group	$Ia\,3d$—O_h^{10}			
Typical ideal formula	$\{Ca_3\}$	$[Al_2]$	(Si_3)	O_{12}
Position of space group	24c	16*a*	24*d*	96*h*
Coordinates	$0\ \frac{1}{4}\ \frac{1}{8}$	000	$0\ \frac{1}{4}\ \frac{3}{8}$	$x\ y\ z$
Point symmetry (local)	222	$\bar{3}$	$\bar{4}$	1
Oxygen coordination	8	6	4	
Polyhedron type	dodecahedron (deformed cube)	octahedron	tetrahedron	

Table 29.11 Refinements of garnet $Y_3Fe_5O_{12}$ crystal structure [112–115]

Lattice constant a = 12.376 Å		
x	y	z
−0.0274 (9)	0.0572 (9)	0.1495 (9)
−0.0270 (4)	0.0569 (5)	0.1505 (5)
−0.0271 (1)	0.0567 (1)	0.1504 (1)

Interatomic angles		
Configuration		Angle, deg
$Fe^{3+}[a]$—O^{2-}—$Fe^{3+}(d)$		125.9 (1)
$Fe^{3+}[a]$—O^{2-}—Y^{3+}*1		101.5 (1)
$Fe^{3+}[a]$—O^{2-}—Y^{3+}*2		104.3 (1)
$Fe^{3+}(d)$—O^{2-}—Y^{3+}*1		123.0 (1)
$Fe^{3+}(d)$—O^{2-}—Y^{3+}*2		93.5 (1)
Y^{3+}—O^{2-}—Y^{3+}		104.5 (1)
$Fe^{3+}[a]$—O^{2-}—$Fe^{3+}[a]$	(4.41)*3	147.2
$Fe^{3+}(d)$—O^{2-}—$Fe^{3+}(d)$	(3.41)*3	86.6
$Fe^{3+}(d)$—O^{2-}—$Fe^{3+}(d)$	(3.68)*3	78.8
$Fe^{3+}(d)$—O^{2-}—$Fe^{3+}(d)$	(3.83)*3	74.7
$Fe^{3+}(d)$—O^{2-}—$Fe^{3+}(d)$	(3.83)*3	74.6

The nearest neighbor interatomic distances					
Ion	Distance, Å				
O^{2-}	$2Y^{3+}$	2.357	(1)	2.436	(1)
		2.017	(1)		
		1.865	(1)		
		2·2.692	(2)	2.789	(2)
		2.837	(2)	2.976	(2)
		2·3.005	(2)	2·3.146	(2)
Y^{3+}		3.46			
		2·3.09		4·3.79	
		4·2.357	(1)	2·2.436	(1)
$Fe^{3+}[a]$		3.46			
		3.46			
		2.017			
$Fe^{3+}(d)$		2·3.09			
		3.46			
		1.865	(1)		

*1 Y^{3+}—O^{2-}, the distance equals 2.436 Å.
*2 The same bond, distance of 2.357 Å.
*3 Values in parentheses are the maximum distances Fe^{3+} (a or d)—O^{2-} in Å.

Table 29.12 Ionic radius R^{3+} (coordination 8), unit cell parameter a and X-ray density d_X of garnet ferrites $R_3Fe_5O_{12}$ [116–117]

Garnet ferrite	R^{3+} ionic radius, Å	a, Å	d_X, g/cm³	Garnet ferrite	R^{3+} ionic radius radius, Å	a, Å	d_X, g/cm³
$La_3Fe_5O_{12}$	1.190	(12.767)*1	(5.67)*1	$Dy_3Fe_5O_{12}$	1.030	12.405	6.61
$Pr_3Fe_5O_{12}$	1.137	(12.646)*1	(5.87)*1	$Y_3Fe_5O_{12}$	1.016	12.376	5.17
$Nd_3Fe_5O_{12}$	1.120	(12.600)*1	(6.00)*1	$Ho_3Fe_5O_{12}$	1.017	12.375	6.77
$Sm_3Fe_5O_{12}$	1.087	12.529	6.23	$Er_3Fe_5O_{12}$	1.004	12.347	6.87
$Eu_3Fe_5O_{12}$	1.073	12.498	6.31	$Tm_3Fe_5O_{12}$	0.991	12.323	6.94
$Gd_3Fe_5O_{12}$	1.061	12.471	6.46	$Yb_3Fe_5O_{12}$	0.982	12.302	7.06
$Tb_3Fe_5O_{12}$	1.044	12.436	6.55	$Lu_3Fe_5O_{12}$	0.972	12.283	7.14

*1 Approximated value.

Table 29.13 Nearest-neighbor interionic distances (in Å) at 673 K from the neutron diffraction data [118]

Garnet constituents	Garnet					
	Tb	Dy	Ho	Er	Tm	Yb
R^{3+} — O^{2-}	2.371 (7)	2.360 (11)	2.362 (5)	2.351 (10)	2.340 (15)	2.336 (6)
R^{3+} — O^{2-}	2.464 (7)	2.441 (11)	2.436 (5)	2.415 (10)	2.424 (15)	2.410 (6)
$Fe^{3+}[a]$ — O^{2-}	2.025 (12)	2.030 (15)	2.018 (9)	2.019 (12)	2.007 (16)	2.025 (8)
$Fe^{3+}[d]$ — O^{2-}	1.862 (12)	1.864 (12)	1.860 (9)	1.868 (12)	1.863 (18)	1.850 (9)

Note: Accuracy of last significant figures is given in parentheses.

Table 29.14 Entering of different ions into $Y_3Fe_5O_{12}$ [Ionic radii R_i (in Å) and maximum entering x_{max} per formula unit with preference for different crystallographic positions are given [119].]

Tetrahedral *d*-positions				Octahedral *a*-positions				Dodecahedral *c*-positions			
Ion	R_i	x_{max}	ICC[*1]	Ion	R_i	x_{max}	ICC[*1]	Ion	R_i	x_{max}	ICC[*1]
Fe^{3+}	0.492	3		Fe^{3+}	0.642	2		Y^{3+}	1.016	3	
B^{3+}	0.11(?)	?		Mn^{4+}	0.530	?		(Mg^{2+})	0.89	?	Si^{4+}
Si^{4+}	0.26	3	Ca^{2+}	(Ge^{4+})	0.530	?	Ca^{2+}	Mn^{2+}	0.96	?	Ge^{4+}
Be^{2+}	0.27	>0.1(?)	Fe^{4+}	(Al^{3+})	0.539	2		Lu^{3+}	0.972	3	
V^{5+}	0.355	1.5	Ca^{2+}	(Te^{6+})	0.56	0.25	Ca^{2+}	Yb^{3+}	0.982	3	
Ge^{4+}	0.390	<3	Ca^{2+}	Ti^{4+}	0.582	≈0.5	Ca^{2+}	Tb^{3+}	1.044	3	
Al^{3+}	0.390	3		(Ga^{3+})	0.610	2		Gd^{3+}	1.061	3	
(Ti^{4+})	0.42	≈0.2	Ca^{2+}	Sn^{4+}	0.708	2	Ca^{2+}	Eu^{3+}	1.073	3	
Ga^{3+}	0.470	3		Co^{2+}	0.745	2	Ge^{4+}	Sm^{3+}	1.087	3	
(Sn^{4+})	0.55	<1.0	Ca^{2+}	Sb^{5+}	0.616	2	Ca^{2+}, Na^{+}	Th^{4+}	1.095	≈0.9	Ca^{2+}
(Co^{2+})	0.58	≈0.5	Ge^{4+}	Cr^{3+}	0.615	≈0.6		Cd^{2+}	1.10	3	Ge^{4+}
Fe^{4+}	(0.585)	0.1	Ca^{2+}	Zn^{2+}	0.740	2	Ge^{4+}	Nd^{3+}	1.120	≈1.8	
(Zn^{2+})	0.60	0.7	F^{-}	Ir^{4+}	0.625	>0.05(?)	Ca^{2+}, Mg^{2+}	Ca^{2+}	1.124	3	V^{5+}
(Ir^{4+})	0.625	?	(Fe^{2+})	Mn^{3+}	0.645	≈0.65(?)		Bi^{3+}	1.132	1.2–1.9	
Fe^{2+}	0.63	1	F^{-}	Fe^{2+}	0.730	≈0.45	Si^{4+}	Pr^{3+}	1.137	≈1.1	
(Ru^{3+})	0.68	?		Ru^{3+}	0.680	>0.02(?)		Ce^{3+}	1.143	≈0.13(?)	
				Ni^{2+}	0.690	≈1.5	Ge^{4+}	Na^{+}	1.18	3	Te^{6+}
				Hf^{4+}	0.710	2	Ca^{2+}	La^{3+}	1.190	0.45	
				Mg^{2+}	0.720	≈1.8	Si^{4+}	Sr^{2+}	1.240	≈1.0	Sn^{4+}
				Cu^{2+}	0.730	<0.2(?)	Ge^{2+}	Pb^{2+}	1.29	1(?)	Ge^{4+}
				Zr^{4+}	0.745	2	Ca^{2+}				
				Sc^{3+}	0.745	≈1.5–1.6					
				In^{3+}	0.792	≈0.92–1.0					
				Mn^{2+}	0.830	≈0.4	Si^{4+}				

[*1] Ions to compensate a charge.

Table 29.15 Curie temperature, compensation temperature [119], and saturation magnetization [120] of rare-earth garnet ferrites at temperature 295 K

Garnet	T_C, K	T_{comp}, K	M_o, $10^3/(4\pi)$ A/m	Garnet	T_C, K	T_{comp}, K	M_s, $10^3/(4\pi)$ A/m
$Y_3Fe_5O_{12}$	558.9		1750	$Ho_3Fe_5O_{12}$	553	132.2	880
$Sm_3Fe_5O_{12}$	573.4		1675	$Er_3Fe_5O_{12}$	566	83	1240
$Eu_3Fe_5O_{12}$	562.4		1170	$Tu_3Fe_5O_{12}$	548	7.6	1397
$Gd_3Fe_5O_{12}$	566	285.5	56	$Yb_3Fe_5O_{12}$	549	<20	1555
$Tb_3Fe_5O_{12}$	568	248.6	198	$Lu_3Fe_5O_{12}$	529.7		1815
$Dy_3Fe_5O_{12}$	553	230.2	376				

Table 29.16 Anisotropy constants for different rare-earth garnet ferrites [119]

Garnet ferrite	T, K	K_1, 10^4 J/m^3	K_2, 10^4 J/m	Garnet ferrite	T, K	K_1, 10^4 J/m^3	K_2, 10^4 J/m
$Y_3Fe_5O_{12}$	295	(−5.7)—(−6.3)	−2.3	$Eu_3Fe_5O_{12}$	293	−38	
	273	−7.8	−0.26		4.2	−400	−156
	77	−22.1	−2.1	$Gd_3Fe_5O_{12}$	320	−4.1	−1.0
	4.2	−24.8	−2.3		80	−44	−3.5
	4.2	−26.3	−1.2		4.2	−241	−3.7
$Sm_3Fe_5O_{12}$	300	(−17)—(−18.2)		$Tb_3Fe_5O_{12}$	300	−8.2	
	77	−1430	2100		80	−760	−7600
	4.2	−4000			78	+480	−4900

Table 29.16 Anisotropy constants for different rare-earth garnet ferrites [119] *(continued)*

Garnet ferrite	T, K	K_1, 10^4 J/m^3	K_2, 10^4 J/m^3	Garnet ferrite	T, K	K_1, 10^4 J/m^3	K_2, 10^4 J/m^3
$Dy_3Fe_5O_{12}$	300	< -5			77	36 (+45)	
	80	−970	214		4.2	9000	50000
$Ho_3Fe_5O_{12}$	300	<-5		$Tm_3Fe_5O_{12}$	293	−5.8 (−11)	0
	80	−800	−270		77	−29.8 (−210)	100
	4.2	−30000		$Yb_3Fe_5O_{12}$	300	−6.1	
$Er_3Fe_5O_{12}$	300	−6.0			80	−38.5	
	110	−22.0			4.2	−6700	

Table 29.17 Fields and constants of magnetic crystallographic anisotropy of Bi–Ca–V garnet ferrites at a temperature of 20°C

Garnet ferrite	M_s, $10^3/(4\pi)$ A/m	K_1/M_s, $10^3/(4\pi)$ A/m	K_2/M_s, $10^3/(4\pi)$ A/m	K_1, 10^2 J/m^3	K_2, 10^2 J/m^3	References
$Bi_{0.8}Ca_{2.2}Fe_{3.9}V_{1.1}O_{12}$	180	−240	−5	−3.44	−0.07	[121]
$Bi_{0.5}Ca_{2.5}Fe_{3.75}V_{1.25}O_{12}$	400	−50	−4	−1.59	−0.1	[121]
$Bi_{0.2}Ca_{2.8}Fe_{3.6}V_{1.4}O_{12}$	630	−22	−2	−1.10	−0.1	[121]
$Ca_3Fe_{3.5}V_{1.5}O_{12}$	750	−16	−1	0.96	−0.06	[122]
$Bi_{0.4}Ca_{2.6}Fe_{3.6}V_{1.3}In_{0.1}O_{12}$	330	−38	−2	−1.0	−0.05	[122]
$Bi_{0.4}Ca_{2.6}Fe_{3.45}V_{1.3}In_{0.25}O_{12}$	80	−55	−4	−0.35	−0.02	[122]
$Bi_{0.2}Ca_{2.8}Fe_{3.5}V_{1.4}Sc_{0.1}O_{12}$	470	−28	−2	−1.05	−0.07	[122]
$Bi_{0.2}Ca_{2.8}Fe_{3.35}V_{1.4}Sc_{0.25}O_{12}$	230	−25	−3	−0.46	−0.05	[122]
$Ca_3Fe_{3.4}V_{1.5}Sc_{0.1}O_{12}$	600	−16	−1	−0.76	−0.05	[122]
$Ca_3Fe_{3.25}V_{1.5}Sc_{0.25}O_{12}$	380	−14	−2	−0.42	−0.06	[122]
$Ca_3Fe_{3.1}V_{1.5}Sc_{0.4}O_{12}$	150	−10	−4	−0.12	−0.05	[122]

Table 29.18 Magnetoelastic and magnetostrictive coefficients of some garnet ferrites

Garnet ferrite	B_1, 10^5 J/m^3		B_2, 10^5 J/m^3		λ_{100}, 10^{-6}		λ_{111}, 10^{-6}		References
	300 K	77 K	300 K	77 K	300 K	77 K	300 K	77 K	
$Y_3Fe_5O_{12}$	3.4	1.7	6.4	12.7	−1.4	−0.7	−2.4	−5.3	[123]
$Y_3Fe_{4.8}Ga_{0.2}O_{12}$	3.0	2.6	6.0	12.5	−1.15	−1.0	−2.1	−4.3	[124]
$Y_3Fe_{4.2}Ga_{0.8}O_{12}$	2.2	2.6	2.86	6.0	−0.85	−1.0	−1.0	−2.1	[124]
$Y_3Fe_{3.66}Ga_{1.34}O_{12}$	0.4	1.75	0.57	2.36	−0.15	−0.67	−0.2	−0.82	[124]
$Y_3Fe_{4.81}Al_{0.19}O_{12}$			—		−0.2	−1.45	−3.45	−6.1	[125]
$Y_3Fe_{4.25}Al_{0.75}O_{12}$					−1.2	−2.2	−2.0	−4.8	[125]
$Y_3Fe_{4.52}Sc_{0.48}O_{12}$	2.3	−4.55	4.0	8.9	−0.91	1.75	−1.4	−3.2	[125]
$Y_3Fe_{4.3}Sc_{0.7}O_{12}$	1.2	−5.5	1.9	4.2	−0.45	2.1	−0.67	−1.45	[125]
$Bi_{0.5}Ca_{2.5}Fe_{3.75}V_{1.25}O_{12}$	4.4	7.6	2.0	3.7	−2.6	−4.5	−1.2	−2.5	[126]
$Cd_3Fe_5O_{12}$					0	7.1	−3.1	−3.9	[127]

Table 29.19 Power law for sublattice magnetization dependence on temperature $M_i(T)/M_i(0\ \mathrm{K}) = D(1 - T/T_C)^{\beta}$ and values of parameters D and β for $Y_3Fe_5O_{12}$ [130–131]

Crystallographic position	β	D	Temperature interval
$[a]$	0.323±0.005	1.18±0.02	$0.65 \leq T/T_C \leq 0.97$
Octahedron	0.40±0.03	1.55±0.20	$0.97 \leq T/T_C \leq 0.9985$
(d)	0.349±0.005	1.09±0.02	$0.65 \leq T/T_C \leq 0.99$
Tetrahedron	0.47±0.03	1.95±0.20	$0.99 \leq T/T_C \leq 0.9997$

Table 29.20 Fields and constants of magnetic crystallographic anisotropy of yttrium garnet ferrites at a temperature of 20°C [121–129]

Garnet ferrite	M_s,	K_1/M_s,	K_2/M_s,	K_1,	K_2,
	$10^3/(4\pi)$ A/m	$10^3/(4\pi)$ A/m		10^2 J/m^3	
$Y_3Fe_5O_{12}$	1750	−42	0	6.0	0
$Y_3Ga_{0.4}Fe_{4.6}O_{12}$	1150	−42	0	3.94	0
$Y_3Ga_{0.6}Fe_{4.4}O_{12}$	850	−45	0	3.05	0
$Y_3Ga_{0.8}Fe_{4.2}O_{12}$	600	−53	0	2.53	0
$Y_3Ga_{1.0}Fe_{4.0}O_{12}$	300	−62	0	1.48	0
$Y_3Ga_{1.2}Fe_{3.8}O_{12}$	100	−96	0	0.76	0
$Y_3In_{0.2}Fe_{4.8}O_{12}$	1850	−22	0	3.24	0
$Y_3In_{0.4}Fe_{4.6}O_{12}$	1820	−12	−1	1.74	0.1
$Y_3In_{0.55}Fe_{4.45}O_{12}$	1750	−7	−1	0.97	0.1
$Y_3Ga_{0.6}In_{0.15}Fe_{4.25}O_{12}$	900	−21	0	1.50	0
$Y_3Ga_{0.6}In_{0.3}Fe_{4.1}O_{12}$	930	−16	−1	1.18	0.05
$Y_3Ga_{0.6}In_{0.48}Fe_{3.92}O_{12}$	950	−5	−1	0.38	0.06
$Y_3Ga_{0.9}In_{0.15}Fe_{3.95}O_{12}$	460	−23	−1	0.84	0.03
$Y_3Ga_{0.9}In_{0.3}Fe_{3.8}O_{12}$	500	−18	−2	0.72	0.05
$Y_3Ga_{0.9}In_{0.48}Fe_{3.62}O_{12}$	570	−6	−2	0.27	0.07
$Y_3Ga_{1.1}In_{0.3}Fe_{3.6}O_{12}$	300	−24	−2	0.57	0.04
$Y_3Ga_{1.1}In_{0.48}Fe_{3.42}O_{12}$	420	−10	−2	0.33	0.06
$Y_3Fe_{4.995}Co_{0.005}O_{12}$	1750	−20	−10	−2.79	−1.39
$Y_3Fe_{4.99}Co_{0.01}O_{12}$	1750	5	−23	0.70	−3.20
$Y_3Fe_{4.97}Co_{0.03}O_{12}$	1750	105	−63	14.63	−8.78
$Y_3Fe_{4.98}Si_{0.02}O_{12}$	1750	-32	−4	−4.46	−0.56
$Y_3Fe_{4.95}Si_{0.05}O_{12}$	1700	-16	−8	−2.17	−1.08

Table 29.21 Coefficients of elastic rigidity of crystals, 10^{10} Pa (10^{11} dyn/cm^2) at a temperature of 300 K

Garnet ferrite	c_{11}	c_1	c_{44}	Ref.
$Y_3Fe_5O_{12}$	26.90	10.77	7.64	[132]
$Eu_3Fe_5O_{12}$	25.10	10.70	7.62	[133]
$Y_3Ga_5O_{12}$	29.03	11.73	9.55	[134]
$Y_3Al_5O_{12}$	33.40	11.12	11.51	[135]
$Y_3Fe_{2.66}Ga_{2.34}O_{12}$	28.47	11.81	8.36	[136]

Table 29.22 Coefficients of elastic rigidity, 10^{10} N/m^2, for garnets with no iron at a temperature of 20°C

Garnet	c_{11}	c_{44}	c_{12}	A_l*1	Ref.
$Y_3Al_5O_{12}$	33.32–33.40	11.50–11.51	11.07–11.12	1.033	[139]
$Y_3Ga_5O_{12}$	29.03	9.547	11.73	1.10	[136]
$Gd_3Ga_5O_{12}$	28.703	9.04	11.601	1.057	[137]
$Gd_3Ga_5O_{12}$ [100]	28.57	9.02	11.49	1.056	[138]
$Gd_3Ga_5O_{12}$ [110]	28.59	9.03	11.49	1.056	[138]
$Gd_3Ga_5O_{12}$ [$\bar{1}$10]	28.51	9.02	11.45	1.057	[138]
$Sm_3Ga_5O_{12}$	28.076	8.604	11.352	1.029	[140]
$Nd_3Ga_5O_{12}$	27.781	8.381	11.155	1.008	[141]

*1 $A_l = 2c_{44}/(c_{11} - c_{12})$ is the coefficient of elastic anisotropy.

Table 29.23 Mean values of microhardness, 10^5 Pa, for planes (110) and (211) and the hardness class for $Y_3Fe_5O_{12}$, $Y_3Ga_5O_{12}$, and $Y_3Al_5O_{12}$ [142–143]

Garnet	Plane		Mean microhardness of a crystal	Hardness class according to the extended 15-point Mohs' hardness scale
	(110)	(211)		
$Y_3Fe_5O_{12}$	{1210±35, 1400±35	1240±50	1230	7.5
$Y_3Ga_5O_{12}$	1450±45	1520±75	1490	8.0
$Y_3Al_5O_{12}$	1900±50	1650±85	1730	8.4

Table 29.24 Typical dislocations in $Y_3Fe_5O_{12}$ detected by an optical-polarization technique [144]

Dislocation axis	[110]					[211]		
Sliding surface	(001)		(110)		(112)	(111)		(311)
Probable direction of the Burger's vector	[110]	[100]	[111]	[100]	[111]	[110]	[110]	[110]
Dislocation type, deg	90	45	35	90	90	90	30	72

Table 29.25 Debye temperature θ_D and heat conductivity κ extrapolated to θ_D for different garnets [145] [Calculated values are in parentheses.]

Garnet	R	θ_D, K	κ, W/(cm·K)
$R_3Al_5O_{12}$	Y	750	0.042
	Gd	(640)	0.045
	Er	630	
	Tm	620	
	Tb	620	
	Lu	(620)	0.043
$R_3Ga_5O_{12}$	Nd	715	
	Yb	715	0.039
	Y	585	0.046
	Gd	520	0.051
$R_3Fe_5O_{12}$	Y		0.038
	Eu		
	Lu		

Table 29.26 Heat conductivity for $Y_3Fe_5O_{12}$ as a function of temperature [145]

T, K	κ, W/(cm·K)
2.5	0.075
6	0.56
10	1.2
21	2.2*1
30	1.80
40	1.12
50	0.63
70	0.41
100	0.26
150	0.150
200	0.106
300	0.074
565	0.038*2

*1 Maximum value.
*2 Extrapolated value.

Table 29.27 Linear thermal expansion factor for polycrystalline garnets $Y_{3-x}Gd_xFe_5O_{12}$ and $Y_3Fe_{5-x}Al_xO_{12}$ [146]

Garnet	Linear thermal expansion factor α, 10^{-6} K^{-1}	
	298 K	623 K
$Y_3Fe_5O_{12}$	8.30	11.0
$Y_{2.24}Gd_{0.76}Fe_5O_{12}$	8.00	11.0
$Y_{1.5}Gd_{1.5}Fe_5O_{12}$	8.10	10.4
$Y_3Fe_{4.7}Al_{0.3}O_{12}$	7.95	10.0
$Y_3Fe_{4.3}Al_{0.7}O_{12}$	8.10	10.0
$Y_3Fe_4Al_1O_{12}$	8.40	11.1
$Y_3Fe_{3.8}Al_{1.2}O_{12}$	8.30	10.5

Table 29.28 Resistivity and dielectric loss factor for $Y_3Fe_5O_{12}$ garnet [147–148]

Measured value		Frequency, GHz	Temperature, K
Polycrystalline sample			
ε'	16.7	9	293
tgδ*1:	$5 \cdot 10^{-5}$	9	293
	$7.2 \cdot 10^{-5}$	9	373
	$2.0 \cdot 10^{-4}$	9	473
	$6.2 \cdot 10^{-4}$	9	573
Single-crystal sample			
ρ, Ω·m	$1.0 \cdot 10^{10}$	0	304
	$1.0 \cdot 10^{7}$	0	373
	$2.6 \cdot 10^{3}$	0	473
	$4.6 \cdot 10^{2}$	0	573

*1 ε'' accuracy is $\pm 3 \cdot 10^{-4}$.

Table 29.29 Typical values for magnetostatic parameters of some garnet ferrite films with cylindrical magnetic domains [119] [h – film thickness; l – characteristic length of a material; H_{col} – field of collapse; σ_W – energy of the domain wall; d_{col} – collapse diameter; K_u – uniaxial anisotropy coefficient; δ_W – the width of domain wall.]

Material composition	$4\pi M_s$,	h,	l,	$\lambda \equiv l/h$	H_{col}, 80 A/m		d_{col}, μm		σ_W,	K_u,	δ_W,
	$10^3/(4\pi)$ A/m	μm	μm		observed	calculated	observed	calculated	10^{-3} J/m^2	10^3 J/m^3	μm
$Gd_{1.5}Eu_{1.5}Fe_{4.5}Al_{0.5}O_{12}$	219	16.0	1.78	0.111	160	111	6.0	8.6	0.68	28.8	0.018
$Gd_{2.34}Tb_{0.66}Fe_5O_{12}$	137	15.0	1.53	0.102	75	72	7.5	7.6	0.23	3.3	0.054
$Er_3Fe_{4.3}Ga_{0.7}O_{12}$	132	4.4	1.50	0.341	32	32	6.0	6.0	0.21	2.8	0.058
$Er_2TbFe_{3.9}Al_{1.1}O_{12}$	136	17.0	1.27	0.075	82	79	7.0	7.0	0.19	2.2	0.067
$Y_{2.7}Sm_{0.3}Fe_{3.8}Ga_{1.2}O_{12}$	128	5.2	0.87	0.167	56	53		3.8	0.12	0.90	0.099
$Y_{1.52}Gd_{0.86}Yb_{0.62}Fe_{4.11}Ga_{0.89}O_{12}$	150	3.6	0.80	0.222		52		3.3	0.14	1.3	0.086
$Y_{1.55}Lu_{0.39}La_{0.06}CaFe_4GeO_{12}$	145	4.1	0.79	0.193	57	56		3.4	0.12	0.59	0.175
$Y_{2.86}La_{0.14}Fe_{3.75}Ga_{1.25}O_{12}$	150	3.8	0.72	0.189	46	58		3.0	0.13	1.0	0.098
$Eu_2ErFe_{4.3}Ga_{0.7}O_{12}$ *1	173	6.0	0.70	0.117		86		3.4	0.17	1.8	0.073
$Y_{1.9}Sm_{0.1}CaFe_4GeO_{12}$	163	4.0	0.66	0.165	70	68		2.9	0.13	0.44	0.249
$Eu_2ErFe_{4.3}Ga_{0.7}O_{12}$ *2	247	18.0	0.64	0.036	182	172	5.0	4.7	0.31	6.0	0.040
$Eu_{1.45}Ca_{1.1}Y_{0.45}Fe_{3.9}Si_{0.6}Ge_{0.5}O_{12}$	218	4.23	0.64	0.151	100	96		2.9	0.25	1.3	0.146
$Y_{2.6}Sm_{0.4}Fe_{3.8}Ga_{1.2}O_{12}$	240	2.8	0.63	0.225	64	83		2.6	0.29	5.2	0.044
$Gd_2LuFe_{4.4}Al_{0.6}O_{12}$	189	9.4	0.53	0.056		119		3.2	0.15	0.47	0.130
$Y_{1.03}Gd_{1.29}Yb_{0.68}Fe_{4.3}Al_{0.7}O_{12}$	175	2.1	0.51	0.243	70	58		2.1	0.12	0.96	0.101
$Er_{1.3}Gd_{0.95}Tb_{0.75}Fe_{4.5}Al_{0.5}O_{12}$	181	11.5	0.35	0.030	140	131	3.0	2.7	0.091	0.52	0.138
$Y_{1.2}Lu_{0.65}Sm_{0.4}Fe_{4.1}(CaGe)_{0.9}O_{12}$	330	3.0	0.35	0.117		163		1.7	0.30	2.3	0.104
$Tm_{2.15}Eu_{0.85}Fe_{4.45}Ga_{0.55}O_{12}$	700	0.67	0.13	0.194	212	268		0.55	0.50	5.35	0.074
$Lu_{1.6}Sm_{0.6}La_{0.2}Fe_{4.4}(GaGe)_{0.6}O_{12}$	784	1.1	0.12	0.109		400		0.59	0.58	7.06	0.065
$Lu_{1.8}Sm_{1.2}Fe_5O_{12}$	1750	4.2	0.05	0.012		1435		0.58	1.25	30.4	0.031
$Tm_{2.15}Sm_{0.85}Fe_5O_{12}$	1378	0.87	0.047	0.054	900	879		2.9	0.71	18.1	0.031

*1 Film on the gadolinium–gallium garnet substrate.
*2 Single crystal.

Table 29.30 Faraday effect ($\lambda = 1.152$ μm) in some garnet ferrites [150] [$\Delta a = a_{substr} - a_{film}$]

Film composition	Δa, 10^{-1}nm	M_s, $10^3/(4\pi)$ A/m	θ_F, 10^2deg/m	θ_F/M_s
$Y_3Fe_5O_{12}$	+0.004	1780 ⊥	+250	0.02
$Bi_{0.1}Y_{2.9}Fe_5O_{12}$	−0.010	1750 ⊥	+150	0.008
$Bi_{0.45}Y_{2.55}Fe_4Ga_1O_{12}$	−0.027	310 ⊥	−340	1.2
$Y_3Fe_{4.75}Sc_{0.25}O_{12}$	−0.021	1780 ∥	+175	0.01
$Y_3Fe_{3.95}Sc_{0.25}Ga_{0.8}O_{12}$	−0.012	620 ∥	+95	0.02
$Y_3Fe_{3.75}Sc_{0.25}Ga_{1.0}O_{12}$	−0.009	400 ∥	+80	0.04
$Y_3Fe_{3.55}Sc_{0.25}Ga_{1.2}O_{12}$	+0.002	180 ∥	+70	0.14
$Gd_{0.6}Y_{2.4}Fe_{4.3}Ga_{0.7}O_{12}$	−0.006	450 ∥	+180	0.16
$Gd_{0.7}Y_{2.3}Fe_{3.8}Ga_{1.2}O_{12}$	−0.013	100 ∥	+140	0.96
$Gd_{0.7}Y_{2.3}Fe_{3.8}Ga_{1.2}O_{12}$	−0.009	30 ∥	+53	3.12
$Yb_{2.6}Pr_{0.4}Fe_4Ga_1O_{12}$	+0.022	280 ∥	−123	0.25
$Yb_{2.3}Pr_{0.7}Fe_4Ga_1O_{12}$	−0.017	260 ∥	−125	0.23
$Yb_2Pr_1Fe_4Ga_1O_{12}$	−0.042	386 ∥	−135	0.12
$(YbPr)_{2.5}Bi_{0.5}Fe_4Ga_1O_{12}$	−0.031	270 ∥	−675	6.25
$(YbPr)_{2.3}Bi_{0.7}Fe_{3.8}Ga_{1.2}O_{12}$	−0.002	150 ∥	−950	40
$(YbPr)_{2.1}Bi_{0.9}Fe_{3.85}Ga_{1.15}O_{12}$	−0.019	220 ∥	−1190	30

Table 29.31 Yttrium garnet sublattice magnetizations (M_s, M_a, and M_d accordingly) along with exchange integrals J_{ij} according to measurements by different methods [151]

Method	$4\pi M_s$(295 K), mT	$\frac{M_s(295\ K)}{M_s(0\ K)}$	$\frac{M_a(295\ K)}{M_a(0\ K)}$	$\frac{M_d(295\ K)}{M_d(0\ K)}$	J_{ad}, cm^{-1}	J_{dd}, cm^{-1}	J_{aa}, cm^{-1}
Nuclear magnetic resonance [186]		0.725	0.89	0.835			
Nuclear magnetic resonance [187]		0.733	0.888	0.837			
Nuclear magnetic resonance [188]		0.730	0.895	0.84			
Neutron diffraction [189]		0.76	0.94	0.88			
Pendulum magnetometer [190]	181.2	0.734					
Pendulum magnetometer [191]	179	0.725					
Magnetic balance [192]	177	0.717			25.36	8.45	11.86
Vibromagnetometric; Magnetostatic modes; Induction [193]	180±1	0.729					
Theory of molecular field [151]	176.7	0.729	0.896	0.84	25.64	7.8	11.0
Scattering of neutrons [191]					27.24	4.38	13.07

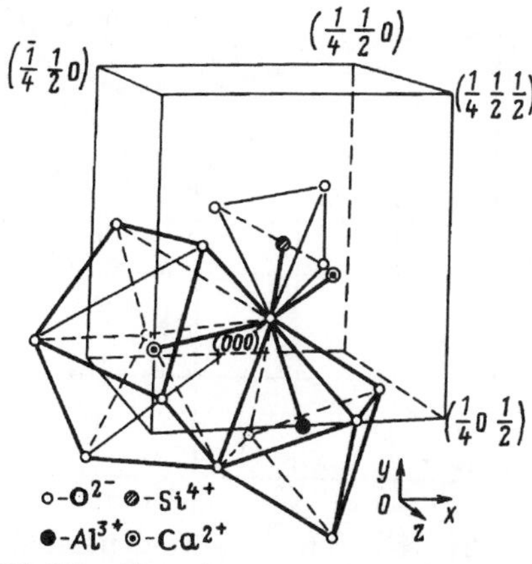

Figure 29.20 Coordination of ions in different sublattices of the garnet $Ca_3Al_2Si_3O_{12}$ structure [152].

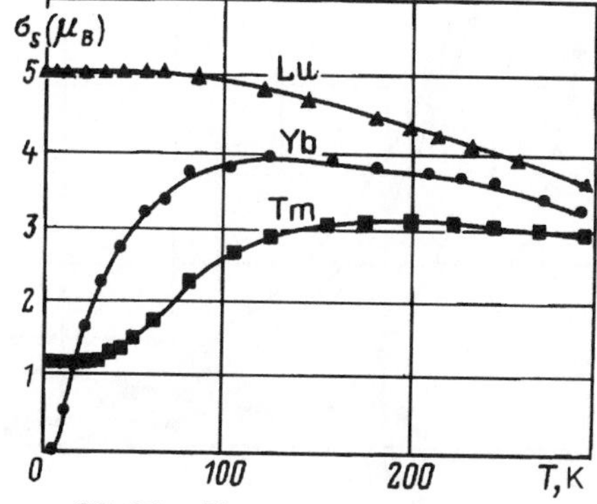

Figure 29.21 Temperature-dependent spontaneous magnetization (in Bohr magnetons) per formula unit in Tm, Yb and Lu garnet ferrites [154].

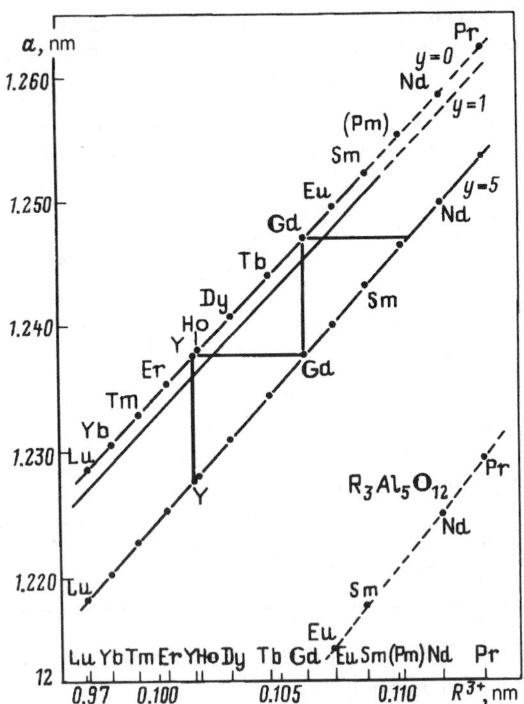

Figure 29.22 Dimensions of an elementary cell in $R_3Fe_{5-y}Ga_yO_{12}$ garnet ferrites; $r^{VI}_{Fe^{3+}} = 6.42 \cdot 10^{-2}$ nm; $r^{VI}_{Ga^{3+}} = 6.17 \cdot 10^{-2}$ nm; $r^{VI}_{Al^{3+}} = 5.40 \cdot 10^{-2}$ nm; $r^{VIII}_{La^{3+}} = 1.192 \cdot 10^{-2}$ nm [153].

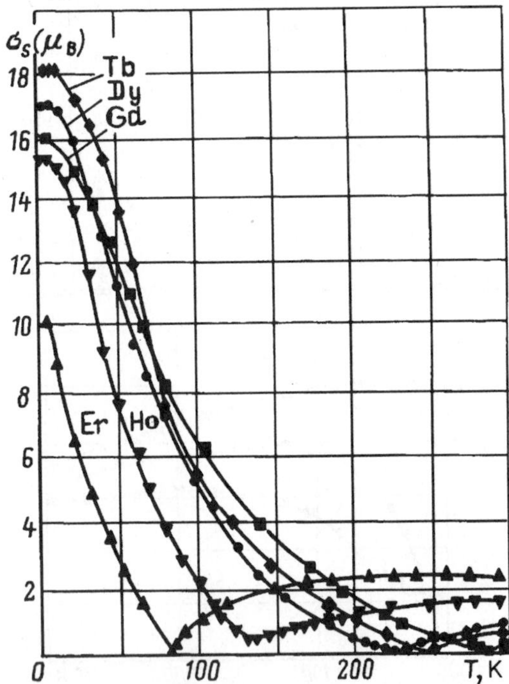

Figure 29.23 Temperature-dependent spontaneous magnetization (in Bohr magnetons) per formula unit in Gd, Tb, Dy, Ho and Er garnet ferrites [154].

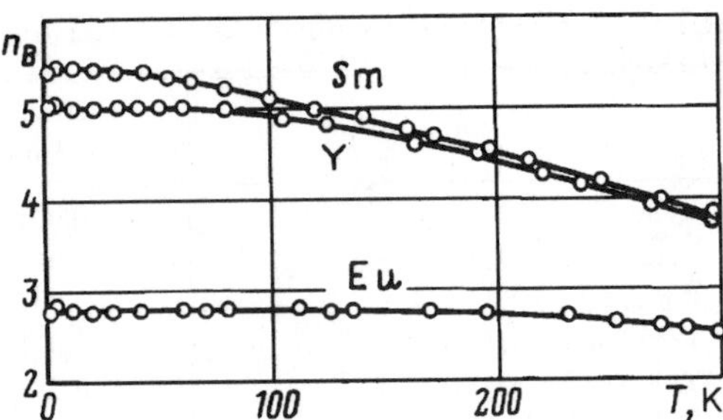

Figure 29.24 Temperature-dependent saturation magnetization (in Bohr magnetons) per formula unit in single-crystal samples of the sphere form, which can rotate easily in external field. Data for yttrium garnet ferrite were obtained on polycrystalline sample [155].

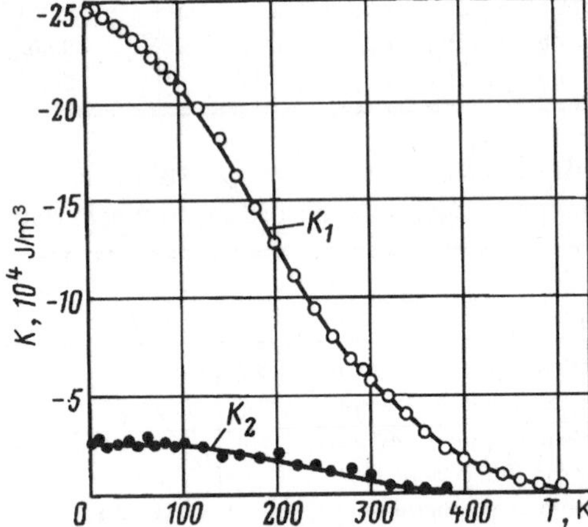

Figure 29.25 Temperature-dependent first and the second constants of magnetic crystallographic anisotropy in $Y_3Fe_5O_{12}$ [156]. The K_2 values at low temperatures strongly depend on admixture concentration (especially on concentration of Fe^{2+}).

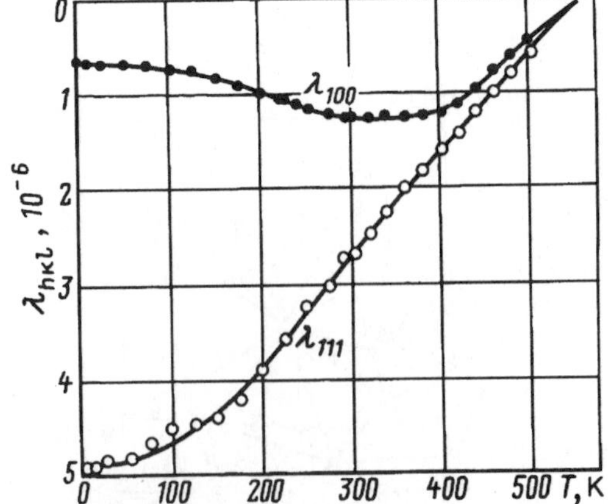

Figure 29.26 Temperature-dependent magnetostriction coefficients λ_{100} and λ_{111} in yttrium garnet ferrite $Y_3Fe_5O_{12}$ [157].

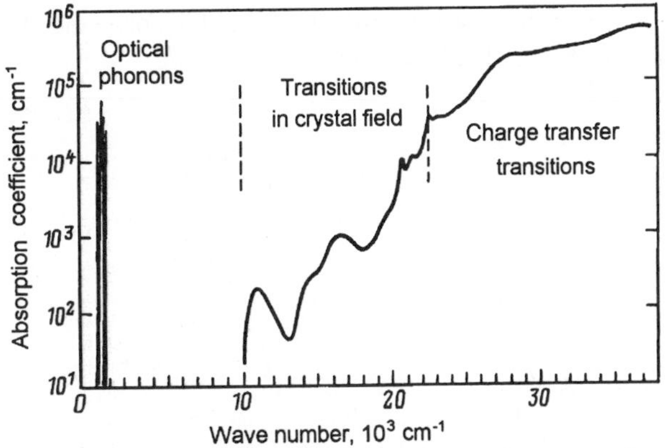

Figure 29.27 Dependence of absorption coefficient (lgα) on wave number in $Y_3Fe_5O_{12}$. In the range from 10,000 cm^{-1} to 40,000 cm^{-1}, data correspond to a temperature of 77 K. The phonon absorption bands are outlined schematically [158].

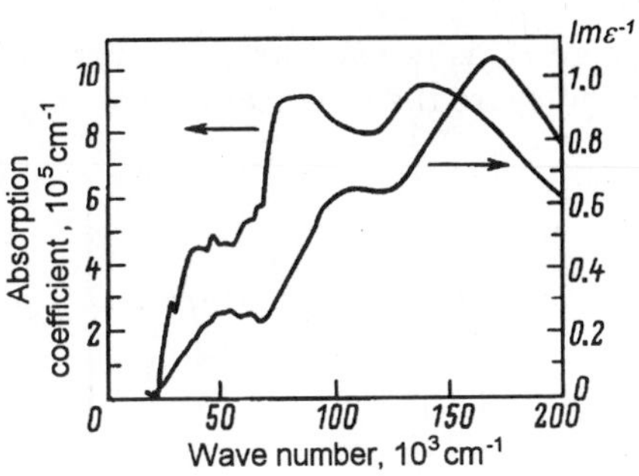

Figure 29.29 Absorption coefficient in $Y_3Fe_5O_{12}$ at high energies [159].

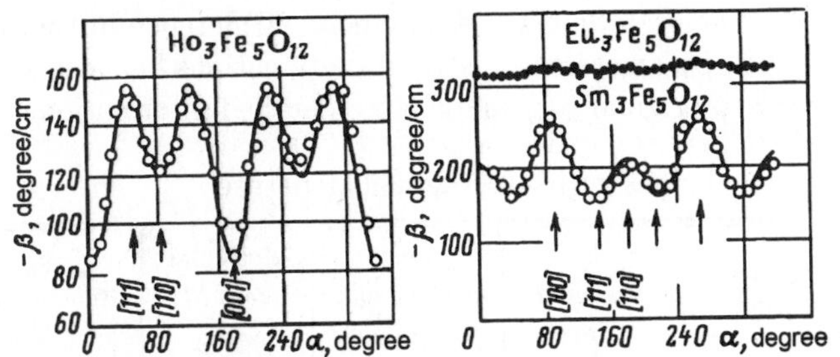

Figure 29.28 Anisotropy of linear magnetic double refraction at a wave-length λ=1.15 μm at T=300 K in holmium, europium, and samarium garnet ferrites [161].

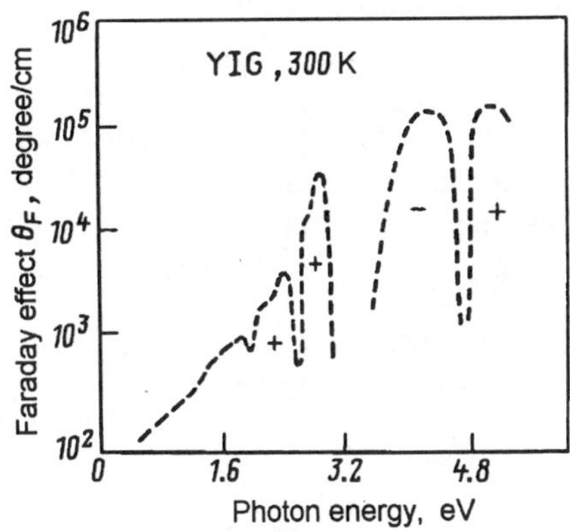

Figure 29.30 Faraday effect in yttrium garnet ferrite $Y_3Fe_5O_{12}$ [160].

29.4 Hexagonal Ferrites

A large group of ferrimagnetic oxides possesses a hexagonal crystal structure. Fig. 29.31 shows the diagram displaying chemical composition of such compounds. In the corners, the BaO, MeO and Fe_2O_3 compounds are placed. The Me symbol denotes the bivalent ion of the first transition group or ions Zn^{2+} and Mg^{2+} as well as combination of such ions (for example, Li^+ and Fe^{3+}). On the line of the diagram connecting BaO and Fe_2O_3, a point is marked that corresponds to the antiferromagnetic barium ferrite $BaFe_2O_4$. The S point in the line MeO – Fe_2O_3 relates the $Me_2Fe_4O_8$ compound with spinel-type cubic crystal structure. The M point is appropriate to the oxide with hexagonal structure and chemical composition $BaFe_{12}O_{19}$=BaO·6Fe_2O_3. The Y point of the diagram corresponds to the compound $Ba_2Me_2Fe_{12}O_{22}$=2(BaO·MeO·3Fe_2O_3).

A great number of compounds with compositions in the lines $M - S$ and $M - Y$ are known. The elementary cell of these compounds can be easily built from the cells of the S, M, and Y structures. In the majority of cases, the Ba^{2+} ion (r = 0.143 nm) can be partially or fully substituted by ions Ca^{2+} (r = 0.106 nm), Sr^{2+} (r = 0.127 nm) or Pb^{2+} (r=0.132 nm), or by the trivalent ions like La^{3+} (r = 0.122 nm).

The crystallographic magnetic anisotropy energy of hexagonal crystals is described by the formula

$$E_a = K_1\sin^2\theta + K_2\sin^4\theta + K_3'\sin^6\theta + K_3\sin^6\cos^6(\varphi - \psi), \quad (29.12)$$

where E_a is the anisotropy energy; K_1, K_2, K'_3, and K_3 are the anisotropy coefficients; and, finally, θ and φ are the spherical coordinates. For the anisotropy fields one has

$$H_\theta^A = \begin{cases} \dfrac{2K_1}{M_s}, & \theta = 0°; \\ -\dfrac{2(K_1+2K_2)}{M_s}, & \theta = 90°; \\ \dfrac{2K_1}{K_2}\dfrac{K_1+2K_2}{M_s}, & \theta = \arcsin\sqrt{\dfrac{K_1}{2K_2}} \end{cases} \tag{29.13}$$

$$H_\varphi^A = 36|K_3|\frac{\sin^4\theta}{M_s}. \tag{29.14}$$

Here H_θ^A is the effective field intensity that is necessary for turning the magnetization vector to these directions where only θ changes; H_φ^A is the field which rotates the magnetization vector at the cone surface.

The hexagonal ferrites' properties are shown in Tables 29.32–29.44 and in Figs. 29.31–29.41.

Some inaccuracies in numerical values of the tabulated parameters are possible because of uncertainties in the composition and physical state of samples under measurement. Difference in thermal treatment history of samples can also change such parameters as distribution of cations between sites, porosity, and so on. That is why one must refer to original papers for more detailed information.

Table 29.32 Some properties of ferromagnetic oxides with a hexagonal structure

Oxide	Symbol	Number of layers in a unit cell	Space group	c parameter, nm	X-ray density d_X, 10^3 kg/m^3	Molecular mass M, a.m.u.	References
$BaFe_{12}O_{19}$	**M**	(10)	$P6_3/mmc$	2.32	5.28	1112	[164]
$Ba_2Fe_4O_8$	**S**				5.24	232	[166]
$BaFe_2Fe_{16}O_{27}$	**W**	$(14)_1$	$P6/mmc$	3.2845	5.31	1575	[165]
$Ba_2Fe_2Fe_{28}O_{46}$	**X**	$(12)_3$	$R\bar{3}m$	8.411	5.29	2686	[165]
$Ba_2Fe_2Fe_{12}O_{22}$	**Y**	$(6)_3$	$R\bar{3}m$	4.3588	5.39	1408	[165]
$Ba_3Fe_2Fe_{24}O_{41}$	**Z**	$(22)_1$	$P6_3/mmc$	5.23	5.33	1520	[165]
$Ba_3Zn_2Fe_{40}O_{65}$		$(17)_2$	$P6_3/mmc$	7.936			[167, 168]
$Ba_4Zn_2Fe_{36}O_{60}$	**M$_2$Y**	$(16)_3$	$R\bar{3}m$	3.81	5.31	3622	[167]
$Ba_4Zn_2Fe_{52}O_{84}$		$(22)_3$	$R\bar{3}m$	15.385			[167, 168]
$Ba_5Zn_2Fe_{64}O_{103}$		$(27)_2$	$P6_3/mmc$	12.576			[167, 168]
$Ba_8Zn_6Fe_{60}O_{104}$		$(28)_3$	$R\bar{3}m$	20.04			[167]
$Ba_{10}Zn_8Fe_{72}O_{126}$		$(34)_3$	$R\bar{3}m$	24.398			[167]
$Ba_{12}Zn_{10}Fe_{84}O_{148}$		$(40)_1$	$R\bar{3}m$	9.584			[167]
$Ba_{14}Zn_{12}Fe_{96}O_{170}$		$(46)_3$	$R\bar{3}m$	33.109			[167]

Table 29.33 Composition and structure of hexagonal ferrites of the Me_2Y_n type, where Me=Mn, Zn [169]

Composition	Sequence of blocks	Number of anion layers	Space group	c parameter, nm	
				experiment	calculation
$Ba_{20}(Mn, Zn)_{18}Fe_{132}O_{236}$	**M(Y)$_4$M(Y)$_5$**	$(64)_3$	$R\bar{3}m$	46.2	46.2
$Ba_{22}(Mn, Zn)_{20}Fe_{144}O_{258}$	**M(Y)$_5$M(Y)$_5$**	$(70)_1$	$P6_3/mmc$	16.86	16.85
$Ba_{24}(Mn, Zn)_{22}Fe_{156}O_{280}$	**M...M(Y)$_{11}$**	$(76)_3$	$R\bar{3}m$	54.93	54.92
$Ba_{24}(Mn, Zn)_{22}Fe_{156}O_{280}$	**M(Y)$_1$M(Y)$_{10}$**	$(76)_1$	$R\bar{3}m1$	18.31	18.31
$Ba_{24}(Mn, Zn)_{22}Fe_{156}O_{280}$	**M(Y)$_4$M(Y)$_7$**	$(76)_1$	$R\bar{3}m1$	18.31	18.31
$Ba_{24}(Mn, Zn)_{22}Fe_{156}O_{280}$	**M(Y)$_5$M(Y)$_6$**	$(76)_3$	$R\bar{3}m$	54.93	54.92
$Ba_{28}(Mn, Zn)_{26}Fe_{180}O_{324}$	**M(Y)$_6$M(Y)$_7$**	$(88)_3$	$R\bar{3}m$	63.67	63.64

Table 29.34 X-ray density d_X and molecular mass M of some oxides with hexagonal crystal structure [5]

Metal	$\mathbf{W}=BaMe_2Fe_{16}O_{27}$		$\mathbf{Y}=Ba_2Me_2Fe_{12}O_{22}$		$\mathbf{Z}=Ba_3Me_2Fe_{24}O_{41}$	
	d_X, 10^3 kg/m^3	M, a.m.u.	d_X, 10^3 kg/m^3	M, a.m.u.	d_X, 10^3 kg/m^3	M, a.m.u.
Mg	5.10	1512	5.14	1346	5.20	2457
Mn	5.31	1573	5.38	1406	5.33	2518
Fe	5.31	1575	5.39	1408	5.33	2520
Co	5.31	1577	5.40	1410	5.35	2522
Ni	5.32	1580	5.40	1414	5.35	2526
Cu	5.36	1590	5.45	1424	5.37	2536
Zn	5.37	1594	5.46	1428	5.37	2539

Table 29.35 Curie temperature and saturation magnetization of $Me_2\mathbf{W}$-type hexaferrites [5]

Metal	T_C, K	σ_s, 10^{-3} kA·m^2/kg		M_s, kA/m	n_B, μ_B at T=0 K	
		0 K	293 K	0 K	experiment	calculation
Mn_2	690	97	59	310	27.4	29.2
Fe_2^{2+}	730	98	78	416	27.4	28
$NiFe^{2+}$	790	79	52	275	22.3	26.4
$ZnFe^{2+}$	700	108	73	382	30.7	31.6
$Ni_{0.5}Zn_{0.5}Fe^{2+}$	720	104	68	362	29.5	29.2

Table 29.36 Curie temperature and saturation magnetization of $Me_2\mathbf{Y}$-type hexaferrites [5]

Metal	T_C, K	σ_s, 10^{-3} kA·m^2/kg		M_s, kA/m	n_B, μ_B at T=0 K	
		0 K	293 K	0 K	experiment	calculation
Mg	550	20	23	120	6.9	2.2
Mn	560	42	31	167	10.6	9.2
Co	610	39	34	183	9.8	7.4
Ni	660	25	24	127	6.3	4.6
Cu		28			7.1	2.6
Zn	400	72	42	227	18.4	20.0

Table 29.37 Curie temperature and saturation magnetization of $Me_2\mathbf{Z}$-type hexaferrites [5]

Metal	T_C, K	σ_s, 10^{-3} kA·m^2/kg		M_s kA/m	n_B, μ_B at T=0 K	
		0 K	293 K	0 K	experiment	calculation
Mg		55			24	26.9
Co	680	69	50	267	31.2	29.8
Ni		54			24.6	26.1
Cu	710	60	46	247	27.2	27.1
Zn	630		58	310		

Table 29.38 Magnetic properties of the SrO·nFe$_2$O$_3$ and PbO·nFe$_2$O$_3$ ferrites [170–172]

Compound		M_s, kA/m	M_r, kA/m	H_c, kA/m	Compound		M_s, kA/m	M_r, kA/m	H_c, kA/m
$SrFe_8O_{13}$	($SrO·4FeO_3$)	259	179	239	$PbFe_8O_{13}$	($PbO·4Fe_2O_3$)	279	159	119
$SrFe_{12}O_{19}$	($SrO·6Fe_2O_3$)	247	183	229	$PbFe_{12}O_{19}$	($PbO·6Fe_2O_3$)	199	143	231
$SrFe_{18}O_{28}$	($SrO·9Fe_2O_3$)	151	120	267	$PbFe_{18}O_{28}$	($PbO·9Fe_2O_3$)	135	80	191

Table 29.39 Anisotropy constant K_3, saturation magnetization M_s, and anisotropy field H_φ^A for some hexagonal ferromagnetic oxides [5]

Compound	T, K	M_s, kA/m	K_3, J/m^3	H_φ^A, A/m
$Ba_2Co_2Fe_{12}O_{22}$	77	210	200	27100
	293	187	80	12300
	443	145	15	2940
$Ba_3Co_{1.5}Fe_{0.5}^{2+}Fe_{24}O_{41}$	293	280	5.5	558
	431	220	0.9	160
$Ba_3Co_{1.92}Fe_{0.08}^{2+}Fe_{24}O_{41}$	293	280	12.0	1270
$Ba_2CoZn_{0.5}Fe_{0.5}^{2+}Fe_{24}O_{41}$	293	290	2.5	239
$Ba_2Zn_{1.5}Fe_{0.5}^{2+}Fe_{12}O_{22}$	293	190	0.6	80

Table 29.40 Anisotropy constants K_1 or $K_1 + 2K_2$, saturation magnetization M_s and anisotropy field H_θ^A for some hexagonal oxides at a temperature of 293 K [5]

Oxide	Symbol	K_1, 10^5 J/m^3	K_1+2K_2, 10^5 J/m^3	M_s, kA/m	H_θ^A, kA/m
$BaFe_{12}O_{19}$*1	**M**	+3.3		380	1350
$BaFe_{18}O_{27}$*1	Fe_2**W**$_2$	+3.0		314	1510
$BaZnFe_{17}O_{27}$	FeZn**W**	+2.4		380	1000
$BaZn_{1.5}Fe_{16.5}O_{27}$	$Fe_{0.5}Zn_{1.5}$**W**	+2.1		380	885
$BaMnZnFe_{16}O_{27}$	MnZn**W**	+1.9		370	811
$BaNi_2Fe_{16}O_{27}$	Ni_2**W**	+2.1		330	1010
$BaNi_{0.5}ZnFe_{16.5}O_{27}$	$ZnFe_{0.5}Ni_{0.5}$**W**	+1.6		350	725
$BaCo_{0.75}Zn_{0.75}Fe_{16.5}O_{27}$	$Fe_{0.5}Co_{0.75}Zn_{0.75}$**W**		−0.4	360	175
$Ba_2Mg_2Fe_{12}O_{22}$	Mg_2**Y**		−0.6	119	800
$Ba_2Ni_2Fe_{12}O_{22}$	Ni_2**Y**		−0.9	127	1110
$Ba_2Zn_2Fe_{12}O_{22}$	Zn_2**Y**		−1.0	227	715
$Ba_2Zn_{1.5}Fe_{12.5}O_{22}$*1	$Fe_{0.5}Zn_{1.5}$**Y**		−0.9	191	756
$Ba_2Co_2Fe_{12}O_{22}$*1	Co_2**Y**		−2.6	185	2230
$Ba_3Co_2Fe_{24}O_{41}$*1	Co_2**Z**		−1.8	270	1030

*1 Measurements were performed on single crystals.

Table 29.41 Magnetic properties of single-crystal $BaFe_{12-2x}Ir_x^{4+}Zn_x^{2+}O_{19}$ hexaferrite with high anisotropy in the basal plane [173]

x	H_a, kA/m (T=300 K)		ΔH, kA/m (T=300 K)	T_C, K	σ_s, A·m^2/kg (T=300 K)
	from magnetization measurements	from ferromagnetic resonance			
0	1350	1350	4.2 at 55 GHz	720	65
0.16	765			635	
0.52	1200	1270	83.5 at 17 GHz	560	65
0.56	1270			550	60
0.60	1600	1750	93.5 at 14 GHz	520	47

Table 29.42 Ferromagnetic resonance line-width for Me_2**Y** single crystals [174]

Single crystal	c parameter, nm	ΔH, A/m (T=300 K)
$Ba_2Zn_2Fe_{12}O_{22}$	4.36	638 (9000 MHz)
$Ba_2(Zn, Mn)Fe_{12}O_{22}$ (3.5% Mn by mass)	4.3564	303 (9000 MHz) 390 (17300 MHz)

Table 29.43 Properties of some waveguides-used hexagonal ferrites in the millimeter wavelength range [175]

Composition	T_{anneal}, K	ρ, 10^3 kg/m^3	Degree of orientation	$\tan\delta$ (f=9.5 GHz)	$\varepsilon'/\varepsilon_H$ (f=9.5 GHz)	M_s, kA/m	H_a, kA/m	T_C, K	f_{res}, GHz
$BaO{\cdot}2NiO{\cdot}xAl_2O_3{\cdot}(8\text{-}x)Fe_2O_3$ *system is denoted as* Ni**W**(x Al)									
Ni_2**W** (0.40 Al)	1573	4.58	0.69				1210	750	49.0
Ni_2**W** (0.60 Al)	1623	4.58	0.88	0.001	15.4	214	1360	730	52.0
Ni_2**W** (0.73 Al)	1623	4.60	0.83	0.003	14.9	195	1430	720	55.0
Ni_2**W** (0.86 Al)	1623	4.63	0.84	0.004	15.1	181	1510	710	58.0
Ni_2**W** (1.00 Al)	1623	4.55	0.86	0.002	14.5	166	1570	690	62.0
$SrO{\cdot}xAl_2O_3{\cdot}(6\text{-}x)Fe_2O_3$ *system is denoted as* Sr**M**(x Al)									
Sr**M** (0.00 Al)	1623	4.91	0.79			334	1510	750	60.0
Sr**M** (0.20 Al)	1573	4.48	0.87	0.001	18.6	263	1600	730	64.0
Sr**M** (0.53 Al)	1573	4.11	0.89	0.001	16.2	168	2010	690	74.0
Sr**M** (0.80 Al)	1573	4.00	0.91				2470	650	86.0
Sr**M** (0.95 Al)	1573	3.95	0.88				2780	630	93.0

Table 29.44 Anisotropy field H_a and $\tan\delta$ for some **M**-type hexaferrites used in millimeter wavelength range [176]

Hexaferrite	H_a, kA/m	$\tan\delta$ (9 GHz)	f_{res}, GHz
$BaZn_{0.3}Ti_{0.3}Fe_{11.4}O_{19}$	1090	0.002	45
$BaAl_{0.3}Fe_{11.7}O_{19}$	1390	0.002	55
$SrNi_{0.3}Ge_{0.3}Al_{0.86}Fe_{9.54}O_{19}$	2170	0.006	82
$SrNi_{0.1}Ge_{0.3}Al_{2.5}Fe_{9.1}O_{19}$	2595	0.008	96

Table 29.45 Magnetic properties of trivalent chrome compounds with halogens [177] [θ_p is the paramagnetic Curie point; $2K_1/M_s$ is the anisotropy field intensity; p_{eff} is the effective magnetic moment.]

Compound	Structure	ρ_x, 10^3 kg/m^3	T_C, K	θ_p, K	n_B, μ_B	M_s, kA/m	$2K_1/M_s$, kA/m (T=1.5 K)	K_1, J/m^3 (T=1.5 K)	p_{eff}, μ_B
CrF_3	D^6_{3d}—$R\bar{3}c$; a=0.52643 nm; α=56.563°		69.8	-124					3.85
$CrCl_3$	D^3_3—$P3_112$; D^5_3—P_212; a=0.6 nm; c=1.73 nm	2.95	16.8	+31		308			3.69
$CrBr_3$	C^2_{3i}—$R\bar{3}$; a=0.626 nm; c=1.82 nm	4.75	35.7	+47	3.0	259	546	$9.4{\cdot}10^4$	3.85
CrI_3	D^3_3—$P3_112$	5.36	68	+70	3.1	214	2260	$3.1{\cdot}10^5$	4.03

Table 29.46 Magnetic properties of $CoMnO_3$, NiMnO, $BiMnO_3$, and $BiCrO_3$ ferrites

Compound	Structure	Cell parameters (T=300 K)	T_C, K	n_B, μ_B	p_{eff}, μ_B	References
$CoMnO_3$	C^2_{3i}—$R\bar{3}$	a=0.5385 nm α=54°31′	120	0.72 (0 K)		[180–182]
$NiMnO_3$	C^2_{3i}—$R\bar{3}$	a=0.5343 nm α=54°39′	120–160	0.76 (0 K)		[180–182]
$BiMnO_3$	perovskite	a=c=0.3935 nm b=0.3989 nm α=γ=91°28′, β=90°58′	103	2 (77 K) 4 (extrapolation)	5	[178, 179]
$BiCrO_3$	perovskite	a=c=0.3906 nm b=0.387 nm α=γ=90°33′, β=89°9′	123			[178, 179]

Table 29.47 Crystallographic and magnetic properties of ferromagnetic fluorides with Fe^{3+}, Cr^{3+}, and Co^{3+} ions [183]

Compound	Cell parameters		T_C, K	n_B, μ_B	M_s, kA/m
	in high-temperature phase	in low-temperature phase		(0 K)	(0 K)
$Na_5Fe_3F_{14}$	a=0.734 nm, c=1.038 nm ρ=3260 kg/m^3	a=7.323 nm, b=0.746 nm c=1.272 nm β=90±0.5°, ρ=3150 kg/m^3	80	5	199
$Na_5Cr_3F_{14}$			<20		
$Na_5Co_3F_{14}$			77< T_C <200		

Table 29.48 Magnetic properties of bivalent europium compounds

Compound	Structure	Cell parameters, nm	T_C, K	θ_p, K	n_B, μ_B (T=0 K)	References
EuF_2	Cubic	0.585	2	−5		[184]
$EuCl_2$	Rhombic	a=0.448, b=0.748 c=0.896		0		[184]
$EuBr_2$	Rhombic	a=0.43, b=0.92 c=1.142		0		[184]
EuI_2	Monoclinic	a=0.762, b=0.823 c=0.788, β=98°	5	+5	7	[184]
EuO	NaCl	0.514	73	+76	6.8	[184]
EuS		0.595	16.5	+19	6.87	
EuSe		0.619	7	+9	6.7	
EuTe		0.66	9.5	−6	6.9	
$Eu_2P_2O_7$	Tetrahedral			−3		[185]
$Eu_3(PO_4)_2$	Rhombohedral			+5	6.7	[185]
Eu_2SiO_4	Rhombic (powder)	a=0.971, b=4.956 c=0.595	7	+7	6.0	[184]
Eu_2SiO_4	Rhombic (single crystal)	a=0.971, b=4.956 c=0.565		+10	6.5	[184]
Eu_3SiO_5	Tetrahedral		4	+19	6.7	[184]
$Eu_3Al_2O_4$				0		[185]
$Eu_3Al_2O_6$	Pseudocubic			10	6.2	
$Eu_5Al_2O_8$				6	5.8	

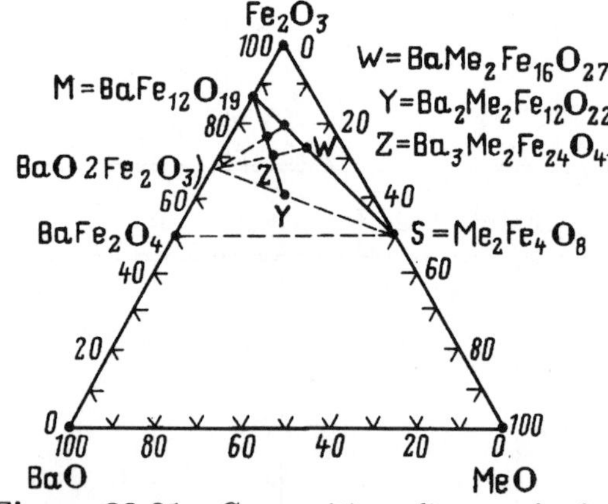

Figure 29.31 Compositions diagram for ferrimagnetic oxides with a hexagonal structure. The Me symbol denotes a bivalent ion (or combination of bivalent ions) [166].

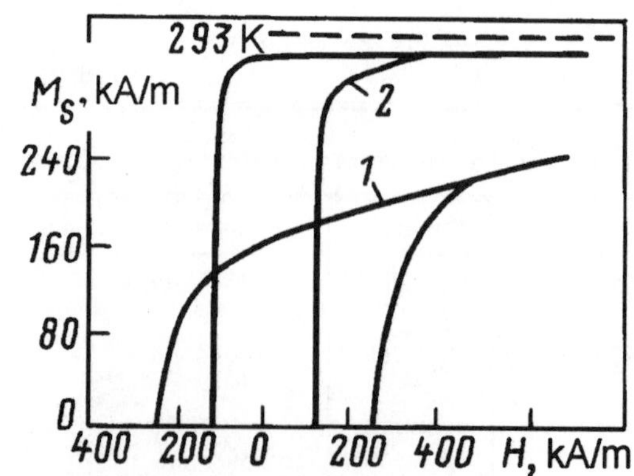

Figure 29.32 Hysteresis loops for $BaFe_{12}O_{19}$ samples [5]: *1* – isotropic sample; *2* – crystallographically textured sample.

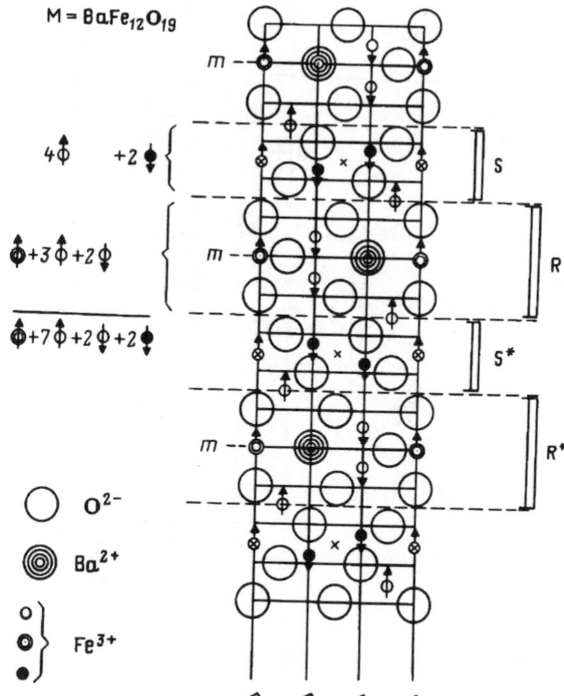

Figure 29.33 Cross–section of magnetoplumbite **M** structure with the c-axis directed vertically [5]: arrows – spin directions; vertical lines – symmetry axes of the third order; crosses – places of symmetry centers; mirror planes denoted by m pass through layers containing barium ions; the displayed structure consists of spinel blocks S separated by blocks with barium ions; stars denote rotation of a given block about c-axis by 180°.

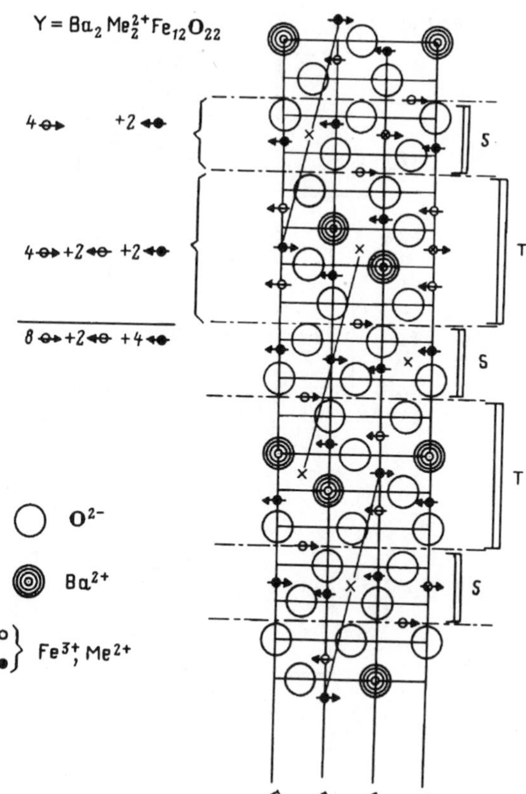

Figure 29.35 Cross–section of the **Y**-type structure with the c axis directed vertically [5]: arrows – spin directions, these being oriented perpendicular to c-axes in this case; vertical lines – symmetry axes of the third order; crosses – places of the symmetry centers; the displayed structure consists of the successively placed blocks **S** and **T**.

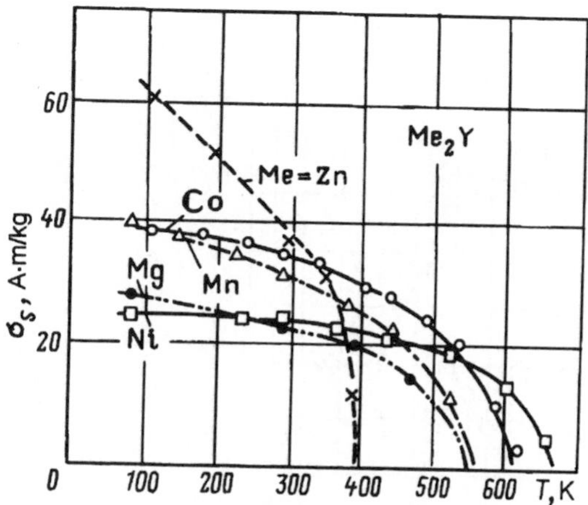

Figure 29.34 Dependence of saturation magnetization σ_s on temperature in compounds with the **Y**-type structure. Measurements were performed on polycrystalline samples in the field of 875 kA/m (11 kOe) [5].

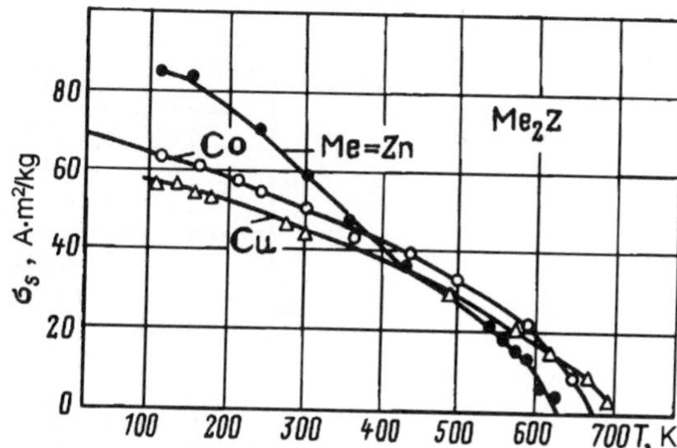

Figure 29.36 Dependence of saturation magnetization σ_s on temperature in compounds with the **Z**-type structure. Measurements were performed on polycrystalline samples in the field of intensity 875 kA/m for Co_2**Z** and Zn_2**Z**, and 1430 kA/m for Cu_2**Z** [5].

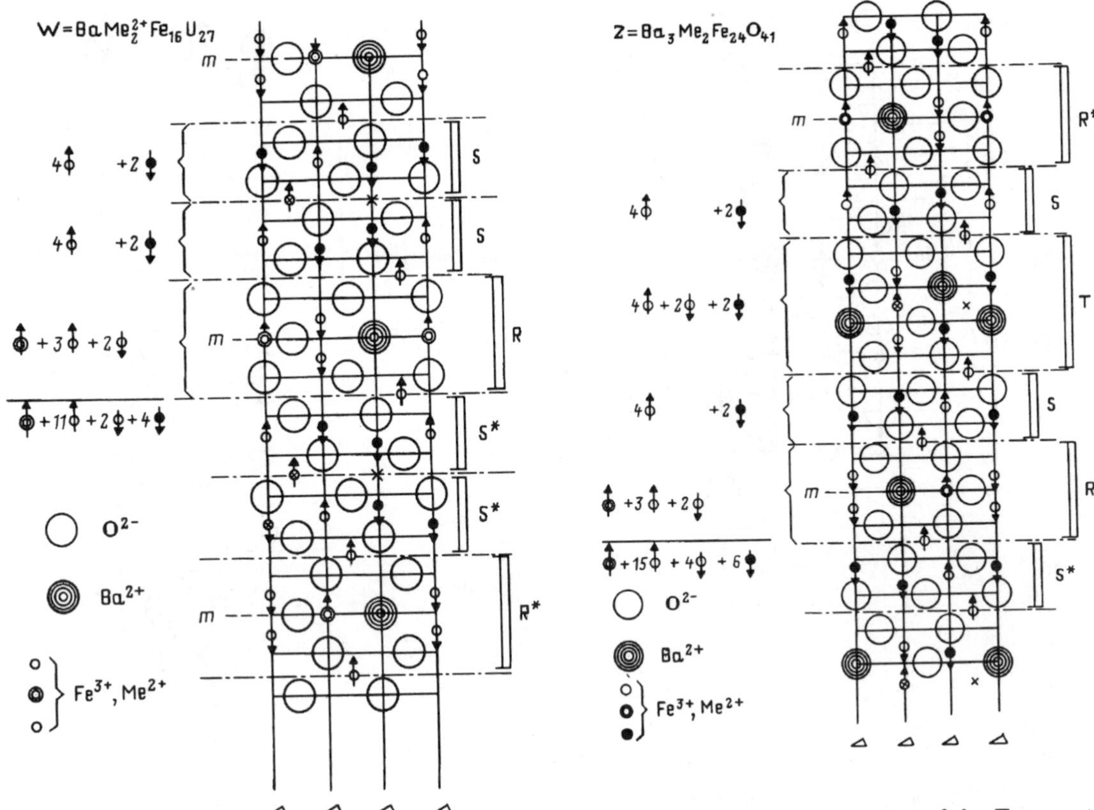

Figure 29.37 Cross–section of the **W**-type structure with the c axis directed vertically [5]: arrows – spin directions; vertical lines – symmetry axes of the third order; stars mark rotation of the given block about the c-axis by 180°; the displayed structure can be considered as the sum of **M**- and **S**-type structures.

Figure 29.38 Cross–section of the **Z**-type structure with the c-axis directed vertically [5]: arrows – spin directions; vertical lines – symmetry axes of the third order; crosses – places of the symmetry centers; stars mark rotation of the given block about the c-axis by 180°; the displayed structure can be considered as the sum of **M**- and **Y**-type structures.

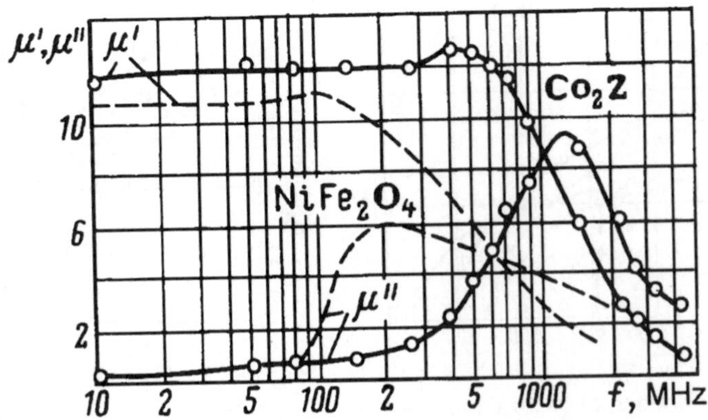

Figure 29.39 Magnetic spectra of polycrystalline sample Co_2**Z** and of spinel $NiFe_2O_4$ (which possesses approximately the same magnetic permeability at low frequencies) [5].

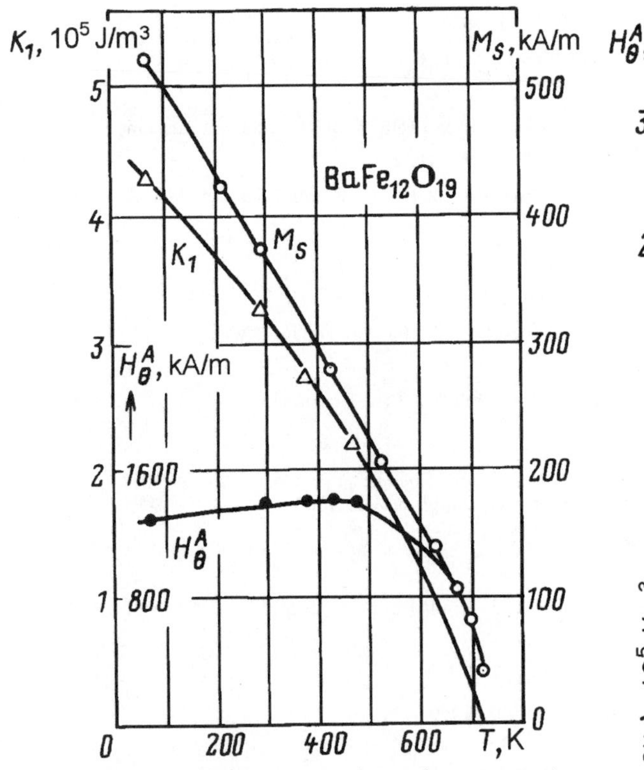

Figure 29.40 Temperature-dependent saturation magnetization M_s, anisotropy constant K_1, and anisotropy field H_θ^A for $BaFe_{12}O_{19}$.

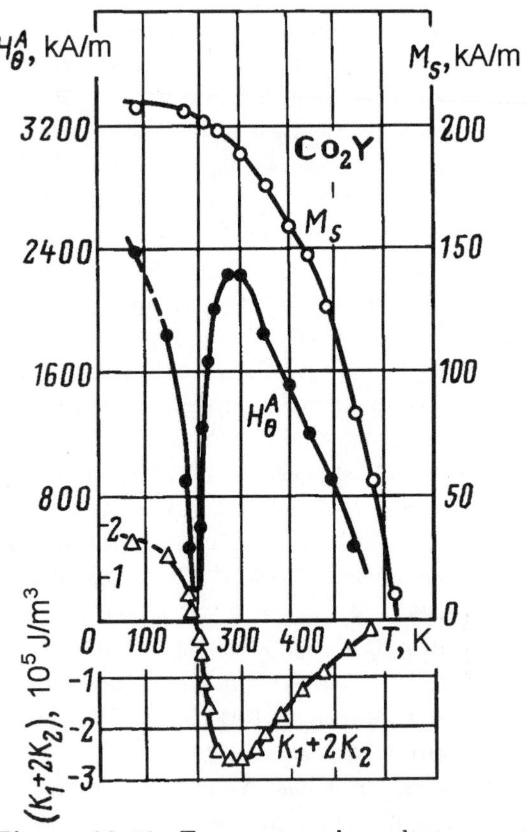

Figure 29.41 Temperature-dependent magnetization M_s, anisotropy constant (K_1+2K_2), and anisotropy field H_θ^A for Co_2**Y** [5].

References

[1] Néel, L., Antiferromagnetics, Izdatel'stvo inostrannoi literatury, Moscow, 1956 (Russian Transl.).

[2] Vonsovskii, S. V., Magnetism, Nauka, Moscow, 1971 (in Russian).

[3] Gurevich, A. G., Magnetic resonance in ferrites and antiferromagnets, Nauka, Moscow, 1973 (in Russian).

[4] Wangsness, R. K., Phys. Rev., 93, 68, 1954.

[5] Smit J., Wijn H., Ferrites, Elsevier, Amsterdam, 1959.

[6] Shull, C. G., Wollen, E. O., Kohler, W. C., Phys. Rev., 84, 912, 1951.

[7] Bickford, L. R., Brownlow, J. M., Penoyer, R. F., Proc. Inst. Electr. Engrs., 104B, Suppl. 5, 238, 1957.

[8] Bickford, L. R., Pappis, J., Stull, J. L., Phys. Rev., 99, 1210, 1955.

[9] Bickford, L. R., Phys. Rev., 78, 449, 1950.

[10] Calhoun, B. A., Phys. Rev., 94, 1582, 1954.

[11] Nathans, R., Proc. Inst. Electr. Engrs., 104B, Suppl. 5, 217, 1957.

[12] Verwey, J. W., Heilmann, E. L., J. Chem. Phys., 15, 174, 1947.

[13] Gorter, E. W., Philips Res. Repts., 9, 295, 1954.

[14] Galt, J. K., Matthias, B. T., Remeika, J. P., Phys. Rev., 79, 391, 1950.

[15] Bozorth, R. M., Tilden, E. F., Williams, A. J., Phys. Rev., 99, 1788, 1955.

[16] Dwight, K., Menyuk, N., Bull. Amer. Phys. Soc., Ser. 2, 3, 41, 1958.

[17] Smith, J., Wijn, H. P. J., Advances in Electronics and Electr. Phys., 6, 83, 1954.

[18] Vager, W. A., Galt, J. K., Merritt, F. R., Phys. Rev., 99, 1203, 1955.

[19] Gendelev, S. Sh., Lapovok, B. L., Rubinshtein, B. E., Fizika tverdogo tela, 5, 3037, 1963 (in Russian).

[20] Sekizawa, H., Sekizawa, H., J. Phys. Soc. Jpn., 17, Suppl. B-1, 380, 1962.

[21] Epstein, D. J., Conf. on Magn. and Magn. Mater., Amer. Institute of Electr. Engng., Boston, MA, 1957, 498.

[22] Van Uitert, L. G., J. Chem. Phys., 24, 306, 1956.

[23] Hastings, J. M., Corliss, L. M., Phys. Rev., 104, 328, 1956.

[24] Dillon, J. F., Geschwind, S., Jaccarino, V., Phys. Rev., 100, 750, 1955.

[25] Gurevich, A. G., Gubler, I. E., Titova, A. G., Fizika tverdogo tela, 3, 19, 1961 (in Russian).

[26] Teale, R. W., J. Appl. Phys., 33, Suppl., 1665, 1962.

[27] Zaveta, K., Svirina, E., Malikova, O., Fizika tverdogo tela, 4, 3593, 1962 (in Russian).

[28] Braun, P. B., Nature, 170, 1123, 1952.

[29] Folen, V. J., J. Appl. Phys., 31, Suppl.3, 1665, 1960.

[30] Schnitzler, A. D., Folen, V. J., Rado, G. T., J. Appl. Phys., 31, Suppl. 3, 348, 1960.

[31] Enz, U., Erzeugung von vetterschaft mit ferriten, Thesis, Zurich, 1955.

[32] Comstock, R. L., Remeika, J. P., J. Appl. Phys., 35, 1018, 1964.

[33] Nielsen, J. W., Lepore, D. A., Zneimer, J., Townsend, G. B., J. Appl. Phys., 33, Suppl. 3, 1379, 1962.

[34] Bertaut, E., Compt. Rend., 230, 213, 1950.

[35] Prince, E., Trueting, R. G., Acta Crystallogr., 9, 1025, 1956.

[36] Weil, L., Bertaut, E., Bochirol, L., J. Phys. et Radium, 11, 208, 1950.

[37] Gorter, E. W., Nature, 165, 798, 1950.

[38] Okamura, T., Kojima, Y., Phys. Rev., 86, 1040, 1952.

[39] Weisz, R. S., Phys. Rev., 96, 800, 1954.

[40] Snoek, J. L., Philips Techn. Rev., 8, 353, 1946.

[41] Okamura, T., Torizuka, Y., Nature, 168, 872, 1951.

[42] Okamura, T., Fujimura, T., Date, M., Phys. Rev., 85, 1041, 1952.

[43] Rabkin, L. I., High frequency ferromagnets, Fizmatgiz, Moscow–Leningrad, 1960 (in Russian).

[44] Bochirol, L., Compt. Rend., 233, 736, 1951.

[45] Prince, E., Phys. Rev., 102, 674, 1956.

[46] Pauthenet, R., Compt. Rend., 230, 1842, 1950.

[47] Guillaud, C., Rev. Modern Phys., 25, 64, 1953.

[48] Tannenwald, P. E., Phys. Rev., 90, 463, 1955.

[49] Torizuka, Y., Sci. Reports Inst. Tohoku Univ., A3, 383, 1951.

[50] Corliss, L. M., Hastings, J. M., Brockman, F. G, Phys. Rev., 90, 1013, 1953.

[51] Bacon, G. N., Roberts, F. F., Acta Crystallogr., 6, 57, 1953.

[52] Jones, G. O., Roberts, F. F., Proc. Phys. Soc. (London), 65B, 390, 1952.

[53] Belson, H. S., Kriessman, C. J., J. Appl. Phys., 30, 170S, 1959.

[54] Rado, G. T., Folen, V. J., Emerson, W. H., Proc. Inst. Electr. Engrs., 104B, Suppl. 5, 198, 1957.

[55] Yager, W. A., Merrit, F. R., Guillaud, C., Phys. Rev., 81, 477, 1951.

[56] Toropov, N. A., Borisenko, A. I., Zhurnal prikladnoi khimii, 88, 1243, 1950 (in Russian).

[57] Inoue, T., Iida, S., J. Phys. Soc. Jpn., 13, 656, 1958.

[58] Chikasumi S., Physics of ferromagnetism, Syokabo, Tokyo, 1980.

[59] Arai, K.-I., Tsuya, N., Ferrites: Proc. Intern. Conf., 1970, p. 51.

[60] Hastings, J. M., Corliss, L. M., Rev. Modern Phys., 25, 114, 1953.

[61] Ferromagnetic materials, A handbook on the properties of magnetically ordered substances, vol. 2, Ed. Wohlfarth, E. P., North Holland Publ. Comp., 1980.

[62] Guillaud, C., Greveaux, H., Compt. Rend., 230, 1458, 1950.

[63] Guillaud, C., J. Phys. et Radium, 12, 239, 1951.

[64] Okamura, T., Kojima, Y., Torizuka, Y., Sci. Reports Inst. Tohoku Univ., A4, 72, 1952.

[65] Okamura, T., Phys. Rev., 85, 690, 1952.

[66] Gorter, E. W., Philips Res. Repts., 9, 295, 1954.

[67] McGuire, T. R., Phys. Rev., 91, 206, 1953.

[68] Carter, A. E., Miles, D. A., Welch, A. J. A., Proc. Inst. Electr. Engrs., 104B, Suppl. 5, 141, 1957.

[69] Gorter, E. W., Philips Res. Repts., 9, 403, 1954.

[70] Maria Neto, J., J. Appl. Phys., 55, No 6, pt. II B, 2338, 1984.

[71] Smolenskii, G. A., Izvestiya Akad. Nauk SSSR, Ser. "Fizika," 16, 728, 1952.

[72] Enz, U., Physica, 24, 609, 1958.

[73] Methfessel, S., Mattis D., Magnetic semiconductors, Springer, Heidelberg, 1968.

[74] Magnetic semiconductors: chalcogenide spinels, Izdatel'stvo MGU, Moscow, 1981 (in Russian).

[75] Nagaev, E. L., Magnetic semiconductor physics, Nauka, Moscow, 1979 (in Russian).

[76] Menyuk, N., Dwight, K., Arnott, R. J., J. Appl. Phys., 37, 1387, 1966.

[77] Austin, I.G., Elwell, D., Contemp. Phys., 11, 455, 1970.

[78] Wojtowicz, P. J., IEEE Trans. Magn., 5, 840, 1969.

[79] Belov, K. P., Koroleva, L. I., Gordeev, I. V., Fizika nizkikh temperatur, 1, 1540, 1975.

[80] Busch, G., Magyar, B., Wachter, P., Phys. Rev. Lett., 23, 438, 1966.

[81] Sato, K., Teranishi, T., J. Phys. Soc. Jpn., 29, 523, 1970.

[82] Jarsen, P. K., Wittekoek, S. W., Phys. Rev. Lett., 29, 1597, 1972.

[83] Shepherd, I. W., Solid State Commun., 8, 1835, 1970.

[84] Hlidek, P., Polivka, V., Proc. 10th Congr. ICO, Prague, 1975, p. 320.

[85] Harbeke, G., Pinch, H. L., Phys. Rev. Lett., 17, 1090, 1966.

[86] Wen, C. P., Hershenov, B., Philipsborn, H., Pinch, H. L., Appl. Phys. Lett., 13, 188, 1968.

[87] Haas, C., IBM J. Res. Develop., 14, 282, 1970.

[88] Minematsu, K., Miyatani, K., Takahashi, T., J. Phys. Soc. Jpn., 31, 123, 1971.

[89] Stoyanov, S. G., Iliev, M. N., Stoyanova, S. P., Solid State Commun., 18, 1389, 1976.

[90] Goodenough, J. B., J. Phys. Chem. Solids, 30, 261, 1969.

[91] Golik, L. L., Grigorovich, S. M., Kun'kova, Z. E., Fizika tverdogo tela, 16, 2151, 1975 (in Russian).

[92] Balkarei, Yu. I., Baru, V. G., Golik, L. L., Mikroelektronika, 5, 475, 1976 (in Russian).

[93] Moser, F., Ahrenkiel, R. K., Carnall, E., J. Appl. Phys., 42, 1449, 1971.

[94] Bongers, P. F., Zanmarchi, G., Solid State Commun., 6, 291, 1968.

[95] Lee, T. H., Coburn, T. J., Gluck, R., Solid State Commun., 9, 1821, 1971.

[96] Ahrenkiel, R. K., Coburn, T. J., Carnall, E., IEEE Trans. Magn., MAG–10, 2, 1974.

[97] Menzer, G., Z. Kristallogr., 69, 300, 1928.

[98] Gibbs, G. V., Smith, J. V., Amer. Minerals, 50, 2023, 1965.

[99] Prandl, W., Z. Kristallogr., 123, 81, 1966.

[100] Abrahams, S. C., Geller, S., Acta Crystallogr., 11, 437, 1958.

[101] Callen, E., J. Appl. Phys., 39, 519, 1968.

[102] Kiryukhin, V. P., Sokolov, V. I., Zhurnal eksperimental'noi i teoreticheskoi fiziki, 51, 428, 1966.

[103] Le Craw, R. C., Comstock, R. L., in: Physical acoustics, vol. 3, Ed. Mason, W. P., Academic Press, New York–London, 1965.

[104] Wickersheim, K. A., in: Magnetism, Eds. Rado, G. and Suhl, H., Academic Press, New York–London, 1963.

[105] Tinkham, H., J. Appl. Phys., 33, Suppl., 1248, 1962.

[106] Wood, D. L., Remeika, J. P., J. Appl. Phys., 38, 1038, 1967.

[107] Dillon, J. F., Jr., J. Phys. et Radium, 20, 374, 1959; J. Appl. Phys., 29, 539, 1958.

[108] Krinchik, G. S., Chetkin, M. V., Zhurnal eksperimental'noi i teoreticheskoi fiziki, 40, 729, 1961; 41, 673, 1962 (in Russian).

[109] Pisarev, R. V., Sinii, I. G., Smolenskii, G. A., Pis'ma ZhETF, 9, 112, 1969; 9, 264, 1969; Zhurnal eksperimental'noi i teoreticheskoi fiziki, 57, 737, 1969 (in Russian).

[110] Dillon, J. F., J. Appl. Phys., 40, 1230, 1969.

[111] Geller, S., Z. Kristallogr., 125, 1, 1967.

[112] Geller, S., Gilleo, M. A., J. Phys. Chem. Solids, 3, 30, 1957.

[113] Euler, F., Bruce, J. A., Acta Crystallogr., 19, 971, 1965.

[114] Bonnet, M., Delapalme, A., Fuess, H., Thomas, M., Acta Crystallogr., B31, 2233, 1975.

[115] Emiraliev, A., Kocharov, A. G., Bakradze, R. V., Kristallografiya, 21, 391, 1976 (in Russian).

[116] Espinosa, G. P., J. Chem. Phys., 37, 2344, 1962.

[117] Strocka, B., Holst, P., Tolksdorf, W., Philips Journ. Res., 33, 186, 1978.

[118] Tcheou, F., Fuess, H., Bertaut, E. F., Solid State Commun., 8, 1745, 1970.

[119] Winkler, G., Magnetic garnets, F, Vieweg und Sohn, Braunschweig/Wiesbaden, 1981.

[120] Yakovlev, Yu. M., Gendelev, S. Sh., Ferrite single crystals in radio electronics, Sovetskoye Radio, Moscow, 1975 (in Russian).

[121] Filippov, V. V., Shil'nikov, Yu. R., Yakovlev, Yu. M., Salyganov, V. I., Synthesis and study of ferromagnetic crystals, CNII "Elektronika," Moscow, 1970, No 9(25), p. 34 (in Russian).

[122] Yakovlev, Yu. M., Shil'nikov, Yu. R., Galaktionova, G. M., Ferrite microwave devices and materials, CNII "Elektronika," Moscow, 1972, p. 188 (in Russian).

[123] Clark, A. E., Desavage, B., Coleman, W., J. Appl. Phys., 34, 1296, 1963.

[124] Petrakovskii, G. A., Smokotin, E. M., Titova, A. G., Fizika tverdogo tela, 9, 2324, 1967 (in Russian).

[125] Smokotin, E. M., Petrakovskaya, E. A., Sablina, K. A., Thin magnetic films, computer and radio technique, vol. 2, Institute for Physics, Krasnoyarsk, 1970 (in Russian).

[126] Mandel, V. S., Smokotin, E. M., Petrakovskii, G. A., Lebed, B. M., Phys. Status Solidi, 30, K111, 1968.

[127] Philips, T. G., White, R. L., Phys. Rev. Lett., 16, 650, 1966.

[128] Yakovlev, Yu. M., Galaktionova, G. M., Burdin, Yu. I., Petrov, R. A., Elektronnaya tekhnika, Ser. 7, No 4 (16), 26, 1968 (in Russian).

[129] Yakovlev, Yu. M., Shil'nikov, Yu. R., Galaktionova, G. M., Salyganov, V. I., Izvestiya Akad. Nauk SSSR, Ser. "Fizika," 35,110, 1971 (in Russian).

[130] Van der Kraan, A. M., Van Loef, J. J., Proc. Int. Conf., Tihany'71: Application of Mössbauer effect, Tihany, 1971, p. 519.

[131] Heller, P., Benedek, G. B., Phys. Rev. Lett., 8, 428, 1962.

[132] Eastman, D. E., J. Appl. Phys., 37, 2312, 1966.

[133] Bateman, T. B., J. Appl. Phys., 37, 2194, 1966.

[134] Spencer, E. G., Denton, R. T., Bateman, T. B., J. Appl. Phys., 34, 3059, 1963.

[135] Alton, W. J., Barlow, A. J., J. Appl. Phys., 38, 3023, 1967.

[136] Petrakovskii, G. A., Izvestiya Akad. Nauk SSSR, Ser. "Fizika," 34, 1052, 1970 (in Russian).

[137] Spencer, E. G., Denton, R. T., Bateman, T. B., J. Appl. Phys., 34, 3059, 1963.

[138] Alton, W. J., Barlow, A. J., J. Appl. Phys., 38, 3023, 1967.

[139] Haussuhl, S., Mateika, D., Z. Naturforsch, 27a, 1522, 1972.

[140] Graham, L. J., Chang, R., J. Appl. Phys., 41, 2247, 1970.

[141] Haussuhl, S., Mateika, D., Tolksdorf, W., Z. Naturforsch, 31a, 390, 1976.

[142] Gendelev, S. Sh., Shcherbak, N, G., Kristallografiya, 10, 708, 1965 (in Russian).

[143] Hergt, R., Gornert, P., Phys. Status Solidi (a), 21, 77, 1974.

[144] Dedukh, L. M., Nikitenko, V. I., Izvestiya Akad. Nauk SSSR, Ser. "Fizika," 34, 1235, 1970 (in Russian).

[145] Slack, G. A., Oliver, D. W., Phys. Rev., 4B, 592, 1971.

[146] Shchelkotunov, V. A., Danilov, V. N., Kalacheva, V. S., Izvestiya Akad. Nauk SSSR, Ser. "Neorg. materialy," 12, 1076, 1976 (in Russian).

[147] Verweel, J., Proc. Inst. Electr. Engrs. B 109, Suppl., 21, 95, 1962.

[148] Bethe, K., Verweel, J., IEEE Trans. Magn., MAG-5, 474, 1969.

[149] Johnson, B., Walton, A. K., Brit. Journ. Appl. Phys., 16, 475, 1965.

[150] Daral, J., Ferrand, B., Geynet, J., IEEE Trans. Magn., MAG–11, No 5, 115, 1975.

[151] Roschmann, P., Hansen, P., J. Appl. Phys., 52, 6257, 1981.

[152] Geller, S., in: Physics of magnetic garnets, Ed. Paoletti, A., North–Holland Publ. Comp., Amsterdam, 1978, p. 1.

[153] Tolksdorf, W., in: Physics of magnetic garnets, Ed. Paoletti, A., North–Holland Publ. Comp., Amsterdam, 1978, p. 521.

[154] Geller, S., Remeika, J. P., Sherwood, R. C., Phys. Rev., 137, 1034, 1965.

[155] Geller, S., Williams, H. J., Sherwood, R. C., Phys. Rev., 131, 1080, 1963.

[156] Hansen, P., Philips Res. Rep. Suppl., No 7, 1, 1970.

[157] Hansen, P., J. Appl. Phys., 45, 3638, 1974.

[158] Scott, G. B., in: Physics of magnetic garnets, Ed. Paoletti, A., North-Holland Publ. Comp., Amsterdam, 1978, p. 445.

[159] Galuza, A. I., Eremenko, V. V., Kirichenko, A. P., Fizika tverdogo tela, 15, 585, 1973 (in Russian).

[160] Wemple, S. H., Blank, S. L., Seman, J. A., Biolsi, W. A., Phys. Rev., B9, 2134, 1974.

[161] Pisarev, R. V., Sinii, I. G., Kolpakova, N. N., Yakovlev, Yu. M., Zhurnal eksperimental'noi i teoreticheskoi fiziki, 60, 2188, 1971 (in Russian).

[162] Geller, S., Williams, H. J., Espinosa, G. P., Sherwood, R. C., Bell Syst. Techn. J., 43, 565, 1964.

[163] Yakovlev, Yu. M., Lebed', B. M., Fizika tverdogo tela, 4, 3654, 1962 (in Russian).

[164] Jonker, G. H., Wijn, H. P. J., Braun, P. B., Philips Techn. Rev., 18, 145, 1956/57.

[165] Braun, P. B., Philips Res. Repts., 12, 491, 1957.

[166] Sitidze, Yu., Sato, H., Ferrites, Mir Publ., Moscow, 1964 (Russian Transl.)

[167] Kohn, J. A., Eckart, D. W., Z. Kristallogr., 119, 454, 1964.

[168] Kohn, J. A., Eckart, D. W., J. Appl. Phys., 35, 968, 1964.

[169] Levine, B. G., Nowlin, C. H., Jones, R. V., Phys. Rev., 174, 571, 1968.

[170] Villers, G., Compt. Rend. Acad. Sci. (Paris), 248, 1974, 1959.

[171] Pauthenet, R., Rimet, G., Compt. Rend. Acad. Sci. (Paris), 249, 565, 1959.

[172] Kojima, H., Sci. Reports Res. Inst. Tohoku Univ., A7, 502, 1955.

[173] Tauber, A., Kohn, J. A., Savage, R. O., J. Appl. Phys., 34, 1265, 1963.

[174] Savage, R. O., Dixon, S., Tauber, A., J. Appl. Phys., 36, 873, 1965.

[175] Taft, D. R., J. Appl. Phys., 35, 776, 1964.

[176] Okazaki, T., Yutaka, H., Electronics and communications in Japan, 57, No 7, 188, 1974.

[177] Dillon, J. F., Kamimura, H., Remeika, J. P., J. Phys. Chem. Solids, 27, 1531, 1966.

[178] Sugawara, F., Iida, H., Syono, Ya., Akimoto, S., J. Phys. Soc. Jpn., 25, 1553, 1968.

[179] Bokov, V. A., Myl'nikova, I. E., Kizhaev, S. A., Fizika tverdogo tela, 7, 3695, 1965 (in Russian).

[180] Bertaut, E. F., Forrat, F., J. Appl. Phys., 29, 247, 1958.

[181] Cloud, W. H., Phys. Rev., 111, 1046, 1958.

[182] Bozorth, R. M., Walsh, D. E., J. Phys. Chem. Solids, 5, 299, 1958.

[183] Knox, K., Geller, S., Phys. Rev., 110, 771, 1958.

[184] Shafer, M. W., McGuire, T. R., J. Appl. Phys., 35, 984, 1964.

[185] Shafer, M. W., J. Appl. Phys., 36, 1145, 1965.

[186] Boyd, E. L., Moruzzi, V. L., Smart, I. S., J. Appl. Phys., 34, 3049, 1963.

[187] Genano, R., Hunt, E., Meyer, H., Phys. Rev., 156, 521, 1967.

[188] Litster, J. D., Benedek, G. B., J. Appl. Phys., 37, 1320, 1966.

[189] Prince, E., J. Appl. Phys., 36, 1845, 1965.

[190] Geller, S., Phys. Rev., 181, 980, 1969.

[191] Zotov, T. D., Sukrovtseva, M. M., Fizika tverdogo tela, 11, 649, 1964 (in Russian).

[192] Andersen, E. E., Phys. Rev., 134, A1581, 1964.

[193] Hansen, P., Roschmann, P., Tolksdorf, W., J. Appl. Phys., 45, 2728, 1974.

[194] Plant, J. S., J. Phys. C: Solid State Phys., 16, 7037, 1983.

30

Galvanomagnetic and Thermomagnetic Phenomena

N.A. Babushkina and V.S. Egorov

30.1 Introduction

Galvanomagnetic and thermomagnetic phenomena show themselves when a conductor carrying an electric current or thermal flux is placed in a magnetic field.

The most distinctive features of the phenomena are associated with the action of magnetic induction **B** on the trajectories of current carriers, which bend because of the Lorentz force and become helices about the **B** as an axis. When the streams of heat and charges flow along **B**, the longitudinal galvano- and thermomagnetic effects (GME and TME) occur. The magnetic field does not change the longitudinal component of the electron momentum, therefore its influence in this case is insignificant and the longitudinal GME and TME for all metals are limited and manifest themselves as a moderate increase (on the order of 1) in electrical or thermal resistance, respectively.

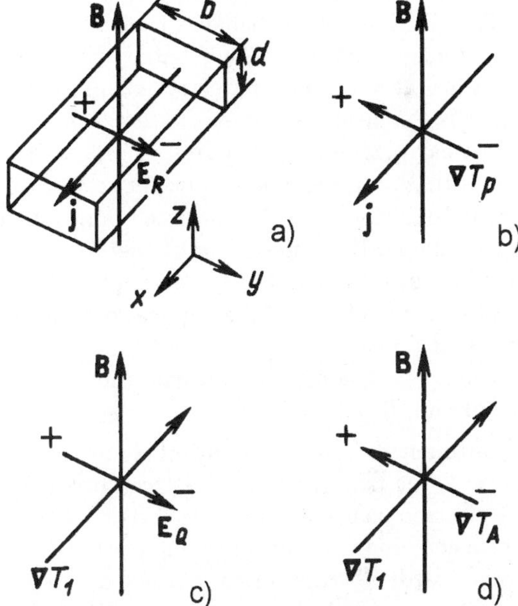

Figure 30.1 Schematic orientations of **B**, **j**, **E**, and ∇T vectors for (*a*) the Hall effect; (*b*) the Ettingshausen effect; (*c*) the Nernst effect; (*d*) the Righi–Leduc effect.

When the streams of heat and electrical charges are perpendicular to **B** ($\mathbf{q} \perp \mathbf{B}$, $\mathbf{j} \perp \mathbf{B}$), the transverse GME and TME appear. In general, the directions of electric field **E** and temperature gradient ∇T do not coincide with the given vectors of electric current density **j** and heat flux density **q** (for definiteness, let the streams **j** and **q** be directed along the x axis and the magnetic field along the z axis).

In a general way, the quantities **j**, **E**, ∇T, and **q** are related to each other by tensor-like formulas

$$\mathbf{j} = \widehat{\sigma}\mathbf{E} - \widehat{\eta}\nabla T, \qquad \mathbf{q} = -T\widehat{\sigma}\widehat{S}\mathbf{E} - \widehat{\kappa}\nabla T, \quad (30.1)$$

where $\widehat{\sigma}$, $\widehat{S}$, $\widehat{\kappa}$, and $\widehat{\eta}$ are the tensors of electrical conductivity, thermal electromotive force (the Seebeck coefficient), thermal conductivity, and thermoelectric current $\widehat{\eta} = \widehat{\sigma}\widehat{S}$, respectively.

0-8493-2861-6/97/$0.00+$.50

If $\mathbf{j} \perp \mathbf{B}$, then:

• there arises an electric field

$$E_R = RjB \tag{30.2}$$

normal to $\mathbf{j}$ and $\mathbf{B}$ (the *Hall effect*), where R is the Hall coefficient (Fig. 30.1a);

• there is a change in electrical resistivity ρ in the direction of $\mathbf{j}$, with $\rho_{xx} = E_x/j_x$ being called the *magnetoresistivity* (MR);

• there arises a temperature gradient

$$\nabla T_P = PjB \tag{30.3}$$

normal to $\mathbf{j}$ and $\mathbf{B}$ (the *Ettingshausen effect*), where P is the Ettingshausen coefficient (Fig. 30.1b).

If $\mathbf{q} \perp \mathbf{B}$, then:

• there arises an electric field

$$E_Q = QB\nabla T_1 \tag{30.4}$$

normal to the original temperature gradient ∇T_1 and $\mathbf{B}$ (the *Nernst effect*), where Q is the Nernst coefficient (Fig. 30.1c);

• there arises a temperature gradient

$$\nabla T_A = AB\nabla T_1 \tag{30.5}$$

normal to the original temperature gradient ∇T_1 and $\mathbf{B}$ (the *Righi - Leduc effect*) where A is the Righi - Leduc coefficient (Fig. 30.1d);

• there is a change in thermal conductivity in the direction of the original heat flow.

The GME and TME behave differently (qualitatively and quantitatively) in the ranges of weak and strong magnetic fields. The boundary between these ranges is determined by a dimensionless quantity $\omega\tau$, where $\omega = eB/m^*c$ is the cyclotron frequency for an electron rotation with an effective mass m^*, and τ is the mean free time between collisions.

In weak magnetic fields ($\omega\tau \leq 1$), $\rho_{xx} \sim B^2$ and $\Delta\rho/\rho \leq 1$ for all metals, where $\Delta\rho = \rho(B) - \rho(0)$, $\rho \cong \rho(0)$; $R = 1/ne$, n is the current carrier density, $e = -1.6 \cdot 10^{-19}$ C is the electron charge.

The condition $\omega\tau \gg 1$ defines the range of strong magnetic fields, where the asymptotic behavior of GME and TME is determined by the *Fermi surface* (FS) topology, differing both for dissimilar metals and for various orientations of monocrystals of the same metal.

The FS is defined as the surface in momentum space, which separates occupied and free electron states at $T = 0$.

The $\omega\tau \geq 1$ condition is usually fulfilled at low temperatures ($T \sim 4$ K) and in clean specimens ($R(300\text{ K})/R(4.2\text{ K}) \sim 10^3$).

Connection of the Fermi surface topology with the galvanomagnetic effects. In the case $\omega\tau \gg 1$, the trajectory of an electron in a magnetic field is described by the equations $\epsilon = const$ (ϵ is the electron energy) and $p_z = const$ (p_z is the momentum projection on the magnetic field direction), which correspond to the line of intersection of the FS by a plane normal to the magnetic field. When an FS is closed, all trajectories in real space make closed orbits similar to those in momentum space and turned around $\mathbf{B}$ by an angle of $\pi/2$. When the FS is a multiply connected infinite surface, apart from the closed orbits there appear open trajectories that correspond to an infinite motion of electrons in real space in the direction turned by an angle of $\pi/2$ with respect to that of open orbits in a momentum space.

When the FS is closed and $\omega\tau \gg 1$, all electrons drift in an electric field $\mathbf{E}$ with a velocity $\mathbf{v}_d = c[\mathbf{E} \times \mathbf{B}]/B^2$, which causes a nondissipative current determining the Hall effect and the nondiagonal conductivity tensor component $\sigma_{xy} = ne/B \equiv \sigma_0/\omega\tau \propto B^{-1}$ ($\sigma_0 = ne^2\tau/m$ is the conductivity in a zero magnetic field). In the electric field direction, the charge is transported by means of a diffusion of orbit centers, i.e. by hopping of electrons to next orbits due to their scattering. This leads to the appearance of a dissipative current and determines the corresponding diagonal conductivity tensor component $\sigma_{xx} \cong \sigma_0/(\omega\tau)^2 \propto B^{-2}$. The electrons on open orbits do not contribute to the Hall conductivity σ_{xy}. Their contribution to conductivity in the direction of open trajectories is the same as that in the absence of any magnetic field, i.e. in this case $\sigma_{xx} \simeq \sigma_0$.

If a lower energy band which must have been entirely filled by electrons overlaps with the adjacent (empty) band, some of the electrons overflow from one band to the other. In this case, the density of empty (hole) states n_2 in the lower band coincides with the density of filled (electron) states n_1 in the upper band. Such a metal is generally called compensated ($n_1 = n_2$). In the first approximation (when $\omega\tau \gg 1$) its Hall conductivity is zero. In the case of a closed FS, it is definitely either electron-like, if it encloses filled states, or hole-like, if it encloses empty ones. When $n_1 = n_2$, all conductivity tensor components are specified by a diffusion of orbit centers, that is $\sigma_{xx} \sim \sigma_{yy} \cong \sigma_0/(\omega\tau)^2 \sim B^{-2}$. (On an open and multiply connected FS there might exist both electron-like and hole-like orbits). The expressions presented for the conductivity tensor components fully describe all the variety of possible asymptotic behaviors in the galvanomagnetic properties of metals.

In the experiment one measures the magnetoresistivity (MR)

$$\rho_{xx} = \frac{E_x}{j_x} = \frac{\sigma_{yy}}{\sigma_{xy}^2 + \sigma_{xx}\sigma_{yy}}, \tag{30.6}$$

and the Hall resistivity

$$\rho_{xy} = \frac{E_y}{j_x} = \frac{\sigma_{xy}}{\sigma_{xy}^2 + \sigma_{xx}\sigma_{yy}} = RB, \tag{30.7}$$

where R is the Hall coefficient which, however, is not constant for all metals. The MR is quantitatively determined by the value of $\omega\tau$ (rather than by the magnetic induction $\mathbf{B}$). This leads to the so-called empiric *Kohler rule*, according to which the MR is described by the magnetic field ($\rho(B)/\rho(0) = f(B/\rho(0))$.

In Figs. 30.3–30.28, principal topological types of the FS are presented, and in Table 30.2 data on the FS topology of metals and a number of alloys can be found. Some comments may be of interest here.

(1) For one-band metals with a closed FS (alkali metals), $\rho_{xx} \sim const\ \sigma_0^{-1}$, the MR does not depend on B and it is usually said that the MR tends to a *saturation* value, while $\rho_{xy} = B/ne$, i.e. the Hall coefficient $R = 1/ne$. In this case the value of R proves to be the same as that in a weak-magnetic-field range $\omega\tau \leq 1$.

(2) For one-band metals with an open FS (noble metals), $\rho_{xx} \sim \sigma_0^{-1}(\omega\tau)^2 \propto B^2$ for those orientations of $\mathbf{B}$ with respect to the crystal axes, when there appear open trajectories along the y axis (perpendicular to the current) in real space, i.e. when $\sigma_{yy} \sim \sigma_0$. For other orientations, ρ_{xx} saturates. In a magnetic field rotation diagram, the MR is small for most orientations of $\mathbf{B}$ and undergoes sharp peaks corresponding to the open trajectories.

(3) For compensated metals ($n_1 = n_2$) with closed FS (beryllium, molybdenum, tungsten, semimetals), $\rho_{xx} \sim \sigma_0^{-1}(\omega\tau)^2 \propto B^2$ for all orientations. A small anisotropy independent of $\mathbf{B}$ is due to the nonspherical shape of the FS. The Hall effect (and hence the Hall coefficient) depends on B, T, and crystal orientation in a complicated way.

(4) For compensated metals with open FS (magnesium, zinc, cadmium, tin, lead, etc.), $\rho_{xx} \sim \sigma_0^{-1}$, i.e. it is independent of B (saturates) for those orientations of $\mathbf{B}$, when there appear open trajectories along the x axis (parallel to the current) in real space, namely, if $\sigma_{xx} \sim \sigma_0$. For other orientations $\rho_{xx} \sim B^2$. In a magnetic field rotation diagram, the MR is large for most orientations of $\mathbf{B}$ and undergoes sharp minima corresponding to the open trajectories.

(5) For multiband noncompensated metals ($n_1 \neq n_2$) with closed or open FS (indium, aluminum, gallium, thallium), the asymptotics of ρ_{xx} are the same as those in §§1 and 2, and $\rho_{xy} = B/ne$, where $n = n_1 - n_2$, except special $\mathbf{B}$ orientations when deviations from compensation due to geometry of the multi-connected FS should be taken into account.

In many multiband metals, a transition of an electron on the FS from one conduction band to another (the so-called magnetic breakdown) becomes possible in the presence of a strong magnetic field. This gives rise to new trajectories and thus considerably affects the galvanomagnetic phenomena.

The FS topology is treated in most detail in [1–3].

Basic theoretical and experimental concepts of the metal magnetoresistance may be found in [4–6]. In Tables 30.3, 30.4 and in Figs. 30.29–30.58, the main data on MR of metals are presented. Note should be taken that at a temperature of 20°C and usually applicable values of magnetic induction $B \sim 1$ T ($\omega\tau \leq 1$) the ratio $\Delta\rho/\rho$ for the majority of metals is very small. For copper, for example, $\Delta\rho/\rho \cong 10^{-4}$ at $B = 2$ T. An exception is presented by semimetals: for example, bismuth display $\Delta\rho/\rho \cong 2$ at $B = 3$ T.

Hall effect. Basic theoretical and experimental data on the Hall effect were given in [7]. The Hall coefficient R in metals may be positive or negative, or even may change its sign as the temperature varies. At high and intermediate temperatures it is practically independent of temperature for the majority of metals. Main data on the Hall effect in metals appear in Table 30.5 and in Figs. 30.59–30.74.

In ferromagnets the electrons are exposed to a magnetic field different from the external one. In this case a special ferromagnetic Hall effect is observed. For ferromagnets, it was experimentally found that the Hall field

$$E_R = \mu_0 j\,(R_0 H + R_1 M)\,, \tag{30.8}$$

where H is the magnetic field strength, μ_0 is the permeability of vacuum, M is the magnetization of a specimen, j is the density of current flowing through the specimen, and, finally, R_0 and R_1 are the ordinary and extraordinary (anomalous) Hall coefficients, respectively.

The above relation with due regard for the equality $B = \mu_0(H + M)$ can be rewritten in the form

$$\mathcal{E} = U_R d/I = R_0 B + \mu_0 R_S M, \tag{30.9}$$

where $\mathcal{E}$ is the experimentally derived Hall voltage U_R normalized to the total current strength I and sample thickness d, $R_S = R_1 - R_0$ is the so-called spontaneous or ferromagnetic Hall coefficient. For the majority of ferromagnetic metals, $R_S \cong R_1$. From this expression one can determine the Hall coefficients by using the experimental dependence $\mathcal{E}(B)$ (Fig. 30.2).

Usually $R_S \gg R_0$ and strongly depends on temperature. With increasing temperature, the spontaneous Hall coefficient R_S first rises, then reaches a maximum at the Curie point, and finally falls. In the paramagnetic temperature range, the Hall effect is described by the relationship

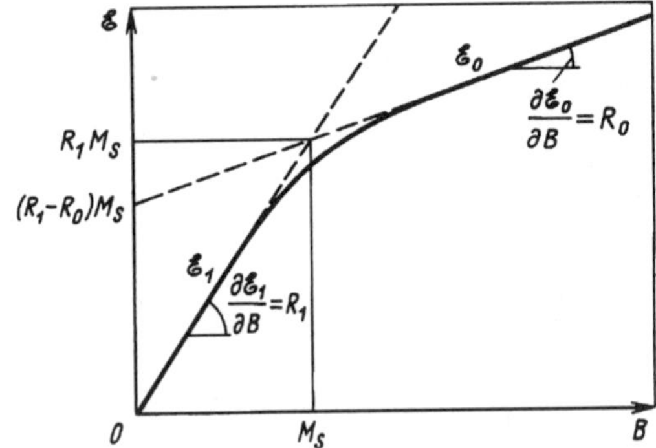

Figure 30.2 Field dependence of the Hall emf E for a ferromagnetic plate. Determination of the ordinary R_0 and extraordinary R_1 Hall coefficients.

$$\mathcal{E} = R^* H, \tag{30.10}$$

where $R^* = R_0 + \chi R_S = R_0 + R_P$, χ is the magnetic susceptibility of a substance, R_P is the paramagnetic Hall–Kikoin coefficient.

Values of the galvanomagnetic coefficients for ferromagnetic metals are presented in Table 30.7.

The thermomagnetic effects in ferromagnets are determined by the formulae:

the Ettingshausen effect

$$\nabla T_P = (P_0 B + \mu_0 P_1 M) j; \tag{30.11}$$

the Righi - Leduc effect

$$\nabla T_A = (A_0 B + \mu_0 A_1 M) \nabla T_1; \tag{30.12}$$

the Nernst effect

$$E_Q = (Q_0 B + \mu_0 Q_1 M) \nabla T_1. \tag{30.13}$$

In ferromagnetic metals, the dependence of ρ on a magnetic field strength also has a number of peculiarities due to the presence of spontaneous magnetization in these substances. In high magnetic fields (when technical saturation of the substance occurs), the resistivity always decreases as the field increases irrespective of the magnetic field direction para to the current.

In tables and figures of the present chapter we used the notation $RRR = \rho(300\ \mathrm{K})/\rho(4.2\ \mathrm{K})$. This parameter characterizes purity of a specimen.

Values of all the galvano- and thermomagnetic coefficients are given below in SI. In order to transform them to another system of units, use must be made of Table 30.1. One should bear in mind that an external magnetic field strength equal to 10 kOe in CGS corresponds to the external magnetic induction in SI, $B \equiv \mu_0 H = 1$ T, where $\mu_0 = 4\pi \cdot 10^{-7}$ H/m $= 1.256 \cdot 10^{-6}$ H/m is the permeability of vacuum.

Table 30.1 Conversion of the galvano- and thermomagnetic coefficients from SI to the "laboratory" system of units

Quantity	Notation	SI	"Laboratory" system
Magnetic induction	B	$\mathrm{Wb/m^2 = V{\cdot}s/m^2}$	10^4 G
Hall coefficient	R	$\mathrm{m^3/C = m^3/(A{\cdot}s)}$	10^{-2} V·cm/(A·G)
Ettingshausen coefficient	P	$\mathrm{K{\cdot}m^3/(V{\cdot}A{\cdot}s)}$	10^{-2} K·cm/(A·G)
Righi–Leduc coefficient	A	$\mathrm{m^2/(V{\cdot}s)}$	10^{-4} $\mathrm{G^{-1}}$
Nernst coefficient	Q	$\mathrm{m^2/(K{\cdot}s)}$	10^{-4} V/(K·G)

30.2 Topology of Fermi Surfaces of Metals

In Table 30.2 and in Figs. 30.3–30.28 data on the topology of Fermi surfaces for various metals are presented.

Table 30.2 Topology of the Fermi surfaces of metals [1]

Metal	Crystal lattice	Topological type and other data on the Fermi surface
Aluminum	fcc	$n_1 \neq n_2$; $B < 3.0$ T – closed; magnetic breakdown at $B > 3.0$ T, Fig. 30.7
Antimony	rhombohedral	$n_1 = n_2$; closed
Arsenic	rhombohedral	$n_1 = n_2$; closed, Fig. 30.12
Beryllium	hcp	$n_1 = n_2$; $B < 5.0$ T – closed; magnetic breakdown at $B > 5.0$ T in the basal plane, Fig. 30.4
Bismuth	rhombohedral	$n_1 = n_2$; closed, Fig. 30.23
Cadmium	hcp	$n_1 = n_2$; open, corrugated cylinder along the [0001] axis, Fig. 30.14
Calcium	fcc	Open (calculation), Fig. 30.6; similar to the Fermi surface of lead (Fig. 30.11)
Cesium	bcc	$n_1 = 1$ electron/atom, $n_2 = 0$; closed, sphere (Fig. 30.3 for potassium)
Chromium	bcc	$n_1 = n_2$; closed, Fig. 30.18; magnetic breakdown along [100] axis in antiferromagnetic chromium at $B > 6.0$ T
Cobalt	fcc	Fig. 30.20
Copper	fcc	$n_1 = 1$ electron/atom, $n_2 = 0$; open, spatial net of corrugated cylinders along [111] axis (primary open directions) and along [110] and [100] axes (secondary open direction), Fig. 30.13
Gadolinium	fcc	Fig. 30.24
Gallium	*1	$n_1 = n_2$; open, corrugated cylinder along the c-axis, Fig. 30.25
Gold	fcc	$n_1 = 1$ electron/atom, $n_2 = 0$; open, spatial net of corrugated cylinders along [111] axis; similar to the Fermi surface of copper, Fig. 30.13
Graphite		Closed, self-intersection, Fig. 30.9
Indium	tetragonal	$n_1 \neq n_2$; closed, similar to the Fermi surface of aluminum
Iron	bcc	$n_1 = n_2$; open, spatial net of corrugated cylinders along [001] axis, Fig. 30.20
Lead	fcc	$n_1 = n_2$; open, spatial net of corrugated cylinders along [111] axis, Fig. 30.11

Table 30.2 Topology of the Fermi surfaces of metals *(continued)*

Metal	Crystal lattice	Topological type and other data on the Fermi surface
Lithium	bcc	$n_1 = 1$ electron/atom, $n_2 = 0$; closed, sphere (Fig. 30.3 for potassium)
Magnesium	hcp	$n_1 = n_2$; open, at $B > 0.5$ T magnetic breakdown in the [0001] plane, similar to beryllium, Fig. 30.5
Mercury	rhombohedral	$n_1 = n_2$; open trajectories are parallel to [100] and [110] axes, Fig. 30.16
Molybdenum	bcc	$n_1 = n_2$; closed, Fig. 30.18
Nickel	fcc	$n_1 = n_2$; open, spatial net of corrugated cylinders along [111] axes; similar to the Fermi surface of copper, Fig. 30.27
Niobium	bcc	$n_1 \neq n_2$; open, net of corrugated cylinders along [001], [110], and [111] axes, Fig. 30.17
Palladium	fcc	$n_1 = n_2$; open, spatial net of corrugated cylinders along [001] axis, Fig. 30.22
Platinum	fcc	$n_1 = n_2$; open, spatial net of corrugated cylinders along [001] axis; similar to the Fermi surface of palladium, Fig. 30.28
Potassium	bcc	$n_1 = 1$ electron/atom, $n_2 = 0$; closed, sphere, Fig. 30.3
Rhenium	hcp	$n_1 = n_2$; open, corrugated cylinder along the [0001] axis; at $B > 3.0$ T magnetic breakdown; additional open trajectories appear along the [0001] and $[10\bar{1}0]$ axes, Fig. 30.19
Rhodium	fcc	Closed (calculation), Fig. 30.21
Rubidium	bcc	$n_1 = 1$ electron/atom, $n_2 = 0$; closed sphere (see Fig. 30.3 for potassium)
Ruthenium	hcp	Open, magnetic breakdown
Scandium	*2	$n_1 \neq n_2$; closed
Silver	fcc	$n_1 = 1$ electron/atom, $n_2 = 0$; open, spatial net of corrugated cylinders along [111] axis similar to the Fermi surface of copper, Fig. 30.13
Sodium	bcc	$n_1 = 1$ electron/atom, $n_2 = 0$; closed, sphere (see Fig. 30.3 for potassium)
Strontium	polymorphic	Open (calculation)
Tantalum	bcc	$n_1 = n_2$; open, spatial net of corrugated cylinders along [001] axis; similar to the Fermi surface of niobium, Fig. 30.17
Thallium	hcp	$n_1 \neq n_2$; open, two corrugated (0001) planes connected by narrow arms along the [0001] axis, Fig. 30.8; magnetic breakdown at $B > 5.0$ T
Tin	tetragonal	$n_1 = n_2$; open, plane net of corrugated cylinders along [010] and [110] axes, Fig. 30.10; magnetic breakdown at $B > 5.0$ T
Titanium	hcp	$n_1 = n_2$; closed
Vanadium	bcc	$n_1 \neq n_2$; open, similar to FS of niobium, Fig. 30.17
Ytterbium	fcc	Closed
Yttrium	hcp	Open (calculation), Fig. 30.26
Zinc	hcp	$n_1 = n_2$; open, corrugated cylinder along the [0001] axis, Fig. 30.15; at $B > 0.25$ T, magnetic breakdown in the basal plane similar to beryllium
Zirconium	hcp	Open (calculation)
AuSn	hexagonal	Open
$AuAl_2$	fcc	Open, similar to the Fermi surface of copper
$AuGa_2$	fcc	Open, similar to the Fermi surface of copper
$AuIn_2$	fcc	Open, similar to the Fermi surface of copper
AgZn	β-brass	Open (calculation)
CuZn	β-brass	Open (calculation)
PtIn	β-brass	Open (calculation)
$MgZn_2$	hexagonal	

*[1] body-centered orthorhombic.

*[2] polymorphic fcc, hcp.

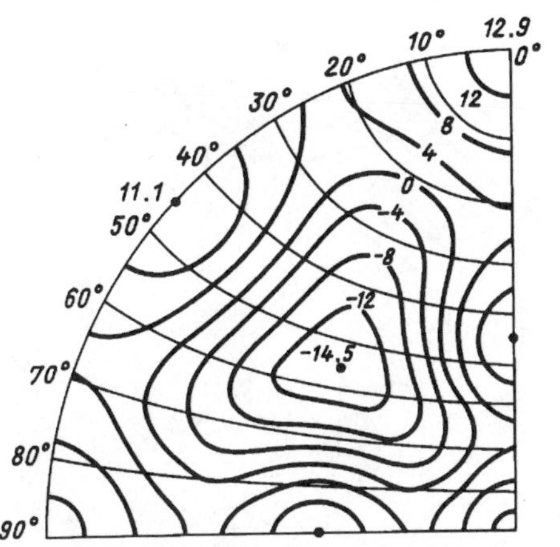

Figure 30.3 Fermi surface of K [2]. Contours of deviations of the Fermi surface from a sphere are drawn in units of $10^4\Delta r/r_1$, where r_1 is the sphere radius (values of $\Delta r/r$ for other alkaline metals are qualitatively the same).

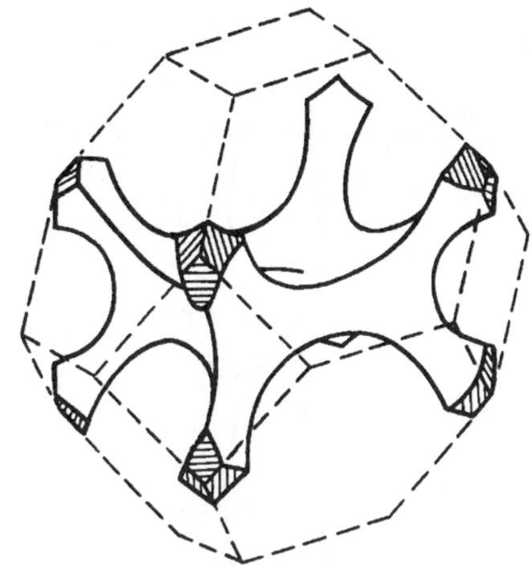

Figure 30.6 Multiply-connected hole Fermi surface of Ca in the first zone (the Harrison model) [2].

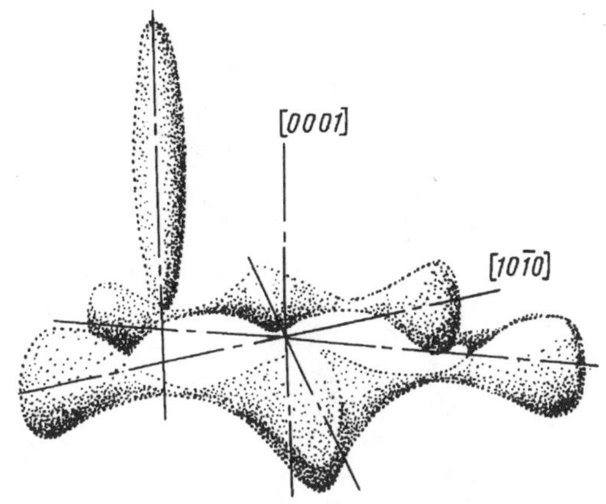

Figure 30.4 Fermi surface of Be (the "cigar" and the "coronet") [2].

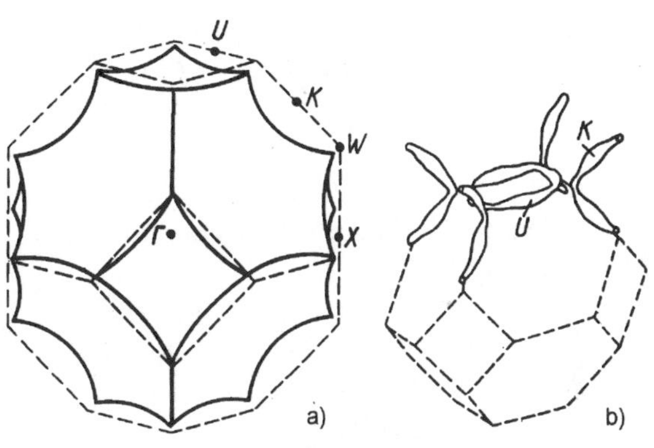

Figure 30.7 (*a*) Hole Fermi surface of Al in the second zone [3], and (*b*) electronic Fermi surface of Al in the third zone [2].

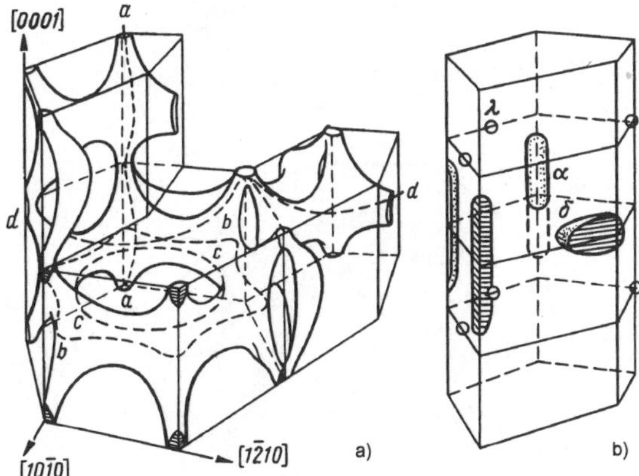

Figure 30.5 Fermi surface of Mg: (*a*) multiply-connected hole surface in the second zone [6]; (*b*) electronic surface in the third and fourth zones [2].

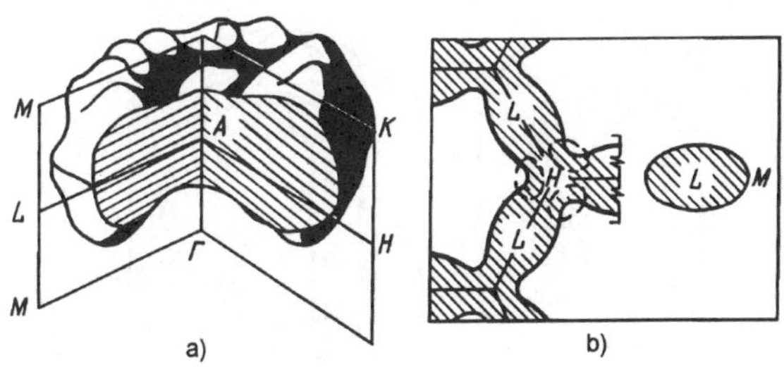

Figure 30.8 (*a*) Fermi surface of Tl in the third zone [2], and (*b*) Fermi surface cross-section by plane AHL in the fourth zone of Tl [2].

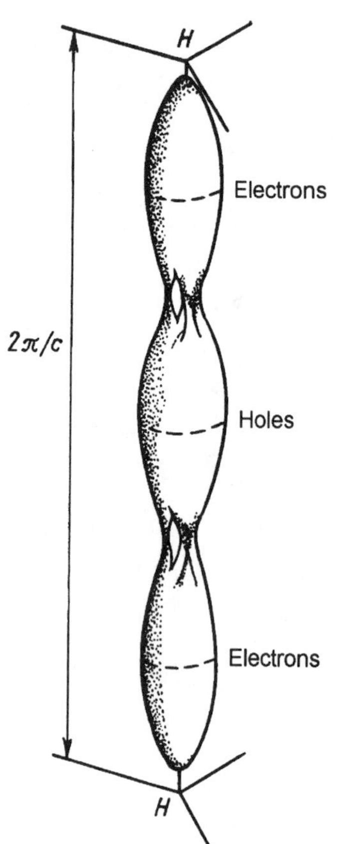

Figure 30.9 Fermi surface of graphite (the McClure model) [2].

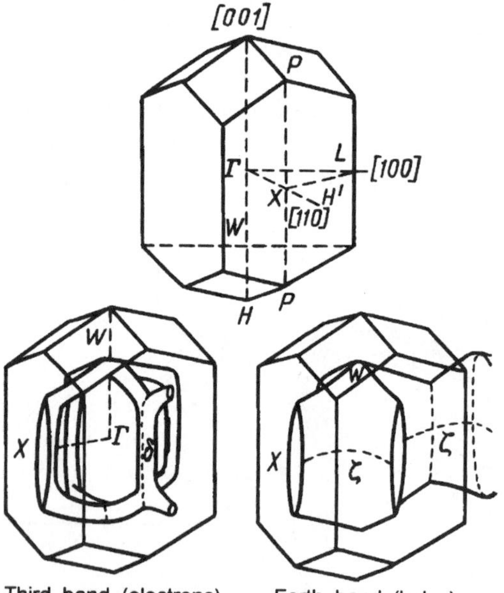

Figure 30.10 Brillouin zone and open hole Fermi surfaces of Sn [1].

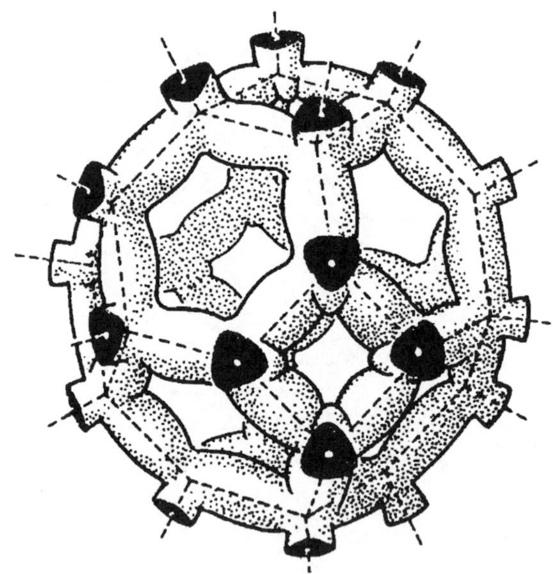

Figure 30.11 Open electron Fermi surface of Pb (the third zone) [1].

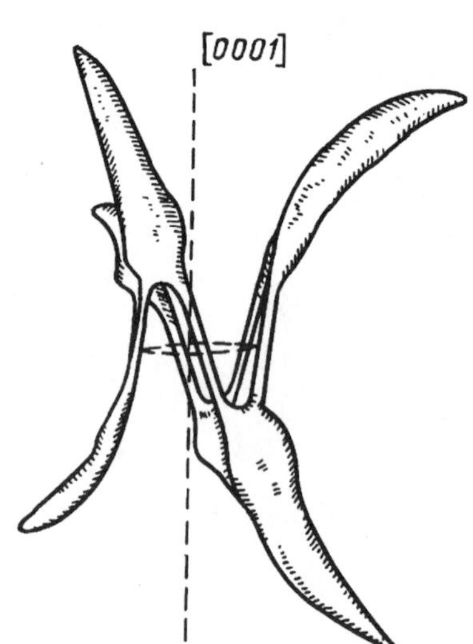

Figure 30.12 Hole Fermi surface of As [2].

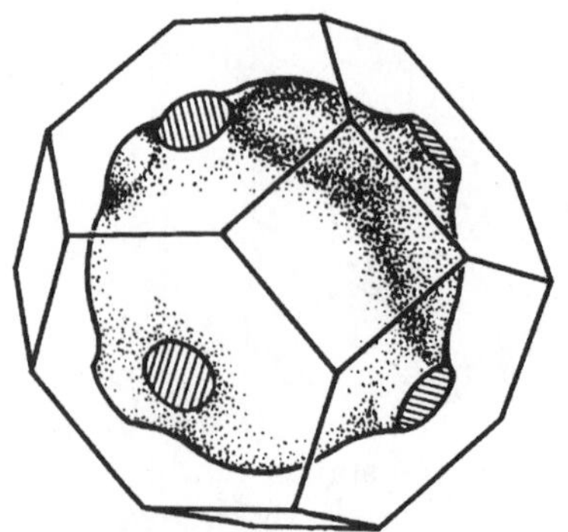

Figure 30.13 Fermi surface of Cu, Au, and Ag [6].

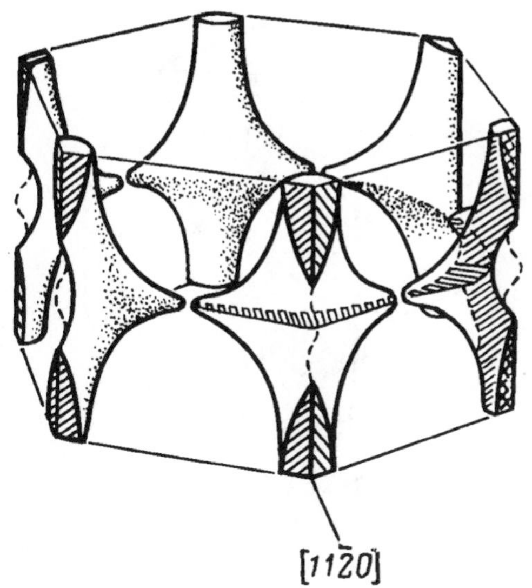

Figure 30.14 Open Fermi surface of Cd [1].

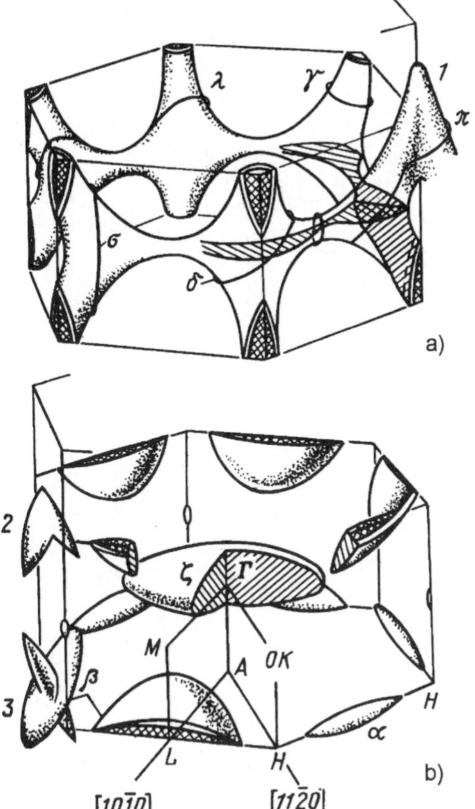

Figure 30.15 Fermi surface of Zn [1]: (*a*) open hole surface in the second zone (pockets in the first zone are shown by the criss-cross shading); (*b*) electron surface in the third and fourth zones (cigars in the fourth zone are criss-cross shaded).

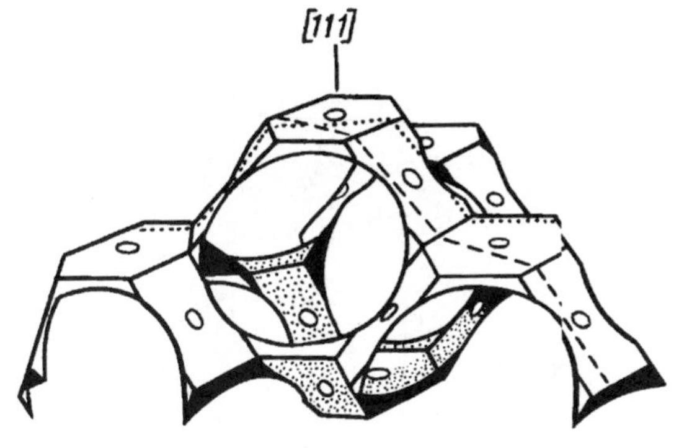

Figure 30.16 Open multiply-connected hole Fermi surface of Hg in the first zone [1].

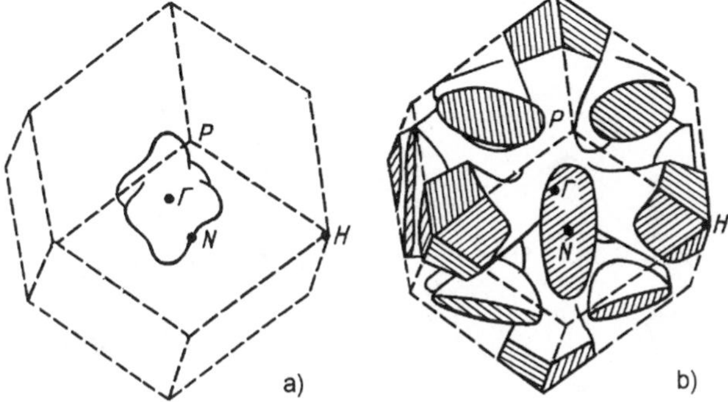

Figure 30.17 Fermi surface of the group V metals V, Nb, and Ta (the Mattheiss model) [2]: (*a*) closed hole surface with the point Γ as a center; (*b*) "toy jungle" consisting of hole tubes and hole ellipsoids in the third zone.

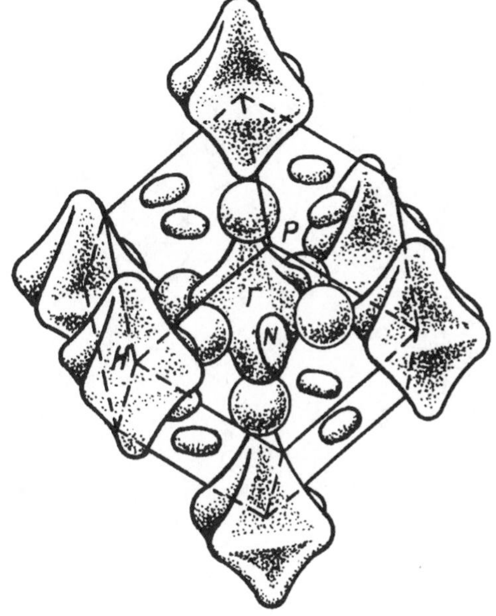

Figure 30.18 Fermi surface of Mo and W metals, and paramagnetic Cr (Lomer model) [2].

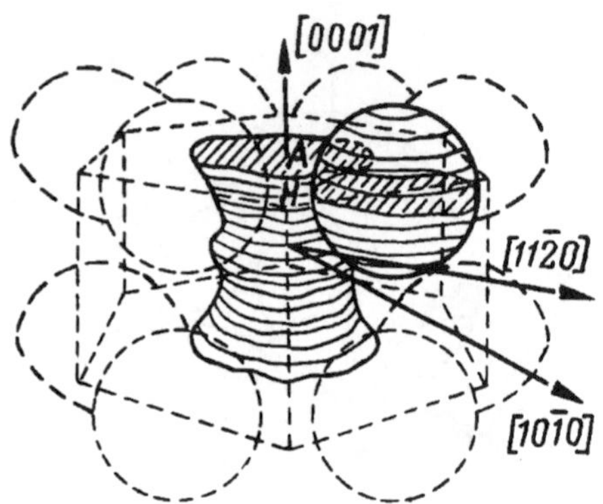

Figure 30.19 Fermi surface of Re: hole surface in the seventh zone (closed, dotted lines) and open electron surface in the eighth zone [2].

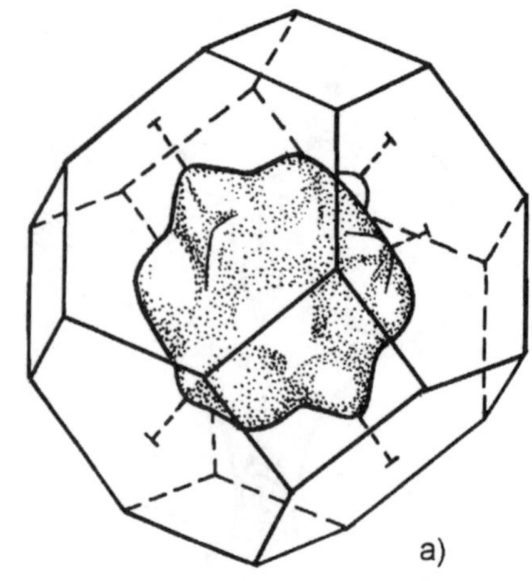

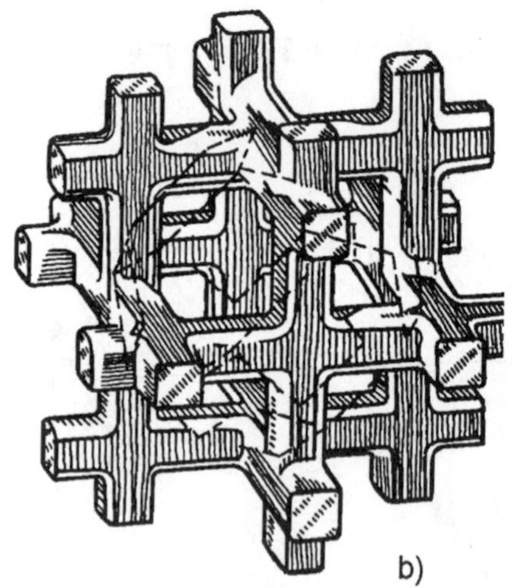

Figure 30.22 Fermi surface of Pd [2]: (*a*) electron surface with the point Γ as a center; (*b*) multiply-connected hole tubes.

Figure 30.20 Theoretical Fermi surface model for Co and Fe [2]: (*a*) hole surface with spin "down"; (*b*) hole surface with spin "up".

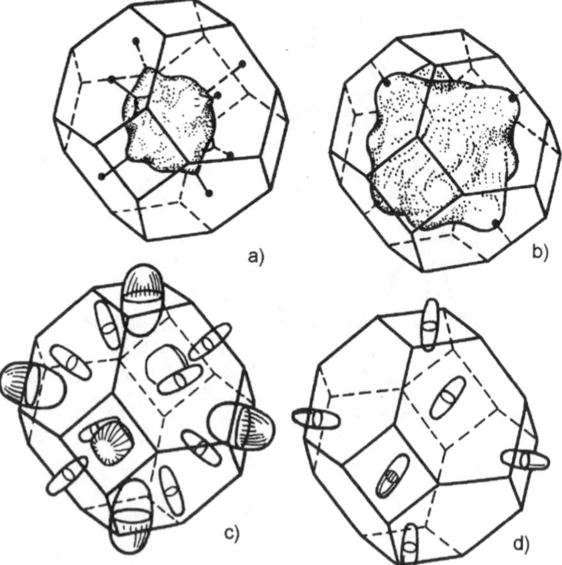

Figure 30.21 Fermi surface model for Rh [2]: (*a*) and (*b*) electronic surfaces; (*c*) and (*d*) hole pockets.

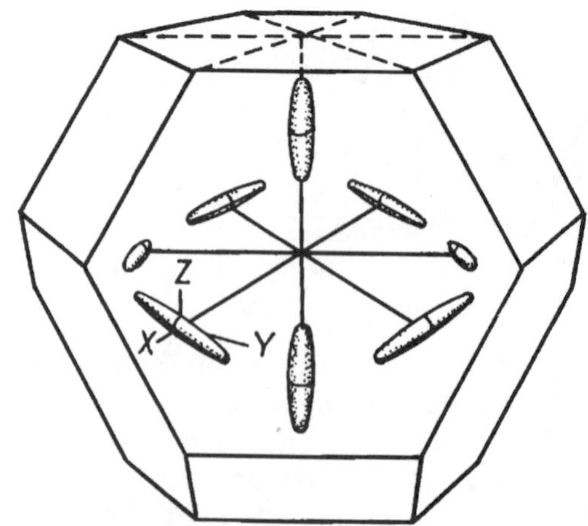

Figure 30.23 Fermi surface of Bi [3].

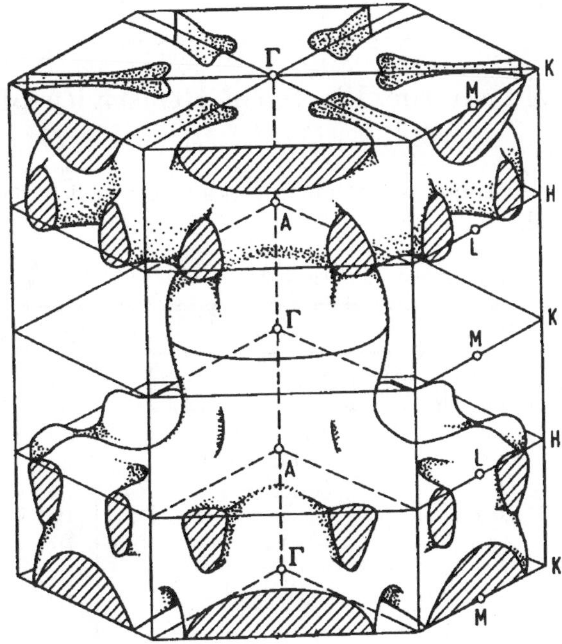

Figure 30.24 Fermi surfaces h_3 and h_4 of Gd [13].

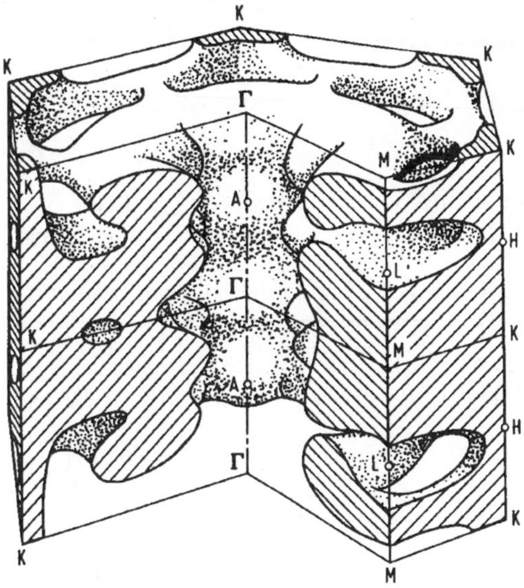

Figure 30.26 Fermi surfaces e_3 and e_4 of Y [13].

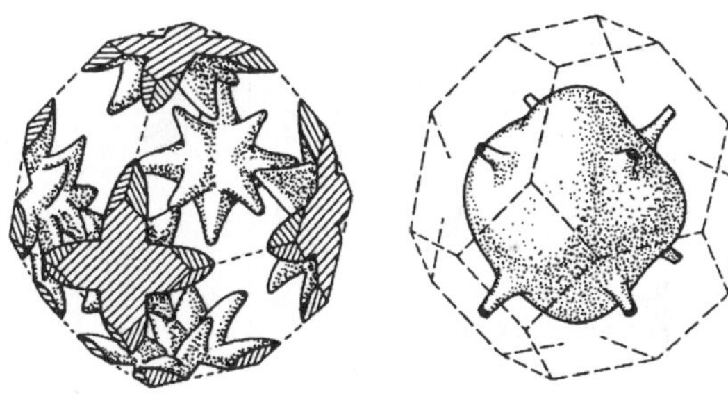

Figure 30.27 Fermi surfaces (*a*) h_5 and (*b*) e_6 of Ni [13].

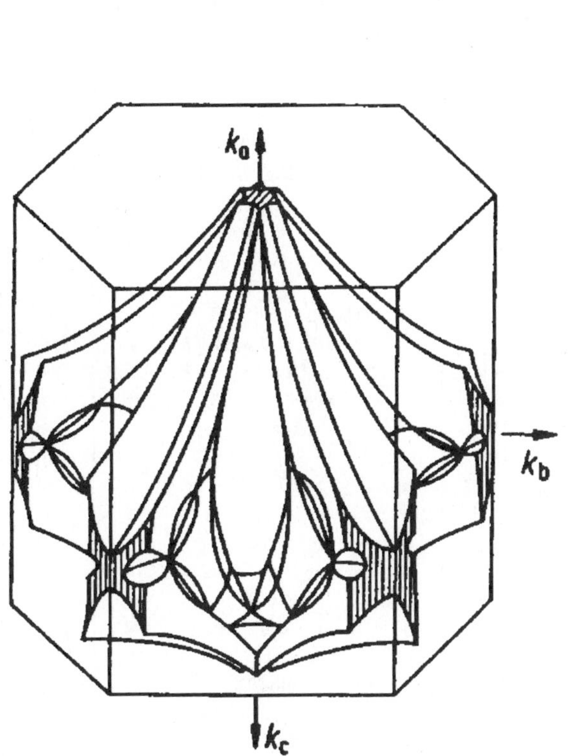

Figure 30.25 Fermi surface h_6 based on modification of free electron model for Ga [13].

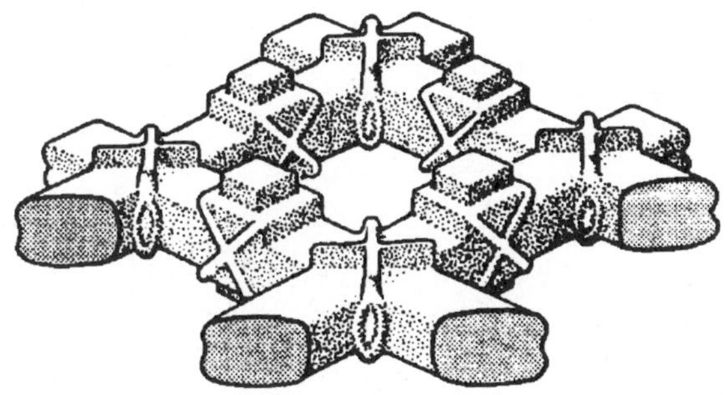

Figure 30.28 Fermi-surface closed region of e_5 at the point Γ for Pt [13].

30.3 The Influence of a Magnetic Field on the Electrical Resistivity of Metals

In Tables 30.3, 30.4 and in Figs. 30.29–30.58, data pertaining to the influence of a magnetic field on the electrical resistivity of various metals are presented.

Table 30.3 Change in electrical resistivity of pure metals in a magnetic field

Material, purity	*RRR*	T, K	B, T	$\Delta\rho/\rho^{*1}$	Ref.
Aluminum (99.999%)	15,000	19.6	4.0	5.0 $(\Delta\rho/\rho_{\perp})$ 2.2 $(\Delta\rho/\rho_{\parallel})$	[8]
	15,000	4.0	4.0	2.4 $(\Delta\rho/\rho_{\perp})$ 1.5 $(\Delta\rho/\rho_{\parallel})$	[8]
	28,200	19.6	2.0	3.1 $(\Delta\rho/\rho_{\perp})$	[8]
	4,000	4.2–70	3.8	See Fig. 30.29	[9]
Antimony polycrystalline		291	30.0	3.5	[11]
		80	30.0	40.0	[11]
Barium	140	20.4	3.32	0.41	[10]
	140	14.0	3.32	3.08	[10]
Beryllium polycrystalline	30	291	30.4	0.66	[11]
	30	78	30.4	2.28	[11]
Beryllium single crystal	40–90	4.2	3-7.0	See Fig. 30.30	[12]
Bismuth polycrystalline		291	30.0	37	[11]
		80	30.0	1360	[11]
Bismuth single crystal		14;4.2		See Fig. 30.31	[125]
Cadmium polycrystalline		291	30.0	0.08	[11]
		78	30.0	0.927	[11]
Cadmium single crystal	200	4.2	1.6	26 $(\Delta\rho/\rho_{\perp})$ 0.05 $(\Delta\rho/\rho_{\parallel})$	[21]
	1000	4.2	1.6	100 $(\Delta\rho/\rho_{\perp})$ 10 $(\Delta\rho/\rho_{\parallel})$	[21]
Cesium polycrystalline		20.4	4.0	0.03	[38]
Chromium polycrystalline		291	30.0	0.03	[11]
		78	30.0	4.36	[11]
Copper polycrystalline		78	30.0	0.429	[11]
Copper single crystal	480–630	4.2–70	3.8	See Fig. 30.29	[24]
Europium single crystal 99.9%		4.2	2.0	0.16	[17]
		4.2	4.0	0.33	[17]
		4.2	8.0	0.58	[17]
Gallium polycrystalline (99.7%)		195	30.0	0.173	[11]
Gallium single crystal (99.999%)	$25 \cdot 10^3$	4.2	2.0	$24 \cdot 10^4$ $(\mathbf{I} \parallel a, \mathbf{B} \parallel c)$	[15]
				$6 \cdot 10^4$ $(\mathbf{I} \parallel a, \mathbf{B} \parallel b)$	[15]
Gold polycrystalline (99.999%)		79	4.0	0.018	[18]
Gold single crystal (99.999%)	1600	4.2	3.0	See Fig. 30.32	[19]
Graphite single crystal (99.995%)		4.2	16.0	$14 \cdot 10^3$ $(\mathbf{B} \parallel c)$	[16]
		4.2	10.0	$9.5 \cdot 10^3$	[16]
		4.2	2.0	$1.5 \cdot 10^3$	[16]
Indium polycrystalline		165	30.0	0.03	[11]
		80	30.0	0.14	[11]
Indium single crystal	12000	4.2	3.5	See Fig. 30.33	[20]
Lead polycrystalline (99.999%)		291	30.0	< 0.01	[18]
		80	30.0	0.05	[18]
	17000	4.2	2.28	417	[35]
	17000	1.86	2.28	4810	[35]
Lead single crystal	10000	4.2	2.4	See Fig. 30.34	[28]
Lithium	985	4.2	0.829	0.978 $(\Delta\rho/\rho_{\perp})$	[23]
	985	77	1.43	0.155 $(\Delta\rho/\rho_{\perp})$	[23]
	985	4.2	1.60	0.975 $(\Delta\rho/\rho_{\parallel})$	[23]
	985	77	1.49	0.07 $(\Delta\rho/\rho_{\parallel})$	[23]
				See Fig.30.35	

Table 30.3 Change in electrical resistivity of pure metals in a magnetic field *(continued)*

Material, purity	*RRR*	*T*, K	*B*, T	$\Delta\rho/\rho$*1		Ref.
Magnesium polycrystalline		291	30.0	0.167		[11]
		78	30.0	2.82		[11]
Magnesium single crystal	230–610	4.2	2.5	See Fig. 30.36		[24]
Molybdenum polycrystalline		195	30.0	0.095		[11]
		78	30.0	0.915		[11]
Molybdenum single crystal	1000	4.2	2.5	See Fig. 30.37		[25]
Neodymium (99.9%)		1.4	0.5	0.04 ($\Delta\rho/\rho_\perp$)	−0.13 ($\Delta\rho/\rho_\parallel$)	[27]
		4.2	0.5	0.02 ($\Delta\rho/\rho_\perp$)	−0.05 ($\Delta\rho/\rho_\parallel$)	[27]
		14.0	0.5	−0.015 ($\Delta\rho/\rho_\perp$)	−0.04 ($\Delta\rho/\rho_\parallel$)	[27]
		14.0	0.5	−0.010 ($\Delta\rho/\rho_\perp$)	−0.02 ($\Delta\rho/\rho_\parallel$)	[27]
Neodymium 99.8%		4.2	8.0	See Fig. 30.38		[28]
Osmium single crystal *2		4.2	8.0	18 ($\varphi = 185°$)		[29]
		4.2	8.0	16 ($\varphi = 170°$)		[29]
		4.2	8.0	4 ($\varphi = 100°$)		[29]
		4.2	8.0	2.7 ($\varphi = 80°$)		[29]
Palladium polycrystalline		78	30.0	0.102		[11]
Palladium single crystal	1730	4.2	1.0	See Fig. 30.39		[25]
Platinum		78	30.0	0.102		[11]
	2400	4.2	2.0	See Fig. 30.40		[24]
Potassium single crystal (99.95%)	3400	4.2	5.5	0.38 at **B** ‖[100]		[22]
	3400	4.2	5.5	0.22 at **B** ‖[110]		[22]
Praseodymium		4.2	8.0	0.04 ($\Delta\rho/\rho_\parallel$)		[17]
Rhenium (99.6%)		80	3.43	0.0196		[30]
Rhenium single crystal	1600	4.2	3.43	See Fig. 30.41		[31]
Rhodium polycrystalline		20.4	3.6	1.546		[32]
		4.2	3.7	1.867		[32]
Rubidium	300	4.2	2.55	0.36 ($\Delta\rho/\rho_\perp$)	0.177 ($\Delta\rho/\rho_\parallel$)	[33]
Ruthenium single crystal	100	4.2	4.0	0.87 (**I** ‖ [1000], **B** ‖ [1120])		[34]
	100	4.2	4.0	0.75 (**I** ‖ [1000], **B** ‖ [1010])		[34]
	100	4.2	4.0	0.59 (**I** ‖ [1120], **B** ‖ [1110])		[34]
	100	4.2	4.0	0.53 (**I** ‖ [1120], **B** ‖ [1000])		[34]
Samarium (99.9%)		4.2	0.5	−2		[17]
		4.2	4.0	5		[17]
		4.2	6.0	12		[17]
		4.2	8.0	17		[17]
Silver polycrystalline		78	30.0	0.376		[11]
	1000	4.2	2.4	See Fig. 30.42		[36]
Sodium polycrystalline		80	30.0	0.07		[11]
	5000	4.2	0.9	0.399		[26]
Tantalum polycrystalline		291	30.0	0.001		[11]
		80	30.0	0.01		[11]
Thallium polycrystalline		80	30.0	0.159		[11]
	100000	4.2	13.0	See Fig. 30.43		[37]
Tin polycrystalline		291	30.0	0.02		[11]
		80	30.0	0.23		[11]
Tin single crystal	10000	4.2	2.0	See Fig. 30.44		[28]
Tungsten		78	30.0	0.938		[11]
Tungsten single crystal	13500	4.2		See Fig. 30.45		[14]
Zinc polycrystalline		291	30.0	0.06		[11]
		78	30.0	0.927		[11]
Zinc single crystal (99.999%)	20000	4.2	1.8	See Fig. 30.46		[39]
Zirconium polycrystalline		195	30.0	0.01		[11]
		80	30.0	0.05		

*1 $\Delta\rho/\rho_\perp$ if **B** ⊥ **I**, $\Delta\rho/\rho_\parallel$ if **B** ‖ **I**. Values of $\Delta\rho/\rho_\perp$ are given in the table except as specially marked.

*2 Specimen axis aligns with [11$\bar{2}$0]; φ is the angle between the direction of **B** and the [11$\bar{2}$0] axis.

Table 30.4 Change in electrical resistivity of ferromagnetic metals in a magnetic field

Metal, purity	*RRR*	T, K	$\mu_0 H$, T	$\Delta\rho/\rho$	Ref.
Cobalt single crystal, whisker	460	4.2	15	See Fig. 30.47	[45]
Cobalt polycrystalline (99.25%)		200–600	0.2	See Fig. 30.48	[46]
		300		$18 \cdot 10^{-4}$	[47]
Dysprosium polycrystalline (99.5%)		93	0.8	-0.02 ($\mathbf{I} \parallel b, \mathbf{B} \parallel a$)	[43]
		60–180	1.5	See Fig. 30.49	[43]
Gadolinium polycrystalline (99.9%)		4.2–350	1.6	See Fig. 30.50	[40]
Gadolinium single crystal		205	1	0.015 ($\mathbf{B} \parallel c, \mathbf{I} \parallel b$)	[40]
		205	1	-0.01 ($\mathbf{B} \parallel b, \mathbf{I} \parallel b$)	[40]
				See Fig. 30.51	[41]
Holmium polycrystalline (99.9%)	16	4–130	2.0	See Fig. 30.52	[42]
Iron single crystal, whisker	260	300	0.1	$9 \cdot 10^{-4}$ ($\Delta\rho/\rho_\parallel$) $-2 \cdot 10^{-4}$ ($\mathbf{I} \parallel [100], \mathbf{B} \parallel [100]$)	[44]
	260	300	0.1	$0.36 \cdot 10^{-2}$ ($\Delta\rho/\rho_\parallel$) $-3 \cdot 10^{-4}$ ($\mathbf{I} \parallel [111], \mathbf{B} \parallel [111]$)	[44]
				See Fig. 30.53, 30.54	
Nickel		300	1.6	$15 \cdot 10^{-3}$ ($\Delta\rho/\rho_\parallel$) $-15 \cdot 10^{-3}$ ($\Delta\rho/\rho_\perp$)	[47]
Nickel polycrystalline		300	0–1.8	See Fig. 30.55	[47]
Nickel single crystal		300–660	1.0	See Fig. 30.56	[48]
Terbium polycrystalline (99.5%)		220	1.5	-0.02 ($\Delta\rho/\rho_\parallel$) -0.018 ($\Delta\rho/\rho_\perp$)	[49]
		80–240	1.4	See Fig. 30.57	[49]

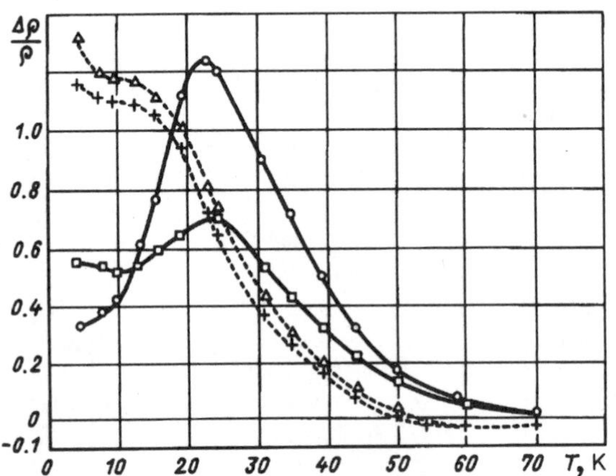

Figure 30.29 Temperature dependence of $\Delta\rho/\rho$ for polycrystalline Al and Cu specimens in a magnetic field $B = 3.8$ T [9]. For Al: ∘ – $RRR = 4000$; □ – $RRR = 1875$; for Cu: △ – $RRR = 600$; + – $RRR = 480$.

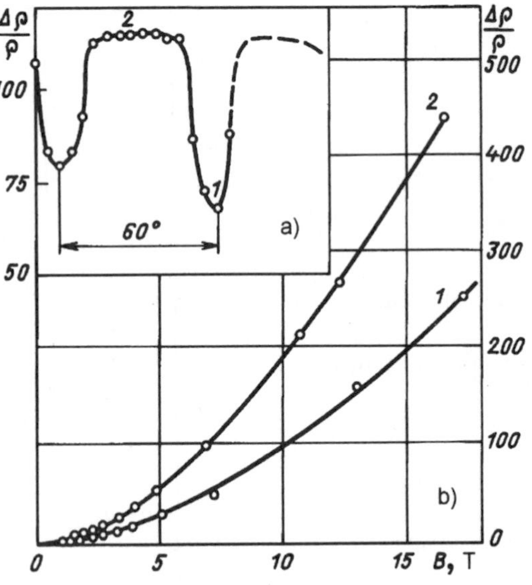

Figure 30.30 (*a*) Anisotropy of $\Delta\rho/\rho$ for a hexagonal Be specimen [12] ($RRR = 88$; $T = 4.2$ K; $B = 7.0$ T; φ is the angle of magnetic field rotation in a plane perpendicular to the electric current through the specimen); (*b*) Dependence of $\Delta\rho/\rho$ for Be on magnetic induction for the directions of minimum (*1*) and maximum (*2*) magnetoresistivity in the rotation diagram (*a*) (see Comment No 3 in the text).

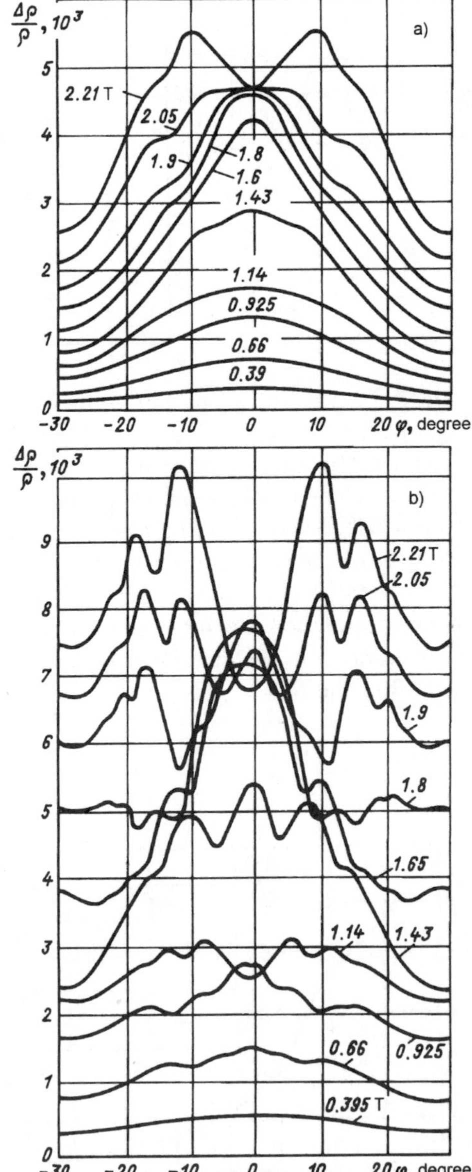

Figure 30.31 Dependence of $\Delta\rho/\rho$ for a Bi single crystal on magnetic field orientation with respect to a binary axis at temperatures of 14 K (*a*) and 4.22 K (*b*) [13] (the principal axis was parallel to the current).

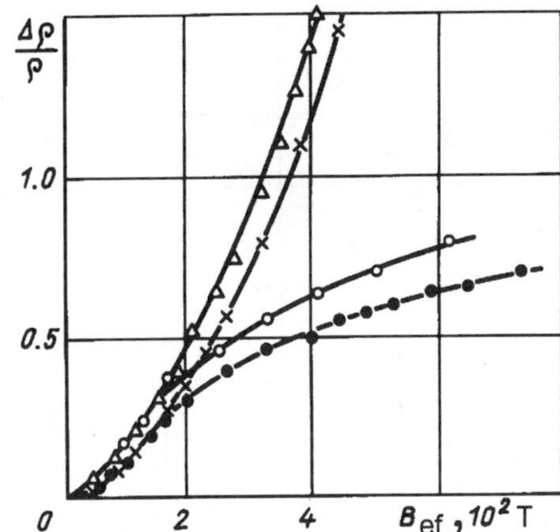

Figure 30.32 Dependence of $\Delta\rho/\rho$ for an Au single crystal on effective magnetic induction $B_{eff} = B \cdot \rho(300\text{ K})/\rho(4.2\text{ K})$ [19] (RRR = 16000): • -in a minimum ($\varphi = 0°$); × – in a maximum ($\varphi = -75°$) of the angular diagram at T=4.2 K; ∘ – in the minimum ($\varphi = 0°$) and Δ – in the maximum ($\varphi = -75°$) at T = 20.4 K; φ is the angle of magnetic field rotation in a plane perpendicular to the current through a specimen (the rotation diagram for Au is similar to that for Ag, see Fig. 30.42; see Comment No 2 in the text).

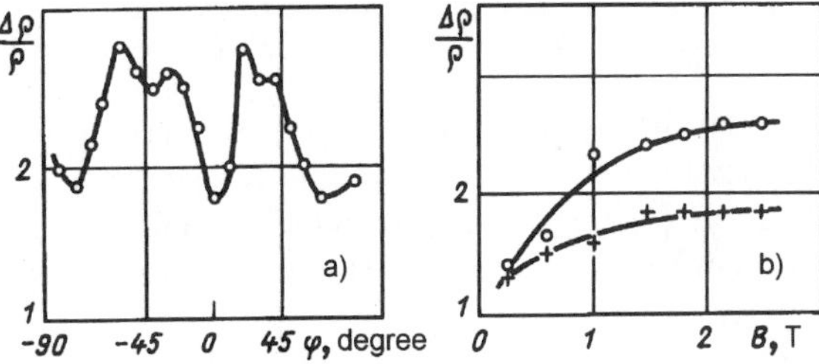

Figure 30.33 (*a*) Anisotropy of $\Delta\rho/\rho$ for an In single crystal [20] (RRR = 12400; T = 4.2 K; B = 2.46 T; φ is the angle of the magnetic field in a plane perpendicular to the electric current through the specimen); (*b*) dependence of $\Delta\rho/\rho$ for In on magnetic induction for the directions of minimum and maximum magnetoresistivity in the rotation diagram (*a*) (see Comment No 1 in the text).

Figure 30.34 (*a*) Polar diagram of $\Delta\rho/\rho$ for a Pb single crystal [28] (RRR = 10000; T = 4.2 K; B = 2.23 T; φ is the angle of magnetic field rotation in a plane perpendicular to the current through a specimen; the crystal axis was parallel to the [111] axis); (*b*) dependence of $\Delta\rho/\rho$ for Pb on magnetic induction for the directions of minimum and maximum magnetoresistivity variations in a polar diagram (*a*) (see Comment No 4 in the text).

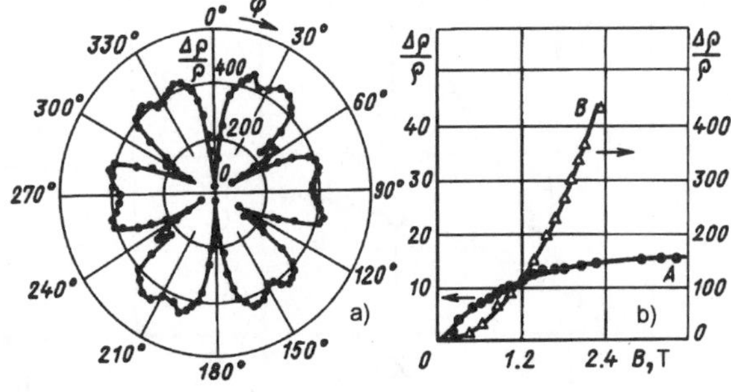

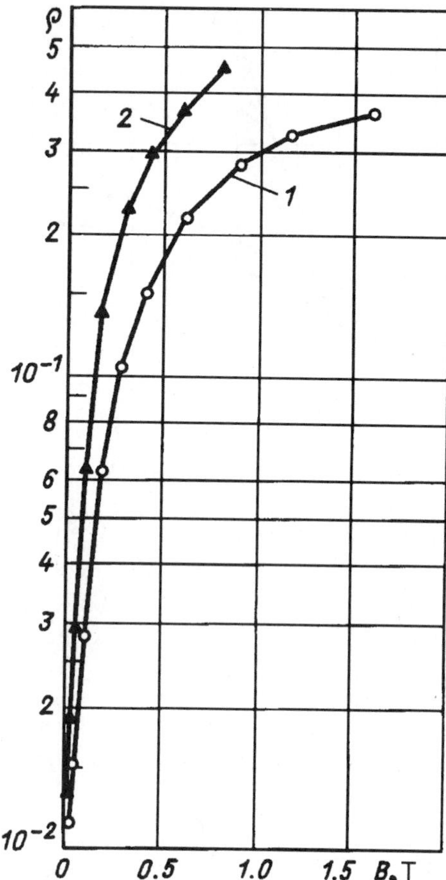

Figure 30.35 Field dependence of electrical resistivity ρ for a polycrystalline Li specimen at 4.2 K [23]: (*1*) **B** is parallel to the current through the specimen; (*2*) **B** is normal to the current.

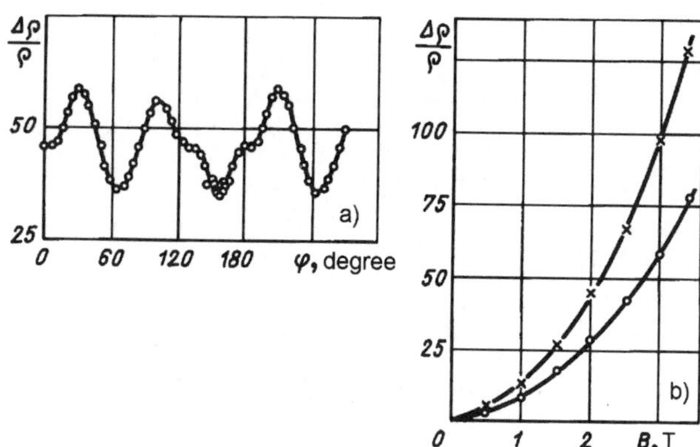

Figure 30.37 (*a*) Anisotropy of $\Delta\rho/\rho$ for a Mo single crystal [25] ($RRR = 1000$; $T = 4.2$ K; $B = 2.3$ T; φ is the angle of magnetic field rotation in a plane perpendicular to the current through the specimen; crystal orientation $\theta = 34°, \xi = 3°$: θ and ξ are the polar and azimuthal angles of the specimen axes with respect to the principal crystal axes); (*b*) dependence of $\Delta\rho/\rho$ for Mo on magnetic induction for the directions of a minimum and a maximum in the angular diagram (*a*) (see Comment No 3 in the text).

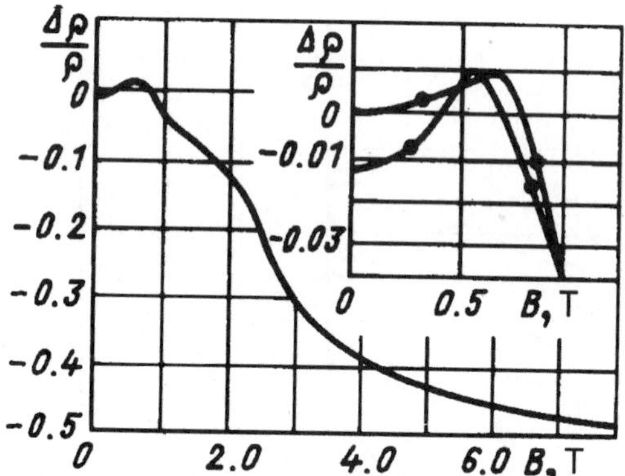

Figure 30.38 Dependence of $\Delta\rho/\rho_{\|}$ for Nd on magnetic induction at 4.2 K [17].

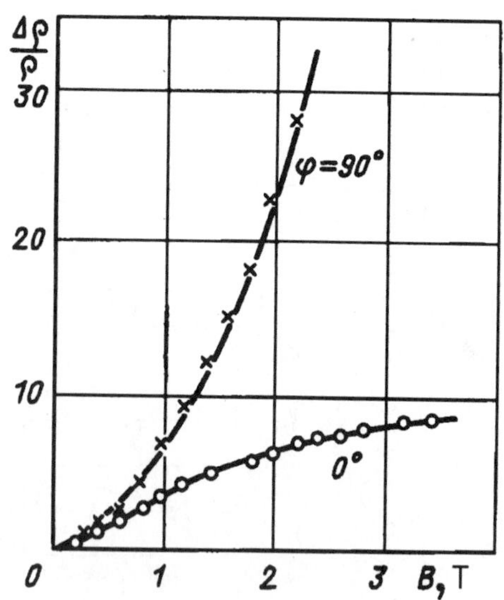

Figure 30.36 Dependence of $\Delta\rho/\rho$ for a Mg single crystal versus magnetic induction for the directions of a maximum ($\varphi = 90°$) and a minimum ($\varphi = 0°$) in the rotation diagram (at 4.2 K) [24]; φ is the angle of a magnetic field rotation in a plane normal to the electric current through the specimen; the specimen axis made an angle of 65° with the [0001] axis (see Comment No 4 in the text).

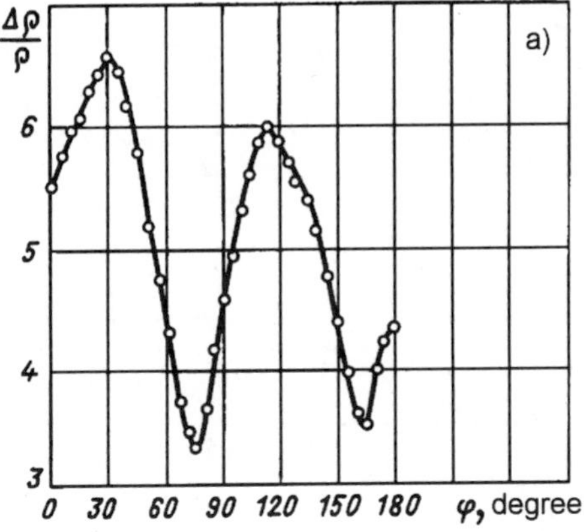

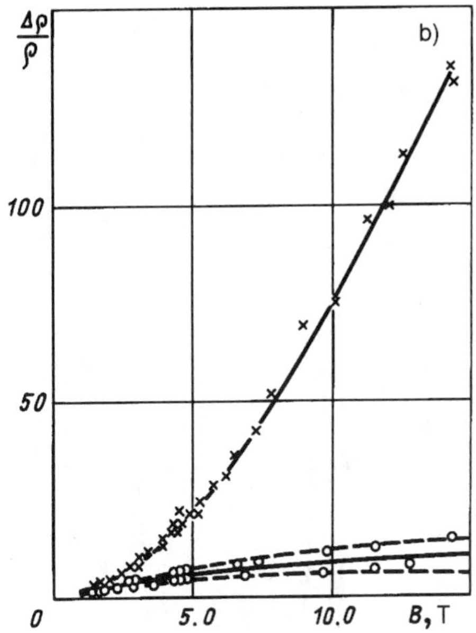

Figure 30.39 (a) Anisotropy of $\Delta\rho/\rho$ for a Pd single crystal [25] (T=4.2 K; B=2.3 T; φ is the angle of magnetic field rotation in a plane perpendicular to the electric current through the specimen; crystal orientation $\theta = 6°, \xi = 27°$: θ and ξ are the polar and azimuthal angles of the specimen axes with respect to the principal crystal axes); (b) dependence of $\Delta\rho/\rho$ on magnetic induction for the directions of a minimum and a maximum in the rotation diagram (a) (see Comment No 4 in the text).

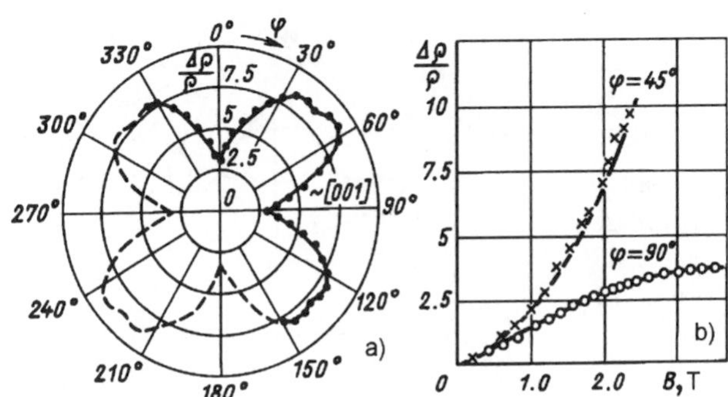

Figure 30.40 (a) Polar diagram of $\Delta\rho/\rho$ for a Pt single crystal [24] ($RRR = 1900$; $B = 2.35$ T; $T = 4.2$ K; φ is the angle of magnetic field rotation in a plane perpendicular to the current through the crystal); (b) dependence of $\Delta\rho/\rho$ for Pt on magnetic induction for the directions of a minimum and a maximum in a polar diagram (a) (see Comment No 4 in the text).

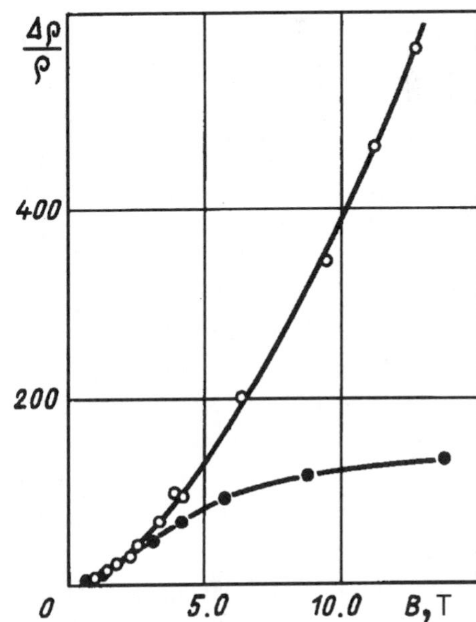

Figure 30.41 Dependence of $\Delta\rho/\rho$ for Re on magnetic induction ($T = 4.2$ K) for the directions of minimum and maximum magnetoresistivity in the rotation diagram [31] (see Comment No 4 in the text).

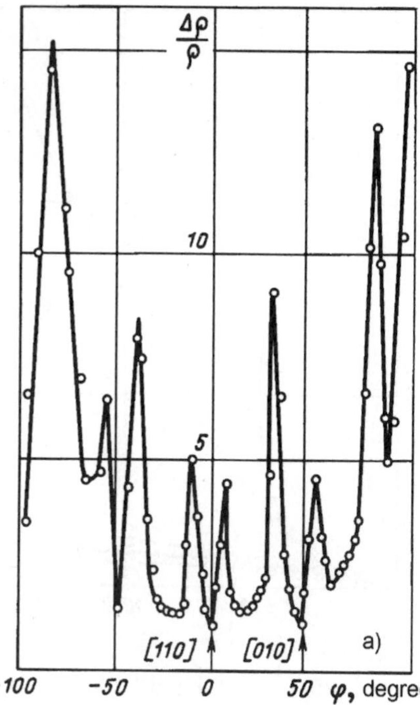

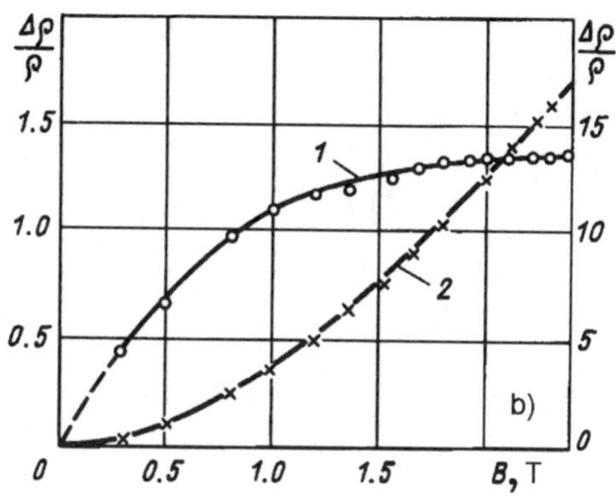

Figure 30.42 (*a*) Anisotropy of $\Delta\rho/\rho$ for an Ag single crystal [36] (RRR = 1,000; T = 4.2 K; B = 2.35 T; φ is the angle of magnetic field rotation in a plane perpendicular to the electric current through the specimen whose axis was parallel to the [001] axis within $\pm 5°$); (*b*) dependence of $\Delta\rho/\rho$ of Ag on magnetic induction for the directions of a minimum ($\varphi = 0°$) and a maximum ($\varphi = 80°$) in the rotation diagram (*a*): *1* – left-side ordinate scale; *2* – right-side ordinate scale (see Comment No 2 in the text).

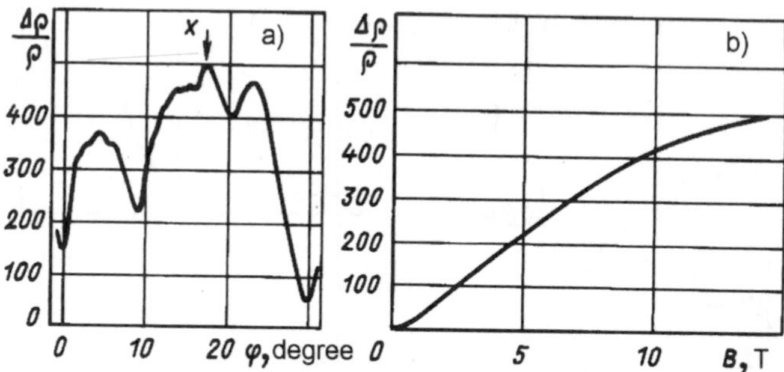

Figure 30.43 (*a*) Dependence of $\Delta\rho/\rho$ for a Tl single crystal on the angle of the magnetic field in the basal plane measured from the $[10\bar{1}0]$ direction (T = 4.2 K; B = 13.4 T [37]); (*b*) $\Delta\rho/\rho$ vs. magnetic induction at 4.2 K in the x direction in the diagram (*a*) (see Comment No 4 in the text).

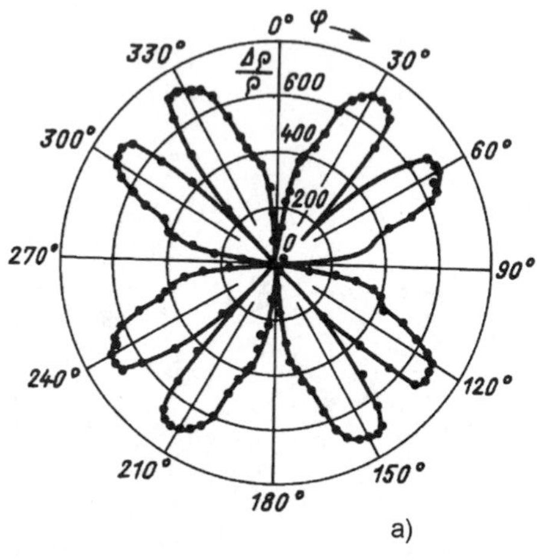

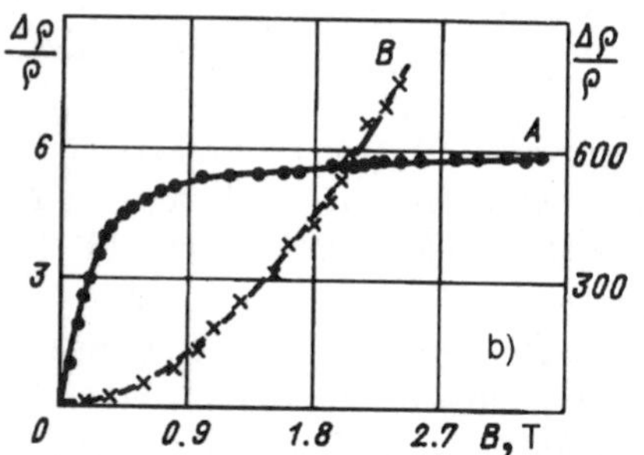

Figure 30.44 (*a*) Polar diagram of $\Delta\rho/\rho$ for a Sn single crystal [28] ($T = 4.2$ K; $B = 2.3$ T; φ is the angle of magnetic field rotation in a plane perpendicular to the electric current through the specimen, $\mathbf{I} \perp \mathbf{B}, \mathbf{I} \parallel [001]$; (*b*) dependence of $\Delta\rho/\rho$ for Sn on magnetic induction for the directions of a minimum ($\varphi = 0°$) and a maximum ($\varphi = 30°$) in a polar diagram (*a*) (see Comment No 4 in the text).

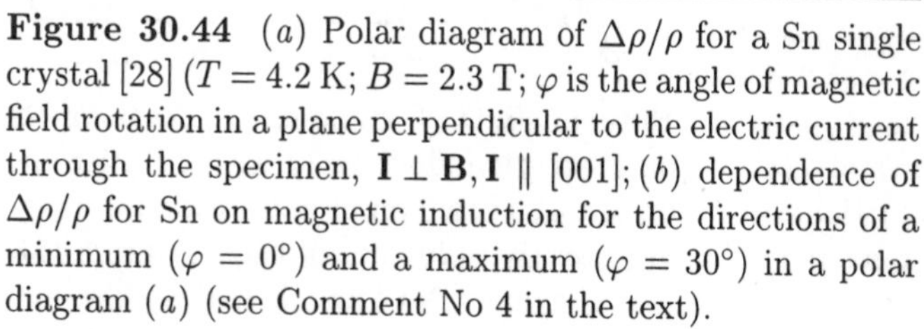

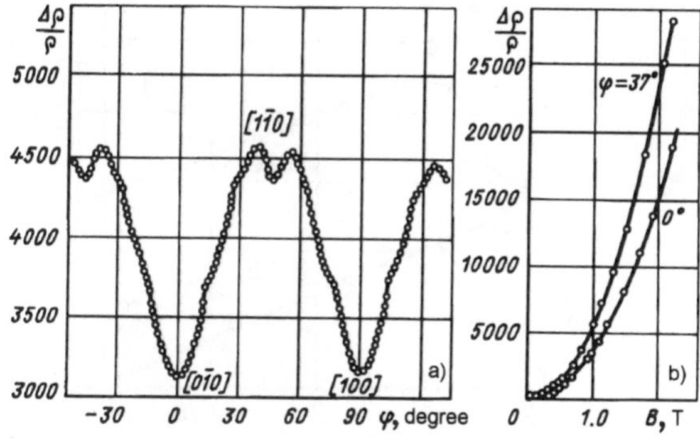

Figure 30.45 (*a*) Anisotropy of $\Delta\rho/\rho$ for a W single crystal [14] ($RRR = 13000$; $T = 4.2$ K, $B = 0.9$ T; the magnetic field rotation plane was perpendicular to the [001] axis; angle φ determining the field direction in the plane was measured from the [010] axis; (*b*) dependence of $\Delta\rho/\rho$ for W on magnetic induction for the directions of maximum ($\varphi = 37°$) and minimum ($\varphi = 0°$) magnetoresistivity in the rotation diagram (*a*) (see Comment No 3 in the text).

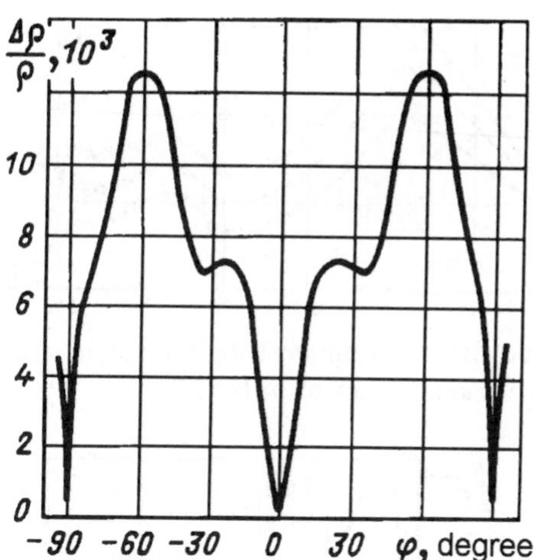

Figure 30.46 Dependence of $\Delta\rho/\rho$ for a Zn single crystal on magnetic field rotation angle [39] ($RRR = 20000$; $T = 4.2$ K; $B = 1.8$ T; **I** $\parallel$ $[11\bar{2}0]$); transverse rotation of **B** in the $(11\bar{2}0)$ plane, **B** $\parallel$ $[10\bar{1}0]$ at $\theta = 0°$ (see Comment No 4 in the text).

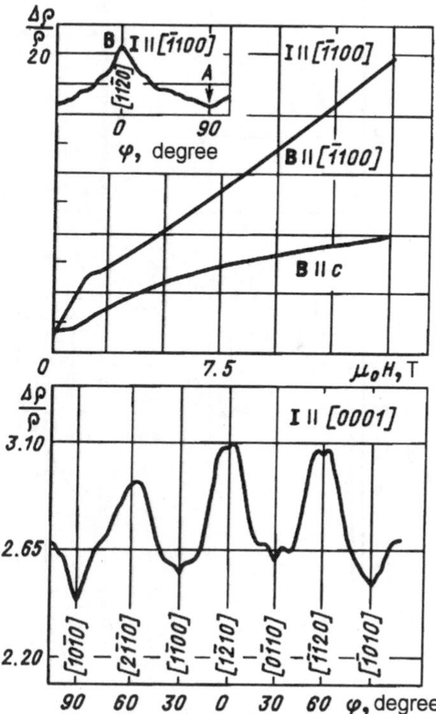

Figure 30.47 Field dependence of $\Delta\rho/\rho_\perp$ for Co single crystals and rotation diagrams [45] ($T = 4.2$ K; $\mu_0 H = 15$ T; $RRR = 204$; φ is the angle of magnetic field rotation in a plane perpendicular to the electric current through the specimen.

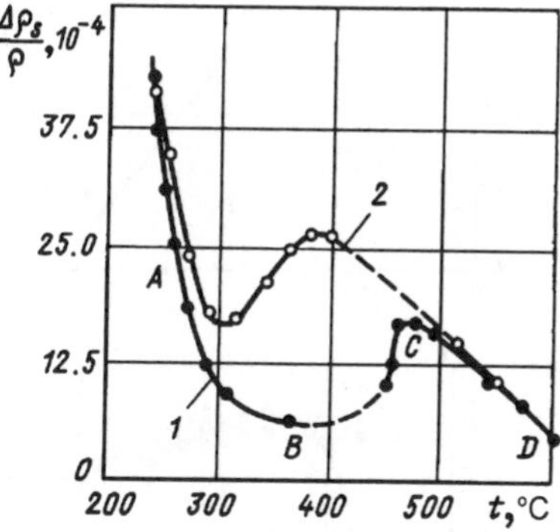

Figure 30.48 Temperature dependence of the saturation value $\Delta\rho_s/\rho$ for Co at (*1*) heating and (*2*) cooling [46].

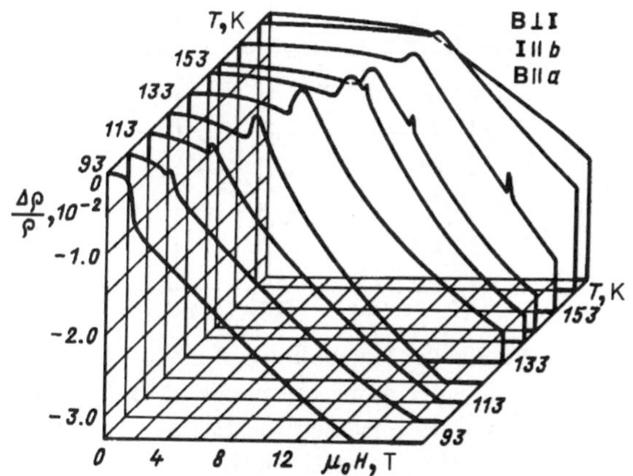

Figure 30.49 Dependence of $\Delta\rho/\rho$ for Dy on $\mu_0 H$ at various temperatures [43].

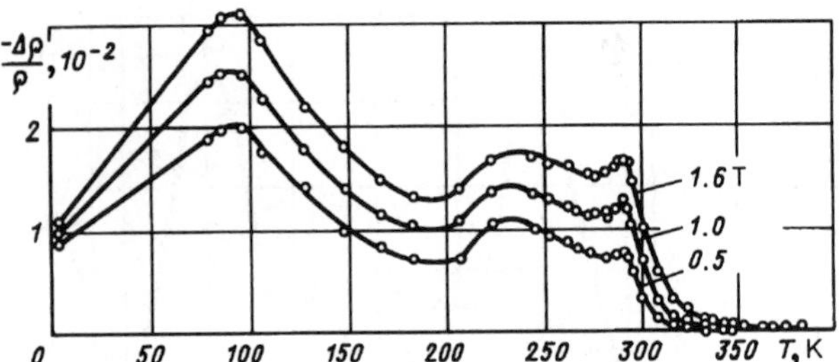

Figure 30.50 Temperature-dependent $\Delta\rho/\rho$ for Gd at for various values of $\mu_0 H$ [40].

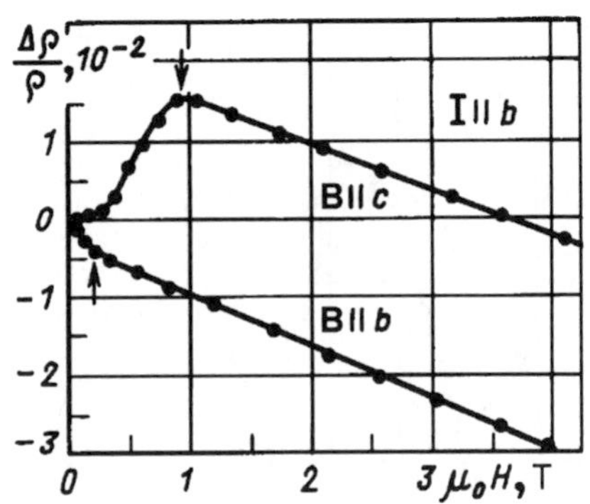

Figure 30.51 Dependence of $\Delta\rho/\rho$ for Gd on $\mu_0 H$ at 205 K [41].

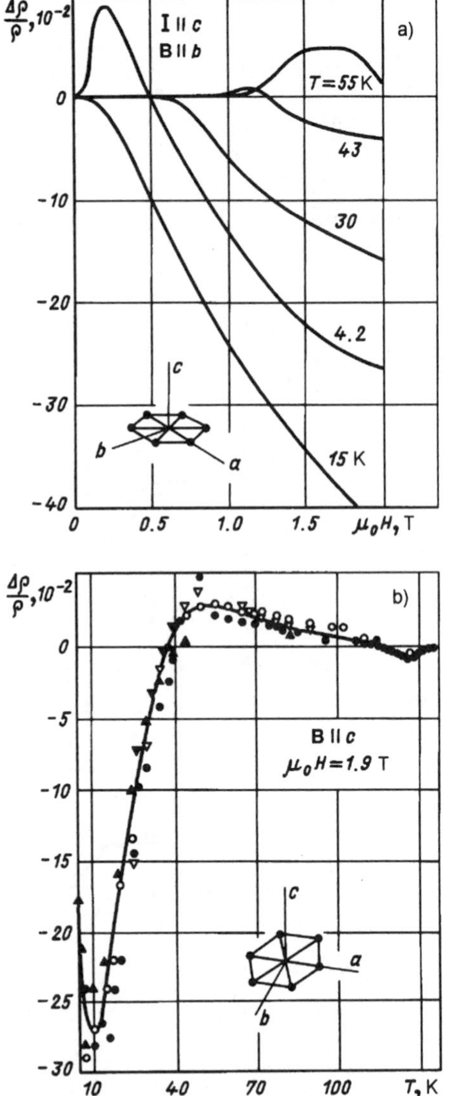

Figure 30.52 (*a*) Dependence of $\Delta\rho/\rho_\perp$ for a Ho single crystal on $\mu_0 H$ along the *c*-axis at **B** ‖ **b** [42] (the same results were obtained for **B** ‖ **a**). (*b*) temperature dependence of $\Delta\rho/\rho$ along the *c*-axis (the symbols used in the graph indicate data from different measuring cycles).

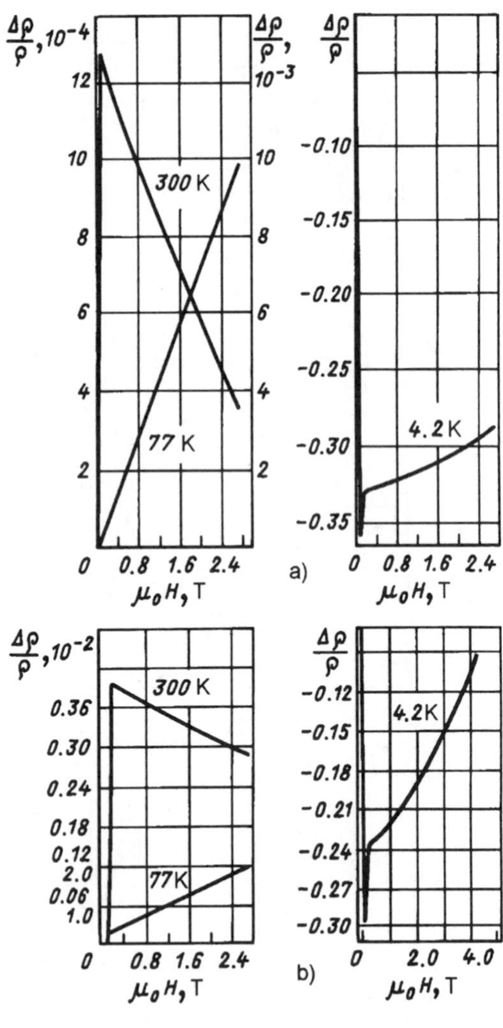

Figure 30.53 Dependence of $\Delta\rho/\rho_\parallel$ for Fe whiskers on $\mu_0 H$ at temperatures of 300, 77, and 4.2 K [44]: (*a*) along the [100] axis **B** ‖ [100] (diameter=240 μm, RRR = 200); (*b*) along the [111] axis **B** ‖ [111] (diameter=300 μm, RRR = 360).

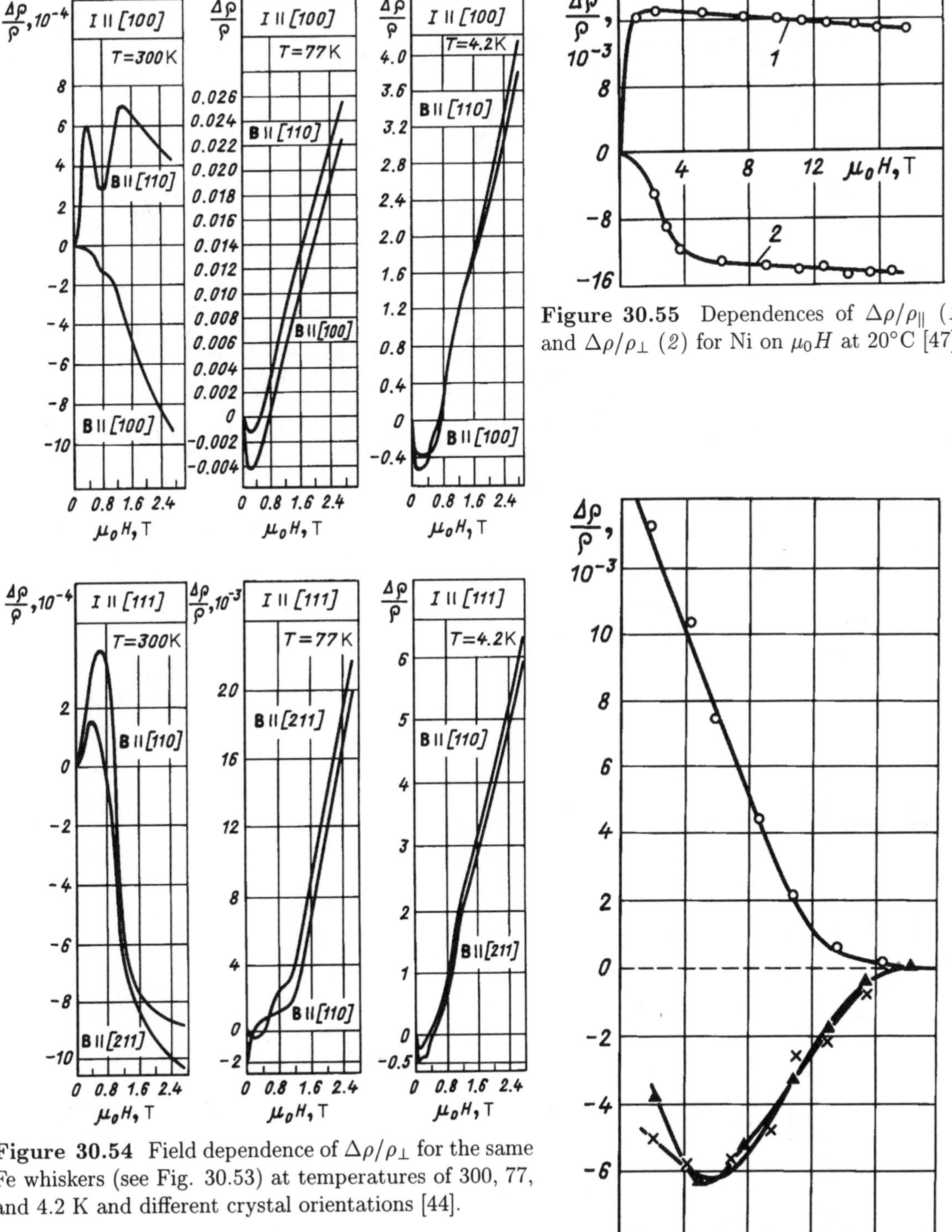

Figure 30.54 Field dependence of $\Delta\rho/\rho_\perp$ for the same Fe whiskers (see Fig. 30.53) at temperatures of 300, 77, and 4.2 K and different crystal orientations [44].

Figure 30.55 Dependences of $\Delta\rho/\rho_\parallel$ (*1*) and $\Delta\rho/\rho_\perp$ (*2*) for Ni on $\mu_0 H$ at 20°C [47].

Figure 30.56 Temperature dependences of the saturation values $\Delta\rho_s/\rho_\parallel$ and $\Delta\rho_s/\rho_\perp$ for Ni [48]: ○ – $\Delta\rho_s/\rho_\parallel$; × – $\Delta\rho_s/\rho_\perp$ ($\theta = 0°$); ▲ – $\Delta\rho_s/\rho_\perp$ ($\theta = 60°$) (θ is the angle between the sample plane and the magnetic field direction).

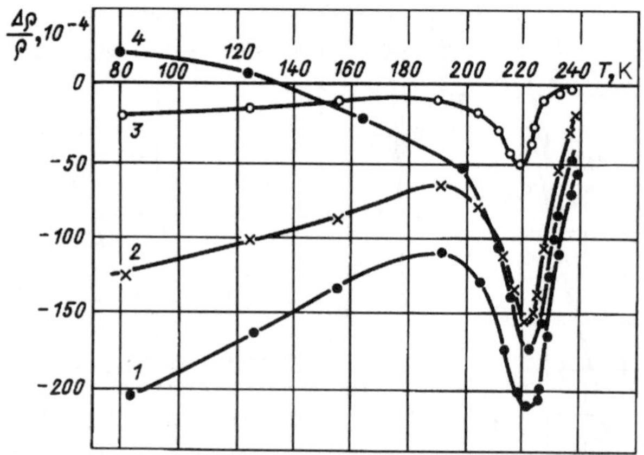

Figure 30.57 Temperature-dependent $\Delta\rho/\rho$ for Tb [49]: *1* – $\Delta\rho/\rho_{\|}, \mu_0 H = 1.5$ T; *2* – $\Delta\rho/\rho_{\|}, \mu_0 H = 0.78$ T; *3* – $\Delta\rho/\rho_{\|}, \mu_0 H = 0.13$ T; *4* – $\Delta\rho/\rho_{\perp}, \mu_0 H = 1.5$ T.

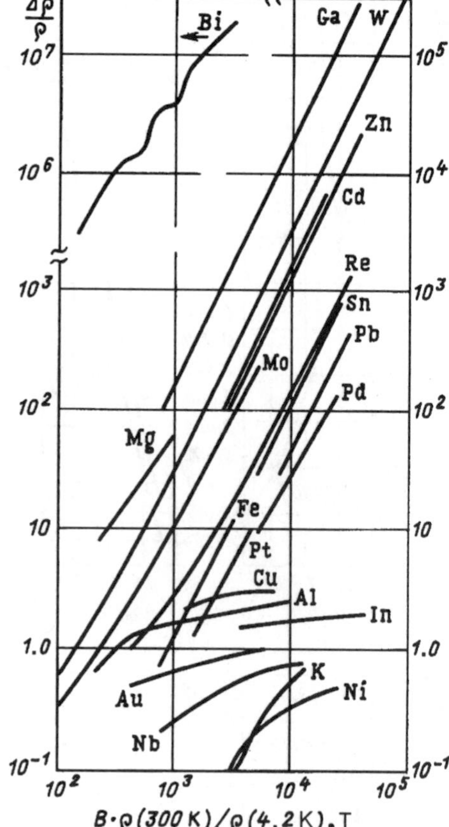

Figure 30.58 Reduced Kohler diagram: dependence of $\Delta\rho/\rho$ on $B \cdot \rho(300\ \mathrm{K})/\rho(4.2\ \mathrm{K})$ for a number of metals [5].

30.4 Hall Coefficients of Metals

In Tables 30.5 and 30.6 and in Figs. 30.59–30.74 data on the Hall coefficients of various metals are presented.

Table 30.5 Hall coefficient of liquid metals

Metal	R, 10^{-11} m^3/(A·s)	T, K	Ref.	Metal	R, 10^{-11} m^3/(A·s)	T, K	Ref.
Aluminum	−3.9	933–1123	[101]	Lead	−4.48	623	[105]
Bismuth	−3.0	558–603	[102]	Mercury	−7.6	303–483	[103]
Cadmium	−7.2	594–673	[103]	Neodymium	< 0		[104]
Cerium	+8.0	873–1223	[106]	Praseodymium	< 0		[104]
Cesium	−(74.6–78.6)	307–372	[107]	Rubidium	−42	308	[99]
Copper	−8.25	135–1423	[101]	Silver	−11.6	1273	[105]
Gallium	−3.9	303–873	[102]	Sodium	−25	371	[102]
Germanium	−3.9	1233–1283	[102]	Thallium	−12.6	500–1000	[102]
Gold	−11.8	1336–1423	[101]	Tin	−4.4	523–593	[103]
Indium	−5.6	429–973	[101]	Uranium	+3.8	273–1473	[104]
Lanthanum	< 0		[104]				

Table 30.6 Hall coefficient of metals

Metal, purity	T, K	B, T	R, 10^{-11} $m^3/(A{\cdot}s)$	Ref.
Aluminum (99.5%), $RRR = 11$	83	0.54	−2.2	[50]
	273	0.54	−3.3	[50]
	573	0.54	−3.9	[50]
	873	0.54	−4.1	[50]
Aluminum 99.9999%, $RRR = 84$–2600	2–80	0.5	See Fig. 30.59	[51]
Beryllium single crystal (99.5%)	290	0.3–2.0	$R_{\mathbf{B}\parallel c_3} = -77$	[52]
Beryllium single crystal (99.99%)	290	0.3–2.0	$R_{\mathbf{B}\parallel c_3} = -76.4$	[53]
	290	0.3–2.0	$R_{\mathbf{B}\perp c_3} = 148$	[53]
			See Fig. 30.60	[53]
Bismuth single crystal			See Fig. 30.61	[55]
Cadmium polycrystalline	90	1.0	10	[65]
	20	1.0	20	[65]
	4.2	1.0	−14	[65]
Cadmium single crystal	297		$R_{\mathbf{B}\perp c_3}$=3.9	[66]
	297		$R_{\mathbf{B}\parallel c_3}$=13.9	[66]
Calcium (99%)	300	0.1	−17.8	[69]
Cerium (99.8%)	293	0.56	+18.1	[70]
Cesium (99.9%)	293		−73.2	[99]
	200		−71.5	[99]
Chromium (99.9%)	287	1.0–2.9	+36.4	[97]
Chromium, $RRR = 15$	8–40	1.0	See Fig. 30.62	[98]
Copper (99.9999%)	300	1.5	−5.3	[57]
	150	1.5	−5.4	[57]
	50	1.5	−6.3	[57]
	40	1.5	−6.8	[57]
	4	1.5	−6.5	[57]
Copper, $RRR = 150$	<80	0.0085–0.51	See Fig. 30.63	[75]
Gold (99.9999%), $RRR = 1957$	300	1.52	−7.3	[57]
	4.2–300		See Fig. 30.64	[57]
Hafnium (99.9%), $RRR = 27.5$	300	1.3	4.2	[56]
	200	1.3	3.0	[56]
	100	1.3	1.9	[56]
	4.2	1.3	−2.6	[56]
Indium (99.9999%), $RRR = 7100$	300		−5.43	[58]
	6–280	$6.3 (\omega\tau > 1)$	16	[58]
			See Fig. 30.65	[59]
Iridium, $\rho(20\ K)/\rho(300\ K) = 0.21$	300		3.18	[60]
	81		3.49	[60]
Lanthanum (99.8%)	20–300	0.56	−8	[70]
Lead (99.9%), $RRR = 6 \cdot 10^3$	293		0.98	[90]
Lead single crystal	4.2	2.3	$R_{\mathbf{B}\parallel[110]}$=43	[91]
Lithium, $RRR = 850$	300	0.95	−13.8	[71]
	4.2	0.95	−14.8	[71]
			See Fig. 30.66	[71]
Lutetium, $RRR = 25$	300	0.55	−5.3	[61]
			See Fig. 30.67	
Magnesium (99.999%)	300	0.4–2.5	−8.3	[72]
	300		$R_{\mathbf{B}\parallel c_3} = -7.8$	[72]
	300		$R_{\mathbf{B}\perp c_3} = -0.8$	[72]
Manganese	297	0.6–2.9	8.44	[73]
α-Manganese	4–300		See Fig. 30.68	[73]
Molybdenum	873	0.54	17.6	[50]
	273	0.54	18.0	[50]
	83	0.54	17.9	[50]
Molybdenum, $RRR = 32000$	4–150	1.25	See Fig. 30.69	[76]

Table 30.6 Hall coefficient of metals *(continued)*

Metal, purity	T, K	B, T	R, 10^{-11} $m^3/(A{\cdot}s)$	Ref.
Neodymium (99.98%)	293	0.56	9.71	[70]
			See Fig. 30.70	[70]
Niobium, $RRR = 13$	300	3.0	8.72	[78]
	77	3.0	8.50	[78]
	4.2	3.0	9.54	[78]
Osmium, $RRR = 12$	300		2.9	[81]
Palladium	300		−7.5	[82]
	115		−6.2	[82]
	4.2		−7.8	[83]
Platinum (99.9%)	300	0.54	−2.2	[50]
	170	0.54	−1.8	[50]
	80	0.54	−2.0	[50]
α–Plutonium	6–438		See Fig. 30.68	[84]
β-Plutonium	100–438		+3	[84]
	7		−90	[84]
γ–Plutonium	400		3	[84]
	7		37	[84]
Potassium	300		−42	[67]
Potassium, $RRR = 1000–8000$	4–100	0.05–0.95	See Fig. 30.71	[68]
Praseodymium (99.9%)	20–300	1.0	See Fig. 30.70	[70]
	300		7.1	[70]
Rhenium, $RRR = 38$	300	1.5	22	[85]
	200	1.5	28	[85]
	100	1.5	32	[85]
Rhodium (99.9%)	300	0.49	5.02	[86]
Rubidium distilled	300		−50	[87]
Rubidium, $RRR = 441–489$		0.95	See Fig. 30.72	[71]
Ruthenium	300	0.49	+22	[88]
Samarium (99.9%)	300	0.55	−2	[61]
Samarium, $RRR = 17.3$	17	0.55	28	[89]
	2.4	0.55	16	[89]
Scandium (99.8%	300	1.3	−6.7	[92]
	4.2	1.3	−3.0	[92]
Silver (99.9999%), $RRR = 918$	300	1.52	−8.98 (See Fig. 30.64)	[57]
Sodium	300	1.0	−13.2	[77]
	78	1.0	−12.4	[77]
	20	1.0	−12.7	[77]
Sodium, $RRR = 1500–6700$	4–100	0.525–0.945	See Fig. 30.73	[68]
Tantalum	300	0.54	10	[94]
Thallium	297	1.7–1.8	2.4	[93]
α–Thorium, $RRR = 18.2$	297	3.0	−11.16	[71]
	77	3.0	−10.35	[71]
	4.2	3.0	−9.7	[71]
Thulium polycrystalline (99.9%)	300	0.55	−15	[61]
Thulium plcr., $RRR = 7.4$			$R_0 = -23.5$	[96]
Tin (99.9%)	294	0.81	−0.22	[79]
Tin, $RRR = 60000$	4.2	0.69	$R_{\mathbf{B}\Vert[00\bar{1}]} = -48$	[80]
Titanium (99.97%) single crystal	300	0.95	$R_{\mathbf{B}\Vert c_3} = 7.1$	[95]
	300	0.95	$R_{\mathbf{B}\perp c_3} = -10.8$	[95]
Tungsten (99.9%)	870	0.54	15.6	[50]
	273	0.54	11.1	[50]
	170	0.54	10.6	[50]
	80	0.54	12.0	[50]
Uranium, $RRR = 12$	273	3.0	3.93	[78]
	77	3.0	4.75	[78]
	4.2	3.0	−0.31	[78]

Table 30.6 Hall coefficient of metals *(continued)*

Metal, purity	T, K	B, T	R, 10^{-11} m^3/(A·s)	Ref.
Vanadium (99.6%), $RRR = 11.5$	20		6.2	[54]
	300		7.9	[54]
Ytterbium	300	0.55	35	[61]
	77	0.55	38	[61]
Ytterbium, $RRR = 17$	20–400	0.20	See Fig. 30.74	[62]
Yttrium single crystal, $RRR = 10$	300	1.0	$R_{\mathbf{B}\perp c_3} = 4.7$	[63]
	300	1.0	$R_{\mathbf{B}\parallel c_3} = -17.2$	[63]
			See Fig. 30.67	[63]
Yttrium single crystal, $RRR = 15$	4–300		See Fig. 30.67	[64]
Zinc (99.9999%) single crystal	297	1.86	$R_{\mathbf{B}\parallel c_3} = 14.4$	[100]
	297	1.86	$R_{\mathbf{B}\perp c_3} = -4.0$	[100]
	77	1.86	$R_{\mathbf{B}\parallel c_3} = 18.5$	[100]
	77	1.86	$R_{\mathbf{B}\perp c_3} = 2$	[100]
Zirconium, $RRR = 27$	300	0–1.5	15.5	[85]
	200	0–1.5	8.0	[85]
	100	0–1.5	4.5	[85]

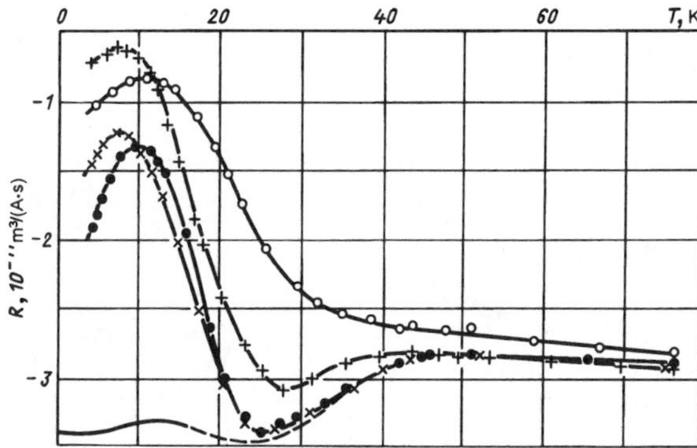

Figure 30.59 Temperature-dependent Hall coefficient for pure Al in a weak magnetic field ($\omega\tau < 1$) [51]: • – $RRR = 2380$ and $6N$; × – $RRR = 2600$ and $6N$; + – $RRR = 1810$ and $5N$; ○ – RRR = 84 and $4N$ ($6N$ corresponds to a 99.9999% purity of material; the lower curve fits calculable model).

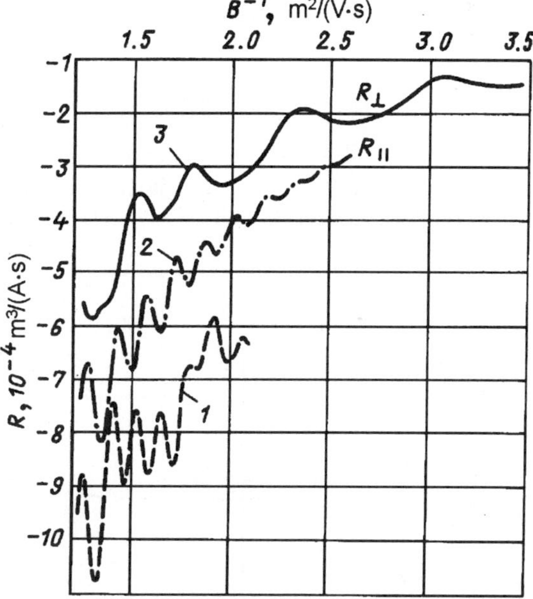

Figure 30.61 Field-dependent Hall coefficients ($R_\perp$ and $R_\parallel$ for a Bi single crystal at 1.4 K [55]: (*1*) angle between the magnetic field and the trigonal axis is 25°, current is parallel to a binary axis; (*2*) magnetic field is parallel to the trigonal axis; (*3*) magnetic field is perpendicular to the trigonal axis.

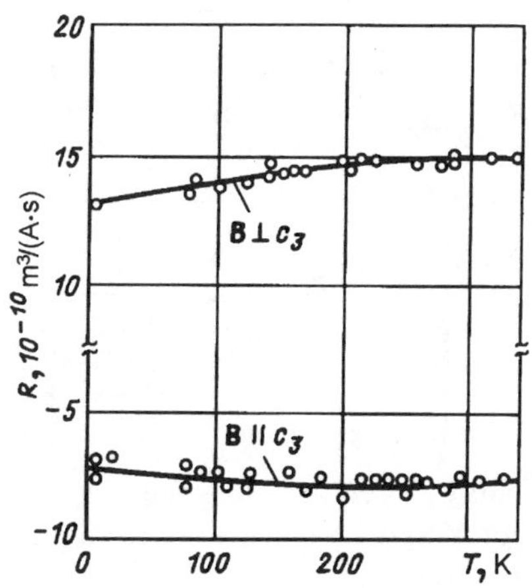

Figure 30.60 Temperature-dependent Hall coefficient for a hexagonal Be single crystal [53].

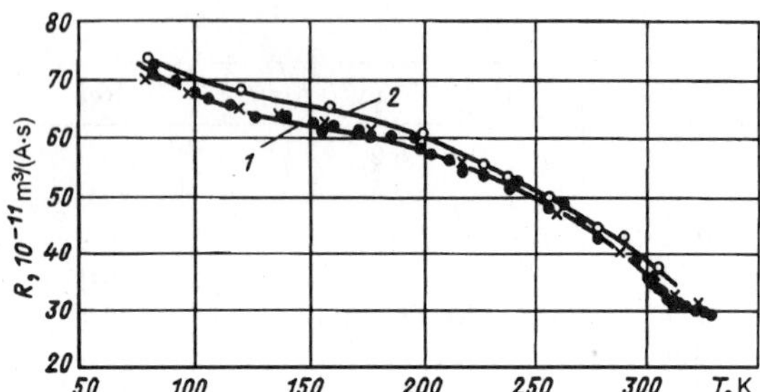

Figure 30.62 Temperature-dependent Hall coefficient for Cr [98]: *1* – before annealing; *2* – after annealing.

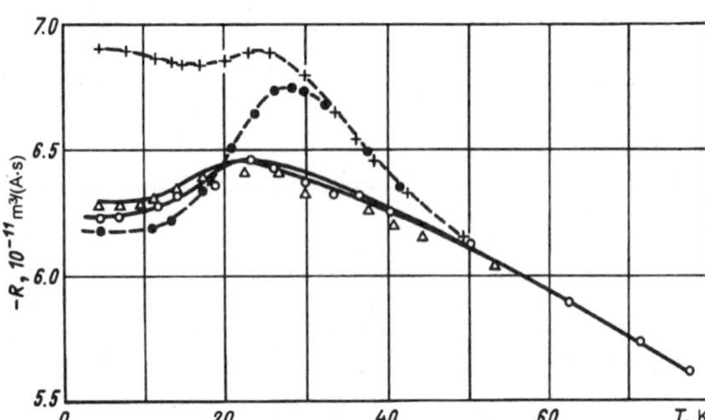

Figure 30.63 Temperature-dependent Hall coefficient for Cu [75]: •, △ – specimen *1*; +, ○ – specimen *2*; - - - $B = 0.5145$ T; —— $B = 0.0085$ T.

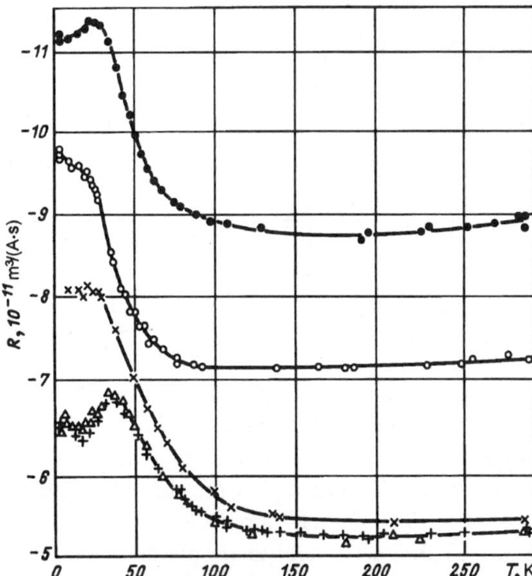

Figure 30.64 Temperature-dependent Hall coefficients for pure (99.9999 %) Ag, Au, and Cu specimens [57]: • – Ag, annealed specimen; ○ – Au, unannealed specimen; × – Cu, unannealed specimen; + – Cu, annealed specimen of 0.929 mm thickness; △ – Cu, annealed specimen of 1.719 mm thickness.

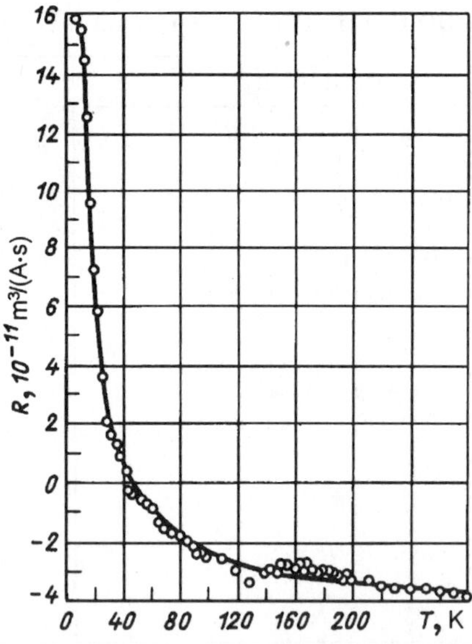

Figure 30.65 Temperature-dependent Hall coefficient for an In single crystal [59] (magnetic field $B = 6.3$ T is oriented along the [101] crystal axis).

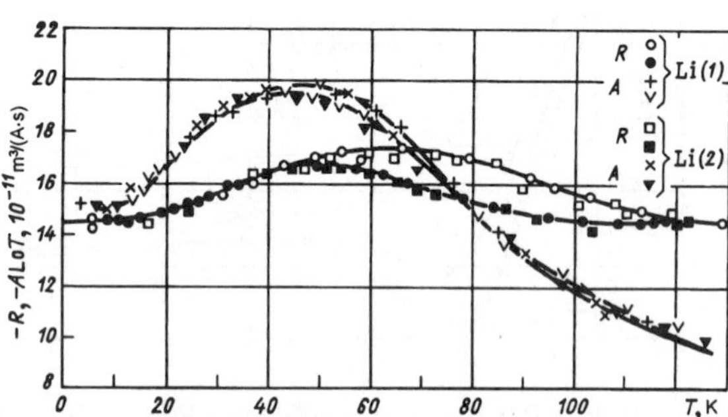

Figure 30.66 Temperature-dependent Hall coefficient *R* and Righi–Leduc coefficient *ALoT* (see footnote to Table 30.9) for Li [71].

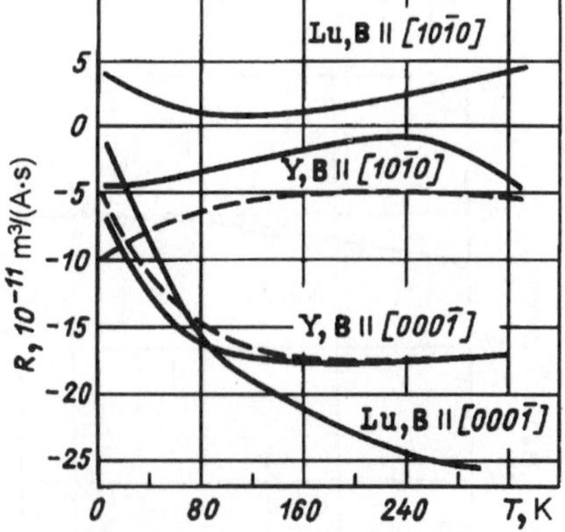

Figure 30.67 Temperature-dependent Hall coefficients for Y and Lu (solid lines from [63], dashed lines from [64]).

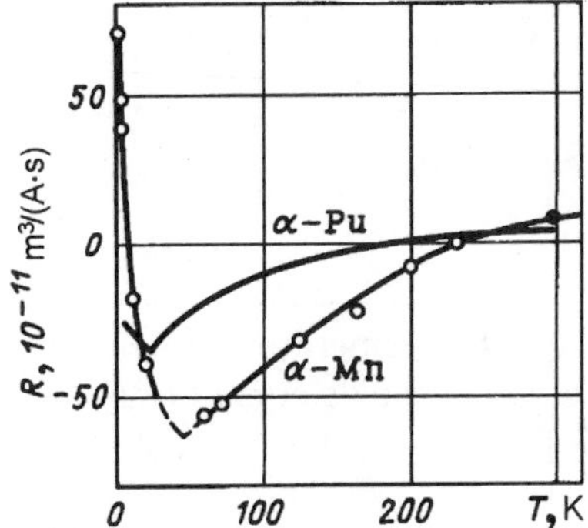

Figure 30.68 Temperature-dependent Hall coefficients for polycrystalline α-Mn [74] and α-Pu [84] specimens.

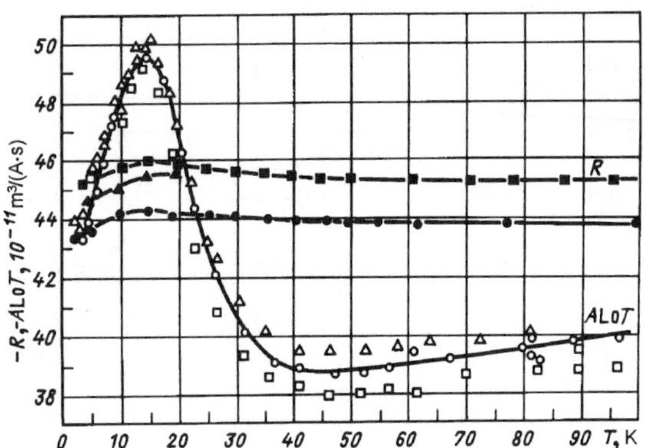

Figure 30.71 Temperature-dependent Hall coefficient R and Righi–Leduc coefficient $ALoT$ (see footnote to Table 30.9) for pure K specimens [68]: ○ - $RRR = 720$; △ - $RRR = 700$; □ - $RRR = 2000$; ● - $RRR = 2900$; ■ - $RRR = 7900$; ▲ - $RRR = 640$; $B = 1$ T.

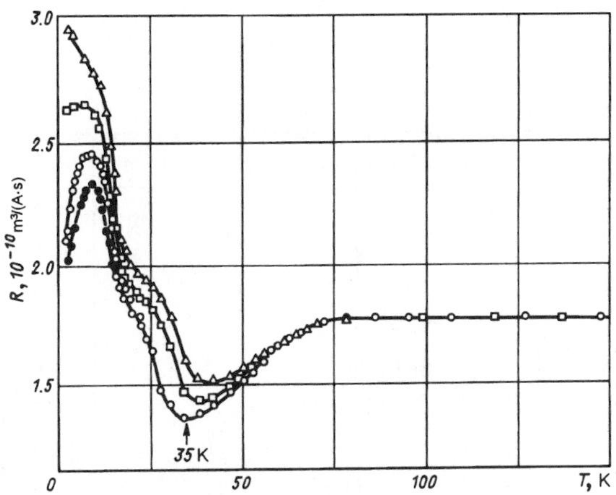

Figure 30.69 Temperature-dependent Hall coefficient for an Mo single crystal [76]: **H** ∥ [100]; ● - $B = 0.25$ T; ○ - $B = 0.45$ T; □ - $B = 0.85$ T; △ - $B = 1.25$ T.

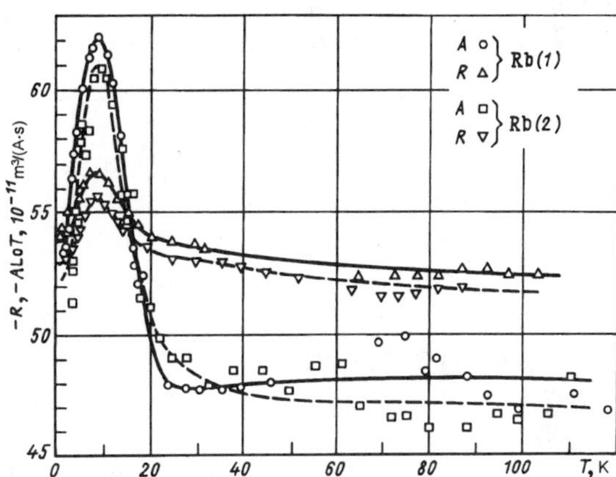

Figure 30.72 Temperature-dependent Hall coefficient R and Righi–Leduc coefficient $ALoT$ (see footnote to Table 30.9) for Rb [71].

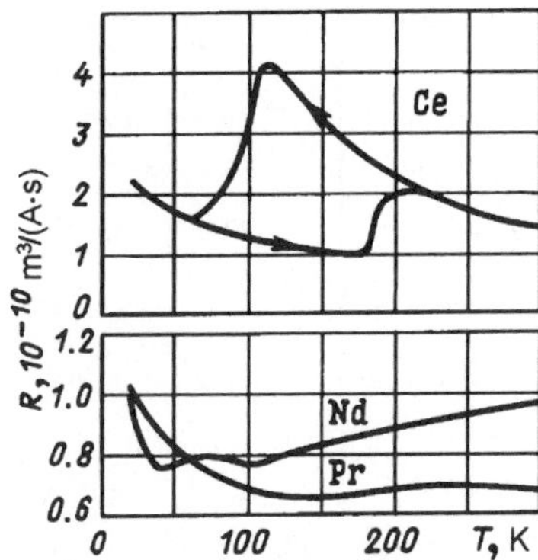

Figure 30.70 Temperature-dependent Hall coefficients for polycrystalline Ce, Nd, and Pr specimens [70].

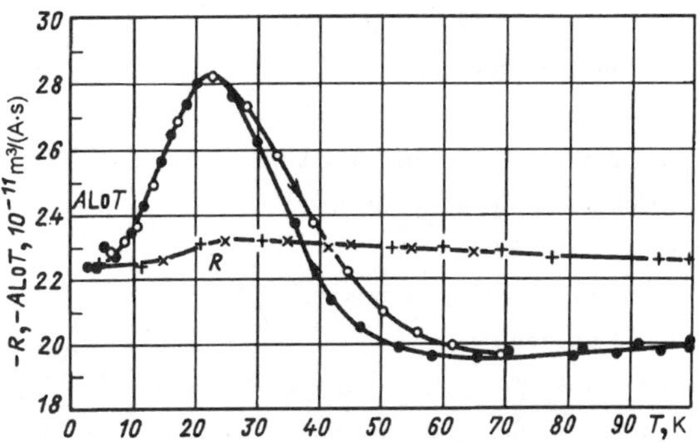

Figure 30.73 Temperature-dependent Hall coefficient R and Righi–Leduc coefficient $ALoT$ (see footnote to Table 30.9) for Na [68]: $RRR = 1500$; ●, + - cooling; ○, × - heating.

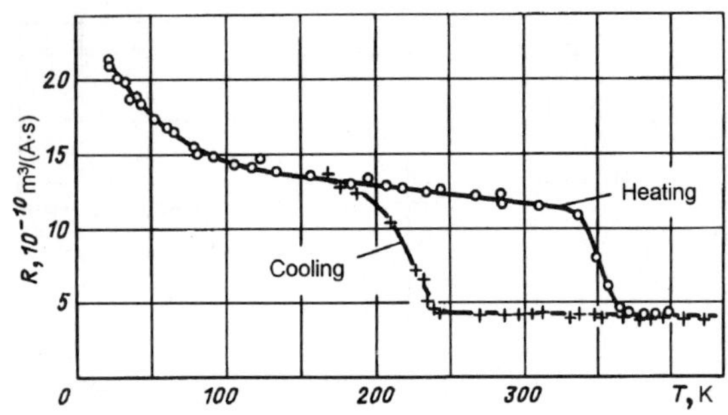

Figure 30.74 Temperature-dependent Hall coefficient of an Yb single crystal ($RRR = 17$; $B = 2.0$ T) [62].

30.5 Hall Coefficients of Ferromagnets and Rare-Earth Metals

In Tables 30.7 and 30.8 and in Figs. 30.75–30.81, data on the Hall coefficients of ferromagnetics and rare-earth metals are presented.

Table 30.7 Hall coefficient of ferromagnetic metals

Metal, purity	T, K	$\mu_0 H$, T	R_0, 10^{-10} m³/(A·s)	R_S, 10^{-10} m³/(A·s)	Ref.
Cobalt (99.99%), $RRR = 60$	293	0–2.8	−0.84	0.14	[109]
Cobalt (99.99%)	273–1400	0.23	See Fig. 30.75		[112]
Iron (99.99%), $RRR = 20$	293	0–3.3	0.2	6.8	[108]
Iron polycrystalline	4.2–300		See Fig. 30.76		[109]
Iron, whiskers, $RRR = 180 - 452$	1–300		See Fig. 30.76		[110]
(99.99%)	1–800		See Fig. 30.77		[111]
Nickel (99.99%), $RRR = 480$	293	0–2.9	−0.56	−5.0	[113]
Nickel (99.99%)	$T > T_C$	0.85	−1.08	−41	[114]
	70–700		See Fig. 30.78		[115]

Table 30.8 Hall coefficient of magnetic rare-earth metals

Metal, purity	T, K	R_0, 10^{-10} m³/(A·s)	R_S, 10^{-10} m³/(A·s)	Ref.
Dysprosium (99.9%), single crystal, $RRR = 10$	$T > T_C$	-5.7^{*1}; -1.0^{*2}	-25.5^{*1}; -32.6^{*2}	[119]
Dysprosium, single crystal, $RRR = 16$	$T < T_C$	See Fig. 30.79		[120]
Erbium, polycrystalline	$T > T_C$	−0.34		[70]
Erbium, single crystal	$T > T_C$	-3.6^{*1}; 0.3^{*2}	-9.4^{*1}; 4.4^{*2}	[124]
Gadolinium (99.9%), polycrystalline, $RRR = 51$	$T > T_C$	−2	−384	[116]
Gadolinium, single crystal, $RRR = 20$	$T > T_C$		-225^{*1}; -452^{*2}	
Gadolinium, $RRR = 40$	$T < T_C$	See Fig. 30.80		[64]
Holmium (99.9%), polycrystalline, $RRR = 12$	$T > T_C$	−2.2	−2.8	[117]
Holmium, single crystal	$T > T_C$	-3.2^{*1}; $+0.2^{*2}$	-13.2^{*1}; $+2.2^{*2}$	[118]
Terbium (99.9%), polycrystalline, $RRR = 46$	$T > T_C$	−4.4	−42	[121]
Terbium, single crystal, $RRR = 20$	$T > T_C$	-4.0^{*1}; -1.1^{*2}	-52^{*1} ; -40^{*2}	[122]
Terbium, single crystal		See Fig. 30.81		[123]

^{*1} **B** $\parallel c_3$.
^{*2} **B** in the basal plane.

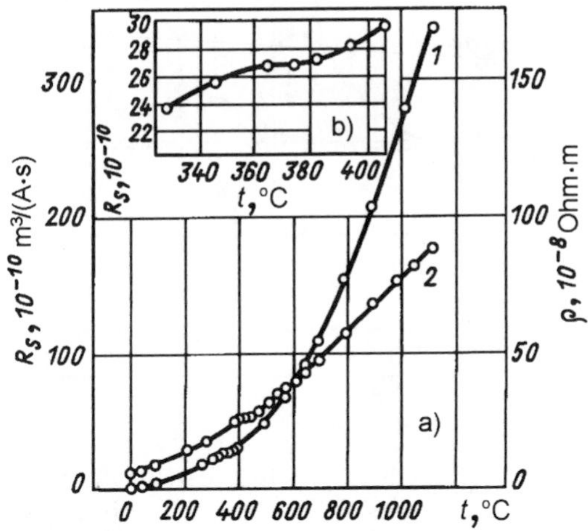

Figure 30.75 (a) Temperature-dependent anomalous Hall coefficient R_S (1) and resistivity ρ (2) for Co [112]; (b) temperature variation of R_S in the phase transition region.

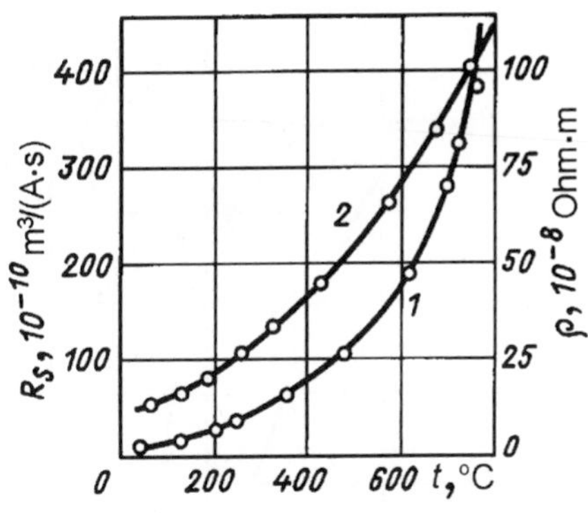

Figure 30.77 Temperature-dependent anomalous Hall coefficient R_S (1) and resistivity ρ (2) of Fe [111].

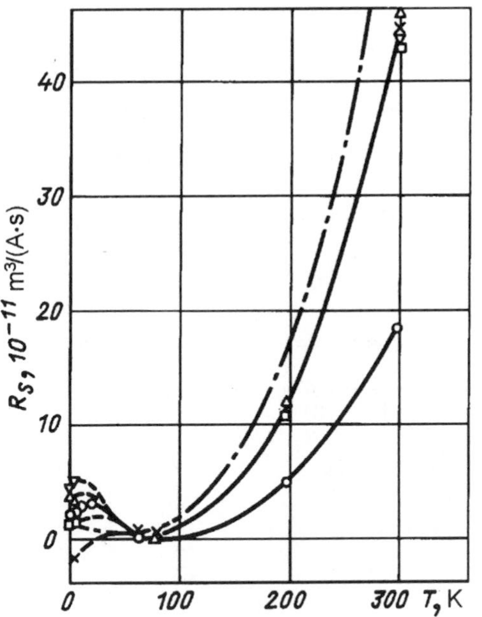

Figure 30.76 Temperature-dependent anomalous Hall coefficient R_S for Fe whiskers [110] (dot-and-dash line for polycrystalline Fe [109]):

Specimen	Direction of **I** and axes of Fe whiskers	Direction of **B**	Direction of Hall emf	RRR
×	[111]	[110]	[112]	196
△	[111]	[110]	[112]	213
o	[100]	[010]	[001]	213
□	[100]	[010]	[001]	452
▽	[100]	[010]	[001]	180

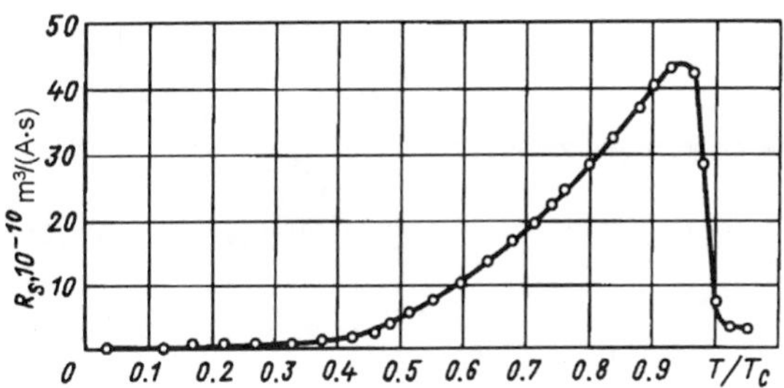

Figure 30.78 Temperature-dependent anomalous Hall coefficient R_S for Ni [115] (T_C is Curie temperature).

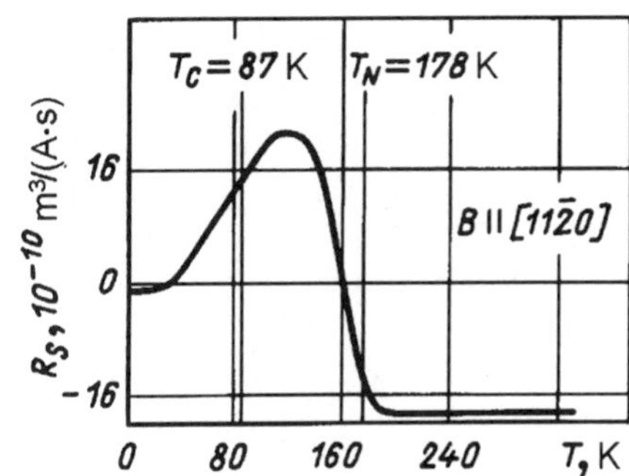

Figure 30.79 Temperature-dependent anomalous Hall coefficient R_S for a Dy single crystal [120].

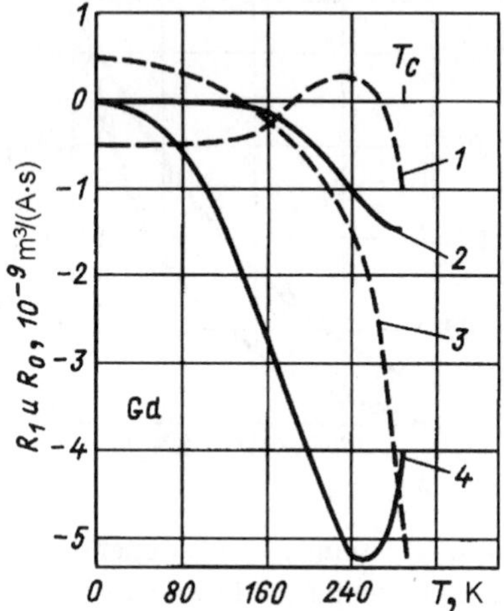

Figure 30.80 Temperature-dependent anomalous R_1 and ordinary R_0 Hall coefficients for a Gd single crystal [64]: (*1*) R_0, **B** in the basal plane; (*2*) R_1, **B** in the basal plane; (*3*) R_0, **B** $\parallel$ **c**; (*4*) R_1, **B** $\parallel$ **c**.

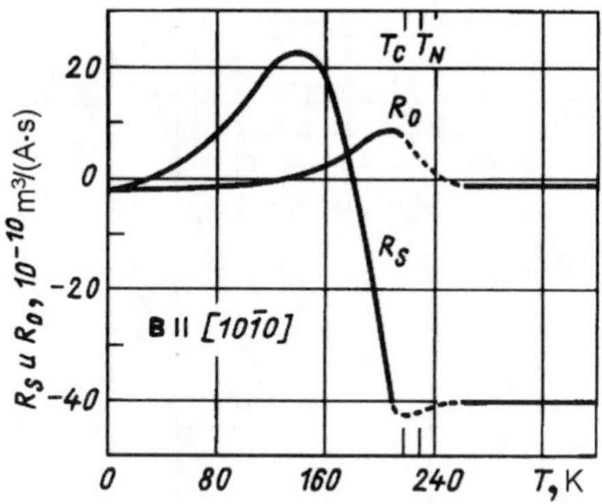

Figure 30.81 Temperature-dependent ordinary R_0 and anomalous R_S Hall coefficients for a Tb single crystal [123].

30.6 Thermomagnetic Coefficients of Metals

In Tables 30.9 and 30.10 and in Figs. 30.82–30.87, data on the thermomagnetic characteristics of metals are presented.

Table 30.9 Thermomagnetic coefficient of metals

Metal, purity	T, K	A, 10^{-3} $m^2/(V{\cdot}s)$	Q, 10^{-8} $m^2/(K{\cdot}s)$	Ref.
Aluminum (99.5%)	313	−0.63	+0.39	[125]
Cadmium	320	+0.89	−1.20	[125]
Copper*3		See Fig. 30.82		[126]
Gold*1		See Fig. 30.82		[126]
Indium		See Fig. 30.83		[127]
Iridium, *RRR* = 23	323	+0.55	−0.05	[125]
	300		−0.44	[81]
Lead	300		−0.05	[125]
Lithium*1		See Fig. 30.66		[71]
Manganese	330		+0.15	[125]
Molybdenum	333		−1.72	[125]
Niobium		See Fig. 30.84		[129]
Osmium, *RRR* = 12	300		−2.32	[81]
Palladium*4	300	−0.487	+3.27	[125]
Platinum, *RRR* = 5.4	323	−0.21		[125]
	300		−1.68	[81]
Potassium*1		See Fig. 30.71		[68]
Rhenium, *RRR* = 447	300		6.4	[81]
Rhodium		See Fig. 30.85		[130]
Rubidium*1		See Fig. 30.72		[71]
Ruthenium		See Fig. 30.85		[130]
Silver*1		See Fig. 30.82		[126]

Table 30.9 Thermomagnetic coefficient of metals *(continued)*

Metal, purity	T, K	A, 10^{-3} m^2/(V·s)	Q, 10^{-8} m^2/(K·s)	Ref.
Sodium*1		See Fig. 30.73		[68]
Tantalum		See Fig. 30.85		[130]
Thallium	333		−0.37	[125]
Tin	330		−0.04	[125]
Tungsten	332	+1.5	−10	[125]
Yttrium*2	293			[128]
Zinc	323	+1.29	−2.40	[125]

*1 For metals with asterisks, the temperature-dependent curves of the Righi–Leduc coefficients A are presented, the values being multiplied by LoT, where Lo $= \pi^2k^2/3e^2 = 2.45 \cdot 10^{-8}$ W·Ω/K^2 is the Lorentz number (k is the Boltzmann constant, e is the electron charge), T is the temperature, to make their comparison with the Hall coefficient more convenient.

*2 $P = 75 \cdot 10^{-4}$ K·m^3/(V·A·s).

*3 $P = -1.446 \cdot 10^{-8}$ K·m^3/(V·A·s).

*4 $P = 16.2 \cdot 10^{-8}$ K·m^3/(V·A·s).

Table 30.10 Thermomagnetic coefficient of ferromagnetic metals

Metal	T, K	B, T	P_1, 10^{-8} K·m^3/(V·A·s)	A_1, 10^{-3} m^2/(V·s)	Q_1, 10^{-8} m^2/(K·s)	Ref.
Cobalt	298	0.95	87	3.77	19	[125]
	300–900				See Fig. 30.86	[132]
Iron	313	2.3	−20.3	3.28	−20.4	[131]
Nickel	313	0.667	55	−49	48.2	[133]
	2–450				See Fig. 30.87	[125]

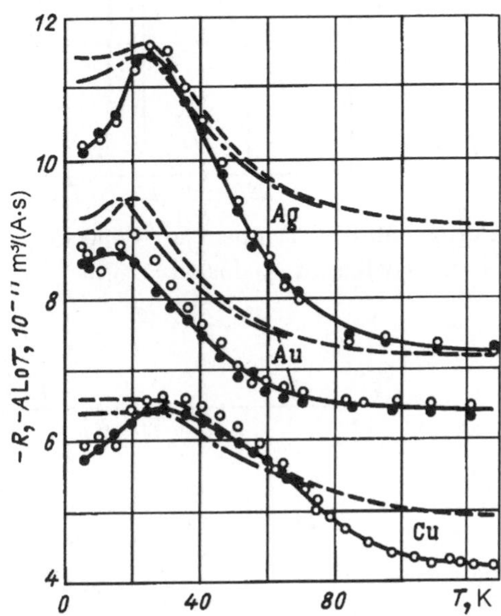

Figure 30.82 Temperature-dependent Hall coefficient R and Righi–Leduc coefficient $ALoT$ (see footnote to Table 30.9) for Ag, Au, and Cu [126]: - - - - R, $B = 0.935$ T; — · — · — R, $B = 0.518$ T; ○ – $ALoT$, $B = 0.935$ T; ● – $ALoT$, $B = 0.518$ T.

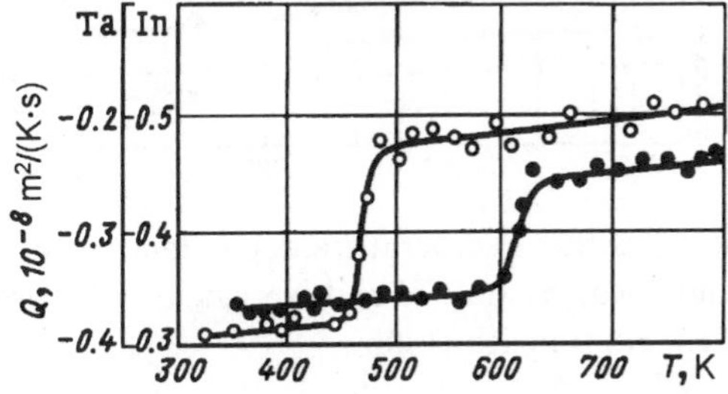

Figure 30.83 Temperature-dependent Nernst coefficients for Tl (●) and In (○) in solid and liquid states [127].

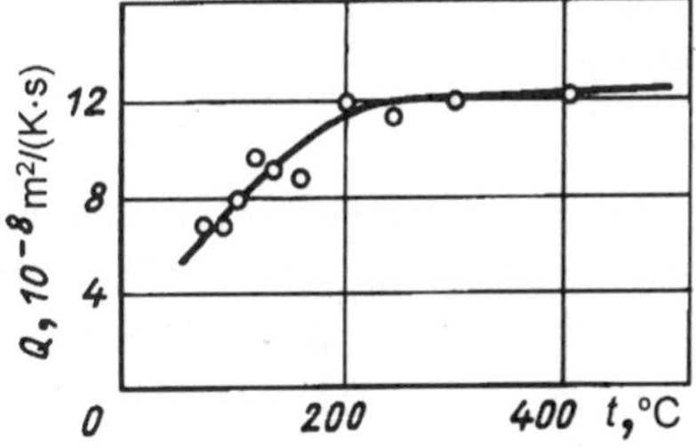

Figure 30.84 Temperature-dependent Nernst coefficient for Nb [129].

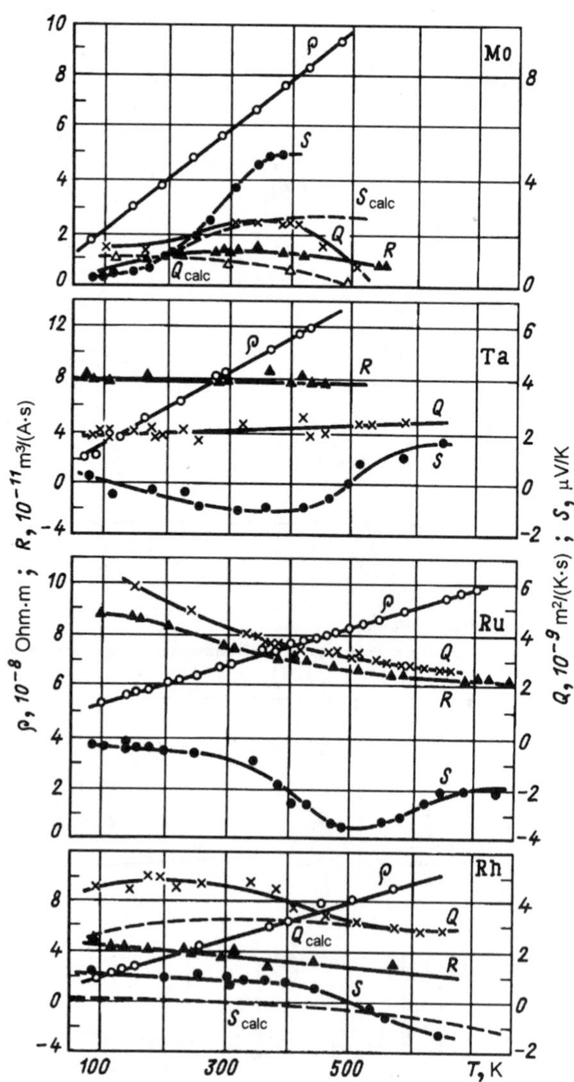

Figure 30.85 Temperature-dependent resistivity ρ, thermopower S, Hall R and Nernst Q coefficients of Mo, Ta, Ru and Rh [130].

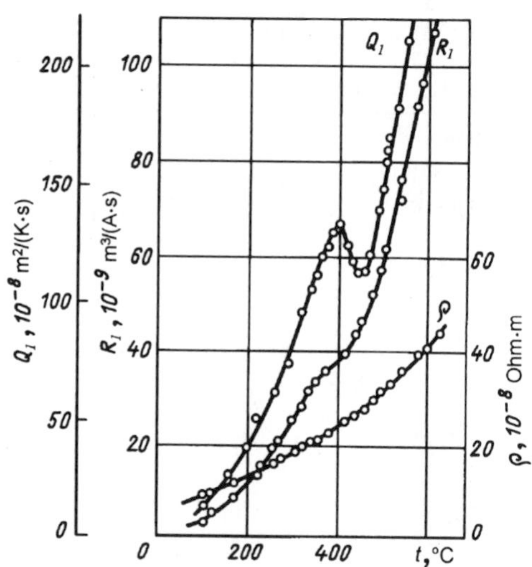

Figure 30.86 Temperature-dependent Hall R_1 and Nernst Q_1 coefficients, and resistivity ρ for Co [132].

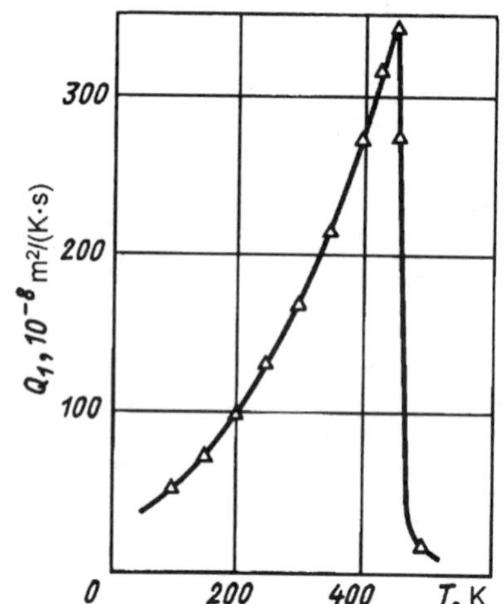

Figure 30.87 Temperature-dependence of Nernst coefficient Q_1 for Ni [125].

References

[1] Gaidukov, Yu. P., Uspekhi fizicheskikh nauk, 100, 449, 1970 (in Russian).

[2] Cracknell, A. P., Wong, K. C., The Fermi surface, Clarendon Press, Oxford, 1973.

[3] Harrison, W. A., Webb, M. B., The Fermi surface, John Wiley, New York–London, 1960.

[4] Lifshits, I. M., Azbel', M. Ya., Kaganov, M. I., Electron theory of metals, Nauka, Moscow, 1971 (in Russian).

[5] Meaden, G. T., Electrical resistance of metals, Plenum Press, New York, 1965.

[6] Fawcett, E., Advances Phys., 13, No 50, 139, 1964.

[7] Hurd, C. M., The Hall effect in metals and alloys, Plenum Press, New York, 1972.

[8] Fickeff, F. R., Phys. Rev., B3, 1941, 1971.

[9] Rocofyllou, E., Papathanassopoulov, C., Physica, B/C 100, 99, 1980.

[10] Justi, E., Kramer, J., Physikal Z., 41, 197, 1940.

[11] Kapitza, P., Proc. Royal Soc. (Lond.), A123, 292, 342, 1929.

[12] Alekseevskii, N. E., Egorov, V. S., Zhurnal eksperimental'noi i teoreticheskoi fiziki, 45, 388, 1963 (in Russian).

[13] Landolt–Börnstein, Numerical data and functional relationships in science and technology, New series, Eds.–in–Chief: Hellwege, K.–H., Madelung, O., Group III: Crystal and Solid State Physics, vol. 13: Metals: Phonon state, electron states and Fermi surfaces, subvol. *c*, Springer–Verlag, Berlin, Heidelberg 1984.

[14] Volkenshtein, N. V., Novoselov, V. A., Startsev, V. E., Fizika metallov i metallovedenie, 22, 175, 1966 (in Russian).

[15] Reed, W. A., Marcus, J. A., Phys. Rev., 126, 1298, 1962.

[16] McClure, P. W., Spry, W. J., Phys. Rev., 165, 809, 1968.

[17] Janos, S., Fener, A., Flachbart, K., Phys. Status Solidi (b), 81, K19, 1977.

[18] Justi, E., Physikal Z., 41, 486, 1940.

[19] Gaidukov, Yu. P., Zhurnal eksperimental'noi i teoreticheskoi fiziki, 37, 1281, 1959 (in Russian).

[20] Volotskaya, V. G., Zhurnal eksperimental'noi i teoreticheskoi fiziki, 45, 49, 1963 (in Russian).

[21] Lazarev, B. G., Nakhimovich, N. M., Parfenova, E. A., Zhurnal eksperimental'noi i teoreticheskoi fiziki, 9, 1169, 1938 (in Russian).

[22] Penz, P. A., Bowers, R., Phys. Rev., 172, 991, 1968.

[23] Gugan, D., Jones, B. K., Helv. Phys. Acta, 36, 6, 1963.

[24] Alekseevskii, N. E., Gaidukov, Yu. P., Zhurnal eksperimental'noi i teoreticheskoi fiziki, 38, 1720, 1960 (in Russian).

[25] Alekseevskii, N. E., Egorov, V. S., Karstens, G. E., Kozak, B. N., Zhurnal eksperimental'noi i teoreticheskoi fiziki, 43, 731, 1962 (in Russian).

[26] Babiskin, J., Siebenmann, P. G., Phys. Kondens. Materia, 9(1/2), 113, 1969.

[27] Nagasawa, H., Phys. Lett., 414, 39, 1972.

[28] Alekseevskii, N. E., Gaidukov, Yu. P., Zhurnal eksperimental'noi i teoreticheskoi fiziki, 36, 446, 1959 (in Russian).

[29] Alekseevskii, N. E., Dubrovin, A. V., Karstens, G. E., Mikhailov, N. N., Zhurnal eksperimental'noi i teoreticheskoi fiziki, 54, 350, 1968 (in Russian).

[30] Aschermann, G., Justi, E., Physikal Z., 43, 207, 1942.

[31] Alekseevskii, N. E., Egorov, V. S., Kozak, B. N., Zhurnal eksperimental'noi i teoreticheskoi fiziki, 44, 1116, 1963 (in Russian).

[32] Schulze, R., Physikal Z., 42, 297, 1941.

[33] MacDonald, D. K. C., Philos. Mag., 2, No 13, 97, 1957.

[34] Bolotin, G. A., Novoselov, V. A., Startsev, V. E., Fizika metallov i metallovedenie, 33, 740, 1972 (in Russian).

[35] Borovik, E. S., Zhurnal eksperimental'noi i teoreticheskoi fiziki, 27, 355, 1954 (in Russian).

[36] Alekseevskii, N. E., Gaidukov, Yu. P., Zhurnal eksperimental'noi i teoreticheskoi fiziki, 37, 672, 1959 (in Russian).

[37] Young, R. C., Phys. Rev., 163, 676, 1967.

[38] Justi, E., Ann. Physik, 3(6), 183, 1948.

[39] Reed, W. A., Brennert, G. F., Phys. Rev., 130, 565, 1963.

[40] Babushkina, N. A., Doklady Akad. Nauk SSSR, Ser. "Fizika," 155, 1290, 1964 (in Russian).

[41] McEven, K. A., Werbber, G. D., Roeland, L. W., Physica, 86-88B, 531, 1977.

[42] Akhavan, M., Blackstead, H. A., Phys. Rev., B13, 1209, 1976.

[43] Akhavan, M., Blackstead, H. A., Phys.Rev., 8, 4258, 1973.

[44] Isin, A., Colemann, R. V., Phys. Rev., 142, 372, 1966.

[45] Colemann, R. V., Morris, R. S., Sellmeyer, D. J., Phys. Rev., B8, 317, 1973.

[46] Bogma, K. K., Zubov, V. V., Fizika metallov i metallovedenie, 20, 135, 1965 (in Russian).

[47] Belov, K. P., Zhurnal eksperimental'noi i teoreticheskoi fiziki, 21, 809, 1951 (in Russian).

[48] Kaul, S. N., J. Phys., F7, 2091, 1977.

[49] Belov, K. P., Nikitin, S. A., Fizika metallov i metallovedenie, 13, 43, 1962 (in Russian).

[50] Frank, V., Appl. Sci. Res., B7, 41, 1958.

[51] Barnard, R. D., Addel Raheim, A.E.E., J. Phys., F10, 2739, 1980.

[52] Borovik, E. S., Zhurnal eksperimental'noi i teoreticheskoi fiziki, 23, 83, 1952 (in Russian).

[53] Shiozaki, J., Phys. Lett., A50, 203, 1974.

[54] Amitin, E. B., Kovalevskaya, Yu. A., Kovdrya, Yu. Z., Fizika tverdogo tela, 9, 905, 1967 (in Russian).

[55] Reynolds, J. M., Hemstreet, H. M., Phys. Rev., 96, 1203, 1954.

[56] Volkenshtein, N. V., Galoshina, E. V., Fizika metallov i metallovedenie, 18, 784, 1964 (in Russian).

[57] Anderson, J. E., Farrel, T., Hurd, C. M., Phys. Rev., 174, 729, 1968.

[58] Cooper, J. N., Cotti, P., Rasmussen, F. B., Phys. Lett., 19, 560, 1965.

[59] Ozimek, E. J., Leisure, R. G., Phys. Lett., A66, 413, 1978.

[60] Gehloff, P. O., Justi, E., Kohler, M., Z. Naturforsch., 5a, 16, 1950; 4a, 561, 1949.

[61] Anderson, G. S., Legvold, S., Spedding, F. H., Phys. Rev., 111, 257, 1959.

[62] Anderson, J. E., Hurd, C. M., Solid State Commun., 11, 1245, 1972.

[63] Volkenshtein, N. V., Galoshina, E. V., Fizika metallov i metallovedenie, 24, 1105, 1967 (in Russian).

[64] Lee, R. S., Legvold, S., Phys. Rev., 162, 431, 1967.

[65] Saeger, K. E., Luck, R., Phys. Kondens. Materia, 9, 91, 1969.

[66] Stringer, J., Hill, J., Huglin, A. S., Philos. Mag., 21, 53, 1970.

[67] Cirles, W., Z. Physik, 147, 481, 1957.

[68] Fletcher, R., Friedman, A. J., Phys. Rev., B8, 5381, 1973.

[69] Frank, V., Jeppesen, O. G., Phys. Rev., 89, 1153, 1953.

[70] Kevane, C. J., Legvold, S., Spedding, F. H., Phys. Rev., 91, 1372, 1953.

[71] Fletcher, R., Phys. Rev., 15, 3602, 1977.

[72] Alty, J. L., Stringer, J., Phys. Status Solidi, 32, 243, 1969.

[73] Foner, S., Phys. Rev., 107, 1513, 1957.

[74] Meaden, G. P. T., Pelloux-Gervais, P., Cryogenics, 7, 161, 1967.

[75] Barnard, R. D., J. Phys., F10, 2251, 1980.

[76] Cherepanov, V. I., Startsev, V. E., Volkenshtein, N. V., Fizika nizkikh temperatur, 5, 1162, 1979 (in Russian).

[77] Justi, E., Ann. Physik, 6, 183, 1948.

[78] Berlincourt, T. G., Phys. Rev., 114, 969, 1959.

[79] Andrewartha, G. G., Evans, E. J., Philos. Mag., 31, 265, 1941.

[80] Kachinskii, V. N., Zhurnal eksperimental'noi i teoreticheskoi fiziki, 43, 1158, 1962 (in Russian).

[81] Nemchenko, V. M., L'vov, S. N., Mal'ko, P. I., Deliev, V. N., Fizika metallov i metallovedenie, 33, 540, 1972 (in Russian).

[82] Plate, H., Phys. Kondens. Materia, 4, 355, 1966.

[83] Schwaller, R., Compt. Rend., 264, 1064, 1967.

[84] Brodsky, M. B., Phys. Rev., 137, A1423, 1965.

[85] Volkov, D. N., Kozlova, T. M., Prudnikov, V. N., Kozis, E. O., Zhurnal eksperimental'noi i teoreticheskoi fiziki, 55, 2103, 1968 (in Russian).

[86] Koster, W., Hagmann, D., Saeger, K. E., Z. Metalkunde, 54, 619, 1963.

[87] Fakidov, I. G., Doklady Akad. Nauk SSSR, 63, No 2, 123, 1948 (in Russian).

[88] Justi, E., Z. Naturforsch, 4, 472, 1949.

[89] Volkenshtein, N. V., Fedorov, G. V., Fizika tverdogo tela, 7, 3213, 1965 (in Russian).

[90] Takano, K., Sato, T., J. Phys. Soc. Jpn., 20, 2013, 1965.

[91] Alekseevskii, N. E., Gaidukov, Yu. P., Zhurnal eksperimental'noi i teoreticheskoi fiziki, 41, 354, 1961 (in Russian).

[92] Volkenshtein, N. V., Galoshina, E. V., Fizika metallov i metallovedenie, 16, 298, 1963 (in Russian).

[93] Smith, A. W., Phys. Rev., 8, 79, 1916.

[94] Krautz, E., Schultz, H., Z. Augewdte Phys., 15, 1, 1963.

[95] Scovil, G. N., Appl. Phys. Lett., 9, 247, 1966.

[96] Volkenshtein, N. V., Fedorov, G. V., Fizika metallov i metallovedenie, 20, 508, 1965 (in Russian).

[97] Foner, S., Phys. Rev., 107, 1513, 1957.

[98] Amitin, E. B., Kovalevskaya, Yu. A., Fizika tverdogo tela, 10, 1884, 1968 (in Russian).

[99] Kendall, P. W., J. Nucl. Mater., 35, 41, 1970.

[100] Lane, Y. S., Huglin, A. S., Stringer, J., Phys. Rev., 135, A1060, 1964.

[101] Busch, G., Guntherodt, H.-J., Phys. Kondens. Materia, 6, 325, 1967; Advances Phys., 16, 651, 1967.

[102] Busch, G., Thieche, Y., Phys. Kondens. Materia, 1, 78, 1963.

[103] Greenfield, A., Phys. Rev., 135, A1589, 1964.

[104] Busch, G., Guntherodt, H.-J., Kunzi, H. U., Phys. Lett., 32A, 376, 1970.

[105] Shackle, P. W., Philos. Mag., 21, 987, 1970.

[106] Busch, G., Guntherodt, H.-J., Kunzi, H. U., Schlapbach, L., Phys. Lett., 31A, 191, 1970.

[107] Fakidov, I. G., Doklady Akad. Nauk SSSR, 63, 123, 1948 (in Russian).

[108] Soffer, S., Dreesden, J. A., Pugh, E. M., Phys. Rev., 140, A668, 1965.

[109] Volkenshtein, N. V., Fedorov, G. V., Zhurnal eksperimental'noi i teoreticheskoi fiziki, 38, 64, 1960 (in Russian).

[110] Dheer, P. N., Phys. Rev., 156, 637, 1967.

[111] Tsonkalas, J. A., Phys. Status Solidi (a), 22, K59, 1974.

[112] Tsonkalas, J. A., Phys. Status Solidi (a), 23, K41, 1974.

[113] Hugnenin, R., Rivier, D., Helv. Phys. Acta, 38, 900, 1965.

[114] Dutta, Roy, S. K., Subrahmanyan, A. V., Phys. Rev., 177, 1133, 1969.

[115] Kaul, S. N., Phys. Rev., B20, 5122, 1979.

[116] Volkenshtein, N. V., Fedorov, G. V., Zhurnal eksperimental'noi i teoreticheskoi fiziki, 50, 1505, 1966 (in Russian).

[117] Volkenshtein, N. V., Fedorov, G. V., Fizika tverdogo tela, 8, 1895, 1966 (in Russian).

[118] Cullen, P. R., Rhyne, J. J., Mancini, F., J. Appl. Phys., 41, 1178, 1970.

[119] Volkenshtein, N. V., Grigorova, I.K., Fedorov, G. V., Zhurnal eksperimental'noi i teoreticheskoi fiziki, 51, 780, 1966 (in Russian).

[120] Rhyne, J. J., Phys. Rev., 172, 523, 1968.

[121] Babushkina, N. A., Fizika tverdogo tela, 7, 3026, 1965 (in Russian).

[122] Fedorov, G. V., Volkenshtein, N. V., Fizika tverdogo tela, 12, 1374, 1970 (in Russian).

[123] Rhyne, J. J., J. Appl. Phys., 40, 1001, 1969.

[124] Rhyne, J. J., O'Connor, M. C., Bull. Amer. Phys. Soc., 14, 306, 1969.

[125] Landolt–Bornstein, Zahlenwerte und funktionen aus physik, chemie, astronomie, 6 Teil, Elektrische eigenschaften, Springer–Verlag, Berlin, Göttingen–Heidelberg, 1959.

[126] Fletcher, R., Friedman, A. J., Stot, M. J., J. Phys., F2, 729, 1972.

[127] Kuvandikov, O. K., Subkhakulov, I., Daminov, A., Fizika metallov i metallovedenie, 44, 670, 1977 (in Russian).

[128] Zeechina, L., Phys. Status Solidi, 42, K153, 1970.

[129] Kuvandikov, O. K., Cherepanov, A. V., Vasil'eva, R. P., Shakerov, V. O., Izvestiya vuzov, Ser. "Fizika," No 8, 128, 1977 (in Russian).

[130] Myasnikova, K. P., Vasilieva, R. P., Cheremushkina, A. V., Izvestiya Acad. nauk SSSR, Ser. "Metally," No 6, 185, 1975 (in Russian).

[131] Butler, E. H., Pugh, E. M., Phys. Rev., 57, 916, 1940.

[132] Cheremushkina, A. V., Vasilieva, R. P., Fizika tverdogo tela, 8, 822, 1966 (in Russian).

[133] Rinder, W., Koch, K. M., Z. Naturforsch. 13a, 26, 1958.

31

Optical Properties of Materials

L.A. Novitskii

31.1 Introduction

The optical characteristics of solid, liquid and gaseous substances in their different states are shown in the present chapter. The analytical dependences allowing us to use these values in practical computations are also represented. The reliability of the optical characteristics (it is usual to list the root-mean-square relative errors of measurements at a confidence level of 0.68) was stated throughout. Some values presented in the tables where the errors are not fixed (these are the cases when the errors of measurements are not shown in the associated publications) should be considered as rough ones.

Optical units are given in accord with the International System of Units (SI).

The index of refraction n is referred to as the ratio of the velocity of the light speed in vacuum to its phase velocity in a given medium; n for the wavelength $\lambda = 546.07$ nm (5460.7 Å) is spoken of as the *principal index of refraction* n_e. The n and n_e values are dimensionless.

Computation of n for any wavelength λ within the specific spectral range is based on dispersion formulas which have similar constructions but different coefficients for materials of various types. For example, the formulas to define n for optical colorless glasses within the wavelength range of 365–1013.9 nm can be written in the following way:

$$n^2 = A_1 + A_2\lambda^2 + A_3\lambda^{-2} + A_4\lambda^{-4} + A_5\lambda^{-6} + A_6\lambda^{-8},$$

where A_1, A_2, ... A_6 are the dispersion formula coefficients depending on glass grades.

To designate the reference-point wavelengths and corresponding indices of refraction, the following literal system of notations is taken in parallel with a digital one:

λ, nm	365.00	404.36	435.83	479.99	486.13	546.07	587.56	589.29	643.85	656.27	706.52
Subscript	i	h	g	F'	F	e	d	D	C'	C	r

The mean dispersion $n_{\lambda_1} - n_{\lambda_2}$ denotes the difference in indices of refraction pertaining to the λ_1 and λ_2 wavelengths.

The principal mean dispersion $n_{F'} - n_{C'}$ is the difference in indices of refraction for the F' and C' wavelengths.

The dispersion coefficient ν_2 is referred to as the ratio of the following type

$$\nu_2 = (n_2 - 1)/(n_1 - n_3),$$

where n_1, n_2, n_3 are the indices of refraction corresponding to the λ_1, λ_2, and λ_3 wavelengths, provided that $\lambda_1 < \lambda_2 < \lambda_3$.

0-8493-2861-6/97/$0.00+$.50

The principal dispersion coefficient ν_e is referred to as the following ratio:

$$\nu_e = (n_e - 1)/(n_{F'} - n_{C'}).$$

The ν_2 and ν_e values are dimensionless.

The refraction index n as a function of the substance temperature T is characterized by the temperature coefficient β defined by the change in n at the wavelength λ as T goes higher by 1 degree according to the absolute temperature scale, i.e.

$$\beta = \Delta n/\Delta T.$$

When $\lambda = \lambda_e$, $\beta = \beta_e = \Delta n_e/\Delta T$ and β_e is said to be the thermooptical constant.

The short-wave transmission boundary $\lambda_{\min}$ is the wavelength on the shorter-wavelength side at which the spectral coefficient of internal transmittance is no less than 0.50 for the materials of 10 mm thick. *The long-wave transmission boundary* $\lambda_{\max}$ is defined in the same way, but for the long–wave range. The $\lambda_{\min}$ and $\lambda_{\max}$ are measured in the units of length.

The radiant transmittance τ equals the ratio of the radiant flux passed through a given body to the radiant flux incident on it.

The internal transmittance τ^* is defined as the radiant flux passed through a given body to its outer surface in relation to the radiation flux entering the body.

The character of light reflection by the surface of any given material depends on the surface finishing quality. Generally, reflection is directional-diffused in nature if the intensity maximum of the reflected light agrees with the direction determined by the reflection law. According to the predominance of reflected light components (direct or diffuse) reflection can be considered either as direct (*coefficient of direct reflection* ρ) or as diffused (*coefficient of diffuse reflection* ρ^*). The surfaces where a diffuse component predominates in a reflected light approximate to some extent to those with directionally independent brightness and with light intensity decreasing in proportion to the cosine of the angle between the normal to the surface and the direction under examination (equal-intensity, or Lambert's surfaces).

The coefficient of absorption α is referred to as the ratio of the radiant flux absorbed by a given body to the radiant flux incident on it.

The τ, τ^*, ρ, ρ^* and α quantities are dimensionless. The relationship between τ, ρ and α is given by the equality

$$\tau + \rho + \alpha = 1.$$

For the radiant flux with a certain wavelength λ the corresponding coefficients are symbolized by τ_λ, τ^*_λ, ρ_λ, ρ^*_λ and α_λ (*spectral coefficients of transmission, internal transmission, reflection and absorption*).

The τ, τ^*, ρ, ρ^* and α values depend on the spectral structure of the radiant flux incident on a given body. In the special case when the radiator simulates the sun radiation, the letter S can be added to the corresponding subscript designation. For example, the absorption coefficient for solar radiation can be written as α_S.

The absorptive index a is referred to as the reciprocal of the distance within which the radiant flux forming a direct beam becomes weaker by a factor of ten. For the radiant flux with a certain wavelength λ the corresponding value can be written as a_λ and referred to as the spectral absorptive index. Quantities a and τ^*, a_λ and τ^*_λ are connected by the expressions

$$\tau^* = 10^{-ad}; \quad \tau^*_\lambda = 10^{-a_\lambda d},$$

where d is the absorbing-layer thickness for a given material (in m).

For convenient comparison of the τ^*_λ values for various materials optically uniform in the direction of propagating radiation, internal radiant transmittance is, as a rule, reduced to the unit length of the absorbing layer. The calculating formula for this reduction takes the form

$$\lg \tau^*_\lambda = d \cdot \lg \tau^*_{\lambda_1},$$

where τ_{λ_1} is the spectral internal transmittance for a unit thickness of absorbing layer.

The τ and τ^* coefficients are connected through the relation

$$\tau^* = \tau/(1-\rho)^2$$

with the same coefficients of reflection at the input and output boundaries of the medium under investigation.

The optical density is said to be the logarithm of the reciprocal of the radiant transmittance, i.e.

$$D = \lg(1/\tau); \quad D^* = \lg(1/\tau^*).$$

Hence,

$$D^* = D + 2\lg(1-\rho).$$

If a radiant flux passes subsequently through the media with the τ_1, τ_2, τ_3, ... radiant transmittance and D_1, D_2, D_3, ... optical densities, then the total transmittance τ and the total optical density D for a set of these media are calculated on the basis of the following formulas:

$$\tau = \tau_1 \cdot \tau_2 \cdot \tau_3 \cdot \ldots; \quad D = D_1 + D_2 + D_3 + \ldots.$$

In a great number of published works the previously specified definition of the adsorptive index is used. Following this way, the absorptive index equals the reciprocal of the distance within which the radiant flux forming a direct beam becomes weaker by a factor of e as a result of absorption in the material (here the natural logarithmic base $e = 2.71828\ldots$). In this case the equations defining τ^* and τ_λ^* take the form

$$\tau^* = e^{-ad}, \quad \tau_\lambda^* = e^{-a_\lambda d}$$

and

$$\ln \tau_\lambda^* = d \ln \tau_{\lambda_1}^*.$$

These cases are described below separately.

The dependence of ρ versus incident angle α_1 and refraction indices n' and n'' of adjacent dielectric media (radiation propagates from the medium with the n' index of refraction into the medium with the n'' index of refraction is determined by the expression

$$\rho = \frac{1}{2}\left[\frac{\sin^2(\alpha_1-\alpha_2)}{\sin^2(\alpha_1+\alpha_2)} + \frac{\tan^2(\alpha_1-\alpha_2)}{\tan^2(\alpha_1+\alpha_2)}\right],$$

where α_2 is the angle of refraction. Angles α_1 and α_2 are connected through the refraction law:

$$n' \sin\alpha_1 = n'' \sin\alpha_2.$$

When a radiant flux falls to a surface normally, then

$$\rho = [(n''-n)/(n''+n')]^2.$$

In the case of optically active materials, the special characteristics described below are additionally used.

The index of refraction for an ordinary ray n_o is the ratio of electromagnetic radiation velocity in vacuum to the phase velocity of an ordinary ray with wavelength λ in an anisotropic medium.

The index of refraction for an extraordinary ray n_e is designated the ratio of the electromagnetic radiation velocity in vacuum to the phase velocity of the extraordinary ray with the wavelength λ in an anisotropic medium. If the propagation of an extraordinary ray with the wavelength λ is considered to be perpendicular to the optical axis of an anisotropic medium (uniaxial anisotropy) or to the bisectrix of an angle between the optical axes (biaxial anisotropy), then n_e is the *principal index of refraction (PIR) for an extraordinary ray.*

One of the most important characteristics of anisotropic bodies is *the birefringence factor* b referred to as the difference between the principal index of refraction for an extraordinary ray in an anisotropic medium and the index of refraction for an ordinary ray, i.e.

$$b = n_e - n_o.$$

The optical stress coefficient B (in Pa^{-1}) is spoken of as the coefficient of proportionality between the birefringence factor b induced by mechanical strains and the difference in the principal normal stresses $\sigma_2 - \sigma_1$ brought about by the strains that arose in the planes coinciding with those of the e-ray and o-ray polarization. If the sample being deformed can be tilted to the uniaxial crystal and the incident parallel light beam is perpendicular to the compressive (tensile) force direction, then the dependence of $n_e - n_o$ on $\sigma_2 - \sigma_1$ takes the following form

$$n_e - n_o = B(\sigma_2 - \sigma_1).$$

The B quantity for the appropriate material makes a function of wavelength and temperature. For the low-class crystal whose optical indicatrix is a triaxial ellipsoid with three unequal unit–perpendicular axes, the indices of refraction in the major-, middle- and minor-axis directions are referred to as the *great (n_G), medium (n_m), and poor (n_p) indices of refraction.*

The relative degree of polarization P_λ is the ratio of intensity difference between the light beams with the wavelength λ polarized orthogonally to their sum. P_λ is a dimensionless quantity.

The total polarization angle (Brewster angle) α_B denotes the angle of incidence at which the reflected light ray is completely polarized in the direction perpendicular to the plane of incidence. α_B is specified by the expression

$$\alpha_B = \arctan n''/n'.$$

For the special case with $n' = 1$, one arrives at

$$\alpha_B = \arctan n''.$$

The specific angle of polarization plane rotation ψ_0 (in angle deg./mm) refers to the angle through which the sample crystal 1 mm in thickness rotates the polarization plane.

The angle of polarization plane rotation ψ by the media with the average thickness of d (along the ray direction) is defined from the expression

$$\psi = \psi_0 d.$$

To describe substances that are optically isotropic under normal conditions but become anisotropic on establishing magnetic or electric fields, special characteristics (Kerr constant, Verdet's constant, etc.) have been applied in Chapter 33 of this handbook.

In studying heat transfer processes, computations of body temperatures and other thermal computations, the optical characteristics pertaining to the thermal radiation of various bodies will continue in use.

The coefficient of thermal radiation ϵ_T is spoken of as the ratio of the thermal radiator luminosity M_e to the luminosity of a black body M_e^0 at the same temperature within the solid angle of 2π.

The spectral coefficient of thermal radiation ϵ_λ is said to be the ratio of the luminosity spectral density $M_{e\lambda}$ for a thermal radiator to the luminosity spectral density of $M_{e\lambda}^0$ of a black body at the same temperature and wavelength within the solid angle of 2π. The ϵ_T and ϵ_λ quantities are dimensionless.

The appropriate calculating formulas have the form

$$M_e = \epsilon_T M_e^0 = \epsilon_T \sigma (T^4 - T_1^4),$$

and

$$M_{e\lambda} = \epsilon_\lambda M_{e\lambda}^0 = \epsilon_\lambda c_1 \lambda^{-5} \left[\exp\left(c_2/\lambda T\right) - 1\right]^{-1},$$

where $\sigma = 5.67051 \cdot 10^{-8}\ \mathrm{W \cdot m^{-2} \cdot K^{-4}}$ is the Stefan–Boltzmann constant; $c_1 = 3.7417749 \cdot 10^{-16}\ \mathrm{W \cdot m^2}$ is the first radiation constant; $c_2 = 0.01438769\ \mathrm{m \cdot K}$ is the second radiation constant; T is the temperature of the radiating surface involved; T_1 is the temperature of the medium being in thermal equilibrium with the surface (in K); the dimensions of M_e and $M_{e\lambda}$ are W/m^2 and W/m^3, respectively.

The connection between ϵ_T and ϵ_λ is established by the following relationship

$$\epsilon_T = \int_0^\infty \epsilon_\lambda M^0_{e\lambda}\, d\lambda / \sigma T^4.$$

In a number of cases, the spectral density of the black body luminosity can be easily written as a function of radiation frequency ν. In this situation

$$M^0_{e\lambda} = \frac{2\pi h\nu^3}{c^2}\left[\exp\left(\frac{h\nu}{kT}\right) - 1\right]^{-1},$$

where $h = 6.62608 \cdot 10^{-34}$ J·s is the Planck's constant, $c = 299792458$ m/s is the speed of light in vacuum; $k = 1.38066 \cdot 10^{-23}$ J/K is the Boltzmann constant; $M^0_{e\lambda}$ dimension fits W·s/m^2.

The ϵ_T coefficient is often used jointly with the absorption coefficient α for computing some heat-transfer processes. In particular, in reckoning the temperature of the materials and coatings irradiated by solar radiation, the ϵ_T and α_S coefficients are widely used. In the event that the heat transfer induced by thermal conductivity and convection may be neglected in comparison with the radiant heat transfer, the temperature of a radiating surface is completely determined by the ϵ_T and α_S quantities and can be calculated with the formula

$$T = (Q\alpha_S/\sigma\epsilon_T)^{1/4},$$

where Q is the specific heat flux of solar radiation at the boundary of the earth's atmosphere in the direction normal to the surface irradiated ($Q \approx 1400$ W/m^2).

Notice that definition of ϵ_T and ϵ_λ by comparing the luminosities of the radiator under examination and the black body within the solid angle of 2π is rather tentative, but commonly used. A comparison can also be made for the other radiation directions but it should be qualified in the corresponding information materials and provided with other symbols. Thus, if the luminosities (or their spectral densities) of real radiators and black bodies are compared out of the solid angle of 2π but in the normal direction to a radiating surface, the coefficients of thermal radiation are denoted as ϵ^n_T and ϵ^n_λ.

When materials are irradiated by the radiant fluxes of great power damage and distortion can occur. In these situations the reliability of materials (so-called optical strength) is characterized by the threshold of surface P (in W/cm^2) and bulk W (in W/cm^2) failures. P and W for a given material are the functions of the flux irradiating this material (continuous or pulse type, radiation wavelength, duration of a pulse), its chemical composition, presence of impurities and inclusions, size of the surface irradiated, production process for the material of samples irradiated, and so on.

A detailed discussion of the above problems will be found in the literature on the following subjects: [1–6] — measurements in the field of modern technology; [7–9, 31, 32] — polarizing materials; [75, 83–85, 89–91] — measurements of the thresholds of surface and bulk failures; [10–13] — optical radiation of various types; [14, 15] — high-temperature measurements; [16] — the light reflection theory; [17] — optical properties of metals. The special handbooks, including [17–24, 47, 97] that contain the additional data on the optical properties of substances, should be consulted.

31.2 Glasses

The optical properties of glasses are presented in Tables 31.1–31.10.

Table 31.1 Optical characteristics of crown glasses [25]

Characteristic	Type of glass										
	Light crown		Crown		Heavy crown						
	Grade of glass										
	ЛК6	ЛК7	K1	K8	TK2	SK5	SK13	SK14	TK16	TK14	TK20
n_e *1	1.472142	1.484608	1.511730	1.518294	1.574860	1.591420	1.594240	1.605880	1.615192	1.615506	1.624702
ν_e	66.64	66.17	61.58	63.87	57.20	61.03	57.97	60.70	58.08	60.33	56.43
n for a laser wavelength, nm:											
488.0	1.47521	1.48777	1.51535	1.52181	1.57926	1.59564	1.59872	1.61023	1.61983	1.61996	1.62955
514.0	1.47371	1.48622	1.51358	1.52009	1.57710	1.59357	1.59653	1.60810	1.61756	1.61778	1.62717
520.8	1.47336	1.48586	1.51316	1.51968	1.57659	1.59308	1.59601	1.60759	1.61702	1.61726	1.62661
530.0	1.47290	1.48538	1.51261	1.51916	1.57593	1.59245	1.59533	1.60694	1.61633	1.61659	1.62589
568.2	1.47121	1.48363	1.51063	1.51722	1.57353	1.59014	1.59288	1.60456	1.61379	1.61416	1.62324
632.8	1.46897	1.48131	1.50802	1.51466	1.57040	1.58709	1.58968	1.60143	1.61049	1.61096	1.61979
647.1	1.46855	1.48088	1.50754	1.51419	1.56982	1.58653	1.58910	1.60085	1.60987	1.61037	1.61916
694.3	1.46732	1.47960	1.50613	1.51279	1.56814	1.58489	1.58738	1.59916	1.60810	1.60864	1.61732
890.0	1.46371	1.47585	1.50212	1.50872	1.56346	1.58020	1.58257	1.59434	1.60313	1.60375	1.61219
1060.0	1.46146	1.47352	1.49976	1.50695	1.56081	1.57745	1.57985	1.59153	1.60032	1.60090	1.60932
β_e, 10^{-7} K^{-1}, within the temperature range, K:											
from 213 to 293	−22	+36	−7.5	+6.0	+15	+17	+9.2	+7.7	+0.70	−1.0	+2.0
from 293 to 393	−7.0	+32	−1.0	+21	+30	+31	+22	+22	+16	+12	+15
τ^*_λ (λ = 560 nm) for d, mm:											
10	0.993	0.991	0.997	0.996	0.995	0.996	0.995	0.995	0.996	0.996	0.996
100	0.932	0.917	0.967	0.958	0.953	0.964	0.954	0.955	0.962	0.958	0.957
B, 10^{-12} Pa^{-1} (λ = 550 nm, T = 293 K)	3.70	3.45		3.20	2.70	2.15			1.80	1.80	1.85
λ_{min}, nm	321	318	324	320	318	344	342	347	347	346	347

Table 31.1 Optical characteristics of crown glasses *(continued)*

Characteristic	Type of glass													
	Superheavy crown			Fluorine crown		*2	Barite crown						Crown-flint	
	Grade of glass													
	SSK2	SSK5	CTK3	FK3	FK5	PSK2	BaK2	БК6	БК8	BaK5	BaK4	БК10	КФ4	KF7
n_e *1	1.625080	1.661510	1.662237	1.466190	1.489150	1.570860	1.542120	1.542136	1.548861	1.558970	1.571260	1.571309	1.520270	1.525770
ν_e	52.83	50.57	57.09	65.47	69.77	62.87	59.37	59.40	62.56	58.28	55.73	55.77	58.71	50.55
n for a laser wavelength, nm:														
488.0	1.63028	1.66727	1.66731	1.46928	1.49218	1.57481	1.54611	1.54612	1.55268	1.56317	1.57575	1.57580	1.52413	1.53034
514.0	1.62772	1.66444	1.66482	1.46777	1.49070	1.57288	1.54415	1.54417	1.55081	1.56111	1.57355	1.57359	1.52224	1.52809
520.8	1.62712	1.66377	1.66424	1.46741	1.49035	1.57242	1.54369	1.54371	1.55037	1.56062	1.57303	1.57307	1.52179	1.52736
530.0	1.62364	1.66291	1.66348	1.46695	1.48989	1.57183	1.54309	1.54311	1.54979	1.56000	1.57235	1.57240	1.52121	1.52688
568.2	1.62352	1.65978	1.66071	1.46525	1.48822	1.56966	1.54091	1.54093	1.54770	1.55770	1.56991	1.56995	1.51910	1.52440
632.8	1.61984	1.65574	1.65708	1.46300	1.48600	1.56680	1.53806	1.53807	1.54494	1.55471	1.56671	1.56676	1.51632	1.52117
647.1	1.61917	1.65500	1.65642	1.46258	1.48558	1.56627	1.53753	1.53755	1.54442	1.55416	1.56612	1.56618	1.51581	1.52058
694.3	1.61722	1.65285	1.65447	1.46136	1.48436	1.56472	1.53599	1.53600	1.54292	1.55255	1.56442	1.56448	1.51431	1.51887
890.0	1.61184	1.64697	1.64903	1.45779	1.48077	1.56024	1.53166	1.53170	1.53861	1.54806	1.55964	1.55974	1.51002	1.51410
1060.0	1.60888	1.64375	1.64594	1.45562	1.47853	1.55756	1.52917	1.52920	1.53604	1.54551	1.55694	1.55709	1.50748	1.51142
β_e, 10^{-7} K^{-1}, within the temperature range, K:														
from 213 to 293	+25	+0.90	−15	−32	−30	+6.6	−2.6		+16	−1.3	+17	+10	+18	+16
from 293 to 393	+39	+17	−2.0	−19	−17	+21	+10		+32	+14	+32	+35	+34	+27
τ^*_λ (λ = 560 nm) for d, mm:														
10	0.995	0.991	0.996	0.996	0.996	0.995	0.996	0.996	0.995	0.997	0.996	0.996	0.996	0.995
100	0.955	0.915	0.961	0.956	0.963	0.956	0.956	0.958	0.956	0.966	0.957	0.957	0.959	0.955
B, 10^{-12} Pa^{-1} (λ = 550 nm, T = 293K)	2.15	1.80	1.50		2.90			2.60	2.80		2.95	2.90	2.90	
λ_{min}, nm	350	383	330	338	319	349	326	322	325	329	343	336	327	363

*1 The indices of refraction are in error of $\pm 1 \cdot 10^{-5}$.

*2 Fluorine heavy crown.

Table 31.2 Optical characteristics of flint glasses [25]

Characteristic	Type of glass														
	*1	Light flint		Flint					Heavy flint						
	Grade of glass														
	LLF8	ЛΦ5	LF7	Φ6	F5	F3	Φ1	F11	SF7	TΦ1	SF2	SF9	TΦ3	TΦ5	TΦ13
n_e	1.535840	1.578326	1.578320	1.607015	1.608130	1.616850	1.616878	1.625070	1.644160	1.652188	1.652210	1.659680	1.723166	1.761712	1.791689
ν_e	44.84	41.03	41.04	37.68	37.88	36.69	36.69	35.69	34.37	33.62	33.61	33.14	29.29	27.32	26.13
n for a laser wavelength, nm:															
488.0	1.54114	1.58457	1.58457	1.61417	1.61527	1.62433	1.62436	1.63288	1.65253	1.66085	1.66088	1.66857	1.73425	1.77426	
514.0	1.53852	1.58149	1.58148	1.61063	1.61174	1.62063	1.62066	1.62901	1.64838	1.65656	1.65658	1.66416	1.72877	1.76802	
520.8	1.53791	1.58076	1.58076	1.60980	1.61091	1.61976	1.61979	1.62811	1.64741	1.65555	1.65558	1.66313	1.72745	1.76657	
530.0	1.53712	1.57983	1.57983	1.60873	1.60985	1.61865	1.61868	1.62695	1.64617	1.65427	1.65429	1.66181	1.72581	1.76470	
568.2	1.53427	1.57647	1.57646	1.60489	1.60602	1.61464	1.61467	1.62278	1.64170	1.64964	1.64966	1.65706	1.71992	1.75806	
632.8	1.53061	1.57215	1.57215	1.59998	1.60113	1.60952	1.60955	1.61747	1.63602	1.64378	1.64379	1.65105	1.71250	1.74971	
647.1	1.52995	1.57137	1.57137	1.59910	1.60024	1.60860	1.60863	1.61651	1.63500	1.64272	1.64273	1.64996	1.71117	1.74822	
694.3	1.52802	1.56912	1.56116	1.59655	1.59770	1.60595	1.60597	1.61374	1.63207	1.63970	1.63970	1.64686	1.70738	1.74397	
890.0	1.52274	1.56303	1.56303	1.58973	1.59090	1.59889	1.59889	1.60631	1.62431	1.63172	1.63172	1.63869	1.69751	1.73299	
1060.0	1.51981	1.55979	1.55978	1.58616	1.58734	1.59523	1.59522	1.60237	1.62036	1.62767	1.62768	1.63455	1.69263	1.72765	
β_e, 10^{-7} K^{-1}, within the temperature range, K:															
from 213 to 293	−21	+25	+14	+30	+21	+23	+31	+18	+27	+18	+22	+33	+49	+62	
from 293 to 393	−6.5	+45	+32	+49	+40	+40	+53	+33	+50	+42	+46	+56	+72	+89	
τ_λ^* (λ = 560 nm) for d, mm:															
10	0.978	0.995	0.995	0.995	0.996	0.995	0.996	0.995	0.996	0.996	0.996	0.996	0.996	0.996	0.996
100		0.954	0.953	0.953	0.961	0.954	0.960	0.955	0.958	0.962	0.963	0.958	0.960	0.960	0.960
B, 10^{-12} Pa^{-1} (λ = 550 nm, T = 293K)	3.30	3.20		2.95			2.90	3.05	2.65	2.50			1.80	1.45	
λ_{min}, nm	381	335	323	343	337	340	346	377	347	365	349	353	372	385	393

*1 Superlight flint.

Table 31.2 Optical characteristics of flint glasses *(continued)*

Characteristic	Type of glass									
	*2	*3	Barite flint					Barite heavy flint		
	Grade of glass									
	СТФ2	BaLF5	БФ1	БФ12	БФ16	БФ24	BaF3	ТБФ3	ТБФ8	BaSF7
n_e	1.955369	1.549830	1.527063	1.629837	1.674385	1.638639	1.585640	1.760210	1.864057	1.705590
ν_e	20.26	53.32	54.68	38.82	47.00	36.50	46.22	40.87	36.40	40.88
n for a laser wavelength, nm:										
488.0	1.97683	1.55436	1.53128	1.63705	1.68073	1.64643	1.59124	1.76846	1.87461	1.71326
514.0	1.96608	1.55213	1.52921	1.63348	1.67761	1.64258	1.58848	1.76439	1.89640	1.70947
520.8	1.96360	1.55161	1.52872	1.63265	1.67688	1.64168	1.58783	1.76343	1.86817	1.70858
530.0	1.96044	1.55093	1.52809	1.63157	1.67593	1.64052	1.58670	1.76220	1.86660	1.70744
568.2	1.94921	1.54847	1.52579	1.62770	1.67250	1.63634	1.58397	1.75776	1.86093	1.70332
632.8	1.93534	1.54527	1.52279	1.62276	1.66809	1.63103	1.58007	1.75207	1.85369	1.69805
647.1	1.93289	1.54468	1.52223	1.62187	1.66728	1.63007	1.57936	1.75104	1.85238	1.69709
694.3	1.92596	1.54298	1.52063	1.61930	1.66496	1.62732	1.57731	1.74808	1.84861	1.69434
890.0	1.90849	1.53827	1.51611	1.61247	1.65864	1.62000	1.57175	1.74020	1.83854	1.68695
1060.0	1.90039	1.53565	1.51352	1.60894	1.65527	1.61623	1.56876	1.73617	1.83330	1.68305
β_e, 10^{-7} K^{-1}, within the temperature range, K:										
from 213 to 293		+0.2	+17	+13	+6	+35	+12	+53	+41	+16
from 293 to 393		+13	+34	+31	+24	+56	+29	+72	+58	+30
τ^*_λ (λ = 560 nm) for d, mm:										
10	0.955	0.996	0.995	0.996	0.996	0.995	0.997	0.980	0.992	0.990
100	0.630	0.962	0.951	0.959	0.963	0.954	0.970	0.817	0.920	0.908
B, 10^{-12} Pa^{-1} (λ = 550 nm, T = 293K)	1.25	2.90	3.15	2.50	1.60	2.60		2.40	2.40	
λ_{min}, nm	450	329	329	347	362	364	340	370	370	390

*2 Superheavy flint.
*3 Barite light flint.

Table 31.3 Coefficients A_i in dispersion formulas for crown glasses*[1] [25]

Grade of glass*[2]	A_1	A_2	A_3	A_4	A_5	A_6
ЛК6	2.1391711	$-9.8913489 \cdot 10^{-3}$	$8.4704778 \cdot 10^{-3}$	$2.8247761 \cdot 10^{-4}$	$-1.9072939 \cdot 10^{-5}$	$9.3359448 \cdot 10^{-7}$
ЛК7	2.1732195	$-9.4960367 \cdot 10^{-3}$	$9.7105457 \cdot 10^{-3}$	$1.0540599 \cdot 10^{-4}$	$-3.8188276 \cdot 10^{-6}$	$5.0116854 \cdot 10^{-7}$
К1	2.249329	$-8.707883 \cdot 10^{-3}$	$1.072569 \cdot 10^{-2}$	$2.718141 \cdot 10^{-4}$	$-1.446565 \cdot 10^{-5}$	$8.846095 \cdot 10^{-7}$
К8	2.2699804	$-9.8250605 \cdot 10^{-3}$	$1.1017203 \cdot 10^{-2}$	$7.6606834 \cdot 10^{-5}$	$1.1616952 \cdot 10^{-5}$	$5.8130900 \cdot 10^{-7}$
ТК2	2.4321820	$-8.2232823 \cdot 10^{-3}$	$1.4772427 \cdot 10^{-2}$	$-1.7487909 \cdot 10^{-5}$	$3.4620141 \cdot 10^{-5}$	$-1.4790076 \cdot 10^{-6}$
SK5	2.487826	$-1.041114 \cdot 10^{-2}$	$1.356371 \cdot 10^{-2}$	$2.208563 \cdot 10^{-4}$	$-3.486038 \cdot 10^{-6}$	$4.105613 \cdot 10^{-7}$
SK13	2.492664	$-8.906364 \cdot 10^{-3}$	$1.481905 \cdot 10^{-2}$	$1.238659 \cdot 10^{-4}$	$1.522812 \cdot 10^{-5}$	$-5.419190 \cdot 10^{-7}$
SK14	2.530211	$-9.687823 \cdot 10^{-3}$	$1.549009 \cdot 10^{-2}$	$-1.944015 \cdot 10^{-4}$	$5.598273 \cdot 10^{-5}$	$-2.679258 \cdot 10^{-6}$
ТК16	2.559113	$-9.8827859 \cdot 10^{-3}$	$1.4212686 \cdot 10^{-2}$	$5.9116562 \cdot 10^{-4}$	$-5.4050675 \cdot 10^{-5}$	$3.2017380 \cdot 10^{-6}$
ТК14	2.5615495	$-1.0649249 \cdot 10^{-2}$	$1.4719854 \cdot 10^{-2}$	$1.6374569 \cdot 10^{-4}$	$7.7935463 \cdot 10^{-6}$	$-1.1965516 \cdot 10^{-7}$
ТК20	2.5854315	$-9.0796080 \cdot 10^{-3}$	$1.6291552 \cdot 10^{-2}$	$1.7763493 \cdot 10^{-4}$	$8.5249807 \cdot 10^{-6}$	$2.5484363 \cdot 10^{-8}$
SSK2	2.582916	$-8.806647 \cdot 10^{-3}$	$1.723626 \cdot 10^{-2}$	$1.944876 \cdot 10^{-4}$	$1.671742 \cdot 10^{-5}$	$-2.215866 \cdot 10^{-7}$
SSK5	2.695150	$-9.406487 \cdot 10^{-3}$	$1.928628 \cdot 10^{-2}$	$2.731062 \cdot 10^{-4}$	$1.195049 \cdot 10^{-5}$	$5.928455 \cdot 10^{-7}$
СТК3	2.7056846	$-1.0711205 \cdot 10^{-2}$	$1.7154336 \cdot 10^{-2}$	$2.8684174 \cdot 10^{-4}$	$-8.5758288 \cdot 10^{-6}$	$1.0797444 \cdot 10^{-6}$
FK3	2.120025	$-8.391972 \cdot 10^{-3}$	$9.205200 \cdot 10^{-3}$	$1.068591 \cdot 10^{-4}$	$3.582447 \cdot 10^{-6}$	$1.571596 \cdot 10^{-7}$
FK5	2.187224	$-8.818712 \cdot 10^{-3}$	$9.930764 \cdot 10^{-3}$	$-1.141033 \cdot 10^{-4}$	$3.011786 \cdot 10^{-5}$	$-1.406416 \cdot 10^{-6}$
PSK2	2.426049	$-1.033777 \cdot 10^{-2}$	$1.297767 \cdot 10^{-2}$	$4.448250 \cdot 10^{-5}$	$1.942162 \cdot 10^{-5}$	$-9.095428 \cdot 10^{-7}$
BaK2	2.336480	$-8.396078 \cdot 10^{-3}$	$1.261039 \cdot 10^{-2}$	$1.454578 \cdot 10^{-4}$	$6.712083 \cdot 10^{-6}$	$-1.378582 \cdot 10^{-7}$
БК8	2.3601309	$-1.0118581 \cdot 10^{-2}$	$1.1761749 \cdot 10^{-2}$	$2.4663544 \cdot 10^{-4}$	$-1.2819425 \cdot 10^{-5}$	$9.3405641 \cdot 10^{-7}$
BaK5	2.386836	$-8.760405 \cdot 10^{-3}$	$1.275121 \cdot 10^{-2}$	$3.488536 \cdot 10^{-4}$	$-1.722515 \cdot 10^{-5}$	$1.018959 \cdot 10^{-6}$
BaK4	2.419484	$-7.993419 \cdot 10^{-3}$	$1.535331 \cdot 10^{-2}$	$-1.168727 \cdot 10^{-4}$	$4.922728 \cdot 10^{-5}$	$-2.107186 \cdot 10^{-6}$
БК10	2.4193499	$-7.6662143 \cdot 10^{-3}$	$1.5535897 \cdot 10^{-2}$	$-1.8549105 \cdot 10^{-4}$	$6.0646835 \cdot 10^{-5}$	$-2.7567697 \cdot 10^{-6}$
КФ4	2.2723952	$-9.3640165 \cdot 10^{-3}$	$1.1794006 \cdot 10^{-2}$	$1.7956625 \cdot 10^{-4}$	$1.2049614 \cdot 10^{-7}$	$3.8219544 \cdot 10^{-7}$
KF7	2.280329	$-7.818331 \cdot 10^{-3}$	$1.444672 \cdot 10^{-2}$	$2.999465 \cdot 10^{-5}$	$3.504189 \cdot 10^{-5}$	$-9.972735 \cdot 10^{-7}$

*[1] With the use of A_i values listed, the λ numerical values in dispersion formulas should be expressed in units of μm; calculation of n with the aid of dispersion formulas yields errors not exceeding $\pm 1 \cdot 10^{-5}$.
*[2] The types of glasses are tabulated in Table 31.1.

Table 31.4 Coefficients A_i in dispersion formulas for flint glasses*[1] [25]

Grade of glass*[2]	A_1	A_2	A_3	A_4	A_5	A_6
LLF8	2.307780	$-9.683781 \cdot 10^{-3}$	$1.382640 \cdot 10^{-2}$	$8.181413 \cdot 10^{-4}$	$-6.495766 \cdot 10^{-5}$	$6.291549 \cdot 10^{-6}$
ЛФ5	2.4246429	$-7.9511272 \cdot 10^{-3}$	$1.9033590 \cdot 10^{-2}$	$3.9095360 \cdot 10^{-4}$	$1.0184703 \cdot 10^{-5}$	$1.7896964 \cdot 10^{-6}$
LF7	2.429964	$-8.661886 \cdot 10^{-3}$	$1.825623 \cdot 10^{-2}$	$5.877478 \cdot 10^{-4}$	$-1.223595 \cdot 10^{-5}$	$2.675158 \cdot 10^{-6}$
Ф6	2.5047749	$-7.8584593 \cdot 10^{-3}$	$2.2129689 \cdot 10^{-2}$	$4.2000253 \cdot 10^{-4}$	$2.3483839 \cdot 10^{-5}$	$1.8712023 \cdot 10^{-6}$
F5	2.508913	$-8.116668 \cdot 10^{-3}$	$2.193760 \cdot 10^{-2}$	$4.351979 \cdot 10^{-4}$	$2.517237 \cdot 10^{-5}$	$1.453801 \cdot 10^{-6}$
F3	2.534317	$-8.648783 \cdot 10^{-3}$	$2.199543 \cdot 10^{-2}$	$7.865099 \cdot 10^{-4}$	$-1.730297 \cdot 10^{-5}$	$4.026666 \cdot 10^{-6}$
Ф1	2.5344898	$-8.9233727 \cdot 10^{-3}$	$2.2149725 \cdot 10^{-2}$	$6.9286357 \cdot 10^{-4}$	$-6.1831815 \cdot 10^{-8}$	$3.0937650 \cdot 10^{-6}$
F11	2.556874	$-9.791700 \cdot 10^{-3}$	$2.407292 \cdot 10^{-2}$	$3.851180 \cdot 10^{-4}$	$3.677796 \cdot 10^{-5}$	$3.744679 \cdot 10^{-6}$
SF7	2.613210	$-9.016895 \cdot 10^{-3}$	$2.437330 \cdot 10^{-2}$	$1.018745 \cdot 10^{-3}$	$-3.079667 \cdot 10^{-5}$	$5.627638 \cdot 10^{-6}$
ТФ1	2.6349371	$-8.5077204 \cdot 10^{-3}$	$2.6185673 \cdot 10^{-2}$	$7.5510993 \cdot 10^{-4}$	$1.5001867 \cdot 10^{-5}$	$3.6330948 \cdot 10^{-6}$
SF2	2.633217	$-7.526166 \cdot 10^{-3}$	$2.720998 \cdot 10^{-2}$	$4.980595 \cdot 10^{-4}$	$4.500670 \cdot 10^{-5}$	$2.281773 \cdot 10^{-6}$
SF9	2.657241	$-8.755496 \cdot 10^{-3}$	$2.652209 \cdot 10^{-2}$	$9.574046 \cdot 10^{-4}$	$-1.245464 \cdot 10^{-5}$	$5.291734 \cdot 10^{-6}$
ТФ3	2.8433701	$-8.5219322 \cdot 10^{-3}$	$3.3994123 \cdot 10^{-2}$	$1.2311168 \cdot 10^{-3}$	$-1.4893638 \cdot 10^{-5}$	$9.3877638 \cdot 10^{-6}$
ТФ5	2.9580175	$-8.2686725 \cdot 10^{-3}$	$3.9383391 \cdot 10^{-2}$	$1.2219807 \cdot 10^{-3}$	$3.1433368 \cdot 10^{-5}$	$8.6507903 \cdot 10^{-6}$
ТФ13	3.0462836	$-6.8088629 \cdot 10^{-3}$	$4.5614337 \cdot 10^{-2}$	$5.5374355 \cdot 10^{-4}$	$1.7210583 \cdot 10^{-4}$	$1.7379726 \cdot 10^{-6}$
СТФ2	3.5505425	$-4.8989192 \cdot 10^{-3}$	$7.3462432 \cdot 10^{-2}$	$1.3716187 \cdot 10^{-3}$	$2.7698053 \cdot 10^{-4}$	$1.7008613 \cdot 10^{-5}$
BaLF5	2.352658	$-6.984001 \cdot 10^{-3}$	$1.519826 \cdot 10^{-2}$	$-1.072165 \cdot 10^{-4}$	$5.019470 \cdot 10^{-5}$	$-1.969284 \cdot 10^{-6}$
БФ1	2.2913292	$-9.5847187 \cdot 10^{-3}$	$1.0929842 \cdot 10^{-2}$	$8.4720003 \cdot 10^{-4}$	$-9.1163748 \cdot 10^{-5}$	$5.5104064 \cdot 10^{-6}$
БФ12	2.5767076	$-7.6835879 \cdot 10^{-3}$	$2.2692139 \cdot 10^{-2}$	$4.1937275 \cdot 10^{-4}$	$2.4633519 \cdot 10^{-5}$	$1.7499673 \cdot 10^{-6}$
БФ16	2.7308360	$-9.2896392 \cdot 10^{-3}$	$2.1405060 \cdot 10^{-2}$	$2.3934677 \cdot 10^{-4}$	$2.7482465 \cdot 10^{-5}$	$2.8808684 \cdot 10^{-7}$

Table 31.4 Coefficients in dispersion formulas for flint glasses *(continued)*

Grade of glasses*2	A_1	A_2	A_3	A_4	A_5	A_6
БФ24	2.6051551	$-1.0884034\cdot10^{-2}$	$2.0201021\cdot10^{-2}$	$1.7331740\cdot10^{-3}$	$-1.4260022\cdot10^{-4}$	$1.0842795\cdot10^{-5}$
BaF3	2.454433	$-8.004484\cdot10^{-3}$	$1.715358\cdot10^{-2}$	$4.065255\cdot10^{-4}$	$-1.841772\cdot10^{-6}$	$1.452719\cdot10^{-6}$
ТБФ3	3.0019055	$-1.0195355\cdot10^{-2}$	$2.5982463\cdot10^{-2}$	$1.2853512\cdot10^{-3}$	$-7.7243846\cdot10^{-5}$	$6.2402164\cdot10^{-6}$
ТБФ8	3.3399592	$-1.2211485\cdot10^{-2}$	$3.8707992\cdot10^{-2}$	$5.7606301\cdot10^{-4}$	$4.4782780\cdot10^{-5}$	$3.5747092\cdot10^{-6}$
BaSF7	2.821595	$-1.018806\cdot10^{-2}$	$2.474911\cdot10^{-2}$	$6.319116\cdot10^{-4}$	$-3.077012\cdot10^{-6}$	$3.961194\cdot10^{-6}$

*1 With the use of A_i values listed, the λ numerical values in dispersion formulas should be expressed in units of μm; calculation of n with the aid of dispersion formulas yields the errors not exceeding $\pm1\cdot10^{-5}$.

*2 The types of glasses are tabulated in Table 31.2.

Table 31.5 Spectral internal transmittances τ_λ and spectral absorptive indices α_λ for quartz optical glasses [25, 85]

λ, nm	Glasses for ultraviolet and visible spectral regions				Glass for a visible spectral region		Glass for visible and infra-red spectral regions		Glass for ultra-violet, visible and infra-red spectral regions	
	КУ-1		КУ-2		КВ		КИ		КУВИ	
	τ^*_λ	α_λ, cm^{-1}	τ^*_λ	α_λ, cm^{-1}	τ^*_λ	α_λ, cm^{-1}	τ^*_λ	α_λ, cm^{-1}	τ^*_λ	α_λ, cm^{-1}
180	0.900	0.246	0.660	0.180					0.845	0.073
190	0.940	0.027	0.755	0.122					0.915	0.038
200	0.960	0.018	0.845	0.073					0.935	0.029
210	0.980	0.009	0.920	0.036	0.030	1.523	0.075	1.125	0.950	0.022
220	0.990	0.004	0.950	0.022	0.220	0.658	0.220	0.658	0.970	0.013
230	>0.999		0.945	0.025	0.565	0.248	0.435	0.362		
240	>0.999		0.920	0.036	0.540	0.268	0.525	0.280		
250	>0.999		0.940	0.027	0.715	0.146	0.585	0.233	0.980	0.009
260	>0.999		0.985	0.007	0.920	0.036	0.800	0.097		
270	>0.999		0.995	0.002	0.995	0.002	0.930	0.032		
280	>0.999		>0.999		>0.999		0.990	0.004	0.985	0.007
300	0.999		>0.999		>0.999		>0.999		0.980	0.009
400	0.999		>0.999		>0.999		>0.999		0.990	0.013
500	0.999		>0.999		>0.999		>0.999		>0.999	0.004
750	0.999		>0.999		>0.999		>0.999		>0.999	
1000	0.999		>0.999		>0.999		>0.999		>0.999	
1385	0.880	0.056	0.960	0.018	0.970	0.013	>0.999		>0.999	
2000	0.999		0.999		0.999		>0.999		>0.999	
2200	0.580	0.237	0.900	0.046	0.925	0.034	>0.999		>0.999	
2300	0.880	0.056	0.955	0.020	0.955	0.020	>0.999		>0.999	
2380	0.980	0.022	0.970	0.013	0.970	0.013	>0.999		>0.999	
2500	0.790	0.102	0.930	0.032	0.930	0.032	>0.999		>0.999	
2720	0.070	>2.5		>2.5	0.030	1.523	>0.999		0.930	0.032
2800		>2.5	0.015	1.824	0.150	0.824	0.955	0.020	0.935	0.029
2900	0.295	0.530	0.720	0.143	0.770	0.114	0.900	0.046	0.910	0.041
3000	0.670	0.174	0.810	0.092	0.810	0.092	0.870	0.060	0.880	0.056
3100	0.785	0.105	0.830	0.081	0.825	0.084	0.850	0.071	0.860	0.066
3300	0.855	0.068	0.860	0.066	0.850	0.071	0.870	0.060	0.875	0.058
3750	0.180	0.745	0.180	0.745	0.180	0.745	0.180	0.745	0.180	0.745
3910	0.185	0.733	0.185	0.733	0.185	0.733	0.185	0.733	0.185	0.733
4000	0.150	0.824	0.150	0.824	0.150	0.824	0.150	0.824	0.150	0.824
4220	0.050	1.301	0.050	1.301	0.050	1.301	0.050	1.301	0.050	1.301
4300	0.070	1.155	0.070	1.155	0.070	1.155	0.070	1.155	0.070	1.155
4400		>2.5		>2.5		>2.5		>2.5	0.060	>2.5*1
4430									0.057*1	>2.5*1
4580									0.030*1	>2.5*1

*1 These values are given for d =4 mm; the remaining values are given for d = 10 mm.

Table 31.6 Optical characteristics of quartz glasses [25, 76]

Characteristic	Numerical value*1					
ε_T at $T = 293\ldots773$ K	0.93					
n for a wavelength of, nm						
488.0	1.46299					
514.0	1.46157					
520.8	1.46123					
530.0	1.46079					
546.1	1.46008					
568.2	1.45918					
632.8	1.45702					
647.1	1.45662					
694.3	1.45543					
890.0	1.45190					
1060.0	1.44968					
ν_e	68.00					
A_1	2.1026513					
A_2, 10^{-3}	-8.5943075					
A_3, 10^{-3}	9.8576238					
A_4, 10^{-4}	-4.4538022					
A_5, 10^{-5}	4.4589827					
A_6, 10^{-6}	-1.9692608					
B, 10^{-12} Pa^{-1}, for a wavelength of 550 nm	3.45					
β, 10^{-7} K^{-1}, for a wavelength, nm:	at a temperature of, K					
	173	203	223	253	293	333
486.13	104	99	100	102	106	112
546.07	101	98	98	100	104	110
589.29	100	96	96	99	102	108
656.27	99	95	95	98	100	107

*1 Obtained by averaging data for quartz glasses of different grades.

Table 31.7 Optical characteristics of electrovacuum glasses, glasses for fiber optics and sitalls at a temperature of 293 K [25, 81, 96] [λ_1= 486.13 nm; λ_2= 656.27 nm.]

Characteristic	Electro-vacuum glasses	Glasses for cores						
	C89-1	TK16	ФВ	BC80	BC586	BC682	BC1851	BC16121
n_e	1.5401	1.6152	1.6291	1.6830	1.5893	1.6855	1.6830	1.6971
$n_{\lambda_1}-n_{\lambda_2}$, 10^{-5}		1050	1757		1422	1298		
λ_{min}, nm		323	334	325	332	325	310	300
τ^*_λ (d=10 mm) for λ of, mm	d=1 mm							
400	0.994	0.992	0.996		0.997	0.992		
600	0.995	0.998	0.999		0.999	0.999		
800	0.993	0.997	0.999		0.999	0.999		
1000	0.980	0.991	0.999		0.999	0.999		
1500	0.974							
2000	0.980							
2500	0.980							
3000	0.950							

Table 31.7 Optical characteristics of electrovacuum glasses, glasses for fiber optics and sitalls at a temperature of 293 K *(continued)*

Characteristic	Glasses for transparent shields							Sitalls		
	BO50	BO73-1	BO488	BO513	BO691	BO695	BO801	CO21	CO115	CO156
n_e	1.4840	1.5152	1.4898	1.5150	1.5091	1.4850	1.5112	1.5531	1.5350	1.5450
$n_{\lambda_1}-n_{\lambda_2}$, 10^{-5}			745	795				1080	1020	1040
λ_{min}, nm	290	310	300	310	300	275	310			
τ^*_λ (d=10 mm) for λ of, mm										
400			0.997	0.994						
600			0.999	0.996						
800			0.999	0.997						
1000			0.999	0.997						

Table 31.8 Spectral coefficients of diffuse reflection ρ^*_λ for light-diffusing optical materials at temperatures of 295 – 300 K [26]

λ, nm	ρ^*_λ						
	Opal glasses				Magnesium oxide		Barium sulfate ($BaSO_4$ compacted powder)
	MC 14		MC 20				
	mirror-finished	clouded	mirror-finished	clouded	MgO ash	MgO compacted powder	
400	0.915	0.892	0.935	0.950	0.977	0.972	0.987
420			0.940	0.950			
440			0.945	0.957			
450	0.945	0.925			0.980	0.980	
460			0.955	0.960	0.980		0.991
480			0.960	0.966	0.980		
490					0.980		
500	0.958	0.942	0.965	0.969	0.980	0.983	0.991
520			0.970	0.974	0.980		
540			0.970	0.975	0.980		
550	0.956	0.945	0.970	0.975	0.980	0.984	
560	0.956	0.945	0.970	0.975	0.980		0.992
570	0.956	0.945	0.970	0.975	0.980		0.992
580	0.956	0.945	0.970	0.975	0.980		0.992
600	0.955	0.945	0.970	0.973	0.980	0.988	0.992
620	0.955	0.945	0.970	0.974	0.980		0.992
640	0.955	0.945	0.965	0.973	0.979		0.992
650	0.955	0.945	0.965	0.975	0.979	0.990	0.992
660	0.955	0.945	0.965	0.975	0.978	0.990	0.992
680	0.955	0.945	0.965	0.971	0.978	0.990	0.992
700	0.955	0.945	0.965	0.970	0.977	0.990	0.992
720	0.955	0.945	0.965	0.971	0.977		0.992
740	0.955	0.945	0.965	0.971	0.976		0.992
750	0.955	0.945	0.960		0.975		0.992
1000			0.935				0.990

Note: The measurement error is no more than ±1%.

Table 31.9 Spectral absorptive indices a_λ and index of refraction n for colored glasses [27]

λ, nm	a_λ, mm^{-1}, for glasses of the grades*																		
	УФС1	УФС6	ФС6	СС2	СС4	СЗС7	СЗС22	СЗС24	ЗС11	ЖЗС6	ЖЗС12	ЖЗС17	ОС5	ИКС6	НС1	НС2	НС8	НС10	ТС10
240	0.32	>3		>3	>3	>3	>3			>3	>3	>3	>3		>3	>3	>3	>3	>3
280	0.025	1.74	0.57	2.73	>3	>3	>3	0.40		>3	>3	>3	>3		>3	>3	>3	>3	>3
320	0.004	0.13	0.084	0.15	0.68	0.75	>3	0.041		1.05	>3	>3	1.77		0.95	>3	>3	>3	>3
350	0.005	0.035	0.032	0.040	0.16	0.086	0.49	0.009		>3	>3	>3	1.24		0.10	0.29	0.75	2.30	0.41
380	0.035	0.104	0.023	0.015	0.050	0.026	0.046	0.004	4.80	>3	>3	>3	1.17		0.011	0.080	0.44	1.32	0.15
420	0.72	>3	0.082	0.010	0.055	0.011	0.009	0.011	1.95	1.21	>3	>3	1.16		0.050	0.14	0.29	0.93	0.43
450	1.07	>3	0.30	0.025	0.15	0.009	0.007	0.011	0.77	0.49	2.33	>3	0.90		0.081	0.29	0.27	0.87	0.67
480	1.33	>3	0.95	0.079	0.49	0.010	0.007	0.009	0.33	0.18	1.27	2.32	0.61		0.073	0.24	0.26	0.85	0.52
520	2.30	>3	1.75	0.24	1.34	0.034	0.028	0.010	0.15	0.056	0.66	1.50	0.36		0.071	0.23	0.27	0.87	0.39
550	2.50	>3	1.55	0.22	1.27	0.079	0.12	0.010	0.21	0.038	0.62	1.43	0.23		0.070	0.22	0.26	0.85	0.37
580	3.50	>3	2.20	0.33	1.86	0.15	0.37	0.013	0.42	0.050	0.80	1.55	0.15		0.072	0.23	0.27	0.88	0.31
620	3.00	>3	3.10	0.37	2.10	0.28	1.09	0.028	0.84	0.094	1.35	2.13	0.10		0.079	0.25	0.27	0.87	0.33
650	1.75	>3	3.40	0.38	2.10	0.39	1.99	0.047	1.12	0.11	1.66	2.43	0.080	5.70	0.076	0.25	0.27	0.86	0.29
680	0.36	2.92	2.70	0.18	1.01	0.48	>3	0.072	1.29	0.092	1.69	2.18	0.070	4.90	0.043	0.13	0.26	0.81	0.15
720	0.080	0.41	0.43	0.009	0.049	0.57	>3	0.12	1.41	0.070	1.69	1.88	0.063	3.70	0.030	0.070	0.25	0.73	0.10
750	0.11	0.30	0.067		0.008	0.61	>3	0.16	1.42	0.050	1.63	1.79	0.060	2.75	0.020	0.060	0.26	0.74	0.12
780	0.15	0.47	0.014		0.005	0.62	>3	0.22	1.43	0.029	1.59	1.87	0.060	2.00	0.024	0.065	0.29	0.79	0.12
840	0.22	0.78	0.014		0.008	0.60	>3	0.34	1.39	0.014	1.52	1.84	0.060	0.99	0.038	0.095	0.35	0.87	0.21
920	0.26	0.94	0.029	0.002	0.016	0.50	>3	0.54	1.22	0.010	1.38	1.64	0.060	0.32	0.044	0.11	0.42	0.98	0.26
1000	0.23	0.99	0.064	0.005	0.037	0.39	>3	0.75	0.99	0.010	1.20	1.48	0.060	0.11	0.050	0.12	0.49	1.10	0.28
1200	0.40	1.51	0.38	0.050	0.27	0.21	>3	0.96	0.50	0.010	0.73	0.96	0.056	0.022	0.040	0.13	0.51	1.06	0.27
1400	0.49	1.41	0.55	0.047	0.36	0.11	2.05	1.01	0.41	0.010	0.44	0.60	0.050	0.017	0.040	0.10	0.36	0.71	0.20
1800	0.43	1.58	0.53	0.043	0.37	0.028	0.45	0.70	0.22	0.010	0.22	0.42	0.032	0.009	0.040	0.090	0.24	0.46	0.18
2460	0.23	1.84	0.17		0.048	0.006	0.15	0.68	0.080	0.010	0.16	0.35	0.030	0.013	0.040	0.080	0.15	0.26	0.15
3000	1.10	0.53	0.37	0.13	0.24	0.11	1.93	0.77	0.11	0.11	0.21	0.55	0.15	0.090	0.13	0.22	0.62	0.63	0.23
n for λ = 589.29 nm	1.540	1.520	1.495	1.520	1.520	1.514	1.535	1.516	1.550	1.522	1.527	1.527	1.523	1.541	1.521	1.523	1.503	1.509	1.523

* УФС is ultraviolet glass; ФС – violet glass; СС – blue glass; СЗС – blue-green glass; ЗС – green glass; ЖЗС – yellow-green glass; ОС – orange glass; ИКС – infra-red glass; НС – neutral glass; ТС – dark glass.

Table 31.10 Refraction indices n, temperature coefficients of refraction index β and dispersion coefficients ν_D for neodymium glasses [75]

Grade of glass	n for λ, μm, at $T = 293$ K				β, 10^{-6} K^{-1}, at $\lambda = 1.06$ μm and $T = 303$ K	$\nu_D = \frac{n_D - 1}{n_F - n_C}$ at $T = 293$ K
	0.488	0.589	0.63	1.06		
Silicate						
ГЛС1		1.534	1.532	1.523	−0.4	57.9
ГЛС2		1.526		1.518	−1.6	57.9
ГЛС3		1.528		1.518		
ГЛС4				1.515		
ГЛС6	1.5531	1.549	1.546	1.518	−5.1	52.2
ГЛС7	1.5609	1.554	1.551	1.516	−4.4	51.2
ГЛС8		1.560		1.538	−4.8	50.6
ГЛС9		1.527		1.542		57.8
Phosphate						
ГЛС21		1.594		1.548	−5.8	58.6
ГЛС22	1.6044	1.596	1.587	1.582	−5.7	58.7
ГЛС23		1.593		1.582		
ГЛС24		1.590		1.582		

31.3 Polarization Materials

Table 31.11 Refraction indices n of Iceland spar (I) and crystalline quartz (II) for ordinary and extraordinary rays at a temperature of 291 K [28, 75]

λ, nm	n_o		n_e (PIR)		λ, nm	n_o		n_e (PIR)	
	I	II	I	II		I	II	I	II
200	1.90284	1.57649	1.64932	1.66232	656	1.65437	1.48459	1.54190	1.55093
214			1.63039	1.64262	670	1.65367	1.48426		
250			1.60032	1.61139	707			1.54049	1.54947
303			1.57695	1.58720	766			1.53907	1.54801
312	1.71425	1.51140			768	1.64974	1.48259		
340			1.56747	1.57738	801	1.64869	1.48216		
405			1.55716	1.56671	845			1.53752	1.54640
410	1.68014	1.49640			905	1.64578	1.48098		
434	1.67552	1.49430			946	1.64480	1.48060		
436			1.55379	1.56322	1000			1.53503	1.54381
467	1.67024	1.49190			1042	1.64276	1.47985		
468			1.55103	1.56037	1080			1.53387	1.54260
480			1.55012	1.55943	1097	1.64167	1.47948		
486	1.66785	1.49074			1159	1.64051	1.47910		
508	1.66527	1.48956			1229	1.63926	1.47870		
509			1.54823	1.55747	1307	1.63789	1.47831		
533	1.66277	1.48841			1396	1.63637	1.47789		
560	1.66046	1.48736			1400			1.52972	1.53826
589	1.65835	1.48640	1.54425	1.55335	1497	1.63457	1.47744		
600			1.54378	1.55288	1530			1.52800	1.53646
628			1.54282	1.55188	1600			1.52703	1.53545
643	1.65504	1.48490			1800			1.52413	1.53242
					2172			1.51799	1.52609

Table 31.12 Temperature coefficients of the refraction indices for Iceland spar (I) and crystalline quartz (II) for ordinary and extraordinary rays at a temperature of 334 K [28, 75]

λ, nm	dn_o/dT, 10^{-5} K^{-1}		dn_e/dT, 10^{-5}, K^{-1}	
	I	II	I	II
467	0.319			
468		0.485		0.681
480		0.499		0.600
508	0.287		0.234	
509		0.514		0.616
589	0.240	0.529	0.213	0.642
643	0.211		0.190	

Table 31.13 Optical characteristics of colorless mica (muscovite) at a temperature of 291 K [29]

λ, nm	n_G	n_m	n_p	τ^*_λ ($d = 0.03$ mm)
589	1.594	1.590	1.561	
1000	1.593	1.586	1.554	0.98
1500	1.586	1.579	1.548	0.92
2000	1.575	1.572	1.544	0.98
2500	1.562	1.562	1.535	0.90
3000				0.95

Table 31.14 Refraction indices n of refraction of sodium nitrate for ordinary and extraordinary rays [29]

λ, nm	n_o	n_e (PIR)
434	1.6126	1.3404
486	1.5998	1.3384
501	1.5968	1.3379
546	1.5899	1.3365
578	1.5860	1.3363
589	1.5848	1.3360
656	1.5791	1.3347
668	1.5783	1.3345

Table 31.15 Refraction indices n of refraction of natural turmaline for ordinary and extraordinary rays at a temperature of 293 K [75]

λ, nm	n_o	n_e (PIR)
477	1.6474	1.6273
488	1.6465	1.6263
497	1.6457	1.6255
502	1.6454	1.6251
515	1.6446	1.6248
532	1.6433	1.6231
633	1.6378	1.6183
1064	1.6274	1.6088

Table 31.16 Refraction indices n of refraction of cadmium sulfide for ordinary and extraordinary rays at a temperature of 293 K [75]

λ, nm	n_o	n_e (PIR)
515	2.726	2.743
520	2.698	2.702
525	2.674	2.675
530	2.649	2.654
535	2.628	2.637
540	2.609	2.622
545	2.594	2.606
550	2.580	2.593
575	2.528	2.545
600	2.493	2.511
625	2.467	2.484
650	2.446	2.463
675	2.427	2.446
700	2.414	2.432
750	2.390	2.409
800	2.374	2.392
850	2.361	2.378
900	2.350	2.368
950	2.341	2.359
1000	2.334	2.352
1050	2.323	2.346
1100	2.324	2.340
1150	2.320	2.336
1200	2.316	2.332
1250	2.312	2.329
1300	2.309	2.326
1350	2.306	2.323
1400	2.304	2.321
1500	2.296	2.312

Table 31.17 Specific angles of polarization plane rotation $psi_0$for crystalline α-quartz at a temperature of 300 K [29, 79]

λ, nm	ψ_0, deg./mm	λ, nm	ψ_0, deg./mm
152	779.9	405	48.9
155	724.0	434	41.9
160	637.0	448	39.2
165	564.5	486	32.8
170	504.5	500	30.8
175	453.5	518	28.6
180	410.5	589	21.7
185	374.0	656	17.3
190	372.5	761	12.6
195	315.5	940	8.14
215	236.0	1342	3.89
279	114.5	2500	0.98
369	60.1	3210	0.52

Table 31.18 Relative degrees of polarization P_λ and spectral coefficients radiant transmittances τ_λ for polarizers [29]

λ, nm	Herapathite polarizer		Polyvinyl polarizer	
	P_λ, %	τ_λ	P_λ, %	τ_λ
400		0.02	100	0.26
450		0.15	100	0.37
500	98.7	0.27	100	0.40
550	99.0	0.32	100	0.37
600	99.0	0.34	100	0.35
650	98.5	0.36	100	0.38
700	98.2	0.36	100	0.42
750	91.5	0.40	100	0.45
800	67.8	0.44	98	0.50
850	41.2	0.50	72	0.59
900	28.2	0.54	32	0.80

Note: The thickness of anisotropic films in polarizers is 0.2 mm; the error of measurement is ±5%.

Table 31.19 Relative degrees of polarization P_λ for iodopolyvinylene films at a temperature of 293 K [75]

λ, μm	P_λ, % for films of the following grades			
	ПТИ–2–А	ПТИ–2–Б	ПТИ–2–В	ПТИ–2–Г
1.0	99.7	99.9	99.9	99.9
1.2	99.9	99.9	99.9	99.9
1.4	99.9	99.9	99.9	99.9
1.6	99.9	99.9	99.9	99.9
1.8	99.9	99.9	99.9	99.9
2.0	99.9	99.9	99.9	99.9
2.2	98.1	99.9	99.9	99.9
2.4	91.0	99.5	99.9	99.9
2.6	68.4	91.6	94.9	98.4
2.7	25.0	84.5	88.8	94.0

31.4 Optical Crystals and Optical Ceramics

Table 31.20 Spectral internal transmittances τ_λ^* for optical crystals of 1 mm thick for an ordinary ray at a temperature of 293 K

λ, μm	AgCl [21]	Al_2O_3[*1] [21]	BaF_2 [21]	CaF_2 [21]	CsI [33]	CuBr [21]	CuCl [21]	Ge [21]	KBr [34]	KCl [21]	KRS-5 [35]	LiF [36]	NaCl [19]	NaF [21]
1.0		0.870	0.996	0.997	0.984		0.878		0.991	0.995	0.819	0.919	0.940	
2.0	0.944	0.900	0.996	0.994	0.984	0.885	0.932	0.548	0.991	0.995	0.841	0.919	0.950	
3.0	0.949	0.910	0.996	0.993	0.984	0.914	0.932	0.700	0.991	0.994	0.846	0.926	0.950	
4.0	0.952	0.870	0.996	0.991	0.981	0.914	0.939	0.707	0.991	0.991	0.846	0.933	0.940	
5.0	0.954	0.770	0.996	0.991	0.981	0.914	0.939	0.714	0.991	0.992	0.846	0.940	0.590	
6.0	0.956	0.520	0.996	0.993	0.981	0.914	0.939	0.714	0.992	0.992	0.846	0.864	0.520	
7.0	0.956	0.500	0.996	0.990	0.981	0.914	0.939	0.707	0.992	0.993	0.846	0.753		
8.0	0.956		0.994	0.975	0.979	0.914	0.939	0.707	0.992	0.994	0.846	0.634		0.960
9.0	0.956		0.993	0.930	0.979	0.914	0.939	0.700	0.992	0.994	0.852			0.950
10.0	0.956		0.987		0.979	0.914	0.939	0.693	0.992	0.995	0.852			0.930
11.0	0.956		0.987		0.977	0.914	0.939	0.678	0.993	0.995	0.852			0.870
12.0	0.956		0.903		0.977	0.914	0.936	0.671	0.993	0.995	0.852			0.780

*1 Sapphire.

Note: The samples are polished disks. The measurement error is ±5%.

Table 31.21 Indices of an ordinary ray refraction n_o for optical crystals at a temperature of 293 K [37, 38, 75, 98–100]

λ, nm	AgCl	Al_2O_3	*1	BaF_2	CaF_2	CdTe	CsI	GaAs	Ge	KBr	KCl	KRS-5	KRS-6	LiF	MgF_2	NaCl	NaF	Si	*2	ZnSe
199					1.4964						1.7172			1.4402		1.7900				
265		1.8336														1.6368				
302		1.8135									1.5463			1.4085	1.4085	1.6068				
340					1.4477					1.6172	1.5271			1.4036		1.5861				
361		1.7945														1.5801				
405		1.7858			1.4415					1.5899	1.5100			1.3985		1.5666			1.8640	
450											1.5021			1.3960					1.8532	
486					1.4370					1.5719	1.4982			1.3948		1.5533				
500		1.7735									1.4971			1.3943		1.5520			1.8450	
546		1.7708	1.7780								1.4929			1.3930		1.5473				2.6620
577		1.7688	1.7694	1.4748								2.6276								
579		1.7687	1.7690	1.4747								2.6251			1.3780					
589		1.7678	1.7681	1.4744	1.4338					1.5600	1.4902	2.6150	2.3367	1.3920		1.5443				2.6113
600			1.7650								1.4900		2.3294	1.3918		1.5431			1.8347	
630				1.4734								2.5720			1.3770				1.8320	
633				1.4733											1.3769				1.8300	2.5902
644		1.7655		1.4730								2.5645								
656				1.4727	1.4325					1.5552	1.4878			1.3908		1.5405				2.5779
694		1.7631		1.4719	1.4319									1.3904						
700		1.7631		1.4718	1.4318						1.4860		2.2892	1.3900					1.8285	
707		1.7630		1.4718										1.3899	1.3760					
750		1.7618		1.4713	1.4311							2.5126	2.2776	1.3894						
767				1.4708	1.4309					1.5502	1.4835			1.3893		1.5366	1.3231			
800				1.4703	1.4306						1.4831		2.2660	1.3889					1.8245	
852		1.7589		1.4698								2.4745								
863				1.4697	1.4299					1.5472	1.4815			1.3883		1.5343				
894		1.7580		1.4694								2.4647								
900				1.4693	1.4297						1.4810		2.2510	1.3879					1.8222	
1000	2.0224	1.7566		1.4686	1.4289	2.8401	1.7576			1.5445	1.4799	2.4474	2.2404	1.3872		1.5322	1.3217		1.8197	2.4801
1014		1.7555		1.4685	1.4288							2.4442								
1050				1.4682	1.4286															
1060				1.4681	1.4285	2.8110				1.5430	1.4790			1.3866	1.3733	1.5314	1.3210		1.8163	
1100				1.4680	1.4283								2.2321	1.3863				3.5435		
1129		1.7534		1.4678				3.4639				2.4309								
1150					1.4280	2.7902									1.3727			3.5313	1.8160	2.4710
1200	2.0158			1.4675	1.4278		1.7530			1.5421	1.4779	2.4258	2.2265	1.3856		1.5302	1.3205	3.5213	1.8152	
1367		1.7494		1.4667				3.4101				2.4138								
1395		1.7489										2.4124						3.4920		
1400					1.4267								2.2176	1.3840					1.8121	

Table 31.21 Indices of an ordinary ray refraction n_o for optical crystals at a temperature of 293 K *(continued)*

λ, nm	AgCl	Al_2O_3	*1	BaF_2	CaF_2	CdTe	CsI	GaAs	Ge	KBr	KCl	KRS-5	KRS-6	LiF	MgF_2	NaCl	NaF	Si	*2	ZnSe
1500	2.0105	1.7465		1.4663	1.4263		1.7494			1.5399	1.4769	2.4089	2.2148	1.3832		1.5284	1.3189			
1529		1.7460		1.4661								2.4066						3.4784		
1600					1.4258								2.2124	1.3824					1.8093	
1699					1.4252								2.2103							
1709		1.7434		1.4654	1.4251							2.4011								
1800		1.7415			1.4249				4.1404				2.2086	1.3806				3.4599	1.8065	
1970		1.7383		1.4647								2.3955								
2000	2.0062	1.7375		1.4647	1.4239		1.7465	3.3510	4.1160		1.4754	2.3951	2.2059	1.3788		1.5268	1.3170	3.4514	1.8035	2.4450
2153		1.7344		1.4641																
2201									4.0900	1.5383			2.2039	1.3767					1.8004	
2325		1.7306		1.4636								2.3908								
2400					1.4217								2.2024	1.3745				3.4382	1.7970	
2500	2.0039			1.4630	1.4211		1.7451		4.0721	1.5374	1.4745	2.3903		1.3733		1.5255	1.3155			
2600					1.4205						1.4742		2.2011	1.3720					1.7935	
2800					1.4192			3.3302			1.4736		2.2001	1.3694					1.7896	
3000	2.0023	1.7115		1.4612	1.4179		1.7444	3.3101	4.0450	1.5368		2.3869	2.1990	1.3666		1.5244	1.3133		1.7855	2.4402
3200														1.3636				3.4259	1.7810	
3244		1.7044		1.4602	1.4161															
3400					1.4149	2.6902			4.0333						1.3554				1.7764	
3422		1.6982		1.4594									2.1984							
3507		1.6950									1.4729		2.1972			1.5230				
3600																			1.7713	
3707		1.6875		1.4588																
3800																			1.7659	
4000	1.9998	1.6748		1.4570	1.4097		1.7434	3.3022	4.0244	1.5357	1.4721	2.3841	2.1956	1.3495		1.5220	1.3085	3.4252	1.7602	2.4350
4255		1.6637			1.4072															
5000	1.9975	1.6179		1.4511	1.3990		1.7427		4.0151	1.5345	1.4704	2.3810	2.1928	1.3267		1.5190	1.3015	3.4220		2.4321
5146		1.6151						3.3001												
5349		1.6020		1.4487																
6000	1.9948			1.4441	1.3856		1.7421			1.5332	1.4684	2.3791	2.1900	1.2975		1.5155	1.2930	3.4202		
7000				1.4352	1.3685				4.0076		1.4660		2.1870	1.2621		1.5110				
8000	1.9885			1.4259	1.3499	2.6801	1.7409			1.5302	1.4633	2.3757	2.1839			1.5066	1.2705	3.4195		2.4180
9000									4.0040		1.4601		2.1805			1.5010				2.4130
10000	1.9803			1.4014			1.7395	3.2702	4.0025	1.5264	1.4570	2.3719	2.1767			1.4949	1.2380	3.4175		2.4071
10600				1.3932		2.6739		3.0820	4.0020	1.5250	1.4550			1.0500			1.2200	3.4173		2.4029
11000								3.0510	4.0019		1.4520		2.1723			1.4881				2.4010
12000	1.9703						1.7378		4.0018	1.5217	1.4481	2.3673	2.1674			1.4801	1.1820	3.4171		2.3941
15000	1.9511						1.7347		4.0014	1.5129	1.4321	2.3592	2.1504			1.4515				

*1 $Al_2O_3:Cr^{3+}$. *2 $Y_3Al_5O_{12}:Nd^{3+}$.

Table 31.22 Spectral radiant transmittances $\tau_{\lambda 1}$ for optical ceramics of 1 mm thick at a temperature of 293 K [20]

λ, μm	$\tau_{\lambda 1}$ for ceramics on the base of					
	Al_2O_3	CdTe	MgF_2	ZnS	ZnSe	Y_2O_3 [39]
1				0.73		
2	0.17	0.65	0.90	0.92	0.28	0.42
3	0.24	0.43	0.92	0.93	0.34	0.45
4	0.29	0.55	0.95	0.93	0.43	0.49
5	0.34	0.46	0.92	0.93	0.53	0.53
6	0.29	0.47	0.94	0.93	0.57	0.51
7	0.068	0.48	0.88	0.93	0.55	0.43
8	0.022		0.44	0.93	0.54	0.29
9			0.20	0.93	0.53	0.13
10				0.91	0.54	0.022
11				0.89	0.54	
12				0.87	0.51	
13				0.83	0.50	
14				0.73	0.50	

Note: The measurement error is ±5%.

Table 31.23 Absorptive indices a_λ for optical ceramics at a temperature of 293 K [75, 78]

λ, μm	a_λ, cm^{-1} for ceramic materials of the grades				
	ПО-2 (on the base of ZnS)	КО-2, КО-2В (on the base of ZnS)	КО-3 (on the base of CaF_2)	КО-4 (on the base of ZnSe)	КО-6 (on the base of CdTe)
3.0			0.16		
4.0			0.12		
5.0	0.40	0.08	0.06		
8.0				0.14	0.24
9.0				0.19	0.25
10.0	0.10	0.10		0.14	0.29
10.6				0.13	0.28
11.0				0.12	0.27

Table 31.24 Refraction indices n for optical ceramics at a temperature of 293 K [75, 77]

λ, μm	n for ceramic materials of the grades				
	ПО-2 (on the base of ZnS)	КО-2, КО-2В (on the base of ZnS)	КО-3 (on the base of CaF_2)	КО-4 (on the base of ZnSe)	КО-6 (on the base of CdTe)
0.26			1.4638		
0.30			1.4540		
0.40			1.4418		
0.50			1.4365		
0.5461				2.6684	
0.5876				2.6250	
0.5893			1.4338		
0.60			1.4336		
0.6563				2.5779	
0.70			1.4318		
0.80	3.3122		1.4306	2.5245	
0.90	3.3007	2.2987		2.5036	

Table 31.24 Refraction indices n for optical ceramics at a temperature of 293 K *(continued)*

λ, μm	n for ceramic materials of the grades				
	ПО-2 (on the base of ZnS)	КО-2, КО-2В (on the base of ZnS)	КО-3 (on the base of CaF_2)	КО-4 (on the base of ZnSe)	КО-6 (on the base of CdTe)
1.0	2.9221	2.2905	1.4289	2.4850	
1.06	2.2881	2.2868			
1.1				2.4796	
1.2	2.2813	2.2799		2.4720	
1.4				2.4615	
1.5				2.4575	2.7394
1.8	2.2673	2.2657		2.4502	
2.0	2.2648	2.2635	1.4239	2.4472	2.7118
2.2				2.4443	
2.4	2.2613	2.2601			
2.5				2.4414	2.6998
2.6				2.4406	
3.0			1.4179	2.4400	2.6930
3.2	2.2562	2.2552			
3.4				2.4371	
3.5				2.4355	2.6889
4.0	2.2520	2.2507	1.4097	2.4350	2.6861
4.5				2.4335	2.6839
4.6				2.4320	
5.0	2.2462	2.2451	1.4006	2.430	2.6824
5.5				2.428	2.6808
6.0	2.2397	2.2386	1.3856	2.427	2.6795
7.0	2.2319	2.2308	1.3693	2.423	2.6771
7.5			1.3596		
8.0	2.2230	2.2219	1.3499	2.418	2.6748
9.0	2.2127	2.2114	1.3268	2.414	2.6722
10.0	2.2006	2.1993	1.3002	2.408	2.6700
10.6	2.1924	2.1912		2.402	
11.0			1.2694	2.401	2.6674
12.0				2.395	2.6646
14.0				2.379	

31.5 Paint and Varnish Materials

Table 31.25 Coefficients of solar radiation absorption α_S for single-layer thermal control coats at temperatures of 293–300 K [40–44]

Type of coat	α_S
Aluminum paint applied on an oxidized aluminum substrate	0.42
Black glossy paint	0.98
Black lusterless paint	0.96–0.99
Fine-grained undercoat on a magnesium substrate	0.94
Titanium dioxide, gray	0.87
Titanium dioxide, white	0.19
White acrylic paint	0.26
White inorganic paint	0.13

Note: The measurement error is ±7%.

Table 31.26 Coefficients of thermal radiation *epsilon*$_T$ for pigmented paints applied on a smooth metallic substrate [45, 80]

Pigment base	Color	T, K	ϵ_T	Pigment base	Color	T, K	ϵ_T
$Al_2O_3 \cdot SiO_2$	White	367–1365	0.92–0.23	Cd	Yellow	293	0.28–0.33
$BaSO_4 \cdot ZnS$		300–1000	0.80–0.42	PbO		115–675	0.90–0.49
CaO		325–815	0.96–0.71	$PbCrO_4$		115–675	0.93–0.59
$MgCO_3$		115–675	0.91–0.89	Cr_2O_3	Green	115–675	0.92–0.67
MgO		115–675	0.91–0.84	Fe_2O_3	Red	115–675	0.91–0.70
$PbCO_3$		115–675	0.93–0.71	$CoO \cdot Cr_2O_3$	Black	673–973	0.74–0.77
ThO_2		117–293	0.90	$CoO \cdot Mn_2O_3$		673–1273	0.87–0.83
Y_2O		115–675	0.90–0.66	$NiO \cdot Cr_2O_3$		973	0.81
ZnO		115–675	0.95–0.91	$NiO \cdot Fe_2O_3$		973	0.80
ZrO_2		115–675	0.95–0.77	CuO		117	0.96
Co_2O_3	Sky-blue	115–675	0.94–0.86	$CoO \cdot Fe_2O_3$		973	0.80
Co_2O_3	Blue	293	0.70–0.80				

Note The measurement error is ±7%.

31.6 Metals and Alloys

Table 31.27 Spectral coefficients of thermal radiation ϵ_λ^n for light metals [46]

Metal	T, K	ϵ_λ^n at λ, μm						
		2	3	4	6	8	10	12
Aluminum	400	0.050	0.041	0.035	0.028	0.025	0.021	0.019
	500	0.054	0.045	0.038	0.032	0.027	0.024	0.021
	600	0.060	0.050	0.044	0.035	0.030	0.026	0.024
	700	0.065	0.055	0.047	0.038	0.033	0.029	0.025
	800	0.070	0.057	0.049	0.042	0.036	0.032	0.029
	900	0.074	0.061	0.053	0.043	0.038	0.034	0.031
Magnesium	400	0.063	0.052	0.044	0.036	0.031	0.027	0.024
	500	0.071	0.058	0.050	0.040	0.035	0.030	0.028
	600	0.085	0.066	0.056	0.045	0.038	0.034	0.032
	700		0.070	0.060	0.048	0.042	0.037	0.035
	800		0.076	0.065	0.053	0.045	0.040	0.037
	900		0.080	0.069	0.056	0.048	0.043	0.040

Note: The samples are polished and unoxidized. The measurement error is ±10%.

Table 31.28 Coefficients of thermal radiation ϵ_T for aluminum [20]

T, K	ϵ_T	T, K	ϵ_T	T, K	ϵ_T	T, K	ϵ_T
50	0.0080	90	0.010	200	0.018	500	0.039
60	0.0085	100	0.011	250	0.021	600	0.046
70	0.0090	120	0.012	300	0.025	700	0.054 [19]
80	0.0095	150	0.013	400	0.032	800	0.062 [19]

Note: The samples are polished. The measurement error is ±10%.

Table 31.29 Spectral coefficients of thermal radiation ϵ_λ^n for refractory metals [18, 75]

Metal	T, K	ϵ_λ^n at λ, μm											
		0.30	0.40	0.50	0.60	0.70	0.80	0.90	1.0	2.0	3.0	4.0	5.0
Chromium [24]	293	0.47	0.40	0.40	0.40	0.43	0.41	0.40	0.40	0.24	0.20	0.21	0.17
Molybdenum	293			0.475		0.480			0.405	0.125	0.082	0.073	0.060
	1000		0.458	0.438	0.417	0.394	0.367	0.333	0.302	0.106	0.063	0.046	0.035
	1200		0.448	0.429	0.410	0.389	0.363	0.333	0.306	0.130	0.081	0.061	0.049
	1600		0.432	0.415	0.397	0.378	0.358	0.333	0.312	0.165	0.108	0.068	0.084
	2000		0.419	0.403	0.387	0.370	0.352	0.333	0.317	0.191	0.133	0.084	0.068
Niobium	1600									0.175	0.151	0.126	0.110
	1800									0.190	0.170	0.145	0.125
	2000									0.205	0.177	0.155	0.135
Rhenium	1810								0.360	0.260			
	2388								0.362	0.275			
	3045								0.365	0.285			
Tantalum	1200		0.525	0.510	0.473	0.421	0.363	0.304	0.262	0.148	0.123	0.108	
	1600		0.516	0.495	0.458	0.412	0.361	0.316	0.281	0.172	0.145	0.128	
	2000		0.507	0.480	0.444	0.405	0.362	0.327	0.299	0.196	0.167	0.148	
	2400		0.498	0.464	0.432	0.399	0.366	0.338	0.317	0.220	0.190	0.168	
Titanium [24]	300									0.35	0.29	0.23	0.18
Tungsten	293									0.070	0.055		0.046
	1200	0.486	0.482	0.474	0.461	0.446	0.428	0.408	0.386	0.186	0.112	0.086	0.078
	1600	0.480	0.479	0.466	0.451	0.436	0.418	0.399	0.378	0.204	0.134	0.108	0.098
	2000	0.474	0.470	0.459	0.441	0.426	0.408	0.390	0.371	0.222	0.157	0.130	0.117
	2600	0.465	0.461	0.447	0.426	0.411	0.394	0.376	0.360	0.248	0.191	0.163	0.146
Vanadium [20]	300									0.16	0.12	0.090	0.080

Note: The samples are polished and unoxidized. The measurement error is ±10%.

Table 31.30 Coefficients of thermal radiation ϵ_T for refractory metals

T, K	V [19]	W [47, 76]	Hf [19]	Mo [18]	Nb [18]	Re [18]	Ta [18]	Ti [19]	Cr [19]	Zr [19]
300		0.032								
1000	0.145	0.105			0.116	0.164	0.132	0.227	0.353	
1100	0.161	0.128		0.105	0.127	0.173	0.141	0.239	0.360	0.204
1200	0.176	0.133	0.284	0.117	0.138	0.181	0.149	0.251	0.372	0.214
1400	0.201	0.164	0.294	0.142	0.158	0.201	0.168	0.274		0.232
1600	0.222	0.195	0.304	0.166	0.178	0.225	0.186	0.297		0.248
1800	0.241	0.223	0.314	0.192	0.195	0.245	0.205	0.316		0.261
2000	0.257	0.249	0.324	0.214	0.212	0.264	0.224			0.272
2200		0.269		0.234	0.228	0.282	0.242			
2400		0.287		0.254	0.244	0.296	0.259			
2600		0.302		0.269		0.309	0.274			
2800		0.314		0.282		0.318	0.288			
3000		0.325					0.300			
3400		0.345								

Note: The sample is polished and unoxidized. The measurement error is ±10%.

Table 31.31 Spectral coefficients of thermal radiation ϵ_λ^n for nonferrous metals and their alloys [46, 75]

Metal	T, K	ϵ_λ^n at λ, μm						
		2	3	4	6	8	10	12
Bronze	273	0.057	0.049	0.043	0.035	0.031	0.029	0.027
	293	0.059	0.051	0.045	0.037	0.033	0.031	0.029
	400	0.070	0.057	0.050	0.041	0.035	0.032	0.029
	600	0.086	0.070	0.062	0.052	0.045	0.040	0.037
	800	0.099	0.082	0.073	0.058	0.051	0.047	0.042
	1000	0.11	0.092	0.080	0.065	0.056	0.052	0.049
	1400	0.13	0.11	0.093	0.078	0.067	0.060	0.056
Cobalt	400	0.074	0.063	0.055	0.044	0.037	0.033	0.030
	500	0.085	0.072	0.062	0.052	0.044	0.039	0.036
	600	0.094	0.080	0.070	0.057	0.050	0.045	0.040
	800	0.11	0.094	0.082	0.067	0.057	0.053	0.048
	1000	0.13	0.11	0.093	0.076	0.066	0.059	0.055
	1400	0.15	0.13	0.11	0.090	0.078	0.071	0.065
	1800	0.17	0.14	0.13	0.10	0.090	0.080	0.074
Constantan [20]	293	0.16	0.14	0.12	0.098	0.085	0.076	0.070
Copper	293	0.033	0.028	0.024	0.019	0.017	0.015	0.014
	400	0.040	0.033	0.027	0.022	0.019	0.017	0.015
	500	0.044	0.037	0.032	0.025	0.022	0.019	0.017
	600	0.048	0.040	0.035	0.028	0.024	0.021	0.019
	800	0.057	0.047	0.040	0.034	0.029	0.025	0.023
	1000	0.064	0.053	0.046	0.037	0.033	0.029	0.027
	1200	0.070	0.059	0.050	0.041	0.035	0.032	0.029
Manganine [20]	293	0.16	0.13	0.11	0.093	0.080	0.072	0.065
Nichrome	400	0.23	0.19	0.17	0.14	0.12	0.11	0.10
	600	0.24	0.20	0.18	0.14	0.13	0.11	0.10
	800	0.25	0.21	0.18	0.15	0.13	0.12	0.11
	1200	0.26	0.22	0.19	0.16	0.14	0.13	0.12
	1600	0.27	0.23	0.20	0.17	0.15	0.13	0.12
Nickel	400	0.083	0.069	0.060	0.048	0.042	0.038	0.034
	500	0.095	0.079	0.069	0.056	0.048	0.043	0.040
	600	0.11	0.088	0.077	0.063	0.054	0.048	0.044
	800	0.12	0.10	0.090	0.074	0.065	0.057	0.053
	1000	0.14	0.12	0.10	0.084	0.073	0.066	0.060
	1400	0.17	0.14	0.12	0.10	0.088	0.078	0.073
	1600	0.18	0.15	0.13	0.11	0.094	0.084	0.077

Note: The samples of nichrome are trimmed; the samples of the remaining metals are polished. The measurement error is ±10%.

Table 31.32 Spectral coefficients of reflection ρ_λ for copper and nickel at temperatures of 291–295 K [28]

λ, nm	ρ_λ		λ, nm	ρ_λ		λ, nm	ρ_λ	
	Cu	Ni		Cu	Ni		Cu	Ni
300	0.250	0.440	800	0.890	0.705	2000	0.955	0.835
500	0.440	0.612	900	0.890	0.710	3000		0.884
589	0.705		1000	0.901	0.725	4000	0.973	0.918
600	0.720	0.650	1100	0.903	0.730	5000	0.968	0.940
700	0.830	0.695	1200	0.905	0.740	10000		0.955

Note: The samples are polished. The incident angle of radiation flux is 0°. The measurement error is ±5%.

Table 31.33 Coefficients of thermal radiation ϵ_T for cobalt, copper and nickel [19, 75]

T, K	ϵ_T			T, K	ϵ_T			T, K	ϵ_T		
	Co	Cu	Ni		Co	Cu	Ni		Co	Cu	Ni
100		0.014		600		0.036	0.099	1100	0.205	0.061	0.156
200		0.023		700	0.125	0.043	0.110	1200	0.225		0.168
300		0.024	0.068	800	0.148	0.050	0.120	1300			0.179
400		0.027	0.078	900	0.157	0.054	0.132	1400			0.188
500		0.031	0.088	1000	0.175	0.058	0.144	1500			0.196

The samples are polished and unoxidized. The measurement error is ±7%.

Table 31.34 Spectral coefficients of thermal radiation ϵ_λ^n for noble metals [46]

Metal	T, K	ε_λ^n at λ, μm						
		2	3	4	6	8	10	12
Gold	400	0.046	0.037	0.032	0.027	0.023	0.020	0.018
	600	0.056	0.047	0.040	0.032	0.027	0.025	0.022
	800	0.065	0.054	0.046	0.038	0.033	0.029	0.027
	1000	0.073	0.060	0.052	0.042	0.036	0.033	0.030
	1200	0.078	0.066	0.057	0.047	0.040	0.036	0.033
Iridium	400	0.072	0.060	0.052	0.042	0.037	0.033	0.030
	600	0.088	0.075	0.065	0.053	0.045	0.041	0.037
	800	0.10	0.085	0.074	0.060	0.052	0.047	0.043
	1000	0.11	0.094	0.082	0.067	0.058	0.053	0.048
	1200	0.12	0.10	0.089	0.074	0.064	0.057	0.053
	1600	0.14	0.12	0.10	0.085	0.074	0.066	0.061
	2000	0.16	0.13	0.11	0.095	0.082	0.074	0.067
	2400	0.17	0.14	0.12	0.10	0.090	0.081	0.075
	2800		0.15	0.14	0.11	0.096	0.087	0.080
Palladium	400	0.095	0.078	0.069	0.057	0.048	0.042	0.039
	600	0.11	0.095	0.083	0.072	0.060	0.052	0.050
	800	0.13	0.11	0.095	0.082	0.069	0.062	0.057
	1000	0.15	0.12	0.11	0.095	0.077	0.068	0.063
	1400	0.17	0.14	0.12	0.11	0.090	0.080	0.075
	1800		0.16	0.14	0.12	0.10	0.091	0.085
	2200		0.18	0.15	0.13	0.11	0.10	0.093
Platinum	400	0.091	0.075	0.067	0.055	0.049	0.042	0.039
	600	0.11	0.093	0.081	0.068	0.059	0.053	0.047
	800	0.13	0.11	0.093	0.077	0.069	0.061	0.057
	1000	0.15	0.12	0.11	0.086	0.075	0.070	0.063
	1400	0.17	0.14	0.12	0.10	0.088	0.079	0.071
	1800		0.16	0.14	0.11	0.10	0.090	0.081
	2000		0.17	0.15	0.12	0.11	0.095	0.087
Rhodium [20]	293	0.10	0.080	0.060	0.042	0.040	0.040	0.040
Silver [20]	293	0.025	0.020	0.020	0.020	0.020	0.015	0.013

Note: The samples are polished and unoxidized. The measurement error is ±10%

Table 31.35 Coefficients of thermal radiation ϵ_T for noble metals [18, 75]

T, K	Au	Ir	Pd	Pt	Rh	Ag
76	0.010			0.030	0.008	0.010
200	0.020					0.016
300	0.025	0.040	0.030	0.032	0.019	0.019
400	0.029					0.022
600	0.038					0.029
800	0.047					0.036
1000	0.056	0.210	0.100	0.128	0.084	0.043
1200	0.065	0.215	0.135	0.149	0.112	
1400		0.216	0.162	0.167	0.133	
1600		0.218	0.179	0.183	0.150	
1800		0.220		0.196	0.163	
2000		0.228			0.178	
2200		0.230			0.183	

Note: The samples are polished and unoxidized. The measurement error is ±7%.

Table 31.36 Spectral coefficients of transmission τ_λ and reflection ρ_λ for gold films with different thicknesses at temperatures from 291 to 295 K [48]

Thickness	τ_λ at λ, nm											
of film, nm	253.6	275.3	296.7	334.1	361.0	404.6	435.8	480.0	508.5	546.1	578.0	643.8
5	0.590	0.583	0.596	0.619	0.635	0.652	0.667	0.690	0.689	0.660	0.618	0.542
10	0.444	0.427	0.438	0.468	0.485	0.506	0.533	0.589	0.616	0.597	0.558	0.454
20	0.238	0.230	0.265	0.274	0.295	0.320	0.349	0.424	0.464	0.433	0.378	0.290
30	0.128	0.121	0.130	0.163	0.180	0.201	0.226	0.293	0.317	0.270	0.226	0.153
40	0.066	0.066	0.072	0.097	0.112	0.130	0.152	0.207	0.220	0.170	0.133	0.084
50	0.033	0.033	0.039	0.057	0.072	0.086	0.104	0.151	0.155	0.111	0.082	0.049
60	0.019	0.019	0.023	0.036	0.046	0.057	0.072	0.107	0.109	0.072	0.053	0.031
70	0.010	0.012	0.014	0.024	0.033	0.042	0.053	0.079	0.081	0.051	0.035	0.022
85	0.010	0.012	0.011	0.018	0.025	0.033	0.041	0.065	0.064	0.037	0.026	0.015
95	0.010	0.010	0.010	0.015	0.021	0.027	0.034	0.054	0.053	0.031	0.021	0.012

Thickness	ρ_λ at λ, nm											
of film, nm	253.6	275.3	296.7	334.1	361.0	404.6	435.8	480.0	508.5	546.1	578.0	643.8
5	0.130	0.135	0.133	0.126	0.129	0.122	0.117	0.109	0.108	0.120	0.139	0.172
10	0.207	0.210	0.207	0.195	0.190	0.181	0.167	0.145	0.145	0.172	0.205	0.273
20	0.272	0.303	0.301	0.289	0.285	0.278	0.259	0.223	0.231	0.300	0.366	0.478
30	0.300	0.323	0.330	0.329	0.335	0.333	0.321	0.286	0.315	0.418	0.493	0.593
40	0.269	0.301	0.316	0.322	0.342	0.352	0.345	0.324	0.369	0.489	0.567	0.660
50	0.253	0.279	0.274	0.294	0.331	0.349	0.346	0.338	0.402	0.530	0.605	0.683
60	0.232	0.259	0.265	0.276	0.320	0.339	0.339	0.338	0.414	0.550	0.622	0.674
70	0.223	0.255	0.261	0.264	0.306	0.320	0.328	0.333	0.422	0.557	0.623	0.675
85	0.218	0.248	0.250	0.253	0.294	0.315	0.322	0.332	0.426	0.569	0.627	0.668
95	0.212	0.248	0.249	0.243	0.256	0.302	0.308	0.327	0.431	0.580	0.629	0.679

Note: The films are deposited by evaporation onto quartz polished substrates at a pressure of $1.33 \cdot 10^{-3}$ Pa. The incident angle of radiant flux is 0°. The measurement error is ±2%.

Table 31.37 Spectral coefficients of reflection ρ_λ for silver films at temperatures from 291 to 295 K [20]

λ, nm	ρ_λ	λ, nm	ρ_λ	λ, nm	ρ_λ
251	0.340	338	0.555	500	0.913
288	0.212	357	0.744	589	0.950
305	0.091	385	0.814	700	0.960
316	0.042	420	0.866	1000	0.970
326	0.146	450	0.905	1100	0.975

Note: The films are deposited by evaporation onto quartz polished substrates at a pressure of $1.33 \cdot 10^{-3}$ Pa. The incident angle of radiant flux is 0°. The measurement error is ±2%.

Table 31.38 Spectral coefficients of reflection ρ_λ for aluminum, palladium and rhodium films at temperatures from 291 to 295 K [20]

λ, nm	ρ_λ			λ, nm	ρ_λ			λ, nm	ρ_λ		
	Al	Pd	Rh		Al	Pd	Rh		Al	Pd	Rh
200		0.25		650	0.82			1100	0.93	0.71	0.85
300	0.80	0.43	0.67	700	0.85	0.68	0.81	1200	0.90	0.72	0.85
400	0.93	0.55	0.75	800	0.82	0.70	0.82	2000			0.91
500	0.93	0.63	0.78	900	0.84	0.70	0.83	3000	0.91		
600	0.90	0.67	0.80	1000	0.90	0.70	0.84	4000	0.92		

Note: The films are deposited by evaporation onto glass substrates at a pressure of $1.33 \cdot 10^{-3}$ Pa. The incident angle of radiant flux is 0°. The measurement error is ±2%.

Table 31.39 Spectral coefficients of thermal radiation ϵ_λ^n for metals with low melting points [46]

Metal	T, K	ϵ_λ^n, at λ, nm						
		2	3	4	6	8	10	12
Antimony	293	0.095	0.079	0.069	0.055	0.047	0.041	0.039
	400	0.11	0.090	0.080	0.066	0.057	0.050	0.046
	500	0.12	0.10	0.088	0.073	0.062	0.055	0.050
	600	0.13	0.11	0.095	0.078	0.070	0.060	0.055
	700	0.14	0.12	0.10	0.084	0.075	0.066	0.060
	800	0.15	0.12	0.11	0.089	0.078	0.072	0.065
	900	0.16	0.13	0.11	0.095	0.084	0.076	0.070
Lead	293	0.11	0.090	0.080	0.065	0.057	0.051	0.048
	400	0.13	0.11	0.093	0.078	0.067	0.061	0.007
	600	0.16	0.13	0.12	0.096	0.084	0.077	0.069
Mercury	273–293		0.17	0.16	0.13	0.11	0.10	0.090
Tin	293	0.083	0.068	0.059	0.049	0.040	0.036	0.031
	400	0.098	0.080	0.070	0.057	0.050	0.045	0.041
	500		0.090	0.080	0.065	0.057	0.050	0.047
Zinc [24]	293	0.051	0.045	0.043	0.039	0.039	0.038	0.037
	500	0.061	0.053	0.050	0.043	0.043	0.043	0.042
	700	0.072	0.060	0.055	0.049	0.047	0.046	0.045

Note: Mercury is purified. The samples of the remaining metals are polished and unoxidized. The measurement error is ±10%.

Table 31.40 Coefficients of thermal radiation ϵ_T^n for ferrous metals [19, 20]

Metal	Condition of the radiating surface	T, K	ϵ_T^n
Cast iron	Polished	473	0.21
	Polished, oxidized at 873 K	473	0.64
		873	0.78
	Rough, oxidized at 1073 K	523	0.95
Iron	Polished	200	0.081
		300	0.101
		500	0.139
		700	0.177
		900	0.216
		1100	0.254
Steel:			
soft	Polished	290	0.10
low-carbon	Polished	300	0.10
constructional	Oxidized	300	0.47
alloyed	Electropolished	290–300	0.13
	Electropolished	400	0.16
	Electropolished	800	0.26
	Electropolished	1200	0.37
	Polished with paste	293	0.19
	Ground with powder	293	0.26–0.31
	Oxidized	300	0.70
	Oxidized	900	0.85
	Oxidized	1100	0.87
	Mirror-finished	290	0.19
	Mirror-finished	300	0.20

Note: The measurement error is ±10%.

Table 31.41 Coefficients of thermal radiation ϵ_T and solar radiation absorption α_S for thermal control coats consisting of metal and dielectric alternating layers [20]

Type of coat	Material of substrate	ϵ_T	α_S
CeO_2–Mo–CeO_2	Molybdenum	0.06	0.90
CeO_2–Mo–CeO_2	Aluminum layer applied on a glass surface	0.07	0.90
CeO_2–Mo–CeO_2	Steel	0.18	0.85
Mo–CeO_2	Aluminum layer applied on a glass surface	0.06	0.90
Mo–SiO_2	The same	0.06	0.90
Ni–SiO_2–Ni–SiO_2	Glass	0.10	0.92
SiO–Al–SiO	Glass	0.08	0.89
SiO_2–Mo–SiO_2	Molybdenum	0.08	0.85
SiO_2–Mo–SiO_2	Aluminum layer applied on a glass surface	0.08	0.85
SiO_2–Mo–SiO_2	Steel	0.15	0.85
ZnS–Al–ZnS	Glass	0.16	
ZnS–Cu–ZnS	Glass	0.11	
ZnS–Ni–ZnS	Glass	0.06	0.85

Note: The measurement error is ±7%.

31.7 Graphite Materials

Table 31.42 Spectral coefficients of reflection ρ_λ and spectral coefficients of thermal radiation ϵ_λ for graphites

λ, μm	Carbon-graphite*1 at $T = 291$ K [49]	Electrographite*2 at $T = 1460$ K [50]	Reactor graphite *3 at $T = 1860$ K [50]
	ρ_λ	ϵ_λ^n	ϵ_λ^n
0.5	0.14	0.86	0.87
1.0	0.16	0.83	0.84
1.5	0.18	0.87	0.87
2.0	0.19	0.87	0.90
2.5	0.21	0.85	0.90
3.0	0.22	0.85	0.89
3.5	0.23	0.82	0.89
4.0	0.25	0.83	0.89
4.5		0.82	0.85
5.0		0.85	0.80

*1 The samples are dense, annealed. The measurement error is ±5%.
*2 The samples are tubular. The surface is unoxidized. The measurement error is ±10%.
*3 The samples are tubular. The surface is porous, unoxidized; $d = 1700$ kg/m^3 at $T = 283$ K. The measurement error is ±10%.

Table 31.43 Coefficients of thermal radiation ϵ_T and spectral coefficients of thermal radiation ϵ_λ^n for graphites ($\lambda = 0.65 - 0.66$ μm)

T, K	Carbon-graphite*1 [49]	Graphite PB*2 [51]	Graphite ГМ3*2 [51]	Reactor graphite*3 [50]		Electrographite*4 [50]	
	ϵ_T	ϵ_T	ϵ_λ^n	ϵ_T	ϵ_λ^n	ϵ_T	ϵ_λ^n
400	0.74						
600	0.76						
800	0.76						
1000	0.77			0.87	0.91	0.86	0.84
1200	0.77	0.76	0.89	0.86	0.90	0.84	0.82
1400	0.78	0.78	0.89	0.86	0.88	0.82	0.80
1600	0.78	0.79	0.89	0.86	0.86	0.80	0.79
1800	0.78	0.80	0.88	0.86	0.85	0.78	0.77
2000	0.78	0.82	0.88	0.86	0.84	0.76	0.76
2200	0.78		0.88	0.86	0.83		
2400	0.78		0.87	0.86	0.81		
2600			0.88	0.85	0.80		
2800				0.84	0.79		

*1 The samples are dense, annealed. The measurement error is ±10%.
*2 The samples are cylindrical; the surface is polished, unoxidized. The measurement errors are ±6–10%.
*3 The samples are tubular. The surface is porous, unoxidized. The measurement errors are ±10–12%.
*4 The samples are tubular. The surface is unoxidized. The measurement error is ±10%.

Table 31.44 Spectral coefficients of reflection ρ_λ for pyrolytic graphites at a temperature of 293 K [49]

λ, nm	ρ_λ at heights of hills on a reflecting surface, μm			
	5–7.5	0.75–1.75	1–1.5	0.1
400	0.190	0.145	0.089	0.075
665	0.210	0.160	0.092	0.080
700	0.215	0.165	0.093	0.082

Table 31.45 Spectral coefficients ϵ_λ^n of thermal radiation for carbon-reinforced plastics and glass-graphites at temperatures from 291 to 293 K [52]

λ, μm	Carbon-reinforced plastic with rough surface	Polished glass-graphite
0.63	0.92	0.81
1.15	0.79	0.73
3.39	0.72	0.66
10.6	0.79	0.50

Note: The measurement error is ±8%.

Table 31.46 Coefficients of thermal radiation ϵ_T for different blacks [20]

T, K	Type of black					
	Acetylene black	Camphor black	Camphor black with water glass	Lamp black	Turpentine black	Carbon black
90		0.930	0.960			
100		0.925	0.958			
120	0.790	0.920	0.955			0.760
140	0.785		0.955			0.805
160	0.777		0.954			0.820
180	0.768		0.953			0.835
200	0.765		0.953		0.950	0.845
220	0.764		0.952		0.950	0.865
240	0.763		0.951		0.951	0.902
260	0.762		0.951	0.942	0.951	0.920
280	0.761		0.950	0.944	0.952	0.940
300	0.760		0.950	0.948	0.952	0.960

Note: The measurement error is ±10%.

31.8 Polymeric Materials

Table 31.47 Optical characteristics of polymeric materials at temperatures from 291 to 293 K [30, 75, 95]

Material	n at λ, nm									ν at $\lambda =$	τ_λ in visible	β, 10^{-5} K^{-1}
	365.0	404.7	486.1	546.1	589.3	630.0	643.9	656.3	1014	589.3 nm	spectral region	at $\lambda = 589.3$ nm
Aminoplast					1.550–1.621							−(8.5–9.0)
Celluloid					1.495–1.520							
Cellulose acetate					1.472–1.502							
Formaldehyde cresylate					1.570–1.650							
Methacrylate					1.501–1.523							
Phenol-formaldehyde					1.540–1.702							

Table 31.47 Optical characteristics of polymeric materials at 291–293 K *(continued)*

Material	n at λ, nm									ν at $\lambda =$	τ_λ in visible	β, 10^{-5} K^{-1}
	365.0	404.7	486.1	546.1	589.3	630.0	643.9	656.3	1014	589.3 nm	spectral region	at $\lambda = 589.3$ nm
Polycarbonate			1.589		1.586			1.581		30.3	0.86 (d=3 mm)	−(11.8–14.3)
Polychloro-styrene		1.651	1.628		1.614			1.608	1.595			−11.0
Polydichloro-styrene		1.651	1.628		1.614			1.608	1.596			−11.0
Polymethyl		1.505	1.496		1.491			1.488	1.482	57.8	0.89–0.92 (d=5 mm)	−(8.5–9.0)
Polystyrene	1.643	1.627	1.604	1.595	1.590	1.587	1.586	1.585	1.573		0.85–0.90 (d=2 mm)	−12.0
Silica gel with polyomethyl-methacrylate						1.466						
Silica gel with polymethyl-methacrylate						1.474						
Styrene					1.610–1.671							
Styrene-acrylonitrile copolymer			1.578		1.567			1.563		36.0	0.88 (d=3 mm)	
Styrene-methyl-methacrylate copolymer			1.592		1.579			1.574		32.2	0.90 (d=3 mm)	
Vinyl acetate					1.473							
Vinyl chloride					1.522–1.531							

Table 31.48 Spectral internal transmittances $\tau^*_{\lambda 1}$ of polymeric materials of 1 mm thick at a temperature of 293 K [20]

λ, μm	*1	*2	*3	*4	*5	*6	*7	*8	*9
0.40	0.0081–0.041	0.0185	0.208	0.563			0.0345		0.0081–0.041
0.50	0.0081–0.041	0.0206	0.208	0.723		0.843	0.0179		0.0081–0.041
0.60	0.0081–0.041	0.0206	0.208	0.723			0.0179		0.0081–0.041
0.70	0.0081–0.041	0.0206	0.249	0.706			0.0345		0.0081–0.041
0.80				0.672					
1.0					0.364	0.936			
1.5					0.364	0.940			
2.0					0.364	0.947		0.0134	
3.0					0.130	0.834		0.0625	
4.0					0.218	0.464		0.0081	
5.0					0.364	0.604		0.0625	

*1 Cellulose acetate.
*2 Cellulose acetobutyrate, white.
*3 Polyamide.
*4 Polycarbonate [53].
*5 Polymethylmethacrylate.
*6 Polytrifluorochloroethylene.
*7 Polyethylene terephthalate.
*8 Teflon (polytetrafluoroethylene).
*9 Ethylcellulose.

Note: The samples are polished disks. The measurement error is ±10%.

Table 31.49 Optical characteristics of plasticized and unplasticized organic glasses [23]

λ, μm	T, K	Plasticized glasses		Unplasticized glasses		λ, μm	T, K	Plasticized glasses		Unplasticized glasses	
		n	$\tau^*_{\lambda 1}$	n	$\tau^*_{\lambda 1}$			n	$\tau^*_{\lambda 1}$	n	$\tau^*_{\lambda 1}$
0.300	291		0.11			0.589	333	1.486	0.90	1.484	0.84
0.320	291		0.53			0.750	291		0.92		0.80
0.340	291		0.76		0.11	0.800	291		0.90		0.92
0.360	291		0.86		0.85	1.00	291		0.90		0.92
0.380	291		0.88		0.89	1.20	291		0.61		0.90
0.400	291		0.89		0.90	1.40	291		0.60		0.60
0.589	294	1.492	0.89	1.489	0.85	1.60	291		0.70		0.69
0.589	318	1.488	0.90	1.486	0.84	2.00	291		0.47		0.45

Table 31.50 Spectral internal transmittances τ^*_λ for the polymeric materials subjected to ultraviolet irradiation [20]

λ, μm	Polycarbonate		Polyethyleneterephthalate	
	before exposure	after exposure	before exposure	after exposure
0.40	0.87	0.63	0.84	0.28
0.45	0.87	0.68	0.85	0.32
0.50	0.88	0.70	0.86	0.38
0.60	0.89	0.76	0.87	0.50
0.70	0.90	0.79	0.87	0.57
0.80	0.90	0.83	0.88	0.62
1.0	0.90	0.84	0.89	0.66

Note: The irradiation simulates the effect of the extraterrestrial solar radiation for 100 h. The samples are polished plates of 30 μm in thickness. The measurements are carried out at 293 K. The measurement error is ±5%.

31.9 Building Materials

Table 31.51 Coefficients of solar radiation absorption α_S for building materials [20]

Material	α_S	Material	α_S	Material	α_S
Alabaster gypsum	0.31	colored with cement paints:		rose	0.52
Asbestos cement:		sky-blue	0.67	light-blue	0.53
grey	0.66	rose	0.56	light-yellow	0.48
white	0.41	light-green	0.66	silicate	0.78
Asbestos slate	0.75	dark-green	0.68	dark-gray	0.75
Brick:		Fiber glass	0.56	cement of the composition:	
white	0.33	Fiber-glass fabric	0.42	1:3	0.66
red	0.48	Foam-plastic insulator,	0.16	1:5	0.64
Cement fibrolite	0.79	Gypsum unpolished	0.25	Porcelain	0.52
Concrete:		Marble:		Roof tile:	
colorless	0.55	white	0.42	brown	0.74
colored with silicate paints:		dark	0.68	red	0.67
white	0.35	Plastering:		Rubberoid	0.93
green	0.59	calcium	0.50	Splint-slab extruded	0.43
sky-blue	0.48	white calcium	0.30	Wood-fiber slab uncolored	0.68
dark green	0.74	adhesive	0.40	and unfinished	

Note: The temperatures of measurements are from 290 to 300 K. The measurement error is ±10%.

Table 31.52 Spectral reflection coefficients ρ_λ for building materials at a temperature of 293 K [20]

Material	ρ_λ for λ, μm										
	0.50	0.75	1.0	2.0	3.0	4.0	5.0	6.0	7.0	8.0	9.0
Concrete:											
unpainted	0.37	0.43	0.42	0.30	0.05	0.06	0.07	0.07	0.32	0.16	0.19
painted with cement blue paint	0.36	0.41	0.51								
painted with cement green paint	0.26	0.28	0.43								
Dry sand cleaned			0.40	0.50	0.54						
Gypsum plastering	0.82	0.85	0.84	0.52	0.08	0.37	0.25	0.12	0.17	0.08	0.06
Porcelain:											
glazed	0.43		0.30	0.25	0.16	0.15	0.03	0.03	0.03	0.05	0.05
unglazed	0.53		0.38	0.33	0.20	0.24	0.08	0.08	0.05	0.10	0.50
White chamotte	0.92		0.95	0.90	0.40	0.62	0.25	0.12	0.08	0.16	0.42
Wood-fiber slabs:											
pressed with polyvinylchloride film	0.08	0.11	0.25								
painted with white enamel	0.71	0.72	0.75								

Note: The measurement error is ±10%.

Table 31.53 Coefficients of thermal radiation ϵ_T^n for building materials [55]

Material	T, K	ϵ_T^n	Material	T, K	ϵ_T^n
Asbestos	293	0.96	Concrete	293	0.92
Asbestos slate	293	0.96	Gray marble polished	293	0.932
Asphalt	298–303	0.95	Gypsum unpolished	293	0.903
Brick:			Roofing paper	293	0.91–0.93
chamotte	293	0.85	Rough calcium plastering	283–363	0.91
	1273	0.75	Rubberoid	293	0.93
	1503	0.59	Window glass:		
fireproof	773–1273	0.65–0.75	lusterless	293	0.96
red unpolished	293	0.932	polished	298–303	0.91
Cement	300	0.54	Wood planed	293	0.8–0.9
Clay roasted	303	0.91	Wood sawdust (coniferous)	298–303	0.96

31.10 Oxides, Borides, Carbides and Nitrides of Refractory Metals

Table 31.54 Coefficients of thermal radiation ϵ_T for refractory oxides [51]

T, K	Al_2O_3	MgO	ZrO_2
1200	0.35	0.33	
1300	0.36	0.28	0.39
1400	0.37	0.28	0.42
1500	0.38	0.30	0.45
1600	0.39	0.34	0.47
1700	0.40	0.38	0.48
1800		0.43	0.49

Note: The measurement error is ±10%.

Table 31.55 Spectral coefficients of thermal radiation ϵ_λ^n for metal oxides and alloys at a wavelength of 0.65 μm [56]

Oxidized material	ϵ_λ^n	Oxidized material	ϵ_λ^n	Oxidized material	ϵ_λ^n
Alumel	0.87	Chromium	0.70	Steel:	
Aluminum	0.30	Cobalt	0.75	carbon	0.85
Beryllium	0.35	Constantan	0.84	stainless	0.80
Cast iron	0.70	Copper	0.70	Thorium	0.50
Chromel:		Iron	0.70	Titanium	0.50
90 Ni/10 Cr	0.87	Magnesium	0.20	Uranium	0.30
80 Ni/20 Cr	0.90	Nickel	0.90	Vanadium	0.70
60 Ni/24 Fe/16 Cr	0.83	Niobium	0.70	Yttrium	0.60
				Zirconium	0.40

Note: The samples are disks with polished and oxidized surfaces. The measurement errors are 10-15%.

Table 31.56 Coefficients of thermal radiation ϵ_T for borides, carbides and nitrides of refractory and rare-earth metals [18]

Material	ϵ_T at a temperature, K									
	1100	1300	1500	1700	1900	2100	2300	2500	2700	2900
Borides:										
gadolinium	0.61	0.62	0.62	0.63	0.64					
hafnium	0.85	0.87	0.89	0.92	0.94					
lanthanum	0.68	0.69	0.69	0.70	0.71					
neodymium	0.56	0.56	0.58	0.58	0.59					
samarium	0.71	0.70	0.69	0.68	0.67					
yttrium	0.63	0.65	0.66	0.67	0.68					
zirconium	0.86	0.88	0.91	0.91	0.95					
Carbides:										
boron	0.84	0.85	0.86	0.87	0.88					
niobium	0.41	0.42	0.43	0.45	0.43	0.44	0.45	0.46	0.47	0.48
tantalum			0.20	0.22	0.24	0.26	0.28	0.30	0.32	
titanium	0.85	0.86	0.87	0.87	0.89					
tungsten			0.20	0.22	0.24	0.26	0.28	0.30	0.32	
zirconium		0.81	0.79	0.77	0.74	0.72	0.70	0.68	0.66	0.64
Boron nitride	0.58	0.59	0.60	0.60	0.60					

Note: The measurement error is ±10%.

31.11 Natural Minerals and Solid Covers

Table 31.57 Refraction index n for white diamonds within the visible spectral range [54, 75]

λ, nm	226	400	450	500	550	600	650	700	750	760
n	2.715	2.465	2.446	2.433	2.423	2.416	2.411	2.407	2.404	2.402

Table 31.58 Refraction indices n for natural precious and semi-precious stones at a wavelength of 589.3 nm [54]

Material	n_o	n_e	n_1–n_2*1 [56]
Beryl	1.571–1.599	1.566–1.590	
Emerald	1.588–1.595	1.581–1.588	0.014
Rock crystal	1.544	1.553	0.013
Sapphire:			
green	1.770–1.779	1.762–1.770	0.018
ruby	1.768–1.778	1.760–1.769	
white	1.768–1.771	1.759–1.761	
Tourmaline	1.669	1.638	0.017
White diamond	2.417		0.044
Material	n_p	n_m	n_G
Malachite		1.66	1.91
Topaz	1.630	1.631	1.638
Turquoise		1.62	1.65

*1 The n_1 values are shown for λ=687.0 μm, and those for n_2 are shown for λ=430.8 μm.

Table 31.59 Spectral coefficients of reflection ρ_λ for various soils [19]

λ, μm	Soapy	Earth			Yellow
	clay	yellow	brown	red	sand
0.4	0.08	0.08	0.08	0.06	0.15
0.5	0.12	0.18	0.12	0.07	0.27
0.6	0.17	0.32	0.17	0.18	0.36
0.7	0.20	0.53	0.20	0.28	0.44
0.8	0.18	0.67	0.21	0.30	0.50
0.9	0.17	0.76	0.23	0.33	0.54
1.0	0.20	0.81	0.20		0.58

Note: For dry loam and dry black soil the mean ρ_λ values are 0.15 and 0.7, respectively, within the wavelength range from 0.4 to 0.7 μm.

Table 31.60 Spectral coefficients of reflection ρ_λ for vegetations of various types [19]

λ, μm	Leaves			Straw	Grass	
	fresh green	dry yellow	dry brown		fresh	dry
0.4	0.10	0.04	0.04	0.10	0.04	0.10
0.5	0.21	0.08	0.08	0.20	0.08	0.18
0.6	0.32	0.22	0.15	0.27	0.19	0.27
0.7	0.40	0.37	0.23	0.33	0.20	0.30
0.8	0.49	0.43	0.38	0.38	0.50	0.35
0.9	0.55	0.48	0.52	0.44	0.69	0.40
1.0	0.58	0.50	0.60	0.48	0.76	0.43

Note: For coniferous needles the mean ρ_λ value is 0.30 within the wavelength range from 0.7 to 1 μm.

31.12 Moon Grounds

Table 31.61 Coefficients of thermal radiation ϵ_T for moon grounds from the Mare Foecunditatis region [58]

T, K	90	120	160	200	240	270	300
ϵ_T	0.9764	0.9763	0.9758	0.9743	0.9706	0.9660	0.9603

Note: The samples have the bulk density of 1900 kg/m^3; the surface of the bulk material is smoothed. The measurement error is $\pm$1%.

Table 31.62 Spectral coefficients of thermal radiation ϵ_λ^n for moon grounds from the Mare Foecunditatis region [58, 59]

λ, μm	ϵ_λ^n		
	I	II	III
2.5	0.82	0.78	0.73
4.0	0.77	0.75	0.72
6.0	0.98	0.86	0.84
8.0	0.97	0.97	0.97
10	0.97	0.97	0.97
12	0.97	0.97	0.97
14	0.98	0.98	0.98

Note: The samples have the following bulk densities, kg/m^3: I — 1400; II — 1600–1700; III — 1900; the surface of bulk material is smoothed.

Table 31.63 Spectral coefficients of diffuse reflection ρ_λ for regolith from different regions of the Moon at temperatures from 293 to 300 K [60]

λ, μm	I	II	III	IV	V
0.30	0.080	0.085	0.075	0.088	0.069
0.50	0.110	0.112	0.095	0.100	0.080
0.70	0.132	0.138	0.112	0.117	0.100
0.90	0.148	0.151	0.131	0.124	0.106
1.2	0.145	0.145	0.117	0.124	0.102
1.5	0.164	0.160	0.140	0.130	0.120
1.8				0.155	

Note: The samples are fine-grained powder; the surface of bulk material is smoothed. The materials are taken from the following regions:
I — The Oceanus Procellarum from the depth of 0.16 m;
II — The Oceanus Procellarum from the depth of 0.33 m;
III — The Mare Tranquillitatis from the depth of 0.10–0.11 m;
IV — The Mare Foecunditatis from the depth of 0.08 m;
V — The Mare Tranquillitatis from the depth of 0.015 m.
The measurement error is ±5%.

31.13 Interferometric Films

Table 31.64 Refraction indices n for interferometric coats [75]

Material	λ, μm	T, K	n	Material	λ, μm	T, K	n	Material	λ, μm	T, K	n
AlF_3	0.55	293	1.38–1.40	MgF_2	0.55	293	1.38	Si	3.0	293	3.42
Al_2O_3	0.55	313	1.59	MgO	0.55	573	1.70	SiO	0.55	293	1.50–1.60
		573	1.63	Na_3AlF_6	0.55	293	1.35	SiO_2	0.55	293	1.47
BiF_3	1.0	293	1.74		0.63		1.34	SrF_3	0.55	293	1.43–1.44
	11.0		1.65	NaF	0.55	293	1.30	Ta_2O_5	0.55	293	2.10–2.30
CaF_2	0.55	293	1.23–1.46	NdF_3	0.55	573	1.61	ThF_4	10.6	293	1.85
CdS	0.55	293	2.40–2.51	Nd_2O_3	0.55	293	1.79	ThO_2	2.2	523	1.86
CdSe	0.55	293	2.40–2.50			533	2.15	TiO	0.55	293	2.20–2.40
CdTe	3.0	293	2.61–3.10	PbF_2	0.30		1.98	TiO_2	0.55	293	2.20–2.40
CeF_3	0.55	573	1.63		0.35	293	1.80	Ti_2O_5	0.55	293	2.20–2.40
GdO_3	0.55	293	1.80		0.55		1.75	Y_2O_3	0.55	523	1.87
Ge	2.0	303	4.01	PbO	0.55	293	2.60	ZnS	0.55	308	2.30
HfO_2	0.55	293	2.02	PbTe	0.55	293	5.60	ZnSe	0.633	293	2.58–2.60
InAs	1.0	293	4.51		5.0		5.00		10.6		2.40
LaF_3	0.55	293	1.59	Sb_2O_3	0.55	293	2.05	ZnTe	1.0	293	2.80
La_2O_3	0.33	293	1.98	Sc_2O_3	0.30	293	1.90	ZrO_2	0.40	293	2.11
LiF	0.55	293	1.40		0.90		1.86		0.55		2.05

31.14 Liquids, Solidified Materials

Table 31.65 Refraction indices n of refraction for some liquids at temperatures from 293 to 298 K

Liquid	λ, nm	n	Ref.	Liquid	λ, nm	n	Ref.
Acetone	546.1	1.3576	[63]	Cyclohexane	546.1	1.4260	[63]
	632.8	1.3542	[63]		632.8	1.4224	[63]
Acid:				Diethyl ether	589.3	1.3526	[66]
acetic	589.3	1.3720	[66]	Ethylsalicylate	589.3	1.523	[62]
acrylic	589.3	1.4224	[30]	Eutenol	589.3	1.540	[58]
hydrochloric	589.3	1.2540	[66]	Glycerin	589.3	1.4370	[64]
isovalerian	589.3	1.4085	[66]	n-Hexane	546.1	1.3742	[63]
methylacetic	589.3	1.4051	[66]		632.8	1.3711	[63]
sulfuric	589.3	1.4290	[66]	Iodomethylene	589.3	1.7559	[64]
valerian	589.3	1.4085	[66]	Methylacetate	589.3	1.450	[61]
vinylacetic	589.3	1.4257	[66]	Methylene iodide	589.3	1.737	[61]
Alcohol:				Methylsalicylate	589.3	1.538	[61]
amyl	589.3	1.4053	[66]	α-Monobromonaphthalene	589.3	1.6588	[60]
ethyl	546.1	1.3612	[63]	Nitrobenzene	546.1	1.5579	[57]
	589.3	1.3611	[66]		589.3	1.55257	[51]
	632.8	1.3583	[63]		632.8	1.5458	[63]
isopropyl	546.1	1.3757	[63]	Oil:			
	632.8	1.3726	[63]	anisic	589.3	1.560	[58]
methyl	546.1	1.3280	[63]	castor	589.3	1.480	[56]
	589.3	1.3265	[56]	cedar	589.3	1.515	[58]
	632.8	1.3253	[63]	cinnamon	589.3	1.602	[58]
Aniline	589.3	1.586	[62]	clove	589.3	1.538	[27]
Benzene	546.1	1.5030	[63]	linseed	589.3	1.485	[29]
	589.3	1.5014	[64]	olive	589.3	1.467	[58]
	632.8	1.4950	[63]	olive oil (high grade)	589.3	1.460	[62]
Benzylbenzoate	589.3	1.568	[62]	paraffin	589.3	1.440	[27]
Bromoform	589.3	1.5980	[64]	poppyseed	589.3	1.463	[29]
Carbon disulfide	546.1	1.6347	[63]	sandal	589.3	1.508	[58]
	589.3	1.620	[58]	sunflower seed	589.3	1.470	[56]
	632.8	1.6185	[67]	turpentine	589.3	1.470	[27]
Carbon tetrachloride	546.1	1.4613	[63]	Paraffin (liquid)	589.3	1.480	[62]
	589.3	1.4601	[58]	Paraldehyde	589.3	1.405	[62]
	632.8	1.4547	[63]	Quinoline	589.3	1.627	[62]
Chlorobenzene	589.3	1.525	[62]	Toluene	546.1	1.4986	[63]
Chloroform	546.1	1.4477	[63]		589.3	1.4980	[56]
	589.3	1.4455	[66]		632.8	1.4901	[63]
	632.8	1.4435	[63]	Turpentine	589.3	1.470	[56]
Cinnamon ethyl	589.3	1.559	[62]	Water + glycerin (1:1)	589.3	1.3981	[65]

Table 31.66 Refraction indices n for distilled water at 293 K [67]

λ, nm	250.0	308.0	359.0	400.0	434.0	486.0	546.1	589.3	632.8	768.0	1000	1250	2000	2600
n	1.3773	1.3569	1.3480	1.3433	1.3403	1.3371	1.3341	1.3330	1.3314	1.3289	1.3247	1.3210	1.290	1.252
λ, nm	3000	3500	3900	4600	5000	6040	7000	8600	10000	11000	12000	12600	13500	14000
n	1.446	1.423	1.353	1.380	1.331	1.312	1.330	1.282	1.212	1.140	1.165	1.280	1.330	1.309

Table 31.67 Spectral absorptive indices a_λ for clear natural waters [67]

λ, nm	390	410	430	450	470	490	510	530	550	570	590	610	650	690
a_λ, m^{-1}	0.038	0.037	0.036	0.037	0.039	0.042	0.054	0.062	0.074	0.094	0.16	0.26	0.38	0.54

Table 31.68 Spectral absorptive indices a_λ, m^{-1} of sea water from different water areas [67]

λ, nm	The Atlantic Ocean	The Pacific Ocean	The Baltic Sea	λ, nm	The Atlantic Ocean	The Pacific Ocean	The Baltic Sea
390	0.032	0.12	2.7	530	0.030	0.060	0.39
410	0.034	0.15	1.9	550	0.034	0.067	0.30
430	0.021	0.15	1.2	570	0.055	0.085	0.25
450	0.018	0.13	0.90	590	0.14	0.17	0.28
470	0.014	0.087	0.83	610	0.24	0.26	0.37
490	0.012	0.064	0.62	650	0.33	0.36	0.41
510	0.018	0.060	0.44	690	0.52	0.56	0.62

Note: The ocean water is taken from the regions of north tradestreams, the water of the Baltic Sea is taken from the area of the Gulf of Riga.
The Caribbean Sea: a_λ (λ=430 nm) = 0.042 m^{-1},
The Sargasso Sea: a_λ (λ=490 nm) = 0.037 m^{-1},
The Mediterranean Sea: a_λ (λ=490 nm) = 0.046 m^{-1},
The Black Sea: a_λ (λ=490 nm) = 0.069 m^{-1}.

Table 31.69 Refraction indices n for optical adhesives and resins for λ=589.3 nm at a temperature of 293 K

Material	n	Material	n	Material	n
Adhesive [68]:				Resin [54]:	
acrylic	1.4861	ОК-72Ф	1.586	dammar	1.515
fir balsam	1.530	ОК-90ПЛ	1.5411	synthetic casein	1.550
balsamic	1.5191	УФ-235М:		copal	1.540
ОК-50П	1.5801	liquid	1.456	orange-colored shellac	1.516
ОК-60	1.5151	solidified	1.462	amber	1.539–1.545

31.15 Gases and Plasma

Table 31.70 Refraction indices n for air at a pressure of 10^5 Pa and a temperature of 288 K [62]

λ, nm	n	λ, nm	n	λ, nm	n	λ, nm	n	λ, nm	n	λ, nm	n
200	1.0003240	250	1.0003012	300	1.0002915	400	1.0002827	600	1.0002769	1400	1.0002734
205	1.0003205	255	1.0002999	310	1.0002901	420	1.0002817	650	1.0002763	1600	1.0002732
210	1.0003175	260	1.0002987	320	1.0002889	440	1.0002809	700	1.0002757	1800	1.0002731
215	1.0003147	265	1.0002976	330	1.0002878	460	1.0002802	750	1.0002753	2000	1.0002730
220	1.0003121	270	1.0002965	340	1.0002868	480	1.0002795	800	1.0002750	2500	1.0002729
225	1.0003098	275	1.0002955	350	1.0002860	500	1.0002789	850	1.0002747	3000	1.0002728
230	1.0003077	280	1.0002946	360	1.0002852	520	1.0002784	900	1.0002745	5000	1.0002727
235	1.0003058	285	1.0002938	370	1.0002845	540	1.0002780	950	1.0002743	7000	1.0002726
240	1.0003041	290	1.0002930	380	1.0002839	560	1.0002776	1000	1.0002741	10000	1.0002726
245	1.0003026	295	1.0002922	390	1.0002833	580	1.0002772	1200	1.0002737	20000–50000	1.0002726

Table 31.71 Refraction indices n of some gases under normal conditions [62]

Gas	λ, nm	n
Acetylene	589.3	1.000606
Ammonia	589.3	1.000375
Argon [63]	546.1	1.0002630
	632.8	1.0002618
Bromine	589.3	1.001125
Carbon dioxide [63]	546.1	1.0004197
	632.8	1.0004174
Carbon oxide [73]	589.3	1.000334
Chlorine	589.3	1.000768
Deuterium	589.3	1.000137
Fluorine	589.3	1.000195
Helium [70]	447.3	1.0002753
	471.5	1.0002745
	492.3	1.0002738
	501.7	1.0002736
	587.7	1.0002719
Hydrogen	589.3	1.000139
Hydrogen bromide	589.3	1.000570
Hydrogen chloride	589.3	1.000444
Hydrogen iodide	589.3	1.000906
Hydrogen sulfide	589.3	1.000619
Krypton (natural) [70]	450.4	1.0002752
	556.4	1.0002724
	565.1	1.0002722
	587.3	1.0002719
	605.8	1.0002716
	645.8	1.0002711
Krypton (^{86}Kr isotope)[71]	556.4	1.0002724
	587.3	1.0002717
	605.8	1.0002716
	760.4	1.0002700
	851.1	1.0002695
	877.9	1.0002694
	893.1	1.0002693
	975.4	1.0002690
Methane	589.3	1.000441
Methyl chloride	589.3	1.000865
Methyl fluoride	589.3	1.000449
Neon [72]	585.4	1.0002719
	607.4	1.0002716
	614.5	1.0002715
	640.4	1.0002711
Nitrogen [63]	546.1	1.0002793
	632.8	1.0002781
Nitrogen lower oxide	589.3	1.000515
Nitrogen oxide	589.3	1.000297
Oxygen [63]	546.1	1.0002531
	632.8	1.0002516
Ozone [63]	589.3	1.000511
Sulfur dioxide	589.3	1.000660
Sulfur hexafluoride	589.3	1.000783
Sulfur trioxide	589.3	1.000737
Tellurium tetrachloride	589.3	1.002600
Water vapor [60]	546.1	1.0002354
	632.8	1.0002337
Xenon	589.3	1.000702

Note: The measurement errors are $\pm(5\text{–}7)\cdot10^{-8}$.

Table 31.72 Refraction indices n for some gases at a liquefaction temperature [74]

Substances	T, K	λ, μm	n
Helium:			
He I	4.2	0.546	1.0206
He II	2.18	0.546	1.0269
Hydrogen	20.4	0.589	1.0974
Nitrogen	78.0	0.589	1.205
Oxygen	92.0	0.589	1.221

Table 31.73 Spectral coefficients α_λ of oxygen absorption in an ultraviolet spectral region [69]

λ, μm	α_λ	λ, μm	α_λ
0.220	0.35	0.250	0.14
0.225	0.29	0.255	0.10
0.230	0.26	0.260	0.09
0.235	0.22	0.265	0.07
0.240	0.19	0.270	0.04
0.245	0.15	0.275	0.02

Note: The temperature is 3000 K, the layer thickness is $5\cdot10^{-2}$ m.

Table 31.74 Thermal radiation coefficients ϵ_T of air at different temperatures and pressures [69]

T, K	ϵ_T at a pressure, Pa			
	10^4	10^5	10^6	$5\cdot10^6$
2000	$2.89\cdot10^{-5}$	$2.91\cdot10^{-4}$	$2.88\cdot10^{-3}$	$1.41\cdot10^{-2}$
3000	$1.58\cdot10^{-5}$	$2.12\cdot10^{-4}$	$2.74\cdot10^{-3}$	$1.66\cdot10^{-2}$
4000	$9.42\cdot10^{-6}$	$2.73\cdot10^{-4}$	$4.30\cdot10^{-3}$	$2.61\cdot10^{-2}$
5000	$1.86\cdot10^{-5}$	$5.93\cdot10^{-4}$	$1.07\cdot10^{-2}$	$5.04\cdot10^{-2}$
6000	$3.02\cdot10^{-5}$	$1.02\cdot10^{-3}$	$2.38\cdot10^{-2}$	$1.33\cdot10^{-1}$
7000	$4.06\cdot10^{-5}$	$1.18\cdot10^{-3}$	$3.51\cdot10^{-2}$	$1.99\cdot10^{-1}$
8000	$1.10\cdot10^{-4}$	$1.46\cdot10^{-3}$	$4.00\cdot10^{-2}$	$2.88\cdot10^{-1}$
9000	$4.14\cdot10^{-4}$	$3.16\cdot10^{-3}$	$5.22\cdot10^{-2}$	$3.52\cdot10^{-1}$
10000	$1.19\cdot10^{-3}$	$7.29\cdot10^{-3}$	$6.89\cdot10^{-2}$	$4.05\cdot10^{-1}$
11000	$2.64\cdot10^{-3}$	$1.39\cdot10^{-2}$	$9.53\cdot10^{-2}$	$4.68\cdot10^{-1}$
12000	$4.39\cdot10^{-3}$	$2.63\cdot10^{-2}$	$1.40\cdot10^{-1}$	$5.09\cdot10^{-1}$
13000	$5.51\cdot10^{-3}$	$4.02\cdot10^{-2}$	$1.63\cdot10^{-1}$	$4.75\cdot10^{-1}$
14000	$5.42\cdot10^{-3}$	$5.25\cdot10^{-2}$	$2.18\cdot10^{-1}$	$6.58\cdot10^{-1}$
15000	$4.52\cdot10^{-3}$	$6.17\cdot10^{-2}$	$2.64\cdot10^{-1}$	$7.34\cdot10^{-1}$
16000	$3.58\cdot10^{-3}$	$6.82\cdot10^{-2}$	$3.14\cdot10^{-1}$	$8.54\cdot10^{-1}$
17000	$2.76\cdot10^{-3}$	$7.04\cdot10^{-2}$	$3.49\cdot10^{-1}$	$8.23\cdot10^{-1}$
18000	$2.24\cdot10^{-3}$	$6.85\cdot10^{-2}$	$3.62\cdot10^{-1}$	$8.56\cdot10^{-1}$
19000	$1.79\cdot10^{-3}$	$6.27\cdot10^{-2}$	$3.65\cdot10^{-1}$	$8.27\cdot10^{-1}$
20000	$1.53\cdot10^{-3}$	$5.37\cdot10^{-2}$	$3.65\cdot10^{-1}$	$8.63\cdot10^{-1}$

Note: The radius of a hemispherical air layer is 0.1 m. The measurement error is ±15%.

Table 31.75 Thermal radiation coefficients ϵ_T of carbon dioxide at different temperatures and pressures [69]

T, K	ϵ_T at a pressure, Pa		
	10^3	10^4	10^5
2000	$4.5 \cdot 10^{-3}$	$3.6 \cdot 10^{-2}$	$8.8 \cdot 10^{-2}$
3000	$1.2 \cdot 10^{-4}$	$3.0 \cdot 10^{-3}$	$3.2 \cdot 10^{-2}$
4000	$2.2 \cdot 10^{-4}$	$2.1 \cdot 10^{-4}$	$3.0 \cdot 10^{-3}$
5000	$2.6 \cdot 10^{-5}$	$2.2 \cdot 10^{-4}$	$1.4 \cdot 10^{-3}$
6000	$6.1 \cdot 10^{-5}$	$8.3 \cdot 10^{-4}$	$3.1 \cdot 10^{-3}$

Note: The radius of a hemispherical gas layer is 0.1 m.

Table 31.76 Coefficients of thermal radiation ϵ_T for hydrogen plasma [69]

T, K	ϵ_T at a pressure, 10^6 Pa						
	20	15	10	7	4	2	1
9000	0.46	0.37	0.27	0.20	0.11	0.05	0.03
10000	0.73	0.65	0.51	0.40	0.26	0.15	0.09
11000		0.85	0.75	0.66	0.50	0.31	0.20
12000			0.92	0.85	0.70	0.51	0.38

Note The radius of a hemispherical plasma layer is 0.5 m. The measurement error is ±20%.

31.16 Optical (Radiant) Strength

Table 31.77 Optical strength of nonmetallic materials [75, 83–90, 92–94]

Material	Irradiator performances		Surface irradiated,	P		W	
	λ, μm	τ^{*1}, s	cm^2	GW/cm^2	J/cm^2	GW/cm^2	J/cm^2
Ammonium deuterophosphate ($NH_4D_2PO_4$)	1.064			$5 \cdot 10^{-1}$			
Ammonium dihydrophosphate ($NH_4H_2PO_4$)	1.064	$12 \cdot 10^{-9}$	$1.3 \cdot 10^{-5}$	$5 \cdot 10^{-1}$		$8.9 \cdot 10^{-1}$	
Barium fluoride (BaF_2)	10.6	$(2.5–3.0) \cdot 10^{-6}$		$(4.0–5.0) \cdot 10^{-2}$			
Cadmium telluride (CdTe)	10.6	$2 \cdot 10^{-9}$			2		
Calcium fluoride (CaF_2)	0.248	$7 \cdot 10^{-8}$	$2.9 \cdot 10^{-4}$			1	
	0.248	$8 \cdot 10^{-9}$	$7.9 \cdot 10^{-3}$		70		
	1.064	$8 \cdot 10^{-9}$				240–280	
Diamond (C)	1.064	$3 \cdot 10^{-8}$	$2.8 \cdot 10^{-7}$	20		160	
Gallium arsenide (GaAs)	0.337	$7 \cdot 10^{-9}$	10^{-6}	10^{-1}			
	10.6	$6 \cdot 10^{-8}$		$3 \cdot 10^{-5}$		10–14	
Germanium (Ge)	10.6	$(2.5–3.0) \cdot 10^{-6}$			30–50	$(2.7–3.5) \cdot 10^{-1}$	
Lithium fluoride (LiF)	0.248	$7 \cdot 10^{-8}$	$8 \cdot 10^{-3}$			300	
	0.6943	$8 \cdot 10^{-9}$				350	
	1.064	$8 \cdot 10^{-9}$				360	
Lithium triborate (LiB_3O_5)	1.064	10^{-7}		25			
Neodymium glass ГЛС22 (phosphate)	1.064	10^{-8}				2	
Neodymium glass ГЛС7 (silicate)	1.064	$5 \cdot 10^{-8}$	$5 \cdot 10^{-4}$	7			
Optical glass K8 (crown)	1.064	$(1.2–3.0) \cdot 10^{-9}$	$2.8 \cdot 10^{-7}$	28		200	
Polymethylmethacrylate	1.064	$12 \cdot 10^{-9}$				40	
Potassium deuterophosphate (KD_2PO_4)	1.064			$5 \cdot 10^{-1}$			
Potassium dihydrophosphate (KH_2PO_4)	1.064	$1.5 \cdot 10^{-3}$			300–700		2100
Potassium chloride (KCl)	0.6943	$8 \cdot 10^{-9}$				80	
	1.064	$8 \cdot 10^{-9}$				70	
	10.6	$(2.5–3.0) \cdot 10^{-6}$	$4 \cdot 10^{-4}$	$(4.0–6.0) \cdot 10^{-2}$			

Table 31.77 Optical strength of nonmetallic materials *(continued)*

Material	Irradiator performances		Surface irradiated,	P		W	
	λ, μm	τ^{*1}, s	cm^2	GW/cm^2	J/cm^2	GW/cm^2	J/cm^2
Quartz crystalline (SiO_2)	0.6943	$7 \cdot 10^{-8}$				230	
Quartzoid glass with polyacrylostyrene	1.064	$12 \cdot 10^{-9}$				60	
Quartzoid glass with polymethylmethacrylate	1.064	$12 \cdot 10^{-9}$				150	
Rubin (Al_2O_3 :Cr^{3+})	0.6943	$2 \cdot 10^{-8}$			80		
Sapphire white (Al_2O_3, Ti)	1.064	$(1.0–3.0) \cdot 10^{-8}$	$2.8 \cdot 10^{-7}$	60		500–600	
Silicon (Si)	2.76	$10^{-6}–10^{-2}$				2.3	
	2.94	$10^{-9}–10^{-2}$				$1.5 \cdot 10^{-1}$	
	10.6	$10^{-12}–10^{-10}$				5.0	
Sodium chloride (NaCl)	0.6943	$8 \cdot 10^{-9}$				150	
	1.064	$8 \cdot 10^{-9}$				120	
	10.6	$(2.5–3.0) \cdot 10^{-6}$	$4 \cdot 10^{-4}$	$(3.0–9.0) \cdot 10^{-2}$			
Thallium bromide–thallium chloride (KRS-6)	10.6	$(2.5–3.0) \cdot 10^{-6}$	$4 \cdot 10^{-4}$	$(3.0–6.0) \cdot 10^{-2}$			
	10.6	10^{-7}				$(4.0–8.0) \cdot 10^{-2}$	
Thallium bromide–thallium iodide (KRS-5)	10.6	$(2.5–3.0) \cdot 10^{-6}$	$4 \cdot 10^{-4}$	$(3.0–6.0) \cdot 10^{-2}$			
Yttrium-alumina garnet ($Y_3Al_5O_{12}$)	0.6943	10^{-8}				65	
	1.064	10^{-8}				200	
Zinc selenide (ZnSe)	10.6	$9.2 \cdot 10^{-8}$		$(2.3–4.6) \cdot 10^{-1}$			

*1 Laser impulse duration.

Note: The P and W values are tentative.

References

[1] Brockhaus ABC der Optik, VEB F.A. Brockhaus Verlag, Leipzig, 1961.

[2] Novitskii, L. A., Teplofizika vysokikh temperatur, 5, 919, 1967 (in Russian).

[3] Novitskii, L. A., Teplofizika vysokikh temperatur, 6, 529, 1968 (in Russian).

[4] Novitskii, L. A., Vdovin, V. G., Fedotov, G. I., Trudy MVTU, No 6, 97, 1974 (in Russian).

[5] Novitskii, L. A., Trushitsyna, A. V., Izmeritel'naya tekhnika, No 7, 46, 1970 (in Russian).

[6] Novitskii, L. A., Trushitsyna, A. V., Varakina, L. P., Pribory i sistemy upravleniya, No 6, 30, 1969 (in Russian).

[7] Zhevandrov, N. D., Light polarization, Nauka, Moscow, 1969 (in Russian).

[8] Shurkliff, W., Polarized light, Cambridge University Press, Cambridge (MA), 1962

[9] Optoelectronic devices for scientific investigations, Ed. Novitskii, L. A., Mashinostroyeniye, Moscow, 1986 (in Russian).

[10] Gurevich, V. Z., Electric infrared radiators, Gosenergoizdat, Moscow–Leningrad, 1963 (in Russian).

[11] Gurevich, M. M., Introduction into photometry, Energiya, Leningrad, 1968 (in Russian).

[12] Novitskii, L. A., Stepanov, B. M., Photometry of fast processes. A handbook, Mashinostroyeniye, Moscow, 1983 (in Russian).

[13] Laboratory optical instruments, Ed. Novitskii, L. A., Mashinostroenie, Moscow, 1979 (in Russian).

[14] Temperature. Its measurement and control in science and industry, Reynold Publ. Corp., New York, 1962.

[15] Kindzneri, V. A., Measurements at high temperatures, Metallurgizdat, Moscow, 1963 (in Russian).

[16] Kizel', V. A., Light reflection, Nauka, Moscow, 1973 (in Russian).

[17] Sokolov, A. V., Optical properties of metals, Fizmatgiz, Moscow, 1961 (in Russian).

[18] Radiating properties of solid materials. A handbook, Ed. Sheindlin, A. E., Energiya, Moscow, 1974 (in Russian).

[19] Kriksunov, L. Z., A handbook of fundamentals of infrared instruments, Sovetskoye radio, Moscow, 1978 (in Russian).

[20] Novitskii, L. A., Stepanov, B. M., Optical properties of materials at low temperatures, Mashinostroyeniye, Moscow, 1980 (in Russian).

[21] Voronkova, E. M., Grechushnikov, B. N., Distler, G. I., Petrov, I. P., Optical materials for infrared devices, Nauka, Moscow, 1965 (in Russian).

[22] A handbook on electrotechnical materials, vol. 2, Eds. Bogoroditskii, N. P. and Pasynkov, V. V., Gosenergoizdat, Moscow - Leningrad, 1960 (in Russian).

[23] Chirkin, V. S., Thermophysical properties of nuclear engineering materials. A handbook, Atomizdat, Moscow, 1968 (in Russian).

[24] Goldsmith, A., Watermann, T. E., Hirschhorn, H. I., Handbook of thermophysical properties of solid materials, vols. 1–4, Pergamon Press, Oxford - London, 1963.

[25] Optical glasses, Catalog USSR-DDR, Mashpriborintorg, Moscow, 1979.

[26] Lagutin, V. I., Likhanov, V. P., Nikonova, E. I., Optiko-Mekhanicheskaya Promyshlennost', No 5, 53, 1984 (in Russian).

[27] Materials for instrument making and automatics. A handbook, Ed. Pyatin, Yu. M., Mashinostroenie, Moscow, 1982 (in Russian).

[28] Landsberg, G. M., Optics, Nauka, Moscow, 1976 (in Russian).

[29] Handbook of designer of optomechanical instruments, Ed. Panov, V. A., Mashinostroenie, Leningrad, 1980 (in Russian).

[30] Mal'tsev, M. D., Karakulina, G. A., Applied optics and optical measurements, Mashinostroenie, Moscow, 1968 (in Russian).

[31] Mustel', E. R., Parygin, V. N., Methods of light modulation and scanning, Nauka, Moscow, 1970 (in Russian).

[32] Baiborodin, Yu. V., Garazha, S. A., Electrooptical effect in crystals and its application to instrument making, Mashinostroenie, Moscow, 1967 (in Russian).

[33] Kruse, P., McGlauchlin, L., McQuistan, R., Elements of infrared technology, London, 1961.

[34] Garbuni, M., Physics of optical phenomena, Energiya, Moscow, 1967 (in Russian).

[35] Hadson, R., Infrared System Engineering, Wiley, New York, 1969.

[36] Hackforth, H. L., Infrared radiation, McGraw–Hill Book Co., New York, 1960.

[37] Peisakhson, I. V., Optics of spectral devices, Mashinostroenie, Moscow, 1970 (in Russian).

[38] Klimkov, Yu. M., Fundamentals of calculating the optoelectronic devices with lasers, Sovetskoye radio, Moscow, 1978 (in Russian).

[39] Smolya, A. V., Tyurina, S. L., Yttrium-oxide-based optical ceramics, in Optophysical measurements, Izdatel'stvo standartov, Moscow, 1977, p. 163 (in Russian).

[40] Rabinovich, G. D., Slobodkin, L. S., Thermoradiation and convectional drying of paint coatings, Nauka i Tekhnika, Moscow, 1962 (in Russian).

[41] Drakin, I. I., Aerodynamic and radiant heating in flight, Oborongiz, Moscow, 1961 (in Russian).

[42] Novitskii, L. A., Teplofizika vysokikh temperatur, 4, 621, 1966 (in Russian).

[43] Novitskii, L. A., Teplofizika vysokikh temperatur, 7, 997, 1969 (in Russian).

[44] Kroshkin, M. G., Physical and engineering fundamentals of space investigations, Mashinostroenie, Moscow, 1969 (in Russian).

[45] Leconte, J., Le spectre infrarouge, Les Presses Universitaires de France, Paris, 1928.

[46] Bramson A. A., Reference tables on infrared radiation of heated bodies, Nauka, Moscow, 1964 (in Russian).

[47] Gutorov, M. M., Fundamentals of illumination engineering and light sources, Energoatomizdat, Moscow, 1983 (in Russian).

[48] Philip, R., J. Phys. et Radium, 20, 535, 1959.

[49] Proc. of the 4th Conf. on Carbon, Pergamon Press, New York, 1960.

[50] Heat exchange, hydrodynamics and thermal properties of substances, Ed. Aladiev, I. T., Nauka, Moscow, 1968 (in Russian).

[51] Petrov, V. A., Radiating power of high-temperature materials, Nauka, Moscow, 1969 (in Russian).

[52] Vlasov, L. V., Liberman, A. A., Samoilov, L. N., Measurement of emissivity of high-temperature materials. Problems of power photometry, Atomizdat, Moscow, 1979 (in Russian).

[53] Mel'nikov, Yu. F., Illumination engineering materials, Vysshaya Shkola, Moscow, 1976 (in Russian).

[54] Smith, G. F., Gemstones, Chapman and Hall, London, 1972.

[55] Blokh, A. G., Basicals of thermal transfer via emission, Gosenergoizdat, Moscow–Leningrad, 1962 (in Russian).

[56] Harrison, T., Radiation pirometry and its underlying principles of radiant heat transfer, New York, 1960.

[57] Kornilov, N. I., Solodova, Yu. P., Jewelry stones, Nedra, Moscow, 1983 (in Russian).

[58] Birckback, R. K., J. Eng.–Mech. Soc. Am., Ser. C: Heat transfer, 94, 72, 1972.

[59] Birckback, R. K., J. Sci. Instr., No 7, 65, 1972.

[60] Lunar soil from the Mare of Fertility, Ed. Vinogradov, A. P., Nauka, Moscow, 1975 (in Russian).

[61] Begunov, B. N., Zakaznov, N. P., A theory of optical systems, Mashinostroenie, Moscow, 1973 (in Russian).

[62] Tables of physical quantities. Handbook, Ed. Kikoin, I. K., Atomizdat, Moscow, 1976 (in Russian).

[63] Hauf, W., Grigull, U., Optical methods in heat transfer, Academic Press, New York, 1970.

[64] Krivovyaz, L. M., Puryaev, D. T., Znamenskaya, M. A., Practices of an optical measurement laboratory, Mashinostroenie, Moscow, 1974 (in Russian).

[65] Voronkov, G. L., Optical emission attenuators, Mashinostroenie, Leningrad, 1980 (in Russian).

[66] Rabinovich, V. A., Khavin, Z. Ya., A concise chemical handbook, Khimiya, Moscow, 1977 (in Russian).

[67] Ivanov, A. P., Physical essentials of hydrooptics, Nauka i Tekhnika, Minsk, 1975 (in Russian).

[68] Manufacturing engineering of optical components, Ed. Semibratov, M. N., Mashinostroenie, Moscow, 1978 (in Russian).

[69] Kamenshchikov, V. A., Plastinin, Yu. A., Nikolaev, V. M., Novitskii, L. A., Radiation properties of gases at high temperatures, Mashinostroenie, Moscow, 1971 (in Russian).

[70] Martin, W. C., J. Opt. Soc. Am., 50, No 2, 174, 1960.

[71] Littlefield, T. A., Nature, 165, 187, 1950.

[72] Burns, J., Adams, G., Longwell, F., J. Opt. Soc. Am., 40, No 5, 340, 1950.

[73] Enokhovich, A. S., A concise handbook on physics, Vysshaya Shkola, Moscow, 1968 (in Russian).

[74] A handbook on physical and technical fundamentals of cryogenics, Ed. Malkov, M. P., Energoatomizdat, Moscow–Leningrad, 1985 (in Russian).

[75] Novitskii, L. A., Yakunin, V. P., Industrial lasers. A handbook, vol. 2, Mashinostroenie, Moscow, 1991 (in Russian).

[76] Galant, V. E., Polukhin, V. N., Urusovskaya, L. N., Optiko-Mekhanicheskaya Promyshlennost', No 7, 50, 1991 (in Russian).

[77] Nazarova, N. A., Optiko-Mekhanicheskaya Promyshlennost', No 7, 3, 1991 (in Russian).

[78] Savushkina, V. N., Mironov, I. A., Volynskaya, S. M., Opticheskii Zhurnal, No 7, 53, 1992 (in Russian).

[79] Kizel', V. A., Burkov, V. I., Crystal gyrotropy, Nauka, Moscow, 1980 (in Russian).

[80] Novitskii, L. A., Polevova, I. F., Zarubezhnaya Tekhnika, No 9, 64, 1964 (in Russian).

[81] Novitskii, L. A., Steklo i Keramika, No 5, 16, 1953 (in Russian).

[82] Vorobiev, V. G., Kruglyakova, M. A., Nikitin, V. A., Optiko-Mekhanicheskaya Promyshlennost', No 2, 23, 1991 (in Russian).

[83] Eshmetieva, E. V., Korolev, V. I., Mesnyankin, E. P., Optiko-Mekhanicheskaya Promyshlennost', No 9, 18, 1991 (in Russian).

[84] Savitskii, A. V., Beisyuk, P. P., Budzulyak, I. V., Fizika i Khimia Obrabotki Materialov, No 3, 44, 1992 (in Russian).

[85] Shelekhov, N. S., Shatilov, A. V., Gusev, G. P., Optiko-Mekhanicheskaya Promyshlennost', No 9, 46, 1991 (in Russian).

[86] Balashov, I. F., Berezin, B. G., Ivanov, B. N., Optiko-Mekhanicheskaya Promyshlennost', No 9, 36, 1991 (in Russian).

[87] Dolotov, S. M., Koldunov, M. F., Manenkov, A. A., Kvantovaya Elektronika, 19, 1134, 1992 (in Russian).

[88] Yashchishin, I. N., Kozii, O. I., Shchavelev, O. S., Optiko-Mekhanicheskaya Promyshlennost', No 2, 71, 1991 (in Russian).

[89] Chmel', A. E., Eron'ko, S. B., Knyazev, S. A., Fizika i Khimia Obrabotki Materialov, No 4, 46, 1992 (in Russian).

[90] Banishev, A. F., Novikova, L. V., Fizika i Khimia Obrabotki Materialov, No 4, 55, 1992 (in Russian).

[91] Goryacheva, T. V., Dernovskii, V. I., Poverkhnost', No 4, 77, 1990 (in Russian).

[92] Bulka, G. R., Butyagin, O. F., Ermakova, G. A., Lazernaya Tekhnika i Optoelektronika, No 1–2, 69, 1992 (in Russian).

[93] Khartsieva, T. N., Lazernaya Tekhnika i Optoelektronika, No 1–2, 76, 1992 (in Russian).

[94] Kondyrev, A. M., Tarasova, Yu. V., Chmel', A. E., Opticheskii Zhurnal, No 10, 78, 1992 (in Russian).

[95] Evstropiev, S. K., Zamoiskaya, L. V., Zgonnik, V. N., Opticheskii Zhurnal, No 10, 53, 1992 (in Russian).

[96] Polukhin, V. N., Ponomareva, V. N., Belikova, N. G., Opticheskii Zhurnal, No 10, 55, 1992 (in Russian).

[97] Zarubina, T. V., Petrovskii, G. T., Opticheskii Zhurnal, No 11, 48, 1992 (in Russian).

[98] Musatov, M. I., Opticheskii Zhurnal, No 11, 83, 1992 (in Russian).

[99] Petrovskii, G. T., Safiulina, S. S., Litova, A. M., Opticheskii Zhurnal, No 11, 32, 1992 (in Russian).

[100] Petrovskii, G. T., Borozdin, S. N., Opticheskii Zhurnal, No 11, 77, 1993 (in Russian).

32

Atomic and Molecular Spectra

A.A. Radzig

32.1 Introduction

The most commonly encountered processes of light emission and absorption in the atomic and molecular media are conditioned by their electron transitions. Thus, they can be subdivided into three types: (1) free–free electron transitions (bremsstrahlung and radiation absorption on electron scattering by atoms and ions, continuous spectrum); (2) bound–free electron transitions (atomic and molecular photoionization, as well as electron photorecombination with ions and neutral particles, continuous spectrum); (3) bound–bound (discrete) electron transitions (line atomic spectrum and molecular band spectrum).

This chapter describes more specifically the discrete spectra of atoms and molecules in the optical range, which come about from the electron transitions within the outer (valence) atomic electron shell and from changes in vibration–rotation states of molecular species. Therefore, information is given in what follows on the excited states of atomic particles along with the parameters of fine-, hyperfine-, and isotope structure in the spectra recorded. In the case of molecular species, we restrict our attention to spectroscopic constants and electronic terms of the most widespread diatomic molecules.

32.2 Atomic Optical Spectra

The line spectra of neutral atoms in the optical range (infrared, visible and ultraviolet, cf. Table 32.1) are conditioned, as a rule, by the transitions of outer (valence) electrons. Figs. 32.1–32.43 display the combined diagrams of atomic energy levels and spectra (often called Grotrian diagrams) for the first 30 elements of the Periodic Table, some atoms with valence electron subshell of the form ns, ns^2, np^5 and np^6 ($n = 5, 6$), and, finally, for the simplest atomic ion–He^+. These atomic systems are of interest as far as the applications of modern atomic spectroscopy are concerned. The diagrams provide a pictorial rendition of excited state distribution over the main electron configurations in atoms and carry information about the most intensive transitions in their optical spectra.

When constructing Grotrian diagrams, we eliminated highly excited Rydberg energy levels and autoionizing levels (resulting from the double electron excitation and placed above the first ionization limit of an atom) from consideration. The position of atomic energy levels (this usually implies the position of the multiplet center of gravity $\bar{T} = \sum T_i g_i / \sum g_i$, where T_i is the multiplet component, g_i is the statistical weight of the i-th sublevel) is determined on the ordinate axis in reverse centimeter. Be-

0-8493-2861-6/97/$0.00+$.50

sides, the figures above the horizontal lines relating to atomic energy levels designate the corresponding excitation energies in electron-volts (1 eV = 8065.54 cm^{-1}).

The energy splittings of multiplet levels, with proper sign characterizing either the normal (+) or inverted (−) multiplet, are enclosed in right-angled frames alongside the lines of energy levels. The primed symbols designate the electron configurations relating to different states of the parent atomic core. When considering the rare gas atoms and atomic iodine, which exhibit the excited state classification according to jl-coupling scheme, the appropriate Grotrian diagrams involve only positions of the top and bottom components of multiplet sublevels. These components are marked with superscribed and subscribed bar, respectively, on the quantum number J of the total atomic angular momenta. The cut-off wavelengths related to the transitions between given multiplet sublevels are indicated in these cases only.

The corresponding absorption oscillator strengths f_{ik} are sometimes given in parentheses alongside the numerical values of transition wavelengths λ. The lower index occasionally added to the values of λ and f_{ik} is used to clarify the identification of the transition multiplet components. This index labels total angular momentum J of the levels (lower, upper, or both).

In the construction of combined energy levels–spectra diagrams in Figs. 32.1–32.43, we resorted to special guides on the Grotrian diagrams [1,2], numerous journal publications on the spectra of elements [3], tables of spectral lines [4,5], and reference books on the transition probabilities (see, for example, [3]). The error in determining the values of T, ΔT_{ik}, λ, and f_{ik} was accounted for in truncating the last significant figures within ±1 for the last figure given.

It should be noted that by convention, all wavelengths in spectroscopy with $\lambda > 200$ nm refer to transitions in air, whereas shorter wavelengths apply to vacuum transitions. Table 32.2 presents the numerical values of wavelength correction $\Delta\lambda$ accounting for light dispersion in air, by use of Edlèn's dispersion formula [3]:

$$\Delta\lambda = \lambda_{\text{vac}} - \lambda_{\text{air}} = \lambda_{\text{vac}}(1 - n^{-1}), \tag{32.1}$$

$$n = 1 + 10^{-4}\left(0.834213 + \frac{2.406030}{1.3 - 10^{-10}\sigma^2} + \frac{0.015997}{0.389 - 10^{-10}\sigma^2}\right), \tag{32.2}$$

where n is the air refraction index for a given wave; σ is a wave number in vacuum, equal to transition energy $\Delta T/h$, cm^{-1}. In practical units, the relation between transition wavelength and energy takes the form:

$$\lambda_{\text{vac}} = 10^8/\Delta T_1 = 12398.5/\Delta T_2, \tag{32.3}$$

here n is expressed in units of 0.1 nm (viz., angstrom), ΔT_1 – in cm^{-1}, and ΔT_2 – in eV.

Table 32.1 Historically established nominal regions of electromagnetic radiation spectrum

Type	Wavelength interval $\Delta\lambda$	Frequency interval $\Delta\nu$	Energy interval $\Delta\epsilon$, eV	Effective temperature interval ΔT, K for black body radiation: $h\nu = 2.82\,kT$
Audio frequencies				
extreme low (ELF)	10^5–10^4 km	3–30 Hz	$1.2\cdot10^{-14}$–$1.2\cdot10^{-13}$	$5.1\cdot10^{-11}$–$5.1\cdot10^{-10}$
super low (SLF)	10^4–10^3 km	30–300 Hz	$1.2\cdot10^{-13}$–$1.2\cdot10^{-12}$	$5.1\cdot10^{-10}$–$5.1\cdot10^{-9}$
infra-low (ILF)	10^3–10^2 km	0.3–3 kHz	$1.2\cdot10^{-12}$–$1.2\cdot10^{-11}$	$5.1\cdot10^{-9}$–$5.1\cdot10^{-8}$
Radio waves				
super long (VLF)	100–10 km	3–30 kHz	$1.2\cdot10^{-11}$–$1.2\cdot10^{-10}$	$5.1\cdot10^{-8}$–$5.1\cdot10^{-7}$
long (LF)	10–1 km	30–300 kHz	$1.2\cdot10^{-10}$–$1.2\cdot10^{-9}$	$5.1\cdot10^{-7}$–$5.1\cdot10^{-6}$
medium frequency (MF)	1000–100 m	0.3–3 MHz	$1.2\cdot10^{-9}$–$1.2\cdot10^{-8}$	$5.1\cdot10^{-6}$–$5.1\cdot10^{-5}$
short (HF)	100–10 m	3–30 MHz	$1.2\cdot10^{-8}$–$1.2\cdot10^{-7}$	$5.1\cdot10^{-5}$–$5.1\cdot10^{-4}$

Table 32.1 Historically established nominal regions of electromagnetic radiation spectrum *(continued)*

Type	Wavelength interval $\Delta\lambda$	Frequency interval $\Delta\nu$	Energy interval $\Delta\epsilon$, eV	Effective temperature interval ΔT, K for black body radiation: $h\nu = 2.82\,kT$
ultra-short				
meter-wave (VHF)	10–1 m	30-300 MHz	$1.2\cdot10^{-7}$–$1.2\cdot10^{-6}$	$5.1\cdot10^{-4}$–$5.1\cdot10^{-3}$
decimeter-wave (UHF)	10–1 dm	0.3–3 GHz	$1.2\cdot10^{-6}$–$1.2\cdot10^{-5}$	$5.1\cdot10^{-3}$–$5.1\cdot10^{-2}$
centimeter-wave (SHF)	10–1 cm	3–30 GHz	$1.2\cdot10^{-5}$–$1.2\cdot10^{-4}$	0.051–0.51
millimeter-wave (EHF)	10–1 mm	30–300 GHz	$1.2\cdot10^{-4}$–$1.2\cdot10^{-3}$	0.51–5.10
submillimeter-wave	1–0.05 mm	0.3–6 THz	$1.2\cdot10^{-3}$–$2.4\cdot10^{-2}$	5.10–102
Infrared radiation				
extreme	1000–40 μm	0.3–7.5 THz	$1.2\cdot10^{-3}$–$3.1\cdot10^{-2}$	5.10–128
far (FIR)	40–6 μm	7.5–50 THz	0.031–0.21	128–850
medium	6–1.5 μm	50–200 THz	0.21–0.83	850–3401
near	1.5–0.77 μm	200–390 THz	0.83–1.61	3401–6626
Visible light				
red	770–622 nm	390–482 THz	1.61–1.99	6626–8203
orange	622–597 nm	482–503 THz	1.99–2.08	8203–8546
yellow	597–577 nm	503–520 THz	2.08–2.15	8546–8842
green	577–492 nm	520–610 THz	2.15–2.52	8842–10370
blue	492–455 nm	610–659 THz	2.52–2.72	10370–11213
violet	455–390 nm	659–769 THz	2.72–3.18	11213–13082
Ultraviolet radiation				
near	390–200 nm	0.769–1.5 PHz	3.18–6.20	13082–25510
far (vacuum - VUV)	200–10 nm	1.5–30 PHz	6.20–124	25510–$5.1\cdot10^{5}$
extreme (XUV)	100–1 nm	3–300 PHz	12.4–1240	$5.1\cdot10^{4}$–$5.1\cdot10^{6}$
X-ray radiation				
ultra-soft (USX)	400–10 Å	$7.53\cdot10^{-3}$–0.3 EHz	31.0–1240	$1.28\cdot10^{5}$–$5.1\cdot10^{6}$
soft	400–2 Å	$7.53\cdot10^{-3}$–1.5 EHz	31.0–6200	$1.28\cdot10^{5}$–$2.6\cdot10^{7}$
hard	2–0.01 Å	1.5–300 EHz	6200–$1.2\cdot10^{6}$	$2.6\cdot10^{7}$–$5.1\cdot10^{9}$
γ-ray radiation	1–10^{-4} Å	$3\cdot10^{18}$–$3\cdot10^{22}$ Hz	$1.2\cdot10^{4}$–$1.2\cdot10^{8}$	$5.1\cdot10^{7}$–$5.1\cdot10^{11}$
Cosmic rays	$< 10^{-3}$ Å	$> 3\cdot10^{21}$ Hz	$> 1.2\cdot10^{7}$	$> 5.1\cdot10^{10}$

Table 32.2 Wavelength correction $\Delta\lambda$, Å to the radiation wavelength λ, Å accounting for light dispersion in air: $\lambda_{\text{vac}} = \lambda_{\text{air}} + \Delta\lambda$, $\lambda_{\text{air}} = \lambda_{\text{vac}} - \Delta\lambda$

λ	$\Delta\lambda$	λ	$\Delta\lambda$	λ	$\Delta\lambda$	λ	$\Delta\lambda$	λ	$\Delta\lambda$	λ	$\Delta\lambda$	λ	$\Delta\lambda$	λ	$\Delta\lambda$
2000	0.648	3200	0.925	4400	1.24	6000	1.66	8200	2.25	10,400	2.85	12,600	3.45	14,800	4.05
2100	0.667	3300	0.950	4500	1.26	6200	1.72	8400	2.31	10,600	2.90	12,800	3.50	15,000	4.10
2200	0.687	3400	0.976	4600	1.29	6400	1.77	8600	2.36	10,800	2.96	13,000	3.56	16,000	4.37
2300	0.708	3500	1.00	4700	1.32	6600	1.82	8800	2.42	11,000	3.01	13,200	3.61	17,000	4.64
2400	0.731	3600	1.03	4800	1.34	6800	1.88	9000	2.47	11,200	3.07	13,400	3.66	18,000	4.92
2500	0.754	3700	1.05	4900	1.37	7000	1.93	9200	2.52	11,400	3.12	13,600	3.72	19,000	5.19
2600	0.777	3800	1.08	5000	1.39	7200	1.98	9400	2.58	11,600	3.18	13,800	3.77	20,000	5.46
2700	0.801	3900	1.10	5200	1.45	7400	2.04	9600	2.63	11,800	3.23	14,000	3.83	30,000	8.18
2800	0.825	4000	1.13	5400	1.50	7600	2.09	9800	2.69	12,000	3.28	14,200	3.88	40,000	10.9
2900	0.850	4100	1.16	5600	1.55	7800	2.15	10,000	2.74	12,200	3.34	14,400	3.94	50,000	13.6
3000	0.875	4200	1.18	5800	1.61	8000	2.20	10,200	2.80	12,400	3.39	14,600	3.99	100,000	27.3
3100	0.900	4300	1.21												

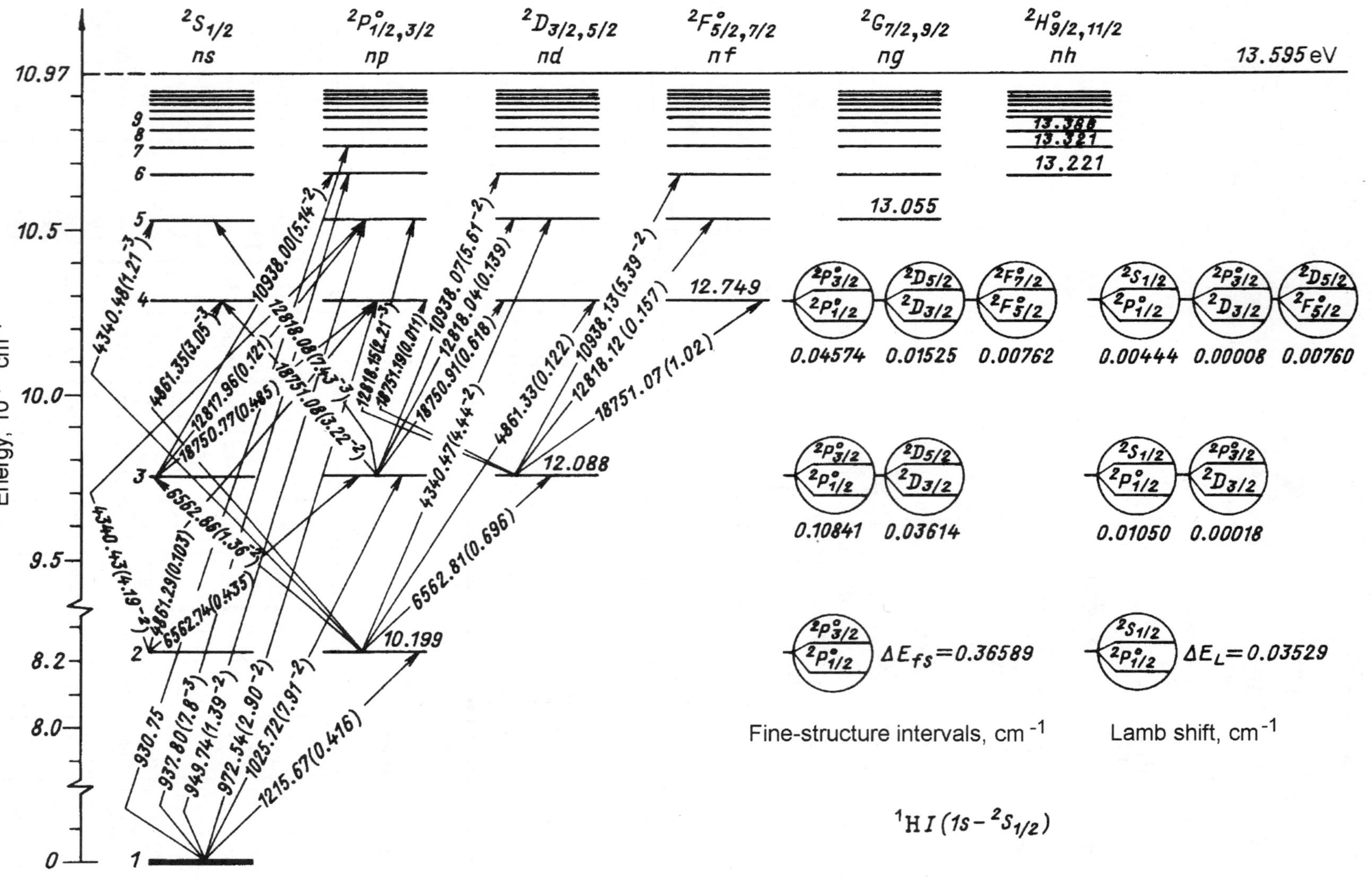

Figure 32.1 Grotrian diagram of atomic hydrogen.

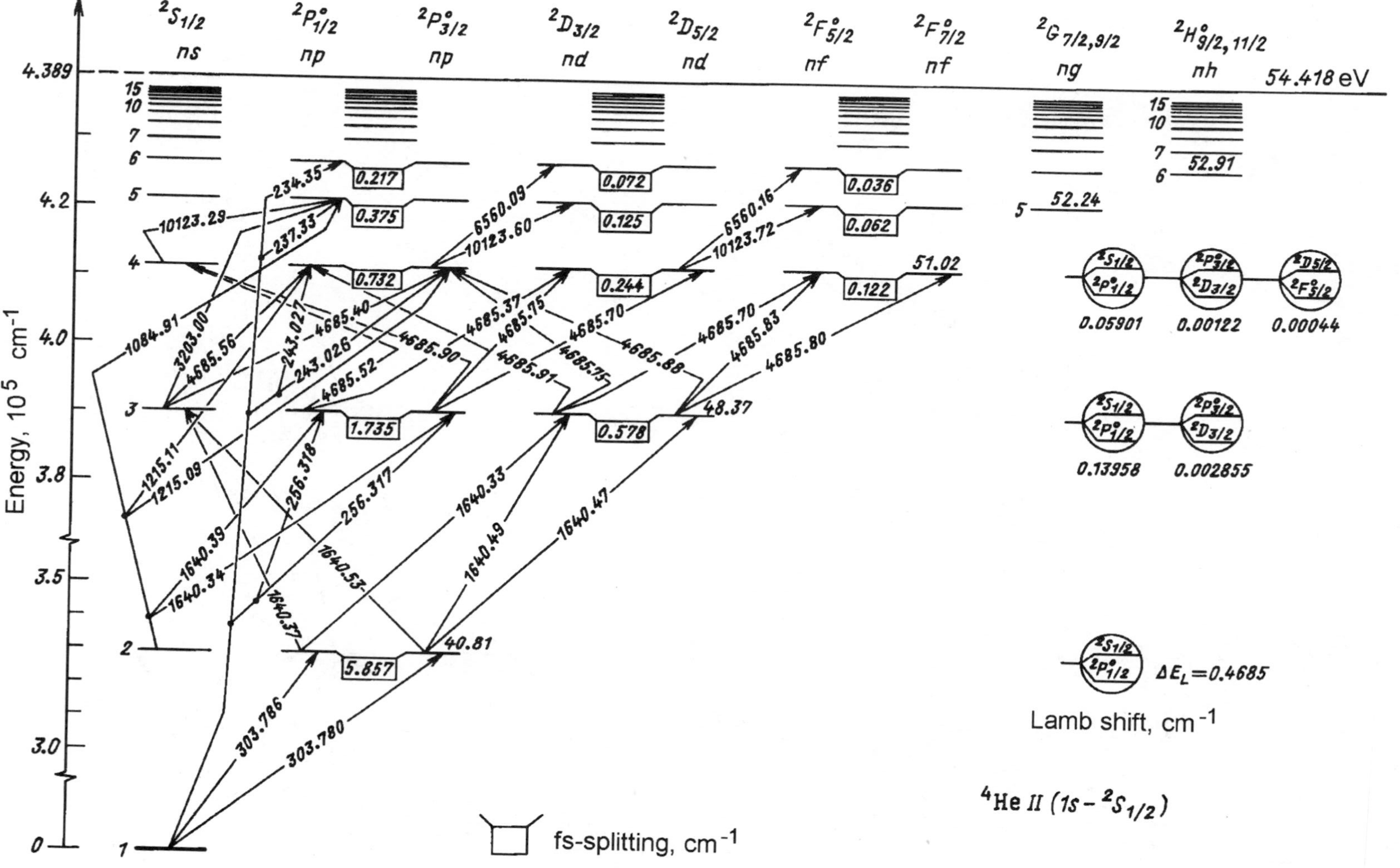

Figure 32.2 Grotrian diagram of single–charged helium ion.

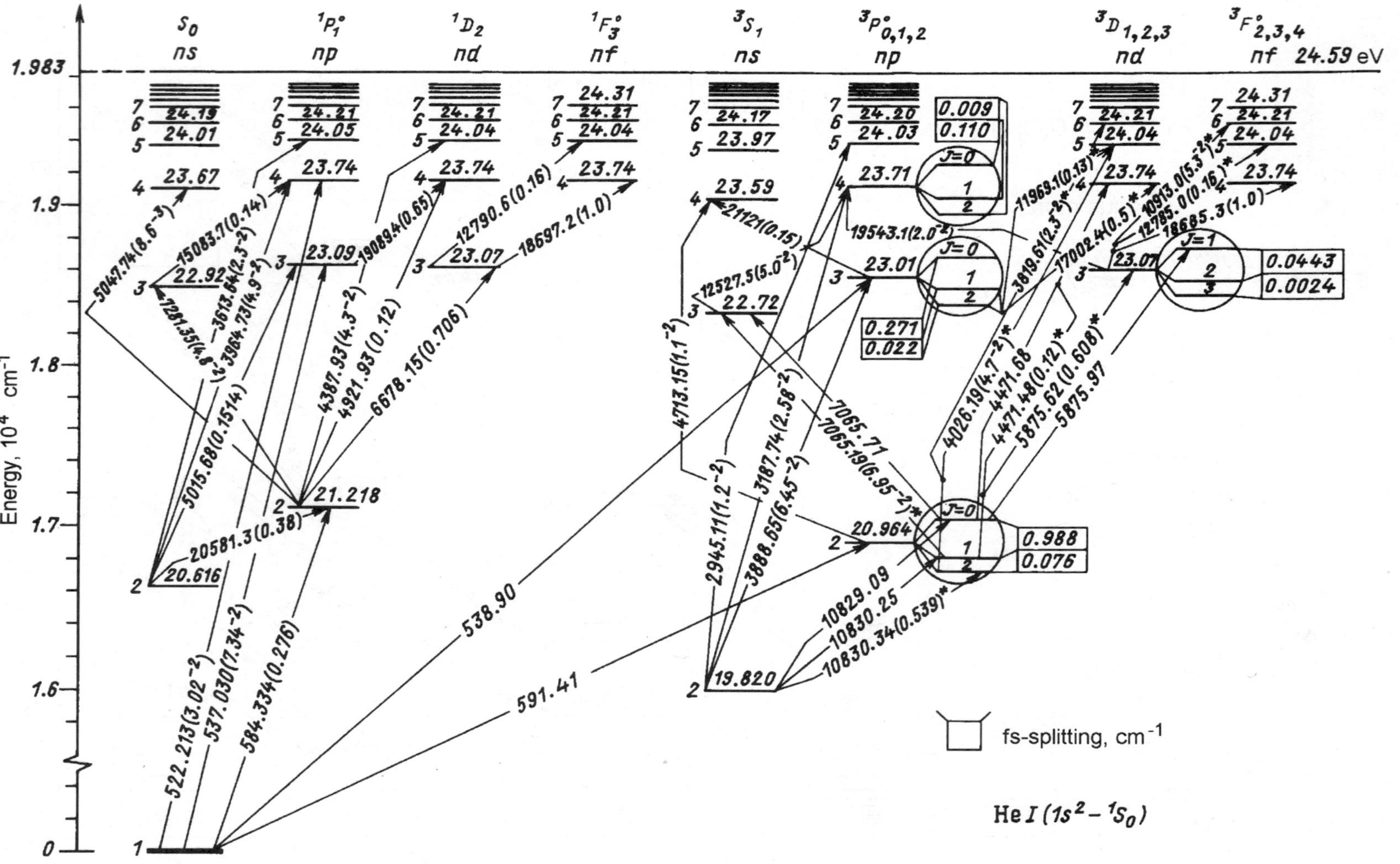

Figure 32.3 Grotrian diagram of atomic helium.

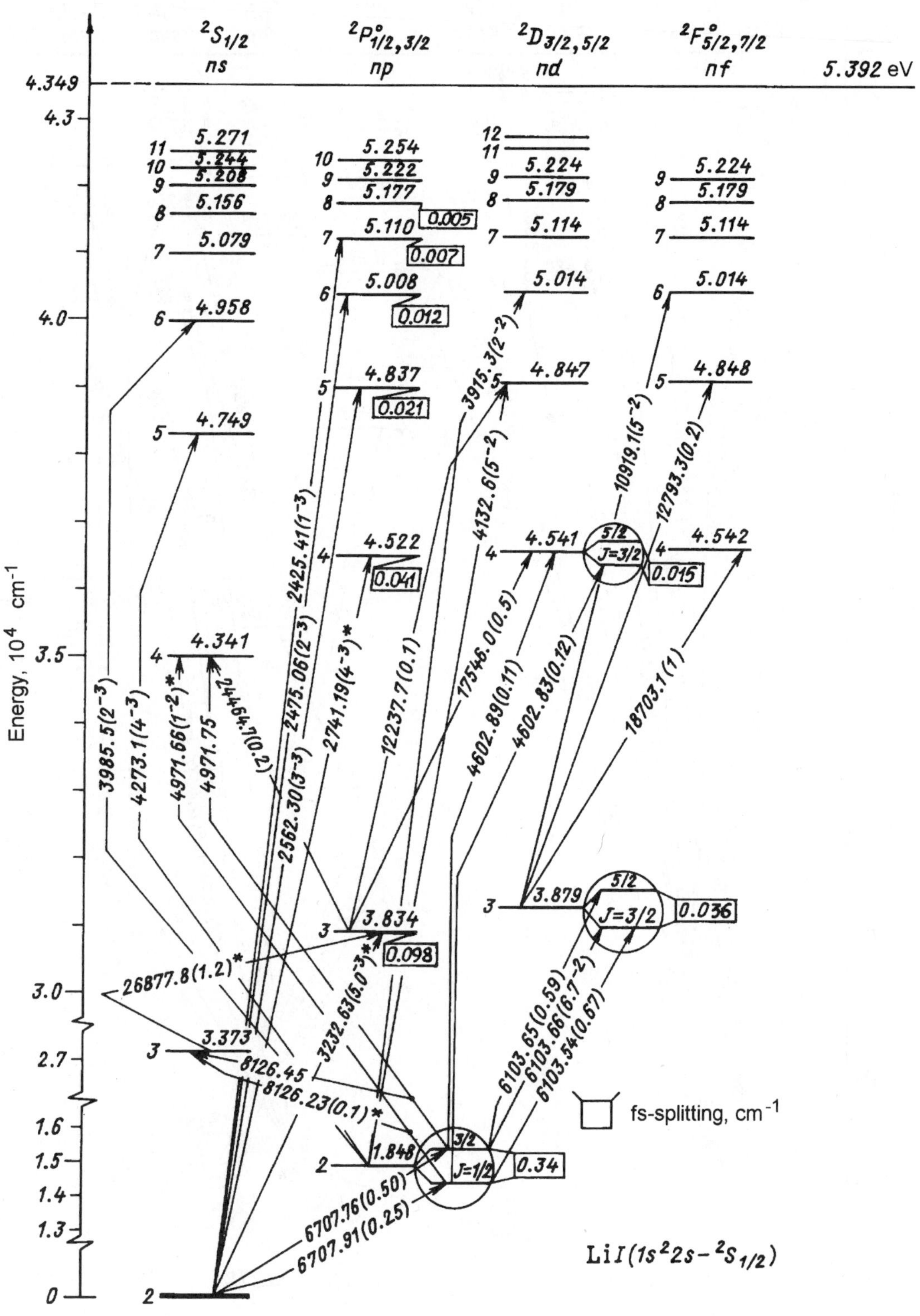

Figure 32.4 Grotrian diagram of atomic lithium.

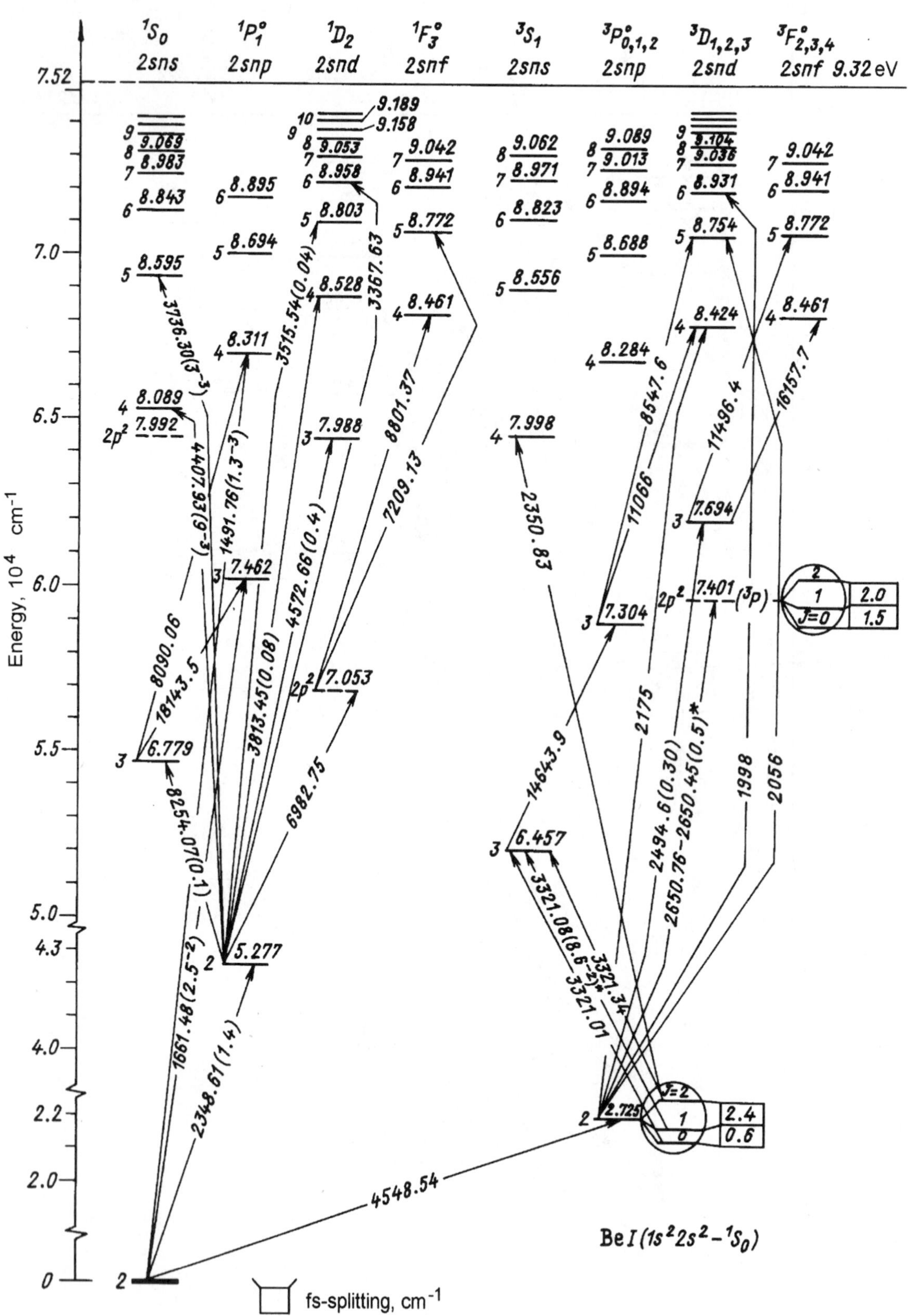

Figure 32.5 Grotrian diagram of atomic beryllium.

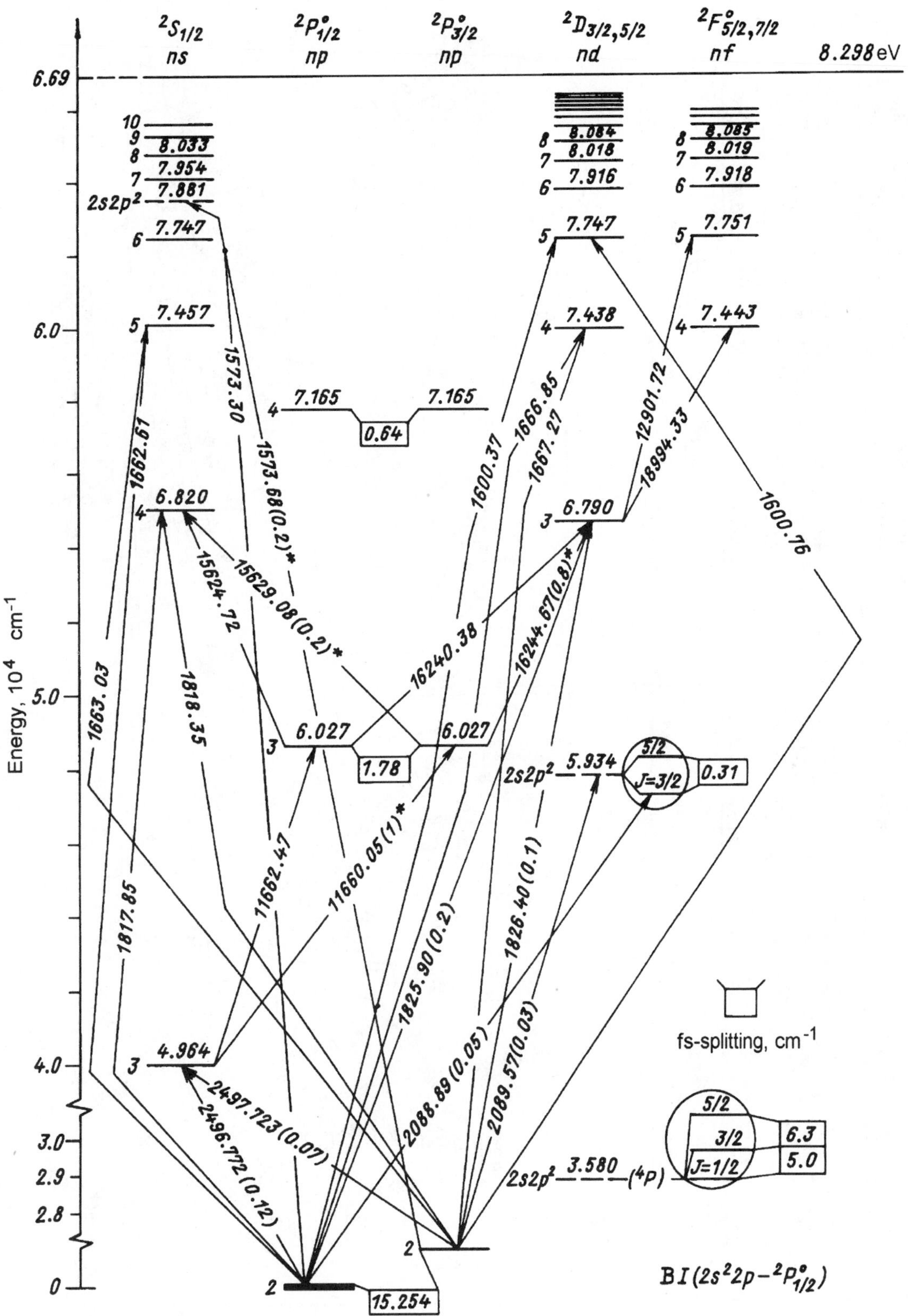

Figure 32.6 Grotrian diagram of atomic boron.

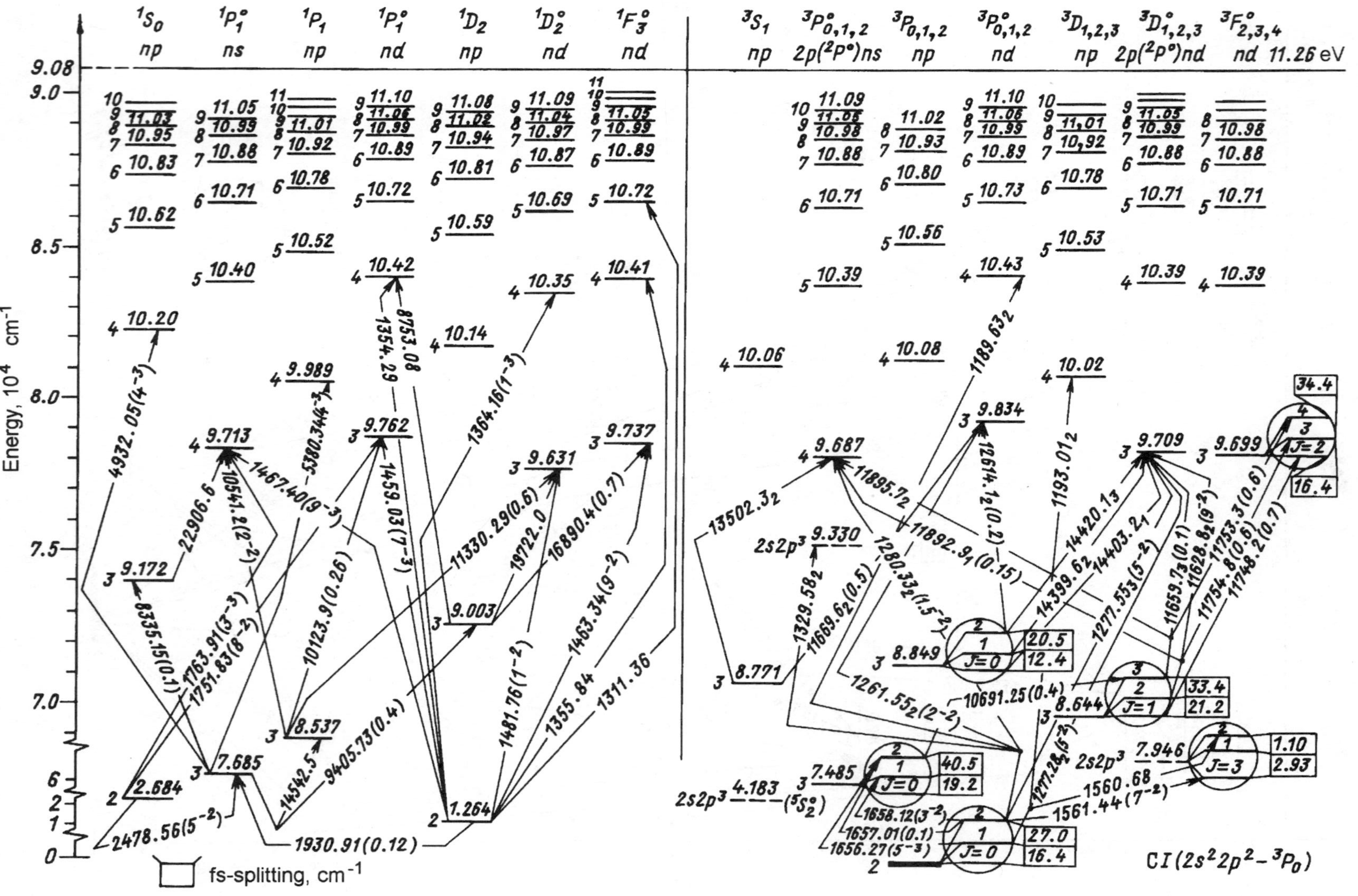

Figure 32.7 Grotrian diagram of atomic carbon.

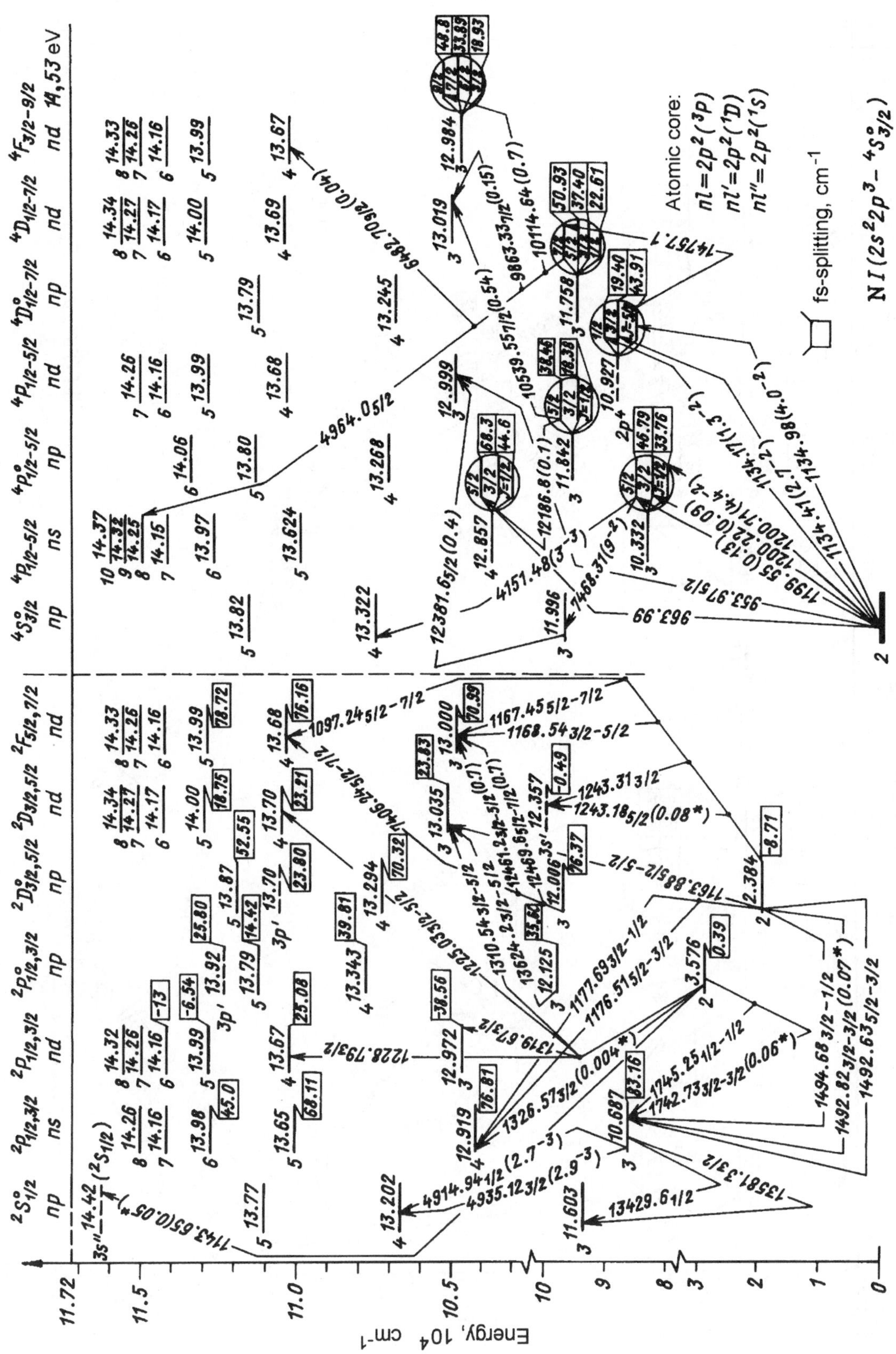

Figure 32.8 Grotrian diagram of atomic nitrogen.

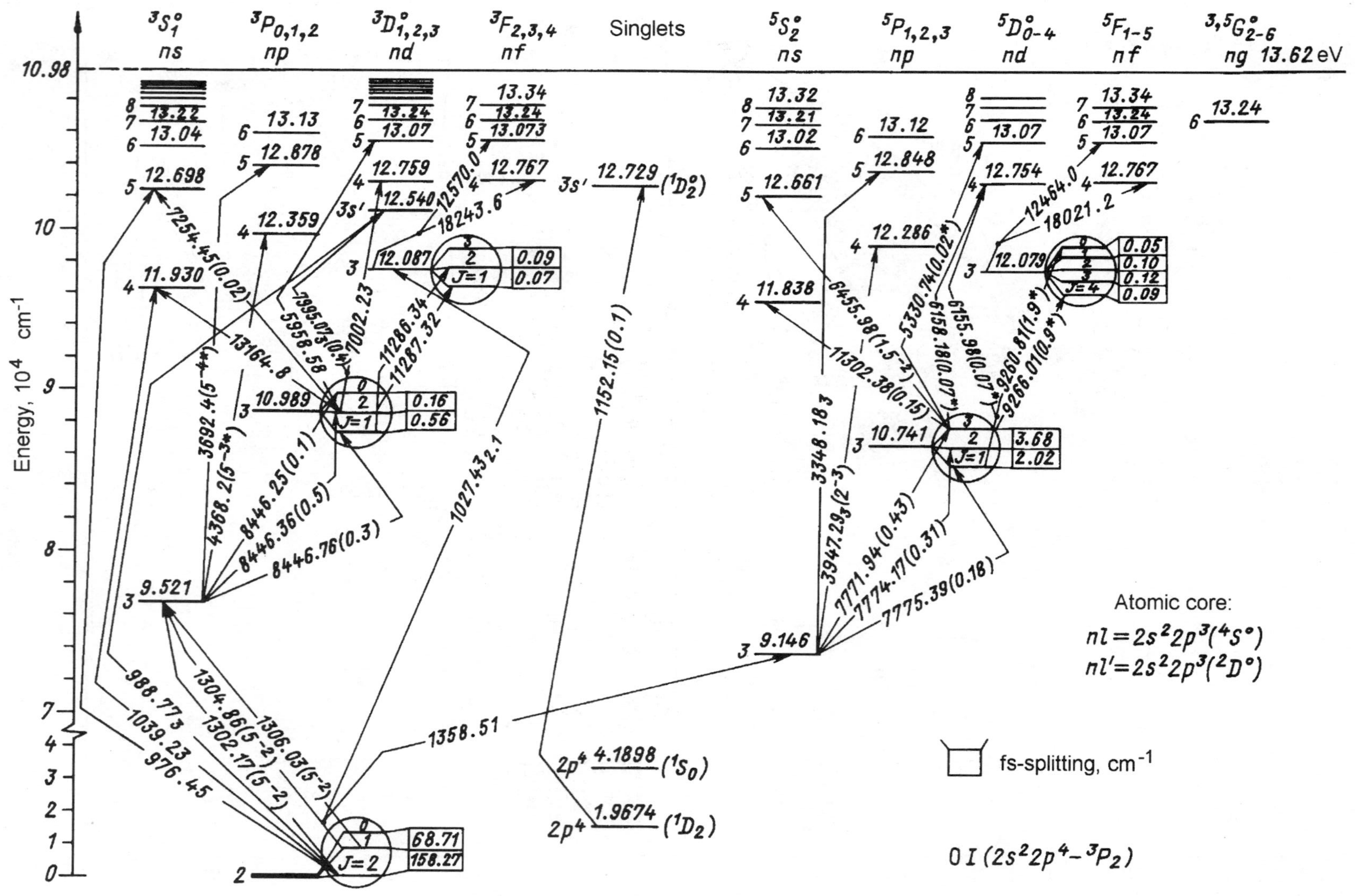

Figure 32.9 Grotrian diagram of atomic oxygen.

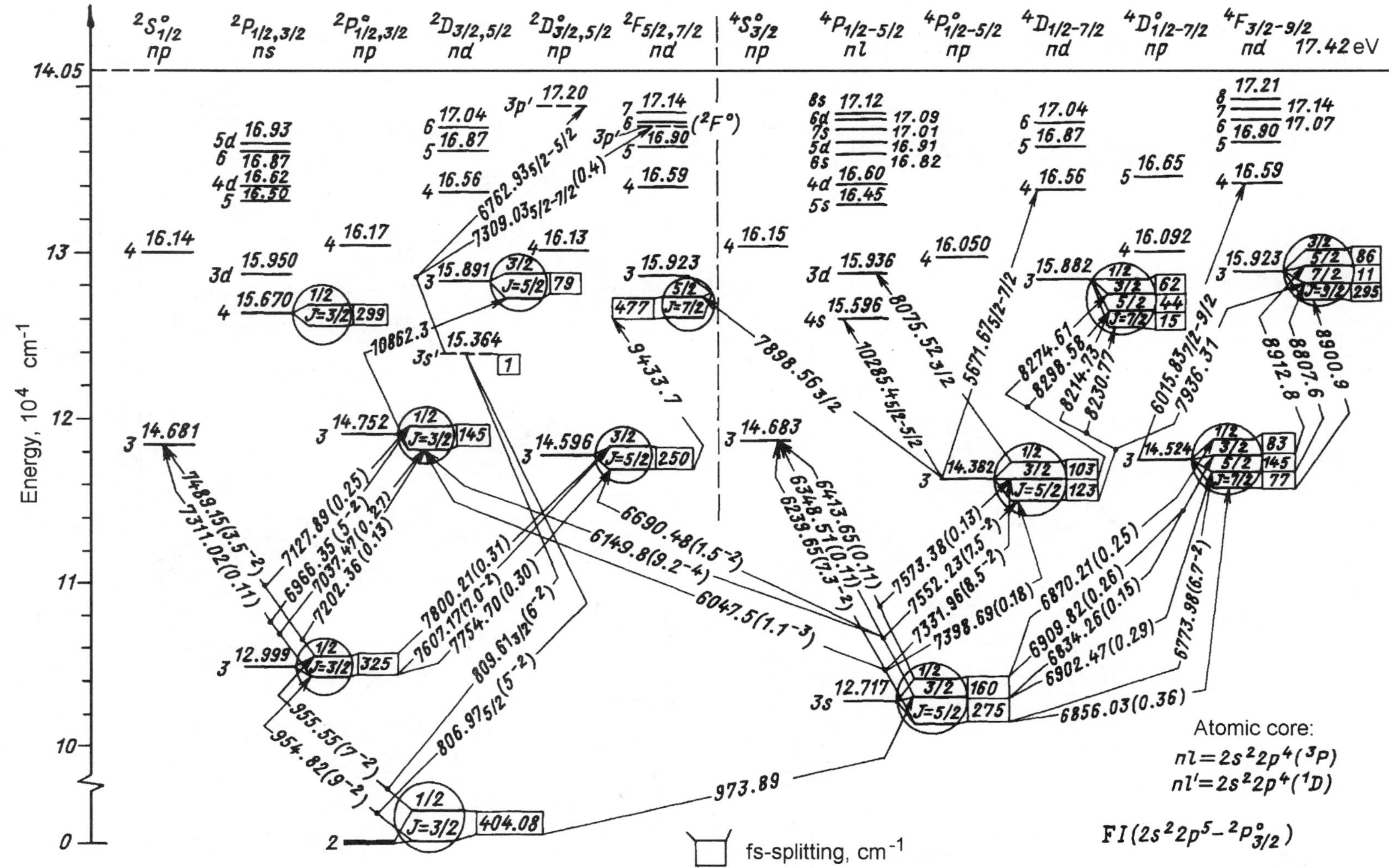

Figure 32.10 Grotrian diagram of atomic fluorine.

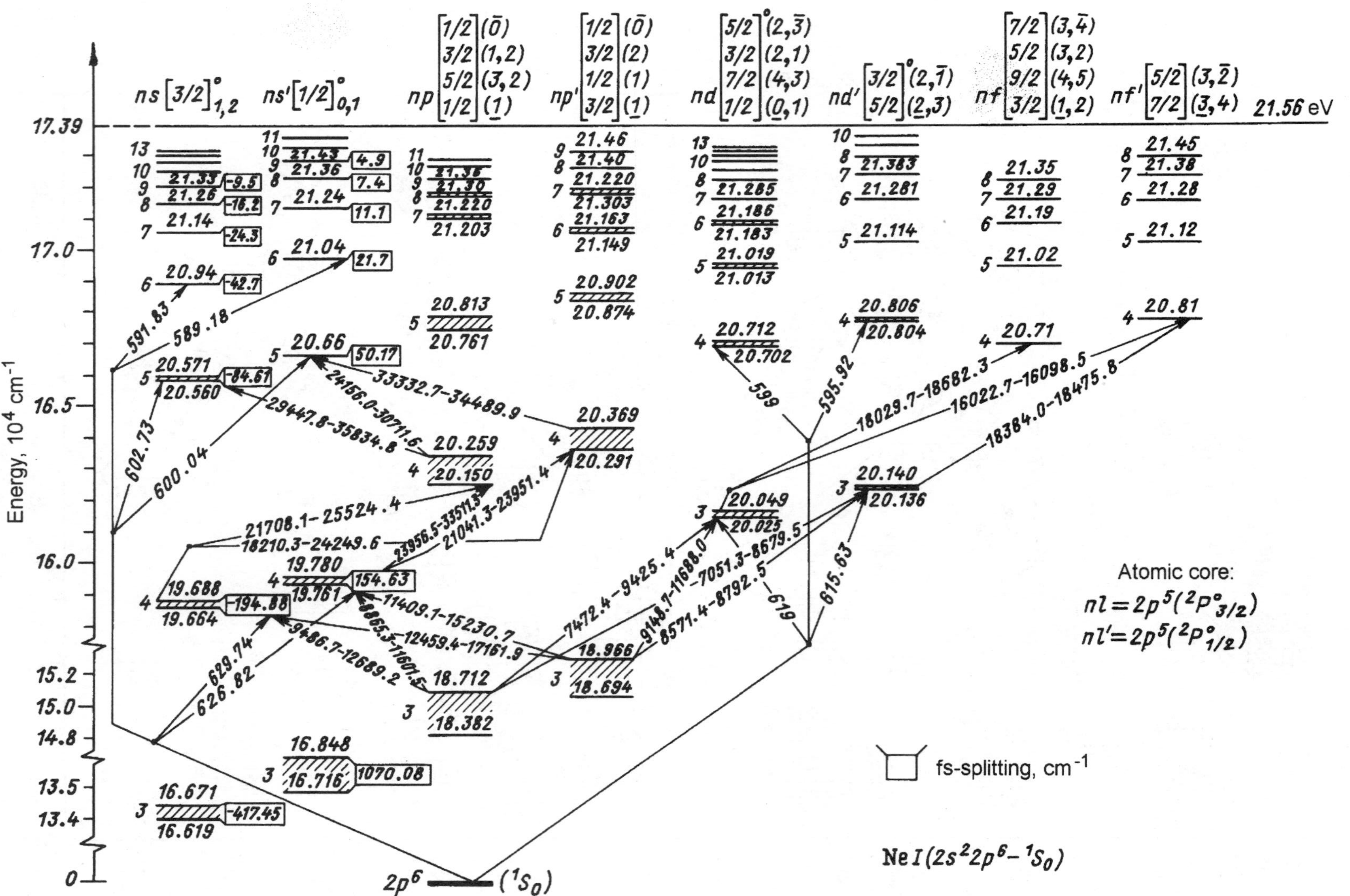

Figure 32.11 Grotrian diagram of atomic neon.

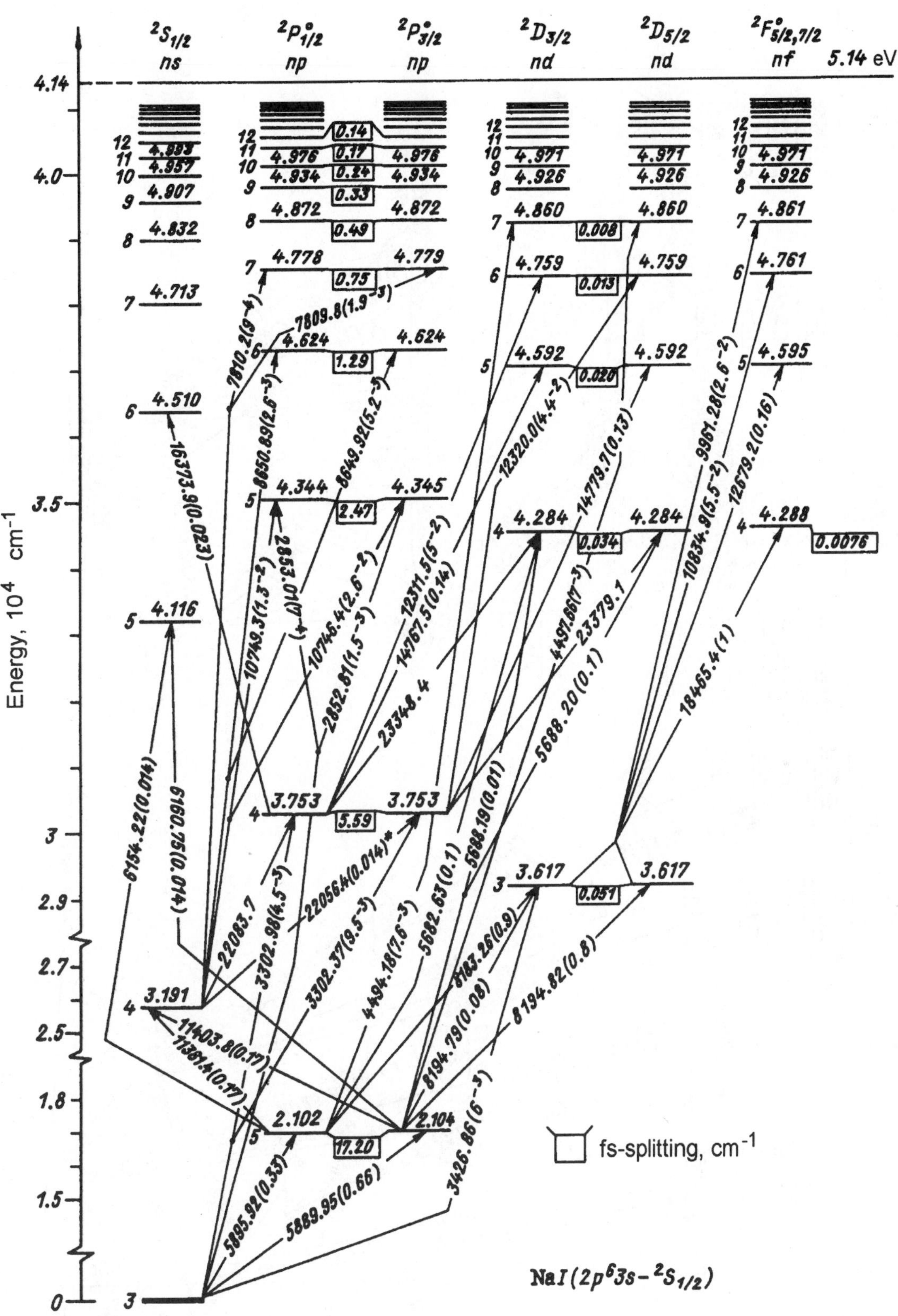

Figure 32.12 Grotrian diagram of atomic sodium.

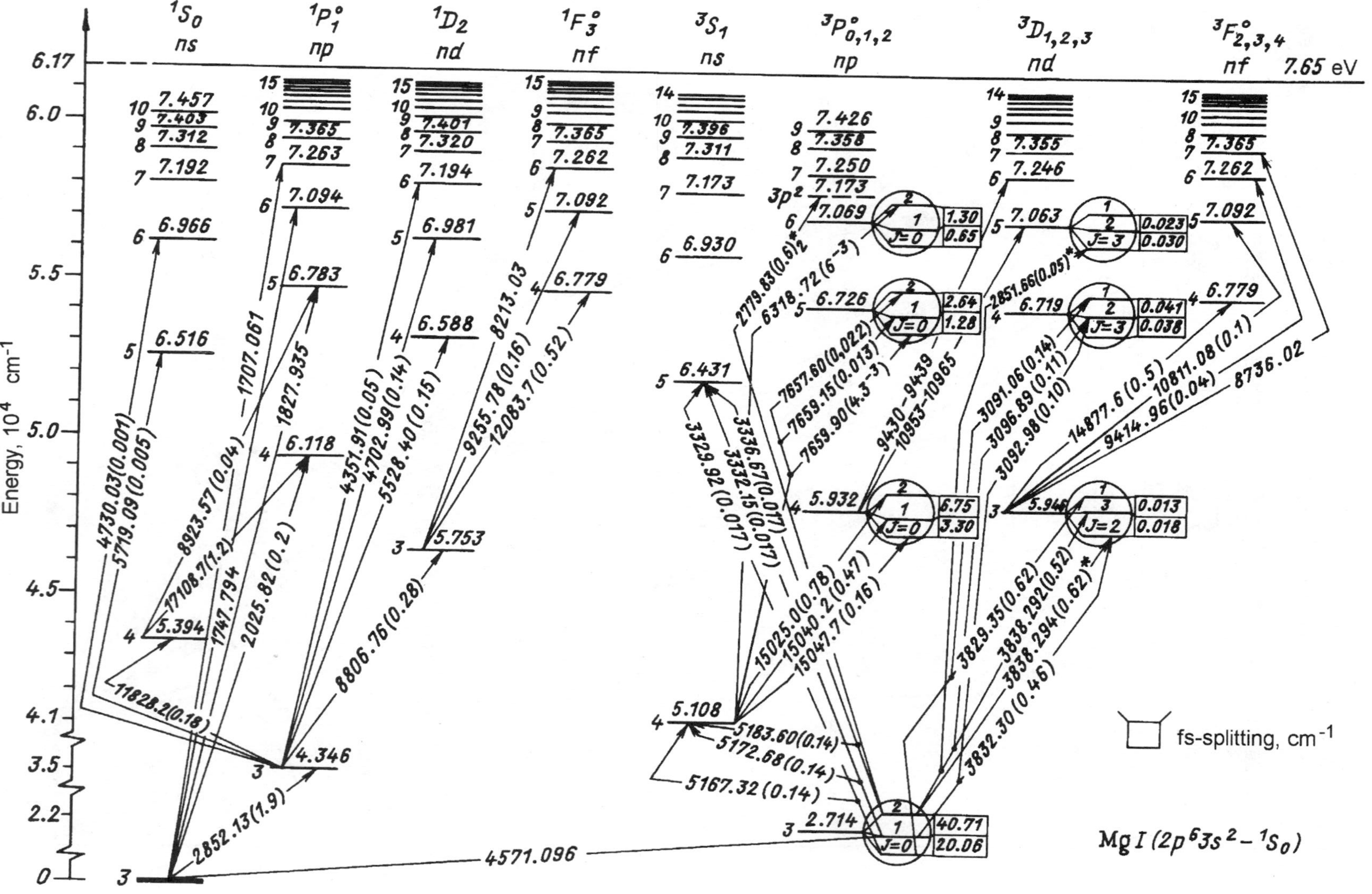

Figure 32.13 Grotrian diagram of atomic magnesium.

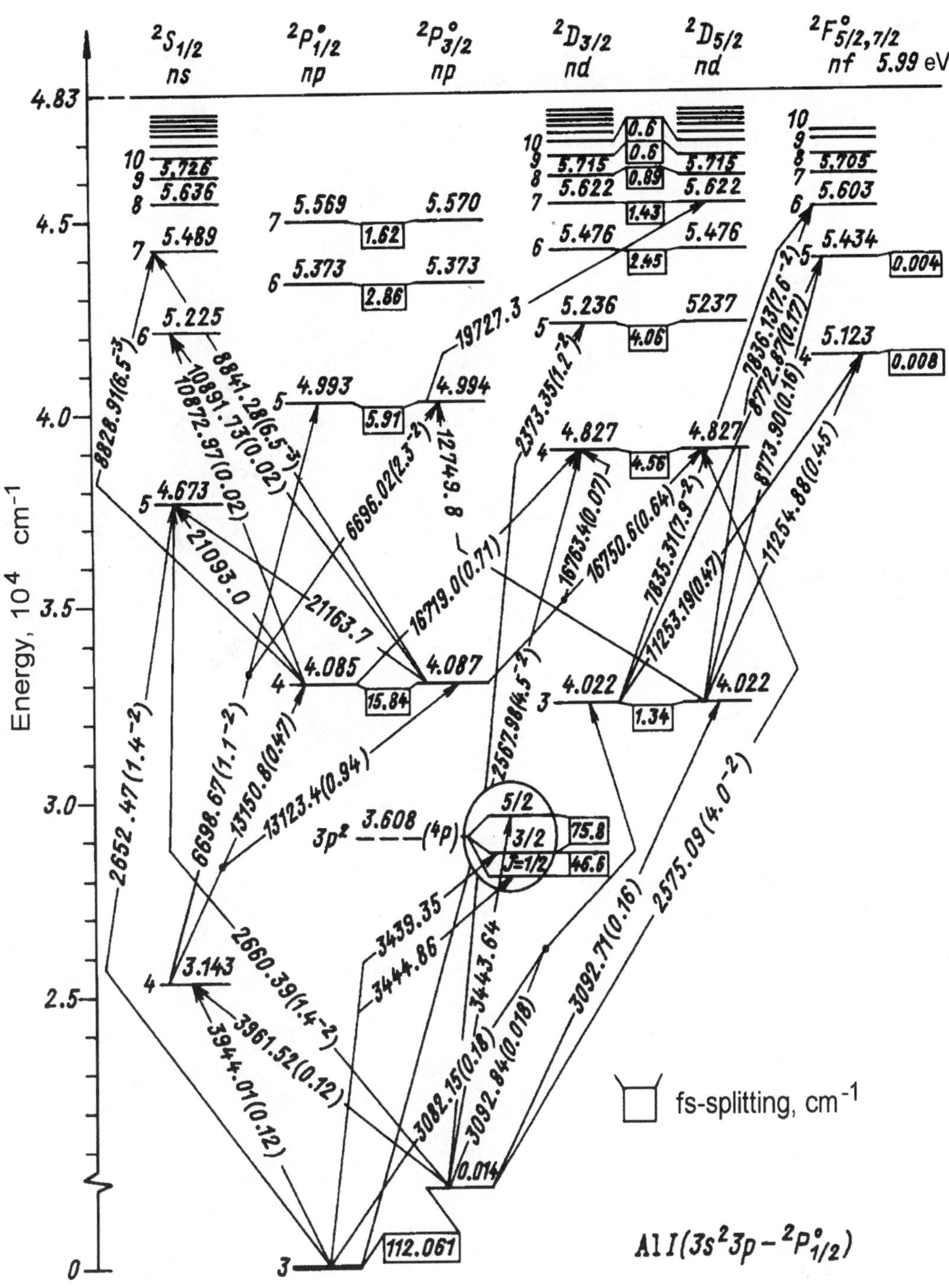

Figure 32.14 Grotrian diagram of atomic aluminum.

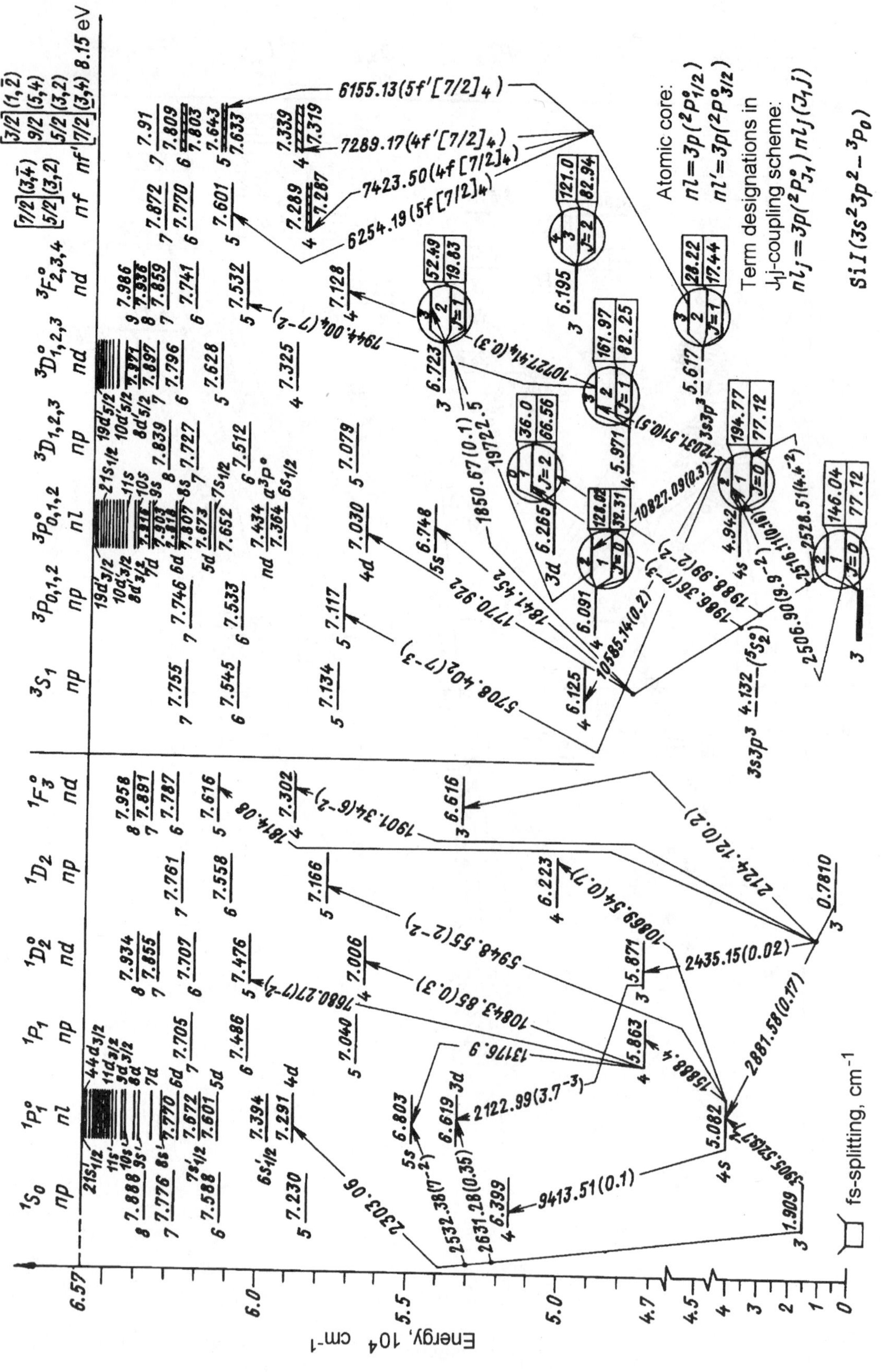

Figure 32.15 Grotrian diagram of atomic silicon.

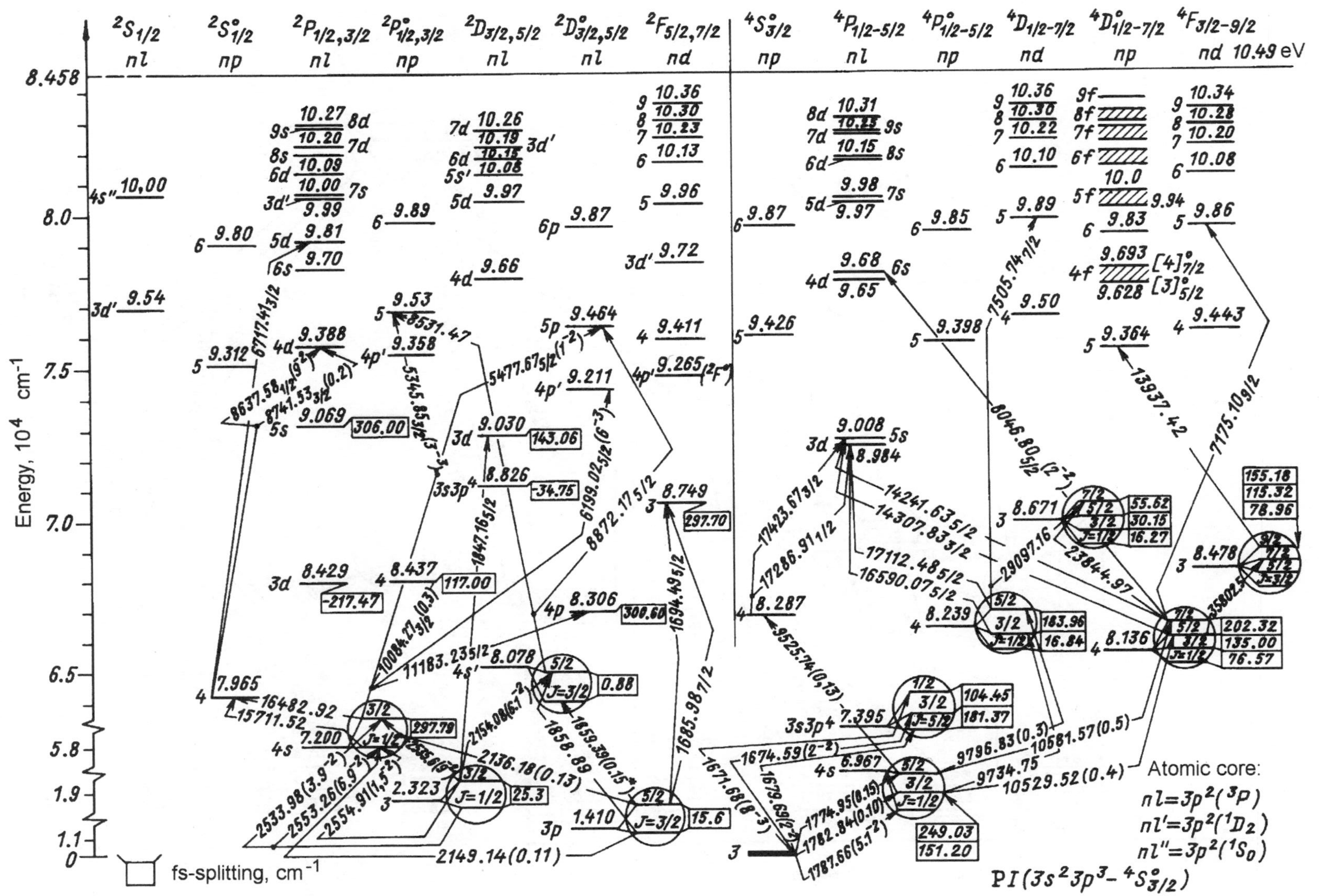

Figure 32.16 Grotrian diagram of atomic phosphorus.

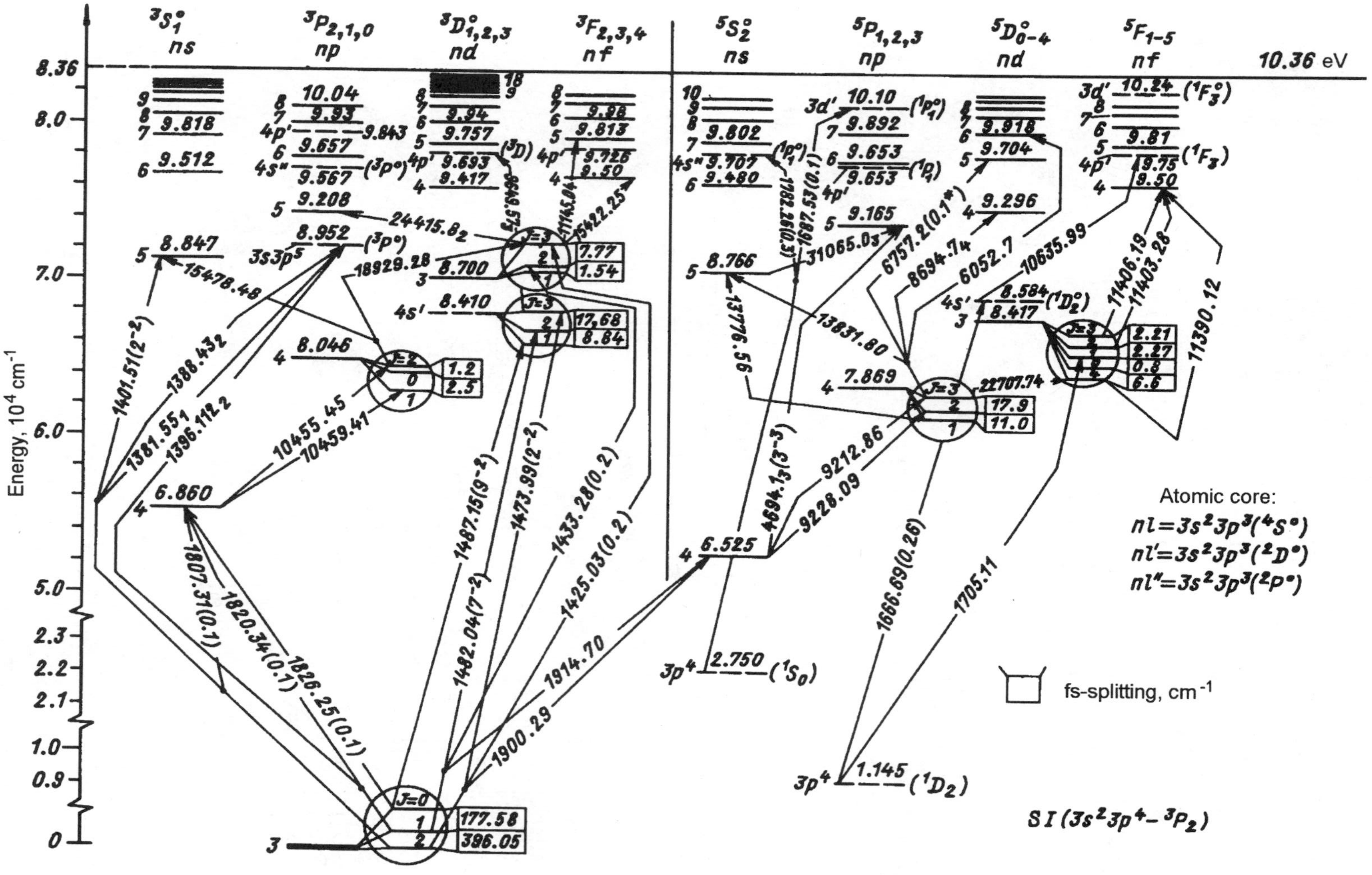

Figure 32.17 Grotrian diagram of atomic sulfur.

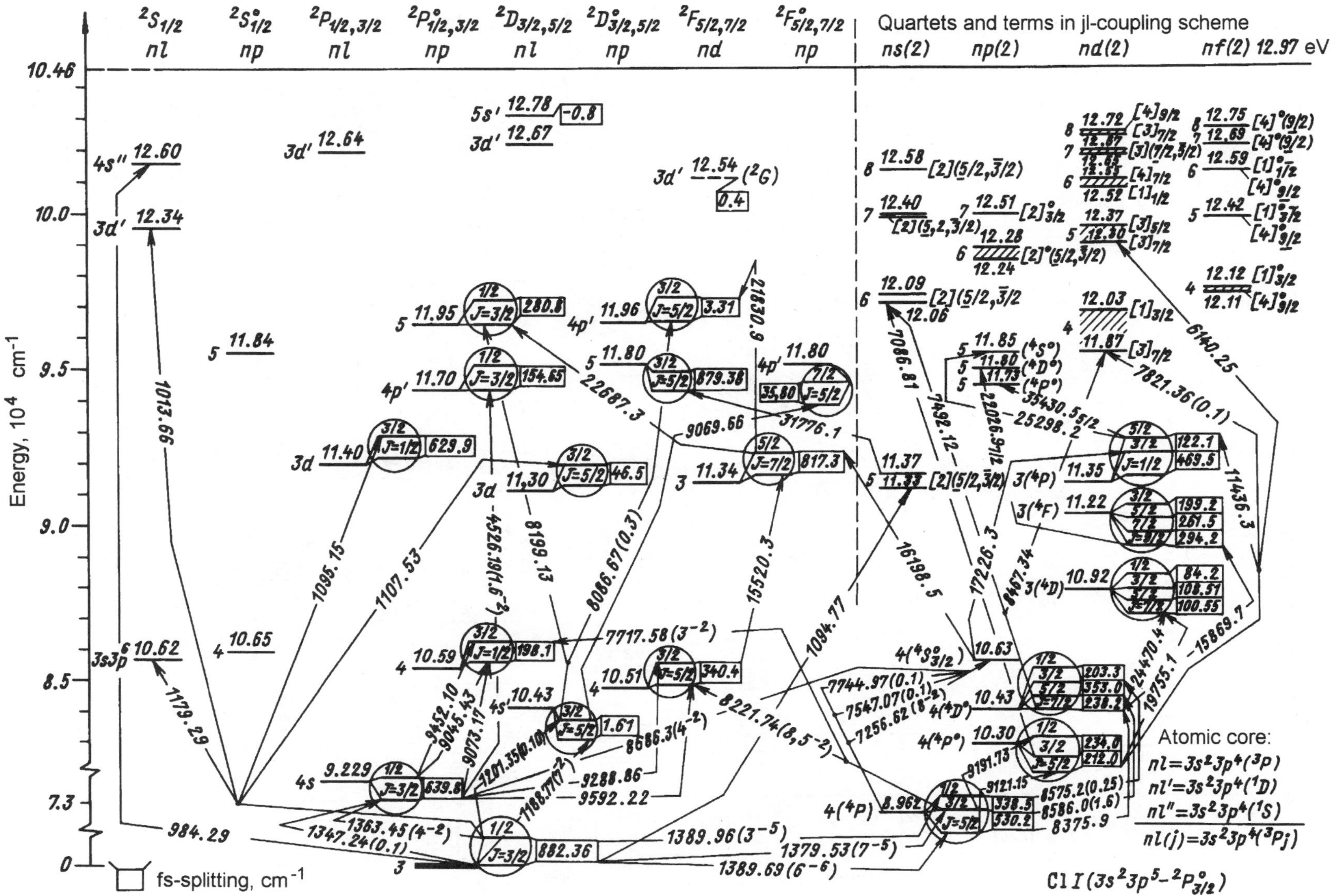

Figure 32.18 Grotrian diagram of atomic chlorine.

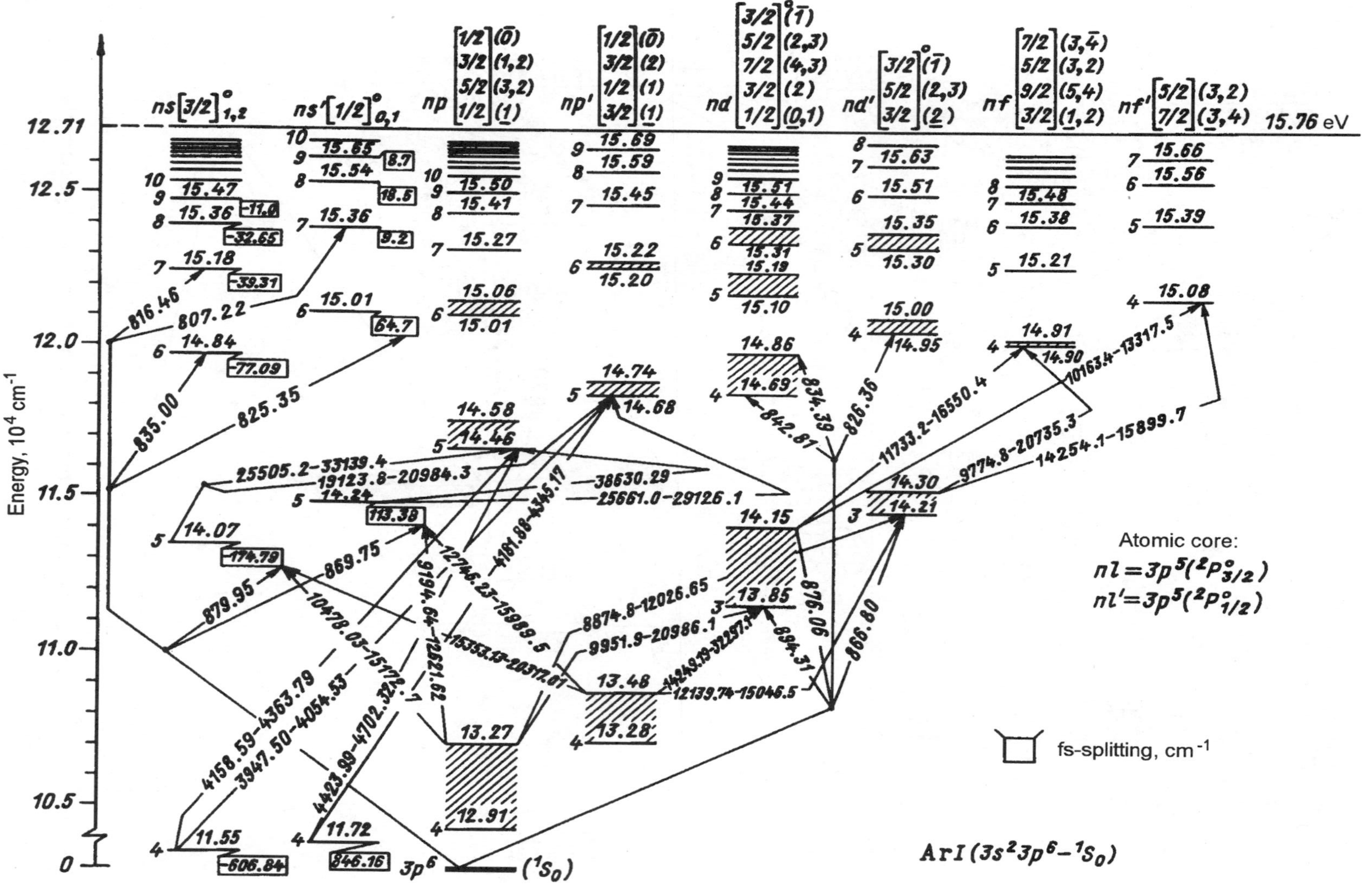

Figure 32.19 Grotrian diagram of atomic argon.

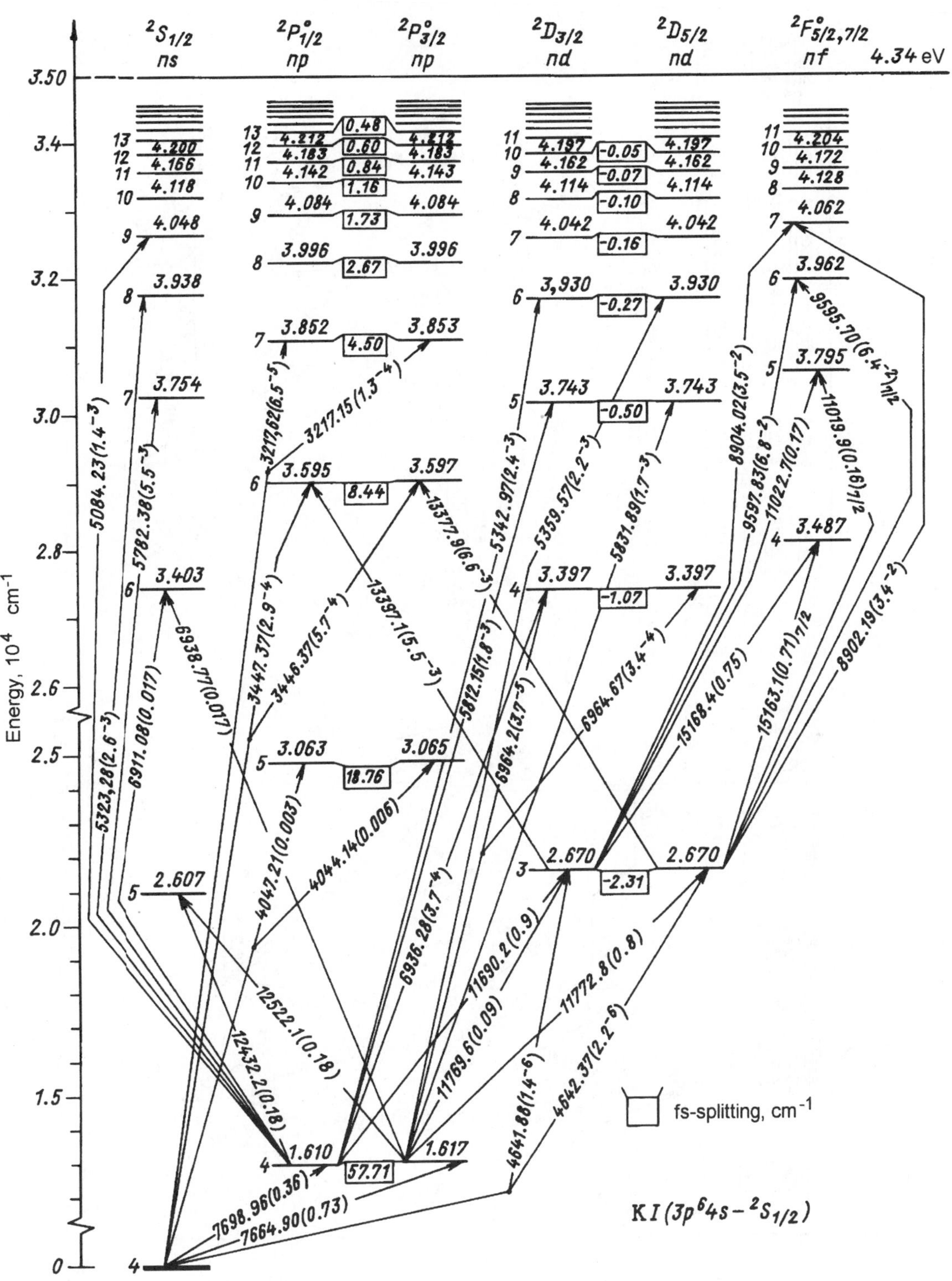

Figure 32.20 Grotrian diagram of atomic potassium.

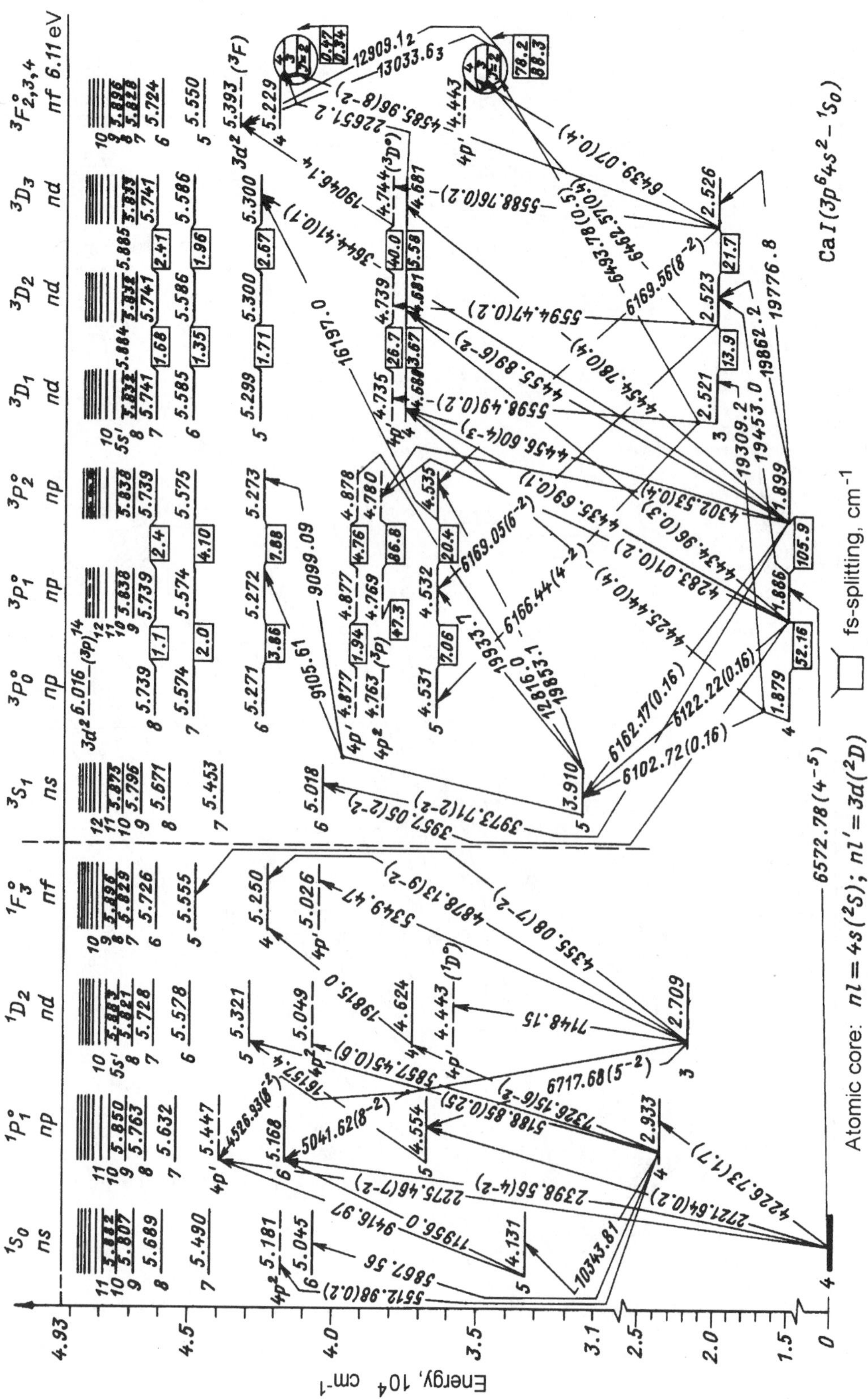

Figure 32.21 Grotrian diagram of atomic calcium.

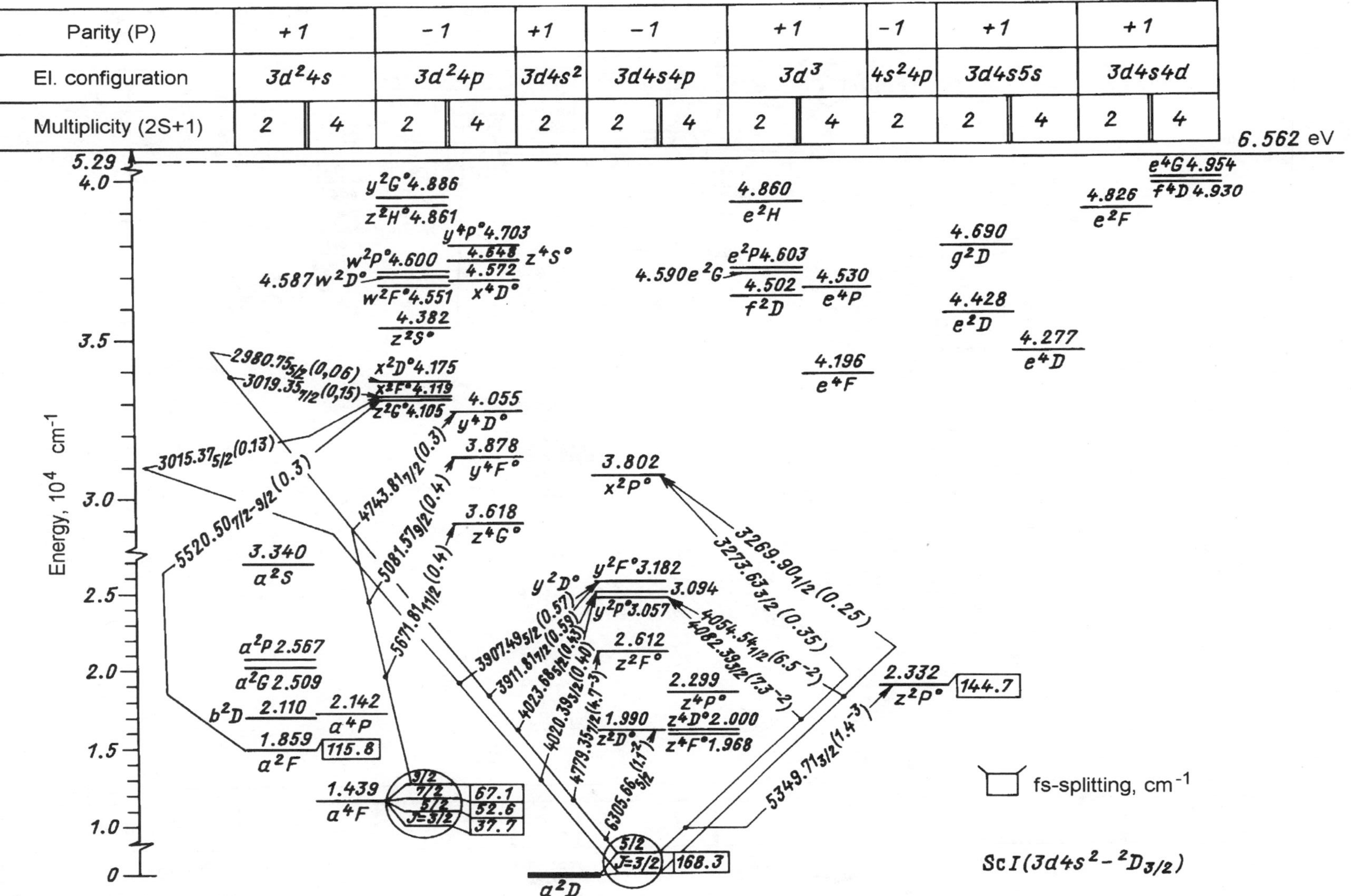

Figure 32.22 Abridged Grotrian diagram of atomic scandium.

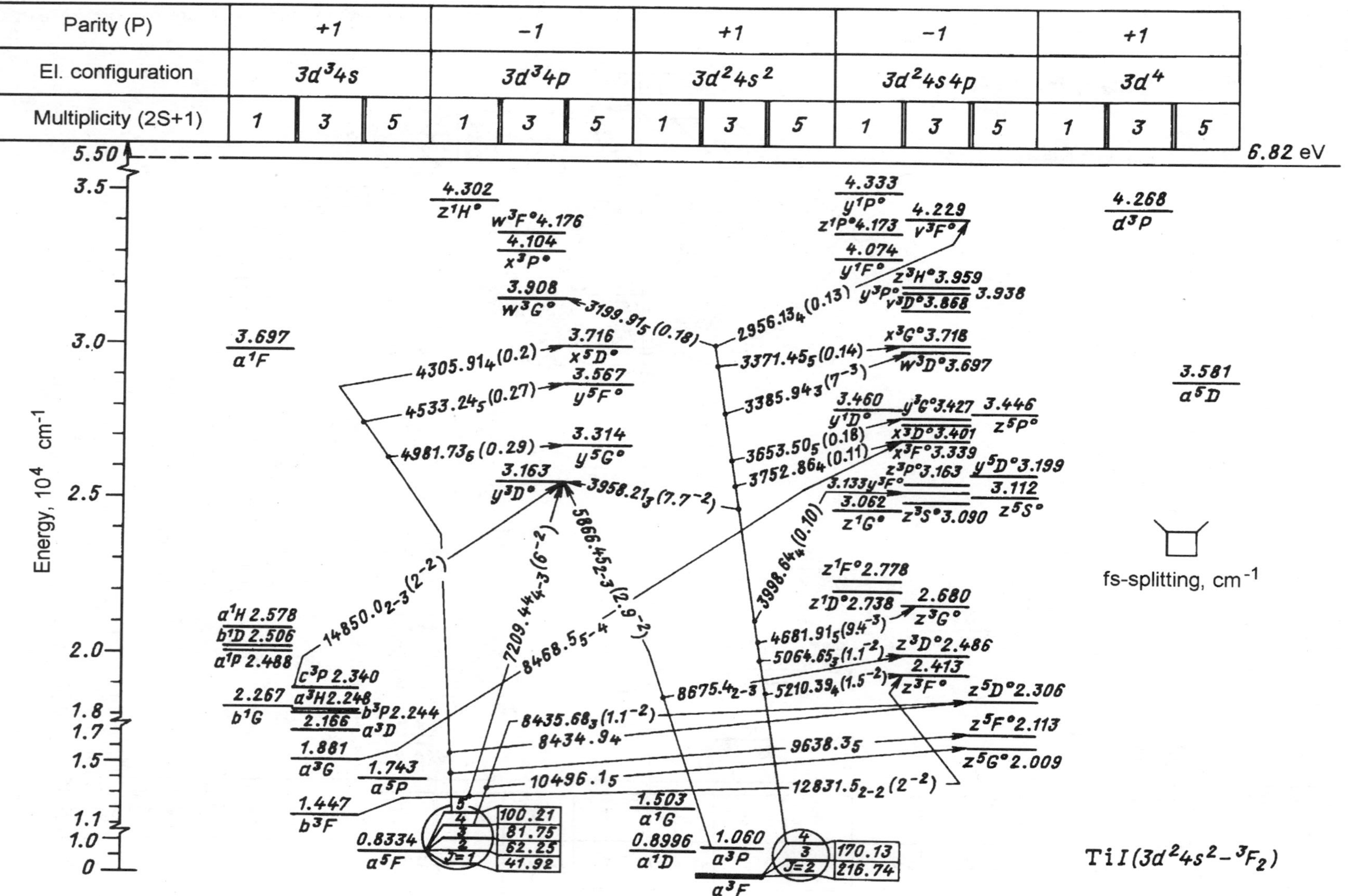

Figure 32.23 Abridged Grotrian diagram of atomic titanium.

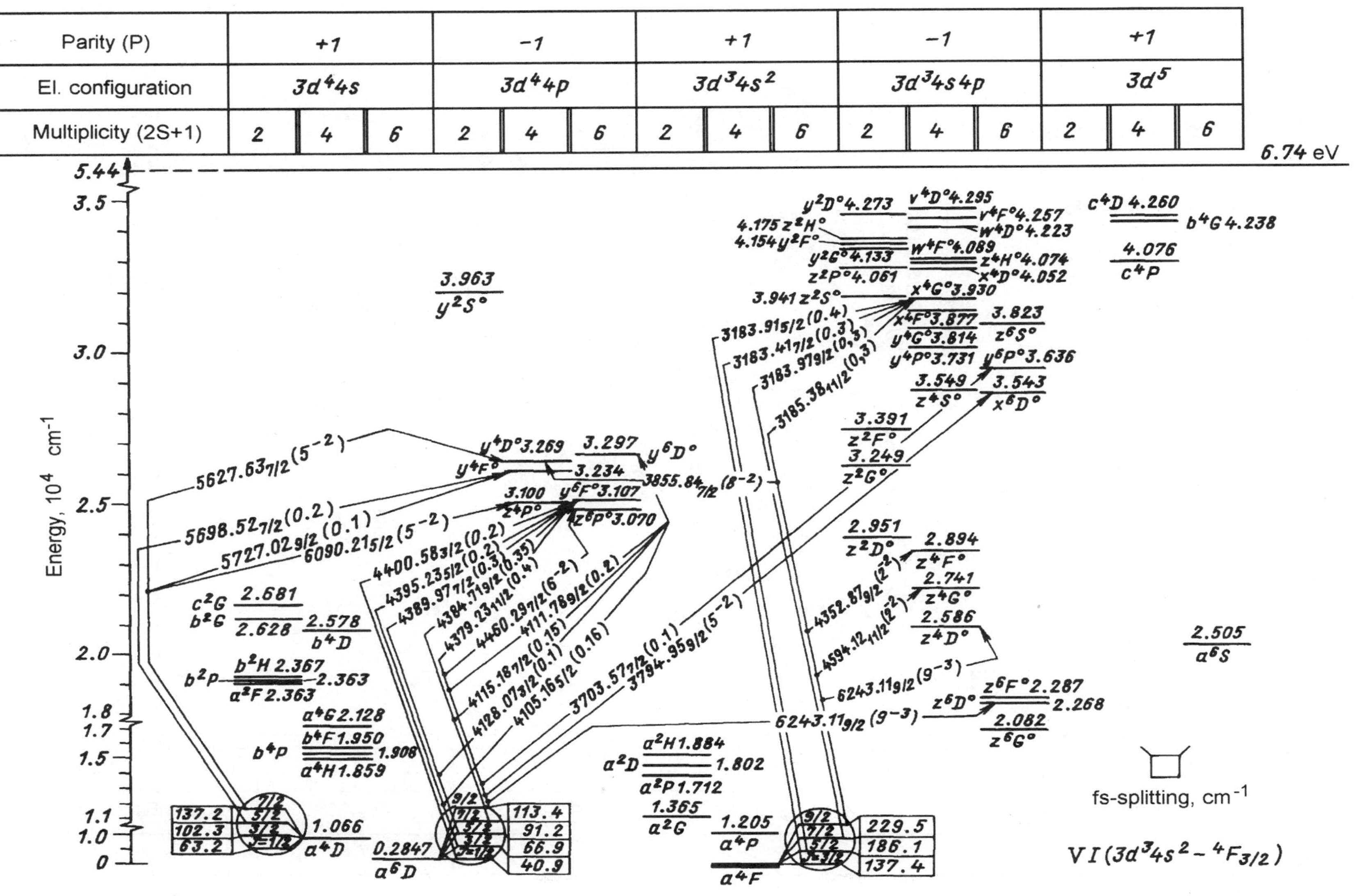

Figure 32.24 Abridged Grotrian diagram of atomic vanadium.

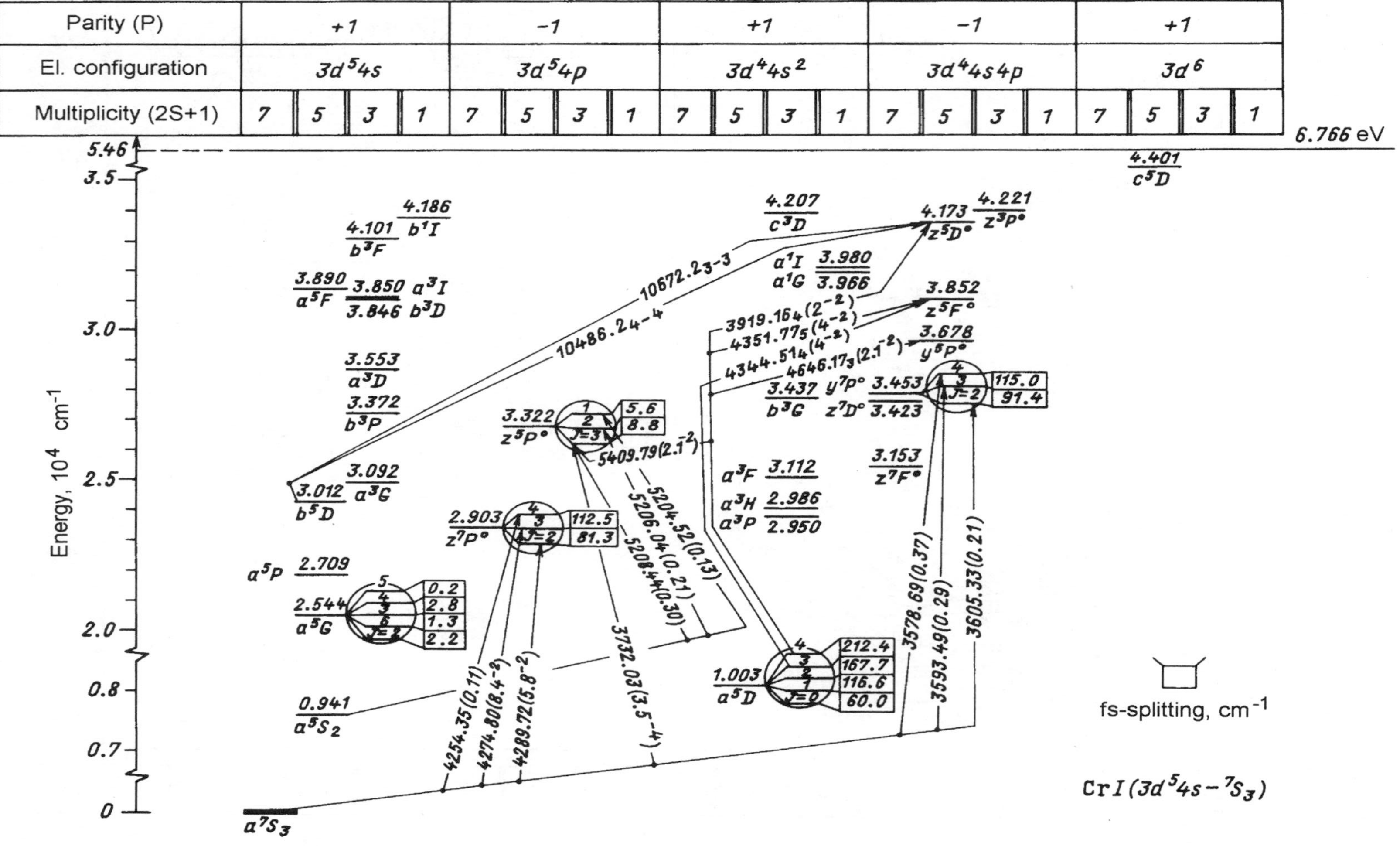

Figure 32.25 Abridged Grotrian diagram of atomic chromium.

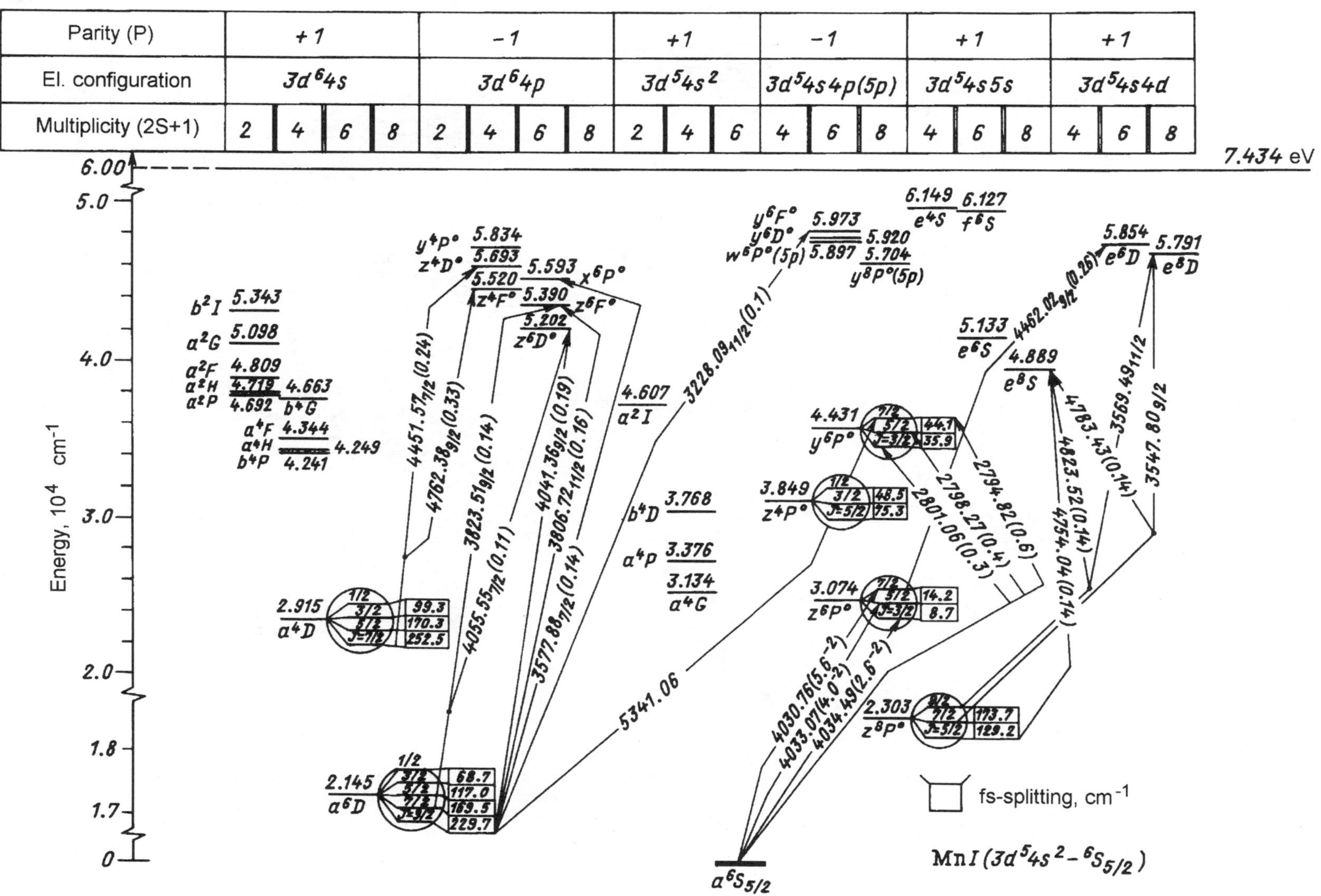

Figure 32.26 Abridged Grotrian diagram of atomic manganese.

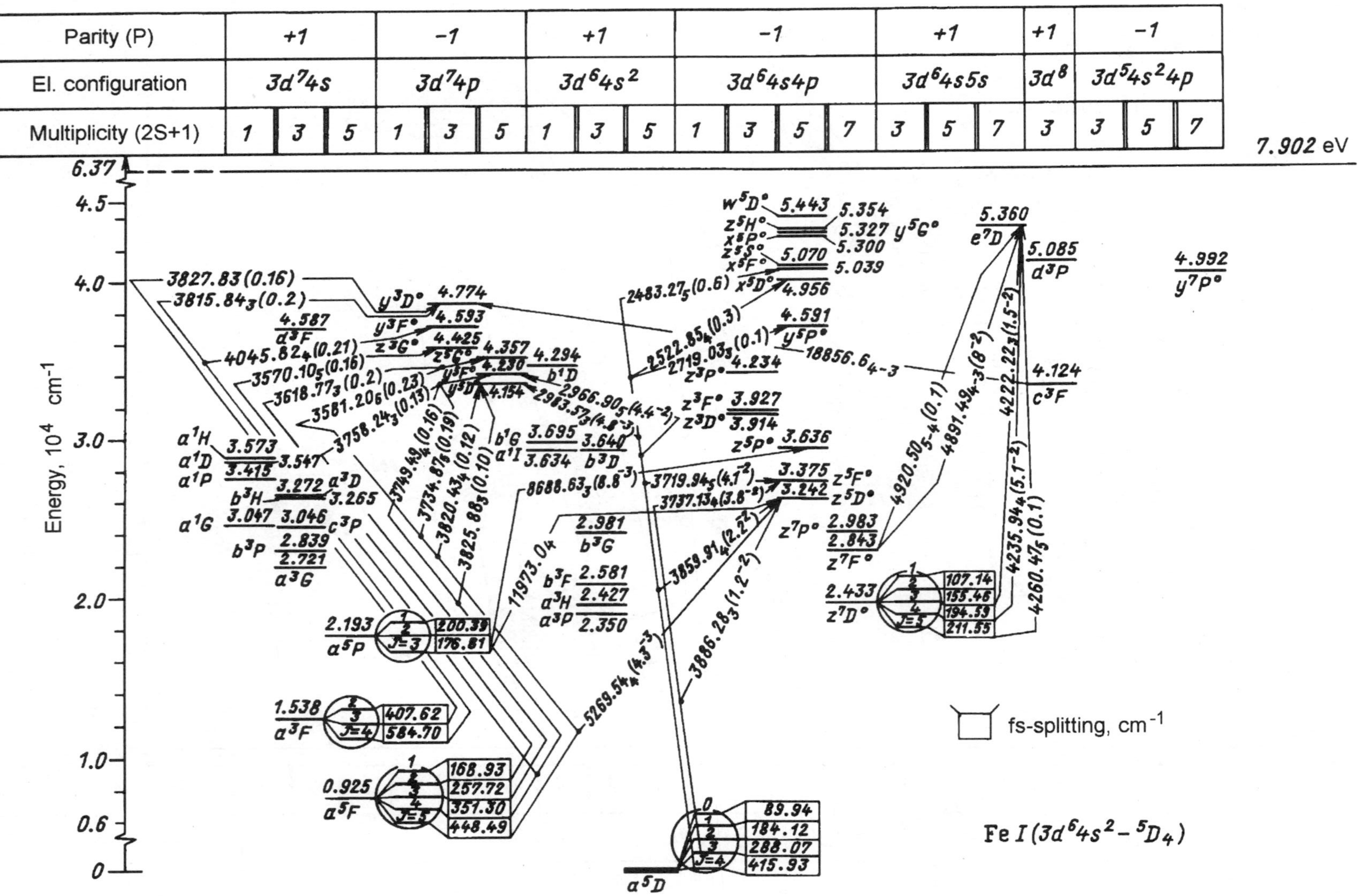

Figure 32.27 Abridged Grotrian diagram of atomic iron.

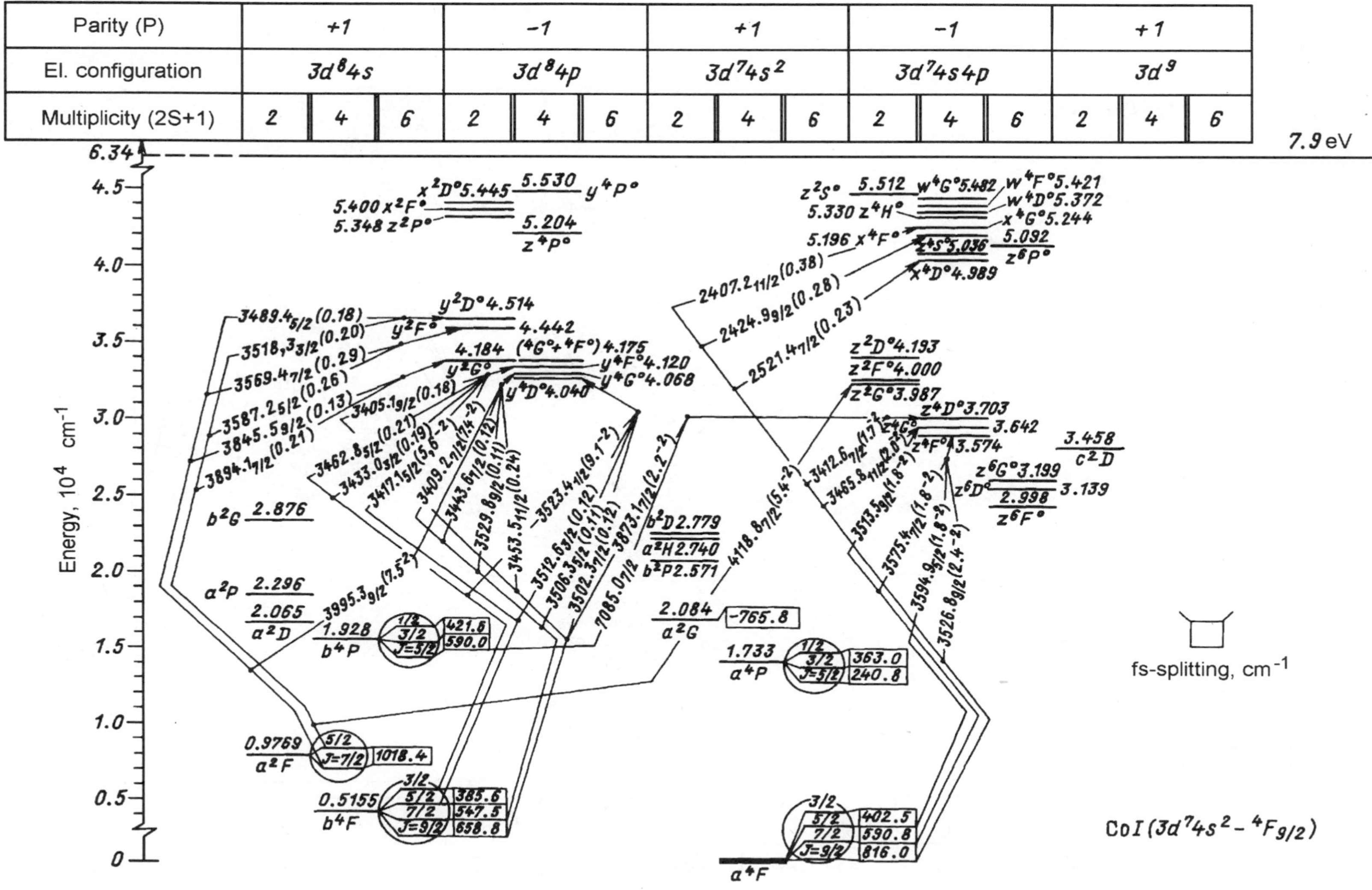

Figure 32.28 Abridged Grotrian diagram of atomic cobalt.

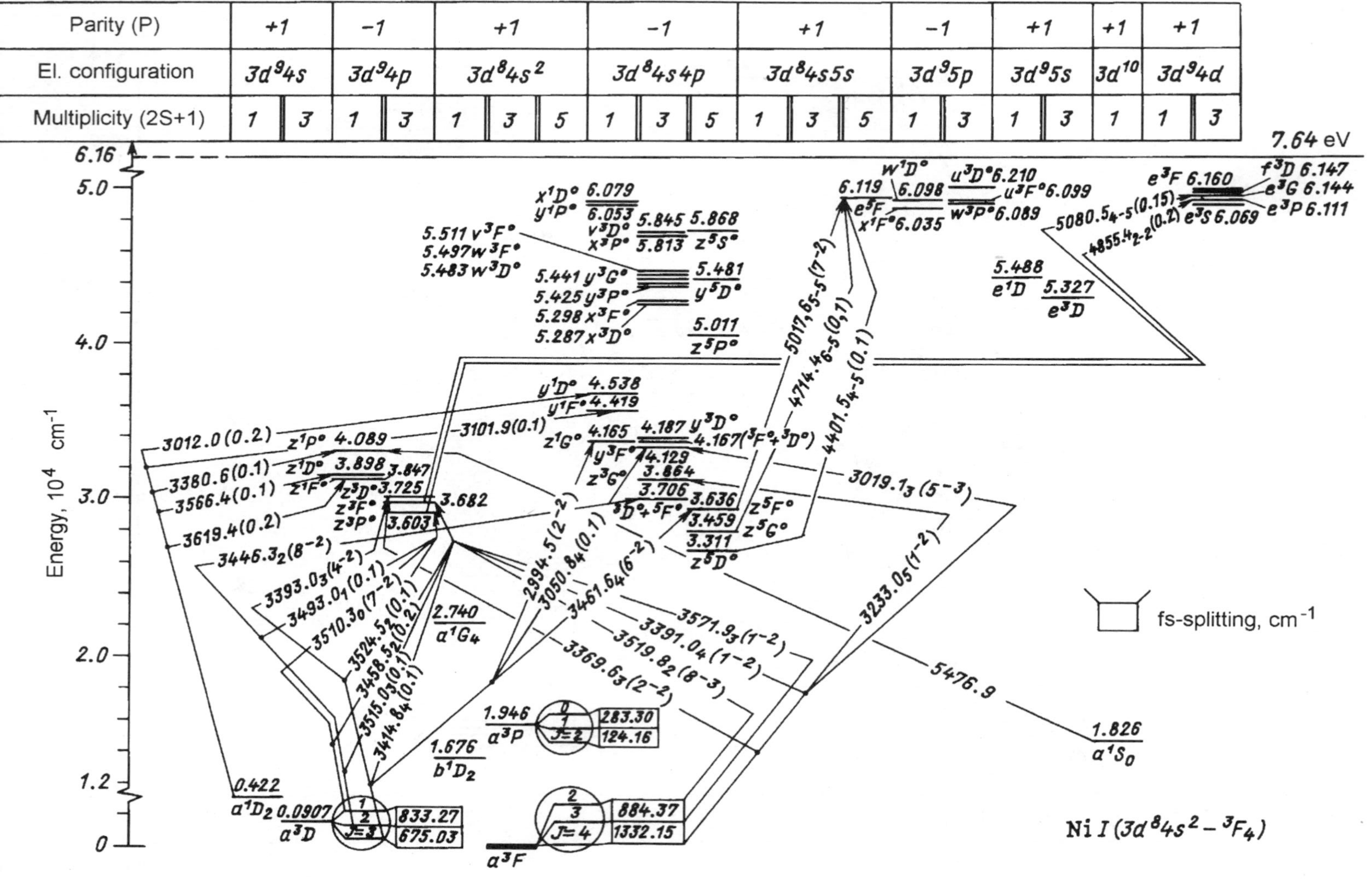

Figure 32.29 Abridged Grotrian diagram of atomic nickel.

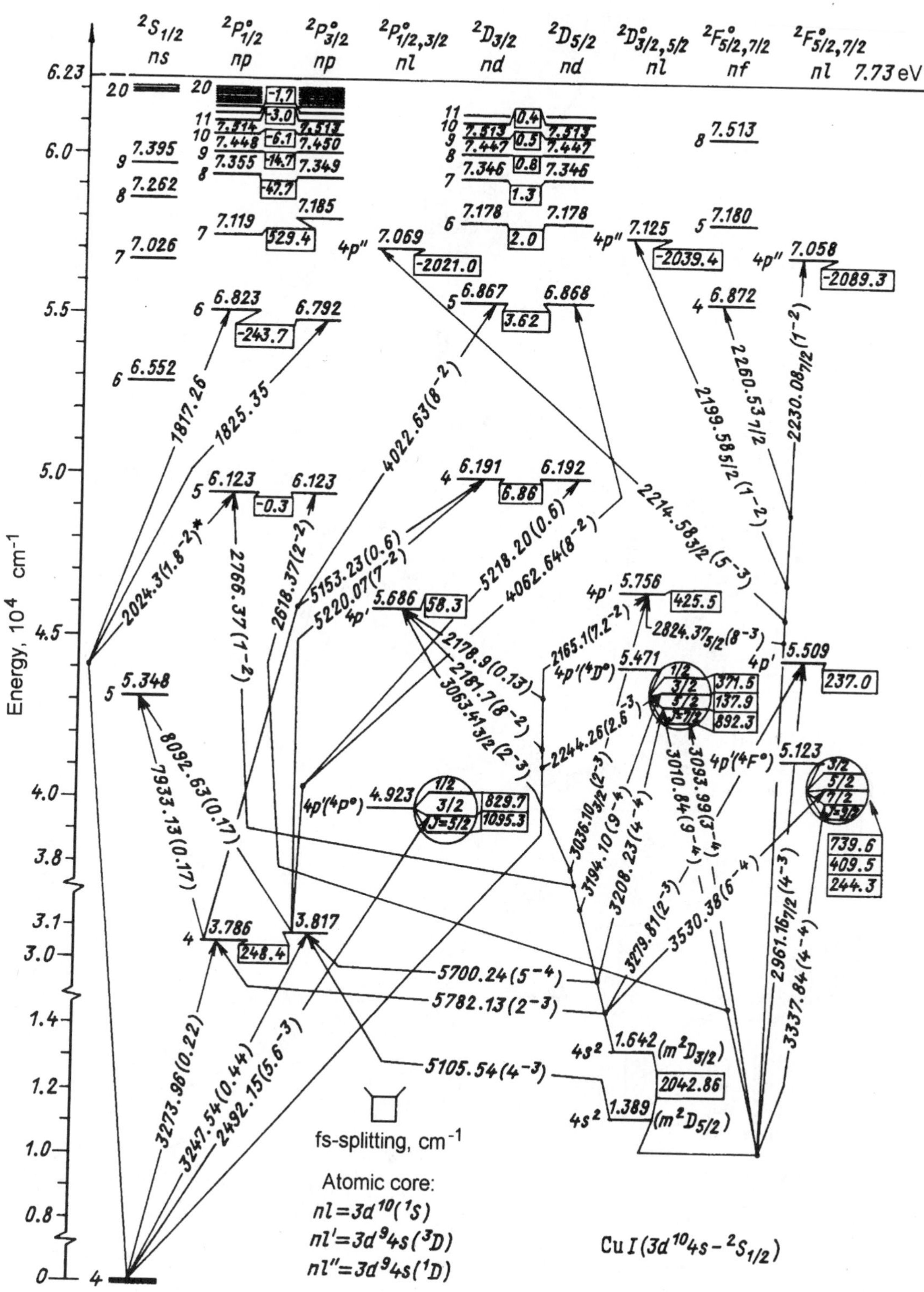

Figure 32.30 Grotrian diagram of atomic copper.

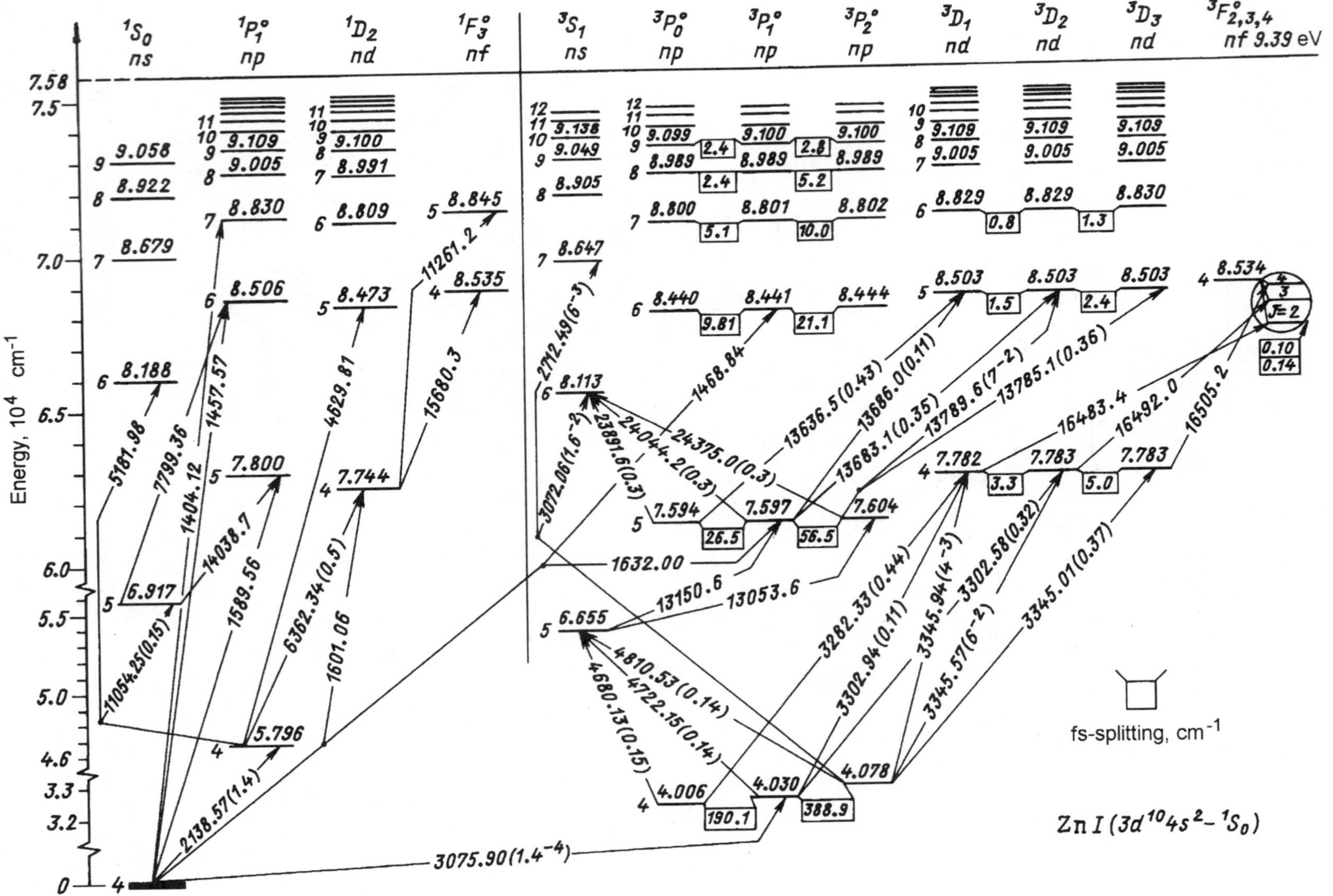

Figure 32.31 Grotrian diagram of atomic zinc.

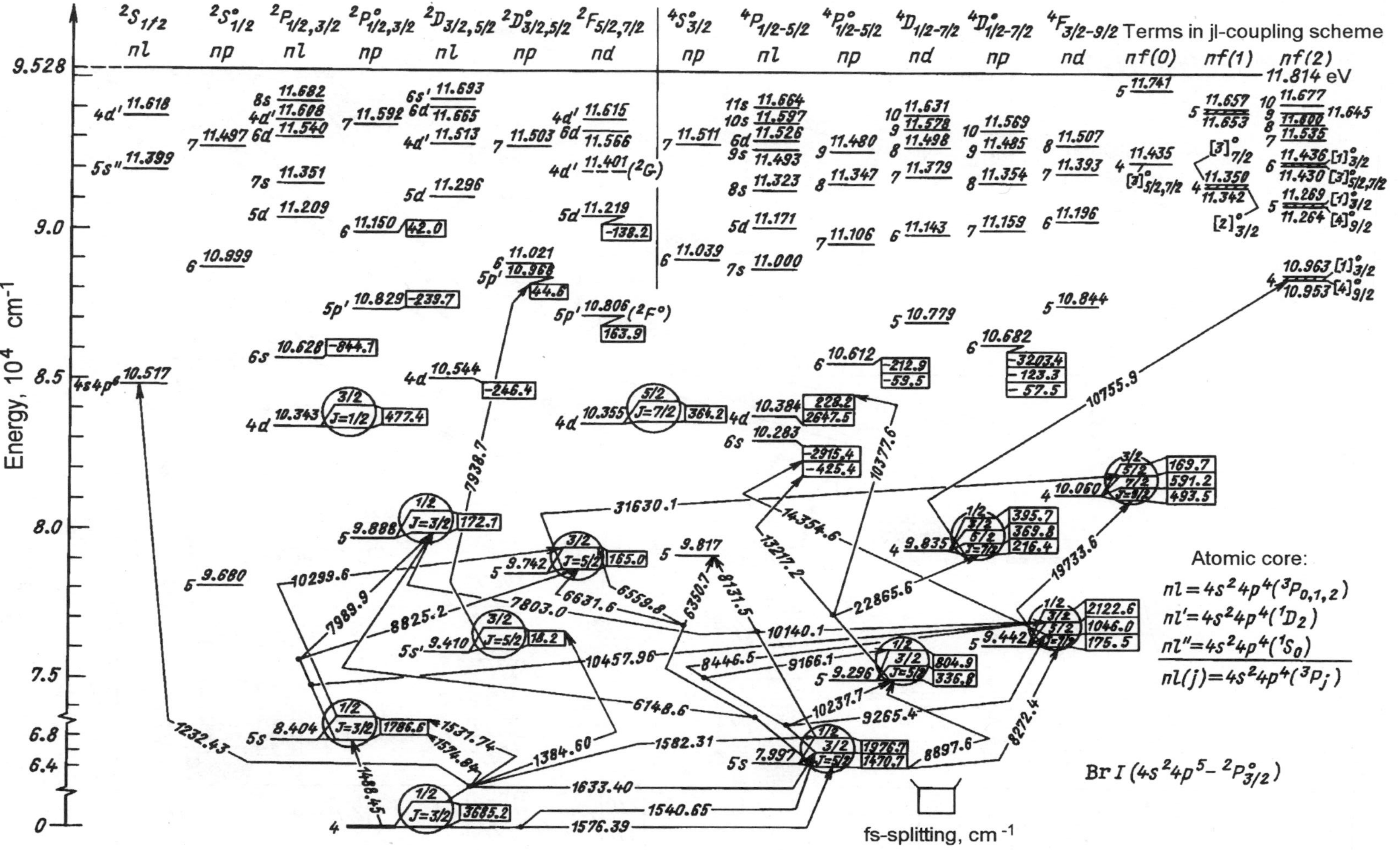

Figure 32.32 Grotrian diagram of atomic bromine.

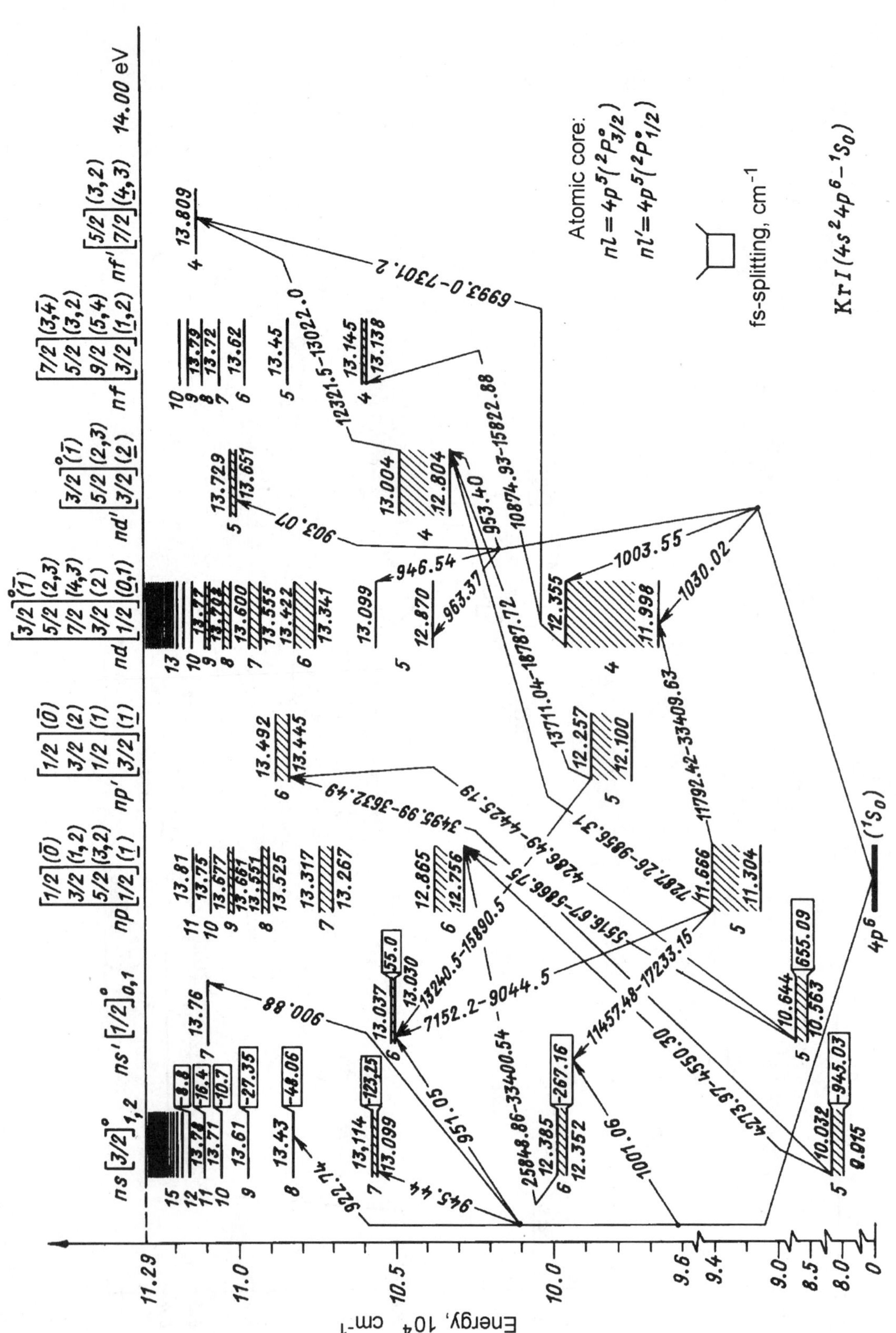

Figure 32.33 Grotrian diagram of atomic krypton.

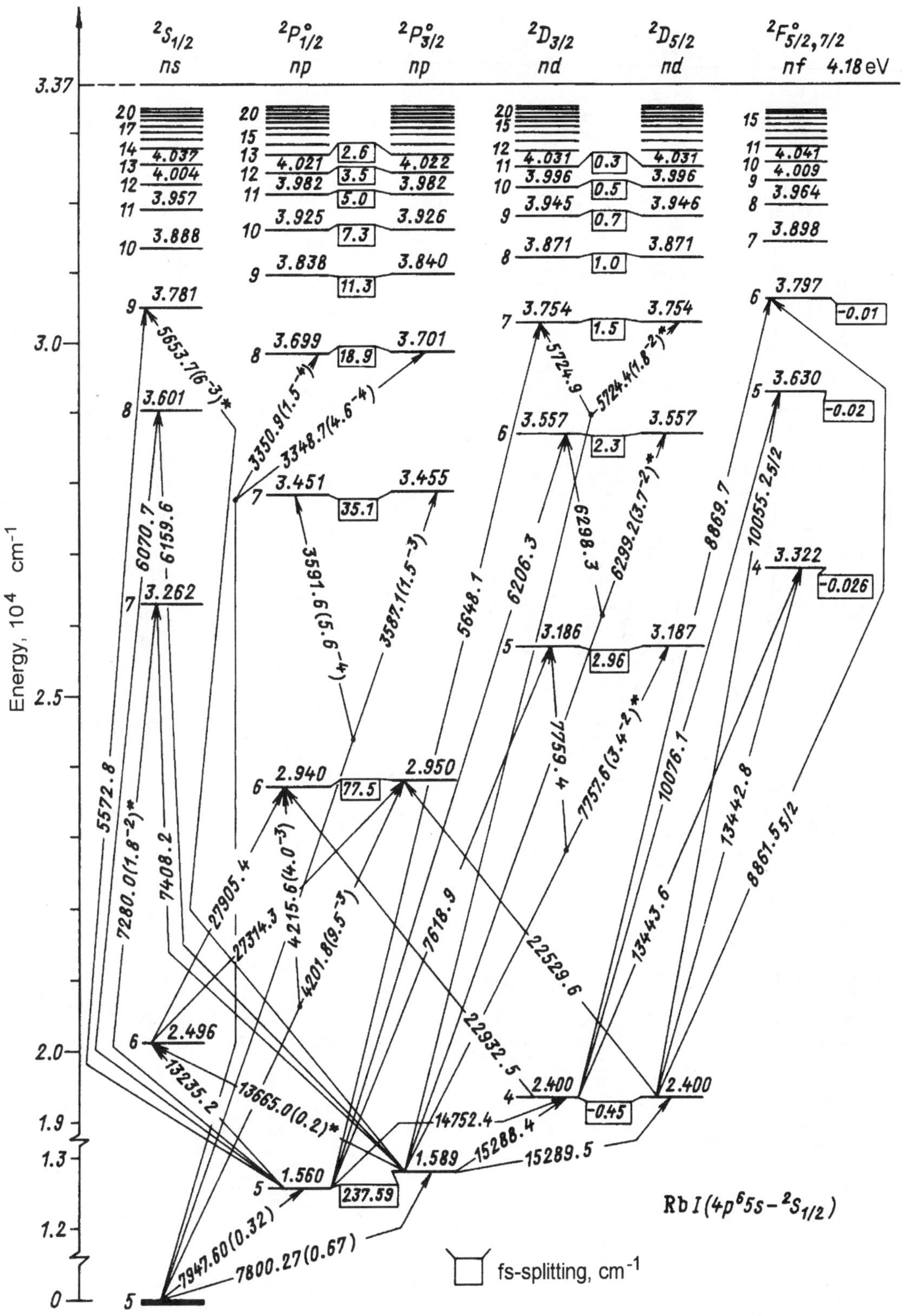

Figure 32.34 Grotrian diagram of atomic rubidium.

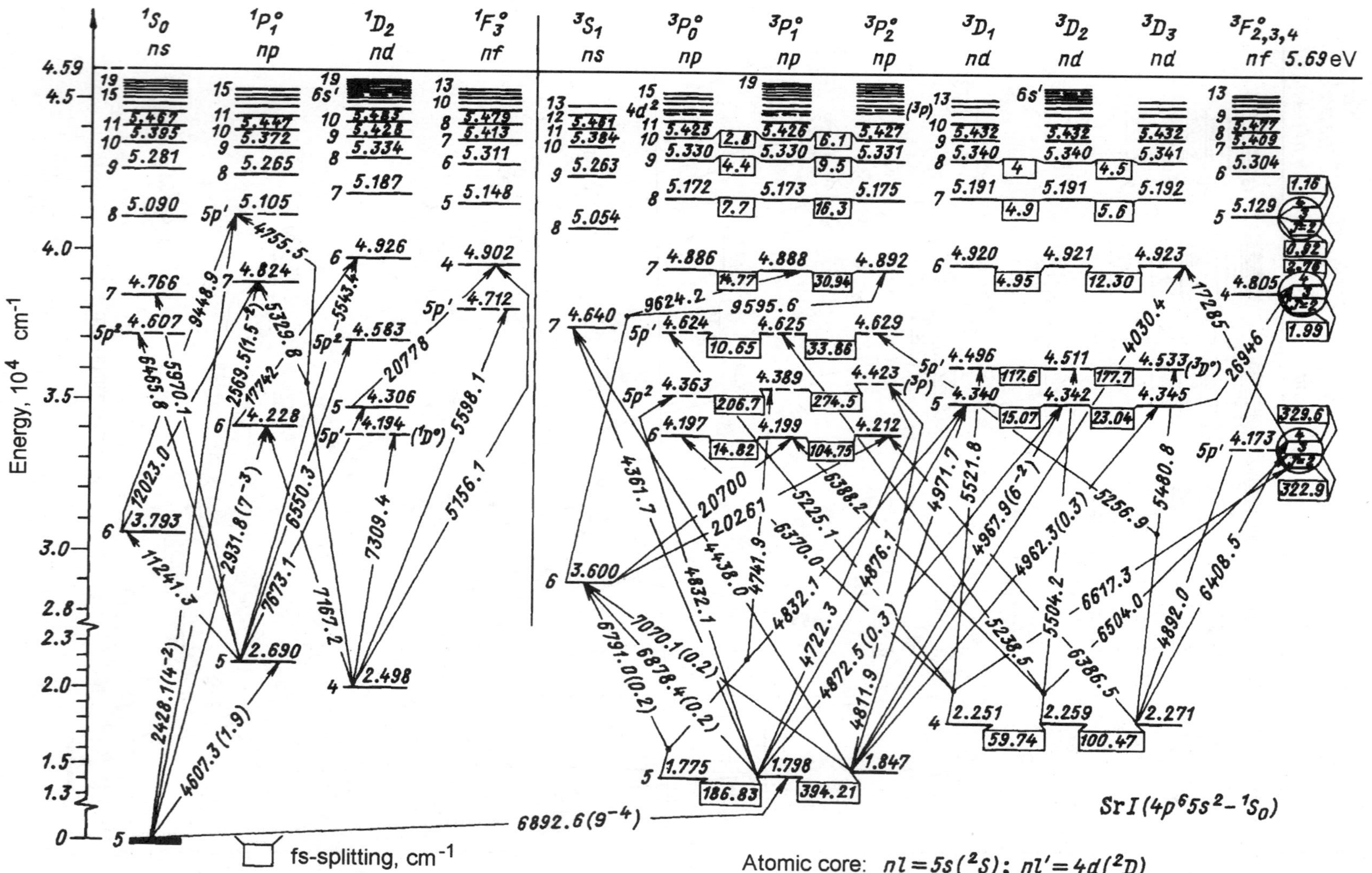

Figure 32.35 Grotrian diagram of atomic strontium.

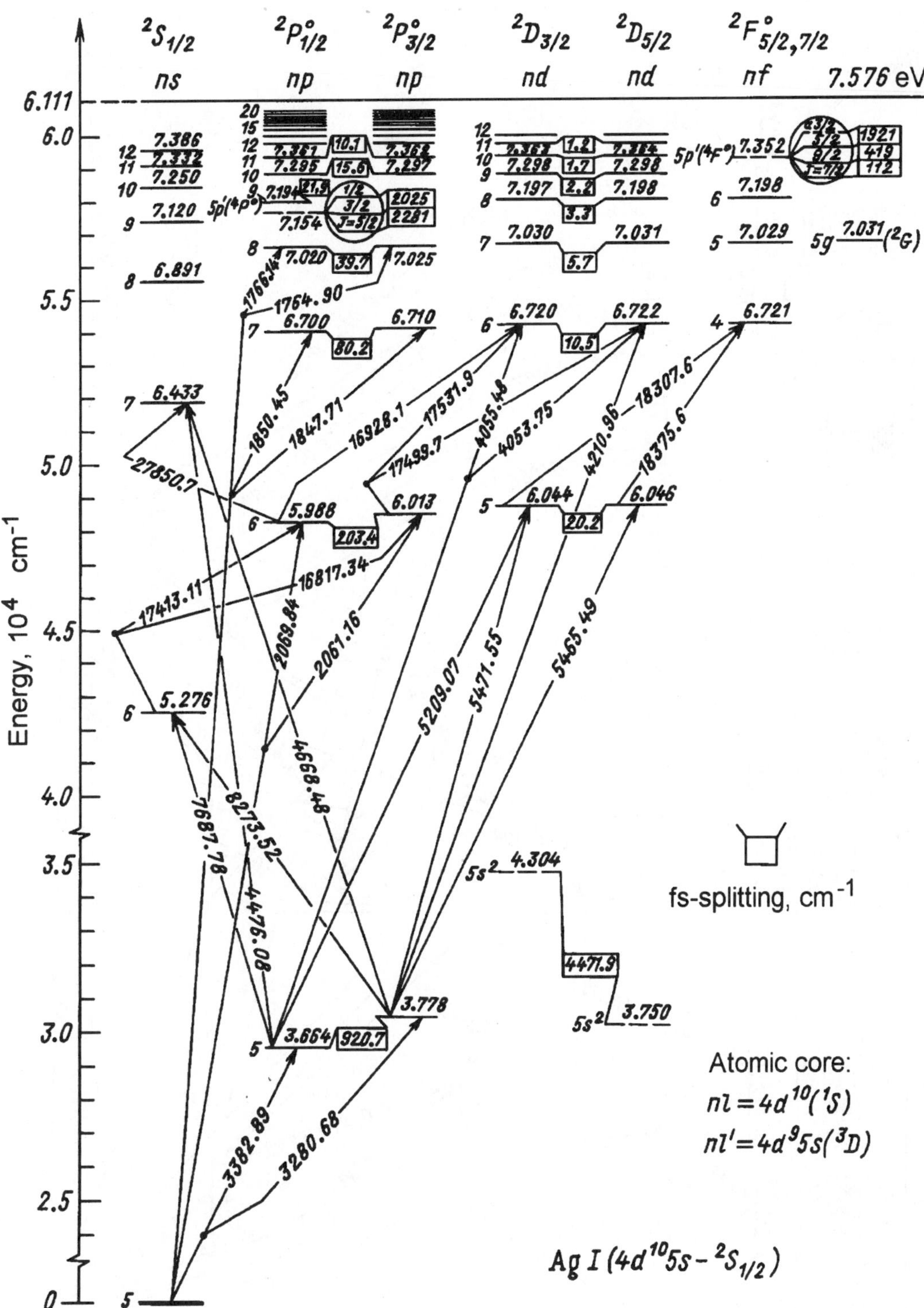

Figure 32.36 Grotrian diagram of atomic silver.

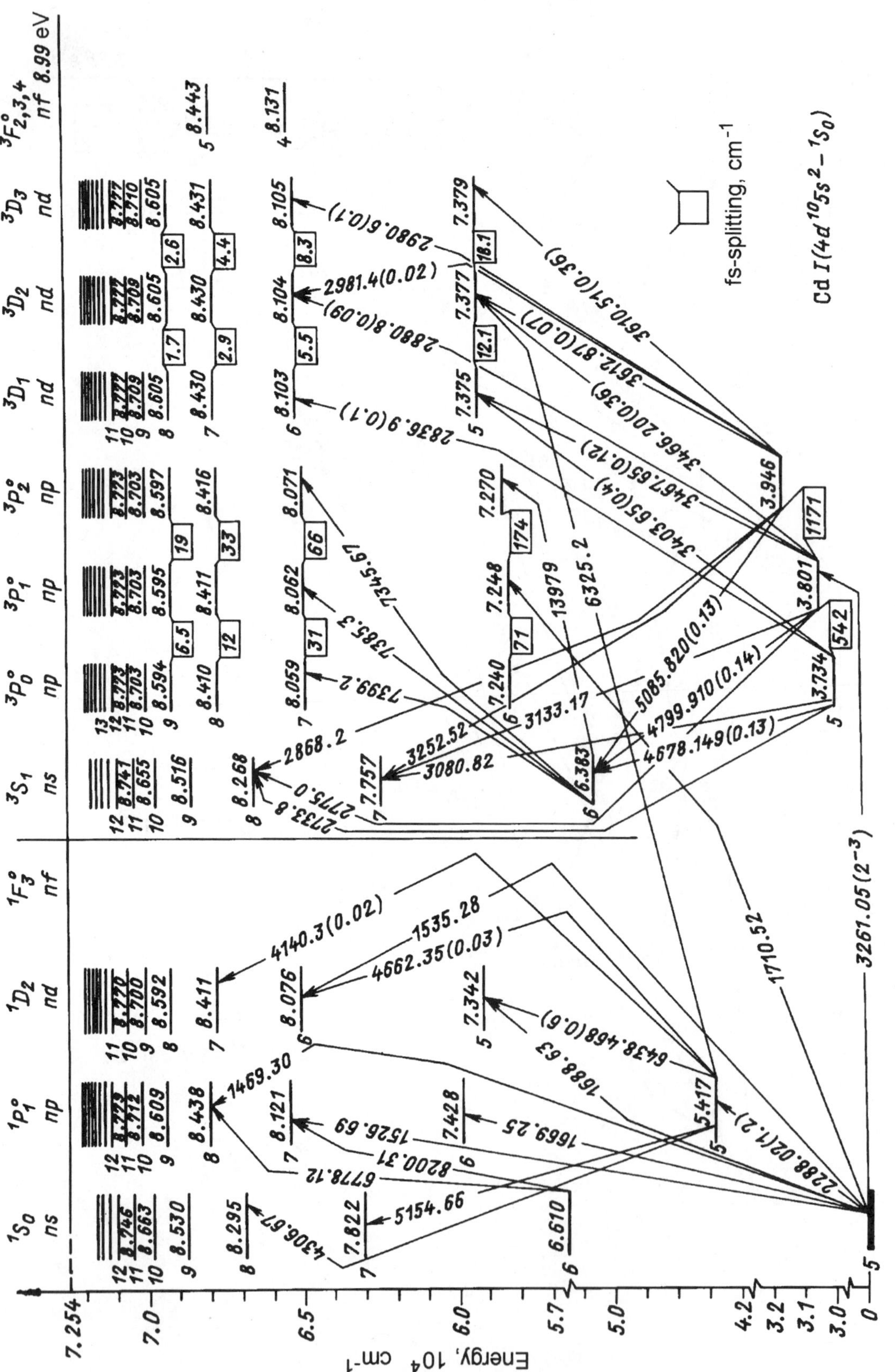

Figure 32.37 Grotrian diagram of atomic cadmium.

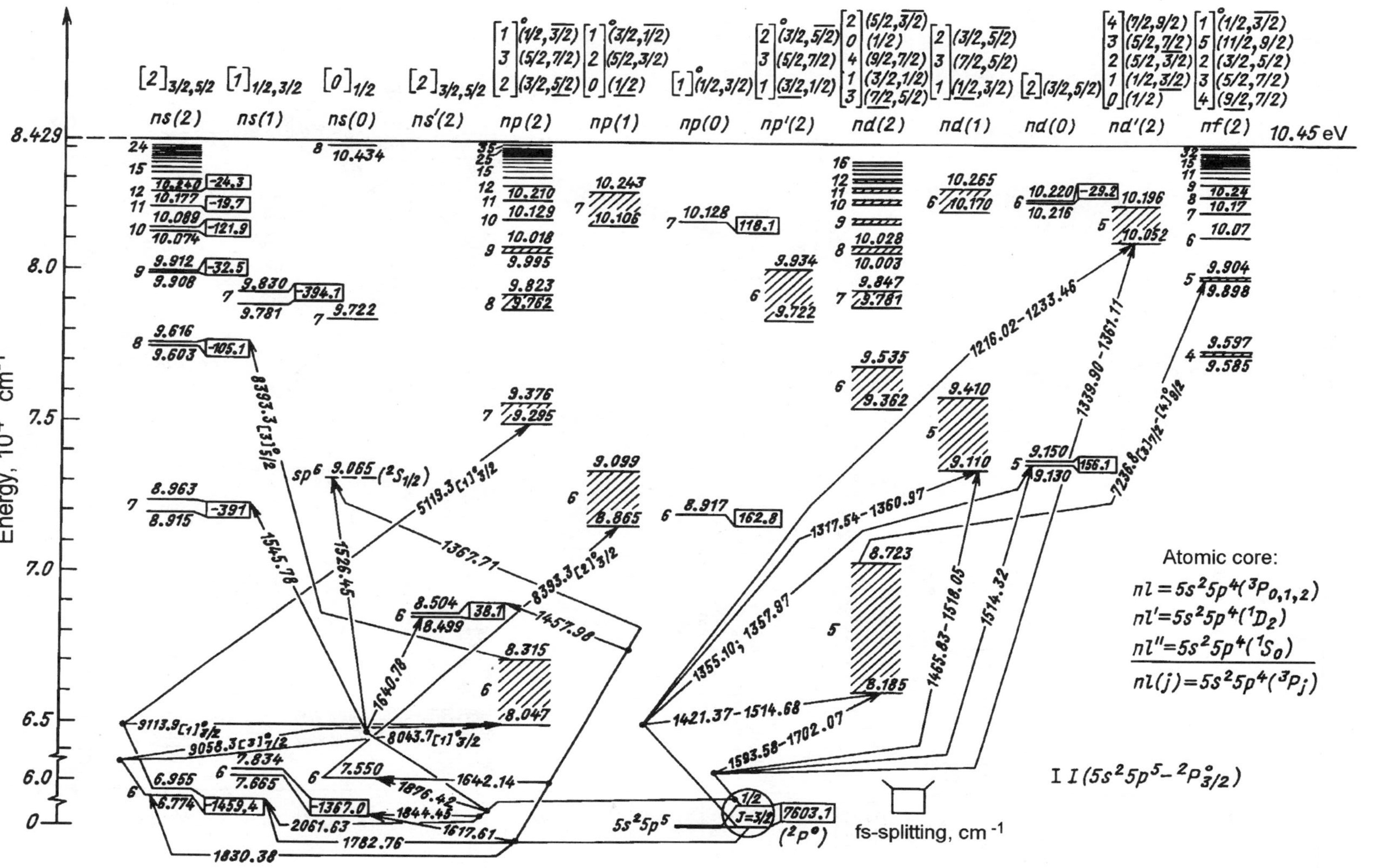

Figure 32.38 Grotrian diagram of atomic iodine.

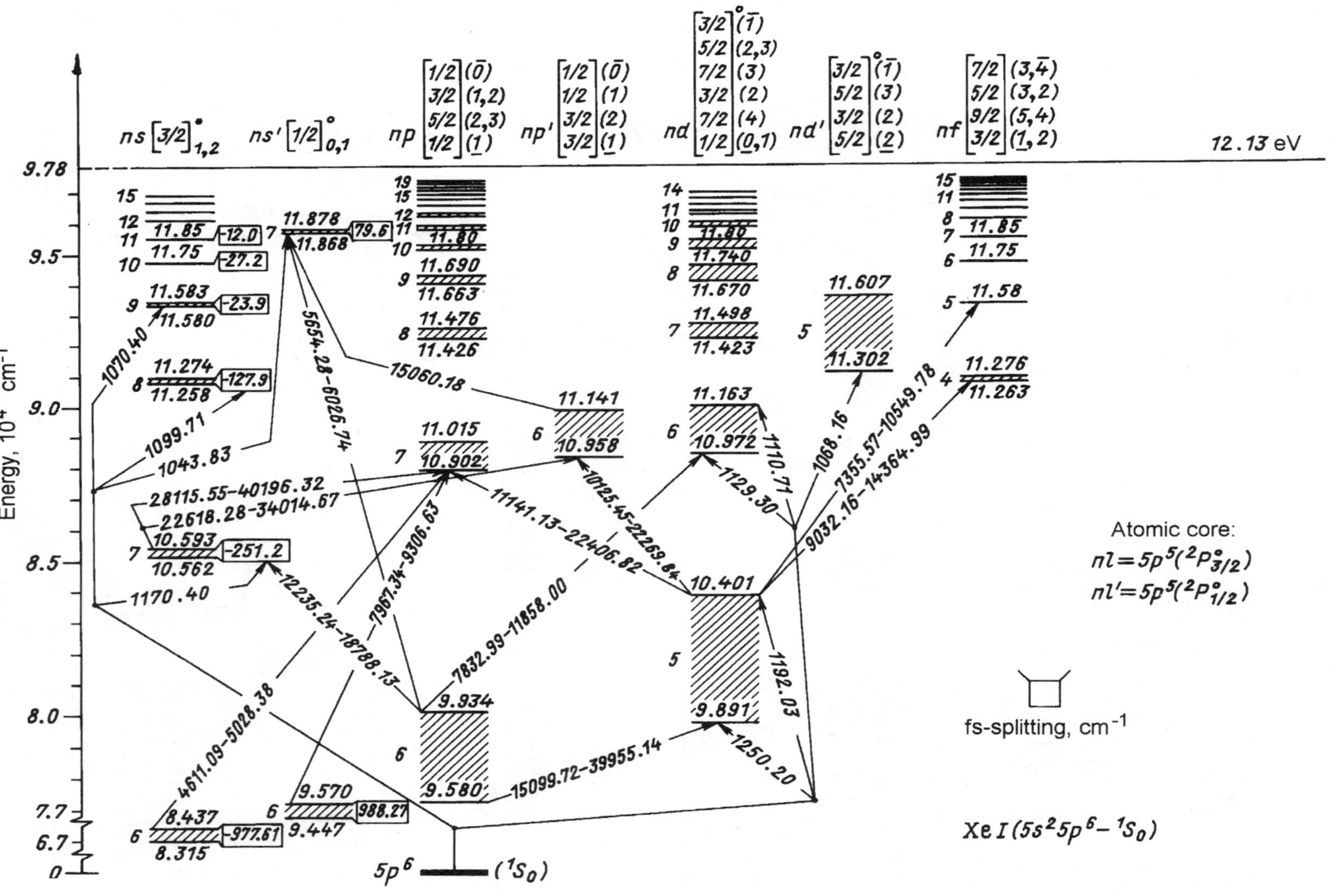

Figure 32.39 Grotrian diagram of atomic xenon.

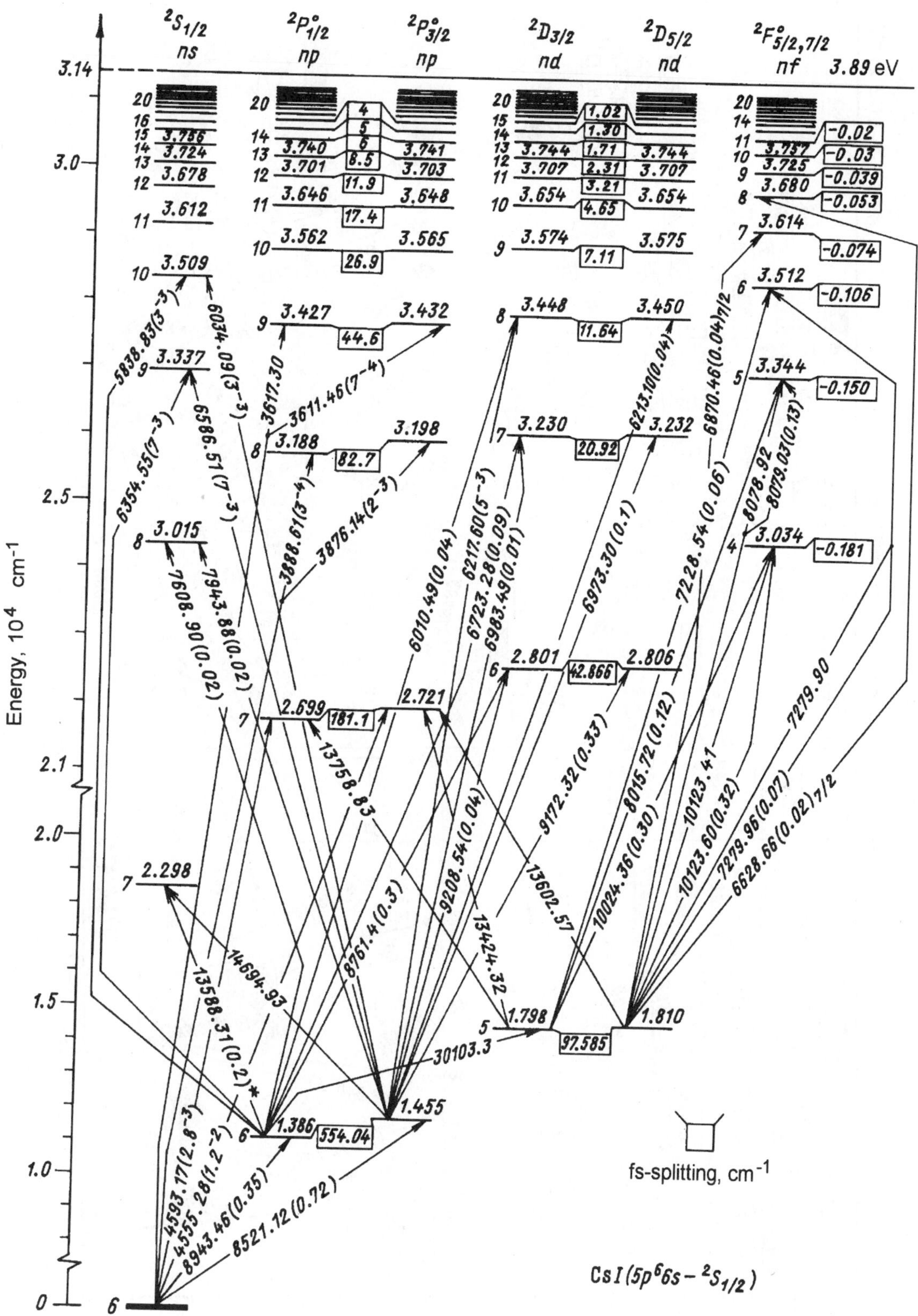

Figure 32.40 Grotrian diagram of atomic cesium.

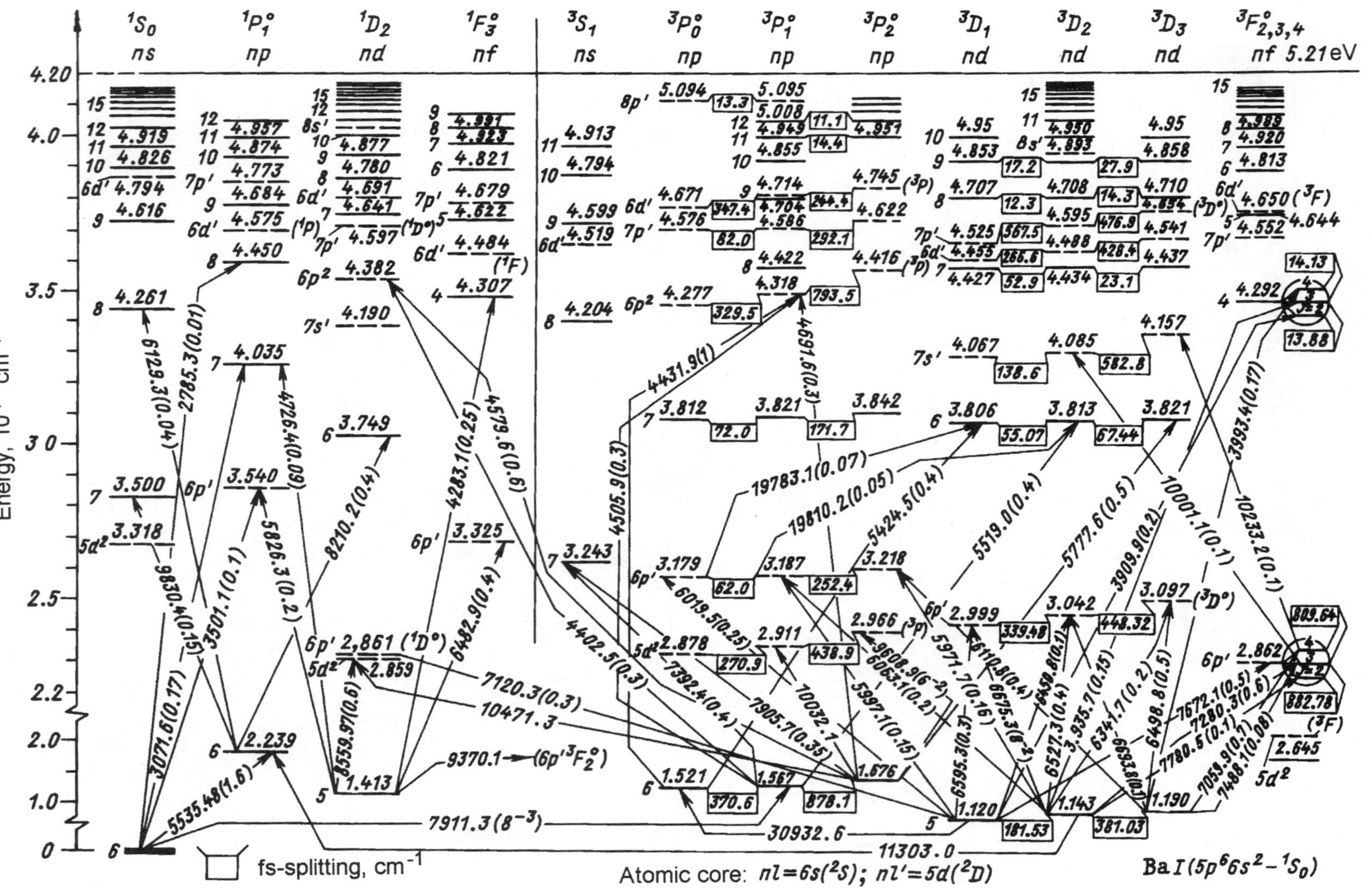

Figure 32.41 Grotrian diagram of atomic barium.

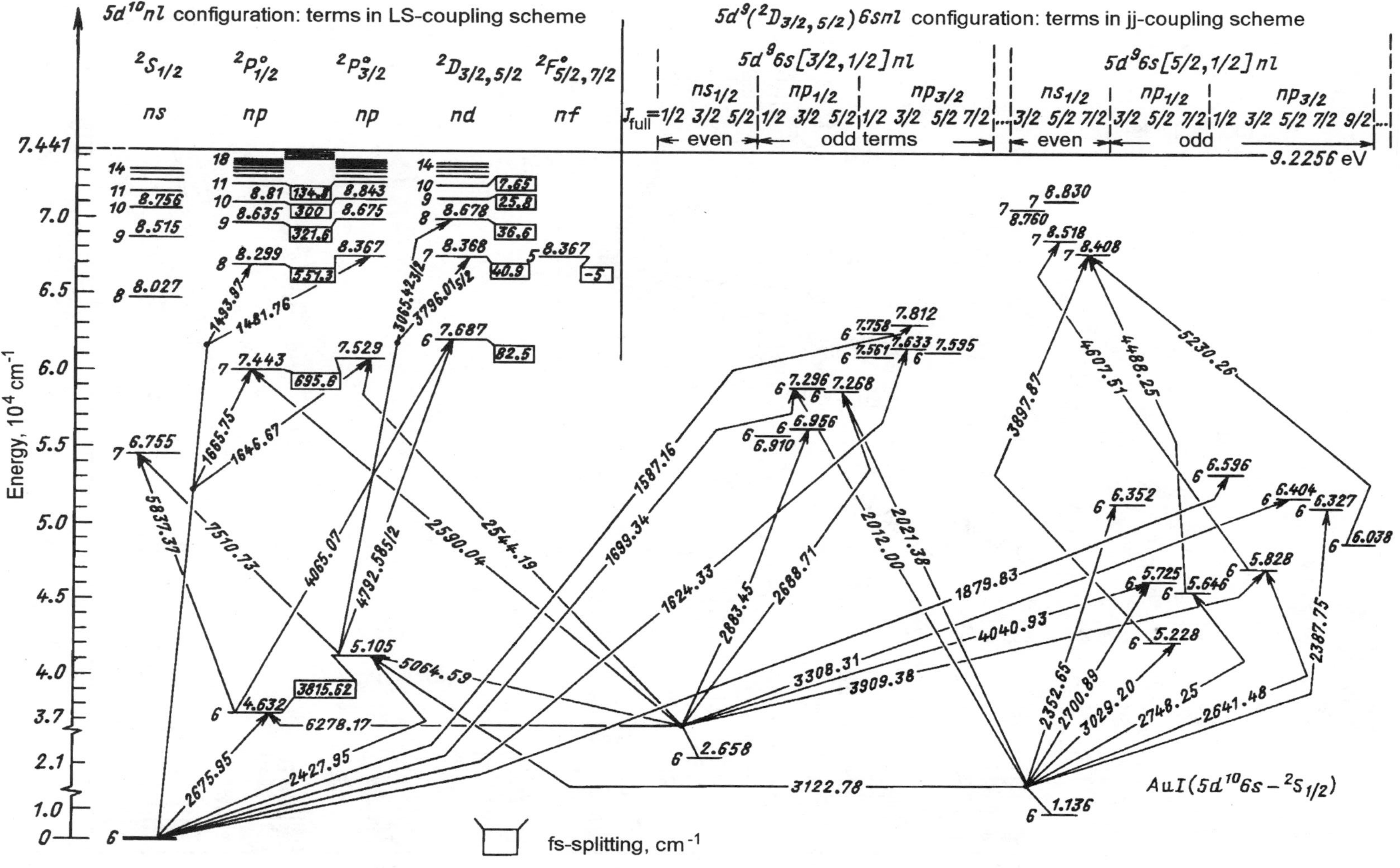

Figure 32.42 Grotrian diagram of atomic gold.

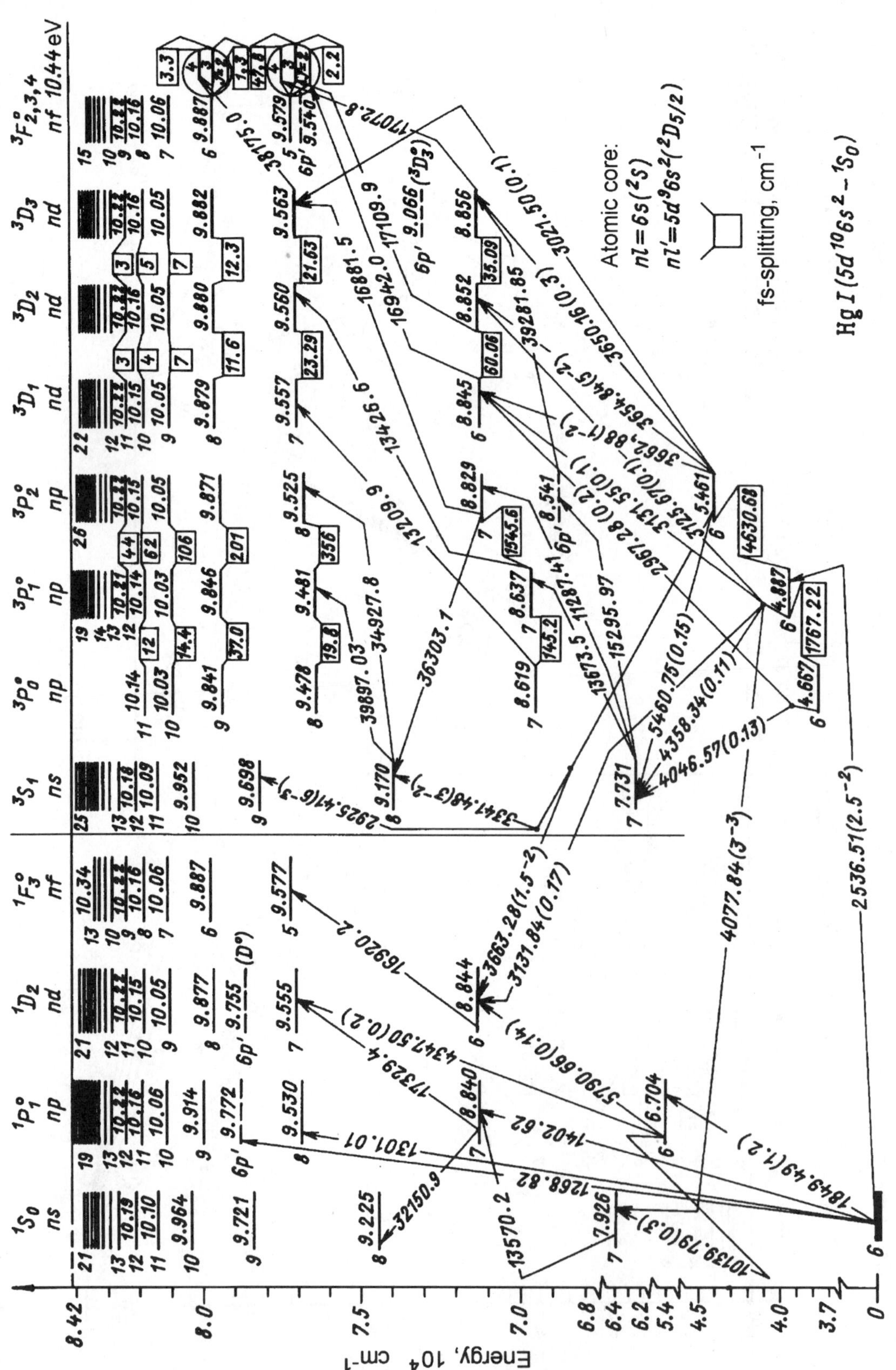

Figure 32.43 Grotrian diagram of atomic mercury.

32.3 Fine-Structure of Atomic Energy Levels

It is common knowledge that classification of many-electron atoms is founded on accounting for some additional perturbations in the self-consistent (effective, central-symmetric) atomic field. They are caused by noncentral electrostatic and relativistic (spin-orbital and spin-spin) interactions of electrons. In nonrelativistic approximation, when only electrostatic interactions are allowed for, the atomic energy levels can be characterized by orbital (L) and spin (S) momenta of electrons and are degenerate within possible directions of **L** and **S** vectors.

Taking into consideration the spin-dependent relativistic interactions, the atomic energy levels in the strict sense must be characterized by the values of conserved total electron angular momentum $\mathbf{J} = \mathbf{L} + \mathbf{S}$, since neither of the individual momenta **L** and **S** is conserved. For relative smallness of relativistic effects as compared to electrostatic electron interactions, the former can be considered according to perturbation theory. Then the atomic energy level with given LS is split into series of components with distinct values of the quantum number $J(|L - S|) < J < L + S)$. As a result, the fine structure of atomic energy levels arises and each sublevel is aligned with a set of quantum numbers LSJ. By accounting for interaction between the electron orbital and spin momenta, the energy difference between adjacent components of the split multiplet term with J and $J - 1$ values of the total electron momenta, respectively, takes the form [6]:

$$\Delta E_J - \Delta E_{J-1} = AJ, \tag{32.4}$$

where A is the constant of spin-orbit interaction depending on electron configuration and, also, on L and S. This formula is termed the Landè interval rule in the framework of LS-coupling scheme, when the fine-structure intervals are small compared to the energy differences between the levels with various L, S.

The constant A can be either positive or negative, and the multiplet is called either normal (minimum value $J = |L - S|$ corresponds to the lowest component) or inverted (maximum value $J = L + S$ corresponds to the lowest component).

The measured values of the splitting energy $\Delta E_{J,J-1}$ for adjacent components of some low-lying atomic multiplets are illustrated in Figs. 32.1–32.43. As a rule, LS-approximation is appropriated for describing not-too-highly excited energy levels of atoms from the top rows and from the middle of the Periodic Table.

The errors in measuring the fine-structure splitting of atomic energy levels were allowed for in rounding off the significant figures, so that they are within the range ± 1 for the last digit given.

32.4 Hyperfine Structure of Atomic Energy Levels

Interaction of the atomic electrons with the nuclear multipole moments (both magnetic and electric) results in splitting of atomic energy levels, which are characterized by the total electron momentum J, into a number of hyperfine components. Each component is defined by a set of four quantum numbers J, I, F, and m_F, where I is the nuclear spin, $\mathbf{F} = \mathbf{J} + \mathbf{I}$ is the total angular momentum of atom ($|J - I| \leq F \leq J + I$), and m_F is the projection of total angular momentum on the quantization axis.

The electron interaction with the nuclear multipole moments of the lowest order (i.e. magnetic dipole and electric quadrupole moments) contributes much to the hyperfine splitting of atomic energy levels. In the first approximation, the angular momenta J and I may be considered as being conserved and then the total energy of an atomic level can be given as a sum [7]:

$$E_F = E_J + E_{M1} + E_{E2} = E_J + \frac{h}{2}AK + \frac{3h}{8}B\frac{K(K+1) - \frac{4}{3}I(I+1)J(J+1)}{I(2I-1)J(2J-1)}; \quad I, J \geq 1, \qquad (32.5)$$

here E_J is the energy of the level unperturbed by the electron interaction with the nuclear moments; E_{M1} is the energy of electron interaction with a nuclear magnetic moment; E_{E2} is the energy of electron interaction with a nuclear quadrupole moment. Besides, the quantum number $K = F(F+1) - I(I+1) - J(J+1)$ and, finally, A and B are the hyperfine splitting constants for the atomic energy levels, the magnetic dipole interaction being always larger than the quadrupole one. The quadrupole splitting constant B is zero for states with electron momentum $J \leq 1/2$ owing to the spherical symmetry of the electron charge distribution.

It is conventional to designate the energy of hyperfine splitting of atomic levels in the form $\Delta\nu(F, F') = \Delta E_{F,F'}/h$, where $\Delta E_{F,F'}$ is the energy separation between two adjacent components with total angular momenta F and $F' = F - 1$, measured in the zero magnetic field. The dependence of the hyperfine splitting $\Delta\nu(F, F')$ on the constants A, B is given as

$$\Delta\nu(F, F') = AF + \frac{3BF(F^2 + 1/2)}{I(2I-1)J(2J-1)}; \quad I, J \geq 1. \qquad (32.6)$$

Measured values of $\Delta\nu(F, F')$ as long as numerical values of A and B constants for some low-excited atomic energy levels are compiled in Table 32.3 [3,7–9]. The errors in determining their values were allowed for when truncating the significant figures, so that they were within the ± 1 range for the last figures quoted.

Table 32.3 Hyperfine splitting of low-lying atomic energy levels

Atomic number Z	Isotope (ground state term), nuclear spin I	Term	Quantum numbers of total angular momentum (F, F')	Hyperfine splitting of levels	
				$\Delta\nu\ (F, F')$; A, B, MHz	$\Delta\ E(F, F')$, 10^{-3} cm^{-1}
1	$^1H(^2S_{1/2})$, 1/2	$1\ ^2S_{1/2}$	(1, 0)	1420.40575	47.3796
		$2\ ^2S_{1/2}$	(1, 0)	177.5568	5.92266
	$^2H(^2S_{1/2})$, 1	$1\ ^2S_{1/2}$	(3/2, 1/2)	327.38435	10.9204
		$2\ ^2S_{1/2}$	(3/2, 1/2)	40.9244	1.36509
	$^3H(^2S_{1/2})$, 1/2	$1\ ^2S_{1/2}$	(1, 0)	1516.70147	50.5917
2	$^3He(^1S_0)$, 1/2	$2\ ^3S_1$	(3/2, 1/2)	6739.701	224.812
		$2\ ^3P_2$	(5/2, 3/2)	6960	232.2
		$2\ ^3P_1$	(3/2, 1/2)	4510	151
	$^3He^+(^2S_{1/2})$, 1/2	$1\ ^2S_{1/2}$	(1, 0)	8665.6499	289.055
3	$^6Li(^2S_{1/2})$, 1	$2\ ^2S_{1/2}$	(3/2, 1/2)	228.20526	7.61211
		$2\ ^2P_{1/2}$		$A = 17.37$	
		$2\ ^2P_{3/2}$		$A = -1.16$	
				$B = -0.1$	
	$^7Li(^2S_{1/2})$, 3/2	$2\ ^2S_{1/2}$	(2, 1)	803.50409	26.80203
		$2\ ^2P_{1/2}$		$A = 45.9$	
		$2\ ^2P_{3/2}$		$A = -3.06$	
				$B = -0.2$	
4	$^9Be(^1S_0)$, 3/2	$2\ ^3P_1$	(5/2, 3/2)	354.44	11.823
			(3/2, 1/2)	202.95	6.7697
		$2\ ^3P_2$	(7/2, 5/2)	435.48	14.526
			(5/2, 3/2)	312.02	10.408
			(3/2, 1/2)	187.62	6.2583

Table 32.3 Hyperfine splitting of low-lying atomic energy levels *(continued)*

Atomic number Z	Isotope (ground state term), nuclear spin I	Term	Quantum numbers of total angular momentum (F, F')	Hyperfine splitting of levels: $\Delta\nu\ (F, F')$; A, B, MHz	Hyperfine splitting of levels: $\Delta\ E(F, F')$, 10^{-3} cm^{-1}
5	$^{10}B(^2P_{1/2})$, 3	$2\ ^2P_{1/2}$	(7/2, 5/2)	429.05	14.312
	$^{11}B(^2P_{1/2})$, 3/2	$2\ ^2P_{1/2}$	(2, 1)	732.15	24.422
		$2\ ^2P_{3/2}$	(3, 2)	222.7	7.428
			(2, 1)	144.0	4.803
			(1, 0)	71	2.37
6	$^{13}C(^3P_0)$, 1/2	$2\ ^3P_1$	(3/2, 1/2)	4.3	0.14
		$2\ ^3P_2$	(5/2, 3/2)	372.6	12.43
7	$^{14}N(^4S_{3/2})$, 1	$2\ ^4S_{3/2}$		$A = 10.45093$	
				$B = 1.3$	
8	$^{17}O(^3P_2)$, 5/2	$2\ ^3P_2$		$A = -219.6$	
		$2\ ^3P_1$		$A = 4.7$	
9	$^{19}F(^2P_{3/2})$, 1/2	$2\ ^2P_{3/2}$	(2, 1)	4020	134
		$2\ ^2P_{1/2}$	(1, 0)	10250	342
10	$^{21}Ne(^1S_0)$, 3/2	$3\ ^3P_2$	(7/2, 5/2)	1034.5	34.51
			(5/2, 3/2)	599.4	19.99
			(3/2, 1/2)	303.9	10.14
11	$^{23}Na(^2S_{1/2})$, 3/2	$3\ ^2S_{1/2}$	(2, 1)	1771.62613	59.09513
		$3\ ^2P_{1/2}$		$A = 94.3$	
		$3\ ^2P_{3/2}$		$A = 18.7$	
				$B = 2.9$	
12	$^{25}Mg(^1S_0)$, 5/2	$3\ ^3P_1$	(7/2, 5/2)	516.1	17.22
			(5/2, 3/2)	350.0	11.7
		$3\ ^3P_2$	(9/2, 7/2)	567.3	18.92
			(7/2, 5/2)	452.3	15.09
			(5/2, 3/2)	329.0	10.97
			(3/2, 1/2)	199.8	6.66
13	$^{27}Al(^2P_{1/2})$, 5/2	$3\ ^2P_{1/2}$	(3, 2)	1506.1	50.24
		$3\ ^2P_{3/2}$	(4, 3)	392	13.1
			(3, 2)	274	9.14
		$4\ ^2S_{1/2}$		$A = 420$	
15	$^{31}P(^4S_{3/2})$, 1/2	$3\ ^4S_{3/2}$		$A = 55.06$	
17	$^{35}Cl(^2P_{3/2})$, 3/2	$3\ ^2P_{3/2}$	(3, 2)	670.0135	22.349
			(2, 1)	355.2210	11.849
			(1, 0)	150.1736	5.009
		$3\ ^2P_{1/2}$	(2, 1)	2074.38	69.19
	$^{37}Cl(^2P_{3/2})$, 3/2	$3\ ^2P_{3/2}$	(3, 2)	555.3043	18.523
			(2, 1)	298.1277	9.944
			(1, 0)	127.4408	4.251
		$3\ ^2P_{1/2}$	(2, 1)	1726.7	57.60
19	$^{39}K(^2S_{1/2})$, 3/2	$4\ ^2S_{1/2}$	(2, 1)	461.71972	15.40132
		$4\ ^2P_{1/2}$		$A = 27.8$	
		$4\ ^2P_{3/2}$		$A = 6.1$	
				$B = 2.8$	
	$^{40}K(^2S_{1/2})$, 4	$4\ ^2S_{1/2}$		$A = -285.73$	
		$4\ ^2P_{1/2}$		$A = -34.5$	
		$4\ ^2P_{3/2}$		$A = -7.5$	
				$B = -3$	
	$^{41}K(^2S_{1/2})$, 3/2	$4\ ^2S_{1/2}$	(2, 1)	254.01387	8.47300
		$4\ ^2P_{1/2}$		$A = 15.2$	
		$4\ ^2P_{3/2}$		$A = 3.4$	
				$B = 3.3$	

Table 32.3 Hyperfine splitting of low-lying atomic energy levels *(continued)*

Atomic number Z	Isotope (ground state term), nuclear spin I	Term	Quantum numbers of total angular momentum (F, F')	Hyperfine splitting of levels: $\Delta\nu$ (F, F'); A, B, MHz	Hyperfine splitting of levels: Δ $E(F, F')$, 10^{-3} cm^{-1}
20	^{41}Ca($^1S_0$), 7/2	$4\ ^3P_1$		$A = -240.7$	
				$B = 4$	
	^{43}Ca($^1S_0$), 7/2	$4\ ^3P_1$		$A = -198.8$	
				$B = 3$	
21	^{45}Sc($^2D_{3/2}$), 7/2	$3\ ^2D_{3/2}$	(5, 4)	1329	44.3
			(4, 3)	1085.8	36.22
		$3\ ^2D_{5/2}$	(6, 5)	635.0	21.18
			(5, 4)	543.8	18.14
			(4, 3)	444.7	14.83
		$4\ ^4F_{3/2}$		$A = 158.5$	
				$B = -5.2$	
		$4\ ^4F_{5/2}$		$A = 154.0$	
				$B = -6.5$	
		$4\ ^4F_{7/2}$		$A = 250.0$	
				$B = -9.1$	
		$4\ ^4F_{9/2}$		$A = 286.0$	
				$B = -15$	
22	^{47}Ti($^3F_2$), 5/2	$3\ ^3F_2$		$A = -85.703$	
				$B = 25.70$	
	^{49}Ti($^3F_2$), 7/2	$3\ ^3F_2$		$A = -85.726$	
				$B = 21.07$	
23	^{51}V($^4F_{3/2}$), 7/2	$3\ ^4F_{3/2}$		$A = 560.07$	
				$B = 3.98$	
		$3\ ^6D_{1/2}$		$A = 751.53$	
		$3\ ^6D_{3/2}$		$A = 405.64$	
				$B = -7.0$	
		$3\ ^6D_{5/2}$		$A = 373.53$	
				$B = -5.0$	
		$3\ ^6D_{7/2}$		$A = 382.37$	
				$B = -2.3$	
24	^{53}Cr($^7S_3$), 3/2	$3\ ^7S_3$	(9/2, 7/2)	371.7	12.40
			(7/2, 5/2)	289.09	9.643
			(5/2, 3/2)	206.50	6.888
25	^{55}Mn($^6S_{5/2}$), 5/2	$3\ ^6S_{5/2}$		$A = -72.4208$	
				$B = -0.018$	
		$4\ ^6D_{9/2}$		$A = 510.3$	
				$B = 132.2$	
		$4\ ^6D_{7/2}$		$A = 458.9$	
				$B = 21.7$	
		$4\ ^6D_{5/2}$		$A = 436.7$	
				$B = -46.8$	
		$4\ ^6D_{3/2}$		$A = 469.4$	
				$B = -65.1$	
		$4\ ^6D_{1/2}$		$A = 882.1$	
26	^{57}Fe($^5D_4$), 1/2	$3\ ^5D_4$		$A = 38.08$	
		$4\ ^5F_5$		$A = 87.25$	
		$4\ ^5F_4$		$A = 78.43$	
		$4\ ^5F_3$		$A = 69.63$	
		$4\ ^5F_2$		$A = 55.99$	

Table 32.3 Hyperfine splitting of low-lying atomic energy levels *(continued)*

Atomic number Z	Isotope (ground state term), nuclear spin I	Term	Quantum numbers of total angular momentum (F, F')	Hyperfine splitting of levels: $\Delta\nu\ (F,F')$; A, B, MHz	Hyperfine splitting of levels: $\Delta\ E(F,F')$, 10^{-3} cm^{-1}
27	^{59}Co($^4F_{9/2}$), 7/2	$3\ ^4F_{9/2}$	(8, 7)	3655	121.9
			(7, 6)	3169.4	105.7
			(6, 5)	2695	89.9
			(5, 4)	2230.6	74.40
			(4, 3)	1774.5	59.19
28	^{61}Ni($^3F_4$), 3/2	$3\ ^3F_4$		$A = -215.04$	
				$B = -56.9$	
29	^{63}Cu($^2S_{1/2}$), 3/2	$4\ ^2S_{1/2}$	(2, 1)	11733.8174	391.398
	^{65}Cu($^2S_{1/2}$), 3/2	$4\ ^2S_{1/2}$	(2, 1)	12568.780	419.250
30	^{67}Zn($^1S_0$), 5/2	$4\ ^3P_2$	(9/2, 7/2)	2418.1	80.66
			(7/2, 5/2)	1855.7	61.90
			(5/2, 3/2)	1312.1	43.77
			(3/2, 1/2)	781.9	26.08
31	^{69}Ga($^2P_{1/2}$), 3/2	$4\ ^2P_{1/2}$	(2, 1)	2677.987	89.328
		$4\ ^2P_{3/2}$		$A = 190.794$	
				$B = 62.522$	
		$5\ ^2S_{1/2}$	(2, 1)	2140	71.3
	^{71}Ga($^2P_{1/2}$), 3/2	$4\ ^2P_{1/2}$	(2, 1)	3402.69	113.50
		$4\ ^2P_{3/2}$	(3, 2)	766.696	25.574
			(2, 1)	445.470	14.859
			(1, 0)	203.043	6.773
		$5\ ^2S_{1/2}$	(2, 1)	2720	90.6
32	^{73}Ge($^3P_0$), 9/2	$4\ ^3P_1$		$A = 15.55$	
				$B = -54.57$	
		$4\ ^3P_2$		$A = -64.427$	
				$B = 111.8$	
33	^{75}As($^4S_{3/2}$), 3/2	$4\ ^4S_{3/2}$		$A = -66.20$	
				$B = -0.53$	
			(3, 2)	819.45	27.33
			(2, 1)	595.12	19.85
36	^{83}Kr($^1S_0$), 9/2	$5p[1/2]_1$		$A = -143.0$	
37	^{85}Rb($^2S_{1/2}$), 5/2	$5\ ^2S_{1/2}$	(3, 2)	3035.732	101.261
		$5\ ^2P_{1/2}$		$A = 120.7$	
		$5\ ^2P_{3/2}$		$A = 25.0$	
				$B = 26.0$	
		$4\ ^2D_{5/2}$		$A = -5$	
		$4\ ^2D_{3/2}$		$A = 7$	
		$6\ ^2S_{1/2}$		$A = 239$	
	^{87}Rb($^2S_{1/2}$), 3/2	$5\ ^2S_{1/2}$	(2, 1)	6834.6826	227.98
		$5\ ^2P_{1/2}$		$A = 406$	
		$5\ ^2P_{3/2}$		$A = 84.9$	
				$B = 12.6$	
		$4\ ^2D_{5/2}$		$A = -17$	
		$4\ ^2D_{3/2}$		$A = 25$	
		$6\ ^2S_{1/2}$		$A = 810$	
39	^{89}Y($^2D_{3/2}$), 1/2	$4\ ^2D_{3/2}$		$A = -57.2$	
		$4\ ^2D_{5/2}$		$A = -28.8$	
42	^{95}Mo($^7S_3$), 5/2	$5\ ^7S_3$		$A = -208.5828$	
				$B = 0.0087$	
		$5\ ^5S_2$		$A = 428.84$	
				$B = 0.028$	
	^{97}Mo($^7S_3$), 5/2	$5\ ^7S_3$		$A = -212.9817$	
		$5\ ^5S_2$		$A = 437.89$	

Table 32.3 Hyperfine splitting of low-lying atomic energy levels *(continued)*

Atomic number Z	Isotope (ground state term), nuclear spin I	Term	Quantum numbers of total angular momentum (F, F')	Hyperfine splitting of levels: $\Delta\nu\ (F, F')$; A, B, MHz	Hyperfine splitting of levels: $\Delta\ E(F, F')$, $10^{-3}\ cm^{-1}$
47	$^{107}Ag(^2S_{1/2})$, 1/2	$5\ ^2P_{3/2}$		$A = -32$	
49	$^{113}In(^2P_{1/2})$, 9/2	$5\ ^2P_{1/2}$	(5, 4)	11,385	379.8
		$6\ ^2S_{1/2}$	(5, 4)	8410	281
	$^{115}In(^2P_{1/2})$, 9/2	$5\ ^2P_{1/2}$		$A = 2281.955$	
		$5\ ^2P_{3/2}$		$A = 242.165$	
		$6\ ^2S_{1/2}$	(5, 4)	8430	281
51	$^{123}Sb(^4S_{3/2})$, 7/2	$5\ ^4S_{3/2}$	(5, 4)	815.6	27.20
			(4, 3)	648.5	21.63
			(3, 2)	484.0	16.1
52	$^{125}Te(^3P_2)$, 1/2	$5\ ^3P_2$		$A = -1010.3$	
		$5\ ^3P_1$		$A = 782.5$	
		$5\ ^1D_2$		$A = -2887.0$	
53	$^{127}I(^2P_{3/2})$, 5/2	$5\ ^2P_{3/2}$	(4, 3)	4226.17	140.97
			(3, 2)	1965.9	65.58
			(2, 1)	737.49	24.60
54	$^{129}Xe(^1S_0)$, 1/2	$6\ ^3P_2$	(5/2, 3/2)	5961.258	198.85
	$^{131}Xe(^1S_0)$, 3/2	$6\ ^3P_2$	(7/2, 5/2)	2693.623	89.850
			(5/2, 3/2)	1608.348	53.649
			(3/2, 1/2)	838.764	27.978
55	$^{133}Cs(^2S_{1/2})$, 7/2	$6\ ^2S_{1/2}$	(4, 3)	9192.63177	306.63342
		$6\ ^2P_{1/2}$		$A = 292$	
		$6\ ^2P_{3/2}$		$A = 50.3$	
				$B = -0.4$	
		$5\ ^2D_{3/2}$		$A = 16.3$	
		$5\ ^2D_{5/2}$		$A = -22$	
		$7\ ^2S_{1/2}$		$A = 550$	
56	$^{135}Ba(^1S_0)$, 3/2	$5\ ^3D_1$		$A = -470$	
				$B = 12$	
		$5\ ^3D_2$		$A = 371$	
				$B = 18$	
		$5\ ^3D_3$		$A = 408$	
				$B = 20$	
		$5\ ^1D_2$		$A = -73.4$	
				$B = 38.7$	
	$^{137}Ba(^1S_0)$, 3/2	$5\ ^3D_1$		$A = -520$	
				$B = 17$	
		$5\ ^3D_2$		$A = 414$	
				$B = 27$	
		$5\ ^3D_3$		$A = 455$	
				$B = 40$	
		$5\ ^1D_2$		$A = -82.2$	
				$B = 59.6$	
57	$^{139}La(^2D_{3/2})$, 7/2	$5\ ^2D_{3/2}$	(5, 4)	737.97	24.62
			(4, 3)	551.98	18.41
			(3, 2)	391.6	13.06
		$5\ ^2D_{5/2}$	(6, 5)	1120.90	37.39
			(5, 4)	912.79	30.45
			(4, 3)	716.29	23.89
			(3, 2)	529.1	17.65
		$6\ ^4F_{3/2}$	(5, 4)	2390.6	79.74
			(4, 3)	1925.5	64.23

Table 32.3 Hyperfine splitting of low-lying atomic energy levels *(continued)*

Atomic number Z	Isotope (ground state term), nuclear spin I	Term	Quantum numbers of total angular momentum (F, F')	Hyperfine splitting of levels: $\Delta\nu\ (F, F')$; A, B, MHz	Hyperfine splitting of levels: $\Delta\ E(F, F')$, 10^{-3} cm^{-1}
		$6\ ^4F_{5/2}$	(6, 5)	1808.9	60.34
			(5, 4)	1503.2	50.14
			(4, 3)	1199.8	40.02
		$6\ ^4P_{1/2}$	(4, 3)	9840.6	328.2
		$6\ ^4P_{3/2}$	(4, 3)	3707.8	123.68
		$6\ ^4P_{5/2}$	(4, 3)	3216.5	107.3
59	^{141}Pr($^4I_{9/2}$), 5/2	$4\ ^4I_{9/2}$		$A = 926.209$	
				$B = -11.88$	
		$4\ ^4I_{11/2}$		$A = 730.393$	
				$B = -11.88$	
		$4\ ^4I_{13/2}$		$A = 613.240$	
				$B = -12.85$	
		$4\ ^4I_{15/2}$		$A = 541.575$	
				$B = -14.56$	
60	^{143}Nd($^5I_4$), 7/2	$4\ ^5I_4$	(15/2, 13/2)	1418	47.3
			(13/2, 11/2)	1257.5	41.95
			(11/2, 9/2)	1084.7	36.18
			(9/2, 7/2)	901.5	30.07
			(7/2, 5/2)	710	23.7
		$4\ ^5I_5$		$A = -153.68$	
				$B = 115.7$	
	^{145}Nd($^5I_4$), 7/2	$4\ ^5I_4$		$A = -121.63$	
				$B = 64.6$	
		$4\ ^5I_5$		$A = -95.53$	
				$B = 61.0$	
62	^{147}Sm($^7F_0$), 7/2	$4\ ^7F_1$		$A = -33.494$	
				$B = -58.692$	
		$4\ ^7F_2$		$A = -41.184$	
				$B = -62.23$	
		$4\ ^7F_3$		$A = -50.240$	
				$B = -33.68$	
	^{149}Sm($^7F_0$), 7/2	$4\ ^7F_1$		$A = -27.611$	
				$B = 16.962$	
		$4\ ^7F_2$		$A = -33.951$	
				$B = 17.99$	
		$4\ ^7F_3$		$A = -41.418$	
				$B = 9.75$	
63	^{151}Eu($^8S_{7/2}$), 5/2	$4\ ^8S_{7/2}$	(6, 5)	120.67	4.025
			(5, 4)	100.29	3.345
			(4, 3)	80.05	2.67
	^{153}Eu($^8S_{7/2}$), 5/2	$4\ ^8S_{7/2}$	(6, 5)	54.04	1.803
			(5, 4)	44.00	1.47
			(4, 3)	35.00	1.17
64	^{155}Gd($^9D_2$), 3/2	$5\ ^9D_2$		$A = 36.575$	
				$B = 179.4$	
		$5\ ^9D_3$		$A = 4.92$	
				$B = -406.67$	
		$5\ ^9D_4$		$A = -6.86$	
				$B = -352.8$	
	^{157}Gd($^9D_2$), 3/2	$5\ ^9D_2$		$A = 47.96$	
				$B = 191.2$	
		$5\ ^9D_3$		$A = 6.45$	
				$B = -433.2$	

Table 32.3 Hyperfine splitting of low-lying atomic energy levels *(continued)*

Atomic number Z	Isotope (ground state term), nuclear spin I	Term	Quantum numbers of total angular momentum (F, F')	Hyperfine splitting of levels	
				$\Delta\nu$ (F, F'); A, B, MHz	$\Delta E(F, F')$, 10^{-3} cm^{-1}
		$5\ ^9D_4$		$A = -9.00$	
				$B = -375.9$	
65	^{159}Tb($^6H_{15/2}$), 3/2	$4\ ^6H_{15/2}$		$A = 673.75$	
				$B = 1449.3$	
		$4\ ^6H_{13/2}$		$A = 682.91$	
				$B = 1167.5$	
		$5\ ^8G_{13/2}$		$A = 532.20$	
				$B = 928.9$	
66	^{161}Dy($^5I_8$), 5/2	$4\ ^5I_8$		$A = -116.232$	
				$B = 1091.57$	
	^{163}Dy($^5I_8$), 5/2	$4\ ^5I_8$		$A = 162.7543$	
				$B = 1152.86$	
		$4\ ^5I_7$		$A = 177.53$	
				$B = 1066.4$	
67	^{165}Ho($^4I_{15/2}$), 7/2	$4\ ^4I_{15/2}$	(9, 8)	7184.8	239.7
			(8, 7)	6540.8	218.2
			(7, 6)	5842.4	194.9
			(6, 5)	5096.3	170.0
			(5, 4)	4309.3	143.7
68	^{167}Er($^3H_6$), 7/2	$4\ ^3H_6$		$A = -120.486$	
				$B = -4552.96$	
69	^{169}Tm($^2F_{7/2}$), 1/2	$4\ ^2F_{7/2}$	(4, 3)	1496.5507	49.920
				$A = -374.13766$	
71	^{175}Lu($^2D_{3/2}$), 7/2	$5\ ^2D_{3/2}$	(5, 4)	2051.2201	68.421
			(4, 3)	345.497	11.524
			(3, 2)	496.578	16.564
		$5\ ^2D_{5/2}$	(6, 5)	1837.570	61.295
			(5, 4)	800.343	26.70
			(4, 3)	161.815	5.398
			(3, 2)	157.73	5.26
			(2, 1)	238.058	7.941
	^{176}Lu($^2D_{3/2}$), 7	$5\ ^2D_{3/2}$		$A = 137.9$	
				$B = 2131$	
		$5\ ^2D_{5/2}$		$A = 104.0$	
				$B = 2624$	
72	^{177}Hf($^3F_2$), 7/2	$5\ ^3F_2$	(11/2, 9/2)	991.792	33.08
			(9/2, 7/2)	477.008	15.91
			(7/2, 5/2)	162.887	5.433
			(5/2, 3/2)	4.864	0.16
	^{179}Hf($^3F_2$), 9/2	$5\ ^3F_2$	(13/2, 11/2)	82.132	2.74
			(11/2, 9/2)	392.848	13.104
			(9/2, 7/2)	541.9104	18.076
			(7/2, 5/2)	558.672	18.635
73	^{181}Ta($^4F_{3/2}$), 7/2	$5\ ^4F_{3/2}$		$A = 509.08$	
				$B = -1012.24$	
		$5\ ^4F_{5/2}$		$A = 313.47$	
				$B = -834.8$	
		$5\ ^4F_{7/2}$		$A = 264.41$	
				$B = -787.5$	
		$5\ ^4F_{9/2}$		$A = 256.62$	
				$B = -650.4$	

Table 32.3 Hyperfine splitting of low-lying atomic energy levels *(continued)*

Atomic number Z	Isotope (ground state term), nuclear spin I	Term	Quantum numbers of total angular momentum (F, F')	Hyperfine splitting of levels: $\Delta\nu$ (F, F'); A, B, MHz	Hyperfine splitting of levels: Δ $E(F, F')$, 10^{-3} cm^{-1}
		$5\ {}^4P_{1/2}$		$A = 884.17$	
		$5\ {}^4P_{3/2}$		$A = 379$	
				$B = -1350$	
74	^{183}W($^5D_0$), 1/2	$5\ {}^5D_1$		$A = 29.12$	
		$6\ {}^7S_3$		$A = 505.6$	
		$5\ {}^5D_2$		$A = 56.3$	
		$5\ {}^5D_3$		$A = 78.0$	
		$5\ {}^5D_4$		$A = 88.3$	
75	^{185}Re($^6S_{5/2}$), 5/2	$5\ {}^6S_{5/2}$		$A = -56.596$	
				$B = 29.635$	
		$5\ {}^4P_{5/2}$		$A = 880.44$	
				$B = 1618.5$	
	^{187}Re($^6S_{5/2}$), 5/2	$5\ {}^6S_{5/2}$		$A = -57.149$	
				$B = 28.05$	
		$5\ {}^4P_{5/2}$		$A = 889.24$	
				$B = 1531.7$	
		$6\ {}^6D_{9/2}$		$A = 2600$	
				$B = 2000$	
77	^{191}Ir($^4F_{9/2}$), 3/2	$5\ {}^4F_{9/2}$	(6, 5)	659.265	21.991
			(5, 4)	189.440	6.319
			(4, 3)	84.050	2.804
	^{193}Ir($^4F_{9/2}$), 3/2	$5\ {}^4F_{9/2}$	(6, 5)	660.090	22.018
			(5, 4)	224.478	7.488
			(4, 3)	33.535	1.119
78	^{195}Pt($^3D_3$), 1/2	$5\ {}^3D_3$		$A = 5702.6$	
		$5\ {}^3D_2$		$A = 2609.6$	
		$6\ {}^3F_4$	(9/2, 7/2)	3820.56	127.4
79	^{197}Au($^2S_{1/2}$), 3/2	$6\ {}^2S_{1/2}$	(2, 1)	6099.320	203.452
		$5\ {}^2D_{5/2}$		$A = 80.24$	
				$B = 1049.8$	
		$5\ {}^2D_{3/2}$		$A = 199.842$	
				$B = 911.077$	
80	^{199}Hg($^1S_0$), 1/2	$6\ {}^3P_2$		$A = 9066.45$	
	^{201}Hg($^1S_0$), 3/2	$6\ {}^3P_2$	(7/2, 5/2)	11,382.629	379.68
			(5/2, 3/2)	8629.522	287.85
			(3/2, 1/2)	5377.49	179.37
		$6\ {}^3D_3$		$A = -2450$	
				$B = 60$	
81	^{203}Tl($^2P_{1/2}$), 1/2	$6\ {}^2P_{1/2}$	(1, 0)	21,105.45	704.0026
		$6\ {}^2P_{3/2}$	(2, 1)	524.0599	17.4808
	^{205}Tl($^2P_{1/2}$), 1/2	$6\ {}^2P_{1/2}$	(1, 0)	21,310.83	710.8534
		$6\ {}^2P_{3/2}$	(2, 1)	530.0765	17.6815
82	^{207}Pb($^3P_0$), 1/2	$6\ {}^1D_2$	(5/2, 3/2)	1524.5	50.85
83	^{209}Bi($^4S_{3/2}$), 9/2	$6\ {}^4S_{3/2}$		$A = -446.94$	
				$B = -304.65$	
		$6\ {}^2D_{3/2}$		$A = -1227$	
				$B = -620$	
		$6\ {}^2D_{5/2}$		$A \simeq 2503$	
				$B = 0$	
		$6\ {}^2P_{1/2}$		$A = 11,268$	
		$6\ {}^2P_{3/2}$		$A = 491.03$	
				$B = 978.64$	

Table 32.3 Hyperfine splitting of low-lying atomic energy levels *(continued)*

Atomic number Z	Isotope (ground state term), nuclear spin I	Term	Quantum numbers of total angular momentum (F, F')	Hyperfine splitting of levels	
				$\Delta\nu\ (F, F')$; A, B, MHz	$\Delta\ E(F, F')$, 10^{-3} cm^{-1}
92	^{235}U$(^5L^\circ_6)$, 7/2	$6\ ^5L^\circ_6$		$A = -60.56$	
				$B = 4104.1$	
		$6\ ^5K^\circ_5$		$A = -68.35$	
				$B = 40.1$	
93	^{237}Np$(^6L_{11/2})$, 5/2	$5\ ^6L_{11/2}$		$A = 778$	
				$B = 645$	

32.5 Isotope Structure of Atomic Spectra

In the spectra of elements involving several isotopes, the splitting of lines into a number of components is observed. Each of the split components characterizes a definite nuclide. This isotope structure occurs due to interaction between electrons and nucleus. The total atomic Hamiltonian in the center-of-mass system allows for the motion of nucleons about the atomic center of mass (normal or Bohr mass effect), the mass-dependent electron-exchange interaction (specific mass effect) and the interaction of valence electrons with extended nuclear-proton charge (field or volume effect). As a result, the observed transition isotope shift of spectral lines $\delta\nu$ for two isotopes with mass numbers A_1 and A_2 ($A = Z + N$, where Z is the number of protons, and N is the number of neutrons) is given by the sum of three parts [10]:

$$\delta\nu^{A_1A_2} = \delta\nu^{A_1A_2}_{BMS} + \delta\nu^{A_1A_2}_{SS} + \delta\nu^{A_1A_2}_{VS}. \tag{32.7}$$

Here, the first term $\delta\nu^{A_1A_2}_{BMS} = \nu_\infty m_e(M_2 - M_1)/M_1M_2$ corresponds to the normal mass shift, which is of particular importance in the light elements ($Z < 30$); ν_∞ is the transition frequency for an infinite nuclear mass; m_e is the electron mass; M_1, M_2 are the nuclear masses. The second term $\delta\nu^{A_1A_2}_{SS}$ complies with the specific mass shift, and, finally, the third term $\nu^{A_1A_2}_{VS}$ is the volume shift between the lines of two isotopes prevailing in the spectra of heavy isotopes ($Z > 60$). In the spectra of medium-weight elements ($Z \sim 20 - 50$), the absolute value of the observed isotope shift is small and difficult to measure accurately.

The transition isotope shift $\delta\nu$ is conditioned by the shifts of the atomic electron terms, so that the relation between $\Delta\nu$ and the isotope shifts of the upper term ($\delta T'$) and the lower one (δT) can be given in the following way:

$$\delta\nu = \delta T' - \delta T. \tag{32.8}$$

By convention, the transition isotope shift is said to be positive if the lines of the heavier isotope are shifted to higher frequencies. This means in the term-shift language that the negative sign of term shift conforms to the case when the energy level of the heavier nuclide lies deeper (its distance to the continuous spectrum edge is larger) than the level of the lighter one. The normal mass shift results in the rise of electron binding energy in the heavier isotope, whereas this energy decreases in the heavier nuclide due to the effect of nuclear finite volume. Thus, the mass and volume nuclear effects are opposite in sign. When the nuclei have magnetic moments and, naturally, the hyperfine structure of the levels can be observed, the level isotope shift should be referred to the centers of gravity of the hyperfine-structure components.

Table 32.4 incorporates the measured isotope shifts $\delta\nu$ of resonance lines for some elements [10–13]. In accordance with the reported experimental errors, the listed numerical values were rounded off and are estimated to be accurate to within an error ± 1 in the last digit quoted.

Table 32.4 Isotope shifts in atomic resonance lines

Atomic number Z	Element (ground-state term)	Transition array	Wavelength λ, Å	Mass numbers of isotopes $A_1 - A_2$	Isotope shift in line $\delta\nu^{A_1A_2}$, 10^{-3} cm^{-1}
1	H ($1s$—$^2S_{1/2}$)	$1\ ^2S_{1/2}$—$2\ ^2P^o$	1215.7	1 — 2	$2.238 \cdot 10^4$
				1 — 3	$2.983 \cdot 10^4$
				2 — 3	$7.477 \cdot 10^3$
2	He ($1s^2$—$^1S_0$)	$1\ ^1S_0$—$2\ ^1P^o$	584.3	3 — 4	$8.8 \cdot 10^3$
3	Li ($2s$—$^2S_{1/2}$)	$2\ ^2S_{1/2}$—$2\ ^2P^o$	6708	6 — 7	351.3
5	B ($2p$—$^2P^o_{1/2}$)	$2p\ ^2P^o$—$3s\ ^2S_{1/2}$	2497	10 — 11	-170
6	C ($2p^2$—$^3P_0$)	$2p^2\ ^1S_0$—$2p3s\ ^1P^o_1$	2478.6	12 — 13	-160
7	N ($2p^3$—$^4S^o_{1/2}$)	$3s\ ^2P_{3/2}$—$3p\ ^2P^o_{3/2}$	8629.2	14 — 15	70
		$3s\ ^4P_{5/2}$—$3p\ ^4P^o_{5/2}$	8216.3	14 — 15	-60
8	O ($2p^4$—$^3P_2$)	$3s\ ^3S_1$—$3p\ ^3P_1$	8446.8	16 — 18	140
10	Ne ($2p^6$—$^1S_0$)	$3s'\ [1/2]^o_1$—$3p\ [5/2]_2$	7173.9	20 — 22	70
		$3s\ [3/2]^o_2$—$3p\ [1/2]_1$	7032.4	20 — 22	50
11	Na ($3s$—$^2S_{1/2}$)	$3s\ ^2S_{1/2}$—$4p\ ^2P^o_{1/2}$	3303.0	23 — 24	24
12	Mg ($3s^2$—$^1S_0$)	$3s^2\ ^1S_0$—$3s3p\ ^1P^o_1$	2852.1	24 — 25	50
				24 — 26	60
18	Ar ($3s^6$—$^1S_0$)	$4s\ [3/2]^o_2$—$4p'\ [3/2]_1$	7147.0	36 — 40	20
		$4s'\ [1/2]^o_1$—$5p\ [1/2]_0$	4510.7	36 — 40	50
19	K ($4s$—$^2S_{1/2}$)	$4s\ ^2S_{1/2}$—$4p\ ^2P^o_{1/2}$	7699.0	39 — 40	4.19
				39 — 41	7.85
		$4s\ ^2S_{1/2}$—$4p\ ^2P^o_{3/2}$	7664.9	39 — 40	4.22
				39 — 41	7.88
20	Ca ($4s^2$—$^1S_0$)	$4s^2\ ^1S_0$—$4s4p\ ^1P^o_1$	4226.7	40 — 42	13.0
				40 — 43	20.4
				40 — 44	25.7
				40 — 48	50.4
		$4s^2\ ^1S_0$—$4s4p\ ^3P^o_1$	6572.8	40 — 41	9.37
				40 — 42	17.0
				40 — 43	26.1
				40 — 44	33.2
				40 — 48	64.1
29	Cu ($3d^{10}4s$—$^2S_{1/2}$)	$4s\ ^2S_{1/2}$—$4p\ ^2P^o_{1/2}$	3274.0	63 — 65	20
30	Zn ($4s^2$—$^1S_0$)	$4s^2\ ^1S_0$—$4s4p\ ^1P^o_1$	2138.6	$\triangle A = 2$ (mean value)	16
31	Ga ($4p$—$^2P^o_{1/2}$)	$4p\ ^2P^o_{1/2}$—$5s\ ^2S_{1/2}$	4033.0	69 — 71	-1.1
		$4p\ ^2P^o_{3/2}$—$5s\ ^2S_{1/2}$	4172.1	69 — 71	-1.3
36	Kr ($4p^6$—$^1S_0$)	$5s\ [3/2]^o_1$—$5p\ [5/2]_2$	8776.7	82 — 84	2
37	Rb ($5s$—$^2S_{1/2}$)	$5s\ ^2S_{1/2}$—$5p\ ^2P^o_{1/2}$	7947.6	85 — 87	2.6
		$5s\ ^2S_{1/2}$—$5p\ ^2P^o_{3/2}$	7800.2	85 — 87	2.6
38	Sr ($5s^2$—$^1S_0$)	$5s^2\ ^1S_0$—$5s5p\ ^1P^o_1$	4607.3	84 — 88	9.0
				86 — 88	4.2
				87 — 88	1.5
		$5s^2\ ^1S_0$—$5s6p\ ^1P^o_1$	2931.8	84 — 88	15.9

Table 32.4 Isotope shifts in atomic resonance lines *(continued)*

Atomic number Z	Element (ground-state term)	Transition array	Wavelength λ, Å	Mass numbers of isotopes $A_1 - A_2$	Isotope shift in line $\delta\nu^{A_1 A_2}$, 10^{-3} cm^{-1}
				86 — 88	7.5
				87 — 88	3.2
		$5s^2\ {}^1S_0$—$5s5p\ {}^3P_1^o$	6892.6	84 — 88	14.1
				86 — 88	6.7
				87 — 88	2.7
40	Zr	$4d^35s\ {}^5F_5$—$4d^35p\ {}^5G_6^o$	4687.8	90 — 92	-12
	($4d^25s^2$—${}^3F_2$)			92 — 94	-7
				94 — 96	-5
47	Ag	$4d^{10}5s\ {}^2S_{1/2}$—$5p\ {}^2P_{1/2}^o$	3382.9	107 — 109	-15
	($4d^{10}5s$—${}^2S_{1/2}$)	$4d^{10}5s\ {}^2S_{1/2}$—$5p\ {}^2P_{3/2}^o$	3280.7	107 — 109	-15
48	Cd	$5s^2\ {}^1S_0$—$5s5p\ {}^3P_1^o$	3261.0	ΔA=2 (mean value)	-15
	($5s^2$—${}^1S_0$)				
49	In	$5p\ {}^2P_{1/2}^o$—$6s\ {}^2S_{1/2}$	4101.8	113 — 115	8.6
	($5p$—${}^2P_{1/2}^o$)	$5p\ {}^2P_{3/2}^o$—$6s\ {}^2S_{1/2}$	4511.3	113 — 115	8.5
54	Xe	$6s\ [3/2]_2^o$—$6p\ [3/2]_2$	8231.6	134 — 136	-3
	($5p^6$—${}^1S_0$)				
55	Cs	$6s\ {}^2S_{1/2}$—$6p\ {}^2P_{3/2}^o$	8521.1	133 — 134	1.2
	($6s$—${}^2S_{1/2}$)			133 — 135	1.2
70	Yb	$6s^2\ {}^1S_0$—$6s6p\ {}^1P_1^o$	3988.0	174 — 176	17
	($4f^{14}6s^2$—${}^1S_0$)			174 — 172	18
71	Lu	$5d6s^2\ {}^2D_{3/2}$—$5d6s6p\ {}^4F_{3/2}^o$	5736.5	175 — 176	-13.1
	($5d6s^2$—${}^2D_{3/2}$)	$5d6s^2\ {}^2D_{5/2}$—$5d6s6p\ {}^4F_{5/2}^o$	6055.0	175 — 176	-13.9
		$5d6s^2\ {}^2D_{5/2}$—$5d6s6p\ {}^4F_{7/2}^o$	5421.9	175 — 176	-13.6
79	Au	$6s\ {}^2S_{1/2}$—$6p\ {}^2P_{1/2}^o$	2675.9	195 — 197	-100
	($5d^{10}6s$—${}^2S_{1/2}$)			193 — 197	-210
				192 — 197	-280
				191 — 197	-320
				190 — 197	-370
80	Hg	$6s^2\ {}^1S_0$—$6s6p\ {}^3P_1^o$	2536.5	198 — 199	-9
	($5d^{10}6s^2$—${}^1S_0$)			198 — 200	-160
				199 — 201	-210
				200 — 202	-180
				202 — 204	-160
81	Tl	$6p\ {}^2P_{3/2}^o$—$7s\ {}^2S_{1/2}$	5350.5	203 — 205	-58.6
	($6s^26p$—${}^2P_{1/2}^o$)				
82	Pb	$6p^2\ {}^3P_0$—$6p7s\ {}^3P_1^o$	2833.1	207 — 208	-47
	($6p^2$—${}^3P_0$)			206 — 208	-75
				204 — 208	-140
				202 — 208	-207

32.6 Spectra of Diatomic Molecules

The molecular band spectra pertaining to light absorption and emission have their origin in radiative transitions between discrete molecular energy levels. A rigorous statement of the problem of determining the energy levels for just the simplest molecule does not lead to exact solution. Mutual effect of electron and nucleus motion, linkage between spin and orbital moments of molecular subsystems must be allowed for approximate methods based on peculiarities of intermolecular interactions.

Due to the pronounced difference in electron and nucleus mass, the velocity of an electron subsystem in a molecule is substantially higher than that of a nuclear subsystem and, hence, the electrons and nuclei contribute variously to the total molecular energy. As this takes place, the possibility exists of separating the problem of energy determination for electrons moving in the nuclei field, from that of determining the energy of nuclear motion itself. Mutual influence of the electron subsystem (marked by relatively high transition rate) and a nuclear subsystem (marked by relatively slow transition rate) in the molecule could be accounted for by the method of successive approximations.

Eventually, the total energy of a diatomic molecule in a given state n is expressible to sufficient accuracy as a sum

$$E_n = T_e + T_{vJ}, \tag{32.9}$$

where T_e is the electronic energy; T_{vJ} is the vibration–rotation energy; v, J are the vibrational and rotational quantum numbers of the molecule, respectively. For not-too-highly excited molecular vibrations, the expansion of T_{vJ} into a power series of v and J could be available, e.g. in the following manner:

$$T_{vJ} = G(v) + F_v(J) = [\omega_e(v+1/2) - \omega_e x_e(v+1/2)^2 + \ldots] + [B_v J(J+1) - D_v J^2(J+1)^2 + \ldots]. \tag{32.10}$$

The vibrational energy $G(v)$ of the molecule in the above expression matches the model of anharmonic oscillator with ω_e equal to the frequency of harmonic oscillations, and $\omega_e x_e$ equal to the anharmonic constant. The rotational energy $F_v(J)$ of the molecule corresponds to the model of a not rigid rotator and accounts for interaction between the molecular vibrations and rotations. So that the rotational constants $B_v, D_v, \ldots$ depend on the level of vibrational excitation v: $B_v = B_e - \alpha_e(v+1/2) + \ldots$, $D_v = D_e + \beta_e(v + 1/2) + \ldots$, here index e labels the equilibrium internuclear distance in diatomics.

With allowance made for molecular energy division into three basic parts, the wave number of transition between two separated states n' and n'' may be written as $\nu = E_{n'} - E_{n''} = T'_e + G' + F' - (T''_e + G'' + F'')$. Then spectra of several kinds appear in the required way: (a) rotational spectra complying with transitions between rotational levels within the confines of steady vibrational and electronic states; (b) vibrational-rotational spectra originating in transitions between various rotational levels of different vibrational states for steady electronic term; (c) electronic spectra related to transitions between vibration-rotation levels of different electronic states. Furthermore, transitions between the sublevels of fine-structure multiplet for a given electronic-vibration-rotation state of the molecule are recorded in the radiowave and microwave spectral regions along with the spectra of electron-spin and nuclear-magnetic resonances occurring in transitions between the Zeeman components of split molecular levels in the magnetic field.

Table 32.5 presents data on the spectroscopic constants of ground state diatomic molecules. In separate columns are: the symbols of the electronic terms, equilibrium parameters r_e, ω_e, $\omega_e x_e$, B_e and α_e[1] of vibration-rotation state, reduced molecular mass $\mu_A = M_1 M_2/(M_1 + M_2)$ for the dominant isotope composition and, finally, dissociation energy D_0° meeting the rupture of a bond in the ground vibrational state, were tabulated. In addition, Figs. 32.44–32.48 show more detailed information concerning the general trend of the electronic terms in a small number of well-studied diatomic molecules H_2, N_2, O_2, NO, and CO.

In selecting numerical data for Table 32.5, we took advantage of spectroscopic tables for diatomic molecules [3,14] and the results from numerous journal publications of the last decade. The errors in the values of spectroscopic parameters included in Table 32.5 have been accounted for when rounding off the significant figures, so that they are within the range ±1 for the last digit given. Estimated uncertainties in the electronic terms of diatomic molecules and ions were indicated immediately alongside the potential curves in Figs. 32.44–32.48.

[1] Asterisked quantities in some columns of Table 32.5 (viz., for r_e, ω_e and B_e) are appropriate to r_0, $\Delta G_{1/2}$ and B_0 molecular parameters.

Table 32.5 Spectroscopic constants of the ground-state diatomic molecules

Molecule	Electronic term	Equilibrium internuclear distance r_e, Å	Vibrational frequency ω_e, cm^{-1}	Anharmonic constant $\omega_e x_e$, cm^{-1}	Rotational constant B_e, cm^{-1}	Rotation-vibration interaction constant α_e, 10^{-3} cm^{-1}	Reduced mass for a dominant isotope composition μ_A, a.m.u.	Dissociation energy D_0°, 10^4 cm^{-1}
Ag_2	$X\ ^1\Sigma_g^+$	2.7	192.4	0.64	0.496	0.19	53.948	1.33
AgBr	$X\ ^1\Sigma^+$	2.393	247.7	0.68	0.0634	0.228	46.424	2.5
AgCl	$X\ ^1\Sigma^+$	2.281	343.5	1.17	0.123	0.6	26.350	2.60
AgF	$X\ ^1\Sigma^+$	1.983	513.45	2.59	0.266	1.92	16.132	2.9
AgH	$X\ ^1\Sigma^+$	1.618	1760	34.1	6.449	201	0.9984	1.84
AgI	$X\ ^1\Sigma^+$	2.545	206.5	0.44	0.0449	0.15	58.025	2.1
AgO	$X\ ^2\Pi_{1/2}$	2.003	490.2	3.1	0.3020	2.5	13.913	1.8
Al_2	$X\ ^3\Sigma_g^-$	2.47	350.0	2.02	0.2054	1.2	13.491	1.5
AlBr	$X\ ^1\Sigma^+$	2.295	378.2	1.33	0.1592	0.86	20.107	3.6
AlCl	$X\ ^1\Sigma^+$	2.130	481.3	1.95	0.2439	1.61	15.230	4.13
AlF	$X\ ^1\Sigma^+$	1.6544	802.3	4.8	0.552	5.0	11.148	5.6
AlH	$X\ ^1\Sigma^+$	1.648	1683	29.1	6.3907	186	0.9715	2.5
AlI	$X\ ^1\Sigma^+$	2.537	316.1	1.0	0.1177	0.559	22.251	3.0
AlO	$X\ ^2\Sigma^+$	1.618	979.2	6.97	0.6414	5.8	10.042	4.23
AlS	$X\ ^2\Sigma^+$	2.029	617.1	3.33	0.2799	1.8	14.633	3.1
AlSe	$X\ ^2\Sigma$		467.6	2.08			20.171	2.8
Ar_2	$X\ ^1\Sigma_g^+$	3.76	25.7*	2.6	0.060	3.7	19.981	0.00848
As_2	$X\ ^1\Sigma_g^+$	2.103	429.6	1.12	0.1018	0.333	37.461	3.2
AsF	$X_1\ ^3\Sigma^-$	1.736	685.8	3.1	0.365	2.4	15.155	3.4
AsH	$X\ ^3\Sigma^-$	1.523	2180	50	7.307	212	0.9944	2.9
AsN	$X\ ^1\Sigma^+$	1.618	1068.5	5.41	0.5455	3.37	11.798	5.2
AsO	$X\ ^2\Pi_{1/2}$	1.624	967.1	4.85	0.4848	3.30	13.181	3.99
AsP	$X\ ^1\Sigma^+$	2.00	604.0	2.0	0.192	0.8	21.914	3.6
AsS	$X_1\ ^2\Pi_{1/2}$	2.017	567.9	1.97	0.1848	0.8	22.409	3.0
Au_2	$X\ ^1\Sigma_g^+$	2.472	190.9	0.420	0.028	0.072	98.483	1.86
AuAl	$X\ ^1\Sigma^+(0^+)$	2.338	333.0	1.16	0.1299	0.67	23.731	2.69
AuBe	$X\ ^2\Sigma^+$	2.060	607.7	3.5	0.4607	4.0	8.6179	2.3
AuH	$X\ ^1\Sigma^+$	1.524	2305.0	43.1	7.240	214	1.0027	2.6
AuMg	$X\ ^2\Sigma^+$	2.443	307.9	1.1	0.1321	0.7	21.381	2.0
B_2	$X\ ^3\Sigma_g^-$	1.589	1051.3	9.4	1.212	14	5.5047	2.42
BBr	$X\ ^1\Sigma^+$	1.89	684.3	3.52	0.489	3.5	9.6615	3.6
BCl	$X\ ^1\Sigma^+$	1.716	839.1	5.11	0.6838	6.46	8.3732	4.5
BF	$X\ ^1\Sigma^+$	1.2626	1402.1	11.8	1.5072*	19.8	6.9702	6.30
BH	$X\ ^1\Sigma^+$	1.232	2367	49.4	12.02	412	0.9233	2.76
BI	$X\ ^1\Sigma^+$	2.14*	575	3.0	0.36*		10.1305	
BN	$X\ ^3\Pi$	1.281	1515	12.3	1.666	25	6.1635	2.6
BO	$X\ ^2\Sigma^+$	1.2048	1885.3	11.7	1.7811	16.5	6.5209	6.68
BS	$X\ ^2\Sigma^+$	1.609	1180.2	6.31	0.7949	6.0	8.1894	4.85
BaBr	$X\ ^2\Sigma^+$	2.8445	193.8	0.41	0.0415	0.122	50.194	3.1
BaCl	$X\ ^2\Sigma^+$	2.683	279.9	0.80	0.0840	0.334	27.895	3.7
BaF	$X\ ^2\Sigma^+$	2.159	469.4	1.83	0.2165	1.16	16.698	4.88
BaH	$X\ ^2\Sigma^+$	2.2317	1168.3	14.5	3.3828	66	1.0005	1.45
BaI	$X\ ^2\Sigma^+$	3.085	152.1	0.27	0.0268	0.066	66.088	3.56
BaO	$X\ ^1\Sigma^+$	1.9397	669.8	2.03	0.3126	1.39	14.333	4.69
BaS	$X\ ^1\Sigma^+$	2.507	379.4	0.884	0.1033	0.32	25.955	3.52
Be_2	$X\ ^1\Sigma_g^+$	2.45	223.4*	20	0.618	20	4.5061	0.067
BeBr	$X\ ^2\Sigma^+$	1.953*	715	3.8	0.546*		8.0885	3.3

Table 32.5 Spectroscopic constants of the ground-state diatomic molecules *(continued)*

Molecule	Electronic term	Equilibrium internuclear distance r_e, Å	Vibrational frequency ω_e, cm^{-1}	Anharmonic constant $\omega_e x_e$, cm^{-1}	Rotational constant B_e, cm^{-1}	Rotation-vibration interaction constant α_e, 10^{-3} cm^{-1}	Reduced mass for a dominant isotope composition μ_A, a.m.u.	Dissociation energy D_0°, 10^4 cm^{-1}
BeCl	$X\ ^2\Sigma^+$	1.797	846.7	4.8	0.7285	6.9	7.1655	3.6
BeF	$X\ ^2\Sigma^+$	1.361	1267	9.1	1.489	17.6	6.1126	5.05
BeH	$X\ ^2\Sigma^+$	1.345	2071.9	48.1	10.274	207	0.9065	1.64
BeI	$X\ ^2\Sigma^+$	2.179*	611.7	1.6	0.422*		8.4146	
BeO	$X\ ^1\Sigma^+$	1.331	1487.3	11.83	1.651	19.0	5.7643	3.7
BeS	$X\ ^1\Sigma^+$	1.742	997.9	6.14	0.7906	6.64	7.0305	3.1
Bi_2	$X\ ^1\Sigma_g^+$	2.661	173.06	0.376	0.0228	0.053	104.49	1.6
BiBr	$X\ 0^+$	2.6095	209.5	0.5	0.0432	0.133	57.285	2.21
BiCl	$X\ 0^+$	2.5	308	1.0	0.09		29.956	2.5
BiF	$X_1\ 0^+$	2.051	513	2.3	0.230	1.5	17.415	2.7
BiH	$X\ ^3\Sigma^-(0^+)$	1.80	1635.7*	32	2.59	54	1.9949	2.0
BiI	$X\ 0^+$	2.800	163.9	0.28	0.0272	0.070	78.957	1.57
BiO	$X_1\ ^2\Pi_{1/2}$	1.93	692	4.3	0.303	2	14.858	2.8
BiS	$X\ ^2\Pi_{1/2}$	2.319	408.7	1.5	0.1128*	0.49	27.730	2.6
Br_2	$X\ ^1\Sigma_g^+(0_g^+)$	2.281	325.32	1.077	0.0821	0.318	39.459	1.589
BrCl	$X\ ^1\Sigma^+$	2.136	444.28	1.84	0.1525	0.77	24.232	1.61
BrF	$X\ ^1\Sigma^+$	1.759	670.75	4.05	0.3558	2.61	15.312	2.055
BrO	$X_1\ ^2\Pi_{3/2}$	1.717	779	6.8	0.4296	3.64	13.299	1.93
C_2	$X\ ^1\Sigma_g^+$	1.2425	1854.7	13.34	1.820	17.6	6.0000	5.01
CCl	$X\ ^2\Pi_{1/2}$	1.645	876.74	5.33	0.6971	6.7	8.9341	3.3
CF	$X\ ^2\Pi_r$	1.272	1308.1	11.1	1.417	18.4	7.3546	4.57
CH	$X\ ^2\Pi_r$	1.120	2858.5	63.0	14.46	534	0.9297	2.79
CN	$X\ ^2\Sigma^+$	1.172	2068.6	13.1	1.8997	17.37	6.4622	6.26
CO	$X\ ^1\Sigma^+$	1.1283	2169.81	13.29	1.93128	17.50	6.8562	8.946
CP	$X\ ^2\Sigma^+$	1.562	1239.7	6.86	0.7986	5.97	8.6491	4.3
CS	$X\ ^1\Sigma^+$	1.535	1285.16	6.50	0.8200	5.92	8.7252	5.93
CSe	$X\ ^1\Sigma^+$	1.676	1035.4	4.9	0.575	3.8	10.433	4.8
Ca_2	$X\ ^1\Sigma^+$	4.278	64.93	1.065	0.04612	0.70	19.981	0.106
CaBr	$X\ ^2\Sigma^+$	2.594	285	0.9	0.0945	0.404	26.529	2.6
CaCl	$X\ ^2\Sigma^+$	2.4368	370.20	1.373	0.1522	0.799	18.650	3.47
CaF	$X\ ^2\Sigma^+$	1.967	581*	2.74	0.338	2.6	12.877	4.4
CaH	$X\ ^2\Sigma^+$	2.002	1298.3	19.1	4.277	97	0.9830	1.4
CaI	$X\ ^2\Sigma^+$	2.829	238.6	0.628	0.0693	0.263	30.392	2.2
CaO	$X\ ^1\Sigma^+$	1.822	732.1	4.8	0.444	3.4	11.423	3.8
CaS	$X\ ^1\Sigma^+$	2.318	462.2	1.78	0.17667	0.84	17.762	2.8
CdH	$X\ ^2\Sigma^+$	1.781*	1337.1		5.323*		0.9990	0.55
Cl_2	$X\ ^1\Sigma_g^+$	1.988	559.7	2.67	0.2440	1.5	17.484	1.9997
ClF	$X\ ^1\Sigma^+$	1.6283	786.15	6.16	0.5165	4.36	12.310	2.111
ClO	$X\ ^2\Pi_i$	1.570	854	5.5	0.6234	6	10.975	2.218
Cr_2	$X\ ^1\Sigma_g^+$	1.679	452.3*	9	0.230	4	25.970	1.26
CrH	$X\ ^6\Sigma$	1.656	1581*	30	6.22	180	0.9886	2.3
CrO	$X\ ^5\Pi_r$	1.61	898	6.7	0.53	5	12.229	3.2
Cs_2	$X\ ^1\Sigma_g^+$	4.648	42.02	0.082	0.0127	0.026	66.453	0.3628
CsBr	$X\ ^1\Sigma^+$	3.072	149.7	0.37	0.036	0.12	49.516	3.36
CsCl	$X\ ^1\Sigma^+$	2.906	214.2	0.73	0.072	0.34	27.685	3.69
CsF	$X\ ^1\Sigma^+$	2.345	352.6	1.61	0.1844	1.18	16.622	4.15
CsH	$X\ ^1\Sigma^+$	2.494	891	12.9	2.710	58	1.0002	1.436

Table 32.5 Spectroscopic constants of the ground-state diatomic molecules *(continued)*

Molecule	Electronic term	Equilibrium internuclear distance r_e, Å	Vibrational frequency ω_e, cm^{-1}	Anharmonic constant $\omega_e x_e$, cm^{-1}	Rotational constant B_e, cm^{-1}	Rotation-vibration interaction constant α_e, 10^{-3} cm^{-1}	Reduced mass for a dominant isotope composition μ_A, a.m.u.	Dissociation energy D_0°, 10^4 cm^{-1}
CsI	$X\ ^1\Sigma^+$	3.315	119.2	0.254	0.0236	0.068	64.918	2.87
Cu_2	$X\ ^1\Sigma_g^+$	2.220	264.5*	1.02	0.1087	0.61	31.465	1.64
CuBr	$X\ ^1\Sigma^+$	2.173	315	0.96	0.1019	0.45	35.011	1.9
CuCl	$X\ ^1\Sigma^+$	2.051	415.3	1.58	0.176	1.00	22.728	3.17
CuF	$X\ ^1\Sigma^+$	1.745	621.55	3.49	0.3794	4.23	14.593	3.6
CuH	$X\ ^1\Sigma^+$	1.463	1941.3	37.5	7.944	256	0.9919	2.3
CuI	$X\ ^1\Sigma^+$	2.338	264	0.6	0.0733	0.284	42.069	2.4
CuO	$X\ ^2\Pi_{3/2}$	1.724	640.2	4.4	0.4445	4.6	12.753	2.3
CuS	$X\ ^2\Pi_{3/2}$	2.054*	414	1.7	0.1884*	1.1	21.201	2.3
F_2	$X\ ^1\Sigma_g^+$	1.412	916.6	11.24	0.8902	13.85	9.4992	1.292
GaBr	$X\ ^1\Sigma^+$	2.352	263.0	0.81	0.0818	0.32	37.221	3.5
GaCl	$X\ ^1\Sigma^+$	2.202	365.3	1.2	0.1499	0.794	23.199	3.97
GaF	$X\ ^1\Sigma^+$	1.774	622.2	3.2	0.3595	2.86	14.893	4.8
GaH	$X\ ^1\Sigma^+$	1.66	1604.5	28.8	6.137	181	0.9933	2.3
GaI	$X\ ^1\Sigma^+$	2.575	216.6	0.5	0.0569	0.19	44.666	2.8
Ge_2	$X\ ^3\Sigma_g^-$	2.44	274		0.078		36.454	2.3
GeF	$X\ ^2\Pi_{1/2}$	1.745	665.7	3.15	0.3658	2.67	15.114	4.0
GeH	$X\ ^2\Pi_r$	1.588	1833.8*	40	6.726	192	0.9939	2.6
GeO	$X\ ^1\Sigma^+$	1.625	986.49	4.47	0.4857	3.08	13.150	5.5
GeS	$X\ ^1\Sigma^+$	2.012	576	1.8	0.1866	0.749	22.319	4.58
GeSe	$X\ ^1\Sigma^+$	2.135	409	1.4	0.0963	0.289	38.401	4.0
GeTe	$X\ ^1\Sigma^+$	2.340	324	0.7	0.0653	0.172	47.113	3.4
H_2	$X\ ^1\Sigma_g^+$	0.7414	4401.21	121.34	60.85	3062	0.5039	3.6118
HD	$X\ ^1\Sigma_g^+$	0.7414	3813.1	91.6	45.65	1990	0.6717	3.6406
HT	$X\ ^1\Sigma_g^+$	0.7414	3597.0	81.68	40.60	1664	0.7554	3.6512
D_2	$X\ ^1\Sigma_g^+$	0.7415	3115.5	61.8	30.44	1079	1.0071	3.6748
DT	$X\ ^1\Sigma_g^+$	0.7414	2845.5	51.4	25.40	822	1.2076	3.6881
T_2	$X\ ^1\Sigma_g^+$	0.7414	2546.5	41.2	20.33	589	1.5080	3.7028
HBr	$X\ ^1\Sigma^+$	1.4144	2649.0	45.22	8.4649	233	0.9954	3.03
HCl	$X\ ^1\Sigma^+$	1.2746	2990.95	52.82	10.593	307.2	0.9796	3.576
HF	$X\ ^1\Sigma^+$	0.9168	4138.3	89.9	20.96	798	0.9571	4.73
HI	$X\ ^1\Sigma^+$	1.6092	2309.01	39.64	6.4264*	169	0.9999	2.463
HfO	$X\ ^1\Sigma$	1.723	974.1	3.23	0.3865	1.72	14.689	6.6
Hg_2	$X\ ^1\Sigma_g^+$	2.92	44	0.5	0.02		100.48	0.089
HgCl	$X\ ^2\Sigma$	2.395	292.6	1.63	0.098	0.5	29.808	0.863
HgH	$X\ ^2\Sigma$	1.766*	1203.2*	120	5.389*		1.0028	0.3020
HoF	X	1.940	615.3	2.60	0.2630	1.4	17.036	4.5
I_2	$X\ ^1\Sigma_g^+$	2.666	214.52	0.609	0.0374	0.114	63.452	1.2333
IBr	$X\ ^1\Sigma^+$	2.469	268.6	0.814	0.0568	0.197	48.659	1.466
ICl	$X\ ^1\Sigma^+$	2.321	384.32	1.51	0.1142	0.53	27.415	1.7366
IF	$X\ ^1\Sigma^+$	1.910	610.2	3.12	0.2797	1.87	16.525	2.32
IO	$X\ ^2\Pi_{3/2}$	1.868	681.5	4.3	0.3403	2.70	14.205	1.8
InBr	$X\ ^1\Sigma^+$	2.543	221	0.65	0.0549	0.19	47.480	3.2
InCl	$X\ ^1\Sigma^+$	2.401	317	1.0	0.1091	0.518	26.810	3.6
InF	$X\ ^1\Sigma^+$	1.985	535.3	2.6	0.2623	1.88	16.303	4.2
InH	$X\ ^1\Sigma^+$	1.838	1476.0	25.6	4.994	143	0.9991	2.0
InI	$X\ ^1\Sigma^+$	2.754	177.08	0.343	0.03687	0.104	60.303	2.8

Table 32.5 Spectroscopic constants of the ground-state diatomic molecules *(continued)*

Molecule	Electronic term	Equilibrium internuclear distance r_e, Å	Vibrational frequency ω_e, cm^{-1}	Anharmonic constant $\omega_e x_e$, cm^{-1}	Rotational constant B_e, cm^{-1}	Rotation-vibration interaction constant α_e, 10^{-3} cm^{-1}	Reduced mass for a dominant isotope composition μ_A, a.m.u.	Dissociation energy D_0°, 10^4 cm^{-1}
K_2	$X\ ^1\Sigma_g^+$	3.925	92.405	0.328	0.0562	0.212	19.482	0.415
KBr	$X\ ^1\Sigma^+$	2.821	213	0.8	0.0812	0.40	26.085	3.2
KCl	$X\ ^1\Sigma^+$	2.667	281	1.3	0.1286	0.790	18.429	3.5
KF	$X\ ^1\Sigma^+$	2.171	428	2.4	0.280	2.34	12.771	4.1
KH	$X\ ^1\Sigma^+$	2.240	986.65	15.84	3.419	94	0.9824	1.429
KI	$X\ ^1\Sigma^+$	3.048	186.5	0.574	0.0609	0.268	29.811	2.7
Kr_2	$X\ ^1\Sigma_g^+$	4.0	24.2	1.3	0.025	1.0	41.956	0.0127
LaF	$X\ ^1\Sigma^+$	2.026*	570*		0.2456*		16.713	5.0
LaO	$X\ ^2\Sigma^+$	1.826	812.7*	2.2	0.353	1.4	14.343	6.6
LaS	$X\ ^2\Sigma^+$	2.355*	456.7	0.96	0.1169*	0.3	25.990	4.8
Li_2	$X\ ^1\Sigma_g^+$	2.673	351.43	2.595	0.6726	7.04	3.5080	0.8342
LiBr	$X\ ^1\Sigma^+$	2.170	563.2	3.5	0.5554	5.64	6.4432	3.5
LiCl	$X\ ^1\Sigma^+$	2.021	643.3	4.50	0.7065	8.01	5.8436	3.9
LiF	$X\ ^1\Sigma^+$	1.564	910.3	7.9	1.345	20.3	5.1238	4.77
LiH	$X\ ^1\Sigma^+$	1.596	1405.6	23.2	7.513	213	0.8812	1.9589
LiI	$X\ ^1\Sigma^+(0^+)$	2.392	496.84	2.84	0.4432	4.10	6.6484	2.86
LiNa	$X\ ^1\Sigma$	2.81	257	1.6	0.396	3.6	5.3755	0.69
LiO	$X\ ^2\Pi_i$	1.7	850	12	1.20	15	4.8768	2.8
LuF	$X\ ^1\Sigma$	1.917	611.8	2.5	0.2676	16	17.137	4.8
LuH	$X\ ^1\Sigma$	1.912	1500	20	4.602	100	1.0021	
LuO	$X\ ^2\Sigma^+$	1.790	842	3.1	0.3588	1.6	14.655	5.8
Mg_2	$X\ ^1\Sigma_g^+$	3.889	51.12	1.64	0.0929	3.78	11.993	0.0404
MgBr	$X\ ^2\Sigma^+$	2.36*	374	1.3	0.164*		18.395	2.7
MgCa	$X\ ^1\Sigma^+$	4.04	60	2	0.069	2	14.989	0.0662
MgCl	$X\ ^2\Sigma^+$	2.199	462.1*	2.1	0.2450	1.6	14.227	2.7
MgF	$X\ ^2\Sigma^+$	1.750	721.6	4.9	0.5192	4.7	10.601	3.8
MgH	$X\ ^2\Sigma^+$	1.730	1495.2	31.9	5.826	186	0.9672	1.1
MgO	$X\ ^1\Sigma^+$	1.7484	784.8	5.3	0.5747	5.32	9.5958	3.0
MgS	$X\ ^1\Sigma^+$	2.142	528.7	2.70	0.2680	1.8	13.704	1.9
MnH	$X\ ^7\Sigma$	1.731	1548	29	5.684	157	0.9897	1.9
MnO	$X\ ^6\Sigma$	1.77	840	4.8	0.43		12.388	3.0
MnS	$X\ ^6\Sigma^+$	2.066	491.1	1.86	0.1954	0.96	20.210	2.3
N_2	$X\ ^1\Sigma_g^+$	1.0977	2358.6	14.32	1.998	17.3	7.0015	7.871
NBr	$X\ ^3\Sigma^-$	1.79	692	4.72	0.44	4	11.893	2.3
NCl	$X\ ^3\Sigma^-$	1.6108	827.96	5.30	0.6498	6.41	9.9990	3.2
NF	$X\ ^3\Sigma^-$	1.317	1141.4	9.0	1.206	14.9	8.0613	2.8
NH	$X\ ^3\Sigma^-$	1.0372	3281	78.3	16.667	647.6	0.94016	2.64
NO	$X\ ^2\Pi_r(\frac{1}{2})$	1.1508	1904.20	14.07	1.6720*	17	7.4664	5.240
NS	$X\ ^2\Pi_r(\frac{1}{2})$	1.494	1219	7.3	0.7696*	6.3	9.7380	4.0
NSe	$X_1\ ^2\Pi_{1/2}$	1.652	956.8	5.6	0.518	4	11.915	3.1
Na_2	$X\ ^1\Sigma_g^+$	3.079	159.12	0.725	0.1547	0.874	11.495	0.5943
NaBr	$X\ ^1\Sigma^+$	2.502	302	1.5	0.1513	0.941	17.803	3.0
NaCl	$X\ ^1\Sigma^+$	2.361	366	2.1	0.2181	1.625	13.871	3.4
NaF	$X\ ^1\Sigma^+$	1.926	536	3.4	0.4369	4.559	10.402	4.3
NaH	$X\ ^1\Sigma^+$	1.889	1176	21.2	4.89	131	0.9655	1.57
NaI	$X\ ^1\Sigma^+$	2.711	258	1.1	0.1178	0.648	19.464	2.4
NaK	$X\ ^1\Sigma^+$	3.59	124.13	0.511	0.0905	0.46	14.459	0.520

Table 32.5 Spectroscopic constants of the ground-state diatomic molecules *(continued)*

Molecule	Electronic term	Equilibrium internuclear distance r_e, Å	Vibrational frequency ω_e, cm^{-1}	Anharmonic constant $\omega_e x_e$, cm^{-1}	Rotational constant B_e, cm^{-1}	Rotation-vibration interaction constant α_e, 10^{-3} cm^{-1}	Reduced mass for a dominant isotope composition μ_A, a.m.u.	Dissociation energy D_0°, 10^4 cm^{-1}
NbO	$X\ ^4\Sigma^-$	1.691	990	3.8	0.432	2	13.646	6.3
Ne_2	$X\ ^1\Sigma_g^+$	3.1*	14*		0.17*	60	9.9962	1.610^{-3}
NiH	$X_1\ ^2\Delta_{5/2}$	1.48	1927*	40	7.70*	230	0.9906	2.5
O_2	$X\ ^3\Sigma_g^-$	1.208	1580.2	12.0	1.446	15.9	7.99746	4.126
OH	$X\ ^2\Pi_i\ (\frac{3}{2})$	0.970	3737.8	84.9	18.91	0.724	0.9481	3.54
P_2	$X\ ^1\Sigma_g^+$	1.893	780.8	2.83	0.3036	1.5	15.487	4.06
PBr	$X\ ^3\Sigma^-(0^+)$	2.171	458.3	1.6	0.1607	0.69	22.244	3.6
PCl	$X\ ^3\Sigma^-$	2.0146	551.38	2.23	0.2529	1.51	16.4251	
PF	$X\ ^3\Sigma^-$	1.590	846.7	4.49	0.567	4.6	11.776	3.7
PH	$X\ ^3\Sigma^-$	1.422	2365	45	8.537	250	0.9761	2.8
PN	$X\ ^1\Sigma^+$	1.491	1337.0	6.9	0.7865	5.54	9.6434	5.1
PO	$X\ ^2\Pi_r$	1.476	1233.3	6.6	0.734	5	10.548	4.92
PS	$X\ ^2\Pi_r$	1.901*	739	3.0	0.2963*		15.733	3.7
Pb_2	$X\ ^3\Sigma_g^-(0_g^+)$	2.93	110.1	0.32	0.0189	0.051	103.99	0.69
PbCl	$X_1\ ^2\Pi_{1/2}$	2.18*	304	0.9	0.119*		29.936	2.5
PbF	$X_1\ ^2\Pi_{1/2}$	2.057	507.3	2.3	0.2287	1.47	17.408	2.9
PbH	$X\ ^2\Pi_{1/2}$	1.839	1564	29.8	4.971	144	1.0030	1.3
PbO	$X\ ^1\Sigma^+$	1.922	721	3.5	0.3073	1.91	14.853	3.1
PbS	$X\ ^1\Sigma^+$	2.287	429.4	1.3	0.1163	0.44	27.712	2.83
PbSe	$X\ ^1\Sigma^+$	2.402	278	0.5	0.0506	0.13	57.732	2.5
PbTe	$X\ ^1\Sigma^+$	2.595	212.0	0.4	0.0313	0.07	79.961	2.1
PdD	$X\ ^2\Sigma^+$	1.529	1446.0	19.6	3.649	81	1.9772	
PtC	$X\ ^1\Sigma$	1.677	1051.1	4.9	0.5304	3.3	11.304	5.1
PtH	$X_1\ ^2\Delta_{5/2}$	1.529	2390	50	7.196	200	1.0026	2.77
PtO	$X\ ^1\Sigma$	1.727	851.1	5.0	0.3822	2.8	14.782	3.2
Rb_2	$X\ ^1\Sigma_g^+$	4.210	57.78	0.139	0.0224	0.055	42.456	0.393
RbBr	$X\ ^1\Sigma^+$	2.945	169.5	0.46	0.0475	0.186	40.903	3.1
RbCl	$X\ ^1\Sigma^+$	2.787	230	0.9	0.0876	0.454	24.769	3.5
RbF	$X\ ^1\Sigma^+$	2.270	375	2	0.2107	1.52	15.525	4.0
RbH	$X\ ^1\Sigma^+$	2.367	937.2	14.17	3.020	73	0.9960	1.5
RbI	$X\ ^1\Sigma^+$	3.177	138.5	0.33	0.0328	0.109	50.873	2.7
RhC	$X\ ^2\Sigma$	1.613	1049.9	4.94	0.603	4.0	10.747	4.85
S_2	$X\ ^3\Sigma_g^-$	1.8894	725.6	2.84	0.2954	1.59	15.986	3.524
SF	$X_1\ ^2\Pi_{3/2}$	1.596	837.6	4.47	0.5552	4.46	11.917	2.7
SH	$X\ ^2\Pi_i$	1.341	2712	60	9.461*	270	0.9770	2.9
SO	$X\ ^3\Sigma^-$	1.481	1149.2	5.6	0.7208	5.74	10.661	4.322
Sb_2	$X\ ^3\Sigma_g^-(0_g^+)$	2.35	270.0	0.59	0.0501		60.948	2.49
SbF	$X_1\ ^3\Sigma^-(0^+)$	1.918	610.2	2.6	0.279	2	16.418	3.6
SbO	$X\ ^2\Pi_r$	1.826	820	4	0.358	2	14.126	3.2
SbP	$X\ ^1\Sigma^+$	2.21	500.1	1.63	0.141	0.5	24.657	2.97
ScCl	$X\ ^1\Sigma^+$	2.23	447	2	0.173	1	19.669	2.7
ScF	$X\ ^1\Sigma^+$	1.788	736	4	0.395	2.7	13.355	5.0
ScO	$X\ ^2\Sigma^+$	1.668*	973.3	4.2	0.5134*	3	11.797	5.6
Se_2	$X_1\ ^3\Sigma_g^-(0_g^+)$	2.166	385.37	0.98	0.0899	0.29	39.958	2.751
SeH	$X_1\ ^2\Pi_{3/2}$	1.47*	2400		7.8*		0.9953	2.6
SeO	$X_1\ ^3\Sigma^-(0^+)$	1.648	914.7	4.5	0.465	3.2	13.327	3.56
SeS	$X_1\ ^3\Sigma^-(0^+)$	2.037	555.6	1.85	0.1793	0.8	22.670	3.0

Table 32.5 Spectroscopic constants of the ground-state diatomic molecules *(continued)*

Molecule	Electronic term	Equilibrium internuclear distance r_e, Å	Vibrational frequency ω_e, cm^{-1}	Anharmonic constant $\omega_e x_e$, cm^{-1}	Rotational constant B_e, cm^{-1}	Rotation-vibration interaction constant α_e, 10^{-3} cm^{-1}	Reduced mass for a dominant isotope composition μ_A, a.m.u.	Dissociation energy D_0°, 10^4 cm^{-1}
Si_2	$X\ ^3\Sigma_g^-$	2.25	511.0	2.0	0.239	1.3	13.988	2.6
SiBr	$X\ ^2\Pi_r$	2.26*	424	1	0.160*		20.655	3.1
SiCl	$X\ ^2\Pi_r$	2.06	535.6	2.17	0.256	1.6	15.542	3.8
SiF	$X\ ^2\Pi_r$	1.601	857.2	4.73	0.5812	4.9	11.315	4.5
SiH	$X\ ^2\Pi_r$	1.520	2042.52	36.06	7.5039	218	0.9728	2.47
SiI	$X_1\ ^2\Pi_{1/2}$	2.4	364	1.2	0.12*		22.923	2.4
SiN	$X\ ^2\Sigma^+$	1.572	1151.4	6.5	0.731	5.6	9.3321	4.2
SiO	$X\ ^1\Sigma^+$	1.510	1241.56	6.0	0.7268	5.04	10.177	6.66
SiS	$X\ ^1\Sigma^+$	1.929	749.5	2.57	0.3035	1.44	14.921	5.2
SiSe	$X\ ^1\Sigma^+$	2.058	580	1.8	0.1920	0.78	20.722	4.5
Sn_2	$X\ ^3\Sigma_g^-(0_g^+)$	2.75	189.7	0.5	0.0385	0.1	59.447	1.53
SnCl	$X_1\ ^2\Pi_{1/2}$	2.36	351	1.1	0.112	0.4	27.073	3.5
SnF	$X\ ^2\Pi_{1/2}$	1.94	583	2.7	0.273	1.4	16.362	4.0
SnO	$X\ ^1\Sigma^+$	1.833	822	3.7	0.3557	2.14	14.112	4.41
SnS	$X\ ^1\Sigma^+$	2.209	487.3	1.36	0.1369	0.506	25.241	3.8
SnSe	$X\ ^1\Sigma^+$	2.326	331	0.74	0.0650	0.17	47.954	3.4
SnTe	$X\ ^1\Sigma^+$	2.523	259	0.5	0.0425	0.10	62.352	3.0
Sr_2	$X\ ^1\Sigma_g^+$	4.5	83		0.019		43.953	0.10
SrBr	$X\ ^2\Sigma^+$	2.735	216.6	0.52	0.05418	0.183	41.585	2.75
SrCl	$X\ ^2\Sigma^+$	2.576	302.6	0.97	0.10156	0.452	25.017	3.4
SrF	$X\ ^2\Sigma^+$	2.075	502	2.3	0.2505	1.55	15.622	4.5
SrH	$X\ ^2\Sigma^+$	2.146	1207.04	17.11	3.6735	80.2	0.9964	1.3
SrI	$X\ ^2\Sigma^+$	2.974	173.8	0.35	0.0367	0.106	51.932	2.3
SrO	$X\ ^1\Sigma^+$	1.920	653.5	4.0	0.3380	2.2	13.533	3.9
SrS	$X\ ^1\Sigma^+$	2.440	388.4	1.3	0.1207	0.44	23.445	2.8
TaO	$X_1\ ^2\Delta_{3/2}$	1.687	1028.7	3.5	0.4028	1.8	14.696	6.6
Te_2	$X_1\ ^3\Sigma_g^-(0_g^+)$	2.558	247.07	0.522	0.03967	0.099	64.953	2.159
TeO	$X_1\ 0^+$	1.825	797.1	4.0	0.355	2.4	14.241	3.1
TeS	$X_1\ 0^+$	2.230	471.2	1.6	0.1322	0.5	25.657	2.8
TeSe	$X_1\ 0^+$	2.37	316.2	0.74	0.062	0.2	48.420	2.4
ThO	$X\ ^1\Sigma$	1.840	895.8	2.4	0.3326	1.3	14.963	7.2
TiO	$X\ ^3\Delta_r$	1.620	1009.0	4.50	0.5354	3.0	11.994	5.5
TiS	$X\ ^3\Delta(1)$	2.083	558.4*	1.9	0.2018	0.9	19.182	3.8
TlBr	$X\ ^1\Sigma^+$	2.618	192.1	0.4	0.0424	0.13	58.014	2.76
TlCl	$X\ ^1\Sigma^+(0^+)$	2.485	284.7	0.88	0.0914	0.40	29.873	3.08
TlF	$X\ ^1\Sigma^+$	2.084	477	2	0.22315	1.50	17.387	3.69
TlH	$X\ ^1\Sigma^+$	1.87	1391	23	4.81	0.15	1.0029	1.6
TlI	$X\ ^1\Sigma^+$	2.814	143*		0.0272	0.07	78.379	2.32
VO	$X\ ^4\Sigma^-$	1.589	1011	4.9	0.5482	3.5	12.173	5.2
Xe_2	$X\ ^1\Sigma_g^+$	4.36	21.1	0.6	0.014		65.194	0.0186
XeCl	$X\ ^2\Sigma^+$	3.2	26	–0.3	0.060		27.641	0.024
YCl	$X\ ^1\Sigma^+$	2.41	381	1	0.116	0.3	25.097	2.8
YF	$X\ ^1\Sigma^+$	1.926	636.3	2.5	0.2904	1.6	15.653	5.0
YO	$X\ ^2\Sigma^+$	1.788	861.46	2.87	0.3889	1.72	13.556	5.9
YbF	$X\ ^2\Sigma^+$	2.016	501.9*	2.2	0.2414*	1	17.128	4.0
YbH	$X\ ^2\Sigma^+$	2.053	1249.5	21.1	3.993	96	1.0020	1.5
ZnH	$X\ ^2\Sigma^+$	1.595	1608	55	6.679	250	0.9922	0.69
ZrO	$X\ ^1\Sigma^+$	1.712	969.8*	4.9	0.4226*	2	13.579	6.3

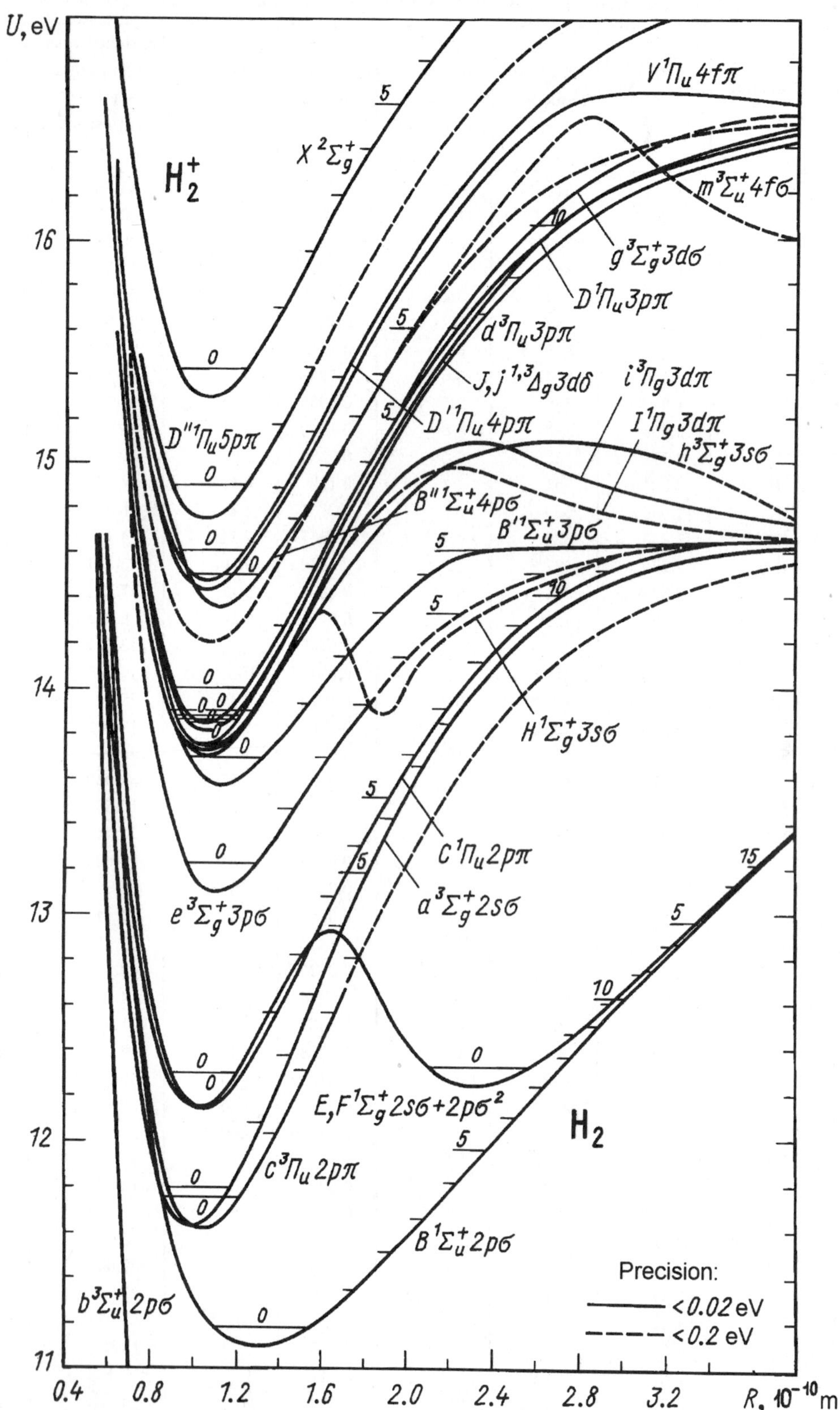

Figure 32.44 The adiabatic electronic terms of H_2 and H_2^+ [15].

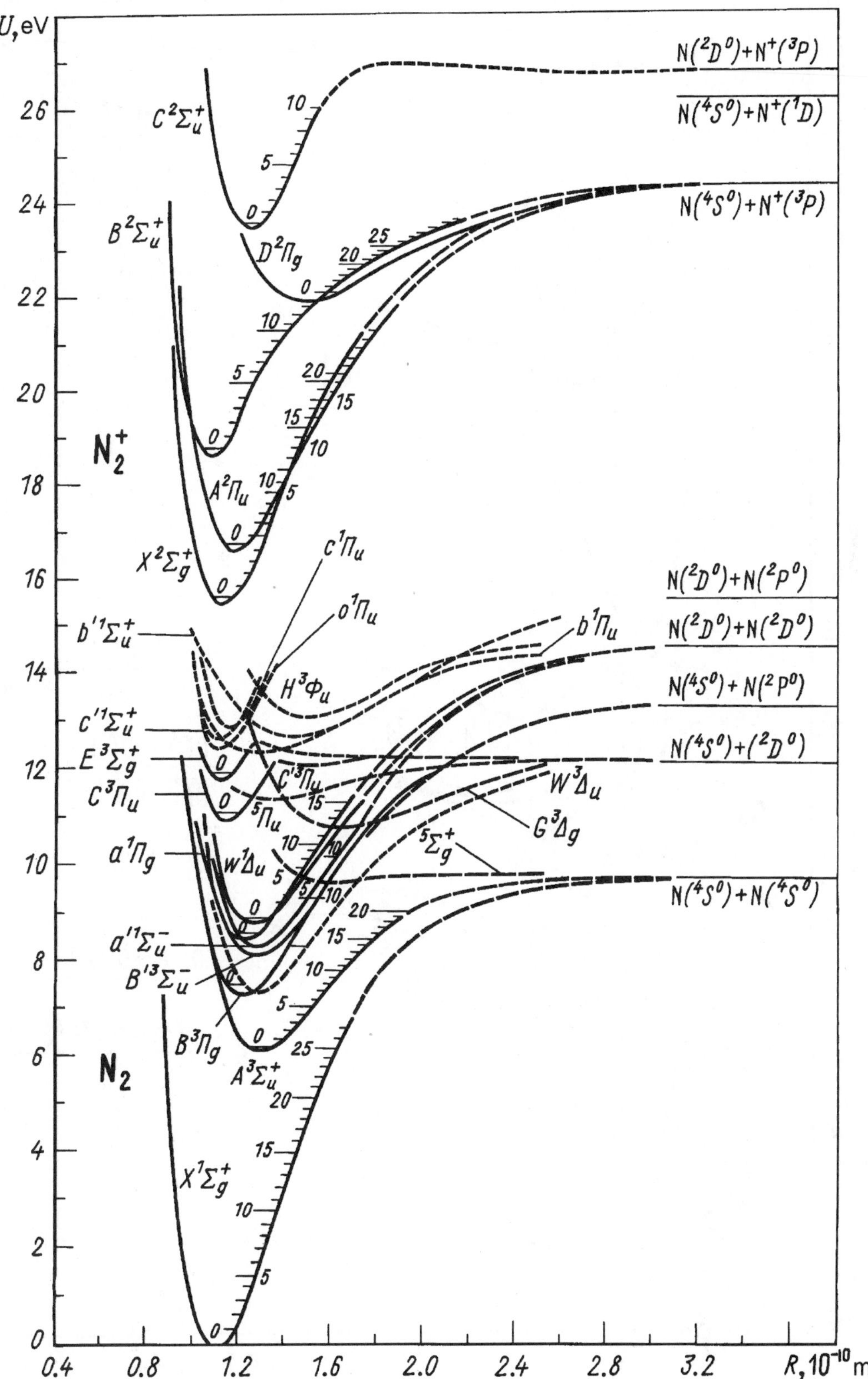

Figure 32.45 The adiabatic electronic terms of N_2 and N_2^+ [17].

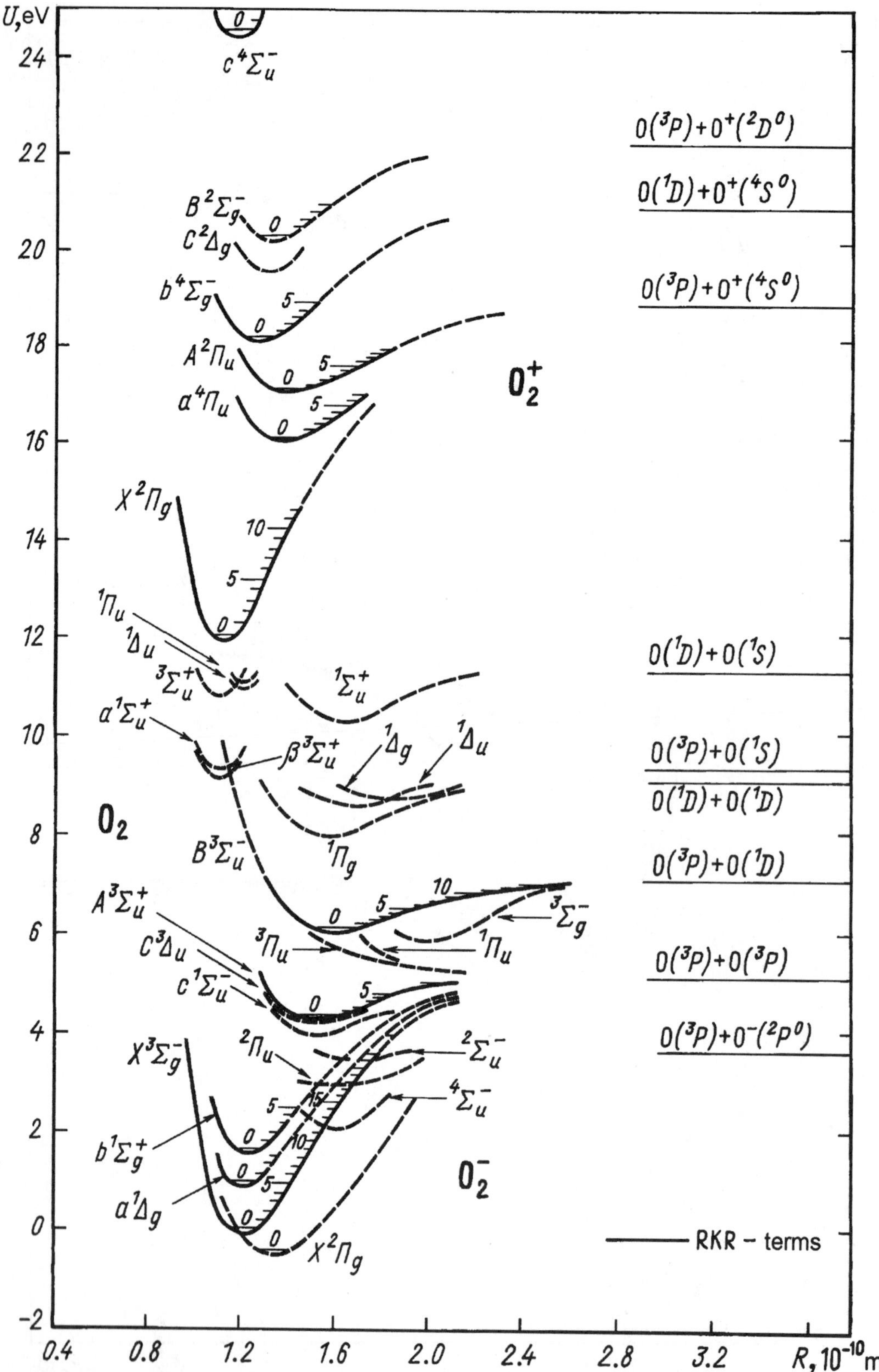

Figure 32.46 The adiabatic electronic terms of O_2, O_2^+ and O_2^- [19].

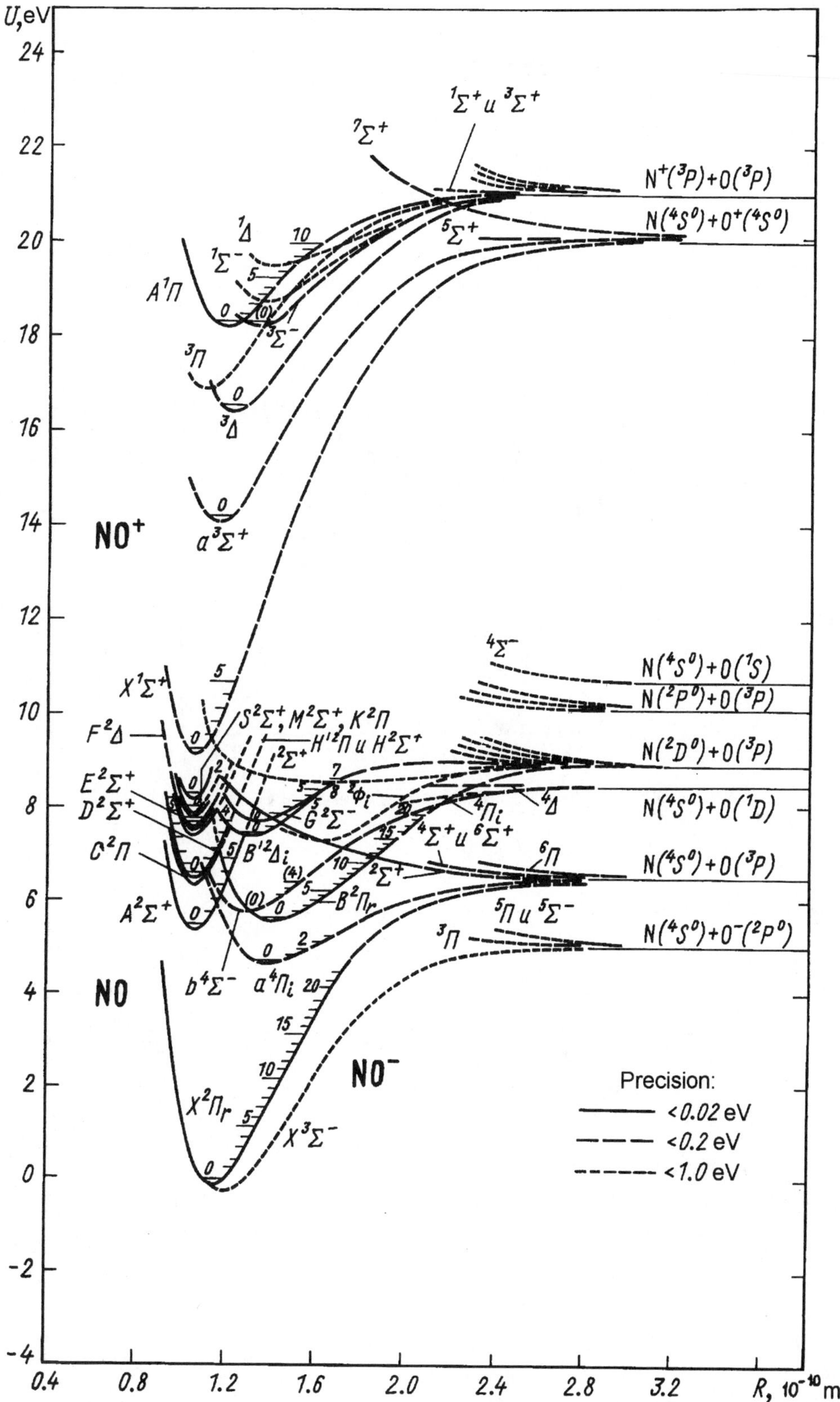

Figure 32.47 The adiabatic electronic terms of NO, NO^+ and NO^- [3,18].

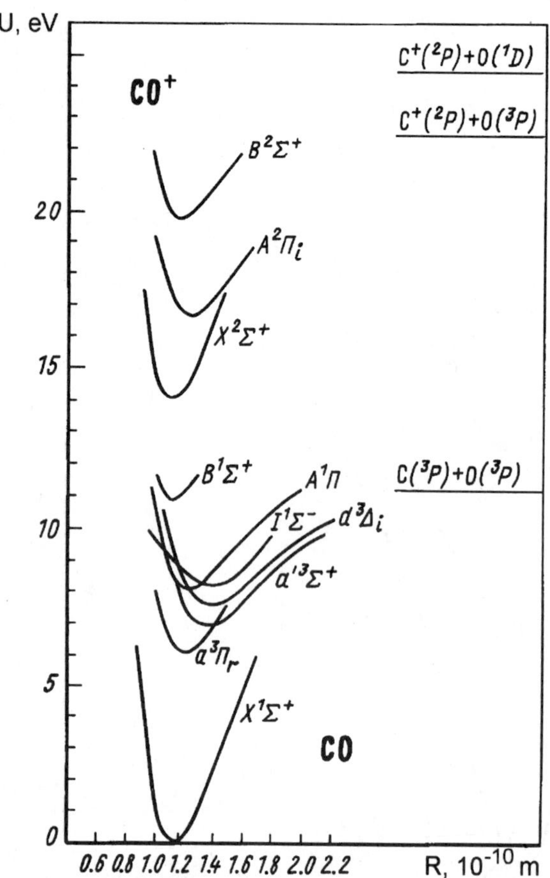

Figure 32.48 The adiabatic electronic terms of CO and CO^+ [3,16].

References

[1] Prokopiev, V. E., Yatsenko, A. S., Grotrian diagrams of neutral atoms (I–III groups, lanthanides, actinides), Preprint No 160 of the Inst. of Automatics and Electronics, Siberian Branch of the USSR Acad. Sci., Novosibirsk, 1981; Energy levels and radiation transitions of neutral atoms (IV–VIII groups), Preprint No 161 of the Inst. of Automatics and Electronics, Siberian Branch of the USSR Acad. Sci., Novosibirsk, 1981 (in Russian).

[2] Bashkin, S., Stoner, J. O., Atomic energy levels and Grotrian diagrams, North-Holland, Amsterdam, 1975 (vol. 1); 1978 (vol. 2); 1981 (vol. 3); 1982 (vol. 4).

[3] Radzig, A. A., Smirnov, B. M., Reference data on atoms, molecules and ions, Springer, Heidelberg, 1985.

[4] Meggers, W. F., Corliss, C. H., Scribner, B. F., Tables of spectral-line intensities, 2nd ed., NBS monograph 145, US Government Printing Office, Washington, 1975 (pt. 1).

[5] Striganov, A. R., Odintsova, G. A., Tables of spectral lines of atoms and ions, Energoizdat, Moscow, 1982 (in Russian).

[6] Condon, E., Shortley, G., The theory of atomic spectra, London, 1935.

[7] Freeman, A. J., Frankel, R. H., Hyperfine interactions, Academic Press, New York, 1967.

[8] Fuller, G. H., J. Phys. Chem. Ref. Data, 5, 835, 1976.

[9] Arimondo, E., Inguscio, M., Violino, P., Rev. Mod. Phys., 49, 31, 1977.

[10] King, W. H., Isotope shifts in atomic spectra, Plenum, New York–London, 1984.

[11] Heilig, K., Bibliography on experimental optical isotope shifts, 1918 through October 1976, Spectrochim. Acta, B32, 1, 1977; Bibliography on experimental optical isotope shifts, November 1976 to October 1981, Spectrochim. Acta, B37, 417, 1982.

[12] Bauche, J., Champeau, R. J., Advances Atom. Mol. Phys., 12, 39, 1976.

[13] Striganov, A. R., Dontsov, Yu. P., Uspekhi fizicheskikh nauk, 55, 315, 1955; Golovin, A. F., Striganov, A. R., Uspekhi fizicheskikh nauk, 93, 111, 1967.

[14] Huber, K.-P., Herzberg, G., Molecular spectra and molecular structure, IV. Constants of diatomic molecules, Van Nostrand Reinhold, New York, 1979.

[15] Sharp, T. E., Atomic data, 2, 119, 1971.

[16] Krupenie, P. H., The band spectrum of carbon monoxide, NSRDS-NBS, NBS, Washington, 1966 (vol. 5).

[17] Lofthus, A., Krupenie, P. H., J. Phys. Chem. Ref. Data, 6, 113, 1977.

[18] Gilmore, F. R., J. Quant. Spectrosc. Radiat. Transfer, 5, 369, 1965.

[19] Krupenie, P. H., J. Phys. Chem. Ref. Data, 1, 423, 1972.

33

Electro-, Magneto-, Elasto- and Nonlinear Optical Materials

A.A. Malyutin, M.E. Brodov, and V.P. Yanovskii

33.1 Introduction

When an electric field is applied to a substance, atomic electrons are perturbed from their equilibrium positions, and polarizability of the matter varies resulting in refractive index changes. In addition to the change of refractive index in the direction of an applied field, it is true for other directions as well. Thus, any material acquires an anisotropy, that is a property common to almost all crystals whose refractive indices are anisotropic even in the absence of external fields. As a result two light waves with orthogonal polarizations propagate variously. Substances that exhibit linear dependence between the refractive index change and the intensity of the applied electric field (*Pockels effect*) are referred to as electrooptical materials and will be described in Section 33.2.

Application of the magnetic field to some materials produces a similar effect, but for waves with left and right circularly polarized states that results in the rotation of the polarization plane of a linearly polarized light (*Faraday effect*). Those materials, which display linear dependence between angle of rotation and the magnetic field intensity, are spoken of as magnetooptical ones (see Section 33.3). Again, there exist materials that produce the same phenomenon with no magnetic field; they are known as displaying natural optical activity. Parameters of this type of materials are compiled in Section 33.6.

Besides, there occur electro- and magnetooptical phenomena that are proportional to a second power of the applied field strength. Parameters of the medium whose susceptibility tensor presents a quadratic function either of the electric field (*Kerr effect*), or of the magnetic field (*Cotton-Mouton and Voigt effects*) are isotropic under ordinary conditions. Only application of the external field makes these materials resemble uniaxial crystals with their optical axes parallel to the field. Some of such materials are considered in Section 33.4. Any material (amorphous or crystalline) under mechanical stress will become optically anisotropic as soon as deformation involves density changes with corresponding variation in refractive index. This phenomenon is known as the *elastooptic* effect. Since an acoustic wave in a medium is a particular type of strain wave, the terms elastooptic effect and acoustooptic effect are often used interchangeably. Elastooptic parameters of materials are presented in Section 33.5.

Light itself represents an electromagnetic wave that at high enough intensity induces changes of the medium optical properties. In nonlinear optical materials, the principle of the light waves superposition is violated. This category of materials includes those with nonlinear response to a single light beam, since it is permissible to regard any light beam as a summation of two or more similar light waves with coincident polarization, frequency and direction. In general, at large values of the electric field

0-8493-2861-6/97/$0.00+$.50

strength the induced optical polarization of the substance can be represented by a series expansion

$$\mathbf{P} = \varepsilon[\chi^{(1)} \cdot \mathbf{E} + \chi^{(2)} \cdot E\mathbf{E} + \chi^{(3)} \cdot E^2\mathbf{E} + \cdots], \tag{33.1}$$

where $\chi^{(1)}$ is the linear susceptibility related to the refractive index n of the medium by

$$\chi^{(1)} = n^2 - 1, \tag{33.2}$$

and $\chi^{(2)}$, $\chi^{(3)}$, etc. are the nonlinear optical susceptibilities of the medium which give rise to a large variety of effects. As an example, if $\mathbf{E}$ and E represent fields of different origin: $\mathbf{E}$ is the low-frequency electric field applied to a medium and E is the optical frequency electric field, then $\chi^{(2)}$ $\chi^{(3)}$ are responsible for Pockels and Kerr effects, respectively. If both fields represent optical frequency fields then $\chi^{(2)}$ describes such phenomena as second harmonic generation, d.c. rectification and parametric generation, while $\chi^{(3)}$ allows for third harmonic generation, two photon absorption and stimulated Rayleigh, Raman and Brillouin scattering. The latter two phenomena relate to the change of the light frequency that is characteristic of the medium itself. They will be addressed in the following Chapter 34. Nonlinear crystal materials described in Section 33.7 do not impose their characteristic frequencies on the light wave.

33.2 Linear Electrooptical Crystals

Linear electrooptic crystal materials exhibit the refraction index change which is proportional to the first power of applied electric field strength E_k, whose highest frequency is below the lattice resonance. This effect is known as the Pockels effect that is widely used to control optical path length in the range from d.c. to microwave frequencies.

The refractive index of a crystal is described by an ellipsoid of dielectric impermeability B_{ij}. By definition

$$B_{ij} = \varepsilon_0 \frac{\partial E_i}{\partial D_j} = \left(\frac{1}{\varepsilon}\right)_{ij}, \tag{33.3}$$

where D_j is the component of the electric induction vector, ε_0 is the vacuum permittivity, and ε is the relative dielectric constant. The relation between the changes of dielectric crystal impermeability and the crystal-applied electric field is written via the tensor of electrooptic coefficients r_{ijk}:

$$\Delta B_{ij} = r_{ijk} E_k, \tag{33.4}$$

where summation over repeating indices is implied, or in a concise form one gets the formula

$$\Delta B_l = r_{lk} E_k, \tag{33.5}$$

in which the indices i, j, k each cover the Cartesian rectangular coordinate axes (X,Y,Z or X_1,X_2,X_3) and $l = (ij)$ refers to six reduced combinations: 1 = (11), 2 = (22), 3 = (33), 4 = (23), 5 = (13), 6 = (12).

The linear electrooptical effect is observed only in acentric crystals. The form of the electrooptic tensor is defined by the point group symmetry of the crystal. In total there exist 21 groups without center of inversion that may have nonvanishing r_{ijk} coefficients. Their reduced matrix forms are listed in Table 33.1.

Table 33.1 Electrooptic matrices

	Group	Electrooptic matrix	Notes		Group	Electrooptic matrix	Notes
Triclinic	1− C_1	$\begin{pmatrix} r_{11} & r_{12} & r_{13} \\ r_{21} & r_{22} & r_{23} \\ r_{31} & r_{32} & r_{33} \\ r_{41} & r_{42} & r_{43} \\ r_{51} & r_{52} & r_{53} \\ r_{61} & r_{62} & r_{63} \end{pmatrix}$	18 elements	Trigonal	1− C_1	$\begin{pmatrix} r_{11} & -r_{22} & r_{13} \\ -r_{11} & r_{22} & r_{13} \\ 0 & 0 & r_{33} \\ r_{41} & r_{51} & 0 \\ r_{51} & -r_{41} & 0 \\ -r_{22} & -r_{11} & 0 \end{pmatrix}$	6 elements
Monoclinic	2− C_2	$\begin{pmatrix} 0 & r_{12} & 0 \\ 0 & r_{22} & 0 \\ 0 & r_{32} & 0 \\ r_{41} & 0 & r_{43} \\ 0 & r_{52} & 0 \\ r_{61} & 0 & r_{63} \end{pmatrix}$	8 elements $2 \parallel X_2$	Trigonal	32− D_3	$\begin{pmatrix} r_{11} & 0 & 0 \\ -r_{11} & 0 & 0 \\ 0 & 0 & 0 \\ r_{41} & 0 & 0 \\ 0 & -r_{41} & 0 \\ 0 & -r_{41} & 0 \end{pmatrix}$	2 elements
Monoclinic	m− C_S	$\begin{pmatrix} r_{11} & 0 & r_{13} \\ r_{21} & 0 & r_{23} \\ r_{31} & 0 & r_{33} \\ 0 & r_{42} & 0 \\ r_{51} & 0 & r_{53} \\ 0 & r_{62} & 0 \end{pmatrix}$	10 elements $m \perp X_2$	Trigonal	$3m$− C_{3v}	$\begin{pmatrix} 0 & -r_{22} & r_{13} \\ 0 & r_{22} & r_{13} \\ 0 & 0 & r_{33} \\ 0 & r_{51} & 0 \\ r_{51} & 0 & 0 \\ -r_{22} & 0 & 0 \end{pmatrix}$	4 elements
Orthorhombic	222− D_2	$\begin{pmatrix} 0 & 0 & 0 \\ 0 & 0 & 0 \\ 0 & 0 & 0 \\ r_{41} & 0 & 0 \\ 0 & r_{52} & 0 \\ 0 & 0 & r_{63} \end{pmatrix}$	3 elements	Hexagonal	6− C_6	$\begin{pmatrix} 0 & 0 & r_{13} \\ 0 & 0 & r_{13} \\ 0 & 0 & r_{33} \\ r_{41} & r_{51} & 0 \\ r_{51} & -r_{41} & 0 \\ 0 & 0 & 0 \end{pmatrix}$	4 elements
Orthorhombic	$mm2$− C_{2v}	$\begin{pmatrix} 0 & 0 & r_{13} \\ 0 & 0 & r_{23} \\ 0 & 0 & r_{33} \\ 0 & r_{42} & 0 \\ r_{51} & 0 & 0 \\ 0 & 0 & 0 \end{pmatrix}$	5 elements	Hexagonal	$\bar{6}$− C_{3h}	$\begin{pmatrix} r_{11} & -r_{22} & 0 \\ -r_{11} & r_{22} & 0 \\ 0 & 0 & 0 \\ 0 & 0 & 0 \\ 0 & 0 & 0 \\ -r_{22} & -r_{11} & 0 \end{pmatrix}$	2 elements
Tetragonal	4− C_4	$\begin{pmatrix} 0 & 0 & r_{13} \\ 0 & 0 & r_{13} \\ 0 & 0 & r_{33} \\ r_{41} & r_{51} & 0 \\ r_{51} & -r_{41} & 0 \\ 0 & 0 & 0 \end{pmatrix}$	4 elements	Hexagonal	622− D_6	$\begin{pmatrix} 0 & 0 & 0 \\ 0 & 0 & 0 \\ 0 & 0 & 0 \\ r_{41} & 0 & 0 \\ 0 & -r_{41} & 0 \\ 0 & 0 & 0 \end{pmatrix}$	1 element
Tetragonal	$\bar{4}$− S_4	$\begin{pmatrix} 0 & 0 & r_{13} \\ 0 & 0 & -r_{13} \\ 0 & 0 & 0 \\ r_{41} & -r_{51} & 0 \\ r_{51} & r_{41} & 0 \\ 0 & 0 & r_{63} \end{pmatrix}$	4 elements	Hexagonal	$6mm$− C_{6v}	$\begin{pmatrix} 0 & 0 & r_{13} \\ 0 & 0 & r_{13} \\ 0 & 0 & r_{33} \\ 0 & r_{51} & 0 \\ r_{51} & 0 & 0 \\ 0 & 0 & 0 \end{pmatrix}$	3 elements
Tetragonal	422− D_4	$\begin{pmatrix} 0 & 0 & 0 \\ 0 & 0 & 0 \\ 0 & 0 & 0 \\ r_{41} & 0 & 0 \\ 0 & -r_{41} & 0 \\ 0 & 0 & 0 \end{pmatrix}$	1 element	Hexagonal	$\bar{6}m2$− D_{3h}	$\begin{pmatrix} 0 & -r_{22} & 0 \\ 0 & r_{22} & 0 \\ 0 & 0 & 0 \\ 0 & 0 & 0 \\ 0 & 0 & 0 \\ -r_{22} & 0 & 0 \end{pmatrix}$	1 element $m \perp X_1$
Tetragonal	$4mm$− C_{4v}	$\begin{pmatrix} 0 & 0 & r_{13} \\ 0 & 0 & r_{13} \\ 0 & 0 & r_{33} \\ 0 & r_{51} & 0 \\ r_{51} & 0 & 0 \\ 0 & 0 & 0 \end{pmatrix}$	3 elements	Cubic	432− O	$\begin{pmatrix} 0 & 0 & 0 \\ 0 & 0 & 0 \\ 0 & 0 & 0 \\ 0 & 0 & 0 \\ 0 & 0 & 0 \\ 0 & 0 & 0 \end{pmatrix}$	0 elements
Tetragonal	$\bar{4}2m$− D_{2d}	$\begin{pmatrix} 0 & 0 & 0 \\ 0 & 0 & 0 \\ 0 & 0 & 0 \\ r_{41} & 0 & 0 \\ 0 & r_{41} & 0 \\ 0 & 0 & r_{63} \end{pmatrix}$	2 elements	Cubic	$23 - T$ $\bar{4}3m - T_d$	$\begin{pmatrix} 0 & 0 & 0 \\ 0 & 0 & 0 \\ 0 & 0 & 0 \\ r_{41} & 0 & 0 \\ 0 & r_{41} & 0 \\ 0 & 0 & r_{41} \end{pmatrix}$	1 element

Measured values of electrooptic coefficients for various types of crystal materials are presented in Tables 33.2 through 33.5 [1]. If r_{lk} have been determined at constant strain (e.g., by making a measurement at frequencies well above acoustic resonances of the sample – the crystal is clamped), the data are indicated by the letter (S) or superscript r_{lk}^S. If r_{lk} have been determined at constant stress (e.g., at frequencies well below the acoustic resonance of the sample – the crystal is free), these data are labelled by the letter (T) or superscript r_{lk}^T. In general, one can write

$$(T)r_{lk} = (S)r_{lk} + p_{ls}d_{ks}, \tag{33.6}$$

where p_{ls} and d_{ks} are the elastooptic and piezoelectric coefficients (see below), respectively. Typical accuracy of the data amounts to ± 15%. Notice that in ferroelectrics and some other materials, the electrooptic coefficients are strongly dependent on temperature, when the latter approaches the critical temperature (T_c) of the phase transition. As a rule, r_{lk} exhibit weak optical wavelength dependence in the range where the crystal is transparent.

One of the most important parameters for crystals used in Pockels cells makes up the half-wave voltage $U_{\lambda/2}$, which is required to produce a phase difference π between two orthogonal polarizations of the radiation propagating through the crystal. In the case of uniaxial crystals, in which the optical axis coincides with the z-axis (or X_3) and the propagation direction is normal to this axis, one has

$$U_{\lambda/2} = \frac{\lambda \cdot 10^{-3}}{n_3^3 r_{33} - n_1^3 r_{13}} = \frac{\lambda \cdot 10^{-3}}{n_3^3 r_c}, \tag{33.7}$$

where $r_c = r_{33} - r_{13}(n_1/n_3)^3$ is measured in nm/V, λ is the light wavelength in nm, and $U_{\lambda/2}$ is in V. According to generally accepted notations, in uniaxial crystals $n_1 = n_o$ and $n_3 = n_e$, respectively, for ordinary and extraordinary light waves.

Table 33.2 Electrooptic parameters of KDP- and ADP-type crystals [Point group symmetry $\bar{4}2m$ above T_c.]

Material	T_c, K	Electrooptic coefficients		Dielectric constants	
		r_{63}, pm/V	r_{41}, pm/V	ε_3	ε_1
KH_2PO_4 (KDP)	123	(T) 9.37	+8.6; $r_{41} < 0$	(T) 21	42
		(S) 8.15; $r_{63} < 0$		(S) 21	44
KD_2PO_4 (DKDP)	222	(T) 26.4	8.8	(T) 50	
		(S) 24.0; $0.93r_{63}^T$		(S) 48	58
KH_2AsO_4 (KDA)	97	(T) 10.9	12.5	(T) 21	54
				(S) 19	53
KD_2AsO_4 (DKDA)	162	(T) 18.2			
RbH_2PO_4 (RDP)	147	(T) 15.5			
		(S) $0.91r_{63}^T$			
RbH_2AsO_4 (RDA)	110	(T) 13.0		(T) 27	41
				(S) 24	39
RbD_2AsO_4 (DRDA)	178	(T) 21.4			
CsH_2AsO_4 (CDA)	143	(T) 18.6			
CsD_2AsO_4 (DCDA)		(T) 36.6			
$NH_4H_2PO_4$ (ADP)	148*1	(T) − 8.5	23.1; $r_{41} < 0$	(T) 15	56
		(S) 4.1		(S) 14	58
$ND_4D_2PO_4$ (DADP)	242*1	(T) 11.9			
$NH_4H_2AsO_4$ (ADA)		(T) 9.2			

*1 Antiferroelectric transition.

Table 33.3 Electrooptic parameters of ABO_3-type compounds

Material (T_c, K)	Symmetry	Electrooptic coefficients r_{l3}, pm/V	r_{lk}, pm/V	λ, nm	Dielectric constants ε_i
$LiNbO_3$ (1470)	$3m$	$(T)\ r_{33} = +32.2$	$(T)\ r_{22} = 6.8$	633	$(T)\ \varepsilon_1 = \varepsilon_2 = 78$
		$(T)\ r_{13} = +10$	$(T)\ r_{51} = 32$	633	$(T)\ \varepsilon_3 = 32$
		$(T)\ r_c = 18$			$(S)\ \varepsilon_1 = \varepsilon_2 = 43$
		$(T)\ r_c = 17$	$(T)\ r_{22} = 5.7$	1150	$(S)\ \varepsilon_3 = 28$
		$(T)\ r_c = 16$	$(T)\ r_{22} = 3.1$	3390	
		$(S)\ r_{33} = +30.8$	$(S)\ r_{22} = 3.4$	633	
		$(S)\ r_{13} = +8.6$	$(S)\ r_{51} = +28$	633	
		$(S)\ r_{33} = 28$	$(S)\ r_{22} = 3.1$	3390	
		$(S)\ r_{13} = 6.5$	$(S)\ r_{51} = 23$	3390	
$LiTaO_3$ (890)	$3m$	$(T)\ r_c = 22$		633	$(T)\ \varepsilon_2 = \varepsilon_1 = 5$
		$(S)\ r_{33} = 30.3$	$(S)\ r_{51} = 20$	633	$(T)\ \varepsilon_3 = 45$
		$(S)\ r_{13} = 7$	$(S)\ r_{22} \approx 1$	633	$(S)\ \varepsilon_2 = \varepsilon_1 = 41$
		$(S)\ r_{33} = 27$	$(S)\ r_{51} = 15$	3390	$(S)\ \varepsilon_3 = 43$
		$(S)\ r_{13} = 4.5$	$(S)\ r_{22} \approx 0.3$	3390	
$BaTiO_3$ (395)	$4mm$	$(T)\ r_c = 108$	$(T)\ r_{51} = 1640$	546	$(T)\ \varepsilon_1 = \varepsilon_3 = 3600$
		$(S)\ r_c = 23$	$(T)\ r_{51} = 820$	546	$(T)\ \varepsilon_3 = 135$
		$(S)\ r_c = 19$		633	$(S)\ \varepsilon_1 = \varepsilon_2 = 2300$
		$(S)\ r_{33} = 28$		633	$(S)\ \varepsilon_3 = 60$
		$(S)\ r_{13} = 8$		633	
$K_3Li_2Nb_5O_{15}$ (693)	$4mm$	$(T)\ r_{33} = 78$		633	$(T)\ \varepsilon_1 = \varepsilon_2 = 309$
		$(T)\ r_{13} = 8.9$		633	$(T)\ \varepsilon_3 = 100$
$Sr_{0.75}Ba_{0.25}Nb_2O_6$ (~330)	$4mm$	$(T)\ r_c = 1410$	$(T)\ r_{51} = 42$	630	$\varepsilon_3 = 3400$ (15 MHz)
		$(T)\ r_{33} = 1340$		633	
		$(T)\ r_{13} = 67$		633	
		$(S)\ r_c = 1090$		633	
$Sr_{0.5}Ba_{0.5}Nb_2O_6$	$4mm$	$(T)\ r_c = 218$		633	$\varepsilon_3 = 450$ (15 MHz)
		(15 MHz)$r_c = 96$		633	
$Sr_{0.25}Ba_{0.75}Nb_2O_6$ (~ 520)	$4mm$	(15 MHz)$r_c = 45$		633	$\varepsilon_3 = 118$ (15 MHz)
$KTa_xNb_{1-x}O_3$ (~ 330)	$4mm$	$(T)\ r_c = 450$	$(T)\ r_{51} = +50$	633	
$PbTiO_3$ (765)	$4mm$	$(S)\ r_{33} = 5.9$		633	$(S)\ \varepsilon_3 = 31$
		$(S)\ r_{13} = 13.8$		633	
$KSr_2Nb_5O_{15}$ (433)	$4mm$ or 4	$(T)\ r_c = 130$		633	$(T)\ \varepsilon_3 = 1000$, $(T)\ \varepsilon_1 = 1200$
$LiIO_3$ (Pyroel.)	6	$(S)\ r_{33} = +6.4$	$(S)\ r_{41} = 1.4$	633	$(T)\ \varepsilon_3 = 554$, $(S)\ \varepsilon_3 = 6.5$
		$(S)\ r_{13} = +4.1$	$(S)\ r_{51} = +3.3$	633	$(T)\ \varepsilon_1 = 65$, $(S)\ \varepsilon_1 = 8$
$Ba_2NaNb_5O_{15}$ (833)	$mm2$	$(T)\ r_c = 34$	$(T)\ r_{42} = 92$	633	$(T)\ \varepsilon_1 = 235$
		$(T)\ r_{33} = 48$	$(T)\ r_{51} = 90$	633	$(T)\ \varepsilon_2 = 247$
		$(T)\ r_{13} = 15$		633	$(T)\ \varepsilon_3 = 5$
		$(T)\ r_{23} = 13$	$(S)\ r_{42} = 75$	633	$(S)\ \varepsilon_1 = 222$
		$(S)\ r_{33} = +29$	$(S)\ r_{51} = 88$	633	$(S)\ \varepsilon_2 = 227$
		$(S)\ r_{23} = 8$		633	$(S)\ \varepsilon_3 = 32$
		$(S)\ r_{13} = 7$		633	

Table 33.4 Electrooptic parameters of AB-type compounds

Material	Symmetry	Electrooptic coefficients r_{lk}, pm/V	λ, nm	Dielectric constants ε_i
CuCl	$\bar{4}3m$	$(T)\ r_{41} = 3.6$	633	$(S)\ \varepsilon = 7.5$
		$(T)\ r_{41} = 3.2$	10600	
		$(S)\ r_{41} = +2.35$	633	
		$(S)\ r_{41} = +2.20$	3390	
CuBr	$\bar{4}3m$	$(T)\ r_{41} = 0.85$	525	

Table 33.4 Electrooptic parameters of AB-type compounds *(continued)*

Material	Symmetry	Electrooptic coefficients r_{lk}, pm/V	λ, nm	Dielectric constants ε_i
ZnO	$6mm$	$(S)\ r_{33} = +2.6,\ (S)\ r_{13} = -1.4$	633	$(S)\ \varepsilon_1 = \varepsilon_2 = 8.15 \approx \varepsilon_3$
		$(S)\ r_{33} = +1.9,\ (S)\ r_{13} = +0.96$	3390	
ZnS	$\bar{4}3m$	$(T)\ r_{41} = 1.2$	400	$(T)\ \varepsilon = 16$
		$(T)\ r_{41} = 2.1$	650	$(S)\ \varepsilon = 12.5$
		$(S)\ r_{41} = 1.6$	633	$(T,S)\ \varepsilon = 8.3$
		$(S)\ r_{41} = 1.4$	3390	
CdS	$6mm$	$(T)\ r_c = 4$	589	$(T)\ \varepsilon_3 = 10.33$
		$(T)\ r_{51} = 3.7$	589	$(T)\ \varepsilon_1 = 9.35$
		$(T)\ r_c = 5.5$	10600	$(S)\ \varepsilon_1 = 9.02$
		$(S)\ r_{33} = 2.4$	633	$(S)\ \varepsilon_3 = 9.53$
		$(S)\ r_{13} = 1.1$	633	
CdSe	$6mm$	$(S)\ r_{33} = 4.3$	3390	$(T)\ \varepsilon_3 = 10.65,\ (S)\ \varepsilon_3 = 10.20$
		$(S)\ r_{13} = 1.8$	3390	$(T)\ \varepsilon_1 = 9.70,\ (S)\ \varepsilon_1 = 9.33$
CdTe	$\bar{4}3m$	$(T)\ r_{41} = 6.8$	3390	
		$(T)\ r_{41} = 6.8$	10600	$(S)\ \varepsilon = 9.4$
		$(T)\ r_{41} = 5.5$	23350	
		$(T)\ r_{41} = 5.0$	27950	
HgS	32	$(S)\ r_{11} = 3.1,\ (S)\ r_{41} = 1.4$	633	
		$(S)\ r_{11} = 4.2,\ (S)\ r_{41} = 2.4$	3390	
ZnS	$6mm$	$(S)\ r_{33} = 1.8$	633	$(T)\ \varepsilon_1 = \varepsilon_3 = 8.7$
		$(S)\ r_{33} = 1.7$	3390	$(S)\ \varepsilon_1 = 8.7$
		$(S)\ r_{13} = 0.9$	633	
ZnSe	$\bar{4}3m$	$(T)\ r_{41} = 2.0$	546	$(T)\ \varepsilon = 9.1$
		$(S)\ r_{41} = 2.0$	633	$(S)\ \varepsilon = 9.1$
		$(T)\ r_{41} = 2.2$	10600	
ZnTe	$\bar{4}3m$	$(T)\ r_{41} = 4.45 - 3.95$	590; 690	$(T)\ \varepsilon = 10.1$
		$(T)\ r_{41} = 1.4$	10600	$(S)\ \varepsilon = 10.1$
		$(S)\ r_{41} = 4.3$	633	
		$(S)\ r_{41} = 3.2$	3390	
GaP	$\bar{4}3m$	$(S)\ r_{41} = (-1.07) - (-0.97)$	560; 3390	$(S)\ \varepsilon = 10 - 12$
GaAs	$\bar{4}3m$	$(S)\ r_{41} = 1.2$	900	$(S)\ \varepsilon = 13.2$
			1080	$(S)\ \varepsilon = 12.3$
		$(S)\ r_{41} = -1.5$	3390	$(T)\ \varepsilon = 12.5$
		$(S{+}T)\ r_{41} = 1.2 - 1.6$	1000; 3000	
		$(T)\ r_{41} = 1.0 - 1.2$	4000; 12000	
		$(T)\ r_{41} = 1.6,\ (S)\ r_{41} = 1.5$	10600	

Table 33.5 Electrooptic parameters of miscellaneous crystals

Material	Symmetry	Electrooptic coefficients r_{lk}, pm/V	λ, nm	Refractive index n_i (λ, nm)
$Bi_4(GeO_4)_3$	$\bar{4}3m$	$(T)\ r_{41} = 1.03$	450 – 620	$n_o = 2.07$
$(CH_2)_6N_4$ (HMT,	$\bar{4}3m$	$(T)\ r_{41} = 0.71 - 0.8$	546	$n_o = 1.591$ (589)
hexamethylenetetramine)		$(T)\ r_{41} = 0.78$	633	$n_o = 1.594$ (633)
		$(S)\ r_{41} < 0.14$	633	
Hauynite (mineral)	$\bar{4}3m$	$(T)\ r_{41} < 0.04$	633	$n_o = 1.496$
$K_2Mg_2(SO_4)_3$	23	$(T)\ r_{41} = 0.40$	546	$n_o = 1.535$ (589)
$(NH_4)_2Cd_2(SO_4)_3$	23	$(T)\ r_{41} = 0.70$	546	$n_o = 1.606$ (589)
$(NH_4)_2Mn_2(SO_4)_3$	23	$(T)\ r_{41} = 0.53$	546	$n_o = 1.57$ (589)
$Tl_2Cd_2(SO_4)_3$	23	$(T)\ r_{41} = 0.37$	546	$n_o = 1.730$ (589)
$K_2Mn_2(SO_4)_3$	23	$(T)\ r_{41} = 2.0$	453 – 642	1.62(450 – 650)
$Rb_2Mn_2(SO_4)_3$	23	$(T)\ r_{41} = 1.9$	453 – 642	1.60(450 – 650)
$Tl_2Mn_2(SO_4)_3$	23	$(T)\ r_{41} = 2.1$	453 – 642	1.80(450 – 650)

Table 33.5 Electrooptic parameters of miscellaneous crystals *(continued)*

Material	Symmetry	Electrooptic coefficients r_{lk}, pm/V	λ, nm	Refractive index n_i (λ, nm)
$K_2Ni_2(SO_4)_3$	23	$(T)\ r_{41} = 1.0$	453 – 642	1.70(450 – 650)
$NaClO_3$	23	$(T)\ r_{41} = 0.4$	589	$n_0 = 1.515$
$NA_3SbS_4{\cdot}9H_2O$	23	$(T)\ n_1^3 r_{41} = 5.66$	420	
		$(T)\ n_1^3 r_{41} = 5.62$	1080	
Sodium uranyl acetate	23	$(T)\ r_{41} = 0.87$	546	$n_0 = 1.507$ (546)
$LiKSO_4$	6	$(T)\ r_c = 1.6$	546	$n_3 \approx n_1 = n_2 = 1.474$ (546)
$LiNaSO_4$	$3m$	$(T)\ r_{22} < 0.02$	546	$n_1 = n_2 = 1.490,\ n_3 = 1.495$
Tourmaline	$3m$	$(T)\ r_{22} = 0.3$	589	$n_1 = n_2 = 1.63$
		$(S)\ r_{33} = r_{13} = 1.7$	633	$n_3 = 1.65$
$NA_3Li(CrO_4)_2{\cdot}6H_2O$	$3m$	$(T)\ r_{22} = 0.92$	500	$n_1 = n_2 = 1.643$ (500)
		$(T)\ r_{22} = 0.82$	520	$n_1 = n_2 = 1.635$ (520)
		$(T)\ r_{22} = 0.77$	600	$n_1 = n_2 = 1.612$ (600)
Ag_3AsS_3 (Proustite)*[1]	$3m$	$(S)\ (n_1^3 r_{13} - n_3^3 r_{33}) = 70$	633	$n_1 = 3.02$ (633)
		$(S)\ n_1^3 r_{22} = 29$	633	$n_3 = 2.74$ (633)
$K_2S_2O_6$	32	$(T)\ r_{11} = 0.26$	546	$n_1 = n_2 = 1.456,\ n_3 = 1.518$
$Cs_2C_4H_4O_6$	32	$(T)\ r_{11} = 1.0$	546	$n_1 = n_2 = 1.546,\ n_3 = 1.546$
$SrS_2O_6{\cdot}4H_2O$	32	$(T)\ r_{11} = 0.1$	546	$n_1 = n_2 = 1.532,\ n_3 = 1.528$
Se*[2]	32	$(S)\ n_1^3 r_{11} = 89$	1150	$n_1 = 2.737(1150),\ n_3 = 3.573$
		$(S)\ r_{11} \sim 2.5$	10600	$n_1 = 2.64(10600),\ n_3 = 3.4$
SiO_2(Quartz)*[3]	32	$(T)\ r_{11} = -0.47$	409 – 605	$n_3 = 1.555$ (546)
		$(T)\ r_{41} = 0.20$	633	$n_1 = n_2 = 1.546$ (546)
		$(S)\ r_{11} = 0.29,\ (S) r_{11} = 0.174$	633	
$(C_6H_{12}O_6)_2NaBr{\cdot}H_2O$	32	$(T)\ r_{11} = 0.1$	546	$n_3 = 1.560$ (546)
AgGaS*[4]	$\bar{4}2m$	$(T)\ r_{63} = 3.0$	633	$n_1 = n_2 = 2.55$ (633)
		$(T)\ r_{41} = 4.0$	633	$n_3 = 2.50$
$Gd_2(MoO_4)_3$*[5]	$\bar{4}2m$	$(T)\ n_1^3 r_{63} = 17$ (450 K)	633	$n_1 = n_2 = 1.528$
	$mm2$	$(T)\ (n_1^3 r_{13} - n_3^3 r_{33}) = 17.5$ (300 K)	633	$n_1 \approx n_2 = 1.848(633)$
				$n_3 = 1.901(633)$
$CdGa_2S_4$	$\bar{4}$	$(T)\ r_{13} = 0.37$	500	$n_1 = n_2 = 2.3$ (500)
		$r_{63} = 3.5$	500	
$(NH_4)_2C_2O_4{\cdot}H_2O$	222	$(T)\ r_{41} = 230$	633	$n_1 = 1.437$ (650)
		$(T)\ r_{52} = 330$	633	$n_2 = 1.547$
		$(T)\ r_{63} = 250$	633	$n_1 = 1.590$
		$(T)\ r \ll 250,\ (S) r_{63} \approx 2$	633	
$NaNO_2$*[6]	$mm2$	$(T)\ r_{22} - (n_3/n_2)^3 r_{32} = 4.1$	546	$n_1 = 1.347$ (546)
		$(T)\ r_{32} - (n_1/n_3)^3 r_{12} = 4.2$	546	$n_2 = 1.415$
		$(T)\ r_{22} - (n_1/n_2)^3 r_{12} = 0.6$	546	$n_2 = 1.661$
		$(T)\ r_{43} = -1.9,\ (T) r_{61} = -3.0$	546	
$C(CH_2OH)_4$	2	$(T)\ r_{52} = 1.45$	460 – 700	$n_1 = 1.528$
		$(T)\ \lvert r_{12} - r_{32}\rvert = 0.7$	460 – 700	$n_2 \approx n_3 = 1.56$
$Ca_2Nb_2O_7$	2	$(T)\ \lvert r_{22} - (n_1/n_2)^3 r_{12}\rvert = 12$	630	$n_1 = 1.97$
		$(T)\ \lvert r_{22} - (n_1/n_2)^3 r_{12}\rvert = 14$	630	$n_2 = 2.16$
		$(T)\ \lvert r_{22} - (n_1/n_2)^3 r_{12}\rvert = 13$	630	$n_3 = 2.17$
		$(S)\ r_{12} = 6.7,\ (S) r_{41} = 2.7$	630	
		$(S)\ r_{22} = 25.5,\ (S) r_{52} < 0.6$	630	
		$(S)\ r_{32} = 6.4,\ (S) r_{63} = 0.9$	630	

*[1] Transparent from 0.6 to 13 μm; $\varepsilon_1 \approx \varepsilon_3 = 20$; $\rho = 10^5\ \Omega{\cdot}$cm.

*[2] Absorption edge approaches ~8 μm; $\varepsilon_1 = 8$.

*[3] $r_{11} < 0$ and $r_{41} > 0$ in left-handed quartz.

*[4] $(S)\ \varepsilon_3 = 14$, $(S)\ \varepsilon_1 = 10$.

*[5] $T_c = 432$ K, $\varepsilon_3 = 8$.

*[6] Author takes X_2 as a polar axis. Transition to mmm occurs at 23 K; $(S)\ \varepsilon_1 = 5,\ \varepsilon_2 = 4,\ \varepsilon_3 = 8$.

33.3 Magnetooptical Materials

Application of an external magnetic field to a paramagnetic or diamagnetic substance causes rotation of the polarization plane of light propagating along the field lines. For such medium, the gyration vector is proportional and parallel to the magnetic field vector **H** (Faraday effect). The relationship between the angle Φ of polarization rotation $\mathbf{\Phi}$ (measured usually in angle minutes) and the applied magnetic field strength H (A/m) takes the form

$$\Phi = VHl\cos\gamma, \tag{33.8}$$

where V is the Verdet constant (angle min./A), l is the length of material, and γ is the angle between the direction of magnetic field and that of light beam propagation. The sense of rotation depends solely on the direction of the magnetic field; reversing the direction of light propagation has no effect on the sense of rotation as soon as the magnetic gyration vector is aligned with **H** rather than with propagation vector **k**. The Faraday effect is one of a few magnetooptical effects that are large enough to be utilized in practice.

For a ferromagnetic material, the magnetic field H in the above relation should be replaced by the magnetization $M(T)$, and the Verdet constant, respectively, by the Kundt constant K(angle deg/T·cm). For these materials values of the Faraday rotation at saturation magnetization $M_s(T)$ per unit length (cm) are usually tabulated for the light travelling parallel to M_s, viz. specific Faraday rotation is of interest:

$$F = K \cdot M_s. \tag{33.9}$$

By virtue of the fact that together with large specific polarization rotation some materials exhibit high absorption of the light, the latter should be taken into account. A material figure of merit accepted for paramagnetic and diamagnetic material is V/α, while for ferromagnetic material $2F/\alpha$, where α is the optical absorption coefficient emerging in the Beer-Lambert law for the light intensity I being diminished on propagating through matter:

$$I = I_0 \exp(-\alpha \cdot l). \tag{33.10}$$

In Tables 33.6 to 33.8 [2] presented below, containing data for some crystals and glasses, the following notations were used:

T_C – Curie temperature
T_p – phase transition temperature
T_N – Néel temperature
T_{comp} – compensation temperature
b.c.c. – body-centered cube
f.c.c. – face-centered cube
h.c.p. – hexagonal close packing
h.t.p. – high-temperature phase
l.t.p. – low-temperature phase
M_s – saturation magnetization
F – specific Faraday rotation
α – absorption coefficient
$2F/\alpha$ – material merit factor

Values of the saturation magnetization M_s are given at 0 K. Verdet constants for some liquids are listed in Tables 33.9 and 33.10 [3].

Table 33.6 Magnetooptical parameters of ferro-, ferri- and antiferromagnetic materials

Material (composition, structure)	Critical temperature, K	M_s, T	F, angle deg./cm	α, cm^{-1}	$2F/\alpha$, angle deg.	T, K	λ, nm
Transition metals							
Iron (b.c.c.)	T_C=1043	0.1735	3.5·10^5	7.6·10^5	0.92	300	546
			5.1·10^5	3.2·10^5	3.1	300	1000
			4.4·10^5	6.5·10^5	1.4	300	500
			6.5·10^5	5·10^5	2.6	300	1000
			7·10^5	4.2·10^5	3.3	300	1500
			7·10^5	3.5·10^5	4.0	300	2000
Cobalt (h.c.p.)	T_C=1390	0.145	3.6·10^5	8.5·10^5	0.85	300	546
			2.9·10^5			300	500
			5.5·10^5	6.1·10^5	1.8	300	1000
			5.5·10^5	4.5·10^5	2.4	300	1500
			4.8·10^5	3.6·10^5	2.7	300	2000
Nickel (f.c.c.)	T_C= 633	0.051	0.99·10^5	8.0·10^5	0.25	300	546
			7.2·10^5	4.2·10^5	3.4	300	4000
			0.8·10^5			300	500
			2.6·10^5	5.8·10^5	0.9	300	1000
			1.5·10^5	4.8·10^5	0.6	300	1500
			1·10^5	4.1·10^5	0.25	300	2000
Nickel (single crystal film[*1]			0.79·10^5			300	546
			0.88·10^5			300	546
			0.97·10^5			300	546
Binary compounds and alloys							
Permalloy (f.c.c., Ni : 82, Fe : 18)	T_C=803	0.085	12·10^5	6·10^5	0.4	300	500
Nickel-Iron (Ni : Fe%)							
100 : 0		0.048	1.2·10^5	7.05·10^5	0.34	300	632.8
90 : 10		0.067	1.6·10^5	7.14·10^5	0.45	300	632.8
80 : 20		0.086	2.2·10^5	7.10·10^5	0.62	300	632.8
70 : 30		0.103	2.7·10^5	7.0·10^5	0.77	300	632.8
60 : 40		0.119	2.9·10^5	7.54·10^5	0.77	300	632.8
50 : 50		0.127	2.8·10^5	8.13·10^5	0.69	300	632.8
40 : 60		0.115	2.2·10^5	8.17·10^5	0.54	300	632.8
30 : 70		0.064		8.13		300	632.8
20 : 80		0.154	3.3·10^5	8.1·10^5	0.81	300	632.8
10 : 90		0.166	3.6·10^5	8.13·10^5	0.88	300	632.8
0 : 100		0.172	3.5·10^5	8.13·10^5	0.86	300	632.8
35 : 65		0.099	2.1·10^5	7.7·10^5	0.55	300	400–700
MnBi Normal (l.t.p.: NiAs)	T_C=639	0.061 (0.060 at 300 K)	4.2·10^5	6.1·10^5	1.4	300	450
			5.0·10^5	5.8·10^5	1.7	300	500
			7.0·10^5	5.1·10^5	2.8	300	600
			7.7·10^5	4.5·10^5	3.4	300	700
			7.6·10^5	4.3·10^5	3.5	300	800
			7.5·10^5	4.2·10^5	3.6	300	900
			7.4·10^5	4.1·10^5	3.6	300	1000

[*1] Data taken along ⟨100⟩, ⟨110⟩ and ⟨111⟩ directions, respectively.

Table 33.6 Magnetooptical parameters of ferro-, ferri- and antiferromagnetic materials *(continued)*

Material (composition, structure)	Critical temperature, K	M_s, T	F, deg/cm	α, cm^{-1}	$2F/\alpha$, angle deg.	T, K	λ, nm
			Binary compounds and alloys				
MnBi	T_C=453	(0.044	3,2·10^5	6.1·10^5	1.0	300	450
Quenched		at	3.3·10^5	5.8·10^5	1.1	300	500
h.t.p.:		300 K)	3.3·10^5	5.1·10^5	1.3	300	600
distorted			3.3·10^5	4.7·10^5	1.4	300	700
NiAs)			3.3·10^5	4.5·10^5	1.4	300	800
			3.2·10^5	4.4·10^5	1.4	300	900
			3.2·10^5	4.4·10^5	1.4	300	1000
MnAs	T_C=313		0.44·10^5	5.0·10^5	0.174	300	500
NiAs			0.49·10^5	4.9·10^5	0.200	300	600
			0.59·10^5	4.6·10^5	0.26	300	700
			0.78·10^5	4.5·10^5	0.34	300	800
			0.62·10^5	4.4·10^5	0.28	300	900
CrTe	T_C=334	0.0081	0.5·10^5	2.0·10^5	0.5	300	550
(NiAs)			0.4·10^5	1.2·10^5	0.7	300	900
			0.4·10^5	0.6·10^5	1.3	300	2500
FeRh	T_p=333		0.9·10^5	3.3·10^5	0.56	348	700
			Ferrites				
YIG	T_N=560	0.020	240	0.069	7·10^3	300	1200
(Garnet)			2400	1500	3.2	300	555
			1750	1350	2.6	300	588
			1250	1400	1.8	300	625
			800	1150	1.4	300	667
			900	670	2.7	300	715
			750	450	3.3	300	770
			175	$<$ 0.06	$>$ 3·10^3	300	1500
GdIG	T_N=564	0.058	−2000	6000	0.6	300	500
(Garnet)	T_{comp}=286		−1050	900	2.3	300	600
			−450	400	2.3	300	700
			−300	100	6	300	800
			−220	230	1.9	300	900
			−80	70	2.3	300	1000
$NiFeO_4$	T_N=858	0.0267	2.0·10^4	5.9·10^4	0.7	300	286
(Spinel)			2.4·10^4	7.4·10^4	0.7	300	330
			−0.75·10^4	16·10^4	0.09	300	400
			−1.0·10^4	10^5	0.2	300	500
			+0.12·10^4	10^4	0.2	300	660
			−120	38	6	300	1500
			+40	32	2.5	300	2000
			+75	15	10	300	3000
			+110	15	15	300	4000
			+110	32	7	300	5000
$CoFe_2O_4$	T_N=793	0.0392	2.75·10^4	12	0.5	300	286
(Spinel)			3.8·10^4	14	0.5	300	330
			3.6·10^4	17	0.4	300	400
			1.3·10^4	13	0.2	300	500
			−2.5·10^4	6	0.8	300	660

Table 33.6 Magnetooptical parameters of ferro-, ferri- and antiferromagnetic materials *(continued)*

Material (composition, structure)	Critical temperature, K	M_S, T	F, angle deg./cm	α, cm^{-1}	$2F/\alpha$, angle deg.	T, K	λ, nm
Ferrites							
$MgFe_2O_4$			−60	100	1	300	2500
			−40	40	2	300	3000
			0	12	0	300	4000
			+30	4	15	300	5000
			+35	6	11	300	6000
			+50	13	8	300	7000
$Li_{0.5}Fe_{2.5}O_4$			−440	150	6	300	1500
			−190	135	3	300	2000
			+10	85	0.2	300	3000
			+85	60	3	300	4000
			+110	44	5	300	5000
			+125	44	6	300	6000
			+135	80	3	300	7000
$BaFe_{12}O_{19}$ (Hexagonal)			−50	−38	3	300	2000
			+75	20	7.5	300	3000
			+130	13	20	300	4000
			+150	20	15	300	5000
			+160	20	16	300	6000
			+165	22	15	300	7000
$BaZn_2Fe_{12}O_{19}$ (Hexagonal)			+90	120	1.5	300	5000
			+80	70	2	300	6000
			+75	65	2	300	7000
			+70	85	2	300	8000
Fluorides							
$RbNiF_3^{*2}$ (Perovskite)	T_N=220	0.0099	360	35	20	77	450
			210	12	35	77	500
			70	10	14	77	600
			−70	30	5	77	700
			310	70	9	77	800
			100	60	3	77	900
			75	25	6	77	1000
$RbNi_{0.75}Co_{0.25}F_3^{*3}$ (Perovskite)	T_N=109		180	9	40	77	600
$RbFeF_3^{*4}$	T_p=102		3400	7	900	82	300
			1600	3	1100	82	400
			950	4.6	410	82	500
			620	1.5	830	82	600
			420	1.2	700	82	700
			300	2.5	240	82	800
FeF_3^{*5}	T_C=365	$3.2\cdot10^{-4}$ (300 K)	670	14	95	300	349
			415	8.2	101	300	404
			180	4.4	82	300	522.5

*2 Measurements along c-axis, which is a magnetic hard axis.
*3 Measurements along c-axis, which is a magnetic easy axis.
*4 Measurements along c-axis ([100] direction at room temperature).
*5 Strong natural birefringence interferes with Faraday effect.

Table 33.6 Magnetooptical parameters of ferro-, ferri- and antiferromagnetic materials *(continued)*

Material (composition, structure)	Critical temperature, K	M_s, T	F, angle deg./cm	α, cm^{-1}	$2F/\alpha$, angle deg.	T, K	λ, nm
Trihalides							
$CrCl_3$ (BiI_3)	T_C=16.8	0.031	2000 −500 −1000	200 300 70	20 3 30	1.5 1.5 1.5	410 450 590
$CrBr_3$ (BiI_3)	T_C=32.5	0.027	$3 \cdot 10^5$ $1.6 \cdot 10^5$	$3 \cdot 10^3$ $1.4 \cdot 10^4$	200 23	 1.5	478 500
CrI_3 (BiI_3)	T_C=68	0.0214	$1.1 \cdot 10^5$ $0.8 \cdot 10^5$	$6.3 \cdot 10^3$ $3 \cdot 10^3$	35 53	1.5 1.5	970 1000
Borates							
$FeBO_3^{*6}$ (Calcite)	T_C=348	0.0009 (300 K)	3200 2300 1100 450	140 40 100 38	45 115 22 24	300 300 300 300	500 525 600 700
Chalcogenides							
EuO (NaCl)	T_C=69	0.189	$-1.0 \cdot 10^5$ $+3.2 \cdot 10^5$ $+5 \cdot 10^5$ $+3.6 \cdot 10^5$ $+0.5 \cdot 10^5$ $+3 \cdot 10^4$ 660	$0.5 \cdot 10^4$ $7.5 \cdot 10^4$ $9.7 \cdot 10^4$ $9.7 \cdot 10^4$ $7.8 \cdot 10^4$ > 0.5 > 1.0	40 8.5 10 7.5 1.3 $\approx 10^5$ ≈ 1300	5 5 5 5 5 20 20	1100 800 700 600 500 2500 10600
EuS (NaCl)	T_C=16.3		$-1.6 \cdot 10^5$ $-9.6 \cdot 10^5$ $+5.5 \cdot 10^5$ $+5.1 \cdot 10^5$	≈ 0 $3.3 \cdot 10^4$ $1.2 \cdot 10^5$ $1.0 \cdot 10^5$	 58 92 10	6 6 6 6	825 690 563 495
EuSe (NaCl)	T_C=7	0.105	$1.45 \cdot 10^5$ $1.17 \cdot 10^5$ $0.95 \cdot 10^5$	80 70 60	3600 3340 3170	4.2 4.2 4.2	750 775 800

Material	M_s T	Intrinsic specific Faraday rotation in angle deg../cm at 300 K for wavelength (nm)						$\alpha(cm^{-1})$ for
		600	800	1000	1200	1400	1600	1250 nm
Orthoferrites*6								
$EuFeO_3$	83							≈38
$GdFeO_3$	94							≈10
$TbFeO_3$	137							≈29
$DyFeO_3$	128							≈40
$HoFeO_3$	91 } ‖ c	8000	2200	1000	800	700	600	≈10
$ErFeO_3$	81							≈15
$TmFeO_3$	140							≈5
$YbFeO_3$	143							12.5
$LuFeO_3$	119							≈5
$SmFeO_3$	84 ‖ a	3400	700	400	300	200	150	≈50
$YFeO_3$	105							≈10
$LaFeO_3$	83							≈10
$PrFeO_3$	71							≈35
$NdFeO_3$	62							≈10

*6 Strong natural birefringence interferes with the Faraday effect.

Table 33.7 Some magnetooptical parameters of paramagnetic materials

Rare earth ions in crystals							
Rare earths	Host material	Valency	T, K	λ, nm	V, angle min./A	α, cm^{-1}	V/α
Eu (3%)	CaF_2	2+	4.2	430	36	11.5	3.1
	CaF_2		4.2	440	28	1.8	15
Eu		2+	300	450	5.7	20	0.28
			300	500	3.3	7	0.47
			300	550	2.0	6	0.34
			300	600	1.4	5.5	0.25
			300	650	1.0	5	0.2
Nd (2.9%)	CaF_2	3+	4.2	426.4	0.24	5	0.048

Perovskites (oxides)					
Material	T, K	λ, nm	V angle min./A	λ, nm at α=$10^4$$cm^{-1}$	λ, nm at α=1cm^{-1}
$SrTiO_3$	298	413	0.98	368	413
		496	0.39		
		620	0.18		
		826	0.083		
$BaTiO_3$	403	427	1.2	380	446
		496	0.48		
		620	0.23		
		826	0.09		
$KTiO_3$	296	352	0.55	327	370
		413	0.24		
		496	0.12		
		620	0.064		
		826	0.028		

Rare-earth aluminum garnets									
Material	T,	Verdet constant, angle min./A at λ, nm							
	K	405	450	480	520	546	578	635	670
TbAlG*[1]	300	−2.797	−1.967	−1.621	−1.306	− 1.146	− 0.989	−0.779	−0.681
	77				−4.304	−3.834	−3.271	−2.523	−2.281
	4.2		−128	−104.9	−81.43	−73.32	−67.57	−60.81	−35.93
	1.45		−252.5	−216.8	−175.0	−157.2	−139.8	−122.5	−117.4
DyAlG	300	−1.559	−1.18	−1.01	−0.838	−0.744	−0.651	−0.516	−0.451
HoAlG	300	−0.891	−0.402	−0.327	−0.421	−0.382	−0.376		0.259
ErAlG	300	−0.237	−0.302	− 0.193	−0.204	−0.197	−0.182	−0.132	−0.112
TmAlG	300	+0.19	+0.129	+0.117	+0.095	+0.087	+0.074	+0.0603	
YbAlG	298	0.361	0.27	0234	0.176	0.167	0.146	0.118	
	77	0.902	0.678	0.604	0.494	0.43	0.379	0.30	

*[1] Absorption coefficient α=0.2–0.6 cm^{-1} at 300 K.

Table 33.8 Magnetooptical parameters of glasses

EUROPIUM GLASS

Composition (mol. %)				Verdet constant V (angle min.·A) at room temperature for wavelength (nm)									
Eu^{3+}	EuO	Al_2O_3	B_2O_3	406	435	450	500	550	600	650	700	750	800
12.9	14.4	15.3	70.2	−0.795	−0.428	−0.354	−0.204	−0.131	−0.095	−0.077	−0.062	−0.053	−0.040
14.7	16.5	11.2	72.0	−0.940	−0.503	−0.411	−0.25	−0.165	−0.113	−0.088	−0.069	−0.058	−0.048
17.9	20.0	13.4	66.2	−1.48	−0.848	−0.66	−0.374	−0.261	−0.187	−0.142	−0.114	−0.05	−0.075
26.7	29.9	11.0	60.0		−1.60	−1.23	−0.657	−0.461	−0.324	−0.24	−0.198	−0.162	−0.138
30.5	34.2	14.8	50.7			−3.2	−1.58	−0.906	−0.601	−0.46	−0.337	−0.283	−0.219

RARE-EARTH BORATE GLASS

Composition[*1]		Verdet constant V (angle min.·A) at room temperature for wavelength (nm)									
R	x	405	436	480	500	520	546	578	600	635	670
La	3.04	0.054	0.045	0.036	0.033	0.029	0.028	0.024	0.023	0.020	0.018
Pr–La	5.44	−0.477	−0.386	−0.290	−0.276	−0.253	−0.224	−0.192	−0.183	−0.161	−0.138
Nd–La	5.41	−0.226	−0.185	−0.151	−0.139	−0.121	−0.118	−0.126	−0.074	−0.070	−0.058
Sm–La	4.97	0.040	0.038	0.031	0.030	0.028	0.024	0.021	0.020	0.018	0.013
Eu–La	4.69	−0.101	−0.075	−0.048	−0.041	−0.036	−0.030	−0.024	−0.020	−0.018	−0.015
Gd–La	4.71	0.040	0.033	0.030	0.028	0.026	0.025	0.023	0.021	0.019	0.016
Tb–La	4.73	−0.643	−0.526	−0.401	−0.362	−0.329	−0.291	−0.258	−0.234	−0.210	−0.178
Dy–La	4.88	−0.548	−0.454	−0.376	−0.343	−0.309	−0.276	−0.242	−0.222	−0.200	−0.173
Ho–La	4.36	−0.338	−0.317	−0.155	−0.165	−0.141	−0.161	−0.131	−0.121		−0.093
Er–La	4.50	−0.117	−0.098	−0.085	−0.103		−0.057	−0.053	−0.050	−0.044	−0.043
Tm–La	4.75	0.075	0.058	0.049	0.043	0.039	0.038	0.029	0.026	0.023	0.020
Yb–La	8.58	0.144	0.118	0.092	0.083	0.075	0.068	0.058	0.054	0.046	0.041
Tb–Pr	4.99	−1.18	−0.987	−0.704	−0.673	−0.614	−0.548	−0.477	−0.437	−0.384	−0.333
Dy–Pr	4.63	−1.07			−0.624	−0.584	−0.519	−0.450	−0.417	−0.364	−0.317
Pr	2.56	−1.06	−0.812	−0.592	−0.603	−0.543	−0.490	−0.420	−0.398	−0.340	−0.305

[*1] Composition for La and Pr: $R_2O_3 \cdot xB_2O_3$; for Tb–Pr and Dy–Pr: $R'_2O_3 \cdot R''_2O \cdot xB_2O_3$; for others: $R_2O_3 \cdot 0.85La_2O_3 \cdot xB_2O_3$.

Table 33.8 Magnetooptical parameters of glasses *(continued)*

RARE-EARTH PHOSPHATE GLASS

Composition ($R_2O_3 \cdot xP_2O_5$)		Verdet constant V (angle min.·A) at room temperature for wavelength (nm)									
R	x	405	436	480	500	520	546	578	600	635	670
La		0.046	0.038	0.030	0.028	0.025	0.023	0.019	0.018	0.016	
Ce	2.67	−0.844	−0.641	−0.460	−0.410	−0.361	−0.318	−0.273	−0.217	−0.217	−0.188
Pr	3.09	−0.562	−0.417	−0.356	−0.388	−0.296	−0.261	−0.229	−0.188	−0.188	−0.166
Nd	2.92	−0.314	−0.263	−0.210	−0.195	−0.171	−0.168	−0.118	− 0.100	−0.100	−0.089
Sm	2.87	0.033	0.030	0.025	0.025	0.021	0.019	0.018	0.015	0.014	0.013
Eu	2.93	−0.031	−0.021	−0.013	−0.008	−0.008	−0.006	−0.005	−0.004	−.0025	−.0025
Gd	3.01	0.023	0.019	0.018	0.015	0.015	0.014	0.014	0.013	0.011	0.011
Tb	2.94	−0.704	−0.575	−0.448	−0.406	−0.371	−0.328	−0.284	−0.259	−0.239	−0.206
Dy	2.51	−0.678	−0.569	−0.451	−0.416	−0.378	−0.337	−0.298	−0.273	−0.247	−0.217
Ho	2.94	−0.376	−0.393	−0.196	−0.192	−0.173	−0.173	−0.149	−0.138	−0.123	−0.106
Er	3.01	−0.175	−0.152	−0.126	−0.139	−0.119	−0.078	−0.075	−0.072	−0.064	−0.055
Tm	2.79	0.024	0.016	0.015	0.011	0.010	0.008	0.006	0.005	0.005	0.009
Yb	3.01	0.109	0.090	0.070	0.063	0.057	0.051	0.045	0.040	0.036	0.030

RARE-EARTH ALUMINA SILICATE GLASS

Composition (mole %)	T, K	Verdet constant V (anglr min.·A) it at room temperature for wavelength (nm)									
		365	404	546	650	700	800	900	1100	1200	1300
Praseodymium											
Pr_6O_{11}(58), SiO_2(25), Al_2O_3(12),	295	−1.31	−0.955	−0.440	−0.299	−0.249	−0.185	−0.141	−0.084	-0.064	-0.044
MgO(4), Sb_2O_3(1)	77.4	−4.30	−3.15	−1.46	−0.980	−0.823	−0.613	−0.466	−0.291	−0.230	−0.173
Dysprosium											
Dy_2O_3(58), SiO_2(25), Al_2O_3(12),	295	−1.11	−0.849	−0.422	−0.290	−0.237					
MgO(4), Sb_2O_3(1)	77.4	−4.20	−3.24	−1.16	−0.109	−0.910					
Terbium											
Tb_4O_7(58), SiO_2(25), Al_2O_3(12),	295	−1.4	−1.08	−0.490	−0.332	−0.276	−0.200	−0.151	−0.093	−0.074	−0.059
MgO(4), Sb_2O_3(1)	77.4	−5.45	−4.02	−1.78	−0.123	−1.01	−0.744	−0.565	−0.353	−0.283	−0.229

Table 33.9 Verdet constant V for some liquids

Liquid	λ, nm	T, K	V, min./(T·cm)
H_2O	546	293	1.54
	578	293	1.37
	589	293	1.307
HCl	589	293	2.24
CCl_4	578	293	1.68
NH_3	578	233	1.73
PCl_3	578	299	3.05
$Ni(CO)_4$	578	190	7.35
Br_2	700	273	5.3
BrH	589	293	3.43
$SiCl_4$	589	289	1.89
HI	589	293	5.13

Table 33.10 Verdet constant V for some organic compounds

Liquid	λ, nm	T, K	V, min./(T·cm)
C_4H_{10}	578	263	1.179
C_5H_{12}	589	293	1.149
C_6H_{12}	589	298	1.240
C_6H_6	578	293	3.10
C_7H_{14}	589	293	1.2946
C_9H_{12}	589	288	2.47
$C_{10}H_{18}$	589	293	1.3785
$C_{12}H_{10}$	589	288	3.82
$C_{13}H_{12}$	589	288	3.22
CH_3I	578	291	3.53
$CHCl_3$	578	291	1.67
CCl_4	578	293	1.68
C_6H_5Cl	589	298	2.891
C_6H_5Br	589	298	3.25
CH_4O	589	299	0.9515
C_3H_6O	589	288.2	1.13
C_4H_5N	589	288.2	2.44
C_6H_7N	589	288	4.14

33.4 Quadratic Electrooptical Materials (Kerr Effect)

All elements in the $\chi^{(2)}$-susceptibility tensor for crystals with center of symmetry and in isotropic substances equal zero. Thus only the quadratic electrooptical effect (or Kerr effect) is observed in such materials, which resemble uniaxial crystals with their optical axes parallel to the applied electric field. The birefringence is measured by Kerr electrooptical constant K, defined as the ratio of the relative difference between two principal refractive indices (n_0 and n_e) to the square of the applied electric field which produces it:

$$K = \{|n_0 - n_e|\}/n_0 E^2. \tag{33.11}$$

The phase delay Δ(rad) between extraordinary and ordinary light rays produced by imposed electric field is given by

$$\Delta = 2\pi K n_0 l E^2/\lambda, \tag{33.12}$$

where l and λ are the optical path length and the light wavelength, respectively, E is the electric field strength (V/m), and K is measured in m^2/V^2. The Kerr constants for gases and liquids are presented in Tables 33.11 and 33.12 [3]. According to these data, the larger molecules with side chains show more pronounced effect. The value of K decreases with increasing temperature. A similar effect is caused by streaming in moving liquid media. These facts suggest that the Kerr effect results largely from alignment of electrically anisotropic molecules. It can attain full magnitude only in electric fields which vary slowly. Nevertheless, some materials display Kerr effect even at optical frequencies.

Table 33.11 Kerr constant K for gases at normal pressure (101.3 GPa) at a wavelength of 546 nm

Gas	T, K	K, 10^{-24} m^2/V^2	Gas	T, K	K, 10^{-24} m^2/V^2
H_2	307.8	0.5	Cl_2	297.2	2.6
O_2	273.2	0.5	CO_2	290.7	1.7
N_2	273.2	0.33	N_2O	329.9	24.0
	295.2	0.16		299.2	3.4

Table 33.11 Kerr constant K *(continued)*

Gas	T, K	K, 10^{-24} m^2/V^2	Gas	T, K	K, 10^{-24} m^2/V^2
HCl	291.2	6.4	$(CH_3)_2O$	291.2	−5.6
H_2S	291.2	1.8	$(C_2H_5)_2O$	335.9	−4.3
NH_3	290.7	3.6	CH_3Cl	291.2	40.5
HCN	293.2	103	C_2H_5Cl	291.2	61.3
SO_2	291.2	−10.4	$CHCl_3$	362.7	−8.5
CH_4	273.2	0.27	CH_2ClCH_2Cl	381.7	5.2
C_2H_6	273.2	0.74	$(CH_3)_2CO$	356.3	5.6
	386.8	6.17	$C_6H_5CH_3$	410.9	10.8
C_3H_8	273.2	0.94	C_6H_5Cl	426.9	41.3
C_2H_4	273.2	1.18	$C_6H_5NO_2$	508.7	162

Table 33.12 Kerr constant K for liquids at a wavelength 546 nm

Liquid	T, K	K, 10^{-24} m^2/V^2	Liquid	T, K	K, 10^{-24} m^2/V^2
O_2	90	1.2	*ortho*-$C_6H_4(CH_3)_2$	293	6.08
N_2	71.4	4.46	$(C_2H_5)_2O$	293	−2.95
CS_2	293	13.4	$CHCl_3$	293	12.9
C_6H_{12}	293	0.252	C_6H_5Cl	293	45.1
C_6H_6	293	1.62	$C_6H_5NO_2$	293	1570

33.5 Elastooptical Materials

The photoelastic effect consists in coupling the external stresses to the optical characteristics of substances and occurs in all their states. The relationship between the change in refractive index and the strain is described by elastooptic constants in a tensor form. In a general way, the change of refraction index indicatrix $\Delta(1/n^2)_{ij}$ is connected with the strain tensor S_{kl} via the fourth rank strain-optical tensor by

$$\Delta(1/n^2)_{ij} = \sum_{k,l=1}^{3} p_{ijkl} S_{kl}. \tag{33.13}$$

Notice that the strain, refraction index and elastooptic coefficients are dimensionless.

Inasmuch as tensors corresponding to strain and indicatrix are symmetric, the summation procedure does not affect the result and the above relation reduces to

$$\Delta(1/n^2)_i = \sum_{j=1}^{6} p_{ij} S_j;\ ij = 1, 2, \ldots, 6, \tag{33.14}$$

where the standard subscript notation is convenient to use: 11→1, 22→2, 33→3, 23 and 32→4, 13 and 31→5, 12 and 21→6. Therefore, the form of the strain-optical tensor can be reduced to a 6·6 p_{ij} matrix of elastooptic coefficients. Fortunately, these 36 coefficients exist solely in the case of triclinic crystals. Under conditions of another point-group symmetries the matrix components vanish or are interrelated. The experimental photoelastic data for some materials are compiled in Table 33.13 [4].

In crystals that are electrooptic in nature (see Section 33.2), mechanical deformation gives rise to an internal electric polarization proportional to the strain. And vice versa, the external electric field generates stresses, and deforms the crystal structure. These substances are termed *piezoelectric*. In most materials used (except for some ferroelectrics like lithium niobate) the piezoelectric effect is small and may be ignored. Where it is not so, the conditions of the photoelastic measurements become important (for lithium niobate, see Table 33.13, the superscript E (p_{ij}^E) denotes measurements at constant electric field).

The terms elastooptic and acoustooptic are interchangeable, since an acoustic wave is a particular type of strain wave. Acoustooptical materials presented in Table 33.14 [4] are closely related to practical devices. Directly measured quantity for acoustooptical materials makes the figure of merit [5]

$$M_2 = n^6 p^2 / \rho v^3, \tag{33.15}$$

where n is the refraction index, p is the strain-optical coefficient (obviously, a crystal cut that maximizes M_2 should be chosen for practical purposes), ρ is the material density, and v is the acoustic velocity. Two alternative expressions

$$M_1 = n^7 p^2 / \rho v, \quad M_3 = n^7 p^2 / \rho v^2 \tag{33.16}$$

are also used in some practical applications. M_1 and M_2 are measured in s^3/g, while M_3 in $cm{\cdot}s^2/g$.

Table 33.13 Photoelastic properties of materials [Crystal syngony and class are marked in headlines. All data are measured at room temperature. Average accuracy runs to p_{ij} is 5%. When the sign of p_{ij} is not given, only the absolute value is known. Acoustooptical figure of merit M_2 has been normalized to melted quartz whose value reaches $0.51 \cdot 10^{-18}$ sec^3/gm.]

Isotropic

Material	λ, nm	p_{11}	p_{12}	M_2
Melted quartz (SiO_2)	630	+0.121	0.270	1.0
As_2S_3-glass	1150	+0.308	0.299	230
Dense flint (SF-4)	630	+0.232	0.256	3.0
Water	630	0.31	0.31	106
$Ge_{33}Se_{55}As_{12}$-glass	1060	0.21	0.21	164
Sb_2O_3	630			18
Various optical glasses	590	0.09–0.24	0.18–0.28	
Dense flint (SchottSF-59)	630	0.27	0.24	12.5
$PbO{:}2Sb_2O_3$	630			18.5
Lucite	630	0.30	0.28	33
Polystyrene	630	0.30	0.31	84

Cubic classes: $\bar{4}3m$, 432, and $m3m$

Material	λ, nm	p_{11}	p_{12}	p_{44}	M_2
GaP	630	−0.151	−0.082	−0.074	29.5
GaAs	1150	−0.165	−0.140	−0.072	69
$Y_3Al_5O_{12}$(YAG)	630	−0.029	+0.009	−0.061	0.048
$Y_3Fe_5O_{12}$(YIG)	1150	0.025	0.073	+0.041	0.22
β-ZnS	630	+0.091	−0.01	+0.075	2.3
Ge	10600	+0.27	+0.235	0.125	540
$ZnAl_2O_4$	630	< 0.009	0.03		0.005
$SrTiO_3$	630	0.15	0.095	0.072	1.1
$Y_3Ga_5O_{12}$	630	0.091	0.019	0.079	0.56
$Bi_4Ge_3O_{12}$	630				3.3
KRS-5 (Thallium bromoiodide)	630				118
Diamond	590	−0.31	−0.03		
		−0.43	+0.19	−0.16	
LiF	590	−0.02	+0.128	−0.064	

Table 33.13 Photoelastic properties of materials *(continued)*

Cubic classes: $\overline{4}3m$, 432 and $m3m$

Material	λ, nm	p_{11}	p_{12}	p_{44}	M_2
MgO	590	+0.32	−0.08		
KBr	590	+0.22	+0.171	−0.026	
KCl	590	+0.17	+0.124		
KI	590	+0.210	+0.169		
NaCi	590	+0.110	+0.153	−0.010	

Cubic classes: 23 and $m3$

Material	λ, nm	p_{11}	p_{12}	p_{13}	p_{44}	M_2
$Ba(NO_3)_2$	630	0.15	0.35	0.29	0.02	15.0
$Bi_{12}GeO_{20}$	630					6.6
$Bi_{12}SiO_{20}$	630					6.0
$Pb(NO_3)_2$	630					17
$NaBrO_3$	590	0.185	0.218	0.213	−0.0139	
$NaClO_3$	590	0.162	0.24	0.20	−0.198	

Trigonal classes: $3m$, 32 and $\overline{3}m$

Material	λ, nm	p_{11}	p_{12}	p_{13}	p_{14}
$LiTaO_3$	630	0.08	0.08	0.09	0.03
α-Al_2O_3	630	0.20	0.078	≈0	
Te	10600	0.155	0.130		
$LiNbO_3(p^E)$	630	−0.02	+0.08	+0.13	−0.08
Ruby (Al_2O_3+0.05% Cr)	630	−0.23	−0.03	+0.02	0.00
α-Quartz (SiO_2)	590	+0.16	+0.27	+0.27	−0.03
		+0.138	+0.25	+0.259	−0.029
$CaCO_3$		+0.095	+0.189	+0.215	−0.006

Material	p_{31}	p_{33}	p_{41}	p_{44}	M_2
$LiTaO_3$	0.09	0.15	0.02	0.02	0.91
α-Al_2O_3	≈0	0.252		0.09	0.22
Te					2920
$LiNbO_3(p^E)$	+0.17	+0.07	−0.15	+0.12	9.0
Ruby (Al_2O_3+0.05% Cr)	−0.04	−0.20	+0.01	−0.10	
α-Quartz (SiO_2)	+0.29	+0.10	−0.047	−0.079	
	+0.258	+0.098	−0.042	−0.0685	
$CaCO_3$	+0.309	+0.178	+0.01	−0.090	

Trigonal classes: 3 and $\overline{3}$

Material	λ, nm	M_2
Li_2WO_4	630	2.5

Hexagonal classes: $\overline{6}m2$, $6mm$, 622, $6/mmm$, 6, $\overline{6}$ and $6/m$

Material	λ, nm	p_{11}	p_{12}	p_{14}	p_{31}	p_{44}	M_2
CdS	630	0.142	0.66		0.041	0.054	8.0
$LiIO_3$	630	0.32		0.03	0.41		8.3

Tetragonal classes: $4mm$, $\overline{4}2m$, 422 and $4/mmm$

Material	λ, nm	p_{11}	p_{12}	p_{13}	p_{31}
TiO_2 (Rutile)	630	−0.011	+0.172	−0.168	−0.096
ADP (Ammonium dihydrogen phosphate)	630	+0.302	+0.246	+0.236	+0.195

Table 33.13 Photoelastic properties of materials *(continued)*

Tetragonal classes: $4mm$, $\bar{4}2m$, 422 and $4/mmm$

Material	λ, nm	p_{11}	p_{12}	p_{13}	p_{31}
KDP (Potassium dihydrogen phosphate)	630	+0.251	+0.249	+0.246	+0.225
$ZrSiO_4$	630	0.06		0.13	0.07
TeO_2	630	0.0074	+0.187	+0.340	+0.090
$Sr_{0.75}Ba_{0.25}Nb_2O_6$	630	0.16	0.10	0.08	0.11
$Sr_{0.5}Ba_{0.5}Nb_2O_6$	630	0.06	0.08	0.17	0.09

Material	p_{33}	p_{44}	p_{66}	M_2
TiO_2(Rutile)	−0.058			2.6
ADP (Ammonium dihydrogen phosphate)	+0.263		0.075	4.2
KDP (Potassium dihydrogen phosphate)	+0.221		0.058	2.5
$ZrSiO_4$	0.09		0.10	2.4
TeO_2	+0.240	−0.17	−0.046	525
$Sr_{0.75}Ba_{0.25}Nb_2O_6$	0.47			26
$Sr_{0.5}Ba_{0.5}Nb_2O_6$	0.23			5.8

Tetragonal classes: 4, $\bar{4}$ and $4/m$

Material	λ, nm	p_{11}	p_{12}	p_{13}	p_{31}	p_{33}	M_2
$PbMoO_4$	630	0.24	0.24	0.255	0.15	0.29	23.7
$CdMoO_4$	630	0.12	0.10	0.13	0.11	0.18	4.5
$PbWO_4$							21.0

Orthorhombic all classes

Material	λ, nm	p_{11}	p_{12}	p_{13}	p_{21}	p_{22}	p_{23}
α-HIO_3	630	0.406	0.277	0.304	0.279	0.343	0.305
$Ca(NbO_3)_2$	630						
$PbCO_3$	630	0.15	0.12	0.16	0.05	0.06	0.21
$Ba_2NaNb_5O_{15}$							
$BaSO_4$	590	+0.21	+0.25	+0.16	+0.34	+0.24	+0.19

Material	p_{31}	p_{32}	p_{33}	p_{44}	p_{55}	p_{66}	M_2
α-HIO_3	0.503	0.310	0.334			0.092	55
$Ca(NbO_3)_2$							1.3
$PbCO_3$	0.14	0.16	0.12				15
$Ba_2NaNb_5O_{15}$	0.17						5–10
$BaSO_4$	+0.27	+0.22	+0.31	+0.002	+0.012	+0.037	

Monoclinic all classes

Material	λ, nm	M_2
Pb_2MoO_5	630	27

33.6 Optical Activity

In some materials even when the light propagation takes place along the optical axis, the plane of a linearly polarized light rotates. This effect, known as natural *optical activity*, is due to the difference between the refractive indices n_l and n_r for the left and right circularly polarized light. Optical activity is observed in many organic substances, crystals and their solutions.

Table 33.14 Some acoustooptical materials

Material	Optical transmission range, nm	λ, nm	n	Acoustic wave polarization and direction	Acoustic velocity v, 10^3 m/s	Optical wave polarization and direction*[1]	M_1	M_2	M_3	Acoustic attenuation at 500 MHz, dB/μs
Melted quartz (SiO_2)	200–4500	633	1.46	long.	5.96	$\perp$	1.0	1.0	1.0	1.8
				shear	3.76		0.12	0.31	0.2	
Water H_2O	200–900	633	1.33	long.	1.5	$\parallel$ or $\perp$	6.1	106	24	75
D_2O	200–1800									
α-HIO_3	300–1800	633	1.98	long. [001]	2.44	$\perp$ [010]	13.6	55	32	0.6
$PbMoO_4$	400–5500	633	2.39	long. [001]	3.66	$\parallel$ or $\perp$ [100]	15.3	23.7	24.9	1.2
$LiNbO_3$	500–4500	633	2.20	long. [$11\bar{2}0$]	6.57		8.3	4.6	7.5	< 0.03
TiO_2	450–5500	633	2.58	long. [$11\bar{2}0$]	7.86	[0001]	7.9	2.6	6.2	
TeO_2	350–5000	633	2.27	long. [001]	4.26	$\perp$ [010]	18.5	22.8	25.6	1
				shear. [110]	0.617	$\parallel$ or $\perp$ [001]	8.8	525	85	3
GaP	600–10000	633	3.31	long. [110]	6.32	$\parallel$	75	29.5	69	< 1
				shear. [100]	4.13	$\parallel$ or $\perp$	17.4	16	25.7	
As_2S_3 glass	600–11000	1150	2.46	long.	2.6	$\perp$	78	230	182	11
$Ge_{33}Se_{55}As_{12}$ glass	1000–14000	1060	2.7	long.	2.52	$\parallel$ or $\perp$	53	164	128	1.8
Ge	2000–20000	10600	4.0	long. [111]	5.50	$\parallel$	1.270	540	1.380	4.2
				shear. [100]	3.51	$\parallel$ or $\perp$	182	190	308	
Te	5000–20000	10600	4.8	long. [$11\bar{2}0$]	2.2	$\parallel$ [0001]	1.320	2.920	3.550	0.8

*[1] The polarization is defined as parallel ($\parallel$) or perpendicular ($\perp$) to the scattering plane formed by the acoustic and optical propagation directions.

The measure of rotary power for solutions can be described by $[\alpha]^t_\lambda$ (in angle deg.·cm^2/g), which is the angle of the polarization plane rotation for the light with wavelength λ after propagation through 1-cm-thick solution layer with optically active material concentration 1 g/cm^3 at a temperature of t°C.

The optical activity of pure substances is defined as

$$[\alpha] = \alpha/\rho l, \tag{33.17}$$

where α is the rotation angle (in angle deg.) for a material of length l (cm) at the density ρ (g/cm^3).

In some cases the, rotary power of solutions can be given by

$$[\alpha] = 100\alpha/\rho l P, \tag{33.18}$$

or by molecular rotation

$$[M] = [\alpha] \cdot M/100, \tag{33.19}$$

where P is the material concentration in g per 100 g of solution, and M is the molecular mass (g). The rotary power for some materials is compiled in Tables 33.15–33.17 [3].

Table 33.15 Optical activity of methyl ethers

Ether	$[\alpha]^{20}_D$, deg.·cm^2/g	$[M]^{20}_D$, deg.·cm^2	Ether	$[\alpha]^{20}_D$, deg.·cm^2/g	$[M]^{20}_D$, deg.·cm^2
$HCO_2C_{10}H_{19}$	−7.952	−14.63	$C_5H_{11}CO_2C_{10}H_{19}$	−6.207	−15.77
$CH_3CO_2C_{10}H_{19}$	−7.942	−15.73	$C_6H_{13}CO_2C_{10}H_{19}$	−5.885	−15.57
$C_2H_5CO_2C_{10}H_{19}$	−7.515	−16.02	$C_7H_{15}CO_2C_{10}H_{19}$	−5.525	−15.58
$C_3H_7CO_2C_{10}H_{19}$	−6.952	−15.69	$C_8H_{17}CO_2C_{10}H_{19}$	−5.316	−15.73
$C_4H_9CO_2C_{10}H_{19}$	−6.555	−15.73			

Table 33.16 Optical activity of ethyltartrate in various solvents

Solvent	Solvent dipole moment, D	$[M]^{20}_{546.1}$, deg.·cm^2	Solvent	Solvent dipole moment, D	$[M]^{20}_{546.1nm}$, deg.·cm^2
s-$C_6H_3(CH_3)_3$	0	0	C_6H_5Cl	1.56	2.30
$C_6H_5OCH_3$	1.16	0.5	C_6H_5CN	3.85	6.79
C_6H_6	0.4	0.74	$C_6H_5NO_2$	3.90	8.06
$C_6H_5CH_3$	0	1.25	C_6H_5COH	2.75	10.77
C_6H_5I	1.25	2.02	Pure ethyltartrate		1.625
C_6H_5Br	1.56	2.18			

Table 33.17 Specific optical activity of crystals

Crystal	Class	λ, nm	α, deg./mm	Crystal	Class	λ, nm	α, deg./mm
$KLiSO_4$	C_6	589	±3.43	HgS	D_3	687	325
SiO_2	D_3	434	41.924	$NaClO_3$	T	556	±1.42
		486.1	32.761	$C_{12}H_{22}O_{11}$	C_2	589	1.6–5.4*[1]
		589.3	21.724	$KNaC_4H_4O_6{\cdot}4H_2O$	D_2	589	−1.4
		656.3	17.320				

*[1] For various directions.

33.7 Nonlinear Crystal Materials

The intensity dependence of the charge polarization in media that do not impose their own characteristic frequencies on the light wave propagated can be written as follows

$$\mathbf{P} = \varepsilon_0 \chi \mathbf{E} + \text{small nonlinear terms.} \tag{33.20}$$

If the electromagnetic wave of a frequency ω propagates through the medium, small nonlinear terms include the second-harmonic component of the charge polarization created by a monochromatic fundamental wave

$$P^{(2\omega)} = \varepsilon_0 \chi^{(2\omega)} : E^{(\omega)} E^{(\omega)}. \tag{33.21}$$

When rewritten via Cartesian components for the special case of second harmonic generation (SHG), this relationship (which describes also d.c. rectification, Pockels effect, etc.) gives

$$P_i^{(2\omega)} = \varepsilon_0 \sum_{j,k=1,2,3} d_{ijk}^{2\omega} \cdot E_j^{(\omega)} \cdot E_k^{(\omega)}, \tag{33.22}$$

where SHG coefficients $d_{ijk}^{2\omega}$ form a third-rank tensor. Inasmuch as the above equation is indifferent to the order in which the two electric field components are taken, only their product being significant, the SHG tensor satisfies the permutation symmetry. This property, which is similar to that of linear electrooptic and elastooptic effects (see Sections 33.2 and 33.5), permits of the contracted form $d_{il}^{2\omega}$, where l substitutes for jk and according to standard subscript notation: 11→1, 22→2, 33→3, 23 and 32→4, 13 and 31→5, 12 and 21→6. In the contracted form, the SHG tensor involves 18 coefficients. The number of non-zero SHG coefficients depends upon the point-group symmetry of the medium. Efficient SHG has been accomplished only in crystals that lack a center of inversion. The contracted form of the SHG tensor for the crystal classes satisfying this criterion is presented in Table 33.18. Using contracted SHG tensors, the latter of above equations can be written in the matrix form

$$\begin{pmatrix} P_x \\ P_y \\ P_z \end{pmatrix} = \varepsilon_0 \begin{pmatrix} d_{11} & d_{12} & d_{13} & d_{14} & d_{15} & d_{16} \\ d_{21} & d_{22} & d_{23} & d_{24} & d_{25} & d_{26} \\ d_{31} & d_{32} & d_{33} & d_{34} & d_{35} & d_{36} \end{pmatrix} \begin{pmatrix} E_x^2 \\ E_y^2 \\ E_z^2 \\ 2E_yE_z \\ 2E_xE_z \\ 2E_xE_y \end{pmatrix}, \tag{33.23}$$

The SHG power $p_{2\omega}$ (in W) generated by a beam of angular frequency ω (in s^{-1}) and power p_ω (in W) and propagating through nonabsorbing medium of the length L (in m) takes the form

$$p_{2\omega} = \frac{2\varepsilon_0^{1/2}\mu_0^{3/2}\omega^2(d_{il}^{2\omega})^2(p_\omega)^2L^2}{\pi r_0^2 n^{2\omega}(n^\omega)^2} f(\Delta k), \tag{33.24}$$

where ε_0 and μ_0 are the free space permittivity and permeability, respectively ($8.85 \cdot 10^{-12}$ F/m and $12.57 \cdot 10^{-7}$ H/m), n^ω and $n^{2\omega}$ are the medium refractive indices at the fundamental and second harmonic wavelengths, $d_{il}^{2\omega}$ is the appropriate SHG coefficient, r_0 is the beam radius, and, finally, $f(\Delta k)$ is the function taking into account the wave vector mismatch between the bound and free harmonic waves in the medium. In the cases corresponding to phase matching , i.e. when $\Delta k = k_{2\omega} - 2k_\omega = 0$ or $v(\omega) = v(2\omega)$, $f(\Delta k) = 1$. In real materials phase velocities $v(\omega)$ and $v(2\omega)$ are not equal due to refractive index dispersion, therefore $\Delta k \neq 0$ and $f(\Delta k) < 1$. The variation of Δk with medium length makes the second harmonic power oscillate. The length of the oscillation period

$$l_{\text{coh}} = \pi / \Delta k \tag{33.25}$$

is known as the *coherence length.* Birefringent crystals offer the peculiar property lying in the fact that, although their nonlinear sources are anchored to the fundamental wave which excites them, in these crystals the direction of harmonic wave polarization may not be aligned with that of fundamental wave. This allows us to exploit birefringence so as to compensate dispersion in certain fortuitous cases. If birefringence exceeds dispersion, two general cases of the phase matching are realized in uniaxial crystals.

In negative uniaxial crystals, the refractive index for an ordinary wave is larger than that for an extraordinary wave, i.e. $n_o > n_e$, and along some direction at an angle θ_1 to the crystal optic axis the equality

$$n_e^{2\omega} = n_o^{\omega} \tag{33.26}$$

holds thus giving type I refractive index matching condition. Simultaneously along direction θ_2 the equality

$$n_e^{2\omega} = \frac{1}{2}\left(n_o^{\omega} + n_e^{\omega}\right) \tag{33.27}$$

holds as well (type II matching condition). In positive uniaxial crystals ($n_o < n_e$), for type I and type II conditions we arrive at equalities

$$n_o^{2\omega} = n_e^{\omega} \tag{33.28}$$

$$n_o^{2\omega} = \frac{1}{2}\left(n_o^{\omega} + n_e^{\omega}\right). \tag{33.29}$$

In biaxial crystals, the phase matching conditions are much more complex.

If one of the above equalities corresponds to $\theta = 90°$, the fundamental wave propagates along a principal axis of the crystal. This phase matching is being termed *noncritical* (NCPM); otherwise, it is termed *critical* (CPM). Achievement of the conditions which are necessary for NCPM, when the second harmonic power exhibits low sensitivity to the beam divergence, is possible for some crystals under temperature tuning.

The second-order nonlinear tensors $d_{ijk}^{2\omega}$ for SHG can be expressed via linear susceptibility tensors χ_{ij} and a new third-rank tensor $\delta_{ijk}^{2\omega}$ (known as Miller's δ-tensor). For lower symmetry crystals (other than triclinic and monoclinic), the relation being discussed can be reduced to

$$d_{ijk}^{2\omega} = \varepsilon_0 \chi_{ii}^{2\omega} \chi_{jj}^{\omega} \chi_{kk}^{\omega} \cdot \delta_{ijk}^{2\omega}, \tag{33.30}$$

where χ_{ij} are dimensionless, $\delta_{ijk}^{2\omega}$ are measured in m^2/C, $d_{ijk}^{2\omega}$ are in the units of (m/V). The following relationships can be employed for converting these coefficients into esu units:

$$d_{ijk}^{2\omega}(\text{esu}) = \frac{3}{4\pi} 10^4 \cdot d_{ijk}^{2\omega}(\text{mks}), \tag{33.31}$$

$$\delta_{ijk}^{2\omega}(\text{esu}) = \frac{4\pi}{3} 10^{-5} \cdot \delta_{ijk}^{2\omega}(\text{mks}). \tag{33.32}$$

SHG coefficients $d_{ijk}^{2\omega}$ and $\delta_{ijk}^{2\omega}$ for nonlinear crystals, together with source wavelengths used in their measurements, and transparency data are incorporated in Table 33.19. In a general way, due to experimental difficulties, the SHG coefficients are determined relative to some well known standard materials, such as quartz (α-SiO_2), KDP, ADP or $LiIO_3$ in the visible part of the spectrum, or, for example, GaAs and GaP when applied to IR radiation.

Table 33.18 Form of SHG-tensor for various crystallographic classes [Crystal singony is marked in the headlines.]

Class	Tensor	Note
	Triclinic system	
$1 - C_1$	$\begin{pmatrix} d_{11} & d_{12} & d_{13} & d_{14} & d_{15} & d_{16} \\ d_{21} & d_{22} & d_{23} & d_{24} & d_{25} & d_{26} \\ d_{31} & d_{32} & d_{33} & d_{34} & d_{35} & d_{36} \end{pmatrix}$	
	Monoclinic system	
$m - C_s$	$\begin{pmatrix} d_{11} & d_{12} & d_{13} & 0 & 0 & d_{16} \\ d_{21} & d_{22} & d_{23} & 0 & 0 & d_{26} \\ 0 & 0 & 0 & d_{34} & d_{35} & 0 \end{pmatrix}$	$m \perp Z$
$m - C_s$	$\begin{pmatrix} d_{11} & d_{12} & d_{13} & 0 & d_{15} & 0 \\ 0 & 0 & 0 & d_{24} & 0 & d_{26} \\ d_{31} & d_{32} & d_{33} & 0 & d_{35} & 0 \end{pmatrix}$	$m \perp Y$ (IRE convention)
$2 - C_2$	$\begin{pmatrix} 0 & 0 & 0 & d_{14} & d_{15} & 0 \\ 0 & 0 & 0 & d_{24} & d_{25} & 0 \\ d_{31} & d_{32} & d_{33} & 0 & 0 & d_{36} \end{pmatrix}$	$2 \parallel Z$
$2 - C_2$	$\begin{pmatrix} 0 & 0 & 0 & d_{14} & 0 & d_{16} \\ d_{21} & d_{22} & d_{23} & 0 & d_{25} & 0 \\ 0 & 0 & 0 & d_{34} & 0 & d_{36} \end{pmatrix}$	$2 \parallel Z$ (IRE convention)
	Orthorhombic system	
$mm2-$ C_{2v}	$\begin{pmatrix} 0 & 0 & 0 & 0 & d_{15} & 0 \\ 0 & 0 & 0 & d_{24} & 0 & 0 \\ d_{31} & d_{32} & d_{33} & 0 & 0 & 0 \end{pmatrix}$	
$222 - D_2$	$\begin{pmatrix} 0 & 0 & 0 & d_{14} & 0 & 0 \\ 0 & 0 & 0 & 0 & d_{25} & 0 \\ 0 & 0 & 0 & 0 & 0 & d_{36} \end{pmatrix}$	
	Tetragonal system	
$4 - C_4$	$\begin{pmatrix} 0 & 0 & 0 & d_{14} & d_{15} & 0 \\ 0 & 0 & 0 & d_{15} & -d_{14} & 0 \\ d_{31} & d_{31} & d_{33} & 0 & 0 & 0 \end{pmatrix}$	
$\bar{4} - S_4$	$\begin{pmatrix} 0 & 0 & 0 & d_{14} & d_{15} & 0 \\ 0 & 0 & 0 & -d_{15} & d_{25} & 0 \\ d_{31} & -d_{31} & 0 & 0 & 0 & d_{36} \end{pmatrix}$	
$4mm-$ C_{4v}	$\begin{pmatrix} 0 & 0 & 0 & 0 & d_{15} & 0 \\ 0 & 0 & 0 & d_{15} & 0 & 0 \\ 0 & 0 & 0 & 0 & 0 & d_{36} \end{pmatrix}$	
$\bar{4}2m-$ D_{2d}	$\begin{pmatrix} 0 & 0 & 0 & d_{14} & 0 & 0 \\ 0 & 0 & 0 & 0 & d_{14} & 0 \\ 0 & 0 & 0 & 0 & 0 & d_{36} \end{pmatrix}$	
$422-$ D_4	$\begin{pmatrix} 0 & 0 & 0 & d_{14} & 0 & 0 \\ 0 & 0 & 0 & 0 & -d_{14} & 0 \\ 0 & 0 & 0 & 0 & 0 & 0 \end{pmatrix}$	
	Trigonal system	
$3 - C_3$	$\begin{pmatrix} d_{11} & -d_{11} & 0 & d_{14} & d_{15} & -d_{22} \\ -d_{22} & d_{22} & 0 & d_{15} & -d_{14} & -d_{11} \\ d_{31} & d_{31} & d_{33} & 0 & 0 & 0 \end{pmatrix}$	
$3m-$ C_{3v}	$\begin{pmatrix} 0 & 0 & 0 & 0 & d_{15} & -d22 \\ -d_{22} & d_{22} & 0 & d_{15} & 0 & 0 \\ d_{31} & d_{31} & d_{33} & 0 & 0 & 0 \end{pmatrix}$	$m \perp X$ (IRE convention)

Table 33.18 Form of SHG-tensor for various crystallographic classes *(continued)*

Class	Tensor	Note
	Trigonal system	
$3m-$ C_{3v}	$\begin{pmatrix} d_{11} & -d_{11} & 0 & 0 & d_{15} & 0 \\ 0 & 0 & 0 & d_{15} & 0 & -d_{11} \\ d_{31} & d_{31} & d_{33} & 0 & 0 & 0 \end{pmatrix}$	$m \perp Y$
$32-D_3$	$\begin{pmatrix} d_{11} & -d_{11} & 0 & d_{14} & 0 & 0 \\ 0 & 0 & 0 & 0 & -d_{14} & -d_{11} \\ 0 & 0 & 0 & 0 & 0 & 0 \end{pmatrix}$	
	Hexagonal system	
$\overline{6}-C_{3h}$	$\begin{pmatrix} d_{11} & -d_{11} & 0 & 0 & 0 & -d_{22} \\ -d_{22} & d_{22} & 0 & 0 & 0 & -d_{11} \\ 0 & 0 & 0 & 0 & 0 & 0 \end{pmatrix}$	
$6-C_6$	$\begin{pmatrix} 0 & 0 & 0 & d_{14} & d_{15} & 0 \\ 0 & 0 & 0 & d_{15} & -d_{14} & 0 \\ d_{31} & d_{31} & d_{33} & 0 & 0 & 0 \end{pmatrix}$	Same as class $4-C_4$
$\overline{6}m2-$ D_{3h}	$\begin{pmatrix} 0 & 0 & 0 & 0 & 0 & -d_{22} \\ -d_{22} & d_{22} & 0 & 0 & 0 & 0 \\ 0 & 0 & 0 & 0 & 0 & 0 \end{pmatrix}$	$m \perp X$ (IRE convention)
$\overline{6}m2-D_{3h}$	$\begin{pmatrix} d_{11} & -d_{11} & 0 & 0 & 0 & 0 \\ 0 & 0 & 0 & 0 & 0 & -d_{11} \\ 0 & 0 & 0 & 0 & 0 & 0 \end{pmatrix}$	$m \perp Y$
$6mm-$ C_{6v}	$\begin{pmatrix} 0 & 0 & 0 & 0 & d_{15} & 0 \\ 0 & 0 & 0 & d_{15} & 0 & 0 \\ d_{31} & d_{31} & d_{33} & 0 & 0 & 0 \end{pmatrix}$	Same as class $4mm-C_{4v}$
$622-$ D_6	$\begin{pmatrix} 0 & 0 & 0 & d_{14} & 0 & 0 \\ 0 & 0 & 0 & 0 & -d_{14} & 0 \\ 0 & 0 & 0 & 0 & 0 & 0 \end{pmatrix}$	Same as class $422-D_4$
	Cubic system	
$23-T$	$\begin{pmatrix} 0 & 0 & 0 & d_{14} & 0 & 0 \\ 0 & 0 & 0 & 0 & d_{14} & 0 \\ 0 & 0 & 0 & 0 & 0 & d_{14} \end{pmatrix}$	
$\overline{4}3m-$ T_d	$\begin{pmatrix} 0 & 0 & 0 & d_{14} & 0 & 0 \\ 0 & 0 & 0 & 0 & d_{14} & 0 \\ 0 & 0 & 0 & 0 & 0 & d_{14} \end{pmatrix}$	
$432-$ O_4	All elements equal zero	

Table 33.19 Second harmonic coefficients [Crystal singony is marked in the headlines.]

Material	Symmetry class	Transparency range, nm	$10^{12} \cdot d_{il}^{2\omega}$, m/V	$10^2 \cdot \delta_{il}^{2\omega}$, m^2/C	λ_1, nm	Ref.
			Monoclinic system			
Lithium sulfate monohydrate, $Li_2SO_4 \cdot H_2O$	$2-C_2$		$d_{22} = 0.46 \pm 0.07$ $d_{23} = 0.33 \pm 0.05$ $d_{34} = 0.29 \pm 0.04$		1064.2	[7]
L-Arginine phosphate (LAP)	$2-C_2$	230–1950	$d_{\text{eff}} \approx (2-3)d_{36}$ (KDP)		1064	[8]

Table 33.19 Second harmonic coefficients *(continued)*

Material	Symmetry class	Transparency range, nm	$10^{12} \cdot d_{il}^{2\omega}$, m/V	$10^2 \cdot \delta_{il}^{2\omega}$, m^2/C	λ_1, nm	Ref.
Monoclinic system						
MAP, $C_{10}H_{11}N_3O_6$	$2-C_2$	500–2000	$d_{21} = d_{16} = 16.7$ $d_{23} = d_{34} = 3.68$ $d_{25} = d_{36} = d_{14} = -0.544$ $d_{22} = 18.4$		1064	[9]
Methyl-nitroaniline (MNA), $NO_2C_6H_3CH_3NH_2$	$m-C_s$	500–2500	$d_{11} = (500 \pm 25\%)$ $\cdot d_{11}(SiO_2) = 168 \pm 42$ $d_{12} = (75 \pm 25\%)$ $d_{11}(SiO_2) = 24.1 \pm 6$		1064	[10]
Potassium malate (KM), COOK-CHOH-$CH_2COOK \cdot 1.5H_2O$	$m-C_s$	240–1300	$d_{31} = 0.61$ $d_{32} = 2.3$		1064	[11]
Potassium tartrate, $K_2C_4H_4O_6 \cdot 1/2H_2O$	$2-C_2$		$d_{21} = 0.13$ $d_{22} = 4.26$ $d_{23} = 0$ $d_{25} = 0.19$	0.68 19.2 0 0.93	694.3	[7]
Sucrose, $C_{12}H_{22}O_{11}$	$2-C_2$	192–1350	$d_{eff} = 0.2d_{\text{eff}}(\text{ADP})$ $d_{\text{eff}} = 0.3d_{\text{eff}}(\text{ADP})$		1060	[12,13]
Triglycine sulfate (TGS), $(NH_2CH_2COOOH)_3 \cdot H_2SO_4$	$2-C_2$		$d_{23} = 0.33$	1.09	694.3	[7]
Orthorhombic system						
Barium sodium niobate, $Ba_2NaNb_5O_{15}$	$mm2$ $-C_{2v}$		$d_{15} = 14.56 \pm 0.73$ $d_{24} = 13.83 \pm 0.73$ $d_{31} = 14.56 \pm 0.73$ $d_{32} = 14.56 \pm 1.46$ $d_{33} = 20 \pm 1.46$	2.35 ± 0.11 2.23 ± 0.11 2.42 ± 0.12 2.41 ± 0.16 4.06 ± 0.30	1064.2	[7]
Cesium titanyl arsenate (CTA), $CsTiOAsO_4$	$mm2$ $-C_{2v}$	350–5300	$d_{31} = 2.1 \pm 20\%$ $d_{32} = 3.4 \pm 20\%$ $d_{32} = 18.1 \pm 10\%$		1064	[14]
COANP, $C_{13}H_{19}N_3O_2$	$mm2$ $-C_{2v}$	470–1500	$d_{31} = 152$ d_{32} =26–34 $d_{33} = 10 \pm 2$		1064	[15]
Gadolinium molybdate, $Gd_2(MoO_4)_3$	$mm2$ $-C_{2v}$		$d_{15} = 2.98 \pm 0.45$ $d_{24} = 2.93 \pm 0.44$ $d_{31} = 2.83 \pm 0.42$ $d_{32} = 2.75 \pm 0.41$ $d_{33} = 0.05 \pm 0.009$	2.44 ± 0.37 2.4 ± 0.36 2.30 ± 0.34 2.24 ± 0.34 0.04 ± 0.007	1064	[7]
Lead niobate, $PbNb_4O_{11}$	$mm2$ $-C_{2v}$		$d_{15} = 6.7 \pm 1.0$ $d_{24} = 6.17 \pm 0.93$ $d_{31} = 7.4 \pm 1.1$ $d_{32} = 6.68 \pm 1.00$ $d_{33} = 10.1 \pm 1.5$	0.84 ± 0.12 0.76 ± 0.11 0.93 ± 0.14 0.83 ± 0.12 1.19 ± 0.18	1064.2	[7]
Lithium formate monohydrate (LFM), $LiCHO_2 \cdot H_2O$	$mm2$ $-C_{2v}$	230–1200	$d_{15} = d_{31} = 0.11 \pm 02$ $d_{24} = d_{32} = (1.27 \pm 0.09)$ $d_{33} = 1.86 \pm 0.11$	1.33 ± 0.24 8.19 ± 0.6 10.03 ± 0.6	1064.2	[7,16]

Table 33.19 Second harmonic coefficients *(continued)*

Material	Symmetry class	Transparency range, nm	$10^{12} \cdot d_{il}^{2\omega}$, m/V	$10^2 \cdot \delta_{il}^{2\omega}$, m^2/C	λ_1, nm	Ref.
Orthorhombic system						
Lithium metagallate, $LiGaO_2$	$mm2$ $-C_{2v}$		$d_{15} = d_{31} = 0.08 \pm 0.009$ $d_{24} = d_{32} = 0.175 \pm 0.019$ $d_{33} = 0.686 \pm 0.07$	0.1 ± 0.01 0.25 ± 0.027 0.88 ± 0.09	1064.2	[7]
Lithium triborate (LBO), LiB_3O_5	$mm2$ $-C_{2v}$	160–3200	$d_{31} = 1.1$ $d_{32} = 1.2$ $d_{33} = 6.5$		1064	[17,18]
Magnesium-barium fluoride, $MgBaF_4$	$mm2$ $-C_{2v}$	185–10000	$d_{31} = 0.0248$ $d_{32} = 0.037$ $d_{33} = 0.01$ $d_{24} = 0.027$		1064	[19]
Meta-dinitrobenzene (MDB)	$mm2$ $-C_{2v}$	400–2200	$d_{31} = 1.78$ $d_{32} = 2.7$ $d_{33} = 0.74$	$\leq 1.61\delta_{33}^{KDP}$ $3\delta_{33}^{KDP}$ $1.8\delta_{33}^{KDP}$	1064	20
Meta-nitroaniline (mNA), $NO_2C_6H_4NH_2$	$mm2$ $-C_{2v}$		$d_{31} = 20$ $d_{32} = 1.6$ $d_{33} = 21$		1064	[21,22]
Polyvinylidene fluoride, $(CH_2CF_2)_N$	$mm2$ $-C_{2v}$		$d_{31} = 0.18$ $d_{32} \cong 0$ $d_{33} \cong 0.36$	1.9 0 3.8	1064	[7]
Potassium dihydrogen phosphate, KH_2PO_4 (Phase below $-150°C$)	$mm2$ $-C_{2v}$		$d_{15} = 0.57 \pm 0.05$ $d_{24} = 0.38 \pm 0.05$ $d_{31} = 0.58 \pm 0.05$ $d_{32} = 0.39 \pm 0.05$ $d_{33} = 0$	3.3 ± 0.3 2.2 ± 0.3 3.4 ± 0.3 2.3 ± 0.3 0	694.3	[7]
Potassium lanthanum nitrate dihydrate (KLN), $K_2La(NO_3)_5 \cdot 2H_2O$	$mm2$ $-C_{2v}$	335–1100	$d_{31} = -1.13 \pm 0.15$ $d_{32} = 1.10 \pm 0.1$ $d_{33} = 0.13 \pm 0.1$		1064	[69]
Potassium niobate, $KNbO_3$	$mm2$ $-C_{2v}$	400–4500	$d_{15} = (2.3 \pm 0.4)d_{33}(LiIO_3)$ $d_{24} = (-2.4 \pm 0.4)d_{33}(LiIO_3)$ $d_{31} = (2.2 \pm 0.2)d_{33}(LiIO_3)$ $d_{32} = (-2.5 \pm 0.2)d_{33}(LiIO_3)$ $d_{33} = (-3.8 \pm 0.2)d_{33}(LiIO_3)$		1064	[23]
Potassium pentaborate tetrahydrate (KB5), $KB_5O_8 \cdot 4H_2O$	$mm2$ $-C_{2v}$	165–1400	$d_{31} = 0.0456$ $d_{32} = 0.0033$		1064	[24,25]
Potassium pentaborate deuterated tetrahydrate (DKB5)	$mm2$ $-C_{2v}$	162.5–1900	$d_{31} = 0.046$ $d_{32} = 0.004$		1064	[25]
Potassium titanyl arsenate (KTA), $KTiOAsO_4$	$mm2$ $-C_{2v}$	350–5300	$d_{31} = 1.6 \pm 0.1$ $d_{32} = 3.3 \pm 0.2$ $d_{31} = 2.8$ $d_{32} = 4.2$ $d_{33} = 16.2$		532 1064	[26] [17]

Table 33.19 Second harmonic coefficients *(continued)*

Material	Symm. class	Transparency range, nm	$10^{12} \cdot d_{il}^{2\omega}$, m/V	$10^2 \cdot \delta_{il}^{2\omega}$, m^2/C	λ_1, nm	Ref.
Orthorhombic system						
Potassium titanyl phosphate (KTP), $KTiOPO_4$	$mm2$ $-C_{2v}$	350 4500	$d_{31} = 1.93 \pm 0.2$ $d_{32} = 3.48 \pm 0.34$ $d_{33} = 8.1 \pm 0.9$ $d_{15} = 6.1$ $d_{24} = 7.6$ $d_{31} = 6.5$ $d_{32} = 5$ $d_{33} = 13.7$		1064 1064	[27,28] [29]
Sodium formate, $NaCHO_2$	$mm2$ $-C_{2v}$	240– 2200	$d_{31} = 0.022$ $d_{32} = -(0.22 \pm 0.11)$ $d_{33} = 0.33 \pm 0.16$ $d_{32} = 0.52 \pm 0.1$		1064 1064	[30] [31]
Sodium nitrite, $NaNO_2$	$mm2$ $-C_{2v}$		$d_{31} = d_{15} = 0.2$ $d_{32} = d_{24} = 0.87$	0.88 15.73	1153	[7]
Terbium molybdate, $Tb_2(MoO_4)_3$	$mm2$ $-C_{2v}$		$d_{15} = 2.87 \pm 0.43$ $d_{24} = 2.9 \pm 0.4$ $d_{31} = 2.6 \pm 0.4$ $d_{32} = 2.52 \pm 0.38$ $d_{33} = 0.12 \pm 0.03$	2.26 ± 0.34 2.33 ± 0.32 2.08 ± 0.32 1.98 ± 0.30 0.08 ± 0.02	1064.2	[7]
Ammonium oxalate monohydrate, $(NH_4)_2C_2O_4 \cdot H_2O$	222 $-D_2$		$d_{14} = 0.43$ $\frac{1}{2}(d_{25} + d_{36})$ $= 0.78 \pm 0.07$	$\delta_{14} = 2.2$	1060	[7]
Barium formate, $Ba(CHO_2)_2$	222 $-D_2$	245– 2200, 4800– 5100	$d_{14} = 0.113$ $d_{25} = 0.117$ $d_{36} = 0.117 - 0.139$		1064	[19]
Cesium triborate (CBO), CsB_3O_5	222 $-D_2$	170– 3000	$d_{14} = 0.648 d_{11}(\text{BBO})$		1064	[32]
Hippuric acid, $C_6H_5CO\text{-}NH(CH_2CO_2H)$	222 $-D_2$		$d_{36} = 2.85$	7.16	694.3	[7]
α-Iodic acid, α-HIO_3	222 $-D_2$	350– 1600	$d_{14} = 7.28 \pm 1.8$ $d_{14} = 4.83 \pm 0.9$ $d_{36} = 4.13 \pm 0.56$	4.43 ± 1.09 3.02 ± 0.56	1064 1152.6 1065	[7] [7] [33]
3-Methyl-4-nitropiridine-1-oxide (POM)	222 $-D_2$	400– 3000	$d_{36} = 10 \pm 15\%$		1064	[34]
5-Nitrouracil(5NU)	222 $-D_2$	410– 2000	$d_{14} = 8.38 \pm 0.13$		1064	[35]
d-Threonine	222 $-D_2$		$d_{14} = 0.45 \pm 0.07$ $d_{25} = 0.49 \pm 0.07$ $d_{36} = 0.46 \pm 0.07$	1.72 ± 0.27 1.86 ± 0.27 1.74 ± 0.27	1064.2	[7]
Trigonal system						
Aluminum phosphate, $AlPO_4$	32 $-D_3$		$d_{11} = 0.41 \pm 0.03$ $d_{14} \cong 0.009$	2.07 ± 0.15 0.046	1058.2	[7]
Benzil, $C_6H_5COCOC_6H_5$	32 $-D_3$		$d_{11} = 4.08 \pm 0.55$	9.36 ± 1.26	1064.2	[7]

Table 33.19 Second harmonic coefficients *(continued)*

Material	Symmetry class	Transparency range, nm	$10^{12} \cdot d_{il}^{2\omega}$, m/V	$10^2 \cdot \delta_{il}^{2\omega}$, m^2/C	λ_1, nm	Ref.
Trigonal system						
Mercury sulfide (cinnabar), α-HgS	32 $-D_3$		$d_{11} = 62.8 \pm 21$ $d_{11} = 50.3 \pm 17$	3.66 ± 1.2 2.93 ± 1	10600	[7]
Potassium dithionate, $K_2S_2O_6$	32 $-D_3$		$d_{11} = 0.087 \pm 0.015$	0.69 ± 0.12	694.3	[7]
Selenium, Se	32 $-D_3$	600–25000	$d_{11} = 97 \pm 25$ $d_{11} = 184 \pm 86$		10600 28000	[36] [37]
Silicon dioxide (α-Quartz), α-SiO_2	32 $-D_3$	150–4500	$d_{11} = 0.335 \pm 5\%$ $d_{11} = 0.009$	$1.49 \pm 5\%$ 0.04	1060 1058.2	[38] [7]
Tellurium, Te	32 $-D_3$	3800–32000	$d_{11} = 650 \pm 30$ $d_{11} = 570 \pm 190$		10600 28000	[38] [39]
β-Barium borate (BBO), β-BaB_2O_4	$3m$ $-C_{3v}$	190–3100	$d_{22} = 4.1 d_{36}(\text{KDP}) = 1.78$ $d_{31} = 0.12$		1064	[40,16]
Lithium niobate, $LiNbO_3$	$3m$ $-C_{3v}$	330–5500	$d_{31} = 6.28 \pm 0.63$ $d_{31} = 5.95$ $d_{33} = 34.4$ $d_{31} = 5.77$ $d_{33} = 33.4$ $d_{33} = 31.8$ $d_{33} = 29.1$	1.18 ± 0.12	632.8 1064 1152 1318 2120	[7,41] [42,43]
Lithium tantalate, $LiTaO_3$	$3m$ $-C_{3v}$		$d_{15} = 2.08 \pm 0.24$ $d_{15} = 1.28 \pm 0.24$ $d_{33} = 19.39 \pm 2.36$	0.48 ± 0.06 0.29 ± 0.06 4.4 ± 0.53	1058.2	[7,41]
Silver arsenic sulfide (proustite), $AgAsS_3$	$3m$ $-C_{3v}$	600 13000	$d_{15} = 15.08 \pm 2.2$ $d_{22} = 24 \pm 4$ $d_{15} = 0.13 d_{36}(\text{GaAs})$ $= 11.3 \pm 2.5$ $d_{22} = 0.2 d_{36}(\text{GaAs})$ $= 18 \pm 2.5$	0.64 ± 0.09 0.91 ± 0.19 1.45 ± 0.29	1152 10600	[7] [44]
Silver antimony sulfide (pyrargyrite), Ag_8SbS_3	$3m$ $-C_{3v}$	700–14000	$d_{15} = 8.4 \pm 2.9$ $d_{22} = 9.2 \pm 2.9$	0.58 ± 0.18 0.55 ± 0.16	10600	[7,36,45] [46]
Thallium arsenic selenide, Tl_3AsSe_3	$3m$ $-C_{3v}$	1260–16500	$d_\Sigma = 3.3 d_\Sigma(\text{pyrargyrite})$ $\pm 30\%$		10600	[45]
Tourmaline	$3m$ $-C_{3v}$		$d_{15} = 0.27 \pm 0.04$ $D_{22} = 0.08 \pm 0.01$ $d_{31} = 0.16 \pm 0.03$ $d_{33} = 0.57 \pm 0.07$	0.70 ± 0.1 0.22 ± 0.03 0.42 ± 0.08 1.58 ± 0.19		[7]
Tetragonal system						
Barium titanate, $BaTiO_3$	$4mm$ $-C_{4v}$		$d_{15} = 17.2 \pm 1.42$ $d_{31} = 17.97 \pm 1.42$ $d_{33} = 6.6 \pm 0.5$ $d_{15} = 19.33 \pm 2.0$ $d_{31} = 17.85 \pm 2.0$ $d_{33} = 7.79 \pm 1.17$	1.98 ± 0.16 2.11 ± 0.17 0.85 ± 0.06 2.23 ± 0.23 2.10 ± 0.23 1.0 ± 0.15	1058 1064.2	[7]

Table 33.19 Second harmonic coefficients *(continued)*

Material	Symmetry class	Transparency range, nm	$10^{12} \cdot d_{il}^{2\omega}$, m/V	$10^2 \cdot \delta_{il}^{2\omega}$, m^2/C	λ_1, nm	Ref.
Tetragonal system						
Lead titanate, $PbTiO_3$	$4mm$ $-C_{4v}$		$d_{15} = 37.9 \pm 5.7$ $d_{31} = 42.8 \pm 6.4$ $d_{33} = 8.5 \pm 1.4$	2.09 ± 0.3 2.39 ± 0.36 0.48 ± 0.08	1064.2	[7]
Potassium lithium niobate, $K_3Li_2Nb_5O_{15}$	$4mm$ $-C_{4v}$	400–5000	$d_{15} = 6.2 \pm 1.1$ $d_{31} = 7.0 \pm 1.5$ $d_{33} = 12.7 \pm 1.8$	1.19 ± 0.2 1.38 ± 0.3 3.14 ± 0.44	1064.2	[7,47]
Potassium sodium barium niobate, $K_{0.8}Na_{0.2}Ba_2Nb_5O_{15}$	$4mm$ $-C_{4v}$		$d_{31} = 12.77$	2.22	1064.2	[7]
Strontium barium niobate, $Sr_{0.5}Ba_{0.5}Nb_2O_6$	$4mm$ $-C_{4v}$		$d_{15} = 6.8 \pm 2.3$ $d_{31} = 4.9 \pm 1.5$ $d_{33} = 12.85 \pm 3.8$	1.07 ± 0.36 0.78 ± 0.24 2.20 ± 0.65	1064.2	[7]
Ammonium dideuterium phosphate (ADDP), $ND_4D_2PO_4$	$\bar{4}2m$ $-D_{2d}$	220–2100	$d_{36} = 0.52 \pm 0.08$	2.88 ± 0.44	694.3	[7,48]
Ammonium dihydrogen phosphate (ADP), $NH_4H_2PO_4$	$\bar{4}2m$ $-D_{2d}$	184–1500	$d_{14} = 0.553 \pm 0.024$ $d_{36} = 0.558 \pm 0.028$ $d_{14} = 0.482 \pm 0.024$ $d_{36} = 0.487 \pm 0.028$ $d_{36} = 0.544 \pm 0.14$ $d_{36} = 0.57 \pm 0.068$ $d_{36} = 0.66 \pm 0.14$	3.19 ± 0.14 3.27 ± 0.16 2.52 ± 0.12 2.57 ± 0.15 3.25 ± 0.84 2.93 ± 0.35 3.65 ± 0.77	1058.2 694.3 1152 632.8 825.0	[7,49]
Beryllium sulfate tetrahydrate, $BeSO_4 \cdot 4H_2O$	$\bar{4}2m$ $-D_{2d}$	187–1300	$d_{36} = 0.24 \pm 0.03$		532	[50]
Cadmium germanium arsenate, $CdGeAs_2$	$\bar{4}2m$ $-D_{2d}$	2400–18000	$d_{36} = (2.6 \pm 0.4)d_{36} =$ 235 ± 38		10600	[51]
Cesium dideuterium arsenate (CDDA), CsD_2AsO_4	$\bar{4}2m$ $-D_{2d}$	270–1660	$d_{36} = 0.402 \pm 0.046$		1064	[52]
Cesium dihydrogen arsenate (CDA), CsH_2AsO_4	$\bar{4}2m$ $-D_{2d}$	250 1430	$d_{36} = 0.25 \pm 0.04$ $d_{36} = 0.402 \pm 0.046$	0.93 ± 0.15	694.3 1060 1064	[7] [52]
Cuprous indium sulfide, $CuInS_2$	$\bar{4}2m$ $-D_{2d}$	1500–13000	$d_{36} = 9.6 \pm 30\%$		10600	[53]
Cuprous thiogallate, $CuGaS_2$	$\bar{4}2m$ $-D_{2d}$	1500–12000	$d_{36} = 14.5 \pm 30\%$		10600	[53]
Potassium dihydrogen arsenate (KDA), KH_2AsO_4	$\bar{4}2m$ $-D_{2d}$	216–1450	$d_{14} = 0.55 \pm 0.02$ $d_{36} = 0.52 \pm 0.03$ $d_{14} = 0.41 \pm 0.05$ $d_{36} = 0.47 \pm 0.05$	2.35 ± 0.08 2.24 ± 0.13 1.58 ± 0.19 1.83 ± 0.19	1058.2 694.3	[7,48]
Potassium dideuterium phosphate (KDDP), KD_2PO_4	$\bar{4}2m$ $-D_{2d}$	200–2000	$d_{14} = 0.5 \pm 0.02$ $d_{36} = 0.5 \pm 0.02$ $d_{14} = 0.46 \pm 0.04$ $d_{36} = 0.5 \pm 0.02$	3.23 ± 0.13 3.24 ± 0.13 2.71 ± 0.23 3.04 ± 0.12	1058.2 694.3	[7,48]

Table 33.19 Second harmonic coefficients *(continued)*

Material	Symmetry class	Transparency range, nm	$10^{12} \cdot d_{il}^{2\omega}$, m/V	$10^2 \cdot \delta_{il}^{2\omega}$, m^2/C	λ_1, nm	Ref.
			Tetragonal system			
Potassium dihydrogen phosphate (KDP), KH_2PO_4	$\bar{4}2m$ $-D_{2d}$	176.5–1500	$d_{14} = 0.49 \pm 0.02$	3.10 ± 0.13	1058.2	[7,49]
			$d_{36} = 0.47$	3.02 ± 0.2		
			$d_{36} = 0.47 \pm 0.03$	2.71 ± 0.17	694.3	
			$d_{36} = 0.47$	2.73		
			$d_{36} = 0.46 \pm 0.07$	3.01 ± 0.46	1152	
			$d_{36} = 0.47 \pm 0.03$	3.01 ± 0.2	1064.2	
			$d_{36} = 0.71$		632.8	[42]
			$d_{36} = 0.63$		1060	
			$d_{36} = 0.62$		1152	
			$d_{36} = 0.60$		1318	
Rubidium dihydrogen arsenate (RDA), RbH_2AsO_4	$\bar{4}2m$ $-D_{2d}$	260–1460	$d_{36} = 0.9d_{36}(\text{KDP})$ $= 0.394 \pm 0.04$	1.27 ± 0.17	694.3	[54,5]
Rubidium dihydrogen phosphate (RDP), RbH_2PO_4	$\bar{4}2m$ $-D_{2d}$	220–1400	$d_{36} = 0.49$	2.83	694.3	[7,56]
			$d_{36} = 1.05 = 0.41 \pm 0.05$		1064.2	[57]
Silver selenagallate, $AgGaSe_2$	$\bar{4}2m$ $-D_{2d}$	800–18000	$d_{36} = 0.368d_{36}(\text{GaAs})$ $= 3.32 \pm 10\%$		10600	[37,58]
Silver thiogallate, $AgGaS_2$	$\bar{4}2m$ $-D_{2d}$	500–1300	$d_{36} = 57.5d_{11}(SiO_2) = 19.3$		1064	[53,59]
			$\cdot d_{36} = (0.154 \pm 30\%)$ $\cdot d_{36}(\text{GaAs}) = 13.9 \pm 4.2$		3390	[60]
			$d_{36} = (0.167 \pm 30\%)$ $\cdot d_{36}(\text{GaAs}) = 15.1 \pm 4.5$	7.17 ± 2.4	10600	
						[7]
Zinc germanium phosphide, $ZnGeP_2$	$\bar{4}2m$ $-D_{2d}$	740–12000	$d_{36} = (0.83 \pm 0.14)$ $\cdot d_{36}(\text{GaAs}) = 75.41 \pm 13$		10600	[37,61]
			$d_{14} = 111 \pm 33$	1.96 ± 0.58		[7]
Cadmium-mercury thiocyanate, $Cd[Hg(SCN)_4]$	$\bar{4} - S_4$		$d_{31} = 7.57 \pm 1.4$	5.3 ± 0.98	1064	[7]
			$d_{36} = 1.76 \pm 0.53$	1.23 ± 0.37		
Mercury thiogallate, $HgGa_2S_4$	$\bar{4} - S_4$	500–13000	$d_{36} = 34.7 \pm 5.2$		1064	[63]
			$d_{31} = 11.6 \pm 1.7$			
			$d_{36} = 80d_{11}(SiO_2)$ $= 26.8 \pm 8$			[63]
			Hexagonal system			
Beryllium oxide, BeO	$6mm$ $-C_{6v}$		$d_{31} = 0.17 \pm 0.01$	0.26 ± 0.015	1064.2	[7]
			$d_{33} = 0.23 \pm 0.01$	0.34 ± 0.01		
Cadmium selenide, CdSe	$6mm$ $-C_{6v}$	750–20000	$d_{33} = 79.6 \pm 4.9$	3.54 ± 0.22	1058	[7,64]
			$d_{15} = 31 \pm 7.5$	2.8 ± 0.67	10600	
			$d_{31} = 28.5 \pm 6.3$	2.56 ± 0.57		
			$d_{33} = 54.5 \pm 12.6$	4.73 ± 1.09		
			$d_{33} = 65.4 \pm 13$		2120	[42]
Cadmium sulfide, CdS	$6mm$ $-C_{6v}$		$d_{15} = 17.0 \pm 1.4$	1.62 ± 0.13	1058.2	[7]
			$d_{31} = 15.6 \pm 0.9$	1.5 ± 0.09		
			$d_{33} = 30.7 \pm 1.9$	2.83 ± 0.17		
			$d_{15} = 28.9 \pm 7.1$	4.88 ± 1.2	10600	
			$d_{31} = 26.4 \pm 6.3$	4.45 ± 1.06		
			$d_{33} = 44.0 \pm 12.6$	7.1 ± 2.0		

Table 33.19 Second harmonic coefficients *(continued)*

Material	Symmetry class	Transparency range, nm	$10^{12} \cdot d_{il}^{2\omega}$, m/V	$10^2 \cdot \delta_{il}^{2\omega}$, m^2/C	λ_1, nm	Ref.
Hexagonal system						
Silicon carbide, SiC	$6mm$ $-C_{6v}$		$d_{15} = 9.1 \pm 1.1$ $d_{31} = 9.83 \pm 1.1$ $d_{33} = 16.4 \pm 1.8$	0.50 ± 0.06 0.54 ± 0.06 0.84 ± 0.09	1064	[7]
Zinc oxide, ZnO	$6mm$ $-C_{6v}$		$d_{15} = 2.32 \pm 0.2$ $d_{31} = 2.1 \pm 0.2$ $d_{33} = 7.0 \pm 0.2$	1.08 ± 0.09 0.98 ± 0.09 3.16 ± 0.09	1058.2	[7]
Zinc sulfide, ZnS	$6mm$ $-C_{6v}$		$d_{33} = 13.5 \pm 0.8$ $d_{15} = 21.4 \pm 8.4$ $d_{31} = 18.85 \pm 6.3$ $d_{33} = 37.3 \pm 12.6$	1.74 ± 0.1 3.42 ± 1.34 3.01 ± 1.0 5.89 ± 1.99	1058 10600	[7]
Gallium selenide, GaSe	$\bar{6}m2$ $-D_{3h}$	650–18000	$d_{22} = (3 \pm 0.6)d_{31}$(CdSe) $= 54 \pm 11$		10600	[65]
Lithium iodate, $LiIO_3$	$6 - C_6$	300–6000	$d_{31} = 7.5 \pm 1.1$ $d_{31} = -8.41 \pm 0.85$ $d_{31} = -5.19 \pm 0.27$ $d_{33} = -5.4 \pm 0.34$ $d_{31} = -4.8 \pm 0.7$ $d_{33} = -3.85 \pm 1.15$ $d_{31} = -4.93 \pm 1.73$ $d_{33} = -4.88 \pm 1.75$ $d_{31} = -6.43$ $d_{33} = -6.41$	6.79 ± 1.0	514.5 6943 1060 1060 1318 2120	[7] [66] [67] [68] [42]
Lithium potassium sulfate, $LiKSO_4$	$6 - C_6$		$d_{31} = 0.38$ $d_{33} = 0.71$		694.3	[7]
Cubic system						
Aluminum antimonide, AlSb	$\bar{4}3m$ $-T_d$		$d_{14} = 49 \pm 36$	0.4 ± 0.29	1058	[7]
Bismuth germanium oxide, $BiGeO_{12}$	$\bar{4}3m$ $-T_d$		$d_{14} = 1.46$	0.47	1064	[7]
Cadmium telluride, CdTe	$\bar{4}3m$ $-T_d$		$d_{14} = 16.7 \pm 6.3$	0.76 ± 0.29	10600	[7]
Cuprous bromide, CuBr	$\bar{4}3m$ $-T_d$		$d_{14} = 7.96 \pm 2.7$ $d_{14} = 15.3 \pm 6.9$	3.73 ± 1.27 5.07 ± 2.28	10600 1064	[7]
Cuprous chloride, CuCl	$\bar{4}3m$ $-T_d$		$d_{14} = 4.19$ $d_{14} = 9.1 \pm 4.1$	2.74 4.64 ± 2.09	10600 1064	[7]
Cuprous iodide, CuI	$\bar{4}3m$ $-T_d$		$d_{14} = 7.96 \pm 2.6$ $d_{14} = 30.2 \pm 13.6$	1.72 ± 0.56 4.65 ± 2.09	10600 1064	[7]
Gallium antimonide, GaSb	$\bar{4}3m$ $-T_d$		$d_{14} = 628 \pm 63$	2.84 ± 0.28	10600	[7]
Gallium arsenide, GaAs	$\bar{4}3m$ $-T_d$		$d_{14} = 137$ $d_{14} = 274 \pm 66$ $d_{36} = 249 \pm 15$ $d_{36} = 173 \pm 28$ $d_{36} = 151 \pm 24$	0.31 1.4 ± 0.34 1.27 ± 0.08 1.70	843-845 1058 2120 10600	[7] [42]

Table 33.19 Second harmonic coefficients *(continued)*

Material	Symmetry class	Transparency range, nm	$10^{12} \cdot d_{il}^{2\omega}$, m/V	$10^2 \cdot \delta_{il}^{2\omega}$, m^2/C	λ_1, nm	Ref.
Cubic system						
Gallium arsenide, GaAs	$\bar{4}3m$ $-T_d$		$d_{14} = 188.5 \pm 19$ $d_{14} = 140 \pm 10$	2.26 ± 0.23 0.72 ± 0.05		[7]
Gallium phosphide, GaP	$\bar{4}3m$ $-T_d$		$d_{14} = 85.6 \pm 14.2$ $d_{36} = 41.2 \pm 2.5$ $d_{14} = 35.0$ $d_{14} = 109$	1.14 ± 0.19 0.55 ± 0.03 0.73 1.9	1058 3390 10600	[7]
Hexamine, $N_4(CH_2)_6$	$\bar{4}3m$ $-T_d$		$d_{14} = 4.73$	15.7	1060	[7]
Indium antimonide, InSb	$\bar{4}3m$ $-T_d$		$d_{14} = 520 \pm 47$	1.8	1058	[7]
Indium arsenide, InAs	$\bar{4}3m$ $-T_d$		$d_{14} = 364 \pm 47$ $d_{14} = 418.9 \pm 12.6$	 3.27 ± 0.1	1058 10600	[7]
Indium phosphide, InP	$\bar{4}3m$ $-T_d$		$d_{14} = 167.0$	0.94	1058	[7]
Zinc selenide, ZnSe	$\bar{4}3m$ $-T_d$		$d_{36} = 31.7 \pm 1.95$ $d_{14} = 78.3 \pm 29.3$	2.12 ± 0.14 7.6 ± 3	1058 10600	[7]
Zinc sulfide, β-ZnS	$\bar{4}3m$ $-T_d$	600–14000	$d_{36} = 24.6 \pm 1.5$ $d_{14} = 30.6 \pm 8.4$	3.2 ± 0.19 5.08 ± 1.38	1058 10600	[7]
Zinc telluride, ZnTe	$\bar{4}3m$ $-T_d$		$d_{36} = 106.7 \pm 6.6$ $d_{14} = 94.6 \pm 9.5$ $d_{14} = 92.1 \pm 33.5$	2.96 ± 0.18 2.62 ± 0.26 4.19 ± 1.5	1058 10600	[7]
Sodium bromate, $NaBrO_3$	23–T		$d_{14} = 0.28$	0.71	694.3	[7]
Sodium chlorate, $NaClO_3$	23–T		$d_{14} = 0.69$	3.43	694.3	[7]

Table 33.20 Refractive indices of nonlinear crystals

Aluminum phosphate ($AlPO_4$)

Wavelength λ, μm	n_o	n_e
0.4	1.5369	1.5465
0.5	1.5287	1.5385
0.6	1.5243	1.5334
0.7	1.5215	1.5301
0.8	1.5192	1.5281
1.0	1.5161	1.5245
1.2	1.5136	1.5223
1.4	1.5112	1.5198
1.6	1.5088	1.5174
1.8	1.5062	1.5145
2.0	1.5034	1.5116
2.2	1.5001	1.5083
2.4	1.4969	1.5048
2.6	1.4928	1.5006

Ammonium dideuterium phosphate ($ND_4D_2PO_4$)

Wavelength λ, μm	n_o	n_e
0.35	1.5414	1.4923
0.53	1.5198	1.4784
0.69	1.5142	1.4737
1.06	1.5088	1.4712

Dispersion equations [70]:

$$n_o^2 = 2.279481 + 1.215870\lambda^2/[\lambda^2 - (7.614168)^2] + 0.10761/[\lambda^2 - (0.115165)^2];$$

$$n_e^2 = 2.151161 + 1.199009\lambda^2/[\lambda^2 - (11.25169)^2] + 0.009652/[\lambda^2 - (0.09855)^2].$$

Ammonium oxalate monohydrate ($(NH_4)_2C_2O_4H_2O$)

Wavelength λ, μm	$n_{\alpha=Z}$	$n_{\beta=Y}$	$n_{\gamma=X}$
0.4471	1.4460	1.5599	1.6119
0.4713	1.4447	1.5561	1.6084
0.4922	1.4435	1.5544	1.6050
0.5016	1.4426	1.5536	1.6037
0.5461	1.4406	1.5493	1.5993
0.5780	1.4391	1.5470	1.5965
0.5876	1.4388	1.5469	1.5952
0.6678	1.4362	1.5426	1.5892
0.7016	1.4352	1.5408	1.5874
1.014	1.4295	1.5312	1.5763
1.129	1.4276	1.5284	1.5728
1.367	1.4235	1.5222	1.5652

Ammonium dihydrogen phosphate, ADP ($NH_4H_2PO_4$)

Wavelength λ, μm	n_o	n_e
0.2138560	1.62598	1.56738
0.2288018	1.60785	1.55138
0.2536519	1.58688	1.53289
0.2967278	1.56462	1.51339
0.3021499	1.56270	1.51163
0.3125663	1.55917	1.50853
0.3131545	1.55897	1.50832
0.3341478	1.55300	1.50313
0.3650146	1.54615	1.49720
0.3654833	1.54608	1.49712
0.3662878	1.54592	1.49698
0.3906410	1.54174	
0.4046561	1.53969	1.49159
0.4077811	1.53925	1.49123
0.4358350	1.53578	1.48831
0.4916036		1.48390
0.5460740	1.52662	1.48079
0.5769590	1.52478	1.47939
0.5790654	1.52466	1.47930
0.6328160	1.52166	1.47685
1.013975	1.50835	1.46895
1.128704	1.50446	1.46704
1.152276	1.50364	1.46666

Accuracy $\approx \pm 3{\cdot}10^{-5}$. Dispersion equations [70]:

$$n_o^2 = 2.301929 + 1.78281\lambda^2/[\lambda^2 - (6.840459)^2] + 0.010569/[\lambda^2 - (0.133602)^2];$$

$$n_e^2 = 2.162273 + 1.823756\lambda^2/[\lambda^2 - (11.31744)^2] + 0.009866/[\lambda^2 - (0.101736)^2].$$

Barium metaborate, BBO (BaB_2O_4) [40]

Wavelength λ, μm	n_o	n_e
0.40466	1.69267	1.56796
0.43583	1.68679	1.56376
0.46782	1.68198	1.56024
0.47999	1.68044	1.55914
0.50858	1.67722	1.55691
0.54607	1.67376	1.55465
0.57907	1.67131	1.55298
0.58930	1.67049	1.55247
0.64385	1.66736	1.55012
0.81890	1.66066	1.55589
0.85212	1.65969	1.54542
0.89435	1.65862	1.54469
1.0140	1.65608	1.54333

Dispersion equations:

$$n_o^2 = 2.7405 + 0.0184/(\lambda^2 - 0.0179) - 0.0155\lambda^2;$$

$$n_e^2 = 2.3730 + 0.0128/(\lambda^2 - 0.0156) - 0.0044\lambda^2.$$

Barium-sodium niobate ($Ba_2NaNb_5O_{15}$)

Wavelength λ, μm	$n_{\alpha=c=Z}$	$n_{\beta=a=Y}$	$n_{\gamma=b=X}$
0.4579	2.2931	2.4266	2.4284
0.4765	2.2799	2.4076	2.4094
0.4880	2.2727	2.3974	2.3991
0.4965	2.2678	2.3903	2.3920
0.5017	2.2649	2.3862	2.3879
0.5145	2.2583	2.3767	2.3786
0.5321	2.2502	2.3655	2.3672
0.6328	2.2177	2.3205	2.3222
1.0642	2.1700	2.2567	2.2580

Dispersion equations:

$$n_\alpha^2 - 1 = 3.6008\lambda^2/[\lambda^2 - (0.17944)^2];$$

$$n_\beta^2 - 1 = 3.9495\lambda^2/[\lambda^2 - (0.20035)^2];$$

$$n_\gamma^2 - 1 = 3.9495\lambda^2/[\lambda^2 - (0.20097)^2].$$

Barium titanate ($BaTiO_3$)

Wavelength λ, μm	n_o	n_e
0.4579	2.5637	2.4825
0.4765	2.5355	2.4605
0.4888	2.5206	2.4487
0.5145	2.4917	2.4255
0.5321	2.4760	2.4128
0.6328	2.4164	2.3637
1.0642	2.3379	2.2970
2.1284	2.2947	2.2593

Dispersion equations:

$$n_o^2 - 1 = 4.239\lambda^2/[\lambda^2 - (0.2229)^2];$$

$$n_e^2 - 1 = 4.0854\lambda^2/[\lambda^2 - (0.2087)^2].$$

Benzil ($(C_6H_5)_2(CO)_2$)

Wavelength λ, μm	n_o	n_e
0.4205	1.737	1.737
0.4358	1.716	1.720
0.4380	1.712	1.718
0.4620	1.694	1.705
0.4860	1.682	1.695
0.5461	1.667	1.684
0.5780	1.660	1.680
0.5893	1.658	1.679
0.6560	1.648	1.672

Dispersion equations:
$n_o^2 - 1 = 1.08 + 0.535\lambda^2/[\lambda^2 - (0.24)^2]$
$+ 0.0150\lambda^2/[\lambda^2 - (0.398)^2]$;
$n_e^2 - 1 = 1.35 + 0.370\lambda^2/[\lambda^2 - (0.24)^2]$
$+ 0.0138\lambda^2/[\lambda^2 - (0.395)^2]$.

Beryllium oxide (BeO), 22.4°C

Wavelength λ, μm	n_o	n_e
0.43	1.73039	
0.44	1.72924	1.74556
0.45	1.72820	1.74447
0.46	1.72725	1.74348
0.47	1.72626	1.74251
0.48	1.72542	1.74162
0.49	1.72460	1.74073
0.50	1.72388	1.74002
0.51	1.72308	1.73918
0.52	1.72249	1.73852
3	1.72177	1.73779
0.54	1.72121	1.73703
0.55	1.72062	1.73644
0.56	1.72006	1.73588
0.57	1.71950	1.73530
0.58	1.71903	1.73477
0.59	1.71856	1.73423
0.60	1.71795	1.73381
0.61	1.71762	1.73322
0.62	1.71710	1.73279
0.63	1.71668	1.73233
0.64	1.71632	1.73191
0.65	1.71589	1.73156
0.66	1.71554	1.73113
0.67	1.71517	1.73075
0.68	1.71482	1.73041
0.69	1.71450	

Dispersion equations:
$n_o^2 - 1 = 1.919087\lambda^2/(\lambda^2 - 0.00727575)$
$+ 3.972323\lambda^2/(\lambda^2 - 199.31087)$;
$n_e^2 - 1 = 1.972142\lambda^2/(\lambda^2 - 0.00748564)$
$+ 17.5787\lambda^2/(\lambda^2 - 779.49122)$.

Beryllium sulfate tetrahydrate ($BeSO_4{\cdot}4H_2O$) [50]

Wavelength λ, μm	n_o	n_e
0.4154	1.4847	1.4431
0.4825	1.4782	1.4379
0.5321	1.4749	1.4348
0.6328	1.4701	1.4315
0.6471	1.4692	1.4312
0.6764	1.4681	1.4304
0.7525	1.4658	1.4292

Dispersion equations:
$n_o^2 = 2.1545 + \frac{0.00835}{\lambda^2 - 0.01606} - 0.03573\lambda^2$;
$n_e^2 = 2.0335 + \frac{0.00806}{\lambda^2 - 0.01354} - 0.01970\lambda^2$.

Bismuth-germanium oxide (Bi_4GeO_{12})

Wavelength λ, μm	n
0.4765	2.142
0.4880	2.135(7)
0.4965	2.131(8)
0.5017	2.128(5)
0.5145	2.123(7)
0.5321	2.115(2)
0.6328	2.086(1)
1.0642	2.044(3)

Dispersion equations:
$n^2 - 1 = 3.08959\lambda^2/(\lambda^2 - 0.01337)$.

Cadmium-germanium arsenide ($CdGeAs_2$)

Wavelength λ, μm	n_o	n_e
5.3	3.5304	3.6209
10.6	3.5046	3.5911

Dispersion equations [71]:
$n_o^2 = 10.1064 + 2.2988\lambda^2/(\lambda^2 - 1.0872)$
$+ 1.6247\lambda^2/(\lambda^2 - 1370)$;
$n_e^2 = 11.8018 + 1.2152\lambda^2/(\lambda^2 - 2.6971)$
$+ 1.6922\lambda^2/(\lambda^2 - 1370)$.

Cadmium-mercury thiocyanate ($Cd[Hg(SCN)_4]$), 27°C

Wavelength λ, μm	n_o	n_e
0.530	2.003(5)	1.792
0.633	1.970	1.753
1.06	1.924(5)	1.728

Cadmium selenide (CdSe)

Wavelength λ, μm	n_o	n_e
0.8	2.6448	2.6607
0.9	2.5826	2.6027
1.0	2.5502	2.5996
1.2	2.5132	2.5331
1.4	2.4929	2.5133
1.6	2.4818	2.5008
1.8	2.4732	2.4930
2.0	2.4682	2.4873
2.2	2.4642	2.4840
2.4	2.4612	2.4798
2.6	2.4590	2.4784
2.8	2.4562	2.4757
3.0	2.4542	2.4741
3.2	2.4532	2.4726
3.4	2.4518	2.4714
3.6	2.4809	2.4702
3.8	2.4498	2.4694
4.0	2.4491	2.4685

Dispersion equations [71]:
$n_o^2 = 4.2243 + 1.768\lambda^2/(\lambda^2 - 0.227)$
$+ 3.12\lambda^2/(\lambda^2 - 3380)$;
$n_e^2 = 4.2009 + 1.8875\lambda^2/(\lambda^2 - 0.2171)$
$+ 3.6461\lambda^2/(\lambda^2 - 3629)$.

Cadmium sulfide (CdS)

Wavelength λ, μm	n_o	n_e
0.5120		2.751
05130		2.743
0.5140		2.737
0.5150	2.743	2.726
0.5160	2.735	2.720
0.5170	2.727	2.714
0.5180	2.718	2.706
0.5190	2.709	2.702
0.5200	2.702	2.698
0.5210	2.700	2.694
0.5220	2.694	2.689
0.5230	2.687	2.685
0.5240	2.681	2.680
0.5250	2.674	2.575
0.5275	2.661	2.665
0.5300	2.649	2.654
0.5325	2.638	2.644
05350	2.628	2.637
0.5375	2.617	2.628
0.5400	2.609	2.622
0.5425	2.602	2.612
0.5450	2.594	2.606
0.5475	2.587	2.600
0.5500	2.580	2.593

Cadmium sulfide (CdS) *(continued)*

Wavelength λ, μm	n_o	n_e
0.5750	2.528	2.545
0.6000	2.493	2.511
0.6250	2.467	2.484
0.6500	2.446	2.463
0.6750	2.427	2.446
0.7000	2.414	2.432
0.7500	2.390	2.409
0.8000	2.374	2.392
0.8500	2.364	2.378
0.9000	2.359	2.368
0.9500	2.341	2.359
1.0000	2.334	2.352
1.0500	2.328	2.346
1.1000	2.324	2.340
1.1500	2.320	2.336
1.2000	2.316	2.332
1.2500	2.312	2.329
1.3000	2.309	2.326
1.3500	2.306	2.323
1.4000	2.304	2.321

Cadmium telluride (CdTe)

Wavelength λ, μm	n
0.903	2.91
1.0	2.84
1.1	2.81
1.0-1.3	2.82
7.0-10.0	2.69
10.0	2.69
14.0	2.69

Dispersion equations:
$n^2 - 1 = 4.68 + 1.53\lambda^2/(\lambda^2 - 0.366)$.

Cesium dideuterium arsenate, CDDA (CsD_2AsO_4) [52]

Wavelength λ, μm	n_o	n_e
0.3472	1.5895	1.5685
0.5321	1.5681	5.5495
0.6943	1.5596	1.5418
1.0642	1.5503	1.5326

Accuracy $\approx \pm 3 \cdot 10^{-4}$. Dispersion equations [70]:
$n_o^2 = 2.40817 + 2.212173\lambda^2/[\lambda^2 - (11.26371)^2]$
$+ 0.015598/[\lambda^2 - (0.138209)^2]$;
$n_e^2 = 2.345809 + 0.651843\lambda^2/[\lambda^2 - (11.28408)^2]$
$+ 0.015141/[\lambda^2 - (0.129754)^2]$.

Cesium dihydrogen arsenate, CDA (CsH_2AsO_4) [52]

Wavelength λ, μm	n_o	n_e
0.3472	1.6027	1.5722
0.5321	1.5733	1.5514
0.6943	1.5632	1.5429
1.0642	1.5516	1.5330

Accuracy $\approx \pm 3 \cdot 10^{-4}$. Dispersion equations [70]:
$n_o^2 = 2.420405 + 1.403336\lambda^2/[\lambda^2 - (7.60422)^2] + 0.016272/[\lambda^2 - (0.134185)^2]$;
$n_e^2 = 2.350262 + 0.685328\lambda^2/[\lambda^2 - (11.28135)^2] + 0.015645/[\lambda^2 - (0.121741)^2]$.

Cesium triborate, CsBO (CsB_3O_5) [32]

Wavelength λ, μm	n_x	n_y	n_z
0.3543	1.5499	1.5849	1.6145
0.4765	1.5370	1.5758	1.6031
0.4880	1.5367	1.5736	1.6009
0.4965	1.5362	1.5716	1.5996
0.5145	1.5349	1.5690	1.5974
0.532	1.5499	1.5849	1.6145
0.5328	1.5370	1.5758	1.6031
1.064	1.5367	1.5736	1.6009

Dispersion equations:
$n_x^2 = 2.2916 + \frac{0.02105}{\lambda^2+0.06525} - 3.1848 \cdot 10^{-5}\lambda^2$;
$n_y^2 = 2.3731 + \frac{0.3437}{\lambda^2+0.1160} - 7.2632 \cdot 10^{-5}\lambda^2$;
$n_z^2 = 2.4607 + \frac{0.03202}{\lambda^2+0.08961} - 5.6332 \cdot 10^{-5}\lambda^2$.

Cuprous bromide (CuBr)

Wavelength λ, μm	n
0.4358	2.336(5)±0.002
0.4678	2.229(0)±0.002
0.4800	2.207(2)±0.002
0.5086	2.171(5)±0.002
0.5461	2.141(1)±0.002
0.5791	2.122(1)±0.002
0.5896	2.117(4)±0.002
0.6438	2.096(9)±0.002
0.7699	2.069(6)±0.004

Cuprous chloride (CuCl)

Wavelength λ, μm	n
0.4047	2.153(5)±0.001
0.4078	2.141(0)±0.001
0.4358	2.072(0)±0.001
0.4678	2.033(6)±0.001
0.4800	2.023(4)±0.001

Cuprous chloride (CuCl) *(continued)*

Wavelength λ, μm	n
0.5086	2.004(2)±0.001
0.5461	1.987(0)±0.001
0.5791	1.976(0)±0.001
0.5896	1.972(6)±0.001
0.6438	1.958(4)±0.001
0.7699	1.941±0.002

Cuprous iodide (CuI)

Wavelength λ, μm	n
0.4358	2.562(1)±0.002
0.4678	2.461(7)±0.002
04800	2.448(5)±0.002
0.5086	2.411(0)±0.002
0.5461	2.372(6)±0.002
0.5791	2.347(5)±0.002
0.5896	2.342(8)±0.002
0.6438	2.315(6)±0.002
0.7699	2.280(2)±0.004

2-Cyclooctilamino-5-nitropyridine, COANP [15]

Wavelength λ, μm	$n_{X=c}$	$n_{Y=a}$	$n_{Z=b}$
0.55	1.681	1.702	1.847

Gadolinium molybdate ($Gd_2(MoO_4)_3$)

Wavelength λ, μm	$n_{\alpha=b=Y}$	$n_{\beta=a=X}$	$n_{\gamma=c=Z}$
0.4579	1.8758	1.8762	1.9342
0.4765	1.8694	1.8699	1.9270
0.4880	1.8659	1.8663	1.9229
0.4965	1.8634	1.8639	1.9201
0.5017	1.8621	1.8625	1.9185
0.5145	1.8588	1.8593	1.9148
0.5321	1.8545	1.8549	1.9102
0.6328	1.8385	1.8390	1.8915
1.064	1.8142	1.8146	1.8637

Dispersion equations:
$n_\alpha^2 - 1 = 2.2450\lambda^2/(\lambda^2 - 0.022693)$;
$n_\beta^2 - 1 = 2.24654\lambda^2/(\lambda^2 - 0.0226803)$;
$n_\gamma^2 - 1 = 2.41957\lambda^2/(\lambda^2 - 0.0245458)$.

Gallium arsenide (GaAs)

Wavelength λ, μm	n	Wavelength λ, μm	n
1.127	3.455	1.550	3.375
1.239	3.425	1.652	3.366
1.377	3.400		

Gallium phosphide (GaP)

Wavelength λ, μm	n	Wavelength λ, μm	n
0.5	3.4595	2.0	3.0379
0.6	3.3495	2.2	3.0331
0.7	3.2442	2.4	3.0296
0.8	3.1830	2.6	3.0271
0.9	3.1430	2.8	3.0236
1.0	3.1192	3.0	3.0215
1.1	3.0981	3.2	3.0197
1.2	3.0844	3.4	3.0181
1.4	3.0646	3.6	3.0166
1.6	3.0509	3.8	3.0159
1.8	3.0439	4.0	3.0137

Gallium selenide (GaSe) [65]

Wavelength λ, μm	n_o	n_e
0.6328	2.993±0.01	2.767±0.01
1.15	2.899±0.01	2.541±0.01
3.39	2.814±0.02	2.462±0.02
5.0	(2.8349)	(2.4611)
10.0	(2.8168)	(2.4597)
15.0	(2.8168)	(2.4527)

Data in parentheses are calculated according to dispersion equations:
$n_o^2 = -0.06 \cdot \lambda^{-4} + 0.526 \cdot \lambda^{-2} + 8.038 - 8.2 \cdot 10^{-4}\lambda^2 - 2.7 \cdot 10^{-6}\lambda^4$;
$n_e^2 = 6.06 + 0.5754/(\lambda^2 - 0.0453) - 1.04 \cdot 10^{-3}\lambda^2$.

Hexamethylenetetramine

Wavelength λ, μm	n	Wavelength λ, μm	n
0.4861	1.5984	0.5780	1.5899
0.5016	1.5953	0.5876	1.5893
0.5461	1.5917	0.6676	1.5856

Hippuric acid ($C_6H_5CO{\cdot}NH(CH_2CO_2H)$)

Wavelength λ, μm	n_α	n_β	n_γ
0.350	1.55	1.61	1.78
0.589	1.5348	1.5921	1.7598
0.700	1.534	1.589	1.755

Indium antimonide (InSb)

Wavelength λ, μm	n	Wavelength λ, μm	n
7.87	4.0	9.01	3.96
8.00	3.99	10.06	3.95

Indium antimonide (InSb) *(continued)*

Wavelength λ, μm	n	Wavelength λ, μm	n
11.01	3.93	16.96	3.86
12.06	3.92	17.85	3.85
12.98	3.91	18.85	3.84
13.90	3.90	19.98	3.82
15.13	3.88	21.15	3.81
15.79	3.87	22.20	3.80

Iodic acid (α-HIO_3)

Wavelength λ, μm	$n_{\alpha=a=X}$	$n_{\beta=c=Z}$	$n_{\gamma=b=Y}$
0.450	1.8798	2.0184	2.0560
0.500	1.8621	1.9930	2.0192
0.5325	1.8547	1.9829	2.0103
0.550	1.8497	1.9787	2.0049
0.600	1.8409	1.9665	1.9922
0.650	1.8452	1.9571	1.9812
0.700	1.8308	1.9505	1.9765
0.800	1.8350	1.9407	1.9672
0.850	1.8223	1.9378	1.9639
0.900	1.8206	1.9347	1.9595
0.950	1.8180	1.9318	1.9564
1.000	1.8147	1.9292	1.9537
1.065	1.8123	1.9275	1.9508
1.100	1.8116	1.9260	1.9484
1.200	1.8086	1.9230	1.9436

Dispersion equations [72]:
$n_x^2 = 2.5761 + 0.6973\lambda^2/[\lambda^2 - (0.2356)^2] - 0.0201\lambda^2$;
$n_y^2 = 2.4701 + 1.2054\lambda^2/[\lambda^2 - (0.2246)^2] - 0.0152\lambda^2$;
$n_z^2 = 2.6615 + 1.1316\lambda^2/[\lambda^2 - (0.2281)^2] - 0.0398\lambda^2$.

Lead niobate ($PbNb_4O_{11}$)

Wavelength λ, μm	$n_{\alpha=a=X}$	$n_{\beta=b=Y}$	$n_{\gamma=c=Z}$
0.4579	2.4754	2.4766	2.5047
0.4765	2.4554	2.4571	2.4845
0.4880	2.4445	2.4465	2.4735
0.4965	2.4371	2.4392	2.466
0.5017	2.4329	2.435	2.4618
0.5145	2.4231	2.4254	2.4518
0.5321	2.4113	2.4137	2.4396
0.6328	2.3644	2.3667	2.3922
1.0642	2.2979	2.301	2.3254

Dispersion equations:
$n_\alpha^2 - 1 = 4.124\lambda^2/[\lambda^2 - (0.202)^2]$;
$n_\beta^2 - 1 = 4.139\lambda^2/[\lambda^2 - (0.2011)^2]$;
$n_\gamma^2 - 1 = 4.246\lambda^2/[\lambda^2 - (0.2014)^2]$.

Lead titanate ($PbTiO_3$)

Wavelength λ, μm	n_o	n_e
0.4880	2.793	2.7744
0.5017	2.7742	2.7574
0.5145	2.7586	2.7431
0.5321	2.7398	2.7260
0.6328	2.6676	2.6594
1.0642	2.5712	2.5692
1.1520	2.5637	2.5623

Dispersion equations:
$n_o^2 - 1 = 5.359\lambda^2/[\lambda^2 - (0.224)^2]$;
$n_e^2 - 1 = 5.365\lambda^2/[\lambda^2 - (0.2170)^2]$.

Lithium formate monohydrate ($LiCHO_2{\cdot}H_2O$)

Wavelength λ, μm	$n_{\alpha=a=X}$	$n_{\beta=b=Y}$	$n_{\gamma=c=Z}$
0.230	1.4256	1.5987	1.6759
0.4579	1.3708	1.4901	1.5308
0.4765	1.3698	1.4883	1.5286
0.4880	1.3692	1.4873	1.5272
0.4965	1.3688	1.4866	1.5264
0.5017	1.3686	1.4862	1.5258
0.5145	1.3680	1.4851	1.5245
0.5321	1.3675	1.4838	1.5229
0.6328	1.3645	1.4784	1.5163
1.0642	1.3593	1.4673	1.5035

Dispersion equations:
$n_\alpha^2 - 1 = 0.8415\lambda^2/[\lambda^2 - (0.0953)^2]$;
$n_\beta^2 - 1 = 1.14106\lambda^2/[\lambda^2 - (0.1183)^2]$;
$n_\gamma^2 - 1 = 1.2454\lambda^2/[\lambda^2 - (0.12496)^2]$.

Lithium-gallium oxide ($LiGaO_3$)

Wavelength λ, μm	$n_{\alpha,\beta=Y,Z}$	$n_{\gamma=X}$
0.41	1.7702	1.8040
0.45	1.7570	1.7895
0.50	1.7466	1.7785
0.55	1.7395	1.7702
0.60	1.7343	1.7615
0.70	1.7268	1.7565
0.80	1.7218	1.7507
0.90	1.7185	1.7475
1.00	1.7160	1.7445
1.20	1.7122	1.7405
1.40	1.7095	1.7372
1.60	1.7070	1.7350
1.80	1.7045	1.7325
2.00	1.7025	1.7303
2.20	1.7005	1.7268
2.40	1.6978	1.7242
2.60	1.6955	1.7225
2.80	1.6925.	1.7200

Lithium iodate ($LiIO_3$) [73]

Wavelength λ, μm	n_o	n_e
0.4047	1.9443	1.7826
0.4358	1.9283	1.7706
0.5086	1.9037	1.7521
0.5461	1.8953	1.7457
0.5791	1.8892	1.7413
0.5896	1.8875	1.7400
0.6438	1.8807	1.7346
1.0140	1.8584	1.7176
1.1286	1.8552	
1.3674	1.8508	1.7122
1.5296	1.8482	1.7101
1.6920	1.8464	1.7089
1.9701	1.8431	1.7072
2.2493	1.8385	1.7050

Dispersion equations:
$n_o^2 = 2.083648 + 1.332068\lambda^2/[\lambda^2 - (0.1879)^2] - 0.00852\lambda^2$;
$n_e^2 = 1.673463 + 1.245229\lambda^2/[\lambda^2 - (0.168)^2] - 0.003641\lambda^2$.

Lithium niobate ($LiNbO_3$)

Wavelength λ, μm	n_o	n_e
0.42	2.4144	2.3638
0.45	2.3814	2.2765
0.50	2.3444	2.2446
0.55	2.3188	2.2241
0.60	2.3002	2.2083
0.70	2.2862	2.1964
0.80	2.2756	2.1874
0.90	2.2598	2.1741
1.00	2.2487	2.1647
1.20	2.2407	2.1580
1.40	2.2291	2.1481
1.60	2.2208	2.1410
1.80	2.2139	2.1351
2.00	2.2074	2.1297
2.20	2.2015	2.1244
2.40	2.1948	2.1187
2.60	2.1882	2.1138
2.80	2.1814	2.1080
3.00	2.1741	2.1020
3.20	2.1663	2.0955
3.40	2.1580	2.0886
3.60	2.1493	2.0814
3.80	2.1398	2.0735
4.00	2.1299	2.0652
4.20	2.1193	2.0564

Temperature-dependent dispersion equations:
$n_o^2 - 1 = 3.913 + \frac{1.173\cdot10^5 + 1.65\cdot10^{-2}T^2}{\lambda^2 - (2.12\cdot10^2 + 2.7\cdot10^{-5}T^2)^2} - 2.78\cdot10^{-8}\cdot\lambda^2$;
$n_e^2 - 1 = 3.5567 + 2.605\cdot10^{-7}T^2 + \frac{0.97\cdot10^5 + 2.7\cdot10^{-2}T^2}{\lambda^2 - (2.01\cdot10^2 + 5.4\cdot10^{-5}T^2)^2} - 2.24\cdot10^{-8}\cdot\lambda^2$.

Lithium sulfate monohydrate ($LiSO_4 \cdot H_2O$)

Wavelength λ, μm	n_α	n_β	n_γ
0.365	1.4771	1.4926	1.5029
0.4047	1.4722	1.4876	1.4980
0.4358	1.4693	1.4849	1.4951
0.4471	1.4686	1.4834	1.4941
0.4713	1.4670		1.4926
0.5016	1.4652	1.4802	1.4905
0.5461	1.4631	1.4782	1.4882
0.5780	1.4619	1.4772	1.4867
0.5876	1.4616	1.4766	1.4866
0.6678	1.4593	1.4743	1.4838
0.7016	1.4585		1.4831
1.014	1.4538	1.4678	1.4777
1.129	1.4525	1.4666	1.4761
1.367	1.4502	1.4636	1.4732
1.530	1.4485		1.4708
1.709	1.4466	1.4588	1.4676

Lithium tantalate ($LiTaO_3$)

Wavelength λ, μm	n_o	n_e
0.45	2.2420	2.2468
0.50	2.2160	2.2205
0.60	2.1834	2.1878
0.70	2.1652	2.1696
0.80	2.1538	2.1578
0.90	2.1454	2.1493
1.00	2.1391	2.1432
1.20	2.1305	2.1341
1.40	2.1236	2.1273
1.60	2.1174	2.1213
1.80	2.1120	2.1170
2.00	2.1066	2.1115
2.20	2.1009	2.1053
2.40	2.0951	2.0993
2.60	2.0891	2.0936
2.80	2.0825	2.0871
3.00	2.0755	2.0799
3.20	2.0680	2.0727
3.40	2.0601	2.0649
3.60	2.0513	2.0561
3.80	2.0424	2.0473
4.00	2.0335	2.0377

Lithium triborate (LiB_3O_5)

Dispersion equations[18]:

$n_x^2 = 2.4542 + \frac{0.01125}{\lambda^2 - 0.01135} - 0.01388\lambda^2;$

$n_y^2 = 2.5390 + \frac{0.01277}{\lambda^2 - 0.01189} - 0.01848\lambda^2;$

$n_z^2 = 2.5865 + \frac{0.01310}{\lambda^2 - 0.01223} - 0.01861\lambda^2.$

Magnesium-barium fluoride ($MgBaF_4$)

Dispersion equations [19]:

$n_X^2 = 2.077 + 0.0076/(\lambda^2 - 0.0079);$

$n_Y^2 = 2.1238 + 0.0086/\lambda^2;$

$n_Z^2 = 2.1462 + 0.00736/(\lambda^2 - 0.0090).$

Mercury sulfide (α-HgS)

Wavelength λ, μm	n_o	n_e
0.62	2.9028	3.2560
0.65	2.8655	3.2064
0.68	2.8384	3.1703
0.70	2.8224	3.1489
0.80	2.7704	3.0743
0.90	2.7383	3.0340
1.00	2.7120	3.0050
1.20	2.6884	2.9680
1.40	2.6730	2.9475
1.60	2.6633	2.9344
1.80	2.6567	2.9258
2.00	2.6518	2.9194
2.20	2.6483	2.9146
2.40	2.6455	2.9108
2.60	2.6433	2.9079
2.80	2.6414	2.9052
3.00	2.6401	2.9036
3.20	2.6387	2.9017
3.40	2.6375	2.9001
3.60	2.6358	2.8987
3.80	2.6353	2.8971
4.00	2.6348	2.8963
5.00	2.6267	2.8863
6.00	2.6233	2.8799
7.00	2.6156	2.8741
8.00	2.6112	2.8674
9.00	2.6066	2.8608
10.00	2.6018	2.8522
11.00	2.5914	2.8434

Dispersion equations [75]:

$n_o^2 = 6.9443 + 0.3665/(\lambda^2 - 0.1351) - 0.0019\lambda^2;$

$n_e^2 = 8.3917 + 0.5405/(\lambda^2 - 0.138) - 0.0027\lambda^2.$

Mercury thiogallate ($HgGa_2S_4$) [76]

Wavelength λ, μm	n_o	n_e
0.5495	2.6592	2.5979
0.6500	2.5796	2.5264
1.0760	2.477	2.432
3.540	2.439	2.398
11.000	2.369	2.329

Dispersion equations:

$n_o^2 = 6.20815221 + 63.70629851/(\lambda^2 - 225) + 0.23698804/(\lambda^2 - 0.09568646);$

$n_e^2 = 6.0090267 + 63.2806592/(\lambda^2 - 225) + 0.21489656/(\lambda^2 - 0.09214633).$

Meta-dinitrobenzene, MDB [20]

Wavelength λ, μm	$n_{X=a}$	$n_{Y=b}$	$n_{Z=c}$
0.436	1.803	1.736	1.508
0.532	1.759	1.698	1.491
0.633	1.738	1.680	1.483
1.064	1.709	1.654	1.471
1.152	1.707	1.652	1.470

Meta-nitroaniline, MNA ($NO_2C_6H_4NH_2$) [21]

Wavelength λ, μm	$n_{X=c}$	$n_{Y=b}$	$n_{Z=a}$
0.5123	1.718	1.750	1.810
0.530	1.705	1.738	1.798
0.6274	1.670	1.709	1.758
1.064	1.631	1.678	1.719
1.54	1.616	1.667	1.700

Dispersion equations [43]:
$n_X^2 = 2.469 + 0.1864\lambda^2/(\lambda^2 - 0.16) - 0.0199\lambda^2$;
$n_Y^2 = 2.6658 + 0.1626\lambda^2/(\lambda^2 - 0.1719) - 0.0212\lambda^2$;
$n_Z^2 = 2.8102 + 0.1524\lambda^2/(\lambda^2 - 0.175) - 0.0294\lambda^2$.

Methyl-(2,4-dinitriphenyl)-amino-2-propanoate, MAP ($C_{10}H_{11}N_3O_6$) [9]

Wavelength λ, μm	n_X	n_Y	n_Z
0.532	1.5568	1.7100	2.0353
1.064	1.5079	1.5991	1.8439

Dispersion equations:
$n_X^2 = 2.1713 + 0.10305\lambda^2/(\lambda^2 - 0.16951) - 0.01667\lambda^2$;
$n_Y^2 = 2.31 + 0.2258\lambda^2/(\lambda^2 - 0.17988) - 0.01886\lambda^2$;
$n_Z^2 = 2.7523 + 0.6079\lambda^2/(\lambda^2 - 0.1606) - 0.05361\lambda^2$.

2-Methyl-4-nitroaniline, MNA ($NO_2C_6H_3CH_3NH_2$) [10]

Wavelength λ, μm	n_X	n_Y
0.532	2.2	
0.6328	2.0±0.1	1.6±0.1
1.064	1.8	

3-Methyl-4-nitropyridine-1-oxide, POM [43]

Wavelength λ, μm	$n_{X=c}$	$n_{Y=a}$	$n_{Z=b}$
0.468	1.6875	1.8085	2.1134
0.509	1.6665	1.7656	2.0279

3-Methyl-4-nitropyridine-1-oxide *(continued)*

Wavelength λ, μm	$n_{X=c}$	$n_{Y=a}$	$n_{Z=b}$
0.532	1.6591	1.7500	1.9969
0.579	1.6487	1.7281	1.9530
0.644	1.6402	1.7091	1.9153
1.064	1.6242	1.6633	1.8287

Dispersion equations:
$n_X^2 = 2.4529 + 0.1641\lambda^2/(\lambda^2 - 0.128)$;
$n_Y^2 = 2.4315 + 0.3556\lambda^2/(\lambda^2 - 0.1276) - 0.0579\lambda^2$;
$n_Z^2 = 2.5521 + 0.7962\lambda^2/(\lambda^2 - 0.1289) - 0.0941\lambda^2$.

5-Nitrouracil [35]

Wavelength λ, μm	$n_{a=X}$	$n_{b=Y}$	$n_{c=Z}$
0.435	1.6351	2.0051	1.7797
0.468	1.6113	1.9737	1.7566
0.480	1.6065	1.9668	1.7500
0.504	1.5958	1.9537	1.7441
0.518	1.5894	1.9411	1.7375
0.546	1.5850	1.9315	1.7242
0.579	1.5787	1.9190	1.7176
0.584	1.5758	1.9135	1.7156
0.636	1.5694	1.9014	1.7070
0.644	1.5670	1.9010	1.7050
1.064	1.5341	1.8517	1.6799
1.320	1.5248	1.8362	1.6719

Dispersion equations:
$n_X = 2.098 + 0.29\lambda^2/(\lambda^2 - 0.0947) - 0.0485\lambda^2$;
$n_Y = 2.39 + 1.033\lambda^2/(\lambda^2 - 0.07) - 0.0549\lambda^2$;
$n_Z = 1.892 + 0.87\lambda^2/(\lambda^2 - 0.0599)$.

Potassium dideuterium phosphate, KDDP (KD_2PO_4)

Wavelength λ, μm	n_o	n_e
0.266	1.5546	1.5085
0.347	1.5278	1.4854
0.355	1.5263	1.4841
0.4047	1.5189	1.4776
0.4078	1.5185	1.4772
0.4358	1.5155	1.4747
0.4916	1.5111	1.4710
0.532	1.5085	1.4690
0.5461	1.5079	1.4683
0.5779	1.5063	1.4670
0.6234	1.5044	1.4656
0.6907	1.5022	1.4639
1.0000	1.4700	1.4400

Potassium dideuterium phosphate *(continued)*

Wavelength λ, μm	n_o	n_e
1.064	1.4928	1.4555
1.315	1.4867	1.4502

Dispersion equations [70]:

$n_o^2 = 2.240921 + 2.246956\lambda^2/[\lambda^2 - (11.26591)^2] + 0.009676/[\lambda^2 - (0.124981)^2]$;

$n_e^2 = 2.126019 + 0.784404\lambda^2/[\lambda^2 - (11.10871)^2] + 0.008578/[\lambda^2 - (0.109505)^2]$.

Potassium dihydrogen arsenate, KDA (KH_2AsO_4)

Wavelength λ, μm	n_o	n_e
0.4861	1.5762	1.5252
0.5460	1.5707	1.5206
0.5893	1.5674	1.5179
0.6563	1.5632	1.5146

Dispersion equations [70]:

$n_o^2 = 2.424647 + 3.742954\lambda^2/[\lambda^2 - (11.26515)^2] + 0.015841/[\lambda^2 - (0.13647)^2]$;

$n_e^2 = 2.262579 + 0.769288\lambda^2/[\lambda^2 - (11.27181)^2] + 0.013461/[\lambda^2 - (0.127145)^2]$.

Potassium dihydrogen phosphate, KDP (KH_2PO_4)

Wavelength λ, μm	n_o	n_e
0.2138560	1.60177	1.54615
0.2288018	1.58546	
0.2446950	1.57228	
0.2464068	1.67105	
0.2536519	1.56631	1.51586
0.2800869	1.55263	1.50416
0.2980628	1.54618	1.49824
0.3021499	1.54433	1.49708
0.3035781		1.49667
0.3125663	1.54117	1.49434
0.3131545	1.54098	1.49419
0.3341478		1.48954
0.3650146	1.52932	1.48432
0.3654833	1.52923	1.48423
0.3662878	1.52909	1.48409
0.3906410		1.48089
0.4046561	1.52341	1.47927
0.4077811	1.52301	1.47898
0.4358350	1.51900	1.47640
0.4916036		1.47254
0.5460740	1.51152	1.46982
0.5769580	1.50987	

Potassium dihydrogen phosphate *(continued)*

Wavelength λ, μm	n_o	n_e
0.5790654	1.50977	1.46856
0.6328160	1.50737	1.46685
1.013975	1.49535	1.46041
1.128704	1.49205	1.45917
1.152276	1.49135	1.45893
1.357070	1.48455	
1.523100		1.455.21
1.529525		1.45512

Dispersion equations:

$n_o^2 = 2.259276 + 0.1008956/[\lambda^2 - (77.26408)^{-1}] + 0.3251305/(0.0025 - \lambda^{-2})$;

$n_e^2 = 2.132668 + 0.008637494/[\lambda^2 - (81.42631)^{-1}] + 0.008069981/(0.0025 - \lambda^{-2})$.

Potassium dithionate ($K_2S_2O_6$)

Wavelength λ, μm	n_o	n_e
0.313	1.480	1.568
0.334	1.475	1.557
0.365	1.470	1.546
0.405	1.465	1.537
0.436	1.463	1.530
0.546	1.456	1.518
0.578	1.455	1.516
1.014	1.448	1.503
1.367	1.446	1.500
1.709	1.444	1.498
2.930	1.436	1.489
3.39	1.430	1.485

Potassium lithium niobate ($K_3Li_2Nb_5O_{15}$)

Wavelength λ, μm	n_o	n_e
0.4500	2.4049	2.2512
0.4750	2.3751	2.2315
0.5000	2.3546	2.2144
0.5250	2.3349	2.2010
0.5324	2.3260	2.1975
0.5500	2.3156	2.1900
0.5750	2.3016	2.1801
0.6000	2.2899	2.1720
0.6250	2.2799	2.1645
0.6500	2.2711	2.1586
0.6750	2.2631	2.1529

Dispersion equations:

$n_o^2 - 1 = 3.708\lambda^2/(\lambda^2 - 0.04601)$;

$n_e^2 - 1 = 3.349\lambda^2/(\lambda^2 - 0.03564)$.

Potassium malate ($COOH\text{-}CHOH\text{-}CH_2COOK\cdot1.5H_2O$)

Dispersion equations [11]:
$n_x^2 = 1.542 + 0.8299/[1 - (0.1419/\lambda)^2]$;
$n_y^2 = 1.470 + 0.7473/[1 - (0.1343/\lambda)^2]$;
$n_z^2 = 1.339 + 0.8519/[1 - (0.1195/\lambda)^2]$.

Potassium niobate ($KNbO_3$) [23]

Wavelength λ, μm	n_x	n_y	n_z
0.4880	2.4190	2.3526	2.2275
0.5145	2.3941	2.3329	2.2116
0.5321	2.3807	2.3224	2.2029
0.6328	2.3291	2.2799	2.1685
1.0642	2.2574	2.2200	2.1196

Dispersion equations:
$n_x^2 = 1 + 3.93281\lambda^2/[\lambda^2 - (0.2118)^2]$;
$n_y^2 = 1 + 3.79361\lambda^2/[\lambda^2 - (0.1969)^2]$;
$n_z^2 = 1 + 3.38362\lambda^2/[\lambda^2 - (0.1857)^2]$.

Potassium-sodium-barium niobate ($K_xNa_{1-x}Ba_2Nb_5O_{15}$), 22°C

Dispersion equations:
$n_o^2 = 3.6680 + 24.681/[(4.3004)^2 - (1.2394/\lambda)^2]$;
$n_e^2 = 2.9198 + 46.737/[(5.1605)^2 - (1.2394/\lambda)^2]$.

Potassium tartrate hemihydrate ($K_2C_4O_6\cdot1/2H_2O$)

Wavelength λ, μm	n_α	n_β	n_γ
0.3650	1.5156	1.5487	1.5630
0.4047	1.5090	1.5409	1.5541
0.4358	1.5049	1.5368	1.5494
0.5461	1.4961	1.5271	1.5384
0.5780	1.4945	1.5253	1.5363
1.014	1.4846	1.5142	1.5238
1.129	1.4832	1.5127	1.5218
1.367	1.4809	1.5102	1.5183

Potassium titanyl phosphate, KTP ($KTiOPO_4$) [78]

Wavelength λ, μm	n_x	n_y	n_z
0.4579	1.7987	1.8093	1.9213
0.4765	1.7923	1.8024	1.9112
0.4880	1.7888	1.7987	1.9058
0.4965	1.7865	1.7961	1.9022
0.5017	1.7851	1.7948	1.9001
0.5145	1.7818	1.7915	1.8952
0.5320	1.7782	1.7874	1.8892

Potassium titanyl phosphate *(continued)*

Wavelength λ, μm	n_x	n_y	n_z
0.5461	1.7755	1.7843	1.8849
0.5770	1.7701	1.7784	1.8767
0.5791	1.7697	1.7780	1.8762
0.6234	1.7636	1.7713	1.8669
0.6328	1.7624	1.7701	1.8652
0.6471	1.7605	1.7681	1.8628
0.6764	1.7574	1.7647	1.8584
0.6943	1.7559	1.7632	1.8559
1.0640	1.7376	1.7430	1.9297
1.1520	1.7356	1.7406	1.8265

Dispersion equations:
$n_x^2 = 1.71543 + 1.2771\lambda^2/[\lambda^2 - (0.18346)^2] - 0.01047\lambda^2$;
$n_y^2 = 1.56353 + 1.44806\lambda^2/[\lambda^2 - (0.18012)^2] - 0.0132\lambda^2$;
$n_z^2 = 1.99703 + 1.3115\lambda^2/[\lambda^2 - (0.2182)^2] - 0.01611\lambda^2$.

Proustite (Ag_3AsS_3)

Wavelength λ, μm	n_o	n_e
0.5876		2.7896
0.6328	3.0190	2.7391
0.6678	2.9804	2.7094
1.0140	2.8264	2.5901
1.1290	2.8067	2.5756
1.3670	2.7833	2.5570
1.530	2.7728	2.5485
1.709	2.7654	2.5423
2.50	2.7478	2.5282
3.56	2.7379	2.5213
4.62	2.7318	2.5178

Dispersion equations [33]:
$n_o^2 = 7.4822 + 0.4635\lambda^2/(\lambda^2 - 0.116) - 0.0016\lambda^2$;
$n_e^2 = 6.3434 + 0.3352\lambda^2/(\lambda^2 - 0.1117) - 0.0007\lambda^2$.

Pyrargyrite (Ag_3SbS_3)

Dispersion equations:
$n_o^2 - 1 = 6.585\lambda^2/[\lambda^2 - (0.4)^2] + 0.1133\lambda^2/[\lambda^2 - (15)^2]$;
$n_e^2 - 1 = 5.845\lambda^2/[\lambda^2 - (0.4)^2] + 0.0202\lambda^2/[\lambda^2 - (15)^2]$.

Quartz (α-SiO_2)

Wavelength λ, μm	n_o	n_e
0.185	1.65751	1.68988
0.198	1.65087	1.66394
0.231	1.61395	1.62555
0.340	1.56747	1.57737
0.394	1.55846	1.56805
0.434	1.55396	1.56339

Quartz (α-SiO_2) *(continued)*

Wavelength λ, μm	n_o	n_e
0.508	1.54822	1.55746
0.5893	1.54424	1.55335
0.7680	1.53903	1.54794
0.8325	1.53773	1.54661
0.9914	1.53514	1.54392
1.1592	1.53283	1.54152
1.3070	1.53090	1.53951
1.3958	1.52977	1.53832
1.4792	1.52865	1.53716
1.5414	1.52781	1.53630
1.6815	1.52583	1.53422
1.7614	1.52468	1.53301
1.9457	1.52184	1.53004
2.0531	1.52005	1.52823
2.3000	1.51561	
2.6000	1.50986	
3.0000	1.49953	1.507
3.5000	1.48451	
4.0000	1.46671	
4.2000	1.4569	
5.0000	1.417	
6.4500	1.274	
7.0000	1.167	

Rubidium dihydrogen arsenate, RDA (RbH_2AsO_4)

Dispersion equations [70]:

$n_o^2 = 2.390661 + 3.487176\lambda^2/[\lambda^2 - (11.25899)^2]$
$+ 0.015513/[\lambda^2 - (0.134582)^2]$;

$n_e^2 = 2.27557 + 0.720099\lambda^2/[\lambda^2 - (11.25304)^2]$
$+ 0.013915/[\lambda^2 - (0.1208)^2]$.

Rubidium dihydrogen phosphate, RDP (RbH_2PO_4)

Wavelength λ, μm	n_o	n_e
0.35	1.5284	1.4969
0.4765	1.5140	1.4861
0.4880	1.5132	1.4832
0.4965	1.5126	1.4827
0.5017	1.5121	1.4825
0.5145	1.5116	1.4820
0.5321	1.5106	1.4811
0.6328	1.4976	1.4775
1.0642	1.4926	1.4700

Dispersion equations:

$n_o^2 - 1 = 1.2068\lambda^2/(\lambda^2 - 0.01539)$;

$n_e^2 - 1 = 1.15123\lambda^2/(\lambda^2 - 0.010048)$.

Selenium (Se)

Wavelength λ, μm	n_o	n_e
1.06	2.790±0.008	3.608±0.008
1.15	2.737±0.008	3.573±0.008
3.39	2.650±0.01	3.460±0.01
10.60	2.64±0.01	3.41±0.01

Silicon carbide (SiC)

Wavelength λ, μm	n_o	n_e
0.4880	2.6916	2.7423
0.5017	2.6837	2.7337
0.5145	2.6771	2.7261
0.5321	2.6689	2.7167
0.6328	2.6351	2.6794
1.0642	2.5830	2.6225

Dispersion equations:

$n_o^2 - 1 = 5.5515\lambda^2/[\lambda^2 - (0.1625)^2]$;

$n_e^2 - 1 = 5.7382\lambda^2/[\lambda^2 - (0.16897)^2]$.

Silver thiogallate ($AgGaS_2$) [53], 20°C

Wavelength λ, μm	n_o	n_e
0.490	2.7148	2.7287
0.500	2.6916	2.6867
0.525	2.6503	2.6239
0.550	2.6190	2.5834
0.575	2.5944	2.5537
0.600	2.5748	2.5303
0.625	2.5577	2.5116
0.650	2.5437	2.4961
0.675	2.5310	2.4824
0.700	2.5205	2.4706
0.750	2.5049	2.4540
0.800	2.4909	2.4395
0.850	2.4802	2.4279
0.900	2.4716	2.4192
0.950	2.4644	2.4118
1.0	2.4582	2.4053
1.1	2.4486	2.3954
1.2	2.4414	2.3881
1.3	2.4359	2.3819
1.4	2.4315	2.3781
1.5	2.4280	2.3745
1.6	2.4252	2.3716
1.8	2.4206	2.3670
2.0	2.4164	2.3637
2.2	2.4142	2.3604
2.4	2.4119	2.3583
2.6	2.4102	2.3567

Silver thiogallate *(continued)*

Wavelength λ, μm	n_o	n_e
2.8	2.4094	2.3559
3.0	2.4080	2.3545
3.2	2.4068	2.3534
3.4	2.4062	2.3522
3.6	2.4046	2.3511
3.8	2.4024	2.3491
4.0	2.4024	2.3488
4.5	2.4003	2.3461
5.0	2.3955	2.3419
5.5	2.3938	2.3401
6.0	2.3908	2.3369
6.5	2.3874	2.3334
7.0	2.3827	2.3291
7.5	2.3787	2.3252
8.0	2.3757	2.3219
8.5	2.3699	2.3163
9.0	2.3663	2.3121
9.5	2.3606	2.3064
10.0	2.3548	2.3012
10.5	2.3486	2.2948
11.0	2.3417	2.2880
11.5	2.3329	2.2789
12.0	2.3266	2.2716
12.5	2.3177	
13.0	2.3076	

Dispersion equations [71]:
n_o^2=3.628+2.1686$\lambda^2/(\lambda^2-0.1003)$+2.1753$\lambda^2/(\lambda^2-950)$;
n_e^2=4.0172+1.5274$\lambda^2/(\lambda^2-0.131)$+2.1699$\lambda^2/(\lambda^2-950)$.

Sodium bromate ($NaBrO_3$)

Dispersion equations:
$$n^2-1=1.3194\lambda^2/[\lambda^2-(0.09)^2] + 0.2357\lambda^2/[\lambda^2-(0.2)^2]-0.0174\lambda^2.$$

Sodium chlorate ($NaClO_3$)

Wavelength λ, μm	n	Wavelength λ, μm	n
0.2310	1.616	0.4862	1.522
0.2573	1.585	0.5173	1.519
0.2748	1.572	0.5892	1.515
0.3256	1.549	0.6563	1.513
0.3404	1.544	0.6867	1.512
0.3467	1.542	0.7188	1.511
0.3611	1.539		

Dispersion equations:
$$n^2-1=1.1825\lambda^2/[\lambda^2-(0.09)^2] + 0.07992\lambda^2/[\lambda^2-(0.185)^2]-0.00864\lambda^2.$$

Sodium formate ($NaCHO_2$) [30]

Wavelength λ, μm	n_X	n_Y	n_Z
0.355	1.4037	1.4867	1.5794
0.532	1.3896	1.4651	1.5472
1.064	1.3814	1.4531	1.5296

Dispersion equations:
$n_X^2=1.2646+0.6381/[1-(0.1101/\lambda)^2]-0.0011\lambda^2$;
$n_Y^2=1.2589+0.8423/[1-(0.1203/\lambda)^2]-0.0005\lambda^2$;
$n_Z^2=1.2515+1.0729/[1-(0.1314/\lambda)^2]-0.0013\lambda^2$.

Sodium nitrite ($NaNO_2$)

Wavelength λ, μm	$n_{\alpha=a=X}$	$n_{\beta=b=Y}$	$n_{\gamma=c=Z}$
0.4358	1.3531	1.4212	1.6900
0.4800	1.3500	1.4166	1.6750
0.5086	1.3484	1.4158	1.6685
0.5461	1.3470	1.4137	1.6620
0.5791	1.3458	1.4122	1.6567
0.5889	1.3455	1.4120	1.6555
06438	1.3442	1.4105	1.6510

Sucrose ($C_{12}H_{22}O_{11}$) [12]

Wavelength λ, μm	n_X	n_Y	n_Z
0.532	1.5404	1.5681	1.5737
1.064	1.5278	1.5552	1.5592

Dispersion equations:
$n_X^2=1.8719+0.4660\lambda^2/(\lambda^2-0.0214)-0.0113\lambda^2$;
$n_Y^2=1.9703+0.4502\lambda^2/(\lambda^2-0.0238)-0.0101\lambda^2$;
$n_Z^2=2.0526+0.3909\lambda^2/(\lambda^2-0.0252)-0.0187\lambda^2$.

Tellurium (Te)

Wavelength λ, μm	n_o	n_e
4.0	4.929	6.372
5.0	4.864	6.316
6.0	4.838	6.286
7.0	4.821	6.270
8.0	4.809	6.257
9.0	4.802	6.253
10.0	4.796	6.246
12.0	4.789	6.237
14.0	4.785	6.230
28	4.7809	6.2263

Dispersion equations [71]:
n_o^2=18.5346+4.3289$\lambda^2(\lambda^2$–3.981)+3.78$\lambda^2(\lambda^2$–11.813);
n_e^2=29.5222+9.3068$\lambda^2(\lambda^2$–2.5766)+9.235$\lambda^2(\lambda^2$–13.521).

Terbium molybdate ($Tb_2(MoO_4)_3$)

Wavelength λ, μm	$n_{\alpha=b=Y}$	$n_{\beta=a=X}$	$n_{\gamma=c=Z}$
0.4579	1.8864	1.8867	1.9433
0.4765	1.8797	1.8800	1.9358
0.4880	1.8760	1.8764	1.9316
0.4965	1.8734	1.8739	1.9288
0.5017	1.8720	1.8724	1.9271
0.5145	1.8687	1.8690	1.9232
0.5321	1.8645	1.8649	1.9185
0.6328	1.8476	1.8482	1.8993
1.0642	1.8222	1.8226	1.8704

Dispersion equations:
$n_\alpha^2 - 1 = 2.27241\lambda^2/(\lambda^2 - 0.023359)$;
$n_\beta^2 - 1 = 2.273955\lambda^2/(\lambda^2 - 0.02333)$;
$n_\gamma^2 - 1 = 2.44301\lambda^2/(\lambda^2 - 0.025133)$.

d-Threonine

Wavelength λ, μm	$n_{\alpha=X}$	$n_{\beta=Y}$	$n_{\gamma=Z}$
0.4579	1.5299	1.6039	1.6125
0.4765	1.5282	1.6017	1.6100
0.4880	1.5272	1.6004	1.6087
0.4965	1.5266	1.5996	1.6077
0.5017	1.5263	1.5991	1.6072
0.5145	1.5254	1.5979	1.6059
0.5321	1.5243	1.5965	1.6043
0.6328	1.5196	1.5998	1.5974
1.0642	1.5114	1.5788	1.5855

Dispersion equations:
$n_\alpha^2 - 1 = 1.273\lambda^2/[\lambda^2 - (0.1032)^2]$;
$n_\beta^2 - 1 = 1.477\lambda^2/[\lambda^2 - (0.1137)^2]$;
$n_\gamma^2 - 1 = 1.497\lambda^2/[\lambda^2 - (0.1169)^2]$.

Tourmaline

Wavelength λ, μm	n_o	n_e
0.4765	1.6474	1.6273
0.4880	1.6465	1.6263
0.4965	1.6457	1.6255
0.5017	1.6454	1.6251
0.5145	1.6446	1.6248
0.5321	1.6433	1.6231
0.6328	1.6378	1.6183
1.0642	1.6274	1.6088

Dispersion equations:
$n_o^2 - 1 = 1.6346\lambda^2/(\lambda^2 - 0.010734)$;
$n_e^2 - 1 = 1.57256\lambda^2/(\lambda^2 - 0.011346)$.

Urea ($CO(NH_2)_2$) [79]

Wavelength λ, μm	n_o	n_e
0.213	1.7308	2.0155
0.266	1.5777	1.7575
0.355	1.5207	1.6580
0.532	1.4939	1.6098
1.064	1.4811	1.5830

Dispersion equations:
$n_o = 2.1678 + 0.0139/(\lambda^2 - 0.0207)$;
$n_e = 2.4917 + 0.0141/(\lambda^2 - 0.0240)$.

Urotropine ($(CH_2)_6N_4$)

Wavelength λ, μm	n	Wavelength λ, μm	n
0.4861	1.5984	0.5780	1.5899
0.5016	1.5953	0.5876	1.5893
0.5461	1.5917	0.6676	1.5856

Zinc germanium phosphide ($ZnGeP_2$)

Wavelength λ, μm	n_o	n_e
0.64	3.5052	3.5802
0.66	3.4756	3.5467
0.68	3.4477	3.5160
0.70	3.4233	3.4885
0.75	3.3730	3.4324
0.80	3.3357	3.3915
0.85	3.3063	3.3593
0.90	3.2830	3.3336
0.95	3.2638	3.3124
1.00	3.2478	3.2954
1.10	3.2232	3.2688
1.20	3.2054	3.2493
1.30	3.1924	3.2346
1.40	3.1820	3.2244
1.60	3.1666	3.2077
1.80	3.1562	3.1965
2.00	3.1490	3.1889
2.20	3.1433	3.1829
2.40	3.1388	3.1780
2.60	3.1357	3.1745
2.80	3.1327	3.1717
3.00	3.1304	3.1693
3.20	3.1284	3.1671
3.40	3.1263	3.1647
3.60	3.1257	3.1632
3.80	3.1237	3.1616
4.00	3.1223	3.1608
4.20	3.1209	3.1595
4.50	3.1186	3.1561
4.70	3.1174	3.1549

Zinc germanium phosphide *(continued)*

Wavelength λ, μm	n_o	n_e
5.00	3.1149	3.1533
5.50	3.1131	3.1518
6.00	3.1101	3.1480
6.50	3.1057	3.1445
7.00	3.1040	3.1420
7.50	3.0994	3.1378
8.00	3.0961	3.1350
8.50	3.0919	3.1311
9.00	3.0880	3.1272
9.50	3.0836	3.1231
10.00	3.0788	3.1183
10.50	3.0738	3.1137
11.00	3.0689	3.1087
11.50	3.0623	3.1008
12.00	3.0552	3.0949

Dispersion equations [80]:
$n_o^2 = 4.4733 + 5.26576\lambda^2(\lambda^2 - 0.13381)$
$+ 1.49085\lambda^2(\lambda^2 - 662.55)$;
$n_e^2 = 4.63318 + 5.34215\lambda^2(\lambda^2 - 0.14255)$
$+ 1.45795\lambda^2(\lambda^2 - 662.55)$.

Zinc oxide (ZnO)

Wavelength λ, μm	n_o	n_e
0.45	2.1058	2.1231
0.50	2.0511	2.0681
0.60	1.9985	2.0147
0.70	1.9735	2.9897
0.80	1.9597	2.9752
0.90	1.9493	2.9654
1.00	1.9435	2.9589
1.20	1.9354	2.9500
1.40	1.9298	2.9429
1.60	1.9257	2.9402
1.80	1.9226	2.9370
2.00	1.9197	2.9330
2.20	1.9173	2.9313
2.40	1.9152	2.9297
2.60	1.9128	2.9265
2.80	1.9100	2.9251
3.00	1.9075	2.9214
3.20	1.9049	2.9186
3.40	1.9022	2.9160
3.60	1.8994	2.9127
3.80	1.8964	2.9101
4.00	1.8891	2.9068

Zinc selenide (ZnSe)

Wavelength λ, μm	n	Wavelength λ, μm	n
0.589	2.61	1.5	2.45
1.0	2.48	2.0	2.44

Zinc sulfide (ZnS)

Wavelength λ, μm	n_o	n_e
0.360	2.705	2.709
0.375	2.637	2.640
0.400	2.560	2.564
0.410	2.539	2.544
0.420	2.522	2.525
0.425	2.511	2.514
0.430	2.502	2.505
0.440	2.486	2.488
0.450	2.473	2.477
0.460	2.459	2.463
0.470	2.448	2.453
0.475	2.445	2.449
0.480	2.438	2.443
0.490	2.428	2.433
0.500	2.421	2.425
0.525	2.402	2.407
0.550	2.386	2.392
0.575	2.375	2.378
0.600	2.363	2.368
0.625	2.354	2.358
0.650	2.346	2.350
0.675	2.339	2.343
0.700	2.332	2.337
0.800	2.324	2.328
0.900	2.310	2.315
1.000	2.301	2.303
1.200	2.290	2.294
1.400	2.285	2.288

Zinc telluride (ZnTe), 25°C

Wavelength λ, μm	n
0.569	3.111
0.577	3.085
0.579	3.079
0.589	3.054
0.600	3.035
0.616	3.005
0.650	2.962
0.700	2.913
0.725	2.893
0.750	2.879
0.760	2.871
0.770	2.866
0.800	2.853
1.000	2.790
1.200	2.758
1.300	2.748
1.400	2.741
1.500	2.734
1.515	2.734
2.060	2.71

33.8 Stimulated Scattering

The harmonic, subharmonic, or mixture frequencies, which are generated in a medium due to $\chi^2 \neq 0$, are determined only by the frequency composition of the applied light beam. The medium responds coherently to the electromagnetic disturbance, but in a passive manner: no detectable exchange of quanta occurs between the light waves and the medium. Nonetheless, several classes of nonlinear optical phenomena exist, in which the optical medium is not a passive element, but imposes its own characteristic frequencies on that of the light. These processes also involve coherent response to the electromagnetic wave, but in these cases, the internal structures of their optically responsive electron systems are modified periodically by internal vibrations, typically mechanical in nature. They accordingly introduce a periodically varying susceptibility into electromagnetic field, thus modulating the scattered light wave. Moreover, the optical electromagnetic field can excite and amplify (stimulate) the vibrations in the medium if its intensity exceeds a threshold value. The excitation of the acoustic vibrations results in *stimulated Brillouin scattering* (SBS); the excitation of the internal molecular vibrations is responsible for the *stimulated Raman scattering* (SRS).

In the simplest case when the frequency of the excited acoustic or molecular vibrations is Ω, and the light wave frequency is ω, the scattered light frequency takes the form

$$\omega' = \omega \pm n\Omega, \tag{33.33}$$

where n is an integer. The minus sign in the above relation refers to the Stokes process (downshift in frequency), while the plus sign refers to the anti-Stokes process (upshift in frequency of the light wave).

If Ω corresponds to SBS (acoustic vibrations), then

$$\Omega_{\mathrm{SBS}} = 2\omega(v_s/c)\sin(\theta/2), \tag{33.34}$$

where v_s and c are the sound and light velocities in the medium, and θ is the observation angle measured from the incident light wave direction. The maximum frequency shift during the SBS occurs in the backward direction ($\theta = \pi$) and typically $\Omega_{\mathrm{SBS}} \approx 10^{10}$ Hz or 1 cm^{-1}.

In the case of SRS (internal molecular vibrations), corresponding Ω_{SRS} depend exclusively on the molecular structure. In a general way, complex molecules possess many different vibrational modes with various frequencies Ω_{SRS}, Ω'_{SRS}, etc. In that event the frequency of a scattered light is given by

$$\omega' = \omega \pm n\Omega_{\mathrm{SRS}} \pm n'\Omega'_{\mathrm{SRS}} \pm \cdots, \tag{33.35}$$

n, n′, ... are integer numbers. Some substances exhibiting SRS are listed in Table 33.21 [81] together with their characteristic frequencies.

Table 33.21 Substances active in stimulated Raman scattering

Substance	Frequency shift, cm^{-1}	Substance	Frequency shift, cm^{-1}
Gases		**Liquids**	
Oxygen	1552	Bromoform	222
Potassium vapor	2721	Tetrachloroethylene	447
Methane	2916	Carbon tetrachloride	460
Deuterium	2991	Ethyl iodide	497
Hydrogen	4155	Hexafluorobenzene	515

Table 33.21 Substances active in stimulated Raman scattering *(continued)*

Substance	Frequency shift, cm^{-1}	Substance	Frequency shift, cm^{-1}
Liquids			
Bromoform	539	*o*-Dichlorobenzene	2202
Trichloroethylene	640	Benzonitrile	2229
Carbon disulfide	656	Acetonitrile	2250
Chloroform	667	1,2-Dimethylaniline	2292
o-Xylene	730	Methylcyclohexane	2817
α-Dimethylphenethylamine	836	Methanol	2831
Dioxane	836	*cis,trans*-1,3-Dimethylcyclohexane	2844
Morpholine	841	Tetrahydrofuran	2849
Triophenol	916	Cyclohexane	2852
Nitromethane	927	*cis*-1,2-Dimethylcyclohexane	2854
Deuterated benzene	944	α-Dimethylphenethylamine	2856
Cumene	990	Dioxane	2856
1,3-Dibromobenzene	990	Cyclohexane	2863
Benzene	992	Cyclohexanone	2863
Pyridine	992	*cis,trans*-1,3-Dimethylcyclohexane	2870
Aniline	997	*cis*-1,4-Dimethylcyclohexane	2873
Styrene	998	Cyclohexane	2884
m-Toluidine	999	Dichloromethane	2902
Bromobenzene	1000	Morpholine	2902
chlorobenzene	1001	2-Octene	2908
Benzonitrile	1002	2,3-Dimethyl-1,5-hexadiene	2910
tert-Butylbenzene	1002	Limonene	2910
Ethylbenzene	1002	*o*-Xylene	2913
Toluene	1004	1-Hexyne	2915
Fluorobenzene	1012	*cis*-2-Heptene	2920
γ-Picoline	1016	Mesitylene	2920
m-Cresol	1029	2-Bromopropane	2920
m-Dichlorobenzene	1030	Acetone	2921
1-Fluoro-2-chlorobenzene	1030	Ethanol	2921
Iodobenzene	1070	Carvone	2922
Benzoyl chloride	1086	*cis*-1,2-Dimethylcyclohexane	2927
Benzaldehyde	1086	Dimethylformamide	2930
Anisole	1097	2-Chloro-2-methylbutane	2931
Pyrrole	1178	2-Octene	2931
Furan	1180	*cis,trans*-1,3-Dimethylcyclohexane	2931
Styrene	1315	*m*-Xylene	2933
Nitrobenzene	1344	1,2-Diethyltartrate	2933
1-bromonaphthalene	1368	*o*-Xylene	2933
1-chloronaphthalene	1368	Piperidine	2933
2-Ethylnaphthalene	1381	1,2-Diethylbenzene	2934
m-Nitrotoluene	1389	2-Chloro-2-methylbutene	2935
Quinoline	1427	1-Bromopropane	2935
bromocyclohexane	1438	Piperidine	2936
Furan	1522	Tetrahydrofuran	2939
Methyl salicylate	1612	Piperidine	2940
Cinnamaldehyde	1624	Cyclohexanone	2945
Styrene	1629	2-Nitropropane	2948
3-Methylbutadiene	1638	1,2-Diethyl carbonate	2955
Pentadiene	1655	1,2-Dichloroethane	2956
Isoprene	1792	*trans*-Dichloroethylene	2956
1-Hexyne	2116	1-Bromopropane	2962

Table 33.21 Substances active in stimulated Raman scattering *(continued)*

Substance	Frequency shift, cm^{-1}	Substance	Frequency shift, cm^{-1}
Liquids		**Solids**	
2-Chloro-2-methylbutane	2962	Quartz	128
α-Dimethylphenethylamine	2967	Lithium niobate	152
Dioxane	2967	α-Sulfur	216
Cyclohexanol	2982	Lithium niobate	248
Cyclopentane	2982	Quartz	466
Cyclopentanol	2982	α-Sulfur	470
Bromocyclopentane	2082	Lithium niobate	628
o-Dichlorobenzene	2982	Calcium tungstate	911
p-Chlorotoluene	2982	Stilbene	997
α-Picoline	2982	Polystyrene	1001
p-Xylene	2988	Calcite	1084
o-Xylene	2992	Diamond	1332
Dibutyl-phthalate	2992	Naphthalene	1380
1,1,1-Trichloroethane	3018	Stilbene	1591
Ethylene chlorohydrin	3022	Triglycine sulfate	2422
Isophorone	3022	Triglycine sulfate	2702
Nitrosodimethylamine	3022	Triglycine sulfate	3022
Propylene glycol	3022	Polystyrene	3054
Cyclohexane	3038		
Styrene	3056		
Benzene	3064		
tert-Butyl benzene	3064		
1-Fluoro-2-chlorobenzene	3084		
Turpentine	3090		
Pseudocumene	3093		
Acetic acid	3162		
Acetonylacetone	3162		
Methyl methacrylate	3162		
γ-Picoline	3182		
Aniline	3300		
Water	3651		

References

[1] Kaminow, I. P., and Turner, E. H., Linear electrooptical materials. in: Handbook of lasers, Ed. Pressley, R.J., CRC Press, Cleveland, 1971, p.447.

[2] Di Chen, Magnetooptical materials. in: Handbook of lasers, Ed. Pressley, R. J., CRC Press, Cleveland, 1971, p.460.

[3] Volkenstein, M. V., Molecular optics, Gostekhizdat, Moscow–Leningrad, 1951 (in Russian).

[4] Pinnow, D. A., Elastooptical materials. in: Handbook of lasers, Ed. Pressley, R.J., CRC Press, Cleveland, 1971, p.478.

[5] Dixon, R. W., Cohen, M. G., Appl. Phys. Lett., 8, 205, 1966.

[6] Handbook of chemist, Khimiya, Moscow–Leningrad, 1965 (in Russian).

[7] Singh, S., Non-linear optical materials, in: Handbook of lasers, Ed. Pressley, R. J., CRC Press, Cleveland, 1971, p.489.

[8] Eimerl, D., Velsko, S., Potopowitz, J., VanUitert, L., IEEE J. Quantum Electron., 25, 179, 1989.

[9] Oudar, J. L., Hierle, R., J. Appl. Phys., 48, 2699, 1977.

[10] Levine, B. F., Bethea, C. G., Thurmond, C. D., Lynch, R. T., Bernstein, J. L., J. Appl. Phys., 50, 2523, 1979.

[11] Schuler, L., Betzler, K., Hesse, H., Opt. Commun., 43, 157, 1982.

[12] Halbout, J.-M., Tang, C. L., IEEE J. Quantum Electron., QE-18, 410, 1982.

[13] Rosker, M. J., Tang, C. L., IEEE J. Quantum Electron., QE–20, 334, 1984.

[14] Cheng, L. T., Cheng, L. K., Bierlein, J. D., Zumsteg, F. C., Appl. Phys. Lett., 63, 2618, 1993.

[15] Gunter, P., Bosshard, Ch., Sutter, K., Chapuis, G., Twieg, R. J., Dobrowolski, D., Appl. Phys. Lett., 50, 486, 1987.

[16] Dmitriev, V. G., Gurzadyan, G. G., Nikogosyan, D. N., Handbook of nonlinear optical crystals, Springer Verlag, Berlin/Heidelberg, 1991.

[17] Chen, C., Wu, Y., Jiang, A., Wu, B., You, G., Li, R., Liu, S., J. Opt. Soc. Amer., B6, 616, 1989.

[18] Kato, K., IEEE J. Quantum Electron., 26, 1173, 1990.

[19] Bechtold, P. S., Haussuhl, S., Appl. Phys., 14, 403, 1977.

[20] Belikova, G. S., Golovei, M. P., Shigorin, V. D., Shipulo, G. P., Opt. Spectrosc., 38, 779, 1975 (in Russian).

[21] Davydov, B. L., Zolin, V. F., Koreneva, L. G., Lavrovsky, E. A., Opt. Spectrosc., 39, 403, 1975.

[22] Kato, K., IEEE J. Quantum Electron. QE–16, 1288, 1980.

[23] Uematsu, Y., Jpn. J. Appl. Phys., 13, 1362, 1974.

[24] Dewey, H. J., IEEE J. Quantum Electron., QE–12, 303, 1976.

[25] Paisner, J. A., Spaeth, M. L., Gerstenberger, D. C., Appl. Phys. Lett., 32, 476, 1978.

[26] Boulanger, B., Feve, J.P., Marnier, G., Menaert, B., Cabirol, X., Villeval, P., Bonnin, C., in: Advanced solid–state lasers and compact blue–green laser digest 1993, vol. 1, (OSA, Washington, DC, 1993), p.35.

[27] Kato, K., IEEE J. Quantum Electron., 27, 1137, 1991.

[28] Kato, K., IEEE J. Quantum Electron., 28, 1974, 1992.

[29] Zumsteg, F. C., Bierlein, J. D., Gier, T. E., J. Appl. Phys., 47, 4980, 1976.

[30] Ito, H., Naito, H., Inaba, H., IEEE J. Quantum Electron., QE–10, 247, 1974.

[31] Kato, K., IEEE J. Quantum Electron., QE–19, 893, 1983.

[32] Yicheng Wu, Sasaki, T., Nakai, S., Yokotani, A., Honggao Tang, and Chuangtian Chen, Appl. Phys Lett., 62, 2614, 1993.

[33] Andrews, R. A., IEEE J. Quantum Electron., 6, 68, 1970.

[34] Zyss, J., Hemla, D. S., and Nicoud, J. F., J. Chem. Phys., 74, 4800, 1981.

[35] Puccetti, G., Perigaud, A., Badan, J., Ledoux, I., Zyss, J., J. Opt. Soc. Am., B10, 733, 1993.

[36] Day, G. W., Appl. Phys. Lett., 18, 347, 1971.

[37] Sherman, G. H., Coleman, P. D., Appl. Phys., 44, 238, 1973.

[38] Levine, B. F., Bethea, C. G., Appl. Phys. Lett., 20, 271, 1972.

[39] Sherman, G. H., Coleman, P. D., J. Quantum Electron., QE–9, 403, 1973.

[40] Eimerl, D., Davis, L., Velsko, S., Graham, E. K., Zalkin, A., J. Appl. Phys., 62, 1968, 1987.

[41] Kuzminov, Yu. S., Lithium niobate and lithium tantalate, Nauka, Moscow, 1975. (in Russian).

[42] Choy, M. M., Byer, R. L., Phys. Rev., B14, 1693, 1976.

[43] Kurtz, S. K., Jerphagnon, J., Choy, M. M., Nonlinear dielectric susceptibility, in: Landolt-Böernstein New Series, Group III; Springer Verlag, Berlin, 1979, vol. 11, pp. 671–743, Jerphagnon, J., Kurtz, S. K., Oudar, J. L., Nonlinear dielectric susceptibilities, in: Landollt–Böernstein, New Series, Group III, Springer–Verlag, Berlin, 1984, vol. 18, pp. 456–506.

[44] Hemla, D. S., Kupecek, Ph. J., Schwartz, S. A., Opt. Commun., 7, 225, 1973.

[45] Feichtner, J. D., Roland, G. W., Appl. Opt., 11, 993, 1972.

[46] Gandrud, W. B., Boyd, G. D., Opt. Commun., 1, 187, 1970.

[47] VanUitert, L. G., Singh, S., Levinstein, H. J., Geisic, J. E., Bonner, W. A., Appl. Phys. Lett., 11, 161, 1967.

[48] Sonin, A. S., Vasilevskaya, A. C., Electrooptic crystals, Atomizdat, Moscow, 1971, 328 p. (in Russian).

[49] Smith, W. L., Appl. Opt., 16, 1798, 1977.

[50] Kato, K., IEEE J. Quantum Electron., 25, 1455, 1990.

[51] Boyd, G. D., Buehler, E., Storz, F. G., Wernick, J. H., IEEE J. Quantum Electron., QE–8, 419, 1972.

[52] Kato, K., IEEE J. Quantum Electron., QE–10, 616, 1974.

[53] Boyd, G. D., Kasper, H., McFee, J. H., IEEE J. Quantum Electron., QE–7, 563, 1971.

[54] Kato, K., Opt Commun., 13, 93, 1975.

[55] Kato, K., IEEE J. Quantum Electron., QE–10, 622, 1974.

[56] Kato, K., Alcock, A. J., Richardson, M. C., Opt. Commun., 11, 5, 1974.

[57] Kato, K., Nakao, S., Jpn. J. Appl. Phys., 13, 1681, 1974.

[58] Boyd, G. D., Kasper, H. M., McFee, J. H., Storz, F. G., IEEE J. Quantum Electron., QE–8, 900, 1972.

[59] Chemla, D. S., Kupecek, P. J., Robertson, D. S., Smith, R. C., Opt. Comm., 3, 29, 1971.

[60] Badikov, V. V., Pivovarov, O. N., Skokov, Yu. V., Skrebneva, O. V., and Trotsenko, N. K., Sov. J. Quantum Electron., 5, 350, 1975.

[61] Boyd, G. D., Bueler, E., Storz, F. G., Appl. Phys. Lett., 18, 301, 1971.

[62] Levine, B. F., Bethea, C. G., Kasper, H., IEEE J. Quantum Electron., QE–12, 367, 1976.

[63] Badikov, V. V., Matveev, I. M., Pshenichnikov, S. M., Rychik, O. V., Trotsenko, N. K., Ustinov N. D., Shcherbakov, S. I., Sov. J. Quantum Electron., 10, 1300, 1980.

[64] Davydov, A. A., Kulevski, L. A., Prochorov, A. M., Saveljev, A. D., and Smirnov, V. V., ZhETF Letters, 15,725, 1972 (in Russian).

[65] Abdullaev, G. B., Kulevski, L. A., Prokhorov, A. M., Saveljev, A. D., Salaev, E. Yu., Smirnov, V. V., ZhETF Letters, 16, 130, 1972 (in Russian).

[66] Pearson, J. E., Evans, G. A., Yariv, A., Opt. Comm., 4, 366, 1972.

[67] Jerphagnon, J., Appl. Phys. Lett., 16, 298, 1970.

[68] Nash, F. R., Bergman, J. G., Boyd, G. D., Turner, E. H., J. Appl. Phys., 40, 5201, 1969.

[69] Ebbers, C. A., DeLoach, L. D., Webb, M., Eimerl, D., Velsko, S. P., Keszer, D. A., IEEE J. Quantum Electron., QE–29, 97, 1993.

[70] Kirby, K. W., Hoeffer, C. S., Deshazer, L. G., Lawrence Livermore National Laboratory Report, 1985.

[71] Bhar, J. C., Appl. Opt., 15, 305, 1976.

[72] Naito, H., Inaba, H., Opto–Electron., 4, 335, 1972.

[73] Umegaki, S., Tanaka, S.–I., Uchiyama, T., Yabumoto, S., Opt. Commun., 3, 244, 1971.

[74] Kabelka, V. I., Piskarskas, A. S., Stabinis, A. Yu., Sher, R. L., Sov. J. Quantum Electron., 5, 255, 1975.

[75] Ayrault, B., Langlois, H., Lecoco–Mayer M. C., Lefin, F., Phys. Status Solidi, 17, 665, 1973.

[76] Badikov, V. V., Matveev, I. N., Panyutin, V. L., Pshenichnikov, S. M., Repyakhova, T. M., Rychik, O. V., Rozenson, A. E., Trotsenko, N. K., Ustinov, N. D., Sov. J. Quantum Electron., 9, 1068, 1979.

[77] Cook, W. R., Hubby, L. M., J. Opt. Soc. Amer., 66, 72, 1976.

[78] Vshivkova, G. D., Maslov, V. A., Polivanov, Yu. N., Chuzavkov, Yu. L., Sov. Solid State Phys., 30, 3550, 1988 (in Russian).

[79] Halbout, J.-M., Blit, S., Donaldson, W., Tang, C. L., IEEE J. Quantum Electron., 15, 1176, 1979.

[80] Bhar, G. C., Samanta L. K., Ghosh, D. K., Das, S., Sov. J. Quantum Electron., 17, 860, 1987.

[81] Johnson, F. M., in: Handbook of lasers, Ed. Pressley, R. J., CRC Press, Cleveland, 1971, p.526.

34

Lasers

A.A. Malyutin[1]

34.1 Introduction

If in an atomic system inverted population of two energy levels E_1 and E_2 ($E_2 > E_1$) occurs, i.e. $N_2/g_2 > N_1/g_1$ (N_2, N_1, g_1, g_2 are population densities and degeneracies of these two levels, respectively), stimulated transitions may prevail over losses, and the radiation with the frequency $\nu = (E_2 - E_1)/h$ will be amplified. To start the generation the medium should be placed inside a resonator, e.g. between a pair of highly reflecting parallel mirrors. Generators utilizing the fundamental physical phenomenon of stimulated emission from such an inverted atomic medium are called *lasers* (acronym for "light amplification by stimulated emission of radiation"). The radiation produced by lasers may exhibit:

* wavelength from soft X-ray to millimeter range;
* high degree of coherence;
* low divergence;
* various temporal parameters, from pulses as short as tens of femtoseconds (10^{-15} s) to continuous operation;
* high energy and/or high power.

Traditionally, lasers are classified according to the type of the active medium: gas, liquid, and solid state lasers. Principles of their operation are presented in detail in many textbooks and special literature. Tables presented in this chapter were compiled on a basis of earlier [1, 2] and contemporary data.

34.2 Gas Lasers

Lasers employing gaseous materials as an active medium emit radiation from the vacuum UV to millimeter wavelengths. Low substance density and high homogeneity facilitate obtaining high quality beams and narrow spectrum bandwidth with this type of lasers, but at the same time that low density does not provide as high population inversion per unit volume as is possible for condensed active materials.

Gas lasers are classified by their medium type as neutral atomic lasers, ion lasers, molecular lasers, excimer lasers, and so on. Laser action can be obtained for noble gases, vapors of metallic or non-

[1]This chapter was compiled with participation of M.E. Brodov and V.P. Yanovskii.

0-8493-2861-6/97/$0.00+$.50

metallic elements. Among more than 90 elements existing in nature, there are few elements for which laser action has not so far been observed in their gaseous phase (see Table 34.1). The only reason is that for some elements either the boiling point is too high (above 2500°C) or available quantities of pure gas are too small (for radioactive elements). Types of gaseous species which can serve as laser mediums are many, but most of them are relatively light, small molecules.

Pumping techniques used in gas lasers.

For pumping of gaseous active materials, various physical mechanisms are used. Among them are direct excitation by electron impact in electric discharges, excitation due to chemical reactions, photodissociation or gas-dynamic processes, and optical and electron beam pumping.

In most gaseous materials the inversion of population is produced in DC, RF, pulsed or hollow cathode *electric discharge.* This technique is widely used for excitation of neutral atoms, ions, stable and unstable molecules. The energy of electrons is transferred directly to active particles (by collisions, recombination), or via the intermediary of another (buffer) species. Very often the discharge in a pure gas is far less efficient than in a mixture. The excitation transfer is much more effective if energy levels of the colliding atoms are coincident. Some transitions in ionized gases are possible due to multiple collisions only.

CW operation of gas lasers is strongly dependent on a gas pressure and a tube diameter product (pD) that determines the average electron energy. For many gas lasers the optimum pD-values were experimentally measured.

Laser performance of a pulsed discharge is not specified by the pD-product only. There are plenty of parameters that determine pulsed laser operation.

Exothermic *chemical reactions* are used for vibrational level excitation of resulting molecules. The population inversion of the molecules persists until the relaxation processes return the system to equilibrium. CW operation is possible but only under permanent delivering of initial chemical materials into the volume where the reaction takes place, simultaneously with removing the reaction products. There exists an optimum speed and temperature of interacting components for each chemical reaction. Some of the chemical lasers use plasmotron preheaters and supersonic nozzles for mixing the components.

Gas-dynamic processes that take place in supersonic nozzles are playing a decisive role for excitation of vibrational molecular levels for some gas lasers.

Though most gas lasers emit a narrow spectrum, a number of infrared gas lasers generate many closely spaced rotational lines of a vibrational transition. If the pressure in the gas discharge is made sufficiently high, a gain profile is formed which allows continuous tuning over a large spectral interval. Transverse discharge (or excitation with *high-energy electrons*) is generally used to achieve a uniform discharge and sufficient gain at high gas pressures.

Some molecules which are bound in excited states but are unstable (and normally do not exist) in their electronic ground states are used as active medium in gas lasers as well. They are usually excited dimers: rare gas molecules (*excimers*) or rare gas halides (*hetero-excimers*) produced by collisions between excited atoms or molecules. Pumping by high-voltage high-current *electron beam* sources or by fast transverse discharge plus high-density original species are necessary to form a sufficient number of *excimers.* Required high gas pressure impedes a uniform discharge in the whole active volume. Preionization by fast electrons or by UV radiation is required to achieve lasing.

As soon as mono- and diatomic gases have very narrow absorption lines, the overlapping with the spectra emitted by radiating sources is not efficient or extremely scarce. As a result, *optical* power can be used efficiently only for pumping of polyatomic gases that have broad molecular absorption bands. Many of these species excited in the middle IR produce lasing in far IR, submillimeter or millimeter range, where *optical* pumping technique is much more fruitful than any kind of electrical excitation.

Some species (including diatomic molecules) pumped in the UV region result in *photo-dissociation* (photolysis) that produces atoms in the excited state which, in turn, may provide laser emission.

One or several pumping techniques were used successfully for excitation of gas laser substances listed in Table 34.1. Laser wavelengths for these gases are given in Table 34.2, respectively either in vacuum or in air. Strong laser lines are printed bold. To distinguish lines of neutral and ionized atoms the latter are marked by Roman figures (II – for singly ionized, III – for doubly ionized, and so on).

In Table 34.2 the generally accepted notations of the molecular spectroscopy are used to designate vibrational (quantum number ν) and rotational (quantum number J) transitions.

In general, the vibrational-rotational transition involves changes in both ν and J within a given electronic state according to the selection rules. Pure rotational transitions are within a given electronic state and a given vibrational state. Quantum numbers ν_1, ν_2, ν_3 for linear triatomic molecules refer to symmetrical stretching, bending and asymmetric stretching modes of vibrations.

More detailed information on gas lasers can be found in [1–7].

Table 34.1 Active mediums for gas lasers (according to their order in Chapter 34)

Neutral and ionized atoms			
He	Ag	In	O
Ne	Au	Tl	S
Ar	Be	C	Se
Kr	Mg	Si	Te
Xe	Ca	Ge	F
Fe	Sr	Sn	Cl
Ni	Ba	Pb	Br
H	Zn	N	I
Na	Cd	P	Mn
K	Hg	As	Sm
Rb	B	Sb	Eu
Cs	Al	Bi	Tm
Cu	Ga	U	Yb
Diatomic molecules			
Xe_2	XeF	Bi_2	HCl
Ar_2	XeO	S_2	DCl
Kr_2	KrO	Te_2	HBr
ArF	HgBr	F_2	DBr
ArCl	HgCl	Br_2	NO
KrBr	H_2	Cl_2	CN
KrCl	D_2	I_2	OH
KrF	HD	ClF	DD

Diatomic molecules			
XeBr	Na_2	HF	CO
XeCl	N_2	DF	
Polyatomic molecules			
Xe_2Cl	OCS	ICN	H_2S
Xe_2Br	CH_2	NH_2	D_2S
Kr_2F	CS_2	NF_2	BF_2
O_3	HCN	FNO	BCl_3
CO_2	DCN	ClNO	SF_6
NO_2	HNC	BrNO	NH_3
C_2O	FCN	H_2O	PH_3
N_2O	ClCN	D_2O	NOCl
S_2O	BrCN	SO_2	
CF_4	CH_2Cl_2	CH_3ClH	C_2H_3Cl
F_3Br	CH_3F	CH_3NH_2	C_2H_3Br
CF_3I	CH_3Cl	CH_2CF_2	C_2H_3CN
C_2H_2	CH_3Br	$C_2H_4[OH]_2$	CH_3OCH_3
C_2H_4	CH_3I	CH_3CH_2F	C_3H_2O
H_2CO	CH_3OH	CH_3CHF	ClO_2
$[H_2CO]_3$	CH_3CN	CH_3CF_3	HCCF
HCOOH	CH_3Ne	C_2H_5Cl	FCN
CH_2F_2		C_2H_5OH	

Table 34.2 Laser transitions in neutral and ionized atoms, diatomic and polyatomic molecules

TRANSITIONS IN SINGLE ATOMS

Active medium: **He**
Pumping: pulsed and cw discharge in He at gas pressure 30–100 Pa; pulsed discharge in gas mixture He (1.5–2 kPa) and H_2 (400 Pa).

wavelengths (in air), μm		
0.706517	2.05813	8.53
0.706521	2.0603	95.763
1.8685	4.60535	216.12
1.9543	4.60567	

Active medium: **Ne** (Fig. 34.1)
Pumping: most atomic transitions are excited in the mixture Ne (1–10 Pa) and He (70 Pa); ion transitions are excited in pulsed discharge at current density 1 kA/cm^2 and gas pressure 0.1–1 Pa.

wavelengths (in air), μm		
IV 0.2018424	IV **0.2065304**	III 0.2180858
IV 0.2022186	III **0.2177705**	V 0.22657

Active medium: **Ne** *(continued)*

wavelengths (in air), μm		
IV 0.2285793	II **0.339320**	0.89886
IV **0.2357980**	II **0.348195**	0.92874
IV 0.2373200	II **0.371309**	1.0295
III **0.2473398**	0.54006	1.0621
III 0.2609982	0.58525	1.0798
III 0.261.34	0.59448	1.0844
III **0.2677918**	0.59393	1.11441
III **0.2678690**	0.60461	1.1177
III 0.2777634	0.61180	1.1390
III 0.2866726	0.61431	1.1409
II 0.3319745	**0.63282**	1.15235
II **0.3323745**	0.63518	1.15259014
II 0.332717	0.64011	1.1601
II 0.332923	0.73048	1.1614
III 0.333114	0.84634	1.17673
II 0.3345446	0.86353	1.1789
II **0.3378256**	0.87717	1.1985
II **0.3392799**	0.88653	1.2066

Table 34.2 *(continued)*

Active medium: **Ne** *(continued)*		
wavelengths (in air), μm		
1.2460	3.3840	8.8528
1.2689	3.3903	9.0871
1.2887	**3.3913**	10.060
1.2912	3.4471	10.978
1.4276	3.4489	11.857
1.4304	3.4750	11.898
1.4321	4.4789	12.831
1.4330	3.5835	13.736
1.4346	3.6515	13.756
1.4368	3.7736	16.634
1.484450	3.9806	16.664
1.486926	4.2171	16.889
1.487247	5.1696	16.943
1.488759	5.3243	17.153
1.489954	5.3249	17.184
1.493623	5.4033	17.800
1.5231	5.4050	17.837
1.7162	5.6652	17.884
1.8210	5.7053	18.392
1.8253	5.7758	20.474
1.8276	5.8844	21.746
1.8304	5.9563	22.830
1.8403	6.7769	25.416
1.8591	6.8865	28.045
1.8597	6.9857	31.544
1.9574	7.3208	31.919
1.9577	7.4201	32.007
2.0350	7.4217	32.507
2.0353	7.4679	33.815
2.1041	7.4779	33.828
2.1708	7.4973	34.543
2.3260	7.5292	34.670
2.3700	7.5674	35.592
2.3951	7.5850	37.221
2.4219	7.6142	41.730
2.4250	7.6440	50.69
2.5393	7.6489	52.40
2.5524	7.6904	53.47
2.7574	7.6994	54.00
2.7819	7.7389	54.10
2.9448	7.7634	55.51
2.9668	7.7794	57.34
2.9805	7.8347	68.31
3.0260	7.8693	72.08
3.0268	7.9406	85.01
3.3173	7.9824	86.93
3.3910	8.0066	88.47
3.3333	8.0599	89.82
3.3353	8.1712	106.00
3.3500	8.3347	124.60
3.3510	8.3472	126.10
3.3804	8.8388	132.80

Table 34.2 *(continued)*

Active medium: **Ar** (Fig. 34.2)
Pumping: almost all transitions, including strong lines 0.351 and 0.364 μm, are excited in cw discharge; typical current for ionized Ar is 30–150 A/cm^2; UV-transitions ($< 0.33\mu$m) are excited at higher current density; IR-transitions (>1.60 μm) are observed in low current cw discharge at gas pressure 7 Pa.

wavelengths (in air), μm		
IV 0.183730	II 0.64831	3.1338
V 0.183733	II 0.6730	3.6312
IV 0.2113982	II 0.68613	3.701.38
IV 0.2248840	0.75030	3.708
IV 0.2513298	II 0.877186	3.71439
IV 0.2621377	0.912297	4.2033
IV 0.2624882	0.965778	4.7138
III 0.2753884	1.0470	4.9148
III 0.2884216	II **1.092344**	4.9199
III 0.2855374	1.1448	4.9496
IV 0.2912924	1.21396	5.02338
IV 0.2926227	1.24028	5.1203
III **0.3002642**	1.27022	5.1205
III 0.302405	1.40948	5.3897
III 0.305484	1.5046	5.4666
III 0.333613	1.6180	5.4680
III 0.334472	1.619395	5.8022
III 0.335849	1.6520	**5.8461**
III 0.351112	1.6941	6.0515
III 0.351418	1.791437	**6.7443**
II 0.357661	**2.0616**	6.9410
III **0.363789**	2.0986	6.942.9
III 0.370520	2.1332	**7.2147**
III 0.379532	2.1534	7.7982
III 0.385829	2.2038	7.8002
III 0.414671	2.2077	7.8042
III 0.418298	2.31339	12.138
II 0.437075	2.3966	12.188
II 0.448181	2.5008	15.032
II 0.454505	2.5504	15.037
II 0.457935	**2.5627**	26.937
II 0.460956	**2.5661**	26.956
II 0.465789	2.6836	
II 0.472686	2.71529	
II **0.476486**	2.7357	
II **0.487986**	2.8195	
II 0.488903	2.8238	
II **0.496507**	2.862	
III 0.499280	2.8776	
II **0.501716**	2.8836	
II 0.506204	2.9273	
II 0.514179	2.9788	
II **0.514532**	3.0454	
II 0.528690	3.0988	
III 0.550220	**3.1325**	

Table 34.2 *(continued)*

Active medium: **Kr**
Pumping: almost all transitions are excited in cw discharge; current density 50–200 A/cm^2 is necessary for excitation of ionized Kr transitions (vacuum UV lines – up to 7–10 kA/cm^2); gas pressure 1–30 Pa.

wavelengths (in air), μm		
IV 0.175641	II 0.616880	3.0528
V 0.183243	III 0.631022	3.1508
IV 0.195027	II 0.631276	3.3401
IV 0.196808	II 0.641661	3.3411
IV 0.2051082	II 0.647088	3.4873
IV 0.2191916	II 0.65100	3.4885
IV 0.2254638	II 0.657012	3.774
IV 0.2338478	II 0.66029	3.956
IV 0.2417843	II 0.676442	3.9573
IV 0.2649357	II 0.687084	4.068
IV 0.2664398	II **0.7435764**	4.142
IV 0.2741380	0.752546	4.3736
III 0.3049704	0.7603	4.3755
III 0.3124363	II 0.79314	4.8760
III 0.3239512	II 0.799322	4.8819
III 0.337496	**0.810433**	4.9983
III **0.3507420**	II 0.828037	4.9999
III 0.356423	II 0.85878	5.1298
III **0.406737**	II 0.86901	5.2985
III **0.413133**	0.88058	5.3004
III 0.415444	0.8929	5.5685
III 0.417179	1.14582	**5.5848**
III 0.422658	1.31775	5.6290
II 0.431781	1.36225	7.0565
II 0.438654	1.44269	7.3605
III 0.444329	1.47648	
II 0.457720	1.4966	
II 0.458285	1.5330	
II 0.461528	1.68533	
II 0.461915	1.68965	
II 0.463386	1.6936	
II 0.465016	1.7843	
II **0.468041**	1.8185	
II 0.469444	1.9211	
II **0.476243**	2.04240	
II 0.476573	**2.1165**	
II **0.482517**	**2.19020**	
II 0.484659	2.2475	
III 0.501645	2.4260	
II 0.502240	**2.52342**	
II 0.512573	2.6260	
II 0.529831	2.6281	
II 0.539865	2.86134	
II **0.568188**	2.8656	
II 0.575298	2.9836	
III 0.603717	2.9870	
II 0.60381	3.0664	

Table 34.2 *(continued)*

Active medium: **Xe**
Pumping: most lines of ionized Xe are excited in cw discharge; current density 70–200 A/cm^2; in pulsed discharge most visible lines are lasing simultaneously; typical gas pressure (1–10)$\cdot 10^{-4}$ Pa; in atomic Xe all lines longer 2.03 μm are excited in cw discharge.

wavelengths (in air), μm		
IV 0.2232442	IV **0.515905**	II 0.926.539
IV **0.2315357**	III 0.523893	II 0.928.854
III 0.247718	IV 0.525637	II **0.969859**
IV 0.2526664	II 0.525992	**0.979970**
III 0.2691939	IV **0.526015**	II 1.063385
IV 0.29837	II 0.526042	II 1.0950
III 0.3079738	II 0.526195	1.36562
IV **0.3246922**	II 0.531387	1.60519
IV 0.3305957	IV 0.534131	1.73254
III 0.330599	IV 0.534331	2.02623
IV **0.3330869**	IV **0.535297**	2.3193
IV 0.334974	IV **0.539461**	2.4825
III 0.3454248	II 0.539525	2.51528
IV 0.348322	III 0.540104	**2.6269**
III 0.359661	II **0.541915**	**2.65146**
IV **0.3645478**	IV 0.549933	2.6601
III 0.366921	III 0.552442	2.6665
III 0.374571	IV 0.55923	2.8590
IV 0.375979	II 0.565938	3.1069
III 0.3780990	II 0.57291	**3.2739**
IV 0.380322	II 0.575103	3.3085
IV 0.397302	II 0.589330	**3.3666**
III 0.405005	IV 0.595565	3.4014
III 0.406048	II **0.597111**	3.4335
III 0.421401	II 0.609361	**3.5070**
III 0.424024	III 0.61766	3.6210
III 0.427259	III 0.623824	3.6509
III 0.428588	II 0.627081	3.6788
IV **0.430577**	IV 0.62865	**3.6849**
III 0.441314	III 0.63435	**3.8666**
III 0.443415	III 0.63435	**3.8940**
IV 0.455862	II 0.652865	**3.9955**
II 0.460303	II 0.66943	4.0196
IV 0.464759	IV 0.669950	4.1516
IV 0.465073	II 0.70723.	4.5381
III 0.467368	III 0.714894	4.5665
III 0.468354	II 0.76186	4.5694
III 0.472357	II **0.782763**	**4.6097**
III 0.474895	II **0.798800**	**5.0230**
II 0.486249	II 0.823162	5.02441
III 0.486946	II 0.833270	**5.3551**
II 0.488730	II 0.840919	5.4735
IV **0.495414**	III 0.85716	5.5739
IV 0.496508	II 0.858251	5.6019
IV 0.500772	II **0.871617**	6.3103
II 0.504492	0.904.539	6.3137
IV 0.515703	II 0.905.930	**7.3147**

Table 34.2 *(continued)*

Active medium: **Xe** *(continued)*		
wavelengths (in air), μm		
7.4294	11.289	12.913
9.0040	11.296	18.500
9.7002	12.263	75.578

Active medium: **Fe**
Pumping: dissociation of $Fe(CO)_5$ in pulsed discharge; pulsed discharge in the mixture of Fe vapor and Ne; photodissociation of $Fe(CO)_5$ by KrF-laser radiation.

wavelengths (in air), μm		
0.360	0.4529	0.563
0.385	0.540	6.847
0.395	0.558	8.4902

Active medium: **Ni**
Pumping: dissociation of $Ni(CO)_4$ in pulsed discharge.

wavelengths (in air), μm		
II 0.79624	II 0.79754	1.3968
1.4550		

Active medium: **H**
Pumping: pulsed discharge in mixture of H_2 (1.3 Pa) and He (470 Pa).

wavelengths (in air), μm		
0.4340	0.4861	1.8751

Active medium: **Na**
Pumping: pulsed discharge in the mixture of Na vapor (0.13–0.4 Pa) and He (0.13–1.3 kPa); photodissociation of sodium halides.

wavelengths (in air), μm		
0.5866	0.5890	1.1382
1.1404		

Active medium: **K**
Pumping: pulsed discharge in the mixture of K vapor (13 Pa) and He (400–700 Pa); photodissociation of vapor by radiation of ruby laser.

wavelengths (in air), μm		
0.4045	1.2523	6.4
0.7665	2.72	7.9
0.7699	3.14	12.5
1.17	3.15	15.95
1.2434	3.16	

Active medium: **Rb**
Pumping: photodissociation of vapor by radiation of ruby lasers; optical pumping by dye lasers.

wavelengths (in air), μm		
0.4210	1.37	2.79
0.7619	1.48	49.68
0.7758	1.53	50.93
0.7800	2.254	
0.7945	2.293	

Table 34.2 *(continued)*

Active medium: **Cs**
Pumping: optical pumping of Cs_2 vapor by 0.388 μm line of He; photodissociation by radiation of UV lasers.

wavelengths (in air), μm		
0.4555	1.376	3.489
0.8521	1.47	3.613
0.8764	2.95	4.220
0.8943	3.010	7.1871
1.01	3.095	
1.36	3.2040	

Active medium: **Cu** (Fig. 34.3)
Pumping: pulsed discharge in the mixture of Cu vapor and He.

wavelengths (in air), μm		
II 0.27032	**0.578213**	II 0.819228
II 0.24858	II 0.72558	II 0.8277
II 0.25063	II 0.73999	II 0.828321
II 0.25905	II 0.740434	II 0.851104
II 0.259	II 0.74382	II 1.7438
II 0.450600	II 0.766470	II 1.7708
II 0.455592	II 0.773868	II 1.8004
II 0.467356	II 0.777874	1.8196
II 0.468199	II 0.780519	1.8228
II 0.485497	II 0.780766	II 1.8448
II 0.490973	II 0.782566	II 1.9154
II 0.493165	II 0.784503	II 1.9328
II 0.501261	II 0.789583	II 1.9479
II 0.502129	II 0.794442	II 1.9712
II 0.505178	II 0.790257	II 2.0006
II 0.506064	II 0.798817	
0.510554	II 0.808858	
0.5700	II 0.809600	

Active medium: **Ag**
Pumping: cw discharge with hollow silver cathode and He or Ne as buffer gas.

wavelengths (in air), μm		
II 0.22434	II 0.84032	II 1.8408
II 0.22774	II 0.87476	II 1.8463
II 0.31807	II 0.8772	II 1.8725
II 0.40859	II 1.3759	II 1.8795
II 0.47884	II 1.5982	II 1.8979
II 0.50273	II 1.6462	1.9370
II 0.64027	II 1.6656	II 1.9714
II 0.80054	II 1.7203	II 1.9823
II 0.82547	II 1.7345	II 2.0796
II 0.8263	II 1.7478	
II 0.83244	II 1.7674	
II 0.83795	1.8380	

Table 34.2 *(continued)*

Active medium: **Au**
Pumping: pulsed discharge in mixture of Au vapor with He; cw discharge with hollow gold cathode and He as buffer gas.

wavelengths (in air), μm		
0.2428	II 0.29182	II 0.69403
II 0.25337	II 0.29594	II 0.75558
II 0.26165	0.3122	II 0.75929
0.2676	II 0.55163	II 0.76005
II 0.28225	II 0.55221	II 0.76067
II 0.28470	II 0.62123	II 0.76351
II 0.28633	0.627818	II 0.82729
II 0.28882	II 0.67014	II 0.88676
II 0.28933	II 0.69029	

Active medium: **Be**
Pumping: pulsed excitation of Be vapor with He or Ne as buffer gas.

wavelengths (in air), μm		
0.4675	0.5272	1.2096

Active medium: **Mg**
Pumping: pulsed and cw discharge in Mg vapor with He, Ne or Ar as buffer gas.

wavelengths (in air), μm		
0.9218	II 2.40415	3.86573
0.9244	II 2.41245	4.20018
1.0952	3.67794	4.36269
1.0915	3.68154	

Active medium: **Ca**
Pumping: pulsed discharge in mixture of Ca vapor with He; hollow cathode discharge; optical pumping by UV radiation.

wavelengths (in air), μm		
0.535	0.6162	II 0.866214
0.586	0.644981	1.9853
0.6102	0.6717	5.5457
0.6122	II **0.854209**	

Active medium: **Sr**
Pumping: pulsed discharge in mixture of Sr vapor with He; hollow cathode discharge.

wavelengths (in air), μm		
0.638075	II 1.091797	6.4567
II 1.033014	3.011	

Active medium: **Ba**
Pumping: pulsed discharge in mixture of Ba vapor with He, Ne, Ar or H_2; hollow cathode discharge.

wavelengths (in air), μm		
II 0.614172	0.712033	**1.5000**
II 0.649690	**1.1303**	1.820

Table 34.2 *(continued)*

Active medium: **Ba** *(continued)*

wavelengths (in air), μm		
1.9017	2.9227	5.0309
2.1568	**3.9578**	5.4798
2.3254	4.0069	5.5636
2.4758	4.330	5.8899
2.5515	4.6706	6.4546
2.5924	4.7156	
2.9057	4.7171	

Active medium: **Zn**
Pumping: cw and pulsed discharge in mixture of Zn vapor with He, Ne or Ar; cw discharge with hollow Zn cathode; dissociation of $Zn(CH_3)_2$ in pulsed discharge.

wavelengths (in air), μm		
II 0.49116	II 0.610253	II 0.77325
II **0.492404**	II 0.747879	II 0.775786
II **0.58944**	II 0.758848	II 1.8308
II 0.6021	II 0.761290	II 5.0848

Active medium: **Cd** (Fig. 34.4)
Pumping: cw or pulsed discharge in mixture of Cd vapor with He or Ne; cw discharge with hollow Cd cathode; dissociation of $Cd(CH_3)_2$ in pulsed discharge.

wavelengths (in air), μm		
II 0.3250	II 0.63601	1.43
II 0.441563	II 0.73269	1.45
II 0.48820	II 0.78443	1.64
II 0.50259	II 0.80669	3.2882
II 0.533749	II 0.85300	13.185
II 0.537804	II 0.88778	14.578
II 0.63548	1.4	

Active medium: **Hg**
Pumping: pulsed discharge in mixture of Hg vapor (0.13 Pa) with He (130 Pa); discharge with hollow cathode; dissociation of $Hg(CH_3)_2$ in pulsed discharge.

wavelengths (in air), μm		
0.3650	II 0.8622	1.5295
III 0.479701	II 0.8677	II 1.5555
III 0.5210	II 0.93968	1.6920
0.5461	II 1.0583	1.6942
II **0.56773**	1.11768	1.7073
II **0.61499**	1.2222	1.71099
III 0.65015	1.2246	1.7329
II 0.7065	1.2545	1.8130
II 0.73466	1.2760	3.93
II 0.74181	1.2981	5.88
II 0.79447	1.3655	5.9817
II 0.85498	1.3675	6.49

Table 34.2 *(continued)*

Active medium: **B** *Pumping:* pulsed discharge in BCl_3 (70Pa).		
wavelengths (in air), μm		
0.345134		

Active medium: **Al** *Pumping:* hollow cathode discharge with He or Ne as buffer gas.		
wavelengths (in air), μm		
0.358744 0.691996	0.704206 0.705656	0.747137

Active medium: **Ga** *Pumping:* pulsed excitation of $Ga(CH_3)_3$ in discharge; photodissociation of GaI.		
wavelengths (in air), μm		
0.4172 6.1460	1.7363	5.7534

Active medium: **In** *Pumping:* pulsed discharge in mixture of In vapor with He or Ne; pulsed dissociation of $In(CH_3)_3$ in discharge.		
wavelengths (in air), μm		
0.4511 2.3779	**0.468082**	1.8732

Active medium: **Tl** *Pumping:* short pulse discharge in mixture of Tl vapor (13 Pa) with Ne or He; dissociation of $Tl(CH_3)_3$ in pulsed discharge; hollow cathode discharge.		
wavelengths (in air), μm		
0.5152 0.53503 0.5949	0.6950 3.8125 5.1059	10.449

Active medium: **C** *Pumping:* lines of ionized C are excited in pulsed discharge in CO_2 or in air; lines of neutral C are excited by cw discharge in CO or CO_2 (1.3 Pa) with He or Ne (250 Pa).		
wavelengths (in air), μm		
0.15482 0.15508 0.164745 0.465016 0.49541	0.51457 0.65780 0.67838 1.0691 1.4540	2.0645 3.4046 3.5155 5.5956

Active medium: **Si** *Pumping:* lines of ionized Si are excited in pulsed discharge due to interaction with walls of the glass tube; lines of neutral Si are excited in pulsed discharge in $SiCl_4$.		
wavelengths (in air), μm		
0.455259 0.456784 0.634724	0.637148 0.667193 1.1984	1.2034 1.5883

Table 34.2 *(continued)*

Active medium: **Ge** *Pumping:* pulsed discharge in Ge vapor with He or Ne; dissociation of $GeCl_4$ in pulsed discharge.		
wavelengths (in air), μm		
0.513175 2.0200	**0.517865**	1.9809

Active medium: **Sn** *Pumping:* pulsed discharge in Sn vapor with He or Ne; cw discharge in mixture of Sn vapor (0.2 Pa) with He or Ne.		
wavelengths (in air), μm		
0.5589 0.579918 0.645350	0.684405 0.6579 1.061	1.062 1.074 4.6146

Active medium: **Pb** *Pumping:* pulsed discharge in Pb vapor with He, Ne or Ar; dissociation of $Pb(CH_4)_4$ in pulsed discharge.		
wavelengths (in air), μm		
0.363954 0.405779 0.4062	**0.53721** 0.72291 3.1738	7.1740 7.9399

Active medium: **N** *Pumping:* lines of ionized N are excited in pulsed discharge in air, N_2 or NH_3 at the pressure 1.3–13 Pa; lines of neutral N are excited in pulsed discharged in N_2 or in mixture of N_2 with He.		
wavelengths (in air), μm		
III 0.336734 IV 0.347867 IV 0.348296 II 0.399501 III 0.409732 III 0.410338 III 0.451088 III 0.451487 II 0.463055	II 0.566663 II 0.567601 **0.567956** 0.8594 0.87274 0.90455 0.93862 0.93921 1.0568	1.0611 1.0623 1.34295 1.35818 1.45423 3.7942 3.8154

Active medium: **P** *Pumping:* pulsed discharge in PF_5 (5 Pa); pulsed discharge in P vapor (0.3–30 Pa) with He or Ne as buffer gas.		
wavelengths (in air), μm		
IV 0.334769 III 0.442208 II 0.602421 II 0.603421 II 0.604325 II 0.608786 II 0.616577	0.667193 0.784563 1.008 1.116 1.119 1.154 1.178	1.571 1.648 1.894 2.060

Table 34.2 *(continued)*

Active medium: **As**

Pumping: pulsed discharge in mixture of As with Ne (13 Pa); cw discharge (hollow cathode) in mixture of As with He; dissociation of $AsCl_3$ in pulsed discharge.

wavelengths (in air), μm		
II 0.538520	1.045	1.463
II 0.549695	1.061	1.8049
II 0.549773	1.124	1.807
II 0.555809	1.1255	1.9750
II 0.56516	1.1519	2.0277
II 0.583790	1.152	2.4460
II 0.617027	1.1521	2.9805
II 0.651174	1.294	5.2865
II 0.710272	1.412	

Active medium: **Sb**

Pumping: pulsed discharge in Sb vapor (0.3 Pa) with He (30 Pa); dissociation of $Sb(CH_3)_3$ in pulsed discharge.

wavelengths (in air), μm		
0.61299	12.0330	

Active medium: **Bi**

Pumping: pulsed discharge in Bi vapor with He or in mixture of $Bi(CH_3)_3$ with He.

wavelengths (in air), μm		
III 0.456084	II 0.571921	II 0.80689
0.4722	III 0.75990	5.3284

Active medium: **V**

Pumping: dissociation of VCl_4 in pulsed discharge.

wavelengths (in air), μm		
2.0195	2.4473	

Active medium: **O**

Pumping: all lines of ionized O are excited in pulsed discharge at the pressure 0.13–13 Pa and current density 0.5–2 kA/cm^2; almost all lines of neutral O are observed in cw discharge in the mixtures O_2–Ar or O_2–Ne.

wavelengths (in air), μm		
V 0.2460	II **0.441488**	0.844680
V 0.278139	III **0.375467**	2.652
III **0.298378**	III **0.375988**	2.890
III 0.304713	**II 0.441697**	4.5607
IV 0.306345	II 0.460552	4.5832
IV 0.338128	II 0.464914	5.981
IV 0.338133	**III 0.559237**	6.8161
IV 0.338554	II 0.66402	6.8161
II **0.374949**	0.672136	6.858
III 0.375426	0.844628	6.8731
II **0.434738**	0.844638	10.400
II **0.435128**	0.844672	

Table 34.2 *(continued)*

Active medium: **S**

Pumping: all lines of ionized S are excited in pulsed high current density discharge in SO_2, SF_6 or H_2S at pressure 1.3–7 Pa; lines of neutral S are excited by cw discharge in S vapor with Ne.

wavelengths (in air), μm		
III **0.2638964**	II **0.532088**	II 0.581935
III 0.3324859	II **0.534583**	0.7725
III 0.3497332	II **0.542874**	1.0455
III **0.3709354**	II **0.543287**	1.0636
II 0.492560	II **0.545388**	1.402
II 0.50116	II 0.547374	1.5422
II 0.501424	II 0.556511	1.6543
II 0.503262	II 0.550990	2.4363
II **0.516032**	II **0.564012**	2.7799
II **0.521962**	II **0.564716**	3.3892

Active medium: **Se**

Pumping: all lines are excited in cw operation; current density 1.5–15 A/cm^2; gas mixture Se (0.7 Pa) and He (0.8–1.1 kPa).

wavelengths (in air), μm		
II 0.446760	**II 0.522751**	II 0.649048
II 0.460434	II 0.525307	II 0.653995
II 0.461877	II 0.525363	II 0.706389
II 0.464844	II 0.527111	II 0.739199
II 0.471823	**II 0.530535**	II 0.767482
II 0.474097	II 0.552242	II 0.772404
II 0.476365	II 0.556693	0.777
II 0.476552	II 0.559116	II 0.779615
II 0.484063	II 0.562313	II 0.783881
II 0.484496	II 0.569788	II 0.830952
II 0.489	II 0.574762	**II 0.92493**
II **0.497566**	II 0.584268	II 0.995515
II **0.499275**	II 0.586627	**II 1.040881**
II **0.506865**	II 0.605596	**II 1.258678**
II 0.509650	II 0.606583	6.3672
II 0.514214	II 0.610196	
II 0.517598	II 0.644425	

Active medium: **Te**

Pumping: pulsed or cw discharge in Te vapor; dissociation of $Te(CH_3)_2$ in pulsed discharge.

wavelengths (in air), μm		
II 0.48429	II 0.57563	II **0.70391**
II 0.50204	II **0.59361**	II 0.78017
II 0.52564	II 0.59726	II 0.79217
II **0.54498**	II **0.59747**	II 0.86046
II 0.54540	III 0.60145	II 0.87338
II **0.54791**	II 0.60823	II 0.88982
II 0.55702	II **0.62307**	II 0.89721
II 0.55764	II 0.62454	II 0.93779
II 0.56405	II 0.63497	0.31720
II **0.56662**	II 0.65851	0.67595
II 0.57081	II 0.66486	0.77856
II **0.57416**	II 0.66761	
II 0.57559	II 0.68851	

Table 34.2 *(continued)*

Active medium: **F**		
Pumping: pulsed discharge in F_2 (3 Pa) or in mixture of CF_4, SF_6, C_2F_6 or NF_3 with He.		
wavelengths (in air), μm		
III 0.275958	0.696635	0.74257
IV 0.282612	0.703745	0.74827
III 0.3121501	0.70394	0.748914
III 0.317413	0.712788	0.75150
II 0.320276	0.71298	0.7552235
II 0.402472	0.720237	0.775470
0.6239651	0.72043	0.780022
0.6348508	0.7310102	1.5900
0.6413651	0.7398688	9.3462

Active medium: **Cl**		
Pumping: UV lines of ionized Cl are lasing in pulsed discharge only (pressure 0.3 Pa); visible lines are excited in cw operation (7 Pa); lines of neutral Cl are excited in HCl or in mixture of Cl_3 with He.		
wavelengths (in air), μm		
III 0.2632686	II 0.413250	II 0.522136
III 0.3191424	II 0.474042	II 0.539216
III 0.3392861	II 0.476871	II 0.609473
III **0.3393444**	II 0.478134	0.9451
III 0.3530016	II 0.489685	1.3859
III 0.3560632	II 0.490483	1.3891
III 0.360210	II 0.491781	1.9755
III 0.361283	II 0.507829	2.0199
III 0.362268	II 0.510310	2.4466
III 0.3720436	II **0.512776**	3.0664
III 0.3748770		

Active medium: **Br**		
Pumping: cw discharge in HBr; pulsed discharge in Br_2 (5 Pa).		
wavelengths (in air), μm		
IV **0.2362465**	II **0.518238**	2.2854
IV 0.2581246	II 0.523826	2.3511
III 0.2787619	II **0.533203**	2.714
II 0.474266	II 0.611756	2.8375
II 0.505463	II 0.616878	

Active medium: **I**		
Pumping: pulsed discharge in pure I at 13 Pa or in mixture with He; pulsed or cw discharge in HI or in mixture of I and He.		
wavelengths (in air), μm		
II 0.448855	II 0.559312	II 0.662235
II 0.453379	II 0.562569	II 0.682523
II 0.467440	II **0.567808**	II 0.690477
II 0.467553	II **0.576072**	II **0.703299**
II 0.493467	II 0.606893	II 0.713879
II 0.498692	II **0.612749**	II 0.761850
II 0.521408	II 0.620486	II 0.773578
II 0.521627	II 0.633997	II **0.817001**
II **0.540736**	II 0.651618	II 0.825381
II 0.5419	II 0.658521	II 0.880428

Table 34.2 *(continued)*

Active medium: **I** *(continued)*		
wavelengths (in air), μm		
II 0.887761	1.553	4.858
0.98	2.5986	**4.8619**
1.01	2.7572	**5.4972**
1.03	3.0360	6.7198
1.06	3.2363	6.902
1.315	3.4296	9.326
1.4542	4.331	

Active medium: **Mn**		
Pumping: pulsed discharge in Mn vapor and He.		
wavelengths (in air), μm		
0.534106	0.553776	1.36267
0.542036	**1.28998**	1.38642
0.547064	1.32938	1.39975
0.551677	1.33190	

Active medium: **Sm**		
Pumping: pulsed discharge in Sm vapor with He, Ne or Ar.		
wavelengths (in air), μm		
1.912	**2.9663**	**4.1368**
2.0482	**3.4654**	**4.8656**
2.6998	**3.5361**	

Active medium: **Eu**		
Pumping: pulsed discharge in Eu vapor with He, Ne or Ar.		
wavelengths (in air), μm		
II 0.6645	1.660	**5.0647**
II 0.9898	**1.7596**	5.2811
II 1.002	**2.5811**	**5.4292**
II 1.016	2.7174	5.7706
II 1.361	4.3202	5.9479
II 1.477	4.6935	6.0576

Active medium: **Tm**		
Pumping: pulsed discharge in Tm vapor with He, Ne or Ar.		
wavelengths (in air), μm		
1.304	1.500	1.973
1.310058	1.637914	1.994160
1.338008	**1.675404**	2.107
1.433973	1.7319	2.384515
1.448509	**1.958443**	

Active medium: **Yb**		
Pumping: pulsed discharge in Yb vapor with He, Ne or Ar.		
wavelengths (in air), μm		
1.0322	II **1.6498**	2.0036
1.2548	1.7454	2.1181
II 1.2714	1.7977	II 2.1480
II 1.3453	II 1.8057	II **2.4377**
1.4280	1.9830	4.8009
1.4787		

Table 34.2 *(continued)*

TRANSITIONS IN DIATOMIC MOLECULES

Active medium: $\mathbf{Xe_2}$ (Fig. 34.5), $\mathbf{Ar_2}$, $\mathbf{Kr_2}$
Pumping: electron beam excitation of high pressure gas

wavelengths (in air), μm	
Xe_2	0.1722
Ar_2	0.1261
Kr_2	0.1457

Active medium: **ArF, ArCl, KrF, etc.**
Pumping: electron beam pumping of the noble gas mixture with halides at high pressure; transversal electric discharge at normal pressure.

molecule	wavelength (in air), μm	molecule	wavelength (in air), μm
ArF	0.1933	XeCl	0.30792
ArCl	0.1750	XeCl	0.30816
KrBr	0.2065	XeF	0.34875
KrCl	0.2229	XeF	0.35091
KrF	0.2481	XeF	0.35097
KrF	0.2484	XeF	0.35114
KrF	0.2485	XeF	0.35365
KrF	0.2495	XeF	0.35354
XeBr	0.2818	XeF	0.483

Active medium: **XeO, KrO**
Pumping: electron beam excitation of the noble gas/oxygen mixture at high pressure.

molecule	wavelength (in air), μm	molecule	wavelength (in air), μm
XeO	0.530	ArO	0.5377
XeO	0.555	ArO	0.557781

Active medium: **HgBr, HgCl**
Pumping: electron beam excitation of the gas at high pressure; $HgBr_2$ dissociation by electric discharge or photodissociation.

molecule	wavelength (in air), μm	molecule	wavelength (in air), μm
HgBr	0.5018	HgBr	0.5042
HgBr	0.5020	HgBr	0.5046
HgBr	0.5023	HgCl	0.55762
HgBr	0.5026	HgCl	0.55835
HgBr	0.5039		

Active medium: $\mathbf{H_2}$ (Fig. 34.6)
Pumping: pulsed discharge in H_2 (13 Pa); electron beam (400 keV) excitation of H_2 (1.3–13 Pa); *para*- and *ortho*-hydrogen lines are labeled by letters *p* and *o*, accordingly.

molecule	wavelength (in air), μm	molecule	wavelength (in air), μm
p	0.109816	*o*	0.144049
p	0.110205	*p*	0.144061

Table 34.2 *(continued)*

Active medium: $\mathbf{H_2}$ *(continued)*

molecule	wavelength (in air), μm	molecule	wavelength (in air), μm
p	0.111515	*p*	0.146017
p	0.111894	*o*	0.146383
p	0.114462	*p*	0.146411
p	0.114862	*p*	0.146841
o	0.115976	*p*	0.148652
p	0.116003	*p*	0.149171
o	0.116136	*o*	0.14942
p	0.116390	*o*	0.149522
o	0.116617	*p*	0.151570
o	0.117436	*o*	0.151867
p	0.117456	*p*	0.151994
o	0.117586	*o*	0.152325
p	0.117830	*p*	0.153494
o	0.118050	*p*	0.154493
o	0.118936	*p*	0.155010
o	0.120497	*o*	0.155345
p	0.120536	*o*	0.15655
o	0.120688	*p*	0.156629
p	0.120929	*p*	0.156644
o	0.121734	*o*	0.156725
p	0.121767	*p*	0.156753
o	0.121900	*o*	0.157199
p	0.121946	*o*	0.15743
p	0.122143	*p*	0.157434
o	0.122358	*o*	0.157739
p	0.122874	*p*	0.157771
o	0.123004	*o*	0.157919
p	0.123230	*o*, *p*	0.157998
p	0.123833	*o*	0.158077
o	0.123956	*p*	0.158110
p	0.124167	*p*	0.158140
p	0.124620	*p*	0.158899
p	0.125202	*o*	0.159131
p	0.126839	*o*, *p*	0.159340
p	0.133856	*o*	0.159606
o	0.134226	*p*	0.159926
p	0.135984	*o*, *p*	0.161033
p	0.136799	*p*	0.161091
p	0.139895	*o*	0.161148
o	0.140264	*p*	0.161165
p	0.160236	*o*	0.161166
o	0.160448	*o*, *p*	0.160829
p	0.160594	*o*	0.160839
o, *p*	0.160623	*p*	0.160844
o	0.160751	*o*	0.160961
p	0.140728	*p*	0.161318
p	0.143262	*p*	0.161485
o	0.143622	*o*	0.16165
p	0.143757		

Table 34.2 *(continued)*

Active medium: $\mathbf{D_2}$
Pumping: pulsed discharge in D_2 (13–400 Pa); for excitation of UV lines short pulse (2.5 ns), high current (several kA/cm^2) discharge.

wavelengths (in vacuum), μm		
0.111336	0.138879	0.16063
0.113770	0.143217	0.160650
0.114757	0.157585	0.160769
0.115650	0.158634	0.160681
0.118811	0.158642	0.160848
0.119753	0.158675	0.160955
0.120640	0.158714	0.161075
0.122800	0.158720	0.161080
0.123556	0.15890	0.161147
0.124239	0.15923	0.161165
0.124831	0.158983	0.161171
0.125329	0.159130	0.161198
0.115840	0.159137	0.161236
0.119005	0.159226	0.161251
0.119940	0.159257	0.161257
0.120821	0.160044	0.161318
0.124412	0.160086	0.161320
0.124997	0.160210	0.161324
0.130363	0.160354	0.161412
0.134590	0.160578	0.161658
wavelengths (in air), μm		
0.827752	0.952367	1.477548
0.944156	0.953005	

Active medium: **HD**
Pumping: pulsed discharge in HD at 13–400 Pa.

wavelengths (in vacuum), μm		
0.113864	0.151359	0.160365
0.114154	0.152989	0.160496
0.115198	0.156201	0.160465
0.117806	0.157136	0.160569
0.118995	0.157242	0.160647
0.119281	0.157267	0.160648
0.120103	0.158008	0.160674
0.121125	0.158085	0.160692
0.122837	0.158185	0.160747
0.124567	0.158253	0.160794
0.125276	0.158305	0.160827
0.130334	0.159378	0.160893
0.135507	0.159524	0.161005
0.140770	0.159713	0.161131
0.148843	0.160233	
wavelengths (in air), μm		
	0.917201	

Active medium: $\mathbf{Na_2}$
Pumping: optical pumping of Na vapor (with 4 kPa of He as buffer gas) at 0.473, 0.659, 0.454, and 0.488 μm.

wavelengths (in air), μm		
0.5250	0.529816	0.5326
0.526333	0.529952	0.534283
0.5279	0.5338	0.5345

Table 34.2 *(continued)*

Active medium: $\mathbf{Na_2}$ *(continued)*

wavelengths (in air), μm		
0.534930	0.5472	0.797474
0.536902	0.5474	0.797657
0.5375	0.5485	0.799091
0.5376	0.5490	0.799660
0.537814	0.5491	0.80011
0.5381	0.549158	0.8002
0.538497	0.5504	0.800840
0.538635	0.5562	0.80154
0.5390	0.5568	0.803650
0.540244	0.5596	0.803931
0.541311	0.784930	0.804447
0.5417	0.786590	0.805366
0.5421	0.789740	0.805611
0.544150	0.789790	0.8066
0.544694	0.791783	0.806943
0.5453	0.792947	0.80715
0.5459	0.793697	0.80859

Active medium: $\mathbf{N_2}$
Pumping: pulsed high current density discharge in gas at 130–500 Pa; electron beam (300–400 keV) excitation at 2.7 kPa; electron beam excitation of high pressure mixture with Ar (Fig. 34.7).

wavelengths (in air), μm		
0.315756	0.3370288	0.3371266
0.315778	0.3370295	**0.3371307**
0.315798	0.3370312	**0.3371366**
0.315803	0.3370381	**0.3371392**
0.315816	**0.3370438**	**0.3371421**
0.315827	0.3370466	**0.3371429**
0.315832	0.3370474	0.3386428
0.315844	0.3370555	0.3575460
0.315853	0.3370562	0.3575798
0.315861	0.3370608	0.3575980
0.315870	0.3370619	0.3576112
0.315874	0.3370665	0.3576194
0.315883	0.3370677	0.3576250
0.315891	0.3370714	0.3576320
0.315900	0.3370726	0.3576571
0.315911	**0.3370749**	0.3576613
0.315919	**0.3370758**	0.3576778
0.3364909	0.3370782	**0.3576899**
0.3365425	**0.3370797**	**0.3576955**
0.3365478	0.3370812	0.3804
0.3366913	0.3370816	0.4058
0.3369541	0.3370826	0.4278
0.3369552	0.3370919	0.7482187
0.3369769	**0.3370986**	0.748274
0.3369823	**0.3371031**	0.7485941
0.3369835	0.3371075	0.7486135
0.3369907	0.3371082	0.7486413
0.3370027	0.3371113	0.7486253
0.3370075	0.3371121	0.7487409
0.3370081	**0.3371135**	0.7488046
0.3370137	**0.3371143**	**0.7488246**
0.3370174	**0.3371172**	0.7489107

Table 34.2 *(continued)*

Active medium: $\mathbf{N_2}$ *(continued)*		
wavelengths (in air), μm		
0.7489626	0.7510923	0.7625709
0.7489809	0.7511592	0.7625770
0.7490096	0.7511799	0.7625812
0.7490317	**0.7512003**	0.7625906
0.7491510	0.7512569	0.7626007
0.7491705	0.7513357	0.7626044
0.7492379	0.7514079	0.7626114
0.7493082	0.7515446	**0.7626180**
0.7493716	0.7515650	0.7626207
0.7493910	0.7517728	**0.7626360**
0.7495086	0.7518013	0.7626560
0.7495465	0.7574329	**0.7626700**
0.7495660	**0.758105**	**0.7626749**
0.7496024	0.758423	0.7626826
0.7497256	0.7586439	0.7627806
0.7497524	0.7587693	0.7628854
0.7497728	0.7589868	0.7629102
0.7498898	0.7591960	0.7630305
0.7499013	0.7593908	0.7631880
0.7499327	0.7594941	0.7632446
0.7499593	0.7597289	0.7633348
0.7499825	0.759870	**0.7633985**
0.7500071	0.7603477	0.7634546
0.7500646	0.7606374	0.7634779
0.7500734	**0.7607626**	**0.7635474**
0.7501056	**0.7608801**	0.7636126
0.7501295	**0.7609853**	0.7636904
0.7501404	0.7613612	0.7637586
0.7501553	0.7610759	**0.7638274**
0.7502139	0.7611082	0.7639571
0.7502729	**0.7611514**	0.7639715
0.7502768	0.7612105	0.7640383
0.7503035	0.7612528	**0.7640794**
0.7503371	0.7613260	0.7641929
0.7503418	0.7615347	0.7642478
0.7503642	0.7616994	0.7644612
0.7503669	**0.7617357**	0.771206
0.7503697	0.7619288	0.7724562
0.7503838	0.7620844	0.7730032
0.7503960	0.7620943	0.7735040
0.7503994	**0.7621161**	0.7739632
0.7504106	0.7622235	0.7743859
0.7504160	0.7622565	0.775270
0.7504184	**0.7622959**	**0.7752354**
0.7504274	0.7623256	0.7753652
0.7504598	0.7623264	0.865331
0.7504768	0.7623311	0.865492
0.7505113	0.7623582	0.866089
0.7505710	0.7623686	0.866256
0.7505903	0.7623918	0.866345
0.7506063	0.7624220	**0.866572**
0.7506356	**0.7624690**	0.86676
0.7508145	**0.7624924**	0.8669223
0.7509890	**0.7625115**	0.866959
0.7510133	**0.7625445**	**0.8671332**

Table 34.2 *(continued)*

Active medium: $\mathbf{N_2}$ *(continued)*		
wavelengths (in air), μm		
0.867554	0.8728430	0.8908878
0.868281	0.8730453	0.8908808
0.8682937	0.8732394	0.8909451
0.868374	**0.8734247**	0.8909527
0.868762	0.8735995	0.8909750
0.869136	**0.8737644**	0.8910132
0.8692580	0.8739162	0.8910480
0.869490	0.8740559	0.8910612
0.8696366	0.8742917	0.8911001
0.8697945	0.884129	0.8911063
0.8698263	0.8845349	0.8911280
0.8699397	0.8846598	0.8911502
0.8700670	0.884758	0.8911538
0.8700684	0.884920	0.8911608
0.8701481	0.885026	0.8911898
0.8701718	0.885261	0.8912139
0.8702541	0.885460	0.8918033
0.8702681	0.8856271	0.8920184
0.8703093	0.885649	0.8922249
0.870331	0.8858470	0.8924223
0.8703457	0.8861195	0.8926099
0.8704549	0.886153	0.8927865
0.8707478	0.886256	0.8929509
0.8710118	0.886278	0.8931019
0.8710273	0.886697	0.8933580
0.8712956	0.886799	0.965389
0.8713533	0.887121	0.965846
0.871450	0.887531	0.967599
0.8715519	0.887918	0.967270
0.871644	0.8880521	0.967758
0.8716718	0.888288	0.967943
0.8717377	0.8884527	0.968061
0.8717970	0.8886204	0.969552
0.8718571	0.8886378	0.969879
0.8718654	0.8887756	1.043588
0.8719537	0.8889111	1.044261
0.8719562	0.8889738	1.044992
0.8719562	0.8890243	1.045519
0.8719791	0.8891133	1.046117
0.8720251	0.8891769	1.046669
0.8720284	0.8892149	1.047195
0.8720308	0.8892940	1.047691
0.8720419	0.8896001	**1.047961**
0.8720848	0.8898930	1.048173
0.8721155	0.8899078	1.048634
0.8721327	0.8901733	**1.049060**
0.8721718	0.8902420	1.049478
0.8721971	0.8902711	1.049873
0.8722007	0.890372	1.050231
0.8722220	0.8904419	1.050512
0.8722341	0.890566	1.050717
0.8722569	0.8906097	1.05083
0.8722836	0.8906649	1.052259
0.8723057	0.8906994	1.052622
0.8726333	0.8907920	1.053093

Table 34.2 *(continued)*

Active medium: $\mathbf{N_2}$ *(continued)*		
wavelengths (in air), μm		
1.053471	1.234631	3.45118
1.230261	3.29372	3.45758
1.231093	3.30076	3.46283
1.231881	3.30665	3.46709
1.232624	3.31149	3.47032
1.233333	3.31526	8.18161
1.233994	3.31801	8.20882

Active medium: $\mathbf{Bi_2}$
Pumping: optical (Ar and dye laser) pumping of the vapor.

wavelengths (in air), μm		
0.5299	0.6550	0.7398
0.6160	0.6809	0.7408
0.6239	0.7006	0.7439
0.6300	0.7013	0.7468
0.6339	0.7292	0.7471
0.6414	0.7301	0.7475
0.6422	0.7335	0.7482
0.6576	0.7364	0.7543
0.6582	0.7366	0.7551
0.6603	0.7376	

Active medium: $\mathbf{S_2}$
Pumping: optical (dye laser) pumping of the vapor; photodissociation by UV radiation.

wavelengths (in air), μm		
0.365	1.0920	1.099
1.086	1.0923	1.100
1.0915	1.0941	1.1587
1.0917	1.0946	

Active medium: $\mathbf{Te_2}$
Pumping: optical pumping of the vapor by Ar-laser 0.4765 μm-line.

wavelengths (in air), μm		
0.5571	0.5721	0.5841
0.5575	0.5724	0.5849
0.5578	0.5766	0.5851
0.5579	0.5767	0.5857
0.5626	0.5773	0.5859
0.5643	0.5780	0.5865
0.5647	0.5783	0.5869
0.5649	0.5784	0.5870
0.5650	0.5785	0.5874
0.5696	0.5786	0.5924
0.5701	0.5787	0.5927
0.5711	0.5790	0.5934
0.5714	0.5793	0.5936
0.5715	0.5794	0.6002
0.5719	0.5797	0.6004
0.5720	0.5798	0.6005

Table 34.2 *(continued)*

Active medium: $\mathbf{Te_2}$ *(continued)*		
wavelengths (in air), μm		
0.6008	0.6170	0.6465
0.6009	0.6204	0.6473
0.6082	0.6278	0.6477
0.6085	0.6287	0.6484
0.6087	0.6288	0.6561
0.6089	0.6295	0.6569
0.6162	0.6379	0.6574
0.6165	0.6381	0.6581
0.6168	0.6388	

Active medium: $\mathbf{F_2}$
Pumping: electric discharge excitation; electron beam excitation of the gas mixture with Ne or He.

wavelengths (in vacuum), μm		
0.15671	0.15748	0.15759

Active medium: $\mathbf{Br_2}$
Pumping: electric discharge excitation of the mixture with Ar; electron beam excitation.

wavelengths (in vacuum), μm		
0.2915	0.61318	0.67456
0.55020	0.61368	0.67506
0.55053	0.63612	0.74582
0.58048	0.63654	0.74638
0.58090	0.63705	0.74641
0.61272	0.67408	0.74704
0.61316	0.67455	

Active medium: $\mathbf{Cl_2}$
Pumping: electron beam excitation of the Cl_2 (300 Pa) and He (1.2 MPa).

wavelengths (in air), μm
0.2580

Active medium: $\mathbf{I_2}$
Pumping: electric discharge excitation of the mixture with Ar; electron beam excitation.

wavelengths (in air), μm		
0.3420	0.5905	0.6258
0.3423	0.5969	0.6330
0.3424	0.6025	0.6352
0.3428	0.6048	0.6490
0.5543	0.6110	0.6511
0.5550	0.617482	0.6592
0.5567	0.617676	0.6645
0.5680	0.617868	0.6936
0.5697	0.617947	0.7114
0.5745	0.618193	0.8144
0.5764	0.618267	0.8358
0.5815	0.618441	0.8578
0.5830	0.618538	0.8804
0.5880	0.6198	0.8813

Table 34.2 *(continued)*

Active medium: $\mathbf{I_2}$ *(continued)*		
wavelengths (in air), μm		
0.9037	1.1347	1.31095
0.9047	1.1349	1.31130
0.9060	1.1350	1.31180
0.9274	1.1454	1.31205
0.9288	1.1464	1.31338
0.9295	1.20359	1.31371
0.9305	1.217	1.31462
0.9518	1.25623	1.31487
0.9545	1.25663	1.3153
0.9555	1.26324	1.3192
0.9766	1.274	1.320
0.9963	1.27883	1.32916
0.9973	1.27945	1.33029
1.0019	1.28177	1.3310
1.0053	1.28210	1.3324
1.0225	1.28433	1.3333
1.0245	1.28478	1.3349
1.0255	1.2870	1.33509
1.0274	1.28722	1.3353
1.0534	1.28754	1.33636
1.0775	1.28931	1.33582
1.0788	1.28972	1.33624
1.1068	1.2925	1.33644
1.1073	1.294	1.338
1.1207	1.304	1.34105
1.1214	1.30509	1.33108
1.1215	1.30536	1.34136
1.1224	1.30620	1.34155
1.1255	1.30748	1.34156
1.1327	1.30864	1.34211
1.1336	1.30890	1.34219

Active medium: **ClF**
Pumping: electron beam excitation of the mixture Ne, Cl_2 and F_2.

wavelengths (in air), μm
0.285

Active medium: **HF**
Pumping: excitation of vibrational levels in chemical reactions; optical pumping for excitation of the vibrational-rotational levels; flash photolysis of the mixture H_2 and F_2.

wavelengths (in vacuum), μm		
Band 1–0	2.6729	3.0582
2.410	2.7075238	3.1125
2.430	2.7440	3.1695
2.450	2.7826	3.2292
2.480	2.8231	3.2919
2.551	2.8657	**Band 2–1**
2.579	2.9103	2.6668
2.6084	2.9573	2.6963
2.6396	3.0064	2.7275

Table 34.2 *(continued)*

Active medium: **HF** *(continued)*		
wavelengths (in vacuum), μm		
2.7604	2.8213	2.9896
2.7952	2.8542	3.026
2.8319	2.8890	3.065
2.8705	2.9257	3.1454
2.9111	2.9644	3.1492
2.9539	3.0052	**Band 5–4**
2.9989	3.8482	3.0982
3.0461	3.0935	3.1350
3.0958	3.1411	3.1640
3.1480	3.1912	3.2151
3.2029	3.2438	3.258
3.2603	3.2991	3.3044
3.2206	**Band 4–3**	**Band 6–5**
Band 3–2	2.9221	3.333
2.7902	2.9549	3.377
Pure rotational transitions		
ν=0	ν=1	ν=2
10.1978	12.2619	10.5819
10.4578	12.7006	10.8117
10.7439	13.1877	13.2211
11.0573	13.7277	14.2881
11.4033	15.0163	19.550
11.7854	18.8010	20.9393
12.2082	20.1337	
12.6781	21.6986	ν=3
13.2009	36.5	11.5408
13.7841	42.4	19.1129
14.4406	50.8	20.3513
15.1744	63.4	21.7885
16.0215	84.4	
18.090	126.5	ν=4
19.350		19.915
20.835		

Active medium: **DF**
Pumping: vibrational transitions are excited during chemical reactions (also initiated by electric discharge), or by photolysis.

wavelengths (in vacuum), μm		
Band 3–1	3.752	3.7651
1.836	3.7901	3.8007
1.844	3.8298	3.8375
1.857	3.8707	3.8757
Band 1–0	3.9133	3.9155
3.493	3.9572	3.9565
3.521	4.0032	3.9995
3.550	4.0502	4.0435
3.581	**Band 2–1**	4.0893
3.612	3.6363	4.1369
3.645	3.6665	4.1862
3.679	3.6923	**Band 3–2**
3.716	3.7310	3.7563

Table 34.2 *(continued)*

Active medium: **DF** *(continued)*		
wavelengths (in vacuum), μm		
Band 3–2	4.0054	3.8817
3.7878	4.0464	3.9145
3.8206	4.0895	3.9487
3.8547	4.1337	3.9843
3.8903	4.1798	4.0212
3.9272	**Band 4–3**	4.0595
3.9654	3.8503	

Active medium: **HCl**
Pumping: excitation during chemical reactions in mixtures of H_2 and Cl_2O or Cl_2 and HI initiated by electric discharge or photolysis; rotational lines are excited by pulsed discharge in mixtures of Cl_2 and CH_3Cl, Cl_2 and CH_3Br, or Cl_2, H_2 and $CClF_3$.

wavelengths (in vacuum), μm		
Band 1–0	**Band 2–1**	**Band 3–2**
$H^{35}Cl$	$H^{35}Cl$	$H^{35}Cl$
3.5728	3.7071	3.8509
3.6026	3.7383	3.8840
3.6337	3.7710	3.9181
3.6660	3.8050	3.9536
3.6996	3.84.01	3.9909
3.7341	3.8768	4.0295
3.7707	3.9149	$H^{37}Cl$
3.8081	$H^{37}Cl$	3.8536
$H^{37}Cl$	3.7098	3.8864
3.6362	3.7408	3.9205
3.6685	3.7735	3.9560
3.7021	3.8074	**Band 4–3**
3.7370	3.8425	$H^{35}Cl$
		4.0054
		4.0399
		4.0759
		4.1135
Pure rotational transitions		
$\nu=0$	27.508	24.5833
$H^{35}Cl$	$H^{37}Cl$	25.7040
13.8720	16.664	$H^{37}Cl$
14.0994	17.997	16.765
14.3434	19.122	18.593
16.2125	$\nu=1$	19.145
16.6085	$H^{35}Cl$	24.6177
17.0304	17.125	$\nu=2$
17.4923	17.575	$H^{35}Cl$
17.9874	18.035	19.183
18.522	18.555	20.9991
20.4106	19.7002	24.3178
21.1556	20.3455	$\nu=3$
21.9706	21.0470	19.783
22.8637	21.8127	19.821
23.8485	22.6514	
26.14.62	23.5705	

Table 34.2 *(continued)*

Active medium: **DCl**
Pumping: excitation due to chemical reaction (pressure respectively 200 and 300 Pa) initiated by pulsed electric discharge.

wavelengths (in vacuum), μm		
Band 2–1	5.1811	5.3562
$D^{35}Cl$	5.2118	5.3889
5.0445	5.2435	5.4577
5.0743	5.2760	5.4935
5.1049	5.3097	5.5304
5.1363	5.3443	$D^{37}Cl$
5.1688	5.3799	5.3629
$D^{37}Cl$	$D^{37}Cl$	5.3956
5.0514	5.1879	5.4295
5.0811	5.2186	**Band 5–4**
5.1118	5.2503	$D^{37}Cl$
5.1431	5.2829	5.5084
Band 3–2	**Band 4–3**	5.5423
$D^{35}Cl$	$D^{35}Cl$	5.5776
5.1511	5.3244	5.6137

Active medium: **HBr**
Pumping: excitation due to chemical reaction (H_2 + Br_2 pressure respectively 200 and 300 Pa) initiated by pulsed electric discharge.

wavelengths (in vacuum), μm		
Band 1–0	4229.5	$H^{81}Br$
$H^{79}Br$	4263.3	4325.5
4017.0	4298.8	4358.5
4047.0	4335.4	4393.1
4078.3	$H^{81}Br$	4430.7
4110.7	4165.8	4465.8
4144.2	4197.5	4504.7
$H^{81}Br$	4263.9	**Band 4–3**
4017.6	4299.4	$H^{79}Br$
4047.5	4335.9	4533.0
4078.8	**Band 3–2**	4569.1
4111.2	$H^{79}Br$	4607.0
4144.8	4325.0	4466.3
4179.6	4357.9	$H^{81}Br$
Band 2–1	4392.5	4533.5
$H^{79}Br$	4428.1	4569.6
4165.3	4465.2	4607.6
4197.0	4504.1	4646.7
Pure rotational transitions		
$\nu=0$	$\nu=1$	31.368
19.399	19.988	32.799
20.360	21.546	40.526
20.896	30.445	$\nu=3$
20.949	31.849	23.436
21.501	33.409	29.786
22.136	$\nu=2$	
30.948	22.226	
32.469	22.855	

Table 34.2 *(continued)*

Active medium: **DBr**

Pumping: excitation due to chemical reaction (pressure respectively 80 and 40 Pa) initiated by pulsed electric discharge.

wavelengths (in vacuum), μm		
Band 3–2	**Band 4–3**	**Band 5–4**
$D^{79}Br$	$D^{79}Br$	$D^{79}Br$
5.8049	6.0209	6.2566
5.8620	6.0529	6.2916
5.8928	6.0858	6.3279
5.9246	6.1200	$D^{81}Br$
5.9573	6.1546	6.2237
$D^{81}Br$	6.1903	6.2581
5.8629	6.2272	6.2932
5.8944	$D^{81}Br$	6.3294
5.9261	6.0225	
5.9590	6.0544	
	6.0873	
	6.1216	
	6.1562	
	6.1918	
	6.2289	

Active medium: **NO**

Pumping: Pulsed discharge in the mixture of NOCl (470 Pa) and He (780 Pa); photodissociation of NOCl or NO in the mixture with NO_2.

wavelengths (in vacuum), μm		
1.1069	6.0267	6.2110
1.2237	6.0324	6.2191
2.6072	6.0386	6.2249
2.6380	6.040	6.2328
5.8462	6.0419	6.2381
5.8549	6.0543	6.2511
5.8584	6.0628	6.2602
5.8706	6.0673	6.2645
5.8789	6.0801	6.2778
5.9036	6.0884	6.2865
5.9083	6.0934	6.2913
5.9423	6.1015	6.2998
5.9546	6.1204	6.3051
5.9550	6.1417	6.3136
5.9632	6.1538	6.3191
5.9673	6.1546	6.3274
5.9756	6.1576	6.3336
5.9799	6.1663	6.3764
5.9882	6.1792	6.3894
5.9931	6.1838	6.3980
6.0010	6.1921	6.4031
6.0054	6.1972	6.4262
6.0192	6.2055	6.4321

Table 34.2 *(continued)*

Active medium: **CN**

Pumping: excitation of electronic and vibrational transitions due to photodissociation or by pulsed discharge in HCN vapor.

wavelengths (in vacuum), μm		
1.09966	1.10485	1.41911
1.09963	1.10445	1.41954
1.09965	1.10521	1.42005
1.09974	1.10603	1.42065
1.09987	1.10689	1.42132
1.10007	1.10726	1.42207
1.10031	1.10782	1.42289
1.10061	1.10879	1.42380
1.10096	1.10981	1.42478
1.10082	1.11090	1.42583
1.10136	1.11200	1.42696
1.10232	1.11321	1.42808
1.10288	1.41830	1.42945
1.10348	1.41849	1.42081
1.10414	1.41876	

Active medium: **OH**

Pumping: photolysis of O_3 and H_2 mixture; pulsed discharge in the mixture of O_3, H_2 and He; rotational transitions are excited by pulsed discharge in the mixture of SF_6, H_2 and O_2.

wavelengths (in vacuum), μm		
2.934.32	14.640	14.655
2.969.99	14.646	14.669
3.078.77	14.662	15.256
3.116.77	15.289	15.274
3.156.97	15.294	18.455
3.326.15	15.313	18.492
3.276.53	18.788	18.502
$\nu=0$	18.828	18.532
12.273	18.849	19.555
12.279	18.878	19.594
12.660	20.050	19.619
12.663	21.480	19.650
13.073	21.570	20.870
13.079	23.140	20.930
13.088	23.260	22.330
13.525	25.110	22.450
13.538	25.280	24.070
13.547	27.470	24.180
13.557	27.710	26.120
14.043	$\nu=1$	26.300
14.059	13.632	$\nu=2$
14.067	13.642	19.273
14.081	14.118	19.321
14.620	14.129	

Table 34.2 *(continued)*

Active medium: **OD** *Pumping*: Pulsed discharge in the mixture of SF_6, H_2 and O_2.		
wavelengths (in vacuum), μm		
ν=0	19.102	19.696
18.121	19.121	19.704
18.138	19.141	20.271
18.590	19.161	20.228
18.603	19.662	20.296
18.624	19.681	30.313

Active medium: **CO** (Fig. 34.8) *Pumping*: excitation of electronic transitions in the high current density discharge; typical pressure for UV lines 8 kPa, for visible lines – 90–250 Pa; excitation of rotational and vibrational lines in the pulsed or cw discharge in the mixture of CO and N_2 (with some He, Xe, Hg), in gas-dynamic expansion and chemical reactions.		
wavelengths (in vacuum), μm		
0.181085	0.519825	0.658285
0.187831	0.519888	0.658805
0.189784	0.519952	0.659287
0.195006	0.558566	0.659729
0.197013	0.558903	**0.660130**
0.450374	0.559213	**0.660438**
0.450508	**0.559498**	**0.660813**
0.450627	**0.559757**	**0.661093**
0.450728	**0.560198**	0.661154
0.450816	**0.560380**	0.661427
0.450888	0.560536	0.661334
0.450947	0.560665	0.661426
0.482082	0.560769	0.661536
0.482290	0.560849	0.661695
0.482483	0.560856	0.661817
0.482659	0.560967	0.662102
0.482820	0.561054	0.662186
0.482964	0.561114	0.662216
0.483091	0.561149	2.3474
0.483202	0.604536	2.3769
0.483297	0.604983	2.4380
0.483602	0.605400	2.4344
0.483638	0.605787	2.4696
0.483658	**0.606464**	2.5019
0.518103	0.606756	2.5350
0.518355	0.607016	2.5689
0.518586	0.607243	2.6036
0.518793	0.607438	2.6392
0.518987	0.607604	2.6756
0.519145	0.607734	2.6886
0.519298	0.607752	2.6914
0.519426	0.607831	2.7129
0.519508	0.607899	2.7262
0.519531	0.608018	2.7290
0.519617	0.608102	2.7319
0.519633	0.608155	2.7511
0.519740	0.608175	2.7647

Table 34.2 *(continued)*

Active medium: **CO** *(continued)*		
wavelengths (in vacuum), μm		
2.7676	**Band 3–2**	4.992099
2.7705	4.846781	5.002121
2.7903	4.856233	5.012268
2.8042	4.865803	5.022539
2.8071	4.875490	5.032938
2.8101	4.885296	5.043462
2.8306	4.895221	5.054117
2.8446	4.905267	5.064899
2.8476	4.915434	5.075812
2.8507	4.925723	5.086856
2.8892	4.936136	5.098033
2.8923	4.946672	5.109343
2.9288	4.957333	5.120787
2.9319	4.968120	5.131252
2.9351	4.979035	5.144084
2.9725	4.990016	5.155938
2.9757	5.001277	5.167931
2.9789	5.012578	5.180064
3.0174	5.023976	5.192338
3.0206	5.035544	5.204755
3.0668	5.047242	5.217312
$^{12}C^{16}O$	5.059073	5.230020
Band 1–0	5.071040	5.242872
4.735872	5.083144	5.255870
4.745130	5.095386	5.269018
4.754501	5.107776	5.282316
4.763984	5.120276	**Band 6–5**
4.773582	**5.132949**	5.008369
4.783295	5.145754	5.017940
4.793123	**Band 4–3**	5.027635
4.803067	4.880759	5.037454
4.813129	4.890016	5.047397
4.823310	4.899391	5.057467
4.833609	4.908883	5.067633
4.844029	4.918494	5.077988
4.854596	4.928238	5.088440
4.865231	4.938978	5.099023
4.876016	4.948052	5.109734
Band 2–1	4.958148	5.120577
4.767821	4.968369	5.131555
4.776892	4.978711	5.142663
4.786076	4.989181	5.153909
4.795373	4.999775	5.165289
4.804785	5.010497	5.176806
4.814312	5.021347	5.188460
4.823954	5.032321	5.200254
4.833714	5.043435	5.21187
4.843591	5.054676	5.224262
4.853586	5.066048	5.236479
4.863700	5.077554	5.248840
4.873935	**Band 5–4**	5.261343
4.884291	4.943828	5.273997
4.894769	4.953240	5.286796
4.905369	4.962772	5.299744
4.916094	4.972425	5.353049
4.926943	4.982220	

Table 34.2 *(continued)*

Active medium: **CO** *(continued)*		
wavelengths (in vacuum), μm		
Band 7–6	5.340935	5.119944
5.074432	5.353322	5.127892
5.312842	5.365859	5.135961
5.326091	5.378545	5.272122
5.339493	5.391382	5.282243
5.084106	5.404372	5.292498
5.094028	5.417516	5.302890
5.104017	5.430815	5.313418
5.114134	5.444270	5.324085
5.124408	5.457884	5.334893
5.134757	5.471632	5.345838
5.145264	5.485591	5.356924
5.155902	5.499688	5.368153
5.166672	**Band 9–8**	5.379526
5.177575	5.108900	5.391045
5.188617	5.100430	5.402707
5.199792	5.092002	5.414516
5.211102	5.201370	5.426477
5.222555	5.211315	5.438584
5.234145	5.221392	5.450840
5.245874	5.231603	5.463249
5.257745	5.241946	5.475810
5.269759	5.252424	5.488524
5.281916	5.263039	5.501394
5.294218	5.273789	5.514421
5.306666	5.284678	5.527606
5.319261	5.295704	5.540950
5.332005	5.306870	5.554455
5.344899	5.318176	5.568122
5.357945	5.329624	**Band 11–10**
5.371143	5.341216	5.122753
5.384494	5.352955	5.129726
5.398001	5.364837	5.136819
5.411665	5.376865	5.144032
Band 8–7	5.389039	5.151336
5.142062	5.401364	5.158822
5.151996	5.413837	5.166400
5.162000	5.426463	5.174100
5.172164	5.439219	5.181924
5.182459	5.452172	5.189871
5.192888	5.465259	5.334601
5.203447	5.478502	5.354901
5.214142	5.491904	5.365340
5.224972	5.505464	5.375920
5.235937	5.519186	5.386640
5.247038	5.553070	5.397499
5.258279	5.547118	5.408504
5.269659	5.561330	5.419652
5.281183	5.575710	5.430943
5.292846	**Band 10–9**	5.442381
5.304651	5.096828	5.453966
5.316600	5.104413	5.465699
5.328694	5.112118	5.477588

Table 34.2 *(continued)*

Active medium: **CO** *(continued)*		
wavelengths (in vacuum), μm		
Band 11–10	5.386899	5.688612
5.489622	5.396031	5.701037
5.501808	5.405298	5.713623
5.514147	5.414703	5.726379
5.526640	5.424245	5.739293
5.539288	5.433926	5.752373
5.552093	5.516474	5.765620
5.565057	5.527444	5.779036
5.578179	5.505651	5.792622
5.591463	5.538558	5.806380
5.604909	5.549827	5.820310
5.618619	5.561252	5.834415
5.632293	5.572816	5.848696
5.646235	5.584536	5.863154
Band 12–11	5.595412	**Band 15–14**
5.179608	5.608443	5.374479
5.186476	5.620630	5.381078
5.193387	5.632975	5.387807
5.207655	5.645485	5.394666
5.214974	5.658148	5.401656
5.222417	5.670974	5.408780
5.229984	5.683961	5.416030
5.237677	5.697113	5.423415
5.245496	5.710431	5.430932
5.429350	5.723916	5.438584
5.439978	5.737568	5.446369
5.450750	5.751390	5.454289
5.461666	5.765384	5.462344
5.472735	5.779551	5.470535
5.483932	5.793891	5.686687
5.495289	5.808408	5.698230
5.506793	5.823103	5.709930
5.518447	**Band 14–13**	5.721786
5.530252	5.354648	5.733804
5.542208	5.362186	5.745984
5.564315	5.369854	5.758322
5.566589	5.377654	5.770827
5.579010	5.385585	5.783496
5.591588	5.393649	5.796340
5.604325	5.401846	5.809343
5.617221	5.410177	5.822516
5.630278	5.418642	5.835859
5.643498	5.427242	5.583870
5.656881	5.435978	5.594893
5.670430	5.444850	5.606068
5.684145	5.583870	5.849374
5.698029	5.594893	5.863063
5.712083	5.606068	5.876926
Band 13–12	5.617391	5.890965
5.343257	5.628872	5.905182
5.351719	5.640505	5.919579
5.360314	5.652298	5.934157
5.369042	5.664243	5.948917
5.377903	5.676348	

Table 34.2 *(continued)*

Active medium: **CO** *(continued)*		
wavelengths (in vacuum), μm		
Band 16–15	6.11397	**Band 22–21**
5.746341	6.12852	6.300749
5.757944	6.14327	6.313813
5.769382	**Band 19–18**	6.327064
5.781150	6.030715	6.340507
5.793071	6.043187	6.354144
5.817413	6.055827	6.367975
5.829830	6.068653	6.382004
5.842415	6.081656	6.410658
5.855168	6.094841	6.425284
5.868096	6.108206	6.440113
5.881189	6.121753	6.455148
5.894455	6.135488	6.470338
5.907895	6.149406	6.485838
5.921510	6.163511	**Band 23–22**
5.935301	6.177806	6.400168
5.949271	6.192291	6.413510
5.963420	6.206969	6.427033
5.977751	6.221841	6.640760
5.992264	**Band 20–19**	6.454686
6.006963	6.097625	6.468812
Band 17–16	6.109989	6.497679
5.842407	6.122525	6.512417
5.854229	6.135248	6.527363
5.866221	6.148152	6.542518
5.878384	6.161240	6.557884
5.890704	6.174509	6.573463
5.903202	6.187966	**Band 24–23**
5.915865	6.201608	6.502437
5.928701	6.215438	6.516055
5.941711	6.229461	6.529870
5.954841	6.243673	6.543890
5.968257	6.258077	6.558117
5.981799	6.272677	6.572552
5.995510	6.287472	6.587194
6.009406	6.302466	6.602046
6.023483	**Band 21–20**	6.617109
6.037744	6.191453	6.632387
6.052189	6.204067	6.647880
Band 18–17	6.216856	6.663590
5.941314	6.229843	7.392748
5.953562	6.243012	**Band 32–31**
5.965936	6.256366	7.441563
5.978508	6.269915	7.457822
5.991256	6.283649	7.474367
6.004170	6.297578	7.491175
6.017267	6.311705	7.508252
6.030541	6.326023	7.525593
6.043400	6.340538	7.543207
6.05760	6.355252	**Band 33–32**
6.07142	6.370176	7.593623
6.08542	6.385283	7.610553
6.09960		7.627777

Table 34.2 *(continued)*

Active medium: **CO** *(continued)*		
wavelengths (in vacuum), μm		
Band 33–32	**Band 10–9**	5.764755
7.645277	5.437800	5.771303
7.663057	5.448674	5.776747
7.681105	5.459684	5.784148
7.734191	5.470881	5.788893
7.751571	5.482117	5.797155
Band 34–33	5.493543	5.798443
7.769230	5.505110	5.801198
7.787173	5.516821	5.810042
7.805408	5.528674	5.810319
7.823924	5.540673	5.813656
Band 35–34	5.552816	5.821796
7.879916	5.500027	5.826267
7.897740	5.510954	5.833703
7.915858	5.522022	5.839055
7.934266	5.533232	5.845769
7.952969	5.544580	5.851981
8.031109	5.556074	5.857987
8.049391	5.567715	5.865079
8.067991	5.574745	5.870367
8.086665	5.579476	5.882903
8.106093	5.585871	5.892408
8.206919	5.591415	5.895604
8.226011	5.597139	5.904387
8.245417	5.603490	5.908465
8.265146	5.608550	5.916519
$^{13}C^{16}O$	5.615709	5.921487
Band 7–6	5.620109	5.928814
5.303284	5.628080	5.934676
5.314516	5.631813	5.941269
5.325877	5.640600	5.948030
5.337376	5.643666	5.953887
Band 8–7	5.651276	5.964829
5.328437	5.653272	5.966669
5.339330	5.655664	5.976868
5.350355	5.662607	5.979617
5.367272	5.667813	5.989075
5.372811	5.674081	5.992728
5.384241	5.680114	6.001439
5.395810	5.685704	6.006059
5.407516	5.685704	6.013970
Band 9–8	5.692566	6.019459
5.377032	5.697477	6.026667
5.387850	5.705170	6.033077
5.398801	5.709398	6.039534
5.409888	5.717932	6.052565
5.421112	5.721472	6.063494
5.4322471	5.729687	6.065767
5.4439721	5.730846	6.075932
5.4556001	5.733698	6.079138
5.4673911	5.741227	6.088540
5.4793111	5.746076	6.092683
5.4913771	5.752912	6.101318
5.5035831	5.758611	6.106404

Table 34.2 *(continued)*

Active medium: **CO** *(continued)*		
wavelengths (in vacuum), μm		
Band 10–9	6.204140	6.301586
6.114263	6.209772	6.310036
6.120296	6.217521	6.323871
6.127383	6.230817	6.337216
6.139855	6.231077	6.349964
6.140673	6.233577	6.350412
6.152366	6.244817	6.352094
6.154138	6.256511	6.363789
6.165049	6.258735	6.366487
6.167780	6.289624	6.377348
6.177904	6.272833	6.381068
6.181597	6.282915	6.391091
6.190935	6.287118	6.405026
6.195595	6.296385	

TRANSITIONS IN POLYATOMIC MOLECULES

Active medium: $\mathbf{Xe_2Cl}$ [8] *Pumping*: electron beam excitation of the mixture Ar, Xe and CCl.
wavelengths (in vacuum), μm
0.518

Active medium: $\mathbf{Xe_2Br}$ [9] *Pumping*: electron beam excitation of the mixture Ar, Xe and $CHBr_3$.
wavelengths (in vacuum), μm
0.440

Active medium: $\mathbf{Kr_2F}$ [9] *Pumping*: electron beam excitation of the mixture Ar, Kr and NF_3.
wavelengths (in vacuum), μm
0.430

Active medium: $\mathbf{O_3}$ *Pumping*: optical, by CO_2 laser radiation.		
wavelengths (in vacuum), μm		
121	163.61	171.5

Active medium: $\mathbf{CO_2}$ *Pumping*: cw discharge in the mixture CO_2, N_2 and He (composition 1:2.5:10, Fig.34.9); excitation in the longitudinal discharge, gas-dynamic laser operation with mixture pump-through; chemical excitation with energy transport via HF or DF molecules; pulsed excitation in transversal discharge.		
wavelengths (in vacuum), μm		
$^{12}C^{16}O_2$	4.3249	4.3612
Band	4.3276	4.3644
102–101	4.3549	4.3677
4.3203	4.3580	4.3711

Table 34.2 *(continued)*

Active medium: $\mathbf{CO_2}$ *(continued)*		
wavelengths (in vacuum), μm		
$^{12}C^{16}O_2$	9.3544134	10.0850408
4.3745	9.3673380	10.0946764
4.3779	9.3805340	10.1046049
4.3814	9.3940033	10.1148262
4.3849	9.4147242	10.1253400
$^{12}C^{18}O_2$	9.4288857	10.1361464
Band	9.4433275	10.1472454
101–100	9.4580515	10.1586374
4.314	9.4730598	10.1703225
4.340	9.4883540	10.1823014
4.354	9.5039361	10.1945745
4.346	9.5198079	10.2071425
4.371	9.5350711	10.2200062
4.377	9.5524275	10.2331666
4.382	9.5691788	10.2466246
4.385	9.5882267	10.2603814
4.392	9.6035727	10.2744384
4.398	9.6212185	10.2887964
$^{12}C^{16}O_2$	9.6391656	10.3034581
Band	9.6574156	10.3184241
001–020	9.6750700	10.3336965
9.0702655	9.6948301	10.3492772
9.0757663	9.7139973	10.3651683
9.0814571	9.7334730	10.3813718
9.0873410	9.7532586	10.3978901
9.0934211	9.7733552	10.4232632
9.0997003	9.7937640	10.4405795
9.1061815	9.8144862	10.4582196
9.1128676	9.8355229	10.4761866
9.1197615	9.8568751	10.4944835
9.1268660	9.8785439	10.5131136
9.1341839	9.9005300	10.5320802
9.1417179	9.9228344	10.5513866
9.1494708	9.9454579	10.5710372
9.1574453	9.9684012	10.5910352
9.1656440	9.9916650	10.6113848
9.1740695	10.0152438	10.6320902
9.1827244	10.0391561	10.6531558
9.1916114	10.0633844	10.6745861
9.2007329	10.0879349	10.6963859
9.2100915	**Band**	10.7185600
9.2196895	**001–100**	10.7411135
9.2295296	9.9985568	10.7640517
9.2396141	10.0049238	10.7873802
9.2499453	10.0115934	10.8111046
9.2605258	10.0185643	10.8352307
9.2713577	10.0258352	10.8597648
9.2824434	10.0334048	10.8847131
9.2937852	10.0412720	10.9100823
9.3053853	10.0494358	10.9358790
9.3172460	10.0578953	10.9221103
9.3293695	10.0666497	10.9887835
9.3417579	10.0756984	11.0159060

Table 34.2 *(continued)*

Active medium: $\mathbf{CO_2}$ *(continued)*		
wavelengths (in vacuum), μm		
Band	**Band**	10.359124
001–100	**002–101**	10.374040
11.0434858	10.146624	10.389256
11.0715308	10.157295	10.404773
11.1000493	10.168257	10.420594
11.1290499	10.179508	10.53097
11.1585415	10.191050	10.54916
11.1885334	10.202833	10.56762
11.2190349	10.215008	10.58646
11.2500559	10.227424	10.60562
Band	10.240133	10.665124
002–021	10.253135	10.685646
9.209171	10.266431	10.706519
9.217773	10.280023	10.727749
9.226615	10.293911	10.749339
9.235699	**10.380097**	10.771295
9.245029	**10.322582**	10.793621
9.254607	10.337367	10.816324
9.264436	10.355455	10.839408
9.274517	10.367847	10.862879
9.284854	10.383545	**Band**
9.295448	10.399550	**004–103**
9.306302	10.415866	10.55376
9.328800	10.458029	10.57170
9.340448	10.475449	10.5900
9.352366	10.493192	10.60858
9.364555	10.511259	**Band**
9.377018	10.529654	**001–110**
9.389757	10.548380	10.591025
9.402774	10.567440	10.789077
9.450554	10.586838	10.890194
9.464848	10.606578	10.900964
9.479432	**10.626664**	10.921469
9.494307	**10.647099**	10.930707
9.509476	**10.667888**	10.942351
9.524939	**10.689036**	10.951486
9.540700	**10.710547**	10.972615
9.556760	10.732425	10.985266
9.573121	10.754676	11.007301
9.589785	10.777305	11.015934
9.606753	10.800317	11.029744
9.624027	10.823718	11.083630
9.641609	10.847513	**Band**
9.655900	10.871709	**011–110**
9.677702	10.896312	10.50816
9.696217	10.921327	10.51001
9.715046	**Band**	10.52029
9.734191	**003–102**	10.52277
9.753653	10.288987	10.53273
9.773433	10.302426	10.54550
9.793533	10.316157	10.54919
9.813954	10.330184	10.55859
	10.344505	10.56284

Table 34.2 *(continued)*

Active medium: $\mathbf{CO_2}$ *(continued)*		
wavelengths (in vacuum), μm		
Band	11.1980	9.06790685
011–110	11.2035	9.07687197
10.57201	11.2235	9.08601437
10.57678	11.2295	9.09533538
10.58575	11.2495	9.10483634
10.60556	11.2545	9.11451859
10.61421	11.2770	9.12438343
10.92146	11.2805	9.13443217
10.93070	**Band**	9.14466611
10.94235	**012–111**	9.15508653
10.95148	10.53907	9.16569469
10.96361	10.54173	9.17649187
10.97261	10.55455	9.18747931
10.98526	10.5639	9.19865824
10.99409	10.58112	9.21002991
11.00730	10.60362	9.22159552
11.01593	10.60885	9.23931022
11.02974	**Band**	9.25136594
11.03813	**101–200**	9.26361978
11.05258	10.51027	9.26361978
11.06069	10.54271	9.27607290
11.07582	**Band ?**	9.22872646
11.08363	13.144	9.30158160
11.09947	13.154	9.31463943
11.10693	13.159	9.32790109
11.12354	13.541	9.34136768
11.13062	13.870	9.35504028
11.14803	14.210	9.36891998
11.15468	14.100	9.38300783
11.17295	14.160	9.39730489
11.17914	14.190	9.41181219
11.19830	16.586	9.42653076
11.20398	16.597	9.44146159
11.22408	17.023	9.45660568
Band	17.029	9.47196399
011–030	17.036	9.48753750
10.9735	17.048	9.50332713
10.9950	17.370	9.51933381
11.0165	17.376	9.53555845
11.0300	17.390	9.55200192
11.0385	$^{12}C^{18}O_2$	9.56866511
11.0535	**Band**	9.58554884
11.0610	**001–020**	9.60265396
11.0760	8.98770094	9.61998126
11.0850	8.99496997	9.63753153
11.1000	9.00240216	**Band**
11.1070	9.00999899	**001–100**
11.1235	9.01776190	10.10434660
11.1315	9.02569233	10.11295490
11.1485	9.03379171	10.12186285
11.1555	9.04206146	10.13107088
11.1736	9.05050297	10.14057947
11.1790	9.05911764	10.15038920

Table 34.2 *(continued)*

Active medium: $\mathbf{CO_2}$ *(continued)*		
wavelengths (in vacuum), μm		
Band	9.78991669	11.10656357
001–100	9.80429565	11.12781897
10.16050076	9.85718656	11.14939871
10.17091494	9.87304135	11.17130639
10.18163264	9.88923095	11.19354572
10.19254484	9.90575779	11.21612059
10.20398263	9.92262428	11.23903501
10.21561722	9.93983281	11.26229313
10.22755988	9.95738572	11.28589928
10.23981204	9.97528536	11.30985791
10.25237519	9.99353402	11.33417365
10.26525095	10.01213398	11.35885125
10.27844106	10.03108747	11.38389568
10.29194733	10.05039672	11.40931204
10.30577171	10.07006390	11.49282306
10.38758981	10.09009116	11.50270688
10.40353724	10.11048062	11.51287542
10.41982107	10.13123435	11.52332884
10.43644425	10.15235439	11.53406738
10.45340986	10.17384275	11.54509140
10.47072112	10.19570139	11.55640135
10.48838137	**Band**	11.56799780
10.50639411	**001–100**	11.57988138
10.52476294	10.58841900	11.59205288
10.54349163	10.60062644	11.60451316
10.56258410	10.61310290	11.61626318
10.58204439	10.62584867	11.63030402
10.60187671	10.63886409	11.64363687
10.62208541	10.65214962	11.65726303
10.64267502	10.66570577	11.67118389
10.66365020	10.67953319	11.68540097
10.68501579	10.69363257	11.69991589
10.70677681	10.70800473	11.71473039
10.72893842	10.72265056	11.72984633
$^{13}C^{16}O_2$	10.73757106	11.81846619
Band	10.75276730	11.83559484
001–020	10.76824046	11.85304268
9.59239947	10.78399183	11.87081258
9.60168999	10.80002276	11.88890760
9.61125719	10.81633473	11.90733089
9.62110404	10.83292931	11.92608578
9.63123346	10.84980818	11.94517571
9.64164836	10.86697309	11.96460427
9.65235161	10.88442594	11.98437521
9.66334607	10.90216870	12.00449241
9.67463455	10.92020346	12.02495992
9.68621987	10.93853242	12.04578194
9.69810480	10.96658256	12.06696282
9.71029210	10.98565737	12.08850708
9.72278448	11.00503495	12.11041941
9.73558466	11.02471801	12.13270467
9.74869533	11.04470940	12.15536788
9.76211913	11.06501204	12.17841426
9.77585872	11.08562903	

Table 34.2 *(continued)*

Active medium: $\mathbf{CO_2}$ *(continued)*		
wavelengths (in vacuum), μm		
Band	9.80651821	11.677
001–020	9.82095385	11.700
9.491.85112	9.83563588	11.723
9.49995782	9.85056572	11.746
9.50826101	9.86574476	11.770
9.51676274	9.88117435	11.794
9.52546505	9.89685583	11.819
9.53436995	9.91279050	11.843
9.54347943	9.92897965	11.868
9.55279546	9.94542451	$^{12}C^{18}O_2$
9.56231997	9.96212632	**Band**
9.57205489	9.97908626	**100–010**
9.58200213	9.99630549	16.596
9.59216355	10.01378513	16.780
9.60254101	10.03152628	16.927
9.61313635	10.04952999	16.970
9.62395138	10.06779729	17.280
9.63498789	10.08632916	17.463
9.64624766	10.10512655	17.596
9.65773242	10.12419038	17.639
9.66944390	10.14352152	17.684
9.68138381	$^{14}C^{16}O_2$	17.730
9.69355383	**Band**	17.775
9.70595561	**001–100**	17.821
9.71859080	11.329	17.915
9.75121048	11.346	17.962
9.76467517	11.364	18.010
9.77838035	11.382	18.053
9.79232753	11.400	

Active medium: $\mathbf{NO_2}$ [6] *Pumping*: gas-dynamic expansion.		
wavelengths (in vacuum), μm		
11.0	16.63	

Active medium: $\mathbf{N_2O}$ *Pumping*: pulsed and cw discharge in the mixture of N_2O, N_2 and He (composition 1:3.5:40, pressure 1.7 kPa); optical pumping by HBr-laser radiation (4.465 μm); gas-dynamic expansion.		
wavelengths (in vacuum), μm		
Band	10.4081	10.5250
001–010	10.4161	10.5337
9.48	10.4242	10.5425
Band	10.4323	10.5513
001–100	10.4405	10.5602
10.3456	10.4570	10.5692
10.3532	10.4653	10.5761
10.3609	10.4737	10.5872
10.3687	10.4821	10.5963
10.3765	10.4906	10.6054
10.3854	10.4991	10.6146
10.3922	10.5077	10.6239
10.4001	10.5163	10.6332

Table 34.2 *(continued)*

Active medium: **N_2O** *(continued)*		
wavelengths (in vacuum), μm		
Band	10.9613	10.59486
001–100	10.9726	10.60382
10.6426	10.9839	10.61282
10.65	10.9953	10.62189
10.6614	11.0067	10.63100
10.6710	11.0182	10.72517
10.6806	11.0298	10.73489
10.6903	11.0415	10.74468
10.6999	**Band**	10.75450
10.7097	**002–101**	10.76939
10.7195	10.40331	10.77434
10.7294	10.41107	10.78435
10.7393	10.41889	10.79441
10.7493	10.42676	10.80453
10.7593	10.43468	10.81471
10.7694	10.44265	10.82495
10.7796	10.45067	10.83524
10.7898	10.45874	10.84560
10.8000	10.46686	10.85601
10.8104	10.47503	10.86648
10.8208	10.48325	10.87701
10.8312	10.49157	10.88760
10.8418	10.49985	10.89825
10.8523	10.50823	10.90896
10.8629	10.51666	10.91973
10.8736	10.52513	10.93056
10.8844	10.53367	10.94145
10.8952	10.54225	10.95241
10.9061	10.55089	10.93642
10.9170	10.55958	10.97450
10.9280	10.56832	10.98564
10.9390	10.57712	10.99684
10.9501	10.58596	

Active medium: **S_2O** [6] *Pumping:* gas-dynamic expansion.		
wavelengths (in vacuum), μm		
12.87	20.57	

Active medium: **OCS** *Pumping:* pulsed discharge in pure OCS, in mixtures: OCS and He, OCS and N_2, OCS and CO, OCS and CO+He; optical pumping by pulsed CO_2-laser.		
wavelengths (in vacuum), μm		
3.428		
Band	8.2595	8.3809
001–100	8.2623	8.3839
8.2388	8.2645	8.3870
8.2416	8.3654	8.3900
8.2439	8.3685	8.3930
8.2518	8.3715	8.3962
8.2543	8.3746	8.3999
8.2571	8.3779	8.4024

Table 34.2 *(continued)*

Active medium: **OCS** *(continued)*		
wavelengths (in vacuum), μm		
Band	8.4146	**Band ?**
001–100	8.4178	18.983
8.4055	8.4213	19.057
8.4065	8.4243	123.0
8.4117		132.0

Active medium: **CH_2** [6] *Pumping:* gas-dynamic expansion.		
wavelengths (in vacuum), μm		
11.0	16.63	

Active medium: **CS_2** *Pumping:* cw discharge in the mixture of CS_2 (13 Pa) and N_2 (250 Pa) pumped through; excitation of N_2 in cw discharge before mixing with CS_2; transversal discharge in CS_2 at normal pressure.		
wavelengths (in vacuum), μm		
11.482	11.524	**11.965**
11.489	11.531	**11.986**
11.596	**11.538**	**12.217**
11.503	**11.545**	**12.241**
11.510	$^{13}C^{32}S_2$	**12.249**
11.517	**11.960**	

Active medium: **HCO** [6] *Pumping:* gas-dynamic expansion.		
wavelengths (in vacuum), μm		
7.15	16.1	

Active medium: **DCO** [6] *Pumping:* gas-dynamic expansion.
wavelengths (in vacuum), μm
9.92

Active medium: **FCO** [6] *Pumping:* gas-dynamic expansion.		
wavelengths (in vacuum), μm		
8.15	11.90	

Active medium: **COS** [6] *Pumping:* gas-dynamic expansion.		
wavelengths (in vacuum), μm		
6.43	8.24	

Active medium: **HCN** [6] *Pumping:* pulsed discharge in various gas mixtures (e.g. CH_4 and NH_3 or CH_4 and N_2); many lines are available at cw operation; gas-dynamic expansion (3.69, 7.19, 7.63 μm) [6].		
wavelengths (in vacuum), μm		
3.69	7.63	96.401
7.19	81.554	98.693

Table 34.2 *(continued)*

Active medium: **HCN** [6]		
wavelengths (in vacuum), μm		
101.257	143.932	310.8870
110.240	138.768	335.1831
112.066	165.150	**336.5578**
113.311	201.059	372.5283
116.132	211.001	538.2
126.164	222.949	545.4
128.629	284.0	676.0
130.838	309.7140	773.5

Active medium: **DCN** *Pumping*: pulsed discharge in the mixture of D_2 and BrCN or CD_4 and ND_3; some lines operate cw; gas-dynamic expansion (4.31 and 12.62 μm) [6].		
wavelengths (in air), μm		
4.31	189.9490	194.7644
12.62	190.0080	204.3872
181.789	194.7027	

Active medium: **NCH** [6] *Pumping*: gas-dynamic expansion.
wavelengths (in vacuum), μm
6.68

Active medium: **FCN** [6] *Pumping*: gas-dynamic expansion.
wavelengths (in vacuum), μm
8.25

Active medium: **ClCN** [6] *Pumping*: gas-dynamic expansion.		
wavelengths (in vacuum), μm		
5.44	6.81	

Active medium: **BrCN** [6] *Pumping*: gas-dynamic expansion.		
wavelengths (in vacuum), μm		
5.38	6.16	

Active medium: **ICN** [6] *Pumping*: gas-dynamic expansion.		
wavelengths (in vacuum), μm		
5.93	6.45	

Active medium: $\mathbf{NH_2}$ [6] *Pumping*: gas-dynamic expansion.
wavelengths (in vacuum), μm
15.4

Active medium: $\mathbf{NF_2}$ [6] *Pumping*: gas-dynamic expansion.		
wavelengths (in vacuum), μm		
18.8	25.4	

Table 34.2 *(continued)*

Active medium: **FNO** [6] *Pumping*: gas-dynamic expansion.
wavelengths (in vacuum), μm
9.28

Active medium: **ClNO** [6] *Pumping*: gas-dynamic expansion.
wavelengths (in vacuum), μm
8.24

Active medium: **BrNO** [6] *Pumping*: gas-dynamic expansion.
wavelengths (in vacuum), μm
6.52

Active medium: $\mathbf{H_2O}$ [6] *Pumping*: pulsed discharge in water vapor at 50−130 Pa; for some lines cw operation; gas-dynamic expansion (4.19 and 4.57 μm) [6].		
wavelengths (in vacuum), μm		
2.28	**27.970755**	48.765
4.19	28.054	49.06
4.57	28.270	49.430
4.77	28.295	53.910
7.095	28.356	55.000
7.206	28.451	55.088
7.287	32.924	56.129
7.299	33.308	57.659
7.392	**33.329**	57.799
7.427	34.60	66.800
7.459	35.017	66.903
7.545	35.383	67.169
7.567	35.833	72.856
7.592	36.606	73.401
7.709	37.848	78.443329
7.711	38.086	79.091010
7.712	39.695	85.564
7.742	40.45	86.301
9.4773	40.638	86.471
9.570	42.51	87.323
11.83	45.517	87.469
11.96	45.91	89.772
16.932	47.244	115.32
23.13	47.39	118.59104
23.365	47.468	120.08
24.966	47.687	220.280
25.162	48.19	350.20
26.595	48.366	
26.660	48.676	

Table 34.2 *(continued)*

Active medium: **D_2O**
Pumping: pulsed discharge in D_2O vapor (27–130 Pa); for some lines cw operation; optical pumping by CO_2-laser.

wavelengths (in vacuum), μm		
26.36	56.830	103.33
33.896	61.182	**107.72019**
35.081	71.944	107.91
36.096	72.427	108.88
36.324	**72.747780**	110.49
36.526	73.337	111.74
37.788	74.341	170.08
37.864	74.526	171.67
39.53	76.305	239
40.994	78.16	263
41.79	83.730	276
48.80	84.111	358.5
50.71	**84.278897**	385
54.73	99.00	

Active medium: **SO_2**
Pumping: pulsed discharge in the mixture of He (500 Pa) and SO_2 (50 Pa); for some lines cw operation; gas-dynamic expansion (3.76 and 4.96 μm) [6].

wavelengths (in vacuum), μm		
3.76	**141.06**	151.35
4.96	142.00	192.80
139.83	149.94	206.53
140.82	151.08	215.27

Active medium: **H_2S**
Pumping: pulsed discharge in H_2S (20 Pa); gas-dynamic expansion (6.58, 6.63 μm) [6].

wavelengths (in vacuum), μm		
6.58	62.6	116.8
6.63	73.54	126.2
33.30	81.45	129.1
37.6000	83.45	130.60
48.70	**87.580**	135.3
52.307	92.0	140.8
55.612	96.4	162.4
60.224	103.8	192.9
61.413	108.8	**225.3**

Active medium: **D_2S** [6]
Pumping: gas-dynamic expansion.

wavelengths (in vacuum), μm		
9.19	9.31	

Active medium: **BF_2** [6]
Pumping: gas-dynamic expansion.

wavelengths (in vacuum), μm		
11.92	26.35	

Table 34.2 *(continued)*

Active medium: **BCl_3**
Pumping: cw discharge in CO_2 with BCl_3.

wavelengths (in air), μm		
18.3	19.4	22.4
18.8	20.2	23.0
19.1	20.6	

Active medium: **SF_6**
Pumping: two photon excitation by CO_2-laser.

wavelengths (in air), μm
15.9005

Active medium: **NH_3**
Pumping: high current density pulsed discharge at 70–130 Pa; optical pumping by N_2O, CO_2 or HF-lasers.

wavelengths (in air), μm		
6.27	**12.689**	64.5
6.69	**12.8115**	64.7274
9.3	12.812	67.19
9.6	**12.851**	67.24
9.7	12.876	72.6
9.9	**12.921**	72.76
10.2	13.031	74.15
10.5	13.114	78.28
10.6	13.124	81.53
10.7	13.145	83.60
11.0	13.176	83.85
11.446	13.269	84.64
11.459	13.218	87.1
11.526	13.331	88.05
11.5547	13.411	88.20
11.721	13.576	88.90
11.80	13.7261	90.50
11.811	13.821	90.93
11.994	15.8782	92.87
12.010	15.9452	112.98
12.078	18.9250	114.29
12.0791	19.5497	119.02
12.1143	25.4744	147.04
12.1558	25.8839	147.15
12.1846	26.1046	147.2
12.245	26.4416	151.49
12.251	26.7068	151.5
12.266	27.8437	155.17
12.280	34.2248	155.28
12.286	35.1573	216.44
12.316	35.5011	223.91
12.348	36.02	225.39
12.520	36.1686	225.07
12.526	49.0356	256.61
12.541	54.45	263.43
12.566	56.8631	280.5
12.591	58.01	281.35
12.631	63.25	281.48

Table 34.2 *(continued)*

Active medium: NH_3 *(continued)*		
wavelengths (in air), μm		
290.4	311.75	15.7
290.9	388	16.0
291.2	404.89	17.8
291.35	$^{15}NH_3$	89.68
291.95	14.3	375.9
301.2	14.6	218.9
306.28	15.2	111.9

Active medium: PH_3 *Pumping*: optical pumping by pulsed CO_2-laser.		
wavelengths (in air), μm		
77.58	90.26	104.4
83.77	97.19	106.04
89.76	97.30	106.05
89.80	102.62	106.09
106.23	140.85	180.54
109.7	145.88	182
116.88	146.07	186.25
117.01	146.34	187.56
121.45	155.07	194.47
129.78	156.34	194.70
129.98	166.73	194.89
130.14	166.79	195.18
135.95	166.84	223.07
136.71	166.87	

Active medium: NOCl *Pumping*: optical pumping by pulsed CO_2-laser.		
wavelengths (in air), μm		
16.4	16.69	16.86
16.52	16.7	16.9
16.57	16.75	16.99

Active medium: CF_4 *Pumping*: optical pumping by pulsed CO_2-laser.		
wavelengths (in air), μm		
15.33	15.74	16.12
15.41	15.76	16.18
15.49	15.77	16.20
15.50	15.84	16.24
15.55	15.85	16.27
15.56	15.91	16.31
15.58	15.94	16.35
15.60	16.00	16.40
15.61	16.03	16.85
15.62	16.07	
15.70	16.10	

Active medium: CF_3Br *Pumping*: optical pumping by pulsed CO_2-laser.		
wavelengths (in air), μm		
823.4	885.2	

Table 34.2 *(continued)*

Active medium: CF_3I *Pumping*: optical pumping by pulsed CO_2-laser.		
wavelengths (in air), μm		
13.54	13.57	13.63

Active medium: C_2H_2 *Pumping*: pulsed electric discharge in the mixture C_2H_2, H_2 and He; optical pumping by CO_2-laser.		
wavelengths (in vacuum), μm		
8.0334	17.45	19.03
8.0340	17.61	19.13
8.0347	17.77	19.21
8.0356	18.67	19.27
8.0380	18.79	19.67
8.0409	18.85	20.01
8.0442	18.96	20.44

Active medium: C_2H_4 *Pumping*: optical pumping by CO_2-laser.		
wavelengths (in air), μm		
10.53	10.98	

Active medium: H_2CO *Pumping*: pulsed discharge at the pressure 7–50 Pa; optical pumping by CO_2-laser.		
wavelengths (in vacuum), μm		
101.9	163.8	D_2CO
119.6	170.2	233
122.8	184.4	245
125.9	HDCO	279
155.1	195	733.5739
157.6	196	752.6807
159.5		

Active medium: $(H_2CO)_3$ *Pumping*: optical pumping by CO_2-laser.		
wavelengths (in vacuum), μm		
384	680	815
433	696	890
460	712	891
512	750	948.9247
619		

Active medium: HCOOH *Pumping*: optical pumping by CO_2-laser.		
wavelengths (in vacuum), μm		
229.39	334.82	392
254.80	334.91	393.6311
278.61	336	394.2
302.08	336.3	396
302.2781	342.74	401
309.23	359.81	403
311.45	368	404.1
319.48	388	405.5848

Table 34.2 *(continued)*

Active medium: **HCOOH** *Pumping:* optical pumping by CO_2-laser.		
wavelengths (in vacuum), μm		
405.75	446.75	669.5308
406	446.8730	670
413	447	742.5723
414	447.58	743
418.51	458.43	744.0503
418.6	458.5229	745
419.55	458.6	761
420	460.51	785
420.26	492	786.1617
421	493.28	HCOOD
428	496	462
432.1093	512.88	925
432.6313	513.2	DCOOD
432.6325	515.1690	381
433.10	518.83	492
435	530	305
437.7	533.6773	526
438	534.5	569
441	534.8	349
445.21	577	790
445.81	580.3872	937
445.8971	580.52	
446.5054	582	

Active medium: $\mathbf{CH_2F_2}$ *Pumping:* optical pumping by CO_2-laser.		
wavelengths (in air), μm		
95.5	194.5	381.8
105.5	202.5	**382.9**
109.3	**214.5**	**394.7**
117.7	227.6	418.1
121.7	**230.1**	432.4
122.4	**235.7**	434.9
122.4	**236.5**	464.5
134.0	255.9	503.6
135.3	261.7	511.3
158.5	270.0	540.8
158.9	**272.2**	567.5
165.8	**287.7**	588.1
165.9	**289.4**	642.5
166.6	293.9	657.2
184.3	298.2	725.1
191.8	326.5	1448.1
193.9	355.2	

Active medium: $\mathbf{CH_2Cl_2}$ *Pumping:* optical pumping by CO_2-laser.		
wavelengths (in air), μm		
249	342	631
254	469	829
258	520	

Table 34.2 *(continued)*

Active medium: $\mathbf{CH_3F}$ *Pumping:* optical pumping by CO_2-laser. CW operation for 496 μm line.		
wavelengths (in air), μm		
9.75	251.91	496.1009
190.3	372.68	541.113
192.78	397.51	541.147
195.0	419	595
196.0	451.903	1221.79
199.14	451.924	
200.3	494	
215.3	496.072	

Active medium: $\mathbf{CH_3Cl}$ *Pumping:* optical pumping by CO_2-laser.		
wavelengths (in air), μm		
227.15	275.09	397.6
236.25	281.67	461.20
240.98	286.79	511.90
250.4	307.65	568.81
254	333.96	870.80
261.03	349.34	943.97
271.29	354	958.25
273.7	364.5	968
275.00	378.57	1886.87
CD_3Cl		
224	383.28	735.12
245	443.26	883.59
246	449.79	1239.47
249	464.76	1990.75
288	480.31	
291.27	519.30	
318	698.55	

Active medium: $\mathbf{CH_3Br}$ *Pumping:* optical pumping by CO_2-laser.		
wavelengths (in air), μm		
245.04	414.98	715.40
264.05	418.31	749.29
279.81	422.78	749.36
294.28	508.48	831.13
311.07	531.06	925.52
311.10	545.21	990.15
311.20	545.39	1310.38
311.21	564.68	1572.64
332.86	585.72	1965.34
333.15	631.93	
352.75	632.00	
380.02	658.53	
407.72	660.70	

Table 34.2 *(continued)*

Active medium: $\mathbf{CH_3I}$ *Pumping*: optical pumping by CO_2-laser.		
wavelengths (in air), μm		
377.45	508.37	583.87
390.53	517.33	639.73
392.48	525.32	670.99
447.1424	529.28	719.30
457.25	542.99	964
459.18	576.17	1063.29
477.87	578.90	1253.738
CD_3I	556.8755	730.3234
272	569.4773	734.2624
301	599.5499	745
390	614.1098	788.48
433.1038	640	895
444.3862	644	918.6101
460.5619	660.5822	953.8799
487.2260	667.2322	981.7094
490.3909	670.0940	1005.3476
523.4061	670.1143	1099.5441
540	691.1292	1549.5048

Active medium: $\mathbf{CH_3OH}$ *Pumping*: optical pumping by CO_2-laser.		
wavelengths (in air), μm		
37.5	164.5076	251.56
40.2	164.77	253.6
42.18	164.7832	254.1
43.4	170.57638	263.7
43.47	171.3	264.6
55.39	185.5	267.4432
58.1	186.03	278.8
60.25	190.3209	280.96
65.1	191.2	290.62
65.6	191.57	292.2
69.70	191.58	292.5
70.511716	191.63	293.78
73.30	193.2	301.9943
77.92	194.01	369.11368
80.3	198.8	386.20
80.6	202.4	390.1
85.59	205.3	392.06871
92.60	206.9	416.5224
92.69	209.89	417.8
96.522394	211.25	451.9
97.48	214.35	469.02330
117.95	218.22	470
118.83409	223.5	471
129.5497	232.85	486.1
133.1196	232.93906	570.56864
151.35	237.6	603.06
159.2	242.4727	614.92
162	242.79	627.34
163.03353	246	694.17
163.9	250.78129	695
164	251.13983	699.42258
164.3		

Table 34.2 *(continued)*

Active medium: $\mathbf{CH_3OH}$ *(continued)*		
wavelengths (in air), μm		
$^{13}CH_3OH$	148.59041	268.57203
85.31729	149.27226	280.21826
86.11179	152.07568	280.23974
103.48081	157.92848	332.6033
115.82324	203.63578	338.9737
118.01314	208.41205	461.3848
146.09739	238.52268	
CD_3OH	266	422
34.8	266.2	435
37.6	267	455
40.1	268	472
41.5	276	480
41.8	277	483
43.9	278	495
49.8	285	498
52.9	286.6	508
60.8	287.4	517
71.0	290.0	551
76.1	297	553
82.1	299	554
86.4	309	583
102.6	310	599
112.3	321	646
128.7	336	648
144.0	346	680
158	350	685
179	351	695
182.4	352	702
184	353	703
191.9	370	711
201	385	722
219.9	386	745
222	398	760
223	407	774
232	409	862
236	410	968
238.3	412	1100
253.2	419	1146
258.7	421	1290
CH_3OD	117	225
46.7	134	229.1
57	134.7	238
69.5	134.8	294.81097
70.3	136	305.72610
100.8	145.66171	330.1
104	212.8	417.1
110	215.37244	
CD_3OD	184	312
35	229	354
41	255	406
78	312	414
119	339	495
150	299	869
165		

Table 34.2 *(continued)*

Active medium: $\mathbf{CH_3OH}$ *(continued)*		
wavelengths (in air), μm		
CH_2DOH	207	322
109	238	364
125	250	374
151	272	396
164	295	468
167	296	616
171	308	
CHD_2OH	238	363
165	260	426
168	346	483
179	355	513

Active medium: $\mathbf{CH_3CN}$ *Pumping*: optical pumping by CO_2-laser.		
wavelengths (in air), μm		
281.18	427.04	704.53
281.96	430.55	713.72
286.88	441.15	741.62
303.54	453.41	854.41
346.32	466.25	1014.89
372.87	480.01	1016.33
380.71	494.74	1086.89
386.41	510.16	1146.83
387.31	561.41	1351.78
388.39	652.68	1814.37
422.14		

Active medium: $\mathbf{CH_3NC}$ *Pumping*: optical pumping by CO_2-laser.		
wavelengths (in air), μm		
280	288	404

Active medium: $\mathbf{CH_3CCH}$ *Pumping*: optical pumping by CO_2-laser.		
wavelengths (in air), μm		
247.89	563.13	675.29
428.87	566.04	757.41
488.88	583.77	798.55
516.77	647.89	1097.11
531.08	649.59	1174.87

Active medium: $\mathbf{CH_3NH_2}$ *Pumping*: optical pumping by CO_2-laser.		
wavelengths (in air), μm		
99.5	159	201
104	164	208
115.5	166	218
118	168	219
126	175	243
134	176	251
139	177	267
141	180	268
143	183	288
147	194	314
148.5	198	347
153		

Table 34.2 *(continued)*

Active medium: $\mathbf{CH_2CF_2}$ *Pumping*: optical pumping by CO_2-laser.		
wavelengths (in air), μm		
288.5	532	884
375.0	554.4	890.0
415	568	890.1
458	663.3	990
464.3	764.1	1020

Active medium: $\mathbf{C_2H_6O_2}$ *Pumping*: optical pumping by CO_2-laser.		
wavelengths (in air), μm		
62.5	132	250
69.1	135	252
70.1	164	262
75.2	169	277
77.4	171	288
90.8	185	290
95.8	189	299
109.1	192	344
117.1	197	358
118	200	388
118.9	231	415
125.8	240	696

Active medium: $\mathbf{CH_3CH_2F}$ *Pumping*: optical pumping by CO_2-laser.		
wavelengths (in air), μm		
206.6	376	519
217.1	378	540.9
226.9	404	593.32
264.7	405	620.4
282.3	405.50	851.9
330.2	462.92	1013
336.7	486	1069
362.1	502.2	1546

Active medium: $\mathbf{CH_3CHF_2}$ *Pumping*: optical pumping by CO_2-laser.		
wavelengths (in air), μm		
458	464	533
755		

Active medium: $\mathbf{CH_3CF_3}$ *Pumping*: optical pumping by CO_2-laser.
wavelengths (in air), μm
379

Active medium: $\mathbf{C_2H_3Cl}$ *Pumping*: optical pumping by CO_2-laser.		
wavelengths (in air), μm		
900	1350	1400
1720		

Table 34.2 *(continued)*

Active medium: C_2H_5OH *Pumping*: optical pumping by CO_2-laser.		
wavelengths (in air), μm		
396		

Active medium: C_2H_3Cl *Pumping*: optical pumping by CO_2-laser.		
wavelengths (in air), μm		
386.0	532	699
424	538	707
445	574	828
487	603	935
507.7	634.4	995
519	638	1041

Active medium: C_2H_3Br *Pumping*: optical pumping by CO_2-laser.		
wavelengths (in air), μm		
283	528.49	784.26
356	553.69	826.94
370	594.72	853.43
396	618.44	900.13
411	624.09	936.15
416	635.35	943.22
419	646	**963.48**
424	**649.42**	985.85
427	**680.54**	989.19
438.5	693.13	990.63
443.5	707.22	1247.59
445	712	1383.88
482.96	724.13	1394.06
490.08	741.11	1614.88
506	780.13	1899.889

Active medium: C_2H_3CN *Pumping*: optical pumping by CO_2-laser.		
wavelengths (in air), μm		
270.6	586	793
489	623	828
503	631	910
550	722	940
574.4	738	1156
578	775	1184
584		

Active medium: CH_3OCH_3 *Pumping*: optical pumping by CO_2-laser.		
wavelengths (in air), μm		
375	480	495
462	492	520

Active medium: C_3H_2O *Pumping*: optical pumping by CO_2-laser.		
wavelengths (in air), μm		
148	156	336
516		

Table 34.2 *(continued)*

Active medium: ClO_2 *Pumping*: optical pumping by CO_2-laser.		
wavelengths (in air), μm		
112	176	204
216	196	

Active medium: HCCF *Pumping*: optical pumping by CO_2-laser.		
wavelengths (in air), μm		
509	1028	

Active medium: FCN *Pumping*: optical pumping by CO_2-laser.		
wavelengths (in air), μm		
308		

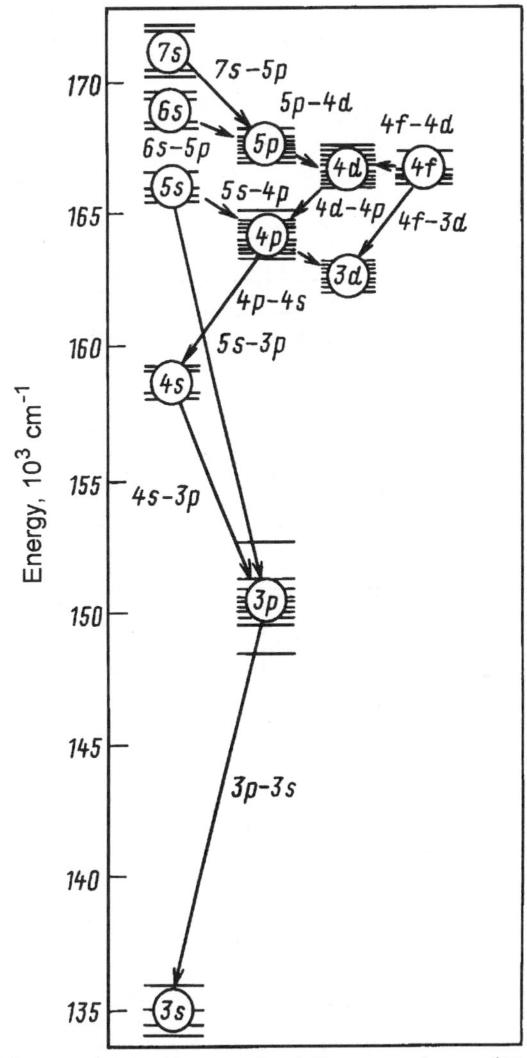

Figure 34.1 Energy–level diagram of neon (Ne), demonstrating principal laser transition groups [2].

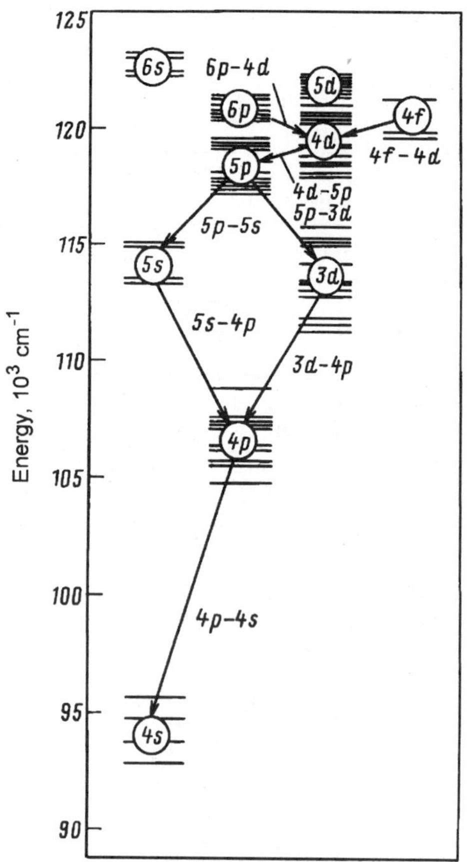

Figure 34.2 Energy-level diagram of argon (Ar), demonstating principal laser transition groups [2].

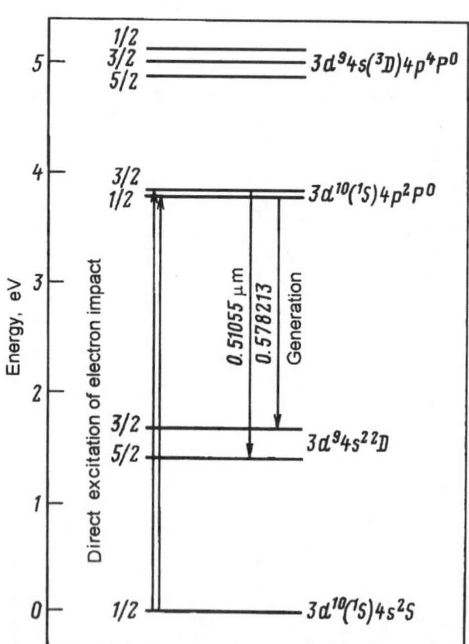

Figure 34.3 Laser transitions in copper (Cu) vapor at 0.5105 and 0.5782 μm [2].

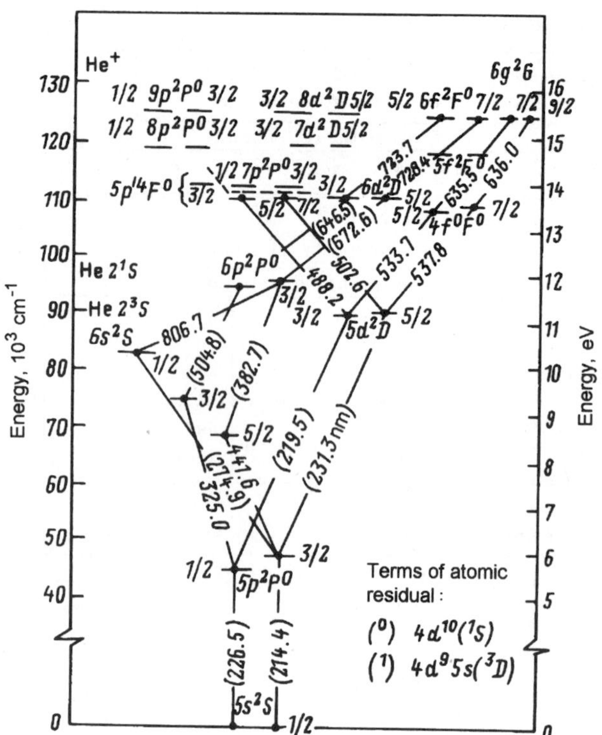

Figure 34.4 Energy–level diagram of singly ionized cadmium (Cd II), showing principal laser transitions (wavelengths given in brackets refer to spontaneous transitions) [3].

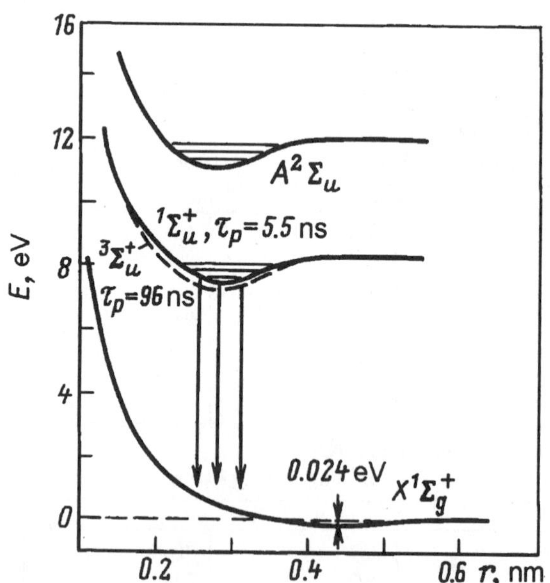

Figure 34.5 Schematic diagram of potential energy and vibrational levels of xenon molecule (Xe_2) [1]. Laser transitions are observed between stable molecular Σ_u^+ – states and unstable ground state ${}^1\Sigma_g^+$ (X denotes the electronic ground state and τ_p – lifetime of upper levels).

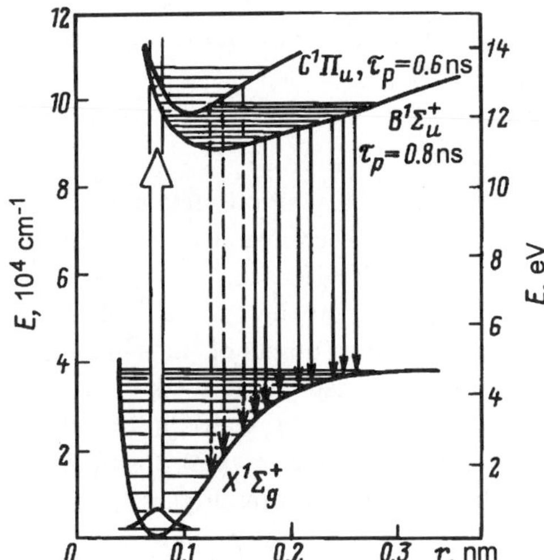

Figure 34.6 Schematic diagram of potential energy and vibrational levels of hydrogen molecule (H_2) [1]. Laser transitions corresponding to the Lyman ($B \to X$) and Werner ($C \to X$) bands are shown with solid and dashed arrows respectively. The wide up-arrow shows transitions from ground vibrational state $X_{\nu=0}$ under electronic excitation, τ_p denotes the lifetime of upper laser levels.

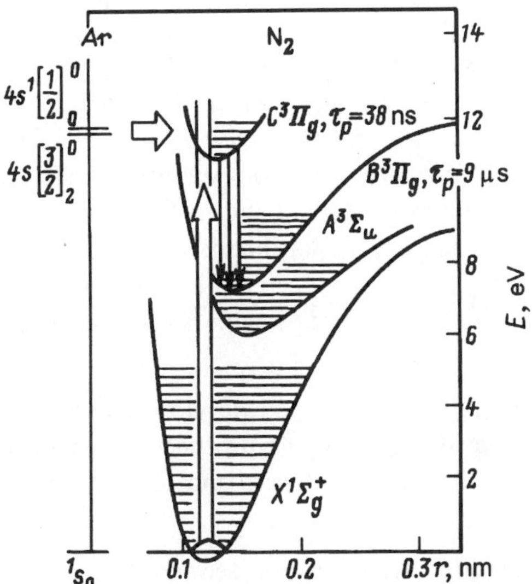

Figure 34.7 Schematic diagram of potential energy and vibrational levels corresponding to laser transitions in N_2+Ar mixture [1]. The wide up-arrow shows transitions from ground vibrational state $X_{\nu=0}$ under electronic excitation. The horizontal arrow denotes energy transfer from excited argon to upper laser level of nitrogen molecule. Solid down-arrows correspond to laser transitions, τ_p denotes lifetimes of laser levels.

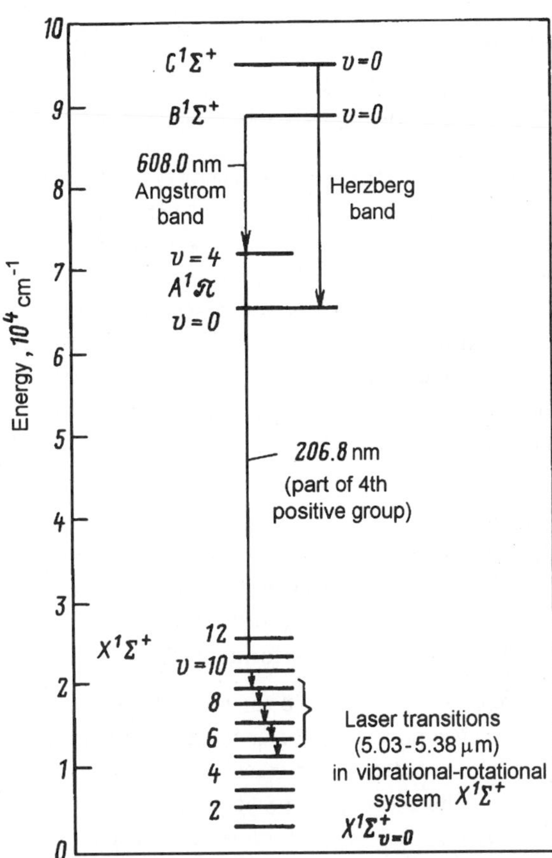

Figure 34.8 Part of the energy-level diagram for CO-molecule, demonstrating the electronic transition corresponding to the Angström band and vibrational-rotational transitions belonging to ground molecular state $X^1\Sigma^+$ [2].

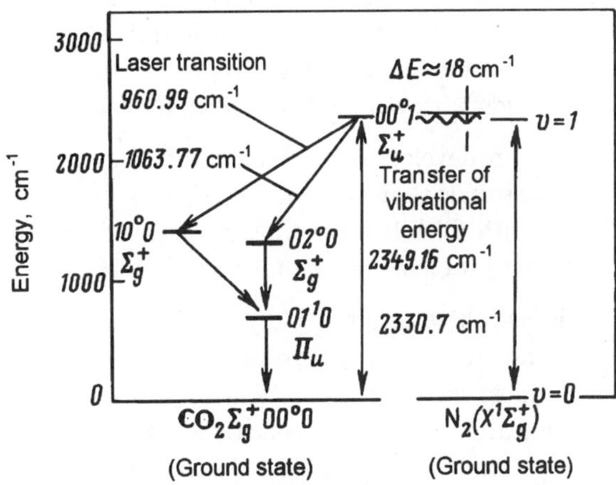

Figure 34.9 Schematic energy-diagram for N_2 + CO_2 mixture [1]. Energy values are given relative to the ground states of N_2 ($X'\Sigma_g^+ v = 0$). Selective excitation of 00° level of the CO_2 molecule due to energy transfer from vibrational level $v = 1$ of N_2 and transitions for CO_2–laser are shown.

34.3 Insulating Crystal Lasers

Insulating crystal laser materials contain ions of rare-earth elements, transition metals or actinide as active species. Though typical active ion concentrations do not exceed several percent, in crystalline solids, unlike gases, active ions are closely packed within a host matrix. As a result the electric field at the ion location has strength and symmetry depending on the type of a host crystal. That is why the same ion placed in different hosts, or even in various points of the same lattice, emits at different wavelengths. The strongest effect is for ion levels corresponding to the outer electrons, as soon as inner shells are shielded. The spectrum of the ion in the crystal is quite different from the spectrum of the free ion; the levels are split, displaced from their original positions, and broadened, that in some cases produce clearly observable vibrational structure (Figures 34.10 and 34.11).

The excitation of the crystalline laser materials is produced with the help of pulsed or CW radiating light sources. Until recently flash and arc lamps were most typical as pump sources for pulsed and continuous modes of operation. They are still used for high-power lasers, but outnumbered little by little by cheaper and more powerful semiconductor diode lasers that provide higher efficiency, better reliability and thermal performance [10, 11].

An active ion can absorb the pump light itself and/or can receive the excitation from sensitizing ions, that increase pumping efficiency. Among various ions Cr^{3+} is used most effectively for sensitization of the crystals containing trivalent ions of Nd, Er and Ho. Ruby (sapphire crystal with Cr^{3+}) laser was the first solid-state laser to be operated successfully. Recent discovery of lasing properties for Cr^{4+} has opened a new page in solid-state laser physics.

Most crystal laser materials use stimulated transitions between electronic levels, and for the majority of ions the wavelength can be changed within only $\approx 1\%$ or a little bit more taking into account a variety of possible hosts and operating temperature variation. Advance in the development of new laser crystal materials has resulted in tunability up to 1/3 of the wavelength for ions exhibiting effective electronic vibrational transitions. Some of these new crystals have a broadband emission spectrum together with good thermal properties, the combination of broad absorption bands and long enough upper-level lifetime to permit both flashlamp and diode pumping. Wide spectrum bandwidth of new laser crystals made possible the reproducible generation of the light pulses as short as several tens of femtoseconds with the peak power at the level above 10^{12} W.

Success in development of tunable lasers and diode pumping has permitted development of those crystal materials that were discovered long ago and were put aside as having no prospect of being used in lasers with flash lamp pumping. Detailed information on laser crystal physics and solid-state laser engineering problems can be found in [12–16].

Table 34.3 lists insulating crystal materials, laser wavelengths and operation temperatures. Laser systems are tabulated according to active (laser) ions. Crystal hosts are given in alphabetical order. In most cases, if the crystal contains sensitizers, corresponding ions are given after host crystal chemical formula; the sensitizer concentration is given in brackets. Concentrations for both an active ion and a sensitizer (if known) are given in weight or atomic percents. Moreover, it may refer to that of the melt or the crystal; the original reference should be traced if it is necessary.

Sometimes, the sensitizing ion is a part of the crystal structure; e.g. the crystal $LiYF_4$ may include nearly equal quantities of Y and Er, in this case, the crystal formula is given as $Li(Y,Er)F_4$. More complicated cases are possible also, e.g. $Li(Y, Er)F_4{:}Tu^{3+}$ where concentration of the second sensitizing ion is rather low (on the order of active ion concentration).

Table 34.3 Wavelengths for insulating crystal lasers

Host crystal	Laser transition	Laser wavelength, nm	Temperature, K
IRON GROUP LASERS			
	Titanium (Ti^{3+}, $3d^1$)		
Al_2O_3 [17]	$^2E \to {}^2T_2$	660–986	300
	Chromium (Cr^{4+}, $3d^2$)		
Mg_2SiO_4 [18]	$^3T_2 \to {}^3A_2$	≈1190–1260	300
$Y_3Al_5O_{12}$ [19]	$^3T_2 \to {}^3A_2$	1340–1560	300
$Y_3Sc_{0.5}Al_{4.5}O_{12}$ [20]	$^3T_2 \to {}^3A_2$	1394–1628	300
$Y_3ScAl_4O_{12}$ [20]	$^3T_2 \to {}^3A_2$	1464–1604	300
$Y_3Sc_{1.5}Al_{3.5}O_{12}$ [20]	$^3T_2 \to {}^3A_2$	1584*1	300
Y_2SiO_5 [21]	$^3T_2 \to {}^3A_2$	1190–1290*2	77
	Chromium (Cr^{3+}, $3d^3$)		
Al_2O_3	$^2E(\overline{E}) \to {}^4A_2$	694.3 (R_1)	300
		693.4	77
	$^2E(2\overline{A}) \to {}^4A_2$	692.9 (R_2)	290
Cr^{52}	$^2E(\overline{E}) \to {}^4A_2$	693.4089	≈70
Cr^{50}		693.4255	≈70
	satellite line	700.9 (N_2)	77
	(Cr^{3+}–Cr^{3+})*3	704.1 (N_1)	77
$Al_2(WO_4)_3$ [22]	$^4T_2 \to {}^4A_2$	810*1	300
$BeAl_2O_4$ [23]	$^4T_2 \to {}^4A_2$	730–800	300
$BeAl_6O_{10}$ [24]	$^4T_2 \to {}^4A_2$	834*1	300
$Be_3Al_2(SiO_6)_3$ [25]	$^4T_2 \to {}^4A_2$	728–809	300
$Gd_3(Sc, Al)_2Al_3O_{12}$ [26]	$^4T_2 \to {}^4A_2$	779–789	300
$Gd_3Sc_2Ga_3O_{12}$ [27, 28]	$^4T_2 \to {}^4A_2$	742–842	300
$Gd_3(Sc, Ga)_2Ga_3O_{12}$ [28]	$^4T_2 \to {}^4A_2$	750–830	300
$KZnF_3$ [29]	$^4T_2 \to {}^4A_2$	785–873	300
$La_3Ga_{5.5}Nb_{0.5}O_{14}$ [30]	$^4T_2 \to {}^4A_2$	≈900–1100	300
$La_3Ga_5SiO_{14}$ [31]	$^4T_2 \to {}^4A_2$	862–1107	300
$La_3Lu_2Ga_3O_{12}$ [32]	$^4T_2 \to {}^4A_2$	830*1	300
$LiCaAlF_6$ [33]	$^4T_2 \to {}^4A_2$	780*1	300
$LiSrAlF_6$ [34]	$^4T_2 \to {}^4A_2$	750–950	300
$LiSr_{0.8}Ca_{0.2}AlF_6$ [35]	$^4T_2 \to {}^4A_2$	750–950	300
$Na_3Ga_2Li_3F_{12}$ [36]	$^4T_2 \to {}^4A_2$	741–841	300
$ScBO_3$ [37]	$^4T_2 \to {}^4A_2$	787–892	300
$ScBeAlO_4$ [34]	$^4T_2 \to {}^4A_2$	792*1	300
$SrAlF_5$ [38]	$^4T_2 \to {}^4A_2$	825–1011	300
$Y_3Al_5O_{12}$	$^2E(\overline{E}) \to {}^4A_2$	687.4	300
$Y_3Sc_2Al_3O_{12}$ [34]	$^4T_2 \to {}^4A_2$	767*1	300
$Y_3(Sc,Ga)_2Ga_3O_{12}$ [39]	$^4T_2 \to {}^4A_2$	758*1	300
$ZnWO_4$ [40]	$^4T_2 \to {}^4A_2$	980–1090	77
	Vanadium (V^{2+}, $3d^3$)		
MgF_2	$^4T_2 \to {}^4A_2$	1121.3	77
$CsCaF_3$ [41]		1240–1330	80
	Ferrum (Fe^{2+}, $3d^3$)		
n-InP [42]	$^5T_2 \to {}^5E$	3530	2

*1 Central wavelength of the tuning range.

*2 Depends on laser pumping wavelength.

*3 Lasing is ascribed to pair centers.

Table 34.3 Wavelengths for insulating crystal lasers *(continued)*

Host crystal	Laser transition	Laser wavelength, nm	Temperature, K
	Cobalt ($Co^{2+}, 3d^7$)		
$KMgF_3$	$^4T_2 \rightarrow {}^4T_1$	1821	77
$KZnF_3$ [43]	$^4T_2 \rightarrow {}^4T_1$	1650–2070	80–200
MgF_2 [44]	$^4T_2 \rightarrow {}^4T_1$	1550–2300	80
ZnF_2	$^4T_2 \rightarrow {}^4T_1$	2165	77
	Nickel ($Ni^{2+}, 3d^8$)		
$CaY_2Mg_2Ge_3O_{12}$ [45]	$^3T_2 \rightarrow {}^3A_2$	1460	80
$KMgF_3$ [46]		1591	77
MgF_2 [45]	$^3T_2 \rightarrow {}^3A_2$	1785–1797	198–240
		1610–1740	80
MgO [47]	$^3T_2 \rightarrow {}^3A_2$	1314.4	77
		1316–1409	77
MnF_2	$^3T_2 \rightarrow {}^3A_2$	1915	77
		1922	77
		1865	20
		1929	85
		1939	85
DIVALENT RARE-EARTH LASERS			
	Samarium ($Sm^{2+}, 4f^6$)		
CaF_2	$5d \rightarrow {}^7F_1$	708.5	20
		708	65–90
		720	>65
		729	>65
SrF_2	$5D_0 \rightarrow {}^7F_1$	696.9	4.2
	Dysprosium ($Dy^{2+}, 4f^{10}$)		
CaF_2	$^5I_7 \rightarrow {}^5I_8$	2360	77
		2358.67	77
SrF_2	$^5I_7 \rightarrow {}^5I_8$	2365.9	77
	Thulium ($Tm^{2+}, 4f^{13}$)		
CaF_2	$^2F_{5/2} \rightarrow {}^2F_{7/2}$	1116	27
TRIVALENT RARE-EARTH LASERS			
	Cerium ($Ce^{3+}, 4f^1$)		
LaF_3 [48]	$5d \rightarrow 4f^{*4}$	≈285	300
$LiLuF_4$ [49]	$5d \rightarrow 4f$	309, 311	300
$LiSrAlF_6$ [50]	$5d \rightarrow 4f$	285 (σ)	300
	$5d \rightarrow 4f$	290 (π)	300
$LiYF_4$ [51]	$5d \rightarrow 4f$	309–325	300
	Praseodymium ($Pr^{3+}, 4f^2$)		
$Ca(NbO_3)_2$	$^1G_4 \rightarrow {}^3H_4$	1040	77
$CaWO_4$	$^1G_4 \rightarrow {}^3H_4$	1046.8	20–90
$GdLiF_4$ [52]	$^3P_0 \rightarrow {}^3F_2$	639	300
$LaBr_3$	$^3P_1 \rightarrow {}^3H_5$	532	
	$^3P_0 \rightarrow {}^3H_6$	621	
	$^3P_0 \rightarrow {}^3F_2$	632	
	$^3P_0 \rightarrow {}^3F_2$	647	

*4 Interconfigurational transition.

Table 34.3 Wavelengths for insulating crystal lasers *(continued)*

Host crystal	Laser transition	Laser wavelength, nm	Temperature, K
	Praseodymium (Pr^{3+}, $4f^2$) *(continued)*		
LaF_3	$^3P_0 \rightarrow {}^3H_6$	598.5	77
$LaCl_3$	$^3P_0 \rightarrow {}^3H_4$	489.2	5.5−14
	$^3P_1 \rightarrow {}^3H_5$	529.8	35
	$^3P_0 \rightarrow {}^3H_6$	616.4	65
		619	
	$^3P_0 \rightarrow {}^3F_2$	645.2	300
$LiYF_4$	$^3P_0 \rightarrow {}^3H_4$	479	300
[52]	$^3P_1 \rightarrow {}^3H_5$	522	300
[53]	$^3P_0 \rightarrow {}^3H_5$	537.8	110
[53]	$^3P_0 \rightarrow {}^3H_6$	607.1	110
[54]	$^3P_0 \rightarrow {}^3F_2$	639.5	300
$LiYF_4$ [53]	$^3P_0 \rightarrow {}^3F_4$	695.4	110
[53]	$^3P_0 \rightarrow {}^3H_6$	719	110
$PrBr_3$	$^3P_0 \rightarrow {}^3F_2$	622	
	$^3P_0 \rightarrow {}^3H_4$	649	300
$PrCl_3$	$^3P_1 \rightarrow {}^3H_5$	489	5.5−14
		531	12
	$^3P_0 \rightarrow {}^3H_6$	617	65
	$^3P_0 \rightarrow {}^3F_2$	620	
[55]	$^3P_0 \rightarrow {}^3H_6$	647	300
PrF_3	$^1G_4 \rightarrow {}^3H_4$	598	16
$SrMoO_4$ [56]	$^3P_0 \rightarrow {}^3H_6$	1040	
$YAlO_3$ [56]	$^3P_0 \rightarrow {}^3F_3$	613.9	110−250
		719.5	110−250
	Neodymium (Nd^{3+}, $4f^3$)		
$Ba_{0.75}Ca_{0.25}Nb_2O_6$	$^4F_{3/2} \rightarrow {}^4I_{11/2}$	1062	295
BaF_2	$^4F_{3/2} \rightarrow {}^4I_{11/2}$	1060	77
BaF_2-LaF_3	$^4F_{3/2} \rightarrow {}^4I_{11/2}$	1054	300
		1053.8	77
		1058	
	$^4F_{3/2} \rightarrow {}^4I_{11/2}$	1054	300
	$^4F_{3/2} \rightarrow {}^4I_{13/2}$	1318.5	300
	$^4F_{3/2} \rightarrow {}^4I_{13/2}$	1328.0	300
	$^4F_{3/2} \rightarrow {}^4I_{13/2}$	1329.0	77
BaF_2-CeF_3	$^4F_{3/2} \rightarrow {}^4I_{11/2}$	1054	300
BaF_2-GdF_3	$^4F_{3/2} \rightarrow {}^4I_{11/2}$	1052.6	300
BaF_2-YF_3	$^4F_{3/2} \rightarrow {}^4I_{11/2}$	1052.1	300
	$^4F_{3/2} \rightarrow {}^4I_{13/2}$	1320	300
$Ba_2MgGe_2O_7$	$^4F_{3/2} \rightarrow {}^4I_{11/2}$	1054.4	300
$Ba_{0.25}Mg_{2.75}Y_2Ge_3O_{12}$	$^4F_{3/2} \rightarrow {}^4I_{11/2}$	1061.5	300
$Ba_2NaNb_5O_{15}$ $(\parallel c)$	$^4F_{3/2} \rightarrow {}^4I_{11/2}$	1061.3	300
$Ba_2ZnGe_2O_7$ $(\parallel c)$	$^4F_{3/2} \rightarrow {}^4I_{11/2}$	1054.4	300
$Bi_2Ge_3O_{12}$	$^4F_{3/2} \rightarrow {}^4I_{11/2}$	1064.25	77
$Bi_4Ge_3O_{12}$	$^4F_{3/2} \rightarrow {}^4I_{11/2}$	1064.4	300
	$^4F_{3/2} \rightarrow {}^4I_{13/2}$	1341.8	300
$CaAl_4O_7$	$^4F_{3/2} \rightarrow {}^4I_{11/2}$	1078.6	300
		1058.95	77
		1065.85	
		1076.55	77
		1077.2	

Table 34.3 Wavelengths for insulating crystal lasers *(continued)*

Host crystal	Laser transition	Laser wavelength, nm	Temperature, K
	Neodymium (Nd^{3+}, $4f^3$) *(continued)*		
	${}^4F_{3/2} \rightarrow {}^4I_{13/2}$	1342	300
		1371	
		1340	77
		1367.5	
$CaAl_{12}O_{19}$	${}^4F_{3/2} \rightarrow {}^4I_{11/2}$	1049.7	300
CaF_2	${}^4F_{3/2} \rightarrow {}^4I_{11/2}$	1045.7	77
CaF_2 (I)	${}^4F_{3/2} \rightarrow {}^4I_{11/2}$	1046.1	300
		1045.7	50
		1046.7	
		1044.8	50
		1050.8	
		1065.0	50
		1048.1	
		1044.8	120
		1066.1	
CaF_2 (II)	${}^4F_{3/2} \rightarrow {}^4I_{11/2}$	1088.5	300
CaF_2-CeF_3	${}^4F_{3/2} \rightarrow {}^4I_{11/2}$	1065.4	300
	${}^4F_{3/2} \rightarrow {}^4I_{13/2}$	1318.5	300
CaF_2-GdF_3	${}^4F_{3/2} \rightarrow {}^4I_{11/2}$	1065.4	300
	${}^4F_{3/2} \rightarrow {}^4I_{13/2}$	1318.5	300
CaF_2-CeO_2	${}^4F_{3/2} \rightarrow {}^4I_{11/2}$	1088.5	300
CaF_2-LaF_3	${}^4F_{3/2} \rightarrow {}^4I_{11/2}$	1064.5	300
	${}^4F_{3/2} \rightarrow {}^4I_{13/2}$	1319.0	300
CaF_2-SrF_2	${}^4F_{3/2} \rightarrow {}^4I_{11/2}$	1036.9	300
CaF_2-SrF_2-BaF_2-YF_3-LaF_3	${}^4F_{3/2} \rightarrow {}^4I_{11/2}$	1053.5	300
	${}^4F_{3/2} \rightarrow {}^4I_{13/2}$	1046.1	300
	${}^4F_{3/2} \rightarrow {}^4I_{11/2}$	1054	300
		1063.2	
	${}^4F_{3/2} \rightarrow {}^4I_{13/2}$	1325.5	77
		1338	
		1360	77
		1327	300
		1337	300
		1385.5	
CaF_2-YF_3-NdF_3	${}^4F_{3/2} \rightarrow {}^4I_{11/2}$	1063.2	300
$Ca_4La(PO_4)_3O$	${}^4F_{3/2} \rightarrow {}^4I_{11/2}$	1061.3	300
$CaLa_4(SiO_4)_3O$	${}^4F_{3/2} \rightarrow {}^4I_{11/2}$	1061	300
$CaGd_4(SiO_4)_3O$	${}^4F_{3/2} \rightarrow {}^4I_{11/2}$	≈1060	300
$Ca(NbO_3)_2$	${}^4F_{3/2} \rightarrow {}^4I_{11/2}$	1060	77
		1061.2	
		1061.5	300
	${}^4F_{3/2} \rightarrow {}^4I_{13/2}$	1338	300
		1342.5	
$CaMoO_4$	${}^4F_{3/2} \rightarrow {}^4I_{11/2}$	1061	300
		1067	
$Ca_5(PO_4)_3F$ (∥a)	${}^4F_{3/2} \rightarrow {}^4I_{11/2}$	1063	300
(∥c)	${}^4F_{3/2} \rightarrow {}^4I_{13/2}$	1334.7	300
		1342.5	77
$CaSc_2O_4$	${}^4F_{3/2} \rightarrow {}^4I_{11/2}$	1072	300
		1075.5	
		1086.8	300

Table 34.3 Wavelengths for insulating crystal lasers *(continued)*

Host crystal	Laser transition	Laser wavelength, nm	Temperature, K
Neodymium (Nd^{3+}, $4f^3$) *(continued)*			
		1073	77
		1086.7	
	${}^4F_{3/2} \rightarrow {}^4I_{13/2}$	1356.5	300
		1387	77
$Ca_3(VO_4)_2$	${}^4F_{3/2} \rightarrow {}^4I_{11/2}$	1067	300
$CaWO_4$	${}^4F_{3/2} \rightarrow {}^4I_{9/2}$	914.5	77
$CaWO_4(Na^+)$	${}^4F_{3/2} \rightarrow {}^4I_{11/2}$	1058.2	300
		1065.2	
		1064.9	77
$CaWO_4(Na^5+)$	${}^4F_{3/2} \rightarrow {}^4I_{11/2}$	1058.7	77
		1060.1	
		1064.9	77
		1065	
		1063.4	77
$CaWO_4$ (‖c, Na^+)	${}^4F_{3/2} \rightarrow {}^4I_{13/2}$	1334	300
		1347.5	
		1331	300
		1345.9	
$CaWO_4$ (⊥c, Na^+)	${}^4F_{3/2} \rightarrow {}^4I_{13/2}$	1337	300
		1339	
		1334.5	77
		1337.2	
		1345.9	77
		1388	
$CaYAlO_4$ [57]	${}^4F_{3/2} \rightarrow {}^4I_{11/2}$	1080.6	300
$Ca_2Y_5F_{19}$	${}^4F_{3/2} \rightarrow {}^4I_{11/2}$	1049.8	300
	${}^4F_{3/2} \rightarrow {}^4I_{13/2}$	1320	77
		1319	300
		1352.5	
$CaY_2Mg_2Ge_3O_{12}$	${}^4F_{3/2} \rightarrow {}^4I_{11/2}$	1058.96	300
$CaY_4(SiO_4)_3$	${}^4F_{3/2} \rightarrow {}^4I_{11/2}$	1067.2	300
CdF_3-CaF_2 [58]	${}^4F_{3/2} \rightarrow {}^4I_{11/2}$	1049.5	300
CdF_2-YF_2	${}^4F_{3/2} \rightarrow {}^4I_{11/2}$	1065.1	300
	${}^4F_{3/2} \rightarrow {}^4I_{11/2}$	1324.5	300
$CeCl_3$ (‖a)	${}^4F_{3/2} \rightarrow {}^4I_{11/2}$	1064.7	300
CeF_3	${}^4F_{3/2} \rightarrow {}^4I_{11/2}$	1041	77
CeF_3	${}^4F_{3/2} \rightarrow {}^4I_{11/2}$	1063.8	90
		1040.4	300
		1063.9	
	${}^4F_{3/2} \rightarrow {}^4I_{13/2}$	1332	300
		1324	77
		1.331	77
		1367.5	
CeP_5O_{14}	${}^4F_{3/2} \rightarrow {}^4I_{11/2}$	1051	300
CsY_2F_7 [59]	${}^4F_{3/2} \rightarrow {}^4I_{11/2}$	1055	300
$GdAlO_3$	${}^4F_{3/2} \rightarrow {}^4I_{11/2}$	1076	300
$Gd_3Ga_5O_{12}$ [60]	${}^4F_{3/2} \rightarrow {}^4I_{11/2}$	1059.1	300
		1060.6	300
		1062.1	300
		1058.4	77
		1059.9	77

Table 34.3 Wavelengths for insulating crystal lasers *(continued)*

Host crystal	Laser transition	Laser wavelength, nm	Temperature, K
Neodymium (Nd^{3+}, $4f^3$) *(continued)*			
	${}^4F_{3/2} \to {}^4I_{13/2}$	1330.7	
		1331.5	
$Gd_3Ga_5O_{12}$:Cr^{3+} [61]	${}^4F_{3/2} \to {}^4I_{13/2}$	1330	300
$GdLiF_4$ [62] (‖c)	${}^4F_{3/2} \to {}^4I_{11/2}$	1047 (π)	300
(⊥c)		1053 (σ)	300
$Gd(MoO_4)_3$	${}^4F_{3/2} \to {}^4I_{11/2}$	1060	135
		1070.1	300
		1060.6	
		1078.9	300
GdO_3	${}^4F_{3/2} \to {}^4I_{11/2}$	1074.1	
		1078.9	77
		1077.6	
GdP_5O_{14}	${}^4F_{3/2} \to {}^4I_{11/2}$	1051	300
$GdScO_3$	${}^4F_{3/2} \to {}^4I_{11/2}$	1085.15	300
$Gd_3Sc_2Al_3O_{12}$	${}^4F_{3/2} \to {}^4I_{11/2}$	1059.95	300
		1059.15	77
		1062	300
		1066	
$Gd_3Sc_2Ga_5O_{12}$ [63]	${}^4F_{3/2} \to {}^4I_{11/2}$	1060	80
		1061	300
		1061.7	420
		1062.1	480
$K_5Bi(MoO_4)_4$	${}^4F_{3/2} \to {}^4I_{11/2}$	1060.4	300
$KGd(WO_4)_2$ (‖[010])	${}^4F_{3/2} \to {}^4I_{11/2}$	1066	300
$KLa(MoO_4)_2$	${}^4F_{3/2} \to {}^4I_{11/2}$	1067.2	300
$KLa(MoO_4)_2$	${}^4F_{3/2} \to {}^4I_{11/2}$	1058.5	77
	${}^4F_{3/2} \to {}^4I_{11/2}$	1058.7	300
$KLu(WO_4)_2$ [64]	${}^4F_{3/2} \to {}^4I_{11/2}$	1071.4	300
	${}^4F_{3/2} \to {}^4I_{13/2}$	1348.2	300
$KY(MoO_4)_2$	${}^4F_{3/2} \to {}^4I_{11/2}$	1066.9	300
$KY(MoO_4)_2$ (‖a)	${}^4F_{3/2} \to {}^4I_{13/2}$	1348.5	300
$KY(WO_4)_2$	${}^4F_{3/2} \to {}^4I_{9/2}$	913.7	77
	${}^4F_{3/2} \to {}^4I_{11/2}$	1068.8	300
	${}^4F_{3/2} \to {}^4I_{13/2}$	1352.5	300
		1354.5	
		1351.5	77
$K_5NdLi_2F_{10}$ [65]	${}^4F_{3/2} \to {}^4I_{11/2}$	1048	77
		1052	300
$KNdP_4O_{12}$	${}^4F_{3/2} \to {}^4I_{11/2}$	1052	300
$K_3Nd(PO_4)_2$	${}^4F_{3/2} \to {}^4I_{11/2}$	1060	300
$K_3(NdLa)(PO_4)_2$	${}^4F_{3/2} \to {}^4I_{11/2}$	1060	300
$K_5Nd(MoO_4)_2$	${}^4F_{3/2} \to {}^4I_{11/2}$	1066	300
KYF_4 [66]	${}^4F_{3/2} \to {}^4I_{11/2}$	1041	300
		1054	300
$LaAlO_3$	${}^4F_{3/2} \to {}^4I_{11/2}$	1080.4	300
$LaBGeO_5$ [67]	${}^4F_{3/2} \to {}^4I_{11/2}$	1048	300
		1071	300
	${}^4F_{3/2} \to {}^4I_{13/2}$	1310	300
		1380	300
$La_2Be_2O_5$ (‖b)	${}^4F_{3/2} \to {}^4I_{11/2}$	1069.8	300

Table 34.3 Wavelengths for insulating crystal lasers *(continued)*

Host crystal	Laser transition	Laser wavelength, nm	Temperature, K
	Neodymium (Nd^{3+}, $4f^3$) *(continued)*		
$La_2Be_2O_5$ (‖X)	$^4F_{3/2} \rightarrow {}^4I_{11/2}$	1079	300
$La_2Be_2O_5$ (‖b)	$^4F_{3/2} \rightarrow {}^4I_{13/2}$	1351.0	300
LaF_3	$^4F_{3/2} \rightarrow {}^4I_{11/2}$	1040.7	300
		1063.3	
		1040.3	77
$LaF_3(\angle cF \approx 20°)$	$^4F_{3/2} \rightarrow {}^4I_{13/2}$	1367.5	300
		1323.5	77
		1367.0	
$LaF_3(\angle cF \approx 73°)$	$^4F_{3/2} \rightarrow {}^4I_{13/2}$	1331.0	300
		1312.5	77
		1330.5	
LaF_3-SrF_2	$^4F_{3/2} \rightarrow {}^4I_{11/2}$	1048.6	300
		1063.5	
	$^4F_{3/2} \rightarrow {}^4I_{11/2}$	1331.5	300
$La_3Ga_5SiO_{14}$ [68]	$^4F_{3/2} \rightarrow {}^4I_{11/2}$	1064.0	300
		1067.0	
		1067.5	77
	$^4F_{3/2} \rightarrow {}^4I_{13/2}$	1373.0	300
$LaNbO_4$	$^4F_{3/2} \rightarrow {}^4I_{11/2}$	1062.4	300
La_2O_3	$^4F_{3/2} \rightarrow {}^4I_{11/2}$	1079	77
La_2O_2S	$^4F_{3/2} \rightarrow {}^4I_{11/2}$	1075	300
$LaMgAl_{11}O_{19}$ [69]	$^4F_{3/2} \rightarrow {}^4I_{11/2}$	1055.2	300
		1081.7	
	$^4F_{3/2} \rightarrow {}^4I_{13/2}$	1376.0	300
$LiGd(MoO_4)_2$ (‖c)	$^4F_{3/2} \rightarrow {}^4I_{11/2}$	1059.9	300
$LiLa(MoO_4)_2$ (‖c)	$^4F_{3/2} \rightarrow {}^4I_{11/2}$	1058.5	300
		1065.8	77
	$^4F_{3/2} \rightarrow {}^4I_{13/2}$	1337.0	300
		1337.5	77
$LiNbO_3$ (⊥c)	$^4F_{3/2} \rightarrow {}^4I_{11/2}$	1084.6	300
$LiNbO_3$ (‖c)		1093.3	300
$LiNbO_3$ (‖c)	$^4F_{3/2} \rightarrow {}^4I_{13/2}$	1387.0	300
$LiNbO_3$ (⊥c)		1374.5	300
$LiNdP_4O_{12}$ (‖a, ‖c)	$^4F_{3/2} \rightarrow {}^4I_{11/2}$	1047.7	300
$Li(Nd,La)P_4O_{12}$	$^4F_{3/2} \rightarrow {}^4I_{11/2}$	1047.7	300
$Li(Nd,Gd)P_4O_{12}$	$^4F_{3/2} \rightarrow {}^4I_{11/2}$	1047.7	300
$LiYF_4$	$^4F_{3/2} \rightarrow {}^4I_{11/2}$	1053.0 (σ)	300
		1047 (π)	
$LuAlO_3$ (‖[112])	$^4F_{3/2} \rightarrow {}^4I_{11/2}$	1067.1	77
		1083.1	120
		1067.5	300
		1075.9	300
		1083.2	
	$^4F_{3/2} \rightarrow {}^4I_{13/2}$	1343.7	300
$Lu_3Al_5O_{12}$	$^4F_{3/2} \rightarrow {}^4I_{9/2}$	947.3	77
	$^4F_{3/2} \rightarrow {}^4I_{11/2}$	1064.25	300
$Lu_3Al_5O_{12}$	$^4F_{3/2} \rightarrow {}^4I_{11/2}$	1060.5	77
	$^4F_{3/2} \rightarrow {}^4I_{13/2}$	1338.2	300
		1353.2	
		1352.5	77

Table 34.3 Wavelengths for insulating crystal lasers *(continued)*

Host crystal	Laser transition	Laser wavelength, nm	Temperature, K
Neodymium (Nd^{3+}, $4f^3$) *(continued)*			
$LuScO_3$	$^4F_{3/2} \rightarrow {}^4I_{11/2}$	1078.5	300
$Lu_3Ga_5O_{12}$	$^4F_{3/2} \rightarrow {}^4I_{11/2}$	1060.9	300
		1062.3	300
		1058.7	77
		1060.25	
		1061.6	77
	$^4F_{3/2} \rightarrow {}^4I_{13/2}$	1331.5	300
$Lu_3Sc_2Al_3O_{12}$	$^4F_{3/2} \rightarrow {}^4I_{11/2}$	1059.9	300
		1059.1	77
	$^4F_{3/2} \rightarrow {}^4I_{13/2}$	1336.0	300
α-$NaCaCeF_6$	$^4F_{3/2} \rightarrow {}^4I_{11/2}$	1065.3	300
	$^4F_{3/2} \rightarrow {}^4I_{13/2}$	1319.0	300
α-$NaCaYF_6$	$^4F_{3/2} \rightarrow {}^4I_{11/2}$	1053.9	300
		1069	
	$^4F_{3/2} \rightarrow {}^4I_{13/2}$	1328.5	300
		1337.5	
		1360.0	300
$5NaF{\cdot}9YF_3$	$^4F_{3/2} \rightarrow {}^4I_{11/2}$	1050.6	300
		1059.5	
	$^4F_{3/2} \rightarrow {}^4I_{13/2}$	1307.0	300
$Na_{0.5}Gd_{0.5}WO_4$	$^4F_{3/2} \rightarrow {}^4I_{11/2}$	1060	77
$NaGdGeO_4$ [70]	$^4F_{3/2} \rightarrow {}^4I_{11/2}$	1061.5	300
	$^4F_{3/2} \rightarrow {}^4I_{13/2}$	1333.4	300
$NaLa(MoO_4)_2$	$^4F_{3/2} \rightarrow {}^4I_{11/2}$	1059.5	300
		1065.3	
$NaLa(MoO_4)_2(\angle cF{=}60°)$	$^4F_{3/2} \rightarrow {}^4I_{13/2}$	1338.0	300
		1344.0	300
		1338.0	77
		1343.0	
$NaLa(WO_4)_2$ (‖c)	$^4F_{3/2} \rightarrow {}^4I_{11/2}$	1063.5	300
	$^4F_{3/2} \rightarrow {}^4I_{13/2}$	1335.5	300
$NaNdP_4O_{12}$	$^4F_{3/2} \rightarrow {}^4I_{11/2}$	1051	300
$Na_3Nd(PO_4)_2$	$^4F_{3/2} \rightarrow {}^4I_{11/2}$	1050	300
$Na_2Nd_2Pb_6(PO_4)_6Cl_2$ [71]	$^4F_{3/2} \rightarrow {}^4I_{11/2}$	1058.5	
		1068.5	
$Na_5Nd(WO_4)_4$	$^4F_{3/2} \rightarrow {}^4I_{11/2}$	1063	300
$NdAl_3(BO_3)_4$	$^4F_{3/2} \rightarrow {}^4I_{11/2}$	1063.5	300
$Nd_xGd_{1-x}(Al_{1-y}Cr_y)_3(BO_3)_4$ [72]	$^4F_{3/2} \rightarrow {}^4I_{11/2}$	1063	
$(NdLa)P_5O_{14}$	$^4F_{3/2} \rightarrow {}^4I_{13/2}$	1051.1	300
$(NdSc)P_5O_{14}$	$^4F_{3/2} \rightarrow {}^4I_{11/2}$	1051	300
$(NdIn)P_5O_{14}$	$^4F_{3/2} \rightarrow {}^4I_{11/2}$	1051	300
$Pb_5Ge_3O_{11}$ [73]	$^4F_{3/2} \rightarrow {}^4I_{11/2}$	1078.9	77
		1079.9	300
$PbMoO_4$	$^4F_{3/2} \rightarrow {}^4I_{11/2}$	1058.6	295
$PbMoO_4$ (⊥c)	$^4F_{3/2} \rightarrow {}^4I_{13/2}$	1334.0	300
		1332.0	77
$Pb_5(PO_4)_3F$	$^4F_{3/2} \rightarrow {}^4I_{11/2}$	1055.1	300

Table 34.3 Wavelengths for insulating crystal lasers *(continued)*

Host crystal	Laser transition	Laser wavelength, nm	Temperature, K
Neodymium (Nd^{3+}, $4f^3$) *(continued)*			
$SrAl_4O_7$	$^4F_{3/2} \rightarrow {}^4I_{11/2}$	1057.6	300
		1082.8	
		1056.6	77
		1056.8	
		1062.7	77
	$^4F_{3/2} \rightarrow {}^4I_{13/2}$	1334.5	300
		1366.5	
		1332.0	77
		1353.0	
$SrAl_{12}O_{19}$	$^4F_{3/2} \rightarrow {}^4I_{11/2}$	1049.1	300
SrF_2	$^4F_{3/2} \rightarrow {}^4I_{11/2}$	1037	295
		1043.7	77
		1044	
SrF_2-CeF_3	$^4F_{3/2} \rightarrow {}^4I_{11/2}$	1059.0	300
	$^4F_{3/2} \rightarrow {}^4I_{13/2}$	1325.5	300
SrF_2-CdF_3	$^4F_{3/2} \rightarrow {}^4I_{11/2}$	1052.8	300
	$^4F_{3/2} \rightarrow {}^4I_{13/2}$	1326.0	300
		1325.0	77
SrF_2-CeF_3-CdF_3	$^4F_{3/2} \rightarrow {}^4I_{11/2}$	1058.9	300
SrF_2-LaF_3	$^4F_{3/2} \rightarrow {}^4I_{11/2}$	1059.7	300
	$^4F_{3/2} \rightarrow {}^4I_{13/2}$	1325.0	300
SrF_2-LuF_3	$^4F_{3/2} \rightarrow {}^4I_{11/2}$	1055.6	300
	$^4F_{3/2} \rightarrow {}^4I_{13/2}$	1320.0	300
SrF_2-YF_3 (10%)	$^4F_{3/2} \rightarrow {}^4I_{11/2}$	1056.7	300
$SrLa_4(SiO_4)_3$	$^4F_{3/2} \rightarrow {}^4I_{11/2}$	1058.6	300
SrF_2-YF_3	$^4F_{3/2} \rightarrow {}^4I_{13/2}$	1322.5	300
		1330.0	
		1332.0	77
$SrMoO_4$	$^4F_{3/2} \rightarrow {}^4I_{11/2}$	1064.3	295
		1057.6	295
		1064.0	77
		1065.2	
		1059	77
		1062.7	
		1061.1	77
$Sr_5(PO_4)_3F$ (‖a)	$^4F_{3/2} \rightarrow {}^4I_{11/2}$	1058.5	300
$Sr_2Y_5F_{19}$	$^4F_{3/2} \rightarrow {}^4I_{11/2}$	1049.3	300
	$^4F_{3/2} \rightarrow {}^4I_{13/2}$	1319.0	300
$YAlO_3$ (‖[112])	$^4F_{3/2} \rightarrow {}^4I_{11/2}$	1079.6	300
		1064.4	300
		1064.05	77
		1072.25	
		1084.2	300
		1091.3	530
$YAlO_3$ (‖c)	$^4F_{3/2} \rightarrow {}^4I_{13/2}$	1339.3	300
		1341.3	
$YAlO_3$ (‖b)		1339.1	77
		1351.4	
$YAlO_3$:Cr^{3+} (0.3%)	$^4F_{3/2} \rightarrow {}^4I_{11/2}$	1064.5	295
		1072.5	300
		1079.5	

Table 34.3 Wavelengths for insulating crystal lasers *(continued)*

Host crystal	Laser transition	Laser wavelength, nm	Temperature, K
Neodymium (Nd^{3+}, $4f^3$) *(continued)*			
		1090.9	300
		1098.9	
	${}^4F_{3/2} \to {}^4I_{13/2}$	1339.1	300
		1341.1	
$Y_3Al_5O_{12}$	${}^4F_{3/2} \to {}^4I_{9/2}$	891	300
		900	
		939	300
		946	
	${}^4F_{3/2} \to {}^4I_{11/2}$	1061.0	77
		1061.2	
		1051.9	300
		1064.2	
		1061.3	300
		1064.0	
		1073.6	300
		1111.9	
		1115.8	300
		1122.5	
		1061.5	300
		1068.2	
		1077.9	300
		1064.15	
	${}^4F_{3/2} \to {}^4I_{13/2}$	1318.7	300
		1333.5	
		1319	300
		1318	
		1338	300
		1358	
[74]		1444	300
	${}^4F_{3/2} \to {}^4I_{15/2}$	1833	233
$YAlO_3$:Cr^{3+}(1%)	${}^4F_{3/2} \to {}^4I_{11/2}$	1061.2	77
		1064.1	300
$YAlO_3$:Lu^{3+} [75]	${}^4F_{3/2} \to {}^4I_{11/2}$	1080	300
$Y_3Ca_5O_{12}$	${}^4F_{3/2} \to {}^4I_{11/2}$	1063.3	300
Y_2O_3	${}^4F_{3/2} \to {}^4I_{11/2}$	1073	77
		1078	77
Y_2O_3-ThO_2-Nd_2O_3	${}^4F_{3/2} \to {}^4I_{11/2}$	1074	300
$Y_3Ga_5O_{12}$	${}^4F_{3/2} \to {}^4I_{11/2}$	1058.9	300
		1060.3	
		1062.05	300
		1058.3	77
		1059.75	77
		1061.4	
YP_5O_{14}	${}^4F_{3/2} \to {}^4I_{11/2}$	1052.5	300
		1051.5	
(Y, Lu)$_3Al_5O_{12}$	${}^4F_{3/2} \to {}^4I_{11/2}$	1064.2	300
		1060.8	77
		1063.6	77
		1072.6	

Table 34.3 Wavelengths for insulating crystal lasers *(continued)*

Host crystal	Laser transition	Laser wavelength, nm	Temperature, K
	Neodymium (Nd^{3+}, $4f^3$) *(continued)*		
$Y_3Sc_2Ga_3O_{12}$	$^4F_{3/2} \rightarrow {}^4I_{11/2}$	1057.5	77
		1058.3	300
		1061.5	300
	$^4F_{3/2} \rightarrow {}^4I_{13/2}$	1358.5	300
$YScO_3$	$^4F_{3/2} \rightarrow {}^4I_{11/2}$	1084.3	300
		1077.0	77
Y_2SiO_5	$^4F_{3/2} \rightarrow {}^4I_{11/2}$	1071.5	300
		1074.2	300
		1078.2	
		1071.0	77
		1078.1	
	$^4F_{3/2} \rightarrow {}^4I_{13/2}$	1358.5	300
		1358.0	77
YVO_4(1%)	$^4F_{3/2} \rightarrow {}^4I_{11/2}$	1069	≈90
		1064.1	300
		1066.4	
	$^4F_{3/2} \rightarrow {}^4I_{13/2}$	1341.5	77
		1342.5	300
ZrO_2-Y_2O_3	$^4F_{3/2} \rightarrow {}^4I_{11/2}$	1060.8	300
	$^4F_{3/2} \rightarrow {}^4I_{13/2}$	1332.0	300
	Samarium (Sm^{3+}, $4f^5$)		
TbF_3 [76]	$^4G_{5/2} \rightarrow {}^6H_{7/2}$	593.2	116
	Europium (Eu^{3+}, $4f^6$)		
Y_2O_3	$^5D_0 \rightarrow {}^7F_2$	611.3	220
YVO_4	$^5D_0 \rightarrow {}^7F_2$	619.3	90
	Terbium (Tb^{3+}, $4f^8$)		
$LiYF_4$:Gd^{3+}(10%)	$^5D_4 \rightarrow {}^7F_5$	544.5	300
	Dysprosium (Dy^{3+}, $4f^9$)		
$Ba(Y_{1.26}Er_{0.7})F_8$	$^6H_{13/2} \rightarrow {}^6H_{15/2}$	3022	77
	Holmium (Ho^{3+}, $4f^{10}$)		
$Ba(Y,Yb)_2F_8$	$^5S_2 \rightarrow {}^5I_8$	551.5	77
$Ba(Y_{1.8}Ho_{0.2})F_8$	$^5F_5 \rightarrow {}^5I_5$	2362	77
		2375	77
		2363	20
		2377	
$BaY_{1.64}Er_{0.03}Tm_{0.03}F_8$	$^5I_7 \rightarrow {}^5I_8$	2710	295
		2065	77
		1074.6	77
		2086.6	
		2055.5	20
		2074	
BaY_2F_8 [77]	$^5I_7 \rightarrow {}^5I_8$	2089.5	300
$Bi_4Ge_3O_{12}$ [78]	$^5I_7 \rightarrow {}^5I_8$	2087	77
CaF_2	$^5I_7 \rightarrow {}^5I_8$	2092	77
	$^5S_2 \rightarrow {}^5I_8$	551.22	77

Table 34.3 Wavelengths for insulating crystal lasers *(continued)*

Host crystal	Laser transition	Laser wavelength, nm	Temperature, K
Holmium ($Ho^{3+}, 4f^{10}$) *(continued)*			
CaF_2:ErF_3(3%), YbF_3(3%), TmF_3(3%)*5	$^5S_2 \to {}^5I_8$	2100	77
CaF_2-ErF_3	$^5I_7 \to {}^5I_8$	2060.0	77
CaF_2-YF_3	$^5I_7 \to {}^5I_8$	2031.8	90
CaF_2-YF_3:(Er^{3+}, Tm^{3+}, Yb^{3+})	$^5I_7 \to {}^5I_8$	2050	100
	$^5I_7 \to {}^5I_8$	2060	298
$CaMoO_4$:Er^{3+}(0.75%)*5	$^5I_7 \to {}^5I_8$	2074.0	77
	$^5I_7 \to {}^5I_8$	2070.7	77
	$^5I_7 \to {}^5I_8$	2055.6	
$Ca(NbO_3)_2$	$^5I_7 \to {}^5I_8$	2047	77
$Ca(PO_4)_3F$:Cr^{3+}(0.3%)*5	$^5I_7 \to {}^5I_8$	2079	77
$CaWO_4$	$^5I_7 \to {}^5I_8$	2046	77
$CaY_4(SiO_4)_3O$:Er^{3+}(37.5%), Tm^{3+}(3.75%)*5	$^5S_2 \to {}^5I_8$	2059	77
	$^5I_7 \to {}^5I_8$	2060	77
$Cd_3Ga_3O_{12}$ [79]	$^5I_6 \to {}^5I_8$	1208.5	110
$ErAlO_3$	$^5S_2 \to {}^5I_5$	1404.0	110
(Er, Lu)AlO_3	$^5I_7 \to {}^5I_8$	2120.5	77
$Er_3Al_5O_{12}$ [77]	$^5I_7 \to {}^5I_8$	2001.0	77
Er_2O_3	$^5I_7 \to {}^5I_8$	2098.5	77
$Er_3Sc_2Al_3O_{12}$:Tm^{3+}	$^5I_7 \to {}^5I_8$	2121	145
Er_2SiO_5	$^5I_7 \to {}^5I_8$	2198.5	110
(Er, Tm, Yb)$_3Al_5O_{12}$	$^5I_7 \to {}^5I_8$	2085	77
$Er_{1.5}Y_{1.5}Al_5O_{12}$	$^5I_7 \to {}^5I_8$	2101.0	77
		2097.9	77
		2091.7	
		≈ 2123	77
$GdAlO_3$	$^5I_7 \to {}^5I_8$	1992.5	90
$Ho_3Al_5O_{12}$	$^5I_7 \to {}^5I_8$	2129.4	90
		2122.4	77
		2097	
HoF_3	$^5I_7 \to {}^5I_8$	2090	77
$Ho_3Ga_5O_{12}$	$^5I_7 \to {}^5I_8$	2113.5	77
	$^5I_7 \to {}^5I_8$	2086	77
(Ho, Lu)$_3Al_5O_{12}$ [80]	$^5I_7 \to {}^5I_8$	2100.5	110
		2125.0	
		2130.0	110
$Ho_3Sc_2Al_3O_{12}$	$^5I_7 \to {}^5I_8$	2117.0	77
		2128.5	
$KGd(WO_4)_2$ [80]	$^5I_7 \to {}^5I_8$	2074.0	110
$KGd(WO_4)_2$ [79]	$^5S_2 \to {}^5I_5$	1398.2	110
$KGd(WO_4)_2$ (‖b)	$^5I_7 \to {}^5I_8$	2934.2	300
$KY(WO_4)_2$ [79]	$^5S_2 \to {}^5I_5$	1390.8	110
[80]	$^5I_7 \to {}^5I_8$	2076.5	110
$KY(WO_4)_2$:Er^{3+},Tm^{3+}	$^5I_7 \to {}^5I_8$	2072.0	300
$KLu(WO_4)_2$ [80]	$^5I_7 \to {}^5I_8$	2079.0	110
$LaNbO_4$ [80]	$^5I_7 \to {}^5I_8$	2072.5	110
$LaNbO_4$:Er^{3+}	$^5I_7 \to {}^5I_8$	2070	90
$LiHoF_4$	$^5F_5 \to {}^5I_7$	979	90
$LiNbO_3$	$^5I_7 \to {}^5I_8$	2078.6	77

*5 Percentage is given for trivalent rare-earth ions co-doping.

Table 34.3 Wavelengths for insulating crystal lasers *(continued)*

Host crystal	Laser transition	Laser wavelength, nm	Temperature, K
Holmium ($Ho^{3+}, 4f^{10}$) *(continued)*			
$LiYF_4$	$^5S_2 \rightarrow {}^5I_7$	749.8	90
	$^5F_5 \rightarrow {}^5I_7$	979.4	90
	$^5S_2 \rightarrow {}^5I_6$	1014.3	90
	$^5S_2 \rightarrow {}^5I_5$	1396.0	300
	$^5I_7 \rightarrow {}^5I_8$	2067.2	90
$LuAlO_3$	$^5I_7 \rightarrow {}^5I_8$	2134.8	90
$Lu_3Al_5O_{12}$	$^5I_7 \rightarrow {}^5I_8$	2102.0	77
$Lu_3Al_5O_{12}:Cr^{3+}$ [81]	$^5I_7 \rightarrow {}^5I_8$	2102.0	110
$Lu_3Al_5O_{12}:Tm^{3+}$ [82]	$^5I_7 \rightarrow {}^5I_8$	2100.0	239
		2100.8	300
$Lu_3Al_5O_{12}:Er^{3+}(2\%),Tm^{3+}(2\%)$	$^5I_7 \rightarrow {}^5I_8$	2102.0	77
$Lu_3Al_5O_{12}$	$^5I_6 \rightarrow {}^5I_7$	2946.0	300
α-$NaCaErF_6$	$^5I_7 \rightarrow {}^5I_8$	2034.5	150
		2031.2	77
		2037.7	
$LiY_{0.5-x-y}Er_{0.5}Tm_xHo_yF_4$ [83]	$^5I_7 \rightarrow {}^5I_8$	2060	
$NaLa(MoO_4)_2:Er^{3+}$	$^5I_7 \rightarrow {}^5I_8$	2050	90
$YAlO_3$ [80]	$^5I_7 \rightarrow {}^5I_8$	2118.5	110
		2130	110
$YAlO_3$ (‖[112])	$^5I_6 \rightarrow {}^5I_7$	2918.0	300
	$^5I_7 \rightarrow {}^5I_8$	2013.2	300
$Y_3Al_5O_{12}$	$^5I_6 \rightarrow {}^5I_7$	2940.3	300
	$^5I_7 \rightarrow {}^5I_8$	2091.4	77
		2097.5	
		2122.3	77
$Y_3Al_5O_{12}:Cr^{3+}(0.5\%)$	$^5I_7 \rightarrow {}^5I_8$	2097.5	77
		2122.3	
$Y_3Al_5O_{12}:Cr^{3+}, Tm^{3+}$ [84]	$^5I_7 \rightarrow {}^5I_8$	2097.4*6	300
$Y_3Al_5O_{12}:Er^{3+}(50\%), Tm^{3+}(6.7\%)$	$^5I_7 \rightarrow {}^5I_8$	2098.2	77
		2122.7	
		2128.8	295
		2130	
$Tm_3Al_5O_{12}$ [77]	$^5I_7 \rightarrow {}^5I_8$	2099.5	
$Y_3Fe_5O_{12}$	$^5I_7 \rightarrow {}^5I_8$	2086	77
$Y_3Fe_5O_{12}:Er^{3+}(5\%), Tm^{3+}(5\%)$	$^5I_7 \rightarrow {}^5I_8$	2089	77
		2107	77
$Y_3Ga_5O_{12}$	$^5I_7 \rightarrow {}^5I_8$	2086	77
		2114	
$Y_3Sc_2Ga_3O_{12}:Cr^{3+}, Tm^{3+}$ [85]	$^5I_7 \rightarrow {}^5I_8$	2086	300
$YVO_4:Er^{3+}, Tm^{3+}$	$^5I_7 \rightarrow {}^5I_8$	2041.2	77
ZrO_2-Er_2O_3	$^5I_7 \rightarrow {}^5I_8$	2115	77
Erbium ($Er^{3+}, 4f^{11}$)			
BaY_2F_8	$^4S_{3/2} \rightarrow {}^4I_{15/2}$	554.0	77
	$^4H_{9/2} \rightarrow {}^4I_{13/2}$	561.7	77
	$^4H_{9/2} \rightarrow {}^4I_{11/2}$	703.7	77
$BaY_2F_8:Yb^{3+}(37.5\%)$	$^4F_{9/2} \rightarrow {}^4I_{15/2}$	670.9	77
		670.0	

*6 Wavelength at a threshold of laser pumping. Above the threshold, lines 2096.3, 2090.0 and 2091.4 nm are also observed.

Table 34.3 Wavelengths for insulating crystal lasers *(continued)*

Host crystal	Laser transition	Laser wavelength, nm	Temperature, K
Erbium ($Er^{3+}, 4f^{11}$) *(continued)*			
$BaEr_2F_8$ [86]	$^4S_{3/2} \rightarrow {}^4I_{9/2}$	1645.5	110
		1735.5	
	$^4S_{3/2} \rightarrow {}^4I_{11/2}$	1232.0	110
	$^4S_{3/2} \rightarrow {}^4I_{13/2}$	842.5	110
		854.3	
	$^4F_{9/2} \rightarrow {}^4I_{11/2}$	1997.5	110
	$^4F_{9/2} \rightarrow {}^4I_{15/2}$	670.0	110
	$^4I_{11/2} \rightarrow {}^4I_{13/2}$	2741.7	110
		2759.5	
		2798.0	110
$BaYb_2F_8$ [87]	$^4F_{9/2} \rightarrow {}^4I_{11/2}$	1965.4	300
	$^4F_{9/2} \rightarrow {}^4I_{13/2}$	1260	300
$Bi_4Ge_3O_{12}$ [78]	$^4S_{3/2} \rightarrow {}^4I_{13/2}$	853	77
	$^4I_{13/2} \rightarrow {}^4I_{15/2}$	1558	77
		1664	
$CaAl_4O_7$	$^4I_{13/2} \rightarrow {}^4I_{15/2}$	1550	77
		1581.5	
CaF_2	$^4I_{13/2} \rightarrow {}^4I_{15/2}$	1617	77
		1529.8	4
	$^4S_{3/2} \rightarrow {}^4I_{13/2}$	845.6	77
		854.8	77
	$^4S_{3/2} \rightarrow {}^4I_{11/2}$	1260	77
	$^4S_{3/2} \rightarrow {}^4I_{9/2}$	1696	77
		1715	
		1726	77
CaF_2-ErF_3 [88]	$^4I_{11/2} \rightarrow {}^4I_{13/2}$	2729.5	300
		2746.0	
		2749.0	300
		2795.5	
		2798.5	300
CaF_2-ErF_3	$^4I_{11/2} \rightarrow {}^4I_{13/2}$	2730.7	300
CaF_2-ErF_3:TmF_3 (0.5%)	$^4I_{11/2} \rightarrow {}^4I_{13/2}$	2690	298
CaF_2-YF_3	$^4S_{3/2} \rightarrow {}^4I_{13/2}$	843.0	77
	$^4I_{13/2} \rightarrow {}^4I_{15/2}$	1544.8	77
		1555.8	
$Ca(NbO_3)_2$	$^4I_{13/2} \rightarrow {}^4I_{15/2}$	1610	77
$CaWO_4$	$^4I_{13/2} \rightarrow {}^4I_{15/2}$	1612	77
$CdaAlO_3$ [89]	$^4S_{3/2} \rightarrow {}^4I_{9/2}$	1657.1	77
$Er_3Al_5O_{12}$ [90]	$^4S_{3/2} \rightarrow {}^4I_{9/2}$	1776.2	110
$Er_3Al_5O_{12}$	$^4I_{11/2} \rightarrow {}^4I_{13/2}$	2937.0	300
$ErAlO_3$ [90]	$^4S_{3/2} \rightarrow {}^4I_{9/2}$	1663.2	110
$(Er, Lu)_3Al_5O_{12}$	$^4I_{11/2} \rightarrow {}^4I_{13/2}$	2939.5	300
$Er_{0.5}Lu_{0.5}AlO_3$	$^4S_{3/2} \rightarrow {}^4I_{9/2}$	1663.1	110
$KEr(WO_4)_2$ [91]	$^4I_{11/2} \rightarrow {}^4I_{13/2}$	2807.0	300
$KGd(WO_4)_2$ (‖c)	$^4I_{11/2} \rightarrow {}^4I_{13/2}$	2722.2	300
		2799.0	
	$^4S_{3/2} \rightarrow {}^4I_{13/2}$	846.8	300
	$^4S_{3/2} \rightarrow {}^4I_{9/2}$	1715.5	300

Table 34.3 Wavelengths for insulating crystal lasers *(continued)*

Host crystal	Laser transition	Laser wavelength, nm	Temperature, K
	Erbium ($Er^{3+}, 4f^{11}$) *(continued)*		
$KGd(WO_4)_2$ (‖b)	$^4I_{11/2} \rightarrow {}^4I_{13/2}$	2799.0	300
	$^4S_{3/2} \rightarrow {}^4I_{13/2}$	846.8	300
	$^4S_{3/2} \rightarrow {}^4I_{9/2}$	1732.5	300
$KLu_{1-x}Er_x(WO_4)_2$ [92]	$^4S_{3/2} \rightarrow {}^4I_{9/2}$	1738.3	300
	$^4S_{3/2} \rightarrow {}^4I_{13/2}$	850	300
	$^4I_{11/2} \rightarrow {}^4I_{13/2}$	2809.2	300
KYF_4 [93]	$^4S_{3/2} \rightarrow {}^4I_{15/2}$	561	300
LaF_3	$^4I_{13/2} \rightarrow {}^4I_{15/2}$	1611.3	300
$LiErF_4$	$^4S_{3/2} \rightarrow {}^4I_{9/2}$	1732	90
$LiErF_4$ [94]	$^4S_{3/2} \rightarrow {}^4I_{9/2}$	1704.2	110
	$^4S_{3/2} \rightarrow {}^4I_{11/2}$	1228.8	110
	$^4S_{3/2} \rightarrow {}^4I_{13/2}$	854.0	110
	$^4F_{9/2} \rightarrow {}^4I_{11/2}$	2000.5	110
	$^4F_{9/2} \rightarrow {}^4I_{13/2}$	2850.0	110
$LiYF_4$	$^4S_{3/2} \rightarrow {}^4I_{13/2}$	850.0	300
[93]	$^4S_{3/2} \rightarrow {}^4I_{15/2}$	551	300
$LuAlO_3$ (‖[112])	$^4S_{3/2} \rightarrow {}^4I_{9/2}$	1667.5	90
$Lu_3Al_5O_{12}$	$^4S_{3/2} \rightarrow {}^4I_{13/2}$	863.25	77
		863.2	300
	$^4I_{13/2} \rightarrow {}^4I_{15/2}$	1652.5	77
		1663.0	
	$^4S_{3/2} \rightarrow {}^4I_{9/2}$	1776.2	300
	$^4I_{11/2} \rightarrow {}^4I_{15/2}$	2940.6	300
		2829.8	
$Lu_3Al_5O_{12}$:Ho^{3+},Tm^{3+}	$^4I_{11/2} \rightarrow {}^4I_{13/2}$	2699.0	300
SrF_2-ErF_3 [95]	$^4I_{11/2} \rightarrow {}^4I_{13/2}$	2728.5	300
		2745.0	
		2793.0	300
$YAlO_3$ (‖[112])	$^4S_{3/2} \rightarrow {}^4I_{13/2}$	849.75	300
		859.4	
		849.65	77
		851.65	
$Y_3Al_5O_{12}$	$^4S_{3/2} \rightarrow {}^4I_{9/2}$	1663.2	300
	$^4I_{112} \rightarrow {}^4I_{13/2}$	2730.9	300
	$^4S_{3/2} \rightarrow {}^4I_{13/2}$	862.75	77
		862.75	300
	$^4I_{13/2} \rightarrow {}^4I_{15/2}$	1632	300
	$^4S_{3/2} \rightarrow {}^4I_{9/2}$	1775.7	300
	$^4I_{11/2} \rightarrow {}^4I_{13/2}$	2830.2	300
		2936.4	
	$^4I_{13/2} \rightarrow {}^4I_{15/2}$	1660.2	
		1645.2	77
$Y_3Al_5O_{12}$:Yb^{3+} (5%)	$^4I_{13/2} \rightarrow {}^4I_{15/2}$	1645.9	295
$Yb_3Al_5O_{12}$	$^4I_{13/2} \rightarrow {}^4I_{15/2}$	1661.5	77
$Yb_{1.5}Er_{1.5}Al_5O_{12}$ [96]	$^4S_{3/2} \rightarrow {}^4I_{9/2}$	1776.0	110
$Yb_{0.75}Er_{0.25}AlO_3$ [96]	$^4S_{3/2} \rightarrow {}^4I_{9/2}$	1663.1	110
(Yb, Er)$_3Al_5O_{12}$ (‖[100])	$^4I_{11/2} \rightarrow {}^4I_{13/2}$	2830.2	300
		2936.4	300
ZrO_2-Er_2O_3 (12%)	$^4I_{13/2} \rightarrow {}^4I_{15/2}$	1620	77

Table 34.3 Wavelengths for insulating crystal lasers *(continued)*

Host crystal	Laser transition	Laser wavelength, nm	Temperature, K
	Thulium ($Tm^{3+}, 4f^{12}$)		
$Bi_4Ge_3O_{12}$ [78]	$^3H_4 \to {}^3H_6$	1850	77
CaF_2-ErF_3	$^3H_4 \to {}^3H_6$	1894	77
		1900	
$CaMoO_4$:Er^{3+}(0.75%)	$^3H_4 \to {}^3H_6$	1911.5	77
		1906.0	
$Ca(NbO_3)_2$	$^3H_4 \to {}^3H_6$	1910	77
$CaWO_4$	$^3H_4 \to {}^3H_6$	1911	77
		1916	
$ErAlO_3$	$^3H_4 \to {}^3H_6$	1872	77
(Er, Lu)AlO_3	$^3H_4 \to {}^3H_6$	1884.5	77
Er_2O_3	$^3H_4 \to {}^3H_6$	1934	77
$Yb_{1.5}Er_{1.5}Al_5O_{12}$	$^3H_4 \to {}^3H_6$	1880	77
		1884	
		2014	85
$GdAlO_3$	$^3H_4 \to {}^3H_6$	1852.9	77
$LiNbO_3$	$^3H_4 \to {}^3H_6$	1853.2	77
$LiYF_4$ [97]	$^3F_4 \to {}^3H_5$	2303.0	77
	$^3H_4 \to {}^3H_6$	1889.0	77
		1909.0	
$Lu_3Al_5O_{12}$	$^3H_4 \to {}^3H_6$	1885.5	77
		2024.0	
$Lu_3Al_5O_{12}$:Cr^{3+} [81]	$^3F_4 \to {}^3H_5$	2342.5	110
α-$NaCaErF_6$	$^3H_4 \to {}^3H_6$	1858.0	77
		1888.5	
SrF_2	$^3H_4 \to {}^3H_6$	1972	77
$YAlO_3$:Cr^{3+} (‖[112])	$^3F_4 \to {}^3H_5$	2274	300
$YAlO_3$:Cr^{3+}(0.75%)	$^3F_4 \to {}^3H_5$	2318	300
		2353	
		2354	300
		2355	
		2274	300
		2318	
$YAlO_3$:Cr^{3+}(0.1%)	$^3F_4 \to {}^3H_5$	2340	300
		2348	300
		2349	
	$^3H_4 \to {}^3H_6$	1856	90
		1883	
		1933.5	90
$YAlO_3$:Er^{3+}(30%)	$^3H_4 \to {}^3H_6$	1861	77
$Y_3Al_5O_{12}$	$^3H_4 \to {}^3H_6$	1883.4	77
		2013.2	85
$Y_3Al_5O_{12}$:Cr^{3+}(0.5%)	$^3H_4 \to {}^3H_6$	2013.2	77
$Y_3Al_5O_{12}$:Cr^{3+}(0.01%)	$^3F_4 \to {}^3H_5$	2324	300
(Yb, Er)$_3Al_5O_{12}$	$^3H_4 \to {}^3H_6$	1885.0	77
		2019.5	
YVO_4	$^3H_4 \to {}^3H_6$	2019	295
[98]	$^3F_4 \to {}^3H_6$	1850–2000	285
ZrO_2-Er_2O_3	$^3H_4 \to {}^3H_6$	1896	77

Table 34.3 Wavelengths for insulating crystal lasers *(continued)*

Host crystal	Laser transition	Laser wavelength, nm	Temperature, K
	Ytterbium ($Yb^{3+}, 4f^{13}$)		
BaF_2 (0.16*7) [99]	$^2F_{5/2} \rightarrow {}^2F_{7/2}$	1024	300
$Bi_4Ge_3O_{12}$	$^2F_{5/2} \rightarrow {}^2F_{7/2}$	1030	77
BaY_2F_8 (1.4*7) [99]	$^2F_{5/2} \rightarrow {}^2F_{7/2}$	1018	300
CaF_2:Nd^{3+}(1–2%)	$^2F_{5/2} \rightarrow {}^2F_{7/2}$	1033.6	120
$Ca_5(PO_4)_3F$ (0.36*7) [99]	$^2F_{5/2} \rightarrow {}^2F_{7/2}$	1043	300
$Ca_4Sr(PO_4)_3F$ [100]	$^2F_{5/2} \rightarrow {}^2F_{7/2}$	1044	300
	$^2F_{5/2} \rightarrow {}^2F_{7/2}$	1108	300
	$^2F_{5/2} \rightarrow {}^2F_{7/2}$	985	300
$Ca_3Sr_2(PO_4)_3F$ [100]	$^2F_{5/2} \rightarrow {}^2F_{7/2}$	1044	300
$Gd_3Ca_5O_{12}$:Nd^{3+}(~2%)	$^2F_{5/2} \rightarrow {}^2F_{7/2}$	1023.2	77
$Gd_3Sc_2Al_3O_{12}$:Nd^{3+}(~1%)	$^2F_{5/2} \rightarrow {}^2F_{7/2}$	1029.9	77
$KCaF_3$ (1.4*7) [99]	$^2F_{5/2} \rightarrow {}^2F_{7/2}$	1031	300
KY_3F_{10} (1.45*7) [99]	$^2F_{5/2} \rightarrow {}^2F_{7/2}$	1011	300
LaF_3 (0.22*7) [99]	$^2F_{5/2} \rightarrow {}^2F_{7/2}$	1009	300
$LiYF_4$ (1.44*7) [99]	$^2F_{5/2} \rightarrow {}^2F_{7/2}$	1020	300
$LiYO_2$ (2.27*7) [99]	$^2F_{5/2} \rightarrow {}^2F_{7/2}$	1020	300
$Lu_3Al_5O_{12}$(~1%)	$^2F_{5/2} \rightarrow {}^2F_{7/2}$	1029.4	77
		1029.7	175
$Lu_3Ga_5O_{12}$:Nd^{3+}(1.5%)	$^2F_{5/2} \rightarrow {}^2F_{7/2}$	1023.0	77
$LuPO_4$ (0.8*7) [99]	$^2F_{5/2} \rightarrow {}^2F_{7/2}$	1011	300
$Lu_3Sc_2Al_3O_{12}$:Nd^{3+}(1.5%)	$^2F_{5/2} \rightarrow {}^2F_{7/2}$	1029.9	77
Rb_2NaYF_6 (1.03*7) [99]	$^2F_{5/2} \rightarrow {}^2F_{7/2}$	1012	300
$ScBO_3$ (1.32*7) [99]	$^2F_{5/2} \rightarrow {}^2F_{7/2}$	1022	300
SrF_2 (0.1*7) [99]	$^2F_{5/2} \rightarrow {}^2F_{7/2}$	1025	300
$Sr_5(PO_4)_3F$ [100]	$^2F_{5/2} \rightarrow {}^2F_{7/2}$	1047	300
$Sr_5(VO_4)_3F$ [100]	$^2F_{5/2} \rightarrow {}^2F_{7/2}$	1044	300
$YAlO_3$ (6.82*7) [99]	$^2F_{5/2} \rightarrow {}^2F_{7/2}$	1014	300
$Y_3Al_5O_{12}$	$^2F_{5/2} \rightarrow {}^2F_{7/2}$	1029.6	77
		1029.3	
$Y_3Al_5O_{12}$:Nd^{3+}(~0.8%)	$^2F_{5/2} \rightarrow {}^2F_{7/2}$	1029.7	200
$Y_3Al_5O_{12}$:Nd^{3+}(~0.8%), Cr^{3+}(0.5%)	$^2F_{5/2} \rightarrow {}^2F_{7/2}$	1029.8	210
$Y_3Ga_5O_{12}$:Nd^{3+}(1.5%)	$^2F_{5/2} \rightarrow {}^2F_{7/2}$	1023.3	77
Y_2SiO_5 (1.83*7) [99]	$^2F_{5/2} \rightarrow {}^2F_{7/2}$	1042	300
(Y, Yb)$_3Al_5O_{12}$	$^2F_{5/2} \rightarrow {}^2F_{7/2}$	1029.3	77
(Yb, Lu)$_3Al_5O_{12}$	$^2F_{5/2} \rightarrow {}^2F_{7/2}$	1029.4	77
	Uranium ($U^{3+}, 5f^3$)		
BaF_2	$^4I_{11/2} \rightarrow {}^4I_{9/2}$	2556	20
CaF_2	$^4I_{11/2} \rightarrow {}^4I_{9/2}$	2613*8	300
		2570*8	300
		2510*9	77
		2440*9	77
		2240*10	77
SrF_2	$^4I_{11/2} \rightarrow {}^4I_{9/2}$	2407	20, 77, 90

*7 Concentration in $10^{20} Yb^{3+}/cm^3$.
*8 This line has been ascribed to U^{3+} ions in tetragonal sites.
*9 This line has been ascribed to U^{3+} ions in orthorhombic sites.
*10 This line has been ascribed to either U^{3+} ions in tetragonal or U^{4+} in trigonal sites.

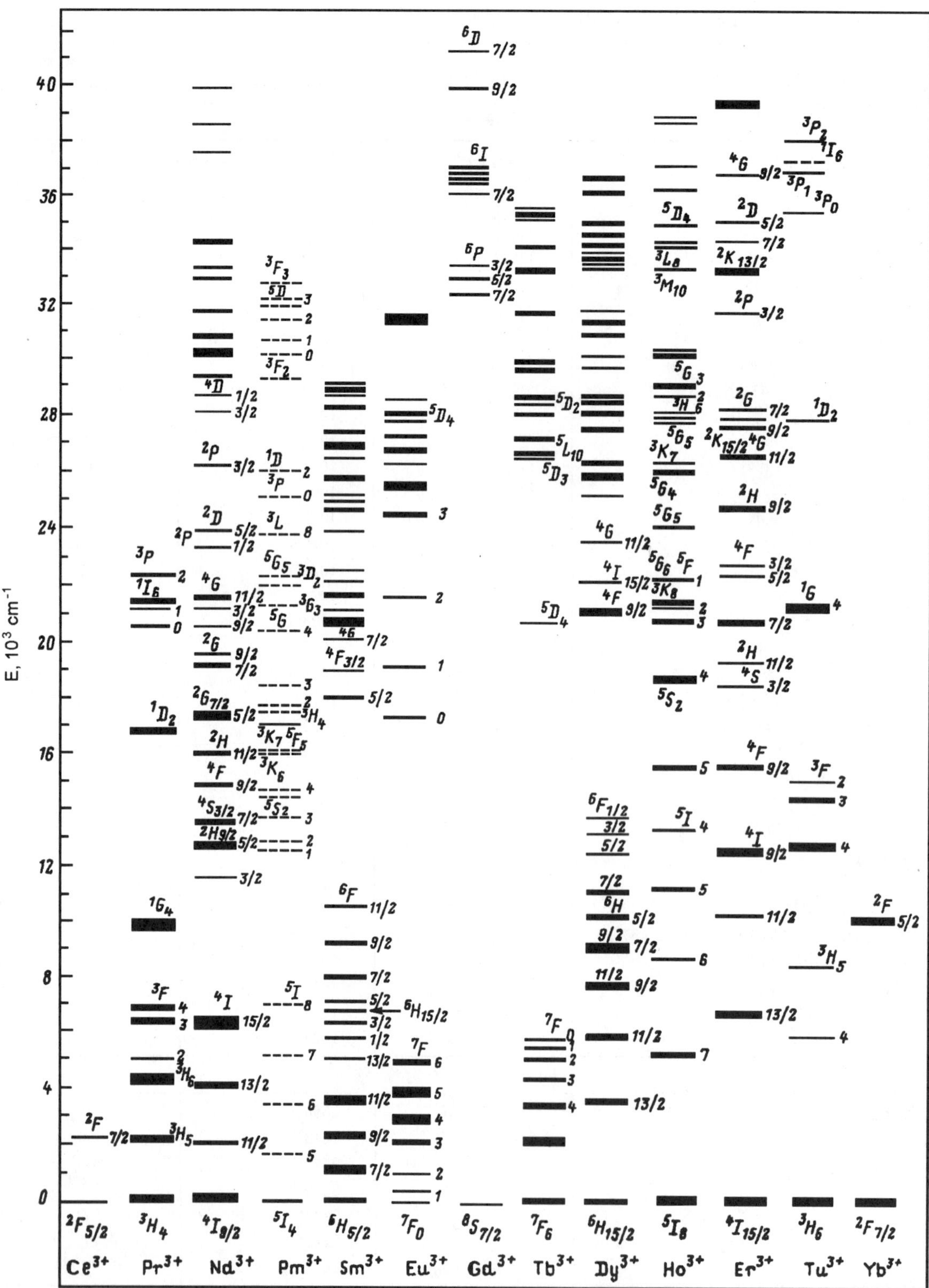

Figure 34.10 Energy levels of trivalent rare-earth ions in crystals [12].

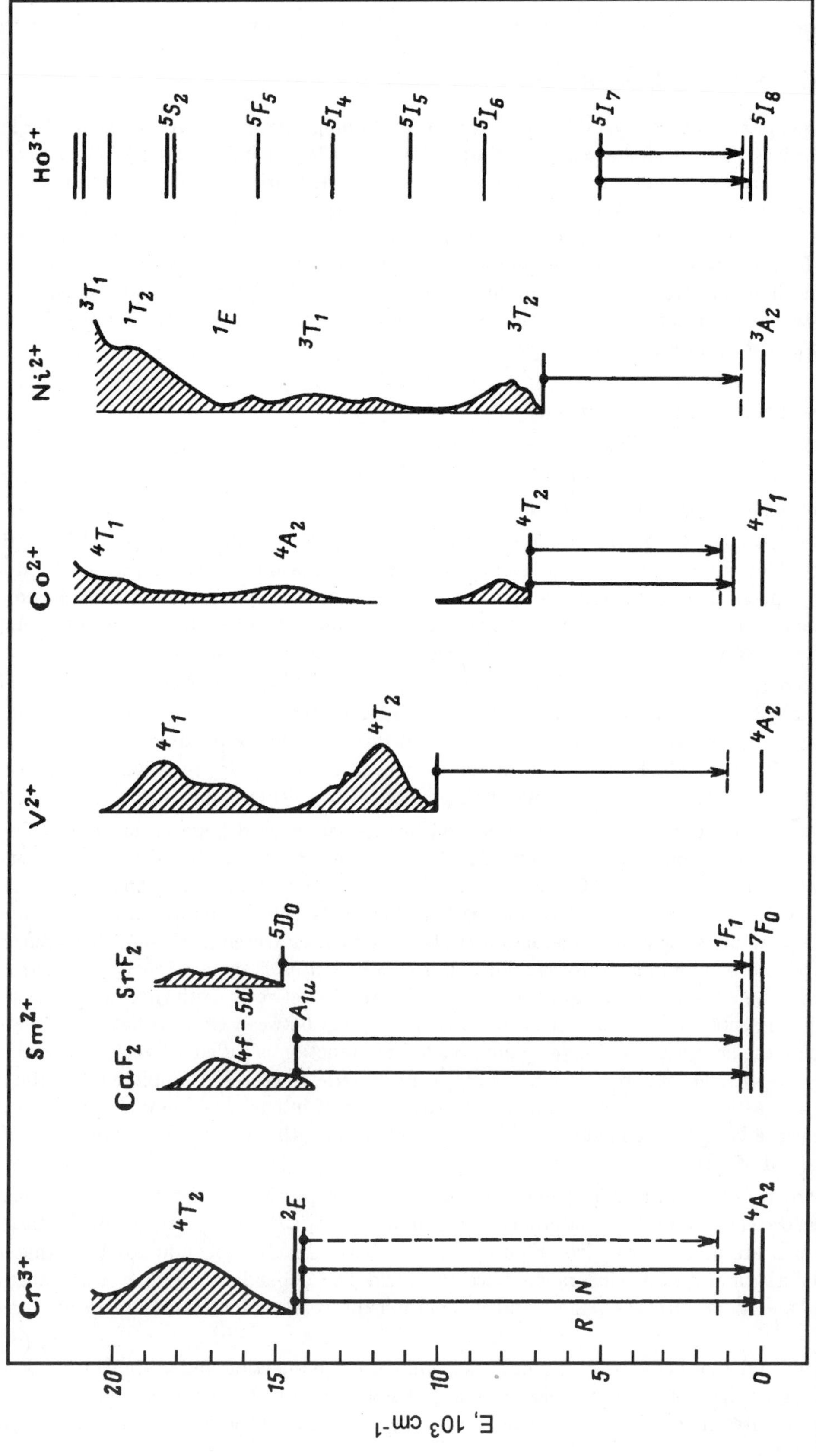

Figure 34.11 Simplified energy–diagrams of insulating crystal lasers employing electronic and electronic–vibrational transitions in some divalent and trivalent ions [12].

34.4 Semiconductor Lasers

In insulating crystal lasers, a crystal is only a matrix containing active ions; in semiconductor lasers a crystal itself plays the role of the active medium, while ions are just impurities supplying electrons in the conduction band and holes in the valence band. The basic principle of semiconductor lasers may be summarized as follows. When an electric current flows in the forward direction through a *p-n* diode, the electrons and holes can recombine within the *p-n* junction emitting the recombination energy in the form of the electromagnetic radiation. The energy difference between the energy levels of electrons and holes is determined by the band gap that can be varied by choosing semiconductor and an impurity concentration. A *p-n* junction is extremely small in size compared to any other laser. Main properties of these lasers are as follows:

* high gain;
* high conversion efficiency of pump energy;
* wide spectrum range tunability;
* possibility of generating ultra-short pulses and high frequency (above 10^{10} Hz) modulated signals;
* poor directionality (spreading out by 5°–30°).

Only direct transitions, i.e. recombination without the emission of a phonon, may result in stimulated emission. This is possible only in so called direct-transition semiconductors (Si and Ge do not belong to this type as there are no direct transitions). The *p-n* junction of the early semiconductor lasers were made from a single material and are known as *homojunctions.* But recent ones employ junctions of two different kinds of semiconductors which are called *heterojunctions.*

Typical semiconductor laser materials are binary compounds or based on multicomponent solid solutions (see Table 34.4). These materials should have the same type of crystal lattice and ability to form a continuous row of solutions with various concentrations. This is possible for ternary alloys, such as $GaAs_{1-x}P_x$, $In_{1-x}Ga_xP$, $Al_xGa_{1-x}As$, and so on. In the $Pb_{1-x}Sn_xSe$ system, the band gap closes at x=0.15 (T=12 K); that makes generating possible at wavelengths up to 32 μm.

The efficiency of the laser material is increased when it is sandwiched between layers of wide-gap materials. For example, comparing GaAs with $Al_xGa_{1-x}As$, one could find that the band gap energy of the latter is greater than that of GaAs, and the refractive index is smaller, while the lattice parameters of $Al_xGa_{1-x}As$ almost do not change with x (Figure 34.12). Therefore, in a *heterojunction* of GaAs and $Al_xGa_{1-x}As$, light and conduction electrons are concentrated in GaAs, whose refractive index is greater, and band gap is narrower. Thus, if a *heterojunction* is made of *p*-type $Al_xGa_{1-x}As$- or *n*-type $Al_xGa_{1-x}As$-layers with a thin active layer of GaAs in between, both light and carriers are confined within the active layer. This leads to a strong coupling between emitted light and pumping current, reducing the laser threshold appreciably. Such a *p-n* junction is called a *double heterojunction* or a *double hetero-structure* (abbreviated in the literature as DH-structure). Complete or partial laser failure may occur as a result of optical damage in the region of intense laser emission. To increase the output power, a laser can be made consisting of many layers with independent confinement of the optical quanta and of carriers.

For a DH-structure laser of (1.2–1.7) μm-wavelength it is not possible to obtain a lattice constant fit with a ternary compound. A quaternary compound $Ga_xIn_{1-x}As_yP_{1-y}$ and similar multicomponent structures are used instead (see Figure 34.13). Many mixed semiconductor laser materials permit wavelength tuning in a wide range since the band gap depends on temperature, magnetic field, pressure, and so on. Review papers and monographs on semiconductor lasers can be found in [101–108].

To excite semiconductor lasers one can employ not only electron injection and avalanche breakdown; electron beam and optical pumping are also usable. In the latter case the active volume can be made 10^4–10^6 times greater than for *p-n* junction laser, with proportional increase of the output power [107].

Table 34.4 Semiconductor laser materials

Compound type	Material	Wavelength, μm	Type of pumping*1
A^3B^5	GaN	0.36	O
	GaAs	0.83–0.92	I,O,E,A
	InP	0.89–0.91	I,O,A
	GaSb	1.5–1.6	I,O,E
	InAs	3.0–3.2	I,O,E
	InSb	4.8–5.3	I,O,E
	(Ga, In)P	0.56–0.9 [103]	I,O
	(Al, Ga)As	0.63–0.9 [103]	I,E
	Ga(As, P)	0.63–0.9 [103]	I,E
	(Al, Ga)(PAs)	0.63–0.9 [103]	I,E
	(In, Ga)As	0.9–3.4 [103]	I,O
	In(P, As)	0.9–3.2	I
	Ga(As, Sb)	0.95–1.6	I
	In(As, Sb)	1.0–5.3	I
	(Al, Ga)Sb	1.2–1.8 [103]	E
	(In, Ga)(P, As)	0.62–3.2	I
	(Al, Ga)(As, Sb)	0.62–1.6	I
A^2B^6	ZnO	0.37	O,E
	Zn	0.33	O,E
	ZnSe	0.46	E
	ZnTe	0.53	E
	CdS	0.49	I,E,A
	CdSe	0.69	I,E
	CdTe	0.79	E
	(Zn, Cd)S	0.33–0.49	I
	C(S, Se)	0.59–0.69	I,E
	(Cd, Hg)Te	3.8–4.1	I
A^4B^6	PbS	4.3	IE
	PbSe	8.5	IE
	PbTe	6.5	IE
	P(S, Se)	4.7–5.5	IOE
	(Pb, Sn)Se	8.5–32	IE
	(Pb, Sn)Te	6.5–32	IE
	(Pb, Ge)Te	4.4–6.5	I
$(A^4,A^2)B^6$	(Pb,Cd)S	3.5	I
A^6	Te	3.7	E
A^3B^6	GaSe	0.59–0.60	OE
	InSe	0.97	E
$A^3_2B^6$	In_2Se	1.6	E
$A^2_3B^5_2$	C_3P_2	2.1	O
$A^2B^4C^6_2$	$CdSnP_2$	1.1	E
	$CdSiAs_2$	0.77	E

*1 Types of pumping are designated as: I – injection, E – electron beam, O – optical, A – avalanche.

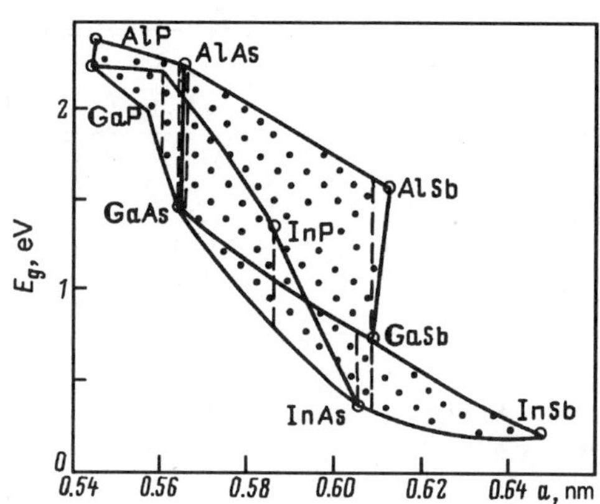

Figure 34.12 The energy band gap versus the crystal lattice parameter for semiconductors and ternary alloys based on A^3B^5 compounds [108].

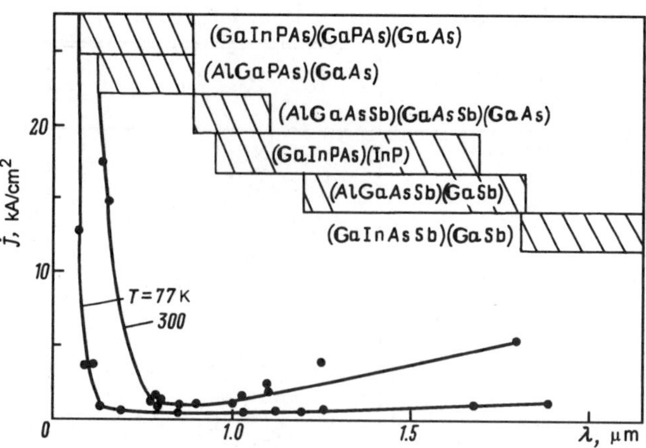

Figure 34.13 The spectrum bandwidth covered with the help of quaternary A^3B^5 compound-based alloys and dependence of the threshold current densities versus wavelength for these heterojunction lasers [103].

34.5 Rare-Earth Liquid Lasers

Rare-earth liquid lasers use rare-earth ions in solutions as an active medium. All these lasers operate in a pulsed mode, using flash lamps for optical pumping. Active ion concentrations in rare-earth liquid materials are typically the same as for solid state lasers. There exist two types of rare-earth liquid solutions:

Metalorganic (or chelate) solutions.

Chelates are liquid solutions in which the active trivalent rare-earth ion is bonded to organic molecules or *ligands*. Laser transition in chelates occurs between the same levels of rare-earth metal ions as for solid state materials:

$Eu^{3+} - {}^5D_0 \rightarrow {}^7F_2$;

$Tb^{3+} - {}^5D_4 \rightarrow {}^7F_5$;

$Nd^{3+} - {}^4F_{3/2} \rightarrow {}^4I_{11/2}$,

while the pumping radiation is converted via ligand absorption bands. As ligands exhibit extremely high absorption (up to 100 cm^{-1}) the generation is possible only in thin active medium layers. Two kinds of chelate compounds are used:

tetrakis chelates – $(RE)^{3+}(ligand^-)_4Q^+$,

and *tris chelates* – $(RE)^{3+}(ligand^-)_3$,

where Q^+ designates a cation. Ligands, cations and solvents most often used in chelate lasers are listed in Table 34.5. Principal parameters of chelate laser liquids are given in Table 34.6.

Inorganic rare-earth solutions.

This type of liquid rare-earth mediums are solutions of rare-earth salts in complex inorganic solvents. These solvents are also known as *aprotic* since all of them lack hydrogen groups. Such materials have no atoms with high vibrational frequencies useful for obtaining high efficiency lasing, though so far trivalent neodymium was used only as an active ion in these materials (see Table 34.7). Intrinsic active ion bands are responsible for absorption of the pumping flash lamp radiation in aprotic lasers.

Despite of attractive lasing properties as high efficiency, low divergence, and narrow spectrum, the aprotic lasers are not studied as intensively as other types of lasers, due to chemical *aggressiveness and toxicity* of their active medium. Most published data on this type of lasers are systematized in [109].

Table 34.5 Ligands, cations and solvents used in rare-earth liquid (metalorganic) lasers

LIGANDS		CATIONS	
Trifluoroacetylacetone	TFA	Piperidinium	Pip
Benzoylacetone	B	Deuterated Piperidinium	DPip
Benzoyltrifluoroacetone	BTF	Pyridinium	Pyr
Deuterated Benzoyltrifluoroacetone	DBTF	Imidazolium	Im
Orthofluorobenzoyltrifluoroacetone	OFBTF	Dimethylammonium	DMA
Methafluorobenzoyltrifluoroacetone	MFBTF	**SOLVENTS**	
Parafluorobenzoyltrifluoroacetone	PFBTF	Methanol	Meth
Orthochlorobenzoyltrifluoroacetone	OCBTF	Ethanol	Eth
Methachlorobenzoyltrifluoroacetone	MCBTF	Propyl alcohol	PA
Parachlorobenzoyltrifluoroacetone	PCBTF	1,4-Dioxane	Diox
Orthobromobenzoyltrifluoroacetone	OBBTF	Acetonitrile	Ace
Methabromobenzoyltrifluoroacetone	MBBTF	Acrylonitrile	Acr
Parabromobenzoyltrifluoroacetone	PBBTF	Dimethylformamide	DMF
Dibenzoylmethane	DBM	Dimethyl sulfoxide	DMSO
Thenoyltrifluoroacetone	TTF	Deuterated Dimethyl sulfoxide	DDMSO
Pentafluoropropionate	PFP	Carbon tetrachloride	CTA
1,10-Phenathrolyne	Phen	Hexafluorobenzene	HFB
Deuterated Tributylphosphate	DTBP	2-Ethoxipropionitrile	EPN
		2-Ethoxiethanole	EE

Table 34.5 *(continued)*

COMBINED SOLVENTS		
solvents	volume combination	solvent number
Eth:Meth	3:1	#1
Eth:Meth:DMF	12:4:1	#2
Eth:Meth:DMF	15:5:2	#3
Eth:Meth:DMF	9:3:2	#4
Eth:DMF:PA	14:5:2	#5
CTA:Acr	9:1	#6
EPN:Ace:EE	2:2:1	#7

Table 34.6 Rare-earth (metalorganic) liquid laser materials

Active complex	Solvent	Concentration, 10^{-3}mol/l	Laser wavelength, μm	Fluorescence lifetime, μs	Transition linewidth, cm^{-1}	Temperature, °C
$Eu^{3+}(B^-)_4Pip^+$	#1	8.7	0.6130	500	21	
$Eu^{3+}(B^-)_4Pip^+$ +	#1 or #2	10	0.6131		21	−150
CH_3COONa		20	0.6114		23	
$Eu^{3+}(B^-)_4Na^+$	#3 or #5	20	0.6111	670	19	−140
$Eu^{3+}(B^-)_4NH_4^+$	#1	15	0.6130			−150
$Eu^{3+}(DBM^-)_4Pip^+$	#4	15	0.6120		16	−145
$Eu^{3+}(DBM^-)_4Pyr^+$	Ace	2.5				+25
$Eu^{3+}(TFA^-)_4NH_4^+$	#1	15	0.6122			−150
	Ace	10	0.6119			−35

Table 34.6 Rare-earth (metalorganic) liquid laser materials *(continued)*

Active complex	Solvent	Concentration, 10^{-3}mol/l	Laser wavelength, μm	Fluorescence lifetime, μs	Transition linewidth, cm^{-1}	Temperature, °C
$Eu^{3+}(TTF^-)_4DMA^+$	Ace	5	0.6125			−35
$Eu^{3+}(BTF^-)_4Im^+$	Ace	7.5	0.6118		69	−35
$Eu^{3+}(BTF^-)_4Pyr^+$	Ace	5	0.6118		107	−40
$Eu^{3+}(BTF^-)_4Pip^+$	Ace	10	0.6119		91	+25
$Eu^{3+}(BTF^-)_4DPip^+$	#7	11.35		700		−105
$Eu^{3+}(DBTF^-)_4Pip^+$	#7	11.35		680		−105
$Eu^{3+}(DBTF^-)_4DPip^+$	#7	11.35		660		−105
$Eu^{3+}(BTF^-)_4DMA^+$	Ace	7.5	0.61175		83	+21
$Eu^{3+}(OFBTF^-)_4DMA^+$	Ace	7.5	0.61167		80	+21
$Eu^{3+}(MFBTF^-)_4DMA^+$	Ace	7.5	0.61178		87	+21
$Eu^{3+}(PFBTF^-)_4DMA^+$	Ace	7.5	0.61173		83	+21
$Eu^{3+}(OCBTF^-)_4DMA^+$	Ace	7.5	0.61174		67	+21
$Eu^{3+}(MCBTF^-)_4DMA^+$	Ace	7.5	0.61171		85	+21
$Eu^{3+}(PCBTF^-)_4DMA^+$	Ace	7.5	0.61172		85	+21
$Eu^{3+}(OBBTF^-)_4DMA^+$	Ace	7.5	0.61174		67	+21
$Eu^{3+}(MBBTF^-)_4DMA^+$	Ace	7.5	0.61173		83	+21
$Eu^{3+}(PBBTF^-)_4DMA^+$	Ace	7.5	0.61171		87	+21
$Tb^{3+}(TFA^-)_3$	Ace or Diox	2.5	0.5470			+25
$Nd^{3+}(PFP^-)_3Phen$	DDMSO	200	1.057		140	+30
$Nd^{3+}(TTF^-)_4Pyr^+$	#6	10				+20
$Nd^{3+}(NO_3^-)_3(DTBP)_3$	Hex or CTA	200	1.054	11	100	+20

Table 34.7 Rare-earth inorganic liquid laser materials

Material composition	Active component concentration, mole/l	Laser wavelength, μm	Fluorescence lifetime, μs	Transition linewidth, cm^{-1}
Nd^{3+}-$SeOCl_2$-$SnCl_4$	0.5	1.056	110	145
	0.1	1.058	83	
	0.3	1.058	230	165
Nd^{3+}-$SeOCl_2$-$SbCl_5$			225	
Nd^{3+}-$SOCl_2$-$GaCl_3$		1.0575	270	102
Nd^{3+}-$POCl_3$-$SnCl_4$	4%*1	1.0525	280	145
	0.86%*1	1.0522	180	100
Nd^{3+}-$SeOCl_2$-$TiCl_3$	3%*1	1.0525	230	
		1.0542	140	
Nd^{3+}-$SeOCl_2$-$AlCl_3$	0.15	1.054	300	
		1.05219		
Nd^{3+}-$SeOCl_2$-$ZrCl_4$	0.3	1.0522	300	145
Nd^{3+}-PBr_3-$AlBr_3$-$SbBr_3$	0.5%*1	1.066	230	113

*1 Concentration is given in weight percents.

34.6 Dye Lasers

Dyes are chemical compounds with complex branching chemical bonds that exhibit strong absorption bands in visible or near UV regions. The scheme of energy levels for a typical *dye* is shown in Fig. 34.14. General features of spectroscopic parameters of a dye are the mirror symmetry of absorption $S_0 \to S_1$ and luminescence bands, as well as partial overlapping of absorption and luminescence spectra (see Fig. 34.15).

The amplification of the radiation in dye solutions occurs due to population inversion between first optically excited singlet state S_1 and vibrational sublevels of the ground state S_0. Taking into account triplet-triplet transition and absorption from excited S_1 state, one can get the gain coefficient of a dye [111]:

$$g = \chi_{S_1 \to S_0}(\nu) \left[\frac{n_{S_1}}{n} \left(1 - \frac{B_{S_1 \to S_i}(\nu)}{B_{S_1 \to S_0}(\nu)} - \frac{n_{S_0}}{n_{S_1}} e^{-\frac{h(\nu_e - \nu)}{kT}} - \frac{n_{T_1}}{n_{S_1}} \cdot \frac{B_{T_1 \to T_j}(\nu)}{B_{S_1 \to S_0}(\nu)} \right) \right], \tag{34.1}$$

where $\chi_{S_1 \to S_0}(\nu) = n(h\nu / c') B_{S_1 \to S_0}(\nu)$ is the maximum gain; c' is the speed of light in the medium (in cm/s); $h\nu$ is the quantum energy (in J); $B_{S_1 \to S_0}$, $B_{S_1 \to S_i}$, $B_{T_1 \to T_j}$ are Einstein coefficients for stimulated emission, singlet and triplet absorption (in J^{-1}·s^{-1}·cm^3), respectively; $\frac{n_{S_0}}{n_{S_1}}$, $\frac{n_{S_1}}{n}$, $\frac{n_{T_1}}{n_{S_1}}$ are relative populations for the states S_0, S_1, T_1; n (in cm^{-3}) is the dye solution concentration.

The amplification is impossible if either

$$B_{S_1 \to S_0}(\nu) \leq B_{S_1 \to S_i}(\nu) \tag{34.2}$$

or when the triplet state population corresponds to the inequality.

$$n_{T_1} / n_{S_1} \geq B_{S_1 \to S_0}(\nu) / B_{T_1 \to T_j}(\nu). \tag{34.3}$$

In a steady state regime the relationship

$$n_{T_1} / n_{S_1} \geq P_{S_1 \to T_1} / P_{T_1 \to S_0} \tag{34.4}$$

is valid where $P_{S_1 \to T_1}$, $P_{T_1 \to S_0}$ are probabilities of corresponding transitions. Therefore, the steady state regime of laser operation is attained only if the condition

$$B_{S_1 \to S_0}(\nu) \geq B_{T_1 \to T_j}(\nu) \cdot P_{S_1 \to T_1} / P_{T_1 \to S_0} \tag{34.5}$$

is fulfilled. Pulsed operation is possible even though the relation (34.5) is violated.

According to (34.5) the effective dye solutions should exhibit high luminescence quantum yield, weak overlapping of the absorption $S_1 \to S_i$, $T_1 \to T_J$ spectrum bands, low population in the triplet state, that is possible at low $P_{S_1 \to T_1}$- and high $P_{T_1 \to S_0}$-probability. One can deduce how important this limitation is from the fact that among many thousands of various dyes the laser effect is observed only for several hundreds if nanosecond pump pulse durations are employed. Microsecond pulses are available from only tens of dyes, while pulses longer than 100 μs are obtained for only several types of dye solutions. The analysis shows that among many different classes of dye lasing conditions the above-mentioned are fulfilled for derivatives of oxazole, oxadiazole, benzene and their analogs – coumarine, rhodamine, oxazine and polimetine dyes.

In Table 34.8 spectral and lasing parameters of most efficient dyes are presented according to [110–113]. Dye lasers can be pumped by incoherent or (more often) by laser sources. A wide spectrum of *dye* luminescence offers broad wavelength tunability in dispersive resonators simultaneously with ultranarrow lines or ultrashort pulses.

Some dyes are used not only as an active medium in lasers, but also as ultrafast shutters for passive Q-switching and mode-locking.

Table 34.8 Spectral and lasing parameters of some most efficient dyes [λ_{abs}^{max}, λ_{lum}^{max}, λ_{osc}^{max} are wavelengths corresponding to maximums of absorption, luminescence and lasing spectra; η is the luminescence quantum yield; $B = B_{S_0 \to S_1}(\lambda_{abs}^{max})$ is the Einstein coefficient at the absorption spectrum maximum; τ_{S_1} is the lifetime of the first excited singlet state S_1; λ_{pmp} is the pump wavelength; λ_{osc} is the lasing wavelength or range; γ is the pump to laser radiation conversion efficiency; $K(\lambda_{pmp})$ is the absorption coefficient at a pump wavelength. *Pump*: FL is the flash lamp pump, Ar and Kr is the pump with argon and krypton continuously working ion lasers at /UV (ultra violet), /VIS (visible), /IR (infrared) or /all lines. *Solvents*: DMF – N,N'-Dimethylformamid; EG – Ethylenglycole; Cycl – Cyclohexane; Meth – Methanol; Bz – Benzyl alcohol; DMSO – Dimethyl sulfoxide; PC – Propylene carbonate; Tol – Toluene; Dichl – Dichlorethane; Eth – Ethanol; Diox – Dioxane.]

Dye	Solvent	λ_{abs}^{max}, nm	λ_{lum}^{max}, (λ_{abs}^{max}) nm	η, %	$B \cdot 10^{-7}$, $J^{-1} \cdot s^{-1} \cdot cm^3$	τ_{S_1}, 10^{-10} s	λ_{pmp}, nm	λ_{osc}, nm	γ, %	$K(\lambda_{pmp})$, cm^{-1}
2,2''-Dimethyl-	Cycl	251	336				248	312–352	14	
-p-terphenyl			(336)							
2,5-Diphenyl-1,3,4-	Diox	284	(334)	65			337	383–395		
oxadiazole (PPD)	Eth	280	(332)	82			337	383–395		
p-Terphenyl	Cycl	275	339				248	322–365	14	
	DMF	284	(345)	30			265	330–360		
	Diox	280	(343)	70			265	330–362		
	Eth	277	(340)	85			265	330–362		
2-Methyl-5-t-butyl-	Diox		(360)				308	346–377	8.5	
-p-quaterphenyl	Cycl	285								
2-Phenyl-5-	Diox	307	361	86			347	370–386	31	7
-(4'-Biphenyl)-							347	372–400		
-1,3,4-oxadiazole	Eth	302	365	21			347	372–384	6	3
(PBD)	Tol	308	365	24			347	374–384	17	20
2-(4-Biphenylyl)-5-	Diox	302	(363)				308	356–386	5.4	
-(4-t-butylphenyl)-			(362)				337	356–390		
-1,3,4-oxadiazole	Cycl		(362)				266	354–388	4.7	
			368							
2,5-Diphenyl-	Tol	306	366	99	4	16	347	360–370	4	14
-1,3-oxazole (PPO)	DMF	305	365	77	5	10	347	372–392	5	16
	Diox	304	364	90	5	11	337	375–387		
							347	378–386	5	6
p-Quaterphenyl	DMF	300	373	68			337	355–390		
	Tol	298	374	69			265	362–390		
							337	362–390		
2,5-Di-(4'-Biphenyl)-	Tol	312	383	47			347	374–386	17	18
-1,3,4-oxadiazole										
(BBD)										
1,4-Diphenyl-	Tol	333	383	23			265	383		
-1,3-butadiene							353	383		
4,4'''-Bis-(2-butylocty-	Diox	313	(388)				308	367–405	11	
loxy)-p-quaterphenyl			(383)				337	354–405		
(BBQ)	Eth/Tol		(392)				355	380–410		
	DMF						FL	389–395		
3,5,3'''',5''''-Tetra-t-	Diox	310	(390)				308	368–402	11	
-butyl-p-			(387)				387	372–412		
-quinquephenyl										

Table 34.8 Spectral and lasing parameters of some most efficient dyes *(continued)*

Dye	Solvent	λ_{abs}^{max}, nm	λ_{lum}^{max}, (λ_{abs}^{max}) nm	η, %	$B \cdot 10^{-7}$, $J^{-1} \cdot s^{-1} \cdot cm^3$	τ_{S_1}, 10^{-10} s	λ_{pmp}, nm	λ_{osc}, nm	γ, %	$K(\lambda_{pmp})$, cm^{-1}
2,5-Di-α-naphtyl-	Diox						337	383–395		
-1,3,4-oxadiazole	Tol	337	392	75			347	383–395	14	20
(αNaND)										
2-(4-Biphenylyl)-6-	Diox	327	(396)				308	386–420	7.3	
-phenylbenzoxazole-			(395)				337	385–420		
-1,3 (PBBO)	Eth		403							
2-(α-naphtyl)-5-	Diox	334	(400)	77			347	396–406	13	23
-phenyl-1,3-oxazole	Eth		400							
(αNPO)										
4,4′-Diphenylstilbene	Diox	340	(406)				308	399–415	11	
			(404)				337	394–416		
	Tol		(408)				355			
	DMF		(409)				FL	406–411		
2,5-Di-(4′-biphenyl)-	Tol	340	412	83			347	404–420	16	19
-1,3-oxazole	Diox	340	406	76			347	404–420		
(BBO)	Eth						347	408–418	29	22
		340	407	96			347	thresh.		4
Stilbene 1	EG	350	(416)				308	405–428	6	
			(417)				337	405–446		
			(410)				Ar/UV	395–435	0.3	
p-Bis(o-	Cycl	350								
Methylstyryl)-	Eth		418							
-benzene	Diox		(423)				308	414–428	8.3	
			(421)				337	412–435		
	Tol		(420)				FL			
Stilbene 3	Eth	350								
	Meth	350	425				308	412–443	8.8	
			424				337	408–457		
			425				355	415–435		
	EG		430				Ar/UV	415–465	0.5	
1,4-Di-(5′-phenyl-	Diox						337	414–426		
-1′,3′-oxazole-2′-							347	416–418	28	18
-yl)-benzene	Tol	360	420	84	8	14	347	416–418	27	22
(POPOP)	DMF	361	426	87	10	11	347	428–430	22	19
1,4-Di-[5′-(4″-	Eth	364	431	86			347	424–432	24	7
Methyl-phenyl)-1′,3-	Diox	364	429	88			347	424–436	34	20
-oxazole-2′-yl]-										
benzene (diMethyl										
POPOP; TOPOT)										
2-[4′-(4″-phenyl-	DMF	368	435	99			347	434–438	17	20
styryl)-phenyl]-										
-5-phenyl-1,3-										
-oxazole										
1,4-Distyryl	DMF	356			8		347	408–422	29	32
benzene							347	408–424	23	13
							347	408–422	14	7

Table 34.8 Spectral and lasing parameters of some most efficient dyes *(continued)*

Dye	Solvent	λ_{abs}^{max}, nm	λ_{lum}^{max}, (λ_{abs}^{max}) nm	η, %	$B \cdot 10^{-7}$, $J^{-1} \cdot s^{-1} \cdot cm^3$	τ_{S_1}, 10^{-10} s	λ_{pmp}, nm	λ_{osc}, nm	γ, %	$K(\lambda_{pmp})$, cm^{-1}
1-Styryl-4-[ω-vynyl--(n-biphenylyl)] benzene	DMF	370			3		347	424–442	30	24
							347	426–430	23	12
							347	430–442	16	6
7-Amino-4-Methyl-coumarin (Coumarin 120)	Eth	354	435							
	Meth		(441)				308	423–462	15	
			(438)				337	418–465		
			(441)				355		11	
			(440)				FL	420–470		
	EG		(450)				Ar/UV	425–475	5	
3,4-DiMethyl-7--hydroxycoumarine	Eth (OH)	366	460	90			347	450–485	15	16
							337	452–483		
							FL	460–480	0.15	30
4,6-DiMethyl-7--ethylaminocoumarin (Coumarin 2)	Meth		(448)				308	432–475	15	
			(444)				337	426–475		
			(448)				355	435–463		
	Eth	366	443				FL	440–458		
	Bz/EG		(450)				Ar/UV	430–480	16	
4-Methyl-7-diethyl-aminocoumarin (Coumarin 1)	Eth	373	450				355	440–475		
		375	(454)	70	5	29	337	442–476		
							347	448–460	25	19
							FL	457–475	0.1	30
	Meth		(456)				308	440–484	18	
3-Carboethoxy-6--hexyl-7-hydroxy-coumarine	Eth (OH)	426	454				337	465–483		
							347	465–485		
							FL	476–484	0.13	30
4-Methyl-7-hydroxy-coumarin (β-Methyl-umbelliferone; coumarin 4)	Eth (OH)	370	455	95	3	42	337	444–473		
							347	446–462	17	19
							FL	456–467	0.1	30
4-Methyl-6-hexyl-7--hydroxycoumarine	Eth (OH)	386	460	90			337	449–478		
							347	450–480	8	16
							FL	458–472	0.15	30
3-Carboethoxy-7--hydroxycoumarine (coumarine 15)	Eth (OH)	416	460	70			347	450–475	16	16
							337	455–471		
							FL	460–480	0.12	30
7-Hydroxycoumarine	Eth(OH) (OH)	375	461	95			337	446–480		
							347	454–474	35	16
							FL	457–465	0.1	30
3-Cyano-6-hexyl-7--hydroxycoumarin	Eth (OH)	444	465				337	490–510		
							347	490–510		
							FL	502–509	0.15	30
3-Acethyl-6-hexyl--7-hydroxycoumarin	Eth (OH)	443	468				347	475–505		
							FL	494–500	0.1	30
3-Phenyl-7-acetoxy-coumarin	Eth (OH)	400	475	76	6	30	337	470–494		
							347	474–494	19	14
							FL	481–495	0.1	30

Table 34.8 Spectral and lasing parameters of some most efficient dyes *(continued)*

Dye	Solvent	λ_{abs}^{max}, nm	λ_{lum}^{max}, (λ_{abs}^{max}) nm	η, %	$B\cdot10^{-7}$, $J^{-1}\cdot s^{-1}\cdot cm^3$	τ_{S_1}, 10^{-10} s	λ_{pmp}, nm	λ_{osc}, nm	γ, %	$K(\lambda_{pmp})$, cm^{-1}
3-Chlor-4-Methyl-7-	Eth	394	476	69	4	28	337	463–498		
-acetoxycoumarin	(OH)						347	476–484	13	22
							FL	474–507	0.1	30
3-Chlor-4-Methyl-7-	Eth	394	476	69	4	28	337	463–498		
-acetoxycoumarin	(OH)						347	476–484	13	22
							FL	474–507	0.1	30
Coumarine 102	Eth	389	465							
		396					347	450–500	20	16
							337	460–500		
							FL	480–500	0.18	30
	Meth		(480)				308	460–510	18	
	Bz/EG		(490)				Ar/UV			
Dianhydrid perylen-		468	486	99	4	119	337	512–530		
3,3,9,10-tetracarbonic							347	516–532	8	7
acid										
2,3,5,6-1H,4H-Tetra-	Eth	396	(491)				308	463–522	5.7	
hydroquinolizino-							FL	477–493		
-<9,9a,1-gh>										
coumarin										
Coumarin 334	Eth	454	496				337	475–508		
							347	475–510		
							FL	513–518	0.15	30
	Meth		(520)				308	506–537	12	
7-Ethylamino-6-	Meth		(500)				308	479–553	16	
-Methyl-4-trifluoro-			(504)				337	478–547		
Methylcoumarin	Eth	395	490				FL	490–510		
Coumarin+	Eth						347	495–515	10	20+0.7
dianhydrid PTK	(OH)						347	505–525	10	20+2
1-Amino-*N*-Methyl-	DMF	454	500	33	3	45	347	502–512	3.5	27
anthrapyridon							347	504–510	4	15
3-(2′-*N*-Methylbenz-	Eth	412	488							
imidazolyl)-7-*N*,*N*-	Meth		(508)				FL	480–540		
-diethylamino-	Meth/		(510)				Kr/VIS	480–555	12	
coumarin	EG									
3-(2′-Benzimidozolyl)	Eth	433	493				FL	517–527		
-7-*N*,*N*-diethylamino-	Benzene		(530)				Ar/VIS	495–570	9	
coumarin										
2,3,5,6-1H,4H-Tetra-	Eth	450	495							
hydro-9-acetylquino-	Meth		(520)				308	506–537	12	
lizino-<9,9a,1-gh>							FL	507–512		
coumarin										
Amidin	DMF	466	540	37	2	71	347	537–548	8	16
							347	540–550	7	28

Table 34.8 Spectral and lasing parameters of some most efficient dyes *(continued)*

Dye	Solvent	λ_{abs}^{max}, nm	λ_{lum}^{max}, (λ_{abs}^{max}) nm	η, %	$B \cdot 10^{-7}$, $J^{-1} \cdot s^{-1} \cdot cm^3$	τ_{S_1}, 10^{-10} s	λ_{pmp}, nm	λ_{osc}, nm	γ, %	$K(\lambda_{pmp})$, cm^{-1}
Anthrapyridon 8	Eth						347	545–575	8	18
2,3,5,6-1H,4H-Tetra-	Eth	423	530							
hydro-8-trifluor-	Meth		(540)				308	522–600	15	
Methylquinolizino-			(537)				337	517–590		
<9,9a,1-gh>coumarin							FL	528–547		
3-(2′-Benzothiazolyl)-	DMSO		(534)				308	515–558	9	
-7-diethylamino-	Eth	458	505				FL	530–539		
coumarin	Bz/EG		(550)				488	520–580	13	
o-(6-Amino-3-imino-	Eth	510	535				FL	551–583		
-3H-xanthen-9-yl)-			(572)				308	547–592	5	
benzoic acid	(acid)	505	529	85	12	71	347	554–567	4	8
(Rhodamine 110)	Meth		(550)				510	528–574	9	
	Meth/EG		(560)				Ar/all	530–600	7	
Disodium	Meth		(540)				308	532–561	9	
Fluorescein	Eth	500	522	61	7	50	337	536–556		
(Uranin)			(550)				355	536–568		
	Meth						FL	549–574		
	Meth/EG		(560)				Ar/all	530–590	7	
o-(6-Ethylamino-3-	Meth		(581)				308	569–608	16	
-ethylimino-2,7-			(480)				532	460–510	12	
dimethyl-3H-xanthen			(585)				510	563–607	20	
-9-yl)benzoic acid	Eth	530	556				337	568–595		
ethylester			(600)				FL	555–620		
(Rhodamin 6G)	Meth/EG		(590)				Ar/all	560–650	25	
Rhodamin G	Eth	530	558	82			347	574–586	20	16
							FL	580–598	0.4	23
o-(6-Ethylamino-3-	Meth		(600)				308	588–644	12	
-diethylimino-3H-			(622)				337	599–650		
xanthe-9-yl)benzoic			(598)				532	583–630	9	
acid	Eth	552	580							
(Rhodamine B)			(591)				510	582–618	21	
			(618)				FL	590–640		
	Meth/EG		(640)				Ar/all	605–675	13	
Kiton Red S	Eth	556	575							
(Sulforhodamin B)	Meth		(605)				308	594–642	12	
			(622)				337	600–646		
			(620)				510	598–645	14	
			(929)				FL	600–650		
	Meth/EG						Ar/all	610–680		
Rhodamine 101	Eth	568								
	Meth		(623)				308	614–672	12	
			(648)				337	623–676		
			(600)				532	590–620	22	
	Meth		(630)				510	607–659	14	
	Meth/EG		(650)				Ar/	620–700	22	
							VIS			

Table 34.8 Spectral and lasing parameters of some most efficient dyes *(continued)*

Dye	Solvent	λ_{abs}^{max}, nm	λ_{lum}^{max}, (λ_{abs}^{max}) nm	η, %	$B \cdot 10^{-7}$, $J^{-1} \cdot s^{-1} \cdot cm^3$	τ_{S_1}, 10^{-10} s	λ_{pmp}, nm	λ_{osc}, nm	γ, %	$K(\lambda_{pmp})$, cm^{-1}
Oxazine 9	Eth	606	630	38	20	35	337	651–689		
(Crezyl violet)							347	652–684	5	7
							FL	660–700	0.25	23
Oxazine 19	Eth	550	640				FL	659–672	0.2	
4-DicyanoMethylene-	Eth	472								
-2-Methyl-6-(p-dime-	DMSO		644							
thylaminostyryl)-4H-			(658)				308	632–690	12	
-pyran			(659)				337	626–703		
	Meth		(642)				532	604–672	24	
			(644)				510	598–677	14	
	DMSO		(655)				FL	610–710	11	
9-Ethylamino-5-ethyl-	Eth	620	650							
imino-10-Methyl-5H-	Meth		(708)				308	660–728	4	
benzo(OH)phenoxazo-			(705)				337	672–727		
nium Perchlorate			(672)				532		20	
(Oxazine 170)			(675)				510	660–712	12	
	EG		(730)				Kr/red	670–740	7	
1,1′-Diphenyl-3,3′-	Eth	618	652	35	63	20	347	660–720		
-diethyl-5,5′-dicarbo-	DMF						347	681–693	8	15
ethoxi-2,2′-imidadi-							347	687–702	14	3
carbocyanin Iodide										
1,1′-Diphenyl-3,3′-	Eth	633	665	40	71	25	347	680–740		
-diethyl-5,5′-di-(benz-	DMF						347	708–720	7	6
oxazole-2″-yl)-2,2′-										
imidadicarbocyanin										
Iodide										
3-Diethylamino-7-di-	Eth	646	670							
ethyliminophenoxazo-			(734)				308	692–768	6	
nium Perchlorate	Meth		(730)				337	692–751		
(Oxazine 1)			(695)				532		18	
	EG		(720)				FL	700–740		
			(720)				Kr/red	695–800	30	
Nile Blue	Eth	635	676				337	686–714		
	Meth		(703)				308	688–747	5	
			(683)				532		18	
			(695)				510	682–730	4	
			(710)				FL	690–750		
	EG		(730)				Kr/red	690–780	7	
1-Ethyl-2-(p-(Dime-	Eth	480								
thylaminophenyl)-	DMSO		(710)				308	670–760	10	
-1,3-butadienyl)-pyri-			(703)				337	675–750		
dinum Perchlorate	Meth		(698)				532	660–730	30	
(Pyridin 1)			(684)				510	661–724	6	
	PC/EG		(710)				Ar/	670–780	20	
							VIS			

Table 34.8 Spectral and lasing parameters of some most efficient dyes *(continued)*

Dye	Solvent	λ_{abs}^{max}, nm	λ_{lum}^{max}, (λ_{abs}^{max}) nm	η, %	$B\cdot10^{-7}$, $J^{-1}\cdot s^{-1}\cdot cm^3$	τ_{S_1}, 10^{-10} s	λ_{pmp}, nm	λ_{osc}, nm	γ, %	$K(\lambda_{pmp})$, cm^{-1}
2-(4-(4-DiMethylami-	Eth	570								
nophenyl)-1,3-butadi-	Meth		(711)				510	703–724	3	
enyl)-3-ethylbenzothi-	PC/EG		(780)				Ar/	700–840	10	
azolium Perchlorate							VIS			
3,3′-Diethyloxatri-	Eth	689	718	53			694	730–740	37	18
carbocyanine Iodide										
(DOTCI)										
Carbazine 122	Eth	655								
	Water		(720)				532		30	
	Meth		(700)				FL	680–740		
	EG		(750)				Kr/red	690–820	20	
1,3,3;1′,3′,3′-Hexame-	Eth	682	720	12	56	7	694	720–740	25	20
thyl-4,5;4′,5′-dibenzo-	EG						694	722–738	32	17
2,2′-indodicarbo-							694	726–748	30	30
cyanine Iodide										
Rhodamine 700	Meth		(723)				308	701–768	11	
	Eth	643					FL	705		
	Meth/EG		(750)				Kr/red	700–810	26	
1,1′-Dibenzyl-4,4′-	Eth	706	730	2	55	1	694	740–760	14	15
-quinocarbocyanine	EG	723	735	16			694	740–760	36	15
Bromide	DMF						694	742–762	30	15
1,1′-Di-(2"-phenyl-	Eth	710	731	3	65	2	694	740–762		
ethylene)-4,4′-	EG	723	735	18	60	13	694	740–762	35	18
-quinocarbocyanine	DMF						694	742–764	33	18
Iodide										
1-Ethyl-4-(p-Di-	Eth	500								
Methylaminophenyl)-	DMSO		(740)				308	695–790	11	
1,3-butadienyl)-pyri-			(743)				337	710–790		
dinium Perchlorate	Meth		(722)				510	687–755	4	
(Pyridin 2)	PC/EG		(720)				Ar/	685–820	18	
							VIS			
3,3′-Diethyl-6,7;6′7′-	Eth	752	780	13	48	12	694	790–825	17	20
-dibezo-10,12-tri-	DMF						694	792–822	30	20
Methylen-11-chlor-										
-2,2′-oxatricarbo-										
cyanine Iodide										
Oxazine 750	DMSO		(777)				308	735–796	6	
	Eth	667	(724)				337	708–780		
	PC/EG		(810)				Kr/red	790–900	12	
3,3′-DiMethyloxatri-	Eth	682	718							
carbocyanine Iodide	DMSO		(792)				308	774–810	4	
(Methyl-DOTCI)			(780)				337	768–820		
			(780)				532			
			(810)				FL			
	EG						Kr/red	745–790		
Dibenzcyanine	Eth	687					694	766–780	10	18

Table 34.8 Spectral and lasing parameters of some most efficient dyes *(continued)*

Dye	Solvent	λ_{abs}^{max}, nm	λ_{lum}^{max}, (λ_{abs}^{max}) nm	η, %	$B{\cdot}10^{-7}$, $J^{-1}{\cdot}s^{-1}{\cdot}cm^3$	τ_{S_1}, 10^{-10} s	λ_{pmp}, nm	λ_{osc}, nm	γ, %	$K(\lambda_{pmp})$, cm^{-1}
3,3-Diethyl-4,5;4′,5′--di-(5″-phenylthio-pheno- 2″,3″-)-2,2′-thiazolodicarbo-cyanine Iodide	Eth DMF	730	780	9	43	6	694 730 694	780–810 780–810 784–814	12 19 17	19 19 19
1,1′-DiMethyl-2,2′-di--(3″,4″-diMethoxiphe-nyl)-5,6;5′,6′-dibenz--4,4′-quinocarbo-cyanine Iodide	Eth	750	793				694	820–854	18	18
1,1′-DiMethyl-2,2′-di--(4″-Methoxiphenyl)--5,6;5′,6′-dibenz-4,4′-quinocarbocyanine Iodide	DMF	768	793	3			694	853–874	35	18
3,3′-Diethyl-4-keto--5-[(3-ethylbenzthia-zolinyliden-2)-1,3--(2,2-diMethyltrime-thyleno)-butenylidem]-7-ethoxy-2,2′-thia-zolinothiacarbocya-nine Bromide	Eth DMF	725	800	0,5	27	0.3	694 730 694	824–866 824–866 824–866	24 32 7	10 10 10
1,1′-DiMethyl-2,2′-di-(3″,4″-Methylene di-oxiphenyl)-5,6;5′,6′--dibenz-4,4′-quino-carbocyanyne Iodide	Eth	768	800	3			694	874–883	35	18
1,3,3;1′,3′,3′-Hexame-thyl-10,12-triMethy-leno-11-chlor-2,2′--indotricarbocyanine Iodide	Eth DMF	742	805	3	22	2	694 750 694	816–844 816–844 816–844	28 40 33	19 19 19
Rhodamine 800	Eth DMSO EG	682	 (810) (795)				 308 Kr/red	 776–823 730–835	 6 22	
3,3′-Diethyl-9,11--(o-phenyleno)-2,2′--thiadicarbocyanine Iodide	Eth DMF	760	815	2	31	3	694 730 694	822–864 822–864 822–864	15 24 23	18 18 18
2-(6-(p-DiMethylami-nophenyl)-2,4-neo-pentylene-1,3,5-hexa-trienyl)-3-Methylben-zothiazolium Perchlorate	Eth DMSO EG Meth PC/EG	585	 (840) (840) (810) (815) (840) (840)				 308 337 532 510 FL Ar/VIS	 810–875 803–875 780–840 793–845 810–860 790–900	 9 15 14 13	

Table 34.8 Spectral and lasing parameters of some most efficient dyes *(continued)*

Dye	Solvent	λ_{abs}^{max}, nm	λ_{lum}^{max}, (λ_{abs}^{max}) nm	η, %	$B{\cdot}10^{-7}$, $J^{-1}{\cdot}s^{-1}{\cdot}cm^3$	τ_{S_1}, 10^{-10} s	λ_{pmp}, nm	λ_{osc}, nm	γ, %	$K(\lambda_{pmp})$, cm^{-1}
1,3,3;1′,3′,3′-Hexa-	Eth	780	844	2	44	2	694	844–890	8	20
-Methyl-4,5;4′,5′-di-	DMF						694	844–890	10	20
benzo-2,2′-indo-	EG						694	844–890	17	20
tricarbocyanine							770	844–890	17	20
Perchlorate										
3,3′-Diethylthiatricar-	Eth	760	815							
bocyanine Iodide	DMSO		(849)				308	828–883	1	
(DTTCI)			(852)				337	834–892		
	Meth		(889)				FL			
	Eth	763	790	34	74	26	694	800–830	22	19
DTTC Bromide	Eth	765					694	820–838	21	18
1,1′,3,3,3′,3′-Hexa-	DMSO		(868)				308	837–905	4	
Methyl-2,2′-			(846)				337	828–891		
indotricarbocy-	Eth	744	770	28	34	31	694	780–826	38	12
anine Iodide	DMSO		(770)				FL			
(HITCI)	DMSO/		(879)				Kr/IR	815–920	10	
	EG		(880)							
1,1′-Diethyl-2,2′-	Eth	823	885	0.7	80	0.4	694	860–920	11	14
quinotricarbocyanine	DMF						820	860–920	32	14
Iodide							694	860–920	19	14
IR 144	Eth	750								
	DMSO		(869)				308	856–879	3	
			(874)				337	862–892		
			(867)				532		6	
			(880)				FL			
IR 125	DMSO	795	(920)				308	890–960	4	
			(918)				337	893–958		
			(913)				532		3	
			(940)				FL			
1,1′,3,3,3′,3′-Hexame-	Eth	780								
thyl-4,4′,5,5′-dibenzo-	DMSO		(932)				308	899–975	4	
-2,2′indotricarbocya-	DMSO/		(920)				Kr/red	880–960	2	
nine Iodide	EG									
3,3′-Diethyl-2,2′-thia-	Eth	865	940	0,3	46	0,3	694	946–980	25	18
tricarbocyanine							820	946–980	49	18
Iodide	DMF						694	946–980	24	18
IR 140	Eth	810	860							
	DMSO		(950)				308	882–985	3	
			(910)				337	900–936		
			(890)				532		5	
			(950)				FL			
	DMSO/		(970)				Kr/Vis	880–	14	
	EG							1010		
IR 26	Dichl	1080								
	Benzene		1290				1064	1200–	4	
								1320		

Table 34.8 Spectral and lasing parameters of some most efficient dyes *(continued)*

Dye	Solvent	λ_{abs}^{max}, nm	λ_{lum}^{max}, (λ_{abs}^{max}) nm	η, %	$B \cdot 10^{-7}$, $J^{-1} \cdot s^{-1} \cdot cm^3$	τ_{S_1}, 10^{-10} s	λ_{pmp}, nm	λ_{osc}, nm	γ, %	$K(\lambda_{pmp})$, cm^{-1}
3,3′-Diethyl-9,11,15, 17-dineopentylene- (5,6,5′,6′-tetra Methoxy)- -selenapentacarbo- cyanine perchlorate	DMSO		(1109)				1064	1076–1147	12	
3,3′-Diethyl-9,11,15, 17-dineopentylene- thiapentacarbocyanine perchlorate	DMSO		(1124)				1064	1102–1148	16	
3,3′-Diethyl-9,11,15, 17-dineopentylene- (5,6,5′,6′-tetraMethoxy)- thiapentacarbocyanine perchlorate	DMSO		(1140)				1064	1107–1285	14	
3,3′-Diethyl-9,11,15, 17-dineopentylene- -(6,7,6′,7′-dibezo)- thiapentacarbocyanine perchlorate	DMSO		(1172)				1064	1151–1216	13	

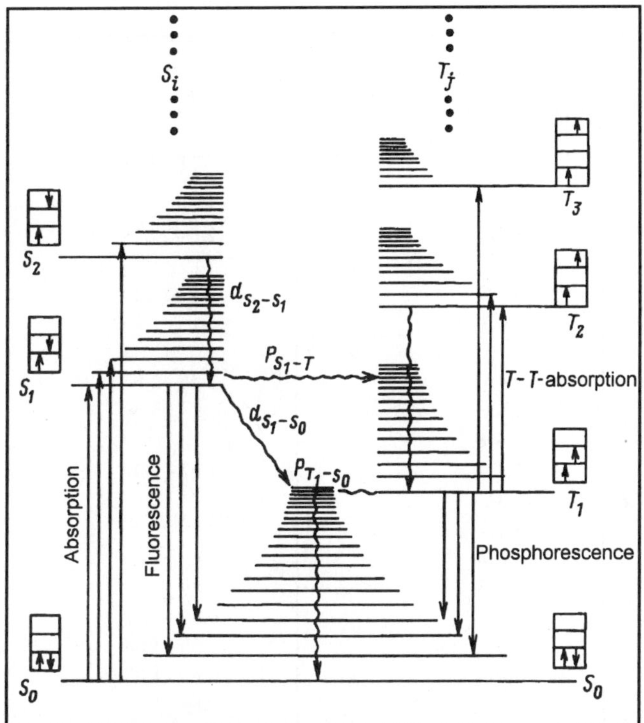

Figure 34.14 Schematic energy–level diagram of singlet (S_i) and triplet (T_j) levels of a complex dye molecule (orientations of outer electron spins are given for each level, respectively).

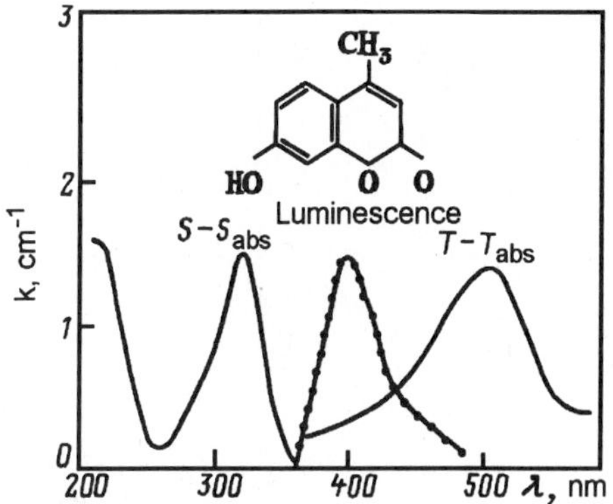

Figure 34.15 Absorption ($S{-}S_{abs}$ and $T - T_{abs}$) and luminescence spectra for some dye (neutral form of the 7–oxy–4–methylcoumarine).

34.7 Color Center Lasers

Color center lasers cover a spectrum range from 0.7 to 3.3 μm. They provide tunability along with narrow bandwidth, operating both in continuous and pulsed regimes.

Basic material for *color center lasers* are ion fluoride and halide crystals of alkaline metals, fluorides of calcium and strontium, and some crystals with impurities.

Color centers are some defects in crystals exhibiting properties of donors or acceptors. These defects may appear when basic material is subjected to ionizing radiation (high-energy electrons, UV-, X-ray, or γ-radiation). Heating basic crystals in vapors of alkaline metals also may result in *color centers* formation.

Color centers can be of different nature. If a single electron is trapped at halide ion vacancy, the imperfection is called *F-center.* If one of the six positive alkali ions that immediately surround the vacancy is foreign (e.g., Na^+ ion in KCl crystal), *F*-center is specified as F_A-*center*, while two foreign ions form F_B-*center.* There is further classification of these F_A- and F_B-centers according to their relaxation behavior. Two *F*-centers aligned along a (110) crystal axis form an F_2-*center*; its single ionized counterpart is known as a F_2^+-*center.* If F_2-*center* captures an electron it is designated as F_2^--*center.*

Stimulated radiation due to transitions between electronic-vibrational levels of color centers leads to highly efficient lasing. Pumping of color center lasers is produced by other laser irradiation with shorter wavelength or with the help of flash lamps. Most color center cw lasers operate at liquid nitrogen temperature. Some color center laser materials are used for passive *Q*-switching, similarly to some dye-laser materials. More detailed information about this type of lasers can be found in [115–118].

Table 34.9 Color center laser materials

Crystal	Type of color center	Pumping wavelength, μm	Tuning range, μm
	Crystals used in CW lasers		
KCl-Li	F_A(II)	0.514	2.3–2.95
		0.647	
	$(F_2^+)_A$	1.32	2.0–2.5
		1.34	
RbCl-Li	F_A(II)	0.647	2.5–3.33
		0.676	
		0.753	
KCl-Na	F_B(II)	0.47	2.25–2.65
		0.53	
		0.568	
	$(F_2^+)_A$	1.32	1.62–1.91
		1.34	
RbCl-Na	F_B(II)	0.647	2.5–2.9
		0.676	
NaF	F_2^+	0.753	0.89–1.0
		0.87	0.99–1.22
		0.9	1.08–1.38
LiF	F_2^+ (77 K)	0.647	0.82–1.07
	(300 K)		0.86–1.0
	F_2^- (300 K)	1.064	1.14–1.2
KF	F_2^+	1.064	1.22–1.5
KCl	F_2^+	1.34	1.6–1.78
NaCl	F_2^+	1.064	1.35–1.6
NaCl-OH$^-$	F_2^+	1.06	1.36–1.77

Table 34.9 Color center laser materials *(continued)*

Crystal	Type of color center	Pumping wavelength, μm	Tuning range, μm
Crystals used in CW lasers *(continued)*			
KCl-OH$^-$	F_2^+	1.32	1.61–1.77
CaF$_2$-Na	$(F_2)_A$ (300 K)	0.61	0.72–0.84
SrF$_2$-Na	$(F_2)_A$ (300 K)	0.694	0.84–0.98
Crystals used in pulsed lasers			
LiF	F_2	0.45	0.665–0.715
		0.53	0.65–0.75
	$F_2 \rightarrow F_2^{+*1}$	0.53	0.84–1.1
	F_2^-	1.06	1.08–1.23
LiF-OH$^-$	F_2^+	0.69	0.84–1.02
NaF-Li	$(F_2^+)_A$	0.69	0.95–1.3
NaF	F_3	0.96	1.15–1.4

*1 Transformation of color center type during laser operation.

References

[1] Handbook on lasers, vol.1, Ed. Prokhorov, A.M., Sov. radio, Moscow, 1978 (in Russian).

[2] Handbook of lasers, Ed. Pressley, R.J., Cleveland, Chemical Rubber Co., 1971.

[3] Beck, R., English, W., and Gurs, K., Table of laser lines in gases and vapors, Springer Series in Optical Sciences, v. 2, 3rd edition, Springer-Verlag, Berlin, 1980.

[4] Douglas, N.G., Millimetre and submillimetre wavelength lasers, Springer Series in Optical Sciences, v.61, Springer-Verlag, Berlin, 1989.

[5] Brederlow, G., Fill, E., and Witte, K.J., The high-power iodine laser, Springer Series in Optical Sciences, vol. 34, Springer-Verlag, Berlin, 1983.

[6] Ablekov, V.K., Denisov Yu.N., and Lyubchenko, F.N., Handbook on gas-dynamic lasers, Mashinostroenie, Moscow, 1982 (in Russian).

[7] Witteman, W.J., The CO_2 laser, Springer Series in Optical Sciences, vol. 34, Springer-Verlag, Berlin, 1987.

[8] Tittel, F.K., Wilson, W.L., Stickel, R.E., et al, Appl. Phys. Lett., 36, 405, 1980.

[9] Tittel, F.K., Marowsky, G., Wilson, W.L., and Williams, R.A., Appl. Phys. Lett., 37, 862, 1980.

[10] Huges, D.W., and Barr, J.R.M., J. Phys. D: Appl. Phys., 25, 563, 1992.

[11] Special issue on diode pumping, IEEE J. Quantum Electron., 28, No. 4, 1992.

[12] Kaminskii, A.A., Mak, A.A., Pashinin, P.P., Handbook on lasers, vol. 1, Ed. Prokhorov, A.M., Sov. radio, Moscow, 1978, pp. 237–324 (in Russian).

[13] Kaminskii, A.A., Laser crystals. Physics and properties, Springer Series in Optical Sciences, v. 14, Springer-Verlag, Berlin, 1981.

[14] Tunable solid-state lasers, Eds. Hammerling, P., Budgor A.B., and Pinto, Springer Series in Optical Sciences, vol. 47, Springer-Verlag, Berlin, 1985.

[15] Tunable solid-state lasers II, Eds. Budgor, A.B., Esterowitz L., and DeShazer, L.G., Springer Series in Optical Sciences, vol. 52, Springer-Verlag, Berlin, 1986.

[16] Koechner, W., Solid-state laser engineering, Springer Series in Optical Sciences, vol 1, Springer-Verlag, Berlin, 1976.

[17] Lacovara, P., Esterovitz, L., and Kokta, M., IEEE J. Quantum Electron., QE–21, 1614-1618, 1985.

[18] Petricevic, V., Gayen, S.K., and Alfano, R.R., Appl. Phys. Lett., 53, 2590, 1988.

[19] Shestakov, A.V., Borodin, N.I., Zhitnyuk, V.A., et al., in: Digest of Conference on Lasers and Electro–Optics (OSA, Washington, D.C., 1991), paper CPDP11.

[20] Kueck, S., Petermann K., and Huber, G., in Advanced Solid–State Lasers, Technical Digest, 1994 (OSA, Washington, D.C., 1994), pp.7–9.

[21] Deka, C., Chai, B.H.T., Shimony, Y., et al., Appl. Phys. Lett., 61, 2141, 1992.

[22] Petermann, K., and Mitzscherlich, P., IEEE J. Quantum Electron., QE–23, 1122, 1987.

[23] Walling, J.C., Laser Focus, 18, 45, 1982.

[24] Alimpiev, A.I., Pestryakov, E.V., Petrov, V.V., et al., Sov. J. Quantum Electron., 18, 323, 1988.

[25] Shand, M.L., and Lay, S.T., IEEE J. Quantum Electron., QE–20, 105, 1984.

[26] Drube, J., Struve, B., and Huber, G., Opt. Comm., 50, 45, 1984.

[27] Zharikov, E.V., Il'ichev N.N., Kalitin, S.P., et al., Sov. J. Quantum Electron, 13, 1274, 1983.

[28] Struve, B., and Huber, G., J. Appl. Phys., 57, 45, 1985.

[29] Durr, U., Brauch, U., Kheirm, W., and Schiller, C., in: Tunable solid-state lasers, Eds. Hammerling, P., Budgor A.B., and Pinto, A., Springer Series in Optical Sciences, vol. 47, Springer-Verlag, Berlin, 1985, pp.20–27.

[30] Kaminskii, A.A., Shkadarevich, A.P., Mill, B.V., et al., Izv. AN, Ser. "Inorg. mater.", 23, 1931, 1987 (in Russian).

[31] Lai, S.T., Chai, B.H.T., Long, M., and Shinn, M.D., IEEE J. Quantum Electron., QE–24, 1922, 1988.

[32] Kaminskii, A.A., Shkadarevich, A.P., Mill, B.V., et al., Izv. AN, Ser."Inorg. mater.", 23, 690, 1987 (in Russian).

[33] Payne, S.A., Chase, L.L., Newkirk, H.W., et al., IEEE J. Quantum Electron., QE–24, 2243, 1988.

[34] Payne, S.A., Chase, L.L., Smith, L.K., et al., J. Appl. Phys., 66, 1051, 1989.

[35] Chai, B.H.T., Letaucheur, J.L., Stalder, M., and Bass, M., Opt. Lett., 17, 1584, 1992.

[36] Caird, J.A., Payne, S.A., Staver, P.R., et al., IEEE J. Quantum Electron., QE–24, 1077, 1988.

[37] Lai, S.T., Chai, B.H.T., Long, M., and Morris, R.C. IEEE J. Quantum Electron., QE–22, 1931, 1986.

[38] Jenssen, H.P., and Lai, S.T., J. Opt. Soc. Am. B, 3, 115, 1986.

[39] Danilov, A.A., Zubenko, D.A., Kalitin, S.P., et al., in: Proc. of General Physics Inst. of Russian Academy of Sciences, v. 26, Moscow, 1990, pp. 5–49 (in Russian).

[40] Kolbe, W., Petermann, K., and Huber, G., IEEE J. Quantum Electron., QE–21, 1596, 1985.

[41] Brauch, U., and Durr, U., Optics Comm., 55, 35, 1985.

[42] Klein, P.B., Furneaux, J.E., and Henry, R.L., Appl. Phys. Lett., 42, 638, 1983.

[43] Kunzel, W., and Durr, U., Appl. Phys., 328, 233, 1982.

[44] Lovold, S., Moulton, P.F., Killinger, D.K., and Mehyuk, N., IEEE J. Quantum Electron., QE–21, 202, 1985.

[45] Moulton, P.F., IEEE J. Quantum Electron., QE–18, 1185, 1982.

[46] Johnson, L.F., Guggenheim, H.J., and Banch, D., Opt. Lett., 8, 371, 1983.

[47] Moulton, P.F., and Mooradian, A., in: Laser spectroscopy IV, Springer Series in Optical Sciences, vol. 21, Springer-Verlag, Berlin, 1979, pp. 584–589.

[48] Ehrlich, D.J., Moulton, P.F. and Oswood, R.M., Opt. Lett., 5, 339, 1980.

[49] Dubinskii, M.A., Semashko, V.V., Naumov, A.K., Abdulsabirov, R.Y., and Korableva, S.L., Laser Physics, 4, 480, 1994.

[50] Marshall, C.D., Payne, S.A., Speth, J.A., and Krupke, W.F., in: Advanced solid-state lasers. Technical digest, 1994 (OSA, Washington, D.C., 1994), pp. 113–115.

[51] Ehrlich, D.J., Moulton, P.F., and Oswood, R.M., Opt. Lett., 4, No. 6, 184–186, 1979.

[52] Sandrock, T., Danger, T., Neumann, E., and Huber, G., Continuous-wave laser action of Pr-doped fluorides at room temperature, in: Advanced solid-state lasers. Technical digest, 1994 (OSA, Washington, D.C., 1994), pp. 175–177.

[53] Kaminskii, A.A., Doklady AN SSSR, 271, 1357, 1983 (in Russian).

[54] Allen, P., Rare-earth modern science and technology, Plenum Press, 1978.

[55] Hegarty, J., and Yen, W.M., J. Appl. Phys., 51, 3545, 1980.

[56] Kaminskii, A.A., Petrosyan, A.G., and Ovanesyan, K.L., Phys. Stat. Solidi, 77, K173, 1983.

[57] Verdun, H.R., and Thomas, L.M., Appl. Phys. Lett., 56, 608, 1990.

[58] Kaminskii, A.A., Agamalyan, N.V., and Denisenko, G.A., Phys. Stat. Solidi, 70, 397, 1982.

[59] Dubinskii, M.A., Khaidukov, N.M., Garipov, I.G., et al., Appl. Opt., 31, 4158, 1992.

[60] Kaminskii, A.A., Osiko, V.V., Sarkisov, S.E., et al, Phys. Stat. Sol., 49, 305, 1978.

[61] Doroshenko, M.E., Osiko, V.V., Sigachev, V.B., and Timoshechkin, M.I., Izv. AN, Ser. "Phys.", 56, 147, 1992 (in Russian).

[62] Zhang, X.X., Villaverde, A.B., Bass, M., et al., J. Appl. Phys., 74, 790, 1993.

[63] Denisov, A.L., Ostroumov, V.G., Saidov, Z.S., et al, J. Opt. Soc. Am. B, 3, 95, 1986.

[64] Kaminskii, A.A., Izv. AN SSSR, Ser. "Inorg. mater.", 15, 2092, 1979 (in Russian).

[65] Lempicki, A., McCollum, B.C., and Chinn, S.R., IEEE J. Quant. Electron., 15, 896, 1979.

[66] Zhang, X.X., Bass, M., Hong, P., Palm, A., and Chai, b.h.t., in: Advanced solid-state lasers. Technical digest, 1994 (OSA, Washington, D.C., 1994), pp. 211–212.

[67] Kaminskii, A.A., Butashin, A.V., Maslyanizin, I.A., et al., Phys. Stat. Sol., 125, 671, 1991.

[68] Kaminskii, A.A., Mill, B.V., and Sil'vestrova, I.M., Izv. AN SSSR, Ser. "Phys.", 47, 1903, 1983 (in Russian).

[69] Garmash, W.M., Kaminskii, A.A., and Polyakov, N.I., Phys. Status Solidi, 75, K111, 1983.

[70] Kaminskii, A.A., Timofeeva, V.A., Agamalyan, N.V. et al, Crystallographya, 27, 522, 1982 (in Russian).

[71] Budin, J.P., Michel, J.C., and Auzel, F., J. Appl. Phys., 50, 641, 1979.

[72] Dianov, E.M., Dmitruk, M.V., and Karasik, A.Ya., Sov. J. Quant. Electron., 10, 1222, 1980.

[73] Kaminskii, A.A., Kursten, G.D., Schultze, D., et al, Doklady AN SSSR, 270, 1373, 1983 (in Russian).

[74] Marling, J., IEEE J. Quantum Electron., QE-14, 56, 1978.

[75] Li, G., Wu, J., Zhuang, J., and Yang, H., J. of Cryst. Growth, 98, 391, 1989.

[76] Kazakov, B.N., Orlov, M.S., Petrov, M.V. et al, Opt. I Spectr., 47, 1217, 1979 (in Russian).

[77] Kaminskii, A.A., Petrosyan, A.G., Fedorov, V.A., et al, Izv. AN SSSR, Ser. "Inorgan. mater.", 17, 1920, 1981 (in Russian).

[78] Kaminskii, A.A., Sarkisov, S.E., and Butaeva, T.I., Phys. Status Solidi, 56, 725, 1979.

[79] Kaminskii, A.A., Fedorov, V.A., Petrosyan, A.G., et al, Izv. AN SSSR, Ser. "Inorgan. mater.", 15, 1494, 1979 (in Russian).

[80] Kaminskii, A.A., Doklady AN SSSR, 260, No.1, 64-67, 1981 (in Russian).

[81] Kaminskii, A.A., Petrosyan, A.G., and Ovanesyan, K.L., Izv. AN SSSR, Ser. "Inorgan. mater.", 19, 1217, 1983 (in Russian).

[82] Rodriguez, W.J., Naranjo, F.L., Barnes, N.P., and Filer, E.D., in: Advanced solid-state lasers. Technical digest, 1994 (OSA, Washington, D.C., 1994), pp. 291–293.

[83] Cockaune, B.J., Cryst. Growth, 54, 407, 1981.

[84] Fan, T.Y., Huber, G., Byer, R.L., and Mitzscherlich, P., IEEE J. Quantum Electron., 24, 924, 1988.

[85] Duczynski, E.W., Huber, G., Ostroumov V.G., and Shcherbakov, I.A., Appl. Phys. Lett., 48, 1562, 1986.

[86] Kaminskii, A.A., Izv. AN SSSR, Ser. "Inorgan. mater.", 18, 482, 1982 (in Russian).

[87] Antipenko, B.M., Mak, A.A., Sinitzin, B.V., et al, Zhurnal technicheskoi fiziki, 52, 521, 1982 (in Russian).

[88] Kaminskii, A.A., Crystallography, 27, 193, 1982 (in Russian).

[89] Mochalov, I.V., Phys. Stat. Solidi (a), 55, 79, 1979.

[90] Kaminskii, A.A., and Petrosyan, A.G., Izv. AN SSSR, Ser. "Inorgan. mater.", 18, 1910, 1982 (in Russian).

[91] Kaminskii, A.A., Pavlyuk, A.A., Butaeva, T.I., et al, Izv. AN SSSR, Ser. "Inorgan. mater.", 15, 541, 1979 (in Russian).

[92] Kaminskii, A.A., Pavlyuk, A.A., Agamalyan, N.V., et al, Izv. AN SSSR, Ser. "Inorgan. mater.", 15, 1496, 1979 (in Russian).

[93] Brede, R., Heumann, E., Huber, G., and Chai, B., in: Advanced solid–state lasers. Technical digest, 1994 (OSA, Washington, D.C., 1994), pp. 187–189.

[94] Kaminskii, A.A., Sarkisov, C.E., Seitranyan, K.B., et al, Izv. AN SSSR, Ser. "Inorgan. mater.", 18, 527, 1982 (in Russian).

[95] Kaminskii, A.A.M, Seitranyan, K.B., Arakelyan, A.Z., et al, Izv. AN SSSR, Ser. "Inorgan. mater.", 18, 1061, 1982 (in Russian).

[96] Kaminskii, A.A., and Petrosyan, A.G., Izv. AN SSSR, Ser. "Inorgan. mater.", 18, 1910, 1982 (in Russian).

[97] Kaminskii, A.A., Izv. AN SSSR, Ser. "Inorgan. mater.", 19, 1388, 1983 (in Russian).

[98] Ohta, K., Saito, H., and Obara, M., J. Appl. Phys., 73, 3149, 1993.

[99] DeLoach, L.D., Payne, S.A., Chase, L.L., et al, IEEE J. Quantum Electron., 29, 1179, 1993.

[100] DeLoach, L.D., Payne, S.A., Smith, L.K., and Krupke, W.F., in: Advanced solid–state lasers. Technical digest, 1994 (OSA, Washington, D.C., 1994), pp. 140–142.

[101] Eliseev, P.G., in: Handbook on lasers, Prokhorov, A.M., Ed., Moscow, Sov. radio, v. 1, 1978, pp. 334-342 (in Russian).

[102] Eliseev, P.G., Semiconductor lasers and transformers. Itogi nauki i techniki, Radiotechnika, vol. 14, part 1, VINITI, Moscow, 1978 (in Russian).

[103] Dolginov, L.M., Eliseev, P.G., and Ismailov, I., Injection radiating devices on multicomponent semiconducting solid solutions. Itogi nauki i techniki, Radiotechnika, vol. 21, VINITI, Moscow, 1980 (in Russian).

[104] Injection lasers. Moscow, Nauka, 1983 (Proc. of P.N. Lebedev Physical Inst. v. 141) (in Russian).

[105] Casey, H.C., and Panish, M., Heterostructure lasers, Academic Press, 1978.

[106] Thompson, G.H.B, Physics of semiconductor laser devices, John Willey & Sons, 1980.

[107] Semiconductors and semimetals, v. 14, Willardson, R.K., and Beer, A.C., Eds., Academic Press, 1979.

[108] Eliseev, P.G., Electron. Promyshlennost', 1980, No. 8-9, p.1980 (in Russian).

[109] Anikiev, Yu.G., Zhabotinskii, M.E., and Kravchenko, V.B., Inorganic liquid lasers, Nauka, Moscow, 1986 (in Russian).

[110] Stepanov, B.I., Rubinov, A.N., and Mostovnikov, V.A., in: Handbook on lasers, vol.1, Ed. Prokhorov, A.M., Sov. radio, Moscow, 1978, pp. 360–379 (in Russian).

[111] Catalogue of active laser mediums based on solutions of organic dyes and related compounds, Ed. Stepanov, B.I., Physics Institute of Belorussia, Minsk, 1977 (in Russian).

[112] Brackmann, U., Lamdachrome laser dyes, Lambda Physics GmbH, Gottingen, 1986.

[113] Kato, K., IEEE J. Quantum Electron., QE–14, 7 (1978); Appl. Phys. Lett., 33, 509, 1978.

[114] Duarte, F.J., High power dye lasers, Springer Series in Optical Sciences, v. 65, Springer–Verlag, 1991.

[115] Mollenhauer, L.F., and Olsen, D.H., J. Appl. Phys., 46, 3109, 1975.

[116] Basiev, T.T., Voron'ko, Yu.K., Mirov, S.B., et al, Izv. AN SSSR, Ser. "Fizika", 46, 1600, 1982 (in Russian).

[117] Feofilov, P.P., and Archangel'skaya, V.A., Izv. AN SSSR, Ser. "Fizika", 45, 302, 1981 (in Russian).

[118] Gellermann, W., J. Phys. Chem. Solids, 52, 249, 1991.

35

X-Radiation

P.M. Imamov

35.1 X-Radiation Generation and Properties

Electromagnetic oscillations with the wavelengths from 10^{-3} to 10 nm are said to be X-radiation (X-rays). The extremely small X-ray wavelengths comparable to interatomic distances in solids and liquids define characteristic features of this radiation. The refractive index of a medium for X-rays is determined by the expression

$$q = 1 - e^2 N\lambda^2/2mc^2\pi, \tag{35.1}$$

where N is the number density of electrons (in 1 cm^3), λ is the wavelength (in cm), e and m are electron charge and mass, respectively. Since $e^2/(mc^2) = 2.8{\cdot}10^{-19}$ cm, $\lambda^2 \approx 10^{-16}$ cm^2, $N \approx 10^{23}$–10^{25} cm^{-3}, the second term in the expression for q is smaller than 10^{-4}. So q is smaller than one, though it differs little from unity for X-rays.

Because of the proximity of q to one it is practically impossible to focus X-rays by means of lenses and prisms. One uses mostly diaphragms and mirrors with total external reflection to form beams in X-ray optics. Diffraction methods of the beam focusing are also employed.

An X-ray electron tube is a source of the X-radiation. Inside the tube electrons emitted by a hot cathode (tungsten thread or spiral) are accelerated by the electric field and traversed to a metallic anode. The energy of electrons is converted into the energy of X-ray photons during the sharp braking of electrons in the anode substance. Then

$$E_{ph} = h\nu = E_1 - E_2, \tag{35.2}$$

where h is the Planck's constant, ν is the radiation frequency, E_1 and E_2 are the energies of an electron just before and after its collision with the anode. The radiation generated consists of bremsstrahlung (braking) and characteristic components. The maximum frequency $\nu_{\max}$ (or the minimum wavelength $\lambda_{\min}$) corresponds to the full stop of electrons ($E_2 = 0$):

$$h\nu_{\max} = \frac{hc}{\lambda_{\min}} = E_1 = eU, \tag{35.3}$$

where U is the accelerating voltage (in kV), $\lambda_{\min} = 1.24/U$ (in nm).

As E_2 can take any value smaller than E_1, the continuous spectrum is limited from the long-wave side only by the absorption of the long-wave radiation in the tube window substance and in the air. The wavelength $\lambda \approx 1.5\lambda_{\min}$ corresponds to the intensity maximum in the continuous spectrum.

Recently the synchrotron, or magneto-braking radiation, arising from the movement of charged particles in a uniform magnetic field, has been used as the source of powerful X-ray radiation. The

0-8493-2861-6/97/$0.00+$.50

synchrotron radiation spectrum fills the interval from infrared to high-energy X-ray radiation practically without breaks. The radiation direction coincides with that of the charged particle instant velocity. The radiation is concentrated within the cone of apex angle $\theta \approx E/(mc^2)$, where E is the charged particle energy, m is its mass, c is the speed of light.

It is recommended to use a subsidiary unit of electron-volt to measure the energy of X-radiation. Electron-volt and its decimal multiples may be used along with the SI units

$$\lambda = \frac{1}{\tilde{\nu}} = \frac{hc}{E}, \quad (\tilde{\nu} = \nu/c);$$

$$E(\lambda = 1 \text{ nm}) = hc/10^{-9} \text{ m} = 1239.8519(32) \text{ nm}, \tag{35.4}$$

$$\lambda(E = 1 \text{ eV}) = hc/1 \text{ eV} = 1239.8519(32) \text{ nm}.$$

Below are given the relations between recommended and subsidiary units kX, Ry, and $\tilde{\nu}$:

$$1 \text{ kX} = 0.100202 \text{ nm}, \quad 1\text{Ry} = 13.605802 \text{ eV},$$

$$\tilde{\nu}(E_1 = 1 \text{ eV}) = 1 \text{ eV}/hc = 8065.479 \text{ cm}^{-1}. \tag{35.5}$$

35.2 Characteristic Spectrum of X-ray Radiation

Characteristic spectrum arises at a certain accelerating voltage U depending on the atomic number Z of anode material. It is easy to explain the generation of the characteristic spectrum on the basis of quantum mechanical considerations.

Electrons accelerated in a tube may "knock out" electrons from the intra-atomic levels of anode atoms. The generation of an electron vacancy transfers the atom to an excited state. The return of the atom into the ground state is accompanied by the liberation of the energy surplus in the form of an X-ray quantum $h\nu = E_1 - E_0$ where E_1 is the energy of an external shell electron and E_0 is the energy of an internal shell electron.

The following designations are used in X-ray spectroscopy [1]. Atomic level terms with the principal quantum numbers equal to 1, 2, 3, 4, 5, and 6 are designated K, L, M, N, O, P, correspondingly. Subscripts on these letters correspond to the different values of the orbital and the total angular momentum according to the following scheme

Atomic electron level	$1s$	$2s$	$2p_{1/2}$	$2p_{3/2}$	$3s$	$3p_{1/2}$	$3p_{3/2}$	$3d_{3/2}$	$3d_{5/2}$	and so on
Level term	K	L_I	L_{II}	L_{III}	M_I	M_{II}	M_{III}	M_{IV}	M_V	and so on

Lines corresponding to the electron transition in an atom into K-, L-, M-, N-shells form K-, L-, M-, N-X-ray series. The lines resulting from the transitions corresponding to such a simple diagram are said to be diagram lines.

The frequency ν of a line (in Hz) changes according to Moseley law when passing from one element to another

$$\nu \approx R_\infty c (Z - \sigma)^2 \left(n_0^{-2} - n_1^{-2}\right), \tag{35.6}$$

where $R_\infty = 109737.3 \text{ cm}^{-1}$ is the Rydberg constant, c is the speed of light, n is the principal quantum number. Here the subscript 0 designates the final electron state and the subscript 1 designates the initial electron state, σ is the screening constant.

If the anode current is kept constant the intensity of the characteristic radiation I_c grows in proportion to $(U - U_c)^{3/2}$ where U_c is the line excitation threshold. The maximum of the ratio of the characteristic radiation intensity to the continuous spectrum intensity is reached at $U = 3U_c$.

Methods of X-ray structure analysis [1], and X-ray spectral analysis [2], X-ray microscopy [3], X-ray diffractometry and topography [4], and so on are based upon the usage of X-rays.

Soft X-ray radiation with the wavelength from 1 to 2.5 nm is used in medicine. In microelectronics technology they produce very large scale integration chips containing 10^5–10^6 elements using photolithography with X-ray masks [5].

35.3 Wavelengths of Basic Lines and Absorption Edges of X-ray Radiation

Table 35.1 Designations of X-ray radiation diagram lines [7]

K-series		*L*-series				*M*-series	
Transition	Line index	Transition	Line index	Transition	Line index	Transition	Line index
K—L_{II}	α_2	L—M_{II}	β_4	L_{II}—O_{IV}	γ_6	M_{III}—N_V	γ
K—L_{III}	α_1	L_I—M_{III}	β_3	L_{III}—M_I	l	M_{IV}—N_{III}	δ
K—M_{II}	β_3	L_I—M_{IV}	β_{10}	L_{III}—M_{II}		M_{IV}—$N_{VI,V,II}$	β
K—M_{III}	β_1	L_I—M_V	β_9	L_{III}—M_{III}	s	M_{IV}—O_{II},O_{III}	η
K—M_{IV}	β_5^{II}	L_I—N_{II}	γ_2	L_{III}—M_{IV}	α_2	M_V—N_{II}	ξ_1
K—M_V	β_5^I	L_I—N_{III}	γ_3			M_V—N_{III}	ξ_2
K—$M_{IV,V}$	β_5	L_I—O_{II}	γ_4'	L_{III}—M_V	α_1	M_V—$N_{II,III}$	ξ
K—N_{II}	β_2^{II}	L_I—O_{III}	γ_4	L_{III}—N_I	β_6	M_V—N_{VI}	α_2
		L_I—$P_{II,III}$	γ_{13}			M_V—N_{VII}	α_1
K—N_{III}	β_2^I	L_{II}—M_I	η	L_{III}—N_{IV}	β_{15}		
K—$N_{II,III}$	β_2	L_{II}—M_{IV}	β_1	L_{III}—N_V	β_2		
K—N_{IV}	β_{4x}	L_{II}—N_I	γ_5	L_{III}—$N_{VI,VII}$	β_7'		
K—N_V	β_4	L_{II}—N_{IV}	γ_1	L_{III}—O_I	β_7		
		L_{II}—O_I	γ_8	L_{III}—$O_{IV,V}$	β_5		

Table 35.2 Wavelengths of diagram lines for the initial *K*-level

Element	Wavelength λ (in nm) for the final level							Absorption edge
	L_I	L_{II}	L_{III}	M_{II}	M_{III}	M_{IV}	M_V	
		α_2	α_1	β_3	β_1	β_5^{II}	β_5^I	
^{3}Li		22.8						22.65
^{4}Be		11.4						11.10
^{5}B		6.76						
^{6}C		4.47						4.368
^{7}N		3.16						3.099
^{8}O		2.362						2.332
^{9}F		1.832						1.800
^{10}Ne		1.461		1.445				1.430
^{11}Na		1.1910		1.1575				1.1569
^{12}Mg		0.9890		0.9521				0.9512
^{13}Al		0.8342	0.8339	0.7961				0.7948
^{14}Si		0.7128	0.7125	0.6753				0.6738
^{15}P		0.6160	0.6157	0.5796				0.5784
^{16}S		0.5375	0.5372	0.5032				0.5019

Table 35.2 Wavelengths of diagram lines for the initial K-level *(continued)*

Element	Wavelength λ (in nm) for the final level							Absorption edge
	L_{I}	L_{II}	L_{III}	M_{II}	M_{III}	M_{IV}	M_{V}	
		α_2	α_1	β_3	β_1	β_5^{II}	β_5^{I}	
^{17}Cl		0.4731	0.4728	0.4403				0.4397
^{18}Ar		0.4195	0.4192	0.3886				0.3871
^{19}K		0.3745	0.3741	0.3454		0.3441		0.3437
^{20}Ca		0.3362	0.3358	0.3090		0.3075		0.3070
^{21}Sc		0.3034	0.3031	0.2780		0.2763		0.2762
^{22}Ti		0.2752	0.2749	0.2514		0.2499		0.2497
^{23}V		0.2507	0.2504	0.2284		0.2270		0.2269
^{24}Cr		0.2294	0.2290	0.2085		0.2071		0.2070
^{25}Mn		0.2106	0.2102	0.1910		0.1897		0.1896
^{26}Fe		0.1940	0.1936	0.1757		0.1744		0.1743
^{27}Co		0.1793	0.1789	0.1621		0.1609		0.1608
^{28}Ni		0.1662	0.1658	0.1500		0.1489		0.1488
^{29}Cu		0.1544	0.1541	0.1392		0.1382		0.1381
^{30}Zn		0.1439	0.1435	0.1295		0.1285		0.1283
^{31}Ga		0.1344	0.1340	0.1208	0.1207	0.1198		0.1196
^{32}Ge		0.1258	0.1254	0.1129	0.1128	0.1119		0.1117
^{33}As		0.1180	0.1176	0.1058	0.1057	0.1049		0.1045
^{34}Se		0.1109	0.1105	0.0993	0.0992	0.0984		0.0980
^{35}Br		0.1044	0.1040	0.0933	0.0932	0.0926		0.0920
^{36}Kr		0.0984	0.0980	0.0879	0.0878	0.0871		0.0866
^{37}Rb		0.0930	0.0926	0.0829	0.0828	0.0822		0.0816
^{38}Sr		0.0879	0.0875	0.0783	0.0782	0.0776		0.0770
^{39}Y		0.0833	0.0829	0.0741	0.0740	0.0735		0.0728
^{40}Zr		0.0790	0.0786	0.0702	0.0701	0.0696		0.0689
^{41}Nb		0.0750	0.0746	0.0666	0.0665	0.0660		0.0653
^{42}Mo		0.0714	0.0709	0.0633	0.0632	0.0627		0.0620
^{43}Tc		0.0679	0.0675	0.0602	0.0601			0.0589
^{44}Ru		0.0647	0.0643	0.0573	0.0572	0.0568		0.0561
^{45}Rh		0.0618	0.0613	0.0546	0.0545	0.0541		0.0534
^{46}Pd		0.0590	0.0585	0.0521	0.0520	0.0516		0.0509
^{47}Ag		0.0564	0.0559	0.0498	0.0497	0.0493		0.0486
^{48}Cd		0.0539	0.0535	0.0476	0.0475	0.0471		0.0464
^{49}In		0.0516	0.0512	0.0455	0.0454	0.0451		0.0444
^{50}Sn		0.0495	0.0491	0.0436	0.0435	0.0432		0.0425
^{51}Sb		0.0475	0.0470	0.0418	0.0417	0.0414		0.0407
^{52}Te		0.0456	0.0451	0.0401	0.0400			0.0390
^{53}I		0.0438	0.0433	0.0385	0.0384			0.0374
^{54}Xe		0.0421	0.0416	0.0369	0.0368			0.0358
^{55}Cs		0.0405	0.0400	0.0355	0.0354			0.0345
^{56}Ba		0.0390	0.0385	0.0342	0.0341	0.0338		0.0331
^{57}La		0.0375	0.0371	0.0329	0.0328	0.0325		0.0318
^{58}Ce		0.0362	0.0357	0.0317	0.0316	0.0313		0.0306
^{59}Pr		0.0349	0.0344	0.0305	0.0304			0.0295
^{60}Nd		0.0336	0.0332	0.0294	0.0293			0.0285
^{61}Pm		0.0325	0.0320	0.0284	0.0283			0.0274
^{62}Sm		0.0314	0.0309	0.0274	0.0273	0.0271		0.0265
^{63}Eu		0.0303	0.0298	0.0264	0.0263			0.0256
^{64}Gd		0.0293	0.0288	0.0255	0.0254	0.0243		0.0247
^{65}Tb		0.0283	0.0279	0.0247	0.0246			0.0238
^{66}Dy		0.0274	0.0269	0.0239	0.0238	0.0236		0.0230
^{67}Ho		0.0265	0.0261	0.0231	0.0230	0.0228		0.0223

Table 35.2 Wavelengths of diagram lines for the initial K-level *(continued)*

Element	Wavelength λ (in nm) for the final level							Absorption edge
	L_I	L_{II}	L_{III}	M_{II}	M_{III}	M_{IV}	M_V	
		α_2	α_1	β_3	β_1	β_5^{II}	β_5^{I}	
^{68}Er		0.0257	0.0252	0.0223	0.0222	0.0221		0.0216
^{69}Tm		0.0249	0.0244	0.0216	0.0215	0.0214		0.0209
^{70}Yb		0.0241	0.0237	0.0209	0.0208	0.0207		0.0202
^{71}Lu		0.0234	0.0229	0.0203	0.0202	0.0200		0.0196
^{72}Hf		0.0227	0.0222	0.0197	0.0196			0.0190
^{73}Ta		0.0220	0.0215	0.0191	0.0190	0.0188		0.0184
^{74}W	0.0216	0.0214	0.0209	0.0185	0.0184	0.0183		0.0178
^{75}Re		0.0208	0.0203	0.0179	0.0178	0.0177		0.0173
^{76}Os		0.0202	0.0197	0.0174	0.0173	0.0172		0.0168
^{77}Ir		0.0196	0.0191	0.0169	0.0168	0.0167		0.0163
^{78}Pt		0.0190	0.0186	0.0164	0.0163	0.0162		0.0158
^{79}Au		0.0185	0.0180	0.0159	0.0158	0.0158		0.0154
^{80}Hg		0.0180	0.0175	0.0155	0.0154	0.0153		0.0149
^{81}Tl		0.0175	0.0170	0.0151	0.0150	0.0149		0.0145
^{82}Pb		0.0170	0.0165	0.0146	0.0145	0.0145		0.0141
^{83}Bi		0.0166	0.0161	0.0142	0.0141	0.0141		0.0137
^{84}Po		0.0161	0.0156	0.0139	0.0138			
^{85}At		0.0157	0.0152	0.0135	0.0134			
^{86}Kn		0.0153	0.0148	0.0131	0.0130			
^{87}Er		0.0149	0.0144	0.0128	0.0127			
^{88}Ra		0.0145	0.0140	0.0124	0.0123			
^{89}Ac		0.0141	0.0136	0.0121	0.0120			
^{90}Th		0.0138	0.0133	0.0118	0.0117	0.0116		0.0113
^{91}Pa		0.0134	0.0129	0.0115	0.0114			
^{92}U		0.0131	0.0126	0.0112	0.0111	0.0110		0.0107

Table 35.3 Wavelengths λ (in nm) for L-series diagram lines [8]

Element	Initial level L_I			L_I-absorption edge	Initial level L_{II}		L_{II}-absorption edge	Initial level L_{III}		L_{III}-absorption edge
	Final level				Final level			Final level		
	$L_{II,III}$	M_{II}	M_{III}		M_I	M_{IV}		M_I	$M_{IV,V}$	
		β_4	β_3		η	β_1		l	$\alpha_{1,2}$	
^{11}Na	37.6				40.71		40.50	40.71		40.50
^{12}Mg	31.6			19.37	25.15		24.93	25.15		25.07
^{13}Al	29.0			14.25	17.14		17.04	17.14		17.04
^{14}Si					13.53		12.30	13.55		12.30
^{15}P					10.38		9.40	10.38		9.40
^{16}S					8.340			8.340		
^{17}Cl					6.733			6.790		
^{18}Ar					5.590			5.530		
^{19}K					4.724		4.210	4.774		4.210
^{20}Ca					4.046	3.594	3.513	4.096	3.633	3.549
^{21}Sc					3.513	3.102		3.559	3.135	
^{22}Ti					3.089	2.705	2.729	3.136	2.742	2.729
^{23}V		2.189			2.734	2.388		2.777	2.425	
^{24}Cr		1.943		1.67	2.430	2.127	1.790	2.478	2.164	2.070
^{25}Mg		1.758			2.185	1.911		2.229	1.945	
^{26}Fe		1.570			1.975	1.726	1.720	2.015	1.759	1.753
^{27}Co		1.427			1.787	1.567	1.562	1.829	1.597	1.592

Table 35.3 Wavelengths λ (in nm) for L-series diagram lines [8] *(continued)*

Element	Initial level L_I			L_I-absorption edge	Initial level L_{II}		L_{II}-absorption edge	Initial level L_{III}		L_{III}-absorption edge
	Final level				Final level			Final level		
	$L_{II,III}$	M_{II}	M_{III}		M_I	M_{IV}		M_I	$M_{IV,V}$	
		β_4	β_3		η	β_1		l	$\alpha_{1,2}$	
^{28}Ni		1.316			1.627	1.427	1.424	1.669	1.456	1.453
^{29}Cu		1.210			1.490	1.305	1.301	1.525	1.334	1.329
^{30}Zn		1.119		1.306	1.368	1.198	1.186	1.402	1.225	1.213
^{31}Ga		1.047		0.952	1.260	1.102	1.083	1.295	1.129	1.110
^{32}Ge		0.964	0.958	0.877	1.161	1.018	0.992	1.194	1.044	1.019
^{33}As		0.893		0.811	1.073	0.941	0.913	1.107	0.967	0.937
^{34}Se		0.832		0.750	0.996	0.874	0.841	1.029	0.899	0.865
^{35}Br		0.777		0.696	0.926	0.813	0.775	0.959	0.837	0.798
^{36}Kr		0.730	0.726	0.647		0.758	0.717		0.782	0.739
^{37}Rb		0.682	0.679	0.601	0.804	0.708	0.664	0.836	0.732	0.686
^{38}Sr		0.640	0.637	0.559	0.752	0.662	0.617	0.784	0.687	0.639
^{39}Y		0.602	0.598	0.522	0.704	0.621	0.576	0.736	0.645	0.596
^{40}Zr		0.567	0.563	0.488	0.661	0.584	0.538	0.692	0.607	0.558
^{41}Nb		0.535	0.531	0.458	0.621	0.549	0.503	0.652	0.573	0.523
^{42}Mo		0.505	0.501	0.430	0.585	0.518	0.472	0.615	0.541	0.491
^{43}Tc				0.406		0.489	0.444		0.511	0.463
^{44}Ru		0.452	0.449	0.384	0.521	0.462	0.418	0.550	0.485	0.437
^{45}Rh		0.429	0.425	0.363	0.492	0.437	0.394	0.522	0.460	0.413
^{46}Pd		0.407	0.403	0.344	0.466	0.415	0.372	0.495	0.437	0.391
^{47}Ag		0.387	0.383	0.326	0.442	0.393	0.353	0.471	0.415	0.370
^{48}Cd		0.368	0.364	0.308	0.419	0.374	0.333	0.448	0.396	0.350
^{49}In		0.351	0.347	0.293	0.398	0.356	0.315	0.427	0.377	0.332
^{50}Sn		0.334	0.341	0.278	0.379	0.339	0.298	0.407	0.360	0.316
^{51}Sb		0.319	0.315	0.264	0.361	0.323	0.283	0.389	0.344	0.300
^{52}Te		0.305	0.301	0.251	0.344	0.308	0.269	0.372	0.329	0.286
^{53}I		0.291	0.287	0.239	0.328	0.294	0.255	0.356	0.315	0.272
^{54}Xe				0.227			0.243		0.302	0.259
^{55}Cs		0.266	0.263	0.217	0.299	0.268	0.231	0.327	0.289	0.247
^{56}Ba		0.256	0.252	0.207	0.286	0.257	0.220	0.314	0.278	0.236
^{57}La		0.245	0.241	0.198	0.274	0.246	0.211	0.301	0.267	0.226
^{58}Ce		0.235	0.231	0.189	0.262	0.236	0.201	0.289	0.256	0.217
^{59}Pr		0.226	0.222	0.181	0.251	0.226	0.193	0.278	0.246	0.208
^{60}Nd		0.217	0.213	0.174	0.241	0.217	0.184	0.268	0.237	0.200
^{61}Pm			0.204	0.167		0.208	0.177		0.228	0.192
^{62}Sm		0.200	0.196	0.160	0.222	0.200	0.169	0.248	0.220	0.185
^{63}Eu		0.193	0.189	0.154	0.213	0.192	0.163	0.239	0.212	0.178
^{64}Gd		0.185	0.182	0.148	0.205	0.185	0.156	0.231	0.205	0.171
^{65}Tb		0.179	0.175	0.142	0.197	0.178	0.150	0.224	0.198	0.165
^{66}Dy		0.172	0.168	0.137	0.190	0.171	0.144	0.216	0.191	0.159
^{67}Ho		0.166	0.162	0.132	0.183	0.165	0.139	0.209	0.185	0.154
^{68}Er		0.160	0.156	0.127	0.176	0.159	0.134	0.202	0.179	0.148
^{69}Tm		0.154	0.151	0.123	0.170	0.153	0.129	0.196	0.173	0.143
^{70}Yb		0.149	0.145	0.118	0.164	0.148	0.124	0.189	0.168	0.139
^{71}Lu		0.144	0.140	0.114	0.158	0.142	0.120	0.184	0.163	0.134
^{72}Hf		0.139	0.135	0.110	0.152	0.137	0.115	0.178	0.158	0.130
^{73}Ta		0.135	0.131	0.106	0.147	0.132	0.111	0.173	0.153	0.126
^{74}W		0.130	0.126	0.102	0.142	0.128	0.107	0.169	0.148	0.122
^{75}Re		0.126	0.122	0.099	0.137	0.124	0.104	0.163	0.144	0.118

Table 35.3 Wavelengths λ (in nm) for L-series diagram lines [8] *(continued)*

Element	Initial level L_I			L_I-absorption edge	Initial level L_{II}		L_{II}-absorption edge	Initial level L_{III}		L_{III}-absorption edge
	Final level				Final level			Final level		
	$L_{II,III}$	M_{II}	M_{III}		M_I	M_{IV}		M_I	$M_{IV,V}$	
		β_4	β_3		η	β_1		l	$\alpha_{1,2}$	
^{76}Os		0.122	0.118	0.096	0.133	0.119	0.100	0.158	0.140	0.114
^{77}Ir		0.118	0.114	0.092	0.128	0.116	0.097	0.154	0.136	0.111
^{78}Pt		0.114	0.110	0.089	0.124	0.112	0.093	0.150	0.132	0.107
^{79}Au		0.111	0.107	0.086	0.120	0.108	0.090	0.146	0.128	0.104
^{80}Hg		0.107	0.103	0.084	0.116	0.105	0.087	0.142	0.125	0.101
^{81}Tl		0.104	0.100	0.081	0.113	0.101	0.084	0.138	0.121	0.098
^{82}Pb		0.101	0.097	0.078	0.109	0.098	0.082	0.135	0.118	0.095
^{83}Bi		0.098	0.094	0.076	0.106	0.095	0.079	0.132	0.115	0.092
^{84}Po		0.097	0.091			0.092		0.128	0.112	
^{85}At			0.088			0.089			0.109	
^{86}Rn			0.085			0.087			0.106	
^{87}Er			0.083			0.084			0.104	
^{88}Ra		0.84	0.080	0.065	0.091	0.081	0.067	0.117	0.101	0.080
^{89}As			0.078			0.079			0.099	
^{90}Th		0.079	0.075	0.061	0.085	0.077	0.063	0.112	0.096	0.076
^{91}Pa		0.077	0.073		0.083	0.074		0.109	0.094	
^{92}U		0.075	0.071	0.057	0.081	0.072	0.059	0.107	0.092	0.072
^{93}Np		0.073	0.069		0.078	0.070		0.104	0.090	
^{94}Pu		0.071	0.067	0.054	0.076	0.068	0.056	0.102	0.088	0.069
^{95}Am		0.069	0.065			0.066		0.100	0.086	

35.4 Widths of X-Ray Radiation Lines

In accordance with the uncertainty principle, energetic levels (terms) have finite widths. This leads to some "straggling" of the energy of quanta belonging to the same spectral line. The line itself is characterized by the finite width comparable with the interdoublet splitting. The magnitudes of the widths of bright lines measured on the half of the height are presented in Table 35.4. These magnitudes are directly determined by the valence zone width.

Table 35.4 Half-height widths $\Delta\lambda$ (in nm) of X-ray bright lines [3]

Element	K-line of the initial level				L-line of the initial level		M-line of the initial level
	α_1	α_2	β_1	β_3	α	β	ξ
^{22}Ti	1.38	1.90					
^{23}V	1.58	2.21					
^{24}Cr	1.96	2.43					
^{25}Mn	2.46	2.96					
^{26}Fe	2.65	3.00					
^{27}Co	2.45	3.12					
^{28}Ni	2.26	3.03					
^{29}Cu	2.31	3.21					

Table 35.4 Half-height widths $\Delta\lambda$ (in nm) of X-ray bright lines [3] *(continued)*

Element	K-line of the initial level				L-line of the initial level		M-line of the initial level
	α_1	α_2	β_1	β_3	α	β	ξ
^{30}Zn	2.44	2.90					
^{31}Ga	2.40	2.55					
^{32}Ge	2.73	2.94					
^{38}Sr	4.5	4.6					1.52
^{39}Y							0.44
^{40}Zr	5.2	5.4			1.94	1.97	0.827
^{41}Nb	5.8	5.4			1.94	2.12	1.24
^{42}Mo	5.86	6.18			1.95	2.21	1.63
^{44}Ru	6.8	6.7			2.55	2.83	2.49
^{45}Rh	7.3	7.2			2.54	2.87	3.77
^{46}Rd	7.8	7.9			2.94	2.97	7.26
^{47}Ag	8.6	8.7			3.16	2.74	10.24
^{48}Cd					2.81	3.50	
^{50}Sn			11.7	11.5			
^{51}Sb	10.6	14.2					
^{52}Te	11.0	14.0					
^{53}I	10.7	15.1					
^{55}Cs	15.0	15.1	17.7	14.6			
^{56}Ba	15.2	18.0					
^{57}La		17.2					
^{58}Ce	12.8	17.0	18.8	17.3			
^{60}Nd	14.4	21.2					
^{62}Sm	23.9	24.2	27.5	24.7			
^{63}Eu		22.2					
^{64}Gd	27.0	26.4					
^{65}Tb		24.6	33.2	33.0			
^{66}Dy	27.9	26.1					
^{67}Ho	30.2	28.8	33.5	34.0			
^{68}Fr		29.5					
^{69}Tm		36.9					
^{70}Yb	40.4	32.5	44.0	37.8			
^{71}Lu	39.3	39.8					
^{72}Hf		38.7					
^{73}Ta	42.4	36.4	44.9	49.3			
^{74}W	43.2	37.4					
^{75}Re	46.4	52.9					
^{76}Os	55.0	50.9					
^{77}Ir	53.0	50.3					
^{78}Pt	58.0	63.3					
^{79}Au	55.5	56.7					
^{80}Hg	64.0	68.5					
^{81}Tl		78.2					
^{82}Pb	65.5	68.7					
^{83}Bi		83.5					
^{90}Th	95.0	92.6					
^{92}U	105.0	107.5					
^{93}Np	98.5	103.9					
^{94}Pu		114.0					
^{95}Am	115.0	129.5					

35.5 Relative Intensities of the Lines

The X-ray line intensity is defined by the oscillator strength and by the frequency of the corresponding transition and also by the statistical weight of the atom level. A calculation of the oscillation strengths comprises a time-taking problem. According to experimental data for the K-series radiation the intensity is defined by the equation $I = \kappa i(U - U_c)^r$ where U_c is the excitation threshold for the series, i is the electron current in a tube, U is the applied voltage, the exponent r equals 1.6–2, κ is an empirical parameter. The relative intensity of a spectrum line is defined by the interlevel transition probability. For the most often used K-series I_{α_1}:I_{α_2}:I_{α_3}= 10:5:2, and λ_α:λ_β =1.09. Values of relative intensities of lines from K- and L-series are presented in Table 35.5 [2, 3].

Table 35.5 K- and L-series relative intensities (in percents of the $K_{\alpha 1}$- and $L_{\alpha 1}$-lines intensities, respectively) [For (Ar–Ca) – in percent of $K_{\alpha 1}$+$K_{\alpha 2}$ doublet and for (K–Co) – in percent of $L_{\alpha 1}$+$L_{\alpha 2}$ doublet intensities, respectively.]

Element	K-series					L-series				
	α_2	β_3	β_1	β_5	β_2	α_1	α_2	l	β_4	$\beta_{2,15}$
^{18}Ar			10.3							
^{19}K			12.2							
^{20}Ca			12.8			100		250		
^{21}Sc	50.7		20.0			100		250		
^{22}Ti	50.2		19.8	0.44		100		165		
^{23}V	50.7		20.5	0.45		100		250		
^{24}Cr	50.5		17.9	0.43		100		77		
^{25}Mn	50.0		22.4	0.34		100		100		
^{26}Fe	50.6		16.7	0.21		100		80		
^{27}Co	50.2		16.0	0.14		100		75		
^{28}Ni	50.7		18.7	0.20						
^{29}Cu	50.7		20.0	0.15						
^{30}Zn	51.3		20.7		0.36					
^{31}Ga	50.7		21.6							
^{32}Ge	51.2		24.0		1.32					
^{33}As	43.2		21.7		0.69					
^{34}Se	52.3		21.0		1.07					
^{35}Br	50.9		22.2		1.73					
^{36}Kr			17.2							
^{37}Rb	51.0		23.9		2.5					
^{38}Sr	52.1		27.4		4.16					
^{39}X	52.3		23.3		3.19					
^{40}Zr	52.4		24.9		3.70	100	12.0	3.5	0.6	2.0
^{41}Nb	52.6		27.9		4.90	100	14.0	4.0	0.7	4.7
^{42}Mo	52.5		26.5		4.30	100	11.0	2.4		5.6
^{44}Ru	51.1		29.3		5.63	100				10.7
^{45}Rh	52.5		27.6		4.60	100				15.2
^{46}Pd	52.6		29.0		6.13	100				16.8
^{47}Ag	52.2		28.4		4.9	100				25.8
^{48}Cd	54.4		29.7		6.42	100				36.0
^{49}In	53.9		29.6		6.47	100				25.2
^{50}Sn	54.1		29.8		5.5	100				27.9
^{51}Sb	51.7		31.0		7.08	100				26.5
^{52}Te	52.7		30.6		7.35					
^{53}I	53.6		30.1		6.0					
^{55}Cs	54.0									
^{56}Ba	56.2		30.4		7.1					

Table 35.5 *K*- and *L*-series relative intensities *(continued)*

Element	*K*-series					*L*-series				
	α_2	β_3	β_1	β_5	β_2	α_1	α_2	l	β_4	$\beta_{2,15}$
^{57}La	56.4					100	11.4			21.2
^{58}Ce	50.8	30.3			8.0	100	9.8	4.7		22.4
^{59}Pr	53.0						11.0			20.4
^{60}Nd	53.2					100	10.4			22.5
^{62}Sm	54.0	32.1			7.5	100	8.5	4.8	0.66	21.7
^{63}Eu	54.9									
^{64}Gd	56.9	31.0			8.7	100	12.5	5.2		19.8
^{65}Tb	56.2					100	11.5	2.64		19.8
^{66}Dy	55.1					100	10.3			19.6
^{67}Ho	56.0	32.0			8.8	100	11.6	5.5	0.57	17.7
^{68}Er	52.2					100	11.5			19.6
^{69}Tm	57.6					100	11.3	3.11	1.08	15.0
^{70}Yb	58.7	32.7			8.9	100	11.9			22.0
^{71}Lu	58.5					100	11.7	3.48		17.9
^{72}Hf	57.1					100	11.6	3.06		19.2
^{73}Ta	55.4	12.5	24.3	0.57	7.8	100	12.0	5.7	1.2	20.2
^{74}W	57.8	13.3	24.8	0.56	8.5	100	11.7	3.38	1.2	22.2
^{75}Re	58.7	12.8	24.6	0.57	8.5	100	11.7		1.8	21.4
^{76}Os	60.2	12.9	25.0	0.69	8.9	100	11.1			26.9
^{77}Ir	57.7	12.7	24.3	0.64	7.8	100	11.5	6.0	1.4	23.8
^{78}Pt	56.3	12.9	24.6	0.69	9.0	100	11.8		1.5	23.2
^{79}Au	58.5	13.2	25.8	0.77	8.8	100	11.5	6.0	1.3	25.9
^{80}Hg	59.5	13.1	26.6	0.72	10.0	100	11.0		1.3	22.9
^{81}Tl	57.4	13.9	25.7	0.65	10.1	100	10.5		1.7	26.8
^{82}Pb	59.6	14.0	26.2	0.74	10.0	100	11.9	4.1	1.7	24.0
^{83}Bi	59.9	13.4	26.8	0.80	10.0	100	11.3	6.3	1.8	24.0
^{90}Tn	61.0	14.0	28.4	0.98	11.8	100	10.7	7.2	1.9	28.4
^{91}Pa						100	10.9			24.0
^{92}U	61.0	14.0	27.8	0.98	12.0	100	10.6	4.5	1.7	26.8
^{93}Np	63.5									
^{94}Pu	63.0					100	10.2		1.7	24.9
^{95}Am	61.9									
^{96}Cm	63.2	11.2	23.0							
^{97}Bk	64.3	12.6	22.2							
^{98}Cf	65.7	13.3	25.9							
^{99}Es	66.6	13.4	24.2							

35.6 The Interaction of X-ray Radiation with Matter

The interaction of X-ray radiation with matter is followed by the secondary radiation arising in the process of the direct pickup of electrons from the atom (photoeffect) and the subsequent detachment of external electrons during filling the internal electronic shells.

The restructuring of electronic shells is accompanied with the emission of X-ray quanta of lower energy (fluorescent radiation) or so called Auger electrons (the secondary photoeffect). The direct interaction of X-ray radiation with external shell electrons leads to the production of Compton electrons. X-ray radiation inelastically scattered by the thermal crystal lattice vibrations, also falls into the category of secondary emission. In some cases the luminescent radiation is observed at X-ray irradiation of crystals. All these processes are responsible for the X-ray absorption and the study of any of them gives information about the properties of the matter under study. As a result of these processes and elastic scattering, the primary beam intensity I_0 decreases exponentially with the thickness t of a traversed substance layer:

$$I = I_0 \exp(-\mu t), \tag{35.7}$$

where μ is the linear attenuation factor (in cm^{-1}). If the substance consists of atoms of the same type it is useful to introduce the so-called mass attenuation factor $\mu_m = \mu/p$ (in cm^2/g), where ρ is the mass density in g/cm^3.

The attenuation factor is approximately proportional to λ^3 and Z^3; with the X-ray wavelength decreasing, the coefficient μ grows down as well. Yet for some wavelength values λ_c the attenuation factor sharply increases (the absorption band edge), and then decreases again with the wavelength decrease according to the same law.

The linear attenuation factor for a composite chemical substance may be expressed in terms of mass concentrations c_i and mass attenuation factors μ_{mi} of elements constituting this composition $\mu = \rho \sum c_i \mu_{mi}$ where the summation is carried out for all the elements entering the complex molecule.

The mass attenuation factor values for K-lines of emitting elements widely used in researches are presented in Table 35.6 [3].

Table 35.6 Values of the mass attenuation factor μ_m (in cm^2/g) for K_α- and K_β-lines of different emitters [8]

Absorbing elements	Cr$K_{\alpha 1}$	Cr$K_{\beta 1}$	Fe$K_{\alpha 1}$	Fe$K_{\beta 1}$	Co$K_{\alpha 1}$	Co$K_{\beta 1}$	Ni$K_{\alpha 1}$	Ni$K_{\beta 1}$	Cu$K_{\alpha 1}$	Cu$K_{\beta 1}$	Mo$K_{\alpha 1}$	Mo$K_{\beta 1}$
^{3}Li	1.77	1.35	1.08	0.816	0.861	0.646	0.689	0.514	0.557	0.414	0.058	0.270
^{4}Be	4.28	3.25	2.62	1.97	2.08	1.56	1.67	1.24	1.35	1.0	0.240	0.223
^{5}B	8.49	6.46	5.20	3.92	4.13	3.10	3.31	2.49	2.67	1.99	0.360	0.290
^{6}C	14.9	11.3	9.10	6.86	7.23	5.42	5.79	4.32	4.67	3.47	0.555	0.415
^{7}N	28.9	18.1	14.6	11.0	11.6	8.70	9.29	6.93	7.50	5.58	0.820	0.626
^{8}O	35.9	27.3	22.0	16.6	17.5	13.10	14.0	10.4	11.3	8.40	1.17	0.884
^{9}F	51.6	39.2	31.6	29.8	25.1	18.80	20.1	15.0	16.2	12.0	1.68	1.20
^{10}Ne	71.3	54.2	43.7	32.9	34.7	26.0	27.8	20.7	22.4	16.7	2.33	1.60
^{11}Na	97.0	74.7	60.7	46.3	48.7	37.0	39.4	30.0	32.1	24.2	3.40	2.18
^{12}Mg	123	94.7	77.0	58.7	61.8	46.9	50.0	37.8	40.8	30.7	4.57	2.95
^{13}Al	153	118	95.8	73.1	76.8	58.4	62.1	47.0	50.7	38.2	5.21	3.68
^{14}Si	187	144	117	89.5	94.1	71.5	76.1	57.6	62.1	46.7	7.12	4.80
^{15}P	226	174	149	108	114	86.3	91.8	69.5	74.9	56.4	8.30	5.73
^{16}S	270	208	196	129	135	103	110	82.9	89.4	67.3	10.3	7.29
^{17}Cl	318	245	199	152	160	121	129	97.8	105	79.4	12.1	8.60
^{18}Ar	372	286	233	178	187	142	151	114	123	92.8	13.5	10.0
^{19}K	407	317	260	201	211	162	172	132	142	108	17.3	11.9
^{20}Ca	462	360	296	228	240	184	196	150	161	123	20.5	13.8
^{21}Sc	521	406	333	258	270	208	221	169	182	139	23.1	16.0
^{22}Ti	585	456	374	289	303	233	248	190	204	156	25.9	18.0
^{23}V	79.9	509	418	323	338	260	276	212	228	174	28.9	20.0
^{24}Cr	90.4	70.1	464	358	376	289	307	235	253	193	32.1	22.2
^{25}Mn	102	79.0	64.6	396	416	320	340	260	280	213	34.8	24.3
^{26}Fe	114	88.5	72.4	55.7	58.5	353	374	287	308	235	39.1	27.8
^{27}Co	127	98.7	80.8	62.2	65.3	50.0	53.2	315	338	258	42.9	32.0
^{28}Ni	141	110	89.8	69.1	72.5	55.6	59.1	45.1	48.5	282	47.0	35.1
^{29}Cu	156	121	99.4	76.5	80.3	61.5	65.4	49.9	53.7	40.8	51.2	39.0
^{30}Zn	173	134	110	84.4	88.6	67.9	72.2	55.1	59.2	45.0	55.7	43.5

Table 35.6 Values of the mass attenuation factor μ_m *(continued)*

Absorbing elements	Cr$K_{\alpha 1}$	Cr$K_{\beta 1}$	Fe$K_{\alpha 1}$	Fe$K_{\beta 1}$	Co$K_{\alpha 1}$	Co$K_{\beta 1}$	Ni$K_{\alpha 1}$	Ni$K_{\beta 1}$	Cu$K_{\alpha 1}$	Cu$K_{\beta 1}$	Mo$K_{\alpha 1}$	Mo$K_{\beta 1}$
^{31}Ga	190	147	121	92.8	97.5	74.7	79.4	60.6	65.1	49.5	60.4	48.0
^{32}Ge	208	162	132	102	107	81.9	87.0	66.4	71.4	54.3	65.5	51.0
^{33}As	228	177	145	0.11	117	89.5	95.1	72.6	78.1	59.3	70.5	55.0
^{34}Se	248	193	158	121	127	97.6	104	79.9	85.1	64.7	75.8	61.0
^{35}Br	270	210	172	132	139	106	113	86.1	92.6	70.4	81.5	66.0
^{36}Kr	293	227	186	143	150	115	122	93.4	100	76.4	87.4	70.0
^{37}Rb	317	246	201	155	163	125	132	101	109	82.6	90.0	77.0
^{38}Sr	342	266	217	167	176	135	143	109	117	89.2	96.0	83.0
^{39}Y	369	286	234	180	189	145	154	118	126	96.1	100	89.0
^{40}Zr	397	308	252	194	204	156	166	127	136	103	16.7	95.0
^{41}Nb	426	331	271	208	219	168	178	136	146	0.11	18.0	100.
^{42}Mo	456	354	290	223	234	180	191	146	157	119	19.3	14.6
^{43}Tc	488	379	310	239	251	192	204	156	168	127	20.7	15.5
^{44}Ru	522	405	332	255	268	205	218	167	179	136	22.0	16.0
^{45}Rh	557	432	354	272	286	219	233	178	191	145	23.5	17.5
^{46}Rd	593	460	377	290	304	233	248	189	204	155	24.5	18.5
^{47}Ag	631	480	401	309	324	248	264	201	217	165	26.6	19.5
^{48}Cd	670	520	426	328	344	264	280	214	230	175	28.3	20.0
^{49}In	711	552	452	348	365	280	298	227	244	186	30.0	21.0
^{50}Sn	754	585	479	369	387	297	315	241	259	197	32.5	22.0
^{51}Sb	798	620	507	390	410	314	334	255	274	208	33.7	23.0
^{52}Te	845	655	537	413	433	332	353	269	290	220	35.6	24.0
^{53}I	892	692	567	436	458	351	375	285	307	233	38.0	25.5
^{54}Xe	786	731	598	460	483	370	394	300	323	245	39.7	26.5
^{55}Cs	816	759	621	478	502	385	409	312	335	255	41.3	28.0
^{56}Ba	599	700	654	508	528	405	430	328	353	268	43.8	24.0
^{57}La	214	730	687	529	555	426	452	345	371	282	45.5	31.0
^{58}Ce	226	517	602	556	583	447	475	363	390	296	48.5	32.0
^{59}Pr	237	188	447	583	612	469	499	381	409	311	50.7	33.5
^{60}Nd	249	197	469	540	534	492	523	399	429	326	53.0	35.0
^{61}Pm	262	207	172	565	399	516	548	418	450	342	55.4	36.5
^{62}Sm	275	218	181	399	418L_{III}	485	470L_{II}	438	471L_I	358	57.9	38.0
^{63}Eu	289	228	190	417	156	500	358	458	410	374	60.6	39.5
^{64}Gd	304	239	200	156	164	353	374	430	429	392	64.5	41.5
^{65}Tb	318	351	209	163	172	369	142	450	324	409	67.3	43.0
^{66}Dy	333	262	219	170	180	140	149	314	339	385	70.3	44.5
^{67}Ho	348	274	229	178	188	146	156	328	129	270	73.4	46.5
^{68}Er	364	287	239	186	197	153	163	126	135	281	76.5	48.5
^{69}Tm	380	299	250	195	205	160	170	131	141	293	79.7	50.0
^{70}Yb	397	312	261	204	214	167	177	137	147	113	85.1	52.0
^{71}Lu	414	326	272	212	224	174	185	143	154	119	86.5	54.0
^{72}Hf	428	335	279	217	228	177	188	145	156	120	84.0	57.0
^{73}Ta	447	350	291	226	238	184	196	151	162	125	88.0	59.0
^{74}W	466	365	304	236	248	192	205	158	170	130	92.1	62.0
^{75}Re	485	380	317	246	259	200	213	164	177	136	96.4	65.0
^{76}Os	506	396	330	256	270	209	222	171	184	141	101	68.0
^{77}Ir	527	418	343	267	281	217	231	178	192	147	105	71.0
^{78}Pt	548	430	357	278	292	226	241	185	200	153	110	74.0
^{79}Au	570	447	372	289	304	235	251	193	208	159	115	77.0
^{80}Hg	593	464	386	300	316	244	261	201	216	166	120	81.0

Table 35.6 Values of the mass attenuation factor μ_m *(continued)*

Absorbing elements	$CrK_{\alpha 1}$	$CrK_{\beta 1}$	$FeK_{\alpha 1}$	$FeK_{\beta 1}$	$CoK_{\alpha 1}$	$CoK_{\beta 1}$	$NiK_{\alpha 1}$	$NiK_{\beta 1}$	$CuK_{\alpha 1}$	$CuK_{\beta 1}$	$MoK_{\alpha 1}$	$MoK_{\beta 1}$
^{81}Tl	616	483	402	312	328	254	271	208	224	172	125	82.0
^{82}Pb	640	501	417	324	341	264	281	216	233	178	130	86.0
^{83}Bi	664	520	433	336	354	274	292	225	242	186	136	89.0
^{84}Po	689	540	449	349	368	284	303	233	251	193	141	93.0
^{85}At	715	560	466	362	381	295	314	242	260	200	118	96.8
^{86}Rn	741	581	483	375	395	306	326	251	270	207	124	100.
^{87}Fr	936	726	602	464	489	375	401	306	330	252	97.8	105.
^{88}Ra	982	764	632	487	514	394	421	321	347	264	101	110.
^{89}Ac	1030	802	663	511	539	414	442	337	364	277	104	100.
^{90}Th	1080	841	696	536	565	434	463	354	382	291	108	76.0
^{91}Pa	1130	881	729	562	593	455	486	371	400	305	115	79.0
^{92}U		923	764	580	621	476	509	388	419	319	124*L*	81.0
^{93}Np		967	800	616	650	499	533	407	439	334	57.6	84.0
^{94}Pu		1010	837	645	681	522	557	426	459	350	60.3	87.0

35.7 Secondary Spectra and Chemical Bond Effects in X-ray Spectroscopy

When X-ray photons with energies $h\nu \geq \epsilon_n$ (where ϵ_n is the electron energy on the n-th level) bombard target atoms the latter may pass into excited states and create vacancies in internal electronic levels. Then the target emits the so-called fluorescent (characteristic) radiation corresponding to the electron transition into the vacant level. In this case the bremsstrahlung (braking radiation) is absent. The maximum wavelength of the primary radiation initiating the excitation of the fluorescent radiation of a given series is said to be a boundary wavelength or an absorption band edge. It may be easily calculated via the expression $h\nu_b = hc/\lambda_b = \epsilon(n,l,i)$ where n, l, and j are the principal, azimuthal, and internal quantum numbers.

The fluorescence yield for different series is given in Table 35.7. For the most part, average experimental yields for singly ionized atoms with vacancies in the K-shell (ω_K), L-shell (ω_L), and M-shell (ω_M) are presented, as well as fluorescence yields for the transfer of the K-shell hole into the L-shell (ω_{KL}) and of the L- shell hole into the M-shell (ω_{LM}).

X-ray fluorescent spectroscopy is widely used for the determination of the contents of different elements in analyzed materials at depths corresponding to 10^4–$3{\cdot}10^5$ atomic layers. For the same goal they use electrons directly picked up from atoms during the X-ray passage through the matter. This method has received the name Electron Spectroscopy for Chemical Analysis (ESCA). It allows one to study 2–10 atomic layers.

The atom transition to the ground state may be accompanied by the emission of an electron, not a photon. Such a transition is called the secondary photoeffect, and corresponding electrons are called Auger electrons. Because the energy spectrum of these electrons is determined by the energy difference of various energy states of the atom, it is the "fingerprint" of a given type of atoms along with the characteristic X-ray emission. The probability of the Auger electron emission for atoms with $Z < 33$ is even higher than the probability of radiation transitions.

Spectra of photo- and Auger electrons escaping the sample are sensitive to the electron structure, chemical bonds, phase composition, and other characteristics of the crystal containing the emitting element.

In crystals of a high degree of perfection a strong space redistribution of the X-ray radiation field takes place that naturally influences the yield intensity of one or another secondary radiation. Such measurements give qualitatively new information [5,9] on the crystal properties in comparison to the information obtained by traditional methods. The method based upon the measurement of the angular dependence of the secondary radiation intensity yield has received the name X-ray standing wave method because the angular dependence features of the secondary radiation yield are determined first of all by generating the standing X-ray wave in the crystal.

If the probability of the secondary radiation yield from the crystal at the depth Z is given by the function $P_A(Z)$ (where A corresponds to one or another type of the secondary emission) the recorded intensity is determined as follows [9]

$$\kappa_A(\Theta) = \int_0^\infty P_A(z)\kappa_A(\Theta, z)dz, \tag{35.8}$$

where

$$\kappa_A(\Theta, z) = \left\{|D_0(\Theta, z)|^2 + |D_h(\Theta, z)|^2 + 2\text{Re}C\epsilon_n(A)D_0(\Theta, z)D_h(\Theta, z)\exp\left[-i\varphi(z) - W(z)\right]\right\}. \tag{35.9}$$

Here D_0, D_h are amplitudes of incident and diffracted waves, respectively, $\varphi(z) = K_hU(z)$ where K_h is the length of the reciprocal lattice vector, U is the displacement of the atomic plane, $\exp[-w(z)]$ is the static Debye–Waller factor, C is the polarization factor, $\epsilon_h = \chi_{ih}/\chi_{i0}$, where χ_{ih} and χ_{i0} are the imaginary parts of polarizability Fourier components (Table 35.8).

First of all, the influence of individual properties of molecules and crystals is felt first by the K_α-line energetic position. At the same time the K_α–line width changes insignificantly in going from one composition to another. Yet the K_α-line width may give the useful information on the symmetry of the nearest surrounding of the atom in the medium. In X-ray spectroscopy the main diagram line is often accompanied by the satellite (weak) lines both from the short-wave side (short-wave satellites) and from the long-wave side (long-wave satellites). The satellites are highly sensitive to the structure of the substance and their sensitivity to the electron or geometrical structure factors often exceeds the sensitivity of main diagram lines. Large numbers of satellites arise as a result of electronic transitions in multiply ionized atoms. As a rule, the short-wave satellites arise as a result of such transitions. For example, a group of K_α -line satellites (α_1, α_2, α_3 and α_4) appear during the interstate $K^1 - L^2$ transitions in doubly ionized atoms.

Some parts of the satellites are situated far from the main line and do not influence its form, but some satellites may significantly influence the form of $K_{\alpha_{12}}$-lines, especially in the ranges of its "tails."

Besides the short-wave satellites, the long-wave satellites of K_α–line may also be observed because during the satellite line emission part of the energy may be carried away by the Auger electron. These satellites, as a rule, have little effect upon the main line form.

One can use the chemical shifts of satellites as well as the main K_α-lines to identify the charge state of the atom under study (Table 35.9).

Table 35.7 Fluorescence yield (in %) per one photoeffect act [8]

K-series		K-series		K-series		K-series		K-series	
Element	ω_K	Element	ω_K	Element	ω_K	Element	ω_K	Element	ω_K
^{4}Be	0.0304	^{11}Na	2.60	^{18}Ar	12.20	^{25}Mn	31.30	^{32}Ge	55.40
^{5}B	0.056	^{12}Mg	3.36	^{19}K	11.50	^{26}Fe	34.20	^{33}As	58.80
^{6}C	0.26	^{13}Al	3.80	^{20}Ca	13.80	^{27}Co	36.60	^{34}Se	59.60
^{7}N	0.60	^{14}Si	4.30	^{21}Sc	19.00	^{28}Ni	41.40	^{35}Br	62.20
^{8}O	0.94	^{15}P	6.00	^{22}Ti	22.10	^{29}Cu	44.30	^{36}Kr	66.0
^{9}F	1.13	^{16}S	8.20	^{23}V	25.30	^{30}Zn	47.90	^{37}Rb	66.9
^{10}Ne	1.82	^{17}Cl	9.55	^{24}Cr	28.30	^{31}Ga	52.80	^{38}Sr	70.2

Table 35.7 Fluorescence yield (in %) per one photoeffect act [8] *(continued)*

K-series		*K*-series		*K*-series		*K*-series		*K*-series	
Element	ω_K	Element	ω_K	Element	ω_K	Element	ω_K	Element	ω_K
^{39}Y	71.1	^{48}Cd	84.0	^{57}La	90.6	^{66}Dy	94.3	^{75}Re	95.9
^{40}Zr	73.0	^{49}In	85.0	^{58}Ce	91.1	^{67}Ho	94.3	^{76}Os	96.1
^{41}Nb	74.8	^{50}Sn	85.9	^{59}Pr	91.5	^{68}Er	94.5	^{77}Ir	96.2
^{42}Mo	76.4	^{51}Sb	86.7	^{60}Nd	92.0	^{69}Tm	94.8	^{78}Pt	96.7
^{43}Tc	77.9	^{52}Te	85.7	^{61}Pm	92.4	^{70}Yb	95.0	^{79}Au	96.4
^{44}Ru	79.3	^{53}I	88.2	^{62}Sm	92.8	^{71}Lu	95.2	^{80}Hg	95.8
^{45}Rh	80.7	^{54}Xe	89.4	^{63}Eu	92.5	^{72}Hf	95.4	^{82}Pb	97.2
^{46}Pd	81.9	^{55}Cs	88.9	^{64}Gd	93.4	^{73}Ta	95.6	^{92}U	97.0
^{47}Ag	83.4	^{56}Ba	90.1	^{65}Tb	93.7	^{74}W	95.7		

L-series			*L*-series			*L*-series			*L*-series		
Element	ω_{KL}	ω_L	Element	ω_{KL}	ω_L	Element	ω_{KL}	ω_L	Element	ω_{KL}	ω_L
^{23}V		0.235	^{51}Sb	7.0	11.9	^{67}Ho	17.0		^{82}Pb	39.5	29.7
^{25}Mn		0.295	^{52}Te	7.3	12.2	^{68}Er	21.0		^{83}Bi	41.4	33.0
^{29}Cu		0.56	^{54}Xe	9.1		^{69}Tm	23.0		^{88}Ra		40.0
^{31}Ga		0.64	^{55}Cs		8.90	^{70}Yb	25.0		^{90}Th		48.8
^{36}Kr		7.50	^{56}Ba	14.8	9.3	^{71}Lu	26.0	29.0	^{91}Pa		50.0
^{37}Rb	1.30	1.10	^{57}La	12.3	11.0	^{72}Hf	29.0	29.0	^{92}K	40.9	53.0
^{39}Y		5.70	^{58}Ce	16.0	16.3	^{73}Ta	28.0	22.5	^{93}Np		49.0
^{40}Zr	3.40		^{59}Pr	12.3	16.7	^{74}W	31.0	29.8	^{94}Pu		56.6
^{41}Nb	2.20		^{60}Nd	16.0	17.0	^{75}Re	30.0		^{95}Cm		53.1
^{42}Mo		6.70	^{61}Pm	18.5		^{76}Os	32.0	34.8		*M*-series	
^{46}Pd	4.70		^{62}Sm	17.0	18.8	^{77}Ir	31.0	30.0	^{76}Os	1.3	
^{47}Ag	4.40	6.59	^{63}Eu	17.0	17.0	^{78}Pt	36.0	32.0	^{79}Au	2.4	2.3
^{48}Cd	5.50		^{64}Gd	18.0	19.8	^{79}Au	28.7	43.0	^{82}Pb	2.6	2.9
^{49}In	6.5		^{65}Tb	19.5	19.4	^{80}Hg	41.0	40.0	^{83}Bi	3.0	3.5
^{50}Sn	6.4		^{66}Dy	21.0		^{81}Tl	44.0	41.0	^{92}U		6.0

Table 35.8 Coefficients of real $\chi_{rh}\cdot 10^6$ and imaginary $\chi_{ih}\cdot 10^6$ parts of the Fourier components of polarizability for some crystals and reflection planes [5]

Crystal	Radiation	(111)		(220)		(400)		(333)	
		χ_{rh}	χ_{ih}	χ_{rh}	χ_{ih}	χ_{rh}	χ_{ih}	χ_{rh}	χ_{ih}
Si	CuK$_\alpha$; (0.154 nm)	8.117	0.2456	9.128	0.3398	7.634	0.3269	4.486	0.2195
	MoK$_\alpha$; (0.071 nm)	1.698	0.0115	1.901	0.0159	1.588	0.0154	0.931	0.0103
	AgK$_\alpha$; (0.056 nm)	1.054	0.0045	1.179	0.0062	0.084	0.0061	0.576	0.0041
Ge	CuK$_\alpha$	17.26	0.601	20.30	0.8282	16.64	0.7938	9.480	0.5295
	MoK$_\alpha$	3.866	0.2449	4.595	0.3462	3.80	0.3337	2.200	0.2243
	AgK$_\alpha$	2.452	0.1041	2.93	0.1437	2.42	0.1380	1.394	0.0931
GaP	CuK$_\alpha$	14.907	0.4651	16.185	0.6349	13.073	0.6349	7.768	0.4671
	MoK$_\alpha$	3.1507	0.1842	3.421	0.1946	2.763	0.1946	1.642	0.1843
	AgK$_\alpha$	1.964	0.0761	2.1326	0.0801	1.7225	0.0801	1.0235	0.0761
GaAs	CuK$_\alpha$	18.024	0.6029	21.544	0.8451	17.671	0.8451	10.043	0.6056
	MoK$_\alpha$	3.800	0.2631	4.553	0.3703	3.735	0.3703	2.123	0.2643
	AgK$_\alpha$	2.375	0.1091	2.839	0.1536	2.328	0.1536	1.323	0.1097
InP	CuK$_\alpha$	17.964	2.2490	18.135	2.425	14.284	2.4254	10.723	2.2507
	MoK$_\alpha$	3.797	0.1154	3.8330	0.1237	3.0192	0.1237	2.2664	0.1155
	AgK$_\alpha$	2.3669	0.0463	2.3895	0.0495	1.8820	0.0495	1.4129	0.0464

Table 35.8 Coefficients of real $\chi_{rh}\cdot 10^6$ and imaginary $\chi_{ih}\cdot 10^6$ parts *(continued)*

Crystal	Radiation	(111)		(220)		(400)		(333)	
		χ_{rh}	χ_{ih}	χ_{rh}	χ_{ih}	χ_{rh}	χ_{ih}	χ_{rh}	χ_{ih}
InAs	CuK_α	19.217	2.0790	22.438	2.4340	18.138	2.4340	11.255	2.0826
	MoK_α	4.0618	0.1968	4.7426	0.2716	3.8340	0.2716	2.3790	0.19760
	AgK_α	2.5321	0.0810	2.9564	0.1115	2.3898	0.1115	1.4830	0.0814
InSb	CuK_α	19.190	2.5722	23.199	3.6301	19.143	3.6301	11.050	2.5843
	MoK_α	4.056	0.1326	4.903	0.1871	4.046	0.1871	2.336	0.1332
	AgK_α	2.528	0.0533	3.057	0.0752	2.522	0.0752	1.456	0.0536

Table 35.9 Shifts of $CrK_{\beta 1,3}$- and $SnL_{\beta 2}$- lines [10]

Ion	Sample	E, eV	Ion	Sample	E, eV
Cr^{3+}	CrF_3	+0.66		Na_2CrO_4	−0.84
	$CrCl_3$	+0.56		$Cu_2Cr_2O_7\cdot 2H_2O$	−0.79
	$CrBr_3$	+0.46		$CaCrO_4$	−0.64
	Cr_2O_3	+0.40		$BaCrO_4$	−0.86
	$CdCr_2Se_4$	+0.77		CrO_3	−0.83
	$CuCr_2S_4$	+0.63	Sn^{2+}	$SnCl_2$	+0.176
	$CoCr_2S_4$	+0.71		SnC_2O_4	+0.144
	$CrCr_2S_4$	+0.64		$SnHPO_4$	+0.098
Cr^{4+}	CrO_2	+0.12		SnS	+0.063
Cr^{5+}	Cr_2O_5	−0.39	Sn^{4+}	$SnCl_4$	−0.085
Cr^{6+}	$K_2Cr_2O_7$	−0.84		$PbSnO_3$	−0.0179
	$Fe_2(CrO_4)_3$	−0.74		SnO_2	−0.180
	K_2CrO_4	−0.91		$CaSnOSiO_4$	−0.198

Note: Positions of $CrK_{\beta 1,3}$ and $SnL_{\beta 2}$ lines in pure Cr and Sn, respectively, are set to 0 eV.

References

[1] Vainshtein, B. K., Modern crystallography, vol. 1, Nauka, Moscow, 1981 (in Russian).

[2] Losev, N.F., Quantitative X-ray fluorescent analysis, Nauka, Moscow, 1969 (in Russian).

[3] X–ray microscopy: Proceedings of the International symposium, Eds. Schmahl, G. and Rudolph, D., Gottingen, Sept. 14–16, 1983, Springer–Verlag, Berlin–Heidelberg–New York–Tokyo, 1984.

[4] Pinsker, Z. G., Dynamical scattering of X–rays in crystals, Springer, Berlin, 1978.

[5] Afanas'ev, A. M., Imamov, R. M., Aleksandrov, P. A., X–ray diffraction diagnostics of submicron layers, Nauka, Moscow, 1989 (in Russian).

[6] Valiev, K. A., Physics of submicron lithography, Plenum Press, New York–London, 1992.

[7] Tables of physical quantities. Handbook, Ed. Kikoin, I. K., Atomizdat, Moscow, 1976 (in Russian).

[8] Blokhin, M. A., Shveitser, I. G., X–ray spectral handbook, Nauka, Moscow, 1982 (in Russian).

[9] Afanas'ev, A. M., Imamov, R. M., Mukhamedzhanov, E. Kh., Cryst. Rev., 3, 157, 1992 (in Russian).

[10] Mazalov, L. N., Treiger, B. A., Zhurnal strukturnoy khimii, 24, 128, 1983.

36

Elementary Particles

V.A. Nikitin

36.1 The Fundamental Particles of Matter

The most important and constant task of physics throughout its entire history has always been to search for the simplest and indivisible particles of matter. The vocation of science is to make use of a minimum number of components and interaction laws between them for classifying and explaining the variety of forms and phenomena encountered in the material world. In the 19th century the atoms of chemical elements claimed to be assigned the role of elementary particles. Soon, however, their composite nature was revealed. Atoms were decomposed into electrons and nuclei, while nuclei, in turn, were caused to undergo fission into protons and neutrons. Thus, at the start of the 20th century, the leadership in the competition for the right to be called elementary was assumed by the electrons, protons and neutrons. They were joined by the photon and (hypothetically, at the beginning) the π-meson, which were assigned the task of carriers of the electromagnetic and nuclear (strong) interactions, respectively. By the middle of the century, owing to the numerous studies carried out in cosmic rays and at accelerators, more than 300 other particles had been discovered, all of which were mutually interrelated through countless production and decay channels. Observation of the interconversion of particles did not, however, lead to straightforward revelation of their composite nature, as in the case of atoms. Thus, for example, the collision of two protons may result in the production of new protons and antiprotons, of pions, and of other particles. Various individual collision events give rise to the production of differing amounts and types of secondary particles. Restrictions are imposed only by conservation laws of energy-momentum and of a series of quantum numbers, such as charge, baryon number, and others. Thus, often several hundred secondary particles are produced in a single event occurring at an accelerator with intersecting proton-antiproton beams with an energy of 1 TeV. The same holds true for decays of particles. Many of them have tens of diverse decay channels, and, here, the secondary particles are apparently not more elementary than the primary ones. This means that particles are not extracted from the interior of other particles, but are produced in the course of interaction. Nature is capable of stamping them in accordance with strictly defined samples. All the particles can be considered elementary from the point of view of the criterion requiring them not to be divisible into component parts. On the other hand, it has been established experimentally that strongly interacting particles, called hadrons, are extended objects of a characteristic size on the order of 10^{-13} cm. The distributions of electric charge, magnetic moment, and optical density have been measured in these particles. They are also observed to exhibit polarizability, i.e. they are deformed in the field of an incident probe wave. The resonance behavior of their interaction with the primary wave is a clear manifestation of their structure. In this connection, the term hadron spectroscopy is used, by analogy with atomic physics.

The dilemma of elementarity and compositeness was solved in a way that has no analogies in the history of physics. A new concept has been formulated, and new objects – quarks – have been in-

0-8493-2861-6/97/$0.00+$.50

troduced that are fundamental fermions of which all hadrons are composed. The novelty consists in that quarks being composite parts of hadrons do not themselves exist in a free state. This nontrivial property is known as quark confinement. The mechanism of confinement can be illustrated with the aid of a potential rising linearly with the distance between a pair of quarks, so that infinite energy is required for freeing a quark. Actually, a nonphysical infinite potential does not appear in the theory. The point is that the strong field acting between quarks polarizes the vacuum. Attempts at pulling quarks apart by a large distance result in virtual quark–antiquark pairs turning into real pairs which form new hadrons. This is what consumes the energy transferred to the quarks, while they themselves remain forever confined constituents of hadrons. Adding the known sextuplet of leptons to the quarks, we obtain the entire list of fundamental fermions, with the aid of which one can build up all known particles and explain many of their properties in the most saving way. At the present-day level of investigations the fundamental fermions themselves are seen to be structureless objects of dimensions smaller than 10^{-16} cm.

36.2 Types of Interaction

There are four known types of fundamental interactions: strong, electromagnetic, weak, and gravity. Quantum field theory deals with particles and forces acting between them from a unique point of view: particles are considered to be field quanta, and forces arise owing to the exchange of field quanta, also. The nature of a force is determined by two parameters: the coupling constant g and the mass m of the quantum serving as a mediator of the coupling. The carriers of interaction are bosons, particles of integer spin. The value of g determines the intensity, with which bosons are emitted and absorbed, and the characteristic interaction potential; m determines the range of the interaction force, $r_0 = \hbar/mc$.

The boson of strong interaction is called the gluon (from the word glue). It is emitted and absorbed by quarks and it "glues" them together in hadrons. Two classes of hadrons are known: mesons and baryons. Mesons consist of quark-antiquark pairs and have integer spin; baryons are composed of three quarks and have half-integer spin. The strong interaction charge exhibited by quarks is such that the total charge of a quark and antiquark equals zero. The three quarks in a baryon are chosen in such a manner, with respect to charge, that together they form a neutral system. Thus, the gluon field is strong in between the quarks in a hadron, and it falls rapidly at the periphery, in accordance with the well-known property of short-range nuclear forces. The gluon possesses zero mass and, therefore, generates a long-range potential of the Coulomb type. But this is true only at small distances $r < 1$ fm (1 fm = 10^{-13} cm), where the polarized vacuum anti-screens (reduces) the strong interaction charge. Thus, at $r = 10^{-17}$ cm the dimensionless strong interaction coupling constant $\alpha_s = g_s/\hbar c = 1/10$. When $r \cong 1$ fm, $\alpha_s \geq 1$ and the potential starts rising linearly, thus confining the quarks in the hadron.

Hadron scattering is characterized by the total cross section $\sigma_{\text{tot}} \approx$ 20–300 mb (1 mb = 10^{-27} cm^2). All hadrons, with the exception of the proton, are unstable. Many of them are understood to be excited states of a quark system. Such a system may emit hadrons and undergo transition to the ground state. This process is due to the strong interaction and proceeds rapidly, i.e. in a nuclear time $\tau \approx 10^{-23}$ s. Accordingly, the excited state, often termed a resonance, has a width $\Gamma = \hbar/\tau =$ 10–200 MeV.

Electromagnetic forces are brought about by the photon, which is emitted and absorbed by the electric charge. The photon has zero mass and generates the well-known long-range Coulomb potential e^2/r. The corresponding dimensionless constant $\alpha_{\text{em}} = e^2/\hbar c = 1/129$ at $r = 10^{-17}$ cm, and $\alpha_{\text{em}} = 1/137$ at $r \geq 1$ fm. The dependence of α_{em} on the distance is also explained by polarization of the vacuum, but in this case the charge is screened by virtual e^+e^- pairs. The closer one is to the "bare" charge, the greater is the interaction force.

The constant α_{em} is two orders of magnitude smaller than α_s, so the cross section characterizing inelastic electromagnetic processes is also two orders of magnitude smaller than the cross section of hadron reactions. The same can be said about the partial widths of particle decays proceeding in the electromagnetic channel.

The weak interaction is mediated by three bosons: W^+, W^-, Z^0. Their masses ~90 GeV. They generate a short–range potential of the form $V(r) = (g_W/r)\exp(-r/r_0)$, $r_0 = 10^{-16}$ cm. Both quarks and leptons take part in weak interactions. The constant $\alpha_W = g_W/\hbar c = 1/27$ at $r = 10^{-17}$ cm.

The range and constant of the weak interaction being small results in the total interaction cross section of the neutrino with matter being small, also. The neutrino is mentioned here because it takes no part in strong and electromagnetic interactions and characterizes pure weak interaction. Thus, for example, $\sigma_{\text{tot}}(\nu_\mu p) = 0.7 \cdot 10^{-38} E_\nu$ cm^2; E_ν is expressed in GeV, $E_\nu = 5 - 220$ GeV. Weak interaction particle decays are characterized by mean lifetimes $\tau = 10^{-13} - 10^{+3}$ s. The large spread of τ values is due to kinematical reasons. In different reactions the laws of energy-momentum conservation impose different limits on the phase volume of the decay products.

The quantities $\alpha_s, \alpha_{\text{em}}, \alpha_W$ depend on the radius or momentum transfer $q = \hbar/r$ at which they are measured. Therefore, they are called running constants. Theory assumes them to compare and the three forces of Nature to join in one at a very high momentum transfer $q = 10^{15}$ GeV/s. No experimental test of this prediction can be implemented, at present, but it is very important for cosmology.

Gravity is transmitted by the graviton, a boson of spin 2 and zero mass. Any mass serves as a source of gravitons, so all particles take part in gravity. The constant $\alpha_g = Gm^2/\hbar c$ is very small. Thus, for instance, in the case of the electron $\alpha_g = 10^{-45}$, which renders impossible, at present, any experimental observation of gravitational effects in particle physics [1, 2].

36.3 Types of Particles

Particles can be classified by the type of interaction in which they take part, and by their quantum numbers.

Hadrons and leptons are the particles distinguished most readily. The former participate in strong interactions, and the latter do not. Three families of leptons are known, (e^-, ν_e), (μ^-, ν_μ), (τ^-, ν_τ), as well as the respective antiparticles. Each family is assigned a conserved quantum number $L_i = 1$, $i = e, \mu, \tau$. For all other particles $L_i = 0$.

In accordance with their spin, particles are divided into fermions (half-integer spin) and bosons (integer spin). This classification is important for at least two reasons: *a*) fermions and bosons obey different statistics, *b*) in quantum field theory fermions play the part of sources of the boson field.

Particles and antiparticles differ by the sign of additive (scalar) quantum numbers (electric charge, lepton number, etc.). If all quantum numbers are zero, then a particle cannot be distinguished from its antiparticle.

In accordance with their spin, hadrons are divided into baryons (half-integer spin) and mesons (integer spin). Baryons are assigned the conserved quantum number $N = 1$. For all other particles $N = 0$. A baryon always differs from its antiparticle. Some mesons coincide with their antiparticles.

The next type of hadrons comprises strange particles. Three main features distinguishing them from nonstrange particles can be indicated. *a*) They are unstable and exhibit lifetimes characteristic of the weak interaction, $\tau \approx 10^{-8}$–10^{-10} s. *b*) They are produced in pairs, which allows assigning them the quantum number of strangeness S, that is conserved in strong and electromagnetic interactions. In each pair produced one of the particles has $S = +1$, and the other has $S = -1$. Weak interactions do not conserve strangeness, so in an average time τ strange particles transform into nonstrange particles. *c*) The mass of the lightest strange baryon is about 170 MeV greater, than the mass of the lightest nonstrange baryon: $m(\Lambda^0) - m(p) = 1115 - 938 = 167$ MeV. In the case of mesons this difference ≈ 350 MeV: $m(K) - m(\pi) = 493 - 140 = 353$ MeV.

Two other types of hadrons can be distinguished in a similar manner: with charm and with beauty. Accordingly, they are carriers of the quantum numbers of charm C and beauty B, that are conserved in strong and electromagnetic interactions. The mass of the lightest charmed baryon Λ_c^+ is 2285 MeV, and its lifetime $\tau = 2 \cdot 10^{-13}$ s. The mass of the lightest baryon with beauty, Λ_b^0, equals 5640 MeV.

Theory foresees the existence of one more type of hadrons with a conserved quantum number T. The mass of the lightest T-particle is expected to be about 130 GeV (mass of the xenon nucleus!).

Now let us list once more, as a separate type of particles, the bosons, which serve as mediators of the fundamental forces of Nature: γ, g (gluon), W^+, W^-, Z^0, and the graviton.

36.4 Conservation Laws and Internal Symmetries of Particles

Energy, momentum, and angular momentum conservation laws are obeyed in all the interactions of particles. These laws follow from the homogeneity and isotropy of space–time, in which physical processes proceed. There also exists a series of quantum numbers pertaining to internal particle symmetries which are conserved with different degrees of rigor.

Exact quantum number conservation laws are obeyed in all types of interactions. These laws include conservation of electric charge Q, of lepton numbers L_i, $i = e, \mu, \tau$, and the baryon number N. (There exist theoretical grounds for believing L and N not to be conserved).

Nonrigorous quantum number conservation laws are obeyed in some types of interactions and violated in others. Such are the conservation laws of strangeness S, charm C, beauty B, space parity P, charge parity C, G-parity, isotopic spin I, and hypercharge Y.

Isotopic spin. Among hadrons there exist groups of particles (multiplets) with identical quantum numbers and close masses ($\Delta m < 3\%$) and differing only in electric charge. Examples are (p, n), (π^+, π^-, π^0), $(\Delta^{++}, \Delta^+, \Delta^0, \Delta^-)$, and others. It is natural to consider a multiplet as representing a sole particle in states of varying charge. Formalization of this observation is achieved by introducing a three-dimensional charge space, where a particle is characterized by a quantum number I, exhibiting properties similar to those of angular momentum. I is termed an isotopic spin. By analogy with ordinary spin, the number of components of the multiplet is $2I + 1$. In the case of the multiplets mentioned above $I = 1/2$, 1, 3/2. By definition, the projection of I onto the z-axis is related to the charge Q of a multiplet member by the formula $Q = I_z + Y/2$. The quantity $Y = N + S + C + B$ is called a hypercharge. A hypothesis has been put forward claiming isotropy of isotopic space, i.e. that particle interactions are invariant relative to rotations in the isotopic space, and that isotopic spin $I^2 = I(I + 1)$ and its projections onto the axes of the reference system are the conserved quantum numbers. The hypothesis of isotopic invariance gives rise to numerous consequences that can be tested. Experimental studies testify to its validity in strong interactions and violation in electromagnetic and weak interactions. The differences in mass between the various multiplet members indicate a measure of the inaccuracy of an isotopic invariance.

The theory of isotopic spin is based on the $SU(2)$ group. The wave function of the lowest multiplet $I = 1/2$ is considered as a two-component spinor, which is an element of the two-dimensional complex space and transforms in it with the aid of unitary two-row matrices of the $SU(2)$ group. The wave functions of the higher multiplets are transformed and classified with respect to the higher representations of this group. Therefore, the $SU(2)$ symmetry is commonly called one of the internal symmetries of elementary particles.

The space parity P of a particle is defined as the eigenvalue of the unitary operator U_P inverting the axes of the reference system: $U_P\psi(r) = \psi(-r)$, $U_P\psi(r) = P\psi(r)$. Since repeated application of the operator U_P transforms the ψ-function into itself, $P^2 = 1$, and $P = \pm 1$. An example illustrating the above is the ψ-function of a particle in a central force field. Solution to the wave equation, in this case, yields spherical functions, the parity of which depends on the orbital momentum l, $P = (-1)^l$. The experiment reveals the quantum number P to be conserved in strong and electromagnetic interactions and not to be conserved in weak interactions. The U_P symmetry points to left and right being indistinguishable in the microcosm, if one excludes weak interactions.

Charge parity C is defined as the eigenvalue of the operator U_C substituting the antiparticle for the particle. It has a definite value only in the case of truly neutral systems, all quantum numbers of which are zero: $Q = N = L = S = B = 0$. C-parity is conserved in strong and electromagnetic interactions and is not conserved in weak interactions. This symmetry signifies equivalence of physical laws in the World and in the Antiworld, if one excludes weak interactions.

G-parity is defined as the eigenvalue of the operator $U_G = U_C \exp(i\pi I_y)$, where I_y is the projection of isospin on the y-axis in isotopic space. The operator $\exp(i\pi I_y)$ rotates the system through π about the y-axis, i.e. it performs the substitution $I_z \to -I_z (p \to n, \pi^+ \to \pi^-$, etc.). G-parity has a definite value for particles with the quantum numbers $N = C = L = B = 0$, and it is conserved only by strong interactions.

Discrete transformations generating conserved quantum numbers P, C, G pertain to internal particle symmetries.

36.5 The Quark Model

The list of particles presented in Section 36.8 includes 500 specimens, not counting the antiparticles! A set of theoretical approaches, termed the Standard model, classifies and describes this variety of particles and processes within the framework of quite a limited number of principles.

An important part of the theoretical scheme is the quark model. Quarks are postulated as particles that are more fundamental than hadrons. The quantum numbers of quarks are chosen to be such that hadrons can be built with them in a way as saving as possible. The quark spin being set equal to 1/2 is clearly because it allows the construction of particles with arbitrary spin. The baryon number $N = 1/3$ (the antiquark has $N = -1/3$). Therefore, baryons are built with three quarks $(q_i q_j q_k)$, while the mesons present quark-antiquark pairs $(q_i \bar{q}_j)$, or quarkonium. The subscript indicates the type of quark or its flavor. At present, the quark model involves six flavors, for which the standard notation u, d, s, c, b, t is adopted. Ordinary flavorless particles are made up of u, d quarks. For example, the proton $p = (uud)$, the neutron $n = (udd)$. Hence one can immediately see the electric charges of the u and d quarks: $2Q_u + Q_d = 1$, $2Q_d + Q_u = 0$, $Q_u = 2/3$, $Q_d = -1/3$, and their isotopic spins $I_u = I_d = 1/2$. The general formula relating all the quantum numbers has the form $Q = I_z + Y/2$; $Y = N + S + C + B + T$. Finally, Y is called the hypercharge.

The quark quantum numbers are presented in Table 36.1.

Since quarks are not observed in the free state, their masses given in Table 36.1 are calculated with the aid of a series of theoretical models. The significant uncertainty in the masses is related to the approximate nature of these models.

Mesons are well described as bound $(q_i \bar{q}_j)$ states with quantum numbers $P = (-1)^{l+1}$, $C = (-1)^{l+s}$. Here, l is the orbital momentum, s is the spin of the $q\bar{q}$ pair. The total momentum of the system, or the spin of the meson, is $J = l + s$. The model forbids the states of $J^{PC} = 0^{--}, 0^{+-}, 1^{-+}, 2^{+-} \ldots$ They can exist as exotic particles with structures that are more complicated, than that of the $q\bar{q}$ pair. The quark composition and the quantum numbers of most (reliably established) mesons are compiled in Table 36.2.

Hadrons composed of u, d, and s quarks group together in multiplets with the same quantum numbers J^{PC} and close masses. This points to the existence of an internal symmetry generalizing the $SU(2)$ isotopic symmetry. The (u, d, s) triplet forms a three-component spinor, which is an element in three-dimensional complex space and is transformed by unitary matrices of the $3 \cdot 3$ main representation of the $SU(3)$ group. Higher representations of this group have dimensions 8 and 10. Meson octets and baryon octets and decuplets, corresponding to the prediction of $SU(3)$ symmetry, have been found among the particles. They are presented in Tables 36.3 and 36.4.

The multiplet includes strange and nonstrange particles of differing masses, which points to a significant violation of $SU(3)$ symmetry. One of the consequences of this violation is a mixing of the neutral members of the η_1 singlet and of the η_8 octet. Their quark functions are constructed on the basis of unit normalization and orthogonalization. The superposition of $u\bar{u}, d\bar{d}, s\bar{s}$ pairs results from their mutual transitions with respect to a strong interaction:

$$\eta_1 = \frac{1}{\sqrt{3}}(u\bar{u} + d\bar{d} + s\bar{s}), \quad \eta_8 = \frac{1}{\sqrt{6}}(u\bar{u} + d\bar{d} - 2s\bar{s}).$$

Owing to the mixing, there actually exists, instead of η_1 and η_8, their superposition with a mixing angle θ:

$$\eta = \eta_s \cos\theta - \eta_1 \sin\theta, \quad \eta' = \eta_s \sin\theta + \eta_1 \cos\theta.$$

Therefore, in reality, instead of meson octets and singlets, as required by exact symmetry, there exist nonets. These nonets are listed in Table 36.3. The quark composition of the $SU(3)$ multiplets is illustrated in Tables 36.5 and 36.6.

The quark model is also successfully applied to describing the interaction dynamics of particles. The mass spectra of the heavy quarkonia $(c\bar{c})$ and $(b\bar{b})$, the jet topology of inelastic processes with large momentum transferred, and so on are well reconstructed.

Certain baryons are built up of three quarks of the same flavor: $\Delta^{++}(uuu)$, $\Omega = (sss)$, and others. To avoid contradicting the Pauli principle (identical fermions cannot be in the same quantum state), a new quantum number is introduced, which is arbitrarily referred to as the color and which can assume 3 values: 1, 2, 3 or yellow, blue, and red. Color also serves as the charge of strong interaction. The colored charge of an antiquark is conjugated with the colored charge of the quark. It assumes the values of antiyellow, antiblue, and antired. When such a choice of quark colors is made, hadrons can be built up as colorless objects of zero strong interaction charge. This is just what Nature does, when it transforms the long-range colored field of quarks (the gluon field) into the short-range field of white hadrons. Now, the quark wave function must be assigned a color index q_α, where $\alpha = 1, 2, 3$. The three-component spinor q is an element of the three-dimensional complex color space, in which the $SU(3)_c$ symmetry group acts. This symmetry has been shown to be exact. Mesons M and baryons B are the $SU(3)_c$ singlets:

$$M = \frac{1}{\sqrt{3}} q\bar{q}_\alpha q^\alpha; \quad B = \frac{1}{\sqrt{6}} q^\alpha q^\beta q^\gamma \epsilon_{\alpha\beta\gamma},$$

where $\epsilon_{\alpha\beta\gamma}$ is a completely antisymmetric tensor of rank three. The three quarks in a baryon form a combination that is antisymmetric with respect to color and does not contradict the Pauli principle [2, 4].

The list of mesons and baryons is presented in Tables 36.7 and 36.8.

Table 36.1 Quantum numbers of quarks

Quark	Q	I	I_3	Spin	N	S	C	B	T	Y	m
u	2/3	1/2	1/2	1/2	1/3	0	0	0	0	1/3	~ 5 MeV
d	−1/3	1/2	−1/2	1/2	1/3	0	0	0	0	1/3	~ 7 MeV
s	−1/3	0	0	1/2	1/3	−1	0	0	0	−2/3	100–300 MeV
c	2/3	0	0	1/2	1/3	0	1	0	0	4/3	1.3–1.7 GeV
b	−1/3	0	0	1/2	1/3	0	0	−1	0	−2/3	4.7–5.3 GeV
t	2/3	0	0	1/2	1/3	0	0	0	1	4/3	176±23 GeV

Table 36.2 Quark composition and quantum numbers of reliably established mesons

$N^{2s+1} \cdot L_J$	J^{PC}	$u\bar{d}, u\bar{u}, d\bar{d}$ $I=1$	$u\bar{u}, d\bar{d}, s\bar{s}$ $I=0$	$c\bar{c}$ $I=0$	$b\bar{b}$ $I=0$	$\bar{s}u, \bar{s}d$ $I=1/2$	$c\bar{u}, c\bar{d}$ $I=1/2$	$c\bar{s}$ $I=0$	$\bar{b}u, \bar{b}d$ $I=1/2$
1^1S_0	0^{-+}	π	η, η'	η_c		K	D	D_s	B
1^3S_1	1^{--}	ρ	ω, ϕ	$J/\psi(1S)$	$\Upsilon(1S)$	$K^*(892)$	$D^*(2010)$	$D_s^*(2110)$	$B^*(5330)$
1^1P_1	1^{+-}	$b_1(1235)$	$h_1(1170), h_1(1380)$			K_{1B}	$D_1(2420)$	$D_{s1}(2536)$	
1^3P_0	0^{++}	$a_0(980)$	$f_0(1400), f_0(975)$	$\chi_{c0}(1P)$	$\chi_{b0}(1P)$	$K_0^*(1430)$			
1^3P_1	1^{++}	$a_1(1260)$	$f_1(1285), f_1(1510)$	$\chi_{c1}(1P)$	$\chi_{b1}(1P)$	K_{1A}			
1^3P_2	2^{++}	$a_2(1320)$	$f_2(1270), f_2'(1525)$	$\chi_{c2}(1P)$	$\chi_{b2}(1P)$	$K_2^*(1430)$	$D_2^*(2460)$		
1^1D_2	2^{-+}	$\pi_2(1670)$							
1^3D_1	1^{--}	$\rho(1700)$	$\omega(1600)$	$\psi(3770)$		$K^*(1680)$			
1^3D_2	2^{--}					$K_2(1770)$			
1^3D_3	3^{--}	$\rho_3(1690)$	$\omega_3(1670), \phi_3(1850)$			$K_3^*(1780)$			
1^3F_4	4^{++}	$a_4(2040)$	$f_4(2050), f_4(2220)$			$K_4^*(2045)$			
2^1S_0	0^{-+}	$\pi(1300)$	$\eta(1295)$	$\eta_c(2S)$		$K(1460)$			
2^3S_1	1^{--}	$\rho(1450)$	$\omega(1390), \phi(1680)$	$\psi(2S)$	$\Upsilon(2S)$	$K^*(1410)$			
3^1S_0	0^{-+}	$\pi(1770)$	$\eta(1760)$			$K(1830)$			
1^3D_2	2^{--}					$K_2(1770)$			
1^3D_3	3^{--}	$\rho_3(1690)$	$\omega_3(1670), \phi_3(1850)$			$K_3^*(1780)$			
1^3F_4	4^{++}	$a_4(2040)$	$f_4(2050), f_4(2220)$			$K_4^*(2045)$			
2^1S_0	0^{-+}	$\pi(1300)$	$\eta(1295)$	$\eta_c(2S)$		$K(1460)$			
2^3S_1	1^{--}	$\rho(1450)$	$\omega(1390), \phi(1680)$	$\psi(2S)$	$\Upsilon(2S)$	$K^*(1410)$			
2^3P_2	2^{++}		$f_2(1810), f_2(2010)$		$\chi_{b2}(2P)$	$K_2^*(1980)$			
3^1S_0	0^{-+}	$\pi(1770)$	$\eta(1760)$			$K(1830)$			

Notes: although the clear signature of associative $t\bar{t}$ quark production in the reaction $\bar{p}p \to t\bar{t} + X$ was observed [5] mesons consisting of quark combination $(t\bar{t})$, $(t\bar{q})$ or $(\bar{t}q)$ are not yet identified.

Table 36.3 Meson nonets

J^{PC}	Nonet members	θ	J^{PC}	Nonet members	θ
0^{-+}	π, K, η, η'	$-(10° - 23°)$	2^{++}	$a_2(1320)$, $K_2^*(1430)$, $f_2'(1525)$, $f_2(1270)$	$28° - 26°$
1^{--}	ρ, $K^*(892)$, ϕ, ω	$39° - 36°$	3^{--}	$\rho_3(1690)$, $K_3^*(1780)$, $\phi_3(1850)$, $\omega_3(1670)$	$29° - 28°$

Table 36.4 Quark composition of hadron octets

Y	I	I_3	Baryons $J^P = 1/2^+$	Quark composition	Mesons $J^P = 0^-$	Quark composition	Mesons $J^P = 1^-$	Quark composition
1	1/2	1/2	p	uud	K^+	$u\bar{s}$	K^{*+}	$u\bar{s}$
1	1/2	−1/2	n	ddu	K^0	$d\bar{s}$	K^{*0}	$d\bar{s}$
0	1	1	Σ^+	uus	π^+	$u\bar{d}$	ρ^+	$u\bar{d}$
0	1	0	Σ^0	uds	π^0	$(u\bar{u} - d\bar{d})/\sqrt{2}$	ρ^0	$(u\bar{u} - d\bar{d})/\sqrt{2}$
0	1	−1	Σ^-	dds	π^-	$d\bar{u}$	ρ^-	$d\bar{u}$
0	0	0	Λ	uds	η, η'	$u\bar{u}, d\bar{d}, s\bar{s}$	ω, φ	$u\bar{u}, d\bar{d}, s\bar{s}$
−1	1/2	1/2	Ξ^0	uss	$\overline{K}^0$	$\bar{d}s$	$\overline{K}^{*0}$	$\bar{d}s$
−1	1/2	−1/2	Ξ^-	dss	K^-	$\bar{u}s$	K^{*-}	$\bar{u}s$

Table 36.5 Baryon octets, decuplets, and singlets

J^P	(D, L_N^P)	s	Octet members				Singlet
$1/2^+$	$(56, 0_0^+)$	1/2	$N(939)$	$\Lambda(1116)$	$\Sigma(1193)$	$\Xi(1318)$	
$1/2^+$	$(56, 0_2^+)$	1/2	$N(1440)$	$\Lambda(1600)$	$\Sigma(1660)$	$\Xi(?)$	
$1/2^-$	$(70, 1_1^-)$	1/2	$N(1535)$	$\Lambda(1670)$	$\Sigma(1620)$	$\Xi(?)$	$\Lambda(1405)$
$3/2^-$	$(70, 1_1^-)$	1/2	$N(1520)$	$\Lambda(1690)$	$\Sigma(1670)$	$\Xi(1820)$	$\Lambda(1520)$
$1/2^-$	$(70, 1_1^-)$	3/2	$N(1650)$	$\Lambda(1800)$	$\Sigma(1750)$	$\Xi(?)$	
$3/2^-$	$(70, 1_1^-)$	3/2	$N(1700)$	$\Lambda(?)$	$\Sigma(?)$	$\Xi(?)$	
$5/2^-$	$(70, 1_1^-)$	3/2	$N(1675)$	$\Lambda(1830)$	$\Sigma(1775)$	$\Xi(?)$	
$1/2^+$	$(70, 0_2^+)$	1/2	$N(1710)$	$\Lambda(1810)$	$\Sigma(1880)$	$\Xi(?)$	$\Lambda(?)$
$3/2^+$	$(56, 2_2^+)$	1/2	$N(1720)$	$\Lambda(1890)$	$\Sigma(?)$	$\Xi(?)$	
$5/2^+$	$(56, 2_2^+)$	1/2	$N(1680)$	$\Lambda(1820)$	$\Sigma(1915)$	$\Xi(2030)$	
$7/2^-$	$(70, 3_3^-)$	1/2	$N(2190)$	$\Lambda(?)$	$\Sigma(?)$	$\Xi(?)$	$\Lambda(2100)$
$9/2^-$	$(70, 3_3^-)$	3/2	$N(2250)$	$\Lambda(?)$	$\Sigma(?)$	$\Xi(?)$	
$9/2^+$	$(56, 4_4^+)$	1/2	$N(2220)$	$\Lambda(2350)$	$\Sigma(?)$	$\Xi(?)$	
			Decuplet members				
$3/2^+$	$(56, 0_0^+)$	3/2	$\Delta(1232)$	$\Sigma(1385)$	$\Xi(1530)$	$\Omega(1672)$	
$1/2^-$	$(70, 1_1^-)$	1/2	$\Delta(1620)$	$\Sigma(?)$	$\Xi(?)$	$\Omega(?)$	
$3/2^-$	$(70, 1_1^-)$	1/2	$\Delta(1700)$	$\Sigma(?)$	$\Xi(?)$	$\Omega(?)$	
$5/2^+$	$(56, 2_2^+)$	3/2	$\Delta(1905)$	$\Sigma(?)$	$\Xi(?)$	$\Omega(?)$	
$7/2^+$	$(56, 2_2^+)$	3/2	$\Delta(1950)$	$\Sigma(2030)$	$\Xi(?)$	$\Omega(?)$	
$11/2^+$	$(56, 4_4^+)$	3/2	$\Delta(2420)$	$\Sigma(?)$	$\Xi(?)$	$\Omega(?)$	

Notes:

D is the dimension of the $SU(6)$-multiplet;

N is the principal excitation quantum number of $q\bar{q}$;
s is the net spin of quarks in $q\bar{q}$ system;

J is the total angular momentum of a particle, i.e. quark spin plus its orbital momentum;

P is the space parity.

Number in parentheses denotes the particle mass in MeV.

Table 36.6 Quark composition of the baryon decuplet

Barions $J^P = 3/2^+$	Y	I	I_3	Quark composition	Barions $J^P = 3/2^+$	Y	I	I_3	Quark composition
Δ^{++}	1	3/2	3/2	uuu	Σ^{*0}	0	1	0	uds
Δ^{+}	1	3/2	1/2	uud	Σ^{*-}	0	1	−1	dds
Δ^{0}	1	3/2	−1/2	udd	Ξ^{*0}	−1	1/2	1/2	uss
Δ^{-}	1	3/2	−3/2	ddd	Ξ^{*-}	−1	1/2	−1/2	dss
Σ^{*+}	0	1	1	uus	Ω^{-}	−2	0	0	sss

Table 36.7 Complete table of mesons

Light mesons without flavor $(S = C = B = 0)$				Strange mesons $(S = \pm 1,\ C = B = 0)$		$c\bar{c}$-mesons	
	$I^G(J^{PC})$		$I^G(J^{PC})$		$I^G(J^{PC})$		$I^G(J^{PC})$
$\bullet\pi^\pm$	$1^-(0^-)$	$\bullet\pi_2(1670)$	$1^-(2^{-+})$	$\bullet K^\pm$	$1/2(0^-)$	$\bullet\eta_c(1S) =$	$0^+(0^{-+})$
$\bullet\pi^0$	$1^-(0^{-+})$	$\bullet\phi(1680)$	$0^-(1^{--})$	$\bullet K^0$	$1/2(0^-)$	$\eta_c(2980)$	
$\bullet\eta$	$0^+(0^{-+})$	$\bullet\rho_3(1690)$	$1^+(3^{--})$	$\bullet K_S^0$	$1/2(0^-)$	$\bullet J/\psi(1S) =$	$0^-(1^{--})$
$\bullet\rho(770)$	$1^+(1^{--})$	$\bullet\rho(1700)$	$1^+(1^{--})$	$\bullet K_L^0$	$1/2(0^-)$	$J/\psi(3097)$	
$\bullet a_0(980)$	$1^-(0^{++})$	$\eta(1760)$	$0^+(0^{-+})$	$\bullet K^*(1410)$	$1/2(1^-)$	$\chi_{c1}(3510)$	
$\bullet\phi(1020)$	$0^-(1^{--})$	$\pi(1770)$	$1^-(0^{-+})$	$\bullet K_0^*(1430)$	$1/2(0^+)$	$\bullet\chi_{c2}(1P) =$	$0^+(2^{++})$
$\bullet h_1(1170)$	$0^-(1^{+-})$	$\pi(1775)$	$1^-(?^{\ -+})$	$\bullet K_2^*(1430)$	$1/2(2^+)$	$\chi_{c2}(3555)$	
$\bullet b_1(1235)$	$1^+(1^{+-})$	$f_2(1810)$	$0^+(2^{++})$	$K(1460)$	$1/2(0^-)$	$\eta_c(2S) =$	$?^{\,?}(?^{\ ?+})$
$f_0(1240)$	$0^+(0^{++})$	$X(1814)$	$1^-(?^{\ ?\ ?})$	$K_2(1580)$	$1/2(2^-)$	$\eta_c(3590)$	
$\bullet a_1(1260)$	$1^-(1^{++})$	$\bullet\phi_3(1850)$	$0^-(3^{--})$	$K_1(1650)$	$1/2(1^+)$	$\bullet\psi(2S) =$	$0^-(1^{--})$
$\bullet f_2(1270)$	$0^+(2^{++})$	$\eta_2(1870)$	$0^+(2^{-+})$	$\bullet K^*(1680)$	$1/2(1^-)$	$\psi(3685)$	
$\bullet f_1(1285)$	$0^+(1^{++})$	$X(1910)$	$0^+(?^{\ ?+})$	$\bullet K_2(1770)$	$1/2(2^-)$	$\bullet\psi(3770)$	$?^{\,?}(1^{--})$
$\bullet\eta(1295)$	$0^+(0^{-+})$	$X(1950)$	$?^{\,?}(?^{?\ ?})$	$\bullet K_3^*(1780)$	$1/2(3^-)$	$\bullet\psi(4040)$	$?^{\,?}(1^{--})$
$\bullet\pi(1300)$	$1^-(0^{-+})$	$\bullet f_2(2010)$	$0^+(2^{++})$	$K(1830)$	$1/2(0^-)$	$\bullet\psi(4160)$	$?^{\,?}(1^{--})$
$a_0(1320)$	$1^-(0^{++})$	$a_4(2040)$	$1^-(4^{++})$	$K_0^*(1950)$	$1/2(0^+)$	$\bullet\psi(4415)$	$?^{\,?}(1^{--})$
$\bullet a_2(1320)$	$1^-(2^{++})$	$a_3(2050)$	$1^-(3^{++})$	$K_2^*(1980)$	$1/2(2^+)$	**b$\bar{\text{b}}$-mesons**	
$h_1(1380)$	$?^-(1^{+\ ?})$	$\bullet f_4(2050)$	$0^+(4^{++})$	$\bullet K_4^*(2045)$	$1/2(4^+)$	$\bullet\Upsilon(1S) =$	$?^{\,?}(1^{--})$

Table 36.7 Complete table of mesons *(continued)*

Light mesons without flavor ($S = C = B = 0$)				Strange mesons ($S = \pm 1,\ C = B = 0$)		$c\bar{c}$-mesons	
	$I^G(J^{PC})$		$I^G(J^{PC})$		$I^G(J^{PC})$		$I^G(J^{PC})$
$\bullet\omega(1390)$	$0^-(1^{--})$	$\eta(2100)$	$0^+(0^{-+})$	$K_2(2250)$	$1/2(2^-)$	$\Upsilon(9460)$	
$\bullet f_0(1400)$	$0^+(0^{++})$	$\pi_2(2100)$	$1^-(2^{-+})$	$K_3(2320)$	$1/2(3^+)$	$\bullet\chi_{b0}(1P) =$	$?\ ^?(0^{++})$
$\rho(1405)$	$1^-(1^{-+})$	$\rho(2110)$	$1^+(1^{--})$	$K_5^*(2380)$	$1/2(5^-)$	$\chi_{b0}(9860)$	
$\bullet f_1(1420)$	$0^+(1^{++})$	$f_2(2150)$	$0^+(2^{++})$	$K_4(2500)$	$1/2(4^-)$	$\bullet\chi_{b1}(1P)$	$?\ ^?(1^{++})$
$f_2(1430)$	$0^+(2^{++})$	$\rho(2150)$	$1^+(1^{--})$	**Charmed mesons**		$\chi_{b1}(9890)$	
$\bullet\eta(1440)$	$0^+(0^{-+})$	$f_2(2175)$	$0^+(2^{++})$	($C = \pm 1$)		$\bullet\chi_{b2}(1P) =$	$?\ ^?(2^{++})$
$\bullet\rho(1450)$	$1^+(1^{--})$	$X(2200)$	$?^?(\text{even}^{++})$	$\bullet D^{\pm}$	$1/2(0^-)$	$\chi_{b2}(9915)$	
$\bullet f_1(1510)$	$0^+(1^{++})$	$f_4(2220)$	$0^+(4^{++})$	$\bullet D^0$	$1/2(0^-)$	$\bullet\Upsilon(2S) =$	$?\ ^?(1^{--})$
$f_2(1520)$	$0^+(2^{++})$	$\rho(2250)$	$1^+(3^{--})$	$\bullet D^*(2010)^{\pm}$	$1/2(1^-)$	$\Upsilon(10023)$	
$f_0(1525)$	$0^+(0^{++})$	$\bullet f_2(2300)$	$0^+(2^{++})$	$\bullet D^*(2010)^0$	$1/2(1^-)$	$\bullet\chi_{b0}(2P) =$	$?\ ^?(0^{++})$
$\bullet f_2'(1525)$	$0^+(2^{++})$	$f_4(2300)$	$0^+(4^{++})$	$\bullet D_1(2420)^0$	$1/2(1^+)$	$\chi_{b0}(10235)$	
$\bullet f_0(1590)$	$0^+(0^{++})$	$\bullet f_2(2340)$	$0^+(2^{++})$	$D_J(2440)^{\pm}$	$1/2(?\ ^?)$	$\bullet\chi_{b1}(2P) =$	$?\ ^?(1^{++})$
$\bullet\omega(1600)$	$0^-(1^{--})$	$\rho_5(2350)$	$1^+(5^{--})$	$\bullet D_2^*(2460)^0$	$1/2(2^+)$	$\chi_{b1}(10255)$	
$X(1600)$	$2^+(2^{++})$	$a_6(2450)$	$1^-(6^{++})$	$D_J^*(2470)^{\pm}$	$1/2(?\ ^?)$	$\bullet\chi_{b2}(2P) =$	$?\ ^?(2^{++})$
$f_2(1640)$	$0^+(2^{++})$	$f_6(2510)$	$0^+(6^{++})$	**Charmed strange**		$\chi_{b2}(10270)$	
$X(1650)$	$1^-(?^?\ ^?)$	$X(3100)$	$?\ ^?(?^?\ ^?)$	**mesons** ($C = S = \pm 1$)		$\bullet\Upsilon(3S) =$	$?\ ^?(1^{--})$
$\bullet\omega_3(1670)$	$0^-(3^{--})$	$X(3250)$	$?\ ^?(?^?\ ^?)$	$\bullet D_s^{\pm}$	$0(0^-)$	$\Upsilon(10355)$	
				$\bullet D_s^{*\pm}$	$?(?\ ^?)$	$\bullet\Upsilon(4S) =$	$?\ ^?(1^{--})$
				$\bullet D_{s1}(2536)^{\pm}$	$0(1^+)$	$\Upsilon(10580)$	
				$D_{sJ}(2564)^{\pm}$	$?(?\ ^?)$	$\bullet\Upsilon(10860)$	$?\ ^?(1^{--})$
				B-mesons ($B = \pm 1$)		$\bullet\Upsilon(11020)$	$?\ ^?(1^{--})$
				$\bullet B^{\pm}$	$1/2(0^-)$		
				$\bullet B^0$	$1/2(0^-)$		
				$\bullet B^*$	$1/2(1^-)$		
• – reliably established particles				B_s^0	$?\ (?\ ^?)$		

Table 36.8 Complete table of baryons

$\bullet p$	P_{11}	$\bullet\Delta(1232)$	P_{33}	$\bullet\Lambda$	P_{01}	$\bullet\Sigma^+$	P_{11}	$\bullet\Xi^0$	P_{11}
$\bullet n$	P_{11}	$\bullet\Delta(1600)$	P_{33}	$\bullet\Lambda(1405)$	S_{01}	$\bullet\Sigma^0$	P_{11}	$\bullet\Xi^-$	P_{11}
$\bullet N(1440)$	P_{11}	$\bullet\Delta(1620)$	S_{31}	$\bullet\Lambda(1520)$	D_{03}	$\bullet\Sigma^-$	P_{11}	$\bullet\Xi(1530)$	P_{13}
$\bullet N(1520)$	D_{13}	$\bullet\Delta(1700)$	D_{33}	$\bullet\Lambda(1600)$	P_{01}	$\bullet\Sigma(1385)$	P_{13}	$\Xi(1620)$	
$\bullet N(1535)$	S_{11}	$\Delta(1750)$	P_{31}	$\bullet\Lambda(1670)$	S_{01}	$\Sigma(1480)$		$\bullet\Xi(1690)$	
$\bullet N(1650)$	S_{11}	$\bullet\Delta(1900)$	S_{31}	$\bullet\Lambda(1690)$	D_{03}	$\Sigma(1560)$		$\bullet\Xi(1820)$	D_{13}
$\bullet N(1675)$	D_{15}	$\bullet\Delta(1905)$	F_{35}	$\bullet\Lambda(1800)$	S_{01}	$\Sigma(1580)$	D_{13}	$\bullet\Xi(1950)$	
$\bullet N(1680)$	F_{15}	$\bullet\Delta(1910)$	P_{31}	$\bullet\Lambda(1810)$	P_{01}	$\Sigma(1620)$	S_{11}	$\bullet\Xi(2030)$	
$\bullet N(1700)$	D_{13}	$\bullet\Delta(1920)$	P_{33}	$\bullet\Lambda(1820)$	F_{05}	$\bullet\Sigma(1660)$	P_{11}	$\Xi(2120)$	
$\bullet N(1710)$	P_{11}	$\bullet\Delta(1930)$	D_{35}	$\bullet\Lambda(1830)$	D_{05}	$\bullet\Sigma(1670)$	D_{13}	$\Xi(2250)$	
$\bullet N(1720)$	P_{13}	$\Delta(1940)$	D_{33}	$\bullet\Lambda(1890)$	P_{03}	$\Sigma(1690)$		$\Xi(2370)$	
$N(1900)$	P_{13}	$\bullet\Delta(1950)$	F_{37}	$\Lambda(2000)$		$\bullet\Sigma(1750)$	S_{11}	$\Xi(2500)$	
$N(1990)$	F_{17}	$\Delta(2000)$	F_{35}	$\Lambda(2020)$	F_{07}	$\Sigma(1770)$	P_{11}		
$N(2000)$	F_{15}	$\Delta(2150)$	S_{31}	$\bullet\Lambda(2100)$	G_{07}	$\bullet\Sigma(1775)$	D_{15}	$\bullet\Omega^-$	
$N(2080)$	D_{13}	$\Delta(2200)$	G_{37}	$\bullet\Lambda(2110)$	F_{05}	$\Sigma(1840)$	P_{13}	$\bullet\Omega(2250)^-$	
$N(2090)$	S_{11}	$\Delta(2300)$	H_{39}	$\Lambda(2325)$	D_{03}	$\Sigma(1880)$	P_{11}	$\Omega(2380)^-$	
$N(2100)$	P_{11}	$\Delta(2350)$	D_{35}	$\bullet\Lambda(2350)$	H_{09}	$\bullet\Sigma(1915)$	F_{15}	$\Omega(2470)^-$	
$\bullet N(2190)$	G_{17}	$\Delta(2390)$	F_{37}	$\Lambda(2585)$		$\bullet\Sigma(1940)$	D_{13}		
$N(2200)$	D_{15}	$\Delta(2400)$	G_{39}			$\Sigma(2000)$	S_{11}	$\bullet\Lambda_c^+$	
$\bullet N(2220)$	H_{19}	$\bullet\Delta(2420)$	$H_{3,11}$			$\bullet\Sigma(2030)$	F_{17}	$\bullet\Sigma_c(2455)$	
$\bullet N(2250)$	G_{19}	$\Delta(2750)$	$I_{3,13}$			$\Sigma(2070)$	F_{15}	$\bullet\Xi_c^+$	
$\bullet N(2600)$	$I_{1,11}$	$\Delta(2950)$	$K_{3,15}$			$\Sigma(2080)$	P_{13}	$\bullet\Xi_c^0$	
$N(2700)$	$K_{1,13}$					$\Sigma(2100)$	G_{17}	Ω_c^0	
• – reliably established particles						$\Sigma(2455)$		$\bullet\Lambda_b^0$	

36.6 Quantum Chromodynamics

The theory of interacting quarks and gluons is termed quantum chromodynamics (based on the Greek word "chromos" meaning color). It has its origin in the invariance principle of the quark field relative to the $SU(3)_c$ transformation group in a color space: $q_\alpha \to q_\beta = S_{\beta\alpha} q_\alpha$, $\alpha, \beta = 1,2,3$ indicate the quark color, where q_α is the quark wave function, i.e. a Dirac spinor; S is the $SU(3)$ group operator – a 3×3 matrix. Its general form is $S = \exp[ig_i(x)\lambda_i/2]$. Here λ_i are the eight Gell-Mann matrices, g_i are eight arbitrary real parameters. If g_i depends on the World point x, then the transformation $q_\alpha \to q_\beta$ is termed a local gauge transformation. The kinetic term in the Lagrangian of the free quark field $\bar{q}\gamma_\mu i\partial_\mu q$ contains a derivative, so it acquires an addition under local gauge transformation. The invariance of the Lagrangian can be retained by a compensating term $L_{\text{int}} = g\bar{q}\gamma_\mu A_\mu q$. It has the same form as the interaction energy of the spinor field q with a vector field A of zero mass. The matrix A is of the general form $A_\mu = A^i_\mu \lambda_i$. Here, the quantities A^i_μ, $\mu = 1, \ldots, 4$, $i = 1, \ldots, 8$ represent potentials of the gluon field. The index μ points to its vector properties, like the photon, and i indicates that there are eight such fields. Thus, the gluon is an octet of the $SU(3)$ group, and it has 8 color states. Like the quark, it carries the strong interaction charge. In this aspect it differs drastically from the photon. The self-action of gluons and the polarization of vacuum results in antiscreening of the strong interaction charge. For this reason, QCD acquires a very important property: $\alpha_s(q) \to 0$ as $q \to \infty \quad (r \to 0)$, which is known as asymptotic freedom meaning that quarks interact weakly at small distances. QCD calculations within perturbation theory well describe reactions involving large momentum transfers. Soft hadron processes ($q < 1\ \text{fm}^{-1}$, $\alpha_s \geq 1$) where perturbation theory cannot be applied remain in the region of phenomenology, although semiquantitative arguments based on QCD are also applied here.

36.7 The Unified Theory of Electromagnetic and Weak Interactions

Gauge invariance of the field serves as an effective means in searching for internal particle symmetries and in applying them to constructing the theory. We have seen above how the requirement of invariance of the quark field with respect to local gauge transformation in a color space leads to the theory of interacting quarks and gluons. The same principle has made possible unification of the electromagnetic and weak interactions in a sole theory. Now let us write down the known fundamental fermions in a symmetric form, as three families:

u	c	t
d	s	b
ν_e	ν_μ	ν_τ
e	μ	τ

The first family involves the (ud) quark pair and the $(\nu_e e)$ lepton pair. The second family includes the (cs), $(\nu_\mu \mu)$ pairs, and the third family, (tb) and $(\nu_\tau \tau)$. The three upper quarks have charges equal to $(+2/3)$, and the three lower ones, to $(-1/3)$. Experiments reveal all the families to participate identically in the weak interaction. This fact is reflected in the structure of the Lagrangian, where the total lepton and quark current includes them in a symmetric form and with identical weights, i.e. with the same coupling constant: $j_W = \bar{e}\nu_e + \bar{\mu}\nu_\mu + \bar{\tau}\nu_\tau + \bar{u}d' + \bar{c}s' + \bar{t}b'$. Here, all the lower quarks are "rotated" (or mixed) with the aid of the Cabbibo-Maskawa matrix $d'_i = V_{ij} \cdot d_j$, where d'_j stands for the lower quark of the j family. The matrix V is unitary. It is expressed through three rotation angles and one common phase, which appears as free parameters of the theory.

The theory of unified electromagnetic and weak interactions is constructed on the base of the $SU(2)\cdot U(1)$ group. The $SU(2)$ group acts in the space of wave functions (doublets) of the leptons and quarks, $(\nu_i l_i)$, $(u_i d_i')$. Here, $i = 1,2,3$ indicates the family number. This group is called the weak isospin group. The unitary operator $U(1)$ transforms the phase. It is quite similar to the gauge transformation in electrodynamics. The requirement of invariance under the $SU(2)$ local gauge transformation generates three vector fields W (in accordance with the amount of generators of the $SU(2)$ group). The $U(1)$ gauge transformation leads, accordingly, to a single vector field B. They have the coupling constants g and g'. Until the W bosons are massless, the above scheme corresponds to nothing in Nature. For them to acquire mass a mechanism was put forward of spontaneous symmetry breaking, under which there appears a condensate of the scalar field in vacuum, and W acquires mass. Two of them, W_1 and W_2, turn out to be charged, while the third, W_3, is neutral. The neutral components of the fields W and B undergo mixing:

$$A = B\cos\theta_W + W_3\sin\theta_W; \qquad Z = -B\sin\theta_W + W_3\cos\theta_W.$$

The field A is interpreted to be a photon, and Z^0 does the neutral boson of weak interaction. The mixing angle θ_W turns out to be a free parameter of the theory. The parameters of the theory are related in the following way:

$$\tan\theta_W = g'/g; \quad e = g\sin\theta_W; \quad G_F = \sqrt{2}g^2/8M_W^2; \quad M_W = \sqrt{\rho}M_{Z_0}\cos\theta_W.$$

The quantity $\rho \approx 1$ and depends on the mass of the t quark. The relationship between electric charge and the weak interaction coupling constant g convincingly testifies in favor of the unification of two quite differing forces of Nature into a sole theory.

The quanta of a scalar field responsible for $SU(2)$ symmetry breaking are termed the Higgs bosons. They have not yet been observed experimentally.

The theory of electroweak interaction explains a great number of properties of electromagnetic and weak processes and sets down a program for further investigations [3].

36.8 Quantum Numbers and the Most Important Particle Decays

Notations:

m – mass	d – electric dipole moment
q – charge	$I^G(J^{PC})$ – quantum number
e – the electron charge	I – isotopic spin
τ – mean lifetime	G – G-parity
Γ – width of resonance	J – spin
μ – magnetic moment	P – parity
μ_B – Bohr's magneton	C – charge parity

Γ_i/Γ – relative partial width of channel i or decay branching ratio for channel i

p (in MeV/c) – maximum momentum of decay products in center-of-mass system of the decaying particle

CL – confidence level in statistical data analysis procedure

Gauge bosons

γ $I(J^{PC}) = 0, 1(1^{--})$

$m < 3 \cdot 10^{-33}$ MeV
$q < 2 \cdot 10^{-32}\ e$
τ = stable

W $J = 1$

$m = 80.22 \pm 0.26$ GeV
$m_{W^+} - m_{W^-} = -0.2 \pm 0.6$ GeV
$\Gamma = 2.12 \pm 0.11$ GeV
Decay channels W^- are charge-conjugate to the decay channels W^+

Decay channels W^+	Γ_i/Γ	p
$e^+\nu$	$(10.5 \pm 0.9)\%$	40300
$\mu^+\nu$	$(10.5 \pm 1.9)\%$	40300
$\tau^+\nu$	$(10.6 \pm 1.6)\%$	40300

Z $J = 1$

$m = 91.173 \pm 0.020$ GeV
$m_Z - m_W = 10.96 \pm 0.26$ GeV
$\Gamma = 2.487 \pm 0.010$ GeV

Decay channels Z	Γ_i/Γ	p
e^+e^-	$(3.345 \pm 0.025)\%$	45600
$\mu^+\mu^-$	$(3.34 \pm 0.04)\%$	45600
$\tau^+\tau^-$	$(3.32 \pm 0.04)\%$	45600
$\ell^+\ell^-$	$(3.337 \pm 0.022)\%$	–
$\nu\bar{\nu}$ (or other nonvisible channels)	$(20.2 \pm 0.4)\%$	45600
Hadrons	$(69.80 \pm 0.33\%$	–
$(u\bar{u} + c\bar{c})/2$	$(13.3 \pm 3.5)\%$	–
$(d\bar{d} + s\bar{s} + b\bar{b})/3$	$(14.4 \pm 2.4)\%$	–
$c\bar{c}$	$(12.6 \pm 2.1)\%$	–
$b\bar{b}$	$(15.2 \pm 1.0)\%$	–

Leptons

ν_e $J = 1/2$

$m < 7.3$ eV, $CL = 90\%$
$\tau/m_{\nu_e} > 300$ c/eV, $CL = 90\%$
$\mu < 1.08 \cdot 10^{-9}\ \mu_B$, $CL = 90\%$

ν_μ $J = 1/2$

$m < 0.27$ MeV, $CL = 90\%$
$\tau/m_{\nu_\mu} > 15.4$ s/eV, $CL = 90\%$
$\mu < 7.4 \cdot 10^{-10}\ \mu_B$, $CL = 90\%$

According to the standard model, the number of light ($m_\nu \ll m_Z$) neutrino types including ν_e, ν_μ, ν_τ must be $N = 3$. Experimental value derived from measured width of Z-boson is $N = 2.99 \pm 0.04$

ν_τ $J = 1/2$

$m < 35$ MeV, $CL = 95\%$
$\mu < 4 \cdot 10^{-6}\ \mu_B$, $CL = 90\%$

e $J = 1/2$

$m = 0.51099906 \pm 0.00000015$ MeV
$\tau > 1.9 \cdot 10^{23}$ years, $CL = 68\%$
$\mu = 1.001159652193 \pm 0.000000000010\ \mu_B$
$d = (-0.3 \pm 0.8) \cdot 10^{-26}\ e\cdot$cm

μ $J = 1/2$

$m = 105.658389 \pm 0.000034$ MeV
$\tau = (2.19703 \pm 0.00004) \cdot 10^{-6}$ s
$\mu = 1.001165923 \pm 0.000000008\ \mu_B$
$d = (3.7 \pm 3.4) \cdot 10^{-19}\ e\cdot$cm
Decay channels μ^+ are charge-conjugate to decay channels μ^-

Decay channels μ^-	Γ_i/Γ	p
$e^-\bar{\nu}_e\nu_\mu$	$\sim 100\%$	53
$e^-\bar{\nu}_e\nu_\mu\gamma$	$(1.4 \pm 0.4)\%$	53
$e^-\bar{\nu}_e\nu_\mu e^+e^-$	$(3.4 \pm 0.4) \cdot 10^{-5}$	53

τ $J = 1/2$

$m = 1784.1^{+2.7}_{-3.6}$ MeV
$\tau = (0.305 \pm 0.006) \cdot 10^{-12}$ s
Decay channels τ^+ are charge-conjugate to decay channels τ^-
$h^\pm \equiv \pi^\pm, K^\pm, h^0 \equiv \pi^0, \gamma$

Decay channels τ^-	Γ_i/Γ	p
(particle)$^-$ $\geq 0\ h^0\nu_\tau$ ("1-prong decay")	$(85.82 \pm 0.25)\%$	–
$\mu^-\bar{\nu}_\mu\nu_\tau$	$(17.58 \pm 0.27)\%$	889
$\mu^-\bar{\nu}_\mu\nu_\tau\gamma$	$(2.3 \pm 1.1) \cdot 10^{-3}$	–
$e^-\bar{\nu}_e\nu_\tau$	$(17.93 \pm 0.26)\%$	892
$h^- \geq 0\ h^0\nu_\tau$	$(50.3 \pm 0.4)\%$	–
$h^-\nu_\tau$	$(12.7 \pm 0.4)\%$	–
$\pi^-\nu_\tau$	$(11.6 \pm 0.4)\%$	887
$K^- \geq 0\ h^0\nu_\tau$	$(1.68 \pm 0.24)\%$	–
$K^-\nu_\tau$	$(6.7 \pm 2.3) \cdot 10^{-3}$	824
$h^- \geq 1\ h^0\nu_\tau$	$(37.6 \pm 0.5)\%$	–
$h^-\pi^0\nu_\tau$	$(24.4 \pm 0.6)\%$	–
$\pi^-\pi^0\nu_\tau$	$(24.0 \pm 0.6)\%$	881
$h^- \geq 2\ \pi^0\nu_\tau$	$(13.2 \pm 0.7)\%$	–
$h^-2\pi^0\nu_\tau$	$(10.3 \pm 0.9)\%$	866
$h^- \geq 3\ \pi^0\nu_\tau$	$(2.7 \pm 0.9)\%$	–
$2h^-h^+ \geq 0\ h^0\nu_\tau$	$(14.06 \pm 0.25)\%$	–

Decay channels τ^- *(continued)*	Γ_i/Γ	p
("3-prong decay")		
$h^-h^-h^+\nu_\tau$	$(8.4 \pm 0.4)\%$	–
$\pi^-\pi^-\pi^+\nu_\tau$	$(5.6 \pm 0.7)\%$	864
$\pi^-\rho^0\nu_\tau$	$(5.4 \pm 1.7)\%$	719
$h^-h^-h^+ \geq 1\ h^0\nu_\tau$	$(5.3 \pm 0.4)\%$	–
$\omega\pi^- \geq 0\ h^0\nu_\tau$	$(1.6 \pm 0.4)\%$	–
$\omega\pi^-\nu_\tau$	$(1.6 \pm 0.5)\%$	713
$K^-h^+h^- \geq 0\ h^0\nu_\tau$	$< 6 \cdot 10^{-3}$	–
$3h^-2h^+ \geq 0\ h^0\nu_\tau$	$(1.11 \pm 0.24) \cdot 10^{-3}$	–
("5-prong decay")		
$3h^-2h^+\nu_\tau$	$(5.6 \pm 1.6) \cdot 10^{-4}$	–
$3h^-2h^+\pi^0\nu_\tau$	$(5.1 \pm 2.2) \cdot 10^{-4}$	–
$4h^-3h^+ \geq 0\ h^0\nu_\tau$	$< 1.9 \cdot 10^{-4}$	–
("7-prong decay")		
$K^0h^- \geq 0\ h^0\nu_\tau$	$(1.30 \pm 0.30)\%$	–
$K^*(892)^- \geq 0\ h^0\nu_\tau$	$(1.43 \pm 0.17)\%$	–
$K^*(892)^0K^- \geq 0\ h^0\nu_\tau$	$(3.2 \pm 1.4) \cdot 10^{-3}$	–
$\overline{K}^*(892)^0\pi^- \geq 0\ h^0\nu_\tau$	$(3.8 \pm 1.7) \cdot 10^{-3}$	–
$K^*(892)^-\nu_\tau$	$(1.42 \pm 0.18)\%$	669

Light flavorless mesons ($S = C = B = 0$)

$\pi^\pm$ $\quad I^G(J^P) = 1^-(0^-)$

$m = 139.5679 \pm 0.0007$ MeV
$\tau = (2.6030 \pm 0.0024) \cdot 10^{-8}$ s
Decay channels π^- are charge-conjugate to the decay channels π^+

Decay channels π^+	Γ_i/Γ	p
$\mu^+\nu_\mu$	$(99.98782 \pm 0.00014)\%$	30
$\mu^+\nu_\mu\gamma$	$(1.24 \pm 0.25) \cdot 10^{-4}$	30
$e^+\nu_e$	$(1.218 \pm 0.014) \cdot 10^{-4}$	70
$e^+\nu_e\gamma$	$(1.61 \pm 0.23) \cdot 10^{-7}$	70
$e^+\nu_e\pi^0$	$(1.025 \pm 0.034) \cdot 10^{-8}$	4
$e^+\nu_e e^+e^-$	$(3.2 \pm 0.5) \cdot 10^{-9}$	70

π^0 $\quad I^G(J^{PC}) = 1^-(0^{-+})$

$m = 134.9743 \pm 0.0008$ MeV
$m_{\pi^\pm} - m_{\pi^0} = 4.5936 \pm 0.0005$ MeV
$\tau = (8.4 \pm 0.6) \cdot 10^{-17}$ s

Decay channels π^0	Γ_i/Γ	p
2γ	$(98.798 \pm 0.032)\%$	67
$e^+e^-\gamma$	$(1.198 \pm 0.032)\%$	67
γ positronium	$(1.82 \pm 0.29) \cdot 10^{-9}$	67
$e^+e^+e^-e^-$	$(3.14 \pm 0.30) \cdot 10^{-5}$	67

η $\quad I^G(J^{PC}) = 0^+(0^{-+})$

$m = 547.45 \pm 0.19$ MeV
$\Gamma = 1.19 \pm 0.11$ keV

Decay channels η	Γ_i/Γ	p
Neutral channels	$(70.8 \pm 0.8)\%$	–
2γ	$(38.9 \pm 0.5)\%$	274
$3\pi^0$	$(31.9 \pm 0.4)\%$	180
$\pi^0 2\gamma$	$(7.1 \pm 1.4) \cdot 10^{-4}$	258
Charged channels	$(29.2 \pm 0.8)\%$	–
$\pi^+\pi^-\pi^0$	(23.6 ± 0.6)	175
$\pi^+\pi^-\gamma$	$(4.88 \pm 0.15)\%$	236
$e^+e^-\gamma$	$(5.0 \pm 1.2) \cdot 10^{-3}$	274
$\mu^+\mu^-\gamma$	$(3.1 \pm 0.4) \cdot 10^{-4}$	253
$\mu^+\mu^-$	$(6.5 \pm 2.1) \cdot 10^{-6}$	253

$\rho(770)$ $\quad I^G(J^{PC}) = 1^+(1^{--})$

$m = 768.1 \pm 0.5$ MeV
$\Gamma = 151.5 \pm 1.2$ MeV
$\Gamma_{ee} = 6.77 \pm 0.32$ keV

Decay channels $\rho(770)$	Γ_i/Γ	p
$\pi\pi$	$\sim 100\%$	358
$\rho(770)^\pm$– decays		
$\pi^\pm\gamma$	$(4.5 \pm 0.5) \cdot 10^{-4}$	371
$\pi^\pm\eta$	$< 8 \cdot 10^{-3}$	144
$\pi^\pm\pi^+\pi^-\pi^0$	$< 2.0 \cdot 10^{-3}$	249
$\rho(770)^0$– decays		
$\pi^+\pi^-\gamma$	$(9.9 \pm 1.6) \cdot 10^{-3}$	358
$\pi^0\gamma$	$(7.9 \pm 2.0) \cdot 10^{-4}$	372
$\eta\gamma$	$(3.8 \pm 0.7) \cdot 10^{-4}$	188
$\mu^+\mu^-$	$(4.60 \pm 0.28) \cdot 10^{-5}$	369
e^+e^-	$(4.44 \pm 0.21) \cdot 10^{-5}$	384

$\omega(783)$ $\quad I^G(J^{PC}) = 0^-(1^{--})$

$m = 781.95 \pm 0.14$ MeV
$\Gamma = 8.43 \pm 0.10$ MeV
$\Gamma_{ee} = 0.60 \pm 0.02$ keV

Decay channels $\omega(783)$	Γ_i/Γ	p
$\pi^+\pi^-\pi^0$	$(88.8 \pm 0.6)\%$	327
$\pi^0\gamma$	$(8.5 \pm 0.5)\%$	379
$\pi^+\pi^-$	$(2.21 \pm 0.30)\%$	365
$\pi^0e^+e^-$	$(5.9 \pm 1.9) \cdot 10^{-4}$	379
$\eta\gamma$	$(4.7^{+2.2}_{-1.8}) \cdot 10^{-4}$	198
$\pi^0\mu^+\mu^-$	$(9.6 \pm 2.3) \cdot 10^{-5}$	349
e^+e^-	$(7.15 \pm 0.19) \cdot 10^{-5}$	391

$\eta'(958)$ $\quad I^G(J^{PC}) = 0^+(0^{-+})$

$m = 957.75 \pm 0.14$ MeV
$\Gamma = 0.198 \pm 0.019$ MeV

Decay channels $\eta'(958)$	Γ_i/Γ	p
$\pi^+\pi^-\eta$	$(44.1 \pm 1.7)\%$	231
$\rho^0\gamma$	$(30.0 \pm 1.4)\%$	171

Decay channels $\eta'(958)$ *(continued)*	Γ_i/Γ	p
$\pi^0\pi^0\eta$	$(20.6 \pm 1.2)\%$	238
$\omega\gamma$	$(3.00 \pm 0.30)\%$	160
$\gamma\gamma$	$(2.17 \pm 0.17)\%$	479
$3\pi^0$	$(1.53 \pm 0.26) \cdot 10^{-3}$	430
$\mu^+\mu^-\gamma$	$(1.06 \pm 0.27) \cdot 10^{-4}$	467

$f_0(975)$ $I^G(J^{PC}) = 0^+(0^{++})$

$m = 974.1 \pm 2.5$ MeV
$\Gamma = 47 \pm 9$ MeV

Decay channels $f_0(975)$	Γ_i/Γ	p
$\pi\pi$	$(78.1 \pm 2.4)\%$	467
$K\overline{K}$	$(21.9 \pm 2.4)\%$	–
$\gamma\gamma$	$(1.19 \pm 0.33) \cdot 10^{-5}$	487

$a_0(980)$ $I^G(J^{PC}) = 1^+(0^{++})$

$m = 982.7 \pm 2.0$ MeV
$\Gamma = 57 \pm 11$ MeV

Decay channels $a_0(980)$	Γ_i/Γ	p
$\eta\pi$	observed	319
$K\overline{K}$	observed	–
$\gamma\gamma$	observed	491

$\phi(1020)$ $I^G(J^{PC}) = 0^-(1^{--})$

$m = 1019.413 \pm 0.008$ MeV
$\Gamma = 4.43 \pm 0.06$ MeV
$\Gamma_{ee} = 1.37 \pm 0.05$ keV

Decay channels $\phi(1020)$	Γ_i/Γ	p
K^+K^-	$(49.1 \pm 0.8)\%$	127
$K_L^0K_S^0$	$(34.4 \pm 0.7)\%$	110
$\rho\pi$	$(12.9 \pm 0.7)\%$	183
$\pi^+\pi^-\pi^0$	$(2.4 \pm 0.9)\%$	462
$\eta\gamma$	$(1.28 \pm 0.06)\%$	362
$\pi^0\gamma$	$(1.31 \pm 0.13) \cdot 10^{-3}$	501
e^+e^-	$(3.09 \pm 0.07) \cdot 10^{-4}$	510
$\mu^+\mu^-$	$(2.48 \pm 0.34) \cdot 10^{-4}$	499
ηe^+e^-	$(1.3 {}^{+0.8}_{-0.6}) \cdot 10^{-4}$	362
$\pi^+\pi^-$	$(8 {}^{+5}_{-4}) \cdot 10^{-5}$	490

$h_1(1170)$ $I^G(J^{PC}) = 0^-(1^{+-})$

$m = 1170 \pm 20$ MeV
$\Gamma = 360 \pm 40$ MeV

Decay channels $h_1(1170)$	Γ_i/Γ	p
$\rho\pi$	observed	311

$b_1(1235)$ $I^G(J^{PC}) = 1^+(1^{+-})$

$m = 1232 \pm 10$ MeV
$\Gamma = 155 \pm 8$ MeV

Decay channels $b_1(1235)$	Γ_i/Γ	p
$\omega\pi$	$\sim 100\%$	349
$\pi^\pm\gamma$	$(1.5 \pm 0.4) \cdot 10^{-3}$	608

$a_1(1260)$ $I^G(J^{PC}) = 1^-(1^{++})$

$m = 1260 \pm 30$ MeV
$\Gamma \sim 400$ MeV

Decay channels $a_1(1260)$	Γ_i/Γ	p
$\rho\pi$	$\sim 100\%$	379
$\pi\gamma$	observed	622

$f_2(1270)$ $I^G(J^{PC}) = 0^+(2^{++})$

$m = 1275 \pm 5$ MeV
$\Gamma = 185 \pm 20$ MeV

Decay channels $f_2(1270)$	Γ_i/Γ	p
$\pi\pi$	$(84.9 {}^{+2.5}_{-1.3})\%$	622
$\pi^+\pi^-2\pi^0$	$(6.9 {}^{+1.5}_{-2.7})\%$	562
$K\overline{K}$	$(4.6 \pm 0.5)\%$	403
$2\pi^+2\pi^-$	$(2.8 \pm 0.4)\%$	559
$\eta\eta$	$(4.5 \pm 1.0) \cdot 10^{-3}$	324
$4\pi^0$	$(3.0 \pm 1.0) \cdot 10^{-3}$	564
$\gamma\gamma$	$(1.39 \pm 0.20) \cdot 10^{-5}$	637

$f_1(1285)$ $I^G(J^{PC}) = 0^+(1^{++})$

$m = 1282 \pm 5$ MeV
$\Gamma = 24 \pm 3$ MeV

Decay channels $f_1(1285)$	Γ_i/Γ	p
4π	$(38 \pm 4)\%$	563
$\rho\pi\pi$	observed	342
$\eta\pi\pi$	$(50 \pm 5)\%$	478
$a_0(980)\pi$	$(37 \pm 7)\%$	233
$K\overline{K}\pi$	$(11.9 \pm 1.4)\%$	308
$\phi\gamma$	$(10 \pm 4) \cdot 10^{-4}$	236
$\gamma\gamma$	$(11 \pm 3) \cdot 10^{-5}$	–

$\eta(1295)$ $I^G(J^{PC}) = 0^+(0^{-+})$

$m = 1295 \pm 4$ MeV
$\Gamma = 53 \pm 6$ MeV

Decay channels $\eta(1295)$	Γ_i/Γ	p
$\eta\pi^+\pi^-$	observed	487
$a_0(980)\pi$	observed	245

$\pi(1300)$ $I^G(J^{PC}) = 1^-(0^{-+})$

$m = 1300 \pm 100$ MeV
$\Gamma = 200 - 600$ MeV

Decay channels $\pi(1300)$	Γ_i/Γ	p
$\rho\pi$	observed	407

Decay channels $\pi(1300)$ *(continued)*	Γ_i/Γ	p
3π	observed	612

$a_2(1320)$ $I^G(J^{PC}) = 1^-(2^{++})$

$m = 1318.2 \pm 0.7$ MeV
$\Gamma = 110 \pm 5$ MeV

Decay channels $a_2(1320)$	Γ_i/Γ	p
$\rho\pi$	$(70.1 \pm 2.7)\%$	420
$\eta\pi$	$(14.5 \pm 1.2)\%$	534
$\omega\pi\pi$	$(10.6 \pm 3.2)\%$	362
$K\overline{K}$	$(4.9 \pm 0.8)\%$	437
$\pi^\pm\gamma$	$(2.7 \pm 0.5) \cdot 10^{-3}$	652
$\gamma\gamma$	$(9.5 \pm 0.9) \cdot 10^{-6}$	659

$\omega(1390)$ $I^G(J^{PC}) = 0^-(0^{--})$

$m = 1394 \pm 17$ MeV
$\Gamma = 229 \pm 40$ MeV

Decay channels $\omega(1390)$	Γ_i/Γ	p
$\rho\pi$	$\sim 100\%$	472

$f_0(1400)$ $I^G(J^{PC}) = 0^+(0^{++})$

$m \sim 1400$ MeV
$\Gamma = 150 - 400$ MeV
$\Gamma_{\gamma\gamma} = 5.4 \pm 2.3$ KeV

Decay channels $f_0(1400)$	Γ_i/Γ	p
$\pi\pi$	$(9.3^{+1.9}_{-1.5})\%$	686
$K\overline{K}$	$(7.5 \pm 0.9)\%$	496
$\eta\eta$	observed	435
$\gamma\gamma$	observed	700

$f_1(1420)$ $I^G(J^{PC}) = 0^+(1^{++})$

$m = 1426.1 \pm 1.6$ MeV
$\Gamma = 56.0 \pm 3.0$ MeV

Decay channels $f_1(1420)$	Γ_i/Γ	p
$K\overline{K}\pi$	$\sim 100\%$	438

$\eta(1440)$ $I^G(J^{PC}) = 0^+(0^{-+})$

$m = 1420 \pm 20$ MeV
$\Gamma = 60 \pm 30$ MeV

Decay channels $\eta(1440)$	Γ_i/Γ	p
$K\overline{K}\pi$	observed	433
$\eta\pi\pi$	observed	566
$a_0(980)\pi$	observed	350
4π	observed	640

$\rho(1450)$ $I^G(J^{PC}) = 1^+(1^{--})$

$m = 1465 \pm 25$ MeV
$\Gamma = 310 \pm 60$ MeV

Decay channels $\rho(1450)$	Γ_i/Γ	p
$\pi\pi$	observed	719
4π	observed	665
e^+e^-	observed	732

$f_1(1510)$ $I^G(J^{PC}) = 0^+(1^{++})$

$m = 1512 \pm 4$ MeV
$\Gamma = 35 \pm 15$ MeV

Decay channels $f_1(1510)$	Γ_i/Γ	p
$K\overline{K}^*(892)$	observed	292

$f_2'(1525)$ $I^G(J^{PC}) = 0^+(2^{++})$

$m = 1525 \pm 5$ MeV
$\Gamma = 76 \pm 10$ MeV

Decay channels $f_2'(1525)$	Γ_i/Γ	p
$K\overline{K}$	$(71.2^{+2.0}_{-2.5})\%$	581
$\eta\eta$	$(27.9^{+2.5}_{-2.0})\%$	529
$\pi\pi$	$(8.2 \pm 1.6) \cdot 10^{-3}$	750
$\gamma\gamma$	$(1.23 \pm 0.22) \cdot 10^{-6}$	763

$f_0(1590)$ $I^G(J^{PC}) = 0^+(0^{++})$

$m = 1587 \pm 11$ MeV
$\Gamma = 175 \pm 19$ MeV

Decay channels $f_0(1590)$	Γ_i/Γ	p
$\eta\eta'(958)$	dominant channel	241
$\eta\eta$	observed	573
$4\pi^0$	observed	735

$\omega(1600)$ $I^G(J^{PC}) = 0^-(1^{--})$

$m = 1594 \pm 12$ MeV
$\Gamma = 100 \pm 30$ MeV

Decay channels $\omega(1600)$	Γ_i/Γ	p
$\rho\pi$	observed	602
$\omega\pi\pi$	observed	564
e^+e^-	observed	797

$\omega_3(1670)$ $I^G(J^{PC}) = 0^-(3^{--})$

$m = 1668 \pm 5$ MeV
$\Gamma = 166 \pm 15$ MeV

Decay channels $\omega_3(1670)$	Γ_i/Γ	p
$\rho\pi$	observed	648
$\omega\pi\pi$	observed	614

$\pi_2(1670)$ $I^G(J^{PC}) = 1^-(2^{-+})$

$m = 1670 \pm 20$ MeV

$\Gamma = 250 \pm 20$ MeV
$\Gamma_{ee} = 1.35 \pm 0.26$ KeV

Decay channels $\pi_2(1670)$	Γ_i/Γ	p
$f_2(1270)\pi$	$(56.2 \pm 3.2)\%$	325
$\rho\pi$	$(31 \pm 4)\%$	649
$f_0(1400)\pi$	$(8.7 \pm 3.4)\%$	212
$K\overline{K}^*(892)$	$(4.2 \pm 1.4)\%$	453
$\gamma\gamma$	$(5.4 \pm 1.1) \cdot 10^{-3}$	835

$\phi(1680)$ $I^G(J^{PC}) = 0^-(1^{--})$

$m = 1680 \pm 50$ MeV
$\Gamma = 150 \pm 50$ MeV

Decay channels $\phi(1680)$	Γ_i/Γ	p
$K\overline{K}^*(892)$	dominant channel	462
$K\overline{K}$	observed	680
e^+e^-	observed	840
$K^0_S K\pi$	observed	619

$\rho_3(1690)$ $I^G(J^{PC}) = 1^+(3^{--})$

$m = 1691 \pm 5$ MeV
$\Gamma = 215 \pm 20$ MeV

Decay channels $\rho_3(1690)$	Γ_i/Γ	p
4π	$(71.1 \pm 1.9)\%$	787
$\pi\pi$	$(23.6 \pm 1.3)\%$	834
$K\overline{K}\pi$	$(3.8 \pm 1.2)\%$	628
$K\overline{K}$	$(1.58 \pm 0.26)\%$	686
$\eta\pi^+\pi^-$	observed	728

$\rho(1700)$ $I^G(J^{PC}) = 1^+(1^{--})$

$m = 1700 \pm 20$ MeV
$\Gamma = 235 \pm 50$ MeV

Decay channels $\rho(1700)$	Γ_i/Γ	p
$\rho\pi\pi$	dominant channel	641
$\rho^0\pi^+\pi^-$	observed	641
$\rho^\pm\pi^\pm\pi^0$	observed	642
$2(\pi^+\pi^-)$	observed	792
$\pi^+\pi^-$	observed	838
$K\overline{K}^*(892)$	observed	479
$\eta\rho$	observed	533
$K\overline{K}$	observed	692
e^+e^-	observed	850

$f_0(1710)$ $I^G(J^{PC}) = 0^+(0^{++})$

$m = 1709 \pm 5$ MeV
$\Gamma = 146 \pm 12$ MeV

Decay channels $f_0(1710)$	Γ_i/Γ	p
$K\overline{K}$	observed	697
$\pi\pi$	observed	843

$\phi(1850)$ $I^G(J^{PC}) = 0^-(3^{--})$

$m = 1854 \pm 7$ MeV
$\Gamma = 87^{+28}_{-23}$ MeV

Decay channels $\phi(1850)$	Γ_i/Γ	p
$K\overline{K}$	observed	785
$K\overline{K}^*(892)$	observed	602

$f_2(2010)$ $I^G(J^{PC}) = 0^+(2^{++})$

$m = 2011^{+60}_{-80}$ MeV
$\Gamma = 202 \pm 60$ MeV

Decay channels $f_2(2010)$	Γ_i/Γ	p
$\phi\phi$	observed	–

$f_4(2050)$ $I^G(J^{PC}) = 0^+(4^{++})$

$m = 2049 \pm 10$ MeV
$\Gamma = 203 \pm 12$ MeV

Decay channels $f_4(2050)$	Γ_i/Γ	p
$\omega\omega$	$(25 \pm 6)\%$	662
$\pi\pi$	$(17.0 \pm 1.5)\%$	1015
$K\overline{K}$	$(6.8^{+3.4}_{-1.8}) \cdot 10^{-3}$	898

$f_2(2300)$ $I^G(J^{PC}) = 0^+(2^{++})$

$m = 2297 \pm 28$ MeV
$\Gamma = 149 \pm 40$ MeV

Decay channels $f_2(2300)$	Γ_i/Γ	p
$\phi\phi$	observed	529

$f_2(2340)$ $I^G(J^{PC}) = 0^+(2^{++})$

$m = 2339 \pm 60$ MeV
$\Gamma = 319^{+80}_{-70}$ MeV

Decay channels $f_2(2340)$	Γ_i/Γ	p
$\phi\phi$	observed	573

Strange mesons ($S = \pm 1$, $C = B = 0$)

$K^\pm$ $I(J^P) = \frac{1}{2}(0^-)$

$m = 493.646 \pm 0.009$ MeV
$\tau = (1.2371 \pm 0.0029) \cdot 10^{-8}$ s
Decay channels K^- are charge-conjugate to the decay channels K^+

Decay channels K^+	Γ_i/Γ	p
$\mu^+\nu_\mu$	$(63.51 \pm 0.19)\%$	236
$e^+\nu_e$	$(1.55 \pm 0.07) \cdot 10^{-5}$	247
$\pi^+\pi^0$	$(21.17 \pm 0.16)\%$	205

Decay channels K^+ (continued)	Γ_i/Γ	p
$\pi^+\pi^+\pi^-$	$(5.59 \pm 0.05)\%$	125
$\pi^+\pi^0\pi^0$	$(1.73 \pm 0.04)\%$	133
$\pi^0\mu^+\nu_\mu$	$(3.18 \pm 0.08)\%$	215
$\pi^0 e^+\nu_e$	$(4.82 \pm 0.06)\%$	228
$\pi^0\pi^0 e^+\nu_e$	$(2.1 \pm 0.4) \cdot 10^{-5}$	206
$\pi^+\pi^- e^+\nu_e$	$(3.91 \pm 0.17) \cdot 10^{-5}$	203
$\pi^+\pi^-\mu^+\nu_\mu$	$(1.4 \pm 0.9) \cdot 10^{-5}$	151
$\mu^+\nu_\mu\gamma$	$(5.50 \pm 0.28) \cdot 10^{-3}$	236
$\pi^+\pi^0\gamma$	$(2.75 \pm 0.15) \cdot 10^{-4}$	205
$\pi^+\pi^0\gamma$ (DE)	$(1.8 \pm 0.4) \cdot 10^{-5}$	205
$\pi^+\pi^+\pi^-\gamma$	$(1.04 \pm 0.31) \cdot 10^{-4}$	125

K^0 $\quad I(J^P) = \frac{1}{2}(0^-)$

50% K_S, 50% K_L

$m = 497.671 \pm 0.031$ MeV

$m_{K^0} - m_{K^\pm} = 4.024 \pm 0.032$ MeV

K^0_S $\quad I(J^P) = \frac{1}{2}(0^-)$

$\tau = (0.8922 \pm 0.0020) \cdot 10^{-10}$ s

Decay channels K^0_S	Γ_i/Γ	p
$\pi^+\pi^-$	$(68.61 \pm 0.28)\%$	206
$\pi^0\pi^0$	$(31.39 \pm 0.28)\%$	209
$\pi^+\pi^-\gamma$	$(1.85 \pm 0.10) \cdot 10^{-5}$	206

K^0_L $\quad I(J^P) = \frac{1}{2}(0^-)$

$m_{K_L} - m_{K_S} = (3.522 \pm 0.016) \cdot 10^{-12}$ MeV

$\tau = (5.17 \pm 0.04) \cdot 10^{-8}$ s

Decay channels K^0_L	Γ_i/Γ	p
$3\pi^0$	$(21.6 \pm 0.8)\%$	139
$\pi^+\pi^-\pi^0$	$(12.38 \pm 0.21)\%$	133
$\pi^\pm\mu^\mp\nu$	$(27.0 \pm 0.4)\%$	216
$\pi^\pm e^\mp\nu$	$(38.7 \pm 0.5)\%$	229
2γ	$(5.70 \pm 0.27) \cdot 10^{-4}$	249
$\pi^0 2\gamma$	$(2.0 \pm 0.5) \cdot 10^{-6}$	231
$(\pi\mu\text{atom})\nu$	$(1.05 \pm 0.11) \cdot 10^{-7}$	216
$\pi^+\pi^-\gamma$	$(4.41 \pm 0.32) \cdot 10^{-5}$	206
$\pi^+\pi^-$	$(2.03 \pm 0.04) \cdot 10^{-3}$	206
$\pi^0\pi^0$	$(9.09 \pm 0.35) \cdot 10^{-4}$	209
$\mu^+\mu^-$	$(7.3 \pm 0.4) \cdot 10^{-9}$	225
$\mu^+\mu^-\gamma$	$(2.8 \pm 2.8) \cdot 10^{-7}$	225
$e^+e^-\gamma$	$(9.1 \pm 0.5) \cdot 10^{-6}$	249

$K^*(892)$ $\quad I(J^P) = \frac{1}{2}(1^-)$

$K^*(892)^\pm$ $\quad m = 891.59 \pm 0.24$ MeV

$K^*(892)^0$ $\quad m = 896.10 \pm 0.28$ MeV

$K^*(892)^\pm$ $\quad \Gamma = 49.8 \pm 0.8$ MeV

$K^*(892)^0$ $\quad \Gamma = 50.5 \pm 0.6$ MeV

Decay channels $K^*(892)$	Γ_i/Γ	p
$K\pi$	$\sim 100\%$	291
$K^0\gamma$	$(2.30 \pm 0.20) \cdot 10^{-3}$	310
$K^\pm\gamma$	$(1.01 \pm 0.09) \cdot 10^{-3}$	309

$K_1(1270)$ $\quad I(J^P) = \frac{1}{2}(1^+)$

$m = 1270 \pm 10$ MeV

$\Gamma = 90 \pm 20$ MeV

Decay channels $K_1(1270)$	Γ_i/Γ	p
$K\rho$	$(42 \pm 6)\%$	71
$K^*_0(1430)\pi$	$(28 \pm 4)\%$	–
$K^*(892)\pi$	$(16 \pm 5)\%$	299
$K\omega$	$(11.0 \pm 2.0)\%$	–

$K_1(1400)$ $\quad I(J^P) = \frac{1}{2}(1^+)$

$m = 1402 \pm 7$ MeV

$\Gamma = 174 \pm 13$ MeV

Decay channels $K_1(1400)$	Γ_i/Γ	p
$K^*(892)\pi$	$\sim 100\%$	401

$K^*(1410)$ $\quad I(J^P) = \frac{1}{2}(1^-)$

$m = 1412 \pm 12$ MeV

$\Gamma = 227 \pm 22$ MeV

Decay channels $K^*(1410)$	Γ_i/Γ	p
$K^*(892)\pi$	$> 40\%$	408
$K\pi$	$(6.6 \pm 1.3)\%$	611

$K^*_0(1430)$ $\quad I(J^P) = \frac{1}{2}(0^+)$

$m = 1429 \pm 6$ MeV

$\Gamma = 287 \pm 23$ MeV

Decay channels $K^*_0(1430)$	Γ_i/Γ	p
$K\pi$	$\sim 100\%$	621

$K^*_2(1430)$ $\quad I(J^P) = \frac{1}{2}(2^+)$

$K^*_2(1430)^\pm$ $\quad m = 1425.4 \pm 1.3$ MeV

$K^*_2(1430)^0$ $\quad m = 1432.4 \pm 1.3$ MeV

$K^*_2(1430)^\pm$ $\quad \Gamma = 98.4 \pm 2.3$ MeV

$K^*_2(1430)^0$ $\quad \Gamma = 109 \pm 5$ MeV

Decay channels $K^*_2(1430)$	Γ_i/Γ	p
$K\pi$	$(49.7 \pm 1.2)\%$	622
$K^*(892)\pi$	$(25.2 \pm 1.7)\%$	423
$K^*(892)\pi\pi$	$(13.0 \pm 2.3)\%$	375
$K\rho$	$(8.8 \pm 0.8)\%$	333
$K\omega$	$(2.9 \pm 0.8)\%$	319

$K^*(1680)$ $\quad I(J^P) = \frac{1}{2}(1^-)$

$m = 1714 \pm 20$ MeV
$\Gamma = 323 \pm 110$ MeV

Decay channels $K^*(1680)$	Γ_i/Γ	p
$K\pi$	$(38.7 \pm 2.5)\%$	779
$K\rho$	$(31.4^{+4.7}_{-2.1})\%$	573
$K^*(892)\pi$	$(29.9^{+2.2}_{-4.7})\%$	615

$K_2(1770)$ $I(J^P) = \frac{1}{2}(2^-)$

$m = 1768 \pm 14$ MeV
$\Gamma = 136 \pm 18$ MeV

Decay channels $K_2(1770)$	Γ_i/Γ	p
$K_2^*(1430)\pi$	dominant channel	282
$K^*(892)\pi$	observed	650
$Kf_2(1270)$	observed	–
$K\phi$	observed	437
$K\omega$	observed	604

$K_3^*(1780)$ $I(J^P) = \frac{1}{2}(3^-)$

$m = 1770 \pm 10$ MeV
$\Gamma = 164 \pm 17$ MeV

Decay channels $K_3^*(1780)$	Γ_i/Γ	p
$K\rho$	$(45 \pm 4)\%$	613
$K^*(892)\pi$	$(27.3 \pm 3.2)\%$	651
$K\pi$	$(19.3 \pm 1.0)\%$	810
$K\eta$	$(8.0 \pm 1.5)\%$	715

$K_4^*(2045)$ $I(J^P) = \frac{1}{2}(4^+)$

$m = 2045 \pm 9$ MeV
$\Gamma = 198 \pm 30$ MeV

Decay channels $K_4^*(2045)$	Γ_i/Γ	p
$K\pi$	$(9.9 \pm 1.2)\%$	958
$K^*(892)\pi\pi$	$(9 \pm 5)\%$	800
$K^*(892)\pi\pi\pi$	$(7 \pm 5)\%$	764
$\rho K\pi$	$(5.7 \pm 3.2)\%$	743
$\omega K\pi$	$(4.9 \pm 3.0)\%$	736
$\phi K\pi$	$(2.8 \pm 1.4)\%$	591
$\phi K^*(892)$	$(1.4 \pm 0.7)\%$	363

Charmed mesons ($C = \pm 1$)

$D^\pm$ $I(J^P) = \frac{1}{2}(0^-)$

$m = 1869.3 \pm 0.5$ MeV
$\tau = (10.66 \pm 0.23) \cdot 10^{-13}$ s
Decay channels D^- are charge–conjugates to the decay channels D^+

Decay channels D^+	Γ_i/Γ	p
Inclusive channels		
e^+X	$(17.2 \pm 1.9)\%$	–
K^-X	$(20.8 \pm 2.8)\%$	–
K^+X	$(5.8 \pm 1.4)\%$	–
$K^0X + \overline{K}^0X$	$(59 \pm 7)\%$	–
Semileptonic channels		
$\overline{K}^0e^+\nu_e$	$(5.5 \pm 1.2)\%$	868
$\overline{K}^0\mu^+\nu_\mu$	$(7.0 \pm 3.0)\%$	865
$K^-\pi^+e^+\nu_e$	$(3.8 \pm 0.9)\%$	863
$\overline{K}^*(892)^0e^+\nu_e$ $\cdot B(\overline{K}^{*0} \to K^-\pi^+)$	$(2.7 \pm 0.4)\%$	720
Hadron channels with K–mesons		
$\overline{K}^0\pi^+$	$(2.6 \pm 0.4)\%$	862
$K^-\pi^+\pi^+$	$(8.0 \pm 0.8)\%$	845
$K^-\pi^+\pi^+$, resonanceless	$(6.7 \pm 0.8)\%$	845
$\overline{K}^0\pi^+\pi^0$	$(8.4 \pm 1.8)\%$	845
$\overline{K}^*(892)^0\pi^+$ $\cdot B(\overline{K}^{*0} \to \overline{K}^0\pi^0)$	$(0.6 \pm 0.2)\%$	712
$\overline{K}^0\rho^+$	$(6.6 \pm 1.7)\%$	680
$K^-\pi^+\pi^+\pi^0$	$(4.9 \pm 1.4)\%$	816
$\overline{K}^*(892)^0\rho^+$ $\cdot B(\overline{K}^{*0} \to K^-\pi^+)$	$(2.7 \pm 1.0)\%$	424
$\overline{K}_1(1400)^0\pi^+$ $\cdot B(\overline{K}_1(1400)^0 \to K^-\pi^+\pi^0)$	$(2.0 \pm 0.5)\%$	390
$\overline{K}^0\pi^+\pi^+\pi^-$	$(6.9 \pm 1.1)\%$	814
$\overline{K}^0a_1(1260)^+$ $\cdot B(a_1(1260)^+ \to \pi^+\pi^+\pi^-)$	$(3.8 \pm 0.9)\%$	290
$\overline{K}_1(1400)^0\pi^+$ $\cdot B(\overline{K}_1(1400)^0 \to \overline{K}^0\pi^+\pi^-)$	$(2.0 \pm 0.5)\%$	390
$K^-\pi^+\pi^+\pi^+\pi^-$	$(6.1 \pm 1.5) \cdot 10^{-3}$	772
$\overline{K}^*(892)^0\pi^+\pi^+\pi^-$ $\cdot B(\overline{K}^{*0} \to K^-\pi^+)$	$(5.1 \pm 1.7) \cdot 10^{-3}$	642
$\overline{K}^0\pi^+\pi^+\pi^-\pi^0$	$(8.7^{+3.5}_{-1.6})\%$	772
$\overline{K}^0\overline{K}^0K^+$	$(2.7 \pm 0.6)\%$	545
$\overline{K}^0K^+$	$(7.3 \pm 1.8) \cdot 10^{-3}$	792
$K^+K^-\pi^+$	$(1.01 \pm 0.13)\%$	744
$\phi\pi^+$ $\cdot B(\phi \to K^+K^-)$	$(3.0 \pm 0.4) \cdot 10^{-3}$	647
$\overline{K}^*(892)^0K^+$ $\cdot B(\overline{K}^{*0} \to K^-\pi^+)$	$(3.1 \pm 0.6) \cdot 10^{-3}$	610
$K^+K^-\pi^+\pi^0$	$(4.0 \pm 0.8) \cdot 10^{-3}$	682
Pion channels		
$\pi^+\pi^+\pi^-$	$(2.8 \pm 0.6) \cdot 10^{-3}$	908
$\eta\pi^+$ $\cdot B(\eta \to \pi^+\pi^-\pi^0)$	$(1.6 \pm 0.5) \cdot 10^{-3}$	848
$3\pi^+2\pi^-$	$(1.5 \pm 1.1) \cdot 10^{-3}$	845

D^0 $I(J^P) = \frac{1}{2}(0^-)$

$m = 1864.5 \pm 0.5$ MeV
$m_{D^\pm} - m_{D^0} = 4.77 \pm 0.27$ MeV
$\tau = (4.20 \pm 0.08) \cdot 10^{-13}$ s
Decay channels $\overline{D}^0$ are charge–conjugates to the decay channels D^0

Decay channels D^0	Γ_i/Γ	p
Inclusive channels		
e^+X	$(7.7 \pm 1.2)\%$	–
μ^+X	$(8.8 \pm 2.5)\%$	–
K^-X	$(46 \pm 4)\%$	–
K^+X	$(3.4^{+0.6}_{-0.5})\%$	–
$K^0X + \overline{K}^0X$	$(42 \pm 5)\%$	–
Half-lepton channels		
$K^-e^+\nu_e$	$(3.31 \pm 0.29)\%$	867
$K^-\mu^+\nu_\mu$	$(2.9 \pm 0.5)\%$	863
$K^-\pi^0e^+\nu_e$	$(1.6^{+1.3}_{-0.5})\%$	861
$\overline{K}^0\pi^-e^+\nu_e$	$(2.8^{+1.7}_{-0.9})\%$	860
$\overline{K}^*(892)^-e^+\nu_e$ $\cdot B(\overline{K}^{*0} \to K^-\pi^+)$	$(1.1 \pm 0.4)\%$	719
$\pi^-e^+\nu_e$	$(3.9^{+2.3}_{-1.2}) \cdot 10^{-3}$	927
Hadron channels with K-mesons		
$\overline{K}^0\pi^0$	$(2.1 \pm 0.5)\%$	860
$K^-\pi^+$	$(3.65 \pm 0.21)\%$	861
$\overline{K}^0\pi^+\pi^-$	$(5.4 \pm 0.5)\%$	842
$\overline{K}^0\rho^0$	$(6.1 \pm 3.0) \cdot 10^{-3}$	677
$K^*(892)^-\pi^+$ $\cdot B(K^{*-} \to \overline{K}^0\pi^-)$	$(3.0 \pm 0.4)\%$	711
$\overline{K}^0\pi^+\pi^-$(resonanceless)	$(1.8 \pm 0.5)\%$	842
$K^-\pi^+\pi^0$	$(11.3 \pm 1.1)\%$	844
$K^-\rho^+$	$(7.3 \pm 1.1)\%$	679
$K^*(892)^-\pi^+$ $\cdot B(\overline{K}^{*-} \to K^-\pi^0)$	$(1.5 \pm 0.2)\%$	711
$K^-\pi^+\pi^+\pi^-$	$(7.5 \pm 0.5)\%$	812
$K^-\pi^+\rho^0$	$(6.4 \pm 0.5)\%$	613
$\overline{K}^*(892)^0\rho^0$ $\cdot B(\overline{K}^{*0} \to K^-\pi^+)$	$(1.0 \pm 0.4)\%$	419
$K^-a_1(1260)^+$ $\cdot B(a_1(1260)^+ \to \pi^+\pi^+\pi^-)$	$(3.7 \pm 0.7)\%$	289
$K_1(1270)^-\pi^+$ $\cdot B(K_1(1270)^- \to K^-\pi^+\pi^-)$	$(3.7 \pm 1.1) \cdot 10^{-3}$	485
$K^-\pi^+\pi^+\pi^-$(no resonance)	$(1.8 \pm 0.5)\%$	812
$\overline{K}^0\pi^+\pi^-\pi^0$	$(10.3 \pm 1.7)\%$	812
$\overline{K}^0\omega$ $\cdot B(\omega \to \pi^+\pi^-\pi^0)$	$(2.2 \pm 0.4)\%$	670
$K_1(1270)^-\pi^+$ $\cdot B(K_1(1270)^- \to \overline{K}^0\pi^-\pi^0)$	$(5.2 \pm 1.6) \cdot 10^{-3}$	485
$K^-\pi^+\pi^+\pi^-\pi^0$	$(3.5 \pm 0.6)\%$	771
$\overline{K}^*(892)^0\pi^+\pi^-\pi^0$ $\cdot B(\overline{K}^{*0} \to K^-\pi^+)$	$(1.1 \pm 0.5)\%$	641
$\overline{K}^0\pi^+\pi^+\pi^-\pi^-$	$(8.5 \pm 1.4) \cdot 10^{-3}$	768
$\overline{K}^0\pi^+\pi^-\pi^0\pi^0(\pi^0)$	$(12.7^{+3.5}_{-2.4})\%$	771
$\overline{K}^0\phi \cdot B(\phi \to K^+K^-)$	$(4.4 \pm 0.6) \cdot 10^{-3}$	520
$\overline{K}^0K^+K^-$	$(5.2 \pm 0.9) \cdot 10^{-3}$	544
$K^0_SK^0_SK^0_S$	$(8.9 \pm 2.5) \cdot 10^{-4}$	538
K^+K^-	$(4.1 \pm 0.4) \cdot 10^{-3}$	791
$K^0\overline{K}^0$	$(1.1 \pm 0.4) \cdot 10^{-3}$	788
$K^0K^-\pi^+$	$(6.4 \pm 1.1) \cdot 10^{-3}$	739
$K^*(892)^+K^-$ $\cdot B(K^{*+} \to K^0\pi^+)$	$(2.3 \pm 0.5) \cdot 10^{-3}$	609
$K^0K^-\pi^+$ (no resonance)	$(2.2 \pm 2.2) \cdot 10^{-3}$	739
$\overline{K}^0K^+\pi^-$	$(4.9 \pm 1.0) \cdot 10^{-3}$	739
$K^+K^-\pi^+\pi^-$	$(2.4 \pm 0.4) \cdot 10^{-3}$	676
$\phi\pi^+\pi^-$ $\cdot B(\phi \to K^+K^-)$	$(1.2 \pm 0.4) \cdot 10^{-3}$	614
$\phi\rho^0$ $\cdot B(\phi \to K^+K^-)$	$(9.0 \pm 2.5) \cdot 10^{-4}$	262
Pionic channels		
$\pi^+\pi^-$	$(1.63 \pm 0.19) \cdot 10^{-3}$	922
$\pi^+\pi^-\pi^0$	$(1.5 \pm 1.0)\%$	907
$\pi^+\pi^+\pi^-\pi^-$	$(7.5 \pm 0.9) \cdot 10^{-3}$	879
$\pi^+\pi^+\pi^-\pi^-\pi^0$	$(1.7 \pm 0.5)\%$	844

$D^*(2010)$ $\quad I(J^P) = \frac{1}{2}(1^-)$

$m = 2010.1 \pm 0.6$ MeV
$m_{D^{*+}} - m_{D^0} = 145.44 \pm 0.06$ MeV
$\Gamma < 1.1$ MeV, $CL = 90\%$
Decay channels $D^*(2010)^-$ are charge-conjugate to the decay channels $D^*(2010)^+$

Decay channels $D^*(2010)^+$	Γ_i/Γ	p
$D^0\pi^+$	$(55 \pm 4)\%$	40
$D^+\pi^0$	$(27.2 \pm 2.5)\%$	39
$D^+\gamma$	$(18 \pm 4)\%$	136

$D^*(2010)^0$ $\quad I(J^P) = \frac{1}{2}(1^-)$

$m = 2007.1 \pm 1.4$ MeV
$m_{D^{*0}} - m_{D^0} = 142.5 \pm 1.3$ MeV
$\Gamma < 2.1$ MeV, $CL = 90\%$
Decay channels $\overline{D}^*(2010)^0$ are charge-conjugate to the decay channels $D^*(2010)^0$

Decay channels $D^*(2010)^0$	Γ_i/Γ	p
$D^0\pi^0$	$(55 \pm 6)\%$	44
$D^0\gamma$	$(45 \pm 6)\%$	138

$D_1(2420)^0$ $\quad I(J^P) = \frac{1}{2}(1^+)$

$m = 2424 \pm 6$ MeV
$\Gamma = 20^{+9}_{-5}$ MeV
Decay channels $\overline{D}_1(2420)^0$ are charge-conjugate to the decay channels $D_1(2420)^0$

Decay channels $D_1(2420)^0$	Γ_i/Γ	p
$D^*(2010)^+\pi^-$	observed	356

$D_2^*(2460)^0$ $\quad I(J^P) = \frac{1}{2}(2^+)$

$m = 2459.4 \pm 2.2$ MeV
$\Gamma = 19 \pm 7$ MeV

Decay channels $\overline{D}_2^*(2420)^0$ are charge-conjugate to the decay channels $D_2^*(2460)^0$

Decay channels $D_2^*(2460)^0$	Γ_i/Γ	p
$D^+\pi^-$	observed	504
$D^*(2010)^+\pi^-$	observed	388

Charmed-strange mesons ($C = S = \pm 1$)

$D_S^\pm$ $\quad I(J^P) = 0(0^-)$

$m = 1968.8 \pm 0.7$ MeV

$m_{D_S^\pm} - m_{D^\pm} = 99.5 \pm 0.6$ MeV

$\tau = (4.50^{+0.30}_{-0.26}) \cdot 10^{-13}$ s

Decay channels D_S^- are charge-conjugate to the decay channels D_S^+

Decay channels D_S^+	Γ_i/Γ	p
Inclusive channels		
K^-X	$(13^{+14}_{-12})\%$	–
K^+X	$(20^{+18}_{-14})\%$	–
$K^0X + \overline{K}^0X$	$(39 \pm 28)\%$	–
Channel without K	$(64 \pm 17)\%$	–
e^+X	$< 20\%$	–
Channels with two K-mesons including ϕ-mes.		
$K^+\overline{K}^0$	$(2.8 \pm 0.7)\%$	851
$K^+K^-\pi^+$	$(3.9 \pm 0.4)\%$	805
$\phi\pi^+$	$(2.8 \pm 0.5)\%$	712
$K^+\overline{K}^*(892)^0$	(2.6 ± 0.5)	683
$K^+K^-\pi^+$ (no resonance)	$(8.1 \pm 3.0) \cdot 10^{-3}$	805
$K^*(892)^+\overline{K}^0$	$(3.3 \pm 0.9)\%$	683
$\phi\pi^+\pi^0$	$(6.7 \pm 3.3)\%$	687
$\phi\rho^+$	$(5.2^{+1.4}_{-1.6})\%$	409
$K^0K^-\pi^+\pi^+$	$(3.3 \pm 1.0)\%$	745
$K^*(892)^+\overline{K}^*(892)^0$	$(5.0 \pm 1.7)\%$	412
$\phi\pi^+\pi^+\pi^-$	$(1.2 \pm 0.4)\%$	640
Other hadron channels		
$\pi^+\pi^+\pi^-$	$(1.2 \pm 0.4)\%$	960
$f_0(975)\pi^+$	$(7.8 \pm 3.2) \cdot 10^{-3}$	735
$\pi^+\pi^+\pi^-$	$(8.0 \pm 3.0) \cdot 10^{-3}$	960
Resonanceless		
$\eta\pi^+$	$(1.5 \pm 0.4)\%$	902
$\eta\rho^+$	$(7.9 \pm 2.1)\%$	727
$\eta'(958)\pi^+$	$(3.7 \pm 1.2)\%$	744
$\eta'(958)\rho^+$	$(9.5 \pm 2.7)\%$	472
Semileptonic channels		
$\phi e^+\nu$	$(1.6 \pm 0.7)\%$	721
$\phi\ell^+\nu$	$(1.4 \pm 0.5)\%$	–

$D_S^{*\pm}$ $\quad I(J^P) = ?(?^?)$

$m = 2110.3 \pm 2.0$ MeV

$m_{D_S^{*\pm}} - m_{D_S^\pm} = 141.5 \pm 1.9$ MeV

$\Gamma < 4.5$ MeV, $CL = 90\%$

Decay channels D_S^{*-} are charge-conjugate to the decay channels D_S^{*+}

Decay channels D_S^{*+}	Γ_i/Γ	p
$D_S^+\gamma$	$\sim 100\%$	137

$D_{S1}(2536)^\pm$ $\quad I(J^P) = 0(1^+)$

$m = 2536.5 \pm 0.8$ MeV

$\Gamma < 4.6$ MeV, $CL = 90\%$

Decay channels $D_{S1}(2536)^-$ are charge-conjugates to the decay channels $D_{S1}(2536)^+$

Decay channels $D_{S1}(2536)^+$	Γ_i/Γ	p
$D^*(2010)^+K^0$	observed	153
$D_S^{*+}\gamma$	observed	390

B-mesons (B = ±1)

$B^\pm$ $\quad I(J^P) = \frac{1}{2}(0^-)$

$m_{B^\pm} = 5278.6 \pm 2.0$ MeV

$\tau = (12.9 \pm 0.5) \cdot 10^{-13}$ s

Decay channels B^- are charge-conjugate to the decay channels B^+

Decay channels B^+	Γ_i/Γ	p
Semileptonic channels		
$\overline{D}^0\ell^+\nu$	$(1.6 \pm 0.7)\%$	–
$\overline{D}^*(2110)^0\ell^+\nu$	$(4.6 \pm 1.0)\%$	–
$\omega\mu^+\nu_\mu$	observed	2580
Channels with D-mesons		
$\overline{D}^0\pi^+$	$(3.8 \pm 1.1) \cdot 10^{-3}$	2308
$\overline{D}^0\rho^+$	$(1.3 \pm 0.6)\%$	2238
$\overline{D}^0\pi^+\pi^+\pi^-$	$(1.1 \pm 0.4)\%$	2289
$\overline{D}^0\pi^+\rho^0$	$(4.2 \pm 3.0) \cdot 10^{-3}$	2209
$\overline{D}^0a_1(1260)^+$	$(5 \pm 4) \cdot 10^{-3}$	2113
$D^*(2010)^-\pi^+\pi^+$	$(2.5 \pm 1.2) \cdot 10^{-3}$	2247
$D^-\pi^+\pi^+$	$(2.5^{+4.8}_{-2.4}) \cdot 10^{-3}$	2299
$\overline{D}^*(2010)^0\pi^+$	$(5.2 \pm 1.5) \cdot 10^{-3}$	2254
$\overline{D}^*(2010)^0\rho^+$	$(1.0 \pm 0.7)\%$	2181
$D^*(2010)^-\pi^+\pi^+\pi^0$	$(1.8 \pm 0.9)\%$	2235
Channels with J/ψ-mesons		
$J/\psi(1S)K^+$	$(7.7 \pm 2.0) \cdot 10^{-4}$	1683
$J/\psi(1S)K^+\pi^+\pi^-$	$(1.1 \pm 0.5) \cdot 10^{-3}$	1612
$J/\psi(1S)K^*(892)^+$	$(1.4 \pm 0.7) \cdot 10^{-3}$	1571
$J/\psi(2S)K^*(892)^+\pi^+\pi^-$	$(1.9 \pm 1.2) \cdot 10^{-3}$	909
In the following channels the sign of the B-meson charge is not defined		
Semileptonic channels		
$e^\pm\nu_e h$	$(10.7 \pm 0.5)\%$	–
$D^*(2010)e\nu_e$	$(7.0 \pm 2.3)\%$	–

Decay channels B^+ *(continued)*	Γ_i/Γ	p
$\mu^\pm \nu_\mu h$	$(10.3 \pm 0.5)\%$	–
Channels with D-mesons		
$D^\pm X$	$(22.7 \pm 3.3)\%$	–
$D^0(\overline{D}^0)X$	$(46 \pm 5)\%$	–
$D^*(2010)^\pm X$	$(26.9 \pm 3.5)\%$	–
$D_s^\pm X$	$(11.5 \pm 2.8)\%$	–
Pairs (DD)	$(6.5 \pm 1.9)\%$	–
Channels with J/ψ or ψ-mesons		
$J/\psi(1S)X$	$(1.12 \pm 0.16)\%$	–
$\psi(2S)X$	$(4.6 \pm 2.0) \cdot 10^{-3}$	–
Channels with K-mesons		
$K^\pm X$	$(85 \pm 11)\%$	–
$K^0(\overline{K}^0)X$	$(63 \pm 8)\%$	–
Channels with baryons		
pX	$(8.2^{+1.4}_{-1.1})\%$	–
ΛX	$(4.2 \pm 0.8)\%$	–
$\Xi^- X$	$(2.8 \pm 1.4) \cdot 10^{-3}$	–
Baryons X	$(7.6 \pm 1.4)\%$	–
$p\bar{p}X$	$(2.50 \pm 0.28)\%$	–
$\Lambda\bar{p}X$	$(2.3 \pm 0.5)\%$	–

B^0 $I(J^P) = \frac{1}{2}(0^-)$

$m_{B^0} = 5278.7 \pm 2.1$ MeV

$|m_{B^0_1} - m_{B^0_2}| = (3.6 \pm 0.7) \cdot 10^{-10}$ MeV

$m_{B^0} - m_{B^\pm} = 0.1 \pm 0.8$ MeV

$\tau = (12.9 \pm 0.5) \cdot 10^{-13}$ s

$\tau_{B^+}/\tau_{B^0} = 0.93 \pm 0.16$

$\Gamma(\mu^- X(\text{via } \overline{B}^0))/\Gamma(\mu^\pm X) = 0.16 \pm 0.04$

Decay channels $\overline{B}^0$ are charge-conjugate to the decay channels B^0

Decay channels B^0	Γ_i/Γ	p
Semileptonic channels		
$D^-\ell^+\nu$	$(1.8 \pm 0.5)\%$	–
$D^*(2010)^-\ell^+\nu$	$(4.9 \pm 0.8)\%$	–
$\pi^-\mu^+\nu_\mu$	observed	2636
Channels with D-mesons		
$D^-\pi^+$	$(3.2 \pm 0.7) \cdot 10^{-3}$	2306
$D^-\rho^+$	$(9 \pm 6) \cdot 10^{-3}$	2236
$D^*(2010)^-\pi^+$	$(3.2 \pm 0.7) \cdot 10^{-3}$	2254
$D^-\pi^+\pi^+\pi^-$	$(8.0 \pm 2.5) \cdot 10^{-3}$	2287
$D^-\pi^+\pi^+\pi^-$ (no resonance)	$(3.9 \pm 1.9) \cdot 10^{-3}$	2287
$D^-\pi^+\rho^0$	$(1.1 \pm 1.0) \cdot 10^{-3}$	2207
$D^-a_1(1260)^+$	$(6.0 \pm 3.3) \cdot 10^{-3}$	2111
$D^*(2010)^-\pi^+\pi^0$	$(1.8 \pm 0.6)\%$	2247
$D^*(2010)^-\rho^+$	$(8 \pm 4) \cdot 10^{-3}$	2181
$D^*(2010)^-\pi^+\pi^+\pi^-$	$(1.41 \pm 0.34)\%$	2234
$D^*(2010)^-\pi^+\rho^0$	$(7 \pm 4) \cdot 10^{-3}$	2151
$D^*(2010)^-a_1(1260)^+$	$(1.8 \pm 0.8)\%$	2051
$D^*(2010)^-\pi^+\pi^+\pi^-\pi^0$	$(4.1 \pm 2.2)\%$	2218
$D^-D_s^+$	$(8 \pm 5) \cdot 10^{-3}$	1812
Channels with J/ψ or ψ-mesons		
$J/\psi(1S)K^0$	$(6.5 \pm 3.1) \cdot 10^{-4}$	1682
$J/\psi(1S)K^+\pi^-$	$(1.0 \pm 0.5) \cdot 10^{-3}$	1652
$J/\psi(1S)K^*(892)^0$	$(1.3 \pm 0.4) \cdot 10^{-3}$	1569
Channels with baryons		
$p\bar{p}\pi^+\pi^-$	$(6.0 \pm 3.0) \cdot 10^{-4}$	2406

B^* $I(J^P) = \frac{1}{2}(1^-)$

$m = 5324.6 \pm 2.1$ MeV

$C\overline{C}$-mesons

$\eta_c(1S)$ $I(J^{PC}) = 0^+(0^{-+})$

$m = 2978.8 \pm 1.9$ MeV

$\Gamma = 10.3^{+3.8}_{-3.4}$ MeV

Decay channels $\eta_c(1S)$	Γ_i/Γ	p
$\eta'(958)\pi\pi$	$(4.1 \pm 1.7)\%$	1319
$\rho\rho$	$(2.6 \pm 0.9)\%$	1276
$K^*(892)^0K^-\pi^+$	$(2.0 \pm 0.7)\%$	1273
$K^*(892)\overline{K}^*(892)$	$(8.5 \pm 3.1) \cdot 10^{-3}$	1193
$\phi\phi$	$(7.1 \pm 2.8) \cdot 10^{-3}$	1086
$K\overline{K}\pi$	$(6.6 \pm 1.8)\%$	1378
$\eta\pi\pi$	$(4.9 \pm 1.8)\%$	1425
$\pi^+\pi^-K^+K^-$	$(2.0^{+0.7}_{-0.6})\%$	1342
$2(\pi^+\pi^-)$	$(1.2 \pm 0.4)\%$	1457
$p\bar{p}$	$(1.2 \pm 0.4) \cdot 10^{-3}$	1157
Radiative decays		
$\gamma\gamma$	$(6^{+6}_{-5}) \cdot 10^{-4}$	1489

$J/\psi(1S)$ $I(J^{PC}) = 0^-(1^{--})$

$m = 3096.93 \pm 0.09$ MeV

$\Gamma = 86 \pm 6$ keV

$\Gamma_{ee} = 5.36 \pm 0.29$ keV

Decay channels $J/\psi(1S)$	Γ_i/Γ	p
h	$(86.0 \pm 2.0)\%$	–
e^+e^-	$(6.27 \pm 0.20)\%$	1548
$\mu^+\mu^-$	$(5.97 \pm 0.25)\%$	1545
Channels with hadron resonances		
$\rho\pi$	$(1.28 \pm 0.10)\%$	1450
$\rho^0\pi^0$	$(4.2 \pm 0.5) \cdot 10^{-3}$	1450
$a_2(1320)\rho$	$(1.09 \pm 0.22)\%$	1126
$\omega\pi^+\pi^+\pi^-\pi^-$	$(8.5 \pm 3.4) \cdot 10^{-3}$	1392
$\omega\pi^+\pi^-$	$(7.2 \pm 1.0) \cdot 10^{-3}$	1435
$K^*(892)^0\overline{K}_2^*(1430)^0$	$(6.7 \pm 2.6) \cdot 10^{-3}$	1005
$\omega K^*(892)\overline{K}$	$(5.3 \pm 2.0) \cdot 10^{-3}$	1098
$\omega f_2(1270)$	$(4.3 \pm 0.6) \cdot 10^{-3}$	1143
$K^+\overline{K}^*(892)^-$	$(5.0 \pm 0.4) \cdot 10^{-3}$	1373
$K^0\overline{K}^*(892)^0$	$(4.2 \pm 0.4) \cdot 10^{-3}$	1371
$\omega\pi^0\pi^0$	$(3.4 \pm 0.8) \cdot 10^{-3}$	1436

Decay channels $J/\psi(1S)$ *(continued)*	Γ_i/Γ	p
$b_1(1235)^{\pm}\pi^{\mp}$	$(3.0 \pm 0.5)\cdot 10^{-3}$	1299
$\omega K^{\pm}K^0_S\pi^{\mp}$	$(2.9 \pm 0.7)\cdot 10^{-3}$	1210
$b_1(1235)^0\pi^0$	$(2.3 \pm 0.6)\cdot 10^{-3}$	1299
$\phi K^*(892)\overline{K}$	$(2.04 \pm 0.28)\cdot 10^{-3}$	969
$\omega K\overline{K}$	$(1.9 \pm 0.4)\cdot 10^{-3}$	1268
$\omega f_0(1710) \to \omega K\overline{K}$	$(4.8 \pm 1.1)\cdot 10^{-4}$	878
$\phi 2(\pi^+\pi^-)$	$(1.60 \pm 0.32)\cdot 10^{-3}$	1318
$\Delta(1232)^{++}\bar{p}\pi^-$	$(1.6 \pm 0.5)\cdot 10^{-3}$	1030
$\omega\eta$	$(1.58 \pm 0.16)\cdot 10^{-3}$	1394
$\phi K\overline{K}$	$(1.48 \pm 0.22)\cdot 10^{-3}$	1179
$\phi f_0(1710) \to \phi K\overline{K}$	$(3.6 \pm 0.6)\cdot 10^{-4}$	714
$p\bar{p}\omega$	$(1.30 \pm 0.25)\cdot 10^{-3}$	769
$\Delta(1232)^{++}\overline{\Delta}(1232)^{--}$	$(1.10 \pm 0.29)\cdot 10^{-3}$	938
$\Sigma(1385)^-\overline{\Sigma}(1385)^+$	$(1.03 \pm 0.13)\cdot 10^{-3}$	692
$\phi\pi^+\pi^-$	$(8.0 \pm 1.2)\cdot 10^{-4}$	1365
$\phi K^{\pm}K^0_S\pi^{\mp}$	$(7.2 \pm 0.9)\cdot 10^{-4}$	1114
$\omega f_1(1420)$	$(6.8 \pm 2.4)\cdot 10^{-4}$	1062
$\phi\eta$	$(6.5 \pm 0.7)\cdot 10^{-4}$	1320
$\Xi(1530)^-\overline{\Xi}^+$	$(5.9 \pm 1.5)\cdot 10^{-4}$	597
$pK^-\overline{\Sigma}(1385)^0$	$(5.1 \pm 3.2)\cdot 10^{-4}$	645
$\omega\pi^0$	$(4.2 \pm 0.6)\cdot 10^{-4}$	1447
$\phi\eta'(958)$	$(3.3 \pm 0.4)\cdot 10^{-4}$	1192
$\phi f_0(975)$	$(3.2 \pm 0.9)\cdot 10^{-4}$	1185
$\Xi(1530)^0\overline{\Xi}^0$	$(3.2 \pm 1.4)\cdot 10^{-4}$	608
$\Sigma(1385)^-\overline{\Sigma}^+$	$(3.1 \pm 0.5)\cdot 10^{-4}$	857
$\phi f_1(1285)$	$(2.6 \pm 0.5)\cdot 10^{-4}$	1032
$\rho\eta$	$(1.93 \pm 0.23)\cdot 10^{-4}$	1398
$\omega\eta'(958)$	$(1.67 \pm 0.25)\cdot 10^{-4}$	1279
$\rho\eta'(958)$	$(1.05 \pm 0.18)\cdot 10^{-4}$	1283
Channels with stable hadrons		
$2(\pi^+\pi^-)\pi^0$	$(3.37 \pm 0.26)\%$	1496
$3(\pi^+\pi^-)\pi^0$	$(2.9 \pm 0.6)\%$	1433
$\pi^+\pi^-\pi^0$	$(1.50 \pm 0.15)\%$	1533
$\pi^+\pi^-\pi^0K^+K^-$	$(1.20 \pm 0.30)\%$	1368
$\pi^+\pi^-K^+K^-$	$(7.2 \pm 2.3)\cdot 10^{-3}$	1407
$K\overline{K}\pi$	$(6.1 \pm 1.0)\cdot 10^{-3}$	1440
$p\bar{p}\pi^+\pi^-$	$(6.0 \pm 0.5)\cdot 10^{-3}$	1107
$2(\pi^+\pi^-)$	$(4.0 \pm 1.0)\cdot 10^{-3}$	1517
$3(\pi^+\pi^-)$	$(4.0 \pm 2.0)\cdot 10^{-3}$	1466
$\Sigma\overline{\Sigma}$	$(3.8 \pm 0.5)\cdot 10^{-3}$	992
$p\bar{p}\pi^+\pi^-\pi^0$	$(2.3 \pm 0.9)\cdot 10^{-3}$	1033
$p\bar{p}$	$(2.16 \pm 0.11)\cdot 10^{-3}$	1232
$p\bar{p}\eta$	$(2.09 \pm 0.18)\cdot 10^{-3}$	948
$p\bar{n}\pi^-$	$(2.00 \pm 0.10)\cdot 10^{-3}$	1174
$\Xi\overline{\Xi}$	$(1.8 \pm 0.4)\cdot 10^{-3}$	818
$n\bar{n}$	$(1.8 \pm 0.9)\cdot 10^{-3}$	1231
$\Lambda\overline{\Lambda}$	$(1.35 \pm 0.14)\cdot 10^{-3}$	1074
$p\bar{p}\pi^0$	$(1.09 \pm 0.09)\cdot 10^{-3}$	1176
$\Lambda\overline{\Sigma}^-\pi^+$	$(1.06 \pm 0.12)\cdot 10^{-3}$	945
$pK^-\overline{\Lambda}$	$(8.9 \pm 1.6)\cdot 10^{-4}$	876
$pK^-\overline{\Sigma}^0$	$(2.9 \pm 0.8)\cdot 10^{-4}$	820
K^+K^-	$(2.37 \pm 0.31)\cdot 10^{-4}$	1468
$\Lambda\overline{\Lambda}\pi^0$	$(2.2 \pm 0.7)\cdot 10^{-4}$	998
$\pi^+\pi^-$	$(1.47 \pm 0.23)\cdot 10^{-4}$	1542
$K^0_SK^0_L$	$(1.08 \pm 0.14)\cdot 10^{-4}$	1466
Radiative channels		
$\gamma\eta_c(1S)$	$(1.3 \pm 0.4)\%$	116
$\gamma\eta\pi\pi$	$(6.1 \pm 1.0)\cdot 10^{-3}$	1486
$\gamma\eta(1440) \to \gamma K\overline{K}\pi$	$(9.1 \pm 1.8)\cdot 10^{-4}$	1223
$\gamma\eta(1440) \to \gamma\gamma\rho^0$	$(6.4 \pm 1.4)\cdot 10^{-5}$	–
$\gamma\rho\rho$	$(4.5 \pm 0.8)\cdot 10^{-3}$	1343
$\gamma\eta'(958)$	$(4.34 \pm 0.34)\cdot 10^{-3}$	1400
$\gamma 2\pi^+2\pi^-$	$(2.8 \pm 0.5)\cdot 10^{-3}$	1517
$\gamma f_4(2050)$	$(2.7 \pm 0.7)\cdot 10^{-3}$	871
$\gamma\omega\omega$	$(1.59 \pm 0.33)\cdot 10^{-3}$	1337
$\gamma\eta(1440) \to \gamma\rho^0\rho^0$	$(1.4 \pm 0.4)\cdot 10^{-3}$	1223
$\gamma f_2(1270)$	$(1.38 \pm 0.14)\cdot 10^{-3}$	1286
$\gamma f_0(1710) \to \gamma K\overline{K}$	$(9.7 \pm 1.2)\cdot 10^{-4}$	1077
$\gamma\eta$	$(8.6 \pm 0.8)\cdot 10^{-4}$	1500
$\gamma f_1(1420) \to \gamma K\overline{K}\pi$	$(8.3 \pm 1.5)\cdot 10^{-4}$	1220
$\gamma f_1(1285)$	$(7.0 \pm 1.8)\cdot 10^{-4}$	1283
$\gamma f'_2(1525)$	$(6.3 \pm 1.0)\cdot 10^{-4}$	1173
$\gamma\phi\phi$	$(4.0 \pm 1.2)\cdot 10^{-4}$	1166
$\gamma p\bar{p}$	$(3.8 \pm 1.0)\cdot 10^{-4}$	1232
$\gamma\eta(2100)$	$(2.9 \pm 0.6)\cdot 10^{-4}$	834
$\gamma\eta(1760) \to \gamma\rho^0\rho^0$	$(1.3 \pm 0.9)\cdot 10^{-4}$	1048

$\chi_{c0}(1P)$ $\qquad I(J^{PC}) = 0^+(0^{++})$

$m = 3415.1 \pm 1.0$ MeV

$\Gamma = 14 \pm 5$ MeV

Decay channels $\chi_{c0}(1P)$	Γ_i/Γ	p
Hadron channels		
$2(\pi^+\pi^-)$	$(3.7 \pm 0.7)\%$	1679
$\pi^+\pi^-K^+K^-$	$(3.0 \pm 0.7)\%$	1580
$\rho^0\pi^+\pi^-$	$(1.6 \pm 0.5)\%$	1609
$3(\pi^+\pi^-)$	$(1.5 \pm 0.5)\%$	1633
$K^+\overline{K}^*(892)^0\pi^-$	$(1.2 \pm 0.4)\%$	1522
$\pi^+\pi^-$	$(7.5 \pm 2.1)\cdot 10^{-3}$	1702
K^+K^-	$(7.1 \pm 2.4)\cdot 10^{-3}$	1635
$\pi^0\pi^0$	$(3.1 \pm 0.6)\cdot 10^{-3}$	1702
Radiative channels		
$\gamma J/\psi(1S)$	$(6.6 \pm 1.8)\cdot 10^{-3}$	303
$\gamma\gamma$	$(4.0 \pm 2.3)\cdot 10^{-4}$	1708

$\chi_{c1}(1P)$ $\qquad I(J^{PC}) = 0^+(1^{++})$

$m = 3510.53 \pm 0.12$ MeV

$\Gamma = 0.88 \pm 0.14$ MeV

Decay channels $\chi_{c1}(1P)$	Γ_i/Γ	p
Hadron channels		
$3(\pi^+\pi^-)$	$(2.2 \pm 0.8)\%$	1683

Decay channels $\chi_{c1}(1P)$ *(continued)*	Γ_i/Γ	p
$2(\pi^+\pi^-)$	$(1.6 \pm 0.5)\%$	1727
$p\bar{p}$	$(8.6 \pm 1.2) \cdot 10^{-5}$	1483
Radiative channels		
$\gamma J/\psi(1S)$	$(27.3 \pm 1.6)\%$	389

$\chi_{c2}(1P)$ $I(J^{PC}) = 0^+(2^{++})$

$m = 3556.17 \pm 0.13$ MeV
$\Gamma = 2.00 \pm 0.18$ MeV

Decay channels $\chi_{c2}(1P)$	Γ_i/Γ	p
Hadron channels		
$2(\pi^+\pi^-)$	$(2.2 \pm 0.5)\%$	1751
$\pi^+\pi^-K^+K^-$	$(1.9 \pm 0.5)\%$	1656
$3(\pi^+\pi^-)$	$(1.2 \pm 0.8)\%$	1707
$\pi^+\pi^-p\bar{p}$	$(3.3 \pm 1.3) \cdot 10^{-3}$	1410
$p\bar{p}$	$(10.0 \pm 1.0) \cdot 10^{-5}$	1510
$\pi^0\pi^0$	$(1.10 \pm 0.28) \cdot 10^{-3}$	1773
Radiative channels		
$\gamma J/\psi(1S)$	$(13.5 \pm 1.1)\%$	430

$\psi(2S)$ $I(J^{PC}) = 0^-(1^{--})$

$m = 3686.00 \pm 0.10$ MeV
$\Gamma = 278 \pm 32$ keV
$\Gamma_{ee} = 2.14 \pm 0.21$ keV

Decay channels $\psi(2S)$	Γ_i/Γ	p
h	$(98.10 \pm 0.30)\%$	–
e^+e^-	$(8.8 \pm 1.3) \cdot 10^{-3}$	1843
$\mu^+\mu^-$	$(7.7 \pm 1.7) \cdot 10^{-3}$	1840
Channels with J/ψ-mesons		
$J/\psi(1S)X$	$(57 \pm 4)\%$	–
$J/\psi(1S)h^0$	$(23.2 \pm 2.6)\%$	–
$J/\psi(1S)\pi^+\pi^-$	$(32.4 \pm 2.6)\%$	477
$J/\psi(1S)\pi^0\pi^0$	$(18.4 \pm 2.7)\%$	481
$J/\psi(1S)\eta$	$(2.7 \pm 0.4)\%$	196
$J/\psi(1S)\pi^0$	$(9.7 \pm 2.1) \cdot 10^{-4}$	527
Hadron channels		
$2(\pi^+\pi^-)\pi^0$	$(3.1 \pm 0.7) \cdot 10^{-3}$	1799
$\pi^+\pi^-K^+K^-$	$(1.6 \pm 0.4) \cdot 10^{-3}$	1726
$\pi^+\pi^-p\bar{p}$	$(8.0 \pm 2.0) \cdot 10^{-4}$	1491
$K^+\overline{K}^*(892)^0\pi^-$	$(6.7 \pm 2.5) \cdot 10^{-4}$	1673
$2(\pi^+\pi^-)$	$(4.5 \pm 1.0) \cdot 10^{-4}$	1817
$\rho^0\pi^+\pi^-$	$(4.2 \pm 1.5) \cdot 10^{-4}$	1751
$\bar{p}p$	$(1.9 \pm 0.5) \cdot 10^{-4}$	1586
Radiative channels		
$\gamma\chi_{c0}(1P)$	$(9.3 \pm 0.8)\%$	261
$\gamma\chi_{c1}(1P)$	$(8.7 \pm 0.8)\%$	171
$\gamma\chi_{c2}(1P)$	$(7.8 \pm 0.8)\%$	127
$\gamma\eta_c(1S)$	$(2.8 \pm 0.6) \cdot 10^{-3}$	639

$\psi(3770)$ $I(J^{PC}) = ?^?(1^{--})$

$m = 3769.9 \pm 2.5$ MeV
$\Gamma = 23.6 \pm 2.7$ MeV
$\Gamma_{ee} = 0.26 \pm 0.04$ keV

Decay channels $\psi(3770)$	Γ_i/Γ	p
$D\overline{D}$	dominant channel	242
e^+e^-	$(1.12 \pm 0.17) \cdot 10^{-5}$	1885

$\psi(4040)$ $I(J^{PC}) = ?^?(1^{--})$

$m = 4040 \pm 10$ MeV
$\Gamma = 52 \pm 10$ MeV
$\Gamma_{ee} = 0.75 \pm 0.15$ keV

Decay channels $\psi(4040)$	Γ_i/Γ	p
e^+e^-	$(1.4 \pm 0.4) \cdot 10^{-5}$	2020
$D^0\overline{D}^0$	observed	777
$D^*(2010)^0\overline{D}^0$	observed	577
$D^*(2010)^0\overline{D}^*(2010)^0$	observed	228

$\psi(4160)$ $I(J^{PC}) = ?^?(1^{--})$

$m = 4159 \pm 20$ MeV
$\Gamma = 78 \pm 20$ MeV
$\Gamma_{ee} = 0.77 \pm 0.23$ keV

Decay channels $\psi(4160)$	Γ_i/Γ	p
e^+e^-	$(10 \pm 4) \cdot 10^{-6}$	2079

$\psi(4415)$ $I(J^{PC}) = ?^?(1^{--})$

$m = 4415 \pm 6$ MeV
$\Gamma = 43 \pm 15$ MeV
$\Gamma_{ee} = 0.47 \pm 0.10$ keV

Decay channels $\psi(4415)$	Γ_i/Γ	p
h	dominant channel	–
e^+e^-	$(1.1 \pm 0.4) \cdot 10^{-5}$	2207

$b\bar{b}$-mesons

$\Upsilon(1S)$ $I(J^{PC}) = ?^?(1^{--})$

$m = 9460.32 \pm 0.22$ MeV
$\Gamma = 52.1 \pm 2.1$ keV
$\Gamma_{ee} = 1.34 \pm 0.04$ keV

Decay channels $\Upsilon(1S)$	Γ_i/Γ	p
$\tau^+\tau^-$	$(2.97 \pm 0.35)\%$	4381
$\mu^+\mu^-$	$(2.48 \pm 0.06)\%$	4729
e^+e^-	$(2.52 \pm 0.17)\%$	4730
Hadron channels		
$J/\psi(1S)X$	$(1.1 \pm 0.4) \cdot 10^{-3}$	–

Decay channels $\Upsilon(1S)$ *(continued)*	Γ_i/Γ	p
Radiative channels		
$\gamma 2(\pi^+\pi^-)$	$(2.5 \pm 0.9) \cdot 10^{-4}$	4720
$\gamma\pi^+\pi^- K^+K^-$	$(2.9 \pm 0.9) \cdot 10^{-4}$	4686
$\gamma\pi^+\pi^- p\bar{p}$	$(1.5 \pm 0.6) \cdot 10^{-4}$	4604

$\chi_{b0}(1P)$ $I(J^{PC}) = ?^?(0^{++})$

$m = 9859.8 \pm 1.3$ MeV

Decay channels $\chi_{b0}(1P)$	Γ_i/Γ	p
$\gamma\Upsilon(1S)$	$< 6\%$	391

$\chi_{b1}(1P)$ $I(J^{PC}) = ?^?(1^{++})$

$m = 9891.9 \pm 0.7$ MeV

Decay channels $\chi_{b1}(1P)$	Γ_i/Γ	p
$\gamma\Upsilon(1S)$	$(35 \pm 8)\%$	422

$\chi_{b2}(1P)$ $I(J^{PC}) = ?^?(2^{++})$

$m = 9913.2 \pm 0.6$ MeV

Decay channels $\chi_{b2}(1P)$	Γ_i/Γ	p
$\gamma\Upsilon(1S)$	$(22 \pm 4)\%$	443

$\Upsilon(2S)$ $I(J^{PC}) = ?^?(1^{--})$

$m = 10.02330 \pm 0.00031$ GeV
$\Gamma = 43 \pm 8$ keV

Decay channels $\Upsilon(2S)$	Γ_i/Γ	p
$\Upsilon(1S)\pi^+\pi^-$	$(18.5 \pm 0.8)\%$	475
$\Upsilon(1S)\pi^0\pi^0$	$(8.8 \pm 1.1)\%$	480
$\tau^+\tau^-$	$(1.7 \pm 1.6)\%$	4683
$\mu^+\mu^-$	$(1.31 \pm 0.21)\%$	5011
e^+e^-	observed	5012
Radiative channels		
$\gamma\chi_{b1}(1P)$	$(6.7 \pm 0.9)\%$	131
$\gamma\chi_{b2}(1P)$	$(6.6 \pm 0.9)\%$	110
$\gamma\chi_{b0}(1P)$	$(4.3 \pm 1.0)\%$	162

$\chi_{b0}(2P)$ $I(J^{PC}) = ?^?(0^{++})$

$m = 10.2320 \pm 0.0006$ GeV

Decay channels $\chi_{b0}(2P)$	Γ_i/Γ	p
$\gamma\Upsilon(2S)$	$(4.6 \pm 2.1)\%$	210
$\gamma\Upsilon(1S)$	$(9 \pm 6) \cdot 10^{-3}$	746

$\chi_{b1}(2P)$ $I(J^{PC}) = ?^?(1^{++})$

$m = 10.2549 \pm 0.0006$ GeV
$m_{\chi_{b1}(2P)} - m_{\chi_{b0}(2P)} = 22.9 \pm 0.6$ MeV

Decay channels $\chi_{b1}(2P)$	Γ_i/Γ	p
$\gamma\Upsilon(2S)$	$(22 \pm 4)\%$	229
$\gamma\Upsilon(1S)$	$(7.9 \pm 1.1)\%$	764

$\chi_{b2}(2P)$ $I(J^{PC}) = ?^?(2^{++})$

$m = 10.26835 \pm 0.00057$ GeV
$m_{\chi_{b2}(2P)} - m_{\chi_{b1}(2P)} = 13.4 \pm 0.4$ MeV
$m_{\chi_{b2}(2P)} - m_{\chi_{b0}(2P)} = 36.4 \pm 0.6$ MeV

Decay channels $\chi_{b2}(2P)$	Γ_i/Γ	p
$\gamma\Upsilon(2S)$	$(19 \pm 4)\%$	242
$\gamma\Upsilon(1S)$	$(7.0 \pm 1.1)\%$	776

$\Upsilon(3S)$ $I(J^{PC}) = ?^?(1^{--})$

$m = 10.3553 \pm 0.0005$ GeV
$\Gamma = 24.3 \pm 2.9$ keV

Decay channels $\Upsilon(3S)$	Γ_i/Γ	p
$\Upsilon(2S)X$	$(10.9 \pm 1.3)\%$	–
$\Upsilon(2S)\pi^+\pi^-$	$(2.1 \pm 0.4)\%$	177
$\Upsilon(2S)\pi^0\pi^0$	$(1.3 \pm 0.4)\%$	190
$\Upsilon(1S)\pi^+\pi^-$	$(4.48 \pm 0.29)\%$	814
$\Upsilon(1S)\pi^0\pi^0$	$(1.8 \pm 0.4)\%$	816
$\mu^+\mu^-$	$(1.81 \pm 0.17)\%$	5177
e^+e^-	observed	–
Radiative channels		
$\gamma\chi_{b2}(2P)$	$(11.4 \pm 0.8)\%$	87
$\gamma\chi_{b1}(2P)$	$(11.3 \pm 0.6)\%$	100
$\gamma\chi_{b0}(2P)$	$(5.4 \pm 0.6)\%$	123

$\Upsilon(4S)$ $I(J^{PC}) = ?^?(1^{--})$

$m = 10.5800 \pm 0.0035$ GeV
$\Gamma = 23.8 \pm 2.2$ MeV
$\Gamma_{ee} = 0.24 \pm 0.05$ keV

Decay channels $\Upsilon(4S)$	Γ_i/Γ	p
e^+e^-	$(1.01 \pm 0.21) \cdot 10^{-5}$	5290

$\Upsilon(10860)$ $I(J^{PC}) = ?^?(1^{--})$

$m = 10.865 \pm 0.008$ GeV
$\Gamma = 110 \pm 13$ MeV
$\Gamma_{ee} = 0.31 \pm 0.07$ keV

Decay channels $\Upsilon(10860)$	Γ_i/Γ	p
e^+e^-	$(2.8 \pm 0.7) \cdot 10^{-6}$	5432

$\Upsilon(11020)$ $I(J^{PC}) = ?^?(1^{--})$

$m = 11.019 \pm 0.008$ GeV
$\Gamma = 79 \pm 16$ MeV
$\Gamma_{ee} = 0.130 \pm 0.030$ keV

Decay channels $\Upsilon(11020)$	Γ_i/Γ	p
e^+e^-	$(1.6 \pm 0.5) \cdot 10^{-6}$	5509

Baryons

Most baryons are observed as resonances among the primary or among the final particles. In the case of nonstrange baryons this is most often the πN-system; in the case of strange baryons it is mainly a KN- or a $H\pi$- system (here H stands for hyperon). The resonance is revealed by partial-wave analysis of the respective reactions. In doing so, the resonance wave is singled out. Its quantum numbers $L_{2I,2J}$ are indicated together with the name of the particle. Thus, for example, the state $N(1675)D_{15}$ is identified as a resonance in the πN-system of mass 1675 MeV, orbital momentum $L = 2$, isotopic spin 1/2, and spin (total angular momentum) 5/2. The resonance occurs at a pion beam momentum $p_{\text{beam}} = 1.01$ GeV/c and has a total production cross section $4\pi\lambda^2 = 15.4$ mb.

N – baryons ($S = 0,\ I = 1/2$)

p $\quad I(J^P) = \frac{1}{2}(\frac{1}{2}^+)$

$m = 938.27231 \pm 0.00028$ MeV
$m(\overline{p})/m(p) = 0.99999998 \pm 0.00000004$
$\mu = 2.79284739 \pm 0.00000006\ \mu_B$
$d = (-4 \pm 6) \cdot 10^{-23}\ e{\cdot}$cm
$|q_p + q_e| < 1.0 \cdot 10^{-21}\ e$ (hydrogen atom charge)
$\tau > 1.6 \cdot 10^{25}$ years (independent of decay channels)
$\tau > 10^{31} - 5 \cdot 10^{32}$ years (depends on assumed decay channels)

n $\quad I(J^P) = \frac{1}{2}(\frac{1}{2}^+)$

$m = 939.56563 \pm 0.00028$ MeV
$m_n - m_p = 1.293318 \pm 0.000009$ MeV
$\tau = 889.1 \pm 2.1$ s
$\mu = -1.9130427 \pm 0.0000005\ \mu_B$
$d < 12 \cdot 10^{-26}\ e{\cdot}$cm, $CL = 95\%$
$q = (-0.4 \pm 1.1) \cdot 10^{-21}\ e$
mean time for the $n\bar{n}$-oscillations $> 1.2 \cdot 10^8$ s, $CL = 90\%$

Decay channels n	Γ_i/Γ	p
$pe^-\bar{\nu}_e$	100%	1.19

$N(1440)P_{11}$ $\quad I(J^P) = \frac{1}{2}(\frac{1}{2}^+)$

$m = 1430 - 1470$ MeV
$\Gamma = 250 - 450$ MeV
$p_{\text{beam}} = 0.61$ GeV/s $\quad 4\pi\lambda^2 = 31.0$ mb

Decay channels $N(1440)$	Γ_i/Γ	p
$N\pi$	60 – 70%	397
$N\pi\pi$	30 – 40%	342
$\Delta\pi$	20 – 30%	143
$N\rho$	< 10%	-
$p\gamma$	0.08 – 0.10%	414
$n\gamma$	0.01 – 0.06%	413

$N(1520)D_{13}$ $\quad I(J^P) = \frac{1}{2}(\frac{3}{2}^-)$

$m = 1515 - 1530$ MeV
$\Gamma = 110 - 135$ MeV
$p_{\text{beam}} = 0.74$ GeV/s $\quad 4\pi\lambda^2 = 23.5$ mb

Decay channels $N(1520)$	Γ_i/Γ	p
$N\pi$	50 – 60%	456
$N\eta$	≈ 0.1%	149
$N\pi\pi$	40 – 50%	410
$\Delta\pi$	15 – 30%	228
$N\rho$	10 – 25%	-
$p\gamma$	0.43 – 0.57%	470
$n\gamma$	0.34 – 0.51%	470

$N(1535)D_{11}$ $\quad I(J^P) = \frac{1}{2}(\frac{1}{2}^-)$

$m = 1520 - 1555$ MeV
$\Gamma = 100 - 250$ MeV
$p_{\text{beam}} = 0.76$ GeV/s $\quad 4\pi\lambda^2 = 22.5$ mb

Decay channels $N(1535)$	Γ_i/Γ	p
$N\pi$	35 – 55%	467
$N\eta$	30 – 50%	182
$N\pi\pi$	5 – 20%	422
$p\gamma$	0.1 – 0.2%	481
$n\gamma$	0.15 – 0.35%	480

$N(1650)S_{11}$ $\quad I(J^P) = \frac{1}{2}(\frac{1}{2}^-)$

$m = 1640 - 1680$ MeV
$\Gamma = 145 - 190$ MeV
$p_{\text{beam}} = 0.96$ GeV/s $\quad 4\pi\lambda^2 = 16.4$ mb

Decay channels $N(1650)$	Γ_i/Γ	p
$N\pi$	60 – 80%	547
$N\eta$	≈ 1%	346
ΛK	≈ 7%	161
$N\pi\pi$	5 – 20%	511
$p\gamma$	0.04 – 0.16%	558

$N(1675)D_{15}$ $\quad I(J^P) = \frac{1}{2}(\frac{5}{2}^-)$

$m = 1670 - 1685$ MeV
$\Gamma = 140 - 180$ MeV
$p_{\text{beam}} = 1.01$ GeV/s $\quad 4\pi\lambda^2 = 15.4$ mb

Decay channels $N(1675)$	Γ_i/Γ	p
$N\pi$	40 – 50%	563
$N\eta$	≈ 1%	374

Decay channels $N(1675)$ *(continued)*	Γ_i/Γ	p
ΛK	$\approx 0.1\%$	209
$N\pi\pi$	$50-60\%$	529
$\Delta\pi$	$50-60\%$	364
$n\gamma$	$0.07-0.12\%$	574

$N(1680)F_{15}$ $I(J^P) = \frac{1}{2}(\frac{5}{2}^+)$

$m = 1675 - 1690$ MeV
$\Gamma = 120 - 140$ MeV
$p_{\text{beam}} = 1.01$ GeV/s $4\pi\lambda^2 = 15.2$ mb

Decay channels $N(1680)$	Γ_i/Γ	p
$N\pi$	$60-70\%$	567
$N\pi\pi$	$30-40\%$	532
$\Delta\pi$	$5-15\%$	369
$N\rho$	$5-15\%$	–
$p\gamma$	$0.21-0.30\%$	578
$n\gamma$	$0.02-0.05\%$	577

$N(1700)D_{13}$ $I(J^P) = \frac{1}{2}(\frac{3}{2}^-)$

$m = 1650 - 1750$ MeV
$\Gamma = 50 - 150$ MeV
$p_{\text{beam}} = 1.05$ GeV/s $4\pi\lambda^2 = 14.5$ mb

Decay channels $N(1700)$	Γ_i/Γ	p
$N\pi$	$5-15\%$	580
ΛK	$0.1-0.3\%$	250
$N\pi\pi$	$85-95\%$	547
$\Delta\pi$	$5-70\%$	385
$p\gamma$	$\sim 0.01\%$	591

$N(1710)P_{11}$ $I(J^P) = \frac{1}{2}(\frac{1}{2}^+)$

$m = 1680 - 1740$ MeV
$\Gamma = 50 - 250$ MeV
$p_{\text{beam}} = 1.07$ GeV/s $4\pi\lambda^2 = 14.2$ mb

Decay channels $N(1710)$	Γ_i/Γ	p
$N\pi$	$10-20\%$	587
$N\eta$	$20-40\%$	410
ΛK	$5-25\%$	264
$N\pi\pi$	$20-50\%$	554
$\Delta\pi$	$10-25\%$	393
$N\rho$	$5-20\%$	48

$N(1720)P_{13}$ $I(J^P) = \frac{1}{2}(\frac{3}{2}^+)$

$m = 1650 - 1750$ MeV
$\Gamma = 100 - 200$ MeV
$p_{\text{beam}} = 1.09$ GeV/s $4\pi\lambda^2 = 13.9$ mb

Decay channels $N(1720)$	Γ_i/Γ	p
$N\pi$	$10-20\%$	594
$N\eta$	$2-6\%$	420
ΛK	$3-10\%$	278
$N\pi\pi$	$>35\%$	561
$\Delta\pi$	$5-15\%$	401
$N\rho$	$25-75\%$	104

$N(2190)G_{17}$ $I(J^P) = \frac{1}{2}(\frac{7}{2}^-)$

$m = 2100 - 2200$ MeV
$\Gamma = 350 - 550$ MeV
$p_{\text{beam}} = 2.07$ GeV/s $4\pi\lambda^2 = 6.21$ mb

Decay channels $N(2190)$	Γ_i/Γ	p
$N\pi$	$10-20\%$	888
$N\eta$	$1-3\%$	790
ΛK	$0.2-0.4\%$	712
$N\pi\pi$	$20-40\%$	868
$N\rho$	$20-40\%$	683

$N(2220)H_{19}$ $I(J^P) = \frac{1}{2}(\frac{9}{2}^+)$

$m = 2180 - 2310$ MeV
$\Gamma = 320 - 550$ MeV
$p_{\text{beam}} = 2.14$ GeV/s $4\pi\lambda^2 = 5.97$ mb

Decay channels $N(2220)$	Γ_i/Γ	p
$N\pi$	$10-20\%$	905
$N\eta$	$0.5-1.0\%$	811

$N(2250)G_{19}$ $I(J^P) = \frac{1}{2}(\frac{9}{2}^-)$

$m = 2170 - 2310$ MeV
$\Gamma = 290 - 470$ MeV
$p_{\text{beam}} = 2.21$ GeV/s $4\pi\lambda^2 = 5.74$ mb

Decay channels $N(2250)$	Γ_i/Γ	p
$N\pi$	$5-15\%$	923
$N\eta$	$1-3\%$	831

$N(2600)I_{1,11}$ $I(J^P) = \frac{1}{2}(\frac{11}{2}^-)$

$m = 2550 - 2750$ MeV
$\Gamma = 500 - 800$ MeV
$p_{\text{beam}} = 3.12$ GeV/s $4\pi\lambda^2 = 3.86$ mb

Decay channels $N(2600)$	Γ_i/Γ	p
$N\pi$	$5-10\%$	1126

Δ-baryons ($S = 0,\ I = 3/2$)

$\Delta(1232)P_{33}$ $I(J^P) = \frac{3}{2}(\frac{3}{2}^+)$

$m = 1230 - 1234$ MeV
$\Gamma = 115 - 125$ MeV
$p_{\text{beam}} = 0.30$ GeV/s $4\pi\lambda^2 = 94.8$ mb

Decay channels $\Delta(1232)$	Γ_i/Γ	p
$N\pi$	$99.3 - 99.5\%$	227
$N\gamma$	$0.56 - 0.66\%$	259

$\Delta(1600)P_{33}$ $I(J^P) = \frac{3}{2}(\frac{3}{2}^+)$

$m = 1550 - 1700$ MeV
$\Gamma = 250 - 450$ MeV
$p_{\text{beam}} = 0.87$ GeV/s $4\pi\lambda^2 = 18.6$ mb

Decay channels $\Delta(1600)$	Γ_i/Γ	p
$N\pi$	$10 - 25\%$	512
$N\pi\pi$	$75 - 90\%$	473
$\Delta\pi$	$50 - 60\%$	301
$N\rho$	$5 - 20\%$	-
$N(1440)\pi$	$20 - 30\%$	74

$\Delta(1620)S_{31}$ $I(J^P) = \frac{3}{2}(\frac{1}{2}^-)$

$m = 1615 - 1675$ MeV
$\Gamma = 120 - 180$ MeV
$p_{\text{beam}} = 0.91$ GeV/s $4\pi\lambda^2 = 17.7$ mb

Decay channels $\Delta(1620)$	Γ_i/Γ	p
$N\pi$	$20 - 30\%$	526
$N\pi\pi$	$70 - 80\%$	488
$\Delta\pi$	$40 - 60\%$	318
$N\rho$	$20 - 35\%$	-
$N\gamma$	$\sim 0.03\%$	538

$\Delta(1700)D_{33}$ $I(J^P) = \frac{3}{2}(\frac{3}{2}^-)$

$m = 1670 - 1770$ MeV
$\Gamma = 200 - 400$ MeV
$p_{\text{beam}} = 1.05$ GeV/s $4\pi\lambda^2 = 14.5$ mb

Decay channels $\Delta(1700)$	Γ_i/Γ	p
$N\pi$	$10 - 20\%$	580
$N\pi\pi$	$80 - 90\%$	547
$\Delta\pi$	$35 - 55\%$	385
$N\rho$	$30 - 50\%$	-

$\Delta(1900)S_{31}$ $I(J^P) = \frac{3}{2}(\frac{1}{2}^-)$

$m = 1850 - 1950$ MeV
$\Gamma = 140 - 240$ MeV
$p_{\text{beam}} = 1.44$ GeV/s $4\pi\lambda^2 = 9.71$ mb

Decay channels $\Delta(1900)$	Γ_i/Γ	p
$N\pi$	$10 - 30\%$	710

$\Delta(1905)F_{35}$ $I(J^P) = \frac{3}{2}(\frac{5}{2}^+)$

$m = 1870 - 1920$ MeV
$\Gamma = 280 - 440$ MeV
$p_{\text{beam}} = 1.45$ GeV/s $4\pi\lambda^2 = 9.62$ mb

Decay channels $\Delta(1905)$	Γ_i/Γ	p
$N\pi$	$5 - 15\%$	713
ΣK	$0.1 - 0.3\%$	415
$N\pi\pi$	$85 - 95\%$	687
$N\rho$	$55 - 95\%$	421
$N\gamma$	$0.01 - 0.05\%$	721

$\Delta(1910)P_{31}$ $I(J^P) = \frac{3}{2}(\frac{1}{2}^+)$

$m = 1870 - 1920$ MeV
$\Gamma = 190 - 270$ MeV
$p_{\text{beam}} = 1.46$ GeV/s $4\pi\lambda^2 = 9.54$ mb

Decay channels $\Delta(1910)$	Γ_i/Γ	p
$N\pi$	$15 - 30\%$	716
$N\pi\pi$	$70 - 85\%$	691
$N\rho$	$5 - 25\%$	426
$N(1440)\pi$	$50 - 70\%$	393

$\Delta(1920)P_{33}$ $I(J^P) = \frac{3}{2}(\frac{3}{2}^+)$

$m = 1900 - 1970$ MeV
$\Gamma = 150 - 300$ MeV
$p_{\text{beam}} = 1.48$ GeV/s $4\pi\lambda^2 = 9.37$ mb

Decay channels $\Delta(1920)$	Γ_i/Γ	p
$N\pi$	$5 - 20\%$	722
ΣK	$1 - 3\%$	431

$\Delta(1930)D_{35}$ $I(J^P) = \frac{3}{2}(\frac{5}{2}^-)$

$m = 1920 - 1970$ MeV
$\Gamma = 250 - 450$ MeV
$p_{\text{beam}} = 1.50$ GeV/s $4\pi\lambda^2 = 9.21$ mb

Decay channels $\Delta(1930)$	Γ_i/Γ	p
$N\pi$	$10 - 20\%$	729

$\Delta(1950)F_{37}$ $I(J^P) = \frac{3}{2}(\frac{7}{2}^+)$

$m = 1940 - 1960$ MeV
$\Gamma = 290 - 350$ MeV
$p_{\text{beam}} = 1.54$ GeV/s $4\pi\lambda^2 = 8.91$ mb

Decay channels $\Delta(1950)$	Γ_i/Γ	p
$N\pi$	$35 - 40\%$	741
ΣK	$0.6 - 0.8\%$	460
$N\pi\pi$	$15 - 40\%$	716

Decay channels $\Delta(1950)$ *(continued)*	Γ_i/Γ	p
$\Delta\pi$	$15-30\%$	574
$N\gamma$	$0.08-0.17\%$	749

$\Delta(2420)H_{3,11}$ $I(J^P) = \frac{3}{2}(\frac{11}{2}^+)$

$m = 2300 - 2500$ MeV
$\Gamma = 300 - 500$ MeV
$p_{\text{beam}} = 2.64$ GeV/s $4\pi\lambda^2 = 4.68$ mb

Decay channels $\Delta(2420)$	Γ_i/Γ	p
$N\pi$	$5-15\%$	1023
ΣK	$0.1-0.9\%$	833

Λ-baryons ($S = -1,\ I = 0$)

Λ $I(J^P) = 0(\frac{1}{2}^+)$

$m = 1115.63 \pm 0.05$ MeV
$\tau = (2.632 \pm 0.020) \cdot 10^{-10}$ s
$\mu = -0.613 \pm 0.004\ \mu_B$
$d < 1.5 \cdot 10^{-16}$ e·cm, $CL = 95\%$

Decay channels Λ	Γ_i/Γ	p
$p\pi^-$	$(64.1 \pm 0.5)\%$	101
$n\pi^0$	$(35.7 \pm 0.5)\%$	104
$n\gamma$	$(1.02 \pm 0.33) \cdot 10^{-3}$	162
$p\pi^-\gamma$	$(8.5 \pm 1.4) \cdot 10^{-4}$	101
$pe^-\bar{\nu}_e$	$(8.34 \pm 0.14) \cdot 10^{-4}$	163
$p\mu^-\bar{\nu}_\mu$	$(1.57 \pm 0.35) \cdot 10^{-4}$	131

$\Lambda(1405)S_{01}$ $I(J^P) = 0(\frac{1}{2}^-)$

$m = 1407 \pm 4$ MeV
$\Gamma = 50.0 \pm 2.0$ MeV

Decay channels $\Lambda(1405)$	Γ_i/Γ	p
$\Sigma\pi$	100%	152

$\Lambda(1520)D_{03}$ $I(J^P) = 0(\frac{3}{2}^-)$

$m = 1519.5 \pm 1.0$ MeV
$\Gamma = 15.6 \pm 1.0$ MeV
$p_{\text{beam}} = 0.39$ GeV/s $4\pi\lambda^2 = 82.8$ mb

Decay channels $\Lambda(1520)$	Γ_i/Γ	p
$N\overline{K}$	$(45 \pm 1)\%$	244
$\Sigma\pi$	$(42 \pm 1)\%$	267
$\Lambda\pi\pi$	$(10 \pm 1)\%$	252
$\Sigma\pi\pi$	$(0.9 \pm 0.1)\%$	152
$\Lambda\gamma$	$(0.8 \pm 0.2)\%$	351

$\Lambda(1600)P_{01}$ $I(J^P) = 0(\frac{1}{2}^+)$

$m = 1560 - 1700$ MeV
$\Gamma = 50 - 250$ MeV
$p_{\text{beam}} = 0.58$ GeV/s $4\pi\lambda^2 = 41.6$ mb

Decay channels $\Lambda(1600)$	Γ_i/Γ	p
$N\overline{K}$	$15-30\%$	343
$\Sigma\pi$	$10-60\%$	336

$\Lambda(1670)S_{01}$ $I(J^P) = 0(\frac{1}{2}^-)$

$m = 1660 - 1680$ MeV
$\Gamma = 25 - 50$ MeV
$p_{\text{beam}} = 0.74$ GeV/s $4\pi\lambda^2 = 28.5$ mb

Decay channels $\Lambda(1670)$	Γ_i/Γ	p
$N\overline{K}$	$15-25\%$	414
$\Sigma\pi$	$20-60\%$	393
$\Lambda\eta$	$15-35\%$	64

$\Lambda(1690)D_{03}$ $I(J^P) = 0(\frac{3}{2}^-)$

$m = 1685 - 1695$ MeV
$\Gamma = 50 - 70$ MeV
$p_{\text{beam}} = 0.78$ GeV/s $4\pi\lambda^2 = 26.1$ mb

Decay channels $\Lambda(1690)$	Γ_i/Γ	p
$N\overline{K}$	$20-30\%$	433
$\Sigma\pi$	$20-40\%$	409
$\Lambda\pi\pi$	$\sim 25\%$	415
$\Sigma\pi\pi$	$\sim 20\%$	350

$\Lambda(1800)S_{01}$ $I(J^P) = 0(\frac{1}{2}^-)$

$m = 1720 - 1850$ MeV
$\Gamma = 200 - 400$ MeV
$p_{\text{beam}} = 1.01$ GeV/s $4\pi\lambda^2 = 17.5$ mb

Decay channels $\Lambda(1800)$	Γ_i/Γ	p
$N\overline{K}$	$25-40\%$	528
$\Sigma\pi$	observed	493
$\Sigma(1385)\pi$	observed	345
$N\overline{K}^*(892)$	observed	-

$\Lambda(1810)P_{01}$ $I(J^P) = 0(\frac{1}{2}^+)$

$m = 1750 - 1850$ MeV
$\Gamma = 50 - 250$ MeV
$p_{\text{beam}} = 1.04$ GeV/s $4\pi\lambda^2 = 17.0$ mb

Decay channels $\Lambda(1810)$	Γ_i/Γ	p
$N\overline{K}$	$20-50\%$	537
$\Sigma\pi$	$10-40\%$	501
$\Sigma(1385)\pi$	observed	356
$N\overline{K}^*(892)$	$30-60\%$	-

$\Lambda(1820)F_{05}$ $I(J^P) = 0(\frac{5}{2}^+)$

$m = 1815 - 1825$ MeV

$\Gamma = 70 - 90$ MeV

$p_{\text{beam}} = 1.06$ GeV/s $4\pi\lambda^2 = 16.5$ mb

Decay channels $\Lambda(1820)$	Γ_i/Γ	p
$N\overline{K}$	55 – 65%	545
$\Sigma\pi$	8 – 14%	508
$\Sigma(1385)\pi$	5 – 10%	362

$\Lambda(1830)D_{05}$ $I(J^P) = 0(\frac{5}{2}^-)$

$m = 1810 - 1830$ MeV

$\Gamma = 60 - 110$ MeV

$p_{\text{beam}} = 1.08$ GeV/s $4\pi\lambda^2 = 16.0$ mb

Decay channels $\Lambda(1830)$	Γ_i/Γ	p
$N\overline{K}$	3 – 10%	553
$\Sigma\pi$	35 – 75%	515
$\Sigma(1385)\pi$	> 15%	371

$\Lambda(1890)P_{03}$ $I(J^P) = 0(\frac{3}{2}^+)$

$m = 1850 - 1910$ MeV

$\Gamma = 60 - 200$ MeV

$p_{\text{beam}} = 1.21$ GeV/s $4\pi\lambda^2 = 13.6$ mb

Decay channels $\Lambda(1890)$	Γ_i/Γ	p
$N\overline{K}$	20 – 35%	599
$\Sigma\pi$	3 – 10%	559
$\Sigma(1385)\pi$	observed	420
$N\overline{K}^*(892)$	observed	233

$\Lambda(2100)G_{07}$ $I(J^P) = 0(\frac{7}{2}^-)$

$m = 2090 - 2110$ MeV

$\Gamma = 100 - 250$ MeV

$p_{\text{beam}} = 1.68$ GeV/s $4\pi\lambda^2 = 8.68$ mb

Decay channels $\Lambda(2100)$	Γ_i/Γ	p
$N\overline{K}$	25 – 35%	751
$\Sigma\pi$	~ 5%	704
$N\overline{K}^*(892)$	10 – 20%	514

$\Lambda(2110)F_{05}$ $I(J^P) = 0(\frac{5}{2}^+)$

$m = 2090 - 2140$ MeV

$\Gamma = 150 - 250$ MeV

$p_{\text{beam}} = 1.70$ GeV/s $4\pi\lambda^2 = 8.53$ mb

Decay channels $\Lambda(2110)$	Γ_i/Γ	p
$N\overline{K}$	5 – 25%	757
$\Sigma\pi$	10 – 40%	711
$\Lambda\omega$	observed	455
$\Sigma(1385)\pi$	observed	589
$N\overline{K}^*(892)$	10 – 60%	524

$\Lambda(2350)H_{09}$ $I(J^P) = 0(\frac{9}{2}^+)$

$m = 2340 - 2370$ MeV

$\Gamma = 100 - 250$ MeV

$p_{\text{beam}} = 2.29$ GeV/s $4\pi\lambda^2 = 5.85$ mb

Decay channels $\Lambda(2350)$	Γ_i/Γ	p
$N\overline{K}$	~ 12%	915
$\Sigma\pi$	~ 10%	867

Σ-baryons ($S = -1,\ I = 1$)

Σ^+ $I(J^P) = 1(\frac{1}{2}^+)$

$m = 1189.37 \pm 0.07$ MeV

$\tau = (0.799 \pm 0.004) \cdot 10^{-10}$ s

$\mu = 2.42 \pm 0.05\ \mu_B$

$\Gamma(\Sigma^+ \to n\ell^+\nu)/\Gamma(\Sigma^- \to n\ell^-\bar{\nu}) < 0.043$

Decay channels Σ^+	Γ_i/Γ	p
$p\pi^0$	$(51.57 \pm 0.30)\%$	189
$n\pi^+$	$(48.30 \pm 0.30)\%$	185
$p\gamma$	$(1.25 \pm 0.07) \cdot 10^{-3}$	225
$n\pi^+\gamma$	$(4.5 \pm 0.5) \cdot 10^{-4}$	185
$\Lambda e^+\nu_e$	$(2.0 \pm 0.5) \cdot 10^{-5}$	71

Σ^0 $I(J^P) = 1(\frac{1}{2}^+)$

J^P undefined; assumed identical to that for Σ^+ and Σ^-

$m = 1192.55 \pm 0.10$ MeV

$m_{\Sigma^-} - m_{\Sigma^0} = 4.89 \pm 0.08$ MeV

$m_{\Sigma^0} - m_\Lambda = 76.92 \pm 0.10$ MeV

$\tau = (7.4 \pm 0.7) \cdot 10^{-20}$ s

Decay channels Σ^0	Γ_i/Γ	p
$\Lambda\gamma$	100%	74
Λe^+e^-	$5 \cdot 10^{-3}$	74

Σ^- $I(J^P) = 1(\frac{1}{2}^+)$

$m = 1197.43 \pm 0.06$ MeV

$m_{\Sigma^-} - m_{\Sigma^+} = 8.07 \pm 0.09$ MeV

$\tau = (1.479 \pm 0.011) \cdot 10^{-10}$ s

$\mu = -1.160 \pm 0.025\ \mu_B$

Decay channels Σ^-	Γ_i/Γ	p
$n\pi^-$	$(99.848 \pm 0.005)\%$	193

Decay channels Σ^- *(continued)*	Γ_i/Γ	p
$n\pi^-\gamma$	$(4.6 \pm 0.6) \cdot 10^{-4}$	193
$ne^-\bar{\nu}_e$	$(1.017 \pm 0.034) \cdot 10^{-3}$	230
$n\mu^-\bar{\nu}_\mu$	$(4.5 \pm 0.4) \cdot 10^{-4}$	210
$\Lambda e^-\bar{\nu}_e$	$(5.73 \pm 0.27) \cdot 10^{-5}$	79

$\Sigma(1385)P_{13}$ $\quad I(J^P) = 1(\frac{3}{2}^+)$

$\Sigma(1385)^+$ $\quad m = 1382.8 \pm 0.4$ MeV
$\Sigma(1385)^0$ $\quad m = 1383.7 \pm 1.0$ MeV
$\Sigma(1385)^-$ $\quad m = 1387.2 \pm 0.5$ MeV
$\Sigma(1385)^+$ $\quad \Gamma = 35.8 \pm 0.8$ MeV
$\Sigma(1385)^0$ $\quad \Gamma = 36 \pm 5$ MeV
$\Sigma(1385)^-$ $\quad \Gamma = 39.4 \pm 2.1$ MeV

Decay channels $\Sigma(1385)$	Γ_i/Γ	p
$\Lambda\pi$	$(88 \pm 2)\%$	208
$\Sigma\pi$	$(12 \pm 2)\%$	127

$\Sigma(1660)P_{11}$ $\quad I(J^P) = 1(\frac{1}{2}^+)$

$m = 1630 - 1690$ MeV
$\Gamma = 40 - 200$ MeV
$p_{\text{beam}} = 0.72$ GeV/s $\quad 4\pi\lambda^2 = 29.9$ mb

Decay channels $\Sigma(1660)$	Γ_i/Γ	p
$N\overline{K}$	10 – 30%	405
$\Lambda\pi$	observed	439
$\Sigma\pi$	observed	385

$\Sigma(1670)D_{13}$ $\quad I(J^P) = 1(\frac{3}{2}^-)$

$m = 1665 - 1685$ MeV
$\Gamma = 40 - 80$ MeV
$p_{\text{beam}} = 0.74$ GeV/s $\quad 4\pi\lambda^2 = 28.5$ mb

Decay channels $\Sigma(1670)$	Γ_i/Γ	p
$N\overline{K}$	7 – 13%	414
$\Lambda\pi$	5 – 15%	447
$\Sigma\pi$	30 – 60%	393

$\Sigma(1750)S_{11}$ $\quad I(J^P) = 1(\frac{1}{2}^-)$

$m = 1730 - 1800$ MeV
$\Gamma = 60 - 160$ MeV
$p_{\text{beam}} = 0.91$ GeV/s $\quad 4\pi\lambda^2 = 20.7$ mb

Decay channels $\Sigma(1750)$	Γ_i/Γ	p
$N\overline{K}$	10 – 40%	486
$\Lambda\pi$	observed	507
$\Sigma\eta$	15 – 55%	81

$\Sigma(1775)D_{15}$ $\quad I(J^P) = 1(\frac{5}{2}^-)$

$m = 1770 - 1780$ MeV
$\Gamma = 105 - 135$ MeV
$p_{\text{beam}} = 0.96$ GeV/s $\quad 4\pi\lambda^2 = 19.0$ mb

Decay channels $\Sigma(1775)$	Γ_i/Γ	p
$N\overline{K}$	37 – 43%	508
$\Lambda\pi$	14 – 20%	525
$\Sigma\pi$	2 – 5%	474
$\Sigma(1385)\pi$	8 – 12%	324
$\Lambda(1520)\pi$	17 – 23%	198

$\Sigma(1915)F_{15}$ $\quad I(J^P) = 1(\frac{5}{2}^+)$

$m = 1900 - 1935$ MeV
$\Gamma = 80 - 160$ MeV
$p_{\text{beam}} = 1.26$ GeV/s $\quad 4\pi\lambda^2 = 12.8$ mb

Decay channels $\Sigma(1915)$	Γ_i/Γ	p
$N\overline{K}$	5 – 15%	618
$\Lambda\pi$	observed	622
$\Sigma\pi$	observed	577

$\Sigma(1940)F_{13}$ $\quad I(J^P) = 1(\frac{3}{2}^-)$

$m = 1900 - 1950$ MeV
$\Gamma = 150 - 300$ MeV
$p_{\text{beam}} = 1.32$ GeV/s $\quad 4\pi\lambda^2 = 12.1$ mb

Decay channels $\Sigma(1940)$	Γ_i/Γ	p
$\Lambda\pi$	observed	639
$\Sigma\pi$	observed	594
$\Sigma(1385)\pi$	observed	460
$\Lambda(1520)\pi$	observed	354
$\Delta(1232)\overline{K}$	observed	410
$N\overline{K}^*(892)$	observed	320

$\Sigma(2030)F_{17}$ $\quad I(J^P) = 1(\frac{7}{2}^+)$

$m = 2025 - 2040$ MeV
$\Gamma = 150 - 200$ MeV
$p_{\text{beam}} = 1.52$ GeV/s $\quad 4\pi\lambda^2 = 9.93$ mb

Decay channels $\Sigma(2030)$	Γ_i/Γ	p
$N\overline{K}$	17 – 23%	702
$\Lambda\pi$	17 – 23%	700
$\Sigma\pi$	5 – 10%	657
$\Sigma(1385)\pi$	5 – 15%	529
$\Lambda(1520)\pi$	10 – 20%	430
$\Delta(1232)\overline{K}$	10 – 20%	498

Σ(2250) $I(J^P) = 1(?^?)$

$m = 2210 - 2280$ MeV
$\Gamma = 60 - 150$ MeV
$p_{\text{beam}} = 2.04$ GeV/s $\quad 4\pi\lambda^2 = 6.76$ mb

Decay channels Σ(2250)	Γ_i/Γ	p
$\Lambda\pi$	observed	842
$\Sigma\pi$	observed	803

Ξ-baryons ($S = -2,\ I = 1/2$)

Ξ^0 $I(J^P) = \frac{1}{2}(\frac{1}{2}^+)$

P undefined; + according to the quark model
$m = 1314.9 \pm 0.6$ MeV
$m_{\Xi^-} - m_{\Xi^0} = 6.4 \pm 0.6$ MeV
$\tau = (2.90 \pm 0.09) \cdot 10^{-10}$ s
$\mu = -1.250 \pm 0.014\ \mu_B$

Decay channels Ξ^0	Γ_i/Γ	p
$\Lambda\pi^0$	100%	135
$\Lambda\gamma$	$(1.06 \pm 0.16) \cdot 10^{-3}$	184
$\Sigma^0\gamma$	$(3.6 \pm 0.4) \cdot 10^{-3}$	117

Ξ^- $I(J^P) = \frac{1}{2}(\frac{1}{2}^+)$

P undefined; + according to the quark model
$m = 1321.32 \pm 0.13$ MeV
$\tau = (1.639 \pm 0.015) \cdot 10^{-10}$ s
$\mu = -0.6507 \pm 0.0025\ \mu_B$

Decay channels Ξ^-	Γ_i/Γ	p
$\Lambda\pi^-$	100%	139
$\Sigma^-\gamma$	$(2.3 \pm 1.0) \cdot 10^{-4}$	118
$\Lambda e^-\bar{\nu}_e$	$(5.5 \pm 0.3) \cdot 10^{-4}$	190
$\Sigma^0 e^-\bar{\nu}_e$	$(8.7 \pm 1.7) \cdot 10^{-5}$	122

Ξ(1530)P_{13} $I(J^P) = \frac{1}{2}(\frac{3}{2}^+)$

$\Xi(1530)^0$ $m = 1531.80 \pm 0.32$ MeV
$\Xi(1530)^-$ $m = 1535.0 \pm 0.6$ MeV
$\Xi(1530)^0$ $\Gamma = 9.1 \pm 0.5$ MeV
$\Xi(1530)^-$ $\Gamma = 9.9 \pm 1.7$ MeV

Decay channels Ξ(1530)	Γ_i/Γ	p
$\Xi\pi$	100%	152

Ξ(1690) $I(J^P) = \frac{1}{2}(?^?)$

$m = 1690 \pm 10$ MeV
$\Gamma < 50$ MeV

Decay channels Ξ(1690)	Γ_i/Γ	p
$\Lambda\overline{K}$	observed	240
$\Sigma\overline{K}$	observed	151
$\Xi^-\pi^+\pi^-$	observed	214

Ξ(1820)D_{13} $I(J^P) = \frac{1}{2}(\frac{3}{2}^-)$

$m = 1823 \pm 5$ MeV
$\Gamma = 24^{+15}_{-10}$ MeV

Decay channels Ξ(1820)	Γ_i/Γ	p
$\Lambda\overline{K}$	dominant channel	400
$\Sigma\overline{K}$	observed	320
$\Xi\pi$	observed	413
$\Xi(1530)\pi$	observed	234

Ξ(1950) $I(J^P) = \frac{1}{2}(?^?)$

$m = 1950 \pm 15$ MeV
$\Gamma = 60 \pm 20$ MeV

Decay channels Ξ(1950)	Γ_i/Γ	p
$\Lambda\overline{K}$	observed	522
$\Sigma\overline{K}$ *and* $\Xi\pi$	observed	–

Ξ(2030) $I(J^P) = \frac{1}{2}(\geq \frac{5}{2}^?)$

$m = 2025 \pm 5$ MeV
$\Gamma = 20^{+15}_{-5}$ MeV

Decay channels Ξ(2030)	Γ_i/Γ	p
$\Lambda\overline{K}$	$\sim 20\%$	589
$\Sigma\overline{K}$	$\sim 80\%$	533
$\Xi\pi$	observed	573
$\Xi(1530)\pi$	observed	421
$\Lambda\overline{K}\pi$	observed	501
$\Sigma\overline{K}\pi$	observed	430

Ω-baryons ($S = -3,\ I = 0$)

Ω^- $I(J^P) = 0\frac{3}{2}^+)$

J^P undefined; $\frac{3}{2}^+$ according to the quark model
$m = 1672.43 \pm 0.32$ MeV
$\tau = (0.822 \pm 0.012) \cdot 10^{-10}$ s
$\mu = -1.94 \pm 0.22\ \mu_B$

Decay channels Ω^-	Γ_i/Γ	p
ΛK^-	$(67.8 \pm 0.7)\%$	211
$\Xi^0\pi^-$	$(23.6 \pm 0.7)\%$	294
$\Xi^-\pi^0$	$(8.6 \pm 0.4)\%$	290

$\Omega(2250)^-$ $I(J^P) = 0(?^?)$

$m = 2252 \pm 9$ MeV
$\Gamma = 55 \pm 18$ MeV

Decay channels $\Omega(2250)^-$	Γ_i/Γ	p
$\Xi^-\pi^+K^-$	observed	531
$\Xi(1530)^0K^-$	observed	437

Charmed baryons ($C = +1$)

Λ_c^+ $I(J^P) = 0\frac{1}{2}^+)$

J undefined; $\frac{1}{2}$ according to the quark model
$m = 2284.9 \pm 0.6$ MeV
$\tau = (1.91 \pm 0.15) \cdot 10^{-13}$ s

Decay channels Λ_c^+	Γ_i/Γ	p
$p\overline{K}^0$	$(1.6 \pm 0.4)\%$	872
$pK^-\pi^+$	$(3.2 \pm 0.7)\%$	822
$p\overline{K}^*(892)^0$	$(8.8 \pm 2.9) \cdot 10^{-3}$	681
$\Delta(1232)^{++}K^-$	$(6.6 \pm 3.0) \cdot 10^{-3}$	709
$p\overline{K}^0\pi^+\pi^-$	$(1.7 \pm 0.6)\%$	753
$pK^-\pi^+\pi^0$	observed	758
$pK^*(892)^-\pi^+$	observed	579
$\Delta(1232)\overline{K}^*(892)$	observed	417
$p\pi^+\pi^-$	$(2.2 \pm 1.3) \cdot 10^{-3}$	926
$p\pi^+\pi^+\pi^-\pi^-$	$(1.2 \pm 0.8) \cdot 10^{-3}$	851
pK^+K^-	$(1.6 \pm 0.9) \cdot 10^{-3}$	615
ΛX	$(27 \pm 9)\%$	-
$\Lambda\pi^+$	$(5.8 \pm 1.6) \cdot 10^{-3}$	863
$\Lambda\pi^+\pi^+\pi^-$	$(2.1 \pm 0.5)\%$	806
$\Sigma^0\pi^+$	$(5.5 \pm 2.6) \cdot 10^{-3}$	824
$\Xi^-K^+\pi^+$	$(4.8 \pm 1.9) \cdot 10^{-3}$	564
Half-lepton channels		
e^+X	$(4.5 \pm 1.7)\%$	-
pe^+X	$(1.8 \pm 0.9)\%$	-
Λe^+X	$(1.2 \pm 0.4)\%$	-
$\Lambda\mu^+X$	$(1.1 \pm 0.7)\%$	-

$\Sigma_c(2455)$ $I(J^P) = 1\frac{1}{2}^+)$

J^P undefined; $\frac{1}{2}^+$ according to the quark model
$\Sigma_c(2455)^{++}$ $m = 2452.7 \pm 0.7$ MeV
$\Sigma_c(2455)^+$ $m = 2452.9 \pm 3.1$ MeV
$\Sigma_c(2455)^0$ $m = 2452.5 \pm 0.8$ MeV

Decay channels $\Sigma_c(2455)$	Γ_i/Γ	p
$\Lambda_c^+\pi$	100%	93

Ξ_c^+ $I(J^P) = \frac{1}{2}(\frac{1}{2}^+)$

$I(J^P)$ undefined; $\frac{1}{2}(\frac{1}{2}^+)$ according to the quark model
$m = 2466.4 \pm 2.1$ MeV
$\tau = (3.0 \pm 1.0) \cdot 10^{-13}$ s

Decay channels Ξ_c^+	Γ_i/Γ	p
$\Lambda K^-\pi^+\pi^+$	observed	786
$\Sigma^+K^-\pi^+$	observed	810
$\Sigma^0K^-\pi^+\pi^+$	observed	734
$\Xi^-\pi^+\pi^+$	observed	850

Ξ_c^0 $I(J^P) = \frac{1}{2}(\frac{1}{2}^+)$

$I(J^P)$ undefined; $\frac{1}{2}(\frac{1}{2}^+)$ according to the quark model
$m = 2472.7 \pm 1.7$ MeV
$m_{\Xi_c^0} - m_{\Xi_c^+} = 6.3 \pm 2.3$ MeV
$\tau = (0.82^{+0.59}_{-0.30}) \cdot 10^{-13}$ s

Decay channels Ξ_c^0	Γ_i/Γ	p
Ξ^-l^+X	observed	-
$\Xi^-\pi^+$	observed	876
$\Xi^-\pi^+\pi^+\pi^-$	observed	818
$pK^-\overline{K}^*(892)^0$	observed	410
Ω^-K^+	observed	876

B-baryons ($B = -1$)

Λ_b^0 $I(J^P) = 0(\frac{1}{2}^+)$

$I(J^P)$ undefined; $0(\frac{1}{2}^+)$ according to the quark model
$m = 5641 \pm 50$ MeV

Decay channels Λ_b^0	Γ_i/Γ	p
$J/\psi(1S)\Lambda$	observed	1756
$pD^0\pi^-$	observed	2383
$\Lambda_c^+\pi^+\pi^-\pi^-$	observed	2336
Λl^-X	observed	-
$\Lambda_c^+l^-X$	observed	-

36.9 Testing the Invariance Principles and Quantum Number Conservation Laws

CPT invariance. The general principles of relativistic quantum field theory lead to the assertion that the equations of motion are invariant with respect to the joint action of the operators of charge conjugation C (transformation of a particle into its antiparticle), inversion of the coordinate axes in space $P(\vec{r} \to -\vec{r})$, and inversion of time $T(t \to -t)$. This is the so-called CPT theorem. CPT invariance of the laws of Nature requires, for example, that the lifetimes as well as the masses of particles and antiparticles be identical. The best test of this consequence is provided by the comparison of the masses of K^0- and $\bar{K}^0$-mesons: $|m(\bar{K}^0) - m(K^0)|/m(K^0) \leq 10^{-18}$, $CL = 90\%$.

P and C invariance. The experiment shows that processes of strong and electromagnetic interactions are invariant with respect to inversion P of the axes of the reference system in space and to charge conjugation C, i.e. they are mirror symmetric and do not alter when antiparticles are substituted for particles. Weak interactions break these symmetries. This is best demonstrated by leptons in particle decays: they are longitudinally polarized, and the helicity of leptons is negative – they are similar to left-handed screws, while the helicity of antileptons is positive, and they look like right-handed screws. Since leptons of opposite helicities are not produced in decays, P and C parities are said to be 100% broken in weak interactions.

CP and T invariance. The CPT theorem signifies that CP and T invariance may be broken simultaneously or obeyed simultaneously. Only one process is known, in which CP invariance does not occur: the K^0-meson decay. Thus, K_L^0 decays either into 3π or into 2π. The former channel is dominant and is CP odd, while the latter occurs in $\sim$ 2% of cases and is CP even. Precisely the above event determines the degree to which CP invariance is violated. Charged nonsymmetry is also observed in the decay of the neutral particle K_L^0: $[\Gamma(K_L^0 \to \pi^- e^+ \nu) - \Gamma(K_L^0 \to \pi^+ e^- \nu)/\text{sum} = 0.333 \pm 0.014\%$. This points to the World and the Antiworld being physically different. Violation of T invariance could be confirmed by the electric dipole moments of particles. The vector $\vec{d} = \vec{r} \cdot e$ (dipole moment) and the pseudovector $\vec{\mu}$ (magnetic moment) would form a T noninvariant system. The most accurately measured dipole moment is the dipole moment of the neutron. Its upper limit amounts to $d_n < 1.2 \cdot 10^{-25}$ e·cm, $CL = 90\%$.

Conservation of lepton numbers. Experimental data and the theory of the electroweak interaction are consistent with conservation of the three lepton numbers L_e, L_μ, L_τ assigned to the three lepton families. Thus, for example, conversion of leptons of different families ($l_i \nleftrightarrow l_j$, $i \neq j$; $i, j = l, \mu, \tau$) is forbidden, and the sum of leptons of a given family $i \sum_k L_i^k$ is conserved in all processes. Here we present some of the most precise data:

$$\Gamma(\mu \to e\gamma)/\Gamma(\mu \to \text{all}) < 5 \cdot 10^{-11}, \quad \Gamma(\mu^- + A \to e^- + A)/\Gamma(\mu^- + A \to \text{all}) < 1 \cdot 10^{-12}$$

(Conversion in a mesoatom)

$$\Gamma(\tau \to \mu\gamma)/\Gamma(\tau \to \text{all}) < 5 \cdot 10^{-4}, \quad (Z, A) \to (Z+2, A) + e^- + e^+, \quad \tau > 1.3 \cdot 10^{24}\ y,$$

$$CL = 68\% \quad \text{for } ^{76}\text{Ge}$$

(Double neutrinoless β-decay).

If the neutrinos in differing families have nonzero and different masses, then formally it is possible to construct a theory (by analogy with the known mixing process of particles and antiparticles in a beam of neutral K^0-mesons, $K^0 \leftrightarrow \bar{K}^0$) with neutrino mixing and oscillations: $\nu_i \leftrightarrow \nu_j, i \neq j$. An experiment for testing this scheme consists in measuring the amount of $\bar{\nu}_e$ in the flux at different distances from the source (the reactor). The disappearance of $\bar{\nu}_e$ could be the result of $\bar{\nu}_e \leftrightarrow \bar{\nu}_\mu$ transitions. In the table presented below this process is denoted as $\bar{\nu}_e \nleftrightarrow \bar{\nu}_e$ (change in the flux). The intensity of $\bar{\nu}_e \leftrightarrow \bar{\nu}_\mu$ transitions (oscillations) is determined by the parameter $\Delta m^2 \sin^2(2\theta)$, where θ

is the mixing angle of the respective neutrinos. If one arbitrarily sets $\sin 2\theta = 1$, then the experiment will yield $\Delta m^2 < 0.0083$ eV^2.

Conservation of quark flavors. In strong and electromagnetic interactions quarks of different flavors (and families)u, d, s, c, b, t do not transform into each other. Accordingly, the quantum numbers S, C, B, T are conserved. The weak interaction violates these laws. There are certain rules, however, that are observed and look as follows.

a) *The* $\Delta S = \Delta Q$ *rule*

This rule relates the change in strangeness and the change in electric charge of hadrons involved in semileptonic decays. Thus, for example, the process $\Sigma^+ \to ne^+\nu_e$, which violates the $\Delta S = \Delta Q$ rule has not been observed, $\Gamma_i/\Gamma_{\text{tot}} < 5 \cdot 10^{-6}$, while the analogous process $\Sigma^- \to ne^-\bar{\nu}_e$ consistent with this rule, is observed and its partial width $\Gamma_i/\Gamma_{\text{tot}} \approx 1 \cdot 10^{-3}$. There exist, also, the rules $\Delta C = \Delta Q$, and $\Delta B = \Delta Q$.

b) *The change of flavor by two units*

The well-known mixing effect $K^0 \leftrightarrow \bar{K}^0$ involves a change of strangeness $\Delta S = 2$. It is due to a second-order weak interaction process: the $(d, \bar{s})$ quark pair twice exchanges W bosons and transforms into the $(\bar{d}, s)$ pair. The data on $K^0 \leftrightarrow \bar{K}^0$ mixing yield directly the mass difference $m(K_S^0) - m(K_L^0) = (3.522 \pm 0.016) \cdot 10^{-12}$ MeV. Mixing has also been found in the $B^0 \leftrightarrow \bar{B}^0 (\Delta B = 2)$. The corresponding mass difference amounts to $|m(B_1^0) - m(B_2^0)| = (0.72 \pm 0.14)\Gamma_B = (3.6 \pm 0.7) \cdot 10^{-10}$ MeV. The mixing of the charmed neutral mesons $D^0 \leftrightarrow \bar{D}^0$ is expected to be small in the standard theory and has not been observed experimentally.

c) *Neutral currents do not change flavor*

Exchanging Z^0-boson does not result in a change in the flavor of quarks. This is confirmed by the absence of the process $K^+ \to \pi^+\nu\bar{\nu}$. It can proceed only in the second order of weak interaction: $u\bar{s} \to W^+ \to u\bar{d};\ \bar{d} \to \bar{d}Z^0 \to \bar{d}\nu\bar{\nu}$. This is also confirmed by the probability of the decay $K_L^0 \to \mu^+\mu^-$ being small and by the absence of the decays $B^0 \to \mu^+\mu^-$, $D^0 \to \mu^+\mu^-$, when $\Gamma_i/\Gamma_{\text{tot}} < 10^{-5}$.

Below a summary is presented of experimental data on tests of quantum number conservation laws and on the fulfillment of C, P, T symmetries [4].

Table 36.9 Testing the quantum number conservation laws and the fulfillment of C, P, T symmetries

Quantity	Value	Tested law of conservation or symmetry
$\pi^0 \to \gamma\gamma\gamma/\text{all}$	$< 3.1 \cdot 10^{-8}$	C
$(e^+e^-)_{J=0} \to 3\gamma/2\gamma$	$< 1 \cdot 10^{-5}$	C
$(e^+e^-)_{J=1} \to 4\gamma/3\gamma$	$< 1 \cdot 10^{-5}$	C
$\eta \to \gamma\gamma\gamma/\text{all}$	$< 5 \cdot 10^{-4}$	C
$\eta \to \pi^0 e^+e^-/\text{all}$	$< 4 \cdot 10^{-5}$	C
$\eta \to \pi^0\mu^+\mu^-/\text{all}$	$< 5 \cdot 10^{-6}$	C
$\eta \to \pi^+\pi^-/\text{all}$	$< 1.5 \cdot 10^{-3}$	P and CP
e electrical dipole moment	$(-3 \pm 8) \cdot 10^{-27}$ e·cm	T and P
μ electrical dipole moment	$(3.7 \pm 3.4) \cdot 10^{-19}$e·cm	T and P
p electrical dipole moment	$(-4 \pm 6) \cdot 10^{-23}$ e·cm	T and P
n electrical dipole moment	$< 1.2 \cdot 10^{-25}$ e·cm	T and P
Λ electrical dipole moment	$< 1.5 \cdot 10^{-16}$ e·cm	T and P
Transverse polarization e^+ in decay of polarized muon $\mu^+ \to e^+\bar{\nu}\nu$	0.007 ± 0.023	T
$K^\pm \to \pi^\pm\pi^+\pi^-$ difference of probabilities/mean	$(0.07 \pm 0.12)\%$	CP
$K^\pm \to \pi^\pm 2\pi^0$ difference of probabilities/mean	$(0.0 \pm 0.6)\%$	CP
$K^\pm \to \pi^\pm\pi^0\gamma$ difference of probabilities/mean	$(0.9 \pm 3.3)\%$	CP
$\lvert\eta_{+-0}\rvert^2 = \Gamma(K_S^0 \to \pi^+\pi^-\pi^0)/\Gamma(K_L^0 \to \pi^+\pi^-\pi^0)$	< 0.12	CP
$\lvert\eta_{000}\rvert^2 = \Gamma(K_S^0 \to 3\pi^0)/\Gamma(K_L^0 \to 3\pi^0)$	< 0.1	CP
$K_L^0 \to \pi^0\nu\bar{\nu}/\text{all}$	$< 7.6 \cdot 10^{-3}$	CP
$\to \pi^0\mu^+\mu^-/\text{all}$	$< 1.2 \cdot 10^{-6}$	CP

Table 36.9 Testing the quantum number conservation laws. *(continued)*

Quantity	Value	Tested law of conservation or symmetry
$\to \pi^0 e^+ e^-/\text{all}$	$< 5.5 \cdot 10^{-9}$	CP
$K_L^0 \to (\pi^-\mu^+\nu - \pi^+\mu^-\bar{\nu})/\text{sum}$	$(0.304 \pm 0.025)\%$	CP (violated)
$\to (\pi^- e^+\nu - \pi^+ e^-\bar{\nu})/\text{sum}$	$(0.333 \pm 0.014)\%$	CP (violated)
$\lvert\eta_{00}\rvert = \lvert A(K_L^0 \to \pi^0\pi^0)/A(K_S^0 \to \pi^0\pi^0)\rvert$	$(2.253 \pm 0.024) \cdot 10^{-3}$	CP (violated)
$\lvert\eta_{+-}\rvert = \lvert A(K_L^0 \to \pi^+\pi^-)/A(K_S^0 \to \pi^+\pi^-)\rvert$	$(2.268 \pm 0.023) \cdot 10^{-3}$	CP (violated)
$\lvert\epsilon'/\epsilon\rvert = (1 - \lvert\eta_{00}/\eta_{+-}\rvert)/3$	$(2.2 \pm 1.1) \cdot 10^{-3}$	CP (violated)
ϕ_{+-} phase η_{+-}	$(46.6 \pm 1.2)°$	CP (violated)
ϕ_{00} phase η_{00}	$(46.6 \pm 2.0)°$	CP (violated)
$(g_{e^+} - g_{e^-})/\text{mean}$, g-gyromagnetic ratio	$(-0.5 \pm 2.1) \cdot 10^{-12}$	CPT
$(g_{\mu^+} - g_{\mu^-})/\text{mean}$, $\frac{\mu}{\mu_B} - 1 = \frac{g-2}{2}$	$(-2.6 \pm 1.6) \cdot 10^{-8}$	CPT
$(\mu_p - \lvert\mu_{\bar{p}}\rvert)/\text{mean}$,$\mu$-magnetic moment	$(-2.6 \pm 2.9) \cdot 10^{-3}$	CPT
$e^+ - e^-$ difference of masses/mean	$< 4 \cdot 10^{-8}$	CPT
$\pi^+ - \pi^-$ difference of masses/mean	$(2 \pm 5) \cdot 10^{-4}$	CPT
$K^+ - K^-$ difference of masses/mean	$(-0.6 \pm 1.8) \cdot 10^{-4}$	CPT
$\lvert K^0 - \overline{K}^0\rvert$ difference of masses/mean	$< 4 \cdot 10^{-18}$	CPT
$p - \bar{p}$ difference of masses/mean	$(2 \pm 4) \cdot 10^{-8}$	CPT
$n - \bar{n}$ difference of masses/mean	$(9 \pm 5) \cdot 10^{-5}$	CPT
$\Lambda - \overline{\Lambda}$difference of masses/mean	$(0.0 \pm 1.1) \cdot 10^{-4}$	CPT
$\Xi^- - \overline{\Xi}^+$ difference of masses/mean	$(1.1 \pm 2.7) \cdot 10^{-4}$	CPT
$\Omega^- - \overline{\Omega}^+$ difference of masses/mean	$(-1 \pm 5) \cdot 10^{-4}$	CPT
$W^+ - W^-$ difference of masses/mean	$(-2 \pm 7) \cdot 10^{-3}$	CPT
$\mu^+ - \mu^-$ difference of lifetimes/mean	$(2 \pm 8) \cdot 10^{-5}$	CPT
$\pi^+ - \pi^-$ difference of lifetimes/mean	$(6 \pm 7) \cdot 10^{-4}$	CPT
$K^+ - K^-$ difference of lifetimes/mean	$(1.1 \pm 0.9) \cdot 10^{-3}$	CPT
$K^\pm \to \mu^\pm\nu$ difference of probabilities/mean	$(-0.54 \pm 0.41)\%$	CPT
$K^\pm \to \pi^\pm\pi^0$ difference of probabilities/mean	$(0.8 \pm 1.2)\%$	CPT
$Z \to e^\pm\mu^\mp/\text{all}$	$< 2.4 \cdot 10^{-5}$	Number of lepton family
$\to e^\pm\tau^\mp/\text{all}$	$< 3.4 \cdot 10^{-5}$	" " "
$\to \mu^\pm\tau^\mp/\text{all}$	$< 4.8 \cdot 10^{-5}$	" " "
$\mu^- \to e^-\nu_e\bar{\nu}_\mu/\text{all}$	$< 1.8 \cdot 10^{-2}$	" " "
$\to e^-\gamma/\text{all}$	$< 5 \cdot 10^{-11}$	" " "
$\to e^-e^+e^-/\text{all}$	$< 1.0 \cdot 10^{-12}$	" " "
$\to e^-\gamma\gamma/\text{all}$	$< 7 \cdot 10^{-11}$	" " "
$\tau^- \to \mu^-\gamma/\text{all}$	$< 5.5 \cdot 10^{-4}$	" " "
$\to e^-\gamma/\text{all}$	$< 2.0 \cdot 10^{-4}$	" " "
$\to \mu^-\pi^0/\text{all}$	$< 8.2 \cdot 10^{-4}$	" " "
$\to e^-\pi^0/\text{all}$	$< 1.4 \cdot 10^{-4}$	" " "
$\pi^+ \to \mu^+\nu_e/\text{all}$	$< 8.0 \cdot 10^{-3}$	" " "
$\pi^0 \to \mu^+e^-/\text{all}$	$< 1.6 \cdot 10^{-8}$	" " "
$K^+ \to \pi^+e^+\mu^-/\text{all}$	$< 7 \cdot 10^{-9}$	" " "
$\to \pi^+e^-\mu^+/\text{all}$	$< 2.1 \cdot 10^{-10}$	" " "
$\to \mu^-\nu e^+e^+/\text{all}$	$< 2 \cdot 10^{-8}$	" " "
$K_L^0 \to e^\pm\mu^\mp/\text{all}$	$< 9.4 \cdot 10^{-11}$	" " "
$D^+ \to \pi e^\pm\mu^\mp/\text{all}$	$< 3.8 \cdot 10^{-3}$	" " "
$B^+ \to \pi^+e^+\mu^-/\text{all}$	$< 6.4 \cdot 10^{-3}$	" " "
$\to \pi^+e^-\mu^+/\text{all}$	$< 6.4 \cdot 10^{-3}$	" " "
$\to K^+e^+\mu^-/\text{all}$	$< 6.4 \cdot 10^{-3}$	" " "
$B^0 \to e^\pm\mu^\mp/\text{all}$	$< 4 \cdot 10^{-5}$	
Neutrino oscillations	$\Delta(m^2)$ for $\sin^2(2\theta) = 1$	
$\bar{\nu}_e \not\to \bar{\nu}_e$	< 0.0083 eV^2	" " "
$\nu_\mu \to \nu_e$	< 0.09 eV^2	" " "
$\bar{\nu}_\mu \to \bar{\nu}_e$	< 0.11 eV^2	" " "

Table 36.9 Testing the quantum number conservation laws. *(continued)*

Quantity	Value	Tested law of conservation or symmetry
$\bar{\nu}_\mu \not\rightarrow \bar{\nu}_\mu$	< 7 eV^2 or > 1200 eV^2	" " "
$\nu_e \rightarrow \nu_\tau$	< 9 eV^2	" " "
Neutrino oscillations	$\sin^2(2\theta)$ for high $\Delta(m^2)$	
$\bar{\nu}_e \not\rightarrow \bar{\nu}_e$	< 0.14	" " "
$\nu_\mu \rightarrow \nu_e$	< 0.0034	" " "
$\bar{\nu}_\mu \rightarrow \bar{\nu}_e$	< 0.004	" " "
$\nu_\mu \rightarrow \nu_\tau$	< 0.004	" " "
$\bar{\nu}_\mu \rightarrow \bar{\nu}_\tau$	< 0.04	" " "
$\mu^-\ {}^{32}S \rightarrow e^+\ {}^{32}Si^*/\mathrm{all}$	$< 9 \cdot 10^{-10}$	Lepton number
$\mu^-\ {}^{127}I \rightarrow e^+\ {}^{127}Sb^{\mathrm{stable}}/\mathrm{all}$	$< 3 \cdot 10^{-10}$	" " "
$\mu^-\ Ti \rightarrow e^+\ Ca/\mathrm{all}$	$< 1.7 \cdot 10^{-10}$	" " "
$\tau^- \rightarrow e^+\pi^-\pi^-/\mathrm{all}$	$< 1.7 \cdot 10^{-5}$	" " "
$\rightarrow \mu^+\pi^-\pi^-/\mathrm{all}$	$< 3.9 \cdot 10^{-5}$	" " "
$K^+ \rightarrow \pi^-e^+e^+/\mathrm{all}$	$< 1.0 \cdot 10^{-8}$	" " "
$\rightarrow \pi^-\mu^+\mu^+/\mathrm{all}$	$< 1.5 \cdot 10^{-4}$	" " "
$\rightarrow \pi^-e^+\mu^+/\mathrm{all}$	$< 7 \cdot 10^{-9}$	" " "
$B^+ \rightarrow \pi^-e^+e^+/\mathrm{all}$	$< 3.9 \cdot 10^{-3}$	" " "
$\rightarrow \pi^-\mu^+\mu^+/\mathrm{all}$	$< 9.1 \cdot 10^{-3}$	" " "
$\rightarrow \pi^-e^+\mu^+/\mathrm{all}$	$< 6.4 \cdot 10^{-3}$	" " "
$\tau_p/BR(p \rightarrow e^+\pi^0)$	$> 550 \cdot 10^{30}$ year	Barion number
$\tau_n/BR(n \rightarrow e^+\pi^-)$	$> 130 \cdot 10^{30}$ year	" " "
$\tau_p/BR(p \rightarrow \mu^+\pi^0)$	$> 270 \cdot 10^{30}$ year	" " "
$\tau_n/BR(n \rightarrow \mu^+\pi^-)$	$> 100 \cdot 10^{30}$ year	" " "
$\tau_p/BR(p \rightarrow e^+K^0)$	$> 150 \cdot 10^{30}$ year	" " "
$\tau_n/BR(n \rightarrow e^+K^-)$	$> 1.3 \cdot 10^{30}$ year	" " "
e mean lifetime	$> 2 \cdot 10^{22}$ year	Charge
$n \rightarrow p\nu\bar{\nu}/\mathrm{all}$	$< 9 \cdot 10^{-24}$	" " "
$K^+ \rightarrow \pi^+\pi^+e^-\bar{\nu}/\mathrm{all}$	$< 1.2 \cdot 10^{-8}$	$\Delta S = \Delta Q$
$\rightarrow \pi^+\pi^+\mu^-\bar{\nu}/\mathrm{all}$	$< 3.0 \cdot 10^{-6}$	" " "
$\Sigma^+ \rightarrow ne^+\nu/\mathrm{all}$	$< 5 \cdot 10^{-6}$	" " "
$\rightarrow n\mu^+\nu/\mathrm{all}$	$< 3.0 \cdot 10^{-5}$	" " "
$K_L^0 \rightarrow \mu^+\mu^-/\mathrm{all}$	$(7.3 \pm 0.4) \cdot 10^{-9}$	Conservation of flavor by neutral current
$\rightarrow e^+e^-/\mathrm{all}$	$< 1.6 \cdot 10^{-10}$	" " "
$\rightarrow \mu^+\mu^-\gamma/\mathrm{all}$	$(2.8 \pm 2.8) \cdot 10^{-7}$	" " "
$\rightarrow e^+e^-\gamma/\mathrm{all}$	$(9.1 \pm 0.5) \cdot 10^{-6}$	" " "
$\rightarrow \pi^0\mu^+\mu^-/\mathrm{all}$	$< 1.2 \cdot 10^{-6}$	" " "
$\rightarrow \pi^0e^+e^-/\mathrm{all}$	$< 5.5 \cdot 10^{-9}$	" " "

References

[1] Okun', L. B., Leptons and quarks, Nauka, Moscow, 1981 (in Russian).

[2] Okun', L. B., Physics of elementary particles, Nauka, Moscow, 1988 (in Russian).

[3] Leader, E., Predazzi, E, Introduction to calibration theories and new physics, Naukova Dumka, Kiev, 1990 (Translation into Russian).

[4] Particle data group. Review of Particle properties, Phys. Rev., D 45, No. 11, pt. 2, 1992; Particle data group. Review of Particle properties, Phys. Rev., D 50, No. 11, pt. 1, 1994.

[5] F.Abe et al. (CDF Collaboration), Phys. Rev. Lett. 74, 2626, 1995; S.Abachi et al. (D0 Collaboration), Phys. Rev. Lett. 74, 2632, 1995.

37

The Nuclear Properties of Elements

V.M. Kulakov

37.1 The Table of Nuclides

Table 37.1 incorporates all reliable information on radioactive and stable nuclides in the order of mass number for every element. The numerical data are arranged in six columns. In Column 1, the symbol of an element (keep in mind that the accepted symbol for the isotopes of the element X is ${}^{A}_{Z}\mathrm{X}_{N}$) and its atomic number Z (i.e. the number of protons) are given. In Column 2, the mass number A ($A = Z + N$, where N is the number of neutrons) is indicated. The symbol m after the mass number denotes an isomeric state. The isomeric states are given only in the case when their half-lives are long enough (>1 s) to be identified independently from the ground state. The symbols m_1 and m_2 denote different isomeric states of the nuclide. In Column 3, the total half-life of the nuclide is given according to the Evaluated Nuclear Structure Data File – ENSDF (in parentheses, the standard deviation of $\Delta T_{1/2}$ is indicated in units of the last significant figures).

In Column 4, the nuclide decay modes are presented. The various decay modes are denoted as α (α-decay), β^- (electron decay), β^+ (positron decay), ϵ (electron capture), β_n, β_p, ϵ_n, ϵ_p (neutron and proton emission following β-decay or electron capture, respectively), IT (isomeric transition through γ-decay or conversion-electron decay), SF (spontaneous fission). If a nucleus shows several decay modes, all of them are indicated in this column. In some cases the decay mode symbol is followed by the branching ratio in parentheses (in %), for example, $\epsilon(60)$. If a nuclide offers several decay modes without indicating the number in parentheses after them, it means that the branching ratio is unknown. The values of characteristic terrestrial isotopic composition (in %) are also given for the stable nuclides. The data were taken from [5] and correspond to the values evaluated principally by the mass–spectrometric method. These data are marked with bold figures. For the radioactive nuclides with long half-lives (therefore they were retained in the Earth's crust), both percent composition and decay modes are indicated; for example, ^{87}Rb | **27.835(13)** β^- | Figures in parentheses following the stable nuclide abundance indicate the standard deviation in the last significant figures. These uncertainties overlap the intervals for both the possible natural variations of isotope composition and the experimental errors.

In Column 5, the energy values (in MeV) are given for the groups of most intensive particles (α, β^-, β^+, n, p) emitted via nuclide decay. The relative intensity of the particle group is listed and the total decay figure (in %) is given in parentheses. The relative intensities of particle groups (in %) are given in slanting brackets only for the concrete decay mode. In the case of continuous β^- and β^+ spectra, the values of the highest boundary energy are given, as a rule. If the β-group with the highest boundary energy possesses a relatively weak intensity, the boundary energies and intensities are listed for one or several β-groups having lower boundary energies.

0-8493-2861-6/97/$0.00+$.50

In Column 6, the energies (in MeV) of the main γ-rays following the nuclide decay are collected. The γ-ray intensity (in %) per the total decay figure is enclosed in parentheses. The relative intensity of γ-rays (in %) is indicated in slanting brackets. If the parentheses after the energy are lacking, it means that the intensity is not precisely defined. The range of γ-ray energies is given in some cases. Thus the symbol 0.511 (an.) means that γ-rays offer the annihilation nature. One, two or three asterisks mark doublet, complex line and set of all complex lines, respectively.

Table 37.1 Table of isotopes [1,2]

Element	A	$T_{1/2}$	Decay type or abundance of stable isotopes	Energy, MeV (relative intensity, %)	
				particle group	γ-rays
$_1$H	1	Stable	**99.985(1)**	–	–
	2	Stable	**0.015(1)**	–	–
	3	12.33(6) y	β^-	0.0186	–
$_2$He	3	Stable	**1.38(3)·10^{-4}**	–	–
	4	Stable	**99.999862(3)**	–	–
	6	0.8081(20) s	β^-	3.508	–
	8	0.122(2) s	β^-, β_n(12)	–	0.98(88)
$_3$Li	6	Stable	**7.5(2)**	–	–
	7	Stable	**92.5(2)**	–	–
	8	0.842(6)s	β^-	13	–
			2α	1.6	–
	9	0.176(2) s	β^-	13.61	–
			n	0.76	–
	11	0.0085 s	β^-, β_n(61)	–	–
$_4$Be	7	53.44(9) d	ϵ	–	0.477(10.3)
	9	Stable	**100**	–	–
	10	1.6(2)·10^6 y	β^-	0.555	–
	11	13.81(8) s	β^-	11.5	2.14(32); 4.67(2.1); 5.85(2.4); 6.79(4.4); 7.99(1.7)
			$\beta^-\alpha$	0.77	
	12	0.0114(5) s	β^-	11.7	–
$_5$B	8	0.769(4) s	β^+	14.1	–
			2α	8.3; 1.6	–
	10	Stable	**19.9(2)**	–	–
	11	Stable	**80.1(2)**	–	–
	12	0.02041(6) s	β^-(~ 100)	13.37	4.43(1.3)
			$\beta^-3\alpha$(1.5)	0.192(1.5)	
	13	0.01736(16) s	β^-, β_n(0.28)	13.44	3.68(7)
	14	0.0161(12) s	β^-	14.0	6.09; 6.73
$_6$C	9	0.1265(9) s	β_p	8.2(60); 1.1(40)	–
	10	19.42(6) s	β^+	1.87	0.511(200, an.); 0.717(100); 1.023(1.7)
	11	20.40(4) min	β^+(> 99)	0.97	0.511(200, an.)
			ϵ(0.19)	–	
	12	Stable	**98.90(3)**	–	–
	13	Stable	**1.10(3)**	–	–
	14	5730(40) y	β^-	0.156	–
	15	2.449(4) s	β^-	9.82(32); 4.51(68)	5.299(68)
	16	0.747(8) s	β_n(> 98.8)	0.79; 1.72	–
$_7$N	12	0.01097(4) s	β^+(~ 100)	16.4	0.511(200, an.); 4.43(2.4)
			3α(~ 3)	0.195	–
	13	9.961(4) min	β^+	1.20	0.511(200, an.)

Table 37.1 Table of isotopes *(continued)*

Element	A	$T_{1/2}$	Decay type or abundance of stable isotopes	Energy, MeV (relative intensity, %) particle group	Energy, MeV (relative intensity, %) γ-rays
$_7$N	14	Stable	**99.634(9)**	–	–
	15		**0.366(9)**	–	–
	16	7.13(2) s	β^-	10.40(26); 4.27(74)	2.75(1); 6.13(69); 7.11(5)
			α(0.0006)	1.7	–
	17	4.169(8) s	β^-	8.68(1.6); 7.81(2.6)	0.87(3); 2.19(0.5)
			n	4.1(95)	
	18	0.63(3) s	β^-	0.40(45); 1.21(45); 1.81(5); 9.4	0.82(59); 1.65(59); 1.98(100); 2.47(41)
$_8$O	13	0.0089(2) s	β_p	6.40/100/; 6.97/24/	–
	14	70.599(22) s	β^+	4.12(0.6); 1.811(99)	0.522(200, an.); 2.312(99)
	15	122.24(16) s	β^+	1.74	–
	16	Stable	**99.762(15)**	–	–
	17		**0.038(3)**	–	–
	18		**0.200(12)**	–	–
	19	26.91(8) s	β^-	4.60; 3.3	0.197(97); 1.37(59)
	20	13.57(10) s	β^-	2.8	1.06(100)
	21	3.4 s	β^-	6.4	1.73; 1.79; 2.80; 3.52
$_9$F	17	60.50(25) s	β^+	1.74	0.511(200, an.)
	18	109.77(5) min	β^+(97)	0.635	0.511(194, an.)
			ϵ(3)	–	
	19	Stable	**100**	–	–
	20	11.03(6) s	β^-	5.41	1.63(100)
	21	4.32(3) s	β^-	5.7; 5.3	0.350/100/; 1.38/13/
	22	4.23(4) s	β^-	11	1.28(100); 2.08(67); 2.17
	23	2.23(14) s	β^-	–	–
$_{10}$Ne	17	0.109(10) s	β_p	3.77; 4.59; 5.12	–
	18	1.67(2) s	β^+	3.42	0.511(200, an.); 1.04(7)
	19	17.40(4) s	β^+	2.22	0.511(200, an.)
	20	Stable	**90.51(9)**	–	–
	21		**0.27(2)**	–	–
	22		**9.22(9)**	–	–
	23	37.24(12) s	β^-	4.38	0.439(33); 1.64(09)
	24	3.38(2) min	β^-	1.99	0.472(100); 0.88(8)
	25	0.602(8) s	β^-	7.3	0.090; 0.980
$_{11}$Na	20	0.446(3) s	β^+	11.4	1.63
			$\beta^+\alpha$(20)	2.14/100/; 2.49/5/; 4.44/21/	–
	21	22.48(3) s	β^+	2.52	0.350(2.3); 0.511(200, an.)
	22	2.602(2) y	β^+(90.6)	1.820(0.05); 0.545	0.511(180, an.); 1.275(100)
			ϵ(9.4)	–	
	23	Stable	**100**	–	–
	24	15.020(7) h	β^-	4.17(0.003); 1.389(100)	1.369(100); 2.754(100)
	24m	0.02018(10) s	IT	–	0.4723
			β^-	6	
	25	59.6(7) s	β^-	3.83	0.39(14); 0.58(14); 0.98(15); 1.61(6)
	26	1.072(9) s	β^-	6.7	1.82(100)
	27	0.304(7) s	β^-, β_n(0.08)	8.0	0.985; 1.698
	28	0.0305(4) s	β^-, β_n(0.6)	13.9	1.47; 2.39

Table 37.1 Table of isotopes *(continued)*

Element	A	$T_{1/2}$	Decay type or abundance of stable isotopes	Energy, MeV (relative intensity, %)	
				particle group	γ-rays
$_{12}$Mg	21	0.122(3) s	ϵ_p	1.77; 1.94	–
	22	3.857(9) s	β^+	3.2	0.58; 0.74
	23	11.317(11) s	β^+	3.03	0.44(9); 0.511(200, an.)
	24	Stable	**78.99(3)**	–	–
	25	Stable	**10.00(1)**	–	–
	26	Stable	**11.01(2)**	–	–
	27	9.462(11) min	β^-	1.75	0.18(0.7); 0.84(70); 1.013(30)
	28	21.07(10) h	β^-	0.9; 0.46	0.031(96); 0.40(30); 0.95(30); 1.35(70)
	29	1.38(13) s	β^-	–	–
$_{13}$Al	23	0.47(3) s	β^+, β_p	–	–
	24	2.066(10) s	β^+(100)	8.5	0.511(200, an.)
			$\alpha(\sim 10^{-2})$	2	–
	24m	0.130(4) s	β^+	–	–
			IT	–	0.439
	25	7.183(12) s	β^+	3.24	0.511(200, an.)
	26	$7.2(3)10^5$ y	β^+(85)	1.17	0.511(170, an.); 1.12(4); 1.81(100)
			ϵ(15)	–	
	26m	6.345(3) s	β^+	3.21	0.511(200, an.)
	27	Stable	**100**	–	–
	28	2.259(9) min	β^-	2.85	1.78(100)
	29	6.56(6) min	β^-	2.40	1.28(94); 2.43(6)
	30	3.60(6) s	β^-	6.3; 5.0	2.23(61); 3.51(39)
	31	0.644(25) s	β^-	7.9; 5.6	1.70; 2.32
$_{14}$Si	25	0.220(3) s	β^+, β_p	–	–
	26	2.210(21) s	β^+	3.83	0.511(200, an.); 0.82(34); 1.62
	27	4.17(1) s	β^+	3.85	0.511(200, an.)
	28	Stable	**92.23(1)**	–	–
	29	Stable	**4.67(1)**	–	–
	30	Stable	**3.10(1)**	–	–
	31	157.3(3) min	β^-	1.48	1.26(0.07)
	32	330(40) y	β^-	0.21	–
$_{15}$P	28	0.2703(5) s	β^+	11.0	0.511(200, an.); 1.78(75); 4.44(10); 7.6(5)
			$\beta^+\alpha$	11.43; 2.10	
	29	4.142(15) s	β+	3.95	0.511(200, an.); 1.28(0.8); 2.43(0.2)
	30	2.498(4) min	β^+	3.24	0.511(200, an.); 2.23(0.5)
	31	Stable	**100**	–	–
	32	14.36(4) d	β^-	1.710	–
	33	25.34(12) d	β^-	0.248	–
	34	12.40(12) s	β^-	5.4	2.13(25); 4.0(0.2)
	35	47.3(7) s	β^-	2.3	1.57
$_{16}$S	29	0.1874 s	β^+, β_p	–	–
	30	1.24(4) s	β^+	5.09(20); 4.42(80)	0.511(200, an.); 0.687(80)
	31	2.584(18) s	β^+	4.42	0.511(200, an.); 1.27(1.1)
	32	Stable	**95.02(9)**	–	–
	33	Stable	**0.75(1)**	–	–
	34	Stable	**4.21(8)**	–	–
	35	87.24(17) d	β^-	0.167	–
	36	Stable	**0.02(1)**	–	–
	37	5.06(1) min	β^-	4.7(10); 1.6(90)	3.09(90)
	38	170.3(7) min	β^-	3.0(5); 1.1	1.88(95)

Table 37.1 Table of isotopes *(continued)*

Element	A	$T_{1/2}$	Decay type or abundance of stable isotopes	Energy, MeV (relative intensity, %) particle group	γ-rays
$_{17}$Cl	32	0.298(2) s	β^+	11.7; 9.9	0.511(200, an.); 2.24(70); 4.29(7); 4.77(14)
			$\beta^+\alpha$(0.01)	2.20	–
	33	2.50(2) s	β^+	4.55	0.511(200, an.); 2.9(0.3)
	34	1.529(4) s	β^+	4.46	0.511(200, an.)
	35	Stable	**75.77(5)**	–	–
	36	3.01(2)·10^5 y	β^-(98.1)	0.714	0.511(0.003, an.)
			ϵ(1.9)	–	
			β^+(0.0012)	–	
	37	Stable	**24.23(5)**	–	–
	38	37.24(5) min	β^-	4.91	1.60(38); 2.17(47)
	38m	0.716(3) s	IT	–	0.671(100)
	39	55.6(2) min	β^-	3.45(7); 2.18(8); 1.91	0.246(44); 1.27(50); 1.52(42)
	40	1.35(2) min	β^-	7.5	1.46/100/; 2.83/100/; 3.10; 5.8
$_{18}$Ar	33	0.18 s	β^+, β_p(34)	–	0.81
	34	0.841(10) s	β^+	5.0	0.67; 3.13
	35	1.781(9) s	β^+	4.94	0.522(200, an.); 1.22(5); 1.76(22)
	36	Stable	**0.337(3)**	–	–
	37	34.8(2) d	ϵ	–	–
	38	Stable	**0.063(1)**	–	–
	39	269(3) y	β^-	0.565	–
	40	Stable	**99.600(3)**	–	–
	41	109.6(4) min	β^-	2.49(0.8); 1.198	1.293(99)
	42	32.9(11) y	β^-	~0.6	–
	43	5.37(6) min	β^-	–	0.74; 0.98; 1.44
	44	11.87(5) min	–	–	1.18; 1.70; 1.89
	45	21.48(15) s	β^-	5.8; 3.2	1.02; 3.71
	46	8(1) s	β^-	–	1.94
$_{19}$K	36	0.340(3) s	β^+	9.9	1.97; 2.21; 2.43
	37	1.23(2) s	β^+	5.14	0.511(200, an.); 2.79(2)
	38	7.636(18) min	β^+	2.68	0.511(200, an.); 2.17(100)
	38m	0.929(3) s	β^+	5.0	0.511(200, an.)
	39	Stable	**93.2581(30)**	–	–
	40	1.277(8)·10^6 y	**0.0117(1)**	–	–
			β^-(89)	1.314	–
			ϵ(11)	–	–
			β^+(0.001)	0.483	–
	41	Stable	**6.7302(30)**	–	–
	42	12.360(3) h	β^-	3.52	0.31(0.2); 1.524(18)
	43	22.3(1) h	β^-	1.82(1); 1.2(3); 0.83	0.373(85); 0.39*(18); 0.59(13); 0.619(81)
	44	22.13(19) min	β^-	5.2	1.156(61); 1.74(8); 2.1**(37); 2.6(7)
	45	17.3(6) min	β^-	4.2; 2.3	0.17; 1.71
	46	107(10) s	β^-	6.4	1.35; 3.70
	47	17.5(3) s	β^-	6.1(1); 4.1	2.0(84); 2.6(15)
	48	6.9(2) s	β^-	–	–

Table 37.1 Table of isotopes *(continued)*

Element	A	$T_{1/2}$	Decay type or abundance of stable isotopes	Energy, MeV (relative intensity, %)	
				particle group	γ-rays
$_{20}$Ca	37	0.175(3) s	β_p	3.10	–
	38	0.447(10) s	β^+	5.6	1.57
	39	0.876 s	β^+	5.49	0.511(an.)
	40	Stable	**96.941(13)**	–	–
	41	1.4(2)·10^5 y	ϵ	–	–
	42	Stable	**0.647(3)**	–	–
	43	Stable	**0.135(3)**	–	–
	44	Stable	**2.086(5)**	–	–
	45	163.8(18) d	β^-	0.252	–
	46	Stable	**0.004(3)**	–	–
	47	4.536(2) d	β^-	1.98(18); 0.67	0.49(5); 0.815(5); 1.308(74)
	48	$> 2 \cdot 10^{18}$ y	**0.187(3)**	–	–
	49	8.716(11) min	β^-	1.95	3.10(89); 4.1(10)
	50	13.9(6) s	β^-	3.1	0.072; 0.257; 1.52; 1.59
$_{21}$Sc	40	0.1823(7) s	β^+, β_p	9.1	0.511(200, an.); 3.75
	41	0.601(12) s	β^+	5.47	–
	42	0.6839(9) s	ϵ	–	0.511(200, an.)
	42m	61.3(7) s	β^+	2.82	0.438(100); 0.511(200, an.); 1.22(100); 1.52(100)
	43	3.891(12) h	β^+	1.20	0.375(22); 0.511(180, an.)
	43m	0.632(8) s	IT	–	–
	44	3.927(8) h	β^+ ϵ(5)	1.47	0.511(188, an.); 1.159(100)
	44m	58.6(1) h	β^+(1.39) IT(98.61)		(0.271(86); 1.02(1.3)
	45	Stable	**100**	–	–
	45m	0.316(9) s	IT	–	0.0124
	46	83.83(2) d	β^-	1.48(0.004); 0.357	0.889(100); 1.120(100)
	46m	18.70(5) s	IT	–	0.142
	47	3.35(2) d	β^-	0.600	0.160(73)
	48	43.7(1) h	β^-	0.65	0.175(6); 0.983(100); 1.040(100); 1.314(100)
	49	57.4(1) min	β^-	2.01	1.76(0.03)
	50	1.708(9) min	β^-	4.2; 3.6	0.520(100); 1.12(100); 1.55(100)
	50m	0.35(1) s	IT	–	0.257
	51	12.4(1) s	β^-	5.0; 4.3	1.44; 1.57; 2.14
$_{22}$Ti	41	0.088(1) s	β_p	2.3/8/; 3.05/17/; 3.68/16/; 4.12/4/; 4.64/50/; 5.30/5/	–
	42	0.199(6) s	β^+	6.0; 5.4	0.611
	43	0.490(20) s	β^+	5.8	–
	44	47.3(12) y	ϵ	–	0.068(90); 0.078(98)
	45	184.8(5) min	β^+	1.04	0.718(0.4); 1.408(0.3)
	46	Stable	**8.0(1)**		
	47	Stable	**7.3(1)**		
	48	Stable	**73.8(1)**		
	49	Stable	**5.5(1)**		
	50	Stable	**5.4(1)**		
	51	5.76(1) min	β^-	2.14	0.320(95); 0.605(1.5); 0.928(5)
	52	1.7(1) min	β^-	–	–
	53	32.7(9) s	β^-	4.8; 3.1	0.101; 0.128; 0.228; 1.676

Table 37.1 Table of isotopes *(continued)*

Element	A	$T_{1/2}$	Decay type or abundance of stable isotopes	Energy, MeV (relative intensity, %)	
				particle group	γ-rays
$_{23}$V	45	0.539(18) s	β^+	–	–
	46	0.4223(2) s	β^+	6.03	0.511(200, an.)
	46m	0.001(1) s	IT	–	0.801
	47	32.6(3) min	β^+	1.89	1.80(0.5)
	48	15.97(4) d	β^+(49)	0.696	0.511(100, an.); 0.945(10);
			ϵ(51)		0.983(100); 1.312(97)
	49	330(15) d	ϵ	–	–
	50	$> 4 \cdot 10^{16}$ y	**0.250(2)**	–	
			ϵ(70)	–	0.783(30); 1.55(70)
			β^-(30)	–	
	51	Stable	**99.750(2)**	–	–
	52	3.75(1) min	β^-	2.47	1.434(100)
	53	1.61(4) min	β^-	2.50	1.006; 1.289
	54	49.8(5) s	β^-	3.3	0.84(100); 0.99(100); 2.21(100)
$_{24}$Cr	45	0.050(5) s	β_p	–	-
	46	0.26(6) s	β^+	–	–
	47	0.460(15) s	β^+	–	–
	48	22.96(3) h	ϵ	–	0.116(98); 0.31(99)
	49	42.09(15) min	β^+	1.54	0.063(14); 0.091(28); 0.153(13)
	50	Stable	**4.345(9)**	–	–
	51	27.704(4) d	ϵ	–	0.320(10)
	52	Stable	**83.789(12)**	–	–
	53	Stable	**9.501(11)**	–	–
	54	Stable	**2.365(5)**	–	–
	55	3.55(3) min	β^-	2.59	–
	56	5.94(10) min	β^-	1.5	0.026; 0.083
$_{25}$Mn	50	0.2832(6) s	β^+	6.61	–
	50m	1.75(3) min	β^+	–	0.227
	51	46.2(1) min	β^+	2.17	–
	52	5.591(3) d	ϵ(66)	–	0.511(67, an.); 0.744(82);
			β^+(34)	0.575	0.935(84); 1.434(100)
	52m	21.1(2) min	β^+(98)	1.63	0.511(193, an.); 1.434(100)
			IT(2)	–	0.378
	53	$3.7(4) \cdot 10^6$ y	ϵ	–	–
	54	312.5(5) d	ϵ	–	0.835(100)
	55	Stable	**100**	–	–
	56	2.5785(6) h	β^-	2.85	0.847(99); 1.811(29); 2.110(15)
	57	1.61(5) min	β^-	2.55	0.014; 0.122; 0.692
	58	65.3(7) s	β^-	–	–
	59	4.6(1) s	β^-	4.8; 4.4	0.473; 0.571; 0.726
	60	1.79(10) s	β^-	–	–
$_{26}$Fe	49	0.075(10) s	β_p	–	–
	51	0.270(6) s	β^+	–	–
	52	8.275(8) h	β^+(56)	0.80	0.165(100); 0.511(112, an.)
			ϵ(44)	–	–

Table 37.1 Table of isotopes *(continued)*

Element	A	$T_{1/2}$	Decay type or abundance of stable isotopes	Energy, MeV (relative intensity, %)	
				particle group	γ-rays
$_{26}$Fe	52m	46(2) s	ϵ, IT	–	–
	53	8.51(2) min	β^+	3.0	0.38(32); 0.511(196, an.)
	53m	2.58(6) min	IT	–	3.041
	54	Stable	**5.8(1)**	–	–
	55	2.7 y	ϵ	–	–
	56	Stable	**91.72(30)**	–	–
	57	Stable	**2.2(1)**	–	–
	58	Stable	**0.28(1)**	–	–
	59	44.496(7) d	β^-	1.57(0.3); 0.475	0.192(2.8); 1.095(56); 1.292(44)
	60	$\sim 3 \cdot 10^5$ y	β^-	0.1	–
	61	5.98(6) y	β^-	2.8	0.13/11/; 0.30/48/; 1.03/98/; 1.20/100/
	62	68(2) s	β^-	2.5	0.506
$_{27}$Co	54	0.19323(14) s	ϵ	–	–
	54m	1.48(2) min	ϵ	–	0.198
	55	17.54(4) h	β^+(81) ϵ(19)	1.50	0.480(12); 0.511(160, an.); 0.930(80); 1.41(13)
	56	78.76(12) d	ϵ(80) β^+(20)	– 1.49	0.511(40, an.); 0.847(100); 1.04(15); 1.24(66); 1.77(15); 2.02(11); 2.60(17); 3.26(13)
	57	270.9(6) d	ϵ	–	0.014(10); 0.122(86); 0.136(11)
	58	70.78(10) d	ϵ(85) β^+(15)	– 4.474	0.511(30, an.); 0.810(99) 0.865(1.4)
	58m	9.15(10) h	IT	–	0.0249
	59	Stable	**100**	–	–
	60	5.271(1) y	β^-	1.48(0.12); 0.314(99)	1.173(100); 1.332(100)
	60m	10.47(4) min	IT(> 99) β^-(0.25)	– 1.55	0.059(2.1); 1.33(0.25)
	61	1.650(5) h	β^-	1.22	0.067(89)
	62	1.50(4) min	β^-	–	–
	62m	13.91(5) min	β^- IT	2.88	1.17(180); 1.47(20); 1.74(19); 2.03(7)
	63	27.4(5) s	β^-	3.6	0.087; 0.982
	64	0.30(3) s	β^-	7.0	0.931; 1.346
$_{28}$Ni	56	6.10(2) d	ϵ	–	0.163(99); 0.276(31); 0.472(35); 0.748(48); 0.812(85); 1.56(14)
	57	36.08(9) h	ϵ(54) β^+(46)	– 0.85	0.127(14); 0.511(92, an.); 1.37(86); 1.89(14)
	58	Stable	**68.27(1)**	–	–
	59	$7.5(13) \cdot 10^4$ y	ϵ	–	–
	60	Stable	**26.10(1)**	–	–
	61	Stable	**1.13(1)**	–	–
	62	Stable	**3.59(1)**	–	–
	63	100.1(20) y	β^+	0.067	–
	64	Stable	**0.91(1)**	–	–
	65	2.520(2) h	β^-	2.13	1.115(16); 1.481(25)
	66	54.6(4) h	β^-	0.20	–
	67	21(1) s	β^-	4.1	0.90*(51); 1.26(15)

Table 37.1 Table of isotopes *(continued)*

Element	A	$T_{1/2}$	Decay type or abundance of stable isotopes	Energy, MeV (relative intensity, %)	
				particle group	γ-rays
$_{28}$Cu	58	3.204(7) s	β^+	7.5	1.45
	59	81.5(5) s	β^+	3.7	0.511(197, an.); 0.879(9); 1.305(11)
	60	23.2(3) min	$\beta^+(93)$	3.92(6); 3.00(18); 2.00	0.511(186, an.); 0.85(15);
			$\epsilon(7)$	–	1.332(80); 1.76(52)
	61	3.408(10) h	$\beta^+(60)$	1.22	0.067(4); 0.284(12); 0.38(3);
			$\epsilon(40)$	–	0.511(120, an.); 1.19(5)
	62	9.74(2) min	β^+	2.91	0.511(195, an.); 0.88(0.3)
	63	Stable	**69.17(2)**	–	–
	64	12.701(2) h	$\epsilon(43)$	–	0.511(38, an.); 1.34(0.5)
			$\beta^-(38)$	0.573	
			$\beta^+(19)$	0.656	
	65	Stable	**30.83(2)**	–	–
	66	5.10(2) min	β^-	2.63	1.039(9)
	67	61.92(9) h	β^-	0.57	0.092*(23); 0.184(40)
	68	31(1) s	β^-	3.5	0.80(17); 1.078(95)
	68m	3.75(5) min	IT(86)	–	–
			β^-	–	–
	69	3.0(1) min	β^-	2.5	0.531; 0.834; 1.007
	70	4.5(10) s	β^-	–	–
$_{30}$Zn	57	0.040(10) s	β_p	1.92; 2.53; 4.57	–
	59	0.1837(23) s	ϵ	–	0.491; 0.914
	60	2.38(5) min	β^+	3.1; 2.5	0.061; 0.273; 0.334; 0.670
	61	89.1(2) s	β^+	4.4	0.48(11); 0.511(198, an.); 0.98(3); 1.64(6)
	62	9.26(2) h	$\epsilon(82)$	–	0.042(20); 0.51**(47); 0.59(22)
			$\beta^+(18)$	0.66	
	63	38.1(3) min	$\beta^+(93)$	2.34	0.511(186, an.); 0.669(8); 0.962(6)
			$\epsilon(7)$	–	
	64	Stable	**48.6(3)**	–	–
	65	244.1(2) d	$\epsilon(98.3)$	–	0.511(3.4, an.); 1.115(49)
			$\beta^+(1.7)$		
	66	Stable	**27.9(2)**	0.327	
	67	Stable	**4.1(1)**	–	–
	68	Stable	**18.8(4)**	–	–
	69	55.6(16) min	β^-	0.90	–
	69m	13.76(2) h	IT(> 99),	–	0.439(95)
			$\beta^-(0.033)$		
	70	Stable	**0.6(1)**	–	–
	71	2.45(10) min	β^-	2.61	0.39(1.3); 0.510(13); 0.92(3); 1.12(153)
	71m	3.94(5) h	β^-	1.46	0.13(9); 0.385(94); 0.495(75); 0.609(95); 0.76(5); 0.99(8)
	72	46.5(1) h	β^-	0.30	0.015(8); 0.145(90); 0.192(10);
	73	23.5(10) s	β^-	4.7	0.216; 0.496; 0.911
	74	95(1) s	β^-	2.3; 2.1	0.057; 0.14; 0.19
	75	10.2(3) s	β^-	5.6	–
	76	5.7(3) s	β^-	3.7	0.08–1.03
	77	1.4(3) s	β^-	4.8	0.189; 0.473
	78	1.47(15)	β^-	5.1	0.182; 0.225; 0.454; 0.636; 0.860
	79	2.63(9) s	β_n	–	–

Table 37.1 Table of isotopes *(continued)*

Element	A	$T_{1/2}$	Decay type or abundance of stable isotopes	Energy, MeV (relative intensity, %)	
				particle group	γ-rays
$_{31}$Ga	62	0.1161(3) s	ϵ	–	–
	63	32.4(5) s	ϵ	–	–
	64	2.630(11) min	β^+	6.05(33); 2.8	0.511(196, an.); 0.80(15); 0.992(43); 1.38(14); 2.18(11); 3.32(18)
	65	15.2(2) min	$\beta^+(> 50)$ $\epsilon(< 50)$	2.24(12); 2.11	0.061(12); 0.115(55); 0.152(10); 0.511(180, an.); 0.75(10)
	66	9.49(8) h	$\beta^+(57)$ $\epsilon(43)$	4.153 –	0.511(114, an.); 1.039(37); 2.183(5); 2.748(25); 4.30(5)
	67	3.261(1) d	ϵ	–	0.388(7)
	68	68.1(3) min	$\beta^+(88)$ $\epsilon(12)$	1.90 –	0.511(176, an.); 1.078(3.5)
	69	Stable	**60.1(2)**	–	–
	70	21.15(5) min	$\beta^-(99.8)$	1.65	0.173(0.16); 1.040(0.5)
	71	Stable	**39.9(2)**	–	–
	72	14.10(2) h	β^-	3.15	0.630(27); 0.835(96); 2.201(26); 2.50*(20)
	73	4.87(3) h	β^-	1.19	0.054(9); 0.295(94); 0.74(6)
	74	8.1(1) min	β^-	2.5	0.60*(100); 0.87*(9); 2.35(45)
	74*m*	9.5(10) s	IT	–	0.0597
	75	2.10(3) min	β^-	3.3	0.58(3)
	76	27.1(2) s	β^-	6	0.546; 0.563; 1.108
	77	13.2(2) s	β^-	5.2	0.459; 0.469
	78	5.09(5) s	β^-	7.5; 5.1	0.567; 0.619; 1.186
	79	3.00(9) s	β^-	6.6	0.09–2.51
	80	1.66(2) s	β^-, β_n	–	1.11
$_{32}$Ge	61	0.040(15) s	–	–	–
	64	63.7(25) s	β^+	3.3; 3.0	0.128; 0.427; 0.667
	65	30.9(7) s	β^+	5.2; 4.6	0.511(197, an.); 0.67(3); 1.72(2)
	66	2.26(5) h	β^+ $\epsilon(38)$	1.1; 0.7	0.046(37); 0.114(22); 0.185(23); 0.27(19); 0.34(19); 0.38(48); 0.47(19); 0.511(124, an.)
	67	18.7(5) min	β^+	3.2; 3.0	0.170*(105); 0.511(170, an.); 1.473
	68	287(6) d	ϵ	–	–
	69	39.05(10) h	$\epsilon(67)$ $\beta^+(33)$	– 1.22	0.511(68, an.); 0.573(13); 0.872(10); 1.107(28)
	70	Stable	**20.5(5)**	–	–
	71	11.8(4) d	ϵ	–	–
	71*m*	0.0202(5) s	IT	–	0.198
	72	Stable	**27.4(6)**	–	–
	73	Stable	**7.8(2)**	–	–
	73*m*	0.499(11) s	IT	–	0.0666
	74	Stable	**36.5(7)**	–	–
	75	82.78(4) min	β^-	1.19	0.199(1.4); 0.265(11)
	75*m*	47.7(7) s	IT(99.97) $\beta^-(0.03)$	–	0.1397
	76	Stable	**7.8(2)**	–	–
	77	11.30(1) h	β^-	2.2	0.21*(61); 0.263(45); 0.368(15); 0.417(25); 0.563(18); 0.73**(14)
	77*m*	52.9(6) s –	$\beta^-(76)$ IT(24)	2.9	0.159(12); 0.215(21)

Table 37.1 Table of isotopes *(continued)*

Element	A	$T_{1/2}$	Decay type or abundance of stable isotopes	Energy, MeV (relative intensity, %) particle group	γ-rays
$_{32}$Ge	78	88(1) min	β^-	0.71	0.277(94); 0.294
	79	19.1(3) s	β^-	4.1	0.109
	79m	39.0(10) s	IT	–	0.186
	80	29.5(4) s	β^-	2.4	0.266; 0.937; 1.564
	81	10.1 s	β^-	~5.6; ~5.3	0.336; 0.737; 0.793
	82	4.6(4) s	β^-	–	1.093
	83	1.9(4) s	–	–	–
	84	1.2(3) s	–	–	–
$_{33}$As	66	0.0958(4) s	ϵ	–	–
	67	42.5(12) s	β^+	5.0; 4.7	0.121; 0.123; 0.244
	68	2.527(13) min	ϵ	–	0.651; 0.762; 1.016; 1.778
	69	15.2(2) min	β^+	2.9	0.23; 0.511(an.)
	70	52.6(3) min	β^+	2.89(6); 2.14	0.511(183, an.); 0.60(23); 0.67(25); 0.75(23); 1.04(78); 1.12(23); 1.71(22)
	71	64.8(7)	ϵ(70) β^+(30)	– 0.81	} 0.175(90); 0.511(60, an.)
	72	26.0(1) h	β^+, ϵ	3.34(17); 2.50	0.511(150, an.); 0.630(8); 0.835(78)
	73	80.30(6) d	ϵ	–	0.054(9)
	74	17.78(3) d	β^-(32) β^+(29) ϵ(39)	1.36 1.54(3); 0.95(26) –	} 0.511(59, an.); 0.596(61); 0.635(14)
	74m	8.0 s	IT	–	0.283
	75	Stable	**100**	–	–
	76	26.32(7) h	β^-	2.97	0.559(43); 0.657(6); 1.22(5)
	77	38.83(5) h	β^-	0.68	0.239(2.5); 0.522(0.8)
	78	90.7(2) min	β^-	4.1	0.614/42/; 0.70/15/; 0.83/8/; 1.31/11/
	79	9.01(15) min	β^-	2.15	0.36(2); 0.43(2); 0.89(1)
	80	15.2(2) s	β^-	6.0	0.66(42); 1.22(4); 1.64(4); 1.77(1.7)
	81	33(2) s	β^-	3.8	0.468; 0.491
	82	21 s	β^-	7.1	0.655; 1.080; 1.731
	83	14.1(11) s	β^-	3.4	0.735; 1.113
	84	5.5(3) s	β^-, β_n(0.1)	–	0.667; 1.455
	85	2.028(12) s	β^-, β_n(23)	–	–
	86	0.9(2) s	β^-, β_n(~4)	–	0.704
	87	0.75(6) s	β^-	–	–
$_{34}$Se	69	27.4(2) s	ϵ; ϵ_p(0.07)	–	0.098; 0.691
	70	41.0(6) min	ϵ	–	–
	71	4.74(5) min	ϵ	–	0.16; 0.511(an.)
	72	8.40(8) d	ϵ	–	0.046(59)
	73	7.15(8) h	β^+(65) ϵ(35)	1.30 –	} 0.066(65); 0.359(99); 0.511(130, an.)
	73m	39.8(13) min	ϵ(27) IT(73)	– –	} 0.0257
	74	Stable	**0.9(1)**	–	–
	75	119.77(1) d	ϵ	–	0.121(17); 0.136(57); 0.265(60); 0.280(25); 0.401(12)
	76	} Stable	**9.0(2)**	–	–
	77	} Stable	**7.6(2)**	–	–
	77m	17.45(10) s	IT	–	0.162(50)
	78	Stable	**23.6(6)**	–	–

Table 37.1 Table of isotopes *(continued)*

Element	A	$T_{1/2}$	Decay type or abundance of stable isotopes	Energy, MeV (relative intensity, %)	
				particle group	γ-rays
$_{34}$Se	79	$\sim 6.5 \cdot 10^4$ y	β^-	0.16	–
	79m	3.91(5)	IT	–	0.096(9)
	80	Stable	**49.7(7)**	–	–
	81	18.5 min	β^-	1.58	0.28**(0.9); 0.56**(0.3); 0.83(0.2)
	81m	57.25 min	IT(> 99)	–	0.103(8)
			β^-(0.058)	–	–
	82	$1.4 \cdot 10^{20}$ y	**9.2(5)**	–	–
	83	22.5(2) min	β^-	1.8	0.22(44); 0.36(69); 1.88(16); 2.29(9)
	83m	70.4(3) s	β^-	3.8	0.35/16/; 0.65/20/; 1.01/100/; 2.02/40/
	84	3.2(2) min	β^-	1.4	0.407
	85	31.7(9) s	β^-	6.2	0.345; 1.427; 3.396
	86	15.3(9) s	β^-	2.6	2.441; 2.660
	87	5.55(20) s	β^-, β_n(0.16)	–	0.243; 0.334; 0.468; 0.573
	88	1.53(6) s	β^-, β_n(0.8)	–	0.159; 0.259; 1.904
	89	0.41(4) s	β^-, β_n(5)	–	–
	91	0.27(5) s	β^-, β_n($\sim$20)	–	–
$_{35}$Br	72	78.6(24) s	ϵ	–	–
	73	3.4(3) min	ϵ	–	–
	74	25.3(3) min	β^+	4.7	0.511(an.); 0.64
	74m	41.5(15) min	β^+	–	0.195
	75	97(2) min	β^+(90)	1.70	0.285; 0.511(an.); 0.62
			ϵ($\sim$ 10)	–	
	76	16.2(2) h	β^+($\sim$ 62)	3.6	0.511(133, an.); 0.559(63); 0.65(19); 1.21(13); 1.86(11)
			ϵ($\sim$ 38)	–	
	77	57.036(6) h	ϵ(99)	–	0.24**(30); 0.52(24); 0.58(7)
			β^+(1)	0.34	
	77m	4.28(10) min	IT	–	0.106
	78	6.46(4) min	β^-($\le$ 0.01)	2.55	0.511(184, an.); 0.614(14)
			ϵ(> 99)	–	
	79	Stable	**50.69(5)**	–	–
	79m	4.864(30) s	IT	–	0.207
	80	17.68(2) min	β^-(92)	2.00	0.511(5, an.); 0.618(7); 0.666(1)
			β^+(2.6)	0.87	
			ϵ(5.7)	–	
	80m	4.42(1) h	IT	–	0.086
	81	Stable	**49.31(5)**	–	–
	82	35.30(2) h	β^-	0.444	0.554(66); 0.619(41); 0.777(83); 1.044(29)
	82m	6.13(8) min	IT(97.6)	–	0.046(0.3); 0.777(0.15)
			β^-(2.4)	3.1	
	83	2.39(2) h	β^-	0.93	0.530(1.4)
	84	31.80(8) min	β^-	4.68	0.88(51); 1.90(18); 3.93(13)
	84m	6.0(2) min	β^-	2.2	0.424; 0.882; 1.463
	85	172(2) s	β^-	2.5	0.802; 0.925
	86	55.0(8) s	β^-	7.1	1.36/39/; 1.56/100/; 2.75/36/
	87	55.69(13) s	β^-	2.6	1.44/100/; 1.85/18/; 2.48/18/; 2.98/25/; 4.19/21/
	88	16.3(3) s	β^-, β_n(6)	–	0.76
	89	4.53(10) s	β^-, β_n(13)	–	0.775; 1.098

Table 37.1 Table of isotopes *(continued)*

Element	A	$T_{1/2}$	Decay type or abundance of stable isotopes	Energy, MeV (relative intensity, %)	
				particle group	γ-rays
$_{35}$Br	90	1.71(14) s	β^-, β_n(23)	–	0.707; 1.362
	91	0.541(5) s	β^-, β_n(9)	–	0.263; 0.803
	92	0.365(7) s	β^-, β_n(16)	–	0.740
$_{36}$Kr	72	17.2(3) s	β^+	3.8	0.163; 0.310; 0.415; 0.577
	73	27.0(12) s	β^+, β_p(0.7)	–	0.178; 0.241; 0.455
	74	11.50(11) min	β^+	2.2; 2.0	0.511(an.)
	75	4.3(1) min	β^+	3.2	0.133; 0.155
	76	14.8(1) h	ϵ	–	0.045; 0.270; 0.316; 0.407
	77	74.4(6) min	β^+(∼ 80) ϵ(∼ 20)	1.86 –	0.130; 0.147
	78	Stable	**0.35(2)**	–	–
	79	35.04(10) h	ϵ(92) β^+(8)	– 0.60	0.398(10); 0.511(15, an.); 0.606(10)
	79*m*	50(3) s	IT	–	0.130
	80	Stable	**2.25(2)**	–	–
	81	2.1(2)·10^5 y	ϵ	–	0.276
	81*m*	13 s	IT	–	0.190(65)
	82	Stable	**11.6(1)**	–	–
	83	Stable	**11.5(1)**	–	–
	83*m*	1.83(2) h	IT	–	0.009; 0.042
	84	Stable	**57.0(3)**	–	–
	85	10.72(2) h	β^-	0.67	0.514(0.4)
	85*m*	4.480(8) h	β(79) IT(21)	0.82 –	0.150(74); 0.305(13)
	86	Stable	**17.3(2)**	–	–
	87	76.31(62) min	β^-	3.8	0.403(84); 0.85(16); 2.57(35)
	88	2.84(3) h	β^-	2.8	0.191(35); 0.85(23); 1.55(14); 2.40(35)
	89	3.07(9) min	β^-	4.0	0.23/85/; 0.51/42/; 0.60/100/; 0.88/65/; 1.12/45/; 1.51/88/
	90	32.32(9) s	β^-	4.4; 2.6	0.120(50); 0.536(48); 1.11(48)
	91	8.57(4) s	β^-	6.4; 3.6	0.109; 0.507; 0.613; 1.109
	92	1.85(1) s	β^-, β_n(0.032)	–	0.142; 0.548; 0.813; 1.219
	93	1.289(12) s	β^-, β_n(1.9)	–	0.253; 0.267; 0.324; 2.350
	94	0.20(1) s	β^-, β_n(6)	–	0.220; 0.359; 0.629
	95	0.78(3) s	β^-	–	–
	97	<0.1 s	–	–	–
$_{37}$Rb	75	17.2(8) s	ϵ	–	0.179
	76	36.8(15) s	β^+	5.2	0.354; 0.423; 2.573
	77	3.70(15) min	ϵ	–	0.067; 0.179; 0.394
	78	17.66(8) min	ϵ	–	0.455; 0.693; 3.438
	79	22.9(5) min	ϵ	–	0.15(73); 0.19(29); 0.511(180, an.)
	80	34(4) s	β^+	4.7	0.511(195, an.); 0.618(39)
	81	4.58(1) h	ϵ(87) β^+(13)	– 1.03	0.446
	81*m*	32 min	β^+, IT	1.4	0.085
	82	1.25(3) min	β^+(96) ϵ(4)	3.15	0.511(192, an.); 0.777(9)
	82*m*	6.2(5) h	ϵ(94) β^+(6)	– 0.78	0.554(66); 0.619(41); 0.777(83)
	83	86.2(1) d	ϵ	–	0.53**(93); 0.79(0.9)

Table 37.1 Table of isotopes *(continued)*

Element	A	$T_{1/2}$	Decay type or abundance of stable isotopes	Energy, MeV (relative intensity, %)	
				particle group	γ-rays
$_{37}$Rb	84	32.87(11) d	ϵ(76)	–	0.511(42, an.); 0.88(74)
			β^+(21)	1.66	
			β^-(3)	0.91	
	84m	20.49(17) min	IT	–	0.216(37); 0.250(65); 0.464(32)
	85	Stable	**72.165(13)**	–	–
	86	18.66(2)	β^-(> 99)	1.78	1.078(8.8)
			ϵ(0.005)		
	86m	1.017(3) min	IT	–	0.556
	87	4.80(13)·10^{10} y	**27.835(13)**	–	–
			β^-	0.274	–
	88	17.8(1) min	β^-	5.3	0.898(13); 1.863(21)
	89	15.2(1) min	β^-	3.92(7); 2.9(5); 1.6	0.66(17); 1.05(75); 1.26(54)
	90	153(3) s	β^-	6.6	0.83*(61); 3.34**(15); 4.13(11); 4.34*(18)
	90m	258(5) s	β^-	–	0.107
			IT	–	
	91	58.4(4) s	β^-	5.8	0.094; 0.346; 2.564; 3.6
	92	4.50(2) s	β^-, β_n(0.012)	–	0.570; 0.815; 2.821
	93	5.8(1) s	β^-, β_n(1.3)	–	0.213; 0.433; 0.986; 1.385
	94	2.69(4) s	β^-, β_n(10)	–	0.837; 1.09; 1.309; 1.578
	95	0.384(6) s	β^-, β_n(8.4)	–	0.204; 0.329; 0.352; 0.681
	96	0.199(3) s	β^-, β_n(13)	–	0.815; 1.037
	97	0.176(6) s	β^-, β_n(27)	–	0.167; 0.585; 0.599
	98	0.114(5) s	β^-, β_n(13)	–	0.144; 2.172
	99	0.076(5) s	β^-	–	0.091; 0.145
$_{38}$Sr	78	~30.6 min	–	–	–
	79	2.25(10) min	ϵ	–	–
	80	106.3(15) min	ϵ	–	0.58
	81	25.5 min	ϵ	–	0.15; 0.19; 0.44
	82	25.0(4) d	ϵ	–	–
	83	32.4(2) h	ϵ(84)	–	0.040(24); 0.38(35); 0.511(32, an.); 0.76(40)
			β^+(16)	1.15	
	84	Stable	**0.56(1)**	–	–
	85	64.84(2) d	ϵ	–	0.514(100)
	85m	67.66(7) min	IT(86)	–	0.150(14); 0.239(85)
			ϵ(14)	–	
	86	Stable	**9.86(1)**	–	–
	87		**7.00(1)**	–	–
	87m	2.81(1) h	IT(> 99)	–	0.388(80)
			ϵ(0.3)	–	
	88	Stable	**82.58(1)**	–	–
	89	50.55(9) d	β^-	1.463	0.91(0.01)
	90	28.6(3) y	β^-	0.546	–
	91	9.52(6) h	β^-	2.67	0.645(15); 0.748(27); 1.025(30)
	92	2.71(1) h	β^-	1.5(10); 0.55	0.44(3); 1.37(90)
	93	7.6(2) min	β^-	2.9	0.60; 0.8; 1.2
	94	78(2) s	β^-	3.5; 2.1	1.42(100)
	95	25.1(2) s	β^-	6.1	0.686; 2.247; 2.717; 2.933
	96	1.06(4) s	β^-	4.4	0.122; 0.809; 0.932
	97	0.40 s	β^-	7.4; 5.2	0.307; 0.652; 0.954; 1.905
	98	0.65(3) s	β^-	5.7	0.037; 0.119; 0.429; 0.445

Table 37.1 Table of isotopes *(continued)*

Element	A	$T_{1/2}$	Decay type or abundance of stable isotopes	Energy, MeV (relative intensity, %)	
				particle group	γ-rays
$_{39}$Y	81	5 min	–	–	–
	82	∼10 min	–	–	–
	83	7.06(8) min	ϵ	–	–
	84	40(1) min	β^+	3.5; 2.24	0.795(100); 0.982(100); 1.041(50)
	85	2.68(5) h	β^+(70)	–	0.231(13); 0.511(140, an.)
			ϵ(30)	1.54	
	85m	4.86(13) h	β^+(55)	–	0.51**(200); 0.92(9)
			ϵ(45)		
	86	14.74(2) h	ϵ(74)	–	0.51*(35); 0.63*(37); 1.077(82);
			β^+(26)	3.15; 2.34	1.16*(35); 1.925(24)
	86m	48(1) min	IT(99.31)	–	0.218(94)
			β(0.69)	–	
	87	80.3(3) h	ϵ(> 90)	–	0.483
			β^+(0.3)	–	
	87m	12.9(4) h	IT(∼98)	–	0.381(74)
			ϵ(∼ 2)	–	
	88	106.60(4) d	ϵ(> 99)	–	0.898(93); 1.836(100);
			β^+(0.2)	0.76	2.734(1)
	89	Stable	**100**	–	–
	89m	16.06(4) s	IT	–	0.909(99)
	90	64.1(1) h	β^-	2.27	–
	90m	3.19(1) h	IT(> 99)	–	0.202(97); 0.482(91)
			β^-(0.0021)	–	
	91	58.51(6) d	β^-	1.545	1.21(0.3)
	91m	49.71(4) min	IT	–	0.555
	92	3.54(1) h	β^-	3.63	0.934(14); 1.40(4.7)
	93	10.1(2) h	β^-	2.89	2.267(6); 0.94(2.3); 1.90(1.8)
	94	19.1(4) min	β^-	5.0	0.56(6); 0.92(43); 1.13(5)
	95	10.3(2) min	β^-	4.4	0.954; 1.324; 2.176; 2.633; 3.577
	96	2.3(1) min	β^-	–	–
	97	1.11(14) s	β^-	6.0; 5.1	–
	98	0.64(3) s	β^-	8.8	1.22; 1.59; 2.94; 4.45
	99	1.5(1) s	β^-, β_n(1)	–	0.122; 0.724
	102	0.27(7) s	β^-	–	–
$_{40}$Zr	81	10 min	–	–	–
	84	5.05(5) min	–	–	–
	85	7.86(4) min	ϵ	–	–
	85m	10.9(3) s	IT	–	0.292
	86	16.5(1) h	ϵ	–	0.028(20); 0.243(96); 0.612(5)
	87	104.0(5) min	β^+	2.10	0.511(an.); 1.2; 2.2
	87m	14.0(2) s	IT	–	0.336; 0.201; 0.135
	88	83.4(3) d	ϵ	–	0.394(97)
	89	78.43(8) h	ϵ(78)	–	0.511(44, an.); 0.91(99)
			β^+(22)	0.90	
	89m	4.18(1) min	IT(94)	–	0.588(87); 1.51(6)
			ϵ(4.7)	–	
			β^+(1.3)	2.40(0.2); 0.89(1.2)	
	90	Stable	**51.45(2)**	–	–
	91	Stable	**11.22(2)**	–	–
	92	Stable	**17.15(1)**	–	–

Table 37.1 Table of isotopes *(continued)*

Element	A	$T_{1/2}$	Decay type or abundance of stable isotopes	Energy, MeV (relative intensity, %)	
				particle group	γ-rays
$_{40}$Zr	93	$1.53(10)\cdot10^6$ y	β^-	0.060	–
	94	Stable	**17.38(2)**	–	–
	95	64.02(4) d	β^-	0.89(2); 0.396	0.724(49); 0.756(49)
	96	$> 3.56\cdot10^{17}$ y	**2.80(1)**	–	–
	97	17.0(2) h	β^-	1.91	0.747(92); 1.148
	98	30.7(4) s	β^-	2.3	–
	99	2.1(1) s	β^-	3.5	0.47; 0.55; 0.59
	102	2.9(2) s	β^-	–	0.535; 0.600
$_{41}$Nb	84	12(3) s	ϵ	–	–
	86	80(12) s	ϵ	–	–
	87	2.60(7) min	β^+	–	0.201; 0.471
	88	14.3(3) min	ϵ	–	0.503; 0.671; 1.06; 1.08
	89	122(4) min	β^+	2.9	0.511(an.); 1.626; 3.577; 3.838
	90	14.60(5) h	β^+, ϵ	1.50	0.142(75); 1.14(97); 2.32(82)
	91	$\sim 1\cdot10^4$ y	ϵ	–	–
	91m	62 d	IT(97)	–	} 0.104(0.5); 1.21(3)
			ϵ(3)	–	
	92	$3.5(3).10^7$ y	ϵ	–	0.561; 0.934
	92m	10.15(2) d	ϵ(> 99)	–	} 0.934(99)
			β^+(0.06)	–	
	93	Stable	**100**	–	}
	93m	13.6(3) y	IT	–	0.030
	94	$2.03(16)\cdot10^4$ y	β^-	0.49	0.702(100); 0.871(100)
	94m	6.26(1) min	IT(> 99)	–	} 0.041; 0.871
			β^-(0.5)	–	
	95	34.97(3) d	β^-	0.160	0.765(105)
	95m	86.6(8) h	IT(97.5)	–	} 0.235
			β^-(2.5)	1.0	
	96	23.35(5) h	β^-	0.7	0.459(28); 0.569(59); 0.778(97); 1.092(49)
	97	72.1(7) min	β^-	1.27	0.665(98)
	97m	60.(1) s	IT	–	0.743(98)
	98	2.86(6) s	β^-	4.6	0.787; 1.024
	98m	51.3(4) min	β^-	3.1	0.720(75); 0.787(100; 1.16(30)
	99	14.3 s	β^-	–	0.098; 0.138
	99m	2.6(2) min	β^-	3.2	0.100/1/; 0.260/1/
	101	7.1(3) s	β^-	4.3	0.158; 0.276; 0.441; 0.480
	103	1.5(2) s	β^-	5.4	0.103; 0.641
	105	1.8(8) s	β^-	–	–
$_{42}$Mo	88	8.2(5) min	β^+	2.5	0.511(an.); 2.69
	90	5.67(5) h	β^+(25)	1.2	} 0.122(71); 0.257(85);
			ϵ(75)	–	0.511(50, an.)
	91	15.49(1) min	β^+	3.44	0.511(an.)
	91m	65.2(8) s	IT(~57)	–	} 0.658(54); 1.21(22); 1.53(15)
			β^+(~ 43)	3.99/15/; 2.78/100/	
	92	Stable	**14.84(4)**	–	–
	93	$3.5(7)\cdot10^3$ y	ϵ	–	
	93m	6.85(7) h	IT(99.88);	–	} 0.264(58); 0.685(100);
			ϵ(0.12)		1.479(100)
	94	Stable	**9.25(2)**	–	–

Table 37.1 Table of isotopes *(continued)*

Element	A	$T_{1/2}$	Decay type or abundance of stable isotopes	Energy, MeV (relative intensity, %)	
				particle group	γ-rays
$_{42}$Mo	95	Stable	**15.92(4)**	–	–
	96	Stable	**16.68(4)**	–	–
	97	Stable	**9.55(2)**	–	–
	98	Stable	**24.13(6)**	–	–
	99	66.02(1) h	β^-	1.23	0.181(7); 0.740(12); 0.780(4)
	100	Stable	**9.63(2)**	–	–
	101	14.6(1) min	β^-	2.23	0.191(25); 0.59(21); 1.02(25)
	102	11.3(2) min	β^-	1.2	0.148; 0.212; 0.224
	103	67.5(15) s	β^-	–	0.424
	104	1.3(3) min	β^-	4.8	0.070
	105	36.7(10) s	β^-	–	0.077; 0.085; 0.148
	106	8.4(5) s	β^-	–	0.054; 0.466; 0.619
	108	1.5(4) s	β^-	–	0.26
$_{43}$Tc	91	3.14(2) min	β^+	5.2	–
	92	4.4(3) min	$\beta^+(\sim 92)$ $\epsilon(\sim 8)$	4.1 –	0.14(67); 0.33(90); 0.511(184, an.); 0.79(95); 1.54(100)
	93	2.75(5) h	$\epsilon(87)$ $\beta^+(13)$	– 0.80	0.511(26, an.); 1.35(65); 1.49(33)
	93m	43.5(10) min	IT(80) $\epsilon(20)$	– –	0.390(63); 2.66(18) –
	94	293(1) min	$\beta^+(11)$ $\epsilon(89)$	0.816 –	0.511(22, an.); 0.702(100); 0.849(100); 0.871(100)
	94m	52(1) min	$\beta^+(66)$ $\epsilon(34)$	2.47	0.511(132, an.); 0.871(91); 1.53(10); 1.87(9)
	95	20.0(1) h	β^+	–	0.768(82); 0.84(11)
	95m	61(2) d	$\epsilon(95)$ $\beta^+(0.42)$ IT(4)	– 0.68 –	0.204(70); 0.584(36); 0.838(27)
	96	4.28(7) d	ϵ	–	0.778(100); 0.81(84); 0.851(100); 1.12(16)
	96m	51.5(10) min	$\epsilon(2)$ IT(98)	– –	0.034; 0.778; 1.200
	97	$2.6(4)\cdot 10^6$ y	ϵ	–	–
	97m	91 d	IT	–	–
	98	$4.2(3)\cdot 10^6$ y	β^-	4.0	0.66(100); 0.76(100)
	99	$2.13(5)\cdot 10^5$ y	β^-	0.292	–
	99m	6.02(3) h	IT(> 99) $\beta^-(\sim 1.10^{-4})$	– –	0.1426(90) –
	100	15.8(1) s	β^-	3.38	0.540; 0.60
	101	14.2(1) min	β^-	1.32	0.307(91); 0.545(8)
	102	5.28(15) s	β^-	4.2	0.475
	102m	4.35(7) min	$\beta^-(\sim 98)$ IT($\sim$ 2)	2.0 –	0.47; 0.63
	103	54.2(8) s	β^-	2.2	0.135/17/; 0.21/10/
	104	18.2(5) min	β^-	4.6	0.36; 0.53; 0.88; 0.89
	105	7.7(2) min	β^-	3.4	0.108; 0.143; 0.159; 0.321
	106	36(1) s	β^-	–	0.270; 1.97; 2.24; 2.79
	107	21.2(2) s	β^-	–	0.10; 0.18
	108	5.17(7) s	β^-	–	0.24; 0.47; 0.71; 0.73; 1.58
	109	1.4(4) s	–	–	–
	110	0.83(4) s	β^-	–	0.241

Table 37.1 Table of isotopes *(continued)*

Element	A	$T_{1/2}$	Decay type or abundance of stable isotopes	Energy, MeV (relative intensity, %) particle group	γ-rays
$_{44}$Ru	92	3.65(5) min	β^+	–	0.135; 0.214; 0.259
	94	51.8(6) min	ϵ	–	0.367; 0.891
	95	1.64(1) h	ϵ(85)	–	0.340(70); 0.511(30, an.); 0.625(13); 1.09(21)
			β^+(15)	1.33	
	96	Stable	**5.52(5)**	–	–
	97	2.9(1) d	ϵ	–	0.215(91); 0.324(8)
	98	Stable	**1.88(5)**	–	–
	99	Stable	**12.7(1)**	–	–
	100	Stable	**12.6(1)**	–	–
	101	Stable	**17.0(1)**	–	–
	102	Stable	**31.6(2)**	–	–
	103	39.35(5) d	β^-	0.70(3); 0.21	0.497(88); 0.610(6)
	104	Stable	**18.7(2)**	–	–
	105	4.44(2) h	β^-	1.87(11); 1.15	0.317*(11); 0.475*(20); 0.67*(16); 0.726(48)
	106	371.63(17) d	β^-	0.039	–
	107	3.75(5) min	β^-	3.2	0.195(14); 0.86(7)
	108	4.55(5) min	β^-	1.3	0.165(28)
	109	35(3) s	β^-	–	0.206; 0.226; 1.93
	110	14.6(10) s	β^-	–	0.096; 0.112
	111	2.2(7) s	β^-	–	–
	112	4.65(14) s		–	–
$_{45}$Rh	95	5.02(10) min	ϵ	–	–
	95m	1.96(4) min	ϵ(12), IT(88)	–	0.543; 0.784
	96	9.90(10) min	β^+	3.3	0.63; 0.68; 0.83
	96m	1.51(2) min	β^+(40), IT(60)	–	0.052; 1.10; 1.70
	97	32 min	β^+	2.1	0.422; 0.840; 0.879
	98	8.7(2) min	β^+	3.5; 2.8	0.65(100)
	98m	3.5(3) min	β^+	–	0.050; 0.745
	99	16 d	ϵ(90)	–	0.511(20, an.); 0.62(20)
			β^+(10)	0.74	
	99m	4.7 h	ϵ	–	0.065; 0.341; 0.618; 1.261
	100	20.8(1) h	ϵ(93)	–	0.511(13, an.); 0.540(88); 0.820(25); 2.37(39)
			β^+(7)	2.62	
	101	3.3(3) y	ϵ	–	0.127(88); 0.198(75); 0.325(11)
	101m	4.34(1) d	ϵ(92.8)	–	0.157; 0.307(83); 0.545(6)
			IT(7.2)	–	
	102	~2.9 y	ϵ	–	0.475; 0.631; 0.697
	102m	207(3) d	ϵ, IT(5)	–	0.475(57); 0.511(25, an.); 0.628
			β^+	1.29	
			β^-(19)	1.25	
	103	Stable	**100**	–	–
	103m	56.12(1) min	IT	–	0.040(0.4)
	104	42.3(4) s	β^-	2.44	0.56(2); 1.24(0.13)
	104m	4.34(5) min	IT(> 99)	–	0.051(47); 0.078(2.5); 0.097(2.6); 0.129
			β^-(0.13)	–	
	105	35.36(6) h	β^-	0.568	0.306(5); 0.319(19)
	105m	45 s	IT	–	0.129
	106	29.80(8) s	β^-	3.54	0.512(21); 0.622*(11)
	106m	130(2) min	β^-	1.7; 0.9	0.140; 0.512; 0.717; 1.046

Table 37.1 Table of isotopes *(continued)*

Element	A	$T_{1/2}$	Decay type or abundance of stable isotopes	Energy, MeV (relative intensity, %) particle group	γ-rays
$_{45}$Rh	107	21.7(4) min	β^-	1.20	0.305(73); 0.390(11)
	108	16.8(5) s	β^-	4.5	0.434(43); 0.51**(10); 0.62(22)
	109	80(2) s	β^-	2.6; 2.3	0.11; 0.18; 0.29; 0.33; 0.43
	110	3.2(2) s	β^-	5.5	0.374; 0.440; 0.797
	112	0.8(1) s	β^-	–	0.349
	113	~0.91 s	β^-	–	0.129
	114	1.68(7) s	β^-	–	0.333
$_{46}$Pd	98	17.7(3) min	β^+	2.3	0.112; 0.663; 0.838
	99	21.4(2) min	β^+	2.2	0.136; 0.264; 0.673
	100	3.63(9) d	ϵ	–	0.074(34); 0.084(49); 0.126(16)
	101	8.47(6) h	ϵ	–	0.296(30); 0.590(24)
			β^+(2.5)	0.78	
	102	Stable	**1.020(12)**	–	–
	103	16.96(2) d	ϵ	–	0.297(0.011); 0.362(0.06); 0.498(0.011)
	104	Stable	**11.14(8)**	–	–
	105	Stable	**22.33(8)**	–	–
	106	Stable	**27.33(5)**	–	–
	107	$6.5(3)\cdot10^6$ y	β^-	0.03	–
	107*m*	21.3(5) min	IT	–	0.215
	108	Stable	**26.46(9)**	–	–
	109	13.46(2) h	β^-	1.028	0.088(5); 0.60(0.03)
	109*m*	4.69(1) min	IT	–	0.189
	110	Stable	**11.72(9)**	–	–
	111	23.4(2) min	β^-	2.2	0.38/5/; 0.60*/13/; 1.4*/8/
	111*m*	5.5(1) h	IT(71)	–	0.172
			β^-(29)	2.0	
	112	21.045(40) h	β^-	0.28	0.019(20)
	113	93(5) s	β^-	–	0.096; 0.222; 0.643; 0.739
	114	2.4(1) min	β^-	–	0.126; 0.136; 0.232; 0.358
	115	41(3) s	β^-	–	0.089; 0.255; 0.343
	116	12.72(44) s	β^-	–	0.115; 0.178
	117	5.0(6) s	β^-	–	–
$_{47}$Ag	98	44.5(12) s	β^+	–	0.571; 0.679; 0.863
	99	1.8 min	β^+	4.2	0.264; 0.806; 0.832
	100	2.3(1) min	β^+	5.4	0.666; 1.694
	101	11.1(3) min	β^+	3.4; 2.7	0.261; 0.588; 0.667; 1.174
	102	12.9(3) min	β^+	2.3	0.56; 0.72; 0.84; 1.74
	102*m*	7.7(5) min	ϵ(51), IT(49)	–	0.009
	103	65.7(7) min	β^+	1.6	0.12*/26/; 0.15/23/; 0.27/34/; 0.511(100, an.)
			ϵ(~ 70)	–	
	103*m*	5.7(3) s	IT	–	0.134
	104	69.2(10) min	β^+	0.99	0.556(84); 0.764(48); 0.854(30)
			ϵ	–	
	104*m*	33.5(20) min	β^+	2.70	0.511(120, an.); 0.556(100)
			ϵ, IT(33)	–	
	105	41.29(7) d	ϵ	–	0.280(32); 0.344**(42)
	105*m*	7.23 min	ϵ, IT(99.7)	–	0.025
	106	24.0(1) min	β^+	1.96	0.511(140, an.)
	106*m*	8.46(10) d	ϵ	–	0.512(86); 0.616(23); 0.717**(31); 0.80**(41); 1.046(29)

Table 37.1 Table of isotopes *(continued)*

Element	A	$T_{1/2}$	Decay type or abundance of stable isotopes	Energy, MeV (relative intensity, %) particle group	γ-rays
$_{47}$Ag	107	Stable	**51.839(5)**	–	–
	107*m*	44.3(2) s	IT	–	0.093(5)
	108	2.37(1) min	β^-(97.5)	1.64	0.434(0.45); 0.511(0.56, an.); 0.632(1.7)
			ϵ(2.2)	–	
			β^+(0.28)	0.90	
	108*m*	127(21) y	ϵ(91)	–	0.079; 0.434(89); 0.614(90); 0.723(90)
			IT(9)	–	
	109	Stable	**48.161(5)**	–	–
	110	24.6(2) s	β^-(99.7)	2.87	0.658(4.5)
			ϵ(0.3)		
	110*m*	249.76(4) d	β^-(98.5)	1.5(0.6); 0.53(31); 0.087	0.658(96); 0.764(23); 0.885(71); 0.937(32); 1.384(21)
			IT	–	
	111	7.45(1) d	β^-	1.05	0.247(1); 0.342(6)
	111*m*	64.8(8) s	IT(99.7)	–	0.060
			β^-(0.3)	–	
	112	3.14(2) h	β^-	3.94	0.617(41); 1.40(5)
	113	5.37(5) h	β^-	2.0	0.12/10/; 0.30/100/; 0.67/17/
	113*m*	68.7(50) s	β^-	1.9; 1.5	0.043; 0.299; 0.316; 0.392
	114	4.6(2) s	β^-	4.9	0.57
	115	20.0(5) min	β^-	3.2	0.14**(12); 0.22**(49); 0.28(13); 1.48(11); 2.12(13)
	116	2.68(1) min	β^-	5.0	0.52; 0.70
	116*m*	10.4(8) s	β^-(~98)	–	0.081; 0.514; 0.706; 1.030
			IT(~ 2)		
	117	72.8(10) s	β^-	–	0.135; 0.338
	118	3.7 s	β^-	–	0.488; 0.677
	118*m*	2.8(3) s	β^-(59), IT(41)	–	0.128
	119	2.1(1) s	β^-	–	0.366; 0.399; 0.626
	120	1.17(5) s	β^-	–	0.506; 0.698
	120*m*	0.32(4) s	β^-(~ 63)	-	0.203; 0.506; 0.698; 0.926
			IT(~37)		
	121	0.8(1) s	β^-	–	0.315; 0.354
	123	0.39(3) s	β^-, β_n	–	–
$_{48}$Cd	100	1.1(3) min	β^+	–	0.124–0.935
	101	1.2(2) min	ϵ, β^+	–	0.098; 0.925; 1.26; 1.72
	102	5.5(5) min	ϵ, β^+	–	0.415; 0.481; 0.505; 1.037
	103	7.3(1) min	ϵ, β^+	–	0.22; 0.511(an.); 0.63; 0.85
	104	57.7(10) min	ϵ, β^+	–	0.084; 0.709
	105	55.5(4) min	ϵ	–	0.308; 0.320; 0.347; 0.433; 0.511(an.); 0.607; 0.962; 1.302; 1.693
			β^+	1.69	
	106	Stable	**1.25(3)**	–	–
	107	6.50(2) h	ϵ(> 99)	–	0.511(0.56, an.); 0.769(0.08); 0.829(0.21)
			β^+(0.28)	0.302	
	108	Stable	**0.89(1)**	–	–
	109	464(1) d	ϵ	–	0.088(4)
	110	Stable	**12.49(9)**	–	–
	111	Stable	**12.80(6)**	–	–
	111*m*	48.6(3) min	IT	–	0.150(30); 0.247(94); 0.396

Table 37.1 Table of isotopes *(continued)*

Element	A	$T_{1/2}$	Decay type or abundance of stable isotopes	Energy, MeV (relative intensity, %)	
				particle group	γ-rays
$_{48}$Cd	112	Stable	**24.13(11)**	–	–
	113	9.3(19)·10^{15} y	**12.22(6)**	–	–
			β^-	–	–
	113*m*	14.1(5) y	β^-(99.9)	0.58	} 0.264
			IT(0.1)	–	
	114	Stable	**28.73(21)**	–	–
	115	53.46(10) h	β^-	1.11	0.262(2); 0.49(10); 0.53(26)
	115*m*	44.6(3) d	β^-	1.62	0.485(0.3); 0.935(1.9); 1.29(0.9)
	116	Stable	**7.49(9)**	–	–
	117	2.49(4) h	β^-	2.23	0.273(31); 0.345(18); 0.434(13); 1.303(19); 1.577(17)
	117*m*	3.36(5) h	β^-	0.67	0.273(18); 0.880(10); 1.24**(11); 1.433(10); 1.998(15)
	118	50.3(2) min	β^-	~0.8	–
	119	2.69(2) min	β^-	–	0.293; 0.343
	119*m*	2.20(2) min	β^-	–	0.146; 1.025; 2.021
	120	50.80(21) s	β^-	–	–
	121	13.5(3) s	β^-	–	0.324; 0.349; 1.041
	126	0.506(15) s	β^-	–	0.260; 0.428
	128	0.94(5) s	–	–	–
$_{49}$In	102	23(4) s	ϵ	–	0.593; 0.777; 0.861
	103	65(7) s	β^+	4.4	0.188; 0.202; 0.720; 0.740
	104	25(6) min	–	–	–
	105	5.1(3) min	β^+	3.9	0.131; 0.260; 0.604
	106	6.2(1) min	β^+	2.0	0.511(an.); 0.63; 1.65; 1.85
	107	32.4(3) min	β^+	2.2	} 0.22(46); 0.511(an.)
			ϵ	–	
	107*m*	50.4(6) s	IT	–	0.678
	108	57 min	β^+	1.29	} 0.150; 0.175; 0.243;
			ϵ	–	0.511(an.); 0.633; 0.872
	109	4.2(1) h	ϵ(94)	–	} 0.205; 0.28**; 0.35**;
			β^+(6)	0.79	0.65**; 0.91**
	109m_1	1.34(7) min	IT	–	0.650
	109m_2	0.21(1) s	IT	–	0.40(20); 0.68(100); 1.04(20); 1.43(77)
	110	66 min	β^+(71)	2.25	} 0.511(142, an.); 0.658(95)
			ϵ(29)	–	0.885; 0.937
	111	2.83(1) d	ϵ	–	0.173(89); 0.247(94)
	111*m*	7.7(2) min	IT	–	0.537
	112	14.4(2) min	β^-(44)	0.66	} 0.511(44, an.); 0.617(6)
			β^+(22)	1.56	
			ϵ(34)		
	112*m*	20.9(2) min	IT	–	0.156(9)
	113	Stable	**4.3(2)**	–	–
	113*m*	1.658(1) h	IT	–	0.392
	114	71.9(1) s	β^-(98)	1.988	} 1.299(0.17)
			ϵ(1.9)	–	
			β^+(0.004)	0.42	
	114*m*	49.51(1) d	IT(96.7)	–	} 0.190(17); 0.558(3.5);
			ϵ(3.3)	–	0.724(3.5)
	115	4.41·10^{14} y	**95.7(2)**	–	–
			β^-	0.48	–

Table 37.1 Table of isotopes *(continued)*

Element	A	$T_{1/2}$	Decay type or abundance of stable isotopes	Energy, MeV (relative intensity, %)	
				particle group	γ-rays
$_{49}$In	115m	4.486(4) h	IT(95)	–	0.336
			β^-(5)	0.83	
	116	14.10(3) s	β^-	3.3	0.434(0.12); 0.95(0.1); 1.293(1.2)
	116m_1	54.15(6) min	β^-	1.00	0.417(36); 1.09(53); 1.293(80); 2.111(20)
	116m_2	2.18(4) s	IT	–	0.164; 0.290
	117	43.8(7) min	β^-	0.74	0.158(87); 0.565(100)
	117m	116.5(7) min	IT(47)	–	0.158(14); 0.315(31)
			β^-(53)	1.78	
	118	5.0(3) s	β^-	4.2	1.230(15)
	119	2.4(1) min	β^-	1.6	0.82(95)
	119m	18.0(3) min	β^-(95)	2.7	0.024; 0.311; 0.91*
			IT(5)	–	
	120	3.2 s	β^-	5.3	1.171(15)
	121	23.1(6) s	β^-	2.5	0.262; 0.657; 0.926
	121m	3.88(10) min	β^-(98.8), IT(1.2)	–	0.314
	122	10.0(5) s	β^-	5	0.99; 1.14
	123	5.98(6) s	β^-	3.3	1.020; 1.131
	123m	47.8(5) s	β^-	4.5	0.126
	124	3.21(6) s	β^-	3.9	0.99/3/; 1.13/10/; 3.21/3/
	125	2.33(4) s	β^-	4.3; 4.1	1.032; 1.335
	126	1.45(22) s	β^-	4.2	0.909; 1.141
	127	1.15(5) s	β^-	5.8; 4.9	1.598
	128	0.9(1) s	β^-	5.0	1.169; 3.520
	129	0.59(2) s	β^-	6.8; 5.5	1.865; 2.119
	130	0.53(5) s	β^-, β_n	–	0.774; 1.221
$_{50}$Sn	106	2.10(15) min	ϵ	–	0.253; 0.387; 0.477
	107	2.90(5) min	ϵ	–	1.129
	108	10.30(8) min	ϵ	–	0.28; 0.42; 0.67
	109	18.0(2) min	ϵ	–	0.335; 0.521; 0.89; 1.12; 1.32; 1.46
			β^+	2.5	
	110	4.11(10) h	ϵ	–	0.283(95)
	111	35.3(8) min	ϵ(73)	–	0.511(54, an.); 0.75(1.1); 1.14(1.8); 1.89(1.0); 1.92
			β+(27)	1.51	
	112	Stable	**0.97(1)**	–	–
	113	115.09(4) d	ϵ	–	0.255(1.8)
	113m	21.4(4) min	IT(91), ϵ(9)	–	0.077(0.6)
	114	Stable	**0.65(1)**	–	–
	115	Stable	**0.36(1)**	–	–
	116	Stable	**14.53(11)**	–	–
	117	Stable	**7.68(7)**	–	–
	117m	13.61(4) d	IT	–	0.158(87); 0.315
	118	Stable	**24.22(11)**	–	–
	119	Stable	**8.58(4)**	–	–
	119m	293.0(13) d	IT	–	0.024(16); 0.090
	120	Stable	**32.59(10)**	–	–
	121	27.06(4) h	β^-	0.383	–
	121m	55(5) y	β^-	0.42	0.037
	122	Stable	**4.63(3)**	–	–
	123	129.2(4) d	β^-	1.42	–
	123m	40.08(7) min	β^-	1.26	0.160

Table 37.1 Table of isotopes *(continued)*

Element	A	$T_{1/2}$	Decay type or abundance of stable isotopes	Energy, MeV (relative intensity, %)	
				particle group	γ-rays
$_{50}$Sn	124	Stable	**5.79(5)**	–	–
	125	9.64(3) d	β^-	2.34	0.811(1.5); 0.904(1.4); 1.068(4); 1.97(0.6)
	125m	9.52(5) min	β^-	2.04	0.325(97)
	126	$\sim 1 \cdot 10^5$ y	β^-	0.3	0.060; 0.067; 0.092
	127	2.10(4) h	β^-	3.2	0.823; 1.096; 1.114
	127m	4.13(3) min	β^-	2.7	0.49(100)
	128	59.1(5) min	β^-	0.80	0.044(7); 0.072(19); 0.50(61); 0.57(22)
	128m	6.5(5) s	IT	–	0.832; 1.169
	129	2.16(4) min	β^-	3.3	0.642; 2.100
	129m	6.7(4) min	IT, β^-	–	–
	130	3.72(11) min	β^-	1.5; 1.1	0.192; 0.780
	130m	1.7(1) min	β^-	–	0.084; 0.145; 0.311; 0.899
	131	61(3) s	β^-	3.4	0.305; 0.450; 1.23
	132	40(1) s	β^-	1.8	0.085; 0.247; 0.340; 0.899; 0.992
	134	1.04(2) s	β^-, β_n(17)	–	–
$_{51}$Sb	109	18.3(5) s	β^+	5.4; 4.4	0.925; 1.062; 1.496
	110	23.0(4) s	β^+	6.9	0.827; 0.985; 1.212; 1.243
	111	75(1) s	β^+	3.3	0.154; 0.489
	112	51.4(5) s	β^+	4.8	0.511(an.); 1.26
	113	6.67(7) min	ϵ	–	0.32; 0.511(an.); 1.03; 1.2*
			β^+	2.42	
	114	3.49(3) min	ϵ	–	0.9; 1.30
			β^+	4.0	
	115	32.1(3) min	ϵ(67)	–	0.499(100); 0.511(67, an.); 0.98(5); 1.24(5)
			β^+(33)	1.5	
	116	15.8(8) min	ϵ(72)	–	0.511(56, an.); 0.93(26); 1.293(85); 2.23(14)
			β^+(28)	2.3	
	116m	60.3(6) min	ϵ(81)	–	0.099(30); 0.140(30); 0.406(36); 0.511(38, an.); 0.545(68); 0.96(75); 1.06(27); 1.293(100)
			β^+(19)	1.16	
	117	2.80(1) h	ϵ(97.4)	–	0.158(87); 0.511(5, an.)
			β^+(2.6)	0.57	
	118	3.6(1) min	ϵ	–	0.511(150, an.); 0.83(0.4); 1.230*(3)
			β^+	2.67	
	118m	5.00(1) h	ϵ(> 99)	–	0.041(29); 0.254(93); 1.049(100); 1.230(100)
			β^+(0.16)	–	
	119	38.1(2) h	ϵ	–	0.024(16)
	120	15.89(4) min	β^+	1.70	0.511(87, an.); 1.171(1.3)
			ϵ	–	
	121	Stable	**57.3(9)**	–	–
	122	2.70(1) d	β^-(97)	2.0; 1.4	0.564(66); 0.686(3.4)
			ϵ(3)	–	
			β^+(0.006)	0.56	
	123	Stable	**42.7(9)**	–	–
	124	60.20(3) d	β^-	2.31	0.603(97); 0.72*(14); 1.691(50)
	124m_1	93(5) s	IT(80)	–	0.505(20); 0.603(20); 0.644(20)
			β(20)	1.19	

Table 37.1 Table of isotopes *(continued)*

Element	A	$T_{1/2}$	Decay type or abundance of stable isotopes	Energy, MeV (relative intensity, %)	
				particle group	γ-rays
$_{51}$Sb	124m_2	20.2(2) min	IT	–	0.035
	125	2.73(3) y	β^-	0.61	0.427(31); 0.463(10); 0.599*(24); 0.634(11)
	126	12.4(1) d	β^-	1.9	0.41; 0.69**
	126m	19.0(3) min	β^-(86), IT(14)	–	0.018
	127	3.85(5) d	β^-	1.5	0.46; 0.68; 0.77
	128	9.01(3) h	β^-	2.0	0.320(83); 0.75*(200)
	129	4.40(1) h	β^-	2.2; 0.6	0.54; 0.81; 0.91
	130	40(1) min	β^-	2.9	0.19; 0.33; 0.82**; 0.94
	130m	6.3(2) min	β^-	3.2; 2.2	0.182; 0.840
	131	23(2) min	β^-	3.0; 1.3	0.64(37); 0.94(48)
	132	4.2(1) min	β^-	3.7	0.104; 0.151; 0.697; 0.974
	133	2.7(1) min	β^-	2.4; 1.2	0.837; 1.096; 1.729; 2.416; 2.755
	135	~2 s	β^-, β_n(20)	–	1.279
	136	0.82(2) s	β^-, β_n(32)	–	–
$_{52}$Te	108	2.1(1) s	α	3.32	–
			ϵ_p	2.6; 3.4; 3.7	–
	109	4.6(3) s	β^+, β_p, α	–	–
	110	18.6(8) s	β^+, α	–	0.108; 0.219; 0.606; 0.895
	111	19.3(4) s	β^+, β_p	–	0.851; 0.881; 1.268; 1.392
	112	2.0(2) min	β^+	–	0.296; 0.373; 0.419
	113	1.7(2) min	β^+	4.7	0.645; 0.814; 1.018; 1.181
	114	15.2(7) min	β^+	–	0.245; 0.727; 0.84; 0.90; 1.897
	115	5.8(2) min	ϵ(~ 20) β^+(~ 80)	– 2.8	} 0.511(160, an.); 0.72(34); 1.28(32); 1.38(32)
	115m	6.7(4) min	ϵ	–	–
	116	2.49(4) h	ϵ	–	0.094; 0.103
	117	62(2) min	ϵ(70) β^+(30)	– 1.81	} 0.511(60, an.); 0.72(65); 0.93(6); 1.78(9)
	117m	0.103(3) s	IT	–	–
	118	6.00(2) d	ϵ	–	–
	119	16.05(5) h	ϵ β^+(5)	– 0.627	} 0.645(85); 0.70(11); 1.76(3.6)
	119m	4.69(4) d	β^+	–	0.153(62); 0.270(25); 1.221(67); 2.09(4)
	120	Stable	**0.096(2)**	–	–
	121	16.78(35) d	ϵ	–	0.508(18); 0.573(80)
	121m	154(7) d	IT(90) ϵ(10)	– –	} 0.212(82); 1.10(3)
	122	Stable	**2.60(1)**	–	–
	123	$> 1 \cdot 10^{13}$ y	**0.908(3)**	–	–
	123m	119.7(1) d	IT	–	0.159(84)
	124	} Stable	**4.816(8)**	–	–
	125		**7.14(1)**	–	–
	125m	58(1) d	IT	–	0.035(7); 0.110(0.3)
	126	Stable	**18.95(1)**	–	–
	127	9.35(7) h	β^-	0.70	0.058(0.01); 0.21*(0.03); 0.360(0.05); 0.417(0.3)
	127m	109(2) d	IT(97.6) β^-(2.4)	–	} 0.059(0.19); 0.088(0.08); 0.67(0.004)

Table 37.1 Table of isotopes *(continued)*

Element	A	$T_{1/2}$	Decay type or abundance of stable isotopes	Energy, MeV (relative intensity, %)	
				particle group	γ-rays
$_{52}$Te	128	$> 8 \cdot 10^{24}$ y	**31.69(2)**	–	–
	129	69.6(3) min	β^-	1.45	0.027(19); 0.455(15)
	129m	33.6(1) d	IT(63) β^-(37)	– 1.60	0.106; 0.69
	130	2.51(27)$\cdot 10^{21}$ y	**33.80(2)**	–	–
	131	25.0(1) min	β^-	2.14	0.150(68); 0.453(16)
	131m	30(2) h	β^-(78) IT(22)	2.46(5); 0.9 –	0.78**(60); 0.85*(31); 1.127(13); 1.206(11)
	132	78.2(8) h	β^-	0.22	0.053(17); 0.230(90)
	133	12.45(28) min	β^-	3.2; 2.8	0.312; 0.408; 1.333
	133m	55.4(4) min	β^-(83) IT(17)	2.4 –	0.432(50); 0.557(35); 0.754(85); 0.91(57)
	134	41.8(8)	β^-	0.7	0.08(13); 0.17(16); 0.204(21); 0.262(19)
	135	18(1) s	β^-	6.0; 5.4	0.267; 0.604; 0.870
	136	20.7(20) s	β^-, β_n(0.7)	–	–
	137	3.5(5) s	β^-, β_n(2.5)	–	0.244
	138	1.4(4) s	β^-, β_n(6)	–	–
$_{53}$I	111	2.5(2) s	α	3.152	0.117; 0.321; 0.341; 0.266
	112	3.42(1) s	β^+, α	–	0.689; 0.787; 0.795; 1.143
	114	2.1(2) s	β^+	6.5	0.682; 0.709; 0.775; 1.091
	115	1.3(2) min	β^+	–	–
	116	2.91(15) s	β^+	6.7	0.540; 0.679
	117	2.3(1) min	β^+	3.5	0.16; 0.34; 0.511(an.)
	118	13.7(5) min	β^+($\sim$ 54) ϵ($\sim$ 46)	5.5 –	0.511(108, an.); 0.55; 0.60; 1.15; 1.34
	118m	8.5(5) min	ϵ, IT	–	0.104; 0.60
	119	19.1(4) min	ϵ(49) β^+(51)	– 2.4	0.26; 0.511(102, an.); 0.78
	120	81.0(6) min	ϵ(54) β^+(46)	– 4.6	0.511(92, an.); 0.56; 0.62; 1.52
	120m	53(4) min	β^+	3.8	0.56; 0.60; 0.61
	121	2.12(1) h	ϵ(91) β^+(9)	– 1.2	0.212(90); 0.32(6); 0.511(18, an.)
	122	3.62(6) min	β^+	3.1	0.511(an.); 0.564; 0.69; 0.78
	123	13.2(1) h	ϵ	–	0.159(83)
	124	4.18(2) d	ϵ(74) β^+(26)	– 2.14	0.511(50, an.); 0.605(67); 0.644(12); 0.73(14); 1.69(14)
	125	60.14(11) d	ϵ	–	0.035(7)
	126	13.02(7) d	ϵ(55) β^-(44) β^+(1.3)	– 1.25 1.13	0.386(34); 0.667(33)
	127	Stable	**100**	–	–
	128	24.99(2) min	β^-(93.6) ϵ(6.4)	2.12 –	0.441(14); 0.528(1.4)
	129	1.57(4)$\cdot 10^7$ y	β^-	0.150	0.040(9)
	130	12.36 h	β^-	1.7(0.4); 1.04	0.419(35); 0.538(99); 0.669(100); 0.743(87)
	130m	9.0(1) min	IT(83), β^-(17)	–	0.048; 0.536

Table 37.1 Table of isotopes *(continued)*

Element	A	$T_{1/2}$	Decay type or abundance of stable isotopes	Energy, MeV (relative intensity, %)	
				particle group	γ-rays
$_{53}$I	131	8.04(1) d	β^-	0.806(0.6); 0.606	0.284(5.4); 0.364(82); 0.637(6.8)
	132	2.30(3) h	β^-	2.12	0.52**(20); 0.67**(144); 0.773(89); 0.955(22)
	132m	83.6(7) min	IT(86), β^-(14)	–	0.175; 0.60; 0.67; 0.77
	133	20.8(1) h	β^-	1.27	0.53(90)
	133m	9 s	IT	–	0.073; 0.647; 0.913
	134	52.6(4) min	β^-	2.43	0.61(18); 0.85(95); 0.89(65); 1.15(10)
	134m	3.69(7) min	IT(98), β^-(2)	–	0.316; 0.847; 0.884
	135	6.61 h	β^-	2.2, 1.4	1.14(37); 1.28(34); 1.46(12); 1.72(19)
	136	84(1) s	β^-	7.0(6); 5.6	0.27(18); 0.39(19); 1.32**(95); 2.3**(19)
	137	24.5(2) s	β^-, β_n(6)	–	0.601; 1.218
	138	6.41(6) s	β^-, β_n(5)	–	0.484; 0.589; 0.875; 2.262
	139	2.30(5) s	β^-, β_n(10)	–	0.528; 0.537; 0.571; 0.848
	140	0.86(4) s	β^-, β_n(14)	–	0.377; 0.458
	141	0.41(8) s	β^-	–	0.192; 0.303; 0.387; 0.579
$_{54}$Xe	112	2.8(2) s	α	3.210	–
	114	10.0(4) s	β^+	–	0.104; 0.162; 0.309; 0.440
	116	56(2) s	β^+	3.3	0.248; 0.311
	117	61(2) s	β^+, β_p(0.003)	–	0.221; 0.295; 0.519; 0.661
	119	5.8(3) min	β^+	–	0.10; 0.23; 0.46
	120	40(1) min	β^+	–	0.055; 0.073; 0.176; 0.76
	121	40.1(20) min	β^+	2.8	0.080; 0.096; 0.132; 0.437; 0.511(an.)
	122	20.1(1) h	ϵ	–	0.060; 0.090; 0.110; 0.148; 0.180; 0.345; 0.417
	123	2.08(2) h	β^+, ϵ	–	0.090; 0.110; 0.149; 0.178; 0.329; 0.511(an.); 0.68; 0.90; 1.10
	124	Stable	**0.10(1)**	–	–
	125	16.9(2) h	ϵ, β^+	–	0.055; 0.188; 0.242
	125m	57(1) s	IT	–	0.075; 0.111; 0.140
	126	Stable	**0.09(1)**	–	–
	127	36.4(1) d	ϵ	–	0.172(22); 0.203(65); 0.375(20)
	127m	69.2(9) s	IT	–	0.125; 0.175
	128	Stable	**1.91(3)**	–	–
	129	Stable	**26.4(6)**		
	129m	8.89(2) d	IT	–	0.040(9); 0.197(6)
	130	Stable	**4.1(1)**	–	–
	131	Stable	**21.2(4)**	–	–
	131m	11.9(1) d	IT	–	0.164(2)
	132	Stable	**26.9(5)**	–	–
	133	5.29(1) d	β^-	0.346	0.081(37)
	133m	2.19(1) d	IT	–	0.233(14)
	134	Stable	**10.4(2)**	–	–
	135	9.083 h	β^-	0.92	0.250(91); 0.61(3)
	135m	15.6 min	IT(> 99) β^-(0.004)	–	0.527
	136	Stable	**8.9(1)**	–	–
	137	3.818(13) min	β^-	4.1	0.455(33)
	138	14.08(8) min	β^-	2.8	0.16/33/; 0.26/100/; 0.42/40/; 1.78/66/; 2.02/58/

Table 37.1 Table of isotopes *(continued)*

Element	A	$T_{1/2}$	Decay type or abundance of stable isotopes	Energy, MeV (relative intensity, %)	
				particle group	γ-rays
$_{54}$Xe	139	39.68(14) s	β^-	5.0	0.18/41/; 0.22/100/; 0.30/57/; 1.15/23/
	140	13.60(10) s	β^-	2.6	0.622; 0.806; 1.315; 1.414
	141	1.72(3) s	β^-, β_n(0.05)	–	0.106; 0.119; 0.909
	142	1.22(2) s	β^-, β_n(0.41)	–	0.54; 0.57; 0.62; 0.66
	144	1.15(20) s	β^-	–	–
	145	0.9(3) s	β^-	–	–
$_{55}$Cs	117	8(2) s	–	–	–
	118	16.4(12) s	β^+, β_p(0.04)	–	–
	120	60.2(15) s	β^+	–	0.332
	121	125.6(14) s	β^+	–	–
	122	4.5(2) min	ϵ	–	–
	123	5.87(5) min	β^+	3.1	0.097; 0.597
	123m	1.60(15) s	IT	–	0.64; 0.95
	124	26.5(15) s	β^+	4.9	0.354; 0.493; 0.915
	125	45(1) min	ϵ(51)	–	} 0.112; 0.511(98, an.)
			β^+(49)	2.05	
	126	1.64(2) min	β^+(82)	3.8	} 0.386(38); 0.511(164, an.); 0.925
			ϵ(18)	–	
	127	6.25(10) h	ϵ(96.5)	–	} 0.125(10); 0.406(72); 0.511(7, an.)
			β^+(3.5)	1.08	
	128	3.62(2) min	β^+(51)	2.9	} 0.441(27); 0.511(110, an.)
			ϵ(49)	–	
	129	32.06(6) h	ϵ	–	0.375(48); 0.416(25); 0.550(5)
	130	29.9 min	β^+	1.97	} 0.54; 0.59
			ϵ	–	
			β(1.6)	0.442	
	131	9.69(1) d	ϵ	–	–
	132	6.475(10) d	ϵ(97)	–	} 0.48*(4); 0.668(99)
			β^+(0.6)	0.40	
			β^-(2)	0.8	
	133	Stable	**100**	–	–
	134	2.062(5) y	β^-	0.662	0.57**(23); 0.605(98); 0.796**(99)
	134m	2.91(1) h	IT	–	} 0.127(14)
			β^-(1)	0.55	
	135	$2.3 \cdot 10^6$ y	β^-	0.21	–
	135m	53(2) min	IT	–	0.781(100); 0.840(96)
	136	13.16(3) d	β^-	0.657(7); 0.341	0.16**(36); 0.340(53); 0.818(100); 1.05(82)
	137	30.0(2) y	β^-	1.176(7); 0.514	0.662(85)
	138	32.2(1) min	β^-	3.40	0.463(23); 1.01(25); 1.426(73); 2.21(18)
	138m	2.90(10) min	IT(75), β^-(25)	–	0.080; 0.463; 1.436
	139	9.27(5) min	β^-	4.2	1.28; 1.42
	140	63.7(3) s	β^-	6.2; 5.6	0.59; 0.88; 1.14; 1.62; 1.85; 2.06; 2.32; 2.72; 3.15
	141	24.94(6) s	β^-, β_n(0.05)	–	0.048; 0.562; 0.589; 1.194
	142	1.80(8) s	β^-, β_n(0.28)	–	0.360; 0.967; 1.326
	143	1.78(3) s	β^-, β_n(1.7)	–	0.196; 0.232; 0.306
	144	1.02(3) s	β^-, β_n(3)	–	0.20; 0.56; 0.64; 0.76
	145	0.59(1) s	β^-, β_n(12)	–	0.11; 0.18; 0.20
	146	0.189(11) s	β^-, β_n(14)	–	–

Table 37.1 Table of isotopes *(continued)*

Element	A	$T_{1/2}$	Decay type or abundance of stable isotopes	Energy, MeV (relative intensity, %)	
				particle group	γ-rays
$_{56}$Ba	117	1.9(2) s	ϵ_p	–	–
	119	5.35(30) s	ϵ_p	–	–
	121	29.7(15) s	β^+, $\beta_p(0.02)$	–	–
	123	2.7(4) min	β^+	–	0.094; 0.116; 0.124
	124	11.9(10) min	ϵ	–	0.17; 0.19; 0.27; 1.22
	125	3.5(4) min	β^+	3.4	0.08; 0.14
	126	100(2) min	β^+	–	0.23/100/; 0.70/33/
	127	12.7(4) min	β^+	2.4	0.07; 0.12; 0.18
	128	2.43(5) d	ϵ	–	0.134; 0.278
	129	2.23(11) h	$\epsilon(94)$	–	} 0.129/26/; 0.181/100/;
			$\beta^+(6)$	1.42	} 0.21**/65/; 1.45/42/
	130	Stable	**0.106(2)**	–	–
	131	11.8(2) d	ϵ	–	0.124**(28); 0.216(19);
	131m	14.6(2) min	IT	–	0.373(13); 0.496**(48)
	132	Stable	**0.101(2)**	–	0.107(40)
	133	10.5(2) y	ϵ	–	0.080**(36); 0.303(14);
	133m	38.9(1) h	IT(> 99)	–	0.356(69)
			$\epsilon(0.011)$	–	0.276(17)
	134	} Stable	**2.417(27)**	–	–
	135	}	**6.592(18)**	–	–
	135m	28.7(2) h	IT	–	0.268(16)
	136	} Stable	**7.854(39)**	–	–
	137	}	**11.23(4)**	–	–
	137m	2.5513(7) min	IT	–	0.662(89)
	138	Stable	**71.70(7)**	–	–
	139	84.6(4) min	β^-	2.3	0.166(23); 1.43(0.4)
	140	12.746(10) d	β^-	1.02	0.030(11); 0.537(34)
	141	18.27(7) min	β^-	3.0	0.193/100/; 0.28/50/; 0.46**/30/;
					0.64/20/
	142	10.6(2) min	β^-	1.7	0.080/30/; 0.26/100/; 0.89/40/;
					1.20/35/
	143	14.5(5) s	β^-	4.2	0.211; 0.799; 0.980; 1.011
	144	11.4(5) s	β^-	2.9; 2.4	0.10; 0.16; 0.39; 0.43
	145	4.31(16) s	β^-	4.9	0.09; 0.38; 0.42
	146	1.91(16) s	β^-	3.9	0.12; 0.14; 0.25
	148	0.47(20) s	β^-	–	0.13; 0.42; 0.55
$_{57}$La	123	17(3) s	β^+	–	0.093
	125	76(6) s	β^+	–	0.068
	126	1.0(3) min	β^+	–	0.256; 0.511(an.)
	127	3.8(5) min	β^+	–	0.056
	128	5.0(3) min	β^+	3.2	0.279; 0.511(an.)
	129	11.6(2) min	ϵ, β^+	2.7; 2.4	0.11; 0.25; 0.28; 0.46
	129m	0.56(5) s	IT	–	0.172
	130	8.7(1) min	ϵ, β^+	–	0.356; 0.45; 0.511(an.); 0.55; 0.72;
					0.81; 0.91; 1.01; 1.19; 1.45; 1.55
	131	59(2) min	$\epsilon(72)$	–	} 0.115(23); 0.364(20);
			$\beta^+(28)$	1.94	} 0.417(20); 0.511(56, an.)
	132	4.8(2) h	β^+	3.7; 3.2	0.47; 0.511(an.); 0.56; 0.66; 1.03;
					1.22; 1.58; 1.91
	133	3.912(8) h	ϵ	–	} 0.511(an.); 0.62; 0.63
			β^+	1.2	}

Table 37.1 Table of isotopes *(continued)*

Element	A	$T_{1/2}$	Decay type or abundance of stable isotopes	Energy, MeV (relative intensity, %)	
				particle group	γ-rays
$_{57}$La	134	6.45(16) min	ϵ(38)	–	0.511(124, an.); 0.605(6)
			β^+(62)	2.7	
	135	19.5 h	ϵ	–	0.481(1.9); 0.588(0.13); 0.87**(0.24)
	136	9.87(3) min	ϵ(67)	–	0.511(66, an.); 0.818(2.5)
			β^+(33)	1.9	
	137	6(2)·10^4 y	ϵ	–	–
	138	1.28(12)·10^{11} y	**0.09(1)**	–	–
			ϵ($\sim$ 68)	–	0.81(30); 1.436(70)
			β^-($\sim$ 32)	0.21	
	139	Stable	**99.91(1)**	–	–
	140	40.272(7) h	β^-	2.175(6); 1.69(15); 1.36	0.329(20); 0.487(40); 0.815(19); 0.923(10); 1.596(96); 2.53(3)
	141	3.93(5) h	β^-	2.43	1.36(2)
	142	92.5(5) min	β^-	4.51	0.65(48); 0.90(9); 1.91(9); 2.41(15); 2.55(11)
	143	14.23(14) min	β^-	3.3	0.62/100/; 0.80/44/; 1.07/26/; 1.17/57/; 1.58/28/; 1.98/35/; 2.56/27/
	144	40.9(4) s	β^-	4.4; 4.1	0.397; 0.541; 0.845
	145	24.8(20) s	β^-	–	0.12; 0.17; 0.36; 0.45; 1.82
	146	8.8(4) s	β^-	5.5; 4.5	0.259; 0.410; 0.503
	147	4.4(5) s	β^-	–	–
	148	1.29(8) s	β^-	–	0.159
$_{58}$Ce	125	11(4) s	β^+	–	–
	128	$\sim$ 6 min	–	–	–
	129	3.5(5) min	–	–	0.080; 0.32; 0.75
	130	25(2) min	–	–	0.13
	131	10(1) min	β^+	–	–
	132	4.2(2) h	ϵ	–	0.18; 0.22
	133	5.40(5) h	ϵ, β^+	–	0.511(an.); 1.8
	134	75.9(9) h	ϵ	–	–
	135	17.6 h	ϵ	–	0.265/100/; 0.300/56/; 0.52**/46/; 0.59**/98/
			β^+($<$ 1)	0.81	
	135m	20 s	IT	–	0.082; 0.150; 0.296
	136	Stable	**0.19(1)**	–	–
	137	9.0(33) h	ϵ($>$ 99)	–	0.446**(2.3); 0.481**(0.06); 0.698(0.04); 0.92**(0.1)
			β^+($\leq$ 0.01)	–	
	137m	34.4(3) h	IT(99.2)	–	0.168(0.4); 0.254(11); 0.762(0.16); 0.825**(0.5)
			ϵ(0.8)	–	
	138	Stable	**0.25(1)**	–	–
	139	137.66(13) d	ϵ	–	0.165(80)
	139m	56.4(5) s	IT	–	0.746(93)
	140	Stable	**88.48(10)**	–	–
	141	32.50(1) d	β^-	0.581	0.145(48)
	142	$> 5 \cdot 10^{16}$ y	**11.08(10)**	–	–
	143	33.0(2) h	β^-	1.39	0.057(11); 0.293(46); 0.668(7); 0.725(8)
	144	284.9(2) d	β^-	0.31	0.080(2); 0.134(11)
	145	2.98(15) min	β^-	2.0	0.063; 0.285; 0.440; 0.724; 1.148
	146	14.2(5) min	β^-	0.7	0.110/20/; 0.142/42/; 0.22/50/; 0.27/12/; 0.32/100/

Table 37.1 Table of isotopes *(continued)*

Element	A	$T_{1/2}$	Decay type or abundance of stable isotopes	Energy, MeV (relative intensity, %)	
				particle group	γ-rays
$_{58}$Ce	147	56.4(12) s	β^-	3.3	0.093; 0.269; 0.374; 0.580
	148	48(1) s	β^-	1.7	0.098; 0.121; 0.292
	151	1.02(6) s	β^-	–	–
$_{59}$Pr	133	6.5(3) min	ϵ	–	0.74; 0.134; 0.361; 0.465
	134	17(2) min	ϵ	–	0.22; 0.30; 0.409; 0.511(an.); 0.639; 0.96
	135	~22 min	β^+	2.5	0.080; 0.22; 0.30; 0.511(an.)
			ϵ	–	
	136	13.1(1) min	$\epsilon(\sim 67)$	–	0.511(66, an.); 0.540; 1.09
			$\beta^+(\sim 33)$	3.5; 3.0	
	137	1.28(3) h	$\epsilon(73)$	–	0.511(54, an.); 0.837
			$\beta^+(27)$	1.7	
	138	1.45(5) min	β^+	–	0.789
	138m	2.1(1) h	$\epsilon(77)$	–	0.298(77); 0.364(9); 0.511(46, an.); 0.79(100); 1.04(100)
			$\beta^+(23)$	1.65	
	139	4.41(4) h	$\epsilon(89)$	–	0.511(18, an.); 1.35(0.5); 1.61(0.3)
			$\beta^+(11)$	1.09	
	140	3.39(1) min	$\epsilon(50)$	–	0.511(100, an.); 1.596(0.3)
			$\beta^+(50)$	2.32	
	141	Stable	**100**	–	–
	142	19.13(4) h	$\beta^-(>99)$	2.16	1.57(3.7)
			$\epsilon(0.016)$		–
	142m	14.6(5) min	IT	–	–
	143	13.58(3) d	β^-	0.933	–
	144	17.28(5) min	β^-	2.99	0.695(1.5); 1.487(0.29); 2.186(0.7)
	144m	7.2(2) min	IT(99.96)	–	0.059
			$\beta^-(0.04)$		
	145	5.98(2) h	β^-	1.80	0.072; 0.68; 0.75; 0.92; 1.05; 1.16
	146	24.07(13) min	β^-	4.1	0.455(77); 0.74(16); 0.78(15); 1.51(27)
	147	13.6(5) min	β^-	2.7; 2.1	0.078(17); 0.127(9); 0.32**(47); 0.56(39); 0.61(10); 0.65(24); 1.26(11)
	148	2.30(3) min	β^-	5.0; 4.7	0.30; 1.36
	149	2.3(2) min	β^-	2.8	0.08; 0.155; 0.325; 0.36; 0.745
	150	6.19(16) s	β^-	–	0.130; 0.723
	151	4.0(7) s	β^-	–	–
$_{60}$Nd	134	8.5(15) min	β^+	–	0.163
	135	12.1 min	ϵ	–	0.204; 0.441; 0.502
	136	50.65(33) min	ϵ	–	0.109; 0.149; 0.575
	137	38.5(15) min	β^+	2.4	0.109; 0.511(an.); 0.55**
	137m	1.60(15) s	IT	–	0.178; 0.286
	138	5.04(9) h	ϵ	–	0.326
	139	29.7(5) min	ϵ	–	0.41; 1.07
	139m	5.5(2) h	IT(12), ϵ	–	0.114/80/; 0.327/50/; 0.511/1400/; 0.73**/210/; 0.82**/70/; 0.983/70/
			β^+	3.1	
	140	3.37(2) d	ϵ	–	–
	141	2.49(3) h	$\epsilon(90)$	–	0.145(0.2); 0.511(6, an.); 1.14(2); 1.30(1)
			$\beta^+(4)$	0.79	
	141m	62.4(9) s	IT(99.97)	–	0.756
			$\epsilon(0.03)$		

Table 37.1 Table of isotopes *(continued)*

Element	A	$T_{1/2}$	Decay type or abundance of stable isotopes	Energy, MeV (relative intensity, %)	
				particle group	γ-rays
$_{60}$Nd	142	Stable	**27.13(10)**	–	–
	143	Stable	**12.18(5)**	–	–
	144	$2.4 \cdot 10^{15}$ y	**23.80(10)**	–	–
			α	1.83	–
	145	$> 1 \cdot 10^{17}$ y	**8.30(5)**	–	–
	146	Stable	**17.19(8)**	–	–
	147	10.98(1) d	β^-	0.9; 0.81	0.091(28); 0.319(3); 0.43**(4); 0.533(13)
	148	Stable	**5.76(3)**	–	–
	149	1.73(1) h	β^-	1.5	0.114(18); 0.210(27); 0.27**(26); 0.541(10)
	150	Stable	**5.64(3)**	–	–
	151	12.44(2) min	β^-	2.3; 1.2	0.118(40); 0.174**(10); 0.256(11); 1.180(9)
	152	11.4(2) min	β^-	1.2; 0.9	0.250; 0.279
	154	40(10) s	β^-	–	0.40; 0.70
$_{61}$Pm	136	107(6) s	ϵ	–	0.374; 0.603; 0.815; 0.858
	137	2.4(1) min	ϵ	–	0.108; 0.178; 0.269; 0.581
	138	3.24(5) min	ϵ	–	0.521; 0.729
	139	4.15(5) min	β^+	3.0	0.368; 0.403; 0.463
	140	9.2(2) s	β^+	5.1	0.717; 0.774; 1.499
	141	20.90(5) min	β^+(57)	2.6	0.195(13); 0.511(114, an.)
			ϵ(43)	–	
	142	40.5(5) s	β^+(∼ 95)	3.78	0.511(190, an.); 1.576
			ϵ(∼ 5)	–	
	143	265(7) d	ϵ	–	0.742(47)
	144	363(14) d	ϵ	–	0.474(45); 0.615(99); 0.695(99)
	145	17.7(4) y	ϵ(> 99)	–	0.067(1.0); 0.072(2.3)
			α($3 \cdot 10^{-7}$)	2.24	
	146	2020(18) d	ϵ(63)	–	0.453(65); 0.75*(65)
			β^-(37)	0.78	
	147	2.6234(2) y	β^-	0.224	–
	148	5.37(1) d	β^-	2.48	0.551(27); 0.914(15); 1.465(23)
	148*m*	41.3(1) d	β^-(95)	1.0; 0.4	0.551(95); 0.630(87); 0.727(36); 0.137
			IT(5)	–	
	149	53.08(5) h	β^-	1.07	0.286(2); 0.58(0.1); 0.85(0.2)
	150	2.68(2) h	β^-	3.05	0.334(71); 0.831(18); 1.165(23); 1.33(22)
	151	28.40(4) h	β^-	1.19	0.17**(18); 0.340(21)
	152	4.1(1) min	β^-	3.5	0.122; 0.841; 0.96*
	152*m*	7.5(2) min	β^-	–	0.120; 0.245
	153	5.4(2) min	β^-	1.65	0.12; 0.18
	154	2.7(1) min	β^-	2.5	0.08; 0.18; 1.44
$_{62}$Sm	134	12(3) s	–	–	–
	137	44(8) s	–	–	–
	138	3.0(3) min	β^+	–	0.05; 0.075
	139	2.57(1) min	β^+	3.6	0.274; 0.306; 0.597
	139*m*	9.5(1) s	IT(93.7)	–	0.155; 0.189
			ϵ(6.3)		
	140	14.82(10) min	β^+	1.9	0.23; 0.14

Table 37.1 Table of isotopes *(continued)*

Element	A	$T_{1/2}$	Decay type or abundance of stable isotopes	Energy, MeV (relative intensity, %)	
				particle group	γ-rays
$_{62}$Sm	141	10.2(2) min	ϵ, β^+	–	0.20; 0.43; 0.78
	141m	22.6(2) min	ϵ(99.69)	–	–
			IT(0.31)		
	142	72.49(5) min	ϵ(∼50)	–	0.15–0.35**; 0.511(100, an.)
			β^+(∼ 50)	1.03	
	143	8.83(2) min	ϵ(52)	–	0.511(100, an.); 1.06
			β^+(48)	2.5	
	143m	66(2) s	ϵ(0.20)	–	0.754
			IT(99.80)		
	144	Stable	**3.1(1)**	–	–
	145	340(3) d	ϵ	–	0.061; 0.485
	146	$1.03(3)\cdot10^8$ y	$< \mathbf{2\cdot10^{-7}}$, α	2.47	–
	147	$1.06(2)\cdot10^{11}$ y	**15.0(2)**	–	–
			α	2.23	–
	148	$8(2)\cdot10^{15}$ y	**11.3(1)**	–	–
			α	1.96	–
	149	$> 1\cdot10^{16}$ y	**13.8(1)**	–	–
	150	Stable	**7.4(1)**	–	–
	151	90(6) y	β^-	0.076	0.022(4)
	152	Stable	**26.7(2)**	–	–
	153	46.7(1) h	β^-	0.80	0.070(5.4); 0.103(28)
	154	Stable	**22.7(2)**	–	–
	155	22.1(2) min	β^-	1.53	0.104(73); 0.246(4)
	156	9.4(2) h	β^-	0.72	0.088(30); 0.166(10); 0.204(20)
	157	8.0(5) min	β^-	2.4	0.20; 0.39
	158	5.51(9) min	β^-	–	0.19; 0.32; 0.36
$_{63}$Eu	139	22(3) s	β^+	–	–
	141	40.0(7) s	β^+	5.0	0.39*; 0.59
	141m	3.3(3) s	ϵ(67), IT(33)	–	0.096
	142	2.4(2) s	β^+	7.0	0.77
	142m	1.22(2) min	ϵ, β^+	4.8	0.77; 1.03
	143	2.63(5) min	β^+	4.1	0.511(an.); 1.11; 1.54; 1.80; 1.91
	144	10.2(1) s	β^+	5.2	0.511(an.); 0.818; 1.660
	145	5.93(4) d	ϵ(99)	–	0.656/30/; 0.894/100/; 1.66/16/
			β^+(1)	1.7(2)	
	146	4.61(24) d	ϵ(96.5)	–	0.511(7, an.); 0.634*(77); 0.749(100)
			β^+(3.5)	2.11(0.14); 1.47(3.3)	
	147	42(1) d	ϵ(99.5)	–	0.122(20); 0.198(24); 0.680(11); 0.957(9); 1.079(9)
			β^+(0.5)	–	
			α(0.002)	2.91	
	148	54.5 d	ϵ(≥99)	–	0.551**(120); 0.62**(90); 0.72**(18)
			β^+(0.13)	0.92	
			$\alpha(9\cdot10^{-7})$	2.63	
	149	93.1(4) d	ϵ	–	0.277/10/; 0.328/10/
	150	12.62(10) h	β^-(90)	1.01	0.334(4); 0.406(3); 0.511(0.8, an.)
			ϵ(9)	–	
			β^+(0.4)	1.24	
	151	Stable	**47.8(5)**	–	–
	152	13.33(4) y	ϵ(73)	–	0.122(37); 0.344(77); 0.965(15); 1.408(22)
			β^-(27)	1.48	
			β^+(0.021)	0.71	

Table 37.1 Table of isotopes *(continued)*

Element	A	$T_{1/2}$	Decay type or abundance of stable isotopes	Energy, MeV (relative intensity, %) particle group	γ-rays
$_{63}$Eu	152m_1	9.32(1) h	β^-(76)	1.88	0.046; 0.122(8); 0.842(13); 0.963(12)
			ϵ(24)	–	
			β^+(0.011)	0.89	
	152m_2	96(1) min	IT	–	0.090; 0.148
	153	Stable	**52.2(5)**	–	–
	154	8.8(1) y	β^-(99.98)	1.85(10); 0.87	0.123(38); 0.724(21); 0.876(12); 1.00**(31); 1.278(37)
			ϵ(0.02)		
	154m	46.0(3) min	IT	–	0.068; 0.101
	155	4.96(1) y	β^-	0.25	0.087(32); 0.105(20)
	156	15.19(6) d	β^-	2.45	0.089(8); 0.812(9); 1.07**(11); 1.15**(14); 1.24**(16)
	157	15.15(4) h	β^-	1.3	0.064(27); 0.37*(14); 0.413(27)
	158	45.9(2) min	β^-	3.4; 2.5	0.080/100/; 0.52**/25/; 0.95**/95/; 1.19/16/
	159	18.7(4) min	β^-	2.6	0.07(42); 0.09(18); 0.15(14); 0.67(21)
	160	50(10) s	β^-	3.9	0.075; 0.17; 0.41; 0.52; 0.82
$_{64}$Gd	142	1.5(3) min	ϵ	–	0.179
	143	39(2) s	ϵ	–	0.20; 0.26; 0.46
	144	4.5(1) min	β^+	3.3	0.333; 0.347
	145	23.9(1) min	ϵ	–	0.511(an.); 0.80/9/; 1.03/10/; 1.75/100/
			β^+	2.3	
	145m	85(3) s	IT(95.3)	–	0.749
			ϵ(4.7)		
	146	48.3(1) d	ϵ, β^+	–	0.078/30/; 0.115**/100/; 0.155/45/
	147	38.1(1) h	ϵ, β^+	–	0.229/150/; 0.39**/85/; 0.64**/70/; 0.77**/60/; 0.932/60/
	148	93(6) y	α	3.183(100)	–
	149	9.4(3) d	ϵ(> 99)	–	0.150(48); 0.299(26); 0.347(25); 0.750(11)
			α(∼ 0.001)	3.01	
	150	1.79(8)·10^6 y	α	2.72(100)	–
	151	120(20) d	ϵ	–	0.0216(3); 0.154(7); 0.175(3); 0.244(7)
			α(∼ $8 \cdot 10^{-7}$)	2.60	
	152	1.08(8)·10^{14} y	0.20(1)	–	–
			α	2.14	–
	153	241.6(2) d	ϵ	–	0.070(2.4); 0.099**(55)
	154	Stable	**2.18(3)**	–	–
	155	Stable	**14.80(5)**	–	–
	156	Stable	**20.47(4)**	–	–
	157	Stable	**15.65(3)**	–	–
	158	Stable	**24.84(12)**	–	–
	159	18.56(8) h	β^-	0.95	0.058(3); 0.363(9)
	160	Stable	**21.86(4)**	–	–
	161	3.7(1) min	β^-	1.7; 1.6	0.102(11); 0.315(25); 0.361(66)
	162	9(1) min	β^-	1.0	0.40; 0.44
$_{65}$Tb	146	23(2) s	ϵ	–	1.08; 1.42; 1.58
	147	1.65(10) h	ϵ, β^+	–	0.305; 0.511(an.); 0.694; 1.152
	148	60(1) min	ϵ, β^+	–	0.511(an.); 0.78; 1.12
	149	4.15(5) h	ϵ(83)	–	0.16; 0.35
			α(∼ 17)	3.95	

Table 37.1 Table of isotopes *(continued)*

Element	A	$T_{1/2}$	Decay type or abundance of stable isotopes	Energy, MeV (relative intensity, %) particle group	Energy, MeV (relative intensity, %) γ-rays
$_{65}$Tb	149m	4.3(1) min	$\epsilon(>99)$	–	0.796
			$\alpha(0.020)$	3.99	
	150	3.27(10) h	$\epsilon(>99)$	–	0.511/100, an./; 0.637/100/; 0.93/35/
			β^+	3.6	
			$\alpha(<0.05)$	3.49	
	151	17.6(1) h	$\epsilon(>99)$	–	0.108(35); 0.252(35); 0.288(32)
			$\alpha(0.009)$	3.41	
	152	17.5(1) h	$\epsilon(80)$	–	0.344/100/; 0.586/14/; 0.779/14/
			$\beta^+(\sim 20)$	2.82	
	152m	4.3(2) min	IT(78); $\epsilon(22)$	–	0.344; 0.411
	153	2.34(1) d	ϵ, β^+	–	0.083**(11); 0.11**(12); 0.212(30)
	154	21.4(5) h	ϵ, β^+	–	0.123; 0.248; 0.347; 0.53**; 0.65**
	155	5.32(6) d	ϵ	–	0.087(37); 0.105(25); 0.180(8)
	156	5.34(9) d	ϵ	–	0.089(17); 0.199(40); 0.535(70); 1.22(29); 1.42(15)
	156m	5.0(1) h	IT	–	0.088
	157	150(30) y	ϵ	–	–
	158	$\sim$ 150 y	$\epsilon(82)$; $\beta^-(18)$	–	0.08; 0.94; 0.96
	158m	10.5(2) s	IT	–	0.110(0.5)
	159	Stable	**100**	–	–
	160	72.3(2) h	β^-	1.74(0.4); 0.86	0.087(12); 0.299(30); 0.879(31); 0.966**(31); 1.178(15)
	161	6.91(2) d	β^-	0.59(10); 0.52	0.026(21); 0.049(19); 0.075(10)
	162	7.7(2) min	β^-	2.4; 1.4	0.180/26/; 0.258/100/; 0.81/44/; 0.89/54/
	163	19.5(3) min	β^-	1.3; 0.8	0.025; 0.235; 0.330; 0.510
	164	3.0(1) min	β^-	2.9; 1.7	0.17; 0.69; 0.75
$_{66}$Dy	148	3.1(1) min	ϵ, β^+	–	0.620
	149	4.6(4) min	β^+	–	0.10; 0.79; 1.78; 1.81
	150	7.17(2) min	ϵ, β^+	–	0.39; 0.511(an.)
			$\alpha(31)$	4.23	
	151	16.9(5) min	β^+, $\epsilon(94)$	–	0.145; 0.511(an.); 0.546
			$\alpha(6)$	4.06	
	152	2.38(2) h	ϵ	–	0.257
			$\alpha(0.09)$	3.65	
	153	6.4(1) h	ϵ	–	0.08**; 0.25**
			$\alpha(0.010)$	3.48	
	154	$\sim 1 \cdot 10^7$ y	α	2.85	–
	155	10.0(3) h	ϵ	–	0.227(68); 0.52**(8); 1.000(6); 1.16**(6)
			$\beta^+(2)$	1.08(0.14); 0.85(2)	
	156	$> 1 \cdot 10^{18}$ y	**0.06(1)**	–	–
	157	8.1(1) h	ϵ	–	0.326(91)
	158	Stable	**0.10(1)**	–	–
	159	144.4(2) d	ϵ	–	0.058(4); 0.348
	160	Stable	**2.34(5)**	–	–
	161	Stable	**18.9(1)**	–	–
	162	Stable	**25.5(2)**	–	–
	163	Stable	**24.9(2)**	–	–
	164	Stable	**28.2(2)**	–	–
	165	2.334(6) h	β^-	1.29	0.095(4); 0.361(1.1)

Table 37.1 Table of isotopes *(continued)*

Element	A	$T_{1/2}$	Decay type or abundance of stable isotopes	Energy, MeV (relative intensity, %)	
				particle group	γ-rays
$_{66}$Dy	165m	1.26(1) min	IT(97.8)	–	} 0.108(3); 0.514(1.8)
			β^-(2.2)	1.04(0.4); 0.89	
	166	81.6(1) h	β^-	0.48(5); 0.40	0.082(12); 0.372(0.5); 0.426(0.5)
	167	6.2 min	β^-	2.0; 1.8	0.25; 0.26; 0.31; 0.57
$_{67}$Ho	150	40(5) s	ϵ	–	–
	151	47(2) s	ϵ(90)	–	–
			α(10)	4.51	–
	152	52.3(5) s	ϵ(94)	–	} 0.614; 0.647
			α(6)	4.45	
	153	9.3(5) min	ϵ	–	} 0.109; 0.162; 0.366
			α(0.1)	3.92	
	154	11.8(5) min	ϵ(> 99)	–	} 0.335; 0.511(an.); 0.873
			α(0.017)	3.93	
	155	48(1) min	ϵ	–	0.092; 0.138; 0.511(an.)
	156	55.6(6) min	β^+	2.9(1); 1.8(18)	0.138/100/; 0.266/99/; 0.367/23/; 0.511(an.)
	157	12.6(6) min	β^+	1.5; 1.2	0.087; 0.152; 0.190; 0.277; 0.341; 0.511(an.)
	158	11.3(4) min	β^+	2.9; 1.3	0.099; 0.218; 0.329; 0.412; 0.52; 0.647; 0.949
	158m	27(2) min	IT(65)	–	} 0.099; 0.218; 0.356; 0.412
			ϵ(35)	–	
	159	33(1) min	ϵ, β^+	–	0.057; 0.080; 0.13; 0.253; 0.309
	159m	8.30(8) s	IT	–	0.206
	160	25.6(3) min	ϵ(> 90)	–	} 0.73; 0.96*
			β^+(~ 0.4)	–	
	160m	5.02(5) h	IT(65)	–	} 0.060; 0.197(20); 0.646(20); 0.729(50); 0.880(26); 0.965**(37)
			$\epsilon + \beta^+$(35)	1.9	
	161	2.48(5) h	ϵ	–	0.026(23); 0.078(15)
	161m	6.7 s	IT	–	0.211(53)
	162	15(1) min	ϵ(95)	–	} 0.081(8); 0.511(9, an.); 1.319
			β^+(5)	1.10	
	162m	68(1) min	IT(61)	–	} 0.185(26); 0.940(13); 1.224(24)
			ϵ(39)	–	
	163	33(23) y	ϵ	–	–
	163m	1.09(3) s	IT	–	0.299
	164	29(1) min	ϵ(58)	–	} 0.073; 0.091
			β^-(42)	0.99	
	164m	37.5(10) min	IT	–	0.037; 0.057
	165	Stable	**100**	–	–
	166	26.80 h	β^-	1.84	0.081(5.4); 1.380(0.9)
	166m	1.20(18)·10^3 y	β^-	0.07	0.184(90); 0.280(30); 0.711(58); 0.711(58); 0.810(60)
	167	3.1(1) h	β^-	0.96; 0.3	0.06 – 0.53
	168	3.0(1) min	β^-	2.2	0.741; 0.821
	169	4.7(1) min	β^-	1.95; 1.2	0.15; 0.68; 0.84; 0.92
	170	42(3) s	β^-	4.0	0.079; 0.812; 1.894; 1.973
$_{68}$Er	151	23(2) s	ϵ	–	–
	152	10.1(2) s	α(~90)	4.80	–
			ϵ(~10)	–	–

Table 37.1 Table of isotopes *(continued)*

Element	A	$T_{1/2}$	Decay type or abundance of stable isotopes	Energy, MeV (relative intensity, %)	
				particle group	γ-rays
$_{68}$Er	153	36(1) s	ϵ(~62)	–	-
			α(~38)	4.67	–
	154	3.75(12) min	α(0.5)	4.15	–
			ϵ(> 99)	–	–
	155	5.3(3)	ϵ(> 99)	–	} 0.110; 0.242; 0.234
			α(~ 0.02)	4.012	
	156	20 min	ϵ	–	0.030; 0.035
	157	25(3) min	ϵ, β^+	–	0.117; 0.386; 0.511(an.); 1.32
	158	2.25(7) h	ϵ	–	} 0.072; 0.250; 0.315; 0.387;
			β^+	0.8	0.511(an.); 0.975
	159	36(1) min	ϵ, β^+	–	0.37 – 2.60
	160	28.6 h	ϵ	–	–
	161	3.24(4) h	ϵ, β^+	–	0.211(9); 0.592(8); 0.826(63)
	162	Stable	**0.14(1)**	–	–
	163	75.0(4) min	ϵ(> 99)	–	} 0.43(0.06); 1.10(0.04)
			β^+(0.004)	0.19	
	164	Stable	**1.60(1)**	–	–
	165	10.36(4) h	ϵ	–	–
	166	} Stable	**33.6(2)**	–	–
	167		**22.95(13)**	–	–
	167m	2.28(3) s	IT	–	0.208(43)
	168	Stable	**26.8(2)**	–	–
	169	9.40(2) d	β^-	0.34	0.008(0.3)
	170	Stable	**14.9(1)**	–	–
	171	7.52(3) h	β^-	1.49(2.3); 1.06	0.112(25); 0.296(28); 0.308(63)
	172	49.3(5) h	β^-	0.89(<10); 0.37	0.407(40); 0.610(40)
	173	1.4(1) min	β^-	–	0.193; 0.199; 0.895
$_{69}$Tm	153	1.59(8) s	α	5.10	–
	154	3.0(2) s	α	5.04	–
	155	39(3) s	α	–	–
	156	19(3) s	α	4.46	–
	157	3.5(3) min	ϵ	–	0.110; 0.348; 0.386; 0.455
	158	4.02(10) min	ϵ	–	0.192; 0.335; 0.628; 1.150
	159	9.0(4) min	ϵ, β^+	2.1	0.038; 0.085; 0.220; 0.271; 0.289
	160	9.2(4) min	ϵ, β^+	–	0.126; 0.264; 0.729
	161	38(4) min	ϵ, β^+	–	0.084; 0.106; 0.112; 0.172; 1.648
	162	21.7(2) min	ϵ, β^+	3.7; 2.1	0.102/20/; 0.236/10/; 0.900
	162m	24.3(17) s	IT(90), ϵ(10)	–	0.192; 0.812
	163	1.81(6) h	ϵ, β^+	–	0.104/8/; 0.240**/5/; 1.4
	164	2.0(1) min	ϵ(50)	–	} 0.091(4); 0.511(100, an.);
			β^+(50)	2.94	1.155
	164m	5.1(1) min	IT(80), ϵ(20)	–	0.208; 0.315
	165	30.06(3) h	ϵ	–	} 0.243(50); 0.297**(35);
			β^+(0.007)	0.30	0.807(15)
	166	7.70(3) h	ϵ(98.2)	–	} 0.081; 0.19*; 0.215; 0.46;
			β^+(~ 2)	1.94	0.60**
	167	9.24(2) d	ϵ	–	0.057(4); 0.208(43); 0.532(2)
	168	93.1(1) d	ϵ(~ 98)	–	} 0.19**(77); 0.448(27);
			β^-(~ 2)	–	0.73**(40); 0.82**(88)
	169	Stable	**100**	–	–

Table 37.1 Table of isotopes *(continued)*

Element	A	$T_{1/2}$	Decay type or abundance of stable isotopes	Energy, MeV (relative intensity, %)	
				particle group	γ-rays
$_{69}$Tm	170	128.6(3) d	β^-(> 99)	0.97	} 0.084(3.3)
			ϵ(0.144)	–	
	171	1.96(1) y	β^-	0.097	0.067
	172	63.6(2) h	β^-	1.88	0.079(5); 0.181(2.2); 1.09(7); 1.39(7); 1.46(7); 1.53(6)
	173	8.24(8) h	β	1.3(2); 0.89	0.399(89); 0.465(8)
	174	5.4(1) min	β^-	1.2	0.176(67); 0.273(85); 0.366(93); 0.50(15); 0.99(89)
	175	15.2(5) min	β^-	2.0	0.51; 0.94
	176	1.9(1) min	β^-	2.8	0.19; 0.38; 1.07
$_{70}$Yb	154	0.42(2) s	α	5.33	–
	155	1.65(15) s	α	5.21	–
	156	24(1) s	α	4.69	–
	157	38.6(10) s	α	4.51	0.164; 0.231
	158	1.38(14) min	ϵ	–	0.074
	159	1.75(20) min	ϵ	–	0.17; 0.18; 0.33; 0.39
	160	4.8 min	ϵ, β^+	–	0.174; 0.216
	161	4.2(2) min	ϵ, β^+	–	0.078; 0.600; 0.631
	162	18.9(2) min	ϵ	–	0.119; 0.163
	163	11.05(25) min	ϵ, β^+	–	0.064; 0.123; 0.860
	164	75.8(17) min	ϵ	–	–
	165	9.9(3) min	ϵ	–	} 0.080; 0.069; 1.09
			β^+	1.6	
	166	56.7(1) h	ϵ	–	0.082(17)
	167	17.5(2) min	ϵ, β^+	–	0.113**(90); 176(15)
	168	Stable	**0.13(1)**	–	–
	169	32.022(8) d	ϵ	–	0.063(45); 0.177(22); 0.198(35)
	169m	46(2) s	IT	–	0.024
	170	} Stable	**3.05(5)**	–	–
	171		**14.3(2)**	–	–
	172		**21.9(3)**	–	–
	173		**16.12(18)**	–	–
	174		**31.8(4)**	–	–
	175	4.19(1) d	β^-	0.466	0.114(1.9); 0.283(3.7); 0.396(6.0)
	176	Stable	**12.7(1)**	–	–
	176m	11.4(5) s	IT	–	0.19; 0.29; 0.39
	177	1.9(1) h	β^-	1.40	0.122(3); 0.151(16); 1.080(5); 1.241(3)
	177m	6.41(2) s	IT	–	0.104(65); 0.228(13)
	178	74(3) min	β^-	0.6	0.348; 0.391
$_{71}$Lu	155	0.07(2) s	α	5.63	–
	156	0.23(3) s	α	5.54	–
	157	5.5(3) s	α	4.996	–
	164	3.1 min	β^+	–	0.124; 0.262; 0.740
	165	12(1) min	ϵ, β^+	–	0.121; 0.132; 0.174; 0.204
	166	2.65(10) min	ϵ, β^+	–	0.102; 0.228; 0.338
	166m_1	1.41(10) min	ϵ(58)	–	} 0.034; 0.102; 0.228; 0.285
			IT(42)	–	
	166m_2	2.12(10) min	ϵ(> 80)	–	1.26; 1.43; 2.10
	167	51.5(10) min	ϵ	–	} 0.030; 0.278; 0.372; 0.402; 0.511(an.); 1.267
			β^+($\sim$ 1)	1.5	

Table 37.1 Table of isotopes *(continued)*

Element	A	$T_{1/2}$	Decay type or abundance of stable isotopes	Energy, MeV (relative intensity, %)	
				particle group	γ-rays
$_{71}$Lu	168	5.3(2) min	ϵ	–	0.087(7); 0.90(10); 0.99(13);
			$\beta^+(\sim 12)$	1.2	
	168m	6.7(4) min	ϵ, β^+	–	0.198; 0.89*; 0.979
	169	34.06(5) h	ϵ	–	0.063; 0.111; 0.191; 0.577
			β^+	1.2	
	169m	160(10) s	IT	–	0.029
	170	2.00(3) d	ϵ	–	0.084(13); 0.98; 1.28; 2.04
			β^+	2.4	
	170m	0.7 s	IT	–	–
	171	8.22(3) d	ϵ	–	0.019(20); 0.668(14); 0.741(68)
			$\beta^+(\sim 0.01)$	–	
	171m	79(2) s	IT	–	0.071(0.2)
	172	6.70(3) d	ϵ	–	0.182(26); 0.81(21); 0.90**(45);
				–	1.09(60)
	172m	3.7 min	IT	–	–
	173	1.37(1) y	ϵ	–	0.079(14); 0.101(7); 0.272(18)
	174	3.31(5) y	ϵ, β^+	–	0.076(6); 1.24(9)
	174m	142(2) d	IT(99.3)	–	0.067; 0.176; 0.273; 0.994
			ϵ(0.7)	–	
	175	Stable	**97.41(2)**	–	–
	176	3.60(16)·10^{10} y	β^-	0.6	0.088(15); 0.202(85); 306(95)
			2.59(2)	–	–
	176m	3.68(1) h	β^-	1.31	0.088; 0.126
	177	6.71(1) d	β^-	0.497	0.113(2.8); 0.208(6.1)
	177m	160.9(3) d	β^-(78)	0.2	0.113(23); 0.208(62); 0.228(37);
			IT(22)	–	0.378(29); 0.418(21)
	178	28.4(2) min	β^-	2.0	0.089; 0.214; 0.326; 0.427
	178m	22.7(4) min	β^-	1.2	0.332
	179	4.59(6) h	β^-	1.35	0.213
	180	5.7(1) min	β^-	2.7; 1.5	0.22; 0.41; 1.11; 1.20
$_{72}$Hf	157	0.110(6) s	α	5.735	–
	158	2.9(2) s	α	5.27	–
	159	5.6(5) s	α	5.095	–
	161	17(2) s	α	4.60	–
	166	6.77(30) min	ϵ, β^+	–	0.079; 0.342; 0.408
	167	2.05(5) min	ϵ, β^+	–	0.315
	168	25.9 min	ϵ	–	0.129; 0.17
			$\beta^+(\sim 2)$	1.7	
	169	3.24(4) min	ϵ	–	0.115; 0.370; 0.493
	170	16.01(13) h	ϵ	–	0.120; 0.165; 0.99; 1.28; 2.03; 2.36;
					2.52; 2.94
	171	12.1(4) h	ϵ	–	0.122; 0.188; 0.29; 0.34; 0.47; 0.66;
					0.86; 1.07
	172	1.87(3) y	ϵ	–	0.024(22); 0.082(10); 0.125**(21)
	173	24.0(5) h	ϵ	–	0.13**(96); 0.30**(52)
	174	2.0(4)·10^{15} y	α	2.50	–
			0.162(2)	–	–
	175	70(2) d	ϵ	–	0.089(3.4); 0.343(85)
	176	Stable	**5.206(4)**	–	–
	177		**18.606(3)**	–	–

Table 37.1 Table of isotopes *(continued)*

Element	A	$T_{1/2}$	Decay type or abundance of stable isotopes	Energy, MeV (relative intensity, %)	
				particle group	γ-rays
$_{72}$Hf	177m_1	1.08(6) s	IT	–	0.113(30); 0.208(81); 0.228(48); 0.378(37)
	177m_2	51.4(5) min	IT	–	0.277; 0.295; 0.327
	178	Stable	**27.297(3)**	–	–
	178m_1	4.0 s	IT	–	0.089; 0.213; 0.326; 0.426
	178m_2	31(1) y	IT	–	0.217; 0.495; 0.574
	179	Stable	**13.629(5)**	–	–
	179m_1	18.68(6) s	IT	–	0.217(94)
	179m_2	25.1(3) d	IT	–	0.12; 0.15; 0.36; 0.45
	180	Stable	**35.100(6)**	–	–
	180m	5.5(1) h	IT	–	0.058(48); 0.215(82); 0.333(93); 0.444(80)
	181	42.4(1) d	β^-	0.41	0.133**(48); 0.346(13); 0.482(81)
	182	9(3)·10^6 y	β^-	0.2	0.271(84)
	182m	61.5(15) min	β^-(54), IT(46)	–	0.224; 0.344; 0.943
	183	64(1) min	β^-	1.5; 1.2	0.46/58/; 0.82/100/
	184	4.12(5) h	β^-	1.1	0.14; 0.18; 0.34
$_{73}$Ta	167	2.9(15) min	ϵ	–	–
	168	2.5(12) min	β^+	–	0.124; 0.262; 0.750
	169	4.9(4) min	ϵ, β^+	–	0.029; 0.154; 0.192
	170	6.76(6) min	ϵ, β^+	–	0.10; 0.22; 0.86; 0.99
	171	23.3(3) min	ϵ	–	0.05; 0.17; 0.50*
	172	36.8(3) min	ϵ, β^+	–	0.092; 0.208; 0.511(an.); 1.109
	173	3.65(5) h	ϵ, β^+	–	0.090**; 0.170**; 0.64; 1.00
	174	1.2(1) h	ϵ, β^+	–	0.091; 0.125; 0.160; 0.205; 0.280; 0.350; 0.511(an.)
	175	10.5(2) h	ϵ	–	0.08; 0.13; 0.21; 0.27; 0.35; 0.45; 0.60; 0.83; 1.2; 1.4; 1.7
	176	8.08(7) h	ϵ, β^+	–	0.088; 0.202
	177	56.6(1) h	ϵ, β^+	–	0.113(6); 0.208(1)
	178	9.31(3) min	ϵ(99) β^+(1)	– 0.89	} 0.093/100/; 0.511/10, an./; 1.10/11/; 1.35**/46/
	179	664.9(42) d	ϵ	–	–
	180	$> 1.0 \cdot 10^{13}$ y	**0.012(2)**	–	–
	180m	8.1(1) h	ϵ(87) β^-(13)	– 1.71	} 0.093(4); 0.103(0.6)
	181	Stable	**99.988(2)**	–	–
	182	115.0(2) d	β^-	0.71(0.3); 0.522	0.068(42); 1.121(34); 1.221(27)
	182m_1	0.28 s	IT	–	–
	182m_2	15.84(10) min	IT	–	0.147(40); 0.172(40); 0.184(20)
	183	5.1(1) d	β^-	0.80; 0.62	0.108(11); 0.161**(17); 0.246**(33); 0.292(11); 0.354(11)
	184	8.7(1) h	β^-	2.64(0.2); 1.76(0.9); 1.19	0.111(21); 0.25(42); 0.30(24); 0.41(71); 0.90**(49)
	185	49(2) min	β^-	1.7	0.175(60)
	186	10.5(5) min	β^-	2.6; 2.2	0.20(74); 0.51(33); 0.61(33); 0.73(48)
$_{74}$W	162	<0.25 s	α	5.538	–
	163	2.5(3) s	α	5.384	–
	164	6.3(5) s	α	5.146	–
	165	5.1 s	α	4.909	–

Table 37.1 Table of isotopes *(continued)*

Element	A	$T_{1/2}$	Decay type or abundance of stable isotopes	Energy, MeV (relative intensity, %)	
				particle group	γ-rays
$_{74}$W	166	16 s	α	4.739	–
	170	4(1) min	ϵ	–	–
	171	9.0(15) min	ϵ	–	–
	172	6.7(10) min	ϵ	–	0.036; 0.458; 0.624
	173	16.5(5) min	ϵ	–	0.050; 0.071; 0.106; 0.365
	174	29 min	ϵ	–	0.035; 0.329; 0.429
	175	34(1) min	ϵ	–	0.26; 0.80; 1.3; 1.6
	176	2.3(1) h	ϵ(> 99)	–	} 0.034; 0.100
			β^+(∼ 0.5)	–	
	176	135(3) min	ϵ	–	0.20; 0.42; 0.62; 0.83; 1.00
	178	21.7(3) d	ϵ	–	–
	179	37.5(5) min	ϵ	–	0.031(22)
	179*m*	6.7(3) min	IT(>99)	–	0.222
			ϵ(0.31)	–	–
	180	$6{\cdot}10^{14}$ y	**0.13(3)**	–	–
	181	121.2(3) d	ϵ	–	0.006(1); 0.136(0.1); 0.152(0.1)
	182	} Stable	**26.3(2)**	–	–
	183		**14.3(1)**	–	–
	184	$> 3 \cdot 10^{17}$ y	**30.67(15)**	–	–
	185	75.1(3) d	β^-	0.429	–
	185*m*	1.67(3) min	IT	–	0.100/16/; 0.13/70/; 0.17/100/
	186	Stable	**28.6(2)**	–	–
	187	23.9(1) h	β^-	1.31(15); 0.63	0.479(23); 0.686(27)
	188	69.4(5) d	β^-	0.349	0.227(0.22); 0.290(0.40)
	189	11.5(3) min	β^-	2.5; 2.0	0.258/100/; 0.417/96/
	190	30.0(15) min	β^-	1.0	0.158; 0.162
$_{75}$Re	170	∼ 7 s	β	–	0.156; 0.306; 0.413
	172	48(12) s	ϵ	–	0.123; 0.254; 0.743
	175	5(1) min	ϵ	–	0.185
	176	5.7(8) min	ϵ	–	0.109; 0.241
	177	14.0(10) min	ϵ	–	0.080*; 0.096; 0.197
	178	13.2(2) min	ϵ	–	0.106; 0.237; 0.939
	179	19.7(5) min	ϵ, β^+	–	0.29; 0.43; 1.68
	180	2.43(6) min	ϵ	–	} 0.11; 0.511(an.); 0.88; 0.90
			β^+	1.8	
	181	20(1) h	ϵ	–	0.365; 0.639
	182	64(5) h	ϵ	–	0.068; 0.100; 1.122; 1.189; 1.23**; 2.0; 2.01
	183	70.0(11) d	ϵ	–	0.046; 0.053; 0.109**; 0.209; 0.246; 0.292
	184	38.0(5) d	ϵ	–	0.111; 0.78**; 0.90**
	184*m*	165(5) d	ϵ(25), IT(75)	–	0.22; 0.25; 0.92
	185	Stable	**37.40(2)**	–	–
	186	90.64(9) h	β^-(99.2)	1.07	} 0.137(9); 0.632(0.03); 0.768(0.035)
			ϵ(7.8)	–	
	186*m*	$2.0{\cdot}10^5$ y	IT	–	0.040; 0.059; 0.099
	187	$5(2){\cdot}10^{10}$ y	β^-	0.003	–
			62.60(2)	–	–
	188	16.98(2) h	β^-	2.12	0.155(10); 0.478(0.6); 0.633(0.9)
	188*m*	18.6(1) min	IT	–	0.092(5); 0.106(10)

Table 37.1 Table of isotopes *(continued)*

Element	A	$T_{1/2}$	Decay type or abundance of stable isotopes	Energy, MeV (relative intensity, %)	
				particle group	γ-rays
$_{75}$Re	189	24.3(4) h	β^-	1.00	0.150*(4); 0.187*(3); 0.218*(10); 0.245(4)
	190	3.1(3) h	β^-	1.8	0.191/10/; 0.392/10/; 0.57/10/; 0.83/3/
	190m	3.2(2) h	β^-(51), IT($\sim$49)	–	0.187; 0.558; 0.569
	191	9.8 min	β^-	1.8	–
	192	16(1) s	β^-	2.5	0.20; 0.29; 0.37; 0.48; 0.57
$_{76}$Os	169	3.2(2) s	α	5.57	–
	170	7.1(5) s	α	5.40	–
	171	8.2(8) s	α	5.24	–
	172	19(2) s	ϵ(> 99) α(< 0.3)	5.11	–
	173	16(5) s	ϵ(99.98) α(0.02)	4.94	–
	174	45(5) s	ϵ(99.98) α(0.02)	4.76	0.118; 0.325
	175	1.4(1) min	ϵ	–	0.125; 0.181
	176	3.0(7) min	ϵ	–	0.776; 1.209; 1.291
	177	3.5(8) min	ϵ	–	0.085; 0.196
	178	5.0(4) min	ϵ	–	–
	179	6.5(5) min	ϵ	–	0.06; 0.22; 0.60; 0.97; 1.33
	180	22(3) min	ϵ	–	0.02
	181	2.7(1) min	ϵ, β^+	–	0.118; 0.145
	182	22(2) h	ϵ	–	0.180/7/; 0.510/10/
	183	13.0(5) h	ϵ, β^+	–	0.114(27); 0.382(90)
	183m	9.9(3) h	ϵ(89), IT(11)	–	0.171; 1.035(6); 1.105**(48)
	184	$> 1 \cdot 10^{17}$ y	**0.02(1)**	–	–
	185	93.6(5) d	ϵ	–	0.646(80); 0.875**(11)
	186	$2(1)\cdot 10^{15}$ y	**1.58(10)**	–	–
			α	2.76	–
	187	Stable	**1.6(1)**	–	–
	188	Stable	**13.3(2)**	–	–
	189	Stable	**16.1(3)**	–	–
	189m	4.8(1) h	IT	–	0.031
	190	Stable	**26.4(4)**	–	–
	190m	9.9(1) min	IT	–	0.187(70); 0.361(94); 0.502(98); 0.616(99)
	191	15.4(1) d	β^-	0.143	0.129(25)
	191m	13.10(5) h	IT	–	0.074
	192	Stable	**41.0(3)**	–	–
	192m	5.9(1) s	IT	–	0.20; 0.30; 0.45; 0.48; 0.57
	193	30.5(4) h	β^-	1.13	0.139(3); 0.28**(2.1); 0.460(3.9); 0.558(2.1)
	194	6.0(2) y	β^-	0.053	0.043(10); 0.078(0.03)
	195	6.5 min	β^-	2	–
	196	34.9(2) min	β^-	0.8	0.126; 0.408
$_{77}$Ir	169	0.4(1) s	α	6.11	–
	171	1.0(3) s	α	5.91	–
	172	1.7(5) s	α	5.81	–
	173	3.0(1) s	α	5.67	–
	174	4(1) s	α	5.4	–

Table 37.1 Table of isotopes *(continued)*

Element	A	$T_{1/2}$	Decay type or abundance of stable isotopes	Energy, MeV (relative intensity, %)	
				particle group	γ-rays
$_{77}$Ir	175	4.5(10) s	α	5.39	–
	176	8(1) s	α	5.12	–
	177	21(2) s	α	5.01	–
	178	22(2) s	ϵ	–	0.132; 0.266; 0.363
	179	4(1) min	ϵ	–	–
	180	1.5(1) min	ϵ	–	0.132; 0.276
	181	5(3) min	ϵ	–	0.05; 0.11; 0.23; 0.32; 1.64
	182	15(1) min	ϵ	–	0.133; 0.278; 0.510; 0.912
	183	57(4) min	ϵ	–	0.24*
	184	3.02(6) h	ϵ	–	0.125/100/; 0.267/200/; 0.392/90/
	185	14(9) h	ϵ	–	0.101; 0.254; 1.67; 1.83
	186	15.8(3) h	ϵ(97)	–	} 0.137(45); 0.297(74); 0.434(35)
			β^+(3)	1.94	
	187	10.5(3) h	ϵ	–	0.18/45/; 0.41/100/; 0.61/45/; 0.98/50/
	188	41.5(5) h	ϵ(> 99)	–	} 0.155(34); 0.633*(29)
			β^+($\sim$ 0.3)	1.66	
	189	13.2(1) d	ϵ	–	0.245(18)
	190	11.78(10) d	ϵ	–	0.187(51); 0.37**(39); 0.40**(39); 0.518(39); 0.56**(72); 0.604(47)
	190m_1	1.2 h	IT	–	0.026
	190m_2	3.2(2) h	ϵ(95)	–	0.175
			IT(5)		
	191	Stable	**37.3(5)**	–	–
	191m	4.94(3) s	IT	–	0.129; 0.171
	192	73.831(8) d	β^-(95.4)	0.67	} 0.296(29); 0.308(30);
			ϵ(4.6)	–	0.317(81); 0.468(49)
	192m_1	1.45(5) min	IT(> 99)	–	} 0.058(0.005); 0.317(0.008);
			β(0.017)	1.5	0.612(0.003)
	192m_2	241(9) y	IT	–	0.155
	193	Stable	**62.7(5)**	–	–
	193m	10.60(11) d	IT	–	0.080
	194	19.15(3) h	β^-	2.24	0.328(10); 0.64*(1)
	194m	171(11) d	β^-	2.3	0.13; 0.32; 0.63
	195	2.5(3) h	β^-	1.1	0.099; 0.211
	195m	3.8(2) h	β^-	1.0	0.10; 0.13; 0.33; 0.37; 0.43; 0.66
	196	52(2) s	β^-	3.2	0.33; 0.36; 0.45; 0.78
	196m	1.40(2) h	β^-	1.2	0.356(94); 0.39(95); 0.44(95); 0.522(99); 0.65(100)
	197	5.8(5)	β^-	2.0	0.50
	198	8(1) s	$\beta-$	3.6	0.407; 0.507
$_{78}$Pt	174	0.7(2) s	α(80)	6.043	–
			$\epsilon+\beta^+$(20)	–	–
	175	2.1(2) s	α($\sim$75)	5.964	0.076
	176	6.33(15) s	α(42)	5.744	–
			$\epsilon+\beta^+$(58)	–	–
	177	11(2) s	α(9)	5.527; 5.435	–
			$\epsilon+\beta$(91)	–	–
	178	21.0(7) s	α(7)	5.458; 5.30	–
			$\epsilon+\beta^+$(93)	–	–

Table 37.1 Table of isotopes *(continued)*

Element	A	$T_{1/2}$	Decay type or abundance of stable isotopes	Energy, MeV (relative intensity, %)	
				particle group	γ-rays
$_{78}$Pt	179	33(4) s	$\alpha(0.27)$	5.15	–
			$\epsilon+\beta^+(>99)$	–	–
	180	52(3) s	$\alpha(\sim0.3)$	5.14	–
			$\epsilon+\beta^+(>99)$	–	–
	181	51(5) s	$\alpha(\sim0.06)$	5.02	–
			$\epsilon+\beta^+(>99)$	–	–
	182	2.6(1) min	$\epsilon(>99)$	–	} 0.136; 0.146; 0.210
			$\alpha(\sim 0.02)$	4.84	
	183	6.6(9) min	$\epsilon(>99)$	–	} 0.119; 0.265*; 0.307
			$\alpha(\sim 0.0013)$	4.73	
	184	17.3(2) min	$\epsilon(>99)$	–	} 0.155; 0.192; 0.548; 0.731
			$\alpha(\sim 0.001)$	4.50	
	185	70.9(24) min	ϵ	–	0.035; 0.63; 1.56
	186	2.0(1) h	$\epsilon(>99)$	–	} 0.67*
			$\alpha(4\cdot10^{-4})$	4.23	
	187	2.35(3) h	ϵ	–	0.106; 0.202; 0.285; 0.709
	188	10.2(3) d	$\epsilon(99)$	–	} 0.140/22/; 0.19**/100/; 0.38/15/
			$\alpha(3\cdot10^{-5})$	3.93	
	189	10.89(11) h	ϵ	–	0.094/120/; 0.141/124/; 0.187/137/; 0.243/100/; 0.56**/230/; 0.61**/180/; 0.722/156/
	190	6(1)·10^{11} y	α	3.18	–
			0.01(1)	–	–
	191	2.9(1) d	ϵ	–	0.36**(5); 0.410(3); 0.539(9)
	192	Stable	**0.79(5)**	–	–
	193	50(9) y	ϵ	–	–
	193m	4.33(3) d	IT	–	0.150
	194	} Stable	**32.9(5)**	–	–
	195		**33.8(5)**	–	–
	195m	4.02(1) d	IT	–	0.099(11); 0.129(1)
	196	Stable	**25.3(5)**	–	—
	197	18.3(3) h	β^-	0.670	0.077(20); 0.191(6)
	197m	94.4(8) min	IT(97)	–	} 0.279(2.6); 0.346(13)
			$\beta^-(3)$	0.737	
	198	Stable	**7.2(2)**	–	–
	199	30.8(4) min	β^-	1.69	0.197(9); 0.32*(8); 0.475*(12); 0.540(24)
	199m	13.6(4) s	IT	–	0.032; 0.392
	200	12.5(3) h	β^-	0.7; 0.6	0.08; 0.14; 0.23; 0.24
	201	2.5(1) min	β^-	2.66	0.15; 0.23; 1.76
$_{79}$Au	176	1.25(30) s	α	6.29; 6.26	–
	177	1.3(4) s	α	6.15; 6.12	–
	178	2.6(5) s	α	5.92	–
	179	7.5(4) s	α	5.84	–
	181	11.3(7) s	$\alpha(1.1)$	5.60; 5.47	–
	182	21(2) s	$\epsilon(>99)$	5.35	0.155; 0.265; 0.787; 0.855
			$\alpha(\sim 0.04)$		
	183	44(2) s	$\alpha(0.3)$	5.34	0.312
	184	53.0(14) s	$\epsilon(>99)$	5.17; 5.11	0.163; 0.273; 0.363
			$\alpha(0.022)$		

Table 37.1 Table of isotopes *(continued)*

Element	A	$T_{1/2}$	Decay type or abundance of stable isotopes	Energy, MeV (relative intensity, %)	
				particle group	γ-rays
$_{79}$**Au**	185	4.3(1) min	$\epsilon(>99)$	5.07	0.243; 0.310; 0.332
			$\alpha(0.09)$		
	186	10.7(5) min	ϵ	–	0.16; 0.22; 0.30; 0.40; 0.76
	187	8.0(4) min	ϵ	–	0.92; 1.33; 1.41
			α	4.69	
	188	8.84(6) min	ϵ, β^+	–	0.25; 0.33; 0.63
	189	28.7(3) min	ϵ	–	0.35; 0.45; 0.71; 0.81
	189m	4.55(10) min	ϵ, β^+	–	0.17; 0.32
	190	42.8(10) min	ϵ, β^+	–	0.29**/100/; 0.60**/5/
	191	3.18(8) h	ϵ, β^+	–	0.14/10/; 0.30/60/; 0.60/10/
	191m	0.92(11) s	IT	–	0.267
	192	4.94(9) h	ϵ	–	0.137; 0.158; 0.296; 0.308; 0.317; 0.612
			$\beta^+(\sim 1)$	2.5	
	193	17.65(15) h	ϵ	–	0.114**(5); 0.18**(11); 0.26*(9)
			β^+	–	
	193m	3.9(3) s	IT(>99),	–	0.258(65)
			$\epsilon(0.03)$	–	
	194	39.5(5) h	$\epsilon(\sim 97)$	–	0.294(12); 0.328(68); 1.469
			$\beta^+(\sim 3)$	1.49	
	195	183(2) d	ϵ	–	0.099(10); 0.129(1)
	195m	30.5(2) s	IT	–	0.261(77)
	196	6.183(10) d	$\epsilon(93)$	–	0.333(25); 0.356(94); 0.426(6)
			$\beta^-(7)$	0.26	
	196m_1	8.1(2) s	IT	–	0.085
	196m_2	9.7(1) h	IT	–	0.148(42); 0.188(32)
	197	Stable	**100**	–	–
	197m	7.8(1) s	IT	–	0.130(8); 0.279(75)
	198	2.696(2) d	β^-	0.962	0.412(95); 0.676(1)
	198m	2.30(4) d	IT	–	0.097; 0.180; 0.204; 0.215
	199	3.139(7) d	β^-	0.46; 0.30	0.158(37); 0.208(8)
	200	48.4(3) min	β^-	2.2	0.368(24); 1.227(23)
	200m	18.7(5) h	$\beta^-(\sim 84)$	0.6	0.256; 0.368; 0.498; 0.579
			IT($\sim$16)	–	
	201	26(1) min	β^-	1.3	0.53; 0.61
	202	28(2) s	β^-	3.5	0.44; 1.12; 1.20; 1.31
	203	53(2) s	β^-	1.9	0.69
	204	40(3) s	β^-	–	0.44; 1.51
$_{80}$Hg	178	0.47(14) s	$\alpha(\sim 84)$,	6.43	–
			$\epsilon(\sim 16)$		
	179	1.09(4) s	$\alpha(\sim 53)$,	6.288	–
			$\epsilon(\sim 47)$		
	180	2.9 s	α	6.120	0.301; 0.381
	181	3.6(3) s	$\epsilon(74)$, $\alpha(26)$	6.006; 5.94; 5.92	0.147
	182	11.2(10) s	$\epsilon(91)$, $\alpha(9)$	5.87; 5.70	0.129; 0.217; 0.413
	183	8.8(5) s	$\epsilon(88)$, $\alpha(12)$	5.905	
	184	30.6(3) s	$\epsilon(98.7)$	5.54	0.156; 0.236; 0.295
			$\alpha(1.3)$		
	185	50(2) s	$\epsilon(<95)$	–	0.222; 0.258
			$\alpha(\geq 5)$	5.64	

Table 37.1 Table of isotopes *(continued)*

Element	A	$T_{1/2}$	Decay type or abundance of stable isotopes	Energy, MeV (relative intensity, %)	
				particle group	γ-rays
$_{80}$Hg	186	1.38(10) min	$\epsilon(>99)$		
			$\alpha(0.016)$	5.09	0.125; 0.27; 0.35; 0.44
	187	3 min	ϵ, α	–	0.175; 0.255; 0.40
	188	3.25(15) min	ϵ	–	0.115; 0.191
			α	4.61	
	189	7.6(1) min	ϵ, β^+	–	0.165; 0.24; 0.32; 0.50
	190	20.0(5) min	ϵ	–	0.14**; 0.17
	191	49(10) min	ϵ	–	0.26**
	191m	50.8(15) min	ϵ	–	–
	192	4.85(20) h	ϵ, $\beta^+(<1)$	–	0.114/30/; 0.157/20/; 0.274/100/
	193	3.80(15) h	ϵ	–	0.187; 0.574; 0.762; 0.855; 1.04; 1.08
	193m	11.8(2) h	$\epsilon(92)$, IT(8)	–	0.141; 0.218; 0.258; 0.574
	194	260(40) y	ϵ	–	–
	195	9.9(5) h	ϵ	–	0.20**; 0.261; 0.59*; 0.780; 0.930; 1.110; 1.172
	195m	41.6(8) h	$\epsilon(50)$, IT(50)	–	0.200(35); 0.261(20); 0.560(20)
	196	Stable	**0.14(10)**	–	–
	197	64.14(5) h	ϵ	–	0.077(18); 0.191(2)
	197m	23.8(1) h	IT(93), $\epsilon(7)$	–	0.134(42); 0.279(7)
	198	Stable	**10.02(7)**	–	–
	199	Stable	**16.84(11)**	–	–
	199m	42.6(2) min	IT	–	0.158(53); 0.375(15)
	200	Stable	**23.13(11)**	–	–
	201	Stable	**13.22(11)**	–	–
	202	Stable	**29.80(14)**	–	–
	203	46.60(2) d	β^-	0.214	0.279(81.5)
	204	Stable	**6.85(5)**	–	–
	205	5.2(1) min	β^-	1.7	0.205
	206	8.15(10) min	β^-	1.3	0.31; 0.65
$_{81}$Tl	184	11(1) s	$\epsilon(98)$, $\alpha(2)$	6.16; 5.99	0.287; 0.340; 0.367
	185m	1.8(2) s	α, IT	5.97	–
	186	45(3) s	$\epsilon(>99)$,		
			$\alpha(\sim 0.006)$	5.77; 5.65	0.36; 0.40
	186m	3 s	IT	–	–
	187	$\sim$ 51 s	ϵ	–	0.127; 0.350
	187m	15.60(12) s	IT, α	5.53	0.16*; 0.25
	188	71(10) s	ϵ, β^+	–	0.413
	189	2.3(2) min	ϵ	–	0.334; 0.942
	190	3.7(3) min	ϵ, β^+	–	0.416; 0.625; 0.731
	191	5.22(16) min	ϵ, β^+	–	0.22; 0.33*; 0.511(an.)
	192	9.6(4) min	ϵ, β^+	–	0.424
	193	21.6(8) min	ϵ, β^+	–	0.158; 0.169; 0.178; 0.187; 0.208; 0.216; 0.247; 0.511(an.)
	193m	2.11(15) min	ϵ, IT	–	0.208; 0.345; 0.586
	194	33.0(5) min	ϵ	–	0.427
	194m	32.8(2) min	ϵ	–	0.097; 0.427; 0.636; 0.749
	195	1.16(5) h	ϵ	–	0.564; 0.885; 1.364
			β^+	1.38	
	195m	3.6(4) s	IT	–	0.383(95)
	196	1.84(3) h	ϵ, β^+	–	0.426; 0.611; 0.635

Table 37.1 Table of isotopes *(continued)*

Element	A	$T_{1/2}$	Decay type or abundance of stable isotopes	Energy, MeV (relative intensity, %)	
				particle group	γ-rays
$_{81}$Tl	196m	1.41(2) h	ϵ(96.2) IT(3.8)	–	0.426; 0.635; 0.695
	197	2.84(4) h	ϵ, β^+	–	0.152; 0.426
	197m	0.54(1) s	IT	–	0.222(40); 0.385(90)
	198	5.3(5) h	ϵ	–	} 0.412(90); 0.65**(40);
			β^+($\sim$ 0.7)	2.4	} 1.20(21); 1.42(24)
	198m	1.87(3) h	ϵ(56), IT(44)	–	0.283(30); 0.412(45); 0.586(35); 0.635(35)
	199	7.42(8) h	ϵ	–	0.158(5); 0.208(12); 0.247(9); 0.455(14)
	200	26.1(1) h	ϵ	–	
			β^+(0.37)	1.44(0.06); 1.07(0.3)	0.368(88); 0.579(10); 1.21**(35)
	201	73.1(2) h	ϵ	–	0.135(2); 0.167(8)
	202	12.23(2) d	ϵ	–	0.439(95); 0.522(0.1)
	203	Stable	**29.524(9)**	–	–
	204	3.78(2) y	β^-(97.4)	0.766	–
			ϵ(2.6)	–	–
	205	Stable	**70.476(9)**	–	–
	206	4.20(2) min	β^-	1.52	–
	206m	3.76(4) min	IT	–	0.216; 0.266; 0.453; 1.021
	207	4.77(2) min	β^-	1.44	0.897(0.16)
	207m	1.33(11) s	IT	–	0.35; 1.00
	208	3.07(2) min	β^-	2.4; 1.80	0.583(86); 2.614(100)
	209	2.20(7) min	β^-	1.99	0.45(100); 1.56(100)
	210	1.30(3) min	β^-	2.3	0.296(80); 0.795(100)
$_{82}$Pb	185	4.1(3) s	α	6.48; 6.40	–
	186	7.9(16) s	α	6.32	–
	187	17.5(36) s	α	6.08	0.34*; 0.39
	188	24.5(15) s	ϵ(97), α(3)	5.98	0.185; 0.758
	189	51(3) s	ϵ(> 99), α($\sim$ 0.4)	5.72	0.27 – 1.11
	190	1.2(1) min	ϵ(> 99), α(0.2)	5.577	0.14; 0.94
	191	1.33(8) min	ϵ(> 99), α(0.013)	5.29	0.937
	192	3.5(1) min	ϵ(> 99), α(0.007)	5.06	0.17; 0.61; 1.20
	193	5.8(2) min	ϵ	–	0.39; 0.72; 0.94
	194	11(2) min	ϵ	–	0.204
	195	15.8(2) min	ϵ	–	0.39*; 0.71; 0.88
	196	37(3) min	ϵ	–	0.192; 0.240; 0.253; 0.367; 0.503
	197	10(2) min	ϵ	–	0.375; 0.386; 0.894
	197m	44.6(9) min	ϵ(81), IT(19)	–	0.085; 0.222; 0.234; 0.386*
	198	2.40(10) h	ϵ	–	0.173(28); 0.290(16); 0.38**(40)
	199	90(10) min	ϵ	–	} 0.353(17); 0.367(80)
			β^+	2.8	}
	199m	12.2(3) min	IT(93), ϵ(7)	–	0.424(20)
	200	21.5(4) h	ϵ	–	0.109; 0.146*; 0.236; 0.26**; 0.290*; 0.450
	201	9.4(2) h	ϵ	–	} 0.330; 0.361; 0.406; 0.585;
			β^+	0.55	} 0.766; 0.907; 0.946

Table 37.1 Table of isotopes *(continued)*

Element	A	$T_{1/2}$	Decay type or abundance of stable isotopes	Energy, MeV (relative intensity, %)	
				particle group	γ-rays
$_{82}$Pb	201m	61 s	IT	–	0.629(51)
	202	$\sim 3 \cdot 10^5$ y	ϵ	–	–
	202m	3.62(3) h	IT(90.5) ϵ(9.5)	–	0.422(90); 0.658(35); 0.787(45); 0.961(90)
	203	52.1(2) h	ϵ	–	0.279(81); 0.401(5)
	203m_1	6.3(2) s	IT	–	0.825(70)
	203m_2	0.48(2) s	IT	–	–
	204	$> 1.4 \cdot 10^{17}$ y	**1.4(1)**	–	–
	204m	67.2(3) min	IT	–	0.375(93); 0.90*(189)
	205	$1.43(14) \cdot 10^7$ y	ϵ	–	–
	206	Stable	**24.1(1)**	–	–
	207	Stable	**22.1(1)**	–	–
	207m	0.805(10) s	IT	–	0.570(98); 1.064(83)
	208	Stable	**52.4(1)**	–	–
	209	3.253(14) h	β^-	0.635	–
	210	22.3(2) y	$\beta^-(> 99)$ $\alpha(1.7 \cdot 10^{-6})$	0.061 3.72	0.047(4)
	211	36.1(2) min	β^-	1.36	0.405(3.4); 0.427(1.8); 0.832(3.4)
	212	10.64(1) h	β^-	0.58	0.239(47); 0.300(3.2)
	213	10.2(3) min	β^-	–	–
	214	26.8 min	β^-	1.03(6); 0.67	0.242(4); 0.295(19); 0.352(36)
$_{83}$Bi	189	<1.5 s	α	6.67	–
	190	5.4(5) s	$\alpha(\sim 90)$	6.45	–
	191	13(1) s	$\alpha(\sim 40)$	6.32	–
	192	42(5) s	$\alpha(\sim 20)$	6.06	–
	193	64(4) s	$\alpha(\sim 60)$	5.90	–
	193m	3.5 s	$\alpha(\sim 25)$	6.48	–
	194	105(15) s	$\epsilon(> 99)$ $\alpha(< 0.2)$	– 5.61	0.280; 0.575; 0.965
	195	170(20) s	$\alpha(< 0.2)$	5.43	–
	195m	90(5) s	$\alpha(4)$	6.11	–
	196	4.6(5) min	ϵ	–	0.372; 0.688; 1.049
	197m	9.5(10) min	$\alpha(0.11)$, $\epsilon(> 99)$	5.77	–
	198	11.85(18) min	ϵ	–	0.20; 0.32; 0.56; 1.06
	198m	7.7(5) s	IT	–	0.248
	199	27(1) min	$\epsilon(> 99)$ $\alpha(\sim 0.01)$	– 5.53	0.425; 0.837; 0.842; 0.946
	199m	24.70(15) min	α	5.484	–
	200	36.4(5) min	ϵ	–	0.245; 0.420; 0.462; 1.027
	200m	31(2) min	ϵ, β^+	–	0.420; 0.462; 1.027
	201	108(3) min	ϵ	–	0.629; 0.786; 0.936; 1.014
	201m	$\sim$ 60 min	ϵ, IT $\alpha(\geq 0.02)$	– 5.28	– –
	202	1.72(5) h	ϵ, β^+	–	0.422; 0.961
	203	11.76(5) h	ϵ, α β^+	– 1.35	0.82**(78); 1.52**(31); 1.87*(35)
	204	11.22(10) h	ϵ	–	0.21**; 0.375; 0.671; 0.91**; 0.98; 1.21**
	205	15.31(4) d	ϵ $\beta^+(0.06)$	– 0.98	0.703(28); 0.988(17); 1.766(27)

Table 37.1 Table of isotopes *(continued)*

Element	A	$T_{1/2}$	Decay type or abundance of stable isotopes	Energy, MeV (relative intensity, %)	
				particle group	γ-rays
$_{83}$Bi	206	6.243(3) d	ε, β^+	–	0.516(46); 0.538(34); 0.803(99); 0.880(72)
	207	38(3) y	ε, β^+	–	0.570(98); 1.064(77); 1.770
	208	3.68(4)·10^5 y	ε	–	2.614(100)
	209	Stable	**100**	–	–
	210	5.013(5) d	β^-(> 99)	1.160	–
			α(1.3 · 10^{-4})	4.69; 4.65	–
	210*m*	3.0(1)·10^5 y	α(99.6)	4.96(58); 4.92(36); 4.57(6)	0.262(45); 0.30(23); 0.34; 0.61
			β^-(0.4)		
	211	2.14(2) min	α(99.72)	6.62(84); 6.28(16)	0.351(14)
			β^-(0.28)	–	
	212	60.55(6) min	β^-(64)	2.25	0.04(2); 0.288(0.5); 0.46**(0.8);
			α(36)	6.09(10); 6.05(26)	0.727(7); 0.785(1.1); 1.620(1.8)
	212m_1	25 min	α(≤ 93), β^-(≥ 7)	–	–
	212m_2	9 min	β^-(≤ 100)	–	–
	213	45.59(6) min	β^-(97.8)	1.39	0.437
			α(2.2)	5.87	
	214	19.9(4) min	β^-(> 99)	3.26	0.609(47); 0.769(5); 1.120(17); 1.238(6); 1.378(5); 1.40**(4); 1.764(17); 2.204(5); 2.445(2)
			α(0.021)	5.45(0.012); 5.51(0.008)	
	215	7 min	β^-	–	–
$_{84}$Po	192	0.034(3) s	α	7.18	–
	193	0.45(15) s	α	6.94	–
	194	0.6(2) s	α	6.85	–
	195	4.5(5) s	α	6.609	–
	195*m*	2.0(2) s	α	6.699	–
	196	5.5(5) s	α	6.520	–
	197	56(3) s	α(90)	6.281	–
	197*m*	26(2) s	α	6.385	–
	198	1.76(3) min	α(70)	6.183	–
			ε(30)	–	–
	199	5.2(1) min	ε(88)	–	0.362; 1.021; 1.034
			α(12)	5.952	
	199*m*	4.2(1) min	ε(61)	–	0.500; 1.002
			α(39)	6.059	
	200	11.5(1) min	ε(86)	–	0.434; 0.671; 0.797
			α(14)	5.863	
	201	15.3(2) min	ε(98.4)	–	0.890; 0.905
			α(1.6)	5.683	
	201*m*	8.9(2) min	IT(40), ε(57)	–	0.412; 0.967
			α(2.9)	5.786	
	202	44.7(5) min	ε(98)	–	0.166; 0.316; 0.689; 0.717;
			α(2)	5.587	0.791
	203	36.7(5) min	ε(∼100)	–	0.215; 0.894; 0.909; 1.091
			α(0.11)	5.384	
	203*m*	1.2(2) min	IT(96), ε(4)	–	0.262; 0.577; 0.905
	204	3.53(2) h	ε(> 99)	–	0.270; 0.884; 1.016
			α(0.6)	5.377	

Table 37.1 Table of isotopes *(continued)*

Element	A	$T_{1/2}$	Decay type or abundance of stable isotopes	Energy, MeV (relative intensity, %)	
				particle group	γ-rays
$_{84}$Po	205	1.80(4) h	$\epsilon(>99)$	–	0.837; 0.850; 0.872; 1.001
			$\alpha(0.5)$	5.22	
	206	8.8(1) d	$\epsilon(94.5)$	–	0.286/35/; 0.338/40/; 0.51**/100/; 0.807/60/; 1.02**/85/
			$\alpha(5.5)$	5.223	
	207	350(4) min	$\epsilon(>99)$	–	0.25/5/; 0.35/4/; 0.41/13/; 0.74/36/; 0.95/84/; 1.15/6/; 1.37/4/; 2.06/1.6/
			$\alpha(0.008)$	5.116	
	207m	2.8(2) s	IT	–	0.26(42); 0.31(40); 0.82(100)
	208	2.898(2) y	$\alpha(>99)$	5.115	0.285(0.003); 0.60**(0.006)
			$\epsilon(0.0018)$	–	
	209	102(5) y	$\alpha(99.74)$	4.881(99)	0.261**(0.4); 0.91(0.5)
			$\epsilon(0.26)$	–	
	210	138.376(2) d	α	5.305(100)	0.803(0.0011)
	211	0.516(3) s	α	7.45(99)	0.570(0.5); 0.90(0.5)
	211m	25.2(6) s	α	8.88(7); 7.28(91)	0.570(92); 1.063(77)
	212	$2.98(3)\cdot10^{-7}$ s	α	8.78(100)	–
	212m	45.1(6) s	α	11.65(97)	0.57(2); 2.61(2.6)
	213	$4.2(8)\cdot10^{-6}$ s	α	8.38	–
	214	$1.64(2)\cdot10^{-4}$ s	α	7.69(100)	–
	215	$1.780(4)\cdot10^{-3}$ s	$\alpha(>99)$	7.38(100)	–
			$\beta^-(0.00023)$	–	–
	216	0.15(1) s	α	6.78(100)	–
	217	<10 s	α	6.55	–
	218	3.05 min	$\alpha(>99)$	6.00(100)	–
			$\beta^-(0.018)$	–	–
$_{85}$At	196	0.3(1) s	α	7.06	–
	197	0.4(1) s	α	6.959	–
	198	4.9(5) s	α	6.755	–
	198m	1.5(3) s	α	6.849	–
	199	7.0(1) s	α	6.643	–
	200	43(2) s	$\alpha(53)$	6.47; 6.42	–
			$\epsilon(47)$	–	–
	201	89(3) s	$\alpha(71)$	6.344	–
			$\epsilon(29)$	–	–
	202	181(3) s	$\epsilon(85)$	–	0.441; 0.570; 0.675
			$\alpha(15)$	6.23(4.3); 6.12(8)	
	203	7.37(20) min	$\epsilon(69)$	–	0.639; 1.002; 1.034
			$\alpha(31)$	6.088	
	204	9.2(2) min	$\epsilon(95.6)$	–	0.425; 0.515; 0.683
			$\alpha(4.4)$	5.951	
	205	26.2(5) min	$\epsilon(90)$	–	0.629; 0.669; 0.719
			$\alpha(10)$	5.902	
	206	29.4(3) min	$\epsilon(99)$	–	0.068(10); 0.396; 0.477; 0.701
			$\alpha(1)$	5.703	
	207	1.80(4) h	$\epsilon(\sim90)$	–	0.301; 0.588; 0.815
			$\alpha(\sim 10)$	5.759	
	208	1.63(3) h	$\epsilon(99.4)$	–	0.18(25); 0.25; 0.66(100)
			$\alpha(0.6)$	5.65	

Table 37.1 Table of isotopes *(continued)*

Element	A	$T_{1/2}$	Decay type or abundance of stable isotopes	Energy, MeV (relative intensity, %)	
				particle group	γ-rays
$_{85}$At	209	5.41(5) h	ϵ(95.9)	–	0.195(23); 0.545(62); 0.780(94)
			α(4.1)	5.647	
	210	8.1(4) h	ϵ(99.82)	–	0.245(79); 1.180(100); 1.436(29); 1.483(48); 1.599(14)
			α(0.18)	5.52(0.05); 5.44(0.05); 5.36(0.06)	
	211	7.214(7) h	ϵ(58.1)	–	0.67
			α(41.9)	5.868	
	212	0.314(2) s	α	7.66(80); 7.60(20)	0.063
	212m	0.119(3) s	α	7.88(20); 7.82(80)	0.063
	213	$1.1(2)\cdot10^{-7}$ s	α	9.08	–
	214	$2\cdot10^{-6}$ s	α	8.78(99)	–
	215	$1\cdot10^{-4}$ s	α	8.026	–
	216	$3\cdot10^{-4}$ s	α	7.80(97); 7.70	–
	217	0.0323(4) s	α(> 99)	7.069	–
			β^-(0.012)	–	–
	218	∼2 s	α(99.9)	6.70(94); 6.65(6)	–
			β^-(0.1)	–	–
	219	0.9(1) min	α(∼97)	6.27	–
			β^-(∼3)	–	–
$_{86}$Rn	200	1.0(2) s	α	6.91	–
	201	7.0(4) s	α	6.72	–
	202	9.85(20) s	α(> 70)	6.64	–
	203	45(3) s	α(65)	6.498	–
			ϵ(35)	–	–
	203m	28(2) s	α	6.548	–
	204	1.24(3) min	α(∼72)	6.417	–
			ϵ(∼28)	–	–
	205	2.83 min	ϵ(77)	–	0.265; 0.465; 0.620
			α(23)	6.263	
	206	5.67(17) min	α(64)	6.260	0.325; 0.387; 0.498
			ϵ(36)	–	
	207	9.3(2) min	ϵ(77)	–	0.345; 0.747
			α(23)	6.133	
	208	24.35(13) min	α(52)	6.141	0.251; 0.287; 0.350; 0.952
			ϵ(48)	–	
	209	28.5(10) min	ϵ(83)	–	0.338; 0.408; 0.689; 0.746
			α(17)	6.039	
	210	2.4(1) h	α(96)	6.040	0.458
			ϵ(4)	–	
	211	14.6(2) h	ϵ(74)	–	0.445(29); 0.680(74); 0.865(18); 0.946(21); 1.13(23); 1.37(38)
			α(26)	5.85(9); 5.78(17)	
	212	24(2) min	α	6.264	–
	213	0.025(2) s	α	8.09	–
	214	$2.7(2)\cdot10^{-7}$ s	α	9.04	–
	215	$2.30(10)\cdot10^{-6}$ s	α	8.67	–
	216	$4.5(5)\cdot10^{-5}$ s	α	8.05	–
	217	$5.4(5)\cdot10^{-4}$ s	α	7.740	–
	218	0.035(5) s	α	7.14(99.8)	0.609(0.2)
	219	3.96(1) s	α	6.82(81); 6.55(11); 6.42(8)	0.272(9); 0.401(5)
	220	55.6(1) s	α	6.29(100)	0.55(0.07)

Table 37.1 Table of isotopes *(continued)*

Element	A	$T_{1/2}$	Decay type or abundance of stable isotopes	Energy, MeV (relative intensity, %)	
				particle group	γ-rays
$_{86}$Rn	221	25(2) min	$\beta(\sim 80)$ $\alpha(\sim 20)$	1.1; 0.8 6.037; 5.788; 5.778	} 0.150; 0.186
	222	3.8235(3) d	α	5.49(100)	0.510(0.07)
	223	43(5) min	β^-	–	–
	224	107(3) min	β^-	–	0.261; 0.266
	225	4.5(3) min	β^-	–	–
	226	6.0(5) min	β^-	–	–
$_{87}$Fr	202	0.34(4) s	α	7.251	–
	203	0.7(3) s	α	7.132	–
	204	2.1(2) s	α	7.028; 6.970	–
	205	3.7(1) s	α	6.916	–
	206	16.0(1) s	$\alpha(85)$ $\epsilon(15)$	6.790 –	} 0.559; 0.575; 0.629
	207	14.8(1) s	$\alpha(93)$ $\epsilon(7)$	6.767 –	– –
	208	59.0(20) s	$\alpha(74)$ $\epsilon(26)$	6.636 –	} 0.325; 0.636; 0.779
	209	50.0(3) s	$\alpha(89)$ $\epsilon(11)$	6.648 –	– –
	210	3.18(6) min	α ϵ	6.543 –	} 0.644; 0.817
	211	3.10(2) min	α, ϵ	6.535	0.281; 0.540; 0.918
	212	20.0(6) min	$\epsilon(56)$ $\alpha(44)$	– 6.42(16); 6.39(17); 6.35(11)	} 0.227; 1.185; 1.274
	213	34.6(3) s	$\alpha(99.45)$ $\epsilon(0.55)$	6.775 –	– –
	214	0.005(2) s	α	8.426; 8.356	–
	214m	0.00335(5) s	α	8.547; 8.477	–
	215	$9.1 \cdot 10^{-8}$ s	α	9.36	–
	216	$70(2) \cdot 10^{-8}$ s	α	9.01	–
	217	$22(5) \cdot 10^{-6}$ s	α	8.315	–
	218	$7(6) \cdot 10^{-4}$ s	α	7.85(93); 7.57	–
	219	0.021(1) s	α	7.313	–
	220	27.4(3) s	$\alpha(99.65)$ $\beta^-(0.35)$	6.68(85); 6.64(13) –	} 0.045; 0.106; 0.162
	221	4.9(2) min	α	6.34(82); 6.12(15)	0.218(14)
	222	14.4(4) min	$\beta^-(> 99)$ $\alpha(0.01–0.1)$	1.8 –	– –
	223	21.8(4) min	$\beta^-(> 99)$ $\alpha(\sim 0.005)$	1.15 5.34	} 0.050(40); 0.080(13); 0.234(4)
	224	2.67(20) min	β^-	2.8; 2.6	0.132; 0.216; 0.837; 1.341
	225	3.9(2) min	β^-	1.6	–
	226	48(1) s	β^-	3.5; 3.2	0.186; 0.254; 1.323
	227	2.4(2) min	β^-	2.4; 1.8	0.090; 0.586
	228	39(1) s	β^-	–	0.474
	229	50(20) s	β^-	–	0.310
$_{88}$Ra	206	0.4(2) s	α	7.272	–
	207	1.3(2) s	α	7.133	–
	208	1.3(2) s	α	7.13	–
	209	4.6(2) s	α	7.010	–

Table 37.1 Table of isotopes *(continued)*

Element	A	$T_{1/2}$	Decay type or abundance of stable isotopes	Energy, MeV (relative intensity, %)	
				particle group	γ-rays
$_{88}$Ra	210	3.7(2) s	α	7.019	–
	211	13(2) s	α	6.911	–
	212	13.0(2) s	α	6.9001	-
	213	2.74(6) min	$\alpha(88)$, $\epsilon(20)$	6.73; 6.62	0.110; 0.215
	214	2.46(3) s	$\alpha(>99)$ $\epsilon(0.059)$	7.136	–
	215	$1.59(9){\cdot}10^{-3}$ s	α	8.699	–
	216	$1.82(10){\cdot}10^{-7}$ s	α	9.349	–
	217	$1.6(2){\cdot}10^{-6}$ s	α	8.99	–
	218	$1.4(2){\cdot}10^{-5}$ s	α	8.39	–
	219	0.01(3) s	α	7.98; 7.68	–
	220	0.023(5) s	α	7.46(99)	0.465(1)
	221	28(2) s	α	6.76(30); 6.67(20); 6.61(34); 6.59(8)	0.091(3.5); 0.151(13); 0.175(2)
	222	38.0(5) s	α	6.56(96); 6.23	0.325(4)
	223	11.434(2) d	α	5.75(9); 5.71(54); 5.61(26); 5.54(9)	0.149**(10); 0.270(10); 0.33**(6)
	224	3.66(4) d	α	5.68(94); 5.45(6)	0.241(3.7)
	225	14.8(2) d	β^-	0.36	0.040(33)
	226	1600(7) y	α	4.78(95); 4.60(6)	0.186(4)
	227	42.2(5) min	β^-	1.31	0.291(4); 0.498(0.6)
	228	5.75(3) y	β^-	0.04	–
	229	4.0(2) min	β^-	1.8	–
	230	93(2) min	β^-	0.8	0.063; 0.072; 0.203; 0.470
$_{89}$Ac	209	0.10(5) s	α	7.59	–
	211	0.25(5) s	α	7.48	–
	212	0.93(5) s	α	7.38	–
	213	0.80(5) s	α	7.36	–
	214	8.2(2) s	$\alpha(\geq 86)$ $\epsilon(\leq 14)$	7.214; 7.082	–
	215	0.17(1) s	$\alpha(99.91)$ $\epsilon(0.09)$	7.604	–
	216	$\sim 3.310^{-4}$ s	α	9.07; 8.99	–
	217	$1.11(3){\cdot}10^{-7}$ s	α	9.65	–
	218	$2.7(4){\cdot}10^{-7}$ s	α	9.21	–
	219	$7(2){\cdot}10^{-6}$ s	α	8.66	–
	220	$2.61(5){\cdot}10^{-2}$ s	α	7.85; 7.68; 7.61	0.134
	221	0.052(2) s	α	7.65; 7.44; 7.38	–
	222	4.2(5) s	α	7.013(93); 6.967	–
	223	2.2(1) min	$\alpha(\sim 99)$ $\epsilon(1)$	6.66(38); 6.65(42); 6.57(13) –	0.082(0.2); 0.096(0.2)
	224	2.9(2) h	$\alpha(\sim 10)$ $\epsilon(\sim 90)$	6.20(3); 6.14(3); 6.04(3) –	0.132(28); 0.217(62)
	225	10.0(1) d	α	5.83(54); 579(28); 5.73(10)*	0.099; 0.150; 0.187
	226	29 h	$\beta(83)$ $\epsilon(17)$ $\alpha(0.006)$	1.1; 0.9 5.34	0.158(32); 0.185(9); 0.230(47); 0.253(11)
	227	21.773(3) y	$\beta^-(98.62)$ $\alpha(1.38)$	0.046 4.95*(1.2); 4.86*(0.18)	0.070; 0.166; 0.190
	228	6.13 h	β^-	2.1; 1.2	0.34**(15); 0.908(25); 0.96**(20)
	229	62.7(5) min	β^-	1.1	0.14*; 0.16; 0.26; 0.57

Table 37.1 Table of isotopes *(continued)*

Element	A	$T_{1/2}$	Decay type or abundance of stable isotopes	Energy, MeV (relative intensity, %) particle group	γ-rays
$_{89}$Ac	230	122(3) s	β^-	2.7	0.455; 0.508; 1.244
	231	7.5(1) min	β^-	2.1	0.185; 0.28; 0.39; 0.71
	232	35(5) s	β^-	–	–
$_{90}$Th	213	0.150(25) s	α	7.69	–
	214	0.125(25) s	α	7.68	–
	215	1.2(2) s	α	7.52; 7.39	–
	216	0.028(2) s	α	7.92	–
	217	$2.52(7)\cdot10^{-4}$ s	α	9.25	–
	218	$1.09(13)\cdot10^{-7}$ s	α	9.67	–
	219	$1.05(3)\cdot10^{-6}$ s	α	9.34	–
	220	$9.7(6)\cdot10^{-6}$ s	α	8.79	–
	221	$1.68(6)\cdot10^{-6}$ s	α	8.47; 8.15	–
	222	$2.8(3)\cdot10^{-6}$ s	α	7.98	–
	223	0.66(1) s	α	7.32; 7.29	–
	224	1.04(5) s	α	7.18(79); 6.91(19)	0.177(9); 0.235(0.4); 0.297(0.3); 0.410(0.8)
	225	8.0(5) min	α(∼ 90)	6.80(8); 6.75(6); 6.50(12); 6.48(39); 6.44(13)	} 0.246(5); 0.322(27); 0.362(5); 0.45(1); 0.49(1)
			ϵ(∼ 10)	–	
	226	30.9 min	α	6.34(79); 6.22(19)	0.111(3.4); 0.242(1.2)
	227	18.718(5) d	α	6.04(23); 5.98(24); 5.76(21); 5.72*(14)	0.050(8); 0.237**(15); 0.31**(8)
	228	1.91313(88) y	α	5.42(73); 5.34(27)	0.084(1.6); 0.132(0.2); 0.167(0.1); 0.214(0.3); 0.239
	229	7340(160) y	α	5.05(5); 4.97**(9); 4.90(10); 4.84(56); 4.81(9)	0.137**(∼3); 0.20*(∼10)
	230	$7.538(30)\cdot10^4$ y	α	4.69(76); 4.62(24)	0.068(0.6); 0.142(0.07)
	231	25.52(1) h	β^-	0.40; 0.30	0.026(2); 0.084**(10)
	232	$14.05(6)\cdot10^9$ y	α	4.01(77); 3.95(23)	–
			100	–	–
	233	22.3(1) min	β^-	1.23	0.029(2.1); 0.087(2.7); 0.171(0.7); 0.453(1)
	234	24.10(3) d	β^-	0.191	0.063*(3.5); 0.093*(4)
	235	6.9(2) min	β^-	–	0.416–0.932
	236	37.1(15) min	β^-	1.1; 1.0	0.11
$_{91}$Pa	216	0.20(4) s	α	7.87; 7.81	–
	217	$4.9(6)\cdot10^{-3}$ s	α	8.33	–
	222	$5.7(5)\cdot10^{-3}$ s	α	8.54; 8.33; 8.21	–
	223	$6.3(10)\cdot10^{-3}$ s	α	8.20; 8.01	–
	224	0.95(15) s	α	7.49	–
	225	1.8(3) s	α	7.25; 7.20	–
	226	1.8(2) min	α(74)	6.86(38); 6.82(34)	–
			ϵ(26)	–	–
	227	38.3(3) min	α(∼ 85)	6.47(43); 6.42**(23); 6.40(8); 6.36(7)	} 0.065**(6); 0.110(2)
			ϵ(∼ 15)	–	
	228	22(1) h	ϵ(∼98)	–	} 0.14(3); 0.20(9); 0.28(5); 0.33(18); 0.41(13); 0.46(32); 0.95(93); 1.57(7); 1.85(4)***
			α(∼ 2)	6.11*(1); 6.08(0.4); 6.03(0.2); 5.80(0.2)	

Table 37.1 Table of isotopes *(continued)*

Element	A	$T_{1/2}$	Decay type or abundance of stable isotopes	Energy, MeV (relative intensity, %)	
				particle group	γ-rays
$_{91}$Pa	229	1.4(4) d	ϵ(99.75)	–	–
			α(0.25)	5.67(0.05); 5.62**(0.07); 5.58(0.10); 5.54(0.03)	–
	230	17.4(5) d	ϵ(90)	–	0.45**(18); 0.91**(24); 0.95(50)
			β^-(10)	0.41	
			α(0.0032)	5.34; 5.32; 5.30	
	231	32760(110) y	α	5.06(11); 5.03(20); 5.01(25); 4.95(23); 4.73(11)	0.027(6); 0.29**(6)
	232	1.31(2) d	β^-	1.3(0.7); 0.32	0.150(12); 0.87**(51); 0.97(40)
	233	27.0(1) d	β^-	0.568(5); 0.257	0.31**(44); 0.34
	234	6.70(5) h	β^-	1.3($\leq$2); 1.13(13); 0.53	0.100(50); 0.126(26); 0.70(24); 0.90(70)***
	234m	1.17(3) min	β^-(99.87)	2.29	0.765(0.30); 1.001(0.60)
			IT(0.13)	–	
	235	24.1(2) min	β^-	1.4	0.128–0.659
	236	9.1(2) min	β^-	3.3	0.64; 0.69; 1.76
	237	8.7(2) min	β^-	2.3	0.090/50/; 0.145/45/; 0.205/55/; 0.330/40/; 0.46/100/; 0.75/50/; 0.87/100/; 0.92/100/
$_{92}$U	226	0.5(2) s	α	7.43	–
	227	1.1(3) min	α	6.87	–
	228	9.1(2) min	α($\geq$ 95)	6.69/70/; 6.60/29/	0.152(0.2); 0.187(0.3); 0.246(0.4)
	229	58(3) min	ϵ($\sim$80)	–	–
			α($\sim$20)		
	230	20.8 d	α	6.36(13); 6.33(4); 6.30(3); 5.89(67); 5.82(32)	0.0.72(0.54); 0.231(0.18)
	231	4.2(1) d	ϵ(> 99)	–	0.084(7); 0.218(1)
			α(0.0055)	5.46	
	232	68.9(4) y	α	5.32(69); 5.26(31)	0.058(0.21); 0.129(0.082)
			SF($0.9\cdot10^{-10}$)	–	
	233	$1.592(2)\cdot10^{10}$ y	α	4.82(84); 4.78(13); 4.73(1.5)	0.029/60/; 0.042/310/; 0.055/68/; 0.097/100/; 0.164/27/; 0.32*/43/
			SF($1.3\cdot10^{-10}$)	–	–
	234	$2.45(2)\cdot10^5$ y	**0.0055(5)**	–	–
			α	4.77(72); 4.72(28)	0.053(0.2)
			SF($1.7\cdot10^{-9}$)	–	–
	235	$7.038(5)\cdot10^8$ y	**0.7200(12)**	–	–
			α	4.58*(8); 4.40(62); 4.36(18); 4.22(6)	0.143(11); 0.185(54); 0.204(5)
			SF($7.2\cdot10^{-9}$)	–	–
	235m	26.1 min	IT	–	0.000076
	236	$2.3416(39)\cdot10^7$ y	α	4.49(74); 4.44(26)	–
			SF($9.6\cdot10^{-8}$)	–	–
	237	6.75(1) d	β^-	0.248	0.026(2); 0.060(36); 0.165(2); 0.208(23)
	238	$4.468(3)\cdot10^9$ y	**99.2745(15)**	–	–
			α	4.20(77); 4.15(23)	
			SF($5.4\cdot10^{-5}$)	–	–
	239	23.50(5) min	β^-	1.29	0.044(4); 0.075(51)
	240	14.1(2) h	β^-	0.36	0.044

Table 37.1 Table of isotopes *(continued)*

Element	A	$T_{1/2}$	Decay type or abundance of stable isotopes	Energy, MeV (relative intensity, %)	
				particle group	γ-rays
$_{93}$Np	227	1.1(3) min	α	–	–
	229	4.0(2) min	α	6.89	–
	230	4.6(3) min	$\alpha(>99)$ $\epsilon(\leq 0.97)$	6.66	–
	231	48.8(2) min	$\epsilon(<99)$ $\alpha(>1)$	6.29	0.264; 0.348; 0.371
	232	14.7(3) min	ϵ	–	0.28; 0.33; 0.82; 0.86**
	233	36.2(1) min	$\epsilon(>99)$ $\alpha(\sim 0.001)$	5.54	–
	234	4.4(1) d	ϵ $\beta^+(\sim 0.05)$	– 0.8	} 0.109; 0.23; 0.25; 0.45; 0.50; 0.75; 0.95; 1.21; 1.56***
	235	396.2(12) d	$\epsilon(>99)$ $\alpha(0.0016)$	– 5.02; 5.00	– –
	236	$1.15(12)\cdot 10^5$ y	$\epsilon(91)$ $\beta^-(9)$	–	} 0.104; 0.160
	237	$2.14(1)\cdot 10^6$ y	α	4.79(51); 4.77(25); 4.65*(9)	0.030(14); 0.086(14); 0.145(1)
	238	2.117(2) d	β^-	1.25	1.01**(42)
	239	2.355(4) d	β^-	0.713(11); 0.437	0.106(23); 0.209(4); 0.228(12); 0.278(14)
	240	65(3) min	β^-	0.89	0.16; 0.25; 0.44; 0.56; 0.60; 0.92; 1.00; 1.16
	240m	7.4(2) min	β^-	2.16	0.56(21); 0.60(13); 0.92**(3); 1.5**(3)
	241	16.0(2) min	β^-	1.3	0.133; 0.174
$_{94}$Pu	232	34.1(7) min	$\epsilon(\geq 80)$ $\alpha(\leq 20)$	– 6.59; 6.54	– –
	233	20.9(4) min	$\epsilon(\sim 100)$ $\alpha(0.12)$	– 6.31	} 0.235; 0.535
	234	8.8(1) h	$\epsilon(94)$ $\alpha(6)$	– 6.20(4); 6.15(1.9)	– –
	235	25.3(10) min	$\epsilon(>99)$ $\alpha(0.003)$	– 5.86	– –
	236	2.851(8) y	α	5.77(69); 5.72(31)	0.048(0.31); 0.109(0.012)
	237	45.3(2) d	$\epsilon(>99)$ $\alpha(0.0033)$	– 5.66/21/; 5.37/79/	} 0.060(5)
	237m	0.18 s	IT	–	0.145(2)
	238	87.74(4) y	α SF$(1.84\cdot 10^{-7})$	5.50(72); 5.46(28) –	$0.099(8\cdot 10^{-3})$; $0.150(1\cdot 10^{-3})$ –
	239	24119(26) y	α SF$(4.4\cdot 10^{-10})$	5.15*(88); 5.10(11.5) –	0.039(0.007); 0.052(0.020); 0.129(0.005); 0.375(0.0012); 0.414(0.0012) –
	240	6537(10) y	α SF$(5.7\cdot 10^{-6})$	5.17(76); 5.12(24) –	$0.65^{**}(2\cdot 10^{-5})$ –
	241	14.4(2) y	$\alpha(2.4\cdot 10^{-3})$ $\beta^-(>99)$	4.90(0.0019); 4.85(0.0003) 0.021	} $0.145(1.6\cdot 10^{-4})$
	242	$3.763(20)\cdot 10^5$ y	α SF$(5.5\cdot 10^{-4})$	4.90(74); 4.86(26) –	–
	243	4.956(3) h	β^-	0.58	0.084(21); 0.381(0.7)

Table 37.1 Table of isotopes *(continued)*

Element	A	$T_{1/2}$	Decay type or abundance of stable isotopes	Energy, MeV (relative intensity, %)	
				particle group	γ-rays
$_{94}$Pu	244	$8.08(10)\cdot10^7$ y	α	4.59; 4.55	–
			SF($1.25\cdot10^{-1}$)	–	–
	245	10.5(1) h	β^-	1.2; 0.9	0.33; 0.56
	246	10.85(2) d	β^-	0.33(10); 0.15	0.044(30); 0.180(10); 0.224(25)
$_{95}$Am	232	55(7) s	–	–	–
	234	2.6(2) min	ϵ, SF	–	–
	237	73.0(10) min	ϵ(> 99)	–	} 0.280; 0.438; 0.474; 0.909
			α(0.025)	6.04	
	238	98(2) min	ϵ(> 99)	–	} 0.36(12); 0.58(29);
			α($1\cdot10^{-4}$)	5.94	0.98*(80); 1.35(76)
	239	11.9(1) h	ϵ(> 99)	–	} 0.209(5); 0.228*(18);
			α(0.01)	5.78	0.278(17)
	240	50.8(3) h	ϵ(> 99)	–	} 0.90(23); 1.00(77)
			α($1.9\cdot10^{-4}$)	5.378	
	241	432.2(5) y	α	5.49(85); 5.44(13)	0.026(2.5); 0.060(36); 0.101**(0.04)
	242	16.02(2) h	β^-(82.7)	0.67	–
			ϵ(17.3)	–	–
	242m	152(7) y	IT(99.52)	–	} 0.049(0.20);
			α(0.48)	5.21(0.41)	0.087(0.036)
			SF($1.6\cdot10^{-8}$)	–	
	243	7380(40) y	α	5.28(88); 5.23(11)	0.044(4); 0.075(50)
			SF($2.2\cdot10^{-8}$)	–	–
	244	10.1(1) h	β^-	0.387	0.099(5); 0.154(19); 0.746(66); 0.900(25)
	244m	~ 26 min	β^-(> 99)	1.50	–
			ϵ(0.036)	–	–
	245	2.05(1) h	β^-	0.91	0.253
	246	39(3) min	β^-	2.10(7); 1.60	0.799(29); 1.07**(65)
	247	22(3) min	β^-	–	0.23; 0.28
$_{96}$Cm	238	2.4(1) h	ϵ(< 90)	–	–
			α(> 10)	6.52	–
	239	~ 2.9 h	ϵ	–	0.188
	240	27(1) d	α	6.29(72); 6.25(28)	–
	241	32.8(2) d	ϵ(99)	–	} 0.475(95); 0.60)
			α(1)	5.94	
	242	162.8(4) d	α	6.11(74); 6.07(26)	0.044(0.041); 0.102(0.004)
			SF($6.8\cdot10^{-6}$)	–	–
	243	28.5(2) y	α(99.76)	6.06*(6); 5.99*(6) 5.78(73); 5.74(11)	} 0.209(4); 0.228(12); 0.278(14)
			ϵ(0.26)	–	
	244	18.10(2) y	α	5.80(77); 5.76(23)	0.043(0.02); 0.100(0.0015); 0.150(0.0013)
			SF($1.3\cdot10^{-4}$)	–	–
	245	8500(100) y	α	5.36(91); 5.31(6)	0.13(5); 0.173(14)
	246	4730(100) y	α	5.39(79); 5.34(21)	–
			SF($2.6\cdot10^{-2}$)	–	–
	247	$1.56(5)\cdot10^7$ y	α	5.27; 4.87	0.278; 0.402
	248	$3.40(4)\cdot10^5$ y	α(91.74)	5.08(82); 5.04(18)	–
			SF(8.26)	–	–
	249	64.15(3) min	β^-	0.9	0.634

Table 37.1 Table of isotopes *(continued)*

Element	A	$T_{1/2}$	Decay type or abundance of stable isotopes	Energy, MeV (relative intensity, %)	
				particle group	γ-rays
$_{96}$Cm	250	$\sim$ 7400 y	SF($\sim$ 65)	–	–
			α($\sim$ 28), β^-($\sim$ 7)		
	251	16.8(2) min	β^-	1.4	0.39; 0.44; 0.53; 0.54
$_{97}$Bk	242	7 min	ϵ	–	–
	243	4.5(2) h	ϵ(99.85)	–	0.755; 0.84; 0.946
			α(0.15)	6.76(0.023); 6.72(0.019); 6.57(0.038); 6.54(0.029); 6.21(0.020)	
	244	4.35(15) h	ϵ(> 99)	–	0.218(100); 0.892(88)
			α(0.006)	6.67(0.003); 6.62(0.003)	
	245	4.94(3) d	ϵ(99.88)	–	0.253(31); 0.39*(3)
			α(0.12)	6.36(0.018); 6.32(0.017); 6.15(0.021); 6.12(0.016); 5.89(0.024)	
	246	1.80(2) d	ϵ	–	0.800(40); 1.07**(12)
	247	1380(250) y	α	5.69(37); 5.53(58)	0.084(40); 0.27(30)
	248	23.7(2) h	β^-(70)	0.9	0.551
			ϵ(30)	–	
	249	320(6) d	β^-(> 99)	0.125	0.32*($3 \cdot 10^{-5}$)
			α(0.0015)	5.42(0.0015); 5.39	
			SF($4.7 \cdot 10^{-8}$)	–	–
	250	3.217(5) h	β^-	1.76(11); 0.73	0.990(47); 1.032(39)
	251	56(2) min	β^-	$\sim$1.0	0.153; 0.178
$_{98}$Cf	240	1.06(15) min	α	7.59	–
	241	3.78(70) min	α	7.342	–
	242	3.68(44) min	ϵ, α	7.392; 7.358	–
	243	10.7(5) min	ϵ($\sim$86)	–	–
			α($\sim$14)	7.17; 7.06	–
	244	19.4(6) min	α	7.213; 7.176	–
	245	43.6(8) min	ϵ($\sim$70)	–	–
			α($\sim$30)	7.137	–
	246	35.7(5) h	α	6.76(78); 6.72(22)	–
			SF		–
	247	3.11(3) h	ϵ(99.96)	–	0.295(1); 0.417; 0.460
			α(0.04)	6.301	
	248	333.5(28) d	α	6.27(82); 6.22(18)	–
			SF	–	–
	249	351(2) y	α	5.81(84); 5.76(4)	0.333(16); 0.388(72)
			SF($5.2 \cdot 10^{-7}$)	–	–
	250	13.08(9) y	α	6.03(85); 5.99(15)	–
			SF($7.7 \cdot 10^{-2}$)	–	–
	251	898(44) y	α	5.85(45); 5.67(55)	0.18; 0.23
	252	2.638(10) y	α(96.91)	6.12(84); 6.08(16)	–
			SF(3.09)	–	–
	253	17.81(8) d	β^-(99.69)	0.27	–
			α(0.31)	5.98	–
	254	60.5(2) d	SF(99.69)	–	–
			α(0.31)	5.834	–
	255	1.9(6) h	β^-	–	–
	256	12.3(12) min	SF	–	–

Table 37.1 Table of isotopes *(continued)*

Element	A	$T_{1/2}$	Decay type or abundance of stable isotopes	Energy, MeV (relative intensity, %)	
				particle group	γ-rays
$_{99}$Es	243	21(2) s	ϵ, α	7.89	–
	244	37(4) s	ϵ(96), α(4)	7.57	–
	245	1.33(15) min	ϵ(60)	–	–
			α(40)	7.73	–
	246	7.7(5) min	ϵ(90)	–	-
			α(10)	7.36	–
	247	4.7(3) min	$\epsilon(\sim 93)$	–	–
			$\alpha(\sim 7)$	7.31	–
	248	27(3) min	ϵ(99.7)	–	–
			α(0.3)	6.87	–
	249	102.2(6) min	ϵ(99.4)	–	} 0.38*; 0.81
			α(0.6)	6.77	
	250	8.6(1) h	ϵ	–	0.303; 0.349; 0.829
	251	33(1) h	ϵ(99.5)	–	} 0.178
			α(0.5)	6.49; 6.46	
	252	471.7(19) d	α(78)	6.64(82); 6.58(13)	} 0.228(0.23); 0.278(0.21);
			ϵ(22)	–	0.40**(1.1)
	253	20.47(3) d	α	6.63(90)	0.387**(0.05)
			SF($8.7\cdot10^{-6}$)	–	–
	254	275.7(5) d	α	6.43(93)	0.063(2); 0.27**(12); 0.31*(0.22)
			SF	–	–
	254m	39.3(2) h	β^-(99.59)	1.13(25); 0.43	} 0.65(31); 0.69**(38)
			α(0.33)	6.382	
			ϵ(0.08)	–	
	255	39.8(12) d	β^-(92)	–	–
			α(8)	6.30; 6.26	–
			SF(0.004)	–	–
	256	7.6 h	β^-	–	0.22; 0.23; 0.86
$_{100}$Fm	243	0.18(6) s	α	8.55	–
	244	0.0037(4) s	SF	–	–
	245	4.2(13) s	α	8.15	–
	246	1.1(2) s	α(92)	8.24	–
			SF(8)	–	–
	247	35(4) s	$\alpha(\geq 50)$	7.93; 7.87	–
			$\epsilon(\leq 50)$	–	–
	248	36(3) s	α(99.9)	7.87; 7.83	–
			SF(0.1)	–	–
	249	2.6(7) min	α	7.53	–
	250	30(3) min	α	7.43	–
	251	5.30(8) h	ϵ(98.2)	–	} 0.453; 0.881
			α(1.8)	6.83; 6.78	
	252	25.39(5) h	SF	–	–
			α	7.04; 7.00	–
	253	3.00(12) d	ϵ(88)	–	} 0.145; 0.272
			α(12)	6.94(9); 6.67(2)	
	254	3.240(2) h	$\alpha(> 99)$	7.20(82); 7.16(17)	–
			SF(0.059)	–	–
	255	20.07(7) h	α	7.02(93); 6.96	0.059*(0.9); 0.081*(1.1)
			SF	–	–
	256	157.6(13) min	SF(91.9)	–	–
			α(8.1)	6.917	–

Table 37.1 Table of isotopes *(continued)*

Element	A	$T_{1/2}$	Decay type or abundance of stable isotopes	Energy, MeV (relative intensity, %)	
				particle group	γ-rays
$_{100}$Fm	257	100.5(2) d	α(99.79)	6.52(94)	0.180(8); 0.242(10)
			SF(0.21)	–	–
	258	0.000380(60) s	SF	–	–
	259	1.5(3) s	SF	–	–
$_{101}$Md	248	7(3) s	ϵ(80), α(20)	8.36; 8.32	–
	249	24(4) s	ϵ($\leq$ 80),	–	–
			α($\geq$20)	8.03	
	250	52(6) s	ϵ(94), α(6)	7.82; 7.75	–
	251	4.0(5) min	ϵ($\geq$90)	–	–
			α($\leq$10)	7.55	
	252	2.3(8) min	ϵ	–	–
	254	10(3) min	ϵ	–	–
	255	27(2) min	ϵ(92)	–	0.430
			α(8)	7.326	
	256	76(4) min	ϵ(90.1)	–	0.400
			α(9.9)	7.205; 7.138	
	257	5.2(5) h	ϵ(90)	–	–
			α(10)	7.063	–
	258	55(4) d	α	6.79; 6.72	–
$_{102}$No	250	0.00025(5) s	α	–	–
	251	0.8(3) s	α	8.68(20); 8.58(80);	–
	252	2.30(22) s	α(73)	8.42; 8.37	–
			SF(27)	–	–
	253	1.7(3) min	α	8.01	–
	254	55(5) s	α	8.10	–
	254m	0.28 s	IT	–	–
	255	3.1(2) min	α	8.12; 8.08; 7.93	–
	256	3.3(2) s	α($\sim$99.7)	8.44; 840	–
			SF(0.3)	–	–
	257	25(2) s	α	8.27(50); 8.22(50)	–
	258	0.0012 s	SF	–	–
	259	60 min	α($\sim$78)	7.53; 7.50	–
			ϵ($\sim$22)	–	–
$_{103}$Lr	255	22(4) s	α	8.43; 8.37	–
	256	28(3) s	α	8.52; 8.43; 8.39	–
	257	0.646(25) s	α	8.86; 8.80	–
	258	4.3(5) s	α	8.65; 8.61; 8.59	–
	259	5.4(8) s	α	8.45	–
	260	180(30) s	α	8.03	–
$_{104}$Db	253	$\sim$ 1.8 s	SF($\sim$50)	–	–
	254	0.0005(2) s	SF	–	–
	256	$\sim$ 0.0005 s	SF	–	–
	257	4.8(3) s	α	9.00; 8.95	0.127
	258	0.011(2) s	SF	–	–
	259	3.1(7) s	α	8.87; 8.77	–
	260	< 0.08 s	SF	–	–
	261	180(30) s	α	8.28	–
$_{105}$Jl	255	$\sim$ 1.2 s ?	SF($\sim$20)	–	–
	257	5.0(12) s	SF($\sim$20), α	9.16	–

Table 37.1 Table of isotopes *(continued)*

Element	A	$T_{1/2}$	Decay type or abundance of stable isotopes	Energy, MeV (relative intensity, %) particle group	γ-rays
$_{105}$Jl	260	1.52(13) s	α(90), SF(10)	9.12; 9.07; 9.04	–
	261	1.8(4) s	α(~75) SF(~25)	8.83	–
	262	34(4) s	α(~40), SF(~60)	8.66; 8.45	–
$_{106}$Rf	259	0.007(3) s	SF(~70); α(?)	–	–
	263	0.8(2) s	α	9.25; 9.06	–
$_{107}$Bh	261	(1–2)$\cdot10^{-3}$ s	SF(~20)	–	–

37.2 The Energy Standards of γ-Rays, α-Particles and Conversion Electrons

The energy standards of radiation are useful in everyday work with spectrometric apparatus both for the spectral interpretation and for spectrometric calibration.

In Table 37.2, the γ-rays' energy standards are compiled [1,2]. E_γ-values comprise the energy range from 12 to 3500 keV and are given in the order of γ–ray energy increasing. Hereinafter, the values in parentheses represent the uncertainties in the last significant figures.

In Table 37.3, the energies and absolute intensities of γ-rays are given [1,2,4] for some nuclides radiating one or several intensive γ-quanta, which are most convinient in calibrating a semiconductor spectrometer. In Table 37.4, the values of the most intensive groups of α-particles are given in the energy range from 2.5 to 7.7 MeV [1,2]. E_α-values are given in the order of increasing energy.

In Table 37.5, the values of the conversion electron energy and intensity are collected in the E_e range from 25 to 2500 keV [2], which are useful for interpreting the conversion electron spectra and calibration of an electron spectrometer.

In Table 37.6, the multipolarities of most intensive γ-transitions and experimental values of the internal conversion coefficients are presented for some nuclides.

Table 37.2 γ-ray energy standards in the E_γ = 12–3500 keV range [1,2]

Nuclide	$T_{1/2}$	E_γ, keV	Nuclide	$T_{1/2}$	E_γ, keV	Nuclide	$T_{1/2}$	E_γ, keV
^{241}Am (NpL_1)*1	432.2 y	11.890(7)	^{137}Cs (Ba$K_{\beta'}$)*1	30.0 y	36.4(1)	^{153}Sm	46.7 h	75.42257(17)
^{57}Co	270.9 d	14.41302(32)	^{183}Ta	5.1 d	46.48502(16)	^{108m}Ag	127 y	79.14(3)
^{241}Am (NpL_β)*1	432.2 y	17.8(1)	^{183}Ta	5.1 d	52.59648(12)	^{133}Ba	10.5 y	80.999(4)
^{241}Am (NpL_γ)*1	432.2 y	20.8(1)	^{133}Ba	10.5 y	53.156(5)	^{153}Sm	46.7 h	83.36765(14)
^{109}Cd (AgK_α)*1	464 d	22.1(1)	^{241}Am	432.2 y	59.537(1)	^{182}Ta	115 d	84.6823(8)
^{109}Cd (AgK_β)*1	464 d	25.0(1)	^{169}Yb	32.022 d	63.12081(4)	^{109}Cd	464 d	88.0341(11)
^{241}Am	432.2 y	26.345(5)	^{182}Ta	115 d	65.72247(14)	^{153}Sm	46.7 h	89.48646(15)
^{137}Cs (BaK_α)*1	30.0 y	32.1(1)	^{182}Ta	115 d	67.74998(12)	^{169}Yb	32.022 d	93.61497(7)
			^{153}Sm	46.7 h	69.67340(10)	^{153}Sm	46.7 h	97.43155(16)
			^{203}Hg (TlK_{α_1})*1	46.6 d	70.8319(8)	^{183}Ta	5.1 d	99.08182(10)
			^{203}Hg (TlK_{α_2})*1	46.6 d	72.8715(9)	^{182}Ta	115 d	100.10652(7)
						^{153}Sm	46.7 h	103.18072(14)
						^{183}Ta	5.1 d	107.93369(12)
						^{169}Yb	32.022 d	109.77988(6)

Table 37.2 γ-ray energy standards in the E_γ = 12–3500 keV range *(continued)*

Nuclide	$T_{1/2}$	E_γ, keV	Nuclide	$T_{1/2}$	E_γ, keV	Nuclide	$T_{1/2}$	E_γ, keV
^{182}Ta	115 d	113.67244(20)	^{110m}Ag	249.8 d	446.812(4)	^{152}Eu	13.33 y	1085.914(13)
^{182}Ta	115 d	116.4201(11)	^{192}Ir	73.83 d	468.07147(27)	^{152}Eu	13.33 y	1089.700(15)
^{169}Yb	32.022 d	118.18996(20)	^{134}Cs	2.062 y	475.36(5)	^{152}Eu	13.33 y	1112.116(17)
^{152}Eu	13.33 y	121.7824(3)	^{7}Be	53.44 d	477.605(3)	^{65}Zn	244.1 d	1115.546(4)
^{157}Co	270.9 d	122.06135(13)	^{192}Ir	73.83 d	484.57797(41)	^{46}Sc	83.83 d	1120.545(4)
^{169}Yb	32.022 d	130.52365(8)	m_0c^2 [*2]	–	511.006(2)	^{182}Ta	115 d	1121.299(14)
^{192}Ir	73.83 d	136.34304(49)	^{134}Cs	2.062 y	563.27(5)	^{182}Ta	115 d	1157.505(15)
^{57}Co	207.9 d	136.47434(30)	^{134}Cs	2.062 y	569.30(3)	^{134}Cs	2062 y	1167.89(6)
^{183}Ta	5.1 d	144.12536(29)	^{152}Eu	13.33 y	586.294(6)	^{60}Co	5.271 y	1173.238(4)
^{182}Ta	115 d	152.43058(24)	^{192}Ir	73.83 d	588.58446(72)	^{56}Co	78.76 d	1175.102(16)
^{182}Ta	115 d	156.38740(29)	^{124}Sb	60.2 d	602.728(12)	^{182}Ta	115 d	1189.051(14)
^{199}Au	3.139 d	158.37945(10)	^{192}Ir	73.83 d	604.41415(47)	^{152}Eu	13.33 y	1212.950(12)
^{183}Ta	5.1 d	160.53005(48)	^{134}Cs	2.062 y	604.68(2)	^{182}Ta	115 d	1221.406(16)
^{133}Ba	10.5 y	160.609(25)	^{192}Ir	73.83 d	612.46504(78)	^{182}Ta	115 d	1231.019(16)
^{183}Ta	5.1 d	161.34799(9)	^{108m}Ag	127 y	614.281(6)	^{56}Co	78.76 d	1238.282(17)
^{183}Ta	5.1 d	162.32522(28)	^{110m}Ag	249.8 d	620.358(3)	^{182}Ta	115 d	1257.421(16)
^{153}Sm	46.7 d	172.85407(15)	^{124}Sb	60.2 d	645.858(12)	^{182}Ta	115 d	1273.735(16)
^{169}Yb	32.022 d	177.21417(9)	^{110m}Ag	249.8 d	657.761(2)	^{22}Na	2.602 y	1274.542(7)
^{182}Ta	115 d	179.39486(24)	^{137}Cs	30.0 y	661.661(3)	^{182}Ta	115 d	1289.158(16)
^{169}Yb	32.022 d	197.95792(9)	^{110m}Ag	249.8 d	677.627(4)	^{152}Eu	13.33 y	1299.124(12)
^{182}Ta	115 d	198.35302(27)	^{110m}Ag	249.8 d	687.010(4)	^{124}Sb	60.2 d	1325.516(21)
^{192}Ir	73.83 d	205.79549(68)	^{152}Eu	13.33 y	688.678(6)	^{60}Co	5.271 y	1332.501(5)
^{199}Au	3.139 d	208.20595(12)	^{110m}Ag	249.8 d	706.680(5)	^{182}Ta	115 d	1342.731(25)
^{183}Ta	5.1 d	209.87220(45)	^{124}Sb	60.2 d	709.320(13)	^{56}Co	78.76 d	1360.25(7)
^{182}Ta	115 d	222.10980(30)	^{124}Sb	60.2 d	713.793(13)	^{134}Cs	2.062 y	1365.17(10)
^{133}Ba	10.5 y	223.116(35)	^{124}Sb	60.2 d	722.789(16)	^{124}Sb	60.2 d	1368.179(30)
^{182}Ta	115 d	229.32197(64)	^{108m}Ag	127 y	722.938(8)	^{24}Na	15.02 h	1368.633(6)
^{228}Th	1.913 y	238.632(2)	^{110m}Ag	249.8 d	744.279(5)	^{182}Ta	115 d	1373.838(24)
^{183}Ta	5.1 d	244.26913(55)	^{110m}Ag	249.8 d	763.949(5)	^{110m}Ag	249.8 d	1384.300(4)
^{152}Eu	13.33 y	244.692(2)	^{152}Eu	13.33 y	778.903(6)	^{152}Eu	13.33 y	1408.011(14)
^{183}Ta	5.1 d	246.06473(23)	^{124}Sb	60.2 d	790.727(16)	^{152}Eu	13.33 y	1457.628(15)
^{169}Yb	32.022 d	261.07865(12)	^{134}Cs	2.062 y	795.78(2)	^{110m}Ag	249.8 d	1475.786(5)
^{182}Ta	115 d	264.07542(30)	^{134}Cs	2.062 y	801.86(3)	^{124}Sb	60.2 d	1488.886(24)
^{133}Ba	10.5 y	276.404(7)	^{152}Eu	13.33 y	810.459(7)	^{110m}Ag	249.8 d	1505.036(5)
^{203}Hg	46.6 d	279.1968(10)	^{110m}Ag	249.8 d	818.032(5)	^{110m}Ag	249.8 d	1562.301(6)
^{183}Ta	5.1 d	291.73096(68)	^{54}Mn	312.5 d	834.848(17)	^{124}Sb	60.2 d	1690.992(26)
^{152}Eu	13.33 y	295.939(8)	^{152}Eu	13.33 y	841.586(8)	^{207}Bi	38 y	1770.237(10)
^{192}Ir	73.83 d	295.95825(13)	^{56}Co	78.76 d	846.772(13)	^{56}Co	78.76 d	1771.351(26)
^{133}Ba	10.5 y	302.858(5)	^{152}Eu	13.33 y	867.388(8)	^{88}Y	106.6 d	1836.063(13)
^{169}Yb	32.022 d	307.73766(13)	^{192}Ir	73.83 d	884.54174(74)	^{56}Co	78.76 d	2015.181(28)
^{192}Ir	73.83 d	308.45689(15)	^{110m}Ag	249.8 d	884.684(5)	^{56}Co	78.76 d	2034.755(29)
^{192}Ir	73.83 d	316.50789(18)	^{46}Sc	83.83 d	889.277(3)	^{124}Sb	60.2 d	2090.962(35)
^{51}Cr	27.7 d	320.08419(42)	^{88}Y	106.6 d	898.042(4)	^{56}Co	78.76 d	2598.458(33)
^{152}Eu	13.33 y	344.275(4)	^{152}Eu	13.33 y	919.401(8)	^{88}Y	106.6 d	2734.087(30)
^{183}Ta	5.1 d	353.99767(36)	^{110m}Ag	249.8 d	937.491(5)	^{24}Na	15.02 h	2754.030(14)
^{133}Ba	10.5 y	356.014(9)	^{124}Sb	60.2 d	968.208(17)	^{56}Co	78.76 d	3201.962(46)
^{152}Eu	13.33 y	367.789(5)	^{152}Eu	13.33 y	1005.279(17)	^{56}Co	78.76 d	3253.416(45)
^{133}Ba	10.5 y	383.859(9)	^{56}Co	78.76 d	1037.840(15)	^{56}Co	78.76 d	3272.990(45)
^{152}Eu	13.33 y	411.115(5)	^{134}Cs	2.062 y	1038.53(5)	^{56}Co	78.76 d	3451.152(47)
^{198}Au	2.696 d	411.80441(15)	^{124}Sb	60.2 d	1045.138(20)	^{56}Co	78.76 d	3547.925(61)
^{108m}Ag	127 y	433.939(4)	^{207}Bi	38 y	1063.662(4)			

[*1] High-intensity X-ray transitions following nucleus decay.

[*2] The energy of annihilation radiation accompanying the positron decay of nuclides.

Table 37.3 Energy and absolute intensity of γ-rays for some sources, convenient for the detector calibration

Nuclide	E_γ, keV	I_γ, %	Nuclide	E_γ, keV	I_γ, %	Nuclide	E_γ, keV	I_γ, %
^{241}Am	26.345(5)	2.47(7)		208.20595(12)	22.1(3)*	^{60}Co	1173.238(4)	99.87(6)
	59.537(1)	36.5(2)	^{203}Hg	279.1968(10)	81.4(2)		1332.501(5)	99.980(9)
^{109}Cd	88.0341(11)	3.75(7)	^{51}Cr	320.08419(42)	9.83(10)	^{88}Y	898.042(4)	93.4(7)
^{57}Co	14.41302(32)	9.6(3)	^{198}Au	411.80441(15)	95.47(8)		1836.063(13)	99.37(2)
	122.06135(13)	85.6(4)	^{137}Cs	661.661(3)	85.4(8)		2734.087(30)	0.72(7)
	136.47434(30)	11.1(3)	^{54}Mn	834.848(17)	99.978(2)	^{24}Na	1368.633(6)	99.994(2)
^{199}Au	158.37945(10)	100*	^{22}Na	1274.542(7)	99.95(3)		2754.030(14)	99.87(2)

* γ-ray relative intensity, %.

Table 37.4 α-particle standard groups in the E_α = 2.5–7.7 MeV range [1,2]

Nuclide	$T_{1/2}$	E_α, MeV	I_α, %	Nuclide	$T_{1/2}$	E_α, MeV	I_α, %
^{146}Sm	$1.03 \cdot 10^8$ y	2.470(6)	100			5.1554(7)	73
^{150}Gd	$1.79 \cdot 10^6$ y	2.719(8)	100	^{240}Pu [3]	6537 y	5.1233(7)	24
^{148}Gd	93 y	3.182787(24)	100			5.1677(7)	76
^{232}Th	$14.05 \cdot 10^9$ y	3.953(8)	23	^{243}Am	7380 y	5.2335(10)	10.6
		4.012(5)	77			5.2754(10)	88
^{238}U	$4.468 \cdot 10^9$ y	4.149(5)	23	^{210}Po	138.38 d	5.30451(7)	100
		4.196(4)	77	^{241}Am	432.2 y	5.44298(13)	12.8
^{235}U	$7.038 \cdot 10^8$ y	4.219(2)	5.7			5.48560(12)	85.2
		4.368(3)	18	^{238}Pu	87.74 y	5.4565(4)	28
		4.400(2)	62			5.49921(20)	72
^{238}U	$2.34 \cdot 10^7$ y	4.445(5)	26	^{243}Cm	28.5 y	5.7415(9)	11
		4.494(3)	74			5.7847(9)	73
^{230}Th	$7.54 \cdot 10^4$ y	4.6210(15)	23.4	^{244}Cm	18.1 y	5.762835(30)	23.3
		4.6875(15)	76.3			5.80496(5)	76.7
^{234}U	$2.45 \cdot 10^5$ y	4.7220(9)	28	^{259}Cf [3]	13.08 y	5.9891(6)	15
		4.7739(9)	72			6.0308(6)	85
^{237}Np [3]	$2.14 \cdot 10^6$ y	4.787(2)	51	^{242}Cm [3]	162.8 d	6.0695(5)	26
^{233}U [3]	$1.59 \cdot 10^5$ y	4.8236(12)	84			6.1129(3)	74
^{241}Pu [3]	14.4 y	4.8530(15)	12	^{252}Cf [3]	2.638 y	6.0757(5)	16
		4.8960(15)	83			6.1183(5)	84
^{242}Pu [3]	$3.76 \cdot 10^5$ y	4.856(2)	26	^{249}Cf [3]	351 y	5.760(1)	4
		4.900(2)	74			5.813(1)	84
^{231}Pa	32760 y	4.9517(8)	22.8			6.1940(7)	2
		5.0141(8)	25.4	^{254}Es	275.7 d	6.4288(15)	93
		5.0297(8)	≤ 20	^{253}Es	20.47 d	6.5916(2)	6.6
		5.0590(8)	11			6.63273(5)	89.8
^{239}Pu	24119 y	5.1046(8)	11.5	^{255}Fm [3]	20.07 h	7.016(2)	93
		5.1429(8)	15.1				

Table 37.5 Conversion electron standard energies E_e in the range 25–2500 keV [2]

Nuclide	$T_{1/2}$	E_e, keV	I_e, %	Nuclide	$T_{1/2}$	E_e, keV	I_e, %
^{199}Au	3.139 d	34.986(7)	2.92(14)	^{192}Ir	73.831 d	217.5634(8)	1.924(14)
		75.273(7)	10.9(5)			230.0621(8)	1.790(25)
		125.099(7)	6.4(3)			238.1131(8)	4.47(14)
		143.536(7)	17.0(8)			302.6280(5)	1.95(6)
		154.813(19)	4.38(19)			389.6767(8)	1.02(4)
^{203}Hg	46.6 d	193.659(5)	16.9(8)	^{198}Au	2.696 d	328.7021(9)	2.87(9)
		263.842(5)	4.35(13)			396.9651(11)	1.02(3)
		275.485(5)	1.06(3)				

Table 37.5 Conversion electron standard energies E_e in the range 25–2500 keV *(continued)*

Nuclide	$T_{1/2}$	E_e, keV	I_e, %	Nuclide	$T_{1/2}$	E_e, keV	I_e, %
^{207}Bi	38 y	481.665(20)	1.55(5)	^{60}Co	5.271 y	1164.906(4)	~0.015
		553.809(20)	0.435(13)			1324.170(5)	~0.011
		975.615(20)	7.04(23)	^{212}Bi	60.55 min	24.510(5)	–
		1047.759(20)	1.78(6)	^{212}Pb	10.64 h	36.153(5)	–
		1059.769(20)	0.587(8)	^{208}Tl	3.07 min	148.099(6)	–
^{137}Cs	30.0 y	624.208(5)	7.64(55)			222.238(6)	–
^{137m}Ba	2.55 min	655.660(5)	1.38(50)			2526.66(10)	–

Table 37.6 Standard values of the internal conversion coefficients (I_e/I_γ) [1]

Nuclide	$T_{1/2}$	E_γ, keV	Multipolarity	I_e/I_γ	Nuclide	$T_{1/2}$	E_γ, keV	Multipolarity	I_e/I_γ
^{109}Gd	464 d	88.0	$E3$	11.2(2)	^{198}Au	2.696 d	411.8	$E2$	0.0302(3)
^{141}Ce	32.50 d	145.4	$M1+0.4\%E2$	0.378(4)	^{137}Cs	30.0 y	661.7	$M4$	0.0916(5)
^{139}Ce	137.66 d	165.9	$M1$	0.2152(33)	^{58}Co	70.78 d	810.8	$E2$	0.000295(10)
^{103}Hg	46.60 d	279.2	$E2+41\%M1$	0.164(2)	^{54}Mn	312.5 d	834.8	$E2$	0.000224(10)
^{113}Sn	115.09 d	391.7	$M4$	0.438(6)	^{65}Zn	244.1 d	1115.5	$M1+16\%E2$	0.0001664(66)

37.3 The Nuclear Quantum Characteristics

In Table 37.7, the experimental values of spins I, magnetic moments μ and electric quadrupole moments Q of the ground and some long-lived metastable states are shown for even–odd, odd–even, and odd–odd nuclei. Table 37.7 does not contain the even–even nuclei, whose ground state spin and magnetic moment equal zero. Values of I, μ and Q are given in the units of $\hbar$ ($\hbar = h/2\pi$, where h is the Planck constant), nuclear magneton μ_N and fm, respectively. The spin values indicated in parentheses, were derived by the indirect route.

Table 37.7 Spins, magnetic moments and electrical quadrupole moments of nuclei [1]

Element	A	$I, \hbar$	μ/μ_N	Q, fm	Element	A	$I, \hbar$	μ/μ_N	Q, fm
$_1$H	1	1/2	+2.7928456(11)	–	$_8$O	15	1/2	0.7189(8)	–
	2	1	+0.8574376(4)	+0.2875(20)		17	5/2	−1.89379(9)	−2.6(3)
	3	1/2	+2.978960(1)	–	$_9$F	17	5/2	+4.7223(12)	10(2)
$_2$He	3	1/2	−2.127624(1)	–		19	1/2	+2.628866(8)	–
$_3$Li	6	1	+0.8220467(6)	−0.0644(7)		20	2	+2.0935(9)	7.0(13)
	7	3/2	+3.256424(2)	−3.66(3)	$_{10}$Ne	19	1/2	−1.887(1)	–
	8	2	+1.6532(8)	2.4(2)		21	3/2	−0.661796(5)	+10.29(75)
$_4$Be	9	3/2	−1.778(9)	+5.3(3)		23	5/2	−1.08(1)	–
$_5$B	8	2	1.0355(3)	–	$_{11}$Na	20	2	+0.3694(2)	–
	10	3	+1.80065(1)	+8.472(56)		21	3/2	+2.38629(10)	−6.0(75)
	11	3/2	+2.688637(2)	+4.196		22	3	+1.746(3)	–
	12	1	+1.00306(15)	1.71(16)		23	3/2	+2.217520(2)	+10.1(8)
	13	3/2	+3.17778(51)	4.78(46)		24	4	+1.6903(8)	–
$_6$C	11	3/2	1.027(10)	3.08(6)		25	5/2	+3.683(4)	+23(8)
	13	1/2	+0.702411(1)	–	$_{12}$Mg	25	5/2	−0.85545(8)	+22
$_7$N	12	1	+0.4573(5)	–	$_{13}$Al	25	5/2	+3.6455(12)	–
	13	1/2	0.32224(35)	–		27	5/2	+3.641504(2)	+14.0(2)
	14	1	+0.4037607(2)	+1.56		28	3	+2.791(1)	–
	15	1/2	−0.2831892(3)	–	$_{14}$Si	29	1/2	−0.55529(3)	–

Table 37.7 Spins, magnetic moments and electrical quadrupole moments *(continued)*

Element	A	$I, \hbar$	μ/μ_N	Q, fm	Element	A	$I, \hbar$	μ/μ_N	Q, fm
$_{15}$P	29	1/2	1.2349(3)	–	$_{27}$Co	55	7/2	+4.822(3)	–
	30	1	–	–		56	4	3.830(15)	–
	31	1/2	+1.13160(3)	–		57	7/2	+4.733(17)	+52(9)
	32	1	−0.2524(3)	–		58	2	+4.044(8)	+22(3)
$_{16}$S	31	1/2	0.48793(8)	–		59	7/2	+4.627(9)	+40.4(40)
	33	3/2	0.643821(1)	−6.4(10)		60	5	+3.799(8)	+44(5)
	35	3/2	±1.00	+4.5(10)		60m	2	+4.40(9)	+30(40)
$_{17}$Cl	35	3/2	+0.8218736(5)	−8.249(2)	$_{28}$Ni	57	3/2	0.88(6)	–
	36	2	+1.28547(5)	−1.80(4)		61	3/2	−0.75002(4)	+16.2(15)
	37	3/2	+0.6841230(5)	−6.493(2)		65	5/2	0.69(6)	–
	38	2	2.05(2)	–	$_{29}$Cu	60	2	+1.219(3)	–
$_{18}$Ar	35	3/2	+0.633(2)	–		61	3/2	+2.14(4)	–
	37	3/2	+0.95(20)	–		62	1	−0.380(4)	–
	39	7/2	−1.3(3)	–		63	3/2	+2.2233(2)	−20.9(3)
$_{19}$K	36	2	(+)0.548(1)	–		64	1	−0.217(2)	–
	37	3/2	+0.20321(6)	–		65	3/2	+2.3817(3)	−19.5(4)
	38	3	+1.3737(10)	–		66	1	−0.282(2)	–
	39	3/2	+0.3914658(4)	+4.9(4)	$_{30}$Zn	63	3/2	−0.28164(5)	+29(3)
	40	4	−1.298099(3)	−6.7(8)		65	5/2	+0.7690(2)	−2.3(2)
	41	3/2	+0.2148699(2)	+6.0(5)		67	5/2	+0.875478(8)	+15.0(15)
	42	2	−1.1425(6)	–	$_{31}$Ga	67	3/2	+1.8507(3)	+19.5
	43	3/2	0.163(2)	–		68	1	0.01175(6)	2.77(14)
	45	3/2	0.1734(4)	–		69	3/2	+2.01659(4)	+16.8
$_{20}$Ca	39	3/2	1.02168(12)	–		71	3/2	+2.56227(2)	+10.6
	41	7/2	−1.594780(9)	–		72	3	−0.13224(2)	+52
	43	7/2	−1.31726(60)	< 23	$_{32}$Ge	69	5/2	0.735(7)	2.4(5)
$_{21}$Sc	41	7/2	5.43(2)	–		71	1/2	+0.547(5)	–
	43	7/2	+4.64(4)	−26(6)		73	9/2	−0.8794669(5)	−17.3(26)
	44	2	+2.56(3)	+10(5)		75	1/2	+0.510(5)	–
	44m	6	+3.88	−19(2)	$_{33}$As	70	4	2.1(2)	–
	45	7/2	4.756483(3)	−22(1)		71	5/2	(+)1.6735(18)	–
	46	4	+3.03(2)	+11.9(6)		72	2	(−)2.1578(22)	–
	47	7/2	+5.34(2)	−22(3)		74	2	−1.597(3)	–
	48	6	–	–		75	3/2	+1.43947(6)	+29
$_{22}$Ti	45	7/2	0.095(2)	1.5(15)		76	2	−0.906(5)	–
	47	5/2	−0.78848(1)	+29(1)	$_{34}$Se	75	5/2	0.67(4)	+100
	49	7/2	−1.10417(1)	+24(1)		77	1/2	+0.534270(8)	–
$_{23}$V	47	3/2	–	–		79	7/2	−1.018(15)	+80
	48	4	1.63(10)	–	$_{35}$Br	76	1	0.5482(1)	27
	49	7/2	4.47(5)	–		79	3/2	+2.106399(4)	+29.3
	50	6	+3.34745(3)	7		80	1	0.5140(6)	19.9(8)
	51	7/2	+5.1514(1)	−5.2(10)		80m	5	+1.3177(6)	+76(3)
$_{24}$Cr	49	5/2	0.476(3)	–		81	3/2	+2.270560(4)	+27(2)
	51	7/2	(−)0.934(5)	–		82	5	+1.6270(5)	+76(3)
	53	3/2	−0.47454(3)	2.2	$_{36}$Kr	83	9/2	−0.970669(3)	+27.0(13)
$_{25}$Mn	51	5/2	3.568(2)	50(10)		85	9/2	1.005(2)	+45(3)
	52	6	+3.0621(14)	+60(8)	$_{37}$Rb	80	1	−0.0834(3)	–
	52m	2	0.0076	–		81	3/2	+2.05(2)	–
	53	7/2	5.024(7)	–		82	5	+1.6434(12)	–
	54	3	+3.2818(13)	–		83	5/2	+1.43(2)	+27(5)
	55	5/2	+3.468716(2)	+35(5)		84	2	−1.297(11)	+0.50(13)
	56	3	+3.2266(2)	–		85	5/2	+1.35303(1)	+27.4(2)
$_{26}$Fe	57	1/2	+0.09044(7)	–		86	2	−1.6920(14)	+20(3)
	59	3/2	0.29(3)	–		87	3/2	+2.75124(1)	+13.2(1)

Table 37.7 Spins, magnetic moments and electrical quadrupole moments of nuclei *(continued)*

Element	A	$I, \hbar$	μ/μ_N	Q, fm	Element	A	$I, \hbar$	μ/μ_N	Q, fm
	88	2	0.508(5)	–	$_{49}$In	109	9/2	+5.53(6)	+89
$_{38}$Sr	87	9/2	−1.09282(65)	15(6)		110	2	+4.365(4)	+37
$_{39}$Y	89	1/2	−0.1374153(3)	–		111	9/2	+5.53(6)	+87
	90	2	−1.630(8)	−15.5(3)		112	1	+2.82(3)	+9.3
	91	1/2	0.1641(8)	–		113	9/2	+5.5289(2)	+84.6
$_{40}$Zr	91	5/2	−1.30362(2)	–		113m	1/2	−0.21074(2)	–
$_{41}$Nb	90	8	4.941(4)	–		114	1	+1.7(4)	–
	93	9/2	+6.1705(3)	−36(7)		114m	5	+4.7(1)	–
	95	9/2	6.123(12)	–		115	9/2	+5.5408(2)	+86.1(45)
	97	9/2	7.3(14)	–		115m	1/2	−0.24398(5)	–
$_{42}$Mo	95	5/2	−0.9142(1)	−1.9(12)		116	1	2.7867(8)	9(2)
	97	5/2	−0.9335(1)	−10.2(39)		116m	5	+4.22(8)	–
$_{43}$Tc	93	9/2	6.15(74)	–		117m	1/2	−0.25174(3)	–
	94	7	5.20(25)	–	$_{50}$Sn	113	1/2	0.880(9)	–
	95	9/2	9.058(140)	–		115	1/2	−0.91883(7)	–
	96	7	+5.37(17)	–		117	1/2	−1.00104(7)	–
	99	9/2	+5.6847(4)	(+)34(34)		119	1/2	−1.04728(7)	–
$_{44}$Ru	97	5/2	0.687(27)	–		119m	11/2	−1.40(8)	21(2)
	99	5/2	−0.6413(51)	+7.6(7)		121	3/2	0.699(7)	8(4)
	101	5/2	−0.7188(60)	+44(4)	$_{51}$Sb	115	5/2	+3.46(1)	−20(4)
	103	5/2	0.67(11)	–		116	3	–	–
	105	(3/2)	<0.3	–		117	5/2	3.43(6)	−30(5)
$_{45}$Rh	101m	9/2	+5.51(9)	–		118	1	2.47(7)	–
	102	(6)	4.11(15)	–		118m	8	2.32(4)	–
	103	1/2	−0.08840(2)	–		119	5/2	+3.45(1)	−21(4)
	103m	7/2	+4.78(10)	–		120	1	2.34(22)	–
	105	(7/2)	+4.428(13)	–		121	5/2	+3.3634(3)	−20(3)
$_{46}$Pd	105	5/2	−0.642(3)	+80(10)		122	2	−1.905(20)	+47(3)
$_{47}$Ag	102m	2	+4.14(25)	–		123	7/2	+2.5498(2)	−26(4)
	103	7/2	+4.47(5)	–		124	3	1.20(2)	–
	104	5	+4.0(1)	–		125	7/2	+2.630(35)	–
	104m	2	+3.7(2)	–		126	(8)	1.28(7)	–
	105	1/2	0.1014(10)	–		127	7/2	2.59(12)	–
	106	1	+2.85(20)	–		128	8	1.31(19)	–
	106m	6	3.71(15)	–	$_{52}$Te	117	1/2	–	–
	107	1/2	−0.113570(20)	–		119	1/2	0.25(5)	–
	108	1	+2.6884(7)	–		119m	11/2	0.95(5)	–
	108m	6	3.580(20)	152(8)		123	1/2	−0.73679(2)	–
	109	1/2	−0.1306905(2)	–		123m	11/2	−1.00(5)	–
	109m	7/2	+4.27(13)	–		125	1/2	−0.88828(3)	–
	110	1	2.7271(8)	–		125m	11/2	−0.93(5)	–
	110m	6	+3.607(4)	165(10)		127	3/2	0.66(5)	–
	111	1/2	−0.146(2)	–		127m	11/2	−0.91(5)	–
	112	2	0.0547(50)	–		129	3/2	0.66(5)	–
	113	1/2	0.159(2)	–		129m	11/2	−1.15(5)	–
$_{48}$Cd	105	5/2	−0.7393(2)	+43(4)		131m	11/2	−1.04(4)	–
	107	5/2	−0.615055(1)	+68(7)	$_{53}$I	125	5/2	3.0(10)	−88.9
	109	5/2	−0.827846(2)	+69(7)		127	5/2	+2.81327(8)	−78.9
	111	1/2	−0.5948856(9)	–		129	7/2	+2.6210(3)	−55.3
	111m	11/2	−1.1051(4)	−85(9)		131	7/2	+2.742(1)	−40(1)
	113	1/2	−0.62306(7)	–		132	4	3.088(7)	9(1)
	113m	11/2	−1.087783(2)	−71(7)		133	7/2	+2.856(5)	−27(1)
	115	1/2	−0.648425(1)	–	$_{54}$Xe	129	1/2	−0.777976(9)	–
	115m	11/2	−1.041034(2)	−54(5)		129m	11/2	−0.847(28)	–

Table 37.7 Spins, magnetic moments and electrical quadrupole moments of nuclei *(continued)*

Element	A	$I, \hbar$	μ/μ_N	Q, fm	Element	A	$I, \hbar$	μ/μ_N	Q, fm
	131	3/2	+0.691861(4)	−12.0(12)		155	5/2	1.93(26)	–
	131m	11/2	−0.80(10)	–	$_{64}$Gd	155	3/2	−0.2591(5)	+159(16)
	133m	11/2	−0.87(12)	–		157	3/2	−0.3398(7)	+203(26)
$_{55}$Cs	125	1/2	+1.41(2)	–		159	3/2	−0.44(3)	–
	127	1/2	+1.46(2)	–	$_{65}$Tb	156	3	1.41(18)	+140(45)
	129	1/2	+1.482(9)	–		157	3/2	2.0(1)	–
	130	1	1.4	–		158	3	+1.758(7)	+270(50)
	131	5/2	+3.543(2)	−62.0(6)		159	3/2	+2.014(4)	118(12)
	132	2	+2.222(7)	+50.8(7)		160	3	+1.702(8)	+300(50)
	133	7/2	+2.582023(9)	−0.3(1)	$_{66}$Dy	153	7/2	−0.72(9)	−15(8)
	134	4	+2.9937(9)	+38.9(3)		155	3/2	−0.34(3)	+94(10)
	134m	8	+1.0978(2)	–		157	3/2	−0.30(3)	+127(14)
	135	7/2	+2.7324(2)	+5.0(2)		161	5/2	−0.4805(51)	+244(17)
	136	5	+3.711(15)	22.5(10)		163	5/2	+0.6726(35)	+257(17)
	137	7/2	+2.8413(4)	+5.1(1)		165	7/2	0.51	280
	138	3	0.48(10)	–	$_{67}$Ho	165	7/2	+4.173(27)	+273(6)
$_{56}$Ba	133	1/2	−0.769(3)	–		166m	(7)	4.1(6)	–
	135	3/2	+0.837943(17)	+18(2)	$_{68}$Er	161	3/2	−0.370(5)	+120(9)
	137	3/2	+0.937365(20)	+28(3)		163	5/2	+0.57(2)	+220(20)
	137m	11/2	–	~ -5		165	5/2	0.66(3)	220(10)
$_{57}$La	137	7/2	+2.695(6)	+26(8)		167	7/2	−0.5665(24)	+282.7(12)
	138	5	+3.7139(3)	+51(9)		169	1/2	+0.515(25)	–
	139	7/2	+2.7832(2)	+22(3)		171	5/2	0.70(5)	240(20)
	140	3	+0.730(15)	+10.3(11)	$_{69}$Tm	163	1/2	0.081(2)	–
$_{58}$Ce	137	3/2	0.91(15)	–		165	1/2	0.139(2)	–
	137m	11/2	0.70(3)	–		166	2	0.092(2)	185(15)
	139	3/2	0.96(20)	–		167	1/2	−0.197(2)	–
	141	7/2	0.970(30)	–		169	1/2	−0.235(3)	–
	143	3/2	~ 1	–		170	1	0.2476(36)	57.4(9)
$_{59}$Pr	141	5/2	4.136(2)	−5.89(42)		171	1/2	0.2303(36)	–
	142	2	+0.234(1)	+2.97(85)	$_{70}$Yb	169	7/2	−0.63(2)	+410(6)
	142m	5	2.2(1)	–		171	1/2	+0.49367(1)	–
	143	7/2	–	–		173	5/2	−0.67989(3)	280(20)
$_{60}$Nd	143	7/2	−1.065(5)	−48.4(20)		175	7/2	0.40(5)	–
	145	7/2	−0.656(4)	−25.3(10)	$_{71}$Lu	171	7/2	2.03(10)	–
	147	5/2	0.578(3)	90(30)		172	4	2.25(10)	–
	149	5/2	0.351(10)	130(30)		173	7/2	2.34(9)	–
$_{61}$Pm	143	5/2	3.78(50)	–		174	(1)	1.94(28)	–
	144	5	1.69(14)	–		174m	(6)	2.34(33)	–
	147	7/2	+2.58(7)	+74(20)		175	7/2	+2.2327(11)	+568(6)
	148	1	1.84(19)	+20(20)		176	7	+3.19(3)	+800(70)
	148m	6	1.82(18)	–		176m	1	+0.318(3)	−239(4)
	149	7/2	3.3(5)	–		177	7/2	+2.239(11)	+551(6)
	151	5/2	1.8(2)	190(30)		177m	23/2	2.75(21)	–
$_{62}$Sm	145	7/2	0.92(6)	–	$_{72}$Hf	175	5/2	0.70(10)	+270(40)
	147	7/2	−0.8148(7)	−18(3)		177	7/2	+0.7935(6)	+450(50)
	149	7/2	−0.6717(7)	+5.2(9)		179	9/2	−0.6409(13)	+510(50)
	151	5/2	0.355(15)	–		179m_2	25/2	7.43(34)	–
	153	3/2	−0.0216(1)	+100(10)		180m	8	+8.7(10)	+440(50)
	155	3/2	–	−90(10)	$_{73}$Ta	181	7/2	2.371	+390
$_{63}$Eu	151	5/2	+3.4717(6)	+115(9)		182	3	2.6(2)	–
	152	3	−1.9414(13)	+316(35)	$_{74}$W	183	1/2	+0.1177847(1)	–
	153	5/2	+1.5330(8)	+294(23)		187	3/2	0.688(21)	–
	154	3	2.005(6)	+390(50)	$_{75}$Re	181	5/2	3.242(65)	–

Table 37.7 Spins, magnetic moments and electrical quadrupolar moments of nuclei *(continued)*

Element	A	$I, \hbar$	μ/μ_N	Q, fm
	183	(5/2)	3.03(11)	–
	184	3	2.499(51)	–
	184m	8	2.86(13)	–
	185	5/2	+3.1871(3)	236(50)
	186	1	+1.739(3)	∼40
	187	5/2	+3.2197(3)	224(50)
	188	1	+1.788(5)	∼40
$_{76}$Os	187	1/2	+0.06465184(6)	–
	189	3/2	0.659933(4)	+91(10)
	193	(3/2)	1.30(19)	–
$_{77}$Ir	191	3/2	+0.1461(6)	78(20)
	191m	11/2	6.026(36)	–
	192	4	+1.880(11)	–
	193	3/2	+0.1591(6)	70(18)
	194	1	0.37(4)	–
$_{78}$Pt	195	1/2	+0.60949(6)	–
	195m	13/2	0.597(15)	–
	197	1/2	0.51(2)	–
$_{79}$Au	190	1	0.066	–
	191	3/2	0.138(7)	–
	192	1	0.0079(11)	–
	193	3/2	0.140(7)	–
	194	1	0.074(4)	–
	195	3/2	0.148(7)	–
	195m	11/2	6.268(31)	–
	196	2	+0.5914(14)	–
	196m_2	12	5.35(20)	–
	197	3/2	+0.145746(9)	59.4(10)
	198	2	+0.5934(4)	–
	198m	(12)	5.55(35)	–
	199	3/2	+0.2715(7)	–
	200m	12	6.10(20)	–
$_{80}$Hg	181	1/2	+0.5071(7)	–
	183	1/2	+0.524(5)	–
	185	1/2	+0.507(4)	–
	187	3/2	−0.593(4)	−50(23)
	191	(3/2)	–	−41(41)
	193	3/2	−0.62757(18)	−86(38)
	193m	13/2	−1.058429(3)	+108(10)
	195	1/2	+0.541475(1)	–
	195m	13/2	−1.044647(3)	+127(11)
	197	1/2	+0.5273741(9)	–
	197m	13/2	−1.027684(3)	+147(13)
	199	1/2	+0.5058851(9)	–
	199m	13/2	−1.014702(3)	+140(42)
	201	3/2	−0.560225(1)	+45.5(40)
	203	5/2	+0.84895(13)	+40(4)
	205	1/2	+0.6010(1)	–
$_{81}$Tl	194	2	0.14(1)	–
	195	1/2	+1.58(4)	–
	196	2	0.07(1)	–
	197	1/2	+1.58(2)	–
	198	2	0.00(1)	–
	198m	7	0.64(7)	–
	199	1/2	+1.60(2)	–
	200	2	0.04(1)	–
	201	1/2	+1.61(2)	–
	202	2	0.06(1)	–
	203	1/2	+1.622257(1)	–
	204	2	0.0908	–
	205	1/2	+1.6382134(7)	–
$_{82}$Pb	207	1/2	0.58219(2)	–
$_{83}$Bi	203	9/2	+4.62(5)	−64(5)
	204	6	+4.28(5)	−41(5)
	205	9/2	4.16(11)	–
	206	6	+4.59(5)	−19(5)
	207	9/2	4.10(2)	−50(15)
	209	9/2	+4.1106(2)	−46
	210	1	−0.0446(1)	+13(1)
$_{84}$Po	205	5/2	∼+0.26	+17
	207	5/2	∼+0.27	+28
	209	1/2	∼+0.77	–
$_{85}$At	211	9/2	–	–
	213	9/2	–	–
$_{89}$Ac	227	3/2	+1.1(1)	+170(20)
$_{90}$Th	229	5/2	+0.46(4)	+430(90)
$_{91}$Pa	231	3/2	2.01(2)	–
	233	3/2	+3.5(8)	−300
$_{92}$U	233	5/2	+0.55	+350
	235	7/2	−0.35	455(9)
$_{93}$Np	237	5/2	+3.14(4)	+410(70)
	238	2	–	–
$_{94}$Pu	239	1/2	+0.203(4)	–
	241	5/2	−0.683(15)	+560(200)
$_{95}$Am	241	5/1	+1.61(3)	+490
	242	1	+0.3878(15)	−276
	243	5/2	+1.61(4)	+490
$_{96}$Cm	243	5/2	0.41	–
	245	7/2	0.5(1)	–
	247	9/2	0.37	–
$_{97}$Bk	249	7/2	2.0(4)	+579
$_{99}$Es	253	7/2	4.70(7)	670(80)
	254m	2	2.90(7)	370(50)

37.4 The Radioactive Series [2]

Figures 37.1–37.4 show the radioactive families of thorium, neptunium, uranium–radium and uranium–actinium, respectively. The chemical symbol of the element, its nuclear mass number and the half-life are indicated as well. Symbols alongside the arrows illustrate the decay type (α, β, IT). If a nuclide decays into two channels, the branching ratio of decay modes is also indicated near the arrows.

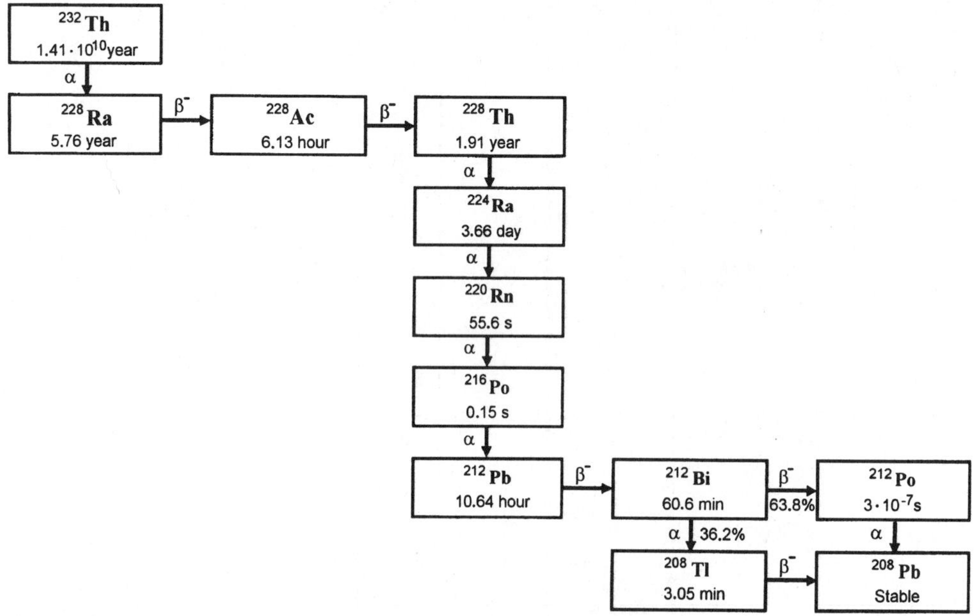

Figure 37.1 Th series ($4n$).

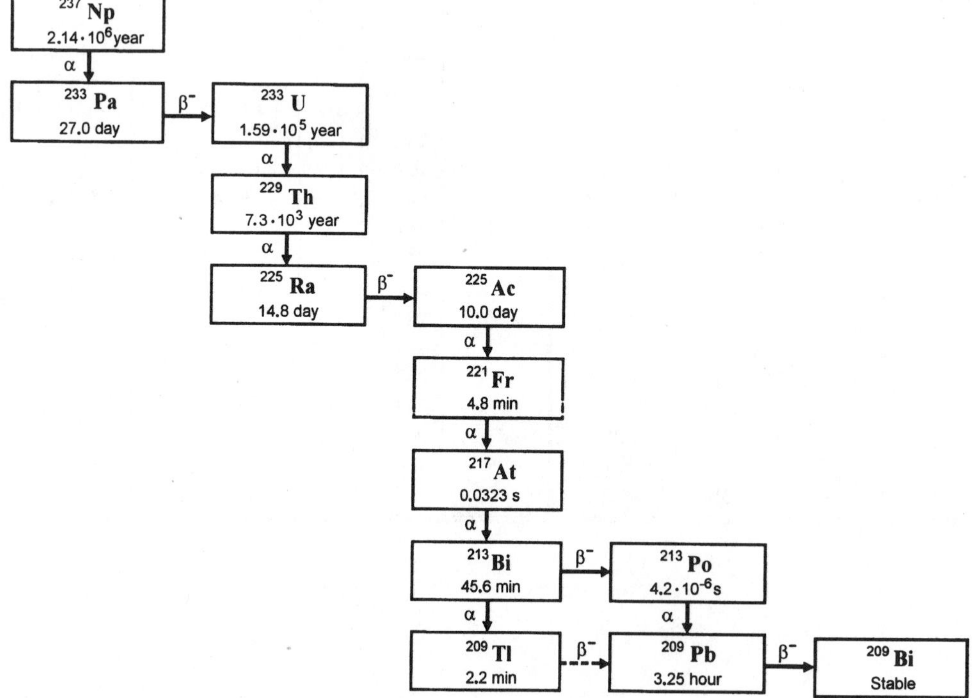

Figure 37.2 Np series ($4n$+1).

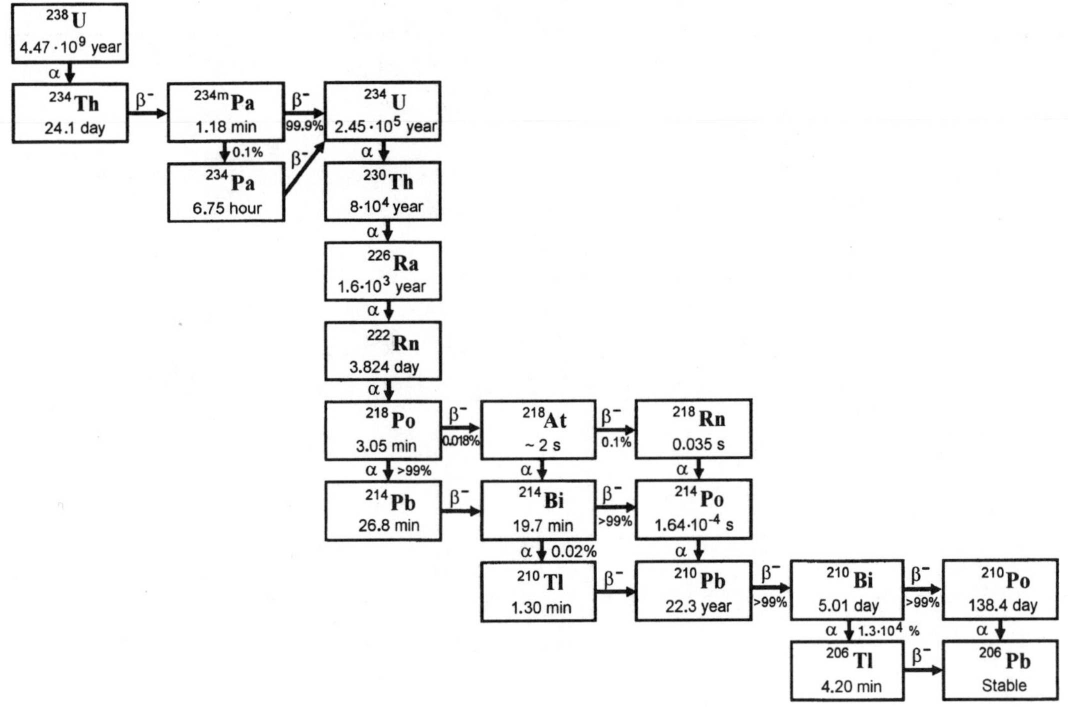

Figure 37.3 U–Ra series ($4n+2$).

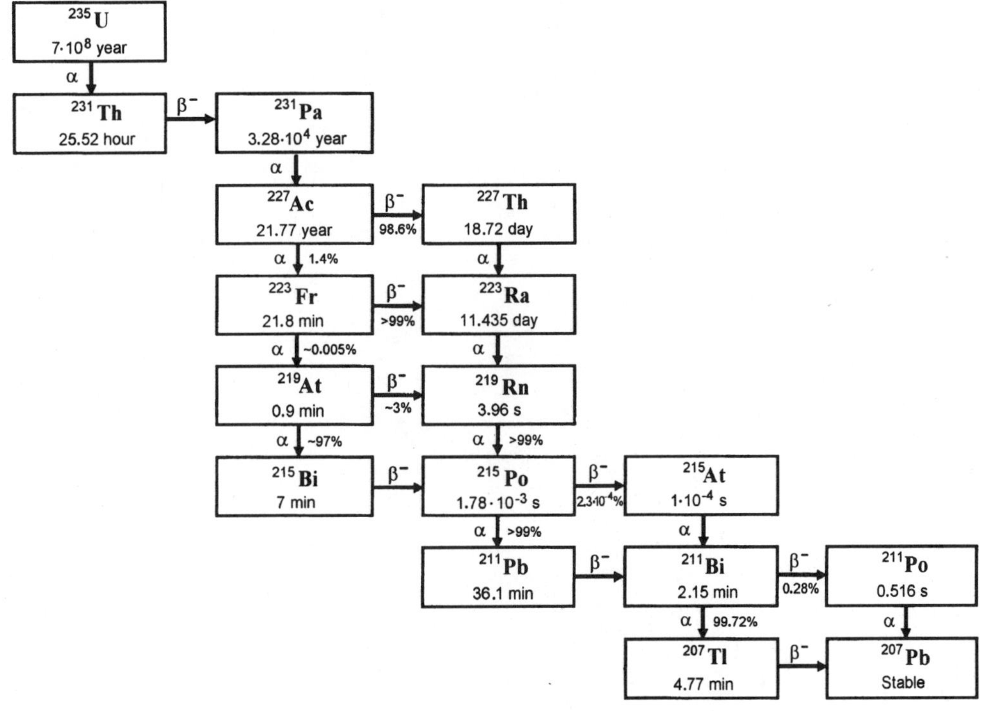

Figure 37.4 U–Ac series ($4n+3$).

References

[1] Table of isotopes, 7th ed., Eds. Lederer, C. M. and Shirley, V. S., John Wiley and Sons, New York, 1978.

[2] Seelmann–Eggebert, W., Pfenning, G., Munzel, H., Klewe–Nebenius, H., Nuklidkarte, 5 Auflage, Gerbash und Sohn Verlag, München, 1981.

[3] Baranov, S. A., Kulakov, V. M., Shatinskii, V. M., Yadernaya fizika, 7, 724, 1969; 14, 1101, 1971 (in Russian).

[4] Greenwood, R. C., Helmer, R. G., Gehrke, R. J., Nucl. Instrum. Methods, 159, 465, 1979.

[5] Holden, N. E., Martin, R. L., Barnes, I. L., Pure Appl. Chem., 56, 675, 1984.

38

Mössbauer Nuclei

S.S. Yakimov and V.M. Cherepanov

38.1 Introduction

The Mössbauer effect or nuclear gamma-resonance (NGR) is referred to as phononless (i.e. without the loss of energy for recoil) radiation or absorption of resonant γ-quanta by atomic nuclei occurring in condensed matter. It is well known [1] that the absorption and emission lines of γ-quanta in free atoms are separated from each other by the twofold recoil energy $E_r = E_0/2Mc^2$, where $E_0 = E_e - E_g$ is the energy of γ-transition between the excited (e) and ground (g) states of nucleus; M is the mass of a nucleus; c is the speed of light. Usually E_r is several orders of magnitude greater than the natural width of a nuclear level $\Gamma_0 = \hbar/\tau$ (here τ is the mean lifetime of the excited nuclear state) and nuclear resonance fluorescence is practically unobservable (even taking account of the additional broadening due to a thermal motion of atoms).

For atoms resided in the solid state samples (for example, in a crystal) all the excitations are collective and they encompass the whole system. When a γ-quantum is emitted with the energy E_r less than the characteristic energy of phonons, the recoil momentum of a separate nucleus ($p = E_0/c$) excites such phonons in the crystal that their total momentum not obligatory coincides with p because some part of it can be transmitted to the crystal as a whole. Due to large mass of a crystal such transfer of momentum is not accompanied by an energy transfer. Though, the energy per one process averaged over a large number of radiation events is equal to E_r. According to the quantum mechanics the energy transferred to the system does not correlate with momentum and can take arbitrary (though quantized) values. Particularly, when it is equal to zero the emission and absorption lines are not shifted and the NGR is possible. The lower temperature and energy E_0 and higher the energy of chemical bonding of atoms in a crystal, the higher the probability of this process f. In the Debye approximation the value of f is determined by the Debye–Waller factor W: $f = \exp(-2W)$, where $W = (3E_r/k\theta_D)[(1/4) + (T/\theta_D)^2 \int_0^{\theta_D/T} x dx/(e^x - 1)]$; θ_D is the Debye temperature.

The high energy resolution of NGR ($\Gamma_0/E_0 \approx 10^{-10} - 10^{-15}$) makes it possible not only to measure very low variations of energy (about 10^{-10} eV) but also to observe the hyperfine structure of nuclear levels that is caused by electric and magnetic electron-nucleus interactions.

The spectrum of radiation traversed through an absorber (which is usually obtained in NGR with the relative motion of a resonance source and absorber with the velocity v) is caused by a change in the γ-quantum energy due to the Doppler effect ($\Delta E = E - E_0 = E_0 v/c$). It is defined by the following expression:

$$\varepsilon(v) = [N(\infty) - N(v)]/[N(\infty) - N_b], \tag{38.1}$$

where $N(\infty)$ is the radiation intensity outside the resonance; $N(v)$ is the intensity at relative velocity v; N_b is the background intensity independent of v, which is determined in a separate experiment.

0-8493-2861-6/97/$0.00+$.50

Analytical expression for the spectrum in the case of a "thin" source (without self-absorption) and of absorber with a single line, provided the energy of transitions in them being equal, may be written (see for example [1]) as follows:

$$\varepsilon(v) = f_S \int_{-\infty}^{\infty} F(E)\{1 - \exp[-f_A n_A \sigma(E)]\}dE, \tag{38.2}$$

where $F(E) = [\Gamma_0/2\pi]\{[E-E_0(1+v/c)]^2+(\Gamma_0/2)^2\}^{-1}$ for the Lorentzian shape of the source emission line; f_S, f_A are the probabilities of emission and absorption of γ-quanta; n_A (cm^{-2}) is the number of resonant nuclei per 1 cm^2; $\sigma(E) = \sigma_0(2\pi c^2\hbar^2/E_0^2)[(E-E_0)^2+(\Gamma_0/2)^2]^{-1}$ is the absorption cross-section for a γ-quanta with the energy E; $\sigma_0 = (2\pi c^2\hbar^2/E_0^2)[(2I_e+1)/(2I_g+1)]/(1+\alpha_T)$ is the cross-section at $E = E_0$; I_g, I_e are nuclear spins in the ground and excited states; α_T is the total coefficient of an internal conversion.

Maximum absorption $\varepsilon(v = 0)$ and the spectrum area are given by the expressions

$$\varepsilon(0) = f_S[1 - \exp(-t_A/2)I_0(\mathrm{i}t_A/2)], \tag{38.3}$$

$$S = \int_{-\infty}^{\infty} \varepsilon(v)dv = (\pi\Gamma_0/2)f_S t_A \cdot \exp(-t_A/2)[I_0(\mathrm{i}t_A/2) + I_1(\mathrm{i}t_A/2)], \tag{38.4}$$

where I_0 and I_1 are the zero- and first-order Bessel functions of the imaginary argument; $t_A = f_A n_A \sigma(E_0)$ is the effective thickness of an absorber.

If the source is not thin, then subject to self-absorption it leads to the formula for $\varepsilon(0)$ of the form [2]:

$$\varepsilon(0) = [f_S/K(t_S)][K(t_S) + K(t_A) - K(t_S + t_A)], \tag{38.5}$$

where $K(x) = x\exp(-x/2)[I_0(\mathrm{i}x/2)+I_1(\mathrm{i}x/2)]$; $x = t_S$, t_A, t_S+t_A; $t_S = f_S n_S \sigma(E_0)$, and the spectrum area does not depend on the form of the radiation line.

The following parameters of the Mössbauer spectra have been considered as the principal ones [3–6].

1. **Isomer (chemical) shift δ of a Mössbauer line** is determined by the difference of nucleus radii in the excited and ground states $\Delta R = R_e - R_g$ and by the difference of electron density on nuclei of the source and absorber ($|\psi_S(0)|^2$ and $|\psi_A(0)|^2$, accordingly):

$$\delta = A\frac{\Delta R}{R}[|\psi_A(0)|^2 - |\psi_S(0)|^2], \tag{38.6}$$

where $A = (4\pi/5)e^2R^2ZS(Z)$; e is the elementary charge; Z is the nuclear charge; $S(Z)$ is the relativistic factor depending on Z and tabulated in [7].

When temperatures of the source and absorber are different the extra shift (which is termed the temperature shift of absorption line due to a relativistic Doppler effect of the second order in v/c) is observed:

$$\delta_T = -E_0 < v^2 > /(2c^2), \tag{38.7}$$

where $< v^2 >$ is the mean squared velocity of a Mössbauer atom.

In the classical limit, which is valid at high temperatures, we have

$$\delta_T = -E_0 \cdot 3kT/(2Mc^2), \tag{38.8}$$

where k is the Boltzmann constant; T is the temperature, and M is the mass of an atom.

2. **Quadrupole splitting Δ of nuclear levels** is caused by the interaction of a nuclear quadrupole moment Q with the nonuniform electric field $q = \mathrm{grad}E$, and is described by the Hamiltonian

$$\mathcal{H} = \frac{eqQ}{4I(2I-1)}\left[3I_z^2 - I(I+1)\right] + \frac{\eta}{2}\left(I_+^2 - I_-^2\right), \tag{38.9}$$

where $I_\pm = I_x \pm \mathrm{i}I_y$. Its eigenvalues are as follows:

$$E_Q = \frac{eqQ}{4I(2I-1)}\left[3m_I^2 - I(I+1)\right](1+\eta^2/3)^{1/2}, \quad m_I = I,\ I-1,\ \ldots,\ -I. \tag{38.10}$$

The asymmetry parameter takes the form $\eta = (V_{xx} - V_{yy})/V_{zz}$, where $|V_{zz}| \geq |V_{xx}| \geq |V_{yy}|$; $V_{xx} = \partial^2 V/\partial x^2$; $V_{yy} = \partial^2 V/\partial y^2$; $V_{zz} = \partial^2 V/\partial z^2$; V is the electric field potential.

In the special case of an axially-symmetrical field gradient, the ^{57}Fe ($I_g = 1/2$, $I_e = 3/2$) nucleus line is split into a doublet with the intercomponent separation of

$$\Delta = E(3/2, 3/2) - E(3/2, 1/2) = eqQ/2, \tag{38.11}$$

where $q = V_{zz}$ is the electric field gradient.

3. **Magnetic splitting of nuclear levels** is caused by hyperfine interaction of the dipole magnetic moment μ of a nucleus with a magnetic field H_n at a nucleus, which is formed by the electrons of the atom, by magnetic moments of nearby atoms as well as by polarized conduction electrons [3 - 6].

Magnetic interaction leads to complete removing of nuclear levels degeneracy, their positions being expressed by the formula

$$E_{m_I} = -\mu_n H_n m_I/I, \tag{38.12}$$

where μ_n is the nuclear magneton; m_I is the magnetic quantum number (selection rules for magnetic dipole transitions are formulated in the following way: $\Delta m_I = 0, \pm 1$).

The distance between the adjacent equidistant sublevels equals $\Delta E_m = g\mu_n H_n$, where g is the gyromagnetic ratio (nuclear g-factor).

The equidistance of sublevels becomes broken with an addition of the electric quadrupole interaction to the magnetic dipole interaction. When $\Delta \ll \Delta E_m$, the nuclear sublevels' positions are determined by the relation

$$E = -\mu_n H_n m_I/I + (-1)^{|m_I|+1/2} eqQ(3\cos^2\theta - 1)/8, \tag{38.13}$$

where θ is the angle between the z component of the electric field gradient and the magnetic field direction.

38.2 Characteristics of Mössbauer Nuclei

Following characteristics of the Mössbauer nuclei are presented in Table 38.1 [8, 9]:
symbol of the nuclide with a mass number;
content of the Mössbauer isotope in a natural mixture;
E_0 – energy of the Mössbauer transition;
$T_{1/2}$ – half-life of the resonant level;
Γ – minimum observable linewidth determined by the formula $\Gamma = 2\ln 2(\hbar/T_{1/2})$;
E_r – recoil energy of a free nucleus;
α_T – total internal conversion coefficient;
σ_0 – total cross-section of a resonant absorption;
I_g, I_e – values of spin and parity in the ground and excited states;
μ_g, μ_e – values of the magnetic moment in the ground and excited nuclear states (expressed in the nuclear magnetons);
Q_g, Q_e – values of the quadrupole moments in the ground and excited nuclear states;
data on the parent isotopes of a Mössbauer nucleus (mass number, type of decay, half-life period; e.c. means the electron capture by a nucleus; β^- denotes beta-decay with the emission of an electron; β^+ labels beta-decay with emission of a positron; α marks alpha-decay with emission of an α-particle).

Table 38.1 Characteristics of the Mössbauer nuclei

Nuclide	Content, %	E_0, keV	$T_{1/2}$, ns	Γ, mm/s	E_r, 10^{-3} eV	α_T	σ_0, 10^{-20} cm^2	I_g	I_e	μ_g	μ_e	Q_g, 10^{-24}cm^2	Q_e, 10^{-24}cm^2	Data on the parent nucleus
^{40}K	0.012	29.56	4.25	2.177	11.73	6.6	28.65	4^-	3^-	−1.298		−0.093		^{39}K(n, γ)(d, p)
^{57}Fe	2.14	14.413	97.81	0.194	1.956	8.21	255.8	$\frac{1}{2}^-$	$\frac{3}{2}^-$	0.09060	−0.1553	0	+0.21	^{57}Co(e.c., 270 d), ^{57}Mn(β^-, 1.6 min)
^{57}Fe	2.14	136.48	8.7	0.230	175.4	0.14	34.50	$\frac{1}{2}^-$	$\frac{5}{2}^-$	0.09060	−0.9785	0		
^{61}Ni	1.19	67.408	5.27	0.770	39.98	0.13	71.16	$\frac{3}{2}^-$	$\frac{5}{2}^-$	−0.7498	+0.478	+0.162	−1.21	^{61}Co(β^-, 99 min), ^{61}Cu(e.c., 3.41 h)
^{67}Zn	4.11	93.317	9150	$32 \cdot 10^{-5}$	69.77	0.89	4.955	$\frac{5}{2}^-$	$\frac{1}{2}^-$	+0.8755	+0.578	+0.17	0	^{67}Ga(e.c., 78 h), ^{67}Cu(β^-, 59 h)
^{73}Ge	7.76	13.263	2953	0.007	1.293	1095	0.761	$\frac{9}{2}^+$	$\frac{5}{2}^+$	−0.8792	−0.094	−0.18		^{78}As(e.c., 80.3 d), ^{73}Ga(β^-, 4.9 h)
^{73}Ge	7.76	68.752	1.86	2.139	34.76	0.8	22.88	$\frac{9}{2}^+$	$\frac{7}{2}^+$					
^{83}Kr	11.55	9.40	147	0.198	0.571	19.6	107.5	$\frac{9}{2}^+$	$\frac{7}{2}^+$	−0.9703	−0.942	+0.253	+0.25	^{83}Br(β^-, 2.39 h), ^{83}Rb(e.c., 86.2 d)
^{99}Tc	0	140.51	0.237	8.215	107.1	0.15	8.621	$\frac{9}{2}^-$	$\frac{7}{2}^+$	+5.6807	+3.7	+0.34		^{99}Mo(β^-, 6.7 h)
^{99}Ru	12.72	89.36	20.5	0.149	43.30	1.5	8.042	$\frac{5}{2}^+$	$\frac{3}{2}^+$	−0.626	−0.285	+0.12	+0.34	^{99}Rh(e.c., 16.1 d), ^{99m}Rh(e.c., 4.7 h)
^{101}Ru	17.07	127.22	0.581	3.701	86.02	0.16	8.688	$\frac{5}{2}^+$	$\frac{3}{2}^+$	−0.68	−0.31			^{101}Rh(e.c., 3 y), ^{101}Tc(β^-, 14.0 min)
^{107}Ag	51.4	93.1	$44.3 \cdot 10^9$	$6.6 \cdot 10^{-11}$	43.4	20	5.3	$\frac{1}{2}^-$	$\frac{7}{2}^+$	−0.1136		0		^{107}Cd(e.c., 6.6 h), ^{107}Pd(β^-, $7 \cdot 10^6$ y)
^{117}Sn	7.61	158.53	0.277	6.230	115.3	0.159	16.80	$\frac{1}{2}^+$	$\frac{3}{2}^+$	+1.0001		0		^{117}Sb(e.c., 2.8 h), ^{117m}In(β^-, 1.93 h)
^{119}Sn	8.58	23.871	17.75	0.647	2.570	5.1	140.3	$\frac{1}{2}^+$	$\frac{3}{2}^+$	−1.0461	+0.633	0	−0.064	^{119m}Sn(245 d), ^{119}Sb(e.c., 38 h)
^{121}Sb	57.25	37.138	3.5	2.105	6.118	11.1	19.55	$\frac{5}{2}^+$	$\frac{7}{2}^+$	+3.3591	+2.47	+0.28	−0.38	^{121m}Sn(β^-, 50 y)
^{125}Te	6.99	35.492	1.481	5.204	5.409	13.6	26.51	$\frac{1}{2}^+$	$\frac{3}{2}^+$	−0.8872	+0.604	0	−0.20	^{125}Sb(β^-, 2.7 y), ^{125}I(e.c., 60 d)
^{127}I	100	57.60	1.91	2.49	14.02	3.78	20.57	$\frac{5}{2}^+$	$\frac{7}{2}^+$	+2.8091	+2.54	−0.79	−0.71	^{127m}Te(β^-, 109 d), ^{127}Xe(e.c., 36.4 d)
^{129}I	0	27.77	16.80	0.586	3.209	5.1	39.01	$\frac{7}{2}^+$	$\frac{5}{2}^+$	+2.6174	+2.797	−0.55	−0.68	^{129m}Te(β^-, 34 d)
^{129}Xe	26.44	39.581	1.01	6.843	6.519	12.3	23.48	$\frac{1}{2}^+$	$\frac{3}{2}^+$	−0.7768	+0.58	0	−0.41	^{129}I(β^-, $1.57 \cdot 10^7$ y), ^{129}Cs(e.c., 2.2 h)

Table 38.1 Characteristics of the Mössbauer nuclei *(continued)*

Nuclide	Content, %	E_0, keV	$T_{1/2}$, ns	Γ, mm/s	E_r, 10^{-3} eV	α_T	σ_0, 10^{-20} cm^2	I_g	I_e	μ_g	μ_e	Q_g, 10^{-24}cm^2	Q_e, 10^{-24}cm^2	Data on the parent nucleus
^{131}Xe	21.18	80.183	0.500	6.823	26.34	1.6	7.403	$\frac{3}{2}^+$	$\frac{1}{2}^+$	+0.6908		−0.12	0	^{131}I(β^-, 8.05 d)
^{133}Cs	100	80.997	6.313	0.535	26.48	1.72	10.28	$\frac{7}{2}^+$	$\frac{5}{2}^+$	−2.5786	+3.44	−0.003		^{133}Ba(e.c., 10.5 y),
														^{133}Xe(β^-, 5.29 d)
^{133}Ba	0	12.29	8.10	2.748	0.610	110	29.18	$\frac{1}{2}^+$	$\frac{3}{2}^+$			0		^{133m}Ba(38.9 h)
^{139}La	99.91	165.85	1.50	1.100	106.2	0.26	5.282	$\frac{7}{2}^+$	$\frac{5}{2}^+$	+2.7781		+0.21		^{139}Ba(β^-, 84.9 min),
														^{139}Ce(e.c., 140 d)
^{141}Pr	100	145.42	1.85	1.017	80.50	0.46	10.57	$\frac{5}{2}^+$	$\frac{7}{2}^+$	+4.162	+2.87	−0.059	+0.3	^{141}Ce(β^-, 32.5 d),
														^{141}Nd(e.c., 2.5 h)
^{145}Nd	8.30	67.25	29.4	0.138	16.74	6.1	3.810	$\frac{7}{2}^-$	$\frac{3}{2}^-$	−0.654		−0.254		^{145}Pm(e.c., 17.7 y),
														^{145}Pr(β^-, 5.98 h)
^{145}Nd	8.30	72.50	0.72	5.241	19.46	4.9	5.917	$\frac{7}{2}^-$	$\frac{5}{2}^-$	−0.654	−0.32	−0.254		
^{145}Pm	0	61.25	2.62	1.705	13.89	6.42	11.72	$\frac{5}{2}^+$	$\frac{7}{2}^+$					^{145}Sm(e.c., 340 d)
^{147}Pm	0	91.03	2.57	1.169	30.26	2.2	6.920	$\frac{7}{2}^+$	$\frac{5}{2}^+$	+2.62	+3.60	+0.71	+0.6	^{147}Nd(β^-, 11.1 d)
^{147}Sm	14.97	122.1	0.80	2.801	54.44	1.0	6.154	$\frac{7}{2}$	$\frac{5}{2}$	−0.813	−0.448	−0.18	−0.31	
^{149}Sm	13.83	22.494	7.12	1.708	1.823	50	7.111	$\frac{7}{2}^-$	$\frac{5}{2}^-$	−0.670	−0.623	+0.058	+0.50	^{149}Eu(e.c., 93 d),
														^{149}Pm(β^-, 53 h)
^{151}Sm	0	65.83	20	0.208	15.40	5.76	8.352							
^{152}Sm	26.72	121.78	1.41	1.593	52.37	1.17	38.01	0^+	2^+	0	+0.84	0	−1.8	^{152}Eu(e.c., 12 y),
														^{149}Pm(β^-, 6m)
^{153}Sm	0	35.842	2.0	3.816	4.507	0.30	146.5	$\frac{3}{2}$	$\frac{3}{2}$					
^{154}Sm	22.71	81.99	3.0	1.112	23.43	5.05	30.08	0^+	2^+	0	+0.79	0	−1.30	Coulomb excitation
^{151}Eu	47.82	21.532	9.7	1.310	1.648	28.6	23.77	$\frac{5}{2}^+$	$\frac{7}{2}^+$	+3.465	+2.587	+1.14	+1.49	^{151}Gd(e.c., 120 d),
														^{153}Sm(β^-, 90 y)
^{153}Eu	52.18	83.365	0.82	4.002	24.38	3.82	9.738	$\frac{5}{2}^+$	$\frac{7}{2}^-$	+1.5294	+1.80	+2.9		^{153}Sm(β^-, 46.7 h),
														^{153}Sm (e.c., 242 d)
^{153}Eu	52.18	97.429	0.21	13.37	33.30	0.42	17.97	$\frac{5}{2}^+$	$\frac{5}{2}^-$	+1.5294	+3.21	+2.9		
^{153}Eu	52.18	103.18	3.9	0.680	37.35	1.78	5.456	$\frac{5}{2}^+$	$\frac{3}{2}^+$	+1.5294	+2.043	+2.9	+1.51	
^{154}Gd	2.15	123.07	1.17	1.900	52.79	1.2	36.71	0^+	2^+	0	0.85	0		^{154}Eu(β^-, 16 y)
^{155}Gd	14.73	60.01	0.134	34	12.47	8.72	10.48	$\frac{3}{2}^-$	$\frac{5}{2}^-$	−0.2584		+1.6		^{155}Eu(β^-, 4.96 y),
														^{155}Tb(e.c., 5.6 d)
^{155}Gd	14.73	86.545	6.33	0.499	25.94	0.43	33.92	$\frac{3}{2}^-$	$\frac{5}{2}^+$	−0.2584	−0.528	+1.6	+0.32	

Table 38.1 Characteristics of the Mössbauer nuclei *(continued)*

Nuclide	Content, %	E_0, keV	$T_{1/2}$, ns	Γ, mm/s	E_r, 10^{-3} eV	α_T	σ_0, 10^{-20} cm^2	I_g	I_e	μ_g	μ_e	Q_g, 10^{-24}cm^2	Q_e, 10^{-24}cm^2	Data on the parent nucleus
^{155}Gd	14.73	105.31	1.168	2.224	38.41	0.26	16.46	$\frac{3}{2}^-$	$\frac{3}{2}^+$	Γ0.2584		1.6	+1.6	
^{156}Gd	20.47	88.966	2.17	1.417	27.23	3.93	31.35	0^+	2^+	0	+0.788	0	Γ2.4	^{156}Eu(β^-, 15 d), ^{156}Tb(e.c., 5.4 d)
^{157}Gd	15.68	54.54	0.187	26.82	10.17	11.9	9.586	$\frac{3}{2}^-$	$\frac{5}{2}^-$	Γ0.3388	Γ0.513	2.0	3.6	^{157}Eu(β^-, 15.2 h)
^{157}Gd	15.68	64.0	460	$9.3\cdot10^{-3}$	14.00	0.97	23.19	$\frac{3}{2}^-$	$\frac{5}{2}^+$	Γ0.3388	Γ0.513	2.0	3.6	
^{158}Gd	24.87	79.51	2.54	1.355	21.48	6.02	27.56	0^+	2^+	0	+0.77	0	+1.3	^{157}Gd(n,γ)
^{160}Gd	21.90	75.26	2.7	1.346	19.00	7.42	25.65	0^+	2^+	0	+0.61	0	+1.34	
^{159}Tb	100	57.995	0.105	44.92	11.35	9.36	10.53	$\frac{3}{2}^+$	$\frac{5}{2}^+$	+2.008	+1.606	+1.3		^{159}Gd(β^-, 18.6 h), ^{159}Dy(e.c., 144.4 d)
^{160}Dy	2.29	86.788	2.037	1.547	25.27	4.5	29.42	0^+	2^+	0	+0.724	0	1.7	^{160}Tb(β^-, 72.1 d)
^{161}Dy	18.88	25.655	28.2	0.378	2.194	2.9	95.31	$\frac{5}{2}^+$	$\frac{5}{2}^-$	Γ0.479	+0.592	+2.35	+2.34	^{161}Tb(β^-, 6.9 d), ^{161}Ho(e.c., 2.45 h)
^{161}Dy	18.88	43.83	0.78	8.002	6.405	4.32	31.92	$\frac{5}{2}^+$	$\frac{7}{2}^+$	Γ0.479	Γ0.140	+2.35	+0.5	
^{161}Dy	18.88	74.577	3.31	1.108	18.54	0.65	6.754	$\frac{5}{2}^+$	$\frac{3}{2}^-$	Γ0.479	Γ0.403	+2.35	+1.4	
^{162}Dy	25.53	80.65	2.27	1.494	21.55	0.55	121.3	0^+	2^+	0	+0.74	0		^{161}Dy(n,γ)
^{164}Dy	28.18	73.392	2.39	1.560	17.63	9.3	22.05	0^+	2^+	0	+0.69	0	Γ1.95	^{164}Ho(e.c., 37 min)
^{165}Ho	100	94.699	0.022	130.1	29.17	3.12	8.277	$\frac{7}{2}^-$	$\frac{9}{2}^-$	+4.12	+4.08	+2.4	+2.6	^{165}Dy(β^-, 2.33 h)
^{164}Er	1.56	91.39	1.47	2.036	27.34	4.18	28.27	0^+	2^+	0	+0.7	0		^{164}Ho(β^-, 37 min)
^{166}Er	33.41	80.557	1.87	1.816	20.98	6.93	23.77	0^+	2^+	0	+0.63	0	Γ1.6	^{166}Ho(β^-, 26.9 h), ^{166}Tm(e.c., 7.7 h)
^{167}Er	22.94	79.322	0.119	29.0	20.22	5.74	7.211	$\frac{7}{2}^+$	$\frac{9}{2}^+$	0.564		2.83		^{167}Tm(e.c., 9.6 d), ^{167}Ho(β^-, 3.1 h)
^{168}Er	27.07	79.80	1.88	1.823	20.35	7.14	23.60	0^+	2^+	0	+0.66	0	Γ2.17	^{168}Tm(e.c., 85 d)
^{170}Er	14.88	79.31	1.90	1.815	19.86	7.33	23.35	0^+	2^+	0	+0.62	0		Coulomb excitation
^{169}Tm	100	8.401	4.0	8.141	0.224	268	25.8	$\frac{1}{2}^+$	$\frac{3}{2}^+$	Γ0.231	+0.534	Γ1.2		^{169}Er(β^-, 9.4 d), ^{169}Yb(e.c., 32 d)
^{170}Yb	3.03	84.253	1.608	2.019	22.41	8.05	19.0	0^+	2^+	0	+0.67	0	Γ2.1	^{170}Tm(β^-, 128.6 d), ^{170}Lu(β^+, 2.0 d)
^{171}Yb	14.31	66.719	0.87	4.713	13.97	13	7.852	$\frac{1}{2}^-$	$\frac{3}{2}^-$	+0.4919	+0.348	0	Γ1.59	^{171}Tm(β^-, 1.92 y), ^{171}Lu(e.c., 8.2 d)
^{171}Yb	14.31	75.875	1.64	2.198	18.07	7.8	14.44	$\frac{1}{2}^-$	$\frac{5}{2}^-$	+0.4919	41.008	0	Γ2.21	

Table 38.1 Characteristics of the Mössbauer nuclei *(continued)*

Nuclide	Content, %	E_0, keV	$T_{1/2}$, ns	Γ, mm/s	E_r, 10^{-3} eV	α_T	σ_0, 10^{-20} cm^2	I_g	I_e	μ_g	μ_e	Q_g, 10^{-24}cm^2	Q_e, 10^{-24}cm^2	Data on the parent nucleus
^{172}Yb	21.82	78.69	1.80	1.931	19.32	8.5	20.79	0^+	2^+	0	+0.664	0	−2.16	^{172}Lu(e.c., 6.7 d), ^{172}Tm(β^-, 63.6 h)
^{174}Yb	31.84	76.469	1.76	2.033	18.04	9.4	20.10	0^+	2^+	0	+0.672	0	−2.14	^{174}Lu(e.c., 3.6 y), ^{174m}Tm(β^-, 5.2 min)
^{176}Yb	12.73	82.13	2.0	1.665	20.57	7.1	22.47	0^+	2^+	0	+0.74	0	−2.24	Coulomb excitation
^{175}Lu	97.41	113.80	0.100	24.04	39.72	2.51	6.728	$\frac{7}{2}^+$	$\frac{9}{2}^+$	+2.22	+4.26	+5.68		^{175}Yb(β^-, 4.19 d), ^{175}Hf(e.c., 70 d)
^{176}Hf	5.20	88.361	1.39	2.227	23.81	5.86	22.84	0^+	2^+	0	+0.53	0	−2.08	^{176m}Lu(β^-, 3.7 h)
^{177}Hf	18.50	112.97	0.5	4.843	38.70	3.0	5.991	$\frac{7}{2}^-$	$\frac{9}{2}^-$	+0.7902	+1.12	+4.5		^{177}Lu(β^-, 6.7 d)
^{178}Hf	27.14	93.174	1.495	1.964	26.18	4.6	25.16	0^+	2^+	0	+0.52	0	−1.95	^{178}Lu(β^-, 28.4 min), ^{178}Ta(e.c., 9.31 min)
^{180}Hf	35.24	93.332	1.50	1.954	25.98	4.7	24.59	0^+	2^+	0	+0.63	0	−1.96	^{180m}Ta(e.c., 8.1 h), ^{180m}Hf(140 d)
^{181}Ta	99.99	6.238	6800	0.0064	0.115	46	167	$\frac{7}{2}^+$	$\frac{9}{2}^-$	+2.356	+5.24	+3.9	+4.4	^{181}W(e.c., 121.2 d), ^{181}Hf(β^-, 42.4 d)
^{181}Ta	99.99	136.25	0.040	50.19	55.05	1.76	5.969	$\frac{7}{2}^+$	$\frac{9}{2}^+$	+2.356	+3.9			
^{180}W	0.14	103.70	1.27	2.077	32.07	3.4	25.62	0^+	2^+	0	+0.52	0	−1.82	^{180m}Ta(β^-, 8.1 h), ^{180}Re(e.c., 2.43 min)
^{182}W	26.41	100.10	1.31	2.086	29.55	3.85	25.17	0^+	2^+	0	+0.51	0	−1.81	^{182}Ta(β^-, 115 d)
^{183}W	14.40	46.484	0.184	31.398	6.338	40	5.523	$\frac{1}{2}^-$	$\frac{3}{2}^-$	+0.1169	−0.10	0	−1.5	^{183}Ta(β^-, 5.1 d), ^{183}Re(e.c., 71 d)
^{183}W	14.40	99.079	0.688	4.013	28.79	4.1	8.178	$\frac{1}{2}^-$	$\frac{5}{2}^-$	+0.1169	+0.92	0	−1.63	
^{184}W	30.64	111.21	1.28	1.922	36.08	2.6	27.40	0^+	2^+	0	+0.58	0	−1.71	^{184}Re(e.c., 38 d), ^{184}Ta(β^-, 8.7 h)
^{186}W	28.41	122.30	1.01	2.215	43.17	1.6	31.46	0^+	2^+	0	+0.621	0	−1.65	^{186}Re(e.c., 90 h), ^{186}Ta(β^-, 10 min)
^{187}Re	62.93	134.24	0.010	203.8	51.73	2.35	40	$\frac{5}{2}^+$	$\frac{7}{2}^+$	+3.2044		+2.2		^{187}W(β^-, 23.9 h)
^{186}Os	1.59	137.16	0.84	2.374	54.29	1.3	28.40	0^+	2^+	0	+0.58	0	−1.5	^{186}Re(β^-, 90 h)
^{188}Os	13.30	155.03	0.695	2.539	68.62	0.8	27.96	0^+	2^+	0	+0.63	0	−1.36	^{188}Re(β^-, 16.8 h), ^{188}Ir(e.c., 41 h)
^{189}Os	16.10	36.22	0.50	15.11	3.726	80	1.151	$\frac{3}{2}^-$	$\frac{1}{2}^-$	+0.6565	+0.22	+0.94	0	^{189}Ir(e.c., 13.3 d), ^{189}Re(β^-, 24.3 h)

Table 38.1 Characteristics of the Mössbauer nuclei *(continued)*

Nuclide	Content, %	E_0, keV	$T_{1/2}$, ns	Γ, mm/s	E_r, 10^{-3} eV	αT	σ_0, 10^{-20} cm^2	I_g	I_e	μ_g	μ_e	Q_g, 10^{-24}cm^2	Q_e, 10^{-24}cm^2	Data on the parent nucleus
^{189}Os	16.10	69.59	1.64	2.397	13.75	8.0	8.420	$\frac{3}{2}^-$	$\frac{5}{2}^-$	+0.6565	+0.986	+0.94	−0.62	
^{189}Os	16.10	95.23	0.23	12.49	25.76	6.7	0.561	$\frac{3}{2}^-$	$\frac{3}{2}^-$	+0.6565		+0.94	0	
^{190}Os	26.40	186.90	0.47	3.114	98.69	0.4	24.66	0^+	2^+	0	+0.63	0	−1.18	^{190}Ir(e.c., 12.1 d),
														^{190}Re(β^-, 3.1 min)
^{191}Ir	37.30	82.398	4.02	0.826	19.08	10.7	1.540	$\frac{3}{2}^+$	$\frac{1}{2}^+$	+0.1453	+0.54	+0.78	0	^{191}Os(β^-, 15.4 d),
														^{191}Pt(e.c., 2.8 d)
^{191}Ir	37.30	129.40	0.089	23.75	47.06	2.88	5.649	$\frac{3}{2}^+$	$\frac{5}{2}^-$	+0.1453	+0.55	+0.78		
^{193}Ir	62.70	73.039	6.30	0.594	14.84	6.5	3.057	$\frac{3}{2}^+$	$\frac{1}{2}^+$	+0.1583	+0.4683	+0.7	0	^{193}Os(β^-, 30.0 h)
^{193}Ir	62.70	138.95	0.080	24.61	54.00	2.26	5.831	$\frac{3}{2}^+$	$\frac{5}{2}^-$	+0.1583		+0.7		
^{195}Pt	33.80	98.857	0.170	16.28	26.90	7.2	6.106	$\frac{1}{2}^-$	$\frac{3}{2}^-$	+0.6060	−0.62	0		^{195}Au(e.c., 183 d),
														^{195}Ir(β^-, 2.8 h)
^{195}Pt	33.80	129.74	0.620	3.401	46.33	1.8	7.426	$\frac{1}{2}^-$	$\frac{5}{2}^-$	+0.6060	+0.854	0		
^{197}Au	100	77.345	1.879	1.883	16.30	4.30	3.858	$\frac{3}{2}^+$	$\frac{1}{2}^+$	+0.1448	+0.416	+0.594	0	^{197}Pt(β^-, 18.3 h),
														^{197}Hg(e.c., 64.1 h)
^{199}Hg	16.84	158.37	2.47	0.699	67.65	0.94	15.08	$\frac{1}{2}^-$	$\frac{5}{2}^-$	+0.506	+1.28	0	+0.7	^{199}Au(β^-, 3.14 d),
														^{199}Tl(e.c., 7.42 h)
^{201}Hg	13.22	32.19	0.20	42.49	2.767	60	0.948	$\frac{3}{2}^-$	$\frac{1}{2}^-$	−0.5567		+0.44	0	^{201}Tl(e.c., 73 h)
^{232}Th	0	49.369	0.345	16.06	5.639	319	1.568	0^+	2^+	0		0		^{236}U(α, $2.4\cdot10^7$ y)
^{231}Pa	0	84.24	41.0	0.079	16.49	1.83	4.751	$\frac{3}{2}^-$	$\frac{5}{2}^-$	+1.98		−1.7	+0.8	^{231}Th(β^-, 25.5 h)
^{234}U	0.057	43.491	0.266	23.65	4.339	780	0.828	0^+	2^+	0		0	−2.8	^{234m}Pa(β^-, 1.17 min),
														^{234}Np(e.c., 4.4 d)
^{236}U	0	45.242	0.235	25.73	4.655	845	0.706	0^+	2^+	0		0	2.96	^{236}Np(e.c., 22 h)
^{238}U	99.27	44.915	0.225	27.07	4.550	660	0.917	0^+	2^+	0	+0.25	+10.5	−3.23	^{242}Pu(α, $3.79\cdot10^5$ y)
^{237}Np	0	59.537	68.3	0.067	8.028	1.12	30.60	$\frac{5}{2}^+$	$\frac{5}{2}^-$	+2.5	+1.3	+4.1	4.1	^{241}Am(α, 458 y),
														^{237}Pu(e.c., 44.63 d)
^{239}Pu	0	57.26	0.101	47.30	7.364	223	0.770	$\frac{1}{2}^+$	$\frac{5}{2}^+$	+0.200		0		^{243}Cm(α, 30 y),
														^{239}Np(β^-, 2.36 d)
^{240}Pu	0	42.824	0.162	39.00	4.100	920	0.724	0^+	2^+		+0.78		−3.3	^{244}Cm(α, 18.1 y),
														^{240}Np(β^-, 63 min)
^{243}Am	0	84.0	2.34	1.392	15.59	0.30	25.9	$\frac{5}{2}^-$	$\frac{5}{2}^+$	+1.58		+4.9	+4.7	^{243}Pu(β^-, 4.96 h),
														^{247}Bk(α, 2.34ns)

38.3 Parameters of Mössbauer Spectra

Positions of the α-Fe spectral lines (at temperatures 298 and 4.2 K, in mm/s) relative to the Pd(^{57}Co) source (at a temperature of 298 K) are presented in Table 38.2. Corresponding values of δ (mm/s) and B_n (T) [10] parameters are illustrated in Figure 38.1 for the Mössbauer absorption spectra of ^{57}Fe nuclei in different materials.

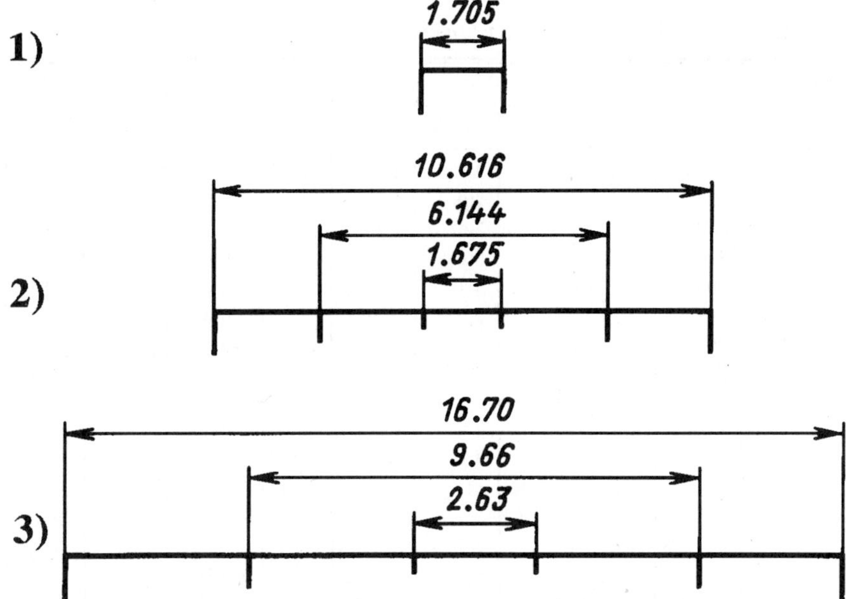

Figure 38.1 Schematics of the Mössbauer absorption spectra for the ^{57}Fe nuclei in different materials at a temperature of 300 K. These materials are used as reference samples for the spectrometer calibration [8]: *1* – sodium nitroprusside $Na_2Fe(CN)_5NO{\cdot}2H_2O$; *2* – α-Fe; *3* – hematite α-Fe_2O_3.

Following parameters are collected in Table 38.3:
formula of the chemical compound (alloy) that contains Mössbauer nuclei in a stable state (absorber);
T_A – temperature of the absorber (if not shown, most probably it was 25°C);
formula of the chemical compound (alloy) that contains radioactive parent nuclei or the chemical symbol of a matrix in which parent isotope atoms were introduced (source);
T_S – temperature of a source;
Γ – width of a separate line in the experimental Mössbauer spectrum;
ε – maximum absorption effect deduced according to the formula $\varepsilon = [N(\infty) - N(v_0)]/N(\infty)$; index 0 appears near the number if the value of ε is given subject to the background;
δ – shift of the observed spectrum; it is counted from the value $v = 0$ and is considered to be positive when a source and an absorber move toward each other (if the sign is not shown it is most probably positive);
Δ – quadrupole splitting;
B_n – magnetic field induction at the atomic nucleus of the Mössbauer isotope in a given substance (absorber);
f_S, f_A – probabilities of the recoil-free emission and absorption of γ-quanta.

Main parameters of the Mössbauer spectra for ^{237}Np nuclei in a number of substances are compiled in Table 38.4 [11].

Table 38.2 α-Fe spectral line positions at helium temperature and at $t = 25°C$ [Cobalt in the matrix of palladium at a temperature of 25°C is chosen as a source.]

T, K	Number of line and its position in the spectrum, mm/s						δ, mm/s	B_n, T
	1	2	3	4	5	6		
298	−5.482	−3.247	−1.013	0.662	2.897	5.134	−0.175	−330.4
4.2	−5.501	−3.210	−0.917	0.801	3.095	5.394	−0.056	−339.0

Table 38.3 Parameters of the Mössbauer spectra

Absorber		Source		Γ,	ε, %	δ,	Δ,	B_n,	f_S	f_A
Formula	T_A, K	Formula	T_S, K	mm/s		mm/s	mm/s	T		
KCI	4,2	KCI(α, p)	4.2	11	16	−0.14				
Fe	298	Pt(Co)	298	0.22	10_0	−0.347	0	33.0		
Fe	300	310(Co)	300	0.39	16_0	+0.090				
Stainless steel 310	300	Cr(Co)	300	0.48	21	+0.05				
α-F_2O_3	300	Cr(Co)	300	0.40		+0.51	0.40	51.7		
$KFeS_2$	298	Pd(Co)	298	0.28	18_0	−0.01	0.50			
$Na_2Fe(CN)_5NO \cdot 2H_2O$	296	Cu(Co)	296			−0.484	1.705			
Ni	80	Ni (Coulomb excitation)	80	3.3	4			9.0	0.09	0.09
ZnO	4.2	MgO(Ga)	4.2	0.006		−0.021				
ZnO	4.2	ZnO	4.2		0.2	0				
Ge	77	Ge (Coulomb excitation)	77	4.4	0.64	+0.11			0.01	
Ru	88	Ru(Rh)	85	0.37	4_0				0.14	
β-Sn	77	SnO_2	77	1.4		+2.55				
Mg_2Sn	300	Mg_2Sn	300	1.0	5					
SnO_2	300	SnO_2	300	1.4	31				0.70	
Te	4.2	Te	4.2	5.2	25		7.8			
TeO_2	300	TeO_2	77	5.2					0.12	0.07
$Na_3H_2{}^{127}IO_6$	20	ZnTe	20	4.1		+1.6				
$^{129}I_2$	100	ZnTe	120	0.87		+0.82				
$Na^{129}I$	77	ZnTe	77			−0.46			0.15	0.29
$^{129}XeF_2$		Nal	4.2	9	2	0	42			
$Na_4^{129}XeO_6$	4.2	Nal	4.2	13	5.4	−0.3				
$^{131}XeF_4$	4.2	$Na_2H_3IO_6$		6.8			6.0			
CsCl	4.2	$BaCl_2 \cdot 2H_2O$	4.2	1.1	0.3_0	−0.01				
Sm_2O_3		Eu_2O_3		4						
^{151}Eu	4.2	Eu_2O_3(Gd)	4.2			−8.1				
$^{151}Eu_2O_3$	77	Eu_2O_3(Gd)	77		25				0.54	0.54
$^{151}EuPd_2Si_2$	292	$^{151}Sm_2O_3$	292	2.41		−8.06	1.23			
$^{155}Gd_2O_3$(60 keV)	80	Sm_2O_3(Eu)	80	4.3						
$^{155}Gd_2O_3$(86 keV)		Sm_2O_3(Eu)	80	1.1	0.56				0.012	0.012
Tb_4O_7	80	Dy_2O_3	80	80	2					
$^{161}Dy_2O_3$	300	Dy_2O_3	300	12					0.10	
$^{161}Dy_2O_3$		Gd_2O_3(Tb)	300	13	6				0.23	
$^{166}Er_2O_3$	20	Ho_2O_3	20	2	21_0	0				
Tm		Er_2O_3	800	30		+5	144	700.0		
Tm_2O_3		Er_2O_3	300	14	4					
$^{170}Yb_2O_3$	20	Tm	20		3.5	+5.7				
$^{171}Yb_2O_3$	4.2	Er_2O_3(Tm)	4.2	11	10		11			
Hf_2O_3		Lu_2O_3		11					0.11	0.04
Ta		W	300	0.26	4.5	+0.9				
TaC	300	W	300	0.06	0.6					

Table 38.3 Parameters of the Mössbauer spectra *(continued)*

Absorber		Source		Γ,	ε, %	δ,	Δ,	B_n,	f_S	f_A
Formula	T_A, K	Formula	T_S, K	mm/s		mm/s	mm/s	T		
^{182}W	22	Ta	22	2.0	44_0	+1.3				
^{183}W	300	Ta	300	40	+0.17					
$^{184}WCl_6$	4.2	W(Re)	4.2			−0.10				
^{186}W	4.2	Re	4.2	4.26						
^{186}Os	10	Re	10	3.3	74					
^{188}Os	15	Re	15	7.5	23_0				0.02	0.11
$Ir(CO)_3Cl_{1.1}$	4.2	^{193}Ir	4.2			−0.04	2.06			
^{191}Ir	91	Os	91	29	1.1				0.07	
^{193}Ir	4.2	Os	4.2			+0.6				
Pt	20	Pt(Au)	20		4					0.13
Pt		Cu(Au)	4.2	20	4					
Au	4.2	Pt	4.2	6	8_0	−1.2			0.34	0.14
$^{199}Hg_2Pt$	4.2	$^{199}Au(Pt)$	4.2	0.37	0.004	+0.1	1.83		0.004	0.004
^{231}Pa	4.2	$^{231}ThO_2$	4.2	4.1	0.73	+0.1	7.8			
$^{232}ThO_2$	25	^{232}Th (Coulomb excitation)	30	9.6		+0.08			0.31	0.35
$^{238}UO_2$	4.2	$^{242}PuO_2$	4.2	45	4	+2		270.0		
$^{239}PuO_2$	4.2	$^{239}NpO_2(ThO_2)$	4.2	5.1	2	0				
$^{240}PuO_2$	4.2	$^{244}CmO_2$	4.2	58	1.2	0.5				
$^{243}AmO_2$	4.2	$^{243}PuO_2$	4.2	22	0.66	0	0	0		
$^{237}NpAl_2$	4.2	$^{241}Am(Th)$	78	1.1	0.04	10				

Table 38.4 ^{237}Np nucleus spectral parameters for some matrices

Absorber	T, K	Source	T, K	δ, mm/s	B_n, T	Δ, mm/s	Γ, mm/s
NpO_2	4.2	$NpO_2(UO_2)$	4.2	0			6
NpO_2	78	$NpO_2(UO_2)$	78	0			3
$NpAl_2$	78	$NpO_2(UO_2)$	78	+5.6	0	0	2
$NpAl_2$	4.2	$NpO_2(UO_2)$	4.2	+6.7	315.0	0.55	
NpO_2	4.2	Am	4.2	+7.5	4.0		4.1
NpO_2	78	Am	4.2		0		3.2
NpO_2	78	Am	78	+7.5	0		2.6
$NpAl_2$	78	Am	4.2		0		3.0
NpO_2	4.2	Th(Am)	4.2	+3.5	4.0		2.0
NpO_2	77	Th(Am)	77	+3.5	0		1.7
Np_3O_8	77	Th(Am)	77	−2.1	0	1.4	
$NpAl_2$	78	Th(Am)	4.2		0		1.1
NpO_2	4.2	Cu(Am)	4.2				10.8
NpO_2	77	$NpO_2(AmO_2)$	77	0			2.5
NpO_2	77	UO_2	77				1.7
NpO_2	77	AmO_2-I	77	+1.1			
NpO_2	77	AmO_2-II	77	+25.2			
$NpCl_4$	4.2	$NpO_2(UO_2)$	4.2	+2.2	260.0	−4.2	
$NpCl_4$	77	$NpO_2(UO_2)$	77	+1.7	0	8.8	
NpO_2	77	Am_2O_3-I	77	+1.8			
NpO_2	77	Am_2O_3-II	77	+28			
NpO_2	4.2	AmF_3	4.2	−44.6			
NpO_2	4.2	$Am_2(C_2O_4)_3 \cdot 6H_2O$	4.2	−40.5			
NpO_2	77	$AmCl_3$	77	−38.5			
NpO_2	77	$AmOCl$	77	−35.6			
NpO_2	4.2	$Am(OH)_4 \cdot H_2O$	4.2	−40.9			

Table 38.4 ^{237}Np nucleus spectral parameters for some matrices *(continued)*

Absorber	T, K	Source	T, K	δ, mm/s	B_n, T	Δ, mm/s	Γ, mm/s
NpO_2	4.2	$K_5[AmO_2(CO_3)_5]_3$	4.2	+25.0			
NpO_2	77	AmSb	4.2	+21.0			
NpO_2	77	Am_3Se_4	4.2	+27		39	
NpO_2	77	Am_3Te_4	4.2	+29		39	
NpO_2	77	Am_2C_3	4.2	−6.5		134	
Np_2C_3	77	Am	4.2	−7.4(NpO_2)	240.0		
α-Np-I	4.2		4.2	+2.0(NpO_2)		29.3	
α-Np-II	4.2		4.2	−0.7(NpO_2)		92.4	
$NpO_2(NO_3)_2 \cdot xH_2O$	4.2	Th(Am)	4.2	−36(NpO_2)	270.0	1.8	
$K_3NpO_2F_5$	4.2	Th(Am)	4.2	−46(NpO_2)	230.0	5.0	
$KNpO_2CO_3$	4.2	Th(Am)	4.2	−12(NpO_2)	640.0	−3.1	
$HNpO_2C_2O_4 \cdot 2H_2O$	4.2	Th(Am)	4.2	−17(NpO_2)	680.0	2.8	
$NpRh_3$	4.2	Th(Am)	4.2	−3.21($NpAl_2$)			1.7
$NpPd_3$	4.2	Th(Am)	4.2	−2.19($NpAl_2$)	335.0		1.7
$NpGe_3$	4.2	Th(Am)	4.2	+0.76($NpAl_2$)			1.7
$NpSn_3$	4.2	Th(Am)	4.2	+1.82($NpAl_2$)	55.0		1.7
NpC	4.2		4.2	−12	479.0	1.1	
NpAs	4.2				470.0		
NpSb	4.2				470.0		
NpP-I	4.2				330.0		
NpP-II	4.2				420.0		
NpN	4.2				270.0		
$RbNpO_4$	4.2	Am	4.2	−52.2		150	3.5
$CsNpO_4$	4.2	Am	4.2	−52.7		138	3.2
Cs_3NpO_5	4.2	Am	4.2	−58.0		81.5	2.7
Li_5NpO_6	4.2	Am	4.2	−63.2		35	3.0
Rb_3NpO_5	4.2	Am	4.2	−59.0		90	2.6
$NpAs_2$	4.2	Am	4.2	+3.0($NpAl_2$)	288.0	23	4.5
NpF_3	4.2	Am	4.2	+41($NpAl_2$)		0.55	
$NpMo_6Se_8$	4.2	Am	4.2	+28($NpAl_2$)			
$NpTe_2$	4.2	Am	4.2	+27.5($NpAl_2$)			
NpS	4.2	Am	4.2	+18.5($NpAl_2$)	170.0		
$NpSb_2$	4.2	Am	4.2	+18($NpAl_2$)			
$NpSe_3$	4.2	Am	4.2	+6($NpAl_2$)			
NpS_3	4.2	Am	4.2	−2($NpAl_2$)			
NpF_4	4.2	Am	4.2	−8($NpAl_2$)			
NpF_5	4.2	Am	4.2	−38($NpAl_2$)			
NpF_6	4.2	Am	4.2	−64($NpAl_2$)			
K_2NpO_4	4.2			−57($NpAl_2$)	109.0	18	
$NpCo_2Si_2$	4.2	Am	4.2	+12.9	287.0		
$[(C_2H_5)_4N]_2NpCl_6$-I	4.2	Th(Am)	4.2	+7.1	110.0	2.7	
$[(C_2H_5)_4N]_2NpCl_6$-II	4.2	Th(Am)	4.2	+7.5	101.0	3.3	
$[(CH_3)_4N]_2NpCl_6$	4.2	Th(Am)	4.2	−6.6	137.0	0.9	
$Cs_2[Np(NO_3)_6]$	4.2	Th(Am)	4.2	−4.0($NpAl_2$)	107.0	−3.3	
$NpRu_2$	4.2		4.2	−17($NpAl_2$)			
$NpIr_2$	4.2			−11($NpAl_2$)	108.0	0	
$NpOs_2$	4.2			−15($NpAl_2$)	76.0		
$NpCo_2$	4.2			−22.5	90.0	−0.4	
$NpNi_2$	4.2			−17.9	230.0	−1	
$NpFe_2$	4.2			−24.0	167.0	−2.8	
$NpMn_2$	4.2			−20.0	40.0		
$NpAl_3$	4.2				263.0		
Np_3S_5-I	4.2	Am	4.2	−5	255.0	2	
Np_3S_5-II	4.2	Am	4.2	+28	260.0	−1	

38.4 The Isomer Shift

Relative isomer shifts δ (in mm/s) for nuclei of the resonance isotope in materials used as reference sources and standard absorbers are presented in Tables 38.5–38.21 [12–13]. The standard absorber employed is shown in parentheses for every isotope.

Table 38.5 Relative isomer shift δ for ^{57}Fe (α-Fe)

Source	Absorber										
	SNP*1	Cr	Stainless steel	SFZ*1	PFZ*1	α-Fe	Rh	Pd	Cu	Pt	α-Fe_2O_3
SNP	0	0.107	0.173	0.188	0.232	0.265	0.354	0.436	0.480	0.612	0.629
Cr	−0.107	0	0.066	0.081	0.125	0.158	0.247	0.329	0.373	0.505	0.522
Stainless steel	−0.173	−0.066	0	0.015	0.059	0.092	0.180	0.263	0.307	0.438	0.456
SFZ	−0.188	−0.081	−0.015	0	0.044	0.076	0.165	0.248	0.292	0.423	0.440
PFZ	−0.232	−0.125	−0.059	−0.044	0	0.033	0.121	0.204	0.248	0.379	0.397
α-Fe	−0.265	−0.158	−0.092	−0.076	−0.033	0	0.089	0.171	0.215	0.347	0.364
Rh	−0.354	−0.247	−0.180	−0.165	−0.121	−0.089	0	0.083	0.127	0.258	0.275
Pd	−0.436	−0.329	−0.263	−0.248	−0.204	−0.171	−0.083	0	0.044	0.175	0.193
Cu	−0.480	−0.373	−0.307	−0.292	−0.248	−0.215	−0.127	−0.044	0	0.131	0.149
Pt	−0.612	−0.505	−0.438	−0.423	−0.379	−0.347	−0.258	−0.175	−0.131	0	0.017
α-Fe_2O_3	−0.629	−0.522	−0.456	−0.440	−0.397	−0.364	−0.275	−0.193	−0.149	−0.017	0

*1 SNP — sodium nitroprusside; SFZ — sodium ferrocianide; PFZ — potassium ferrocianide.

Table 38.6 Relative isomer shift δ for ^{99}Ru (Rh)

Source	Absorber		
	RuO_2	Ru	Ru(Rh)
RuO_2	0	0.25	0.26
Ru	−0.25	0	0.006
Ru(Rh)	−0.26	−0.006	0

Table 38.7 Relative isomer shift δ for ^{121}Sb (InSb)

Source	Absorber			
	β-Sn	InSb	$Ni_{21}Sn_2B_6$	SnO_2
β-Sn	0	2.6	4.3	11.1
InSb	−2.6	0	1.67	8.55
$Ni_{21}Sn_2B_6$	−4.3	−1.67	0	6.88
SnO_2	−11.1	−8.55	−6.88	0

Table 38.8 Relative isomer shift δ for ^{119}Sn (SnO_2)

Source	Absorber								
	SnO_2	Me_2SnF_2	V(Sn)	Pd(Sn)	Pd_3Sn	Mg_2Sn	α-Sn	β-Sn	SnTe
SnO_2	0	1.29	1.58	1.59	1.76	1.86	2.02	2.54	3.46
Me_2SnF_2	−1.29	0	0.29	0.30	0.47	0.57	0.73	1.26	2.17
V(Sn)	−1.58	−0.29	0	0.01	0.18	0.28	0.44	0.96	1.88
Pd(Sn)	−1.59	−0.30	−0.01	0	0.17	0.27	0.43	0.95	1.87
Pd_3Sn	−1.76	−0.47	−0.18	−0.17	0	0.10	0.26	0.78	1.70
Mg_2Sn	−1.86	−0.57	−0.28	−0.27	−0.10	0	0.16	0.682	1.60
α-Sn	−2.02	−0.73	−0.44	−0.43	−0.26	−0.16	0	0.52	1.44
β-Sn	−2.54	−1.26	−0.96	−0.95	−0.78	−0.682	−0.52	0	0.92
SnTe	−3.46	−2.17	−1.88	−1.87	−1.70	−1.60	−1.44	−0.92	0

Note: The isomer shifts for SnO_2, $CaSnO_3$, and $BaSnO_3$ are experimentally identical.

Table 38.9 Relative isomer shift δ for ^{125}Te (ZnTe)

Source	Absorber					
	β-TeO_3	Cu(Sb)	Cu(I)	ZnTe	SnTe	PbTe
β-TeO_3	0	1.07	1.11	1.14	1.2	1.31
Cu(Sb)	−1.07	0	0.04	0.07	0.13	0.24
Cu(I)	−1.11	−0.04	0	0.03	0.09	0.19
ZnTe	−1.14	0.07	−0.03	0	0.06	0.17
SnTe	−1.2	−0.13	−0.09	−0.06	0	0.11
PbTe	−1.31	−0.24	−0.19	−0.17	−0.11	0

Table 38.10 Relative isomer shift δ for ^{127}I (CuI)

Source	Absorber				
	ZnTe	CsI	NaI	KI	CuI
ZnTe	0	0.12	0.14	0.16	0.28
CsI	−0.12	0	0.02	0.04	0.16
NaI	−0.14	−0.02	0	0.02	0.14
KI	−0.16	−0.04	−0.02	0	0.13
CuI	−0.28	−0.16	−0.14	−0.13	0

Table 38.11 Relative isomer shift δ for ^{129}I (CuI)

Source	Absorber				
	NaI	KI	CuI	CsI	ZnTe
NaI		0.01	0.05	0.08	0.45
KI	−0.01	0	0.04	0.07	0.44
CuI	−0.05	−0.04	0	0.03	0.41
CsI	−0.08	−0.07	−0.03	0	0.38
ZnTe	−0.45	−0.44	−0.41	−0.38	0

Table 38.12 Relative isomer shift δ for ^{149}Sm (SmF_3)

Source	Absorber			
	EuF_3	Eu_2O_3	SmF_3	Sm_2O_3
EuF_3	0	0.01	0.01	0.04
Eu_2O_3	−0.01	0	0.00	0.03
SmF_3	−0.01	0.00	0	0.03
Sm_2O_3	−0.04	−0.03	−0.03	0

Table 38.13 Relative isomer shift δ for ^{151}Eu (EuF_3)

Source	Absorber					
	EuS	SmF_3	EuF_3	$EuF_3 \cdot 2H_2O$	Eu_2O_3	Sm_2O_3
EuS	0	11.45	11.50	11.53	12.52	12.56
SmF_3	−11.50	−0.05	0	0.038	1.024	1.06
EuF_3	−11.45	0	0.05	0.09	1.07	1.11
$EuF_3 \cdot 2H_2O$	−11.53	−0.09	−0.038	0	0.98	1.02
Eu_2O_3	−12.52	−1.07	−1.024	−0.98	0	0.04
Sm_2O_3	−12.56	−1.11	−1.06	−1.02	−0.04	0

Table 38.14 Relative isomer shift δ for ^{153}Eu (EuF_3)

Source	Absorber			
	Eu_2O_3	Sm_2O_3	EuF_3	EuS
Eu_2O_3	0	0.25	1.2	15.2
Sm_2O_3	−0.25	0	0.9	15.0
EuF_3	−1.2	−0.9	0	14.0
EuS	−15.2	−15.0	−14.0	0

Table 38.15 Relative isomer shift δ for ^{155}Gd (GdF_3)

Source	Absorber			
	Pd(Eu)	Gd	$SmAl_3$(Eu)	GdF_3
Pd(Eu)	0	0.07	0.53	0.66
Gd	−0.07	0	0.46	0.59
$SmAl_3$(Eu)	−0.53	−0.46	0.0	0.14
GdF_3	−0.66	−0.59	−0.14	0

Table 38.16 Relative isomer shift δ for ^{161}Dy (DyF_3)

Source	Absorber					
	DyF_3	GdF_3(Tb)	Gd_2O_3(Tb)	Dy_2O_3	Gd(Tb)	Dy
DyF_3	0	0.2	0.5	0.67	2.26	2.9
GdF_3(Tb)	−0.2	0	0.3	0.5	2.1	2.7
Gd_2O_3(Tb)	−0.5	−0.3	0	0.1	1.7	2.4
Dy_2O_3	−0.67	−0.5	−0.1	0	1.6	2.21
Gd(Tb)	−2.26	−2.1	−1.7	−1.6	0	0.62
Dy	−2.9	−2.7	−2.4	−2.21	−0.62	0

Table 38.17 Relative isomer shift δ for ^{170}Yb ($YbAl_2$)

Source	Absorber						
	YbB_6	Yb	$YbAl_2$	TmB_{12}	$TmAl_2$	$YbAl_3$	Tm
YbB_6	0	0.18	0.19	0.22	0.26	0.30	0.35
Yb	−0.18	0	0.02	0.04	0.09	0.12	0.17
$YbAl_2$	−0.19	−0.02	0	0.02	0.07	0.10	0.15
TmB_{12}	−0.22	−0.04	−0.02	0	0.05	0.08	0.13
$TmAl_2$	−0.26	−0.09	−0.07	−0.05	0	0.03	0.08
$YbAl_3$	−0.30	−0.12	−0.10	−0.08	−0.03	0	0.05
Tm	−0.35	−0.17	−0.15	−0.13	−0.08	−0.05	0

Table 38.18 Relative isomer shift δ for ^{181}Ta (Ta)

Source	Absorber		
	Mo(W)	W	Ta
Mo(W)	0	21.8	22.6
W	−21.8	0	0.835
Ta	−22.6	−0.835	0

Table 38.19 Relative isomer shift δ for ^{193}Ir (Ir)

Source	Absorber			
	Pt	Ir	Os	V(Os)
Pt	0	0.645	1.183	2.34
Ir	−0.645	0	0.539	1.70
Os	−1.183	−0.539	0	1.16
V(Os)	−2.34	−1.70	−1.16	0

Table 38.20 Relative isomer shift δ for ^{197}Au (Au)

Source	Absorber	
	Au	Pt
Au	0	1.23
Pt	−1.23	0

Table 38.21 Relative isomer shift δ for ^{237}Np ($NpAl_2$)

Source	Absorber			
	Am	Th(Am)	NpO_2	$NpAl_2$
Am	0	4.0	7.5	13.8
Th(Am)	−4.0	0	3.5	9.8
NpO_2	−7,5	−3.5	0	6.3
$NpAl_2$	−13.8	−9.8	−6.3	0

References

[1] Chemical applications of Mössbauer spectroscopy, Eds. Goldanskii, V. I. and Herber, R. H., Academic Press, New York – London, 1968.

[2] Bykov, G. M., Fam Zui Khien, Zhurnal eksperimental'noi i teoreticheskoi fiziki, 43, 909, 1962.

[3] Wertheim, G. K., Mössbauer effect. Principles and applications, Academic Press, New York – London, 1964.

[4] Gibb, T. C., Principles of Mössbauer spectroscopy, Chapman & Hall, London, 1977.

[5] Mössbauer effect methodology, Ed. Gruverman, I. J., Plenum Press, New York, 1965 – 1971 (v. I–IX).

[6] Mössbauer spectroscopy, Eds. Dickson, D. P. E. and Berry, F. J., Cambridge University Press, Cambridge, 1968.

[7] Shirley, D. A., Rev. Mod. Phys., 36, 339, 1964.

[8] Muir, A. H., Ando, K. J., Coogan, H. M., Mössbauer effect data index 1958 – 1965, Interscience Publ., New York – London – Sydney, 1966.

[9] Stevens, J. G., Stevens, V. E., Mössbauer effect data index, covering the 1976 literature, Plenum Press, New York, 1978.

[10] Violett, C. E., Pipkorn, D. N., J. Appl. Phys., 42, 4339, 1971.

[11] Kalvius, G. M., Wagner, F. E., Potzel, W., J. de Phys. Colloq. C6, 97, 657, 1976.

[12] Mössbauer effect reference and data journal, 1978 - 1987.

[13] Stevens, J. G., Gettis, W. L., Isomer shift reference scales, Int. Conf. Applied Mössbauer Effect, Jaipur, India, 1981.

39

Nuclear Reactions

V.P. Rudakov

39.1 Introduction

At present, two classes of microobjects are conventionally distinguished in nuclear physics: atomic nuclei and elementary particles. Atomic nuclei are objects composed of protons and neutrons with mass numbers of two or higher. All other microobjects pertain to elementary particles. If special refinements are not required, then both atomic nuclei and elementary particles are said to be microparticles or, simply, particles.

A nuclear reaction is a process of an elementary particle interaction with a nucleus or of nuclei with each other. A nuclear reaction is usually written down in the form

$$A + a \to B + b \tag{39.1}$$

or, briefly, as

$$A(a, b)B. \tag{39.2}$$

Such a notation indicates that at the initial stage of the reaction (in the initial-state channel) there are two particles A and a whose interaction results in the production of the particles B and b at the final stage (the final-state channel). These particles may be the same ones that interacted in the primary channel (elastic scattering); they may be the same particles as in the initial state but in different intrinsic states (inelastic scattering); and, finally, they may be other particles, and their number may, generally, equal more than two.

Each nuclear reaction is described by two main characteristics: the probability of its occurrence and its energetics, i.e. the amount of absorbed or released energy.

The probability of a nuclear reaction is described by its *cross section.* A square meter or any part of it (for instance, 10^{-30} m^2 = 1 fm^2) can serve as a unit of cross section. Previously, the barn – a unit not belonging to any system – was used for measuring the cross section (1 b = 10^{-28} m^2).

The differential reaction cross section is a quantity proportional to the probability for a given particle (reaction product) to depart from the reaction vertex at a given angle (relative to the movement direction of the particle initiating the reaction) with a given energy. This quantity is denoted as $d\sigma/d\Omega$ and is expressed in units of m^2/sr. The integral of the differential cross section over the total solid angle yields the *total cross section.*

The cross section is ultimately determined by the properties of forces responsible for the interaction between the particles and at present can be computed exactly only for elastic scattering in a Coulomb field (Rutherford scattering). The cross sections of all other reactions are computed within the frames of various models and are actually only estimates.

0-8493-2861-6/97/$0.00+$.50

The bibliography relevant to studies of specific nuclear reactions and schemes of nuclear levels are published periodically in the journals *Nuclear Data Sheets* and *Nuclear Physics*.

39.2 The Rutherford Scattering Cross Section

The center-of-mass differential Rutherford cross section $d\sigma/d\Omega$ (in $\mathrm{fm}^2/\mathrm{sr}$) for scattering of a non-relativistic particle of mass m, and charge ze on a nucleus of mass M and charge Ze is calculated by the formula

$$(d\sigma/d\Omega)_{\mathrm{CMS}} = 0.1296 \left(\frac{zZ}{E}\right)^2 \left(\frac{m+M}{M}\right)^2 \csc^4(\theta/2),\ \mathrm{fm}^2/\mathrm{sr}, \tag{39.3}$$

where E (in MeV) is the energy of the incident particle in the laboratory reference system (LS); θ is the scattering angle in the center-of-mass system (CMS).

The same cross section will be obtained in the LS, if $\csc^4(\psi/2) - 2(m/M)^2 + \ldots$ is substituted for $\csc^4(\theta/2)$, where the angle ψ is read in the LS. The next term in the expansion will be on the order of $(m/M)^4$.

39.3 The Energy of a Nuclear Reaction

The energetics relationships for a nuclear reaction are defined by the energy and momentum conservation laws.

When a reaction $A(a,b)B$ occurs, the quantity

$$Q = [(M_A + Ma) - (M_B + M_b)]c^2, \tag{39.4}$$

where M_i are the masses of the particles participating in the reaction and c is the speed of light, is said to be the *reaction energy*.

Usually *mass defects* ΔM are used, instead of masses, for computing Q:

$$Q = (\Delta M_A + \Delta M_a) - (\Delta M_B + \Delta M_b). \tag{39.5}$$

The mass defect is the quantity $\Delta M = M - A$ where M is the actual mass of the particle (atom), A is the so-called mass number, i.e. the total number of nucleons (protons and neutrons) in the atomic nucleus. If M is expressed in atomic mass units (a.m.u.) and A is assigned the same unit, then ΔM is also expressed in a.m.u. One a.m.u. represent 1/12 of the ^{12}C nuclide mass and equals $1.6605655 \cdot 10^{-27}$ kg. For calculations of reaction energies it is more convenient to express ΔM in kilo-electronvolts: a.m.u. = 931501.59 keV.

Employing the mass defects, one can handle numbers that are many times smaller than the nuclear masses or the binding energies.

Mass defects are presented in Table 39.1.

Table 39.1 Mass defects $\Delta M = M - A$ (in keV) [1] [N is the number of neutrons; Z is the number of protons; $A = N + Z$ is the mass number; s indicates values resulting from interpolation or extrapolation based on available data.]

N	Z	A	Element	Mass excess, keV	N	Z	A	Element	Mass excess, keV
1	0	1	n	8071.431 (39)	10	5	15	B	29530 (s)
0	1		H	7289.034 (23)	9	6		C	9873.2 (8)
1	1	2	H	13135.84 (4)	8	7		N	101.514 (36)
2	1	3	H	14949.94 (5)	7	8		O	2855.4 (7)
1	2		He	14931.32 (5)	6	9		F	17660 (s)
3	1	4	H	25920 (500)	11	5	16	B	38000 (s)
2	2		H	2424.94 (4)	10	6		C	13693 (16)
1	3		Li	25130 (300)	9	7		N	5681.6 (23)
4	1	5	H	33790 (800)	8	8		O	−4737.02 (4)
3	2		He	11390 (50)	7	9		F	10692 (14)
2	3		Li	11680 (50)	6	10		Ne	24110 (140)
4	2	6	He	17597.0 (35)	12	5	17	B	45270 (s)
3	3		Li	14087.3 (8)	11	6		C	21060 (s)
2	4		Be	18375 (6)	10	7		N	7870 (15)
5	2	7	He	26111 (30)	9	8		O	−809.9 (8)
4	3		Li	14908.2 (9)	8	9		F	1951.66 (18)
3	4		Be	15770.1 (9)	7	10		Ne	16478 (26)
2	5		B	29940 (100)	12	6	18	C	25370 (s)
6	2	8	He	31609 (12)	11	7		N	13274 (30)
5	3		Li	20946.9 (9)	10	8		O	−783.03 (30)
4	4		Be	4941.76 (10)	9	9		F	872.5 (7)
3	5		B	22921.9 (13)	8	10		Ne	5319 (5)
2	6		C	35085 (25)	7	11		Na	25320 (s)
6	3	9	Li	24954.8 (20)	13	6	19	C	34430 (s)
5	4		Be	11348.0 (4)	12	7		N	15600 (300)
4	5		B	12416.1 (10)	11	8		O	3331.4 (27)
3	6		C	28912.1 (39)	10	9		F	−1487.33 (13)
7	3	10	Li	33830 (250)	9	10		Ne	1750.9 (6)
6	4		Be	12607.6 (6)	8	11		Na	12930 (12)
5	5		B	12051.7 (5)	13	7	20	N	22200 (s)
4	6		C	15702.9 (7)	12	8		O	3799 (8)
3	7		N	39500 (s)	11	9		F	−17.1 (6)
8	3	11	Li	40940 (120)	10	10		Ne	−7043.0 (5)
7	4		Be	20176 (6)	9	11		Na	6844 (7)
6	5		B	8667.9 (4)	8	12		Mg	17568 (27)
5	6		C	10650.0 (11)	14	7	21	N	26950 (s)
4	7		N	25230 (100)	13	8		O	8120 (80)
8	4	12	Be	25030 (40)	12	9		F	−47 (7)
7	5		B	13369.5 (13)	11	10		Ne	−5733.1 (11)
6	6		C	0.0 (0.0)	10	11		Na	−2185.8 (7)
5	7		N	17338 (1)	9	12		Mg	10912 (16)
4	8		O	32070 (260)	14	8	22	O	9490 (220)
9	4	13	Be	34900 (s)	13	9		F	2826 (30)
8	5		B	16562 (4)	12	10		Ne	−8026.1 (5)
7	6		C	3125.038 (18)	11	11		Na	−5184.0 (7)
6	7		N	5345.6 (9)	10	12		Mg	−394.1 (19)
5	8		O	23105 (10)	9	13		Al	18210 (s)
10	4	14	Be	40970 (s)	15	8	23	O	17950 (s)
9	5		B	23657 (30)	14	9		F	3350 (170)
8	6		C	3019.922 (24)	13	10		Ne	−5155.1 (21)
7	7		N	9863.444 (23)	12	11		Na	−9529.6 (8)
6	8		O	8008.3 (5)	11	12		Mg	−5470.6 (15)
5	9		F	33610 (s)	10	13		Al	6768 (25)

Table 39.1 Mass defects $\Delta M = M - A$ (in keV) [1] *(continued)*

N	Z	A	Element	Mass excess, keV	N	Z	A	Element	Mass excess, keV
15	9	24	F	8650 (s)	17	15		P	−24304.7 (8)
14	10		Ne	−5949 (10)	16	16		S	−26015.1 (6)
13	11		Na	−8417.5 (8)	15	17		Cl	−13329 (8)
12	12		Mg	−13930.6 (7)	14	18		Ar	−2210 (130)
11	13		Al	−52 (4)	21	12	33	Mg	4130 (s)
10	14		Si	10740 (120)	20	13		Al	−9370 (s)
16	9	25	F	12840 (s)	19	14		Si	−20570 (50)
15	10		Ne	−2150 (90)	18	15		P	−26336.9 (21)
14	11		Na	−9357 (7)	17	15		S	−26585.9 (8)
13	12		Mg	−13190.8 (11)	16	17		Cl	−21003.0 (9)
12	13		Al	−8912.9 (11)	15	18		Ar	−9385 (30)
11	14		Si	3824 (10)	21	13	34	Al	−4150 (s)
16	10	26	Ne	−190 (s)	20	14		Si	−20250 (s)
15	11		Na	−6888 (23)	19	15		P	−24550 (50)
14	12		Mg	−16212.4 (9)	18	16		S	−29931.25 (28)
13	13		Al	−12207.6 (10)	17	17		Cl	−24438.3 (8)
12	14		Si	−7143.1 (31)	16	18		Ar	−18379.2 (30)
11	15		P	11260 (s)	15	19		K	−1480 (s)
17	10	27	Ne	6670 (s)	22	13	35	Al	−840 (s)
16	11		Na	−5630 (80)	21	14		Si	−15040 (s)
15	12		Mg	−14585.0 (14)	20	15		P	−24940 (80)
14	13		Al	−17194.3 (7)	19	16		S	−28846.27 (21)
13	14		Si	−12385.3 (15)	18	17		Cl	−29013.73 (10)
12	15		P	−590 (s)	17	18		Ac	−23048.9 (16)
17	11	28	Na	−1130 (120)	16	19		K	−11169 (20)
16	12		Mg	−15016.4 (22)	22	14	36	Si	−12670 (s)
15	13		Al	−16848.2 (8)	21	15		P	−20770 (s)
14	14		Si	−21491.2 (6)	20	16		S	−30665.9 (15)
13	15		P	−7159.5 (38)	19	17		Cl	−29521.77 (21)
12	16		S	4190 (120)	18	18		Ar	−30231.32 (28)
18	11	29	Na	2660 (150)	17	19		K	−17426 (8)
17	12		Mg	−10750 (50)	16	20		Ca	−6650 (270)
16	13		Al	−18212 (5)	23	14	37	Si	−7010 (s)
15	14		Si	−21893.7 (8)	22	15		P	−19010 (s)
14	15		P	−16949.3 (18)	21	16		S	−26908 (30)
13	16		S	-3160 (50)	20	17		Cl	-31761.76 (13)
19	11	30	Na	8380 (300)	19	18		Ar	−30947.9 (6)
18	12		Mg	−9790 (s)	18	19		K	−24799.4 (14)
17	13		Al	−15890 (40)	17	20		Ca	−13164 (39)
16	14		Si	−24431.7 (9)	23	15	38	P	−14560 (s)
15	15		P	−20204.5 (27)	22	16		S	−26862 (12)
14	16		S	−14062.0 (31)	21	17		Cl	−29798.0 (4)
13	17		Cl	4840 (s)	20	18		Ar	−34715.0 (8)
20	11	31	Na	10610 (s)	19	19		K	−28802.0 (16)
19	12		Mg	−3900 (s)	18	20		Ca	−22060 (9)
18	13		Al	−15100 (100)	17	21		Sc	−4460 (s)
17	14		Si	−22948.7 (10)	24	15	39	P	−12300 (s)
16	15		P	24439.5 (6)	23	16		S	−23000 (s)
15	16		S	−19044.1 (16)	22	17		Cl	−29803 (18)
14	17		Cl	−7070 (50)	21	18		Ar	−33241 (5)
21	11	32	Na	16410 (s)	20	19		K	−33806.2 (8)
20	12		Mg	−2890 (s)	19	20		Ca	-27282 (5)
19	13		Al	−11290 (s)	18	21		Sc	−14080 (s)
18	14		Si	−24092 (7)	24	16	40	S	−22240 (s)

Table 39.1 Mass defects $\Delta M = M - A$ (in keV) [1] *(continued)*

N	Z	A	Element	Mass excess, keV	N	Z	A	Element	Mass excess, keV
23	17		Cl	−27540 (500)	23	24		Cr	−34618 (25)
22	18		Ar	−35040.2 (7)	22	25		Mn	−22920 (s)
21	19		K	−33535.2 (8)	29	29	48	K	−32220 (500)
20	20		Ca	−34846.8 (8)	28	20		Ca	−44216 (4)
19	21		Sc	−20527 (4)	27	21		Sc	−44498 (6)
18	22		Ti	−9040 (230)	26	22		Ti	−48487.7 (14)
25	16	41	S	−18100 (s)	25	23		V	−44472.8 (31)
24	17		Cl	−27400 (160)	24	24		Cr	−42818 (7)
23	18		Ar	−33067.7 (9)	23	25		Mn	−29170 (s)
22	19		K	−35559.7 (9)	29	20	49	Ca	−41286 (5)
21	20		Ca	−35138.5 (9)	28	21		Sc	−46555 (4)
20	21		Sc	−28643.5 (16)	27	22		Ti	−48558.7 (14)
19	22		Ti	−15780 (40)	26	23		V	−47956.9 (16)
25	17	42	Cl	−24420 (s)	25	24		Cr	−45329.0 (26)
24	18		Ar	−34420 (40)	24	25		Mn	−37613 (24)
23	19		K	−35022.8 (13)	23	26		Fe	−24470 (160)
22	20		Ca	−35543.9 (14)	30	20	50	Ca	−39572 (8)
21	21		Sc	−32120.7 (16)	29	21		Sc	−44539 (16)
20	22		Ti	−25122 (6)	28	22		Ti	−51432.1 (26)
19	23		V	−8020 (s)	27	23		V	−49219.3 (16)
26	17	43	Cl	−23140 (60)	26	24		Cr	−50258.0 (14)
25	18		Ar	−31980 (70)	25	25		Mn	−42625.7 (17)
24	19		K	−36588 (10)	24	26		Fe	−34430 (s)
23	20		Ca	−38405.4 (14)	30	21	51	Sc	−43220 (20)
22	21		Sc	−36185.0 (24)	29	22		Ti	−49733.0 (28)
21	22		Ti	−29324 (7)	28	23		V	−52199.1 (14)
20	23		V	−18020 (s)	27	24		Cr	−51447.8 (14)
26	18	44	Ar	−32271 (20)	26	25		Mn	−48239.8 (14)
25	19		K	−35807 (39)	25	26		Fe	−40228 (17)
24	20		Ca	−41466.0 (14)	24	27		Co	−27230 (s)
23	21		Sc	37810.7 (24)	31	21	52	Sc	−40140 (s)
22	22		Ti	−37546.2 (31)	20	22		Ti	−49469 (10)
21	23		V	−23850 (s)	29	23		V	−51438.9 (17)
20	24		Cr	−13500 (130)	28	24		Cr	−55415.3 (15)
27	18	45	Ar	−29730 (60)	27	25		Mn	−50704.2 (24)
26	19		K	−36611 (11)	26	26		Fe	−48332 (12)
25	20		Ca	−40809.6 (15)	25	27		Co	−34230 (s)
24	21		Sc	−41066.5 (14)	31	22	53	Ti	−46890 (100)
23	22		Ti	−39004.0 (28)	30	23		V	−51863 (25)
22	23		V	−31879 (27)	29	24		Cr	−55283.7 (16)
21	24		Cr	−19460 (150)	28	25		Mn	−54687.4 (15)
28	18	46	Ar	−29730 (70)	27	26		Fe	−50944.2 (22)
27	19		K	−35420 (16)	26	27		Co	−42640 (18)
26	20		Ca	−43138.2 (37)	25	28		Ni	−29410 (180)
25	21		Sc	−41755.6 (15)	32	22	54	Ti	−45330 (s)
24	22		Ti	−44122.7 (14)	31	23		V	−49930 (100)
23	23		V	−37070.8 (18)	30	24		Cr	−56931.3 (16)
22	24		Cr	−29461 (30)	29	25		Mn	−55554.3 (18)
21	25		Mn	−12470 (s)	28	26		Fe	56251.4 (14)
28	19	47	K	−35698 (8)	27	27		Co	−48009.6 (19)
27	20		Ca	−42342.9 (37)	26	28		Ni	−39210 (s)
26	21		Sc	−44330.5 (24)	32	23	55	V	−49010 (s)
25	22		Ti	−44931.0 (15)	31	24		Cr	−55106.3 (17)
24	23		V	−42001.1 (17)	30	25		Mn	−57710.0 (16)

Table 39.1 Mass defects $\Delta M = M - A$ (in keV) [1] *(continued)*

N	Z	A	Element	Mass excess, keV	N	Z	A	Element	Mass excess, keV
29	26		Fe	−57478.6 (15)	32	32		Ge	−47550 (s)
28	27		Co	−54023.9 (16)	37	27	64	Co	−59791 (20)
27	28		Ni	−45334 (11)	36	28		Ni	−67097.9 (16)
26	29		Cu	−31530 (s)	35	29		Cu	−65423.0 (17)
33	23	56	V	−46210 (s)	34	30		Zn	−66001.2 (19)
32	24		Cr	−55265 (30)	33	31		Ga	−58836 (8)
31	25		Mn	−56908.8 (17)	32	32		Ge	−54430 (250)
30	26		Pe	−60604.1 (14)	37	28	65	Ni	−65124.5 (17)
29	27		Co	−56036.7 (24)	36	29		Cu	−67261.5 (20)
28	28		Ni	−53902 (11)	35	30		Zn	−65909.6 (20)
27	29		Cu	−38500 (s)	34	31		Ga	−62653.8 (22)
33	24	57	Cr	−52790 (s)	33	32		Ge	−56410 (100)
32	25		Mn	−57487 (8)	32	33		As	−47150 (s)
31	26		Fe	−60179.0 (14)	38	28	66	Ni	−66021 (19)
30	27		Co	−59342.4 (15)	37	29		Cu	−66256.7 (21)
29	28		Ni	−56099 (7)	36	30		Zn	−68898.3 (16)
28	29		Cu	−47620 (s)	35	31		Ga	−63723.3 (34)
27	30		Zn	−32630 (130)	34	32		Ge	−61621 (13)
34	24	58	Cr	−52050 (s)	33	33		As	−51520 (s)
33	25		Mn	−56210 (100)	39	28	67	Ni	−63470 (90)
32	26		Fe	−62151.8 (16)	38	29		Cu	−67305 (8)
31	27		Co	−59844.0 (18)	37	30		Zn	−67879.6 (16)
30	28		Ni	−60224.3 (15)	36	31		Ga	−66878.5 (18)
29	29		Cu	−51661.7 (28)	35	32		Ge	−62450 (50)
28	30		Zn	−42260 (s)	34	33		As	−56650 (s)
34	25	59	Mn	−55478 (30)	33	34		Se	−47080 (s)
33	26		Fe	−60661.4 (26)	39	29	68	Cu	−65390 (50)
32	27		Co	−62226.4 (15)	38	30		Zn	−70006.3 (17)
31	28		Ni	−61152.9 (15)	37	31		Ga	−67085.2 (21)
30	29		Cu	−56352.2 (21)	36	32		Ge	−66972 (12)
29	30		Zn	−47590 (s)	35	33		As	−58770 (s)
35	25	60	Mn	−52950 (s)	34	34		Se	−54170 (s)
34	26		Fe	−61437 (10)	40	29	69	Cu	−65940 (70)
33	27		Co	−61646.6 (15)	39	30		Zn	−68417.0 (18)
32	28		Ni	−64470.2 (15)	38	31		Ga	−69321.5 (31)
31	29		Cu	−58343.3 (25)	37	32		Ge	−67096 (4)
30	30		Zn	−54184 (11)	36	33		As	−63120 (40)
35	26	61	Fe	−59010 (70)	35	34		Se	−56300 (40)
34	27		Co	−62897.0 (17)	34	35		Br	−45620 (s)
33	28		Ni	−64219.1 (15)	41	29	70	Cu	−63390 (110)
32	29		Cu	−61980.7 (20)	40	30		Zn	−69559.9 (34)
31	30		Zn	−56580 (200)	39	31		Ga	−68905.2 (32)
30	31		Ga	−47750 (s)	38	32		Ge	−70561.4 (18)
36	26	62	Fe	−58930 (50)	37	33		As	−64339 (20)
35	27		Co	−61504 (14)	36	34		Se	−61590 (200)
34	28		Ni	−66745.4 (15)	35	35		Br	−51140 (s)
33	29		Cu	−62796 (5)	41	30	71	Zn	−67324 (10)
32	30		Zn	−61169 (10)	40	31		Ga	−70141.5 (26)
31	31		Ga	−51770 (s)	39	32		Ge	−69905.8 (21)
37	27	63	Co	−61850 (19)	38	33		As	−67893 (4)
36	28		Ni	−65512.6 (115)	37	34		Se	−63090 (s)
35	29		Cu	−65578.5 (15)	36	35		Br	−56490 (s)
34	30		Zn	−62211.1 (22)	35	36		Kr	−46500 (s)
33	31		Ga	−56690 (100)	42	30	72	Zn	−68134 (6)

Table 39.1 Mass defects $\Delta M = M - A$ (in keV) [1] *(continued)*

N	Z	A	Element	Mass excess, keV	N	Z	A	Element	Mass excess, keV
41	31		Ga	−68591.0 (28)	48	31	79	Ga	−62810 (170)
40	32		Ge	−72582.6 (20)	47	32		Ge	−69570 (150)
39	33		As	−68232 (7)	46	33		As	−73720 (50)
38	34		Se	−67894 (12)	45	34		Se	−75920.6 (39)
37	35		Br	−58930 (s)	44	35		Br	−76070.0 (36)
36	36		Kr	−53870 (s)	43	36		Kr	−74439 (9)
43	30	73	Zn	−65030 (200)	42	37		Rb	−70860 (110)
42	31		Ga	−69730 (40)	41	38		Sr	−65460 (s)
41	32		Ge	−71293.5 (19)	49	31	80	Ga	−59530 (s)
40	33		As	−70949 (4)	48	32		Ge	−69430 (310)
39	34		Se	−68209 (11)	47	33		As	−72060 (300)
38	35		Br	−63670 (220)	46	34		Se	−77761.3 (35)
37	36		Kr	−56980 (140)	45	35		Br	−75891.0 (36)
44	30	74	Zn	−65670 (140)	44	36		Kr	−77897 (11)
43	31		Ga	−68020 (100)	43	37		Rb	−72190 (23)
32	32		Ge	−73422.1 (19)	42	38		Sr	−70390 (s)
41	33		As	−70859.7 (25)	49	32	81	Ge	−66340 (s)
40	34		Se	−72212.7 (26)	48	33		As	−72640 (100)
39	35		Br	−65295 (15)	47	34		Se	−76391.0 (37)
38	36		Kr	−62020 (100)	46	35		Br	−77976 (6)
45	30	75	Zn	−62460 (s)	45	36		Kr	−77707 (18)
44	31		Ga	−68560 (200)	44	37		Rb	−75445 (35)
43	32		Ge	−71856.1 (24)	43	38		Sr	−71460 (50)
42	33		As	−73033.9 (23)	50	32	82	Ge	−65790 (s)
41	34		Se	−72169.0 (24)	49	33		As	−70190 (s)
40	35		Br	−69159 (20)	48	34		Se	−77586 (10)
39	36		Kr	−64160 (s)	47	35		Br	−77498 (6)
38	37		Rb	−57510 (600)	46	36		Kr	−80591 (6)
46	30	76	Zn	−62550 (170)	45	37		Rb	−76213 (20)
45	31		Ga	−66440 (150)	44	38		Sr	−75999 (9)
44	32		Ge	−73213.5 (25)	43	39		Y	−67910 (s)
43	33		As	−72290.6 (23)	50	33	83	As	−69950 (220)
42	34		Se	−75259.2 (25)	49	34		Se	−75410 (32)
41	35		Br	−73103 (15)	48	35		Br	−79025 (15)
40	36		Kr	−69100 (200)	47	36		Kr	−79984.6 (39)
39	37		Rb	−60610 (270)	46	37		Rb	−78987 (32)
47	30	77	Zn	−58910 (s)	45	38		Sr	−76737 (30)
46	31		Ga	−66410 (s)	44	39		V	−72440 (s)
45	32		Ge	−71214.3 (26)	51	33	84	As	−66160 (s)
44	33		As	−73915.7 (37)	50	34		Se	−75942 (18)
43	34		Se	−74606.1 (25)	49	35		Br	−77759 (26)
42	35		Br	−73241.5 (38)	48	36		Kr	−82431.9 (36)
41	36		Kr	−70236 (30)	47	37		Rb	−79752 (4)
40	37		Rb	−65110 (120)	46	38		Sr	−80641 (4)
38	38		Sr	−57960 (280)	45	39		Y	−73692 (30)
48	30	78	Zn	−58080 (320)	44	40		Zr	−71440 (s)
47	31		Ga	−63680 (200)	51	34	85	Se	−72570 (s)
46	32		Ge	−71760 (70)	50	35		Br	−78670 (100)
45	33		As	−72740 (70)	49	36		Kr	−81471.8 (37)
44	34		Se	−77031.5 (26)	48	37		Rb	−82158.8 (33)
43	35		Br	−73458 (5)	47	38		Sr	−81095 (7)
42	36		Kr	−74150 (8)	46	39		Y	−77835 (12)
41	37		Rb	−67090 (180)	45	40		Zr	−73130 (s)
40	38		Sr	−63850 (s)	52	34	86	Se	−70860 (s)

Table 39.1 Mass defects $\Delta M = M - A$ (in keV) [1] *(continued)*

N	Z	A	Element	Mass excess, keV	N	Z	A	Element	Mass excess, keV
51	35		Br	−75960 (400)	57	36	93	Kr	−64920 (s)
50	36		Kr	−83263 (5)	56	37		Rb	−72920 (170)
49	37		Rb	−82737.7 (32)	55	38		Sr	−80280 (150)
48	38		Sr	−84512.1 (28)	54	39		Y	−84227 (20)
47	39		Y	−79239 (10)	53	40		Zr	−87116.7 (28)
46	40		Zr	−77940 (s)	52	41		Nb	−87209.0 (29)
45	41		Nb	−69340 (s)	51	42		Mo	−86803 (4)
46	34		Se	−77761.3 (35)	50	43		Tc	−83610 (5)
52	35	87	Br	−74210 (s)	49	44		Ru	−77310 (s)
51	36		Kr	−80707 (5)	57	37	94	Rb	−69460 (s)
50	37		Rb	−84595.7 (30)	56	38		Sr	−78960 (70)
49	38		Sr	−84868.9 (27)	55	39		Y	−82382 (12)
48	39		Y	−83007.2 (30)	54	40		Zr	−87263.9 (31)
47	40		Zr	−79430 (80)	53	41		Nb	−86367.1 (30)
46	41		Nb	−74430 (s)	52	42		Mo	−88412.3 (34)
53	35	88	BBr	−71090 (s)	51	43		Tc	−84156 (6)
52	36		Kr	−79686 (14)	50	44		Ru	−82571 (13)
51	37		Rb	−82602 (12)	58	37	95	Rb	−66550 (310)
50	38		Sr	−87910.6 (27)	57	38		Sr	−75140 (90)
49	39		Y	−84298 (4)	56	39		Y	−81233 (20)
48	40		Zr	−83621 (10)	55	40		Zr	−85663.4 (34)
47	41		Nb	−76420 (s)	54	41		Nb	−86786.5 (25)
46	42		Mo	−72920 (s)	53	42		Mo	−87712.1 (24)
53	36	89	Kr	−76790 (60)	52	43		Tc	−86013 (8)
52	37		Rb	−81717 (13)	51	44		Ru	−83452 (12)
51	38		Sr	−86203 (4)	50	45		Rh	−78340 (150)
50	39		Y	−87695.3 (30)	59	37	96	Rb	−62770 (s)
49	40		Zr	−84859.5 (32)	58	38		Sr	−73070 (140)
48	41		Nb	−80621 (19)	57	39		Y	−78430 (100)
47	42		Mo	−75220 (s)	56	40		Zr	−85444.7 (37)
54	36	90	Kr	−75180 (70)	55	41		Nb	−85608 (5)
53	37		Rb	−79570 (60)	54	42		Mo	−88794.9 (24)
52	38		Sr	−85934.7 (38)	53	43		Tc	−85821 (6)
51	39		Y	−86480.7 (33)	52	44		Ru	−86075 (9)
50	40		Zr	−88764.6 (29)	51	45		Rh	−79633 (13)
49	41		Nb	−82654 (5)	59	38	97	Sr	−69080 (s)
48	42		Mo	−80167 (6)	58	39		Y	−76280 (130)
47	43		Tc	−70970 (s)	57	40		Zr	−82954.2 (37)
55	36	91	Kr	−71770 (110)	56	41		Nb	−85611.6 (31)
54	37		Rb	−77970 (40)	55	42		Mo	−85544.5 (24)
53	38		Sr	−83666 (5)	54	43		Tc	−87224 (5)
52	39		Y	−86349.5 (35)	53	44		Ru	−86070 (100)
51	40		Zr	−87892.5 (29)	52	45		Rh	−82560 (100)
50	41		Nb	−86636.9 (38)	51	46		Pd	−77760 (s)
49	42		Mo	−82199 (12)	60	38	98	Sr	−67380 (s)
48	43		Tc	−75980 (200)	59	39		Y	−73190 (S)
56	36	92	Kr	−69150 (220)	58	40		Zr	−81291 (20)
55	37		Rb	−75120 (200)	57	41		Nb	−83530 (6)
54	38		Sr	−82892 (34)	56	42		Mo	−88115.4 (24)
53	39		Y	−84822 (16)	55	43		Tc	−86434 (6)
52	40		Zr	−88456.1 (28)	54	44		Ru	−88226 (6)
51	41		Nb	−86448.1 (34)	53	45		Rh	−83168 (12)
50	42		Mo	−86807 (4)	52	46		Pd	−81270 (s)
49	43		Tc	−78936 (26)	60	39	99	Y	−71500 (220)

Table 39.1 Mass defects $\Delta M = M - A$ (in keV) [1] *(continued)*

N	Z	A	Element	Mass excess, keV	N	Z	A	Element	Mass excess, keV
59	40		Zr	−77890 (100)	61	44		Ru	−85938 (6)
58	41		Nb	−82346 (16)	60	45		Rh	−87855 (6)
57	42		Mo	−85969.5 (24)	59	46		Pd	−88422 (5)
56	43		Tc	−87326.2 (25)	58	47		Ag	−87075 (10)
55	44		Ru	−87619.8 (26)	57	48		Cd	−84336 (11)
54	45		Rh	−85517 (10)	56	49		In	−79340 (s)
53	46		Pd	−82112 (23)	55	50		Sn	−73090 (s)
52	47		Ag	−76510 (S)	63	43	106	Tc	−80030 (s)
60	40	100	Zr	−76600 (200)	62	44		Ru	−86333 (10)
59	41		Nb	−79960 (130)	61	45		Rh	−86372 (10)
58	42		Mo	−86189 (6)	60	46		Pd	−89913 (5)
57	43		Tc	−86018.8 (27)	59	47		Ag	−86929 (6)
56	44		Ru	−89221.6 (26)	58	48		Cd	−87131 (6)
55	45		Rh	−85592 (20)	57	49		In	−80586 (31)
54	46		Pd	−85230 (15)	56	50		Sn	−76990 (s)
53	47		Ag	−77930 (400)	55	51		Sb	−66190 (s)
61	40	101	Zr	−73050 (s)	64	43	107	Tc	−79510 (s)
60	41		Nb	−78950 (100)	63	44		Ru	−83710 (300)
59	42		Mo	−83516 (6)	62	45		Rh	−86860 (40)
58	43		Tc	−86327 (24)	61	46		Pd	−88371 (6)
57	44		Ru	−87951.6 (29)	60	47		Ag	−88404 (6)
56	45		Rh	−87410 (18)	59	48		Cd	−86987 (7)
55	46		Pd	−85428 (18)	58	49		In	−83500 (150)
54	47		Ag	−81330 (s)	57	50		Sn	−78400 (s)
53	48		Cd	−75530 (s)	56	51		Sb	−70400 (s)
61	41	102	Nb	−76360 (s)	64	44	108	Ru	−83820 (600)
60	42		Mo	−83562 (21)	63	45		Rh	−85020 (600)
59	43		Tc	−84600 (s)	62	46		Pd	−89523 (5)
58	44		Ru	−89100.5 (29)	61	47		Ag	−87602 (6)
57	45		Rh	−86777 (7)	60	48		Cd	−89251 (6)
56	46		Pd	−87925 (9)	59	49		In	−84100 (80)
55	47		Ag	−82330 (50)	58	50		Sn	−81900 (s)
54	48		Cd	−79430 (s)	57	51		Sb	−72400 (s)
53	49		In	−70130 (s)	56	52		Te	−65320 (s)
62	41	103	Nb	−75410 (s)	65	44	109	Ru	−80810 (s)
61	42		Mo	−80610 (s)	64	45		Rh	−85110 (s)
60	43		Tc	−84910 (100)	63	46		Pd	−87606 (5)
59	44		Ru	−87261.4 (30)	62	47		Ag	−88722 (4)
58	45		Rh	−88024 (4)	61	48		Cd	−88540 (5)
57	46		Pd	−87478 (9)	60	49		In	−86524 (10)
56	47		Ag	−84800 (50)	59	50		Sn	−82620 (s)
55	48		Cd	−80600 (140)	58	51		Sb	−76120 (s)
54	49		In	−74100 (s)	57	52		Te	−67470 (s)
62	42	104	Mo	−80500 (s)	65	45	110	Rh	−82930 (100)
61	43		Tc	−82700 (s)	64	46		Pd	−88335 (20)
60	44		Ru	−88099 (6)	63	47		Ag	−87456 (4)
59	45		Rh	−86952 (4)	62	48		Cd	−90349 (4)
58	46		Pd	−89400 (5)	61	49		In	−86409 (30)
57	47		Ag	−85150 (30)	60	50		Sn	−85834 (16)
56	48		Cd	−83850 (s)	59	51		Sb	−77430 (s)
55	49		In	−75850 (s)	58	52		Te	−71760 (s)
54	50		Sn	−71150 (s)	66	45	111	Rh	−82530 (s)
63	42	105	Mo	−77140 (s)	65	46		Pd	−86030 (50)
62	43		Tc	−82540 (200)	64	47		Ag	−88226 (5)

Table 39.1 Mass defects $\Delta M = M - A$ (in keV) [1] *(continued)*

N	Z	A	Element	Mass excess, keV	N	Z	A	Element	Mass excess, keV
63	48		Cd	−89254 (4)	62	55		Cs	−66850 (s)
62	49		In	−88405 (11)	70	48	118	Cd	−86707 (20)
61	50		Sn	−85941 (8)	69	49		In	−87450 (300)
60	51		Sb	−80840 (s)	68	50		Sn	−91653.6 (35)
59	52		Te	−73470 (70)	67	51		Sb	87967 (5)
66	46	112	Pd	−86326 (26)	66	52		Te	−87671 (24)
65	47		Ag	−86620 (29)	65	53		I	−81370 (s)
64	48		Cd	−90577.9 (34)	64	54		Xe	−78070 (s)
63	49		In	−88000 (7)	63	55		Cs	−68670 (s)
62	50		Sn	−88658 (6)	71	48	119	Cd	−84230 (300)
61	51		Sb	−81740 (100)	70	49		In	−87730 (18)
60	52		Te	−77550 (s)	69	50		Sn	−90066.7 (35)
59	53		I	−67440 (s)	68	51		Sb	−89483 (12)
67	46	113	Pd	−83640 (s)	67	52		Te	−87189 (13)
66	47		Ag	−87040 (20)	66	53		I	−83820 (100)
65	48		Cd	−89050.3 (35)	65	54		Xe	−78830 (160)
64	49		In	−89372 (5)	64	55		Cs	−72530 (s)
63	50		Sn	−88332 (5)	63	56		Ba	−64530 (s)
62	51		Sb	−84443 (32)	72	48	120	Cd	−83981 (30)
61	52		Te	−78540 (s)	71	49		In	−85700 (100)
60	53		I	−71440 (s)	70	50		Sn	−91101.8 (35)
67	47	114	Ag	−85160 (140)	69	51		Sb	−88421 (8)
66	48		Cd	−90019.6 (33)	68	52		Te	−89404 (21)
65	49		In	−88576 (5)	67	53		I	−84000 (200)
64	50		Sn	−90560 (4)	66	54		Xe	−82050 (280)
63	51		Sb	−84870 (50)	65	55		Cs	−73640 (320)
62	52		Te	−82190 (s)	64	56		Ba	−69050 (s)
61	53		I	−73070 (s)	72	49	121	In	−85842 (28)
60	54		Xe	−67090 (s)	71	50		Sn	−89201.8 (35)
68	47	115	Ag	−84910 (100)	70	51		Sb	−89588.4 (35)
67	48		Cd	−88093 (8)	69	52		Te	−88508 (15)
66	49		In	−86541 (8)	68	53		I	−86140 (40)
65	50		Sn	−90035.1 (37)	67	54		Xe	−82350 (110)
64	51		Sb	−87005 (20)	66	55		Cs	−77150 (s)
63	52		Te	−82420 (50)	65	56		Ba	−70570 (410)
62	53		I	−76620 (s)	73	49	122	In	−83600 (150)
61	54		Xe	−68700 (s)	72	50		Sn	−89946 (4)
69	47	116	Ag	−82620 (s)	71	51		Sb	−88323.3 (36)
68	48		Cd	−88717.6 (37)	70	52		Te	−90304 (4)
67	49		In	−88253 (8)	69	53		I	−86160 (40)
66	50		Sn	−91526.1 (36)	68	54		Xe	−85160 (S)
65	51		Sb	−86930 (40)	67	55		Cs	−78010 (S)
64	52		Te	−85370 (110)	66	56		Ba	−74260 (S)
63	53		I	−77610 (170)	74	49	123	In	−83440 (40)
62	54		Xe	−73270 (260)	73	50		Sn	−87821 (4)
61	55		Cs	−62630 (s)	72	51		Sb	−89217.5 (38)
70	47	117	Ag	−82240 (100)	71	52		Te	−89165.5 (38)
69	48		Cd	−86416 (13)	70	53		I	−87970 (100)
68	49		In	−88944 (9)	69	54		Xe	−85290 (100)
67	50		Sn	−90398.9 (35)	68	55		Cs	−80890 (s)
66	51		Sb	−88654 (18)	67	56		Ba	−75390 (s)
65	52		Te	−85164 (35)	75	49	124	In	−81100 (90)
64	53		I	−80850 (110)	74	50		Sn	−88240 (5)
63	54		Xe	−74480 (s)	73	51		Sb	−87613.4 (38)

Table 39.1 Mass defects $\Delta M = M - A$ (in keV) [1] *(continued)*

N	Z	A	Element	Mass excess, keV	N	Z	A	Element	Mass excess, keV
72	52		Te	−90518.3 (38)	75	55		Cs	−86863 (12)
71	53		I	−87361 (5)	74	56		Ba	−87303 (12)
70	54		Xe	−87450 (140)	73	57		La	−81600 (s)
69	55		Cs	−81530 (480)	81	50	131	Sn	−77480 (s)
68	56		Ba	−78750 (s)	80	51		Sb	−82100 (s)
76	49	125	In	−80500 (300)	79	52		Te	−85201 (5)
75	50		Sn	−85902 (5)	78	53		I	−87451 (5)
74	51		Sb	−88252 (5)	77	54		Xe	−88421 (5)
73	52		Te	−89019 (4)	76	55		Cs	−88066 (8)
72	53		I	−88841 (5)	75	56		Ba	−86726 (19)
71	54		Xe	−87110 (40)	74	57		La	−83770 (100)
70	55		Cs	−84040 (40)	73	58		Ce	−79470 (s)
69	56		Ba	−79460 (250)	82	50	132	Sn	−76390 (220)
77	49	126	In	−77900 (120)	81	51		Sb	−79610 (200)
76	50		Sn	−86024 (12)	80	52		Te	−85213 (21)
75	51		Sb	−86402 (32)	79	53		I	−85706 (21)
74	52		Te	−90066 (4)	78	54		Xe	−89286 (5)
73	53		I	−87911 (6)	77	55		Cs	−87175 (23)
72	54		Xe	−89162 (8)	76	56		Ba	−88453 (10)
71	55		Cs	−84330 (140)	75	57		La	−83740 (50)
70	56		Ba	−52560 (s)	74	58		Ce	−82340 (s)
78	49	127	In	−77170 (130)	82	51	133	Sb	−78980 (210)
77	50		Sn	−83600 (100)	81	52		Te	−82930 (70)
76	51		Sb	−86704 (7)	80	53		I	−85902 (31)
75	52		Te	−88285 (5)	79	54		Xe	−87662 (9)
74	53		I	−88980 (5)	78	55		Cs	−88089 (8)
73	54		Xe	−88316 (6)	77	56		Ba	−87569 (9)
72	55		Cs	−86206 (21)	76	57		La	−85570 (s)
71	56		Ba	−82760 (100)	75	58		Ce	−82170 (s)
70	57		La	−77760 (s)	74	59		Pr	−77970 (s)
79	49	128	In	−74340 (250)	83	51	134	Sb	−73870 (s)
78	50		Sn	−83440 (150)	82	52		Te	−82670 (s)
77	51		Sb	−84730 (150)	81	53		I	−83970 (60)
76	52		Te	−88992.3 (39)	80	54		Xe	−88125 (7)
75	53		I	−87734 (5)	79	55		Cs	−86909 (8)
74	54		Xe	−89861.2 (16)	78	56		Ba	−88968 (8)
73	55		Cs	−85935 (6)	77	57		La	−85268 (31)
72	56		Ba	−85482 (20)	76	58		Ce	−84770 (s)
71	57		La	−78680 (s)	75	59		Pr	−78470 (s)
80	49	129	In	−73120 (180)	83	52	135	Te	−77600 (250)
79	50		Sn	−80640 (130)	82	53		I	−83796 (29)
78	51		Sb	−84630 (22)	81	54		Xe	−86506 (11)
77	52		Te	−87007 (4)	80	55		Cs	−87665 (9)
76	53		I	−88505 (4)	79	56		Ba	−87870 (7)
75	54		Xe	−88697.5 (20)	78	57		La	−86670 (12)
74	55		Cs	−87563 (24)	77	58		Ce	−84550 (100)
73	56		Ba	−85116 (19)	76	59		Pr	−80990 (140)
72	57		La	−81120 (s)	75	60		Nd	−76290 (s)
81	49	130	In	−70080 (s)	84	52	136	Te	−74830 (s)
80	50		Sn	−80380 (130)	83	53		I	−79430 (100)
79	51		Sb	−82380 (80)	82	54		Xe	−86425 (8)
78	52		Te	−87348 (5)	81	55		Cs	−86358 (8)
77	53		I	−86897 (10)	80	56		Ba	−88906 (7)
76	54		Xe	−89881.1 (16)	79	57		La	−86040 (70)

Table 39.1 Mass defects $\Delta M = M - A$ (in keV) [1] *(continued)*

N	Z	A	Element	Mass excess, keV	N	Z	A	Element	Mass excess, keV
78	58		Ce	−86500 (40)	83	59		Pr	−83790 (6)
77	59		Pr	−81400 (60)	82	60		Nd	−85949 (5)
76	60		Nd	−79190 (70)	81	61		Pm	−81060 (60)
84	53	137	I	−76720 (200)	80	62		Sm	−78978 (16)
83	54		Xe	−82215 (22)	79	63		Eu	−71480 (s)
82	55		Cs	−86560 (7)	88	55	143	Cs	−68360 (s)
81	56		Ba	−87733 (7)	87	55		Ba	−74010 (s)
80	57		La	−87130 (s)	86	57		La	−78310 (80)
79	58		Ce	−85910 (s)	85	58		Ce	−81610 (6)
78	59		Pr	−83210 (s)	84	59		Pr	−83065 (6)t
77	60		Nd	−79410 (s)	83	60		Nd	−84000 (5)
76	61		Pm	−74210 (s)	82	61		Pm	−82959 (7)
85	53	138	I	−71730 (s)	81	62		Sm	−79511 (11)
84	54		Xe	−80030 (s)	80	63		Eu	−74410 (50)
83	55		Cs	−82770 (s)	79	64		Gd	−68510 (s)
82	56		Ba	−88273 (7)	89	55	144	Cs	−63930 (s)
81	57		La	−86524 (7)	88	56		Ba	−72030 (s)
80	58		Ce	−87565 (13)	87	57		La	−74930 (s)
79	59		Pr	−83128 (16)	86	58		Ce	−80431 (6)
78	60		Nd	−82030 (s)	85	59		Pr	−80750 (6)
77	61		Pm	−75030 (s)	84	60		Nd	−83746 (5)
85	54	139	Xe	−75750 (90)	83	61		Pm	−81416 (7)
84	55		Cs	−80630 (70)	82	62		Sm	−81964 (6)
83	56		Ba	−84925 (7)	81	63		Eu	−75636 (30)
82	57		La	−87231 (6)	80	64		Gd	−71940 (s)
81	58		Ce	−86966 (8)	90	55	145	Cs	−61720 (s)
80	59		Pr	−84854 (13)	89	56		Ba	−67820 (s)
79	60		Nd	−82050 (50)	88	57		La	−72920 (s)
78	61		Pm	−77500 (210)	87	58		Ce	−77120 (90)
77	62		Sm	−72300 (450)	86	59		Pr	−79625 (11)
86	54	140	Xe	−73180 (260)	85	60		Nd	−81430 (5)
85	55		Cs	−77240 (250)	84	61		Pm	−81270 (6)
84	56		Ba	−83285 (12)	83	62		Sm	−80656 (6)
83	57		La	−84320 (6)	82	63		Eu	−77936 (16)
82	58		Ce	−88081 (6)	81	64		Cd	−72940 (s)
81	59		Pr	−84693 (8)	80	65		Tb	−66240 (s)
80	60		Nd	−84220 (40)	90	56	146	Ba	−65560 (s)
79	61		Pm	−78180 (110)	89	57		La	−69460 (s)
78	62		Sm	−75480 (s)	88	58		Ce	−75760 (120)
87	54	141	Xe	−69000 (140)	87	59		Pr	−76840 (100)
86	55		Cs	−75000 (100)	86	60		Nd	−80923 (5)
85	56		Ba	−79980 (60)	85	61		Pm	−79442 (8)
84	57		La	−83008 (31)	84	62		Sm	−80984 (8)
83	58		Ce	−85438 (6)	83	63		Eu	−77111 (11)
82	59		Pr	−86018 (6)	82	64		Gd	−75910 (s)
81	60		Nd	−84203 (10)	81	65		Tb	−67810 (s)
80	61		Pm	−80470 (40)	90	57	147	La	−67540 (s)
79	62		Sm	−75910 (60)	89	58		Ce	−72240 (T)
78	63		Eu	−69880 (100)	88	59		Pr	−75440 (200)
88	54	142	Xe	−66050 (170)	87	60		Nd	−78144 (5)
87	55		Cs	−70950 (130)	86	61		Pm	−79040 (5)
86	56		Ba	−77820 (100)	85	62		Sm	−79265 (5)
85	57		La	−80018 (9)	84	63		Eu	−77535 (8)
84	58		Ce	−84535 (6)	83	64		Gd	−75207 (26)

Table 39.1 Mass defects $\Delta M = M - A$ (in keV) [1] *(continued)*

N	Z	A	Element	Mass excess, keV	N	Z	A	Element	Mass excess, keV
22	65		Tb	−70510 (s)	91	62		Sm	−72557 (5)
81	66		Dy	−64210 (s)	90	63		Eu	−73363 (6)
90	58		Ce	−70710 (s)	89	64		Gd	−73119 (6)
89	59	148	Pr	−72510 (s)	88	65		Tb	−71329 (9)
88	60		Nd	−77407 (5)	87	66		Dy	−69155 (8)
87	61		Pm	−76870 (10)	86	67		Ho	−64954 (34)
86	62		Sm	−79335 (5)	85	68		Er	−60310 (s)
85	63		Eu	−76235 (22)	84	69		Tm	−53870 (s)
84	64		Gd	−76268 (6)	83	70		Yb	−47210 (s)
83	65		Tb	−70640 (80)	93	61	154	Pm	−68450 (100)
82	66		Dy	−67770 (s)	92	62		Sm	−72454 (5)
91	58	149	Ce	−67470 (s)	91	63		Eu	−71726 (7)
90	59		Pr	−71310 (200)	90	64		Gd	−73704 (5)
89	60		Nb	−74374 (5)	89	65		Tb	−70240 (50)
88	61		Pm	−76063 (6)	88	66		Dy	−70392 (12)
87	62		Sm	−77135 (5)	87	67		Ho	−64635 (s)
86	63		Eu	−76439 (7)	86	68		Er	−62440 (s)
85	64		Gd	−75131 (7)	85	69		Tm	−54530 (s)
84	65		Tb	−71434 (16)	84	70		Yb	−50050 (s)
83	66		Dy	−67530 (s)	94	61	155	Pm	−67100 (s)
82	67		Ho	−61530 (s)	93	62		Sm	−70196 (5)
91	59	150	Pr	−68680 (s)	92	63		Eu	−71825 (6)
90	60		Nd	−73682 (6)	91	64		Gd	−72071 (5)
89	61		Pm	−73550 (80)	90	65		Tb	−71256 (15)
88	62		Sm	−77049 (5)	89	66		Dy	−69157 (13)
87	63		Eu	−74756 (11)	88	67		Ho	−66055 (24)
86	64		Gd	−75765 (11)	87	68		Er	−62057 (27)
85	65		Tb	−71098 (11)	86	69		Tm	−56459 (s)
84	66		Dy	−69140 (s)	85	70		Yb	−50450 (s)
83	67		Ho	−62040 (s)	84	71		Lu	−42600 (s)
82	68		Er	−57940 (s)	94	62	156	Sm	−69368 (14)
92	59	151	Pr	−67440 (s)	93	63		Eu	−70083 (11)
91	60		Nd	−70945 (6)	92	64		Gd	−72536 (5)
90	61		Pm	−73386 (11)	91	65		Tb	−70098 (7)
89	62		Sm	−74574 (5)	90	66		Dy	−70527 (9)
88	63		Eu	−74650 (5)	89	67		Ho	−65410 (s)
87	64		Gd	−74168 (9)	88	68		Er	−63930 (s)
86	65		Tb	−71608 (8)	87	69		Tm	−56940 (80)
85	66		Dy	−68601 (27)	86	70		Yb	−53060 (s)
84	67		Ho	−63440 (s)	85	71		Lu	−43810 (s)
83	68		Er	−58200 (s)	95	62	157	Sm	−66860 (200)
82	69		Tm	−50800 (s)	94	63		Eu	−69465 (16)
92	60	152	Nd	−70146 (31)	93	64		Gd	−70825 (5)
91	61		Pm	−71290 (130)	92	65		Tb	−70767 (6)
90	62		Sm	−74761 (5)	91	66		Dy	−69425 (9)
89	63		Eu	−72884 (5)	90	67		Ho	−66890 (50)
88	64		Gd	−74703 (6)	89	68		Er	−63090 (s)
87	65		Tb	−70853 (16)	88	69		Tm	−58490 (s)
86	66		Dy	−70116 (8)	87	70		Yb	−53270 (s)
85	67		Ho	−63710 (80)	86	71		Lu	−46470 (s)
84	68		Er	−60410 (s)	85	72		Hf	−38960 (s)
83	69		Tm	−51810 (s)	95	63	158	Eu	−67240 (80)
93	60	153	Nd	−67360 (s)	94	64		Gd	−70691 (5)
92	61		Pm	−70760 (100)	93	65		Tb	−69475 (6)

Table 39.1 Mass defects $\Delta M = M - A$ (in keV) [1] *(continued)*

N	Z	A	Element	Mass excess, keV	N	Z	A	Element	Mass excess, keV
92	66		Dy	−70410 (6)	93	70		Yb	−59170 (s)
91	67		Ho	−66433 (8)	92	71		Lu	−54370 (s)
90	68		Er	−65030 (s)	91	72		Hf	−48770 (s)
89	69		Tm	−58430 (s)	90	73		Ta	−42370 (s)
88	70		Yb	−55530 (s)	89	74		W	−34850 (s)
87	71		Lu	−47230 (s)	99	65	164	Tb	−62110 (150)
86	72		Hf	−42220 (s)	98	66		Dy	−65967 (6)
96	63	159	Eu	−65930 (50)	97	67		Ho	−64937 (6)
95	64		Gd	68562 (6)	96	68		Er	−65940 (6)
94	65		Tb	−69536 (6)	95	69		Tm	−61978 (21)
93	66		Dy	−69171 (6)	94	70		Yb	−60880 (s)
92	67		Ho	−67318 (11)	93	71		Lu	−54580 (s)
91	68		Er	−64390 (100)	92	72		Hf	−51280 (s)
90	69		Tm	−60190 (s)	91	73		Ta	−43080 (s)
89	70		Yb	−55290 (s)	90	74		W	−38040 (s)
88	71		Lu	−49490 (s)	99	66	165	Dy	−63611 (6)
87	72		Hf	−42800 (s)	98	67		Ho	−64896 (6)
97	63	160	Eu	−63540 (s)	97	68		Er	−64518 (6)
96	64		Gd	−67943 (5)	96	69		Tm	−62924 (6)
95	65		Tb	−67840 (6)	95	70		Yb	−60161 (21)
94	66		Dy	−69774 (6)	94	71		Lu	−56160 (s)
93	67		Ho	−66388 (16)	93	72		Hf	−51260 (s)
92	68		Er	−66052 (29)	92	73		Ta	−45360 (s)
91	69		Tm	−60130 (60)	91	74		W	−38670 (s)
90	70		Yb	−57550 (s)	100	66	166	Dy	−62583 (8)
89	71		Lu	−49930 (s)	99	67		Ho	−63067 (6)
88	72		Hf	−45750 (s)	98	68		Er	−64921 (6)
87	73		Ta	−35780 (s)	97	69		Tm	−61874 (13)
97	64	161	Gd	−65507 (6)	96	70		Yb	−61582 (9)
96	65		Tb	−67466 (6)	95	71		Lu	−56100 (160)
95	66		Dy	−68056 (6)	94	72		Hf	−53480 (s)
94	67		Ho	−67203 (6)	93	73		Ta	−46100 (s)
93	68		Er	−65197 (12)	92	74		W	−41480 (s)
92	69		Tm	−61680 (s)	100	67	167	Ho	−62316 (21)
91	70		Yb	−57400 (s)	99	68		Er	−63286 (6)
90	71		Lu	−52080 (s)	98	69		Tm	−62537 (6)
89	72		Hf	−46130 (s)	97	70		Yb	−60583 (7)
88	73		Ta	−38840 (s)	96	71		Lu	−57450 (70)
98	64	162	Cd	−64360 (120)	95	72		Hf	−53150 (s)
97	65		Tb	−65760 (70)	94	73		Ta	−47950 (s)
96	66		Dy	−68181 (6)	93	74		W	−41950 (s)
95	67		Ho	−66047 (7)	92	75		Re	−34650 (s)
94	68		Er	−66335 (6)	101	67	168	Ho	−60270 (100)
93	69		Tm	−61540 (60)	100	68		Er	−62985 (6)
92	70		Yb	−59340 (s)	99	69		Tm	−61306 (6)
91	71		Lu	−52340 (s)	98	70		Yb	−61565 (7)
90	72		Hf	−48760 (s)	97	71		Lu	−57100 (80)
89	73		Ta	−39710 (s)	96	72		Hf	−55100 (s)
88	74		W	−34130 (s)	95	73		Ta	−48400 (s)
98	65	163	Tb	−64680 (50)	94	74		W	−44500 (s)
97	66		Dy	−66382 (6)	93	75		Re	−35700 (s)
96	67		Ho	−66379 (6)	102	67	169	Ho	−58793 (21)
95	68		Er	−65168 (7)	101	68		Er	60917 (6)
94	69		Tm	−62770 (s)	100	69		Tm	−61269 (6)

Table 39.1 Mass defects $\Delta M = M - A$ (in keV) [1] *(continued)*

N	Z	A	Element	Mass excess, keV	N	Z	A	Element	Mass excess, keV
99	70		Yb	−60361 (7)	99	75		Re	−43580 (s)
98	71		Lu	−57881 (26)	98	76		Os	−39620 (s)
97	72		Hf	−54530 (100)	97	77		Ir	−30890 (s)
96	73		Ta	−50030 (s)	96	78		Pt	−24930 (s)
95	74		W	−44890 (s)	106	69	175	Tm	−52290 (50)
94	75		Re	−38130 (s)	105	70		Yb	−54691 (6)
93	76		Os	−30550 (s)	104	71		Lu	−55159 (5)
103	67	170	Ho	−56100 (200)	103	72		Hf	−54548 (10)
102	68		Er	−60104 (6)	102	73		Ta	−52350 (s)
101	69		Tm	−59791 (6)	101	74		W	−49450 (s)
100	70		Yb	−60759 (6)	100	75		Re	−45150 (s)
99	71		Lu	−57319 (21)	99	76		Os	−39710 (s)
98	72		Hf	−56120 (s)	98	77		Ir	−33160 (s)
97	73		Ta	−50120 (s)	97	78		Pt	−25640 (s)
96	74		W	−46920 (s)	96	79		Au	−17160 (s)
95	75		Re	−38920 (s)	107	69	176	Tm	−49590 (s)
94	76		Os	−33530 (s)	106	70		Yb	−53490 (6)
103	68	171	Er	−57714 (6)	105	71		Lu	−53381 (5)
102	69		Tm	−59205 (6)	104	72		Hf	−54567 (6)
101	70		Yb	−59302 (6)	103	73		Ta	−51470 (100)
100	71		Lu	−57821 (6)	102	74		W	−50570 (s)
99	72		Hf	−55300 (s)	101	75		Re	−44970 (s)
98	73		Ta	−51600 (s)	100	76		Os	−41810 (s)
97	74		W	−46900 (s)	99	77		Ir	−33840 (s)
96	75		Re	−41100 (s)	98	78		Pt	−28540 (s)
95	76		Os	−34160 (s)	97	79		Au	−18400 (s)
94	77		Ir	−26180 (s)	107	70	177	Yb	−50986 (6)
104	68	172	Er	−56491 (13)	106	71		Lu	−52382 (5)
103	69		Tm	−57380 (11)	105	72		Hf	−52879 (6)
102	70		Yb	−59250 (6)	104	73		Ta	−51721 (7)
101	71		Lu	−56726 (6)	103	74		W	−49720 (s)
100	72		Hf	−56330 (s)	102	75		Re	−46120 (s)
99	73		Ta	−51410 (s)	101	76		Os	−41620 (s)
98	74		W	−48810 (s)	100	77		Ir	−35820 (s)
97	75		Re	−41510 (s)	99	78		Pt	−29350 (s)
96	76		Os	−36840 (s)	98	79		Au	−21190 (s)
95	77		Ir	−27320 (s)	97	80		Hg	−12650 (s)
105	68	173	Er	−53730 (200)	108	70	178	Yb	−49660(50)
104	69		Tm	−56226 (31)	107	71		Lu	−50300 (40)
103	70		Yb	−57546 (6)	106	72		Hf	−52434 (6)
102	71		Lu	−56871 (6)	105	73		Ta	−50520 (100)
101	72		Hf	−55270 (s)	104	74		W	−50430 (100)
100	73		Ta	−52370 (s)	103	75		Re	−45770 (210)
99	74		W	−48470 (s)	102	76		Os	−43350 (s)
98	75		Re	−43370 (s)	101	77		Ir	−36270 (s)
97	76		Os	−37410 (s)	100	78		Pt	−31630 (s)
96	77		Ir	−29910 (s)	99	79		Au	−22410 (s)
95	78		Pt	−21790 (s)	98	80		Hg	−15930 (s)
105	69	174	Tm	−53850 (50)	108	71	179	Lu	−49110 (40)
104	70		Yb	−56940 (s)	107	72		Hf	−50462 (6)
103	71		Lu	−55562 (6)	106	73		Ta	−50347 (8)
102	72		Hf	−55830 (8)	105	74		W	−49283 (17)
101	73		Ta	−51980 (80)	104	75		Re	−46590 (50)
100	74		W	−50080 (s)	103	76		Os	−42890 (s)

Table 39.1 Mass defects $\Delta M = M - A$ (in keV) [1] *(continued)*

N	Z	A	Element	Mass excess, keV	N	Z	A	Element	Mass excess, keV
102	77		Ir	−37890 (s)	110	75		Re	−43802 (7)
101	78		Pt	−32010 (s)	109	76		Os	−42787 (7)
100	79		Au	−24750 (s)	108	77		Ir	−40290 (s)
99	80		Hg	−16800 (s)	107	78		Pt	−36490 (s)
109	71	180	Lu	−46680 (70)	106	79		Au	−31730 (s)
108	72		Hf	−49779 (6)	105	80		Hg	−26140 (s)
107	73		Ta	−48914 (13)	104	81		Tl	−19110 (s)
106	74		W	−49624 (8)	103	82		Pb	−11740 (s)
105	75		Re	−45829 (31)	113	73	186	Ta	−38600 (60)
104	76		Os	−44220 (s)	112	74		W	−42498 (7)
103	77		Ir	−37930 (s)	111	75		Re	−41910 (7)
102	78		Pt	−34120 (s)	110	76		Os	−42987 (7)
101	79		Au	−25630 (s)	109	77		Ir	−39156 (21)
100	80		Hg	−19860 (s)	108	78		Pt	−37830 (s)
109	72	181	Hf	−47403 (6)	107	79		Au	−31690 (s)
108	73		Ta	−48425 (6)	106	80		Hg	−28350 (s)
107	74		W	−48237 (9)	105	81		Tl	−19860 (s)
106	75		Re	−46440 (s)	104	82		Pb	−14330 (s)
105	76		Os	−43410 (s)	113	74	187	W	−39893 (7)
104	77		Ir	−39340 (s)	112	75		Re	−41205 (7)
103	78		Pt	−34060 (s)	111	76		Os	−41208 (7)
102	79		Au	−27640 (s)	110	77		Ir	−39710 (s)
101	80		Hg	−20790 (s)	109	78		Pt	−36810 (s)
110	72	182	Hf	−45990 (50)	108	79		Au	−32870 (s)
109	73		Ta	−46417 (6)	107	80		Hg	−28060 (s)
108	74		W	−48228 (6)	106	81		Tl	−21930 (s)
107	75		Re	−45430 (s)	105	82		Pb	−14940 (s)
106	76		Os	−44580 (s)	114	74	188	W	−38657 (7)
105	77		Ir	−38980 (s)	113	75		Re	−39006 (7)
104	78		Pt	−35980 (s)	112	76		Os	−41125 (7)
103	79		Au	−28180 (s)	111	77		Ir	−38323 (13)
102	80		Hg	−23210 (s)	110	78		Pt	−37788 (11)
111	72	183	Hf	−43269 (32)	109	79		Au	−32490 (s)
110	73		Ta	−45279 (12)	108	80		Hg	−29880 (s)
109	74		W	−46347 (6)	107	81		Tl	−22290 (s)
108	75		Re	−45791 (10)	106	82		Pb	−17500 (s)
107	76		Os	43490 (s)	115	74	189	W	−35470 (200)
106	77		Ir	−40090 (s)	114	75		Re	−37970 (11)
105	78		Pt	−35630 (s)	113	76		Os	−38978 (7)
104	79		Au	−30010 (s)	112	77		Ir	−38480 (s)
103	80		Hg	−23690 (s)	111	78		Pt	−36570 (s)
102	81		Tl	−15830 (s)	110	79		Au	−33410 (s)
112	72	184	Hf	−41480 (60)	109	80		Hg	−29210 (s)
111	73		Ta	−42821 (27)	108	81		Tl	−24020 (s)
110	74		W	−45687 (7)	107	82		Pb	−17860 (s)
109	75		Re	−44191 (9)	106	83		Bi	−9870 (s)
108	76		Os	−44233 (7)	116	74	190	W	−34220 (360)
107	77		Ir	−39510 (250)	115	75		Re	−35520 (200)
106	78		Pt	−37210 (s)	114	76		Os	−38699 (7)
105	79		Au	−30220 (s)	113	77		Ir	−36700 (200)
104	80		Hg	−26040 (s)	112	78		Pt	−37318 (21)
103	81		Tl	−16900 (s)	111	79		Au	−32876 (26)
112	73	185	Ta	−41360 (21)	110	80		Hg	−30960 (80)
111	74		W	−43370 (7)	109	81		Tl	−24160 (310)

Table 39.1 Mass defects $\Delta M = M - A$ (in keV) [1] *(continued)*

N	Z	A	Element	Mass excess, keV
108	82		Pb	−20220 (s)
107	83		Bi	−10850 (s)
116	75	191	Re	−34343 (12)
115	76		Os	−36388 (7)
117	77		Ir	−36698 (7)
113	78		Pt	−35698 (16)
112	79		Au	−33870 (50)
111	80		Hg	−30480 (70)
110	81		Tl	−25670 (210)
109	82		Pb	−20230 (s)
108	83		Bi	−13050 (s)
116	76	192	Os	−35875 (7)
115	77		Ir	−34826 (7)
114	78		Pt	−36283 (7)
113	79		Au	−32768 (17)
112	80		Hg	−31970 (s)
111	81		Tl	−25590 (s)
110	82		Pb	−22290 (s)
109	83		Bi	−13670 (s)
117	76	193	Os	−33387 (8)
116	77		Ir	−34519 (6)
115	78		Pt	−34458 (7)
114	79		Au	−33360 (s)
113	80		Hg	−31020 (s)
112	81		Tl	−27020 (s)
111	82		Pb	−22070 (s)
110	83		Bi	−15560 (s)
109	84		Po	−8310 (s)
118	76	194	Os	−32417 (7)
117	77		Ir	−32514 (6)
116	78		Pt	−34765 (6)
115	79		Au	−32256 (16)
114	80		Hg	−32206 (26)
113	81		Tl	−26810 (s)
112	82		Pb	−23810 (s)
111	83		Bi	−15980 (s)
110	84		Po	−10810 (s)
119	76	195	Os	−29690 (500)
118	77		Ir	−31692 (31)
117	78		Pt	−32802 (6)
116	79		Au	−32572 (6)
115	80		Hg	−31050 (50)
114	81		Tl	−27850 (210)
113	82		Pb	−23550 (s)
112	83		Bi	−17680 (210)
111	84		Po	−11060 (s)
119	77	196	Ir	−29440 (60)
118	78		Pt	−32652 (6)
117	79		Au	−31162 (9)
116	80		Hg	−31846 (10)
115	81		Tl	−27350 (s)
114	82		Pb	−25150 (s)
113	83		Bi	−17760 (s)
112	84		Po	−13210 (s)
111	85		At	−4050 (s)
120	77	197	Ir	−28430 (200)
119	78		Pt	−30431 (6)
118	79		Au	−31150 (6)
117	80		Hg	−30735 (21)
116	81		Tl	−28330 (s)
115	82		Pb	−24630 (s)
114	83		Bi	−19410 (s)
113	84		Po	−13230 (s)
112	85		At	−6030 (s)
121	77	198	Ir	−25520 (300)
120	78		Pt	−29921 (20)
119	79		Au	−29591 (6)
118	80		Hg	−30964 (6)
117	81		Tl	−27500 (80)
116	82		Pb	−25900 (s)
115	83		Bi	−19300 (s)
114	84		Po	−15070 (s)
113	85		At	−6670 (310)
121	78	199	Pt	−27420 (25)
120	79		Au	−29104 (6)
119	80		Hg	−29557 (6)
118	81		Tl	−28080 (220)
117	82		Pb	−25280 (90)
116	83		Bi	−20610 (s)
115	84		Po	−15050 (s)
114	85		At	−8470 (210)
122	78	200	Pt	−26600 (s)
121	79		Au	−27300 (50)
120	80		Hg	−29514 (6)
119	81		Tl	−27060 (10)
118	82		Pb	−26160 (s)
117	83		Bi	−20460 (s)
116	84		Po	−16740 (s)
115	85		At	−8670 (s)
114	86		Rn	−3740 (s)
123	78	201	Pt	−23740 (110)
122	79		Au	−26400 (100)
121	80		Hg	−27672 (6)
120	81		Tl	−27185 (16)
119	82		Pb	−25327 (35)
118	83		Bi	−21410 (s)
117	84		Po	−16410 (s)
116	85		At	−10520 (s)
115	86		Rn	−3950 (s)
123	79	202	Au	−23860 (200)
122	80		Hg	−27356 (6)
121	81		Tl	−25988 (18)
120	82		Pb	−25942 (11)
119	83		Bi	−21040 (s)
118	84		Po	−17780 (s)
117	85		At	−10520 (s)
116	86		Rn	−5880 (s)
115	87		Fr	−3160 (310)

Table 39.1 Mass defects $\Delta M = M - A$ (in keV) [1] *(continued)*

N	Z	A	Element	Mass excess, keV	N	Z	A	Element	Mass excess, keV
124	79	203	Au	−22980 (s)	127	82		Pb	−17624 (5)
123	80		Hg	−25277 (6)	126	83		Bi	−18268 (5)
122	81		Tl	−25769 (6)	125	84		Po	−16373 (7)
121	82		Pb	−24794 (10)	124	85		At	−12888 (9)
120	83		Bi	−21600 (50)	123	86		Rn	−8994 (35)
119	84		Po	−17360 (90)	122	87		Fr	−3760 (s)
118	85		At	−11970 (s)	121	88		Ra	1970 (s)
117	86		Rn	−6000 (s)	120	89		Ac	9120 (s)
116	87		Fr	1230 (210)	129	81	210	Tl	−9251 (13)
125	79	204	Au	−20200 (300)	128	82		Pb	−14738 (5)
124	80		Hg	−24703 (6)	127	83		Bi	−14801 (5)
123	81		Tl	−24353 (6)	126	84		Po	−15963 (5)
122	82		Pb	−25117 (6)	125	85		At	−11976 (12)
121	83		Bi	−20820 (s)	124	86		Rn	−9608 (12)
120	84		Po	−18250 (s)	123	87		Fr	−3640 (s)
119	85		At	−11970 (s)	122	88		Ra	610 (s)
118	86		Rn	−7770 (s)	121	89		Ac	8860 (s)
117	87		Fr	870 (s)	129	82	211	Pb	−10491.9 (38)
116	88		Ra	6280 (s)	128	83		Bi	−11865 (6)
125	80	205	Hg	−22299 (8)	127	84		Po	−12444 (5)
124	81		Tl	−23837 (5)	126	85		At	−11653 (9)
123	82		Pb	−23777 (6)	125	86		Rn	−8761 (11)
122	83		Bi	−21070 (9)	124	87		Fr	−4220 (50)
121	84		Po	−17576 (35)	123	88		Ra	780 (90)
120	85		At	−12960 (s)	122	89		Ac	7400 (s)
119	86		Rn	−7600 (s)	130	82	212	Pb	−7562 (6)
118	87		Fr	−1040 (s)	129	83		Bi	−8135 (6)
117	88		Ra	5980 (s)	128	84		Po	−10381 (5)
126	80	206	Hg	−20955 (21)	127	85		At	−8625 (6)
125	81		Tl	−22269 (5)	126	86		Rn	−8666 (7)
124	82		Pb	−23795 (5)	125	87		Fr	−3690 (s)
123	83		Bi	−20033 (12)	124	88		Ra	−110 (s)
122	84		Po	−18190 (11)	123	89		Ac	7180 (s)
121	85		At	−12730 (s)	131	82	213	Pb	−3140 (s)
120	86		Rn	−8970 (s)	130	83		Bi	−5243 (11)
119	87		Fr	−1180 (s)	129	84		Po	−6663 (7)
118	88		Ra	3960 (s)	128	85		At	−6589 (13)
126	81	207	Tl	−21041 (6)	127	86		Rn	−5706 (11)
125	82		Pb	−22463 (5)	126	87		Fr	−3556 (1)
124	83		Bi	−20058 (8)	125	88		Ra	290 (35)
123	84		Po	−17150 (11)	124	89		Ac	6170 (s)
122	85		At	−13310 (50)	123	90		Th	12240 (s)
121	86		Rn	−8690 (90)	132	82	214	Pb	−185.3 (33)
120	87		Fr	−2650 (s)	131	83		Bi	−1209 (12)
119	88		Ra	3700 (s)	130	84		Po	−4479 (5)
127	81	208	Tl	−16768 (6)	129	85		At	−3389 (6)
126	82		Pb	−21759 (5)	128	86		Rn	−4328 (11)
125	83		Bi	−18879 (5)	127	87		Fr	−965 (13)
124	84		Po	−17475 (6)	126	88		Ra	90 (13)
123	85		At	−12640 (s)	125	89		Ac	6140 (s)
122	86		Rn	−9560 (s)	124	90		Th	10870 (s)
121	87		Fr	−2770 (s)	132	83	215	Bi	1710 (100)
120	88		Ra	1930 (s)	131	84		Po	−540.5 (37)
128	81	209	Tl	−13650 (15)	130	85		At	−1262 (7)

Table 39.1 Mass defects $\Delta M = M - A$ (in keV) [1] *(continued)*

N	Z	A	Element	Mass excess, keV	N	Z	A	Element	Mass excess, keV
129	86		Rn	−1179 (10)	132	91		Pa	22330 (19)
128	87		Fr	309 (13)	137	87	224	Fr	21710 (s)
127	88		Ra	2531 (12)	136	88		Ra	18813 (6)
126	89		Ac	5950 (50)	135	89		Ac	20219 (8)
125	90		Th	10870 (90)	134	90		Th	19993 (18)
133	83	216	Bi	5970 (s)	133	91		Pa	23798 (20)
132	84		Po	1769 (6)	138	87	225	Fr	23790 (s)
131	85		At	2237 (6)	137	88		Ra	21987.3 (36)
130	86		Rn	245 (11)	136	89		Ac	21626 (12)
129	87		Fr	2975 (14)	135	90		Th	22303 (11)
128	88		Ra	3285 (11)	134	91		Ra	24320 (21)
127	89		Ac	7980 (s)	139	87	226	Fr	27460 (330)
126	90		Th	10390 (s)	138	88		Ra	28665.7 (33)
133	84	217	Po	5960 (s)	137	89		Ac	24301.0 (38)
132	85		At	4382 (12)	136	90		Th	23189 (6)
131	86		Rn	3649 (8)	135	91		Pa	26029 (12)
130	87		Fr	4307 (15)	134	92		U	27186 (34)
129	88		Ra	5881 (13)	140	87	227	Fr	29580 (100)
129	89		Ac	8701 (15)	139	88		Ra	27185 (20)
127	90		Th	12141 (36)	138	89		Ac	25850.0 (31)
134	84	218	Po	8354.6 (33)	137	90		Th	25806.3 (37)
133	85		At	8099 (13)	136	91		Pa	26832 (10)
132	86		Rn	5212 (5)	135	92		U	28880 (s)
131	87		Fr	7050 (6)	140	88	228	Ra	28941 (5)
130	88		Ra	6644 (14)	139	89		Ac	28895 (5)
129	89		Ac	10837 (16)	138	90		Th	26758 (6)
128	90		Th	12362 (16)	137	91		Pa	28870 (9)
134	85	219	At	10530 (80)	136	92		U	29221 (21)
133	86		Rn	8830.7 (37)	141	88	229	Ra	32720 (s)
132	87		Fr	8617 (9	140	89		Ac	30720 (150)
131	88		Ra	9377 (14)	139	90		Th	29580.9 (34)
130	89		Ac	11560 (16)	138	91		Pa	29887 (13)
129	90		Th	14470 (24)	137	92		U	31201 (11)
135	85	220	At	14200 (s)	136	93		Np	33758 (29)
134	86		Rn	10599 (6)	142	88	230	Ra	34560 (s)
133	87		Fr	11470 (8)	141	89		Ac	33760 (s)
132	88		Ra	10263 (15)	140	90		Th	30861.3 (29)
131	89		Ac	13747 (17)	139	91		Pa	32165.5 (37)
130	90		Th	14663 (23)	138	98		U	31607 (6)
135	86	221	Rn	14380 (s)	137	93		Np	35232 (24)
134	87		Fr	13265 (12)	142	89	231	Ac	35910 (100)
133	88		Ra	12957 (9)	141	90		Th	33812.2 (30)
132	89		Ac	14518 (18)	140	91		Pa	33423.1 (31)
131	90		Th	16934 (14)	139	92		U	33780 (50)
136	86	222	Rn	16370.0 (33)	138	93		Np	35626 (13)
135	87		Fr	16338 (21)	143	89	232	Ac	39150 (s)
134	88		Ra	14312 (6)	142	90		Th	35447.2 (23)
133	89		Ac	16617 (7)	141	91		Pa	35934 (12)
132	90		Th	17197 (16)	140	92		U	34597 (6)
131	91		Pa	21959 (35)	139	93		Np	37290 (s)
136	87	223	Fr	18382.3 (37)	138	94		Pu	38362 (23)
135	88		Ra	17234.8 (37)	143	90	233	Th	38732.3 (24)
134	89		Ac	17825 (9)	142	91		Pa	37487.1 (24)
133	90		Th	19256 (17)	141	92		U	36914.7 (33)

Table 39.1 Mass defects $\Delta M = M - A$ (in keV) [1] *(continued)*

N	Z	A	Element	Mass excess, keV	N	Z	A	Element	Mass excess, keV
140	93		Np	38010 (s)	147	95		Am	55462.7 (27)
139	94		Pu	40042 (23)	146	96		Cm	54801.5 (24)
144	90	234	Th	40612 (4)	145	97		Bk	57800 (s)
143	91		Pa	40349 (5)	144	98		Cf	59332 (33)
142	92		U	38142.6 (24)	149	94	243	Pu	57752.5 (35)
141	93		Np	39951 (9)	148	95		Am	57170.1 (32)
140	94		Pu	40342 (8)	147	96		Cm	57177.4 (26)
139	95		Am	44460 (s)	146	97		Bk	58685 (6)
145	90	235	Th	44150 (s)	145	98		Cf	60910 (s)
144	91		Pa	42320 (100)	144	99		Es	64800 (s)
143	92		U	40916.4 (24)	150	94	244	Pu	59803 (5)
142	93		Np	41039.5 (26)	149	95		Am	59878.6 (31)
141	94		Pu	42160 (6)	148	96		Cm	58449.6 (23)
140	95		Am	44650 (s)	147	97		Bk	60646 (21)
145	91	236	Pa	45540 (200)	146	98		Cf	61465 (6)
144	92		U	42442.0 (23)	145	99		Es	65970 (s)
143	93		Np	43426 (10)	151	94	245	Pu	63157 (30)
142	94		Pu	42889 (6)	150	95		Am	61897.3 (36)
141	95		Au	46020 (s)	149	96		Cm	61001.3 (29)
140	96		Cm	47890 (s)	148	97		Bk	61811 (5)
146	91	237	Pa	47640 (50)	147	98		Cf	63377 (6)
145	92		U	45388.7 (25	146	99		Es	66380 (s)
144	93		Np	44869.3 (23)	145	100		Fm	70020 (s)
143	94		Pu	45087 (6)	152	94	246	Pu	65290 (50)
142	95		Am	46640 (s)	151	95		Am	64920 (50)
141	96		Cm	49170 (s)	150	96		Cm	62616.0 (34)
147	91	238	Pa	51270 (300)	149	97		Bk	64020 (s)
146	92		U	48307.0 (22)	148	98		Cf	64096.2 (31)
145	93		Np	47452.6 (23)	147	99		Es	67930 (s)
144	94		Pu	46160.8 (24)	146	100		Fm	70131 (36)
143	95		Am	48417 (32)	152	95	247	Am	67130 (s)
142	96		Cm	49398 (31)	151	96		Cm	65530 (5)
141	97		Bk	54280 (s)	150	97		Bk	65484 (6)
147	92	239	U	50572.2 (22)	149	98		Cf	66150 (s)
146	93		Np	49306.4 (30)	148	99		Es	68550 (31)
145	94		Pu	48585.1 (24)	147	100		Fm	71540 (s)
144	95		Am	49389 (5)	153	95	248	Am	70490 (s)
143	96		Cm	51090 (s)	152	96		Cm	67389 (6)
142	97		Bk	54280 (s)	151	97		Bk	67990 (s)
148	92		U	52712 (5)	150	98		Cf	67243 (31)
147	93	240	Np	52210 (60)	149	99		Es	70220 (s)
146	94		Pu	50122.8 (23)	148	100		Fm	71891 (21)
145	95		Am	51443 (20)	147	101		Md	77000 (s)
144	96		Cm	51712 (6)	153	96	249	Cm	70748 (8)
143	97		Bk	55710 (s)	152	97		Bk	69848.0 (35)
142	88		Cf	58030 (s)	151	98		Cf	69721.6 (30)
148	93		Np	54310 (100)	150	99		Es	71116 (7)
147	94	241	Pu	52953.0 (23)	149	100		Fm	73500 (s)
146	95		Am	52932.2 (23)	148	101		Md	77260 (s)
145	96		Cm	53606 (6)	154	96	250	Cm	72986 (12)
144	97		Bk	56100 (s)	153	97		Bk	72950 (5)
143	98		Cf	59190 (s)	152	98		Cf	71169.8 (35)
149	93	242	Np	57250 (s)	151	99		Es	73170 (s)
148	94		Pu	54715.0 (23)	150	100		Fm	74069 (31)

Table 39.1 Mass defects $\Delta M = M - A$ (in keV) [1] *(continued)*

N	Z	A	Element	Mass excess, keV	N	Z	A	Element	Mass excess, keV
149	101		Md	78600 (s)	152	103		Lr	90250 (s)
154	97	251	Bk	75250 (s)	157	99	256	Es	87260 (s)
153	98		Cf	74130 (5)	156	100		Fm	85481 (8)
152	99		Es	74503 (8)	155	101		Md	87420 (s)
151	100		Fm	76000 (s)	154	102		No	87801 (40)
150	101		Md	79030 (s)	153	103		Lr	91820 (s)
155	97	252	Bk	78530 (s)	157	100	257	Fm	88588 (10)
154	98		Cf	76031 (6)	156	101		Md	89040 (s)
153	99		Es	77150 (s)	155	102		No	90223 (31)
152	100		Fm	76822 (33)	154	103		Lr	92970 (s)
151	101		Md	80500 (s)	153	104		Db	95950 (s)
150	102		No	82862 (26)	157	101	258	Md	91820 (s)
155	98	253	Cf	79299 (10)	156	102		No	91520 (s)
154	99		Es	79012.4 (35)	155	103		Lr	94820 (s)
153	100		Fm	79346 (5)	154	104		Db	96550 (s)
152	101		Md	81240 (s)	157	102	259	No	94026 (11)
151	102		No	84330 (s)	156	103		Lr	96000 (s)
156	98	254	Cf	81342 (12)	155	104		Db	98500 (s)
155	99		Es	81992 (6)	157	103	260	Lr	98140 (s)
154	100		Fm	80899 (5)	156	104		Db	99230 (s)
153	101		Md	83390 (s)	155	105		Jl	103650 (s)
152	102		No	84729 (34)	157	104	261	Db	101250 (s)
156	99	255	Es	84080 (s)	156	105		Jl	104460 (s)
155	100		Fm	83796 (5)	157	105	262	Jl	106040 (s)
154	101		Md	84880 (s)	157	106	263	Rf	110310 (s)
153	102		No	86870 (s)					

39.4 The Threshold of a Nuclear Reaction

If the reaction energy $Q < 0$, then the reaction consumes energy, and for such a reaction to start the incident particle must have an energy exceeding the reaction threshold:

$$E_{\text{thresh}} = |Q|(M_1 + M_2)/M_2, \tag{39.6}$$

where M_1 and M_2 are the masses of particles in the input reaction channel.

39.5 The Coulomb Barrier

Charged particles that are to participate in a reaction must first overcome electrostatic repulsion between their charges – the so-called Coulomb barrier.

The Coulomb barrier height of a nucleus with a charge $Z_1 e$ and a radius R_1 for a particle of a charge $Z_2 e$ and a radius R_2 is given by the expression

$$E_{\text{Coul}} = Z_1 Z_2 e^2/(R_1 + R_2), \tag{39.7}$$

where $R_1 + R_2 = R_0(A_1^{1/3} + A_2^{1/3})$; R_0 is a constant that is usually set to about 1 fm; A_1 and A_2 are the mass numbers of the interacting nuclei.

Fig. 39.1 shows Coulomb barrier heights for protons and α-particles calculated by the above equation for R_0 equal to 1.0, 1.2, and 1.4 fm. The curves are plotted as functions of Z_1 and are drawn smoothly through the points calculated for the most abundant isotopes of given A_1 and Z_1 values [2].

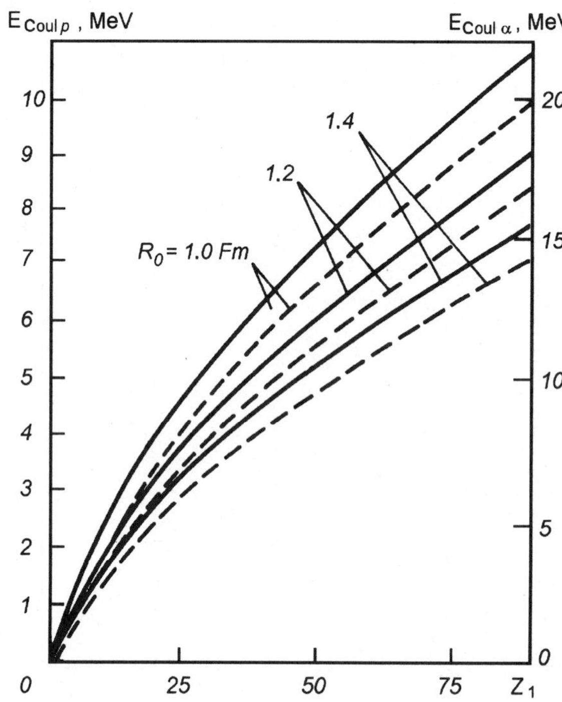

Figure 39.1 Coulomb barrier heights for protons (solid lines) and α-particles (dotted lines).

39.6 Kinematics of the Nuclear Reactions

The kinematics of nuclear reactions is described by the relationships between energies of the particles participating in the nuclear reaction, angles and solid-angles in laboratory (LS) and center-of-mass (CMS) systems.

Below are given some important nonrelativistic equations (the definitions cf. in Fig.39.2, primed energies relate to CMS):

$$\sqrt{E_3} = \frac{\sqrt{M_1 M_3 E_1}}{M_3 + M_4} \cos\psi \left(1 \pm \left\{ \frac{1 + \frac{M_4}{M_3}}{\cos^2\psi} \left[\frac{M_4}{M_1}\left(1 + \frac{Q}{E_1}\right) - 1 \right] \right\}^{1/2} \right), \tag{39.8}$$

$$Q = \frac{M_3 + M_4}{M_4} E_3 - \frac{M_4 - M_1}{M_4} E_1 - \frac{2\sqrt{M_1 M_3 E_1 E_3}}{M_4} \cos\psi. \tag{39.9}$$

Let us introduce the following notations: ψ and ξ are escape angles in LS, θ and φ are the same angles in CMS, E_i and E are energies in LS and CMS, respectively;

$$Q = (M_1 + M_2 - M_3 - M_4)\, c^2, \tag{39.10}$$

$$E_T = E_1 + Q = E_3 + E_4, \tag{39.11}$$

$$A = \frac{M_1 M_4 (E_1/E_T)}{(M_1 + M_2)(M_3 + M_4)}, \tag{39.12}$$

$$B = \frac{M_1 M_3 (E_1/E_T)}{(M_1 + M_2)(M_3 + M_4)}, \tag{39.13}$$

$$C = \frac{M_2 M_3}{(M_1 + M_2)(M_3 + M_4)}\left(1 + \frac{M_1 Q}{M_2 E_T}\right) = \frac{E_4'}{E_T}, \tag{39.14}$$

$$D = \frac{M_2 M_4}{(M_1 + M_2)(M_3 + M_4)}\left(1 + \frac{M_1 Q}{M_2 E_T}\right) = \frac{E_3'}{E_T}, \tag{39.15}$$

(Note that $A + B + C + D = 1$ and $AC = BD$.)
Then LS energy of light product equals

$$\frac{E_3}{E_T} = B + D + 2(AC)^{1/2}\cos\theta = B\left[\cos\psi \pm \left(\frac{D}{B} - \sin^2\psi\right)^{1/2}\right]^2 \tag{39.16}$$

(use only plus sign unless $B > D$, in which case $\psi_{\max} = \sin^{-1}(D/B)^{1/2}$).
LS energy of heavy product:

$$\frac{E_4}{E_T} = A + C + 2(AC)^{1/2}\cos\phi = A\left[\cos\xi \pm \left(\frac{C}{A} - \sin^2\xi\right)^{1/2}\right]^2 \tag{39.17}$$

(use only plus sign unless $A > C$, in which case $\xi_{\max} = \sin^{-1}(C/A)^{1/2}$).
LS angle of heavy product:

$$\sin\xi = \left(\frac{M_3 E_3}{M_4 E_4}\right)^{1/2}\sin\psi. \tag{39.18}$$

CMS angle of light product:

$$\sin\theta = \left(\frac{E_3}{E_T D}\right)^{1/2}\sin\psi. \tag{39.19}$$

Intensity or solid-angle ratio for light product:

$$\frac{\sigma(\theta)}{\sigma(\psi)} = \frac{\sin\psi d\psi}{\sin\theta d\theta} = \frac{\sin^2\psi}{\sin^2\theta}\cos(\psi - \theta) = \frac{(AC)^{1/2}\left(D/B - \sin^2\psi\right)^{1/2}}{E_3/E_T}. \tag{39.20}$$

Intensity or solid-angle ratio for heavy product:

$$\frac{\sigma(\phi)}{\sigma(\xi)} = \frac{\sin\xi d\xi}{\sin\phi d\phi} = \frac{\sin^2\xi}{\sin^2\phi}\cos(\phi - \xi) = \frac{(AC)^{1/2}\left(C/A - \sin^2\xi\right)^{1/2}}{E_4/E_T}. \tag{39.21}$$

Intensity or solid-angle ratio for associated particles in LS system:

$$\frac{\sigma(\xi)}{\sigma(\psi)} = \frac{\sin\psi d\psi}{\sin\xi d\xi} = \frac{\sin^2\psi\cos(\theta - \psi)}{\sin^2\xi\cos(\phi - \xi)}. \tag{39.22}$$

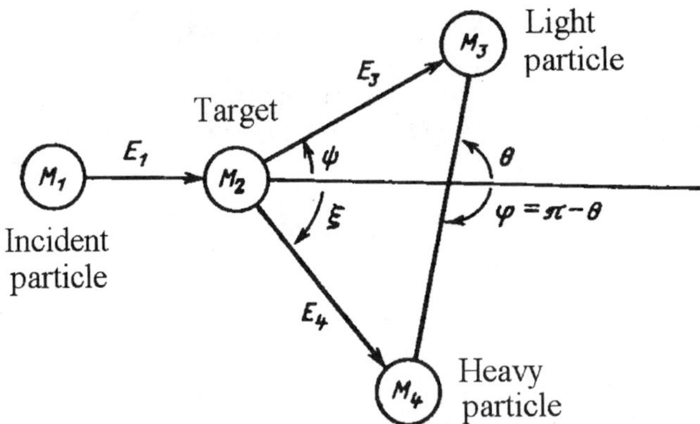

Figure 39.2 Kinematic diagram of nuclear reaction. M_i and E_i are masses and energies of participating particles; ψ and ξ are scattering angles in LS; θ and φ are the same angles in CMS.

39.7 Thermonuclear Reactions

In this section examples are presented of nuclear reactions (which play a fundamental role in the histories of mankind and of the Universe). These are thermonuclear reactions that underlie thermonuclear weapons and future energetics, and may also serve as energy sources for stars.

The basic thermonuclear reaction is the reaction between the deuteron D and the triton T which results in the production of a helium nucleus ^{4}He and a neutron n:

$$\mathrm{T} + \mathrm{D} \to {}^4\mathrm{He}(3.52) + \mathrm{n}(14.07) + 17.59 \text{ MeV}. \qquad (39.23)$$

This reaction has a low Coulomb barrier (see section 39.5), a very large cross section (see Fig. 39.3) and a very high reaction energy (see section 39.3). The figures in the brackets are the energies (in MeV) of the particles produced in the reaction if the particles in the input channel of the reaction have zero energies (see section 39.6). One can see that the neutron carries away the greater part of the reaction energy. For utilization of this energy, application of a lithium blanket has been proposed:

$$^6\mathrm{Li} + \mathrm{n} \to {}^4\mathrm{He} + \mathrm{T} + 4.78 \text{ MeV}. \qquad (39.24)$$

This reaction results in reproduction of the tritium.

Two other important thermonuclear reactions are the following ones:

$$\mathrm{D} + \mathrm{D} \to {}^3\mathrm{He}(0.82) + \mathrm{n}(2.45) + 3.27 \text{ MeV}, \qquad (39.25)$$

$$\mathrm{D} + \mathrm{D} \to \mathrm{T}(1.01) + \mathrm{p}(3.02) + 4.03 \text{ MeV}. \qquad (39.26)$$

The cross sections of these reactions are presented in Figs. 39.4 and 39.5.

One more thermonuclear reaction with a low Coulomb barrier is as follows:

$$\mathrm{T} + \mathrm{T} \to {}^4\mathrm{He}(1.25) + 2\mathrm{n}(5.04) + 11.33 \text{ MeV}. \qquad (39.27)$$

The cross section of this reaction is presented in Fig. 39.6.

The main disadvantage of the reactions above mentioned is the difficulty in utilization of the energy carried off by neutrons. From this point of view, one can consider reactions in which neutrons are not produced:

$$^6\mathrm{Li} + \mathrm{p} \to {}^3\mathrm{He}(2.30) + {}^4\mathrm{He}(1.72) + 4.02\ \mathrm{MeV}, \tag{39.28}$$

$$^9\mathrm{Be} + \mathrm{p} \to {}^4\mathrm{He}(1.28) + {}^6\mathrm{Li}(0.85) + 2.13\ \mathrm{MeV}, \tag{39.29}$$

$$^{11}\mathrm{B} \to 3^4\mathrm{He}(2.90) + 8.70\ \mathrm{MeV}. \tag{39.30}$$

These reactions exhibit much higher Coulomb barriers and require higher heating of the plasma. The cross sections for these reactions are presented in Figs. 39.7 and 39.8.

The composition of matter in the Universe is as follows: about 70% hydrogen, about 30% helium, and about 1% heavier elements. Hence, if nuclear reactions are the energy sources of stars they have to be reactions involving the burning of hydrogen. The only feasible reaction which produces a deuteron from two protons is the exotic reaction

$$\mathrm{p} + \mathrm{p} \to \mathrm{d} + \mathrm{e}^- + \nu_e. \tag{39.31}$$

Estimation of the cross section of this reaction for proton energies of about 1 MeV, yields a value of about 10^{-23} b. This means that processes in stars involving this reaction, are very slow.

The chain of nuclear reactions (referred to as the proton cycle) which produces ^{4}He nucleus from four protons and results in the reaction energy of about 26.7 MeV is as follows:

$$\mathrm{p} + \mathrm{p} \to \mathrm{d} + \mathrm{e}^- + \nu_e. \tag{39.32}$$

$$\mathrm{p} + \mathrm{d} \to {}^3\mathrm{He} + \gamma, \tag{39.33}$$

$$^3\mathrm{He} + {}^3\mathrm{He} \to {}^4\mathrm{He} + 2\mathrm{p} \tag{39.34}$$

or

$$^4\mathrm{He} + {}^3\mathrm{He} \to {}^7\mathrm{Be} + \gamma, \tag{39.35}$$

$$^7\mathrm{Be} + \mathrm{e}^- \to {}^7\mathrm{Li} + \nu_e, \tag{39.36}$$

$$^7\mathrm{Li} + \mathrm{p} \to 2\,{}^4\mathrm{He} \tag{39.37}$$

or

$$^7\mathrm{Be} + \mathrm{p} \to {}^8\mathrm{B} + \gamma, \tag{39.38}$$

$$^8\mathrm{B} \to {}^8\mathrm{Be} + \mathrm{e}^+ + \nu_e, \tag{39.39}$$

$$^8\mathrm{Be} \to 2\,{}^4\mathrm{He}. \tag{39.40}$$

Another possible nuclear reaction chain leading to the same result ($4\mathrm{p} \to {}^4$ He + 26.7 MeV) is spoken of as the carbon cycle because ^{12}C nucleus plays there the role of a catalyzer:

$$^{12}\mathrm{C} + \mathrm{p} \to {}^{13}\mathrm{N} + \gamma, \tag{39.41}$$

$$^{13}\mathrm{N} \to {}^{13}\mathrm{C} + \mathrm{e}^- + \nu_e, \tag{39.42}$$

$$^{13}\mathrm{C} + \mathrm{p} \to {}^{14}\mathrm{N} + \gamma, \tag{39.43}$$

$$^{14}\mathrm{N} + \mathrm{p} \to {}^{15}\mathrm{O} + \gamma, \tag{39.44}$$

$$^{15}\mathrm{O} \to {}^{15}\mathrm{N} + \mathrm{e}^- + \nu_e, \tag{39.45}$$

$$^{15}\mathrm{N} + \mathrm{p} \to {}^{12}\mathrm{C} + {}^4\mathrm{He}. \tag{39.46}$$

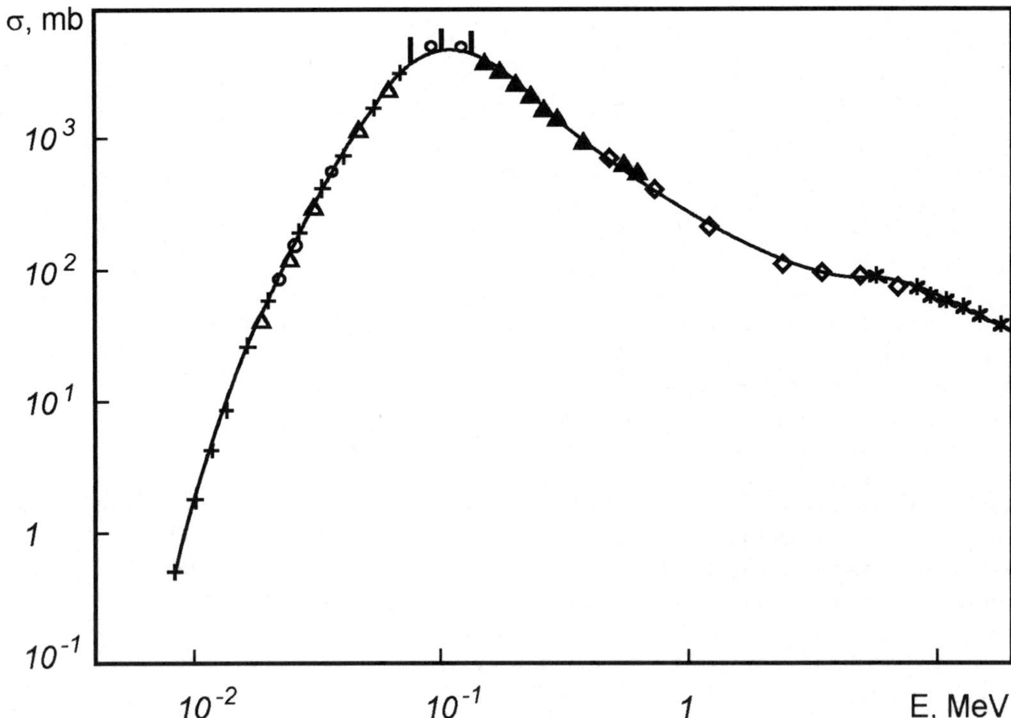

Figure 39.3 Excitation function for the T(d, n)^{4}He reaction (the total cross section for the T(d, n)^{4}He reaction versus the deuteron energy). The points indicate the results of measurements. The solid line is the best fit to the experimental data [3].

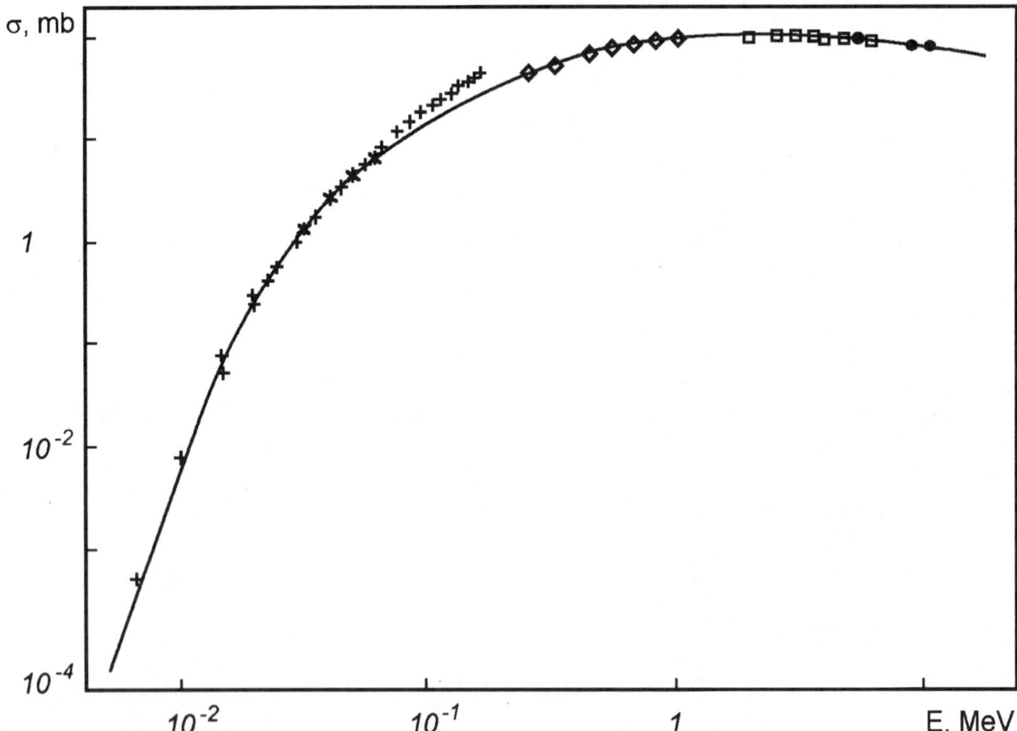

Figure 39.4 Excitation function for the D(d, n)^{3}He reaction. The notations are the same as in Fig. 39.3.

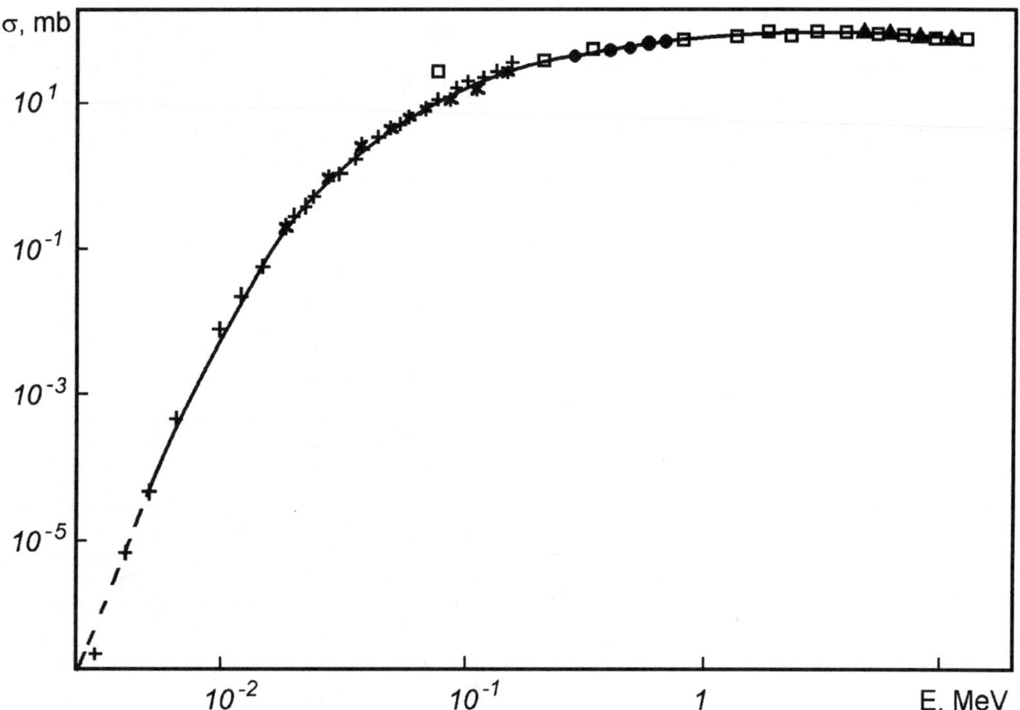

Figure 39.5 Excitation function for the D(d, p)T reaction. The notations are the same as in Fig. 39.3.

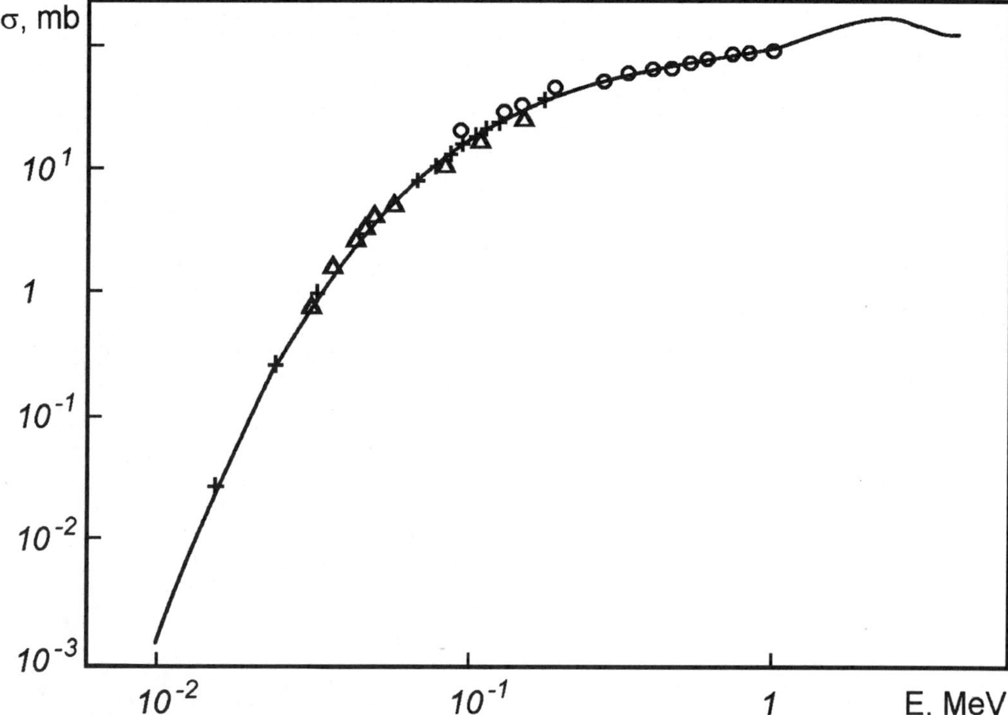

Figure 39.6 Excitation function for the $T(T, 2n)^4He$ reaction. The notations are the same as in Fig. 39.3.

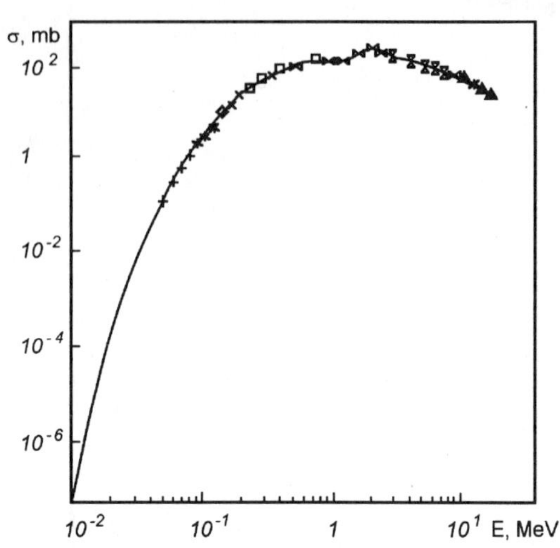

Figure 39.7 Excitation function for the ^{6}Li(p, α)^{3}He reaction. The notations are the same as in Fig. 39.3.

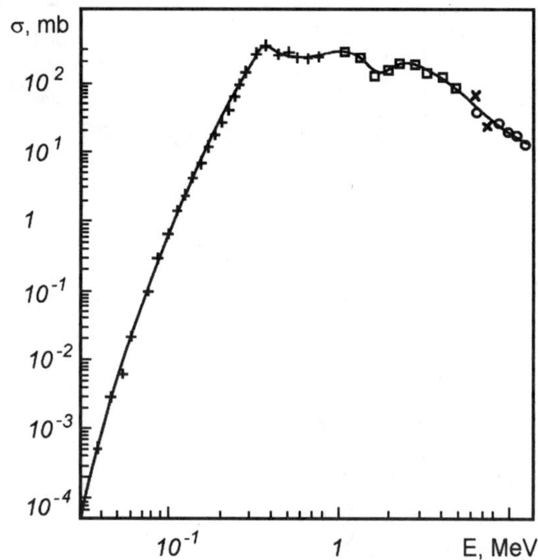

Figure 39.8 Excitation function for the ^{9}Be(p, α)^{6}Li reaction. The notations are the same as in Fig. 39.3.

References

[1] Wapstra, A.N., Bos, K., Atomic data and nuclear data tables, vol.19, 177, 1977.

[2] Marion, J.B., Young, F.C., Nuclear reaction analysis: graphs and tables, North Holland Publ. Comp., Amsterdam, 1968.

[3] Abramovich, S.N., Gugovsky, B.J., Gerebtsov, V.A., and Zvenigorodsky, A.G., Yaderno-fizicheskie konstanti termoyadernogo sinteza, Moscow, 1989 (in Russian).

40

Nuclear Fission

A.I. Obukhov and I.S. Grigoriev

40.1 Introduction

Fission of an atomic nucleus denotes the process of its decay into two (more rarely, three and four) nuclei (fission fragments) comparable in mass. Nuclear fission was first observed in irradiating uranium nuclei with neutrons [1], then spontaneous uranium fission was discovered [2]. For the nuclei with mass number $A > 100$, the fission reaction is exothermic because the binding energy per nucleon is larger in nuclei–fragments as opposed to a fissile nucleus. The energy released in the course of nuclear fission is given out as the kinetic energy of fragments, the energy carried away by neutrons, γ-quanta, β-particles, and antineutrinos accompanying the fission process.

Excitation energy can be imparted to a nucleus either via irradiating with γ-quanta or bombarding with neutrons and high-energy charged particles which transfer a part of their energy to the nucleus in inelastic collisions. Of fundamental importance is the process of nuclear fission by the neutron capture.

The threshold energy of γ-quanta-induced fission, i.e. the energy of the lowest-energy γ-quanta still able of inducing fission, is a direct measure of the minimal excitation energy necessary for fission onset. The threshold energy of neutron-induced fission is smaller than that of photofission of a compound nucleus by the binding energy of a neutron and a target nucleus; therefore, some threshold-energies of the neutron-induced fission are negative. This means that the fission process is feasible on the capture of thermal and cold neutrons by nuclei. The latter are called the well-fissile nuclei (^{233}U, ^{235}U, ^{239}Pu, ^{241}Pu).

Apart from the nuclear fission under the action of the above excitation mechanisms, nuclear fission is also possible without apparent external influences onto the nucleus. This process is called spontaneous nuclear fission. It is commonly supposed that in nonexcited nuclei (considered as small drops) there occur oscillations with a 10^{-20}–10^{-21} s period and amplitude of 0.1–0.2 nuclear radius. The fission barrier restrains spontaneous disintegration of a nucleus; however, with a huge number of oscillations, the barrier may happen to be overcome through tunneling. Lifetimes of nuclei with respect to spontaneous fission change from 10^{21} y for uranium and thorium isotopes to milliseconds for nuclei with charges $Z = 104$–107.

Fission fragments of a heavy nucleus in most cases have different masses. For instance, in uranium fission, the mass of a light fragment equals 90–100 a.m.u., whereas that of a heavy fragment, 130–140 a.m.u. The velocities of fragments are on the order of 10^9 cm/s. Their momenta are equal in magnitude and opposite in sign. The velocity of fragments reaches 0.9 of its final value in a time of about 10^{-20} s at a distance between them on the order of 10^{-11} cm, when they are still in the lowest electron shell of the atom.

The fission fragments find themselves originally in excited states; their total excitation energy ranges to about 20 MeV. In the first place, the fragments throw off superfluous neutrons (fission prompt

0-8493-2861-6/97/$0.00+$.50

neutrons) in 10^{-14}–10^{-17} s. The average excitation of a fragment upon emitting neutrons equals half the neutron binding energy and amounts to 6–7 MeV for both the fragments. This energy is emitted in the form of γ-quanta in 10^{-9}–10^{-14} s.

At the beginning, the fragments possess a positive charge from 10 to 20 electron charges because some orbital electrons are "shaken off" in the course of fission. Passing through matter, the fragments lose their energy by ionization of the medium. Along the path of the two fragments, about $5 \cdot 10^6$ ion pairs are created. The time of fragments slowing down in air is on the order of 10^{-9} s; in dense media, 10^{-12} s. In this time, they cover the distance of about 2 cm in air; and about 10^{-3} cm, in dense media.

The charge of the retarded fragments differs widely from that of stable nuclides with the same mass. This difference reduces in the series of β-decays (as a rule, three or four decays for each fragment).

In particular cases, the β-decay of fission products gives rise to highly excited states of a daughter nucleus for which the neutron emission is possible. Since the decay constant with respect to emission of this "delayed" neutron much exceeds the β-decay constant, the neutron emission follows the β-decay of a parent nucleus almost instantly, and the time dependence of the delayed neutrons intensity is the same as that for the β-decay of the parent nucleus.

More detailed information on the nuclear fission can be found in [3–9].

40.2 Nuclear Fission Barriers

The probability of nuclear fission (within the liquid–drop nuclear model) at low excitation energies, $E \leq 6$ MeV, was connected with the potential barrier penetration whose dependence on the nucleus deformation δ nearby the barrier top was assumed to be roughly parabolic:

$$U(\delta) = B_{\max} - \text{const}(\delta - \delta_0)^2,$$

where δ_0 is the deformation corresponding to the maximum height $B_{\max}$ of the barrier.

Recent discovery of a variety of new phenomena (spontaneously fissile isomers, broad subbarrier resonances, groups of narrow resonances with a large partial fissionability, etc.) has led to the idea of a more complicated double-humped structure of the barrier with maxima of the height B_A and B_B at deformations $\delta_A \simeq 0.4$ and $\delta_B \simeq 0.8$, respectively. The saddle between maxima lies at $\delta_{II} \simeq 0.6$, and its bottom settles by $E_{\rm II}$ above the ground state [10].

Table 40.1 shows the double-humped barrier parameters of heavy nuclei ($Z \geq 88$), whereas Table 40.2, the values of $B_{\max}$ for nuclei with $Z \leq 85$ and with a single-humped fission barrier. The internal barrier B_A of the nuclei from Th up to Fm amounts to 5–6 MeV.

Fission barriers of nuclei with $Z < 85$ enhance with decreasing Z^2/A. For still lighter nuclei the liquid–drop model predicts the growth of the fission barrier, its maximum value in the neighborhood of molybdenum nuclei, and then its decrease down to zero when $Z^2/A \to 0$.

Table 40.1 Fission barriers B_A and B_B and the depth of the second potential well $E_{\rm II}$ (in MeV) [10] [In all tables, errors are given in parentheses.]

Fissile nucleus	$B_{\rm A}$	$E_{\rm II}$	$B_{\rm B}$	Fissile nucleus	$B_{\rm A}$	$E_{\rm II}$	$B_{\rm B}$
^{228}Ra	8.0 (0.5)		8.5 (0.5)	^{231}Th	6.0 (0.1)	<5.8	6.1 (0.3)
^{226}Ac	6.0 (0.6)		7.7 (0.3)	^{232}Th	5.8 (0.2)	≪4.5	6.2 (0.2)
^{227}Th	5.9 (0.3)		6.6 (0.3)	^{233}Th	6.3 (0.2)	<6.2	6.3 (0.2)
^{228}Th	6.2 (0.3)		6.5 (0.3)	^{234}Th	6.1 (0.2)		6.5 (0.2)
^{229}Th			6.5 (0.3)	^{231}Pa	5.9 (0.2)		5.9 (0.3)
^{230}Th	6.1 (0.2)		6.5 (0.3)	^{232}Pa	6.1 (0.3)	<5.7	6.2 (0.2)

Table 40.1 Fission barriers B_A and B_B and the depth of the second potential well E_{II} *(continued)*

Fissile nucleus	B_A	E_{II}	B_B	Fissile nucleus	B_A	E_{II}	B_B
^{233}Pa	6.1 (0.3)		6.1 (0.3)	^{245}Pu	5.6 (0.2)		5.0 (0.2)
^{232}U	5.2 (0.2)		5.1 (0.3)	^{237}Am		2.4±0.2	
^{234}U	5.6 (0.2)		5.5 (0.2)	^{238}Am		2.6±0.2	
^{235}U	5.9 (0.2)	2.5 ±0.3	5.6 (0.2)	^{239}Am	6.2 (0.3)	2.4±0.2	
^{236}U	5.6 (0.2)	2.3 ±0.2	5.5 (0.2)	^{240}Am	6.5 (0.2)	3.0±0.2	5.2 (0.3)
^{237}U	6.1 (0.2)	2.5 ±0.4	5.9 (0.2)	^{241}Am	6.0 (0.2)	2.2±0.2	5.1 (0.3)
^{238}U	5.7 (0.2)	2.6 ±0.1	5.7 (0.2)	^{242}Am	6.5 (0.2)	2.9±0.2	5.4 (0.3)
^{239}U	6.3 (0.2)	1.9 ±0.3	6.1 (0.2)	^{243}Am	5.9 (0.2)	2.3±0.2	5.4 (0.3)
^{240}U	5.7 (0.2)		5.5 (0.2)	^{244}Am	6.3 (0.2)	2.8±0.4	5.4 (0.3)
^{234}Np	5.5 (0.2)		5.1 (0.2)	^{245}Am	5.9 (0.2)		5.2 (0.3)
^{235}Np	5.5 (0.2)		5.2 (0.2)	^{247}Am	5.5 (0.2)		
^{236}Np	5.8 (0.2)		5.6 (0.2)	^{241}Cm	6.3 (0.3)	2.1±0.3	4.3 (0.5)
^{237}Np	5.7 (0.2)	2.8±0.3	5.4 (0.2)	^{242}Cm	5.8 (0.4)		4.0 (0.5)
^{238}Np	6.1 (0.2)	2.3±0.3	6.0 (0.2)	^{243}Cm	6.4 (0.3)	1.9±0.3	
^{239}Np	5.9 (0.2)		5.4 (0.2)	^{244}Cm	5.8 (0.2)		4.3 (0.3)
^{232}Pu	5.3 (0.4)			^{245}Cm	6.2 (0.2)	2.1±0.3	
^{234}Pu	5.8 (0.7)			^{246}Cm	5.7 (0.2)		4.2 (0.3)
^{235}Pu		2.6±0.4	5.1 (0.4)	^{247}Cm	6.0 (0.2)		
^{236}Pu			4.5 (0.4)	^{248}Cm	5.7 (0.2)		
^{237}Pu		2.8±0.2		^{249}Cm	5.6 (0.2)		
^{238}Pu	5.5 (0.2)	2.7±0.2	5.0 (0.2)	^{250}Cm	5.3 (0.2)		
^{239}Pu	6.2 (0.2)	2.6±0.2	5.5 (0.2)	^{249}Bk	6.1 (0.2)		
^{240}Pu	5.6 (0.2)	2.4±0.3	5.1 (0.2)	^{250}Bk	6.1 (0.2)		4.1 (0.3)
^{241}Pu	6.1 (0.2)	1.9±0.3	5.4 (0.2)	^{250}Cf	5.6 (0.3)		
^{242}Pu	5.5 (0.2)		5.1 (0.2)	^{253}Cf	5.4 (0.3)		
^{243}Pu	5.9 (0.2)	1.7±0.3	5.2 (0.2)	^{250}Es	6.7		
^{244}Pu	5.4 (0.2)		5.0 (0.2)	^{255}Fm	5.7		

Table 40.2 Fission barrier heights B_{max} for nuclei from ^{213}At to ^{153}Tb [11]

Compound nucleus	B_{max}, MeV	Compound nucleus	B_{max}, MeV	Compound nucleus	B_{max}, MeV
^{213}At	17.3	^{200}Hg	23.5–24.6	^{190}Os	25.6
^{212}At	18.6	^{199}Hg	23.1–24.7	^{188}Os	24.4
^{212}Po	19.6	^{198}Hg	21.4–22.7	^{187}Os	24.6–24.8
^{211}Po	20.5–20.6	^{197}Hg	21.1–22.7	^{186}Os	24.1
^{210}Po	21.2	^{196}Hg	19.7–21.2	^{185}Os	24.0
^{209}Po	21.1	^{198}Au	23.7–24.9	^{185}Re	26.2
^{208}Po	19.9	^{197}Au	23.4–25.1	^{184}Re	26.3
^{207}Po	19.3	^{196}Au	22.5–23.9	^{181}Re	19.0 [12]
^{210}Bi	24.2–24.4	^{195}Au	20.8–22.9	^{184}W	28.3
^{208}Bi	23.6–24.1	^{194}Au	20.1–21.7	^{183}W	28.3
^{207}Bi	22.8–22.9	^{196}Pt	24.9–26.3	^{182}W	27.4
^{206}Bi	22.4	^{194}Pt	22.8–24.4	^{181}W	26.9–27.0
^{208}Pb	27.4	^{193}Pt	22.7–24.2	^{180}W	26.5–26.6
^{207}Pb	26.9–27.0	^{192}Pt	21.4–23.1	^{179}W	25.8
^{206}Pb	25.3	^{191}Pt	20.8–22.5	^{179}Ta	28.6
^{205}Pb	24.6	^{191}Ir	23.2	^{173}Lu	30.5
^{204}Pb	23.2–23.8	^{190}Ir	22.6–22.9	^{173}Yb	33.4
^{201}Tl	23.1	^{189}Ir	22.0–22.1	^{170}Yb	30.6
^{200}Tl	22.8	^{188}Ir	22.2	^{153}Tb	28.9 [12]

40.3 Spontaneous Fission

Spontaneous fission half-lives of nuclei in the ground and isomer states are presented in Tables 40.3 and 40.4.

In Table 40.3, the spontaneous fission half-lives are given for the ground state nuclei of isotopes of three natural and seventeen synthesized elements. The half-life period diminishes by about 31 orders for nuclei from Th up to Db, and then it changes slightly. Average numbers of prompt neutrons and kinetic energies are there presented for pair fragments formed in spontaneous fission. Besides, Table 40.3 contains information on a new type of radioactivity, viz. spontaneous disintegration with emission of ^{14}C type fragments for francium and radium nuclei, and ^{24}Ne for uranium nuclei. In these cases, instead of $T_{1/2}$, given is the portion ϵ of decays into heavy fragments as opposed to the emission of α-particles. For these cases the E_k column presents the kinetic energies of the fragments.

Table 40.4 contains half-lives of spontaneous fission from an isomer state localized within the second potential well for nuclei from U to Bk. As compared to the spontaneous fission from the ground state in the first potential well, the spontaneous fission from an isomer state shows considerably smaller life-times, which is due to the size and width of the external barrier B_B being much smaller.

Table 40.3 Spontaneous fission half-lives $T_{1/2}$ of ground state nuclei, mean numbers of prompt neutrons $\bar{\nu}_{pr}$, and average kinetic energies E_k of fission fragments

Nuclide	A	$T_{1/2}$ [13–17]	$\bar{\nu}_{pr}$ [18]	E_k, MeV [19]	Nuclide	A	$T_{1/2}$ [13–17]	$\bar{\nu}_{pr}$ [18]	E_k, MeV [19]
^{87}Fr	221	$^{6}_{14}$C+$^{207}_{81}$Tl;		31.3 [35]	^{95}Am	232	10^2 y [22]		
		$\epsilon \leq 4.4 \cdot 10^{-12}$				234	10^4 y [22]		
^{88}Ra	221	$^{14}_{6}$C+$^{207}_{82}$Pb;		32.4 [35]		240	10^{14} y [22]		
		$\epsilon \leq 4.4 \cdot 10^{-12}$				241	$1.06\ (3) \cdot 10^{14}$ y		
	222	$^{14}_{6}$C+$^{208}_{82}$Pb;		33.0 [35]		242	$8.8\ (33) \cdot 10^{11}$ y	2.42(14)*1 [20]	
		$\epsilon = 3.7(6) \cdot 10^{-10}$		33.0 [35]		243	$3.35\ (31) \cdot 10^{13}$ y		
	223	$^{14}_{6}$C+$^{209}_{82}$Pb;		31.9 [35]		244	10^{14} y [22]		
		$\epsilon = 6.1(10) \cdot 10^{-10}$		31.9 [35]	^{96}Cm	240	$1.9 \cdot 10^6$ y		
	224	$^{14}_{6}$C+$^{210}_{82}$Pb;		30.5 [35]		242	$6.5\ (6) \cdot 10^6$ y	2.538 (12)	
		$\epsilon = 4.3(12) \cdot 10^{-11}$				244	$1.344\ (2) \cdot 10^7$ y	2.696 (10)	181.5(10)
^{90}Th	230	$1.5\ (15) \cdot 10^{17}$ y	1.24(15)*1 [20]			246	$1.81\ (4) \cdot 10^7$ y	2.950 (14)	184.2 (5)
	232	$> 10^{21}$ y	2.13 (20)[21]			248	$4.11\ (4) \cdot 10^6$ y	3.157 (15)	182.5 (9)
^{91}Pa	230	$> 10^{16}$ y [20]				250	$1.13\ (5) \cdot 10^4$ y	3.17 (8)	179.1 (27)
	231	$1.1 \cdot 10^{16}$ y			^{97}Bk	246	$2 \cdot 10^{13}$ y [22]		
^{92}U	232	$7.8\ (60) \cdot 10^{13}$ y				249	$1.864\ (87) \cdot 10^9$ y	3.395 (26)	
	232	$^{24}_{10}$Ne+$^{208}_{82}$Pb;		32.8 [36]	^{98}Cf	246	$1.83 \cdot 10^3$ y	3.14(9)	196.6 (20)
		$\epsilon = 2.0(5) \cdot 10^{-12}$				248	$7 \cdot 10^3$ y		189.3 (30)
	233	$1.2\ (3) \cdot 10^{17}$ y				249	$6.98\ (15) \cdot 10^{10}$ y		
	234	$1.42\ (8) \cdot 10^{16}$ y	1.63 (15)*1			250	$1.70\ (7) \cdot 10^4$ y	3.51 (4)	187.0 (5)
	235	$9.8\ (28) \cdot 10^{18}$ y				252	85.38 (39) y	3.7675 (40)	186.3 (10)
	236	$2.43\ (13) \cdot 10^{16}$ y	1.89 (5)			254	0.1656 (5) y	3.83 (5)	188.3 (20)
	238	$8.08\ (26) \cdot 10^{15}$ y	1.98 (3)	168.2(11)	^{99}Es	253	$6.42\ (22) \cdot 10^5$ y		191.3(30)
^{93}Np	228	10^8 y [22]				254	$> 2.5 \cdot 10^7$ y		
	237	10^{18} y				255	$2.44\ (14) \cdot 10^3$ y		
	239	$> 5 \cdot 10^{12}$ y [20]			^{100}Fm	242	$0.82 \cdot 10^{-3}$s		
^{94}Pu	236	$3.5\ (10) \cdot 10^9$ y	2.21(18)			244	$3.3 \cdot 10^{-3}$s		
	238	$4.77\ (13) \cdot 10^{10}$ y	2.24 (8)			246	15s		199.6 (40)
	239	$7.8\ (16) \cdot 10^{15}$ y				248	$3.6 \cdot 10^4$s		198.7 (40)
	240	$1.15\ (4) \cdot 10^{11}$ y	2.17 (1)	180.4 (3)		250	10 y		
	241	$> 3 \cdot 10^{15}$ y				252	115 y		
	242	$6.84\ (8) \cdot 10^{10}$ y	2.153 (19)	179.96(20)		254	228 (1) d	3.98 (19)	194.9 (10)
	244	$6.56\ (32) \cdot 10^{10}$ y	2.30 (19)	181.8(10)		255	$\sim 10^4$ y		

Table 40.3 Spontaneous fission half-lives $T_{1/2}$ of ground state nuclei *(continued)*

Nuclide	A	$T_{1/2}$ [13–17]	$\bar{\nu}_{pr}$ [18]	E_k, MeV [19]	Nuclide	A	$T_{1/2}$ [13–17]	$\bar{\nu}_{pr}$ [18]	E_k, MeV [19]
	256	2.6 h	3.74 (18)	198.2 (5)		256	$7 \cdot 10^{-3}$ s		
	257	131 (3) y	3.796 (14)	197.6		257	47 s		
	258	0.38 ms		238 [15]		258	13 (3)$\cdot 10^{-3}$ s		
	259	1.5 s		245.2 (60)		259	3 s		
^{101}Md	255	$\geq 5 \cdot 10^{-2}$ y				260	0.020–0.08 s		199
	256	$\geq 10^{-2}$ y					[16, 34]		
	257	≥ 6 d				261	70 s		
	259	95 min		203.9 (14)		262	$47 \cdot 10^{-3}$ s		
^{102}No	250	$0.25 \cdot 10^{-3}$ s			^{105}Jl	257	≥ 8 s		
	252	~ 8 s	4.15 [23]	202.4		260	~ 15 s		
	254	~ 24 h				261	~ 7 s		
	256	10^3 s				262	47 s		
	258	$1.2 \cdot 10^{-3}$ s			^{106}Rf	259	> 2 s		
^{103}Lr	256	$\sim 10^5$ s				260	$7 \cdot 10^{-3}$ s		
^{104}Db	254	$5 \cdot 10^{-4}$ s				263	~ 1 s		
	255	~ 4.5 s			^{107}Bh	261	10^{-2} s		

*[1] The value is extrapolated from the one measured with thermal or fast neutrons.

Table 40.4 Spontaneous fission half-lives $T_{1/2}$ of nuclei in isomer states [22, 24] [In the cases when two values of $T_{1/2}$ are given for a nuclide, they refer to different isomer states.]

Nucleus	$T_{1/2}$, s	Nucleus	$T_{1/2}$, s	Nucleus	$T_{1/2}$, s	Nucleus	$T_{1/2}$, s
^{236}U	$1.16 \cdot 10^{-7}$	^{239}Pu	$8 \cdot 10^{-6}$	^{238}Am	$3.5 \cdot 10^{-5}$	^{242}Cm	$1.8 \cdot 10^{-7}$
^{238}U	$\geq 10^{-9}$	^{240}Pu	$3.4 \cdot 10^{-9}$	^{239}Am	$1.6 \cdot 10^{-7}$		$5 \cdot 10^{-11}$
	$1.95 \cdot 10^{-7}$	^{241}Pu	$3 \cdot 10^{-8}$	^{240}Am	$9 \cdot 10^{-4}$	^{243}Cm	$4.2 \cdot 10^{-8}$
^{228}Np	60		$2.1 \cdot 10^{-5}$	^{241}Am	$1.5 \cdot 10^{-6}$	^{244}Cm	$> 1 \cdot 10^{-7}$
^{237}Np	$4 \cdot 10^{-8}$	^{242}Pu	$5 \cdot 10^{-8}$	^{242}Am	$1.4 \cdot 10^{-2}$		$\leq 5 \cdot 10^{-12}$
^{235}Pu	$3 \cdot 10^{-8}$		$3.6 \cdot 10^{-9}$	^{243}Am	$5.5 \cdot 10^{-6}$	^{245}Cm	$1.3 \cdot 10^{-8}$
^{236}Pu	$3.4 \cdot 10^{-8}$	^{243}Pu	$6 \cdot 10^{-8}$	^{244}Am	$1.0 \cdot 10^{-3}$	^{242}Bk	$9.5 \cdot 10^{-9}$
^{236}Pu	$3.7 \cdot 10^{-11}$	^{244}Pu	$3.8 \cdot 10^{-10}$	^{245}Am	$6.4 \cdot 10^{-7}$		$6 \cdot 10^{-7}$
^{237}Pu	$1.1 \cdot 10^{-6}$	^{245}Pu	$9 \cdot 10^{-8}$	^{246}Am	$7.3 \cdot 10^{-5}$	^{244}Bk	$8.2 \cdot 10^{-7}$
	$1.1 \cdot 10^{-7}$	^{232}Am	85	^{240}Cm	$5.5 \cdot 10^{-8}$	^{245}Bk	$2 \cdot 10^{-9}$
^{238}Pu	$6 \cdot 10^{-9}$	^{234}Am	150		$1 \cdot 10^{-11}$		
	$5 \cdot 10^{-10}$	^{237}Am	$5 \cdot 10^{-9}$	^{241}Cm	$1.5 \cdot 10^{-8}$		

40.4 Cross Sections of Nuclear Fission

In Tables 40.5 and 40.6, the cross sections σ_f are presented for nuclear fission induced by thermal neutrons. The values with asterisks are recommended for the energy $E_n = 0.0253$ eV ($v = 2200$ m/s); E_f is the average energy of fission neutrons. The values marked by the letter r were obtained for the distribution of neutrons over the reactor. Figures 40.1 – 40.4 show the dependences $\sigma_f(E_n)$ for main fissile nuclei and ^{238}U.

In Table 40.6, collected are the cross sections of thermal-neutron fission. With growing energy of neutrons, the fission cross sections of fissile nuclei such as ^{233}U, ^{235}U, ^{239}Pu, ^{241}Pu are first decreasing undergoing several resonance peaks and at a neutron energy of 1–5 MeV they form the first plateau (Figs. 40.1, 40.3, and 40.4). Easy fission of nuclei of the second group: ^{232}Th, ^{238}U, ^{240}Pu, ^{242}Pu, and some others, starts only at a certain energy above E_f^{thresh}. Their fission cross sections first grow abruptly, and then reach saturation, the first plateau. When the neutron energy exceeds 5 MeV, the fission cross sections of the two groups of nuclei behave in a similar manner: at neutron energies from about 8 MeV to 14 MeV, the cross sections form the second plateau, then the third plateau, etc.

In precision measurements, an intermediate structure was observed in the fission cross sections of "threshold" nuclei, like ^{236}U, ^{238}U, ^{237}Np, ^{240}Pu, and ^{242}Pu, whereas for ^{230}Th and others, subbarrier resonances were found. An example of the intermediate structure in the fission cross section of ^{238}U nuclei is drawn in Fig. 40.2. These phenomena have been interpreted within the two-hump model [3, 14]. Basic characteristics of nuclei fissile in interaction of ^{233}U, ^{235}U, ^{239}P, and ^{241}Pu with thermal neutrons are presented in Table 40.6. The fission cross sections of main nuclei fissile in interaction with 0.1–20 MeV neutrons can be found in Ref. [3].

Table 40.5 Cross sections of thermal-neutron fission of nuclei σ_f and numbers of secondary neutrons per fission act $\bar{\nu}_{pr}$

Nucleus-target	σ_f, 10^{-28} m^2 [18]	$\bar{\nu}_{pr}$ [6, 18]	Nucleus-target	σ_f, 10^{-28} m^2 [18]	$\bar{\nu}_{pr}$ [6, 18]
^{223}Ra	0.7(3)		^{236}Pu	170(35)	
^{226}Ra	<5·10^{-5}		^{237}Pu	2455(295)	
^{228}Ra	<2		^{238}Pu	17.9(4)*	2.900(20)
^{227}Ac	<2.9·10^{-4}		^{239}Pu	748.0(20)* [25]	2.874(7)
^{227}Th	202(13)		^{240}Pu	0.056(30)	2.921(12)
^{228}Th	<0.3		^{241}Pu	1011.1(62)* [25]	
^{229}Th	30.8(15)	2.08(2)	^{242}Pu	0.0007* [26]	
^{230}Th	<1.2·10^{-3}		^{243}Pu	196(16)	
^{232}Th	2.5·10^{-6}	2.12(10)	^{241}Am	3.20(9)*	
^{233}Th	15(2)		^{242}Am^m	6950(280)*	3.260(24)
^{234}Th	<0.01		^{242}Am^g	2100(200)	
^{230}Pa	1500(250)		^{243}Am	0.1983(43)	3.214(38)
^{231}Pa	0.020(1)		^{244}Am^m	1600(300)	
^{232}Pa	700(100)r		^{244}Am^g	2300(300)	
^{233}Pa	<0.1		^{242}Cm	< 5	
^{234}Pa^m	<500		^{243}Cm	617(20)*	3.430(40)
^{234}Pa^g	<5000		^{244}Cm	1.04(20)*	
^{230}U	25(10)		^{245}Cm	2145(58)*	3.717(60)
^{231}U	400(300)		^{246}Cm	0.14(5)*	
^{232}U	76.8(48)*		^{247}Cm	81.9(44)	3.79(15)
^{233}U	529.1(12)* [25]	2.480(7)	^{248}Cm	0.37(5)	
^{234}U	<0.65		^{250}Bk	960(150)	
^{235}U	582.6(11)* [25]	2.407(7)	^{249}Cf	1642(33)*	4.06(4)
^{236}U	0.07	2.338(22)	^{251}Cf	4895(250)	
^{237}U	<0.35		^{252}Cf	32(4)	
^{238}U	(4–11)·10^{-6}		^{253}Cf	1300(240)	
^{239}U	14(3)		^{254}Es^m	1826(80)	
^{234}Np	900(300)		^{254}Es^g	1966(160)	
^{236}Np	2500(150)	3.12(14)	^{255}Fm	3360(170)	
^{237}Np	0.0215(24)*	2.525(16)	^{257}Fm	2950(160)	
^{238}Np	2088(30)				

Table 40.6 Constants of fissile nuclei interaction with neutrons of velocity v = 2200 m/s [25]

Characteristic	^{233}U	^{235}U	^{239}Pu	^{241}Pu
σ_γ, 10^{-28} m^2	45.5(7)	98.3(8)	269.3(2.2)	358.2(5.1)
σ_f, 10^{-28} m^2	529.1(1.2)	582.6(1.1)	748(20)	1011.1(6.2)
σ_{abs}, 10^{-28} m^2	574.7(1.0)	680.9(11)	1017.3(2.9)	1369.4(7.7)
ν_{pr}[6]	2.480(7)	2.407(7)	2.874(7)	2.921(12)
ν_{delay}	0.0066(30)	0.01659(55)	0.00622(32)	0.01563(160)
ν_{abs}	2.4933(39)	2.4251(34)	2.8768(57)	2.9369(73)
$\bar{E}_n$, MeV	2.0627(248)	2.0266(220)	2.0990(232)	2.0858(232)

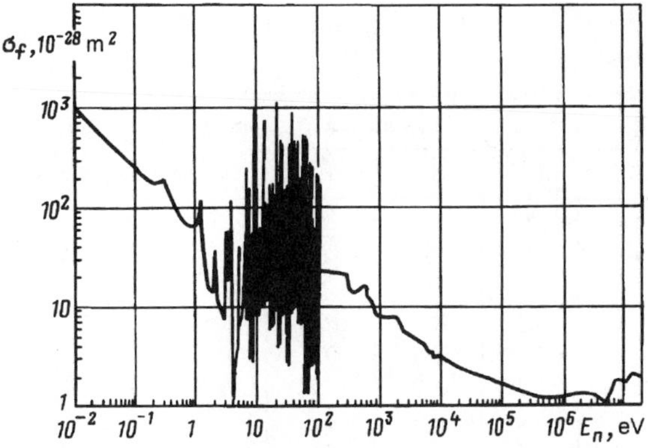

Figure 40.1 Dependence of the fission cross section of ^{235}U on neutron energy [27].

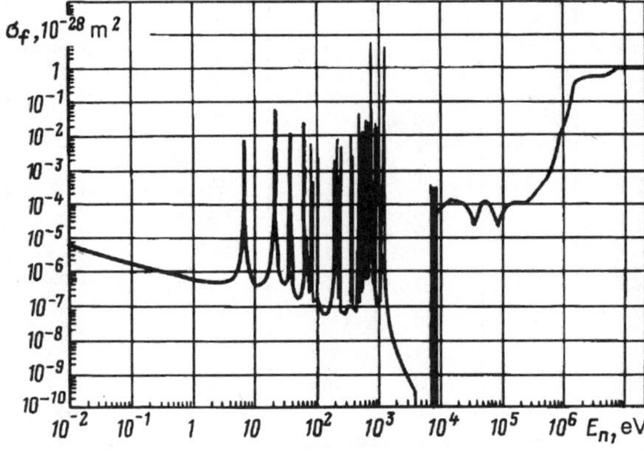

Figure 40.2 Dependence of the fission cross section of ^{238}U on neutron energy [27].

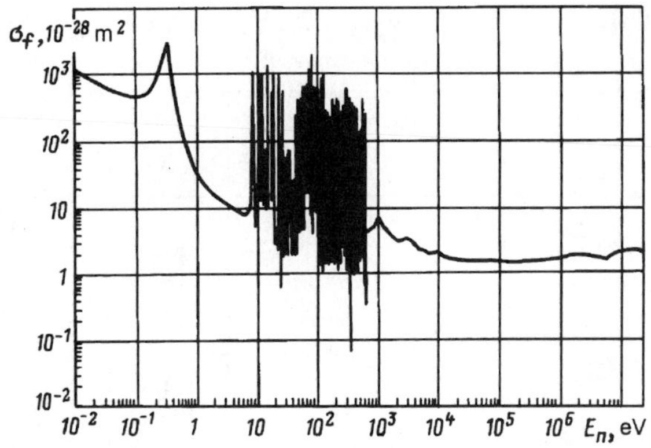

Figure 40.3 Dependence of the fission cross section of ^{239}Pu on neutron energy [27].

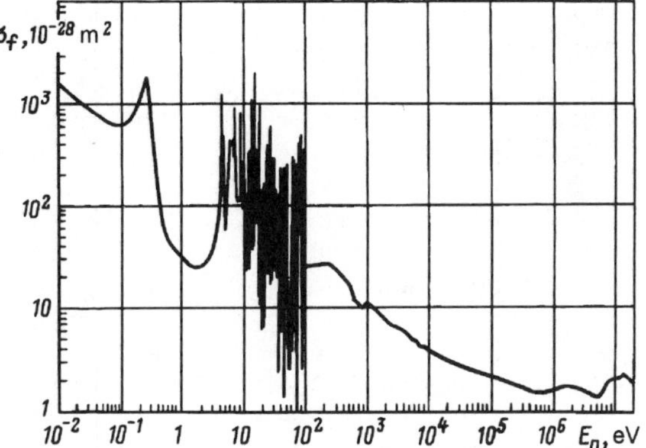

Figure 40.4 Dependence of the fission cross section of ^{241}Pu on neutron energy [27].

40.5 Energy Released in Fission

The total energy released in the process of nuclear fission is a sum of the following components:

kinetic energy of the fission fragments E_k;
energy of prompt γ-quanta $E_{\gamma\text{prompt}}$;
energy of fission neutrons E_f;
energy of β-radiation of fission products E_β;
energy of γ-radiation of fission products E_γ;
energy carried away by antineutrinos in β-decays $E_{\bar{\nu}}$.

The energy carried away by antineutrinos is not given out as thermal energy; therefore, about 196 MeV are released per an act of thermal-neutron ^{235}U fission, i.e. about $3.2 \cdot 10^{10}$ fissions occur per second at the power of 1 watt. In fission of 1 kg of ^{235}U released is the energy of $8 \cdot 10^{13}$ J.

Energies released in fission of various nuclei are presented in Tables 40.7, 40.8, and Figure 40.5.

Table 40.7 Energy (in MeV) released in nuclear fission (average value per one fission) [The energy carried away by delayed neutrons, $E_{n,\mathrm{del}} \approx 10$ keV, is here omitted.]

Nucleus-target	Particle and its energy, MeV	Prompt radiation			Delayed radiation			E_{tot}	References
		E_k	E_n	E_γ	E_β	E_γ	$E_{\bar\nu}$		
^{197}Au	^{16}O,135	156	80	12	3	4	5		[9]
^{209}Bi	^{22}Ne,175	171	90	12	3	4	5		[9]
^{226}Ra	p, 12	159	20	6	5	5	7		[9]
^{232}Th	n, 3.35	161.8	4.7	6.1	8.1	7.86	10.9	199.46	[7]
	n, 14.0	161.8	7.4	6.1	7.2	6.99	9.7	199.14	[7]
^{233}U	n, thermal	171.5	4.9	7.6	5.1	4.9	6.8	200.8	[7]
	n, 0.5	169.4	4.9	7.6	5.0	4.9	6.8	198.6	[7]
	n, 14.0	169.4	7.5	7.6	4.3	4.1	5.7	198.6	[7]
	α, 30	176	36	8	3	4	5		[9]
^{235}U	n, thermal	172.7	4.8	6.9	6.4	6.2	8.6	205.2	[7]
	n, 0.5	169.8	4.8	6.9	6.4	6.2	8.6	202.8	[7]
	n, 14.0	169.8	7.3	6.9	5.3	5.2	7.1	201.7	[7]
^{236}U	n, 2.82	170.4	5.3	6.3	6.9	6.7	9.3	204.6	[7]
^{238}U	n, 3.10	170.3	5.5	6.3	8.2	8.0	11.0	209.3	[7]
	n, 14.0	170.3	7.1	6.3	7.1	6.9	9.5	207.2	[7]
^{237}Np	n, 2.37	173.0	6.1	6.4	5.5	5.4	7.4	203.9	[7]
^{239}Pu	n, thermal	178.6	5.9	7.8	5.3	5.1	7.1	209.8	[7]
	n, 0.5	176.1	5.9	7.8	5.2	5.1	7.0	207.2	[7]
	n, 14.0	176.1	7.5	7.8	4.3	4.2	5.7	205.5	[7]
^{240}Pu	n, 2.39	179.0	6.2	6.5	5.7	5.6	7.7	207.7	[7]
^{241}Pu	n, thermal	175.4	6.0	7.9	6.5	6.4	8.8	210.9	[7]
	n, 0.5	175.6	6.0	7.9	6.5	6.3	8.7	211.0	[7]
^{242}Pu	n, 2.32	176.8	4.6	6.5	6.6	6.4	8.9	209.9	[7]
^{252}Cf	Spontaneous fission	190.0	7.4	8.4	6.0	5.9	8.1	225.8	[7]
^{256}Fm	Spontaneous fission	198	32	8	8	8	12		[9]

Table 40.8 Mean total kinetic energy of the fragments of nuclear fission induced by thermal neutrons [19]

Nucleus-target	E_k, MeV
^{229}Th	162.7(5)
^{231}Pa	165.1(5)
^{232}U	169.3(5)
^{237}Np	171.0(7)
^{238}Pu	178.1(5)
^{241}Am	180.0(4)
^{243}Am	177.5(6)
^{245}Cm	184.5(5)
^{249}Cf	189.4(5)
^{251}Cf	185.0(27)
^{254}Es	194.6(5)
^{255}Fm	195.7(29)

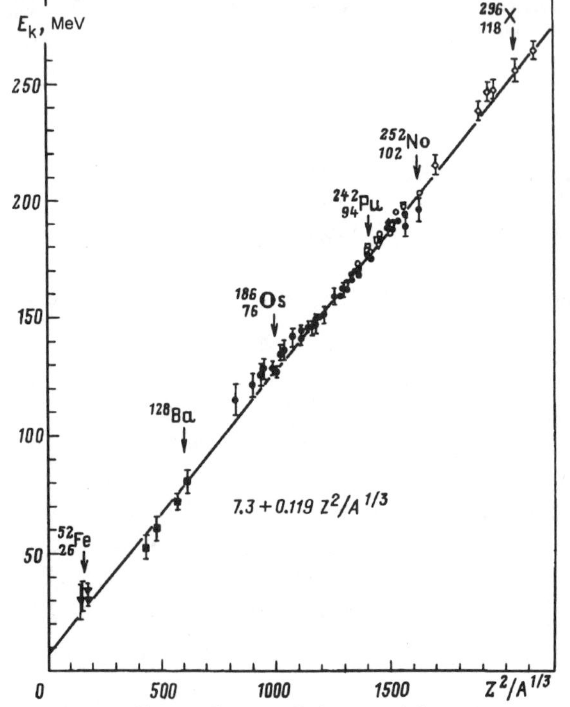

Figure 40.5 Dependence of the total kinetic energy E_k of fission fragments on $Z^2/A^{1/3}$ of a fissile nucleus [28].

40.6 Products of Nuclear Fission

Direct yield of nuclear fission fragments is referred to as the independent yield. The total yield (Fig. 40.6) is a sum of the independent yield and the part resulting from the chain of β–decays.

The ranges of the fission fragments are presented in Table 40.9.

Table 40.9 Ranges of nuclear-fission fragments $\bar{R}$ (in mg/cm^2) during retardation in various media [6] [$\bar{R}_l$ is the range of a light fragment, whereas $\bar{R}_h$, of a heavy fragment.]

Retarding medium	^{235}U+$n_{\rm thermal}$			^{252}Cf, spontaneous fission	
	$\bar{R}$	$\bar{R}_l$	$\bar{R}_h$	$\bar{R}_l$	$\bar{R}_h$
Aluminum		4.17	3.22	4.16	3.44
Argon [30]	4.123	4.45	3.84		
Air	2.16 cm	2.45 cm	2.03 cm	2.32 cm*2	1.90 cm*2
Carbon		3.09	2.48		
Collodion	20 μm [29]				
Copper	5.08				
Helium [30]	1.252	1.38	1.14		
Iron	4.22				
Lavsan		17.8 μm*1	14.0 μm*1		
Nickel	4.40	5.53	4.30	5.52	4.57
Silver		7.33	5.50	7.28	5.82
Uranium	6.7 μm [29]	10.64*2	8.44*2		
Uranium dioxide	9.9 μm [29]				
Uranosouranic mixed oxide	13.7 μm [29]				
Water	21 μm [29]				
Zirconium		6.12	4.78		

*1 Ranges of fragments in nuclear photofission, $E_{\max} = 20$ MeV.
*2 Ranges of ^{99}Mo and ^{140}Ba fragments, $\bar{R}_l$ and $\bar{R}_h$, respectively.

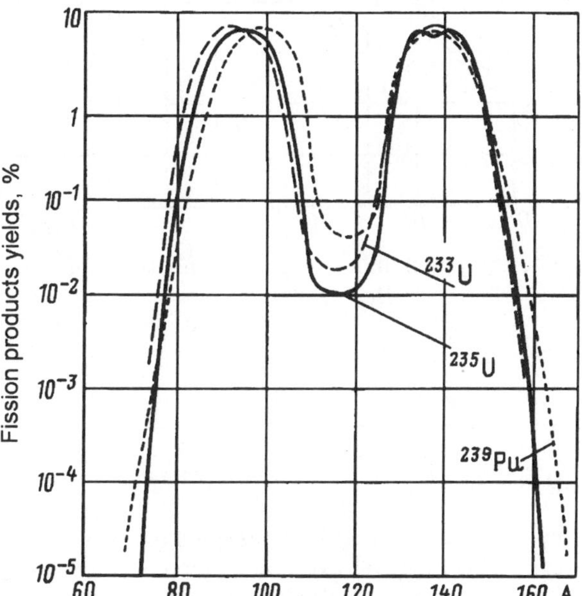

Figure 40.6 Total yields of fission products formed in thermal-neutron fission of ^{233}U, ^{235}U, and ^{239}Pu (the whole curve is normalized to 200 per cent) [20].

40.7 Fission Neutrons

Energy spectra of prompt fission neutrons are similar for different nuclei. In the simplest form, within the experimental error, the fission-neutron spectra can be represented by the Maxwellian distribution

$$N(E) = \left(2/\sqrt{\pi\theta^3}\right)\sqrt{E}\exp(-E/\theta),$$

where E is the neutron energy in the laboratory system of coordinates, and the parameter θ is connected with the average energy of the spectrum $\bar{E}$: $\theta = 2\bar{E}/3$ (Table 40.10).

The average number of prompt neutrons $\bar{\nu}_{\text{prompt}}$ arising in nuclear fission tends to increase with energy E_n of neutrons inducing fission (Figure 40.7). This dependence can be represented by the expression $\bar{\nu}(E_n) = \bar{\nu}_0 + aE_n$, where $\bar{\nu}_0$ is the average number of prompt neutrons produced in thermal-neutron fission. The parameter $a \simeq 0.15$ weakly depends on the neutron energy.

Table 40.10 Average energies $\bar{E}$ of the energy distributions of prompt neutrons for different nuclei

Nucleus	Type of fission	$\bar{E}$, MeV	Nucleus	Type of fission	$\bar{E}$, MeV
^{233}U	by thermal neutrons	2.0627 (248) [25]	^{240}Pu	spontaneous	1.86 [20]
^{235}U	by thermal neutrons	2.0266 (220) [25]	^{241}Pu	by thermal neutrons	2.0858 (232) [25]
^{239}Pu	by thermal neutrons	2.0990 (232) [25]	^{252}Cf	spontaneous	2.1409

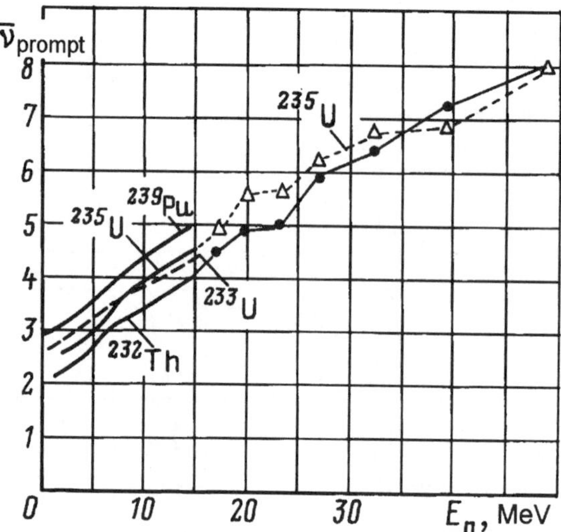

Figure 40.7 The average number of prompt neutrons of fission of ^{232}Th [21, 31], ^{233}U [20], ^{235}U [23, 31], ^{239}Pu [23, 32].

40.8 Prompt γ-Radiation

The excitation energy of fission fragments that remains upon emission of prompt neutrons, is usually equal to 3–4 MeV per fragment. This fragment excitation is removed by emission of prompt γ-quanta. The radiation process occurs in a time of 10^{-9}–10^{-14} s after the neutron emission.

The spectrum of γ-quanta and total energies $E_{\gamma,\text{tot}}$ of prompt γ-quanta produced in fission of some nuclei are presented in Tables 40.11 and 40.12.

Table 40.11 The spectrum of prompt γ-quanta accompanying thermal-neutron fission of ^{235}U [20]
[$N(E_\gamma)$ is the number of γ-quanta emitted when fission occurs in the energy range 0.1 MeV wide, $E_{eff} = E_\gamma N(E_\gamma)$.]

E_γ, MeV	$N(E_\gamma)$	E_{eff}, MeV/fission	E_γ, MeV	$N(E_\gamma)$	E_{eff}, MeV/fission	E_γ, MeV	$N(E_\gamma)$	E_{eff}, MeV/fission
0.1	0.176	0.0176	2.7	0.0409	0.1104	5.2	0.00430	0.0224
0.2	0.815	0.1630	2.8	0.0369	0.1033	5.3	0.00399	0.0217
0.3	0.697	0.2091	2.9	0.0330	0.0957	5.4	0.00371	0.0200
0.4	0.661	0.2644	3.0	0.0298	0.0894	5.5	0.00341	0.0188
0.5	0.662	0.3110	3.1	0.0268	0.0831	5.6	0.00314	0.0176
0.6	0.553	0.3318	3.2	0.0244	0.0781	5.7	0.00290	0.0165
0.7	0.474	0.3318	3.3	0.0219	0.0723	5.8	0.00264	0.0153
0.8	0.408	0.2464	3.4	0.0198	0.0673	5.9	0.00243	0.0143
0.9	0.353	0.3177	3.5	0.0181	0.0634	6.0	0.00223	0.0134
1.0	0.310	0.310	3.6	0.0165	0.0594	6.1	0.00204	0.0124
1.1	0.272	0.2992	3.7	0.0150	0.0555	6.2	0.00188	0.0117
1.2	0.240	0.2880	3.8	0.0136	0.0517	6.3	0.00172	0.0108
1.3	0.205	0.2665	3.9	0.0126	0.0491	6.4	0.00157	0.0100
1.4	0.180	0.2520	4.0	0.0116	0.0464	6.5	0.00139	0.0090
1.5	0.158	0.2370	4.1	0.0106	0.0435	6.6	0.00128	0.0084
1.6	0.139	0.2224	4.2	0.00985	0.0414	6.7	0.00115	0.0077
1.7	0.125	0.2125	4.3	0.00908	0.0391	6.8	0.00103	0.0070
1.8	0.113	0.2034	4.4	0.00830	0.0365	6.9	0.000916	0.0063
1.9	0.102	0.1938	4.5	0.00764	0.0344	7.0	0.000833	0.0058
2.0	0.113	0.2260	4.6	0.00704	0.0324	7.1	0.000731	0.0052
2.1	0.0818	0.1718	4.7	0.00649	0.0305	7.2	0.000629	0.0045
2.2	0.0726	0.1597	4.8	0.00604	0.0290	7.3	0.000547	0.0040
2.3	0.0651	0.1497	4.9	0.00556	0.0272	7.4	0.000467	0.0034
2.4	0.0579	0.1390	5.0	0.00519	0.0260	7.5	0.000388	0.0029
2.5	0.0512	0.1280	5.1	0.00470	0.0240	7.6	0.000308	0.0023
2.6	0.0457	0.1188						

Table 40.12 The total energy $E_{\gamma,tot}$ of prompt γ-quanta formed under fission of some nuclei [20]
[$\bar{N}_\gamma$ is the average number of γ-quanta per fission, E_n is the energy of neutrons inducing a fission.]

Nucleus	E_n, MeV	$\bar{N}_\gamma$	$E_{\gamma,tot}$, MeV	Nucleus	E_n, MeV	$\bar{N}_\gamma$	$E_{\gamma,tot}$, MeV
^{235}U	Thermal	7.4 (8)	7.2 (8)	^{238}U	2.8		7.5 (11)
	2.8		7.5 (11)		14.7		7.5 (11)
	14.7		7.5 (11)	^{252}Cf	Spontaneous fission	10.3	8.2

40.9 Delayed Emission of Fission Products

β- and γ-emission of fission products. β- and γ-emission of fission products is the result of decay of more than 200 radionuclides; therefore, as a rule, joint β- or γ- emission is considered at the point time t after fission averaged over many fission acts. The time dependence of the decay rate (in MeV·s^{-1}·fission^{-1}) of the fission products can be approximated as follows:

$$A_\beta(t) \approx 1.4t^{-1.2}; \quad A_\gamma(t) \approx 8.2t^{-1.2},$$

where t (in s) is the time upon fission. These expressions are valid within the range 10 s $< t < 10^7$ s within $\pm$20% accuracy.

The fuel (uranium) and fission products in an active reactor are intensively bombarded with neutrons during a campaign lasting from 1 to 3 y. Prompt and thermal neutrons induce, in these nuclides, numerous nuclear reactions that give rise to new nuclides, specifically, to transuranium elements up to americium and curium. The new nuclide set formed by the end of the reactor campaign contains more than 500 radionuclides and determines subsequent radiation characteristics of the irradiated fuel. In time, the short-lived radionuclides transform into various stable isotopes, and the activity of the irradiated fuel becomes increasingly dependent on long-lived actinoids such as ^{238}Pu with the half-life $T_{1/2}$ = 87.7 y, ^{241}Am ($T_{1/2}$ = 432 y), ^{239}Pu ($T_{1/2}$ = 24,100 y), etc. Some characteristics of the irradiated fuel are presented in Tables 40.13 and 40.14.

Table 40.13 Integral radiation characteristics of the irradiated fuel (actinoids+fission products) [37]

I. Water-moderated reactor WWER–1000 upon the campaign of 1,014 days duration including two scheduled outages for 48 days each. The total weight of the fuel equals 80 tons (the uranium weight is 70 tons); the initial enrichment amounts to 4.4 per cent; the rated thermal power is 3.2 GW; the depth of depletion is 40.5 GW·d/t.

Endurance after outage	Activity, Bq	Energy release, MW			Mean energy, keV/particle		Contribution of actinides, %
		α	β	γ	β	γ	
0	$7.7 \cdot 10^{20}$	0.14	109	97	925	756	
10 s	$6.7 \cdot 10^{20}$	0.14	77	76	753	694	
1 min	$5.8 \cdot 10^{20}$	0.14	54.9	60	625	642	
10 min	$4.5 \cdot 10^{20}$	0.14	33.3	40	494	563	
1 h	$3.2 \cdot 10^{20}$	0.14	18.6	25	391	521	
6 h	$2.4 \cdot 10^{20}$	0.14	11.0	15	308	430	
1 d	$1.8 \cdot 10^{20}$	0.14	7.1	10	259	396	
5 d	$1.0 \cdot 10^{20}$	0.14	4.1	6.1	272	463	
30 d	$4.5 \cdot 10^{19}$	0.13	2.2	2.4	331	575	
120 d	$2.1 \cdot 10^{19}$	0.095	1.3	0.87	411	595	4.26
1 y	$8.7 \cdot 10^{18}$	0.046	0.68	0.24	509	584	4.81
3 y	$3.3 \cdot 10^{18}$	0.021	0.18	0.11	394	613	6.78
10 y	$1.4 \cdot 10^{18}$	0.020	0.05	0.036	281	615	19.2
30 y	$7.8 \cdot 10^{17}$	0.020	0.03	0.019	324	602	30.5
100 y							66.4

II. Graphite-water-moderated reactor RBMK–1000 upon the campaign lasting 1,100 days in the regime of sustained overloads. The total weight of the fuel is 217 tons (that of uranium is 192 tons); enrichment amounts to 2 per cent; depletion equals 24.9 GW·d/t; thermal power is 3.2 GW.

Endurance after outage	Activity, Bq	Energy release, MW			Mean energy, keV/particle	
		α	β	γ	β	γ
0	$7.8 \cdot 10^{20}$	0.31	103	91	864	719
10 s	$6.8 \cdot 10^{20}$	0.31	75	73	712	659
1 min	$6.0 \cdot 10^{20}$	0.31	54.6	58	595	607
10 min	$4.7 \cdot 10^{20}$	0.31	33.8	38	473	531
1 h	$3.4 \cdot 10^{20}$	0.31	18.8	25	373	493
6 h	$2.6 \cdot 10^{20}$	0.31	11.4	15	297	409
1 d	$2.0 \cdot 10^{20}$	0.31	7.7	11	258	379
5 d	$1.1 \cdot 10^{20}$	0.31	4.7	6.3	319	528
30 d	$5.1 \cdot 10^{19}$	0.28	2.7	2.6	360	570
120 d	$2.6 \cdot 10^{19}$	0.19	1.7	1.0	442	589
1 y	$1.2 \cdot 10^{19}$	0.082	0.96	0.34	513	586
3 y	$4.9 \cdot 10^{18}$	0.024	0.28	0.16	396	616
10 y	$2.2 \cdot 10^{18}$	0.024	0.08	0.058	291	615

Table 40.14 Activity A and concentration C of main long-lived nuclides in the irradiated 1t-fuel WWER–1000 and RBMK–1000 reactors upon operating periods of 1014 and 1100 days, respectively, at nominal thermal power 3.2 GW for endurances of 0.5 and 10 years [37]

Fission products					
Reactor type		WWER–1000 Depletion 40.5 GW·d/t		RBMK–1000 Depletion 24.9 GW·d/t	
Nuclide	$T_{1/2}$	Activity A, Bq		Activity A, Bq	
		$T_{\rm end}$=0.5 y	$T_{\rm end}$=10 y	$T_{\rm end}$=0.5 y	$T_{\rm end}$=10 y
^{85}Kr	10.7 y	$2.3\cdot10^{14}$	$1.2\cdot10^{14}$	$3.2\cdot10^{14}$	$1.7\cdot10^{14}$
^{90}Sr	28.5 y	$1.4\cdot10^{15}$	$1.1\cdot10^{15}$	$2\cdot10^{15}$	$1.6\cdot10^{15}$
^{90}Y	61.4 h	$1.4\cdot10^{15}$	$1.1\cdot10^{15}$	$2\cdot10^{15}$	$1.6\cdot10^{15}$
^{91}Y	58.5 d	$8.1\cdot10^{15}$	~ 0	$2\cdot10^{15}$	~ 0
^{95}Zr	64 d	$1.1\cdot10^{16}$	~ 0	$3.7\cdot10^{15}$	~ 0
^{95}Nb	35 d	$2.2\cdot10^{16}$	~ 0	$7.4\cdot10^{15}$	~ 0
^{103}Ru	39.3 d	$2.3\cdot10^{15}$	~ 0	$1.2\cdot10^{15}$	~ 0
^{103m}Rh	56.1 min	$2.3\cdot10^{15}$	~ 0	$1.2\cdot10^{15}$	~ 0
^{106}Ru	368 d	$5\cdot10^{15}$	$7.3\cdot10^{12}$	$1\cdot10^{16}$	$1.5\cdot10^{13}$
^{106}Rh	29.9 s	$5\cdot10^{15}$	$7.3\cdot10^{12}$	$1\cdot10^{16}$	$1.5\cdot10^{13}$
^{134}Cs	2.06 y	$8.1\cdot10^{14}$	$3.3\cdot10^{13}$	$3.4\cdot10^{15}$	$1.4\cdot10^{14}$
^{137}Cs	30.2 y	$1.6\cdot10^{15}$	$1.25\cdot10^{15}$	$2.8\cdot10^{15}$	$2.3\cdot10^{15}$
^{137m}Ba	2.56 min	$1.5\cdot10^{15}$	$1.2\cdot10^{15}$	$2.7\cdot10^{15}$	$2.2\cdot10^{15}$
^{141}Ce	32.5 d	$1.7\cdot10^{15}$	~ 0	$5.7\cdot10^{14}$	~ 0
^{144}Ce	284 d	$2.4\cdot10^{16}$	$5\cdot10^{12}$	$1.6\cdot10^{16}$	$3.5\cdot10^{12}$
^{147}Pm	2.62 y	$4.3\cdot10^{15}$	$3.5\cdot10^{14}$	$3.9\cdot10^{15}$	$3.2\cdot10^{14}$
^{151}Sm	87 y	$1.1\cdot10^{13}$	$1.0\cdot10^{13}$	$4\cdot10^{12}$	$3.7\cdot10^{12}$
^{154}Eu	8.5 y	$4.9\cdot10^{13}$	$2.2\cdot10^{13}$	$1.9\cdot10^{14}$	$9.3\cdot10^{13}$
^{155}Eu	4.96 y	$5\cdot10^{13}$	$1.3\cdot10^{13}$	$1.3\cdot10^{14}$	$3.6\cdot10^{13}$

Actinoids									
Reactor type		WWER–1000 Depletion 40.5 GW·d/t				RBMK–1000 Depletion 24.9 GW·d/t			
Nuclide	$T_{1/2}$	$T_{\rm end}$=0.5 y		$T_{\rm end}$=10 y		$T_{\rm end}$=0.5 y		$T_{\rm end}$=10 y	
		C, g/t	A, Bq	C, g/t	A, Bq	C, g/t	A, Bq	C, g/t	A, Bq
^{235}U	$7.04\cdot10^{8}$ y	$\sim$ 12300	$9.8\cdot10^{8}$	12300	$9.8\cdot10^{8}$	$\sim$ 2940	$2.3\cdot10^{8}$	2940	$2.3\cdot10^{8}$
^{236}U	$2.34\cdot10^{7}$ y	5730	$1.4\cdot10^{10}$	$\sim$ 5730	$1.4\cdot10^{10}$	2610	$6.2\cdot10^{9}$	$\sim$ 2610	$6.2\cdot10^{9}$
^{238}U	$4.47\cdot10^{9}$ y	929000	$1.1\cdot10^{10}$	929000	$1.1\cdot10^{10}$	962000	$1.1\cdot10^{10}$	962000	$1.1\cdot10^{10}$
^{238}Pu	87.7 y	126	$8\cdot10^{13}$	122	$7.7\cdot10^{13}$	68.6	$4.3\cdot10^{13}$	68.4	$4.3\cdot10^{13}$
^{239}Pu	24120 y	5530	$1.3\cdot10^{13}$	5530	$1.3\cdot10^{13}$	2630	$6\cdot10^{12}$	$\sim$ 2630	$6\cdot10^{12}$
^{240}Pu	6537 y	2420	$2\cdot10^{13}$	2420	$2\cdot10^{13}$	2190	$1.8\cdot10^{13}$	$\sim$ 2190	$1.8\cdot10^{13}$
^{241}Pu	14.4 y	1470	$5.6\cdot10^{15}$	960	$3.7\cdot10^{15}$	713	$2.7\cdot10^{15}$	453	$1.7\cdot10^{15}$
^{241}Am	432 y	71.6	$9.1\cdot10^{12}$	616	$7.8\cdot10^{13}$	35.7	$4.5\cdot10^{12}$	293	$3.7\cdot10^{13}$
^{242}Pu	$3.76\cdot10^{5}$ y	582 y	$8.4\cdot10^{10}$	$\sim$ 582	$8.4\cdot10^{10}$	508	$7.4\cdot10^{10}$	$\sim$ 508	$7.4\cdot10^{10}$
^{242m}Am	152 y	0.264	$9.5\cdot10^{10}$	0.253	$9\cdot10^{10}$	0.34	$1.2\cdot10^{11}$	0.32	$1.1\cdot10^{11}$
^{242}Am	16.2 h	$3.16\cdot10^{-6}$	$9.5\cdot10^{10}$	$3.03\cdot10^{-6}$	$9.2\cdot10^{10}$	$4\cdot10^{-6}$	$1.2\cdot10^{11}$	$3.9\cdot10^{-6}$	$1.2\cdot10^{11}$
^{242}Cm	163 d	6.1	$7.5\cdot10^{14}$	$6.2\cdot10^{-4}$	$7.6\cdot10^{10}$	5.2	$6.3\cdot10^{14}$	$7.9\cdot10^{-4}$	$9.6\cdot10^{10}$
^{243}Am	7380 y	120	$8.8\cdot10^{11}$	$\sim$ 7380	$8.8\cdot10^{11}$	73.8	$5.5\cdot10^{11}$	$\sim$ 73.8	$5.5\cdot10^{11}$
^{243}Cm	28.5 y	0.24	$4.6\cdot10^{11}$	0.194	$3.7\cdot10^{11}$	0.21	$4\cdot10^{11}$	0.16	$3\cdot10^{11}$
^{244}Cm	18.1 y	45.7	$1.4\cdot10^{14}$	31.7	$9.5\cdot10^{13}$	8.1	$24\cdot10^{13}$	5.7	$1.7\cdot10^{13}$
^{245}Cm	8500 y	2.2	$1.4\cdot10^{10}$	$\sim$ 2.2	$1.4\cdot10^{10}$	0.19	$1.2\cdot10^{9}$	$\sim$ 0.19	$1.2\cdot10^{9}$
^{246}Cm	4730 y	0.33	$3.7\cdot10^{9}$	$\sim$ 0.33	$3.7\cdot10^{9}$	0.044	$5\cdot10^{8}$	$\sim$ 0.044	$5\cdot10^{8}$

Delayed neutrons. Neutrons emitted by excited fission fragments are conditionally divided into groups in accordance with the half-lives of progenitors. As a rule, six groups are distinguished (see, for instance, Figure 40.8 and Table 40.15). Usually, the first β-active fission fragments are referred to as the progenitors of delayed neutrons, whereas the products of the fragment decay are called the emitters of delayed neutrons. The half-lives and yields of neutrons for one and the same fissile nucleus weakly depend on the energy of a neutron (up to $E \simeq 5$ MeV) inducing the fission. The total yields of delayed neutrons are presented in Table 40.16 and Figure 40.9.

Table 40.15 The relative yield I_{rel} of delayed neutrons in groups per one fission act when fission of uranium and plutonium isotopes is induced by thermal neutrons [20]

Nuclide	Group number	$T_{1/2}$, s	I_{rel}	Nuclide	Group number	$T_{1/2}$, s	I_{rel}
^{233}U	1	55.00(54)	0.086	^{239}Pu	1	54.28(234)	0.035
	2	20.57(38)	0.299		2	23.04(1.67)	0.298
	3	5.00(21)	0.252		3	5.60(40)	0.211
	4	2.13(20)	0.278		4	2.13(24)	0.326
	5	0.615(242)	0.051		5	0.618(213)	0.086
	6	0.277(47)	0.034		6	0.257(45)	0.044
	Total yield		1.000		Total yield		1.000
^{235}U	1	55.72(128)	0.033	^{241}Pu	1	54.0(10)	0.009
	2	22.72(71)	0.219		2	23.2(5)	0.233
	3	6.22(23)	0.196		3	5.6(6)	0.176
	4	2.30(9)	0.395		4	1.97(10)	0.397
	5	0.610(83)	0.11		5	0.43(4)	0.185
	6	0.230(25)	0.042		6		
	Total yield		1.000		Total yield		1.000

Table 40.16 The number of delayed neutrons per 100 fissions and mean energy of a delayed neutron $\bar{E}_{n,del}$ [32] [Letters (t), (r), and number (14) stand for nuclear fission induced by thermal, reactor neutrons and neutrons with 14 MeV energy, respectively; the abbreviation (sf) denotes spontaneous fission of a nucleus.]

Nucleus	N_n	$\bar{E}_{n,del}$, keV
^{232}Th (r)	4.76 (0.34)	424.6
^{232}Th (14)	3.03 (0.29)	457.9
^{233}U (t)	0.845 (0.066)	407.7
(r)	0.916 (0.089)	394.8
(14)	0.708 (0.095)	389.4
^{235}U (t)	1.77 (0.081)	415.8
(r)	1.98 (0.18)	517.6
(14)	0.978 (0.097)	400.8
^{236}U (r)	2.26 (0.19)	424.0
^{238}U (r)	3.51 (0.27)	421.9
(14)	2.69 (0.21)	428.5
^{237}Np (r)	1.28 (0.13)	418.5
^{239}Pu (t)	0.769 (0.058)	419.8
(r)	0.724 (0.009)	412.9
(14)	0.387 (0.062)	383.2
^{240}Pu (r)	0.923 (0.108)	416.6
^{241}Pu (t)	1.58 (0.13)	428.1
(r)	1.49 (0.16)	426.7
^{242}Pu (r)	1.41 (0.14)	420.0
^{252}Cf (sf)	0.690 (0.092)	409.8

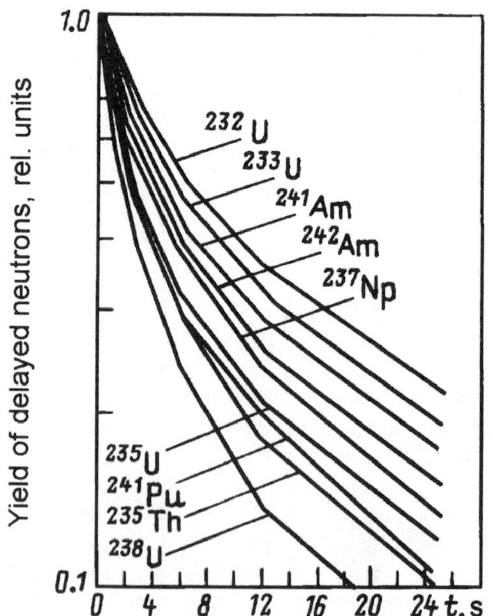

Figure 40.8 Time-dependent yield of delayed neutrons for various fissile nuclei [9].

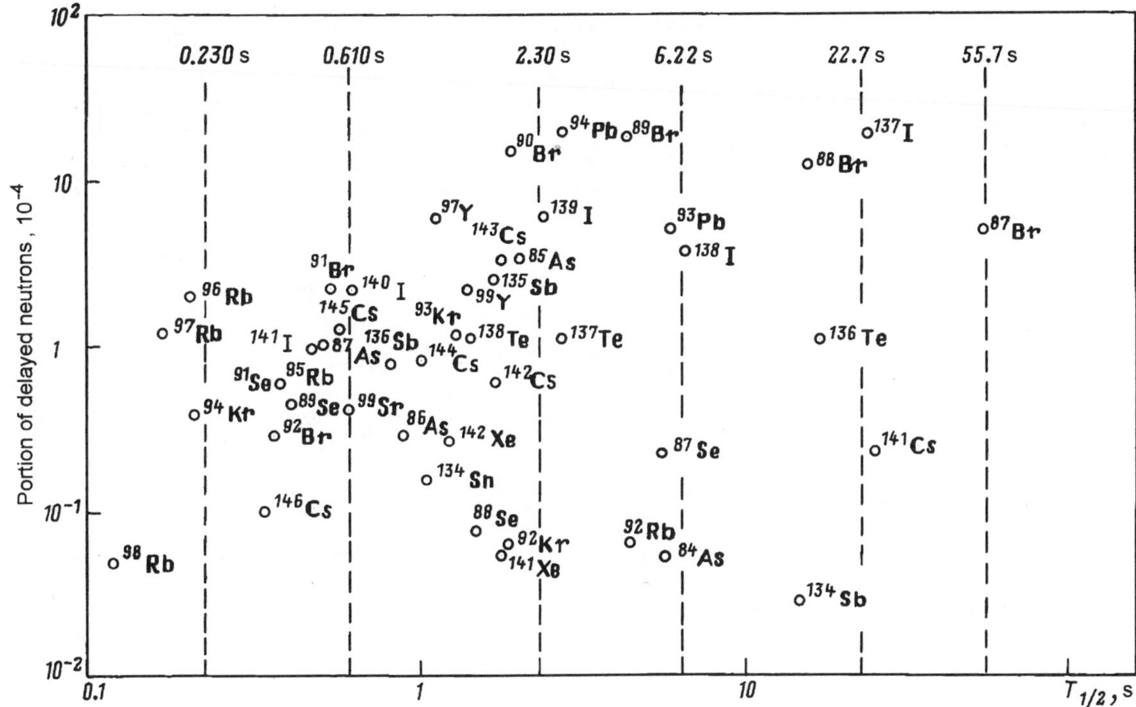

Figure 40.9 Characteristics of emitters of delayed neutrons [9].

References

[1] Hahn, O., Strassmann, F., Naturwissenschaften, 27, 11, 1939.

[2] Flerov, G. N., Petrzhak, K. A., Zhurnal eksperimental'noi i teoreticheskoi fiziki, 10, 1013, 1940.

[3] Obukhov, A. I., Perfilov, N. A., Uspekhi fizicheskikh nauk, 92, 621, 1967 (in Russian).

[4] Hyde, E.K., The nuclear properties of the heavy elements, Prentice Hall Inc., Englewood Cliffs, New Jersey, 1965.

[5] Vandenbosch, R., Huisenga, J. R., Nuclear fission, Academic Press, New York–London, 1973.

[6] Gorbachev, V. M., Zamyatin, Yu. S., Lbov, A. A., Interaction of radiations with nuclei of heavy elements and nuclear fission, Atomizdat, Moscow, 1976 (in Russian).

[7] Michaudon, A., Basic physics of the fission process. Nuclear fission and neutron-induced fission cross sections, Ed. Michaudon, A., Pergamon Press, Oxford, 1981.

[8] Pik-Pichak, G. A., Nuclear fission, in: Fizicheskaya entsiklopedia, vol.1, Sovetskaya Entsiklopedia, Moscow, 1988, p. 578 (in Russian).

[9] Gangrskii, Yu. P., Dalkhsuren, B., Markov, B. N., Nuclear fission fragments, Energoatomisdat, Moscow, 1986 (in Russian).

[10] Bjornholm, S., Lynn, J. E., Rev. Mod. Phys., 52, 725, 1980.

[11] Ignatyuk, A. V., Smirenkin, G. N., Itkis, M. G., Fizika elementarnykh chastits i atomnogo yadra, 16, 709, 1985 (in Russian).

[12] Plasil, F., Awes, T. C., Cheynis, B., Phys. Rev. C, 29, 1145, 1984.

[13] Proposed recommended list of heavy elements radionuclide decay data, Pt.1, Ed. Lorenz, A., INDC (NDC) - 149/NE, IAEA, Vienna. 1983.

[14] Brown, E., Firestone, R. B., in: Tables of radioactive isotopes, Ed. Shirley, V. C., A. Wiley-Interscience Publ., New York, 1986.

[15] Tuli, J. K., Nuclei properties, in: Handbook on nuclear activation data technical report, Ser. No 273, IAEA, Vienna, 1987, p. 3.

[16] Sommerville, L. P., Nurmia, M. J., Nitschke, J. M., Phys. Rev. C, 31, 1801, 1985.

[17] Druzhinin, A. A., Polynov, V. N., Korochkin, A. N., Atomnaya energiya, 56, 68, 1985 (in Russian).

[18] Mughabhab, S. F., Neutron cross sections, BNL–325, vol. 1, pt. B, Academic Press, New York, 1984, pt. B.

[19] Vorobieva, V. G., Kuz'minov, B. D., Voprosy atomnoi nauki i tekhniki, Ser. Yadernye konstanty, No 19, 16, 1975; No 2, 27, 1985 (in Russian).

[20] Tables of physical quantities. Handbook, Ed. Kikoin, I. K., Atomizdat, Moscow, 1976 (in Russian).

[21] Malinovskii, V. V., Taras'ko, M. Z., Kuz'minov, B. D., Voprosy atomnoi nauki i tekhniki, Ser. Yadernye konstanty, No 1, 24, 1985; No 2, 36, 1985; Atomnaya energiya, 58, 430, 1985 (in Russian).

[22] Flerov, G. N., in: Isotope production, Eds. Bochkarev, V. V., Brezhnev, N. E. and Kulish, E. E., Atomizdat, Moscow, 1973, p. 19 (in Russian).

[23] Malinovskii, V. V., Vorobieva, V. G., Kuz'minov, B. D., Voprosy atomnoi nauki i tekhniki, Ser. Yadernye konstanty, No 5, 19, 1983 (in Russian).

[24] Metag, V., Comments Nucl. Particle Phys., 10, No 2, 79, 1981.

[25] Devadeenam, M., Stehn, J. R., Ann. Nucl. Energy, 11, 375, 1984.

[26] Kon'shin, V. A., Nuclear physical constants of fissile nuclei. Handbook, Energoatomisdat, Moscow, 1984 (in Russian).

[27] Asami, T., Graphs of evaluated neutron cross sections in JENDL–2, JAERI-H-84-052, Japanese Nuclear Data Committee, Toaki, Ibaraki, 1984.

[28] Viola, V. E., Kwiatkowski, K., Walker, M., Phys. Rev., C 31, 1550, 1985.

[29] Steinberg, M., Nucleonics, 21, No 8, 151, 1963.

[30] Rustichelli, F., Z. Physik, 262, 211, 1973.

[31] Howe, R. E., Nucl. Sci. Engng., 86, 157, 1984.

[32] England, T. R., Wilson, W. B., Scheuter, R. E., Mann, F. M., Nucl. Sci. Engng., 85, 139, 1983.

[33] Broder, D. L., Popkov, K. K., Rubanov, S. N., Small-size shield of reactors, Atomizdat, Moscow, 1967 (in Russian).

[34] Druin, V. A., Bochev, B., Korotkin, Yu. S., Atomnaya energiya, 43, 155, 1977 (in Russian).

[35] Price, P. B., Stevenson, J. D., Barwick, S. W., Phys. Rev. Lett., 54, 297, 1985.

[36] Barwick, S. W., Price, P. B., Stevenson, J. D., Phys. Rev., C 31, 1984, 1985.

[37] Radiation parameters of irradiated nuclear fuel. Handbook, Eds. Kolobashkin, V. M., Rubtsov, P. M., Ruzhanskii, P. A., Sidorenko, V. D., Energoatomizdat, Moscow, 1983 (in Russian).

[38] Rose, H.J., Jones, G.A., Nature, 307, 245, 1984.

41

The Penetration of Neutrons through Matter

S.V. Marin

41.1 Introduction

In this Chapter, the physical constants of neutron interaction with the nuclei of matter in a wide energy region from 0.0253 eV to about 20 MeV are presented. This list of nuclear-physical constants is far from being a full list of all the existing data. Therefore, to obtain more information, e.g. on the parameters of resonances, angular and energy distributions of secondary neutrons or on other data, one should appeal to particular reference books or to evaluated neutron data libraries. They contain the recommended values of nuclear–physical constants in the form admitting their periodical reexamination and applicable to calculations in solving a wide class of problems. In particular, ENDF/B [1], ENDL [2], UKNDL [3], KEDAK [4] are among the best known libraries of evaluated nuclear data. To simplify the search of needed information, one should apply to the bibliographical list of neutron data, CINDA - Computer Index of Neutron Data [5] accepted in 1965 as the basic document for international exchange of bibliographical information. A detailed description of nuclear reaction mechanisms involving neutrons can be found in the literature devoted to the nuclear and neutron physics [6–11, 45]. The solution to neutron transport problems is best described in the literature on the physics and technology of nuclear reactors [12–16].

41.2 Basic Characteristics of the Neutron

When the process of neutron transport in matter is investigated, it is to be taken into account that the time characteristic of neutron interaction with nuclei is much shorter that its half-life. Therefore, the neutron in that case is considered to be a long-lived and stable particle.

It is well known that the neutron in a free state is radioactive. Its basic characteristics are presented below:

rest mass $m_n = 1.6749286(10) \cdot 10^{-27}$ kg [17];
energy equivalent of the neutron rest mass $E^0_{m_n} = m_n c^2 = 939.5731(27)$ MeV [11];
charge $|q_n| < 10^{-21}|q_e|$, q_e being the electron charge [18];
decay scheme $n_0 \to p^+ + e^- + \tilde{\nu}$;
half-life $T_{1/2} = 10.13(9) - 10.69(13)$ min [11];
boundary energy of β-spectrum $E_{\text{bound}} = 782.43(4)$ keV [11];
spin $S_n = \pm 1/2$ [11];
magnetic moment $\mu_n = 1.91304275(45)\mu_N$ [17];

0-8493-2861-6/97/$0.00+$.50

Compton wavelength $\lambda_{C,n} = h/m_n c = 1.3195909(22) \cdot 10^{-15}$ m [17].

Upper estimation of the neutron electric-dipole moment d_n obtained in experiments with ultracold neutrons gives $d_n/e < 6 \cdot 10^{-27}$ [11]. It is commonly supposed that the neutron is either an electrically neutral particle or a particle with a negligible (of an order of $10^{-19} q_e$) electric charge.

As an elementary particle the neutron manifests wave properties, i.e. the particle is associated with a wave of the length λ defined by the de Broglie relation

$$\lambda = h/p = 2.86 \cdot 10^{-11}/\sqrt{E_n},$$

where h is the Planck's constant; p is the neutron momentum; E_n is the neutron energy (in eV). The relation is valid when the relativistic correction is small (Figure 41.1). Deviation from the straight line in this Figure is observed at energies above 10^8 eV and it is due to high-energy relativistic effects. The wave properties show themselves most clearly at low energies; for instance, at $E_n = 0.0253$ eV, the value of λ is comparable with the atomic size (Table 41.1). The neutrons of those energies diffract on a crystal lattice like the X-rays [19].

The energy of neutrons determines the type of their interaction with medium nuclei (Figure 41.2). When analyzing the data drawn in Figure 41.2, one should take into account that the boundaries of energy regions are symbolic and do overlap.

Table 41.1 Some characteristics of neutrons of different energies [8]

Group	Energy, eV	Temperature, K	Velocity, m/s	Wavelength, m
Ultracold ($< 10^{-7}$ eV)	10^{-7}	$1.1 \cdot 10^{-3}$	0.44	$0.9 \cdot 10^{-8}$
Cold ($10^{-7} - 10^{-2}$ eV)	10^{-3}	11.6	$4.37 \cdot 10^{2}$	$9.04 \cdot 10^{-10}$
Thermal (0.01 – 0.1 eV)	0.0253	293	2200	$1.80 \cdot 10^{-10}$
Resonance (0.1 – 50 eV)	1.0	$1.16 \cdot 10^{4}$	$1.38 \cdot 10^{4}$	$2.86 \cdot 10^{-11}$
Slow (50 – 500 eV)	100	$1.16 \cdot 10^{6}$	$1.38 \cdot 10^{5}$	$2.86 \cdot 10^{-12}$
Intermediate ($500 - 10^{5}$ eV)	10^{4}	$1.16 \cdot 10^{8}$	$1.38 \cdot 10^{6}$	$2.86 \cdot 10^{-13}$
Fast ($10^{5} - 10^{7}$ eV)	10^{6}	$1.16 \cdot 10^{10}$	$1.38 \cdot 10^{7}$	$2.86 \cdot 10^{-14}$
High-energy ($10^{7} - 10^{9}$ eV)	10^{8}	$1.16 \cdot 10^{12}$	$1.28 \cdot 10^{8}$	$2.79 \cdot 10^{-15}$
Relativistic ($> 10^{9}$ eV)	10^{10}	$1.16 \cdot 10^{14}$	$2.99 \cdot 10^{8}$	$1.14 \cdot 10^{-16}$

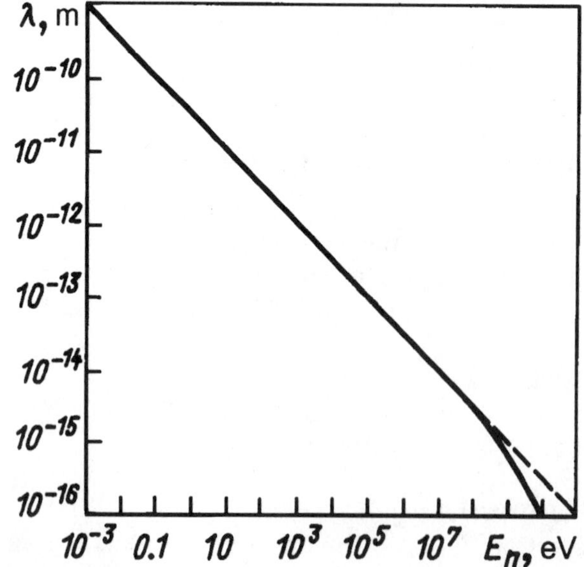

Figure 41.1 Dependence of the neutron wave length on energy [8].

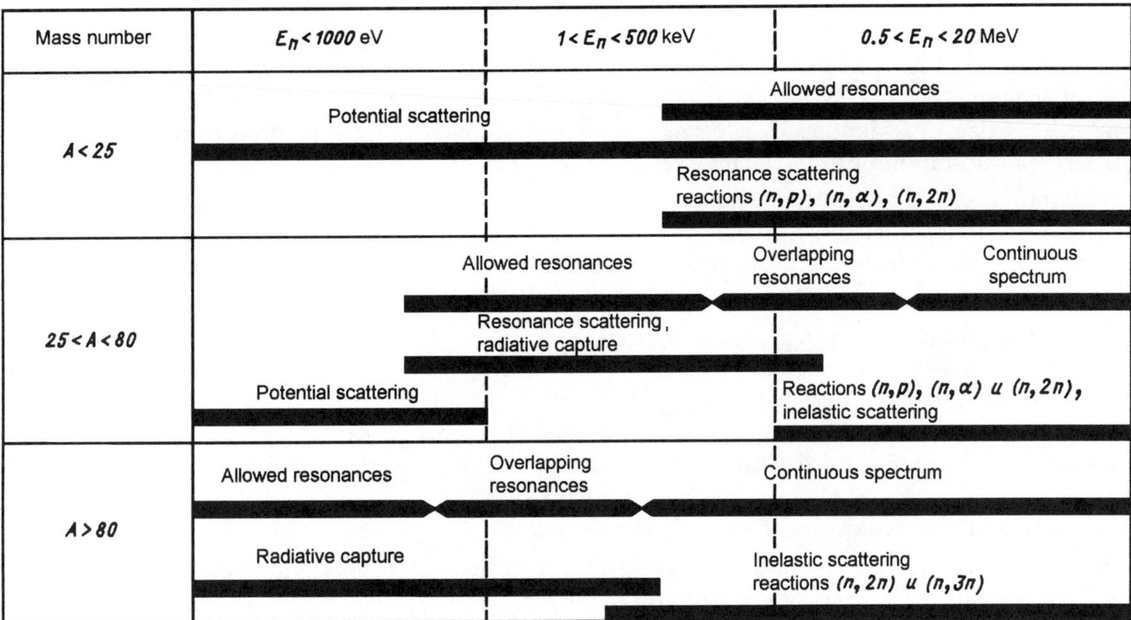

Figure 41.2 Classification of nuclear reactions involving neutrons [9].

41.3 Neutron Cross Sections

Neutron induced nuclear reactions are described by the following notation [20]:

$$n + X \to Y + b + Q,$$

which means the neutron interaction with a nucleus X resulting in the production of a nucleus Y and a particle b. The thermal effect, or the reaction energy Q, is the difference of particle masses before and after reaction:

$$Q = E_b + E_Y - E_n,$$

where E_Y and E_b are the kinetic energies of the reaction products.

When $Q > 0$, the reaction is exoergic and proceeds at any neutron kinetic energy. When $Q < 0$, the reaction is endoergic and cannot occur unless the neutron energy exceeds a certain value called the threshold energy of reaction, $E_{\mathbf{thresh}}$:

$$E_{\mathbf{thresh}} = |Q|$$

in the center–of–mass system (CMS); and

$$E_{\mathbf{thresh}} = |Q|(M + 1)/M$$

in the laboratory system of coordinates (LS), where M is the mass of a nucleus X expressed in terms of the neutron mass.

The microscopic cross section σ_i of a reaction of the i-th type is regarded as the number of events of the same type i per unit time, per one nucleus of matter and divided by the number of particles that fall onto the surface of unit area per unit time. The calculations are, as a rule, based on the quantity

$$\Sigma_i = \rho\sigma_i,$$

where Σ is given in cm^{-1}; ρ is the number density of nuclei in a substance(in cm^{-3}); σ_i is the cross section (in cm^2). In the literature on neutron physics, there sometimes happens an out-of system unit of the cross section, barn, that is expressed via the SI unit by the relation 1 barn $= 10^{-28}m^2$. The macroscopic cross section Σ_i is considered as the probability for a neutron to exhibit an act of scattering or absorption within the interval of 1 cm-length.

The total result of interaction can be represented either as a sum of the effects of elastic scattering and inelastic interactions or as a sum of elastic and inelastic effects of scattering and the effects of neutron absorption. Inelastic interaction comprises inelastic scattering and absorption of neutrons.

We present the symbolic notation (Figure 41.3):

total cross section, $\sigma_{\text{tot}} = \sigma_{n,n} + \sigma_x = \sigma_s + \sigma_{\text{abs}}$;
scattering cross section, $\sigma_s = \sigma_{n,n} + \sigma_{n,n'}$;
inelastic cross section, $\sigma_x = \sigma_{n,n'} + \sigma_{\text{abs}}$;
absorption cross section, $\sigma_{\text{abs}} = \sigma_{n,f} + \sigma_{n,2n} + \sigma_{n,3n} + \sigma_{n,np} + \sigma_{n,\gamma} + \sigma_{n,p} + \sigma_{n,\alpha}\cdots$;
cross section of the processes resulting in the neutron emission, $\sigma_{em} = \sigma_{n,n'} + 2\sigma_{n,2n} + 3\sigma_{n,3n} + \nu\sigma_{n,f} + \sigma_{n,np} + \ldots$;
cross section of the processes with neutron production, $\sigma_p = \sigma_{n,n'} + \sigma_{em}$.

Here $\sigma_{n,n}$ and $\sigma_{n,n'}$ are cross sections of elastic and inelastic scattering of neutrons; $\sigma_{n,2n}$ and $\sigma_{n,3n}$ are cross sections of the reactions $(n, 2n)$ and $(n, 3n)$; $\sigma_{n,f}$ is the fission cross section; $\sigma_{n,\gamma}$ is the cross section of radiative capture; $\sigma_{n,p}$, $\sigma_{n,d}$, $\sigma_{n,\alpha}$, $\sigma_{n,np}$ are cross sections of reactions (n,p), (n,d), (n,α), (n,np), respectively; ν is the average yield of neutrons in fission.

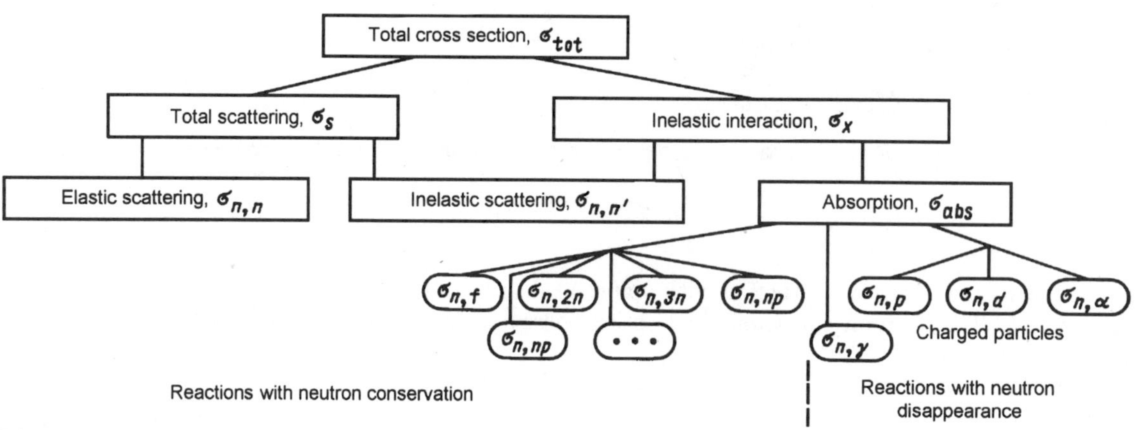

Figure 41.3 The scheme of connection of neutron cross sections [26].

41.4 Mechanism of Nuclear Reactions Involving Neutrons

Nuclear reactions occurring in collisions of neutrons with nuclei are rather diverse and depend on the individual properties of colliding particles and the energy of their relative motion. The whole set of nuclear reactions can be conditionally divided into two groups, the reactions with formation of a compound nucleus and direct nuclear reactions. A system formed out of an absorbed neutron and a nucleus–target and being in a highly excited state is called the compound nucleus. The lifetime of a compound nucleus amounts to about 10^{-17} s, and the excitation energy is a sum of the kinetic energy and binding energy of the absorbed neutron. The excitation energy of a compound nucleus is distributed over a large number of degrees of freedom.

In accordance with the modern ideas of nuclear reactions, it is assumed that the process occurs in two stages, production of a compound nucleus and its decay into the reaction products, i.e.

$$n + X \to C \to Y + b,$$

where C is a compound nucleus formed at the intermediate stage. Its decay depends upon the energy, angular momentum, and parity, and is independent of the way the compound nucleus forms.

However, the process is feasible when a neutron colliding with a single nucleon of the nucleus–target leaves the nucleus without interacting with other nucleons with a great probability. A process of that sort is called the direct reaction. Unlike the nuclear reaction with formation of a compound nucleus, when a large number of degrees of freedom is excited, a small number of degrees of freedom gets excited in the direct nuclear reaction. When energies of incident neutrons are smaller than 20 MeV, the probability of that process is small.

There are also reactions that are intermediate in nature between direct reactions and reactions with formation of a compound nucleus. The nucleus can decay before the energy imported by the captured particle is distributed between all the nuclear nucleons (pre-equilibrium decay).

A compound nucleus can decay in several ways: with emission of a neutron of the same energy as the absorbed energy (elastic or resonance scattering); with emission of one or several γ-quanta (radiative capture); with emission of charged particles or neutrons when the excitation energies are large enough.

A distinguishing feature of the nuclear reactions with formation of a compound nucleus at low neutron energies (lower than 1 MeV) is the presence of resonances in the energy dependence of cross sections. Resonance scattering arises from the nucleus interior, whereas potential scattering arises from its surface. In this energy region, the most important are the processes of elastic scattering and radiative capture of neutrons. Other processes are either forbidden energetically or characterized by a low probability. Inelastic scattering is impossible if the energy of an incident neutron is smaller than the nucleus excitation energy (for nuclei with intermediate mass numbers, the excitation energy, as a rule, exceeds several tens of MeV). Reactions $(n,p), (n,a)$ and others are of low probability owing to the Coulomb barrier that hinders the emission of low-energy charged particles.

When the neutrons of thermal and resonance energies interact with the nuclei of heavy nuclides, the most important are elastic scattering and radiative capture, whereas for some heavy nuclides, fission. If the neutron energy is higher than 1 MeV, there are possible other nuclear reactions such as inelastic scattering, reactions with emission of charged particles, etc.

The process of elastic scattering of a neutron is considered to consist of two parts, pure resonance scattering with formation of a compound nucleus and potential scattering when a neutron does not penetrate the nucleus interior but reflects from its surface. Resonance and potential scatterings are coherent and interfere with each other. According to the Breit–Wigner formula for elastic scattering one has

$$\sigma_s(E) = \pi \bar{\lambda}^2 \frac{\Gamma_n^2}{(e - E_R)^2 + (\Gamma/2)^2} + 4\pi a^2 - 4\pi \bar{\lambda} \Gamma_n a \frac{E - E_R}{(E - E_R)^2 + (\gamma/2)^2},$$

where $\bar{\lambda} = \lambda/2\pi$ is the reduced de Broglie wavelength; Γ and Γ_n are the total and neutron widths of resonances ($\Gamma = \Gamma_n + \Gamma_\gamma$ is the level width, a sum of partial widths, the neutron width Γ_n and radiative width Γ_γ; the ratios Γ_n/Γ and Γ_γ/Γ are the probabilities of elastic scattering with formation of a compound nucleus and radiative capture, respectively; the resonance widths are taken at half-height of the cross section in a resonance and are expressed in energy units); E_R is the resonance energy; a is the nucleus effective radius. The first term describes resonance scattering; the second – the potential scattering; the third – the interference of potential and resonance scattering processes. It is seen that when $E_R > E$ the interference term becomes negative, which corresponds to the decrease in the cross section owing to the interference [9, 12].

The total cross section of neutron scattering is often considered as a sum of the coherent and noncoherent cross sections [8]:

$$\sigma_s = \sigma_{\mathrm{s,coh}} + \sigma_{\mathrm{s,noncoh}}.$$

41.5 Cross Sections of Nuclear Reactions for Thermal Neutrons

In Tables 41.2–41.4, listed are the cross sections of nuclear reactions for neutrons of thermal energies (0.0253 eV) determined either experimentally or by comparing the results of different authors.

The first column contains the symbol of an element and mass number. When the mass number is not indicated, the cross section is given for a natural mixture of isotopes of a given element. The second column shows its content in the natural mixture for a long-lived nucleus–target; for the radioactive nucleus, the half-life. For some nuclides, both the quantities are reported. The third column presents concentrations of nuclei; the fourth, cross sections of reactions: $\sigma_{n,\gamma}, \sigma_{n,\alpha}, \sigma_{n,p}, \sigma_{\text{abs}}$. The cross sections of nuclear reactions $\sigma_{n,\alpha}, \sigma_{n,p}, \sigma_{\text{abs}}$ differ from the values of $\sigma_{n,\gamma}$ given in this column in that they are accompanied by indication of the nuclear reaction type. For instance, the entries $70.5(1.9)^{(\text{abs})}$ and $940(4)^{(n,\alpha)}$ refer to the cross sections of reactions σ_{abs} and $\sigma_{n,\alpha}$, respectively. In some cases, given are the cross sections of an isomer state production for which the half-life or/and identification of the state is shown in brackets: m is the nucleus in a metastable state; and g, in the ground state. In the fifth–seventh columns, we list the cross sections of scattering and its components $\sigma_{s,\text{coh}}$ and $\sigma_{s,\text{noncoh}}$.

The last column of Tables 41.3 and 41.4 contains the total cross section and the fission cross section, respectively.

41.6 Plots of Total Cross Sections versus Energy

Figures 41.4–41.32 show the energy dependences of total cross sections for neutron interaction with nuclei ${}^1_1$H, ${}^3_1$H, ${}_5$B, ${}_6$C, ${}_9$Be, ${}_{26}$Fe, ${}_{49}$In, ${}_{79}$Au, ${}_{92}$U, hydrogen in light water, and deuterium in heavy water. Solid curves are the result of estimating and handling full experimental data set and employing the relevant theoretical models. For the total list of experimental data, see [9]. In Figure 41.19 qualitative cross section description is given for the energy range from 0.1 to 3 MeV because of the complex resonance structure of the real dependence.

Table 41.2 Absorption and scattering cross sections for neutrons with energy 0.0253 eV (elements with atomic number from 1 to 60) [21]

Symbol, mass number		Contents in natural mixture, % or (and) $T_{1/2}$ [22]	Number density of nuclei ρ, 10^{24} cm^{-3}	$\sigma_{n,\gamma}$, 10^{-28} m^2	σ_s, 10^{-28} m^2	$\sigma_{s\ \text{coh}}$, 10^{-28} m^2	$\sigma_{s\ \text{noncoh}}$, 10^{-28} m^2
H	1	99.985	$5.37 \cdot 10^{-5}$	0.3326(7)	20.491(14)	0.439 (3)	20.052 (14)
	2	0.0148		$0.519\ (7) \cdot 10^{-3}$	3.390 (12)	2.484 (5)	0.906 (13)
He			$2.68 \cdot 10^{-5}$		0.76 (1)		
	3	0.000138		$0.031\ (9) \cdot 10^{-3}$	3.10 (13)	2.48 (4)	0.62 (12)
	4	99.999862			0.76 (1)		
Li			0.0464	0.0448 (30) $70.5\ (1.9)^{(\text{abs})}$	0.95 (4)		0.25 (4)
	6	7.52		0.0385 (30) $940\ (4)^{(n,\alpha)}$	0.75 (2)	0.45 (8)	0.30 (8)
	7	92.48		0.0454 (3)	0.97 (4)	0.51 (1)	0.46 (4)
Be	7	53.30 d		$48\ 000\ (9000)^{(n,p)}$			
	9	100	0.1236	$7.6\ (8) \cdot 10^{-3}$	6.151 (5)		0.004 (1)
B			0.136	0.10 (4)	4.27 (7)	3.01 (5)	1.26 (3)
	10	19.8		0.5 (2) $3873\ (9)^{(\text{abs})}$	2.23 (6)	0.1180 (3)	2.11 (8)
	11	80.2		$5.5\ (3.3) \cdot 10^{-3}$	4.84 (4)	4.69 (3)	0.15 (3)

Table 41.2 Absorption and scattering cross sections for neutrons with energy 0.0253 eV *(continued)*

Symbol, mass number	Contents in natural mixture, % or (and) $T_{1/2}$ [22]	Number density of nuclei ρ, 10^{24} cm^{-3}	$\sigma_{n,\gamma}$, 10^{-28} m^2	σ_s, 10^{-28} m^2	$\sigma_{s\ coh}$, 10^{-28} m^2	$\sigma_{s\ noncoh}$, 10^{-28} m^2
C		0.08023	3.50 (7)$\cdot 10^{-3}$	4.740 (5)		0.3 (1)$\cdot 10^{-3}$
12	98.89		3.53 (7)$\cdot 10^{-3}$	4.746 (2)		
13	1.11		1.37 (4)$\cdot 10^{-3}$	4.19 (12)	4.16 (12)	0.029 (10)
14	5730 y		< 0.001			
N		5.38$\cdot 10^{-5}$	74.7 (7.3)$\cdot 10^{-3}$	10.03 (8)	9.60 (4)	0.45 (11)
		1.90 (3)$^{(abs)}$				
14	99.63		75.0 (7.5)$\cdot 10^{-3}$	10.05 (12)		0.46 (5)
15	0.366		0.024 (8)$\cdot 10^{-3}$	4.59 (5)		3.5$\cdot 10^{-3}$
O		5.38$\cdot 10^{-5}$	0.19 (2)$\cdot 10^{-3}$	3.761 (6)	3.761 (6)	
16	99.760		0.190 (19)$\cdot 10^{-3}$	3.761 (6)		
17	0.038		0.538 (65)$\cdot 10^{-3}$	3.61 (6)		0.007 (2)
			0.235 (10)$^{(abs)}$			
18	0.202		0.16 (1)$\cdot 10^{-3}$			
F 19	100	5.39$\cdot 10^{-5}$	9.6 (5)$\cdot 10^{-3}$	3.641 (10)	3.641 (10)	0.40 (2)$\cdot 10^{-3}$
Ne		2.69$\cdot 10^{-5}$	39 (4)$\cdot 10^{-3}$	2.415 (10)		0.015 (5)
20	90.51		37 (4)$\cdot 10^{-3}$	2.47 (3)		
21	0.27		0.666 (110)	5.1 (3)		
			< 1.5$^{(n,\alpha)}$			
22	9.22		45.5 (6.0)$\cdot 10^{-3}$	1.705 (9)		
Na 23	100	0.02541	0.530 (5)	3.025 (2)	1.54 (2)	1.55 (3)
			0.400 (30)			
Mg		0.04310	0.063 (3)	3.4140 (24)		0.046 (6)
24	78.99		0.051 (5)	3.74 (4)		
25	10.0		0.190 (30)		1.54 (12)	0.023
26	11.01		0.0382 (8)	2.83 (17)		
27	9.46 min		0.07 (2)			
Al 27	100	0.06024	231 (3)$\cdot 10^{-3}$	1.4134 (10)		9.8 (6)$\cdot 10^{-3}$
Si		0.04996	0.171 (3)	2.0437 (17)		0.009 (2)
28	92.23		0.177 (5)	1.992 (6)		
29	4.67		0.101 (14)			
30	3.10		0.107 (2)	2.49 (4)		
P		0.03539	0.172 (6)	3.134 (10)		0.006 (4)
S		0.03950	0.52 (1)	0.9787 (50)	0.9733(50)	5.4 (7)$\cdot 10^{-3}$
			8 (4)$\cdot 10^{-3(n,\alpha)}$			
32	95.02		0.53 (4)	0.9432 (21)		
			7 (4)$\cdot 10^{-3(n,\alpha)}$			
33	0.75		0.35 (4)	2.8 (7)	2.7 (2)	
			2 (1)$\cdot 10^{-3(n,p)}$			
			190(80)$\cdot 10^{-3(n,\alpha)}$			
34	4.21		227 (5)$\cdot 10^{-3}$		1.45 (3)	
Cl		5.45$\cdot 10^{-5}$	33.1 (3)	15.8 (2)		4.9 (2)
			0.37 (2)$^{(n,p)}$			
			33.5 (3)$^{(abs)}$			
35	75.77		43.6 (4)	20.6 (3)	16.32 (25)	4.24 (25)
			0.489 (14)$^{(n,p)}$			
			0.08 (4)$\cdot 10^{-3(n,\alpha)}$			
36	3.00$\cdot 10^{5}$ y		< 10.0			
37	24.23		0.433 (6)	1.15 (5)	<0.08$\cdot 10^{-3}$	

Table 41.2 Absorption and scattering cross sections for neutrons with energy 0.0253 eV *(continued)*

Symbol, mass number	Contents in natural mixture, % or (and) $T_{1/2}$ [22]	Number density of nuclei ρ, 10^{24} cm^{-3}	$\sigma_{n,\gamma}$, 10^{-28} m^2	σ_s, 10^{-28} m^2	$\sigma_{s\ coh}$, 10^{-28} m^2	$\sigma_{s\ noncoh}$, 10^{-28} m^2
Ar		$2.68 \cdot 10^{-5}$	0.675 (9)	0.647 (3)		0.209 (2)
36	0.337		5.2 (5)	73.7 (4)		
			$5.5\ (1) \cdot 10^{-3\,(n,\alpha)}$			
37	35.0 d		$69\ (14)^{(n,p)}$			
			$1970\ (330)^{(n,\alpha)}$			
38	0.063		0.8 (2)	1.5 (1.5)		
39	269 y		600 (300)			
40	99.60		0.660 (10)	0.40 (2)		
K		0.01325	2.1 (1)	2.04 (10)		0.37 (10)
39	93.258		2.1 (2)			0.34 (10)
			$4.3\ (5) \cdot 10^{-3\,(n,\alpha)}$			
40	0.0117		30 (8)			
	$1.28 \cdot 10^9$ y		$4.4\ (3)^{(n,p)}$			
			$0.39\ (3)^{(n,\alpha)}$			
41	6.730		1.46 (3)			0.7 (2)
Ca		0.02329	0.43 (2)	2.93 (4)		0.025 (3)
40	96.94		0.41 (2)	3.01 (8)		
			$2.5(1.1) \cdot 10^{-3\,(n,\alpha)}$			
41	$1.0 \cdot 10^5$ y		4.0			
42	0.647		0.680 (70)	1.2 (2)		
43	0.135		6.2 (6)			
44	2.09		0.88 (5)			
45	165 d		15.0			
46	0.0035		0.74 (7)			
48	0.187		1.09 (14)			
Sc 45	100	0.03349	27.2 (2)	22.4 (4)	18.1 (3)	4.3 (4)
Ti		0.05670	6.09 (13)	4.09 (3)	1.34 (2)	2.75 (4)
46	8.2		0.59 (18)	2.78 (24)		
47	7.4		1.7 (2)	3.1 (2)	1.3 (2)	1.8 (3)
48	73.7		7.84 (250	4.1 (2)		
49	5.4		2.2 (3)	0.7 (3)	0.08 (4)	
50	5.2		0.179 (3)	3.7 (3)		
V		0.07050	5.08 (4)	4.8 (1)	0.017 (1)	
50	0.25		60 (40)	7.5 (1.0)		
51	99.75		4.9 (1)	4.8 (2)		
Cr		0.0801	3.07 (8)	3.38 (1)	1.63 (1)	1.75 (1)
50	4.35		15.9 (2)		2.41 (6)	
52	83.79		0.76 (6)	2.96 (2)		
53	9.50		18.2 (1.5)	7.78 (20)	2.10 (3)	5.68 (19)
54	2.36		0.36 (4)	2.54 (10)		
Mn 55	100	0.08145	13.3 (2)	2.2 (2)	1.65 (2)	0.6 (2)
Fe		0.08487	2.56 (3)	11.35 (3)		0.38 (3)
54	5.8		2.25 (18)	2.17 (10)		
56	91.72		2.59 (14)	12.46 (49)		
57	2.2		2.48 (30)			
58	0.28		1.28 (5)			
Co	100	0.0890	37.18 (6)	6.00 (6)	0.96 (3)	5.04 (8)
60	5.271 y		2.0 (2)			
60m	10.5 min		58.0 (8.0)			

Table 41.2 Absorption and scattering cross sections for neutrons with energy 0.0253 eV *(continued)*

Symbol, mass number	Contents in natural mixture, % or (and) $T_{1/2}$ [22]	Number density of nuclei ρ, 10^{24} cm^{-3}	$\sigma_{n,\gamma}$, 10^{-28} m^2	σ_s, 10^{-28} m^2	$\sigma_{s\ \mathrm{coh}}$, 10^{-28} m^2	$\sigma_{s\ \mathrm{noncoh}}$, 10^{-28} m^2
Ni		0.08980	4.49 (16)	17.8 (4)	13.0 (3)	4.8 (2)
58	68.27		4.6 (3)	25.3 (4)		
60	26.10		2.9 (2)	0.98 (7)		
61	1.13		2.5 (8)	9.0 (1.0)	7.1 (1)	1.9 (3)
62	3.59		14.5 (3)	9.1 (4)		
64	0.91		1.52 (3)	0.0014 (3)		
65	2.52 h		22.4 (2.0)			
Cu		0.8493	3.78 (2)	7.78 (3)	7.28 (1)	0.50 (4)
63	69.20		4.50 (20)	5.1 (2)	5.1 (2)	0.006 (1)
64	12.70 h		< 6000			
65	30.80		2.17 (3)	14.1 (5)	13.7 (5)	0.36 (5)
66	5.10 min		135 (10)			
Zn		0.06572	1.11 (2)	4.08 (3)	4.00 (3)	0.075 (7)
64	48.6		0.76 (2)	3.9 (3)		
			0.015 (1)·$10^{-3 (n,\alpha)}$			
65	244.1 d		250 (150)			
66	27.9		0.85 (20)	4.9 (3)		
67	4.1		6.8 (8)			
			0.006 (4)·$10^{-3 (n,\alpha)}$			
68	18.8		0.72 (4)[^{69m}Zn]	5.4 (3)		
			1.0 (1)[^{69g}Zn]			
			<0.020·$10^{-3 (n,\alpha)}$			
70	0.62		83.0 (5.0)·10^{-3}[^{71g}Zn]			
			8.7 (5)·10^{-3}[^{71m}Zn]			
Ga		0.05105	2.90 (10)	6.50 (20)		
69	60.1		1.68 (7)			
71	39.9		4.71 (23)			
Ge		0.04530	2.3 (2)	8.37 (6)		
70	20.5		3.43 (20)	8.8 (8)		
72	27.4		0.98 (9)	7.5 (7)		
73	7.8		15 (2)			
74	36.5		0.51 (8)	6.1 (4)		
76	7.8		0.15 (2)			
As 75	100	0.04606	4.5 (1)	5.43 (3)	5.37 (2)	0.06 (1)
Se		0.0343	11.7 (2)	8.56 (1.0)	8.22 (10)	0.34 (10)
74	0.87		51.8 (1.2)			
76	9.0		85 (7)	18.4 (3)		
77	7.6		42 (4)	8.43 (16)		
78	23.5		0.43 (2)	8.40 (2)		
80	49.6		0.61 (45)	6.95 (6)		
82	9.2		0.039 (3)	5.0 (2)		
	1.4·10^{20} y		[70 s, ^{83m}Se]			
			0.0052 (4)			
			[22.5 min, ^{83g}Se]			
Br		0.02351	6.9 (2)	6.1 (2)		0.1 (4)
79	50.69		11.0 (7)			
81	49.31		2.7 (2)			

Table 41.2 Absorption and scattering cross sections for neutrons with energy 0.0253 eV *(continued)*

Symbol, mass number		Contents in natural mixture, % or (and) $T_{1/2}$ [22]	Number density of nuclei ρ, 10^{24} cm^{-3}	$\sigma_{n,\gamma}$, 10^{-28} m^2	σ_s, 10^{-28} m^2	σ_s coh, 10^{-28} m^2	σ_s noncoh, 10^{-28} m^2
Kr			$2.67 \cdot 10^{-5}$	25 (1)	7.50 (13)		
	78	0.356		0.17 (2)			
				[50 s, ^{79m}Kr]			
				6.2 (9)[$^{79m+g}$Kr]			
	80	2.27		11.5 (5)			
	82	11.6		28.20			
	83	11.5		180 (30)			
	84	57.0		0.110 (15)			
	85	10.7 y		1.66 (2)			
	86	17.30		0.003 (2)			
				[76 min, ^{87g}Kr]			
Rb			0.01078	0.38 (4)	6.4 (2)	6.25 (4)	
	85	72.17		0.48 (1)		6.2 (1)	
	87	27.83		0.120 (30)		6.6 (1)	
		$4.8 \cdot 10^{10}$ y					
Sr			0.01746	1.28 (6)	10 (1)		
	84	0.56		0.87 (7)			
	86	9.80		1.04 (7)			
	87	7.00		16.3			
	88	82.60		0.058 (4)			
	89	50.5 d		0.42 (4)			
	90	28.8 y		0.9 (5)			
Y	89	100	0.03733	1.28 (2)	7.67 (6)		0.13 (2)
	90	64.1 h		<6.5			
	91	58.5 d		1.4 (3)			
Zr			0.03732	0.185 (3)	6.40 (4)	6.2 (2)	0.15 (3)
	90	51.50		0.011 (5)	5.3 (3)		
	91	11.20		1.24 (25)	10.7 (6)	10.0 (6)	0.7 (1)
	92	17.10		0.220 (60)			
	93	$1.5 \cdot 10^6$ y		$1.3 < \sigma_{n\gamma} < 4.0$			
	94	17.40		0.0499 (24)	6.1 (4)		
	96	2.80		0.0229 (1)	6.6 (4)		
Nb	93	100	0.05445	1.15 (5)	0.37 (7)	6.32 (4)	$6.20 (6) \cdot 10^{-3}$
	94	$2 \cdot 10^4$ y		14.9 (1.0)			
				[35 d, ^{95g}Nb]			
				0.6 (1)			
				[87 h, ^{95m}Nb]			
	95	35 d		< 7.0			
				[23 h, ^{96g}Nb			
Mo			0.06403	2.55 (5)	5.59 (4)		0.27 (12)
	92	14.80		0.019			
	94	9.30		0.015			
	95	15.90		14.0 (5)			
	96	16.70		0.5 (2)			
	97	9.60		2.1 (5)			
	98	24.10		0.130 (6)			
	100	9.60		0.199 (3)			
Tc	99	$2.14 \cdot 10^5$ y		20 (1)[15.8 s, ^{100g}Tc]			

Table 41.2 Absorption and scattering cross sections for neutrons with energy 0.0253 eV *(continued)*

Symbol, mass number		Contents in natural mixture, % or (and) $T_{1/2}$ [22]	Number density of nuclei ρ, 10^{24} cm^{-3}	$\sigma_{n,\gamma}$, 10^{-28} m^2	σ_s, 10^{-28} m^2	$\sigma_{s\ coh}$, 10^{-28} m^2	$\sigma_{s\ noncoh}$, 10^{-28} m^2
Ru			0.07270	2.56 (13)	6.5 (1)		
	96	5.50		0.29 (2)			
	98	1.86		< 8.0			
	99	12.7		7.1 (1.0)			
	100	12.6		5.0 (6)			
	101	17.0		3.4 (9)			
	102	31.6		1.21 (7)			
	104	18.7		0.32 (2)			
	105	4.44 h		0.39 (6)			
	106	367 d		0.146 (45)			
Rh	103	100	0.07263	145 (2)[$^{104m+g}$Rh]			
	104	42.3 s		40 (30)			
	105	35.4 h		11000 (3000) [29.8 s, ^{106g}Rh] 5000 (1000) [130 min, ^{106m}Rh]			
Pd			0.06906	6.9 (4)	4.2 (2)		0.091 (9)
	102	1.00		3.4 (3)			
	104	11.00		0.6 (3)			
	105	22.20		20 (3)	5 (6)		
	106	27.30		0.292 (29) [6.5·10^6 y, ^{107g}Pd] 0.013 (2) [213 s, ^{107m}Pd]	5.1 (6)		
	107	6.5·10^6 y		1.8 (2)			
	108	26.70		8.3 (5) [13.5 h, ^{109g}Pd] 0.183 (33) [4.7 min, ^{109m}Pd]			
	110	11.80		0.190 (30) [23.4 min, ^{111g}Pd] 0.037 (6) [5.5 h, ^{111m}Pd]			
Ag			0.05857	63.3 (4)	5.08 (3)		0.55 (4)
	107	51.83		37.6 (1.2)[$^{108m+g}$Ag]	7.44 (9)		0.12 (3)
	109	48.17		91.0 (1.0)[$^{110m+g}$Ag]	2.55 (6)		0.32 (5)
	110m	249.9 d		82 (11)			
Cd			0.04635	2520 (50)$^{(abs)}$	5.6 (6)		
	106	1.25		~ 1			
	108	0.89		1.1 (3)			
	109	464 d		700 (100) 0.05$^{(n,\alpha)}$			
	110	12.50		11 (1)[$^{111m+g}$Cd]			
	111	12.80		24 (3)	5 (1)		
	112	24.10		2.2 (5)	7 (1)		
	113g	12.20 9·10^{15} y		20600 (400)			
	114	28.70		0.30 (2) [53.4 h, ^{115g}Cd] 0.036 (7) [44.8 d, ^{115m}Cd]	6 (1)		

Table 41.2 Absorption and scattering cross sections for neutrons with energy 0.0253 eV *(continued)*

Symbol, mass number	Contents in natural mixture, % or (and) $T_{1/2}$ [22]	Number density of nuclei ρ, 10^{24} cm^{-3}	$\sigma_{n,\gamma}$, 10^{-28} m^2	σ_s, 10^{-28} m^2	σ_s coh, 10^{-28} m^2	σ_s noncoh, 10^{-28} m^2
116	7.50		0.050 (8)			
			[2.40 h, ^{117g}Cd]			
			0.025 (10)			
			[3.4 h, ^{117m}Cd]			
In		0.03818	193.8 (1.5)	2.45 (20)	1.95 (10)	0.5 (1)
113	4.3		12.0 (1.1)	3.75 (7)		0.000037
115	95.7		202 (2)	2.6 (1)		0.6 (1)
	$5.1 \cdot 10^{14}$ y					
Sn		0.03703	0.626 (9)	4.909 (6)	4.887 (3)	0.022 (5)
112	1.01		0.30 (4)			
			[21 min, ^{113m}Sn]			
			0.71 (10)			
			[115.1 d, ^{113g}Sn]			
114	0.67		0.115 (30)	4.6 (5)		
115	0.38		30 (7)			
116	14.80		0.140 (30)		4.26 (15)	
117	7.75		2.3 (5)		5.2 (4)	
118	24.3		0.220 (50)	4.26 (15)		
119	8.6		2.2 (5)		4.55 (38)	
120	32.4		0.140 (30)	5.17 (16)		
			[27.1 h, ^{121g}Sn]			
			0.001 (1)			
			[55 y, ^{121m}Sn]			
122	4.56		0.180 (20)			
			[40.1 min, ^{123m}Sn]			
			0.001 (1)			
			[129 d, ^{123g}Sn]			
124	5.64		0.130 (5)	4.41 (30)		
			[9.5 min, ^{125m}Sn]			
			0.004 (2)			
			[9.62 d, ^{125g}Sn]			
Sb		0.03076	5.1 (1)	4.2 (1)	4.03 (2)	0.17 (12)
121	57.3		5.9 (2)[$^{122m+g}$Sb]			
123	42.7		4.1 (1)[^{124g}Sb]			
Te		0.02841	4.7 (1)		3.74 (6)	0.2 (2)
120	0.091		2.0 (3)[^{121g}Te]			
122	2.5		3.4 (5)			
123	0.89		418 (30)			
			0.046 (6)$\cdot 10^{-3(n,\alpha)}$			
124	4.60		6.8 (1.3)	3.8 (4)		
125	7.00		1.55 (16)		4 (4)	
126	18.70		1.04 (15)			
128	31.70		0.215 (8)			
130	34.50		0.002 (1)			
	$2 \cdot 10^{21}$ y		[30 h, ^{131m}Te]			
			0.27 (6)			
			[25 min, ^{131g}Te]			

Table 41.2 Absorption and scattering cross sections for neutrons with energy 0.0253 eV *(continued)*

Symbol, mass number	Contents in natural mixture, % or (and) $T_{1/2}$ [22]	Number density of nuclei ρ, 10^{24} cm^{-3}	$\sigma_{n,\gamma}$, 10^{-28} m^2	σ_s, 10^{-28} m^2	$\sigma_{s\ \mathrm{coh}}$, 10^{-28} m^2	$\sigma_{s\ \mathrm{noncoh}}$, 10^{-28} m^2
I		0.02340	894 (90)[13 d, ^{126g}I]			
126	13 d		5960			
127	100		6.2 (2) [25 min, ^{128g}I]		3.54 (3)	$\sim$ 0.0
129	$1.6{\cdot}10^7$		18 (2)			
			[9.2 min, ^{130m}I]			
			9(1)			
			[12.36 h, ^{130g}I]			
130	12.36 h		18 (3) [8.04 d, ^{131g}I]			
131	8.04 d		80 (50) [2.28 h, ^{132g}I]			
Xe		$2.68{\cdot}10^{-5}$	23.9 (1.2)	4.30 (2)		
124	0.096		165 (20) [$^{125m+g}$Xe]			
125	17 h		$<0.03^{(n,\alpha)}$			
126	0.09		3.5 (8) [$^{127m+g}$Xe]			
127	36.41 d		$<0.01^{(n,\alpha)}$			
128	1.92		$<$ 8.0[$^{129m+g}$Xe]			
129	26.4		21 (5)			
130	4.1		$<$ 26.0[$^{131m+g}$Xe]			
131	21.2		85 (10)			
132	26.9		0.45 (6) [$^{133m+g}$Xe]			
133	5.25 d		190 (90)			
134	10.4		0.265 (20)[$^{135m+g}$Xe]			
135	9.10 h		2.65 (11)$\cdot 10^6$			
136	8.9		0.26 (2)			
Cs 133	100	0.00847	29 (1.5)[$^{134m+g}$Cs]			0.22
134	2.062 y		140 (12)			
135	$3.0{\cdot}10^6$ y		8.5 (5)[13.1 d, ^{136g}Cs]			
137	30.17 y		0.110 (33)[32.2 min, ^{138g}Cs]			
Ba		0.01535	1.2 (1)	3.42 (4)		
130	0.106		11.3 (1.0)			
132	0.101		6.5 (8)			
			[10.7 y, ^{133g}Ba]			
			0.5			
			[38.9 h, ^{133m}Ba]			
134	2.417		2.0 (1.6)			
135	6.592		5.8 (9)			
136	7.854		0.4 (4)			
137	11.23		5.1 (4)			
138	71.70		0.360 (36)			
139	82.9 min		6.2 (1.6)			
140	12.79 d		1.6 (3)			
La		0.02667	8.97 (5)	10.13 (22)	8.64 (10)	1.49 (20)
138	0.089		57.2 (5.7)			
	$1.1{\cdot}10^{11}$ y					
139	99.911		8.93 (4)	10.13 (22)	8.64 (10)	1.49 (20)
140	40.3 h		2.7 (3)			
Ce		0.02966	0.63 (4)	4.7 (3)		
136	0.19		0.95 (25)			
			[34.4 h, ^{137m}Ce]			
			6.3 (1.5)			
			[9.0 h, ^{137g}Ce]			

Table 41.2 Absorption and scattering cross sections for neutrons with energy 0.0253 eV *(continued)*

Symbol, mass number	Contents in natural mixture, % or (and) $T_{1/2}$ [22]	Number density of nuclei ρ, 10^{24} cm^{-3}	$\sigma_{n,\gamma}$, 10^{-28} m^2	σ_s, 10^{-28} m^2	$\sigma_{s\ coh}$, 10^{-28} m^2	$\sigma_{s\ noncoh}$, 10^{-28} m^2
138	0.254		0.015 (5) [56 s, ^{139g}Ce]			
			1.1(3) [137.2 d, ^{139g}Ce]			
139	137.2 d		500			
140	88.50		0.57 (4)	2.83 (11)		
141	35.2 d		29 (3)			
142	11.08		0.95 (5)	3.7 (2)		
143	33.0 h		6.0 (7)			
144	284.9 d		1.0 (1)			
Pr 141	100	0.02769	1.5 (3)[$^{142m+g}$Pr]	2.54 (6)		
142	19.2 h		20 (3)			
143	13.58 d		90 (10)			
Nd		0.02914	50.5 (2.0)	16.0 (1.0)		11 (2)
142	27.20		18.7 (7)	7.7 (5)		
143	12.20		325 (10) 17.4 (1.6)$\cdot 10^{-3(n,\alpha)}$	80 (2)		
144	23.80 $2.1\cdot 10^{15}$ y		3.6 (3)	1.0 (2)		
145	8.30		4.2 (2) $<0.1\cdot 10^{-3(n,\alpha)}$			
146	17.20		1.4 (1) [11 d, ^{147g}Nd]			
147	11 d		440 (150)			
148	5.76		2.5 (2) [1.73 h, ^{149g}Nd]	4.0 (5)		
150	5.64		1.2 (2) [12.4 min, ^{151g}Nd]	3.5 (5)		

Table 41.3 Absorption and scattering cross sections for neutrons with energy 0.0253 eV (elements with atomic number from 61 to 89) [24]

Symbol, mass number	Contents in natural mixture, % or/and $T_{1/2}$ [22]	Number density of nuclei ρ, 10^{24} cm^{-3}[23]	$\sigma_{n,\gamma}$, 10^{-28} m^2	σ_s, 10^{-28} m^2	σ_{tot}, 10^{-28} m^2
Pm	5.53 y		8400 (1680)		
147	2.6234 y		181 (7)		205 (7)
148m	41.3 d		22000 (2500)		
148	5.37 d		2000 (1000)		
149	53.1 h		1400 (300)		
151	28.4 h		<700		
Sm		0.0311	5800 (100)		
144	3.30		0.7		
145	340 d		~110		
147	15.10 $1.06\cdot 10^{11}$ y		64 (5)	11(7)	75 (6)
148	11.30 $8\cdot 10^{15}$ y		2.7 (6)		

Table 41.3 Absorption and scattering cross sections for neutrons with energy 0.0253 eV *(continued)*

Symbol, mass number		Contents in natural mixture, % or/and $T_{1/2}$ [22]	Number density of nuclei ρ, 10^{24} cm^{-3}[23]	$\sigma_{n,\gamma}$, 10^{-28} m^2	σ_s, 10^{-28} m^2	σ_{tot}, 10^{-28} m^2
Sm	149	13.90		41000 (2000)		
	150	7.4		102 (5)		133 (8)
	151	90 y		15000 (1800)		
	152	26.60		206 (6)		
	154	22.60		5.5 (1.1)		
Eu			0.0207	4600 (100)	8.0 (1.0)	
	151	47.9		9200 (100) 9.0 (2.0)·$10^{-6(n,\alpha)}$		
	152	13 y		2300 (1000)		
	153	52.1		390 (30)	8.0 (2)	
	154	8.5 y		1500 (400)		
Gd			0.0305	49000 (1000)		
	152	0.2 1.1·10^{14} y		1100 (100) 7.0·$10^{-3(n,\alpha)}$		
	154	2.10		85 (12)		
	155	14.80		61000 (500)	60	
	156	20.6		1.5 (1.2)		
	157	15.7		254 (2)·10^3	1011	
	158	24.80		2.5 (5)		
	160	21.80		0.77 (2)		
	161	3.7 min		31000 (12000)		
Tb	159	100	0.0316	25.5 (1.1)	20 (2)	
	160	72.1 d		525 (100)		
Dy			0.0317	930 (20)	100 (10)	1030 (25)
	156	0.057		33 (3) <9·$10^{-3(n,\alpha)}$		
	158	0.10		43 (6) <6·$10^{-3(n,\alpha)}$		
	160	2.3		61 (6) <0.3·$10^{-3(n,\alpha)}$		
	161	19.0		585 (30) <3·$10^{-5(n,\alpha)}$	22 (1)	
	162	25.5		180 (20)	2.5 (8)	
	163	24.9		130 (10) <2·$10^{-5(n,\alpha)}$	9.7 (4)	140 (10)
	164	28.1		2700 (75)	347 (30)	
	165	2.33 h		3900 (300)		
Ho	165	100	0.0320	66.5 (3.3) <2·$10^{-5(n,\alpha)}$	9.4 (2)	
Er			0.0172	162 (8)	11.0 (8)	
	162	0.14		19 (2) <11·$10^{-3(n,\alpha)}$	1.8 (2.0)	
	164	1.56		13.2 <1.2·$10^{-3(n,\alpha)}$	12.1 (7)	
	166	33.4		35 (3) <7·$10^{-5(n,\alpha)}$	19.1 (1.0)	
	167	22.9		670 (30) <7·$10^{-5(n,\alpha)}$	7.9 (8)	
	168	27.1		1.95 (5) <9·$10^{-5(n,\alpha)}$	15.0 (8)	
	170	14.90		5.7 (2)	15.0 (8)	
	171	7.52 h		280 (30)		

Table 41.3 Absorption and scattering cross sections for neutrons with energy 0.0253 eV *(continued)*

Symbol, mass number		Contents in natural mixture, % or/and $T_{1/2}$ [22]	Number density of nuclei ρ, 10^{24} cm^{-3}[23]	$\sigma_{n,\gamma}$, 10^{-28} m^2	σ_s, 10^{-28} m^2	σ_{tot}, 10^{-28} m^2
Tm	169	100	0.0331	103 (3) $<1{\cdot}10^{-5(n,\alpha)}$	12 (2)	
	170	128.6 d		92 (4)		
	171	1.92 y		4.5 (2)		
Yb			0.0244	36.6 (2.0)	25.0 (8)	
	168	0.135		3470 (100) $<4{\cdot}10^{-3(n,\alpha)}$		
	170	3.1		10 (1) $<4{\cdot}10^{-5(n,\alpha)}$		
	171	14.4		50 (4) $<4{\cdot}10^{-5(n,\alpha)}$		
	172	21.9		1.3 (8) $<3{\cdot}10^{-5(n,\alpha)}$		
	173	16.2		19 (2)		
	174	31.60		65 (5) $<2{\cdot}10^{-5(n,\alpha)}$		
	176	12.7		2.4 (2)		
Lu			0.0335	77 (3)	8.0 (2.0)	84.7 (5)
	175	97.39		23.4 (2.0) $<6{\cdot}10^{-5(n,\alpha)}$		
	176	2.61 $3.6{\cdot}10^{10}$ y		2100 (50) $<2{\cdot}10^{-3(n,\alpha)}$		
Hf			0.0449	102 (2)	8 (2)	110 (2)
	174	0.16 $2{\cdot}10^{15}$ y		390 (55)		
	176	5.2		38 (6)		
	177	18.60		365 (20)		372 (23)
	178	27.1		86 (7)		91 (6)
	179	13.7		45 (5)		51 (6)
	180	35.2		12.6 (7)		
Ta	180	0.0123		700 (200)		
	181	99.9877	0.0553	10.3 (2.5)$\cdot 10^{-3}$ [15.8 min, ^{182m}Ta] 21.0 (7) [115 d, ^{182g}Ta]	6.2 (6)	27.2 (2)
	182	115 d		8200 (600)		
W			0.0619	18.5 (5)		
	180	0.13		3.5 [121 d, ^{181}W]		
	182	26.3		20.7 (5)		
	183	14.3		10.2 (3)		
	184	30.7		1.8 (2)		
	186	28.6		37.8 (1.5)		
	187	23.9 h		64 (10)		
Re			0.0950	88 (4)	11.3 (5)	
	185	37.40		112 (3)		
	187	62.60 $4{\cdot}10^{10}$ y		74 (4)		
	188	16.9 h		<2.0		

Table 41.3 Absorption and scattering cross sections for neutrons with energy 0.0253 eV *(continued)*

Symbol, mass number	Contents in natural mixture, % or/and $T_{1/2}$ [22]	Number density of nuclei ρ, 10^{24} cm^{-3}[23]	$\sigma_{n,\gamma}$, 10^{-28} m^2	σ_s, 10^{-28} m^2	σ_{tot}, 10^{-28} m^2
Os		0.0734	15.3 (7)		
184	0.018		3000 (150) $<1{\cdot}10^{-2(n,\alpha)}$		
186	1.582 $2{\cdot}10^{15}$ y		$<1{\cdot}10^{-4(n,\alpha)}$		
187	1.60		336 (17) $<1{\cdot}10^{-4(n,\alpha)}$		344 (12)
188	13.3		4.3 (1.0) $<3{\cdot}10^{-5(n,\alpha)}$		
189	16.1		23 (4) $<1{\cdot}10^{-5(n,\alpha)}$		
190	26.4		13.0 (3)		
192	41.0		2.0 (1)		
193	30.2 h		1540		
Ir		0.0782	426 (4)	14 (2)	440 (4)
191	37.30		924 (53)		
192	74.20 d		1100 (400)		
193	62.7		112.5 (7.5)		
193m	10.6 d		>0.035		
Pt		0.0661	10.0 (2)	11.2 (1.0)	
190	0.013 $6.0{\cdot}10^{11}$ y		150 (150) $<8{\cdot}10^{-3(n,\alpha)}$		
192	0.78		<14.0 $<2{\cdot}10^{-4(n,\alpha)}$		
194	32.9		1.2 (9) $<5{\cdot}10^{-6(n,\alpha)}$		
195	33.8		27 (2) $<5{\cdot}10^{-6(n,\alpha)}$		
196	25.3		0.74 (8)		
198	7.2		3.7 (2)		
199	30.8 min		15 (10)		
Au 197	100	0.0590	98.85 (9)		
198	2.695 d		25100 (370)		
199	3.14 d		30 (15)		
Hg		0.0408	375 (5)		
196	0.15		120 (13) [23.8 h, ^{197m}Hg] 3080 (200) [64.1 h, ^{197g}Hg]		
198	10.0		1.9		
199	16.8		2000 (1000)		
200	23.1		<60.0		
201	13.2		<60.0		
202	29.8		4.9 (1)		
204	6.90		0.43 (10)		
Tl		0.0351	3.4 (5)	9.7 (4)	13.1 (7)
203	29.5		11.0 (5)		
204	3.77 y		21.6 (2.0)		
205	70.5		0.10 (3)		

Table 41.3 Absorption and scattering cross sections for neutrons with energy 0.0253 eV *(continued)*

Symbol, mass number		Contents in natural mixture, % or/and $T_{1/2}$ [22]	Number density of nuclei ρ, 10^{24} cm^{-3}[23]	$\sigma_{n,\gamma}$, 10^{-28} m^2	σ_s, 10^{-28} m^2	σ_{tot}, 10^{-28} m^2
Pb			0.0329	0.170 (2)	11.4 (2)	11.6 (2)
	204	1.42		0.661 (70)		
	206	24.1		30.5 (8)$\cdot 10^{-3}$		
	207	22.1		709 (10)$\cdot 10^{-3}$		
	208	52.30		487 (30)$\cdot 10^{-6}$		
	210	22.3 y		0.5 (5)		
Bi	209	100	0.0280	33 (4)$\cdot 10^{-3}$		
	210*m*	3.5$\cdot 10^6$ y		54 (5)$\cdot 10^{-3}$		
Rn	220	55.6 s		<0.2		
	222	3.8235 d		0.72 (7)		
Ra	223	11.435 d		130 (20) 0.7 (3)$^{(n,f)}$		
	224	3.64 d		12.0 (5)		
	226	1600 y		11.5 (1.5) <1$\cdot 10^{-4(n,f)}$		
	228	5.76 y		36 (5) <2.0$^{(n,f)}$		
Ac	227	21.773 y		515 (35) <2$\cdot 10^{-3(n,f)}$		

Table 41.4 Cross sections of fission and capture, scattering and total cross sections for neutrons with energy 0.0253 eV (elements with atomic number from 90 to 100)

Symbol, mass number		Contents in natural mixture, % or/and $T_{1/2}$ [22]	Number density of nuclei ρ, 10^{24} cm^{-3} [23]	$\sigma_{n,\gamma}$, 10^{-28} m^2 [25]	σ_s, 10^{-28} m^2 [24]	σ_f, 10^{-28} m^2 [25]	σ_{tot}, 10^{-28} m^2 [24]
Th	227	18.718 d				200(20)	
	228	1.1931 y		123 (15)		<0.3	
	229	7340 y		54 (6)		30.5 (3.0)	84.5 (6.7)
	230	8.0$\cdot 10^4$ y		23.2 (6)	13 (14) [26]	<1.2$\cdot 10^{-3}$	
	232	100 1.41$\cdot 10^{10}$ y	0.0299	7.40 (8)	12.67 (8)	39 (4)$\cdot 10^{-6}$	20.07 (11)
	233	22.3 min		1500 (100)		15 (2)	
	234	24.10 d		1.8 (5) [24]		<0.01 (24) 1500 (250) [24]	
Pa	230	17.7 d					
	231	3.28$\cdot 10^4$ y	0.0402	201 (20)		0.012 (6)	211 (4)
	232	1.31 d		760 (100)		700 (100)	
	233	27.0 d		21 (6) [1.17 min, ^{234m}Pa] 20 (3) [6.67 h, ^{234g} Pa]		<0.1	55 (3)
	234*m*	1.175 min				<500	
	234	6.75 h				<5000	
U	230	20.8 d				25 (10)	
	231	41.2 d				400 (300)	
	232	72 y		73.1 (1.5)	14.7	75.2 (4.7)	163 (10)
	233	1.592$\cdot 10^5$ y	0.0484	45.5 (7) [27]	8.2 (2.0)	529.1 (1.2) [27]	574.7 (1.0) [27]

Table 41.4 Cross sections of fission and capture, scattering and total cross sections for neutrons with energy 0.0253 eV *(continued)*

Symbol, mass number	Contents in natural mixture, % or/and $T_{1/2}$ [22]	Number density of nuclei ρ, 10^{24} cm^{-3} [23]	$\sigma_{n,\gamma}$, 10^{-28} m^2 [25]	σ_s, 10^{-28} m^2 [24]	σ_f, 10^{-28} m^2 [25]	σ_{tot}, 10^{-28} m^2 [24]
U 234	0.0054		100.0 (1.5)	12.4	< 0.65	112.4
	$2.45 \cdot 10^5$ y					
235	0.720	0.0479	98.3 (8) [27]	14.0 (5) [27]	582.6 (1.1) [27]	694.9 (1.1) [27]
	$7.038 \cdot 10^8$ y					
236	$2.342 \cdot 10^7$ y		5.2 (3)			
237	6.75 d		380 (100)		<0.35	411 (138)($^{(abs)}$)
238	99.275	0.0473	2.71 (2)	8.90 (16)	$5.3 \cdot 10^{-6}$	11.60 (16)
	$4.468 \cdot 10^9$ y					
239	23.5 min		22 (5)	14 (3)		
Np 234	4.4 d				900 (300) [24]	
235	396 d		1600 (200) [24]			
			[22.5 h, ^{236m}Np]			
			184 (4) [24]			
			[$1.29 \cdot 10^6$ y, ^{236g}Np]			
236	$1.29 \cdot 10^6$ y				2500 (150)	
237	$2.14 \cdot 10^6$ y		169 (3)		0.019 (3)	
238	2.117 d				2070 (30)	
239	2.35 d		31(6)		<1.0	
			[7.5 min, ^{240m}Np]			
			14 (14)			
			[65 min, ^{240g}Np]			
Pu 236	2.85 y				162 (30)	
237	45.4 d				2200 (400)	
238	87.74 y		547 (20)	564 (20)	16.5 (5)	588 (20)
239	$2.410 \cdot 10^4$ y		269.3 (2.2) [27]	7.7 (5)	741.7 (2.0) [27]	1017.3 (2.9) [27]
240	6570 y	287.0 (1.4)	1.54 (9)	0.05 (5)	291.1 (1.4)	
241	14.4 y		358.2 (5.1) [27]	11 (1)	1011.1 (6.2) [27]	1369.4 (7.7) [27]
242	$3.76 \cdot 10^5$ y	0.0492	18.5 (4)	8.0 (2)	<0.2	26.5 (5)
243	4.956 h		87 (13)		180 (30)	
244	$8.1 \cdot 10^7$ y		1.7 (1) [24]			
245	10.5 h		150 (30) [24]			
Am 241	433 y		$0.10\ (5) \cdot 10^{-3}$[22]		3.14 (10)	
			[13 s, ^{242m2}Am]			
			83.6 (2.6)			
			[152 y, ^{242m1}Am]			
			752 (20)			
			16.01 y, ^{242g}Am]			
			835.6 (20)			
242m	152 y		1100 (1100)		6900 (400)	8000 (800)$^{(abs)}$
243	7370 y		79.0 (4.0)		0.20 (11)	85 (4)
			75.2 (1.8)			
			[26 min, ^{244m}Am]			
			4.1 (2)			
			[10.1 h, ^{244g}Am]			
244m	26 min				1600 (300)	
244	10.1 h				2300 (300)	

Table 41.4 Cross sections of fission and capture, scattering and total cross sections for neutrons with energy 0.0253 eV *(continued)*

Symbol, mass number		Contents in natural mixture, % or/and $T_{1/2}$ [22]	Number density of nuclei ρ, 10^{24} cm^{-3} [23]	$\sigma_{n,\gamma}$, 10^{-28} m^2 [25]	σ_s, 10^{-28} m^2 [24]	σ_f, 10^{-28} m^2 [25]	σ_{tot}, 10^{-28} m^2 [24]
Cm	242	162.8		20 (10)		<5.0	
	243	28.5 y		131 (10)		609 (25)	825 (125)$^{(abs)}$
	244	18.11 y		13.5 (2.0)	8 (3)	1.0 (2)	23 (3)
	245	$8.5 \cdot 10^3$ y		350 (30)		2030 (60)	2375 (100)
	246	$4.7 \cdot 10^3$ y		1.3 (3)		0.15 (7)	
	247	$1.60 \cdot 10^7$ y		59.6		80.7	
	248	$3.5 \cdot 10^5$ y		2.9 (3)		0.37 (7)	
	249	65 min		1.6 (8) [24]			
Bk	249	321.4 d		1800 (100)			1300 (300)$^{(abs)}$
	250	3.22 h				960 (150)	
Cf	249	351.0 h		450 (30)		1660 (50)	
	250	13.1 y		1750 (250)		< 350.0	
	251	900 y		2850 (290)		4800 (480)	7150 (350)$^{(abs)}$
	252	2.64 y		20.4 (1.5)		32(4)	
	253	17.8 d		12.0 (2.0)		1100 (220)	
Es	253	20.47 d		155 (20) [39.3 h, ^{254m}Es] <3.0 [276 d, ^{254g}Es]			
	254m	39.3 h		1.3 [39.3 h, ^{254m}Es]		1840 (80)	
	254	276 d		40.0 [38.3 d, ^{255}Es]		2900 (110)	
	255	38.3 d		43 (10)		3400 (170) [28]	
Fm	254	3.24 h		76 (76)			
	255	20.1 h		26 (3)			
	256	2.63 h		45 (45)			
	257	100.5 d				2950 (160) [24]	6100 (600)$^{(abs)}$

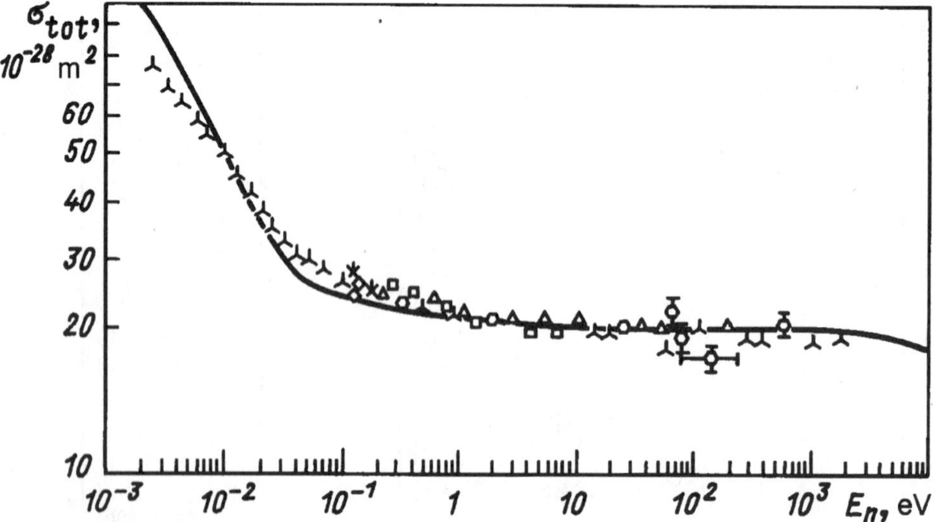

Figure 41.4 Dependence of the total cross section of neutron interaction with hydrogen nuclei on the neutron energy ($10^{-3} < E_n < 10^4$ eV) [29].

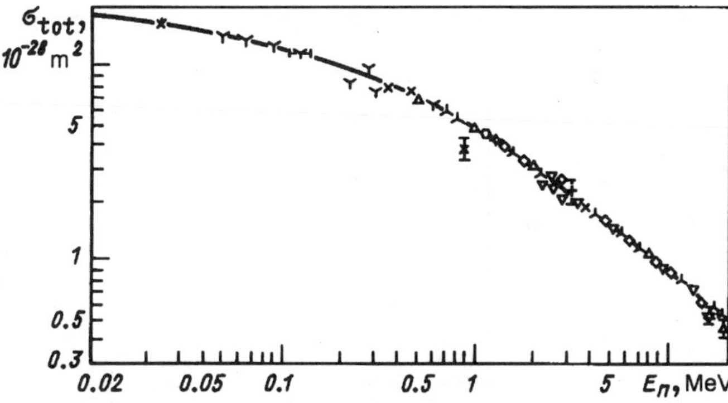

Figure 41.5 The neutron–energy dependence of the total cross section of neutron interaction with hydrogen nuclei ($2 \cdot 10^{-2} < E_n < 20$ MeV) [29].

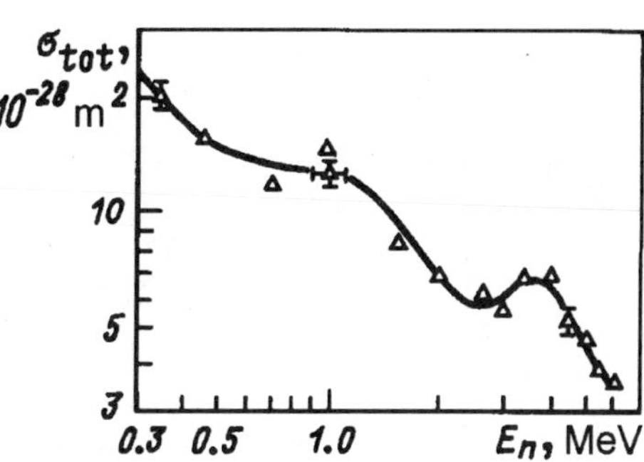

Figure 41.7 The neutron–energy dependence of the total cross section of neutron interaction with hydrogen nuclei in water ($0.3 < E_n < 7$ MeV) [29].

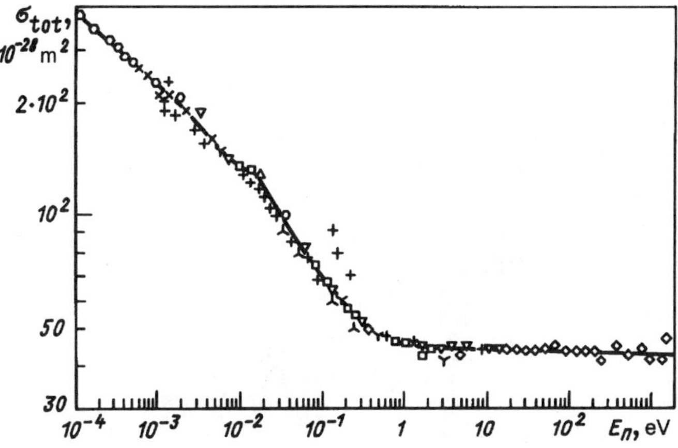

Figure 41.6 The neutron–energy dependence of the total cross section of neutron interaction with hydrogen nuclei in water ($10^{-4} < E_n < 2 \cdot 10^3$ eV) [29].

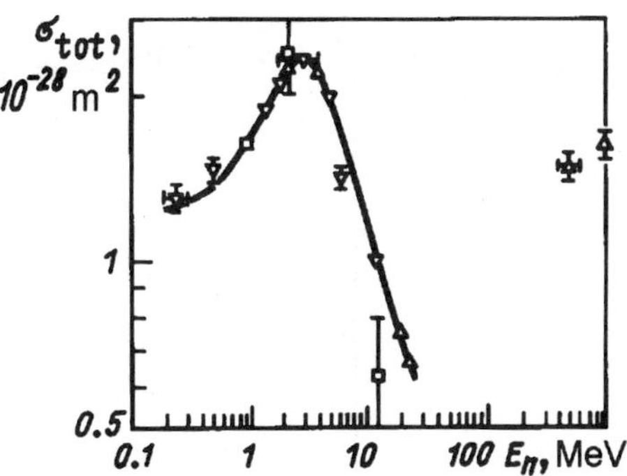

Figure 41.8 The neutron–energy dependence of the total cross section of neutron interaction with tritium nuclei ($0.2 < E_n < 12$ MeV) [29].

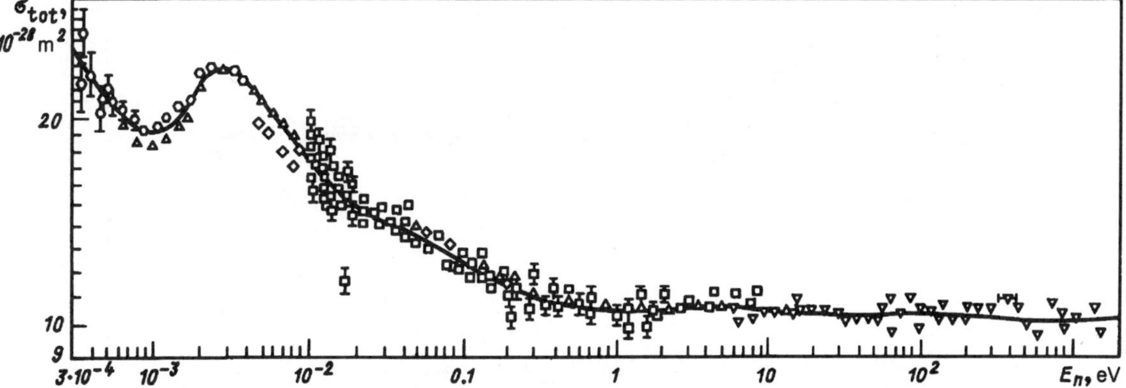

Figure 41.9 The neutron–energy dependence of the total cross section of neutron interaction with deuterium nuclei in heavy water ($3 \cdot 10^{-4} < E_n < 2 \cdot 10^3$ eV) [29].

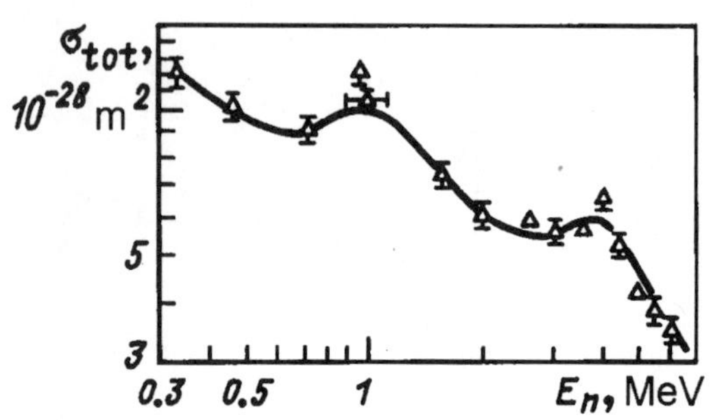

Figure 41.10 The neutron–energy dependence of the total cross section of neutron interaction with deuterium nuclei in heavy water ($0.3 < E_n < 7$ MeV) [29].

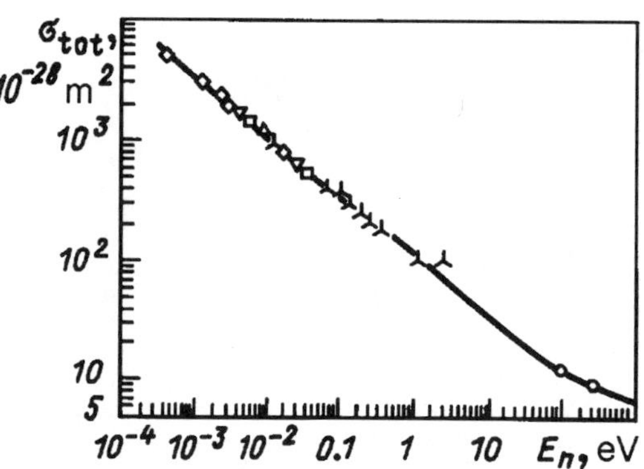

Figure 41.11 The neutron–energy dependence of the total cross section of neutron interaction with boron nuclei ($4 \cdot 10^{-4} < E_n < 10^3$ eV) [29].

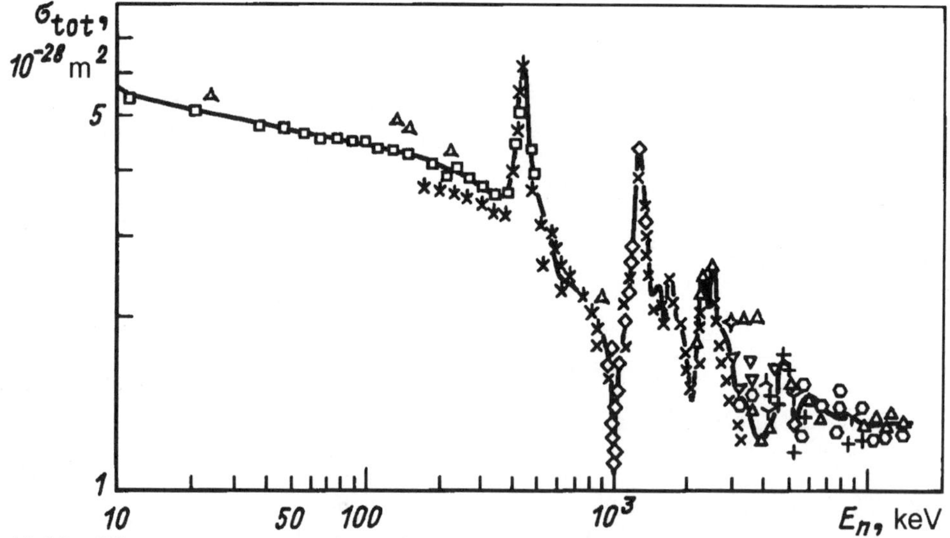

Figure 41.12 The neutron–energy dependence of the total cross section of neutron interaction with boron nuclei ($10 < E_n < 2 \cdot 10^4$ keV) [29].

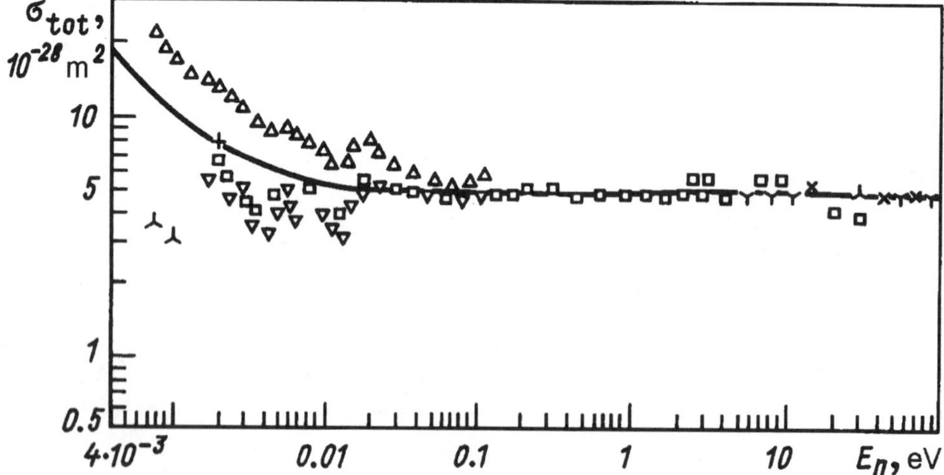

Figure 41.13 The neutron–energy dependence of the total cross section of neutron interaction with carbon nuclei ($4 \cdot 10^{-3}10 < E_n < 100$ eV) [29].

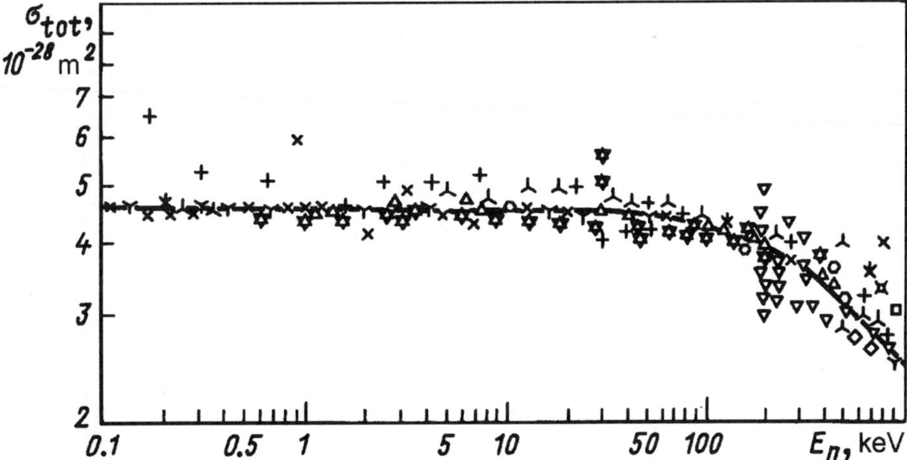

Figure 41.14 The neutron–energy dependence of the total cross section of neutron interaction with carbon nuclei (0.1 keV< E_n < 1 MeV) [29].

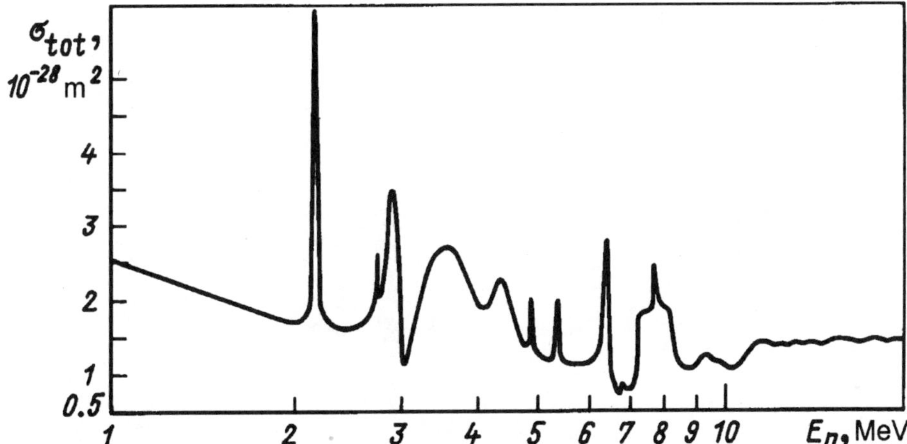

Figure 41.15 The neutron–energy dependence of the total cross section of neutron interaction with carbon nuclei (1 MeV < E_n < 20 MeV) [29].

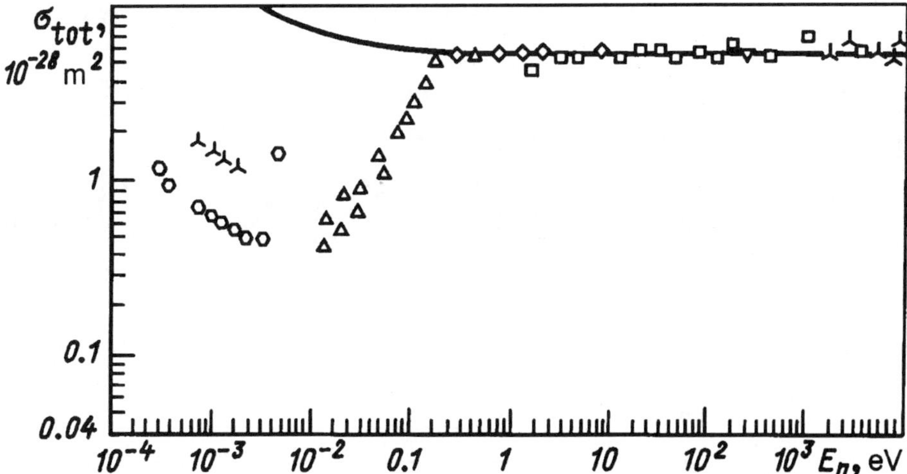

Figure 41.16 The neutron–energy dependence of the total cross section of neutron interaction with beryllium nuclei (10^{-3} < E_n < $10^{.4}$ eV) [29].

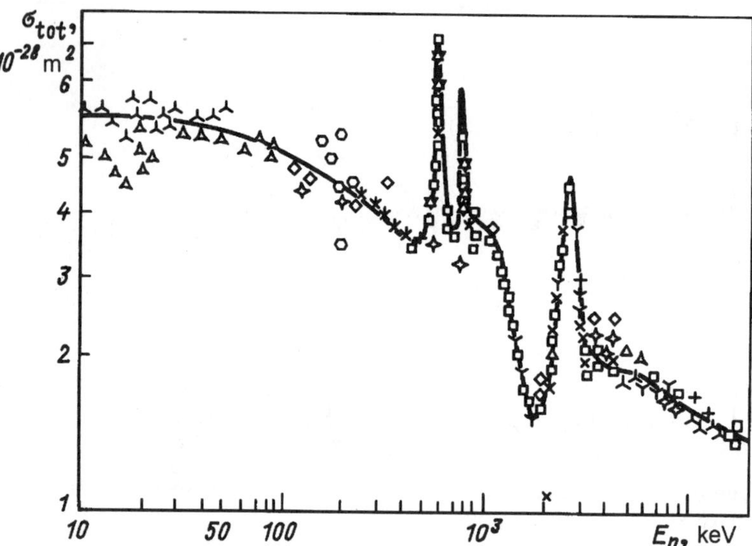

Figure 41.17 The neutron–energy dependence of the total cross section of neutron interaction with beryllium nuclei (10 keV< E_n < 30 MeV) [29].

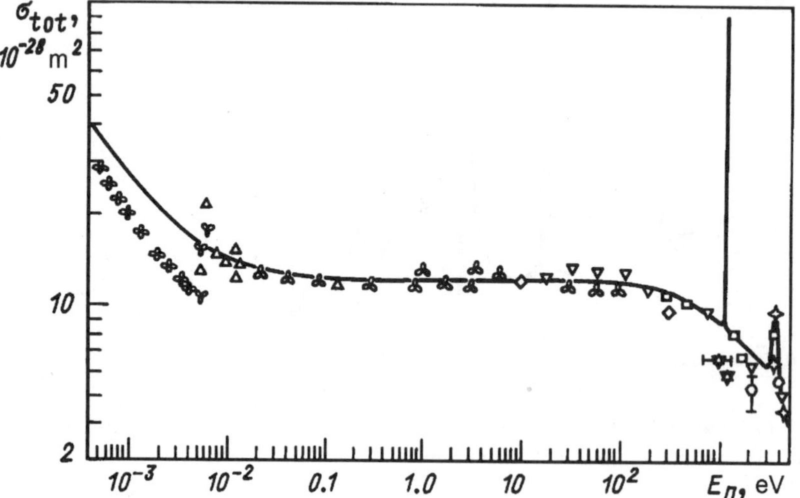

Figure 41.18 The neutron–energy dependence of the total cross section of neutron interaction with iron nuclei ($10^{-3} < E_n < 5 \cdot 10^3$ eV) [29].

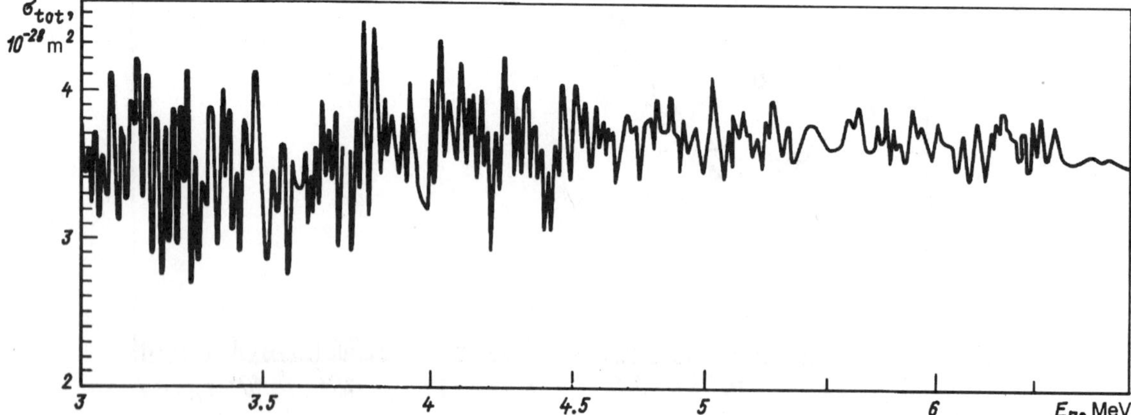

Figure 41.19 The neutron–energy dependence of the total cross section of neutron interaction with iron nuclei ($3 < E_n < 7$ MeV) [29].

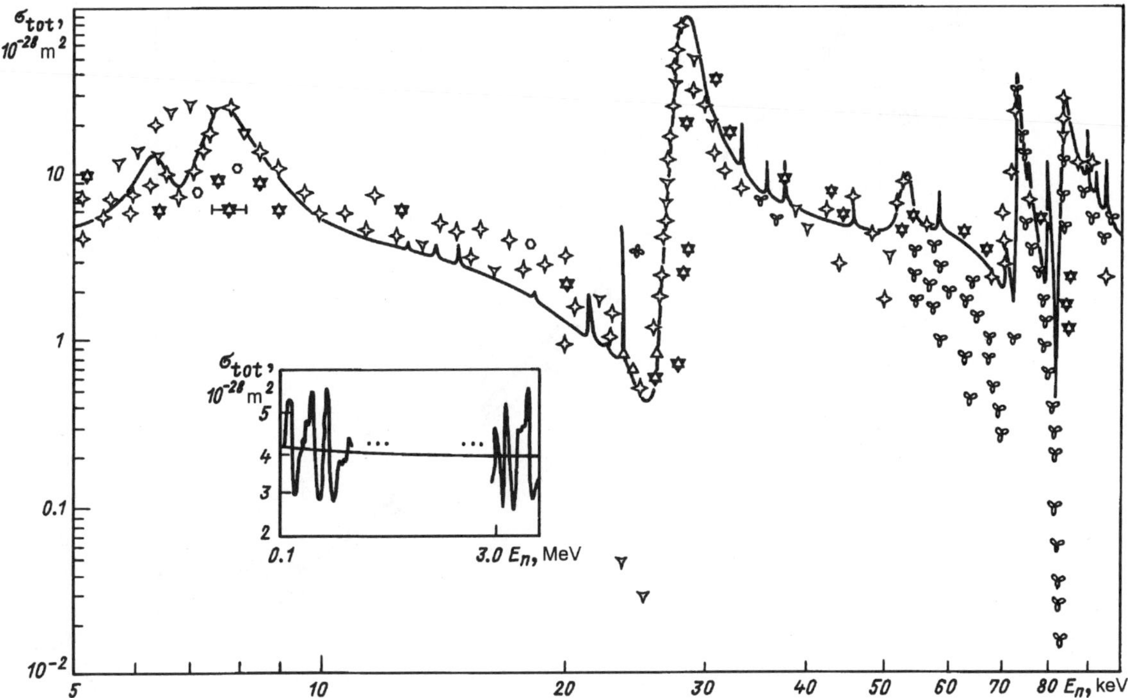

Figure 41.20 The neutron–energy dependence of the total cross section of neutron interaction with iron nuclei (5 keV< E_n < 3 MeV) [29].

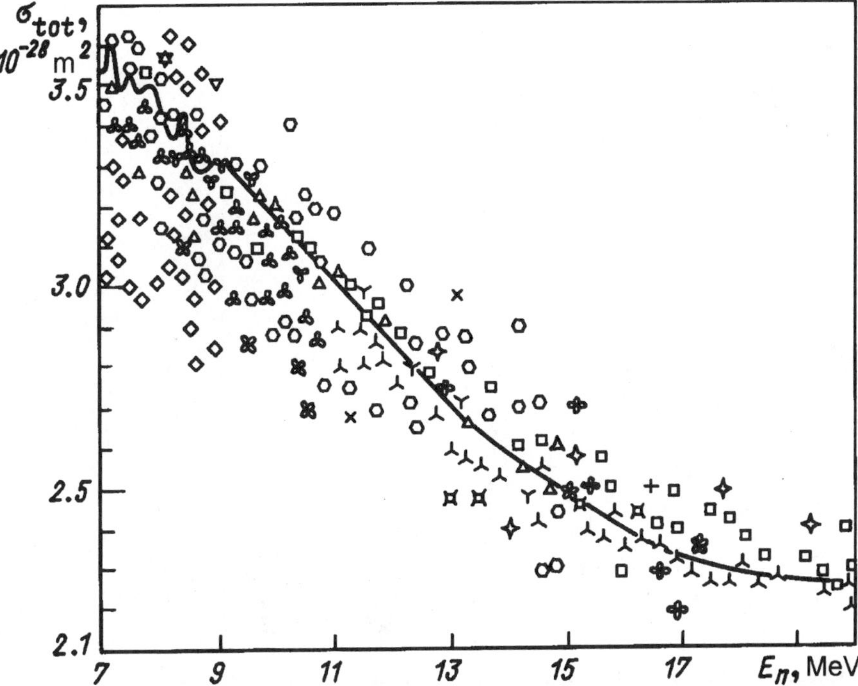

Figure 41.21 The neutron–energy dependence of the total cross section of neutron interaction with iron nuclei (7< E_n < 20 MeV) [29].

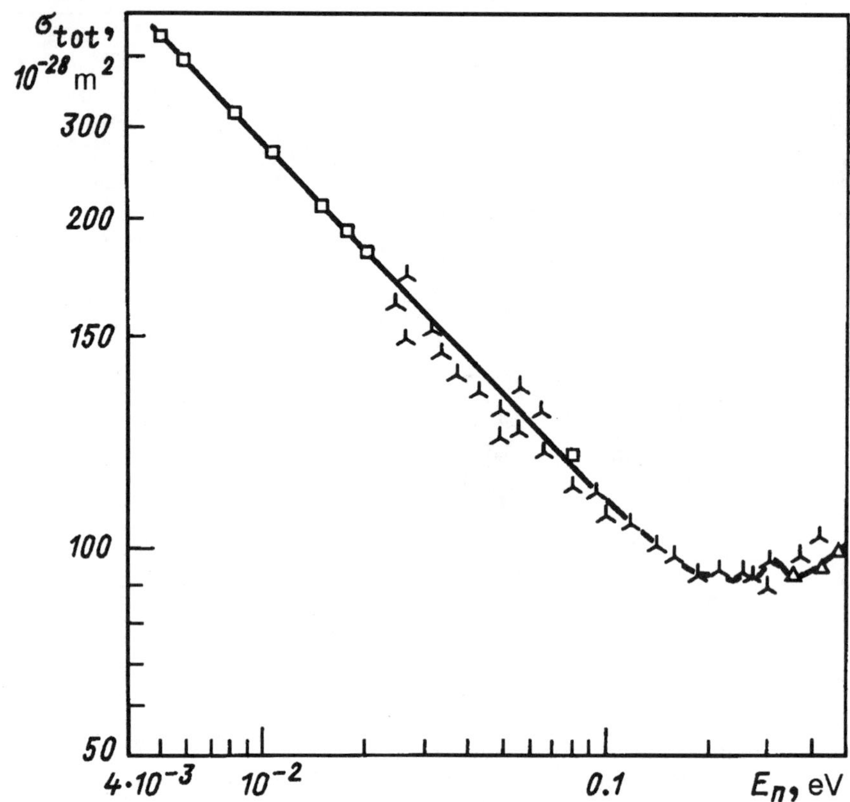

Figure 41.22 The neutron–energy dependence of the total cross section of neutron interaction with indium nuclei ($5 \cdot 10^{-3} < E_n < 0.5$ eV) [29].

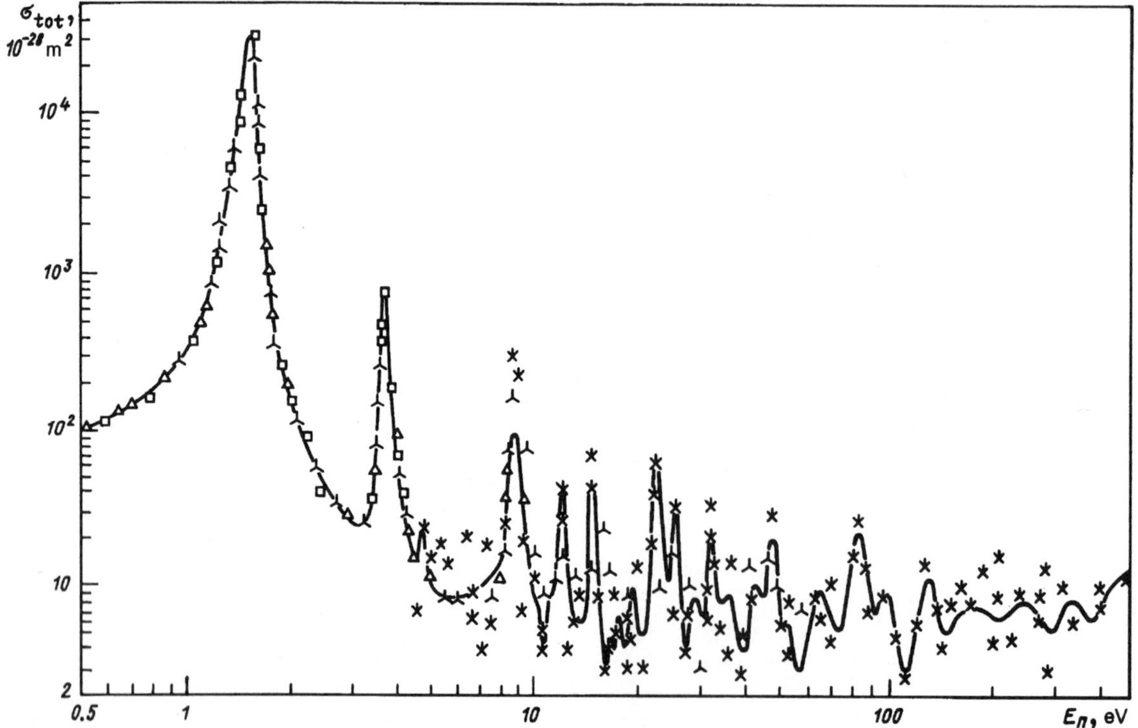

Figure 41.23 The neutron–energy dependence of the total cross section of neutron interaction with indium nuclei ($0.5 < E_n < 500$ eV) [29].

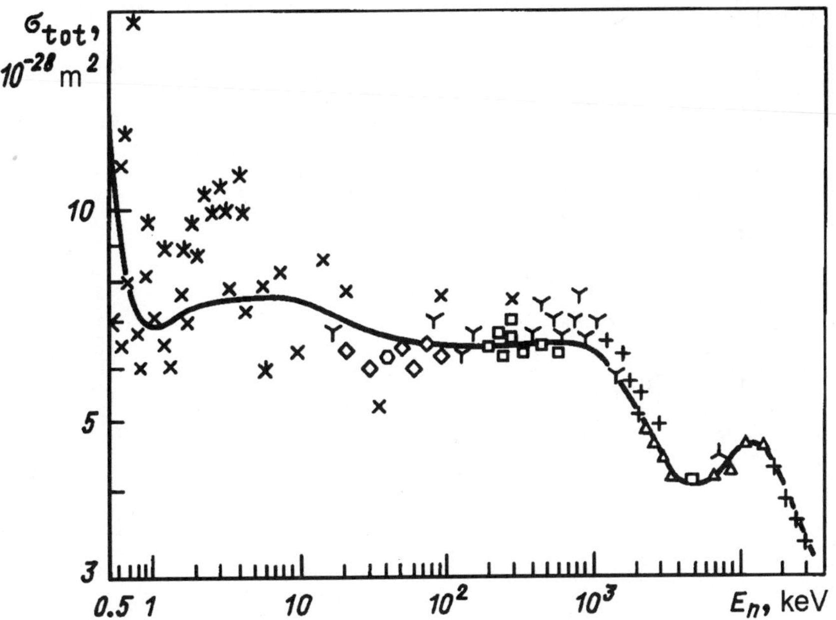

Figure 41.24 The neutron–energy dependence of the total cross section of neutron interaction with indium nuclei (0.5 keV $< E_n <$ 30 MeV) [29].

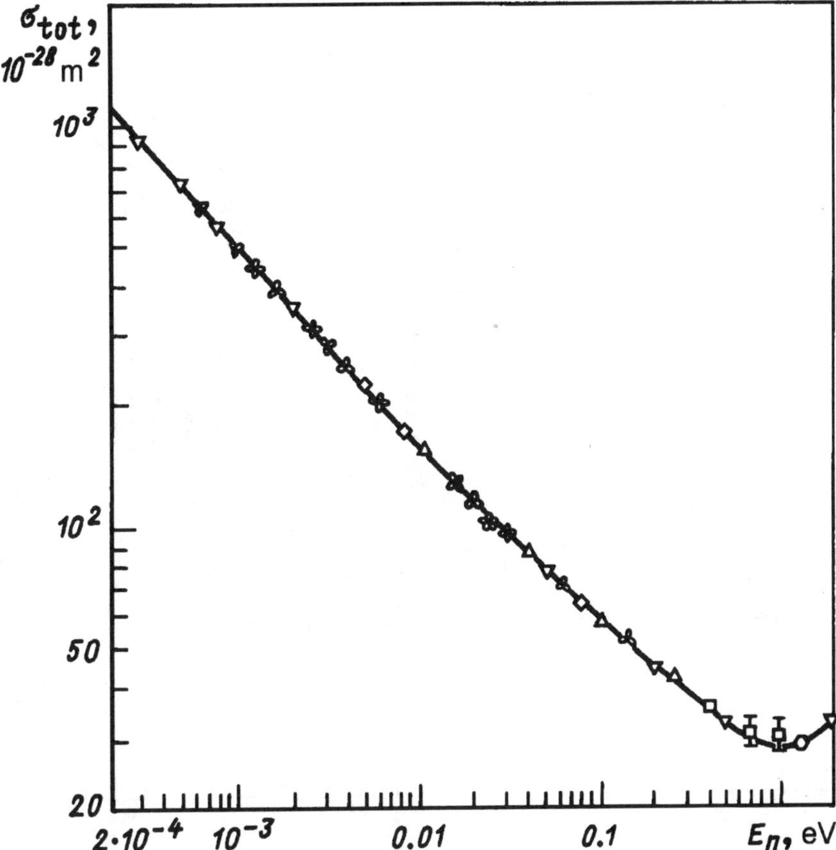

Figure 41.25 The neutron–energy dependence of the total cross section of neutron interaction with gold nuclei ($10^{-4} < E_n <$ 2 eV) [29].

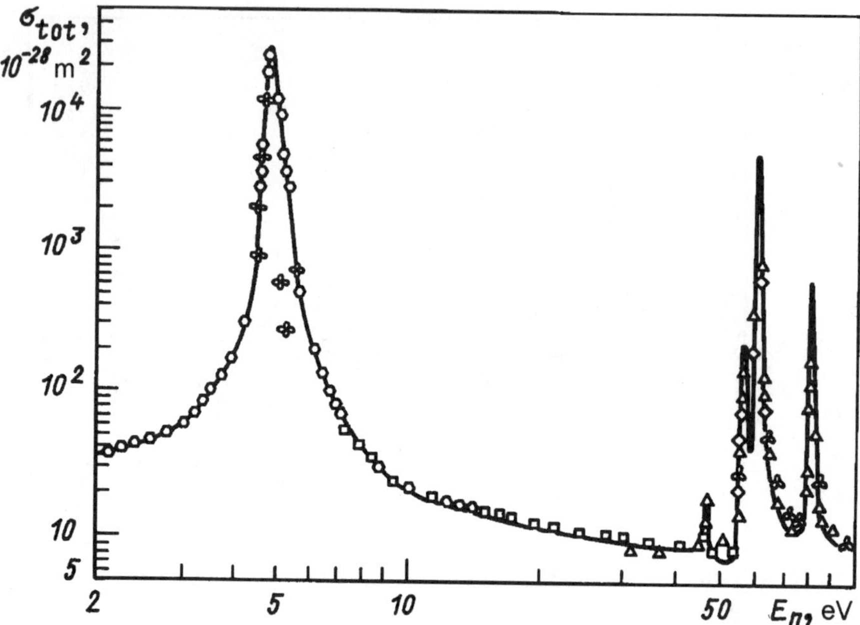

Figure 41.26 The neutron–energy dependence of the total cross section of neutron interaction with gold nuclei ($2 < E_n < 100$ eV) [29].

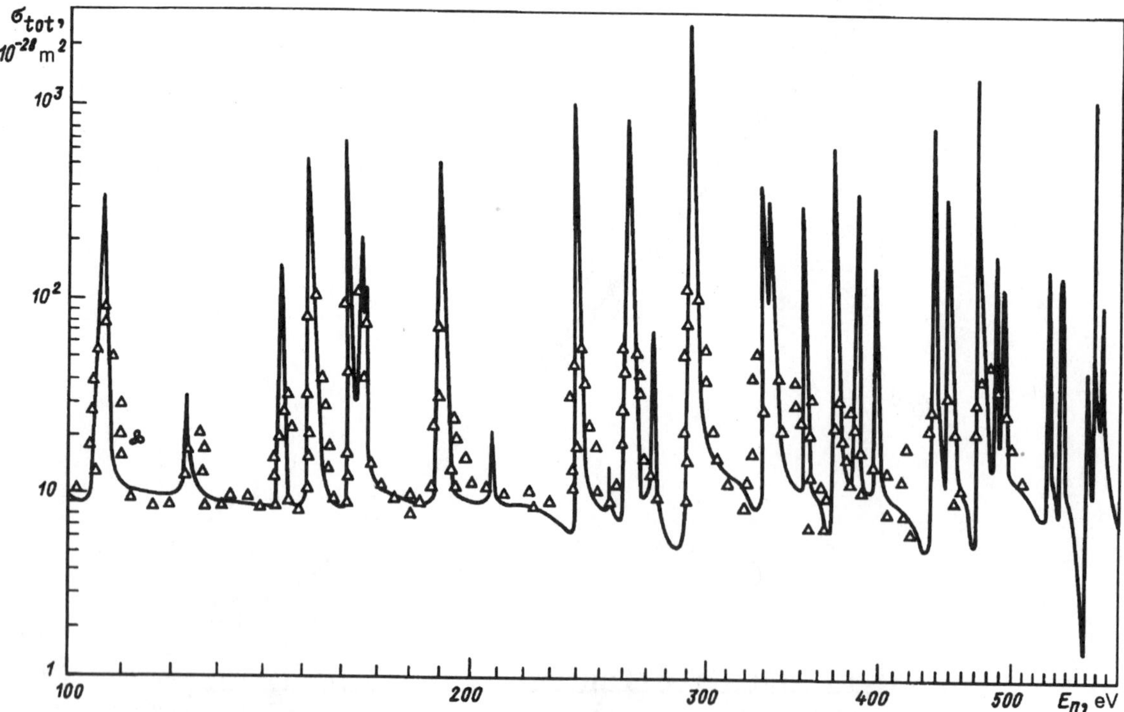

Figure 41.27 The neutron–energy dependence of the total cross section of neutron interaction with gold nuclei ($100 < E_n < 600$ eV) [29].

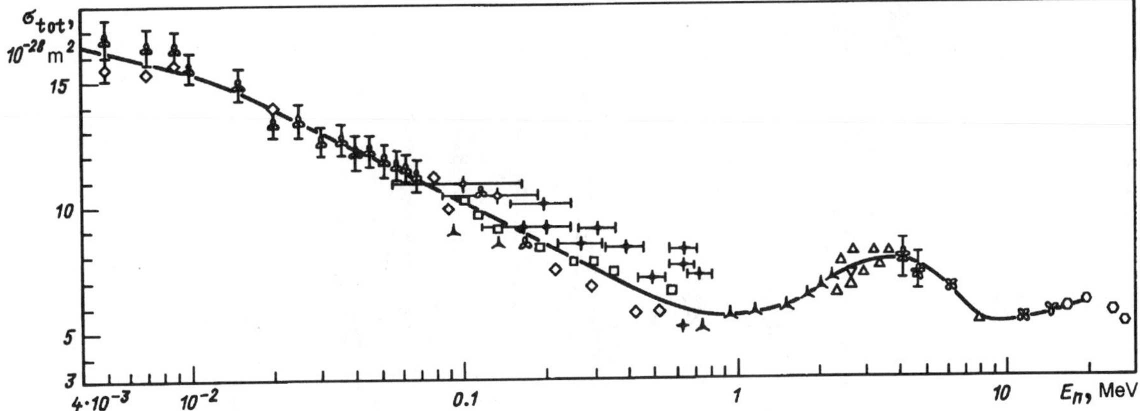

Figure 41.28 The neutron–energy dependence of the total cross section of neutron interaction with gold nuclei (4 keV$< E_n <$ 30 MeV) [29].

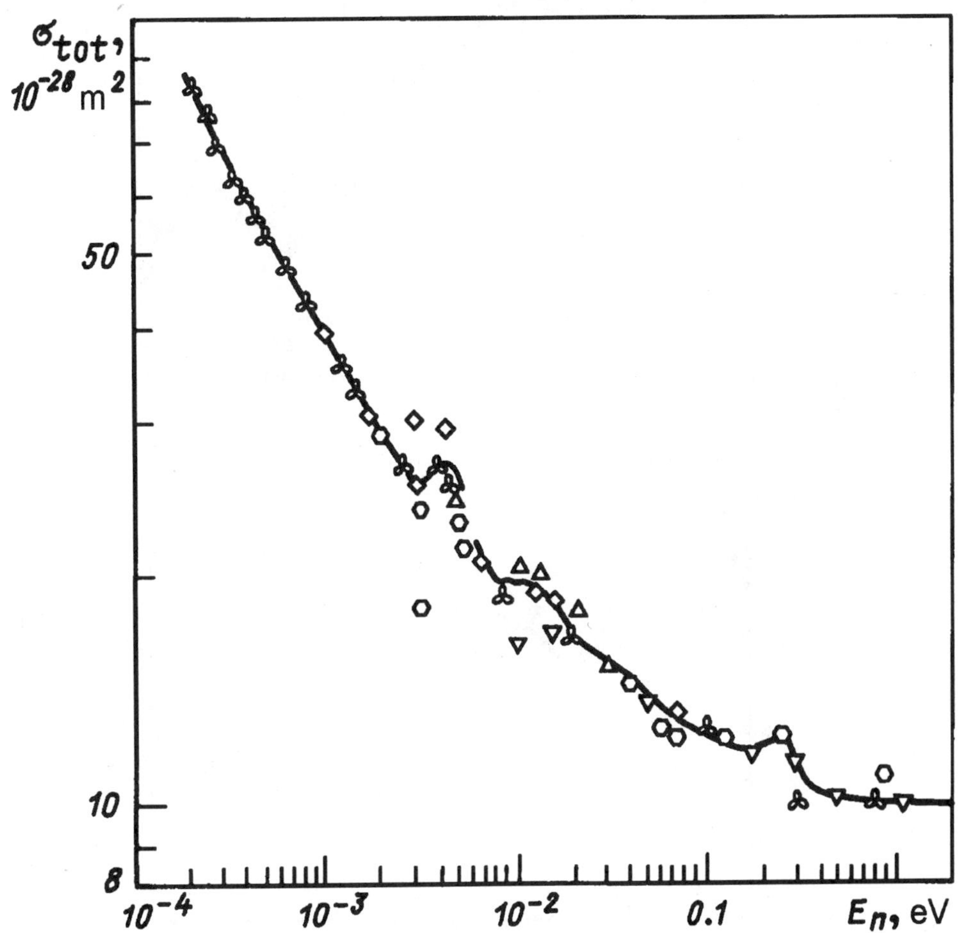

Figure 41.29 The neutron–energy dependence of the total cross section of neutron interaction with uranium nuclei ($10^{-4} < E_n <$ 2 eV) [29].

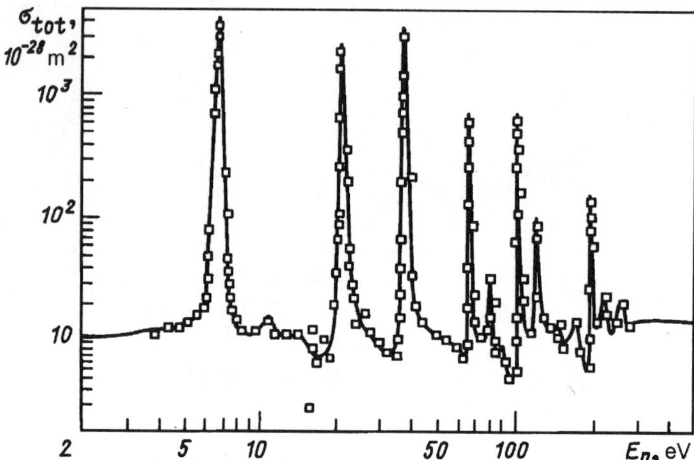

Figure 41.30 The neutron–energy dependence of the total cross section of neutron interaction with uranium nuclei ($2 < E_n < 500$ eV) [29].

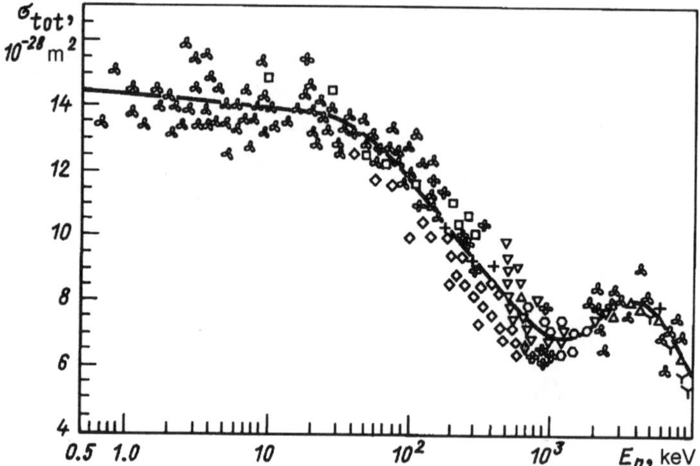

Figure 41.31 The neutron–energy dependence of the total cross section of neutron interaction with uranium nuclei (0.5 keV$< E_n < 10$ MeV) [29].

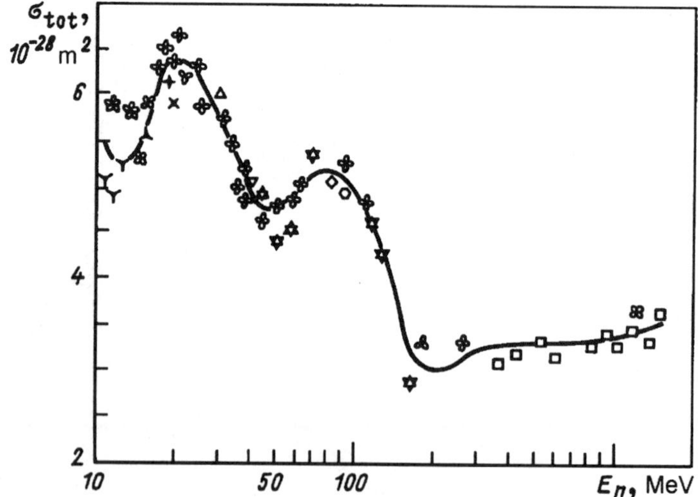

Figure 41.32 The neutron–energy dependence of the total cross section of neutron interaction with uranium nuclei ($10 < E_n < 10^3$ MeV) [29].

41.7 Resonance Integrals

The quantity

$$I_i = \int_{E_{\min}}^{E_{\max}} \sigma_i(E) \frac{dE}{E}$$

is called the resonance integral, where I_i is taken in 10^{-28}m^2, $E_{\min}$ and $E_{\max}$ (in eV) are the lower and upper limits of the energy spectrum whose values depend on the experimental conditions; i is the index of a process ($i = \gamma$ stands for radiative capture; $i = f$, for fission, etc.); $\sigma_i(E)$ is the energy–dependent cross section of an ith process. The contribution of a high-energy region to $\sigma_i(E)$ is, as a rule, negligible; therefore, it is assumed that $E_{\max} \to \infty$.

When the energy dependence of a nuclear reaction is known in analytic form, the resonance integral may be calculated. Basic methods of determining resonance integrals embrace measurements in the neutron field formed out of the thermal Maxwell spectrum with an appropriate temperature T, K and the epithermal $1/E$–spectrum with the lower limit μkT, where μ varies with the nuclear–reactor type; k is the Boltzmann constant. For a heavy-water reactor, $\mu \approx 5$; for a graphite reactor, $\mu \approx 3$, therefore, at $T = 293.6$ K, μkT equals 0.126 and 0.076 eV, respectively.

The quantity

$$I_1' = \int_{\mu kT}^{\infty} \left[\sigma(E) - g(T)\sigma_0\sqrt{\frac{E_0}{E}}\right] \frac{dE}{E}$$

is called the excess resonance integral, where $g(T)$ is the Westcott factor or g–factor that determines the departure of the cross section $\sigma(E)$ energy dependence from the law $1/\sqrt{E}$; $E_0 = 0.0253$ eV and σ_0 is the cross section at E_0. Since the cadmium cut-off energy E_{Cd} is higher than μkT, the integration is carried out over two energy regions:

$$I_1' = \int_{\mu kT}^{E_{\text{Cd}}} \left[\sigma(E) - g(T)\sigma_0\sqrt{\frac{E_0}{E}}\right] \frac{dE}{E} + \int_{E_{\text{Cd}}}^{\infty} \left[\sigma(E) - g(T)\sigma_0\sqrt{\frac{E_0}{E}}\right] \frac{dE}{E} =$$
$$[\Delta I - \Delta I(1/v)] + [I - I(1/v)] = \Delta I' + I',$$

where I' is the epicadmium resonance integral without the part dependent on $1/v$; $\Delta I'$ is the part that cut off by the cadmium filter, depends on the neutron temperature and is small for thermal energies (see Table 41.5). The contribution of the term $\Delta I'$ is to be taken into account for those nuclei whose resonance peaks lie below E_{Cd} (^{113}Cd, ^{151}Eu, ^{176}Lu, ^{182}Ta, ^{191}Ir, ^{231}Pa, ^{239}Pu, etc.).

The data on epicadmium resonance integrals refer either to I' or to I which contains the $(1/v)$–dependent part. This part is determined from the expression

$$I(1/v) = I - I' = \int_{E_{\text{Cd}}}^{\infty} g(T)\sigma_0\sqrt{\frac{E_0}{E}}\frac{dE}{E} \approx 2g(T)\sigma_0\sqrt{\frac{E_0}{E_{\text{Cd}}}}.$$

The values of the resonance integrals are given in Tables 41.6 and 41.7. For some elements, further information on the value of the epicadmium excess resonance integral I' is presented in column 5. Besides, for the majority of elements, the standard sample and value of its resonance integral are indicated which have been used in processing the experimental data. Also, the data from Refs. [6, 21, 44] are recommended.

Table 41.5 Specification of resonance integrals

Symbol	Meaning	Integration limits	Comments
I'	Epicadmium excess resonance integral	$E_{Cd} - \infty$	Cadmium-filter method
$I(1/v)$	Resonance integral part dependent on $1/v$	$E_{Cd} - \infty$	
I	Epicadmium resonance integral including $(1/v)$-dependent part		$I = I' + I(1/v)$
I'_1	Excess resonance integral	$\mu kT - \infty$	Method without cadmium filter
$\Delta I'$	Excess resonance integral at $E < E_{Cd}$	$\mu kT - E_{Cd}$	$\Delta I' = I'_1 - I'$

Table 41.6 Resonance integrals [28]

Symbol and	mass number	Half-life of the reaction product	I_γ, 10^{-28} m^2	Standard	E_{Cd}, eV or I'
H	1		0.1489		0.5
	2	12.346 y	$6.298 \cdot 10^{-4}$		0.5
He	3		2390 (10)		0.5
Li			28	Au (1558)	0.4
	6		425.4		0.5
	7	844 ms	0.01756		0.5
Be	9	$1.6 \cdot 10^6$ y	$4\ (1) \cdot 10^{-3}$		0.5
B			341 (2)		0.5
	10		1722 (5)		0.5
	11	0.0203 s	0.0757		0.5
C			$1.5\ (2) \cdot 10^{-3}$		0.5
	12		$1.5\ (2) \cdot 10^{-3}$		0.5
N			0.90 (5)		0.5
	14		0.90 (5)		0.5
O			$0.31\ (4) \cdot 10^{-3}$		0.5
	16		$0.27\ (3) \cdot 10^{-3}$		0.5
F	19	11.0 s	$17.6\ (3.0) \cdot 10^{-3}$		0.5
Na	23	0.02 s, ^{24m}Na			
		15.03 h, ^{24g}Na			
		$^{24m+g}$Na	0.360 (93)	Au (1558)	0.5
Mg			0.076 (20)	Au (1556)	$I' = 0.045(20)$
	24		0.030 (4)		0.5
Al	27	2.246 min	0.175 (5)	Au (1551)	0.55
Si			0.5	Au (1558)	0.4
	28		0.078(5)		0.5
P	31	14.3 d	0.08 (2)		0.5
S			0.6	Au (1558)	0.4
	32		1.672		0.5
Cl			12.8 (1.7)	Li (32.2)	0.5
	35	$3.0 \cdot 10^5$ y	17 (2)		0.5
Ar			0.42 (5)		0.5
	40	1.83 h	0.41 (3)		0.5
K			1.0 (1)		0.5
	39		0.9 (1)		0.5
Ca			0.20 (2)		0.5
		$1.3 \cdot 10^5$ y	0.18 (3)		
	40	18.7 s, ^{48m}Sc			
Sc	45	84.0 d, ^{48g}Sc			0.5
	51	$^{46(m+g)}$Sc	12 (1)	Au (1560)	0.5
Ti			3.8 (9)	Li (32.2)	0.5
	48		3.7 (3)		0.5

Table 41.6 Resonance integrals [28] *(continued)*

Symbol and mass number		Half-life of the reaction product	I_γ, 10^{-28} m^2	Standard	E_{Cd}, eV or I'
V			2.7 (1)		0.5
	51	3.75 min	3.53 (1.40)	Au (1558)	0.5
Cr			1.7 (2)		0.5
	50	27.7 d	12.5 (3.24)	Au (1558)	0.5
	52		0.60 (5)		0.5
Mn	55	2.58 h	15.6 (5)	Au (1560)	0.5
Fe			1.4 (2)		0.5
	56		1.4 (2)		0.5
Co	59	10.5 min, ^{60m}Co	39.7 (4.3)	Au (1551)	0.55
		5.272 y, ^{60g}Co	31.4 (4.8)	Au (1551)	0.55
		$^{60(m+g)}$Co	71.1 (1.8)	Au (1551)	0.55
Ni			2.2 (2)		0.5
	58	7.5·10^4 y	2.2 (2)		0.5
Cu			3.2 (3)		0.5
	63	12.7 h	5.3 (1)	Au (1560)	0.5
Zn			2.3 (3)		0.5
	64	244 d	1.73 (8)	Au (1560)	0.5
Ga			18.7 (1.5)		0.5
	69	21.1 min	15.6 (1.5)		0.5
Ge			6.1 (1.0)		0.5
As	75	26.4 h	75 (3)	Au (1560)	0.5
Se			9.6 (1.2)	Li(32.2)	0.5
	80	57.3 min, ^{81m}Se	0.50 (2)	Au, Co	0.55
		18 min, ^{81g}Se			
		$^{81(m+g)}$Se	1.7 (2)		0.5
Br			90 (6)		0.5
	79	4.42 h, ^{80m}Br	34.5 (4.0)		0.5
		18 min ^{80g}Br	92.0 (10.0)		0.5
		$^{80(m+g)}$Br	132.5 (10)		0.5
Kr			53 (7)		0.5
	84	4.48 h, ^{85m}Kr	6.03		$I' = 6$
		10.76 y, ^{85g}Kr	2.01		$I^1 = 2$
		$^{85(m+g)}$Kr	2.7 (7)		0.5
Rb			6.0 (5)		0.5
	85	1.02 min, ^{86m}Rb	1.16 (3)	Au, Co	0.55
		18.7 d, ^{86g}Rb			
		$^{86(m+g)}$Rb	7.5 (5)		0.5
Sr			11 (2)		0.5
	88	50.5 d	0.05 (2)		0.5
Y	89	3.19 h, ^{90m}Y	0.88 (8)	Au, Co	0.55
		64.1 h, ^{90g}Y			
		$^{90(m+g)}$Y	1.0 (2)		0.5
Zr			1.10 (15)		0.5
	90		0.20 (3)		0.5
Nb	93	6.2 min, ^{94m}Nb	6.56 (1.96)	Au (1558)	0.5
		2.0·10^4 y, ^{94g}Nb			
		$^{94(m+g)}$Nb	8.5 (5)		0.5
Mo			22 (2)		0.5
Tc	99	15.8 s	340 (20)		0.5
Ru			42 (4)		0.5
	102	39.35 d	4.1 (4)		0.5

Table 41.6 Resonance integrals [28] *(continued)*

Symbol and	mass number	Half-life of the reaction product	I_γ, 10^{-28} m^2	Standard	E_{Cd}, eV or I'
Rh	103	4.4 min, ^{104m}Rh	89 (7)	Au (1549)	$I' = 78(7)$
		42 s, ^{104g}Rh	1111 (74)	Au (1549)	$I' = 1054(74)$
		$^{104(m+g)}$Rh	1100 (50)		0.5
Pd			90 (5)		0.5
	106	21.3 s, ^{107m}Pd			
		6.5·10^6 y, ^{107g}Pd			
		$^{107(m+g)}$Pd	5.73 (57)		0.5
Ag			747 (20)		0.5
	107	127 y, ^{108m}Ag	1.26 (19)	Co (75)	$I' = 3.0(5)$
		2.41 min, ^{108g}Ag			
		$^{108(m+g)}$Ag	95 (4)	Au (1558)	$I' = 79(4)$
	109	250.4 d, ^{110m}Ag	71.7 (3.6)	Co (75)	$I' = 15.5(8)$
		24.6 s, ^{110g}Ag	1112 (68)	Au, Co	0.55
		$^{110(m+g)}$Ag	1450 (40)		0.5
Cd			102 (2)		0.5
In			3200 (50)		0.5
	115	2.2 s, ^{116m2}In			
		54 min, ^{116m1}In			
		14 s, ^{116g}In	690 (45)	Au	1.3
		$^{116(m1+m2)}$In	2114 (23)	Au, Co	0.55
		$^{116(g+m1+m2)}$In	3300 (850)		
Sn			8.5 (2.0)		0.5
Sb			175 (10)		0.5
	121	4.2 min, ^{122m}Sb			
		2.7 d, ^{122g}Sb			
		$^{122(m+g)}$Sb	230 (10)	Au (1560)	0.5
Te			54 (3)		0.5
I	125		13730 (2000)		0.5
	127	24.99 min	150 (30)	Au (1560)	0.5
Xe	129		250 (25)		0.5
Cs	133	2.89 h, ^{134m}Cs	29.2 (6.2)	Au (1550)	0.55
		2.05 y, ^{134g}Cs	359 (90)	Au (1550)	0.55
		$^{134(m+g)}$Cs	415 (15)		0.5
Ba			7.5 (1.0)		0.5
	138	83.3 min	0.4 (2)	Au (1560)	0.5
La			19 (1)	Au (1558)	0.5
	139	40.22 h	12.5 (4)	Au (1560)	0.5
Ce			3.0 (8)		0.5
	140	32.5 d	0.43 (2)	Au (1550)	0.55
Pr	141	19.2 h	14.1 (2)		0.5
Nd			45 (5)		0.5
Pm	147	41.3 d, ^{148m}Pm	1026 (280)		0.5
		5.37 d, ^{148g}Pm	1274 (66)		0.5
		$^{148(m+g)}$Pm	3220	Au (1558)	
Sm			1400 (200)		0.5
Eu			6320 (869)	Au (1551)	0.2
	153	8.5 y	3414 (197)	Au (1551)	0.55
Gd			390 (10)		0.5
Tb	159	72.1 d	400 (24)	Au (1558)	$I' = 390(24)$
Dy			1600 (100)		0.5

Table 41.6 Resonance integrals [28] *(continued)*

Symbol and mass number		Half-life of the reaction product	I_γ, 10^{-28} m^2	Standard	E_{Cd}, eV or I'
Ho	165	1200 y, ^{166m}Ho			
		27.2 h, ^{166g}Ho	660 (30)	Au (1550)	0.5
		$^{166(m+g)}$Ho	700 (20)		0.5
Er			740 (10)		0.5
Tm	169	0.004 ms, ^{170m}Tm			
		130 d, ^{170g}Tm			
		$^{(m+g)}$Tm	1548 (56)	Au (1550)	0.55
Yb			182 (10)		0.5
Lu			900 (50)		0.5
	175	3.69 h, ^{176m}Lu	523 (57)	Au (1551)	0.55
Hf			2000 (100)		0.5
Ta			720 (25)	Au (1575)	0.5
	181	16.5 min ^{182m}Ta	0.415 (110)	Au (1551)	0.55
		115.1 d, ^{182g}Ta	717 (58)	Au (1551)	0.55
		$^{182(m+g)}$Ta	717 (58)	Au (1551)	0.55
W			352 (30)		0.5
Os			209	Au (1558)	0.5
Ir			2250 (200)		0.5
	193	171 d, ^{194m}Ir			
		19 h, ^{194g}Ir	1370 (150)		0.5
		$^{194(m+g)}$Ir	1386 (110)	Au (1550)	0.55
Pt			140 (6)		0.5
Au	197	2.7 d	1560 (40)		0.5
Hg			73 (10)		0.5
Tl			12 (1)		0.5
	205	4.19 min, ^{206m}Tl	0.7 (2)		0.5
Pb			0.16 (4)		0.5
	207		0.4 (2)		0.5
Bi	209	$3.5{\cdot}10^6$ y, ^{210m}Bi			
		5.01 d, ^{210g}Bi			
		$^{210(m+g)}$Bi	0.19 (3)		0.5
Ra	226	41.2 min	222 (15)		0.5
Ac	227	6.13 h	1017 (103)	Co (70)	0.5

Table 41.7 Resonance integrals of fissile elements [25]

Symbol and mass number		Half-life	g_γ (300 K)	I_γ, 10^{-28} m^2	g_f (300 K)	I_f, 10^{-28} m^2
Th	228	1.913 y		1000		
	229	7300 y	1.043	1000 (180)	1.025	464 (70)
	230	80000 y	1.013	1010 (30)		
	232	$1.4{\cdot}10^{10}$ y	0.995	85 (3)		0.619
	233	22.3 mon		400 (100)		
Pa	231	32760 y	1.20	470 (100)		
	233	27.0 d	0.980	895 (30)		
U	232	72 y	0.973	280 (15)	0.976	350 (100)
	233	159200 y	1.022	140 (6)	0.998	764 (3)
	234	244600 y	0.989	645 (70)		
	235	$7.038{\cdot}10^8$ y	0.981	144 (6)	0.980	275 (5)
	236	$2.342{\cdot}10^7$ y	1.002	365 (20)		
	237	6.75 d		1200 (200)		
	238	$4.468{\cdot}10^9$ y	1.002	278 (5)		2.03

Table 41.7 Resonance integrals of fissile elements [25] *(continued)*

Symbol and mass number		Half-life	g_γ (300 K)	I_γ, 10^{-28} m^2	g_f (300 K)	I_f, 10^{-28} m^2
Np	237	$2.14 \cdot 10^6$ y	0.952	660 (50)		6.9
	238	2.117 d				880 (70)
Pu	236	2.85 y		197 [28]		960 [28]
	238	87.74 y	0.956	162 (15)	0.956	23 (5)
	239	24110 y	1.131	190 (20)	1.065	310 (10)
	240	6553 y	1.028	8260 (250)		
	241	14.7 y	1.04	162 (8)	1.046	570 (17)
	242	$3.76 \cdot 10^5$ y	1.010	1280 (50)		4.7 (4.7)
	243	4.956 h		265 (60)		540 (140)
Am	241	432.6 y	0.994	1400 (90)	1.014	22 (2)
	241	432.6 y		1190 (80), (*g*)		
	241	432.6 y		220 (15), (*m*)		
	242*g*	16.01 h		300		
	242*m*	141 y	1.104	230 (100)	1.100	1900 (300)
	243	7380 y	1.013	2050 (100)		10 (60)
	243	7380 y		110 (10), (*g*)		
	243	7380 y		1940 (100), (*m*)		
Cm	242	162.8 d	0.927	150 (40)		
	243	28.5 y		215 (20)		1550 (200)
	244	18.11 y	1.001	625 (50)	0.998	19 (2)
	245	8500 y	0.936	104 (8)	0.942	790 (40)
	246	4700 y	1.005	117 (8)	1.006	12.2
	247	$1.6 \cdot 10^7$ y	1.002	500 (75)	0.995	750 (100)
	248	$3.5 \cdot 10^5$ y	1.002	265 (25)		14 (2)
Bk	249	321.4 d		1400 (700)		
Cf	249	351 y		660 (120)		1900 (100)
	250	13.1 y		8300 (4000)		
	251	900 y		1590 (70)		5400 (800)
	252	2.84 y		43 (4)		110 (20)
	253	17.8 d		12 (2)		2000 (500)
Es	253	20.47 d		7300 (400)		
	253*g*	20.47 d		4300 (220)		
	253*m*	20.47 d		3000 (180)		
	254	276 d				2200 (100)
Fm	257	100.5 d		5000 [28]		

41.8 Nuclear Reaction $(n, 2n)$

The nuclear reaction $(n, 2n)$ belongs to threshold reactions, and when the neutron energy exceeds the threshold energy by several MeV, this reaction, as a rule, proceeds with great probability. The $(n, 2n)$ reaction threshold is approximately equal to the neutron binding energy in a nucleus–target system; therefore, it occurs only through the interaction with neutrons of energies above 8 MeV. An exception is the reaction on beryllium, for which the threshold is around 2 MeV.

The cross sections of nuclear reaction $(n, 2n)$ for various nuclides averaged over the neutron spectrum of ^{235}U fission and the energy dependences of the cross section for nuclear reaction $(n, 2n)$ for ^{9}Be, ^{56}Fe, ^{127}In, ^{208}Pb, ^{232}Th, ^{238}U are given in Table 41.8 and Figures 41.33–41.38.

The cross section of nuclear reaction $(n, 2n)$ at energies from about 14 to 15 MeV can be estimated via the following formula:

$$\sigma_{(n,2n)} = \begin{cases} (1000 + 7.5\,A)\left(7.8\,\dfrac{N-Z}{A} - 0.234\right), & \dfrac{N-Z}{A} \le 0.13; \\ (1000 + 7.5\,A)\left(0.65 + \dfrac{N-Z}{A}\right), & \dfrac{N-Z}{A} > 0.13, \end{cases}$$

where $\sigma_{(n,2n)}$ is given in 10^{-31} m^2; N and Z are the number of neutrons and protons in a nucleus–target system with the mass number A. The calculated values are in agreement with the experimental data within 10–15 per cent for nuclei with parameters $(N - Z)/A > 0.06$. When $(N - Z)/A < 0.06$, the calculated cross sections are much larger than the experimental cross sections because the formula has been derived without accounting for the competing reaction (n, np) [30].

Table 41.8 Cross sections of the reactions induced by neutrons with energies about 14.5 MeV and neutrons of the fission spectrum [30, 31]

Nucleus-target	Nuclear reaction	Residual nucleus	Half-life of reaction product	Reaction energy, MeV	Cross section at $E = 14.5$ MeV, 10^{-31} m^2	Cross section averaged over fission spectrum of ^{235}U, 10^{-31} m^2
Li-6	(n,p)	He-6	0.808 s	−2.73	8.6 (2)	4.18
	(n,α)	H-3	12.34 y	4.78	25 (3)	
	$(n,2n)$			−3.7	72 (5)	0.158
	(n,np)			−4.65	170	
	$(n,n\alpha)$	H-2	Stable	−1.47	400	
	(n,t)	He-4		4.785	10	
Li-7	$(n,2n)$	Li-6		−7.25	22	0.37
	(n,np)	He-6	0.808 s	−10	105	
	$(n,n\alpha)$	H-3	12.34 y	−2.47	340	
	(n,d)	He-6	0.808 s	−7.76	10	
	(n,t)			−3.42	55	
	$(n,3n)$			−12.92	0.2	
	$(n,2n\alpha)$	H-2	Stable	−8.72	33	
Be-9	(n,γ)	Be-10	$2.7\cdot10^{6}$ y	6.81	0.001	
	(n,p) [31]	Li-9	0.175 s	−12.836	< 4	10^{-3}
	(n,α)	He-6	0.808 s	−0.6022	10 (1)	32.8 (3.8)
	$(n,2n)$			−1.665	524 (25)	144 (6)
	(n,t)	Li-7	Stable	−10.439	20	
B-10	$(n,2n)$			−8.4352	27	0.18
	(n,t)			0.2318	94 (20)	23.8
	(n,np)	Be-9	Stable	−6.59	75	
B-11	(n,p)	Be-11	13.57 s	−10.726	3.3 (7)	10^{-3}
	(n,α)	Li-8	0.84 s	−6.633	30.5 (3)	0.14 (7)
	$(n,2n)$	B-10	Stable	−11.456	19	0.008
	(n,t)	Be-9		−9.559	15	
C-12	(n,p)	B-12	0.02 s	−12.588	0.19	$0.26\cdot10^{-3}$
	(n,α)	Be-9	Stable	−5.7016	80 (20)	0.37
	$(n,2n)$	C-11	20.38 min	−18.723	0	$4.2\ (1.4)\cdot10^{-7}$
	$(n,n\alpha)$			−7.37	190	
N-14	(n,p)	C-14	5730 y	0.626	80 (15)	
	(n,α)	B-11	Stable	−0.157		91
	$(n,2n)$	N-13	9.97 min	−10.554	7.3 (1)	$0.94\cdot10^{-3}$
	(n,d)	C-13	Stable	−5.326	49	
	(n,t)	C-12		−4.015	29	
	(n,np)	C-13		−7.55	46 (13)	
	$(n,2\alpha)$	Li-7		−2.62	32	

Table 41.8 Cross sections of the reactions induced by neutrons *(continued)*

Nucleus-target	Nuclear reaction	Residual nucleus	Half-life of reaction product	Reaction energy, MeV	Cross section at $E = 14.5$ MeV, 10^{-31} m^2	Cross section averaged over fission spectrum of ^{235}U, 10^{-31} m^2
O-16	(n,p)	N-16	7.11 s	−9.638	41 (3)	0.019 (1)
	(n,α)	C-13	Stable	−2.215	103 (20)	11.3
	$(n,2n)$	O-15	122 s	−15.669	0	5.3 (2.4)·10^{-6}
	(n,d)	N-15	Stable	−9.903	150	
	(n,np)	N-15		−12.11	15	
F-19	(n,d) [31]	O-18		−5.77	23	
	(n,p)	O-19	26.9 s	−4.036	20 (2)	0.83 (2)
	(n,α)	N-16	7.11 s	−1.523	33 (5)	15.1 (2)
	$(n,2n)$	F-18	109.8 min	−10.431	57 (10)	7.3 (7)·10^{-3}
	(n,t)	O-17	Stable	−7.557	10	
Ne-20	(n,p)	F-20	11 s	−6.244	92	0.078
	(n,α)	O-17	Stable	−0.588	14	12
	$(n,2n)$	Ne-19	17.22 s	−16.866	0	10^{-3}
Na-23	(n,γ) [31]	Na-24	15 h	6.96	0.24	
	(n,p)	Ne-23	37.24 s	−3.596	44 (4)	1.43 (2)
	(n,α)	F-20	11 s	−3.866	150 (10)	0.53 (2)
	$(n,2n)$	Na-22	2.6 y	−12.418	44 (4)	2.2 (2)·10^{-3}
	(n,np)	Ne-22	Stable	−8.79	18	
Mg-24	(n,γ) [31]	Mg-25		7.33	0.25 (E=4MeV)	
	(n,p)	Na-24	15 h	−4.732	186 (15)	1.48 (82)
	(n,α)	Ne-21	Stable	−2.553	63	1.8
	$(n,2n)$	Mg-23	11.327 s	−16.531	0	0.002
Al-27	(n,γ) [31]	Al-28	2.24 min	7.73	0.56	
	(n,p)	Mg-27	9.46 min	−1.828	74 (5)	3.86 (25)
	(n,α)	Na-24	15 h	−3.132	118 (5)	0.705 (40)
	$(n.2n)$	Al-26	7.38 ·10^5 y	−13.058	7	5·10^{-3} a
	$(n,2n)$	Al-26m	6.35 s		0	
	(n,np)	Mg-26	Stable	−8.27	50	
Si-28	(n,γ) [31]	Si-29		8.47	0.45	
	(n,p)	Al-28	2.243 min	−3.860	260 (25)	6.4 (8)
	(n,α)	Mg-25	Stable	−2.653	11	0.56
	$(n,2n)$	Si-21	4.11 s	−17.177	0	10^{-3}
	(n,np)	Al-27	Stable	−11.59	27	
P-31	(n,γ) [31]	P-32	14.3 d	7.94	0.34	
	(n,p)	Si-31	2.62 h	−0.709	83 (5)	35.5 (2.7)
	(n,α)	Al-28	2.243 min	−1.944	110 (10)	1.9 (6)
	$(n,2n)$	P-30	2.499 min	−12.307	12.5 (3)	1.09 ·10^{-3}
	(n,d)	Si-30	Stable	−5.073	15	
	$(n$, He$)$	Al-29	6.52 min	−13.086	0.013	
	(n,np)	Si-30	Stable	−7.30	100	
S-32	(n,γ) [31]	Si-33		8.63	0.54	
	(n,p)	P-32	14.3 d	−0.928	230 (10)	66.8 (3.7)
	(n,α)	P-29	Stable	1.526	68 (10)	43.6
	$(n,2n)$	P-31	2.61 s	−15.088	0	0.63·10^{-5}
	(n,t)	P-30	2.499 min	−12.689	0	1.06·10^{-5}
	(n,np)	P-31	Stable	−8.86	78	

Table 41.8 Cross sections of the reactions induced by neutrons *(continued)*

Nucleus-target	Nuclear reaction	Residual nucleus	Half-life of reaction product	Reaction energy, MeV	Cross section at $E = 14.5$ MeV, 10^{-31} m^2	Cross section averaged over fission spectrum of ^{235}U, 10^{-31} m^2
Cl-35	(n,p)	S-35	88 d	0.615	110 (10)	78 (23)
	(n,α)	P-32	14.3 d	0.938	117 (10)	8.8 (4.6)
	$(n,2n)$	Cl-34	1.525 s	−12.646	9 (1)	$0.79{\cdot}10^{-3}$
	$(n,2n)$	Cl-34m	32.06 min		6 (1)	$0.51{\cdot}10^{-3}$
Ar-40	(n,p)	Cl-40	1.32 min	−6.72	18	0.01
	(n,α)	S-37	5.06 min	−2.486	11.3 (2)	0.11
	$(n,2n)$	Ar-39	269 y	−9.871	570	0.15
	(n,np)	Cl-39	56.2 min	−12.51	1.7	
K-39	(n,p)	Ar-39	269 y	0.217	179 (60)	82.2
	(n,α)	Cl-36	$3.01{\cdot}10^5$ y	1.363	115 (30)	8.0 (3)
	$(n,2n)$	K-38	7.71 min	−13.085	4 (1)	$0.37{\cdot}10^{-3}$
	$(n,2n)$	K-38m	0.9256 s		2 (4)	
	(n,np)	Ar-38	Stable	−6.38	180	
	$(n,n\alpha)$	Cl-35		−7.22	30	
Ca-40	(n,p)	K-40	$1.28{\cdot}10^9$ y	−0.529	470 (30)	77
	(n,α)	Ar-37	35.06 d	1.748	320	13 (6)
	$(n,2n)$	Ca-39	0.86 s	−15.643	0	$3{\cdot}10^5$
	(n,t)	K-38	7.71 min	−12.933	0.027	
	(n,np)	K-39	Stable	−8.33	200	
	$(n,n\alpha)$	Ar-36		−7.04	23	
Sc-45	(n,p)	Ca-45	163 d	0.526	58 (6)	15 (12)
	(n,α)	K-42	12.36 h	−0.395	55 (5)	0.182 (12)
	$(n,2n)$	Sc-44	3.99 h	−11.321	340 (20)	0.04
	$(n,2n)$	Sc-44m	2.44 d		116 (15)	0.012
	$(n$, He$)$	K-43	22.6 h	−11.341	0.0086	
	$(n,2p)$	K-44	22.15 min	−9.65	0.21	
Ti-48	(n,p)	Sc-48	43.8 h	−3.208	66 (5)	0.3 (18)
	(n,α)	Ca-45	163 d	−2.033	31 (8)	0.013 (6)
	$(n,2n)$	Ti-47	Stable	−11.628	550	0.016
	(n,np)	Sc-47	3.4 d	−11.45	10	0.0013
V-51	(n,p)	Ti-51	5.8 min	−1.676	33 (3)	0.456 (23)
	(n,α)	Sc-48	43.8 h	−2.055	16 (2)	0.022 (3)
	$(n,2n)$	V-50	Stable	−11.052	660 (50)	0.21
	$(n,n\alpha)$	Ti-47		−10.29	2	0.000087
Cr-52	(n,p)	V-52	3.76 min	−3.196	102 (20)	1.09 (8)
	(n,α)	Ti-49	Stable	−1.211	40 (4)	0.083
	$(n,2n)$	Cr-51	27.7 d	−12.041	357 (30)	0.033
	(n,np)	V-51	Stable	−10.5	30	0.005
Mn-55	(n,p)	Cr-55	3.55 min	−1.806	45 (10)	1.2
	(n,α)	V-52	3.76 min	−0.626	29 (2)	0.11 (3)
	$(n,2n)$	Mn-54	312.3 d	−10.224	809 (35)	0.244 (15)
	$(n$, He$)$	V-53	1.55 min	−12.709	0.8 (3)	
Fe-56	(n,p)	Mn-56	2.579 h	−2.918	110 (10)	1.03 (75)
	(n,α)	Cr-53	Stable	0.321	40	0.397(12)
	$(n,2n)$	Fe-55	2.72 y	−11.203	540 (40)	0.0754
	(n,t)	Mn-54	312.3 d	−11.931	0.045	
	(n,np)	Mn-55	Stable	−10.19	35	0.0051

Table 41.8 Cross sections of the reactions induced by neutrons *(continued)*

Nucleus-target	Nuclear reaction	Residual nucleus	Half-life of reaction product	Reaction energy, MeV	Cross section at $E = 14.5$ MeV, 10^{-31} m^2	Cross section averaged over fission spectrum of ^{235}U, 10^{-31} m^2
Co-59	(n,p)	Fe-59	45.1 d	−0.783	60 (10)	1.42 (14)
	(n,α)	Mn-56	2.579 h	0.320	29 (2)	0.143 (10)
	$(n,2n)$	Co-58	70.78 d	−10.46	707 (70)	0.40 (4)
	$(n,2n)$	Co-58m	9.15 h		380	
	(n,He)	Mn-57	1.61 min	−11.47	0.0046	
Ni-58	(n,α)	Co-58	70.78 d	0.395	374 (30)	108.5 (5.4)
	(n,p)	Co-58m	9.15 h		200	35.4 (2.2)
	(n,α)	Fe-55	2.72 a	2.89	120 (15)	3.0 (9)
	$(n,2n)$	Ni-57	36.16 h	−12.203	30 (3)	5.77 (31)$\cdot10^{-3}$
	(n,t)	Co-56	78.76 d	−11.073	0.092	
	(n,np)	Co-57	270.9 d	−8.18	400	0.21
	$(n,n\alpha)$	Fe-54	Stable	−6.39	30	0.0014
Cu-63	(n,p)	Ni-63	100.1 a	0.716	120 (30)	9.8
	(n,α)	Co-66	5.27 a	1.715	35 (8)	0.50 (56)
	(n,α)	Co-66m	10.47 min		25	
	$(n,2n)$	Cu-62	9.74 min	−10.854	551 (30)	0.12 (12)
	$(n,\text{ He})$	Co-61	1.65 h	−9.528	0.113 (4)	
Zn-64	(n,p)	Cu-64	12.71 h	0.208	176 (20)	29.9 (1.6)
	(n,α)	Ni-61	Stable	3.867	57.5	1.1
	$(n,2n)$	Zn-63	38.1 min	−11.856	178 (15)	0.017
	(n,t)	Cu-62	9.74 min	−10.08	0.086 (23)	
Ga-69	(n,p)	Zn-69	55.6 min	−0.124	34 (3)	1.5
	(n,p)	Zn-69m	14 h		25 (2)	0.496 (73)
	(n,α)	Cu-66	5.1 min	2.584	18 (2)	0.2
	$(n,2n)$	Ga-68	68 min	−10.31	945 (50)	0.227
As-75	(n,p)	Ge-75	82.78 min	−0.406	19.2 (2)	0.45 (15)
	(n,p)	Ge-75m	48.3 s		18	
	(n,α)	Ga-72	14.1 h	1.205	11.6 (1)	7.1$\cdot10^{-3}$
	$(n,2n)$	As-74	17.78 d	−10.243	1061 (40)	0.33 (2)
	(n,He)	Ga-73	4.86 h	−10.15	0.0035	
Se-80	(n,p)	As-80	16.5 s	−5.22	7.2	3.8$\cdot10^{-3}$
	(n,α)	Ge-77	11.3 h	−0.95	2.6 (4)	1.2$\cdot10^{-3}$
	$(n,2n)$	Se-79	6.5$\cdot10^4$ y	−9.896	1132 (60)	0.432
	$(n,2n)$	Se-79m	3.91 min		90	
Br-79	(n,p)	Se-79	6.5$\cdot10^4$ y	0.641	31	0.87
	(n,p)	Se-79m	3.91 min		10	
	(n,α)	As-76	26.32 h	1.859	12.7 (1.5)	0.031
	$(n,2n)$	Br-78	6.46 min	−10.693	974 (50)	0.204
Kr-84	(n,p)	Br-84	31.8 min	−3.92	8 (1.5)	9.3$\cdot10^{-3}$
	(n,α)	Se-81	18.5 min	−0.4	4.7	1.5$\cdot10^{-3}$
	$(n,2n)$	Kr-83	Stable	−10.518	1290	0.33
Rb-85	(n,p)	Kr-85	10.71 y	0.0955	18.3	0.26
	(n,p)	Kr-85m	4.84 h		5 (5)	
	(n,α)	Br-82	35.3 h	0.991	5.9 (1)	5.3$\cdot10^{-3}$
	$(n,2n)$	Rb-84	32.77	−10.6	1123 (100)	0.37 (1)
	$(n,2n)$	Rb-84m	20.5 min		350	

Table 41.8 Cross sections of the reactions induced by neutrons *(continued)*

Nucleus-target	Nuclear reaction	Residual nucleus	Half-life of reaction product	Reaction energy, MeV	Cross section at $E = 14.5$ MeV, 10^{-31} m^2	Cross section averaged over fission spectrum of ^{235}U, 10^{-31} m^2
Sr-88	(n,p)	Rb-88	17.8 min	−4.522	15 (2)	$3.8{\cdot}10^{-3}$
	(n,α)	Kr-85	10.71 y	−0.788	6	$5.9{\cdot}10^{-3}$
	$(n,2n)$	Sr-87	Stable	−11.113	1200	0.14
	$(n,2n)$	Sr-87m	2.805 h		318 (30)	0.0451
Y-89	(n,p)	Sr-89	50.55 d	−0.707	24.6 (3)	0.31 (6)
	(n,α)	Rb-86	18.66 d	0.699	5.4 (1)	$8.3{\cdot}10^{-3}$
	$(n,2n)$	Y-88	107.15 d	−11.468	966 (100)	0.156 (11)
Zr-90	(n,p)	Y-90	64.1 h	−1.506	45 (3)	0.38 (2)
	(n,p)	Y-90m	3.19 h		9.1	
	(n,α)	Sr-87	Stable	1.75	14	0.014
	(n,α)	Sr-87m	2.805 h		4.1 (3)	
	$(n,2n)$	Zr-89	78.43 h	−11.983	768 (30)	0.076 (1)
	$(n,2n)$	Zr-89m	4.18 min		86 (8)	
	(n,t)	Y-88	107.15 d	−11.352	0.041	
Nb-93	(n,p)	Zr-93	$1.53{\cdot}10^{6}$ y	0.719	35	1.0
	(n,α)	Y-90	64.1 h	4.914	9.5 (5)	0.0974 (68)
	(n,α)	Y-90m	3.19 h		5	0.0267 (17)
	$(n,2n)$	Nb-92	$1.2{\cdot}10^{8}$ y	−8.826	1375 (70)	1.04
	$(n,2n)$	Nb-92m	10.13 d		482 (35)	0.475 (32)
	$(n,$He$)$	Y-91	58.51 d	−7.719	0.0031	
Mo-98	(n,p)	Nb-98	2.86 s	−3.82	11 (4)	0.015
	(n,α)	Zr-95	64.05 d	3.202	6.5 (1)	0.014 (2)
	$(n,2n)$	Mo-97	Stable	−8.642	1370	1.3
Tc-99	(n,p)	Mo-99	66.02 h		12 (3)	0.10
	(n,α)	Nb-96	23.35 h		7 (1)	0.065
	$(n,2n)$	Tc-98	$4.2{\cdot}10^{6}$ y		1230 (120)	1.02
	(n,nd)	Nb-95	34.97 d		1.3 (2)	
Ru-102	(n,p)	Tc-102	5.28 s	−3.72	17.3	0.012
	(n,p)	Tc-102m	4.35 min		5.7	
	(n,α)	Mo-99	66.02 h	2.502	5 (1)	0.007
	$(n,2n)$	Ru-101	Stable	−9.216	1390	0.802
Rh-103	(n,p)	Ru-103	39.35 d	0.0198	16 (1)	0.107 (6)
	(n,α)	Tc-100	15.8 s	3.48	11 (2)	0.016
	$(n,2n)$	Rh-102	207 d	−9.31	1325 (100)	0.729
	$(n,2n)$	Rh-102m	2.89 y		380	0.715
	$(n,$He$)$	Tc-101	14.2 min	−8.55	0.016 (7)	
Rd-106	(n,p)	Rh-106	29.9 s	−2.758	23 (6)	0.04
	(n,p)	Rh-106m	132 min		5.7	
	(n,α)	Ru-103	39.35 d	2.998	5.6 (7)	$7{\cdot}10^{-3}$
	$(n,2n)$	Pd-105	Stable	−9.561	1400	$6.11{\cdot}10^{-1}$
Ag-107	(n,p)	Pd-107	$6.5{\cdot}10^{6}$ y	0.747	41	0.47
	(n,p)	Pd-107m	21.3 s		15	
	(n,α)	Rh-104	42.3 s	482	10.8	0.025
	$(n,2n)$	Ag-106	23.96 min	−9.551	1260 (120)	0.601
	$(n,2n)$	Ag-106m	8.41 d		400	
Cd-114	(n,p)	Ag-114	4.52 s	−4.22	5 (2)	0.003
	(n,α)	Pd-111	22 min	1.66	0.65 (1)	$5{\cdot}10^{-4}$
	(n,α)	Pd-111m	5.5 h		0.13	
	$(n,2n)$	Cd-113	Stable	−9.041	1500	1.07
	$(n,2n)$	Cd-113m	13.6 y		860	

Table 41.8 Cross sections of the reactions induced by neutrons *(continued)*

Nucleus-target	Nuclear reaction	Residual nucleus	Half-life of reaction product	Reaction energy, MeV	Cross section at $E = 14.5$ MeV, 10^{-31} m^2	Cross section averaged over fission spectrum of ^{235}U, 10^{-31} m^2
In-115	(n,p)	Cd-115	53.46 h	0.668	15 (5)	0.041
	(n,p)	Cd-115m	44.6 d		7	
	(n,α)	Ag-112	3.12 h	2.68	2.5 (5)	10^{-3}
	$(n,2n)$	In-114	71.9 s	−9.029	1710 (80)	1.07
	$(n,2n)$	In-114m	49.5 d		1262 (100)	0.761
	(n,He)	Ag-113	5.37 h	−9.34	0.007	
Sn-120	(n,p)	In-120	3.08 h	−4.82	4.3 (7)	0.001
	(n,α)	Cd-117	2.4 h	0.96	2.6	10^{-4}
	$(n,2n)$	Sn-119	Stable	−9.104	1560	1.03
Sb-121	(n,p)	Sn-121	27.06 h	0.395	9.1	0.16
	(n,α)	In-118	5 s	3.51	3.6	$4{\cdot}10^{-4}$
	$(n,2n)$	Sb-120	15.89 min	−9.248	1580 (100)	0.846
	$(n,2n)$	Sb-120m	5.76 d		610	
Te-130	(n,p)	Sb-130	6.33 min	−4.22	1.8 (3)	0.002
	(n,p)	Sb-130m	40.9 min		0.6	
	(n,α)	Sn-127	2.1 h	1.810	0.4 (1)	$2.0{\cdot}10^{-4}$
	$(n,2n)$	Te-129	69.6 min	−8.413	1700 (20)	1.80
	$(n,2n)$	Te-129m	33.5 d		1000	
	(n,He)	Sn-128	59.3 min	−10.797	0.015 (8)	
I-127	(n,p)	Te-127	9.35 h	0.09	9 (3)	0.068
	(n,p)	Te-127m	109 d		6 (2)	0.013 (1)
	(n,p)	Te-127g	9.35 h			0.009 (5)
	(n,α)	Sb-124	60.2 d	4.279	1.4 (2)	0.003
	$(n,2n)$	I-126	12.93 d	−9.139	1496 (100)	1.05 (65)
Xe-132	(n,p)	I-132	2.3 h	−2.798	3.0 (5)	0.012
	(n,α)	Te-129	69.6 min	3.372	1.7	$7{\cdot}10^{-4}$
	$(n,2n)$	Xe-131	Stable	−8.936	1670	1.068
	$(n,2n)$	Xe-131m	11.9 d		770	
Cs-133	(n,p)	Xe-133	5.245 d	0.355	11.3 (2)	0.081
	(n,p)	Xe-133m	2.191 d		4.8 (8)	
	(n,α)	I-130	12.36 h	4.448	1.3 (3)	0.0033 (8)
	(n,α)	I-130m	9 min		0.54 (8)	
	$(n,2n)$	Cs-132	6.475 d	−8.979	1603 (100)	0.992
	(n,d)	Xe-132	Stable	−3.87	0.9	
	(n,He)	I-131	8.04 d		0.0032	
Ba-138	(n,p)	Cs-138	33.41 min	−4.62	3.0 (5)	$5{\cdot}10^{-4}$
	(n,α)	Xe-135	9.083 h	3.875	2.6 (3)	0.0019(3)
	(n,α)	Xe-135m	15.65 min		0.55	
	$(n,2n)$	Ba-137	Stable	−8.612	1720	1.71
	$(n,2n)$	Ba-137m	2.55 min		1250	
La-139	(n,p)	Ba-139	84.9 min	−1.478	4.8 (4)	$4{\cdot}10^{-3}$
	(n,α)	Cs-136	12.98 d	4.817	1.8 (3)	$2{\cdot}10^{-3}$
	$(n,2n)$	La-138	Stable	−8.778	1710	1.40
Ce-140	(n,p)	La-140	40.22 h	−2.984	6.5 (5)	$5{\cdot}10^{-3}$
	(n,α)	Ba-137	Stable	5.338	11.5 (1)	$3{\cdot}10^{-3}$
	$(n,2n)$	Ce-139	139.3 d	−9.203	1750 (70)	1.32
	$(n,2n)$	Ce-139m	56.2 s		963 (120)	

Table 41.8 Cross sections of the reactions induced by neutrons *(continued)*

Nucleus-target	Nuclear reaction	Residual nucleus	Half-life of reaction product	Reaction energy, MeV	Cross section at $E = 14.5$ MeV, 10^{-31} m^2	Cross section averaged over fission spectrum of ^{235}U, 10^{-31} m^2
Pr-141	(n,p)	Ce-141	32.5 d	0.201	9 (1)	0.035
	(n,α)	La-138	Stable	6.146	3.2	$7 \cdot 10^{-3}$
	$(n,2n)$	Pr-140	3.39 min	−9.397	1660 (200)	1.1
Nd-142	(n,p)	Pr-142	19.13 h	−1.381	14 (2)	0.042
	(n,α)	Ce-139	139.3 d	6.642	7.1 (8)	0.01
	$(n,2n)$	Nd-141	2.42 h	−9.813	1701 (120)	0.627
	$(n,2n)$	Nd-141m	62.1 s		600 (50)	
Sm-152	(n,p)	Pm-152	3.8 min	−2.62	3.7 (4)	$5 \cdot 10^{-3}$
	(n,α)	Nd-149	1.73 h	5.275	1.7 (2)	$8 \cdot 10^{-4}$
	$(n,2n)$	Sm-151	90 y	−8.267	1855 (150)	2.34
Eu-153	(n,p)	Sm-153	46.44 h	−0.02	6 (1)	0.015
	(n,α)	Pm-150	2.68 h	5.83	2.2 (3)	10^{-3}
	$(n,2n)$	Eu-152	13.2 y	−8.555	1950 (200)	2.15
	$(n,2n)$	Eu-152$m1$	9.3 h		500 (100)	
	$(n,2n)$	Eu-152$m2$	96 min		70	
Gd-158	(n,p)	Eu-158	45.9 min	−2.65	2.5	$4 \cdot 10^{-3}$
	(n,α)	Sm-155	22.1 min	5.16	1.4	$3 \cdot 10^{-4}$
	$(n,2n)$	Gd-157	Stable	−7.931	1900	3.76
Tb-159	(n,p)	Gd-159	18.6 h	−0.168	4.7 (7)	0.01
	(n,α)	Eu-156	15.19 d	6.215	1.8	$8 \cdot 10^{-4}$
	$(n,2n)$	Tb-158	150 y	−8.136	1800 (120)	3.02
	$(n,2n)$	Tb-158m	10.5 s		450 (65)	
Dy-164	(n,p)	Tb-164	2.9 min	−2.56	2.8 (5)	$3 \cdot 10^{-3}$
	(n,α)	Gd-161	3.6 min	5.207	1.2	$2 \cdot 10^{-4}$
	$(n,2n)$	Dy-163	Stable	−7.655	1950	4.88
Ho-165	(n,p)	Dy-165	2.334 h	−0.513	3.2	$4 \cdot 10^{-3}$
	(n,α)	Tb-162	7.7 min	6.460	1.5	$6 \cdot 10^{-4}$
	$(n,2n)$	Ho-164	29 min	−7.989	2000 (200)	3.49
	$(n,2n)$	Ho-164m	37.5 min		1200 (200)	
Er-166	(n,p)	Ho-166	27 h	−1.077	4.5 (7)	0.022
	(n,α)	Dy-163	Stable	7.094	2	10^{-3}
	$(n,2n)$	Er-165	10.36 h	−8.474	1960 (150)	2.25
Tm-169	(n,p)	Er-169	9.3 d	0.431	4.6	0.014
	(n,α)	Ho-166	27 h	7.44	1.8	10^{-3}
	$(n,2n)$	Tm-168	93.1 d	−8.06	2071 (100)	3.43
Yb-174	(n,p)	Tm-174	5.4 min	−2.28	3.5 (1.0)	$3 \cdot 10^{-3}$
	(n,α)	Er-171	7.52 h	6.414	1.2 (2)	$2 \cdot 10^{-4}$
	$(n,2n)$	Yb-173	Stable	−7.469	2020	5.67
Lu-175	(n,p)	Yb-175	4.19 d	0.314	3.7 (5)	$8 \cdot 10^{-3}$
	(n,α)	Tm-172	63.6 h	7.867	1.6	10^{-3}
	$(n,2n)$	Lu-174	3.31 y	−7.659	2030 (200)	4.79
	$(n,2n)$	Lu-174m	142 d		630 (50)	
Hf-180	(n,p)	Lu-180	5.7 min	−2.52	1.9	10^{-3}
	(n,α)	Yb-177	1.9 h	6.856	1.1	$2 \cdot 10^{-4}$
	$(n,2n)$	Hf-179	Stable	−7.388	2080	6.88
	$(n,2n)$	Hf-179m	18.68 s		600	

Table 41.8 Cross sections of the reactions induced by neutrons *(continued)*

Nucleus-target	Nuclear reaction	Residual nucleus	Half-life of reaction product	Reaction energy, MeV	Cross section at $E = 14.5$ MeV, 10^{-31} m^2	Cross section averaged over fission spectrum of ^{235}U, 10^{-31} m^2
Ta-181	(n,p)	Hf-181	42.4 d	−0.240	4.5 (5)	$1 \cdot 10^{-3}$
	(n,α)	Lu-178	28.4 min	7.41	1.4	$3 \cdot 10^{-4}$
	$(n,2n)$	Ta-180	10^{13} y	−7.644	2090 (100)	4.96
	$(n,2n)$	Ta-180m	8.1 h		1110 (100)	
	$(n,$He$)$	Lu-179	4.59 h	−6.56	0.0034	
W-184	(n,p)	Ta-184	8.7 h	−2.248	4.0 (1.0)	$2 \cdot 10^{-3}$
	(n,α)	Hf-181	42.4 d	7.369	1.2 (2)	$2\ (5) \cdot 10^{-4}$
	$(n,2n)$	W-183	Stable	−7.411	2100	6.55
	$(n,2n)$	W-183m	5.15 s		1600	
	(n,np)	Ta-183	5.1 d		0.7 (2)	
Re-187	(n,p)	W-187	23.9 h	−0.529	4.3 (5)	$2 \cdot 10^{-3}$
	(n,α)	Ta-184	8.7 h	7.102	1.2	$1 \cdot 10^{-4}$
	$(n,2n)$	Re-186	90.64 h	−7.371	1700 (200)	10 (6)
	$(n,$He$)$	Ta-185	49 min	−6.6	0.004 (3)	
Os-192	(n,p)	Re-192	16 s	−3.19	1.3	$3 \cdot 10^{-4}$
	(n,α)	W-189	11.5 min	5.24		$1 \cdot 10^{-4}$
	$(n,2n)$	Os-191	15.4 d	−7.559	2120 (150)	5.40
Ir-193	(n,p)	Os-193	31.5 h	−0.350	3.8 (5)	$1 \cdot 10^{-3}$
	(n,α)	Re-190	3.1 min	6.64	1.1	$1 \cdot 10^{-4}$
	$(n,2n)$	Ir-192	74.02 d	−7.772	2048 (150)	3.71
Pt-195	(n,p)	Ir-195	2.5 h	−0.153	2.0 (5)	$2 \cdot 10^{-3}$
	(n,α)	Os-192	Stable	8.711	1.1	$4 \cdot 10^{-4}$
	$(n,2n)$	Pt-194	Stable	−6.124	2190	18.5
Au-197	(n,p)	Py-197	18.3 h	0.036	2.1 (2)	$2 \cdot 10^{-3}$
	(n,α)	Ir-194	19.15 h	6.979	0.35 (5)	$1 \cdot 10^{-4}$
	$(n,2n)$	Au-196	6.18 d	−8.08	2254 (200)	3.0 (3)
	$(n,2n)$	Au-196m	9.7 h		230	
Hg-202	(n,p)	Au-202	29 s	−2.72	1.6	$4 \cdot 10^{-4}$
	(n,α)	Pt-199	31 min	5.706	0.9	$1 \cdot 10^{-4}$
	$(n,2n)$	Hg-201	Stable	−7.756	2160	4.45
Tl-205	(n,p)	Hg-205	5.2 min	−0.747	1.9 (2)	$5 \cdot 10^{-4}$
	(n,α)	Au-202	29 s	5.68	0.85	$1 \cdot 10^{-4}$
	$(n,2n)$	Tl-204	3.78 y	−7.541	2006 (200)	5.61
Pb-208	(n,p)	Tl-208	3.07 min	−4.211	1.0 (5)	$1 \cdot 10^{-4}$
	(n,α)	Hg-205	5.2 min	6.186	0.8	$1 \cdot 10^{-4}$
	$(n,2n)$	Pb-207	Stable	−7.368	2300	6.18
	$(n,2n)$	Pb-207m	0.8 s		1282	
Bi-209	(n,p)	Pb-209	3.31 h	0.135	0.8 (2)	$1 \cdot 10^{-3}$
	(n,α)	Tl-206	4.2 min	9.634	0.8 (3)	$3 \cdot 10^{-4}$
	$(n,2n)$	Bi-208	$3.68 \cdot 10^{5}$ y	−7.453	2261 (100)	5.92
Th-232	(n,γ)	Th-233	22.2 min	4.79	5.0	
[31]	$(n,2n)$	Th-231	25.5 h	−6.43	1160	
	(n,α)	Ra-229		8.08	4.6	
	$(n,3n)$	Th-230		−11.6	850	
U-238	(n,γ)	U-239	23.5 min	4.80	1	
[31]	$(n,2n)$	U-237	6.75 d	−6.14	700-900	
	(n,p)	Pa-238		−3.18	1.5	
	(n,t)	Pa-236	12 min	−5.10	0.02	
	(n,α)	Th-235		9.07	1.5	
	$(n,3n)$	U-236	$2.39 \cdot 10^{7}$ y	−11.27	500	
	(n,np)	Pa-237	39 min	−7.66	0.23	

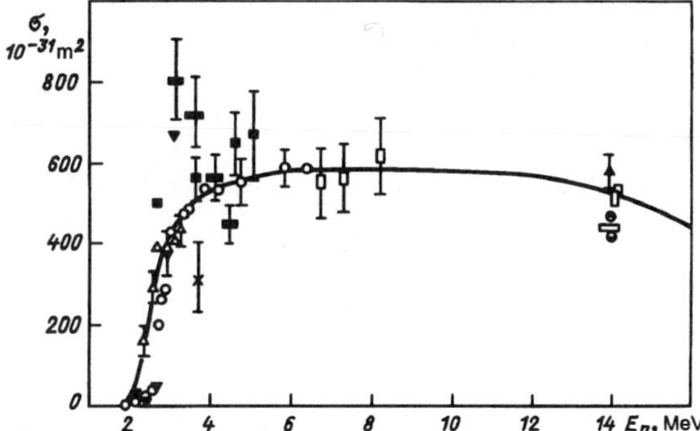

Figure 41.33 Dependence of the cross section of nuclear reaction ^{9}Be(n, 2n)^{8}Be on the neutron energy [30].

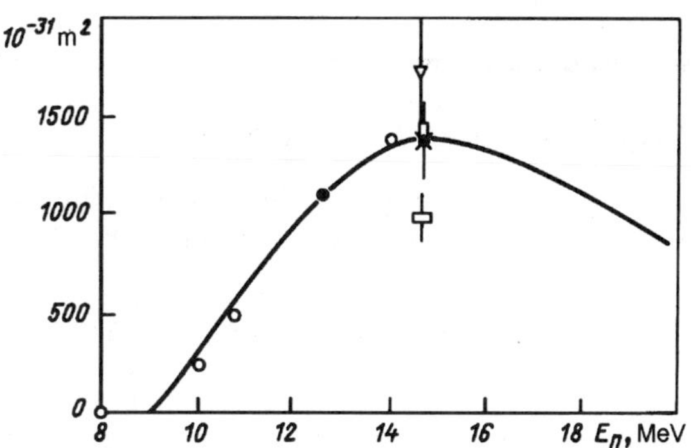

Figure 41.36 Dependence of the cross section of nuclear reaction ^{208}Pb(n, 2n)^{207}Pb on the neutron energy [30].

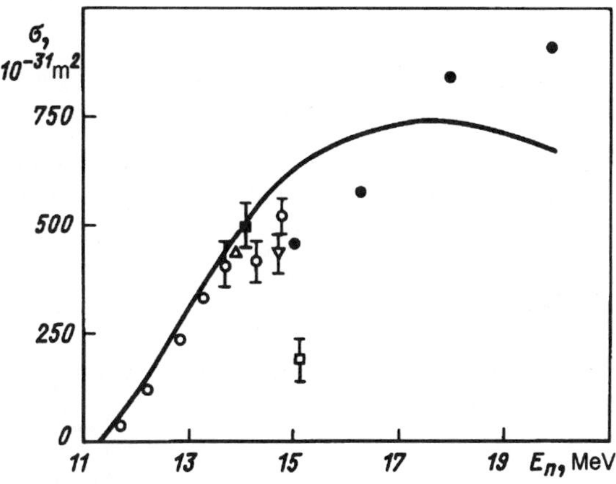

Figure 41.34 Dependence of the cross section of nuclear reaction ^{56}Fe(n, 2n)^{55}Fe on the neutron energy [30].

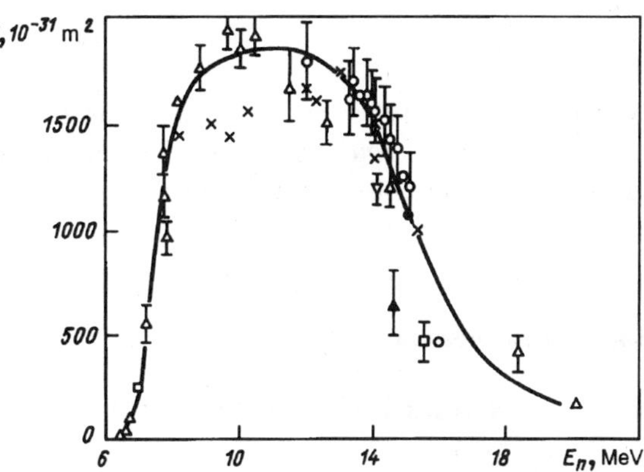

Figure 41.37 Dependence of the cross section of nuclear reaction ^{232}Th(n, 2n)^{231}Th on the neutron energy [30].

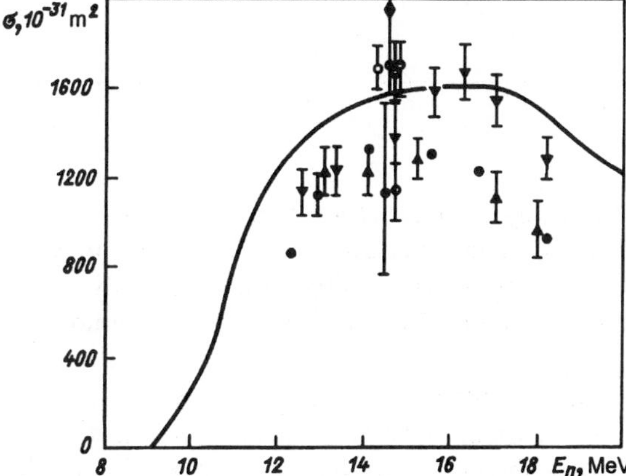

Figure 41.35 Dependence of the cross section of nuclear reaction ^{127}I(n, 2n)^{126}I on the neutron energy [30].

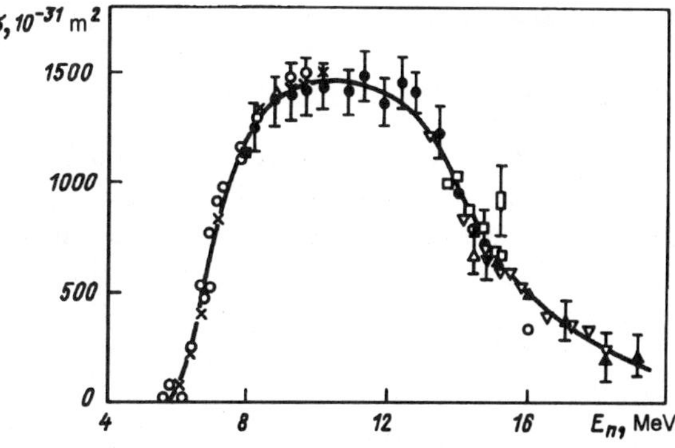

Figure 41.38 Dependence of the cross section of nuclear reaction ^{238}U(n, 2n)^{237}U on the neutron energy [30].

41.9 Nuclear Reaction (n, p)

The cross section of the reaction (n,p) in the region of neutron energies from about 14 MeV to 15 MeV can be estimated by the formula

$$\sigma_{n,p} = 43.5(A^{1/3}+1)\exp\left\{0.083\sqrt{A}\left[-50\frac{N-Z-1}{A} + 0.58\frac{Z-1}{A^{1/3}} - 3.26\right]\right\},$$

(where $\sigma_{n,p}$ is in 10^{-31} m^2 units) which gives the results in agreement with experimental data within 25–percent error for most nuclei [30].

41.10 Nuclear Reaction (n, α)

The formula

$$\sigma_{n,\alpha} = 17.84\exp[-33(N-Z)/A],$$

(where $\sigma_{n,\alpha}$ is in 10^{-31} m^2 units) is proposed for evaluating the cross section of the nuclear reaction involving neutrons with energies from about 14 to 15 MeV and produces results compatible with experiment within 40–percent error [30, 31].

41.11 Nuclear Reaction (n, t)

The formula

$$\sigma_{n,t} = 4.52(A^{1/3}+1)\exp[-10(N-Z)/A]$$

(where $\sigma_{n,t}$ is in 10^{-31} m^2 units) is proposed for evaluating the cross section of the reaction (n,t) involving neutrons with energies from about 14 to 15 MeV and produces results compatible with experiment within 70–percent error.

41.12 Activation Detectors

Many elements possess a large activation cross section and give radioactive elements with half-lives convenient for measurements. The measurements of fast neutrons spectra by an activation method are based on employing those materials as detectors whose cross sections differ from zero only above a certain threshold energy E_{thresh}. Activation methods of the neutron spectroscopy are expounded in detail in [32–36, 44].

The characteristics of activation and threshold detectors are listed in Tables 41.9 and 41.10. The following notation is used there: T_n is the temperature of neutrons; thermal, resonance, and fast stand for the energy regions of neutrons; for the resonance region shown in parentheses is the value of the threshold energy; γ_{sp} denotes the method of gamma-spectroscopy; β_{sp}, $4\pi\beta$, γ–γ, etc. relate to the methods of measuring the induced activation.

The characteristics of detectors employed in spectroscopy on assemblies with a source of 14 MeV-neutrons are drawn in Figures 41.39 and 41.40 [43].

Table 41.9 A brief characteristic of detectors of various types [33]

Element or element and type of reaction (for threshold detectors)	Detector material	Medium temperature, K	Application region	Thickness, 10^{-3} kg/m^2	Density of neutron flux, (neutrons/s·m^2)	Method of activity measurements
Na	Tablets (pressing)	1000	Resonance (2950 eV)	150–2000	10^9–10^{14}	γ_{sp}, β_{sp}
V	Foil	1300	$1/v$	50	10^9–10^{14}	$4\pi\beta$
Cr	Foil	800	Resonance, $1/v$	10–400	10^{12}–10^{14}	
Mn	Ceramics	1300	Thermal, resonance (337 eV)	3–1000	10^9–10^{18}	γ_{sp}, β_{sp}
	Alloy	650				
	Polyethylene	350				
Co	Foil (electrolysis)	900	Thermal, resonance (132 eV)	0.5–700	10^{13}–10^{18}	γ_{sp}, γ–γ
	Alloy	650				
	Polyethylene	350				
Cu	Foil	800	Resonance (580 eV)	50–300	10^9–10^{18}	γ_{sp}, β_{sp}
Ag	Alloy	650	Resonance (5.2 eV)	0.5–10	10^{12}–10^{18}	γ_{sp}
In	Foil	400	Thermal, resonance (1.44 eV)	0.5–10	10^7–10^{18}	γ_{sp}, β_{sp}
	Alloy	650				
La	Ceramics	1300	Resonance (73.5 eV)	5–200	10^{10}–10^{18}	γ_{sp}, β_{sp}
	Alloy	650				
	Polyethylene	350				
Sm	Ceramics	1300	Resonance (8.0 eV)	1.0–20	10^{10}–10^{18}	γ_{sp}, β_{sp}
	Alloy	650				
	Polyethylene	350				
Eu	Ceramics	1300	T_n	0.3–800	10^9–10^{14}	γ_{sp}, β_{sp}
	Alloy	650				
	Polyethylene	350				
Dy	Ceramics	1300	T_n, thermal, resonance	3–40	10^9–10^{18}	γ_{sp}, β_{sp}
	Alloy with aluminum	650				
	Polyethylene	400				
Lu	Ceramics	1300	T_n, thermal	1.3–800	10^9–10^{18}	γ_{sp}, β_{sp}
	Alloy	650				
	Polyethylene	350				
W	Foil (rolling, electrolysis)	1300	Resonance (18.8 eV)	0.5–400	10^{10}–10^{18}	γ_{sp}, β_{sp}
	Alloy	650				
	Polyethylene	350				
Au	Foil (rolling)	900	Thermal	0.5–600	10^9–10^{16}	$\gamma_{sp}, 4\pi\beta$
	Alloy	650				
Mg (n,p)	Foil	650	Fast, > 7 MeV	1.0–10	10^{12}	γ_{sp}
	Ceramics	1300				
Al (n,p)	Foil	800	Fast, > 4 MeV	1.0–10	10^{13}	γ_{sp}
	Ceramics	1300				

Table 41.9 A brief characteristic of detectors of various types [33] *(continued)*

Element or element and type of reaction (for threshold detectors)	Detector material	Medium temperature, K	Application region	Thickness, 10^{-3} kg/m^2	Density of neutron flux, (neutrons/s·m^2)	Method of activity measurements
S (n, p)	Tablets (pressing)	400	Fast, > 2 MeV	10–30	10^{11}	β_{sp}
^{56}Fe(n, p)	Foil (pressing)	1200	Fast, > 3.7 MeV	1–10	10^{12}	γ_{sp}, β_{sp}
Ni (n, p)	Foil	1200	Fast, > 3.0 MeV	1–10	10^{12}	γ_{sp}
^{65}Cu$(n, 2n)$	Foil	800	Fast, > 9.0 MeV	1–10	10^{13}	γ_{sp}, β_{sp}
^{64}Zn(n, p)	Foil	800	Fast, > 3.0 MeV	1–10	10^{12}	γ_{sp}
Rh (n, n')	Fool	1300	Fast, > 0.8 MeV	0.1–10	10^{11}	$x, 4\pi\beta$–x
In (n, n')	Foil	400	Fast, > 1.4 MeV	0.1–10	10^{11}	γ_{sp}
	Ceramics	1300				
Er (n, γ)	Ceramics	1300	T_n	0.5–40	10^9–10^{18}	γ_{sp}
	Polyethylene	350				
Ir (n, γ)	Foil	1300	T_n, thermal, resonance	3–100	10^{12}–10^{18}	γ_{sp}
	Alloy	650				

Table 41.10 Characteristics of threshold detectors [32]

Nuclear reaction		E_{thresh}, MeV [35]	Half-life of reaction product	E_γ, MeV	Yield of gamma-quanta per decay	E_{eff}, MeV	σ_{eff}, 10^{-34} m^2
^{103}Rh (n, n')	^{103m}Rh	0.1	56.1 min	0.020	1.00	0.8	935 (45)
^{115}In (n, n')	^{115m}In	0.32	4.5 h	0.333	0.46	1.15	985 (10)
^{238}U $(n$, f)		0.5				1.6	974 (68)
^{232}Th $(n$, f)		1.0				1.6	143 (36)
^{64}Zn (n, p)	^{64}Cu	0.96	12.78 h	0.511	0.38	2.6	158 (8)
^{32}S (n, p)	^{32}P	0.92	14.3 d	$E_\beta = 1.71$	1.0	2.6	232 (37)
^{58}Ni (n, p)	^{58}Co	1.0	71.3 d	0.511	0.30	2.6	416 (37)
^{54}Fe (n, p)	^{54}Mn	1.1	312.6 d	0.835	1.0	3.0	398 (36)
^{27}Al (n, p)	^{27}Mg	1.8	9.46 min	0.84	0.70	4.5	51.3 (4.5)
^{90}Zr (n, p)	^{90}Y	3.5	64.4 h	$E_\beta = 2.27$	1.0	6.2	15.8 (1)
^{56}Fe (n, p)	^{56}Mn	2.9	2.58 h	0.847	0.99	6.6	65.3 (4.8)
^{24}Mg (n, p)	^{24}Na	4.9	15.0 h	1.369	1.0	7.2	143 (9)
^{59}Co (n, α)	^{56}Mn	5.0	2.58 h	0.847	0.99	7.1	13.7 (1.2)
^{27}Al (n, α)	^{24}Na	4.8	15.0 h	1.369	1.0	7.4	75.5(3.2)
^{232}Th $(n, 2n)$	^{231}Th	6.5	25.64 h	$E_\beta = 2.30$	1.0	7.6	1620(90)
^{65}Cu $(n, 2n)$	^{64}Cu	10.0	12.78 h	0.511	0.38	11.2	670 (100)
^{55}Mn $(n, 2n)$	^{54}Mn	10.4	312.6 d	0.835	1.0	11.7	540 (120)
^{19}F $(n, 2n)$	^{18}F	11.0	109.8 min	0.511	1.94	12.8	50 (8)
^{63}Cu $(n, 2n)$	^{62}Cu	11.0	9.76 min	0.511	1.95	12.9	620 (70)
^{58}Ni $(n, 2n)$	^{57}Ni	12.5	36.0 h	1.37	0.86	13.7	38.0 (4.4)
^{64}Zn $(n, 2n)$	^{63}Zn	12.5	38.4 min	1.15	0.49	14.8	690 (40)

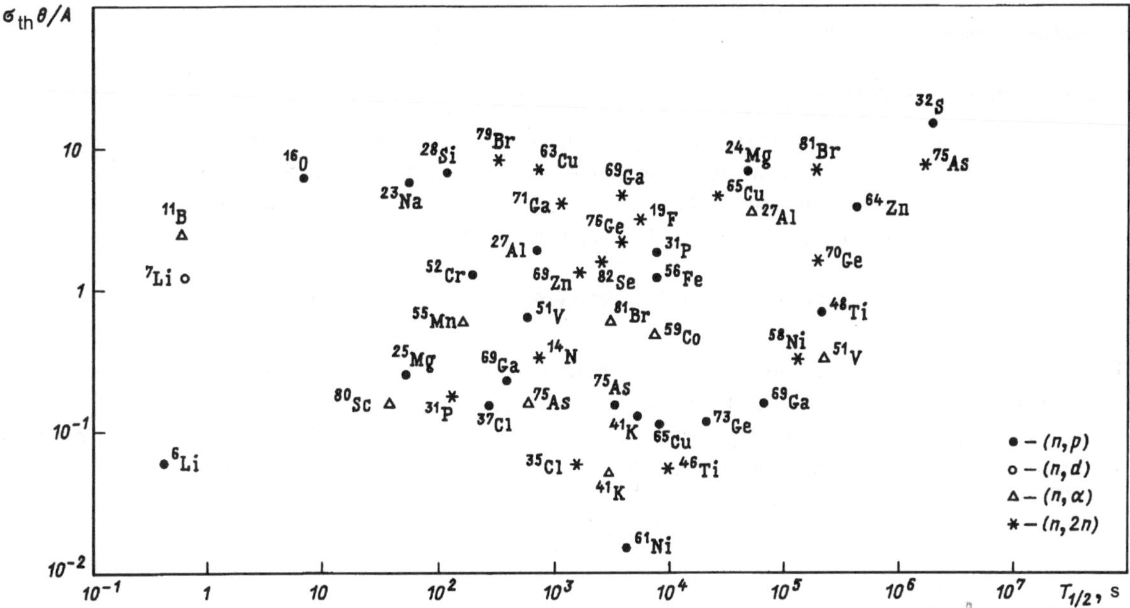

Figure 41.39 Values of the half-life and quantity $\sigma_{\text{thresh}}\theta/A$ for threshold nuclear reactions ($1 \leq A \leq 85$) [43], where σ_{thresh} is the cross section of a threshold nuclear reaction at the neutron energy 14 MeV, 10^{-31} m^2; θ is the nuclide abundance in relative units.

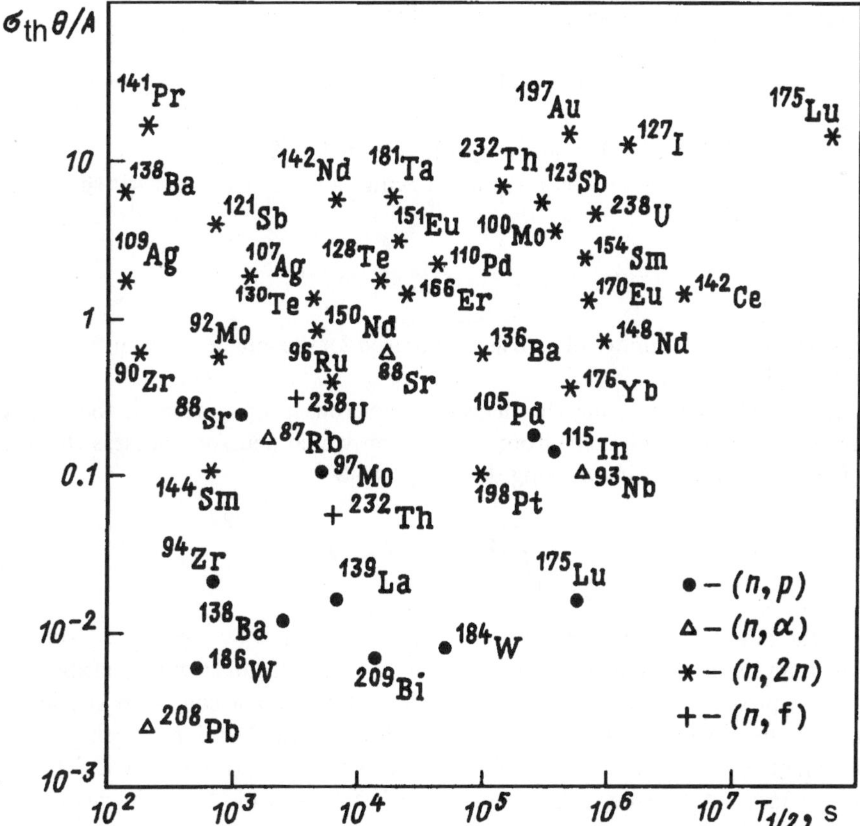

Figure 41.40 Values of the half-life and quantity $\sigma_{\text{thresh}}\theta/A$ for threshold nuclear reactions ($86 \leq A \leq 238$) [43]. The notations are the same as in Fig. 41.39.

41.13 Neutron Slowing Down

In the elementary theory of neutron slowing down, it is assumed that the neutron slows down solely by elastic collisions with medium nuclei. The neutron in these collisions loses part of its energy which transforms into the kinetic energy of the recoil nucleus. Upon collision, the energy of a scattered neutron in the LS-system is determined by the relationship [38]

$$E/E' = (1/2)[1 + \alpha + (1 - \alpha)\cos\theta], \quad \alpha = [(A-1)/(A+1)]^2,$$

where E and E' are neutron energies before and after collision, respectively; A is the mass of a nucleus–target; θ is the neutron scattering angle in the CMS.

The scattering angle φ in the LS-system (the angle between vectors of neutron velocities prior to and upon collision) and the scattering angle θ in the CMS are connected as follows:

$$cos\varphi = 1 + A\cos\theta/\sqrt{(1 + 2A\cos\theta + A^2)}.$$

Here and above the neutron mass is set unity.

In elastic collisions the neutron energy changes from $\alpha E'$ to E'.

The average logarithmic energy loss per one collision with a nucleus of mass A is determined by the formula

$$\xi = \overline{(\ln E' - \ln E)} = 1 + (A-1)^2/2A - \ln[(A-1)/(A+1)]$$

and does not depend on energy. When $A \gg 1$, the following relation

$$1/\xi = A/2 + 1/3 + 1/18A$$

holds valid. At $A = 2$, the error in this quantity is less than 1% [37].

The mean number of collisions N necessary for slowing down a neutron with the initial energy E' to energy E is defined by

$$N = (1/\xi)\ln(E'/E).$$

Table 41.11 reports the mean number of collisions required for slowing a neutron from 2 MeV down to the thermal energy (0.0253 eV).

One of the quantities characterizing the degree of neutrons propagation in the course of slowing down is the mean square of the distance from a source on which neutrons possess the energy E. The mean square of the migration during slowing down equals [38]

$$\overline{r^2} = \int r^2 W(r,\tau)dr = 2n\tau,$$

where $W(r,\tau)$ is the function describing the neutron field generated by a point-like source of monoenergetic neutrons in an infinitely homogeneous medium. It is a solution to the equation $\partial W(r,\tau)/\partial\tau = \Delta W(r,\tau)$ with the initial condition $W(r,0) = \delta(r); n = 1,2,3$ for a source in the form of a plane, string, and point, respectively. The quantity $\sqrt{\tau}$ is called the moderation length; and τ, the age of neutrons. The quantities τ are experimentally determined from activation of a detector at various distances from a source. As a rule, τ is measured with the help of the ^{115}In detector which has a resonance at $E = 1.46eV$.

The addition $\Delta\tau$ due to neutron moderation from 1.46 eV to E_c is small as compared to τ and can be approximately calculated by the formula [38]

$$\Delta\tau = \int\limits_{E_c}^{1.46\ \mathrm{eV}} \frac{dE}{3\xi \sum_s \sum_{\mathrm{tr}} E},$$

where Σ_{tr} is the transport macroscopic cross section; it replaces the total cross section in computations (the so called transport approximation): $\Sigma_{\mathrm{tr}} = \Sigma_{\mathrm{tot}} + \Sigma_s(1 - \overline{\cos\varphi}), \overline{\cos\varphi}$ is the mean cosine of the scattering angle in the LS-system.

The $\Delta\tau$ obtained by this formula in the region where it is necessary to take account of the chemical bond's influence on the process of energy transfer from neutrons to nuclei of a moderator, is smaller than the actual value. It is impossible to compute the age τ by the above formula for a mixture of a light moderator with heavy nuclei since in this case the inelastic collisions and anisotropy of scattering on heavy nuclei at high energies become significant.

The time of neutron slowing down to the age τ is defined by

$$T(\tau) = \int\limits_0^u T_s(u')\frac{du'}{\bar{\xi}},$$

where $T_s(u) = 1/v(u)\Sigma_s(u)$ is the mean time between two collisions of a neutron; $du/\bar{\xi}$ is the average number of collisions in the interval of moderation du. Since for nuclei with mass numbers $A > 1$ in the energy region up to 100 keV, $1/\Sigma_{\mathrm{tr}} \simeq$ const, the energy dependence of the neutrons age is logarithmic [13]:

$$\tau(E) = \frac{1}{3\xi \sum_{\mathrm{tr}} \sum_s} \int\limits_E^{E_0} \frac{dE'}{E'} = \frac{1}{3\xi \sum_{\mathrm{tr}} \sum_s} \ln \frac{E_0}{E}.$$

For the neutron sources with $E_0 \le 100$ keV the time of neutron slowing down:

$$T(E) = \frac{1}{\xi \sum_s} \int\limits_E^{E_0} \frac{dE'}{v(E')E'} = \frac{2}{\xi \sum_s}\left(\frac{1}{v} - \frac{1}{v_0}\right).$$

For low energies, i.e. $E \ll E_0$

$$T(E) \approx 2/\xi v(E) \sum_s .$$

In Tables 41.12 and 41.13, the values are presented for the age of neutrons from a source with the fission spectrum during their moderation to the In resonance energy (1.46 eV) and for τ of some moderators.

Table 41.11 Properties of some moderators [8]

Moderator	1-α	ξ	$\overline{\cos\varphi}$	Mean number of collisions	A
$^1_1$H	1.00	1.00	0.67	18.2	1
$^2_1$H	0.889	0.725	0.33	25.1	2
He	0.640	0.425	0.167	43	4
Li	0.438	0.268	0.095	68	7
Be	0.360	0.209	0.074	87	9
C	0.284	0.158	0.056	115	12
O	0.222	0.120	0.042	152	16
U	0.016	0.00838	0.0028	2172	238

Table 41.12 Parameters characterizing the properties of moderators [13]

Moderator	τ, cm^2 [38]	$T, 10^{-5}$ s
H_2O	26.48 (32)	1
D_2O	111 (1)	5
C	282.5 (1.8)	15
Be	86.6 (2.4)	7
BeO	92.0 (1.5)	7.8

Table 41.13 The age of neutrons for some moderators [39]

Moderator (density, g/cm^3)	τ, cm^2	Neutron source (E_0, MeV)	Detector or threshold energy E_{thresh}, eV
H_2O	5.48 (15)	Sb-γ-Be (0.025)	In
	13.9 (2)	Na-γ-Be (0.970)	In
	27.86 (1)	With fission spectrum	In
	34.6 (2.2)	Reaction $D(d,n)$ (2.6)	In
	48.5	Ra-α-Be	In
	57.3 (2.0)	Po-α-Be	In
	150.0 (6.0)	Reaction $T(d,n)$ (14.1)	In
	30.3 (1.5)	With fission spectrum	0.025
	55	Ra-α-Be	Dy
	62 (5)	Po-α-Be	0.025
50% H_2O+50% Al	76.8	With fission spectrum	In
50% H_2O+50% Zr	81.7	The same	0.025
50% H_2O+50% Pb	72.8	Ra-α-Be	In
50% H_2O+50% Fe	57.9	The same	In
50% H_2O+50% Zr	92	Po-α-Be	0.025
D_2O	109 (3)	With fission spectrum	In
	120	The same	0.025
C	142–147	Sb-γ-Be (0.025)	In
	312.5 (5)	With fission spectrum	In
	380	Ra-α-Be	In
Be (1.85)	80.2 (2)	With fission spectrum	In
Be (1.78)	120 (23)	Ra-α-Be	In
Be (1.84)	97.2	With fission spectrum	0.025
BeO (2.96)	93.4 (4.7)	The same	In
BeO	105 (10)		0.025

41.14 Neutron Diffusion

The quantity $L = \sqrt{D/\Sigma_a}$ is called the diffusion length; here D is the diffusion coefficient; Σ_a is the absorption cross section. The diffusion coefficient D is connected with the diffusion constant D_0 by the formula $D_0 = Dv$, where v is the velocity. The diffusion length, up to a numerical factor, characterizes the mean distance between the point of neutron production and the point of its absorption, i.e. $\overline{r^2} = 6L^2$, where r is the straight-line distance from the point of neutron production to the point of its capture.

The quantity t_T defined by the formula $t_T = 1/(v\Sigma_a)$ where v is the velocity of a neutron is called the diffusion time or the lifetime of thermal neutrons.

The diffusion length L, diffusion coefficient D, mean logarithmic energy loss ξ, and the quantity $1 - \bar{\mu}$, where $\bar{\mu}$ is the mean cosine of the scattering angle, are listed in Table 41.15 for some elements. The time of thermalization t_{th} (Table 41.16) is defined as the rate of approaching the average energy of neutrons to the equilibrium energy, i.e.

$$\overline{E} - (2/3)kT = \text{const}\,\exp(-t/t_{th}).$$

In experiments with pulse neutron sources of, the decay constant of the asymptotic density of the neutron flux is connected with diffusion properties of a medium and geometrical parameter B^2 (in cm^2)

by the formula

$$\lambda_0 = v\sum\nolimits_a + D_0B^2 - CB^4 + FB^6,$$

where D is the diffusion coefficient; v, the neutron velocity; Σ_a, the absorption cross section; $D_0 = Dv$; C is the coefficient of diffusion cooling; F is a numerical factor. The quantities specifying the diffusion properties of neutron moderators are given in Table 41.17.

Table 41.14 Lifetime of thermal neutrons [9]

Moderator	Density, g/cm^3	Lifetime, t_T, 10^{-4} s
H_2O	1.00	2.13
D_2O	1.10	1300
C	1.6	129
Be	1.85	36.5
BeO	2.96	76.0
Paraffin	0.87	1.78
Organic glass	1.18	2.32
$ZrH_{1.7}$	3.48	2.66

Table 41.15 Diffusion characteristics of some substances [38]

Medium	$1-\bar{\mu}$ [21]	ξ [21]	L, cm	D, cm
H_2O			2.69 (2)	0.1423 (19)
D_2O			147 (2)	0.84 (1)
Be	0.9259	0.209	24.4 (13)	0.487 (5)
BeO			36.5 (3)	0.627 (7)
C	0.9440	0.158	56.4 (6)	0.828 (8)
Na	0.9710	0.0845	17.7 (1)	3.755 (6)
Mg	0.9722	0.0811	30.4 (8)	2.229 (2)
Al	0.9754	0.0723	16.7 (1)	3.418 (7)
Ca [23]	0.9833	0.0492	21.8	4.87
Cd [23]	0.9940	0.0178	0.0949	1.03
Mn [23]	0.9878	0.0359	1.30	1.80
Fe	0.9881	0.0353	1.22 (1)	0.288 (1)
Co [23]	0.9887	0.0335	0.40	0.541
Cu [23]	0.9896	0.0309	1.30	0.552
Zr	0.9927	0.0218	13.1 (1)	1.195 (9)
Nb [23]	0.9928	0.0214	4.42	1.23
Mo [23]	0.9931	0.0207	2.08	0.749
Ag [23]	0.9938	0.0184	0.508	0.953
Pb	0.9968	0.0096	13.3 (1)	0.875 (16)
Bi	0.9968	0.0095	39.20 (24)	1.263 (5)
Th		0.0086	1.78 (1)	0.601 (4)
U		0.0084	1.20 (1)	0.453 (4)
Pu [23]		0.0083	0.0708	0.286

Table 41.16 Thermalization time in some moderators [40]

Moderator	t_{th}, 10^{-6} s
H_2O (77 K)	55.5 (1.5)
H_2O (300 K)	5.8 (0.6)
C	185 (45)
Be	28
BeO	67

Table 41.17 Diffusion parameters of moderators [41]

Medium (density, g/cm^3)	B^2, cm^{-2}	λ_0, s^{-1}	L, cm	D_0, 10^5 cm^2/s	C, 10^4 cm^4/s [F, cm^6/s]
H_2O	0.02–0.42		2.80 (3)	0.368 (4)	0.4137 (688)
D_2O				1.96 (7)	3.72 (50)
C					
(1.70)	1.76–18.9	74.7 (0.6)	53.8 (3)	2.14 (1)	340 (30)
(1.64)			52.4 (1.4) [42]	2.086 (81) [42]	390 (25) [42]
Be					
(1.85)	3.4–74	277 (8)	20.8 (3)	1.235 (13)	28 (3)
(1.79)			21.8 (3) [42]	1.233 (7) [42]	31.3 (8) [42]
BeO					
(2.96)		132 (3)	29.9 (3)	1.178 (3)	38.5 (8)
(2.79)			28.9 (5) [42]	1.548 (9) [42]	46.3 (2.4) [42]
Zr [40]		3765 (89)		10.579 (32)	21.2 (3.5) [5300 (14)]
Benzene [40]		2886 (111)		0.4865 (137)	1.387 (385)
Paraffin		5860 (70)	2.13 (4)	0.266 (5)	0.120 (50)

Table 41.17 Diffusion parameters of moderators [41] *(continued)*

Medium (density, g/cm^3)	B^2, cm^{-2}	λ_0, s^{-1}	L, cm	D_0, 10^5 cm^2/s	C, 10^4 cm^4/s [F, cm^6/s]
Polyethylene		5900 (90)	2.12 (3)	0.265 (6)	0.260 (80)
Dowtherm A (Diphenyl – 26.8 %, oxydiphenyl – 73.2 %)		2870 (40)	4.14 (7)	0.492 (6)	1.19 (21)
Organic glass		4300		0.34	
$ZrH_{1.7}$	0.03–0.39	3757 (98)		0.579 (32)	21.2 (3.5) 50 (14)
Lucite			3.14 (3)		
Diphenyl ($T = 77°$C)	0.0816–0.28		4.82 (7)	0.5437 (117)	0.985 (55)
n-Heptane ($T = 17.5°$C)	0.11–1.5	4950 (120)	2.59 (6)	0.3313 (12)	0.480 (29)
Monoisopropylbiphenyl ($T = 30°$C)	0.0945–0.2952			0.3780 (33)	0.125 (43)

References

[1] ENDF-102. Data formates and procedures for the evaluated nuclear data file, ENDF, revised by Garber, D., Dunford, C., Pearlstein, S., BNL-NCS-50496, New York, Upton, 1975.

[2] Hawerton, R. J., Plechaty, E. F., Cullen, D. E., The LLL evaluated nuclear data library (ENDL): evaluation techniques, graphical displays and descriptions of individual evaluations, UCRL–50400, Lawrence Livermore Laboratory, 1971 (v. 4).

[3] Parker, K., The Aldermaston nuclear data library as at May 1963, AWRE 0-70/63, Aldermaston, England, 1963.

[4] Woll, D., Card image format of the Karlsruhe evaluated nuclear data file KEDAK, Karlsruhe report KFK 800, Institut fur Angewandte Kernphysik Kernforschungszentrum, Karlsruhe, 1968.

[5] CINDA 84 (1981 - 1984), The index to literature and computer files on microscopic neutron data, Published on behalf of USA National Nuclear Data Center, USSR Nuclear Data Center, NEA Data Bank IAEA Nuclear Data Section, IAEA, Vienna, 1984.

[6] Radiative neutron capture, A Handbook, Eds. Belanov, T. S., Ignatyuk, A. V., Pashchenko, A. B., Plyaskin, V. I., Energoatomisdat, Moscow, 1986 (in Russian).

[7] Blatt, J., Weisskopf, V., Theoretical nuclear physics, New York, 1952.

[8] Curtiss, L.F., Introduction to neutron physics, D.Van Nostrand Co., Princeton, New Jersey, Toronto, London, New York, 1962

[9] Beckurts, K.H. and Wirtz, K., Neutron physics, Springer–Verlag, Berlin–New York, 1964.

[10] Sitenko, A. G., Theory of nuclear reactions. Higher school students' guide, Energoatomisdat, Moscow, 1983 (in Russian).

[11] Aleksandrov, Yu. A., Basic properties of neutrons, 2-nd ed., Energoizdat, Moscow, 1982 (in Russian).

[12] Weinberg, A., Wigner, E., The physical theory of neutron chain reactors, Chicago, 1959.

[13] Feinberg, S. M., Shikhov, S. B., Troyanskii, V. B., Theory of nuclear reactors. Higher school students' guide, Atomizdat, Moscow, 1980 (v. 1) (in Russian).

[14] Computing methods in reactor physics, Eds. Greenspan, H., Kelber, C.N. and Okrent, D., Gordon and Breach Sci. Publ., New York, London, Paris, 1968.

[15] Nikolaev, M. N., Bazazyants, N. O., Elastic neutron scattering anisotropy, Atomizdat, Moscow, 1972 (in Russian).

[16] Antsipov, G. V., Kon'shin, V. A., Sukhovitskii, E. Sh., Nuclear constants for plutonium isotopes, Nauka i tekhnika, Minsk, 1982 (in Russian).

[17] Symbols, units and nomenclature in physics, Document U.I.P. (1978), Int. Union of Pure and Appl. Phys., S.U.N. Commission.

[18] Particle data group, Review of particle properties, Rev. Mod. Phys., 56, S122, 1984.

[19] Neutron physics, Springer tracts in modern physics, vol. 80, No 1, Springer–Verlag, Berlin, 1977.

[20] Medvedev, Yu. A., Stepanov, B. M., Trukhanov, G. Ya., Nuclear physical constants of interactions with elements entering into the atmosphere and crust composition. Handbook, Energoizdat, Moscow, 1981 (in Russian).

[21] Mughabhab, S. F., Divadeenam, M., Holden, W. E., Neutron cross section, vol. 1, pt. A, Academic Press, New York, 1981.

[22] Nuclear wallet cards, Ed. Shirley, V. S., Lederer, C. M., NBS, Office of Stanford Reference Data, New York, 1979.

[23] Gordeev, I. V., Kardashev, D. A., Malyshev, A. V., Nuclear physical constants. Handbook, Atomizdat, Moscow, 1963 (in Russian).

[24] Mughabhab, S. F., Garder, D. I., Neutron cross sections, 3-d ed., vol. 1, BNL - 325, Brookhaven National Laboratory Associated Universities, Inc., 1973.

[25] Abagyan, L. P., Yudkevich, M. S., Evaluated neutron data for thermal reactor calculations, Voprosy atomnoi nauki i tekhniki, Ser. "Yadernye constanty", vol. 4(43), CNIIAtominform, Moscow, 1981, p.24.

[26] Gorbachev, V. M., Zamyatin, Yu. S., Lbov, A. A., Interaction of radiations with nuclei of heavy elements and nuclear fission, Atomizdat, Moscow, 1976 (in Russian).

[27] Nuclear data standards for nuclear measurements, Technical report series No 227, IAEA, Vienna, 1983.

[28] Gryntakis, E. M., Kim, J. I., J. Radioanal. Chem., 76, 341, 1983.

[29] Garber, D. I., Kinsey, R. R., Neutron cross sections. Curvers, 3-d ed., BNL–325, vol. 2, Brookhaven National Laboratory Associated Universities, Inc., 1976.

[30] Bychkov, V. M., Manokhin, V. N., Pashchenko, A. B., Plyaskin, V. I., Cross–section of neutron–induced threshold reactions. Handbook, Energoizdat, Moscow, 1982 (in Russian).

[31] Tsukada, K., Table of nuclear reactions and subsequent radioactive decays induced by 14–MeV neutrons, JAERI 1252, Japan Atomic Energy Research Institute, 1977.

[32] Lapenas, A. A., Neutron spectra measurements by the activation technique, Zinatne, Riga, 1975 (in Russian).

[33] Lomakin, S. S., Petrov, V. I., Samoilov, P. S., Neutron radiometry by the activation technique, Energoatomizdat, Moscow, 1983 (in Russian).

[34] Kramer–Ageev, E. A., Troshin, V. S., Tikhonov, E. G., Activation techniques of neutron spectrometry, Atomizdat, Moscow, 1976 (in Russian).

[35] Bondars, Kh. I., Zagryadskii, V. A., Novikov, V. M., Chuvilin, D. Yu., On the selection of the optimal set of activation detectors for neutron spectrometry in assemblies with an external source of 14–MeV neutrons, Preprint No 3798/4, Institute of Atomic Energy, Moscow, 1983 (in Russian).

[36] Handbook of nuclear activation cross sections, Technical report series No 156, IAEA, Vienna. 1974.

[37] Usachev, L. N., Bobkov, Yu. G., Perturbation theory and laying out an experiment in the problem of nuclear data for reactors, Atomizdat, Moscow, 1980 (in Russian).

[38] Galanin, A. D., Introduction to the theory of thermal neutron nuclear reactor, Higher school students' guide, Energoatomisdat, Moscow, 1984 (in Russian).

[39] Tables of physical quantities. Handbook, Ed. Kikoin, I. K., Atomizdat, Moscow, 1976 (in Russian).

[40] Spectra of slow neutrons. Collection of papers, Atomizdat, Moscow, 1971 (in Russian).

[41] Advances in nuclear science and technology, vol. 3, Eds. Greebler, P., Henley, E.J., Academic Press, New York–London, 1966.

[42] Zhezherun, I. F., Experimental study of certain problems of physics of nuclear reactors with crystal decelerators, Phys. Math. Doctoral theses, 1974 (in Russian).

[43] Barbier, M., Induced radioactivity, North–Holland Publ. Comp., Amsterdam–London, 1969.

[44] Handbook on nuclear activation data, Technical report series No 273, IAEA, Vienna, 1987.

42

Penetration of Ionizing Radiation through Matter

V.P. Rudakov

42.1 Passage of Heavy Charged Particles through Matter

The heavy (i.e. heavier than electron) charged particles passing through matter lose their energy E mainly due to ionization and excitation of the atoms. The energy loss is expressed in terms of the specific energy loss dE/dx or $(1/\rho)dE/dx$, where ρ is the density of matter (in mg/cm^3), with x denoting the particle coordinate in the bulk. The specific energy loss is also referred to as the stopping power of matter.

The energy loss of a heavy particle can be calculated with the Bethe formula

$$\frac{dE}{dx} = \frac{2\pi e^4 Z z^2 n}{m_e v^2}\left[\ln\frac{2m_e v^2 Q_{\text{max}}}{I^2(1-\beta^2)} - 2\beta^2 - \delta - U\right], \tag{42.1}$$

where e and m_e are the electron charge and mass, respectively; n is the number density of atoms (in cm^{-3}) and Z is their atomic number (number of atomic protons); v and z are the velocity and charge of an incident particle, respectively; $\beta = v/c$ (c is the speed of light); I is the average ionization potential of atoms in the matter; Q_{max} is the maximum energy transferred by the incident particle to atomic electron; U is the correction factor accounting for the electron binding energies in K- and L-atomic electron shells, and, finally, d is the correction factor connected with a so-called density effect. In the energy region of the incident particles 1–100 MeV, the corrections U and d can be disregarded.

The range R of the particle with initial energy E is given by the formula

$$R = \int_0^E \frac{\rho dE}{dE/dx}, \tag{42.2}$$

the unit of the range is mg/cm^2.

The heavy charged particle with 1–100 MeV energy loses only a small part of its energy in each act of interaction with atomic electrons and therefore the beam of the monoenergetic charged particles passing through matter does not practically change its intensity until the end of the range. Statistical fluctuations of the energy losses result in small (1 – 2%) straggling of the ranges. The escape of the heavy charged particles from the beam due to multiple scattering can be estimated using the formulae given below in Section 42.2.

From Bethe's formula one can see that the stopping power of a particular material for a given ionizing particle depends only on charge and velocity of this particle. Thus, for example, if the proton range

0-8493-2861-6/97/$0.00+$.50

in some material is known, one can calculate the deuteron and triton ranges in the same material by using the following relationship

$$R_m(E) = \frac{m}{m_p} R_p \left(\frac{m_p}{m} E\right), \tag{42.3}$$

where R_p, m_p and $(m_p/m)E$ are the proton range, mass and energy, respectively; R_m, m and E are the deuteron (or triton) range, mass and energy. If α-particle range is known, using this relation one can calculate the range of $^3He^+$, and so on.

In general, if the range of some particle with charge Z and mass m is known one can calculate approximately the range of any other particle with charge Z_1 and mass m_1 by employing the relationship

$$R_{m_1}(E) = \frac{m_1}{m} \frac{Z^2}{Z_1^2} R_m \left(\frac{m}{m_1} E\right) F, \tag{42.4}$$

where E and $(m/m_1)E$ are the energies of the particles with masses m and m_1, respectively; F is the correction factor, which differs noticeably from unit only at high energies. This relation does not take into account the charge exchange processes at low energies and therefore can be used only for energies above 5 MeV. In Tables 42.1 – 42.20, the ranges and stopping powers for some light ions (from hydrogen to neon) in different substances are given. The more complete tables of ranges have been published in [1–3].

Table 42.1 Mass stopping power of different substances (in MeV·mg^{-1}·cm^2) for hydrogen ions $^1_1H^+$ [1] [E_m is the energy per mass unit of incident particle; E is a total energy; 1 u=1 a.m.u.]

E_m, MeV/u	Be	C	Al	Ti	Ni	Ge	Zr	Ag	Eu	Ta	Au	U	E, MeV
0.0125	0.348	0.288	0.211	0.140	0.114	0.103	0.096	0.089	0.056	0.048	0.044	0.038	0.0126
0.0160	0.394	0.326	0.239	0.159	0.129	0.117	0.108	0.100	0.064	0.055	0.050	0.043	0.0161
0.0200	0.441	0.364	0.267	0.178	0.144	0.131	0.121	0.112	0.071	0.061	0.056	0.048	0.0202
0.0250	0.497	0.411	0.301	0.200	0.163	0.148	0.137	0.126	0.080	0.069	0.063	0.054	0.0252
0.0320	0.559	0.463	0.339	0.225	0.183	0.167	0.154	0.142	0.091	0.078	0.072	0.061	0.0322
0.0400	0.611	0.509	0.371	0.248	0.202	0.184	0.170	0.157	0.100	0.086	0.079	0.068	0.0403
0.0500	0.655	0.549	0.400	0.268	0.220	0.200	0.184	0.171	0.108	0.094	0.087	0.075	0.0504
0.0600	0.683	0.578	0.419	0.282	0.232	0.211	0.195	0.180	0.116	0.101	0.093	0.080	0.0605
0.0700	0.699	0.598	0.430	0.292	0.241	0.219	0.202	0.188	0.121	0.105	0.098	0.084	0.0705
0.0800	0.707	0.613	0.437	0.298	0.247	0.224	0.207	0.192	0.125	0.109	0.101	0.087	0.0806
0.0900	0.709	0.622	0.440	0.303	0.251	0.228	0.210	0.195	0.128	0.112	0.104	0.090	0.0907
0.1000	0.706	0.628	0.440	0.305	0.253	0.230	0.211	0.198	0.130	0.114	0.106	0.091	0.1008
0.1250	0.688	0.635	0.433	0.305	0.255	0.231	0.212	0.199	0.133	0.116	0.109	0.094	0.1260
0.1600	0.651	0.632	0.415	0.297	0.250	0.226	0.208	0.195	0.133	0.117	0.109	0.095	0.1612
0.2000	0.605	0.614	0.391	0.284	0.240	0.218	0.200	0.187	0.130	0.114	0.107	0.094	0.2016
0.2500	0.551	0.580	0.362	0.268	0.227	0.206	0.189	0.177	0.125	0.110	0.104	0.090	0.2519
0.3200	0.488	0.530	0.327	0.246	0.210	0.192	0.175	0.164	0.117	0.104	0.098	0.086	0.3225
0.4000	0.431	0.478	0.295	0.225	0.193	0.177	0.161	0.151	0.110	0.098	0.092	0.080	0.4031
0.5000	0.376	0.423	0.263	0.203	0.176	0.162	0.147	0.137	0.102	0.090	0.085	0.075	0.5039
0.6000	0.331	0.373	0.236	0.185	0.161	0.148	0.135	0.125	0.094	0.084	0.079	0.070	0.6047
0.7000	0.298	0.334	0.216	0.170	0.149	0.137	0.124	0.116	0.088	0.078	0.074	0.066	0.7055
0.8000	0.271	0.302	0.199	0.159	0.139	0.129	0.117	0.109	0.082	0.074	0.070	0.062	0.8062
0.9000	0.249	0.275	0.186	0.149	0.131	0.122	0.110	0.102	0.078	0.071	0.067	0.059	0.9070
1.0000	0.231	0.253	0.174	0.141	0.124	0.115	0.104	0.097	0.075	0.067	0.064	0.057	1.0078
1.2500	0.196	0.210	0.151	0.124	0.111	0.103	0.093	0.086	0.068	0.061	0.058	0.052	1.2597
1.6000	0.163	0.171	0.129	0.108	0.096	0.090	0.081	0.075	0.060	0.054	0.051	0.046	1.6125
2.0000	0.138	0.144	0.111	0.094	0.084	0.079	0.071	0.066	0.053	0.049	0.046	0.041	2.0156
2.5000	0.115	0.121	0.095	0.082	0.074	0.069	0.062	0.058	0.047	0.043	0.041	0.037	2.5195

Table 42.1 Mass stopping power of different substances for hydrogen ions $^1_1H^+$ *(continued)*

E_m, MeV/u	Be	C	Al	Ti	Ni	Ge	Zr	Ag	Eu	Ta	Au	U	E, MeV
3.2000	0.095	0.100	0.079	0.069	0.063	0.059	0.053	0.049	0.041	0.038	0.036	0.032	3.2250
4.0000	0.079	0.084	0.067	0.059	0.054	0.051	0.046	0.043	0.036	0.033	0.031	0.029	4.0312
5.0000	0.066	0.071	0.057	0.051	0.047	0.044	0.040	0.037	0.031	0.029	0.028	0.025	5.0395
6.0000	0.057	0.061	0.050	0.045	0.041	0.039	0.035	0.032	0.028	0.026	0.025	0.023	6.0468
7.0000	0.050	0.054	0.044	0.040	0.037	0.035	0.032	0.029	0.025	0.023	0.022	0.021	7.0546
8.0000	0.045	0.049	0.040	0.036	0.034	0.032	0.029	0.027	0.023	0.021	0.021	0.019	8.0624
9.0000	0.044	0.045	0.037	0.033	0.031	0.029	0.027	0.025	0.022	0.020	0.019	0.018	9.0702
10.0000	0.038	0.041	0.034	0.031	0.029	0.027	0.025	0.023	0.020	0.019	0.018	0.017	10.078
11.0000	0.035	0.038	0.032	0.029	0.027	0.025	0.023	0.022	0.019	0.018	0.017	0.016	11.086
12.0000	0.033	0.036	0.030	0.027	0.025	0.024	0.022	0.020	0.018	0.017	0.016	0.015	12.094

E_m, MeV/u	H	He	N	O	Ne	Ar	Kr	Xe	Rn	Mylar	$(CH_2)_n$	Water	E, MeV
0.0125	0.792	0.338	0.224	0.210	0.187	0.123	0.076	0.054	0.035	0.283	0.360	0.275	0.0126
0.0160	0.879	0.371	0.249	0.234	0.208	0.141	0.087	0.062	0.041	0.318	0.405	0.306	0.0161
0.0200	0.956	0.392	0.272	0.256	0.229	0.159	0.099	0.072	0.048	0.353	0.449	0.333	0.0202
0.0250	1.048	0.416	0.299	0.282	0.255	0.183	0.117	0.085	0.062	0.394	0.502	0.367	0.0252
0.0320	1.156	0.439	0.325	0.309	0.283	0.212	0.138	0.104	0.072	0.441	0.562	0.403	0.0322
0.0400	1.244	0.452	0.347	0.334	0.307	0.240	0.159	0.122	0.087	0.481	0.614	0.435	0.0403
0.0500	1.324	0.466	0.368	0.358	0.333	0.268	0.181	0.141	0.102	0.518	0.660	0.465	0.0504
0.0600	1.399	0.479	0.388	0.375	0.355	0.292	0.198	0.156	0.115	0.545	0.696	0.489	0.0605
0.0700	1.459	0.494	0.405	0.393	0.373	0.311	0.212	0.168	0.124	0.566	0.721	0.511	0.0705
0.0800	1.516	0.509	0.420	0.408	0.388	0.326	0.224	0.178	0.132	0.582	0.742	0.531	0.0806
0.0900	1.561	0.524	0.433	0.422	0.402	0.339	0.233	0.186	0.138	0.595	0.756	0.549	0.0907
0.1000	1.610	0.540	0.488	0.434	0.413	0.350	0.242	0.192	0.143	0.604	0.768	0.565	0.1008
0.1250	1.719	0.574	0.474	0.457	0.436	0.368	0.254	0.202	0.150	0.621	0.790	0.597	0.1260
0.1600	1.837	0.613	0.503	0.484	0.460	0.383	0.263	0.209	0.155	0.633	0.805	0.635	0.1612
0.2000	1.919	0.637	0.522	0.501	0.474	0.385	0.262	0.208	0.154	0.631	0.800	0.658	0.2016
0.2500	1.962	0.649	0.527	0.507	0.473	0.374	0.253	0.198	0.146	0.613	0.777	0.669	0.2519
0.3200	1.943	0.634	0.516	0.492	0.450	0.345	0.230	0.180	0.131	0.576	0.732	0.654	0.3225
0.4000	1.839	0.607	0.482	0.458	0.415	0.310	0.205	0.159	0.116	0.528	0.673	0.612	0.4031
0.5000	1.678	0.558	0.433	0.412	0.369	0.271	0.179	0.140	0.102	0.472	0.602	0.553	0.5039
0.6000	1.501	0.503	0.381	0.362	0.324	0.237	0.157	0.122	0.090	0.416	0.534	0.488	0.6047
0.7000	1.325	0.455	0.335	0.318	0.285	0.208	0.139	0.108	0.080	0.370	0.475	0.430	0.7055
0.8000	1.169	0.410	0.297	0.283	0.253	0.185	0.125	0.098	0.073	0.331	0.426	0.381	0.8052
0.9000	1.039	0.370	0.265	0.253	0.225	0.166	0.113	0.089	0.066	0.300	0.384	0.340	0.9070
1.0000	0.938	0.339	0.242	0.230	0.205	0.152	0.104	0.083	0.062	0.274	0.351	0.308	1.0078
1.2500	0.753	0.282	0.201	0.192	0.170	0.128	0.090	0.072	0.055	0.227	0.288	0.254	1.2597
1.6000	0.583	0.228	0.165	0.158	0.141	0.108	0.077	0.063	0.049	0.184	0.230	0.205	1.6125
2.0000	0.460	0.188	0.140	0.133	0.120	0.093	0.068	0.056	0.044	0.153	0.189	0.169	2.0156
2.5000	0.359	0.154	0.118	0.112	0.103	0.082	0.061	0.051	0.040	0.128	0.155	0.140	2.5195
3.2000	0.278	0.124	0.098	0.094	0.087	0.070	0.053	0.045	0.036	0.105	0.125	0.114	3.2250
4.0000	0.223	0.101	0.082	0.079	0.074	0.061	0.047	0.039	0.032	0.088	0.104	0.095	4.0312
5.0000	0.182	0.083	0.069	0.067	0.063	0.052	0.040	0.034	0.028	0.074	0.087	0.079	5.0390
6.0000	0.155	0.071	0.060	0.058	0.055	0.046	0.036	0.030	0.025	0.064	0.075	0.069	6.0468
7.0000	0.137	0.062	0.053	0.051	0.049	0.041	0.032	0.027	0.023	0.057	0.066	0.061	7.0546
8.0000	0.122	0.055	0.048	0.046	0.044	0.037	0.029	0.025	0.021	0.051	0.060	0.055	8.0624
9.0000	0.111	0.050	0.044	0.042	0.040	0.034	0.027	0.023	0.019	0.047	0.054	0.050	9.0702
10.0000	0.102	0.046	0.040	0.039	0.037	0.031	0.025	0.021	0.018	0.043	0.050	0.046	10.078
11.0000	0.094	0.043	0.038	0.036	0.034	0.029	0.023	0.020	0.017	0.040	0.046	0.043	11.086
12.0000	0.087	0.040	0.035	0.034	0.032	0.027	0.022	0.019	0.016	0.037	0.043	0.040	12.094

Table 42.2 Range (in mg/cm^2) of hydrogen ions $^1_1H^+$ for different substances [1] [E_m is the energy per mass unit of incident particle; E is a total energy.]

E_m, MeV/u	Be	C	Al	Ti	Ni	Ge	Zr	Ag	Eu	Ta	Au	U	E, MeV
0.0125	0.059	0.070	0.102	0.158	0.193	0.216	0.231	0.246	0.373	0.421	0.449	0.491	0.0126
0.0160	0.068	0.082	0.118	0.181	0.222	0.247	0.265	0.283	0.430	0.488	0.521	0.575	0.0161
0.0200	0.078	0.093	0.134	0.205	0.251	0.279	0.300	0.320	0.489	0.556	0.595	0.661	0.0202
0.0250	0.089	0.106	0.151	0.231	0.283	0.315	0.338	0.362	0.554	0.632	0.678	0.757	0.0252
0.0320	0.102	0.122	0.173	0.264	0.323	0.360	0.386	0.414	0.635	0.727	0.781	0.877	0.0322
0.0400	0.116	0.139	0.196	0.298	0.365	0.406	0.436	0.468	0.719	0.824	0.886	0.999	0.0403
0.0500	0.132	0.158	0.222	0.337	0.413	0.458	0.493	0.529	0.815	0.935	1.006	1.139	0.0504
0.0600	0.147	0.175	0.246	0.374	0.458	0.507	0.546	0.586	0.905	1.038	1.117	1.268	0.0605
0.0700	0.161	0.193	0.270	0.409	0.500	0.554	0.597	0.641	0.990	1.136	1.223	1.391	0.0705
0.0800	0.175	0.209	0.293	0.443	0.541	0.600	0.646	0.694	1.072	1.230	1.324	1.509	0.0806
0.0900	0.190	0.226	0.316	0.476	0.582	0.644	0.694	0.746	1.152	1.321	1.423	1.623	0.0907
0.1000	0.204	0.242	0.339	0.510	0.622	0.688	0.742	0.797	1.230	1.410	1.519	1.734	0.1008
0.1250	0.240	0.281	0.397	0.592	0.721	0.798	0.861	0.924	1.421	1.629	1.754	2.006	0.1260
0.1600	0.293	0.337	0.480	0.709	0.861	0.952	1.029	1.103	1.686	1.931	2.077	2.378	0.1612
0.2000	0.357	0.402	0.580	0.848	1.025	1.133	1.227	1.314	1.993	2.280	2.448	2.804	0.2016
0.2500	0.444	0.486	0.714	1.031	1.241	1.371	1.486	1.591	2.389	2.728	2.926	3.351	0.2519
0.3200	0.580	0.614	0.919	1.306	1.564	1.726	1.874	2.006	2.973	3.387	3.628	4.151	0.3225
0.4000	0.756	0.774	1.179	1.649	1.964	2.164	2.355	2.520	3.685	4.188	4.481	5.122	0.4031
0.5000	1.007	0.998	1.542	2.122	2.511	2.761	3.010	3.222	4.642	5.262	5.624	6.422	0.5039
0.6000	1.293	1.253	1.947	2.642	3.110	3.412	3.727	3.992	5.676	6.421	6.855	7.814	0.6047
0.7000	1.615	1.539	2.394	3.210	3.761	4.119	4.506	4.828	6.789	7.666	8.174	9.301	0.7055
0.8000	1.971	1.857	2.881	3.824	4.461	4.877	5.344	5.726	7.976	8.990	9.577	10.881	0.8062
0.9000	2.359	2.207	3.405	4.480	5.207	5.682	6.235	6.684	9.230	10.385	11.054	12.541	0.9070
1.0000	2.780	2.589	3.967	5.177	5.997	6.533	7.178	7.697	10.546	11.848	12.602	14.275	1.0078
1.2500	3.967	3.685	5.524	7.088	8.153	8.851	9.746	10.461	14.095	15.789	16.769	18.935	1.2597
1.6000	5.944	5.552	8.056	10.148	11.584	12.529	13.827	14.869	19.647	21.933	23.269	26.189	1.6125
2.0000	8.643	8.132	11.435	14.165	16.068	17.317	19.155	20.629	26.796	29.798	31.589	35.443	2.0156
2.5000	12.661	11.975	16.361	19.938	22.471	24.137	26.772	28.843	36.865	40.850	43.256	48.349	2.5195
3.2000	19.456	18.433	24.518	29.373	32.867	35.196	39.120	42.169	52.948	58.480	61.830	68.810	3.2250
4.0000	28.827	27.268	35.569	42.008	46.704	49.898	55.475	59.875	74.010	81.494	86.028	95.428	4.0312
5.0000	42.846	40.387	51.855	60.461	66.794	71.210	79.124	85.476	104.075	114.385	120.447	133.199	5.0390
6.0000	59.307	55.720	70.767	81.732	89.870	95.609	106.176	114.724	138.207	151.568	159.272	175.591	6.0468
7.0000	78.125	73.180	92.199	105.704	115.830	122.961	136.444	147.441	176.200	192.816	202.300	222.377	7.0546
8.0000	99.234	92.701	116.062	132.276	144.562	153.185	169.769	183.485	217.769	237.963	249.315	273.414	8.0624
9.0000	122.542	114.237	142.282	161.361	175.962	186.185	206.058	222.734	262.856	286.831	300.174	328.493	9.0702
10.0000	147.989	137.754	170.795	192.866	209.966	221.870	245.277	265.037	311.387	339.290	354.688	387.461	10.078
11.0000	175.544	163.200	201.546	226.713	246.486	260.164	287.342	310.256	363.240	395.146	412.594	450.213	11.086
12.0000	205.152	190.534	234.483	262.868	285.442	300.978	332.155	358.304	418.272	454.277	473.872	516.684	12.094

Table 42.2 Range (in mg/cm^2) of hydrogen ions ${}^1_1H^+$ for different substances [1] *(continued)*

E_m, MeV/u	H	He	N	O	Ne	Ar	Kr	Xe	Rn	Mylar	$(CH_2)_n$	Water	E, MeV
0.0125	0.022	0.056	0.089	0.095	0.109	0.174	0.292	0.403	0.562	0.068	0.052	0.067	0.0126
0.0160	0.026	0.065	0.104	0.111	0.126	0.200	0.334	0.463	0.651	0.080	0.061	0.079	0.0161
0.0200	0.030	0.076	0.119	0.127	0.144	0.227	0.377	0.522	0.739	0.092	0.070	0.091	0.0202
0.0250	0.035	0.088	0.136	0.146	0.165	0.256	0.423	0.585	0.829	0.105	0.081	0.105	0.0252
0.0320	0.041	0.104	0.159	0.169	0.191	0.292	0.478	0.659	0.933	0.122	0.094	0.123	0.0322
0.0400	0.048	0.122	0.182	0.194	0.218	0.327	0.533	0.730	1.033	0.139	0.107	0.142	0.0403
0.0500	0.056	0.144	0.210	0.223	0.249	0.367	0.529	0.807	1.139	0.159	0.123	0.164	0.0504
0.0600	0.063	0.166	0.237	0.251	0.279	0.403	0.645	0.875	1.232	0.178	0.138	0.185	0.0605
0.0700	0.070	0.186	0.263	0.277	0.306	0.436	0.694	0.937	1.316	0.196	0.152	0.206	0.0705
0.0800	0.077	0.206	0.287	0.302	0.333	0.468	0.740	0.995	1.394	0.214	0.166	0.225	0.0806
0.0900	0.083	0.226	0.311	0.326	0.358	0.498	0.784	1.050	1.469	0.231	0.179	0.244	0.0907
0.1000	0.090	0.245	0.333	0.350	0.383	0.527	0.827	1.103	1.541	0.248	0.193	0.262	0.1008
0.1250	0.105	0.290	0.388	0.406	0.442	0.597	0.928	1.231	1.712	0.289	0.225	0.305	0.1260
0.1600	0.125	0.349	0.460	0.481	0.521	0.691	1.065	1.402	1.943	0.345	0.269	0.362	0.1612
0.2000	0.146	0.414	0.539	0.563	0.607	0.796	1.219	1.596	2.204	0.409	0.319	0.424	0.2016
0.2500	0.172	0.492	0.635	0.663	0.714	0.929	1.414	1.844	2.541	0.490	0.383	0.500	0.2519
0.3200	0.208	0.602	0.770	0.804	0.867	1.125	1.707	2.218	3.051	0.609	0.477	0.607	0.3225
0.4000	0.251	0.732	0.932	0.974	1.053	1.372	2.079	2.695	3.706	0.755	0.592	0.735	0.4031
0.5000	0.308	0.906	1.153	1.206	1.311	1.721	2.606	3.372	4.635	0.957	0.750	0.908	0.5039
0.6000	0.372	1.096	1.402	1.468	1.604	2.120	3.209	4.144	5.692	1.185	0.928	1.103	0.6047
0.7000	0.444	1.307	1.684	1.765	1.936	2.575	3.892	5.021	6.885	1.442	1.129	1.323	0.7055
0.8000	0.525	1.540	2.004	2.101	2.312	3.089	4.657	6.002	8.209	1.730	1.353	1.572	0.8062
0.9000	0.616	1.799	2.364	2.479	2.736	3.665	5.505	7.082	9.662	2.050	1.602	1.852	0.9070
1.0000	0.718	2.084	2.762	2.897	3.206	4.301	6.433	8.251	11.237	2.403	1.877	2.164	1.0078
1.2500	1.019	2.902	3.909	4.102	4.560	6.119	9.045	11.508	15.585	3.418	2.673	3.067	1.2597
1.6000	1.554	4.298	5.856	6.140	6.845	9.138	13.298	16.756	22.446	5.153	4.050	4.620	1.6125
2.0000	2.336	6.248	8.522	8.934	9.948	13.167	18.863	23.567	31.173	7.563	5.992	6.792	2.0156
2.5000	3.582	9.216	12.465	13.075	14.479	18.947	26.709	33.053	43.166	11.178	8.953	10.083	2.5195
3.2000	5.827	14.351	19.060	19.995	21.953	28.283	39.129	47.972	61.844	17.293	14.045	15.702	3.2250
4.0000	9.081	21.609	28.071	29.413	32.030	40.662	55.340	67.335	85.883	25.701	21.140	23.481	4.0312
5.0000	14.110	32.706	41.483	43.369	46.857	58.666	78.649	94.977	119.798	38.231	31.808	35.137	5.0390
6.0000	20.125	45.939	57.172	59.674	64.112	79.413	105.264	126.386	157.996	52.904	44.369	48.838	6.0468
7.0000	27.062	61.206	75.032	78.221	83.712	102.808	135.109	161.429	200.471	69.628	58.705	64.477	7.0546
8.0000	34.874	78.444	94.976	88.900	105.595	128.788	168.137	200.041	247.074	88.323	74.781	81.953	8.0624
9.0000	43.542	97.563	116.955	121.671	129.683	157.403	204.352	242.228	297.738	108.953	92.552	101.233	9.0702
10.0000	53.047	118.429	140.896	146.465	155.891	188.297	243.626	287.921	352.412	131.476	111.983	122.269	10.078
11.0000	63.367	140.982	166.715	173.205	184.154	221.721	285.865	337.042	411.094	155.833	133.023	144.996	11.086
12.0000	74.495	165.192	194.371	201.846	214.428	257.523	330.985	389.490	473.713	181.985	155.646	169.377	12.094

Table 42.3 Mass stopping power of different substances (in MeV·mg^{-1}·cm^2) for helium ions $^4_2$He$^+$ [1] [E_m is the energy per mass unit of incident particle; E is a total energy.]

E_m, MeV/u	Be	C	Al	Ti	Ni	Ge	Zr	Ag	Eu	Ta	Au	U	E, MeV
0.0125	0.877	0.726	0.532	0.354	0.287	0.261	0.241	0.223	0.142	0.121	0.112	0.096	0.0500
0.0160	0.993	0.821	0.602	0.400	0.325	0.295	0.273	0.253	0.161	0.137	0.126	0.108	0.0640
0.0200	1.110	0.918	0.673	0.447	0.363	0.330	0.305	0.282	0.180	0.154	0.141	0.121	0.0801
0.0250	1.241	1.026	0.752	0.500	0.406	0.368	0.341	0.316	0.201	0.172	0.158	0.135	0.1001
0.0320	1.403	1.163	0.852	0.566	0.461	0.419	0.387	0.358	0.227	0.195	0.180	0.154	0.1281
0.0400	1.554	1.294	0.945	0.631	0.514	0.469	0.432	0.400	0.254	0.219	0.201	0.173	0.1601
0.0500	1.704	1.428	1.040	0.697	0.571	0.520	0.479	0.444	0.282	0.245	0.226	0.194	0.2001
0.0600	1.819	1.539	1.115	0.751	0.617	0.561	0.518	0.480	0.308	0.267	0.247	0.212	0.2402
0.0700	1.906	1.631	1.173	0.795	0.657	0.596	0.551	0.511	0.328	0.287	0.266	0.229	0.2802
0.0800	1.972	1.709	1.219	0.832	0.689	0.624	0.576	0.536	0.347	0.305	0.282	0.243	0.3202
0.0900	2.020	1.773	1.254	0.863	0.715	0.649	0.598	0.557	0.365	0.319	0.296	0.255	0.3602
0.1000	2.054	1.827	1.280	0.887	0.737	0.668	0.614	0.575	0.379	0.332	0.307	0.265	0.4003
0.1250	2.094	1.932	1.317	0.927	0.774	0.702	0.645	0.604	0.405	0.354	0.331	0.286	0.5003
0.1600	2.077	2.018	1.323	0.948	0.797	0.723	0.663	0.622	0.424	0.372	0.349	0.304	0.6404
0.2000	2.010	2.040	1.299	0.944	0.799	0.723	0.664	0.622	0.431	0.380	0.357	0.312	0.8005
0.2500	1.901	1.999	1.248	0.922	0.784	0.711	0.651	0.610	0.431	0.381	0.357	0.312	1.0007
0.3200	1.745	1.896	1.170	0.881	0.752	0.686	0.626	0.585	0.419	0.372	0.349	0.307	1.2808
0.4000	1.587	1.762	1.086	0.827	0.712	0.651	0.594	0.555	0.404	0.359	0.338	0.296	1.6010
0.5000	1.424	1.601	0.996	0.770	0.667	0.613	0.558	0.520	0.384	0.343	0.322	0.284	2.0013
0.6000	1.268	1.428	0.904	0.708	0.617	0.568	0.515	0.480	0.359	0.321	0.302	0.269	2.4016
0.7000	1.148	1.288	0.832	0.657	0.575	0.530	0.480	0.448	0.338	0.302	0.285	0.253	2.8018
0.8000	1.050	1.171	0.773	0.616	0.541	0.500	0.452	0.422	0.320	0.288	0.271	0.241	3.2021
0.9000	0.972	1.074	0.724	0.581	0.512	0.475	0.429	0.399	0.306	0.275	0.260	0.232	3.6023
1.0000	0.905	0.991	0.682	0.552	0.487	0.452	0.409	0.380	0.293	0.264	0.250	0.223	4.0026
1.2500	0.775	0.830	0.598	0.490	0.436	0.407	0.366	0.340	0.267	0.241	0.228	0.204	5.0033
1.6000	0.649	0.680	0.512	0.428	0.382	0.357	0.322	0.297	0.238	0.216	0.204	0.183	6.4042
2.0000	0.548	0.572	0.442	0.374	0.336	0.316	0.283	0.262	0.212	0.194	0.183	0.165	8.0052
2.5000	0.459	0.481	0.379	0.325	0.294	0.277	0.247	0.230	0.189	0.172	0.163	0.148	10.007
3.2000	0.378	0.399	0.318	0.276	0.252	0.237	0.212	0.196	0.164	0.150	0.142	0.129	12.808
4.0000	0.316	0.336	0.270	0.237	0.217	0.205	0.184	0.170	0.144	0.132	0.125	0.114	16.010
5.0000	0.264	0.283	0.229	0.203	0.187	0.176	0.159	0.147	0.126	0.115	0.110	0.100	20.013
6.0000	0.228	0.246	0.200	0.178	0.164	0.156	0.141	0.130	0.111	0.103	0.098	0.090	24.016
7.0000	0.202	0.218	0.178	0.159	0.147	0.140	0.127	0.117	0.101	0.093	0.089	0.082	28.018
8.0000	0.181	0.196	0.161	0.145	0.134	0.127	0.116	0.107	0.093	0.086	0.082	0.076	32.021
9.0000	0.165	0.179	0.147	0.133	0.123	0.117	0.107	0.099	0.086	0.080	0.076	0.071	36.023
10.0000	0.152	0.165	0.136	0.123	0.114	0.109	0.099	0.092	0.080	0.074	0.072	0.066	40.026
11.0000	0.141	0.153	0.127	0.115	0.107	0.102	0.093	0.086	0.075	0.070	0.068	0.062	44.029
12.0000	0.132	0.143	0.118	0.108	0.100	0.096	0.087	0.082	0.071	0.066	0.064	0.059	48.031

Table 42.3 Mass stopping power of different substances (in MeV·mg^{-1}·cm^2) for helium ions ${}^4_2\mathrm{He}^+$ [1] *(continued)*

E_m, MeV/u	H	He	N	O	Ne	Ar	Kr	Xe	Rn	Mylar	$(CH_2)_n$	Water	E, MeV
0.0125	1.994	0.851	0.564	0.530	0.471	0.311	0.190	0.135	0.089	0.713	0.907	0.692	0.0500
0.0160	2.214	0.934	0.627	0.590	0.525	0.354	0.220	0.156	0.104	0.802	1.020	0.771	0.0640
0.0200	2.408	0.989	0.684	0.644	0.576	0.400	0.250	0.182	0.121	0.889	1.131	0.840	0.0801
0.0250	2.617	1.038	0.746	0.703	0.636	0.456	0.292	0.213	0.154	0.985	1.254	0.916	0.1001
0.0320	2.904	1.102	0.817	0.777	0.710	0.532	0.346	0.261	0.181	1.107	1.412	1.013	0.1281
0.0400	3.167	1.152	0.882	0.849	0.782	0.611	0.405	0.310	0.220	1.224	1.562	1.106	0.1601
0.0500	3.442	1.211	0.956	0.930	0.865	0.698	0.470	0.366	0.265	1.346	1.716	1.208	0.2001
0.0600	3.723	1.274	1.033	0.999	0.944	0.777	0.528	0.416	0.305	1.450	1.851	1.302	0.2402
0.0700	3.977	1.347	1.103	1.071	1.016	0.847	0.577	0.459	0.339	1.542	1.966	1.394	0.2802
0.0800	4.229	1.420	1.171	1.138	1.084	0.910	0.624	0.497	0.369	1.623	2.068	1.482	0.3202
0.0900	4.451	1.493	1.235	1.204	1.146	0.967	0.665	0.532	0.394	1.695	2.155	1.565	0.3602
0.1000	4.685	1.572	1.303	1.262	1.201	1.018	0.703	0.559	0.415	1.757	2.235	1.642	0.4003
0.1250	5.227	1.745	1.442	1.389	1.327	1.121	0.772	0.615	0.457	1.888	2.403	1.816	0.5003
0.1600	5.863	1.955	1.604	1.546	1.466	1.222	0.838	0.667	0.495	2.021	2.568	2.025	0.6404
0.2000	6.376	2.117	1.734	1.665	1.574	1.279	0.870	0.691	0.510	2.096	2.660	2.188	0.8005
0.2500	6.764	2.239	1.816	1.748	1.632	1.290	0.872	0.684	0.503	2.114	2.679	2.306	1.0007
0.3200	6.947	2.268	1.844	1.760	1.610	1.234	0.823	0.642	0.469	2.061	2.617	2.337	1.2808
0.4000	6.775	2.237	1.776	1.688	1.529	1.141	0.755	0.587	0.428	1.947	2.479	2.254	1.6010
0.5000	6.355	2.112	1.640	1.561	1.396	1.027	0.677	0.529	0.385	1.786	2.280	2.094	2.0013
0.6000	5.742	1.926	1.459	1.385	1.238	0.905	0.600	0.468	0.343	1.593	2.044	1.869	2.4016
0.7000	5.110	1.756	1.292	1.227	1.098	0.802	0.537	0.418	0.309	1.427	1.834	1.659	2.8018
0.8000	4.540	1.593	1.152	1.098	0.982	0.718	0.486	0.379	0.282	1.287	1.652	1.480	3.2021
0.9000	4.056	1.446	1.036	0.987	0.878	0.647	0.442	0.349	0.259	1.170	1.500	1.328	3.6023
1.0000	3.676	1.329	0.949	0.900	0.802	0.594	0.409	0.326	0.242	1.073	1.374	1.209	4.0026
1.2500	2.971	1.113	0.792	0.756	0.672	0.504	0.354	0.286	0.216	0.895	1.137	1.003	5.0033
1.6000	2.316	0.907	0.655	0.626	0.561	0.428	0.307	0.250	0.193	0.730	0.914	0.814	6.4042
2.0000	1.830	0.750	0.556	0.530	0.479	0.372	0.272	0.223	0.175	0.610	0.751	0.674	8.0052
2.5000	1.432	0.616	0.470	0.447	0.413	0.327	0.242	0.202	0.161	0.510	0.617	0.557	10.007
3.2000	1.121	0.494	0.392	0.374	0.347	0.281	0.214	0.178	0.143	0.420	0.501	0.456	12.808
4.0000	0.891	0.403	0.329	0.316	0.297	0.243	0.186	0.157	0.127	0.352	0.415	0.380	16.010
5.0000	0.728	0.331	0.277	0.266	0.251	0.208	0.162	0.137	0.112	0.296	0.347	0.318	20.013
6.0000	0.621	0.282	0.240	0.231	0.219	0.183	0.143	0.121	0.100	0.256	0.299	0.274	24.016
7.0000	0.546	0.248	0.213	0.206	0.194	0.163	0.128	0.110	0.091	0.228	0.265	0.243	28.018
8.0000	0.489	0.221	0.192	0.186	0.175	0.148	0.116	0.100	0.083	0.205	0.238	0.219	32.021
9.0000	0.443	0.201	0.175	0.169	0.160	0.135	0.107	0.092	0.077	0.187	0.217	0.200	36.023
10.0000	0.407	0.186	0.162	0.156	0.148	0.125	0.099	0.085	0.071	0.172	0.199	0.184	40.026
11.0000	0.376	0.172	0.151	0.146	0.138	0.116	0.092	0.079	0.066	0.160	0.185	0.171	44.029
12.0000	0.350	0.161	0.141	0.136	0.129	0.109	0.087	0.075	0.062	0.149	0.172	0.160	48.031

Table 42.4 Range (in mg/cm^2) of helium ions ${}^4_2He^+$ for different substances [1] [E_m is the energy per mass unit of incident particle; E is a total energy.]

E_m, MeV/u	Be	C	Al	Ti	Ni	Ge	Zr	Ag	Eu	Ta	Au	U	E, MeV
0.0125	0.085	0.099	0.142	0.219	0.267	0.307	0.337	0.371	0.583	0.685	0.746	0.874	0.0500
0.0160	0.100	0.117	0.166	0.255	0.310	0.356	0.390	0.428	0.673	0.790	0.860	1.007	0.0640
0.0200	0.114	0.135	0.190	0.292	0.356	0.406	0.444	0.478	0.764	0.897	0.976	1.143	0.0202
0.0250	0.131	0.155	0.218	0.333	0.406	0.462	0.505	0.552	0.867	1.017	1.107	1.296	0.1001
0.0320	0.152	0.180	0.252	0.385	0.470	0.532	0.580	0.634	0.996	1.167	1.269	1.486	0.1281
0.0400	0.174	0.206	0.288	0.438	0.534	0.603	0.658	0.718	1.127	1.320	1.435	1.679	0.1601
0.0500	0.198	0.235	0.327	0.498	0.607	0.684	0.745	0.812	1.275	1.490	1.620	1.895	0.2001
0.0600	0.221	0.262	0.365	0.553	0.674	0.758	0.825	0.898	1.410	1.645	1.788	2.090	0.2402
0.0700	0.243	0.287	0.400	0.604	0.737	0.827	0.900	0.979	1.535	1.789	1.944	2.271	0.2802
0.0800	0.263	0.311	0.433	0.654	0.796	0.892	0.971	1.055	1.654	1.924	2.090	2.441	0.3202
0.0900	0.283	0.334	0.465	0.701	0.853	0.955	1.039	1.128	1.766	2.053	2.229	2.602	0.3602
0.1000	0.303	0.356	0.497	0.747	0.908	1.016	1.105	1.199	1.874	2.176	2.362	2.756	0.4003
0.1250	0.351	0.409	0.574	0.857	1.041	1.162	1.263	1.369	2.129	2.467	2.675	3.118	0.5003
0.1600	0.418	0.480	0.680	1.006	1.218	1.358	1.477	1.597	2.466	2.852	3.086	3.591	0.6404
0.2000	0.496	0.559	0.802	1.175	1.419	1.579	1.718	1.854	2.840	3.277	3.539	4.110	0.8005
0.2500	0.599	0.658	0.959	1.389	1.672	1.858	2.022	2.178	3.304	3.802	4.099	4.751	1.0007
0.3200	0.752	0.802	1.191	1.700	2.036	2.259	2.461	2.647	3.964	4.547	4.892	5.656	1.2808
0.4000	0.945	0.977	1.475	2.075	2.474	2.738	2.986	3.209	4.743	5.423	5.825	6.718	1.6010
0.5000	1.211	1.216	1.860	2.577	3.055	3.372	3.682	3.955	5.769	6.564	7.039	8.099	2.0013
0.6000	1.510	1.481	2.283	3.120	3.680	4.052	4.430	4.758	6.838	7.773	8.323	9.551	2.4016
0.7000	1.842	1.777	2.745	3.708	4.353	4.782	5.236	5.622	7.989	9.060	9.687	11.088	2.8018
0.8000	2.207	2.103	3.244	4.337	5.072	5.560	6.095	6.544	9.207	10.418	11.126	12.709	3.2021
0.9000	2.604	2.460	3.779	5.006	5.833	6.381	7.004	7.521	10.485	11.842	12.633	14.402	3.6023
1.0000	3.031	2.848	4.349	5.714	5.635	7.245	7.961	8.549	11.821	13.328	14.205	16.164	4.0026
1.2500	4.228	3.955	5.920	7.642	8.811	9.584	10.553	11.339	15.403	17.304	18.411	20.866	5.0033
1.6000	6.211	5.826	8.460	10.711	12.251	13.272	14.645	15.759	20.971	23.466	24.929	28.141	6.4042
2.0000	8.906	8.403	11.835	14.723	16.729	18.054	19.967	21.512	28.111	31.320	33.237	37.383	8.0052
2.5000	12.910	12.232	16.742	20.475	23.109	24.850	27.556	29.696	38.143	42.332	44.862	50.241	10.007
3.2000	19.669	18.656	24.857	29.860	33.450	35.850	39.838	42.951	54.141	59.869	63.338	70.594	12.808
4.0000	28.981	27.434	35.838	42.415	47.199	50.459	56.090	60.545	75.070	82.738	87.383	97.044	16.010
5.0000	42.903	40.463	52.011	60.740	67.151	71.624	79.576	85.969	104.928	115.401	121.564	134.555	20.013
6.0000	59.245	55.686	70.787	81.859	90.060	95.848	106.434	115.008	138.814	152.317	160.110	176.642	24.016
7.0000	77.926	73.019	92.063	105.656	115.831	123.000	136.481	147.486	176.530	193.264	202.825	223.087	28.018
8.0000	98.882	92.397	115.752	132.035	144.354	153.004	169.563	183.268	217.797	238.083	249.497	273.752	32.021
9.0000	122.021	113.778	141.783	160.909	175.528	185.766	205.590	222.233	262.558	286.598	299.989	328.433	36.023
10.0000	147.287	137.127	170.092	192.190	209.289	221.197	244.530	264.234	310.742	338.683	354.114	386.980	40.026
11.0000	174.648	162.394	200.627	225.799	245.552	259.221	286.299	309.135	362.231	394.145	411.613	449.291	44.029
12.0000	204.051	189.540	233.337	261.704	284.239	299.753	330.801	356.851	416.882	452.868	472.467	515.303	48.031

Table 42.4 Range (in mg/cm^2) of helium ions ${}^4_2He^+$ for different substances [1] *(continued)*

E_m, MeV/u	H	He	N	O	Ne	Ar	Kr	Xe	Rn	Mylar	$(CH_2)_n$	Water	E, MeV
0.0125	0.032	0.081	0.123	0.132	0.149	0.237	0.409	0.596	0.917	0.097	0.075	0.095	0.0500
0.0160	0.038	0.097	0.146	0.156	0.176	0.277	0.474	0.688	1.056	0.115	0.089	0.113	0.0640
0.0200	0.045	0.113	0.170	0.181	0.204	0.318	0.540	0.780	1.194	0.133	0.103	0.133	0.0801
0.0250	0.053	0.132	0.197	0.210	0.236	0.364	0.613	0.879	1.337	0.154	0.120	0.155	0.1001
0.0320	0.062	0.158	0.232	0.247	0.277	0.420	0.699	0.996	1.501	0.180	0.140	0.183	0.1281
0.0400	0.073	0.186	0.269	0.286	0.319	0.475	0.783	1.107	1.659	0.207	0.162	0.213	0.1601
0.0500	0.085	0.219	0.312	0.330	0.367	0.536	0.874	1.225	1.822	0.238	0.186	0.247	0.2001
0.0600	0.096	0.251	0.352	0.371	0.411	0.590	0.954	1.327	1.963	0.267	0.208	0.279	0.2402
0.0700	0.106	0.282	0.390	0.410	0.452	0.639	1.027	1.419	2.087	0.294	0.229	0.309	0.2802
0.0800	0.116	0.311	0.425	0.446	0.490	0.685	1.093	1.503	2.200	0.319	0.249	0.336	0.3202
0.0900	0.125	0.338	0.458	0.480	0.526	0.727	1.156	1.580	2.305	0.343	0.268	0.363	0.3602
0.1000	0.134	0.364	0.490	0.513	0.560	0.768	1.214	1.654	2.404	0.366	0.286	0.388	0.4003
0.1250	0.154	0.425	0.563	0.588	0.639	0.861	1.350	1.824	2.633	0.421	0.329	0.445	0.5003
0.1600	0.179	0.500	0.654	0.684	0.739	0.981	1.523	2.042	2.927	0.493	0.386	0.518	0.6404
0.2000	0.206	0.579	0.750	0.783	0.844	1.109	1.711	2.278	3.245	0.570	0.447	0.594	0.8005
0.2500	0.236	0.671	0.863	0.900	0.969	1.264	1.940	2.569	3.640	0.665	0.522	0.683	1.0007
0.3200	0.277	0.795	1.016	1.060	1.142	1.486	2.271	2.991	4.217	0.799	0.627	0.804	1.2808
0.4000	0.323	0.937	1.193	1.246	1.346	1.756	2.677	3.513	4.932	0.959	0.753	0.943	1.6010
0.5000	0.385	1.121	1.428	1.493	1.620	2.126	3.238	4.232	5.919	1.174	0.922	1.128	2.0013
0.6000	0.451	1.320	1.687	1.765	1.925	2.542	3.867	5.038	7.022	1.412	1.107	1.330	2.4016
0.7000	0.525	1.538	1.979	2.073	2.269	3.013	4.573	5.944	8.255	1.678	1.314	1.558	2.8018
0.8000	0.608	1.778	2.308	2.418	2.655	3.541	5.357	6.951	9.614	1.973	1.545	1.814	3.2021
0.9000	0.701	2.042	2.674	2.803	3.087	4.128	6.222	8.052	11.096	2.300	1.799	2.099	3.6023
1.0000	0.805	2.331	3.079	3.227	3.564	4.774	7.164	9.239	12.695	2.658	2.078	2.416	4.0026
1.2500	1.109	3.156	4.236	4.444	4.931	6.608	9.801	12.526	17.083	3.682	2.881	3.328	5.0033
1.6000	1.645	4.556	6.189	6.487	7.222	9.636	14.065	17.789	23.963	5.422	4.262	4.885	6.4042
2.0000	2.426	6.504	8.851	9.277	10.321	13.660	19.623	24.590	32.679	7.829	6.201	7.054	8.0052
2.5000	3.668	9.461	12.779	13.403	14.836	19.419	27.441	34.042	44.628	11.431	9.151	10.333	10.007
3.2000	5.900	14.568	19.340	20.286	22.270	28.705	39.795	48.882	63.208	17.513	14.216	15.922	12.808
4.0000	9.134	21.781	28.294	29.645	32.283	41.006	55.903	68.123	87.095	25.868	21.267	23.652	16.010
5.0000	14.129	32.801	41.613	43.505	47.007	58.885	79.051	95.574	120.776	38.311	31.861	35.228	20.013
6.0000	20.100	45.939	57.189	59.692	64.139	79.483	105.475	126.758	158.699	52.879	44.322	48.830	24.016
7.0000	26.987	61.095	74.919	78.105	83.596	102.708	135.103	161.546	200.864	69.481	58.564	64.355	28.018
8.0000	34.742	78.207	94.718	98.633	105.319	128.499	167.891	199.876	247.128	88.040	74.522	81.703	32.021
9.0000	43.437	97.188	116.539	121.239	129.234	156.809	203.844	241.758	297.426	108.521	92.165	100.844	36.023
10.0000	52.785	117.905	140.308	145.857	155.254	187.581	242.838	287.126	351.710	130.883	111.457	121.730	40.026
11.0000	63.031	140.299	165.946	172.408	183.319	220.770	284.780	335.901	409.980	155.069	132.350	144.298	44.029
12.0000	74.083	164.342	193.410	200.852	213.383	256.324	329.587	387.987	472.165	181.040	154.816	168.510	48.031

Table 42.5 Mass stopping power of different substances (in MeV·mg^{-1}·cm^2) for lithium ions ${}^7_3Li^+$ [1] [E_m is the energy per mass unit of incident particle; E is a total energy.]

E_m, MeV/u	Be	C	Al	Ti	Ni	Ge	Zr	Ag	Eu	Ta	Au	U	E, MeV
0.0125	1.355	1.121	0.821	0.546	0.444	0.402	0.373	0.345	0.219	0.188	0.172	0.148	0.0877
0.0160	1.533	1.268	0.929	0.618	0.502	0.455	0.422	0.390	0.248	0.212	0.195	0.167	0.1123
0.0200	1.714	1.418	1.039	0.691	0.561	0.509	0.472	0.436	0.277	0.237	0.218	0.187	0.1403
0.0250	1.917	1.586	1.162	0.772	0.627	0.569	0.527	0.488	0.310	0.265	0.245	0.209	0.1754
0.0320	2.167	1.796	1.315	0.874	0.711	0.647	0.597	0.552	0.351	0.301	0.277	0.238	0.2245
0.0400	2.407	2.004	1.464	0.976	0.796	0.726	0.669	0.619	0.394	0.339	0.312	0.268	0.2806
0.0500	2.654	2.224	1.620	1.085	0.889	0.810	0.747	0.692	0.439	0.382	0.352	0.302	0.3508
0.0600	2.854	2.415	1.749	1.179	0.969	0.880	0.813	0.754	0.483	0.420	0.387	0.333	0.4210
0.0700	3.015	2.579	1.856	1.258	1.039	0.943	0.872	0.809	0.520	0.455	0.421	0.362	0.4911
0.0800	3.144	2.724	1.943	1.327	1.098	0.995	0.919	0.855	0.554	0.486	0.449	0.387	0.5613
0.0900	3.244	2.847	2.014	1.385	1.148	1.043	0.960	0.894	0.586	0.512	0.475	0.410	0.6314
0.1000	3.322	2.954	2.070	1.435	1.192	1.081	0.994	0.929	0.613	0.536	0.497	0.428	0.7016
0.1250	3.437	3.171	2.162	1.522	1.271	1.152	1.059	0.992	0.665	0.581	0.544	0.470	0.8770
0.1600	3.467	3.370	2.210	1.582	1.330	1.206	1.107	1.038	0.707	0.621	0.582	0.508	1.1226
0.2000	3.412	3.463	2.204	1.602	1.355	1.228	1.126	1.056	0.732	0.646	0.606	0.529	1.4032
0.2500	3.290	3.460	2.160	1.596	1.356	1.231	1.128	1.056	0.745	0.659	0.618	0.540	1.7540
0.3200	3.103	3.371	2.080	1.566	1.337	1.221	1.113	1.040	0.745	0.661	0.621	0.546	2.2451
0.4000	2.908	3.229	1.989	1.516	1.305	1.194	1.088	1.017	0.740	0.658	0.619	0.543	2.8064
0.5000	2.703	3.037	1.890	1.461	1.266	1.162	1.058	0.987	0.730	0.650	0.611	0.539	3.5080
0.6000	2.497	2.812	1.781	1.395	1.215	1.118	1.015	0.946	0.707	0.632	0.596	0.529	4.2096
0.7000	2.321	2.605	1.683	1.329	1.163	1.072	0.971	0.905	0.683	0.611	0.577	0.512	4.9112
0.8000	2.164	2.413	1.594	1.270	1.114	1.031	0.932	0.869	0.660	0.593	0.559	0.497	5.6128
0.9000	2.031	2.244	1.513	1.214	1.070	0.993	0.896	0.834	0.640	0.575	0.543	0.484	6.3144
1.0000	1.911	2.092	1.440	1.165	1.028	0.955	0.863	0.802	0.619	0.557	0.527	0.471	7.0160
1.2500	1.665	1.784	1.284	1.053	0.937	0.873	0.787	0.731	0.574	0.517	0.489	0.438	8.7700
1.6000	1.412	1.480	1.116	0.932	0.832	0.778	0.701	0.647	0.519	0.470	0.444	0.398	11.226
2.0000	1.204	1.257	0.972	0.823	0.740	0.694	0.622	0.576	0.467	0.426	0.402	0.363	14.032
2.5000	1.019	1.067	0.840	0.721	0.653	0.613	0.548	0.509	0.418	0.380	0.361	0.328	17.540
3.2000	0.843	0.890	0.709	0.617	0.561	0.528	0.474	0.438	0.366	0.335	0.318	0.289	22.451
4.0000	0.708	0.754	0.605	0.532	0.487	0.458	0.413	0.381	0.322	0.295	0.281	0.256	28.064
5.0000	0.593	0.636	0.514	0.455	0.419	0.396	0.357	0.330	0.283	0.258	0.247	0.226	35.080
6.0000	0.513	0.552	0.449	0.400	0.369	0.350	0.316	0.292	0.250	0.231	0.221	0.203	42.096
7.0000	0.453	0.489	0.400	0.358	0.331	0.314	0.284	0.263	0.228	0.209	0.201	0.185	49.112
8.0000	0.407	0.440	0.361	0.325	0.301	0.286	0.260	0.240	0.209	0.192	0.185	0.170	56.128
9.0000	0.370	0.401	0.330	0.298	0.276	0.263	0.239	0.221	0.193	0.178	0.171	0.158	63.144
10.0000	0.340	0.368	0.304	0.276	0.256	0.244	0.222	0.206	0.179	0.166	0.160	0.148	70.160
11.0000	0.315	0.341	0.283	0.257	0.239	0.227	0.207	0.193	0.168	0.157	0.151	0.139	77.176
12.0000	0.293	0.318	0.264	0.241	0.224	0.214	0.195	0.182	0.159	0.148	0.143	0.132	84.192

Table 42.5 Mass stopping power of different substances (in MeV·mg^{-1}·cm^2) for lithium ions $^7_3Li^+$ [1] *(continued)*

E_m, MeV/u	H	He	N	O	Ne	Ar	Kr	Xe	Rn	Mylar	$(CH_2)_n$	Water	E, MeV
0.0125	3.080	1.314	0.871	0.818	0.727	0.480	0.294	0.209	0.138	1.101	1.401	1.069	0.0877
0.0160	3.420	1.442	0.969	0.912	0.810	0.547	0.339	0.242	0.160	1.239	1.576	1.190	0.1123
0.0200	3.719	1.527	1.057	0.995	0.890	0.618	0.385	0.281	0.188	1.373	1.746	1.298	0.1403
0.0250	4.042	1.604	1.152	1.086	0.983	0.705	0.451	0.329	0.238	1.522	1.936	1.415	0.1754
0.0320	4.483	1.701	1.261	1.200	1.097	0.822	0.534	0.402	0.280	1.709	2.180	1.565	0.2245
0.0400	4.904	1.783	1.366	1.315	1.211	0.946	0.627	0.480	0.341	1.896	2.418	1.713	0.2806
0.0500	5.362	1.886	1.489	1.448	1.348	1.087	0.732	0.570	0.413	2.096	2.673	1.882	0.3508
0.0600	5.842	1.999	1.621	1.567	1.481	1.219	0.829	0.652	0.479	2.275	2.905	2.043	0.4210
0.0700	6.290	2.130	1.744	1.694	1.607	1.340	0.913	0.726	0.536	2.438	3.110	2.204	0.4911
0.0800	6.742	2.263	1.867	1.815	1.727	1.451	0.995	0.793	0.589	2.588	3.297	2.363	0.5613
0.0900	7.148	2.398	1.983	1.933	1.840	1.552	1.067	0.854	0.632	2.722	3.461	2.513	0.6314
0.1000	7.576	2.542	2.107	2.041	1.942	1.646	1.136	0.905	0.671	2.842	3.614	2.656	0.7016
0.1250	8.581	2.864	2.367	2.280	2.179	1.839	1.267	1.009	0.750	3.100	3.945	2.981	0.8770
0.1600	9.788	3.263	2.678	2.581	2.448	2.039	1.399	1.114	0.826	3.374	4.286	3.381	1.1226
0.2000	10.822	3.593	2.942	2.826	2.671	2.171	1.477	1.173	0.866	3.557	4.514	3.714	1.4032
0.2500	11.707	3.875	3.143	3.026	2.825	2.233	1.510	1.184	0.870	3.659	4.638	3.992	1.7540
0.3200	12.353	4.032	3.280	3.130	2.864	2.194	1.464	1.142	0.834	3.664	4.654	4.155	2.2451
0.4000	12.413	4.098	3.255	3.093	2.801	2.091	1.383	1.076	0.784	3.567	4.542	4.130	2.8064
0.5000	12.058	4.007	3.113	2.962	2.650	1.949	1.285	1.004	0.731	3.389	4.326	3.973	3.5080
0.6000	11.310	3.794	2.871	2.729	2.438	1.783	1.183	0.983	0.675	3.138	4.027	3.681	4.2096
0.7000	10.332	3.551	2.612	2.482	2.221	1.622	1.085	0.845	0.624	2.886	3.709	3.354	4.9112
0.8000	9.356	3.283	2.373	2.263	2.023	1.479	1.001	0.781	0.580	2.652	3.404	3.051	5.6128
0.9000	8.474	3.020	2.164	2.063	1.834	1.353	0.923	0.729	0.542	2.444	3.134	2.775	6.3144
1.0000	7.762	2.807	2.003	1.901	1.693	1.254	0.864	0.688	0.511	2.265	2.902	2.552	7.0160
1.2500	6.382	2.390	1.701	1.624	1.443	1.082	0.760	0.614	0.464	1.922	2.441	2.153	8.7700
1.6000	5.043	1.975	1.426	1.363	1.221	0.932	0.668	0.543	0.421	1.590	1.989	1.772	11.226
2.0000	4.024	1.649	1.223	1.164	1.054	0.818	0.598	0.491	0.386	1.341	1.652	1.482	14.032
2.5000	3.175	1.365	1.042	0.992	0.915	0.724	0.537	0.448	0.356	1.130	1.368	1.235	17.54
3.2000	2.481	1.102	0.874	0.834	0.775	0.627	0.477	0.397	0.318	0.938	1.118	1.017	22.451
4.0000	1.996	0.902	0.738	0.708	0.665	0.544	0.417	0.351	0.284	0.790	0.931	0.851	28.064
5.0000	1.634	0.743	0.621	0.508	0.564	0.467	0.363	0.307	0.252	0.665	0.779	0.713	35.080
6.0000	1.396	0.635	0.539	0.519	0.491	0.410	0.320	0.272	0.224	0.576	0.673	0.616	42.096
7.0000	1.227	0.557	0.479	0.461	0.436	0.367	0.288	0.246	0.203	0.511	0.595	0.547	49.112
8.0000	1.097	0.496	0.431	0.416	0.393	0.332	0.261	0.224	0.186	0.460	0.534	0.492	56.128
9.0000	0.993	0.451	0.393	0.379	0.359	0.303	0.239	0.206	0.172	0.418	0.485	0.448	63.144
10.0000	0.910	0.415	0.362	0.350	0.331	0.280	0.221	0.190	0.159	0.385	0.445	0.412	70.160
11.0000	0.839	0.385	0.337	0.325	0.307	0.260	0.206	0.177	0.148	0.356	0.412	0.382	77.176
12.0000	0.779	0.359	0.315	0.304	0.287	0.243	0.193	0.166	0.139	0.332	0.384	0.357	84.192

Table 42.6 Range (in mg/cm^2) of lithium ions ${}^7_3Li^+$ for different substances [1] [E_m is the energy per mass unit of incident particle; E is a total energy.]

E_m, MeV/u	Be	C	Al	Ti	Ni	Ge	Zr	Ag	Eu	Ta	Au	U	E, MeV
0.0125	0.091	0.105	0.149	0.228	0.276	0.320	0.353	0.390	0.614	0.724	0.790	0.937	0.0877
0.0160	0.108	0.125	0.176	0.268	0.324	0.375	0.411	0.454	0.713	0.840	0.917	1.084	0.1123
0.0200	0.125	0.145	0.203	0.310	0.374	0.430	0.472	0.519	0.815	0.960	1.047	1.236	0.1403
0.0250	0.144	0.168	0.234	0.356	0.431	0.493	0.540	0.593	0.930	1.094	1.193	1.407	0.1754
0.0320	0.167	0.196	0.273	0.414	0.502	0.572	0.625	0.685	1.075	1.263	1.376	1.621	0.2245
0.0400	0.192	0.225	0.313	0.474	0.575	0.653	0.712	0.779	1.222	1.434	1.562	1.838	0.2806
0.0500	0.219	0.258	0.358	0.541	0.657	0.743	0.809	0.884	1.388	1.626	1.770	2.080	0.3508
0.0600	0.245	0.288	0.399	0.602	0.731	0.825	0.898	0.980	1.538	1.799	1.957	2.298	0.4210
0.0700	0.269	0.316	0.438	0.660	0.801	0.902	0.982	1.070	1.677	1.957	2.129	2.498	0.4911
0.0800	0.292	0.343	0.475	0.714	0.866	0.974	1.060	1.154	1.807	2.107	2.290	2.685	0.5613
0.0900	0.313	0.368	0.510	0.766	0.929	1.043	1.134	1.235	1.930	2.247	2.442	2.861	0.6314
0.1000	0.335	0.392	0.545	0.815	0.989	1.109	1.206	1.312	2.047	2.381	2.586	3.029	0.7016
0.1250	0.387	0.449	0.627	0.934	1.131	1.266	1.377	1.494	2.322	2.694	2.923	3.418	0.8770
0.1600	0.458	0.524	0.739	1.092	1.319	1.473	1.603	1.735	2.679	3.102	3.358	3.919	1.1226
0.2000	0.539	0.606	0.866	1.268	1.528	1.704	1.854	2.003	3.068	3.544	3.830	4.460	1.4032
0.2500	0.644	0.708	1.027	1.487	1.786	1.989	2.165	2.334	3.543	4.082	4.402	5.116	1.7540
0.3200	0.797	0.851	1.259	1.797	2.151	2.389	2.603	2.803	4.202	4.825	5.195	6.020	2.2451
0.4000	0.984	1.021	1.535	2.162	2.576	2.854	3.113	3.349	4.958	5.676	6.101	7.050	2.8064
0.5000	1.235	1.246	1.897	2.633	3.122	3.450	3.767	4.050	5.913	6.746	7.242	8.348	3.5080
0.6000	1.505	1.486	2.279	3.125	3.688	4.066	4.445	4.776	6.890	7.843	8.405	9.663	4.2096
0.7000	1.796	1.745	2.685	3.641	4.278	4.707	5.151	5.535	7.900	8.972	9.601	11.012	4.9112
0.8000	2.110	2.025	3.113	4.181	4.895	5.374	5.889	6.326	8.945	10.138	10.836	12.403	5.6128
0.9000	2.444	2.327	3.565	4.746	5.538	6.068	6.657	7.151	10.025	11.340	12.109	13.833	6.3144
1.0000	2.801	2.651	4.041	5.336	6.207	6.789	7.455	8.009	11.140	12.580	13.421	15.302	7.0160
1.2500	3.786	3.560	5.333	6.922	7.996	8.712	9.587	11.303	14.085	15.850	16.879	19.170	8.7700
1.6000	5.391	5.076	7.390	9.407	10.782	11.699	12.901	13.883	18.595	20.841	22.158	25.061	11.226
2.0000	7.549	7.139	10.092	12.620	14.368	15.528	17.162	18.489	24.312	27.129	28.811	32.461	14.032
2.5000	10.727	10.178	13.986	17.184	19.431	20.921	23.184	24.983	32.273	35.868	38.036	42.665	17.540
3.2000	16.050	15.237	20.377	24.576	27.575	29.584	32.858	35.424	44.873	49.681	52.588	58.696	22.451
4.0000	23.345	22.115	28.890	34.412	38.347	41.030	45.590	49.207	61.270	67.597	71.426	79.418	28.064
5.0000	34.219	32.291	41.612	48.727	53.931	57.561	63.935	69.066	84.591	93.110	98.124	108.717	35.080
6.0000	46.969	44.168	56.261	65.202	71.805	76.461	84.889	91.721	111.029	121.911	128.197	141.553	42.096
7.0000	61.546	57.693	72.863	83.772	91.915	97.648	108.335	117.065	140.460	153.864	161.529	177.795	49.112
8.0000	77.913	72.827	91.364	104.373	114.191	121.081	134.173	145.010	172.689	188.866	197.979	217.365	56.128
9.0000	96.006	89.546	111.719	126.952	138.567	146.699	162.344	175.480	207.690	226.803	237.462	260.122	63.144
10.0000	115.792	107.830	133.888	151.447	165.005	174.444	192.836	208.369	245.422	267.589	279.846	305.969	70.160
11.0000	137.251	127.647	157.836	177.806	193.446	204.266	225.596	243.585	285.804	311.088	324.942	354.839	77.176
12.0000	160.349	148.972	183.531	206.012	223.836	236.107	260.555	281.068	328.736	357.218	372.746	406.695	84.192

Table 42.6 Range (in mg/cm^2) of lithium ions ${}^7_3Li^+$ for different substances [1] *(continued)*

E_m, MeV/u	H	He	N	O	Ne	Ar	Kr	Xe	Rn	Mylar	$(CH_2)_n$	Water	E, MeV
0.0125	0.034	0.087	0.130	0.138	0.156	0.245	0.421	0.618	0.961	0.103	0.080	0.101	0.0877
0.0160	0.041	0.104	0.155	0.165	0.186	0.290	0.496	0.720	1.116	0.123	0.096	0.121	0.1123
0.0200	0.049	0.123	0.182	0.193	0.217	0.336	0.568	0.823	1.269	0.144	0.112	0.143	0.1403
0.0250	0.057	0.145	0.213	0.226	0.253	0.387	0.649	0.934	1.429	0.167	0.130	0.168	0.1754
0.0320	0.068	0.174	0.252	0.268	0.299	0.450	0.746	1.065	1.613	0.197	0.154	0.200	0.2245
0.0400	0.080	0.205	0.294	0.311	0.347	0.512	0.841	1.190	1.791	0.227	0.178	0.234	0.2806
0.0500	0.094	0.243	0.343	0.361	0.401	0.580	0.943	1.322	1.975	0.262	0.205	0.272	0.3508
0.0600	0.106	0.279	0.387	0.407	0.450	0.641	1.031	1.436	2.131	0.294	0.230	0.307	0.4210
0.0700	0.117	0.312	0.428	0.450	0.495	0.695	1.112	1.538	2.269	0.324	0.253	0.340	0.4911
0.0800	0.128	0.344	0.467	0.490	0.537	0.746	1.186	1.630	2.394	0.352	0.275	0.371	0.5613
0.0900	0.138	0.375	0.504	0.527	0.576	0.792	1.254	1.715	2.509	0.378	0.296	0.400	0.6314
0.1000	0.148	0.403	0.538	0.563	0.613	0.836	1.317	1.795	2.616	0.403	0.316	0.427	0.7016
0.1250	0.170	0.468	0.616	0.644	0.698	0.937	1.463	1.978	2.863	0.462	0.362	0.489	0.8770
0.1600	0.196	0.548	0.714	0.745	0.804	1.063	1.647	2.209	3.174	0.538	0.422	0.566	1.1226
0.2000	0.223	0.630	0.813	0.849	0.914	1.196	1.842	2.454	3.505	0.619	0.485	0.645	1.4032
0.2500	0.255	0.724	0.929	0.968	1.041	1.356	2.077	2.752	3.909	0.716	0.562	0.736	1.7540
0.3200	0.295	0.848	1.081	1.128	1.214	1.577	2.407	3.174	4.485	0.850	0.668	0.857	2.2451
0.4000	0.341	0.986	1.253	1.308	1.412	1.839	2.802	3.681	5.179	1.005	0.790	0.992	2.8064
0.5000	0.398	1.159	1.474	1.540	1.670	2.187	3.328	4.356	6.107	1.207	0.948	1.165	3.5080
0.6000	0.458	1.339	1.709	1.787	1.946	2.564	3.898	5.086	7.106	1.422	1.116	1.349	4.2096
0.7000	0.523	1.530	1.965	2.057	2.248	2.977	4.517	5.881	8.187	1.656	1.298	1.549	4.9112
0.8000	0.595	1.736	2.247	2.353	2.579	3.430	5.191	6.745	9.353	1.909	1.495	1.768	5.6128
0.9000	0.673	1.959	2.557	2.678	2.943	3.926	5.921	7.675	10.605	2.185	1.710	2.009	6.3144
1.0000	0.760	2.200	2.894	3.032	3.341	4.465	6.707	8.665	11.939	2.483	1.943	2.273	7.0160
1.2500	1.010	2.878	3.846	4.032	4.466	5.973	8.875	11.368	15.547	3.326	2.603	3.023	8.7700
1.6000	1.444	4.012	5.427	5.687	6.321	8.425	12.328	15.631	21.120	4.735	3.721	4.284	11.226
2.0000	2.069	5.572	7.558	7.921	8.802	11.647	16.779	21.076	28.099	6.662	5.274	6.021	14.032
2.5000	3.054	7.918	10.676	11.195	12.385	16.217	22.983	28.577	37.581	9.520	7.615	8.523	17.540
3.2000	4.813	11.941	15.843	16.617	18.240	23.531	32.713	40.266	52.215	14.311	11.604	13.025	22.451
4.0000	7.346	17.591	22.858	23.948	26.085	33.168	45.333	55.339	70.929	20.856	17.128	19.081	28.064
5.0000	11.247	26.198	33.261	34.774	37.585	47.133	63.413	76.781	97.236	30.575	25.403	28.122	35.080
6.0000	15.906	36.449	45.413	47.403	50.952	63.203	84.029	101.111	126.825	41.941	35.125	38.735	42.096
7.0000	21.280	48.275	59.249	61.771	66.134	81.327	107.149	128.256	159.727	54.896	46.238	50.849	49.112
8.0000	27.336	61.640	74.712	77.803	83.100	101.469	132.756	158.192	195.859	69.390	58.701	64.398	56.128
9.0000	34.066	76.482	91.774	95.480	101.800	123.606	160.869	190.942	235.189	85.406	72.497	79.365	63.144
10.0000	41.456	92.705	110.387	114.758	122.176	147.702	191.404	226.468	277.697	102.917	87.604	95.720	70.160
11.0000	49.492	110.269	130.495	135.582	144.187	173.733	224.299	264.722	323.398	121.886	103.990	113.420	77.176
12.0000	58.174	129.156	152.070	157.926	167.804	201.663	259.498	305.639	372.249	142.288	121.639	132.440	84.192

Table 42.7 Mass stopping power of different substances (in MeV·mg^{-1}·cm^2) for beryllium ions $^9_4Be^+$ [1] [E_m is the energy per mass unit of incident particle; E is a total energy.]

E_m, MeV/u	Be	C	Al	Ti	Ni	Ge	Zr	Ag	Eu	Ta	Au	U	E, MeV
0.0125	1.800	1.489	1.091	0.725	0.589	0.534	0.495	0.458	0.291	0.249	0.229	0.196	0.1127
0.0160	2.036	1.684	1.234	0.821	0.666	0.605	0.560	0.518	0.329	0.282	0.259	0.222	0.1442
0.0200	2.276	1.883	1.380	0.917	0.745	0.676	0.626	0.579	0.368	0.315	0.290	0.248	0.1802
0.0250	2.545	2.105	1.542	1.026	0.833	0.756	0.700	0.648	0.412	0.352	0.325	0.278	0.2253
0.0320	2.876	2.384	1.745	1.160	0.944	0.859	0.792	0.733	0.466	0.400	0.368	0.316	0.2884
0.0400	3.199	2.664	1.946	1.298	1.058	0.965	0.889	0.823	0.523	0.450	0.414	0.356	0.3605
0.0500	3.538	2.966	2.160	1.447	1.186	1.080	0.996	0.922	0.585	0.509	0.469	0.403	0.4506
0.0600	3.820	3.233	2.341	1.578	1.297	1.177	1.088	1.009	0.646	0.562	0.518	0.446	0.5407
0.0700	4.050	3.465	2.493	1.690	1.396	1.266	1.171	1.087	0.698	0.611	0.566	0.486	0.6309
0.0800	4.236	3.670	2.618	1.788	1.479	1.340	1.238	1.152	0.746	0.654	0.605	0.521	0.7210
0.0900	4.381	3.846	2.720	1.871	1.550	1.409	1.297	1.208	0.791	0.692	0.642	0.553	0.8111
0.1000	4.494	3.996	2.800	1.940	1.613	1.462	1.344	1.257	0.829	0.725	0.672	0.580	0.9012
0.1250	4.653	4.293	2.926	2.060	1.721	1.560	1.434	1.343	0.900	0.787	0.736	0.636	1.1265
0.1600	4.692	4.561	2.991	2.141	1.800	1.633	1.498	1.406	0.957	0.840	0.788	0.688	1.4420
0.2000	4.633	4.702	2.993	2.176	1.841	1.667	1.529	1.434	0.994	0.877	0.823	0.718	1.8024
0.2500	4.508	4.742	2.960	2.187	1.859	1.687	1.545	1.447	1.021	0.903	0.847	0.740	2.2530
0.3200	4.327	4.701	2.900	2.184	1.865	1.702	1.551	1.450	1.038	0.922	0.866	0.761	2.8839
0.4000	4.142	4.598	2.833	2.159	1.859	1.700	1.550	1.448	1.054	0.938	0.881	0.773	3.6049
0.5000	3.947	4.435	2.760	2.133	1.849	1.697	1.546	1.441	1.065	0.949	0.893	0.787	4.5061
0.6000	3.734	4.205	2.663	2.085	1.816	1.673	1.518	1.414	1.057	0.945	0.891	0.791	5.4073
0.7000	3.547	3.981	2.572	2.032	1.777	1.638	1.484	1.384	1.044	0.934	0.882	0.782	6.3085
0.8000	3.374	3.762	2.485	1.980	1.737	1.608	1.453	1.354	1.029	0.924	0.872	0.775	7.2098
0.9000	3.222	3.560	2.401	1.925	1.697	1.575	1.421	1.323	1.015	0.912	0.862	0.768	8.1110
1.0000	3.079	3.371	2.320	1.877	1.656	1.538	1.390	1.292	0.998	0.898	0.849	0.759	9.0122
1.2500	2.766	2.962	2.132	1.749	1.557	1.450	1.307	1.213	0.953	0.859	0.812	0.727	11.265
1.6000	2.410	2.526	1.904	1.590	1.420	1.327	1.195	1.104	0.885	0.801	0.758	0.680	14.420
2.0000	2.091	2.183	1.688	1.430	1.285	1.205	1.080	1.001	0.810	0.739	0.699	0.630	18.024
2.5000	1.789	1.875	1.475	1.267	1.146	1.077	0.963	0.894	0.735	0.668	0.634	0.575	22.530
3.2000	1.491	1.575	1.254	1.091	0.993	0.934	0.838	0.775	0.647	0.592	0.562	0.510	28.839
4.0000	1.256	1.337	1.073	0.943	0.864	0.814	0.733	0.676	0.572	0.524	0.499	0.454	36.049
5.0000	1.053	1.129	0.913	0.809	0.745	0.703	0.633	0.586	0.502	0.458	0.439	0.401	45.061
6.0000	0.912	0.980	0.797	0.711	0.656	0.622	0.561	0.519	0.445	0.410	0.393	0.361	54.073
7.0000	0.804	0.869	0.709	0.635	0.587	0.558	0.505	0.467	0.404	0.372	0.357	0.328	63.085
8.0000	0.721	0.781	0.640	0.576	0.533	0.507	0.461	0.426	0.370	0.341	0.328	0.302	72.098
9.0000	0.656	0.710	0.584	0.528	0.489	0.466	0.424	0.392	0.342	0.316	0.303	0.280	81.110
10.0000	0.602	0.651	0.538	0.488	0.452	0.431	0.392	0.364	0.318	0.295	0.284	0.262	90.122
11.0000	0.557	0.603	0.500	0.455	0.422	0.402	0.366	0.341	0.298	0.277	0.267	0.246	99.134
12.0000	0.519	0.562	0.467	0.426	0.395	0.378	0.344	0.321	0.281	0.261	0.252	0.232	108.15

Table 42.7 Mass stopping power of different substances (in MeV·mg^{-1}·cm^2) for beryllium ions $^9_4\mathrm{Be}^+$ [1] *(continued)*

E_m, MeV/u	H	He	N	O	Ne	Ar	Kr	Xe	Rn	Mylar	$(CH_2)_n$	Water	E, MeV
0.0125	4.090	1.745	1.156	1.086	0.965	0.637	0.390	0.277	0.183	1.463	1.861	1.420	0.1127
0.0160	4.541	1.915	1.287	1.210	1.076	0.727	0.450	0.321	0.213	1.645	2.093	1.581	0.1442
0.0200	4.939	2.028	1.403	1.322	1.182	0.821	0.512	0.372	0.249	1.824	2.319	1.723	0.1802
0.0250	5.367	2.130	1.530	1.442	1.305	0.936	0.598	0.436	0.316	2.020	2.571	1.879	0.2253
0.0320	5.950	2.258	1.673	1.593	1.455	1.091	0.708	0.534	0.372	2.268	2.893	2.077	0.2884
0.0400	6.518	2.370	1.815	1.747	1.609	1.257	0.833	0.638	0.453	2.520	3.214	2.276	0.3605
0.0500	7.150	2.514	1.985	1.931	1.797	1.449	0.976	0.760	0.551	2.795	3.564	2.510	0.4506
0.0600	7.818	2.676	2.170	2.097	1.983	1.635	1.110	0.873	0.641	3.045	3.888	2.734	0.5407
0.0700	8.450	2.861	2.343	2.276	2.159	1.800	1.226	0.975	0.720	3.275	4.177	2.961	0.6309
0.0800	9.084	3.050	2.516	2.445	2.327	1.956	1.340	1.068	0.793	3.487	4.443	3.183	0.7210
0.0900	9.655	3.239	2.679	2.611	2.486	2.097	1.441	1.153	0.854	3.677	4.675	3.394	0.8111
0.1000	10.248	3.438	2.850	2.761	2.626	2.226	1.537	1.224	0.907	3.844	4.889	3.592	0.9012
0.1250	11.618	3.877	3.204	3.087	2.950	2.490	1.715	1.367	1.015	4.196	5.341	4.035	1.1265
0.1600	13.248	4.417	3.625	3.493	3.314	2.760	1.893	1.507	1.118	4.567	5.802	4.576	1.4420
0.2000	14.694	4.878	3.995	3.837	3.627	2.948	2.005	1.592	1.176	4.830	6.129	5.043	1.8024
0.2500	16.043	5.310	4.307	4.147	3.872	3.061	2.069	1.622	1.193	5.014	6.355	5.470	2.2530
0.3200	17.225	5.623	4.573	4.364	3.993	3.059	2.041	1.592	1.163	5.109	6.490	5.794	2.8839
0.4000	17.679	5.836	4.635	4.405	3.989	2.978	1.969	1.533	1.116	5.080	6.468	5.882	3.6049
0.5000	17.609	5.851	4.546	4.325	3.870	2.846	1.877	1.466	1.068	4.949	6.318	5.802	4.5061
0.6000	16.912	5.673	4.293	4.080	3.646	2.666	1.768	1.380	1.009	4.693	6.022	5.505	5.4073
0.7000	15.792	5.427	3.992	3.794	3.395	2.479	1.659	1.291	0.954	4.411	5.669	5.126	6.3085
0.8000	14.584	5.118	3.700	3.528	3.153	2.306	1.560	1.217	0.904	4.134	5.307	4.755	7.2098
0.9000	13.444	4.792	3.433	3.272	2.910	2.146	1.464	1.157	0.859	3.877	4.972	4.403	8.1110
1.0000	12.505	4.522	3.227	3.062	2.728	2.021	1.392	1.109	0.824	3.649	4.675	4.111	9.0122
1.2500	10.598	3.968	2.825	2.697	2.397	1.798	1.262	1.019	0.770	3.192	4.054	3.576	11.265
1.6000	8.604	3.369	2.433	2.326	2.083	1.590	1.140	0.927	0.718	2.713	3.394	3.023	14.420
2.0000	6.988	2.865	2.123	2.022	1.830	1.421	1.038	0.852	0.670	2.329	2.870	2.574	18.024
2.5000	5.577	2.397	1.831	1.742	1.608	1.272	0.944	0.786	0.626	1.986	2.403	2.169	22.530
3.2000	4.390	1.950	1.546	1.476	1.371	1.110	0.844	0.702	0.563	1.659	1.978	1.800	28.839
4.0000	3.542	1.601	1.309	1.256	1.181	0.966	0.741	0.624	0.504	1.402	1.652	1.510	36.049
5.0000	2.903	1.319	1.104	1.062	1.001	0.830	0.645	0.545	0.447	1.181	1.383	1.267	45.061
6.0000	2.478	1.127	0.958	0.921	0.873	0.728	0.569	0.484	0.398	1.023	1.194	1.094	54.073
7.0000	2.177	0.988	0.849	0.819	0.774	0.651	0.512	0.437	0.361	0.907	1.056	0.970	63.085
8.0000	1.946	0.880	0.765	0.738	0.697	0.588	0.463	0.397	0.330	0.815	0.947	0.872	72.098
9.0000	1.759	0.799	0.696	0.672	0.636	0.538	0.424	0.364	0.304	0.741	0.860	0.793	81.110
10.0000	1.610	0.735	0.641	0.619	0.586	0.495	0.391	0.336	0.282	0.681	0.788	0.729	90.122
11.0000	1.484	0.681	0.595	0.575	0.544	0.460	0.364	0.313	0.262	0.630	0.729	0.676	99.134
12.0000	1.377	0.634	0.556	0.537	0.508	0.429	0.341	0.294	0.246	0.587	0.678	0.630	108.15

Table 42.8 Range (in mg/cm^2) of berillium ions ${}^{9}_{4}Be^{+}$ for different substances [1] [E_m is the energy per mass unit of incident particle; E is a total energy.]

E_m, MeV/u	Be	C	Al	Ti	Ni	Ge	Zr	Ag	Eu	Ta	Au	U	E, MeV
0.0125	0.085	0.097	0.137	0.210	0.252	0.295	0.325	0.360	0.566	0.669	0.730	0.868	0.1127
0.0160	0.101	0.116	0.163	0.248	0.298	0.346	0.381	0.421	0.660	0.779	0.850	1.009	0.1448
0.0200	0.117	0.136	0.189	0.287	0.346	0.400	0.438	0.483	0.758	0.893	0.974	1.154	0.1802
0.0250	0.135	0.157	0.219	0.332	0.400	0.460	0.503	0.554	0.868	1.002	1.114	1.318	0.2253
0.0320	0.158	0.185	0.256	0.388	0.469	0.536	0.585	0.642	1.006	1.183	1.289	1.522	0.2884
0.0400	0.182	0.213	0.295	0.445	0.539	0.613	0.669	0.733	1.148	1.348	1.468	1.731	0.3605
0.0500	0.208	0.245	0.338	0.510	0.617	0.699	0.762	0.834	1.307	1.532	1.667	1.963	0.4506
0.0600	0.233	0.273	0.378	0.568	0.689	0.778	0.848	0.926	1.451	1.697	1.847	2.172	0.5407
0.0700	0.256	0.300	0.414	0.623	0.754	0.851	0.926	1.011	1.583	1.849	2.011	2.362	0.6309
0.0800	0.277	0.325	0.450	0.675	0.816	0.920	1.001	1.091	1.706	1.989	2.163	2.539	0.7210
0.0900	0.298	0.349	0.483	0.724	0.876	0.986	1.072	1.168	1.823	2.123	2.307	2.707	0.8111
0.1000	0.319	0.372	0.516	0.771	0.933	1.048	1.140	1.241	1.935	2.250	2.444	2.866	0.9012
0.1250	0.368	0.427	0.595	0.884	1.068	1.197	1.302	1.414	2.195	2.548	2.764	3.236	1.1265
0.1600	0.435	0.498	0.701	1.033	1.246	1.394	1.517	1.643	2.534	2.934	3.177	3.711	1.4420
0.2000	0.512	0.575	0.821	1.200	1.444	1.612	1.755	1.896	2.903	3.354	3.624	4.224	1.8024
0.2500	0.611	0.671	0.973	1.407	1.688	1.881	2.048	2.209	3.350	3.860	4.163	4.841	2.2530
0.3200	0.754	0.804	1.188	1.695	2.027	2.253	2.455	2.644	3.963	4.551	4.900	5.682	2.8839
0.4000	0.924	0.960	1.440	2.027	2.414	2.677	2.920	3.142	4.652	5.326	5.726	6.621	3.6049
0.5000	1.147	1.159	1.762	2.447	2.900	3.208	3.503	3.766	5.503	6.282	6.742	7.777	4.5061
0.6000	1.382	1.368	2.094	2.875	3.392	3.743	4.091	4.398	6.352	7.233	7.753	8.191	5.4073
0.7000	1.629	1.588	2.439	3.313	3.894	4.287	4.691	5.042	7.209	8.192	8.769	10.065	6.3085
0.8000	1.890	1.821	2.795	3.762	4.407	4.842	5.305	5.700	8.079	9.162	9.797	11.223	7.2098
0.9000	2.163	2.067	3.164	4.223	4.931	5.409	5.932	6.374	8.961	10.144	10.836	12.391	8.1110
1.0000	2.450	2.328	3.546	4.698	5.469	5.988	6.573	7.063	9.856	11.140	11.890	13.571	9.0122
1.2500	3.222	3.041	4.560	5.942	6.872	7.497	8.246	8.863	12.167	13.705	14.603	16.605	11.265
1.6000	4.445	4.196	6.127	7.835	8.996	9.773	10.771	11.590	15.603	17.508	18.625	21.095	14.420
2.0000	6.054	5.734	8.141	10.230	11.668	12.627	13.947	15.024	19.865	22.196	23.585	26.611	18.024
2.5000	8.389	7.967	11.003	13.584	15.389	16.590	18.373	19.796	25.715	28.618	30.364	34.110	22.530
3.2000	12.265	11.650	15.656	16.967	21.320	22.898	25.417	27.399	34.890	38.676	40.960	45.783	28.839
4.0000	17.552	16.634	21.891	26.095	29.126	31.193	34.644	37.388	46.773	51.660	54.612	60.800	36.049
5.0000	25.417	23.996	31.029	36.449	40.399	43.152	47.914	51.753	63.643	70.115	73.925	81.994	45.061
6.0000	34.639	32.586	41.624	48.366	53.326	56.820	63.069	68.138	82.764	90.945	95.675	105.743	54.073
7.0000	45.188	42.373	53.637	61.803	67.878	72.152	80.035	86.477	104.061	114.067	119.795	131.968	63.085
8.0000	57.040	53.333	67.036	76.722	84.010	89.122	98.746	106.715	127.401	139.416	146.192	160.624	72.098
9.0000	70.155	65.451	81.789	93.088	101.678	107.691	119.165	128.800	152.770	166.913	174.809	191.616	81.110
10.0000	84.508	78.716	97.872	110.858	120.858	127.818	141.286	152.660	180.143	196.501	205.557	224.876	90.122
11.0000	100.089	93.104	115.259	129.997	141.508	149.471	165.072	178.229	209.464	228.085	238.300	260.359	99.134
12.0000	116.873	108.600	133.931	150.492	163.592	172.608	190.475	205.467	240.660	261.605	273.037	298.040	108.15

Table 42.8 Range (in mg/cm^2) of berillium ions ${}^{9}_{4}Be^{+}$ for different substances [1] *(continued)*

E_m, MeV/u	H	He	N	O	Ne	Ar	Kr	Xe	Rn	Mylar	$(CH_2)_n$	Water	E, MeV
0.0125	0.031	0.081	0.120	0.127	0.143	0.224	0.385	0.566	0.883	0.095	0.074	0.093	0.1127
0.0160	0.038	0.097	0.144	0.153	0.171	0.267	0.454	0.663	1.030	0.114	0.089	0.112	0.1442
0.0200	0.045	0.115	0.169	0.180	0.201	0.311	0.524	0.761	1.176	0.134	0.105	0.133	0.1802
0.0250	0.054	0.136	0.199	0.211	0.236	0.360	0.601	0.867	1.328	0.156	0.122	0.157	0.2253
0.0320	0.064	0.164	0.237	0.251	0.280	0.420	0.695	0.993	1.505	0.185	0.145	0.188	0.2884
0.0400	0.076	0.194	0.277	0.293	0.326	0.480	0.786	1.113	1.676	0.215	0.168	0.220	0.3605
0.0500	0.089	0.231	0.324	0.341	0.378	0.545	0.884	1.240	1.852	0.248	0.194	0.257	0.4506
0.0600	0.100	0.265	0.366	0.385	0.425	0.603	0.969	1.349	2.002	0.278	0.218	0.291	0.5407
0.0700	0.111	0.297	0.406	0.426	0.468	0.655	1.045	1.466	2.134	0.307	0.240	0.322	0.6309
0.0800	0.122	0.328	0.443	0.464	0.508	0.703	1.115	1.535	2.253	0.333	0.261	0.351	0.7210
0.0900	0.131	0.356	0.478	0.500	0.545	0.748	1.180	1.616	2.363	0.359	0.281	0.373	0.8111
0.1000	0.140	0.383	0.510	0.533	0.580	0.789	1.241	1.691	2.465	0.382	0.300	0.405	0.9012
0.1250	0.151	0.445	0.584	0.610	0.661	0.885	1.379	1.865	2.699	0.438	0.344	0.464	1.1265
0.1600	0.186	0.521	0.677	0.706	0.762	1.005	1.554	2.084	2.994	0.510	0.400	0.537	1.4420
0.2000	0.212	0.598	0.771	0.804	0.866	1.131	1.739	2.317	3.308	0.587	0.461	0.612	1.8024
0.2500	0.241	0.687	0.880	0.917	0.986	1.281	1.960	2.597	3.688	0.678	0.533	0.697	2.2530
0.3200	0.279	0.802	1.022	1.065	1.146	1.487	2.267	2.990	4.224	0.803	0.631	0.809	2.8839
0.4000	0.320	0.928	1.179	1.230	1.327	1.726	2.626	3.451	4.857	0.945	0.742	0.933	3.6049
0.5000	0.372	1.082	1.375	1.436	1.556	2.036	3.095	4.053	5.683	1.124	0.883	1.087	4.5061
0.6000	0.424	1.239	1.579	1.651	1.796	2.363	3.590	4.687	6.551	1.311	1.030	1.247	5.4073
0.7000	0.479	1.401	1.797	1.880	2.052	2.713	4.116	5.162	7.469	1.510	1.184	1.416	6.3085
0.8000	0.538	1.572	2.031	2.126	2.328	3.090	4.676	6.081	8.439	1.721	1.348	1.599	7.2098
0.9000	0.603	1.754	2.284	2.392	2.625	3.496	5.273	6.840	9.462	1.946	1.524	1.796	8.1110
1.0000	0.672	1.948	2.555	2.676	2.945	3.928	5.904	7.636	10.533	2.185	1.711	2.008	9.0122
1.2500	0.868	2.480	3.302	3.461	3.827	5.111	7.605	9.756	13.363	2.846	2.229	2.596	11.265
1.6000	1.199	3.344	4.506	4.722	5.241	6.980	10.236	13.004	17.610	3.919	3.080	3.557	14.420
2.0000	1.665	4.506	6.095	6.387	7.090	9.381	13.554	17.063	22.813	5.356	4.238	4.851	18.024
2.5000	2.389	6.230	8.386	8.793	9.723	12.739	18.113	22.575	29.781	7.456	5.958	6.763	22.530
3.2000	3.669	9.159	12.148	12.740	13.986	18.065	25.198	31.087	40.437	10.944	8.862	9.968	28.839
4.0000	5.505	13.254	17.232	18.053	19.672	25.049	34.344	42.011	53.999	15.688	12.866	14.357	36.049
5.0000	8.327	19.481	24.757	25.884	27.991	35.151	47.423	57.521	73.030	22.718	18.851	20.897	45.061
6.0000	11.697	26.894	33.547	35.019	37.658	46.774	62.333	75.117	94.429	30.939	25.883	28.573	54.073
7.0000	15.585	35.452	43.558	45.415	48.644	59.889	79.063	94.761	118.238	40.313	33.924	37.339	63.085
8.0000	19.971	45.131	54.756	57.026	60.931	74.476	97.607	116.440	144.404	50.810	42.950	47.151	72.098
9.0000	24.849	50.888	67.123	69.839	74.485	90.521	117.984	140.178	172.912	62.418	52.950	58.000	81.110
10.0000	30.210	67.658	80.627	83.823	89.267	108.002	140.136	165.950	203.749	75.122	63.909	69.865	90.122
11.0000	36.045	80.410	95.226	98.943	105.248	126.901	164.020	193.726	236.932	88.895	75.807	82.716	99.134
12.0000	42.353	94.135	110.904	115.180	122.410	147.197	189.598	223.458	272.429	103.720	88.631	96.537	108.15

Table 42.9 Mass stopping power of different substances (in MeV·mg^{-1}·cm^2) for boron ions $^{11}_{5}B^{+}$ [1] [E_m is the energy per mass unit of incident particle; E is a total energy.]

E_m, MeV/u	Be	C	Al	Ti	Ni	Ge	Zr	Ag	Eu	Ta	Au	U	E, MeV
0.0125	2.222	1.838	1.347	0.896	0.727	0.660	0.611	0.566	0.360	0.308	0.283	0.242	0.1376
0.0160	2.514	2.080	1.524	1.013	0.823	0.747	0.692	0.640	0.407	0.348	0.320	0.274	0.1761
0.0200	2.811	2.325	1.703	1.133	0.920	0.835	0.773	0.715	0.455	0.389	0.358	0.307	0.2202
0.0250	3.142	2.600	1.904	1.266	1.028	0.933	0.865	0.800	0.508	0.435	0.401	0.343	0.2752
0.0320	3.551	2.943	2.155	1.433	1.166	1.060	0.978	0.905	0.575	0.493	0.455	0.390	0.3523
0.0400	3.940	3.281	2.397	1.599	1.304	1.189	1.095	1.014	0.645	0.555	0.510	0.439	0.4404
0.0500	4.341	3.638	2.650	1.776	1.455	1.325	1.222	1.132	0.718	0.624	0.575	0.494	0.5504
0.0600	4.668	3.950	2.860	1.928	1.584	1.439	1.330	1.233	0.789	0.686	0.634	0.545	0.6605
0.0700	4.932	4.218	3.035	2.058	1.700	1.542	1.426	1.323	0.850	0.744	0.689	0.592	0.7706
0.0800	5.145	4.458	3.180	2.172	1.797	1.628	1.504	1.399	0.906	0.795	0.735	0.633	0.8807
0.0900	5.317	4.667	3.300	2.271	1.881	1.710	1.574	1.465	0.960	0.840	0.779	0.672	0.9908
0.1000	5.457	4.852	3.400	2.356	1.958	1.775	1.632	1.527	1.006	0.881	0.816	0.704	1.1009
0.1250	5.689	5.249	3.578	2.519	2.104	1.907	1.753	1.642	1.100	0.963	0.900	0.778	1.3761
0.1600	5.830	5.667	3.716	2.661	2.237	2.029	1.862	1.746	1.189	1.044	0.979	0.855	1.7614
0.2000	5.861	5.948	3.786	2.753	2.329	2.109	1.935	1.814	1.257	1.109	1.041	0.909	2.2018
0.2500	5.806	6.108	3.813	2.817	2.394	2.173	1.990	1.864	1.315	1.163	1.090	0.953	2.7521
0.3200	5.672	6.163	3.802	2.863	2.445	2.232	2.034	1.901	1.361	1.209	1.135	0.998	3.5229
0.4000	5.506	6.112	3.766	2.870	2.470	2.260	2.060	1.924	1.401	1.247	1.171	1.028	4.4036
0.5000	5.309	5.966	3.713	2.870	2.487	2.283	2.079	1.938	1.433	1.277	1.201	1.058	5.5045
0.6000	5.091	5.733	3.631	2.843	2.476	2.280	2.070	1.928	1.442	1.289	1.215	1.078	6.6054
0.7000	4.894	5.494	3.549	2.804	2.452	2.261	2.048	1.909	1.441	1.288	1.217	1.079	7.7063
0.8000	4.707	5.248	3.466	2.762	2.423	2.243	2.028	1.889	1.435	1.289	1.217	1.081	8.8072
0.9000	4.540	5.017	3.383	2.713	2.392	2.219	2.003	1.864	1.431	1.286	1.214	1.083	9.9081
1.0000	4.379	4.795	3.300	2.670	2.356	2.188	1.977	1.838	1.419	1.277	1.208	1.079	11.009
1.2500	4.015	4.300	3.096	2.538	2.260	2.105	1.898	1.761	1.384	1.248	1.179	1.056	13.761
1.6000	3.577	3.749	2.825	2.359	2.108	1.969	1.774	1.639	1.314	1.190	1.125	1.009	17.614
2.0000	3.159	3.297	2.550	2.160	1.941	1.821	1.632	1.512	1.224	1.117	1.056	0.951	22.018
2.5000	2.741	2.872	2.260	1.941	1.756	1.650	1.476	1.369	1.125	1.024	0.972	0.881	27.522
3.2000	2.309	2.439	1.942	1.689	1.538	1.447	1.297	1.200	1.002	0.916	0.870	0.790	35.229
4.0000	1.955	2.082	1.671	1.469	1.345	1.267	1.141	1.053	0.891	0.816	0.777	0.707	44.036
5.0000	1.644	1.763	1.425	1.263	1.163	1.097	0.989	0.915	0.784	0.715	0.685	0.626	55.045
6.0000	1.424	1.531	1.245	1.110	1.024	0.971	0.876	0.810	0.694	0.641	0.614	0.564	66.054
7.0000	1.255	1.356	1.107	0.992	0.916	0.871	0.788	0.729	0.631	0.580	0.557	0.512	77.063
8.0000	1.125	1.218	0.998	0.898	0.831	0.791	0.719	0.664	0.577	0.532	0.511	0.471	88.072
9.0000	1.022	1.106	0.910	0.822	0.762	0.725	0.660	0.611	0.532	0.492	0.472	0.437	99.081
10.0000	0.936	1.013	0.838	0.760	0.704	0.671	0.611	0.567	0.494	0.458	0.441	0.408	110.09
11.0000	0.865	0.937	0.776	0.706	0.655	0.625	0.569	0.530	0.463	0.430	0.415	0.383	121.10
12.0000	0.804	0.871	0.724	0.660	0.613	0.586	0.534	0.498	0.435	0.405	0.391	0.361	132.11

Table 42.9 Mass stopping power of different substances (in $MeV \cdot mg^{-1} \cdot cm^2$) for boron ions ${}^{11}_{5}B^+$ [1] *(continued)*

E_m, MeV/u	H	He	N	O	Ne	Ar	Kr	Xe	Rn	Mylar	$(CH_2)_n$	Water	E, MeV
0.0125	5.050	2.155	1.427	1.341	1.192	0.786	0.482	0.342	0.226	1.806	2.297	1.751	0.1376
0.0160	5.607	2.365	1.589	1.495	1.329	0.897	0.556	0.396	0.263	2.031	2.584	1.952	0.1761
0.0200	6.098	2.504	1.732	1.632	1.460	1.014	0.632	0.460	0.308	2.252	2.863	2.128	0.2202
0.0250	6.628	2.630	1.889	1.781	1.611	1.156	0.739	0.539	0.390	2.495	3.175	2.320	0.2752
0.0320	7.347	2.788	2.066	1.967	1.797	1.347	0.875	0.659	0.459	2.801	3.572	2.564	0.3523
0.0400	8.029	2.919	2.236	2.152	1.982	1.548	1.026	0.786	0.558	3.104	3.959	2.804	0.4404
0.0500	8.772	3.085	2.435	2.369	2.205	1.778	1.198	0.933	0.676	3.429	4.373	3.079	0.5504
0.0600	9.553	3.269	2.651	2.563	2.423	1.993	1.356	1.067	0.784	3.721	4.751	3.341	0.6605
0.0700	10.288	3.484	2.853	2.771	2.628	2.191	1.493	1.187	0.877	3.988	5.086	3.605	0.7706
0.0800	11.035	3.705	3.056	2.970	2.827	2.376	1.628	1.297	0.964	4.236	5.397	3.867	0.8807
0.0900	11.717	3.931	3.251	3.168	3.017	2.545	1.749	1.399	1.036	4.462	5.674	4.119	0.9908
0.1000	12.444	4.175	3.461	3.352	3.189	2.703	1.867	1.486	1.102	4.668	5.936	4.362	1.1009
0.1250	14.206	4.741	3.918	3.775	3.607	3.045	2.097	1.671	1.242	5.131	6.530	4.934	1.3761
0.1600	16.461	5.488	4.504	4.340	4.117	3.430	2.352	1.873	1.390	5.674	7.209	5.685	1.7614
0.2000	18.590	6.172	5.055	4.854	4.589	3.729	2.537	2.014	1.488	6.111	7.754	6.380	2.2018
0.2500	20.664	6.840	5.547	5.341	4.987	3.942	2.665	2.089	1.536	6.458	8.185	7.046	2.7522
0.3200	22.583	7.372	5.995	5.722	5.235	4.011	2.676	2.087	1.525	6.699	8.508	7.596	3.5229
0.4000	23.499	7.758	6.161	5.856	5.302	3.958	2.617	2.037	1.484	6.752	8.598	7.818	4.4036
0.5000	23.686	7.871	6.115	5.818	5.205	3.828	2.525	1.971	1.437	6.657	8.498	7.804	5.5045
0.6000	23.057	7.734	5.853	5.563	4.971	3.635	2.411	1.881	1.376	6.398	8.210	7.505	6.6054
0.7000	21.790	7.488	5.508	5.235	4.684	3.421	2.289	1.782	1.317	6.086	7.822	7.083	7.7063
0.8000	20.346	7.140	5.161	4.922	4.398	3.216	2.177	1.698	1.262	5.767	7.403	6.634	8.8072
0.9000	18.945	6.752	4.838	4.611	4.100	3.024	2.064	1.631	1.211	5.463	7.006	6.204	9.9081
1.0000	17.787	6.432	4.590	4.356	3.881	2.874	1.980	1.577	1.171	5.191	6.649	5.848	11.009
1.2500	15.386	5.761	4.102	3.916	3.480	2.610	1.833	1.480	1.118	4.634	5.885	5.191	13.761
1.6000	12.771	5.001	3.611	3.453	3.091	2.359	1.692	1.376	1.065	4.026	5.038	4.487	17.614
2.0000	10.557	4.327	3.208	3.055	2.764	2.147	1.568	1.288	1.012	3.519	4.335	3.889	22.018
2.5000	8.542	3.672	2.805	2.669	2.463	1.948	1.446	1.205	0.958	3.042	3.681	3.322	27.522
3.2000	6.796	3.019	2.394	2.285	2.122	1.718	1.307	1.087	0.872	2.569	3.062	2.786	35.229
4.0000	5.515	2.493	2.039	1.955	1.838	1.504	1.153	0.971	0.785	2.182	2.572	2.351	44.036
5.0000	4.532	2.059	1.723	1.659	1.563	1.295	1.007	0.851	0.698	1.844	2.159	1.978	55.045
6.0000	3.871	1.760	1.496	1.439	1.363	1.138	0.889	0.755	0.622	1.598	1.866	1.709	66.054
7.0000	3.398	1.543	1.326	1.278	1.209	1.016	0.799	0.682	0.563	1.416	1.648	1.514	77.063
8.0000	3.034	1.373	1.193	1.151	1.087	0.917	0.722	0.619	0.514	1.272	1.477	1.361	88.072
9.0000	2.740	1.245	1.084	1.047	0.990	0.837	0.660	0.567	0.473	1.154	1.339	1.235	99.081
10.0000	2.504	1.143	0.997	0.963	0.911	0.771	0.609	0.523	0.438	1.059	1.226	1.134	110.09
11.0000	2.306	1.057	0.925	0.893	0.845	0.714	0.566	0.487	0.408	0.979	1.133	1.050	121.10
12.0000	2.136	0.984	0.862	0.833	0.788	0.666	0.529	0.455	0.382	0.911	1.052	0.977	132.11

Table 42.10 Range (in mg/cm^2) of boron ions $^{11}_{5}B^+$ for different substances [1] [E_m is the energy per mass unit of incident particle; E is a total energy.]

E_m, MeV/u	Be	C	Al	Ti	Ni	Ge	Zr	Ag	Eu	Ta	Au	U	E, MeV
0.0125	0.082	0.093	0.130	0.199	0.237	0.278	0.307	0.341	0.535	0.633	0.691	0.824	0.1376
0.0160	0.097	0.111	0.155	0.236	0.282	0.329	0.361	0.400	0.627	0.740	0.808	0.961	0.1761
0.0200	0.113	0.130	0.181	0.274	0.329	0.381	0.418	0.461	0.722	0.851	0.929	1.103	0.2202
0.0250	0.131	0.152	0.210	0.318	0.382	0.440	0.481	0.530	0.829	0.977	1.065	1.263	0.2752
0.0320	0.154	0.179	0.247	0.373	0.449	0.514	0.562	0.617	0.965	1.135	1.237	1.463	0.3523
0.0400	0.177	0.206	0.285	0.429	0.517	0.590	0.644	0.706	1.104	1.297	1.413	1.668	0.4404
0.0500	0.203	0.238	0.328	0.493	0.595	0.676	0.737	0.806	1.261	1.478	1.610	1.898	0.5504
0.0600	0.227	0.266	0.367	0.551	0.666	0.754	0.821	0.897	1.404	1.643	1.788	2.105	0.6605
0.0700	0.250	0.293	0.404	0.606	0.732	0.827	0.900	0.982	1.536	1.794	1.951	2.295	0.7706
0.0800	0.272	0.318	0.439	0.657	0.794	0.895	0.974	1.062	1.659	1.935	2.103	2.472	0.8807
0.0900	0.293	0.342	0.473	0.707	0.854	0.961	1.045	1.139	1.776	2.069	2.246	2.640	0.9908
0.1000	0.313	0.365	0.506	0.754	0.911	1.024	1.114	1.212	1.887	2.197	2.384	2.800	1.1009
0.1250	0.363	0.420	0.585	0.867	1.046	1.173	1.276	1.386	2.148	2.495	2.705	3.172	1.3761
0.1600	0.429	0.490	0.690	1.015	1.223	1.369	1.489	1.613	2.485	2.879	3.115	3.643	1.7614
0.2000	0.505	0.566	0.808	1.178	1.416	1.582	1.721	1.860	2.844	3.288	3.550	4.142	2.2018
0.2500	0.599	0.657	0.952	1.375	1.649	1.838	2.001	2.159	3.272	3.772	4.066	4.733	2.7522
0.3200	0.733	0.783	1.155	1.647	1.967	2,188	2.384	2.568	3.847	4.421	4.758	5.522	3.5229
0.4000	0.891	0.926	1.387	1.954	2.326	2.580	2.814	3.028	4.485	5.138	5.522	6.391	4.4036
0.5000	1.094	1.109	1.682	2.337	2.770	3.065	3.346	3.598	5.262	6.011	6.450	7.447	5.5045
0.6000	1.306	1.297	1.981	2.723	3.213	3.547	3.877	4.168	6.028	6.869	7.361	8.477	6.6054
0.7000	1.527	1.493	2.288	3.113	3.660	4.032	4.411	4.742	6.792	7.723	8.267	9.498	7.7063
0.8000	1.756	1.698	2.602	3.508	4.112	4.521	4.952	5.321	7.557	8.577	9.171	10.517	8.8072
0.9000	1.994	1.913	2.924	3.910	4.569	5.014	5.498	5.908	8.325	9.432	10.077	11.534	9.9081
1.0000	2.241	2.137	3.253	4.219	5.033	5.514	6.051	6.503	9.098	10.291	10.986	12.553	11.009
1.2500	2.897	2.743	4.114	5.377	6.225	6.796	7.472	8.032	11.061	12.471	13.291	15.131	13.761
1.6000	3.915	3.703	5.417	6.951	7.991	8.689	9.572	10.300	13.919	15.634	16.637	18.864	17.614
2.0000	5.226	4.957	7.059	8.903	10.169	11.016	12.161	13.099	17.393	19.455	20.680	23.361	22.018
3.0000	7.099	6.749	9.355	11.595	13.155	14.196	15.713	16.929	22.088	24.609	26.120	29.379	27.522
3.2000	10.171	9.668	13.044	15.861	17.856	19.196	21.296	22.955	29.361	32.581	34.519	38.631	35.229
4.0000	14.329	13.588	17.947	21.467	23.995	25.719	28.553	30.811	38.706	42.792	45.256	50.442	44.036
5.0000	20.491	19.355	25.105	29.577	32.825	35.087	38.947	42.063	51.920	57.249	60.384	67.044	55.045
6.0000	27.704	26.073	33.392	38.899	42.937	45.778	50.801	54.880	66.877	73.542	77.396	85.620	66.054
7.0000	35.955	33.730	42.790	49.410	54.320	57.772	64.073	69.225	83.536	91.629	96.264	106.135	77.063
8.0000	45.234	42.310	53.279	61.090	66.950	71.057	78.722	85.069	101.808	113.474	116.930	128.568	88.072
9.0000	55.513	51.808	64.842	73.916	80.797	85.610	94.725	102.378	121.692	133.025	139.359	152.858	99.081
10.0000	66.777	62.217	77.463	87.862	95.849	101.406	112.085	121.103	143.173	156.245	163.489	178.960	110.09
11.0000	79.021	73.524	91.128	102.902	112.077	118.422	130.777	141.196	166.215	181.065	189.220	206.844	121.10
12.0000	92.230	85.719	105.822	119.031	129.455	136.629	150.768	162.631	190.765	207.444	216.556	236.498	132.11

Table 42.10 Range (in mg/cm^2) of boron ions $^{11}_{5}B^+$ for different substances [1] *(continued)*

E_m, MeV/u	H	He	N	O	Ne	Ar	Kr	Xe	Rn	Mylar	$(CH_2)_n$	Water	E, MeV
0.0125	0.030	0.077	0.113	0.120	0.135	0.211	0.362	0.533	0.832	0.090	0.070	0.088	0.1376
0.0160	0.036	0.093	0.137	0.145	0.162	0.253	0.429	0.627	0.974	0.109	0.085	0.107	0.1761
0.0200	0.043	0.110	0.162	0.171	0.192	0.295	0.497	0.722	1.116	0.128	0.100	0.127	0.2202
0.0250	0.052	0.131	0.191	0.202	0.226	0.343	0.572	0.825	1.265	0.150	0.118	0.150	0.2752
0.0320	0.062	0.159	0.228	0.242	0.269	0.402	0.664	0.949	1.438	0.179	0.140	0.181	0.3523
0.0400	0.073	0.189	0.268	0.283	0.314	0.461	0.754	1.067	1.606	0.208	0.163	0.213	0.4404
0.0500	0.086	0.225	0.314	0.331	0.365	0.526	0.850	1.192	1.780	0.241	0.189	0.249	0.5504
0.0600	0.098	0.259	0.356	0.374	0.412	0.584	0.939	1.300	1.929	0.271	0.212	0.283	0.6605
0.0700	0.109	0.291	0.396	0.415	0.455	0.636	1.011	1.397	2.060	0.299	0.235	0.314	0.7706
0.0800	0.119	0.321	0.432	0.453	0.495	0.684	1.082	1.485	2.179	0.326	0.256	0.343	0.8807
0.0900	0.129	0.350	0.467	0.489	0.533	0.728	1.147	1.567	2.289	0.351	0.275	0.371	0.9908
0.1000	0.138	0.377	0.500	0.523	0.568	0.770	1.208	1.643	2.392	0.376	0.294	0.397	1.1009
0.1250	0.158	0.439	0.575	0.600	0.649	0.866	1.347	1.818	2.627	0.432	0.339	0.456	1.3761
0.1600	0.184	0.514	0.666	0.695	0.749	0.985	1.520	2.035	2.920	0.503	0.395	0.528	1.7614
0.2000	0.209	0.590	0.758	0.790	0.850	1.108	1.700	2.261	3.226	0.578	0.453	0.601	2.2018
0.2500	0.237	0.674	0.862	0.898	0.965	1.251	1.911	2.530	3.589	0.665	0.522	0.683	2.7522
0.3200	0.272	0.783	0.996	1.038	1.115	1.445	2.199	2.898	4.092	0.782	0.615	0.789	3.5229
0.4000	0.311	0.899	1.141	1.190	1.283	1.666	2.532	3.325	4.678	0.913	0.718	0.903	4.4036
0.5000	0.357	1.040	1.320	1.378	1.492	1.949	2.960	3.875	5.432	1.077	0.846	1.044	5.5045
0.6000	0.404	1.181	1.504	1.572	1.709	2.244	3.407	4.446	6.215	1.246	0.978	1.188	6.6054
0.7000	0.453	1.326	1.698	1.776	1.937	2.556	3.875	5.048	7.033	1.422	1.116	1.339	7.7063
0.8000	0.506	1.476	1.904	1.993	2.179	2.888	4.368	5.681	7.887	1.608	1.260	1.500	8.8072
0.9000	0.562	1.635	2.125	2.224	2.438	3.241	4.888	6.342	8.777	1.804	1.413	1.671	9.9081
1.0000	0.622	1.802	2.358	2.470	2.714	3.615	5.433	7.029	9.702	2.011	1.574	1.854	11.009
1.2500	0.788	2.254	2.993	3.136	3.464	4.620	6.877	8.830	12.107	2.572	2.014	2.354	13.761
1.6000	1.063	2.973	3.994	4.184	4.639	6.173	9.066	11.031	15.638	3.465	2.723	3.152	17.614
2.0000	1.443	3.920	5.289	5.542	6.147	8.131	11.770	14.840	19.879	4.636	3.666	4.208	22.018
2.5000	2.024	5.303	7.127	7.472	8.529	10.826	15.429	19.263	25.472	6.321	5.046	5.742	27.522
3.2000	3.039	7.625	10.110	10.601	11.638	15.047	21.045	26.010	33.918	9.086	7.348	8.282	35.229
4.0000	4.482	10.845	14.108	14.780	16.110	20.540	28.238	34.601	44.585	12.816	10.497	11.734	44.036
5.0000	6.693	15.722	20.002	20.914	22.626	28.453	38.482	46.751	59.491	18.323	15.185	16.857	55.045
6.0000	9.329	21.521	26.877	28.058	30.188	37.544	50.145	60.514	76.320	24.753	20.685	22.861	66.054
7.0000	12.370	28.216	34.709	36.191	38.782	47.803	63.232	75.881	94.854	32.087	26.976	29.718	77.063
8.0000	15.804	35.793	43.476	45.281	48.401	59.223	77.750	92.852	115.339	40.304	34.042	37.400	88.072
9.0000	19.627	44.224	53.168	55.323	59.024	71.798	93.721	111.457	137.682	49.402	41.879	45.902	99.081
10.0000	23.834	53.460	63.765	66.298	70.624	85.517	111.105	131.683	161.883	59.372	50.480	55.213	110.09
11.0000	28.420	63.482	75.238	78.180	83.183	100.369	129.874	153.510	187.959	70.195	59.830	65.313	121.10
12.0000	33.384	74.283	87.576	90.957	96.689	116.341	150.003	176.908	215.894	81.862	69.922	76.189	132.11

Table 42.11 Mass stopping power of different substances (in MeV·mg^{-1}·cm^2) for carbon ions $^{12}_{6}C^+$ [1] [E_m is the energy per mass unit of incident particle; E is a total energy.]

E_m, MeV/u	Be	C	Al	Ti	Ni	Ge	Zr	Ag	Eu	Ta	Au	U	E, MeV
0.0125	2.558	2.116	1.550	1.031	0.837	0.760	0.704	0.651	0.414	0.354	0.326	0.279	0.1500
0.0160	2.894	2.394	1.754	1.166	0.947	0.859	0.796	0.737	0.468	0.401	0.368	0.316	0.1920
0.0200	3.236	2.677	1.961	1.304	1.059	0.961	0.890	0.824	0.524	0.448	0.412	0.353	0.2400
0.0250	3.618	2.993	2.192	1.458	1.184	1.074	0.995	0.921	0.585	0.501	0.462	0.395	0.3000
0.0320	4.088	3.388	2.481	1.650	1.342	1.220	1.126	1.042	0.662	0.568	0.523	0.449	0.3840
0.0400	4.559	3.797	2.773	1.850	1.509	1.376	1.267	1.173	0.746	0.642	0.591	0.508	0.4800
0.0500	5.042	4.226	3.078	2.062	1.690	1.539	1.419	1.314	0.834	0.725	0.668	0.574	0.6000
0.0600	5.455	4.616	3.342	2.253	1.852	1.681	1.554	1.441	0.922	0.802	0.740	0.637	0.7200
0.0700	5.806	4.967	3.573	2.423	2.001	1.815	1.679	1.558	1.000	0.875	0.811	0.697	0.8400
0.0800	6.108	5.293	3.775	2.578	2.133	1.933	1.786	1.661	1.076	0.944	0.872	0.751	0.9600
0.0900	6.365	5.587	3.951	2.718	2.252	2.047	1.885	1.754	1.150	1.006	0.932	0.804	1.0800
0.1000	6.587	5.856	4.104	2.844	2.364	2.142	1.970	1.843	1.215	1.063	0.985	0.850	1.2000
0.1250	6.991	6.450	4.397	3.095	2.585	2.344	2.155	2.018	1.352	1.183	1.106	0.956	1.5000
0.1600	7.286	7.082	4.644	3.325	2.796	2.536	2.327	2.183	1.486	1.305	1.224	1.068	1.9200
0.2000	7.407	7.518	4.785	3.479	2.943	2.665	2.445	2.292	1.589	1.402	1.316	1.148	2.4000
0.2500	7.402	7.786	4.860	3.592	3.052	2.770	2.537	2.377	1.677	1.482	1.390	1.215	3.0000
0.3200	7.294	7.925	4.889	3.681	3.144	2.870	2.616	2.444	1.750	1.555	1.459	1.283	3.8400
0.4000	7.140	7.926	4.884	3.721	3.204	2.930	2.671	2.496	1.817	1.617	1.519	1.333	4.8000
0.5000	6.950	7.810	4.860	3.757	3.256	2.989	2.722	2.537	1.876	1.672	1.572	1.385	6.0000
0.6000	6.692	7.537	4.773	3.738	3.255	2.998	2.721	2.535	1.895	1.695	1.597	1.418	7.2000
0.7000	6.456	7.247	4.682	3.698	3.235	2.982	2.701	2.519	1.901	1.699	1.606	1.423	8.4000
0.8000	6.229	6.944	4.587	3.655	3.206	2.967	2.683	2.500	1.899	1.706	1.610	1.431	9.6000
0.9000	6.025	6.658	4.490	3.601	3.174	2.945	2.658	2.474	1.899	1.706	1.612	1.437	10.800
1.0000	5.828	6.382	4.392	3.553	3.136	2.912	2.631	2.446	1.889	1.700	1.607	1.436	12.000
1.2500	5.382	5.764	4.150	3.403	3.029	2.822	2.544	2.361	1.855	1.672	1.581	1.415	15.000
1.6000	4.844	5.077	3.826	3.195	2.854	2.667	2.403	2.219	1.779	1.611	1.523	1.366	19.200
2.0000	4.327	4.515	3.492	2.958	2.657	2.493	2.235	2.071	1.676	1.529	1.446	1.303	24.000
2.5000	3.799	3.981	3.132	2.690	2.434	2.286	2.045	1.898	1.560	1.419	1.347	1.222	30.000
3.2000	3.240	3.423	2.725	2.371	2.158	2.030	1.820	1.684	1.406	1.286	1.221	1.109	38.400
4.0000	2.770	2.950	2.368	2.081	1.906	1.795	1.617	1.492	1.262	1.156	1.101	1.002	48.000
5.0000	2.347	2.516	2.034	1.802	1.660	1.566	1.412	1.306	1.119	1.021	0.978	0.893	60.000
6.0000	2.041	2.194	1.784	1.591	1.468	1.392	1.256	1.161	0.995	0.919	0.880	0.808	72.000
7.0000	1.804	1.948	1.590	1.425	1.317	1.252	1.132	1.048	0.907	0.833	0.800	0.736	84.000
8.0000	1.619	1.752	1.436	1.293	1.196	1.137	1.034	0.955	0.830	0.766	0.735	0.678	96.000
9.0000	1.472	1.592	1.310	1.183	1.097	1.044	0.950	0.879	0.767	0.708	0.680	0.629	108.00
10.0000	1.348	1.459	1.206	1.094	1.013	0.966	0.879	0.816	0.712	0.660	0.636	0.587	120.00
11.0000	1.245	1.349	1.118	1.017	0.943	0.900	0.819	0.763	0.666	0.619	0.598	0.551	132.00
12.0000	1.158	1.254	1.042	0.950	0.883	0.843	0.768	0.717	0.626	0.584	0.563	0.519	144.00

Table 42.11 Mass stopping power of different substances (in MeV·mg^{-1}·cm^2) for carbon ions ${}^{12}_{6}C^{+}$ [1] *(continued)*

E_m, MeV/u	H	He	N	O	Ne	Ar	Kr	Xe	Rn	Mylar	$(CH_2)_n$	Water	E, MeV
0.0125	5.814	2.481	1.643	1.544	1.372	0.905	0.555	0.394	0.260	2.079	2.645	2.019	0.1500
0.0160	6.455	2.722	1.829	1.721	1.529	1.033	0.640	0.456	0.303	2.338	2.975	2.247	0.1920
0.0200	7.020	2.883	1.994	1.879	1.681	1.167	0.728	0.529	0.354	2.592	3.296	2.449	0.2400
0.0250	7.630	3.028	2.175	2.050	1.855	1.331	0.851	0.620	0.449	2.872	3.655	2.670	0.3000
0.0320	8.459	3.210	2.379	2.265	2.069	1.550	1.007	0.759	0.528	3.225	4.113	2.952	0.3840
0.0400	9.291	3.378	2.587	2.490	2.294	1.792	1.187	0.910	0.646	3.591	4.582	3.245	0.4800
0.0500	10.188	3.583	2.829	2.752	2.561	2.065	1.391	1.083	0.785	3.983	5.079	3.577	0.6000
0.0600	11.163	3.820	3.098	2.995	2.831	2.330	1.584	1.247	0.916	4.348	5.551	3.904	0.7200
0.0700	12.113	4.102	3.359	3.262	3.094	2.580	1.758	1.397	1.033	4.695	5.989	4.245	0.8400
0.0800	13.100	4.398	3.628	3.526	3.356	2.820	1.933	1.540	1.144	5.028	6.406	4.591	0.9600
0.0900	14.027	4.706	3.892	3.793	3.611	3.046	2.094	1.675	1.241	5.342	6.792	4.931	1.0800
0.1000	15.021	5.040	4.178	4.047	3.850	3.263	2.253	1.793	1.330	5.635	7.166	5.265	1.2000
0.1250	17.456	5.826	4.815	4.639	4.432	3.742	2.577	2.053	1.526	6.305	8.024	6.063	1.5000
0.1600	20.573	6.859	5.628	5.424	5.146	4.286	2.940	2.341	1.737	7.091	9.009	7.105	1.9200
0.2000	23.495	7.800	6.388	6.135	5.800	4.713	3.206	2.546	1.881	7.723	9.800	8.063	2.4000
0.2500	26.341	8.719	7.071	6.809	6.357	5.025	3.397	2.663	1.959	8.233	10.434	8.981	3.0000
0.3200	29.040	9.480	7.710	7.358	6.732	5.158	3.442	2.684	1.960	8.614	10.941	9.768	3.8400
0.4000	30.475	10.061	7.990	7.594	6.876	5.133	3.394	2.642	1.924	8.757	11.150	10.139	4.8000
0.5000	31.007	10.303	8.004	7.616	6.814	5.011	3.305	2.581	1.881	8.714	11.125	10.216	6.0000
0.6000	30.311	10.167	7.695	7.313	6.535	4.778	3.170	2.473	1.809	8.411	10.793	9.867	7.2000
0.7000	28.745	9.878	7.266	6.905	6.180	4.513	3.020	2.350	1.737	8.029	10.318	9.330	8.4000
0.8000	26.923	9.448	6.829	6.513	5.820	4.256	2.880	2.247	1.669	7.632	9.797	8.779	9.6000
0.9000	25.142	8.961	6.420	6.119	5.441	4.014	2.739	2.164	1.607	7.251	9.298	8.234	10.800
1.0000	23.673	8.560	6.109	5.797	5.165	3.825	2.635	2.099	1.559	6.909	8.850	7.783	12.000
1.2500	20.623	7.722	5.498	5.249	4.664	3.498	2.457	1.984	1.498	6.212	7.888	6.959	15.000
1.6000	17.295	6.773	4.890	4.676	4.186	3.195	2.292	1.863	1.443	5.452	6.822	6.076	19.200
2.0000	14.457	5.926	4.393	4.183	3.785	2.940	2.148	1.763	1.386	4.819	5.936	5.325	24.000
2.5000	11.839	5.090	3.887	3.699	3.414	2.700	2.005	1.669	1.328	4.216	5.102	4.604	30.000
3.2000	9.538	4.238	3.360	3.208	2.979	2.412	1.834	1.526	1.224	3.605	4.298	3.911	38.400
4.0000	7.814	3.533	2.889	2.770	2.605	2.131	1.634	1.376	1.113	3.093	3.644	3.332	48.000
5.0000	6.468	2.939	2.459	2.368	2.231	1.849	1.438	1.214	0.997	2.632	3.081	2.823	60.000
6.0000	5.548	2.523	2.144	2.062	1.953	1.631	1.274	1.083	0.892	2.291	2.674	2.449	72.000
7.0000	4.883	2.217	1.905	1.837	1.737	1.460	1.148	0.980	0.810	2.034	2.368	2.176	84.000
8.0000	4.366	1.975	1.716	1.656	1.564	1.320	1.038	0.890	0.740	1.830	2.126	1.958	96.000
9.0000	3.945	1.793	1.561	1.507	1.426	1.206	0.950	0.816	0.681	1.662	1.928	1.778	108.00
10.0000	3.606	1.646	1.436	1.387	1.312	1.110	0.877	0.754	0.631	1.524	1.766	1.633	120.00
11.0000	3.320	1.522	1.331	1.285	1.216	1.028	0.815	0.701	0.587	1.409	1.631	1.511	132.00
12.0000	3.074	1.416	1.241	1.198	1.134	0.959	0.762	0.656	0.549	1.311	1.514	1.407	144.00

Table 42.12 Range (in mg/cm^2) of carbon ions ${}^{12}_{6}C^{+}$ for different substances [1] [E_m is the energy per mass unit of incident particle; E is a total energy.]

E_m, MeV/u	Be	C	Al	Ti	Ni	Ge	Zr	Ag	Eu	Ta	Au	U	E, MeV
0.0125	0.075	0.085	0.119	0.181	0.216	0.255	0.282	0.313	0.491	0.581	0.635	0.758	0.1500
0.0160	0.089	0.102	0.143	0.216	0.258	0.302	0.332	0.368	0.576	0.681	0.744	0.886	0.1920
0.0200	0.104	0.120	0.167	0.252	0.302	0.350	0.385	0.425	0.665	0.785	0.857	1.019	0.2400
0.0250	0.121	0.140	0.194	0.293	0.351	0.406	0.444	0.490	0.766	0.903	0.985	1.169	0.3000
0.0320	0.143	0.165	0.229	0.345	0.414	0.476	0.520	0.572	0.893	1.051	1.146	1.357	0.3840
0.0400	0.164	0.191	0.264	0.398	0.478	0.547	0.597	0.655	1.024	1.203	1.311	1.549	0.4800
0.0500	0.189	0.221	0.304	0.457	0.551	0.627	0.684	0.749	1.171	1.373	1.495	1.764	0.6000
0.0600	0.212	0.247	0.341	0.512	0.617	0.700	0.763	0.834	1.304	1.526	1.661	1.957	0.7200
0.0700	0.233	0.272	0.375	0.562	0.678	0.767	0.836	0.913	1.426	1.666	1.812	2.133	0.8400
0.0800	0.253	0.296	0.408	0.609	0.735	0.830	0.904	0.986	1.539	1.795	1.952	2.295	0.9600
0.0900	0.272	0.318	0.439	0.655	0.789	0.891	0.968	1.056	1.645	1.916	2.083	2.447	1.0800
0.1000	0.291	0.339	0.468	0.698	0.841	0.948	1.030	1.123	1.747	2.032	2.208	2.592	1.2000
0.1250	0.335	0.387	0.539	0.799	0.962	1.082	1.176	1.278	1.981	2.299	2.494	2.925	1.5000
0.1600	0.394	0.449	0.631	0.929	1.118	1.253	1.363	1.478	2.276	2.636	2.854	3.339	1.9200
0.2000	0.459	0.515	0.733	1.070	1.285	1.438	1.564	1.692	2.588	2.991	3.232	3.771	2.4000
0.2500	0.540	0.593	0.857	1.240	1.485	1.658	1.804	1.949	2.955	3.406	3.675	4.279	3.0000
0.3200	0.654	0.700	1.030	1.471	1.756	1.956	2.130	2.297	3.445	3.959	4.264	4.951	3.8400
0.4000	0.787	0.821	1.226	1.730	2.058	2.287	2.493	2.685	3.983	4.564	4.909	5.684	4.8000
0.5000	0.957	0.974	1.472	2.051	2.429	2.692	2.938	3.162	4.633	5.294	5.685	6.567	6.0000
0.6000	1.133	1.130	1.721	2.371	2.798	3.093	3.379	3.635	5.269	6.007	6.442	7.423	7.2000
0.7000	1.316	1.292	1.975	2.694	3.168	3.494	3.821	4.110	5.901	6.714	7.192	8.268	8.4000
0.8000	1.505	1.462	2.234	3.020	3.540	3.898	4.267	4.589	6.533	7.419	7.938	9.109	9.6000
0.9000	1.701	1.638	2.498	3.351	3.916	4.303	4.716	5.071	7.165	8.122	8.683	9.946	10.800
1.0000	1.903	1.822	2.769	3.686	4.297	4.713	5.170	5.559	7.798	8.827	9.428	10.781	12.000
1.2500	2.439	2.317	3.471	4.549	5.270	5.760	6.330	6.807	9.401	10.605	11.310	12.885	15.000
1.6000	3.262	3.094	4.526	5.823	6.698	7.291	8.029	8.642	11.713	13.164	14.016	15.905	19.200
2.0000	4.311	4.097	5.840	7.385	8.442	9.153	10.101	10.882	14.493	16.223	17.252	19.504	24.000
2.5000	5.793	5.514	7.656	9.514	10.803	11.668	12.910	13.911	18.206	20.299	21.555	24.264	30.000
3.2000	8.193	7.795	10.537	12.846	14.476	15.574	17.271	18.618	23.888	26.527	28.116	31.492	38.400
4.0000	11.405	10.823	14.325	17.178	19.219	20.614	22.878	24.688	31.108	34.416	36.411	40.617	48.000
5.0000	16.125	15.240	19.809	23.391	25.984	27.791	30.841	33.308	41.232	45.491	48.001	53.336	60.000
6.0000	21.621	20.359	26.123	30.493	33.687	35.936	39.873	43.073	52.627	57.905	60.963	67.489	72.000
7.0000	27.887	26.173	33.259	38.475	42.332	45.044	49.951	53.967	65.278	71.640	75.291	83.068	84.000
8.0000	34.921	32.678	41.210	47.329	51.905	55.114	61.055	65.977	79.129	86.683	90.956	100.073	96.000
9.0000	42.705	39.870	49.967	57.042	62.392	66.136	73.175	79.085	94.186	103.004	107.942	118.468	108.00
10.0000	51.232	47.750	59.521	67.599	73.786	78.093	86.316	93.260	110.448	120.581	126.208	138.227	120.00
11.0000	60.500	56.310	69.865	78.984	86.071	90.974	100.466	108.470	127.890	139.370	145.686	159.335	132.00
12.0000	70.501	65.543	80.991	91.197	99.229	104.760	115.603	124.700	146.479	159.344	166.385	181.788	144.00

Table 42.12 Range (in mg/cm^2) of carbon ions $^{12}_{6}C^{+}$ for different substances [1] *(continued)*

E_m, MeV/u	H	He	N	O	Ne	Ar	Kr	Xe	Rn	Mylar	$(CH_2)_n$	Water	E, MeV
0.0125	0.027	0.070	0.103	0.109	0.122	0.192	0.330	0.487	0.762	0.082	0.064	0.080	0.1500
0.0160	0.033	0.085	0.125	0.133	0.148	0.231	0.392	0.574	0.894	0.100	0.078	0.097	0.1920
0.0200	0.040	0.102	0.148	0.157	0.176	0.271	0.456	0.663	1.026	0.118	0.092	0.116	0.2400
0.0250	0.048	0.121	0.175	0.186	0.207	0.315	0.526	0.760	1.166	0.139	0.108	0.138	0.3000
0.0320	0.057	0.147	0.210	0.223	0.248	0.371	0.612	0.876	1.329	0.165	0.129	0.167	0.3840
0.0400	0.068	0.175	0.248	0.262	0.290	0.426	0.696	0.987	1.486	0.192	0.151	0.197	0.4800
0.0500	0.080	0.209	0.291	0.306	0.338	0.487	0.787	1.104	1.649	0.223	0.175	0.231	0.6000
0.0600	0.091	0.241	0.330	0.347	0.382	0.541	0.866	1.205	1.788	0.252	0.197	0.262	0.7200
0.0700	0.101	0.270	0.367	0.385	0.422	0.589	0.936	1.294	1.909	0.278	0.218	0.291	0.8400
0.0800	0.110	0.298	0.401	0.420	0.458	0.633	1.000	1.376	2.019	0.302	0.237	0.318	0.9600
0.0900	0.119	0.325	0.433	0.452	0.493	0.673	1.060	1.451	2.120	0.325	0.255	0.343	1.0800
0.1000	0.127	0.349	0.463	0.483	0.525	0.711	1.115	1.520	2.213	0.347	0.273	0.367	1.2000
0.1250	0.146	0.403	0.529	0.552	0.597	0.797	1.239	1.676	2.424	0.397	0.312	0.420	1.5000
0.1600	0.168	0.471	0.610	0.636	0.685	0.902	1.392	1.867	2.681	0.460	0.361	0.483	1.9200
0.2000	0.190	0.536	0.690	0.719	0.773	1.008	1.548	2.063	2.946	0.525	0.412	0.547	2.4000
0.2500	0.214	0.609	0.779	0.811	0.871	1.131	1.729	2.293	3.258	0.600	0.471	0.617	3.0000
0.3200	0.244	0.701	0.892	0.930	1.000	1.296	1.975	2.607	3.686	0.700	0.550	0.707	3.8400
0.4000	0.276	0.799	1.015	1.058	1.141	1.483	2.256	2.968	4.181	0.810	0.637	0.803	4.8000
0.5000	0.315	0.917	1.165	1.216	1.316	1.719	2.614	3.427	4.811	0.947	0.745	0.921	6.0000
0.6000	0.354	1.035	1.318	1.377	1.496	1.965	2.984	3.902	5.462	1.088	0.854	1.041	7.2000
0.7000	0.395	1.154	1.478	1.546	1.685	2.223	3.372	4.400	6.139	1.234	0.968	1.166	8.4000
0.8000	0.438	1.278	1.648	1.725	1.885	2.497	3.779	4.922	6.843	1.387	1.087	1.298	9.6000
0.9000	0.484	1.409	1.830	1.915	2.098	2.787	4.206	5.466	7.576	1.548	1.213	1.439	10.800
1.0000	0.534	1.546	2.021	2.116	2.324	3.093	4.653	6.029	8.334	1.718	1.345	1.589	12.000
1.2500	0.669	1.915	2.539	2.660	2.936	3.913	5.832	7.499	10.297	2.176	1.704	1.997	15.000
1.6000	0.892	2.496	3.349	3.508	3.887	5.170	7.603	9.684	13.153	2.898	2.277	2.643	19.200
2.0000	1.196	3.254	4.386	4.594	5.093	6.737	9.767	12.333	16.548	3.835	3.032	3.488	24.000
2.5000	1.655	4.348	5.839	6.121	6.764	8.868	12.661	15.831	20.971	5.168	4.124	4.701	30.000
3.2000	2.448	6.162	8.169	8.565	9.404	12.166	17.048	21.102	27.570	7.327	5.922	6.685	38.400
4.0000	3.563	8.650	11.258	11.794	12.858	16.410	22.605	27.739	35.811	10.210	8.354	9.352	48.000
5.0000	5.256	12.368	15.774	16.493	17.850	22.472	30.454	37.047	47.231	14.429	11.946	13.277	60.000
6.0000	7.265	16.804	21.011	21.936	23.611	29.398	39.339	47.534	59.984	19.327	16.136	17.857	72.000
7.0000	9.574	21.888	26.959	28.113	30.138	37.189	49.278	59.203	74.127	24.896	20.914	23.158	84.000
8.0000	12.177	27.631	33.604	35.003	37.429	45.845	60.283	72.068	89.655	31.125	26.270	28.881	96.000
9.0000	15.072	34.016	40.944	42.607	45.474	55.369	72.377	86.157	106.576	38.015	32.205	35.320	108.00
10.0000	18.257	41.008	48.966	50.915	54.255	65.754	85.537	101.468	124.896	45.562	38.716	42.369	120.00
11.0000	21.728	98.594	57.651	59.910	63.763	76.997	99.745	117.991	144.635	53.755	45.793	50.014	132.00
12.0000	25.487	56.772	66.993	69.585	73.989	89.090	114.986	135.707	165.787	62.589	53.435	58.249	144.00

Table 42.13 Mass stopping power of different substances (in MeV·mg^{-1}·cm^2) for nitrogen ions $^{14}_{7}N^+$ [1] [E_m is the energy per mass unit of incident particle; E is a total energy.]

E_m, MeV/u	Be	C	Al	Ti	Ni	Ge	Zr	Ag	Eu	Ta	Au	U	E, MeV
0.0125	2.972	2.458	1.801	1.198	0.973	0.882	0.818	0.756	0.481	0.412	0.378	0.324	0.1750
0.0160	3.362	2.781	2.038	1.355	1.100	0.998	0.925	0.856	0.544	0.466	0.428	0.367	0.2240
0.0200	3.759	3.110	2.278	1.515	1.230	1.116	1.034	0.957	0.608	0.521	0.478	0.410	0.2801
0.0250	4.202	3.477	2.547	1.694	1.375	1.248	1.156	1.070	0.680	0.582	0.536	0.458	0.3501
0.0320	4.749	3.936	2.881	1.916	1.559	1.418	1.308	1.210	0.769	0.660	0.608	0.522	0.4481
0.0400	5.296	4.410	3.222	2.149	1.753	1.598	1.472	1.363	0.867	0.746	0.686	0.590	0.5601
0.0500	5.843	4.898	3.567	2.390	1.958	1.784	1.644	1.523	0.967	0.840	0.774	0.665	0.7001
0.0600	6.306	5.337	3.864	2.604	2.141	1.944	1.797	1.665	1.067	0.927	0.856	0.736	0.8402
0.0700	6.700	5.731	4.123	2.795	2.309	2.094	1.938	1.798	1.154	1.010	0.936	0.804	0.9802
0.0800	7.038	6.099	4.350	2.971	2.458	2.227	2.058	1.914	1.240	1.088	1.005	0.866	1.1202
0.0900	7.331	6.435	4.551	3.131	2.594	2.357	2.171	2.020	1.324	1.158	1.074	0.926	1.2603
0.1000	7.589	6.748	4.729	3.277	2.724	2.468	2.270	2.123	1.400	1.225	1.135	0.979	1.4003
0.1250	8.096	7.470	5.092	3.585	2.994	2.714	2.495	2.337	1.566	1.370	1.281	1.107	1.7504
0.1600	8.557	8.317	5.454	3.905	3.283	2.978	2.732	2.563	1.745	1.533	1.437	1.254	2.2405
0.2000	8.862	8.994	5.725	4.162	3.521	3.189	2.925	2.742	1.901	1.677	1.574	1.374	2.8006
0.2500	9.030	9.498	5.929	4.382	3.723	3.380	3.095	2.899	2.046	1.808	1.696	1.482	3.5008
0.3200	9.058	9.841	6.071	4.571	3.904	3.564	3.248	3.035	2.173	1.931	1.812	1.594	4.4810
0.4000	8.958	9.944	6.127	4.669	4.019	3.676	3.352	3.131	2.279	2.028	1.906	1.673	5.6012
0.5000	8.759	9.843	6.125	4.735	4.104	3.767	3.430	3.197	2.364	2.107	1.981	1.746	7.0015
0.6000	8.476	9.546	6.046	4.734	4.123	3.797	3.446	3.210	2.400	2.146	2.022	1.796	8.4018
0.7000	8.209	9.215	5.953	4.703	4.114	3.792	3.435	3.203	2.417	2.161	2.042	1.810	9.8021
0.8000	7.947	8.860	5.852	4.664	4.091	3.786	3.424	3.189	2.423	2.177	2.054	1.826	11.202
0.9000	7.710	8.520	5.745	4.608	4.062	3.769	3.401	3.166	2.430	2.183	2.063	1.839	12.603
1.0000	7.478	8.188	5.635	4.559	4.023	3.736	3.375	3.139	2.423	2.181	2.062	1.843	14.003
1.2500	6.943	7.435	5.353	4.389	3.908	3.640	3.281	3.046	2.393	2.157	2.039	1.825	17.504
1.6000	6.286	6.589	4.966	4.146	3.704	3.461	3.118	2.880	2.309	2.090	1.976	1.773	22.405
2.0000	5.646	5.892	4.557	3.860	3.468	3.254	2.916	2.702	2.187	1.996	1.887	1.700	28.006
2.5000	4.986	5.225	4.111	3.531	3.194	3.001	2.684	2.491	2.047	1.862	1.768	1.603	35.007
3.2000	4.281	4.522	3.600	3.132	2.851	2.682	2.405	2.225	1.858	1.699	1.613	1.465	44.810
4.0000	3.682	3.921	3.147	2.766	2.533	2.386	2.150	1.983	1.677	1.536	1.463	1.331	56.012
5.0000	3.138	3.364	2.719	2.409	2.219	2.094	1.887	1.746	1.496	1.365	1.308	1.194	70.015
6.0000	2.742	2.948	2.397	2.138	1.973	1.869	1.687	1.560	1.337	1.234	1.182	1.086	84.018
7.0000	2.433	2.628	2.145	1.922	1.776	1.688	1.527	1.414	1.223	1.124	1.079	0.993	98.021
8.0000	2.191	2.372	1.944	1.750	1.619	1.540	1.400	1.293	1.124	1.036	0.995	0.918	112.02
9.0000	1.998	2.162	1.779	1.606	1.489	1.418	1.290	1.194	1.041	0.961	0.923	0.854	126.03
10.0000	1.835	1.986	1.641	1.489	1.379	1.315	1.197	1.111	0.968	0.898	0.865	0.799	140.03
11.0000	1.699	1.841	1.525	1.388	1.278	1.228	1.118	1.042	0.909	0.845	0.816	0.752	154.03
12.0000	1.583	1.714	1.425	1.299	1.207	1.153	1.050	0.980	0.856	0.798	0.769	0.710	168.04

Table 42.13 Mass stopping power of different substances (in MeV·mg^{-1}·cm^2) for nitrogen ions $^{14}_{7}N^{+}$ [1] *(continued)*

E_m, MeV/u	H	He	N	O	Ne	Ar	Kr	Xe	Rn	Mylar	$(CH_2)_n$	Water	E, MeV
0.0125	6.754	2.881	1.909	1.794	1.594	1.052	0.645	0.457	0.302	2.415	3.072	2.347	0.1750
0.0160	7.498	3.162	2.125	1.999	1.777	1.200	0.744	0.530	0.352	2.716	3.456	2.610	0.2240
0.0200	8.155	3.349	2.317	2.182	1.952	1.355	0.845	0.615	0.411	3.012	3.829	2.845	0.2801
0.0250	8.863	3.517	2.527	2.381	2.155	1.546	0.988	0.721	0.522	3.336	4.246	3.102	0.3501
0.0320	9.826	3.729	2.763	2.631	2.403	1.801	1.170	0.882	0.614	3.746	4.778	3.429	0.4481
0.0400	10.792	3.924	3.006	2.893	2.664	2.081	1.379	1.057	0.751	4.172	5.322	3.769	0.5601
0.0500	11.807	4.152	3.278	3.189	2.968	2.394	1.612	1.256	0.910	4.616	5.886	4.145	0.7001
0.0600	12.907	4.417	3.582	3.462	3.273	2.693	1.832	1.441	1.059	5.027	6.418	4.513	0.8402
0.0700	13.977	4.733	3.876	3.764	3.570	2.977	2.028	1.612	1.192	5.417	6.910	4.898	0.9802
0.0800	15.095	5.068	4.180	4.063	3.867	3.249	2.227	1.775	1.318	5.794	7.382	5.290	1.1202
0.0900	16.155	5.420	4.482	4.369	4.159	3.509	2.412	1.929	1.429	6.152	7.823	5.679	1.2603
0.1000	17.306	5.807	4.814	4.662	4.435	3.759	2.596	2.066	1.532	6.492	8.256	6.067	1.4003
0.1250	20.214	6.747	5.576	5.372	5.133	4.333	2.984	2.378	1.767	7.302	9.293	7.022	1.7504
0.1600	24.160	8.055	6.610	6.370	6.043	5.034	3.452	2.749	2.040	8.328	10.580	8.344	2.2405
0.2000	28.110	9.332	7.643	7.339	6.939	5.639	3.836	3.046	2.250	9.240	11.725	9.647	2.8006
0.2500	32.135	10.637	8.627	8.307	7.755	6.131	4.144	3.249	2.389	10.044	12.730	10.957	3.5008
0.3200	36.060	11.771	9.574	9.137	8.359	6.405	4.274	3.333	2.434	10.697	13.586	12.129	4.4810
0.4000	38.233	12.622	10.024	9.528	8.627	6.440	4.258	3.315	2.414	10.986	13.988	12.720	5.6012
0.5000	39.078	12.985	10.088	9.598	8.587	6.315	4.165	3.252	2.370	10.982	14.020	12.875	7.0015
0.6000	38.389	12.877	9.745	9.262	8.276	6.052	4.014	3.132	2.291	10.652	13.669	12.496	8.4018
0.7000	36.552	12.561	9.239	8.781	7.858	5.739	3.840	2.988	2.209	10.210	13.121	11.865	9.8021
0.8000	34.352	12.055	8.714	8.310	7.426	5.431	3.675	2.868	2.130	9.738	12.500	11.201	11.202
0.9000	32.174	11.468	8.216	7.831	6.963	5.136	3.505	2.769	2.057	9.279	11.899	10.537	12.603
1.0000	30.373	10.983	7.838	7.438	6.627	4.908	3.381	2.694	2.000	8.684	11.355	9.985	14.003
1.2500	26.604	9.962	7.093	6.771	6.017	4.512	3.169	2.559	1.932	8.013	10.176	8.977	17.504
1.6000	22.444	8.789	6.346	6.068	5.432	4.146	2.974	2.418	1.872	7.076	8.854	7.885	22.405
2.0000	18.866	7.733	5.733	5.459	4.940	3.837	2.803	2.301	1.809	6.289	7.747	6.949	28.006
2.5000	15.539	6.680	5.102	4.855	4.481	3.544	2.631	2.191	1.743	5.533	6.697	6.043	35.007
3.2000	12.601	5.599	4.439	4.238	3.935	3.186	2.423	2.016	1.617	4.763	5.678	5.166	44.810
4.0000	10.386	4.696	3.840	3.682	3.462	2.832	2.172	1.828	1.479	4.110	4.843	4.428	56.012
5.0000	8.648	3.930	3.288	3.165	2.983	2.472	1.923	1.624	1.333	3.519	4.120	3.775	70.015
6.0000	7.454	3.389	2.881	2.771	2.624	2.191	1.711	1.455	1.198	3.077	3.593	3.291	84.018
7.0000	6.586	2.991	2.570	2.478	2.343	1.969	1.549	1.321	1.092	2.744	3.194	2.935	98.021
8.0000	5.909	2.673	2.323	2.241	2.117	1.786	1.405	1.205	1.001	2.477	2.887	2.650	112.02
9.0000	5.355	2.434	2.119	2.046	1.936	1.637	1.290	1.108	0.925	2.256	2.617	2.414	126.03
10.0000	4.908	2.241	1.955	1.888	1.786	1.510	1.193	1.026	0.859	2.075	2.403	2.223	140.03
11.0000	4.529	2.077	1.816	1.754	1.659	1.403	1.112	0.956	0.801	1.923	2.225	2.062	154.03
12.0000	4.203	1.936	1.697	1.639	1.550	1.311	1.042	0.896	0.751	1.792	2.070	1.924	168.04

Table 42.14 Range (in mg/cm^2) of nitrogen ions $^{14}_{7}N^{+}$ for different substances [1] [E_m is the energy per mass unit of incident particle; E is a total energy.]

E_m, MeV/u	Be	C	Al	Ti	Ni	Ge	Zr	Ag	Eu	Ta	Au	U	E, MeV
0.0125	0.073	0.083	0.116	0.176	0.209	0.247	0.272	0.303	0.474	0.561	0.614	0.734	0.1750
0.0160	0.088	0.100	0.139	0.210	0.250	0.293	0.322	0.358	0.559	0.661	0.721	0.861	0.2240
0.0200	0.103	0.118	0.163	0.246	0.293	0.341	0.375	0.414	0.647	0.763	0.833	0.992	0.2801
0.0250	0.120	0.138	0.191	0.287	0.342	0.396	0.434	0.479	0.747	0.880	0.960	1.141	0.3501
0.0320	0.141	0.163	0.225	0.338	0.405	0.466	0.509	0.560	0.874	1.028	1.120	1.328	0.4481
0.0400	0.163	0.189	0.261	0.391	0.469	0.537	0.586	0.644	1.004	1.180	1.285	1.520	0.5601
0.0500	0.188	0.219	0.301	0.451	0.542	0.617	0.673	0.737	1.151	1.349	1.469	1.734	0.7001
0.0600	0.211	0.245	0.338	0.505	0.608	0.691	0.752	0.823	1.284	1.503	1.635	1.928	0.8402
0.0700	0.232	0.270	0.372	0.556	0.669	0.758	0.826	0.902	1.407	1.643	1.787	2.105	0.9802
0.0800	0.252	0.294	0.405	0.604	0.727	0.822	0.894	0.976	1.521	1.774	1.928	2.269	1.1202
0.0900	0.272	0.316	0.436	0.649	0.781	0.882	0.960	1.046	1.628	1.896	2.060	2.422	1.2603
0.1000	0.291	0.337	0.466	0.693	0.833	0.940	1.023	1.114	1.730	2.012	2.185	2.567	1.4003
0.1250	0.335	0.387	0.538	0.795	0.956	1.075	1.169	1.271	1.966	2.282	2.475	2.903	1.7504
0.1600	0.394	0.449	0.630	0.926	1.112	1.247	1.357	1.470	2.261	2.619	2.835	3.317	2.2405
0.2000	0.458	0.513	0.730	1.064	1.276	1.429	1.554	1.681	2.568	2.968	3.207	3.743	2.8006
0.2500	0.536	0.589	0.850	1.228	1.469	1.642	1.787	1.929	2.923	3.369	3.635	4.233	3.5008
0.3200	0.644	0.690	1.013	1.447	1.726	1.924	2.095	2.259	3.387	3.893	4.193	4.870	4.4810
0.4000	0.769	0.803	1.197	1.689	2.008	2.233	2.435	2.622	3.890	4.458	4.795	5.555	5.6012
0.5000	0.927	0.945	1.425	1.987	2.353	2.609	2.847	3.064	4.493	5.135	5.515	6.374	7.0015
0.6000	1.089	1.089	1.655	2.282	2.693	2.979	3.254	3.501	5.080	5.794	6.215	7.165	8.4018
0.7000	1.257	1.238	1.889	2.579	3.033	3.348	3.661	3.938	5.662	6.444	6.904	7.941	9.8021
0.8000	1.430	1.393	2.126	2.878	3.374	3.717	4.070	4.376	6.240	7.089	7.587	8.712	11.202
0.9000	1.609	1.555	2.368	3.180	3.718	4.088	4.480	4.817	6.817	7.731	8.267	9.476	12.603
1.0000	1.793	1.722	2.614	3.485	4.064	4.461	4.893	5.261	7.394	8.373	8.946	10.236	14.003
1.2500	2.279	2.171	3.251	4.268	4.947	5.410	5.945	6.393	8.847	9.986	10.653	12.145	17.504
1.6000	3.021	2.871	4.202	5.417	6.235	6.791	7.477	8.047	10.932	12.294	13.093	14.868	22.405
2.0000	3.962	3.771	5.380	6.817	7.798	8.461	9.335	10.056	13.425	15.036	15.994	18.095	28.006
2.5000	5.283	5.034	6.999	8.715	9.904	10.703	11.839	12.756	16.736	18.670	19.830	22.339	35.007
3.2000	7.409	7.055	9.552	11.668	13.157	14.164	15.703	16.927	21.769	24.188	25.644	28.743	44.810
4.0000	10.237	9.721	12.887	15.481	17.333	18.601	20.640	22.271	28.126	31.134	32.948	36.777	56.012
5.0000	14.369	13.587	17.687	20.920	23.254	24.882	27.610	29.816	36.987	40.828	43.092	47.909	70.015
6.0000	19.153	18.044	23.183	27.102	29.961	31.974	35.472	38.317	46.907	51.635	54.376	60.230	84.018
7.0000	24.584	23.083	29.369	34.021	37.454	39.868	44.208	47.759	57.873	63.540	66.795	73.734	98.021
8.0000	30.658	28.700	36.235	41.667	45.721	48.564	53.797	58.131	69.834	76.530	80.323	88.419	112.02
9.0000	37.359	34.891	43.773	50.128	54.748	58.052	64.229	69.414	82.796	90.580	94.944	104.253	126.03
10.0000	44.678	41.656	51.974	59.190	64.529	68.316	75.510	81.582	96.755	105.669	110.624	121.215	140.03
11.0000	52.615	48.985	60.832	68.839	75.048	79.346	87.626	94.607	111.691	121.757	127.303	139.290	154.03
12.0000	61.160	56.874	70.337	79.273	86.290	91.125	100.559	108.473	127.573	138.822	144.987	158.473	168.04

Table 42.14 Range (in mg/cm^2) of nitrogen ions $^{14}_{7}N^{+}$ for different substances [1] *(continued)*

E_m, MeV/u	H	He	N	O	Ne	Ar	Kr	Xe	Rn	Mylar	$(CH_2)_n$	Water	E, MeV
0.0125	0.027	0.069	0.100	0.106	0.119	0.186	0.318	0.469	0.733	0.080	0.063	0.077	0.1750
0.0160	0.033	0.084	0.122	0.129	0.144	0.224	0.379	0.555	0.863	0.097	0.076	0.095	0.2240
0.0200	0.039	0.100	0.145	0.154	0.171	0.263	0.442	0.643	0.994	0.116	0.090	0.114	0.2801
0.0250	0.047	0.119	0.172	0.182	0.203	0.308	0.512	0.739	1.133	0.136	0.107	0.136	0.3501
0.0320	0.057	0.145	0.207	0.219	0.244	0.363	0.597	0.854	1.295	0.163	0.128	0.164	0.4481
0.0400	0.067	0.173	0.244	0.258	0.286	0.419	0.681	0.965	1.452	0.190	0.149	0.194	0.5601
0.0500	0.079	0.207	0.288	0.303	0.334	0.479	0.772	1.082	1.615	0.221	0.173	0.228	0.7001
0.0600	0.090	0.239	0.327	0.344	0.378	0.533	0.851	1.184	1.754	0.250	0.196	0.260	0.8402
0.0700	0.100	0.269	0.364	0.382	0.418	0.582	0.922	1.274	1.876	0.276	0.216	0.289	0.9802
0.0800	0.110	0.297	0.398	0.417	0.455	0.626	0.987	1.355	1.986	0.301	0.236	0.316	1.1202
0.0900	0.119	0.324	0.430	0.450	0.489	0.667	1.048	1.431	2.088	0.324	0.254	0.341	1.2603
0.1000	0.127	0.349	0.460	0.481	0.522	0.706	1.104	1.501	2.183	0.346	0.272	0.365	1.4003
0.1250	0.146	0.405	0.528	0.550	0.595	0.792	1.229	1.658	2.395	0.397	0.312	0.419	1.7504
0.1600	0.168	0.471	0.608	0.634	0.683	0.897	1.381	1.850	2.653	0.460	0.361	0.483	2.2405
0.2000	0.189	0.535	0.687	0.716	0.769	1.002	1.535	2.043	2.914	0.523	0.411	0.545	2.8006
0.2500	0.212	0.606	0.773	0.805	0.864	1.121	1.710	2.265	3.215	0.596	0.468	0.613	3.5008
0.3200	0.241	0.693	0.881	0.918	0.986	1.277	1.943	2.562	3.621	0.690	0.543	0.698	4.4810
0.4000	0.271	0.785	0.995	1.038	1.118	1.451	2.205	2.899	4.082	0.794	0.624	0.788	5.6012
0.5000	0.308	0.894	1.134	1.184	1.280	1.671	2.537	3.325	4.667	0.921	0.724	0.897	7.0015
0.6000	0.344	1.002	1.275	1.332	1.446	1.897	2.880	3.764	5.268	1.050	0.825	1.007	8.4018
0.7000	0.381	1.112	1.423	1.488	1.620	2.135	3.236	4.222	5.890	1.185	0.929	1.122	9.8021
0.8000	0.421	1.226	1.579	1.652	1.803	2.386	3.609	4.700	6.536	1.325	1.039	1.244	11.202
0.9000	0.463	1.345	1.744	1.825	1.998	2.651	3.999	5.197	7.205	1.472	1.154	1.373	12.603
1.0000	0.507	1.470	1.919	2.009	2.204	2.930	4.406	5.710	7.895	1.627	1.274	1.509	14.003
1.2500	0.631	1.805	2.388	2.502	2.758	3.674	5.476	7.043	9.675	2.042	1.600	1.879	17.504
1.6000	0.831	2.329	3.119	3.267	3.616	4.807	7.072	9.013	12.251	2.693	2.116	2.462	22.405
2.0000	1.104	3.009	4.048	4.240	4.698	6.212	9.012	11.388	15.295	3.533	2.793	3.219	28.006
2.5000	1.513	3.984	5.344	5.602	6.187	8.112	11.592	14.507	19.239	4.722	3.766	4.301	35.007
3.2000	2.215	5.590	7.408	7.767	8.526	11.034	15.479	19.176	25.085	6.635	5.360	6.059	44.810
4.0000	3.197	7.781	10.128	10.610	11.568	14.770	20.372	25.021	32.341	9.173	7.501	8.406	56.012
5.0000	4.680	11.051	14.080	14.723	15.937	20.076	27.242	33.168	42.337	12.865	10.645	11.842	70.015
6.0000	6.428	14.897	18.640	19.461	20.952	26.106	34.977	42.297	53.439	17.130	14.293	15.824	84.018
7.0000	8.430	19.303	23.795	24.815	26.609	32.858	43.592	52.411	65.698	21.957	18.433	20.337	98.021
8.0000	10.678	24.263	29.534	30.765	32.905	40.334	53.095	63.521	79.107	27.336	23.059	25.366	112.02
9.0000	13.170	29.760	35.852	37.311	39.831	48.531	63.506	75.649	93.672	33.267	28.168	30.908	126.03
10.0000	15.904	35.762	42.739	44.443	47.369	57.446	74.803	88.792	109.399	39.745	33.757	36.959	140.03
11.0000	18.876	42.258	50.175	52.145	55.510	67.074	86.969	102.941	126.301	46.761	39.817	43.505	154.03
12.0000	22.087	49.245	58.157	60.410	64.246	77.406	99.990	118.077	144.373	54.308	46.346	50.542	168.04

Table 42.15 Mass stopping power of different substances (in MeV·mg^{-1}·cm^2) for oxygen ions $^{16}_{8}O^+$ [1] [E_m is the energy per mass unit of incident particle; E is a total energy.]

E_m, MeV/u	Be	C	Al	Ti	Ni	Ge	Zr	Ag	Eu	Ta	Au	U	E, MeV
0.0125	3.228	2.671	1.956	1.301	1.056	0.959	0.888	0.822	0.522	0.447	0.411	0.352	0.1999
0.0160	3.652	3.021	2.213	1.472	1.195	1.085	1.005	0.930	0.591	0.506	0.465	0.398	0.2559
0.0200	4.083	3.378	2.475	1.646	1.336	1.213	1.124	1.039	0.661	0.565	0.520	0.445	0.3199
0.0250	4.565	3.777	2.767	1.840	1.494	1.356	1.256	1.162	0.739	0.632	0.582	0.498	0.3999
0.0320	5.159	4.276	3.130	2.082	1.694	1.540	1.421	1.315	0.836	0.717	0.660	0.567	0.5118
0.0400	5.754	4.791	3.500	2.334	1.904	1.736	1.599	1.480	0.941	0.810	0.745	0.640	0.6398
0.0500	6.363	5.334	3.885	2.603	2.133	1.942	1.791	1.659	1.053	0.915	0.843	0.725	0.7997
0.0600	6.890	5.830	4.222	2.846	2.339	2.124	1.963	1.820	1.165	1.013	0.935	0.804	0.9597
0.0700	7.347	6.284	4.521	3.065	2.532	2.297	2.125	1.971	1.266	1.108	1.026	0.882	1.1196
0.0800	7.748	6.714	4.789	3.271	2.706	2.452	2.265	2.107	1.365	1.197	1.106	0.953	1.2796
0.0900	8.103	7.113	5.030	3.461	2.867	2.606	2.399	2.233	1.464	1.280	1.187	1.024	1.4395
0.1000	8.423	7.489	5.248	3.637	3.023	2.739	2.519	2.356	1.553	1.359	1.260	1.086	1.5995
0.1250	9.075	8.373	5.707	4.018	3.356	3.042	2.797	2.620	1.755	1.535	1.435	1.241	1.9994
0.1600	9.710	9.438	6.189	4.431	3.726	3.379	3.101	2.909	1.980	1.739	1.631	1.423	2.5592
0.2000	10.171	10.323	6.571	4.777	4.041	3.660	3.358	3.147	2.181	1.925	1.807	1.577	3.1990
0.2500	10.478	11.022	6.880	5.084	4.321	3.922	3.591	3.364	2.374	2.098	1.968	1.720	3.9988
0.3200	10.635	11.555	7.128	5.367	4.583	4.184	3.814	3.564	2.552	2.267	2.128	1.871	5.1184
0.4000	10.635	11.806	7.274	5.543	4.772	4.364	3.979	3.717	2.706	2.408	2.262	1.986	6.3980
0.5000	10.525	11.828	7.360	5.689	4.931	4.526	4.122	3.842	2.841	2.532	2.381	2.098	7.9975
0.6000	10.274	11.571	7.328	5.738	4.998	4.602	4.177	3.891	2.909	2.601	2.451	2.176	9.5970
0.7000	10.004	11.230	7.255	5.731	5.013	4.621	4.186	3.903	2.945	2.633	2.488	2.205	11.196
0.8000	9.716	10.832	7.155	5.702	5.001	4.629	4.185	3.899	2.962	2.662	2.511	2.232	12.796
0.9000	9.445	10.438	7.038	5.645	4.976	4.617	4.167	3.878	2.977	2.675	2.527	2.252	14.395
1.0000	9.172	10.043	6.912	5.592	4.935	4.583	4.140	3.850	2.972	2.675	2.530	2.260	15.995
1.2500	8.532	9.137	6.578	5.394	4.802	4.473	4.033	3.743	2.941	2.651	2.506	2.243	19.994
1.6000	7.743	8.116	6.116	5.107	4.562	4.263	3.841	3.547	2.844	2.575	2.434	2.183	25.592
2.0000	6.978	7.282	5.632	4.770	4.286	4.021	3.604	3.340	2.703	2.467	2.332	2.101	31.990
2.5000	6.196	6.492	5.108	4.338	3.969	3.729	3.335	3.095	2.544	2.314	2.196	1.992	39.987
3.2000	5.360	5.662	4.508	3.922	3.570	3.358	3.011	2.786	2.326	2.128	2.020	1.835	51.184
4.0000	4.646	4.947	3.971	3.490	3.196	3.010	2.712	2.501	2.116	1.938	1.846	1.680	63.980
5.0000	3.988	4.275	3.456	3.062	2.820	2.661	2.398	2.219	1.901	1.735	1.662	1.517	79.975
6.0000	3.502	3.765	3.061	2.731	2.519	2.388	2.155	1.993	1.708	1.577	1.509	1.387	95.970
7.0000	3.118	3.368	2.749	2.463	2.276	2.164	1.957	1.812	1.567	1.441	1.383	1.273	111.96
8.0000	2.814	3.046	2.497	2.247	2.080	1.977	1.798	1.660	1.443	1.331	1.278	1.178	127.96
9.0000	2.569	2.779	2.288	2.066	1.915	1.823	1.659	1.535	1.338	1.235	1.187	1.098	143.95
10.0000	2.361	2.556	2.112	1.916	1.774	1.692	1.540	1.430	1.246	1.155	1.113	1.029	159.95
11.0000	2.186	2.368	1.962	1.786	1.656	1.580	1.438	1.340	1.169	1.087	1.050	0.967	175.94
12.0000	2.036	2.205	1.833	1.672	1.552	1.483	1.351	1.261	1.102	1.026	0.990	0.913	191.94

Table 42.15 Mass stopping power of different substances (in $MeV \cdot mg^{-1} \cdot cm^2$) for oxygen ions ${}^{16}_{8}O^{+}$ [1] *(continued)*

E_m, MeV/u	H	He	N	O	Ne	Ar	Kr	Xe	Rn	Mylar	$(CH_2)_n$	Water	E, MeV
0.0125	7.337	3.130	2.074	1.949	1.731	1.143	0.700	0.497	0.328	2.624	3.338	2.547	0.1999
0.0160	8.146	3.435	2.309	2.171	1.930	1.304	0.808	0.576	0.382	2.951	3.754	2.835	0.2559
0.0200	8.860	3.638	2.517	2.371	2.121	1.472	0.918	0.668	0.447	3.272	4.160	3.091	0.3199
0.0250	9.629	3.821	2.745	2.587	2.341	1.679	1.074	0.783	0.567	3.625	4.612	3.370	0.3999
0.0320	10.674	4.051	3.002	2.858	2.611	1.956	1.271	0.958	0.667	4.069	5.190	3.725	0.5118
0.0400	11.724	4.263	3.265	3.143	2.894	2.261	1.498	1.148	0.815	4.532	5.782	4.095	0.6398
0.0500	12.859	4.522	3.570	3.473	3.232	2.607	1.756	1.367	0.991	5.027	6.410	4.514	0.7997
0.0600	14.101	4.826	3.914	3.783	3.576	2.943	2.001	1.575	1.157	5.493	7.012	4.931	0.9597
0.0700	15.326	5.190	4.250	4.128	3.915	3.264	2.224	1.768	1.307	5.941	7.577	5.371	1.1196
0.0800	16.617	5.579	4.602	4.473	4.257	3.577	2.452	1.954	1.451	6.379	8.127	5.823	1.2796
0.0900	17.857	5.991	4.955	4.829	4.597	3.878	2.666	2.133	1.579	6.801	8.647	6.278	1.4395
0.1000	19.208	6.445	5.342	5.175	4.923	4.172	2.881	2.293	1.700	7.205	9.163	6.733	1.5995
0.1250	22.658	7.562	6.250	6.021	5.753	4.857	3.345	2.665	1.980	8.184	10.416	7.871	1.9994
0.1600	27.416	9.141	7.501	7.228	6.857	5.712	3.917	3.119	2.315	9.450	12.006	9.469	2.5592
0.2000	32.262	10.710	8.772	8.424	7.964	6.472	4.402	3.496	2.582	10.605	13.457	11.072	3.1990
0.2500	37.290	12.343	10.010	9.639	8.999	7.114	4.809	3.770	2.773	11.655	14.771	12.714	3.9988
0.3200	42.341	13.821	11.241	10.728	9.815	7.520	5.018	3.913	2.858	12.560	15.953	14.242	5.1184
0.4000	45.391	14.985	11.900	11.311	10.242	7.645	5.056	3.935	2.866	13.043	16.607	15.101	5.3980
0.5000	46.957	15.603	12.122	11.533	10.319	7.588	5.005	3.908	2.848	13.196	16.847	15.471	7.9975
0.6000	46.534	15.609	11.813	11.227	10.032	7.335	4.866	3.796	2.777	12.912	16.569	15.147	9.5970
0.7000	44.543	15.307	11.259	10.700	9.576	6.993	4.679	3.642	2.691	12.442	15.989	14.458	11.196
0.8000	41.998	14.739	10.653	10.160	9.079	6.639	4.493	3.506	2.604	11.905	15.282	13.694	12.796
0.9000	39.414	14.048	10.065	9.593	8.530	6.292	4.293	3.392	2.520	11.367	14.576	12.908	14.395
1.0000	37.256	13.471	9.615	9.124	8.129	6.020	4.147	3.304	2.454	10.873	13.928	12.248	15.995
1.2500	32.694	12.242	8.716	8.322	7.394	5.546	3.894	3.144	2.375	9.848	12.505	11.032	19.994
1.6000	27.644	10.825	7.816	7.474	6.691	5.107	3.663	2.978	2.306	8.715	10.905	9.712	25.592
2.0000	23.317	9.558	7.085	6.747	6.105	4.742	3.464	2.844	2.236	7.772	9.574	8.589	31.990
2.5000	19.307	8.300	6.339	6.032	5.567	4.403	3.269	2.722	2.166	6.875	8.320	7.508	39.987
3.2000	15.777	7.010	5.558	5.306	4.927	3.989	3.034	2.524	2.024	5.964	7.109	6.469	51.184
4.0000	13.103	5.924	4.844	4.646	4.368	3.574	2.740	2.307	1.866	5.186	6.111	5.587	63.980
5.0000	10.990	4.994	4.178	4.023	3.791	3.142	2.443	2.063	1.693	4.472	5.236	4.797	79.975
6.0000	9.520	4.329	3.680	3.539	3.352	2.798	2.186	1.858	1.531	3.931	4.589	4.203	95.970
7.0000	8.440	3.832	3.294	3.175	3.002	2.524	1.985	1.694	1.399	3.516	4.094	3.761	111.96
8.0000	7.590	3.433	2.983	2.879	2.719	2.294	1.805	1.548	1.286	3.181	3.695	3.403	127.96
9.0000	6.886	3.129	2.725	2.631	2.489	2.105	1.659	1.425	1.190	2.901	3.365	3.104	143.95
10.0000	6.315	2.883	2.515	2.429	2.298	1.943	1.535	1.320	1.105	2.670	3.092	2.860	159.95
11.0000	5.828	2.673	2.337	2.257	2.135	1.805	1.430	1.230	1.030	2.474	2.863	2.653	175.94
12.0000	5.407	2.491	2.183	2.108	1.994	1.686	1.340	1.153	0.966	2.306	2.663	2.474	191.94

Table 42.16 Range (in mg/cm^2) of oxygen ions $^{16}_{8}O^+$ for different substances [1] [E_m is the energy per mass unit of incident particle; E is a total energy.]

E_m, MeV/u	Be	C	Al	Ti	Ni	Ge	Zr	Ag	Eu	Ta	Au	U	E, MeV
0.0125	0.075	0.084	0.117	0.176	0.208	0.247	0.273	0.304	0.474	0.562	0.614	0.736	0.1999
0.0160	0.090	0.101	0.141	0.212	0.251	0.295	0.324	0.360	0.561	0.663	0.725	0.866	0.2559
0.0200	0.106	0.120	0.166	0.249	0.295	0.345	0.378	0.419	0.662	0.770	0.840	1.002	0.3199
0.0250	0.123	0.141	0.194	0.291	0.346	0.402	0.440	0.486	0.756	0.891	0.971	1.156	0.3999
0.0320	0.146	0.167	0.230	0.345	0.411	0.474	0.518	0.570	0.887	1.044	1.138	1.350	0.5118
0.0400	0.168	0.195	0.267	0.400	0.478	0.549	0.598	0.657	1.023	1.202	1.309	1.550	0.6398
0.0500	0.194	0.225	0.309	0.462	0.554	0.632	0.689	0.755	1.176	1.378	1.500	1.773	0.7997
0.0600	0.218	0.253	0.348	0.519	0.623	0.709	0.771	0.844	1.314	1.538	1.673	1.974	0.9597
0.0700	0.240	0.279	0.384	0.572	0.687	0.779	0.848	0.926	1.442	1.684	1.831	2.158	1.1196
0.0800	0.262	0.304	0.418	0.622	0.746	0.845	0.919	1.003	1.560	1.819	1.977	2.328	1.2796
0.0900	0.282	0.327	0.450	0.668	0.802	0.907	0.986	1.076	1.671	1.945	2.113	2.486	1.4395
0.1000	0.301	0.349	0.481	0.714	0.856	0.966	1.050	1.144	1.775	2.064	2.241	2.635	1.5995
0.1250	0.347	0.399	0.554	0.818	0.981	1.104	1.201	1.305	2.017	2.340	2.538	2.978	1.9994
0.1600	0.406	0.462	0.648	0.950	1.139	1.279	1.390	1.507	2.316	2.682	2.903	3.399	2.5592
0.2000	0.470	0.527	0.748	1.089	1.304	1.460	1.588	1.718	2.624	3.031	3.275	3.825	3.1990
0.2500	0.548	0.602	0.867	1.251	1.495	1.671	1.818	1.963	2.974	3.428	3.699	4.309	3.9988
0.3200	0.654	0.701	1.026	1.465	1.746	1.947	2.120	2.286	3.428	3.941	4.245	4.932	5.1184
0.4000	0.774	0.810	1.204	1.699	2.019	2.246	2.448	2.637	3.915	4.488	4.827	5.595	6.3980
0.5000	0.925	0.945	1.422	1.984	2.349	2.605	2.843	3.060	4.491	5.135	5.516	6.379	7.9975
0.6000	1.079	1.082	1.640	2.264	2.671	2.956	3.228	3.474	5.047	5.758	6.178	7.127	9.5970
0.7000	1.236	1.222	1.859	2.542	2.990	3.302	3.610	3.884	5.593	6.369	6.825	7.856	11.196
0.8000	1.398	1.367	2.081	2.822	3.309	3.648	3.992	4.294	6.135	6.973	7.465	8.577	12.796
0.9000	1.565	1.518	2.306	3.104	3.630	3.994	4.375	4.705	6.673	7.572	8.099	9.290	14.395
1.0000	1.737	1.674	2.536	3.389	3.953	4.432	4.760	5.119	7.211	8.170	8.732	9.999	15.995
1.2500	2.189	2.091	3.128	4.117	4.774	5.225	5.739	6.172	8.563	9.671	10.319	11.774	19.994
1.6000	2.878	2.741	4.011	5.183	5.970	6.507	7.161	7.708	10.498	11.813	12.585	14.303	25.592
2.0000	3.749	3.574	5.102	6.480	7.417	8.052	8.881	9.568	12.807	14.352	15.271	17.291	31.990
2.5000	4.967	4.739	6.594	8.230	9.358	10.119	11.190	12.057	15.858	17.702	18.808	21.202	39.987
3.2000	6.914	6.589	8.932	10.933	12.337	13.288	14.728	15.876	20.468	22.755	24.131	27.067	51.184
4.0000	9.483	9.012	11.963	14.398	16.132	17.320	19.214	20.732	26.244	29.067	30.768	34.367	63.980
5.0000	13.209	12.499	16.291	19.303	21.471	22.985	25.499	27.536	34.235	37.809	39.916	44.407	79.975
6.0000	17.497	16.493	21.218	24.844	27.483	29.342	32.547	35.156	43.128	47.496	50.031	55.451	95.970
7.0000	22.346	20.992	26.740	31.021	34.172	36.389	40.345	43.586	52.916	58.124	61.117	67.505	111.96
8.0000	27.753	25.992	32.852	37.828	41.532	44.131	48.882	52.819	63.565	69.689	73.160	80.579	127.96
9.0000	33.709	31.495	39.552	45.259	49.555	52.563	58.155	62.848	75.085	82.175	86.156	94.653	143.95
10.0000	40.209	37.502	46.835	53.306	58.240	61.678	68.172	73.653	87.481	95.574	100.080	109.714	159.95
11.0000	47.254	44.008	54.698	61.961	67.578	71.469	78.928	85.215	100.740	109.856	114.886	125.760	175.94
12.0000	54.840	51.012	63.137	71.224	77.559	81.926	90.409	97.525	114.840	125.007	130.587	142.791	191.94

Table 42.16 Range (in mg/cm^2) of oxygen ions $^{16}_{8}O^+$ for different substances [1] *(continued)*

E_m, MeV/u	H	He	N	O	Ne	Ar	Kr	Xe	Rn	Mylar	$(CH_2)_n$	Water	E, MeV
0.0125	0.027	0.070	0.101	0.107	0.119	0.185	0.316	0.466	0.729	0.081	0.063	0.078	0.1999
0.0160	0.033	0.085	0.123	0.131	0.145	0.225	0.379	0.555	0.863	0.099	0.077	0.096	0.2559
0.0200	0.040	0.102	0.147	0.156	0.174	0.266	0.444	0.645	0.998	0.118	0.092	0.116	0.3199
0.0250	0.048	0.122	0.175	0.186	0.206	0.312	0.517	0.745	1.141	0.139	0.109	0.138	0.3999
0.0320	0.058	0.149	0.212	0.224	0.249	0.370	0.605	0.865	1.309	0.167	0.131	0.168	0.5118
0.0400	0.069	0.179	0.251	0.265	0.293	0.427	0.693	0.980	1.473	0.195	0.153	0.199	0.6398
0.0500	0.082	0.214	0.296	0.311	0.343	0.491	0.787	1.102	1.643	0.228	0.179	0.235	0.7997
0.0600	0.093	0.248	0.337	0.354	0.388	0.547	0.870	1.208	1.787	0.258	0.202	0.268	0.9597
0.0700	0.104	0.279	0.376	0.394	0.430	0.597	0.944	1.302	1.914	0.285	0.224	0.298	1.1196
0.0800	0.113	0.308	0.411	0.430	0.469	0.643	1.011	1.386	2.028	0.311	0.244	0.326	1.2796
0.0900	0.123	0.335	0.444	0.464	0.504	0.685	1.073	1.464	2.134	0.335	0.263	0.352	1.4395
0.1000	0.131	0.361	0.475	0.496	0.538	0.725	1.130	1.537	2.231	0.357	0.281	0.377	1.5995
0.1250	0.150	0.418	0.544	0.567	0.613	0.814	1.259	1.698	2.449	0.409	0.322	0.432	1.9994
0.1600	0.173	0.485	0.626	0.652	0.702	0.920	1.413	1.892	2.710	0.473	0.372	0.496	2.5592
0.2000	0.194	0.550	0.705	0.734	0.788	1.025	1.567	2.085	2.971	0.537	0.422	0.559	3.1990
0.2500	0.217	0.619	0.790	0.822	0.882	1.143	1.740	2.305	3.269	0.609	0.479	0.626	3.9988
0.3200	0.245	0.705	0.895	0.932	1.001	1.295	1.968	2.596	3.666	0.701	0.551	0.709	5.1184
0.4000	0.274	0.794	1.006	1.048	1.129	1.464	2.222	2.922	4.113	0.801	0.630	0.796	6.3980
0.5000	0.309	0.898	1.139	1.188	1.284	1.674	2.539	3.329	4.672	0.923	0.725	0.901	7.9975
0.6000	0.343	1.001	1.272	1.329	1.441	1.888	2.863	3.744	5.240	1.045	0.821	1.005	9.5970
0.7000	0.378	1.104	1.411	1.475	1.604	2.111	3.198	4.174	5.825	1.171	0.919	1.113	11.196
0.8000	0.415	1.211	1.557	1.628	1.776	2.346	3.547	4.622	6.429	1.303	1.022	1.227	12.796
0.9000	0.455	1.322	1.711	1.790	1.958	2.593	3.911	5.085	7.054	1.440	1.129	1.347	14.395
1.0000	0.496	1.438	1.874	1.961	2.150	2.853	4.290	5.563	7.697	1.584	1.241	1.474	15.995
1.2500	0.611	1.749	2.311	2.420	2.665	3.545	5.285	6.803	9.353	1.970	1.544	1.818	19.994
1.6000	0.797	2.236	2.989	3.130	3.462	4.597	6.767	8.633	11.745	2.575	2.024	2.359	25.592
2.0000	1.049	2.865	3.849	4.031	4.463	5.898	8.564	10.831	14.563	3.353	2.650	3.060	31.990
2.5000	1.427	3.764	5.044	5.286	5.836	7.650	10.942	13.707	18.198	4.448	3.547	4.058	39.987
3.2000	2.070	5.235	6.934	7.269	7.978	10.325	14.502	17.983	23.551	6.200	5.006	5.667	51.184
4.0000	2.962	7.226	9.405	9.852	10.741	13.720	18.948	23.293	30.145	8.506	6.952	7.801	63.980
5.0000	4.299	10.175	12.970	13.561	14.682	18.505	25.143	30.640	39.159	11.836	9.787	10.898	79.975
6.0000	5.866	13.622	17.057	17.809	19.177	23.910	32.077	38.823	49.111	15.658	13.057	14.468	95.970
7.0000	7.653	17.556	21.658	22.587	24.227	29.938	39.767	47.852	60.055	19.967	16.753	18.497	111.96
8.0000	9.654	21.971	26.767	27.884	29.833	36.593	48.227	57.743	71.992	24.756	20.871	22.974	127.96
9.0000	11.869	26.856	32.383	33.703	35.988	43.879	57.480	68.522	84.938	30.028	25.412	27.900	143.95
10.0000	14.297	32.186	38.498	40.036	42.682	51.795	67.512	80.194	98.903	35.780	30.375	33.273	159.95
11.0000	16.935	37.953	45.100	46.873	49.908	60.342	78.312	92.754	113.908	42.008	35.755	39.084	175.94
12.0000	19.786	44.156	52.186	54.211	57.665	69.515	89.873	106.192	129.952	48.709	41.551	45.331	191.94

Table 42.17 Mass stopping power of different substances (in MeV·mg^{-1}·cm^2) for flourine ions $^{19}_{9}F^+$ [1] [E_m is the energy per mass unit of incident particle; E is a total energy.]

E_m, MeV/u	Be	C	Al	Ti	Ni	Ge	Zr	Ag	Eu	Ta	Au	U	E, MeV
0.0125	3.408	2.819	2.066	1.374	1.115	1.012	0.938	0.868	0.551	0.472	0.434	0.372	0.2375
0.0160	3.856	3.190	2.337	1.554	1.262	1.145	1.061	0.981	0.624	0.534	0.491	0.421	0.3040
0.0200	4.311	3.566	2.613	1.737	1.411	1.280	1.186	1.097	0.698	0.597	0.549	0.470	0.3800
0.0250	4.820	3.987	2.921	1.943	1.577	1.431	1.326	1.227	0.780	0.667	0.615	0.526	0.4749
0.0320	5.446	4.514	3.305	2.198	1.788	1.626	1.500	1.388	0.882	0.757	0.697	0.598	0.6079
0.0400	6.074	5.058	3.695	2.464	2.010	1.833	1.689	1.563	0.994	0.855	0.787	0.676	0.7599
0.0500	6.767	5.672	4.131	2.768	2.268	2.066	1.904	1.764	1.120	0.973	0.896	0.770	0.9499
0.0600	7.350	6.220	4.504	3.036	2.495	2.265	2.094	1.941	1.243	1.081	0.998	0.858	1.1399
0.0700	7.862	6.725	4.838	3.280	2.710	2.458	2.274	2.110	1.355	1.185	1.098	0.943	1.3299
0.0800	8.319	7.208	5.141	3.512	2.905	2.632	2.432	2.262	1.465	1.285	1.188	1.023	1.5198
0.0900	8.727	7.660	5.417	3.727	3.088	2.806	2.584	2.405	1.576	1.379	1.279	1.102	1.7098
0.1000	9.100	8.091	5.670	3.929	3.266	2.960	2.722	2.546	1.678	1.469	1.361	1.174	1.8998
0.1250	9.882	9.117	6.125	4.375	3.654	3.313	3.045	2.853	1.911	1.672	1.563	1.352	2.3747
0.1600	10.685	10.386	6.810	4.876	4.100	3.718	3.412	3.201	2.179	1.914	1.794	1.566	3.0397
0.2000	11.313	11.481	7.308	5.313	4.495	4.071	3.734	3.501	2.426	2.141	2.010	1.754	3.7996
0.2500	11.781	12.392	7.735	5.717	4.858	4.409	4.038	3.783	2.669	2.359	2.212	1.934	4.7495
0.3200	12.090	13.135	8.103	6.102	5.210	4.757	4.335	4.052	2.901	2.577	2.419	2.127	6.0794
0.4000	12.195	13.538	8.341	6.356	5.472	5.005	4.563	4.262	3.103	2.761	2.594	2.277	7.5992
0.5000	12.162	13.668	8.505	6.574	5.698	5.231	4.763	4.440	3.283	2.926	2.751	2.424	9.4990
0.6000	11.924	13.430	8.505	6.660	5.801	5.341	4.848	4.516	3.377	3.019	2.845	2.526	11.399
0.7000	11.647	13.074	8.446	6.672	5.836	5.380	4.873	4.544	3.429	3.066	2.897	2.568	13.299
0.8000	11.339	12.642	8.350	6.655	5.837	5.402	4.885	4.551	3.457	3.106	2.931	2.605	15.198
0.9000	11.047	12.208	8.232	6.602	5.820	5.400	4.873	4.536	3.482	3.128	2.955	2.634	17.098
1.0000	10.749	11.769	8.100	6.553	5.783	5.370	4.852	4.512	3.483	3.135	2.965	2.649	18.998
1.2500	10.046	10.758	7.745	6.351	5.654	5.267	4.748	4.407	3.462	3.121	2.951	2.641	23.747
1.6000	9.176	9.618	7.248	6.052	5.407	5.052	4.552	4.204	3.370	3.051	2.885	2.587	30.397
2.0000	8.330	8.693	6.723	5.694	5.116	4.800	4.303	3.987	3.227	2.945	2.783	2.508	37.996
2.5000	7.459	7.816	6.149	5.282	4.778	4.489	4.015	3.726	3.062	2.786	2.644	2.398	47.495
3.2000	6.520	6.888	5.484	4.771	4.343	4.085	3.663	3.389	2.830	2.588	2.457	2.232	60.794
4.0000	5.706	6.077	4.877	4.287	3.926	3.697	3.331	3.073	2.600	2.380	2.268	2.063	75.992
5.0000	4.945	5.300	4.285	3.796	3.496	3.299	2.974	2.751	2.357	2.151	2.061	1.881	94.990
6.0000	4.372	4.701	3.822	3.409	3.145	2.981	2.691	2.488	2.133	1.968	1.884	1.731	113.99
7.0000	3.912	4.226	3.450	3.091	2.857	2.715	2.456	2.274	1.967	1.808	1.735	1.597	132.99
8.0000	3.544	3.837	3.145	2.830	2.620	2.491	2.264	2.091	1.818	1.676	1.610	1.484	151.98
9.0000	3.245	3.511	2.890	2.609	2.419	2.303	2.095	1.939	1.690	1.560	1.500	1.387	170.98
10.0000	2.988	3.235	2.673	2.424	2.245	2.141	1.949	1.810	1.577	1.462	1.409	1.302	189.98
11.0000	2.770	3.002	2.487	2.263	2.099	2.002	1.823	1.698	1.482	1.378	1.330	1.226	208.98
12.0000	2.583	2.797	2.325	2.120	1.969	1.881	1.713	1.600	1.397	1.302	1.255	1.158	227.98

Table 42.17 Mass stopping power of different substances (in MeV·mg^{-1}·cm^2) for flourine ions $^{19}_{9}F^+$ [1] *(continued)*

E_m, MeV/u	H	He	N	O	Ne	Ar	Kr	Xe	Rn	Mylar	$(CH_2)_n$	Water	E, MeV
0.0125	7.746	3.305	2.189	2.057	1.828	1.206	0.739	0.525	0.346	2.770	3.524	2.689	0.2375
0.0160	8.600	3.627	2.437	2.292	2.038	1.376	0.853	0.608	0.403	3.115	3.963	2.994	0.3040
0.0200	9.353	3.841	2.657	2.503	2.239	1.555	0.969	0.705	0.472	3.454	4.392	3.263	0.3800
0.0250	10.165	4.034	2.898	2.731	2.471	1.773	1.133	0.827	0.599	3.827	4.869	3.558	0.4749
0.0320	11.269	4.276	3.169	3.017	2.756	2.066	1.342	1.011	0.704	4.296	5.479	3.933	0.6079
0.0400	12.378	4.500	3.447	3.318	3.056	2.387	1.581	1.212	0.861	4.785	6.104	4.323	0.7599
0.0500	13.674	4.808	3.796	3.693	3.437	2.772	1.867	1.454	1.053	5.346	6.816	4.800	0.9499
0.0600	15.043	5.148	4.175	4.035	3.815	3.139	2.135	1.680	1.234	5.860	7.481	5.261	1.1399
0.0700	16.402	5.554	4.548	4.417	4.190	3.493	2.380	1.892	1.398	6.358	8.109	5.748	1.3299
0.0800	17.840	5.990	4.941	4.802	4.571	3.841	2.632	2.098	1.558	6.848	8.725	6.252	1.5198
0.0900	19.232	6.452	5.336	5.201	4.951	4.177	2.871	2.297	1.701	7.324	9.312	6.761	1.7098
0.1000	20.752	6.963	5.772	5.591	5.318	4.508	3.113	2.478	1.837	7.785	9.900	7.275	1.8998
0.1250	24.673	8.235	6.805	6.557	6.265	5.289	3.642	2.902	2.157	8.912	11.342	8.570	2.3747
0.1600	30.169	10.059	8.254	7.954	7.546	6.286	4.311	3.432	2.547	10.399	13.212	10.420	3.0397
0.2000	35.883	11.912	9.756	9.369	8.858	7.199	4.896	3.888	2.872	11.795	14.967	12.314	3.7996
0.2500	41.926	13.877	11.255	10.837	10.118	7.998	5.407	4.239	3.117	13.104	16.608	14.295	4.7495
0.3200	48.133	15.712	12.779	12.195	11.158	8.549	5.705	4.449	3.249	14.278	18.135	16.190	6.0794
0.4000	52.050	17.183	13.646	12.971	11.745	8.767	5.797	4.513	3.286	14.956	19.043	17.317	7.5992
0.5000	54.262	18.031	14.008	13.327	11.924	8.769	5.783	4.516	3.291	15.249	19.468	17.877	9.4990
0.6000	54.009	18.116	13.711	13.030	11.644	8.514	5.648	4.406	3.224	14.986	19.230	17.580	11.399
0.7000	51.858	17.821	13.108	12.458	11.149	8.142	5.448	4.240	3.133	14.485	18.665	16.833	13.299
0.8000	49.015	17.201	12.433	11.857	10.596	7.749	5.244	4.092	3.039	13.894	17.836	15.982	15.198
0.9000	46.098	16.431	11.771	11.220	9.977	7.359	5.021	3.968	2.947	13.294	17.048	15.097	17.098
1.0000	43.659	15.787	11.267	10.692	9.526	7.055	4.860	3.872	2.876	12.741	16.322	14.353	18.998
1.2500	38.494	14.414	10.262	9.796	8.706	6.529	4.585	3.702	2.796	11.595	14.724	12.989	23.747
1.6000	32.760	12.828	9.263	8.857	7.929	6.052	4.341	3.530	2.732	10.328	12.923	11.509	30.397
2.0000	27.833	11.409	8.458	8.054	7.288	5.661	4.135	3.395	2.669	9.278	11.429	10.253	37.996
2.5000	23.244	9.992	7.631	7.262	6.703	5.301	3.935	3.278	2.607	8.277	10.017	9.039	47.495
3.2000	19.193	8.527	6.761	6.454	5.994	4.853	3.691	3.071	2.426	7.255	8.648	7.869	60.794
4.0000	16.095	7.277	5.950	5.706	5.365	4.389	3.365	2.834	2.292	6.370	7.506	6.862	75.992
5.0000	13.626	6.192	5.180	4.988	4.701	3.895	3.029	2.558	2.100	5.545	6.492	5.947	94.990
6.0000	11.886	5.404	4.594	4.418	4.185	3.493	2.729	2.320	1.911	4.907	5.729	5.247	113.99
7.0000	10.592	4.809	4.133	3.985	3.767	3.167	2.491	2.125	1.756	4.413	5.137	4.720	132.99
8.0000	9.560	4.324	3.758	3.626	3.425	2.890	2.274	1.950	1.620	4.006	4.654	4.286	151.98
9.0000	8.698	3.953	3.442	3.323	3.144	2.658	2.095	1.800	1.503	3.664	4.251	3.921	170.98
10.0000	7.992	3.649	3.184	3.074	2.908	2.459	1.943	1.671	1.398	3.379	3.913	3.619	189.98
11.0000	7.386	3.387	2.962	2.860	2.706	2.288	1.813	1.559	1.306	3.136	3.628	3.362	208.98
12.0000	6.858	3.160	2.769	2.674	2.529	2.139	1.700	1.462	1.225	2.925	3.378	3.139	227.98

Table 42.18 Range (in mg/cm^2) of fluorine ions $^{19}_{9}F^+$ for different substances [1] [E_m is the energy per mass unit of incident particle; E is a total energy.]

E_m, MeV/u	Be	C	Al	Ti	Ni	Ge	Zr	Ag	Eu	Ta	Au	U	E, MeV
0.0125	0.081	0.090	0.125	0.187	0.219	0.260	0.287	0.321	0.497	0.589	0.644	0.773	0.2375
0.0160	0.098	0.109	0.151	0.225	0.265	0.313	0.344	0.382	0.592	0.700	0.765	0.916	0.3040
0.0200	0.115	0.130	0.179	0.266	0.314	0.368	0.403	0.447	0.692	0.817	0.891	1.064	0.3800
0.0250	0.135	0.153	0.210	0.313	0.371	0.431	0.471	0.521	0.806	0.950	1.036	1.234	0.4749
0.0320	0.160	0.183	0.251	0.373	0.443	0.512	0.558	0.614	0.952	1.119	1.219	1.449	0.6079
0.0400	0.186	0.214	0.292	0.434	0.517	0.594	0.648	0.711	1.102	1.294	1.409	1.670	0.7599
0.0500	0.215	0.248	0.339	0.504	0.601	0.687	0.748	0.820	1.272	1.490	1.621	1.917	0.9499
0.0600	0.241	0.279	0.382	0.567	0.678	0.772	0.840	0.919	1.425	1.666	1.812	2.140	1.1399
0.0700	0.266	0.308	0.421	0.626	0.748	0.850	0.924	1.010	1.566	1.828	1.986	2.343	1.3299
0.0800	0.289	0.335	0.459	0.680	0.814	0.923	1.003	1.094	1.697	1.977	2.147	2.529	1.5198
0.0900	0.311	0.360	0.494	0.732	0.876	0.991	1.077	1.174	1.818	2.115	2.297	2.703	1.7098
0.1000	0.333	0.384	0.528	0.781	0.934	1.056	1.147	1.249	1.932	2.246	2.437	2.866	1.8998
0.1250	0.383	0.440	0.608	0.895	1.069	1.207	1.311	1.425	2.192	2.542	2.756	3.235	2.3747
0.1600	0.447	0.508	0.710	1.039	1.238	1.396	1.517	1.645	2.513	2.908	3.146	3.685	3.0397
0.2000	0.516	0.577	0.818	1.188	1.413	1.591	1.730	1.871	2.839	3.279	3.541	4.137	3.7996
0.2500	0.598	0.657	0.944	1.360	1.615	1.815	1.974	2.132	3.209	3.697	3.987	4.648	4.7495
0.3200	0.710	0.761	1.111	1.584	1.879	2.105	2.291	2.471	3.686	4.235	4.561	5.302	6.0794
0.4000	0.834	0.875	1.296	1.828	2.163	2.416	2.632	2.836	4.191	4.804	5.167	5.992	7.5992
0.5000	0.990	1.014	1.521	2.122	2.503	2.787	3.039	3.272	4.786	5.472	5.877	6.799	9.4990
0.6000	1.148	1.154	1.745	2.409	2.833	3.146	3.434	3.696	5.356	6.111	6.556	7.567	11.399
0.7000	1.309	1.298	1.969	2.693	3.159	3.500	3.825	4.115	5.914	6.735	7.217	8.312	13.299
0.8000	1.474	1.445	2.195	2.978	3.485	3.853	4.214	4.533	6.466	7.350	7.869	9.046	15.198
0.9000	1.644	1.598	2.424	3.265	3.811	4.204	4.604	4.951	7.013	7.960	8.514	9.772	17.098
1.0000	1.818	1.757	2.656	3.554	4.138	4.557	4.994	5.371	7.559	8.566	9.156	10.491	18.998
1.2500	2.275	2.179	3.256	4.290	4.968	5.450	5.984	6.436	8.926	10.084	10.761	12.285	23.747
1.6000	2.968	2.832	4.144	5.362	6.171	6.739	7.414	7.980	10.872	12.238	13.040	14.828	30.397
2.0000	3.838	3.664	5.233	6.657	7.616	8.282	9.131	9.837	13.177	14.773	15.722	17.812	37.996
2.5000	5.044	4.818	6.711	8.390	9.539	10.330	11.418	12.303	16.200	18.092	19.225	21.687	47.495
3.2000	6.954	6.633	9.005	11.043	12.462	13.439	14.890	16.051	20.723	23.050	24.449	27.441	60.794
4.0000	9.451	8.987	11.949	14.409	16.148	17.356	19.248	20.768	26.334	29.181	30.896	34.533	75.992
5.0000	13.035	12.341	16.113	19.128	21.286	22.806	25.295	27.314	34.023	37.593	39.698	44.193	94.990
6.0000	17.128	16.154	20.816	24.417	27.023	28.873	32.022	34.587	42.510	46.839	49.352	54.734	113.99
7.0000	21.728	20.422	26.055	30.277	33.369	35.559	39.421	42.585	51.798	56.922	59.870	66.171	132.99
8.0000	26.836	25.145	31.829	36.707	40.321	42.872	47.484	51.306	61.856	67.846	71.246	78.520	151.98
9.0000	32.443	30.326	38.136	43.704	47.876	50.812	56.215	60.749	72.703	79.602	83.482	91.771	170.98
10.0000	38.548	35.968	44.977	51.263	56.034	59.373	65.624	70.898	84.347	92.188	96.561	105.919	189.98
11.0000	45.156	42.070	52.351	59.379	64.791	68.556	75.711	81.741	96.780	105.582	110.446	120.966	208.98
12.0000	52.262	48.631	60.257	68.056	74.141	78.352	86.467	93.273	109.989	119.774	125.154	136.920	227.98

Table 42.18 Range (in mg/cm^2) of fluorine ions $^{19}_{9}F^+$ for different substances [1] *(continued)*

E_m, MeV/u	H	He	N	O	Ne	Ar	Kr	Xe	Rn	Mylar	$(CH_2)_n$	Water	E, MeV
0.0125	0.029	0.076	0.108	0.114	0.126	0.196	0.331	0.487	0.760	0.087	0.068	0.084	0.2375
0.0160	0.036	0.093	0.133	0.140	0.156	0.239	0.399	0.583	0.905	0.107	0.084	0.104	0.3040
0.0200	0.044	0.111	0.159	0.168	0.187	0.284	0.471	0.682	1.052	0.128	0.100	0.125	0.3800
0.0250	0.052	0.134	0.190	0.201	0.223	0.335	0.551	0.792	1.209	0.152	0.119	0.150	0.4749
0.0320	0.064	0.164	0.231	0.244	0.270	0.399	0.649	0.925	1.395	0.182	0.144	0.183	0.6079
0.0400	0.076	0.197	0.274	0.289	0.319	0.463	0.747	1.053	1.576	0.214	0.169	0.218	0.7599
0.0500	0.090	0.236	0.325	0.341	0.375	0.534	0.852	1.189	1.765	0.251	0.197	0.258	0.9499
0.0600	0.103	0.273	0.371	0.389	0.426	0.596	0.943	1.306	1.925	0.284	0.223	0.295	1.1399
0.0700	0.114	0.308	0.413	0.432	0.472	0.652	1.025	1.409	2.066	0.314	0.247	0.328	1.3299
0.0800	0.125	0.340	0.452	0.473	0.514	0.703	1.099	1.502	2.191	0.342	0.269	0.359	1.5198
0.0900	0.135	0.370	0.489	0.510	0.553	0.749	1.167	1.587	2.305	0.369	0.290	0.388	1.7098
0.1000	0.145	0.398	0.522	0.545	0.590	0.793	1.230	1.667	2.413	0.394	0.310	0.415	1.8998
0.1250	0.165	0.461	0.598	0.623	0.672	0.890	1.371	1.844	2.651	0.451	0.354	0.475	2.3747
0.1600	0.190	0.534	0.687	0.715	0.768	1.005	1.539	2.054	2.934	0.520	0.409	0.545	3.0397
0.2000	0.213	0.603	0.771	0.803	0.861	1.118	1.704	2.261	3.215	0.588	0.462	0.612	3.7996
0.2500	0.237	0.677	0.862	0.897	0.961	1.243	1.888	2.495	3.531	0.664	0.523	0.683	4.7495
0.3200	0.267	0.767	0.972	1.012	1.086	1.404	2.127	2.801	3.948	0.761	0.599	0.771	6.0794
0.4000	0.297	0.859	1.087	1.133	1.219	1.579	2.391	3.139	4.413	0.865	0.681	0.861	7.5992
0.5000	0.333	0.967	1.224	1.277	1.379	1.795	2.718	3.560	4.990	0.991	0.779	0.969	9.4990
0.6000	0.368	1.072	1.361	1.421	1.540	2.015	3.051	3.985	5.573	1.117	0.877	1.076	11.399
0.7000	0.404	1.178	1.503	1.570	1.707	2.243	3.393	4.425	6.170	1.245	0.978	1.187	13.299
0.8000	0.441	1.286	1.652	1.727	1.882	2.482	3.748	4.881	6.785	1.379	1.082	1.302	15.198
0.9000	0.481	1.399	1.809	1.891	2.066	2.734	4.118	5.352	7.420	1.519	1.191	1.425	17.098
1.0000	0.524	1.517	1.974	2.065	2.261	2.997	4.503	5.837	8.073	1.665	1.305	1.554	18.998
1.2500	0.640	1.832	2.415	2.529	2.783	3.697	5.509	7.091	9.747	2.056	1.611	1.901	23.747
1.6000	0.827	2.321	3.097	3.243	3.583	4.755	6.999	8.930	12.152	2.664	2.093	2.445	30.397
2.0000	1.079	2.950	3.957	4.143	4.583	6.054	8.793	11.126	14.966	3.440	2.719	3.145	37.996
2.5000	1.453	3.840	5.140	5.386	5.944	7.789	11.149	13.974	18.568	4.525	3.608	4.133	47.495
3.2000	2.084	5.284	6.994	7.332	8.045	10.414	14.642	18.170	23.821	6.244	5.039	5.713	60.794
4.0000	2.950	7.217	9.395	9.840	10.730	13.712	18.961	23.329	30.226	8.484	6.929	7.785	75.992
5.0000	4.236	10.054	12.824	13.409	14.521	18.316	24.922	30.398	38.899	11.688	9.657	10.765	94.990
6.0000	5.732	13.344	16.725	17.463	18.811	23.475	31.540	38.208	48.398	15.337	12.778	14.172	113.99
7.0000	7.427	17.076	21.091	21.997	23.602	29.194	38.836	46.774	58.781	19.425	16.285	17.995	132.99
8.0000	9.318	21.247	25.917	27.001	28.897	35.480	46.827	56.117	70.057	23.948	20.174	22.223	151.98
9.0000	11.403	25.847	31.204	32.479	34.692	42.340	55.539	66.266	82.245	28.911	24.450	26.862	170.98
10.0000	13.683	30.853	36.948	38.427	40.980	49.776	64.962	77.229	95.363	34.315	29.111	31.909	189.98
11.0000	16.158	36.261	43.139	44.839	47.757	57.791	75.091	89.008	109.434	40.156	34.157	37.358	208.98
12.0000	18.829	42.072	49.777	51.714	55.023	66.384	85.920	101.596	124.464	46.432	39.586	43.210	227.98

Table 42.19 Mass stopping power of different substances (in MeV·mg^{-1}·cm^2) for neon ions $^{20}_{10}Ne^+$ [1] [E_m is the energy per mass unit of incident particle; E is a total energy.]

E_m, MeV/u	Be	C	Al	Ti	Ni	Ge	Zr	Ag	Eu	Ta	Au	U	E, MeV
0.0125	3.548	2.935	2.150	1.430	1.161	1.054	0.976	0.903	0.574	0.491	0.452	0.387	0.2499
0.0160	4.014	3.320	2.432	1.618	1.314	1.192	1.104	1.022	0.649	0.556	0.511	0.438	0.3199
0.0200	4.487	3.712	2.720	1.809	1.469	1.333	1.235	1.142	0.726	0.621	0.571	0.490	0.3998
0.0250	5.017	4.150	3.041	2.022	1.642	1.490	1.380	1.277	0.812	0.695	0.640	0.547	0.4998
0.0320	5.669	4.699	3.440	2.288	1.861	1.692	1.562	1.445	0.918	0.788	0.726	0.623	0.5397
0.0400	6.323	5.265	3.846	2.565	2.092	1.908	1.758	1.627	1.035	0.890	0.819	0.704	0.7997
0.0500	7.043	5.904	4.300	2.881	2.361	2.150	1.982	1.836	1.165	1.013	0.933	0.802	0.9996
0.0600	7.679	6.498	4.705	3.171	2.607	2.367	2.188	2.028	1.299	1.129	1.042	0.896	1.1995
0.0700	8.242	7.050	5.072	3.439	2.840	2.577	2.384	2.211	1.420	1.243	1.151	0.989	1.3994
0.0800	8.749	7.581	5.407	3.693	3.055	2.768	2.558	2.379	1.541	1.352	1.249	1.076	1.5994
0.0900	9.207	8.081	5.715	3.932	3.258	2.961	2.726	2.538	1.663	1.455	1.349	1.163	1.7993
0.1000	9.630	8.526	6.000	4.158	3.456	3.132	2.880	2.694	1.776	1.554	1.440	1.242	1.9992
0.1250	10.534	9.719	6.625	4.664	3.896	3.531	3.246	3.041	2.037	1.782	1.666	1.441	2.4990
0.1600	11.502	11.179	7.331	5.249	4.413	4.002	3.673	3.445	2.346	2.060	1.932	1.686	3.1987
0.2000	12.301	12.484	7.946	5.777	4.887	4.426	4.061	3.806	2.638	2.328	2.185	1.908	3.9984
0.2500	12.946	13.617	8.500	6.282	5.338	4.845	4.437	4.157	2.933	2.593	2.431	2.125	4.9980
0.3200	13.431	14.592	9.002	6.778	5.788	5.284	4.816	4.501	3.223	2.863	2.687	2.363	6.3974
0.4000	13.664	15.169	9.346	7.122	6.131	5.608	5.112	4.776	3.477	3.094	2.907	2.551	7.9968
0.5000	13.728	15.427	9.600	7.421	6.432	5.904	5.376	5.011	3.706	3.302	3.106	2.736	9.9960
0.6000	13.495	15.199	9.626	7.537	6.565	6.045	5.487	5.111	3.821	3.417	3.220	2.859	11.995
0.7000	13.214	14.833	9.582	7.570	6.621	6.104	5.529	5.155	3.890	3.478	3.287	2.913	13.994
0.8000	12.894	14.376	9.495	7.568	6.637	6.143	5.555	5.175	3.931	3.532	3.333	2.962	15.994
0.9000	12.589	13.912	9.381	7.524	6.632	6.154	5.554	5.169	3.968	3.565	3.368	3.002	17.993
1.0000	12.275	13.440	9.250	7.483	6.605	6.133	5.541	5.152	3.978	3.580	3.286	3.025	19.992
1.2500	11.526	12.343	8.886	7.287	6.487	6.043	5.447	5.056	3.972	3.581	3.386	3.030	24.990
1.6000	10.587	11.097	8.363	6.983	6.239	5.829	5.252	4.850	3.889	3.521	3.328	2.985	31.987
2.0000	9.664	10.085	7.800	6.607	5.936	5.569	4.992	4.625	3.744	3.416	3.229	2.909	39.984
2.5000	8.704	9.121	7.176	6.164	5.576	5.238	4.686	4.439	3.574	3.251	3.086	2.799	49.980
3.2000	7.661	8.092	6.443	5.605	5.103	4.800	4.304	3.982	3.325	3.041	2.886	2.622	63.974
4.0000	6.747	7.186	5.767	5.069	4.642	4.371	3.939	3.633	3.074	2.814	2.682	2.439	79.968
5.0000	5.885	6.309	5.100	4.519	4.162	3.927	3.539	3.274	2.805	2.560	2.453	2.239	99.960
6.0000	5.233	5.626	4.574	4.080	3.764	3.568	3.220	2.978	2.552	2.356	2.255	2.072	119.95
7.0000	4.704	5.082	4.148	3.717	3.435	3.265	2.954	2.734	2.365	2.174	2.087	1.921	139.94
8.0000	4.279	4.632	3.797	3.417	3.163	3.007	2.734	2.525	2.195	2.024	1.944	1.792	159.94
9.0000	3.932	4.255	3.502	3.162	2.931	2.791	2.539	2.350	2.049	1.891	1.817	1.681	179.93
10.0000	3.634	3.933	3.250	2.948	2.730	2.603	2.369	2.200	1.918	1.778	1.713	1.583	199.92
11.0000	3.378	3.660	3.033	2.760	2.560	2.441	2.223	2.071	1.807	1.680	1.622	1.495	219.91
12.0000	3.159	3.420	2.843	2.593	2.408	2.300	2.095	1.956	1.709	1.592	1.535	1.416	239.90

Table 42.19 Mass stopping power of different substances (in MeV·mg^{-1}·cm^2) for neon ions $^{20}_{10}Ne^+$ [1] *(continued)*

E_m, MeV/u	H	He	N	O	Ne	Ar	Kr	Xe	Rn	Mylar	$(CH_2)_n$	Water	E, MeV
0.0125	8.063	3.440	2.279	2.141	1.903	1.256	0.770	0.546	0.360	2.883	3.668	2.799	0.2499
0.0160	8.951	3.775	2.537	2.386	2.121	1.433	0.888	0.632	0.420	3.342	4.125	3.116	0.3199
0.0200	9.736	3.998	2.766	2.605	2.331	1.618	1.009	0.734	0.491	3.595	4.572	3.397	0.3998
0.0250	10.581	4.199	3.016	2.843	2.572	1.846	1.180	0.860	0.623	3.983	5.069	3.703	0.4998
0.0320	11.730	4.451	3.299	3.141	2.869	2.150	1.397	1.053	0.733	4.472	5.704	4.094	0.6397
0.0400	12.884	4.684	3.588	3.454	3.181	2.485	1.646	1.262	0.896	4.981	6.354	4.500	0.7997
0.0500	14.233	5.005	3.952	3.844	3.578	2.885	1.944	1.514	1.097	5.564	7.095	4.997	0.9996
0.0600	15.716	5.378	4.362	4.216	3.985	3.280	2.230	1.755	1.289	6.122	7.816	5.496	1.1995
0.0700	17.194	5.823	4.768	4.631	4.392	3.662	2.495	1.983	1.466	6.665	8.501	6.026	1.3994
0.0800	18.763	6.299	5.196	5.050	4.807	4.039	2.768	2.206	1.638	7.202	9.176	6.575	1.5994
0.0900	20.289	6.807	5.630	5.487	5.224	4.406	3.029	2.423	1.795	7.727	9.825	7.133	1.7993
0.1000	21.960	7.368	6.108	5.916	5.628	4.770	3.294	2.622	1.944	8.238	10.476	7.698	1.9992
0.1250	26.302	8.778	7.254	6.989	6.678	5.638	3.882	3.094	2.299	9.500	12.091	9.136	2.4990
0.1600	32.474	10.827	8.885	8.562	8.122	6.766	4.640	3.695	2.742	11.194	14.221	11.216	3.1987
0.2000	39.016	12.952	10.608	10.187	9.631	7.827	5.324	4.227	3.123	12.825	16.274	13.390	3.9984
0.2500	46.070	15.249	12.368	11.909	11.118	8.789	5.942	4.658	3.426	14.399	18.250	15.708	4.9980
0.3200	53.471	17.455	14.196	13.548	12.396	9.497	6.337	4.942	3.610	15.861	20.146	17.986	6.3974
0.4000	58.319	19.253	15.290	14.533	13.159	9.823	6.495	5.056	3.682	16.757	21.337	19.402	7.9968
0.5000	61.248	20.352	15.811	15.043	13.459	9.898	6.528	5.098	3.715	17.213	21.974	20.179	9.9960
0.6000	61.122	20.502	15.516	14.746	13.177	9.635	6.391	4.986	3.648	16.960	21.763	19.896	11.995
0.7000	58.833	20.218	14.871	14.133	12.648	9.237	6.180	4.810	3.555	16.433	21.119	19.097	13.994
0.8000	55.737	19.560	14.138	13.483	12.049	8.812	5.963	4.653	3.456	15.800	20.282	18.174	15.994
0.9000	52.534	18.725	13.415	12.786	11.370	8.387	5.722	4.522	3.358	15.151	19.428	17.205	17.993
1.0000	49.858	18.028	12.867	12.210	10.878	8.057	5.550	4.422	3.284	14.550	18.639	16.391	19.992
1.2500	44.165	16.538	11.774	11.241	9.988	7.491	5.261	4.240	3.208	13.303	16.893	14.902	24.990
1.6000	37.799	14.802	10.687	10.219	9.149	6.983	5.009	4.073	3.153	11.917	14.911	13.280	31.987
2.0000	32.292	13.237	9.812	9.344	8.455	6.568	4.797	3.939	3.097	10.764	13.260	11.895	39.984
2.5000	27.125	11.661	8.905	8.475	7.822	6.186	4.593	3.825	3.043	9.659	11.690	10.549	49.980
3.2000	22.550	10.019	7.944	7.583	7.042	5.702	4.336	3.608	2.893	8.524	10.160	9.245	63.974
4.0000	19.031	8.604	7.036	6.747	6.344	5.190	3.979	3.351	2.710	7.532	8.875	8.114	79.968
5.0000	16.218	7.369	6.166	5.936	5.595	4.636	3.606	3.045	2.499	6.599	7.726	7.079	99.960
6.0000	14.225	6.468	5.498	5.287	5.008	4.181	3.266	2.776	2.287	5.873	6.856	6.280	119.95
7.0000	12.736	5.783	4.970	4.791	4.530	3.808	2.995	2.555	2.112	5.306	6.177	5.675	139.94
8.0000	11.543	5.221	4.537	4.378	4.135	3.489	2.745	2.354	1.955	4.837	5.620	5.175	159.94
9.0000	10.540	4.790	4.171	4.027	3.810	3.222	2.539	2.182	1.821	4.440	5.151	4.752	179.93
10.0000	9.718	4.436	3.871	3.738	3.536	2.990	2.363	2.031	1.700	4.108	4.758	4.401	199.92
11.0000	9.007	4.130	3.612	3.488	3.300	2.790	2.211	1.901	1.592	3.824	4.425	4.100	219.91
12.0000	8.387	3.864	3.386	3.270	3.093	2.616	2.078	1.788	1.498	3.577	4.131	3.838	239.90

Table 42.20 Range (in mg/cm^2) of neon ions $^{20}_{10}Ne^+$ for different substances [1] [E_m is the energy per mass unit of incident particle; E is a total energy.]

E_m, MeV/u	Be	C	Al	Ti	Ni	Ge	Zr	Ag	Eu	Ta	Au	U	E, MeV
0.0125	0.079	0.087	0.121	0.181	0.212	0.253	0.279	0.312	0.483	0.573	0.626	0.753	0.2499
0.0160	0.096	0.107	0.147	0.220	0.257	0.305	0.335	0.373	0.577	0.683	0.745	0.894	0.3199
0.0200	0.114	0.127	0.175	0.260	0.306	0.360	0.394	0.437	0.676	0.798	0.871	1.041	0.3998
0.0250	0.133	0.151	0.207	0.307	0.362	0.423	0.462	0.511	0.790	0.930	1.014	1.210	0.4998
0.0320	0.158	0.181	0.247	0.367	0.434	0.503	0.549	0.604	0.935	1.100	1.198	1.424	0.6397
0.0400	0.184	0.211	0.289	0.428	0.509	0.586	0.638	0.701	1.085	1.274	1.387	1.645	0.7997
0.0500	0.213	0.246	0.336	0.498	0.593	0.679	0.739	0.811	1.255	1.470	1.600	1.893	0.9996
0.0600	0.240	0.277	0.379	0.562	0.670	0.764	0.831	0.910	1.409	1.647	1.791	2.116	1.1995
0.0700	0.265	0.306	0.418	0.620	0.740	0.842	0.915	1.001	1.550	1.808	1.965	2.319	1.3994
0.0800	0.288	0.333	0.456	0.675	0.806	0.915	0.994	1.085	1.680	1.957	2.125	2.505	1.5994
0.0900	0.310	0.358	0.491	0.726	0.867	0.983	1.067	1.164	1.801	2.095	2.274	2.678	1.7993
0.1000	0.331	0.382	0.525	0.775	0.925	1.047	1.137	1.239	1.914	2.224	2.414	2.840	1.9992
0.1250	0.381	0.437	0.604	0.888	1.059	1.195	1.297	1.410	2.171	2.517	2.729	3.204	2.4990
0.1600	0.444	0.504	0.704	1.029	1.225	1.378	1.497	1.623	2.485	2.876	3.111	3.645	3.1987
0.2000	0.511	0.571	0.808	1.174	1.395	1.566	1.701	1.841	2.802	3.236	3.495	4.084	3.9984
0.2500	0.591	0.648	0.930	1.340	1.588	1.781	1.935	2.090	3.158	3.638	3.924	4.575	4.9980
0.3200	0.696	0.747	1.089	1.553	1.839	2.057	2.237	2.413	3.612	4.151	4.470	5.198	6.3974
0.4000	0.814	0.854	1.263	1.783	2.107	2.351	2.558	2.757	4.089	4.687	5.041	5.849	7.9968
0.5000	0.960	0.985	1.474	2.058	2.425	2.698	2.939	3.165	4.645	5.312	5.706	6.605	9.9960
0.6000	1.107	1.115	1.682	2.325	2.733	3.032	3.307	3.560	5.176	5.907	6.338	7.319	11.995
0.7000	1.256	1.248	1.890	2.590	3.036	3.361	3.670	3.949	5.694	6.486	6.952	8.011	13.994
0.8000	1.410	1.385	2.100	2.854	3.337	3.687	4.031	4.336	6.205	7.056	7.556	8.692	15.994
0.9000	1.566	1.526	2.311	3.119	3.639	4.012	4.390	4.723	6.711	7.620	8.152	9.362	17.993
1.0000	1.727	1.673	2.526	3.385	3.941	4.338	4.751	5.110	7.214	8.179	8.744	10.025	19.992
1.2500	2.147	2.061	3.077	4.062	4.704	5.158	5.660	6.089	8.471	9.574	10.220	11.675	24.990
1.6000	2.781	2.658	3.889	5.042	5.804	6.337	6.968	7.502	10.251	11.544	12.304	14.001	31.987
2.0000	3.572	3.415	4.879	6.220	7.118	7.741	8.530	9.190	12.347	13.850	14.743	16.714	39.984
2.5000	4.662	4.458	6.216	7.787	8.857	9.593	10.598	11.420	15.081	16.851	17.911	20.218	49.980
3.2000	6.379	6.089	8.277	10.171	11.483	12.387	13.718	14.787	19.145	21.306	22.604	25.389	63.974
4.0000	8.607	8.190	10.905	13.176	14.774	15.883	17.608	18.998	24.154	26.780	28.360	31.720	79.968
5.0000	11.786	11.165	14.599	17.361	19.330	20.717	22.971	24.805	30.973	34.240	36.166	40.287	99.960
6.0000	15.395	14.526	18.744	22.024	24.389	26.066	28.902	31.216	38.456	42.391	44.678	49.580	119.95
7.0000	19.429	18.270	23.340	27.164	29.955	31.930	35.391	38.231	46.602	51.235	53.904	59.612	139.94
8.0000	23.890	22.395	28.382	32.779	36.026	38.317	42.433	45.848	55.386	60.775	63.838	70.397	159.94
9.0000	28.768	26.902	33.869	38.865	42.598	45.223	50.028	54.062	64.822	71.002	74.432	81.923	179.93
10.0000	34.060	31.793	39.799	45.418	49.670	52.645	58.185	62.860	74.915	81.913	85.820	94.188	199.92
11.0000	39.770	37.066	46.171	52.432	57.237	60.580	66.901	72.230	85.660	93.487	97.819	107.191	219.91
12.0000	45.893	42.719	52.984	59.909	65.294	69.021	76.169	82.167	97.042	105.717	110.492	120.938	239.90

Table 42.20 Range (in mg/cm^2) of neon ions ${}^{20}_{10}Ne^+$ for different substances [1] *(continued)*

E_m, MeV/u	H	He	N	O	Ne	Ar	Kr	Xe	Rn	Mylar	$(CH_2)_n$	Water	E, MeV
0.0125	0.028	0.074	0.104	0.110	0.122	0.189	0.320	0.471	0.735	0.084	0.066	0.081	0.2499
0.0160	0.035	0.091	0.129	0.136	0.151	0.231	0.387	0.566	0.879	0.104	0.082	0.101	0.3199
0.0200	0.043	0.109	0.155	0.164	0.182	0.276	0.458	0.664	1.024	0.125	0.098	0.122	0.3998
0.0250	0.051	0.132	0.186	0.197	0.218	0.327	0.538	0.774	1.181	0.149	0.117	0.147	0.4998
0.0320	0.063	0.162	0.227	0.240	0.265	0.391	0.636	0.906	1.366	0.180	0.141	0.180	0.6397
0.0400	0.075	0.195	0.271	0.285	0.314	0.456	0.733	1.034	1.547	0.212	0.167	0.215	0.7997
0.0500	0.089	0.235	0.321	0.337	0.371	0.527	0.839	1.170	1.737	0.248	0.195	0.255	0.9996
0.0600	0.102	0.272	0.368	0.385	0.421	0.589	0.930	1.288	1.897	0.281	0.221	0.292	1.1995
0.0700	0.113	0.307	0.410	0.429	0.467	0.645	1.012	1.391	2.038	0.312	0.245	0.326	1.3994
0.0800	0.124	0.339	0.449	0.469	0.510	0.696	1.086	1.484	2.163	0.340	0.267	0.357	1.5994
0.0900	0.134	0.369	0.485	0.506	0.549	0.742	1.154	1.569	2.277	0.366	0.288	0.385	1.7993
0.1000	0.144	0.397	0.519	0.541	0.585	0.786	1.216	1.648	2.384	0.392	0.308	0.412	1.9992
0.1250	0.164	0.459	0.594	0.618	0.666	0.882	1.355	1.823	2.620	0.448	0.352	0.471	2.4990
0.1600	0.188	0.530	0.681	0.709	0.761	0.995	1.520	2.030	2.898	0.516	0.405	0.540	3.1987
0.2000	0.211	0.598	0.763	0.794	0.851	1.105	1.680	2.232	3.171	0.582	0.458	0.605	3.9984
0.2500	0.234	0.669	0.850	0.885	0.948	1.225	1.858	2.457	3.476	0.656	0.516	0.674	4.9980
0.3200	0.262	0.754	0.955	0.994	1.067	1.378	2.085	2.747	3.873	0.748	0.589	0.757	6.3974
0.4000	0.291	0.841	1.064	1.108	1.192	1.543	2.334	3.067	4.311	0.846	0.666	0.843	7.9968
0.5000	0.324	0.942	1.192	1.243	1.342	1.746	2.641	3.460	4.850	0.964	0.758	0.944	9.9960
0.6000	0.357	1.040	1.320	1.377	1.492	1.950	2.950	3.856	5.393	1.080	0.849	1.043	11.995
0.7000	0.390	1.138	1.451	1.516	1.647	2.162	3.268	4.264	5.948	1.200	0.942	1.146	13.994
0.8000	0.425	1.239	1.589	1.661	1.808	2.384	3.597	4.687	6.518	1.324	1.039	1.253	15.994
0.9000	0.462	1.343	1.734	1.813	1.979	2.616	3.939	5.123	7.105	1.453	1.140	1.336	17.993
1.0000	0.501	1.452	1.886	1.973	2.159	2.859	4.294	5.570	7.707	1.588	1.245	1.485	19.992
1.2500	0.608	1.741	2.292	2.399	2.638	3.502	5.219	6.723	9.246	1.947	1.526	1.805	24.990
1.6000	0.779	2.189	2.916	3.052	3.370	4.470	6.582	8.405	11.446	2.503	1.967	2.302	31.987
2.0000	1.008	2.760	3.697	3.871	4.280	5.651	8.213	10.401	14.005	3.209	2.536	2.939	39.984
2.5000	1.346	3.566	4.768	4.995	5.510	7.220	10.344	12.977	17.262	4.191	3.340	3.832	49.980
3.2000	1.913	4.863	6.434	6.743	7.398	9.579	13.482	16.748	21.982	5.735	4.626	5.251	63.974
4.0000	2.686	6.588	8.577	8.983	9.795	12.523	17.338	21.353	27.699	7.735	6.313	7.101	79.968
5.0000	3.827	9.105	11.618	12.148	13.157	16.606	22.624	27.622	35.392	10.576	8.732	9.744	99.960
6.0000	5.145	12.005	15.057	15.722	16.940	21.154	28.459	34.508	43.766	13.792	11.484	12.746	119.95
7.0000	6.633	15.279	18.887	19.699	21.142	26.171	34.858	42.022	52.873	17.378	14.560	16.101	139.94
8.0000	8.283	18.921	23.101	24.068	25.766	31.660	41.837	51.180	62.721	21.329	17.956	19.793	159.94
9.0000	10.098	22.922	27.701	28.834	30.807	37.628	49.416	59.009	73.324	25.646	21.675	23.828	179.93
10.0000	12.074	27.262	32.680	33.990	36.258	44.074	57.584	68.513	84.695	30.330	25.717	28.203	199.92
11.0000	14.213	31.935	38.030	39.531	42.114	51.000	66.337	78.691	96.855	35.378	30.077	32.913	219.91
12.0000	16.514	36.942	43.750	45.455	48.375	58.405	75.668	89.539	109.805	40.786	34.755	37.955	239.90

42.2 Multiple Scattering of Charged Particles Passing through Matter

The charged particle passing through matter changes the direction of its motion due to interaction with atomic electrons. The angle of deflection for a single interaction act is very small. But statistical addition of the deflection angles in the course of many interactions could result in divergence of the original parallel beam after passing some distance in matter. The angular distribution of the beam, i.e. the dependence of the beam density on the angle with respect to the beam axis, is approximately Gaussian one and takes the form

$$F = F_0 \exp(-\theta^2/\theta_0^2), \tag{42.5}$$

where F is the beam density at an angle θ; F_0 is the beam density at an angle 0° with respect to the beam axis; θ_0 is the angle at which the beam density has fallen to $1/e$ of its value at 0°. The angle θ is called the angle of multiple scattering, and angle θ_0 is used for description of beam divergence.

The most recent and most complete theory of multiple scattering is that given in [4, 5]. The parameters of this theory are as follows:

$$\chi_c = 0.1569\frac{Z(Z+1)z^2t}{A(pv)^2}, \qquad \text{and} \quad b = \ln\left[2730(Z+1)Z^{1/3}z^2t/(A\beta^2)\right] - 0.1544, \tag{42.6}$$

where z and Z are the atomic numbers of the incident particle and the scatterer, respectively; A is the atomic weight (mass number) of the scatterer; t is the thickness of the scatterer in g/cm^2; pv is the momentum–velocity product of the incident particle in MeV; $\beta = v/c$ (c is the speed of light); $(pv)^2 = (E^2 + 2EMc^2)\beta^2$ and $\beta^2 = 1 - (1 + E/Mc^2)^{-2}$; E and Mc^2 are the incident particle energy and rest mass (in MeV), respectively.

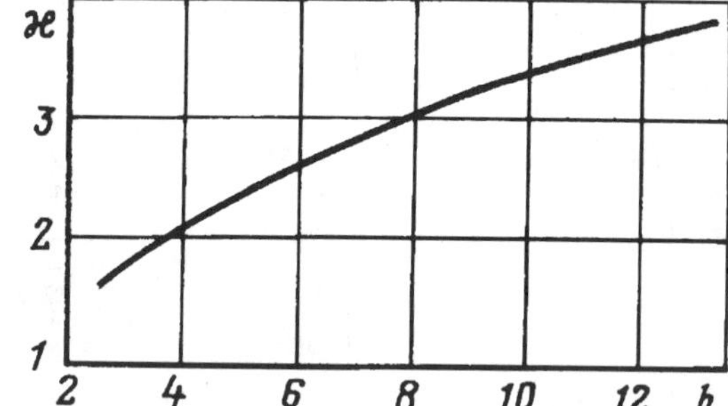

Figure 42.1 Dependence of κ on b (κ and b are the parameters used to compute the angle of multiple scattering).

In order to find the angle of multiple scattering θ_0, at which the beam density has fallen to $1/e$ of its value at 0°, there is a need

(1) to calculate χ_c and b from the above expressions;

(2) to read the value of κ from the graph in Fig. 42.1 in accordance with calculated value of b;

(3) to compute $\theta = \kappa\chi_c$ (in rad).

42.3 Passage of Electrons through Matter

The electrons passing through matter lose their energy mainly due to ionization and excitation of atoms and to bremsstrahlung. The probability for the incident electron to lose a substantial part of its initial energy in a single act of interaction with atomic electron and to escape the beam due to large angle scattering can be quite significant. Therefore there is no concept of average range for electrons in matter and one can speak only about the maximum depth of penetration (or extrapolated range). In order to estimate the extrapolated range of electrons one can use the empirical relationships. For example, the following empirical electron range in aluminum has been proposed [6]:

$$R = 412E^n, \quad 0.01\,\text{MeV} \le E \le 3\,\text{MeV}, \tag{42.7}$$

where $n = 1.265 - 0.0954 \ln E\,(\text{MeV})$ and $R(\text{mg/cm}^2) = 530E - 106$ for 3 MeV $\le E \le$ 20 MeV.

The graph in Fig. 42.2 is based on these relationships. Since the extrapolated electron range (expressed in mg/cm^2) varies only slightly with the atomic number Z of the medium, this graph is approximately correct for any stopping material if the following correction is included:

$$R_X = R_{Al} \frac{(Z/A)_{Al}}{(Z/A)_X}, \tag{42.8}$$

where R and Z/A are the electron range and ratio of the atomic number to the mass number for Al and element X, respectively.

Figure 42.3 shows the graph of specific energy loss for electrons in silicon, because of its importance in charged particle detection [7]. Units of keV/μm have been chosen to simplify the calculations for silicon detectors. The value of dE/dx for other materials with not too different Z may be obtained by multiplying the value of dE/dx from the graph by the ratio of the density of the material to the density of silicon (2.33 g/cm^3).

The electrons originating in β-decay of radioactive nuclei possess a continuous energy spectrum with maximum energy E_{max}. The beam of β-particles passing through matter is attenuated approximately according to the exponential law

$$F = F_0 \exp(-\mu_m x), \tag{42.9}$$

where x is the thickness of an absorber (in g/cm^2); μ_m is the mass absorption coefficient for electrons (in cm^2/g), which can be approximately expressed as $\mu_m = 15.5E^{-1.41}$ cm^2/g [8].

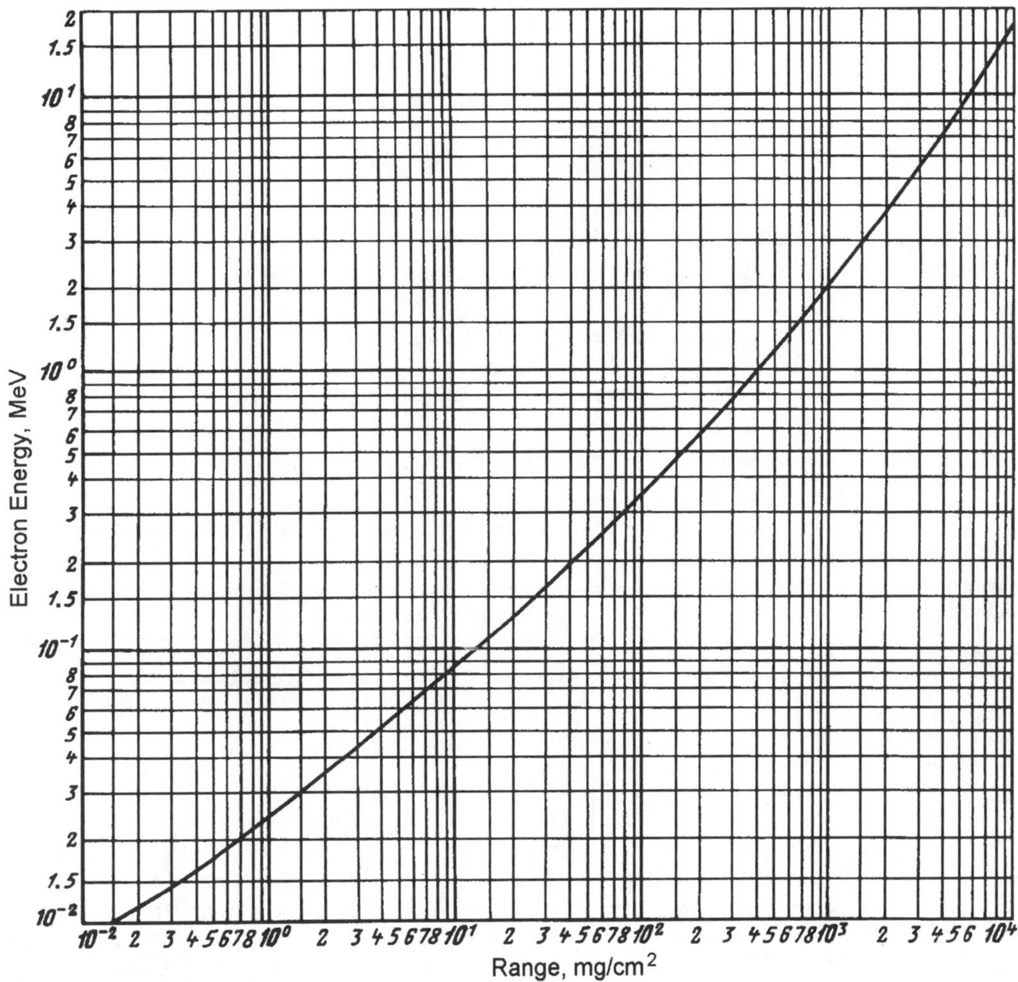

Figure 42.2 The electron range in aluminum.

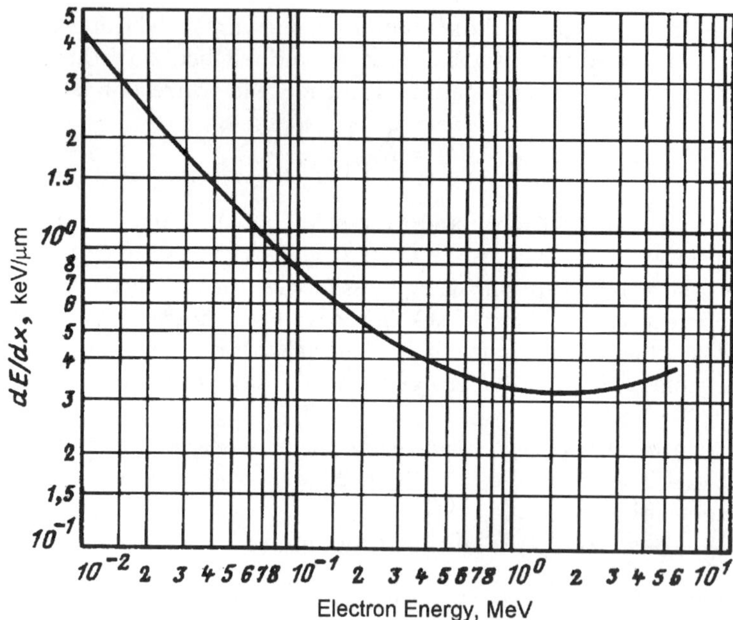

Figure 42.3 Specific energy losses in silicon.

42.4 Penetration of the γ-Rays through Matter

When γ-rays penetrate matter, three main processes take place: the photoelectric effect, Compton scattering, and pair production. Each of these processes results in removal of the photon from the beam. Therefore, the attenuation of a narrow monoenergetic γ-ray beam is given by the exponential relationship

$$F = F_0 \exp(-\mu_m x), \tag{42.10}$$

where x is the thickness of an absorber (in g/cm^2); μ_m is the mass attenuation coefficient (in cm^2/g).

Mass attenuation coefficients for various materials are given in Fig. 42.4 [9]. Tables 42.21–42.24 present the thickness of the shield as a function of attenuation rate and γ-ray energy [8].

Table 42.21 Thickness of lead shielding (in mm) as a function of the attenuation rate and γ-ray energy (wide beam) [8]

Attenuation rate	Energy of γ-rays, MeV															
	0.1	0.2	0.3	0.4	0.5	0.6	0.7	0.8	0.9	1.0	1.5	2	3	4	6	10
2	1	2	3	4	5	7	8	10	11.5	13	17	20	21	20	16	13.5
5	2	4	6	9	11	15	19	22	25	28	38	43	46	45	38	30
10	3	5.5	9	13	16	21	26	30.5	35	38	51	59	65	64	55	42
30	3.5	7	11.5	17	23	30	36.5	43	49.5	55	73	85	93	92	80	63
100	5	10	16	23	30	38.5	47	55	63	70	96.5	113	122	121	109	87
500	6.5	14	22	31	40	51	61	72	82	92	129	150	163	161	149	119
10^3	7	15	24	33	44	57	69.5	81	92	102	141	165	160	178	165	133
$5 \cdot 10^3$	9	19	30	42	55	70	85	99	112	124	170	198	219	217	203	166
10^4	10.5	21	33	45.5	59	75	91	106	120	133	183	213	236	234	220	180
$5 \cdot 10^4$	11.5	23.5	37	52	69	87	105	123	140	156	214	247	263	272	258	215
10^5	11.5	24.0	38	54	72	92	111	130	148	165	227	262	289	289	275	229

Table 42.22 Thickness of iron shielding (in cm) as a function of the attenuation rate and γ-ray energy (wide beam) [8]

Attenuation	Energy of γ-rays, MeV															
rate	0.1	0.2	0.3	0.4	0.5	0.6	0.7	0.8	0.9	1.0	1.5	2.0	3.0	4	6	10
2	0.7	1.2	1.7	2.2	2.5	2.7	2.9	3.1	3.2	3.3	3.6	3.9	4.4	4.5	4.6	3.4
5	1.4	2.5	3.4	4.1	4.8	5.1	5.5	5.7	6.1	6.4	7.4	8.1	8.9	9.4	9.6	8.0
10	1.9	3.5	4.6	5.6	6.3	6.8	7.3	7.7	8.1	8.5	10.0	11.0	12.2	12.6	13.2	11.4
30	2.4	4.5	6.2	7.9	8.5	9.2	9.8	10.4	10.9	11.4	13.6	15.1	17.0	17.7	18.8	17.0
100	3.4	6.1	8.1	9.6	10.8	11.7	12.5	13.2	13.9	14.5	17.3	19.5	22.1	23.3	25.0	23.1
500	4.4	7.7	10.1	12.0	13.7	14.9	16.0	17.0	17.9	18.7	22.3	25.0	28.8	30.6	32.7	31.2
10^3	4.5	8.2	11.0	13.2	15.0	16.3	17.5	18.6	19.6	20.5	24.4	27.5	31.7	33.7	36.0	34.6
$5{\cdot}10^3$	5.6	10.1	13.4	15.8	17.7	19.3	20.7	22.0	23.2	24.3	29.4	33.3	38.2	40.7	43.2	42.2
10^4	6.8	11.5	14.2	17.1	19.0	20.7	22.3	23.6	24.9	26.0	31.3	35.5	40.9	43.7	46.5	45.2
$5{\cdot}10^4$	8.6	13.8	17.0	19.6	21.8	23.6	25.2	26.9	29.4	29.9	35.9	40.8	47.2	50.4	55.0	53.0
10^5	10.0	15.8	18.2	20.8	23.0	24.9	26.7	28.4	30.0	31.5	38.0	43.2	50.0	53.4	58.3	56.1
10^6	12.8	17.9	21.4	24.2	26.7	28.9	31.2	33.3	35.2	37.0	44.7	50.6	58.8	63.3	69.0	67.0
10^7	15.0	20.3	24.3	27.6	30.5	33.2	35.8	38.1	40.2	42.5	51.3	57.9	67.5	73.1	79.4	78.0

Table 42.23 Thickness of concrete shielding (in cm) as a function of the attenuation rate and γ-ray energy (wide beam) [8]

Attenuation	Energy of γ-rays, MeV															
rate	0.1	0.2	0.3	0.4	0.5	0.6	0.7	0.8	0.9	1.0	1.5	2.0	3.0	4	6	10
2	4.7	7.6	9.9	11.3	12.3	12.4	12.4	12.6	12.7	12.9	13.6	14.1	15.3	16.4	18.8	18.8
5	5.6	11.0	15.5	18.8	21.1	21.8	22.3	22.6	23.0	23.5	25.8	28.2	32.9	35.2	38.7	39.9
10	8.2	14.6	19.7	23.7	25.8	26.8	27.6	28.4	29.1	29.9	34.0	37.6	43.4	47.5	51.5	54.0
30	8.5	16.4	22.8	27.7	32.9	34.8	36.4	37.8	39.2	40.5	46.5	51.6	59.9	65.7	71.5	78.1
100	11.5	21.1	48.9	35.2	39.9	43.0	45.3	47.2	48.8	50.5	58.3	66.7	77.5	84.5	95.1	105.1
500	13.8	24.6	35.2	43.9	50.5	54.5	57.3	58.8	62.5	64.6	74.8	84.5	101.0	110.4	124.4	139.7
10^3	15.5	28.2	39.2	48.1	55.2	59.2	62.5	65.3	76.8	70.4	81.7	92.7	110.9	120.9	137.9	155.0
$5{\cdot}10^3$	18.8	33.1	45.6	56.4	65.7	70.0	74.0	77.0	80.2	82.8	97.4	110.9	132.7	146.8	166.7	186.7
10^4	21.8	35.2	48.5	60.3	69.3	74.7	79.1	82.9	85.2	89.2	104.5	118.6	143.2	156.7	179.0	201.3
$5{\cdot}10^4$	23.3	42.3	56.4	68.6	78.1	83.4	88.7	93.4	97.9	102.1	120.4	136.2	164.9	181.6	206.6	233.6
10^5	30.5	50.5	64.6	75.1	82.8	83.3	93.5	98.1	102.5	106.6	126.6	144.4	173.8	191.4	218.4	248.9
10^6	49.3	66.4	79.8	89.9	97.4	103.7	109.2	114.1	119.5	124.4	149.8	171.4	205.4	225.4	260.6	295.8
10^7	64.0	84.9	15.7	130.7	110.3	117.6	123.6	130.0	136.2	142.0	170.8	194.9	236.0	259.4	299.4	340.5

Table 42.24 Thickness of water shielding (in cm) as a function of the attenuation rate and γ-ray energy (wide beam) [8]

Attenuation	Energy of γ-rays, MeV															
rate	0.1	0.2	0.3	0.4	0.5	0.6	0.7	0.8	0.9	1.0	1.5	2.0	3.0	4	6	10
2	18	27	30	30	29	28	27	27	26	26	28	33	37	38	40	45
5	27	27	42	44	46	46	47	47	48	48	53	59	69	74	85	98
10	35	47	52	55	57	58	60	61	62	63	71	79	92	101	116	135
30	45	58	65	69	73	75	77	79	81	82	96	107	124	140	163	192
100	53	70	79	84	88	92	95	98	101	104	120	136	160	182	212	253
500	66	85	96	104	110	116	120	124	128	132	153	173	206	234	277	334
10^3	72	92	104	113	118	124	129	134	138	143	167	189	226	257	306	368
$5{\cdot}10^3$	81	105	120	130	138	147	154	160	165	170	198	225	271	309	368	446
10^4	91	114	128	139	147	155	162	168	175	181	212	241	290	331	396	480
$5{\cdot}10^4$	103	128	144	156	167	175	183	190	196	205	241	276	334	383	460	554
10^5	110	135	152	164	175	184	192	201	204	216	255	292	353	404	487	587
10^6	128	156	176	192	205	215	224	233	243	253	299	342	415	479	571	695
10^7	150	178	200	217	232	243	255	265	277	288	344	393	478	554	657	807

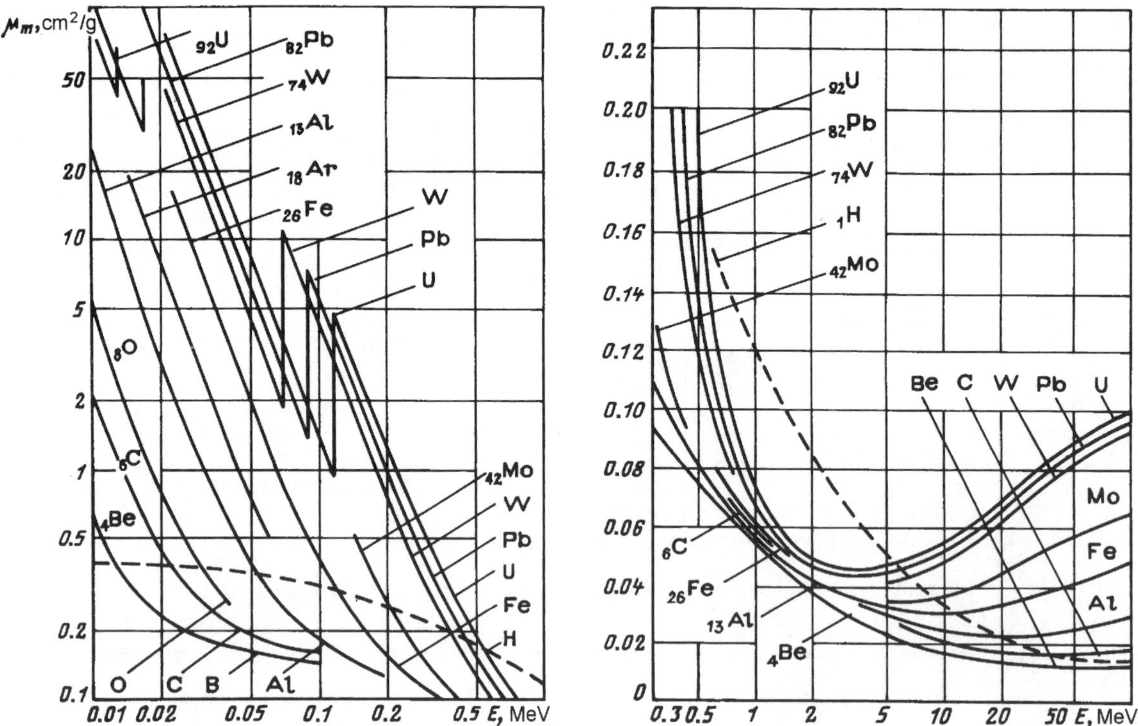

Figure 42.4 Mass attenuation coefficient for a narrow γ-ray beam in different substances as a function of γ-ray energy.

42.5 Relative Biological Effectiveness and Tolerance Levels of Various Radiations

The gray unit (Gy) is an SI measure of the absorbed dose of ionizing radiation. The substance of 1 kg mass therewith gets energy of 1 J via the absorbed radiation.

The röntgen presents the out–of–SI unit of exposure of X- or γ-radiations. Radiation exposure equals 1 R, if 1 CGS unit of ion charge of each (opposite) sign appears in 1 cm^3 of dry air under exposition. As the weight of 1 cm^3 dry air comprises 0.001293 g and 1 unit of charge in the CGS system corresponds to $2.08 \cdot 10^9$ ions, then 1 R = $2.58 \cdot 10^{-4}$ C/kg.

For those radiation modes and energies, which necessitate $\simeq$34 eV in producing an ion pair in air, it follows that 1 Gy $\simeq$114 R. This relationship is valid for X-ray and γ-ray radiations, as well as for electrons with energy $\leq$3 MeV. However, the relationship derived is invalid for α-radiation and fission fragments.

The relative biological effectiveness is different for various radiations. The coefficient of relative biological effectiveness (RBE) of different radiations characterizes the intensity of biological impact as compared with γ-radiation provided equal absorbed dose.

The RBE unit determines a biological action of X-radiation with energy 200 keV creating 1 hundred pairs of ions along the radiation path in water. RBE of different biological reactions variously depends on linear transfer of energy. Since the radiation impact on the human body is an object of quantitative control, the reglemented RBE was introduced for checking the whole irradiated human body.

The radiation exposure is expressed in terms of equivalent dose D_{eq}, i.e. energetic dose of radiation is multiplied by its RBE coefficient. D_{eq} is measured in the units of sievert, which serves as a biological equivalent of gray unit.

The natural human irradiation due to cosmic rays and terrestrial radioisotopes is estimated to be close to 0.8 mSv per year (0.09 R/y $\simeq$ 10 μR/h). The professional exposure is stated legislatively. It amounts to 0.05 Sv/y$\simeq$ 28 μSv/h (5 R/y or 3 mR/h) for Russia and is calculated on the basis of a 36-h operating week [11].

Coefficients of relative biological effectiveness (CRBE) for different ionizing radiations are tabulated in Tables 42.25, 42.26. The effectivness of Pb-shielding is illustrated in Figure 42.5.

Table 42.25 Coefficient of relative biological effectiveness (CRBE) of different ionizing radiations for chronicling irradiation of the whole body [10]

Radiation	CRBE	Radiation	CRBE
γ-rays	1	Neutrons of energy 5 keV	2.5
X-rays	1	20 keV	5
Electrons and positrons	1	100 keV	8
α-particles ($E \leq 1$ MeV)	10	500 keV	10
Protons ($E \leq 1$ MeV)	10	1 MeV	10.5
Recoil heavy nuclei	20	5 MeV	7
Thermal neutrons	3	10 MeV	6.5

Table 42.26 Coefficient of relative biological effectiveness (CRBE) for the radiations with various linear transfer of energy (LTE) [10]

LTE in water, keV/μm	3.5	3.5–7.0	7.0–23	23–53	53–175
CRBE	1	1–2	2–5	5–10	10–20

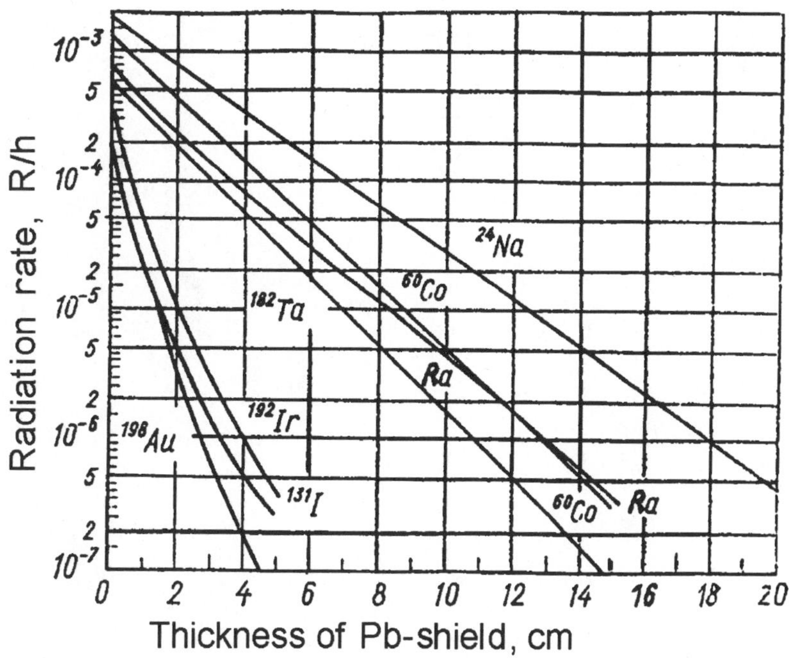

Figure 42.5 Radiation rate at the distance 1 m from different γ-ray sources of 1 mCi intensity as a function of the lead shield thickness.

References

[1] Northclifte, L. C., Schilling, R. F., Nuclear data tables, A7, 233, 1970.

[2] Nemets, O. F., Gofman, Yu. V., A Handbook on nuclear physics, Naukova Dumka, Kiev, 1975 (in Russian).

[3] Pucherov, N. N., Romanovskii, S. V., Chesnokova, T. D., Tables of mass stopping power and range of charged particles in 1–100 MeV energy interval, Naukova Dumka, Kiev, 1976 (in Russian).

[4] Nigam, B. P., Sandersan, M. K., Wu, T.-W., Phys. Rev., 115, 491, 1959.

[5] Marion, I. B., Zimmerman, B. A., Nucl. Instrum. and Methods, 51, 93, 1967.

[6] Katz, L., Penfold, A. S., Rev. Mod. Phys., 24, 28, 1952.

[7] Marion, I. B., Young, F. C., Nuclear reaction analysis: Graphs and tables, North-Holland Publ. Comp., Amsterdam, 1968.

[8] Aglintsev, K. K., Kodukov, V. M., Applied dosimetry, Gosatomisdat, Moscow, 1962 (in Russian).

[9] Price, B.T., Horton, C.C., Spinney, K.T., Radiation shielding, Pergamon Press, London–New York–Paris, 1957.

43

Cosmic Rays

V.S. Ptuskin

43.1 The Sources of Cosmic Rays

Cosmic rays (CR) are high-energy elementary particles and atomic nuclei of cosmic origin. The particles with energies from 10^6 to 10^{20} eV were observed in CR. Particles produced as a result of interaction of primary CR with air atomic nuclei in the Earth's atmosphere are also classified as cosmic rays.

The main fraction of CR coming to the Earth's atmosphere is of galactic origin (galactic CR). Supernovae, supernova remnants and neutron stars are considered to be the sources of these particles [1, 2]. Some fraction of CR, mostly with energies 10^6–10^9 eV, comes to the Earth from the Sun. The solar cosmic rays are accelerated at chromospheric flares and during other active processes on the Sun [3]. As supposed, particles with the highest observed energies $E > 10^{17}$–10^{19} eV are of extragalactic origin. They might be accelerated in active galaxies [2]. Jupiter magnetosphere supplies the interplanetary medium with electrons with energies $E < 3{\cdot}10^7$ eV [4]. So called anomalous CR were observed at energies 10^6–10^8 eV. These nuclei are accelerated in the outer part of the heliosphere near the external boundary of the supersonic solar wind [5]. The integral flux density of CR is measured in the units of $\mathrm{s^{-1}{\cdot}m^{-2}{\cdot}sr^{-1}}$. The particle energy E is measured in electron-volt, and the unit of energy per nucleon $\varepsilon = E/A$ (in eV/nucleon) is also used for nuclei (here A is the nucleus atomic number). The *rigidity* of a charged particle with a momentum p (in kg·m/s), and a charge Ze (in C) equals to $R = pc/Ze$; usually R is measured in volts. The particle momentum (in eV/c) (c is the speed of light) numerically equals rigidity in volts for a particle of unit charge, i.e. $Z = 1$. What follows is a summary of basic data on CR [6]:

CR flux density beyond the region of the Earth's magnetic field action, $\mathrm{s^{-1}{\cdot}m^{-2}}$	$(3–6){\cdot}10^3$
Number density of primary CR in the interstellar medium, $\mathrm{m^{-3}}$	10^{-4}
Energy density of primary CR in the interstellar medium, $\mathrm{J/m^3}$	$1.6{\cdot}10^{-13}$
Rate of ion pairs production by CR at sea level (on average), $\mathrm{s^{-1}{\cdot}m^{-3}}$	$1.6{\cdot}10^6$
Total power of CR reaching the Earth's surface, W	$4{\cdot}10^8$

43.2 Galactic Cosmic Rays

Given below are the parameters of the galactic model of cosmic ray origin [2]:

0-8493-2861-6/97/$0.00+$.50

Total power of the galactic CR sources, W	$3{\cdot}10^{33}$
Height of the galactic CR halo (the region of the Galaxy where CR are confined), kpc*1	3–30
Time of CR confinement in the Galaxy*2, year	10^7–10^8
Average matter thickness traversed by CR in the Galaxy*2, kg/m^2	100

*1 1 pc = $3.0857{\cdot}10^{16}$ m.

*2 For $\varepsilon = 2$ GeV/nucleon; the time decreases with energy at $\varepsilon > 2$ GeV/nucleon.

CR flux density as a function of particle energy is shown in Figs. 43.1–43.4. The power law dependence on energy is the characteristic feature of $I(E)$. The energy spectrum observed on Earth varies with the level of the solar activity (see section 43.3).

Galactic cosmic rays contain a small fraction of antiprotons with the flux density $(3.8 \pm 0.95){\cdot}10^{-2}$ s^{-1}·m^{-2}·sr^{-1}·GeV^{-1} in the energy range $4.7 \leq E \leq 11.3$ GeV at the Earth's atmosphere boundary [12]. This value is larger than expected from the production of antiprotons due to interactions of CR nuclei with the interstellar gas.

Detailed data on nucleus composition in CR are obtained only for energies less than 30 GeV/n [13,14].

The flux density of different nuclei with the energies of 600–1000 MeV/nucleon at the boundary of the atmosphere measured in September 1977 is as follows [15]:

Table 43.1

Element	I, s^{-1}·m^{-2}·sr^{-1}	Element	I, s^{-1}·m^{-2}·sr^{-1}	Element	I, s^{-1}·m^{-2}·sr^{-1}	Element	I, s^{-1}·m^{-2}·sr^{-1}
He	54.08 (1.10)	Ne	0.228 (5)	Si	0.202 (4)	Cr	0.0240 (15)
C	1.52 (3)	Na	0.51 (2)	S	0.0467 (19)	Fe	0.142 (4)
N	0.42 (1)	Mg	0.282 (6)	Ar	0.0205 (12)	Ni	0.0074 (9)
O	1.41 (3)	Al	0.056 (2)	Ca	0.0323 (15)		

Relatively high abundance of Li, Be, B, ^{2}H, ^{3}He, and the nuclei with charges $21 \leq Z \leq 25$ as well as other rare-in-nature elements and isotopes (see Figs. 43.5, 43.6) is attributed to their secondary origin. The secondary nuclei are generated by fragmentation of heavier nuclei interacting with the interstellar gas. The fraction of the secondary nuclei falls at high energy (Fig. 43.7) due to the corresponding decrease of the CR confinement time in the Galaxy. Table 43.2 shows the source abundances of the primary nuclei corrected for fragmentation.

Table 43.2 Relative nuclei content for galactic [19] and solar [20] primary cosmic rays [The conventional content of Si nuclei is equal 100.]

Element	Galactic CR	Solar CR	Element	Galactic CR	Solar CR
H	77000 (9000)		Al	14 (3)	9.4 (3.8)
He	11500 (1000)	41600 (2000)	Si	100	100
C	465 (40)	270 (51)	S	14(3)	23 (12)
N	20 (10)	80 (22)	Ar	4 (2)	4.1 (2)
O	525 (50)	625 (76)	Ca	7 (2)	7.9 (2)
Ne	62 (5)	83 (23)	Fe	92 (12)	106 (39)
Na	9 (3)	8.7 (3.2)	Ni	4.8 (0.6)	5.2 (2.6)
Mg	110 (7)	124 (25)			

Radioactive isotope ^{10}Be is used for the determination of CR age in the Galaxy. In the interstellar medium the fraction of non-decayed isotope ^{10}Be in CR constitutes 0.22 ± 0.04 at an energy of about $\varepsilon = 400$ MeV/nucleon [21].

Fig. 43.8 illustrates the deviation of the angular distribution of galactic cosmic rays from the isotropic distribution.

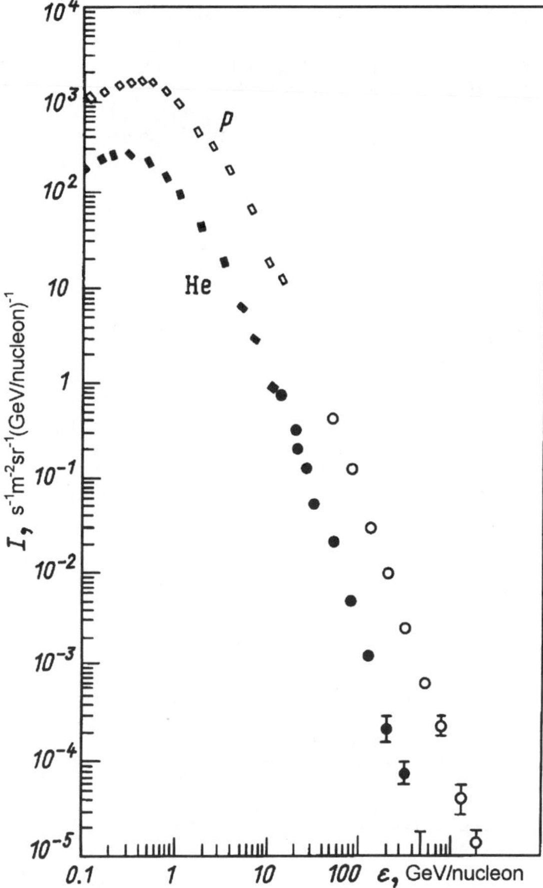

Figure 43.1 Energy spectra of protons and helium in CR [7].

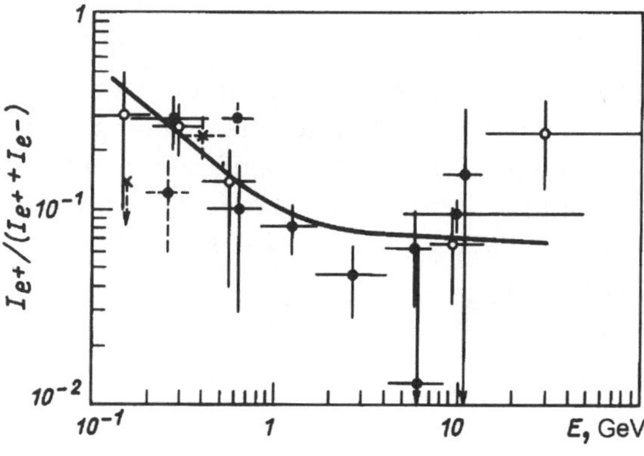

Figure 43.2 Fraction of relativistic positrons in the total flux density of electrons and positrons in CR [9, 10]. Most of these positrons are produced in the decays of π^+-mesons which were born in the interactions of CR nuclear component with the interstellar gas nuclei.

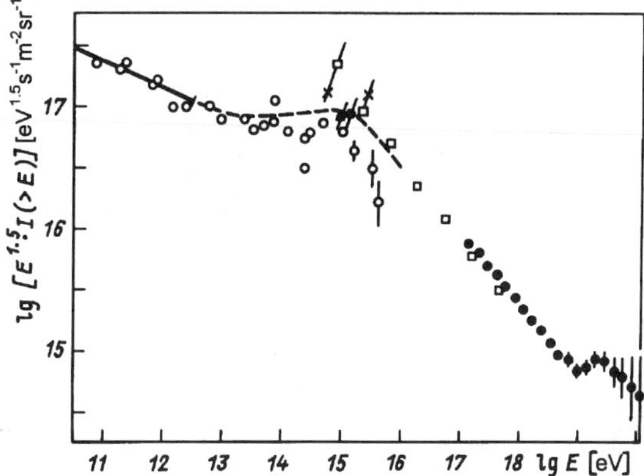

Figure 43.3 Spectrum of high energy CR [8].

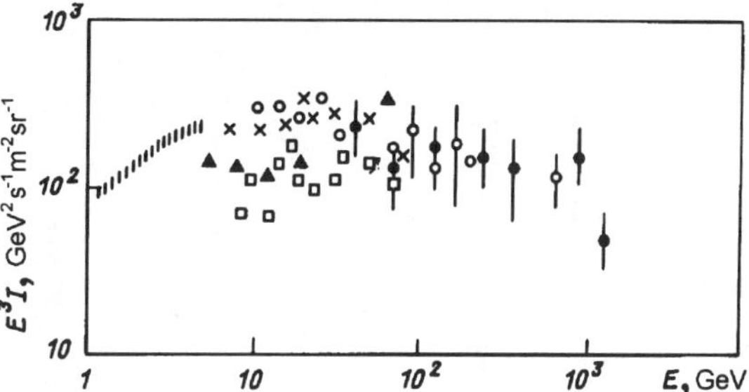

Figure 43.4 Energy spectrum of CR electrons [9, 10].

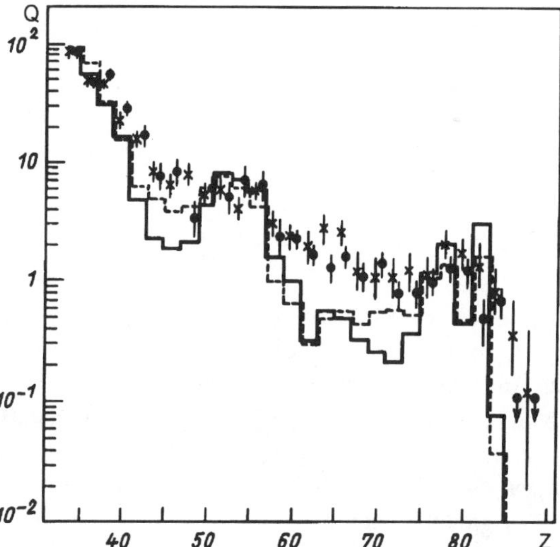

Figure 43.5 Relative abundance Q of elements with the nucleus charge $Z \geq 30$ (concentration of the Fe nuclei is set equal to 100). The circles and crosses are experimental data; the solid line is for the Solar system abundance; dashed line shows expected CR composition under the assumptions that CR source composition coincides with the Solar system composition and the average matter thickness traversed by cosmic rays in the interstellar gas equals 55 kg/cm^2 [17].

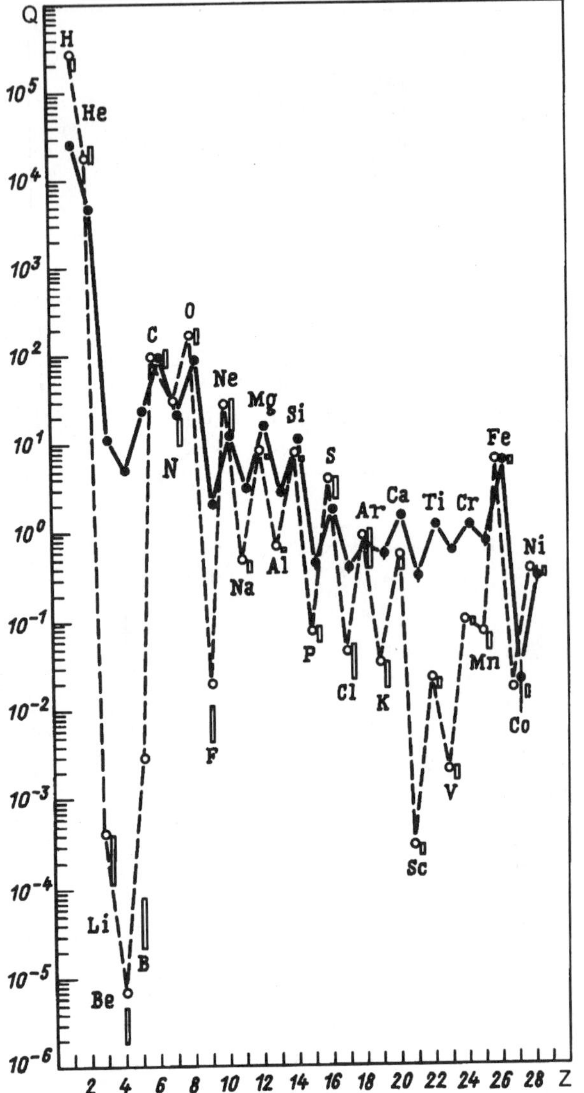

Figure 43.6 Relative abundance Q of elements from H to Ni; conventional content of C nuclei equals 100 [16]. The full circles are experimental data for the nuclei with energies 70–280 MeV/nucleon, the open circles represent average Solar system abundances, and the rectangles show abundances for the nearby galactic region.

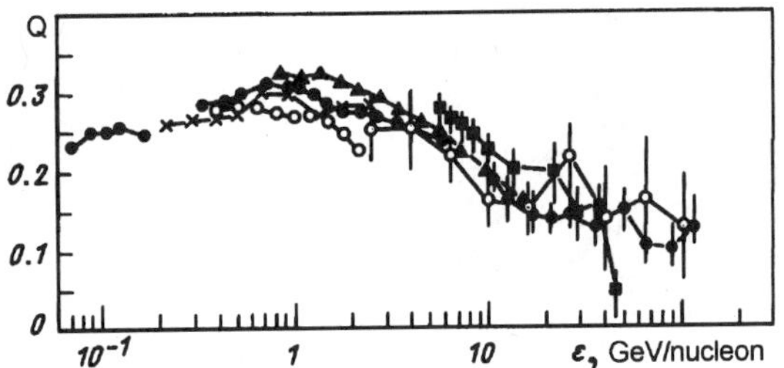

Figure 43.7 The ratio of fluxes B and C nuclei as a function of energy [18].

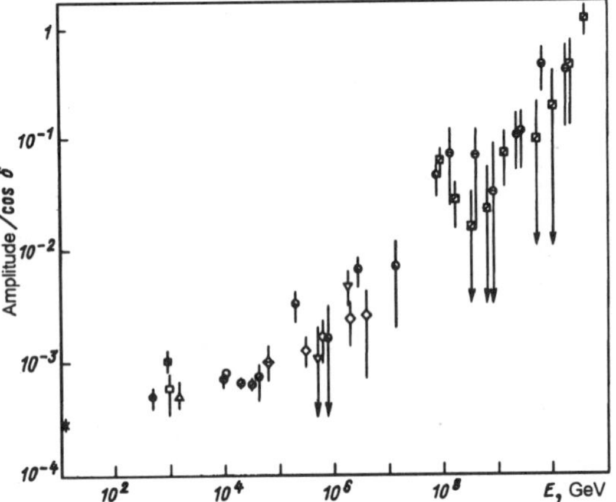

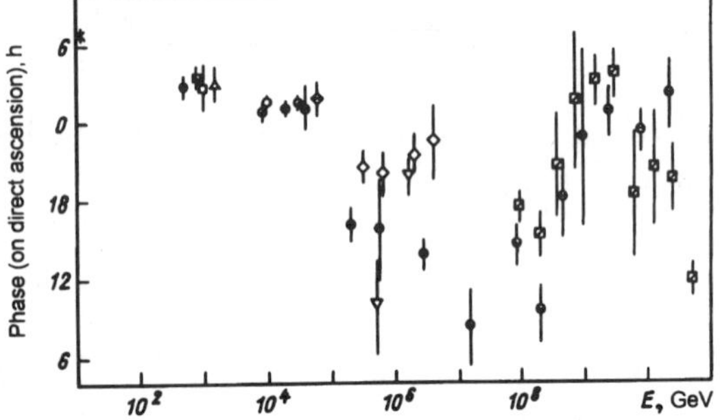

Figure 43.8 CR anisotropy according to harmonic analysis of CR detectors' counting rates vs. sidereal time [22]. The amplitude and phase of the first harmonic are shown; δ is the geographical latitude where the observations were performed.

43.3 Modulation of Galactic Cosmic Rays in the Interplanetary Medium

The presence of regular and random magnetic fields of the solar origin in the interplanetary medium results in CR modulation, i.e. variations of its intensity in time and space [23–25] (see class IIIa in Table 43.3, and Figures 43.9, 43.10). Space craft measurements up to the distance of more than

40 a.u. from the Sun (1 a.u. = $1.496 \cdot 10^{11}$ m) reveal the radial gradient of cosmic ray intensity in the interplanetary medium near the ecliptic plane of about $3 \cdot 10^{-2}$ (a.u.)$^{-1}$ at the maximum of solar activity and $1.5 \cdot 10^{-2}$ (a.u.)$^{-1}$ at the minimum of solar activity for energies $\varepsilon \approx 1$ GeV/nucleon [26]. The decrease of CR intensity in the internal regions of the Solar system results from the interactions of CR with the regular and random magnetic fields carried by the solar wind (the plasma flow from the solar atmosphere). The characteristic scale of the region where modulations of galactic CR occurs is estimated as 60–120 a.u.

Table 43.3 Summary of the main classes of intensity variations for CR with energies below 100 GeV [23] [$\delta I/I$ is the observed amplitude variation.]

Class	Type of CR variations	$\delta I/I$, %			Nature of the variations
		Muon component	Neutron component	In the stratosphere and beyond the boundary of the atmosphere	
I. Meteorological variations	Seasonal	5	1		Variation of the absorption and decay of secondary particles in the Earth's atmosphere due to the change of the meteorological conditions (temperature, pressure) above the point of observation
	Diurnal (mask analogous variations of class III)	0.1–0.2 %/K			
	Temperature effect	0.1–0.2 %/K	0.02 %/K		
	Barometric effect	$\sim 10^{-3}$ %/Pa	$7 \cdot 10^{-3}$ %/Pa		
II. Variation of the geomagnetic cut-off threshold	Increase during the main phase of magnetic storm against a generally reduced background	3	10	Direct observation of the geomagnetic threshold variation	Influence of the geomagnetic field variations on the CR trajectories and the geomagnetic cut-off threshold
	Solar-day-periodical (local source)	1	3		Asymmetrical variations of the geomagnetic cut-off threshold
IIIa. Modulation effects due to solar wind	11-years- and 22-years-periodical	6	20	100	Scattering of galactic CR by solar wind inhomogeneities. Variation of the solar activity. Inversion of the solar global magnetic field
	27-day-periodical	0.5	1–2	10	Asymmetrical stream of inhomogeneous plasma from the Sun and sector structure of the interplanetary magnetic field
	Solar-day-periodical	0.2–0.3	0.5	2	CR anisotropy caused by the action of regular and random electromagnetic fields in the interplanetary space
	Forbush effect (decrease of CR intensity at the time of magnetic storm)	10	30	50	Interaction of CR with the interplanetary shock waves

Table 43.3 Summary of the main classes of intensity variations for CR *(continued)*

Class	Type of CR variations	$\delta I/I$, %			Nature of the variations
		Muon component	Neutron component	In the stratosphere and beyond the boundary of the atmosphere	
IIIb. Generation of fast particles on the Sun	Large increases of CR intensity due to powerful solar chromospheric flares (amplitudes are given for the outburst of February 23, 1956)	400	5000	Thousands times higher than the normal level	Particle acceleration in the region of a solar flare with subsequent propagation from the Sun
	Small increases associated with the ordinary chromospheric flares	0.1–0.2	1	Tens or hundreds times higher than the normal level	
IIIc. Anisotropy and intensity variations of galactic CR	Possible anomalous long-term increases of CR intensity by a factor 10^3 on a time scale more than 10^3 years				Local galactic supernovae at a distance of about 20 pc from the Sun
	Sidereal-day variations	0.1 at solar activity minimum			CR leakage out of the Galaxy, peculiar motion of the Sun, CR fluxes from the nearest individual galactic sources

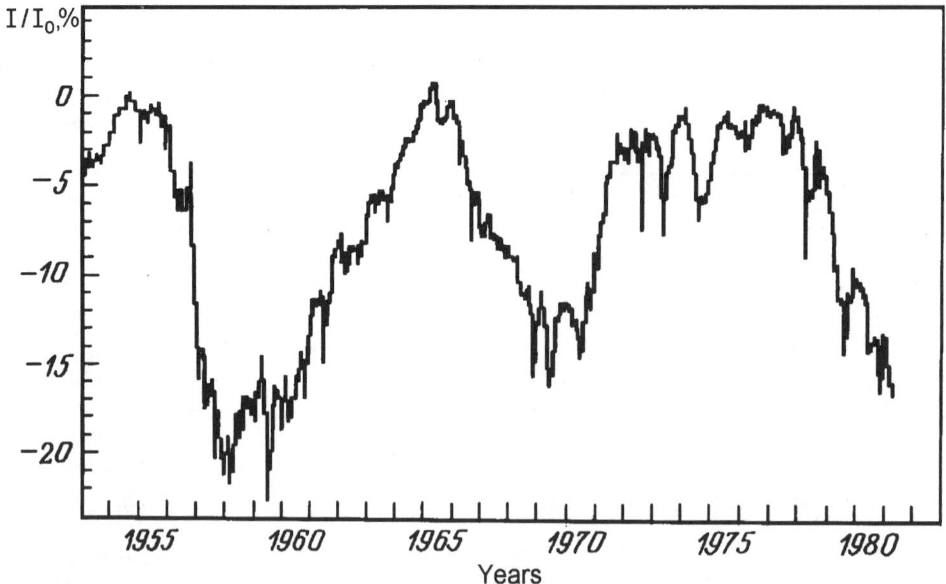

Figure 43.9 The variations of CR flux density averaged over 27 days as observed with the neutron monitor at Climax (geographical latitude 39.37°, longitude 253.82°). The mean energy of the primary CR beyond the atmosphere approximately equals 6 GeV/nucleon. The flux density is normalized to the minimum of the solar activity in 1954.

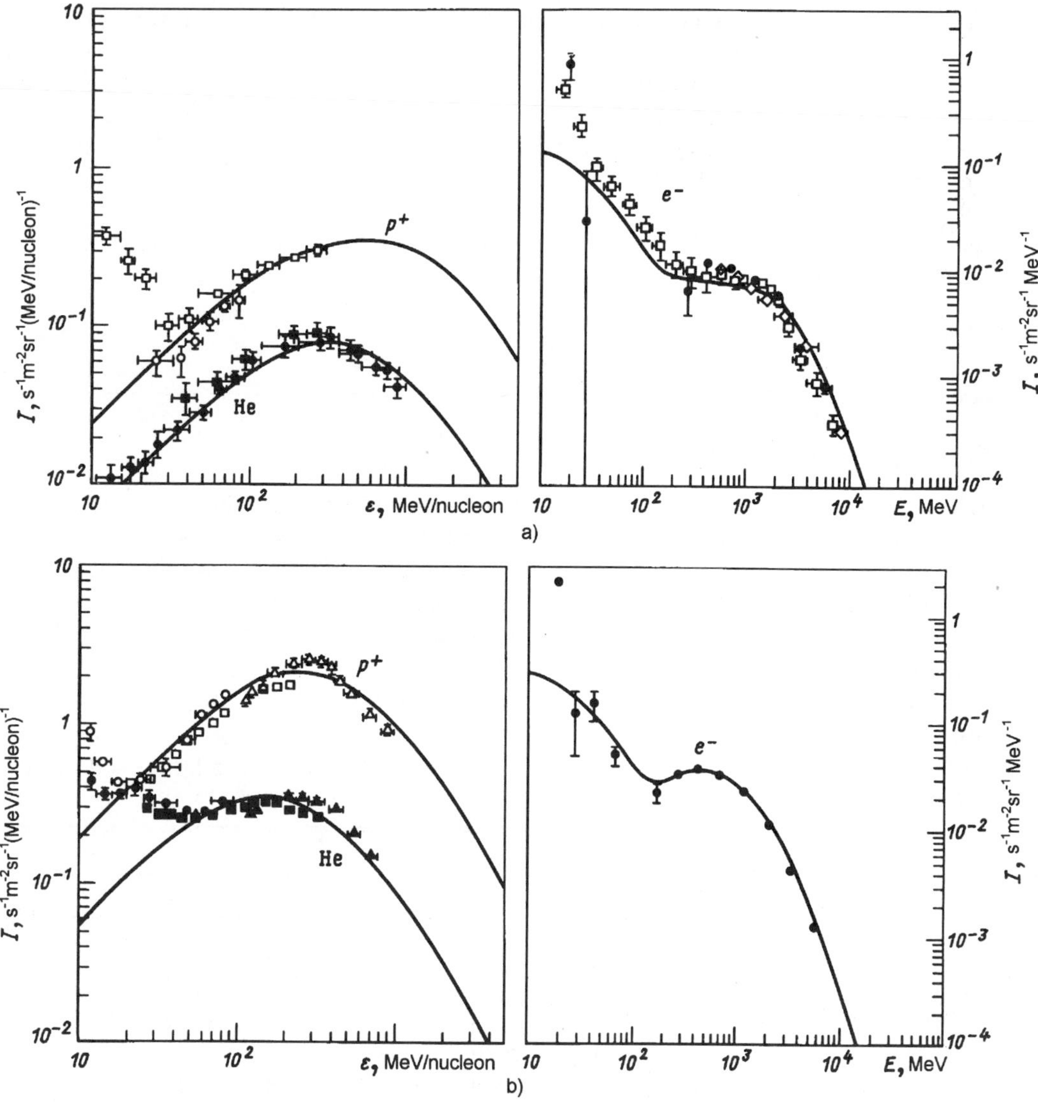

Figure 43.10 CR energy spectra at the periods of maximum 1969 (*a*) and minimum 1977 (*b*) of the solar activity [27].

43.4 Solar Cosmic Rays

Protons constitute the main fraction of solar cosmic rays. For the powerful solar flares the flux ratio of protons and helium nuclei approximately equals 50 at the energies of a few tens MeV/nucleon.The total energy of accelerated particles in the flare ranges up to 10^{25} J. The charge composition of accelerated particles, the total energy and the form of the energy spectrum vary significantly from one flare to another. The differential particle spectrum vs. momentum has a power law (with the typical power law exponent being of 3–7) or exponential form. The flux of the solar CR usually reaches its maximum value at the Earth's orbit in 8–16 hours after the flare starts and then falls substantially in 30–32 hours (see [3, 28, 29] for detail).

CR flares accompanied by the emission of particles with energies more than 450 MeV penetrating to the Earth's surface via production of secondary particles were registered 54 times during the period of observations 1942–1984. These flares, as a rule, were observed at the time of the solar activity variation but not at the moments of the solar activity maximum. The proton flares with the particle energy in the range $10 < E < 100$ MeV were recorded at the upper atmosphere much more often. The proton flares at the Sun are accompanied by the emission of fast electrons ($E > 30$ keV, the flux density up to 10^8 (s·cm^2·sr)$^{-1}$).

Averaged (over the time) CR power of the Sun amounts to 10^{17} W.

The variations of CR intensity on Earth resulting from the generation of fast particles at the Sun are listed in Table 43.3 (class IIIb).

43.5 Geomagnetic Effects

The terrestrial magnetic field changes the initial trajectories of CR particles [30]. As a result the effect of geomagnetic cut off occurs: for a given geomagnetic latitude and direction only those particles can reach the surface of the Earth whose rigidity R is above some threshold cut-off rigidity (see Table 43.4). The value of the rigidity R (in eV) determines the particle trajectory in the magnetic field. The radius r (in m) of the circular motion of the charged particles in the uniform magnetic field B (in T) equals r=3.34·10$^{-9}R/B$.

The variations of CR intensity of geomagnetic origin are indicated in the Table 43.3 (class II).

Table 43.4 Magnetic cut-off rigidities (in GeV) for the vertical directions [31] [δ is the geomagnetic latitude and ϕ is the geomagnetic longitude (in angle deg.)]

δ	ϕ, deg											
	0	30	60	90	120	150	180	210	240	270	300	330
75	0.1	0.2	0.3	0.3	0.3	0.3	0.2	0.1	0.0	0.0	0.0	0.0
60	1.1	1.5	1.7	1.8	2.0	2.2	1.8	1.1	0.4	0.2	0.2	0.6
45	4.7	5.4	5.8	6.2	6.6	7.0	5.6	4.2	2.4	1.3	1.4	3.0
30	11.2	12.1	13.3	14.4	14.4	13.6	11.9	9.9	9.4	4.6	4.2	8.9
15	14.5	15.5	16.6	17.5	17.0	15.9	14.9	14.1	13.0	9.8	9.3	13.2
0	14.0	14.8	16.1	17.0	16.8	16.2	15.8	15.2	14.6	13.6	13.1	13.7
−15	11.0	11.5	12.4	13.0	13.4	13.7	14.2	14.4	14.2	13.7	13.0	12.3
−30	6.8	6.1	5.7	5.1	5.3	6.0	8.0	9.5	12.1	12.3	11.4	10.3
−45	3.7	2.7	2.1	1.4	1.1	1.6	2.8	4.3	5.9	9.3	8.6	6.5
−60	1.9	1.1	0.6	0.2	0.1	0.2	0.5	1.2	2.3	3.9	4.2	3.3
−75	0.6	0.3	0.1	0.0	0.0	0.0	0.0	0.2	0.6	0.9	1.2	1.0

43.6 Cosmic Rays in the Earth's Atmosphere

The total thickness of the atmosphere is equivalent to about 13 nuclear path lengths and 27 electromagnetic radiation lengths for incoming cosmic fast particles. That is why the primary cosmic rays with sufficiently high energies produce the branching chain of interactions. Arising from these interactions π^+, π^--mesons and partly K-mesons decay and generate muons and neutrinos which constitute the penetrating component of the secondary cosmic rays. (Muons with energies less than 10^{12} eV lose their energy mainly due to the ionization of the medium. Their path length is approximately proportional to the energy and amounts to 1 km in the ground. The number of muons with higher energies attenuate according to (approximately) exponential law and penetrate into the ground to 3–5 km.) The pions which have no time to decay along with the nucleons continue the cascade of interactions until the generated particles reach an energy on the order of 10^9 eV. These particles constitute a nuclear secondary component of CR. The decay $\pi^0 \to 2\gamma$ gives rise to the electron-photon shower which constitutes a secondary electron-photon component of CR. The photons knock electrons out from their parent atoms and generate e^+e^- pairs which in their turn produce photons through the radiative deceleration and annihilation (see [23, 32, 33], Table 43.5, and Figures 43.11–43.13).

The showers of secondary particles in the Earth's atmosphere produced by the primary CR with energies more than 10^{14} eV are called the extensive air showers.

Table 43.5 Distribution of energy dissipation at latitude 50° [32]

Process of dissipation	Power, 10^{-8} W/(m^2·sr)
Ionization in the atmosphere	120
Remainder at the see level	6
Nuclear spallation	24
Neutrino	37
Total amount	187

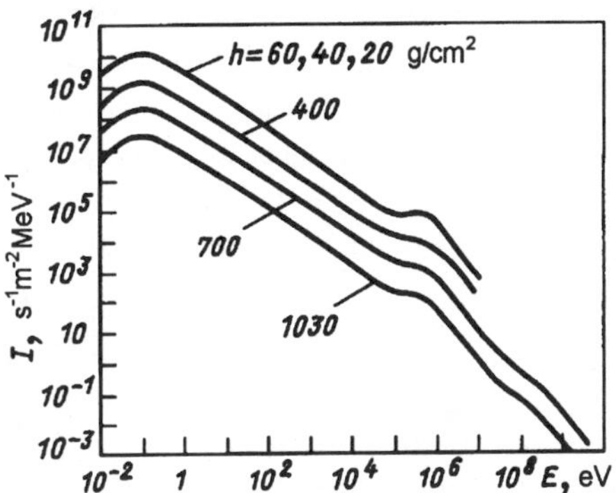

Figure 43.11 The equilibrium neutron flux density versus energy at different depths in the atmosphere for geomagnetic latitude 44° [32]. The neutron flux at the Earth's surface in the energy range 1–10^7 eV is subject to large variations due to the changes of soil features and other factors which are difficult to account for.

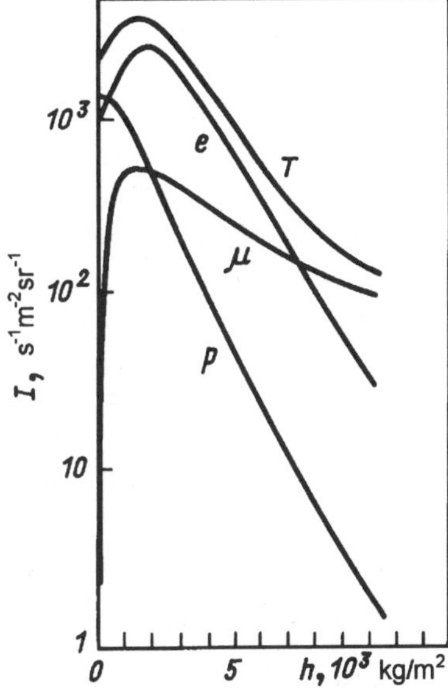

Figure 43.12 Vertical flux density of CR versus atmospheric depth measured at Saskatoon (geographical latitude 60.5°, longitude 311.9°) at the year of the solar activity minimum: T is the total intensity; e is for electrons; μ is for muons, and p is for protons [34].

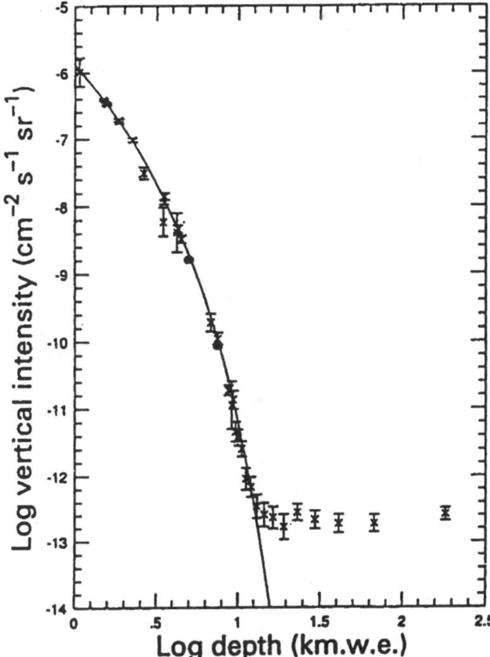

Figure 43.13 Relation between muon flux density and ground thickness (depth) [33].

References

[1] Ginzburg, V.L., Syrovatskii, S.I., The origin of cosmic rays, Pergamon Press, Oxford, 1964.

[2] Berezinskii, V. S., Bulanov, S. V., Dogiel, V. A., Ginzburg, V. L., Ptuskin, V. S., Astrophysics of cosmic rays, North Holland Publ. Comp., Amsterdam, 1990.

[3] Dorman, L. I., Venkatesan, D., Space Science Rev., 63, 4, 1993.

[4] Eraker, J. H., Astrophys. J., 257, 862, 1982.

[5] Garcia-Munoz, M., Pyle, K. R., Sympson, J. A., Astrophys. J., 274, L93, 1983.

[6] Allen, C. W., Astrophysical quantities, Athlone Press, Univ. of London, London, 1963.

[7] Webber, W. R., Lezniak, J. A., Astrophys. Space Sci., 30, 361, 1974.

[8] Linsley, J., IAU Symp. No 94: Origin of Cosmic Rays, Eds. Setti, G., Spada, G., Wolfendale, A., D. Reidel Publ. Comp., Dordrecht–Boston–London, 1980, p. 53.

[9] Muller, D., Tang, J., Proc. 17th Intern. Cosmic Ray Conf., vol. 2, Paris, 1981.

[10] Nishimura, J., Fujii, M., Taira, T., Astrophys. J., 238, 394, 1980.

[11] Tan, L. C., Ng, L. K., Proc. 17th Intern. Cosmic Ray Conf., vol. 2, Paris, 1981, p. 202.

[12] Golden, R. L., Horan, S., Mauger, B. G., Phys. Rev. Lett., 43, 1196, 1979.

[13] Waddington C. J., Cosmic abundances of matter, AIP, New York, 1989.

[14] Sympson, J. A., Ann. Rev. Nucl. Part. Sci., 33, 323, 1983.

[15] Webber, W. R., Astrophys. J., 255, 329, 1982.

[16] Meyer, P., IAU Symp. No 94: Origin of Cosmic Rays, Eds. Setti, G., Spada, G., Wolfendale, A., D. Reidel Publ. Comp., Dordrecht–Boston–London, 1980, p. 7.

[17] Israel, M. H., Proc. 17th Intern. Cosmic Ray Conf., Paris, 12, 53, 1981.

[18] Webber, W. R., in: Composition and Origin of Cosmic Rays, Ed. Shapiro, M. M., D. Reidel Publ. Comp., Dordrecht–Boston–London, 1983, p. 25.

[19] Mewaldt, R. A., Rev. Geophys. Space Phys., 21, 295, 1983.

[20] Meyer, J., Proc. 17th Intern. Cosmic Ray Conf., Paris, 3, 145, 1981.

[21] Lukasiak, A., Ferrando, P., McDonald, F. B., Webber, W. R., Proc. 23rd Intern. Cosmic Ray Conf., Calgary, 1, 523, 1993.

[22] Linsley, J., Proc. 18th Intern. Cosmic Ray Conf., Bangalore, 13, 135, 1983.

[23] Dorman, L. I., Cosmic rays, North Holland Publ. Comp., Amsterdam, 1974.

[24] Quenby, J. J., Space Sci. Rev., 37, 201, 1984.

[25] Toptygin, I. N., Cosmic rays in interplanetary magnetic fields, D. Reidel Publ. Comp., Dordrecht, 1985.

[26] McKibben, R. B., Pyle, K. R., Sympson, J. A., Astrophys. J., 254, L23, 1982.

[27] Evenson, P., Garcia-Munoz, M., Meyer, P., Astrophys. J., 275, L15, 1983.

[28] Shea, M. A., Smart, D. F., Solar Phys., 127, 297, 1990.

[29] Forman, M., Ramaty, R., Zweibel, E. G., in: The Physics of the Sun, vol. 2, Ed. Sturrock, P. A., D. Reidel Publ. Comp., Dordrecht, 1986, p.249.

[30] Dorman, L. I., Smirnov, V. S., Tyasto, M. I., Cosmic rays in the Earth magnetic field, Nauka, Moscow, 1971 (in Russian).

[31] Shee, M. A., Smart, D. F., J. Geophys. Res., 72, 2021, 1967.

[32] Hayakawa, S., Cosmic ray physics, Wiley Interscience, New York–London–Sydney–Toronto, 1969.

[33] Gaisser, T. K., Cosmic rays and particle physics, Cambridge Univ. Press, Cambridge, 1990.

[34] Komori, H., J. Phys. Soc. Jpn., 17, 457, 1962.

44

Physics of the Earth

I.A. Maslov

44.1 General Characteristics of the Earth

44.1.1 Figure of the Earth

The term "real Earth's figure" is used in reference to the shape of the Earth's physical surface, i.e. surface of the continents, oceans, seas and lakes [1]. In this context some estimates of the surface areas may be of interest:

Earth's area	$5.1 \cdot 10^{14}$ m^2
Global mean sea area	$3.61 \cdot 10^{14}$ m^2
Continents' area	$1.49 \cdot 10^{14}$ m^2
Glaciers and the land under ice	$1.62 \cdot 10^{13}$ m^2
Sea Ice zones	
in March	$1.75 \cdot 10^{13}$ m^2
in September	$2.84 \cdot 10^{13}$ m^2

In scientific investigations and in solving practical problems, the geoid surface is widely considered, i.e. the level surface of the gravitational potential when describing the Earth's figure.

For practical use, a generalized and rather simple mathematical approximation involving the ellipsoid of revolution (oblate spheroid) is also adopted. Its parameters are obtained from the best fitting to geoid (quasigeoid if more rigorous).

In some cases along with a purely geometric concept of the Earth's ellipsoid, the notion of the Normal Earth is employed. Its mass equals that of the real Earth, and its surface comprises the ellipsoid of revolution.

The WGS-84 ellipsoid (World Geodetic System, 1984), employed in the U.S. since 1984, is characterized by the following parameters [2]:

Geocentric gravitational constant, GM (subject to atmospheric mass)	$3\,986\,005 \cdot 10^8$ m$^3 \cdot$s^{-2}
Gravitational constant of atmosphere, GM_a	$3.5 \cdot 10^8$ m$^3 \cdot$s^{-2}
Major semi-axis, a	6 378 137 m
Reverse oblateness, $1/\alpha$	298.257
Gravitational acceleration at the equator, γ_e	9.780 m$\cdot$s^{-2}
Gravity potential on the geoid surface, W_0	6263686 m$^2 \cdot$s^{-2}

In USSR, the ellipsoid of Krasovskii was adopted in 1946 as the Earth's figure. Its parameters are [3]:

Equatorial radius, a	6 378 2455 m
Polar radius, b	6 356 863 m
Reverse oblateness, f^{-1}	298.3

0-8493-2861-6/97/$0.00+$.50

Surface area of ellipsoid,	$5.10 \cdot 10^8$ km^2
Volume of ellipsoid	$1.083 \cdot 10^{12}$ km^3
Mean radius of the isometric sphere, R	6 371 110 m
Mean density of the Earth, ρ	5.517 g/cm^3
Mass of the Earth with the mean density ρ,	$5.975 \cdot 10^{27}$ g

In the case of the Krasovskii ellipsoid, the dependence between increments of geodetic latitude dB and longitude $dL\cos B$ and that of arcs of meridian dx and parallel dy takes the form [3]:

$dL\cos B$, dB	$1''$	$1'$	$1°$
dx, dy	30.9 m	1.85 km	111 km.

44.1.2 Structure of the Earth

The closeness of the Earth's figure to the ellipsoid of revolution suggests that the Earth's substance is in hydrodynamic equilibrium with respect to the forces acting on it (attracting and centrifugal forces), i.e. the Earth behaves under the influence of long-term forces as a liquid body [4, 5]. Estimation of the Earth's viscosity gives 10^{22} Pa·s.

At the same time, the shear modulus determined for the Earth as a whole as it relates to short-term actions (earthquakes, tides, motion of atmospheric masses, etc.), amounts to about $15 \cdot 10^{10}$ N·m^{-2}. Thus, the Earth is a viscoelastic body with a relaxation time of $\tau \approx 10^{10}$ s.

As a rule, the interior structure of the Earth is judged by the seismic observation of elastic waves propagation, these originating from earthquakes and from high-power nuclear explosions.

Elastic properties of the Earth's interior change abruptly at definite depths, and they change smoothly within the layers between these boundaries. The most important boundaries are the Mohorovičič discontinuity at a depth of 10 to 70 km and the Weichert–Gutenberg discontinuity at a depth of 2900 km, which abruptly refracts longitudinal elastic waves and does not transmit transverse waves. These surfaces divide the Earth into three main zones, namely the crust, mantle, and the core. The crust possesses the highest rigidity; the mantle is characterized by high viscosity; and the core is in nearly liquid state and responds only to the longitudinal waves by changing its volume. There are also less pronounced boundaries within the three main zones of the earth.

The inner structure of the Earth is estimated also on the basis of the known mass and moment of inertia and by studying the elastic waves caused by earthquakes. It was discovered that the density in the center of the Earth is $\rho_c \geq 12.2$ g/cm^3 and that at a depth of 2900 km the core is separated from the upper layers by an abrupt density change of about 4 g/cm^3. The stepwise variation of density with the depth may be caused both by the modification of the rock composition and by the change of its phase state [6]. The crust of continents is from 3 to 10 times thicker than that of oceans. The continental crust thickness is different at the platforms (30–40 km) and in geosynclines (40–80 km). In the regions of the highest mountains of the Pamirs and the Himalayas it ranges from 70 to 80 km. The lower boundary of the crust (the Mohorovičič discontinuity M) forms the roots of the mountains in such regions. They penetrate the mantle deeper (30–40 km) than in the platform flatland. The ocean crust is relatively thin (about 4–8 km). The M surface occurs there at a depth of 10 to 15 km. The difference between the M surface depths on continents and in oceans reaches approximately 20–50 km. The average value of the continental crust density runs to about 2.7–2.8 g/cm^3, and under oceans it is about 2.9 g/cm^3. The upper mantle density ranges from 3.3 to 3.4 g/cm^3. The mantle surface on the continents forms cavities, whereas in the oceans it forms giant hills. The Earth's crust of oceans and continents differs in the velocity of the elastic waves propagation. There are no layers with the velocity of longitudinal waves of 6 km/s in the ocean crust, which value is typical for the crust of continents.

Two major layers are recognized in the Earth's crust. They are the sedimentary layer, which lies nearly horizontally, and the consolidated (or crystalline) layer. The velocity of elastic waves in sedimentary rocks varies widely but in the thick layer it is usually less than 5 km/s. In the upper part of the continental consolidated crust, it is close to 6 km/s and in the lower part, close to 7 km/s; under the oceans this velocity ranges from 6.5 to 7 km/s.

The Earth's interior structure is described by models that characterize the changes in density, pressure, gravitational acceleration, seismic waves velocity and in other parameters with the depth. Classical models are spherically symmetrical.

It is customary to account for differences in the plutonic structure of the Earth by constructing the Earth's models of oceanic and continental type along with the mean Earth model, which shows the combination of the previous ones (Table 44.1).

The outer shells of the Earth involve:

atmosphere (a gas shell, restricted underneath by the solid or water surface and smoothly transforming into the outer space above);

hydrosphere (a water shell, partly covering the rigid earth body)

Earth's mass	$5.977 \cdot 10^{24}$ kg
Mass of the atmosphere	$5.3 \cdot 10^{18}$ kg
Mass of the Ocean	$1.4 \cdot 10^{21}$ kg
Mass of water in rocks	$2 \cdot 10^{20}$ kg
Mass of ice on the Earth	$2.2 \cdot 10^{19}$ kg
Mass of water vapor in atmosphere	$1.3 \cdot 10^{16}$ kg.

Table 44.1 Physical parameters of the Earth's models [7]

Radius, km	Depth, km	Density, g/cm^3	Compressional wave velocity, km/s	Shear wave velocity, km/s	Bulk modulus, 10^8 Pa	Shear modulus, 10^8 Pa	Pressure, 10^8 Pa	Acceleration due to gravity, cm/s^2
0	6371.0	13.012	11.241	3.565	14 237	1653	3632.4	0
1217.1	5153.9	12.704	11.091	3.439	13 625	1502	3288.7	436.2
1217.1	5153.9	12.139	10.258	0.000	12 773	0	3288.7	436.2
3485.7	2885.3	9.909	8.002	0.000	6 345	0	1354.0	1069.3
3485.7	2885.3	5.550	13.732	7.243	6 582	2911	1354.0	1069.3
5701.0	670.0	5.377	10.928	6.114	3 045	1639	239.1	1001.2
5701.0	670.0	4.077	10.038	5.417	2 480	1220	239.1	1001.2
5951.0	420.0	3.768	9.554	5.052	2 157	961	141.1	997.6
Oceanic Earth								
5951.0	420.0	3.553	8.949	4.789	1 758	815	141.1	997.6
6360.0	11.0	3.305	7.900	4.550	1 150	684	2.2	983.5
6360.0	11.0	2.850	6.400	3.700	647	390	2.2	983.5
6366.0	5.0	2.850	6.400	3.700	647	390	0.6	983.5
6366.0	5.0	1.500	2.000	1.000	40	15	0.6	983.5
6367.0	4.0	1.500	2.000	1.000	40	15	0.4	982.9
6367.0	4.0	1.030	1.500	0.000	23	0	0.4	982.9
6371.0	0.0	1.030	1.500	0.000	23	0	0.0	982.0
Continental Earth								
5951.0	420.0	3.553	9.135	4.816	1 865	823	140.7	997.6
6336.0	35.0	3.320	8.020	4.690	1 161	730	9.7	984.2
6336.0	35.0	2.920	6.500	3.750	686	410	9.7	984.2
6351.0	20.0	2.320	6.500	3.750	686	410	5.3	983.3
6351.0	20.0	2.720	5.800	3.450	483	323	5.3	983.3
6371.0	0.0	2.720	5.800	3.450	483	323	0.0	981.6
Mean Earth's model								
5951.0	420.0	3.553	8.967	4.806	1 762	820	141.1	997.6
6352.0	19.0	3.310	7.934	4.654	1 127	716	4.8	983.7
6352.0	19.0	2.902	6.500	3.750	681	408	4.8	983.7
6357.0	14.0	2.902	6.500	3.750	681	408	3.3	983.4
6357.0	14.0	2.802	6.500	3.550	537	353	3.3	983.4
6368.0	3.0	2.802	6.000	3.550	537	353	0.3	982.6
6368.0	3.0	1.030	1.500	0.000	23	0	0.3	982.6
6371.0	0.0	1.030	1.500	0.000	23	0	0.0	981.9

44.2 Lithosphere

44.2.1 Structure and properties

The Earth's lithosphere consists of sedimentary and crystalline rocks. Within the lithosphere two markedly different regions are recognized [8]: the deep oceanic region ($268{\cdot}10^6$ km^2 area, mean depth below sea level is 4.5 km; the layer thickness equals 6 km) and the continental shield region ($105{\cdot}10^6$ km^2 area, mean height over sea level is 0.75 km; the layer thickness equals 35.05 km). In addition transition regions are discriminated: the region of young folded series ($42{\cdot}10^6$ km^2 area) and the suboceanic region ($93{\cdot}10^6$ km^2 area). The land area of $2{\cdot}10^6$ km^2 is generally formed of volcanic islands.

Table 44.2 Average chemical composition of the lithosphere [6]

Compound	Mass, 10^{18} kg	Content, %
SiO_2	13 050	55.2
Al_2O_3	3 629	15.3
CaO	2 082	8.8
FeO	1 381	5.8
MgO	1 234	5.2
Na_2O	682	2.9
Fe_2O_3	661	2.8
K_2O	452	1.9
TiO_2	385	1.6
P_2O_5	62.0	0.3
MnO	42.6	0.2

Table 44.3 Main physical properties of rocks [10]

Property	Limiting values for rocks	
	magmatic and metamorphic	sedimental
Density, g/cm^3	2.5–3.3	1.9–2.9
Porosity, %	up to 10	up to 40
Magnetic susceptibility, 10^{-9} m^3/kg	100–4000	up to 600
Residual magnetization, 10^{-3} A/m	up to 320	
Young's modulus, 10^{11} g/cm^2	5–30	2.5–12
Poisson's ratio	0.21–0.28	0.23–0.27
Compressional wave velocity, km/s	4.9–10.1	0.5–5.9
Resistivity, Ohm·m	400–100 000	0.3–5000
Heat conductivity, W/(cm·K)	0.001–0.03	0.01–0.03
Specific heat, J/(g·K)	0.5–1.17	0.67–0.96

Table 44.4 The abundance of chemical elements in the Earth's crust [11,12]

*1	Element	*2	*1	Element	*2	*1	Element	*2	*1	Element	*2
1	Hydrogen	1.00	22	Titanium	0.61	44	Ruthenium	$5{\cdot}10^{-6}$	66	Dysprosium	$7.5{\cdot}10^{-4}$
2	Helium	10^{-6}	23	Vanadium	0.02	45	Rhodium	10^{-6}	67	Holmium	10^{-4}
3	Lithium	$5{\cdot}10^{-3}$	24	Chromium	0.03	46	Palladium	$5{\cdot}10^{-6}$	68	Erbium	$6.5{\cdot}10^{-4}$
4	Beryllium	$4{\cdot}10^{-4}$	25	Manganese	0.10	47	Silver	$5{\cdot}10^{-6}$	69	Thulium	10^{-4}
5	Boron	$5{\cdot}10^{-3}$	26	Iron	4.20	48	Cadmium	$5{\cdot}10^{-4}$	70	Ytterbium	$8{\cdot}10^{-4}$
6	Carbon	0.35	27	Cobalt	$2{\cdot}10^{-3}$	49	Indium	10^{-5}	71	Lutecium	$1.7{\cdot}10^{-4}$
7	Nitrogen	0.04	28	Nickel	0.02	50	Tin	$8{\cdot}10^{-3}$	72	Hafnium	$4{\cdot}10^{-4}$
8	Oxygen	49.13	29	Copper	0.01	51	Antimony	$5{\cdot}10^{-5}$	73	Tantalum	$2.4{\cdot}10^{-5}$
9	Fluorine	0.08	30	Zinc	0.02	52	Tellurium	10^{-6}	74	Tungsten	$7{\cdot}10^{-3}$
10	Neon	$5{\cdot}10^{-7}$	31	Gallium	10^{-4}	53	Iodine	10^{-4}	75	Rhenium	10^{-7}
11	Sodium	2.40	32	Germanium	$4{\cdot}10^{-4}$	54	Xenon	$3{\cdot}10^{-9}$	76	Osmium	$5{\cdot}10^{-6}$
12	Magnesium	2.35	33	Arsenic	$5{\cdot}10^{-4}$	55	Cesium	10^{-3}	77	Iridium	10^{-6}
13	Aluminum	7.45	34	Selenium	$8{\cdot}10^{-5}$	56	Barium	0.05	78	Platinum	$2{\cdot}10^{-5}$
14	Silicon	26.00	35	Bromine	10^{-3}	57	Lanthanum	$6.5{\cdot}10^{-4}$	79	Gold	$5{\cdot}10^{-7}$
15	Phosphorus	0.12	36	Krypton	$2{\cdot}10^{-8}$	58	Cerium	$2.9{\cdot}10^{-3}$	80	Mercury	$5{\cdot}10^{-6}$
16	Sulfur	0.10	37	Rubidium	$8{\cdot}10^{-3}$	59	Praseodymium	$4.5{\cdot}10^{-4}$	81	Thallium	10^{-5}
17	Chlorine	0.20	38	Strontium	0.035	60	Neodymium	$1.7{\cdot}10^{-3}$	82	Lead	$1.6{\cdot}10^{-3}$
18	Argon	$4{\cdot}10^{-4}$	39	Yttrium	$5{\cdot}10^{-3}$	62	Samarium	$7{\cdot}10^{-4}$	83	Bismuth	10^{-5}
19	Potassium	2.35	40	Zirconium	0.025	63	Europium	$2{\cdot}10^{-5}$	90	Thorium	10^{-3}
20	Calcium	3.25	41	Niobium	$3.2{\cdot}10^{-3}$	64	Gadolinium	$7.5{\cdot}10^{-4}$	92	Uranium	$4{\cdot}10^{-4}$
21	Scandium	$6{\cdot}10^{-4}$	42	Molybdenum	10^{-5}	65	Terbium	10^{-4}			

*1 Atomic number. *2 Abundance, mass. %.

44.2.2 Gravity

The gravitational field on the Earth's surface is determined by the geopotential and by its first and second derivatives [9]. We introduce these values in the orthogonal coordinate system: x, y are directed to the North and to the East, respectively, and z is directed downwards along the vertical line. The geopotential W is the sum of the attracting potential of the Earth's mass (gravitational potential) and the potential of the centrifugal forces due to rotation of the Earth. Its value is expressed in joules.

For its first-order derivative (gravitational acçeleration) $g_z = dW/dz$ (see Table 44.5) and for the second-order ones (W_{xx}, $W_{yx}, \ldots$) special names of units were introduced. For g we have 1 gal = 1 cm/s^2 and for the second derivatives we have 1 eötvös = E = 10^{-9} s^{-2}.

The values of the second-order derivatives of the geopotential on the surface of the Krasovskii ellipsoid are defined by the formulas (in units of E):

$$W_{yy} - W_{xx} \equiv W_\Delta = 5.1(1 + \cos 2\varphi); \quad W_{xy} = 0;$$

$$W_{yz} = 0; \quad W_{xz} = 8.1 \sin 2\varphi; \quad W_{zz} = -3086(1 + 0.00071 \cos 2\varphi).$$

Tidal periodic change of g value [9] on the surface of the Earth determined by the Moon influence is up to $2.49 \cdot 10^{-4}$ cm/s^2. The amplitude of the solar tidal influence reaches $9.6 \cdot 10^{-5}$ cm/s^2. Thus, the total influence of the Moon and the Sun on g amounts to $3.45 \cdot 10^{-4}$ cm/s^2. The amplitude of tidal change in g depends on the latitude of the observation point (see Table 44.6).

Table 44.5 Dependence of gravitational acceleration g on latitude of the place φ (angle deg.) at the surface of the Krasovskii ellipsoid [9]

φ	g, cm/s^2	φ	g, cm/s^2	φ	g, cm/s^2
0	978.0300				
5	978.0692	35	979.7299	65	982.2853
10	978.1856	40	980.1659	70	982.6061
15	978.3757	45	980.6159	75	982.8664
20	978.6338	50	981.0663	80	983.0584
25	978.9521	55	981.5034	85	983.1759
30	979.3212	60	982.2853	90	983.2155

Table 44.6 Main tide forming components [13]

Type of tide	Latitude factor	Amplitude factor	Classification	Period, h
Long-period	$(1 - 3\sin^2\varphi)/4$	0.156	M_f	327.84
24-hour	$\sin\varphi \cdot \cos\varphi$	0.377	O_1	25.82
		0.176	P_1	24.07
		0.531	K_1	23.93
12-hour	$\cos^2\varphi/2$	0.174	N_2	12.66
		0.908	M_2	12.42
		0.424	S_2	12.00
		0.115	K_2	11.97

44.2.3 Seismicity

Earthquakes consist of the vibrations of the Earth as a result of the seismic waves propagation from some source of seismic energy [14,43,44].

Elastic waves in the Earth [16]. Propagation of elastic strains during earthquakes shows the wave character. Longitudinal (P) and transverse (S) volume waves as well as the surface Rayleigh (R) waves (in which particle oscillations occur in the plane perpendicular to the surface and passing through the direction of seismic ray) and the surface Love (L) waves are usually considered.

Magnitude of an earthquake M is the characteristic of the origin energy expressed as the decimal logarithm of the largest ground oscillation amplitude, the one being recorded in the passage of the seismic wave (of this or that type). The relation between the earthquake's magnitude and intensity (power) depends on the distance between the origin (hypocenter) and the registration point on the surface of the Earth (particularly, the epicenter). The lower the depth of the origin, the higher the intensity of the surface shaking at the same value of a magnitude. As a rough estimate of the earthquake's intensity I in epicenter for average depth of source in the Earth's crust, the following dependence is true:

$$I = 1.7\,M - 2.2.$$

This quantity is corrected with the use of mean empirical dependences of earthquakes' intensity on the epicentral distance.

Three scales of magnitude are widely used. They involve the Richter scale (local) M_L; the one determined through the volume longitudinal waves m or m_b; and the one determined through the surface waves M_S. Some differences between scales exist, which depend on the earthquake energy. The relationship between the elastic waves energy E (in J) and magnitude M_S may be expressed by the equation:

$$\log E = 5.24 + 1.44\, M_S.$$

In 1935, G. Richter accepted as a standard earthquake one that can be detected with the amplitude of 1 μm on the seismogram at a distance of 100 km by a seismograph with a static magnification factor of 2800, and with period of 0.8 s and damping 0.8. It was experimentally found that the energy of an earthquake accepted as the one with the unit energy (and with zero magnitude) corresponds to emission of approximately 10^5 J of elastic wave energy. The 0.5 increase in the earthquake magnitude conforms to the change of energy in the origin by an order of magnitude. The most severe earthquake does not exceed a magnitude of 8.5, which corresponds to 10^{20} J of energy in the origin.

The intensity (power) of an earthquake is determined according to the extent of the destruction of buildings, to the character of the Earth's surface change, and to the data on the feelings experienced by people. There exist a lot of different earthquake intensity scales. In Russia and in a number of European countries, the magnitude scales GSSSD 6949-52 and MSK-64 are accepted. Below the approximate relationship between the magnitude and intensity of an earthquake is presented.

Magnitude according to the Richter scale	4.0		5.0		6.0		7.0		8.0	
Origin depth, km	3	5–10	5	10	10	20	15	30	25	40
Intensity according to MSK-64, number	7	6	8	7	8–9	7–8	9–10	8–9	10–9	9–10

Microseismic activity [17, 18]. Planetary microscopic vibrations of the Earth's surface (microseisms of the first kind) are detected in the range of frequencies from 0.03 to 100 Hz, depending on the place, observation conditions and frequency. Their amplitudes amount to 10 μm for low-frequency vibrations and decrease to 10^{-3} μm in very still regions for frequencies of about 1 Hz.

Microseisms of the second kind are caused by the surface sources (transport, industrial plants, surf and others). They possess periods from 0.001 to 0.1 s, depend on the ground conditions and rapidly decrease with the distance from the source. The displacement amplitudes lie in the range from 10^{-4} to $5{\cdot}10^{-2}$ μm.

Table 44.7 Scale of earthquake

Number	Name	Displacement, mm	Acceleration, mm/s^2	Characteristics
1	Insignificant		2.5	Soil vibrations are detected by instruments (microseisms)
2	Very light		2.5–5.0	Sometimes are felt by people
3	Light		5–10	Are felt by a few people
4	Moderate	0.5	5–10	Are noticed by many people; vibrations of windows and doors are possible
5	Rather severe	0.5–1	25–230	Sway of hanging things, creak of floors, crumbling of whitewash
6	Severe	1.1–2	50–100	Light damage to buildings: narrow cracks in plaster, cracks in stoves, etc.
7	Very severe	2.1–4	100–250	Substantial damage to buildings: fall of plaster, narrow cracks in walls
8	Destructive	4.1–8	250–500	Some destruction of buildings: large cracks in walls, fall of cornices and chimney
9	Devastating	8.1–16	500–1000	Collapse of some buildings: fall of walls, floors, and roofings
10	Annihilating	16.1–32	1000–2500	Collapse of many buildings, cracks in soil of about one meter in thickness
11	Disastrous	32	2500–5000	Numerous cracks on the earth surface, large collapses in mountains
12	Catastrophic		>5000	Large-scale changes in relief

Table 44.8 Energy and magnitudes of earthquakes

Magnitude, m	Energy, J	Comment
0	$1.7 \cdot 10^5$	For very weak quakes data are not reliable
1	$4.8 \cdot 10^6$	
2	$1.3 \cdot 10^8$	
2.5		The weakest earthquakes that can be felt
3	$3.6 \cdot 10^9$	
4	$1.0 \cdot 10^{11}$	
5	$2.7 \cdot 10^{12}$	The weakest earthquakes that can cause damage to buildings
6.0	$4.4 \cdot 10^{13}$	Earthquake in Skoplje, 1963
7.0	$2.1 \cdot 10^{15}$	Earthquake in Inaigahua, 1968
7.8		Buller earthquake, 1929
7.9		Earthquake in Hocks-Bay, 1931
8.3		Earthquake in San Francisco, 1906
8.4		Chile, 1960; Alaska, 1964
8.9		The most strong registered seismic quakes: Colombia, 1906; Sanriku, 1933

Table 44.9 Earthquakes' statistics over 1918–1964

Magnitude m	Number of quakes per 10 years	Energy per 10 years, 10^{16} J	Magnitude m	Number of quakes per 10 years	Energy per 10 years, 10^{16} J
8.5–8.9	3	156	7.0–7.4	149	58
8.0–8.4	11	113	6.5–6.9	560	41
7.5–7.9	31	80	6.4–6.8	2100	30

44.2.4 Geomagnetism

The magnetic field of the Earth is characterized by the vector of intensity $\mathbf{T}$. The projections of this vector on the axes of orthogonal system of coordinates form components of geomagnetic field: Z, X, and Y are the vertical, northern, and eastern components, accordingly. Horizontal component $H = \sqrt{X^2 + Y^2}$ is also often used.

The angle between horizontal direction of H and geographic meridian is spoken of as the declination D, and the angle I between the intensity vector and horizontal plane is called inclination.

Constant magnetic field [19, 20, 21]. The Earth possesses its own magnetic field. Its main component is the field of magnetic dipole or of uniformly magnetized sphere. The axis of the dipole is inclined to the rotation axis of the Earth at an angle of 10.5°. The magnetic moment of the dipole equals $8.3 \cdot 10^{25}$ cgsm.

The North Magnetic Pole (the point on the Earth's surface where the intensity vector is directed vertically downwards) is at $\varphi = 73°$ north latitude, $\lambda = 100°$ west longitude. The South Magnetic Pole (where the same vector is directed vertically upwards) is at $\varphi = 68°$ south latitude, $\lambda = 143°$ east longitude.

Geomagnetic poles (the points of intersection of magnetic dipole axis with the day's surface) do not coincide with the magnetic poles.

For the period of year 1960, their coordinates were $\varphi = 78.3°$ north latitude, $\lambda = 68.3°$ west longitude for the North, and $\varphi = 78.3°$ south latitude, $\lambda = 111.7°$ east longitude for the South Geomagnetic Pole.

The magnetic field intensity at the magnetic pole reaches 51.74 A/m, and that at the magnetic equator is 27.86 A/m.

The parameters of a constant magnetic field vary at its surface within the following ranges:

T: (49.3–58.11) A/m,

H: (0–32.64) A/m,

Z: (0–32.64) A/m.

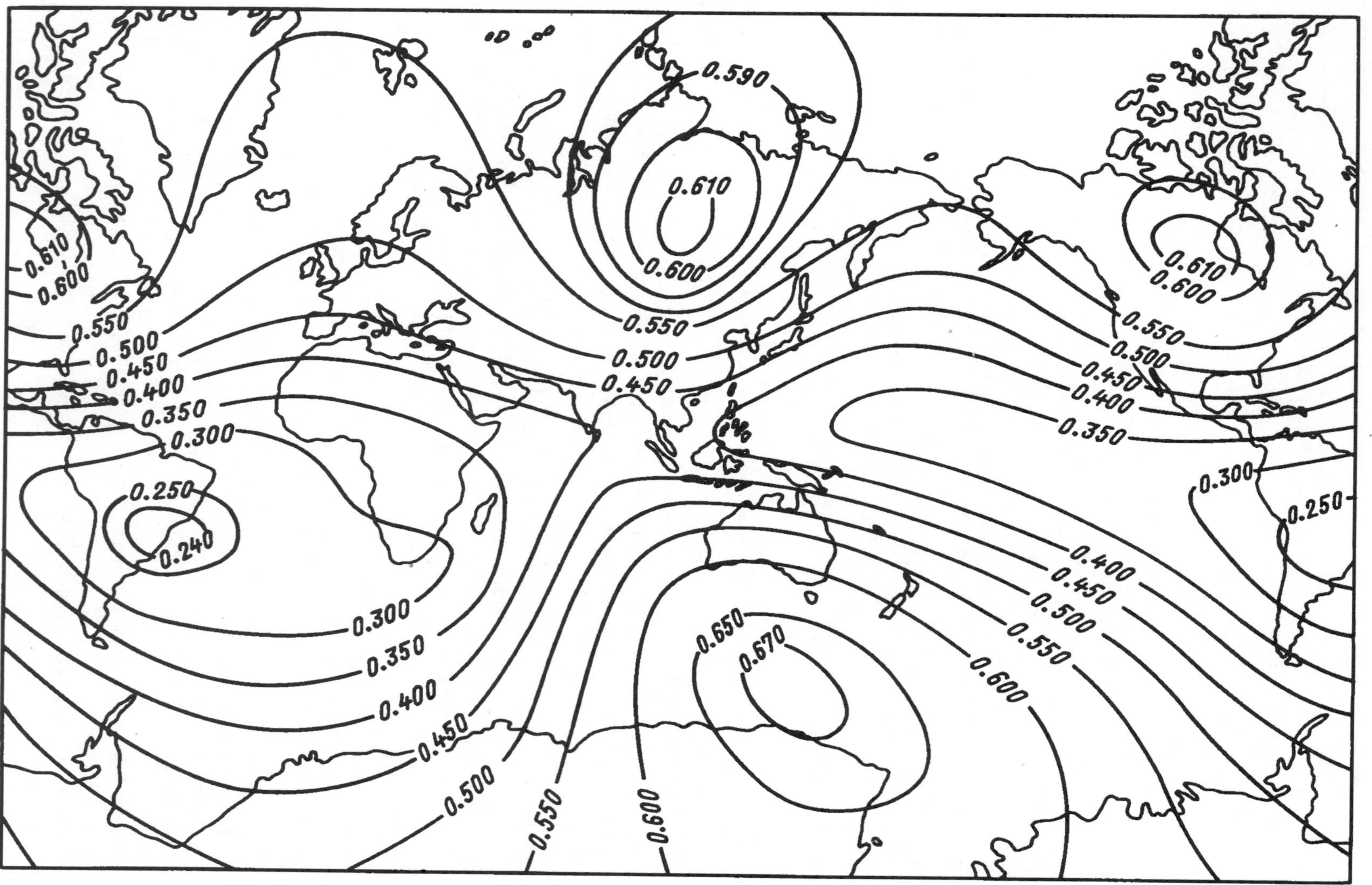

Figure 44.1 The map of normal geomagnetic field intensity T_n, 10^{-4}T; epoch of year 1980 [21].

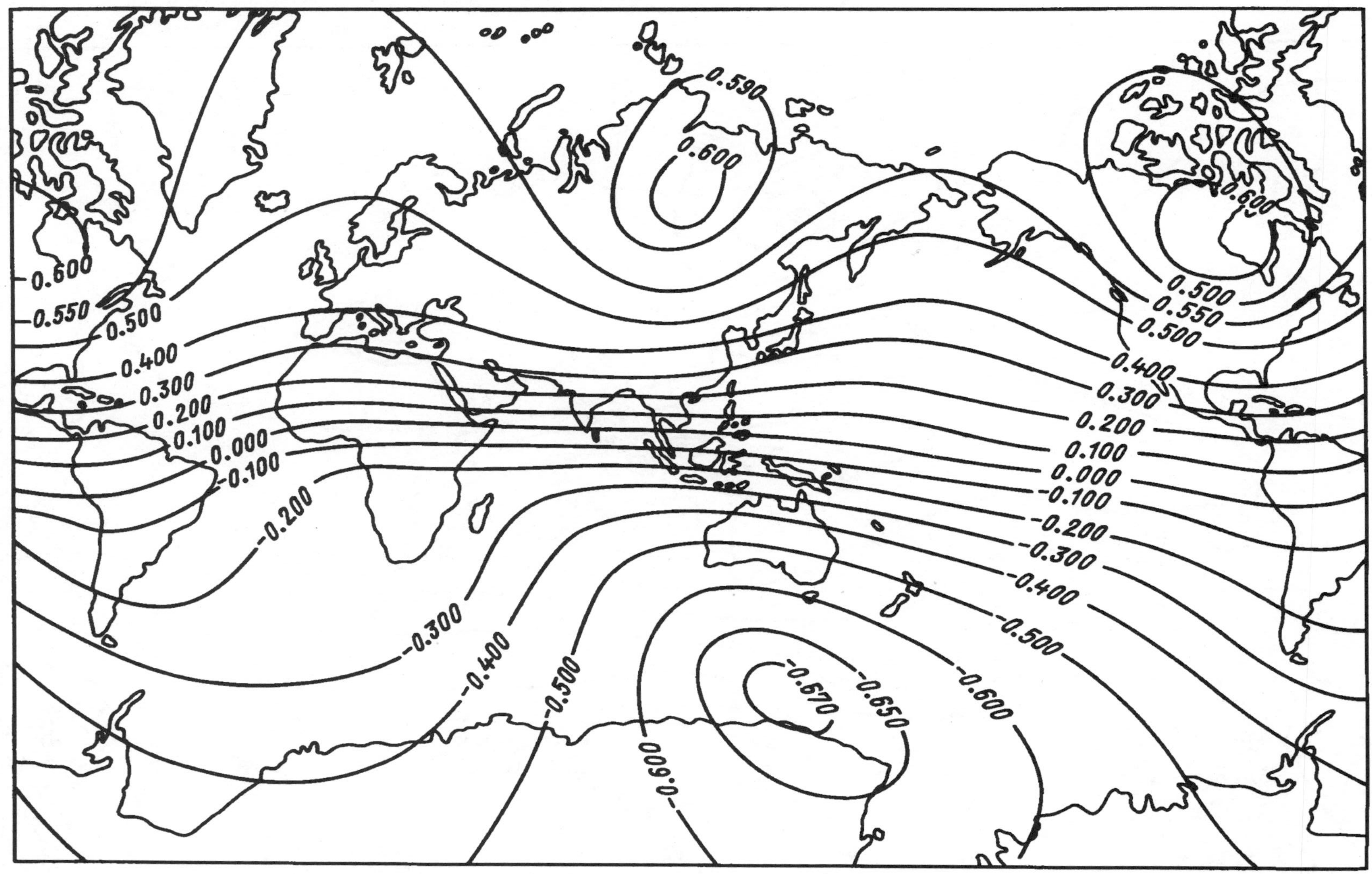

Figure 44.2 The map of vertical component of normal geomagnetic field intensity Z_n, 10^{-4}T; epoch of year 1980 [21].

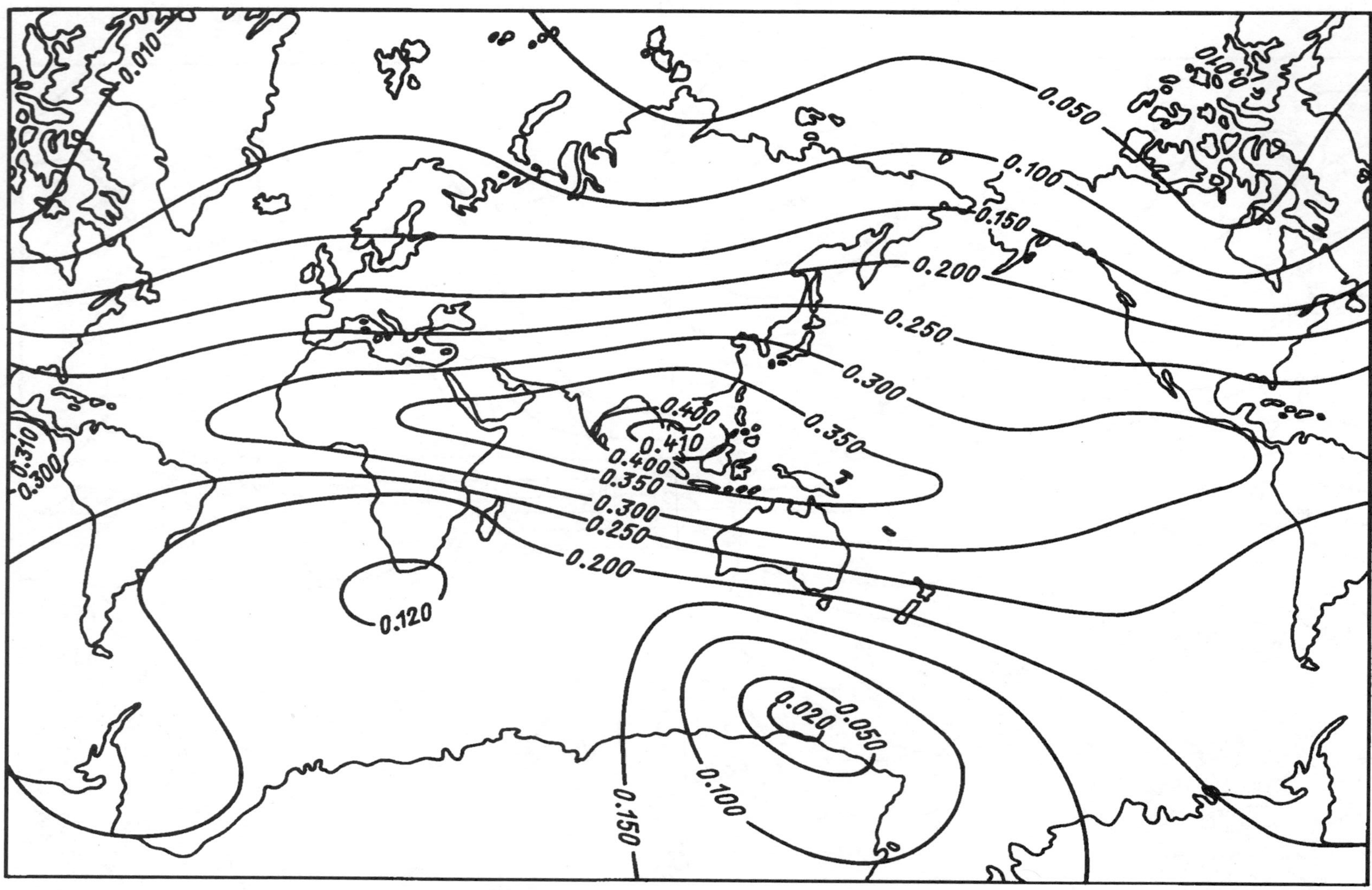

Figure 44.3 The map of horizontal component of normal geomagnetic field intensity H_n, 10^{-4}T; epoch of year 1980 [21].

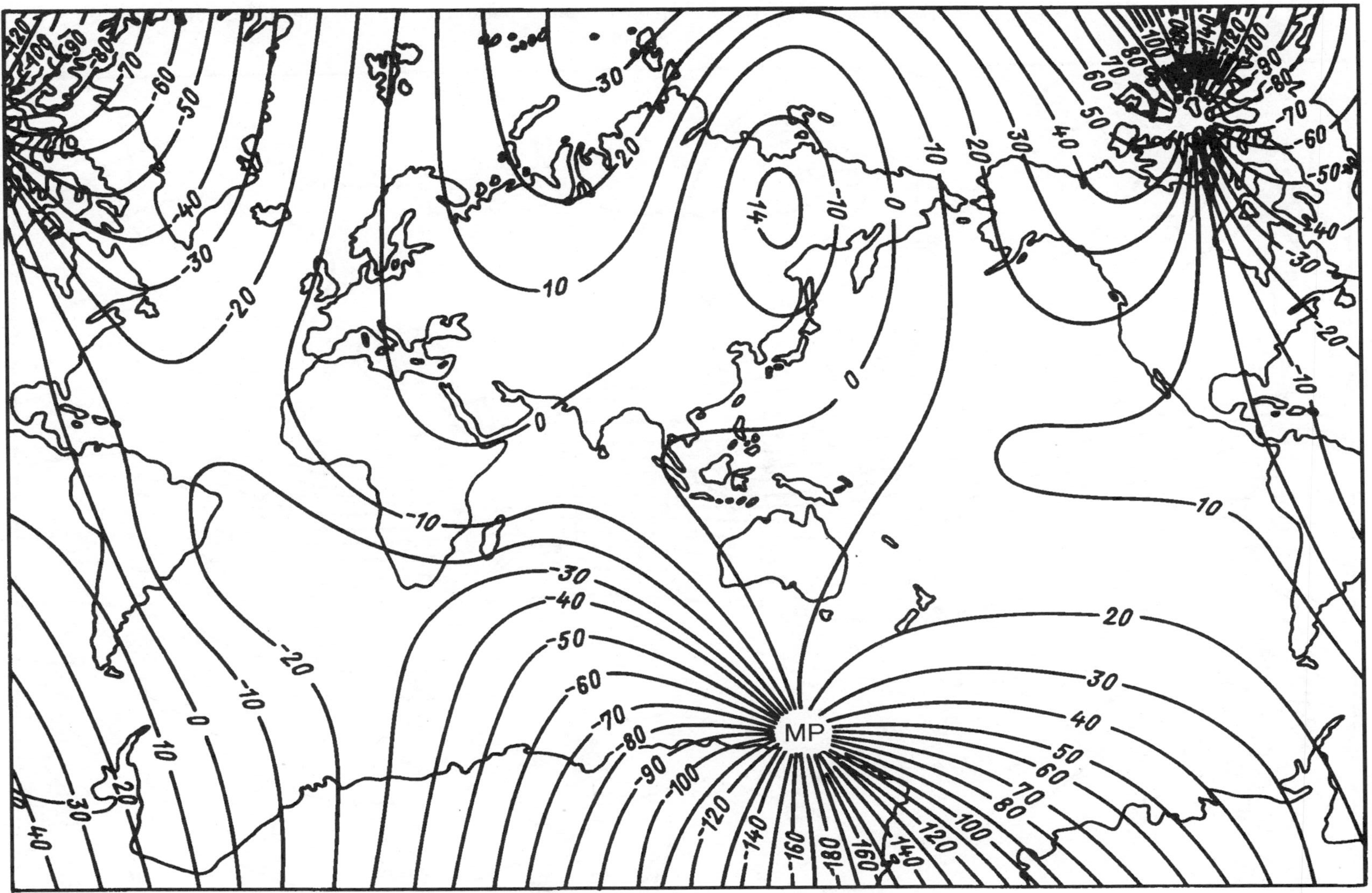

Figure 44.4 The map of magnetic declination values, angle deg.; epoch of year 1980 [21].

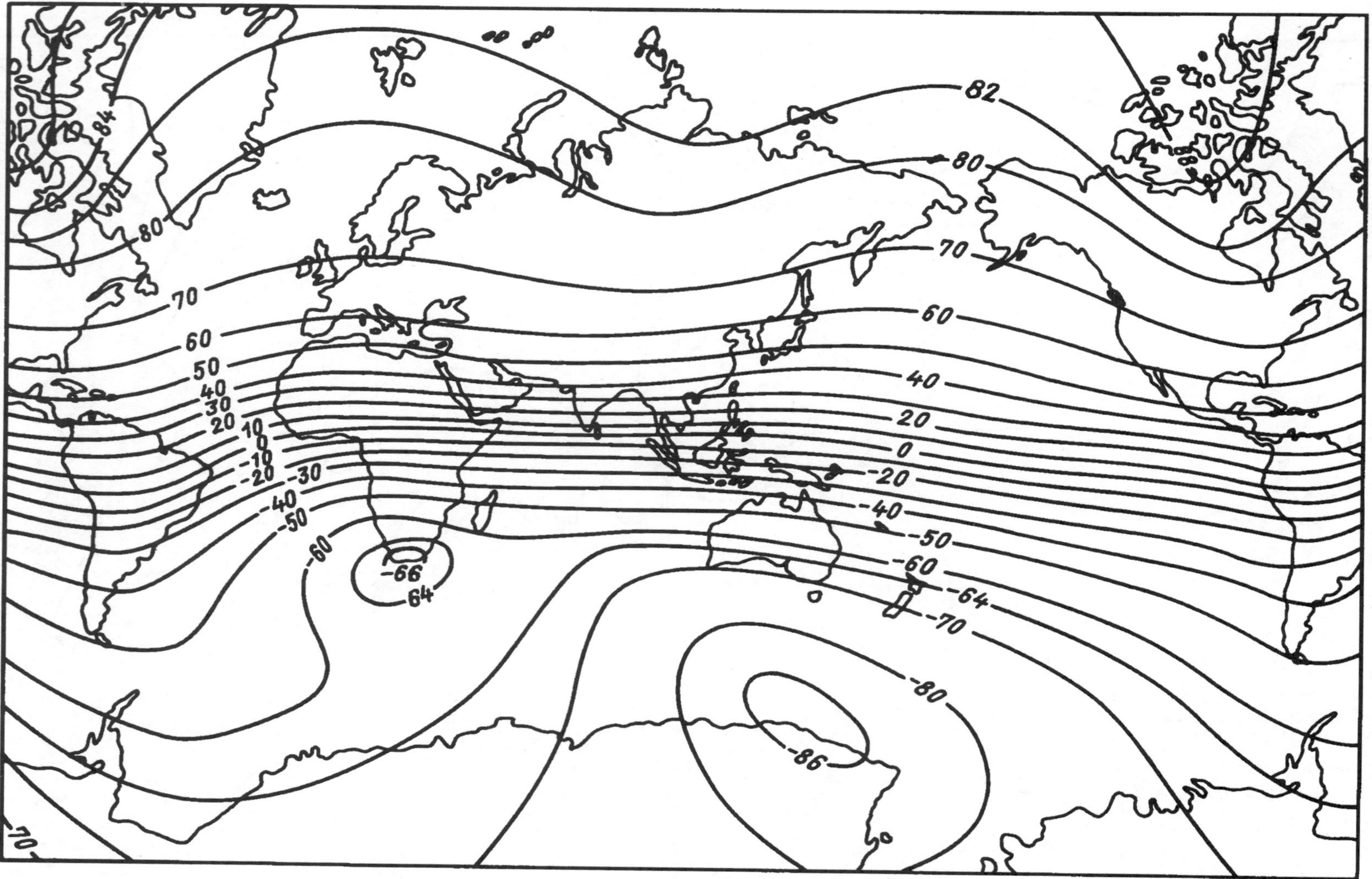

Figure 44.5 The map of magnetic inclination normal values, angle deg.; epoch of year 1980 [21].

The features common to magnetic fields of continents and oceans are as follows [21]:

the presence of large-block structure of anomalous magnetic field (with the dimensions of different character magnetic field regions over several hundred kilometers);

the presence of minimum in the spectrum of anomalous magnetic field from 400 to 4000 km;

the presence of regions with pronounced field anisotropy.

The value of magnetic field anomaly near the surface of continents varies from several tenths of an ampere per meter to several amperes per meter and more (near the outcrops of platform base at the surface).

Geographic distribution of constant magnetic field of the Earth (with consideration for magnetic anomalies that occupy the area from a number of kilometers to whole continents) is represented by the maps pertinent to a definite epoch (Figures 44.1–44.5).

Secular variations of a geomagnetic field. The mean values of the geomagnetic field elements change with time. The comparison of the field elements for the years 1885 and 1950 leads to a conclusion that the total magnetic moment of the Earth decreases at an approximate rate of $7{\cdot}10^{-4}$ over the course of a year.

Varying magnetic field of the Earth. Periodic variations [22]. All the periodic variations of a magnetic field of the Earth have their sources outside of the Earth. These variations are classified according to the length of a period.

The solar-daily variations (caused by the diurnal motion of the Earth), Moon-daily, annual, cyclic with 11-years-period variations (connected with the change of solar activity) and others are singled out. The amplitudes of all the periodic variations (except for solar-daily) account for units of the angle minutes of declination and the thousandths of A/m in magnetic field intensity (see Table 44.10).

Nonperiodic variations of the Earth's magnetic field. Magnetic storms. Magnetic storms and total magnetic activity are caused by the interaction of corpuscular radiation of the Sun with the constant magnetic field of the Earth. Magnetic storms (i.e. abrupt and irregular in form variations of the Earth's magnetic field) start simultaneously over the whole Earth's magnetic globe and they have a tendency to repeat within 27 days. The field is changed in its intensity and direction by several percent over a period of a few hours to several days.

The Earth's telluric currents [23, 24]. The currents and a varying magnetic field comprise the connected phenomena. The telluric current density j for different parts of the Earth's surface are approximately the same: $j = 2$ A/km^2. The intensity of telluric currents grows from low latitudes to high latitudes. At low latitudes, the magnetic field intensity does not usually exceed several tens of millivolts per kilometer. In the polar countries, the field intensity can reach units and even tens of volts per kilometer. Telluric currents are strongest during magnetic storms.

Table 44.10 Solar-diurnal variations of a geomagnetic field (in $0.3{\cdot}10^{-3}$ A/m) [22]

Component of a field intensity	Years of low magnetic activity				Years of high magnetic activity			
	medium latitudes		high latitudes		medium latitudes		high latitudes	
	still days	disturbed days	still days	disturbed days	still days	disturbed days	still days	disturbed days
Horizontal H	7–40	18–58	44–80	90–402	8–74	25–109	27–77	50–460
Vertical Z	4–16	9–36	20–90	103–305	6–26	14–128	15–70	128–315
The total vector modulus	6–26	11–64	20–65	150–300	8–35	13–104	35–110	230–275

44.2.5 Geothermics

Interior temperature of the Earth [25]. Near the surface of the Earth the temperature of soil and of shallow rocks is determined by the balance of heat obtained from the Sun and that radiated to atmosphere. The water and air shells of the Earth play the role of a thermal regulator. The mean penetration depth of the diurnal variations of the soil temperature (depending on its properties and geographic conditions) varies from 35 to 100 cm. Delay time of the extrema appearance runs on the average 2–3 hours for every 10 cm of depth.

The penetration depth of annual temperature variations in low latitudes comes to about 5–10 m, and in medium and high latitudes it is 8–24 m, reaching 30 m. The penetration depth of the secular variations equals more than 50 m and survives for a long time due to retardation of the thermal wave phase with the depth. The permafrost (penetrating in some places up to several hundreds of meters) is a relic of the glacial period, which finished several tens of thousands years ago. Observations in mines and bore holes reveal a slow increase in the temperature with the depth. In California at a depth of about 2800 m, the temperature reaches 400 K and in the test pits at the North Caucasus a temperature of about 430 K was detected at a depth of 3200 m. The rate of the temperature change with the depth is characterized by geothermic gradient or by the reciprocal value of geothermic step. The value of dT/dh varies from 0.1 to 0.01 K/m. At the bottom of the oceans the mean value of dT/dh is on the order of 0.08 K/m for the Pacific Ocean and about 0.04 K/m for the North Atlantic. Thermal conductivity λ of continental rocks runs on the average to 2–2.5 W/mK.

Heat transport budget of the Earth [20]. The mean heat flux through the continents amounts to 0.06 W/m^2. The lowest value of the heat flux (about 0.037 W/m^2) was reached in the ancient Precambrian shields; the highest one (up to 0.092 W/m^2) is observed in the regions of modern volcanism (excluding the sources themselves). The heat flux in the ocean regions varies from 0.04 W/m^2 in the deep-sea trench to 0.3 W/m^2 near the rift zones of the mid-oceanic ridges. The mean heat flux through the ocean bottom is estimated at 0.1 W/m^2. The mean heat flux through the Earth's surface comes to $\geq$ 0.062 W/m^2. Total thermal losses of the Earth range up to about $4{\cdot}10^{13}$ W. Radioactive elements constitute the permanent source of the interior heat of the Earth (see Table 44.11).

The cooling of the plutonic layers of the Earth is slow at present. Thermally active are the upper layers of the Earth (down to a depth of 100–200 km) especially in the regions with the sour, volcanic, and magmatic rocks.

Heat balance of the Earth [27]. The Earth absorbs annually 7030 MJ/m^2 with 4690 MJ/m^2 of them absorbed by the Earth's surface and 2340 MJ/m^2 by the Earth's atmosphere. The losses by long-wave effective radiation reach 4670 MJ/m^2; by evaporation it loses 2470 MJ/m^2. The heat conduction to atmosphere by the turbulent heat transfer is 540 MJ/m^2.

Table 44.11 Mean content (in 10^{-6} g/t) of radioactive elements in the main types of sedimental rocks and quantity A of energy released [26]

Rocks	^{235}U	^{232}Th	^{40}K	A, 10^{-13} J/(cm^2·s)
Sands and aleurolite	2.3	5.9	$16{\cdot}10^3$	11.30
Clay and shale	2.9	10.9	$20{\cdot}10^3$	16.7
Carbonates	2.1	2.2	$7{\cdot}10^3$	5.44
Average				12.56

44.3 Hydrosphere

Hydrologic cycle. The total amount of water in the annual rainfalls for the whole of the Earth's globe is estimated as $5{\cdot}10^{17}$ kg, which is 40 times more than the total content of water in the atmosphere (see Table 44.12). By current data (with an accuracy of 10%), the total quantity of the rainfall in the regions of the world's ocean gives rise on the average to a layer of 102 cm/year, which corresponds to 370 000 km^3 of water. Evaporation from the ocean surface amounts to 113 cm/year (or 407 000 km^3 of water) and the river sink to oceans appears to be 10.3 cm/year (or 37 000 km^3).

For the land surface the following values (in cm/year) were found (with allowance made for the sink into ocean): 70.0 (102 000 km^3) in rainfall, 44.6 (64 900 km^3) in evaporation, and 25.4 (37 000 km^3)

in sink. There are only about 8000 km^3 of rainfall at the dry land parts that have no sink to oceans. The same figure holds for evaporation. For the Earth's globe as a whole, the annual layer of rainfall (which is equal to the annual evaporation) comes to 92.8 cm/year (472 000 km^3). Only 22% of this value falls to the dry land, and 78% to the ocean. Evaporation from the dry land ranges up to about 14% of the total value. The rest 86% evaporates from the surface of oceans.

The greater part of the water (90%) is held in the lower layer of the atmosphere (up to 8–10 km) and it comprises 0.3–0.4% of the atmospheric air mass in this layer. The main part of water in the atmosphere is in the vapor form (about 95%). The cloud particles portion (water droplets and ice crystals) measures less than 5% of the atmospheric water.

Table 44.12 Distribution of surface water resources over the Earth [33]

Water resource	Volume, km^3	Mass, kg
Oceans and seas	$1336 \cdot 10^6$	$1.3 \cdot 10^{21}$
Polar ice	$3.5 \cdot 10^6$	$3.5 \cdot 10^{18}$
Lakes	$2.5 \cdot 10^5$	$2.5 \cdot 10^{17}$
Underground lakes	$2.5 \cdot 10^5$	$2.5 \cdot 10^{17}$
Rivers	$5 \cdot 10^4$	$5 \cdot 10^{16}$
Bogs	$6 \cdot 10^3$	$6 \cdot 10^{15}$
Snow cover	250	$2.5 \cdot 10^{14}$
Total resources	$\sim 1340 \cdot 10^6$	$(1.8\text{–}2.7) \cdot 10^{18}$

Table 44.13 Most important substances dissolved in sea, lake and river water

Substances dissolved in sea water			Substances dissolved in lake and river water		
Ion	Concentration, g/10^3 kg	Total mass, 10^{15} kg	Ion	Concentration, g/10^3 kg	Total mass, 10^{12} kg
Cl^-	19 360	30 976	CO_2	51.2	1178
Na^{+1}	10 770	17 232	CO^{+2}	29.8	680
SO_4^{-2}	2701	4 321	CO_4^{-2}	17.7	407
Mg^{+2}	1298	2 077	SiO_2	17.1	393
Ca^{+2}	408	653	Cl^-	8.3	191
K^{+1}	387	619	Na^{+1}	8.4	193
HCO^{-1}	128	205	Mg^{+2}	5.0	155
Br^-	66	106	$(Al, Fe)_2O_3$	4.0	92
H_3BO_3	27	43	K^{+1}	3.1	71
Sr^{+2}	14	22	NO_3^{-1}	1.3	30

Table 44.14 Solubility (in cm^3/l) of gases in sea and fresh water

	In fresh water at temperature, °C				In sea water ($S = 35‰$) at temperature, °C			
	0	10	20	30	0	10	20	30
N_2	18.18	14.60	12.24	10.98	14.04	11.72	10.18	9.08
O_2	10.29	8.02	6.57	5.57	8.04	6.41	5.35	4.50
Ar	0.54	0.42	0.35	0.30	0.41	0.31	0.25	0.21
CO_2	0.52	0.36	0.26	0.20	0.44	0.31	0.23	0.18

44.3.1 The Ocean [26]

The area of the world's ocean accounts for $361.3 \cdot 10^6$ km^2 (70.8% of the Earth's surface); the water volume equals 1338.5 km^3 (see Table 44.15) with the mean depth approaching 3705 m.

Table 44.15 Main morphometric features of oceans

Ocean	Area		Volume		Mean depth, m
	10^6 km^2	Percentage of the World Ocean, %	10^6km^3	Percentage of the World Ocean, %	
Pacific	178.68	49.4	707.1	52.8	3957
Atlantic	91.66	25.4	330.1	24.7	3601
Indian	76.17	21.1	284.7	21.3	3736
Arctic	14.79	4.1	16.7	1.2	1129

The ocean bottom shows three distinct steps: continental shelf, continental slope, and the bed of the ocean or sea. The continental shelf (shelf or continental plateau) represents the continuation of the continental surface and occupies about 7.5% of the ocean area. The mean width of the continental shelf comprises 78 km, but near the shores of Africa it is practically absent and near the north shores of Asia its width would run to about several hundred kilometers.

From the ocean side the continental shelf is restricted by the edge (its depth being from 20 to 500 m with the average value of 133 m).

The continental slope is steeply lowered from the edge to the bed of the ocean or sea and it forms about 15.5% of the ocean area. Its lower boundary traverses a depth of 2000–4000 m and the width varies from several kilometers to several hundred kilometers. The mean slope angle measures about 3.5° and its maximum value covers 40–45°.

The bed of the ocean or sea makes a central (presumably plain) part of the bottom. It occurs more than 2000–4000 m beneath the surface and occupies about 77% of the world's ocean area. When the ocean surface interacts with atmosphere, the wind sea-way appears (see Table 44.16).

Table 44.16 The scale of the power (number) of wind sea-way and the depths of surface wave propagation [32]

Sea-way, number	Characteristics	Wave dimensions			Depth of disturbance, m
		Height, m	Length, m	Period, s	
0	Absent	0	0	0	0
I	Calm	up to 0.25	up to 5.0	up to 2.0	up to 2.6
II	Moderate	0.25–0.75	5–15	2–3	2.6–10.3
III	Substantial	0.75–1.25	15–25	3–4	10.3–19
IV	Substantial	1.25–2.0	25–40	4–5	19.2–34
V	Rough	2.0–3.5	40–75	5–7	33.7–70
VI	Rough	3.5–6.0	75–125	7–9	69.9–127
VII	Very rough	6.0–8.5	125–170	9–11	127.3–183
VIII	Very rough	8.5–11.0	170–220	11–12	182.6–245
IX	Exceptional	>11.0	>220	>12	>245.3

Internal waves. Internal waves appear at the interface of water layers with different density. They influence the variability parameters of hydrophysic fields.

Short-period (1–10 min) internal waves propagate by groups. Waves with a period from 10 min to 5 hours are with the lengths from hundred meters to several kilometers. Their velocities measure a fraction of m/s and amplitudes range from 10 to 20 m.

The Rossby waves (with a length of thousands of kilometers and with a period up to a few tens days), tsunami (with a period from 2 to 40 min, the propagation velocity $C_{ts} = \sqrt{gH}$, where H is the depth of a basin, the height from 1 to 2 m, and the length from 20 to 40 km), as well as the tides and inertial oscillations are also observed in the ocean.

Temperature field of the ocean [28]. The mean temperature of the ocean comes to 3.8°C. The highest temperature of 33°C is found in August at the surface of the Persian Gulf. The lowest values are observed under the ice in the polar regions, where the temperature is close to the freezing point. At a salinity of 33, 34 and 35‰ the freezing point equals −1.80, −1.85, and −1.91°C, respectively.

Thermal balance of the ocean [31]. The mean value of solar radiation absorbed by the ocean surface constitutes on the average 175 W/m^2. The total emission from the surface (the back radiation) is constant and adds up to 65 W/m^2. The water warm up by solar radiation strongly decreases with the depth of the layer. For example, with the absorbed energy of 4186.6 J/cm^2, the distribution of the temperature increase in the layers of 1 m in thickness for the pure ocean water looks as follows:

The depth of a layer, m	0–1	1–2	5–6	10–11	20–21
Extent of warm-up in a layer, K	6.24	0.61	0.236	0.104	0.04

44.3.2 Physical properties of the ocean water [35]

Chemical composition (Table 44.17, see also Tables 44.13, 44.14). The salt composition of the ocean mass is regulated by the solubility, by the pulling down of rainfall from continents, by the processes of exchange with atmosphere and with sediments at the bottom, as well as by the vital activity of the sea organisms.

There appear the following salts in 1 liter of sea water (in g): 25.518 of NaCl; 3.305 of $MgSO_4$, 2.447 of $NaCl_2$; 1.141 of $CaCl_2$; 0.725 of KCl; 0.202 of $NaHCO_3$; 0.083 of NaB_2 and oxides of boron.

The water pH value ranges usually between 7.8 and 8.3 in the surface layers. In the river estuaries, the pH value climbs up to 9.0.

Table 44.17 Mean chemical composition of the ocean water at $T = 5°C$ and chlorinity 19%

Element	Content, %	Element	Content, %	Element	Content, %	Element	Content, %
Oxygen	85.94	Silicium	$>2 \cdot 10^{-4}$	Lead	$4 \cdot 10^{-7}$	Krypton	$2.8 \cdot 10^{-8}$
Hydrogen	10.80	Fluorine	$1.3 \cdot 10^{-4}$	Manganese	$4 \cdot 10^{-7}$	Bismuth	$2 \cdot 10^{-8}$
Chlorine	1.898	Argon	$6.1 \cdot 10^{-5}$	Selenium	$4 \cdot 10^{-7}$	Neon	$1.1 \cdot 10^{-8}$
Sodium	1.056	Rubidium	$2 \cdot 10^{-5}$	Cesium	$2 \cdot 10^{-7}$	Cobalt	10^{-8}
Magnesium	$1.272 \cdot 10^{-1}$	Lithium	$1 \cdot 10^{-5}$	Uranium	$1.5 \cdot 10^{-7}$	Silver	10^{-8}
Sulfur	$8.84 \cdot 10^{-2}$	Phosphorus	$1 \cdot 10^{-5}$	Molybdenum	$5 \cdot 10^{-8}$	Xenon	$9.4 \cdot 10^{-9}$
Calcium	$4.00 \cdot 10^{-2}$	Iodine	$5 \cdot 10^{-6}$	Gallium	$5 \cdot 10^{-8}$	Scandium	$4 \cdot 10^{-9}$
Potassium	$3.80 \cdot 10^{-2}$	Barium	$5 \cdot 10^{-6}$	Thorium	$<5 \cdot 10^{-8}$	Mercury	$3 \cdot 10^{-9}$
Bromine	$6.5 \cdot 10^{-3}$	Arsenic	$1.5 \cdot 10^{-6}$	Nickel	$3 \cdot 10^{-8}$	Helium	$5.2 \cdot 10^{-10}$
Carbon	$3.0 \cdot 10^{-3}$	Zinc	$1 \cdot 10^{-6}$	Vanadium	$3 \cdot 10^{-8}$	Gold	$5 \cdot 10^{-10}$
Nitrogen	$1.7 \cdot 10^{-3}$	Aluminum	$1 \cdot 10^{-6}$	Cerium	$3 \cdot 10^{-8}$	Radium	$(0.2\text{–}3) \cdot 10^{-10}$
Strontium	$1.33 \cdot 10^{-3}$	Iron	$1 \cdot 10^{-6}$	Yttrium	$3 \cdot 10^{-8}$		
Boron	$4.6 \cdot 10^{-4}$	Copper	$6 \cdot 10^{-7}$	Lanthanum	$3 \cdot 10^{-8}$		

Radioactivity of the ocean water [35]

Table 44.18 Content of natural radionuclides in upper layers (5–100 m) of the ocean water

Radionuclide	Concentration, $g \cdot cm^{-3}$	Specific radioactivity, $Bq \cdot cm^{-3}$	Total quantity, 10^9 kg	Total radioactivity, 10^{-3}Ci
^{40}K	$4.5 \cdot 10^{-8}$	$1.2 \cdot 10^{-2}$	63 000	460 000
^{87}Rb	$8.4 \cdot 10^{-8}$	$2.2 \cdot 10^{-4}$	118 000	8 400
^{238}U	$2.0 \cdot 10^{-9}$	$1.0 \cdot 10^{-4}$	2 800	3 800
^{235}U	$1.5 \cdot 10^{-11}$	$3.0 \cdot 10^{-6}$	21	110
^{232}Th	10^{-11}	$1.0 \cdot 10^{-7}$	14	8strut
^{226}Ra	$3.0 \cdot 10^{-16}$	$3.0 \cdot 10^{-5}$	$4.2 \cdot 10^{-4}$	1 100

Table 44.19 Artificial radioactivity in the ocean (γ-sources)

Radionuclide	Half-life $T_{1/2}$, years	Energy, MeV	Specific activity range, $Bq{\cdot}cm^{-3}$	
			open ocean	places of sinks
^{137}Cs	30	0.67	$5{\cdot}10^{-7}$–10^{-5}	10^{-5}–10^{-2}
^{60}Co	5.3	1.17–1.33		10^{-7}–10^{-4}
$^{103}Ru+^{106}Ru$	1.0	0.51	$5{\cdot}10^{-7}$–$2{\cdot}10^{-5}$	10^{-5}–10^{-2}
^{51}Cr	0.08	0.32		10^{-4}–10^{-1}
^{65}Zn	0.7	1.2		10^{-5}–10^{-2}
$^{95}Zr+^{95}Nb$	0.17–0.1	0.75	10^{-6}–10^{-5}	

The salinity field. The ocean salinity (without considering the epeiric seas and estuarine regions) ranges from 31 to 38‰. The average salinity at the surface is 34.73‰. The extremal values are observed in the near-surface layer. Maximum values were found in the epeiric seas: the salinity is more than 39‰ in the Mediterranean Sea; it is up to 41‰ in the Red Sea. The minimum values (up to 3‰) were measured in the Baltic Sea and in the river estuaries.

Owing to the high ionizing power of water, the salts dissolved in it are in the ion-dispersed form. Relative quantity of the dissolved substances in 1 liter of sea water is called salinity. It is expressed in terms of pro mille and denoted as S, ‰. There exists a relationship between S and the chlorinity (the content of chlorine in the sea water):

$$S = 0.03 + 1.805\,\mathrm{Cl}.$$

The mean salinity of the sea water is $S = 34.85$‰, and the mean chlorinity Cl = 19.37‰.

The density and the electric conductivity of the sea water depend on temperature, salinity and pressure (Tables 44.20, 44.21). The density ρ in the ocean varies from 1.0757 to 0.9960 g/cm^3.

For mean values of the density distribution, the density field is steadily stratified, the density continually growing with the depth.

The values of the sea surface density, which depend on temperature, are within the range from 0.9960 to 1.0283 g/cm^3.

Table 44.20 Density (in g/cm^3) of sea water as a function of temperature and salinity

t,°C	S, ‰		
	32	35	38
0	1.0257	1.0281	≈1.031
14	1.0239	1.0262	≈1.028
28	1.0202	1.0224	1.0247

Table 44.21 Electric conductivity (in 10^{-5} S/m) of sea water as a function of temperature and salinity

t,°C	S, ‰			
	10	20	30	40
0	923	1747	2528	3276
15	1378	2594	3740	4834
25	1712	3214	4626	5967

The specific heat of the ocean water is close to 4.18 J/(g·K) and it decreases with a rise in salinity, temperature and pressure. At atmospheric pressure, $t = 0$°C, and $S = 0$‰ it reaches 4.22 J/(g·K); at $t = 30$°C, and $S = 40$‰ it equals 3.87 J/(g·K). Specific heat of water at $t = 0$°C and $S = 34.85$‰ decreases from 3.93 J/(g·K) at the surface to 3.89 J/(g·K) at a depth of 1 km, and to 3.64 J/(g·K) at a depth of 10 km.

Heat conduction. The decisive role in heat transfer through the ocean is played by the turbulent heat conduction under frictional and convective water stirring.

Molecular heat conductivity is very low. For example, at $t = 17.5$°C it comprises:

S, ‰	0	10	20	30
λ, W/(m·K)	0.581	0.569	0.565	0.557.

Water viscosity η represents the combination of the molecular and turbulent components. Molecular viscosity of sea water at $S = 40$‰ takes the following values:

t, °C	0	10	20	30
η, 10^{-3} Pa·s	1.92	1.4	1.07	0.86

Turbulent viscosity is far in excess of the molecular one and is primarily responsible for water frictional stirring.

The surface tension α of the ocean water decreases with temperature and increases with a rise in saltiness S and chlorinity Cl. The dependence of α, mN/m on t, °C and Cl, ‰ takes the form

$$\alpha = 75.611 - 0.144t + 0.0399\mathrm{Cl}.$$

Sound velocity in the ocean [37]. The most simple formula for calculating the sound velocity c, m/s in the ocean water as a function of t, °C, and S, ‰ is written in the following way:

$$c = 1450 - 4.206t - 0.036t^2 + 1.137S - 35.$$

The calculation error is minimum at t = 10°C and does not exceed 1.5 m/s. The sound velocity increases with t, S, and the depth. When S grows by 1‰ or the depth grows by 100 m, the sound velocity rises by 1.2 and 1.6 m/s, respectively.

Optical properties of the ocean water [35].

Table 44.22 Optical properties of the clean ocean water at t = 20°C

λ, nm	n	$\sigma(90°)$, 10^{-3} m^{-1}	σ, 10^{-3} m^{-1}	ϵ, 10^{-3} m^{-1}	κ, 10^{-3} m^{-1}	$\Lambda = \sigma/\epsilon$
250	1.377	2.00	32.0	220	190	0.15
300	1.359	0.96	15.0	55	40	0.27
320	1.354	0.74	12.0	32	20	0.38
350	1.349	0.51	8.2	20	12	0.41
400	1.343	0.30	4.8	11	6	0.44
420	1.342	0.25	4.0	9	5	0.44
440	1.340	0.20	3.2	7	4	0.46
460	1.339	0.17	2.7	5	2	0.54
480	1.337	0.14	2.2	5	3	0.44
500	1.336	0.12	1.9	8	6	0.24
520	1.335	0.10	1.6	16	14	0.10
530	1.335	0.095	1.5	23	22	0.065
540	1.335	0.088	1.4	30	29	0.047
550	1.334	0.082	1.3	36	35	0.036
560	1.334	0.076	1.2	40	39	0.030
580	1.333	0.066	1.1	75	74	0.015
600	1.333	0.058	0.93	200	200	0.0046
620	1.332	0.051	0.82	240	240	0.0034
640	1.332	0.045	0.72	270	270	0.0027
660	1.331	0.040	0.64	310	310	0.0021
680	1.331	0.035	0.56	380	380	0.0015
700	1.330	0.031	0.50	600	600	0.0008
740	1.329	0.025	0.40	2250	2250	0.0002
750	1.329	0.024	0.39	2620	2620	0.0001
760	1.329	0.022	0.35	2560	2560	0.0001
800	1.328	0.018	0.29	2020	2020	0.0001

Note: n is the refraction coefficient; σ is the light scattering coefficient; γ is the scattering angle; ϵ is the attenuation factor; κ is the absorption coefficient; Λ is the luminous transmittance of water medium; $\sigma = -(1/\Phi)(d\Phi_\sigma/dl)$; $\epsilon = (1/\Phi)(d\Phi_\epsilon/dl)$; $\kappa = -(1/\Phi)(d\Phi_\kappa/dl)$, where Φ is the flux of parallel monochromatic radiation; dl is the thickness of the irradiated layer with the volume dV; $d\Phi_\sigma$, $d\Phi_\epsilon$, $d\Phi_\kappa$ are the scattered, attenuated and absorbed elementary radiant fluxes, respectively, on passing through the volume dV.

Table 44.23 Dependence of the total molecular scattering coefficient σ_M (in 10^{-3}m^{-1}) of the ocean water on wavelength λ, nm

λ	σ_M	λ	σ_M	λ	σ_M	λ	σ_M	λ	σ_M
250	4.0	420	4.9	520	2.0	580	1.4	680	6.9
300	1.9	440	4.0	530	1.9	600	1.15	700	6.2
320	1.5	460	3.3	540	1.7	620	1.00	740	4.9
360	1.0	480	2.7	550	1.6	640	0.89	760	4.3
400	0.59	500	2.4	560	2.0	660	0.79	800	3.6

Table 44.24 Light refraction coefficient in the ocean water as a function of wavelength λ and salinity S

λ,	S, ‰			
nm	0	10	20	30
667.8	1.33087	1.33271	1.33452	1.33726
587.6	1.33305	1.33491	1.33675	1.33951
501.6	1.33635	1.33824	1.34011	1.34293
447.2	1.33945	1.34138	1.34329	1.34616

Physical properties of the sea ice [36]. The freezing point of ocean water t_f,°C decreases with a rise in salinity S as

$$t_f = -0.003 - 0.0527\,S - 0.0000004\,S^3.$$

Dependences of t_f and the temperature θ of the maximum density on the salinity S of the ocean water can be viewed from Table 44.25. Generally, there is no definite freezing point for the sea water. With the general water salinity of 33‰, the ice formation starts at −1.8°C. But some sea water remains between the ice crystals. In it, individual salts are crystallized at lower temperatures. Only at −5.5°C is the formed brine solidified. The saltiness and the portion of air in ice determine the density of the sea ice (Table 44.26).

Table 44.25 Dependence of the freezing point t_f, °C and highest density temperature θ, °C on the salinity S, ‰ of the ocean water [37]

S	0	5	10	15	20	24.695	25	30	35	40
t_f	0	−0.3	−0.5	−0.8	−1.1	−1.332	−1.35	−1.6	−1.9	−2.2
θ	3.98	2.9	1.9	0.8	0.3	−1.332	−1.4	−2.5	−3.5	−4.5

Table 44.26 Sea ice density ρ, g/cm^3 dependence on the salinity and air content

Volume content	S, ‰			
of air, %	0	10	20	30
0	0.918	0.925	0.934	0.942
3	0.890	0.898	0.906	0.914
6	0.863	0.871	0.851	0.859
9	0.835	0.843	0.851	0.859

44.4 Atmosphere

44.4.1 Structure of the atmosphere [33]

Atmosphere is not uniform. Its properties change most significantly along the vertical line. Along this direction atmosphere can be divided into a number of layers according to composition, temperature regime, and electric properties. The most pronounced differences are in the temperature distribution.

The layer neighboring the earth (troposphere) is characterized by a temperature decreasing with the height (about 6 K/km) and it ends with tropopause at a height of 7 km at the pole and of 17 km at the equator. The latter is overlain by the stratosphere. Temperature rises in this region from approximately 200 K in tropopause to 280 K in stratopause (at a height of 50 km). Higher one finds the mesosphere. The temperature in this layer decreases with the height and is about 170–180 K at approximately 85 km (mesopause).

The three named layers (troposphere, stratosphere, and mesosphere) are characterized by the constant gas composition and have the common term homosphere.

Atmospheric temperature grows anew starting from 85 km, because of absorption of the ultraviolet radiation of the Sun. The mean temperature gradient equals 20 K/km up to a height of 150 km. Then the growth is slowed and it stops at a height of 300 km. This region of atmosphere is called thermosphere and ends with thermopause, the one being at a height of 350–450 km in the daytime and dropping down to a height of 200–250 km at night. Thermosphere and the spacious layer above it (metasphere) took a common name of heterosphere. Atomic oxygen prevails at the height of about 750 km and helium prevails at a height of 1500 km. Gas separation stops at the height of several thousand kilometers by the transition to the hydrogen atmosphere.

Together with the atmosphere classification based on the temperature structure, other stratifications also exist. Pronounced electromagnetic phenomena are observed in the region from 80 to 500 km (which is called the ionosphere), and in the region above this, where the effect of magnetic field is prevailing (which is called the magnetosphere). At such heights particles with high energy can overcome the gravitational field of the Earth. That is why this region is also spoken of as the exosphere.

Stratification of atmosphere according to the degree of ionization [39]. Observations of the propagation of radio waves have shown that gases in atmosphere are ionized. Four more or less pronounced and regularly observed layers are known at present. They are referred to as the D, E, F_1, and F_2 layers.

The E and F_2 layers are continuous and cover all the Earth's globe. The D and F_1 layers appear only at a definite time of the day and of the year. In addition, in the regions of continuous layers E and F_2 sporadic layers appear from time to time. They make separate clouds with high concentration of ions and electrons. The lower boundary of the ionosphere coincides with the beginning of the D layer. The number of electrons in 1 cm^3 reaches several thousands. The D layer reflects long (several kilometers) waves. At the inclined incidence it partly reflects and noticeably absorbs short electromagnetic waves (30–100 m) and strongly absorbs the waves with the length from 100 to 500 m. At the heights from 85–90 to 130–140 km the E layer is situated. It represents the permanently existing ionization region with the electron concentration maximum (up to $2{\cdot}10^5$ cm^{-3}) at a height of 120–130 km. At night the concentration of electrons decreases to $5{\cdot}10^3$ cm^{-3}. The E layer reflects and noticeably absorbs waves longer than 10 m, and at the inclined incidence it reflects more short waves (15–10 m) in the daytime. At the level of the maximum electron concentration in the E layer, the lower boundary of the aurora borealis is placed.

The highest concentrations of electrons are observed at the height of 200–500 km in the regions of F_1 and F_2 layers. The F_1 layer is formed only in summer in the daytime hours, normally at the height of 180–220 km. Maximum electron concentration in the F_1 layer ranges from $(2\text{–}5){\cdot}10^5$ cm^{-3}. The F_1 layer substantially influences the short wave propagation. Maximum electron concentration in the F_2 layer reaches about several millions of electrons in 1 cm^3. The highest concentration zone is at a height of 200–400 km. The state of the F_2 layer strongly affects the wave propagation in the 10 to 200 m interval. Above the F_2 layer maximum, the ion and electron concentrations decrease very weakly with the height, and at a height of 2000–3000 km they reach the concentration of the interplanetary gas (10^3–10^2 cm^{-3}).

The upper atmosphere ionization is strongly determined by the influence of the Sun. The degree of ionization changes during the day, with seasons, and the phase of solar activity. Ionization is also strongly affected by the bombardment of atmosphere with the particles originating from the Sun. They cause magnetic storms and aurora borealis. The E region corresponds hypothetically to the oxygen dissociation region ($O_2 \rightarrow O+O$), and the D layer to ionization of O_2 near the first ionization potential. Maximums of the F_1 and F_2 regions are situated approximately at a height of 200 and 272 km, respectively. During the night, the F_1 and F_2 layers merge together forming one ionization layer. The D layer disappears at night, and the E layer is noticeably dissipated. At the height up to 200 km, atmosphere is stratified into 11 layers due to temperature variations. The common feature of these layers is a linear change of molecular temperature T_m with the geopotential height Φ. Geopotential (Φ) and geometric (Z) heights are related by the equation $\Phi = rZ/(r + Z)$, where r is the mean radius of the Earth.

Molecular temperature T_m is connected with the kinetic one T_0 by the relationship

$$T_M = T_0 M_0 / M_z,$$

where M_0 and M_z are the molecular masses of the air at sea level and at the height under study, respectively.

Barometric formula. A simplified formula can be used for determining the heights' difference between two points Z_2 and Z_1 with the pressures P_2 and P_1 in them:

$$Z_2 - Z_1 = 18400(1 + 0.00366\bar{t})\lg(P_1/P_2).$$

This formula was derived for the dry air and for a constant (not depending on latitude and altitude) value of gravitational acceleration $g_0 = 980.665\ \text{cm/s}^2$.

According to numerous data (which were obtained by direct and indirect methods), the characteristics of the mean (or standard) atmosphere were deduced (Tables 44.27–44.30) [40].

Table 44.27 Composition of the atmosphere and molar mass

Gas	Volume percentage, %	Molar mass M, kg/kmol
Nitrogen N_2	78.084000	28.01340
Oxygen O_2	20.94760	31.99880
Argon Ar	0.934000	39.94800
Carbon dioxide CO_2	0.031400*1	44.00995
Neon Ne	$1.818 \cdot 10^{-6}$	20.18300
Helium He	$524.0 \cdot 10^{-6}$	4.00260
Krypton Kr	$114.0 \cdot 10^{-6}$	88.00000
Xenon Xe	$8.7 \cdot 10^{-6}$	131.30000
Hydrogen H_2	$50.0 \cdot 10^{-6}$	2.01594
Nitrous oxide N_2O	$50.0 \cdot 10^{-6}$*1	44.01280
Methane CH_4	$200 \cdot 10^{-6}$	16.04303
Ozone O_3	In summer up to $7.0 \cdot 10^{-6}$*1	47.99820
	In winter up to $2.0 \cdot 10^{-6}$*1	
Sulfuric anhydride SO_2	Up to $100 \cdot 10^{-6}$*1	64.06280
Nitrogen peroxide NO_2	Up to $2.0 \cdot 10^{-6}$*1	46.00550
Iodine I_2	Up to $1.0 \cdot 10^{-6}$*1	253.80800
Air	100	28.96442*2

*1 Gas composition can substantially differ depending on place and time of measurement.
*2 Calculated according to the equation of state of an ideal gas.

Table 44.28 Parameters of the atmosphere at mean sea level

Parameter	Notation	Value
Sound velocity, m/s	a_c	340.294
Gravitational acceleration, m/s^2	g_c	9.80665
Pressure height scale, m	H_{pc}	8434.5
Mean free path of air particles, m	l_c	$66.328 \cdot 10^{-9}$
Molar mass, kg/kmol	M_c	28.964420
Particle concentration, m^{-3}	n_c	$25.471 \cdot 10^{24}$
Pressure, Pa	P_c	101325.0
Kelvin temperature, K	T_c	288.15
Mean velocity of air particles, m/s	v_c	458.94
Specific weight, N/m^3	γ_c	12.013
Kinematic viscosity, m^2/s	ν_c	$14.607 \cdot 10^{-6}$
Dynamic viscosity, Pa·s	μ_c	$17.894 \cdot 10^{-6}$
Heat conductivity, W/(m·K)	λ_c	$25.343 \cdot 10^{-3}$
Air particles collision frequency, s^{-1}	ω_c	$6.9193 \cdot 10^{9}$
Density, kg/m^3	ρ_c	1.2250

Table 44.29 Height distribution of the molecular mass and temperature

Height, km	Molecular mass, a.m.u.	Temperature, K		Molecular temperature gradient, K/m
		molecular	kinetic	
0	28.966	288.15	288.15	−0.00651122
11	28.966	216.66	216.66	0
25	28.966	216.66	216.66	0.00276098
46	28.966	247.00	247.00	0
54	28.966	274.00	274.00	−0.00349544
80	28.966	185.00	185.00	0
95	28.966	185.00	185.00	0.00500000
110	28.934	257.64	257.36	0.00801741
120	28.727	335.00	332.24	0.02345357
150	28.107	1010.00	980.05	0.01987408
160	27.900	1199.40	1155.26	0.00308461
170	27.700	1228.71	1175.00	0.00308461
180	27.476	1257.93	1193.20	0.00308461
190	27.245	1287.06	1210.60	0.00308461
200	27.000	1316.10	1226.80	0.00308461

Table 44.30 Dependences of P/P_c, ρ/ρ_c and $(\rho/\rho_c)^{1/2}$ ratios, sound velocity, viscosity, and heat conductivity in atmosphere on the geometric height

Geometric height, m	P/P_c	ρ/ρ_c	$\sqrt{\rho/\rho_c}$	Sound velocity a, m/s	Viscosity		Heat conductivity λ, 10^{-2} W/(m·K)
					dynamic μ, 10^{-5} Pa·s	kinematic ν, m^2/s	
−2000	1.26112	1.20666	1.09848	347.888	1.8515	1.2525–5	2.6359
−1500	1.19117	1.15218	1.07340	346.005	1.8361	1.3009–5	2.6106
−1000	1.12441	1.09966	1.04862	344.111	1.8206	1.3516–5	2.5853
−500	1.06073	1.04889	1.02416	342.208	1.8050	1.4048–5	2.5598
0	1.00000	1.00000	1.00000	340.294	1.7894	1.4607–5	2.5343
500	9.42130–1	9.52876–1	9.76154–1	338.370	1.7737	1.5195–5	2.5087
1000	8.87010–1	9.07477–1	9.52616–1	336.435	1.7579	1.5813–5	2.4830
2000	7.84618–1	8.21676–1	9.06464–1	332.532	1.7260	1.7147–5	2.4314
3000	6.92042–1	7.42248–1	8.61538–1	328.584	1.6938	1.8628–5	2.3795
4000	6.08541–1	6.68854–1	8.17835–1	324.589	1.6612	2.0275–5	2.3273
5000	5.33415–1	6.01166–1	7.75349–1	320.545	1.6282	2.2110–5	2.2747
6000	4.66002–1	5.38866–1	7.34075–1	316.452	1.5949	2.4162–5	2.2218
7000	4.05677–1	4.81648–1	6.94008–1	312.306	1.5612	2.6461–5	2.1687
8000	3.51854–1	4.29113–1	6.55144–1	308.105	1.5371	2.9044–5	2.1152
9000	3.03979–1	3.81276–1	6.17475–1	303.848	1.4926	3.1957–5	2.0614
10000	2.61533–1	3.37559–1	5.80999–1	299.532	1.4577	3.5251–5	2.0072
15000	1.19534–1	1.58983–1	3.98727–1	295.069	1.4216	7.2995–5	1.9518
20000	5.45699–2	7.25793–2	2.69405–1	295.069	1.4216	1.5989–4	1.9518
30000	1.18137–2	1.50286–2	1.22591–1	301.709	1.4753	8.0134–4	2.0345
40000	2.83388–3	3.26176–3	5.71119–2	317.189	1.6009	4.0067–3	2.2313
50000	7.87354–4	8.38264–4	2.89528–2	329.799	1.7037	1.6591–2	2.3954
60000	2.16714–4	2.52797–4	1.58996–2	315.073	1.5837	5.1141–2	2.2041
70000	5.15261–5	6.76150–5	8.22283–3	297.061	1.4377	1.7358–1	1.9765
80000	1.03871–5	1.50678–5	3.88172–3	282.538	1.3208	7.1558–1	1.7987

Note: Notation of the form 5.15261–5 means $5.15261 \cdot 10^{-5}$.

44.4.2 Radiational balance in the atmosphere [41,42]

Energy coming from the Sun is accepted in the 0.2 to 4 μm wavelength range. About 40% of this energy falls into the visible part of the spectrum, i.e. 0.4–0.67 μm (Table 44.31).

Table 44.31 Spectral intensity S_0 (in W·cm^{-2}·μm) of solar radiation at the upper boundary of atmosphere as a function of λ, nm [35]

λ	S_0	λ	S_0	λ	S_0	λ	S_0
0.22	0.0057	0.39	0.1098	0.56	0.1695	1.60	0.0244
0.23	0.0067	0.40	0.1429	0.57	0.1712	1.80	0.0159
0.24	0.0063	0.41	0.1751	0.58	0.1715	2.00	0.0103
0.25	0.0070	0.42	0.1747	0.59	0.1700	2.20	0.0079
0.26	0.0130	0.43	0.1639	0.60	0.1666	2.40	0.0064
0.27	0.0232	0.44	0.1810	0.62	0.1602	2.60	0.0048
0.28	0.0222	0.45	0.2006	0.64	0.1544	2.80	0.0039
0.29	0.0482	0.46	0.2066	0.66	0.1486	3.00	0.0031
0.30	0.0514	0.47	0.2033	0.68	0.1427	3.20	0.0023
0.31	0.0689	0.48	0.2074	0.70	0.1369	3.40	0.0017
0.32	0.0830	0.49	0.1950	0.72	0.1314	3.60	0.0013
0.33	0.1059	0.50	0.1942	0.75	0.1235	3.80	0.0011
0.34	0.1074	0.51	0.1882	0.80	0.1107	4.00	0.0009
0.35	0.1093	0.52	0.1833	0.90	0.0889	4.50	0.0006
0.36	0.1068	0.53	0.1842	1.00	0.0746	5.00	0.0004
0.37	0.1181	0.54	0.1783	1.20	0.0484	6.00	0.0002
0.38	0.1120	0.55	0.1725	1.40	0.0366	7.00	0.0001

The mean energy flux per unit time (solar constant S) from the Sun at the radius of the mean Earth's orbit equals $S = 1.376$ kW/m^2. The mean energy sink falling on the unit Earth's area per 1 s from the Sun measures $(1/4)S = 344$ W/m^2 with the season variations of 3.5%.

Albedo (R) is the integrated ratio of the light flux reflected into all the directions to the one incident on the reflecting surface of the Earth (Table 44.32). Its value is close to 30% and the cloud albedo contribution (75%) is judged to be principal. The value of 15% is characteristic of the dry land. In the deserts $R = 0.2$–0.3 and for the ice cover it is $R = 0.4$–0.6. The mean radiant flux reflected by the Earth's surface comes to $Q_R = (S/4) \cdot R \approx 100$ W/m^2. The mean flux absorbed by the Earth's surface equals $Q_{\text{abs}} = (S/4) \cdot (1 - R) \approx 240$ W/m^2.

Table 44.32 Albedo of different surfaces [33]

Surface	Albedo, %	Surface	Albedo, %	Surface	Albedo, %
Woody flora	10–18	Clay blue:		Plowed field:	
Chernozem:		dry	23	dry	8–12
dry	14	wet	16	wet	5–7
wet	8	Yellow sand:		Rye and wheat	10–25
Serozem:		dry	35	Grass:	
dry	25–30	gray	18–23	fresh	26
wet	10–12	river	43	dry	19

Absorption of radiation energy in the atmosphere (Table 44.33) [33]. The main parts in absorption of radiation energy in atmosphere are played by oxygen, ozone, carbon dioxide, water vapor, and dust. As a whole, atmosphere absorbs 17–25% of the solar radiation. Oxygen possesses absorption bands mainly in the ultraviolet range of the spectrum. In the visible range it absorbs in the A band with the center close to 0.76 μm, and in the B band with the center close to 0.69 μm. But absorption in these bands is not strong and weakly influences the attenuation of radiation.

Ozone is formed as a result of solar radiation absorption ($\lambda < 242.0$ nm) at the heights from 10 to 60 km with the absorption maximum at a height of 22 km. The main absorption bands of ozone are also in the ultraviolet range. The spectrum observed near the Earth breaks close to the 300.0 nm wavelength. In this spectral region, solar radiation absorption by ozone comes to about 2–3% of the total integral flux.

The most important absorption band of carbon dioxide falls on the wide band from 12.9 to 17.1 μm, which is in the maximum of the atmosphere thermal radiation. Water vapor is also of great importance for radiation absorption in atmosphere (Table 44.34). This is determined not only by its high content, but also by a great number of lines and bands in its spectrum. The most important of them are the ones in the infrared part of the spectrum.

In the visible, two bands are mainly responsible for strong absorption. They involve $\lambda \sim$ 730–685 nm band and the "rain" band within 606–585 nm.

Table 44.33 Mean radiation balance of the northern hemisphere at mean cloud conditions [43]

Components of radiation balance	Energy flux, mW/cm^2	Components of radiation balance	Energy flux, mW/cm^2
Insolation with short-wavelength radiation at the upper atmosphere boundary	35	Radiation absorption by the Earth's surface:	
		total	16.5
Long-wave radiation (total)	22.6	direct solar radiation	7.8
Radiation absorption in atmosphere:		transmitted by clouds	5.0
total	6	scattered	3.7
by ozone	0.9	Effective radiation by the Earth's surface:	
by water vapor and dust	4.5	heat radiation	39.9
by clouds	0.6	atmosphere backscattering	33.6
Reflection and scattering of radiation into outer space:		effective radiation	2.1
		Heat radiation of troposphere:	
total	123	absorbed by troposphere	38
by atmosphere	2.4	self-radiation	53.4
by clouds	8.4	Heat emission into outer space:	
by the Earth's surface	1.5	by the Earth's surface	1.9
		by troposphere	19.7
		by stratosphere	0.98

Table 44.34 Water vapor absorption bands

Band	α	β	ρ	σ_r	Ψ	Ω	ω_1	ω_2	x		y	
Center of band λ, μm	0.72	0.82	0.93	1.13	1.38	1.86	2.01	2.05	2.68	3.2–4.0	4.0	4.9

44.4.3 Electric phenomena in the atmosphere [33]

Ions in the atmosphere. As a result of ionization of gases in atmosphere, the primary (molecular) ions and stable complexes of 10–15 molecules (light ions) are formed. When light ions attach to aerosol particles, larger ions (heavy and ultraheavy) are formed. Intermediate ions are also detected, their nature being not clear yet.

Fog droplets and cloud elements exhibit the size of 10^{-4}–10^{-3} cm and can have a charge. But usually they cannot be said to be ions. As a rule, every ion carries one elementary charge if its radius is below 10^{-6} cm.

Table 44.35 Main groups of atmospheric ions [39]

Ion group	Mobility, cm^2/(V·s)	Ion radius, nm
Light	10^{-2}–1	0.66–8
Intermediate	10^{-3}–10^{-2}	8–25
Heavy	10^{-3}–$2.5 \cdot 10^{-3}$	25–55
Ultraheavy	$\leq 2.5 \cdot 10^{-3}$	> 55

Table 44.36 Mean number of ions appearing in 1 cm^3 of air for 1 s [39]

Air mass position	Ionizer			Sum
	Radioactive elements' radiation		Cosmic radiation	
	of soil	of air		
Over dry land	4.0	4.6	1.5–1.8	10.1–10.4
Over ocean			1.5–1.8	1.5–1.8

Table 44.37 Concentration of cosmogenous nuclides in the surface layer of air [28]

Nuclide	^{14}C	^{7}Be	^{3}H	^{25}S	^{32}P	^{22}Na
Specific activity, Ci/m^3	$5{\cdot}10^{-13}$	$(2\text{–}32){\cdot}10^{-14}$	10^{-13}–10^{-14}	10^{-15}	$2{\cdot}10^{-16}$	10^{-17}

Among the numerous known ionizers, the most important for the lower atmospheric layers are the radiations from radioactive substances, which are contained in crust and in atmosphere (Tables 44.35–44.37), as well as cosmic rays. Cosmic radiation is a main ionizer above the oceans.

Electric field in the atmosphere. The vertical component of electric field in the atmosphere nearly always exceeds substantially its horizontal components, the fact correlated with the negative charge of the Earth. The mean surface density of an electric charge of the Earth is $dQ/dS = -1.15 \cdot 10^{-13}$ C·cm^{-2}. The total charge of the Earth equals $Q = -5.7 \cdot 10^5$ C. These values are deduced under the assumption that the mean vertical gradient of the electric potential near the Earth's surface is equal to 130 V/m.

Electric charges of sediments. Particles of sediments of all kinds carry electrical charges (Table 44.38), which appear in a number of groups of charging processes. Among them occur collisions of polarized particles, capture of the air ions by the sedimental particles, spraying of the water droplets, charging under the aggregate state change.

The charge of separate particles varies widely. The number of positively charged particles is on the average 1.5 times higher than the number of negatively charged particles. This ratio varies from 1 : 1 to 3 : 1. At the same time, the mean negative charge per droplet ($1.3{\cdot}10^{-12}$ C) is greater than the positive one ($1.1{\cdot}10^{-12}$ C). The droplets of the steady rain are charged to the potential of 0.5–10 V; those of the thunderstorm shower are charged up to 300 V with the mean value close to 40 V.

Table 44.38 Electric charge Q of sediments of different origin and the current i caused by them

Character of sediments	Q_{av}, C	Q_{max}, C	i_{av}, A	i_{max}, A
Steady rain	10^{-15}–10^{-14}	$5{\cdot}10^{-13}$	$5{\cdot}10^{-15}$	$5{\cdot}10^{-14}$
Shower rain	10^{-13}–10^{-12}	$5{\cdot}10^{-12}$	10^{-14}	$5{\cdot}10^{-12}$
Hail	10^{-11}	10^{-10}	–	10^{-11}
Snow	10^{-12}–10^{-11}	$5{\cdot}10^{-11}$	10^{-14}	$5{\cdot}10^{-13}$

References

[1] Gill, A., Atmosphere–ocean dynamics, vol. 1, Academic Press, New York, 1982, p. 361.

[2] White, H. L., The world geodetic system 1984, vol. 1, Proc. 4th International Geodet. Symposium Satellite Position, Austin, Texas, 28 April - 2 May, 1986, p. 93.

[3] Morozov, V. P., A course of spheroidal gravimetry, Nedra, Moscow, 1979 (in Russian).

[4] Gutenberg, B., Physics of the Earth's interior, Academic Press, New York, 1959.

[5] Jeffreys, H., The Earth, its origin, history, and physical constitution, 4th ed., Cambridge University Press, Cambridge, 1959.

[6] Kosminskaya, I. P., Vestnik Akad. Nauk SSSR, 2, 51, 1965 (in Russian).

[7] Zharkov, V. N., Internal structure of Earth and planets, Nauka, Moscow, 1978 (in Russian).

[8] Crust of the Earth (A symposium), Ed. Poldervaart, A., The Geological Soc. of America, Baltimore, 1955.

[9] Veselov, K. E., Sagitov, M. U., Gravimetric reconnaissance, Nedra, Moscow, 1968 (in Russian).

[10] Berch, F., in: Crust of the Earth (A symposium), Ed. Poldervaart, A., The Geological Soc. of America, Baltimore, 1955.

[11] Fersman, A. E., Geochemistry, vols. 1–4, ONTI, Moscow, 1933-1939 (in Russian).

[12] Cherdyntsev, V. V., Abundance of chemical elements, Gostekhizdat, Moscow, 1956 (in Russian).

[13] Hendershott, M. C., Munk, W. H., Ann. Rev. Fluid Mech., 2, 205, 1970.

[14] Riznichenko, Yu. V., Problems of seismology, Nauka, Moscow, 1985 (in Russian).

[15] Savarenskii, E. F., Kirnos, D. P., Fundamentals of seismology and seismometry, Gostekhizdat, Moscow, 1954 (in Russian).

[16] Iaby, G.A., Earthquake, Heinemann, New Zealand, 1980.

[17] Proskuryakova, T. A., Rykunov, L. N., International geophysical year, vol. 5, Nauka, Moscow, 1963 (in Russian).

[18] Helton, B. C., Jonson, D. P., Proc. JBB, 50, 2328, 1962.

[19] Yanovskii, B. M., Geomagnetism, Gostekhizdat, Moscow, 1953 (in Russian).

[20] Ocean geophysics, oceanology, vol. 2, Ed. Monin, A. S., Nauka, Moscow, 1979 (in Russian).

[21] Pochtarev, V. I., The normal terrestrial magnetic field, Nauka, Moscow, 1984 (in Russian).

[22] A handbook on the variable terrestrial magnetic field, Gidrometeoizdat, Leningrad, 1954 (in Russian).

[23] Kraev, A. P., Essentials of geoelectricity science, Isdatel'stvo LGU, Leningrad, 1950 (in Russian).

[24] Vinogradov, P. A., Geologiya i geofizika, No 12, 111, 1963.

[25] Lyubimova, E. A., Terrestrial tectonosphere, Nauka, Moscow, 1978 (in Russian).

[26] Jecobs, J. B., The Earth interior, in: Encyclopedia of physics, vol. 47, Ed. Flügge, E.S., Springer-Verlag, Berlin, 1956.

[27] Budyko, M. I., Kondratiev, K. Ya., Kosmicheskiye issledovaniya, 2, 62, 1964 (in Russian).

[28] A handbook on hudroacoustics, Sudostroenie, Leningrad, 1988 (in Russian).

[29] Brown, R. D., Toward a more dynamic geoid, Proc. 9th Intern. Symposium Earth Tides, New York, August 17–22, 1981, Stuttgart, 1983, p. 453.

[30] Sound transmission through a fluctuating ocean, Ed. Flatte, S., Cambridge University Press, Cambridge, 1979

[31] Gill, A., Atmosphere-ocean dynamics, vol. 1, Academic Press, New York, 1982, p.51.

[32] Elizarov, A. A., Oceanological essentials of fishing, Isdatel'stvo LGU, Leningrad, 1983 (in Russian).

[33] Tverskoi, P. N., A course of meteorology, Gidrometeoizdat, Leningrad, 1962 (in Russian).

[34] Bruevich, S. V., The element composition of the World ocean water, vol. 1–2, Trudy instituta okeanologii Akad. Nauk SSSR, Nauka, Moscow, 1948 (in Russian).

[35] Ocean physics. Ocean hydrophysics, Ed. Monin, A. S., Nauka, Moscow, 1978 (in Russian).

[36] Dietrich, G., Kalle, K., Allgemeine meereskunde. Eine einfuhrung in die ozeanographie, Gebruder Bornstraeger, Berlin, Nikolassee, 1957.

[37] Gusev, A. M., Fundamentals of oceanology, Isdatel'stvo MGU, Moscow, 1983 (in Russian).

[38] List, R. J., Smithsonian meteorological tables, 6th ed., Smithson Misc. Collection, No 114, Smithsonian Institute, Washington, 1951.

[39] Nicolet, M., Aeronomy, Mir, Moscow, 1964 (in Russian).

[40] Standard atmosphere. Parameters. GSSD 4401–81, Izdatel'stvo standartov, Moscow, 1981 (in Russian).

[41] Kondratyev, K. Ya., Radiation in the atmosphere, Academic Press, New York, 1969.

[42] Stephens, G. K., Campbell, G. G., Vonder Haar, T. H., Earth radiation budgets, J. Geophys. Res., 86, 9737, 1981.

[43] Seismic zoning of the USSR territory, Nauka, Moscow, 1980 (in Russian).

[44] Nikonov, A. A., Earthquakes, Znanie, Moscow, 1984 (in Russian).

45

Astronomy and Astrophysics

Yu.E. Lyubarskii and R.A. Syunyaev

45.1 Some Astronomical Units and Constants [1]

The following units are most frequently used in the field:

astronomical unit, 1 a.u. = $1.495989(1) \cdot 10^{11}$ m;

parsec, 1 pc = $3.085778 \cdot 10^{16}$ m;

light year, 1 ly = $9.4606030 \cdot 10^{15}$ m;

tropical year (between subsequent equinoxes), 1 trop. yr. = 31556926 s = 365.24219 d;

Jansky (unit of the spectral flux density), 1 Jy= 10^{-26} W/(m$^2 \cdot$ Hz).

Stellar magnitudes [1–3]. The ratio of illuminances E_1 and E_2, produced by two stars, is connected with their stellar magnitudes m_1 and m_2 by a relationship

$$E_2/E_1 = 10^{0.4(m_1 - m_2)}. \tag{45.1}$$

Absolute stellar magnitude M is equal to the visual stellar magnitude m of the object located at a distance r, when it would be placed at a distance of 10 pc:

$$M = m + 5 - 5 \lg r, \tag{45.2}$$

where r is the distance in pc.

As the energy received from the object is always measured within a finite wavelength range, one marks visual stellar magnitudes with indexes pointing to the spectral band of measurements. As a standard, one takes a three-color photometrical system UBV, based on three standard spectral bands: ultraviolet (U), blue (B), and visual (V) (Fig. 45.1). The color of a star is characterized by the difference between stellar magnitudes measured in different bands, for example, $B - V$ or $U - B$. A star belonging to the spectral class A0 has $U - B = B - V = 0$. At present, the UBV system is expanded to cover the infrared band (Table 45.1).

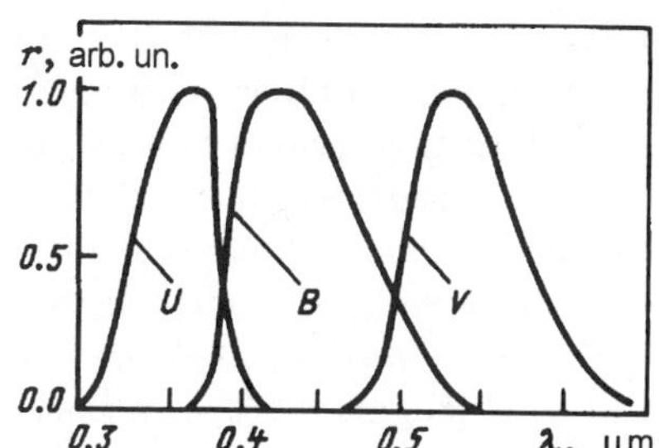

Figure 45.1 Curves of transparency for U, B, and V filters [3].

0-8493-2861-6/97/$0.00+$.50

Table 45.1 System of standard photometrical bands [3]

Band	Effective wavelength, μm	Effective bandwidth, μm	Flux density corresponding to m=0, Jy	Band	Effective wavelength, μm	Effective bandwidth, μm	Flux density corresponding to m=0, Jy
U	0.36	0.04	1880	J	1.25	0.3	1770
B	0.44	0.10	4440	K	2.2	0.6	630
V	0.550	0.08	3810	L	3.5	0.9	310
R	0.700	0.21	3010	M	5.0	1.1	180
I	0.88	0.22	2430	N	10.4	6.0	43

Bolometric magnitude m_b is the stellar magnitude corresponding to the total energy flux from the object. The luminosity of a $M_b = 0$ star comes to $2.97 \cdot 10^{28}$ W. At the Earth's atmosphere boundary, the incident flux density from a $m_b = 0$ star is $2.48 \cdot 10^{-8}$ W/m^2, and the illuminance measures $2.54 \cdot 10^{-6}$ lx.

Celestial coordinates [2]. Main points and coordinate systems on the celestial sphere are as follows.

Northern and southern world poles denote the points of intersection of the celestial sphere with an extension of the Earth's rotational axis in the northern and southern directions.

Zenith and nadir denote the points of intersection of the celestial sphere with an extension of the plumb line at the point of observations upward and downward, respectively.

Celestial equator is termed a large circle made by intersection of the celestial sphere with a plane of the terrestrial equator.

Ecliptic is an apparent path of the Sun about the celestial sphere. It is inclined at an angle $\epsilon = 23°27'$ to the plane of the celestial equator.

Point of spring (vernal) equinox, Υ, denotes a point at which the ecliptic crosses the celestial equator, and which is passed by the Sun moving from the southern to the northern hemisphere.

Galactic equator is referred to as a circle made by intersection of the celestial sphere with an extension of the galactic plane. It is inclined at $62°36'$ to the plane of the celestial equator.

Northern galactic pole is termed a point of intersection of the celestial sphere with an extension of the rotational axis of the Galaxy in the northern direction. Its equatorial coordinates are $\alpha = 12^h 49^m, \delta = 27°24'$.

Galactic center has equatorial coordinates $\alpha = 17^h 42^m 24^s$, $\delta = -28°55'$.

Equatorial coordinate system (Fig. 45.2). *Declination* δ of a celestial object denotes an angle expressed in degrees and counted from the celestial equator toward the star along the *circle of declinations* (a large circle passing through world poles and the star). The declination is positive for objects in the northern hemisphere. *Right ascension* α of a celestial object is an angular distance measured in hours eastward from the vernal equinox along the celestial equator.

Ecliptical coordinate system (Fig. 45.2). *Astronomical latitude* β of a celestial object is termed an angle expressed in degrees, which is measured between the ecliptic and the object along the *astronomical latitude circle* (a large circle passing through poles of ecliptic and the object). Astronomical latitude is positive to the north of the ecliptic. *Astronomical longitude* λ is termed an angle (in degrees) measured along the ecliptic through the south to the east between the spring equinox point and a point of intersection of the ecliptic with the circle of astronomical latitude passing through the object.

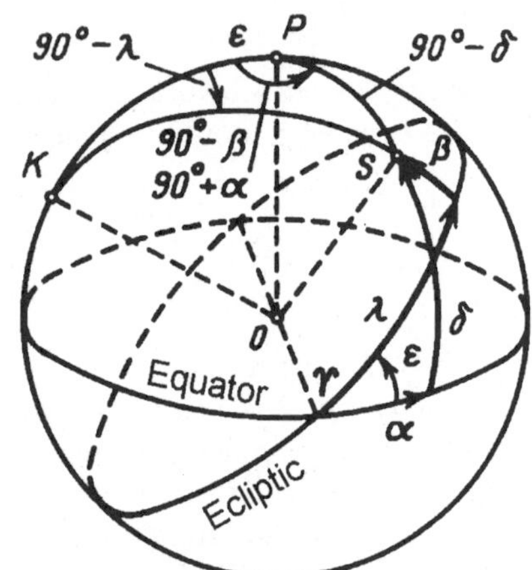

Figure 45.2 The main points and coordinates on the celestial sphere: K, P, S, O are the pole of ecliptic, the northern pole of the world, location of the object, and location of the observer, respectively [2].

Galactic coordinate system. Galactic latitude b of a celestial object is an angle expressed in degrees and measured along the *galactic latitude circle* (a large circle passing through the galactic poles and the object) between the galactic equator and the object. Galactic latitude is positive to the north of the galactic equator. *Galactic longitude l* of an object is termed an angle expressed in degrees and measured along the galactic equator from the galactic center in the direction through the South to the East up to a point of intersection with the galactic latitude circle passing through the object.

The following astronomical symbols have found wide use:

☉	Sun	♁, ⊕	Earth	⛢	Uranus
☾	Moon	♂	Mars	♆	Neptune
☿	Mercury	♃	Jupiter	♇	Pluto
♀	Venus	♄	Saturn	☄	Comet
♈	Aries, 0°	♌	Leo, 120°	♐	Sagittarius, 240°
♉	Taurus, 30°	♍	Virgo, 150°	♑	Capricorn, 270°
♊	Gemini, 60°	♎	Libra, 180°	♒	Aquarius, 300°
♋	Cancer, 90°	♏	Scorpio, 210°	♓	Pisces, 350°

45.2 The Sun

Basic characteristics of the Sun are as follows [1] (see also Figs. 45.3–45.5):

radius, $R_\odot = 6.9599(7)\cdot 10^{10}$ cm;
mass, $M_\odot = 1.989(1)\cdot 10^{33}$ g;
mean density, $\rho = 1.409$ g/cm^3;
sidereal rotational period (relative to remote stationary stars at a latitude $\varphi = 17°$), 25.38 d;
synodic rotational period (relative to the Earth), $26.75 + 5.7\sin^2\varphi$ d;
equatorial inclination to the plane of ecliptic, $7°15'$;
acceleration of the free fall at the surface, $2.740\cdot 10^4$ cm/s^2;
mean magnetic induction near the poles at minimal number of the spots, $(1\text{–}2)\cdot 10^{-4}$ T;
angular diameter seen at average distance from the Earth, $31'59.26''$.

Location of the Sun in the Galaxy is described in the following way [1]:

distance to the galactic center equals (10 ± 0.8) kpc;
distance to the north of the galactic plane is (8 ± 12) pc.

The Sun moves relative to nearby stars with a velocity of 15.5 km/s in the direction $\alpha = 17^h40^m$, $\delta = +21°$ [4]; rotational velocity about the galactic center is 250 km/s.

The Sun as a star is characterized by the following parameters [1]:

$$
\begin{aligned}
m_V &= -26.74; & M_V &= +4.83;\\
m_B &= -26.09; & M_B &= +5.48;\\
m_U &= -25.96; & M_U &= +5.61;\\
m_b &= -26.82; & M_b &= +4.75;
\end{aligned}
$$

its spectral refers to G2V; the effective temperature reaches 5770 K; the age comes to $5\cdot 10^9$ y.

Structure of the solar atmosphere is understood by reference to Fig. 45.6.

Photosphere designates a layer with a depth of about 500 km, in which continuum radiation is produced with a spectrum similar to the blackbody spectrum. Narrow absorption lines called Fraunhofer lines are superimposed on this spectrum (Table 45.2).

Chromosphere designates a transition region between the photosphere and the corona with a width of order 10^4 km, and emitting in lines observed during solar eclipses.

Corona corresponds to an upper portion of the solar atmosphere going over immediately into the interplanetary medium. A high temperature (about 10^6 K) of the corona is sustained by the energy generated on dissipation of the magnetic fields going up from the photosphere, and on dissipation of sound and Alfvén waves excited by convection in the photosphere. The electrons are distributed over the corona according to the law [7]:

$$N_e = 10^8(0.036/r^{1.5} + 1.55/r^6 + 2.99/r^{16}. \tag{45.3}$$

Here N_e is the electron number density (in cm^{-3}), r is the distance from the solar center (in $R_\odot$). Continuum emission of the corona is due to scattering of the Sun photons by the electrons. Strong forbidden lines of highly ionized heavy elements are also observed (Table 45.3). The corresponding transitions are forbidden due to selection rules in the dipole approximation, so their upper states are metastable. In ordinary conditions they are quenched in collisions, but in a low-density medium the collisions are rare and deexcitation is accompanied by a forbidden photon emission. The emissive ability of the corona is determined by the emission measure, $ME = \int N_e^2 dV$, with its standard value of $4.4 \cdot 10^{49}$ cm^{-3}. The total luminous flux of the corona beyond 1.3 $R_\odot$ at maximal and minimal number of the spots comprises a $1.3 \cdot 10^{-6}$ and $0.8 \cdot 10^{-6}$ fraction, respectively, of the total flux from the Sun [1].

Vertical oscillations of the solar atmosphere with a 5 min period are observed. Their horizontal scale equals 5000 km and amplitude is about 0.4 km [5]. Some data provide evidence for oscillations of the whole solar atmosphere or its significant part with a period of 160 min [8]. Characteristics of numerous nonstationary formations in the solar atmosphere are compiled in Table 45.4.

Emission of the Sun is exemplified by the following quantities [1]:

total luminosity, $L_\odot = 3.826(8) \cdot 10^{26}$ W;

radiant flux from the surface unit, $6.284 \cdot 10^7$ W/m^2;

luminous intensity, $2.84 \cdot 10^{27}$ cd;

illuminance produced by the Sun beyond the terrestrial atmosphere at a mean distance of the Earth from the Sun, 127 000 lx;

solar constant (the total power of radiation that falls onto a unit square located outside the terrestrial atmosphere at a mean distance of the Earth from the Sun), 1373(20) W/m^2.

The solar spectrum is shown in Figs. 45.7–45.10. The major part of the emission comes from the photosphere. In a short-wavelength region ($\lambda \leq 100$ nm), the spectrum consists of emission lines of numerous ions originating in the chromosphere and corona.

Radioemission of the Sun in a quiet state is caused by thermal emission of the corona. Luminance temperature at wavelengths $\lambda > 1$ m is equal to the electron temperature in the corona (about 10^6 K). At shorter wavelengths, the corona becomes transparent and its luminance temperature decreases. Thermal emission of the chromosphere dominates at $\lambda \leq 1$ cm. A slow-varying thermal radiation (S-component) with the intensity well correlated with the solar spots area is also observed. Characteristics of different types of a strongly nonstationary radiation produced by a burst-like energy release in active regions of the solar atmosphere are listed in Tables 45.5, 45.6.

Neutrino flux from the Sun is equal to 2.1(3) solar neutrino units [6]. One solar neutrino unit corresponds to 10^{-36} neutrino absorptions per second per one Cl atom in the reaction $\nu + {}^{37}\mathrm{Cl} \rightarrow e^- + {}^{37}\mathrm{Ar}$. As the energy threshold for this reaction is $E_\nu = 0.814$ MeV, only the most energetic neutrinos may be observed, which are generated inside the Sun in the reaction ${}^8\mathrm{B} \rightarrow {}^8\mathrm{Be}^* + e^+ + \nu$. The major part of the neutrino flux is generated in the reaction $p + p \rightarrow {}^2\mathrm{H} + e^+ + \nu$; energy of these neutrinos, however, $E_\nu < 0.420$ MeV, and they are not recorded.

Solar cycle [5]. The level of solar activity is characterized by the Wolf number

$$W = 10g + f, \tag{45.4}$$

where g is a number of spot groups; f is a number of spots. Average duration of the spot-formation cycle (cycle of solar activity) comprises 11.04 y (Fig. 45.11). Long-term periods were observed without formation of spots. The last such period, the Mounder minimum, lasted from 1645 to 1715.

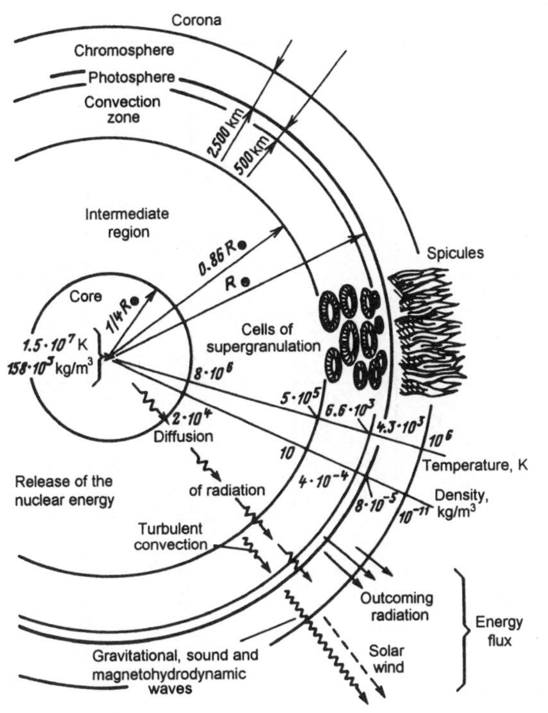

Figure 45.3 Structure of the Sun (without a scale kept properly) [5].

Table 45.2 The strongest Fraunhofer lines in the solar spectrum [5]

Wavelength, nm	Equivalent linewidth* W, nm	W/λ, 10^{-4}	Absorbing atom or ion
358.1209	0.2144	5.99	Fe I
371.9947	0.1664	5.34	Fe I
373.4874	0.3027	9.45	Fe I
373.7141	0.1071	4.28	Fe I
374.5574	0.1202	4.59	Fe I
374.9495	0.1907	5.78	Fe I
375.015	0.1388	4.30	H I
375.8245	0.1647	4.97	Fe I
377.063	0.1860	6.21	H I
379.790	0.3463	10.85	H I
382.0436	0.1712	5.12	Fe I
382.5891	0.1519	4.21	Fe I
383.2310	0.1685	6.00	Mg I
383.539	0.2362	7.19	H I
383.8302	0.1920	6.41	Mg I
385.9922	0.1554	4.00	Fe I
388.905	0.2346	7.22	H I
393.3682	2.0253	48.74	Ca II
396.8492	1.5467	34.35	Ca II
397.0076	0.3076	7.76	H I ε
410.1748	0.3133	7.46	H I δ
434.0475	0.2855	6.59	H I γ
486.1342	0.3680	7.50	H I β
656.2808	0.4020	6.49	H I α

* Equivalent linewidth is a width of the continuum part (adjacent to the line) whose energy is equal to that absorbed in the line.

Table 45.3 Some emission lines of the corona in the visible and near infrared ranges [5]

Transition	Ion	Wavelength, nm	Equivalent linewidth, nm	Transition	Ion	Wavelength, nm	Equivalent linewidth, nm
$^2P_{3/2} \to ^2P_{1/2}$	Ni XVI	360.1	0.13	$^3P_1 \to ^3P_0$	Ni XV	670.2	0.12
$^2P_{3/2} \to ^2P_{1/2}$	Fe XIV	530.3	2.00	$^3P_1 \to ^3P_0$	Fe XIII	1074.7	5.00
$^1D_2 \to ^3P_2$	Fe XIII	338.8	1.00	$^3P_1 \to ^3P_2$	Fe XI	789.2	0.60
$^3P_2 \to ^3P_1$	Fe XIII	107.98	3.00	$^2P_{1/2} \to ^2P_{3/2}$	Fe X	637.4	0.50

Table 45.4 Characteristics of formations in the solar atmosphere [1,3,5]

Formation	Dimensions, km	Lifetime	Temperature, K
Solar spots	2000–100000	1 d–10 months	4200 (shadow) 5700 (penumbra)
Granules	1000	10 min	T(granules) – T(intergranular spacing) = 300 K
Cells of supergranulation	20000–40000	20 h	
Spicules	1000 (horizontal) 7000 (vertical)	10 min	$(1–4) \cdot 10^4$
Faculae	5000	10 d	T(facula) – T(photosphere) = 1000 K
Protuberances	30000 (height) 20000 (length) 5000 (width)	2 months	$(6–8) \cdot 10^3$

Table 45.5 Mean characteristics of radiobursts [10]

Characteristics	Microwave bursts			Bursts in the meter band							
							Noise storms				
	pulse-like	gradual	IV μ type	IVdm type	IV$mF/mA$ type	IVmB type	continual	I type	II type	III type	IV type
Frequency band, MHz	≥ 1000			~ 150–1500	≤ 200		≤ 500		≤ 200	≤ 500	≤ 200
Rate of bursts at maximal solar activity	≤ 1 h^{-1}	≥ 1 d^{-1}	~1–10 per month; very intensive bursts occur at the solar activity maximum				~1–3 per week	$\geq 10^3$ h^{-1}	0.5 d^{-1}	3 h^{-1} (by groups)	1 h^{-1}
Frequency range, MHz	~500–50000		~ 5000–50000	1000	> 200	≤ 200	< 500	≥ 4	~ 5	50–100	
Duration	≤ 10 min	≥ 10 min	from 10 min to several hours			up to a few hours	hours or days	< 1 s (groups of ~ 0.5 min)	5–30 min	≤ 10 s (1 min groups)	$\leq$1 min
Luminance temperature, K	$\sim 10^6$–10^{10}	$\leq 10^6$	$\leq 10^9$	10^7–10^8		$\leq 10^{10}$		$\geq 10^{10}$	$\leq 10^8$– 10^{11}		
Circular polarization	absent or partial		partial	partial, strong	weak	very strong			very weak	partial	
Linear polarization	weak		partial	10%				very weak		unreliable data	
Angular size of the source, deg.	1–4	1–2	2–4	3–5	6–12	2–6	2–10	1–6	6–$\geq$ 30	3–12	

Table 45.6 Typical values of the total energy E (in J) and power F (in W), released by different channels for large and small solar bursts [3]

Channel	The largest bursts		Subburst	
	E	F	E	F
Radiation:				
soft X-ray and UV	$(3–5)\cdot10^{24}$	$(3–5)\cdot10^{21}$	$(1–3)\cdot10^{22}$	$(1–3)\cdot10^{20}$
optical continuum	$(1–3)\cdot10^{24}$	$(1–3)\cdot10^{21}$		
in H_α line	$(1–3)\cdot10^{23}$	$(1–3)\cdot10^{20}$	10^{19}	$3\cdot10^{16}$
hard X-ray	$(3–5)\cdot10^{19}$	$(3–5)\cdot10^{16}$	$(1–3)\cdot10^{17}$ *	$(1–3)\cdot10^{15}$
in γ-range	$(1–3)\cdot10^{18}$	$(1–3)\cdot10^{15}$		
in radioband	$\sim10^{17}$	$\sim10^{14}$	10^{15}	10^{13}
Accelerated particles:				
electrons ($\geq$ 20 keV)	$(3–5)\cdot10^{24}$	$(3–5)\cdot10^{21}$	10^{20} *	10^{18}
protons ($\geq$ 20 MeV)	$(1–3)\cdot10^{24}$	$(1–3)\cdot10^{21}$		
Hydrodynamical plasma motions:				
interstellar ejection and shock waves	$(1–3)\cdot10^{25}$			
motion beyond chromosphere	10^{25}	10^{22}	10^{22}	10^{19}

* Absent for majority of subbursts.

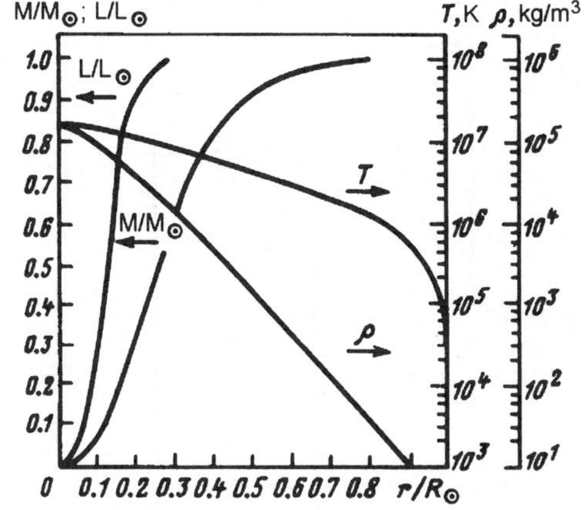

Figure 45.4 Model of the internal structure of the Sun [5].

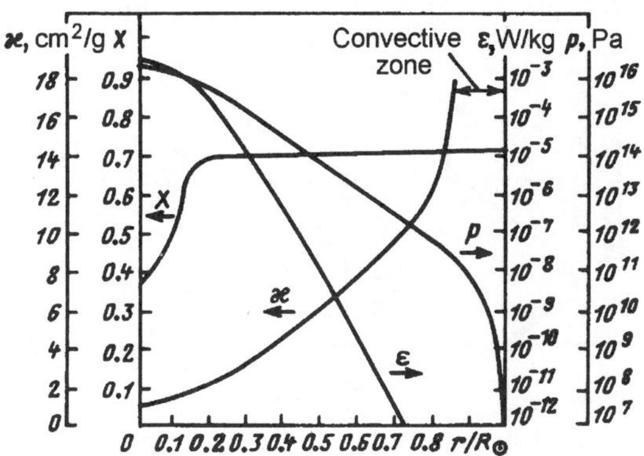

Figure 45.5 Properties of the solar matter (X is the mass fraction of hydrogen, κ is the opacity, ε is the energy output rate, p is the pressure) [5].

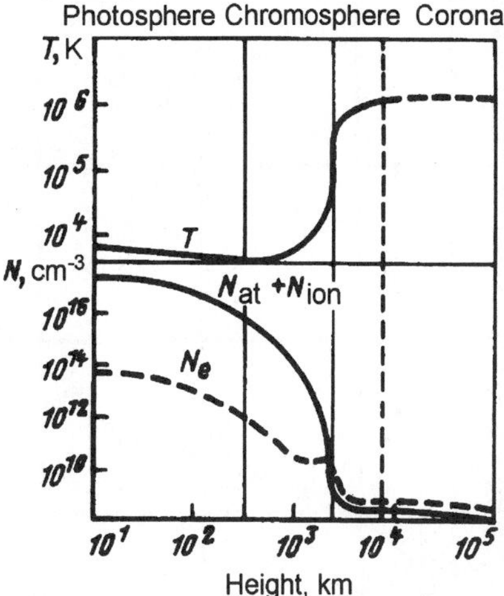

Figure 45.6 Temperature T and concentrations of electrons N_e, ions N_{ion}, and neutral atoms N_{at} in the solar atmosphere. Height is counted from the level of the optical thickness equal to unity at a wavelength of 0.5 μm [5].

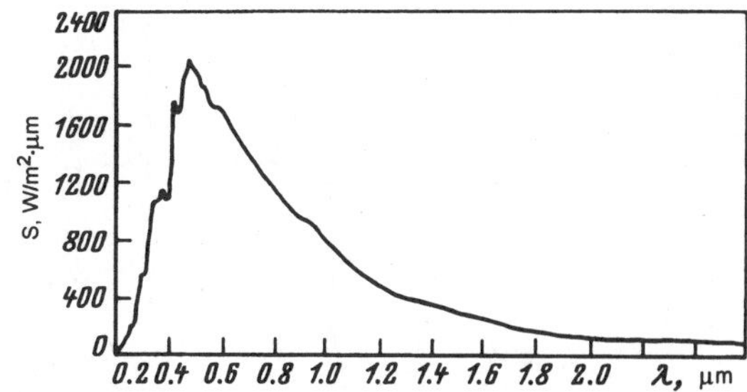

Figure 45.7 Spectrum of the Sun (λ=0.2–2.6 μm) [5].

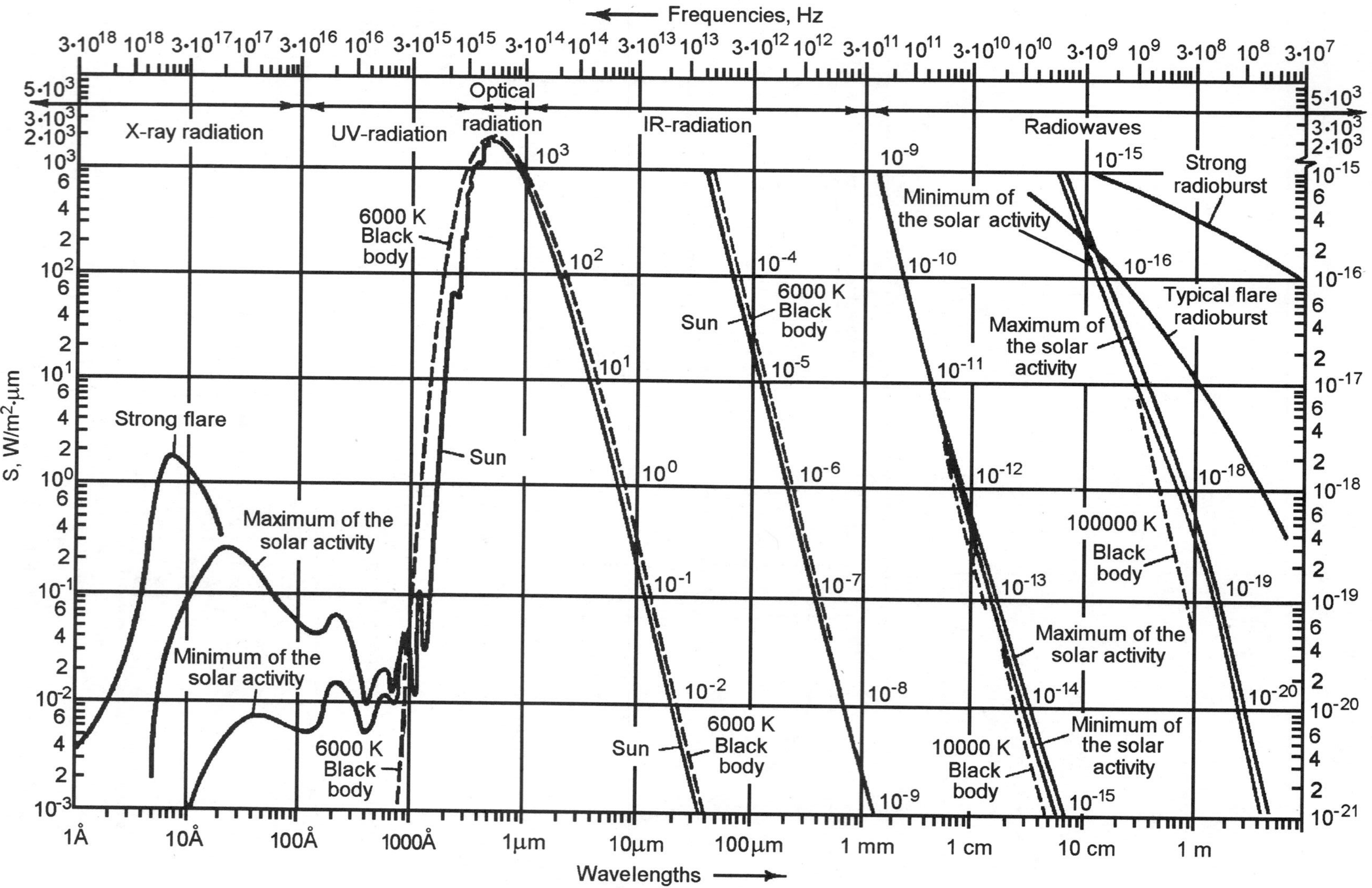

Figure 45.8 Spectrum of the Sun [3].

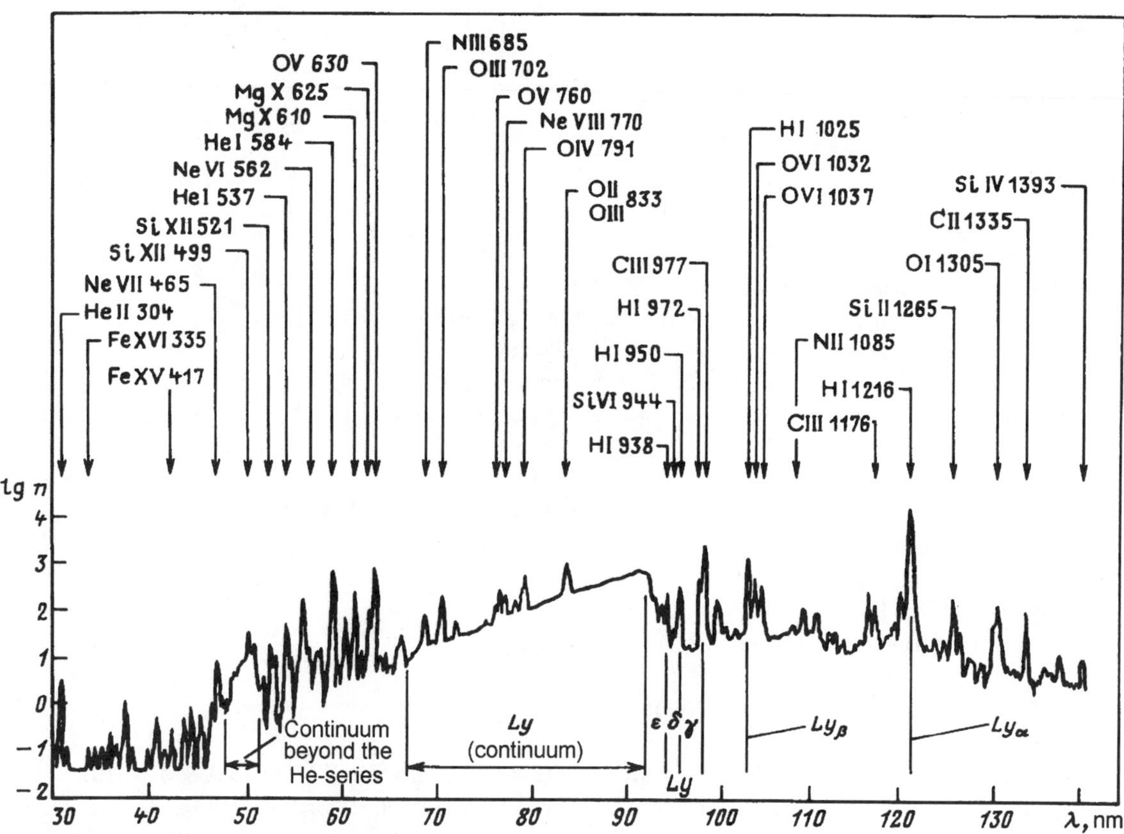

Figure 45.9 Spectrum of the Sun in the ultraviolet range [5]: n is the number of pulses per 0.08 s.

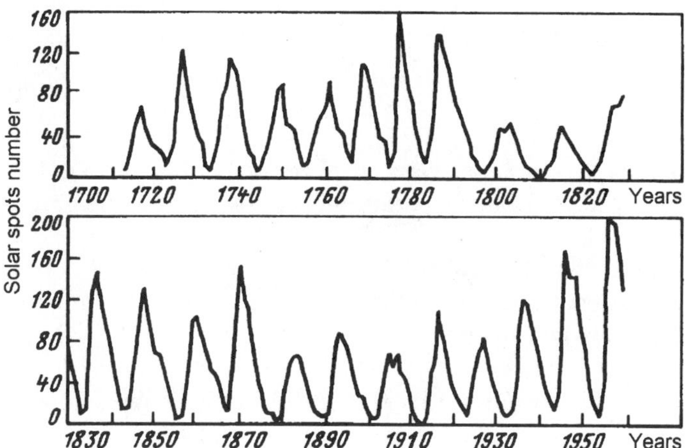

Figure 45.10 Variation of the Wolf's number with time [5].

45.3 Planets and Satellites, and Interplanetary Medium [1]

Basic information on the planets of the solar system (see also Tables 45.7 – 45.8, Figs. 45.11, 45.12) can be shown as follows:

mass of the Earth, $M_{\oplus} = 5.976(4) \cdot 10^{27}$ g;

radius of the Earth's equator, $R_\oplus$=6378 km (about the Earth's figure see Chapter 44);
total mass
of the solar system planets is 447.8 $M_\oplus$;
of the planet satellites is 0.12 $M_\oplus$;
of asteroids is $3 \cdot 10^{-4}$ $M_\oplus$;
of meteorite and comet material is 10^{-9} $M_\oplus$;
of the solar planet system is 448.0 $M_\oplus$;

The Moon's parameters [1]:
average distance from the Earth is 384 401(1) km;
minimal and maximal distances from the Earth are 356400 – 406700 km;
sidereal period (relative to remote stars) is 27.322 days;
synodic period (between two subsequent new moons) is 29.531 days;
inclination angle of the orbit to the ecliptic is $5°8'43''$, variations of $\pm 9'$ are observed with a period of 173 days;
eccentricity of the orbit is 0.0549;
angle of the equator inclination to the ecliptic is $1°32'30''$, and to the orbit, $6°41'$;
average radius is 1738.2 km;
mass is $7.350 \cdot 10^{22}$ kg;
average density is $3.341 \cdot 10^3$ kg/m^3;
acceleration of free fall at the surface is 1.622 m/s^2;
escape velocity at the surface is 2.38 km/s;
visual stellar magnitude at the full Moon is -12.73;
albedo is 0.067.

Parameters of the planets' satellites and rings are presented in Tables 45.9 and 45.10.

Asteroids [3, 15]. About 3000 asteroids (Table 45.11) with well-defined orbits exist, with 98% of them moved between Mars's and Jupiter's orbits, thus forming a belt of asteroids. Parameters of their orbits are:
major semiaxes range 2.2–3.2 a.u.;
periods of orbiting are 3–9 years with lags near 4.0, 4.8, and 5.9 y, i.e. 1/3, 2/5 and 1/2 of Jupiter's period;
eccentricities are 0–0.2;
inclination to ecliptic is 5–15°.

Comets [3, 16]. A source of comets in the solar system is the Oort cloud remote at 10^4–10^5 a.u. from the Sun. The cloud contains about 10^{11} comet cores. Comets' parameters are (see also Table 45.12):
size of a core is 0.5–20 km;
size of the comet's head is 10^4–10^6 km at the distance of about 1 a.u. from the Sun;
tail is 10^6–10^7 km in length at a distance of about 1 a.u. from the Sun;
distance from the Sun at which the tail appears is 4–6 a.u.;
mass is 10^{11}–10^{16} kg.

Interplanetary medium. Parameters of the solar wind nearby the Earth's orbit are as follows [3, 18]:
velocity, 400–700 km/s;
temperature, $5{\cdot}10^4$–$5{\cdot}10^5$ K;
magnetic induction, 10^{-8}–10^{-9} T (Fig. 45.14);
number density, 1–10 cm^{-3};
mass flux, 10^{11}–10^{13} g/s;
kinetic energy flux, 10^{19} W.

Heliopause (a boundary between the solar wind and the interstellar medium, Fig. 45.15) is located at about 200 a.u. from the Sun [19].

The solar wind impinges on the Earth's atmosphere in a manner illustrated by Fig. 45.16. The boundary of magnetosphere–magnetopause passes through the subsolar point at the distance of 9–11 $R_\oplus$ from the Earth. Magnetic force lines binding the Earth and the solar wind are formed by the reconnection of the magnetic field lines in the magnetospheric region between the Sun and the Earth. They are carried away by the wind forming the magnetospheric tail $\approx 30R_\oplus$ in diameter. Partially, the plasma of a solar wind flows into magnetosphere, forming the plasma mantle with a particle concentration of 0.1–5 cm^{-3} and magnetic field of (2–3)$\cdot 10^{-8}$ T. The difference of electric potentials transverse the magnetospheric tail is 10–100 kV. Magnetic fields in northern and southern parts of the magnetospheric tail are oppositely directed. At a distance of $\approx 100R_\oplus$, reconnection of the magnetic force lines takes place and then magnetic force lines of the solar wind separate those of the Earth. The magnetic lines with plasma kept by them drift towards the Earth, flow about both Earth's sides and come back to the subsolar side of magnetopause. The total cycle takes about 3–6 hours during which an energy of 10^{10}–10^{11} J is injected into magnetosphere. At a distance up to 4–5 $R_\oplus$, magnetic force lines of the Earth, with the cooled plasma of ionospheric origin kept by them, rotate together with the Earth to form plasmosphere. Concentration of particles in the latter is 10^2–10^3 cm^{-3}; magnetic field intensity is 10^{-7}–10^{-5} T.

Size of the ionized hydrogen zone around the Sun is 5.5 and 20 a.u. along the direction of the Sun motion and in the opposite direction, correspondingly [20]; in the direction perpendicular to the Sun motion, the size of the zone runs to 10 a.u. The Sun moves with respect to the interstellar medium with a velocity of 20 km/s in the direction of $\alpha = 252°$, $\delta = -15°$.

Interstellar dust forms a disk of about 3 a.u. in radius in the plane of the ecliptic [3]; mass of the dust grains is $10^{-3} - 10^{-5}$ g. The total mass of the dust in the solar system is 10^{19}–10^{20} g. Inclination angles of the dust grain orbits to the ecliptic do not exceed 30–40°.

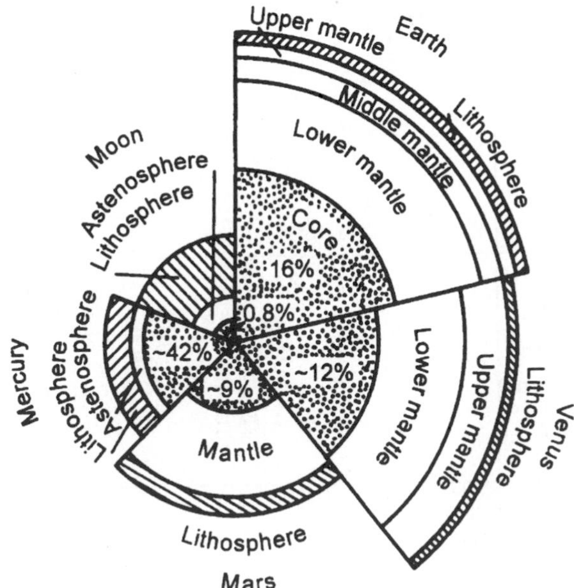

Figure 45.11 Internal structure of the planets of the Earth's group. The relative volumes of the cores are given in percent [13].

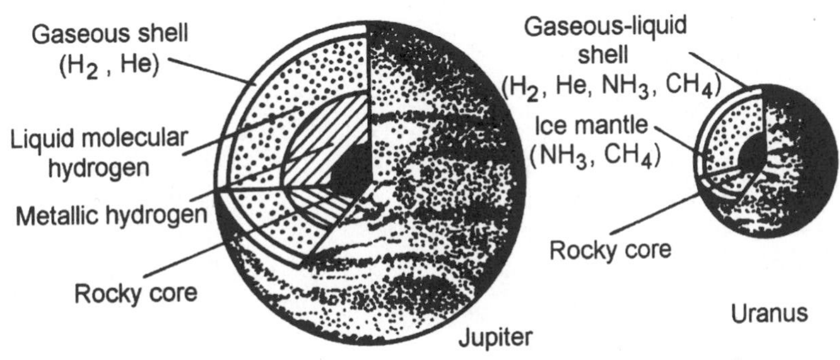

Figure 45.12 Models of the Jupiter and Uranus internal structure [13].

Table 45.7 Characteristics of the planets and their orbits [11]

Planet	Major orbital semiaxis, a.u.	Orbital period	Orbital inclination to ecliptic, deg.	Orbital eccen-tricity	Rota-tional period	Inclination of the rotational axis to the orbital plane, deg.	Magnetic moment [12], A·m^2	Mass, $M_\oplus$	Radius, $R_\oplus$	Mean density, g/cm^3	Acceleration of free fall at the surface, cm/s^2	Albedo*
Mercury	0.387	87.969 d	7.00	0.206	58.7 d	$\sim$0	(2–6)·10^{25}	0.055	0.382	5.4	370	0.06
Venus	0.723	224.701 d	3.39	0.007	243 d	−2	<3·10^{24}	0.815	0.949	5.3	887	0.75
Earth	1.00	365.256 d		0.017	23.9 h	23.5	8·10^{28}	1.00	1.00	5.5	980	0.36
Mars	1.52	1.88 y	1.85	0.093	24.6 h	24.0		0.107	0.553	3.9	370	0.24
Jupiter	5.20	11.9 y	1.30	0.049	9.92 h	3.1	1.5·10^{33}	318	11.2	1.3	2490	0.50
Saturn	9.54	29.5 y	2.49	0.056	10.7 h	29	4.7·10^{31}	95.2	9.45	0.7	1050	0.76
Uranus	19.2	84.0 y	0.77	0.047	23.9 h	−82.1		14.6	4.10	1.2	852	0.62
Neptune	30.1	165 y	1.77	0.009	17.8 h	28.8		17.2	3.88	1.6	1117	0.50
Pluto	39.4	248 y	17.2	0.249	6.39 d	$\geq$50		$\sim$0.002	$\sim$0.24	0.8	$\sim$300	0.09

* Portion of the light reflected by the whole planet surface in reference to the incident light.

Table 45.8 Main characteristics of the planet atmospheres [13] [See also Fig. 45.13]

Characteristics	Earth	Mercury	Venus	Mars	Jupiter*
Chemical composition (volume content,%)	N_2 (78); CO (10^{-5}); O_2 (21); CH_4 (10^{-4}); Ar (0.93); H_2 (5·10^{-5}); H_2O (0.1–1); Ne (10^{-3}); CO_2 (0.03); He (10^{-4})		CO_2 (95); HCl (4·10^{-5}); N_2 (3–5); HF (10^{-6}); Ar (0.01); O_2 (<5·10^{-4}); H_2O (0.1–1); SO_2 (10^{-5}); CO (3·10^{-3}); H_2S (8·10^{-3})	CO_2 (95); N_2 (2–3); Ar (1–2); H_2O (10^{-3}–10^{-1}); O_2 (0.1–0.4)	H_2 (87); HCl (10^{-5}); He (12.8); C_2H_6 (4·10^{-2}); H_2O (10^{-4}); C_2H_2 (8·10^{-3}); CH_4 (7·10^{-2}); PH_4 (4·10^{-5}); NH_3 (2·10^{-2}); CO (2·10^{-7})
Mean molecular mass, a.m.u.	28.97		43.2	43.5	2.25
Temperature at the surface (for average latitudes), K	T_{max} =310 T_{min} =240		735 735	270 200	135 –
Mean pressure nearby the surface, Pa	10^5	<2·10^{-9}	9·10^6	6·10^2	5·10^7
Mean density nearby the surface, g/cm^3	1.27·10^{-3}	<10^{-17}	61·10^{-3}	1.2·10^{-5}	10^{-4}

* Jupiter has no solid surface. The data are presented for the lower boundary of the stratosphere.

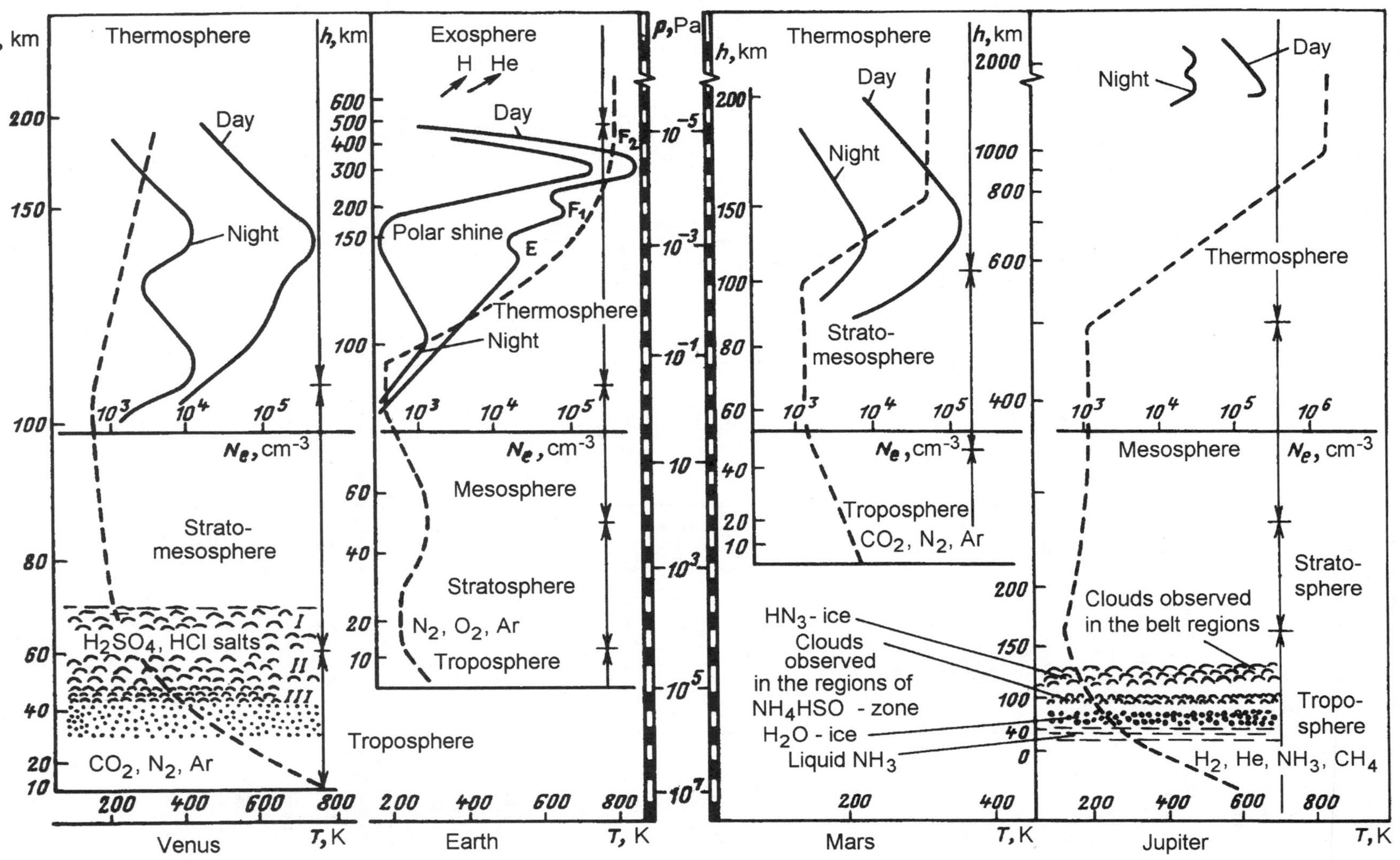

Figure 45.13 Structure of Venus, Earth, Mars, and Jupiter atmospheres. The profiles of temperature T (dashed lines) and electron concentration N_e (solid lines) are shown. The height h above the planet's surface and the pressure p are plotted along the vertical axis [13].

Table 45.9 Characteristics of planetary satellites and their orbits [11]

	Satellite	Major orbital semiaxis, km	Orbital period, d	Eccentricity	Orbital inclination to the ecliptic, deg.	Mass, M_{Moon}	Radius, km
	Satellites of Mars						
	Phobos	9.380	0.319	0.018	1.0	$1.3 \cdot 10^{-7}$	13 (1)
	Deimos	23.500	1.26	0.002	2.0	$2.7 \cdot 10^{-8}$	7.5 (1)
	Satellites of Jupiter						
	(Adrasthea)	128000	0.297	~0	~0		~20
	(Methys)	127000	0.295	~0	~0		~20
	Amaltea	181000	0.489	0.003	0.4		135 (1)
	Theba	221000	0.670	0	0		40
Galileo's satellites	Io	422000	1.77	small, variable	0.0	1.2	1816
Galileo's satellites	Europe	671000	3.55	small, variable	0.0	0.66	1563
Galileo's satellites	Ganimed	1070000	7.16	0.001*1	0.2	2.0	2638
Galileo's satellites	Kallisto	1880000	16.7	0.01	0.2	1.5	2410
	Leda	11100000	240	0.146*1	26.7		~5
	Gamalia	11500000	251	0.158*1	27.6		~90
	Lysiphoia	11700000	260	0.130*1	29.0		~10
	Elara	11700000	260	0.207*1	24.8		~40
	Ananke	20700000	617*2	0.17*1	33		~10
	Karma	22400000	692*2	0.21*1	16		~15
	Paciphea	23300000	735*2	0.38*1	35		~20
	Synopa	23700000	758*2	2.38*1	27		~15
	Satellites of Saturn						
	(Athlas)	138000	0.602	0.002	0.3		20 (1)
	S27	139000	0.613	0.004	0.0		70 (1)
	S26	142000	0.629	0.004	0.1		55 (1)
	Epimetheus	151000	0.694	0.009	0.3		70 (1)
	Yanus	152000	0.695	0.007	0.1		110 (1)
	Mimas	186000	0.942	0.020	1.5	~0.0005	196
	Encelad	238000	1.37	0.004	0.0	~0.001	255
	Tephia	295000	1.89	0.000	1.1	~0.01	530
	Telesta	295000	1.89				17 (1)
	Kalliposo	295000	1.89				17 (1)
	Diona	377000	2.74	0.0022	0.0	0.014	560
	S6 (Diona B)	377000	2.74	0.005	0.2		18 (1)
	Rhea	527000	4.52	0.001	0.4	0.034	765
	Titan	1220000	16.0	0.029	0.3	1.8	2575
	Hyperion	1480000	21.3	0.104	0.4		205 (1)
	Yapeth	3560000	79.3	0.028	14.7*1	0.026	730
	Pheba	13000000	550*2	0.163	30		110
	Satellites of Uranus						
	Miranda	130000	1.41	0.000	~0	0.0005	~160
	Ariel	192000	2.52	0.003	~0	0.03	~705
	Umbriel	267000	4.14	0.004	~0	0.01	~580
	Titania	438000	8.71	0.002	~0	0.04	~845
	Oberon	586000	13.5	0.001	~0	0.04	~760
	Satellites of Neptune						
	Triton	355000	5.88*2	0.000	20	0.8	1600
	Nereida	5560000	360	0.75	28	$2 \cdot 10^{-8}$	470
	Satellite of Pluto						
	(Charon)	~17000	6.39	~0		~0.02	~400

*1 Variable orbital eccentricity.

*2 Reversal rotation.

Note: in parentheses are not commonly accepted names.

Table 45.10 Characteristics of rings around planets [13, 14]

Planet	Outer radius, km	Width, km	Thickness, km	Number of rings
Jupiter	126000	600	1	1
Saturn	137000	60000	1–2	5
Uranus	56000	variable		9

Table 45.11 Parameters of some asteroids and their orbits [1,15]

Name	Radius, km	Mass, kg	Rotational period	Orbital period, d	Major orbital semiaxis, a.u.	Eccentricity	Orbital inclination to the ecliptic, deg.
Cerera	500	$1.2{\cdot}10^{21}$	9 h 05 min	1681	2.766	0.079	10.6
Pallada	304	$2.3{\cdot}10^{19}$	10 h	1684	2.768	0.235	34.8
Junona	123		7 h 13 min	1594	2.668	0.256	13.0
Vesta	269	$2.4{\cdot}10^{19}$	5 h 20 min	1325	2.362	0.088	7.1
Heba	100		7 h 17 min	1380	2.426	0.203	14.8
Iris	104		7 h 07 min	1344	2.386	0.230	5.5
Hygia	225		18 h	2042	3.151	0.099	3.8
Eunomia	136		6 h 05 min	1569	2.643	0.185	11.7
Psychea	125		4 h 18 min	1826	2.923	0.135	3.1
Icarus	0.7		2 h 16 min	408	1.078	0.827	23.0

Table 45.12 Characteristics of some short-periodical comets [1,16]

Comet	Perihelion passage		Period,	Orbital inclination	Eccentricity	Perihelion	Major orbital
	the last, y	number of returns	y	to the ecliptic, deg.		distance, a.u.	semiaxis, a.u.
Halley	1986	30	76.1	162	0.97	0.59	17.8
Enke	1994	56	3.31	12	0.85	0.34	2.21
Tempel-2	1994	19	5.26	12	0.55	1.37	3.0
Olbers	1956	3	69	45	0.93	1.20	16.8
Krommelin	1984	5	27.9	29	0.92	0.74	9.2
Pons-Bruks	1954	3	71	74	0.96	0.78	17.2

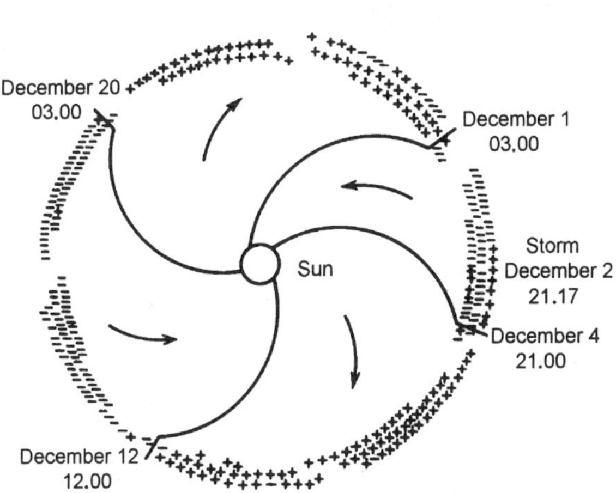

Figure 45.14 The sectoral structure of interplanetary magnetic field in the ecliptic plane. (Minuses and pluses show the field oriented toward the Sun and in the opposite direction, respectively [18].)

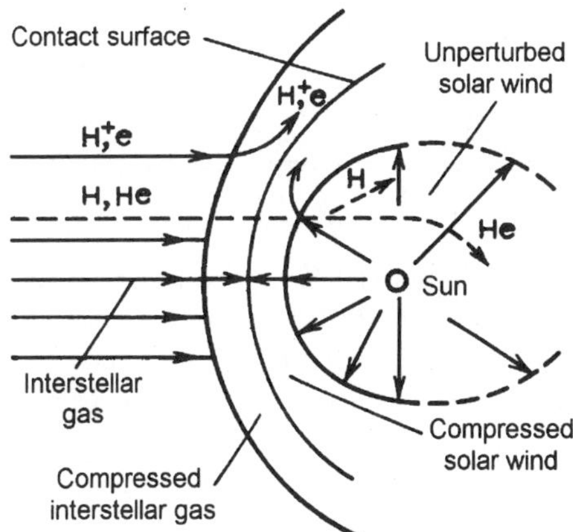

Figure 45.15 Interaction between the solar wind and interstellar matter; solid lines follow the trajectories of protons and electrons of interstellar gas, dashed lines are the trajectories of neutral atoms, thick lines denote external and internal shock waves [19].

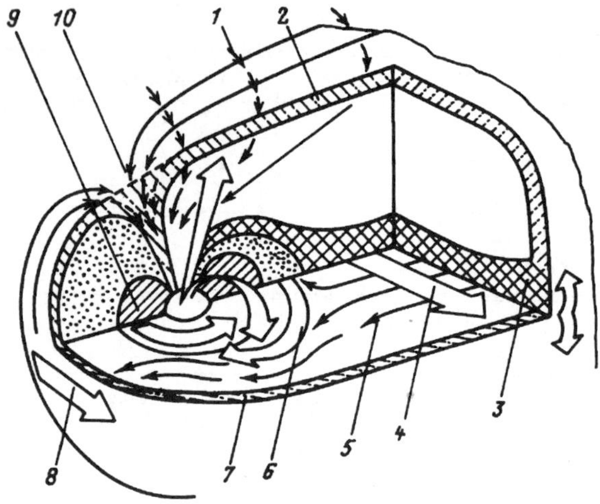

Figure 45.16 The solar wind flowing around the Earth's magnetosphere [17]:
1 – interplanetary magnetic field; *2* – mantle of plasma; *3* – plasma layer; *4* – current across the tail; *5* – plasma convection; *6* – ring current; *7* – magnetopause; *8* – current in the magnetopause; *9* – plasmosphere (rotates with the Earth); *10* – cusp.

45.4 Stars

Basic spectral classes (types) of stars are presented in Table 45.13 that follows.

Table 45.13 Spectral classification of stars [1–3]

Spectral class	Characteristics of the class	Surface temperature, K	Color of the star (B–V)
O	hot stars with absorption lines of He II	30000–50000	blue (-0.3^m)
B	absorption lines of He I (H lines enhance towards class A)	12000–30000	slightly blue-white (0.0–0.3^m)
A	H lines reach maximum intensity and then attenuate; lines of Ca II enhance	7600–11000	white (0.2–0.0^m)
F	CaII lines enhance, H lines attenuate, metal lines are developing	6000–7600	yellow-white (0.6–0.3^m)
G	intensive lines of Ca II and other metals; H lines attenuate	5000–6000	yellow (0.8–0.6^m)
K	intensive lines of metals; absorption bands of CH and CN appear	4000–5000	orange (1.4–0.8^m)
M	intensive bands of TiO	2500–4000	red (2.0–1.4^m)

Additional spectral classes to those listed above are:

S distinguished by the presence of the ZrO absorption bands in their spectra and related to K-stars by their physical parameters;

N, R distinguished by the presence of absorption bands of carbon, carbon oxide, and cyan molecular compounds; they correspond to basic spectral classes K and G, respectively;

Q Novae;

W hot Wolf-Rayet stars with wide emission lines.

Each spectral class is divided into 10 subclasses and is identified by a number 0 to 9 (moving down in temperature) placed after the letter. Roman numeral following the type designation denotes a luminosity class of the star: I — supergiants, II — bright giants, III — giants, IV — subgiants, V — main sequence, VI — subdwarfs, VII — white dwarfs (Tables 45.14, 45.15).

Hertzsprung-Russel diagram [21] (Fig. 45.17) relates luminosity of the stars to their spectral classes. Each type of star is into a definite zone in this diagram. The majority of stars belong to the *main sequence.* These are the stars with a thermonuclear reaction H→He being a source of their energy. Minimal mass needed for the thermonuclear reaction onset in the stellar interior comes to

0.085 $M_\odot$ [22]. The main reaction in the stars with mass $M \leq M_\odot$ is a *pp*-cycle (see Chapter 39). Under the conditions typical for interior of a star, the rate of energy release (in W/kg) in the *pp*-reaction takes the form

$$\epsilon_{pp} = 2.50\rho X^2 T_6^{-2/3} \exp\left(-33.8T_6^{-1/3}\right), \tag{45.5}$$

where ρ is the density, g/cm^3; X is the hydrogen mass fraction; T_6 is the temperature of the matter in 10^6 K. In the stars with mass $M > M_\odot$, the double CNO-cycle serves as a basic energy source (see again Chapter 39) with the rate of energy release

$$\epsilon_{\text{CNO}} = 9.5 \cdot 10^{24} \rho X X_{\text{CN}} T_6^{-2/3} \exp\left(-152.3T_6^{-1/3}\right), \tag{45.6}$$

where X_{CN} is the mass fraction of carbon and nitrogen.

Once hydrogen in the core is depleted, the hydrogen burning begins in the shell surrounding the core, followed by the subsequent burning of helium, carbon, and some other elements. At these stages, the size and luminosity of the star increase, resulting in star displacement to the top right part of the Hertzsprung-Russel diagram. Stars with the shell source of energy lie in the *red giants* region. Stars of moderate masses (near $M_\odot$) with helium burning in their core fall on a *horizontal branch.* At late evolutionary stages, the stars lose intensively their masses. Once the nuclear fuels in the star are depleted, the remnant of the star may become a white dwarf, a neutron star or a black hole depending on its mass.

Characteristics of different types of stars are presented in Table 45.16 and in Figs. 45.18, 45.19.

Binary and multiple stars. Among a hundred stellar systems there are 30 single stars, 47 binary systems (94 companions), and 23 multiple stars (81 companions) [1].

Variable stars. In addition to *eclipsing variable stars* making binary systems whose brightness varies owing to periodical eclipses of one companion by another, there exist many other types of physically variable stars. The most numerous among them are *pulsating stars* (Table 45.17).

A period-luminosity relationship exists for classical Cepheids [2]:

$$M_V = -1.67 - 2.54 \lg P, \quad M_B = -1.33 - 2.25 \lg P, \tag{45.7}$$

where M_V and M_B are the absolute stellar magnitudes in optical and blue spectral ranges, respectively; P is a pulsating period in days. This dependence is used in distance determination of nearby galaxies.

Cataclysmic variables (Table 45.18) make close binaries (their orbital periods range from 1 to 10 hours) with one of the companions being a white dwarf and another being a normal star [24,25]. An instability induced by accretion (mass transfer from a normal star to a white dwarf) results in flashes of optical radiation. In the case of classical Novae, it is a thermal instability that leads to a thermonuclear explosion of hydrogen accumulated on the white dwarf's surface due to accretion. Cataclysmic variables are the X-ray sources (luminosity ranges from 10^{24} to 10^{26} W in the steady state).

Parameters of some cataclysmic variables are listed in Tables 45.19, 45.20.

The stars described just below relate to other types of variable stars.

T *Taurus stars* [27, 28] denote stars of mass 0.5–3 $M_\odot$ that have not yet reached the main sequence. These stars undergo irregular variations of their brightness up to a few stellar magnitudes; they correspond to spectral classes from M to G with strong emission lines; their luminosities fall in the range of 0.3–30 $L_\odot$; mass loss rate is 10^{-7}–10^{-9} $M_\odot$/yr.

UV *Cetus stars* [1, 29] are said to be flaring dwarf stars of spectral class M with masses 0.1–0.5 $M_\odot$; flares occur irregularly with a characteristic frequency of about 1 per day and with a brightness growth taking about 1 min; flare lasts about 20 min, and its total energy comprises 10^{25} J as of order.

R *Corona Borealis stars* [1] are referred to as supergiants of F–K and R spectral classes; they undergo nonperiodical decrease of brightness by 1–9^m lasting from tens to a few hundred days.

Symbiotic type stars [30] denote close binaries composed of the cold and hot stars, chaotically varying their brightness by 1–3^m during about one year (sometimes during tens of years).

Peculiar stars [3]. Some of them are outlined below.

Wolf-Rayet stars denote highly luminous (of order $10^6 L_\odot$) stars with very bright and broad emission lines; they are distinguished by the presence in their spectra of both the lines of highly ionized elements ($T \approx 10^5$ K) and a rather low-temperature continuum ($T \approx (1\text{–}2) \cdot 10^4$ K). They are massive (about 10 $M\odot$) stars at the final stages of strong stellar evolution rapidly loosing their masses (10^{-4}–10^{-5} $M\odot$/year) in the form of a strong stellar wind. About 300 of these objects are known in our Galaxy and in the nearby galaxies.

Magnetic stars possess very strong (10^{-2}–1 T) magnetic fields. All of them have an abnormal chemical composition (apparently only in the outer layers): abundance of helium is low, whereas abundance of heavy elements (Si, Cr, Mn, Sr, Eu, Gd, etc.) is abnormally high. The element excess grows in average with increase of the element atomic number and reaches 10^4–10^6 for rare-earth elements. Magnetic stars fall on the main sequence within B–F range of spectral classes; 10–15% of the stars from this part of the diagram belong to them.

Supernovae [31]. Evolution of the star may end as a supernova explosion; that is why explosion of a star results either in complete destruction of the star and dispersion of its material or in the gravitational collapse of the stellar core and ejection of its outer layers. A neutron star or a black hole may be formed after the collapse; binding energy of the core, $\sim 10^{46}$ J, is carried away by neutrinos in $\sim$ 100 s. The matter ejected in the explosion moves away with a velocity of 10,000–20,000 km/s, carrying away an energy of $10^{44} - 10^{45}$ J. An expanding shell is formed, whose radiation is observed as the supernova explosion (Fig. 45.20). Luminosity of a supernova reaches 10^{35}–10^{37} W at the brightest stage. Radiation is provided by the thermal energy stored in the shell, as well as by the decay of radioactive elements (products of the explosion). The element ^{56}Ni is mostly formed, which is converted into ^{56}Co (a half-life period is 6.1 day) and then into ^{56}Fe (78.8 day).

Rate of supernova explosions runs to approximately 0.05 per year in our Galaxy. Most of these explosions are not observed due to absorption of the radiation by the interstellar dust. During all the history of astronomical observations, only four supernovae have been indicated in the Galaxy: in 1006, 1054, 1572, and in 1604. Approximately 500 supernova explosions have occured in other galaxies.

The richest information has been gleaned from the supernova explosion occurring in the nearest galaxy, the Large Magellanic Cloud, on February 23, 1987 (Fig. 45.21) [7]. At first, a neutrino flash was detected: 20 neutrinos with energies from 6 to 39 MeV were recorded for 12 s. The total energy of the explosion of this supernova amounts to $(1.4 \pm 0.6) \cdot 10^{44}$ J, mass of the ejected shell is 10–15 $M_\odot$. At the stage of the brightness decrease, radiation of the shell was provided by radioactive decay of 0.075 $M_\odot$ of ^{56}Co. Gamma-lines of ^{56}Co were measured at energies 0.843 and 1.238 MeV, as well as softer X-ray continuum radiation resulting from multiple Compton scattering of gamma quanta by electrons of the shell (Fig. 45.22). In the old photoplates at the supernova location, a blue B3 supergiant was revealed. Its luminosity and radius reached $4 \cdot 10^{31}$ W and $3 \cdot 10^{10}$ m, correspondingly.

Final stages of stellar evolution [33]. Once a star has lost (ejected) a part of its mass during evolution (this may result from a supernova explosion) and its nuclear fuel has been depleted, the fate of the star depends on the mass of a collapsing remnant. *White dwarfs* denote stars in which the gravity force is balanced by the pressure of a degenerated electron gas. Their radiation is provided by the thermal energy stored in their interiors. White dwarf mass cannot exceed a value of $M = 1.46(2/\mu)^2 M_\odot$ (Chandrasekhar limit), where $\mu = A/Z$ is the molecular mass per one electron (for elements from He to Fe, $\mu \approx 2$). White dwarfs' radii are (10^6–10^7) m (Fig. 45.23), their luminosity equals $(10^{-2}\text{–}10^{-4})L_\odot$, and central densities are of the order 10^9 kg/m^3. The Zeeman line splitting evidences for the magnetic field induction of $B = 10^2$–10^3 T for some white dwarfs.

Within stars' mass range $0.1M_\odot < M < (2\text{–}3)M_\odot$, the equilibrium state constitutes a neutron star (Fig. 45.24). Typical parameters of neutron stars are as follows: radius gets to about 10 km, B reaches up to 10^6–10^9 T, gravitational red shift at the surface is $\Delta\lambda/\lambda \approx 10\%$, the central density is (10^{17}–10^{19}) kg/m^3, the moment of inertia is of order 10^{38} kg$\cdot$ m^2. The upper limit of neutron stars' mass is determined by an unknown equation of state at superhigh densities, but this limit does not certainly exceed 3.2 $M_\odot$. Stellar remnants of larger masses inevitably collapse into *black holes.*

Rotating neutron stars with superstrong magnetic fields may appear as *radiopulsars* [35, 36], i.e. powerful sources of the strictly periodical radiopulses whose period coincides with a rotational period of the neutron star (Table 45.21). Radioemission exhibits a power-like spectrum (Fig. 45.25). The energy source for pulsars is the rotational energy of a neutron star, so the periods of all the pulsars increase. More than 500 pulsars are now known.

Compact stars entering the close binary systems may manifest themselves as *X-ray sources* [33]. An acceleration serves as an energy source in such stars, that is a mass transfer from the normal to the compact star. Luminosity L (in W) of accreting sources is connected with the accretion rate $\dot{m}$, $10^{-8} M_\odot$/year by the relationship

$$L = 6 \cdot 10^{31} \xi \dot{m}; \qquad (45.8)$$

$\xi \approx 10^{-4}$ for white dwarfs, $\xi = 0.1$–0.2 for neutron stars, and $\xi = 0.06$–0.4 for disk-accreting black holes (depending on the angular momentum of the black hole). The complete classification of X-ray sources does not exist yet. Selected types of sources were specified. Once accretion onto a neutron star with a magnetic field $B > 10^6$ T occurs, the matter falls into a region of magnetic poles of the star. X-ray emission of the neutron star is modulated by its axial rotation. Such sources are named *X-ray pulsars.* There are 20 known X-ray pulsars with periods ranging from 0.1 to 1000 s.

Bursters are the X-ray sources in which bursts of X-ray radiation occur nonperiodically with time intervals from a few hours to one day against the background of a stationary X-ray emission with a luminosity of about 10^{29} W. Growth and decay of the burst last correspondingly 1 s and 3–100 s; the maximal luminosity may be of order 10^{31} W. About 30 such sources have been known. Bursts result from thermonuclear explosions of the matter accumulated on the surface of a neutron star during accretion.

Transient (nova-like) sources are the systems where accretion is not permanent, thus the source appears and disappears from time to time within intervals from a few months to a few years. It may be connected with an ellipticity of the orbit of a relativistic companion of a binary or by pulsations of the normal companion resulting in large variations of accretion rate. Occasionally, X-ray pulsars and bursters are among transients.

Parameters of some of the X-ray sources are listed in Tables 45.22, 45.23; Figs. 45.26–45.28 demonstrate their typical spectra.

Cosmic gamma-bursts are characterized by the following parameters [44] (see also Fig. 45.29):
the total energy recorded during the full time of the burst is $3 \cdot 10^{-9}$–$1.5 \cdot 10^{-5}$ J/m^2;
energy flux density at the maximum is $5 \cdot 10^{-8}$–$4 \cdot 10^{-7}$ J/(m^2·s);
burst duration is 0.1–100 s;
number of separate peaks (pulses) in the burst is 1–5; sometimes it may be more;
period of the first pulse rise is 10^{-3}–3 s.

Table 45.14 Characteristics of some bright stars [1]

Star	Visual stellar magnitude	Absolute visual stellar magnitude	Spectral and luminosity classes	Distance, pc
Polaris (α Ursae Minoris)	2.3*	−4.6	F8I	240
Achernar (α Eridani)	0.48	−2.2	B5IV–V	39
Algol (β Persei)	2.2*	−0.3	B8V	32
Aldebaran (α Tauri)	0.85	−0.7	K5III	21
Capella (α Aurigae)	0.08	−0.6	G8+F	14
Rigel (β Orionis)	0.11	−7.0	B8I	250
Betelgeuse (α Orionis)	0.8*	−6	M2I	200
Canopus (α Carinae)	−0.73	−4.7	F0I	60
Sirius (α Canis Majoris)	−1.45	1.41	A1V	2.7
Procyon (α Canis Minoris)	0.35	2.65	F5IV	3.5
Spica (α Virginis)	0.96	−3.4	B1V	80
α Centauri	−0.1	4.3	G2V	1.33
Arcturus (α Bootis)	−0.06*	−0.2	K2*p*III	11
Antares (α Scorpii)	1.0	−4.7	M1I	130
Vega (α Lyrae)	0.04	0.5	A0V	8.1
Altair (α Aquilae)	0.77	2.3	A7V	5.0

* Variable star.

Table 45.15 Characteristics of the stars from different spectral classes [2]

Star	Spectral and luminosity classes	Effective temperature, K	Luminosity, $L_\odot$	Radius, $R_\odot$	Mass, $M_\odot$
α Scorpii A (Antares)	M0I	3300	34000	530	19
α Bootis (Arkturus)	K2III	4000	130	26	4.2
η Orionis	B1V	23000	13000	7.2	13.7
α Canis Majoris A (Sirius A)	A1V	9700	61	2.4	3.3
Barnard star	M5V	3000	0.015	0.50	0.38
α Canis Majoris B (Sirius B)	A5VII	8200	$2.6{\cdot}10^{-3}$	$2.6{\cdot}10^{-2}$	0.96

Table 45.16 Mass, radius, luminosity, and mean density depending on the spectral class of the star [1]

Spectral	$\lg(M/M_\odot)$			$\lg(R/R_\odot)$			$\lg(L/L_\odot)$			$\lg\bar{\rho}$ (g/cm^3)		
class	SG	G	MS	SG	G	MS	SG	G	MS	SG	G	MS
O5	+2.2		+1.6			+1.25			+5.7			−2.0
B0	+1.7		+1.25	+1.3	+1.2	+0.87	+5.4		+4.3	−2.1		−1.2
B5	+1.4		+0.81	+1.5	+1.0	+0.58	+4.8		+2.9	−2.9		−0.78
A0	+1.2		+0.51	+1.6	+0.8	+0.40	+4.3		+1.2	−3.5		−0.55
A5	+1.1		+0.32	+1.7		+0.24	+4.0		+1.3	−3.8		−0.26
F0	+1.1		+0.23	+1.8		+0.13	+3.9		+0.8	−4.2		−0.01
F5	+1.0		+0.11	+1.9	+0.6	+0.08	+3.8		+0.4	−4.5		+0.03
G0	+1.0	+0.4	+0.04	+2.0	+0.8	+0.02	+3.8	+1.5	+0.1	−4.9	−1.8	+0.13
G5	+1.1	+0.5	−0.03	+2.1	+1.0	−0.03	+3.8	+1.7	−0.1	−5.2	−2.4	+0.20
K0	+1.1	+0.6	−0.11	+2.3	+1.2	−0.07	+3.9	+1.9	−0.4	−5.7	−2.9	+0.25
K5	+1.2	+0.7	−0.16	+2.6	+1.4	−0.13	+4.2	+2.3	−0.8	−6.4	−3.4	+0.38
M0	+1.2	+0.8	−0.33	+2.7		−0.20	+4.5	+2.6	−1.2	−6.7	−4	+0.4
M2	+1.3		−0.41	+2.9		−0.3	+4.7	+2.8	−1.5	−7.2		+0.7
M5			−0.67			−0.5		+3.0	−2.1			+1.0
M8			−1.0			−0.9			−3.1			+1.8

Note: SG – supergiant, G – giant, MS – main sequence.

Table 45.17 Pulsating variable stars [23]

Type of the star	Period	Typical period	Population	Spectral class	Absolute stellar magnitude
RR Lyra	1.5–24 h	0.5 d	II	A2–F2	0.0–1.0^m
Classical Cepheid variable	1–50 d	5–10 d	I	F6–K2	(−0.5)–(−6^m)
W Virginis	2–45 d	12–20 d	II	F2–G6 (?)	0–(−3^m)
RV Taurus	20–150 d	75 d	II	G, K	−3^m
Red semiregular variables	100–200 d	100 d	I and II	(K), M, R,	(−1)–(−3^m)
Long-periodical variables	100–700 d	270 d	I and II	Me, Re, Ne, Se	(+1)–(−2^m)
β Cephei stars (β Canis Majoris)	4–6 h	5 h	I	B1–B2	(−3.5)–(−4.5^m)
Dwarf Cepheides and δ Scuti variables	1–3 h	2 h	I	A2–F5	(+2)–(+3^m)
Beat Cepheides (double periodical)	1–7 d	2 d	I (?)	F0–G0 (?)	(−1)- (−3^m) (?)
White dwarf variables (ZZ Cetus stars)	200–1000 s	500 s (?)	I (?)	A5–F5 (?)	(+10)–(+15^m) (?)

Table 45.18 Types of cataclysmic variables

Type	Amplitude of stellar magnitude variations during burst	Luminosity at maximum, Wt	Burst total energy, J	Burst duration, d	Period between bursts
Classical novae	9–(>14^m)	10^{31}	10^{38}–10^{39}	50–5000	~10^3 y
Recurrent novae	7–9^m	10^{31}	10^{35}–10^{36}	10–100	10–100 y
Dwarf novae:					
U Gem type	2–6^m	10^{27}	10^{31}–10^{32}	10	50–500 d
Z Cam type	2–5^m	10^{28}	10^{31}–10^{32}	10	10–50 d

Table 45.19 Selected galactic novae [1]

Star	Year of burst	Stellar magnitude			Absolute stellar magnitude at maximum	t_3*, d	Orbital period [26]
		before burst	at maximum	after burst			
T Aur	1891	>13.	4.0	14.8	−6.2	120	4 h 20 min
GK Per 2	1901	13.6	0.2	13.2	−8.3	12	45 h 20 min
V603 Aql	1918	10.6	−1.1	10.9	−8.4	7	3 h 20 min
DQ Her	1934	14.3	1.4	13.8	−6.2	105	4 h 39 min

* t_3 – time of a brightness decline after maximum by 3^m.

Table 45.20 Selected recurrent novae [1]

Star	Year of burst	Stellar magnitude		Absolute stellar magnitude at maximum	t_3*, d	Orbital period [26]
		at maximum	at minimum			
T Cr B	1866, 1946	2.1	10.6	−8.1	6	230 d
RS Oph	1898, 1933, 1958	4.3	11.6	−8.5	10	
WZ Sge	1913, 1946	7.3	15.9	−7.1	33	81.5 min
U Sco	1866, 1906, 1936	8.9	17.6	−7.6	6	

* t_3 – time of a brightness decline after maximum by 3^m.

Table 45.21 Characteristics of some radiopulsars [36]

Pulsar	Period P, s	Spin down time, $P/\dot{P}$, 10^6 y	Equivalent pulse width*1 at 400 MHz, ms	Flux density at 400 MHz, mJy	Distance, kpc	Dispersion measure, pc/cm^3	Energy loss rate*2, 10^{28} W
PSR 1937+24 [35]	$1.5578{\cdot}10^{-3}$	150	0.125*3		2.5		40
PSR 0531+21*4	0.0332	0.0024	1.9	480	2.0	56.7	4600
PSR 1913+16*5	0.0590	212	10	12	6.1	167	$2{\cdot}10^{-2}$
PSR 0833−45*6	0.0892	0.022	1.7	5000	0.5	69.0	67
PSR 0950+08	0.2530	34	9.5	500	0.1	2.9	$5{\cdot}10^{-3}$
PSR 0823+26	0.5306	10	6.0	41	0.7	19.4	$4{\cdot}10^{-3}$
PSR 0031−07	0.9429	74	42	27	0.4	10.8	$2{\cdot}10^{-4}$
PSR 1133+16	1.1879	10	30	350	0.1	4.8	$9{\cdot}10^{-4}$
PSR 1919+21*7	1.3373	32	25	56	0.5	12.4	$2{\cdot}10^{-4}$
PSR 0525+21	3.7454	3.0	75	93	2	50.9	$3{\cdot}10^{-4}$
PSR 1845−19	4.3081	5.8	66	15	0.7	19.1	10^{-4}

*1 Pulse energy relative to maximal flux density.
*2 Rotational energy loss rate $I\Omega\dot{\Omega}$ determined for a characteristic inertial moment of the neutron star I=10^{38} kg·m^2.
*3 Pulse width at half maximum.
*4 Pulsar in the Crab Nebula. Pulsing radiations in optical, X-ray and γ-bands are observed.
*5 Enters into close binary system, P_{orb}=27906.98 s.
*6 Vela pulsar. Pulsing radiations in optical and γ-bands are observed.
*7 The first pulsar discovered.

Table 45.22 Characteristics of some X-ray pulsars [40, 41]

Source	X-ray luminosity, 10^{30} W	Pulsation period P, s	Relative rate of period changes $\dot{P}/P$, y^{-1}	Binary orbital period, d	Distance, kpc
SMC X-1	50	0.71	$-6{\cdot}10^{-4}$	3.89	50
Hercules X-1	2.5	1.24	$-3{\cdot}10^{-6}$	1.7	5
Centaurus X-3	8–16	4.84	$-3{\cdot}10^{-4}$	2.087	8
X Perseus	$4{\cdot}10^{-4}$	835	$-2{\cdot}10^{-4}$	>40	0.35
Vela X-1	0.8–0.25	283	$-2{\cdot}10^{-5}$	8.97	1.4

Table 45.23 X-ray sources as candidates for black holes

Source	X-ray luminosity, 10^{30}W	Mass, $M_\odot$	Normal star of the binary system	Binary orbital period, d	Distance, kpc
Cygnus X-1 [38]	4–6	8–11	O9.7 I	5.6	2.5
LMC X-1 [39]	20	>3	O7	3.9	50
LMC X-3 [39]	20	7–14	B3	1.7	50

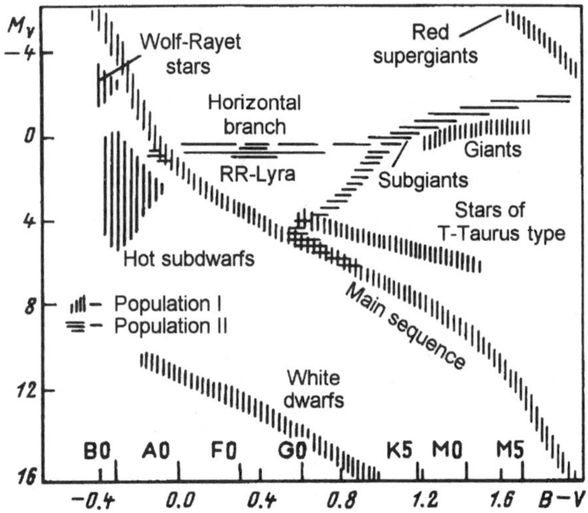

Figure 45.17 The Hertzsprung-Russel diagram [2].

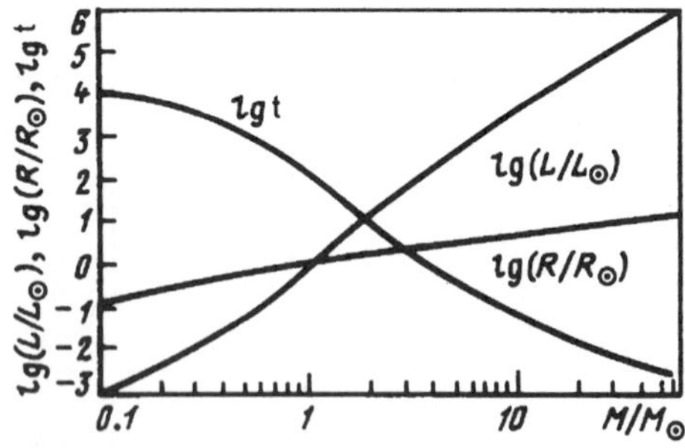

Figure 45.18 Dependence of the luminosity L and radius R of a main sequence star on its mass. The time $t = 10^9$ years corresponds to the main sequence lifetime of the star.

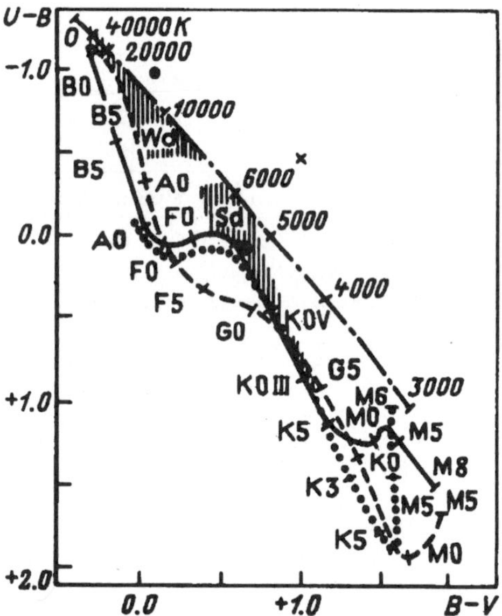

Figure 45.19 Relation between color indices $U-B$ and $B-V$ for main sequence stars (the solid line), supergiants (the dotted line), giants (dots). The hatched regions are occupied by white dwarfs (Wd) and subdwarfs (Sd). The dot-and-dash line corresponds to the black body radiation. Locations of the Sun ($\odot$), the quasar 3C273 ($\bullet$), and the X-ray source Cygnus X-1 ($\times$) are shown [3].

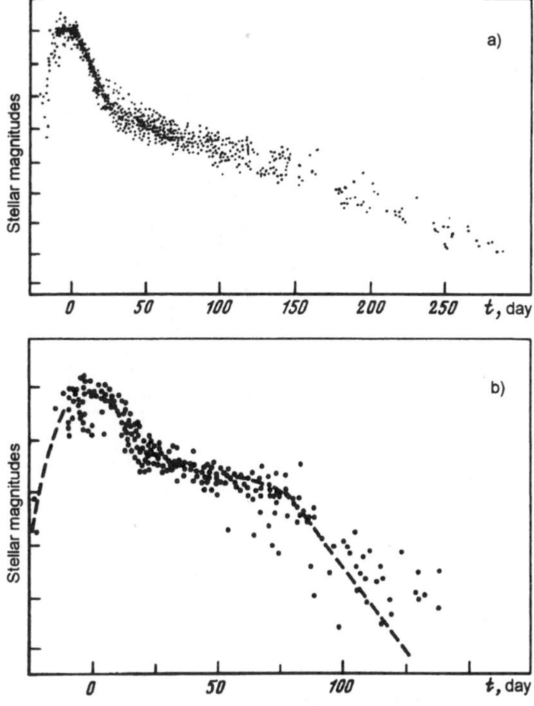

Figure 45.20 Combined brilliance curves of type I (**a**) and II (**b**) supernovae [32].

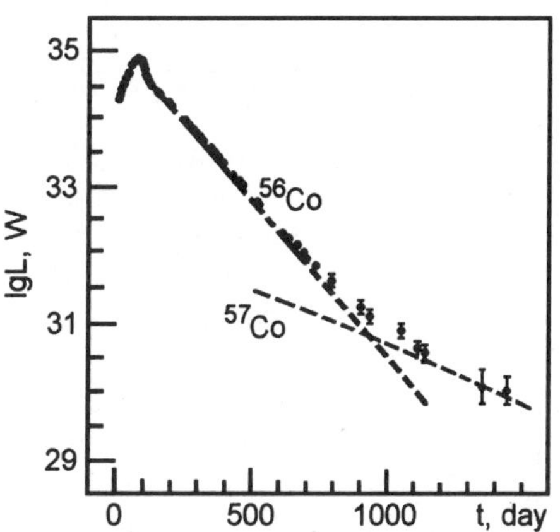

Figure 45.21 Dependence of the Supernova 1987a luminosity on time [7]. The dashed lines show the energy release resulting from radioactive decay of Co.

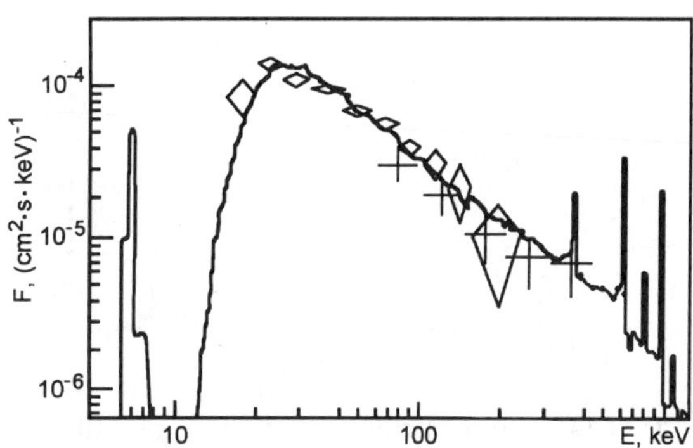

Figure 45.22 X-ray spectrum of the Supernova 1987a in 320 days after the explosion [9]. The polygons and crosses are the measured values, the solid line fits theoretical calculation of the spectrum generated during γ-radiation of radioactive Co transition across the supernova shell.

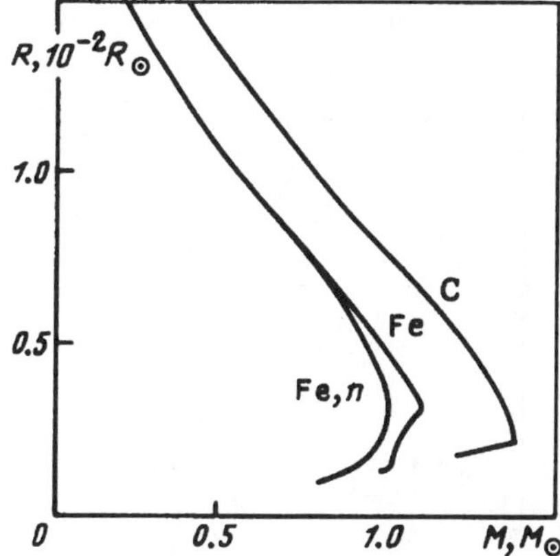

Figure 45.23 Mass-radius relation for white dwarfs. Calculations are shown for white dwarfs composed of carbon, iron and iron with account for neutronization [34].

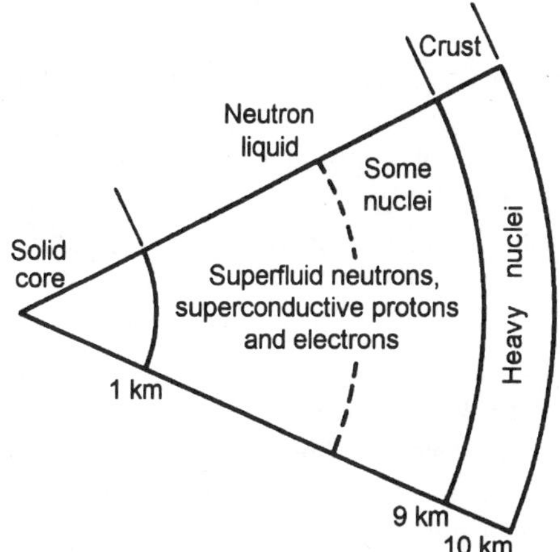

Figure 45.24 Schematics of a neutron star [35].

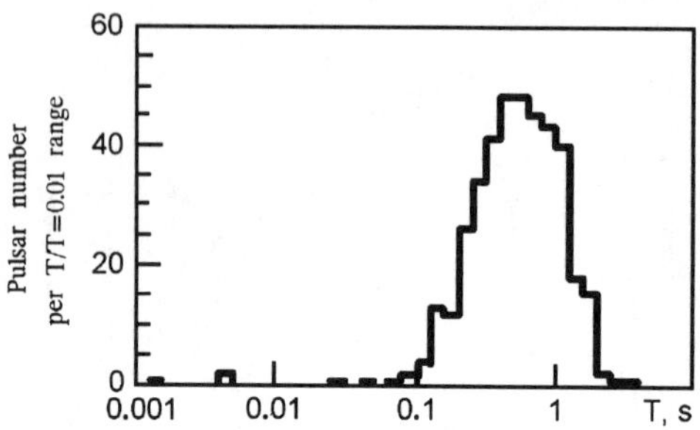

Figure 45.25 Distribution of the radiopulsars periods [90].

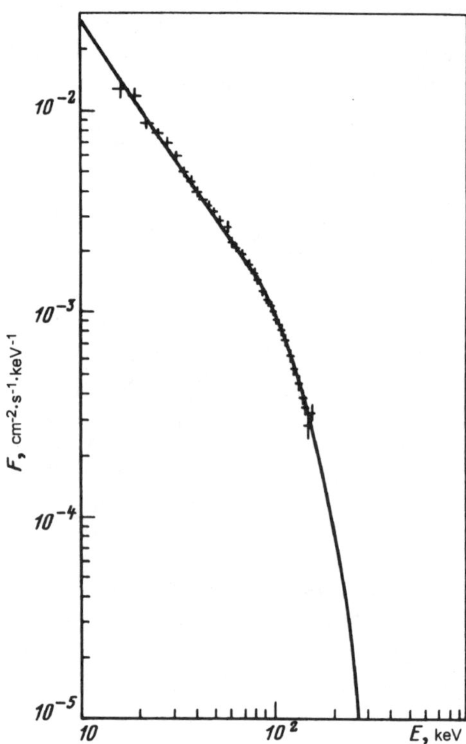

Figure 45.26 X-ray spectrum of the black hole candidate Cygnus X-1: the solid line represents the spectrum of a plasma layer with the temperature $T = 27$ keV and optical half-thickness relative to Thompson scattering $\tau = 2$ [42].

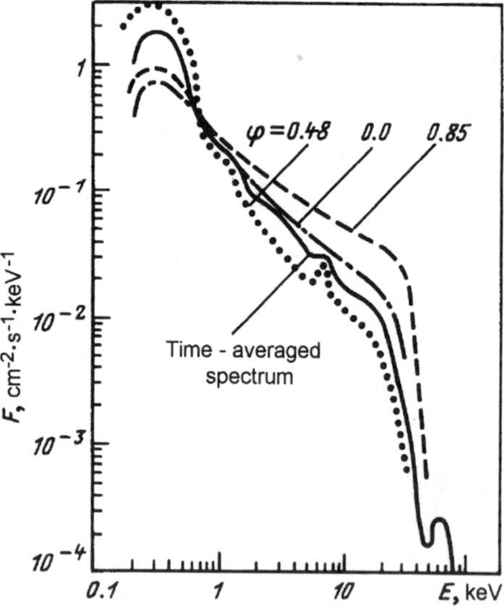

Figure 45.27 Spectrum of the X-ray pulsar Herculus X-1 at different phases Φ of the pulsation period. The spectral feature near $E = 7$ keV is a result of iron fluorescence. The feature near $E = 50$ keV corresponds to the cyclotron frequency of electrons in the magnetic field $B = 5 \cdot 10^8$ T [43].

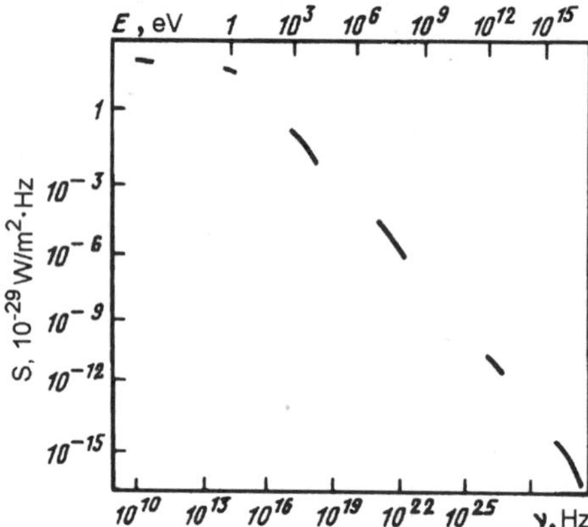

Figure 45.28 Cygnus X-3 spectrum. The source reveals its activity in all spectral ranges from radio- to gamma-radiation of ultrahigh energy; it enters a close binary system (with an orbital period 4.8 h) located at 10 kpc from the Sun [45].

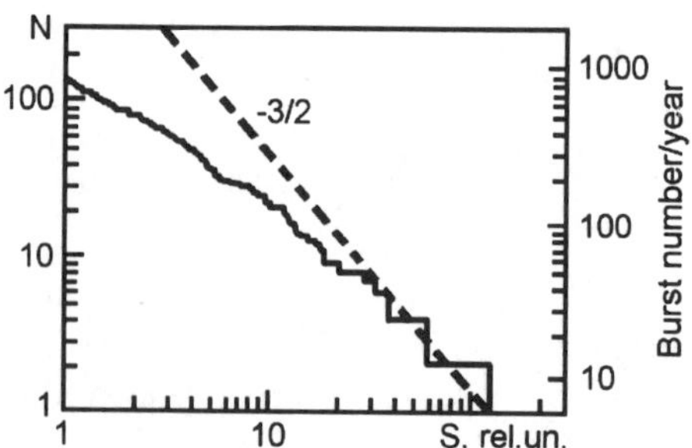

Figure 45.29 Distribution of gamma-bursts by maximal intensity S. The dashed line corresponds to the law $N(>S) \propto S^{-3/2}$, which would be observed for a uniform and isotropic space distribution of gamma-sources [89].

45.5 The Galaxy

The Galaxy parameters [1,3,46,47]. Our Galaxy constitutes a radiating disk involving stars. It belongs to the class of spiral galaxies. The galactic disk thickens at its center forming a bulge with a compact galactic core inside. A flat component of the galactic disk – a thin layer of the interstellar gas where new young stars are forming – may be discerned. The disk is surrounded by a spheroidal halo consisting of old weakly luminous stars. It follows from dynamic considerations based on the analysis of the rotational curve (Fig. 45.30) and stability, that the Galaxy has to be surrounded by a corona containing the main part of a system mass. The corona is not observed directly and hence it has to be composed of a dark matter – stars of low masses and low luminosity, "dead" stellar remnants, neutrinos with non-zero rest mass (if there are any).

Parameters of the Galaxy are listed below (see also Figs. 45.30–45.36, and Table 45.24):

the total luminosity equals $5 \cdot 10^{36}$ W;

number of stars with absolute visual magnitudes less than 16.5^m is $(6\text{–}8) \cdot 10^{10}$;

period of the Sun orbiting about the galactic core is $2.5 \cdot 10^8$ year;

age of the Galaxy is about 10^{10} year;

contribution to the energy density in the galactic disk from:

radiation of stars — $0.7 \cdot 10^{-13}$ W/m^3;

turbulent gas motions — $0.5 \cdot 10^{-13}$ J/m^3;

relic radiation — $0.4 \cdot 10^{-13}$ J/m^3;

cosmic rays — $1.6 \cdot 10^{-13}$ J/m^3;

magnetic field — about 10^{-13} J/m^3;

the total density of matter in the central plane of the Galaxy in the Sun's neighborhood determined by the virial theorem [48] is $0.14 M_\odot/\text{pc}^3 = 8.8 \cdot 10^{-21}$ kg/m^3;

density of the observed stellar matter in the solar neighborhood is $0.046 M_\odot/\text{pc}^3 = 3.0 \cdot 10^{-21}$ kg/m^3;

density of the interstellar gas in the solar neighborhood reaches $0.03 M_\odot/\text{pc}^3 = 1.2 \cdot 10^{-21}$ kg/m^3.

Types of stellar populations [1,3]. Stars and other objects in the Galaxy are divided into two types. Population I consists of the objects forming a flat subsystem of the Galaxy: stars from the galactic disk, interstellar gas and dust, diffusive nebulae, open galactic clusters. Age of population I objects does not exceed 10^{10} years. Population II involves old stars (their age is about 10^{10} years) with a spherical spatial distribution; these stars are characterized by low metallicity and high space velocities. Globular clusters relate to population II.

Stellar clusters [1,3,50,51]. The total number of clusters in the Galaxy comes to about 20,000.

Stellar associations denote groups of a few tens or a few hundreds of young stars. Size of such a group may be about a few tens parsecs. They are connected with star formation regions. The following types of associations may be distinguished: O-associations, where hot O and B stars are concentrated, and T-associations with T Taurus variables included. About 70 O-associations and 25 T-associations are known.

Open stellar clusters (Table 45.25) are said to be groups of a few hundred or a few thousand stars. Their mass is about $10^3 M_\odot$, and size is 1–10 pc. These clusters belong to population I. About 1000 open stellar clusters are known. Abundance of heavy elements in stellar associations and open stellar clusters is close to the solar abundance. Open clusters and associations concentrate toward the galactic plane.

Globular clusters (Table 45.26) are referred to as stable formations of 10^5–10^7 stars. The total number of globular clusters in the Galaxy is about 500. Their age is $(10\text{–}18) \cdot 10^9$ years, abundance of heavy elements is 10–100 times lower than the solar one. Globular clusters form a spheroidal system and concentrate towards the center of the Galaxy.

Interstellar gas [52, 53]. Several regions can be distinguished in this situation (see Table 45.27):

(1) giant molecular clouds (about 4000) containing almost half of the total gaseous mass;

(2) regions of neutral hydrogen (H I);

(3) zones of ionized hydrogen (H II), of which the emission nebulae surrounding hot stars are comprised;

(4) corridors of a hot rarefied gas (coronal gas). Average density of the interstellar gas over the galactic disk is 0.5–0.7 cm^{-3}. Thickness of the gas layer equals 200–300 pc and it increases up to a few kiloparsecs to the periphery of the Galaxy (beyond 12–15 kpc).

Average electron concentration is approximately 0.03 cm^{-3}. A bulk of gas is concentrated in the galactic spiral arms. A significant part of this gas appears in the molecular state (Table 45.28). The total mass of gas in the Galaxy is $(5\text{–}10)\cdot 10^9 M_\odot$. Mass of the neutral hydrogen determined by radiation at $\lambda = 21$ cm, is $(2\text{–}4)\cdot 10^9 M_\odot$. Clouds of the interstellar gas, in addition to orbiting about the galactic center, show chaotical motions with velocities about 10 km/s. The mass distribution of the clouds is amenable to a certain law: number of clouds dN with masses from M to $M + dM$ obeys

$$dN \propto M^{-3/2} dM. \tag{45.9}$$

Interstellar magnetic field is characterized by the induction over a range $(2\text{–}7) \cdot 10^{-10}$ T.

Interstellar gas effectively absorbs ionizing radiation; this makes it difficult to perform the measurements in the UV and soft X-ray spectral bands (Fig. 45.38).

Nebulae and supernova remnants. *Planetary nebulae* (Table 45.29) [56] exhibit spherical shells surrounding very hot ($T \approx 3 \cdot 10^4\text{–}10^5$ K) stars, the UV-radiation of which forms a source of atomic excitations in the nebula. The planetary nebula originates when a star at the late stage of evolution ejects its outer layers, thus forming an expanding shell.

Diffusive nebulae may be of three types. *Emissive nebulae* (H II-zones) arise around hot stars, the UV-radiation of which ionizes surrounding interstellar gas. In *reflecting nebulae* the interstellar dust is illuminated by bright stars of a later spectral type (cooler than B2). Relativistic electrons and gas heated by a shock wave exhibit sources of radiation in *supernova remnants* (Figs. 45.39, 45.40).

Radius R (in pc) of an H II-zone depends on the medium density N (in cm^{-3}) and on the type of the exciting star according to the law [1]:

$$R = S_0 N^{-2/3}, \tag{45.10}$$

where S_0 may be determined from the spectral class of the star:

class	O5	O8	B0	B2	B5	A0
S_0	100	65	35	15	3	1

Owing to a low density of the gas, H II-zones intensively radiate in forbidden lines (Table 45.30).

Parameters of several diffusive nebulae are listed in Tables 45.31, 45.32.

Interstellar masers [59]. Maser radiation may be generated in separate dense formations of the interstellar gas located near strong sources of excitation. Masers corresponding to the rotational transitions in OH, H_2O, SiO, and CH_2OH molecules were observed (Table 45.33). Two different types of masers are distinguished: those connected with large cold stars radiating in the IR spectral range (Table 45.34), and those connected with H II regions surrounding young hot stars (Table 45.35).

Interstellar dust and absorption of stellar light [1, 61]. Parameters of the interstellar dust and characteristics of light absorption in the interstellar medium read as follows:

mass of a dust grain is about 10^{-13} g;

size of a dust grain is $10^{-5}\text{–}10^{-4}$ cm;

concentration of dust grains is of order 10^{-12} cm^{-3};

density of absorbing material in the interstellar space is 10^{-22} kg/m^3 $= 0.0015 M_\odot/\text{pc}^3$;

absorption of the starlight near the galactic plane in the vicinity of the Sun is defined by a value $A_V = 1.9$ stellar magnitudes/kpc.

Absorption depends on the wavelength and this leads to reddening of the starlight, which is characterized by a color excess $E(\lambda_1, \lambda_2)$ that is equal to the difference in absorption at the indicated wavelengths λ_1, λ_2, expressed in the terms of stellar magnitudes m (Fig. 45.41). Centers of the B- and V-bands are agreed to be standard values for λ_1, λ_2. The corresponding color excess is denoted by E_{B-V}. To get absolute value of absorption from the color excess one uses a scaling factor R:

$$R = A_V / E_{B-V}. \tag{45.11}$$

In the average $R = 3.1$, but in dark cloud regions it may be equal to 5–6 (Table 45.36).

Mass ratio of the dust and the gas in different clouds is the same and comprises approximately 10^{-2}. Relation of the number of atoms on the line of sight N_H (in cm^{-2}) to the color excess E_{B-V} is given as

$$N_H = 5.9 \cdot 10^{21} E_{B-V}. \tag{45.12}$$

Gaseous dust complexes and star formation [3, 52]. A significant part of interstellar gas in spiral arms is accumulated in the gaseous dust complexes of tens or a few hundred parsecs in size. Dense ($n \approx 10^3$ cm^{-3}) and cold ($T \approx 10$ K) molecular clouds are embedded to these complexes, where gas is condensed into stars (Fig. 45.42). Young hot stars form around themselves *compact H II-zones* [62] of 0.1–1 pc in size, with a density of 10^3–10^6 cm^{-3}, and a mass of 10^{-2}–1 $M_\odot$. These zones are surrounded in turn by an opaque shell of dust and gas. Such shells reemit a short-wave radiation of the central star into IR-radiation (luminosity covers 10^2–10^5 $L_\odot$). In young stars, an intensive mass outflow is observed, in which dense ($n \approx 10^4$ cm^{-3}) formations are created. The latter move with velocities about 100 km/s; these are *Herberg-Aro objects* [63]. Their temperatures are of the order of 10^4 K, and luminosity (mainly emission lines) is 1–10^3 $L_\odot$.

Masers are connected with star formation regions. Star formation rate in the Galaxy comprises $4M_\odot$/year. Masses of newly formed stars are distributed according to the law: number of stars dN in a mass interval from M to $M + dM$ obeys

$$dN = \psi(M) dM, \tag{45.13}$$

where $\psi(M) = AM^{-3.5}$ (the Salpeter function). This approximation is valid in a mass range $0.3\ M_\odot < M < 50 M_\odot$.

The galactic center [64, 65]. Interstellar absorption in the direction of the galactic center exceeds 27^m, hence the galactic center can be observed only in radio, IR-, X-ray, and gamma-ray spectral ranges. A spheroid of stars with a mass of 10^{10} $M_\odot$ and a gaseous disk consisting of molecular and atomic hydrogen and rotating with a velocity of 200 km/s are located in the galactic center (Fig. 45.43). The central extended H II-zone looks like a spheroid of 150 pc in radius and about 10^6 $M_\odot$ in mass.

Immediately in the center of the Galaxy, the radiosource Sagittarius A West (Sgr AW) is placed. Its size is less than 10^{13} m, and its luminosity equals $3{\cdot}10^{26}$ W. The total IR-luminosity of the dust in the central region of 1 pc in radius is $2{\cdot}10^6 L_\odot$. For keeping the gas ionized in the central region and for heating the dust responsible for the IR-radiation, the power of the central ionizing source must be as high as $(1\text{–}3) \cdot 10^7 L_\odot$. Analysis of the gas velocity distribution shows that a mass of about 10^6 $M_\odot$ is concentrated in the central region of 1 pc in size.

The supernova remnant Sagittarius A East (Sgr AE) is offset 10 pc from the center of the Galaxy. This remnant is moving away from the center with a velocity of 40 km/s. The radiosource Sagittarius B2 (Sgr B2) is a molecular cloud about 30 pc in size, with a mass up to $3{\cdot}10^6 M_\odot$. The center is surrounded by a molecular ring approximately 200 pc in radius expanding with a velocity of 140 km/s and rotating with a velocity of 50 km/s. Its mass is of the order of 10^7 $M_\odot$. In the Sgr B2 cloud compact H II-zones are observed, which are masers radiating in lines of hydroxyl and water vapor. In the central zone, several X-ray sources have been discovered. One of them coincides with the Sgr AW. Its luminosity is $1.5{\cdot}10^{28}$ W in the energy range $E = 0.5$–4.5 keV. In the spectrum of another source, $1E1740$-294, a gamma-line of 511 keV corresponding to $e^+ e^-$ annihilation was observed. Power radiated in this line varies during the time about 1/2 year and attains $2{\cdot}10^{30}$ W. The linewidth is less than $(1.6^{+0.9}_{-1.6})$ keV. The total luminosity of the galactic center reaches $3{\cdot}10^{31}$ W in the energy range 10 keV–10 MeV.

Table 45.24 Dimensions a_0, mass M and oblateness ε (semiaxes relation) of the Galaxy subsystems

Subsystem	ε	a_0, kpc	M, $10^{10}M_\odot$
Core	0.6	0.005	0.009
Bulge	0.6	0.21	0.442
Disc*	0.1	4.62	7.68
	0.45	1.026	−0.379
Flat*	0.02	6.4	1.0
	0.025	5.12	−0.64
Halo	0.3	1.9	1.2
Corona	1	75	110

* A density decrease is observed in the central part of the flat subsystem and in the disc, so an additional "negative mass" is introduced for describing these systems.

Table 45.25 Characteristics of some open clusters

Cluster	Distance, pc	Lower mass limit, $\lg(M/M_\odot)$	$\lg N_5^*$	Core radius, pc	Corona radius, pc	Age, 10^8 y
Pleiades	134	2.6	2.2	2.6	8.1	0.5
Hyades	46	2.2	1.8	3.2	8.0	5
Praesepe	174	2.5	2.0	2.8	6.9	5
M67	830	3.0	2.7	1.8	12.6	33

* N_5 is a number of stars brighter than 5^m.

Table 45.26 Characteristics of some globular clusters

Cluster	Core diameter, pc	Distance, kpc	Integral visual stellar magnitude	Mass, $10^4 M_\odot$
M5	12	8.5	5.9	6
M13	11	7.7	5.9	30
M92	10	10	6.5	14
M71	5	4.5	8.3	

Table 45.27 Typical parameters of the main structural components of interstellar gas in spiral arms of the Galaxy [53]

Phase	Temperature, K	Density, cm^{-3}	Mass, $M_\odot$	Size, pc	Part of the occupied volume
Coronal gas	$\sim 5{\cdot}10^5$	~0.003			~0.5
H II zones of low density	$\sim 10^4$	~3			~0.01
H I warm regions	$\sim 10^3$	~1			~0.01
Intercloud medium	$\sim 10^4$	~0.1			~0.5
H I clouds	~80	~10	~100	~10	~0.01
Dark clouds	~10	$\sim 10^3$	~300	~1	$\sim 10^{-5}$
Great globules	~10	$\sim 10^4$	~20	~0.3	$3{\cdot}10^{-9}$
H II regions	$\sim 10^4$	~30	~300	~10	$\sim 10^{-4}$
Giant molecular clouds	~20	~300	$\sim 3{\cdot}10^5$	~40	$\sim 3{\cdot}10^{-4}$
Consolidations in molecular clouds	~6	$\sim 10^5$	~100	~0.5	

Table 45.28 Some interstellar molecules [52, 55]

Molecule	Characteristic wavelength, cm	$\lg(N/N_H)$	Molecule	Characteristic wavelength, cm	$\lg(N/N_H)$
H_2		0	H_2O	1.4	−7
CH	9.6	−8	C_2H	0.34	−6
OH	18	−8	HCN	0.34	−6
CN	0.27	−8	H_2S	0.18	−8
CO	0.26	−4	OCS	0.25	−8
SiO	0.23	−7	NH_3	1.3	−6
CS	0.20	−7	H_2CO	6.2	−8
SO	0.30	−7	H_2CS	11	−10
	0.22		CH_2NH	5.7	−10
SiS	0.33	−7	HCOOH	18	−10
	0.27		CH_3C_2H	0.35	−9

Table 45.29 Some characteristics of planetary nebulae [1]

Nebula	Flux density in H_β line, 10^{-15} W/m^2	Temperature, 10^3 K	Central star temperature, 10^3 K	$\lg N$, cm^{-3}	Diameter, pc
IC 418	800	12	36	4.1	0.09
NGC 3242	300	14	50	3.0	0.1
NGC 6572	800	11	50	4.0	0.05
NGC 6720 ("Ring")	320	10	90	3.0	0.2
NGC 7009 ("Saturn")	280	12	50	4.0	0.08
NGC 7662	250	14	60	3.9	0.06

Nebula	Mass, $M_\odot$	Flux density at a frequency of 1 GHz, Jy	Radioemission spectral index*	Expanding rate, km/s	Distance, pc
IC 418	0.04	0.66	+0.9	0	1500
NGC 3242	0.04	0.90	0.0	20	800
NGC 6572	0.10	0.24	+2.0	4	900
NGC 6720 ("Ring")	0.17	0.44	+0.1	19	700
NGC 7009 ("Saturn")	0.09	0.52	+0.8	19	700
NGC 7662	0.07	0.07	0.0	25	900

* Spectral index $d\lg(\text{intensity})/d\lg(\text{frequency})$.

Table 45.30 Transitions probabilities A of spontaneous emission for some forbidden lines observed in nebulae [57]

Transition	O III		N II		O I	
	Wavelength, nm	A, s^{-1}	Wavelength, nm	A, s^{-1}	Wavelength, nm	A, s^{-1}
$^3P_2$—$^1D_2$	500.684	0.021	658.34	0.0030	630.023	0.0069
$^3P_1$—$^1D_2$	495.891	0.0071	654.81	0.00103	636.388	0.0022
$^3P_0$—$^1D_2$	493.10	$1.9\cdot10^{-6}$	652.74	$4.2\cdot10^{-7}$	639.2	$1.1\cdot10^{-6}$
$^1D_2$—$^1S_0$	436.321	1.6	575.48	1.08	557.735	1.28

Table 45.31 Characteristics of some emissive (E) and reflective (C) nebulae [1]

Nebula	Type	Distance, pc	Diameter, pc	Gas mass, $M_\odot$	Density, cm^{-3}	Radiation flux density in H_α, 10^{-6} W/(m^2 sr)	Radiowave flux density (λ=20 cm), Jy	Class of the exciting star
In stellar cluster Pleiades (M45)	C	126	1.5					B7
In Orion constellation (M42)	E	460	5	300	600	13	440	O8
Horsehead	CE	350	3	0.6	25			B1
Tarantula	E	$5\cdot10^4$	250	10^5				O class stars
Rosette	E	$1.1\cdot10^3$	15	$9\cdot10^3$	30	1.8	300	O6
Lagoon (M8)	E	$1.2\cdot10^3$	9	10^3	80	7	380	O5
North America	CE	700	20	$8\cdot10^3$	15	0.8	510	A2
Cocoon	C	$1.6\cdot10^3$	2	7	70			B1
Trifid (M20)	E	10^3	4	150	100	6	30	O7

Table 45.32 Characteristics of supernova remnants [58]

Name of the remnant (year of burst)	Distance, kpc	Diameter, pc	Age up to 1980, y	Radiowave		X-ray		Estimate of mean expanding rate, km/s
				Flux density (ν=1000 MHz), Jy	Spectral index	Luminosity*, 10^{28} W	Temperature, 10^6 K	
Cassiopeia A (1680)	3	3.5	300	3100	−0.8	33	15 and 60	5500
Kepler's supernova (1604)	10	6.6	376	20	−0.6	30		
Brahe's supernova (1572)	5	10.7	408	52	−0.6	17	6 and 40	5000
Supernova 1181	8	12	799	35	−0.1	not found		
Crab Nebula (1054)	2	3	926	1000	−0.2	310		1200
Supernova 1006	4	40	974	25	−0.6	0.2	20	1800
Supernova 185	2.5	28	1795	33	−0.4	17	6	600
IC 443	1.5	20	3400	180	−0.5	1	17	800
Puppis A	2.2	20	5000	145	−0.4	60	7	700
Vela X	0.5	30	13000	1800	−0.3	5	4	500
Cygnus loop	0.8	30	20000	160	−0.5	16	3	400
HB21	1.1	35	>35000	225	−0.4	1	<2	<200

* In 0.2–10 keV range.

Table 45.33 Observed maser transition parameters

Molecule	Transition	Frequency ν, MHz	Wavelength λ, cm	A, s^{-1}
OH	$^2\Pi_{3/2}$ $J=3/2$ $F=1\to 2$	1612.231	18.6	$1.29\cdot 10^{-11}$
	$^2\Pi_{3/2}$ $J=3/2$ $F=1\to 1$	1665.402	18.0	$7.11\cdot 10^{-11}$
	$^2\Pi_{3/2}$ $J=3/2$ $F=2\to 2$	1667.359	18.0	$7.11\cdot 10^{-11}$
	$^2\Pi_{3/2}$ $J=3/2$ $F=2\to 1$	1720.530	17.4	$9.42\cdot 10^{-12}$
	$^2\Pi_{3/2}$ $J=5/2$ $F=2\to 2$	6030.747	5.0	$1.53\cdot 10^{-9}$
	$^2\Pi_{3/2}$ $J=5/2$ $F=3\to 3$	6035.092	5.0	$1.57\cdot 10^{-9}$
	$^2\Pi_{1/2}$ $J=1/2$ $F=0\to 1$	4660.242	6.4	$1.08\cdot 10^{-9}$
	$^2\Pi_{1/2}$ $J=1/2$ $F=1\to 0$	4765.562	6.3	$3.86\cdot 10^{-10}$
	$^2\Pi_{3/2}$ $J=7/2$ $F=4\to 4$	13441.371	2.2	$9.26\cdot 10^{-9}$
H_2O	$6_{16}\to 5_{23}$	22235.080	1.35	$1.91\cdot 10^{-9}$
SiO	$^1\sum \nu=1,\ J=1\to 0$	43122.03	0.70	$3.00\cdot 10^{-6}$
	$\nu=1,\ J=2\to 1$	86243.27	0.35	$2.87\cdot 10^{-5}$
	$\nu=1,\ J=3\to 2$	129363.12	0.23	$1.04\cdot 10^{-4}$
	$\nu=2,\ J=1\to 0$	42820.48	0.70	$2.93\cdot 10^{-6}$
CH_3OH	$J=4$ $k=2\to 1$	24933.468	1.20	$8.40\cdot 10^{-8}$
	5 $k=2\to 1$	25959.080	1.20	$8.74\cdot 10^{-8}$
	6 $k=2\to 1$	25018.123	1.20	$8.98\cdot 10^{-8}$
	7 $k=2\to 1$	25124.873	1.20	$9.21\cdot 10^{-8}$
	8 $k=2\to 1$	25294.411	1.20	$9.48\cdot 10^{-8}$

Table 45.34 Characteristics of masers connected with the infrared stars

Parameter	H_2O	OH	SiO
Number of observed transitions	1	3	4
Number of known objects	~50	~50	16
Linewidth, km/s	1–2	1–2	0.5–2
Kinetic temperature*1, K	400–1600	400–1600	250–3500
Number of spectral details	1–10	2–10*2	1–10
Range of speeds, km/s	5–50	5–80	2–15
Polarization	No	Small	No(?)
Details' lifetime, s	$>10^7$	$>10^7$	
Spot size, cm	10^{14}	10^{15}	$<10^{16}$
Luminance temperature, K	10^{11}–10^{12}	10^9–10^{11}	$>10^3$
Size of sources group, cm	10^{15}	$2\cdot 10^{15}$	
Power, W*3	10^{17}–10^{21}	10^{17}–10^{21}	10^{19}–10^{20}

*1 Determined by linewidth assuming no line narrowing and no large-scale gas motions.
*2 More for supergiants.
*3 Under isotropic radiation assumption.

Table 45.35 Characteristics of masers connected with H II regions

Parameter	H_2O	OH	SiO	CH_3OH
Number of observed transitions	1	9	4	4
Number of known objects	~50	~50	1*1	1*1
Linewidth, km/s	0.5–2	0.1–1	2	0.5
Kinetic temperature, K*2	100–1500	4–400	3500	150

Table 45.35 Characteristics of masers connected with H II regions *(continued)*

Parameter	H_2O	OH	SiO	CH_3OH
Number of spectral details	1–100	1–50	5	3
Range of speeds, km/s	1–300	1–30	25	4
Polarization, %	0–10 (linear)	0–100 (linear) 0–100 (circular)	0–20 (circular)	No
Details' lifetime, s*3	10^6–10^7	10^7–10^8		
Spot size, cm	10^{13}–10^{14}	10^{14}–10^{15}	$<10^{16}$	$<10^{16}$
Luminance temperature, K	10^{13}–10^{15}	10^{12}–10^{13}	$>10^{13}$	$<10^3$
Size of the sources group, cm	10^{16}–10^{17}	10^{16}–10^{17}		10^{17}
Power, W*4	10^{20}–10^{26}	10^{20}–10^{23}	10^{22}	10^{20}

*1 The only known source – Orion A.
*2 Assuming no line narrowing and no large-scale gas motions.
*3 There are some cases when the characteristic time is less.
*4 Under isotropic radiation assumption.

Table 45.36 Statistical properties of the dust clouds [61]

Characteristic	Standard cloud	Large cloud
Mean E_{B-V} for the cloud, E_0	0.061 (6)	0.29 (6)
k – number of clouds at a distance of 1 kpc	6.2 (3)	0.8 (2)
Selective absorption per 1 kpc, kE_0	0.38 (5)	0.23 (1)

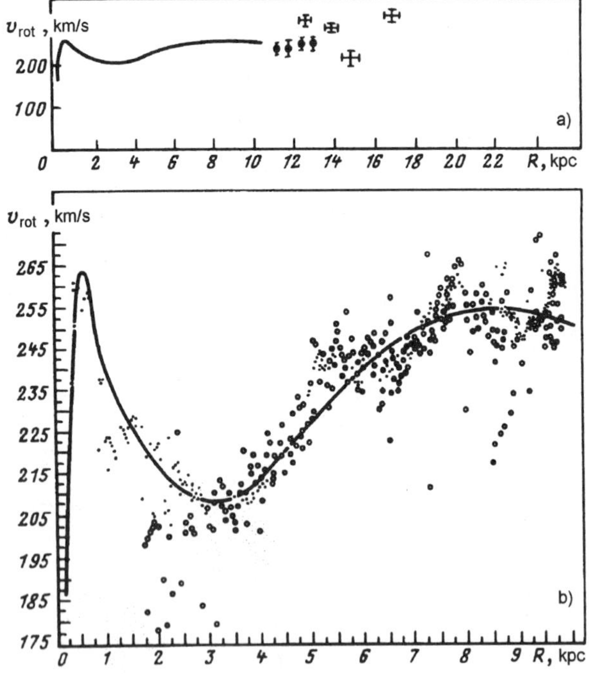

Figure 45.30 Rotational curve of the Galaxy (a) (b – a zoomed scale) [46].

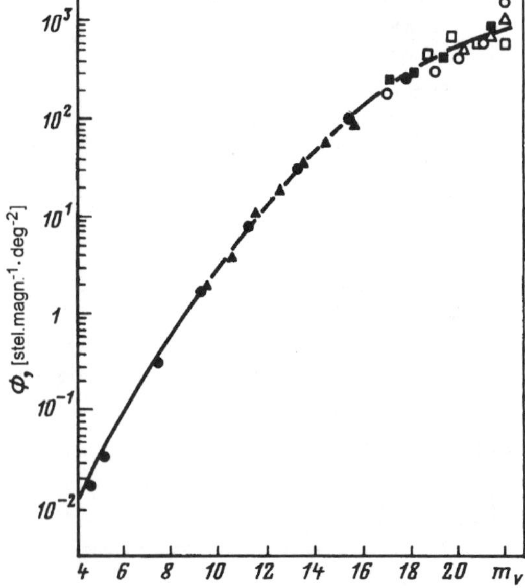

Figure 45.31 Distribution of the stars by their visual stellar magnitudes in the Galaxy pole direction [49].

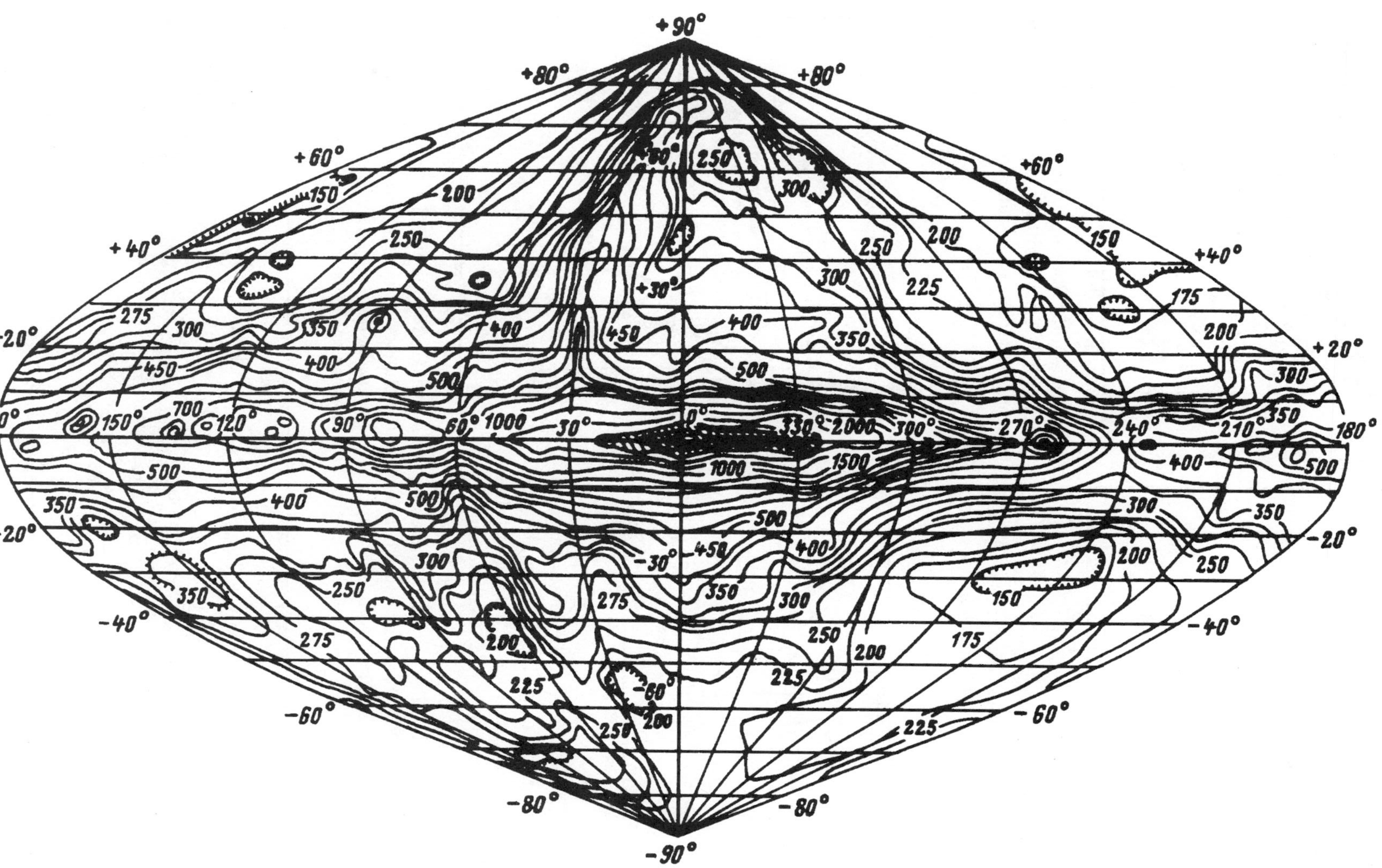

Figure 45.32 Brightness distribution over the sky at 150 MHz frequency in the galactic coordinates: numbers in the isophotes show the luminance temperature of the radiation, K [54].

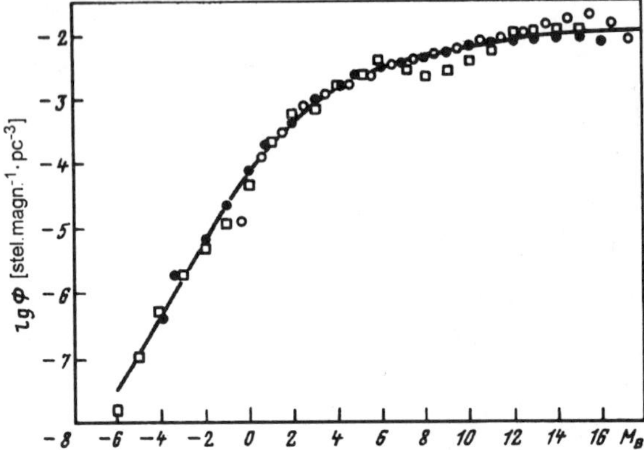

Figure 45.33 Luminosity function of stars in the Galaxy $\Phi(M_B)$ (distribution of stars by absolute stellar magnitudes in the B-band) [49].

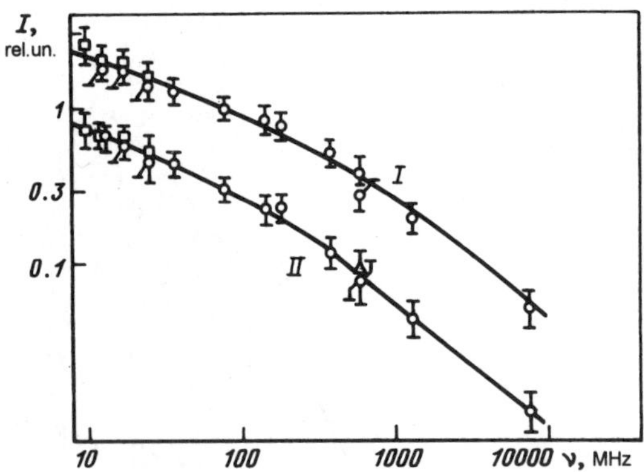

Figure 45.35 Radiospectrum of the Galaxy: I – radiation of high-latitude regions in the anti-center direction; II – radiation of spiral arms [54].

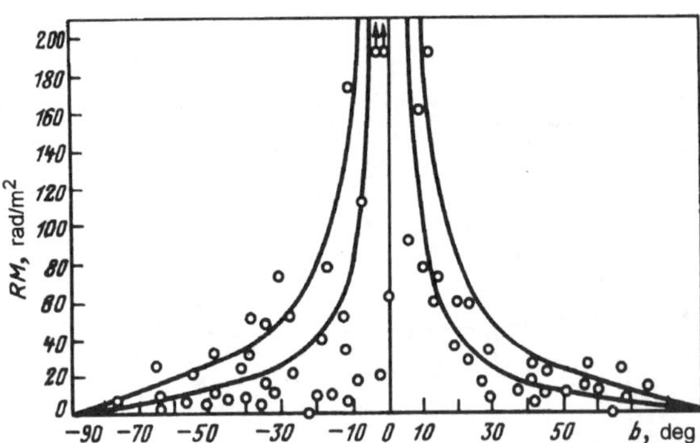

Figure 45.34 Dependence of the rotational measure RM of extragalactical radiosources on the galactic latitude. Faraday rotation of the polarization plane of the radiation $\Delta\Phi = RM\lambda^2$ enables one to determine $RM = 0.81 \int_0^l B_{\|} N_e dl$ (rad/m^2), where $B_{\|}$ is a component of the magnetic induction along the line of sight, 10^{-10} T; N_e is the concentration of electrons, cm^{-3}, l is the distance to the source, pc [54].

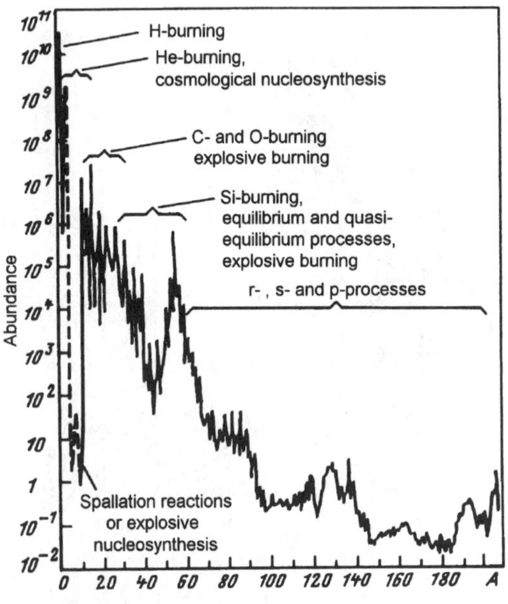

Figure 45.36 Chemical abundance: A is the mass number; nuclear cycles in stars are shown that may be responsible for features in the chemical abundance curve [2].

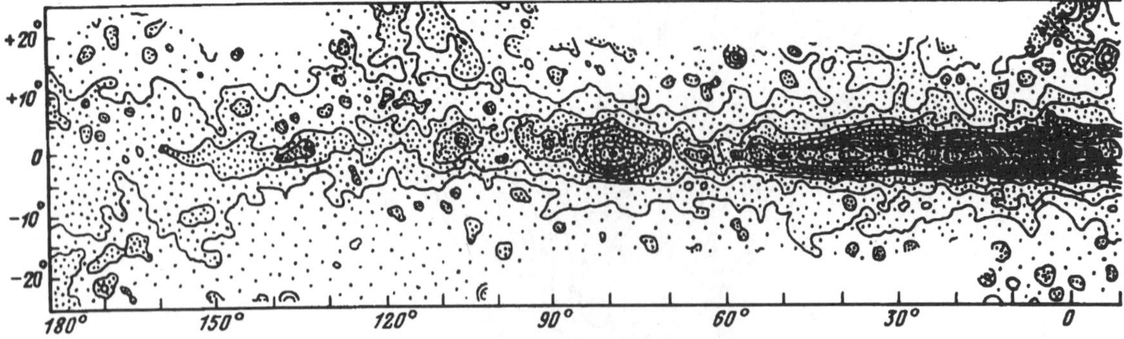

Figure 45.37 γ-radiation distribution (70 MeV–5 GeV) over sky in the galactic coordinates [54].

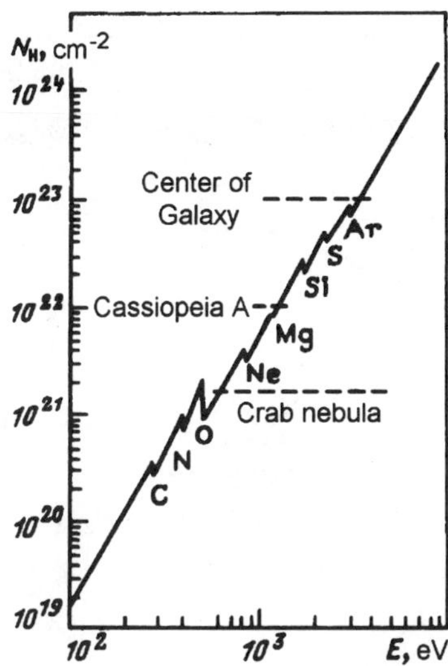

Figure 45.38 Absorption of the X-ray radiation by the interstellar gas. Shown is the number of hydrogen atoms N_H along the line of sight that gives unit optical thickness for a given photon energy E [52].

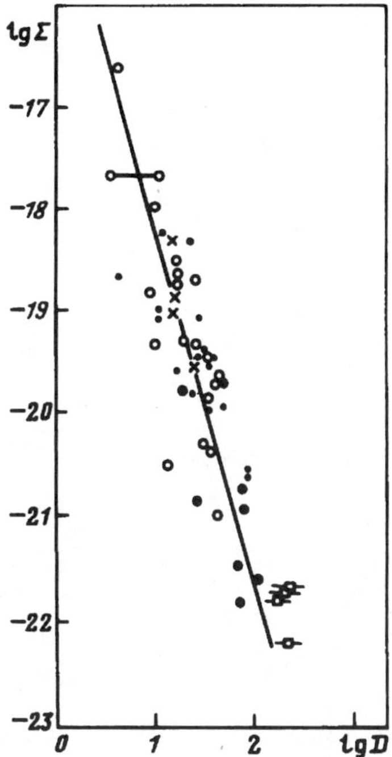

Figure 45.40 Dependence of the radiobrightness Σ, W/(m^2·Hz·sr) of supernova remnants at frequency 1 GHz on their diameters D, pc [59].

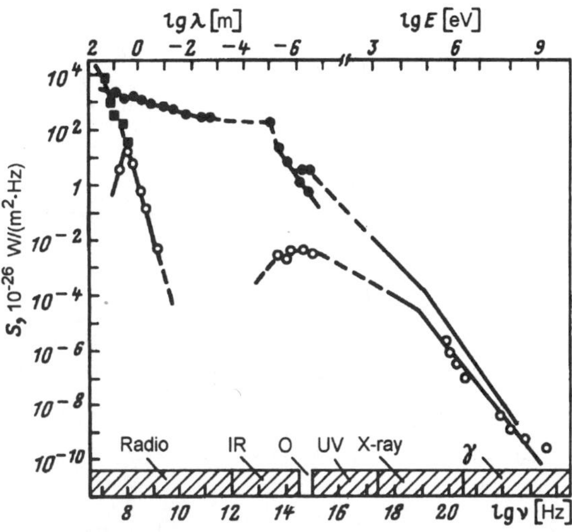

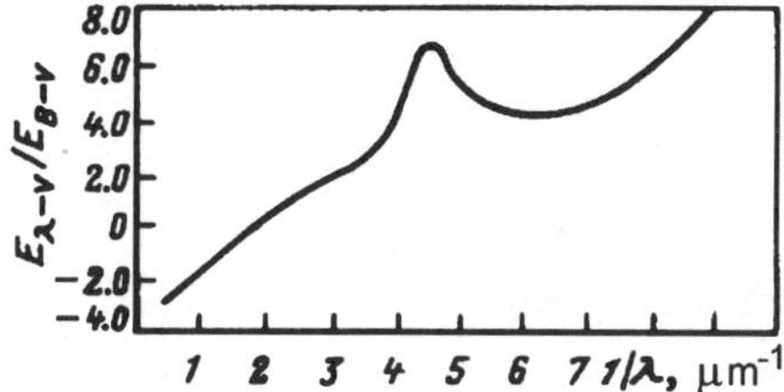

Figure 45.41 The normalized curve of the interstellar absorption [52].

Figure 45.39 Spectra of the Crab nebula (•), of its pulsar (○) and of a compact radiosource in the nebula center (□) [32]. Energy ranges are the same as in Fig. 45.8.

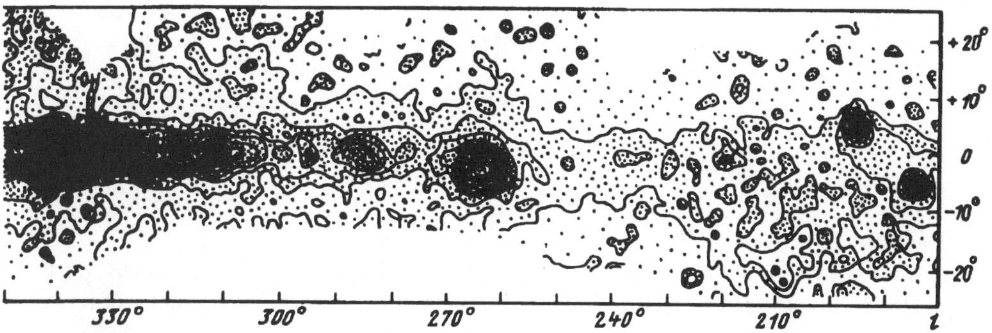

Figure 45.37 *(continued)*

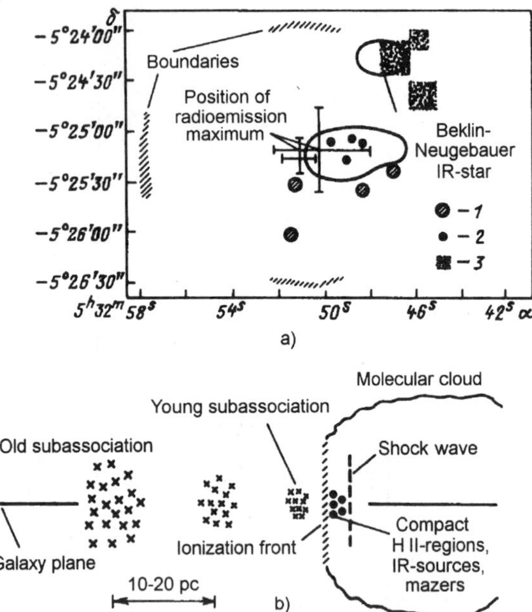

Figure 45.42 Scheme of the star formation origin in the Orion nebula (a), and a typical scheme of the large star formation with the running across wave of star formation (b) [53]: *1* – compact H II-zones, *2* – stars in the Orion Trapecium; *3* – H_2O-masers; IR-nebulae are enclosed by the solid line.

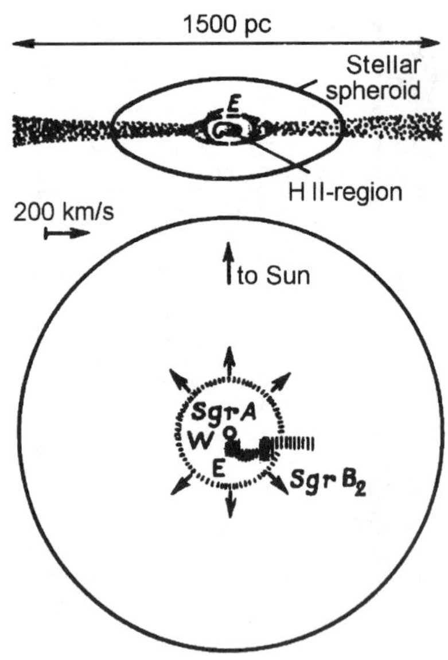

Figure 45.43 The center of the Galaxy. Locations of radiosources Sagittarius B2 and Sagittarius A (with W and E companions) and an expanding molecular ring are shown [64].

45.6 Galaxies and the Universe

Structure and classification of galaxies [66, 67]. The majority of galaxies are composed of two main components: a massive disc and a spheroidal component. Dependence of luminosity of the spheroidal component I on the radius r follows the law

$$\lg I(r)/I_1(0) \approx (r/r_e)^{1/4}, \tag{45.14}$$

where r_e is an effective radius. Luminosity of the disc component is distributed over the disc plane according to the relationship:

$$I(r)/I_2(0) \approx \exp(-r/r_0), \tag{45.15}$$

where $r_0 \approx$ 2–6 kpc, and the luminosity distribution over the plane perpendicular to the disc plane obeys:

$$I(z)/I_3(0) \approx \cosh^{-2}(z/z_0), \tag{45.16}$$

where $z_0 \approx$ 0.6–0.8 kpc.

Elliptical galaxies (denoted by En, where numerals $n = 0, 1, \ldots, 7$ correspond to compression of the galaxy $\epsilon = (a - b)/a$, $n = 10\epsilon$) are practically composed of the spheroidal component only. Their masses and luminosity range from 10^5 to $10^{13} M_\odot$, and from 10^4 to $10^{12} L_\odot$, respectively. Almost 25% of all the galaxies belongs to this type.

In *spiral galaxies* (which are denoted by Sa, Sb, Sc in line with the stage of spirals development), a disc component dominates in addition to a spheroidal component, which is always present. Spiral galaxies with a bar form a separate class (SBa, SBb, SBc). Fraction of the galaxies of these two types is nearly 50%. Their masses and luminosity are of order 10^8 to $10^{12} M_\odot$, and 10^8 to $10^{11} L_\odot$, respectively. The lens-shaped (SO) galaxies represent a class intermediate between the elliptical and spiral galaxies. 20% of the total number of galaxies belongs to this former type.

Irregular galaxies IrI and IrII are divided into two types corresponding to two types of stellar population (I and II, respectively). Their masses and luminosity do not exceed the limits of $10^{10} M_\odot$ and $10^{10} L_\odot$, respectively.

Parameters of different types of galaxies are listed in Table 45.37. Luminosity distribution of galaxies takes the form:

$$\Phi(L)dL = \Phi^*(L/L^*)^{-1.25} \exp(-L/L^*)d(L/L^*), \tag{45.17}$$

where $\Phi^* = 1.5 \cdot 10^{-3}$ Mpc^{-3}; absolute stellar magnitude $M_B^* = -21$ corresponds to the luminosity L^*. Density of galaxies luminosity $\rho_V = 10^8\ L_\odot$ Mpc^{-3}. The nearest galaxies are combined into a gravitationally bounded system, called the Local group (Table 45.38). Characteristics of some other galaxies are presented in Table 45.39.

Active galaxies and quasars [3]. Active galaxies are characterized by:

(1) presence of a compact core with a luminosity of 10^{34}–10^{41} W;

(2) a rapid (months and days) variability of radiation;

(3) a nonthermal spectrum with excessive fluxes in radio-, IR- and X-ranges (Fig. 45.44);

(4) broad (10^3–10^4 km/s) emission lines (Fig. 45.45);

(5) peculiar appearance evidenced by an explosion and jet-like out-bursts.

The following main types of active galaxies exist.

Seyfert galaxies denote spirals (as a rule) with small and abnormally bright cores. Width of emission lines of Seyfert galaxies of type I and of type II are 1000–3000 km/s and 500–1000 km/s, respectively. IR and optical luminosity lie in a range of 10^{35}–10^{39} W. Seyfert galaxies of type I have comparable X-ray luminosity. In the energy range $E = 1$–100 keV, the spectra usually follow a power law.

Quasars are point-like sources with luminosity 10^{36}–10^{41} W in IR, optical and X-ray spectral ranges. Widths of the emission lines are 4000–6000 km/s. Approximately 1% of quasars are radiosources of 10^{39} W in power. Quasars are the most powerful objects in the observable Universe (Fig. 45.46). The most distant of known quasars has a red shift $z = 4.04$.

Lacertids (BL Lacertae objects) are sources similar to quasars. They are characterized by the absence of emission lines and by a strong (10–30%) variable polarization of radiation.

Radiogalaxies denote (as a rule) elliptical galaxies with a radioluminosity of 10^{35}–10^{38} W. Most of them possesses a binary structure: two radioemitting clouds are symmetrically spaced at a distance of a few megaparsecs from the central source. Their radiospectrum usually follows a power law (Fig. 45.47).

Figs. 45.48–45.50 demonstrate luminosity and spectral index distribution for the active galaxies.

Galaxy clusters [74]. Distribution of galaxies in space is strongly nonuniform. Correlation function for galaxies

$$\xi(R) = < N(r)N(r+R) - N^2(r) > / < N^2(r) >, \tag{45.18}$$

where N is the number density of galaxies, is expressed as [75]: $\xi(R) = (R_0/R)^\gamma$, $R_0 = (4.23 \pm 0.52)h^{-1}$ Mpc; γ=1.77(4). Here $h = H/50$, where H is the Hubble constant measured in km/(s·Mpc). This function fits the observable data for $R < 20h^{-1}$ Mpc.

A significant fraction of galaxies is concentrated in clusters (Fig. 45.51). Typical cluster masses are 10^{12}–$10^{15}\ M_\odot$; they contain a few hundreds or a few thousands of galaxies (Table 45.40). Rich clusters are those containing within a distance of $3h^{-1}$ Mpc from the cluster center not less than 50 galaxies with a brightness ranging from m_3 to $m_3 + 2$, where m_3 is the stellar magnitude of the third galaxy by the brightness. Density of galaxies in the central parts of rich clusters is distributed by King's law:

$$N(r) = N_0(1 + r^2/r_c^2)^{-3/2}, \tag{45.19}$$

where r_c=0.25(4) h^{-1} Mpc. The entire diameter of the clusters reaches 3–5 Mpc. Clusters are filled with gas at a temperature of the order of 10^8 K and a density of about 10^{-3} sm^{-3}. Bremmstrahlung X-ray radiation of the gas is observed in this case. The correlation function of clusters in the interval $5\ h^{-1} < R < 100\ h^{-1}$ Mpc takes the form [76]

$$\xi(R) = (R_0/R)^{\gamma}, \quad \gamma = 1.6, \quad R_0 = 50h^{-1}\text{Mpc}. \tag{45.20}$$

In greater scales, superclusters are also observed. Its typical size is about 30 Mpc, and its mass is of the order of 10^{16} $M_{\odot}$. A typical supercluster contains two-three rich clusters [77].

Counts of sources [79]. In the Euclidian space uniformly filled with sources, a number of sources $N(S)$ with the flux density higher than S, obeys the law

$$N \propto S^{-3/2} \tag{45.21}$$

(in the differential form, $dN/dS \propto S^{-5/2}$). Deviations of the counting results from this law are explained by evolution of the sources, as well as by cosmological effects (Figs. 45.52–45.54).

Background radiation (Fig. 45.55) is contributed from distant galaxies; hot intergalaxy gas, as well as the relict radiation (microwave background radiation), remained after the early hot stages of the evolution of the Universe. Parameters of the background radiation in different spectral ranges are presented in Table 45.41.

Microwave background radiation [83] (named the relic radiation) features a spectrum close to that of a black body with a temperature of T=2.7 K (Fig. 45.56) and the following characteristics:

density of energy is 0.25 eV/cm^3;

velocity of the Sun's motion relative to the microwave background is 300–400 km/s in the direction $\alpha = 12^h$, $\delta = 0°$;

dipole anisotropy of the brightness arising as a result of the Sun's motion, $\delta T/T = 1.3 \cdot 10^{-3} \cos\theta$, where θ is the angle between the direction of motion of the Sun and the line of sight;

quadrupole anisotropy of brightness [83] was discovered to be $\delta T = 13 \pm 4\ \mu$K [84].

The limits for small-scale brightness fluctuations are shown in Fig. 45.57.

Cosmology. The basic parameters of our Universe look now as follows:

the Hubble constant H = 50–100 km/(s·Mpc);

a critical density of the matter (when the Universe is closed) is ρ_c=4.7·$10^{-30}h^2$ g/m^3, where h=H/50 km/(s·Mpc);

density of the matter observed is $\rho \approx 10^{-31}$ g/cm^3;

baryon to photon ratio N_b/N_{ph} = (3–10) · 10^{-10} [85];

baryon density parameter $\Omega_b = \rho_b/\rho_c$, 0.01$\leq \Omega_b \leq$0.14 [86];

density of the relic neutrinos (of three sorts) is 450 cm^{-3} [88];

cosmological limitation on the total neutrinos' masses (of all sorts) is $\Sigma m_\nu c^2 < 200$ eV [88];

age of the Universe in years $t = (1.96/h)f(\Omega) \cdot 10^{10}$, where the function $f(\Omega)$ is shown in Fig. 45.58. Here Ω is the density of any kind of matter expressed in units of the critical density ρ_c.

Expansion of the Universe results in the red shift of the radiation of the distant sources: the farther distance, the more red shift (Fig. 45.59). Therefore, the location of sources is characterized by the quantity $z = (\lambda - \lambda_0)/\lambda_0$, where λ_0 and λ are the wavelengths radiated by the source and detected by an observer, correspondingly. Fig. 45.60 illustrates a non-Euclidian character of the Universe. It demonstrates that the angular size of a body with a fixed size never vanishes with a rise in the distance to the body, and begins increasing after a definite minimal value is achieved.

At the early stages of evolution, the Universe was filled with a hot plasma, which cooled down during the expansion. Approximately in 500 s after the beginning of expansion, when the temperature fell down to 10^9 K, nuclear reactions that sustained a balance between neutrons and protons stopped. At the end of this stage, an intensive synthesis of elements was going on (Fig. 45.61); in particular, the most part of the observed helium was synthesized over this period. Approximately 200,000 years later, at a temperature T= 4000 K, recombination of the hydrogen occurred (helium recombined somewhat earlier), and after that the Universe became practically transparent for filling thermal radiation. Owing to this fact, a growth of perturbations in the substance density becomes possible (radiation pressure prevented it before the recombination). This growth eventually led to formation of gravitationally bound bodies: clusters, galaxies, etc. (Fig. 45.62).

Table 45.37 Mass–luminosity ratio and mass portion of the neutral hydrogen for galaxies of different types [48, 67]

Type of galaxy	$(M/L)/(M_\odot/L_\odot)$	M_{HI}/M	Type of galaxy	$(M/L)/(M_\odot/L_\odot)$	M_{HI}/M
E	20–40	~0	Sb	10	0.05
SO	10	0.005	Sc	<10	0.07
Sa	10	0.03	IrI	<10	0.2

Table 45.38 Population of the Local group of galaxies [3]

Family	Constitution of the family showing the constellation in which the galaxy is located	Type	Distance, kpc	Absolute stellar magnitude	Mass, $10^6 M_\odot$
Galaxies	Galaxy (center in Sagittarius)	Sb	10	−21	250000
	Large Magellanic Cloud (Tucan)	Ir	52	−18	14000
	Small Magellanic Cloud (Dorado)	Ir	71	−16	5000
	Fornax	Ep	188	−13	20
	Sculptor	Ep	84	−12	3
	Leo I	Ep	220	−11	4
	Draco	Ep	76	−9	0.1
	Ursa Minor	Ep	67	−9	0.1
	Leo II	Ep	220	−9	1
	Pegasus	Ep	170	−9	
	Orion	Ep	80	−7	
	Capricornus	Ep	70	−6	
	Ursa Major	Ep	120	−6	
	Ursa Major	Ep	130		
	Sextant C	Ep	140		
	Serpens	Ep	30		
	Carinae	Ep	170		
Andromeda nebulae	Andromeda (M31, NGC 224)	Sb	690	−22	360000
	Triangulum (M33, NGC 598)	Sc	720	−19	20000
	Andromeda (M32)	E2	690	−16	2600
	Andromeda (NGC 205)	SB0	690	−15	2000
	Cassiopeia (NGC 185)	E3	690	−15	100
	Cassiopeia (NGC 147)	E5	690	−14	150
	Andromeda I	Ep	690	−11	1
	Andromeda II (Pisces)	Ep	690	−11	1
	Andromeda III	Ep	690	−11	1
	Andromeda IV	Ir	690	−11	10
	Pisces	Ir	690	−9	10
Periphery of the Local group	Cassiopeia (IC 10)	Ir	1250	−17	15000
	Cetus (IC 1613)	Ir	770	−15	400
	Sagittarius (NGC 6822)	Ir	600	−15	1500
	Centus (Wolf-Lundmark-Mellot galaxy)	Ir	1300	−14	300
	Sextant A	Ir	1300	−14	1000
	Leo A	Ir	1100	−13	400
	Capricornus	Ir	1000	−11	30
	Virgo	Ir	1000	−11	40
	Sculptor	Ir	1400	−10	10
	Sagittarius	Ir	500	−9	10

Table 45.39 Some bright galaxies [1]

Galaxy	Type	Diameter, kpc	Distance, Mpc	Absolute stellar magnitude	Mass*, $\lg(M/M_\odot)$
NGC 55	Sc	12	2.3	−19.9	10.5
M81	Sb	16	3.2	−20.9	11.2
M82	IrII	7	3	−19.6	10.5
NGC 3115	E7	5	4	−19.3	10.9
M87	E1	13	13	−21.7	12.6
M104 (Sombrero)	Sa	8	12	−22	11.7
Centaurus A (NGC 5128)	E0p	15	4.4	−20	11.3
M51 (Whirlpool)	Sc	9	3.8	−19.7	10.9
M83	SBc	12	3.2	−20.6	

* Mass of the observed matter. There are arguments that the whole mass of some galaxies is greater, as they are surrounded by massive invisible coronas.

Table 45.40 Some clusters of galaxies

Cluster	Number of the observed galaxies [1]	Distance*, Mpc [2]	Angular size, deg. [2]	Optical luninosity, $10^{11}L_\odot$ [2]	Radial velocity dispersion, km/s [78]	Gas temperature, keV [78]	X-ray luminosity (2–10 keV), 10^{37} W [78]
Perseus (A 426)	500	120	4	10	1400	6.8	12.3
Virgo	2500	23	12	12	700	2.2	0.3
Coma Berenices (A 1656)	800	135	4	50	900	8.8	7.6

* For H=50 km/(s·Mpc).

Table 45.41 Energy density and photon number density for the background radiation in different ranges [82]

Range	Radiant energy density, eV/cm^3	Photon number density, cm^{-3}
Long-wave radioemission	$\sim10^{-7}$	~1
Microwave background radiation	~0.25	~400
Infrared emission	$\sim10^{-2}$	~1
Optical emission	$\sim3\cdot10^{-3}$	$\sim10^{-3}$
Long-wave X-ray emission ($E<1$ keV)	$\sim(10^{-4}$–$10^{-5})$	$3\cdot10^{-7}$–$3\cdot10^{-8}$
Shor-twave X-ray emission ($E>1$ keV)	$\sim10^{-4}$	$\sim3\cdot10^{-9}$
γ-ray emission:		
$E=1$–6 MeV	$\sim3\cdot10^{-5}$	$\sim10^{-11}$
$E>10$ MeV	10^{-5}	10^{-12}

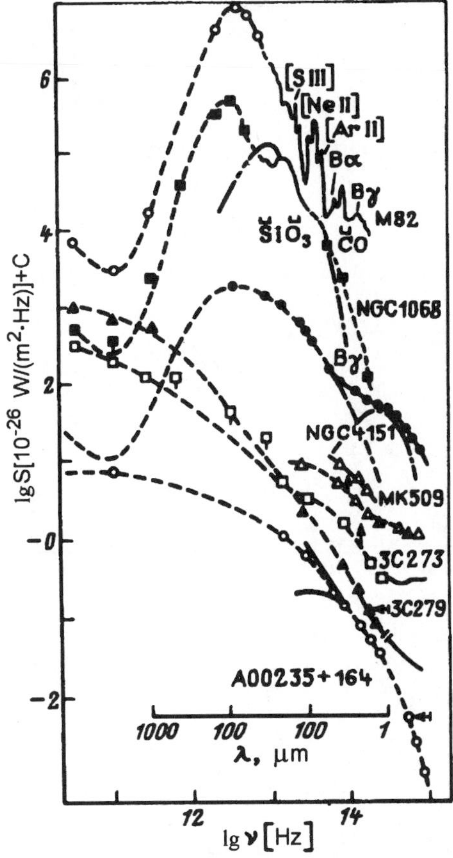

Figure 45.44 Spectra of active galaxies in IR-, optical and radio-bands. For the sake of space the spectra are shifted along the ordinate axis (each by a definite value C). Presented are the spectra of: the "exploding" galaxy M82 ($C = 3.77$), the Seyfert II galaxy NGC1068 ($C = 3.07$), Seyfert I galaxies NGC4151 ($C = 2.5$) and MK509 ($C = 1.77$), the quasars 3C273 ($C = 1$) and 3C279 ($C = 2$), and the Lacertid A00235+164 ($C = 0$) [68].

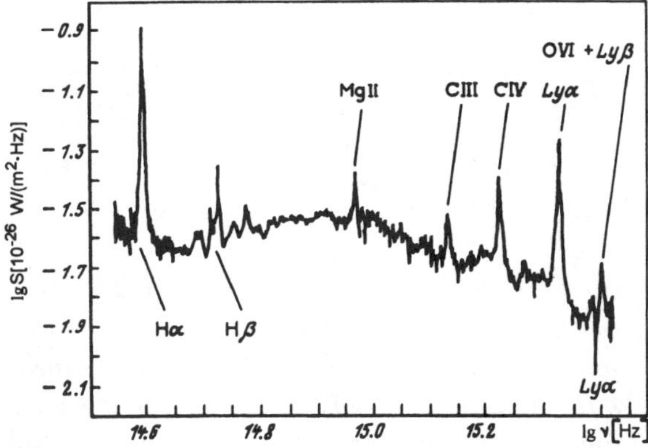

Figure 45.45 Spectrum of the quasar 3C273 in the optical and ultraviolet ranges [69].

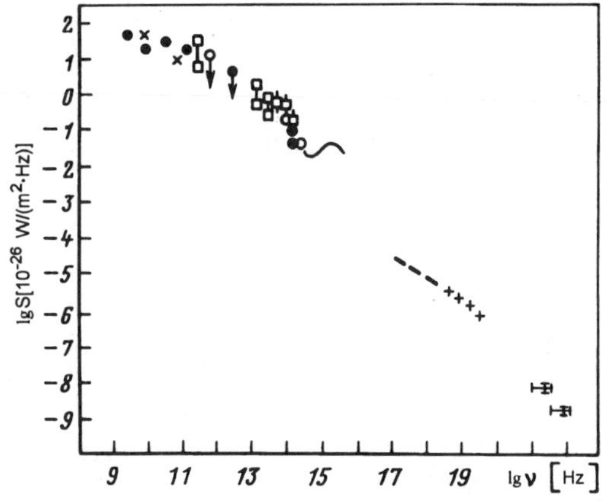

Figure 45.46 Spectrum of the quasar 3C273 in a spectral range from radio to gamma diapason [70].

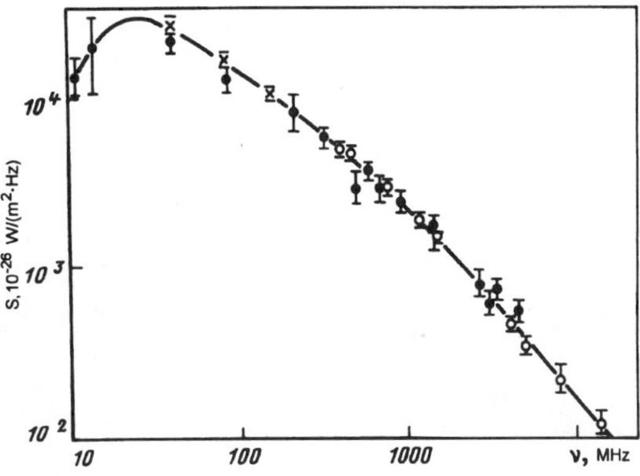

Figure 45.47 The radiogalaxy Cygnus A spectrum [54].

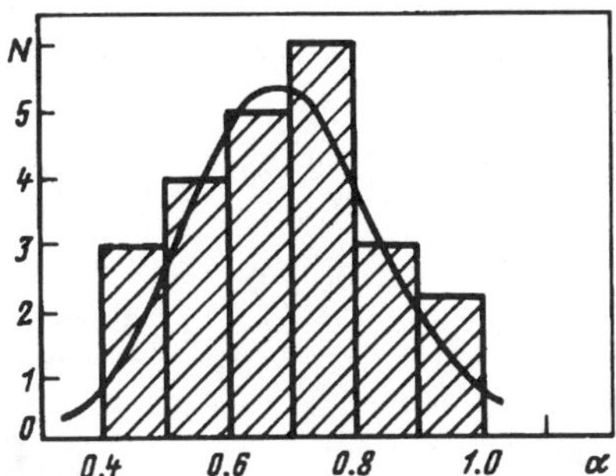

Figure 45.48 Distribution of the spectral indices α of the active galaxies in the X-ray range: $\alpha = -d\lg(\text{intensity})/d\lg(\text{frequency})$; N is the number of galaxies [71].

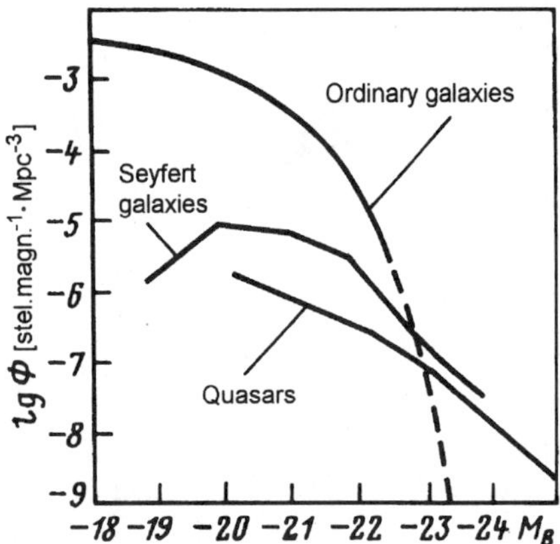

Figure 45.49 Luminosity function Φ of active galaxies. The absolute stellar magnitudes in the B-band is plotted along the abscissa axis [72].

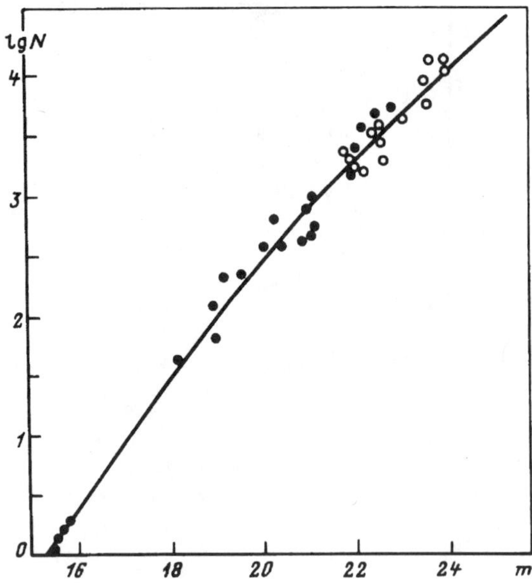

Figure 45.52 Number of galaxies $N(m)$ brighter than a visual stellar magnitude m by a square degree [81].

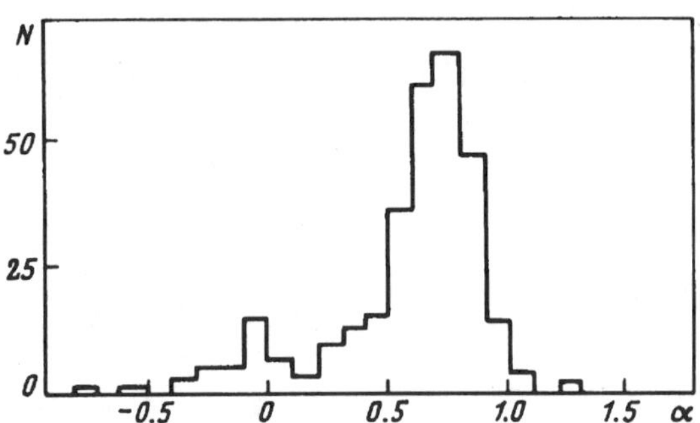

Figure 45.50 Distribution of spectral indices α of radiogalaxies in a frequency range from 178 to 1400 MHz [73].

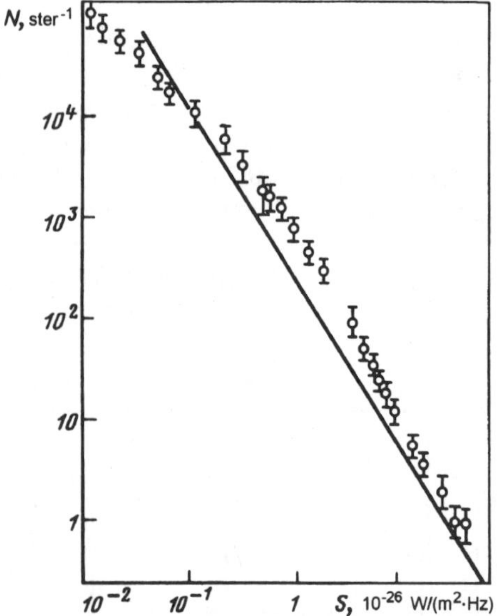

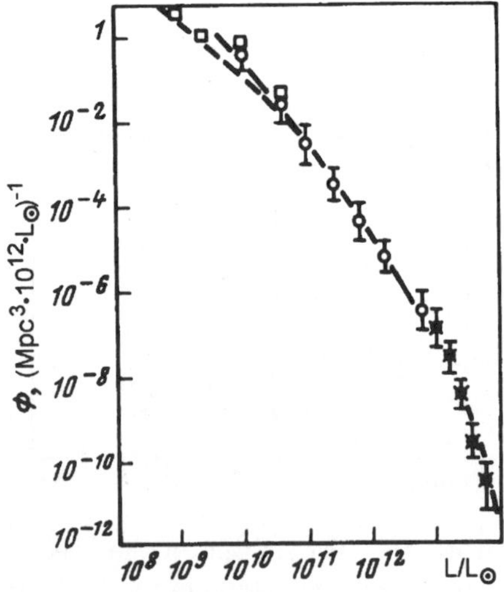

Figure 45.51 Luminosity functions of clusters and groups of galaxies [3].

Figure 45.53 Counts of radiosources at a frequency of 408 MHz: the solid line fits the law $S^{-3/2}$. Deviations from this law for large values of S mean that in the past the sources were either more numerous or/and they were brighter; flattening of the count curve for small values of S is due to cosmological effects [79].

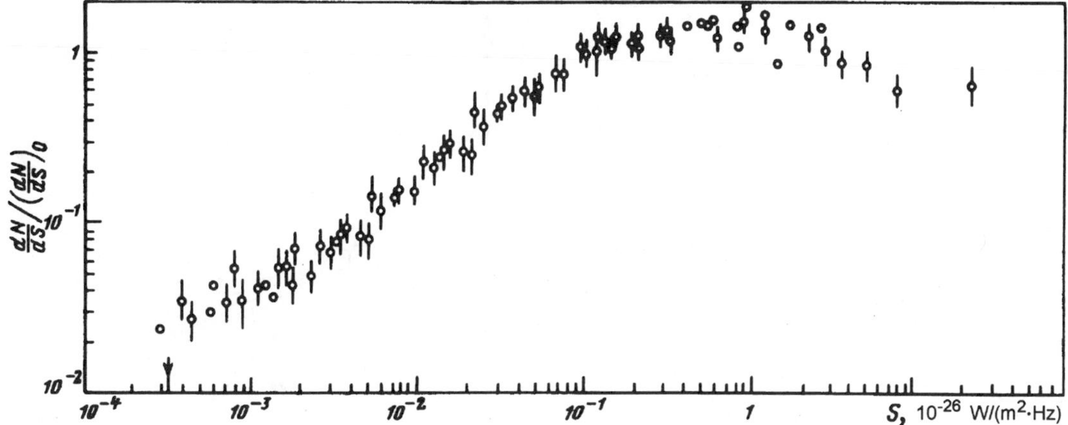

Figure 45.54 Counts of radiosources at $\lambda = 21$ cm: the curve dN/dS (in $\mathrm{sr}^{-1}\cdot\mathrm{Jy}^{-1}$) was normalized to the law $(dN/dS)_0 = 225\ S^{-5/2}$. Deviations from this law for large values of S are evidence for decreasing of the radiosource number at the present time. Decrease at small S is connected with a non-Euclidian nature of our world [80].

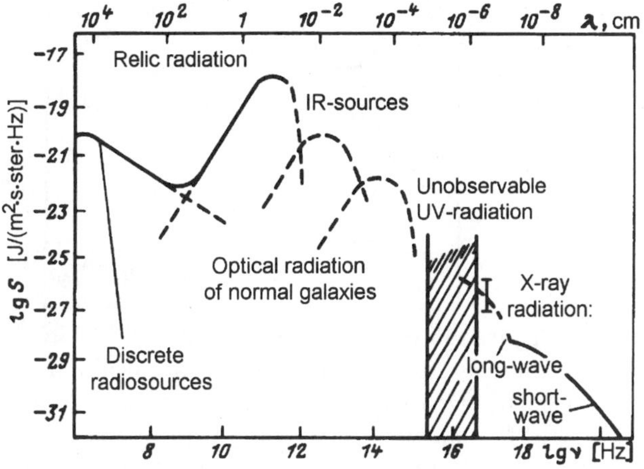

Figure 45.55 Spectrum of the electromagnetic background radiation in the Universe: the solid and the dashed lines represent measured values and the theoretical estimations, respectively [82].

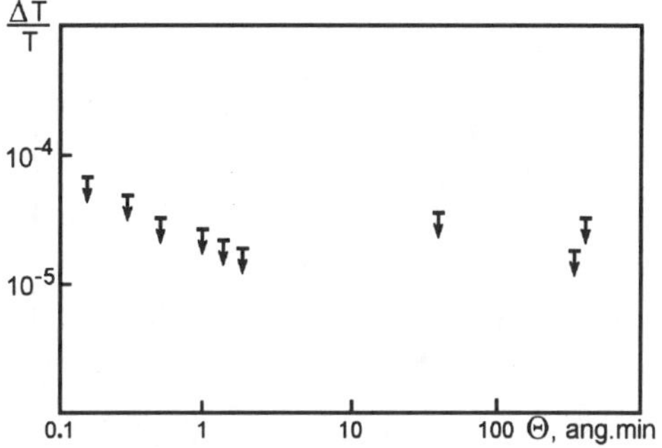

Figure 45.57 Limits for temperature fluctuations of the microwave background radiation in different angular scales [91, 92].

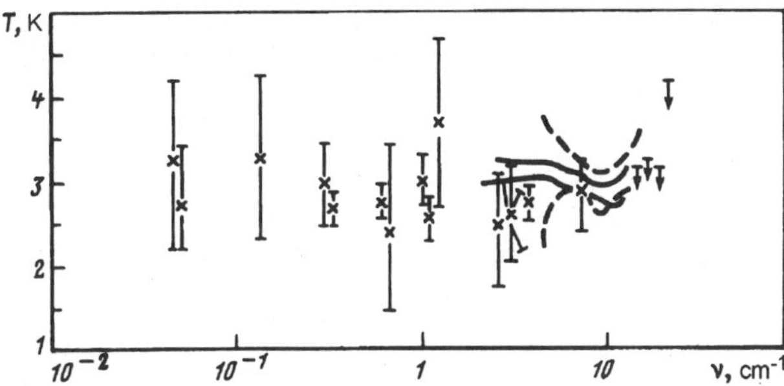

Figure 45.56 Measured values of the luminance temperature T of the microwave background radiation at various wavelengths [83].

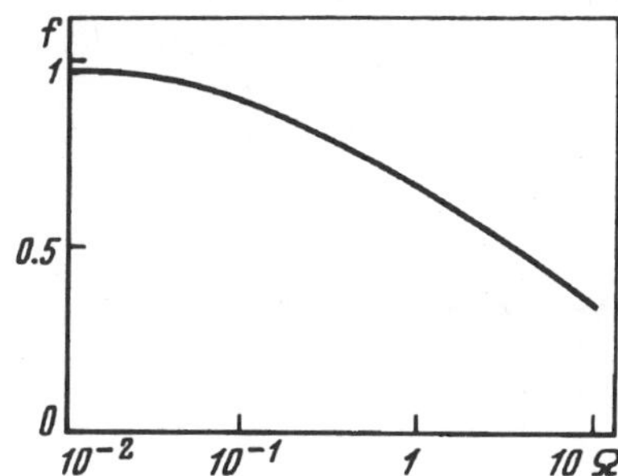

Figure 45.58 Function $f(\Omega)$ entering the expression for the age of the Universe [85].

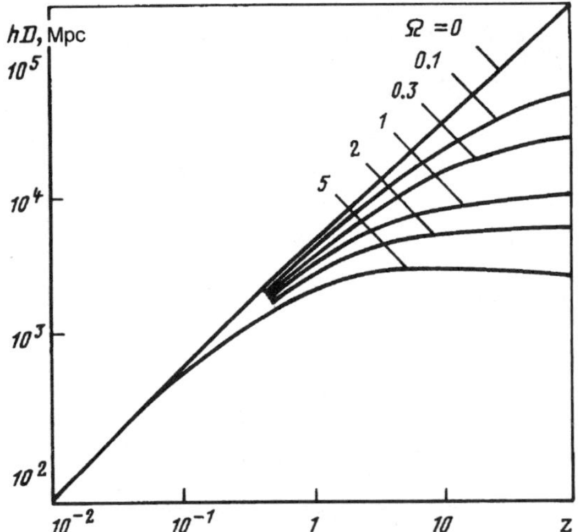

Figure 45.59 Dependence of the distance D to a source on its red shift z for different values of $\Omega = \rho/\rho_c$. The distance is defined in such a way that the flux density is proportional to D^{-2} [85].

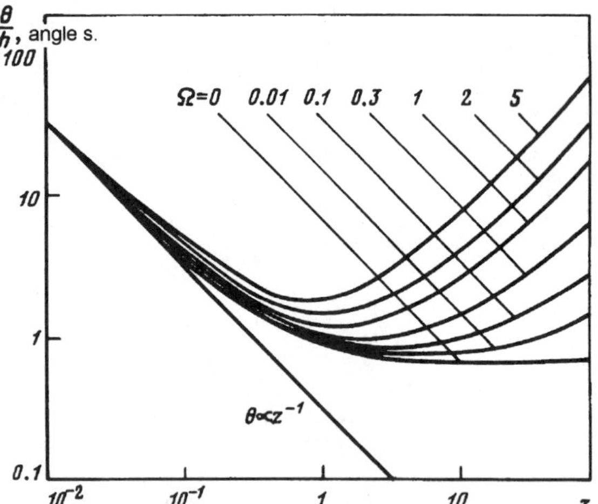

Figure 45.60 Dependence of the angular size θ of a 10 kpc-across source on its red shift z [85].

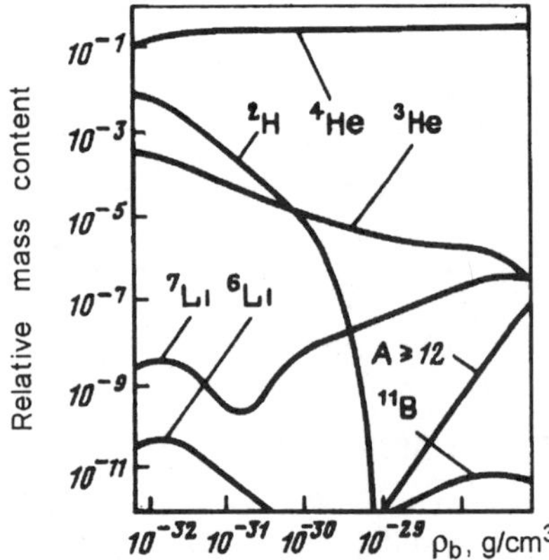

Figure 45.61 Dependence of the relative content of elements originated in the cosmic nucleosynthesis on the present baryon density. The calculations were performed for the present temperature of the microwave background radiation at 2.7 K [87]. The measurements give the following values for the relative mass content: $^2H/^1H \geq 10^{-5}$, $^7Li/^1H \approx 10^{-10}$, $(^2H + ^3He)/^1H \leq 10^{-4}$ [86].

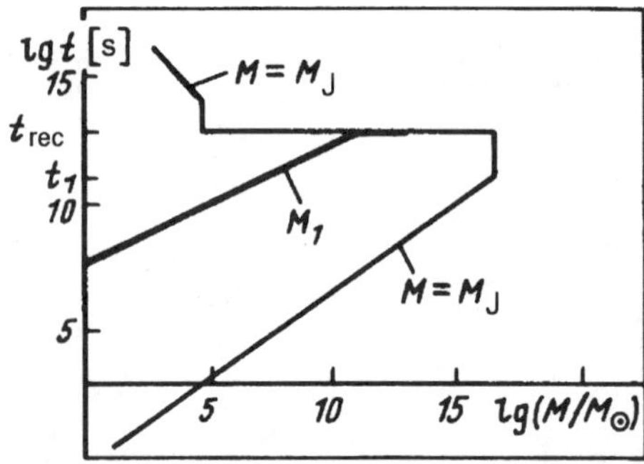

Figure 45.62 Gravitational instability in the Universe. Time counted from the beginning of the Universe expansion and mass of perturbation ($\rho\lambda^3_{\text{pert}}$) are plotted correspondingly along y- and x-axes. The moment t_1 when the matter density and the radiation density are matched and the moment t_{rec} of the hydrogen recombination are pointed out; M_J is the Jeans mass (i.e. the minimal mass at which the gravitational instability begins); M_1 is the maximum mass of perturbations which decayed to that moment due to radiative viscosity and thermal conductivity; $\Omega = 1$ [85].

References

[1] Allen, C. W., Astrophysical quantities, Althlone Press, London, 1973.

[2] Leng, K. R., Astrophysical formulae, Springer–Verlag, Berlin, 1974.

[3] Physics of space, A concise encyclopedia, Sovetskaya entsiklopedia, Moscow, 1986 (in Russian).

[4] Kulikovskii, P. G., Stellar astronomy, Nauka, Moscow, 1978 (in Russian).

[5] Gibson, E. G., The quiet Sun, Scientific and Technical Information Office, NASA, Washington, 1973.

[6] Bahcall, J. N., Huebner, W. F., Lubow, S. H., Rev. Mod. Phys., 54, 767, 1982.

[7] Shklovskii, I. S., Physics of the solar corona, Fizmatgiz, Moscow, 1962 (in Russian).

[8] Vorontsov, S. V., Zharkov, V. N., Uspekhi fizicheskikh nauk, 134, 675, 1981.

[9] The solar output and its variation, Ed. White, O., Colorado Associated University Press, Boulder, Colorado, 1977.

[10] Kruger, A., Introduction to solar radio astronomy and radio physics, D.Reidel Publishing Company, Dordrecht–Boston–London, 1979.

[11] Mercury, XII, No 4, 118, 1983.

[12] Russel, C. T., Adv. Space Res., 1, 257, 1981.

[13] Marov, M. Ya., Planets of the solar system, 2nd ed., Nauka, Moscow, 1986 (in Russian).

[14] Ip, W. H., Space Sci. Rev., 26, 39, 1980.

[15] Simonenko, A. N., Meteorites as fragments of asteroids, Nauka, Moscow, 1979 (in Russian).

[16] Churyumov, K. I., Comets and their observation, Nauka, Moscow, 1980 (in Russian).

[17] Stern, D. P., Ness, N. F., Ann. Rev. Astron. Astrophys., 20, 139, 1982.

[18] Akasofu, S. I., Ann. Rev. Astron. Astrophys., 20, 117, 1982.

[19] Baranov, V. B., Comm. Astrophys., IX, 75, 1981.

[20] Kurt, V. G., in: Astrophysics and space physics, Ed. Syunyaev, R. A., Nauka, Moscow, 1982 (in Russian).

[21] Martynov, D. Ya., A course of general astrophysics, 3rd ed., Nauka, Moscow, 1979 (in Russian).

[22] Grossman, A. S., Hays, D., Grabaske, H. C., Astron. Astrophys., 30, 95, 1974.

[23] Cox, J. P., Theory of stellar pulsations, Princeton University Press, Princeton, 1980.

[24] Bath, G. T., Quart. Royal Astron. Soc., 19, 442, 1978.

[25] Robinson, E. L., Ann. Rev. Astron. Astrophys., 14, 119, 1976.

[26] Gorbatskii, V. G., Nova-like and nova stars, Nauka, Moscow, 1974 (in Russian).

[27] Rydgren, A. E., Strom, S. E., Strom, B. H., Astrophys. J. Suppl., 30, 307, 1976.

[28] Cohen, M., Phys. Rep., 116, 173, 1984.

[29] Gershberg, R. E., Small-mass flaring stars, Nauka, Moscow, 1978 (in Russian).

[30] Boyarchuk, A. A., Science and technology developments, Ser. "Astronomiya", 22, 83, VINITI, Moscow, 1983 (in Russian).

[31] Trimble, V., Rev. Mod. Phys., 54, 1183, 1982.

[32] Shklovskii, I. S., Supernova stars, 2nd ed., Nauka, Moscow, 1976 (in Russian).

[33] Shapiro, S. L., and Teukolsky, S. A., Black holes, white dwarfs and neutron stars, Wiley-Interscience Publication, John Wiley and Sons, New York, 1983.

[34] Van Horn, H. M., Phys. Today, 32, 23, 1979.

[35] Smith, F. G., Pulsars, W. H. Freeman and Company, San Francisco, 1977.

[36] Manchester, R. N., and Taylor, J. H., in: Pulsars, W.H.Freeman and Company, San Francisco, 1977.

[37] Backer, D. C., Kulkarni, S. R., Heiles, C., Nature, 300, No 5893, 615, 1982.

[38] Liang, H. P., Nolan, P. L., Space Sci. Rev., 38, No 314, 353, 1984.

[39] White, N. E., Advances Space Res., 3, No 10–12, 9, 1984.

[40] Borner, G., Phys. Rep., 60, 153, 1980.

[41] White, N. E., Swank, J. H., Holt, S. S., Astrophys. J., 270, 711, 1983.

[42] Syunyaev, R. A., Trumper, J., Nature, 279, 506, 1979.

[43] Holt, S. S., McCray, R., Ann. Rev. Astron. Astrophys., 20, 323, 1982.

[44] Mazets, E. P., Golenetskii, S. V., in: Astrophysics and space physics, Ed. Syunyaev, R. A., Nauka, Moscow, 1982 (in Russian).

[45] Vladimirskii, B. M., Gal'per, A. M., Luchkov, B. I., Uspekhi fizicheskikh nauk, 145, 255, 1985 (in Russian).

[46] Marochnik, L. S., Suchkov, A. A., Galaxy, Nauka, Moscow, 1984 (in Russian).

[47] Kinematics, dynamics and structure of the Milky Way, Ed. Shuter, W. L. H., D.Reidel, Dordrecht, 1983.

[48] Faber, S. M., Gallagher, J. S., Ann. Rev. Astron. Astrophys., 17, 135, 1979.

[49] Bachall, J. N., Soneira, R. M., Astrophys. J. Suppl., 44, 73, 1980.

[50] Efremov, Yu. N., Science and technology developments, Ser. "Astronomiya", 27, 102, VINITI, Moscow, 1985 (in Russian).

[51] Khlopov, P. N., Star clusters, Nauka, Moscow, 1981 (in Russian).

[52] Kaplan, S. A., Pikel'ner, S. B., Physics of interstellar medium, Nauka, Moscow, 1979 (in Russian).

[53] Bochkarev, N. G., in: Stars and stellar systems, Ed. Martynov, D. Ya., Nauka, Moscow, 1981, p. 265 (in Russian).

[54] Longair, M. S., High energy astrophysics, Cambridge University Press, Cambridge, 1981.

[55] Turner, B. E., in: Galactic and extragalactic radio astronomy, Eds. Verschur, G. L., and Kellermann, K. I., Springer–Verlag, Berlin, 1974, p.199.

[56] Kostyakova, E. B., Physics of planetary nebulas, Nauka, Moscow, 1982 (in Russian).

[57] Sobolev, V. V., A course of theoretical astrophysics, 3rd ed., Nauka, Moscow, 1985 (in Russian).

[58] Pskovskii, Yu. P., in: Stars and stellar systems, Ed. Martynov, D. Ya., Nauka, Moscow, 1981, p. 88 (in Russian).

[59] Lozinskaya, T. A., Science and technology developments, Ser. "Astronomiya", 22, 33, VINITI, Moscow, 1983 (in Russian).

[60] Moran, J. M., in: Frontiers of astrophysics, Ed. Avrett, E. H., Harvard University Press, Cambridge, 1976.

[61] Spitzer, L., Jr., Physical processes in the interstellar medium, Wiley-Interscience Publication, John Wiley and Sons, New York, 1978.

[62] Habing, H. J., Israel, F. P., Ann. Rev. Astron. Astrophys., 17, 345, 1979.

[63] Schwartz, R. D., Ann. Rev. Astron. Astrophys., 21, 209, 1983.

[64] Oort, J. H., Ann. Rev. Astron. Astrophys., 15, 295, 1977.

[65] The galactic center, Eds. Riegler, G. R., and Blandford, R. D., American Institute of Physics, New York, 1982.

[66] Zasov, A. V., Science and technology developments, Ser. "Astronomiya", 18, 3, VINITI, Moscow, 1981 (in Russian).

[67] Tayler, R. G., Galaxies: structure and evolution, Wykeham Publications (A member of the Taylor & Francis Group), London, 1987.

[68] Rieke, G. H., Lebofsky, M. J., Ann. Rev. Astron. Astrophys., 17, 447, 1979.

[69] Puetter, R. C., Burbridge, E. M., Smith, H. E., Astrophys. J., 257, 487, 1982.

[70] Ulrich, M. H., Space Sci. Rev., 28, 89, 1981.

[71] Mushotsky, R. F., Advances Space Res., 3, No 10–12, 157, 1984.

[72] Meurs, E. J., Wilson, A. A., Astron. Astrophys., 136, 206, 1984.

[73] Pacholczyk, A. G., Radio galaxies, Pergamon Press, Oxford, 1977.

[74] Bahcall, N., Ann. Rev. Astron. Astrophys., 15, 505, 1977.

[75] Peebles, P. J. E., The large-scale structure of the Universe, Princeton University Press, Princeton, 1980.

[76] Klypin, A. A., Kopylov, A. I., Pis'ma v astronomicheskii zhurnal, 9, 75, 1983 (in Russian).

[77] Oort, J. H., Ann. Rev. Astron. Astrophys., 21, 373, 1983.

[78] Mushotsky, R. F., Serlemitsos, P. J., Smith, B. W., Astrophys. J., 225, 21, 1978.

[79] Longair, M., Uspekhi fizicheskikh nauk, 99, 229, 1969 (in Russian).

[80] Windhorst, R., Ph.D. Thesis, University of Leiden, 1984.

[81] Karachentsev, I. O., Kopylov, A. I., Pis'ma v astronomicheskii zhurnal, 3, 246, 1977 (in Russian).

[82] Longair, M. S., Syunyaev, R. A., Uspekhi fizicheskikh nauk, 105, 41, 1971 (in Russian).

[83] Weiss, R., Ann. Rev. Astron. Astrophys., 18, 489, 1980.

[84] Partridge, R. B., Phys. Scripta, 21, 624, 1980.

[85] Zel'dovich, Ya. B., Novikov, I. D., Universe structure and evolution, Nauka, Moscow, 1975 (in Russian).

[86] Yang, J., Turner, M. S., Steigman, G., Astrophys. J., 281, 493, 1984.

[87] Davis, M., in: Frontiers of astrophysics, Ed. Avrett, E. H., Harvard University Press, Princeton, 1976.

[88] Zel'dovich, Ya. B., Syunyaev, R. A., Pis'ma v astronomicheskii zhurnal, 6, 451, 1980 (in Russian).

Subject Index

The most effective way to use this Index is to look for the pertinent phenomenon, class of substance, or general concept. The reference given for each index term is the inclusive pages of the respective definition, figure or table. References to pages are shown by the letter p*; those to figures and tables, by the letters* f *and* t*, respectively.*

G

L

O